WITHDRAWN

3 8008 00276 9320

AF616209

F2 057°
20 NM
FRANKFURT FIR
Hugelsheim
BADEN-BADEN/OOS
KARLSRUHE
0:10
BECKER
BUSY
DIM
OFF
ENT
CURSOR
MDE
PLN
F1
F2
SQL
MDE
STO
COM
NAV
ADF
XPDR
CODE
TMR
IDT
REC
BFO
SBY
ON
ALT

Jane's

AVIONICS

Edited by Chris Johnson

Seventeenth Edition

1998-99

ISBN 0 7106 1544 2

"Jane's" is a registered trade mark

In the USA and its dependencies

Jane's Information Group Inc, 1340 Braddock Place, Suite 300, Alexandria, Virginia 22314-1651, USA

British Library Cataloguing-in-Publication Data.

A catalogue record for this book is available from the British Library.

Printed and bound in Great Britain by Biddles Ltd, Guildford and King's Lynn

Contents

How to use *Jane's Avionics*

At the beginning of this edition of *Jane's Avionics* we have added some extra pages to help the reader. The first is this 'How to use *Jane's Avionics*' which gives an overview of what is contained in each section and an explanation of the three categories into which the various entries have been divided.

The reader will also find New entries in this edition, and Entries deleted from this edition, which are designed to enable the reader to quickly find out what is new and what has been removed.

These pages, together with the Foreword and Contents, are printed on coloured paper to make it easier for readers to find the pages they need.

Jane's Avionics contains information on airborne electronic equipment for both military and civil aircraft. The criteria for inclusion is that the equipment should have a significant electronic content, be in development, production or service and be operated in manned aircraft. Avionics equipment designed specifically for unmanned aircraft is covered in *Jane's Unmanned Aerial Vehicles and Targets*, although those items which are carried by both manned and unmanned aircraft will continue to be found in *Jane's Avionics*.

Jane's Avionics is divided into 12 main sections based on the function of the systems concerned. These sections are: Communications; Maritime operations; Radar GPWS and TCAS systems; Electro-optics; Electronic warfare; Data handling; Data recording; Navigation and nav/attack; Flight management and control; Cockpit displays, instruments and indicators; Head-up displays, helmet-mounted displays and weapon aiming sights, and Stores management.

Some systems spread across more than one function and could be placed in more than one section as, for instance, is the case with nav/comm radios. The aim has been consistency and all examples of a particular type of system have been placed in one section.

Within each main section entries are arranged alphabetically by country of manufacture. For each country, entries are attributed to a contractor, and contractors are listed alphabetically. The entries for each contractor are also listed alphabetically. Entries are structured as follows:

Title.

Narrative: contains information on function, development history, and technical description.

Specification: lists the main technical parameters and dimensions.

Operational status: contains information on development/production/service status.

Contractor: details of the prime contractor.

In addition to the main text, *Jane's Avionics* contains three indexes: a contractors index; an alphabetical index; and an index that correlates entries to contractors.

The Contractors index contains details of the name, address and, where available, the telephone/fax number of each contractor represented in the book.

The Alphabetical index lists every entry in alphabetical order.

The Manufacturers index lists entries alphabetically by contractor.

To help users of this title evaluate the published data, *Jane's Information Group* has divided entries into three categories:

● ***VERIFIED*** The editor has made a detailed examination of the entry's content and checked its relevance and accuracy for publication in the new edition to the best of his ability. In 50 per cent of cases the contractor has also verified the accuracy of the text.

● ***UPDATED*** During the verification process, significant changes to content have been made to reflect the latest position known to *Jane's* at the time of publication.

● ***NEW ENTRY*** Information on new equipment and/or systems appearing for the first time in this title.

Since 1995 all new pictures have been dated with the year of publication. All new pictures this year are dated 1998. Some are followed by seven-digit numbers for ease of identification by our image library.

CD-ROM *Jane's Avionics* is included in two Jane's CD-ROMs: *Jane's Defence Equipment Library*; *Jane's Transport Library*.

Internet An electronic index of the full text of this publication is now available at Jane's homepage on the Internet. Access to the index is available through WWW or ftp at the following addresses:

http://www.janes.com/defence/resources/wordset.html
ftp://ftp.janes.com/janes/index

email address Material for *Jane's Avionics* can be sent to the following email address: jav@janes.co.uk

We hope these measures increase the value of this title to our thousands of customers worldwide. If you have comments or suggestions for further improvements, the Publisher would be pleased to receive them.

This edition of *Jane's Avionics* contains 305 new entries.

Glossary of acronyms and abbreviations

A	Ampère
AAA	Anti-Aircraft Artillery
AAAA Arrays	Advanced Architectures for Airborne
AACMI	Autonomous Air Combat Manoeuvring Installation
AACS	Airborne Advanced Communications System
AADS	Airborne Active Dipping Sonar
AAED	Advanced Airborne Expendable Decoy
AAICP	Air-to-Air Interrogator Control Panel
AAIP	Analogue Autoland Improvement Programme
AAU	1. Audio Amplifier Units 2. Audio Amplifier Unit
AAW	Anti-Air Warfare
AAWWS	Airborne Adverse Weather Weapon System
ABCCC	Airborne Battlefield Command and Control Centre
ABICS	Ada Based Integrated Control System
ABIS	All-Bus Instrumentation System
ABL	AirBorne Laser
AC	Alternating Current
ACAS	Airborne Collision Avoidance System
ACARS	Automatic Communications And Reporting System
ACC	1. Avionics Computer Control 2. Axis Controlled Carrier
ACCS	Airborne Computing and Communications System
ACE	1. Actuator Control Electronics 2. Advanced Communication Engine 3. Autonomous Combat (maneouvres) Evaluation
ACEM	Aerial Camera Electro-optical Magazine
ACI	Armament Control Indicator
ACIDS	Automated Communications and Intercom Distribution System
ACIS	1. Armament Control Indicator Set 2. Advanced Cabin Interphone System
ACLS	Automatic Carrier Landing System
ACM(I)	Air Combat Manoeuvring Installation
ACMS	1. Aircraft Condition Monitoring System 2. Armament Control and Monitoring System 3. Avionics Control and Management System
ACNIP	Auxiliary (or Advanced) Communication Navigation and Identification Panel
ACP	1. Advanced Controller Processor 2. Armament Control Panel 3. Audio Converter Processor
ACP(C)	Automatic Communications Processor (Control)
ACS	Armament Control System
ACT	Airborne Crew Trainer
ACTIVE	Advanced Control Technology for Integrated Vehicles
ACU	1. Adaptive Control Unit 2. Airborne Computer Unit 3. Annotation Control Unit 4. Antenna Control Unit 5. Audio Control Unit 6. Auxiliary Control Unit
ADAAPS	Aircraft Data Acquisition, Analysis and Presentation System
ADACS	Airborne Digital Automatic Collection System
ADAD	Air Defence Alerting Device
ADAS	1. Airborne Data Acquisition System 2. Auxiliary Data Annotation Set
ADAU	Auxiliary Data Acquisition Unit
ADC	1. Advanced Data Controller 2. Air Data Computer
ADDS	Advanced Digital Dispensing System
ADECS	Advanced Digital Engine Control System
ADELE	Alerte Détection Et Localisation des Emetteurs
ADELT	Automatically Deployed Emergency Locator Transmitter
ADF	Automatic Direction-Finder
ADI	1. Attitude Director Indicator 2. Azimuth Display Indicator
ADID	Aircraft Data Interface Device
ADIRS	Air Data and Inertial Reference System
ADIRU	Air Data Inertial Reference Unit
ADLGP	Advanced Datalink for Guided Platforms
ADM	1. Advanced Development Model 2. Air Data Module
Ad-Me	Advanced Metal Evaporated
ADOCS	Advanced Digital Optical Control System
ADR	Accident Data Recorder
ADS	1. Audio Distribution System 2. Automatic Dependent Surveillance 3. Airlifter Defense Systems
ADS-B	Automatic Dependent Surveillance Broadcast
ADSU	Air Data Sensor Unit
ADTU	Auxiliary Data Transfer Unit
ADU	1. Air Data Unit 2. Annotation Display Unit
ADVCAP	Advanced Capability
ADVICE	Acoustic Data Vessel Identification, Classification and Explanation
AEC	Automatic Exposure Control
AEGIS	Airborne Early warning/Ground environment Integration Segment
AELS	Airborne Electronic Library System
AERA	Automated En-Route Air traffic control
Aero-H	Aeronautical High-gain antenna
Aero-I	Aeronautical Intermediate-gain antenna
AES	Aeronautical Earth Station
AESOP	Airborne Electro-optical Special Operations Payload
AETMS	Airborne Electronic Terrain-Mapping System
AEU	Airborne Electronics Unit
AEW	Airborne Early Warning
AF	Audio Frequency
AFA	Audio Frequency Amplifier
AFC	Automatic Frequency Control
AFCAS	Automatic Flight Control Augmentation System
AFCS	Automatic Flight Control System
AFDAS	Aircraft Fatigue Data Analysis System
AFDS	Automatic Flight Director System
AFH	Advanced Fibre Heater
AFIS	Airborne Flight Information System
AFMCS	Advanced Flight Management Computer System
AFMS	Automatic Flight Management System
AFSAT	Air Force Satellite
AFSATCOM	Air Force Satellite Communications
AFTI	Advanced Fighter Technology Integration
AFV	Armoured Fighting Vehicle
AGC	Automatic Gain Control
AGES	Air-to-Ground Engagement Simulator
AGINT	Advanced GPS Inertial Navigation Technology
AGL	Above Ground Level
AGPPE	Advanced General Purpose Processor Element
AGRA	Automatic Gain Ranging Amplifier
AGREE	Advisory Group on the Reliability of Electronic Equipment (US)
AGTV	Active Gated Television
AGU	AIRLINK Gateway Unit
AHEAD	Attitude, Heading and Rate of Turn Indicating System
AHIP	Army Helicopter Improvement Programme
AHRS	Attitude and Heading Reference System
AI	Air Interception
AIBU	Advanced Interference Blanker Unit
AICS	Airborne Integrated Communications System
AIDC	Aero Industry Development Centre
AIDS	Aircraft Integrated Data Suite (or System)
AIG	ACAS Implementation Group
AIM	1. Air Intercept Missile 2. Air traffic control beacon/IFF/Mk XII system
AIME	Autonomous Integrity Monitored Extrapolation
AIMES	Avionics Integrated Maintenance Expert System
AIMS	1. Advanced Integrated MAD System 2. Air traffic control radar beacon, IFF, Mk 12 transponder System 3. Aircraft Information Management System 4. Aircraft Integrated Monitoring System
AINS	1. Aided Inertial Navigation System 2. Airborne Inertial Navigation System
AIP	Australian Industrial Participation
AIPT	Advanced Image Processing Terminal
AIRSTAR	Airborne Surveillance and Target Acquisition Radar
AIT	Advanced Intelligence Tape
AIU	1. Armament Interface Unit 2. Automatic Ignition Unit
AJB	Audio Junction Box
AKU	Avionic Keyboard Unit
ALARMS	Airborne Laser Radar Mine Sensor
ALAT	French Army Light Aviation Corps
ALE	Automatic Link Establishment
ALFS	Airborne Low Frequency Sonar
ALLTV	All Light Level Television
ALRAD	Airborne Laser Rangefinder And Designator
AM	Amplitude Modulation
AMAC	Airborne Multi-Application Computer
AMC	Advanced Micro-electronics Converter
AMCS	Airborne Missile Control System
AME	1. Amplitude Modulation Equivalent 2. Angle Measuring Equipment
AMHMS	Advanced Magnetic Helmet-Mounted Sight
AMICS	Adaptive Multidimensional Integrated Control System
AMIDS	Advanced Missile Detection System
AMIMU	Advanced Multisensor Inertial Measurement Unit
AMLCD	Active Matrix Liquid Crystal Display
AMMCS	Airborne Multiservice/Multimedia Communications System
AMP	Advanced Modular Processor
AMPA	Advanced Mission Planning Aid
AMRAAM	Advanced Medium-Range Air-to-Air Missile
AMRS	Advanced Maintenance Recorder System
AMS	1. Airborne Maintenance Subsystem 2. Avionics Management System
AMSAR	1. Airborne Multirole Steerable Array Radar 2. Airborne Multirole multifunction Solid-state Active array Radar
AMSL	Above Mean Sea Level
AMSS	Aeronautical Mobile Satellite Service
AMU	1. Audio Management Unit 2. Auxiliary Memory Unit 3. Avionics Management Unit
ANC	Active Noise Cancellation
ANDVT	Advanced Narrowband Digital Voice Terminal
ANMI	Air Navigation Multiple Indicator
ANSI	American National Standards Institute
ANVIS	Aviator's Night Vision Imaging System

AOA	1. Airborne Optical Adjunct 2. Angle Of Attack
AOC	Assumption Of Control (message)
AOCM	Advanced Optical CounterMeasures
AOPT	Advanced Optical Position Transducer
APALS	Autonomous Precision Approach and Landing System
APAR	Advanced Phased-Array Radar
A/PDMC	Aircraft Products Data Management Computer
APHIDS	Advanced Panoramic Helmet Interface Demonstrator System
APIRS	Aircraft Piloting Inertial Reference System
APR	Auto Power Reserve
APS	1. Adaptive Processor System 2. Aircraft Position Sensor 3. Altitude Position Sensor
APSP	Advanced Programmable Signal Processor
APU	Auxiliary Power Unit
AQL	Advanced Quick Look
ARBS	Angle Rate Bombing Set
ARCADS	Armament Control And Delivery System
ARDS	Airborne Radar Demonstrator System
ARI	Azimuth Range Indicator
ARIA	Advanced Range Instrumentation Aircraft
ARIC	Airborne Radio and Intercom Control
ARIES	Airborne Recorder for IRLS and EO Sensors
ARINC	Aeronautical Radio Incorporated
ARJS	Airborne Radar Jamming System
ARM	Anti-Radiation Missile
ARPA	Advanced Research Projects Agency (US)
ARPS	Advanced Radar Processing System
ARSA	Advisable Radar Service Area
ARTCC	Air Route Traffic Control Centre
ARTI	Advanced Rotorcraft Technology Integration
ARWS	Advanced Radar Warning System
AS	Anti-Spoofing
ASAC	Airborne Surveillance Airborne Control
ASAP	Airborne Shared Aperture Programme
ASARS	1. Airborne Search And Rescue System 2. Advanced Synthetic Aperture Radar System
ASAS	Airborne Separation Assistance System
ASC	Airborne Strain Counter
ASCB	1. Aircraft Standard Communications Bus 2. Avionics Standard Communications Bus
ASCII	American Standard Code for Information Interchange
ASCOT	Aerial Survey Control Tool
ASCTU	Air Supply Controller/Test Unit
ASDC	Armament Signal Data Converter
ASE	AutoStabilisation Equipment
ASETS	Airborne Seeker Evaluation Test Set
ASGC	Airborne Surveillance Ground Control
ASI	ARINC Standards Interface
ASIC	Application Specific Integrated Circuit
ASIT	Adaptive Surface Interface Terminals
ASMD	Anti-Ship Missile Defence
ASMU	Avionics System Management Unit
ASNI	Ambient Sea Noise Indication
ASP	1. Advanced Signal Processor 2. Airborne Signal Processor 3. Aircraft Systems Processor
ASPIS	Advanced Self-Protection Integrated Suite
ASPJ	Airborne Self-Protection Jammer
ASPRO	Associative Processor
ASR	Advanced Special Receiver
ASSG	Acoustic Sensor Signal Generator
AST	Airborne Surveillance Testbed
ASTAR	Airborne Search Target Attack Radar
ASTE	Advanced Strategic and Tactical Expendables
ASTOR	Airborne StandOff Radar
ASU	Acoustic Simulation Unit
ASUW	Anti-Surface Warfare
ASV	Anti-Surface Vessel
ASVW	Anti-Surface Vessel Warfare
ASW	Anti-Submarine Warfare
ASWAC	Airborne Surveillance Warning And Control
ATA	1. Actual Time of Arrival 2. Advanced Tactical Aircraft
ATAC	1. Applied Technology Advanced Computer 2. Air Transportable Acoustic Communication
ATAL	Appereillage de Télévision sur Aéronef Léger
ATARS	Advanced Tactical Air Reconnaissance System
ATC	Air Traffic Control
ATCRBS	Air Traffic Control Radar Beacon System
ATDS	Airborne Tactical Data System
ATE	Automatic Test Equipment
ATF	Advanced Tactical Fighter
ATH	Autonomous Terminal Homing
ATHS	Automatic Target Handover System
ATIRCM	Advanced Threat InfraRed CounterMeasures
ATIS	Automatic Terminal Information Service
ATLANTIC	Airborne Targeting Low-Altitude Navigation Thermal Imaging and Cueing
ATLIS	Automatic Tracking Laser Illumination System
ATM	Air Traffic Management
ATN	Aeronautical Telecommunications Network
ATOPS	Advanced Transport Operations System
ATR	1. Air Transport Racking 2. Automatic Target Recognition
ATRJ	Advanced Threat Radar Jammer
AVAD	Automatic Voice Alert Device
AVDAS	Airborne Video Data Acquisition System
AVE	Airborne Vehicle Equipment
AVICS	Air Vehicle Interface and Control System
AVTR	Airborne VideoTape Recorder
AWACS	Airborne Warning And Control System
AWADS	Adverse Weather Aerial Delivery System
AWARE	Advanced Warning of Active Radar Emissions
BBU	Battery Back-up Unit
BCSV	Bearing Compartment Scavenge Value
BDHI	Bearing, Distance and Heading Indicator
BDI	Bearing Distance Indicator
BER	BIT Error Rate
BFDAS	Basic Flight Data Acquisition System
BFO	Beat Frequency Oscillation
BIT	Built-In Test
BITE	Built-In Test Equipment
BPI	Bits Per Inch
BPSK	Bi-Phase Shift Keyed
B-RNAV	Basic Area NAVigation
BSIN	Bus System Interface Unit
BSP	Barra Side Processor
BT	Bathythermograph
BTH	Beyond The Horizon
BTU	Basic Terminal Unit
BVCU	Bleed Valve Control Unit
BVR	Beyond Visual Range
C^{3}CM	Command, Control and Communications CounterMeasures
C^{3}I	Command, Control, Communications and Intelligence
C^{4}I	Command, Control, Communications, Computers and Intelligence
C/A	Coarse Acquisition
CAA	Civil Aviation Authority (UK)
CAB	Common Avionics Baseline
CACTCS	Cabin Air Conditioning and Temperature Control System
CAD	Computer-Aided Design
CADC	Central Air Data Computer
CAINS	Carrier Aircraft Inertial Navigation System
CAMBS	Command Activated MultiBeam Sonobuoy
CAMEL	Cartridge Active Miniature Electromagnetic
CARA	Combined Altitude Radar Altimeter
CARABAS	Coherent All Radio Band Sensing
CASS	1. Command Active Sonobuoy System 2. Consolidated Automatic Support System
CAT	CATegory
CBL	Control By Light
CC	Countermeasures Computer
CCD	Charge Coupled Device
CCG	Communication Control Group
CCIL	Continuously Computed Impact Line
CCIP	Continuously Computed Impact Point
CCIR	Comité Consultatif International des Radiocommunications
CCRP	Continuously Computed Release Point
CCS	Conformal Countermeasures System
CCTWT	Coupled Cavity Travelling Wave Tube
CCU	1. Cockpit Control Unit 2. Common Control Unit 3. Communication Control Unit
CDC	Cabin Display Computer
CDI	Course Deviation Indicator
CDIRRS	Cockpit Display of InfraRed Reconnaissance System
CDM	Collaborative Decision Making
CDNU	Control and Display Navigation Unit
CDR	Critical Design Review
CDTI	Cockpit Display of Traffic Information
CDU	1. Cockpit Display Unit 2. Control and Display Unit
CEAT	Centre d'Essais Aeronautique de Toulouse
CEDAM	Combined Electronic Display And Map
CEP	Circular Error of Probability
CEV	Centre d'Essais de Vol
CFAR	Constant False Alarm Rate
CFD	Chaff and Flare Dispensing
CFDCU	Chaff and Flare Dispenser Control Unit
CFDIU	Centralised Fault Display Interface Unit
CFDS	Centralised Fault Display System
CFIT	Controlled Flight Into Terrain
CG	Centre of Gravity
CGCC	Centre of Gravity Control Computer
CHAALS	Communications High-Accuracy Airborne Location System
CHT	Cylinder Head Temperature
CI	Control Indicator
CI-F	Control Indicator - Front
CIP	Common Integrated Processor
CIRCE	Cossor Interrogation and Reply Cryptographic Equipment
CIRTEVS	Compact InfraRed Television System
CI-S	Control Indicator - Side
CIS	1. Commonwealth of Independent States 2. Control Indicator Set
CIT	Combined Interrogator/Transponder
CITS	Centrally Integrated Test System
CIU	1. Cockpit Interface Unit 2. Control Interface Unit
CLASS	Coherent Laser radar Airborne Shear Sensor
CLDP	Convertible Laser Designation Pod
CLDS	Cockpit Laser Designation System
CM	COMSEC Module
CMD	1. Colour Multipurpose Display 2. CounterMeasures Dispenser
CMDR	Card Maintenance Data Recorder
CMDS	CounterMeasures Dispensing System
CMF	Central Maintenance Function
CMLSA	Commercial Microwave Landing System Avionics
CMOS	Complementary Metal Oxide Silicon
CMP	Central Maintenance Panel
CMRA	Cruise Missile Radar Altimeter
CMRS	Crash/Maintenance Recorder System
CMS	Computer Module System
CMT	Cadmium Mercury Telluride
CMUP	Conventional Mission Upgrade Programme
CMWS	Common Missile Warning System
C-Nite	Cobra Nite
CN2H	Conduit Nuit, second-generation, Helicopters
CNC	Communication/Navigation equipment Controls
CNI	Comunications, Navigation and Identification
CNI-MS	CNI Management System
CNIU	Communications/Navigation Interface Unit
CNS	Communication, Navigation and Surveillance
CODAR	Correlation Detection and Ranging
COIL	Chemical Oxygen Iodine Laser
COMAC	Cockpit Management Computer
COMED	Combined Map and Electronic Display
COMINT	Communications Intelligence
COMPASS	Compact MultiPurpose Advanced Stabilised System
COMSEC	Communications Security
COTIM	Compact Thermal Imaging Module
COTS	Commercial-Off-The-Shelf
CP	Computer Processor
CPA	Cabin Public Address
CPC	Cabin Pressure Controller
CPCS	Cabin Pressure Control System
CPDLC	Controller-Pilot DataLink Communications
CPM/P	Command Post Modem/Processor
CPR	Covert Penetration Radar
CPS	Covert Penetration System
CPU	Central Processing Unit
CPU-F	Control Panel Unit - Front
CPU-S	Control Panel Unit - Side
CPVR	Crash Protected Video Recorder
CR	Countermeasures Receiver
CRE	Communications Radar Exciter
CRPA	Controlled Reception Pattern Antenna
CRT	Cathode Ray Tube
CS	Communications Subsystem

CSA	Control Stick Assembly
CSAS	Command and Stability Augmentation System
CSC	1. Communication System Controller 2. Compass System Controller
CSCG	Communications System Control Group
CSD	Common Strategic Doppler
CSMU	Crash Survivable Memory Unit
CSS	1. Computer Support System 2. Complimentary Satellite System
CSU	1. Communications Switching Unit 2. Control Status Unit 3. Crew Station Unit
CSVR	Crash Survivable Voice Recorder
CT	1. Control Transmitter 2. Crew member Terminal
CTAS	Chrysler Technologies Airborne Systems
CTC	Cabin Temperature Controller
CTS	Central Tactical System
CTT/H-R	The Commanders' Tactical Terminal/ Hybrid-Receive only
CTU	Control Terminal Unit
CU	Control Unit
CUGR	Cargo Utility GPS Receiver
CV	Aircraft carrier
CVFDR	Cockpit Voice and Flight Data Recorder
CVR	Cockpit Voice Recorder
CVR/DFDR	Cockpit Voice Recorder/Digital Flight Data Recorder
CW	Continuous Wave
CWI	Continuous Wave Illuminator
CWS	Control Wheel Steering
DADC	Digital Air Data Computer
DAFCS	Digital Automatic Flight Control System
DAFD	Digital Autopilot/Flight Director
DAFICS	Digital Automatic Flight Inlet Control System
DAI	DCMS Audio Interface
DAIS	Digital Avionics Information System
DAIRS	Distributed-Architecture InfraRed Sensor
DAMA	Demand Assigned Multiple Access
DAMS	Drum Auxiliary Memory Sub-unit
DAPU	Data Acquisition and Processing Unit
DAR	1. Direct Access Recorder 2. Drone Anti-Radar
DARPA	Defense Advanced Research Projects Agency (US) (now ARPA)
DARTS	Digital Airborne Radar Threat Simulator
DARU	Data Acquisition and Recording Unit
DASH	Display And Sight Helmet
DASS	Defensive Aids SubSystem
DAT	Digital Audio Tape
DAU	1. Data Acquisition Unit 2. Digital Amplifier Unit
dB	decibel(s)
DBC	DCMS Bus Coupler
DBI	DCMS Bus Interface
dBm	decibels $\times 10^{-3}$
DC	Direct Current
DCC	Digital Computer Complex
DCI	DCMS Crew member Interface
D/CM	Diagnostic and Condition Monitoring
DCMS	Digital Communication Management System
DCPS	Data Collection and Processing System
DCSU	Dual Crew Station Unit
DCU	1. Data Collection Unit 2. Digital Computer Unit
DDI	1. Data Display Indicator 2. Digital Display Indicator
DDIC	Digital Display Indicator Control
DDPS	Digital Display Processing System
DDU	Disk Drive Unit
DDVR	Displayed Data Video Recorder
DEC	Digital Electronic Control
DECS	Digital Engine Control System
DECU	Digital Engine Control Unit
DED	Data Entry Display
DEEC	Digital Electronic Engine Control
DEFCS	Digital Electronic Flight Control System
DEMON	Demodulation Of Noise
DES	Data Encryption Standard
DEU	Display Electronics Unit
DF	Direction-Finding
DFAD	Digital Feature Analysis Data
DFDAU	Digital Flight Data Acquisition Unit
DFDR	Digital Flight Data Recorder
DFGC	Digital Flight Guidance Computer
DFIR	Deployable Flight Incident Recorder
DFLCC	Digital Flight Control Computer
DFU	Deployable Flotation Unit
DG	Directional Gyro
DGA	Displacement Gyro Assembly
DGNS	Differential Global Navigation System
DGPS	Differential GPS
DIANE	Détection Identification Analyse des Nouveaux Emetteurs
DICASS	Directional Command Activated Sonobuoy System
DICU	Display Interface Control Unit
DID	Data Insertion Device
DIFAR	Directional acoustic Frequency Analysis and Recording
DIFM	Digital Instantaneous Frequency Measurement
DII	DCMS Interphone Interface
DIL	Digital Integrated Logic
DIM	Dispense Interface Microprocessor
DIRCM	Directional InfraRed CounterMeasures
DITACS	Digital Tactical System
DITS	Digital Information Transfer System
DITU	De-Icer Timer Unit
DIU	Data Interface Unit
DLMS	Digital Land Mass System
DLPP	DataLink Pre-Processor
DLS	Data Loader System
DLT	Digital Linear Tape
DMA	Direct Memory Access
DMDG	Digital Map Display Generator
DME	Distance Measuring Equipment
DME-P	Distance Measuring Equipment - Precision
DMG	Digital Map Generator
DMM	Data Management Module
DMS	Data Multiplexer Sub-unit
DMU	1. Data Management Unit 2. Digital Master Unit
DOA	Direction Of Arrival
DoD	Department of Defense (US)
DOLE	Detection Of Laser Emissions
DOLRAM	Detection Of Laser, Radar And Millimetric threats
DOP	Digital Onboard Processor
DOS	Disk Operating System
DP	Display Processor
DPCM	Digital Pulse Code Modulation
DPG	Data Processor Group
DP/MC	Display Processor/Mission Computer
DPRAM	Dual-Port Random Access Memory
DPS	Data Processing System
DPSK	Digital Phase Shift Keying
DPU	Digital Processing Unit
DRAM	Dynamic Random Access Memory
DRC	Data Recording Cartridge
DRD	Digital Radar Display
DRFM	Digital Radio Frequency Memory
DRU	Data Retrieval Unit
DSCS	Defense Satellite Communication System
DSDC	Digital Signal Data Converter
DSP	1. Day Surveillance Payload 2. Digital Signal Processing 3. Digital Signal Processor
DSS	Data Storage Set
DSU	1. Data Storage Unit 2. Digital Switch Unit 3. Dynamic Sensor Unit
DSUR	Data Storage Unit Receptacle
DTC	Data Transfer Cartridge
DTD	Data Transfer Device
DTE	Data Transfer Equipment
DTED	Digital Terrain Elevation Data
DTF	Digital Tape Format
DTIU	Data Transfer Interface Unit
DTM	Data Transfer Module
DTM/D	Digital Terrain Management and Display
DTN	Data Transfer Network
DTS	1. Data Terminal Set 2. Data Transfer System 3. Digital Terrain System
DTU	1. Data Transfer Unit 2. Display Terminal Unit
DU	Display Unit
DUCK	DCMS Universal Configuration Key
DV/A	Doppler Velocimeter/Altimeter
DVI	Direct Voice Input
DVOF	Digital Vertical Obstruction File
DVRS	Display Video Recording System
DVS	Doppler Velocity Sensor
EADI	Electronic Attitude Director Indicator
EAMED	Eventide Airborne Multipurpose Electronic Display
EAP	1. Emergency Audio Panel 2. Experimental Aircraft Programme
EAR	Electronically Agile Radar
EAROM	Electrically Alterable Read-Only Memory
EARS	ECI Airborne Relay System
EASU	Engine Analyser and Synchrophase Unit
EATCHIP	European Air Traffic Control Harmonisation and Integration Programme
EATMS	European Air Traffic Management System
EAU	Engine Analyser Unit
ECA	Electronic Control Amplifier
ECAC	European Civil Aviation Conference
ECAM	Electronic Centralised Aircraft Monitor
ECB	Electronic Control Box
ECCM	Electronic Counter-CounterMeasures
ECDU	Enhanced Control and Display Unit
ECIPS	Electronic Combat Integrated Pylon System
ECL	Emitter Coupled Logic
ECM	Electronic CounterMeasures
ECNI	Enhanced Communications, Navigation and Identification
ECOP	Electronic Co-Pilot
ECP	Engineering Change Proposal
ECR	Electronic Combat and Reconnaissance
ECS	Environmental Control System
ECU	1. Electronics Control Unit 2. Environmental Control Unit 3. Exercise Control Unit
EDAU	Extended Data Acquisition Unit
EDC	Error Detection and Correction
EDIU	Engine Data Interface Unit
EDM	Engine Data Multiplexer
EDTS	Expanded Data Transfer System
EDU	1. Electronic Display Unit 2. Engine Diagnostic Unit
EEC	Electronic Engine Controls
EEMS	Electrostatic Engine Monitoring System
EEPROM	Electrically Erasable Programmable Read-Only Memory
EEZ	Economic Exclusion Zone
EFCS	Electronic Flight Control System
EFCU	Electrical Flight Control Unit
EFDARS	Expansible Flight Data Acquisition and Recording System
EFIS	Electronic Flight Instrumentation System
EFMCS	Enhanced Flight Management Computer System
EGAC	Enhanced General Avionics Computer
EGNOS	European Geostationary Navigation Overlay System
EGT	Exhaust Gas Temperature
EHF	Extra High Frequency
EHSI	Electronic Horizontal Situation Indicator
EIA	Electronic Industries Association (US)
EICAS	Engine Indication and Crew Alerting System
EID	Emitter Identification
EIS	Electronic Instrument System
EIU	1. Electronic Interface Unit 2. Engine Interface Unit
EL	ElectroLuminescent
ELAC	Elevator and Aileron Computer
ELB	Emergency Locator Beacon

For more information see pages 602, 603 + 604.

ELF	Electronic Location Finder
ELINT	Electronic Intelligence
ELIOS	ELINT Identification and Operating System
ELIPS	Electronic Integrated Protection Shield
ELMS	Electrical Load Management System
ELS	1. Electronic Library System 2. Emitter Location System
ELT	Emergency Locator Transmitter
E-MAGR	Enhanced-Miniaturised Airborne GPS Receiver
EM	ElectroMagnetic
EMC	ElectroMagnetic Compatibility
EMD	Engineering Model Derivative
EMDU	Enhanced Main Display Unit
EMI	ElectroMagnetic Interference
EMMU	Engine Monitor Multiplexer Unit
EMP	ElectroMagnetic Pulse
EMS	Entry Monitor System
EMSC	Engine Monitoring System Computer
EMTI	Enhanced Moving Target Indicator
EMU	Engine Monitoring Unit
EMux	Electronic Multiplexing
EO	Electro-Optic
EOB	Electronic Order of Battle
EOCM	Electro-Optical CounterMeasures
EOIVS	Electro-Optical/Infrared Viewing System
EOSS	Electro-Optic Sensor System
EOVS	Electro-Optical Viewing System
EPAD	Electrically Powered Actuator Design
EPI	Engine Performance Indicator
EPLD	Electronic Programmable Logic Device
EPMS	Electrical Power Management System
EPR	Engine Pressure Ratio
EPROM	Erasable Programmable Read-Only Memory
EPRT	Engine Pressure Ratio Transmitter
EQAR	Expanded Quick Access Recorder
ERAPS	Expandable Reliable Acoustic Path Sonobuoy
EROM	Erasable Read-Only Memory
ERP	Effective Radiated Power
ERS	Electronic Resource System
ERWE	Enhanced Radar Warning Equipment
E/SA	Embedded and Special Application
ESC	Engine Supervisory Control
ESG	Electrostatically Suspended Gyro
ESM	Electronic Support Measures
ESP	1. Expandable System Programmer 2. Expendable Signal Processor
ESS	Exercise Support System
ESU	Electronic Storage Unit
ETA	Estimated Time of Arrival
ETE	Estimated Time En route
ETIPS	ElectroThermal Ice Protection System
ETMP	Enhanced Terrain Masked Penetration
ETPU	Engine Transient Pressure Unit
EU	Electronics Unit
EUROCAE	European Organisation for Civil Aviation Electronics
EVM	Engine Vibration Monitor
EVS	Enhanced Vision Sensor
EW	Electronic Warfare
EWAAS	End-state WAAS (wide area augmentation system)
EWMS	Electronic Warfare Management System
EWMU	EW Management Unit
EWPI	Electronic Warfare Prime Indicator
EXCAP	Expanded Capability
FAA	Federal Aviation Administration (US)
FAADC^{3}I	Forward Area Air Defence Command, Control and Intelligence
FAC	1. Flight Augmentation Computer 2. Forward Air Controller
FACS	Fully Automatic Compensation System
FACTS	FLIR Augmented Cobra TOW Sight
FADEC	Full Authority Digital Engine Control
FAF	Final Approach Fix
FAFC	Full Authority Fuel Control
FAM	Final Approach Mode
FAMIS	Full Aircraft Management/Inertial System
FANS	Future Air Navigation System
FAP	Final Approach
FBL	Fly-By-Light
FBS	Fly-By-Speech
FBW	Fly-By-Wire
FCC	Flight Control Computer
FCDC	Flight Control Data Concentrator
FCMC	Flight Control and Monitoring Computer
FC/NP	Fire Control/Navigation Panel
FCPC	Flight Control Primary Computer
FCS	Flight Control System
FCSC	Flight Control Secondary Computer
FCU	Flight Control Unit
FDAMS	Flight Data Acquisition Management System
FDAU	Flight Data Acquisition Unit
FDEP	Flight Data Entry Panel
FDIU	Flight Data Interface Unit
FDM	Frequency Division Multiplex
FDMU	Flight Data Management Unit
FDP	Flight Data Panel
FDR	Flight Data Recorder
FDR/FA	Flight Data Recorder/Fault Analyser
FDS	Flight Director System
FDT	Flight-Deck Terminals
FDU	Flight Data Unit
FET	Field Effect Transistor
FEWSG	Fleet Electronic Warfare Support Group
FFSP	Full Function Signal Processor
FFT	Fast Fourier Transform
FGCP	Flight Guidance Control Panel
FINAS	Ferranti Inertial Nav/Attack System
FIRAMS	Flight Incident Recorder and Aircraft Monitoring System
FIRMU	Flight Incident Recorder Memory Unit
FIS	Flight Information Service
FL	Foot-Lambert
FLAG	Four-mode Laser Gyro
FLAGSHIP	Four-mode Laser Gyro Software Hardware Implemented Partitioning
FLASH	Folding Light Acoustic Systems for Helicopters
FLIR	Forward-Looking InfraRed
Flops	Floating point operations per second
FM	Frequency Modulated
FMA	Flight Mode Annunciator
FMC	1. Flight Management Computer 2. Forward Motion Compensation
FMCS	1. Fatigue Monitoring and Computing System 2. Flight Management Computer System
FMCW	Frequency Modulated Continuous Wave
FMGC	Flight Management and Guidance Computer
FMGS	Flight Management and Guidance System
FMICW	Frequency Modulated Interrupted Continuous Wave
FMS	Flight Management System
FMU	Flight Management Unit
FOAEW	Future Organic Airborne Early Warning
FOC	Full Operational Capability
FOV	Field Of View
FPD	Flat-Panel Display
FPMU	Fuel Pump Monitoring Unit
FQIS	Fuel Quantity Indication System
FRPA	Fixed Reception Pattern Antenna
FRP	Federal (US) Radionavigation Plan
FRS	Fighter/Reconnaissance/Strike
FSA/CAS	Fuel Savings Advisory and Cockpit Avionics System
FSAS	Fuel Saving Advisory System
FSC	Fuel Savings Computer
FSD	Full-Scale Deflection
FSK	Frequency Shift Keying
FSRS	1. Flight Safety Recording System 2. Frequency Selective Receiver System
FSS	Flight Service Station
ft	feet
FT	Fault Tolerant
FT-ADIRS	Fault Tolerant Air Data Inertial Reference System
FT-ADIRU	Fault Tolerant Air Data Inertial Reference Unit
FTC	Fast Time Constant
FTI	Fixed Target Indication
FTIT	Fan Turbine Inlet Temperature
FTRG	Fleet Tactical Readiness Group
G	Giga = 1,000,000,000
GaAs	Gallium Arsenide
GAS	Global positioning Adaptive Antenna System
GATM	Global Air Traffic Management
GATS	GPS-Aided Targeting System
GBIB	Ground-Based Integrity Broadcast
GCA	Ground Collision Avoidance
GCAS	Ground Collision Avoidance System
GCR	Ground Clutter Reduction
GDE	Graphics Differential Engine
GDP	Graphics Drawing Processor
GEM	GPS Embedded Module
GEMMA	GEC MPA Mission Avionics
GEN-X	Generic Expendable
GES	Ground Earth Station
GGP	GPS Guidance Package
GHz	Giga Hertz
GIC	GPS Integrity Channel
GIC	GPS/WAAS integrity channels (GPS global positioning system) (WAAS wide area augmentation system)
GIG	GPS integration guidelines
GIT	General Interface Terminal
GLINT	Gated Laser Illuminator for Night Television
GLONASS	Global Orbital Navigation Satellite System
GMR	Ground Mapping Radar
GMT	Greenwich Mean Time
GMTI	Ground Moving Target Indication
GNC	General Navigation Computer
GNLS	GPS Navigation and Landing System
GNSS	Global Navigation Satellite System
GNSSU	Global Navigation Satellite Sensor Unit
GP&C	Global Positioning and Communication
GPCDU	General Purpose Control Display Unit
GPIN	Global Positioning laser Inertial Navigation
GPIRS	Global Positioning/Inertial Reference System
GPIRU	Global Positioning Inertial Reference Unit
GPS	Global Positioning System
GPSSU	Global Positioning System Sensor Unit
GPVI	Graphics Processor Video Interface
GPWS	Ground Proximity Warning System
GRE	Ground Readout Equipment
GS	GroundSpeed
GSDI	GroundSpeed and Drift Indicator
GSE	Ground Support Equipment
GSM	GPS Sensor Module
GTAR	GEC Thomson Airborne Radar
GTRE	Gas Turbine Research Establishment (UK)
h	hour(s)
HAC	Anti-tank variant of Tiger helicopter
HACLCS	Harpoon Airborne Command, Launch and Control System
HAD	Hybrid Analogue Digital
HADAS	Helmet Airborne Display And Sight
HADS	Helicopter Air Data System
HAINS	High-Accuracy Inertial Navigation System
HAP	Escort variant of Tiger helicopter
HAPS	Helicopter Acoustic Processing System
HARM	High-speed Anti-Radiation Missile
HBR	High Bit Rate
HDD	Head-Down Display
HDDR	Head-Down Display Recorder
HDG	Heading
HDS	Hard Disk Subsystem
HDTV	High-Definition TV
HER	Harsh Environment Recorder
HERALD	Helicopter Equipment for Radar And Laser Detection
HEU	HUD Electronics Unit
HF	High Frequency
HFAC	Helicopter Flight Advisory Computer
HFDU	High-Frequency Data Unit
HgCdTe	Mercury Cadmium Telluride
HGS	Holographic Guidance System
HHTI	Hand-Held Thermal Imager
HHUD	Holographic Head-Up Display
HIADC	High-Integration Air Data Computer
HIBIRD	Helicopter Identification By InfraRed Detection
HIBRAD	Helicopter Identification By Radar Detection
HICU	HIPSS Interface Control Unit
HIDAS	Helicopter Integrated Defensive Aids System
HIDEC	Highly Integrated Digital Engine Control
HIDSS	Helmet Integrated Display Sight System
HIPAS	High-Performance Active Sonar
HIPSS	Helicopter Integrated Power and Switching System
HIRES	High Resolution
HIRF	High-Intensity Radiated Field
HIRNS	Helicopter InfraRed Navigation System
HIRS	Helicopter InfraRed System
HISAR	Hughes Integrated Surveillance And Reconnaissance system
HIT	Hughes Improved Terminal
HITMORE	Helicopter Installed Television Monitor Recorder
HLD	Head Level Display
HLWE	Helicopter Laser Warning Equipment
HMCU	Hydraulic Monitoring Computer Unit
HMD	Helmet-Mounted Display
HMDD	Helmet-Mounted Display Device
HMFU	HydroMechanical Fuel Unit
HMU	HydroMechanical Unit
HNVS	Helicopter Night Vision System
HOCAS	Hands-On Collective And Stick
HOFIN	Hostile Fire Indicator
HOPS	Helmet Optical Position Sensor
HOS	Helitow Observation System

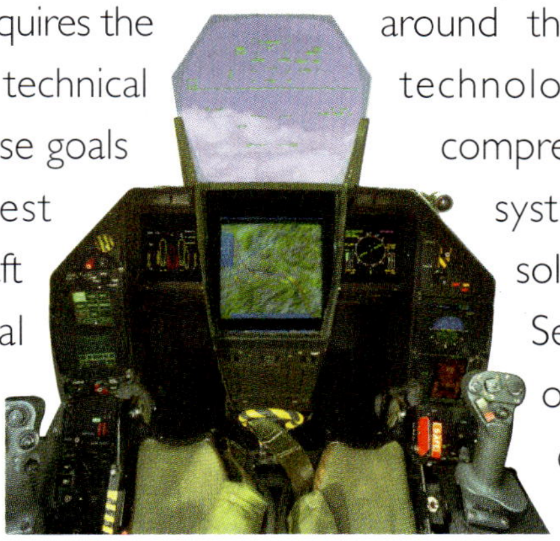

GLOSSARY

HOTAS Hands-On Throttle And Stick
HOWLS Hostile Weapons Location System
HP High Pressure
HPA High-Power Amplifier
HPAG High-Power Amplifier Group
HRP Headset Receptacle Panel
HSDB High-Speed DataBus
HSI Horizontal Situation Indicator
HSIC High-Speed Integrated Circuit
HSSL Helicopter Self-Screening Launcher
HTC Hover Trim Control
HTNS Helicopter Tactical Navigation System
HTTB High-Technology TestBed
HUD Head-Up Display
HUDC Head-Up Display Computer
HUDWAC Head-Up Display and Weapon Aiming Computer
HUMC Health and Usage Monitoring Computer
HUMS 1. Health and Usage Monitoring System
2. Health and Usage Monitoring and Sensing
HVPSU High-Voltage Power Supply Unit
Hz Hertz

IAC Integrated Avionics Computer
IAD Integrated Antenna Detector
IAHFR Improved Airborne High-Frequency Radio
IAHFR/NOE Improved Airborne High-Frequency Radio Nap-Of-the-Earth
IAM Initial Approach Mode
IAMS Integrated Armament Management System
IAPS Integrated Avionics Processing System
IAS Indicated Air Speed
IB Interconnecting Box
I/C Interface and Control
ICAAS Integrated Controls and Avionics for Air Superiority
ICAO International Civil Air traffic Organisation
ICAP Increased Capability
ICCP Integrated Communications Control Panel
ICDU 1. Integrated Control and Display Unit
2. Intelligent Control Display Unit
ICE Improved Combat Efficiency
ICMS Integrated CounterMeasures System
ICNIA Integrated Communications Navigation Identification Avionics
ICNIS Integrated Communications Navigation Identification Set
ICP Integrated Control Panel
ICS 1. InterCommunications System (or set)
2. Internal Countermeasures Set
ICSM Integrated Conventional Stores Management
ICSM/GPS Integrated Conventional Stores Management/Global Positioning System
ICU 1. Interface Computer Unit
2. Interface Converter Unit
3. Interstation Control Unit
ICW Interrupted Continuous Wave
ID Identification
IDACS Integrated Digital Audio Control System
IDAP Integrated Defense Avionics Program
IDAS Integrated Design Automation System
IDECM Integrated Defensive Electronic CounterMeasures
IDF Instantaneous Direction-Finding
IDM Inductive Debris Monitor
IDP Imagery Display Processor
IDS 1. Infrared Detection Set
2. InterDiction Strike
IEEE Institute of Electrical and Electronic Engineers (US)
IEU Interface Electronics Unit
IEW Integrated Electronic Warfare
IEWCS Intelligence and Electronic Warfare Common Sensor
IF Intermediate Frequency
IFF Identification Friend or Foe
IFFCP Identification Friend or Foe Control Panel
IFM Instantaneous Frequency Measurement
IFM/SHR IFM SuperHeterodyne Receiver
IFMU Integrated Flight Management Unit
IFOG Interferometric Fibre Optic Gyro
IFR Instrument Flight Rules
IGAC Israeli General Avionics Computer
IHADSS Integrated Helmet And Display Sighting System
IHDTV Intensified High-Definition TeleVision
IHEWS Integrated Helicopter Electronic Warfare Suite
IHU Integrated Helmet Unit
IHUMS Integrated Health and Usage Monitoring System
IIDS Integrated Instrumentation Display System
IIS Infrared Imaging System
IJMS Interim JTIDS Message System
ILS Instrument Landing System
IMA Integrated Modular Avionics
IMC 1. Instrument Meteorological Conditions
2. Image Motion Compensation
IMSS Integrated MultiSensor System
IMU Inertial Measurement Unit
in inch(es)
INACP Integrated Navigation Aids Control Panel
INEWS Integrated Electronic Warfare System
INS Inertial Navigation System
INU Inertial Navigation Unit
I/O Input/Output
IOC 1. Initial Operational Capability
2. Input Output Computer
I/P Identification Position
IP Intermediate Pressure
IPADS Improved Processing And Display System
IPCS Intelligent Power Control System
IPEC Inflight Passenger Entertainment and Communications
IPT Intelligent Power Terminal
IPU Interface Processor Unit
IR InfraRed
IRCCD InfraRed Charge Coupled Device
IRCM InfraRed CounterMeasures
IRFIS Inertial Referenced Flight Inspection System
IRIS 1. InfraRed Imaging Subsystem
2. Integrated Radar Imaging System
IRLS InfraRed LineScanner
IRMS Integrated Radio Management System
IRP Interphone Receptacle Panel
IRS Inertial Reference System
IRST(S) InfraRed Search and Track (System)
IRU Inertial Reference Unit
IR/UV InfraRed/Ultraviolet
IRV InfraRed Vision
IRVAT InfraRed Video Automatic Tracking
ISA Instruction Set Architecture
ISAHRS Improved Standard Attitude and Heading Reference System
ISAR Inverse Synthetic Aperture Radar
ISB Independent Sideband
ISBA Inertial Sensor Based Avionics
ISC Intercommunications Set Control
ISDS IRCM Self-Defence System
ISIS Integrated Strike and Interception System
ISLS Interrogation Side-Lobe Suppression
ISO International Standardisation Organisation
ISS Integrated Sensor System
ISU Intercommunications Set control Unit
ITAR Integrated Terrain Access and Retrieval system
I-TOW Improved Tube-launched Optically tracked Wire-guided (missile)
IUMS Integrated Utilities Management System
IVMMS Integrated Vehicle Mission Management System
IVSC Integrated Vehicle Subsystem Controls
IVSI Instantaneous Vertical Speed Indicator
IWAAS Initial WAAS (wide area augmentation system)

JAR Joint Airworthiness Requirement
JASS Joint Airborne SIGINT System
JAST Joint Advanced Strike Technology
JDAM Joint Direct Attack Munition
JIAWG Joint Integrated Avionics Working Group (US)
Joint STARS Joint Surveillance and Target Attack Radar System (also JSTARS)
JPALS Joint Precision Approach Landing System
JPATS Joint Primary Aircraft Training System
JPO Joint Program Office (US)
JSAF LBSS Joint SIGINT Avionics Family Low-Band SubSystem
JTIDS Joint Tactical Information Distribution System
JTW Joint Targeting Workstation

k 1,000
K Kelvin
KAPS Kollsman Auto-schedule Pressurisation System
kbit kilobit
KBU KeyBoard Unit
kbyte kilobyte
kg kilogram
KFD Key Fill Device
kHz kilohertz
kips Thousand instructions per second
kops Thousand operations per second
kt Nautical miles per hour (knot)
KTD Key Transfer Device

L1 GPS carrier frequency (1227.6 MHz)
L2 GPS carrier frequency (1575.42 MHz)
LAAP Low-Altitude AutoPilot
LAAS Local Area Augmentation System
LAASH LITEF Analytical Air data System for Helicopters
LAAT Laser Augmented Airborne TOW
LADGNSS Local Area Differential Global Navigation Satellite System
LADGPS Local Area Differential Global Positioning System
LAEO Low-Altitude Electro-Optical
LAHRS Kearfott Low-cost Altitude Heading Reference System
LAIRS 1. Light Aircraft Reconnaissance System
2. Loral Advanced Imaging Radar System
LAMPS Light Airborne MultiPurpose System
LAN Local Area Network
LANA Low-Altitude Night Attack
LANE Low-Altitude Navigation Equipment
LANTIRN Low-Altitude Navigation and Targeting InfraRed for Night
LASS Low-Altitude Surveillance System
LASTE Low-Altitude Safety and Target Enhancement
LCC Leadless Ceramic Chip-carrier
LCD Liquid Crystal Display
LCF Low Cycle Fatigue
LCINS Low-Cost Inertial Navigation System
LCLU Landing Control Logic Unit
LCM Lance-Cartouches Modulaire
LCOS(S) Lead Computing Optical Sight (System)
LCTAR Le Centre Thomson d'Applications Radar
LCWDS Low-Cost Weapon Delivery System
LDT/SCAM Laser Detector and Tracker/Strike Camera
LDU Launcher Decoder Unit
LEA Leurre Electromagnetique Actif
LECOS Light Electronic Control System
LED Light Emitting Diode
LED-RHA Light Emitting Diode-Recording Head Assembly
LESM Lightweight Electronic Support Measures
LF Low Frequency
LGCIU Landing Gear Computer and Interface Unit
LH Light Helicopter
LHN LITEF Helicopter Navigation
LHR LITEF Helicopter Reference
LICAS Low-Intensity Conflict Aircraft System
LIDAR Light Detection And Ranging
LIMAR Laser Imaging And Ranging
LINS Laser Inertial Navigation System
LIP Limited Installation Programme
LISCA Leurre Infrarouge a Signature et Cinematique Adaptee
LITDL Link 16 Interoperable Tactical Data Link
LLLTV Low-Light Level Television
LLTV Low-Light Television
LNA Low-Noise Amplifier
LO Local Oscillator
LOCUS Laser Obstacle Cable Unmasking System
LOFAR Low frequency Omnidirectional acoustic Frequency Analysis and Recording
LORAN Long-Range Aid to Navigation
LORES Low Resolution
LOROP Long-Range Oblique Photography
LOS Line Of Sight
LOSI Line Of Sight Indicator
LP Low Pressure
LPC Linear Predictive Coding
LPCBA Low-Pressure Compressor Bleed Actuator
LPD Low Probability of Detection
LPHUD Low-Profile Head-Up Display
LPI Low Probability of Interception
LQA Link Quality Analysis
LRCU Landing Roll-out Control Unit
LRM Line-Replaceable Module
LRMTS Laser Ranger and Marked Target Seeker
LRU Line-Replaceable Unit

LSB	Lower SideBand
LSI	Large Scale Integration
LST	Laser Spot Tracker
LTD/R	Laser Target Designator/Ranger
LTR	Loop Transfer Recovery
LVDT	Linear Variable Differential Transformer
LWA	Laser Warning Analyser
LWF	LightWeight Fighter
LWR	Laser Warning Receiver
m	metre
M	1,000,000 or mega
M-ADS	Modified Automatic Dependent Surveillance
MACC	Multi-Application Control Computer
MACS	Multiple Application Control System
MAD	Magnetic Anomaly Detector
MADAR	1. Maintenance Analysis, Detection And Recording 2. Malfunction Detection, Analysis and Recording
MADC	1. Micro Air Data Computer 2. Miniature Air Data Computer
MADGE	Microwave Aircraft Digital Guidance Equipment
MAESTRO	Modular Avionics Enhancement System Targeted for Retrofit Operations
MAG	Micromachined Accelerometer Gyro
MAGR	Miniature Airborne GPS Receiver
MAHRU	Microflex Attitude and Heading Reference Unit
MAP	Missed Approach Point
MAP	Modular Airborne Processor
MARA	Modular Architecture for Real-time Applications
MARE	Miniature Analogue Recording Electronics
MARS	1. Modular Airborne Recording System 2. Multi-Application Recorder/ Reproducer System
MASTER	Military Aircraft Satcoms Terminal
MASPA	Minimum Aviation System Performance Standards
MATSS	Mobile Aerostat Tracking and Surveillance System
MAW	1. Missile Approach Warner 2. Mission Adaptive Wing
MAWS	Missile Approach Warning System
mb	millibar
MBAT	MultiBeam Array Transmitter
Mbit	Megabit
Mbyte	Megabyte
MC	Mission Computer
MCCP	Main Communications Control Panel
MCDU	Multifunction Control and Display Unit
MCE	Modular Control Equipment
Mcops	Million complex operations per second
MCT	1. Mercury Cadmium Telluride 2. Metal oxide semiconductor Controlled Thyristor
MCU	1. Management Control Unit 2. Master Control Unit 3. Missile Control Unit 4. Modular Concept Unit 5. Multifunction Control Unit
MCW	Modulated Continuous Wave
MDF	Mission Data File
MDG	Map Display Generator
MD(G)T	Mission Data (Ground) Terminal
MDI	Multifunction Display Indicator
MDL	Mission Data Loader
MDP	1. Maintenance Data Panel 2. Modular Display Processor
MDRI	Multipurpose Display Repeater Indicator
MDT	Mission Data Terminal
MDTC/P	Mega Data Transfer Cartridge with Processor
MDTS	Mission Data Transfer System
MEA	Minimum En route Altitude
MEC	Modular Electronics Concept
MEECN	Minimum Essential Emergency Communications Network
MERLIN	Modular Ejection-Rated Low-profile Imaging for Night
MESA	Minimum Emergency Safe Altitude
METR	Multiple Emitter Targeting Receiver
MFCD	MultiFunction Colour Display
MFD	MultiFunction Display
MFD(S)	MultiFunction Display (System)
MFHD	MultiFunction Head-down Display
MFMS	Military Flight Management System
MGR	Miniature GPS Receiver
MHDD	Multifunction Head-Down Display
MHz	megahertz
MIAMI	Microwave Ice Accretion Measurement Instrument
MICNS	Modular Integrated Communications and Navigation System
Micro-AIDS	Micro-Aircraft Integrated Data System
MIDA	Message Interchange Distributed Application
MIDS	Multifunction Information Distribution System
MIGITS	Miniature Integrated GPS/INS Tactical System
MIL-SPEC	Military Specification
MILSTAR	Military Strategic and Tactical Relay
MIL-STD	Military Standard
MIMU	Multisensor Inertial Measurement Unit
min	Minute(s)
MIPS	Million Instructions Per Second
MIRLS	Miniature InfraRed Linescan System
MIRTS	Modular InfraRed Transmitting System
MLI	Mid-Life Improvement
MLPRF	Modular Low-Power Radio Frequency
MLS	Microwave Landing System
MLU	1. Mid-Life Update 2. Monitor and Logic Unit
MLV	Memory Loader and Verifier
mm	millimetre(s)
MM	Mission Manager
MMC	Mission Management Computer
MMIC	Monolithic Microwave Integrated Circuit
MMLSA	Military Microwave Landing System Avionics
MMMS	MultiMission Management System
Mmo	Maximum operating Mach number
MMP	Maintenance Monitoring Panel
MMR	MultiMode Receiver
MMRS	MultiMission Radar System
MMS	Mast-Mounted Sight
MMSS/R	Mobile Mass Storage System Model R
MMW(R)	MilliMetric Wave (Radar)
MNOS	Metal Nitride Oxide Silicon
MNPS	Minimum Navigation Performance Standards
MNS	Modular Navigation System
MOA	Minimum Operating Altitude
MODAR	Modular Aviation Radar
MODAS	Modular Data Acquisition System
MODIR	Modulated InfraRed jammer
MONOHUD	Monocular Head-Up Display
Mops	Million operations per second
MOPS	Minimum Operating Performance Standard
MOS	Metal Oxide Silicon
MOSFET	Metal Oxide Silicon Field Effect Transistor
MOSP	Multimission Optronic Stabilised Payload
MP	Mission Planner
MPA	Maritime Patrol Aircraft
MPCD	MultiPurpose Colour Display
MPD	MultiPurpose Display
MPHD	MultiPurpose Head-Down
MPM	MultiPurpose Modem
MPPU	MultiPurpose Processor Unit
MPS	Mission Planning Subsystem
MQLF	Mobile Quick-Look Facility
MR	Maritime Reconnaissance
MRAALS	Marine Remote Area Approach Landing System
MRAAM	Medium-Range Air-to-Air Missile
mrad	milliradian
MRLG	Monolithic Ring Laser Gyro
MRT	1. Miniature Receive Terminal 2. MultiRole Turret
MRTI	MultiRole Thermal Imager
MRTU	Multiple Remote Terminal Unit
MRU	1. Maintenance Recorder Unit 2. Mobile Reporting Unit
ms	millisecond
m/s	metres per second
MSA	Minimum Safe Altitude
MSAR	Miniature Synthetic Aperture Radar
MSCADC	Miniature Standard Central Air Data Computer
MSD	1. Map Storage Display 2. Mass Storage Device
MSF	Mission Support Facility
MSIP	1. Multi(national) Staged Improvement Programme 2. MultiStaged Improvement Programme
MSIS	Multisensor Stabilised Integrated System
MSP	Mission System Processor
MSPS	Modular Self-Protection System
MSR	1. Marconi Secure Radio 2. Modular Strain Recorder 3. Modular Survivable Radar
MSS	Maritime Surveillance System
MSTAR	Moving and Stationary Target Acquisition and Recognition
MSU	1. Maintenance Station Unit 2. Mass Storage Unit 3. Mode Selector Unit
MSWS	MultiSensor Warning System
MTAS	Millimetric Target Acquisition System
MTBF	Mean Time Between Failures
MTBR	Mean Time Between Repairs
MTBUR	Mean Time Between Unscheduled Repairs
MTI	Moving Target Indicator
MTIS	Modular Thermal Imaging Sight
MTL	Magnetic Tape Loader
MTTR	Mean Time To Repair
MTU	Magnetic Tape Unit
MUST	Multimission UHF Satcom Transceiver
Mux	Multiplexer
mV	millivolt
MVP	Military VME Processor
mW	milliwatt
MWR	Microwave Radiometer
NAS	Navigation Attack System
NASA	National Aeronautics and Space Administration (US)
NASH	Navigation and Attack System for Helicopters
NAT	National Air Traffic
NATO	North Atlantic Treaty Organisation
nav/com	navigation and communication
NavHARS	Navigation, Heading and Attitude Reference System
NavWASS	Navigation and Weapon Aiming SubSystem
NBC	Nuclear, Biological and Chemical
NCS	Network Control Station
NCU	Navigation Computer Unit
ND	Navigation Display
NDB	Non-Directional Beacon
Nd:YAG	Neodymium/Yttrium Aluminium Garnet
NEACP	National Emergency Airborne Command Post
Ng	Gas Generator RPM
Ni/Cd	Nickel/Cadmium
NIDJAM	Navigation/Identification Deception Jammer
NILE	NATO Improved Link 11
NIS	NATO Identification System
NITE-OP	Night Imaging Through Electro-Optics
NIU	Navigation Interface Unit
n mile(s)	Nautical mile(s)
NMOS	Negative Metal Oxide Semiconductor
NMS	Navigation Management System
NMU	Navigation Management Unit
NOCUS	North Continental US
NOE	Nap-Of-the-Earth
NOR	Logic circuit usable as either AND/OR
NOTAR	No Tail Rotor
NOVRAM	Non-Volatile Random Access Memory
NPR	No Power Recovery
NPRM	Notice of Proposed Rule Making
NPU	Navigation Processor Unit
NQIS	Navigation Quality Inertial Sensor
NRT	Near Real-Time
ns	nanoseconds
NSA	National Security Agency (US)
NSIU	Navigation Switching Interface Unit
NSP	Night Surveillance Payload
NSSL	National Severe Storms Laboratory (US)
NTDS	Naval Tactical Data System
NTS	Night Targeting System
NV	1. Night Vision 2. Non-Volatile
NVG	Night Vision Goggles
NVG/HUD	Night Vision Goggles/Head-Up Display
NV/HUD	Night Vision/Head-Up Display
NVIS	Night Vision Imaging Systems
NVM	Non-Volatile Memory
NWDS	Navigation and Weapon Delivery System
OADS	Omnidirectional Air Data System
OAS	Offensive Avionics System
OASYS	Obstacle Avoidance System
OAT	Outside Air Temperature
OBEWS	OnBoard Electronic Warfare Simulator
OBI	Omni-Bearing Indicator
OBTEX	OnBoard Targeting EXperiments
OCU	Optical Control Unit
ODIN	Operational Data Interface
ODS	1. Operational Debrief Station 2. Optical Disk System
ODU	Optical Display Unit
OEM	Original Equipment Manufacturer
OFP	Operational Flight Programme
OHU	Optical Head Unit
OMMS	Oxygen Mask-Mounted Sight
OMS	1. Onboard Maintenance System 2. Operational Management System

OMT	Onboard Maintenance Terminal
OOOI	Out/Off/On/In
OQAR	Optical Quick Access Recorder
OQPSK	Offset Quadrature Phase Shift Keyed
ORS	Offensive Radar System
OSA	Operational Support Aircraft
OSC	Optical Sensor Converter
OSF	Optronique Secteur Frontal
OSU	Omega/VLF Sensor Unit
OTAR	Over The Air Re-keying
OTCIXS	Officer in Tactical Command Information eXchange Subsystem
OTH(T)	Over The Horizon (Target)
OTIS	Optronic Tracking and Identification System
OTPI	On-Top Position Indicator
OWL	Obstacle Warning Ladar
OWL/D	Optical Warning, Location and Detection
OWS	Obstacle Warning System
P	Precision
P^{3}I	PrePlanned Product Improvement
PA	1. Pilot's Associate 2. Public Address
PAC	1. Precision Attitude Control 2. Public Address set Control
PACA/IAGSS	Precision Attitude Control Augmentation/Improved Air-to-Ground Sight System
PACIS	Pilot Aid and Close-In Surveillance
PAL	Phase Alternation Line
PAM	Pulse Amplitude Modulation
PAMIR	Passive Airborne Modular InfraRed
PAR	Power Analyser and Recorder
PAS	Performance Advisory System
PAT	Pilot Access Terminal
PAWS	Passive Airborne Warning System
PBDI	Position, Bearing and Distance Indicator
PBIL	Projected Bomb Impact Line
PC	1. Personal Computer 2. Printed Circuit 3. Pulse Compression
PCB	Printed Circuit Board
PCM	1. Power Converter Module 2. Pulse Code Modulation
PCSB	Pulse Coded Scanning Beam
PCU	Pilot's Control Unit
PDC	Programme Development Card
PDC/PMM	Programme Development Card/ Performance Monitor Module
PDES	Pulse Doppler Elevation Scan
PDNES	Pulse Doppler Non-Elevation Scan
PDR	Programmable Digital Radio
PDS	Portable Data Store
PDU	Pilot's Display Unit
PE	Processing Element
PENETRATE	Pass no Enhanced Navigation with Terrain Referenced Avionics
PEO	Program Executive Officer
PEP	Peak Envelope Power
PERP	Peak Effective Radiated Power
PFCS	Primary Flight Control (or Computer) System
PFD	Primary Flight Display
PFD/ND	Primary Flight Display/Navigation Display
PFIS	Portable Flight Inspection System
PI	Process (or Programme) Instruction
PICC	Processor Interface Controller and Communication
PIDS	Pylon Integrated Dispenser Station
PILOT	Pod Integrated Localisation, Observation, Transmission
PIN	Positive-Intrinsic-Negative (semiconductor)
PIRATE	Passive InfraRed Airborne Track Equipment
PISA	Pilot's Infrared Sighting Ability
PITS	Passive Identification and Targeting System
PIU	1. Processor Interface Unit 2. Pylon Interface Unit
PLB	Personal Locator Beacon
PLGR	Precision Lightweight GPS Receiver
PLL	Phase-Locked Loop
PLRS	Precision Location Reporting System
PLS	Personnel Location System
PLSS	Precision Location Strike System
PLU	Programme Load Unit
PMA	Projected Map Assembly
PMAWS	Passive Missile Approach Warning System
PMF	Processeur Militaire Français
PMM	Performance Monitor Module
PNVS	Pilot Night Vision Sensor
PODS	Portable Data Store
POET	Primed Oscillator Expendable Transponder
PPI	Plan Position Indicator
PPM	1. PreProcessor Module 2. Programmable Processing Modules
pps	pulse per second
PPS	1. Photovoltaic Power System 2. Precise Positioning Service (GPS)
PRF	Pulse Recurrence Frequency
PRI	Pulse Repetition Interval
PRIDE	Pulse Recognition Interval De-interleaving
PROM	Programmable Read-Only Memory
PRU	Performance Reference Unit
PSC	Performance Seeking Control
PSCS	Photographic Sensor Control System
PSI	Pounds per Square Inch
PSK	Phase Shift Keyed
PSP	Programmable Signal Processor
PSU	1. Power Supply Unit 2. Power Switching Unit
PTC	Pack Temperature Controller
PTM	Pressure Transducer Module
PTMU	Pressure and Temperature Measurement Unit
PTP	Programmable Touch Panel
PTT	Press To Transmit
PULSE	Precision Up-shot Laser Steerable Equipment
PVD	ParaVisual Director
PVI	Pilot-Vehicle Interface
PVS	Pilot's Vision System
P/Y	GPS precision code
QAR	Quick Access Recorder
QDM	Magnetic heading to runway
QEC	Quadrantal Error Correction
QPSK	Quadrature Phase Shift Keyed
QR	Quadrant Receiver
RA	1. Relay Assembly 2. Resolution Advisory
RADC	Rome Air Development Center (US)
RAE	Royal Aerospace Establishment (UK)
RAF	Royal Air Force (UK)
RAIM	Receiver Autonomous Integrity Monitoring
RAM	Random Access Memory
RAMS	1. Racal Avionics Management System 2. Removable Auxiliary Memory Set
RAMU	Removable Auxiliary Memory Unit
RAN/RAWS	Royal Australian Navy Role Adaptable Weapons System
RAPPORT	Rapid Alert Programmed Power management Of Radar Targets
RA/TA	Resolution Advisory/Traffic Advisory
RA/VSI	Resolution Advisory/Vertical Speed Indicator
RC	Resistance Capacitance
RCS	Radar Cross-Section
RCU	1. Remote-Control Unit 2. Rudder Control Unit
RDAS	Reconnaissance Data Annotation Set
RDC	Remote Data Concentrator
RDI	Radar Doppler à Impulsions
RDM	Radar Doppler Multifunction
RDP	Range Doppler Profile
RDU	Remote Display Unit
REACT	Rain Echo Attenuation Compensation Technology
REU	Remote Electronics Unit
REWTS	Responsive Electronic Warfare Training System
RF	Radio Frequency
RFA	Royal Fleet Auxiliary
RFI	Radio Frequency Interference
RFP	Request For Proposals
R/FPU	Recorder/Film Processor Unit
RFTDL	RangeFinder Target Designator Laser
RFU	Radio Frequency Unit
RGB	Red, Green, Blue
RGS	Recovery Guidance System
RHA	Recording Head Assembly
RHWR	Radar Homing and Warning Receiver
RIMS	Replacement Inertial Measurement System
RIS	Reconnaissance Interface System
RISC	Reduced Instruction Set Computer
RLG	Ring Laser Gyro
RMI	1. Radio Magnetic Indicator 2. Remote Magnetic Indicator
RMPA	Replacement Maritime Patrol Aircraft
RMR	Remote Map Reader
RMS	1. Reconnaissance Management System 2. Root Mean Square
RNav	Area Navigation
RNP/ANP	Required Navgiation Performance/ Actual Navigation Performance
RNS	Radar Navigation System
RO	Range Only
RODS	Ruggedised Optical Data System
ROM	Read-Only Memory
ROMAG	Remote Map Generator
RPA	Rotocraft Pilot's Associate
RPG	Receiver Processor Group
rpm	revolutions per minute
RPU	Receiver Processor Unit
RPV	Remotely Piloted Vehicle
RRU	Remote Readout Unit
RSC	Remote Switching Control
RSIP	Radar System Improvement Programme
R/T	Receiver/Transmitter
RTA	Receiver-Transmitter-Antenna
RTCA	Radio Technical Commission for Aeronautics (US)
RTCA FFSC	Radio Technical Commission For Astronautics' Free Flight Select Committee (US)
RTD	Real-Time Display
RTIC	Real-Time Information into Cockpit
RTMM	Removable Transport Media Module
RTT	Radio Telemetry Theodolite
RTU	Radio Tuning Unit
RVDT	Rotating Variable Differential Transformer
RWR	Radar Warning Receiver
RWS	Range-While-Search
s	second(s)
SA	1. Selective Availability 2. Situation Awareness
SAAHS	Stability Augmentation and Attitude Hold System
SAARU	Secondary Attitude and Air data Reference Unit
SA/AS	Selective Availability/Anti-Spoofing
SAC	Strategic Air Command (US)
SACT	Signal Acquisition Conditioning Terminal
SADANG	Système Acoustique D'Atlantique Nouvelle Génération
SAFCS	Standard Automatic Flight Control System
SAFFIRE	Synthetic Aperture Fully Focused Imaging Radar Equipment
SAFIS	Semi-Automatic Flight Inspection System
SAHRS	Standard Attitude and Heading Reference System (Kearfott)
SAIRS	Standardised Advanced InfraRed Sensor
SAM	1. Situation Awareness Mode 2. Surface-to-Air Missile
SAMIR	Systeme d'Alérte Missile Infra Rouge
SAMSON	Special Avionics Mission Strap-On (pod)
SAR	1. Search And Rescue 2. Signal Acquisition Remote 3. Synthetic Aperture Radar
SARPs	Standard And Recommended Practices
SARSAT	Search And Rescue Satellite Aided Tracking
SASS	Small Aerostat Surveillance System
SAT	1. Situational Awareness Technology 2. Static Air Temperature
SATCOM	Satellite Communications
SATURN	Second generation of Anti-jam Tactical UHF Radios for NATO
SAU	1. Safety and Arming Unit 2. Signal Acquisition Unit
SAW	Surface Acoustic Wave
SAWS	Silent Attack Warning System
SBC	Single Board Computer
SC	Single Card
SCADC	Standard Central Air Data Computer
SCAR	Sistemi de Control de Armamento
SCAT	Speed Command of Attitude and Thrust
SCAT-1	Special Category 1
SCD	Signal Command Decoder
SC-DDS	Sensor Control-Data Display Set
SCDL	Surveillance and Control DataLink
SCI	Serial Communication Interface
SCNS	Self-Contained Navigation System
SCP	Single Card Processing
SCS	Sidewinder Control System
SCT	Single-Channel Transponder
SCU	1. Station Control Unit 2. Stores Control Unit 3. Supplemental Control Unit
SDC	1. Signal Data Computer 2. Signal Data Converter 3. Situation Display Console
SDCS	Satellite Data Communications System
SDCU	Smoke Detection Control Unit
SDP	Signal Data Processor
SDS	Satellite Data System
SDU	1. Satellite Data Unit 2. Smart Display Unit
SEAD	Suppression of Enemy Air Defences
SEAFAC	Systems Engineering Avionics Facility
SEC	Spoiler and Elevator Computer
SECU	Spoilers Electronic Control Unit

SELCAL	Selective Calling
SEM	Standard Electronic Module
SEMA	1. Smart ElectroMechanical Actuator
	2. Special Electronic Mission Aircraft
SEMC	Standard Electronic Memory Cartridge
SENAP	Signal emulation of aero-navigation and landing
SEOS	Stabilised Electro-Optical System
SEP	Spherical Error Probability
SETS	Severe Environment Tape platform
SEU	Sight Electronics Unit
SFCC	Slat Flap Control Computer
SFDP	Smart Flat-Panel Display
SFDR	Standard Flight Data Recorder
SGU	Signal Generator Unit
SHARP	1. Standard Hardware Acquisition and Reliability Programme
	2. Strapdown Heading and Attitude Reference Platform
SHF	Super High Frequency
SHIP	(FLAG) Software/Hardware Implemented Partitioning
SHR	SuperHeterodyne Receiver
SHUD	Smart Head-Up Display
SICAS	Secondary Surveillance Radar (SSR) Improvements And Collision Avoidance Systems
SID	1. Sensor Image Display
	2. Standard Instrument Departure
SIF	Selective Identification Facility
SIGINT	Signals Intelligence
SIMOP	Simultaneous Operation (of collocated RF sets)
SIMS	Signal Identification Mobile System
SINCGARS	Single-Channel Ground/Air Radio System
SIRFC	Suite of Integrated RF Countermeasures
SIS	Superhet IFM Subsystem
SIT	Silicon Intensified Target
SITREP	Situation Report
SIU	1. Sensor Interface Unit
	2. Sidewinder Interface Unit
SKE	Station-Keeping Equipment
SLAM	Standoff Land Attack Missile
SLAMMR	Side-Looking Airborne Modulated Multimission Radar
SLAR	Side-Looking Airborne Radar
SLIR	Side-Looking InfraRed
SLOS	Stabilised Long-range Observation System
SMA	Surface Movement Advisor
SMDU	Strapdown Magnetic Detector Unit
SMEU	Switchable Main Electronic Unit
SMP	Stores Management Processor
SMS	Stores Management System
SMT	Surface Mount Technology
SMU	Systems Management Unit
SMWP	Standby Master Warning Panel
S/N	Signal/Noise ratio
	Stress against number of alternating load cycles to failure
SOCUS	South Continental US
SOF	Special Operations Force
SOJ	StandOff Jamming
SOLL	Special Operations Low Level
SOTAS	StandOff Target Acquisition System
SPASYN	Space Synchro
SPEES	Système Pour l'Elevation de l'Endommagement Structural
SPEW	Small Platform Electronic Warfare (system)
SPEWS	Self-Protection Electronic Warfare System
SPHERIC	System for Protection of Helicopters by Radar and Infrared Countermeasures
SPI	Short Pulse Insertion
SPILS	Spin Prevention and Incidence Limiting System
SPN/GEANS	Standard Precision Navigator/Gimballed Electrostatic Aircraft Navigation System
SPO	System Program Office (US)
SPRITE	Signal Processing In The Element
SPS	Standard Positioning Service (GPS)
SPT	Signal Processing Tools
SPU	1. Signal Processing Unit
	2. Stores Power Unit
SRA	1. Shop Repair Assembly
	2. Shop Replaceable Assembly
SRAM	Static Random Access Memory
SRC	Surveillance Radar Computer
SRFCS	Self-Repairing Flight Control System
SRL	Systems Research Laboratories
SRPC	Strain Range Pair Counter
SRS	1. Sonobuoy Reference System
	2. Superhet Receiver Subsystem
	3. Survival Radio Set
SRU	1. Scanner Receiver Unit
	2. Shop Replaceable Unit
SS	System Status
SSB	Single SideBand
SSCVFDR	Solid-State Combined Voice and Flight Data Recorder
SSCVR	Solid-State Cockpit Voice Recorder
SSFDR	Solid-State Flight Data Recorder
SSICA	Stick Sensor and Interface Control Assembly
SSID	Solid-State Ice Detector
SSQAR	Solid-State Quick Access Recorder
SSR	Secondary Surveillance Radar
SSTI	Stabilised Steerable Thermal Imager
SSU	Sensor Surveying Unit
STANAG	Standard NATO Agreement
STAR	1. Signal Threat Analysis and Recognition
	2. Standard Terminal Arrival Route
STARS	1. Small Tethered Aerostat Relocatable System
	2. Stand-off Target Attack Radar System
START	Solid-state Angular Rate Transducer
STC	1. Sensitivity Time Control
	2. Supplemental Type Certificate (US)
	3. Swept Time Constant
STCA	Short-Term Conflict Alert
STEVI	Sperry Turbine Engine Vibration Indicator
STIRS	Strapdown Inertial Reference System
STIS	Stabilised Thermal Imaging System
STORMS	Stores Management System
STP	Status Test Panel
STR	Sonar Transmitter/Receiver
STRAP	Straight Through Repeater Antenna Performance
STS	Support and Test Station
STTE	Special-to-Type Test Equipment
STU	Satellite Terminal Unit
SUAWACS	Soviet Union Airborne Warning And Control System
SUM	Structural Usage Monitor
SWIP	Systems Weapons Improvement Programme
T^{2}A	Total Terrain Avionics
TA	Traffic Advisory
TACAMO	Take Charge And Move Out
TACAN	Tactical Air Navigation
TACCO	Tactical Co-ordinator
TACDS	Threat Adaptive Countermeasures Dispensing System
TACNAVMOD	Tactical Navigation Modification
TADIXS	Tactical Data Information exchange Subsystem
TADIXS-B	Tactical Data Information eXchange System Broadcast
TADS	Target Acquisition Designation Sight
TADS/PNVS	Target Acquisition and Designator Set/ Pilot Night Vision Sensor
TADS/PNVS	Targeting Air Data System/Pilot's Night Vision System
TAIMS	Three-Axis Inertial Measurement System
TANS	Tactical Air Navigation System
TAOM	Tactical Air Operations Module
TARPS	Tactical Aircraft Reconnaissance Pod System
TAS	True AirSpeed
TAT	Total Air Temperature
TAWS	Terrain Avoidance and Warning System
TBCP	TeleBrief Control Panel
TBMS	Tactical Battlefield Management Subsystem
TCA	Terminal Control Area
TCAS	Traffic alert and Collision Avoidance System
TCCP	Take Command Control Panel
TCS	Television Camera Set
TDM	1. Tactical Data Modem
	2. Time Division Multiplex
TDMA	Time Division Multiple Access
TDMS	Tactical Data Management System
TDS	Tactical Data System
TDU	Test Display Unit
TED	Transferred Electron Device
TEMS	Turbine Engine Monitoring System
TEORS	Tactical Electro-Optical Reconnaissance System
TERCOM	Terrain Contour Matching
TEREC	Tactical Electronic Reconnaissance
TERPROM	Terrain Profile Matching
TESS	Threat Emitter Simulator System
TEVI	Turbine Engine Vibration Indicator
TEWS	Tactical Electronic Warfare System
TF	Terrain-Following
TF/TA^2	Terrain-Following, Terrain-Avoidance, Threat-Avoidance
TFEL	Thin Film ElectroLuminescent
TFPRT	Thin Film Platinum Resistant Thermometer
TFR	Terrain-Following Radar
TFT	Thin Film Transistor
TFTS	Terrestrial Flight Telecommunication System
TIALD	Thermal Imaging And Laser Designator
TIBS	Tactical Information Broadcast Service
TICM	Thermal Imaging Common Modules
TIFS	Total In-Flight Simulator
TILS	Tactical Instrument Landing System
TINS	Thermal Imaging Navigation Set
TIP	Technical Improvement Programme
TIRRS	Tornado InfraRed Reconnaissance System
TIS	Traffic Information Service
TISEO	Target Identification System Electro-Optical
TIU	Time Insertion Unit
TJS	Tactical Jamming System
TLR	Target Locating Radar
TM	Transverse Magnetic
TNR	Tornado Nose Radar
TODS	Tactical Optical Disk System
TOF	Trigger-On-Failure
TOPMS	Take-Off Performance Monitoring System
TOW	Tube-launched Optically tracked Wire-guided (missile)
TP	Tactics Planner
TPU	Transmitter Processing Unit
T/R	Transmitter/Receiver
TRAAMS	Time Reference Angle of Arrival Measurement System
TRAC-A	Total Radiation Aperture Control-Antenna
TRANSEC	Transmission Security
TRF	Tuned Radio Frequency
TRIXS	Tactical Reconnaissance Intelligence Exchange Service
TRN	Terrain Reference Navigation
TRSB	Time Reference Scanning Beam
TRU	Transmitter/Receiver Unit
TSD	Tactical Situation Display
TSFC	Thrust Specific Fuel Consumption
TSO	Technical Service Order (US)
TSPJ	Tornado Self-Protection Jammer
TSSAM	Tri-Service Standoff Attack Missile
TTC	Tape Transport Cartridge
TTFF	Time To First Fix
TTL	Transistor/Transistor Logic
TTS	Time To Station
TTTS	Tanker Transport Training System
TTU	Triplex Transducer Unit
TWMS	Tactical Weapons Management System
TWS	Track-While-Scan
TWT	Travelling Wave Tube
TV	Television
UAC	Universal Avionics Computer
UAV	Unmanned Aerial Vehicle
UCS	Utilities Control System
UDL	The Universal DataLink
UFCD	Up-Front Control Display
UFCP	Up-Front Control Panel
UFD	Up-Front Display
UFDR	Universal Flight Data Recorder
UHF	Ultra High Frequency
UHU	Multipurpose escort/anti-armour variant of Tiger helicopter
UKIRCM	United Kingdom InfraRed CounterMeasures
UK MoD	United Kingdom Ministry of Defence
ULAIDS	Universal Locator Airborne Integrated Data System
URR	Ultra Reliable Radar
USAF	United States Air Force
USB	Upper SideBand
USN	United States Navy
USTS	UHF Satellite Tracking (or Terminal) System
UTM	Universal Transverse Mercator
UV	UltraViolet
V	Volt
Vl	Relative velocity of flow on the underside of an aerofoil
VADR	Voice And Data Recorder
VAMP	VHSIC Avionics Modular Processor
VATS	Video Augmented Tracking System
VAWS	Voice Alarm Warning System
VBM	Volatile Bulk Memory
VCR	Video Cassette Recorder
VCS	Video Camera System
VDA	Versatile Drone Autopilot
VDL	VHF (very high frequency) DataLink
VDM	Visual Display Module
VDU	Visual Display Unit

VEMD Vehicle and Engine Management Display
VFR Visual Flight Rules
VG Vertical Gyro
VGG Visual Graphics Generator
V/H Velocity/Height (ratio)
VHF Very High Frequency
VHSIC Very High-Speed Integrated Circuit
VID Virtual Image Display
VIEWS Vibration Indicator Engine Warning System
VIGIL Vinten Integrated Infrared Linescan
VISTA Variable stability In-flight Simulator Test Aircraft
VITS Video Image Tracking System
VLA Vertical Line-Array
VLAD Vertical Line-Array DIFAR
VLC Very Low Clearance
VLF Very Low Frequency
VLLR Very Light Laser Rangefinder
VLSI Very Large Scale Integration
Vmo Maximum permitted operating speed
VMS Vehicle Management System
VMU 1. Velocity Measuring Unit
2. Voice Message Unit
VNav Vertical Navigation
Vne Never to be exceeded speed
Vocoder Voice encoder/decoder
VOGAD Voice-Operated Gain Adjustment Device
VOR VHF Omnidirectional Range
VOR/Loc VHF Omnidirectional Range/Locator
VORTac VHF Omnidirectional Range/Tacan
VOS Voice-Operated Switch
Vox Voice keying or activation
VPU Voice Processor Unit
VPU(D) Voice Processor Unit (with Data mode)
Vr Radial velocity
VRD Virtual Retinal Display
Vref Typical approach speed
VRU Vertical Reference Unit
Vs Stalling speed
VSD Video Symbology Display
VSI Vertical Speed Indicator
VSRA Vertical/Short take-off and landing Research Aircraft
V/STOL Vertical and Short Take-Off and Landing
VSVA Variable Stator Vane Actuator
VSW Verification SoftWare
VSWR Voltage Standing-Wave Ratio
VTA Voice Terrain Advisory
VTAS Visual Target Acquisition System
VTM Voltage Tunable Magnetron
VTO Volumetric Top Off
VTOL Vertical Take-Off and Landing
VTR VideoTape Recorder
V/UHF Very/Ultra High Frequency

W Watt
WAAS Wide Area Augmentation System
WAC Weapon Aiming Computer
WACCS Warning and Caution Computer System
WAGS Windshear Alert and Guidance System
WAN Wide Area Network
WASP Weasel Attack Signal Processor
WBC Weight and Balance Computer
WBS Weight and Balance System
WCCS Wireless Communication and Control System
WCMS Weapons Control and Management System
WCP Weapon Control Panel
WDA Weather Display Adaptor
WDC Weapon Delivery Computer
WDIP Weapon Data Input Panel
WDNS Weapon Delivery and Navigation System
WEAC Weapons Computer
WFOV Wide Field Of View
WIU Weapon Interface Unit
WNC Weapons and Navigation Computer
WPU Weapon Processing Unit
WRA Weapon-Replaceable Assembly
WSIP Weapons System Improvement Programme
WSW WindShear Warning
WSW/RGS WindShear Warning/Recovery Guidance System
WTC Windshield Temperature Controller

YIG Yttrium Indium Garnet

ZTC Zone Temperature Controller

The Electromagnetic Spectrum

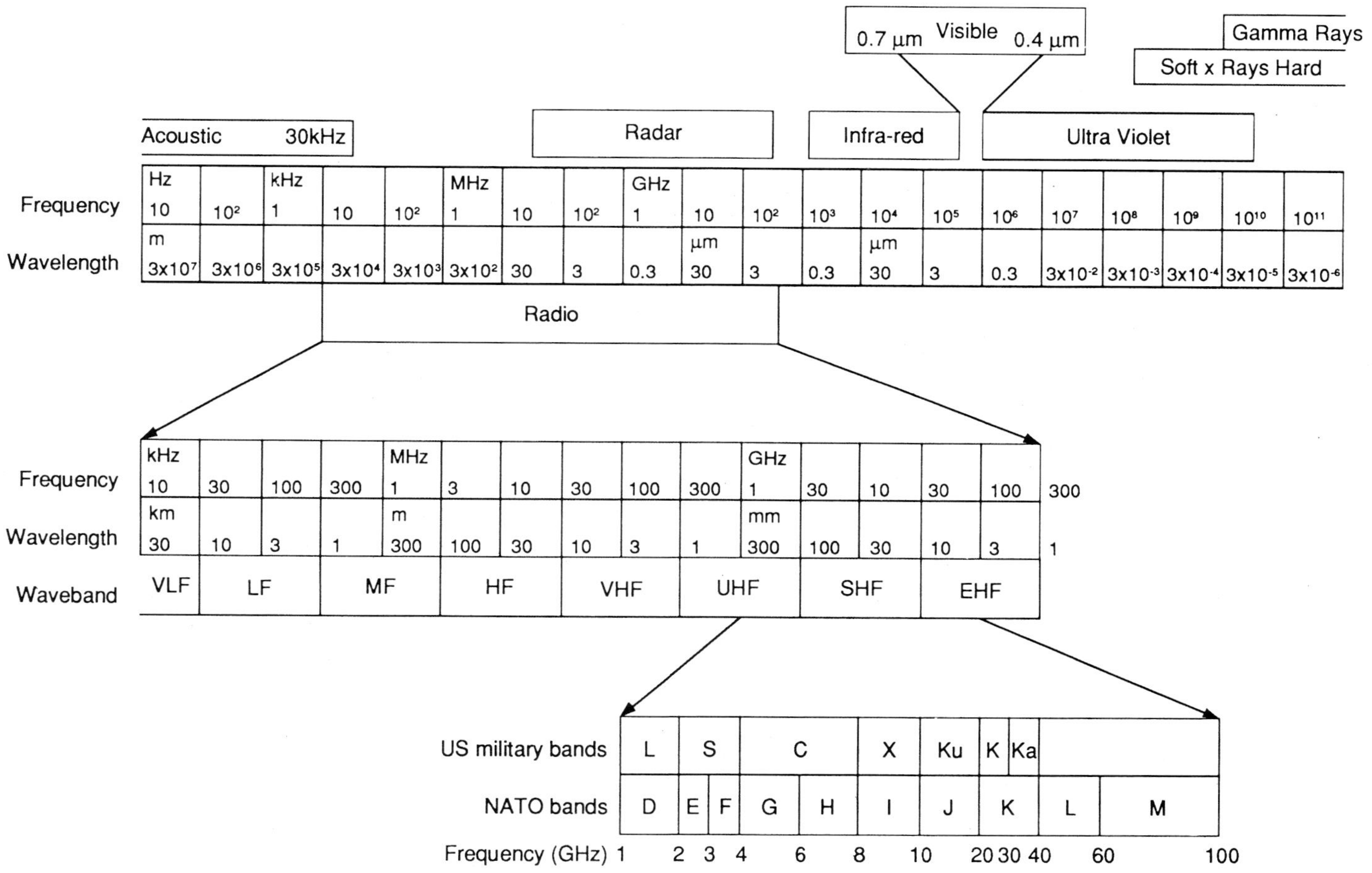

The Joint Electronics Type Designation System (JETDS)

The Joint Electronic Type Designation System (JETDS) is an unclassified US military system designed to identify, by a series of numbers and letters, the use of an equipment and its major components. The system was formerly known as the AN system. A typical example of the system would be:

AN/FPS-5A

In this example the prefix AN indicates that the type number has been assigned in the JETDS system. It does not necessarily mean that the army, navy or air force uses the equipment.

The next letter (F) gives the type of installation where the equipment is used such as fixed, mobile, shipborne and so on. See Table A for a complete listing of installations. The following letter (P) indicates the type of equipment (radar, teletype, radio and so on). See Table B for a complete listing of equipment types. The letter before the dash (S) gives the purpose of the equipment such as transmitting, receiving, detecting and so on. See Table C for a complete listing of equipment purposes. The number after the dash (5) is the model number. The final letter (A) gives notification when a model has been modified but can be interchanged with, or substituted for, the equipment in its original state. If the modified system cannot be interchanged with, or substituted for the original, a new type designation is assigned.

Table A Installation

A	Piloted aircraft
B	Underwater mobile, submarine
C	Air transportable (inactive)
D	Pilotless carrier
F	Fixed, ground
G	Ground, general
K	Amphibious
M	Ground, mobile
P	Pack or portable
S	Water surface craft
T	Ground, transportable
U	General utility
V	Ground, vehicular
W	Water, surface and underwater
Z	Piloted and pilotless vehicle combination

Table B Type of Equipment

A	Infrared
B	Pidgeon (inactive)
C	Carrier (wire)
D	Radiac
E	Nupac (inactive)
F	Photographic
G	Telegraph or teletype
I	Interphone and public address
J	Electromechanical
K	Telemetering
L	Countermeasures
M	Meteorological
N	Sound in air
P	Radar
Q	Sonar and underwater sound
R	Radio
S	Special types or combinations of types
T	Telephone (wire)
V	Visual and visible light
W	Armament
X	Facsimile or television
Y	Data processing

Table C Purpose of Equipment

A	Auxiliary assemblies (inactive)
B	Bombing
C	Communications
D	Direction-finder or reconnaissance/ surveillance
E	Ejection or release
G	Fire control or searchlight directing
H	Recording or reproducing
K	Computing
L	Searchlight control (inactive)
M	Maintenance and test assemblies
N	Navigation aids
P	Reproducing (inactive)
Q	Special, or combination of purposes
R	Receiving, passive detecting
S	Detecting, or range and bearing search
T	Transmitting
W	Control (automatic flight control or remote control)
X	Identification and recognition
Y	Surveillance (search, detect and tracking) and control (fire control, air control)

New entries in this edition

Entry	Contractor
Communications	
Canada	
CMA-2200 intermediate gain satcom antenna system	Canadian Marconi Company
France	
JET-SAT 97	Dassault Electronique
DLP DataLink Processor for NATO interoperability	Rockwell-Collins France
VIP communications suite	Rockwell-Collins France
SAVIB69 audio management system	TEAM, Telecommunications, Electronique, Aeronautique et Maritime
SELCAL airborne selective calling system	TEAM, Telecommunications, Electronique, Aeronautique et Maritime
SIB31/45/54/66/73/85 audio management systems	TEAM, Telecommunications, Electronique, Aeronautique et Maritime
TC 20 control unit	Thomson-CSF Communications
12000 VHF AM/FM radio	Thomson-CSF Communications
12100 VHF FM radio	Thomson-CSF Communications
THOMRAD 6000 V/UHF ECCM transceiver series	Thomson-CSF Communications
TRA 2020 VHF/UHF radio	Thomson-CSF Communications
International	
MCS-7000 aeronautical satellite communications system	Honeywell Inc, Business and Commuter Aviation/Racal Avionics Ltd
TRA 6032/XT 621 P1 V/UHF	Thomson-CSF Communications/
SATURN airborne transceiver	Rohde & Schwarz GmbH & Co KG
Israel	
ASARS-G Airborne Search and Rescue System with GPS & relay	Tadiran Spectralink Ltd
South Africa	
TR 2800 airborne HF transceiver	Grinel Comms
ACR 500 series air/ground V/UHF transceivers	Reutech Defence Industries (RDI) (Pty) Ltd
Sweden	
JAS 39 Gripen radio communication system	CelsiusTech Electronics AB
UK	
Scorpio 2000 series datalink systems	Caledonian Airborne Systems
VLT 15-10-6MK3/VLR 15-16-1 helicopter video downlink	ECS Enterprise Control Systems Limited
Advanced Digital Radio ADR + VHF radio	ITT Defence Ltd
NavSymm DR5-96S UHF differential datalink system	Navstar Systems Ltd
NavSymm DR5-RDS VHF differential datalink system	Navstar Systems Ltd
Airborne real-time datalink	Racal Avionics Limited
STR Satellite Transceiver	Racal Avionics Limited
Eurofighter voice control system	Smiths Industries Aerospace
USA	
GLOBALink/CNS™	ARINC Inc/Magellan Systems
7522 series VHF comm radio tuning panel	AVTECH Corporation
ST-800S/L series wideband microwave transmitter	AYDIN TELEMETRY
RTA-44D VHF Data Radio (VDR)	Chelton Avionics Inc, Wulfsberg Electronics Inc
CDR-3100 series LF-HF receivers	Cubic Communications Inc
PA-5050A 1kW power amplifier	Cubic Communications Inc
T-4180 LF-HF digital signal processor (DSP) exciter	Cubic Communications Inc
Airborne microwave transmission system	FLIR Systems Inc
GMA 340 audio panel	Garmin International Inc
IFPS Intra-Formation Positioning System	Lockheed Martin Federal Systems, Owego
COM 211 VHF/UHF transceiver	Narco Avionics Inc
95S-1A direct conversion receiver	Rockwell Collins
AN/ARC-230/HF-121C high performance radio system	Rockwell Collins
VHF-900/VHF-900B transceivers	Rockwell Collins
VLF/LF High Power Transmit Set (HTPS)	Rockwell Collins
SAMS-1000 audio management system	Sigtronics Corporation
Sigtronics aircraft intercoms	Sigtronics Corporation
Telelink helicopter datalink	Teledyne Controls, Business and Commuter Avionics
Telephonics Communications Management System (TCOMSS)	Telephonics Corporation
Unilink air-to-ground two-way datalink	Universal Avionics Systems Corporation
Maritime operations	
France	
TRES Tactical Radar ESM System	Thomson-CSF Radars Contre-Mesures
Italy	
ATR42 Maritime Patrol Mission System (MPMS)	Alenia Difesa, Avionic Systems and Equipment Division
RFAS	
Sea Dragon maritime surveillance mission system architecture	Leninetz Holding Company
Streege maritime surveillance mission suite architecture	Leninetz Holding Company
UK	
Mission recording system	Normalair-Garrett Ltd
USA	
AN/ARR-78(V) ASCL Advanced Sonobuoy Communications Link	GEC-Marconi Hazeltine Corporation
Radar, GPWS and TCAS systems	
Canada	
SSAR Spotlight Synthetic Aperture Radar for CP-140 aircraft	Lockheed Martin Canada
France	
GCAS Ground Collision Avoidance System	Dassault Electronique
NRAI-11/IDEE 1 Mk XII interrogator-decoder	Thomson-CSF Communications
TSB 2500 Combined IFF Interrogator and Transponder (CIT)	Thomson-CSF Communications
TSC 2050 IFF Mk XII/Mode S diversity transponder	Thomson-CSF Communications
ORB 37 radar system	Thomson-CSF Radars Contre-Mesures
RC 400 compact multimission multitarget radar	Thomson-CSF Radars Contre-Mesures
Romeo II obstacle avoidance radar	Thomson-CSF Radars Contre-Mesures
Germany	
MIDAS Microwave Imaging DLR/Dasa Airborne System	Daimler-Benz Aerospace, Defense and Civil Systems
International	
ENR European Navy Radar for the NH 90 NFH	Alenia Difesa/Daimler-Benz/Thomson-CSF
IFF for Eurofighter 2000	Daimler-Benz Aerospace AG/Italtel/ Raytheon Systems Limited
Italy	
Creso airborne battlefield surveillance radar	Alenia Difesa, Avionic Systems and Equipment Division
RFAS	
SPPZ ground proximity and warning systems	AeroPribor-Voskhod Joint Stock Company
NIT side-looking airborne radar	Leninetz Holding Company
VEGA-M airborne surveillance radar	Moscow Scientific Institute of Instrument Engineering MNIIP
IFF 60P system	All-Russian Joint Stock Company Nizhegorodskaya Tearmarka
SRZO-KR airborne interrogator-transponder	Kazan Scientific Research Institute of Radio-Electronics/RadioPribor
Topaz multifunction fire-control radar	Phazotron Scientific & Production Company
IFF 6201R/6202R and 6231R/6232R	Production Association RadioPribor
South Africa	
PA-5429 radar altimeter	Plessey South Africa Limited
XBT-2000 X-band radar transponder	Plessey South Africa Limited
UK	
IFF 4800 transponder	Raytheon Systems Limited, Electronic Systems Division
USA	
Stormscope WX-500 weather mapping system	BFGoodrich Aerospace Avionics Systems
GTX 320 IFF transponder	Garmin International Inc
RCZ-852 diversity Mode-S transponder	Honeywell Inc, Business & Commuter Aviation Systems
XS-950 Mode-S ATDL (Air Transport DataLink) transponder	Honeywell Inc, Business & Commuter Aviation Systems
TCAS II (TCAS 2000) Traffic alert and Collision Avoidance System	Honeywell Inc, Business & Commuter Aviation Systems
HG9550 LPI radar altimeter system	Honeywell Inc, Sensors and Guidance Products
Sea Vue (SV) surveillance radar	Raytheon Systems Company
Electro-optics	
Australia	
LRTS Long Range Tactical Surveillance sensor	British Aerospace Australia
Ranger 600 eye-safe laser rangefinder	British Aerospace Australia

Entry	Contractor
Canada	
P-3C Advanced Imaging Multispectral System (AIMS)	Wescam Inc
France	
JADE night vision goggles	SFIM Industries
AP 40 panoramic film camera	Thomson-CSF Optronique
Chlio-S multisensor airborne FLIR	Thomson-CSF Optronique
Damocles multimode multifunction laser designator pod	Thomson-CSF Optronique
NAVFLIR navigation and attack pod	Thomson-CSF Optronique
Presto/Desire reconnaissance pod	Thomson-CSF Optronique
TMV 632 airborne laser spot tracker and rangefinder	Thomson-CSF Optronique
Germany	
Hellas helicopter obstacle warning system	Daimler-Benz Aerospace AG, Defense and Civil Systems
IRLS InfraRed Linescanner Systems	Honeywell Regelsysteme GmbH
VOS 80C digital video colour camera	Zeiss-Eltro Optronic GmbH (ZEO)
International	
Litening airborne laser designator and navigation pod	Rafael Missile Division/Northrop Grumman, Electronics & Systems Integration Division
Israel	
DSP-1 Dual Sensor Payload	Controp Precision Technologies Ltd
LRFD Laser RangeFinder Designator systems	ELOP Electro-Optics Industries Ltd
Topaz electro-optical targeting and surveillance system	Rafael Missile Division
Toplite multisensor payload	Rafael Missile Division
POP Plug-in Optronic Payload	TAMAM Division, Electronics Group, Israel Aircraft Industries
RFAS	
GEO-NV-III-TV day/night tracking system	Geophizika-NV
GEO-NVG-III Night vision Goggles	Geophizika-NV
LRLS airborne Laser Radar Landing System	State Research Institute of Aviation Systems (GosNIIAS)
AMORS Airborne Multifunction Optical Radar System	State Research Institute of Aviation Systems (GosNIIAS)
A-84 panoramic aerial camera	Zenit Foreign Trade Firm, State Enterprise P/C S.A. Zverev Krasnogorsky Zavod
AC-707 spectrozonal aerial camera	Zenit Foreign Trade Firm, State Enterprise P/C S.A. Zverev Krasnogorsky Zavod
AK-108Ph vertical and oblique aerial camera	Zenit Foreign Trade Firm, State Enterprise P/C S.A. Zverev Krasnogorsky Zavod
Sweden	
SEOS 200 helicopter observation system	Saab Dynamics AB
TopEye survey system	Saab Survey Systems
Turkey	
ASELFLIR-200 second generation gyroscope stabilised airborne FLIR	Aselsan Inc, Microelectronics, Guidance and Electro-optics Division
USA	
SAFIRE thermal imaging system AN/AAQ-22	FLIR Systems Inc
Star SAFIRE thermal imaging system AN/AAQ-22	FLIR Systems Inc
ULTRA 6000 dual-camera 24 hour imaging system	FLIR Systems Inc
EOSDS Electro-Optical Surveillance and Detection Systems	Northrop Grumman Electronic Sensors and Systems Division
Colour video HUD cameras	Photo-Sonics Inc
AN/AAQ-26 infrared detecting set	Raytheon Systems Company
AN/AAQ-27 (3FOV) imaging infrared system	Raytheon Systems Company
ATFLIR Advanced Targeting Forward-Looking InfraRed system	Raytheon Systems Company
DB-110 Dual-Band reconnaissance system	Raytheon Systems Company
CA-236 36-inch E-O LOROP camera	Recon/Optical Inc
Electronic warfare	
Canada	
Electronic warfare training system for CT-133 aircraft	Lockheed Martin Canada
France	
DDM missile launch detector	Matra BAe Dynamics/SAGEM
LCM (Lance-Cartouches Modulaire) modular cartridge dispenser	Matra BAe Dynamics
LEA (Leurre Electromagnetique Actif) active radar decoy	Matra BAe Dynamics/Thomson-CSF Radars Contre-Mesures
LISCA (Leurre Infrarouge a Signature et Cinematique Adaptee) smart IR decoy	Matra BAe Dynamics/SNPE
Germany	
AN/ALQ-119GY/ALR 68	Daimler-Benz Aerospace AG, Defense and Civil Systems
LWR Laser Warning Receiver	Daimler-Benz Aerospace AG, Defense and Civil Systems
Towed decoy	Daimler-Benz Aerospace AG, Defense and Civil Systems
International	
ERWE II Enhanced Radar Warning Equipment	Daimler-Benz Aerospace AG, Defense and Civil Systems/Litton Applied Technology
NRWE Tornado New Radar Warning Equipment	Daimler-Benz Aerospace AG, Defense and Civil Systems/Litton Applied Technology
Israel	
EL/L-8233 Integrated Self-Defence System (ISDS)	Elta Electronics Industries Ltd
CDF-3001 airborne V/UHF COMINT/DF system	Tadiran Electronic Systems Ltd
Italy	
ARWE Advanced Radar Warning Equipment	Elettronica S.p.A.
ELT/553(V)-2 airborne pulse and CW jammer	Elettronica S.p.A.
RFAS	
20 SP M-01 airborne flare dispenser system	Aviaavtomatika/Joint Stock Company PRIBOR
Irtysh EW system on Su-39 Strike Shield	Central Scientific Institute for Radiotechnical Measurement TSNIITI, Omsk
YB-3A flare dispenser system	Vympel State Machine Building Design Bureau
Sweden	
EWS 39 - Gripen integrated EW suite	Ericsson Saab Avionics AB
UK	
DRFM TG Digital Radio-Frequency Memory Techniques Generator	Racal Radar Defence Systems Ltd
USA	
AR-900 ESM system	ARGOSystems Inc
CS-6700 ACES Automatic processing ESM System	Condor Systems Inc
AN/ALR-91(V)3 series threat warning systems	Litton Applied Technology
AN/APR-39(V) Radar Warning Receiver (RWR)	Lockheed Martin Fairchild Systems
AN/APR-39A(V) threat warning systems	Lockheed Martin Fairchild Systems
Nimrod MRA4 Defensive Aids Sub-System (DASS)	Lockheed Martin Fairchild Systems
AN/ALQ-184(V)9 combined ALQ-184 ECM pod and ALE-50 towed decoy system	Raytheon Systems Company
AN/ALE-38/41 dispenser system	Tracor Aerospace Inc
WTSS Wideband Tactical Surveillance System	TRW Systems Integration Group
AN/ALQ-504 airborne VHF/UHF communications signals intercept/ DFJamming/Deception system	Zeta
Data handling	
France	
Airborne data loader ARINC 615/603	SFIM Industries
DFDAU-ACMS Digital Flight Data Acquisition Unit-Aircraft Condition Monitoring System	SFIM Industries
DMU-ACMS Data Management Unit - Aircraft Condition Monitoring System	SFIM Industries
FDAU-ACMS Flight Data Acquisition Unit - Aircraft Condition Monitoring System	SFIM Industries
FDIU Flight Data Interface Unit	SFIM Industries
RFAS	
SVS series of digital air data computers	AeroPribor Voskhod Joint Stock Company
C300 programmed signal processor	Leninetz Holding Company
BTsVM-386 airborne digital computer	Ramenskoye Design Company AO RPKB
South Africa	
AM1000 distributed architecture data acquisition and processing system	Analysis, Management & Systems (Pty) Ltd
UK	
Advanced Memory Unit AMU	Smiths Industries Aerospace
Displays and Mission Computer DMC	Smiths Industries Aerospace
USA	
Modular Mission & Display Processor MDP	Astronautics Corporation of America
HG1140 multirole air data computer	Honeywell Inc, Sensor and Guidance Products
Integrated Core Processor ICP	Raytheon Systems Company
Data recording	
France	
MONITAIR flight data recorder	MONIT'AIR
EVS 925 video tape recorder	SFIM Industries

Entry	Contractor
SSCVR solid-state cockpit voice recorder	SFIM Industries/Dassault Electronique/ TEAM
Germany	
MDFDR Multiple Dislocated Flight Data Recorder system	Daimler-Benz Aerospace, Defense and Civil Systems
International	
AH-64D Apache Longbow HUMS	Stewart Hughes Ltd/Base 10/The Boeing Company
Helicopter Flight Data Recording/ Health and Usage Monitoring System (FDR/HUMS)	Teledyne Controls/Stewart Hughes Ltd
Norway	
EE 235 solid-state recorder	EIDEL Eidsvoll Electronics AS
RFAS	
KARAT integrated monitoring and flight data recording system	GosNIIAS State Research Institute of Aviation Systems/JSC Pribor Design Bureau Aviaavtomatika
UK	
AE6100HW cassette data recorder	Avalon Electronics Ltd
USA	
DCRsi 120 Digital Cartridge Recording system	Ampex Corporation Data Systems Division
PAR 1000 Mini-HUMS	Avionics Specialties Inc
Health and Usage Management Systems (HUMS) for helicopters	BFGoodrich Aerospace, Aircraft Integrated Systems
AN/AQH-9 mission recorder system	DRS Technologies Inc, DRS Precision Echo Inc
AN/AQH-11 high density mission recorder system	DRS Technologies Inc, DRS Precision Echo Inc
AN/USH-42 ( ) mission recorder/ reproducer set	DRS Technologies Inc, DRS Precision Echo Inc
DCMR-24 Digital Cassette Mission Recorder	DRS Technologies Inc, DRS Precision Echo Inc
DCMR-100 Digital Cassette Mission Recorder	DRS Technologies Inc, DRS Precision Echo Inc
Model FA2100 recorder family	L.3 Communication Corp, Aviation Recorders
Colour Cockpit TV Sensor CCTVS	Lockheed Martin Fairchild Systems
Model 32HE variable speed digital recorder	Metrum-Datatape Inc
GenHUMS	Smiths Industries Aerospace
NuHums™ Health and Usage Monitoring System	SPS Signal Processing Systems
T'AIMS (Teledyne Aircraft Integrated Monitoring System)	Teledyne Controls
Navigation and nav/attack	
Canada	
CMA-3000 Single Unit Navigator (SUN)	Canadian Marconi Company
CMA-3012 Global Navigation Satellite Sensor Unit (GNSSU)	Canadian Marconi Company
France	
COSPAS-SARSAT emergency locator transmitter - A06 range	CEIS TM - LCD Division
IPG-120F GPS receiver	Rockwell-Collins France
Mercator digital map generator	SAGEM SA, Defence & Security Division
Meghas™ new-generation avionics suite for helicopters	Sextant Avionique/SFIM Industries
Nadir 1000 integrated navigation and mission management system	Sextant Avionique
NASH Night Attack System for Helicopters	Sextant Avionique
Stratus and Totem 3000 flight systems	Sextant Avionique
TLS 2020 MultiMode Receiver MMR	Thomson-CSF Communications
TLS 2030 MultiMode Receiver MMR	Thomson-CSF Communications
TLS 2040 MultiMode Receiver MMR	Thomson-CSF Communications
Germany	
NFS-3000 series navigation management system	Daimler-Benz Aerospace AG, Defense and Civil Systems
EuroNav III task management system	EuroAvionics Navigationssysteme GmbH & Co
H-764G Embedded GPS/INS	Honeywell Regelsysteme GmbH
u-INS/GPS	Honeywell Regelsysteme GmbH
International	
ILS-85 Instrument Landing System	NIIAO Institute of Aircraft Equipment, Moscow/Ukraine Research Institute of Radio Equipment, Kiev
Tiger combat helicopter avionics system	Sextant Avionique/Daimler-Benz Aerospace AG/LITEF GmbH/Nord-Micro Electronik Feinmechanik AG/ Rohde & Schwarz GmbH & Co KG/ VDO-Luftfahrgerate Werk GmbH
Israel	
3D GPS attitude determination receiver	Rokar International Ltd
GPS NAVPOD rugged receiver	Rokar International Ltd
GPS SWIFT high velocity, high acceleration, receiver	Rokar International Ltd
NTS/NTS A - Night Targeting System	TAMAM Division, Electronics Group, Israel Aircraft Industries
TN-90Q/G compact inertial navigation system	TAMAM Division, Electronics Group, Israel Aircraft Industries
RFAS	
DME/P-85	All-Russia Research Institute of Radio Equipment/NIIAO Institute of Aircraft Equipment
VOR-85	All-Russia Research Institute of Radio Equipment/NIIAO Institute of Aircraft Equipment
INS-85 Inertial Navigation System	AviaPribor
GINS-3 gravimetric navigation system	AviaPribor
I-21 inertial navigation system	AviaPribor
CH-3301 GLONASS/GPS airborne receiver	NAVIS
INS-80 Inertial Navigation System	Ramenskoye Design Company AO RPKB
NS BKV-95 integrated navigation system	Ramenskoye Design Company AO RPKB
SINUS integrated navigation, flight management and display system	Ramenskoye Design Company AO RPKB
Turkey	
LN100GT EGI embedded GPS inertial navigation system	Aselsan Inc, Microelectronics, Guidance and Electro-optics Division
UK	
NavSymm Sharpe XR6-12 channel GPS receiver	Navstar Systems Ltd
CDU/IN/GPS	Racal Avionics Ltd
LWCCU LightWeight Common Control Unit	Racal Avionics Ltd
USA	
GNC 250XL GPS/Comm system	Garmin International
IFR/VFR avionics stacks	Garmin International
HT1000 GNSS navigation management system	Honeywell Inc, Business & Commuter Aviation Systems/Trimble Navigation Ltd, Avionics Products
KN-4071 Attitude Heading Reference System (AHRS)	Kearfott Guidance & Navigation Corporation
NAV 122D and NAV 122D/GPS self-contained NAV receiver indicators	Narco Avionics Inc
AHS-3000 Attitude Heading System	Rockwell Collins
APR-4000 GPS approach sensor	Rockwell Collins
Flight management and control	
Canada	
Contaminant & Fluid Integrity Measuring System (C/FIMS)	AlliedSignal Aerospace Canada
Germany	
Air data systems	Nord-Micro Elektronik Feinmechanik AG
International	
Aria - EFIS-95 airborne integrated avionics system	NIIAO Institute of Aircraft Equipment/ AlliedSignal Aerospace
RFAS	
ASShU-334 fly-by-wire flight control system	AviaPribor
ASUU-96 automatic control and stability augmentation system	AviaPribor
EDSU-77 fly-by-wire flight control system	AviaPribor
SVS-V1 helicopter air data computer system	AviaPribor
VSUP-85 flight control computer system	AviaPribor
VSUPT-334 flight and thrust control computer system for Tu-334	AviaPribor
VSUT-85 thrust control computer system	AviaPribor
Aircraft Systems Control System (ASCS) - Electronic Flight Engineer	GosNIIAS State Research Institute of Aviation Systems
Antonov-70 airborne information system	Leninetz Holding Company
EFIS-85 electronic flight instrument system	NIIAO Institute of Aircraft Equipment/ Ulyanovsk Instrument Design Office
FCS-85 flight control system	NIIAO Institute of Aircraft Equipment/ Moscow Institute of Electromechanics and Automatics
FILS Fault Isolation and Localisation System	NIIAO Institue of Aircraft Equipment
FMS-85 flight management system	NIIAO Institute of Aircraft Equipment
705-6 attitude heading reference system	Ramensky Instrument Engineering Plant
STR7-4 fuel quantity and flow metering system	TekhPribor State Enterprise
SUITS-10 fuel management and indicating system	TekhPribor State Enterprise
USA	
Fire Detection Suppression system (FiDS)	BFGoodrich Aerospace Aircraft Integrated Systems
Fuel Quantity Indicating Systems (FQIS)	BFGoodrich Aerospace Aircraft Integrated Systems

Entry	Contractor
EQUIX^{SM} integrated avionics architecture for JSF	Honeywell Inc, Defense Avionics Systems
FMS-6000 Flight Management System	Rockwell Collins
UNS-1K Flight Management System (FMS)	Universal Avionics Systems Corporation

Cockpit displays, instruments and indicators
Czech Republic

Entry	Contractor
Altimeters	Mikrotechna Praha a.s
Artificial horizon LUN 1205.XX-8	Mikrotechna Praha a.s
Integrated artificial horizon LUN 1208.XX-8	Mikrotechna Praha a.s
Germany	
Eurogrid European geographic information display system	Daimler-Benz Aerospace AG, Defense and Civil Systems
Cockpit warning system	Diehl GmbH & Co Luftfahrt Elektronik
RFAS	
VM type altimeters	AeroPribor-Voskhod Joint Stock Company
AGB-96, AGB-98 and AGB-100 horizon gyros	AviaPribor
AGR-29 and AGR-81 standby horizon gyros	AviaPribor
Airspeed and altitude indicators	AviaPribor
IM-3, IM-5, IM-6, IGM multifunction displays	AviaPribor
IRM-1 radio compass indicator	AviaPribor
SEI-85 and KISS-1-1M multifunction displays	ElectroPribor, Kazan Plant
Tachometer indicators	ElectroPribor, Kazan Plant
Temperature gauges	ElectroPribor, Kazan Plant
MIKBO series of compact integrated avionic systems	NIIAO Institute of Aircraft Equipment
MFI multifunctional active-matrix liquid-crystal display	Ramenskoye Design Company AO RPKB
PKP-72 and PKP-77 flight directors and PNP-72 compass	Ramenskoye Instrument Engineering Plant
Switzerland	
Altimeters, types 3A and 3H, 2 in	Revue Thommen AG
Encoding altimeters, types 3A and 3H, 3 in	Revue Thommen AG
MACH/Airspeed Indicators (MAI), type 5, 3 in	Revue Thommen AG
Vertical Speed Indicator (VSI), type 4A16, 3 in	Revue Thommen AG
UK	
Airborne radar indicator unit and track while scan system	Caledonian Airborne Systems Ltd
SAHIS Standby Attitude, Heading and rate of turn Indicating System	Ferranti Technologies Ltd
Angle of attack system	Ferranti Technologies Ltd
Attitude indicators - FH series	Ferranti Technologies Ltd
DEWD Dedicated Electronic Warfare Display	Meggitt Avionics
Observer™ moving map task management system	Skyforce Avionics
Ruggedised flat panel LCD aircraft monitors	Skyforce Avionics
Skymap II™ and Tracker II™	Skyforce Avionics
Aircraft moving coil indicators	Weston Aerospace
USA	
8 × 10 in colour AMLCD Display Unit	AlliedSignal Inc, Commercial Avionics Systems
MFD 5200 colour AMLCD multifunction display	Arnav Systems Inc
Airborne multifunction CRT display	Astronautics Corporation of America
Colour AMLCD MultiFunction Displays (MFDs)	Astronautics Corporation of America
Colour AMLCD multifunction indicators	Avionic Displays Corporation
AIM 1100 3 in self-contained attitude indicator	BFGoodrich Aerospace Avionics Systems
Eagle-19 colour AMLCD	dpiX, a Xerox New Enterprise Business
LSZ-860 lightning sensor system	Honeywell Inc Business & Commuter Aviation Systems
Primus 2000XP integrated avionics cockpit	Honeywell Inc Business & Commuter Aviation Systems
Control Display System (CDS) for OH-58D Kiowa Warrior	Honeywell Inc Defense Avionics Systems
Radar Control Display Unit (RCDU) for UK E-3D	Honeywell Inc Defense Avionics Systems
EDM-700 family of Engine Data Management systems	JP Instruments
KN-0001 Attitude Director Indicator ADI	Kearfott Guidance and Navigation Corporation
Full-function display/processing system	Litton Guidance & Control Systems
Liquid Crystal Crew Display Unit (LCCDU)	Palomar Products Inc
NeoAV Model 550 EFIS Electronic Flight Instrument System	Rogerson Kratos, a Rogerson Aircraft Corporation Subsidiary
NeoAV IIDS Integrated Instrument Display Systems	Rogerson Kratos, a Rogerson Aircraft Corporation Subsidiary
Primary Flight Display Subsystem for the S-92 Helibus	Sanders, a Lockheed Martin Company
TAMMAC Tactical Aircraft Moving Map Capability	Smiths Industries Aerospace

Head-up displays, helmet-mounted displays and weapon aiming sights
France

Entry	Contractor
Strix and Viviane day/night sights	SFIM Industries
International	
HUD 2020 and HUD 2022 Head-Up Displays	GEC-Marconi Avionics Ltd, Mission Avionics Division/Honeywell Inc Business and Commuter Aviation Systems/Honeywell Inc Air Transport Systems
Virtual Retinal Display (VRDTM) Technology in helmet-Mounted displays	Microvision Inc/Saab AB/Ericsson Saab Avionics
UK	
HMSS Helmet-Mounted Sighting System	GEC-Marconi Avionics Ltd, Mission Avionics Division, Rochester
USA	
Helmet-Mounted Cueing System	Honeywell Inc Sensor and Guidance Products

Stores management
USA

Entry	Contractor
Weapons control and management system for Nimrod 2000	Smiths Industries Aerospace

Addenda

Communications
Denmark

Entry	Contractor
TT-30224A Inmarsat-C aeronautical capsat	Thrane & Thrane A/S
TT-3608F Aero-C capsat printer unit	Thrane & Thrane A/S
TT-5000 series Inmarsat Aero-I system	Thrane & Thrane A/S

Cockpit displays, instruments and indicators

Entry	Contractor
Cockpit control panels	Gables Engineering Inc

Head-up displays, helmet-mounted displays and weapon aiming sights
Ukraine

Entry	Contractor
SURA Helmet-Mounted Target Designation System (HMTDS)	Arsenal Central Design Office

Entries deleted from this edition

Entry	Contractor
Communications	
France	
Chameleon L11/L16 tactical datalink server	Rockwell-Collins France
MCU-2201F modem control unit	Rockwell-Collins France
THOM'RAD 2000 control unit	Thomson-CSF Communications
THOM'RAD 2000 VHF AM/FM radio	Thomson-CSF Communications
THOM'RAD 2000 VHF FM radio	Thomson-CSF Communications
THOM'RAD 2000 VHF/UHF radio	Thomson-CSF Communications
International	
Satcom conformal antenna subsystem	AlliedSignal Commercial Avionics Systems/Dassault Electronique
Terrestrial Flight Telecommunication System TFTS	Rohde and Schwarz GmbH & Co KG/ Mors SA
THOM'RAD 6000 VHF/UHF secure radio communication system	Thomson-CSF Communications/Rohde and Schwarz GmbH & Co KG
Israel	
IS-10 digital image processing and communication system	Elbit Ltd
Advanced DataLinks for Guided Platforms (ADLGP)	Tadiran/Spectralink Ltd
RFAS	
Orlan aircraft transmitter/receiver	Aviaexport
R-800 radio set	Aviaexport
South Africa	
C-Band video/data transceiver	Plessey Tellumat
UK	
AD980 central suppression unit	GEC-Marconi Electro-Optics Ltd, Sensors Division
Cordless cabin telephone system	GEC-Marconi Electro-Optics Ltd, Sensors Division
Terrestrial flight telephone system	GEC-Marconi Electro-Optics Ltd, Sensors Division
RT 150 personal locator beacon	Graseby Dynamics Ltd
MCA 6010 satellite communications radio antenna	Racal Avionics Ltd
Satfone single channel voice/data system	Racal Avionics Ltd
SDR 2000 satellite data radio	Racal Avionics Ltd
USA	
Passenger address/cabin interphone system	AVTECH Corporation
VHF COMM radio tuning panel	AVTECH Corporation
PA805S power amplifier	Aydin Vector
VSD-100 airborne subcarrier discriminator	Aydin Vector
AT-101 HF antenna tuning system	Collins Avionics & Communications
IFM-101/AM-7189 VHF/FM power amplifier	Collins Avionics & Communications
MX-11641/ARC(244D-1) low noise amplifier/diplexer	Collins Avionics & Communications
37R-2 VHF communications antenna	Collins Avionics & Communications
490S-1 HF antenna tuning unit	Collins Avionics & Communications
490T-1 HF antenna coupler	Collins Avionics & Communications
CPL-920D digital antenna coupler	Collins Avionics & Communications
PAC-230 HF antenna coupler	Collins Avionics & Communications
Model 1150 Advanced Cabin Interphone System (ACIS)	Hughes Aircraft Company, Sensors and Communications Systems
Advanced Data Controller	Motorola Government & Systems Technology Group
LSSC-100/200/300 series satcom and line of sight terminals	Motorola Government & Systems Technology Group
LST-5C UHF satcom and line of sight transceiver	Motorola Government & Systems Technology Group
Proteus/URC-200 VHF and UHF multiband transceiver	Motorola Government & Systems Technology Group
AS-3521/ARN Controlled Reception Pattern GPS Antenna (CRPA-2)	Raytheon E-Systems, ECI Division, St Petersburg
AS-3822A/URN Fixed Reception Pattern GPS Antenna (FRPA-3)	Raytheon E-Systems, ECI Division, St Petersburg
FDM 1500 frequency division multiplexer	SCI Systems Inc
Frequency Shift Keyed converter	SCI Systems Inc
ACU-150D HF antenna coupler	Sunair Electronics Inc
Sure-Comm intercommunication system	Telephonics Corporation
Jetfone TD-3000 airborne telephone system	Trimble Navigation
TX 10 VHF transceiver	Trimble Navigation
Maritime operations	
Canada	
AN/AQA-801 Barra Side Processor	Computing Devices Canada
AQQ-T501	Computing Devices Canada

Entry	Contractor
UK	
A627 airborne acoustic processor	Ultra Electronics Limited, Sonar and Communication Systems
A628 speech processor	Ultra Electronics Limited, Sonar and Communication Systems
USA	
AN/ASA-82 anti-submarine warfare display	Lockheed Martin Display Systems
AN/ASQ-164 control indicator set	Lockheed Martin Display Systems
Radar, GPWS and TCAS systems	
Canada	
LN66 radar system	Canadian Marconi Company
China, People's Republic	
F-8 fire-control radar	China Leihua Electronic Technology Research Institute
France	
AHV-8 radio altimeter	Thomson-CSF Communications
AHV-530 radio altimeter	Thomson-CSF Communications
AHV-530A radio altimeter	Thomson-CSF Communications
AHV-540 radio altimeter system	Thomson-CSF Communications
AHV-550 radio altimeter	Thomson-CSF Communications
AHV-2550 digital radio altimeter	Thomson-CSF Communications
NRAI-10A/SB20 Mk XII IFF interrogator	Thomson-CSF Communications
NRAI-11/SB13 MkXII interrogator-decoder	Thomson-CSF Communications
SB 25 combined IFF interrogator and transponder	Thomson-CSF Communications
SC 2050 IFF Mk XII diversity transponder	Thomson-CSF Communications
Germany	
ATC 2000 R-(2) transponder	Becker Avionic Systems
International	
Ground Collision Avoidance System (GCAS)	Dassault Electronique/Collins Avionics and Communications
South Africa	
PT-1500A radar altimeter	Plessey Tellumat South Africa Ltd
UK	
Millimetric Target Acquisition System (MTAS)	British Aerospace Defence Ltd
USA	
Mode S datalink transponder	Honeywell Inc Business & Commuter Aviation Systems
Primus 450 and 650 weather radars	Honeywell Inc Business & Commuter Aviation Systems
Primus 500 ColoRadar	Honeywell Inc Business & Commuter Aviation Systems
Primus 708A radar	Honeywell Inc Business & Commuter Aviation Systems
Primus 800 radar	Honeywell Inc Business & Commuter Aviation Systems
Primus 870 turbulence detection weather radar	Honeywell Inc Business & Commuter Aviation Systems
TCAS II Traffic alert and Collision Avoidance System	Honeywell Inc Business & Commuter Aviation Systems
AN/APQ-113/-144/-161/-163/-165/ -169 attack radars	Lockheed Martin Ocean Radar and Sensor Systems
ASARS-1 Advanced Synthetic Aperture Radar	Lockheed Martin SAR Imaging Systems
Loral Advanced Imaging Radar System (LAIRS)	Lockheed Martin SAR Imaging Systems
Miniature Synthetic Aperture Radar (MSAR)	Lockheed Martin SAR Imaging Systems
AN/APQ-156 radar for the A-6E	Norden Systems Inc
AN/APS-130 mapping radar	Norden Systems Inc
AIRSTAR surveillance airborne target acquisition radar	Sanders Inc, a Lockheed Martin Company
AN/APQ-( ) terrain-following radar	Texas Instruments Inc
AN/APQ-99 terrain-following radar for the RF-4C	Texas Instruments Inc
AN/APQ-171 radar	Texas Instruments Inc
AN/APS-127 radar	Texas Instruments Inc
Transponder test set	Telephonics Corporation
Sentinel II Airborne Traffic alert and Collision Avoidance System (TCAS)	Trimble Navigation
Electro-optics	
International	
Lite Pod	Rafael Missiles Division/Northrop Grumman Electronics & Systems Integration Division/Sargent Fletcher Inc

Entry	Contractor
UltraMedia airborne camera system	Broadcast & Surveillance Systems Ltd/ FLIR Systems Inc
Israel	
ANVIS 7 – *entry moved to International section, Head-up displays, helmet-mounted displays and weapon aiming sights*	Elbit Ltd
Rangefinder Target Designator Laser	El-Op Electro-Optics Industries Ltd
Litening targeting and navigation pod	Rafael Missiles Division
Day Surveillance Payload (DSP)	Tamam Precision Instrument Industries
Night Surveillance Payload (NSP)	Tamam Precision Instrument Industries
Stabilised Long-range Observation System (SLOS)	Tamam Precision Instrument Industries
Italy	
CIRTEVS infrared television system	Ottico Meccanica Italiana SpA
UK	
Ultra 4000 airborne stabilised system	Broadcast & Surveillance Systems Ltd
Ultra 5000 airborne surveillance system	Broadcast & Surveillance Systems Ltd
Airborne laser designator	GEC-Marconi Electro-Optics Ltd, Sensors Division, Basildon
USA	
A/A24Q-1(V) Photographic Sensor Control System (PSCS)	Fairchild Defense OSC
AN/ASQ-90 reconnaissance management system	Fairchild Defense OSC
AN/ASQ-172A reconnaissance system	Fairchild Defense OSC
AN/ASQ-197 Sensor Control-Data Display Set (SC-DDS)	Fairchild Defense OSC
AN/AYA-10 reconnaissance management system	Fairchild Defense OSC
Reconnaissance Data Anotation Set (RDAS)	Fairchild Defense OSC
Autonomous Landing Guidance (ALG)	FLIR Systems Inc
Situation awareness FLIR	FLIR Systems Inc
ULTRA 3000 compact stabilised thermal imaging system	FLIR Systems Inc
UltraMedia aerial camera systems	FLIR Systems Inc
UltraMedia-RS lightweight gyrostabilised camera system	FLIR Systems Inc
AN/AVS-6 Aviator's Night Vision Imaging System (ANVIS)	ITT Defense Electro-Optical Products Division
F4949 Aviator's Night Vision Imaging System (ANVIS)	ITT Defense Electro-Optical Products Division
Laser target designator/rangefinder for F/A-18	Litton Systems Inc, Laser Systems Division
AN/AAQ-9 Infrared detecting set	Texas Instruments Inc, Defense Systems and Electronics Group
Falcon Eye FLIR	Texas Instruments Inc, Defense Systems and Electronics Group
OR-89/AA forward-looking infrared system	Texas Instruments Inc, Defense Systems and Electronics Group
OR-5008/AA Forward-Looking InfraRed system (FLIR)	Texas Instruments Inc, Defense Systems and Electronics Group
AN/AVS-7 ANVIS/HUD – *entry moved to International section, Head-up displays, helmet-mounted displays and weapon aiming sights*	Tracor Inc
Electronic warfare	
International	
MAWS Missile Approach Warning System	Daimler-Benz Aerospace AG, Defense and Civil Systems/Litton Applied Technology
Israel	
SPS-2100 self-protection system	Elisra Electronic Systems Ltd
ACS-500 Automatic VHF/UHF COMINT System	Tadiran Systems Ltd
Owl ELINT and ESM system	Tadiran Systems Ltd
Passive Identification and Targeting System (PITS)	Tadiran Systems Ltd
RAS-1B ELINT and ESM system	Tadiran Systems Ltd
RAS-2A ELINT and ESM system	Tadiran Systems Ltd
RDF-500 Radar Direction Finder	Tadiran Systems Ltd
Skyjam computerised jamming system	Tadiran Systems Ltd
TDF-500 VHF/UHF automatic direction-finding system	Tadiran Systems Ltd
Italy	
ELT/553 self-protection jammer	Elettronica S.p.A.
RFAS	
Landfish	
Mak	
Miass	
SG-1 radar warning receiver	

Entry	Contractor
South Africa	
PD 1000 panoramic display unit	Grinaker System Technologies
R1000 VHF/UHF radio monitoring receiver	Grinaker System Technologies
SR1000 VHF/UHF fast scanning receiver	Grinaker System Technologies
Sweden	
AQ-31 deception jamming pod	CelsiusTech Electronics AB
AQ-800 noise jamming pod	CelsiusTech Electronics AB
AQ-861 noise and deception jammer	CelsiusTech Electronics AB
USA	
APR-39 VIKING III threat warning system	Litton Applied Technology
AN/ALQ-131 receiver/processor	Lockheed Martin Electronic Defense Systems
AN/APR-38 Control Indicator Set for the F-4G Wild Weasel	Lockheed Martin Electronic Defense Systems
AN/APR-43 receiver	Lockheed Martin Electronic Defense Systems/Tracor Aerospace Electronic Systems Inc
AN/APR-38/47 radar warning receiver	Lockheed Martin Federal Systems
AN/APR-50 Radar Warning Receiver	Lockheed Martin Federal Systems
AN/ALE-39 chaff dispenser	Lockheed Martin Tactical Defense Systems
AN/ALQ-76 jamming pod	McDonnell Douglas Aerospace
F-4G Wild Weasel electronic warfare aircraft	McDonnell Douglas Aerospace
EA-6B Prowler Electronic Warfare aircraft	Northrop Grumman Corporation
EF-111A Raven Electronic Warfare aircraft	Northrop Grumman Corporation
AN/ALQ-101 noise/deception jamming pod	Northrop Grumman Corporation
Airborne Radar Jamming System (ARJS)	Northrop Grumman Corporation
AN/AAR-44(V) infrared warning system	Raytheon E-Systems, Goleta Division
AN/APR-39(V)1 radar warning receiver	Raytheon E-Systems, Goleta Division
Compact signal intelligence workstation	Raytheon E-Systems, Goleta Division
S200 self-defense jammer	Rodale Electronics Inc/Ericsson Saab Avionics AB
AN/ALQ-149 communications countermeasures system	Sanders, a Lockheed Martin Company
OnBoard Electronic Warfare Simulator (OBEWS)	Sanders, a Lockheed Martin Company
Data handling	
Canada	
Interface Converter Unit (ICU)	Computing Devices Canada
France	
Type 30 Pressure and Temperature Measurement Units (PTMU)	Sextant Avionique
Type 40 Pressure Measurement Unit (PMU)	Sextant Avionique
Type 120 air data computer	Sextant Avionique
Data analogue converter	Sextant Avionique
Israel	
ACE-3 computer	Elbit Ltd
ACE-4 computer	Elbit Ltd
Memory Loader Verifier (MLV)	Elisra Electronic Systems Ltd
EL/S-9000 computer	Elta Electronics Ltd
Intelligent Data Transfer Cartridge (DTC) system	Rada Electronic Industries Ltd
RMB 2000 advanced MIL-SPEC memory board	Rada Electronic Industries Ltd
Rover RD-220 MIL-SPEC terminal and RD-220PC MIL-SPEC computer	Rada Electronic Industries Ltd
Rover RD-286 MIL-SPEC computer	Rada Electronic Industries Ltd
Italy	
Mission Data Transfer System (MDTS)	Nardi Sistemi Elettronica SpA
RFAS	
CBC-85 digital air data computer	Aviaexport
UK	
53-020-01 Digital Signal Data Converter (DSDC)	GEC-Marconi Avionics Ltd, Rochester
USA	
Fibre optic High-Speed DataBus (HSDB)	Harris Corporation
MIL-STD-1553 EMUX system for B-1B	Harris Corporation
ARINC 706 Digital Air Data Computer (DADC)	Honeywell Inc Air Transport Systems
PG 1152AC03 Air Data Module (ADM)	Honeywell Inc Air Transport Systems
PG 1152AC04 Air Data Module (ADM)	Honeywell Inc Air Transport Systems
Pressure Transducer Module (PTM)	Honeywell Inc Air Transport Systems
LG1189 Engine Pressure Ratio Transmitter (EPRT)	Honeywell Inc Business & Commuter Aviation Systems

ENTRIES DELETED FROM THIS EDITION

Entry	Contractor
LG1197 Engine Pressure Ratio Transmitter (EPRT)	Honeywell Inc Business & Commuter Aviation Systems
MIL-STD-1553 terminals	ILC Data Device Corporation
Enhanced General Avionics Computer (EGAC)	Litton Guidance & Control Systems
TDY-750 advanced standard computer	Litton Guidance & Control Systems
TDY-750EV enhanced advanced standard computer	Litton Guidance & Control Systems
Data transfer equipment	Lockheed Martin Advanced Recorders
Mass Storage Unit (MSU)	Lockheed Martin Advanced Recorders
Remote terminal	Lockheed Martin Advanced Recorders
STAR single board computer	Sanders, a Lockheed Martin Company
Data recording	
Canada	
CMA-3030 Cockpit Voice Recorder (CVR)	Canadian Marconi Company
Poland	
ATM-QR3 Aircraft Integrated Monitoring Systems (AIMS)	ATM Inc
Sweden	
AMOT engine data acquisition unit	CelsiusTech Electronics AB
SSC airborne camera system	CelsiusTech Electronics AB
USA	
Engine Analyser Unit (EAU)	Ametek Aerospace Products
DCRsi Digital Cassette Recording system	Ampex Corporation Data Systems Division
ELI-101A engine monitor	BFGoodrich Aerospace Avionics Systems
Disk Drive Unit	Computing Devices International
MCU 110 and 111 Management Control Units	Hamilton Standard Division of UTC
SDC-100 Signal Data Computer	Hamilton Standard Division of UTC
AN/AXQ-15 Helicopter Installed Television Monitor Recorder (HITMORE)	Lockheed Martin Fairchild Systems
AN/USQ-85 Turbine Engine Monitoring System (TEMS)	Northrop Grumman Corporation
AN/ASH-27 signal data recorder	Solaris Systems
Magnetic mass memory products	Solaris Systems
Model 885 intensified CCD video camera	Videospection Inc
Model 891 miniature CCD video camera	Videospection Inc
Model 2091 high-resolution CCD video camera	Videospection Inc
Model 2531-B cockpit colour television camera	Videospection Inc
Model 4087 air-to-air tanker video system	Videospection Inc
Navigation and nav/attack	
Canada	
CMA-734 Omega/VLF systems	Canadian Marconi Company
CMA-764-1 GPS/Omega/VLF sensor system	Canadian Marconi Company
CMA-771 GPS/Omega VLF systems	Canadian Marconi Company
CMA-2000 microlander microwave landing system	Canadian Marconi Company
CMA-3000 Helicopter Tactical Navigation System (HTNS)	Canadian Marconi Company
Portable Flight Inspection System (PFIS)	Litton Systems Canada Ltd
France	
Avionique Nouvelle (AN) series suite for helicopters	Sextant Avionique/SFIM Industries
Nadir 10 integrated navigation and mission management system	Sextant Avionique
Navigation and attack system for helicopters	Sextant Avionique
Ring Laser Gyro (RLG) inertial reference units	Sextant Avionique
Strapdown Heading and Attitude Reference Platform (SHARP)	SFIM Industries
8900 VOR/ILS	Thomson-CSF Communications
TDM-709 distance measuring equipment	Thomson-CSF Communications
TLS-2000 integrated airborne landing system	Thomson-CSF Communications
Germany	
VOR/ILS NAV 900 system	Becker Avionic Systems
International	
M-ADS Modified Automatic Dependent Surveillance system	Kongsberg Aerospace/Racal Avionics Ltd
Euronav GPS system	Hughes Aircraft Company/Acatel SEL AG/Elmer SpA
Israel	
Low-Cost Weapon Delivery System (LCWDS)	Elbit Ltd

Entry	Contractor
Nightsight integrated avionics system	Elbit Ltd
Update weapon delivery and navigation system for F-4	Elbit Ltd
WDNS 391 weapon delivery systems	Elbit Ltd
GPS Sky receiver	Rokar International Ltd
Aided Inertial Navigation Systems (AINS)	Tamam Precision Instrument Industries
Cobra laser night attack system	Tamam Precision Instrument Industries
UK	
STR 2100 series Fixed Reception Pattern Antenna (FRPA)	Cossor Electronics Ltd
STR 2200 series controlled reception pattern antenna	Cossor Electronics Ltd
STR 2300 series antenna electronics unit	Cossor Electronics Ltd
STR 2400 series antenna control unit	Cossor Electronics Ltd
Solid-State Angular Rate Transducer (START)	GEC-Marconi Avionics, Rochester
RNS 5000 area navigation system	Racal Avionics Ltd
Deltafix LR differential GPS receiver	Racal Survey Ltd
Euronav III moving map and task management system	Skyquest Aviation
USA	
Multiport NDB-2 Navigation Data Bank	AlliedSignal Commercial Avionics Systems
Navision 1000/2000 navigation management systems	Arnav Systems Inc
R-40/R-60 Loran C receivers	Arnav Systems Inc
Locator Loran receiver	Azure Technology
Long Ranger Loran C system	Azure Technology
Long Ranger Plus Loran C receiver	Azure Technology
ELF V Electronic Location Finder	Cubic Defense Systems
MLZ-900 Microwave Landing System (MLS) receiver	Honeywell Inc Business & Commuter Aviation Systems
IEC 9001 GPS Navigation and Landing System (GNLS)	Interstate Electronics Corporation
KH-0002 Low-cost Attitude Heading Reference System (LAHRS)	Kearfott Guidance and Navigation Corporation
SKH-4212 ring laser gyro Attitude Motion Sensor Set (AMSS)	Kearfott Guidance and Navigation Corporation
SKN-2416 inertial navigation system	Kearfott Guidance and Navigation Corporation
SKN-2440 High-Accuracy Inertial Navigation System (HAINS)	Kearfott Guidance and Navigation Corporation
SKN-2443 High-Accuracy Inertial Navigation System (HAINS)	Kearfott Guidance and Navigation Corporation
EC-10X cartographic GPS navigator	Magellan Systems Corporation
Nav 1000M5 GPS receiver	Magellan Systems Corporation
SkyNav 5000 GPS receiver	Magellan Systems Corporation
Escort II nav/com receiver	Narco Avionics Inc
NCS 812 nav/com/DME system	Narco Avionics Inc
NS 801/NS 800 Rnav systems	Narco Avionics Inc
AN/ARN-130 Tacan	NavCom Defense Electronics Inc
XR5-M GPS receiver	NavSymm
NAS-21 astro/inertial navigation system	Nothrop Grumman Corporation
GPS navigator	Trimble Navigation Ltd, Avionics Products
TNL 1000 GPS receiver	Trimble Navigation Ltd, Avionics Products
TNL 7880 Airborne GPS/Omega/VLF navigation system	Trimble Navigation Ltd, Avionics Products
TNL 7900 Airborne GPS/Omega/VLF navigation systems	Trimble Navigation Ltd, Avionics Products
TNL 8000 airborne GPS sensor/navigator	Trimble Navigation Ltd, Avionics Products
Flight management and control	
Canada	
Electro-Thermal Ice Protection Systems (ETIPS)	AlliedSignal Aerospace Canada
France	
BR 1260/BR 2060 analogue control units	ELECMA, the Electronics Division of SNECMA
CP 1654 airborne engine life monitoring unit	ELECMA, the Electronics Division of SNECMA
RM 1627 Full Authority Digital Engine Control (FADEC) unit	ELECMA, the Electronics Division of SNECMA
RN 1763 and 1764 Full Authority Digital Engine Control (FADEC) units	ELECMA, the Electronics Division of SNECMA
RM 1901/RN 2288 Full Authority Digital Engine Control (FADEC) units	ELECMA, the Electronics Division of SNECMA
RN 1993 Full Authority Digital Engine Control (FADEC) unit	ELECMA, the Electronics Division of SNECMA

Entry	Contractor
RN 2151 and RN 2185 hardened Full Authority Digital Engine Control (FADEC) units	ELECMA, the Electronics Division of SNECMA
TE 1374 Adour engine electronic fuel dipping timer box	ELECMA, the Electronics Division of SNECMA
VT 1153/1160/1840 Adour engine control amplifiers	ELECMA, the Electronics Division of SNECMA
AP 725 autopliot	Sextant Avionique
Centre of Gravity Control Computer (CGCC)	Sextant Avionique
Elevator and Aileron Computer (ELAC) for the A320	Sextant Avionique
Engine Interface Unit	Sextant Avionique
Germany	
APU Control for the A319/A320/A321/A330/A340	Bodenseewerk Gerattechnik GmbH/BGT
Digital Engine Control Unit (DECU)	Bodenseewerk Gerattechnik GmbH/BGT
Thrust control computer for the Airbus A300-600 and A310	Bodenseewerk Gerattechnik GmbH/BGT
Versatile Electronic Control Box	Bodenseewerk Gerattechnik GmbH/BGT
Cabin Pressure Control System (CPCS)	Nord-Micro Elektronik Feinmechanik AG
India	
Engine Electrical Monitoring System (EEMS)	Hindustan Aeronautics Ltd
International	
Digital Engine Control Unit for the EJ2000 engine	Dornier GmbH/ENOSA/Tecnost SpA
EF 2000 Stick Sensor and Interface Control Assembly (SSICA)	GEC-Marconi Avionics/BGT/GF-Sistemi Avionici SpA/ENOSA
Engine Monitoring Unit (EMU) for the EF 2000	ENOSA/GEC-Marconi Avionics/Microtecnica
FADEC co-operative programmes	ELECMA/General Electric Aircraft Engines/Lockheed Martin Control Systems
FADEC for the MTR 390 turboshaft engine	ELECMA/BGT
FADEC for the PZL-10W turboshaft engine	ELECMA/PZL-HYDRAL SA
Fuel Savings Advisory System (FSAS) for KC-135	GEC-Marconi Avionics/Lear Astronics Corp
Onboard Maintenance Terminal (OMT)	Computing Devices International/Sextant Avionique
Slat/Flap Control Computer (SFCC) for the Airbus A330/A340	AlliedSignal Commercial Avionics Systems/BGT
Israel	
HALO advanced helicopter avionics	Elbit Ltd
Advanced Brake Controller (ABC) for F-16	RSL Electronics Ltd
AS-3000 Diagnostic Engine Start System Controller (DESSC) for F-16	RSL Electronics Ltd
AS-3100 Digital Generator Control Units (DGCU) for F-4	RSL Electronics Ltd
RS-404 Digital Temperature Control Amplifier (DTCA) for J79 engine	RSL Electronics Ltd
Fuel Asymmetry Caution Unit (FACU) for F-15	RSL Electronics Ltd
Italy	
FQG-28 Fuel Quantity Gauging system	Ottico Meccanica Italiana SpA
Japan	
Yaw damper system for the T-4	Japan Aviation Electronics Industry Ltd
Fly-by-wire actuator and controller	Shimadzu Corporation
UK	
Turbine engine vibration monitoring system	British Aerospace (Systems & Equipment) Ltd
Intelligent Power Control System (IPCS)	GEC-Marconi Avionics
Full Flight Regime Autothrottle for the Boeing 747	GEC-Marconi Avionics
Slat/Flap Control Computer (SFCC) for the Airbus A310 and A300-600	GEC-Marconi Avionics
Slat/Flap Control Computer (SFCC) for the Airbus A319, A320 and A321	GEC-Marconi Avionics
Spoiler Electronics Control Unit (SECU) for the Canadair Regional Jet (RJ)	GEC-Marconi Avionics
Yaw damper controller for the T-45A	GEC-Marconi Avionics
Digital Engine Control System (DECS) for vectored thrust power plants	Lucas & Smiths Industries Controls Ltd
Actuators for the BR700 engine	Lucas Varity Aerospace
AE2100 D3 Full Authority Digital Engine Control (FADEC)	Lucas Varity Aerospace
Automatic Ignition Unit (AIU)	Lucas Varity Aerospace
Auxiliary and emergency power control unit	Lucas Varity Aerospace
CUE-400 main engine control unit	Lucas Varity Aerospace
DECU-500 Digital main Engine Control Unit	Lucas Varity Aerospace
Dedicated alternator for the RB211-524G/H	Lucas Varity Aerospace
Engine transient pressure unit	Lucas Varity Aerospace
ESC-102 Engine Supervisory Control	Lucas Varity Aerospace
FADEC for the Trent engine	Lucas Varity Aerospace
FAFC 2000 Full Authority Fuel Control	Lucas Varity Aerospace
Flap control unit	Lucas Varity Aerospace
Generic variable stator vane actuation system for the RB211 engine	Lucas Varity Aerospace
GMA2100/GMA3007 FADEC system	Lucas Varity Aerospace
Power management unit	Lucas Varity Aerospace
SDS-400 engine control computer	Lucas Varity Aerospace
Steering control unit	Lucas Varity Aerospace
T406 FADEC system	Lucas Varity Aerospace
Auxiliary power unit controller	Normalair-Garrett Ltd
Aircraft nose wheel steering controller	Normalair-Garrett Ltd
Emergency power supply	Page Aerospace Ltd
BR710 Full Authority Digital Engine Control (FADEC)	RoSEC
EEC2000 FADEC	Smiths Industries Aerospace
Engine limiters	Smiths Industries Aerospace
Health and Usage Monitoring System (HUMS)	Smiths Industries Aerospace
Radiation pyrometer	Smiths Industries Aerospace
'Smart' throttle actuators	Smiths Industries Aerospace
Temperature monitor units	Smiths Industries Aerospace
SN700 series yaw damper computer	Smiths Industries Aerospace
Engine control accessories	Ultra Electronics, Controls Division
Flaps computer for the Avro regional jet	Ultra Electronics, Controls Division
Hub Integrated Power and Switching System (HIPSS)	Ultra Electronics, Controls Division
Hydraulic supply circuit Electronic Control Unit (H-ECU)	Ultra Electronics, Controls Division
Landing Gear Computer and Interface Unit (LGCIU)	Ultra Electronics, Controls Division
Landing gear computer modules for the EF 2000	Ultra Electronics, Controls Division
Propeller electronic controller	Ultra Electronics, Controls Division
Remote Data Concentrators (RDC)	Ultra Electronics, Controls Division
Rotortuner 1000/2000	Ultra Electronics, Helitune Division
Rudder Control Unit (RCU)	Ultra Electronics, Controls Division
Timer/Monitor Control Unit (TMCU)	Ultra Electronics, Controls Division
UltraQuiet Active Tuned Vibration Attenuators (ATVA)	Ultra Electronics, Noise and Vibration Systems
T55 DECU Digital Electronic Control Unit	Vosper Thorneycroft Controls Ltd
USA	
EGPWS: Enhanced Ground Proximity Warning System	AlliedSignal Commercial Avionics Systems
Digital electronic control for the GE38/CFE738 engine family	AlliedSignal Engine Systems & Accessories
Engine Monitor Multiplex Unit (EMMU) for the RB211-524	AlliedSignal Engine Systems & Accessories
Fuel handling unit for the CF6-80C2 FADEC engine	AlliedSignal Engine Systems & Accessories
Supplemental Control Unit (SCU) for the PW4000 Engine	AlliedSignal Engine Systems & Accessories
Digital engine control for TFE109 turbofan engine	AlliedSignal Engine Systems & Accessories
Electronic control unit for the Garrett GTC 36-200 auxiliary power unit	AlliedSignal Engine Systems & Accessories
Engine performance reserve controller for the TFE731 engine	AlliedSignal Engine Systems & Accessories
Engine power trim system	AlliedSignal Engine Systems & Accessories
Engine synchroniser	AlliedSignal Engine Systems & Accessories
Full Authority Digital Engine Control for the Garrett TFE731 engine (FADEC)	AlliedSignal Engine Systems & Accessories
Full authority engine control for the Garrett ATF-3 engine	AlliedSignal Engine Systems & Accessories
Full authority engine control for the Garrett GTCP 36-50, -55 and -100 auxiliary power units	AlliedSignal Engine Systems & Accessories
Full authority engine control for the Garrett CTCP 36-150 auxiliary power unit	AlliedSignal Engine Systems & Accessories
Full authority engine control for the Garrett GTCP 660-4 auxiliary power unit	AlliedSignal Engine Systems & Accessories
Full authority engine control for the Garrett TPE331 engine	AlliedSignal Engine Systems & Accessories

ENTRIES DELETED FROM THIS EDITION

Entry	Contractor
Integrated engine computer for TPE331-14 power plant	AlliedSignal Engine Systems & Accessories
Propeller synchrophaser	AlliedSignal Engine Systems & Accessories
Conductive fuel gauging systems	Ametek Aerospace Products
Fuel gauging systems	Ametek Aerospace Products
Liquid level measurement system	Ametek Aerospace Products
Mass fuel flow indicating system	Ametek Aerospace Products
Electronic ballast	AVTECH Corporation
Windshield Temperature Controller (WTC)	AVTECH Corporation
Fuel gauging and fuel level sensing systems	BFGoodrich Aerospace Aircraft Integrated Systems
PS-834 lightweight power supply	BFGoodrich Aerospace Avionics Systems
PS-855 emergency power supply	BFGoodrich Aerospace Avionics Systems
Special purpose moitoring systems	BFGoodrich Aerospace Avionics Systems
Oil debris detection and emergency lubrication control system	BFGoodrich Aerospace Miltary Fuel & Integrated Systems Division
KT Series of Helicopter Health and Usage Monitors (HUMS)	BFGoodrich Aerospace, Technology Integration Inc
Yaw damper	Century Flight Systems Inc
AIC 12 air inlet control for the F-15	Hamilton Standard Division of UTC
ASCT100 Air Supply Controller/Test Unit (ASCTU)	Hamilton Standard Division of UTC
CPC22 cabin pressure controller	Hamilton Standard Division of UTC
CPC32 cabin pressure controller	Hamilton Standard Division of UTC
CPC100 cabin pressure controller	Hamilton Standard Division of UTC
CTC111 zone temperature controller	Hamilton Standard Division of UTC
CTC114 pack temperature controller	Hamilton Standard Division of UTC
CTC129 pack temperature controller	Hamilton Standard Division of UTC
CTC130 zone temperature controller	Hamilton Standard Division of UTC
EC300/200TC-1 Cabin Temperature Controller (CTC)	Hamilton Standard Division of UTC
EDM110 and EDM112 PMUX propulsion multiplexers	Hamilton Standard Division of UTC
EEC90 engine supervisory control	Hamilton Standard Division of UTC
EEC103 engine supervisory control	Hamilton Standard Division of UTC
EEC104 full authority engine control for the Boeing 757	Hamilton Standard Division of UTC
EEC106 Digital Electronic Engine Control (DEEC)	Hamilton Standard Division of UTC
EEC118 Multiple Application Control System (MACS)	Hamilton Standard Division of UTC
EEC131 engine control for the PW4000 turbofan engine	Hamilton Standard Division of UTC
EEC132 multiple application control	Hamilton Standard Division of UTC
EEC150 full authority engine control	Hamilton Standard Division of UTC
EEC153 APU control for the PW901A	Hamilton Standard Division of UTC
EEC160 electronic control for the PW545 engine	Hamilton Standard Division of UTC
EEC206 electronic engine control for the PW206 engine	Hamilton Standard Division of UTC
Engine Data Interface Unit (EDIU)	Hamilton Standard Division of UTC
EPR102 Engine Pressure Ratio transmitter	Hamilton Standard Division of UTC
FCC110-1 stabilator control for the AH-64 Apache	Hamilton Standard Division of UTC
Flight management system PMS 500	Hamilton Standard Division of UTC
PSC100/101/102 solid-state propeller synchrophaser	Hamilton Standard Division of UTC
PSC103 digital propeller synchrophaser	Hamilton Standard Division of UTC
Air/Ground System (AGS)	Honeywell Inc, Air Transport Systems
Automatic flight control system for the AS 350/355	Honeywell Inc, Business & Commuter Aviation Systems
Automatic flight control system for the Bell 212 and 412	Honeywell Inc, Business & Commuter Aviation Systems
Flight Management System (FMS) for General Aviation	Honeywell Inc, Business & Commuter Aviation Systems
Digital fully fly-by-wire system for JAS-39	Lear Astronics Corporation
Fuel savings advisory system for the KC-135	Lear Astronics Corporation
Commercial FADECs	Lockheed Martin Control Systems
Airspeed select features for SCAT autopower systems	Safe Flight Instrument Corporation
Speed Command of Attitude and Thrust (SCAT) system	Safe Flight Instrument Corporation
Electrostatic Engine Monitoring System (EEMS)	Smiths Industries Aerospace
Accelerometer-controlled yaw damper	S-TEC Corporation

Cockpit displays, instruments and indicators

Entry	Contractor
France	
ALEV 3 laser airspeed measurement system	Sextant Avionique
Cabin pressure indicator	Sextant Avionique
AC 6445 data entry panel and time display	SFIM Industries
AC 6620 cockpit display unit	SFIM Industries
ISG 80 attitude indicator	SFIM Industries
Germany	
Lighting dimmer	Diehl GmbH & Co Luftfahrt Elektronik
Warning system	Diehl GmbH & Co Luftfahrt Elektronik
Beamstar colour multifunction display	ESW-Extel Systems Wedel
Israel	
Fighter display system	Elbit Ltd
MultiFunction Colour Displays (MFCD)	Elbit Ltd
NTS display	Elbit Ltd
UK	
C2G-gyrosyn compass system	British Aerospace (Systems & Equipment) Ltd
CL 11 directional gyro	British Aerospace (Systems & Equipment) Ltd
GM9-gyro magnetic compass system	British Aerospace (Systems & Equipment) Ltd
Synchro repeater Type A	British Aerospace (Systems & Equipment) Ltd
EuroNav III moving map display	Skyquest Aviation
USA	
C-17A main battery charger	Aerospace Avionics
F-16 battery charger/bus control system	Aerospace Avionics
VG-204 Vertical Gyro	BFGoodrich Aerospace Avionics Systems
VG-208 Vertical Gyro	BFGoodrich Aerospace Avionics Systems
Liquid Crystal Displays (LCDs)	Collins Avionics & Communications
Starship 1 integrated avionics system	Collins Avionics & Communications
AD-300 series Attitude Director indicators	Honeywell Inc Business & Commuter Aviation Systems
AD-500/550 Attitude Director indicator	Honeywell Inc Business & Commuter Aviation Systems
AD-600/650 Attitude Director indicator	Honeywell Inc Business & Commuter Aviation Systems
AD-800 Attitude Director indicator	Honeywell Inc Business & Commuter Aviation Systems
ARINC 700 symbol generator	Honeywell Inc Business & Commuter Aviation Systems
EDZ-800 Electronic Flight Instrument System (EFIS) for general aviation	Honeywell Inc Business & Commuter Aviation Systems
GH-14 attitude director indicator	Honeywell Inc Business & Commuter Aviation Systems
HAI 5 Heading and Attitude Indicator	Honeywell Inc Business & Commuter Aviation Systems
HZ-6F attitude director indicator	Honeywell Inc Business & Commuter Aviation Systems
LSZ-850 lightning sensor system	Honeywell Inc Business & Commuter Aviation Systems
RD-450 Horizontal Situation Indicator (HSI)	Honeywell Inc Business & Commuter Aviation Systems
RD-550 Horizontal Situation Indicator (HSI)	Honeywell Inc Business & Commuter Aviation Systems
RD-650 Horizontal Situation Indicator (HSI)	Honeywell Inc Business & Commuter Aviation Systems
Air Navigation Multiple Indicator (ANMI) for the F-15	Honeywell Inc Defense Avionics Systems
Monochrome multifunction display	Honeywell Inc Defense Avionics Systems
AN/AVA-1 electronic attitude display for the A-6	Kaiser Electronics
AN/AVA-12 attitude display for the F-14	Kaiser Electronics
Kollsman Auto-schedule Pressurisation System (KAPS)	Kollsman Inc
CN-998B/ASN-43A Military directional gyro	Litton Special Devices
DCU-201 Digital Convertor Unit	Litton Special Devices
DG-9100 slaved directional gyro system	Litton Special Devices
RVG-801 Remote Vertical Gyros	Litton Special Devices
Avionics Management Unit (AMU)	Sanders, a Lockheed Martin Company
COM-NAV/Breaker Panel (CNBP)	Sanders, a Lockheed Martin Company
Control and Display System for the C-130J	Sanders, a Lockheed Martin Company

Head-up displays, helmet-mounted displays and weapon aiming sights

Entry	Contractor
France	
Strix sight for armed reconnaissance and anti-tank helicopters	SFIM Industries
Viviane day and night sights for helicopters	SFIM Industries
International	
Head-up displays for corporate aircraft	GEC-Marconi Avionics/Honeywell Inc Defense Avionics Systems

Entry	Contractor
HUD 2020	Honeywell Inc Business & Commuter Aviation Systems/GEC-Marconi Avionics Ltd, Mission Avionics Division
Israel	
AN/AVS-7 ANVIS/HUD - *moved to International section*	Elbit Ltd
UK	
Alpha Sight system	GEC-Marconi Avionics Ltd, Mission Avionics Division, Rochester
USA	
AN/ASG-26A Lead-computing optical sight	Lockheed Martin Ocean Radar and Sensor Systems
ANVIS/HUD system - duplicate entry	Tracor Inc

Stores management systems

Entry	Contractor
Israel	
Sidewinder Control System (SCS)	Elbit Ltd

ADMINISTRATION

Director, Defence Business Unit: *Alan Condron*
Publisher: *Karen Heffer*
Managing Editor: *Simon Michell*
Database Manager/Data Administrator: *Ruth Simmance*
Editorial: *Geoff Vince*

EDITORIAL OFFICE

Jane's Information Group Limited, Sentinel House, 163 Brighton Road,
Coulsdon, Surrey CR5 2YH, UK
Tel: (+44 181) 700 37 00 Fax: (+44 181) 700 37 88
email: jav@janes.co.uk

SALES OFFICES

Send enquiries to: *Tony Kingham – International Sales Manager*
(Europe, Russian Federation and Associated States (CIS), Africa, Middle East)

Jo Moon (Scandinavia, Far East, UK)
Jane's Information Group Limited, UK address as Editorial Office
Tel: (+44 181) 700 37 59 Fax: (+44 181) 763 10 06

Send USA enquiries to: *Robert Loughman – Vice-President Product Sales*
Jane's Information Group Inc, 1340 Braddock Place, Suite 300, Alexandria,
Virginia 22314-1651, USA
Tel: (+1 703) 683 37 00 Fax: (+1 703) 836 00 29 Telex: 6819193

Send Asia enquiries to: *David Fisher*
Jane's Information Group Asia, 60 Albert Street, #15-01 Albert Complex,
Singapore 189969
Tel: (+65) 336 64 11 Fax: (+65) 336 99 21

Send Australia/New Zealand enquiries to: *Pauline Roberts*
Jane's Information Group, PO Box 3502, Rozelle,
New South Wales 2039, Australia
Tel: (+61 2) 85 87 79 00 Fax: (+61 2) 85 87 79 01

Advertisement Sales Manager: *Richard West*
Jane's Information Group, Sentinel House, 163 Brighton Road, Coulsdon,
Surrey CR5 2YH, UK
Tel: (+44 181) 700 37 39 Fax: (+44 181) 700 37 44
email: richard.west@janes.co.uk

Australia: *Richard West*
(see Advertisement Sales Manager)

Austria and Germany (South): *Nathalie Rosmorduc*
Jane's Information Group, Sentinel House, 163 Brighton Road, Coulsdon,
Surrey CR5 2YH, UK
Tel: (+44 181) 700 39 61 Fax: (+44 181) 700 37 44
email: nathalie.rosmorduc@janes.co.uk

Benelux: *Nathalie Rosmorduc*
Jane's Information Group (see Austria and Germany (South))

Brazil: *Richard West*
(see Advertisement Sales Manager)

Czech Republic and Slovakia: *Nathalie Rosmorduc*
Jane's Information Group (see Austria and Germany (South))

France: *Patrice Février*
Jane's Information Group – France, BP 418, 35 avenue MacMahon,
F-75824 Paris Cedex 17
Tel: (+33 1) 45 72 33 11 Fax: (+33 1) 45 72 17 95
email: patrice.fevrier@wanadoo.fr

Germany (North): *Richard West*
(see Advertisement Sales Manager)

Greece and Turkey: *Richard West*
(see Advertisement Sales Manager)

India and Pakistan: *Richard West*
(see Advertisement Sales Manager)

Ireland: *Nathalie Rosmorduc*
Jane's Information Group (see Austria and Germany (South))

Israel: *Oreet Ben-Yaacov*
Oreet International Media, 15 Kinneret Street, IL-51201 Bene Berak
Tel: (+972 3) 570 65 27 Fax: (+972 3) 570 65 26
email: oreetimc@netvision.net.il

Italy and Switzerland: Ediconsult Internazionale Srl
Piazza Fontane Marose 3, I-16123 Genova, Italy
Tel: (+39 10) 58 36 84 Fax: (+39 10) 56 65 78
email: ediconsult@iol.it

Japan: Richard West
(see Advertisement Sales Manager)

Korea, South: *Young Seoh Chinn*
JES Media International, 6th Floor Donghye Building, 47-16 Myungil-Dong,
Kangdong-Gu, Seoul 134-070
Tel: (+82 2) 481 34 11 Fax: (+82 2) 481 34 14
email: jesmedia@unitel.co.kr

Poland: *Nathalie Rosmorduc*
Jane's Information Group (see Austria and Germany (South))

Russian Federation and Associated States (CIS): *Simon Kay*
23 Church Road West, Crowthorne, Berkshire RG45 7NQ, UK.
Tel: (+44 1344) 777123 Fax: (+44 1344) 775885
email: crowkay@msn.com

Scandinavia: *Gillian Thompson*
The Falsten Partnership, 11 Chardmore Road, Stamford Hill,
London N16 6JA, UK
Tel: (+44 181) 806 23 01 Fax: (+ 44 181) 806 81 37
email: falsten@dial.pipex.com

Singapore: *Richard West*
(see Advertisement Sales Manager)

South Africa: *Richard West*
(see Advertisement Sales Manager)

Spain: *Michael Andrade*
Via Exclusivas SL, Modesto Lafuente 4, E-28010 Madrid
Tel: (+34 91) 448 76 22 Fax: (+34 91) 446 01 98
email: via@varenga.com

Thailand: *Richard West*
(see Advertisement Sales Manager)

UK – Midlands and Wales: *Nathalie Rosmorduc*
Jane's Information Group (see Austria and Germany (South))

UK – North and Scotland: *Richard West*
(see Advertisement Sales Manager)

UK – South: *Richard West*
(see Advertisement Sales Manager)

USA, Mid Atlantic and Midwest; New England and Canada:
Kimberley S Hanson
Global Media Services Inc, 1227 Shaker Drive, Suite 100, Herndon,
Virginia 20170, USA
Tel: (+1 703) 406 88 00 Fax: (+1 703) 406 81 08
email: kim@globlmedia.com

Southern USA: *Kristin Schulze*
Global Media Services Inc, 5370 East Bay Drive, Suite 104, Clearwater,
Florida 33764, USA
Tel: (+1 813) 524 77 41 Fax: (+1 813) 524 75 62
email: kristin@globlmedia.com

Western USA and Western Canada: *Richard Ayer*
Global Media Services Inc, 127 Avenida del Mar, Suite 2A, San Clemente,
California 92672, USA
Tel: (+1 949) 366 84 55 Fax: (+1 949) 366 92 89
email: ayercomm@earthlink.net

Administration:
USA and Canada: *Maureen Nute – Advertising Production Manager*
Jane's Information Group Inc, 1340 Braddock Place, Suite 300, Alexandria,
Virginia 22314-1651, USA
Tel: (+1 703) 683 37 00 Fax: (+1 703) 836 00 29
email: nute@janes.com

UK and Rest of World: *Fay Lenham*
Jane's Information Group Limited, Sentinel House, 163 Brighton Road,
Coulsdon, Surrey CR5 2YH, UK
Tel: (+44 181) 700 37 42 Fax: (+44 181) 700 38 59

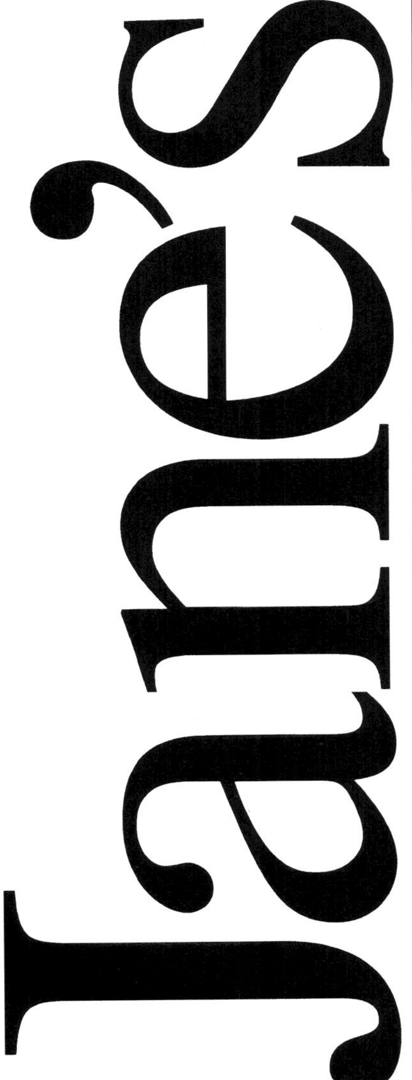

Users' Charter

This publication is brought to you by Jane's Information Group, a global company with more than 100 years of innovation and an unrivalled reputation for impartiality, accuracy and authority.

Our collection and output of information and images is not dictated by any political or commercial affiliation. Our reportage is undertaken without fear of, or favour from, any government, alliance, state or corporation.

We publish information that is collected overtly from unclassified sources, although much could be regarded as extremely sensitive or not publicly accessible.

Our validation and analysis aims to eradicate misinformation or disinformation as well as factual errors; our objective is always to produce the most accurate and authoritative data.

In the event of any significant inaccuracies, we undertake to draw these to the readers' attention to preserve the highly valued relationship of trust and credibility with our customers worldwide.

If you believe that these policies have been breached by this title, you are invited to contact the Editor.

A copy of Jane's Information Group's Code of Conduct for its editorial teams is also available from the Editor.

INVESTOR IN PEOPLE

A two-seat Super Hornet, designated F/A-18F, launching AIM-120 Advanced Medium-Range Air-to-Air Missile (AMRAAM) *1998*/0018934

Foreword

What is new in the 1998-99 edition of *Jane's Avionics?*

This edition of *Jane's Avionics* contains over 2,700 entries. There are over 300 new entries (with a significant number being from the Russian Federation), over 430 new photographs, 950 updated entries, 700 entries verified by the editor with assistance from the relevant contractor, and 750 entries verified by the editor based on best data available to Jane's Information Group. Notwithstanding the most welcome assistance of the contractors in checking some of the entries marked verified, readers are cautioned that the responsibility for the data presented in *Jane's Avionics* is Jane's alone, representing the best endeavour of the editor at the time of going to press (early in 1998), subject to space and other constraints.

Again this year there has been pressure to constrain the growth in the size, weight and cost of *Jane's Avionics*. Avionics continues to be a very dynamic area of development, with: many new technical and operational developments; many new items of equipment; many significant modifications and upgrades to existing systems; and many new companies entering the business of avionics equipment design, development and manufacture. These factors have together contributed to the large number of new entries included in this edition, which represent a 12 per cent increase on the 1997-98 edition. However, publishing policy dictates against an increase in size of this magnitude each year and, accordingly, it has again been necessary to delete certain entries relating to equipment still in service but no longer in production, and to be yet more rigorous in limiting *Jane's Avionics* to the main aspects of permament fit avionic equipment concerned principally with the functions of piloting and navigation. The following types of equipment, that have appeared in earlier editions, have had to be deleted for 1998-99:

1. within the communications section: antennas, unless they include major avionic content, or have particular merit in terms of new technology or application;
2. within the flight management systems section: power system controls, which although now almost universally electronic, are not primarily avionic, but principally electronic analogues of earlier electro-mechanical control systems; Full Authority Digital Engine Controls (FADECs), which, again, although having significant electronic content, are principally only part of engine automation; control computers used as part of aircraft systems such as flaps, brakes and so on, as opposed to computers that form part of the main avionic functions concerned with aircraft piloting and navigation;
3. within the instruments section: gyros, unless they are an integral part of a display.

For reasons of space, it has been decided not to include any of the burgeoning array of In-flight Passenger Entertainment and Communications (IPEC) services; however, dual-use flight crew and passenger communication services have been included when they have significant importance as flight crew systems.

Mergers and takeovers

Publishing timescales in relation to company merger and take-over activities continue to present difficulties in providing latest data concerning affiliation of products to manufacturers. At the time of going to press, the formal status of various proposed mergers/takeovers remained uncertain, including the Lockheed Martin/Northrop Grumman and Dassault Electronique/Thomson-CSF situations, and accordingly relevant entries appear under their formal affiliations at the time. Similarly, the GEC-Marconi takeover of Tracor occurred too late to be reflected in the entry details of *Jane's Avionics*.

The Raytheon merger with Hughes' defense operations occurred during the production cycle, as did the formation of Raytheon Systems Company with its five major business segments: Defense Systems; Sensors and Electronic Systems; C^3 Systems; Intelligence, Information and Aircraft Integration Systems; and Training and Services. However information on the detailed allocation of individual items of avionic equipment to the new segments was not available, and although it appears probable that the vast majority of items relevant to this publication will be included within the Sensors and Electronic Systems Segment, there are some items that may be within the remit of the C^3 Segment, and the Intelligence, Information and Aircraft Integration Systems Segment. Accordingly, all items are shown under the overall contractor allocation of Raytheon Systems Company, at least for this year.

Avionic developments

Functional integration of avionic systems continues to be the main trend throughout the avionics industry, with navigation, flight management, communications and display systems being ever more

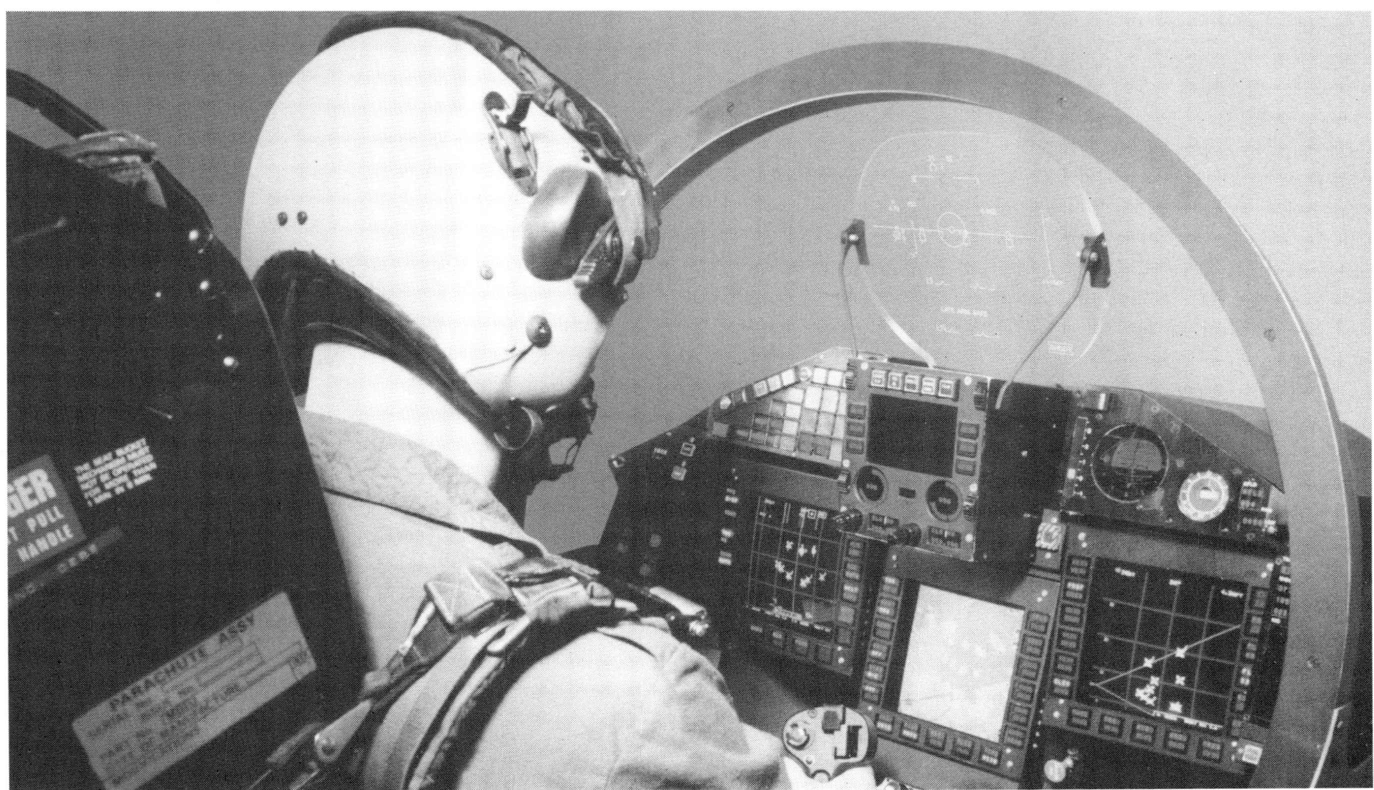

The Eurofighter cockpit illustrates the principles of a modern fighter aircraft cockpit, featuring an array of colour multifunction head-down displays with soft-key control, head-up display, helmet-mounted display, threat and other warning displays, and hands-on-throttle and stick controls **1998**/0018931

The flight deck of the New Generation Boeing 737-700 aircraft (left) and, on the right, a computer-generated image of the Airbus-proposed A3XX airliner. Both aircraft are designed to be operated in the future CNS/ATM environment **1998**/0018932/0016824

highly integrated. There are a number of operational and technical imperatives driving this trend. The continuing increase in computing power and the continuing miniaturisation of electronic components are driving forces, as is the continuing progress in the development of flat-panel colour Active-Matrix Liquid-Crystal Display (AMLCD) technology. The convergence of these developments has made it possible to develop ever larger MultiFunction Colour Display (MFCD) units that can be used for all avionics purposes, with the result that modern aircraft flight decks feature anything from one to 10 large AMLCDs, offering almost unlimited flexibility and redundancy. This capability has been harnessed most notably by Airbus to produce cockpit avionics that are so similar across the whole range of its aircraft that crew cross-training between types has been significantly reduced, and crewing flexibility much improved, offering significant savings in operating costs. This policy also produces significant potential flight safety benefits accruing from the similarity of aircraft cockpit layout, and the familiarity it provides to flight crews.

The ever increasing integration of avionic control and display functions outlined above matches well the capabilities required to implement the Future Air Navigation System/Air Traffic Management (FANS/ATM) and Communications Navigation Surveillance/Aeronautical Telecommunications Network (CNS/ATN) systems, and the requirements for transition to Global Positioning System-based (GPS-based) navigation and landing guidance. Implementation is now approaching, at least in the USA, where the US Federal Aviation Administration (FAA) has published its plan for the transition to GPS-based navigation and landing, together with the development of the Wide Area Augmentation System (WAAS) operational system and Local Area Augmentation System (LAAS) avionic architecture. It is planned that the Initial WAAS (IWAAS) capability will be achieved in 1998, with a build up of services throughout the period 2000-2010. At the same time the FAA plans to run down conventional navigation services, based on Omega, Loran-C, VOR/DME, TACAN and ILS/MLS.

The FAA plan is fine as a national policy, and as far as it goes, but it fails to address a number of widely held concerns. Amongst these are:

1. continuing concern about the vulnerability of the GPS system to intentional and unintentional interference if GPS is to become the sole means of navigation and landing;
2. continuing concern amongst many nations, other than the USA, about the total dependence of all civil aviation on the US-owned GPS system;
3. concern about commitment of the worldwide aviation community to funding the necessary WAAS/LAAS and CNS/ATN ground-based infrastructure;
4. concern about commitment on the part of all aircraft operators worldwide to funding and fitting WAAS/LAAS - CNS/ATN compatible avionics. Although regulatory pressures, added to financial benefits, may be adequate to induce most large civil operators to adopt CNS/ATM procedures, the response of many small civil operators, and many military forces, is less certain, with serious implications for air safety;
5. that ever greater dependence on the computer, as it becomes yet more dominant in data management and aircraft control, can result in the pilot losing situational awareness, with potentially catastrophic results.

Two military transport aircraft that are being updated to operate in the Global Air Transport Management (GATM) era, showing: on the left, the cockpit of the PACER CRAG Block 20 C/KC-135 aircraft (updated by Rockwell Collins), with displays for windshear radar and enhanced ground proximity warning system, and (right) the cockpit of the Lockheed Martin C-130J-30, the latest derivative of the Hercules transport aircraft, which in addition to a dual embedded GPS/INS system features twin head-up displays **1998**/0018929/0018930

Air safety

Together, the above factors lead directly to the major topic of air safety, which very correctly continues to be a priority concern. Fortunately, recent technical developments provide real gains in improving civil aircraft safety; three notable examples are:

1. Enhanced Ground Proximity Warning System (EGPWS): EGPWS is designed to provide greater warning of the possible Controlled Flight Into Terrain (CFIT) incidents that continue to be a major cause of accidents; EGPWS is already being adopted by many of the major airline operators ahead of the formal requirement for fitment being ordered by various regulatory bodies, and the indications at the time of going to press were that the FAA intends to issue a Notice of Proposed Rule Making (NPRM) in the near future to require the fitment of EGPWS (or Terrain Avoidance and Warning Systems [TAWS] as they are increasingly being called) into all passenger carrying jet and turbine aircraft with six or more seats, with a requirement for new aircraft to be fitted from December 2000 and all affected aircraft to be retrofitted by December 2003;
2. Head-Up Displays (HUDs): HUDs have been used for targeting and other functions in military aircraft for many years, but they are now being introduced into civil aircraft to aid the pilot in landing during adverse weather conditions and at airports with limited landing aids; again, HUD fitment is in process by a number of airlines, and widespread adoption seems probable for both safety and economic reasons;
3. Clear Air Turbulence (CAT) warning: CAT warning, as part of the weather radar system, is being fitted increasingly, to minimise the potential serious damage to aircraft and passengers associated with serious CAT conditions.

Acknowledgements

I am most grateful to the many contractors who replied to my request for help in checking the data presented on their products in the previous edition of *Jane's Avionics* (1997-98); comments and new data supplied have been most useful in preparing the 1998-99 edition.

I am also most grateful to the many staff of Jane's Information Group who have again guided me through the publishing process, particularly: Belinda Cunningham, who dealt with all the mailing activities; my copy editor, Geoff Vince, for his unceasing patience and help in preparing text and illustrations, Sharon Jackson, Sarah Erskine, Violet Carrett and Christine Varndell, who have been responsible for all the word processing.

Chris Johnson
Editor, *Jane's Avionics*
Fleet

May 1998

COMMUNICATIONS

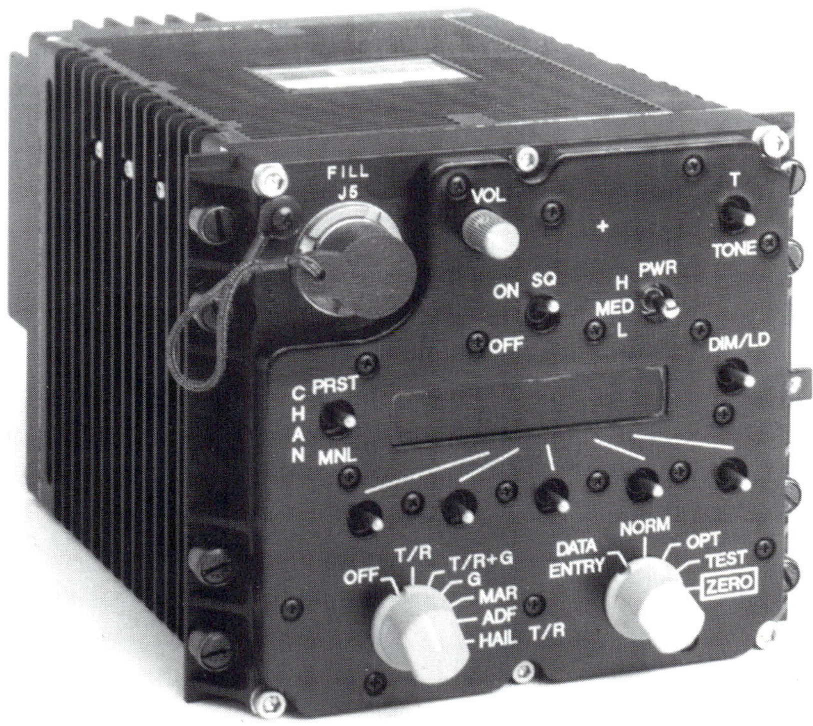

The Thomson-CSF Communications Transceivers Airborne TRA 6030 (above) and TRA 6020 (below) of the THOMRAD 6000 V/UHF ECCM transceiver series
1998/0011853/0011854

COMMUNICATIONS

CANADA

AMT-100 aeronautical mobile terminal

The AMT-100 aeronautical mobile terminal provides air-to-ground and ground-to-air voice communications to and from any location on the globe via the Inmarsat satellite system. The unit communicates through aeronautical standard ground stations operated by Inmarsat signatories.

The AMT-100 is an all-in-one system providing a single-voice channel with direct dial capability for automatic global operation. The system, consisting of a transceiver, high-power amplifier and non-obtrusive high-gain antenna, has been designed for corporate aviation aircraft.

The transceiver assembly handles all protocols, performs voice and data modulation and handles frequency conversion and Doppler correction. It also contains the antenna control processor. The high-power amplifier is a reliable solid-state unit designed for a wide range of aircraft installations. The antenna assembly is a high-gain tailfin-mounted steerable antenna which provides full azimuth and elevation coverage without gaps. The antenna may be pointed using aircraft navigation system inputs.

Specifications

Dimensions:
(transceiver) 381 × 190.5 × 317.5 mm
(amplifier) 127 × 190.5 × 317.5 mm
(diplexer/LNA) 431.8 × 50.8 × 198.1 mm
(antenna drive assembly) 355.6 × 63.5 × 190.5 mm
Weight:
(transceiver) 14.5 kg
(amplifier) 7.7 kg
(diplexer/LNA) 2.72 kg
(antenna drive assembly) 3.17 kg
Power supply: 28 V DC

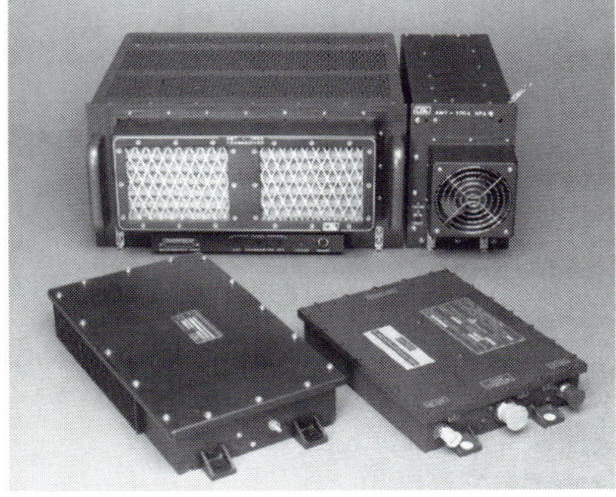

The AMT-100 provides voice communications with the public telephone network via the Inmarsat satellite system

Operational status

The first AMT-100-equipped Canadair Challenger was in service by the end of 1992. Full access approval was gained from Inmarsat in August 1993. Systems upgrades to include fax, data and multichannel capability are under development.

Contractor

CAL Corporation.

VERIFIED

ADT-200A aeronautical data terminal

The ADT-200A aeronautical data terminal has been designed principally as an aeronautical transportation fleet management tool, utilising satellite communications. The communications system is based on Inmarsat Standard C communications format and is capable of two-way communications consisting of position reporting, general messaging and the sending of coded messages. The fleet communications control is performed through a network control centre which allows communications with the individual ADT-200A terminals. Position information is derived from a GPS receiver and encoded along with a terminal identification code in the transmit signal.

The ADT-200A is compliant with all applicable mechanical and electrical airborne regulations. The system consists of a compact keyboard/display unit, modular transceiver and antennas.

The keyboard/display unit is easily mounted in the cockpit or elsewhere in the aircraft to alert the operator to incoming messages. It enables the operator to read messages under any lighting condition and to respond via a keyboard.

The transceiver houses the transmitter, GPS position locator and power supply unit. The transceiver provides a link between the operator's keyboard/display unit and the antenna, sends and receives messages and transmits the position of the aircraft as determined by the GPS.

The antennas are rugged, lightweight and weather resistant with a durable aerodynamic cover. They are used for both L-band satellite data communications and for receiving position signals from the GPS system. Once installed, the antennas are automatically aligned with the satellite.

Specifications

Dimensions:
(transceiver) 190.5 × 210.8 × 324.1 mm
(MDS antenna) 25.4 × 127 mm diameter
(GPS antenna) 19.1 × 76.2 mm diameter
(keyboard/display) 267 × 140 × 51 mm
Weight:
(transceiver) 7.5 kg
(MDS antenna) 0.6 kg
(GPS antenna) 0.1 kg
(keyboard/display) 0.9 kg
Power supply: 22-32 V DC
Frequency:
(transmit) 1,626.5-1,660.5 MHz
(receive) 1,530-1,559 MHz
Channel spacing: 5 kHz

Contractor

CAL Corporation.

VERIFIED

JS-100A aeronautical satcom system

CAL has developed the JS-100A Satcom system specifically for use on smaller aircraft. The unit has been engineered to minimise impact on airframe and performance.

The system consists of a transceiver assembly, high-power amplifier and antenna assembly. The transceiver assembly handles all protocols, performs voice and data modulation and handles frequency conversion and Doppler correction. It also contains the antenna control processor. The high-power amplifier is a reliable solid-state unit designed for a wide range of aircraft installations. The antenna assembly is a high-gain steerable antenna which provides full azimuth and elevation coverage without gaps. The antenna may be pointed using aircraft navigation system inputs or a signal strength measurement from the transceiver.

Specifications

Dimensions:
(transceiver) 317.5 × 317.5 × 190.5 mm
(high-power amplifier) 127 × 317.5 × 195.6 mm
(antenna) 1,371.6 × 203.2 × 200.7 mm
Weight:
(transceiver) 12.25 kg
(high-power amplifier) 5.44 kg
(antenna) 15.42 kg
Power supply: 115 V AC, 400 Hz or 28 V DC

Contractor

CAL Corporation.

VERIFIED

AN/ASH-503 voice message system

The Canadian Marconi AN/ASH-503 voice message system is a single-box device using advanced microprocessor and memory technology to store and reproduce high-quality digitally produced speech messages.

The AN/ASH-503 Micro VMS can store 15 three-word messages or 13 seconds of continuous speech. Each system incorporates extensive self-test facilities. High-quality speech reproduction is ensured by using a 30 kHz digitising sampling rate. Any type of voice or language can be stored and reproduced. Each message can be preceded by an alerting tone and repeated as often as necessary. The messages can be sorted in order of priority so that, in the event of coincidental inputs, the most important message is reproduced first.

Specifications

Dimensions: 64 × 65 × 67 mm
Weight: 0.45 kg
Power supply: 28 V DC, 2.5 W

Operational status

In production and in service.

Contractor

Canadian Marconi Company.

UPDATED

CMA-2102 high-gain satcom antenna system

The CMA-2102 high-gain satcom antenna system is designed to support Inmarsat's Aero-H satellite communications service, which provides aircraft with simultaneous two-way, digital voice and real-time data communications capability for flight crew, cabin crew and passengers.

The CMA-2102 is a single, phased-array, electronically-steered, top-mounted, high-gain antenna, with a single Beam Steering Unit (BSU) and Diplexer/Low Noise Amplifier (D/LNA) which provides hemispherical coverage in an extremely reliable installation. Since the CMA-2102 is mounted on the top of the aircraft, its coverage pattern does not suffer from keyholes/blindspots, and installation is greatly simplified. The system conforms to ARINC Characteristic 741.

The CMA-2102 covers 360° in azimuth and −5 to +90° in elevation and operates over the frequency range 1,525 to 1,660.5 MHz.

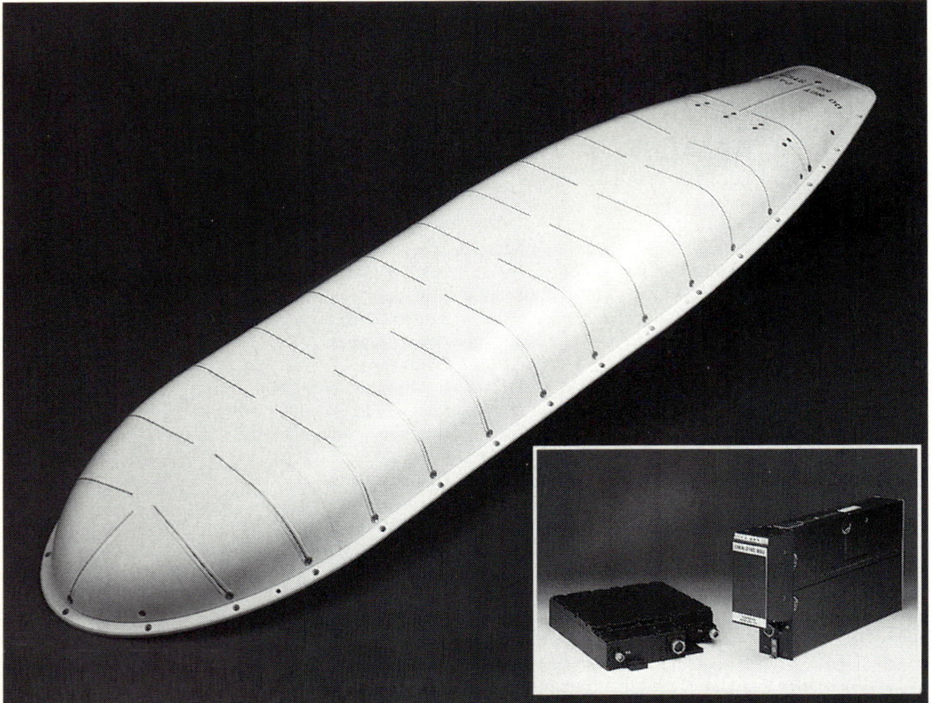

The CMA-2102 high-gain satcom antenna system with (inset) the low noise amplifier (left) and beam steering unit (right) ***1998***/0018149

Specifications

Gain: between 12 dBiC and 17 dBiC over 90% of Inmarsat hemisphere; minimum of 9 dBiC over 100% of Inmarsat hemisphere
Axial ratio: less than 6 dB for all steering angles and all frequencies of operation within coverage region
Dimensions:
(BSU) 2 MCU
(antenna) 1,702 × 470 × 126 mm
(D/LNA) 281 × 197 × 50 mm
Weight:
(BSU) 2.7 kg
(antenna) 27.9 kg
(D/LNA) 3 kg
Power consumption:
(D/LNA) 6 W
(BSU/Antenna) 46 W

Operational status

Over 600 CMA-2102 systems delivered, with selections/orders for as many more to 36 major airlines and VIP/military operators, including American Airlines, Air France, Air Canada; Cathay Pacific, Japan Airlines, KLM; Lufthansa, Quantas, Singapore Airlines; Saudi Arabian Airlines, Swissair, United Airlines and several other airlines. Commissioned on the following aircraft types (TCs and/or STCs); Boeing (B707/727/737/747/757/767/777); DC-10; MD-11 and MD-90; and Airbus (A300/319/320/321/330/340). Other installations include E-4B, P-3C, and F-27/28.

Contractor

Canadian Marconi Company.

UPDATED

CMA-2200 intermediate-gain satcom antenna system

The CMA-2200 intermediate-gain satcom antenna system design supports the requirements of the new generation Inmarsat Aero-I satellite communication service. The Inmarsat-3 satellite provides aircraft with telephony, fax and real-time data communications and has been developed to meet the communication requirements of short-/medium-haul regional and corporate jet operators.

The CMA-2200 Intermediate Gain Antenna (IGA) is a derivative of the proven technology and architecture of the CMA-2102 High-Gain Antenna (HGA). The CMA-2200 is a top-mounted, linear array, electronically steered antenna, which conforms to ARINC Characteristic 761.

The CMA-2200 system consists of the Antenna and Diplexer/Low Noise Amplifier (D/LNA). The traditional Beam Steering Unit (BSU) has been eliminated, the steering of the IGA being performed directly by the terminal equipment provided by leading manufacturers.

Specifications

Coverage: >95% of the Inmarsat hemisphere
Frequency:
(receive) 1,525.0-1,559.0 MHz
(transmit) 1,626.5-1,660.5 MHz
Gain: 8.5 dBiC typical; 6.0 dBiC min
Dimensions:
(D/LNA) 280 × 197 × 50 mm
(IGA) 759 × 97 × 109 mm
Weight:
(D/LNA) 3.0 kg
(IGA) 2.7 kg
Power Consumption:
(D/LNA) 6 W
(IGA) 7 W

Operational status

Inmarsat's rigorous AERO-I IGA testing was completed in August 1997. The Canadian Marconi Company claim that CMA-2200 is the first Aero-I Antenna subsystem to be Inmarsat-approved. CMA-2200 black-label production shipsets started deliveries in January 1998. The CMA-2200 system has been selected for the Honeywell/Racal MCS-3000/6000 and MCS-7000 satcom systems, with certification scheduled for May 1998.

Contractor

Canadian Marconi Company.

NEW ENTRY

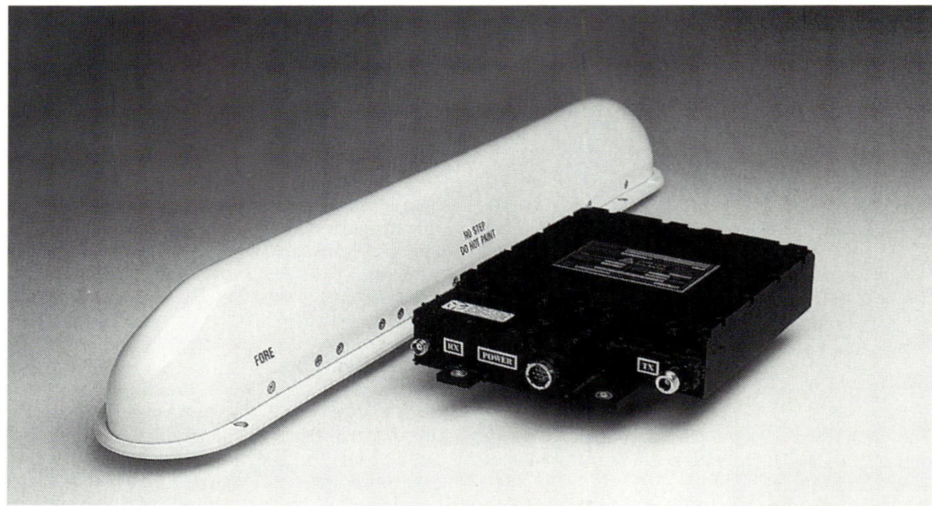

The CMA-2200 intermediate-gain satcom antenna system with the antenna unit (left) and diplexer/low noise amplifier (right) ***1998***/0018148

CHINA, PEOPLE'S REPUBLIC

CT-3 airborne radio

The CT-3 is an airborne AM VHF radio set covering the 100 to 150 MHz frequency range. It has 601 channels, 20 of which may be preset. Channel spacing is 83.333 kHz. When a VHF antenna with a travelling wave coefficient of more than 0.4 is used and the radio is operated at 14 W, communication range is over 120 km at a flight altitude of 3,300 ft, 230 km at 16,500 ft and 350 km at 33,000 ft. Maximum permissible altitude is 82,000 ft.

Contractor

China National Electronics Import & Export Corporation.

VERIFIED

JDF-2HF SSB transceiver

JDF-2HF is a solid-state automatic tuning SSB airborne radio set designed for fighter/interceptors, bombers, transport aircraft and helicopters. Operating in the 2 to 11.999 MHz range, the transceiver's modes are A3J USB and LSB, A2J USB CW and LSB CW, and A3 AM. The full set consists of a transceiver, an antenna tuner and a preset control box. Power output ranges from 15 to 50 W.

Contractor

China National Electronics Import & Export Corporation.

VERIFIED

FRANCE

JET-SAT 97

The JET-SAT 97 is designed to operate under the Inmarsat 3 satellite spot beams, offering cockpit and cabin voice, fax and data services. This fully integrated system provides five channels (four voice/fax/PC data and one packet data), and meets FANS operational requirements. It comprises one IGA (Intermediate Gain Antenna) and two LRUs (Line Replaceable Units).

Specifications

Dimensions:
2 LRUs each of 4 MCU
(antenna) 650 × 336 × 140 mm
Weight: 18 kg

Dassault Electronique JET-SAT 97 **1998**/0011852

Contractor

Dassault Electronique.

NEW ENTRY

Syracuse II airborne terminal

Dassault Electronique has developed an airborne terminal to work in conjunction with the Syracuse II satellite military telecommunications system. The terminal will enable an aircraft to communicate at any time with other terminals on land or at sea within the coverage of the satellite. It is designed for long-range aircraft such as the Atlantique 2 maritime patrol aircraft, KC-135 in-flight refueller and military transport aircraft. Weight of the system is expected to be about 135 kg.

The terminal employs two electronically scanned active antennas, each with several hundred active elements. The two transmit/receive antennas, mounted in the fuselage of the aircraft, have no significant protrusions and hence no adverse effect on drag. Faulty elements will be inhibited individually and it is estimated that satisfactory communications will be possible with up to 10 per cent of the elements defective. The system will therefore have a high tolerance to component failure, giving it a high MTBF.

Syracuse antennas installed on a C-160 Transall aircraft

Operational status

An electronically scanned active antenna was successfully flight-tested in 1991.

Contractor

Dassault Electronique.

VERIFIED

Airborne Multiservice/Multimedia Communication System (AMMCS)

The Airborne Multiservice Multimedia Communication System (AMMCS) provides voice, computer data, telex and facsimile air-to-ground communications facilities over different media such as satellite, HF, VHF or UHF radio.

Message handling facilities allow editing, retrieving, modifying and encryption/decryption of messages. The AMMCS provides automatic store-and-forward message services including logging, routeing and relaying. It supports a wide variety of message types and formats including telefax, telex, and email. Multiple compression algorithms allow message transmissions to be speeded up. Message routeing is made automatically using different criteria such as urgency, medium availability, cost and priority.

The reliable transmission of long messages, such as high-quality telefax or error free computer data files, over HF implies some specific modems, datalink protocols and transmission management protocols.

AMMCS integrates the Rockwell MDM-2501, multi-waveform adaptive high-speed modem. MDM-2501 offers 12 different standard waveforms, ensuring interoperability on different media. Among these waveforms is the MIL-STD-188-110A single-tone waveform which allows the efficient transmission of data up to 2,400 bits/s with forward error correction on HF between 300 and 3,000 Hz. When this waveform is used, the automatic multipath compensation provided by this modem overcomes the data rate limitations usually imposed on HF channels.

The AMMCS uses a multimedia FED-STD-1052 link protocol. It can operate over half- or full-duplex channels. This type of robust protocol is mandatory for HF links when propagation conditions may vary drastically and the link can be interrupted for variable periods of time. This protocol supports the automatic adaptability of system parameters to optimise the transmission performance in terms of throughput and bit error rate.

The AMMCS features functions which automatically adapt its parameters during a transmission depending on the type of link, type of service, required performance and propagation characteristic variations to optimise the performance in terms of bit error rate, required transmit power, transmission delay and throughput. These functions automatically select the best HF propagating channel and change the frequency when conditions degrade. The system uses the MIL-STD-188-141A/FED-STD-1045A ALE protocol on HF.

The AMMCS may present different levels of automation. In a fully automated system, direct dialling is offered to aircraft users for voice communications. The system selects the appropriate media and uses the dialling transcription to perform automatic selective call and link establishment procedures. A similar procedure applies to telex, data or faxes queued in the message handling system file for transmission. AMMCS uses the routeing information contained in the message envelopes to establish the appropriate link automatically and to manage the message transmission.

The AMMCS system is integrated into a high-impact magnesium alloy case laptop PC featuring 486/33 MHz Intel processor, up to 500 Mbyte hard disk, high view angle touch panel colour screen and mouse or trackerball interface. A TEMPEST version is available.

Contractor

Rockwell-Collins France.

VERIFIED

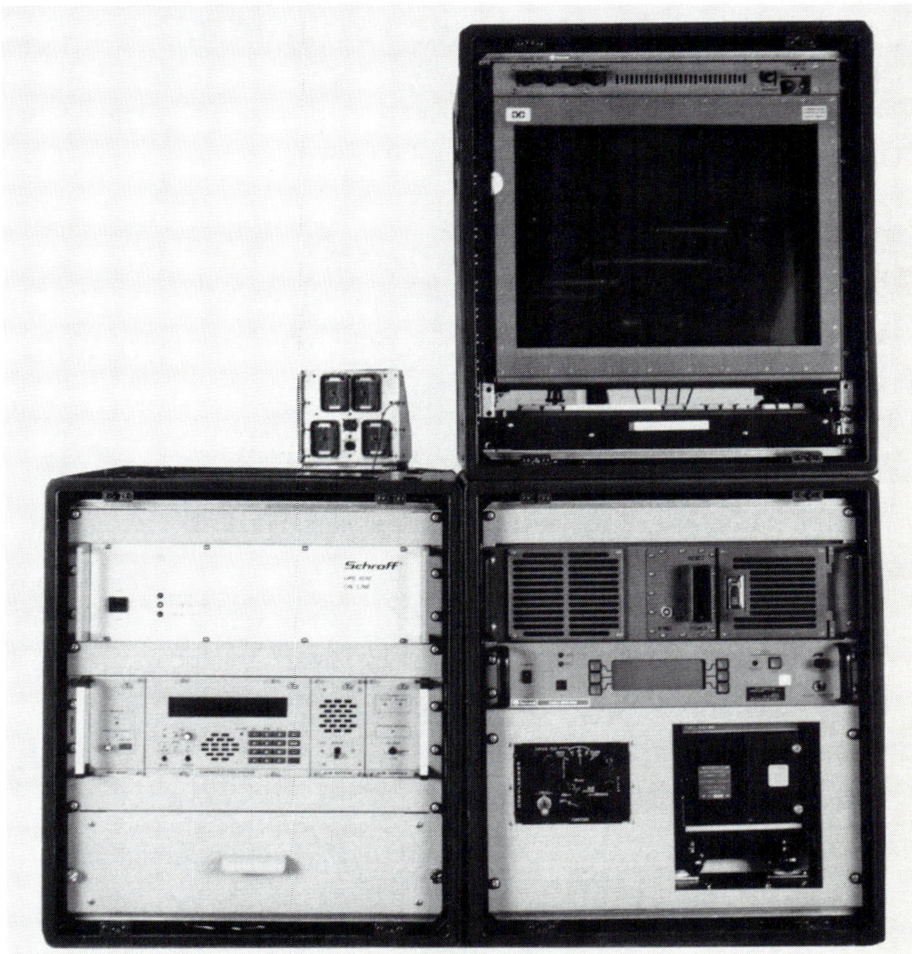

DLP DataLink Processor **1998**/0011851

DLP DataLink Processor system for NATO interoperability

The DLP (DataLink Processor) is an advanced tactical datalink server that supports TADIL-A (Link 11A), TADIL-B (Link 11B) and TADIL-J (Link 16) for NATO interoperability. It receives, transmits, processes and

ADLP-100F Airborne DataLink Processor
1998/0011850

displays tactical data in accordance with STANAG 5511 and/or STANAG 5516 message specification.

The DLP has the capability to simultaneously process data provided by two or more networks (that is Link 11/Link 16, 2 × Link 11).

Depending upon the nature of the application, the DLP can be configured as a tactical datalink front-end processor only, or as a tactical datalink processor together with graphic display and keyboard/trackball.

The DLP provides the necessary external interfaces including: NTDS/ATDS, MIL-STD-1553B, Ethernet and serial interfaces.

The DLP product family includes equipment designed for aircraft, ground station (fixed/transportable) and shipboard applications. It uses commercial workstations for benign environments, 19 in ruggedised VME-based equipment for ground and shipboard and ATR VME-based equipment for airborne environments.

Operational status

In production since 1996. In operation with the French Navy, French Air Force and UK Royal Navy.

ADLP-100F Airborne DataLink Processor for NATO interoperability

The ADLP-100F is part of Rockwell-Collins France DataLink Processor (DLP) product line. It is an advanced tactical datalink processor specifically designed for Link 11 airborne applications including both fixed- and rotary-wing platforms.

The ADLP supports the capability to receive, process, transmit and display tactical data in accordance with STANAG 5511 message specification for NATO interoperability. Depending upon the nature of the application, the ADLP-100F can be installed on board the platform as a front end processor connected to the onboard mission system through a MIL-STD-1553B interface or equivalent, or as a standalone subsystem together with a display and keyboard/trackball.

The ADLP-100F is a VME-based system packaged in an ATR format for easy installation on board the platform. It provides an ATDS interface to a KG-40(A) crypto and a 1553B interface or equivalent to the onboard mission system or sensors.

The ADLP-100F hardware is qualified for both fixed- and rotary-wing aircraft.

Operational status

Selected by ECF to equip the Super Puma and Panther helicopters with Link 11 for a foreign navy; qualified for the French Navy ATL2 aircraft.

Contractor

Rockwell-Collins France.

NEW ENTRY

ETC-40X0F centralised control system

The ETC-40X0F provides frequency and mode control of any radio communication and radio navigation equipment in the aircraft. It is a fully modular architecture system allowing easy adaptation to the avionics configuration of the aircraft. Main features are tandem seat operation, compatibility with any type of radio com/nav equipment, flexibility, modularity, colour display, easy installation and BITE.

ETC-40X0F basic configuration is composed of two EDU-40X0F control and display units and one MPU-40X0F bus concentrator.

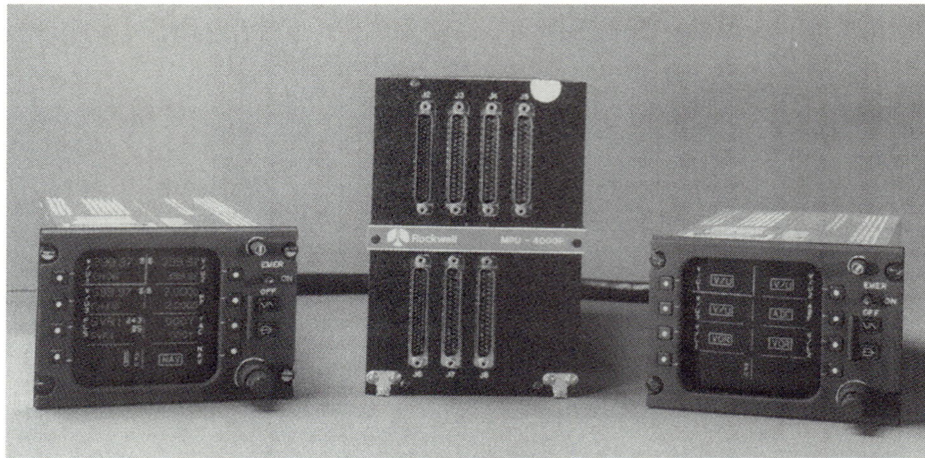

The Rockwell-Collins France ETC-40X0F centralised control system consists of two EDU-40X0F control and display units and the MPU-40X0F bus concentrator (centre)
1998/0011849

ETC-40X0F can control and display ARINC 429 as well as any other type of equipment. ARINC 429-equipped systems are directly linked to the EDU-40X0F, and non-ARINC 429 radios are linked through dedicated interfaced boards, within MPU-40X0F.

The ETC-40X0F allows the operator a global display of all com/nav equipment in the aircraft. For each piece of equipment, essential information is displayed throughout three levels of pages. Top level pages display active and preset frequencies. Technical parameters, such as mode, channel and test, are available on two additional pages. The system provides simultaneous access to all avionics equipment through each of the two EDU-40X0F units.

The MPU-40X0F unit houses the ARINC 429 bus concentrator and interface boards for non-ARINC 429 equipment. ETC-40X0F is also offered with colour LCD technology and embedded GPS.

Specifications

Dimensions:
(EDU-40X0F) 146 × 242 × 105 mm
(MPU-40X0F) 322 × 124 × 193 mm
Weight:
(EDU-40X0F) 3.6 kg
(MPU-40X0F) 4.5 kg
Power supply: 28 V DC, 4 A max
Radio interfaces: 6 ARINC 429, 12 non- ARINC 429
Programmable channels: 16 per radio
Altitude: up to 55,000 ft
NVG compatible

Operational status

In production for Eurocopter Cougar Mk 1, Mk 2, Fennec, Raytheon Hawker 800 XP and Lockheed Martin C-130 aircraft.

Contractor

Rockwell-Collins France

UPDATED

MCU-2202F data terminal set control

The MCU-2202F data terminal set control has been designed for the control of the Collins MDM-2202 Link 11 modem. It is a processor-controlled unit which provides configuration and control instructions to the data terminal set and provides address storage and sequencing instructions for control of the network roll call operation.

Specifications

Dimensions: 146 × 105 × 242 mm
Weight: 3.6 kg
Power supply: 28 V DC
Temperature range: −20 to +70°C
Altitude: up to 55,000 ft

Operational status

In production for aircraft of the French Navy.

Contractor

Rockwell-Collins France.

VERIFIED

VIP communications suite

The VIP communications suite provides voice, computer data, telex (TTY) and fax air-to-ground communications over different media such as satellite, HF or UHF radio.

VIP message handling facilities allow users to edit, store, retrieve and modify messages, and encrypt or decrypt them off line. The system provides an automatic store-and-forward capability including logging, routeing and relaying. It supports a variety of message types and formats. Multiple compression algorithms speed up message transmission, a useful capability when operating in poor propagation conditions. Routeing is on the basis of criteria such as urgency, medium availability, cost and priority.

The VIP communications suite integrates the Rockwell MDM-2501, multiwaveform adaptive high-speed modem. This offers a dozen different standard waveforms, including the MIL-STD-188-110A single-tone waveform for speeds up to 2,400 bits/s with forward error correction on 300 to 3,000 Hz HF channels. This waveform uses automatic multipath compensation.

The system uses the FED-STD-1052 multimedia link protocol. It can operate over half- or full-duplex channels. The protocol supports automatic adaptivity to optimise transmission performance in terms of throughput and bit error rate.

The VIP communications suite also features MIL-STD-188-141A and FED-STD-1045A ALE.

The system has different levels of automation. In a fully automatic configuration the VIP communications suite will direct dial, choosing the appropriate media and performing SELCALL and ALE functions.

The VIP communications suite is integrated into a high-impact magnesium alloy case and is based on the 486 Intel processor with up to 500 Mbytes of hard disk. A TEMPEST version is available.

Contractor

Rockwell-Collins France.

NEW ENTRY

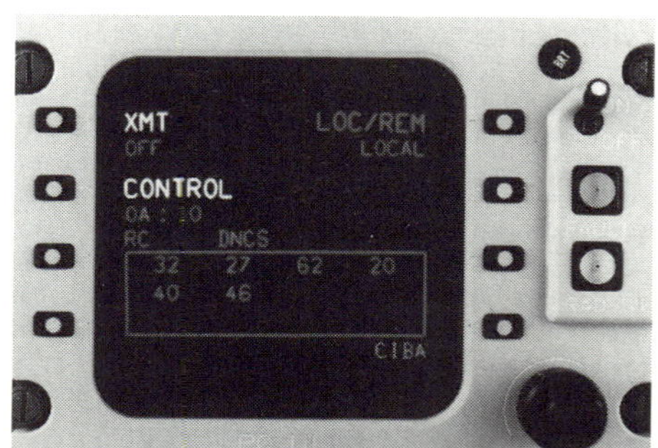

The Rockwell-Collins France MCU-2202F data terminal set control

Telemir infrared communication system

Telemir uses an infrared beam for air-to-air, omni-directional air-to-ground, ground-to-air and ground-to-ground communications. The airborne equipment consists of an optical head (mounted on top of the tailfin) and a processing unit. It is extremely difficult to jam.

The system is used by a carrier-based aircraft for the reception of navigational updating and reference data such as altitude, location and speed from the ship's inertial navigation system for aligning its own INS. A new version is now available with a MIL bus 1553 datalink.

Operational status

In service in French Navy Super Etendard aircraft and integrated into the naval version of Rafale.

Contractor

SAGEM SA, Defence and Security Division.

VERIFIED

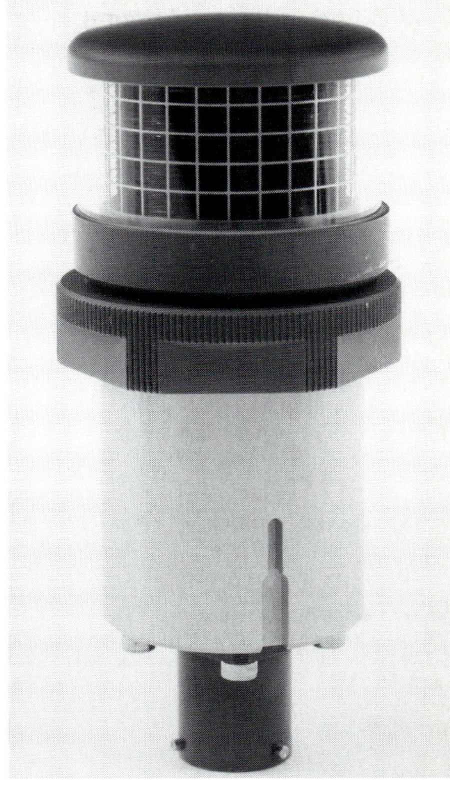

The Telemir optical head is fitted on top of the tailfin for infrared communications (Rafale version)
1997/0001173

EVR 716/750 enhanced VHF data radio

The Sextant Avionique EVR 716/750 (Enhanced VHF Radio) is a new VHF AM transceiver designed to comply with 25 kHz and 8.33 kHz channel spacing in the VHF band: it is immune to FM radio signals in accordance with ICAO annex 10

The EVR 716 offers 25 kHz and 8.33 kHz channel spacing for voice communications and 25 kHz channel spacing data transmission capability with an external modem (MSK/CSMA 2.4 kbps).

The EVR 750 VHF Data Radio (VDR) offers, in addition to the EVR 716, ARINC 750 data modes compatible with the ATN protocols (31.5 kbps) with internal modems.

The EVR 716 complies with ARINC 716, and the EVR 750 with ARINC 750. The two systems share a common hardware configuration, and the EVR 716 can be upgraded to the EVR 750 capability by software only.

Specifications

Dimensions: 3 MCU
Weight: 4.8 kg
Power supply: 28 V DC
Temperature: −15 to +70°C

ARINC specifications

ARINC 716
ARINC 750

Operational status

The EVR 716 is in production; the full EVR 750 standard is still in development.

Sextant Avionique EVR 716/750 enhanced VHF data radio ***1997***/0001172

British Airways has selected the EVR 716 for its fleet to meet January 1999 regulatory requirements for 8.33 kHz channel spacing with the requirement to upgrade to the full EVR 750 VHF Data Radio (VDR) standard when available to meet future data transmission requirements.

Contractor

Sextant Avionique.

UPDATED

Onboard image transmission system

The onboard image transmission system comprises an image compressor for installation in the aircraft and an image decompression device installed in a ground station.

The system is capable of acquiring a standard 625-line video signal in real time. After acquisition, the image is digitised into 256 lines of 256 columns each with 8 bits per column. The data reduction rate is sufficient to allow transmissions down a telemetry link with a bandwidth of around 1 MHz, at a rate of approximately 2 Mbytes/s. The system can transmit composite video images consisting of low- and medium-frequency signals, for landscaped sequences, or high-frequency signals, mainly for superimposed synthetic messages. The signals are processed in different channels.

After processing, the binary data at the CPR output is formatted into a continuous message with a perfectly cyclical constant structure compatible with the digitising process. The IRIG standard PCM output message is coded NRZ-L. If no useful data is available for transmission to the DCPR, the above message structure is transmitted at a sufficient transition rate to constitute a clock pulse upstream of the DCPR.

Contractor

SFIM Industries.

VERIFIED

SAVIB69 audio management system

SAVIB69 audio management system provides control of the aircraft's external and internal communications systems. It includes the ADAV aural warning function with 265 synthetic voice messages. SAVIB69 comprises: one audio management unit CTA 3488 (including voice alert); one audio control panel (BCA 3487) for each cockpit, one fuselage ground connection; and one telebriefing connection.

Each control panel provides an automatic transmit/receive function on the interphone channel; the ADAV aural warning system memory capacity of 265 messages corresponds to a global alert time of 5 mins. The system permits the selection of: up to four communication channels with volume control; up to six navigation channels with volume control; four listening channels without volume control; one telebriefing connection; one double track audio recording system; 24 discrete signals corresponding respectively to one alert.

Each audio control panel transmits the status of its selector switches to the audio management unit via ARINC 429 busses. Each audio management unit incorporates up to two independent audio signal processors. The ADAV mode provides the crew with a prioritised listening monitor on a fixed frequency.

SAVIB69 is connected to the aircraft MIL-STD-1553B databus.

Specifications

BCA

Bandwidth: 300-5,000Hz ±3 dB
Dimensions: 65(H) × 146(W) × 110(D) mm
Weight: 0.7 kg
Power: 17-32 V DC, 10 W
MTBF: 10,000 flight hours

CTA

Bandwidth: 300-5,000Hz ± 3 dB
Dimensions: 194(H) × 60(W) × 320(D) mm
Weight: 3 kg
Power: 17-32 V DC, 20 W
MTBF: 4,500 flight hours

Operational status

Designed for Rafale, both single-seat and two-seat versions.

Contractor

TEAM (Télécommunications, Electronique, Aéronautique et Maritime).

NEW ENTRY

SAVIB69 audio management system ***1998***/0011848

SELCAL airborne selective calling system

The SELCAL system is a selective calling device which enables a ground operator to call an aircraft without the pilot having to constantly monitor a particular frequency. The SELCAL system comprises: a signal processing unit; a code selection unit; visual and audio annunciators.

The SELCAL system can monitor any HF or VHF radio receiver installed on an aircraft. On receipt of a signal it lights a channel indicator and sounds an alarm. The system comprises five radio communication channels operating simultaneously and continually. The code selection panel is recommended as part of both the SIB54 (A320) and the SIB73 (A330340) systems.

Options: code selection panel BC2065; SELCAL call signal indicator BC2066.

Specifications

Standards: ARINC 714 and ARINC 531
Approvals: TSO C59 according to DO 160C and DO 178A

Operational status

SELCAL model 2253 is currently installed on the following aircraft: Aerospatiale/Aeritalia ATR42; Airbus Industrie A300, A310; Boeing 737-500, 747-400, 757, 767, 777, MD-80; British Aerospace BAe 146; Fokker F-100. The SC2539 model is currently installed on Boeing 737 first-generation aircraft.

Contractor

TEAM (Télécommunications, Electronique, Aéronautique et Maritime).

NEW ENTRY

SELCAL airborne selective calling system ***1998***/0011847

SIB31/45/54/66/73/85 audio management systems

The SIB family of audio management systems provides crew members with an interphone link (in multi-crew aircraft) and enables each one to operate the following facilities:

SIB31: radio communications; radio navigation; emergency warning system
SIB45: radio communications; radio navigation; emergency warning system
SIB54: radio communications; radio navigation; interphone; public address
SIB66: radio communications; radio navigation; interphone
SIB73: radio communications; radio navigation; interphone; public address; emergency warning system

SIB31 audio management system ***1998***/0011846

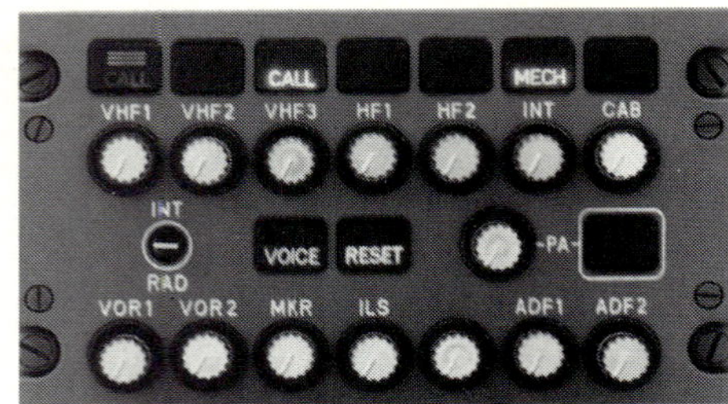

SIB73 audio management system ***1998***/0011845

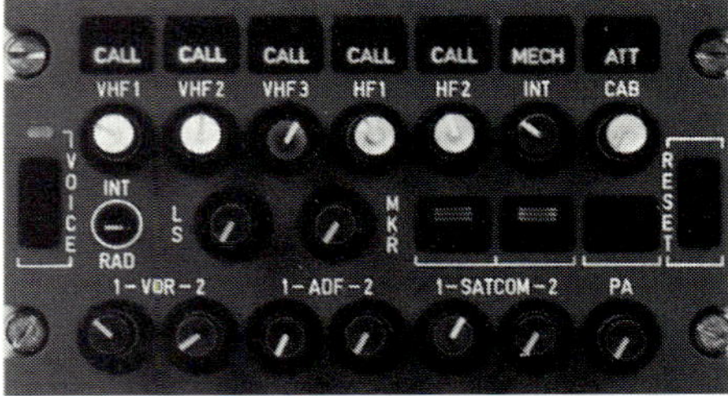

SIB85 SATCOM audio management system ***1998***/0011844

SIB85 SATCOM: radio communications (including satcom); radio navigation; interphone; public address; emergency warning system.

The systems comprise the following units:
SIB31: junction box BJ1977; three to five main audio control panels CP1976
SIB45: junction box BJ2620A; three to five main audio control panels CP2618A
SIB54: up to five audio control panels ACP2788; 1 × audio management unit AMU2790; 1 × SELCAL code selection panel BC2065
SIB66: up to four audio control panels BC3438; 1 × ground crew panel PC2748; 1 × SELCAL code selection panel BC2065
SIB73: up to five audio control panels ACP2788AC01; 1 × audio management unit AMU3514; 1 × SELCAL code selection panel BC2065; 2 × speakers HP3520
SIB85 SATCOM: up to five audio control panels; ACP2788AF01; 1 × audio management unit AMU3514; 1 × SELCAL code selection panel BC2065; 2 × speakers HP3520.

Facilities provided include:
SIB31: 4 × Tx/Rx selective tunable; 4 × navigation channels tunable; 3 × listening channels
SIB45: 6 × Tx/Rx selective tunable; 8 × navigation channels tunable; 3 × listening channels
SIB54: up to five audio channels; up to 12 navigation channels; one each public address/intercom/cabin channel
SIB66: six audio channels; six navigation channels; 2 × interphone channels; 1 × general call channel
SIB73: up to six audio channels; up to 11 navigation channels; one each public address/intercom/cabin channel
SIB85 SATCOM: up to seven audio channels (including 2 × satcom channels); up to six navigation channels; one each public address/intercom/cabin channels.

Operational status

Systems have been installed on the following aircraft:
SIB31: Dauphin SA365C; Ecureuil AS355F; Puma SA330J; Super Puma SA332L1 helicopters
SIB45: Super Puma SA332 helicopters
SIB54: Airbus A320 aircraft
SIB66: Transall aircraft of the French Air Force Transport Air Command (COTAM)
SIB73: Airbus A330/340 aircraft
SIB85 SATCOM: Airbus A330/340 aircraft equipped with VOICE SATCOM from the cockpit.

Contractor

TEAM (Télécommunications, Electronique, Aéronautique et Maritime).

NEW ENTRY

3527 HF/SSB radio

The 3527 HF/SSB radio comprises the 3527F transceiver, 3527H remote control, 3596A antenna coupler, 3597A digital preselector and the 3598A FSK modem. The preselector filter enables the simultaneous operation of two 3527H units on the same aircraft, one for transmission and the other for reception.

The fully solid-state transmitter supplies a modulated peak power of 400 W. The transceiver is driven by a digital synthesiser tuned for setting to 280,000 channels spaced at 100 Hz intervals. The antenna coupler provides tuning with wire antennas from 10 to 30 m long. The FSK modem handles messages of 50 to 100 bauds.

The comprehensive fault identification BITE increases maintenance efficiency and the modular construction minimises repair time by fast access to all circuit modules.

Specifications
Power supply: 115 V AC, 400 Hz, 3 phase
Power output: 400 W PEP, 200 W average (SSB)
Modes: USB, data (USB), CW, Link 11
Frequency stability: $\pm 5 \times 10^7$
Number of channels: 280,000
Operational specification: STANAG 5035, AIR 7304

Operational status
In production.

Contractor
Thomson-CSF Communications.

VERIFIED

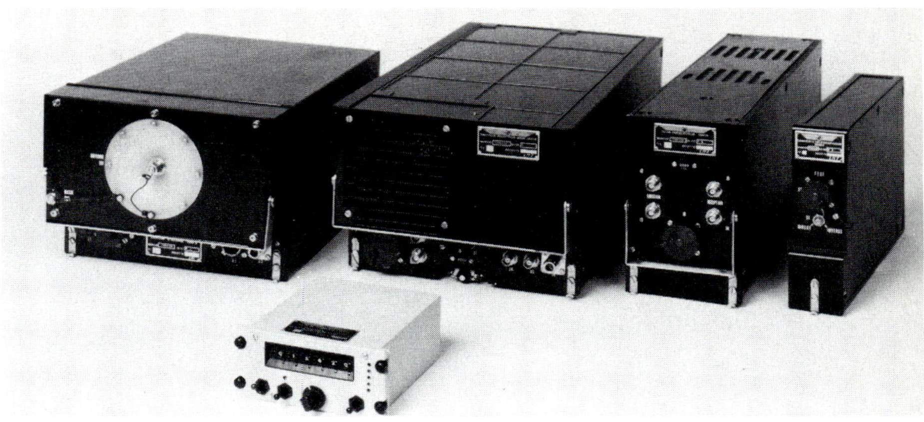

The Thomson-CSF 3527 HF/SSB radio

4600-E (TRAP-138A) VHF AM radio

Thomson-CSF's 4600-E (designated TRAP-138A for military use) is a VHF transmitter/receiver designed primarily for light military aircraft and helicopters for which it provides both air-to-air and air-to-ground communication facilities. Secondary applications include vehicular and fixed ground stations.

The 4600-E covers the VHF band from 118 to 143.975 MHz, in which it provides either 520 or 1,040 channels at intervals of 50 or 25 kHz, generated by means of frequency synthesis techniques. It offers either A1 or A3 transmission modes and has a maximum transmitted carrier wave power output of more than 20 W.

Frequency stability is guaranteed over a temperature range of −40 to +70°C, although the equipment will operate over a rather wider temperature range.

Operational status
No longer in production. This equipment is widely used by the French armed forces in a wide range of fixed-wing aircraft and helicopters. It has also proved a successful export and is in service with the armed forces of Argentina and Spain.

Contractor
Thomson-CSF Communications.

VERIFIED

6800 series UHF AM transceivers

The 6800 UHF AM 5 W airborne remote-controlled transceivers meet severe electrical environmental conditions and are suitable for use in many types of fixed-wing aircraft and helicopters. The equipment provides air-to-air and air-to-ground communications in the 225 to 400 MHz range and has 7,000 channels at 25 kHz spacing. Operating mode is A3E.

The 6800 family can operate with homing and is capable of voice scrambling, frequency hopping and radio relay. Different control units have been developed to facilitate installation in various aircraft. An integrated independent Guard receiver and BIT are provided.

There are three versions of the transceiver: the 6857-01 ¼ ATR unit which has a peripheral flange for mounting on the aircraft structure, the 6857-20 ¼ ATR rack mounting unit and the 6857-21 ¼ ATR rack-mounted unit with Guard receiver. Weight is 4.2 kg.

Operational status
No longer in production. In service with the French Air Force and Army on the Xingu, TB 31 Epsilon, AS 355 and AS 332.

Contractor
Thomson-CSF Communications.

VERIFIED

BER 8500/8700 transceivers family

BER 8500/8700 is a family of transceivers covering the VHF and UHF bands from 100 to 173.975 MHz and 225 to 400 MHz in A3 and A1D (A9) AM modes and in F3 and F1D (F9) FM modes. ECCM versions include Have Quick II embedded operation.

The family consists of the BER 8523 covering VHF and UHF AM/FM with Guard receiver, the BER 8524 VHF and UHF AM/FM with Guard receiver and ECCM, the BER 8751 UHF AM/FM and the BER 8752 UHF AM/FM with ECCM.

The ECCM capability is characterised by TRANSEC provided by embedded Have Quick II circuits; COMSEC provided by an external cypher unit (the family is compatible with KY 58 in both NRZ and diphase operation) and SCP 5000 compatibility through direct connection for external TDP 5000 processor.

BER 8500/8700 transceivers are controlled by one or several units of the ECCM VHF/UHF control systems, including the BCA 1217 main VHF/UHF or UHF control unit, DBC 1317 secondary VHF/UHF or UHF control unit or DMC 1317 and DMCB 1317 loading modules. These modules allow control of the transceivers, display of the mode, frequency and alarm and loading of ECCM parameters. The BER 8523 and BER 8751 can be controlled by the TC20 control unit.

Specifications
Dimensions:
(BER 85xx) 57 × 202.5 × 493 mm
(BER 87xx) 57 × 202.5 × 455 mm
Weight:
(BER 85xx) 8.5 kg
(BER 87xx) 7.5 kg
Power supply: 28 V DC or 115 V AC 400 Hz
(transmit) 180 W
(receive) 35 W
Frequency:
(BER 85xx) 100-173.975 MHz and 225-400 MHz
(BER 87xx) 225-400 MHz

Operational status
In service with the French Air Force and several other air forces.

Contractor
Thomson-CSF Communications.

VERIFIED

ERA-8500 VHF/UHF radio

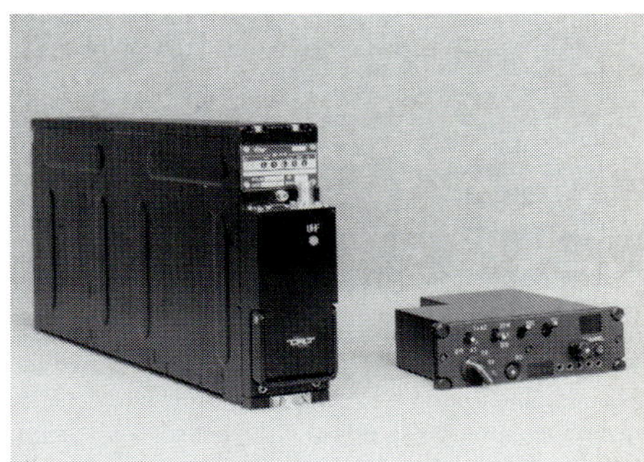

The ERA-8500 VHF/UHF radio

The ERA-8500 transceiver covers the VHF and UHF bands from 100 to 173.975 MHz and 225 to 400 MHz in A3 and A9 AM modes and F1, F3 and F9 FM modes at 25 kHz spacing. It consists of the BER 8500 V/UHF transceiver and the BCA 11XX or BCA 12XX control unit. Its tuning time is compatible with very fast frequency-hopping modes. Power output is 15 W in AM and 20 W in FM.

The ERA-8500 is designed for use by high-performance military aircraft and is offered with a Have Quick II module incorporated in the transceiver. It operates between −55 and +90°C at altitudes up to 100,000 ft.

The ERA-8500 transceiver is coupled with the TDP-5000 radio processing unit to form the SCP-500 integrated ECCM radio communication system.

Specifications
Dimensions:
(BER 8500 V/UHF R/T unit) 60 × 202.5 × 493 mm
(BCA 11XX main control unit) 146 × 76 × 173 mm
(BCA 12XX auxiliary control unit) 146 × 47 × 173 mm
Weight:
(BER 8500 V/UHF R/T unit) 8.5 kg
(BCA 11XX main control unit) 1.3 kg
(BCA 12XX auxiliary control unit) 1 kg
Power supply: 115 V AC, 400 Hz or 28 V DC
(transmit) 180 W
(receive) 35 W

Operational status
In service.

Contractor
Thomson-CSF Communications.

UPDATED

ERA-8700 UHF radio

The ERA-8700 transceiver is limited to the UHF band between 225 and 400 MHz at 25 kHz spacing. It consists of the BER 8700 UHF transceiver and the BCA 11XX or BCA 12XX control unit.

The equipment is available with the Have Quick II frequency-hopping mode.

Specifications
Dimensions:
(BER 8700 UHF R/T unit) 60 × 202.5 × 455 mm
(BCA 11XX main control unit) 146 × 76 × 173 mm
(BCA 12XX auxiliary control unit) 146 × 47 × 173 mm
Weight:
(BER 8700 UHF R/T unit) 7.5 kg
(BCA 11XX main control unit) 1.3 kg
(BCA 12XX auxiliary control unit) 1 kg

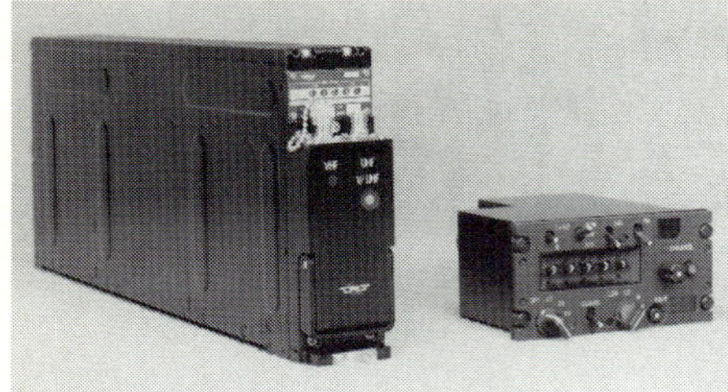

The Thomson-CSF ERA-8700 UHF radio

Power supply: 115 V AC, 400 Hz or 28 V DC
(transmit) 180 W
(receive) 35 W

Operational status
In service.

Contractor
Thomson-CSF Communications.

UPDATED

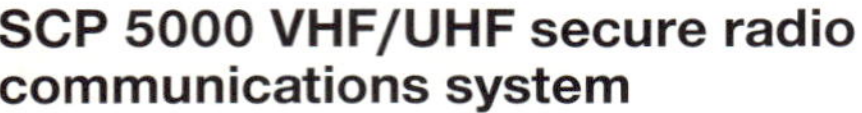

SCP 5000 VHF/UHF secure radio communications system

The SCP 5000 secure radio communications system is designed for military platforms. It provides voice and data transmission services protected against the electronic warfare threat such as jamming, eavesdropping, intrusion and localisation.

The system features pseudo-random fast frequency hopping, time variable unsigned synchronisation dwell, voice and data cyphering and use of the complete UHF band. The Have Quick II mode is available as an option.

The SCP 5000 offers fixed-frequency plain analogue voice, KY 58-compatible fixed-frequency cyphering voice, frequency-hopping plain or cyphered voice, point-to-point data transmission to transmit mission data and inter-weapons system co-operation data and data transmission over a TDMA network to allow participants to exchange tactical information between aircraft or between command and control centres and aircraft.

The SCP 5000 airborne system comprises one or two BER 8500 or 8700 V/UHF or UHF transceivers, a TDP 5000 processor to control the frequency hopping, two BCA 1217 control boxes and an optional BVA 500 frequency display.

The TDP 5000 processor is connected to the aircraft system via the avionics bus or a dedicated link. It manages the frequency-hopping procedures; communicates cyphering/decyphering; manages data transmission, message formatting, link or network management and error detection and correction; and interfaces with the platform for voice digitisation and bus coupling.

The Thomson-CSF SICOP-500 advanced ECCM radio system

Specifications
Dimensions:
(BER 8500) 57 × 202.5 × 493 mm
(BER 8700) 57 × 202.5 × 455 mm
(TDP-5000) 57 × 193.5 × 380 mm
(BCA 1217) 146 × 47.6 × 148 mm
(BVA 500) 41.5 × 41.5 × 125 mm
Weight:
(BER 8500) 8.5 kg
(BER 8700) 7.5 kg
(TDP-5000) 7 kg
(BCA 1217) 1.2 kg
(BVA 500) 0.5 kg
Power supply:
(BER 8500/8700) 28 V DC or 115 V AC, 400 Hz
(TDP-5000, BCA 1217, BVA 500) 28 V DC
Frequency: 100-156 and 225-400 MHz
Channel spacing: 25 kHz

Operational status
In production for the Mirage 2000 for the French Air Force and the export version Mirage 2000-5.

Contractor
Thomson-CSF Communications.

VERIFIED

SICOP-500 integrated radio communication system

SICOP-500 is an integrated ECCM radio communication system designed for air force applications. It operates in the VHF and UHF bands and capabilities include plain data and voice, cipher voice, jam-resistant voice and data, ground-to-air, air-to-air and air-to-ground communications.

The jam-resistant operation is achieved by using frequency-hopping techniques for voice and data transmissions, while error correction is provided for data transmissions. For voice transmissions the ciphering can operate both in jam-resistant and fixed-frequency modes.

The SICOP-500 comprises an ERA-8500 VHF/UHF transceiver, ERA-8700 UHF transceiver (see earlier items) and TDP-500 ECCM radio processor unit. It is designed for use by high-performance military aircraft in a jamming environment. It operates between −55 and +90°C at pressure altitudes up to 96,000 ft.

Specifications
Dimensions:
(BER 8500 V/UHF R/T unit) 57 × 202.5 × 493 mm
(BER 8700 UHF R/T unit) 57 × 202.5 × 455 mm
(BCA 1111 main control unit) 146 × 76 × 173 mm
(BCA 1211 auxiliary control unit) 146 × 47 × 173 mm
(TDP-500 radio processing unit) 57 × 193 × 380 mm
(BCSA-500 radio processing control unit) 146 × 38 × 100 mm
(BMD-500 security unit) 80 × 60 × 160 mm
Weight:
(BER 8500 V/UHF R/T unit) 8.5 kg
(BER 8700 UHF R/T unit) 7.5 kg
(BCA 1111 main control unit) 1.3 kg
(BCA 1211 auxiliary control unit) 1 kg
(TDP-500 radio processing unit) 6 kg
(BCSA-500 radio processing control unit) 0.5 kg
(BMD-500 security unit) 0.8 kg

Operational status
In service on Dassault Mirage 2000 and various other platforms.

Contractor
Thomson-CSF Communications.

VERIFIED

The SCP 5000 has been selected for the French Air Force Mirage 2000 and the export Mirage 2000-5

TC 20 control unit

The TC 20 control unit is a multipurpose VHF/UHF control unit which controls all the Thomson-CSF Communications fixed-frequency transceivers.

It can control two transceivers simultaneously. Programming allows the automatic adaptation of the control unit to the transceiver configuration. Mode controls provided include frequency setting from 30 to 400 MHz, 800 preset channels in four fields, Guard emergency control and Guard receiver, squelch, power and secure mode control. Preparation of the next frequency or channel to be used is possible on the second line of the display for an instantaneous transfer to the active line by a one-touch button. It is NVG-compatible.

Specifications
Dimensions: 57 × 146 × 120 mm
Weight: 1.25 kg
Power supply: 28 V DC

Operational status
In production.

Contractor
Thomson-CSF Communications.

UPDATED

12000 VHF AM/FM radio

The 12000 is a remote-controlled VHF transceiver designed for air traffic and air-to-ship communications in the 100 to 173 MHz frequency range with a channel spacing of 12.5 kHz. The AM/FM radio can operate with direction-finding equipment and sonobuoys in anti-submarine warfare and is capable of radio relay and voice scrambling. It has basic output powers of 15 and 3 W, with a 6 mW discrete capability. An integrated Guard receiver function is incorporated. The equipment's interface could be either ARINC 410 or 429. The transceiver is suitable for combat aircraft, helicopters and maritime surveillance aircraft. The transceiver can be controlled by the TC 20 control unit.

Specifications
Dimensions:
(transceiver) 197 × 61 × 375 mm
Weight:
(transceiver) 5 kg
Power supply: 22.5-31.5 V DC
Frequency:
(transmission) 100-156.975 MHz
(reception) 100-173.5 MHz

Operational status
In service with the French Navy and Air Force.

Contractor
Thomson-CSF Communications.

UPDATED

The 12000 VHF AM/FM radio

12100 VHF FM radio

The 12100 radio is a tactical VHF FM remote-controlled airborne transceiver designed specifically for air-to-ground communications between military aircraft and ground forces.

It covers 30 to 88 MHz. In addition to the standard military FM mode, it provides interoperability with security forces, with a 12.5 kHz step and lower frequency deviation mode.

The Series is capable of KY 58 cyphered voice, external homing and radio delay. Nominal power is 15 W, with a reduced power output of 3 W.

The transceiver can be controlled by the TC 20 control unit.

Specifications
Dimensions:
(transceiver) 200 × 100 × 340 mm
Weight:
(transceiver) 5 kg

Operational status
In service with the French Air Force and Navy on the AS 355.

Contractor
Thomson-CSF Communications.

UPDATED

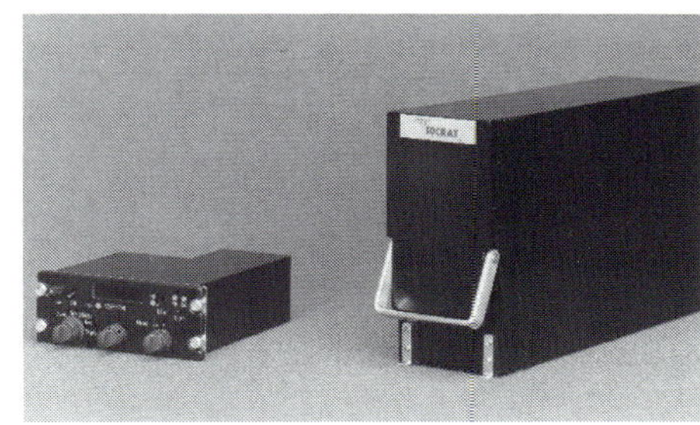

The 12100 VHF FM radio

TRA 2020 VHF/UHF radio

The main features of the TRA 2020 are a frequency range of 100 to 400 MHz, Guard receiver, clear or KY 58 secure voice, homing/direction-finding output, discrete power and remote control. The main criteria which guided the design were reliability, maintainability, lightness, low power consumption and the ability to adapt to any military platform.

The TRA 2020 architecture is centred around two printed circuit boards using SMC technology and carrying all power supply, transmission, synthesiser and main receiver functions and optional Guard receiver and control interface functions. The easily adaptable mechanical configuration allows installation on any type of platform. The optional control interface is compatible with MIL-STD-1553 or ARINC 429 busses. In the ARINC 429 versions, it is controlled by the TC 20 control unit.

Specifications
Dimensions: 356 × 57.15 × 193 mm
Weight: <4.5 kg
Power supply: 28 V DC
(VHF) 118-156 MHz
(UHF) 225-400 MHz
Channel spacing: 12.5 kHz

Status
In production.

Contractor
Thomson-CSF Communications.

UPDATED

The TRA 2020 VHF/UHF radio and its associated TC 20 control unit (right) **1998**/0011843

THOMRAD 6000 V/UHF ECCM transceiver series

The THOMRAD 6000 product line is designed for voice and data communication and provides a full tri-service interoperability in ECCM mode. It covers the VHF and UHF bands from 30 to 400 MHz and operates in AM and FM modes. All transceivers include the channel spacing at 8.33 kHz in accordance with the new ICAO requirements.

ECCM protection is provided by use of the NATO SATURN mode, or a proprietary embedded fast frequency-hopping and ciphering capability that uses modern digital modulation and synthesiser technology. The provision of embedded COMSEC and TRANSEC modules reduce weight and save installation and integration effort. The THOMRAD 6000 product line features, in addition, the NATO Have Quick II mode and can be associated with external crypto devices.

Data transmission mode is available for point-to-point datalink operation using Time Division Multiple Access protocol. Use of the datalink mode, with fast frequency hopping and error detection and correction procedures, ensures both transmission reliability and minimal detectability. THOMRAD 6000 radios are designed to transmit and receive Link 11 and Link 22.

The THOMRAD 6000 product line includes two airborne versions (TRA 6030 and TRA 6020), ground and shipborne versions (TRG 6030), and a manpack

The Thomson-CSF Communications Transceivers Airborne TRA 6020 (left) and TRA 6030 (right), of the THOMRAD 6000 V/UHF ECCM transceiver series **1998**/0011853/0011854

version (TRM 6020). Together they represent a fully integrated and interoperable communication system for all air/ground, air/air and surface/surface communications.

Particular versions are available for specific requirements.

Specifications

Dimensions:
(TRA 6030) 90 × 194 × 320.5 mm
(TRA 6020) 127 × 120 × 183 mm
Weight:
(TRA 6030) 7 kg
(TRA 6020) 5 kg
Power: 28 V DC

Operational status

All versions in production. Selected for Rafale, Mirage 2000, E2C, Cougar and Panther.

Contractor

Thomson-CSF Communications.

NEW ENTRY

TRC 9600 VHF/FM secure radio communication system

The TRC 9600 is a VHF/FM military airborne transceiver designed for military platforms with a high ECCM protection level ensuring reliable communications in a dense electronic warfare situation. It is the airborne version of the PR4G system, fully interoperable with manpack and vehicular versions and covers the frequency range 30 to 88 MHz at 25 kHz spacing to give 2,320 channels.

The Thomson-CSF TRC 9600 VHF/FM secure radio communication system comprises the TRC 9610 transceiver unit (left) and the TRC 9620 control unit (right)

The TRC 9600 provides communications protected from interception, direction-finding, jamming, listening-in and spoofing. The lightweight and compact 10 W transceiver embodies frequency hopping, free channel search and high-security digital encryption functions.

Its light weight, small size and high reliability result from the use of advanced and well-proven technology such as powerful and fast microprocessors of the new HCMOS 68000 family, VLSI circuits, wide use of surface mount components and the use of proximity filters for co-site operation.

Frequencies are generated by a digital synthesiser with an ultra-rapid acquisition time, driven by a high-stability oscillator. Seven channels can be memorised and are stored for more than a year. The output power can be 10 W, 5 W or 0.5 W depending on the requirement. In the analogue fixed-frequency mode, the TRC 9600 is directly interoperable with existing VHF/FM sets and is provided with a noise squelch.

The system consists of the TRC 9610 transceiver unit, TRC 9620 control unit, shockmount, antenna with logic converting unit and power supply. Accessories include a KY 58 encryption unit, relay cable, aircraft interphone system and fill device.

Specifications

Dimensions:
(control unit) 145 × 76 × 160 mm
(transceiver unit) 125 × 196 × 340 mm
Weight: 8 kg
Power supply: 28 V DC

Contractor

Thomson-CSF Communications.

UPDATED

GERMANY

AR 3202 VHF transceiver

The AR 3202 VHF transmitter/receiver is a member of Becker's 3000 Series Prime Line avionic systems which feature microprocessor control. It provides 760 channels in the VHF band which extends from 118 to 136.975 MHz for civil aircraft. This range can be extended to 144 or 152 MHz for military aircraft. The system features a non-volatile memory and solid-state switches which eliminate mechanical contacts, giving increased ruggedness and reliability. Opto-electronic switching is employed for frequency selection. Frequency generation and display are microprocessor-controlled. The display comprises two liquid crystal presentations, one of which indicates the active channel while the other shows a preselected frequency.

The complete system is housed in a single compact unit, which complies with ARINC standards. It does not require the remote boxes, interconnecting cables, or external cooling of older designs.

Additionally the set can be operated using the CU 3202 remote controller. This is particularly useful in a tandem-seat trainer where any frequency selection made in one cockpit is displayed in the other.

The transmitter output power is 20 W and the AR 3202 is suitable for both fixed-wing aircraft and helicopters.

Specifications

Dimensions: 47.5 × 146 × 225 mm
Weight: 1.3 kg
Frequency ranges:
Standard: 118.0-136.975 MHz
Option 1: 118.0-144.0 MHz
Option 2: 118.0-152.0 MHz
Channel spacing: 25 kHz
Power supply: 25-30 V DC; 3.5 A (transmit), 0.24 A (standby). Certified to FAA TSO C37c, TSO C38c and ICAO Annex 10 specifications.

Operational status

In production and in service.

Contractor

Becker Avionic Systems.

UPDATED

AR 3209 VHF transceiver

There are two versions of the AR 3209 transceiver, the AR 3209-(09) 5 W system and the AR 3209-(11) 10 W system.

General features include: storage memory for up to 20 of 760 channels; front panel adjustment of squelch, side tone and intercom; two sunlight-readable LCDs.

The AR 3209 is an ideal retrofit for the Becker COM 2000 and is cable, plug and form/fit compatible.

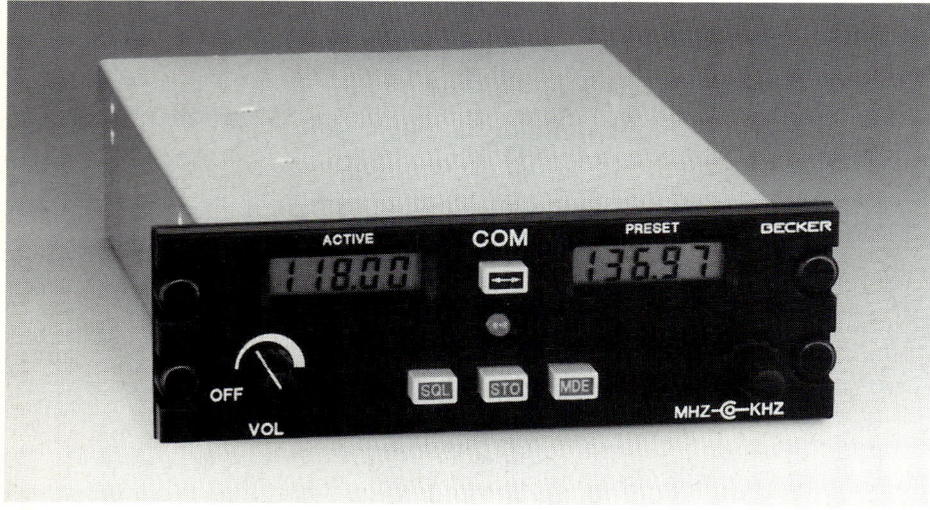

Becker Avionic Systems AR 3209 ***1997***/0001174

Specifications

Dimensions: 146 × 47.5 × 229 mm
Weight: 1.2 kg
Radio frequency: 118.0-136.975 MHz
Channel spacing: 25 kHz
Memory channels: 20
Power supply: AR 3209-(09): 13.75 V DC
AR 3209-(11): 27.5 V DC
Certifications: TSO C37d, TSO C38d, and ICAO Annex 10

Operational status

In production.

Contractor

Becker Avionic Systems.

UPDATED

AR 4201 VHF-AM transceiver

The small lightweight AR 4201 offers 760 channels and is certified for use in VFR and IFR equipped aircraft. It is ideal for installation in gliders and home-built and small single-engined aircraft due to its limited power requirement and 57 mm round format.

The equipment has a transmit power of 5 to 7 W. A standby frequency and 99-channel memory are available and can be easily programmed and recalled. The AR 4201 features intercom, panel lighting, voltage indicator, an RF input, automatic test routines, a serial interface and an optional temperature indication.

Two dynamic and two standard microphones can be connected without any alterations. The display shows the active frequency and either the standby frequency, the memory channel used or the supply voltage and also the external temperature (with the optional temperature sensor).

Using the RS-232 interface, all functions can be remotely operated. With this standard feature, the AR 4201 can be integrated into future Becker Systems.

Specifications

Dimensions: 192 × 60.6 × 60.6 mm
Weight: 0.67 kg

Becker Avionic Systems AR4201 VHF-AM transceiver ***1997***/0001176

Power supply: 12.4-15.1 V DC;
(transmit) <2.5 A
(standby) <0.07 A
Frequency range: 118.0-136.975 MHz.
Channel spacing: 25 kHz
Memories: 99
Certifications: TSO C37d, TSO C38d, and ICAO Annex 10

Contractor
Becker Avionic Systems.

UPDATED

AS 3100 audio selector and intercommunication system

A member of the Becker 3000 Series avionic systems, the AS 3100 controls four transmitters and up to six receivers. By addition of an auxiliary unit, a further six receiver units can be added to the audio chain. In different versions the AS 3100 is capable of either voice-operated switch or push-to-talk operation with all other stations, voice filter for ADF and navigation systems, connection of various types of microphone and emergency operation and providing redundancy for the transmitter/receiver operation. The intercom amplifier has a common bus connecting up to six cabin and three cockpit stations. A cockpit voice recorder output is incorporated.

The system is of modular construction and may be tailored to precise customer requirements. It is suitable for fixed-wing aircraft and helicopters. To achieve maximum adaptability, a full range of sub-units has been developed to extend the function of the main AS 3100 controller to a full cabin communication and passenger entertainment system. An intercom amplifier permits communication between passengers and crew in noisy aircraft such as helicopters, and a service station allows communication between the crew and flight attendants as well as a public address facility. A tape player, used in conjunction with the public address amplifier, is the basis of passenger entertainment through headsets or loudspeakers. The public address amplifier includes one mono or stereo amplifier and a double-tone gong which operates when activated by the fasten seat belts or no smoking signs.

An external jack box, which is normally installed in the wheel well or any other readily accessible location, permits communication between cockpits or flight deck and ground crew during starting and departure checks.

All units operate from a 28 V DC supply.

Specifications
Dimensions:
(main control unit) 38 × 146 × 35 mm
(auxiliary unit) 29 × 146 × 26 mm
(cassette player) 57 × 146 × 170 mm
(service station) 210 × 66 × 115 mm
(public address amplifier) 129 × 45 × 245 mm
(external jack box) 117 × 80 × 80 mm
Weight:
(main control unit) 0.8 kg
(auxiliary unit) 0.2 kg
(cassette player) 1 kg
(service station) 1 kg
(public address amplifier) 0.8 kg
(external jack box) 0.6 kg

Operational status
In production and in service.

Contractor
Becker Avionic Systems.

UPDATED

COM 5200 series VHF communication systems

The COM 5200 Systems flexible architecture allows the VHF communication radios installation to be customised to provide the required combination of output power and cost. Changes in output power can be made later without extensive additional installation costs. The system is intended for installations where minimum panel space is to be used. It utilises a lightweight CU 5209 control unit, which fits into a standard 2.25 in (57 mm) round instrument panel cut-out, and is only 2.5 in (63.5 mm) deep. Lightweight remote transceiver units, with output powers from 5 to 20 W are mated with this control head, to complete the systems and can be installed at any convenient place in the aircraft.

The CU 5209 control unit uses a high-contrast, double line LCD display, which is readable in bright sunlight.

Both active and standby frequencies are displayed, and can be transferred by a single stroke of the 'Flip-Flop' button. Up to 99 preset frequencies can be entered from the front panel and stored in non-volatile memory. A BITE automatically checks and monitors the system, to ensure proper operation and to facilitate maintenance. All systems will be JTSO certified for either VFR or IFR use in all types of fixed-wing and rotary-wing aircraft, to most ICAO requirements for VHF radio. The COM 5200 systems can be combined with other Becker Compact-Line avionics systems, such as VOR/ILS navigation receivers and transponders, which

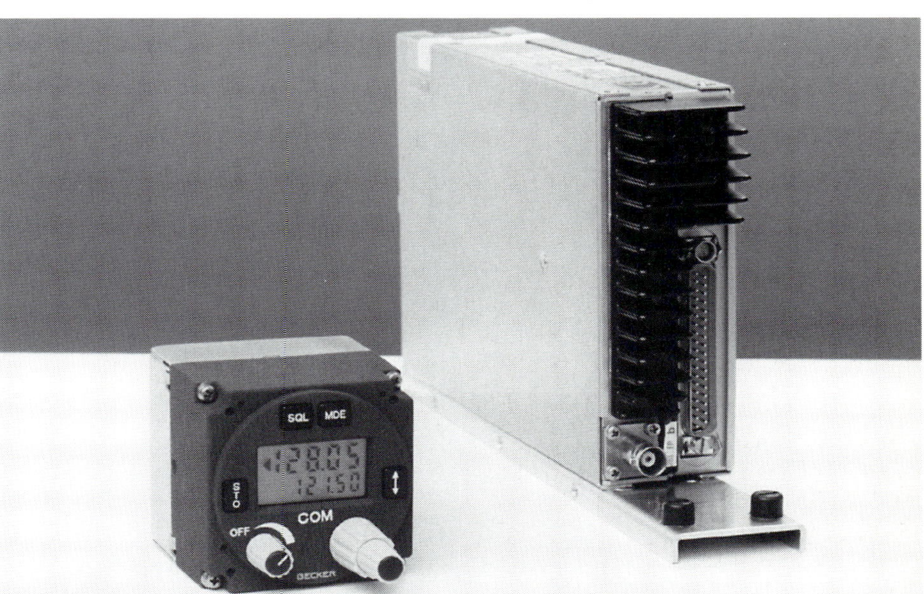

COM 5200 series VHF communications systems ***1997***/0001175

Members of the Becker AS 3100 family

have similar control units. The remote transceiver units can be controlled by other types of CDU or FMS devices and can be integrated into flight management systems.

Specifications

CU 5209 for RT 3209 Control unit
CU 5202 for RT 3202
RT 3209-(11) ≥10 W (CW) transceiver
RT 3202 ≥16 W (CW) transceiver
Frequency range: 118.000-136.975 MHz
Channel number: 760
Channel spacing: 25 kHz
Dimensions excluding connector: CU 5209
CU 5209 and CU 5202: 61.3 × 61.3 × 62 mm
RT 3202: 134 × 50 × 265 mm
RT 3209: 134 × 50 × 243 mm
Power supply: 28 V DC;
Transmit:
(RT 3202) ≤4 A
(RT3209-(11)) ≤2.5 A
Standby: ≤0.25 A

Contractor

Becker Avionic Systems.

VERIFIED

ODIN Operational Data INterface

The purpose of the Operational Data INterface (ODIN) is to establish air-to-air and air-to-ground data transfer communication via already existing voice communication channels in the HF and VHF/UHF frequency ranges. The mode of communication is half-duplex, so that ODIN can only transmit or receive information at any given time. When data transfer is not taking place normal voice communication is possible.

The ODIN system consists of two LRUs: the electronics unit and the control panel. The initiation of message transfer via a radio channel and the related transmission mode are controlled by the control panel. Acknowledge, broadcast, interrogate and automatic modes for message transfer are available.

ODIN features a 68000 communication processor and an ADSP 2100 modem processor plus a MIL-STD-1553 interface. The data transfer rate is 2,400 bauds for VHF and UHF and 300 bauds for HF.

Operational status

In service with Panavia IDS and ECR Tornado aircraft.

Contractor

LITEF GmbH.

VERIFIED

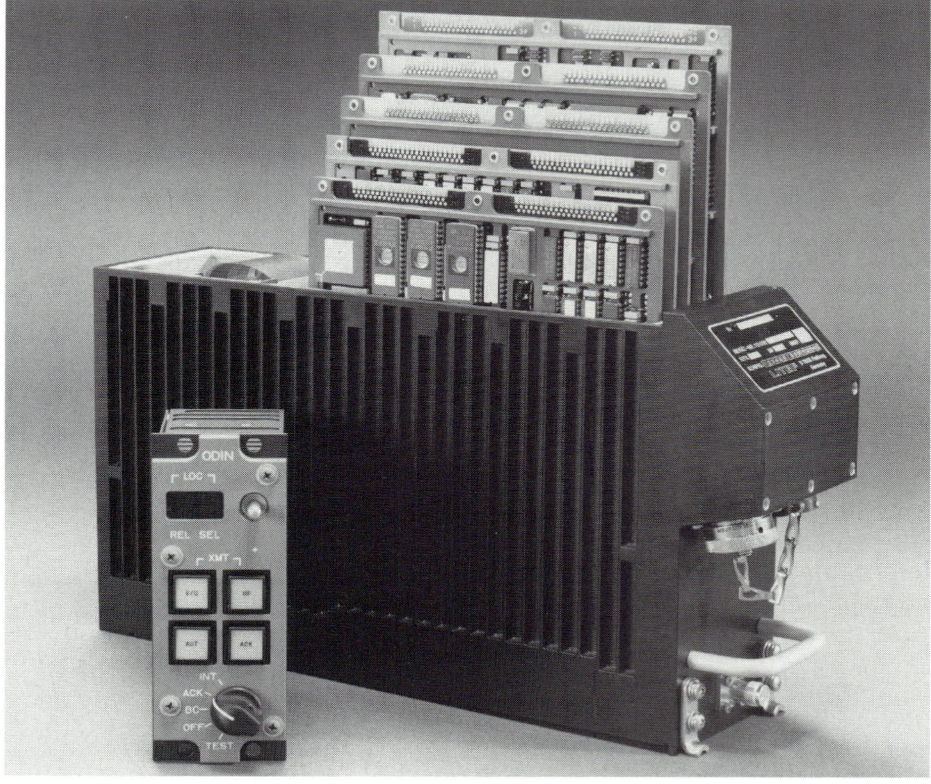

The LITEF Operational Data INterface (ODIN) is employed in Panavia Tornado aircraft

610 series VHF and UHF radios

The Rohde and Schwarz 610 family of VHF and UHF radio communication systems comes in two basic versions: a single panel-mount cockpit unit or a remotely controlled transmitter/receiver with a panel-mounted control unit. The controllers for the VHF and UHF variants are electrically and mechanically identical. They permit parallel operation of both a VHF and a UHF transmitter/receiver from a single controller and/or the operation of a single transmitter receiver from two control units. All systems in the 610 series provide not only radio telephonic communication but also incorporate a 16 kbit baseband data transmission and ADF facilities.

Technical specifications of the VHF and UHF variants are virtually identical. Each type has a transmitter power output of 10 W at normal supply voltage of 28 V, or 1 W carrier wave with a reduced emergency power supply of 16 V. Frequency range of the VHF systems is from 100 to 155.975 MHz and that of the UHF equipments from 225 to 399.975 MHz. Guard receivers cover the emergency channels of 121.5 and 243 MHz respectively. Frequency setting increments are 25 kHz in each case and the channel spacing is also 25 kHz, although in the case of UHF systems the spacing is optionally adaptable to 50 kHz. Up to 30 channels, plus the Guard channel, may be preselected and a remote frequency/channel indicator is an optional accessory.

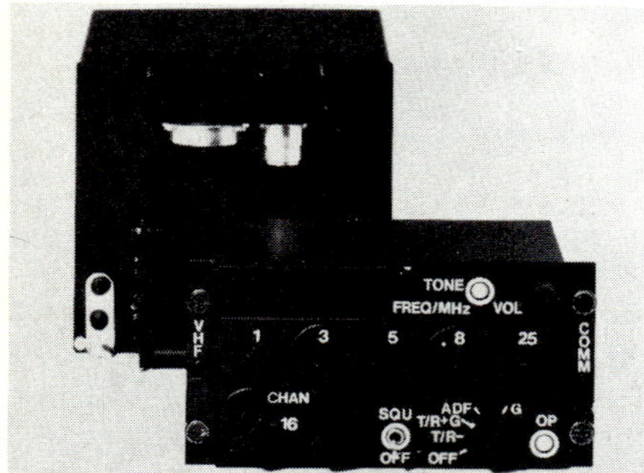

The Rohde and Schwarz 610 VHF/UHF radio

All 610 series transmitter/receivers are compatible with one another and the various modules have precisely designed interfaces to permit easy replacements to be made without need for adjustment. The systems are suitable for retrofitting and Rohde and Schwarz has produced replacement kits with tailored adaptor trays for aircraft such as the The Boeing Company F-4F and RF-4E. A range of Special-to-Type Test Equipment (STTE) is available, with first line test sets which can isolate faults down to module level and full-scale automatic test equipments for base repair facilities.

Specifications

Dimensions:
(XU 610 VHF cockpit version) 127 × 124 × 165 mm
(XU 611 VHF remote-controlled) 127 × 124 × 165 mm
(XD 610 UHF cockpit version) 127 × 124 × 165 mm
(XD 611 UHF remote-controlled) 127 × 124 × 165 mm
(GB 600 remote-control unit) 146 × 76 × 110 mm
Weight:
(XU 610 VHF cockpit version) 5.2 kg
(XU 611 VHF remote-controlled) 4.6 kg
(XD 610 UHF cockpit version) 5.2 kg
(XD 611 UHF remote-controlled) 4.6 kg
(GB 600 remote-control unit) 1.9 kg

Operational status

In production.

Contractor

Rohde and Schwarz GmbH and Co KG.

VERIFIED

Have Quick 610 VHF/UHF ECCM radios

The Have Quick airborne radio is based on the Rohde and Schwarz 610 Series of radios. As a retrofit for the German Air Force The Boeing Company F-4F and RF-4E Phantom aircraft, the radio includes UHF Have Quick ECCM and its hardware is compatible with Have Quick II. The retrofit adaptor provides an extension to a combined VHF/UHF radio. The radio complies with STANAG 4246 and is interoperable with Rohde and Schwarz Secos 400 Have Quick ground-based or shipborne radios.

Operational status

In service in German Air Force F-4F and RF-4E aircraft.

Contractor

Rohde and Schwarz GmbH and Co KG.

VERIFIED

HF 510 radio system

The HF 510 is an airborne system for air-to-air and air-to-ground communication using AME and SSB to transmit voice and data during nap of the earth flying under non-line of sight and extreme environmental conditions over the whole operational area. Suitable for both helicopters and fixed-wing aircraft, the system includes types for voice and data or voice only and consists of 100 W or 400 W HF transceivers, antenna tuning units and control units.

The system features include high-density integration, a high degree of modularity and built-in speech processor, test programme and silent tuning.

The XK 510 HF data transceiver, especially developed for the German Army Eurocopter PAH-2

Tiger helicopter, has passive and active channel analysis, automatic channel selection from a preset frequency list, adaptive reaction and error detection and correction. Modes are fixed channel, adaptive or ECCM.

Specifications

Dimensions:
(GB 510 control unit) 146 × 229 × 165 mm
(XK 510 D1 transceiver) 223 × 194 × 320 mm
(XK 516 S1 transceiver) 255 × 194 × 320 mm
(FK 510 antenna tuning unit) 123 × 194 × 320 mm
(FK 511 antenna tuning unit) 255 × 194 × 320 mm
Weight:
(GB 510 control unit) 4.5 kg
(XK 510 D1 transceiver) 13.5 kg
(XK 516 S1 transceiver) 19 kg
(FK 510 antenna tuning unit) 5 kg
(FK 511 antenna tuning unit) 5 kg
Power supply:
(XK 510 and 515) 28 V DC
(XK 516) 115 V AC, 400 Hz, 3 phase
Power output:
(XK 510) 100 W PEP CW
(XK 515) 150 W PEP
(XK 516) 400 W PEP
Frequency: 2-29.999 MHz
Channel spacing: 100 kHz
Preset channels: 100 or any number with computer control or cockpit management

Operational status

In production for German Army Eurocopter PAH-2 helicopters.

Contractors

Rohde and Schwarz GmbH and Co KG.
Daimler-Benz Aerospace AG, Defense and Civil Systems.

UPDATED

Secos 610 VHF/UHF ECCM radios

Secos 610 secure radios are designed for airborne and land-mobile use and are based on the Rohde and Schwarz 610 Series of radios. The Secos 610 features digital encryption, medium-speed frequency hopping with collision-free operation of a large number of networks and interoperability with the ground-based or shipboard Secos 400. The frequency range for ECCM operations is 225 to 400 MHz. There is an additional fixed-channel plain language AM option covering the 100 to 156 MHz VHF band. Channel spacing is 25 kHz.

The equipment consists of transceivers, control units and a key entry device. Control is through a compact cockpit transceiver; there are also remote-controlled versions. One common unit controls both the UHF and optional VHF. Power supply is 28 V DC and power output is 15 W on FM. MIL-STD-1553B-controlled versions are also available.

Operational status

In service in F/A-18 aircraft.

Contractor

Rohde and Schwarz GmbH and Co KG.

VERIFIED

The Rohde and Schwarz Secos 610 UHF ECCM transceiver

XK 401 HF/SSB radio

The XK 401 has been developed jointly by Rohde and Schwarz with Siemens AG for the trinational Panavia Tornado, with the aim of providing reliable air-to-air and air-to-ground communication over long ranges.

The system covers the HF band from 2 to 29.999 MHz in 100 Hz increments, providing more than 280,000 channels, of which any 11 are preselectable. All channels are individually selectable by means of decade switches. Operational modes are upper sideband, A3J (duplex) and continuous wave. Transmitter power output is 400 W peak power for modulated transmission and 100 W in carrier wave mode transmission.

The XK 401 comprises three basic units: the XK 401 transmitter/receiver, the GB 401 remote controller and the VK 241 power amplifier. Additionally, two optional units, the FK 241 antenna tuner and a control frequency selector, are available and are recommended for optimum operation. The system is of modular construction and solid-state components are used.

The transmitter/receiver section uses digital synthesis frequency-generation techniques and the synthesiser itself is said to possess outstandingly good noise characteristics. Two intermediate frequencies, 72.03 MHz and 30 kHz, are used for both transmission and reception paths. The receiver section has automatic squelch which operates if the HF/SSB level exceeds an adjustable threshold. In the transmitter section, the lower sideband is suppressed by mechanical filters. Built-in test equipment is incorporated.

The remote controller, which may be up to 50 m from the other major units, contains all necessary system controls as well as storage facilities for the 11 preselectable channels.

Two identical amplifier modules, with a common output matching circuit, form the power amplifier section, this duplication being applied in the interests of reliability. In normal operation both modules are in use and, in the event of one module failing, the only consequence is reduction of output power rather than total power output breakdown. Thermal and mismatch overload protection circuits, which include open and short-circuit protection, are provided. Heat dissipated by the amplifier section is extracted by an external ventilator.

Use of the optional antenna tuner and the control frequency selector considerably enhances system performance. The former unit matches the base impedance of the antenna to transmitter/receiver output impedance and, during reception periods, acts as a preselector. The control frequency selector is used to control the digital tuning information. Average tuning time for the system is less than half a second, but use of the control frequency selector unit enables tuning information to be stored for all preselected channels and the tuning time is thus further reduced. No power is radiated during tuning, resulting in radio silence being maintained during such an operation.

Specifications

Dimensions:
(controller) 146 × 86 × 165 mm
(transmitter/receiver) 124 × 194 × 319 mm
(power amplifier) 257 × 194 × 319 mm
Weight:
(controller) 1.8 kg
(transmitter/receiver) 11.2 kg
(power amplifier) 17.2 kg

Operational status

In service in the Panavia Tornado.

Contractor

Rohde and Schwarz GmbH and Co KG.

VERIFIED

XT 3000 VHF/UHF radio

The XT 3000 is a combined VHF/UHF air-to-air and air-to-ground transmitter/receiver covering the frequency ranges 100 to 162 MHz and 225 to 400 MHz, in both FM and AM modes for voice and data. Frequency increments are spaced at 25 kHz intervals but the channel spacing is switchable by increments of 25, 50 or 100 kHz as required. Transmitter output is 10 W. Up to 28 operational channels, together with the international distress frequencies of 121.5 and 243 MHz, may be preselected on remote-control units. The system incorporates built-in test equipment and inputs for special-to-type and automatic test equipment.

The XT 3000 comprises a single VHF/UHF transmitter/receiver, VHF and UHF amplifiers, two control units for remote operation and a channel/ frequency indicator. It meets MIL-E-5400 and MIL-STD-810B, -461, -462, -463, -781B and VG 95211.

Specifications

Dimensions: ½ ATR short plus two ¼ ATR short units
Weight: 30 kg

Operational status

In production and in service.

Contractor

Rohde and Schwarz GmbH and Co KG.

VERIFIED

XT 3011 UHF radio

The XT 3011 represents what may be regarded as the UHF section of the XT 3000 system, with which it is almost identical. Exceptions are the lack of a VHF transmitter/receiver and appropriate amplifier module and the inclusion of an integrated but independent Guard receiver.

Frequency range is from 225 to 400 MHz with a frequency spacing of 25 kHz. Channel spacing is normally 50 kHz but this may be modified to give 25 kHz increments. Modulation mode is AM only.

The system is operable from either one or two remote-control positions and there are three different

The Rohde and Schwarz XT 3000 VHF/UHF radio with, from left to right (top), the UHF power amplifier, transmitter/receiver and VHF power amplifier. Two control units and a remote frequency indicator are shown at the bottom

types of controller available. Provision is also made for a remote frequency/channel indicator.

Specifications
Dimensions:
(transceiver) ½ ATR short controllers in accordance with MIL-STD-25212
Weight: 7.5 kg

Operational status
In service.

Contractor
Rohde and Schwarz GmbH and Co KG.

VERIFIED

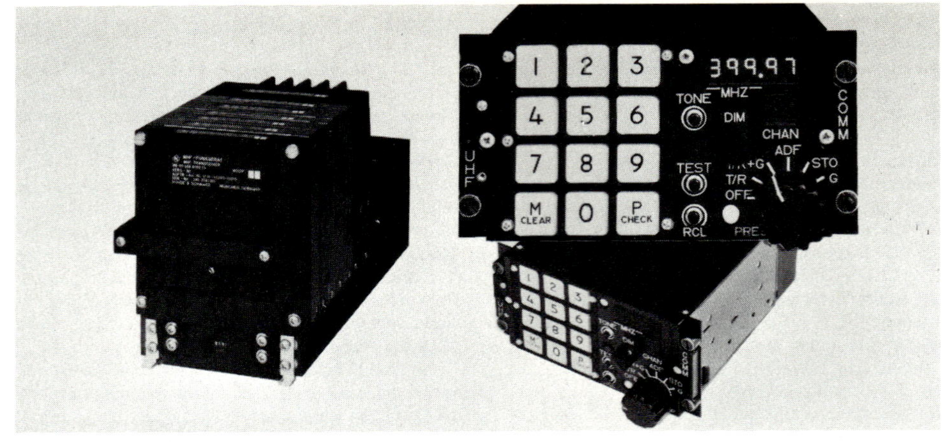

The Rohde and Schwarz XT 3011 UHF transceiver with control units

---

INDIA

Intercom - Audio Management Unit (AMU) 1301A

The Intercom AMU 1301A is designed for use in fighter aircraft, transports and helicopters to cater for up to five crew members in a net with call and conference facilities. It provides management control of five transceivers, three receivers and five audio warning signals. Additional features include: VOS for hands-free operation; dual redundancy.

Specifications
Dimensions:
(junction box) 120 × 120 × 250 mm
(station box) 76 × 146 × 210 mm
Weight:
(junction box) 1.4 kg
(station box) 1.5 kg
Power supply: 27.5 V DC, 2 A
Bandwidth: 200 Hz to 2.5 kHz

Contractor
Hindustan Aeronautics Ltd.

UPDATED

Hindustan's AMU 1301A Intercom ***1996***

Audio Management Unit (AMU) 1303A

AMU 1303A is designed for use in fighter and trainer aircraft. It provides management control of two transceivers, three receivers and seven audio warning signals. Additional features include: a telebrief facility; voice-operated switch for hands-free operation.

Specifications
Dimensions:
(junction box) 60 × 175 × 180 mm
(station box) 67 × 146 × 180 mm
Weight:
(junction box) 1.1 kg
(station box) 1.2 kg
Power supply: 27.5 V DC, 1 A
Bandwidth: 300 Hz to 3.5 kHz

Contractor
Hindustan Aeronautics Ltd.

UPDATED

COM 32XA HF/SSB communication system

The all-solid-state modular construction HF/SSB communication system is designed for air-to-air and air-to-ground communication on 2.06 to 29.999 MHz (COM 326A) or 2.5 to 23.5 MHz (COM 327A). Channel spacing is 100 kHz and the system has seven in-flight programmable preset channels and instant manual selection of any channel.

The salient features of the system include solid-state design with software-controlled advanced automatic antenna matching, high-stability frequency synthesiser, high-power solid-state amplifier with protection against high VSWR, high-performance receiver with front panel configurable logic, in-flight programming of channnel and modes and BITE.

Specifications
Dimensions:
(receiver/exciter) 190 × 194 × 320 mm
(power amplifier) 190 × 194 × 320 mm
(ATU for COM 327 and 325) 150 × 180 × 321 mm
(control unit) 146 × 124 × 102 mm
Weight:
(receiver/exciter) 11.5 kg
(power amplifier) 15 kg
(ATU for COM 327 and 325) 7.5 kg
(control unit) 1.5 kg
Power supply:
(COM 326A) 200 V AC, 400 Hz, 3 phase
(COM 325/327/328/329A) 115 V AC, 400 Hz, single phase
27.5 V DC
(receiver) 100 W DC
(transmitter) 850 W AC 100 W DC
Temperature range: −40 to +55°C
Altitude: up to 65,000 ft

Operational status
The COM 325A is fitted on the Hawker Siddeley 748, COM 326A on the Jaguar, COM 327A on the An-32, COM 328A on the Do 228 and the COM 329A on the Il-38.

Contractor
Hindustan Aeronautics Ltd.

VERIFIED

COM 105A VHF transceiver

The all-solid-state modular construction COM 105A VHF transceiver provides air-to-air and air-to-ground voice telephony communication in the VHF band between 118 and 136 MHz at 25 kHz spacing, giving 720 channels. Twenty preset channels are available.

Specifications
Dimensions:
(transceiver) 135 × 145 × 247 mm
(control unit) 104 × 82 × 95.5 mm
Weight:
(transceiver) 4.3 kg
(control unit) 0.5 kg
Power supply: 27.5 V DC
(transmit) 108 W
(receive) 55 W
Temperature range: −40 to +55°C
Altitude: up to 65,000 ft

Contractor
Hindustan Aeronautics Ltd.

VERIFIED

COM 150A UHF transceiver

The COM 150A UHF transceiver provides 7,000 channels at 25 kHz spacing between 225 and 399.975 MHz for radio telephony (A3) transmissions. There are four preset channels.

Specifications
Dimensions: 121 × 174 × 294 mm
Weight: 6.5 kg
Power output: 5 W
Temperature range: −55 to +55°C
Altitude: up to 70,000 ft

Contractor
Hindustan Aeronautics Ltd.

VERIFIED

UHF communication system - COM 1150A

The AM transceiver COM 1150A is designed with hybridised circuits for improved reliability, reduced volume and weight, providing A3 communication in VHF band.

Specifications

Dimensions:
(transceiver unit) 124 × 178 × 250 mm
(control unit) 80 × 80 × 107 mm
Weight:
(transceiver unit) 5 kg
(control unit) 0.3 kg
Operating frequency: 225-399.975 MHz
Channel spacing: 25 kHz
No of preset channels: 10
Type of transmission: A3
Receiver sensitivity: 97.0 dBm
Antenna impedance: 52±2 ohms
Audio output: 100 mW into 150/600 ohms
Power output: 5 W (nominal)
Microphone: EM type (150 ohms)

Contractor

Hindustan Aeronautics Ltd.

UPDATED

HF-SSB communication transceiver - COM 1330A

COM 1330A transceiver provides long-range communication for transport aircraft, ground attack fighters and helicopters. It is designed with state-of-the-art techniques featuring: microprocessor-based antenna tuning for different types of wire, cap and rod antennae, frequency synthesis, 100 W solid-state power amplifier and hybrid circuits. It provides a separate audio channel for SELCAL decoding.

Specifications

Dimensions:
(receiver/exciter) 190 × 194 × 320 mm
(PA-ATU) 190 × 194 × 320 mm
(controller) 146 × 124 × 102 mm
Weight:
(receiver/exciter) 8 kg
(PA-ATU) 12.5 kg
(controller) 0.8 kg
Frequency: 2-26.9999 MHz
Frequency stability: ±0.8 ppm
Modes of operation: CW, MCW, AM, USB & ISB
Sensitivity:
(AM mode) −101 dBm
(SSB mode) −110 dBm
Power output: 100 W PEP
Tuning time: 18s
Preset channels: 9
Input power: 22-31 V DC, nominal 27.5 V
Audio ouput: >50 mW into 600 ohms

Contractor

Hindustan Aeronautics Ltd.

UPDATED

VUC 201A VHF/UHF system

The all-solid-state modular construction VUC 201A is a standard air-to-air and air-to-ground radio, offering 2,240 channels between 100 and 155.975 MHz and 7,000 channels between 225 and 399.975 MHz, at 25 kHz spacing. The Guard channel can be tuned between 238 and 248 MHz and 19 channels can be preset. It includes BITE.

Specifications

Dimensions: 160 × 155 × 357 mm
Weight: 13 kg
Power supply: 27.5 V DC, 100 W Rx, 550 W Tx
Power output: 10 W VHF, 20 W UHF
Temperature range: −55 to +55°C
Altitude: up to 70,000 ft

Contractor

Hindustan Aeronautics Ltd.

VERIFIED

INTERNATIONAL

MCS 3000/6000 aeronautical satellite communications system

Honeywell and Racal Avionics are teamed for the development, manufacture and marketing of a multichannel satellite communications system for commercial aircraft, fully compatible with ARINC 741 and (beginning early in 1998) ARINC 761 for Aero-I.

The Honeywell/Racal MCS 3000/6000 systems provide a three- or six-channel full-duplex voice and data communications capability supporting such functions as Airline Communications And Reporting System (ACARS), Automatic Dependent Surveillance (ADS) and flight deck and passenger telephone and fax communications between an aircraft and the ground.

The MCS 6000 airborne terminal comprises a Satellite Data Unit (SDU), Radio Frequency Unit (RFU) and High-Power Amplifier (HPA), and may be interfaced to a variety of high-gain phased-array antenna subsystems and voice/data communications devices.

The SDU performs the functions of system controller, data modulation and demodulation, data synchronisation and decoding and voice coding/decoding.

The RFU performs the functions of down converting the received L-band (NATO D-band) signals to a lower frequency for input to the digital processing circuits and up converting the modem output signals to the L-band (NATO D-band) transmit frequency for each operational channel. The RFU operates in full-duplex mode, simultaneously supporting both receive and transmit functions at all times.

The HPA is a linear power amplifier which provides the gain to generate the required output power. The output power of the HPA is under the control of the SDU which receives data from the ground station commanding an increase or decrease in output power to maintain the satellite signal at a satisfactory level. The HPA is available in either an ARINC 741 or ARINC 761 configuration.

The system operates at L-band (NATO D-band) frequencies. Signals are relayed via the Inmarsat space segment satellites, linking in to the ground telecommunications network through a series of dedicated Ground Earth Stations (GES). The satellite and GES networks combine to provide a worldwide communications service.

MCS 3000/6000 systems support both air-to-ground and ground-to-air communications. The systems are capable of supporting 9.6 kbytes/s voice, 4.8 kbytes/s fax, 2.4 kbytes/s PC modem and 10.5 kbytes/s packet data services for ARINC 741 applications; the systems will support 4.8 kbytes/s voice, 2.4 kbytes/s fax, 2.4 kbytes/s PC/modem and 1.2 kbytes/s packet data services for ARINC 761 (Aero-I) applications. Typical applications for these services break down into the areas of passenger services, airline operational and administrative services and air traffic control.

6-channel MCS-6000 system

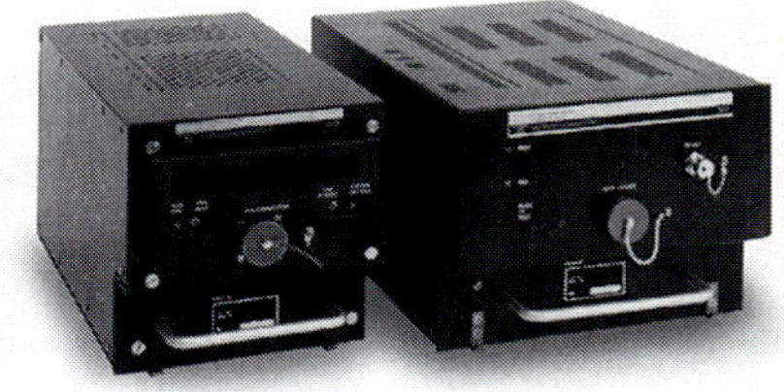

3-channel MCS-3000 system

Honeywell/Racal MCS 3000/6000 aeronautical satellite communications system showing both the six-channel MCS-6000 system (radio frequency unit (left), satellite data unit (centre) and high-power amplifier (right)); and the three-channel MCS-3000 system (satellite data unit (left) high-power amplifier (right))
***1997**/0002115*

Passenger services include telephone, fascimile, PC and value added data services such as catalogue sales, hire car reservations and duty free sales.

Airline operational and administrative services include the ACARS datalink supporting engineering, operational and cabin management functions.

Air traffic control uses include Automatic Dependent Surveillance aircraft position reporting.

MCS 3000/6000 operates in full accordance with Inmarsat specifications and type approval has been received on all major wide-body aircraft types.

Aero-I is an upgraded capability of the MCS 3000/6000 to allow narrow-body aircraft with intermediate-gain antennas to utilise the new Inmarsat third-generation Aero-I spot-beam services. As an extension to the Aero-I upgrade, existing high-gain MCS equipment can be upgraded, via a 'Service Bulletin'; this takes advantage of 'evolved Aero-H' services, which provide a similar range of passenger global communications for telephone, fax and pc-date, as Aero-I. Using 'evolved Aero-H' in spot-beam coverage allows the operator to benefit from reduced service charges and 4.8 kbytes/s voice CODECs (digital transmission).

Specifications

Dimensions:
(SDU) 6 MCU; weight 10.73 kg
(RFU) 4 MCU; weight 7.68 kg
ARINC 741 (HPA) 8 MCU; weight 12.86 kg (Aero-H)
ARINC 761 (HPA) 4 MCU: weight 7.05 kg (Aero-I)
Power supply: 115 V AC, 400 Hz; or 28 V DC
Output power: 60 W typical
Frequency:
(transmit) 1,626.5-1,660.6 MHz
(receive) 1,530-1,553 MHz

Operational status

Over 1,200 MCS systems are on order from over 57 major airlines. Installations have been completed on all major wide-body aircraft types, such as the Boeing 747, 767, 777, MD-11, MD-90, Airbus A310, A320, A330 and A340 and top of the range executive aircraft such as the Gulfstream IV and V, Challenger 603 and 604 and Falcon 900.

Integration has been completed with all major antenna, ACARS and passenger telephone equipment vendors.

In March 1996, the Racal/Honeywell team was named as the US government's provider of choice for multichannel SATCOM using the MCS-3000/6000 'systems, featuring STU-III secure voice, access to Microsoft-Mail, 9600 bps fax and other capabilities, in conjunction with the Tecom T-4000 High-Gain Antenna System and the Honeywell CM-250 Communications Management Units.' The MCS-6000 is also fitted to the Advanced Range Instrumentation Aircraft (ARIA) RC-135s.

System upgrades available in May 1998 include new MCS-3000i/+ and MCS-6000i/+ variants which, in addition to supporting Aero-H, will be able to support Aero-I and Aero-H+.

The Aero-H+ system will use the same high-gain antenna as Aero-H but, with the capability of using spot-beam satellites, it offers a potentially lower service cost. The Racal/Honeywell Aero-I systems operate in the spot-beams of the new-generation Inmarsat-3 satellites. All services offered with Aero-H are available on Aero-H+ and Aero-I.

The Racal/Honeywell team is preparing to certify the

Canadian Marconi CMA-2200 antenna as its selected Aero-I antenna and Supplemental Type Certification (STC) flight testing, in Honeywell's Citation III aircraft, is anticipated in May 1998. First commercial certification is planned to be on a Boeing 737-800.

Contractors

Honeywell Inc, Business and Commuter Aviation Systems.
Racal Avionics Ltd.

UPDATED

MCS-7000 aeronautical satellite communications system

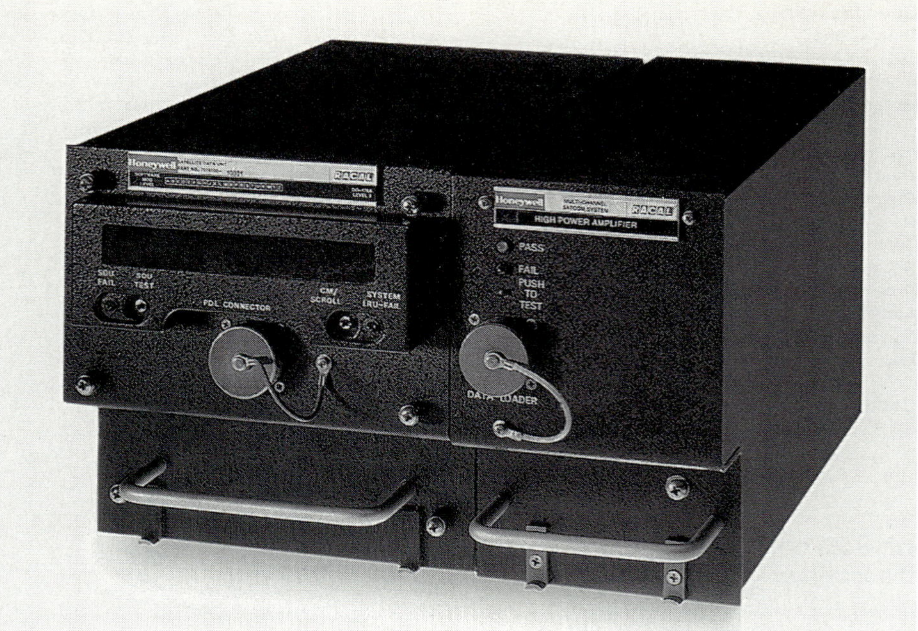

The Honeywell/Racal MCS-7000 aeronautical satellite communications system showing Satellite Data Unit (left) and High Power Amplifier (right) ***1998***/0018147

Honeywell and Racal Avionics have launched the MCS-7000 as their latest-generation enhanced satellite communications system for commercial airliners and business jet aircraft, based on their existing MCS-3000 and MCS-6000 systems. The new MCS-7000 provides up to seven channels of voice/data communications and the capability of Aero-H, Aero-H+ or Aero-I (spot-beam) services, depending on the High Power Amplifier (HPA) and antenna configuration. The MCS-7000 satcom systems consist of only two units, a 6 MCU-size Satellite Data Unit (SDU) and a 4 MCU-size High Power Amplifier (HPA) with embedded Beam Steering Unit functionality.

The Aero-H+ system will use the same high gain as Aero-H but, with the capability of using spot-beam satellites, it offers a potentially lower cost. The Aero-I system operates in spot beams of the new generation Inmarsat-3 satellites. Spot beams have lower power requirements and therefore smaller, lower power HPAs and smaller Intermediate-Gain Antennas can be used. The selected antenna is the Canadian Marconi CMA-2200.

All services currently offered with Aero-H systems are available on Aero-H+ and Aero-I including: cockpit voice (allowing instantaneous communication with operations, maintenance and air traffic control); passenger telephony, passenger fax, news and weather broadcasts; interactive passenger services and a PC data capability.

MCS-7000 will be the standard production system available from April 1999 and it provides several additional features, including an optional internal PBX phone system with digital handsets. For aircraft which are not equipped with an Inertial Reference System (IRS), a new Signal Conditioning Unit (SCU) supports satcom operation and a proprietary interface supports the imminent Complementary Satellite Systems, including Low Earth Orbit (LEO) and Medium Earth Orbits (MEO). Maximum flexibility is provided for operators to choose Inmarsat services for the cockpit (safety/ATC services) and an alternative service for passenger requirements (voice/fax).

Company data also uses the terminology MCS-7000+ and MCS-7000i when describing this system. It is also claimed that, in addition to supporting existing Inmarsat Aero-H, Aero-H+ and Aero-I, the MCS-7000 system has growth to support the Complementary Satellite Systems (CSS).

Operational status

MCS-7000 will be the standard production system available from April 1999.

Contractors

Honeywell Inc, Business and Commuter Aviation Systems.
Racal Avionics Ltd.

NEW ENTRY

MIDS-FDL Multifunctional Information and Distribution System — Fighter DataLink

GEC-Marconi Hazeltine Corporation and Rockwell Collins have entered into a joint venture to qualify, build and sell Link 16 Multifunctional Information Distribution System (MIDS) Fighter DataLink (FDL) terminals for the US Air Force F-15 aircraft. Features include: full Link 16 interoperability -TADIL-J and IJMS; 50 W transmit power and <1 W LPI mode; dual antenna transmit and receive to provide the following capabilities: position, location and identification messages; relative and geodetic navigation; adaptable host interface processing.

Specifications

Performance characteristics:
Pseudo random frequency hopping
51 frequencies — 969 to 1,206 MHz
Frequency hop rate — 13us
238 kbps Data Rate/expandable to 2 Mbps
Jam resistant and crypto secure
128 possible nets
Dimensions: ¾ ATR form factor
Weight: Maximum 20 kg (option dependant)
Power supply: 115 V AC, 400 Hz, 350 W average

Operational status

Contracts have been awarded by the US Air Force to integrate the MIDS-FDL onto F-15 C/D and F-15E aircraft.

Contractors

GEC-Marconi Hazeltine Corporation.
Rockwell Collins.

UPDATED

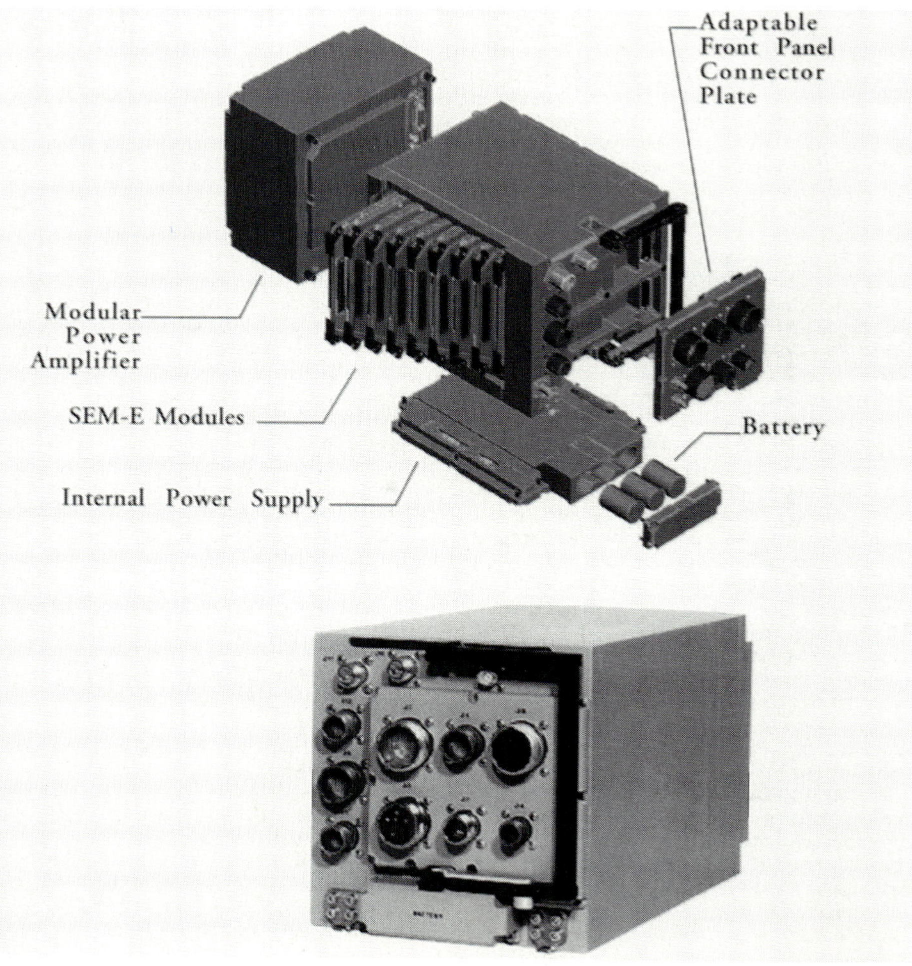

MIDS-FDL Fighter DataLink
1998/0011842

MIDS-LVT Multifunctional Information and Distribution System — Low Volume Terminal (MIDS-LVT)

The Multifunctional Information and Distribution System (MIDS-LVT) is under development by a joint venture consisting of GEC-Marconi Hazeltine Corporation (USA), Thomson-CSF (France), MID (Italy), Siemens (Germany) and ENOSA (Spain). These companies operate through the prime contractor, MIDSCO headed in New Jersey, USA. An engineering manufacturing development contract was awarded in March 1994 by the International Program Office.

MIDS is to become the standard NATO Interoperable data communications system, implementing Link 16 protocols under STANAGs 4175 and 5516. The MIDS terminal will also implement the interim JTIDS message standard, to ensure compatibility with the JTIDS Class 1 terminals which are still in use. It serves a nodeless network with a fast data exchange rate. Also voice transmission capability is provided.

While maintaining full interoperability with the family of JTIDS terminals, MIDS is a new-generation design that will satisfy a broad range of Link 16 applications. To allow its use in the latest and future aircraft designs, such as EFT 2000, the AMX, Tornado, Rafale and the F/A-18, the MIDS terminal has been reduced in size, weight, cost and power consumption to less than half that of the present JTIDS Class 2 terminal. It retains, however, all of the JTIDS capabilities, including three-dimensional receiver coverage, 200 W transmit power, complete Tacan capability, full relative/geodetic navigation, precise self-identification and powerful anti-jam capabilities. MIDS-LVT is configured in a single main terminal box and a standoff remote power supply. It will have a MIL-STD-1553 bus interface, allowing standard interoperation with designated aircraft systems, plus a high-speed optical databus (3910) and X.25 and ethernet interface capabilities. An exhaustive built-in-test (BIT) function eases the failure identification and location. Minimum maintenance effort has to be spent by the modular and interchangeable construction.

The MIDS terminal is designed around multiprocessor architecture, with industry standard VME and IEEE 486 busses offering a truly open architecture. This gives maximum flexibility and additional growth potential. Receiver/transmitter modules incorporate the latest MMIC technology.

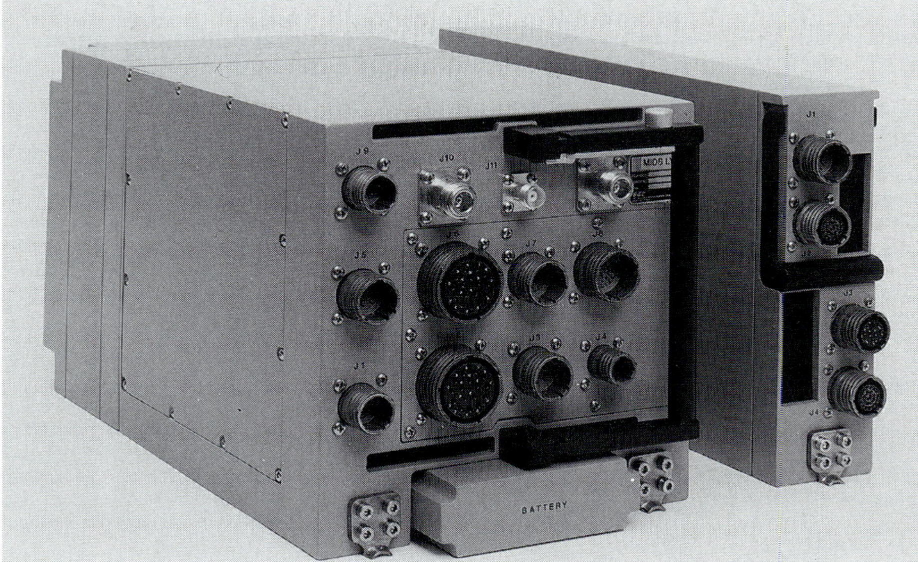
The MIDS terminal will incorporate the same functional capabilities as the JTIDS Class 2 terminal, with significant decreases in size, weight, power and cost ***1995***

Operational status

Contracted for 440 US Navy F/A-18 C/D aircraft.

Contractors

GEC-Marconi Hazeltine Corporation
Thomson-CSF Communications
Siemens AG Defence Electronics
MID(Marconi-Italtel-Defence)
ENOSA

UPDATED

Satcom conformal antenna subsystem

Dassault Electronique produces an antenna subsystem for aeronautical communications in the L-band, to meet the specific needs of the airline and general aviation industry. It is compliant with ARINC 741 and meets DO-160C.

The equipment is composed of two side-mounted conformal antennas, two Beam-Steering Units (BSUs) located inside the aircraft and the associated Diplexer/Low-Noise Amplifier (D/LNA) assemblies.

In order to provide a Satcom system with no operational limitations, an antenna with optimised RF characteristics has been designed. A large number of radiating elements arranged in an optimum pattern offers a better combination of high-gain and low-sidelobe levels. The thin profile of the High-Gain Antenna (HGA) results in a negligible drag penalty of less than 0.02 per cent of total aircraft drag and the antenna subsystem is adaptable to all high-gain Satcom avionics subsystems available or under development.

Specifications

Dimensions:
(High-Gain Antenna ×2) 566.4 × 495.3 × 7.6 mm
(Beam-Steering Unit ×2) 342.9 × 261.6 × 88.9 mm
Weight:
(High-Gain Antenna ×2) 7.5 kg
(Beam-Steering Unit ×2) 8.5 kg
(Diplexer/LNA ×2) 3 kg

Operational status

The antenna has received Inmarsat multichannel access approval with no restrictions.

The antenna system is certified on Airbus A300, A310, A330 and A340; Boeing 707, 737-300, 747-400, 767, MD-11, L-1011 and MD-80 aircraft, and the Falcon 900. More than 400 Satcom antenna systems have been ordered by over 20 major airlines.

Contractors

AlliedSignal Commercial Avionics Systems.
Dassault Electronique.

UPDATED

TRA 6032/XT 621 P1 V/UHF SATURN airborne transceiver

The TRA 6032 (French identity) or XT 621 P1 (German identity) is an airborne secure radiocommunication system designed for clear and encrypted voice and data transmission, in simplex or half duplex mode in the 100 to 156 MHz and 225 to 400 MHz frequency range.

TRA 6032 uses SATURN as the main transmission security technique which provides the highest level of protection against severe jamming, direction-finding and deception. Communication security is performed with standard NATO crypto devices.

It offers downwards interoperable voice modes such as AM and FM in fixed frequency and first-generation Have Quick I and II ECCM techniques.

The TRA 6032 is capable of data transmission, in fixed frequency as well as in SATURN mode and is particularly adapted to Link 11 and NATO Improved Link Eleven (NILE) or Link 22 transmission.

The TRA 6032 is a 3 MCU transceiver compliant with ARINC 600 and is designed for the rotary-wing environment. Its Guard receiver monitors the 121.5 MHz and 243 MHz international distress frequencies and the 156.525 MHz GMDSS frequency.

Specifications

Dimensions: 90 × 194 × 320 mm
Weight: 7.6 kg
Power supply: 28 V DC

Operational status

In preproduction. Selected for Tiger and NH 90.

Contractors

Thomson-CSF Communications.
Rohde and Schwarz Gmbh and Cc KG.

NEW ENTRY

XK 516D HF airborne voice/data radio

The XK 516D airborne radio is designed for use in commercial aircraft. The system provides conventional voice and high-speed data air-to-ground, ground-to-air and air-to-air communications over long distances. The data communication is suitable for aircraft operational and administrative communications, as well as air traffic communications. The XK 516D is a joint development by Rohde and Schwarz and AlliedSignal Aerospace.

The XK 516D consists of the XK 516D1 transceiver and the FK 516 antenna coupler. The data modules which provide the high-speed data function are fully integrated within the transceiver. The voice/data therefore fits within the space of a conventional radio and additional space for the data capability is not needed.

The functioning of the equipment is controlled by the integrated test system in which a number of functions are continuously monitored. After the test routine has been triggered, any faulty module will be located and indicated. BITE results are reported to the onboard CFDS/CMC system via two ARINC 429 busses. Interfaces to the central maintenance systems of Airbus, Boeing and The Boeing Company aircraft are implemented in the radio, featuring one part number for nearly all aircraft types.

The XK 516D is designed to meet the requirements of ARINC 719 for the voice function and ARINC 753/635 for the data function. The integrated data communication capability meets the specifications of ARINC 753 and 635 and high-speed data communication up to 1,800 bits/s user rate is provided.

To provide full compatibility between existing and new equipment and aircraft wiring, multiwire serial interface to ARINC 753, conventional ARINC 719 control lines and a single wire coaxial interface between transceiver and the antenna coupler are available. This provides interchangeability between existing voice transceivers and the coupler.

The antenna coupler is a digitally tuned coupler with tuning times, typically of less than three seconds. The learn mode can provide even shorter tuning times of several hundred milliseconds.

Contractors

AlliedSignal Commmercial Avionics Systems.
Rohde and Schwarz GmbH and Co KG.

VERIFIED

ISRAEL

Airborne Advanced Communications System (AACS)

The Airborne Advanced Communications System (AACS) provides secure airborne communications, allowing the use of radio and navigation receivers and monitoring of alarm signals on the aircraft. A microcomputerised Communications Switching Unit (CSU) mixes and routes individual volume-controlled audio signals to the Intercommunications Set Control (ISC) unit. The CSU circuits include an audio switching matrix, radio interface and control circuitry that can accommodate up to 16 operators and 30 radio transceivers, receivers and alarm tones.

Specifications

Dimensions:
(CSU) 193.55 × 241.3 × 406.4 mm
(ISC) 95.2 × 146 × 127 mm
Weight:
(CSU) 13.6 kg
(ISC) 1.13 kg

Contractor

ECI Telecom Ltd.

VERIFIED

The Airborne Advanced Communications System (AACS) showing the communication switching unit (rear) and the intercommunications set control (front)

Airborne Radio and Intercom Control system (ARIC)

The Airborne Radio and Intercom Control (ARIC) system is a communications control system designed to meet the requirements of a wide range of military helicopters and fixed-wing aircraft. It provides secure communications for up to ten crew stations, allowing use of intercom and radio equipment and monitoring of navigation and audible alarm signals.

The ARIC system features a remote transmitter selection switch capability. This feature, which enables remote transmitter selection and keying from a special switch mounted on the stick, is for helicopters requiring Hands-On Collective And Stick (HOCAS) capabilities during nap of the earth flight.

Each station is able to access five radio transmitters, six navigation receivers, one intercom network and four alarm signals. There are two modes for remote transmitter selection: direct selection and sequential selection. Extremely low cross-talk allows mixing of secure and clear communications channels and the system has Vox capability for operation in high-noise environments. It features independent crew stations, independent selection and volume control for each channel, very low intrinsic noise, NVG-compatible front panel and sunlight-readable and NVG-compatible optional remote status indicator.

Specifications

Dimensions: 169.2 × 37 × 36.2 mm
Weight: 1.09 kg
Power supply: 22-32 V DC
Reliability: >6,000 h MTBF

Contractor

ECI Telecom Ltd.

VERIFIED

ECI Airborne Relay System (EARS)

The ECI Airborne Relay System (EARS) is designed to meet the demands of the modern integrated battlefield. It serves as an automatic radio relay system for voice and data communications and provides reliable communications through wide aerial coverage, communications in different frequency bands and a long-distance link.

The system operates in the frequency bands 30 to 87.975 MHz VHF/FM, 116 to 151.975 MHz VHF/AM and 225 to 400 MHz UHF/AM. It is simple and quick to install on aircraft such as the Boeing 707, 727 and 737, Lockheed Martin C-130B, C-130H and P-3C and the Fokker F27 and F28.

Contractor

ECI Telecom Ltd.

VERIFIED

ICU-12 internal communication control system

The ICU-12 is a 12-channel audio control system which provides secure internal communication for up to ten crew stations, allowing use of intercom and radio equipment and monitoring of navigation and audible alarm signals. It is a lightweight and compact high-performance system intended for helicopter, transport aircraft and light aircraft installations. The system comprises crew station Audio Control Units (ACUs) and a central Audio Junction Box (AJB). A full system contains 10 ACUs and one AJB.

Each crew station is able to access remotely up to 12 channels consisting of two intercom nets, seven radio transceivers, one radio navigation channel and two alarm tones. The ICU-12 features extremely low cross-talk and allows mixing of secure and non-secure communication channels. The crew stations are independent of one another, avoiding total system breakdown due to single station failure, and independent volume control is provided for each input channel.

Specifications

Dimensions:
(ACU) 146 × 95.5 × 143 mm
(AJB) 150 × 101 × 180 mm
Weight:
(ACU) 1.5 kg
(AJB) 2.3 kg
Power supply: 22-32 V DC
Reliability: 4,000 h MTBF

Contractor

ECI Telecom Ltd.

VERIFIED

The ICU-12 internal communications system showing the audio junction box (left) and two audio control units (right)

Giga-Links digital communication systems

Giga-Links is a family of digital communication products for the transmission of high data rates generated by wideband imaging systems. Typical applications include airborne observation sensors such as high-resolution electro-optical cameras and thermal imaging scanners, as well as high-rate ELINT sensors.

Giga-Links operates at microwave frequencies and uses a modular concept; a single-channel implementation for rates of up to 150 Mbytes and parallel channels for higher rates. Compression techniques are used to reduce the data rate when frequency bandwidth is constrained or when higher rate sensors are used.

The channelised airborne transmitter consists of frequency sources, QPSK modulators and solid-state power amplifiers. Error correcting codes are used to improve the link performance. A compact and lightweight design is used to meet the specific airborne requirements.

The associated ground station consists of a sensitive receiving chain with auto-tracking capability.

For those applications requiring data compression, dedicated units are added at the transmitting and receiving ends, using an adaptive compression algorithm implemented in VLSI. The sensor output data rate is reduced by up to a factor of four, while maintaining the excellent image quality required for reconnaissance applications. An extensive monitoring and test capability is included in both the aircraft and ground installation.

Contractor

Elisra Electronic Systems Ltd.

VERIFIED

ARC-740 UHF secure radio

The ARC-740 provides UHF communication for air-to-air, air-to-ground and ground-to-ground use for plain speech and secure speech, using frequency hopping at 10 hops/s and data transmissions at 2.4 kbits/s. The frequency band from 225 to 399.975 MHz is covered by 7,000 channels at 25 kHz steps in both AM and FM modes; any channel can be manually selected. There are also ECCM-protected normal and high-powered modes. There are 99 preset channels and a continuous watch is kept on the Guard frequency.

Specifications

Dimensions:
(transceiver) 250 × 127 × 177 mm
(power amplifier) 190 × 170 × 145 mm
(control box) 143 × 146 × 124 mm
Weight:
(transceiver) 7 kg
(power amplifier) 4.9 kg
(control box) 1.8 kg

Contractor

Elta Electronics Industries Ltd.

VERIFIED

The Elta ARC-740 UHF secure radio

ASARS Airborne Search And Rescue System

The Tadiran Spectralink Airborne Search and Rescue System is used in the rescue of downed aircrew by an airborne platform, special forces pick up, or drop zone marking. It facilitates rapid rescue under adverse conditions, and is designed for use both in combat and peacetime. Both covert and standard beacon mode operation are possible, as well as voice communication.

ASARS consists of the ARS-700 airborne system and the PRC-434A radio transponder. The system utilises advanced range measurement technology, enabling accurate and quick survivor location in bad weather and difficult or hostile terrain.

First pass pick up is made possible by the accurate azimuth and range measurements provided by the lightweight airborne system. A beacon mode supplements the rescue radio's special transponder capabilities, and downed aircrew survival is enhanced by the fallback voice communication capability on every channel.

Short burst-type interrogations (600 ms) and special modulation techniques eliminate the dangers of repeated or continuous transmissions being picked up by enemy forces. Ten preprogrammed frequency channels, out of 3,000 operating frequencies, further enhance security. Up to a million call codes allocated to every radio transponder enable selective interrogation, minimising the number of transmissions. Several survivors can be handled by one airborne system, with each one clearly identified. If the selective call code of the transponder radio is unknown, it can be automatically retrieved by the airborne system via secure algorithms, enabling the successful completion of the rescue mission.

The transponder radio is activated automatically upon bail-out, and it can be easily operated with one hand by a disabled survivor. The radio is exceptionally rugged, compact and lightweight, and simple to use. It is powered by a durable lithium battery that provides a long operating life.

The transponder radio is compatible with both the Cubic AN/ARS-6V and the Tadiran ARS-700, and operates over the 225 to 300 MHz frequency range, with an effective range of 200 km. The beacon mode and default voice capabilities are compatible with other rescue systems and techniques.

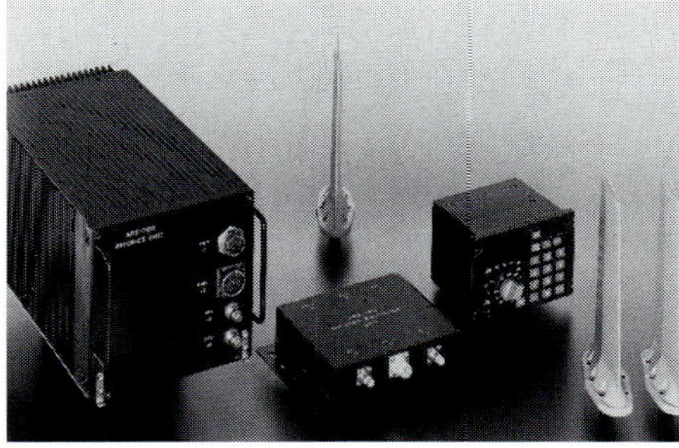

The Tadiran Spectralink ARS 700 airborne search and rescue system

ASARS can be installed in all types of fixed-wing aircraft and helicopters, without the need for any aircraft modification. It is easy both to install and remove; re-installation takes less than 30 minutes. The compact control unit mounts in the standard radio slot on the instrument panel. System parameters are easily programmed using a standard PC interface.

DF accuracy is ±2.5° RMS. Range measurement accuracy is 50 m.

Specifications

ARS 700 airborne system
Dimensions:
(avionics unit) 190 × 195 × 380 mm
(control display unit) 100 × 127 × 130 mm
(antenna switching unit) 50 × 150 × 150 mm
(remote display unit) 76 × 76 × 50 mm
Weight: 18 kg
Power supply: 28 V DC, 10 A
Power output: 20 W peak (20 W, 2 W, 0.4 W PEP)

PRC 434 radio transponder
Dimensions: 186 × 74 × 38 mm
Weight: 0.9 kg
Power supply:
(lithium battery) 15 h at 1:10 T/R ratio
Power output: 1 W RMS

Operational status

Selected by the French, Italian and Turkish air forces.

Contractor

Tadiran Spectralink Ltd.

VERIFIED

ASARS-G Airborne Search And Rescue System with GPS & relay

ASARS-G maintains all the existing facilities of ASARS; in addition, it includes: a complete survivor position locating system, with GPS navigation receiver, and full navigation support, designated ARS-700G; a secure data channel relay capability; voice channel reporting on both VHF and UHF; remote activation of embedded GPS features by the airborne units (ARS-7000G & ARS-434R); a simple operator interface that turns the PRC-434 rescue radio into a complete airborne relay system, designated ARS-434R.

Specifications

(for survivor rescue radio PRC-434G)
Frequency range: 225-299.975 + 121.5 MHz
Channels: 3,000
Programmed channels: 10 + 2 guard
Position accuracy: GPS P-code or C/A code
Modulation:
(voice or swept tone) AM
(narrowband data) FSK
(transponder) OOK, PSK for ASARS
Activation: manual or automatic upon ejection
Remote operation: beacon mode; navigaton mode; data transfer mode
Transmit power:
(beacon in UHF) 2 W peak
(voice in UHF) 0.75 W CW
(121.5 MHz) 0.1 W peak
adaptable power management for data transfer

Contractor

Tadiran Spectralink Ltd.

NEW ENTRY

ITALY

AN/ARC-150(V) UHF AM/FM transceiver

The AN/ARC-150(V) designation represents a family of small, high-performance, lightweight, airborne UHF transceivers manufactured by Elmer and based on a Magnavox design as well as Elmer's own research and development. It has produced many versions, including the 10 W panel-mounted ARC-150(V)10, the 10 W remote-controlled ARC-150(V)2 and the 30 W remote-controlled ARC-150(V)8.

Elmer has also developed a series of control panels and frequency/channel repeaters to meet specific installation requirements on different aircraft and helicopters. A feature of this family is 'slice' assembly, which simplifies maintenance and facilitates growth.

A series of mounting adaptors has been developed and produced to allow the basic ARC-150 transceivers to replace older UHF radios such as the AN/ARC-51BX, AN/ARC-109, AN/ARC-52 and AN/ARC-552 without any mechanical or electrical modification to the aircraft.

A version of the ARC-150(V) with ECCM capabilities has also been developed using the frequency-hopping Have Quick technique. This capability is easily implemented by the substitution of the synthesiser slice and by minor changes on the control panel unit. No mechanical or electrical modifications are required on the aircraft. The modified radio retains the normal non-hopping mode.

The ARC-150(V)8 is a development of the basic 30 W ARC-150(V) transceiver incorporating the FM modulation capability by means of an additional slice. This facility makes the unit particularly suitable for use with data, frequency-shift keying and secure voice modems.

It has been demonstrated to be fully compatible with the Vinson KY-58 system both in AM and FM and in the diphase and baseband modes. It has also been successfully used as a main component of an airborne system for UHF satellite communication.

Specifications

Dimensions:
(10 W panel-mounted RT-1136) 146 × 124 × 193 mm
(10 W remote RT-1051) 127 × 120 × 183 mm
(30 W remote RT-1073) 127 × 120 × 291 mm
Weight:
(10 W panel-mounted RT-1136) 4.3 kg
(10 W remote RT-1051) 3.7 kg
(30 W remote RT-1073) 6 kg
Power supply: 28 V DC
Power output: 10 W (30 W for the AN/ARC-150(V)8 model)
Frequency: 225-400 MHz
Channel spacing: 25 kHz
Preset channels: 20 using electronic memory (MNOS)
Frequency accuracy: 2 kHz

Guard receiver: 243 MHz
Operating modes: AM voice, ADF, homer, secure voice/data, ECCM

Operational status

In production and in service with Italian and other armed forces. Over 1,000 units have been delivered and installed on a wide range of aircraft and helicopters including the Panavia Tornado, Aermacchi MB-339, Aeritalia G91Y and F-104S fixed-wing aircraft, and Agusta A 109, Agusta-Bell 212, Agusta-Sikorsky SH-3D and HH-3F and other helicopters.

Contractor

Elmer SpA.

VERIFIED

RV-4/213/A VHF/FM transceiver

The RV-4/213/A VHF/FM equipment is an airborne transceiver suitable for light aircraft and helicopters. It is in effect an airborne version of the Italian Army RV-3/213/V which has been repackaged for installation in small aircraft. It is compatible with, and equivalent to, the Thomson-CSF TRAP-113 and is intended to replace equipment of the ARC-131 type. Tuning is fully automatic and digital over the frequency range 26 to 72 MHz in 50 kHz steps. Modes of operation are F3E FM voice and homing and RF power output is 20 W.

Specifications

Dimensions: 320 × 92 × 194 mm
Weight: 6 kg

Operational status

In production and in service.

Contractor

Elmer SpA.

VERIFIED

SRT-170/EB HF/SSB transceiver

Following research and development into HF propagation, Elmer has designed and produced an advanced HF/SSB system for aircraft and helicopters. The standard system comprises a 100 W transceiver and a tuner directly connected to an advanced design HF loop antenna.

The transceiver is a scaled-down version of the SRT-470F 400 W transceiver developed by Elmer for the Panavia Tornado programme, with a slightly modified receiver/exciter and a newly designed 100 W RF power amplifier which are all contained within a ⅜ ATR short enclosure. The tuner and loop antenna were jointly developed with British Aerospace Dynamics Group.

Extensive ground and flight tests over the last three years have used ground mockups and helicopters such as the Aerospatiale/Westland Gazelle, Sikorsky SH-3D, Agusta-Bell 212 and Agusta A 109, A 129 and CH-47.

The loop antenna is designed to achieve a high angle of incidence in the skywave mode to improve communication during helicopter operation near the ground and over unfavourable terrain. Attention has been given to defining the best geometric configuration and installation so as to optimise ground wave and near vertical incidence skywave propagation. Extensive field-strength measurements conducted over the HF range have demonstrated a significant improvement over the current wire antennas.

The SRT-170/EB comprises an SP-648/SM control panel, SP-649/SM receiver/exciter, SP-480/E power amplifier, SP-1036 mounting tray and AN16C tuner and loop antenna.

Its main features are all-solid-state technology with extensive use of miniaturisation techniques, frequency synthesis, fully automatic tuning, free air convection cooling, low power consumption, remote control of operating functions by means of a serial bit stream and incorporated BIT facilities for continuous and interrupted self-diagnosis. There is optional compatibility with a MIL-STD-1553B databus, channel presetting by associated control panel and NATO Link 11 compatibility. The SRT-170 has SIMOP (SIMultaneous OPeration of collocated RF sets) capability with the addition of ancillary pre- and post-selector equipments. The tuner is mounted externally on the helicopter or aircraft surface, to reduce electromagnetic compatibility problems, and directly connected to the antenna; there is no limitation in distance between power amplifier and tuner.

Specifications

Dimensions:
(receiver/exciter) ⅜ ATR short
(power amplifier) ⅜ ATR short
(control panel) 146 × 85.7 × 125 mm
Weight:
(receiver/exciter) 5.1 kg
(power amplifier) 6.2 kg
(control panel) 0.9 kg
(tuner) 4.1 kg
Power supply: 28 V DC
(transmit) 450 W
(receive) 90 W
RF power output: 100 W PEP and average
Frequency: 2-30 MHz
Number of channels: 280,000
Antenna: loop
Modes: A2J (CW), A3H (AME), A3J (USB, LSB)
Tuning: automatic and digital at 100 Hz minimum step
Tuning time: 2 s typical (including tuner)

Operational status

In current production and in service with Italian and other armed forces, with more than 300 units delivered.

Contractor

Elmer SpA.

VERIFIED

SRT-194 VHF/AM radio

Elmer's SRT-194 is a VHF/AM transmitter/receiver covering the 108 to 156 MHz band with channel spacings of 25 kHz. Up to 20 channels, plus an additional Guard channel, are preselectable. The system provides air-to-air and air-to-ground communication in AM mode, together with modulated carrier wavetone and retransmit facilities; ADF and homer functions are also optionally available by the addition of appropriate ancillary equipment. The system comprises a transmitter/receiver unit with a remote controller; an additional channel/frequency selector can also be provided.

A modular 'slice' style of construction is employed and new functions can be incorporated by simply adding appropriate slices. Construction is all-solid-state and miniaturisation techniques are extensively employed. Power consumption is low and the thermal design is claimed to be extremely efficient, so that no forced-air cooling supply is required. The system is compatible with a wide range of interphone systems and identical pin-to-pin connections with Elmer UHF systems render the SRT-194 readily interchangeable with these without change to aircraft wiring or mountings. Transmitter output power is 10 W minimum. The equipment conforms to MIL-E-5400 Class II modified specification.

Specifications

Dimensions:
(control panel) 83 × 146 × 181 mm
(transmitter/receiver) 228 × 127 × 127 mm
(channel/frequency indicator) 83 × 33 × 154 mm
Weight:
(control panel) 1.4 kg
(transmitter/receiver) 4.3 kg
(channel/frequency indicator) 0.5 kg

Operational status

In production and in service.

Contractor

Elmer SpA.

VERIFIED

SRT-470F HF/SSB transceiver

The SRT-470F was designed to meet the HF/SSB requirements of the Panavia Tornado for air-to-air and air-to-ground communications and consists of a transmitter/receiver, a power amplifier/antenna coupler and a panel-mounted control unit. The transceiver is fully solid state with advanced circuit techniques such as wideband RF amplification. The antenna coupler is completely automatic and matches the power amplifier output to a notch antenna which is an integral part of the aircraft structure. The equipment operates over the frequency range 2 to 30 MHz with 280,000 channels at 100 Hz spacing and has a power output of 400 W PEP.

Specifications

Dimensions:
(transceiver) 199 × 261 × 319 mm
(power amplifier/antenna coupler) 199 × 94 × 319 mm
(controller) 86 × 146 × 125 mm

Operational status

In service with Italian Air Force Panavia Tornado aircraft.

Contractor

Elmer SpA.

VERIFIED

SRT-651 VHF/UHF AM/FM transceiver

The SRT-651 is an all-solid-state, compact, lightweight airborne transceiver covering the 30 to 400 MHz frequency range and using the most advanced techniques in the area of large-scale integration components and microprocessors. The system comprises a CP-1200 control panel, RT-651 receiver/transmitter and optional ID-1151 channel/frequency repeater.

Conceived and designed for airborne use, the SRT-651 is also suitable for a wide range of applications where space and weight are limited. The flexible modular 'slice' design permits easy assembly and disassembly and allows expansion of the operating functions. The RF power rating can be easily upgraded to 30 W by changing the transmitter slice. Different interface standards are available, including an Elmer serial data interface using control panel CP-1200 or MIL-STD-1553B or ARINC 429.

BIT facilities can be implemented by using different plug-in cards for the interface slice. The test result is automatically monitored on the control panel display, which gives a direct identification of the faulty slice.

The SRT-651 has facilities for ECCM including frequency-hopping ('Have Quick' has been implemented) and spread spectrum pseudo-noise modulation. It can be operated with a wide variety of ancillaries including homer indicators, VHF/UHF

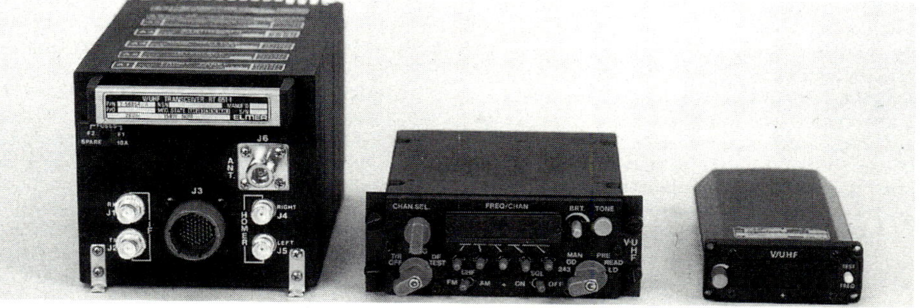

The Elmer SRT-651 VHF/UHF transceiver

direction-finders, UHF emergency beacons, KY-58 secure voice modems in diphase and baseband modes, Elmer SP-1212 ECCM spread spectrum voice and data modem, NATO multitone Link 11 data modem and FSK modems. Various mountings can be provided to retrofit the SRT-651 directly into existing VHF and UHF installations.

Specifications

Dimensions:
(RT-651) 126.7 × 120.6 × 224 mm
(CP-1200) 146 × 57.1 × 150 mm
Weight:
(RT-651) 3.5 kg
(CP-1200) 1.4 kg
Power supply: 28 V DC
(transmit) 150 W max
(receive) 25 W max
Power output:
(AM) 10 W
(FM) 15 W
Frequency: 30-88 MHz VHF/FM, 108-156 MHz VHF/AM, 156-174 MHz VHF/FM and 225-400 MHz UHF AM/FM
Channel spacing: 25 kHz in all bands
Guard receiver: 40.5, 121.5, 156.8 and 243 MHz automatically selected
Preselected channels: 20 using electronic memory

Operational status

In production for Italian, Brazilian and other armed forces and for installation on Alenia/Aermacchi/ Embraer AMX aircraft.

Contractor

Elmer SpA.

VERIFIED

The Elmer SRT-653 VHF/UHF transceiver is installed on Italian Air Force Tornado aircraft

SRT-653 VHF/UHF transceiver

Designed to meet a requirement of the Italian Air Force for its version of the Panavia Tornado, the SRT-653 VHF/UHF transceiver comprises the RT-1051 UHF transceiver, SP-1047 VHF transceiver, SP-1203 adaptor/power supply, SP-1204 mounting tray, CP-1001 control panel and ID-1150 frequency repeater.

The RT-1051 UHF transceiver belongs to the basic ARC-150(V) family. The SP-1047 VHF transceiver has been derived by Elmer from this system and maintains its general characteristics.

Specifications

Dimensions:
(ID-1150) 86 × 50 × 159 mm
(SP-1047/RT-1051 on tray) 265 × 155 × 408 mm
(CP-1001) 146 × 95.2 × 165.5 mm
Weight:
(ID-1150) 0.35 kg
(SP-1047 and RT-1051 on tray) 13.5 kg
(CP-1001) 1.6 kg
Power supply: 115 V AC, 400 Hz, 3 phase
(transmit) 320 VA max
(receive) 140 VA max
Power output: 10 W
Frequency: 108-156 and 225-400 MHz
Channel spacing: 25 kHz
Preset channels: 17 + 2 Guard channels

Operational status

In service on Italian Air Force Panavia Tornado aircraft.

Contractor

Elmer SpA.

VERIFIED

MIDA airborne data terminal

The Message Interchange Distributed Application (MIDA) is a digital datalink used for data communication between manned or unmanned aircraft and the ground control station. It is a spread spectrum bidirectional microwave link structured on a time division basis.

MIDA includes a video processing section composed of an ADPCM codec for video data bandwidth compression and an automatic lock follower unit performing the automatic tracking of the target selected by the command and control station.

MIDA is able to provide both downlink and uplink facilities. Downlink includes codified monochrome video data from a 625/50 CCIR standard camera output or uncodified monochrome video data at a low frame rate or MTI, ESM, SAR surveillance radar data and telemetry and aircraft status data on altitude, airspeed, heading and failures. Uplink includes remote command data for camera platform steering commands, target designation, aircraft flight profile and mission updating; and link service data for codified or uncodified video mode selection, surveillance data retransmission requests, acknowledgement procedures and so on. DME capability is also provided by MIDA.

MIDA operates in the D- or J-bands. It consists of the airborne data terminal which has two LRUs: the spread spectrum transceiver unit and the antenna unit. Besides the microwave, IF and baseband sections, the transceiver unit contains the processing units which perform the video signal bandwidth compression and target automatic tracking. The airborne data terminal is interfaced via ARINC 429 or MIL-STD-1553 to the airborne main computer.

The Marconi *MIDA airborne data terminal*

Contractor

Marconi SpA.

VERIFIED

PRT 403 airborne transmitter

The PRT 403 comprises the transmitter and a high-gain rotating aerial to which the transmitter is connected by a waveguide feed. The equipment is completely solid state including the modulator, except for the power amplifier which employs a klystron tube. It is used for transmitting mid-course guidance information to a missile.

Both the logic data handling section and the RF units employ LSI, thin film and microstrip hybrid circuits, multilayer and crossover thick film hybrid circuits, allowing extremely small weights and dimensions.

The main functions of the equipment are periodically checked by BITE which is initiated by the system computer. Major defects and performance degradation are detected and displayed to the operator in real time. A major failure to the power unit and its auxiliary circuits instantaneously operates automatic protection circuits.

The equipment is designed and manufactured for operation in the MIL-E-5400 Class 1B environment.

Specifications

Dimensions: 257.1 × 193.5 × 497 mm
Weight: 25 kg
Power supply: 115 V AC, 400 Hz, 3 phase

Contractor

Marconi SpA.

VERIFIED

JAPAN

Direction-finding and receiving system

The direction-finding and receiving system is designed to detect SHF frequencies between 6.4 and 7.1 GHz used for video transmissions from ground stations and employs a directional antenna to track the transmission automatically. The system operates up to a maximum aircraft speed of 155 kt and a maximum altitude of 20,000 ft as an emergency communications system.

The system consists of separate direction-finding and receiving antennas, a DF signal processor, DF control panel and monitor on which is displayed the angle of ground transmissions relative to the aircraft to an accuracy of ±5°. One or more optional antenna directing systems may be added, to enable the aircraft to relay transmissions.

Contractor

Tokimec Inc Electronics Systems Division.

UPDATED

The Tokimec direction-finding and receiving system with the receiving and DF antennas mounted underneath the ventral surface of an S-76 helicopter just aft of the nosewheel. An antenna directing system is also mounted on each side of the cabin

Satellite communication system

Toshiba has developed a satellite communication system which enables air-to-ground international telephone calls from commercial airliners via the Inmarsat satellite service. This was the first type-approved multichannel operation aeronautical satellite communications system in the world. The onboard system consists of radio transceiver equipment for communication with the Inmarsat satellites via a separate antenna and terminal equipment that includes cordless telephones and a control display unit. For passenger use, up to three telephones can be installed in an aircraft and one of the three telephone channels can be used in the cockpit. A 600 bits/s data channel is available for ACARS data communication for automatic dependent surveillance.

Toshiba has developed LSIs to achieve the component miniaturisation necessary for light compact equipment suitable for installation in aircraft.

The radio transceiver consists of a high-power amplifier, radio frequency unit and satellite data unit. The equipment meets ARINC 741 international standards for electronic equipment used in commercial airlines. It can receive and send digital voice signals at 9,600 bits/s, allowing up to three digital voice lines and one 600 bits/s digital data line to be connected to the system.

Specifications

Dimensions:
(high-power amplifier) 8 MCU
(radio frequency unit) 4 MCU
(satellite data unit) 6 MCU

Operational status

The equipment has received Supplemental Type Certification approval for the Boeing 747-400 from the FAA; certification for use in other aircraft is pending. The equipment is used by All Nippon Airways.

Contractor

Toshiba Corporation.

VERIFIED

NETHERLANDS

Link-Y Mk 2 datalink

The Link-Y Mk 2 datalink system exchanges system track data, management data, status data and commands between shipborne, land-based and airborne units participating in the datalink network.

Link-Y Mk 2 allows the establishment of a fleetwide common tactical database through the exchange of real-time and non real-time track data. By exchanging management data, such as IFF/SIF reports, information difference reports and conflict reports, a complete common tactical picture is established.

The exchange of commands and status data, such as force disposition orders, weapon readiness reports and force engagement status reports, is an indispensable tool for engagement planning.

Engagement execution is supported through the exchange of commands, such as weapon doctrine orders, engagement orders and hold-fire orders.

Link-Y Mk 2 allows a maximum of 31 units to participate in the network. Reporting normally takes place using a TDMA method. To each unit, one or more time slots are automatically allocated by the net control station. In radio silence mode all units maintain silence but urgent messages are transmitted by means of a single report.

The Link-Y Mk 2 terminal can be set in the Link-Y Mk 1 mode, resulting in complete interoperability with older Link-Y units which lack the automatic slot allocation and encryption facilities.

The system operates on HF, VHF and UHF frequencies. Effective range is 925 km for HF and line of sight for VHF and UHF.

Specifications

Dimensions:
(Link-Y terminal) 57.2 × 194 × 318 mm
(optional control unit) 146 × 66 × 68 mm
Weight:
(Link-Y terminal) 4 kg
(optional control unit) 0.4 kg
Power supply: 28 V DC
Data rates: 300, 600, 1,200, 2,400 and 4,800 bits/s

Operational status

In production for several navies.

Contractor

Hollandse Signaalapparaten BV, Signaal Special Products.

VERIFIED

Vesta-VC datalink

The Vesta transponder receiver system can easily be extended with a voice channel datalink. This makes it suitable for data transfer and over-the-horizon targeting for naval vessels or ground control stations by using the helicopter. The target data is received from the helicopter sensors.

The datalink is established via an existing communications voice channel. The Vesta transponder switches the available radio in the helicopter from voice to data, after which the data is transmitted via the voice channel in a frequency shift keyed signal.

A receiver extractor and datalink unit on the base station then converts this data to an 'own position' reference by triangulation of the helicopter position and target data, and interfaces the data handling system (the weapon control system) with the radar displays. Vesta is compatible with all airborne and base system interfaces.

Operational status

In production for several navies.

Contractor

Hollandse Signaalapparaten BV, Signaal Special Products.

VERIFIED

SOUTH AFRICA

TR 2800 airborne HF transceiver

The TR 2800 is a new-generation 100 W airborne HF transceiver. Its features include: frequency-hopping capabilities for enhanced ECCM performance, compatible with the TR 250/390 mobile base station HF transceivers; selective calling with channel enhancement.

Optionally, the AT 2820 HF-NVIS-ECCM loop ATU can be supplied that includes 400 W linear amplifiers, antenna tuning units and built-in computer control.

Contractor

Grintek Comms, Grintek Electronics Limited.

NEW ENTRY

TR 2800 airborne HF transceiver
***1998**/0011841*

ACA 340 V/UHF airborne radio

The ACA 340 multimode radio is an airborne tactical radio. In addition to the full-frequency coverage required by modern air forces, the ACA 340 has advanced ECCM features such as frequency hopping, data transmission and encryption. It is capable of operating in extreme environmental and electronic conditions.

The control of the radio has been designed to be integrated into existing avionics, or stand-alone controls can be supplied.

Specifications

Dimensions: 124 × 385 × 193 mm
Weight: <8.26 kg
Power supply: 28 V DC
Frequency: 30-400 MHz
Channel spacing: 25 kHz
Modulation: AM, FM, data, binary FM secure speech
RF output: 10 W AM (80% modulation) 15 W FM
Preset channels: 99
Temperature range: −40 to +71°C
Altitude: up to 70,000 ft

Contractor

Reutech Defence Industries (RDI) (Pty) Ltd.

UPDATED

ACA 340V/UHF airborne radio ***1998**/0011840*

ACR 160 maritime band radio

The ACR 160 VHF airborne panel-mount transceiver has been designed to fulfil the most demanding requirements in new and existing airborne platforms. It operates in the 156 to 162 MHz maritime frequency band and complies with the CH00-CH28 and CH60-CH88 international marine frequency allocations.

The ACR 160 has a capacity of 57 preset channels which can be selected via the user-friendly front panel. Both active and standby channels are simultaneously displayed and transfer is effected via a single push-button. It is capable of operating in extreme environmental and electronic conditions and conforms to RTCA DO-160B.

Specifications

Power supply: 27.5 V DC, 2 A (typical)
Frequency: 156-162.025 MHz
Channel spacing: 25 kHz
Temperature range: −20 to +55°C
Altitude: up to 50,000 ft

Contractor

Reutech Defence Industries (RDI) (Pty) Ltd.

VERIFIED

ACR 210 airborne radio

The ACR 210 VHF transceiver has been designed to be installed in new aircraft and for retrofit. Its dimensions correspond to the ARINC standard for control equipment.

The transceiver is equipped with a wideband transmitter for operation in the 118 to 136.975 MHz frequency band and has two frequency displays which are microprocessor-controlled. In addition, the ACR 210 can be controlled via a dual RS-485 bus.

The ACR 210 can be adapted for tandem operation with a TAN 210 tandem controller.

Specifications

Power supply: 27.5 V DC, 3.8 A (typical)
Frequency: 118-136.975 MHz
Channel spacing: 25 kHz
Temperature range: −20 to +55°C
Altitude: up to 50,000 ft

Contractor

Reutech Defence Industries (RDI) (Pty) Ltd.

VERIFIED

ACR 500 series air/ground V/UHF transceivers

The ACR 500 series is RDI's latest air/ground V/UHF transceiver family. The 500 system comprises: two airborne transceivers; the ACR 500 V/UHF tactical airborne transceiver, and the ACR 520 VHF tactical airborne transceiver; two ground-based transceivers; the GBF 500 UHF power-agile filters transceiver and the GBR 500 V/UHF ground-based transceiver.

The ACR 500 and ACR 520 transceivers employ frequency-hopping and software encryption to ensure communications security, and very high rate direct sequence spread spectrum techniques for 'own probability of detection' communications, and to reduce the probability of interception (ACR 500 only).

High-quality vocoded speech transmission (essentially digitised speech) is used to ensure speech quality.

Other features include: bandwidth efficient/4DQPSK data mode, and noise resistant DSP squelch.

ACR 500 V/UHF transceiver

The ACR 500 is a multimode/multiband transceiver operating over the 30 to 420 MHz range. It also incorporates a fully synthesised auxiliary receiver with analogue and digital modes, to allow simultaneous voice and data reception on two frequencies, to improve situational awareness.

ACR 500 V/UHF tactical airborne transceiver
***1998**/0011839*

Specifications

(ACR 500 V/UHF transceiver)
Frequency range: 30-420 MHz
Channel spacing: 12.5 kHz; 25 kHz
Modulation formats: AM, FM (WB/NB); SSB (USB); CPFSK Data (binary); Pi/4 DQPSK
RF output, max: 32.4 W PEP AM; 20 W FM, SSB
Collocation: full operation 5% off frequency
ECCM: fast frequency hopping, direct sequence spread spectrum
Encryption: Cat A, built in
Audio capabilities: analogue voice, CVSD, high-quality vocoder
Data capabilities: high-speed, addressable, network function
Auxiliary receiver: built-in
Power: 28 V DC
Dimensions: 124 × 385 × 193 mm
Weight: <9.5 kg

ACR 520 VHF transceiver

The ACR 520 operates in the 30 to 88 MHz portion of the VHF band, its capability for communications with

ACR 520 VHF tactical airborne transceiver
***1998**/0011838*

ground forces, using the GBR 500 receiver, and SSB for extended range operation.

Specifications

(ACR 520 VHF transceiver)
Frequency range: 30-88.975 MHz
Channel spacing: 12.5 kHz; 25 kHz
Modulation formats: AM, FM (WB/NB); SSB (USB); CPFSK Data (binary); Pi/4 DQPSK
RF output, max: 25 W PEP SSB; 15 W FM
ECCM: fast frequency hopping
Encryption: Cat A, built in
Audio capabilities: analogue voice, CVSD, high-quality vocoder
Data capabiities: high-speed, addressable, network function
Power: 28 V DC
Dimensions: ½ ATR short
Weight: <6 kg

Contractor

Reutech Defence Industries (RDI) (Pty) Ltd.

NEW ENTRY

SWEDEN

AMR 345 VHF/UHF transceiver

The AMR 345 is a small VHF/UHF AM/FM panel-mounted radio for speech and data communications, covering the frequency range 104 to 162 and 223 to 408 MHz, with 25 kHz channel spacing. It has an ECCM capability and is designed to operate in a secure voice system. Transmitter output power is 10 W in the AM mode and 15 W in FM.

The transceiver memory has a flexible storage capacity for up to 1,000 preset channels. These channels can be used for immediate access to all frequencies at all airbases. The keyboard is used to select the preset frequencies, as well as for manual setting. A BIT facility performs a go/no go test in the aircraft.

An autonomous system is formed by the transceiver, a 28 V power supply, a headset and an antenna. It is also operable with other systems in the aircraft by means of a serial data control link. The microprocessor control allows easy adaptation of the transceiver (including the front panel function) to any system configuration. Preset channels and data back-up are stored in an electrically reprogrammable non-volatile memory.

The transceiver is hardened to ensure full operation after EMP exposure. ECCM capability is provided by use of an external ECCM control unit which can be designed in accordance with customer requirements.

The transceiver is designed to meet applicable MIL standards.

Specifications

Dimensions:
(transceiver) 230 × 146 × 76 mm
Weight: 3.5 kg
Frequency accuracy: 5 ppm (−40 to +70°C)

Operational status

In production. The AMR 345 is included in the communication system for the JAS 39 Gripen aircraft for the Royal Swedish Air Force, and has also been selected for re-equipping the Saab 105 (SK 60), Saab J32 Lansen, Saab SK 35 Draken, C-130 Hercules, the Eurocopter Super Puma helicopter and the CASA C-212 transport aircraft. The radio is expected to find additional applications in maritime and ground communication systems.

Contractor

CelsiusTech Electronics AB.

VERIFIED

The AMR 345 VHF/UHF transceiver dual command version

AMR 345 dual and multicommand VHF/UHF airborne radio system

The CelsiusTech Electronics AMR 345 dual and multicommand system is designed for trainers, transport aircraft and helicopters. It consists of the AMR 345 VHF/UHF AM and FM speech and data transceiver and the AMR 349 control units. Performance characteristics are as for the single AMR 345 VHF/UHF transceiver described above.

It features easy handling between instructor and trainee without any priority switches, the latest command being valid, extensive non-volatile reprogrammable memories for up to 1,000 channels and BIT.

The AMR 345 is designed to be operated under the worst flying conditions with a minimum of time and attention required by the pilots. In the trainer installation the trainee has full command over the frequency selection, as in front-line aircraft, and the instructor is fully informed on what the trainee is doing and can at any time take control himself.

The AMR 345 is reprogrammable in the aircraft by means of a fill gun.

In the manual frequency mode a keyboard gives access to any frequency in the VHF or UHF bands. AM or FM is selected by a separate switch on the panel. In the preset mode, a keyboard gives direct access to 1,000 channels and can be used for immediate access to different services at all available airbases. The control features may be adjusted according to customer requirements via software modifications and change of key-tops. One version of the equipment includes VHF and UHF Guard channels.

Specifications

Dimensions:
(transceiver) 76 × 146 × 230 mm
(control unit) 76 × 146 × 40 mm
Weight:
(transceiver) 3.5 kg
(control unit) 0.35 kg
Power supply: 16-32 V DC
Power output:
(AM) 10 W
(FM) 15 W
Frequency: 104-162 and 223-408 MHz
Channel spacing: 25 kHz

Operational status

Dual and multicommand radio systems are fitted in the Pilatus PC-7 and PC-9, Raytheon Hawker 125-800, Gates LearJet and Falcon F20.

Contractor

CelsiusTech Electronics AB.

VERIFIED

The AMR 345 panel-mounted VHF/UHF transceiver - part of the communication system for JAS 39 Gripen aircraft

JAS 39 Gripen radio communication system

The JAS 39 Gripen communication system is a dual-transceiver installation for simultaneous voice and data communication on the VHF and UHF frequency bands. The system can also operate as a relay link, and there is provision for adding a third Command and Control (C^2) data receiver.

The system provides both clear analogue voice and encryption modes, together with ECCM capabilities to ensure jam resistant communication for both voice and data. The Aircraft DataLink (ADL) operates in time sharing mode and enables continuous exchange of data between several aircraft, presented to the pilot via the main computer.

Functions and features included in the system comprise:

(a) The secure voice function is mechanised by use of a vocoder for narrowband digitising, and by applying encryption and error correction coding. The vocoder is specially designed for operation, including voice recognition, in noisy environments, as found in the Gripen aircraft cockpit

(b) A ground telecommunication amplifier function is incorporated to provide for analogue (voice) and digital (voice and data) telebrief operation whilst on the ground to eliminate the need for unnecessary free-space transmissions

(c) A tone generator and speech synthesiser function is provided to provide warning and alert signals as well as voice messages to the pilot

(d) A data transfer unit provides for loading: pre-coded frequency channels; crypto and ECCM keys; mission data; time synchronisation; and other requirements for the aircraft avionics system pre-flight briefing and post-flight analysis activities. During 1997, a more modern data transfer unit has been produced with application to other air and ground platforms, which uses 40 Mbytes of memory expansion and additional interfaces including MIL-STD-1553B and Ethernet.

JAS 39 Gripen radio communication system **1998**/0011837

Operational status

In production for JAS 39 A and B versions, and for the Saab 340 (popularly named Argus), and the S100B airborne radar aircraft. In a modified version, it forms part of the Swedish Air Force Tactical Radio System (TARAS). It is also planned as a baseline for the communication system in the JAS 39 Gripen export system.

Contractor

CelsiusTech Electronics AB.

NEW ENTRY

UNITED KINGDOM

CONTRAN® - VHF radio anti-blocking system

CONTRAN® is a dual-function avionics system, the primary purpose of which is to prevent simultaneous or conflicting VHF radio transmissions between aircraft and Air Traffic Control.

CONTRAN® operates by monitoring any transmissions which are in progress on a specific frequency as a pilot tries to initiate a transmission. It will then, if necessary, inhibit the attempted transmission and make the pilot aware that it has done so. The pilot can then retransmit after the preceding transmission has been completed.

The unit's secondary purpose is to guard against unintentional transmissions by alerting the flight crew to an onboard 'stuck microphone' which is then automatically terminated if appropriate.

Specifications

Standards: FAA TSO-C122, TSO-C128; RTCA DO207, DO209; Eurocae ED67, ED68
Size: 1 MCU Standard ARINC 600 Mounting
Weight: <1.9 kg

Operational status

CONTRAN® has been ordered by Britannia Airways Ltd for its entire fleet of Boeing 757/767 aircraft.

Deliveries commenced in 1996.

Contractor

British Aerospace Systems and Equipment.

VERIFIED

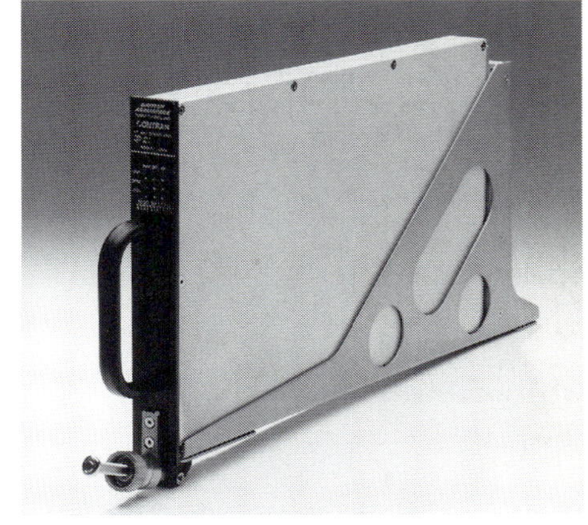

CONTRAN-VHF radio system **1997**/0001179

Scorpio 2000 series datalink systems

Scorpio 2000 is one of a series of expandable datalink systems which accept host aircraft and designated target global position data from the radar/navigation system databus and translates, encodes and up-converts them for transmission via radio link to ground station.

Typically, the radar used with the system is the AlliedSignal RDR-1500B, and the message data can be secured to meet user requirements. Options include: GPS internal (or external DGPS) interfaces, flat panel display and keyboard control.

Specifications

Radio link: VHF, power to suit requirements
Data interface/speed: ARINC 419, 9,600 baud
Weight: 2.4 kg
Power: 28 V DC

Contractor

Caledonian Airborne Systems Ltd.

NEW ENTRY

Scorpio 2000 series datalink, with AlliedSignal RDR-1500B radar display **1998**/0011835

Series 7-202 HF airborne receiving system

The Series 7-202 HF receiving system comprises a Type 7-202-1 HF receiver and Type 19-197 active antenna. The system is designed to provide worldwide reception of transmissions.

The Type 7-202-1 is a fully synthesised HF receiver capable of receiving transmissions anywhere in the long, medium or short wavebands. The Type 19-197 is a high-efficiency low-noise active antenna providing continuous coverage of the long, medium and short wave and VHF/FM broadcast bands. The antenna is protected against low-level lightning strikes.

The receiving system is fully automatic in operation and, once the desired service has been selected, it will automatically search out and tune to the best available transmission anywhere in the world. A database programmed into the receiver contains details of the coverage areas of all the available transmitters together with their frequencies and times of transmissions. The receiver, by interrogating the aircraft's navigational computer, obtains latitude and longitude, together with time of day. Using this information, the receiver tunes through available transmissions and, using its inbuilt signal quality measuring circuits and algorithm, determines which transmission offers the best quality and tunes the receiver to the appropriate frequency. The resulting audio is then output for distribution over the aircraft audio system. The inbuilt microprocessor continuously monitors the quality of the received signal and, should the quality deteriorate for any reason, the radio will automatically retune to any better quality transmission which may be available.

The receiver may be programmed with the worldwide coverage information for two different services which are selected by a single control line. Programming of the receiver's database to define the transmissions available, together with their broadcast times and frequencies, is accomplished with a user-friendly database program incorporating a graphical user interface. This program runs on an IBM PC and, once completed, may be downloaded directly into the receiver, enabling any operator to adapt the receiver to meet the specific needs of passengers and routes. A trace program built into the receiver records the selected frequencies and received signal strengths, thus enabling fine tuning of the database to ensure optimum performance.

Specifications

Weight:
(antenna) 2.5 kg
(receiver) 1.8 kg
Power supply: 28 V DC, 500 mA max
Frequency: 150 kHz-108 MHz
Channel spacing: 5 kHz and 9 kHz

Contractor

Chelton (Electrostatics) Ltd.

VERIFIED

805 Series UHF transceivers

The 805 is a fully synthesised UHF transceiver covering the frequency range 225 to 399.975 MHz. The state-of-the-art design provides both AM and FM communications, together with a dedicated guard receiver to monitor the 243 MHz distress frequency. When used with the Chelton 715-7 control unit and a suitable UHF antenna, such as the Chelton type 16-1 or any other Chelton multiband antenna, the 805-1 offers a complete AM/FM communications system suitable for connection to the aircraft audio system. The 805-2 variant provides a datalink capability.

The use of software variables in the design permits the implementation of user-specific requirements through simple reprogramming. This ability, coupled with the inbuilt growth facilities, enables the design to be readily configured for a wide variety of applications addressing interfaces such as microphone sensitivity and audio output levels. These variations are indicated by a suffix to the transceiver part number. A typical system comprises the following units: 805-1 UHF transceiver; 715-7 control unit; 715-7S slave control unit; 27601 mounting tray; optional remote fill gun.

The 715-7 control unit provides control of the transceiver's operating mode and frequency. Selection of the operating frequency may be either by direct entry of the desired frequency or by recalling a stored channel. Up to 100 frequencies can be stored in the transceiver. Each stored channel can be assigned a separate transmit and receive frequency for half-duplex operation and compatibility with satellite communication systems. Alternatively the transceiver can be controlled by a simple command set supplied via an RS-422 databus.

For users who wish to change the stored frequencies on a regular basis a remote fill gun is available. A computer program running on a PC-compatible computer permits the desired frequencies to be assigned to any channel number. The database is then down-loaded to the 715-7 controller in a few seconds via an infrared link using the remote fill gun. This enables a complete fleet of aircraft to be updated in a matter of minutes without the need to remove any equipment from the aircraft.

Chelton 805 Series UHF transceivers ***1996***

For installations requiring dual-control units a slave control unit, the 715-7S, is offered. When the main control unit is in control, the slave control units acts as a remote readout unit. Control of the system can be passed from the main control unit to the slave control unit and the main control unit will then act as the remote readout unit. Control units can be provided with lighting options, including NVG compatibility, to match cockpit specifications.

The datalink communications transceiver variant, 805-2, is rated to provide a 100 per cent transmit duty cycle.

805 series transceivers are compatible with Chelton 930 Series direction-finders.

Specifications

Transmitter
Power output: 10 W nominal
Frequency response:
(voice) 300 Hz-3 kHz at +3 dB
(data) 100 Hz-10 kHz at +3 dB
Sidetone level: 5.0 Vrms into 150 ohms

Main receiver
Sensitivity: <3.0 μV for 10 dB (S+N)/N at 30% AM
<1.5 μV for 10 dB SINAD at +3 kHz FM
Audio response:
(AM voice) 300 Hz-3kHz at +3 dB
(FM voice) 300 Hz-3 kHz at +3 dB
(adf) 100 Hz-6 kHz at +3 dB
(AM data) 100 Hz-10 kHz at +3 dB
(FM data) 100 Hz-10 kHz at +3 dB

Guard receiver
Sensitivity: <3.0 V for 10 dB (S+N)/N at 30% AM
<1.5 V for 10 dB SINAD at +3 kHz FM

Contractor

Chelton (Electrostatics) Ltd.

VERIFIED

ACCS 3100 audio management unit

The ACCS 3100 Audio Management Unit (AMU) provides complete integration of audio selection, aural warnings generation and speech recognition functions within a single unit. Selections and control are normally provided via the MIL-STD-1553 avionics databus from an integrated avionics suite, although a dedicated control panel can also be offered.

The AMU implements all the intercommunications functions required on the next generation of military aircraft such as the EF 2000 or for avionic upgrades to existing aircraft. Intercommunication is provided between radios, navaids, pilot, co-pilot, ground crew, recorders and the avionics system status outputs which activate aural warnings. Digital signal processing techniques effect noise tracking voice-operated switching for hands-free intercom, and noise filtering and voice-operated gain adjustment devices ensure audio signals routed to transmitters are of optimum clarity and modulation.

Aural warnings include voice warnings as well as attentions and tones, as commanded by external status inputs from dedicated discretes or via a MIL-STD-1553 databus. Voice warnings storage for up to 200 messages is currently provided. Digitised voice output of status and informatory data can also be activated by direct voice input commands from the aircrew.

Direct voice input uses speech recognition algorithms within the AMU to ensure accurate performance despite ambient cockpit noise and stress-induced voice variations. A programmable 600 word vocabulary, together with a dynamic syntax pointer responsive to avionics system status, gives sufficient flexibility to meet most demanding applications.

The AMU uses a powerful Ada-programmed 68020 processor, running BIT as well as normal operating programs; a non-volatile maintenance memory is available to store fault data. The AMU databus interface is compatible with MIL-STD-1553B, STANAG 3838 and STANAG 3910 fibre optic databus protocols.

Specifications

Dimensions: 380 × 200 × 124 mm
Weight: 7.7 kg
Power: 100 W
Processing: 30 Mips

Operational status

In service. Also selected as the communications and audio management unit for the EF 2000.

Contractor

Computing Devices Company Ltd.

VERIFIED

VLT 15-10-6MK3/VLR 15-16-1 helicopter video downlink

The video link has been designed for use as a helicopter-to-ground system, either for surveillance or outside broadcast use. The VLT 15-10-6MK3 transmitter is a multichannel 10 W unit. Any 16 frequencies within a 180 MHz bandwidth may be pre-programmed within the band 2.2 to 2.6 GHz. The VLR 15-16-1 is its matching 16 channel receiver.

A 3.5 GHz version is also available, and the 2.5 GHz unit may be upgraded to 3.5 GHz standard.

Contractor

ECS Enterprise Control Systems Limited.

NEW ENTRY

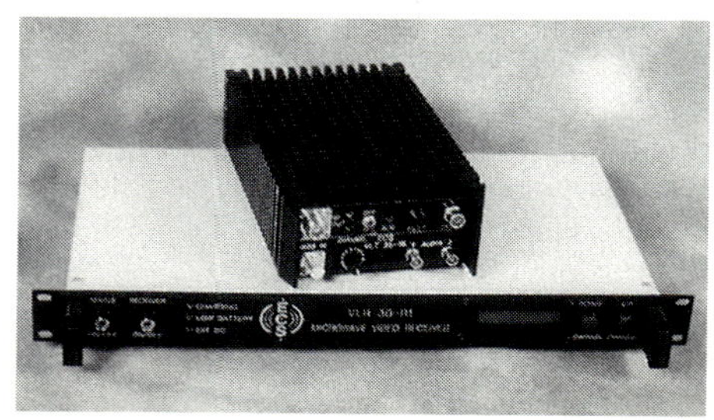

VLT 15-10-6 MK3/VLR 15-16-1 helicopter video downlink
1998/0011834

PTR 1721 V/UHF radio

The PTR 1721, a combined VHF and UHF radio covering the UHF band from 225 to 400 MHz and the VHF band from 100 to 156 MHz, is designed for all types of military aircraft. The ability to communicate on VHF in addition to UHF offers increased flexibility to air forces which, on occasion, may need to operate from civil airfields equipped with VHF facilities only. The PTR 1721 has been chosen for the RAF Panavia Tornado aircraft.

The PTR 1721 offers up to 9,240 channels, 2,240 on VHF and 7,000 on UHF. Channel spacing in the VHF band is at 25 kHz and standard spacing in UHF is at 50 kHz although an optional interval of 25 kHz spacing is available. Frequencies are selectable directly from the remote-control unit which also provides preselection of up to 17 channels with the addition of the UHF and VHF Guards at the international distress frequencies of 243 and 121.5 MHz respectively. Frequency synthesis techniques ensure good frequency stability and channel selection characteristics.

The system is entirely solid-state in construction and, wherever possible, employs conventional technology in the interests of reliability enhancement. The receiver is varactor-tuned and the frequency synthesiser is compared against a single reference oscillator.

A sealed case houses the transmitter/receiver unit and access to the modules is gained by removal of the sides of the case, to which the modules themselves are attached. Heat is conducted from the modules via the chassis, which acts as a heatsink, to the sides of the case and hence to external air. Forced-air cooling is directed through the mounting tray installation which also acts as a vibration insulator. Forced-air cooling may be dispensed with in less demanding aircraft environments.

Switching for dual-control operation is external to the system and in two-seat aircraft identical control units are installed at each crew position. Options include an antenna lobe switch for azimuth homing requirements and a UHF antenna switch for automatic direction-finder operation.

Recently, the company has introduced an optional modification which is available for both in-service and new radios. This modification provides a frequency-hopping capability which gives a significant improvement in communications performance when the radio is operated in a jamming environment.

The environmental temperature range extends from −40 to +70° C ambient and the normal operational altitude is to a maximum of 50,000 ft, although operation is possible for short periods at altitudes of up to 70,000 ft.

Specifications

Dimensions:
(controller) 94 × 145 × 175 mm
(transmitter/receiver) 125 × 194 × 339 mm
Weight:
(controller) 1.8 kg
(transmitter/receiver) 11.25 kg

Operational status

In production and in service in Royal Air Force Panavia Tornado aircraft.

Contractor

GEC-Marconi Electro-Optics Limited, Airadio Division, Portsmouth.

VERIFIED

PTR 1741 VHF/AM radio

The PTR 1741 equipment may be regarded as the VHF counterpart of the GEC-Marconi PTR 1751 system, both units being members of the same family of GEC-Plessey military airborne radios. Covering the frequency range from 100 to 155.975 MHz, the PTR 1741 provides 2,240 channels at 25 kHz separation. If desired, however, channel spacing can be set at steps of 50 kHz to ease interface with older equipment which does not have the same frequency stability as the new-generation systems. As in the case of the PTR 1751, the PTR 1741 is available with a choice of manual or manual/preset control unit; however, only a single power output level of 10 W is available.

The PTR 1741 shares a number of common features with the PTR 1751 with regard to build standard, operating range, options and so on. Both are suitable for dual UHF/VHF installation, their electrical interfaces and mechanical dimensions being identical. A single controller may be used to operate the two transceivers in such a dual installation.

Specifications

Dimensions:
(manual controller) 146 × 48 × 108 mm
(preset controller) 146 × 95 × 108 mm
(transmitter/receiver) ½ ATR short × 160 mm
Weight:
(manual controller) 0.7 kg
(preset controller) 1.4 kg
(transmitter/receiver) 5 kg
(Guard receiver module) 0.3 kg

Operational status

No longer in production but still in service.

Contractor

GEC-Marconi Electro-Optics Limited, Airadio Division, Portsmouth.

VERIFIED

PTR 1751 UHF/AM radio

The PTR 1751 is a lightweight UHF/AM radio designed for all types of military fixed-wing aircraft and helicopters. It provides 7,000 channels in the frequency band 225 to 399.975 MHz at a channel spacing of 25 kHz and is available with either 10 or 20 W transmitter outputs. Channel spacing of 50 kHz is available as an option.

Both versions comprise a single transmitter/receiver unit with either a manual controller or an optional manual and preset controller. Each controller provides full selection of the range of 7,000 channels together with control of the built-in test functions; the manual/preset unit additionally permits preselection of up to 30 channels, plus Guard frequency, through incorporation of a non-volatile memory store. A remote frequency and channel indicator is also available.

Options include continuous monitoring on the 243 MHz international distress frequency via a separate Guard receiver module which plugs directly into the main transmitter/receiver chassis, an external homing unit and a wideband secure speech facility which requires no additional interface equipment.

Recently the company has introduced an optional modification which is available for both in-service and new radios. This modification provides a frequency-hopping capability which gives a significant improvement in communications performance when the radio is operated in a jamming environment.

The PTR 1751 conforms generally to DEFSTAN-07-55 and operates satisfactorily over a temperature range from −35 to +70° C. It is of all-solid-state modular construction with high reliability as a principal design aim, and GEC-Marconi Radar and Defence Systems claims a current MTBF of 800 hours.

Specifications

Dimensions:
(transmitter/receiver)
(10 W version) ½ ATR short × 160 mm high
(20 W version) ½ ATR medium × 160 mm high
(manual controller) 146 × 48 × 108 mm
(preset controller) 146 × 95 × 108 mm
Weight:
(transmitter/receiver)
(10 W version) 5 kg
(20 W version) 6.7 kg
(preset controller) 1.4 kg
(manual controller) 0.7 kg
(Guard receiver module) 0.3 kg

Operational status

In production and service.

Contractor

GEC-Marconi Electro-Optics Limited, Airadio Division, Portsmouth.

VERIFIED

AD120 VHF/AM radio

Originally designed for civil aviation applications by the King Radio Corporation in the USA, GEC-Marconi has requalified this VHF/AM system for military roles and, manufacturing under licence from King, markets the system under the designation AD120. It is installed in most UK military fixed-wing aircraft and helicopters and has also been supplied to a number of overseas customers.

Covering the frequency band 108 to 137 MHz, the AD120 provides channel spacing of 25 kHz. Services provided are double sideband AM voice communication. Power output can be varied between 10 and 20 W. The system is an all-solid-state design of modular construction and is designed for easy installation and maintenance. It is also exceptionally simple to operate. The system's standard remote controller possesses only five controls: an on-off switch, volume control, a test button and two rotary switches for frequency selection. Tuning is instantaneous. Automatic squelch and gain control eliminate the need for manual adjustment.

The self-test facility may be used during operation as a confidence check.

Specifications

Dimensions:
(transmitter/receiver) 60 × 127 × 315 mm
(controller) 146 × 47 × 90 mm
Weight:
(transmitter/receiver) 2.3 kg
(controller) 0.6 kg

Operational status

In production and service. More than 1,200 systems currently in service.

Contractor

GEC-Marconi Electro-Optics Ltd, Sensors Division, Basildon.

VERIFIED

The GEC-Marconi AD120 VHF radio

AD190 (ARC-340) VHF/FM radio

The GEC-Marconi AD190 radio, a company designation for the ARC-340 system, is a tactical VHF/FM equipment designed specifically for air-to-ground communication between military aircraft and ground forces. It is in service with the UK armed forces, in conjunction with Army Clansman VHF equipment, and is in operation in a wide range of fixed-wing aircraft and helicopters in many parts of the world.

The system provides clear and secure speech communication, data transmission, automatic rebroadcast for extending the range of tactical communications and homing facilities. The homing mode may be used simultaneously with a communications channel without mutual interference. The range capability of the homing facility is the same as the communications range.

The AD190 covers the VHF band from 30 to 75.975 MHz at selectable channel spacing of either 25 or 50 kHz increments. Tuning is silent and instantaneous and transmitter output power is selectable at either 1 or 20 W.

Multistation collocated operation is possible with a maximum of three systems in the same aircraft. This is subject to the proviso that antennas must be sited at least three feet apart and frequency separation of 3.5 per cent for three systems or 3 per cent for two systems must be maintained. The AD190 incorporates diagnostic BITE.

Specifications
Dimensions:
(controller) 146 × 67 × 85 mm
(transmitter/receiver) 385 × 197 × 125 mm
Weight:
(controller) 0.9 kg
(transmitter/receiver) 9.2 kg

Operational status
In production and service. 900 systems in service.

Contractor
GEC-Marconi Electro-Optics Ltd, Sensors Division, Basildon.

VERIFIED

AD3400 multimode secure radio system

The GEC-Marconi AD3400 provides, in a single transmitter/receiver, coverage of the entire airborne line of sight frequency band from 30 to 400 MHz with channel increments of 25 kHz. Up to 20 channels may be preselected in this range.

In effect, the system provides the equivalent of four radios in a single unit since it covers the following bands and modes: VHF FM for tactical close support, the civil ATC VHF AM band, the civil and maritime VHF FM bands and the military UHF AM and FM bands. Its other major feature is that it provides secure communications for both speech and data transmission. For speech transmission in a secure mode, an encryption unit is used and for data security the appropriate modem is connected to the radio equipment.

Separate, continuously operating Guard receivers are incorporated and homing and ADF facilities are also available with suitable keying and antenna installations.

Construction of the system features 'slice' techniques and Large-Scale Integrated (LSI) circuitry. The system is based on Intel 8085 microprocessors and uses distributed processing. The extensive use of LSI and proven components provides a high measure of reliability. Serviceability is further enhanced by the comprehensive BITE facility which can provide rapid fault diagnosis, particularly since intermittent fault data is stored in a non-volatile memory for maintenance purposes.

The system is convection cooled and protected by a sensing device which progressively reduces power output at high temperatures to prevent transmitter damage. The transmitter is also protected from damage caused by short or open circuits at the transmitter output.

The AD3400 has been designed for ease of installation with particular attention to retrofit requirements in older aircraft in which it is often possible to fit two of these systems in the space formerly occupied by a single radio.

Specifications
Dimensions:
(AA34024 controller) 57 × 146 × 152 mm
(AA34001 transmitter/receiver) 194 × 125 × 256 mm
(AA34601-1 encryption unit) 194 × 58 × 320 mm
Weight:
(AA34024 control unit) 1.3 kg
(AA34001 transmitter/receiver) 6.5 kg
(AA34601-1 encryption unit) 4.5 kg

Operational status
In production.

Contractor
GEC-Marconi Electro-Optics Ltd, Sensors Division, Basildon.

VERIFIED

AD3430 UHF datalink radio

The GEC-Marconi AD3430 is a UHF Link 11 compliant radio which operates in the 225 to 399.975 MHz band. It is tunable in 25 kHz increments and provides transmission and reception of clear and secure speech using amplitude or frequency modulation. Secure speech operation is provided in association with a BID 250 encryptor. The capability for providing transmission and reception of Link 11 data using frequency modulation is provided in accordance with all the general radio and UHF specific Link 11 radio requirements of MIL-STD-188-203-1A.

The equipment can deliver maximum power outputs of 20 W in AM operation and 60 W in FM operation with a transmit duty cycle of 100 per cent. The primary control of the radio is via an ARINC 429 serial data highway.

Specifications
Dimensions: 365 × 194 × 124 mm
Weight: 9.5 kg

Operational status
The AD3430 is in production in support of the GEC-Marconi Communications subsystem for the Royal Navy EH 101 helicopter.

Contractor
GEC-Marconi Electro-Optics Ltd, Sensors Division, Basildon.

VERIFIED

AD3500 V/UHF radio

The AD3500 was selected in 1986 as the standard tactical V/UHF radio on the Royal Air Force's British Aerospace Harrier GR. Mk 7 aircraft. The radio covers the standard VHF and UHF frequency bands, in both AM and FM, with 30 channels being preset and automatic or manual selection between AM and FM operation. The system also offers both clear and secure speech and advanced capabilities in an ECCM environment. The radio has a MIL-STD-1553 interface.

Specifications
Dimensions:
(RX/TX) 241 × 127 × 124 mm
(control unit) 152 × 146 × 57 mm
Weight:
(RX/TX) 4.5 kg
(control unit) 1.3 kg
Power supply: 28 V DC
Power output:
(AM) 10 W
(FM) 15 W

Operational status
In service in Royal Air Force British Aerospace Harrier GR. Mk 7 aircraft and GKN Westland Helicopters Sea King Mk IV.

Contractor
GEC-Marconi Electro-Optics Ltd, Sensors Division, Basildon.

VERIFIED

Secure speech communication system

The GEC-Marconi secure speech communication system has been developed to enable strategic and intelligence information to be passed between aircraft and ships without risk of interception. Messages can be transmitted in the complete confidence that only the recipient can understand the contents. The system has been designed to allow integration with existing communications control equipment in naval and air force aircraft with minimum aircraft modification.

System development has involved the design and manufacture of A and B model LRUs, system test rigs and special-to-type test equipment, and commissioning of the aircraft rigs. The production programme has been built on a combination of modern control methods and advanced automatic test procedures to ensure the total set of units for each aircraft comes together as a proven system. All aspects of overall system cost, including those of ownership, installation, crew training, maintenance and support have been reduced by employing a modular concept. This concept allows system configuration to be cost-effective, giving maximum flexibility to meet customer requirements with a built-in provision for expansion.

For a complete clear and secure multistation multi-radio communications fit, the development system has been selected by Loral for the Royal Navy Merlin production helicopter. The system uses microprocessor technology to control the routeing of audio paths. MIL-STD-1553 and ARINC 429 data highways are used to transfer control commands between the various elements of the subsystem. The system also incorporates BIT and circuit redundancy features.

Operational status
In production.

Contractor
GEC-Marconi Electro-Optics Ltd, Sensors Division, Basildon.

VERIFIED

Units of the GEC-Marconi secure speech communication system

Advanced airborne speech recogniser for ASR 1000

GEC-Marconi developed the Model ASR 1000 speech recogniser to meet the UK Ministry of Defence's requirements for an advanced airborne speech recogniser. Its development has followed an evolutionary path after successful flight trials by the Defence Evaluation Research Agency of other GEC-Marconi speech recognisers such as SR-128 and Macrospeak in Tornado, Buccaneer and BAe 1-11 aircraft and the Wessex helicopter. The system incorporates the advanced technology of the Macrospeak commercial recogniser with additional enhancements to meet the DERA's specific requirements for accurate performance in the noisy environment of fast jet aircraft, where lower signal-to-noise ratios are normal. A complementary ground-based speech template preparation facility uses advanced statistical techniques to produce robust templates for better performance when the pilot is under stress and other cases where the variable nature of speech has proved to be a problem in the past.

The ASR 1000 is a speaker-dependent 1000-word continuous recogniser suitable for two-crew operation and performs its tasks in real time. It features automatic gain control with a dynamic noise mask and can implement a true finite state syntax for each crew member. It is contained in a ½ ATR case but the technology, when proved, will be customised to meet customers' specific requirements for integration in advanced avionics and other critical areas.

Applications will include its use by aircrew to control communications, navigation and weapon delivery systems; it will also be used for low bit-rate communications.

Specifications

Dimensions: ½ ATR short
Vocabulary size: 1,000 words

Operational status

In service.

Contractor

GEC-Marconi Radar and Defence Systems Ltd.

UPDATED

GEC-Marconi speech recognition unit

Helicopter secure speech system

GEC-Marconi has developed the helicopter secure speech system to provide high-quality secure voice encrypted communications for military helicopters, to interface with the existing avionics and be interoperable with the current secure ground-based communications system. The equipment is easily fitted for specific missions or roles, or may be used as a permanent fixed installation.

The system comprises the Digital Master Unit (DMU) with integral VHF FM ARC-340 radio and encryption equipment. Multiple user access is provided within the aircraft by means of user selector boxes over which the pilot has overall control via the Master Control Unit (MCU). A crew member may therefore operate over a secure net within the aircraft or to the ground if required. Features of the system include full air certification to military specification, transilluminated controls with a night vision facility and a rebroadcast and intercom facility.

Specifications

Dimensions:
(DMU) 420 × 540 × 237 mm
(MCU) 147 × 97 × 60 mm
(USB) 147 × 97 × 60 mm
(controller) 153 × 203 × 205 mm
Weight:
(DMU) 24.3 kg
(MCU) 0.6 kg
(USB) 1.1 kg
(controller) 2.6 kg

Operational status

In production.

Contractor

GEC-Marconi Radar and Defence Systems Ltd.

UPDATED

MarCrypDix Air secure speech system

MarCrypDix Air is a secure speech encryption equipment developed specifically for installation in civil and military aircraft. It is designed to interface with any airborne digital communication system and provides high-grade speech and data encryption. There is storage for eight different key variables or code settings, fully automatic synchronisation with zero error extension and negligible range degradation.

The key variables are generated by a key management unit and are inserted either manually, using a hexadecimal keyboard, or may be generated automatically within the key management unit using a true random noise generator. Transfer of the key variables from the management unit to the encryption unit is accomplished by a key fill gun. This is a battery operated, pocket-sized device with an internal memory system, data interfacing being performed optically. Key variables stored within the gun can be transmitted to any number of encryption units. Provision is also made for the instant erasure of all information stored within the gun's memory. The number of key variables is 2^{128}, with a key variable length of 128 bits, plus a further 16 bits set uniquely for each customer.

The equipment is self-monitoring and will inhibit transmission should any fault occur in the equipment, thus ensuring that sensitive information cannot be transmitted with less than complete protection.

Although MarCrypDix Air has been designed for the airborne environment, the same unit, suitably mounted, can be employed in a ground role for ground-to-air or ship-to-air communication.

Specifications

Dimensions: 193.5 × 320.5 × 58 mm
Weight: 4 kg

Operational status

In production and in service.

Contractor

GEC-Marconi Radar and Defence Systems Ltd.

UPDATED

Voice encoder/decoder

GEC-Marconi has developed a channel voice encoder/decoder (vocoder) equipment designed to operate in environments of high acoustic noise and high bit error rates, such as those encountered in tactical military roles. The 2,400 bits/s vocoder incorporates a robust pitch extractor, regenerator and so on. According to GEC-Marconi the system provides improved resistance to acoustic noise and transmission errors over other systems such as linear predictive vocoders. The pitch extraction technique is less sensitive to acoustic noise and forward error correction, and median smoothing/majority vote methods applied to the pitch information result in reduced sensitivity to transmission errors.

Good communication has been achieved between operators in high acoustic noise and transmission conditions. Subjective tests, using standard helmets and face-mask microphones, have been carried out in a noise chamber where operators were subjected to levels of 120 dB (typical fighter noise) and 115 dB (fighter bomber level), said to be representative of high-speed low-level flight. Communication could be maintained at these noise levels for random transmission error levels of up to 2.5 per cent.

The system is almost completely digital in character and is based upon an Intel 8085 microprocessor chip combined with a special purpose digital filter bank. It is now being manufactured as a four-board pack.

Specifications

Dimensions: 51 × 176 × 181 mm
Weight: 1 kg

Operational status

In production as a four-board pack.

Contractor

GEC-Marconi Radar and Defence Systems Ltd.

UPDATED

The GEC-Marconi four-board vocoder pack

Advanced Digital Radio ADR + VHF radio

The ITT Defence Ltd software programmable ADR + VHF radio has been evolved from the proven Single Channel Ground Airborne Radio System (SINGARS); it uses the proven functionality with the benefit of an all-digitial signal processing architecture, and features:

(a) Embedded GPS position and user information in data transmissions, enabling accurate monitoring of friendly force positions for situational awareness
(b) New frequency-hopping packet data waveform with forward error correction, reduced on-air transmit time, reduced transmission overhead, and improved message throughput
(c) Improved channel access algorithm which combines voice and packet data on a common net, with packet throughput rates and minimal impact on voice communication
(d) Bowman Internet Communications Controller (BICC) with a programmable interface which converts protocols, providing a seamless flow of data across the battlefield and facilitates horizontal and vertical integration of command and control.

Waveform improvements have been implemented to improve simultaneous voice and data transmission.

The BICC has been developed to interface both the UK Bowman battlefield communications system and the US Force XXI digitisation programme.

The ADR + VHF radio interfaces directly with the main battlefield Bowman radios and the compatible ITT-enhanced portable VHF radio.

Specifications

Frequency range: 30.00-87.975 MHz
Modulation: Gaussian Minimum Shift Keying (GMSK)
Channels: 2,320
Channel spacing: 25 kHz
Modes of operation: fixed frequency; frequency hopping; free channel search
Preset channel operation: 6 single-channel; 6 frequency hopping; 6 free channel search
Data rate: 30 kbps
Transmit power: 16 W (adjustable)
Power: 28 V DC
Dimensions: 139.7 × 195 × 104 mm
Weight: 3.73 kg

Contractor

ITT Defence Ltd.

NEW ENTRY

Advanced Digital Radio ADR + VHF radio **1998**/0011833

NavSymm DR5-96S UHF differential datalink system

The NavSymm DR5-96S differential datalink system is designed to work with the NavSymm Sharpe XR6 GPS receiver to provide a complete solution to the differential GPS requirement.

The NavSymm DR5-96S operates at 9,600 bps, enabling the receiver to accept RTCM SC-104 (version 2.1) messages and return positional information down the same datalink. It can be synchronised to GPS time so that mobile location data can be retrieved and returned at scheduled times.

The NavSymm DR5-96S can also be used to transmit raw measurement data in real-time positioning systems in order to achieve accuracies down to a few centimetres. The radio is tolerant of noisy RF environments, and is available with two software-selectable, predefined, factory-set frequencies in the FCC licence-free band of 450-470 MHz.

Specifications

RF range: 450-470 MHz; 2 programmable synthesised channels
Channel spacing: 12.5 or 25 kHz
Transmit power: 2 W
Transmit data rate: 9,600 or 4,800 bps
Operating mode: half duplex
Dimensions: 175 × 80 × 57 mm
Weight: 1 kg
Power: 10-32 V DC; 1.2 W receiver; 13.5 W transmit

Operational status

Available

Contractor

Navstar Systems Ltd.

NEW ENTRY

NavSymm DR5-96S II UHF differential datalink system **1998**/0011832

NavSymm DR5-RDS VHF differential datalink system

By receiving GPS differential corrections from Differential Corrections Services provided over the Radio Data System (RDS) the NavSymm DR5-RDS is able to provide DGPS accuracy over a wide area without establishing an RTCM correction transmitter.

The RDS is a system of transmitting data over an inaudible FM sub-carrier with normal communications.

The unused capacity of some RDS channels is used to transmit DGPS correction data. The NavSymm DR5-RDS decodes these signals and outputs them in standard RTCM SC 104 form to the GPS receiver.

Specifications

Frequency range: 87.5-107.9 MHz, 100 kHz steps, agile PLL
Scanning performance: 10 to 40 seconds to detect all the RDS stations of the FM band
Dimensions: 175 × 78 × 74 mm
Weight: 0.77 kg
Power: 11-32 V DC; 0.6 W

Operational status

Available

Contractor

Navstar Systems Ltd.

NEW ENTRY

NavSymm DR5-RDS VHF differential datalink system **1998**/0011831

Automatic Voice Alert Device (AVAD)

The Automatic Voice Alert Device (AVAD) stores prerecorded human speech in digitised form and operates under microprocessor control to assemble messages from words or phrases held in a vocabulary store. Since the voice is not synthesised, it can be either male or female in any language to provide the appropriate degree of stress and urgency for any situation. Racal says that high-quality reproduction ensures that the voice remains recognisable and intelligible under all conditions.

Message priority order, repetition rate and volume are programmable to individual requirements to provide the maximum information to aircrew without disrupting their primary tasks. Racal claims that existing discrete audio warning systems can be replaced by AVAD with a minimum of installation cost and complexity.

The units in the current AVAD range are the V694, V695 and V697.

V694

The AVAD V694 is designed to integrate and control audio alerts keyed by signals from system sensors and push-buttons. The message format is completely flexible and the system conforms to ARINC 577 audible warning systems and ARINC 726 flight warning computer systems for civil aviation applications.

Up to 4 minutes of prerecorded speech can be encoded and stored as a vocabulary of messages, phrases, words and tones. Messages can be constructed using words and phrases drawn from the unit memory under software control. This allows the total duration of all messages to exceed the vocabulary by a considerable amount. Up to 15 message channels are based on positive or zero volt keying. Four channels can respond to variable DC voltage inputs, offering a sensitivity of up to 256 logic switching steps across the input range. A stabilised 5 V DC output is available for use as a reference for an external potentiometer. Typical variable readouts might be aircraft altitude, electrical system voltages or engine pressure ratios.

Unit operation and vocabulary store may be defined by the user and program software allows control of each message cycle. Each message has normal, regrade and test priority values assigned to it. The output sequence is then controlled by the regrade and test inputs.

There are two audio outputs, one for a telephone or headset and the other for driving a loudspeaker direct. A variant can be provided where the loudspeaker drive output is removed and replaced by an auxiliary audio input. This auxiliary input facilitates the summing of existing aircraft audio warnings with the single AVAD output for routeing into the aircraft audio integration/intercommunication control system. This AVAD configuration is particularly suitable for helicopter use. The nominal level of output can be adjusted by individual potentiometers offering a 20 dB range, with the relative volume of each message or message format under software control. The complete message format, including repetition rate, pauses, message inhibit/de-inhibit and tones, may be similarly controlled. A key input is provided which inhibits any message in progress without affecting other keyed messages. A message will remain inhibited until its sensor input is removed and reapplied. Alternatively, the inhibit may be removed after a specified time if the warning message keyline remains activated during this period.

Full test/reset facilities provide fast checking of unit operation. Correct operation of the microprocessor and amplifier circuitry and control program status are continuously self-monitored. In the event of a fault condition, both audio outputs are automatically inhibited and a fail output is provided to activate an indicator to warn of unit failure.

V695
The V695 is a smaller seven-channel unit with microprocessor control. It offers all the programming flexibility of the V694 but has a storage capacity of 13.5 seconds. An auxiliary audio input is provided.

V697
The V697 provides a single output for applications in which only one warning such as 'fire' or 'pull up' is required. A second keyline may be used to provide a system enable/disable input. The V697 can be enabled by either positive or zero volt keying and the resultant warning output can be either a continuous or one message cycle per key event.

All the above units provide a bandwidth of 100 Hz to 4 kHz. Audio distortion is less than 4 per cent. Volume is preset and adjustable by 20 dB. The V694 is also programmable to different levels for each message and variation during messages.

Specifications

Dimensions:
(V694) 165 × 115 × 51 mm
(V695) 134 × 61 × 32 mm
(V697) 110 × 50 × 28 mm
Weight:
(V694) 0.95 kg
(V695) 0.2 kg
(V697) 0.14 kg
Power supply: 28 V DC
Temperature range: −40 to +70°C

Operational status

In production and in worldwide service with a variety of fixed- and rotary-wing aircraft including Royal Air Force Jaguar fixed-wing aircraft and Puma and all Sea King helicopters.

Contractor

Racal Acoustics Ltd.

VERIFIED

RA800 Integrated Digital Audio Control System (IDACS)

Designed to satisfy the present and envisaged future requirements for audio management, the RA800 Integrated Digital Audio Control System (IDACS) can be installed in a variety of military and civil fixed-wing aircraft and helicopters with a minimum of custom engineering to integrate and control all the audio signals.

Control, interfacing and amplification is provided for all audio equipment on the aircraft, together with a sophisticated and flexible intercom network for the crew. Cross-talk and interference are reduced to a minimum and control is exercised over MIL-STD-1553B or ARINC 429 databus links. A single communications audio management unit can support up to eight major crew stations or intercom nets, providing interfacing and control for up to eight transmitter/receivers, eight receivers, eight unswitched audios and four recorder outputs. Options include voice and tone alerting, Tempest speech security and power supply for active noise reduction headsets or helmets. Built-in test routines and dual separate power supplies ensure increased reliability and failure damage protection.

The RA800 'Light' secure Communications Control Systems (CCS) is a development of the basic system. It is a fully digital system in which advanced signal processing ensures extremely high signal quality and integrity. Communications within the system are via fibre optic cables which eliminate many of the EMC and security difficulties associated with conventional aircraft systems.

The Communications Audio Management Unit (CAMU) forms the core of the system and provides the control and secure communications switching using digital signal processing. The optical connections between the CAMU and the crew stations permit good routeing redundancy with low weight. Additionally, an analogue fallback is incorporated into the system to ensure 'minimum facilities' operation in the event of digital system failure or battle damage.

The RA 800 'Light' offers full compatibility with warning and management systems and is designed to interface with HUMS (Health and Usage Monitoring Systems) to enhance overall maintainability and flight safety. The system can incorporate AVAD, an automatic crew alerting device which provides warnings in natural speech, and active noise reduction headsets for improved speech intelligibility and safety in high noise environments.

Operational status

In service on BAe Hawk aircraft. Ordered for EH 101 and RAF Chinook helicopters, and for SAAF C-130B aircraft.

Contractor

Racal Acoustics Ltd.

UPDATED

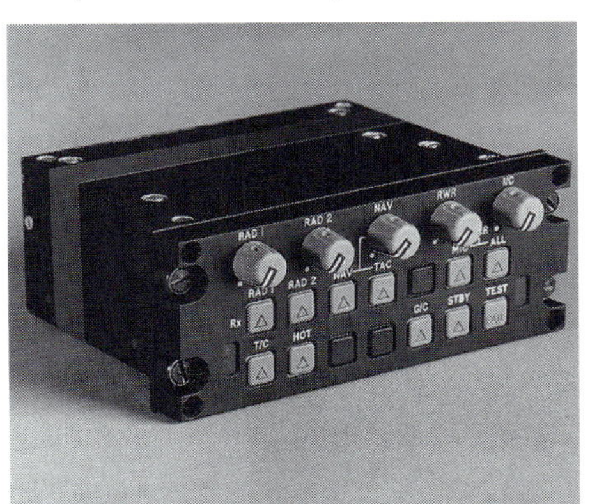

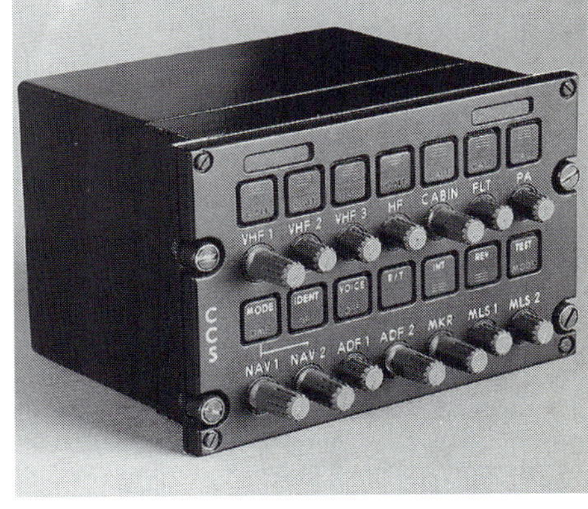

The two units of Racal's RA800 Integrated Digital Audio Control System **1996**

Six-Ninety series audio amplifiers

Racal's six-ninety series of audio amplifiers has been developed to meet the requirements of aircraft intercommunication, microphone conversion and amplification, audio isolation, telephone and loudspeaker drive. Units in the range are said to be able to satisfy the onboard requirements of any type of aircraft. Inputs at telephone, electromagnetic, dynamic or carbon levels can be accommodated to provide operational flexibility with corresponding outputs to suit impedances of most current telephones, headsets or loudspeakers.

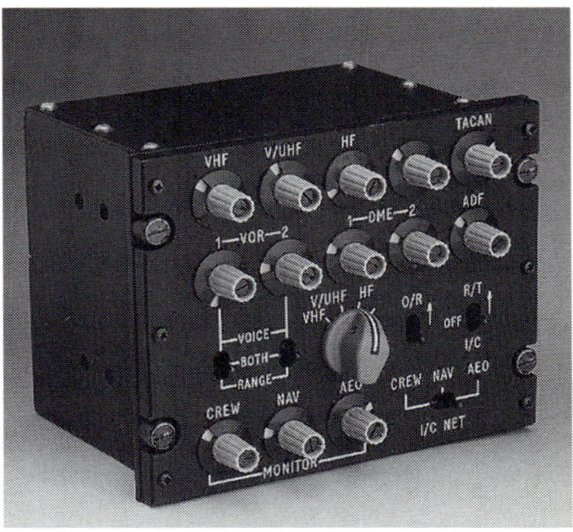

Racal Six-Ninety series station box **1996**

A special version, the A697 Cockpit Voice Recorder (CVR) summing amplifier is designed to meet all possible needs in aircraft having to satisfy the mandatory CAA specification No 11 requirement for CVR installations. All audio signals monitored by the pilot, co-pilot and a third crew member, together with hot mike signals in all control column switch modes, can be recorded on channels of a CVR in association with a separate 'area' microphone.

The A697 can also provide the interface between aircraft audio systems from different manufacturers. Electromagnetic, dynamic or carbon microphones may be employed and input may be made to any one of several approved CVR systems.

In the A6912, a development of the A697, the third crew member audio summing circuitry has been replaced by circuitry enabling helicopter rotor speed to be recorded. Signals relating to the rotor speed are taken from a tachometer, encoded in the A6912 and recorded on one channel of the CVR.

Operational status

In production.

Contractor

Racal Acoustics Ltd.

VERIFIED

Airborne real-time datalink

Racal Avionics has integrated its satcom and navigation/mission management products to provide a complete real-time datalink package which includes both the ground segment and the airborne Line Replaceable Units (LRUs). The airborne LRUs utilised are the: multifunction Control Display Navigation Unit (CDNU) as the operator interface and the Satellite TRansceiver (STR) as the communications system. The primary function of the CDNU is to provide aircraft navigation and subsystem control. It is also the airborne datalink end system, providing Automatic Dependent Surveillance (ADS) and Controller to Pilot DataLink Communication (CPDLC), which also allows free text messages and pre-formatted messages to be sent to a variety of ground destinations. It displays messages received from the ground and allows the user to reply to those messages. The CDNU also has an Emergency ADS Mode.

ADS reports are initiated from the ground, via an ADS contract, and include: present position, altitude, velocity, and aircraft intent. All ADS contracts are transparent to the pilot.

The STR equipment which supports packet data services — Data 2 (ACARS) and Data 3 (ATN FANS X.25) communications over the worldwide Inmarsat satellite system, comprises two LRUs: the combined Satellite Data Unit (SDU)/Radio Frequency Unit (RFU) and a High Power Amplifier (HPA).

The Inmarsat system utilises a digital format for real-time full-duplex data communications at up to 1,200 bps. Being a low-gain system, the STR operates via a 0 dBic omnidirectional antenna and therefore requires no other navigation input for beam-steering purposes.

The ground terminal allows the operator to initiate ADS contracts with numerous aircraft and displays the requested information on a map and status window. The terminal also allows text messages to be sent to selected aircraft and displays messages received from aircraft. Any aircraft that initiates an Emergency ADS message is highlighted on the map with an audio alert.

Contractor

Racal Avionics Limited.

NEW ENTRY

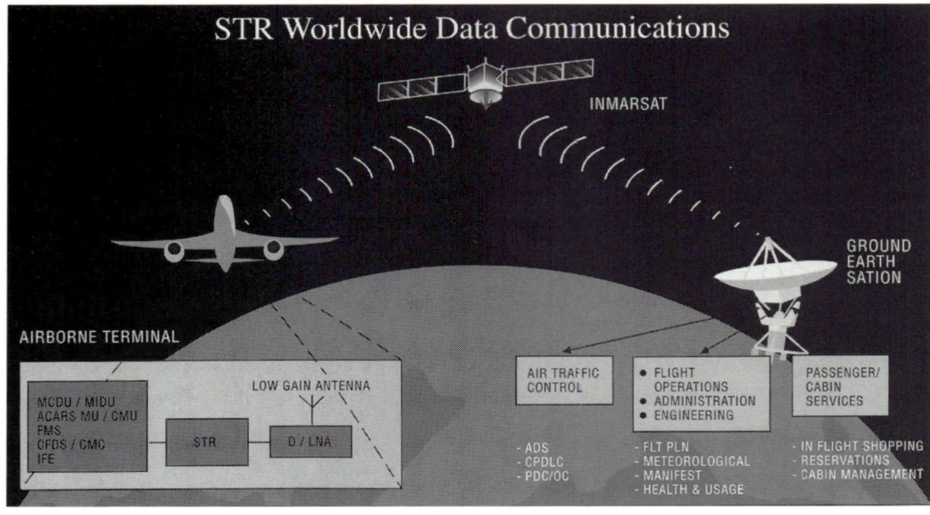

STR worldwide data communications ***1998***/0011830

M-ADS helicopter satellite communications system

Racal Avionics has been selected to supply a small, lightweight satellite communications system to Kongsberg Aerospace. The system will form part of a Modified Automatic Dependent Surveillance (M-ADS) system currently under development by Kongsberg Aerospace for installation in helicopters operated by the Norwegian Helikopter Service.

The system will transmit position, waypoint and groundspeed information from air to ground via the Inmarsat satellite system, providing a means of accurate position reporting. This information, which will be automatically transmitted at regular intervals, will provide pinpoint accuracy and keep air traffic controllers up to date with the aircraft's position and intentions.

Specifications

SDU: 2 MCU weighing <4.5 kg
HPA: 2 MCU weighing <4.5 kg
ADSU: 2 MCU weighing <3 kg

Operational status

The M-ADS system is being evaluated in Eurocopter Super Puma AS332LI and Sikorsky S-61N helicopters, with Chinook, Super Puma Mk II, AS365N-2 Dauphin and Bell 214ST helicopters to follow.

Contractor

Racal Avionics Ltd.

UPDATED

STR Satellite TRansceiver

The STR Satellite TRansceiver is a single channel satcom system which supports packet data services — Data-2 (ACARS, AFIS, A622 FANS) and Data-3 (ATN FANS, X-25 cabin) communications — over worldwide Inmarsat satellite system.

The STR is an Inmarsat Aero 'L' Class 1 system, which meets the integrity and real-time communication requirements necessary to support Air Traffic Services, both today and in the future, while simultaneously supporting operational, administrative and passenger datalink services.

Applications typically supported by the STR are:

(a) Operational/fleet management services: flight planning data, departure times, ETAs, position reports, manifest and engineering data
(b) Communications/Navigation Surveillance (CNS) services: FANS applications including: Automatic Dependent Surveillance (ADS), Controller to Pilot DataLink Communications (CPDLC) and Pre-Departure and Oceanic Clearance (PDC/OC) via A622 FANS, and via ATN FANS as a growth option
(c) Passenger and cabin management data services — X-25 based.

The STR utilises the proven technology of the Racal/Honeywell MCS 3000/6000 multichannel systems, reconfigured into two compact LRUs: a Satellite Data Unit/Radio Frequency Unit (SDU/RFU) and High Power Amplifier (HPA). The system interfaces to an omni-directional low-gain antenna and to a variety of cockpit and cabin communication management devices including: ACARS Mus/CMUs, AFIS, FMS, and In-Flight Entertainment (IFE).

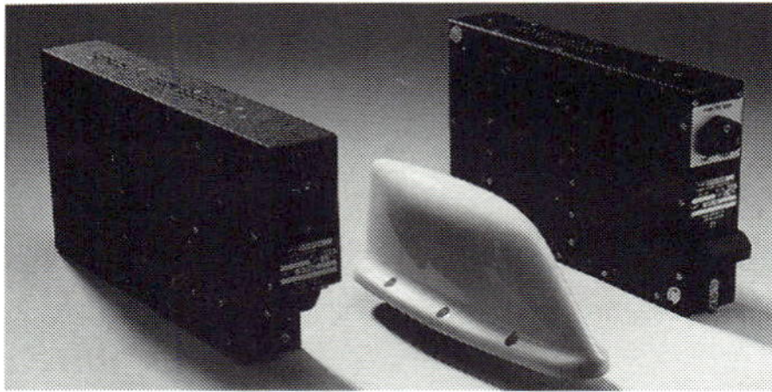

Racal Avionics STR Satellite TRansceiver and omnidirectional antenna ***1998***/0011829

Specifications

Data rate: 600, 1,200 bps
Dimensions: each unit; 2 MCU
Weight: each unit; <4.5 kg
Power: 115 V AC, 400 Hz, <275 V A

Operational status

In production

Contractor

Racal Avionics Limited.

NEW ENTRY

BCC306 helicopter VHF/FM transmitter/receiver

The BCC306 is a development of the Clansman RT351 VHF receiver/transmitter in service with the British Army for vehicle and manpack use and is intended for tactical use between airborne and ground units. The BCC306 provides air-to-ground and air-to-air voice communications over the 30 to 76 MHz frequency range, giving 1,841 channels at 25 kHz spacing. The equipment functions in the single-frequency simplex mode and uses F3E narrowband modulation. It is simple to operate with channel selection and all other control functions via a remote-control unit.

The system consists of the transmitter/receiver, remote-control unit, power supply unit, antenna tuning unit and an aircraft interface unit, with an overall weight of 10.37 kg. The transmitter/receiver is an RT351 equipment from which all the controls have been removed and resited on a remote-control unit near the pilot.

The tuning unit is capable of tuning the 1.7 m whip antenna over the complete frequency range. The unit does not require frequency setting or antenna impedance information, antenna matching and tuning being carried out automatically following operation of the appropriate switch on the control unit.

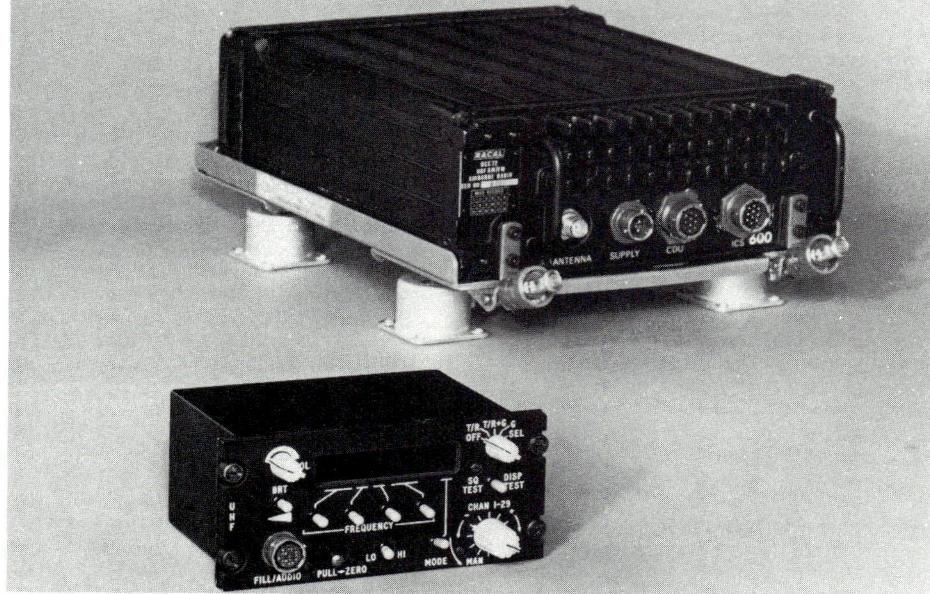

The Racal-Radio BCC 72 transceiver and BCC 584B control

Operational status

In operational service with the UK and other armed forces. Although developed originally for installation in helicopters, especially the Westland Wessex, the equipment has also been fitted in various types of light aircraft.

Contractor

Racal Radio Ltd.

VERIFIED

Jaguar-U (BCC 72) UHF radio

The Racal Jaguar-U system is a range of radio communications equipment operating in the UHF band, available for airborne, shipborne and land use. The airborne version, BCC72, is designed to operate over the frequency range 225 to 400 MHz and provides fixed frequency or frequency-hopping FM with selectable encryption, and fixed frequency AM compatible with current operational systems. It provides 7,000 channels at 25 kHz spacing, with 30 programmable channels (including a Guard channel on the distress frequency of 243 MHz) with flexibility to select clear, secure, fixed and hopping modes.

Design of Jaguar-U is based on experience gained in the Jaguar-V systems and employs the same method of medium-speed frequency hopping to protect against interception, direction-finding and jamming. To simplify frequency management, the 225 to 400 MHz band has been divided into 13 sub-bands, allowing the co-sited operation of multiple radio systems without interference. However, the radio can also be programmed to use any one of three larger hop bands. In either mode, orthogonal hop sets are available to assist frequency management. Large numbers of nets can operate in the same frequency bands and individual bands can be barred to avoid jamming or to protect other fixed frequency stations. Selective communication can be performed within an individual net or, conversely, a radio can be selectively barred from a net should it be captured. Once the radios have been programmed with hop codes and frequencies they synchronise automatically without the need for time of day input. Communications security is also enhanced by the use of either a built-in encryption unit using a second keystream generator or any external 16 kbits/s system in fixed or frequency-hopping modes

The airborne BCC 72 system consists of a rack-mounted transceiver and a panel-mounted control/display unit. The transceiver provides output powers of either 10 mW or 15 W on FM and 40 mW or 40 W PEP on AM, output level being selected on the controller. Two control/display units are available, depending on the aircraft requirements. The BCC 584B controller gives full manual selection of any of the 7,000 channels plus selection of the 30 programmable channels, hopping or fixed mode selection, low- or high-power selection and frequency/mode display by light-emitting diodes.

Specifications

Dimensions:
(transceiver) 230 × 90 × 350 mm
(controller BCC 584B) 145 × 66 × 104 mm
Weight:
(transceiver) 6.5 kg
(controller BCC 584B) 0.8 kg

Operational status

In production and in service with a number of users.

Contractor

Racal Radio Ltd.

VERIFIED

Eurofighter voice control system

Smiths Industries Aerospace has been selected by Computing Devices, Hastings Ltd and endorsed by Eurofighter GmbH to provide the direct voice input system for the Eurofighter programme. A design and development contract has been awarded to the company's Defence Systems division, Cheltenham, covering a Speech Recognition Module (SRM) for installation in the communications audio management unit to be supplied for Eurofighter by Computing Devices.

It is claimed that Eurofighter will be the first production aircraft to incorporate interactive voice technology as a standard fit. Some 25 different cockpit functions will be controlled by voice command through a system containing a 200 word vocabulary developed by Smiths Industries Aerospace. Activities such as calling up screen displays, selecting radio channels and frequencies and other similar functions will be controlled simply though verbal instructions from each individual pilot, even when flying under the most extreme operating conditions.

Each pilot will be required to enrol his voice in a ground-based PC station in advance of each flight. The system is capable of recognising that pilot's individual voice patterns at all times and can cope with physical strains which are placed on the vocal chords through the effects of *g* forces and when flying at high speeds or avoiding enemy aircraft under combat conditions.

Leading up to the Eurofighter application, prototype voice systems developed by Smiths Industries were successfully trialled in a number of aircraft including the F-16, F-18 and the AV-8B/Harrier. These led to the development of an interactive voice module designed specifically for the AV-8B/Harrier mission computer. The trials proved the potential of such systems to reduce pilot workload, enhance situation awareness and improve aircraft safety.

The development of voice technology has benefited from the R&D at the Speech Research Unit of the Defence Evaluation and Research Agency (DERA) Malvern, and application development in the DERA Speech Systems Group, Farnborough.

As a result of this co-operation, a contract has also been signed for Smiths Industries to equip a GKN Westland Lynx helicopter with an SRM to evaluate performace in rotary-wing aircraft.

Operational status

Smiths Industries' SRM contract for Eurofighter includes the provision of hardware to support the aircraft development programme, with options covering the future production phase. Smiths Industries will supply a number of major avionic systems for Eurofighter, including cockpit displays and control equipment.

Contractor

Smiths Industries Aerospace.

NEW ENTRY

A628 speech processor

The A628 speech processor provides clear recognisable speech by reducing the effects of background noise and enhancing speech quality. It automatically adjusts to the acoustic environment, and is particularly suitable for vocoders and harsh environments. High extraneous noise levels at the transmitter reduce the quality of speech received over radio circuits. This is a particular problem when vocoders are used because the resulting digitised signal to the radio transmitter often provides an unusable signal at the receiver's decoder.

The A628 is able to separate noise and voice signals and, using active techniques, cancels noise picked up by the microphone before it reaches the radio transmitter. Not only does the speech processor reduce noise from the transmitted audio signal but it also enhances the speech waveform at the receiver. This ensures that received speech is much easier to understand. The A628 is of particular importance to operators of HF radio systems which are typically noisier than other RF links. A further feature of the A628 is its ability to adapt to the acoustic environment, providing the precise amount of processing to ensure recognisable speech.

The speech processor unit fits between the operator's microphone and the radio system.

Contractor

Ultra Electronics Limited, Sonar and Communication Systems.

VERIFIED

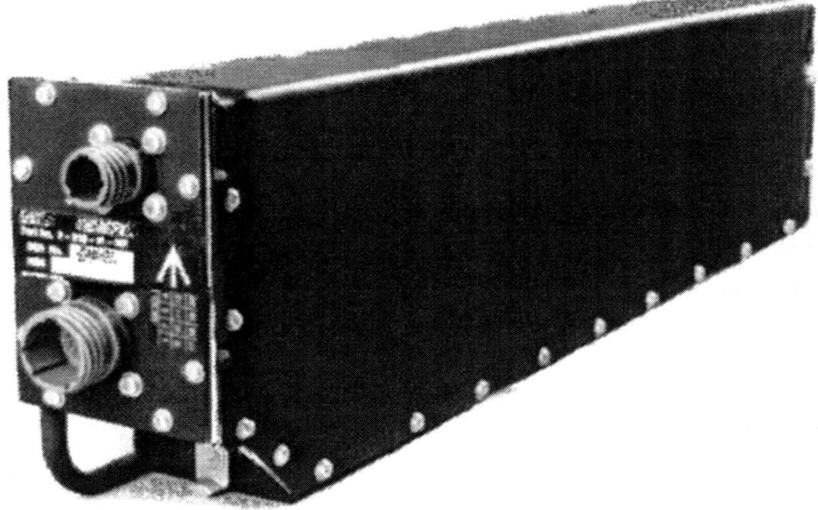

Ultra Electronics' A628 Speech Processor ***1997***/0002117

T618 Link 11 Data Terminal Set

The T618 Data Terminal Set (DTS) is a small lightweight unit suitable for helicopter, fixed-wing and other applications. It provides all Link 11 modem and network control functions defined by MIL-STD-188-203-1A, STANAG 5511 and ADat P-11 in picket and network control modes.

As a modem, the DTS converts the Link 11 data into audio tones suitable for transmission over radio circuits. As a network controller, it provides error corrections in USB, LSB and Diversity modes and roll call management functions in the net.

The T618 includes features which are of great importance to Link 11 operators. It provides an outlet to display the performance of every station in the network in real time; the link monitor displays the received signal quality on both upper and lower sidebands. The percentage of time each picket responds to interrogation and the analysis of the data being received is also displayed for each picket unit.

For the system maintainer, the T618 contains a BIT which isolates malfunctions to card level. In addition, various loopback modes provide signal paths which help to isolate difficulties in system configuration, including multiple stations. The combination of these controls and the link quality assessment provide a high visibility in the system operation.

The single-tone Link 11 system enhancement is available with the T618. Provided as a switchable option, it will enable the DTS to be backwards compatible with current conventional Link 11 modes.

Ultra Electronics Link 11 datalink processor (left) and Data Terminal Set (right) ***1998***/0011828

The DTS is also programmable for operations in combinations of TDMA network protocols and frequency hopping.

The T618 DTS can be upgraded easily to provide NATO Improved Link 11 (NILE) performance and the interim Link 11 improvements.

Specifications

Dimensions: ¼ ATR short
Weight: 3.6 kg
Power supply: 115 V AC, <30 W
Interfaces: serial data ATDS, radio interfaces as defined by MIL-STD-188-203-1A
Computer control: ARINC 429, MIL-STD-1553B, RS-232C as options
Temperature range: −40 to +55°C
Reliability: 7,000 h MTBF

Operational status

In serial production. Developed for, and in use on, the Merlin ASW helicopter.

Contractor

Ultra Electronics, Sonar and Communication Systems.

VERIFIED

T619 Link 11 datalink processor

The T619 datalink processor acts as a Link 11 processor. It uses up to date technology to provide a small lightweight unit suitable for helicopter, fixed-wing and other applications.

The datalink processor is compliant with STANAG 5511 and OPSEC 411. It provides database management, track correlation, gridlock, message assembly/disassembly, transmit and receive filtering, operator control and monitoring. All functions are automated where practical to reduce the operator workload to a minimum.

The unit consists of a single processor card and an interface card, both of which are VME-based. A spare slot is available to double processor throughput and memory capability or to provide additional functions such as to integrate a single card cryptographic device. For this purpose the datalink processor has the appropriate qualifications and is equipped with all the required front panel controls.

Specifications

Dimensions: ½ ATR short
Weight: <10 kg
Power supply: 115 V AC, 40 Hz, <40 W
Interface: MIL-STD-1553B
Temperature range: −40 to +55°C
Reliability: 6,000 h MTBF

Operational status

In serial production. In use on the Merlin ASW helicopter.

Contractor

Ultra Electronics, Sonar and Communication Systems.

VERIFIED

UNITED STATES OF AMERICA

AN/ARC-199 HF radio

The AlliedSignal AN/ARC-199 is a solid-state HF communications system providing 280,000 channels in the frequency band 2 to 30 MHz. It incorporates power management and low probability of intercept with selectable power output of 4, 40 or 150 W PEP. Twenty channels can be preselected for instantaneous recall and use. Automatic recognition of incoming messages by a selective addressing system is used.

The radio is microprocessor-controlled and contains a dual MIL-STD-1553B databus interface. Remote control can be provided through this interface, as well as through a dedicated keyboard/CRT display controller. The equipment can operate with both voice and data formats and can be used 'in clear' or with encryption devices. The CRT display is compatible with NVG. The antenna coupler is able to tune a variety of antennas throughout the frequency range, including long open wires, grounded wires, whips and shunt antennas.

Specifications

Weight: 14.5 kg

Operational status

In service on US Army helicopters.

Contractor

AlliedSignal Commercial Avionics Systems.

VERIFIED

ARINC Communications And Reporting System (ACARS)

The ARINC Communications and Reporting System (ACARS) is an advanced bidirectional datalink communication system operating in the VHF band. The datalink allows automatic transfer of operational data and maintenance reports between onboard systems and the user's ground-based computer. Additional

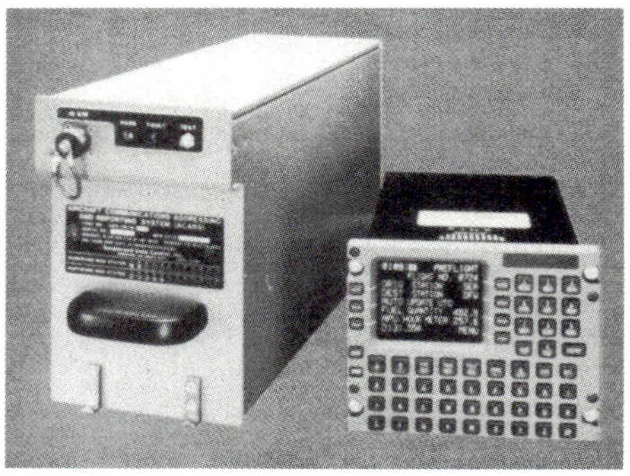

The ACARS with the management unit on the left and the control display unit on the right

messages may be generated utilising a menu-driven control display unit.

The Mk II datalink system provides a 50 per cent growth capacity in memory and processing. The system's software and database are loadable via a portable/airborne data loader without the need for removal from the aircraft. Downlink data contents, trigger selection and uplink display formatting can be modified.

Specifications

Dimensions:
(management unit) 194 × 320.5 × 123.2 mm
(control display unit) 114.3 × 158.8 × 146.1 mm
Weight:
(management unit) 5.45 kg
(control display unit) 3 kg
Power supply: 115 V AC, 400 Hz, single phase, 70 W

Operational status

In service with many Airbus and Boeing aircraft.

Contractor

AlliedSignal Commercial Avionics Systems.

VERIFIED

Datalink system

The datalink system is an ACARS (ARINC Communications Addressing and Reporting System) digital link used for data management between aircraft and the airline's base. Messages can be downlinked through the VHF communications transceiver to a central processing system which determines the registration and flight number of the aircraft and the nature and destination of the message, and then processes the message. Information can also be uplinked to the aircraft in the same manner. Messages are entered manually using the control unit keyboard. The datalink system can also interface with appropriate aircraft systems for automatic reporting of data, including the weather radar indicator, for extended display of information.

Optional items include a dedicated hard-copy printer, a management terminal for use by the cabin crew and a 20 page display memory. Provisions have been made for additional ARINC 429 input/output ports, checklist and multifunction display interface and HF modem.

The family of AlliedSignal datalink products includes the MUA-45A management unit, CNA-45A and CNA-45B control units, CDA-45B cabin management terminal, PTA-45A and PTA-45B data printers, AUA-45A data acquisition unit and HDA-45A hand-held terminal.

The MUA-45A management unit controls all datalink functions on the aircraft. It interfaces with and processes data between all onboard systems and the ground. Datalink messages are sent via the VHF communication transceiver using ARINC/SITA or company assigned frequencies. The MUA-45A tunes the desired frequency for data communication and also acts as a control head for the radio in the voice mode. The MUA-45A has a self-contained GMT clock and a large expandable memory.

The CNA-45A and CNA-45B control units provide pilot interface to the management unit. The CNA-45A has a single line of 16 characters. However, when the datalink is interfaced with a digital weather radar indicator the indicator's CRT offers a full-page colour display of messages, including uplinked weather maps and turbulence plots, rather than just the co-ordinates.

The CNA-45B features a 4 in diagonal display which accommodates 16 lines of 22 characters each, providing a full page of data on a dedicated screen. Both control units utilise an easy-to-read menu entry system, with a full alphanumeric keyboard with positive action keys. They have full-function capability and free-text formatting for unlimited message transmission.

The CDA-45B cabin management terminal enables flight attendants to send and receive messages directly, without interfering with the flight crew. It offers a reliable, economical and easy-to-use method of record keeping and communications concerning cabin activities.

The PTA-45B data printer provides printed copy of uplinked messages, flight plans and graphics on receipt of a hard-copy request from the control unit. The data printer can print any datalink message displayed on the radar indicator. The PTA-45B is an ARINC 740 multiport printer.

The AUA-45A data acquisition unit provides the interface to gather information from any aircraft analogue or digital sensor and processes it into ACARS format for downlinking. Information can be provided automatically or upon request from either the pilot or the ground. The AUA-45A also accepts uplinks from the datalink for various user-defined functions including uplink requests for data, parameter changes or format changes for downlinks.

The HDA-45A hand-held terminal provides a point-of-sale terminal for duty free and audio/video sales. It provides currency conversions, issues receipts, records sales and delivers telex message capability to the passenger's seat. The magnetic strip reader enables credit card transactions and retrieval of coded boarding pass information.

Operational status

In production. The system is widely used.

Contractor

AlliedSignal Commercial Avionics Systems.

VERIFIED

The MUA-45 datalink being set prior to a Delta Air Lines flight

Units of the AlliedSignal MUA-45 datalink set include (left to right) the CNA-45B control unit, AUA-45A data acquisition unit, CDA-45B cabin management terminal, HDA-45A hand-held terminal, MUA-45A management unit and PTA-45B data printer

HF datalink

The HF datalink service is an extension of the VHF Aircraft Communications Addressing and Reporting System (ACARS). It provides an aeronautical data communication link beyond the line of sight limitations of VHF, with the potential for worldwide coverage.

The avionics system consists of an airborne HF Data Unit (HFDU) incorporating an HF modem and a datalink processor which implements the air-ground link and network access protocols. The system makes use of the aircraft's existing HF radio and antenna. Data transmission rates are comparable to those of low data rate Satcom.

The HFDU provides interface between the ACARS MU and HF and VHF radios. In VHF mode, it is transparent to the MU and VHF radios. In the HF mode, there is no operator involvement. The HFDU performs automatic search and selection of HF frequency, controls tuning and keying of the HF radio and employs the HF modem to send or receive ACARS messages. The 4 MCU enclosure provides HF modem and airborne datalink processor, power supply plus backplane and two spare slots for enhancements such as ATN.

Operational status

The HF datalink has been certified on an American Airlines Boeing 767-300 aircraft.

Contractor

AlliedSignal Commercial Avionics Systems.

VERIFIED

KHF-950 HF radio

The KHF-950 has coverage up to 29.999 MHz and offers 280,000 frequencies at 100 Hz spacing. The system operates in USB, LSB and AM modes. Transmitter output in each SSB mode is 150 W PEP and 35 W average over the full frequency range.

The KHF-950 employs synthesised frequency generation techniques and uses microprocessor control for easy in-flight operation. It possesses a non-volatile memory which allows preselection storage for up to 99 channels and their appropriate modes but, in addition, allows the user to tune manually to any other frequency within the covered bandwidth without disturbance to the stored presets.

Operation may be either simplex, for normal air traffic or similar communication, or semi-duplex which permits patch-through into public utility telephone circuits. Provision is also made for a SELCAL facility and the dedicated circuits enable continuous SELCAL monitoring to be maintained without having to select the AM mode.

A feature of the KHF-950 is its automatic antenna tuning capability, an operation carried out by simply keying the microphone. AlliedSignal claims that the system will operate satisfactorily on antennas only 10 ft long. It will also tune to fixed-rod aerials and towel-rail antennas used on helicopters. AlliedSignal has developed equipment to facilitate operation from shunt and notch antenna systems.

The KHF-950 system comprises the KCU-951 all-digital remote controller, the KAC-952 power amplifier/antenna coupler and the KTR-953 transmitter/receiver. The remote controller (or frequency selector) is designed for panel mounting and presents channel, frequency and mode selection data on a self-dimming gas discharge numeric display.

An optional controller is the KFS-954 which is also panel-mounted and measures only 14.5 cm^2. This unit contains storage for all 176 ITU maritime radiotelephone channels plus additional preselected simplex air traffic or conventional airborne communication channels. By preprogramming the ITU channels, it is possible for the operator to call any radiotelephone station without having to select the separate transmit and receive channels manually. The operator merely selects the radiotelephone mode and the required channel. The KHF-950 also interfaces with teletype and facsimile systems and a dual installation equipment allows dual-frequency reception from a single antenna.

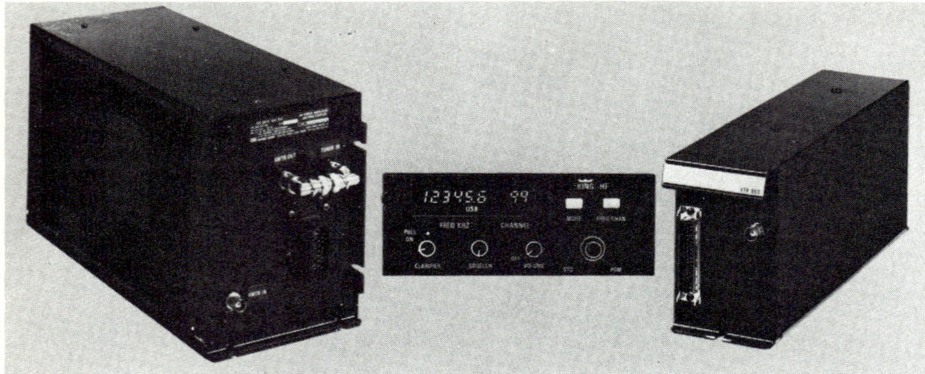

The AlliedSignal KHF-950 radio in dual configuration

Specifications
Weight: 9.16 kg

Operational status
In production and in service. The system has been widely adopted and has been installed in aircraft such as Canadair Challenger, Gulfstream G III, Lear 55, Citation III and Falcon 50. The KHF-950 has also been selected by the US Army for helicopters.

Contractor
AlliedSignal Commercial Avionics Systems.

VERIFIED

KHF-970 HF radio

The KHF-970 is a derivative of the KHF-950 HF/SSB airborne radio which has been adapted specifically for military use. Selection of the system for nap of the earth helicopter communications, an application for which HF is now recognised as being superior to VHF, was announced by the US Army in 1982. A surface vehicle version for military use has been designated KVR-980.

The KHF-970 covers the HF band from 2 to 29.999 MHz, providing 280,000 channels at 100 Hz increments. Transmitted output power is 150 W. Preset channel selection from a non-volatile memory is also provided but, unlike the KHF-950, all channels and frequencies are selected on a keyboard and displayed on a CRT.

Other features include scanning of preset channels and provision for multiple selective addressing, frequency link analysis and for automated communications operating instructions.

Operational status
In production.

Contractor
AlliedSignal Commercial Avionics Systems.

VERIFIED

KHF-990 HF radio

The KHF-990 HF radio is a helicopter system which draws on technology used in the company's airborne KHF-950 and marine KMC-95 systems. It provides 280,000 channels in the 2 to 30 MHz HF band at frequency increments of 100 Hz. Modulation is in SSB mode and transmitted output power is 150 W PEP.

The KHF-990 has been optimised for helicopter operation. It uses the miniature KFS-594 controller, a KAC-992 combined antenna coupler/probe antenna and a remotely located KTR-993 receiver/exciter/power amplifier. This combination provides a fully capable yet lightweight system.

The KFS-594 controller provides access to 176 permanently programmed ITU marine radiotelephone channels and to 19 programmable channels which may be selected or retuned by the pilot. The KAC-992 is an automatic, digital antenna coupler which is self-contained in the end of a probe antenna system. It may be mounted externally or internally with only the probe portion of the antenna protruding from the aircraft.

The KTR-993 receiver/exciter/power amplifier can be mounted in any convenient location within the helicopter with no restrictions on proximity to the other two units. It meets the TSO requirements for explosion-proof, drip-proof and salt-spray categories.

Specifications
Weight: 9.9 kg

Operational status
In production.

Contractor
AlliedSignal Commercial Avionics Systems.

VERIFIED

KMA 24 and KMA 24H audio control systems

The KMA 24 and KMA 24H are compact, lightweight systems for the integrated control of a number of radio communication and navigation systems. Designed principally for the general aviation sector, they are single box units for panel mounting.

The KMA 24 can control up to three transmitter/receivers and six receivers, including an internal marker beacon receiver for which it contains an automatically dimmed three light presentation. In the 24H version, the internal marker beacon facility is replaced by a five station hot microphone intercom and associated volume control. The 24H's marker switches control audio from an external marker beacon receiver. This system also features switching which gives the pilot's microphone priority over that of the co-pilot.

Both units provide transmitter/receiver and receiver outputs to speakers, headphones or both. A separate headphone isolation amplifier maintains constant noise-free volume levels even when several receivers are keyed simultaneously. When a microphone is keyed, all receivers are automatically muted to eliminate feedback. Both systems are offered in a choice of eight configurations with a number of different optional facilities.

Specifications
Dimensions:
(KMA 24) 172 mm length behind panel
(KMA 24H) 173 × 33 × 159 mm
Weight: 0.77 kg

Operational status
In production and in service.

Contractor
AlliedSignal Commercial Avionics Systems.

VERIFIED

KTR 908 VHF transceiver

The KTR 908 is a remote-mounted airborne VHF communications transceiver with a standard frequency range of 118 to 135.975 MHz and channel spacing of 25 kHz. An optional extension to 151.975 MHz is available. Power output is 20 W from a 28 V DC power supply and storage of active and standby frequencies is provided. It is operated from a cockpit-mounted KFS598 controller which requires 57 mm panel space.

Specifications
Dimensions:
(transceiver) 147 × 45 × 299 mm
Weight:
(transceiver) 1.6 kg

Operational status
In operational service with the US Army and National Guard.

Contractor
AlliedSignal Commercial Avionics Systems.

VERIFIED

KTR 909 UHF transceiver

The all-solid-state KTR 909 transceiver operates in the 225 to 399.975 MHz range in 25 kHz increments and is capable of 10 W of transmitter power. Weighing only 2.09 kg, the compact system includes the KTR 909 remote-mounted transceiver and the Gold Crown III KFS 599A control head. The KFS 599A is offered in two versions: with standard gas discharge display or with ANVIS NVG-compatible display.

The KFS 599A can be tuned by dialling in the desired operating frequency or by selecting any of the 20 user-programmable channels.

The KTR 909 is capable of operating with tandem KFS 599A control heads designated as master and slave, which is ideally suited for training applications. Additionally, one version of the KTR 909 can be tuned via compatible radio management systems, such as the AlliedSignal FMS 555 or FMS systems with frequency management.

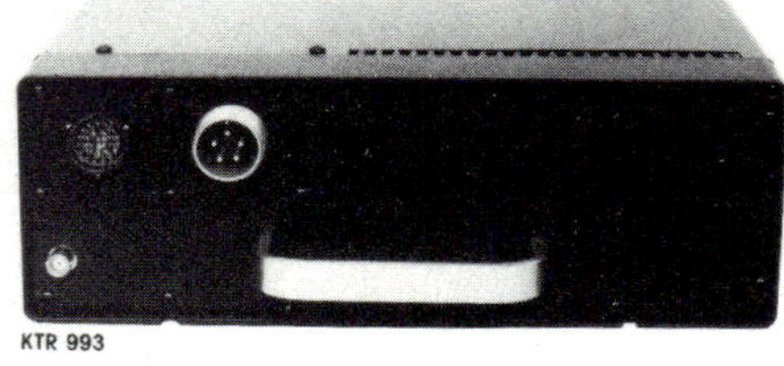

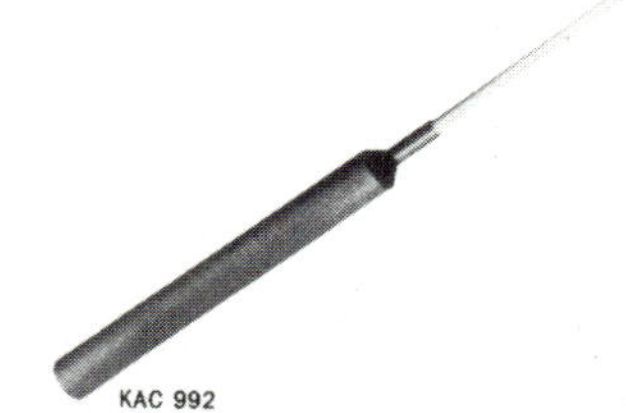

The AlliedSignal KHF-990 radio for helicopters

Dual-monitoring capability allows the KTR 909 to monitor either the main receiver, the Guard receiver or both simultaneously.

The KTR 909 offers 1,000 Hz tone modulation used in DF operations. In addition, it provides an ADF mode which allows the unit to perform the tuning function for remote ADF systems. This function, when interfaced with peripheral DF equipment, permits an ADF indicator to be used as a bearing indicator for DF operations.

Operational status
In service.

Contractor
AlliedSignal Commercial Avionics Systems.

VERIFIED

KY 96A and KY 97A VHF communications transceivers

The KY 96A and KY 97A VHF communications transceivers are identical in all respects except that the KY 96A operates at 28 V, while the KY 97A operates at 14 V. Frequency coverage is from 118 to 136.975 MHz at 25 kHz spacing.

Both the active and a standby frequency are displayed on the illuminated liquid crystal display and the set is switched between the two by a single button push. Up to nine channels can be programmed into the non-volatile memory. The transceivers feature audio levelling, so that weak signals are automatically amplified and strong signals are muted, and are equipped with an audio amplifier to drive a speaker for those installations not equipped with an audio panel.

Specifications
Dimensions: 33 × 158.8 × 266.7 mm
Weight: 1.32 kg
Power output: 5 W min
Frequency: 118-136.975 MHz
Temperature range: −20 to +55°C

Contractor
AlliedSignal Commercial Avionics Systems.

VERIFIED

The KY 96A VHF communications transceiver

KY 196A and KY 197A VHF transceivers

During 1987, AlliedSignal updated the KY 196 and 197 transceiver with the addition of some new features, designating the new transceivers KY 196A and KY 197A.

The KY 196A is a compact lightweight panel-mounted transmitter/receiver particularly suitable for light aircraft. It covers the VHF band from 118 to 136.975 MHz in which range 760 channels are provided; channels are selectable at increments of 25 or 50 kHz. One of the features of the update is the expansion of the frequency coverage by 1 MHz at the top end of the range.

A principal feature of the system is that a second frequency, in addition to the one in use, may be stored for immediate selection. Both the operating and standby frequencies are presented on a self-dimming gas discharge display in a window on the front panel. When the standby channel is selected the former operating channel is entered into the standby store. Non-volatile storage, provided by an electrically alterable read-only memory chip, ensures both frequencies remain stored when the power supply is off or disconnected. No separate memory power supply is required.

Solid-state construction is employed throughout. The system is microprocessor-controlled and digital synthesis techniques are used for frequency generation. A MOSFET RF amplifier and mixer stage is used to provide clear signal reception.

Other improvements in the upgrade include a bigger selection knob, lighted push-buttons, pilot programmable lighting and dimming levels and a facility which detects a stuck microphone and stops transmission after 2 minutes.

Specifications
Dimensions: 159 × 33 × 267 mm
Weight: 1.45 kg
Power supply:
(KY 196A) 28 V DC
(KY 197A) 14 V DC
Power output:
(KY 196A) 16 W
(KY 197A) 10 W
Temperature range: −20 to +55°C

Operational status
In production and in service.

Contractor
AlliedSignal Commercial Avionics Systems.

VERIFIED

RMS 555 radio management system

The RMS 555 offers high-technology features and performance in a compact RMU 556 control/display unit, driven by a ¼ ATR dwarf KDA 557 data adaptor. Utilising a 3.6 in diagonal CRT display, the RMU 556 provides a multitude of different pages, including the normal active frequency, memory and diagnostic pages.

Line item push-buttons provide quick access to each of the frequencies to be selected. Positive detent concentric knobs are used to provide secure input of frequency, channel or codes.

In a dual RMS installation, the pilot and co-pilot can each tune all radios in the system, including the cross-side radios. Each RMS is capable of handling 17 different pieces of equipment via an ARINC 429 databus, including three VHF comms, dual navs, dual Tacans, dual ADFs, dual DMEs, dual ATCRBs or Mode S transponders, dual MLS and TCAS.

Standby frequencies for comm, nav and ADF allow for flip-flop tuning. The memory capability of the RMS 555 system also allows the pilot to store up to 20 pilot-programmable frequencies for each VHF comm, 10 for each ADF and 10 for each Tacan.

Specifications
Dimensions:
(data adaptor) 99 × 85.9 × 231 mm
(control/display unit) 152.4 × 62.6 × 231 mm
(configuration module) 53.3 × 40.5 × 38.6 mm
Weight:
(data adaptor) 1.87 kg
(control/display unit) 1.46 kg
(configuration module) 0.046 kg
Power supply: 28 V DC, 1.25 A

Contractor
AlliedSignal Commercial Avionics Systems.

VERIFIED

Satellite Data Communications System (SDCS)

The Satellite Data Communications System (SDCS) uses the INMARSAT network to send and receive information worldwide. Interfaced to AFIS, it provides worldwide communication capabilities.

It provides two-way unrestricted message forwarding to another SDCS-equipped aircraft, the Global Data Center, a fax machine, an auto answer terminal and other service providers such as BASEOPS International, Air Routing International, Jeppesen Dataplan, Universal Weather and Aviation and MEDLINK. It also enables provision of worldwide weather information and flight planning/flight plan filing.

Specifications
Weight:
(HPA/LNA) 2.04 kg
(antenna) 0.4536 kg
(SCU) 2.72 kg
Power supply: 27.5 V DC
(transmit) 4.5 A
(receive) 0.5 A

Contractor
AlliedSignal Commercial Avionics Systems.

VERIFIED

VCS 40 VHF communication system

The VCS 40 VHF communication system is an all-digital transceiver employing microprocessor-controlled circuitry. It incorporates a full complement of AlliedSignal Series III performance features including 20 W solid-state transmitter, continuous transmit capability at reduced power, white-on-black dichroic liquid crystal frequency display, 760-channel operation at 25 kHz spacing over 118 to 136.975 MHz, optional 1,360 channel extension up to 151.975 MHz, built-in SELCAL and ACARS capability and automatic self-test and diagnostics.

The VCS 40 consists of the VC 401B digital transceiver and the panel-mounted CD 402B control/display unit.

Specifications
Dimensions:
(transceiver) 104 × 101.6 × 333.6 mm
(control/display unit) 63.5 × 80 × 63.5 mm
Weight:
(transceiver) 2.78 kg
(control/display unit) 0.27 kg
Power supply: 18-33 V DC
(transmit) 4.5 A
(receive) 0.5 A
Altitude: up to 55,000 ft

Contractor
AlliedSignal Commercial Avionics Systems.

VERIFIED

GLOBALink/CNS™

ARINC and Magellan jointly market GLOBALink/CNS™, a low-cost communications, navigation, and surveillance service designed for regional and commuter airlines and business and general aviation.

The system consists of:

(a) ARINC air/ground datalink network
(b) Global Positioning System (GPS) augmentation
(c) Magellan CNS-12™, the Magellan-integrated avionics unit.

Over 4,600 aircraft now use the ARINC Aircraft Communications Addressing and Reporting System (ACARS). ACARS is used for Air Traffic Control (ATC), Airline Operational Control (AOC), and Airline Administrative Communications (AAC). With an end-to-end operational availability consistently higher than 99.99 per cent, ACARS provides data communications to and from the addressed aircraft. Communications are automatically and accurately transmitted between flight crews and ground personnel. ACARS provides full coverage at over 300 airports, including every major US airport.

ARINC plans to substantially increase the number of ground stations in the network, further increasing the coverage available at low altitudes.

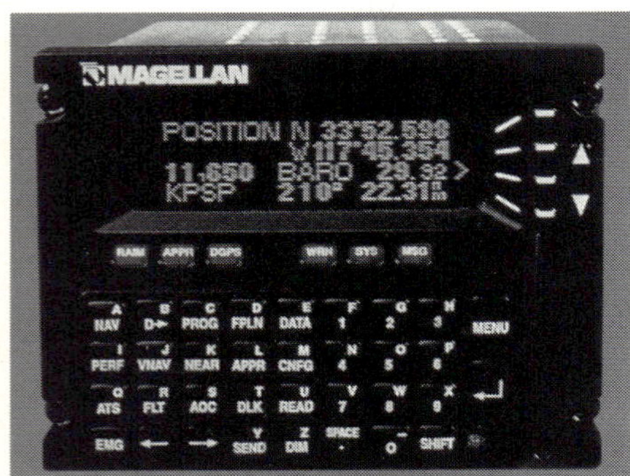

Front panel of the Magellan CNS-12, part of the GLOBALink/CNS™ system ***1998***/0011856

The present ACARS ground station network provides continuous en route coverage to all aircraft operating above 20,000 ft anywhere over the continental United States.

Almost full coverage is also provided in Canada, Jamaica, Central America, and Mexico. In addition, stations are also located in Bermuda, San Juan, Hawaii, Guam, and Alaska.

Satellite navigation and GPS now allow aircraft to navigate anywhere in the world with a single system. Aircraft equipped with the CNS-12 can access any ARINC Local Area Differential Ground Station (LADGS). LADGS provides the augmentation signals that contain integrity and differential corrections needed for precision approaches. Highly reliable integrity monitoring systems ensure full protection for instrument flight. The unit is also upgradable for Wide-Area Augmentation System (WAAS) operation.

The CNS-12 combines the components necessary to transmit and receive data – VHF radio, ACARS Management Unit (MU), and the Control and Display Unit (CDU) – with a GPS receiver, and is suitable for space- and weight-constrained aircraft and helicopters.

Main characteristics include:

(a) two-way VHF DataLink (VDL) with a frequency range of 118 to 136.975 MHz

(b) dedicated dGPS navigation datalink receiver within the VOR frequency range

(c) a total of 12 channels, of which 2 are for WAAS.

Specifications

Dimensions: 146 × 114.3 × 215
Weight: approx 4.55 kg
Display: Four line, 20 characters per line

Contractors

ARINC Inc.
Magellan Systems.

NEW ENTRY

5163-1 SELCAL decoder

The purpose of the SELective CALling (SELCAL) system is to permit exclusive calling of individual aircraft over normal radio channels that link the ground station to that aircraft. The system operates with HF and VHF ground-to-air transmitters and receivers and does not interfere with the normal operation of communications except when the SELCAL is performing its calling function.

Each SELCAL-equipped aircraft is assigned a four letter identifier which is used when the ground station wishes to contact it. The SELCAL decoder is designed to respond only to the identifier for which it is set. When the identifier is received by the aircraft, the decoder actuates a signal indicator in the form of a lamp, bell, chime or any combination of these. Typically, the signal will be annunciated on an audio control panel microphone select switch.

The 5163-1 SELCAL decoder is a rack-mounted, 16 tone decoder designed for the ICAO and ARINC standard system. It may be installed in non-pressurised and non-temperature controlled locations on aircraft up to altitudes of 55,000 ft. The unit is housed in a 1 MCU ARINC 600 package.

Operational status

In service.

Contractor

AVTECH Corporation.

VERIFIED

7522 series VHF comm radio tuning panel

The AVTECH 7522 series VHF comm radio tuning panel updates ARINC 500 (analogue) and ARINC 700 (digital) series radios to provide the new 8.33 kHz frequency spacing standard, required in order to meet certain European ATC requirements after 1 January 1999. The new panel has two connectors: a 55-pin connector used exclusively for ARINC 500 series radios, and a 24-pin connector for ARINC 700 series radios. The 500 series radios are tuned with a modified 2 × 5 coding. The 700 series radios are tuned via an ARINC 429 databus. This approach allows airlines with mixed fleets (500 and 700 series radios) to stock only one type of tuning panel. The current range of 760 channels at 25 kHz spacing is increased to 2,280 channels at 8.33 kHz spacing.

Operational status

TSO approved September 1997. Selected by a number of airlines to meet new European standards.

FAA STC approval was granted in January 1998 for Rockwell Collins VHF-700B, VHF-900B (digital) and 618M-5 (analogue) radios with AVTECH's 7522-1-2 control panels.

Contractor

AVTECH Corporation.

UPDATED

AVTECH 5060-1 audio selector panels for the de Havilland DHC-8

AvFax airborne facsimile machine

AvFax is a commercial transport-qualified facsimile terminal. It has been designed and tested to meet all the applicable DO-160C and FAR Part 25.853 requirements for cabin equipment. Categories include shock, vibration, temperature, altitude, EMC, flammability and toxicity.

Key features of the AvFax include the availability of multiple system interfaces for GMIS, Honeywell/Racal, Collins and GTE equipment; and support for fax transmission or reception via Satcom or NATS at 2.4, 4.8 and 9.6 kbits/s.

Operational status

In production for several international carriers.

Contractor

AVTECH Corporation.

VERIFIED

The AvFax airborne facsimile machine

Audio selector panels

The AVTECH audio selector panel provides complete selection and volume control of audio communications for cockpit crews of transport aircraft. These panels are custom-designed and manufactured specifically for each aircraft application, whether supplied as standard equipment by the airframe manufacturer or specified as buyer-furnished equipment by the purchasing airline.

The panels are designed to minimise cross-talk and spurious noise, while maximising fidelity and reliability.

Operational status

In production and in service with Boeing 727, 737, 757, 767, MD-80 and MD-90 aircraft and a number of other aircraft types, including the Beech T34C, Bell 204A, Canadair CL-41, Cessna Citation and Citation Jet, de Havilland DHC-8, Embraer EMB-312 and Learjet 20, 30 and 50 Series.

Contractor

AVTECH Corporation.

VERIFIED

Cockpit Voice Recorder (CVR) audio mixer

The Cockpit Voice Recorder (CVR) audio mixer is designed to comply with FAA regulations mandating cockpit voice recorders for all multi-engined turbine-powered aircraft that require two crew members and seat six or more passengers. The CVR audio mixer sums and routes audio signals from microphones, headphones and speakers to the cockpit voice recorder. It also provides hot mic biasing, adjustable channels for balancing audio levels and paired pins to make installation easy and economical. Both single- and dual-station units are available. The systems are qualified to TSO C50c requirements.

Operational status
In service in various business aircraft.

Contractor
AVTECH Corporation.

VERIFIED

Digitally Controlled Audio System (DCAS)

At the heart of each DCAS is the Digital Signal Processor (DSP). DSP circuitry gives flight crews crisp drift-free audio communications with few electronic parts and dramatically increased reliability over analogue systems, extending system flexibility through software add-on features.

AVTECH DSP systems consist of a Remote Electronics Unit (REU), plus an Audio Control Panel (ACP) for each user. In operation, control switch selections are made at the ACP, then multiplexed and sent to the REU. All analogue audio signals are also sent to the REU where they are filtered and converted to digital signals. The resulting digital audio is selected, amplified, filtered and summed in the DSP circuit according to control selections. After processing, the digital audio signal is converted back to analogue for distribution to the user. All processing is done at microprocessor speeds resulting in real-time audio communication with crisp digitally processed quality. Advanced features include a digital control bus between each flight deck ACP, BITE to monitor and report digitally on system operational integrity and advanced data reporting via an ARINC 629 interface to other aircraft systems. Jack panels, headsets, microphones and speakers provide audio signal input/output, completing the system.

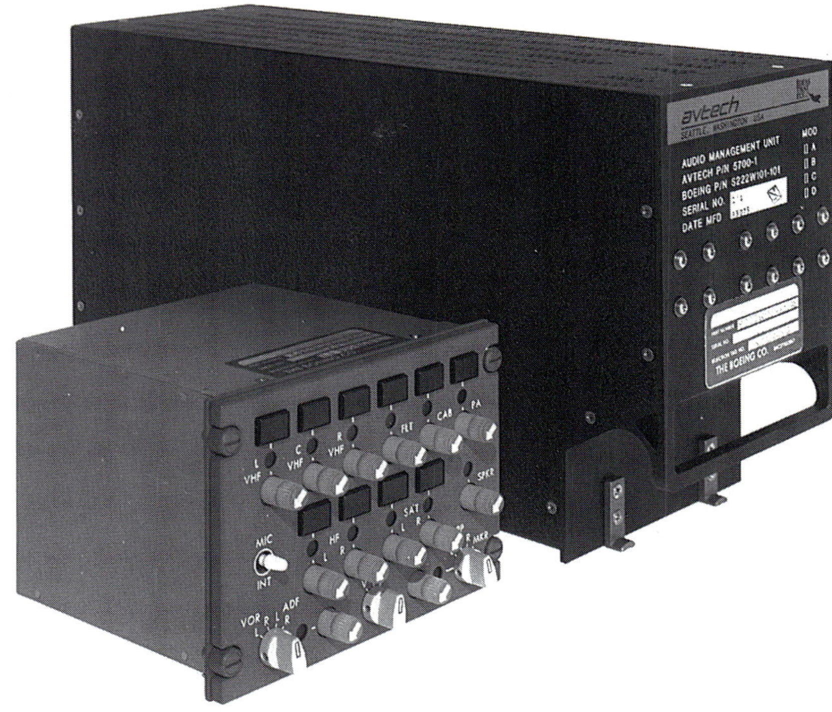

The digitally controlled audio system for the Boeing 777 **1995**

Operational status
In service, widely used.

Contractor
AVTECH Corporation.

UPDATED

RCC-200-SPM airborne command control receiver

The RCC-200-SPM airborne command control receiver is a dual-conversion superheterodyne UHF receiver capable of operating in L/S or TV bands with a 100 MHz tuning range. It is designed for operation in aerospace environments where size, weight and power consumption are critical. The RCC-200-SPM incorporates advanced design features which rely heavily on the use of integrated circuit technology to enhance the overall performance.

Performance specifications meet the requirements of the majority of airborne telemetry and command control applications. Additional features can be incorporated to meet specific customer requirements.

Specifications
Dimensions: 152.4 × 82.6 × 52.5 mm
Weight: 1.19 kg
Power supply: 24-36 V DC
Temperature range: −40 to +70°C

Contractor
AYDIN TELEMETRY.

UPDATED

RCC-210 series command control receivers

The RCC-210 series UHF command control receivers are solid-state FM receivers designed for operation in aerospace environments where size, weight and power consumption are critical. Models are available with input frequency ranges of 1,400 to 1,550 MHz, 1,700 to 1,850 MHz and 2,200 to 2,400 MHz.

Specifications
Weight: 0.48 kg
Power supply: 28 V DC, 220 mA

Contractor
AYDIN TELEMETRY.

UPDATED

RCC-500 command control receiver

The RCC-500 command control receiver is a multipurpose FM receiver specifically designed to meet the stringent electrical, environmental and reliability requirements of missile terminal flight, while providing for a variety of optional configurations determined by the addition of external units.

When combined with an AYDIN TELEMETRY Series 294-bit synchroniser, the RCC-500 receiver may be used to recover data, or for command and control when integrated with the TDC-100 Series command tone decoder.

The RCC-500 receiver (with external decoder) meets the requirements of Range Commanders Council document 3134-80. An ultra-high Q preselector and high IF allow the use of a single heterodyne receiver front end which provides a more reliable and cost-effective approach over others. The receiver front end has sufficient dynamic range to meet the susceptibility requirements of RCC document 313-80 and MIL-STD-461 and -462.

Contractor
AYDIN TELEMETRY.

UPDATED

ST260 series UHF transmitters

ST260 series UHF transmitters are designed for video and telemetry datalink operation in tactical weapons such as standoff attack missiles and laser-guided bombs.

The ST260 is completely solid state. It incorporates a mode selectable dual-power output of 40 W in high-power mode and 2 W in low power with other power outputs available as an option. The ST260 contains a power line regulator, ensuring uniform performance over the entire allowable input voltage range and compliance with power line conducted susceptibility and interference requirements. An output circulator internal to the unit allows operation into any load impedance, including short and open circuits.

The ST260 provides an output frequency stability of ±0.003 per cent, frequency response of 10 Hz to 6 MHz ±1.5 dB and deviation sensitivity of ±6 MHz.

Synthesised frequency selection is available in 1 MHz steps over a 100 MHz band.

Specifications
Volume: 869 cm^3
Weight: 1.45 kg

Contractor
AYDIN TELEMETRY.

UPDATED

ST-800S/L series wideband microwave transmitter

AYDIN TELEMETRY's ST-800S/L series microwave transmitters are designed for highly reliable operation in the severe environmental flight conditions of missiles, space vehicles and aircraft, where size and weight efficiency are critical. These solid-state, crystal stabilised, true FM telemetry transmitters can accommodate various modulation formats such as standard analogue pre-emphasised video, TTL (Transistor Transistor Logic), differential TTL and fully isolated differential TTL (opto-coupled).

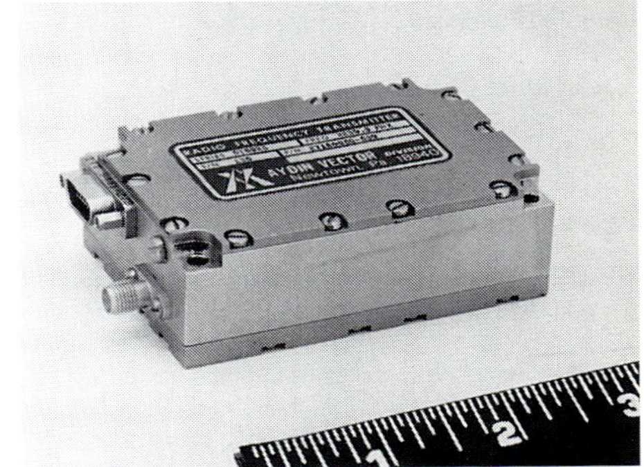

ST-800S/L series wideband microwave transmitter
1998/0011827

Available in 2, 5 and 10 W minimum power output, the ST-800S/L measures 50.8 × 76.2 × 20.3 mm, excluding connectors and weighs 0.2 kg. It operates in the frequency range of 2,200 to 2,400 MHz and 1,435 to 1,540 MHz.

Contractor

AYDIN TELEMETRY.

NEW ENTRY

ST4160S S-band airborne transmitter

The ST4160S forward error-corrected 175 W dual-frequency PSK S-band (NATO E/F-band) airborne transmitter is rack mountable and operates on raw 28 V aircraft power. Convolutional encoding, high output of 160 W minimum and RS-422 bus compatibility are prime features of the ST4160S. Modulation is PCM/PM linear phase and data rate is 700 kbits/s convolutional encoded or 1,400 kbits/s non-convolutional encoded.

A typical application of the airborne ST4160S transmitter is preparing and relaying digitised SAR (Synthetic Aperature Radar) video imagery and annotated data via a microwave link to a ground station. The electronics in the data network use scrambling and convolution encoding to enhance the transmission and reception process significantly.

Principal components of the ST4160S transmitter are the AYDIN TELEMETRY T-700 transmitter and four PA-440 50 W power amplifiers plus additional electronics.

Specifications

Dimensions: 482.6 × 304.8 × 482.6 mm
Weight: 38.56 kg
Temperature range: −10 to +50°C
Frequency stability: ±0.003 %

Contractor

AYDIN TELEMETRY.

UPDATED

T-300 series airborne UHF transmitter

The T-300 series is a subminiature solid-state crystal-stabilised UHF/FM transmitter capable of transmitting wideband telemetry and digital multiplex signals. It is designed for extremely reliable operation in the severe environmental flight conditions associated with missiles, space vehicles or aircraft.

The T-300 operates at 2,200 to 2,400 MHz, with a frequency stability of ±0.002 per cent. Power output is 5 W. The T-300 series meets IRIG-106-93 standards.

Specifications

Dimensions: 63.5 × 38.1 × 19 mm
Power supply: 28 V DC ±4 V
Temperature range: −20 to +70°C

Contractor

AYDIN TELEMETRY.

UPDATED

AIRLINK antenna system for satcoms

AIRLINK low- and high-gain antenna systems are designed for use with INMARSAT satellite communications.

The high-gain antenna system uses two conformal, electronically steered, phased-arrays in a side-mounted architecture. This configuration yields superior coverage with minimal aerodynamic drag penalties. The high-gain antenna system is fully approved for multichannel data, voice and data applications.

The low-gain antenna system consists of a single-blade antenna and is used for low-speed data applications. It is ideally suited as a back-up for the high-gain system.

Ball has recently introduced the AIRLINK Gateway Unit (AGU) which provides the digital signal processing necessary for operating with INMARSAT's circuit mode data channel. This channel provides users with a host of applications at the 9.6 kbits/s rate and can also provide secure satellite communications.

Specifications

Dimensions:
(antenna array) 407 × 813 × 9.5 mm
(beam-steering unit) 89 × 264 × 343 mm
(diplexer/low-noise amplifier) 51 × 198 × 282 mm
(high-power amplifier) 193 × 257 × 925 mm
(AIRLINK Gateway Unit) 7 MCU
Weight:
(antenna array) 7.1 kg
(beam-steering unit) 8.4 kg
(diplexer/low-noise amplifier) 3 kg
(high-power amplifier) 20 kg
(AIRLINK) Gateway Unit) 13.5 kg
Power supply: 115 V AC, 400 Hz, single phase
Frequency: 1,530-1,559 MHz, 1,626.5-1,660.5 MHz

Operational status

Selected by United Airlines and British Airways for Boeing 777 aircraft and by Scandinavian Airline System for Boeing 767-300s. Also selected by the United States government for VIP/SAM fleet.

The Ball AIRLINK antenna system showing the conformal array (above) and beam-steering unit (left), diplexer/low-noise amplifier (centre) and high-power amplifier (right)

Contractor

Ball Aerospace and Technologies Corp.

VERIFIED

Boeing phased-array antenna

The Boeing Company has developed an electronically steered high-performance, low-profile phased-array communication antenna that will help revolutionise mobile satellite communication by increasing the data flow by thousands of times over current capabilities. This will make possible high data-rate retrieval, in-flight entertainment and many other high-bandwidth applications that have been unavailable on mobile platforms hitherto.

For commercial carriers or military aircraft, the phased-array communication antenna offers the ability to provide operators with more information options.

In use, an antenna beam is directed to acquire a BSS satellite emitting signals from a stationary orbit above the Earth's equator. The antenna beam location is controlled electronically to acquire and then track the satellite of interest. Once the antenna beam locks on to the satellite, broadcast signals are brought aboard the aircraft where they are decoded by the receiver and distributed to operators (or passengers).

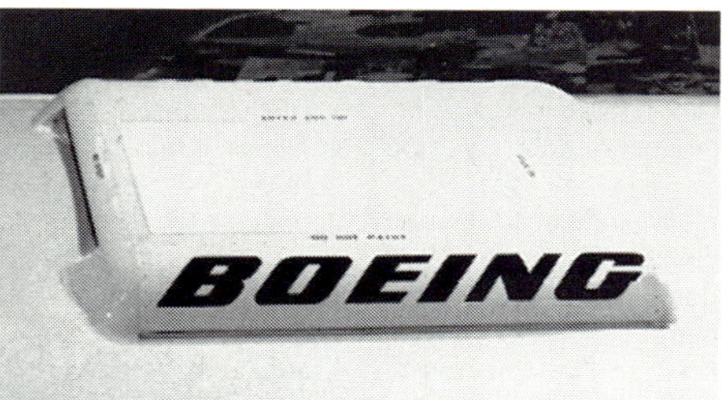

The Boeing phased-array antenna **1998**/0011826

The 1,500+ element antenna measures approximately 2 ft by 3 ft and is 1 in thick. Unlike conventional, mechanically steered antennas that are bulky and slow to switch between satellites, the Boeing phased-array antenna steers beams electronically, permitting instantaneous connections between satellites and mobile platforms.

In June 1996, Boeing flight-tested a prototype phased-array antenna. The tests demonstrated the antenna system's ability to automatically acquire and track broadcast service satellites and display video data on board the aircraft while in flight.

Also in June 1996, the antenna was installed on a Boeing C-135 US Air Force avionics testbed aircraft to support a series of Joint Warrior Interoperability Demonstration (JWID '96 and '97) exercises.

In November 1996, the antenna was installed on a private 757 business jet. With the antenna on board, the aircraft is able to receive live television as well as business data.

Operational status

In development.

Contractor

The Boeing Company.

UPDATED

C-1000 communications management controller

The C-1000 is a fully frequency-agile control unit which provides thumbwheel control of all available Flexcomm channels and provides for the storage of up to 30 preset channels for simplex or semi-duplex operation. Any of the channels may be changed by the operator. However, this capability may be disabled if the operator so desires. It also provides control of CTCSS tones in both receive or transmit. The C-1000 has edge lighting, using either 5 or 28 V. When used with a complete Flexcomm system it automatically selects the appropriate radiotelephone unit depending on the desired frequency.

The system also contains full discrete switches for use with external devices such as antennas and DTMF coders.

The EAROM memory chips remember programmed channel information indefinitely without external power.

Specifications

Dimensions: 146 × 76 × 191 mm
Weight: 1.2 kg

Operational status

In production and in service.

Contractor

Chelton Avionics Inc, Wulfsberg Electronics Division.

UPDATED

C-5000 communications management controller

Designed to operate with the existing Flexcomm, RT-9600 and RT-7200 radios and future products, the C-5000 communications management controller will operate multiple radios simultaneously and performs the same functions as the C-1000, with the addition of a number of new features.

The C-5000 includes channel identification via alphanumeric or frequency display, simultaneous control of up to three RT systems, built-in dual-tone multifrequency, built-in dual-independent microphone inputs which allow independent operation of two RTs, multiple control head installation capability, simulcast, relay and repeater modes, programme scan and a 350-channel memory. Each channel has individual transmit and receive frequencies and CTCSS selection and control.

The operator can monitor main and Guard receiver audio from all transceivers simultaneously and the C-5000 has independent volume control of the active or selected RT system and secondary RT systems. The communications management controller is easy to operate, with vacuum fluorescent display, which is easily readable in bright sunlight, and optional NVG-compatibility.

Future growth designs incorporate an interface for voice encryption and also allow for the installation of commercially available digital squelch protocols.

Operational status

Available.

Contractor

Chelton Avionics Inc, Wulfsberg Electronics Division.

UPDATED

Flexcomm communication system

Flexcomm, as the name implies, is a flexible communication system providing very wide frequency coverage capable of operating from the C-1000 or C-5000 controller. This provides for thumbwheel control of frequency as well as 30 programmable preset channels.

A complete Flexcomm system consists of the C-1000/5000 control, the RT-30 (30 to 50 MHz FM), RT-138 (138 to 174 MHz FM) and the RT-406F (406 to 512 MHz FM). Selection of frequency at the control chooses the appropriate radio unit. A single mount may be installed and the FM radio units may be interchanged. It is only necessary to connect the proper antenna to the radio unit. A total of 24,910 channels is available with all FM radiotelephone units.

Each FM set may carry an optional Guard receiver. In addition to straight simplex operation, semi-duplex (talk on one frequency, listen on another) may be used for operation with repeaters. All UHF international marine frequencies are covered.

Flexcomm II consists of the RT-5000 transceiver and the C-5000 communications controller.

Specifications

Weight: 6.81 kg
Frequency: 29.7-960 MHz
Channels: 688,640

Operational status

In production and in service. Widely used in helicopters.

Contractor

Chelton Avionics Inc, Wulfsberg Electronics Division.

UPDATED

RT-30 VHF/FM radio

The RT-30 FM transceiver uses a digital frequency synthesiser to provide FM communications over the frequency range 29.7 to 49.99 MHz. There are no band-spread limitations; the receiver can operate at one frequency extreme with the transmitter at the other.

Fully solid state, it also provides 32 subaudible CTCSS tones. An available system includes a single channel Guard receiver operating anywhere in the band. Separate audio inputs and outputs are provided for use with external CTCSS tones, tone bursts and DTMF encoders.

The RT-7200 VHF radio with the C-722 controller

Specifications

Dimensions: 111 × 266 × 127 mm
Weight: 3.4 kg
Power output: 10 W continuous
Frequency: 29.7-49.99 MHz
Temperature range: −40 to +60°C

Operational status

In production and in service.

Contractor

Chelton Avionics Inc, Wulfsberg Electronics Division.

UPDATED

RT-138F VHF/FM transceiver

The RT-138F uses a digital frequency synthesiser to provide FM communications over the frequency range 138 to 173.975 MHz. There are no band-spread limitations; the receiver can operate at one frequency extreme with the transmitter at the other. Of fully solid-state construction, it also provides 32 subaudible CTCSS tones. A single-channel Guard receiver is available which can operate anywhere in the band.

Separate audio inputs and outputs are provided for use with external CTCSS tones, tone bursts, DTMF encoders, voice scramblers, data and so on.

Specifications

Dimensions: 111 × 266 × 127 mm
Weight: 3.4 kg
Power output: 10 W continuous
Frequency: 138-173.975 MHz
Temperature range: −40 to +50°C

Operational status

In production and in service.

Contractor

Chelton Avionics Inc, Wulfsberg Electronics Division.

UPDATED

RT-406F UHF/FM transceiver

The RT-406F uses a digital frequency synthesiser to provide FM communications over the frequency band 406 to 512 MHz. There are no band-spread limitations; the receiver can operate at one frequency extreme and receive at the other.

Fully solid state, the system also provides 32 sub-audible CTCSS tones and a single-channel Guard receiver is available operating anywhere in the band.

Separate audio inputs and outputs are provided for use with the external CTCSS tones, tone bursts, DTMF encoders, voice scramblers, data and so on.

Specifications

Dimensions: 111 × 266 × 127 mm
Weight: 3.4 kg
Power output: 10 W continuous
Frequency: 406-512 MHz
Temperature range: −40 to +60°C

Operational status

In production and in service.

Contractor

Chelton Avionics Inc, Wulfsberg Electronics Division.

UPDATED

RT-7200 VHF/FM radio

The RT-7200 transmitter/receiver is an air-to-ground communications system suitable for a wide range of fixed-wing aircraft or helicopters. It provides FM operation over a choice of 7,200 channels within the band 138 to 174 MHz at either 25 or 50 kHz increments but can be tuned to the much lower incremental value of 5 kHz. Like other equipment in the range, the RT-7200 provides a semi-duplex facility with automatic push-to-talk tuning.

In simplex operation, frequency selection may be made on an individual dial-up basis using conventional thumbwheel controls; a digital readout on the C-722 remote controller confirms the selection. Alternatively, up to 15 channels are available on a programmable preselection basis. Power output is operator selectable at either 1 or 10 W.

The system has an optional two-channel built-in Guard receiver which conforms to the same general specification as the main equipment's receiver section. There is no frequency separation restriction if the optional Guard receiver is fitted.

Automatic signal-to-noise squelch, with manual override, is provided and a separate input is included for encoder, voice scrambler, data or other systems. Construction is all-solid-state and digital synthesis is employed for frequency generation.

Specifications

Dimensions: 127 × 320 × 127 mm
Weight: 4.21 kg
Power output: 1 or 10 W selectable
Frequency: 138-174 MHz
Channels: 7,200
Channel spacing: 25 or 50 kHz (5 kHz tunable)

Operational status

In production and in service.

Contractor

Chelton Avionics Inc, Wulfsberg Electronics Division.

UPDATED

RT-9600F VHF/FM radio

The RT-9600F is virtually identical to the RT-7200 system in its general technical specification. The major difference is in its frequency coverage and channel capacity. It covers the frequency range 150 to 174 MHz but provides a total of 9,600 channels in this band. Normal frequency spacing is 25 or 50 kHz but the RT-9600F (also known as the AN/ARC-513) is capable of tuning to a finer 2.5 kHz incremental spacing.

The C-962A controls are edge-lit, using 5 or 28 V DC supply for lighting. Unlike the Flexcomm units, the RT-7200 and the RT-9600F have an additional AM/IF detector to provide an output for a DF or ADF system.

Specifications

Dimensions: 127 × 320 × 132 mm
Weight: 4.21 kg
Frequency: 150-174 MHz
Channels: 9,600
Channel spacing: 25 or 50 kHz (2.5 kHz tunable)

Operational status

In production and in service.

Contractor

Chelton Avionics Inc, Wulfsberg Electronics Division.

UPDATED

RTA-44A VHF transceiver

The RTA-44A is designed to be a highly reliable VHF transceiver for sharp clear voice communications. Many of the RTA-44A's performance capabilities come from a single microprocessor which is utilised in the computer architecture of the transceiver for high firmware reliability, a minimum parts count and reduced power usage. The microprocessor also provides the desired flexibility for offset operation and power-on channel tuning.

The transmitter consists of linear amplifier stages connected by microstrip impedance matching sections. These eliminate conventional interstage coupling capacitors and inductors along with their inherent failures and misadjustments.

The receiver features a programmable RF attenuator controlled by an off-channel signal sensor which automatically varies the attenuation to maintain a maximum desired/undesired signal input to the preselector. This makes the rejection of signals possible, even when an onboard transmitter isolated by 30 dB is being operated.

The microprocessor reduces maintenance time by monitoring pertinent modes throughout the transceiver, allowing a thorough fault analysis and fault isolation routine to be conducted down to subassembly/module level. Failed subassemblies are identified by an alphanumeric code displayed on the HEX LED display.

Three dedicated buffers and an 8-bit analogue-to-digital converter are utilised in the computer architecture to monitor all events external to the microcomputer for BITE signal processing.

Specifications

Weight: 4.1 kg
Frequency: 118-137.975 MHz
Number of channels: 800 (25 kHz spacing)

Operational status

In production.

Contractor

Chelton Avionics Inc, Wulfsberg Electronics Division.

UPDATED

RTA-44D VHF Data Radio (VDR)

The RTA-44D VHF Data Radio (VDR) was part of AlliedSignals' new Quantum Line of Communications, Navigation, and Surveillance (CNS) Equipment. Avionics Yearbook does not know its status in this programme following sale of the Wulfsberg Electronics Division to Chelton Avionics, Inc. It meets all ARINC 716 Series specifications and is fully interchangeable with the RTA-44A VHF Comm and other ARINC 716 Series radios.

The VDR provides VHF voice and data communication between onboard aircraft systems, to other aircraft, and to ground-based systems. It operates as a standard double sideband AM analogue voice transceiver, and also as a data capable transceiver. Depending on the selected data mode, the unit performs transceiver, modem, and/or link layer functions.

In an Aviation VHF Packet Communications System environment, the VDR is either a simple transceiver with an analogue interface or a link layer bridge for the VHF subnetwork. In an ACARS environment, the unit is either a simple transceiver with an analogue interface to the ACARS management unit or a Minimum Shift Keying modem.

Features of the RTA-44D VDR include:

(a) Standard VHF COM (ARINC 716) functions; voice and analogue data processing accomplished by Digital Central Processing (DSP)
(b) VHF data radio (ARINC 750) functions: high-speed D8PSK modem; transmitter capable of D8PSK transmission; receiver capable of D8PSK reception
(c) Cooling meets ETOPS 180-min requirements

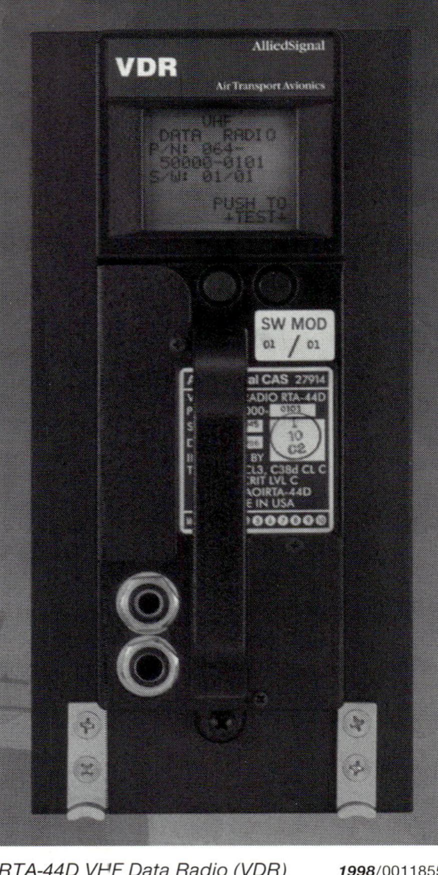

RTA-44D VHF Data Radio (VDR) **1998**/0011855

(d) Complies with DO-178B software requirements
(e) Meets ICAO Annex 10 requirements
(f) Meets HIRF requirements
(g) Meets DO-160C lightning protection requirements
(h) Meets 200 ms power interrupt transparency requirements

Specifications

Dimensions: 324.0 × 90.9 × 194.0 mm (L × W × H)
Weight: 5.0 kg
Power: 27.5 V DC
Form factor: 3 MCU per ARINC Specifications 600

Contractor

Chelton Avionics Inc, Wulfsberg Electronics Division.

NEW ENTRY

CDR-3100 series LF-HF receivers

The CDR-3100 family of receivers incorporates digital signal processing microchip technology and is designed for communications, surveillance and direction-finding tasks on board ships and aircraft.

Operation is menu driven and little or no operator training is required. The receivers have digitally tuned preselectors and 250 programmable channels. There are 51 selectable synthesised IF bandwidths from 100 Hz to 18 kHz, and the system constantly performs built-in-test checks. When a problem is detected, a message on the unit's display screen identifies the defective module.

The CDR-3100 series of LF-HF receivers comprises the following units:

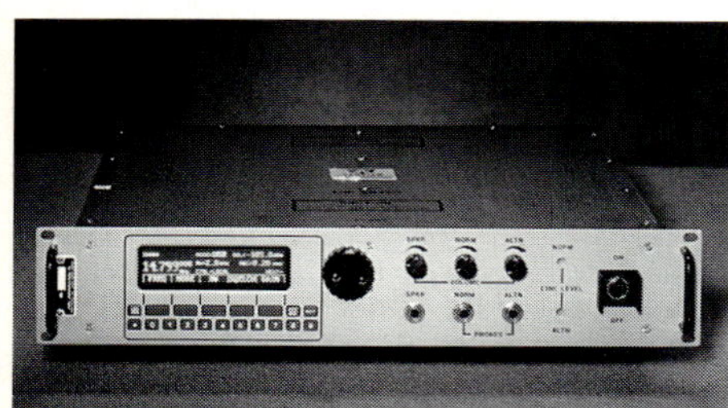

The Cubic Communications CDR-3150 Receiver **1998**/0011825

(a) CDR-3120 dual remote receiver – this receiver is operated remotely and does not have a front panel. Optional FSK demodulators and four-pole pre-selectors are mounted externally
(b) CDR-3130 dual surveillance receiver – in addition to the displays, soft keys, and key pads, each receiver has a phone/speaker jack. Optional FSK demodulators and four-pole preselectors are mounted externally
(c) CDR-3138 dual DF receiver – the chassis contains a master and slave receiver
(d) CDR-3150 single 19 in rack mount receiver – includes separate speaker jack and two phone jacks on the front panel. One phone jack for USB, LSB, CW, FM, and the other phone jack is for LSB during ISB operation. Optional FSK demodulator and four-pole preselector are integral
(e) CDR-3180 half-rack surveillance receiver – has front panel display keypad and controls including phone/speaker jack. Optional FSK demodulator and four-pole preselector are external
(f) RCU-3100 remote-control unit – controls up to 100 receivers via RS-232 and RS-422 databus.

Operational status

This product is currently in production.

Contractor

Cubic Communications Inc.

NEW ENTRY

PA-5050A 1 kW power amplifier

The model PA-5050A solid-state amplifier features continuous-duty operation and is designed to operate with the T-4180 HF DSP exciter, capable of LSB, USB, ISB, AM, FM, CW and FMfax operating modes. Included in the PA-5050A is the PS-7130A power supply consisting of five power supply modules. This integrated design concept minimises the effect of single point failures by providing gradual degradation of system power output.

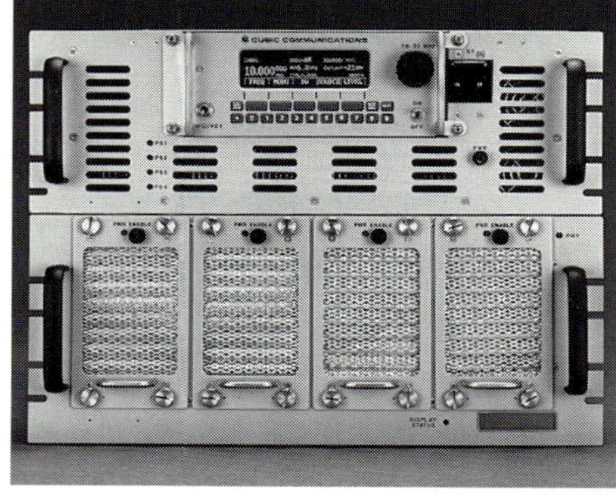

The Cubic Communications PA-5050A Power Amplifier **1998**/0011823

Specifications
Frequency range: 1.6-30 MHz
Power output: 1 kW peak
Dimensions: 355.3 × 482.2 × 538.1 mm
Weight: 34 kg

Operational status
This product is currently in production.

Contractor
Cubic Communications Inc.

NEW ENTRY

T-4180 LF-HF Digital Signal Processing (DSP) exciter

The T-4180 DSP exciter incorporates digital signal processing microchip technology designed to provide greater linearity and spectral purity. It is intended for use with the PA-5050A 1 kW power amplifier, and operates at 1.6 to 30 MHz.

The T-4180 exciter is available in a half-rack chassis. The unit is capable of local or remote-control operation. In addition, operation is menu-driven and little or no operator training is required. The exciter has a digitally tuned IF filter and 250 programmable channels. Operating modes include LSB, USB, ISB, AM, AM, CW, FSK and FMfax. BITE constantly troubleshoots the receiver and, when a problem is detected, a message on the unit's display screen identifies the module which should be removed and replaced.

Operational status
This product is currently in production.

Contractor
Cubic Communications Inc.

NEW ENTRY

AN/URQ-34 anti-jam tactical datalink

The AN/URQ-34 is an anti-jam tactical datalink transmission system operating in the Ku (NATO J-band) frequency band, built to support US Army and US Air Force real-time combat sensors. It was originally developed for the StandOff Target Acquisition System (SOTAS) and was delivered and deployed for special projects after SOTAS was discontinued.

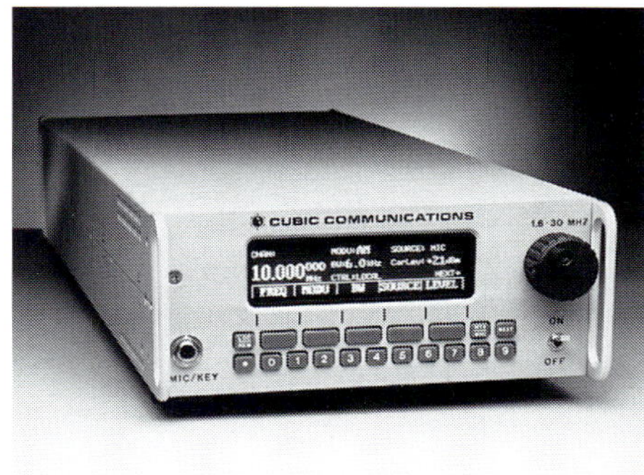

The Cubic Communications T-4180 DSP exciter
1998/0011822

The AN/URQ-34 uses fast frequency-hopping technology to defeat the hostile ECM threat specified for the Modular Integrated Communications and Navigation System (MICNS) and has a demonstrated range of 240 km. Input of digital data can be in clear or encrypted and the data rate is 25 to 100 kbytes nominal. It meets all applicable MIL, MICNS and SOTAS standards for environment, nuclear hardening, ballistic hardening, MTBF and MTTR. The MTBF is given as between 480 and 1,680 hours. The AN/URQ-34 makes extensive use of BITE, giving fault location to module level.

Operational status
In service with the US Army and US Air Force as the Joint Surveillance Target Attack Radar Systems (J-STARS) datalink.

Using a combination of US Army and company records and funds, Cubic has begun a System Improvement Program (SIP) to reduce Size, Weight And Power (SWAP) and cost of the AN/URQ-34 by 50 per cent.

Contractor
Cubic Defense Systems.

UPDATED

Airborne microwave transmission systems

FLIR Systems Inc airborne microwave transmission systems are designed to meet the requirements of government forces (police, customs, SAR) and TV companies to transmit high data rate TV and FLIR pictures from helicopters to ground stations; they are compatible with most surveillance systems, including all of those manufactured by Broadcast and Surveillance Systems Ltd (BSS) and FLIR Systems Inc.

Two models are available: the medium-range system, capable of transmitting good quality live pictures from air to ground up to 56 km; the long-range system, which can achieve 96 km.

The system comprises: a 1 W RF transmitter, 20 or 30 W ERP power amplifier, 4 dBi circularly polarised omnidirectional antenna actuator, CAA-certified to lower below the helicopter during flight to provide unobstructed 360° coverage. Pilot controls allow full control of the antenna system.

The system is normally supplied for operation around a nominal frequency of 2.4 GHz, but all bands from 0.6 to 7.5 GHz are possible.

Contractor
FLIR Systems Inc.

NEW ENTRY

FLIR Systems Inc airborne microwave transmission system, showing controllable antenna
1998/0011836

GMA 340 audio panel

The GMA 340 audio panel provides separate front panel volume controls for the pilot, co-pilot and passengers. The GMA 340 is capable of full audio stereo output at 14 and 28 V, and can deliver full power to 4 or 8 ohm speakers without the need for voltage converters or dropping resistors. The GMA 340 also includes a six position intercom, each with its own VOX circuit. The GMA 340 actively de-emphasises cabin noise to improve intelligibility. The display features LED push-button controls.

Specifications
Audio panel: 3 × transceiver inputs; 5 × receiver inputs; 2 × unswitched inputs
Intercom: 6 × positions: pilot, co-pilot, 4 × passengers
Headphones: 3 × stereo: pilot, co-pilot, passengers
Speakers outputs: 2
Marker beacon receiver: RF crystal-controlled at 75 MHz
Dimensions: 172.7 × 159.77 × 33 mm
TSO compliance: TSO-C50c; TSO-C35d

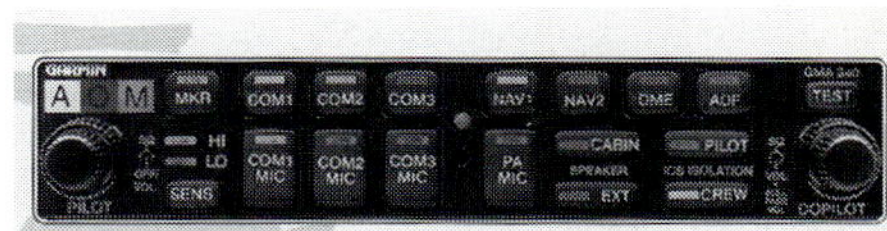

Garmin GMA 340 audio panel ***1998***/0011821

Contractor
Garmin International Inc

NEW ENTRY

Channelised Programmable Digital Radio (PDR)

The modular Programmable Digital Radio (PDR) can implement virtually any narrowband or wideband function from 2 MHz to 2 GHz for comms, nav, ident, landing systems or datalinks. A common module transceiver pair of receiver/processor and exciter/processor is the basic core of the system with RF input and digital data output. The common modules can be programmed to provide any function at system turn-on or during a mission or flight segment. Multisystem simultaneous operation is achieved by adding one or more receiver modules and exciter modules to the system. The architecture associated with the PDR offers simplistic system fault detection and isolation. In addition, system functional redundancy is a part of the basic design via software reconfiguration of a lower priority function automatically or the automatic reprogramming of a spare module. The PDR is designed for both military and commercial applications.

Operational status
The PDR is currently in development. The core common modules have been developed and demonstrated in 1995. The existing form factor utilises an ARINC standard packaging. However, a transition to SEM-E and VME is expected in the near future.

Contractors
GEC-Marconi Hazeltine Corporation
Northrop Grumman Corporation, Electronic Sensors and Systems Division, Baltimore.

UPDATED

Joint Tactical Information Distribution System (JTIDS)

The Joint Tactical Information Distribution System (JTIDS) is a US joint service command and control support system providing secure jam-resistant communications and embedded navigation and identification for land, sea and air platforms. Using Time Division Multiple Access (TDMA) technology, it provides high-capacity networking among diverse airborne and surface users. It allows all stations to share an integrated awareness of the combat situation for friendly forces as well as detected threats in real time, using the tri-service multinational message catalogue TADIL-J. It can thus provide a language-independent method to ensure co-ordinated operations on a multinational battlefield. The data received can be displayed in both symbology and

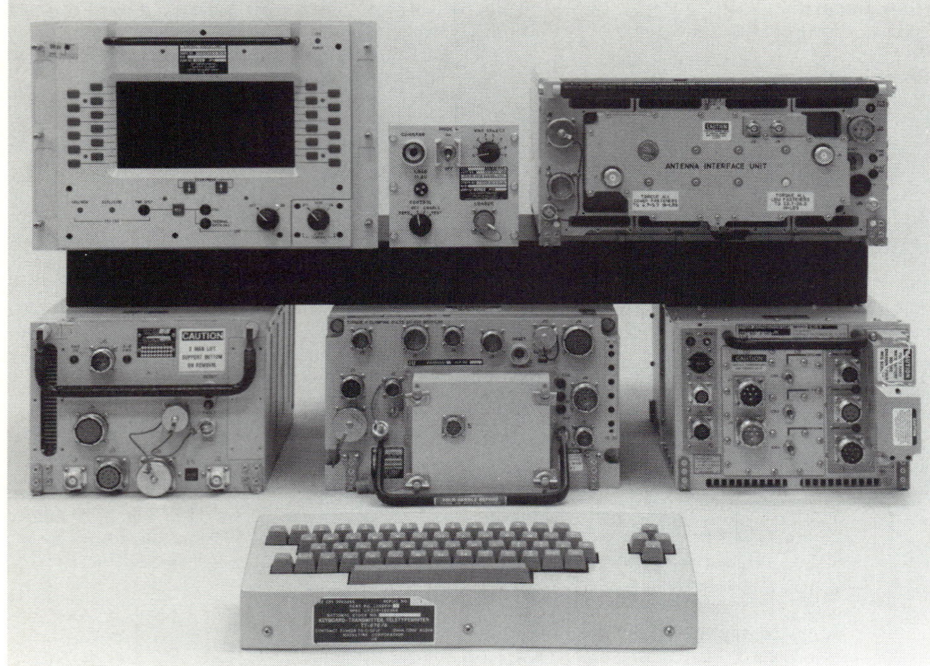

GEC-Marconi Hazeltine Class 2H JTIDS equipment

language of the host nation's platform. Thus, data received by a British system would use English as the language for a given message; an Italian system could display the same message in Italian.

JTIDS operates on 51 frequencies in the 960 to 1,215 MHz frequency band, sharing with Tacan but strictly avoiding the IFF transponder frequencies which are also in the band. The JTIDS TDMA scheme breaks up time into 7.8125 ms time slots and allocates these slots to users, based on projected traffic demand. Users employ their time slots to transmit while listening during all other times, thus enabling relays of opportunity, to maximise the probability of message reception. Participants routinely inject information into the network through their regular broadcast slots without necessarily knowing who needs the information, using a broadcast-oriented architecture. These broadcast slots might include identification and location data, as well as host processor-generated traffic such as target acquisition tracks and platform status information on weapons and fuel status or equipment readiness.

JTIDS is designed to survive the highest levels of enemy radio electronic combat. The links are encrypted with the latest approved crypto devices. Jam resistance is achieved by multiple techniques through fast hopping over all frequencies, direct sequence spread spectrum of the waveform and Reed-Solomon forward error correction. The result of the signal processing gains from this combination of techniques means that the JTIDS omnidirectional radiation pattern can offer as much signal improvement as if it were being transmitted by a highly directional antenna, such as parabolic dish, without the difficulties of beam pointing or the limitations of single path links.

The initial emphasis in the development of JTIDS hardware was on Class 1 terminals developed from 1974. First deliveries took place in 1977 and have resulted in terminals on board the Boeing E-3A AWACS aircraft, in ground terminals in the Royal Air Force IUKADGE system and in transportable shelters designated as Adaptive Surface Interface Terminals (ASIT) used by the US Army and US Air Force ground control facilities. The terminal is also used in the NATO Air Defense Ground Environment. The Class 1 terminal comprises rack-mounted components occupying almost 0.26 m^3 and weighing approximately 192 kg.

Operational status

The operational fielding of JTIDS has begun, primarily with navy and army (FAAD) platforms. Various multiservice demonstrations, notably the All-Services Combat Identification Evaluation Team (ASCIET) exercises, have shown the superior capabilities of Link 16 systems such as JTIDS. There have also been exercises (and combat experience during Desert Storm) which proved interoperability among US, British and French JTIDS equipment. The Japanese AWACS will be equipped with JTIDS capability and the Netherlands air defense ground forces will also soon get JTIDS terminals. Other nations continue to show interest in receiving Link 16 capabilities. All the members of the JTIDS equipment family are now in production.

JTIDS Class 2 terminal

The Class 2 terminal, for fighter aircraft installation, began its development early in 1980. It was designed by GEC-Marconi Hazeltine Corporation in a leader-follower arrangement, with Rockwell Collins as the subcontractor providing RF design expertise.

The Class 2 terminal, designated by the USA as the AN/URC-107(V)1, AN/URC-107(V)6, AN/URC-107(V)8, AN/URC-107(V)10, provides the basic JTIDS functions of high-capacity data communications, embedded TACAN functionality, dual-channel/dual-mode integral secure voice, dual-grid relative and geodetic navigation, GPS interface, PPLI self-identification messages, crypto-secure and jam-resistant connectivity and system status monitoring and reporting. Since JTIDS operates in the Tacan and IFF frequency band, extensive interference protection circuits are built into the equipment to prevent inadvertent interference with those systems. Circuits continuously monitor the JTIDS radio operation and can shut down transmission in the event of any out of specification emission. During peacetime activities, the terminal operation maintains restricted duty cycles, further limiting the potential for interference with civilian facilities. The terminal incorporates SRU/LRU BIT hardware and software, achieving 98 per cent fault detection and 95 per cent fault isolation.

The Class 2 terminal comprises two boxes occupying approximately 0.045 m^3 and weighing 57 kg. One box houses the receiver/transmitter circuits; the other is a dual LRU component. The dual LRUs are the digital processor section, standard for all terminals and applications, and the interface unit which is customised for each host platform. The digital processor section carries out all signal processing and digital computing functions, as well as advanced position location and tactical air navigation. The key

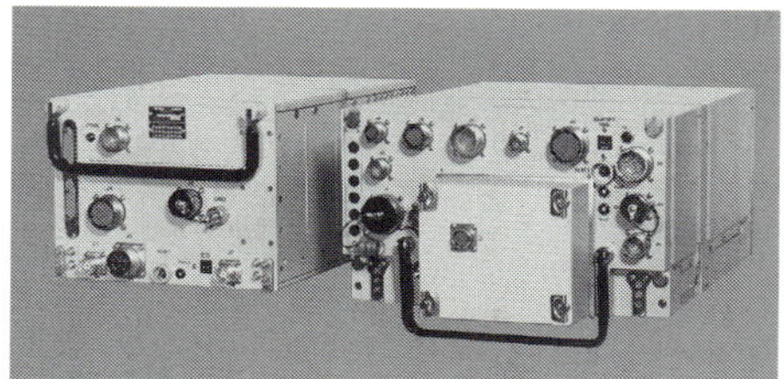

The Class 2 JTIDS terminal for the US Air Force F-15 with the receiver/transmitter (left) and the data processor group (right) **1995**

JTIDS information shown on the colour display of an F-15. The aircraft is shown at the centre of the display, concentric circles indicate distance in nautical miles. Round symbols indicate friendly aircraft, rectangles are unknowns and triangles are enemy aircraft

output of the system is the display of information passed over the JTIDS network, allowing network users to share a common situational awareness. In the fighter terminal, the JTIDS interface feeds the multifunction display which is capable of selective control by the pilot and automatic display under instructions in preloaded mission files. Selected platforms can receive precise direction from a central control point or can support local decision making based on the shared awareness. In other installations, the JTIDS data can be displayed on host processor displays, providing real-time depiction of the combat situation.

Specifications

Dimensions:
(data processing group) 324 × 193 × 485 mm
(receiver/transmitter) 257 × 193 × 395 mm
Weight:
(data processing group) 36 kg
(receiver/transmitter) 25.8 kg
Power supply: 120/208 V AC, 50/60/400 Hz or 240/280 V DC, 1,400 W
Power output:
(TDMA) 200 W
(Tacan) 500 W
Data rate: 238 kbits/s
Range: (normal) 557 km, (available) 928 km

Operational status

The final Class 2 production contract was recently awarded. Production is expected to continue until at least the year 2000.

Terminals are currently being integrated into US Air Force aircraft: F-15, Rivet Joint, Joint STARS, and Airborne Battlefield Command/Control Centers (ABCCC) and the ground-based Modular Control Equipment (MCE). They are also being fitted into the US Navy's F-14 aircraft and the US Marine Corps' Air Defense Command Post (ADCP). They are also operational in many UK Tornado aircraft. Successful testing of Class 2 terminals on US Navy terminals on US Navy submarines has shown the usefulness of JTIDS for the entire fleet.

JTIDS Class 2H Terminal

The Class 2H terminal, derived from the Class 2 to replace the Class 1, occupies 0.15 m^3 and weighs 155 kg, while giving the additional capabilities of increased throughput (115 kbits/s v 28.8 kbits/s), relative navigation and Tacan functionality, interoperability with TADIL-J message protocols and increased system functionality such as an increased quantity of crypto variables and over-the-air initialisation and rekey.

The Class 2H terminal is the high-power 1,000 W output transceiver version of the basic Class 2 terminal. It currently has three configurations. The first (AN/URC-107(V)4 and AN/URC-107(V)5) is an airborne system with receiver/transmitter, data processor group, high-power amplifier/antenna interface group, control monitor set and power conditioner and crypto loading devices. The other two configurations are for land-based use (AN/URC-107(V)9) and for surface ships (AN/URC-107(V)7). The Class 2H possesses all the functionalities of the basic Class 2 and is totally interoperable.

Specifications

Volume: 0.15 m³
Weight: 155 kg
Power supply: 120/208 V AC, 50/60/400 Hz, 3 phase
Power output:
(TDMA high power) 950 W
(TDMA low power) 235 W
(Tacan) 500 W
Data rate: up to 238 kbits/s
Range: (normal) 557 km, (available) 928 km
Message types: TADIL-J or interim JTIDS message standard
Interface: MIL-STD-1553B

Operational status

Class 2H terminals are currently being integrated into US Naval surface ships and E-2C Hawkeye aircraft, US Marine Corps Tactical Air Operations Modules (TAOM) and upgrades to US Air Force E-3 AWACS aircraft. They are also fitted to British and French AWACS aircraft.

JTIDS Class 2M terminals have been ordered by the US Air Force for JTIDS operations

JTIDS Class 2M terminal, designated AN/GSQ-240, is a single-box variant of the basic JTIDS terminal, designed for land-based operations.

Contractor

GEC-Marconi Hazeltine Corporation.

UPDATED

Alpha 12 VHF radio

The Genave Alpha 12 is a panel-mounted VHF/AM transmitter/receiver for light aircraft. It has a low power consumption, particularly suiting it to aircraft with a limited electrical generation capacity such as gliders, certain agricultural aircraft or home-builts. It provides 12 channels in the band 118 to 135.975 MHz at a channel spacing of 25 kHz. Transmitter power output is a nominal 4 W carrier, with 3.3 W minimum.

Features include a MOSFET, track-tuned front end and crystal intermediate frequency filtering. A light-emitting diode is incorporated to act as a transmit indicator.

Specifications

Dimensions: 63 × 165 × 254 mm
Weight: 1.81 kg

Operational status

In production and in service.

Contractor

Genave Inc.

VERIFIED

Alpha 100 VHF radio

The Genave Alpha 100 is a panel-mounted VHF/AM transmitter/receiver providing 100 channels in the band 118 to 127.9 MHz at a channel spacing of 100 kHz. Transmitter power output is 8 W peak power, 2 to 3 W carrier. Construction is fully solid state and the receiver section is of the double conversion, superheterodyne type and is crystal controlled. Facilities include a manually adjustable squelch disable and automatic gain control. Frequencies are selected by means of a dual-knob selector with digital readout and a light-emitting diode is employed as a transmit indicator.

The system has a low power requirement, in common with other Genave equipment, making it suitable for aircraft with little or no electrical generation capacity.

Specifications

Dimensions: 165 × 63 × 228 mm
Weight: 1.82 kg

Operational status

In production and in service.

Contractor

Genave Inc.

VERIFIED

The Alpha 720 transmitter/receiver with microphone

Alpha 720 VHF radio

The Genave Alpha 720 is a panel-mounted VHF/AM transmitter/receiver providing 720 channels in the 118 to 135.975 MHz band at a channel separation of 25 kHz. Transmitter output power is 4 W nominal. It is designed for the general aviation and light aircraft market.

The system is a single crystal unit using digital phase-locked synthesis techniques for frequency generation. Construction is fully solid state, with extensive employment of integrated circuitry. Features include a transformerless series modulator in the transmitter section, a single conversion receiver and field effect transistor front end and mixer circuitry. Facilities include automatic squelch disable and active impulse noise limitation to reduce external interference effects. Channel selection is performed by use of a dual-control frequency selector knob on the equipment's front casing; the selection is confirmed by a dimmable incandescent readout display.

Like many Genave products, the Alpha 720 is a low power consumption system suitable for aircraft with limited electrical power. A variant, the Man-Pack, designed for portable use, is produced for gliders, home-builts and agricultural aircraft without electrical systems.

Specifications

Dimensions: 63 × 165 × 254 mm
Weight: 1.81 kg

Operational status

In production and in service.

Contractor

Genave Inc.

VERIFIED

AN/USQ-130(V) MX-512PA Link-11/TADIL-A data terminal

The AN/USQ-130(V) Link-11/TADIL-A data terminal set is designed to provide all required modem and network control functions in a Link-11/TADIL-A system using either HF or UHF radio equipment. The equipment meets the data terminal set requirements of MIL-STD-188-203-1A and may be operated as a picket or net control station in a TADIL-A net. As a net control station, the AN/USQ-130(V) accepts addresses from the tactical data computer or from a separate control panel.

The equipment provides all the modes of Link-11/TADIL-A systems including net control or picket, high- and low-data rate, net test, net synchronisation, short broadcast, long broadcast and full-duplex (for single station system tests and sidetone verification). Doppler correction circuits which operate independently on both sidebands are operator-selectable. The AN/USQ-130(V) also operates in the Improved Link-11 Waveform (ILEW) mode.

The AN/USQ-130(V) can be externally controlled by a computer over a MIL-STD-188-114, RS-232C-compatible asynchronous control interface, or 1553 databus. The AN/USQ-130(V) may also be controlled from a separate remote-control panel using menus standard to the MX-512P DTS family.

The set is programmable. All modem, network control and link monitoring functions are performed digitally in microprocessors using a modular, multiprocessor architecture. Selection of the conventional Link 11 or ILEW is made over the remote-control interface.

The single-tone waveform for Link-11 provides improved performance in HF Link-11 networks on an SSB HF channel. Single-tone Link-11 uses an eight-phase modulated 1,800 Hz tone. Adaptive equalisation is used to demodulate the signal under the severe multipath conditions typical of HF propagation paths. Error detection and correction codes are used to provide enhanced message throughput.

The AN/USQ-130(V) provides, as an option, a 2,400 bits/s, full-duplex, RS-232C satellite-wireline interface which transmits and receives compatible Link-11 data in digital form. Link-11 data may be sent over satellite, wireline, or other tactical circuits. The AN/USQ-130(V) can be operated in either the digital mode, the conventional mode, or in a mixed mode (gateway), where some pickets operate in the digital transmission mode and some in the conventional HF or UHF mode.

The unit provides, as another option, link quality analysis indicators which include multipath spread, fading bandwidth, net cycle time since last reply, and tone power spectrum for each participating unit in the network. Using these indicators, an operator can troubleshoot equipment failures and configuration set up problems in the net and determine when HF propagation problems require a change of radio frequency. BITE provisions in the AN/USQ-130(V) include loop-back functions which verify operation of the system.

The data terminal set is compatible with ATR short, measures 193 × 57 × 32 mm, and weighs 4.6 kg. It is powered from 24 to 32 V DC and meets certain requirements of MIL-HDBK-217E, MIL-E-5400T, MIL-STD-188 and MIL-STD-1553B.

Specifications

Dimensions:
Size: ¼ ATR-short
Height: 19.3 cm
Width: 5.7 cm
Depth: 32.0 cm
Weight: 4.6 kg

Power:
24 to 32 V DC
28 W

Reliability:
MTBF: Over 10,000 h per MIL-HDBK-217F at 50°C AIC

Contractor

General Atronics Corporation.

VERIFIED

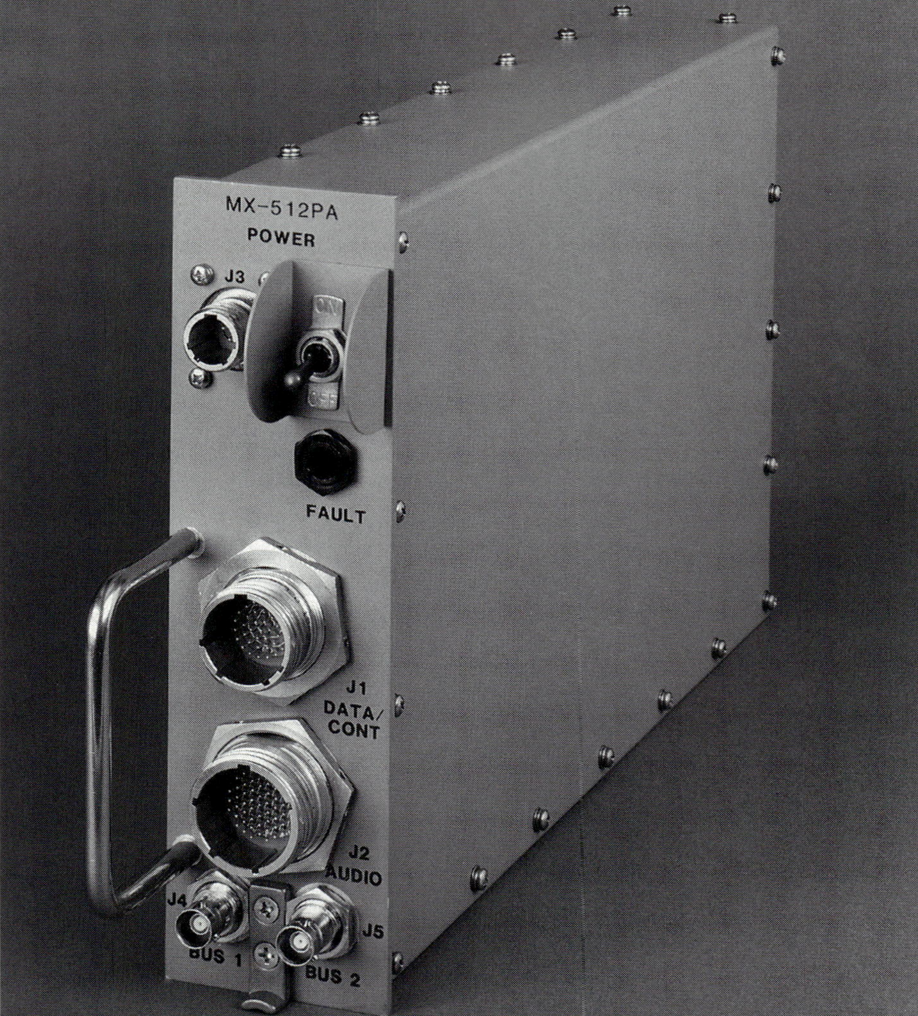

AN/USQ-130(V) MX-512PA Link-11/TADIL-A data terminal ***1997***/0002119

GA-540 TADIL-A/Link-11 Serial DataLink Translator (SDLT)

The GA-540 Serial DataLink Translator (SDLT) provides the interface in a TADIL-A/Link-11 system between a serial encryption device and a tactical data system processor. The GA-540 serial interface to the encryption device meets the Airborne Tactical Data System (ATDS) interface requirements of MIL-STD-188-203-1A (Appendix D-2). All data buffering, timing and ATDS handshaking are performed automatically. The GA-540 connects to the tactical data system processor using one of the following interfaces: RS-422/RS-423 conditioned diphase; transformer-coupled conditioned diphase; RS-422/RS-423 synchronous; RS-423 asynchronous; VME bus.

The GA-540 is offered as a 19 in rack-mountable unit, as a ¼-ATR-short airborne unit, or as a 6U-VME card. The rack-mounted and ATR units include a power supply, BIT indicator, and panel-mounted connectors. The VME card set can be hosted in any VME operating environment.

Specifications

Dimensions:
(size): ¼ ATR-short
(height): 19.3 cm
(width): 5.7 cm
(depth): 32.0 cm
Power:
24 to 32 V DC
15 W
Reliability: 78,000 h

Contractor

General Atronics Corporation.

VERIFIED

GA-540 TADIL-A/Link-11 Serial DataLink Translator (SDLT)
1997/0002120

Laser transmitter

GTE Government Systems is developing an improved airborne laser transmitter for aircraft-to-submarine communications, under contract to the US Navy. Experiments conducted during the past few years with a blue-green laser system have demonstrated the feasibility of communicating through cloud with a submerged submarine at significant depths, and the existing high-performance equipment is being upgraded to operate with a high area coverage rate in various weather conditions and water types. Laser communication systems offer a variety of benefits for use with submerged submarines compared with conventional radio communications, including less distortion as the beam travels through water, higher data rates, greater security and less susceptibility to jamming.

It is intended that the airborne transmitter will provide nearly instantaneous communication between carrier-based aircraft and submerged submarines in tactical situations. In addition, advanced versions of the transmitter would enable a space platform to communicate both tactical and strategic messages over large areas.

Lasers are also being developed for air-to-air applications under the Have Lace Laser Airborne Communications Experimental programme. GTE is producing two terminals, mainly from off-the-shelf equipment, which include laser transmitters and receivers as well as acquisition and tracking equipment.

Operational status

In development. The blue-green laser transmitter was mounted in a Lockheed P-3C Orion aircraft for experimental flights. The Have Lace terminals were evaluated on board a Boeing KC-135 testbed aircraft. GTE is also developing submarine optical receivers based on atomic resonance filter techniques.

Contractor

GTE Government Systems Electronic Defense Sector.

VERIFIED

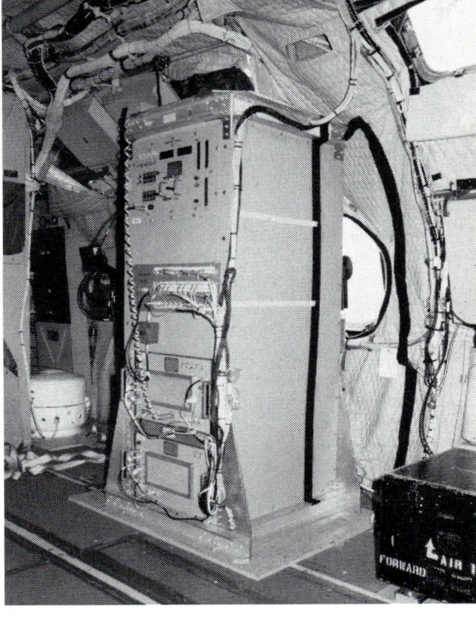

The GTE air-to-air laser communication system

AN/ASW-54(X) Link 16 Interoperable Tactical DataLink (LITDL)

The AN/ASW-54(X) Link 16 Interoperable Tactical DataLink (LITDL) multinet TDMA architecture supports mixed force operations and provides a real-time tactical situational awareness capability to tactical forces.

Multilink flexibility ensures connectivity to existing datalink systems and an innovative gateway systems architecture interoperates with the broad area command and control JTIDS net.

LITDL contains an airborne datalink for tactical aircraft and UAVs. The transponder mode, for operation with a command and control terminal, also contains a fighter-to-fighter situational awareness mode. The integrated self-contained unit consists of a UHF transceiver, programmable modem, MIL-STD-1553 interfaces and a military power supply. BIT is incorporated in the unit.

Specifications

Dimensions: 128.3 × 180.6 × 273 mm
Weight: 6.23 kg
Power supply: 28 V DC
Frequency: 300-324.9 MHz
Channels:
(FSK) 250 100 kHz channels,
(anti-jamming) 250 orthogonal channels

Contractor

Harris Corporation.

VERIFIED

AN/USQ-86(V) datalink

The AN/USQ-86(V) is a modular integrated communications and navigation system which provides a nuclear and ballistic-hardened anti-jam datalink for air-to-ground applications. Modules of the system, which has been developed under the auspices of the US Army's Electronics Research and Development Command, can be used with a variety of manned aircraft and remotely piloted vehicles.

In the manned aircraft role the system can be used as part of a standoff target radar system wideband sensor configuration for remote-control airborne intercept, to compromise enemy communications and to locate hostile command and control facilities.

The key element in all these facilities is the secure air-to-ground datalink. This is achieved by a combination of spread spectrum modulation techniques, message coding, signal processing and advanced antenna techniques.

Operational status

In late development.

Contractor

Harris Corporation.

VERIFIED

AN/ZSW-1 weapon control datalink

The AN/ZSW-1 weapon control datalink is flight qualified on the F-15E for use with the GBU-15 and AGM-130 standoff guided weapons. It provides real-time target acquisition and battle damage information at long standoff range and is resistant to jamming in all RF environments. The AN/ZSW-1 is fully MIL-qualified and is adaptable to other aircraft and missile configurations.

Operational status

In production.

Contractors

Harris Corporation.
Raytheon Systems Company.

UPDATED

Multifunction/multiband antenna subsystem

The tactical airborne antenna array with multiband and multifunction antennas is integrated into a single aperture structure and configured so that the entire structure can be conformally mounted on the aircraft fuselage. An additional feature is the electronics system which is needed to interface the aperture to radios and also provides anti-jam adaptive processing. It covers VHF, UHF and L-, S- and C-bands (NATO D, E/F and G/H bands).

The switching electronics is a highly integrated array of RF switches, power dividers and receive amplifiers which provide multi-use selection of each major antenna band. The resource manager is a general purpose processor with interfaces to a radio via mission avionics bus and intercoms. Its function is to manage antenna resources in order to optimise coverage and minimise interference.

Specifications

Weight: 58.1 kg
Reliability: 1,800 h MTBF

Contractor

Harris Corporation.

VERIFIED

Primus II radios

Honeywell introduced the Primus II series of radios in April 1987; they are aimed at the business aircraft and regional airliner segment of the market. The system incorporates VLSI technology and digital bus tuning and control, together with centralised radio management and a digital audio system.

Primus II radios are controlled and display their information via a radio system bus which is the Sperry-developed Avionics Standard Communications Bus (ASCB) formatted for radios. The system comprises the

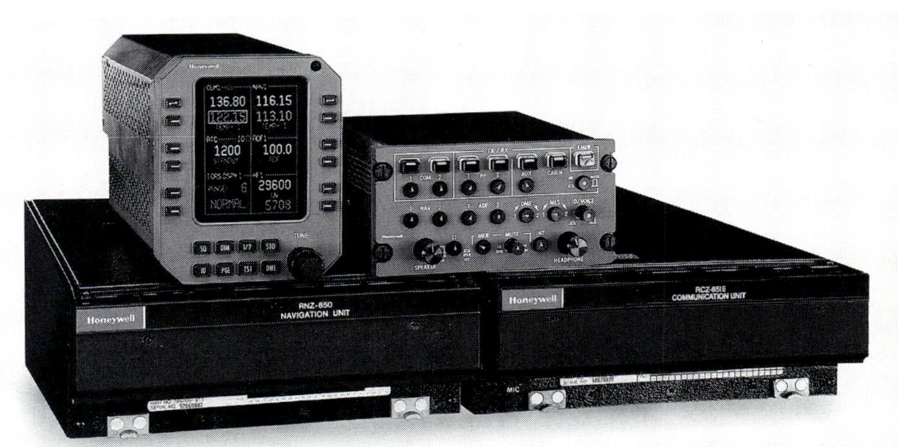

The Honeywell Primus II radio system showing the RM-850 radio management unit (top left), AV-850 audio control panel (top right), RNZ-850 navigation unit (bottom left) and RCZ-851E communication unit (bottom right)
1997/0001185

full-colour RM-850 series radio management unit, RNZ-850 series integrated navigation unit, RCZ-851E series integrated communications unit and AV-850 audio control unit.

The radio management unit provides control over operating modes, frequencies and codes for all units in the system. Five dedicated windows support the com, nav, transponder, ADF and MLS functions. The unit also provides BITE control and readout.

The integrated navigation unit contains a VHF navigation receiver, DME transceiver modules and ADF receiver module. The VHF navigation receiver houses the functions of VOR/localiser receiver, glide slope receiver and marker beacon receiver to provide an ILS that meets Cat II low-approach requirements. The DME transceiver is a six-channel scanning DME that simultaneously tracks selected DME channels and two preselect navigation frequencies. It meets the initial approach mode accuracy requirements of the P-DME specification with an accuracy of better than 100 ft and can also operate on W, X, Y or Z DME channels. The extended range ADF module can receive low-frequency NDBs below 200 kHz as well as the marine emergency band of 2.181 to 2.183 MHz.

The integrated communications unit incorporates separate VHF communication transceiver and transponder modules. Optional transponders include Mode A/C, Mode S and Mode S with diversity. The communication transceiver operates across the entire 118 to 152 MHz frequency range, but for civil use the upper limit can be reduced to 136 MHz.

The MLS receiver operates as an extension of the integrated navigation unit and can interface with standard digital and analogue outputs.

The digital audio system receives digitised audio from the other units via one high-speed digital bus from each side, providing immunity to noise and virtual elimination of cross-talk. It can control 16 or more audio signals and multiple audio panels may be installed in the aircraft. It is available in three- or four-row versions, with a variety of layouts.

Operational status

In production and in service.

Contractor

Honeywell Inc Business & Commuter Aviation Systems.

VERIFIED

ETICS Embedded Tactical Internet Control System for OH-58D Kiowa Warrior helicopter

ETICS is an embedded digital messaging system that communicates on the tactical internet for both the fire support (advanced field artillery tactical data system (AFTADS)) and the command and control (Appliqué) nets. With ETICS, the OH-58D pilots can communicate tactical information directly and quickly with other air and ground battlefield systems.

ETICS features Variable Message Format (VMF), message parsing, VMF message display and situational awareness reporting and display. It incorporates a COTS operating system that is compliant with the US Army Technical Architecture, as well as reuse of government source code for message parsing.

Operational status

In operational evaluation; developed by Honeywell in co-operation with the US Army Kiowa Warrior Project Management Office and Bell Helicopter Textron Incorporated - manufacturers of the Kiowa Warrior helicopter.

Contractor

Honeywell Inc Defense Avionics Systems.

VERIFIED

AN/ARC-201 VHF/FM transceiver (SINCGARS-V)

The AN/ARC-201 SINCGARS-V is an airborne VHF/FM frequency-hopping radio and is an all-solid-state equipment for use in helicopters, light observation aircraft and fighters. It provides single-channel and frequency-hopping modes. A six-channel non-volatile preset memory is incorporated for single-channel and ECCM modes. An interface and controls for the AM-7189A/ARC 50 W amplifier are incorporated.

The AN/ARC-201 is available in panel-mounted, dedicated-remote and 1553B multiplex bus-remote configurations. It is interoperable with the current VHF/FM radios in the single-channel mode, and with the Single Channel Ground and Airborne Radio Subsystem (SINCGARS) VRC-87 to VRC-92 and manpack PRC-119 ground radios in the frequency-hopping mode. Electroluminescent lighting is provided on the front panel, compatible with the use of NVGs.

The radio has module and component commonality with the ground SINCGARS communication equipment, with extensive use of LSI circuitry and microprocessors being made for high reliability. A BIT function isolates faults to the module level with 90 per cent confidence. A data rate adaptor interfaces the radio with data devices for data communication. An automatic single-channel cueing capability in the ECCM modes allows a single-channel user to alert members of an ECCM net. Internal and external COMSEC can be used to provide secure communications in voice and data modes.

The ARC-201 can be supplied as the RT-1476 single-unit panel-mounted radio, as the RT-1477 remote radio with the C-11466 remote-control unit or as the RT-1478 remote bus radio.

The ITT AN/ARC-201 VHF/FM transceiver, with RT-1477 remote radio and C-11466 remote-control unit

The ITT RT-1478 remote radio operates with a MIL-STD-1553B databus

The ITT RT-1476 panel-mounted radio

Specifications

Dimensions:
(RT-1476) 146 × 104 × 239 mm
(RT-1477) 127 × 102 × 237 mm
(C-11466) 146 × 76 × 132 mm
(RT-1478) 127 × 102 × 259 mm
Weight:
(RT-1476) 3.1 kg
(RT-1477) 2.9 kg
(C-11466) 1.0 kg
(RT-1478) 3.1 kg
Power output: 10 W
Frequency: 30-87.975 MHz
Channel spacing: 25 kHz
Channels: 2,320

Operational status

In production. Contracted requirements for the US Army continue until 2002.

Contractor

ITT Aerospace, Communications Division, Fort Wayne.

VERIFIED

AN/ARC-51 UHF radio

The Lapointe AN/ARC-51 is a military airborne radio covering the UHF band from 225 to 400 MHz in which it provides 3,500 channels. The system, intended mainly for high-performance aircraft, is available in a number of versions designated ARC-51A, ARC-51AX, ARC-51B and ARC-51BX. These, together with a wide range of controllers, give users the flexibility to assemble a number of configurations to suit particular installations. ARC-51 combinations are also used in surface vehicles and in other land applications.

All variants operate in AM/DSB mode and have a transmitted power output of 20 W. They provide azimuth homing facilities, when used in conjunction with suitable indicator equipment, and can be used for automatic rebroadcast purposes.

Although many different types of controller are available, a typical ARC-51 installation would provide preselection of up to 20 channels together with manual selection of any of the total of 3,500 frequencies covered. An independent Guard receiver provides simultaneous continuous monitoring of the international UHF distress frequency of 243 MHz. Other installations permit the dual control of one or more transmitter/receivers from more than one crew position, and miniature frequency indicator displays are available for use in situations where space is limited.

A feature of this system is the hermetically sealed pressurised container which allows full operational performance in unpressurised avionics bays at aircraft altitudes up to 70,000 ft when used in conjunction with a forced-air cooling supply. This sealed case assembly also renders the system particularly suitable for operation in situations where dust or water contamination could otherwise be expected, such as in desert vehicles or high-humidity tropical environments. The system's normal operating temperature range is from −54 to +71°C.

Specifications

Dimensions: 429 × 222 × 171 mm
Weight: 14.5 kg

Operational status

Variants of the ARC-51 are in service with the Belgian, Canadian, German, Indian, Italian, Thai and US armed forces.

Contractor

Lapointe Industries.

VERIFIED

Communication Control Group (CCG)

The Communication Control Group (CCG) was developed for the US Navy S-3B ASW aircraft. It provides intercom system control between crew stations, air-to-air and air-to-ground communications control, data communications via datalink and TEMPEST secure/clear isolation. The system comprises a Communication System Controller (CSC), two control indicators and four system control indicators.

The CSC provides the interface between the radios and the intercom system, as well as other audio and digital system interfaces. In response to display units, the CSC connects and configures radios and other communications equipment. The CSC maintains communication configuration and control menus for the display units, provides control of all modes of frequency selection for the HF and VHF/UHF radios in addition to the sonobuoy receiver and the On-Top Position Indicator (OTPI). Complete communication control is provided to multiple crew stations as well as access to all radios. The CSC provides control of UHF radio channelisation frequencies and the antenna switching unit. It can also control switching for secure voice equipment.

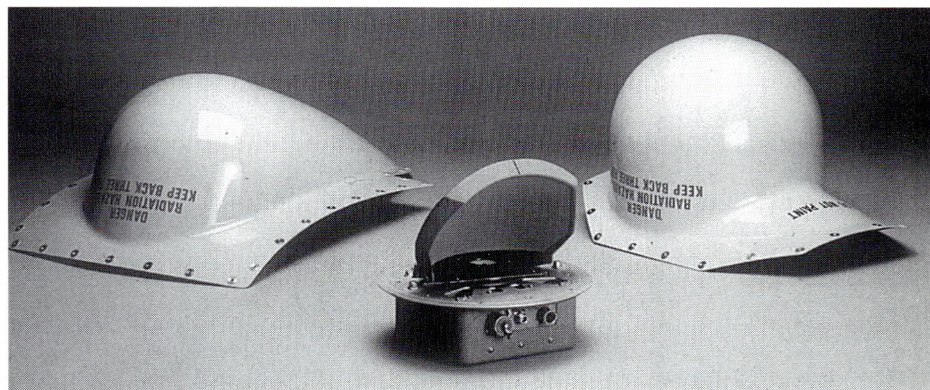

The Litton AN/ARQ-44 datalink antenna and radomes for the SH-60B LAMPS II helicopter **1995**

The control indicators work in conjunction with the CSC to provide management of communications configuration and mode control, provision of navigation initialisation and update capability, provision of armament system status and mode control, provision of acoustic sensor status and updating capability and display of mission avionics diagnostics and BIT information.

The Communication Control Group showing the communication system controller, two control indicators and four radio system control indicators **1995**

Specifications

Dimensions:
(CSC) 178 × 350 × 477 mm
(control indicator) 122 × 146 × 240 mm
(radio system control indicator) 169 × 146 × 2.1 mm
Weight:
(CSC) 23.63 kg
(control indicator ×2) 4.09 kg
(radio system control indicator ×4) 6.4 kg
Power supply:
(CSC) 115 V AC, 400 Hz, 150 W 28 V DC, 15 W
(control indicator) 115 V AC, 400 Hz, 30 W,
5 V AC, 400 Hz, 15 W, 28 V DC, 24 W
Reliability:
(CSC) 2,837 h MTBF
(control indicator) 10,501 h MTBF
(radio system control indicator) 12,039 h MTBF

Operational status

In production for the US Navy S-3B ASW aircraft.

Contractor

Litton Guidance & Control Systems.

VERIFIED

EC-130E airborne TV and radio broadcasting station

The latest version of psychological operations airborne TV and radio broadcasting EC-130E aircraft is capable of broadcasting TV in full colour in any format worldwide. Aircraft modification kits and installations include VHF and UHF antennas housed in 6 ft diameter, 23 ft long (1.829 × 7.01 m) pods under each wing dedicated to higher-frequency TV and rotatable to alter polarities; VHF low-frequency TV antennas mounted on each side of the tailplane; upgraded transmitters; equipment for formatting TV signal to worldwide standards and increased output power from 1 to 10 kW. Two retractable trailing-wire antennas provide both HF and VHF AM omnidirectional radio broadcast coverage.

The aircraft are upgraded versions of EC-130E Volant Solo aircraft.

Operational status

In service on US Air National Guard EC-130E aircraft.

Contractor

Lockheed Martin Aircraft and Logistics Centers.

UPDATED

IFPS Intra-Formation Positioning System

Lockheed Martin Federal Systems, Owego is developing an IFPS Intra-Formation Positioning System for use by US Air Force Operations Command on its MH-53J helicopters and MC-130H tanker aircraft during covert tanker operations.

Four separate types of message will be used on the AN/ARC-164 low-probability of intercept (LPI) radio to correlate position with: aircraft in the same formation (A-Net datalink); aircraft in other formations (E-Net datalink); refuelling rendezvous (R-Net datalink); secure voice communications.

A cockpit display shows relative positions, velocities and altitudes of other aircraft on the IFPS link.

Operational status

In development.

Contractor

Lockheed Martin Federal Systems, Owego.

NEW ENTRY

COM 2II VHF/UHF transceiver

The COM 211 was designed for aircraft operations requiring both VHF and UHF capability in one box. It features: frequency memory channels, flip-flop frequency transfer, non-volatile memory, acoustic audio levelling, seamless transition between VHF and UHF, and active and standby frequency display.

Specifications

Frequency range:
(VHF) 118.000-136.975 MHz, in 25 kHz steps
(UHF): 225.000-399.975 MHz, in 25 kHz steps
Memory channels: 10
Output power:
(VHF) 8 W nominal
(UHF) 10 W nominal
Dimensions: 160.3 × 34.3 × 273.7 mm
Weight: 1.32 kg
Power:
(receive) 28 V DC 0.5 A
(transmit) 6.0 A

Operational status

Development complete.

Contractor

Narco Avionics Inc.

NEW ENTRY

The Narco Avionics COM 211 VHF/UHF transceiver ***1998***/0011819

COM 810/811 series VHF radios

Narco's COM 810 and COM 811 models are solid-state microprocessor-controlled communication systems designed principally for light and general aviation aircraft. Each covers the VHF band from 118 to 136.975 MHz in which it provides 760 channels, two of which are preselectable; one is for active and the other for standby use. The 810 and 811 systems are essentially similar except that the former is designed for operation from a 13.75 V DC supply and the latter from a 27.5 V DC supply.

The active and standby frequencies are presented on LED displays which are automatically dimmed during darkness by a built-in photocell circuit. The legend XMT is illuminated when the microphone is keyed for transmission.

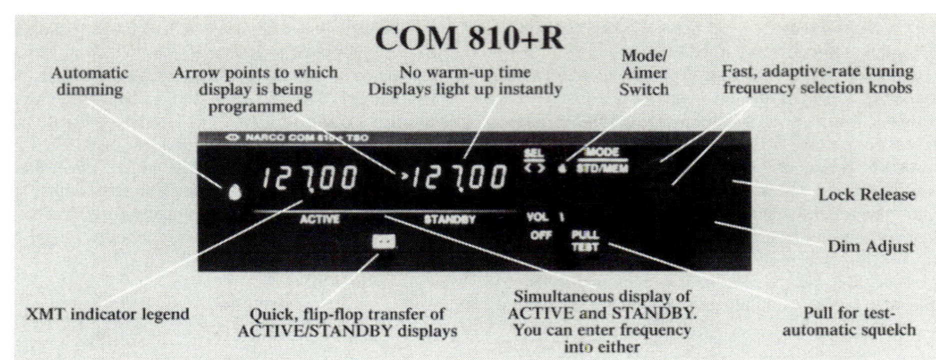

COM 810+R 'Direct Replacement' COM Transceiver ***1997***/0001186

New frequencies may be entered in either the active or standby positions when desired; an illuminated arrow indicates which section has been selected for new frequency entry. Frequency selection is completed by use of a concentric tuning control, the outer part of which makes frequency readout changes at the rate of 1 MHz per detent and the inner part providing kHz changes at 25 kHz per detent. Clockwise rotation increases the numerical value of the frequency selection and counter-clockwise rotation decreases it. A transfer switch is used to exchange selected frequencies between active and standby modes.

An optional feature is a connection which enables the last entered frequencies to be retained in the system memory when the radio is inactive. This requires a trickle current of 0.1 mA from the aircraft's battery. If this circuit is not connected, then the radio automatically retunes to the 121.5 MHz internationally designated emergency frequency in the active mode and to the 121.9 MHz ground control frequency in standby mode the next time it is switched on. In the event of a display failure, the radio automatically reverts to these frequencies and may be retuned to the desired channel by counting the detent clicks of the tuning control.

Built-in automatic squelch control, deactivated by use of a pull/test switch, maintains audio silence until a signal is received. Automatic audio-levelling in both transmitter and receiver allows all signals to be heard at the same level regardless of modulation. A built-in 10 W amplifier, provision for multiple audio inputs and intercommunication facilities are also included.

Latest in the range is the TSO'd COM 810+R, which directly replaces all Narco COMs from the COM 11 to the COM 120 and the COM 810/811 and COM 810+/811+.

Specifications

Dimensions:
(transmitter/receiver) 159 × 38 × 279 mm
Weight:
(transmitter/receiver) 1.3 kg
(mounting tray) 0.34 kg
Frequency: 118-136.975 MHz
Channels: 760
Power output: 8 W nominal

Operational status

In production and in service.

Contractor

Narco Avionics Inc.

VERIFIED

CP 136 and CP 136M audio control panels

Belonging to the Centerline range, the CP 136 and the slightly larger 136M variant provide fully solid-state control of all radios, headsets and loudspeakers using push-button control and LED selection display. The units provide 10 W across 4 ohms to speakers or 50 MW to 600 ohms headphones.

Specifications

Dimensions: 28 × 159 × 213 mm
Weight:
(CP 136) 0.82 kg
(CP 136M) 0.91 kg
Power supply: 13.75 or 27.5 V DC

Operational status

CP 136 no longer in production; the CP 136M is still being produced.

Contractor

Narco Avionics Inc.

VERIFIED

AN/AIC-28(V)1, 2, 3 and 4 audio distribution systems

The AN/AIC-28(V)1 audio distribution system provides COMSEC-compatible internal communications among aircrew and direct access communications between them and personnel in support of aircraft mission and maintenance activities. The Audio Distribution System (ADS) consists of up to 27 subscriber control panels located at the various crew stations, two central switching units to route audio traffic and a programming, display and test panel to control, monitor and test the ADS.

The key to achieving ADS secure communication performance lies in the microminiature LSI design which provides extremely low levels of cross-talk (greater than 100 dB between communication channels). Interlock circuitry prevents any subscriber from transmitting simultaneously on a clear and a secure or classified channel. The interlock also inhibits all direct access transmissions from a subscriber station initiating a public address announcement.

Any subscriber control panel or station can signal any other subscriber station for a private conversation via a four-channel selective intercom. Other subscribers can be added to form progressive conferences.

Access to radios by subscribers is accomplished remotely from the programming, display and test panel. Subscribers are provided with panel controls for selection of monitor and transmit functions and controls for adjusting the receive volume level for up to four radios at a time. The ADS retains all selective interphone and radio access operations in the event of an aircraft power failure.

The programming display and test panel enables a control operator to programme selectively the direct access radio and intercom nets to the various subscribers. It also displays the current programme status by visual display of the channels assigned to each subscriber. The programming display and test system isolates faults down to a replaceable primary unit level.

The central switching unit interfaces directly with the system's subscriber stations and serves as an audio distribution matrix. It consists of an array of audio multiplexer devices with appropriate station and channel input/output audio buffers and station data buffers. All audio switching and mixing is performed within the central switching unit so that, irrespective of the total number of channels a subscriber chooses to monitor simultaneously, composite earphone audio is conveyed from the central switching unit to the subscriber station via a single twisted cable pair. Conversely, when a subscriber transmits, the microphone audio is conveyed on a single twisted pair to the central switching unit which distributes it to other subscriber stations and to radios on the channels on which he is transmitting. The two central switching units operate synchronously in a master/slave relationship.

Four separate types of subscriber station panels can operate with the system: a flight deck audio panel, a mission audio panel, a special audio channel and maintenance audio panel.

All flight deck and mission audio panels can initiate public address announcements which are broadcast over all headsets and loudspeakers in the aircraft. When in public address mode, all direct access radio transmission from the originating subscriber station is inhibited and when originating from the flight deck the speaker nearest the originating station is muted. All-weather 127 mm and 203 mm speakers are provided. The public address amplifier is packaged for standard air transport racking. The package contains controls and an output power meter for testing and adjusting the

120 W main amplifier, a 16 W auxiliary amplifier and the speakers. Provision is made for the incorporation of recorders to log all two-way external communications.

The system has a built-in self-test capability at both system level and at subscriber station level. Replacement of primary units is possible during mission operations. A semi-automatic test station is available to facilitate ground maintenance and repair of primary units.

System features include the automatic assignment of nets on the selective intercom together with logic to decode call digits and determine which nets are available to the respective subscribers; capability of subscribers to transmit on more than one communication net; direct access to as many as 22 radios with additional access from flight crew stations to seven navigation monitors, and capability for distributing audible alarm signals to subscribers in response to sensors and switches external to the ADS.

The AN/AIC-28(V)2 and AN/AIC-28(V)3 audio distribution systems are identical to the AN/AIC-28(V)1 system except for an increased number of subscriber audio stations and direct access radios available for mission operations. These changes required expansion of the programming display and test panel and the central switching unit.

The AN/AIC-28(V)4 tanker audio distribution system provides secure communications between crew members and radio equipment. The system consists primarily of sunlight-readable refuelling audio panels located at stations for the boom operator, observers and loadmaster; flight deck audio panels and their associated clear-secure panels, located at the aircraft commander's, pilot's, flight engineer's and navigator's station; central switching units to route the audio traffic; a radio select panel to programme access to radios, and an ADS test panel to test and monitor operation of the system.

The system audio traffic is divided into clear and secure communications. The central switching unit interlock circuitry prevents compromise of secure communications, such as transmitting simultaneously on a clear and secure channel. Any subscriber panel can be used to signal other subscribers for private or conference conversations, or can be used to make public address announcements, which are broadcast over the headsets and loudspeakers.

Operational status

In service. The AN/AIC-28(V)1, 2 and 3 are no longer in production.

Contractor

Palomar Products Inc.

VERIFIED

AN/AIC-29(V)1 intercommunication system

The AN/AIC-29(V)1 intercommunication system provides secure internal communications between helicopter crew members and direct access between crew members and radios and/or security equipment for external communications. The system can accommodate up to 15 transmit and 18 receive radio channels and provides greater than 100 dB cross-talk isolation between a transmit and any other transmit or receive channel. The system consists of six crew station units, a single maintenance station unit and a communications switching unit. The crew station units are NVG-compatible. The communications switching unit performs switching and mixing of audio channels in accordance with digital data multiplexed from each crew station unit. The switching unit's response to the multiplexed data depends on the communication plan programmed into the system.

The system provides emergency back-up intercom and radio communication selection which bypasses the communications switching unit for audio transmit and receive functions.

System features include the capability for distributing audible alarms to crew stations in response to sensors/switches external to the intercommunication system. The system also provides interface with the MIL-STD-1553 databus, two-way chime call capability and built-in test circuits.

Operational status

In production.

Contractor

Palomar Products Inc.

VERIFIED

AN/AIC-30(V)1 and (V)2 intercommunication systems

The AN/AIC-30(V)1 intercommunication system provides communications among helicopter crew members and direct access to radio communications in support of mission activities. The system includes a microprocessor-controlled communications switching unit which is controlled by multiplexed data from six crew station units during normal system operation. Interlock circuitry prevents a crew member from transmitting on a clear channel while receiving on a secure channel. Cross-talk isolation between any two channels is greater than 126 dB at 1 kHz.

An intercom back-up call channel, which bypasses the switching unit matrix, may be activated at the crew station units. This back-up channel may be constantly monitored using the crew station unit master volume control. In addition, two cockpit crew station units are provided with a direct back-up interface to two radio transceivers. The intercom back-up call channel is also used for extended ground communication operations with the switching unit power removed. Two audio frequency amplifiers provide volume control for communications at the maintenance stations.

The AN/AIC-30(V)2 intercommunication system is an enhancement to the AN/AIC-30(V)1 intercommunication system for use in larger aircraft. The AN/AIC-30(V)2 system increases the number of crew stations and audio frequency amplifiers, modifies the communications switching unit, expands the radio channels and hard-wired back-up channels and provides dual-headset and microphone capability at each crew station unit.

The communications switching unit modification includes added circuit cards and wiring for multiplexing control and audio access transmit and receive. A fourth connector is included on the front panel of the switching unit to accommodate the additional audio traffic.

The crew station unit front panel changes reflect the additional radios and navaids utilised for the system. Each crew station unit provides dual-headset capability. The cross-talk isolation between transmit and receive channels is greater than 126 dB at 1 kHz. The crew station unit also provides intercom back-up circuits and is NVG-compatible. The three audio frequency amplifiers provide headset volume control at the maintenance or remote crew stations.

Operational status

In service. The AN/AIC-30(V)1 is no longer in production.

Contractor

Palomar Products Inc.

VERIFIED

AN/AIC-32(V)1 intercommunication system

The AN/AIC-32(V)1 intercommunication system provides secure internal communications between crew members and direct access to mission radios and security equipment for external communications. The system's primary units consist of a communication control unit, four flight deck crew station units, five mission area crew station units, eight maintenance station units and one maintenance control unit.

Audio traffic from all station units is controlled by the communication control unit during normal system operations. A back-up operating mode, initiated from selected crew station units, bypasses the communication control unit to provide hard-wired access to a set of predetermined radios. A back-up intercom network, integrated with the call function on all station units, is also provided for emergency intercommunication between crew members.

A test switch on each crew station unit permits preflight verification of audio and lamp indicator circuitry and digital interface with the communication control unit. A public address system, accessed from the flight crew station, allows announcements over the loudspeakers and headsets. Auxiliary control units, located at selected mission area crew station units, expands the total system direct access capability to 30 transmit and 36 receive channels.

Operational status

In production and in service.

Contractor

Palomar Products Inc.

VERIFIED

AN/AIC-34(V)1 and (V)2 intercommunication systems

The AN/AIC-34(V)1 secure intercommunication system is a programmable microprocessor-controlled modular audio and digital communication distribution system designed for use on board fixed- and rotary-wing aircraft or in ground-based C^3 shelters. The AN/AIC-34(V)1 provides internal communications between crew members as well as crew member access to mission radios and communication security equipment for external communications. Designed in accordance with MIL-E-5400, it is qualified to MIL-STD-810 for air transport, MIL-STD-461 for EMI and NACSIM 5100 for TEMPEST compliance.

The AN/AIC-34(V)1 intercommunication system comprises a Communication Control Unit (CCU), five Crew Station Units (CSU), 16 Control Display Units (CDU), an Emergency Audio Panel (EAP), three Maintenance Station Units (MSU) and 26 jack boxes.

The CCU is a microprocessor-controlled modular audio switching and control unit that provides up to 32 crew stations with access to as many as 48 receive channels and 30 transmit channels. It also accepts up to 70 binary discrete inputs and provides up to 122 switched relay outputs for discrete control and crypto switching.

The CSUs and CDUs are the crew member interfaces to the communication control unit and communication assets. The functions of channel select switches and volume controls on the CSU front panel are firmware dependent and are reconfigurable to meet the needs of specific missions or system requirements. The CDU allows operator assignment of communication assets and displays communication system operational and BIT status on an integral eight-colour CRT display. Both units are capable of operating with two headsets when operated in a monaural mode or can operate with one headset when operated in a binaural mode. Monaural or binaural operating modes are configured by jumpers at the unit connectors in the aircraft wiring.

The MSU provides intercom access for maintenance personnel and ground crew. Each communication control, control display and maintenance station unit is connected to a jack box. This allows switching of multiple microphone inputs and provides for switching of audio from an auxiliary source or attached unit to various headset or speaker interfaces.

The EAP provides an operator with the capability to select multiple levels of degraded or emergency back-up modes of operation to provide for continued operation in the event of hardware failures or battle damage.

The AN/AIC-34(V)2 secure intercommunication set is a subset of the AN/AIC-34(V)1 system, and is used in airborne military applications where the crew complement is significantly less than that of the AIC-34(V)1 but the requirement to access numerous radio channels and the need for the ability to reconfigure the operating modes of the onboard communication assets are similar. The AN/AIC-34(V)2 intercommunication set is designed to be compliant with the requirements of MIL-E-5400, MIL-STD-810, MIL-STD-461 and NACSIM 5100.

The set comprises a CCU, five CSUs, two CDUs, an EAP, three MSUs and eight jack boxes.

The CCU is the same as that used in the AN/AIC-34(V)1 except that it contains an additional set of relay circuit cards to provide for greater flexibility and expanded switching modes of the communication assets. The firmware resident in this CCU is also unique to the requirements of the AN/AIC-34(V)2.

The CSU, CDU, MSU, EAP and jack boxes are identical to those used in the AN/AIC-34(V)1 except for the CSU front panel switch legends.

Specifications

Dimensions:
(CCU) 497.6 × 257 × 193.7 mm
(CSU) 165 × 146 × 152.4 mm
(CDU) 165.1 × 146 × 152.4 mm
(EAP) 106.7 × 146 × 47.8 mm
(MSU) 78.7 × 160 × 107.9 mm
(jack box) 150.5 × 127.5 × 43.2 mm
Weight:
(CCU) 19.8 kg
(CSU) 2.7 kg
(CDU) 3.6 kg
(EAP) 0.45 kg
(MSU) 1.03 kg
(jack box) 0.34 kg

Operational status

In service with the US Navy.

Contractor

Palomar Products Inc.

VERIFIED

AN/AIC-38(V)1 and AN/AIC-40(V)1 intercommunication systems

The AN/AIC-38(V)1 and AN/AIC-40(V)1 secure intercommunication systems are programmable microprocessor-controlled modular audio and digital communication distribution systems designed for use in both fixed- and rotary-wing aircraft or in ground-based C³ shelters. The AN/AIC-38(V)1 and AN/AIC-40(V)1 provide internal communications between crew members as well as access to mission radios and communications security equipment for external communications. Designed in accordance with MIL-E-5400, the systems are qualified to MIL-STD-810D for air transport, MIL-STD-461 for EMI and NACSIM 5100 for TEMPEST.

The systems consist of a Communication Control Unit (CCU), Crew Station Units (CSU), a Digital Switch Unit (DSU) and Emergency Audio Panel (EAP).

The CCU is a microprocessor-controlled modular audio switching and control unit that provides up to 32 crew stations with access to up to 48 receive channels and 30 transmit channels. It also accepts up to 70 binary discrete inputs and provides up to 122 switched relay outputs for discrete control and crypto switching. The communications connectivity plan resident within the CCU may be reconfigured in real time by a higher-order controller via MIL-STD-1553B databus, or by operator control via a CSU.

The CSU provides crew member interface to communication assets. Channel select keyswitches and volume controls on the CSU are firmware-controlled and are reconfigurable to meet the needs of specific missions or system requirements.

The DSU, under CCU control, routes eight bidirectional channels of digital data and control lines to data terminals and radios or communication security equipment. Using the EAP, multiple levels of emergency back-up operation are operator selectable for continued operation after hardware failures or battle damage.

Operational status

In production and in service.

Contractor

Palomar Products Inc.

VERIFIED

AN/AIC-39(V)1 intercommunication system

The AN/AIC-39(V)1 secure intercommunication system is a programmable microprocessor-controlled modular audio and digital communication distribution system designed for use on board both fixed-wing aircraft and helicopters or in ground-based C³ shelters. The AN/AIC-39(V)1 provides internal communications between crew members as well as crew member access to mission radios and communications security equipment for external communications. Designed in accordance with MIL-E-5400, the AN/AIC-39(V)1 is qualified to MIL-T-5422 and MIL-STD-810D for air transport, MIL-STD-461 for EMI and NACSIM 5100 for TEMPEST.

The system consists of a Communication Control Unit (CCU), Crew Station Units (CSU), Dual Crew Station Units (DCSU), Audio Amplifier Units (AAU), Auxiliary Control Units (ACU), an Emergency Audio Panel (EAP) and secure jack box.

The CCU is a microprocessor-controlled modular audio switching and control unit that provides up to 32 crew stations with access to up to 48 receive channels and 30 transmit channels. It also accepts up to 70 binary discrete inputs and provides up to 122 switched

Crew station unit for the AN/AIC-34, AN/AIC-38, AN/AIC-39 and AN/AIC-40 intercommunication sets

relay outputs for discrete control and crypto switching. The communications connectivity plan resident within the CCU may be reconfigured in real time by a higher-order controller, via MIL-STD-1553B databus, or by operator control via a CSU or DCSU.

The CSU and DCSU provide crew member interface to communication assets. Channel select keyswitched and volume controls on the CSU and DCSU are firmware controlled and are reconfigurable to meet the needs of specific missions or system requirements. DCSUs are identical to CSUs except that the DCSUs are equipped with additional audio circuitry to support a subordinate ACU.

The ACU used in conjunction with the DCSU, provides limited access to communication assets at remote crew positions. The ACU operates with a subset of the functions provided by the DCSU.

The AAUs provide a binaural headset and microphone interface to the CCU, but have no annunciators or indicators.

The EAP allows operator selection of multiple levels of emergency back-up operation. The back-up capability permits continued operation after hardware failures or battle damage.

Operational status

In production and in service.

Contractor

Palomar Products Inc.

VERIFIED

Advanced Narrowband Digital Voice Terminal (ANDVT)

The Advanced Narrowband Digital Voice Terminal (ANDVT) provides half-duplex secure voice and data communications for a variety of military tactical applications, including shipboard, land-based and airborne. Typical user modes include secure voice, data and signalling, point-to-point and modem processor only.

The CV-3591 Basic Terminal Unit (BTU) provides voice and modem processing by using two similar signal processors. Communications security is achieved by encoding and decoding the digital data to and from the voice processor external data device.

In the standard terminal configuration, the KYV-5 COMSEC Module (CM) is a front panel plug-in unit.

Specifications

Dimensions:
(BTU) 193.8 × 124.7 × 337.8 mm
(CM) 158.75 × 123.95 × 76.2 mm
(MPU/VPU) 157.5 × 124.7 × 75.4 mm
(interface unit) 146 × 69.85 × 285.75 mm
Weight:
(BTU) 9.89 kg
(CM) 1.63 kg
(MPU/VPU) 1.27 kg
(interface unit) 2.13 kg
Temperature range: −46 to +95°C
Altitude: up to 70,000 ft
Reliability: 2,000 h MTBF

Contractor

Raytheon Systems Company.

UPDATED

AN/ARC-114A VHF radio

The AN/ARC-114A is a military airborne radio which covers the VHF band from 30 to 75.95 MHz and provides 920 channels at a spacing of 50 kHz. The system also incorporates an independent Guard receiver to monitor a single, pretuned Guard channel in the frequency range 40 to 42 MHz.

A compact lightweight single container unit, the ARC-114A is designed primarily for helicopters, although its high-altitude performance up to 50,000 ft means it is suitable for higher-flying fixed-wing aircraft.

The ARC-114A operates in FM mode and can transmit either voice or data signals. Transmitter power output is 10 W. Facilities include azimuth homing and a rebroadcast capability is also available when the system is used in conjunction with a suitable transmitter/receiver.

The system is of all-solid-state construction and employs crystal-controlled digital synthesis techniques for frequency generation. In receive mode, varactor and band selection tuning methods are employed and for transmit mode the frequency selection is accomplished by indirect synthesis and digital tuning techniques.

Specifications

Dimensions: 105 × 146 × 215 mm
Weight: 3.1 kg
Temperature range: −32 to +55°C

Operational status

No longer in production. In service in helicopters of some six forces around the world. The ARC-114A has been produced in quantity for the US Army which procured over 5,000 systems for service in its Bell UH-1H and Sikorsky UH-60 utility helicopters.

Contractor

Raytheon Systems Company.

UPDATED

AN/ARC-115A VHF radio

The AN/ARC-115A broadly has the same range of facilities as the company's AN/ARC-114A system but provides, within the frequency band 116 to 149.975 MHz, 1,360 channels at a spacing of 25 kHz. Like the ARC-114A, it has a rebroadcast facility when operated in conjunction with suitable equipment. It also has an azimuth homing capability.

Specifications

Dimensions: 124 × 146 × 235 mm
Weight: 3.26 kg

Operational status

No longer in production. In service with the UH-1H helicopters of the US Army.

Contractor

Raytheon Systems Company.

UPDATED

AN/ARC-164 UHF radio

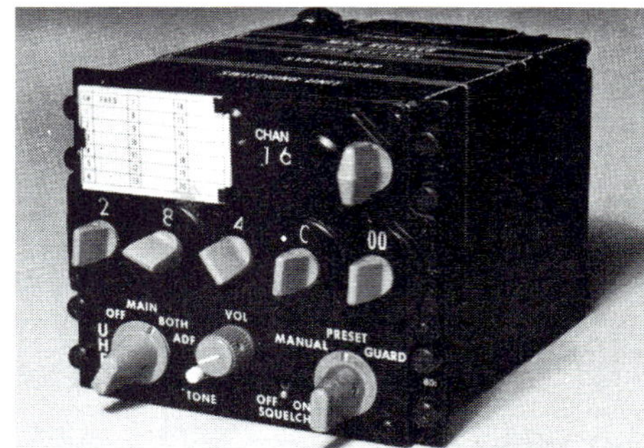

The ARC-164 UHF radio

The ARC-164 is the basic member of a family of radio communications equipment and subvariants, each designed for particular applications yet with a high degree of commonality.

The basic ARC-164 covers the UHF band, providing 7,000 channels over the range 225 to 400 MHz in 25 kHz increments. Any 20 channels may be preselected.

A fully solid-state system, the ARC-164 is distinguished by its 'slice' module construction in which a series of modules, connected by a flexible harness, are simply bolted together to form the desired electronic configuration. A typical simple system would comprise transmitter, receiver, guard receiver and synthesiser. The control unit may either form part of this consolidated package or be remotely located. The modular approach allows growth capability, extra modules being added as required. A range of optional facilities, such as data transmission, secure speech and ECCM capability, is available by the addition of the appropriate slices.

A number of directly connected or remote-control units are produced for the ARC-164. These include: a simple frequency-selection controller; a 32-channel preset control with LED readout of the selected channel; a 20-channel preset unit with provision for two-cockpit take-control; a microprocessor controller with 400 UHF and VHF, AM or FM, preset channels, liquid crystal channel and frequency readout display and the capability of controlling up to four systems simultaneously. Additional remote frequency/channel indicators are available. Also available is a variety of mounting trays to suit differing installations for new types of aircraft and for the updating of older aircraft equipment.

A remote ARC-164 radio compatible with MIL-STD-1553B databus operation has been developed under contract to the US Army and is in production and service. Panel-mounted radios and certain controls can be furnished with ANVIS Green A lighting compatible with NVG in accordance with MIL-STD-85762. These features can also be obtained by retrofitting appropriate radios and controls. A Low Probability of Intercept and Detection (LPI/LPD) version of the ARC-164, which allows covert UHF communications, is known as StealthComm.

StealthComm waveform features include a hybrid of several LPI/LPD techniques: hybrid direct sequence and frequency hopping; feature suppression modulation techniques; receive sensitivity enhancements; 60 dB of power control.

StealthComm has enhanced data capability from 16 to 80 to 120 kbytes.

All StealthComm features can be incorporated in existing ARC-164 radios through mod kits; or new ARC-164/LPD radios can be installed as direct replacements form-fit for existing ARC-164s.

Specifications

Dimensions:
(transmitter/receiver)
(10 W version) ½ ATR × 178 mm
(30 W version) ½ ATR × 374 mm
(controller) ½ ATR × 83 mm
Weight:
(transmitter/receiver)
(10 W version) 3.7 kg
(30 W version) 6.8 kg
(controller) 2 kg
Frequency: 225-400 MHz
Channels: 7,000
Channel spacing: 25 kHz
Power output: 10 W (standard), 30 W (uprated)
Reliability: 2,000 h MTBF demonstrated

Operational status

In production and service. More than 42,000 ARC-164s have been produced to date.

The system is fitted to a wide range of US Air Force aircraft including the F-16 fighter. Total US Air Force orders exceed 18,000 sets.

The ARC-164 equips the Royal Navy's Sea King helicopters and Sea Harrier aircraft and the Hawk, Jaguar and other Royal Air Force aircraft. Further British Aerospace Hawk aircraft, notably those delivered to Kenya, are also fitted with the ARC-164, as are some BAe Strikemaster strike/trainers.

Contractor

Raytheon Systems Company.

UPDATED

The ARC-164 (seen on the right-hand side of the cockpit below the main instrument panel) installed in the British Aerospace Hawk

AN/ARC-181 TDMA radio terminal

The AN/ARC-181 Time-Division Multiple Access (TDMA) terminal is a secure, jam-resistant radio terminal for airborne surveillance, command and control centres developed for US and NATO E-3A AWACS aircraft, Hawk missile batteries and NATO Air Defence Ground Environment centres under the Airborne Early Warning/Ground Environment Integration Segment (AEGIS) programme. The terminal comprises a communications processor, transmitter, receiver, high-power amplifier and control and display panel.

Part of the JTIDS programme, the terminals provide the channel for continuous communications exchange, resulting in a constantly updated information pool which is available to all network members. Spread spectrum, data interleaving and frequency-hopping techniques give enhanced data and jam-resistance capabilities.

Operational status

First production terminals went into service with NATO Boeing E-3A aircraft and ground stations in 1983.

Contractor

Raytheon Systems Company.

UPDATED

AN/ARC-187 UHF radio

The ARC-187 is a further development of the company's ARC-164 US Air Force standard system which has been adapted to meet a US Navy requirement for a low-cost terminal designed to operate with communication satellites, principally the US Navy's FltSatCom. It covers the UHF band from 225 to 400 MHz, in which range it provides 7,000 channels at increments of 25 kHz. Up to 20 channels may be preselected.

Operating modes include AM with a secure speech facility and FM and FSK data transmission in both analogue and digital form. ECCM capability is incorporated internally in the receiver/transmitter.

The system, which is remotely controlled, uses standard ARC-164 'slice' modules including a modified synthesiser section designed for compatibility with communications satellite data rate requirements.

A new control for the ARC-187 has been developed, which provides for compatibility with Satcom and MIL-STD-1553B databus modes of operation. This control incorporates ANVIS Green A lighting in accordance with MIL-STD-85762 for compatibility with Gen III NVG. A Satcom modem for the ARC-187 is under development.

Specifications

Dimensions:
(controller) 132 × 147 × 124 mm
(transmitter/receiver) 440 × 153 × 143 mm
Weight:
(controller) 2 kg
(transmitter/receiver) 7.4 kg
Frequency: 225-400 MHz
Channels: 7,000
Channel spacing: 25 kHz
Power output: 30 W (AM), 100 W (FM/FSK)

Operational status

In service in US Navy P-3C aircraft.

Contractor

Raytheon Systems Company.

UPDATED

AN/ARC-195 VHF radio

The AN/ARC-195 radio may be considered as a variant of the UHF AN/ARC-164, with which it has a component commonality of 93 per cent. It is available in 10 W and 30 W output versions. The major differences between the two systems consist of some component value changes and the substitution in the AN/ARC-195 of a synthesiser with a frequency standard appropriate to the VHF section of the radio frequency spectrum.

The AN/ARC-195 covers the VHF band from 116 to 156 MHz, providing 1,750 channels at a frequency separation of 25 kHz. In other respects it is almost identical to its UHF counterpart. Control units are also almost identical and certain units from the AN/ARC-164 range of controllers may be used for combined UHF/VHF operation.

Operational status
In production and in service.

Contractor
Raytheon Systems Company.

UPDATED

AN/ARC-222 SINCGARS radio

The AN/ARC-222 SINCGARS radio is the replacement for the AN/ARC-186 and is designed for air-to-ground and air-to-air communications. It includes SINCGARS-V capability and covers the frequency ranges 30 to 87.975 MHz in VHF FM, 108 to 151.975 MHz in VHF AM (108 to 115.975 MHz receive only) and 152 to 174 MHz FM for the International Maritime Band.

It consists of a control unit and a receiver/transmitter.

Specifications
Dimensions:
(control unit) 57.4 × 127 × 95.8 mm
(receiver/transmitter) 120.6 × 127 × 174.6 mm
Weight:
(control unit) 0.73 kg
(receiver/transmitter) 4.67 kg
Power supply: 28 V DC, 4.5 A max
Frequency:
30-87.975 MHz (FM)
108-151.975 MHz (AM) (108-115.975 MHz receive only)
152-174 MHz (International Maritime Band)

Contractor
Raytheon Systems Company.

UPDATED

AN/AXQ-14 datalink

The AN/AXQ-14 is a two-way communication datalink to guide the GBU-15 glide bomb. It provides a video and command link between the command aircraft and the weapon, enabling the systems operator to remain in the control loop while the weapon is being directed to its target. In effect, the datalink permits a command authority similar to a fly-by-wire system, in which the operator can transmit guidance instruction from launch to impact. Alternatively, he may select any one of a number of autonomous weapon control modes, including an override mode which permits target updating or redesignation as required.

The extended weapon control capability conferred by the datalink contributes to weapon system performance in terms of standoff range and operational utility. Target acquisition is deferred until the weapon, rather than the command aircraft, is closer to the target. Tactically, the aircraft can leave the target zone immediately after launch.

The AN/AXQ-14 system comprises three major elements: a datalink pod mounted on the command aircraft, a datalink control panel used in conjunction with an existing display within the aircraft and a weapon datalink module mounted on the rear of the weapon itself.

The pod is an aerodynamically shaped container mounted on a standard stores carriage strongpoint on the fuselage centreline or on an underwing station, according to aircraft type. It contains four LRUs comprising an electronics section incorporating all radio frequency generating and receiving equipment, a demultiplexer to decode all aircraft command and pod control signals, an encoder and antenna controls; a phase-scanned array for weapon tracking in normal operation; a forward horn antenna that provides additional coverage; and a mission tape recorder which maintains a permanent record of weapon video data. The pod is suitable for high-performance aircraft, is certificated for operation at speeds in excess of M1.0 and is also compatible with high- and low-altitude operations. There is said to be no compromise of aircraft performance attributable to carriage of the pod.

The AN/AXQ-14 datalink pod mounted inboard of a GBU-15 glide bomb, under the fuselage of a US Air Force F-15 ***1995***

Used in conjunction with an existing display system, the aircraft control panel acts as the interface between the weapon system operator and the weapon guidance system. The panel accepts signal inputs from the aircraft as well as from its own controls, and formats these into discrete commands as required via the datalink. Although the panels are tailored to the individual requirements of the aircraft type and intended customer usage, each unit accepts the standard configurations of the GBU-15/AGM-130 datalink and the pod.

Attached to the aft of the GBU-15/AGM-130 weapon is the ultimate component in the datalink chain, the weapon datalink module. This simultaneously transmits video from the weapon's seeker-head and processes incoming command signals from the aircraft to the weapon. Heading changes during the weapon's flight are effected through discrete command signals. Dual-analogue command channels enable the operator to slew the weapon in pitch and yaw during approach to the target.

Digital techniques are employed in the AN/AXQ-14 system and the transmitter is of all-solid-state construction. The system's electronically phase-scanned antenna array provides the datalink with high-rate tactical manoeuvring capability. A comprehensive range of test equipment is provided, including a flight checkout unit for testing aircraft cables from the pod connection point, an aircraft simulator unit which permits functional checks of the control panel and a weapon simulator unit for test of the aircraft pod and isolating faults down to LRU level. Used together, these two latter units permit full system functional checkout.

Two primary launch modes are envisaged for operation of the GBU-15 weapon and AN/AXQ-14 control combination: low-altitude penetration and high-altitude standoff. It is claimed that use of the datalink has improved weapon delivery accuracy over non-link weaponry in various profiles from airborne platforms such as the US Air Force's B-52, F-4 and F-15 aircraft; and US FMS F-16 aircraft. The system is also said to be compatible with the A-4, A-7 and F/A-18 aircraft. Potential weapon applications include Harpoon, Maverick and cruise missiles.

Operational status
In production and in service.

Contractor
Raytheon Systems Company.

UPDATED

AN/URQ-33(V) Class 1 JTIDS terminal

The Joint Tactical Information Distribution System (JTIDS) uses frequency hopping, spread spectrum, automatic relay and other high-technology techniques to provide data and voice communications which are highly resistant to jamming.

The AN/URQ-33(V) Class 1 JTIDS terminal is used on board the US Air Force and NATO E-3 AWACS and for a number of ground-based applications for communicating information on command and control, surveillance, intelligence, force status, target assignments, warnings and alerts, weather and logistics.

JTIDS uses a computer-controlled Time Division Multiple Access (TDMA) technique in which information is transmitted in short bursts lasting only a fraction of a second. Bursts are synchronised by computer with bursts from other users, to allow simultaneous transmission on the network without causing interference. The JTIDS burst is spread in frequency, encoded and hopped across a number of frequencies in a split second, making it hard to intercept and almost impossible to jam. The receiver selects pertinent data by means of software filtering.

JTIDS communications are automatically relayed by other terminals and this both extends the range beyond the line of sight and provides another layer of defence against jammers. It also enables terminals to provide users with position and navigation data, without the need for extra equipment, by means of the highly accurate message time of arrival measurement which can be converted to range between the transmitter and the receiver.

JTIDS is broadcast in the 960 to 1,215 MHz frequency range. The system consists of a radio set control, transceiver processor unit, high-power amplifier, high-power amplifier power supply, low-power amplifier power supply and antenna coupler. It also includes a general purpose digital computer, programmed to perform most of the communications tasks and interface with the host platform's computer.

Contractor
Raytheon Systems Company.

UPDATED

C-1282AG remote-control unit

The C-1282AG Remote-Control Unit (RCU) provides the capability to control and monitor remotely the operation of the RT-1273AG Multimission UHF Satcom Transceiver (MUST) or the MD-1269A MultiPurpose Modem (MPM) over an asynchronous MIL-STD-188-114A balanced interface. Frequency and mode selection for the AN/ARC-171 and preset selection for the AN/WSC-3 are provided by the RCU via the MPM.

The C-1282AG RCU allows full control of all MUST and MPM radio and modem features, including frequency, data rate, modulation type, modem emulation mode, transmit power level and BIT. Menu and Arrow keys allow the operator to scroll through various menus to view or update radio or modem

The Raytheon Systems Company C-1282AG remote-control unit

configurations. The Preset switch allows access to each of the eight radio and modem preset configurations. The volume knob may be used to attenuate plain text receive audio out of the radio or modem.

The C-1282AG RCU has an illuminated panel designed to MIL-P-7788E which provides a menu-driven keypad/display interface. Display intensity is adjustable and tracks the externally controlled edge-lit panel brightness. The panel and display are designed for upgrade to NVG compatibility.

Specifications

Dimensions: 146 × 76.2 × 133.35 mm
Weight: 0.91 kg
Power supply: 19-30 V DC, 0.5 A max
Temperature range: −45 to +55°C
Altitude: up to 30,000 ft

Contractor

Raytheon Systems Company.

UPDATED

CA-657 VHF/AM radio

The CA-657 is another VHF variant of the ARC-164 UHF system and its derivative, the VHF ARC-195. Its frequency band coverage, however, is somewhat broader than the ARC-195 since it covers from 100 to 159.975 MHz, in which range it provides 2,400 channels. Microprocessor memory control systems used with the CA-657 permit alternative preselection of up to either 20 or 30 channels. Channel separation is at intervals of 25 kHz and the transmitter output power is 10 W.

Construction of the system is based upon that of the ARC-164 and a high degree of component commonality exists.

Operational status

In production and in service.

Contractor

Raytheon Systems Company.

UPDATED

Commanders' Tactical Terminal/ Hybrid-Receive only (CTT/H-R)

The Commanders' Tactical Terminal/Hybrid-Receive Only (CTT/H-R) is a multichannel, multifunction terminal used to receive real-time intelligence reports from a variety of sources. It allows the tactical user to receive data from the Tactical Reconnaissance Intelligence Exchange Service (TRIXS) or the Tactical Information Broadcast Service (TIBS), while simultaneously receiving the Tactical Receive equipment and related APplications (TRAP) and Tactical Data Information eXchange System-Broadcast (TADIXS-B). This capability provides the user with a significant increase in access to real-time intelligence data and offers the flexibility to utilise whatever resources are available in the theatre.

CTT/H-R is the result of an evolutionary development that began with the full-duplex CTT field terminals and has continued with the development of a CTT/TEREC terminal, a multichannel CONSTANT SOURCE receiver system, a TIBS interface unit and a TIBS receive only unit. The CTT/H-R combines the functions of all these systems into a single receive-only terminal.

The CTT/H-R is packaged into a single unit and uses open architecture VME technology. Some of the key VME modules used include embedded COMSEC modules based on CTIC and Ricebird crypto chips, multiple 680X0 microprocessor modules and a MIL-STD-1553 interface. Additional modules include a single-board UHF receiver/synthesiser and a single-board modem using TMS320 processors. The CTTR/H-R is qualified for installation in fixed-wing aircraft and helicopters, as well as vehicles and ships. It uses Ada software and has a download capability for changes or upgrades.

The Raytheon Systems Class 1 JTIDS terminal on board a US Air Force E-3A AWACS aircraft

Specifications

Dimensions: 190.5 × 266.7 × 520.7 mm
Weight: 19.5 kg
Power supply: 100-132 V AC, 47-140 Hz, 180 W or 200-264 V AC, 47-66 Hz, 180 W
Frequency: 225-400 Hz
Channel spacing: 5.25 kHz
Temperature range: −45 to +55°C
Altitude: up to 30,000 ft

Operational status

In production.

Contractor

Raytheon Systems Company.

UPDATED

CV-3670/A digital speech processor

The CV-3670/A is an airborne digital, linear predictive data converter which provides digitised speech output at a data rate of 2.4 kbits/s, permitting narrowband system operation on standard voice quality circuits. The output can be multiplexed with other databit streams to allow simultaneous voice and data transmissions. Independent data clocks, provided by the data terminal to the transmitter analyser and receiver synthesiser of the speech processor, are used to synchronise the equipment to the receiver and transmitter of the data terminal.

The CV-3670/A is a remotely operated unit packaged in a ½ ATR box, with voice input/output via the aircraft intercommunication system. It is operated by use of the C-10085/A controller/indicator.

Contractor

Raytheon Systems Company.

UPDATED

E-SAT 300A satellite receiver

The E-SAT 300A is an all-digital satellite communication system designed for aircraft applications. Airborne telephone calls and data messages are automatically transmitted via the INMARSAT network to ground stations. Two telephone calls can be made simultaneously, while a third channel is available for relaying data messages. The system is designed to accommodate growth to eight simultaneous telephone channels.

The steered high-gain single-helix antenna features small size and weight and minimises the number of LRUs. The design eliminates the keyhole and is mechanically steerable through 360° in azimuth and −30 to +90° in elevation.

Specifications

Dimensions:
(satellite data unit) 9 MCU
(radio frequency unit) 12 MCU
(Class A high-power amplifier) 8 MCU
(antenna control unit) 6 MCU
(system power supply) 8 MCU
(radome) 374.7 × 431.8 × 2,794 mm
Weight: (total system) 99.79 kg
Power: 1.3 kVA
Frequency:
(receive) 1,530-1,559 MHz
(transmit) 1,626.5-1,660.5 MHz
Reliability: 31,500 h MTBF

Contractor

Raytheon Systems Company.

UPDATED

Have Quick System

Have Quick provides the user with an effective air-to-air, air-to-ground and ground-to-air jam-resistant UHF voice communication capability that will allow operations in a jamming environment.

The Have Quick system consists of an ECCM modification to selected airborne and ground-based radios, which gives them a frequency-hopping capability. Part of the strength of the system comes from the use of channels in an apparently random manner so that no pattern is evident to the external observer. Jamming is consequently more difficult.

The frequency-hopping scheme is implemented by storing a pattern of the frequencies to be used for a given day within every Have Quick radio and utilising this pattern according to the time of day. For every time slot in the day, where each time slot is a small part of a second, there is a specific frequency which must be used for a given communications net, whether it is transmitting or receiving. This frequency changes pseudorandomly from one time slot to the next. Thus, Have Quick terminals require some means to store the frequency pattern for channel use on a given day and also an accurate clock to control the times at which the pattern is consulted.

The Have Quick radio retains the normal non-hopping mode where it uses any one of the 7,000 channels available in the 225 to 400 MHz UHF communication band. The use and operation of the radio in normal mode is essentially unchanged from present-day procedures.

If jamming is encountered, the Have Quick radios can switch over to the ECCM mode and continue their communication. In order to permit this switch over, the radios must be suitably primed so that they will be synchronised in the ECCM mode; this is usually done before take-off.

The Have Quick system also has a capability termed multichannel or break-in operation. This permits a Have Quick radio to receive two simultaneous transmissions on the same net while avoiding the beat note which typically prevents the listener from understanding either transmission. The technique is implemented automatically in the transmitter where it is recognised whether or not the net is already in use. If so, the transmitter side-steps by one 25 kHz channel. Since the receiver is set to wideband mode, the second signal is received in addition to the original. This multichannel capability may be selected by the operator whenever the system is in ECCM mode.

Operational status

Have Quick is in service in a number of military aircraft. Both the AN/ARC-164 and the AN/ARC-171 have been modified for Have Quick.

Contractor

Raytheon Systems Company.

UPDATED

MD-1269A multipurpose modem

The MD-1269A multipurpose modem is an advanced digital signal processor-based, full-duplex 70 MHz modem for UHF Satcom and line of sight communications links. It simplifies existing communications systems and terminals by combining the important modulation modes of six currently used modems into a single unit.

Digital data shaping provides spectral containment and permits 2,400 bps BPSK and 4,800 bps OQPSK operation over a 5 kHz satellite channel. The MD-1269A also includes baseband interfaces with audio equipment, data terminals or encryption devices to provide narrowband AM and FM voice, TADIL A (link 11), TADIL B, TADIL C (Link 4), KY-57/58 (Vinson AM and FM), KY-65/75 (Parkhill), KG-84 and ANDVT compatibility.

The MD-1269A has a standard 70 MHz RF interface and can be used with any 70 MHz compatible receiver/transmitter in the non-DAMA modes. The 70 MHz interface is synthesised to provide 5 kHz channelisation capability with the R/T operation on 25 kHz channels. For DAMA operation, the R/T must be compatible with the stricter requirements for DAMA operation. The modem is compatible with the US Navy 25 kHz TDMA-1 DAMA and the US Air Force 5 kHz USTS DAMA. All non-DAMA capabilities are preserved in the DAMA version of the MD-1269A for backward compatibility.

The MD-1269A includes synchronous and asynchronous control/data ports for various local or remote-control options. All control and data ports can be configured as balanced or unbalanced MIL-STD-188-114A interfaces to a host terminal controller or C-1282AG control head. The entire modem is contained in a ½ ATR airborne package and is ruggedised to withstand airborne, shipboard or tactical environments.

Specifications

Dimensions: 123.95 × 193.55 × 497.84 mm
Weight: 11.34 kg
Power supply: 115 or 230 V AC, 47-440 Hz, 50 W max
Temperature range: −40 to +55°C
Altitude: up to 30,000 ft

Contractor

Raytheon Systems Company.

UPDATED

MX 42000 UHF Satcom datalink

The MX 42000 UHF Satcom datalink is suitable for a wide range of manned and unmannned airborne applications, as well as ground-based ones. It has been adapted from the MXF-420 which has been selected by the US Army for the Joint Services Multiband AN/PCS-5 programme.

The MX 42000 features embedded, variable ratio video/image compression, an embedded ViaSat DAMA modem to provide full 5 and 25 kHz capability and embedded COMSEC. It is also available with multiband VHF/UHF capability in addition to its UHF Satcom capability and can be operated as a voice/data communications relay.

Specifications

Volume: 0.0112 m³
Weight: 9.98 kg
Power output: 300 W (with 100 W amplifier)

Contractor

Raytheon Systems Company.

UPDATED

MXF-400 Series V/UHF communication system

The MXF 400 family of VHF/UHF multiband multifunction communication systems uses a modular technology common to ground and airborne systems.

The MXF-400 modules are adapted to manpack, vehicular and airborne applications through the use of interfaces proven on the PACER SPEAK, ARC-164, ARC-187 and ARC-222.

Raytheon Systems is developing the MXF-430 and AN/PSC-5 receiver-transmitters into a form, fit and function replacement for the RT-1319 in the VRC-83 and GRC-206. The MXF-440 will provide a complete UHF/VHF LOS/UHF Satcom communication system, including Have Quick II, COMSEC and DAMA, within the footprint of the ARC-187. The MXF-450 provides UHF/VHF LOS communication with Have Quick II and COMSEC in a shell that is form, fit and functionally interchangeable with an RT-1504/ARC-164 or RT-1614/ARC-164.

Raytheon Systems is also studying the following developments to the MXF-400 family: SATURN UHF ECCM for NATO applications is an option on current contracts; 8.3 kHz channel spacing in VHF-AM for ICAO ATC requirements in European airspace; TDMA in VHF-AM for FAA ATC requirements in US airspace; Link 4 and Link 11 datalink capability; SINCGARS; expansion of band coverage to 512 MHz with continuous tone code subaudible system; video compression imagery.

Operational status

In development.

Contractor

Raytheon Systems Company.

UPDATED

RT-1273AG DAMA UHF Satcom transceiver

The RT-1273AG UHF Satcom transceiver is an advanced full-duplex transceiver which upgrades and simplifies existing communications systems by combining modem and transceiver functions into one ATR unit. It incorporates the functionality of the Digital Signal Processor (DSP) based on the MD-1269A multipurpose modem which provides interoperability with all existing and planned UHF satellite communication equipment. The DSP architecture is designed to allow new waveforms to be programmed and incorporated as software updates. This integrated Satcom system reduces cabling, hardware size, weight and power consumption.

The RT-1273AG transceiver section features an integral 100 W transmitter and a highly sensitive receiver for UHF line of sight or satellite communication. Transmitter power output is variable from 1 to 100 W for optimum link performance when using either low- or high-gain Satcom antennas. A high-performance synthesiser provides independent 5 kHz channel selection of receive and transmit frequencies from 225 to 400 MHz for full-duplex operation.

The RT-1273AG modem section is based on the MD-1269A multipurpose modem. The modem section features a DSP-based architecture that offers multiple modulation modes and data rates. DSP-generated AM and FM is provided for low-distortion line of sight communications. Data rates from 75 bits/s to 38.4 bits/s are selectable for binary FSK, BPSK, Shaped BPSK, QPSK, Offset QPSK, or Shaped Offset QPSK operation. Doppler frequency offsets up to 1 kHz can be tracked by the modem to allow airborne operation. To maximise data throughput while minimising adjacent channel interference, sidelobes of the transmit signal are significantly reduced by advanced digital signal processing techniques that shape the modulated waveform. This permits 2.4 kbps BPSK and 4.8 kbps OQPSK operation over a 5 kHz satellite channel in compliance with JCS requirements.

The Demand Assigned Multiple Access (DAMA) waveforms are processed internally, allowing the RT-1273AG to participate in either 5 or 25 kHz DAMA networks with no additional equipment. Two simultaneous full-duplex baseband ports are available in the DAMA modes. Provisions for adding embedded COMSEC and digital voice have been designed into the equipment.

The RT-1273AG includes synchronous and asynchronous control/data ports for various local and remote-control options. Both ports can be configured as balanced or unbalanced MIL-STD-188-114A interfaces to a host terminal controller or control head.

Specifications

Dimensions: 256.5 × 193 × 497.8 mm
Weight: 22.68 kg max
Power supply: 115 or 230 V AC, 50/60/400 Hz, single phase
115 V AC, 400 Hz, 3 phase
28 V DC
Frequency: 225-399.995 MHz
Temperature range: −45 to +55°C
Altitude: up to 30,000 ft

Contractor

Raytheon Systems Company.

UPDATED

SADL Situational Awareness DataLink

The SADL Situational Awareness DataLink integrates air force close air support aircraft with the digitised battlefield via the US Army Enhanced Position Location Reporting System (EPLRS). SADL provides fighter-to-fighter, air-to-ground and ground-to-air data communications that are robust, secure, jam-resistant, and contention free. With its inherent automatic and on-demand position and status reporting for situational awareness, SADL provides an effective solution to the air-to-ground combat identification problem.

Fighter-to-Fighter Operation

The SADL radio is integrated with aircraft avionics over the 1553 multiplex databus, providing the pilot with cockpit displays of data from other SADL-equipped aircraft as well as EPLRS-equipped aircraft and ground units. SADL is capable of fighter-to-fighter network operation without reliance on ground-based EPLRS network control. Fighter positions, radar targets, and ground target positions are shared relative to the fighter's own inertial navigation system or global positioning system. Based on the number of fighters on the network, the data capacity of the network can be customised from the cockpit. Automated fighter-to-fighter relay and adaptive power control capabilities ensure connectivity, jam resistance, and reduced probability of undesired detection.

Air-to-Ground Mode

In the air-to-ground mode, the pilot commands the SADL radio to synchronise with a specific ground network based on encryption keys. The fighter's radio then returns to sharing fighter-to-fighter data while recording ground positions from the ground EPLRS network. EPLRS tracks the fighter and provides the fighter position and altitude for ground-to-air combat identification. At the beginning of an air-to-ground attack, the pilot uses a switch on the control stick to request a view of the five closest friendly EPLRS-netted positions shown with Xs on both the head-up display and the multifunction displays. The pilot decides whether to fire based on the proximity of friendly positions to the target for effective air-to-ground combat ID. Friendly air defence units also have a picture of friendly air tracks to minimise fratricide.

Operational status

The Joint Interoperability Test Command at Fort Huachuca, in conjunction with the USAF 422nd Test and Evaluation Squadron, successfully conducted gateway interoperability tests connecting the Joint Tactical Information Distribution System (JTIDS) to SADL. Tactical awareness displays were reliably exchanged in both directions between two SADL-equipped F-16s and a JTIDS Class 2-equipped F-15 aircraft. SADL's update rates and message structure permitted exchange of TADIL-J message sets for displays of relative fighter positions, radar targets, aircraft fuel status, weapon loads and additional information. In addition to the 1,816 EPLRS radios fielded to divisions of the Contingency Force, EPLRS received a full-scale production award in 1997 for a multiyear 1997-98 buy of 2,100 radios. A further 535 SADL radios have been ordered by the US Air National Guard for F-16 Block 25/30 and A-10 aircraft assigned to the close air support role.

Contractor

Raytheon Systems Company.

UPDATED

95S-1A direct conversion receiver

Designed for communications and surveillance use, the Collins 95S-1A direct conversion receiver spans the 5 kHz to 2,000 MHz range in 1 Hz steps. The remotely tunable receiver uses state-of-the-art Direct Conversion Receiver DSP technology to attain low-noise, spurious-free, high-performance reception of AM, FM, SSB and ISB signals from 100 Hz to 300 KHz bandwidth.

Direct conversion architecture uses a single mixer with a single oscillator tuned to the desired frequency, converting a signal directly to baseband – the zero frequency IF signal. Rockwell Collins claims that this technology reduces spurious signal generation, and that it improves reliability due to lower part counts.

Software upgrades are installed to internal flash memory via the serial control port.

Operational status

In production and in service.

Contractor

Rockwell Collins.

NEW ENTRY

Rockwell Collins 95S-1A direct conversion receiver
1998/0011824

618M-3 and 618M-3A VHF radio

The 618M-3 radio and its variant, the 618M-3A, are air transport-category VHF transceivers of 25 W nominal transmitter output. They were developed by Rockwell Collins as retrofit replacements for the earlier 618M-1 radio and for similar ARINC 546 and 566 systems as a response to the introduction of 25 kHz spacing between channels for VHF air traffic control communication.

Coverage of the VHF band by the 618M-3 is from 118 to 135.975 MHz, while the 618M-3A version coverage extends from 116 to 151.975 MHz. The former provides 720 channels, the latter 1,440 channels, in each case in 25 kHz incremental steps. Besides basic voice communications, each system also possesses data and SELCAL facilities.

Frequency generation is supplied from a digitally controlled frequency standard. Low component density and use of solid-state circuitry is a feature and heat-sinking with cooling vanes assists in maintaining low transmitter temperatures. All of these factors result in improved reliability and the calculated MTBF is greater than 4,000 hours.

Speech compression ensures good intelligibility irrespective of the user's voice characteristics or microphone technique and overmodulation is avoided by input signal amplitude limiting. Carrier-to-noise, carrier override squelch control and automatic gain control are also incorporated.

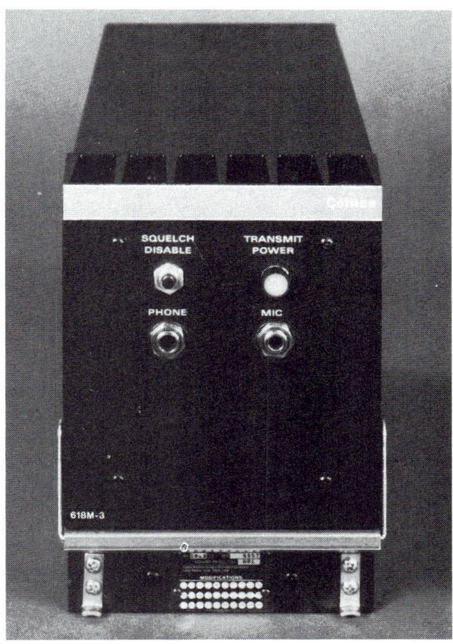

The Rockwell Collins 618M-3 HF radio

Specifications

Dimensions: ½ ATR short
Weight: 4.53 kg
Temperature range:
(with cooling air) −54 to +55°C
(without cooling air) −54 to +30°C

Operational status

In production and in service.

Contractor

Rockwell Collins.

UPDATED

628T-1 HF/SSB radio

The 628T-1 is an HF/SSB transceiver for long-range transport aircraft operating on overwater routes or other areas over which reliable extended range communications are required. It provides USB, AM voice and data communications on any of 24,200 channels, at 1 kHz increments in the 2.8 to 26.999 MHz band. Tuning is automatically controlled through a remote-control unit. Nominal transmission power is 200 W peak power in SSB or 100 W average in compatible AM.

This all solid-state system uses digital synthesis techniques for frequency generation. High stability is maintained through temperature compensation of the frequency standard.

Mechanical design is aimed at maximising maintainability. The transmitter/receiver is housed in a case with hinged tray and fold-out doors for easy accessibility. Plug-board circuitry is used extensively and bench-testing is simplified through the provision of a built-in test connector at the rear of the casing. Special provision is made for the replacement of the power amplifier transistors without the necessity of removing the entire power amplifier board.

Transmitter cooling is achieved through a heatsink and filtered forced air flow, while the receiver section relies on conventional convection cooling and is not dependent on a cool air supply.

An AM SELCAL facility is provided by means of a special audio output through which SELCAL signals are monitored irrespective of the selected operating mode. Options include automatic antenna tuning couplers, to permit antenna performance optimisation over the frequency spectrum covered by the 628T-1, and a 999W-1/A1 adaptor unit which permits interchangeability with the Rockwell Collins 618T-2/5 transceivers without disturbance to aircraft wiring, racks, connectors, antenna couplers or frequency selector.

Specifications

Dimensions: ¾ ATR short
Weight: 13.6 kg

Operational status

In production and in service.

Contractor

Rockwell Collins.

UPDATED

628T-2 HF radio

The 628T-2 is based on the 628T-1 and to some extent it is also based on the earlier 618T-2/5 equipment which is still to be found on many transport aircraft. It uses much of the operational and design experience derived from these systems. The 628T-2, however, was designed to offer certain advantages over the earlier equipment, principally higher power output and extended coverage of the HF band, hence a greater number of channels and more options in terms of operating modes.

Transmitter output power is 400 W peak power and full coverage of the HF band, from 2 to 30 MHz, permits use of up to 280,000 channels at 100 Hz increments or 28,000 channels at a separation of 1 kHz.

Operational modes include USB, LSB, AM, carrier wave and data. Full 400 W peak output is available in the sideband modes with 125 W average in compatible AM and 125 W nominal in continuous wave mode.

In mechanical and electronic design, the 628T-2 is very similar to the 628T-1 although Rockwell Collins has incorporated a number of improvements with regard to sensitivity, cross-modulation elimination, intermediate frequency translation and heat dissipation in the power amplifier stage. All are aimed at extension of performance and enhancement of reliability.

Specifications

Dimensions: 6 MCU to ARINC 600
Weight: 12.72 kg

Operational status
In production and in service.

Contractor
Rockwell Collins.

UPDATED

628T-3 and 628T-3/A HF radios

Like the 628T-2, the 628T-3 and 628T-3/A are based on the 628T-1 design and represent the latest standard in this series. The principal difference between the 628T-1/2 and the 628T-3( ) is the 28 V DC power requirement of the latter. The principal difference between the 628T-3 and the 628T-3/A variant is the narrower bandwidth intermediate frequency filtration of the latter, resulting in heightened selectivity over part of the operating spectrum and slight differences in audio response.

As in the case of the 628T-2, each system provides full coverage of the 2 to 30 MHz HF band and offers 280,000 channels at 100 Hz increments. There is, however, no option for a smaller number of channels at 1 kHz separation; transmitter output is 200 W peak power.

Operating modes are voice, and voice and data in both upper and lower sidebands, with compatible AM and carrier wave in USB only. SELCAL facilities are as those of the 628T-1.

Specifications
Dimensions: ¾ ATR short
Weight: 11.36 kg

Operational status
In production and in service.

Contractor
Rockwell Collins.

UPDATED

714E HF control units

The 714E control units used in conjunction with the 618T series transmitter/receivers provide remote selection of the available 28,000 channels. Frequencies are indicated in a direct reading digital display and can be selected in 1 kHz increments throughout the 2 to 29.9999 MHz range. Frequency selection is accomplished by rotating four knobs until the desired frequency appears in the window. A function selector and radio frequency sensitivity adjustment are included.

The 714E-2 control unit may be used either in new installations or as a replacement for the 614C-2 control unit in retrofit applications.

The 714E-3 is used with equipment operated in the continuous wave or data mode.

Specifications
Dimensions: 146 × 65 × 159 mm
Weight: 0.9 kg

Operational status
In service.

Contractor
Rockwell Collins.

UPDATED

Airborne Integrated Communications System (AICS)

The Airborne Integrated Communications System (AICS) provides a complete communications system covering from HF- up to, and including, L-band, while achieving significant reductions in size, power and weight compared with traditional LRUs. Communication links include multiple channels of VHF and UHF with ECCM, high-frequency ECCM, Satcom, Link 11, Link 16 and a digital modem.

AICS is designed to reduce aircrew workload by automating time-consuming tasks associated with the control of conventional communication equipment and present information to the operator to improve situation awareness.

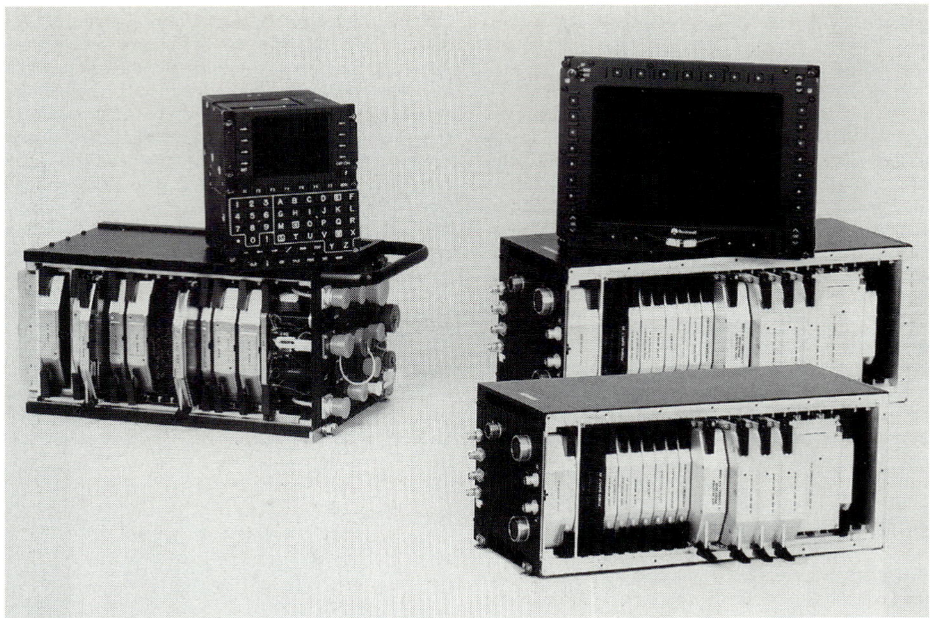

AICS is a modular avionics package which provides a complete communication system **1995**

The system combines air-cooled SEM-E modular RF and digital packaging with a colour graphic map display, mission planning station, digital map databases and communications management software. The four major subsystems that make up AICS are modular communication, COMSEC, intercom and information management and display.

Operational status
Rockwell Collins has been awarded a contract valued at nearly US$3 million for an AICS demonstrator for the UK Defence Evaluation Research Agency.

Contractor
Rockwell Collins.

UPDATED

AN/ARC-171(V) UHF radio

The Rockwell Collins AN/ARC-171(V) system comprises a family of UHF communications radios, each of which covers the band 225 to 399.975 MHz with 7,000 channels at increments of 25 kHz. In each case, an independent Guard channel covering the emergency frequency of 243 MHz is provided. Up to 20 of the normal operational channels may be preselected.

Members of the family provide a very wide range of transmission modes which include AM voice, AM secure voice, FM voice and data, FSK data and automatic rebroadcast. Certain versions also include an ADF navigation facility. With one exception with a full-duplex transmitter/receiver, all models use the same basic chassis configuration and the modification required to convert one version to another is accomplished by replacing circuit cards or complete modules. All equipment in the ARC-171(V) series may be remotely controlled and differing control units are employed according to the variant in use or the role in which it is employed. For example, in AM-only versions a simplified controller, covering functions common only to AM modes, is used, but a more comprehensive unit is available for use with those variants which provide AM/FM/FSK facilities. An alternative type of controller is provided for use with a satellite communications version.

Specifications
Dimensions: 406 × 241 × 170 mm
Weight: 15.88 kg

Operational status
No longer in production. The ARC-171(V), together with the Raytheon Systems Company ARC-164, is a standard system selected for US Air Force service and is installed in a variety of aircraft, particularly those with a satellite communication requirement. Certain configurations are intended primarily for use on board US Air Force Boeing E-3A Sentry AWACS aircraft.

Contractor
Rockwell Collins.

UPDATED

AN/ARC-182 VHF/UHF radio

The Rockwell Collins ARC-182 is a combined VHF/UHF military communications system designed for all types of fixed-wing aircraft and helicopters. Small and light enough to be especially attractive for installation in the lighter aircraft classes, it covers the frequency bands from 30 to 88 MHz in FM, 116 to 156 MHz in AM, 156 to 174 MHz in FM and for the UHF band 225 to 400 MHz in both AM and FM modes. Additionally, a receive-only facility covering the band 108 to 116 MHz is provided for navigation purposes. Channel spacing throughout the range is at 25 kHz intervals. A total of 11,960 channels, up to 28 of which may be preselected, is available and Guard channel coverage on the 243 MHz emergency frequency is also provided. A Guard precedence mode is activated by single-switch selection.

Developed for a US Navy requirement, the ARC-182 is also in service with the US Air Force. The provision of a comprehensive communication system, which also covers the marine band, renders the equipment useful in a number of tactical roles, especially those associated with extended economic zone maritime patrol. Later additions include a facility to scan up to five channels, allowing monitoring of multiple communication nets, and an added satellite communications ability which allows transoceanic flights with line of sight radio.

Specifications
Dimensions: 165 × 146 × 124 mm
Weight: 4.54 kg

Operational status
In continuous production and service. Rockwell Collins has been awarded annual US Navy contracts with options that to date total more than 7,000 sets. These contracts include receiver-transmitters and associated mounts, controllers, filters, high-power amplifiers and so on for US Navy fixed-wing aircraft and helicopters and ground-transportable applications. The contracts also contain foreign military sales arrangements for aircraft destined for Australia, Norway, Spain, and other countries.

Contractor
Rockwell Collins.

UPDATED

AN/ARC-182(V) high-power/ frequency-agile transceiver system

A variant of the AN/ARC-182 is a frequency-agile system that is designed for simultaneous operation of several transceivers in airborne applications. The system permits operation of transmitters and receivers frequency spaced as close as 5 per cent, with antenna isolation as small as 20 dB in the 225 to 400 MHz band in both single channel and frequency-hopping modes.

The equipment consists of the RT-1250A/ARC-182 transceiver, the F-1556/ARC UHF high frequency-agile filter, the AM-7177A/ARC UHF high-power amplifier, the MT-6330/ARC mounting tray and the C-10319A/ XN-3/4 controller.

The system provides enhanced AN/ARC-182 performance in UHF at 30 W while permitting standard ARC-182 operation in the VHF band. The high-power amplifier is wideband, while the electronically tuned filter used in both transmit and receive allows medium-speed frequency hopping. In addition, a synchronous filter capable of providing high-energy pulsed emitter protection is provided.

Specifications

Dimensions: 188.9 × 241.3 × 444.5 mm
Weight: 14.78 kg

Contractor

Rockwell Collins.

UPDATED

AN/ARC-186(V)/VHF-186 VHF AM/FM radio

The Rockwell Collins AN/ARC-186(V)/VHF-186 is a tactical VHF AM/FM radio communications system designed for all types of military aircraft.

The basic ARC-186(V) is a solid-state 10 W system of modular construction which provides 4,080 channels at 25 kHz spacing. The 1,760 FM channels are contained in the 30 to 88 MHz band and 2,320 AM channels within the range 108 to 152 MHz. A secure speech facility can be used in both AM and FM modes and the equipment is compatible with either 16 or 18 kbit secure systems in diphase and baseband operation.

Up to 20 channels may be programmed for preselection on the ground or in the air. Preselection is accomplished through incorporation of a non-volatile NMOS memory which continues to retain data in the event of a loss of power supply. Two dedicated channel selector switch positions cover the FM and AM emergency channel frequencies of 40.5 and 121.5 MHz respectively.

Either panel mounting with direct control through an integral controller or remote mounting with an identical control panel presentation is possible. A half-size remote controller which contains the same control functions is also available and a typical configuration in a two-seat aircraft would comprise a full panel mount in the pilot's cockpit with a half-size controller at the crew position. In these dual-control configurations, a manual take-control switch provides full communications control for either crew member.

Conversion from panel to remote control is made by removing the panel controller and replacing it with a plug-in serial control receiver module. A typical conversion is said to require less than 5 minutes. Frequency displays on both types of controller are immune to fade-out during periods of low voltage.

Circuitry of the ARC-186(V) is of modular design. Seven module cards are held in place by the body chassis or card cage and are electrically interconnected by a planar card in which all hard wiring has been virtually eliminated. All radio frequency lines in the interconnecting planar card have been buried to minimise electromagnetic interference. Individual module cards are readily removable and may be replaced in the field to reduce fault finding and repair time.

Current options for the ARC-186(V) include AM/FM homing facility, but growth capability has been designed into the equipment from the outset and possible future developments could include SELCAL, burst data and target hand-off. One present simple modification, carried out by replacement of the decoder module in the remote transceiver, permits the radio to be directly connected to a MIL-STD-1553 digital databus and it is claimed that the system will be equally compatible with avionics suites of future generation equipment. An additional possibility is the uprating of transmitter output power.

US Air Force testing has demonstrated a MTBF in excess of 9,000 hours.

A principal design objective for the ARC-186 series equipment was that it should be capable of easy retrofit in existing installations. Since the system is considerably smaller than the equipment it is designed to replace, this is accomplished by use of plug-in adaptor trays which permit rapid replacement without disturbance to existing aircraft wiring harnesses.

The Rockwell Collins AN/ARC-186 military VHF system in both direct and remote-control configurations

Specifications

Dimensions:
(remote-mounted transmitter/receiver) 127 × 165 × 123 mm
(panel-mounted transmitter/receiver) 146 × 165 × 123 mm
(half-size remote-control) 146 × 95 × 57 mm
Weight:
(transmitter/receiver) 2.95 kg
(remote-control) 0.79 kg
(FM homing module) 0.45 kg

Operational status

In production and in service. More than 28,000 sets have been delivered worldwide. It has been selected by the US Air Force as the standard equipment for all aircraft requiring VHF AM/FM capability. Aircraft to be equipped include the A-10, C-130 and F-16.

Contractor

Rockwell Collins.

UPDATED

AN/ARC-190(V) HF radio

The Rockwell Collins AN/ARC-190(V) is a military HF transmitter/receiver designed as a replacement for a number of earlier HF systems in a US Air Force modernisation programme. The system is therefore particularly suited to retrofit applications as well as for installation as original equipment in a wide range of aircraft, such as the B-1B, B-52, C-5, C-25, C-130, C-141, F-15, F-16, H-53, KC-10, RF-4C and VC/KC-135. As well as retrofit kits for 618T systems, a MIL-STD-1553 system is available. The system can be used in either 1553 or non-1553 applications. A selective calling (SELCAL) AM detector is available in some versions of the ARC-190.

It covers the 2 to 30 MHz band in which it provides 280,000 channels, any 30 of which are preselectable, in incremental steps of 100 kHz. Operational modes include USB, LSB, AME and CW. Data transmission facilities are also available in USB and LSB modes and in these modes the system is able to operate with audio-frequency shift keying or multitone modems.

The system is remotely controlled and dual control of the radio is possible from two crew stations. Serial data control is applied between each of the major units in the system and this is said to render it adaptable to future requirements such as SELCAL and remote frequency management. Transmitter power output is 400 W. The ARC-190 is supplied with AC or DC interface power, as well as MIL-STD-1553 control scheme.

Construction of the ARC-190(V) is all-solid-state. The full system includes an antenna coupler which ensures compatibility with military cap, shunt, wire, whip or probe antenna systems. An F-1535 bandpass filter unit is available to provide added selectivity and overload protection for improved receiver performance in a strong signal environment. In the transmit mode it provides additional filtering to the exciter RF output. The system operates to pressure altitudes up to 70,000 ft over temperatures ranging from −55 to +71°C.

Rockwell Collins supplies the US Air Force with an automatic communications processor for the ARC-190(V) which automatically selects the optimum HF frequency after the operator has selected the station to be called. The processor also features anti-jam modes. After test and evaluation, production of the processor commenced in 1988.

Specifications

Dimensions:
(controller) 114 × 146 × 66 mm
(transmitter/receiver) 480 × 257 × 194 mm
(antenna coupler) 545 × 211 × 189 mm
Weight:
(controller) 0.68 kg
(transmitter/receiver) 22.68 kg
(antenna coupler) 10.89 kg

The Rockwell Collins AN/ARC-190(V) HF radio ***1995***

Operational status

In production and in service in US Air Force aircraft such as the B-1B, B-52, C-5, C-25, C-130, C-141, F-15, F-16, H-53, KC-10, RF-4C, VC/KC-135, C-17 and US Navy E-6A.

Contractor

Rockwell Collins.

UPDATED

AN/ARC-210(V) multimode integrated communications system

The Rockwell Collins AN/ARC-210(V) multimode integrated communications system was derived from the AN/ARC-182 system to provide multimode voice and data communications in either normal or jam-resistant modes through software reconfiguration. The RT-1556 transceiver is capable of establishing two-way communication links over the 30 to 400 MHz frequency range within tactical aircraft environments. There are currently nine different variants of the transceiver, but a great deal of commonality is retained between these.

The ARC-210 meets the 8.33 kHz European ATC channel spacing requirements.

Rockwell Collins has produced the RT-1794(C)/ARC-210 which incorporates all the features of the basic ARC-210, plus embedded Satcom/DAMA, COMSEC (KG-84, KY-58, ANDVT, RGV-11), CTIC, MIL-STD-188-220A JVMFa, Link 4A and is software reprogrammable via the Advanced Memory Loader Verifier (AMLV) feature, also embedded.

The US Navy's F/A-18C/D and E/F, MV-22, C-2 and EA-6B, and the US Air Forces' C-17A, CV-22 and HH-60G are on contract to receive this unit. A fully fuctional auxiliary remote control, the C-12571/ARC-210 is available, or the system may be integrated into aircraft Control Display Units (CDUs) via MIL-STD-1553B databus.

The transceiver is the nucleus of the multimode communication system which includes an appliqué for Have Quick, Have Quick II and SINCGARS-V waveforms. In addition, the AN/ARC-210 has been demonstrated to provide Have Quick IIA ECCM and for Link 4A and Link II data communications. The system will also provide Satcom wideband and narrowband operation. Along with the AM-7525/-7526 UHF high-power amplifier and MX-11641 low-noise amplifier/diplexer, the ARC-210 provides a flexible Satcom terminal and complies with MIL-STD-188-181/182/183 DAMA requirements. Maritime, land-mobile, ATC and ADF are included modes of operation. The system is controllable by a MIL-STD-1553B databus and includes a remote controller for manual operation, a remote indicator and a family of broadband and electronically tunable antennas.

Rockwell Collins was placed on contract on 16 August 1995 to incorporate the Secretary of Defence's acquisition streamlining initiative into the ARC-210 communications system.

Planned growth capabilities include: Have Quick IIA and SATURN capability, SINCGARS Improvement Program (SIP) – data rate adaptor, USA ATC DAMA (digital voice) and Differential GPS datalink.

The Rockwell Collins AN/ARC-210(V) communications system **1996**

The AN/ARC-217(V) HF transceiver is designed for nap of the earth communications. It consists of (left to right): control unit, receiver/transmitter and antenna coupler

Specifications

Dimensions: 127 × 142.2 × 248.9 mm
Weight: 5.44 kg

Operational status

In production for US Navy (F/A-18 C/D, AH-1W, AV-8B, CH-46E, C/MH-53E, KC-130 F/R/T, UH-1N, EA-6B), US Air Force, US Marine Corps, US Army and US Air National Guard platforms plus Canadian, Finnish, Italian, Spanish and Swiss forces. Over 4,400 systems have been ordered in total.

Contractor

Rockwell Collins.

UPDATED

AN/ARC-217(V) HF system

The AN/ARC-217(V) is a lightweight tactical HF airborne transceiver system designed for both rotary- and fixed-wing applications. When installed with a simple retrofit kit, the AN/ARC-217(V) is a direct replacement for the AN/ARC-199, but provides the additional features of embedded ECCM, automatic link establishment and data transmission. It also has a MIL-STD-1553B control interface.

The primary requirement was that the system should establish communication links 24 hours a day with ground stations on 90 per cent of the attempts and that 90 per cent of each message be accurately transmitted and received.

Operational status

The AN/ARC-217(V) has a wide application for a variety of US military combat helicopters and fixed-wing aircraft. Also procured for the presidential helicopter fleet.

Contractor

Rockwell Collins.

UPDATED

AN/ARC-220 HF tactical communications system

The AN/ARC-220 is designed to provide the US Army with an easy to use standard airborne HF tactical communications system for helicopters, to overcome propagation problems of VHF and UHF line of sight systems and eliminate reliance on strategic satellite channels. The AN/ARC-220 provides secure or unsecured voice and data communications in the 2-30 MHz frequency band. It also features Automatic Link Establishment and Electronic Counter CounterMeasures capabilities (ECCM - also designated EP (Electronic Protection)). Advanced HF features will permit communication via near vertical incidence skywave, making possible reliable propagation in nap of the earth profiles. The AN/ARC-220's capabilities will benefit helicopters, such as the AH-64 Apache, CH-47 Chinook, UH-60 Black Hawk and OH-58 Kiowa flying nap of the earth profiles.

The AN/ARC-220 will replace the ARC-199 and provide the US Army with a full digital signal processing radio that reduces the parts count by more than 50 per cent. It also provides many additional capabilities such as embedded Automatic Link Establishment (ALE), ECCM, standard 110A data modem, flash memory for field reprogrammability and a spare card slot for future growth. The primary requirement for the ARC-220 is to provide simplified, reliable, long-range, low-altitude nap of the earth communications with a high degree of link connectivity on the battlefield.

The communications system consists of a receiver/transmitter, antenna coupler and control display unit.

Operational status

In service with US Army helicopters.

Contractor

Rockwell Collins.

UPDATED

AN/ARC-230/HF-121C high-performance radio system

The Rockwell Collins AN/ARC-230/HF-121C high-performance radio system is designed for military voice and data HF applications requiring 400 W operation. Compliant with the requirements of MIL-STD-188-

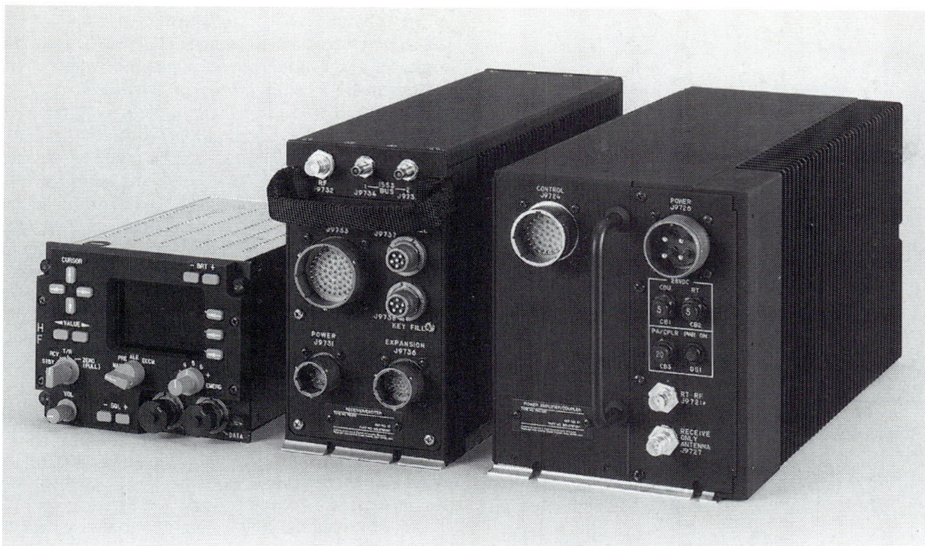

The AN/ARC-220 is an HF nap of the earth communications system **1995**

203-1A for Link 11/TADIL A and MIL-E-5400, this 400 W PEP/and average power radio provides maximum performance for airborne applications. The AN/ARC-230/HF-121C has been optimised for tacitcal digital data communications and SIMultaneous OPeration (SIMOP) of multiple radio sets with minimum frequency and antenna separation. Embedded Automatic Link Establishment (ALE) and ECCM capabiities are also available.

The basic radio set is partitioned into a receiver-exciter with integral prepost-selector on a mounting shelf, and a power amplifier-power supply on a mounting shelf. The receiver-exciter and/or prepost-selector can be used independently of the power amplifier. Serial control via RS-232 or MIL-STD-1553B is available.

Operational status

In production and operation in commercial and military aircraft in the USA and other countries. It has been selected as the standard HF radio by many aircraft manufacturers. The radio set has received military nomenclature designation for, AN/ARC-153, 157, 191(V), 207(V), 229, & 230, AN/URC-91, and 97(V) by the United States and AN/ARC-512 for other nations.

Contractor

Rockwell Collins.

NEW ENTRY

AN/ASC-15B communications central

The AN/ASC-15B communications central, referred to as a command console, functions as an airborne and ground command post, providing tactical voice communications in both secure and non-secure modes. This highly mobile communications combat command centre provides NATO and US tri-service forces interoperability during all types of military operations and special missions.

The AN/ASC-15B can be operated from a UH-60A or UH-1H helicopter, or removed and configured for ground operation. It provides HF plus VHF and UHF communications in AM and FM modes, channel scanning of four V/UHF preset channels in each AN/ARC-182 radio, automatic retransmission in VHF and UHF bands and UHF satellite communications.

The AN/ASC-15B consists of an AN/ARC-174 HF transceiver, three AN/ARC-182 V/UHF transceivers, two AM-7189A IFM power amplifiers, an MX-931B/URC repeater, an AM-7402 Satcom power amplifier and two C-11128 ECCM (HQ) controls.

The ASC-15B has been modified with three ARC-210(V) radios to replace the ARC-182s presently installed. It was given the nomenclature AN/ASC-15C after radio upgrading.

Specifications

Weight: 129.28 kg
Power supply: 28 V DC
Power output:
(2-30 MHz) 100 W PEP
(30-400 MHz) 15 W FM, 10 W AM
Frequency: 2-30 MHz and 30-400 MHz
Modes:
(2-30 MHz) HF/SSB, AME and CW
(30-400 MHz) V/UHF AM and FM

Operational status

In service with the US Army.

Contractor

Rockwell Collins.

UPDATED

AN/URC-138(V)1(C) Link 16 Low Volume Terminal (LVT)

The AN/URC-138(V)1(C) Information Distribution System provides anti-jam protected, encrypted, high throughput data distribution. Because it is low cost, it makes JTIDS participation affordable. The small size and low weight of the AN/URC-138(V)1(C) terminal makes it suitable for a broad variety of tri-service platforms.

The AN/URC-138(V)1(C) terminal is waveform, message format and network compatible with existing JTIDS (TADIL J) Link 16 systems. The terminal provides JTIDS interoperability between the US tri-services and NATO forces.

JTIDS provides situation awareness by providing threat, target and friendly ID, position and status information among participating platforms in near real time, with anti-jam security without any voice communication.

It provides full stacked net capacity, up to 128, and full JTIDS data throughput. The system can automatically exchange information from a variety of platform sensors. This can include functions such as IR and optics scan, target identification and steering commands. Real-time data updates can also be used to provide landing cues.

In addition to robust data communication, the AN/URC-138(V)1(C) terminal also provides two voice ports to enable secure voice communication in a jamming environment.

Specifications

Operational characteristics
Net participation:
TDMA
128 nets maximum
128 time slots/sec/net
Message catalog: J-Series messages as defined in STANAG 5516
Frequency operating range: 969-1206 MHz

The AN/ASC-15B command and control console is designed for helicopters such as the UH-60 Black Hawk

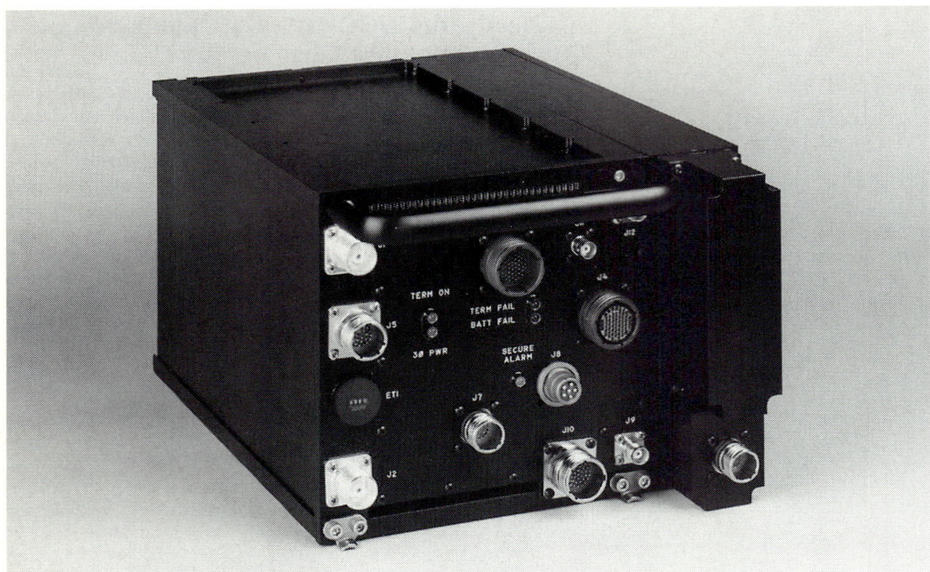

AN/URC 138(V)1(C) Link 16 Low Volume Terminal (LVT) ***1997***/0001183

Informational data rates: 28.8 to 238 kbs
Anti-jam: Frequency hopping, forward error detection/correction
Range (max): 400 n miles

Electrical characteristics
RF power output (max): 200 W
I/O data/voice:
Data: MIL-STD-1553
Voice port 1: LPC-10 (2.4 kbs)
Voice port 2: CVSD (16 kbs)
Antenna: JTIDS compatible
Primary power: 120 V AC 3ø 400 Hz, 750 W

Physical characteristics
Dimensions: 318 × 256.3 × 191 mm
Weight (max): 20 kg
Packaging: SEM-E module form factor cooling, forced air mounting

Contractor
Rockwell Collins.

UPDATED

Automatic communications processor

The automatic communications processor and associated ARC-190 400 W airborne radio provide a system that automatically scans multiple frequencies, selects the best frequency on which to make a call and automatically repeats the call until contact is confirmed. The system also provides an anti-jam frequency-hopping capability for effective ECCM.

The CP-2024 Automatic Communications Processor (ACP) and the C-11814/ARC-190(V) Automatic Communication Processor Control (ACPC) operate together as a microprocessor-based remote-control subsystem which can be added to existing AN/ARC-190(V) radio systems to automate and simplify HF radio operation. These units are completely interoperable with MIL-STD-188-141A ALE Rockwell Collins Selscan commercial air and ground units and FED-STD-1045.

The ACP combines receive scanning and selective calling under microprocessor control to monitor up to 100 preset channels for incoming ALE calls. Link quality analysis circuits measure and store signal-to-noise and bit error rate characteristics of received ALE signals for use by automatic frequency selection algorithms.

Selective calling addresses and preset channels may be programmed by the user from the ACPC front panel or from a remote ASCII terminal. All presets are stored in non-volatile memory for power-off retention.

The ACP provides frequency control of the associated ARC-190(V) HF radio in order to monitor multiple frequencies by scanning multiple preset channels chosen from a total of up to 100 stored simplex or half-duplex preset channels. Incoming ALE calls are answered automatically and the calling station's address is displayed to the user. Positive squelch is automatically broken whenever contact is established in response to an incoming call or as a result of an outgoing call. The system also provides a standard selective calling (SELCAL) capability when used with the AN/ARC-190 RT-1341(V)6, RT-1341(V)7 or RT-1341(V)8 radios.

Outgoing calls can be initiated on a station-to-station or net broadcast basis.

Specifications
Dimensions:
(ACP) 198.6 × 121.9 × 495.8 mm
(ACPC) 66.5 × 146 × 106.7 mm
Weight:
(ACP) 9.53 kg
(ACPC) 1.81 kg
Power supply: 115 V AC, 400 Hz, 110 W, 28 V DC, 25 W

Operational status
In service on C-5, C-20, C-25, C-27, C-130, C-141, KC-10 and VC-135 aircraft.

Contractor
Rockwell Collins.

UPDATED

Automatic link establishment for HF

The 309M-1 Automatic Link Establishment (ALE) processor and 514A-13 control operate together as a microprocessor-based remote-control subsystem which can be added to existing Rockwell Collins HF radio systems to automate and simplify operation.

The 309M-1 ALE processor combines receive scanning and selective calling under microprocessor control to monitor up to 100 preset channels for incoming ALE calls. Link quality analysis circuits measure and store signal-to-noise and bit error rate characteristics of received ALE signals for use by automatic frequency selection algorithms. The ALE processor automatically mutes the receive audio output from the HF radio while scanning to eliminate distracting HF background noise and irrelevant channel activity.

When an automatic call is placed, the operator selects the preset ALE address of the individual station or net to be contacted and initiates the call. Automatic channel selection algorithms choose the calling channel from the list of channels currently being scanned. Automatic channel selections are made according to the order in which the candidate channels are ranked.

The channels actually scanned, and the choice of which of the multiple self-addresses are valid at any time, are determined by the scan list or lists selected. Multiple scan lists may be selected simultaneously, resulting in a combined list of channels for scanning purposes. The unique flexibility provided by selectable scan lists allows the 309M-1 to participate in multiple networks simultaneously.

Specifications
Dimensions:
(309M-1) 198.6 × 121.9 × 318 mm
(514A-13) 66.5 × 146 × 106.7 mm
Weight:
(309M-1) 4.5 kg
(514A-13) 1.36 kg
Power supply: 28 V DC
(309M-1) 30 W
(514A-13) 25 W

Contractor
Rockwell Collins.

UPDATED

CP-1516/ASQ Automatic Target Hand-off System

The CP-1516/ASQ Automatic Target Hand-off System (ATHS) is a battlefield mission management system used in conjunction with a control and display unit and up to four standard HF, VHF or UHF radios to provide a tactical Command Control, Communications and Information (C^3I) network. The digital communication network can provide for stores management, target handovers and other similar functions to be passed to airborne, artillery and ground forces in short radio bursts which are difficult for the enemy to detect or jam.

The CP-1516 features a recall capability for 12 previously received messages and allows the transmitting of preformatted messages or free-text

The Rockwell Collins CP-1516/ASQ automatic target hand-off system and control/display unit

messages using an alphanumeric keyboard. Non-volatile memory in the unit retains all critical information in the event of a power loss. In addition, the CP-1516 maintains the current status of up to 10 active airborne missions and two preplanned missions.

Various control/display unit options are available for data entry and display. The CP-1516 is fully compatible with the AH-64 Apache data entry panel and TADS/PNVS display, the Bell OH-58D control/display and mast-mounted sight display and combat helicopter control/display unit.

Modern electronic battlefield systems including SINCGARS, E-PLRS/JTIDS hybrid (PJH), Tacfire communications, COMSEC and all MIL-STD-1553 avionics, including digitally generated map displays, are completely compatible with the CP-1516.

The CP-1516, which was first demonstrated in May 1985, is suitable for use in conjunction with the Rockwell Collins CMS-80 avionics management system and is used on the US Army's OH-58D and JOH-58 helicopters.

The computer within the CP-1516 incorporates 8 kbytes of RAM, 2 kbytes of EAROM and 196 kwords of program memory. The system is compatible with MIL-STD-1553A and B databusses.

Specifications

Dimensions: 136 × 165 × 203 mm
Weight: 4.5 kg
Power: (DC) 40 W max

Operational status

Over 750 units have been delivered.

Contractor

Rockwell Collins.

UPDATED

CP-2228/ASQ Tactical Data Modem (TDM-200)

The CP-2228/ASQ Tactical Data Modem (TDM-200) is an upgraded and improved version of the CP-1516/ASQ ATHS. TDM-200 was developed to provide additional capabilities to meet the more stringent environmental requirements of fighter and close air support aircraft, while also meeting the needs for future datalink applications. TDM-200 is form, fit and functionally compatible with the CP-1516/ASQ ATHS.

TDM-200 is capable of transmitting and receiving FSK from baud rates of 75 to 1,220 and digital data from 75 to 16,000 bits/s. The higher frequency operation dramatically reduces transmission time, thus making it more difficult to detect and jam. It has four ports and up to four modems which can simultaneously transmit or receive messages.

The computer within TDM-200 is a Z80180 with 256 kwords program memory with potential growth to 1 Mwords. The computer also incorporates 64 k RAM and 8 k non-volatile memory. Mission data and operational flight program data may be programmed via the MIL-STD-1553 databus or a digital data loader.

Specifications

Dimensions: 136 × 165 × 203 mm
Weight: 4.5 kg
Power: 50 W max

Operational status

The TDM-200 is fully developed and has completed qualification testing for the US Marine Corps AV-8B Harrier II aircraft.

Contractor

Rockwell Collins.

UPDATED

CP-2378 (C) JTIDS datalink terminal

The CP-2378(C) low-cost JTIDS terminal provides Link 16 situational awareness to the naval aircraft to which it is fitted, by displaying friendly and hostile threat information in real time in the cockpit.

It is being developed and funded by Rockwell Collins and the UK Ministry of Defence. Rockwell Collins claim that it is one-fifth the cost and one-third the size of standard Class 2 JTIDS terminals, but that it provides much the same capability, including integrated communication, navigation and identification systems for distributing tactical information within a theatre of operations.

The CP-2378(C) terminal is interoperable with all JTIDS (TADIL J) Link 16 systems, including waveforms, message format and networks. It provides compatibility and interoperability between US and NATO forces. In addition, the system can automatically exchange information from a variety of platform sensors, including infrared and optics scan, target identification and steering commands. It uses SEM-E modular architecture and Ada software.

The Rockwell Collins CP-2378(C) JTIDS datalink terminal ***1997***/0002118

Operational status

Seven units manufactured; four units being used for installation testing on Sea Harrier F/A-2 aircraft and Sea King AEW. Mk 7 helicopters by the Royal Navy; three units being used for further testing at Rockwell Collins.

Contractor

Rockwell Collins.

UPDATED

Datalink systems

Rockwell Collins datalink systems are designed for commercial and military transports and conform to ARINC 724 and 597. Their associated control panels have full alphanumeric keyboards with a two-line, 32-character light-emitting diode display. Information can also be displayed on a compatible weather radar screen.

These datalinks can control VHF radios, act as a checklist memory and provide the interface between flight and performance management systems and ground-based data transmitters. The operator can change the software and the units have comprehensive built-in test facilities and a continuous GMT clock facility.

The DLC-800 is a menu-driven control/display designed to be used with the Rockwell Collins DL-700 datalink systems. The unit is completely compatible with other datalink management units which use the ARINC 429 databus and can be retrofitted into existing installations. Airline programmable software enables an airline to create distinctive symbols, display formats and messages for use on the DLC-800. The unit has a high-resolution infrared touch-input system and a wide viewing angle for easy reading by both pilots. The display is easily readable in direct sunlight. Key actions can be highlighted by using inverse characters in the form of dark on light background. Messages can flash on the screen to attract attention.

Management units are available in two formats. The DLM-700 management unit is in accordance with ARINC 724, 597 and 600. The 597A-1 is in accordance with ARINC 404A and 597.

Specifications

Dimensions:
(control unit) 114 × 146 × 120 mm
(DLM-700 management unit) 4 MCU
(579A-1 management unit) ⅜ ATR short
Weight:
(control unit) 1.2 kg
(DLM-700 management unit) 4 kg
(597A-1 management unit) 3.6 kg
Temperature range: −15 to +70°C
Altitude: 55,000 ft

Operational status

In production.

Contractor

Rockwell Collins.

UPDATED

DLM-700B datalink system

The DLM-700B datalink management unit is designed in accordance with ARINC 724B for all analogue and digital aircraft, including the Airbus A320, Boeing 747-400 and MD-11. The unit is part of the Rockwell Collins Airline Communications And Reporting Systems (ACARS) product line. Growth capacity is provided for additional memory or ARINC 429 interfaces. Compatibility with the Collins DL-700 and DL-500 systems can be added.

Message decoding in the DLM-700B is enhanced through the use of a Collins MSK demodulator which removes distortion from the VHF communications transceiver. A frequency management function provides automatic frequency changes without any pilot action. A channel management function, for use with Category B service providers, allows the DLM-700B to select the optimum ground station. This automatically provides fast message response time, as well as reduced message congestion in ground networks.

A communications statistics package monitors air-to-ground link performance; data is gathered on every flight leg. A message store and forward function allows flight plans, weather and ATIS information to be held for display upon demand. A flight profile report provides the crew with details on departure and arrival times, cruise times, engine data and other items. Custom data, such as frequency tables, delay codes and pilot alterable messages, can be stored in the system. Data can be loaded on the bench or in the aircraft. An optional floating point maths package is available for data processing. Using this package, complex engine reports from any analogue ARINC 717 or 573 equipped aircraft can be prepared.

Specifications

Dimensions: 125 × 193 × 325 mm
Weight: 5 kg
Power supply: 115 V AC, 400 Hz, single phase, 80 W optional 28 V DC battery, 75 mA max

Temperature range: −20 to +70°C
Altitude: up to 55,000 ft

Operational status
More than 2,300 units in service.

Contractor
Rockwell Collins.

UPDATED

DLM-900 datalink system

The Rockwell Collins DLM-900 is a next-generation datalink management unit specifically designed to provide an affordable and seamless transition as the global datalink system evolves from ACARS to ATN. The unit provides easy upgrade via software data load to add Aeronautical Telecommunications Network (ATN), Future Air Navigation System (FANS) and other future capabilities as they come online. The DLM-900 is certifiable to the DO-160C and DO-178B, Level D and Level C standards required for future ATN and FANS operations. Software update costs are minimised by providing isolated partitions for essential (Level C) and non-essential (Level D) functions as well as aircraft-specific and aircraft-independent functions. A powerful user application development capability allows airlines to perform their own datalink non-essential application maintenance and modifications. This unit was designed with sufficient processor throughput, memory capacity and interface capabilities to host future functionalities and provides a spare full-depth card slot to accommodate growth.

Specifications
Dimensions: 4 MCU per ARINC 600
Weight: 5.5 kg
Power supply: 115 V AC, 400 Hz, with 28 V DC optional
Temperature range: −40 to +55°C
Altitude: up to 55,000 ft

Operational status
In production in 1997 for 10 airlines.

Contractor
Rockwell Collins.

UPDATED

HF-121/121B/121C (AN/ARC-191, AN/ARC-207, AN/ARC-512) radio

The Rockwell Collins HF-121/121B/121C, known under the JETDS nomenclature as the AN/ARC-191, AN/ARC-207 and AN/ARC-512, is a high-reliability airborne communication system which covers the full HF-band from 2 to 30 MHz, in which range it provides 280,000 channels. The system comprises elements from two other Rockwell Collins airborne HF systems, the AN/ARC-153 and the AN/ARC-157, both produced for the US Navy. The resultant configuration complies with MIL-E-5400.

The HF-121 family can transmit and receive both data and voice signals and operates in USB, LSB, ISB and AME modes. Transmitted power output level is selectable at either 100, 500 or 1,000 W. The system operates over temperatures from −54 to +55°C with short-term operation up to 70°C. Maximum operational pressure altitude is 26,000 ft.

An HF-121C with 100 W, 200 W and 400 W selectable outputs and embedded Automatic Link Establishment (ALE), a pre/post-selector and modem is also available. The system includes a DSP receiver/exciter, a solid-state power amplifier and digital antenna couplers. The maximum operational pressure altitude is 50,000 ft and the operating temperature ranges from −54 to +55°C.

Operational status
In production and in service in US Navy and US Air Force aircraft.

Contractor
Rockwell Collins.

UPDATED

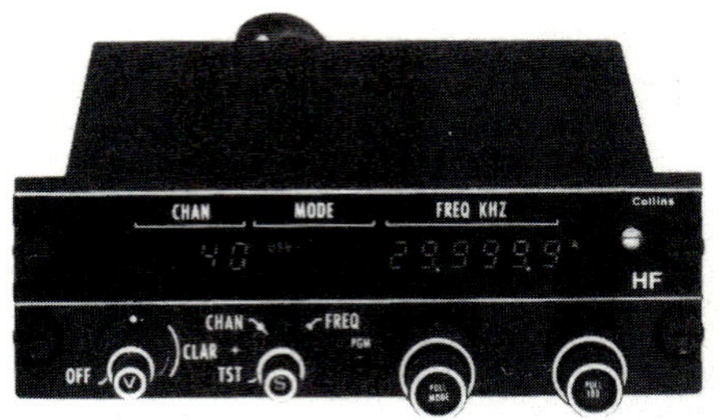

The Rockwell Collins HF-230 HF radio and ITU radio-telephony transceiver

HF-230 HF radio

The Rockwell Collins HF-230 radio is for use in fixed-wing aircraft and helicopters. It provides 280,000 channels at 100 Hz channel spacing between 2 and 29.9999 MHz. All 176 ITU radio-telephony channels are preprogrammed, giving phone-patch capability over very long ranges wherever this facility is available. Lower sideband operation is possible for international or maritime communications.

The system comprises a TCR-230 transceiver, PWR-230 power amplifier and a range of antenna couplers. A DSA-220 adaptor permits two such systems to be operated in the same aircraft.

The radio features 40 pilot programmable channels and, when selected, channel number and frequency are displayed.

The new CTL-230 display unit forms part of the HF-230 and features gas-discharge symbology.

An automatic probe antenna coupler, the PAC-230, is available for helicopter applications.

Specifications
Weight: 11.1 kg
Power output: 100 W PEP
Temperaure range: −55 to +70°C
Altitude: up to 55,000 ft (with pressurised antenna coupler)

Operational status
In production and in service.

Contractor
Rockwell Collins.

UPDATED

HF-9000 series HF radios

The HF-9000 series is a series of lightweight HF radios developed to meet the HF communications requirements of commercial business jets and military aircraft ranging from helicopters to high-performance fighters.

The initial emphasis was on the development of a system for light fixed- and rotary-wing tactical aircraft. That system, designated the HF-9000 and the AN/ARC-217(V) in its military form, is now in production.

Additional capabilities have been added to the HF-9000 family to include MIL-STD-188-141A Automatic Link Establishment (ALE) and MIL-STD-188-148 ECCM capabilities. The AN/ARC-217 version employs modular design combined with fibre optics, microprocessor technology, digital synthesisers and couplers and MIL-STD-1553B or ARINC 429 control.

The basic communications equipment includes an HF receiver/transmitter with a 175 W HF power amplifier/antenna coupler. A 200 W version is also available. System design is such that it can be configured with a control unit for panel mounting in the cockpit or for a MIL-STD-1553B control system. All control and status information transferred between the transceiver and the power amplifier/coupler is transferred through a small fibre optic cable, permitting fast exchange of large amounts of data between the two units.

The system can be operated in simplex or duplex modes over the frequency range from 2 to 29.9999 MHz in 100 Hz increments in both Upper SideBands and Lower SideBands (USB, LSB) voice and data, Amplitude Modulation Equivalent (AME) and Continuous Wave (CW) modes, with growth to HF datalink operations. Up to 99 programmable preset channels can be stored in a non-volatile memory and each memory channel can store separate receive and transmit modes and frequencies. The transceiver uses a direct digital frequency synthesiser for rapid frequency changes with microprocessor control to improve stability. The antenna coupler is designed to permit rapid tuning of a wide range of antennas in a variety of aircraft.

Specifications
Dimensions:
(control unit) 66 × 145 × 152 mm
(transmitter/receiver) 193 × 127 × 317 mm
(coupler) 193 × 89 × 317 mm
Weight:
(control unit) 1.2 kg
(transmitter/receiver) 5.5 kg
(coupler) 4.0 kg

Operational status
In service with Gulfstream IV aircraft, and with the Royal Australian Air Force as a replacement for the 618T HF radio on aircraft such as the C-130 Hercules. A total of approximately 2,000 sets has been sold.

Contractor
Rockwell Collins.

UPDATED

HFS-700 HF radio

The Rockwell Collins HFS-700 is an HF transmitter/receiver intended for service aboard transport aircraft which have a long-range communication requirement. Designed and developed in accordance with ARINC 719, the HFS-700 draws largely upon operational experience gained with the company's earlier 628T-1 and 618T-2/5 systems.

Covering the full HF band from 2 to 30 MHz, the HFS-700 provides 28,000 channels at increments of 1 kHz and uses ARINC 429 serial channel selection. Operating modes are USB, LSB, AM equivalent, data and CW. Nominal transmitted power outputs are 400 W PEP in SSB modes, 125 W in compatible AM and 125 W in CW. A special audio output permits SELCAL monitoring to continue in all mode settings.

The system is of all solid-state construction and extensive use is made of CMOS and linear integrated circuits. The digital frequency generation synthesiser is locked to a highly accurate reference standard.

Features include a Digital Information Transfer System (DITS) tuning interface, new installation concept connectors and cooling with a blower. Particular attention has been paid to control of excessive temperature during abnormal operating conditions and a special design of heatsink has been employed to maintain output transistors within derated temperature limits. Additionally, a dissipation detector reduces drive to lower levels under high-dissipation conditions.

Mechanical design features include plug-in boards with hinged tray and fold-out doors to allow easy access to components during servicing without use of card extenders.

Specifications
Dimensions: 6 MCU per ARINC 719600
Weight: 12.7 kg
Temperature range: −55 to +70°C
Altitude: up to 40,000 ft

Operational status
In production and in service.

Contractor
Rockwell Collins.

UPDATED

HFS-900D HF data radio

The HFS-900D is designed to provide the air transport industry with next-generation HF voice and data communications service. It provides all traditional HF modes of operation, including USB, LSB, CW, AME and analogue data, with ARINC 635/753 standard High Frequency DataLink (HFDL) modes compatible with the ATN network. The HFS-900D provides the means to process, transmit and receive data and analogue voice and is designed to operate on frequencies spaced 100 Hz apart in the 2-30 MHz band. For compatibility with existing ARINC 719 installations, the HFS-900D provides Single SideBand (SSB) voice, Amplitude Modulated Equivalent (AME), Continuous Wave (CW), Selective Calling (SELCAL) and analogue data functions. The unit contains an internal data modem and controller, and voice transmission is compatible with current SSB HF transceivers. Data transmission is compatible with ground HF transmitting and receiving systems using conventional HF transceivers and ARINC 635 compliant modems and controllers. The HFS-900D is ARINC 634 system design guideline compliant and offers improved built-in test capability and software dataload capability in accordance with ARINC 615. It is compatible with a new high-speed digital antenna coupler (CPL-920D) and a new ARINC 753 coupler interface to improve communication with the coupler. The HFS-900D offers HFDL through two different upgrade paths, a service bulletin/upgrade kit for 628T-2A, HFS-700, HFS-900 transceivers, or a new HF-900D data radio manufactured with the service bulletin.

Specifications
Dimensions: 6 MCU
Weight: 12 kg
Power supply: 115 V AC, 3 phase
Frequency: 2-29.9999 MHz
Channels: 280,000 in 100 Hz increments
Emissions:
(receive) AM, USB/LSB/AME, data, CW
(transmit) USB/LSB/AME, data, CW
Temperature range: −55 to +70°C
Altitude: up to 50,000 ft

Operational status
In production October 1997.

Contractor
Rockwell Collins.

UPDATED

ICS-150 intercommunications set

The Rockwell Collins ICS-150 intercommunications set is a fully militarised aircraft audio system which provides selectable channels of communications between aircraft crew stations. It also provides communications between each crew station and various transceivers, receivers and warning systems.

This set is particularly applicable for aircraft with multiple crew stations, a variety of communication, navigation and warning receivers or transceivers and stringent requirements for cross-talk isolation, electromagnetic interference and nuclear hardening.

The set delivered to the US Air Force for its B-1B aircraft provides an audio interface for up to eight crew stations, five ground crew/maintenance stations, eight separate avionics transceivers and 10 receivers.

An InterCommunication Set (ICS) is made up of one central control unit, up to eight crew station units and up to five maintenance station units.

Each crew station unit provides 10 receive-monitor control functions such as on/off and volume and transmit selection control of ICS, plus up to six other transmit functions. The crew station also provides master volume, hot mic, all call and LRU test facilities.

The central control unit is equipped with secure interlock capability to prevent secure communications from being heard on non-secure transmissions.

This system has high channel isolation and provision has been made for the future incorporation of a COMSEC switch capability. The COMSEC switch will permit a single speech encryption device to be switched between several radios.

The ICS-150 was designed to meet all military requirements, from parts utilisation to qualification testing.

A central mixing architecture requires very few interconnect lines between each crew station and the central control unit. Three twisted/shielded pairs of wires are used for microphone audio, headset audio and serial control data.

The ICS-150 central control unit has redundant input regulators and separate line regulators in each module to prevent any single point failure from causing system failure. An additional back-up mode is provided in case there is structural damage to the central control unit. This back-up mode provides for a direct connection between the pilot and co-pilot and two separate transceivers.

Specifications
Dimensions:
(central control unit) 124 × 193 × 497 mm
(crew station unit) 146 × 95 × 112 mm
(maintenance station unit) 117 × 91 × 99 mm
Weight:
(central control unit) 8.3 kg
(crew station unit) 0.9 kg
(maintenance station unit) 0.23 kg
Power supply: 28 V DC

Operational status
In production and in service. Applications include US Air Force B-1B, B-2 and C-135C, US Army Special Operations Force aircraft and Royal Australian Navy helicopters.

Contractor
Rockwell Collins.

UPDATED

Joint Tactical Information Distribution System (JTIDS)

The first AN/URC-107(V) Joint Tactical Information Distribution System (JTIDS) Class 2 terminal manufactured by Rockwell Collins was delivered to the US Air Force in September 1984 for testing aboard an F-15 fighter. This terminal, built under the US Air Force 'leader-follower' concept, was the culmination of 3½ years full-scale development by Singer (now GEC-Marconi Hazeltine Corporation) and Collins. Under this 'leader-follower' arrangement, GEC-Marconi Hazeltine Corporation led the design of the data processor group and Rockwell Collins developed the receiver/transmitter. Rockwell Collins now manufactures the complete JTIDS Class 2 line for US Air Force, Navy, Army and NATO applications.

Operational status
Collins was awarded an initial US$42 million contract for low-rate production of JTIDS for the F-15. Deliveries began in 1992. A second contract for US$59 million was awarded in August 1991 for JTIDS for F-14, E-2 and E-3 aircraft for delivery in 1993. A third contract for US$43.3 million for Class 2 and Class 2H terminals for US ships and F-14D and E-2C aircraft was awarded in November 1992, with delivery between May 1994 and January 1995. Rockwell Collins has been awarded 60 per cent of the DoD full-rate terminal production requirement (US$11.7 million) for completion by September 1999. In January 1997, a contract worth US$196 million was awarded to Rockwell Collins for up to 230 (minimum 40) Class 2/2H terminals from US Air Force; contract to be completed March 2001.

Contractor
Rockwell Collins.

UPDATED

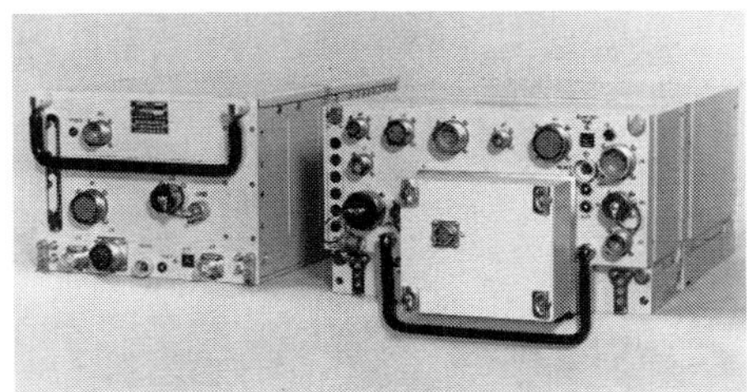

The Rockwell Collins/GEC-Marconi Hazeltine JTIDS Class 2 terminal for the US Air Force F-15

Miniature Receive Terminal (MRT)

The Miniature Receive Terminal (MRT) forms part of the minimum essential emergency communications network that provides secure VLF links between the National Command Authority and Strategic Air Command bombers. It includes the Datametrics quarter-page black and white printer.

Operational status
The MRT was developed for the B-52H and is in service on the B-1B.

Contractor
Rockwell Collins.

UPDATED

OG-187/ART-54 VLF/LF transmitter

The OG-187/ART-54 is a self-contained 200 kW VLF/LF transmitter designed to meet transmission requirements over the 17 to 60 kHz frequency range in 10 kHz increments.

The OG-187/ART-54 is an all solid-state transmitter designed to operate in the minimum shift keying, frequency shift keying, continuous shift keying and frequency shift continuous wave modes. The transmitter can be controlled locally or remotely via a standard serial bus. Automatic tuning is achieved in 10 seconds maximum. Constant surveillance tuning ensures matching antenna impedances regardless of the environmental effects after deployment.

Specifications
Dimensions: 1,874 × 1,990 × 1,450 mm
Weight: 1,545 kg
Power supply: 115/200 V AC, 400 Hz, 3 phase, 252 kVA
28 V DC, 8 A

Contractor
Rockwell Collins.

UPDATED

OG-188/ARC-96A VLF/LF transmitter

The OG-188/ARC-96A is a self-contained 100 kW VLF/LF transmitter designed to meet the requirements for the US Air Force World Wide Airborne Command Post mission. Operating over the 17 to 60 kHz frequency range in 10 kHz increments, it provides 100 kW to the dual trailing-wire antenna aboard the EC-135 aircraft.

The OG-188/ARC-69A is an all solid-state transmitter designed to operate in the minimum shift keying, frequency shift keying, continuous shift keying and frequency shift continuous keying modes. The transmitter can be controlled locally or remotely via a standard serial bus. Constant surveillance tuning ensures matching antenna impedances regardless of the environmental effects after deployment.

Specifications
Dimensions: 1,750 × 1,340 × 1,570 mm
Weight: 1,091 kg
Power supply: 115/200 V AC, 400 Hz, 3 phase, 119 kVA
28 V DC, 8 A

Contractor
Rockwell Collins.

UPDATED

RT-1379A/ASW transmitter/receiver/processor

The RT-1379A/ASW is an AN/ARC-182 derivative design providing a 5 kbits/s half-duplex or simplex RF datalink using the TADIL-C message protocol and modulation. As currently configured, the radio covers the 225 to 400 MHz UHF band with 25 kHz channel spacing and is compatible with a number of US Navy data systems. These include the naval tactical data system, airborne tactical data system, AN/SPN 10/42 automatic carrier landing system, AN/TPQ-10/27 precise course direction system and the inertial navigation system.

The RT-1379A/ASW interfaces with the mission computer on either of two (redundant) MIL-STD-1553B multiplex buses. Jumpers in the aircraft wiring harness determine the unique multiplex address assigned to the radio.

TADIL-C address assignment is via five octal encoded switches under a front protective cover. The radio's address is normally selected on the flight line before a mission. The last three (least significant) octal address digits can be changed by commands from the mission computer at any time, causing the radio to assume a new TADIL-C address.

Among the types of information which can be handled are two-way transfer of target information, aircraft vectoring data, INS update data, landing system data and general data reporting of aircraft status.

In general, any data which is available on the aircraft multiplex bus can be transmitted by the radio. It can be modified to communicate using a message protocol other than TADIL-C format and can accommodate other data rates up to 16 kbits/s. The radio can also be made to operate on any channelised frequency between 30 and 400 MHz, and voice communications capability can be added.

Specifications

Dimensions: 135.9 × 127 × 270.5 mm
Weight: 4.9 kg

Operational status

The RT-1379A/ASW radio is in operational service on US Navy F/A-18 Hornet aircraft.

Contractor

Rockwell Collins.

UPDATED

RTU-4200 series Radio Tuning Units

The RTU-4200 Radio Tuning Unit (RTU) provides centralised control of VHF comms, VOR, ILS, DME, ADF, transponder and TCAS. Integration with a Flight Management System (FMS) also allows tuning of radios via either the FMS control/display unit or the RTU, with the tuned frequency always appearing on the RTU display. Each RTU can control all radio sensors. It also has the capability to store 20 preset comm and 20 preset nav frequencies.

Brightness of the active matrix liquid crystal display can be controlled by the aircraft's master dimming control bus or by crew member adjustment on the RTU. Parallax compensation is provided and control settings and radio diagnostics are stored in non-volatile memory for availability after power shutdown.

Operational status

The RTU-4200 is standard equipment on Gulfstream IV SP business jet aircraft. The RTU-4210 has been selected by Cessna for the Citation X business jet. The RTU-4280 has been selected by Gulfstream for the Gulfstream V business jet. The RTU-4200 has been certified aboard the Sikorsky S-76B helicopter.

Contractor

Rockwell Collins.

UPDATED

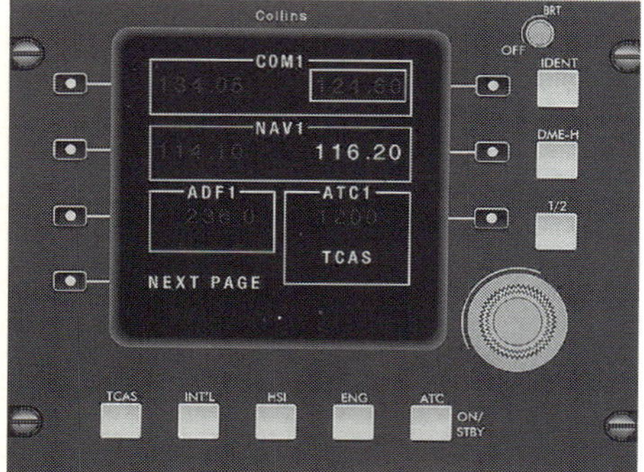

The RTU-4280 has been selected for the Gulfstream V **1995**

SAT-900/901 satellite communications systems

The SAT-900/901 satellite communications systems operate at L-band through the INMARSAT organisation space segment. Both systems support a low-speed digital link for the transmission of ACARS/AIRCOM datalink messages when out of range of terrestrial VHF stations. Rockwell Collins Satcom systems are currently operational with the DLM-700B and management units from other commercial datalink equipment suppliers.

In addition to low-speed data, the SAT-901 can support a digital voice channel for use by either the cockpit or cabin. The selection of the single channel for voice or data is made via the MCDU or MIDU in the cockpit. Cabin communications require the use of the optional CTI-901 cabin telecommunications interface unit. The CTI-901 provides the industry standard CEPT E1 interface between the SDU-901 and the cabin telephone system.

The SAT-900/901 systems are compatible with all ARINC 741 low- and high-gain antenna subsystems. A low-gain antenna can be used with either system to support low-speed data communications, but a high-gain antenna subsystem is required with the SAT-901 for voice communications.

Each system is made up of a Satellite Data Unit (SDU), Radio Frequency Unit (RFU), High-Power Amplifier (HPA) and ARINC 741 antenna subsystem. The SAT-900 consists of an SDU-900, RFU-900 and HPA-900. The SAT-901 uses the same RFU and HPA as the SAT-900, along with the voice capable SDU-901.

The SAT-900 is compatible with INMARSAT Data-1 protocols; the SAT-901 is compatible with INMARSAT Data-2 and Voice-2 protocols.

Specifications

Dimensions:
(SDU-901) 190 × 193 × 325 mm
(RFU-900) 125 × 193 × 325 mm
(HPA-900) 125 × 193 × 325 mm
Weight:
(SDU-901) 8.84 kg
(RFU-900) 8.3 kg
(HPA-900) 6.8 kg
Power supply: 115 V AC, 400 Hz
Power output: 40 W continuous
Frequency:
(receive) 1,530-1,559 MHz
(transmit) 1,625.5-1,660.5 MHz
Synthesiser step size: 2.5 kHz
Doppler correction: within 20 Hz over ± 2 kHz
Modulation type: A-BPSK

Operational status

No longer in production. Both the US FAA and the Australian CAA have certified the SAT 900 system. Northwest Airlines ordered the system for its Boeing 747s. United Airlines and Japan Airlines also ordered it for the Boeing 747-400.

Contractor

Rockwell Collins.

UPDATED

SAT-906 satellite communications system

The SAT-906 is the Rockwell Collins industry standard satellite communications system. It is fully compliant with ARINC 741 and 746 and provides six high-quality voice and data channels for cockpit and cabin use. It provides low- and high-speed data, cockpit voice and passenger PC, fax and telephone services when mated with a high-gain antenna subsystem.

The SAT-906 consists of the SDU-906, RFU-900 and HPA-901A. The SDU-906 provides the interface to all other aircraft systems, and includes modems, codecs and protocol support for communication with ground earth stations.

The RFU-900 consists of a wideband L-band (NATO D-band) to IF down-converter for receive operation and a wideband IF to L-band up-converter for transmit operation. The RFU operates in full-duplex mode, receiving L-band signals in the range 1,530 to 1,559 MHz.

The HPA-901A is a linear amplifier that provides the necessary intermodulation performance characteristics for multichannel operation. At 63 W, the HPA-901A is claimed by Rockwell Collins to be the "most powerful in the industry and the only one capable of full 6-channel operation with typical (12 dB) high-gain antenna performance".

The SAT-906 is compatible with Inmarsat Data-2, Data-3 and Voice-3 protocols and provides enhanced en route navigation and communications functionality, including Automatic Dependent Surveillance (ADS).

In secure mode, a STU-III Secure Telephone Unit is used in conjunction with the CIU-906 to pass encrypted multichannel voice, fax, and data through the SAT-906 system.

A Cabin Interface Unit (CIU-906) is available, which serves as an onboard switchboard capable of providing secure communications to the ground. CIU-906 can accommodate up to 10 six-wire telephones and four ports for two-wire devices, such as cordless phones, faxes and PC modems.

Aero-I

In 1997, Rockwell Collins launched the Aero-I system as a derivative of the SAT-906 system. Aero-I incorporates many features and benefits of the SAT-906 and provides new features unique to Aero-I, making satellite communication services accessible to a broader range of aircraft operators, Aero-I is specifically designed for air transport, military and business and regional applications by minimising costs using access to the Inmarsat 3 spot beam coverage. Aero-I is well suited to operators seeking a global, economical, single-channel, SATCOM datalink capability. Aero-H will remain the system of choice for customers requiring continuous, full-capabiity, multichannel global satellite service.

Specifications

Dimensions:
(SDU-906) 6 MCU
(RFU-900) 4 MCU
(HPA-901) 8 MCU

The Sabreliner 65 satellite communications demonstration aircraft showing one of the 4 in square satellite antennas aft of the aircraft door near the top of the fuselage

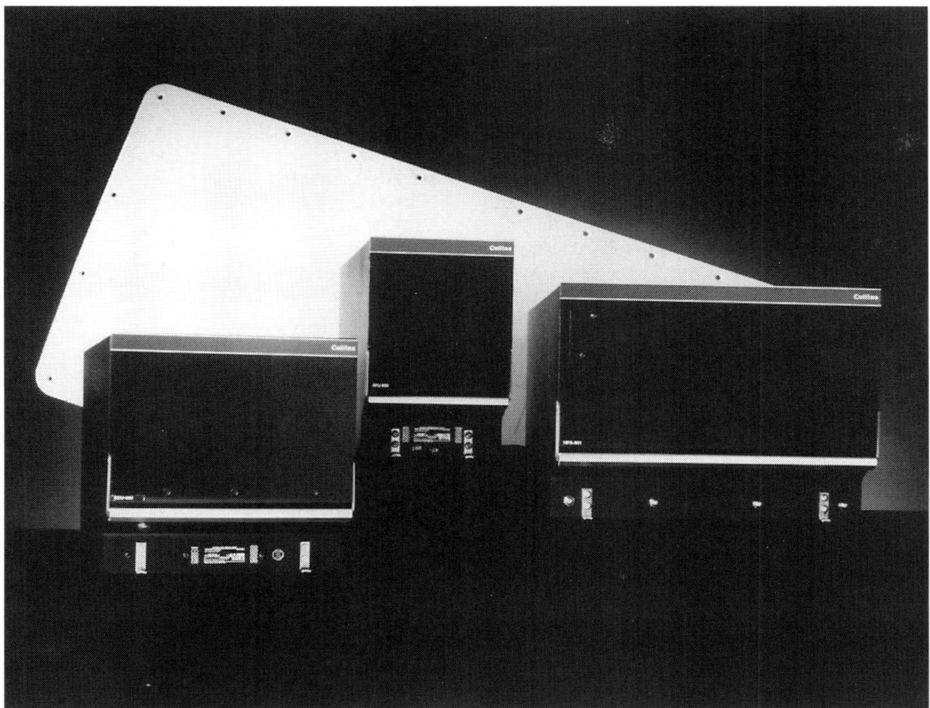

The Rockwell Collins Satcom 906 communications system showing (left to right) the SDU-906 satellite data unit, the RFU-900 Radio Frequency Unit and the HPA 901 High-Power Amplifier ***1997***/0001184

Weight:
(SDU-906) 12.7 kg
(RFU-900) 8.3 kg
(HPA-901) 17.24 kg
Power supply:
(SDU-906) 115 V AC, 400 Hz, 180 W
(RFU-900) 60 W
(HPA-901) <300 W

Operational status

SAT-906 Aero-H first certified in December 1993, on A310. Since then, 20 airlines have ordered a total of more than 500 systems. Aero-I availability is linked to the availability of Inmarsat Aero-I services.

Contractor

Rockwell Collins.

UPDATED

TACAMO II communications system

The TAke Charge And Move Out (TACAMO) II system provides airborne VLF communications links with the US Navy strategic submarine fleet. The system is a manned communications relay link to strategic forces, normally passing messages one way from the national command to submarines and other strategic forces.

At present, a complete communications centre in the Boeing E-6A TACAMO II aircraft allows simultaneous receive and transmit throughout the frequency range VLF to UHF. The system receives multiple frequency low-level signals, while simultaneously transmitting at high power in a stressed environment. The VLF power amplifier provides amplification of the signal to 200 kW power and automatic tuning of the signal to the dual trailing-wire antenna system. This latter consists of two antennas, one nearly 1,500 m long and the other more than 8,500 m. Only the short wire is charged, the energy reradiating off the longer wire, the length of which varies with the frequency in use. The transmitted signal to the submarine is vertically polarised, with the E-6A aircraft flying in a continuous tight turn. This allows most of the antenna system to hang vertically from the aircraft.

Operational status

In service on E-6A TACAMO II aircraft of the US Navy.

Contractor

Rockwell Collins.

UPDATED

VHF-21/22/422 Pro Line II VHF Radios

Designed primarily for general aviation aircraft of all types, the Rockwell Collins VHF-21/22/422 transmitter receivers are remotely controlled, rack-mounted sets with 20 W transmitter output. They are available in two versions: the A equipment covering the VHF band from 118 to 136.975 MHz and the B variant from 118 to 151.975 MHz. Channel spacing is 25 kHz. Units with broad receiver bandwidths are available.

The radios use digital synthesis frequency generation techniques and are of all solid-state construction. They provide automatic carrier and phase noise squelch and automatic gain control and are designed to drive cabin audio systems of all types. Principal attractions are low weight, compactness and the low power consumption of 6.5 A during transmission. Consequently they require no forced-air supply and electronic section cooling is carried out by a combination of heatsink and convective air flow.

The VHF-21 can directly replace the earlier VHF-20 series radios. The VHF-422 is compatible only with CSDB or ARINC 429 controls.

Either hard or soft mounting may be used and all connections are made through a single connector on the rear of the casing.

Specifications

Dimensions: 84 × 86 × 355 mm
Weight: 2.1 kg

Operational status

In production and in service.

Contractor

Rockwell Collins.

UPDATED

VHF-700/VHF 700B transceivers

The Rockwell Collins VHF-700 is a VHF AM transmitter/receiver designed principally for air transport use which covers the band 118 to 135.975 MHz with channel spacings of 25 kHz. Typical transmitted power output is 30 W.

The system, which has been designed and developed to ARINC 716 specifications, has a number of advanced features including a new receiver design which eliminates mutual interference between equipment even when up to three transmitter/receivers are used in the same aircraft. A complete end-to-end self-test facility, which checks 99 per cent of all critical components, is built-in. The ARINC 429 serial databus is also tested to ensure that the system is receiving valid information from the remote tuning selector and centralised ARINC 604 fault monitoring capability.

The VHF-700 is microprocessor-controlled, of all solid-state construction and employs a temperature compensated crystal oscillator as the reference standard for its digital frequency generation synthesiser.

VHF-700B

The VHF-700B adds 8.33 kHz channel spacing capability to the VHF-700 range.

Specifications

Dimensions: 3 MCU short
Weight: 4 kg
Temperature range: −55 to +71°C
Altitude: up to 50,000 ft

Operational status

The VHF-700 is widely used on Boeing and Airbus airliners. The VHF-700B, first offered in 1997, has been selected by Quantas Airways to replace existing analogue VHF equipment on its Boeing 747-200/300 aircraft in time for the January 1999 European requirements.

Rockwell Collins has received US FAA supplemental type certification for the VHF-700B radio with the AVTECH 7522-1-2 control panel.

Contractor

Rockwell Collins.

UPDATED

VHF-900/VHF-900B transceivers

The Rockwell Collins VHF-900 hardware was developed for Boeing CAT III GPS autoland procedures. The VHF-900B transceiver has been designed to meet the January 1999 European requirement for 8.33 kHz channel spacing. It has been selected by Quantas Airways to retrofit its fleet of Boeing 747-400 aircraft.

VHF-900 hardware is also being evaluated by Rockwell Collins as the basis of a system to provide graphical weather data to aircraft flight decks, using a VHF radio datalink operating at 31.5 kbps, at which speed a weather data bitmap file can be uplinked in 3 to 5 seconds (the same file, uncompressed, would require 15 minutes on an ACARS link).

Operational status

Rockwell Collins has received US FAA supplemental type certification for the VHF-900B radio with the AVTECH 7522-1-2 control panel.

Contractors

Rockwell Collins

NEW ENTRY

VLF/LF High Power Transmit Set (HPTS)

The HPTS system consists of a Very Low Frequency/Low Frequency (VLF/LF) 200 kW solid-state power amplifier and dual-trailing wire antenna system. It is designed to improve the reliability of systems that provide survivable communications links from the US Navy's E-6A TACAMO aircraft to the US strategic forces.

Operational status

In service and in production for the US Navy E-6A TACAMO aircraft.

Contractor

Rockwell Collins.

NEW ENTRY

VP-110 voice encryption device

The VP-110 is a voice encryption device designed for use with airborne radio communications systems. The unit is packaged in a ½ ATR short unit and requires 28 V DC. A companion unit, the VP-100, performs the same function for fixed-station radios. Although aimed primarily at HF radios, the system will work equally well on both VHF and UHF narrowband equipment and ordinary telephone lines with the addition of a TA-110 adaptor.

The equipment is intended for a range of uses such as law enforcement, business, diplomatic, government agency and selected military voice transmission applications. It has been designed for the encryption of sensitive transmissions.

The system eliminates all syllabic content in the encrypted mode while retaining clear voice quality and recognition. For transmission, the voice is converted into analogue signals and divided into low- and high-band frequency ranges. It is then encoded and transmitted in a random mode with regard to time and frequency.

Public keying is provided, enabling private conversations between two stations without prior manual exchange of a recognition code. In this method of communication, the operator selects the mode and the two units exchange a set of numbers, using a complex mathematical algorithm, which in effect establishes a signature for connecting private conversations. Eight codes, or key variables, can be entered in the unit microprocessors, providing 10×7^{19} code possibilities.

Rockwell Collins produces two versions of the VP-110: one for the US market and one for export. The former uses a Data Encryption Standard (DES) algorithm while the latter is provided with a Rockwell-developed algorithm. An Over-The-Air Rekeying (OTAR) option is available.

Operational status
In production and in service.

Contractor
Rockwell Collins.

UPDATED

AN/AGC-9(V) communications management terminal

The SCI AN/AGC-9(V) communications management terminal is a modular compact-distributed microprocessor-controlled message processing system currently in service on the US Navy's P-3 aircraft. The AN/AGC-9 is a multipurpose message handling system which includes word processing, control and message storage. The AGC-9 has a high-speed rotary printer which provides hard-copy output at a speed in excess of 1,500 characters/s. The AN/AGC-9 provides the P-3 crew with complete message centre control capabilities. It can be configured for multiple I/O ports to interface with telephone/teletypewriters, the naval tactical data system, MIL-STD-1553B, RS-232 and other interfaces.

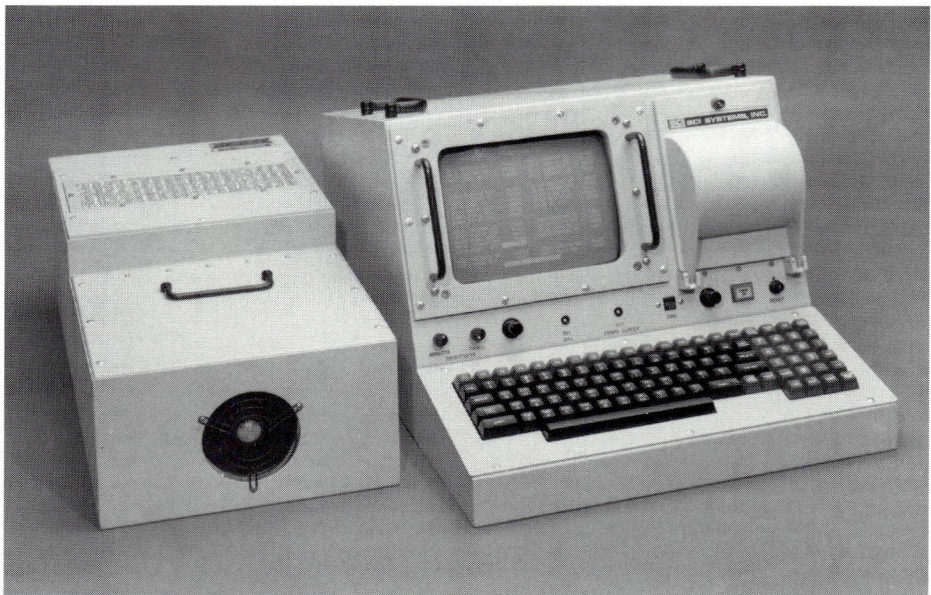
The SCI AN/AGC-9(V) communications management terminal

Specifications
Dimensions:
(terminal) 431.8 × 317.5 × 546 mm
(electronic assembly) 571.5 × 279.4 × 193.8 mm
Weight:
(terminal) 19.5 kg
(electronic assembly) 16.3 kg
Power supply: 115 V AC, 60 or 400 Hz
Interfaces: TTY, NTDS, MIL-STD-1553B, RS-232 plus others

Operational status
In service with the US Navy Lockheed P-3 Orion.

Contractor
SCI Systems Inc.

VERIFIED

Auxiliary Communications, Navigation and Identification Panel for the AV-8B

The Auxiliary Communications, Navigation and Identification Panel (ACNIP) is an integral part of the communications, navigation and identification system used on the AV-8B and TAV-8B Harrier.

When interfaced with the other components of the aircraft communications system, the ACNIP performs audio amplification, control inhibit and distribution functions, generates audio warning messages in response to discrete serial and/or analogue inputs, provides code, mode, remote variable load, baseband/diphase, and control zeroing functions for two KY-58 secure communications units and controls functions of the identification system such as zeroisation and emergency operation. Additionally, the ACNIP provides logic-controlled push-to-talk switch closures and switch functions for landline telephone communications, control for ground crew communications and a hot mic capability for the operator. BIT circuitry detects 98 per cent of all electrical component failures.

The ACNIP controls a non-volatile EEPROM for the storage of the code and mode operating parameters of the secure speech units. On initial power-up, the ACNIP will update the KY-58 units to the operating modes as selected before power-down.

An LCD module, backlit and with variable illumination level located on the front panel of the ACNIP, provides a visual readout of the functional status of the KY-58 units, the operating mode, code and type of cypher used by each unit being displayed.

Operational status
In service with the AV-8B.

Contractor
SCI Systems Inc.

VERIFIED

Integrated radio control panel

The integrated radio control panel is designed to provide the pilot and/or co-pilot with the means to manage all communications equipment on board a commercial or military aircraft. Two interface units are available: MIL-STD-1553 for military applications and ARINC 429 for commercial aircraft.

The primary display medium is a CRT which displays mode and frequency information. In addition to the CRT, several dedicated controls are provided for those functions which require immediate access. The unit has a non-volatile memory for power-off data retention.

Specifications
Dimensions: 203 × 146 × 200 mm
Weight: approx 4.5 kg

Contractor
SCI Systems Inc.

VERIFIED

Intercommunications set for the V-22 Osprey

The full V-22 intercommunications set consists of a Communications Switching Unit (CSU) with up to six Intercom Set Control (ISC) stations, two Audio Frequency Amplifier (AFA) assemblies, a Cabin Public Address (CPA) amplifier and four cabin PA speakers.

It provides simultaneous intercom for multiple crew stations, five channels of radio transmission and

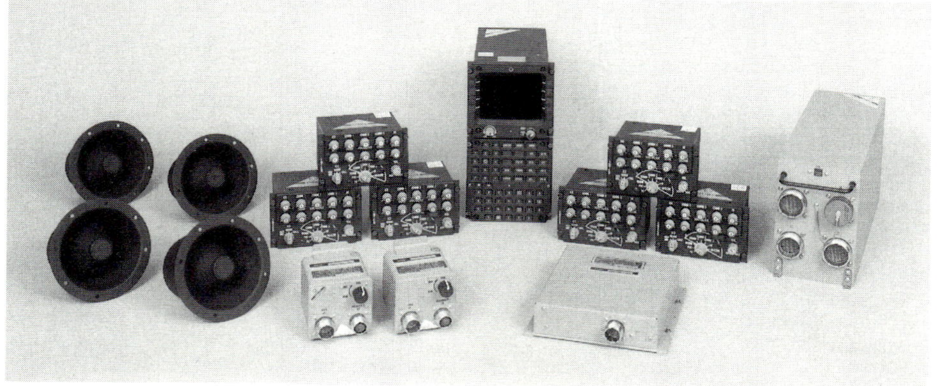
The intercommunications set for the V-22 Osprey

reception for each ISC, and reception of four composite navaids, five warning tones, one IFF and one radar warning receiver on two ISC panels. All audio switching and mixing functions are software-controlled and interfaces to clear and secure communications equipment are provided. There is also a digital message device I/O channel. The electroluminescent panel provides high-contrast NVG-compatible lighting.

The CSU contains the switching circuitry, logic circuit and a large portion of the audio circuitry required for ICS operation. In addition, it is the interconnect unit for the ICS system components and peripheral devices. The CSU provides impedance matching, audio reporting and push-to-talk functions between the crew station ICS panels and the aircraft radios. Radio selection status is provided to the aircraft control and display subsystem through the CSU.

Each intercommunication set is essentially the individual crew station control panel, on which the frequencies, sources and signal levels are selected for monitoring and/or transmission. The audio frequency amplifier permits the selection of intercom facilities only.

The system is based on a low-power CMOS microprocessor and large-scale integration circuitry. Extensive built-in testing has been designed in for simplified maintenance. Options with the system include a digitised message card, RS-232, ARINC 429 and MIL-STD-1553 interfaces and an incandescent light panel.

Operational status
In service and in production for the V-22 Osprey.

Contractor
SCI Systems Inc.

VERIFIED

C³SAT 2000 High-Power Amplifier (HPA)

The C³SAT High-Power Amplifier (HPA) is a linear multi-carrier amplifier. As part of the System 2000, it receives L-band signals from the 2020 Satellite Terminal Unit (STU) (see item below), amplifies the signals and provides the required output to the antenna subsystem.

The HPA has power management capability to minimise average power consumption. Its output gain is automatically controlled by the STU through an ARINC 429 bus. The HPA provides the antenna subsystem with RF power, necessary to transmit simultaneous voice or data messages through the high-gain antenna. The ARINC 429 bus is also used to relay the status of the HPA through the STU to the cockpit. Parameters include BIT outputs, total power output, number of carriers being transmitted, output VSWR and internal temperature.

The System 2000 HPA is used as part of an INMARSAT commissioned multicarrier Aeronautical Earth Station (AES) to provide reliable worldwide telephone and data communications for commercial airlines and corporate aviation customers.

Specifications
Dimensions: 194.1 × 257.8 × 318.5 mm
Weight: 13.83 kg
Power supply: 115 V AC, 400 Hz, 250 W

Operational status
Access-approved on Boeing 727, 747-400, Gulfstream II, III, IV, Canadair Challenger 601 and WC135.

Contractor
Sextant Electronics Inc.

VERIFIED

C³SAT 2020 Satellite Terminal Unit (STU)

The C³SAT 2020 Satellite Terminal Unit (STU) provides reliable worldwide multichannel telephone and data communications for commercial airlines and corporate aviation customers while using the INMARSAT network of satellites and ground stations.

Each channel within the system provides unrestricted voice and/or data communications. The STU is manufactured in accordance with ARINC 741 requirements and it is compatible with all ARINC 741 compliant antenna subsystems, ARINC 746 compliant cabin communications systems and ARINC 724 ACARS equipment.

Future growth and enhancements are being developed and increased channel capacity and a fax capability are available.

Specifications
Dimensions: 194.1 × 322.33 × 318.6 mm
Weight: 16.78 kg
Power supply: 115 V AC, 400 Hz, 250 W max
Frequency:
(transmit) 1,625.6-1,660.5 MHz
(receive) 1,530-1,559 MHz
Channel spacing: 17.5 kHz

Contractor
Sextant Electronics Inc.

VERIFIED

SAMS-1000 audio management system

SAMS-1000 is an audio management system designed for use with up to six radios and five navigation receivers, in addition to siren, intercom and public address functions. Each operator can independently monitor any or all radios. SAMS-1000 includes a 'hot microphone' or push-to-talk intercom; voice-operated intercom is an option. SAMS-1000 is designed for use in airborne law enforcement, and search and rescue operations.

Contractor
Sigtronics Corporation

NEW ENTRY

Sigtronics aircraft intercoms

Sigtronics Corporation makes a range of aircraft intercom systems:

SAS-440/-640 auto squelch panel-mounted intercom series
These intercoms virtually eliminate the need to constantly readjust the squelch during flight. The SAS-440 has radio priority to assure that the only voice heard by air traffic control is that of the crew. SAS-440 supports four headsets. SAS-640 supports six headsets. TSO approval was pending in mid-1997.

SDB-800 dual audio panel intercom
Effectively, a dual SPA-400/-600 installation can support two pilots, and up to eight headsets.

SPA-400/600 intercom series
An industry standard intercom for many years, the SPA-400/600 has radio priority, and a pilot fail-safe feature, which ensures that the pilot will always hear the radios, even if the intercom is set to off.

SPA-400N/-600N intercom series
Specially designed version of the SPA-400/-600 series for very high noise cockpits; helicopters; warbirds and ultralights.

STN-400/-600 stereo intercom series
A full stereo version of the SPA-400/-600 series.

Contractor
Sigtronics Corporation.

NEW ENTRY

C-10382/A Communication System Control (CSC) set

The primary function of the Communication System Control (CSC) set is to provide the pilot with integrated, centralised control of data transferring capability, power switching, mode selection, operating frequencies, interconnections and signal flow routes of the aircraft's CNI equipment. The CSC provides for highly efficient operation of communications by integrating these primary CNI systems controls into the aircraft's advanced avionics architecture and also into a single convenient easy-to-operate pilot's control panel.

On the US Navy's F/A-18 aircraft, the controls and displays of the cockpit control panel are engineered to optimise pilot control of the CNI equipment. The control panel is positioned to allow the pilot to keep his eyes focused straight ahead, with only the pertinent information and controls he needs presented in his field of view.

A redundant MIL-STD-1553 multiplex bus provides connection between the CSC and the AN/AYK-14(V) mission computer for the flow of information and control. The mission computer provides CNI control signals, BIT commands and information for the control panel's alphanumeric display. In return, the CSC transmits equipment status, received CNI data, operating options and BIT response to the mission computer. Dedicated serial digital lines interface the control panel with the CSC.

To process data to and from the CSC, mission computer, control panel and CNI equipment, the CSC interfaces serial digital signals, discrete signals, analogue signals, synchro signals, avionic multiplex bus signals and audio signals. As the CSC microcomputer processes at least 1,300 parameters/s, it controls and processes the data required, leaving 40 per cent of real time available for growth.

Operational status
In service on the US Navy F/A-18 aircraft.

Contractor
Smiths Industries Aerospace.

VERIFIED

TEC LINE VHF-251A communications receiver

The TEC LINE VHF-251A 760-channel communications receiver is a direct enhanced replacement for the VHF-250/251/251S and 251E communications transceivers.

Features include: 760 channels in the 118 to 136.975 MHz band; 10 W power; non-volatile 10 frequency memory; failed display emergency mode; stuck-microphone protection.

Specifications
FAA TSO: C37c, C38c, DAO-160c
Dimensions: front panel: 79.25 × 66.29 × 43.69 mm
chassis: 80.77 × 67.41 × 316.2 mm
Weight: 1.75 kg

Contractor
S-TEC Corporation.

VERIFIED

ASB-500 HF/SSB radio

The Sunair ASB-500 and the associated ACU-150D (see item above) have been specifically designed for aircraft and helicopters requiring a large number of operational frequencies but where space and weight are limiting factors. It covers the HF band and provides USB, AM and optionally LSB modes of operation.

The radio comprises a transceiver, remote control and automatic antenna coupler. The controller is panel-mounted and includes a six-digit LED frequency display and illuminated status and antenna coupler tuning monitors. The coupler is solid-state and tunes extremely quickly with a 10-channel last-tuned memory. The radio provides 100 W output power and is certificated to the relevant FCC and FAA TSOs.

Specifications

Dimensions:
(transceiver) 123.8 × 193.7 × 39.4 mm
(controller) 146 × 127 × 66.7 mm
(antenna coupler) 177 × 152 × 305 mm
Weight:
(transceiver) 6.6 kg
(controller) 0.8 kg
(antenna coupler) 3.9 kg
Power supply: 27.5 V DC
(receive) 2.7 A
(transmit) 13 A
Frequency: 2-17.9995 MHz
Power output: 100 W PEP
Channels: 32,000 at 500 Hz spacing
Temperature range: −46 to +55°C
Altitude: up to 30,000 ft

Sunair ASB-500/ACU-150D HF radio set

Operational status

The ASB-500 and ACU-150D are available for general aviation and helicopter applications.

Contractor

Sunair Electronics Inc.

VERIFIED

ASB-850A HF/SSB radio

Sunair's ASB-850A is a multipurpose synthesised military HF transceiver. It is a particularly light and compact system designed for light fixed-wing aircraft and helicopters operating in tactical roles. The system is remotely controlled from a miniature panel-mounted unit which contains an LED frequency selection display. It operates in USB, LSB and AME modes.

The ASB-850A is of all solid-state construction and is claimed to be of exceptionally robust design and manufacture.

A new high-speed automatic antenna coupler is incorporated within the system and this unit can tune the antenna to the frequency selected in 1 second or less for initial tuning and in a matter of milliseconds for tuning to the last 10-tuned channels memory.

Specifications

Dimensions:
(transceiver) 470 × 124 × 241 mm
(amplifier/antenna coupler) 448 × 152 × 235 mm
(controller) 172 × 146 × 64 mm
Weight:
(transceiver) 7.9 kg
(amplifier/antenna coupler) 7.9 kg
(controller) 0.6 kg
Power supply: 27.5 V DC
(receive) 2 A
(transmit) 17 A
Frequency: 2-29.9999 MHz
Channels: 280,000 at 100 Hz spacing
Temperature range: −46 to +71°C
Altitude: up to 30,000 ft

Operational status

In service. Production ended in 1994. A spares and support service is continuing as long as is possible and practical.

Contractor

Sunair Electronics Inc.

VERIFIED

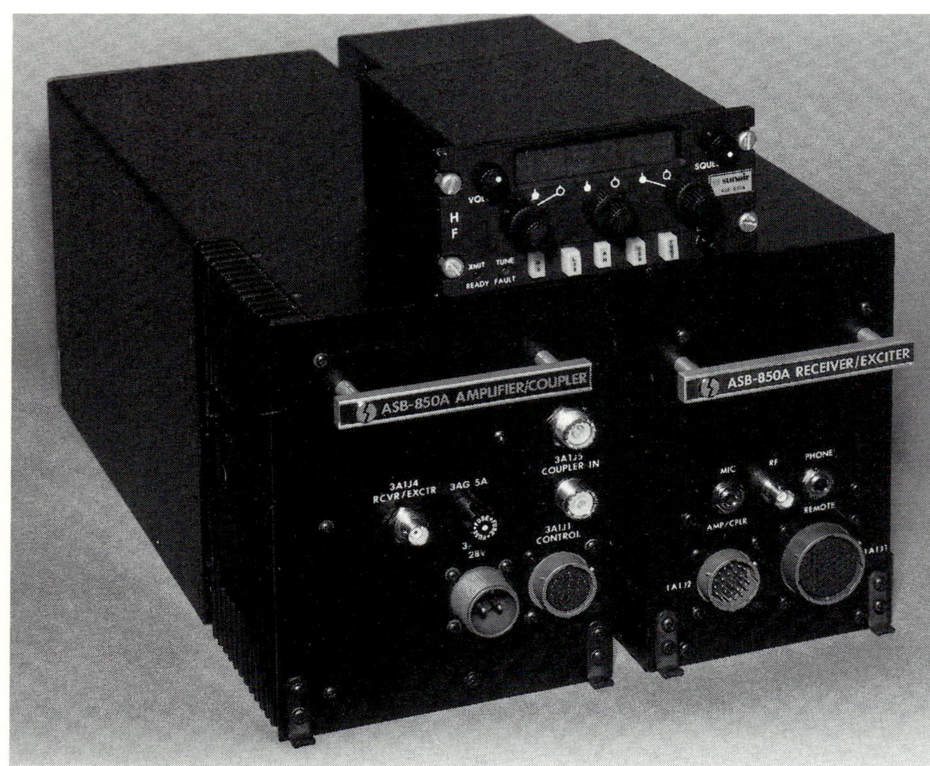

Sunair ASB-850A HF/SSB airborne transceiver

TeleLink helicopter datalink

The Teledyne TeleLink helicopter datalink supports a variety of bidirectional communication media, including: ACARS, satellite and telephone. Capabilities include: flight operations data exchange; GPS-based position reporting; uplink/downlink of FDR/HUMS data; custom messaging. Interfaces include: ARINC 429, 739, CSDB & RS 232/422 I/O.

Contractor

Teledyne Controls, Business and Commuter Avionics.

NEW ENTRY

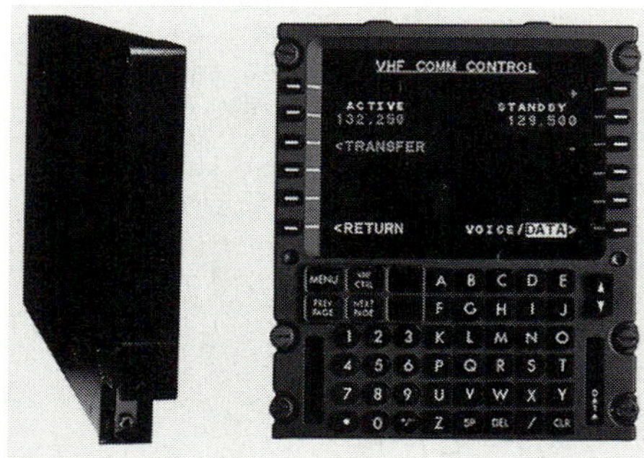

TeleLink helicopter datalink ***1998***/0011820

TeleLink TL-608 datalink system

The TeleLink TL-608 digital datalink system meets the needs of the corporate or regional airline pilot by providing a variety of features and versatile connectivity options in a small package. Numerous bidirectional communications media are supported, including: VHF ACARS, satellite, airborne telephone and other, to meet the communication requirements of the flight crew and the passengers.

Although derived from ACARS, the TL-608 is designed to take advantage of emerging datalink technologies and open architecture software to provide the operator with a choice of service providers and communications media.

Specifications

Dimensions: 1 MCU: 25.4 × 194 × 386.8 mm
Weight: 1.8 kg
Power: 28 V, 10 W
Inputs: ARINC 429 receivers, RS-232/422 receivers, discretes, VHF modem, telephone modem
Outputs: ARINC 429 transmitters, RS-232/422 transmitters, discretes, VHF modem, telephone modem
Interfaces: FMS, CDU, DAU, printer, dataloader, maintenance terminal, laptop PC
Communications media: VHF radio, airborne telephone, Satcom, Mode S

Operational status

Selected by Bombardier Inc as the standard option Communications Management Unit (CMU) for the Global Express aircraft, where it will interface with the Honeywell flight management system and Satcom to provide data and voice coverage worldwide.

Contractor

Teledyne Controls, Business & Commuter Avionics.

VERIFIED

AN/AIC-37(V) Digital Communication Management System

The AN/AIC-37(V) Digital Communication Management System (DCMS) is a digital, distributive, TEMPEST certified intercommunication system designed to manage the communications assets of airborne platforms. It integrates and manages all the navigation aids and alert/warning tones, as well as providing for the real-time control and switching of modems and encryption devices between transceivers. Both frequency management and a MIL-STD-1553B interface are offered. In addition, the AN/AIC-37(V) has no single point failure, contains BIT to card level (eliminating the need for intermediate level maintenance) and is field reconfigurable and expandable. A voice-activated switch, capable of operating in high-noise environments, and 25 intercom nets are standard features. Individual volume control for all assets is provided, plus a variety of front panel configurations ranging from push-buttons to a full display.

The AN/AIC-37(V) operates on a redundant 10 mbits/s databus. This permits up to 84 non-blocking channels of communication between crew stations and external radio systems, and a means to interface with an external computer through a redundant MIL-STD-1553B bus. External voice and data transmissions can be made on encrypted or plain networks. The system can deny access to any crew position not authorised to receive secure information.

When operators communicate, the crew terminal digitises the audio into an allocated data time slot with a destination address to the selected interface. The destination unit then converts the digital information back into baseband audio for application to the specific asset. Similarly, when a radio or remote communication device receives audio, it is digitised and made available to any of the operators that have been enabled to receive the audio. The entire system is digitally reconfigurable, enabling asset assignment based on the operational scenario. All channels within the system can be selected, combined and monitored.

Specifications

Dimensions:
(DCI) 171.5 × 127 × 204.7 mm
(DAI) 152.4 × 127 × 242.8 mm
(DBI) 152.4 × 127 × 242.8 mm
(DII) 123.9 × 127 × 86.1 mm
(DUCK) 19.1 × 80.8 × 132.1 mm
(DBC) 57.1 × 111 × 57.1 mm
Weight:
(DCI) 4.11 kg, (DAI) 4.14 kg
(DBI) 3.64 kg, (DII) 1.02 kg
(DUCK) 0.25 kg, (DBC) 0.34 kg
Power supply: 28 V DC
Temperature range: –54 to +55°C
Altitude: up to 50,000 ft

Elements of the digital communication management system, which are installed in the EP-3E electronic intelligence aircraft

Operational status

In production and operational on aircraft such as US Navy EP-3, ES-3 and P-3C, US Air Force E-8 JSTARS and the US President's Air Force 1. Also selected for the Norwegian Air Force P-3 upgrades and the German ATL-1 upgrade.

Contractor

Telephonics Corporation.

UPDATED

C-11746(V) communication system control unit

The C-11746 communication system control unit is designed for secure TEMPEST crew intercommunication and radio transmit and receive control in high-noise airborne applications.

The unit provides individual on/off and receive level control of radios and navigation receivers, voice-operated switching for hands-free intercom control and a remote select capability for HOTAS/HOCAS operation. A single unit can handle five transmit/receive radios, six navigation receivers, four auxiliary inputs and two intercom buses. Units are available with MIL-L-85762A NVG-compatible or standard edge-lit front panels.

Specifications

Dimensions: 66.675 × 127 mm
Weight: 1.18 kg
Power supply: 28 V DC, 7 W
Environmental: MIL-C-58111
TEMPEST: NACSIM 5100

Operational status

In service on the AH-64 Apache helicopter.

Contractor

Telephonics Corporation.

VERIFIED

Integrated Radio Management System (IRMS)

The Integrated Radio Management System (IRMS) is a communications management system providing control of aircraft radios, radar transponders and intercom. It provides total communications back-up in the event of battle damage through redundant control panels and centralised control units. The system comprises two Communication equipment Control Units (CCU), two Communication/Navigation equipment Controls (CNC), seven Intercommunication Set control Units (ISU), three Public Address set Controls (PAC), six Interphone Receptacle Panels (IRP) and nine Headset Receptacle Panels (HRP).

The CCU is the centralised 1750 processor-based MIL-STD-1553B bus controller unit which provides control of all audio, digital and analogue signal processing. It interfaces to the CNC, ICS and various radio equipments, via a MIL-STD-1553B interface bus. The CNC is used to tune navigation and communication radios, display radio frequencies and modes, IFF modes and VOR/ILS course selection. ICS provides microphone, PTT, VOX, audio and volume selection for all audio sources. ISU enables intercom/radio, talk/listen or PA selection. PAC selects speakers and volume levels. IRP and HRP provide microphone preamplification and headset impedance matching.

Specifications

Dimensions:
(CCU ×7) 198 × 191 × 498 mm
(CNC ×2) 95 × 184 × 178 mm
(ICS ×7) 124 × 146 × 152 mm
(ISU ×2) 67 × 146 × 89 mm
(PAC ×3) 38 × 136 × 102 mm
(IRP ×6) 64 × 102 × 38 mm
(HRP ×9) 38 × 127 × 102 mm
Weight:
(CCU) 13.8 kg, (CNC) 3.14 kg
(ICS) 2.3 kg, (PAC) 0.41 kg
(IRP) 0.23 kg, (HRP) 0.34 kg
Power supply:
(CNC, ICS, PAC) 28 V DC
(ISU, IRP, HRP) 16 V DC
Temperature range: –54 to +71°C

The Telephonics Integrated Radio Management System (IRMS)

Altitude: up to 50,000 ft
Reliability: 3,300 h MTBF

Operational status

In production and in service. The IRMS is installed and fully operational on the US Air Force C-17A transport aircraft.

Contractor

Telephonics Corporation.

VERIFIED

Joint STARS Interior Communications System (ICS)

The Interior Communications System (ICS) is utilised on the E-8A aircraft as the communication system for the US Air Force/US Army Joint STARS. It is a fully distributed digital communications system, supporting 96 full-duplex channels for simultaneous non-blocking secure and non-secure operations. The ICS interfaces with various receiver/transmitters, secure speech devices, communication, navigation and identification receivers via General Interface Terminals (GIT), as well as supporting the interface with all mission crew members and flight deck personnel.

The ICS is fully modular and other crew members and GITs may be added as the system grows or the mission requirements change, with no changes to the existing system hardware. At the present time, there are six Flight Deck Terminals (FDT), 18 Crew member Terminals (CT) and five GITs with five channels each. Each of the mission operator CTs has a single display and keyboard. The keyboard provides radio selection or selection of the mission net, preset conference net, progressive/selective conference net, call and a telephone dialling net. The FDTs are identical to the CTs except that they contain an additional display and control keys for radio frequency and parameter selection. The GITs are digitally configurable for input and output levels, thereby allowing one design to accommodate many different peripheral devices without changing design or adjustments. Each GIT contains the input/output and control for five full-duplex audio channels. Once the system has been configured, all information within all of the units is retained in non-volatile memory storage.

When operators communicate, the terminals digitise audio in an allocated data time slot with a destination address to the selected interface, which then converts the digital information back into baseband audio for application to the specific asset. Likewise, when a radio or remote communication device receives audio, this audio is digitised and made available to any of the operators that have been enabled to receive the audio. The entire system is digitally reconfigurable, enabling asset assignment based on the operational scenario. All channels within the system can be selected, combined, monitored and individually volume-controlled by any operator.

Battle damage that renders a single CT, FDT or GIT terminal, or multiple terminals, inoperable does not impact on system operation or affect the operation of the other system components. Each of the units is self-contained and utilises microprocessors that transfer function and allow continued operation.

Specifications

Dimensions:
(CT ×18) 171.5 × 127 × 177.8 mm
(FDT ×6) 247.7 × 127 × 177.8 mm
(GIT ×5) 142.7 × 171.5 × 190.5 mm
(single TAP) 47.8 × 85.1 × 85.1 mm
(dual TAP) 47.8 × 161.3 × 85.1 mm
Weight:
(CT) 3.98 kg
(FDT) 4.45 kg, (GIT) 4.05 kg
(single TAP) 0.25 kg, (dual TAP) 0.43 kg
Power supply: 28 V DC
Temperature range: −54 to +71°C
Altitude: up to 50,000 ft

Operational status

In service on the E-8A Joint STARS aircraft.

Contractor

Telephonics Corporation.

VERIFIED

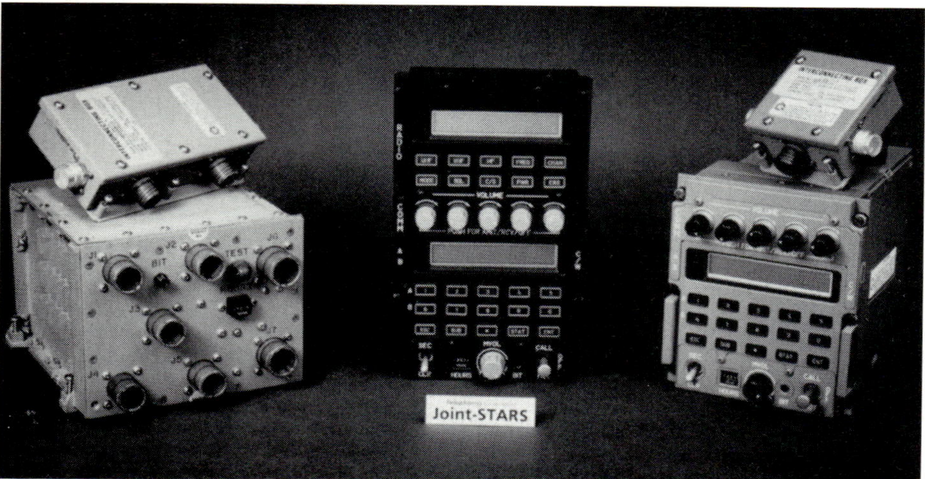

The US Air Force/US Army Joint STARS Interior Communications System (ICS)

OK-374/ASC Communication System Control Group (CSCG)

The Communication System Control Group (CSCG) is an airborne communication and control system utilised on the LAMPS III SH-60B helicopter. It provides centralised microprocessor-regulated access and control of the various internal and external communication, navigation, voice encryptors and antenna selections via the Relay Assembly (RA) for the four main crew members, while providing for two additional maintenance-only positions utilising the Interconnecting Box (IB).

The CSCG is designed to meet the requirements of NACSIM 5100, MIL-E-5400, MIL-STD-461, MIL-STD-1553A and other applicable standards for airborne communication systems operating in severe helicopter environments. It provides control of communications and intercom configurations based on manual control or serial digital data instructions from an AN/AYK-14 airborne general purpose digital computer. The manual control panel switch settings of the four Remote Switching Controls (RSC) and the Control Indicator (CI) are transferred over a serial digital datalink to the central Audio Converter Processor (ACP). The ACP then performs the required functions such as tuning the radios, selecting audio for presentation to the crew members or selecting ICOM sets.

The CSCG provides integrated control via the CI of the HF radio, two UHF radios, datalink, IFF interrogator and transponder, direction-finder group, voice encryptors and antenna selections. It also provides external aircraft computer control of sonobuoy receivers, sonobuoy command via UHF radio, radio modes and frequencies and other supervisory functions, plus external radio communications access via the two UHF clear or secure radios, the clear or secure HF radio and the secure datalink. Internal ICOM communications are provided over a common ICS net and two separate conference nets: one for the pilot and co-pilot and the other for the sensor operator and the observer. Both PTT and VOX are provided for the ICOM nets and common ICS net access is also provided at two other positions called hoist and maintenance. The system distributes various warning tones such as stabilator warning, radar altimeter warning, helo threat warning and IFF Mode 4.

The Communication System Control Group (CSCG) for the LAMPS III SH-60B helicopter

A manual back-up mode is provided, enabling selected crew members to have access to radios and ICOM in the event of system failure.

Specifications

Dimensions:
(ACP) 185 × 334 × 524 mm
(CI) 372 × 146 × 165 mm
(RSC ×4) 124 × 146 × 165 mm
(RA) 81 × 147 × 112 mm
(IB) 47 × 94 × 54 mm
Weight:
(ACP) 22.3 kg
(CI) 6.4 kg, (RSC) 7.9 kg
(RA) 0.9 kg, (IB) 0.2 kg
Power supply: 115/200 V AC, 400 Hz, 3 phase, 200 VA
Temperature range: −40 to +71°C
Altitude: up to 15,000 ft
Reliability: 1,200 h MTBF

Operational status

In service in the LAMPS III SH-60B helicopter.

Contractor

Telephonics Corporation.

VERIFIED

STARCOM intercommunication system

STARCOM is a high-intelligibility audio communication system that meets the intercom needs of a wide variety of airborne and ground-based applications. The system is designed to provide secure communications capability in high-noise environments.

The baseline STARCOM system consists of Communications System Controls (CSC) and an Audio Distribution Unit (ADU). It can accommodate five transmit/receive radios, six navigation receivers, two intercom channels and controlled and uncontrolled audio warning signals. Radio and nav receive channels can be individually monitored and controlled for level. Voice-operated switching, with an adjustable threshold for hands-free intercom control, and a remote select capability for HOTAS/HOCAS operation are standard features. Up to 10 CSCs can be interconnected through the ADU. MIL-L-85762A NVG-compatible or standard front panel lighting is available.

Various additions to the baseline system can be installed to support specific applications.

Operational status

STARCOM has been installed on AH-1W, AH-6, AS 565 MA, CE-144A, CH-47D ACMS, CH-146, MH-47D, MH-47E, MH-60G, MH-60K, MH-60L, OH-58D and UH-1N helicopters, and C-130 and P-3C aircraft. Also selected for the SH-2G(A), SH-2G(NZ), LCAC upgrade, and UH-60Q.

Contractor

Telephonics Corporation.

UPDATED

Telephonics Communications Management System (TCOMSS)

The Telephonics TCOMSS is a fibre optic, open architecture, VME-based system that uses COTS hardware to provide secure digital audio and full digital control from all operator positions. TCOMSS operates on a 100 Mbps fibre optic network that supports full-duplex, non-blocking communication/management of up to 738 audio sources (crew members, external transceivers, navigation aids and encryption devices). Selectable bandwidths support audio and digital data distribution from narrow to wideband signals. System features include: unlimited conferencing, individual volume control, binaural (dichotic) audio, SIMOP, VOX, radio relay, frequency management, data transmission and real-time modem and encryption device switching. Because TCOMSS is modular, the system can be upgraded with options: auditory localisation (3D audio), multilevel security, combined or separate red/black busses as well as interfaces with FDDI, ATM, Fibre Channel, MIL-STD-1553, RS-232, and RS-422.

TCOMSS consists of three module types: the Audio Control Subsystem (ACS); Audio Control Panel (ACP); and the jack box. The ACS provides the main interface for the TCOMSS. It links the fibre optic rings with operators and communications assets. A plug-in module interfaces with the platform databus. ACS modules can be arranged in single or mulitple configurations to support varied platforms.

The ACP is the interface between the operator and the system. Using an RS-422 interface, the ACP enables the operator to manage all assets assigned to a particular crew position. This includes: independent radio transmit and receive selection and control for all assigned radios; programmable intercom/conference net selection and control; VOX; selectable dichotic capability; clear/secure selection; radio relay; and crypto and modem control.

The jack box is available for certain applications, typically larger platforms, to provide digital audio out to the headset jack. The jack box provides the proper controls and interfaces to support an airborne system. In smaller TCOMSS applications, the ACS incorporates the jack box functions.

When operators communicate, the audio is digitised and placed into an allocated time slot with a destination address within the ACS. The ACS then routes the audio to the appropriate address; either another jack box for interphone audio or for conversion to analogue for radio transmission. Similarly, when a radio or remote communication device receives audio, it is digitised and made available to any of the operators that have been enabled to receive the audio. The entire system is digitally reconfigurable, enabling asset assignment based on the operational scenario. All channels within the system can be selected, combined and monitored.

Specifications

Dimensions:
ACS 190.5 × 194.0 × 376.7 mm
ACP 146.0 × 171.4 × 127.0 mm
J-Box 165.1 × 50.8 × 127.0 mm
Weight:
ACS 9.98 kg
ACP 2.04 kg
J-Box .91 kg

Operational status

Selected for the UK Nimrod MRA4 programme and the NATO AWACS upgrade.

Contractor

Telephonics Corporation.

NEW ENTRY

Wireless Communications and Control System (WCCS)

The Wireless Communications and Control System (WCCS) is a wireless FM communications system which provides voice-operated hands-free full-duplex party line operation for up to six cordless headset users. The system provides highly intelligible communications in a 115 dBSPL environment through the use of a special enhanced noise-cancelling microphone and high-noise attenuation headset. Communication between users occurs within an aircraft through the use of an internally installed leaky coaxial line antenna or externally through a UHF blade antenna. Additional aircraft communications flexibility is provided by a hard-wired two-way audio interface to the aircraft interphone system. In addition to the communications capability which the system provides, it also supports a wireless hand-held remote Control Transmitter (CT) which is used to control cargo and retrieval winches.

The WCCS comprises a Receiver/Transmitter radio (RT), remote-control cargo winch controller, headset rack, up to six cordless headsets and the battery charger.

The RT, which is housed within the headset rack, is the repeater which provides full-duplex operation between users. Remote operation is achieved by using the RT in the portable battery-powered mode. Direct communications between cordless headset users can be accomplished by operating the system in the half-duplex mode, independent of the RT. The simultaneous use of multiple WCCS systems within the same operating range of each other without cross-talk interference is achieved by the use of seven frequency groups and 32-tone squelch codes.

The CT provides variable speed bidirectional cargo winch capability and two single-speed bidirectional retrieval winch capabilities.

Specifications

Dimensions:
(RT) 191 × 203 × 152 mm
(CT) 217 × 84 × 121 mm
(headset) 254 × 226 × 216 mm
(charger) 255 × 229 × 84 mm
(rack) 1,118 × 483 (stowed), 978 (extended) × 178 mm
Weight:
(RT) 2 kg
(CT) 0.55 kg, (headset) 1.23 kg
(charger) 1.9 kg, (rack) 7.73 kg
Power supply: 28 V DC, 28 W
Frequency: 410-420 MHz
Temperature range: −20 to +71°C
Altitude: up to 50,000 ft
Reliability: 5,165 h MTBF

Operational status

In production and in service in the US Air Force C-17A transport aircraft.

Contractor

Telephonics Corporation.

VERIFIED

ProCom 4 aircraft intercom

The Telex ProCom 4 aircraft intercom is compact and offers optional panel-mounting configurations. It provides noise-free voice-activated communications for pilot, co-pilot and up to three passengers. It has provisions for optional connection of a music/auxiliary source and tape recorder.

The ProCom 4 is equipped with a master squelch control and individual squelch circuits with trimmers for each user. With the ProCom 4 only the microphone of the person talking is hot, resulting in less noise. In addition, using separate squelch trimmers solves adjustment problems caused by such things as different ambient noise levels at different microphones throughout the cockpit, several types of microphone being used for different user voice levels. In the event of an intercom failure, the pilot can still use the radio.

Specifications

Weight: 0.28 kg
Power supply: 12-28 V DC, 125 mA at 28 V

Contractor

Telex Communications Inc.

VERIFIED

AN/USC-42 UHF Satcom and line of sight communication set

The AN/USC-42 Miniaturised Demand Assigned Multiple Access (Mini-DAMA) set is a down-sized member of the TD-1271 terminal family. It achieves interoperability with the US Navy's TD-1271B/U multiplexer and AN/WSC-3 and the AFSATCOM system. The Mini-DAMA will function in nine operational modes. Among them is 25 kHz Satcom; here the system will support Navy TDMA-1 network operations and non-TDMA communications. On 25 kHz line of sight channels, it will support short-range tactical communications. On 5 kHz UHF Satcom channels, it will interoperate with Navy non-TDMA communications, US Air Force DAMA network operations, US Air Force non-TDMA communications and AFSATCOM network operations. Product improvement growth paths exist for embedding AFSATCOM IIR and Have Quick IIA capabilities.

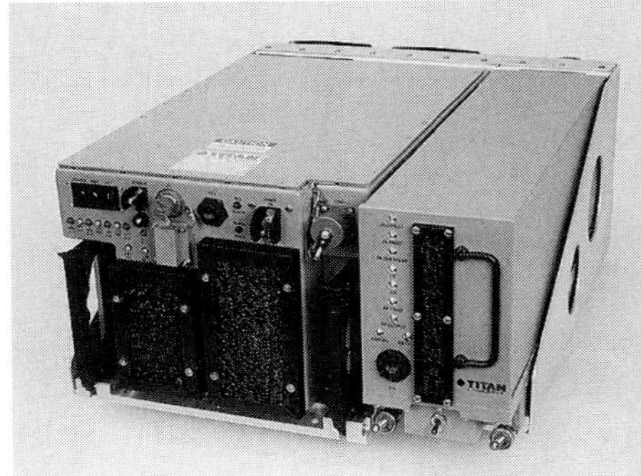

AN/USC-42(V)3 Mini-DAMA Satcom Terminal
1996

The US Navy's FLTBDCST, CUDIXS/NAVMACS, SSIXS, OTCIXS, secure voice TACINTEL, TADIXS A and ORESTES will use Mini-DAMA for data exchange. Mini-DAMA modem/receiver/transmitters will come in two configurations: 483 mm rack for ship and shore installations and as a 1 ATR-long package for aircraft.

Principal components include an integrated modem/receiver/transmitter and a separately housed power amplifier. The airborne version will contain a remote operation display/entry panel. Operations will be either half- or full-duplex through the Mini-DAMA embedded radio or through a 70 MHz IF interface to an external receiver/transmitter. The AN/USC-42(V)3 Mini-DAMA is configured for airborne platforms.

Specifications

Frequency: 225-399.995 MHz
Channel spacing: 5 kHz or multiples of 5 kHz
Temperature range: −32 to +55°C
Reliability: 2,000-4,000 h MTBF

Operational status

In production and in service. Preplanned product improvement (P[3]I) growth paths exist for embedding COMSEC, vocoding, AFSATCOM IIR, Have Quick IIA and SAFEVET. OTCIXS II is embedded in all terminals.

Contractor

Titan Linkabit.

VERIFIED

LSM-1000 UHF DAMA Satcom modem

The LSM-1000 is a federated (free-standing) modem which is capable of implementing Demand Assigned Multiple Access (DAMA) MIL-STD-188-181, -182, and -183 waveforms when used in conjunction with UHF Satcom radios. It can be operated in either full- or half-duplex applications and includes an embedded CTIC chip to implement the encrypted Orderwire Channel operation. The LSM-1000 is configured as a ½ ATR extra-short, low-cost modem in a rugged airborne chassis, providing growth capability for embedding Vocoder and COMSEC functions plus two spare SEM-E card slots. The user configures the modem using a selected platform control device.

Control of the modem may be accomplished via a MIL-STD-1553B interface, a standard RS-232 or RS-422 computer serial interface, or with a dedicated remote Control Indicator (CI). Key loading is accomplished with DS-102-compatible fill devices using the front panel keyfill interface.

The LSM-1000 is compatible with external KG-84A and KY-57-58 COMSEC equipment and Advanced Narrowband Digital Voice Terminals (ANDVTs). The LSM-1000 provides various serial and discrete interfaces including radio control and status to support system applications. It is also available as SEM-E modules for embedded modem applications. The LSM-1000 is prewired to accept an embedded SEM-E Vocoder/COMSEC module available as a plug-in option. A companion mounting tray is also available.

The Titan Linkabit MD-1035B/A UHF dual modem

Operational status
Currently available.

Contractor
Titan Linkabit.

VERIFIED

MD-1035B/A UHF dual voice/data modem

The MD-1035B/A UHF dual modem was designed to meet a US Air Force Electronic Systems Division requirement for communication over both current and anticipated satellite systems with widely varying characteristics. It provides a variety of modulation/demodulation, convolutional or block-coding error control and multiple access options in a single package. Communication and network control functions are performed by a flexible, multistack microcomputer which permits demodulation of virtually any digital signalling scheme through software changes. It can interface with a number of RF systems including those operating in frequency bands above the nominal UHF range.

Features include AFSAT 1 and AFSAT 2 (US Air Force Satellite) modulation schemes, dedicated interleaving, error control coding/decoding, input/output and network control and BIT facilities which permit fault identification down to card level. Incorporated firmware changes provide an additional channel for use with the Single Channel Transponder (SCT), 2,400 bits/s for secure voice, coding and interleaving to reduce scintillation mitigation, probing for SCT report-back, 1,200 bits/s and demultiplexing for fleet broadcast and 2,400 bits/s data for tactical operations.

The dual modem upgrade programme is extending the equipment capability further, to operate in a MILSTAR UHF network and to receive the SCT AFSAT 1 and 2 type downlinks. The upgraded modem will receive modified versions of the AFSAT 2 signalling from the SDS and DSCS 3 SCT and AFSAT 1 channel 1.5 from the DSCS 3 SCT. This programme includes development of a MILSTAR payload and command post simulator to allow DMU testing prior to the availability of the MILSTAR system.

Specifications
Dimensions:
(control indicator) 173 × 147 × 175 mm
(telegraph modem) 198 × 127 × 362 mm
(electrical equipment) 98 × 135 × 435 mm
Weight:
(control indicator) 2.54 kg
(telegraph modem) 8.63 kg
(electrical equipment) 1.59 kg

Operational status
In production and in service. The MD-1035B/A is installed in strategic force elements of the US Air Force in aircraft such as the B-52, B-1B, KC-10, EC-135G and RC-135G. The system is the basis for the Titan Linkabit airborne command post and US Navy attack submarine modem/processor.

Contractor
Titan Linkabit.

VERIFIED

Jet Call

Jet Call is an ARINC ground-to-air selective calling system utilising thumbwheel coding with four buttons in full view. Sixteen available tones provide over 10,000 possible combinations and take about 10 seconds to set. The Jet Call uses no wire jumpers or remote switches. Two or five decoder channels are available to handle up to three VHF comms and two HF transceivers. The Jet Call is TSO'd to FAA C059. It uses all solid-state circuitry and switched capacitor filters with high inherent stability and has low power consumption and easy installation, using standard Mil type D connectors with insertable and removable pins.

Specifications
Dimensions: 127 × 57 × 321 mm
Weight: 1.45 kg

Operational status
In service.

Contractor
Trimble Navigation Ltd, Avionics Products.

UPDATED

TX 760D communications transceiver

The all solid-state TX 760D communications transceiver features a 5 W transmitter and includes voice-activated intercom capability. It has 760 channels over the frequency range 118 to 136.975 MHz (including 10 memory channels), a planar gas discharge display, dual displays for active and standby, single-knob tuning and digital frequency synthesis.

Specifications
Dimensions: 79.4 × 41.3 × 290.8 mm
Weight: 0.68 kg
Power supply: 13.75 V DC
(standby) 0.325 A
(transmit) 2 A

Contractor
Trimble Navigation Ltd, Avionics Products.

UPDATED

UniLink air-to-ground two-way datalink

UniLink is designed to be interfaced with and controlled through Universal Avionics Systems Corporation's colour flat-panel Flight Management Systems, which include the UNS-1B plus, UNS-1C, UNS-1Csp, and UNS-1D. The UniLink menu software integrates with the UNS FMS and provides access for sending and receiving data and graphics. Flight plans can be uplinked through UniLink and loaded directly into the FMS. Position reports and other data from the FMS can be automatically downlinked.

UniLink has been designed to support several communications media including VHF, telephony and Satcom. Other media, such as HF, will be added as they become supported by the Aeronautical Telecommunications Network (ATN).

UniLink has been designed to support all ACARS message types, including triggered events such as OOOI (Out, Off, On, In) and planned interface with Digital Flight Data Acquisition Units (DFDAUs).

The UniLink module is available as a model UL-600 housed in a 1-MCU sized unit, supporting single-, dual- or triple-FMS installations; it will also be available as a separate PCB for the UNC-1C, UNS-1Csp, and UNS-1D FMS.

Universal Avionics Systems Corporation claims that the combined UNS-1 and UniLink suite fulfils the evolving Communications/Navigation/Surveillance (CNS) routing and communication requirements in the future Aircraft Traffic Management (ATM) system.

Specifications
(UL-600)
Dimensions: 1-MCU
Weight: 1.47 kg
Power: 28 V DC, 5 W
Memory: 2 Mbytes FLASH memory; 1 Mbyte SRAM; 32-bit controller
Interfaces: VHF modem: 1 input/1 output; telephony; 1 input/1 output; ARINC 429: 8 input/3 output; RS 422/232: 8 input/8 output; RS 232 diagnostics/load port: 1 input/1 output; discretes: 16 input/16 output; configuration module 1 input/2 output

Contractor
Universal Avionics Systems Corporation

NEW ENTRY

UniLink air-to-ground two-way datalink
1998/0011857

Integrated Communications Navigation Identification Avionics (ICNIA)

The Integrated Communications Navigation Identification Avionics (ICNIA) radio terminal concept, relied on the use of rapidly emerging advanced Radio Frequency (RF) and digital technologies to provide pilots with more Communications, Navigation, Identification (CNI) availability and flexibility than ever before. The ICNIA effort demonstrated that the pilot can do the same CNI functions as individual radios with the same performance, but in a smaller integrated terminal; a terminal capable of maintaining operation through multiple failures or battle damage while responding to the pilot's needs for different priorities of the CNI functions throughout his mission. Savings were projected for a 45 to 50 per cent reduction in size and weight with an accompanying 50 to 55 per cent reduction in life cycle cost. The ICNIA programme was a Tri-service programme, Air Force led, with the Army and Navy as participating services.

The functions implemented by ICNIA were in the 2 MHz to 2 GHz range and included:

Joint Tactical Information Distribution System (JTIDS)
Enhanced Position Locating and Reporting System (EPLRS)
Have Quick
Single Channel Ground-to-Air Radio System (SINCGARS)
Global Positioning System (GPS)
Tactical Air Navigation (TACAN)
Mark 12 Interrogate
Mark 12 Transpond
Microwave Landing System (MLS)
Instrument Landing System (ILS)
VHF Omnidirectional Range (VOR)
Traffic Collision Avoidance System (TCAS)
HF
UHF
VHF
Mode S
Link 4
Link 11
Fleet Satellite Communications (FLTSATCOM).

The ICNIA terminal configurations were defined as the Army terminal (ADM-1), the Air Force terminal (ADM-2) and the Full Function terminal (ADM-3 and ADM-4). The ADM-1 terminal provided reception, transmission, processing and control of a tailored set of CNI functions for laboratory demonstration and later flight demonstrated on an Army UH-60 Black Hawk helicopter. ADM-2 was used to test Ada software activities to support the Advanced Tactical Fighter (ATF), and was delivered in place at the contractor's facility as an RF test bench and to support module repair. The ADM 3/4 terminals were delivered to Wright Laboratory's Avionics Directorate for further laboratory testing, demonstration, and system development. The ADM-4 terminal was also used to demonstrate the Navy's functions of Link 4, Link 11 and FLTSATCOM.

ICNIA made use of redundancy, resource-sharing, extensive Built-In Test (BIT), and Very High-Speed Integrated Circuit (VHSIC) technology to produce an architecture capable of detecting/isolating fault to the LRM-level and dynamically reconfiguring to improve operation availability. The ICNIA software also employed a modular concept. As new functions or requirements became known, new software could be added to the ICNIA system to implement the new function or requirements, that is, ICNIA was software reprogrammable. This allowed flexibility to the ICNIA system that could not be realised with a federated or black box implementation. The implication is that new functions can be added to a weapon system that has an ICNIA system via a software upgrade and no new hardware modifications. This saves significant cost in Group A modifications (the cost of installing new equipment) to the weapon system. This was demonstrated by the addition of the Navy functions.

ICNIA technology was transitioned to the F-22 Advanced Tactical Fighter and formed the foundation the F-22's CNI suite.

Operational status

With ICNIA technology transitioned to the F-22, the next step is to transition this technology to the retrofit aircraft market, such as the F-15 and F-16. A new programme, called the Integrated CNI Subsystem (ICNIS) program, is planned to address technical and risk issues involved with transitioning this technology to retrofit platforms.

Contractor

US Air Force Material Command, Wright Laboratory.

VERIFIED

Wideband secure voice and data equipment

Whittaker secure communications equipment consists of the Voice Processor Unit (VPU) or Voice Processor Unit with Data mode (VPUD), Remote-Control Unit (RCU), Key Fill Device (KFD) and Key Transfer Device (KTD). The equipment is easily installed in aircraft, ground stations and ships and readily interfaces with existing UHF/VHF AM/FM military transceivers with X-mode capability. Modifications are available for operation with frequency-hopping radios.

Electrical and mechanical interfaces ensure direct replacement for US KY-58 crypto. Cryptographical keys can be erased from memory by the operator and an automatic alarm security feature continuously tests for proper secure operation. Key usage accountability is monitored by a non-resettable usage counter in the KFD. Key memory retention is provided during power-off or transient conditions. Resynchronisation occurs within half a second in late entry or temporary loss situations. A remote over-the-air rekey option is available and there is a retransmission capability for relay link operation.

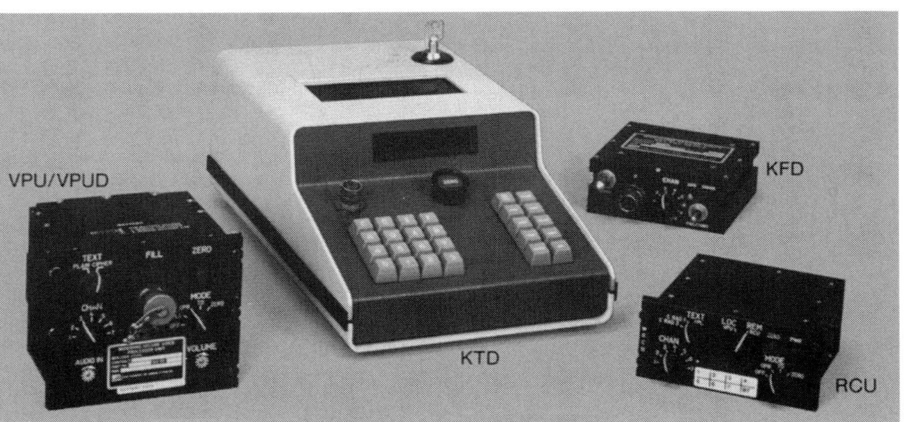

Whittaker secure communications equipment consists of (left to right) the VPU/VPUD, KTD, KFD and RCU

Specifications

Dimensions:
(VPU/VPUD) 146.1 × 124 × 121.9 mm
(RCU) 146.1 × 66.8 × 121.9 mm
(KFD) 129.5 × 58.4 × 149.9 mm
(KTD) 254 × 152.4 × 350.5 mm
Weight:
(VPU/VPUD) 2.36 kg
(RCU) 1.09 kg
(KFD) 1.18 kg
(KTD) 8.92 kg
Power supply:
(VPU/VPUD, RCU) DC per MIL-STD-704A
(KFD) battery self-power
(KTD) 115/230 V AC, 50-400 Hz
Temperature range:
(KTD) −40 to +55°C
(VPU/VPUD, RCU, KFD) −54 to +71°C
Altitude:
(KTD) up to 15,000 ft
(VPU/VPUD, RCU, KFD) up to 70,000 ft
Reliability:
(RCU) 60,000 h MTBF
(VPU/VPUD) 7,000 h MTBF
(KFD) 38,000 h MTBF
(KTD) 19,000 h MTBF

Contractor

Whittaker Electronic Systems.

VERIFIED

TCDL Tactical Common DataLink programme

The US Defense Airborne Reconnaissance Office (DARO), in conjunction with the US Defense Advanced Research Projects Agency (DARPA), who are the contracting and technical agency, has selected three contractor teams for Phase I of the TCDL programme.

The TCDL programme is a multiple phase, multiple award programme to develop a family of CDL interoperable digital links to support both unmanned and manned airborne reconnaissance platforms including: Outrider, Predator, Reef Point, Rivet Joint, Joint STARS, Airborne Reconnaissance Low (ARL) and others. The TDCL will support air-to-surface transmission of radar, imagery, video and other sensor information at ranges up to 200 km. The TDCL will operate with existing CDL systems operating at the 10.71 Mbps return link and the 200 Kbps command link data rates.

Operational status

Phase I, during which the teams will develop designs for airborne and surface TCDL terminals for tactical unmanned vehicle applications, this will last six months. DARO and DARPA will then select two teams to continue into Phase II, an 18 month prototype development and demonstration effort.

Contractor Teams

US DARO and US DARPA contracting three teams.

Team 1: Harris Corporation, Melbourne; GEC-Marconi Hazeltine Corporation, Fort Wayne; TSI Telsyn Inc, Columbia.

Team 2: Lockheed Martin, Salt Lake City; Rockwell Collins, Cedar Rapids.

Team 3: Motorola, Scotsdale; Raytheon Systems Company, Falls Church; Cubic Defense Systems, San Diego.

UPDATED

SENSORS

Maritime operations
Radar, GPWS and TCAS systems
Electro-optics
Electronic warfare

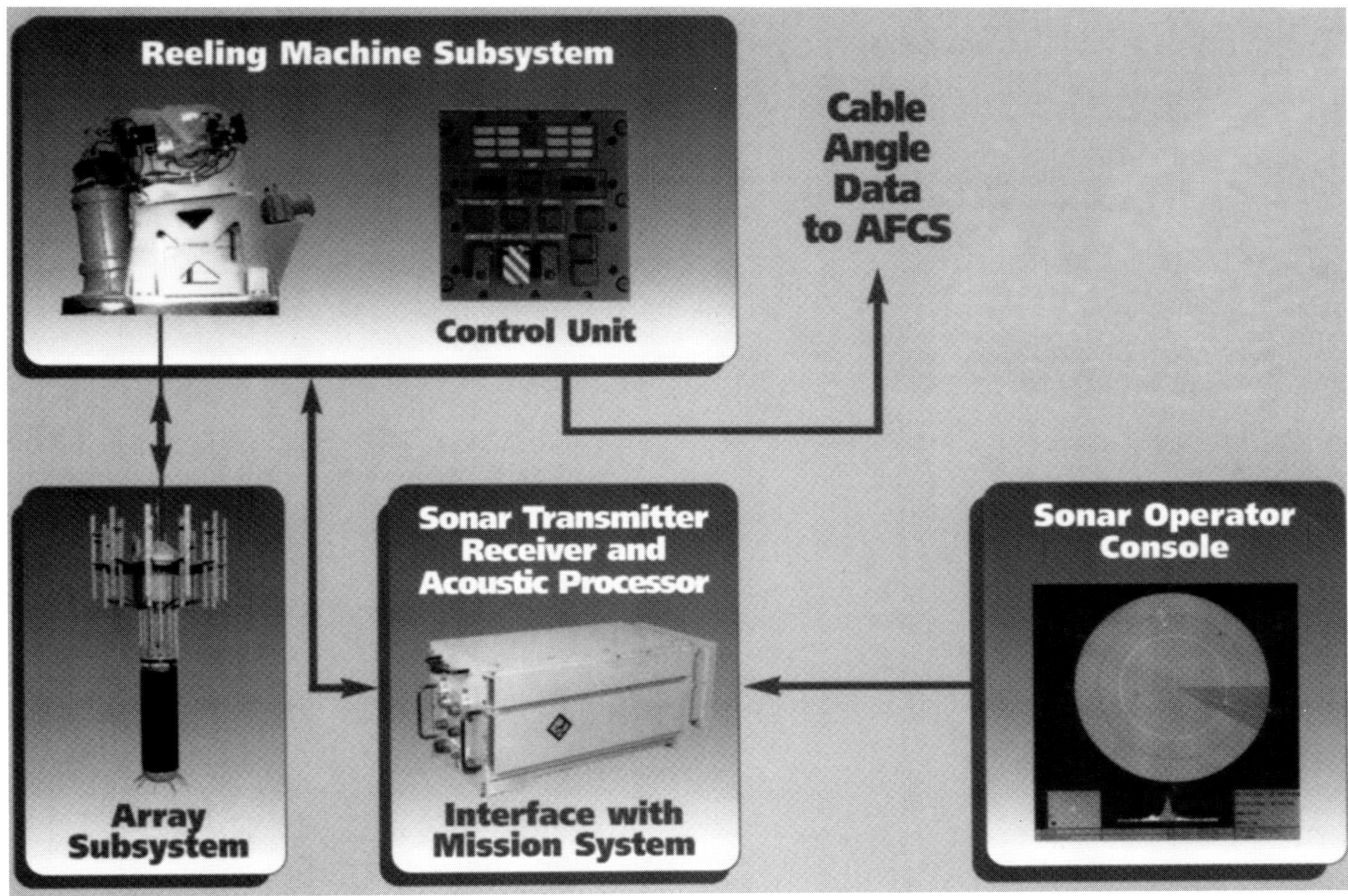

Thomson Marconi Sonar FLASH Folding Light Acoustic System for Helicopters showing the major system components: Reeling machine subsystem with Control unit; Array subsystem; Sonar transmitter receiver and Acoustic processor; Sonar operator console **1998**/0011863

MARITIME OPERATIONS

AUSTRALIA

Marine surveillance system

The ASTA low-level airborne marine surveillance system includes a 360° Litton Canada AN/APS-504V(5) search radar, a FLIR Systems 2000G infrared detection system with recorder and a comprehensive avionics suite with DME, Omega/VLF and an optional inertial navigation system and a two-axis autopilot. These systems are integrated and operate together to detect, track and identify targets and geographically locate them with date, time, latitude and longitude co-ordinates. The system, fitted to a suitable platform, provides a surveillance package that can carry out detection and identification, often without the crew of the target being aware of being under surveillance.

The Litton AN/APS-504(V)5 features digital display subsystems, a track-while-scan mode capable of tracking up to 20 targets simultaneously, coherent pulse compression, frequency agility and videotape recording. Detection of a 2 m^2 target in Sea State 3 conditions at ranges up to 65 km is claimed.

Signals provided to the FLIR Systems 2000G by the radar and the long-range navigation system give it the capability to identify marine targets on a 24 hour basis and record these targets on video.

The cockpit of a Searchmaster aircraft fitted with the ASTA marine surveillance system

Operational status

The ASTA marine surveillance system was developed for the Searchmaster N22S Series 2. This aircraft is in service with the US Customs Service in its drug interdiction programme.

Contractor

AeroSpace Technologies of Australia Ltd.

VERIFIED

CANADA

AN/ASA-64 Magnetic Anomaly Detector (MAD)

The AN/ASA-64 MAD identifies and marks local distortions in the earth's magnetic field induced by the presence of submarines. The operator is alerted by visual and aural alarms, thereby reducing the level of experience needed to operate the system. As the system does not require constant monitoring, the operator can devote more time to other sensors.

CAE has completed a product improvement programme in support of the AN/ASA-64 submarine anomaly detector originally built for the US Navy P-3C Orion maritime patrol aircraft. The improved version has increased processing power. A variant for helicopters is proposed.

Specifications

Dimensions:
(control unit) 102 × 146 × 90 mm
(ID-1559 processor) 229 × 150 × 153 mm
Weight:
(control unit) 0.68 kg
(ID-1559 processor) 3 kg

CAE AN/ASA-64 Magnetic Anomaly Detector (MAD)

Power supply: 115 V AC, 20 W
Environmental: MIL-E-5400

Operational status

In production and in service.

Contractor

CAE Electronics Ltd.

VERIFIED

AN/ASA-65(V) nine-term compensator

CAE developed the AN/ASA-65(V) semi-automatic Magnetic Anomaly Detector (MAD) compensator to improve the effectiveness of MAD on aircraft with only manual compensation for aircraft interference with the earth's magnetic field. Previously, fixed-permalloy strips and copper coils were mounted in the MAD boom to create induced and eddy-current fields equal and opposite to those caused by the aircraft. These compensators had to be custom-designed for each individual aircraft, took a long time to adjust on flight test and did not cater for the changes which take place during the aircraft's life. CAE also says that new, more sensitive MAD equipment needs greater precision than fixed compensators can provide.

The AN/ASA-65(V) compensates for permanent interference after only five minutes' flying, compared with about an hour needed for manual compensators. The system allows for manoeuvre interferences after 30 to 45 minutes, improving MAD detection range, especially when frequent manoeuvres are performed and conditions are turbulent. This compares favourably with manual compensation procedures which traditionally took 90 minutes.

The all solid-state AN/ASA-65(V) is compatible with all current MAD systems. Internal patch connectors are used to adjust the system for the aircraft concerned.

Specifications

Dimensions:
(control indicator) 229 × 146 × 165 mm
(electronic control amplifier) 197 × 149 × 346 mm
(magnetometer assembly) 152 mm cube
(coil assembly) 89 mm cube
Total weight: 13.4 kg
Power supply: 115 V AC, 100 W
28 V DC or AC, 10 W for panel lamps
Figure of merit: $<$1 gamma
Max compensation field: 50 gamma on each side of aircraft
Reliability: $>$1,800 h MTBF

Operational status

In production and in service. The nine-term compensator is used by the US Navy P-3C Orion and S-3A Viking ASW aircraft.

Contractor

CAE Electronics Ltd.

VERIFIED

AN/ASQ-504(V) Advanced Integrated MAD System

The AN/ASQ-504(V) Advanced Integrated MAD System (AIMS) is an inboard system for helicopters, fixed-wing aircraft and lighter-than-air platforms. This fully automatic system improves detection efficiency while reducing significantly the operator's workload. For helicopter installation, the detecting head is mounted inboard the aircraft, thus providing 'on-top' contact when over a target by eliminating the time delay inherent in a towed detecting head system.

The AN/ASQ-504(V) system combines sensitivity and accuracy with ease of operation, eliminates aircraft generated interference, reduces geological and solar interference and provides automatic contact alert both visually and audibly. Detection data, via the control indicator or an avionics bus interface, allows the operator to determine if the aircraft is within target acquisition range.

AIMS eliminates the hazard associated with towed systems in a helicopter application. It also allows surveillance and manoeuvrability at higher speed, thereby increasing patrol range and detectability and reducing the incidence of false alarms. When used with dipping sonar, transition between the systems can be performed quickly and effectively.

AIMS comprises a 0.005 nanotesla optically pumped magnetometer, vector magnetometer, amplifier computer and control indicator. It can operate independently or it can accept and execute commands from common control/display units via a MIL-STD-1553 digital databus.

Specifications

Dimensions:
(control indicator) 190 × 145 × 145 mm
(amplifier computer) 193 × 257 × 559 mm
(vector magnetometer) 152 × 152 × 152 mm
(detecting head) 178 × 813 mm
Weight: 23 kg
Power supply: 108/118 V AC, 380/420 Hz, single-phase, 200 VA
Sensitivity: 0.01 gamma (in flight)
Feature recognition: automatic target detection; visual and audible operator alert
Environmental: MIL-E-5400

Operational status

In production. In July 1987, CAE announced a C$38 million contract to supply 242 ASQ-504(V) AIMS systems to equip Sea King helicopters of the British Royal Navy and Nimrod MR. Mk 2 aircraft of the Royal Air Force. Deliveries started in early 1989 and continued until late 1991. The system has also been ordered by several other countries for various helicopter and fixed-wing applications.

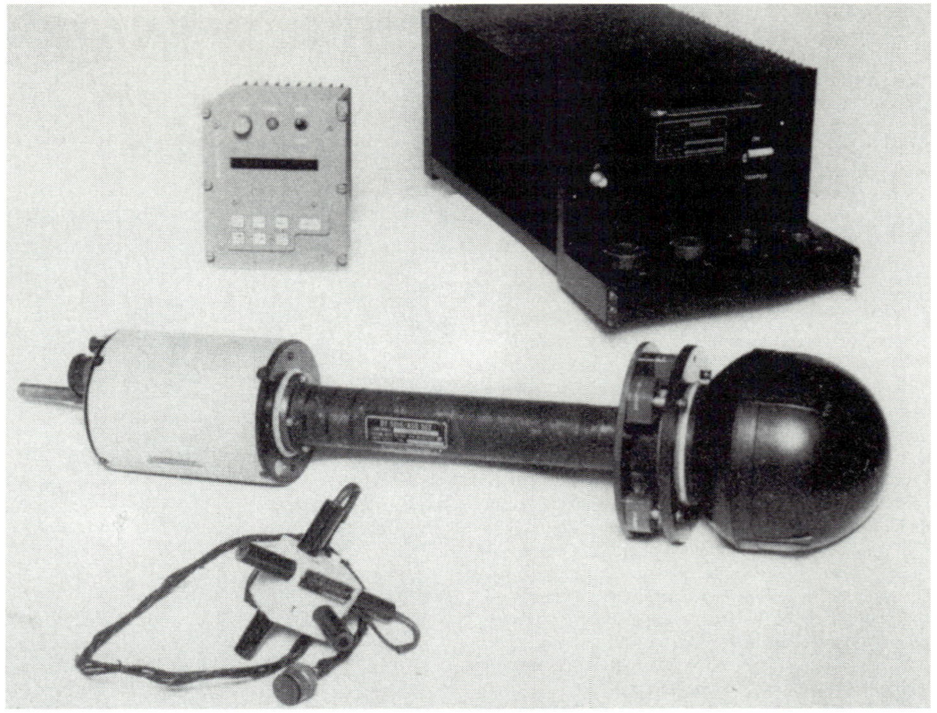

The CAE AN/ASQ-504(V) Magnetic Anomaly Detector (MAD)

Contractor

CAE Electronics Ltd.

VERIFIED

AN/UYS-503 ASW acoustic processor

The AN/UYS-503 is a small, lightweight acoustic processing system designed for use in a variety of airborne, surface and subsurface surveillance platforms. It employs a processor architecture that exploits the advantages of high-density digital technology and, consequently, has remained ahead of the ever changing threat to meet the detection, localisation and attack challenges posed by the latest generation of nuclear submarines, and diesel submarines operating in shallow waters. It functions as a complete system by providing all the required input signal conditioning, signal processing and analysis, post-detection processing and control and display processing.

A typical maritime patrol aircraft configuration would feature concurrent processing of 32 or 64 sonobuoys, while a helicopter configuration could include processing for eight or 16 sonobuoys and a low-frequency dipping sonar. Unique algorithms provide acoustic data fusion functions that combine available data to compute fixes and automatically track targets of interest. Other features of the AN/UYS-503 include proprietary algorithms that provide a consistent and reliable detection and localisation capability for broadband swathes and emissions and a new colour capability that is an intrinsic part of the signal processing and greatly reduces operator workload while enhancing detection and tracking performance.

Sensors processed include, analogue or digital: OMNI, DIFAR/VLAD, VLAD, DICASS, CAMBS,CODAR, BARRA, dipping sonar bathythermal, ambient noise.

Specifications

Dimensions: 262 × 396 × 246 mm
Weight: 27 kg (16 DIFAR system)
Frequency: fullband DIFAR
Input channels: any standard sonobuoy receiver
Control input: MIL-STD-1553B, RS-232C or RS-422
Tactical data output: MIL-STD-1553B, RS-232C, RS-422 or other as specified
Video output: RS-343 composite video colour or monochrome

Operational status

Currently in service with ASW aircraft of the Australian, Canadian, Swedish, UK and US defence forces. Also in service with the Japanese Maritime Self-Defence Force.

Contractor

Computing Devices Canada Ltd.

VERIFIED

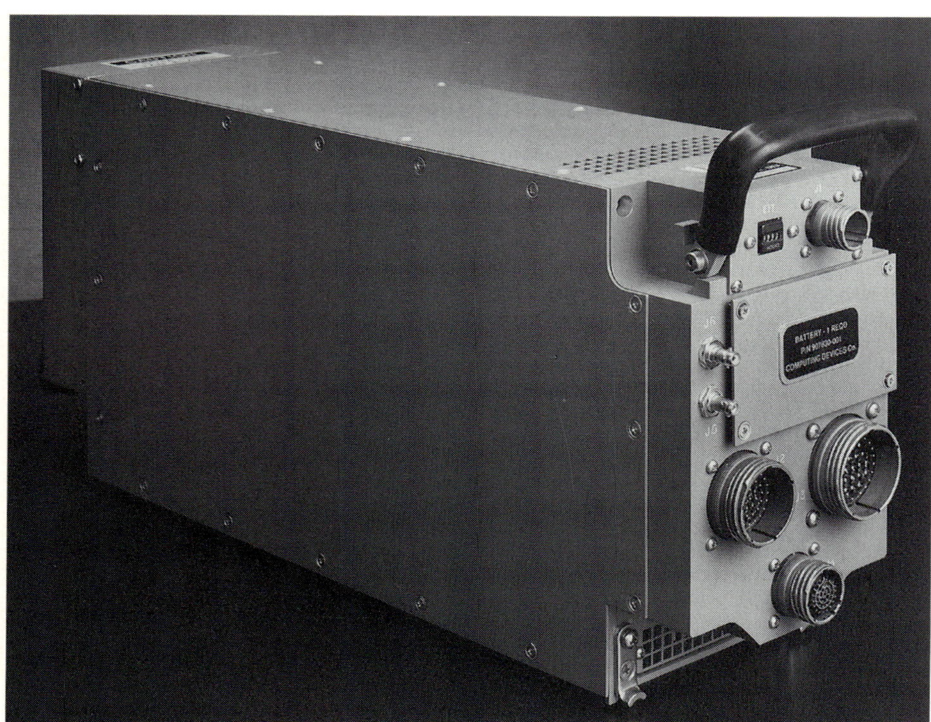

The Mission System Processor (MSP) for the ASW-503 mission data management system **1995**

ASW-503 mission data management system

The ASW-503 is a flexible mission data handling system designed to support maritime operations by airborne, surface and subsurface surveillance. It provides the necessary tactical processing, database management, control and display functions to integrate and manipulate a suite of sensor, navigation, communication and stores/armament subsystems for the effective conduct of operations.

In the ASW-503 individual subsystem interfaces are accommodated without modification to the subsystem. All integration activities are contained within the ASW-503, allowing the procurement of off-the-shelf systems without incurring high integration costs.

The heart of the ASW-503 is the Mission System Processor (MSP) and its associated software developed in Ada. It is a ½ ATR multiprocessor unit which includes all the electronics needed to support complete mission system functionality. Each MSP is capable of driving two independent workstations or a single workstation comprising an Integrated Control Panel (ICP) and two high-resolution colour displays. All subsystems are controlled through the programmable ICP and information is displayed on a high-resolution colour VDU with 1,280 × 1,024 pixels. Tactical data relating to the current mission is held in a central database and is used to develop a map representation referred to as the Tactical Situation Display (TSD).

In a typical MPA configuration, three MSPs are used to drive and control six displays and control panels. This results in six universal workstations, with each station being able to perform any operational role selected by the operator. In this system one MSP is designated the master and acts as bus controller and maintains the central tactical database. Communication between the MSPs and ICPs is achieved through a high-speed serial channel.

Specifications

Dimensions:
(MSP) 132 × 262 × 546 mm
(ICP) 615 × 101 × 316 mm
Weight:
(MSP) 15.4 kg
(ICP) 12.4 kg

Contractor

Computing Devices Canada Ltd.

VERIFIED

Tactical Data Management System

The Tactical Data Management System (TDMS) and APS-504(V)5 radar together form a high-performance integrated radar sensor and tactical display system. The APS-504(V)5 is an airborne I/J-band search radar, primarily used to detect and track sea surface targets in the presence of clutter. The TDMS provides a radar display with an advanced digital map, overlay and data interface to control and enhance the operation of the radar. FLIR video can also be displayed. The TDMS/radar combination provides surveillance of maritime activity within and beyond the 370 km (200 n miles) Economic Exclusion Zone (EEZ).

The TDMS is a VME-based data management system which contains application software for mission and flight management, provision of tactical aids and maintenance of tabular data for presentation to the operator. In addition to a high-resolution colour display, the operator interface includes a keyboard, trackerball, radar rotary controls and keypad panels. The TDMS interfaces to navigation systems such as the LTN-92. Using the TDMS to integrate radar and navigation data, the operator develops tactical and navigation plots to fit the mission requirements. From simple missions to complex tactical operations, the TDMS provides an essential aid to tactical decision making.

Ground support for the TDMS is provided by the Mission Support Facility (MSF). This PC-based facility is used for preflight mission planning, data entry and post-flight mission analysis. Preflight data is transferred to the TDMS on a high-capacity optical disk. Data collected and stored during flight on the optical disk can be offloaded to the MSF for post-flight mission analysis.

The APS-504(V)5 radar is the most advanced of the Litton APS-504 family of airborne search radars. It employs a TWT-based transmitter with wideband frequency agility, high-ratio pulse compression, scan-to-scan integration and digital signal processing to enhance the detection of surface targets, including targets with radar cross-sections as small as 1 m^2 in Sea State 3. The APS-504(V)5 can be configured to meet various installation and performance requirements. It has been installed in aircraft ranging from small twin-engined turboprops such as the Beech 200 to larger aircraft such as the Boeing 737.

Specifications

Weight:
(TDMS) 105 kg, varying with configuration
Power supply:
(TDMS) 115 V AC, 400 Hz, single phase, 28 V DC

Operational status

In production.

Contractor

Litton Systems Canada Ltd.

VERIFIED

FRANCE

SDF-123F sonobuoy direction-finder

The SDF-123F sonobuoy direction-finder provides relative bearing information from any sonobuoy signal in the 136-174 MHz frequency range through a companion FM receiver. It consists of two LRUs: the ANT-123F static rotating antenna and the MPU-123F monitoring and processing unit. The system is fully solid state, uses a standard sonobuoy FM receiver and has On-Top Position Indication (OTPI).

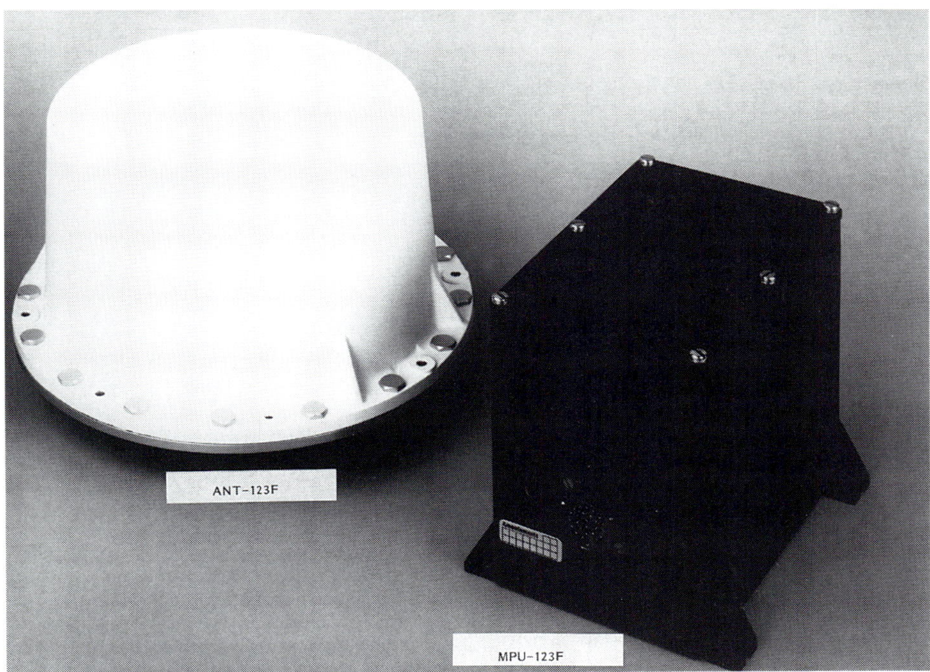

The ANT-123F is a directional rotating antenna, made of eight elements which are switched according to a specific pattern in order to simulate a low-speed antenna rotation. This creates a frequency modulation of the incoming VHF signal, the relative phase of which contains the bearing information.

The MPU-123F processing unit is the heart of the SDF-123F system and supplies power to the system, provides signal generation for antenna control, processes the demodulated AF signal from the companion FM receiver and generates the bearing signal. Output is available in ARINC 407 (three-wire synchro) or (optionally) ARINC 429 format.

Specifications

Dimensions:
(MPU-123F) 135 × 166 × 146 mm
(ANT-123F) 130 × 250 mm diameter
Weight:
(MPU-123F) 1.5 kg
(ANT-123F) 2.1 kg
Power supply: 26 V AC, 8 VA
27.5 V DC, 1 A

Operational status

In production for Eurocopter ASW helicopters, Pakistani ATLI and Chilean Navy P3 aircraft.

Contractor

Rockwell-Collins France.

UPDATED

SDF-123F sonobuoy direction finder units, showing the antenna unit ANT-123F (left) and processing unit MPU-123F (right)
1998/0011862

Mk 3 Magnetic Anomaly Detector

The Magnetic Anomaly Detector (MAD) Mk 3 is specifically designed for use both on fixed-wing aircraft and inboard on helicopters to detect the presence of a submersible by measuring the disturbance to the earth's magnetic field. It is an integral, digital, solid-state airborne system which becomes operational at switch on without any warm-up time. Target parameters are automatically delivered in real time on a CRT control and display unit.

The system consists of three separate units: a detection unit, a computer and a control/display unit. A MIL-STD-1553B databus interface card is included in the system. A graphic recorder is available as an option.

The sensor operates on the nuclear magnetic resonance principle and uses the precession of protons in a liquid, the frequency of which, measured by the pick-up coils, is proportional to the magnetic field to be measured. Compensation is employed to eliminate from the received signal all disturbances created by the magnetic element in the aircraft and their movement in the earth's field. The compensator uses a 16-term model representing the magnetic components in the aircraft. The MAD Mk 3 includes rapid aircraft identification modes.

The target is detected and located automatically. This is a fundamental role, allowing the MAD system to give a high detection probability with a very low false alarm rate in extracting the target signal from the background noise. The Sextant Avionique MAD system is based on a mathematical comparison between the current MAD signal and an analytical model of the target signals, as opposed to the more conventional use of threshold detection in several frequency bands. All computing tasks are performed by the Alpha 732 proprietary 1 Mops digital computer operating in Pascal.

Specifications

Dimensions:
(detection element) 1,250 × 125 mm diameter
(computer) ½ ATR
(control/display unit) 190 × 146 × 162 mm
Weight:
(detection element) 5.5 kg
(computer) 10 kg
(control/display unit) 3.5 kg

Operational status

In production for the French Navy Dassault Atlantique 2 maritime patrol aircraft, and in development for the Italian Navy EH 101 ASW helicopter.

Contractor

Sextant Avionique.

VERIFIED

The Sextant Avionique MAD probe on the Dassault Aviation Atlantique 2

Mk 3 towed magnetometer for helicopters

The Mk 3 towed magnetometer version of Sextant Avionique's MAD equipment is intended for ASW helicopters. It operates in exactly the same way as the system described previously and, in order to eliminate disturbances created by the helicopter, the detector probe is placed in a streamlined housing which is towed at the end of a 70 m cable. The digital computer measures the signals from the sensor and transforms them into suitable formats for the graphic recorder at the operator's station.

The housing assembly includes the detection probe and an electronic unit, combining to produce a nuclear oscillator whose Larmor frequency is a function of the magnetic field exerted on the probe. The geometry of the probe is chosen so that its position with respect to the magnetic field vector can be ignored.

Maintainability is assisted by in-flight checking of the towed magnetometer housing, computer, control unit and recorder. The ground-test points are easily accessible and the subassemblies are plug-in units.

Specifications

Background noise: typical deviation 0.006 gamma
Measurement range: 25,000-70,000 gamma
Sensitivity for relative field output: 1, 2, 5, or 10 gamma for 100 mm stylus deviation
Paper speed: 6, 75, or 300 mm/min

Streamlined body
Dimensions: (length) 1,300 mm × (diameter) 160 mm
Weight: 16 kg

Computer
Dimensions: 346 × 124 × 194 mm
Weight: 7.5 kg
Power supply: 200 V AC 400 Hz, 3 phase, 100 W
28 V DC, 5 W

Control unit
Dimensions: 165 × 150 × 57 mm
Weight: 1 kg

Recorder
Dimensions: 190 × 150 × 190 mm
Weight: 4 kg
Power supply: 115 V AC 400 Hz, 50 W
28 V DC, 20 A

Winch and cradle
Dimensions: 1,330 × 350 × 755 mm
Weight: 44 kg
Power supply: 27 V DC, 40 A

Operational status

Helicopter version in production and in service.

Contractor

Sextant Avionique.

VERIFIED

AMASCOS multisensor system

AMASCOS (Airborne Maritime Situation Control System) is designed for building up and updating tactical situations in real time and as a decision aid for operators. It is a family of maritime systems, which uses a modular approach to system design, and can be integrated on any type of fixed-wing aircraft or helicopter.

The three versions of AMASCOS – AMASCOS 100, AMASCOS 200 and AMASCOS 300 – correspond to the broad categories of mission requirement ranging from simple maritime surveillance to anti-surface and anti-submarine warfare. The typical AMASCOS configuration integrates Thomson-CSF equipment such as radar, FLIR, sonics, MAD and communications, but its modular architecture makes it possible to tailor each system to a specific requirement.

AMASCOS 100
AMASCOS 100 is a lightweight configuration which weighs less than 250 kg. It includes radar and FLIR plus a tactical computer and is suited for a wide range of missions, such as EEZ surveillance, search and rescue and law enforcement. AMASCOS 100 is suitable for fitment to light turboprop aircraft or carrier-based helicopters with an operating crew of one to two.

AMASCOS 200
AMASCOS 200 adds ESM equipment to the AMASCOS 100. It is suitable for anti-surface warfare and can be extended to provide an anti-surface warfare capability. AMASCOS 200 is suitable for fixed- and rotary-wing aircraft of the 8 ton class, with two or three operators.

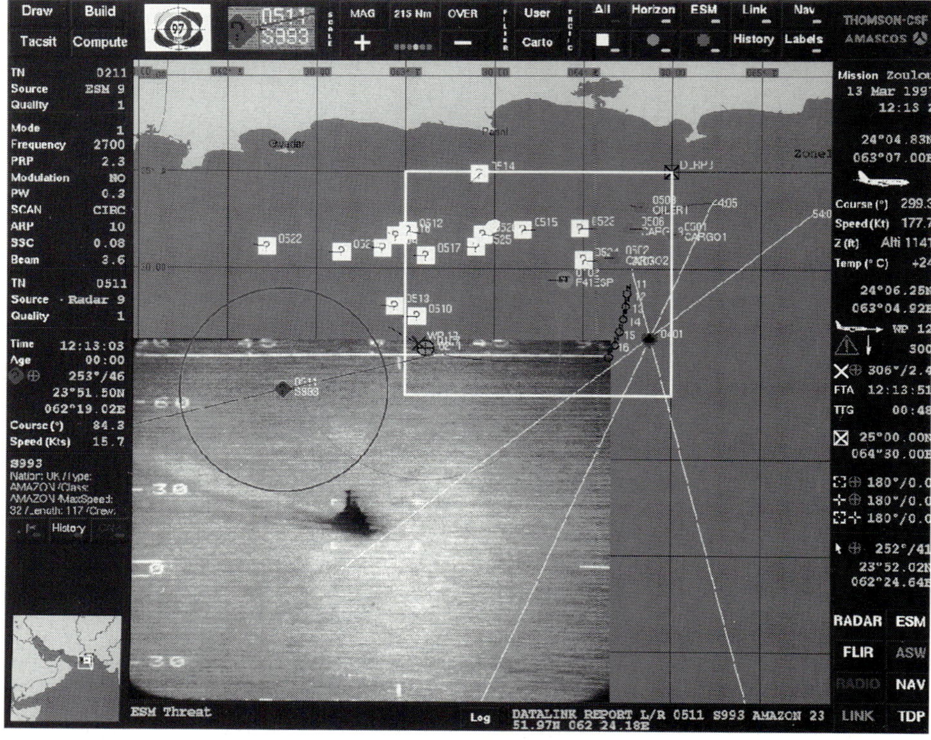

AMASCOS display

1998/0011861

AMASCOS 300

The AMASCOS 300 is the most versatile version and is likely to comprise the Ocean Master radar, Nadir Mk II inertial GPS, Chlio FLIR, DR 3000 ESM, Link W datalink, Sadang 1000 sonobuoys, HS 312S dipping sonar and MAD Mk III. This version is suitable for both anti-surface and anti-submarine warfare missions and is also suitable for command and control assignments. The heart of the system is a dedicated tactical computer which collates and processes data from different sensors and other onboard equipment. AMASCOS 300 is suitable for installation on any maritime patrol aircraft of over 10 tons with a crew of three or more operators.

Operational status

In 1996, the Indonesian Navy chose AMASCOS for its six NC-212 maritime patrol aircraft (Ocean Master radar and Chlio FLIR) and its three NBO 105 helicopters (Ocean Master radar). System fits also include the Sextant Avionique Gemini navigation computer.

Contractor

Thomson-CSF Radars Contre-Mesures.

UPDATED

TRES Tactical Radar ESM System

TRES is a combined Tactical Radar and ESM System designed for fitment to naval helicopters to provide wide area maritime surveillance, early warning and weapon control. The system integrates the Varan sea surveillance radar and the DR 2000 Dalia ESM systems.

Contractor

Thomson-CSF Radars Contre-Mesures.

NEW ENTRY

TRES Tactical Radar ESM System for naval helicopters
1998/0011860

DUAV-4 helicopter sonar

The DUAV-4 is an active/passive directive sonar designed for submarine surveillance and location (azimuth, distance and radial speed). It is specially designed for use on board light ship-based helicopters such as the Lynx. It may also be fitted on small surface vessels.

The DUAV-4 differs from conventional sonars in its signal processing system, which is designed to give improved detection, especially in severe reverberation conditions such as in shallow waters. The sonar can be operated in either the active or passive mode. True bearing, range and radial speed are measured in the active mode; true bearing only in the passive mode.

A combined display unit permits surveillance display, for initial detection, or plotting display, for precise azimuth determination. Total weight, including the electronic rack, cables and dome, is 250 kg.

Operational status

The DUAV-4 is in service with the French Navy, Royal Netherlands Navy and several other navies. Over 90 systems have been produced. In some French Navy Lynx helicopters, the DUAV-4 will be replaced by the HS12.

Contractor

Thomson Marconi Sonar SAS.

VERIFIED

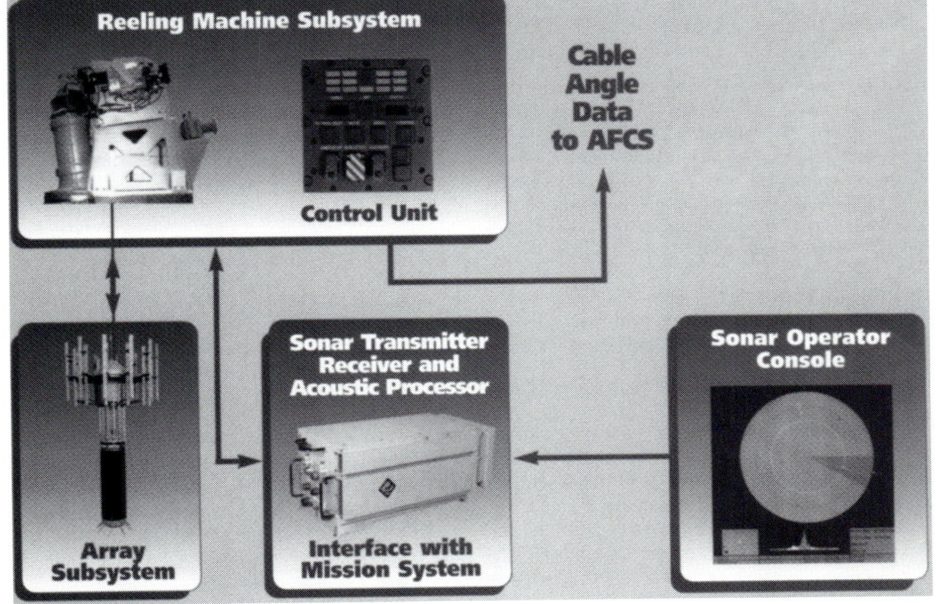

FLASH dipping sonar system components ***1998***/0011863

FLASH dipping sonar

FLASH is Thomson Marconi Sonar's most recent generation of dipping sonar, updating the earlier HS12 and HS312 (1980's technology) and DUAV-4 (1970's technology) systems.

FLASH (Folding Light Acoustic System for Helicopters) is a new low-frequency helicopter dipping sonar with sonobuoy processing capabilities developed by Thomson Marconi Sonar for a number of applications. It operates at five frequencies below 5 kHz, with 24 preformed beams. The winch is capable of 750 m immersion depth at a speed up to 10 m per second.

The system offers an FM mode to enable detection of low-speed targets in shallow waters and strong reverberation conditions. With its light weight and volume, the FLASH can be fitted on board light ASW helicopters such as the Super Lynx, and on a wide variety of medium to large helicopters.

Operational status

Thomson Marconi Sonar supplies FLASH (designated Sonar 2089 in the UK) for the long-range dipping sonar requirement of the Royal Navy's EH 101 Merlin helicopter. This system combines the Thomson Marconi Sonar SAS low-frequency array and winch with the Thomson Marconi Sonar Ltd acoustic processor designated the AQS 950. Thomson Marconi Sonar has also teamed with Hughes Aircraft Company for the US Navy's ALFS requirement in which configuration it is designated AN/AQS-22. Selected by the UAE Naval Forces for the Cougar helicopter.

The FLASH array and reeling machine have already been evaluated in trials for the US Navy. The FLASH array, reeling machine, transmitter and new acoustic processor are now being evaluated in flight trials by the

French Navy for the NFH 90 programme. Acoustic in-water testing has been conducted in France, the UK, and the USA. The system is currently in production.

Contractor

Thomson Marconi Sonar SAS.

UPDATED

HS 12 helicopter sonar

The HS 12 is an active/passive panoramic helicopter version of the SS 12 small ship sonar and uses the same electronics as that version. It has similar capabilities for operation in shallow or noisy waters and has a system weight of 230 kg, making it suitable for installation on light helicopters such as the Lynx. The HS 12 transducer is lowered and raised by a hydraulic winch at high speed.

Operation in CW and FM modes is possible and digital signal processing is employed by the system's microprocessor. Automatic tracking of two targets and transmission of elements to an external equipment, such as a plotting table, are provided.

The system operates on 13 kHz in the active mode and in the 7 to 12 kHz range passively. A total of 12 preformed beams is employed, giving 30° sectors, with a maximum range of about 10 km. The display consists of four quadrants, these being obtained by processing adjacent beams. The operator can select a CW mode which provides target range and Doppler. FM processing can also be selected and a sector mode is provided.

Specifications

Weight: 230 kg

Operational status

In series production for French and foreign navies. Fitted to Sea King helicopters.

Contractor

Thomson Marconi Sonar SAS.

UPDATED

HS 312 ASW system

The HS 312 is an acoustic system for helicopters incorporating the facilities of the HS 12 system and the SADANG acoustic processor. The equipment functions in both passive and active CW and MF modes and has a longer range than the HS 12. The acoustic subsystem can be fitted to process sonobuoys simultaneously with the dipping sonar.

Only a single operator is required and the light weight and compactness of the system mean that the HS 312 can be fitted to any type of light- or medium-size helicopter. Performance of the basic components has been improved by integration of the processing units, integration of a standardised keyboard and the use of only one display screen.

Operational status

In series production for a foreign navy. Designed for Super Puma and Cougar Naval helicopters.

Contractor

Thomson Marconi Sonar SAS.

UPDATED

LAMPARO processing equipment

Derived from SADANG equipment, LAMPARO airborne sonar signal processing and display systems are designed as modern digital equipments for ASW fixed- and rotary-wing aircraft.

The processing capabilities of the systems are based on a single processing unit in a light and compact package, for use by a single ASW operator. The same wide range of facilities and processing modes is available. The system is biased towards the processing of omnidirectional passive and active buoys.

Signals are displayed on a CRT in TV format and a hard copier can be connected for permanent recording. An on-top position indicator is used in conjunction with a radio compass for homing on the sonobuoy. Weight of the complete system is 70 kg.

Operational status

In operational service.

Contractor

Thomson Marconi Sonar SAS.

VERIFIED

Upgraded DUAV-4 helicopter sonar

The upgraded DUAV-4 (DUAV-4 UPG) differs from the DUAV-4 in its acoustic processor which is now based on the SADANG new-generation acoustic processor. The system offers more powerful signal and data processing, as well as an improved man/machine interface.

The DUAV-4 UPG also offers specific software tools and operator aids designed to meet the requirement for detection in shallow, rocky bottom waters.

Operational status

The upgraded DUAV-4 sonar has been supplied to the Royal Swedish Navy.

Contractor

Thomson Marconi Sonar SAS.

VERIFIED

INTERNATIONAL

Airborne Active Dipping Sonar

The Airborne Active Dipping Sonar (AADS) system consists of the sonar array, reeling machine, sonar transmitter/receiver and signal and display processor. It is designed to be fully compatible with all types of maritime helicopters in service today.

The sonar transducer contains the acoustic projector for sonic pulse generation and an expandable array of receiving hydrophones. The projector resonates at the frequency and duration of the pulse generated by the sonar transmitter. It is capable of producing maximum acoustic source levels at any depth below 6 m. Centre frequencies are selectable for optimal acoustic performance in different environments. The expandable receiving array provides a large acoustic aperture for increased sensitivity and directivity. The expanding mechanism is designed to fail-safe standards to ensure safe retrieval of the array.

Rapid deployment and retrieval of the sonar transducer is provided by the lightweight reeling machine. Layering and stowage of the required length of the extra strong cable has been tested with an MTBF of 4,440 cycles. Replacement of cable and assembly can be accomplished during the refuelling cycle.

The Sonar Transmitter/Receiver (STR) generates the transmit waveforms and provides the output power to drive the acoustic projector. It also processes the signal returns from the transducer receive array. The STR is located in the helicopter cabin and provides local control functions for the dipping sonar subsystem, including the reeling machine, the acoustic transducer and the interfaces between the dipping sonar system and the signal processor.

Contractors

Raytheon Systems Company.
Thomson Marconi Sonar SAS.

UPDATED

AN/AQS-22 Airborne Low-Frequency Sonar

The AN/AQS-22 Airborne Low-Frequency Sonar (ALFS) is a US Navy project for a dipping sonar which will largely replace the AN/AQS-13F. It is scheduled to be fitted to SH-60F and LAMPS SH-60B helicopters. No technical details have been released, except that the associated processor will be the UYS-2 manufactured by AT & T. The sonar will use the expandable sonar array and reeling machine subsystem of the Thomson Marconi Sonar FLASH system.

Operational status

Trials of the FLASH array and reeling machine have been completed for the US Navy, together with acoustic in-water trials.

Contractors

Raytheon Systems Company.
Thomson Marconi Sonar SAS.

UPDATED

AQS 960 acoustic processor

The AQS 960 proposal for the UK RMPA programme is developed from the AQS 903. Capabilities include: full sonobuoy processing, new active DICASS sonobuoy networks, automated classification and tactical management. Options include GPS and integrated crew-training capabilities.

Contractors

Thomson Marconi Sonar Ltd.
Thomson Marconi Sonar SAS.
Digital System Resources.

VERIFIED

P-3C upgrade program for RAAF

The Royal Australian Air Force (RAAF) P-3C upgrade program comprises two major elements: Project Air 5140 and Project Air 5276. Project Air 5140 is managed by British Aerospace Australia. It comprises: replacement of the ESM kit with the Elta ALR-2001 'semi-SIGINT suite' to give the P-3C a true electronic surveillance capability. This change involves significant reallocation of the operational tasking for Sensor Station Three. The ESM and infrared detection system operation is transferred from Sensor Station Three to a new Sensor Station Four. The new Sensor Station Four is equipped with a BARCO ruggedised RGDS 651/EX terminal unit with a 19 in monitor. About 75 per cent of the fleet has already been modified with the ALR-2001 ESM; the project is due for completion in 1998.

Project Air 5276 is managed by Raytheon Systems Company; it brings aircraft up to AP-3C standard; it consists of updating the avionics system and associated support facilities for the RAAF's 18 P-3C aircraft, together with structural work. Items included in the update include: Lockheed Martin's Tactical Defence Systems Data Management System (DMS-2000); Elta's EL/M-2022A(V)3 Synthetic Aperture Radar (SAR)/Inverse Synthetic Aperture Radar (ISAR); Computing Devices Canada's acoustic processor; Honeywell's fully integrated navigation suite; Hughes Defense Communications ICS, HF/VHF/UHF communications system.

Contractors

British Aerospace Australia.
Raytheon Systems Company.
Lockheed Martin Tactical Defense Systems.

UPDATED

ISRAEL

EL/M-2022A maritime surveillance radar

The EL/M-2022A multimode maritime surveillance radar is designed for fixed- or rotary-wing aircraft. The main missions for the radar are maritime target surveillance, ASW, search and rescue and economic zone control. Modular hardware design, software/remote control and a flexible avionic interface ensure easy installation in a variety of airborne platforms. The radar features multiple Track-While-Scan (TWS) for up to 100 targets, expand and freeze capabilities and sector or full-scan coverage. Optional features include integral IFF compatibility and air-to-air detection.

Operational modes consist of long-range sea surveillance, small target detection for periscopes based on a high-resolution waveform, high-resolution mapping based on Doppler beam-sharpening, moving target indication, navigation and weather capabilities, and SAR/ISAR/range signature classification.

The EL/M-2022A consists of an ultra-low sidelobe planar-array antenna, two-axis electric drive system and coherent TWT-based transmitter. The receiver/processor features a wide dynamic range receiver and programmable signal processor.

Three basic configurations are available: EL/M-2022A(V)1, EL/M-2022A(V)2 and EL/M-2022A(V)3.

The radar is suitable for both fixed- and rotary-wing aircraft.

EL/M-2022A(V)1 is a lightweight radar for small airborne platforms such as helicopters and UAVs with medium-range detection and a 50 target TWS capability. Typical detection ranges in Sea State 3 are 130 km against a small ship. Weight of the EL/M-2022A (V)1 is 65 kg.

LRUs of the EL/M-2022A(V)2 maritime surveillance radar

EL/M-2022A(V)2 is a version with a long-range periscope detection capability. Typical detection ranges in Sea State 3 are 55 km on a periscope and 130 km on a small ship. The EL/M-2022A(V)2 weighs 86-95 kg.

EL/M-2022A(V)3 is a version with long-range classification, periscope detection and a 100 target TWS capability. Detection ranges in Sea State 3 are 55 km on a periscope and 130 km on a small ship. The EL/M-2022A(V)3 weighs 95-103 kg.

Operational status

The EL/M-2022A(V)3 is in production for various customers. The radar has been selected for the Australian P-3C upgrade programme.

Contractor

Elta Electronics Industries Ltd.

UPDATED

ITALY

ATR42 Maritime Patrol Mission System (MPMS)

Alenia Difesa has been awarded a contract for the development and integration of a Maritime Patrol Mission System (MPMS) for the ATR42 aircraft for the Italian customs service (Guardia di Finanza); this aircraft has also been given the designation SAR42. Equipment details are not known, but it is believed to include a podded radar and secure datalink.

Contractor

Alenia Difesa, Avionic Systems and Equipment Division.

NEW ENTRY

RUSSIAN FEDERATION AND ASSOCIATED STATES (CIS)

Sea Dragon maritime surveillance mission system architecture

The Sea Dragon maritime surveillance mission system architecture is a proposal offered by the Leninetz Holding Company of St Petersburg. The roles proposed include: Anti-Submarine Warfare (ASW); maritime surveillance; Anti-Surface Warfare (ASuW); Search And Rescue (SAR); surface environment monitoring. The mission suite comprises: radar; electro-optics, Magnetic Anomaly Detector (MAD); acoustic system; Electronic Support Measures (ESM); and Mission Control Computer Unit (MCCU).

The radar system proposed provides: long-range detection of surface vessels in rough sea states and against precipitation and jamming environments; detection of small targets and emergency beacons in high sea states; air-to-air detection against a background of sea clutter; target acquisition and tracking, with range heading and velocity data; Synthetic Aperture Radar (SAR) and Inverse SAR (ISAR) modes.

The electro-optic surveillance system combines InfraRed (IR) and Low Light Level TV (LLLTV) cameras. There are two video channels, and features include: high resolution and selectable field of view IR and TV images; 360 degree coverage; 3-D gyrostabilised gimbal; autotrack controlled by the Mission Control Computer Unit (MCCU).

The MAD system comprises the sensitive quantum-mechanical magnetometer and provides: real-time data processing; use of target detection and aircraft-generated interference compensation combined algorithms with a high degree of automation; high accuracy of the beam range and beam passage time determination; reliable magnetic and electromagnetic interference clutter protection.

The acoustic system provides: automatic submarine detection and localisation by means of active and passive sonobuoys with target motion parameter measurement; operation near ice edges; underwater acoustic conditions reconnaissance and contacts prediction; multichannel receiving and processing capability.

The ESM provides detection of radar emissions with multi-octave frequency coverage in dense signal environments; monopulse measurements; analysis of frequency and modulation data; emitter identification and multitarget tracking.

The Sea Dragon mission system is centrally controlled by the Mission Control Computer Unit (MCCU), which provides: three identical operator positions with access to all processing functions, via multiplex databuses. Each operator station is equipped with two multicolour 13 in liquid crystal displays and multifunction consoles. Each operator has a digital map display and the ability to perform data fusion from all sensors, together with access to aircraft navigation and flight plan data. Laser disk memory is used to support system requirements.

Contractor

Leninetz Holding Company.

NEW ENTRY

Streege maritime surveillance mission system architecture

The Streege maritime surveillance mission system architecture is a proposal offered by the Leninetz Holding Company of St Petersburg. The roles proposed include: maritime surveillance including infrared, optical and radio frequency bands imaging; search and rescue; surface environment monitoring; frontier and Exclusive Economic Zone (EEZ) monitoring.

The mission suite comprises: Synthetic Aperture Radar (SAR) and Inverse SAR (ISAR) radar imaging; infrared and Low Light Level TV (LLLTV) cameras mounted in fully stabilised platforms; data processing and multifunction display systems; datalinks to co-operating ground and sea-based platforms. System growth options include: Magnetic Anomaly Detectors (MAD); gas analysing and radiation monitoring systems.

Contractor

Leninetz Holding Company.

NEW ENTRY

SOUTH AFRICA

Pelican TACCO system

The Pelican TACCO system provides for the integration of sensor and datalink information, the compilation of tactical picture, and the overall management of Maritime Patrol Aircraft (MPA) operations. It assists the Tactical Co-ordinator (TACCO) with the co-ordination and control of the aircraft's missions, including surveillance and intelligence gathering, vectored attacks on surface ships, anti-submarine warfare and search and rescue operations. The system also facilitates mission planning and debriefing. The basic configuration of the Pelican system allows integration with the following aircraft subsystems: navigation, ESM, acoustics, radar, sonobuoy launcher, and datalink controller. Other integrations can be added to match customer requirements.

Contractor

Altech Defence Systems.

VERIFIED

SWEDEN

SSC Maritime Surveillance System

The Maritime Surveillance System (MSS) was developed for the Swedish coastguard and has been in operation for 20 years. It is used for maritime surveillance tasks including: Exclusive Economic Zone (EEZ) protection; oil pollution detection and assessment; fishing activities monitoring; border patrol.

The MSS comprises:

(a) Sensor data package that includes: SLAR, IR/UV scanner, MicroWave Radiometer (MWR), camera(s), FLIR (optional), FLAR interface (optional), user-defined sensors (optional);

(b) Sensor Data Processor (SDP) that processes data from the selected sensor set;

(c) Data Management Unit (DMU) that provides high-performance image timing data and position information, display processing, sensor control, mission control and data reporting and datalink administration.

The current version of the MSS system is the MSS5000. MSS5000 has been improved over the previous versions of the MSS in a large number of ways, the main ones being: real-time resampling of all sensor data to georeferenced image; real-time display of all sensor images together with the digital map; Side-Looking Airborne Radar (SLAR) swath with 160 km wide coverage; new improved graphical user interface and image display windows.

The SLAR, manufactured by Ericsson, is designed exclusively for airborne maritime surveillance. The SLAR is an imaging radar, which produces map-like images, suitable for superimposition on a digital map. Its main applications are surveillance of sea traffic, oil spill detection, fishery protection, search and rescue and sea ice mapping.

The IR/UV scanner, manufactured by Daedalus, is used to obtain high-resolution imagery of accident sites. It is ideal for mapping oil spills. The infrared and ultraviolet channel images are both resampled and presented in real time on top of the digital map.

The MWR, manufactured by Ericsson, is a scanning radiometer, used for detailed mapping of oil spills. It measures the thickness of the oil on the water surface, and estimates the total oil volume. The MWR image can be superimposed on the other sensor images in the digital map.

The camera(s) are used for high-quality, high-resolution photographic evidence. Both hand-held and vertically mounted installations are possible. Each frame is automatically annotated with relevant time and position data.

All sensors of the MSS5000 can be operated single handed. The operator can also monitor resulting images in real time, and superimpose selected images on the digital map. Image interpretation is supported by online image analysis tools. This gives an excellent overview, as well as detail when needed.

Selected mission report text and images can be transmitted to the headquarters on the ground or to a coastguard ship to support, for instance, oil spill clean-up operations.

Operational status

MSS systems, or selected subsystems thereof, have been delivered to coastguard organisations in China, Germany, India, Netherlands, Norway, Poland, Portugal, Sweden, UK and the USA. They have been installed in aircraft such as CASA-212, Cessna 402B and 404, Fairchild Dornier Do 28, Fairchild Dornier Merlin, Turbolet L410 and Y12.

Contractor

Swedish Space Corporation.

VERIFIED

UNITED KINGDOM

System 7 homing systems

The System 7 is a versatile building block homing system capable of providing both broadband and emergency guard channel homing. In its simplest form, a System 7 installation comprises a homing indicator, an antenna feed unit, an antenna system and the aircraft receiver. When an independent self-contained emergency guard channel homing system is required, the System 7 uses a Chelton 7-28 Series two-channel receiver in place of the aircraft receiver. It is also possible to interface both the aircraft receiver and the 7-28 Series receiver with the System 7, thus providing broadband plus guard channel homing. The System 7 interfaces with all AM receivers, including the ARC116, ARC159, ARC164, ARC182, ARC186, PTR1751 and VHF20, and also with those FM equipments having AM facilities.

System 7 comprises a number of sub-units which include:

Series 7-24 homing indicator units. These are self-contained homing directors incorporating all necessary electronic control circuits, voltage regulators and computer reference amplifiers packaged within an 89 mm housing. The indicators have a variable sensitivity control located on the front panel and are edge-lit.

Series 7-27 indicator feed units. These comprise the electronic circuitry otherwise packaged within the 7-24 homing indicator units and are intended to feed existing aircraft navigation indicators or flight directors.

Series 7-25 antenna feed units. These contain the necessary antenna phasing and switching circuits to couple antennas into the homing system. Unlike the sub-units listed above, the choice of antenna feed unit is dependent upon the frequency band required.

Series 7-28 twin-channel emergency guard receivers. These small receivers are designed to provide a completely self-contained homing system at VHF and UHF distress frequencies.

Series 7-60 complete homing adaptors. These are combinations of Series 7-27 indicator feed units and Series 7-25 antenna feed units packaged in one small module.

Specifications

Weight: less than 1 kg
Power: 28 V DC, 500 mA max

Operational status

More than 700 systems are in service with military and commercial operators worldwide.

Contractor

Chelton (Electrostatics) Ltd.

VERIFIED

Series 700 sonobuoy and VHF homing receivers

Series 700 receivers are frequency synthesised dual-bandwidth VHF receivers designed to form part of a system to provide port and starboard and fore and aft VHF homing with automatic on-top indication. The dual bandwidth permits reception of 25 kHz spaced signals when in the narrowband mode or 375 MHz spaced sonobuoy signals in the wideband mode. An audio output is available to provide aural port-starboard homing and recognition of amplitude modulated signals.

The Series 700 was developed from, and embodies technology and subsystems used in the Chelton System 7 VHF/UHF homing systems, of which more than 700 are in service for worldwide military and civil operations. A complete system comprises four LRUs plus antennas, and features frequency synthesised dual-bandwidth operation, up to 99 preset channels with full OTPI interface, low weight and small size.

Specifications

Type 700 receiver
Weight: 1.325 kg
Power supply: 28 V DC, 500 mA
Frequency:
(Type 700-1) 100.00-173.500 MHz,
(Type 700-2) 117.975-173.975 MHz

Type 710 control unit
Weight: 0.98 kg
Power supply: 22-31.5 V DC, < 1.0 A
Frequency:
(Type 710-1) 100.00-159.975 MHz,
(Type 710-2) 117.975-173.975 MHz
Channel mode frequency:
(Type 710-1) 136.00-173.500 MHz,
(Type 710-2) 136.00-173.500 MHz

Operational status

Chelton 700 receivers have been ordered by Bristow Helicopters, the Helicopter Services (Navy), the Royal Navy, the Royal Norwegian Navy and the Royal Swedish Navy.

Contractor

Chelton (Electrostatics) Ltd.

UPDATED

Series 730/930 sonobuoy homing and channel occupancy system

The Series 730 99-channel sonobuoy homing and channel occupancy system operates over the frequency range 136.000 to 173.500 MHz and has homing or channel occupancy independent modes of operation.

In the homing mode, the system supplies azimuthal information so that the aircraft may be flown overhead a sonobuoy. Additional information indicates whether the sonobuoy is ahead or astern of the homing aircraft. This information is further decoded to give an indication of the instant the aircraft passes overhead the sonobuoy.

In the channel occupancy mode, the receiver measures the received signal strength to show when a selected channel is clear for the reception of sonobuoy signals.

The system comprises eight LRUs: the Type 730-1 radio receiver, Type 731-1 homing/channel occupancy switch, Type 732-1 azimuth switch coupler, Type 733-1 fore and aft coupler, Type 734-1 receiver mounting tray, azimuth antenna assembly, fore and aft antenna assembly and channel occupancy antenna assembly.

The heart of the system is the radio receiver which interfaces the system to the aircraft via a dual-redundant MIL-STD-1553B databus.

In the channel occupancy mode, the radio receiver provides the necessary control signals to disable the azimuth and fore and aft switch couplers. The receiver cycles through channels as designated by the mission computer and provides information on the databus channel status, together with received signal strength.

In the homing mode, the radio receiver provides the necessary control signals of the azimuth and fore and aft switch couplers to switch the RF input of the radio receiver between the antenna assemblies.

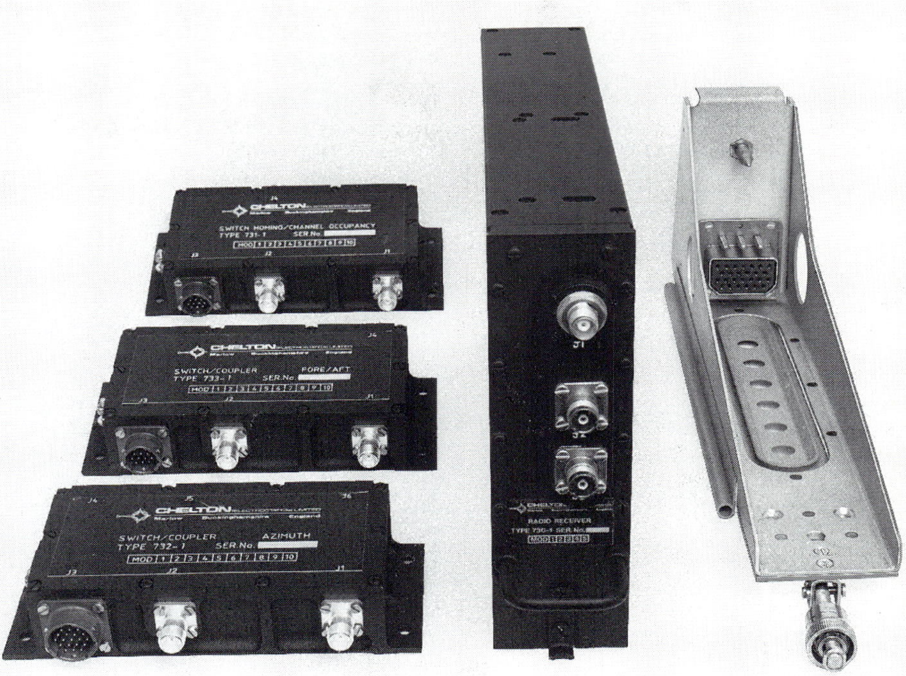

The Chelton 730 sonobuoy homing and channel occupancy system comprises (left, top to bottom) the Type 731-1 homing/channel occupancy switch, the Type 733-1 fore and aft switch coupler, the Type 732-1 azimuth switch coupler, (centre) the Type 730-1 radio receiver and (right) the Type 734-1 receiver mounting tray

BITE is capable of diagnosing 95 per cent of all defects to LRU level.

Chelton's experience with the Series 730 system has led to the development of the Series 930 VHF/UHF direction-finding systems. These provide 360° bearing information in analogue or digital format to accuracies better than 5° and use high reliability, stepper motor driven, rotating antennas and enhanced software.

Specifications

Weight:
(all LRUs) 8 kg

Operational status

The 730 has been selected for the Westland/Agusta EH 101 Merlin ASW helicopter.

Contractor

Chelton (Electrostatics) Ltd.

UPDATED

ASN-900 series tactical processing systems

The ASN-900 series tactical processing systems provide facilities to correlate and process data for display from the wide variety of sensors and navigation systems installed in modern maritime patrol aircraft, ASW/ASV helicopters and AEW aircraft. The system enables the tactical co-ordinator to display data in an easily assimilated form and assist in the solution of complex navigation, intercept and attack problems. ASN-902 and 924 systems, which are based on AQS-902/920 hardware, have a flexible design which makes them readily adaptable for use as the central element in an integrated mission management system. It can replace the variety of individual sensor system control and display units with common integrated units, providing flexibility of operation. Standard ARINC 419, 429 and MIL-STD-1553B data interfaces allow installation as original equipment or as a retrofit.

Systems integration considerably improves the efficiency and flexibility of any mission avionics suite, minimises the weight of combinations of multiple sensors and simplifies logistic and training problems with common control units and multipurpose displays.

GEC-Marconi Avionics can offer various levels of systems integration, extending to a totally integrated mission management system. The company has carried out systems integration for the advanced Sea King Mk 42B. This programme incorporates an AQS-920 Series acoustic processor and, as its core element, an ASN-902 tactical processing system.

The ASN-902 system, in service in Sea King Mk 42B helicopters, has a monochrome display, whereas the ASN-924, which is in production, has a colour display to enhance the information presented to the tactical co-ordinator. A new series, designated ASN-990, is being developed. These systems are based on power PCs and high-resolution colour displays with picture-in-picture capability.

Operational status

The ASN-902 tactical processing system is in service in the Sea King Mk 42B. The ASN-924 is in production.

Contractor

GEC-Marconi Avionics Limited, Mission Avionics Division, Rochester.

VERIFIED

The GEC-Marconi Avionics integrated mission avionics suite in a Sea King Mk 42B

AA34030 sonobuoy command transmitter

The AA34030 sonobuoy command transmitter is a variant of the GEC-Marconi, Basildon AD 3400 VHF/UHF transmitter/receiver which has been in full production for a number of years. This variant has been developed in conjunction with GEC-Marconi Avionics, Rochester for the AQS-903 acoustic processing system and is currently in service with several air forces around the world.

The equipment is a UHF transmitter which has been specifically developed to provide the RF downlink between an airborne acoustic processing system and active sonobuoys. The transmitter provides AM or FM analogue or digital communications over the frequency range 282 to 292 MHz with channel increments of 25 kHz, although for normal sonobuoy operations 50 kHz is used. Control of the transmitter is via an ARINC 429 digital data highway and a number of discrete inputs. Control information is normally provided by the acoustic processing system, although a discrete control unit can be provided.

Specifications

Dimensions: 194 × 125 × 256 mm
Weight: 6 kg max

Operational status

In production for the Royal Navy EH 101 Merlin helicopter.

Contractor

GEC-Marconi Electro-Optics Limited, Sensors Division, Basildon.

VERIFIED

Mission recording system

Normalair-Garrett's mission recording system is designed for use in ASW operations. It uses Ampex Corporation DCRSi digital cassette recording technology.

Operational status

In production for the EH 101 Merlin HM Mk 1 ASW helicopter.

Contractor

Normalair-Garrett Ltd.

NEW ENTRY

NGL sonobuoy dispenser

The sonobuoy dispenser forms part of an automatic system for the deployment of sonobuoys during ASW operations.

The Normalair-Garrett Ltd sonobuoy dispenser is a 10-station rotary device for the carriage and release, by gravity, of various sonobuoys in any combination of the standard 'A', 'F' or 'G' sizes. It can be remotely controlled from a tactical mission console or the stores management system and can be used to release a predetermined pattern of sonobuoys.

The dispenser, although primarily designed for a helicopter application, can also be used for fixed-wing operations, where the addition of a pressurisation valve and associated safety interlocks are required to ensure that sonobuoys cannot be dropped above aircraft pressurisation levels.

The composite dispenser is supported by a base structure which is attached to the floor of the aircraft by means of receptacles in the aircraft seat rails. Additionally, the dispenser is braced at the top, by a top-steady arrangement which attaches to a spigot on the top cover.

NGL sonobuoy dispensing system **1997**/0002126

Specifications

Weight: 34.5 kg, excluding the aircraft/dispenser interface unit and top steady bracket.
Power Consumption: Drive mechanism: 115/200 V, 3 phase, 400 Hz AC, 300 W-28 V DC, 8 A.
Release solenoid: 28 V DC-13 A for 160 ms

Operational status

In production for EH 101 Merlin HM Mk 1 ASW helicopter. Selected for Nimrod MRA4.

Contractor

Normalair-Garrett Ltd.

UPDATED

Acoustic Data Vessel Identification, Classification and Expert (ADVICE) subsystem

The Acoustic Data Vessel Identification, Classification and Expert (ADVICE) subsystem assists the acoustic operator in interpreting the ever-increasing quantity and quality of data output by the prime acoustic system. ADVICE separates the data into vessel types, providing a detailed classification of possible targets in an easily assimilable graphic presentation. The classification includes the allocation of frequency lines to component parts of the vessel and, where a definitive classification is not possible, a list of possibilities is produced with an associated confidence level. The operator may choose to input his own match of signals to vessel attributes and this data will be used by ADVICE to reclassify the vessel. ADVICE will then assign a level of confidence in that reclassification.

ADVICE can be embedded either in the prime acoustic processor on board the aircraft or installed as separate hardware.

Contractor

Thomson Marconi Sonar Limited.

VERIFIED

AQS 901 acoustic processing systems

The Thomson Marconi Sonar AQS 901 acoustic processing system is installed in both the Royal Air Force Nimrod MR.2 Mk 2 and the Royal Australian Air Force Lockheed P-3C Orion maritime patrol aircraft and has been in service since 1979. The AQS 901 is a very powerful airborne processor that is able to handle data from all types of sonobuoys in the NATO inventory. This includes advanced sonobuoys such as Barra, CAMBS and VLAD.

Every aircraft contains two AQS 901 systems, each with two CRT displays and two electrographic chart recorders (hard-copy display). A fifth display, for system management activities, shows processed and tabular data from either system. Control and interrogation of processed data is achieved using a mix of keyboard, on-screen menus, rollerball and cross-wire cursor.

A series of software and hardware updates over the past 10 years has enabled the AQS 901 to continue to fully satisfy current ASW requirements.

In 1995 the RAF's Nimrod fleet were further upgraded when its AQS 901 systems were updated to include advanced colour processing.

Operational status

In service. AQS 901 equips over 30 Royal Air Force Nimrod MR.2 Mk 2s and 20 Royal Australian Air Force P-3C Orions. Deliveries were completed in autumn 1986. Software and hardware product improvements will continue for several years.

The AQS 901 installation in an RAF Nimrod MR.2 Mk 2

Contractor

Thomson Marconi Sonar Limited.

VERIFIED

AQS 902/AQS 920 series acoustic processors

The AQS 902 and its export variants, the AQS 920 series, have been designed for ASW helicopters, maritime patrol aircraft and small ships. This range of lightweight and versatile systems will handle data from all current and projected NATO inventory sonobuoys and will interface with a wide range of 31 or 99 RF channel sonobuoy receivers. An AQS 902/920 series can process and display up to eight buoys or a combination of sonobuoys and dipping sonar simultaneously. The modular structure of the AQS 902/920 series allows systems to be tailored to specific customer requirements. The signal processor unit converts received sonobuoy signals into digital form, carries out the necessary filtering and analysis, and processes the data into a suitable form for display. It is packaged in a 1 ATR(S) box. Colour acoustic data is displayed on a CRT or electrographic paper chart recorder. Either or both may be specified, but the CRT display requires a post-processor unit to prepare data for presentation and provides many additional operator aiding facilities. The operator uses a key panel and rollerball or stiff stick cursor control device to specify processing modes and display formats, measure and extract information and access the wide range of automatic and semi-automatic system facilities.

The AQS 902G/DS installation in a Royal Navy Sea King Mk 6

The AQS 902/920 offers numerous processing and display facilities including simultaneous passive and active processing, wide and narrowband spectral analysis, broadband correlation, CW and FM active modes, passive auto alerts, Doppler fixing, calculation of target range and speed, bathythermal processing and ambient noise measurement. Additionally, the system provides automatic tracking and calculation of line frequency, target bearing and range and signal-to-noise ratio.

Thomson Marconi Sonar claims that a unique feature of these systems is the acoustic localisation plot which presents the operator with a geographical plot of sonobuoy bearing and range information.

Operational status

Configurations of this system are in service with the Royal Navy, the Royal Swedish Navy, the Indian and Italian navies. Further variants have been sold to Grumman for S-2(T) Turbo Tracker aircraft.

More than 150 AQS 902G/DS systems in service with the Royal Navy's Sea King Mk 5 and Mk 6 ASW helicopters. Software and hardware product improvements will continue for several years.

Contractor

Thomson Marconi Sonar Limited.

VERIFIED

AQS 903/AQS 930 series acoustic processing system

The Thomson Marconi Sonar AQS 903 and its export variants, the AQS 930 series, are compact acoustic processing systems which are designed for use in all types of ASW helicopters. AQS 903/930 series systems can process and display all current and projected sonobuoy types, as well as dipping sonar.

AQS 903/930 series systems make extensive use of distributed processing, and incorporate the latest component and custom chip technology to provide optimum performance. Modularity enables systems to be configured to specific customer requirements for processing between 8 and 32 sonobuoys. Full 360° surveillance is provided with sector coverage offering additional classification and/or tracking analysis to allow smooth transition from one mission phase to another, irrespective of the buoy type in use. System control is achieved by using a simple control panel with multifunction keys and a cursor control device. All analysis and display cues are in plain language, requiring a minimum of operator training.

AQS 903/930 systems will interface (via either ARINC 419/429 or MIL-STD-1553B buses) to other avionics systems responsible for sonobuoy launching, navigation, tactical tracking and weapon release.

The Royal Navy EH 101 Merlin ASW helicopter

Operational status

Selected for the Royal Navy EH 101 Merlin helicopter. A production order for 44 systems has been received, and initial production deliveries have been made.

Contractor

Thomson Marconi Sonar Limited.

VERIFIED

ASW crew trainer

Thomson Marconi Sonar's Anti-Submarine Warfare (ASW) crew trainers (ACTs) provide fully dynamic, real-time ASW training for maritime patrol aircraft and ASW helicopter crews. ACT systems offer comprehensive acoustic, tactical and crew co-operation and co-ordination training by simulating sonobuoy, dipping sonar, target and ocean environmental data. Data can be controlled, processed and displayed on the aircraft's acoustic and tactical systems in the same way as real data. Training is therefore achieved without the need to deploy sonobuoys or dipping sonar, or use co-operating submarines and surface vessels, thus substantially reducing costs and increasing training opportunities. There are two basic variants of ACT.

ACT 1 is a software-based system developed specifically for the Thomson Marconi Sonar AQS 901 acoustic processing systems now in service on Royal Air Force Nimrod MR.2 Mk 2 and Royal Australian Air Force P-3C Orion maritime patrol aircraft. It comprises an Exercise Control Unit (ECU) and the computer programme which is downloaded into the AQS 901 from a Magnetic Tape Unit (MTU). The crew member acting as the exercise controller uses the ECU to set the target's initial start position, course, speed and depth and, if desired, controls subsequent target manoeuvres and ocean conditions. Otherwise the facilities and functions necessary to simulate fully dynamic ASW scenarios are automatic.

ACT 2 is a hardware-based system which is far more flexible than ACT 1 and can be interfaced with any acoustic processor resident in the platform or, by means of a VHF radio link and an ACT ground station. ACT 2 can simulate the characteristics of any sonobuoy, dipping sonar, target and ocean environment specified by the user. It comprises an Acoustic Simulation Unit (ASU) and Exercise Control Unit (ECU). The ASU, which executes the ASW scenarios and generates synthetic passive and active acoustic data, can be configured to simulate any number of independent sonobuoy and dipping sonar channels. The ECU is an intelligent control terminal with an integral keyboard and display. In addition to performing all exercise control functions, the battery-supported ECU is used for the temporary storage of exercise scenario data.

ACT 2 systems currently available can simulate up to 32 independent data channels from any combination of LOFAR, DIFAR, Barra, Ranger, CAMBS, DICASS and bathythermal sonobuoys, and can generate simultaneously the acoustic signature of up to three independent targets. Simulation of specific dipping sonars and other types of sonobuoys can be easily accommodated within the existing hardware.

Unlike ACT 1, there are no limitations with ACT 2 on the number and variety of exercise scenarios for training. These are created and maintained by each customer, using an Exercise Support System (ESS). The ECU and ESS functions can be performed using a single/common IBM PC-compatible computer, which can also be used as a classroom trainer and for post-flight debriefs.

Operational status

ACT 1 is in service with the Royal Air Force and Royal Australian Air Force. ACT 2 has completed development and is in full production; deliveries have been made to several customers.

Contractor

Thomson Marconi Sonar Limited.

VERIFIED

CRISP Compact Reconfigurable Interactive Signal Processor

Thomson Marconi Sonar Limited has developed 'CRISP', a Compact, Reconfigurable Interactive Signal Processor utilising the Series 5 technology, which is based on many years experience of providing powerful signal processors to navies around the world.

This version of CRISP is now available to support the surface, subsurface and airborne processing community, with a compact portable mission support system.

A high-resolution colour graphics display (available as a flat screen) with pop-up menus and trackermouse roll and click selection allows interactive and user-friendly operation.

Using modular boards already in operational naval service, CRISP brings the following advances to Mission Support Systems for the processing of airborne sonars: simultaneous broadband and narrowband analysis; digital DIFAR processing; frequency range from infrasonic to intercept; very fast time replay; short-term event capture and analysis; rapid software reconfigurability; commonality with RN and RAF equipment.

Operational status

In service with the Royal Navy, Royal Air Force and Canadian Forces.

Contractor

Thomson Marconi Sonar Limited.

VERIFIED

TMS 2000 series acoustic processing system (typical passive display) ***1997***/0001187

TMS 2000 series acoustic processing system

The TMS 2000 is an advanced and powerful airborne acoustic processor. It uses a cost-effective and totally open architecture based on rugged, state-of-the-art, off-the-shelf hardware. The TMS 2000 uses existing and proven algorithms which have been developed as part of the AQS 900 and SADANG series of processors. They have been rehosted and enhanced with new colour broadband, data fusion, and multistatic processing techniques.

The system is designed for use by one to four operators, with each operator having access to all the processed data as graphical data fusion plots, localisation plots, acoustic formats or tabular information. The system can support multiple high-resolution colour monitors.

All current and projected NATO sonobuoy types are processed and sonobuoy data may be received from either traditional analogue receivers or from new digital receivers.

System control uses the latest techniques of pull-down or pop-up menus, augmented by on-screen editing and fast dedicated keys. The system can be easily interfaced to other avionics systems via any industry standard or MIL-STD-1553B interface.

Passive functions

The TMS 2000 provides a range of advanced processing detection modes, which cover a full range of target noise characteristics including narrowband frequency lines, broadband signals, swaths, transients and DEMON signals. Each channel has two main processes, plus up to six auxiliary processes which are tuned to the target characteristics and environmental conditions. Computer assistance is provided to link database information to observed signal data to aid target classification. DOUBLE DEMON processing ensures full coverage of both the carrier noise bandwidth and modulating spectrum to achieve permanent detection of transient cavitation phenomena.

Active functions

The TMS 2000 has the capability to process up to 16 DICASS sonobuoys simultaneously in the multistatic mode as well as the monostatic mode, and so satisfies typical brown shallow water operational requirements. Typical coverage is 200 square nautical miles with 16 DICASS sonobuoys.

Specifications

Sonobuoy capability: 8, 16, or 32 sonobuoy channels; compatible with all NATO sonobuoys
Passive processing: up to 32 DIFAR per unit; 2 main and up to 6 auxiliary processes for each channel; manual/automatic noise nulling; cardioid processing; computer-assisted classification
Active processing: 16 DICASS multistatic/monostatic modes; ping-to-ping integration, auto-handling of pulse sequences; reverberation suppression and stationary artefact removal
Displays: passive and active colour enhancement; passive and active geographic energy plots

Operational status

Under development. Ground-based prototype systems have been delivered to the UK Ministry of Defence for evaluation and airborne versions have been flight-trialed on board the RAF's Nimrod MR.2 Mk 2.

Contractor

Thomson Marconi Sonar Limited.

UPDATED

Type 2069 sonar

The Type 2069 dunking sonar sensor

The Type 2069 sonar helicopter dunking system is installed in Sea King helicopters. It provides full 360° coverage and is understood to be effective at ranges up to 7,300 m. Full azimuth coverage is provided in stepped fashion over 90° arcs progressively, but manual control allows the operator to concentrate on any particular sector. The system is programmed to undertake an automatic search. The operator is provided with audio, visual Doppler and visual sector sonar information, and close contact maintenance is provided for the tracking of nearby targets and those at greater depths.

The sonar may be employed for either surveillance or attack control, or for both simultaneously. Pulse length and detection range settings are operator selectable to optimise the system according to sea conditions and the tactical situation.

The Type 2069 sonar is the latest upgrade of the 195M employing new solid-state transmitters and with a longer cable. The sonar transducer has been re-engineered to provide a greater operating depth and is now integrated with the AQS 902 G/DS acoustic system.

The total system is the first in-service sonar with simultaneous control, processing and display of sonobuoys and sonar. It is installed in the Royal Navy Sea King Mk 6.

Operational status

Type 2069 is now in service with the Royal Navy and several other navies.

Contractor

Thomson Marconi Sonar Limited.

VERIFIED

AQS 970 acoustic processor

The Ultra AQS 970 airborne acoustic processor has been selected for the Nimrod MRA4 programme and is being developed through collaboration between Ultra Electronics and Computing Devices of Canada. The CDC UYS-503 system is already in service with the defence forces of Australia, Canada, Sweden and the US. The AQS 970 interfaces with the Tactical Command System (TCS) and the Sonobuoy Receiving System (SRS) in fixed-wing Maritime Patrol Aircraft (MPA) and Anti-Submarine Warfare (ASW) helicopters, providing signal conditioning, signal and data processing, display generation and control functions.

The Ultra AQS 970 is designed to meet the detection, localisation and attack requirements posed by the latest-generation nuclear and diesel submarines operating in open ocean and littoral water environments. The modular architecture of the AQS 970 and its ample processing power and memory capacity enable it to readily accommodate new buoy types and processing techniques without hardware modification.

The system employs a scalable hardware and software architecture, configurable as 16, 32 or 64 identical processing channels, providing full 64-buoy processing capability (subject to buoy RF channel limitations) for all NATO variants of the following existing or planned sonobuoy types:

Passive search and localisation
- DIFAR
- LOFAR
- VLAD
- CODAR (DIFAR/omnipairs)
- HIDAR (digital DIFAR)
- BARRA

Active search and localisation
- CAMBS
- DICASS
- RO
- SR(SA)903 Active sonobuoy search system

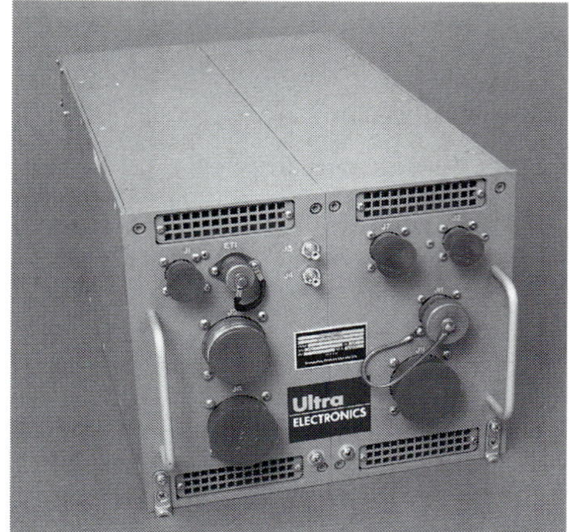

Ultra Electronics AQS 970 acoustic processor
1998/0011859

Special purpose
Ambient Noise Monitoring (ANM)
BATHY
Built-in Acoustic Test Signal Generator (ATSG) and Tactical Acoustic Trainer (TAT)

Processing and display capabilities include: bearing coherent processing and effective colour-coded displays; Energy map and acoustic situation displays; narrowband, broadband/swath, diesel, DEMON, and transient detection processes; Variable band CODAR (using DIFAR buoys); Adaptive interference rejection/ anti-jam processing (DIFAR, BARRA, CAMBS)

Operator aids include: Gram markers and harmonic dividers, CPA analysis, Lloyd's mirror analysis, display magnify/freeze/zoom/pan, cross-hair and hook, adjustable thresholds and integration times, concurrent multiresolution processing, rapidly selectable operator-defined display formats and combinations.

General aides provided include: frequency/bearing/ range trackers, synchronised audio, range prediction, data fusion and contact tracking, auto alerts.

Operational status

Ultra Electronics is providing the integrated acoustic system for Nimrod MRA4.

Contractor

Ultra Electronics, Sonar and Communication Systems.

UPDATED

ARR 970 sonobuoy telemetry receiver

The ARR 970 16-channel and 32-channel sonobuoy receiver system has been designed to satisfy the requirements of current Anti-Submarine Warfare (ASW) aircraft while providing optional facilities to meet future requirements. It amplifies and demodulates FM sonobuoy transmissions and provides the output to the AQS 970 acoustic processor for analysis and display. The ARR 970 provides RF and acousitic performance as well as lightweight and compact size.

Both 16-channel and 32-channel models are available, with options for an On-Top Position Indicator (OTPI) function, and channel scanning receivers sonobuoy reference system.

Ultra Electronics ARR 970 sonobuoy telemetry receiver 32- and 16-channel configurations
1998/0011858

Specifications

16-channel system
Dimensions: 545 × 215 × 345 mm
Weight: 33 kg
Power supply: 115 V, 3 phase, 400 Hz, MIL-STD-704A: 300 VA
RF channels:
99+ test (375 kHz spacing, standard)
505 (75 kHz spacing, optional)
Audio interfaces: balanced, 2 Vrms for 75 kHz peak deviation
IF bandwidth: 240 kHz, standard
Control interfaces: MIL-STD-1553B, RS-422

32-channel system
Dimensions: 545 × 328 × 345 mm
Weight: 39 kg
Power supply: 115 V, 3 phase, 400 Hz, MIL-STD-704A: 450 VA
RF channels:
99+ test (375 kHz spacing, standard)
505 (75 kHz spacing, optional)
Audio interfaces: balanced, 2 Vrms for 75 kHz peak deviation
IF bandwidth: 240 kHz, standard
Control interfaces: MIL-STD-1553B, RS-422

Operational status

Selected for use on Nimrod MRA4.

Contractor

Ultra Electronics, Sonar and Communication Systems.

UPDATED

The Ultra R605 sonobuoy receiver
1995

R605 sonobuoy receiving set

The R605 sonobuoy receiver is in use on the Royal Navy EH 101 Merlin ASW helicopter. The receiver amplifies and demodulates four simultaneous FM sonobuoy transmissions and provides the output to the acoustic processor for analysis and display. It provides coverage of 99 RF channels per receiver module and is mechanically interchangeable and connector compatible as a direct replacement for the AN/AAR-75.

For applications where the receiving sets must be mounted at a distance from the receiving antenna, a remote preamplifier is provided. MIL-STD-1553B control is available. Control functions include RF channel selection, sonobuoy type for optimal bandwidth control, RF level reporting and self-test initiation and reporting.

The R605 receiver features a large RF dynamic range and high sensitivity over a broad base bandwidth. In addition, the receiver provides a wide, linear phase, audio output with close uniformity between each of the four audio channels. Digital sonobuoy data undergoes matched filter detection and clock regeneration in the receiver. Differentially driven clock and data signals are available for each of the four receiver outputs at any of the data rates and code functions.

Specifications

Dimensions: 159 × 351 × 232 mm
Weight: 10 kg
Power supply: 115 V AC, 400 Hz, 3 phase, 100 VA max
Frequency: 136-173.5 MHz
Channels: 99 RF, 4 acoustic
Interfaces: MIL-STD-1553, RS-422, ARINC 429
Reliability: >2,000 h MTBF

Operational status

In serial production. In service on Royal Air Force Nimrod MR.2 Mk 2 and Royal Australian Air Force P-3C Orion aircraft, and selected for the Royal Navy EH 101 Merlin helicopter.

Contractor

Ultra Electronics, Sonar and Communication Systems.

VERIFIED

R612 On-Top Position Indicator (OTPI)

The 99-channel R612 On-Top Position Indicator (OTPI) offers flexibility and is designed to operate with a variety of direction-finding antennas and various control options, including RS-422, and MIL-STD-1553B via the R605 receiver or a manual control box. The OTPI produces an adequate signal strength indication when the RF signal input level is sufficient for reliable direction-finder operation. The design is compatible with all known sonobuoys and incorporates BIT facilities.

Specifications

Dimensions: 126 × 135 × 78 mm
Weight: 2.16 kg
Power supply: 115 V AC, 400 Hz, single phase, 15 VA
Interfaces: RS-422, MIL-STD-1553B
Reliability: >3,000 h MTBF

Contractor

Ultra Electronics, Sonar and Communication Systems.

VERIFIED

R613 sonobuoy receiver

The R613 receiver is an improved version of the T843 receiver, which it supersedes. The receiver amplifies and demodulates FM sonobuoy transmissions and provides an output to the acoustic processor for analysis and display. It is installed in Royal Navy Sea King Mk 6 ASW helicopters. The receiver operates on four simultaneous channels and has a full 99-channel capability, but without the high level of sophisticated BITE provided with other Ultra receivers. A dedicated control unit and separate RF amplifier for increased signal-to-noise performance complete this receiver.

Specifications

Dimensions: 194 × 385 × 232 mm
Weight: 11 kg
Power supply: 115 V AC, 400 Hz, 3 phase, 200 VA max
Frequency: 136-173.5 MHz
Channels: 99 RF, 4 acoustic
Reliability: >1,100 h MTBF

Operational status

In service in Royal Navy Sea King Mk 6 ASW helicopters.

Contractor

Ultra Electronics, Sonar and Communication Systems.

VERIFIED

UNITED STATES OF AMERICA

AN/AQS-13 sonar for helicopters

The AN/AQS-13 is a helicopter dunking system, the -13B and -13F models being in production. It is one of a series of equipments which began with the AN/AQS-10 in 1985. The AN/AQS-13B is a long-range active scanning sonar which detects and maintains contact with underwater targets through a transducer lowered into the water from a hovering helicopter. Opening or closing rates can be accurately determined and the system also provides target classification information.

The AN/AQS-13B has significant advantages in operation and maintenance over earlier systems. To aid the operator, some electronic functions have been automated. Maintenance has been simplified by eliminating all internal adjustments and adding BITE circuits. These advantages were brought about by the use of the latest electronic circuits and packaging techniques, which also reduced system size and weight.

To enhance detection capability in shallow water and reverberation limited conditions, while eliminating false alarms from the video display, AlliedSignal developed an Adaptive Processor Sonar (APS) for the system. The APS is a completely digital processor employing fast Fourier transform techniques to provide narrowband analysis of the uniquely shaped CW pulse transmitted in the APS mode. The display retains the familiar PPI readout of target range and bearing, but APS adds precise digital readout of the radial component of target Doppler. With APS, processing gains of greater than 20 dB with zero false alarm rates have been measured for target Dopplers under 0.5 kt.

The AN/AQS-13E was the first system to integrate APS and sonobuoy processing in a common processor, the sonar data computer. Improvements to this system led to the AN/AQS-13F.

The higher energy transmitted with the longer pulse APS mode, combined with the narrowband analysis, also substantially improves the figure of merit in the non-reverberant conditions typical of deep water operations. Measured processing gains for APS under ambient wideband noise limited conditions exceed 7 dB.

The AN/AQS-13F has been designed to provide rapid tactical response against the most advanced submarine threats. It is a sister equipment to the AN/AQS-18 (see next item) and is identical in many respects. A new transducer, when lowered to depths of up to 450 m, permits instantaneous range improvements of over 100 per cent compared to previous systems. Very high-speed reeling allows a dip to maximum depth to be completed in approximately 3 minutes. The powerful omnidirectional transducer providing 216±1 dB source level is integrated with a sensitive directional receiver array providing azimuth resolution in a small rugged unit. The sonar data computer offers digital matched filter processing for 200 ms and 700 ms sonar pulses, as well as sonobuoy control and processing. The azimuth and range indicator and receiver provides a video display for the operator.

Specifications

Weight:
(13A) 373 kg
(13B) 282 kg
(13F) 280 kg
Frequency: 9.25 to 10.75 kHz
Sound pressure level:
(13B) 113 dB
(13F) 216 dB
Range scales: 1, 3, 5, 8, 12, 20 n miles (0.9, 2.7, 4.6, 7.3, 11, 18.3 km)
Operational modes:
(13A) active 3.5 or 35 ms, MTI, APS, passive, voice communications, key communications
(13F) active 3.5 or 35 ms rectangular pulse, 200 ms or 700 ms shaped pulse, MTI, passive 500 Hz (bandwidth 9 to 11 kHz), SSB voice communications on 8 kHz
Visual outputs:
(13A) range and bearing
(13B) range, range rate, bearing and operator verification
Audio output:
(13A) single channel with gain control
(13B) dual channel with gain control plus constant level to aircraft intercom
Recorder operation: bathythermograph, range, aspect, MAD self-test
Operating depth:
(13F) 1,450 ft at 50 ft hover

Operational status

The AN/AQS-13 is widely used by US forces and 1,000 sets have been ordered or supplied for the helicopters of 15 foreign navies in Asia, Europe, the Middle East and South America. The AN/AQS-13F was selected by the US Navy and is now in operation on the SH-60F carrier-based ASW helicopter.

Contractor

AlliedSignal Ocean Systems.

VERIFIED

AN/AQS-18 sonar for helicopters

The AN/AQS-18 is a helicopter long-range active scanning sonar. The system detects and maintains contact with underwater targets through a transducer lowered into the water from a hovering helicopter. Active echo-ranging determines a target's range and bearing and opening or closing rate relative to the aircraft. Target identification information is also provided.

The AN/AQS-18 is an advanced version of earlier dunking sonars made by AlliedSignal and includes digital technology, improved signal processing and improved operator displays. The system consists of a small high-density transducer with a high sink and retrieval rate, a built-in multiplex system to permit use of a single conductor cable, a 330 m cable and compatible reeling machine and a lightweight transmitter built into the transducer package. The Adaptive Processor Sonar (APS), which provides enhanced performance in shallow water areas, is an integral part of the system.

The AN/AQS-18 offers a number of improvements over earlier dipping sonars. These include increased transmitter power output to give longer range, high-speed dip cycle time and reductions in weight of all units.

The APS increases detection capability in shallow water and limited reverberation conditions, while eliminating false alarms from the video display. The APS is a digital processor which uses fast Fourier transform techniques to provide narrowband analysis of the uniquely shaped CW pulse transmitted in the APS mode. The PPI display retains the normal readout of target range and bearing.

The APS processing gain improvement over the normal AN/AQS-18 analogue processing is 20 dB for a 2 kt target and 15 dB for a 5 kt Doppler target. The higher energy transmitted with the longer pulse APS mode, combined with the narrowband analysis, also improves operation in the non-reverberant conditions more typical of deep water. In general, the gain improvement above a speed of 10 kt exceeds 7 dB.

The latest version is the AN/AQS-18(V) which is available with both 300 and 450 m length cables.

Specifications

Weight: 252 kg plus 13.3 kg for APS
Frequency: 9 23, 10, 10.77 kHz
Sound pressure level: 217 dB/μPa/yd (0.9 m)
Range scales: 1,000, 3,000, 5,000, 8,000, 12,000, 20,000 yds
Modes: 3.5 or 35 ms pulse (energy detection) and 200 or 700 ms pulse (narrowband analysis)
Visual outputs: range, range rate, bearing, operator verification
Audio output: dual channel with gain control plus constant level to aircraft intercom
Recorder operation: bathythermograph, range, ASPECT, MAD, BITE
Operating depth: 330 m

Operational status

In production and in service with the German, Greek, Itailian, Japanese, Portuguese, Spanish, Taiwanese and US navies on Lynx, SH-3, SH-60J and S-70C(M)-1 helicopters.

Contractor

AlliedSignal Ocean Systems.

VERIFIED

AN/AQS-18(V) dipping sonar system

The AN/AQS-18(V) is the export version of the AN/AQS-13F helicopter dipping sonar. These systems employ a transducer which is lowered into the water from a hovering helicopter to detect and maintain contact with underwater targets. Active echo-ranging determines target range and bearing, and opening and closing rate relative to the helicopter. Target classification indications are also provided.

The AN/AQS-13F/18(V) Series of dipping sonars is specifically designed for ASW helicopters where great mobility is required for fast reaction. AN/AQS-13F/18(V) equipped helicopters are well-suited for redetection of contacts, target localisation and weapon delivery against shallow and deep water threats. ASW helicopters are often required to search areas which are difficult for other sensor platforms, including shallow water with high noise areas, coastal regions, constrained passages, high-density shipping lanes and areas of concentrated naval activity.

Features of the Series include high source level to provide long-range, shallow water signal processing for high reverberation areas, capability to control, process and display sonobuoys in a single integrated system, high-speed reeling machine to achieve maximum depth and retrieval within 3 minutes and adaptable reeling machine designs compatible with a wide range of helicopters.

AN/AQS-13F/18(V) systems include digital technology, improved signal processing, extensive use of hybrid integrated circuits and improved operator displays.

The AN/AQS-18(V) deployed from a German Navy Westland Lynx Mk 88 helicopter ***1995***

AN/AQS-13G/18A

The AN/AQS-13G/18A is the latest stage in evolutionary development of the system. The improved 'dry end' of the system provides 14 dB improvement over earlier versions, thus providing search rates which are more than four times greater.

Specifications

Weight: 275 kg total
Frequency: 9.23, 10.003 and 10.744 kHz
Operating depth: 440 m
Sound pressure level: 217 ±1 dB/µPa/yd (0.9 m)
Range scales: 0.9, 2.7, 4.6, 7.3, 11, 18 km
Modes:
(active) 3.5, 35 ms pulses, 200, 700 ms shaped pulses (passive), (communicate) SSB at 8 kHz
Raise speed: 6.7 m/s average
Lower speed: 4.9 m/s
Water exit speed: 1.5 m/s

Contractor

AlliedSignal Ocean Systems.

UPDATED

AQS-18A dipping sonar system

The AQS-18A dipping sonar system represents a development of the medium frequency AN/AQS-18(V) dipping sonar. The dome control, reeling machine and transducer of the AN/AQS-18(V) have been interfaced with a powerful digital processor, control unit and colour display. The advanced processing brings greater performance through high-resolution digital processing, greater contact memory space and the flexibility to increase the number of sonar beams, type and length of pulses and menus of operator displays.

The AQS-18A has additional pulse lengths of 1.6, 3.2 and 4 seconds. The longer pulses put more energy on the target and provide higher Doppler resolution for maximum performance in high reverberation shallow water conditions. An FM mode is available for extremely low Doppler target detection and maximum range resolution. The total ASW system improvement of the AQS-18A is 14 dB over current systems and can provide more than four times the area search rate of the AN/AQS-18(V).

The AQS-18A has spare processing power and space for additional processing features such as computer-aided detection and classification, multisensor target fusion, embedded training and performance prediction, based on environmental data collected during past or current missions. MIL-STD-1553 databus protocol facilitates integration with other aircraft subsystems and components. BIT eases support and boosts availability. The new weapon-replaceable assemblies have a significantly higher MTBF than current systems.

Specifications

Weight: 265 kg total
Frequency: 9.23, 10.003, 10.774 kHz
Sound pressure level: 217 ±1 dB/µPa/yd (0.9 m)
Range scales: 0.9, 2.7, 4.6, 7.3, 11, 18, 29 km
Operating depth: 440 m
Modes:
(active) 3.5, 35 ms pulses, 0.2, 0.7, 1.6, 3.2, 4 s shaped; 0.625 s FMs (passive), (obstacle avoidance), (communicate) SSB at 8 kHz
Raise speed: 6.7 m/s average
Lower speed: 4.9 m/s
Water exit speed: 1.5 m/s

Operational status

The AQS-18A is in production and has been delivered to the Italian Navy.

Contractor

AlliedSignal Ocean Systems.

VERIFIED

Low Frequency Active Dipping Sonar (LFADS)

LFADS is similar in size and weight to the mid-frequency AN/AQS-18(V). LFADS has been installed on SH-3, EH 101 and SH-60 helicopters. The cable interface, sonar processor, sonar control unit, display, reeling machine and control devices are common between the 10 kHz AQS-18 and the 1.38 kHz LFADS systems.

The LFADS is capable of depths up to 440 m, and has figure-of-merit sufficient to achieve convergence zone detections in deep water, and transmission/ receive characteristics optimised for extremely long ranges in shallow water. The low-frequency capability designed into LFADS using proprietary transducer and beam-forming technology allows multiple boundary interactions and reduced reverberation contamination of the received signals. Use of high-resolution Doppler processing and shaped pulses achieves detection of targets even at speeds as low as 1 kt. Extended duration FM pulses are available to detect the near zero Doppler target as well.

Specifications

Operating depth: 440 m
Operating frequencies: 1.33-1.43 kHz
Source level: 220 dB/µPa/yd
Range scales: 1, 3, 5, 10, 15, 20, 30, 40, 60, 80, 100, 120 kYd; 0.9, 2.7, 4.6, 9, 14, 18, 27, 37, 55, 73, 91, 110 km
Operational modes: Active-CW to 10 s pulses (FM to 5 s pulses); passive; UQC; raytrace
Raise speed (avg): 4.3 m/s
Lower speed (max): 4.9 m/s
Water exit speed: 1.5 m/s
Seating speed: 0.6 m/s
System weight: (stand-alone)
(Proc. & Display) 59 kg
(R/M & Cable) 95 kg
(Xducer) 141 kg
(Total weight) 295 kg

Operational status

LFADS has been demonstrated in differing water conditions, including the Timor and Mediterranean Seas and Vestfjorden Fjord. Results of these tests indicate that LFADS can even outperform shipborne systems against diesel-electric submarines.

Contractor

AlliedSignal Ocean Systems.

VERIFIED

MPAvionics system

ARGOSystems' MPAvionics is an adaptable, readily customised MPA mission system that integrates operational control and display of information from the wide range of sensors, communications, and navigation equipment used in maritime patrol missions. In addition to MPA, the MPAvionics system can serve as the basis of an integrated SIGINT system, Airborne Early Warning (AEW) system, ground- or ship-based support station, or any other complex sensor information collection system with selection of the required sensors.

The heart of the MPAvionics system is the ruggedised Distributed Processing and Display System (DPDS). The DPDS is a streamlined mission command and control centre that processes and displays integrated subsystem data and controls the mission surveillance radar, IFF interrogator, InfraRed (IR) sensor, ESM, acoustic, navigation, GPS, communications, and datalink subsystems. It replaces dedicated sensor control and display units with integrated controls to minimize the number of operator actions required to control individual sensors. The DPDS provides a complete picture of the real-time tactical situation during MPA missions on the tactical plot, a map overlaid with radar, video, and contact information.

The DPDS records detailed aircraft, Built-In Test (BIT) status, and contact data for post-mission analysis; enables premission data preparation; and integrates with voice and datalinks to transmit critical information to supporting platforms or to a ground-based MPA operations centre. The DPDS also provides aircraft navigation references for mission sensors and displays.

The DPDS consists of one or more workstations (consoles), linked by an Ethernet Local Area Network (LAN). Consoles are modular in design and have an open and distributed VMEbus architecture that allows the user to specify a desired number of consoles and to allocate subsystem display and control among consoles to meet mission and geographic needs. This approach ensures that technological advances in both hardware and software can be readily integrated into the system as the state of the art evolves.

Each DPDS console comprises RISC-based display and input/output processors and multifunctional displays, a keyboard and trackball, and two electroluminescent Programmable Entry Panels or PEPs, displaying touchscreen buttons. The touchscreen buttons map to customisable menus and submenus, designed specifically to control the sensors and subsystems in the system and to control the tactical plot and manage content and status information.

The DPDS has 11 multiwindow displays. Up to four windows can be used for real-time video displays or playback of prerecorded video. The display format permits operators to simultaneously display multiple windows of the tactical situation plot, contact and status information, and radar, IR, and ESM videos. All windows can be moved, sized, opened, or closed as desired, enabling the operator to optimise the displays to meet the precise demands of any tactical situation. Display data can be recorded on a video recorder and added to data stored on the mission loader/recorder to provide a complete record of the mission which can be used as documentary evidence to establish legal liability for violations of territorial waters. All console displays are loaded from a master mission database maintained in the input/output processor. Removal of the mission loader/recorder enhances security for sensitive mission data and permits easy exchange of data files at a ground station.

A typical DPDS configuration allocates mission responsibilities to functional operators via consoles for a Tactical Controller (TACCO), one or more Sensor Operators (SENSOs), and an ASW operator. Each operator is assigned primary responsibility for the operation and control of certain sensor subsystems, aircraft systems, and other control functions, including tactics and weapons. Each console in the DPDS provides integrated graphical user interface controls for all subsystem control and display functions, and can be configured for individual operator responsibility or particular mission. All DPDS consoles are essentially identical in hardware and software, except for placement of manual equipment controls, which may vary by console.

The DPDS uses a real-time operating system, based on commercial off-the-shelf standard, and an Ada development tool set to provide a flexible mission application development environment and enable successful performance of time-critical operations, such as sensor interfacing and co-ordination. DPDS software can be executed on multiple consoles in the aircraft or on a ground-based console to provide simulation and debugging capabilities and permit development and testing of mission data on any operator or ground-based workstation. The modular structure allows for reuse of existing software when the system must be modified to address evolving geographic or mission needs.

A ground-based MPA Operations Centre supports operational and organisational activities by providing the means to classify, store, and disseminate data for mission analysis and maintenance of national database resources. ARGOSystems provides integrated logistics support, including software maintenance and development training programs.

In fielded MPAvionics systems, the DPDS has been interfaced with the following sensors and aircraft subsystems:

Litton Systems Canada AN/APS-504 (V)5 search radar;
Raytheon Systems Company AN/APS-134(LW) search radar;
Raytheon Systems Limited Model 3500 IFF interrogator;
GEC-Marconi Sensors TICM II MRT-S infrared detector;
FSI-FLIR-2000 infrared detector;
ARGOSystems AR-900 ESM system;
Litton LTN-92 Inertial Navigation System (INS);
Sextant Avionique AOC-31 Air Data Computer (ADC).

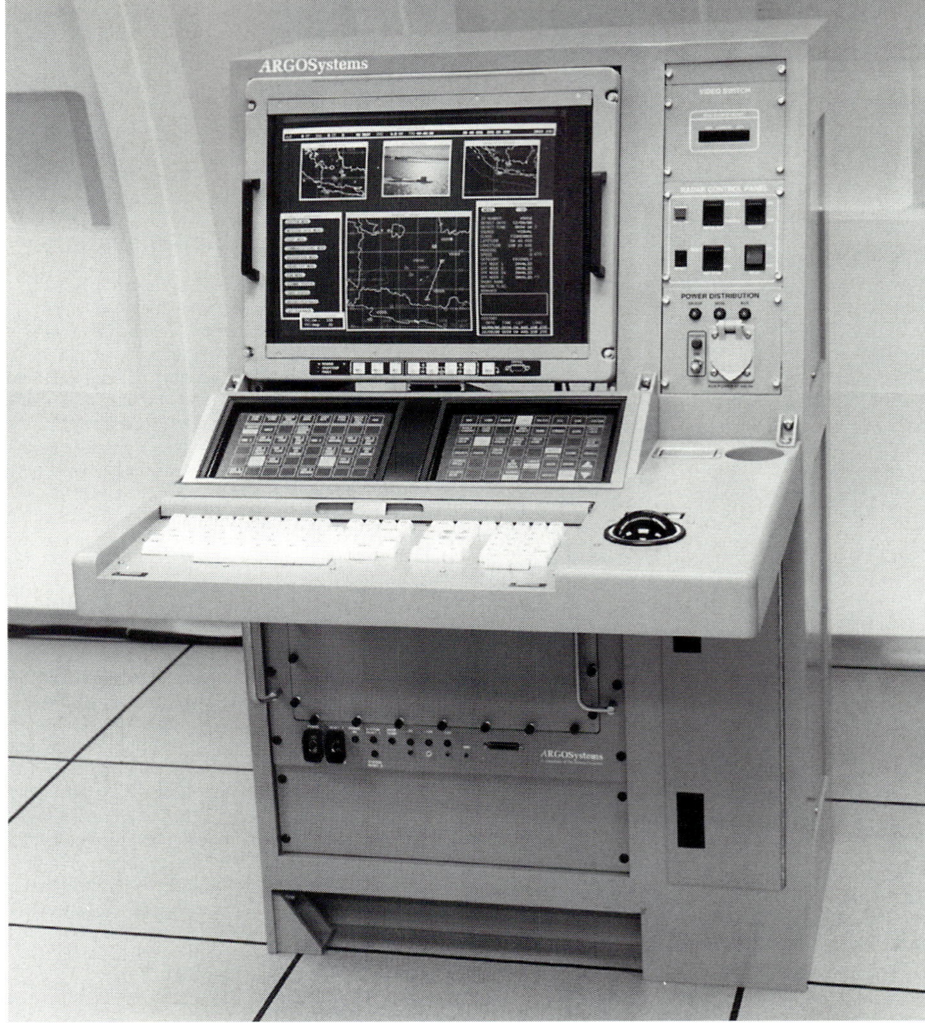

ARGOSystems Inc MPAvionics ***1997***/0001188

Operational status

The MPAvionics mission system is a derivative of deployed systems on the Boeing 737 Surveiller and other operational platforms. It is currently in service with the Indonesian Air Force in the 737 Surveiller MPA. The DPDS has been demonstrated worldwide on the IPTN CN-235 MPA under MPA mission conditions.

Contractor

ARGOSystems (a subsidiary of The Boeing Company).

VERIFIED

Tactical Data System (TDS)

The Rockwell Collins Tactical Data System (TDS) is installed on the Royal Australian Navy's Role Adaptable Weapons System (RAN/RAWS), the Sikorsky S-70B-2 Seahawk helicopter. With this system the Seahawk can perform ASW missions with three crew members, do all sensor processing on board the aircraft, locate, classify and prosecute targets autonomously and transfer target data automatically via digital datalink.

The TDS consists of a tactical display unit, a horizontal situation video display, a multifunction keyboard for communication navigation and identification management, an advanced mission adaptable computer, a data loader that simplifies the preflight mission preparation and a datalink to transfer target and mission data.

The S-70B-2 with the TDS has the capability to assist the crew in making tactical decisions autonomously and each crew member has access to virtually all tactical data. That implies greater situational awareness, so that the entire mission can be flown by three crewmen. System functions include navigation, flight plan management, datalink, tactical display and databus controller.

The tactical display unit is a multifunction display which provides an integrated pictorial representation of processed sensor information to the tactical co-ordinator and sensor operator. As it is raster driven, there is no limit to the amount of information that can be displayed. Two identical, yet independent, 8 × 8 in (203 × 203 mm) CRTs display standard naval combat data system symbology and various special display icons in the presentation of the tactical plot. Crew members use multifunction slew controllers to select symbols to provide access to the various display functions and to designate displayed data to the system. There are no complicated or extensive sets of buttons and switches to master and during a mission the crew can concentrate on the tactical display, not on managing a host of individual subsystems.

A powerful mission adaptive computer links all the aircraft sensors and displays via a MIL-STD-1553B multiplex databus management system. A very efficient functionally redundant integrated design places specific emphasis on maintaining a consistently high mission success rate through graceful degradation.

The Rockwell Collins Tactical Data System (TDS) is in service on the Royal Australian Navy's Sikorsky S-70B-2 Seahawk helicopters ***1995***

Operational status

In service on the Royal Australian Navy Sikorsky S-70B-2 Seahawk helicopter.

Contractor

Rockwell Collins.

UPDATED

AN/ARS-3 sonobuoy reference system

The AN/ARS-3 is a major part of the Update II improvements to the US Navy's P-3C ASW aircraft. Comprising 10 blade antennas and a receiver-converter, the system provides for the passive detection and location of sonobuoys transmitting on 31 standard channels. The system uses the P-3C's Univac CP-901 computer for processing and display. It uses the aircraft's inertial navigation system to determine the geographical position of the buoys. In conjunction with the CP-901, the AN/ARS-3 can locate a faulty module 95 per cent of the time. Unlike the AN/ARS-2, the system needs no cooling air.

Specifications

Dimensions: 279 × 546 × 311 mm
Weight:
(including antennas) 35.8 kg
(receiver-converter only) 24.5 kg
Power supply: 115 V AC, 150 W
Temperature range: −54 to +71°C
Cooling air: none
Vibration: MIL-T-5422E
Environmental: MIL-E-5400
Reliability: 1,000 h MTBF

Operational status

In service on P-3C Orions flown by the US Navy, Australia, Japan and Netherlands.

Contractor

Cubic Defense Systems Inc.

VERIFIED

AN/ARS-4 sonobuoy reference system

The AN/ARS-4 is a modified version of the AN/ARS-2 designed for the S-3B ASW aircraft. It is retrofitted into all S-3 aircraft in the US Navy inventory. The unit's updated electronics provide improved performance, reliability and maintainability, as well as the ability to locate sonobuoys operating on the 99 sonobuoy channels.

Specifications

Dimensions: 222 × 229 × 483 mm
Weight: (receiver-converter only) 15.9 kg
Power supply: 115 V AC, 150 W
Temperature range: −54 to +71°C
Cooling air: 0.24 kg/m
Vibration: MIL-T-5422E
Environmental: MIL-E-5400
Reliability: (SBR only) 2,862 h MTBF

US Navy S-3B Viking ASW aircraft are equipped with Cubic AN/ARS-4 sonobuoy reference systems

Operational status

In service in US Navy S-3 aircraft.

Contractor

Cubic Defense Systems Inc.

VERIFIED

AN/ARS-5 sonobuoy reference system

Designed for the P-3C Update III, Cubic's AN/ARS-5 is an improved version of the AN/ARS-3. It has 99-channel capability as well as advanced design and built-in expansibility for such additional capabilities as anti-jamming, communications intelligence and search and rescue. The AN/ARS-5 incorporates the latest advances in self-test antenna radiation and built-in test equipment; the key difference between the ARS-5 and the other systems in the series is that it incorporates an embedded computer, while the others rely on computing within the aircraft.

Specifications

Dimensions: 279 × 546 × 311 mm
Weight:
(including antennas) 35.8 kg
(receiver-converter only) 24.5 kg
Power supply: 115 V AC, 150 W
Temperature range: −54 to +71°C
Cooling air: none
Vibration: MIL-T-5422E
Environmental: MIL-E-5400
Reliability: 1,000 h MTBF

Operational status

In production for the US Navy's P-3C Update III programme.

Contractor

Cubic Defense Systems Inc.

VERIFIED

Stand-alone Sonobuoy Reference System (SRS)

Under an independent research and development programme, Cubic is developing a stand-alone Sonobuoy Reference System (SRS) that can allow virtually any fixed-wing aircraft or helicopter to have full SRS capabilities. This unit will incorporate such additional features as anti-jamming, DME, COMINT and search and rescue. Other features include self-calibration and built-in fault isolation for easier maintenance.

This stand-alone capability can be incorporated into either the AN/ARS-4 or the AN/ARS-5 chassis and can interface with any aircraft databus. Using its own embedded computer system, the system can operate in both active and passive modes with sonobuoys operating on up to 99 channels.

Performance specifications for the stand-alone SRS are identical to those for the AN/ARS-5 system.

Contractor

Cubic Defense Systems Inc.

VERIFIED

AN/AKT-22(V)4 telemetry data transmitting set

The AN/AKT-22(V)4 system relays up to eight channels of sonobuoy data from an ASW helicopter to a ship. It comprises the T-1220B transmitter-multiplexer, the C-8988A control indicator, an AS-3033 antenna and a TG-229 actuator. The sonobuoy signals are received by dual AN/ARR-75 radio receiving sets and passed into the transmitter-multiplexer. The control indicator has four trigger switches, each of which disables two data channels. Composite trigger tones are brought into the multiplexer separately via the control indicator and combined with the sonic data channels and the single voice channel. The resulting FM signal is used to modulate the transmitter.

The data transmitting set has a DIFAR operating mode in which the extra voice channel is inoperative and the composite FM modulating signal is disconnected from the transmitter input. Two DIFAR sonobuoy transmissions enter the transmitter-multiplexer on dedicated channels. After conditioning in an amplifier-adaptor, the DIFAR signal is split into two components: DIFAR A and DIFAR B. The A signal is conditioned in a low-pass filter, while the B signal drives a variable-cycle oscillator centred at 70 kHz, and then passes through a bandpass filter. The two filter outputs are combined linearly, and the resulting composite modulation signal drives the transmitter. A switch on the controller indicator controls whether the normal sonic or composite DIFAR signals are transmitted.

The AS-3033 antenna has two sections: a VHF element for receiving the sonobuoy signals and a UHF part which sends the multiplexed data down to the ship. The ship receives the information on an AN/SKR telemetric data receiving set.

Specifications

Power supply: 115 V AC, 3-phase, at 0.85 A, 1.5 A, and 1 A respectively
Warm-up time: <1 min in standby mode < 15 min under environmental extremes
Operating stability: > 100 h for continuous or intermittent operation
Frequency: 2,200-2,290 MHz, 1 of 20 switch-selectable E-L-band channels
Multiplexer inputs:
(a) 8 sonar data channels (7 with 10-2,000 Hz bandwidth; 1 with 10-2,800 Hz), at 0.16-16 V
(b) 4 sonar trigger channels, 26-38 kHz, at 1-3 V
(c) 1 voice channel, 300-2,000 Hz bandwidth, at 0-0.25 V
(d) 2 composite DIFAR channels, 10-2,000 Hz bandwidth at 3-6 V or 10.6 V
Channel phase correlation: difference in phase delay between any 2 passive data channels < 1° (10-500 Hz)

Operational status

In service with US Navy.

Contractor

Flightline Electronics Inc.

VERIFIED

AN/ARN-146 On-Top Position Indicator (OTPI)

The AN/ARN-146 radio receiving set is a 99-RF channel On-Top Position Indicator (OTPI) which is used on board ASW rotary- and fixed-wing aircraft to provide bearing and on-top position indication of deployed sonobuoys. When used in conjunction with a suitable ADF system, the AN/ARN-146 enables an operator to locate and verify the position of sonobuoys which are operating on any of 99 RF channels.

The AN/ARN-146 has a modular solid-state design for use with computer control, either via RS-422 directly or when connected with the AN/ARR-84 sonobuoy receiver. The AN/ARN-146 is form and fit interchangeable with the R-1651/ARA and R-1047 A/A OTPI receivers. It is compatible with ARA-25, ARA-50 and OA-8697/ARD ADF antenna systems.

The AN/ARN-146 set consists of the R-2330/ARN-146 receiver and the C-11699/ARN-146 radio set control. The R-2330/ARN-146 is a VHF AM receiver which receives signals from sonobuoys operating on any of 99 RF channels in the frequency range from 136 to 174 MHz. This receiver houses all the electronics, including internal transfer circuits for the switching of RF, baseband, power and phase compensation circuits for sharing the DF system between the OTPI and an associated UHF receiver system for the ADF. The R-2330/ARN-146 also contains a PLL synthesised local oscillator, digital address decoder, voltage tuned BP filters, AGC and BIT circuitry.

The C-11699/ARN-146 radio set control is an optional manual control box which is used in place of the RS-422 bus or the MIL-STD-1553B dual bus of the AN/ARR-84. This control box provides the capability to apply power to the AN/ARN-146 system to select any one of 99 RF channels, to activate the BIT circuitry and to display adequate signal strength and results of the BIT via the adequate signal strength indicator.

Specifications

Dimensions:
(radio receiver) 125.8 × 134.6 × 77.47 mm
(radio set control) 57.15 × 127 × 146 mm
(mounting plate) 15.2 × 144.8 × 77.47 mm
Weight:
(radio receiver) 2.15 kg
(radio set control) 0.57 kg
(mounting plate) 0.27 kg
Power supply: 115 V AC, 400 Hz, single phase, 15 VA
28 V DC, 20 W
Frequency: 136-173.5 MHz (99 channels)
Sonobuoy compatibility: DIFAR, LOFAR, Ranger, BT, CASS, DICASS, VLAD, CAMBS, Barra, ERAPS, HLA, ATAC, SAR
Reliability: 3,000 h MTBF

Operational status

In production for aircraft of the US Navy.

Contractor

Flightline Electronics Inc.

VERIFIED

AN/ARR-72 sonobuoy receiver system

The AN/ARR-72 sonobuoy receiver system is used on the P-3C patrol aircraft, in conjunction with acoustic signal processors and a digital computer. The AN/ARR-72 system receives, amplifies and demodulates FM signals transmitted by deployed sonobuoys in the 162.25 to 173.5 MHz VHF band. The AN/ARR-72 is compatible with the AN/SSQ-36, 41, 50, 53 and 62 sonobuoys. The receiver is in five parts: AM-4966 preamplifier, CH-169 receiver, SA-1065 audio assembly, C-7617 control indicator and SG-791 Acoustic Sensor Signal Generator (ASSG) which performs diagnostic functions. The AN/ARR-72 system is compatible with LOFAR, CODAR, BT, RO, CASS and DICASS equipment.

The AN/ARR-72 is a dual-conversion superheterodyne VHF receiver system. Radio frequency signals are received at the dual-aircraft VHF blade antennas, amplified by the system's AM-4966 dual RF amplifiers and passed on to the CH-169 31-channel receiver assembly. Within this assembly, a multicoupler distributes the preamplified RF to the 31 fixed tuned receivers, where it is further amplified and demodulated to provide baseband audio and RF level signals.

US Navy SH-60 LAMPS Mk III helicopters have Flightline Electronics AN/ARR-72 sonobuoy receivers

Each of the 31 receiver channels contains a channelised first converter, an IF filter, a second converter and a discriminator/amplifier. The first converter contains a crystal-controlled local oscillator and mixed/IF circuit. The plug-in discriminator/amplifier provides RF level and FM signal detection. The optional phaselock discriminator is directly interchangeable with the discriminator/amplifier module. The first converter, second converter and discriminator/amplifier assemblies are plug-in units and, excepting the local oscillator crystals, are identical for the 31 channels.

The SA-1605 audio assembly includes 19 audio switching and amplifier cards and two audio power supply regulators. The audio assembly accepts the baseband and RF level outputs of the 31 receivers and outputs them to the computer and processing equipment. Each of the 19 audio channels contains a 31 by 1 switching matrix to select the output of a given receiver. This receiver signal is then amplified and provided through two individually buffered outputs to the processing equipment. Selection of a particular receiver channel may be accomplished by the digital computer or the C-7617 dual-channel control indicator. RF level for the selected receiver is displayed on the corresponding control-indicator meter.

The SG-791 ASSG is the BIT for the receiver system. RF test signals are generated for application to the preamplifier, the multicoupler or to a radiating antenna. Internal circuits generate simulated signals for the testing of normal LOFAR, extended LOFAR, range only and BT processing equipment. External modulation inputs are provided to accept modulation from devices such as the target generator in the demultiplexer of the AN/AQA-7(V) DIFAR equipment, thus allowing end-to-end checks of sophisticated ASW equipment.

The ASSG contains a redesigned multifrequency oscillator to provide for generation of the 31 individual RF frequencies. This replaces a previously utilised crystal turret. This synthesiser reduces the channelling time from 10 seconds to less than a second.

Specifications

Weight:
(preamplifier) 0.9 kg
(receiver) 26.53 kg
(audio) 17.69 kg
(ASSG) 8.71 kg
(8 control indicators) 14.51 kg
Total weight: 68.36 kg

Power supply: 115 V AC, 400 Hz, single phase, 300 W
28 V DC, 280 mA for panel lighting
18 V DC, 250 mA for ASSG annunciators
Frequency: (31 channels) 162.25-173.5 MHz at 0.375 MHz spacing
Noise figure: 5 dB max (3.5 dB is typical)
IF rejection: 66 dB min (>100 dB typical)
Image rejection: 66 dB min (>100 dB typical)
High audio level: 16 VRMS at ±75 kHz deviation
Standard audio level: 2 VRMS at ±75 kHz deviation
Crosstalk: >54 dB min
Output isolation: 60 dB min
Audio frequency response: ±1 dB from −20 Hz to 20 kHz; ±6 dB from 5 Hz to 40 kHz
Audio distortion: −20 Hz to 5 Hz <3% 5-18 kHz <5%
Controlled audio phase characteristics: −100 Hz to 3 kHz ±5°
Audio noise: 10-300 Hz 2 mV
301-2,400 Hz 3 mV
Operational stability: 500 h
Operating life: 20,000 h
Reliability: 500 h MTBF including ASSG and 10 dual-channel control indicators

Operational status

In production and in service with P-3C aircraft and Sikorsky SH-60 LAMPS Mk III helicopters.

Contractor

Flightline Electronics Inc.

VERIFIED

AN/ARR-75 sonobuoy receiving set

The AN/ARR-75 sonobuoy receiving set is a 31-RF channel FM receiver designed for ASW fixed-wing aircraft and shipboard applications. Independent receiver modules provide four simultaneous demodulated audio outputs, each capable of selecting one of 31 RF input channels.

The AN/ARR-75 receiving set is composed of two units. The first of these units is the OR-75/ARR-75 receiver group assembly, which consists of the PP-6551/ARR-75 power supply, four R-1717/ARR-75 receiver modules and the CH-670/ARR-75 chassis. The receiver group assembly contains the majority of the system's electronics.

The second receiving set unit is the C-8658/ARR-75 or C-10429/ARR-75 radio set control. This control unit

Modules of the Flightline Electronics AN/ARR-72 sonobuoy receiver (from left): preamplifier, acoustic sensor signal generator, receiver, audio assembly and control indicator

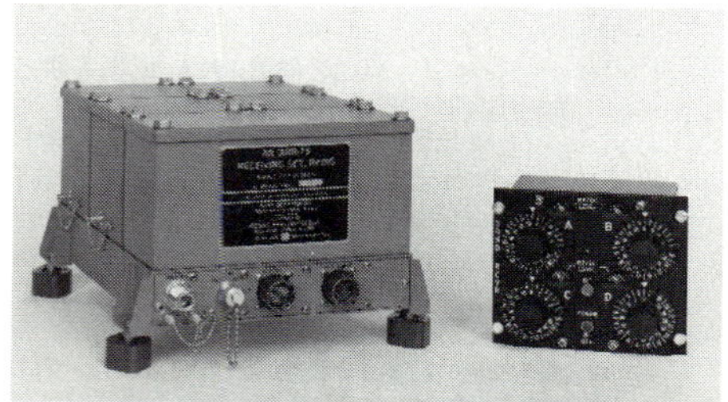

The Flightline Electronics AN/ARR-75 sonobuoy receiver

provides independent RF channel selection and signal strength monitoring of any one of the 31 RF channels for each of the four receiver modules.

Specifications

Dimensions:
(receiver) 203 × 186 × 305 mm
(controller) 122 × 145 × 71 mm
Weight:
(receiver) 9.8 kg
(controller) 1.1 kg
Power supply: 115 V AC, 400 Hz, 3 phase, 75 VA max 27 V(C-8658) or 5 V(C-10429), 5.6 W for lighting at either 27 V (C-8658) or 5 V (C-10429)
Frequency: 162.25-173.5 MHz (31 channels)
Noise figure: 5 dB at 50 ohms
Specifications: MIL-STD-461, 462, 463, 781, MIL-E-5400, MIL-R-81681, AR-5, 8, 10, 34
Operating life: 20,000 h
Reliability: 1,500 h MTBF

Operational status

In production and in service on the LAMPS Mk I, LAMPS Mk III and SH-3H helicopters.

Contractor

Flightline Electronics Inc.

VERIFIED

AN/ARR-84 sonobuoy receiver

The AN/ARR-84 is designed to receive signals from all current and planned future US and allied sonobuoys. This receiver has four acoustic channels capable of receiving signals from up to four deployed sonobuoys simultaneously on any of 99 RF channels. The AN/ARR-84 is installed on a variety of ASW platforms, including rotary- and fixed-wing aircraft as well as large surface combatants and fast patrol craft. This receiver is a form and fit replacement for its predecessor, the AN/ARR-75 31-channel sonobuoy receiver.

The AN/ARR-84's major assemblies include a power supply, four identical receiver modules and the electrical equipment chassis. The power supply contains the input/output assembly for MIL-STD-1553B and the FSK printed circuit board, allowing digital sonobuoy data extraction.

Additional features of the AN/ARR-84 include dual-selectable IF bandwidths for optimum performance with a wide variety of sonobuoy types. Outstanding AM rejection and mechanical vibration immunity severely reduce spurious returns due to propeller/rotor multipath and platform vibration. Low susceptibility to conducted and radiated energy prevents desensitisation and allows the AN/ARR-84 to be utilised near strong onboard emitters and shipboard search radars.

Flightline also provides an optional radio set control, which is intended for use when an online computer is not available and which provides power control and BIT operation for the receiver group. RF channel selection, sonobuoy type selection and RF level readout is provided independently for each of the four receivers in the group. Operator input is effected through a multifunction keypad, with signal strength provided as a histogram display and entry readback/failure data provided by a 16-character message display.

The Flightline Electronics AN/ARR-84 99-RF channel sonobuoy receiver shown with the AN/ARN-146 OTPI (on the right)

Specifications

Dimensions: 190.5 × 381 × 254 mm
Weight:
(receiver) 11.34 kg
(optional control box) 1.81 kg
Power supply: 115 V AC, 400 Hz, 3 phase, 100 VA
Frequency: 136-173.5 Hz (99 channels)
Sonobuoy compatibility: DIFAR, LOFAR, Ranger, BT, DICASS, VLAD, CAMBS, Barra, ERAPS, HLA, ATAC, SAR
Environmental: MIL-E-5400 Class 1B
Reliability: 1,500 h MTBF

Operational status

In production and in service with aircraft of the US Navy.

Contractor

Flightline Electronics Inc.

VERIFIED

AN/ARR-502 sonobuoy receiver

The AN/ARR-502 sonobuoy receiver is intended for a wide variety of maritime patrol aircraft and ASW helicopters. The standard configuration system provides for 16 acoustic receiver channels and an OTPI function in a lightweight package. However, the system's modularity allows it to be easily reconfigured to any specific application. The receiver module is a plug-in assembly containing advanced circuitry and software selectable tuning flexibility, allowing for reception of thousands of RF channels down to 5 kHz channel spacing with the current standard being 99. This hybridised receiver module weighs only 0.36 kg and also contains AM demodulation capability.

The AN/ARR-502 features extremely low acoustic receiver baseband noise floor, active mixer technology for unsurpassed third order intermodulation and spurious response performance, threshold extending FM detector with programmable characteristics for sonobuoy type optimisation, progressive AGC with programmable time constants to improve multipath, propeller artefact and countermeasure rejection. It also has latest-generation surface acoustic wave IF filtering for improved phase and delay response to provide the best bearing accuracy for multiplexed buoys and best data error rate for digital buoys, high-resolution self-calibrating signal strength indication and high-resolution local oscillator tuning which can be used to optimise performance with off-tuned sonobuoys and for operating in a countermeasure environment.

The system is capable of providing acoustic, On-Top Position Indicator (OTPI) and SRS receivers in the same package and has optional dual-antenna input with each receiver capable of independently using either antenna through autonomous or external control. The OTPI receiver is compatible with AN/ARA-25, AN/ARA-48/50 or DF-301E antenna systems. Independent or simultaneous operation from MIL-STD-1553B databus or manual control units is possible, as is operation with either internal or antenna collocated preamplifiers. Variable receiver configurations allow 16, 20 or 40 acoustic channel packaging options in a single package.

Reliability and maintainability features are fully solid state with no mechanical relays and self-test hardware and software perform complete sensitivity and signal-to-noise ratio tests on receiver modules. All self-test failure data is logged in non-volatile memory and no special test equipment is needed. The system has full ATE compatibility, with test connectors on modules. Receiver module design is common to any type of analogue, digital, OTPI, scanning, aural monitoring or SRS receiver.

Specifications

Dimensions: 406.4 × 304.8 × 203.2 mm
Weight: 21.77 kg
Reliability: >1,000 h MTBF

Contractor

Flightline Electronics Inc.

VERIFIED

R-1651/ARA On-Top Position Indicator (OTPI)

The solid-state modular construction R-1651 radio receiving set is a 31-channel On-Top Position Indicator (OTPI) which is being used on board ASW rotary- and fixed-wing aircraft to provide bearing information and on-top position indication of deployed sonobuoys. This OTPI, when used in conjunction with a suitable ADF system, enables the operator to locate and verify the position of deployed sonobuoys operating on any of 31 VHF RF channels.

The R-1651/ARA is form and fit interchangeable with the R-1047 A/A OTPI. It is compatible with the ARA-25 and ARA-50 antennas, and it is also available in a modified configuration which is compatible with the DF-3010E antenna system. The R-1651/ARA is controlled by the optional C-3840/A control box.

The R-1651/ARA is a single-conversion, 31-channel superheterodyne VHF receiver which is designed to receive signals from deployed sonobuoys over the frequency range of 162.25 to 173.5 MHz.

The antenna input is switched by a relay to the receiver or the UHF receiver, on external command. The RF signals pass through bandpass filters, are amplified and converted to an IF of 25 MHz. After IF amplification and detection, the resultant audio signal is amplified to the level required by the ADF system. The AGC level is developed at the detector, amplified and used for gain control of the IF and RF stages and to operate the adequate signal strength circuits.

The local oscillator provides one of 31 possible LO signals. The same number of crystals is appropriately selected using external controlled code lines containing a 6-bit binary code. Line receivers, decoding gates and matrix drivers connect one of the 31 crystals to the local oscillator circuits.

The optional C-3840/A radio receiver control is used to switch the antenna system, by energising an external coaxial relay, from the UHF/ADF mode of operation to the sonobuoy OTPI mode of operation to select any one of the 31 RF channels and to indicate adequate signal strength when the R-1651/ARA receives an RF signal of −86 dBm or greater.

Specifications

Dimensions: 76.2 × 124.46 × 134.6 mm
Weight: 1.59 kg
Power supply: 115 V AC, single phase, 15 VA
28 V DC, 20 W
Frequency: 162.25-173.5 MHz (31 channels)
Altitude: up to 35,000 ft
Temperature range: −54 to +55°C
Operating life: 5,000 h
Reliability: 1,000 h MTBF

Operational status

In production and in service in the US Navy P-3C Updates I and III and the Sikorsky SH-60 LAMPS Mk III.

Contractor

Flightline Electronics Inc.

VERIFIED

AN/ARR-78(V) Advanced Sonobuoy Communication Link (ASCL)

ASCL is the US Navy standard sonobuoy receiver and is deployed on the Orion P-3C Update III and Viking S-3B ASW aircraft. The P-3C Update III employs two ASCLs in a 32-acoustic-channel expansion configuration (CHEX). ASCL operates on both types of aircraft in conjunction with standard VHF antennas. The following units comprise a typical ASCL configuration.

RF Preamplifier

The AM-6875 is an optimised high-performance, low-noise VHF preamplifier that provides amplification and prefiltering of received RF signals.

Radio Receiver

The R-2033 (P-3C, Update III) or the R-2066 (S-3B) contains 20 fully synthesized receiver modules (16 acoustic and 4 auxiliary) and one each of the

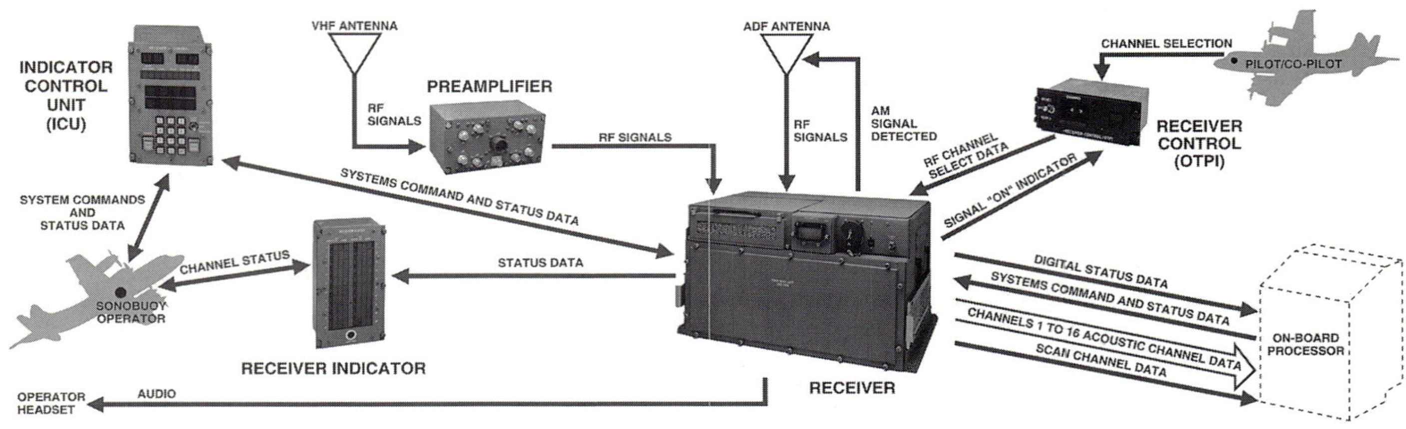

Typical ASCL system ***1998***/0011865

following modules: RF/ADF amplifier multicoupler, reference oscillator, I/O Proteus digital channel (P-3C), I/O Manchester digital channel (S-3B), I/O processor, clock generator, BITE and DC power supply module. Each single-conversion module includes mixer conversion, frequency synthesised local oscillator, demodulator, and output interface circuits.

Each of the acoustic receiver modules processes FM/analogue signals at any of the 99 channels in the extended VHF band. Each of the four auxiliary receiver modules processes FM/analogue signals at any of the 99 channels and provides one channel for selection and processing of the On-Top Position Indicator (OTPI) signals, two channels for the operator to monitor acoustic information, and one channel to monitor the RF signal level in any of the RF channels. Common receiver modules are interchangeable.

BITE circuits provides comprehensive end-to-end evaluation of each receiver from the VHF preamplifiers to the receiver output interface circuits. BITE is initiated automatically by the computer (such as Proteus) and/or by the operator through the Indicator Control Unit (ICU). Performance status is displayed on the ICU and routed to the computer.

Indicator Control Unit (ICU)
The C-10126 ICU provides the operator with a means for manual control of each receiver channel frequency assignment, receiving mode, and self-test. It also displays status of the receiving set and operator entry information.

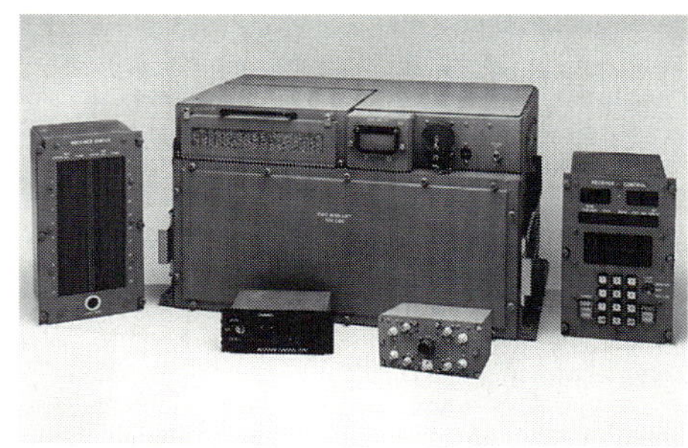

AN/ARR-78(V) Advanced Sonobuoy Communication Link (ASCL) ***1998***/0011864

Receiver Indicator (Receiver Status)
The ID-2086 Indicator continuously displays the control mode setting, the RF channel number, and the received signal level for each receiver.

Receiver Control (OTPI)
The C-10127 Receiver Control Unit provides the operator with control over the OTPI.

Specifications

Frequency: extended VHF
Receiver: 20 (16 acoustic/4 auxiliary)
Channels: 99 per receiver
Audio Output:
(analog) 2 V rms balanced
(monitor) 50 mW, 300 ohms
Power: 115 V AC±10%, 380 to 440 Hz, 3-phase, 500 W; 18 to 32 V DC, 7 W; 26.5 V AC±10%, 400 Hz, 50 W

Contractor

GEC-Marconi Hazeltine Corporation.

NEW ENTRY

Magic Lantern for SH-2G helicopter

Magic Lantern uses a blue-green laser and camera array to scan the water below the helicopter from surface level down to keel depth. The system correlates multiple scans to identify mines. It is able to sweep the entire 'upper column' at and below the ocean surface, and is cleared for day and night operations. Accurate navigation data from the GPS enables the SH-2G crew with Magic Lantern to locate mines precisely.

The podded Magic Lantern system replaces the Magnetic Anomaly Detector (MAD) on a strengthened hardpoint on the right side of the helicopter. The Magic Lantern system display in the aft cabin of the SH-2G shows the Sensor Operator (SENSO) mine detection symbology and real-time video imagery of suspected mine contacts. The ASN-150 Tactical Navigation system (TACNAV) provides signals to the HSI in the cockpit enabling the pilots to fly predetermined search patterns. A Tactical Decision Aid (TDA) has been developed to aid aircrews in determining search patterns and analysing post-mission data. A miniaturised airborne GPS is part of the SH-2G modifications to provide EOD crews with precise position data.

The SH-2G Super Seasprite advanced glass cockpit developed by Kaman Aerospace and Litton Guidance & Control Systems ***1997***/0002197

Operational status

Two systems are in service with the US Navy Reserve (HSL-94), a third was to be added before the end of 1997. Kaman Aerospace is contracted to modify five more helicopters for the US Navy Reserve.

Contractor

Kaman Aerospace Corporation.

VERIFIED

SH-2G Super Seasprite Integrated Tactical Avionics System (ITAS)

Working with Litton Guidance & Control Systems, Kaman has created an advanced glass cockpit which contains a highly automated Integrated Tactical Avionics System (ITAS), enabling a crew of two to fly the aircraft and manage its multimission equipment suite. The ITAS is a low-risk avionics system designed specifically to meet the requirements of the Royal Australian Navy.

The Kaman-Litton ITAS, driven by two mission data processors, integrates the input of radar, thermal imager and electronic protection measures for manageable cockpit presentations. It enables the Super Seasprite crew to attack targets with a variety of anti-ship missiles. Electronic Flight Instrumentation

System (EFIS), engine and transmission data, tactical plots, and sensor imagery are posted on any of the SH-2G's four colour multifunction displays.

The new glass cockpit retains the under-glare shield caution/advisory panels introduced originally on the SH-2G but eliminates all electromechanical instruments except for back-up airspeed and altitude gauges and a standby compass.

Operational status

Selected as 'preferred tenderer' by Royal Australian Navy for 11 helicopter requirement.

Contractors

Kaman Aerospace Corporation.
Litton Guidance & Control Systems.

VERIFIED

AN/UYS-1 signal processor

The UYS-1 is a fully programmable, high-performance signal processor capable of processing a wide variety of sensor data from analogue or digital inputs. Programmable software and modular hardware features have enabled the UYS-1 to satisfy processing requirements for 16 different US weapon systems.

A total of 160 S-3A Viking aircraft were modified, beginning in 1987, as part of the programme that denotes the S-3B standard. The UYS-1 also equips the P-3C Update III standard aircraft, deliveries of which to the US Navy began in 1984: this standard is now in fleet use and it is planned that 167 Orions will be modified. The same processor has also been installed in the US Navy's SH-60B Lamps III helicopters, with a total planned requirement of 204.

Specifications

Dimensions: 1,455 × 594 × 284 mm
Weight: 109 kg

Operational status

In production since 1981.

Contractor

Lockheed Martin Federal Systems.

VERIFIED

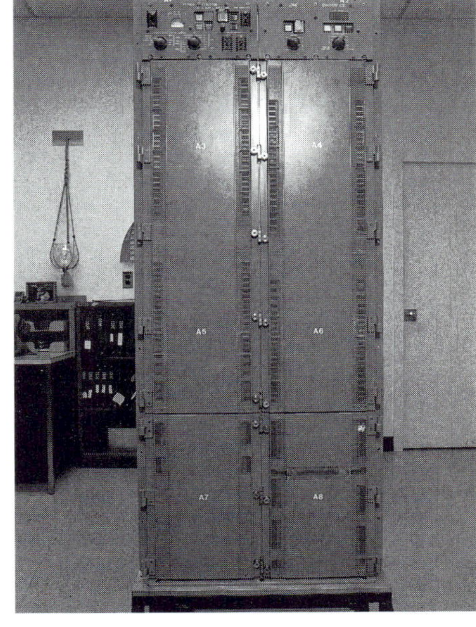

The Lockheed Martin Federal Systems AN/UYS-1 signal processor

AN/AYA-8C data processing system

Lockheed Martin has been producing the data processing system for the US Navy P-3C Orion anti-submarine warfare aircraft since 1968. It provides the interface between the CP-901 central computer and the aircraft systems and so constitutes a major part of the P-3C's mission avionics.

The system is connected to all the crew stations on the aircraft: tactical co-ordinator, non-acoustic sensor, navigation/communications, acoustic sensor and flight deck. In addition, it communicates directly with the radar interface unit, armament/ordnance system, navigation systems, sonar receiver, magnetic anomaly detector, infrared detection system, electronic support measures, sonobuoy reference system and Omega. A few systems, such as a digital magnetic tape system and the datalink, go directly into the CP-901.

The P-3C's data processing activities are divided into four logic units, forming separate boxes in which all electronic operations are conducted in a combination of digital and analogue formats. Various keysets and panels complete the system hardware.

Logic Unit 1 interfaces between the central computer and four types of peripheral information system: manual data entry, system status, sonobuoy receiver and auxiliary readout display.

The manual entry subsystem provides the communication between the various operator stations and the central computer. Each operator has a panel of illuminated switches and indicators by which he communicates with the central computer. System status for the navigation and magnetic anomaly detector is received and stored by part of Logic Unit 1, then transmitted to the central computer. The computer receives the status information on demand, or when any status changes. Finally, the auxiliary readout display logic interfaces between the central computer and the auxiliary displays at the tactical co-ordinator and nav/com stations. The radar, sonar antenna, infrared detection system and electronic support measures interfaces are also achieved by Logic Unit 1.

Logic Unit 2 is concerned with communicating between the central computer and the navigation and armament/ordnance systems. This logic unit transmits Doppler and inertial navigation data to the computer, and instructions to launch search-and-kill stores.

Logic Unit 3 controls the CRT displays provided for the tactical co-ordinator, sensor operators and pilot. The tactical co-ordinator and pilot displays can generate characters, vectors and conics, while the sensor displays use characters and vectors only.

Logic Unit 4 is mainly an expansion unit comprising two items: the Data Multiplexer Sub-unit (DMS) and the Drum Auxiliary Memory Sub-unit (DAMS). These provide extra input/output capacity and memory capacity respectively.

The DMS can service four input and output peripherals, as selected by the central computer. One output channel presents characters, vectors and conics for the auxiliary sensor display, and one input/output channel is used for the aircraft's Omega navigation system, a command launch system for the Harpoon anti-ship missile and sonobuoy reference system.

The DAMS was incorporated to give an additional 393,216 words of memory to the computer, so that the operational program could be expanded to accommodate extra functions and equipment.

Various keysets and control panels allow access to the central computer via the data processing system. A universal keyset allows the transfer of information between the computer and the nav/com operator. The pilot uses his own keyset for controlling the information presented on his CRT display for entering navigation stabilisation data, dropping weapons and flares and entering information on visual contacts. The ordnance panel displays the commands which the computer has given to the ordnance operator concerning status and position of the search stores, such as sonobuoys, which are available for deployment. Finally, there is an armament/ordnance test panel which monitors the output from the data processing system Logic Unit 2 to those systems.

Operational status

In production for P-3C Orion.

Contractor

Lockheed Martin Ocean, Radar and Sensor Systems.

VERIFIED

HIgh Performance Active Sonar (HIPAS)

Lockheed Martin is developing the HIgh Performance Active Sonar (HIPAS) to meet the US Navy's Airborne Low Frequency Sonar (ALFS) requirements. HIPAS is a heliborne dipping system that provides both active and passive detection and tracking of submarines. Designed for installation on the SH-60F carrier-based inner zone and SH-60B LAMPS III helicopters, as well as the HV-22 Osprey tilt-rotor aircraft, HIPAS fits in the same space, weight and power envelope as present sonar systems.

Operational status

Under development.

Contractor

Lockheed Martin Ocean, Radar and Sensor Systems.

UPDATED

AN/ASQ-212 mission processing system

The AN/ASQ-212 system consists of the CP-2044 computer and several interconnection devices which comprise a form, fit and function replacement for the AN/ASQ-114 computer, data analysis logic units and the signal data converter. The extended memory upgrade of the AN/ASQ-114 computer is transferred to the CP-2044 and used both for global and secondary memory. The CP-2044 incorporates Motorola 680030 processors to provide a throughput ranging from 10 to 25 Mips, which is 30 times greater than the current system in the P-3C Update I/III aircraft at a fraction of the current size, weight and power requirements. In the full Update III Ada implementation, less than 50 per cent of the CP-2044 minimum throughput and memory capacity is utilised.

The CP-2044 VME bus open architecture can be configured with additional processing, memory and input/output modules to meet the requirements of new subsystems such as GPS and Satcom, and of processing intensive functions such as sensor post-processing and data fusion.

Initially designed for retrofit into P-3C Update I/III aircraft, the AN/ASQ-212 can be easily tailored to the requirements of other P-3C configurations as well as new aircraft.

Operational status
The AN/ASQ-212 system has been developed for the US Navy under a two-phase programme that began in September 1989. The first production systems were installed in Navy test aircraft and training facilities beginning in May 1993, at a rate of four systems per month. Plans called for 150 operational systems to be installed beginning in October 1993. In addition, two foreign P-3C operators took delivery of the AN/ASQ-212 beginning in 1993 and several other countries are also considering use of the system.

Contractor
Lockheed Martin Tactical Defense Systems.

VERIFIED

P-3C upgrade programmes

US Navy P-3C Anti-surface warfare Improvement Program (AIP)

The AIP effort involves system upgrades to the P-3C Update III aircraft to improve its capability to support a range of naval operations including: ASuW; over-the-horizon targeting; and command, control, communications and intelligence activities.

Major upgrades include non-acoustic sensor enhancements (radar, ESM, IR detection system); survivability improvements; data management system upgrades; and a communications upgrade (ICS, VHF, tactical communications, UHF, SATCOM).

Equipment to be fitted in the programme includes: the AN/APS-137B(V)5; the AN/AVX-1 Cluster Ranger electro-optical sensor and a missile approach warner. Provision for Link 16 to be added is a preplanned product improvement.

Operational status
The exact funding status of the programme is uncertain, but it could be applied to any number required of the 146 aircraft that are already modified to Update III standard. It is understood that the P3-AIP fleet update is required by 2005, and that it is expected to be in use until the new Large Land-Based Aircraft (LLBA) becomes available between, approximately, the years 2015 and 2020.

Contractors
Lockheed Martin Tactical Defense Systems.

Royal Australian Air Force (RAAF) P-3C Upgrade Program

See International section for details of this programme.

Royal Norwegian Air Force (RNoAF) P-3C Upgrade Improvement Program (UIP)

The RNoAF P-3C UIP will upgrade four aircraft with a variant of the US Navy's AIP tailored for Norwegian requirements; upgrades include improvements to: non-acoustic sensors (IRDS, new ESM, APS-137 ISAR/SAR radar); SATCOM and new ICS; new DMS.

Operational status
Modification of the four aircraft was planned to begin during 1997, with final aircraft delivery in 1999.

Contractor
Lockheed Martin Tactical Defense Systems.

UPDATED

AN/AQS-14 sonar

The AN/AQS-14 sonar equips the US Navy Sikorsky MH-53E helicopter; first deliveries took place in 1984 and the system has been in service since June 1986. To date 32 systems have been delivered. Northrop Grumman has delivered eight additional AN/AQS-14 Mod 2 systems with a new airborne electronics console which has greatly improved system performance. The AQS-14 saw extensive use in the Red Sea during 1984 and in the Gulf during 1987/88 and 1990/91 during Operations Desert Shield and Desert Storm.

Used for minehunting duties, towed from helicopters, hovercraft or small surface vessels, the AQS-14 is a side-looking multibeam sonar with electronic beam-forming, all-range focusing and an adaptive processor. The underwater vehicle is 3 m long and has an active control system which allows it to be towed at a selected distance above the seabed or under the surface. The vehicle is controlled by an operator in the helicopter, and it is connected by a non-magnetic cable. The operator, assisted by computer-ordered detection and classification, has a real-time sonar display on which he can mark targets of interest. A tape recorder allows the recording, classification, position logging and review of data concerning mines and similar objects.

Operational status
In production and in service with the US Navy on MH-53E ASW helicopters and multimission air cushion vehicles.

Contractor
Northrop Grumman Corporation, Electronic Sensors and Systems Division.

VERIFIED

AN/AQA-7 Sonobuoy Processor

The AN/AQA-7 is the standard processor aboard the US Navy's P-3A/B and C (Update II) ASW aircraft. The system works in conjunction with the AN/SSQ-53B DIFAR sonobuoy, the SSQ-62B DICASS and SSQ-77A VLAD buoys. A new development programme, Improved Processing And Display System (IPADS), will significantly enhance the capabilities of the sensor station operators and the tactical co-ordinators on P-3A and P-3B aircraft.

Operational status
In service in the P-3. The system is in service with five foreign navies and some 1,325 AN/AQA-7s have been delivered.

The Royal Netherlands Navy P-3C Orion uses the AN/ASQ-81(V)1

Contractor
Raytheon Systems Company.

UPDATED

AN/ASQ-81(V) Magnetic Anomaly Detection system (MAD)

The AN/ASQ-81(V) Magnetic Anomaly Detection (MAD) system was developed for US Navy use in the detection of submarines from an airborne platform. The system operates on the atomic properties of optically pumped metastable helium atoms to detect variations of intensity in the local magnetic field. The Larmor frequency of the sensing elements is converted to an analogue voltage which is processed by bandpass filters before it is displayed to the operator.

Four configurations of the AN/ASQ-81(V) are available: two for use within an airframe and two for towing behind an aircraft. The US Navy uses the AN/ASQ-81(V)1 in the land-based P-3C Orion, where it is housed in a tail sting. The AN/ASQ-81(V)3 is installed in the carrier-based S-3 Viking aircraft, where it is extended on a boom.

The AN/ASQ-81(V)2 is a towed version employed by the US Navy on Sikorsky SH-3H and Kaman SH-2D helicopters. It is also in service with other countries, including the Netherlands for use on the Westland Lynx, Japan for use on the Mitsubishi HSS-2, and with forces employing the Hughes 500D helicopter.

The second towed version, the AN/ASQ-81(V)4, is used by the US Navy on the Sikorsky SH-60B LAMPS III helicopter.

All versions of the AN/ASQ-81(V) have the same C-6983 detecting set control, AM-4535 amplifier and power supply unit. The AN/ASQ-81(V)1 and 3 use a DT-323 magnetic detector, while the AN/ASQ-81(V)2 and 4 have a TB-623 magnetic detecting towed body. The towed version is controlled by the C-6984 reel control, which works the RL-305 magnetic detector launching and reeling machine.

Operational status
In production and in service.

Contractor
Raytheon Systems Company.

UPDATED

AN/ASQ-208 magnetic anomaly detection system

The AN/ASQ-208 is a derivative of the AN/ASQ-81 magnetometer that operates on the atomic properties of optically pumped metastable helium atoms to detect variations in total magnetic field intensity. The AN/ASQ-208 is a digital system which incorporates microprocessor technology to achieve aircraft compensation, multiple channel filtered display and threshold processing. Two system configurations are available: one for inboard installations and one for towed installations.

The system features a high-sensitivity helium sensor and choice of inboard or towed configuration. Control implementation is by MIL-STD-1553B or Manchester databus for online operation or dedicated offline control unit.

Enhanced signal recognition is provided by three-channel filtered data for optimum operator display and automatic detection with range/confidence estimate. Automatic aircraft compensation is provided by electronic compensation of aircraft interfering terms and an automatic figure of merit estimator.

The AN/ASQ-208(V) is configured for minimum impact on existing aircraft or new aircraft installations and utilises existing AN/ASQ-81(V) aircraft wiring. It requires a single cable change for the vector sensor.

System performance enhancements include shallow water detection, elimination of dedicated Magnetic Anomaly Detection (MAD) compensation flights and automatic detection.

Operational status

In initial production and in service.

Contractor

Raytheon Systems Company.

UPDATED

LAMPS MK III datalink (Hawk Link)

The LAMPS MK III datalink system provides full duplex, secure and highly reliable communications between airborne and shipboard platforms. The system, better known as the Hawk Link, was designed specifically to enable the sensors and weapons of the SH-60B Seahawk, LAMPS MK III helicopter to function as integrated subsystems of the Navy's Light Airborne MultiPurpose System (LAMPS) MK III Weapon System. The Hawk Link multiplies SH-60B based processing of airborne sensor information by the power of shipboard processing capabilities through parallel operations, increasing the speed and accuracy of target/threat classification and/or localisation. Airborne sensors become, through the Hawk Link, an extension of the ship's sensor suite.

The Hawk Link system consists of two subsystems, the AN/ARQ-44 Airborne Data Terminal and AN/SRQ-4 Shipboard Data Terminal. The AN/ARQ-44 is functionally responsible for communicating SH-60B Seahawk sensor data to its parent ship over the wide data bandwidth digital downlink, and receiving command and control data from the parent ship over the narrower data bandwidth digital uplink. The AN/SRQ-4 is linked with shipboard computers via an NTDS slow interface, hands-off Anti-Ship Surveillance and Targeting (ASST) sensor information (radar video and tactical plot data) to shipboard mission control consoles, and hands-off Anti-Submarine Warfare (ASW) sensor information to the ship's sonar signal processing system.

Hawk Link features include: full duplex, digital communication; wide-bandwidth downlink; downlink error correction; high fade margins; up to four wide plus four narrow acoustic channels; communication security (KG-45); full Mil design with online BIT; directional (jam resistant and LPI); narrow-bandwidth uplink; uplink error detection; high-intelligibility voice com; search and track radar video; advanced sensor interfaces, interoperability; LPI upgrades under consideration.

AN/ARQ-44 Airborne Data Terminal Specification:

Multiplexer/demultiplexer, TD-1254/ARQ-44, analogue and digital sensor interfaces, 1553A interface, 18 kg.

Radio, receiver/transmitter, RT-1275/ARQ-44, FM FSK modulation, 30 kg.

Directional antenna (2 each), AS-3273/ARQ-44, Azimuth steerable, 3.9 kg each (with radome).

Operational status

The Hawk Link is operational in the US Navy's SH-60B Seahawk helicopters, and AN/SQQ-89(V) Surface ASW Combat System equipped ships including DDG-51 'Arleigh Burke' Class and DD-963 'Spruance' Class Destroyers, CG-47 'Ticonderoga' Class Cruisers, FFG-7 'Oliver Hazard Perry' Class Frigates. In addition, the Spanish Navy selected the Hawk Link to support its SH-60B Seahawks and LAMPS MK III Frigates.

Contractor

Sierra Technologies, Inc.

VERIFIED

Tactical navigation modification suite

Developed for the US Navy's Orion P-3B fleet, the Smiths Industries Tactical Navigation Modification (TACNAVMOD) suite incorporates modern displays, computer architecture and a MIL-STD-1553B databus. Also included are ESM display integration, hybrid acoustics processing, sensor integration and advanced navigation integration processing. TACNAVMOD comprises 10 LRUs, plus two high-resolution raster displays and additional search/track sensors; other facilities can be added as options.

Operational status

In service. No longer in production.

Contractor

Smiths Industries Aerospace.

VERIFIED

RADAR, GPWS and TCAS SYSTEMS

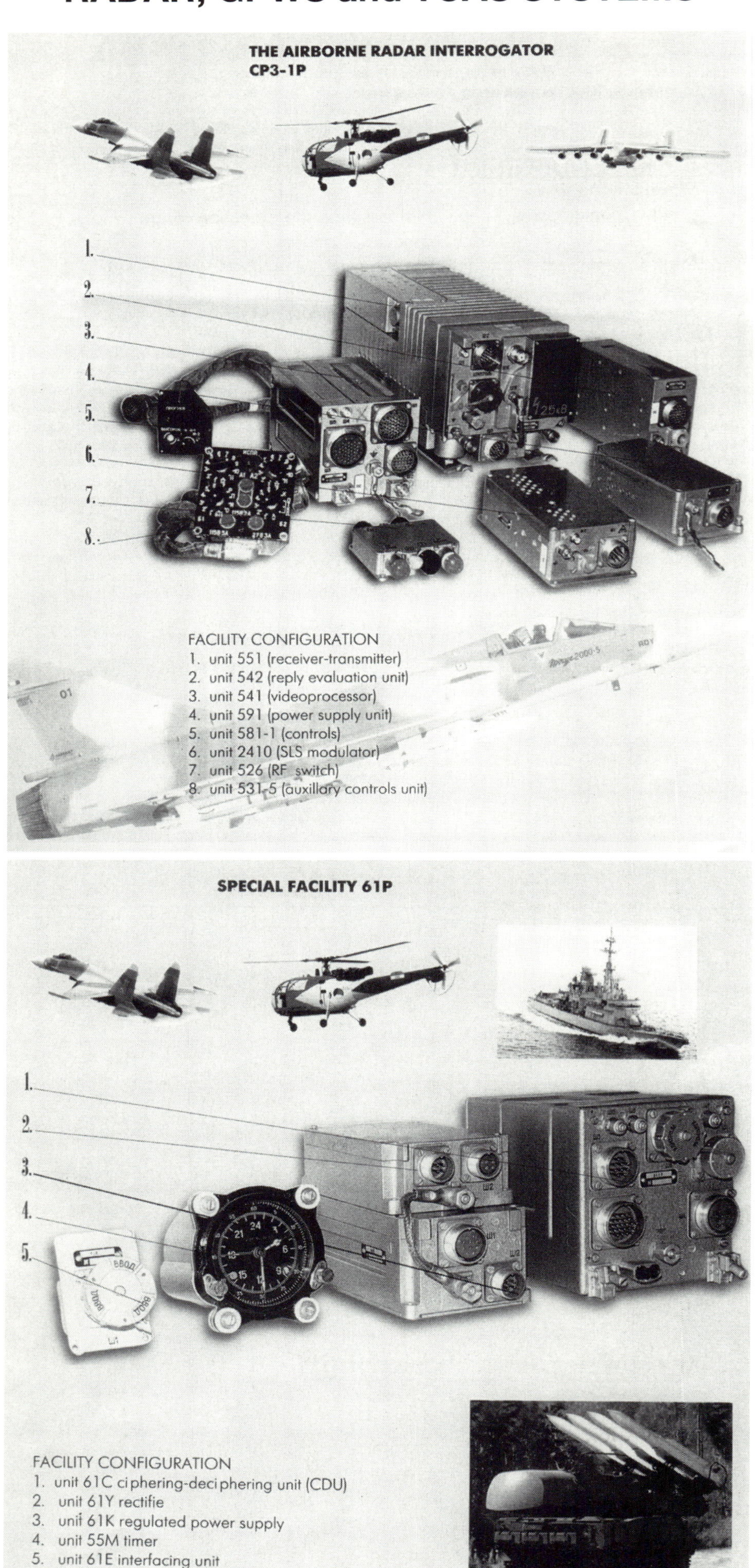

The All-Russian TSC Nizhegorodskaya Yamarka 60P IFF system showing the CP3-1P airborne interrogator system and 61P cipher system **1998**/0011880/0011881

CANADA

Polaris

CAL Corporation is developing a polarimetric version of the SAFFIRE system called Polaris. The Polaris system includes a microstrip dual-polarisation antenna, with the associated waveguide switches to select horizontal or vertical transmit polarisation. This antenna is designed to provide a minimum of 20 dB isolation between orthogonal polarisation channels. An additional receiver provides the capability of simultaneously receiving two orthogonal polarisation returns. The four returns, HH, HV, VH and VV, are used to synthesise the effect of transmitting in any polarisation and receiving in any polarisation state. This polarisation preprocessing enhances the detectability of targets such as ships at sea, manmade objects in foliage, different types of terrain and vegetation and oil slicks. The parameters for synthesis are operator chosen, using a simple menu or graphic interface.

Polaris options include a high-resolution colour video display and a high-density tape recorder.

Operational status

In development.

Contractor

CAL Corporation Ltd.

VERIFIED

SAFFIRE

The Synthetic Aperture Fully Focused Imaging Radar Equipment (SAFFIRE) produces map-like displays with range and azimuth resolutions of 37.5 m, in swaths of 25, 50 or 100 km and at scales of 1:1,000,000 or 1:125,000, overlaid with a latitude and longitude grid and alphanumeric annotation. A selection of single- or dual-sided operation, horizontal or vertical beam polarisation, a variable number of looks and STC (Swept Time Constant) are available.

The system has two fixed-mounted dual-polarised microstrip printed circuit antennas positioned along the underside of the aircraft fuselage, each 40 cm high and either 5 m, 2.5 m or 1.23 m in length. A high peak power, coherent-on-reception transceiver provides coherent signals to the digital processor which performs multilook registration, yaw compensation, clutter lock and autofocus processing. The radar data is recorded with a choice of dry processed paper, film or 8 mm format high-density digital tape recorder and can be sent to a video monitor unit or real-time downlink. SAFFIRE interfaces with the host aircraft inertial navigation system, GPS receiver or radar altimeter and conditioned electrical power supply.

The transmitter produces a peak power of 200 kW, a pulsewidth of 230 ns, a frequency of 9,250 MHz and a PRF between 550 and 1,100 Hz slaved to the aircraft groundspeed. The system sensitivity provides detection of a 2 m^2 target at the maximum range. At maximum range the synthetic aperture length consisting of 64 elements, presummed over three to six radiated pulses, is 37.5 m, the same as the aperture spacing. The SAFFIRE integrated sidelobe ratio is −15 dB.

SAFFIRE operates at altitudes between 5,000 and 20,000 ft, at a velocity of between 150 and 300 kt and with a squint angle between ±15°.

Specifications

Dimensions:
(transmitter) 444 × 482 × 610 mm
(receiver) 266 × 482 × 610 mm
(processor) 266 × 482 × 610 mm
(display and control) 266 × 482 × 610 mm
(recorder) 310 × 482 × 508 mm
(antenna) 400 × 5,000, 2,500 or 1,250 mm
Weight:
(transmitter) 60 kg
(receiver) 15 kg
(processor) 45 kg
(display and control) 15 kg
(recorder) 57 kg
(antenna each) 20, 10 or 5 kg
Power supply:
(transmitter plus receiver) 115 V AC, 400 Hz, 3 phase, 1,200 W and 28 V DC, 6 A
(processor) 115 V AC, 400 Hz, single phase, 500 W
(display and control) 28 V DC, 2 A
(recorder) 115 V AC, 400 Hz, single phase, 225 W

Operational status

Completing development.

Contractor

CAL Corporation Ltd.

VERIFIED

SLAR 100 Side-Looking Airborne Radar

The first SLAR 100 system was installed in a modified de Havilland Dash 7 operated by the Canadian Atmospheric Environment Service, which uses the aircraft for monitoring ice flows and other reconnaissance duties. The equipment entered service at the end of 1985 and has been joined by a second set, fitted to a Lockheed L188 Electra.

The basic SLAR consists of a control unit, hard-copy film recorder, transmitter/receiver, central processor unit and 5.18 m antennas. The basic SLAR can be upgraded to a synthetic aperture fixed focus radar by integrating an options processor containing digital downlink interface and/or tape drive interface, Doppler beam-sharpening processor, moving target indicator, 2.44 m antennas and constant false alarm rate.

The SLAR 100 has a maximum range of 100 km on either side of the aircraft, producing map-like displays in swaths of 25, 50 or 100 km, at scales of 1:1,000,000, 1:500,000, 1:250,000 or 1:125,000 overlaid with a latitude and longitude grid.

The system has two antennas positioned along the underside of the fuselage for the Dash 7 application, each 5.28 m long and 40 cm high; equally the antennas could be mounted in an external pod. A magnetron transmitter is used. The radar imagery is combined in a central processor with aircraft attitude and navigation data before being recorded on a roll of thermally developed black and white film. The radar data can also be displayed in the aircraft or datalinked to a ground station.

The SLAR 100 radar in the Dash 7 would normally operate at between 5,000 and 10,000 ft. Aircraft roll angle must be maintained within ±4° and yaw to within ±15°.

The transmitter operates with a peak power of 200 kW, a PRF of 800 Hz, a pulsewidth of 0.23 μs and at a frequency of 9,250 MHz. In the SLAR 100 the along track range resolution is proportional to target range, being 7.8 m/km. Across track the range resolution is constant at 37.5 m.

Specifications

Dimensions:
(transmitter) 444 × 482 × 584 mm
(central processor) 265 × 482 × 559 mm
(control unit) 265 × 482 × 406 mm
(antenna) 404 × 5,285 mm
Weight:
(transmitter) 60 kg
(central processor) 29 kg
(control unit) 22 kg
(antenna (each)) 36.5 kg
(total system) 267 kg

Operational status

In service in Boeing Canada DHC-7 Dash 7 aircraft operated by the Canadian government.

Contractor

CAL Corporation Ltd.

VERIFIED

The CAL Corporation SLAR 100 is fitted in the Boeing Canada DHC-7 Dash 7 for use over the Northern Territory in all-weather conditions

SLAR 300 Side-Looking Airborne Radar

The SLAR 300 is a real aperture side-looking non-coherent airborne imaging radar with a 100 km range on either side of the aircraft which produces a 200 km swath. The radar's high sensitivity is due to the use of a 250 kW peak power magnetron transmitter operating at I/J-band, although a G-band version can be provided. The SLAR 300 operates with a vertical or horizontal polarisation, which is selectable if a dual-polarisation antenna is used. The system works with either fixed or gimballed antennas mounted on both sides of the aircraft.

The CAL Corporation lightweight low-cost high-efficiency modular dual-polarised microstrip antenna can be used with this system, where size and weight are primary considerations. Output devices include a film recorder, downlink, digital tape recorder and video display.

SLAR 300 is the basic model of an upgradeable family of radars. Options such as synthetic aperture processing, polarimetric processing and moving target indicator processing may be added after delivery to upgrade this system.

Specifications

Range: 100 km each side
Range resolution: 37.5 m
Azimuth resolution: 7.5 m/km of range
Swath width: 25, 50 or 100 km
Swath offset: 0, 25, 50 or 75 km
Aircraft groundspeed: 150-330 kt
Altitude: 5,000-20,000 ft
Max squint angle: ±15°

Contractor

CAL Corporation Ltd.

VERIFIED

AN/APS-140(V) radar

AN/APS-140(V) is the US designation for a configuration of the AN/APS-504(V)5 radar (see later item). This configuration includes back-to-back radar and ESM parabolic antennas, a three-axis pedestal and a MIL-STD-1553B control interface.

Operational status

In production.

Contractor

Litton Systems Canada Ltd.

VERIFIED

AN/APS-503 radar

The AN/APS-503 is a lightweight radar, with operational parameters suitable for maritime helicopters. It consists of five units: a transmitter/receiver, pedestal with parabolic antenna, PPI type azimuth range indicator, radar control unit and range bearing unit.

The transmitter operates in the I-band, with pulsewidth and PRF optimised for maritime reconnaissance and a 50 kW peak power. The receiver has a solid-state local oscillator with logarithmic response and sensitivity time control and fast time constant.

The antenna is a 61 × 46 cm parabola, providing a 4° beam in azimuth and a 5° beam in elevation. The antenna is stabilised for pitch and roll and has a manual tilt control. Top-, belly- and nose-mounted antenna versions are available. A pilot's bright display is available as an option.

The range and bearing unit positions a range strobe and an azimuth marker on the PPI, at the same time transmitting this information to ancillary equipment such as navigation systems.

Specifications

Transmitter frequency: 9.2-9.4 GHz, fixed
Power output: 50 kW peak
Pulsewidth: 0.5 μs
PRF: 400 Hz
Receiver noise: 8 dB
Antenna: 61 × 46 cm parabolic
Gain: 30 dB
Beamwidth: 4° azimuth; 5° elevation
Sidelobes: −20 dB
Polarisation: horizontal
Scan rate: 30 rpm
Stabilisation: automatic compensation for pitch and roll to ±20°
Tilt control: ±8°

Operational status

No longer in production but still in service with Canadian Forces' Sea King helicopters. There have been 50 AN/APS-503 radars produced.

Contractor

Litton Systems Canada Ltd.

VERIFIED

AN/APS-504 radar

The AN/APS-504 airborne search radar was developed for the Canadian Forces' Tracker aircraft. It is a development of the AN/APS-503 radar and is designed for maritime search operations. It uses a 100 kW peak power magnetron at I/J-band with two transmitter pulsewidths: 0.5 and 2.4 μs. The AN/APS-504 includes a two-axis antenna unit with parabolic antenna, transmitter/receiver unit, analogue PPI display unit and radar control unit.

The Litton AN/APS-504(V)5 radar installed in a Beech 200 aircraft

Operational status

No longer in production. There have been 40 AN/APS-504 radars produced, for two customers.

Contractor

Litton Systems Canada Ltd.

VERIFIED

AN/APS-504(V) series radar

The APS-504(V) Series of airborne search radars has been designed primarily for maritime surveillance applications. They can be installed in either a fixed- or rotary-wing aircraft. In addition to coastal and offshore surveillance missions, these radars can be used for weather avoidance, low-resolution land mapping and navigation.

The APS-504(V)2 is the commercial version of the AN/APS-504 which was developed specifically for Canadian Forces' Tracker aircraft. It uses the same 100 kW peak power I/J-band magnetron and transmitter pulsewidths of 0.5 and 2.4 μs. The system consists of a two-axis antenna unit with parabolic antenna, transmitter/receiver unit, analogue PPI display unit and radar control unit.

The APS-504(V)3 was developed to include an improved transmitter/receiver and a digital signal processor and scan converter that produce a ground-stabilised PPI display in a high-resolution 875-line video format. Navigation and cursor data are overlayed on the non-fading radar video, which can also be recorded and played back. Several sizes of flatplate antennas are available, mounted on a two-axis pedestal.

The APS-504(V)5 is the most advanced of the APS-504 family. It employs a TWT-based transmitter with wideband frequency agility, high-ratio pulse compression, scan-to-scan integration and digital signal processing to enhance the detection of sea surface targets, including targets with radar cross-sections as small as 1 m² in Sea State 3. The APS-504(V)5 can be configured to meet various installation and performance requirements.

Specifications
(APS-504(V)5)

Weight:
(with 3-axis pedestal) 180 kg
Power supply: 115 V AC, 400 Hz, 3 phase, 1.5 kVA
28 V DC, 15 A
Frequency: 8.9-9.4 GHz (16 frequencies)
Peak power: 6.6 kW
Antenna: flatplate and parabolic (various sizes available)
Scan rate: 7.5-120 rpm automatically selected
Display: RS-343 875-line video
Range scales: 5.5-370 km

Operational status

The APS-504(V)2 is no longer in production. There were 53 of this version produced.

The APS-504(V)3 is still in production. There have been 25 of this version produced.

The APS-504(V)5 is still in production. It has been installed in aircraft ranging from twin-engined turboprops such as the Beech 200 to larger aircraft such as the Boeing 737. More than 65 of this version have been produced.

Contractor

Litton Systems Canada Ltd.

VERIFIED

SSAR Spotlight Synthetic Aperture Radar for CP-140 aircraft

Lockheed Martin Canada's Spotlight SAR is an adaptation of the AN/APS-506 multimode radar for the Canadian Forces CP-140 Aurora aircraft. Spotlight SAR is being developed as an affordable real-time airborne reconnaissance and surveillance imaging radar, and in particular to provide exceptionally long standoff range detection and classification of ships and submarine periscopes.

Spotlight SAR is based on the enhancement of conventional radar components: receiver, transmitter antenna and related control units to provide detailed cross-section images in three modes: Spotlight, Strip Map and Range Doppler Profiling/Inverse SAR (RDP/ISAR), whilst retaining the original conventional radar modes. In the Spotlight Mode, the aircraft's lateral motion is used to form a long synthetic aperture. Resolution (<1m) is sufficient to establish fine details of naval and ground-based targets from long standoff ranges.

The Strip Mode provides an endless strip of high resolution, real-time imagery along the aircraft's flight path, and as much as 60° off track.

Employing enhanced ISAR techniques, Range Doppler Profiling provides a continuous series of snapshots (frames) of moving targets for analysis and classification.

Spotlight SAR has potential both for military and civilian SAR missions.

Operational status

Lockheed Martin Canada, collaborating with the Canadian Department of National Defence has defined and designed SSAR to provide the operator interface. Lockheed Martin Canada is leading a team comprising: Applied Analytics, Array Systems Computing, the IMP Group and Raytheon Canada for development of the Spotlight SAR system, for the Canadian Forces Bombardier CP-140 Aurora aircraft.

Contractor

Lockheed Martin Canada.

NEW ENTRY

IRIS synthetic aperture radar

Available for both civil and military applications, the Integrated Radar Imaging System (IRIS) is an airborne synthetic aperture radar reconnaissance system.

A major advantage afforded by the IRIS is real-time tactical operation. The full-resolution, full-swath, onboard processing unique to the IRIS enables the simultaneous viewing of reconnaissance images in the air and, via downlink, on the ground. In a battlefield or other active military situation, this capability allows the integration of the airborne reconnaissance system with a ground-based, tactical command and communications system to co-ordinate and control ground forces and air strike support. Intelligence derived from the IRIS reconnaissance images on the position of opposing forces can be used by field commanders to direct ground operations and airborne support within seconds of acquisition of the reconnaissance data.

Two operational modes are provided, with the capability of switching inflight from a wide swath to a high-resolution mode in seconds. Images are produced from distances of up to 100 km and altitudes of up to 49,000 ft, providing standoff range capability.

The IRIS consists of three segments: an airborne segment, a transportable ground segment and a precision analysis facility. The standard equipment for these segments is augmented by options for image display, storage and interpretation. Drawing on these options, the IRIS can be configured for a variety of user requirements.

The airborne segment consists of an imaging synthetic aperture radar with onboard digital processing, downlink transmission, high-density magnetic tape data recording and image production capabilities. Images are processed in real time and displayed on a video monitor and hard-copy paper strip. The airborne segment has been packaged in compact, rugged modules for application in high-performance executive-size turboprop and jet aircraft. In addition, independent processors are included for simultaneous fixed and moving target imaging. Moving and fixed targets can be displayed independently or superimposed in colour in a single image.

The transportable ground segment consists of a downlink receiver, digital data processor and tactical workstation. The tactical workstation is rugged, transportable and inexpensive - ideal for rapid deployment in critical reconnaissance areas. It features high-density magnetic tape data recording, continuous paper strip printing, continuous image and frame image video display capabilities.

The precision analysis facility provides radar interpreters with the ability to analyse reconnaissance data received via downlink or recorded on board the aircraft and the transportable ground segment. The facility can archive and retrieve multiple data sets from the airborne and transportable segments, as well as maps and other interpretation aids entered at the facility. The precision analysis facility allows for in-depth interpretation of imagery, change detection, production of map mosaics and preparation of precision hard-copy film products.

Specifications

Weight: 400 kg
Frequency: 9,375 MHz
Average transmitter power: 160 W
Platform speed: 550 kt (max)
Antenna length: 1.0 m standard, 1.4 m high gain

Operational status

In production.

Contractor

MacDonald Dettwiler and Associates Ltd.

VERIFIED

CHINA, PEOPLE'S REPUBLIC

204 Radar

It is the first I/J-band monopulse airborne interceptor radar, successfully developed by Leihua for all weather F-8 fighters. It has search, acquisition and tracking capabilities and can be used for attack on flying targets with gun, rockets and missiles in association with an onboard fire control computer and optical gunsight.

It provides good anti-interference, high reliability and easy maintenance.

Specifications

Detection range: 29 km
Scan range:
(azimuth) ±38°
(elevation) −12 to +24°
Volume: 0.145 m³
Weight: 145 kg
Frequency: I/J-band

Operational status

The 204 radar was confirmed through final evaluation in 1984 and put into mass production.

204 Radar **1996**

Contractor

China Leihua Electronic Technology Research Institute.

VERIFIED

698 side-looking radar

The I/J-band 698 side-looking radar is designed specifically for detection of periscopes and ships. The radar features coherent moving target detection, slotted feed double parabolic reflector antenna, parametric amplifier, high-stability local oscillator, coherent receiver, IF log amplifier and digital filter. Detection ranges are quoted as 60 km against ships and 17 km against a periscope.

Specifications

Detection range: periscope 17 km; ship 60 km
Display ranges:
(transversal) 60 km (normal)
(longitudinal) 30 km (searching)
High resolution:
(transversal) 300 m (searching)
(longitudinal) 50 m
Operational altitude: 50-500 m (searching)
Volume: 0.8 m
Weight: 230 kg

Operational status

In service.

Contractor

China Leihua Electronic Technology Research Institute.

UPDATED

Colour weather radar

The colour weather radar designed for civil aircraft is now under development and testing, with assistance from Rockwell International Corporation Collins General Aviation Division. The radar system incorporates a variety of new technologies such as the narrowed waveguide planar slotted array, high-reliabillty transmitter, miniaturised logic-controlled receiver, and digital colour indicator. The transmitter employs a pulse-modulated magnetron. It also has ground-mapping features.

Specifications

Weather avoidance range: >500 km (for strong thunderstorm areas)
Scan range:
(azimuth) ±60°
(elevation) ±15°
Antenna stabilisation range:
(roll) ±30°
(pitch) ±10°
Weight: 20 kg (no cables)
Antenna beamwidth: <6°

Contractor

China Leihua Electronic Technology Research Institute.

VERIFIED

CWI illuminator

The continuous wave illuminator, now used for F-8 fighters, is designed to perform semiactive radar guidance of air-to-air medium-range interception missiles when operated in combination with airborne radar.

Operational status

It is in production with an annual output of 30 to 50 sets. There are a series of CWIs, including CWI-A, CWI-B, CWI-C and CWI-D, available for different types of aircraft.

Specifications

Frequency: I/J-band
Radiation power: 200 W
FM noise: LFM<−99 dB/Hz/10 KHz
AM noise: LAM<115 dB/KHz/10 KHz
Weight: 40 kg
Volume: 0.035 m^3

Contractor

China Leihua Electronic Technology Research Institute.

VERIFIED

JL-7 fire-control radar

The multifunction JL-7 fire-control radar for the F-7C aircraft is designed to search, detect and track airborne targets and carry out air-to-ground ranging. It can be used for attack on air or ground targets using missiles, guns or bombs in association with a gunsight or HUD.

Specifications

Volume: 0.12 m^3
Weight: 115 kg
Frequency: J-band
Range:
(detection) 27.8 km
(track) 18.5 km
Coverage:
(azimuth) ±35°
(elevation) −13 to +17°
Altitude: 2,300-65,000 ft

Operational status

In production and in service with Chinese Air Force F-7C aircraft.

Contractor

China Leihua Electronic Technology Research Institute.

VERIFIED

JL-7 fire-control radar 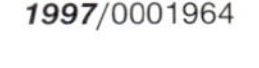 ***1997***/0001964

JL-7 fire-control radar installed on the F-7C aircraft  ***1997***/0001965

JL-10A ***1997***/0001191

JL-10A airborne radar

The JL-10A airborne radar is designed for the fighter requirements of medium-range omnidirectional attack, close-range dogfight, look-up and look-down, and surface moving target attack over land and sea. It is the first airborne full-wave pulse Doppler radar, with high, medium and low PRF, produced in China. The JL-10A is a highly digitised system which uses a slotted array antenna, signal exciter and sophisticated signal processor.

Specifications

Range: 59.3 km look-up, 53.7 km look-down (5 m^2 target)
Tracking range: 29.6 km
Range resolution: 150 m
Range accuracy: 15 m
Reliability: >70 h MTBF

Operational status

Development and flight trials have been completed successfully.

Contractor

China Leihua Electronic Technology Research Institute.

VERIFIED

Shenying multimode airborne radar

The Shenying multimode airborne radar is a coherent I/J-band pulse Doppler system which will provide fighters with capabilities of medium-range dogfight, look-up and look-down weapon delivery, and ground or sea moving target attack.

The antenna is a flat plate slotted array which features low sidelobes and full azimuth and elevation monopulse operation. A gridded TWT transmitter is employed, which operates at low, medium and high PRFs. The receiver, with low-noise front end, consists of two channels which provide the monopulse sum and multiplexed difference channels (azimuth and elevation). Shenying incorporates digital signal/data processors to handle radar mode control, to conduct the built-in test, and to perform radar signal and data processing.

Operational status

Under development.

Specifications

Detection range:
(search) 80 km (max); 54 km (look-down)
(tracking) 40 km (look-up); 32 km (look-down)
Search angle:
(azimuth) ±60°
(elevation) ±60°
Frequency: I/J-band

Contractor

China Leihua Electronic Technology Research Institute.

VERIFIED

PL-7 fire-control radar

The PL-7 is a lightweight, monopulse, J-band fire-control radar developed for the Chinese Air Force and designed for use in fighter aircraft. It has air-to-air and air-to-ground modes and can operate in conjunction with a fire-control computer, IFF, head-up display or aiming sight.

The radar consists of 18 LRUs located in the nose and cockpit. It has five modes in air-to-air: search (from 400 m to 30 km, and through ±45°); manual acquisition; boresight; attack/track (up to 15 km, through ±45° and down to 2,300 ft altitude) and transponder. The three air-to-ground modes are: slant range, attack and acquisition.

A horizontally polarised antenna is used, with a 3.4° beamwidth in azimuth, 5.6° in elevation and a 30 dB gain.

Specifications

Volume: 0.23 m^3
Weight: 115 kg
Peak power: 75 kW
Reliability: 50 h MTBF

Contractor

China National AeroTechnology Import and Export.

VERIFIED

Model 265 radar altimeter

The Model 265 is a nanosecond pulse radar altimeter. It can be used on various types of civil or military high-speed jet aircraft, helicopters and unmanned aerial vehicles. It outputs high-precision and continuous true height signals to navigation, autopilot and fire-control systems.

The Model 265 features a large altitude range, high accuracy, high reliability and stability, small volume and the use of the front edge of the tracking pulse to measure the shortest range between the aerial and the ground, without interference. The system incorporates a self-test function and provides warning signals at preset heights.

Specifications

Weight: 3 kg
Altitude: 0-4,920 ft
Accuracy: ±3 ft or 3% of true height
Resolution: ±0.5 ft or 0.1% of height

Operational status

Model 265 radar altimeters have been used on the J7-IIM and J8-II fighters, Y-12 transport aircraft, the J-8 trainer and the CKIE low-altitude pilotless target.

Contractor

China National Electronics Import & Export Corporation.

VERIFIED

CZECH REPUBLIC

SO-69/ICAO airborne transponder

The SO-69/ICAO airborne transponder is a reconstructed and updated version of the Soviet SO-69 system. It complies with ICAO design standards. The encoder unit Š-ICAO and control unit O-ICAO can be modified to support IFF Mk10 modes 1 and 2.

The new transponder can easily be retrofitted in aircraft equipped with the SO-69 transponder. It weighs less than the existing SO-69 equipment and consumes less power.

Test equipment designated ZZ-69/ICAO is available to support the system.

Specifications

Dimensions/Weight:
Transmitter-receiver unit PV-ICAO: 412 × 64 ×160 mm/5.5 kg
Encoder unit Š-ICAO: 436 × 86 × 160 mm/2.3 kg
Control unit O-ICAO: 71.5/98 × 112 × 59 mm/0.35 kg
Communication unit BK-ICAO: 217 × 133 × 59 mm/1.1 kg
Frame S0-69: 436 × 155 × 216 mm/2.4 kg
Receiver frequency: 1,030 MHz (bandwidth 6 MHz)
Transmitter frequency: 1,090 MHz
Transmitter power: 250 to 500 W
Modes: A, A/C, B
Power supply: 27 V DC and 115 V AC/400 Hz

Operational status

The SO-69/ICAO transponder has already been installed in MiG-23, MiG-29, Su-22 and Su-25 aircraft.

Contractor

Elektrotechnika-Tesla Kolin, a.s.

UPDATED

SO-69/ICAO airborne transponder **1997**/0002198

FRANCE

Anemone radar

Anemone was intended for the modernised Super Etendard. The Anemone search, tracking and ranging radar developed under the prime management of Dassault Electronique with the participation of Thomson-CSF has the following functions: air-to-surface (main function); air-to-ground (ranging and ground-mapping); and air-to-air.

It consists of a nosecone (antenna and circuitry), an aircraft/radar interface unit and radar controls.

The Anemone radar operates in the I/J-band with frequency agility. A wideband monopulse flat slotted array antenna with low-level sidelobes is provided, and reinforced ECCM is incorporated. In the air-to-surface mode, the system enables a surface target to be detected and tracked. Targets are detected in the search mode with an elevation angle automatically adjusted as a function of the selected range angle. Following target designation and lock on, the change to track-while-scan or continuous tracking is automatic. In the air-to-air mode the radar allows linear scan, search with semi-automatic acquisition, and continuous tracking.

The Dassault Electronique Anemone radar

Operational status

The Anemone has been fitted to the modernised Super Etendard of the French Navy. It is currently in production and is being fitted to the Super Etendard aircraft as part of an upgrade programme for the French Navy.

The Dassault Electronique 3300 single-box transponder

Contractors

Dassault Electronique.
Thomson-CSF Radars Contre-Mesures.

VERIFIED

Antilope V radar

The Antilope V radar was designed for the Mirage 2000N. Its basic functions are terrain-following, ground-mapping, interlace (terrain-following and ground-mapping), air-to-air and air-to-surface. Essential characteristics of Antilope V are:

(a) J-band transmission providing high ground reflectivity
(b) High-speed vertical scanning of the antenna
(c) Asymmetric antenna, providing accurate localisation of obstacles in the path of the aircraft
(d) An antenna of the flat slotted-array type producing a weighted polar diagram with very low level diffuse sidelobes
(e) Receiver with a wide dynamic range and image-frequency suppression
(f) Image sharpening of the monopulse type with compression in the elevation plane for highly accurate determination of the height of obstacles
(g) Real-time radar data processing
(h) Continuous and automatic test system
(i) Protection against reception by antenna sidelobes.

Radar information is displayed on a head-up display and on a three-colour multimode CRT head-down display, as well as being sent to the navigation and

Antilope V terrain-following and navigation radar

weapon system. The system can provide terrain-following commands at 300 ft (91 m) and 600 kt, computing a preset obstacle clearance height, and with a preselected *g* level.

Operational status

Dassault Electronique is the design leader on this programme which is shared with Thomson-CSF. The Mirage 2000N radar has been in service since 1987. First deliveries of an upgraded version of Antilope V for the Mirage 2000D began in late 1992.

Contractors

Dassault Electronique.
Thomson-CSF Radars Contre-Mesures.

VERIFIED

DAV air-to-air warning and surveillance radar

The DAV warning and surveillance system for helicopters (developed by Dassault Electronique), detects low- and medium-altitude air threats, including hovering helicopters; it provides accurate co-ordinate data at standoff range. After detection, DAV sorts the threats into aircraft and helicopter categories and identifies the helicopters by their type.

The radar is attached to the rotor, so it rotates at rotor speed, thereby delivering data at a high refresh rate over 360° in azimuth.

DAV delivers a number of data items: target designation, classification, identification, warning, target tracking, and air-to-air missile fire control. It provides real-time comprehensive air-to-air situation awareness information.

DAV can provide information to an FAAD (Forward Area Air Defence) system. As an FAAD extension, it provides a capability for warning, attack and destruction of low-flying targets beyond the action ranges of the ground-to-air defence facilities.

The system has been designed for all-weather operation. Its location accuracy is such that it delivers accurate azimuth, elevation, range and radial speed data on targets for air-to-air missile engagement. It also allows very short response times for gun firing either by the carrier helicopter or other attack helicopters.

DAV features a multitargeting autotracking capability. This E/F-band radar's radiated power endows it with high-detection probability performance up to 9 km while remaining mostly covert. It is equipped with a frequency-hopping transmitter.

Radar integration into the helicopter's system is via a 1553B bus.

Specifications

Dimensions:
(antenna module) 420 × 900 mm diameter
Weight:
(aerial unit) 60 kg
(interface) 5 kg
Frequency: E/F-band
Transmitter: solid-state, frequency-agile
Antenna/radome: rotational velocity same as that of platform helicopter main rotor
Coverage: 360° azimuth; 24° elevation
Receiver: superheterodyne
Doppler analysis: FFT processing
Target recognition: fixed-wing and rotary-wing aircraft
Helicopter identification: from 'library' signatures
Target designation information: azimuth, elevation, range and speed

Operational status

The first volume production radar for the Tiger HAP helicopter was scheduled to be available by the end of 1997 but current status is uncertain. The US Army has already been evaluating the radar.

Contractor

Dassault Electronique.

UPDATED

Ground Collision Avoidance System (GCAS)

Dassault Electronique is aiming its new GCAS system at the airline Ground Proximity Warning System (GPWS) and Enhanced Ground Proximity Warning System (EGPWS) market. GCAS incorporates a worldwide digital terrain database (like EGPWS), and an alert algorithm that includes aircraft performance potential by incorporating attitude and aircraft type-specific and configuration data (for example, true aircraft flap settings, actual aircraft weight, and actual engine performance capabilities). Audio and visual caution and warning messages are similar to those found in GPWS. GCAS is designed as a near form/fit replacement for existing GPWS systems and to be linked to aircraft navigation or GPS data. GCAS reverts to basic GPWS mode if a reliable navigation signal is lost. The digital GCAS can be operated without a display, a feature intended to encourage retrofits on older aircraft, although it is intended to provide display data on appropriate weather radar or EFIS screens, using ARINC 453 protocols. An upgraded 3D display, for future aircraft, such as Airbus A3XX, is planned.

GCAS calculates aircraft time to clear obstacles, rather than time to impact. In making its calculations, GCAS demands no more than 0.5*g* manoeuvre, 75 per cent airframe load limit and 90 per cent take-off power.

Operational status

The system has been under evaluation at the French Flight Test Centre in an Airbus A300 simulator since 1995, and on board a Dassault Falcon 2000 since 1997. Initial certification in the Falcon 2000 is expected in July 1998 under European Joint Airworthiness Authority (JAA) rules. FAA/JAA certification for Airbus and Boeing transports is anticipated for spring 1999.

Dassault Electronique is teaming with Teledyne Controls for marketing, support and repair of the GCAS product in the USA.

Contractor

Dassault Electronique.

NEW ENTRY

Horizon radar on AS 532 UL Cougar helicopter **1997**/0001192

Horizon battlefield surveillance system

The Horizon system (Hélicoptère d'Observation Radar et d'Investigation sur Zone) has been developed and manufactured for the French armed forces/l'Armee de l'Air for tactical intelligence data gathering. It is a development of the earlier Orchidee concept evaluation system.

The Horizon system is an I-band long-range ground surveillance system which can detect, from a stand-off position, in all weather conditions, day and night, moving objects over large areas up to 200 km distance, including: wheeled and tracked vehicles; helicopters (moving and hovering); aircraft; ships. Each moving object is detected, localised and automatically analysed and classified. Up to 20,000 km^2 can be surveyed every 10 seconds. Airborne activities are controlled from an operator panel in the Cougar helicopter. Collected date is transmitted to a ground station over the secure, all-digital, frequency hopping, Agatha datalink, up to a distance of 150 km.

Each Horizon system comprises: one fully equipped ground station with Agatha datalink; two AS 532 UL Cougar helicopters, each equipped with Target MTI radar unit and operator console, navigation and communication equipment and secure Agatha datalink.

The Target radar is an all-digitial, frequency agile, I/J-band, Doppler MTI radar. It combines mechanical and electronic scanning and provides instantaneous panoramic surveillance of a sector bounded by the following operating parameters: 60 and 90° sectors at scanning rates of 2, 4 or 8° per second, independently of helicopter course/heading and speed.

Each Horizon system is able to work alone, or with other Horizon systems, or as part of a larger C3I system.

The Horizon system is also proposed to NATO, within the framework of an Alliance Ground Surveillance acquisition programme.

System elements are provided by the following contractors: Eurocopter International: Cougar helicopter; Thomson CSF Radars Contre-Mesures: Target MTI radar; Dassault Electronic: Agatha datalink and ground station.

Specifications

Target radar
Range:
(clear weather) 200 km
(rain/cloud) 150 km
Resolution:
(range) 40 m
(velocity) 2 m/s
Scan sector: 360°
Scan rate: 2.4 or 8°/s
Agatha datalink
RF: J-band
Data rate: up to 0.5 Mbyte/s
Range: up to 150 km

Operational status

In 1991, a demonstrator system was used in Desert Storm operations by the French Daguet division and by the American XVIIIth Army Corps.

Two systems, each comprising two helicopters and one ground station, have been delivered to the French armed forces.

Contractors

Dassault Electronique.
Eurocopter International.
Thomson-CSF Radars Contre-Mesures.

UPDATED

RDN 85-B Doppler velocity sensor

The RDN 85-B is a single-box Doppler velocity sensor designed for use in helicopters, said to have good operating characteristics over calm seas. The radar interfaces with an ARINC 429 digital databus and along and across track velocities are also transmitted as DC signals for display by a hover meter and for coupling to an autopilot. There is BITE for in-flight and ground system checkout.

Specifications

Dimensions: 437 × 437 × 130 mm
Weight: <10 kg
Power supply: 28 V DC, 30 W
Frequency: 13.325 ±20 MHz
Transmitter power: 30 MHz
Velocity range: −50 to +350 kt
Altitude: up to 20,000 ft
Accuracy: 0.15% or 0.12 kt
Transmitter: Gunn diode oscillator
Reliability: >3,200 h MTBF

Operational status

In production for the French Navy export search and rescue Puma and Dauphin helicopters, French Army Super Pumas and the Fennec helicopter for Singapore. More than 300 units have been ordered.

Contractor

Dassault Electronique.

VERIFIED

RDN 2000 Doppler velocity sensor

The RDN 2000 J-band Doppler velocity sensor has been designed for light, medium and heavy helicopters, where autonomous navigation and assistance for the pilot is needed for missions of all types over land and sea. It employs FM/CW techniques and digital signal processing to provide accurate ground velocity data. The lightweight single unit includes an antenna, transmitter/receiver, power supply and signal processor.

Coupled with a computer and an AHRS or INS, the RDN 2000 provides a stealthy autonomous navigation system which is insensitive to countermeasures. In addition, it can operate with a local jammer.

The system provides high performance when flying over land or sea, especially over calm water, and it has an automatic land/sea transition capability. This allows autopilot operation irrespective of flying conditions.

Specifications

Dimensions: 437 × 240 × 80 mm
Weight: 4.2 kg

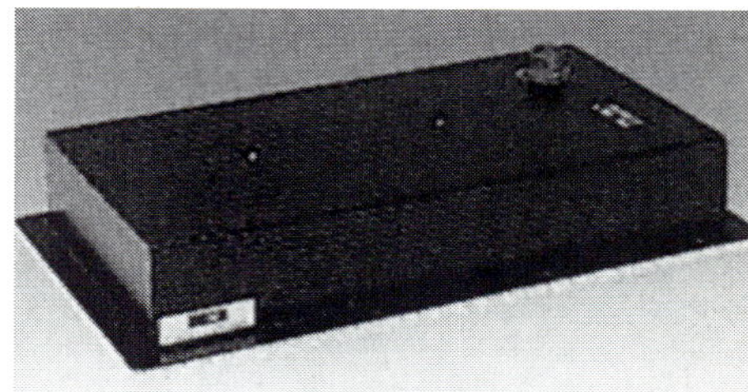

RDN 2000 Doppler velocity sensor **1998**/0010925

Power supply: 28 V DC, 25 W
Transmit power: 50 mW
Accuracy (95%):
0.15% of Vt or 0.1 m/s (along track)
0.22 % of Vt or 0.1 m/s (across track)
0.15% of Vt or 0.2 m/s (vertical)
Reliability: 8640 h MTBF

Operational status

First deliveries were due to begin in 1997; status uncertain.

Contractor

Dassault Electronique.

UPDATED

RBE2 airborne radar

The RBE2 airborne radar is being developed by GIE (Groupement d'Intérêt Economique) which includes Thomson-CSF Radars Contre-Mesures, as leader, and Dassault Electronique. The venture is shared on the basis of two-thirds for Thomson-CSF Radars Contre-Mesures and one-third for Dassault Electronique.

The RBE2 is the first of a new class of airborne radars using the Thomson-CSF Radant passive electronic scanning process. It is a multirole radar with large air-to-air and air-to-surface capabilities designed for the Rafale ACT and ACM.

For the air defence role the RBE2 is able to carry out all air defence functions including search in look-up and look-down modes, identification, automatic multi-target tracking and dogfight. Detection is optimised by means of automatic waveform selection of high, medium or low pulse recurrence frequencies. Multitarget tracking capability is improved by the use of a two-plane electronically scanned antenna which can track targets whatever their relative position in space to provide the RBE2 radar with its simultaneous multimode capability.

For the air-to-ground role the RBE2 provides Rafale with all-weather deep strike capability, as well as close support and battlefield interdiction, by means of automatic Terrain-Following/Terrain-Avoidance/Threat-Avoidance (TF/TA2), high-resolution mapping, fixed-target search and tracking, ground moving target search and tracking and air-to-ground ranging. The simultaneous operation of TF/TA2 and air-to-air modes bring vital capabilities in terms of self-defence while flying penetration missions and of ground clearance in air-to-air missions.

In addition to air defence and air-to-ground roles the RBE2 is optimised for shipping strike missions with long-range detection, multitarget tracking and target recognition and assessment.

With its low observable profile the RBE2 has important features such as full integration in the multisensor weapon system and real-time interworking with the EW suite.

Operational status

In production for Rafale ACT and ACM aircraft - first delivery 1997.

Contractor

GIE Radar ACT/ACM-Rafale.

UPDATED

The RBE2 two-plane electronically scanned multimode radar **1995**

GIE RBE2 radar for the Dassault Rafale ACT and ACM **1997**/0001193

AHV-9T radio altimeter

The AHV-9T is designed for the Panavia Tornado. Low- and high-altitude bands can be used, the variable output being available from a microprocessor-based receiver unit. The system features CMOS integrated electronics, stripline technology antenna and built-in fault detection capability.

Specifications

Dimensions:
(indicator) 61 × 61 × 158 mm
(transmitter/receiver unit) 109 × 154 × 324 mm
(antenna) 78 × 88 × 33 mm
Weight:
(indicator) 0.9 kg
(transmitter/receiver unit) 4.5 kg
(antenna) 0.2 kg
Power supply: 115 V AC, 400 Hz, 70 VA
Altitude: up to 50,000 ft
Accuracy: 1 ft ±2%
RF power: 60 MW (FM/continuous wave)
Outputs: 5 digital, 3 analogue

Operational status

In service in German Panavia Tornados.

Contractor

Thomson-CSF Communications.

VERIFIED

AHV-12 radio altimeter

Designed for use on the Dassault Mirage 2000 and Atlantique 2, the AHV-12 radio altimeter incorporates recent digital developments. Its main features are wide altitude range, high-accuracy and high-integrity levels.

The Thomson-CSF Communications AHV-12 digital radio altimeter

Specifications

Typical installation
Dimensions:
(antenna) circular, rectangular or small size
(transmitter/receiver unit) 193 × 90 × 315 mm
(indicators) ARINC 429 (digital), ARINC 552 (analogue)

Weight:
(transmitter/receiver unit) 5 kg
Power supply: 115 V AC, 400 Hz, 45 VA
Altitude: up to 70,000 ft
Accuracy: 1 ft ±1%

Operational status

In production and service in Mirage 2000 variants, Atlantique 2 and US reconnaissance aircraft.

Contractor

Thomson-CSF Communications.

VERIFIED

AHV-16 radio altimeter

The AHV-16 microprocessor-based radio altimeter has been built with a reprogrammable memory and is designed for commuter and military aircraft. It has both digital and analogue outputs to ARINC 552 and ARINC 429 standards. It can interface with existing digital and analogue avionic equipment and with electro-mechanical instruments as well as electronic flight instrument systems.

Specifications

Dimensions:
(indicator) 3 ATI
(transmitter/receiver unit) 230 × 90 × 90 mm
Weight:
(indicator) 1.2 kg
(transmitter/receiver unit) 2.0 kg
Power supply: 28 V DC
(R/T unit) 15 W,
(indicator) 8 W
Acccuracy: 1 ft ±2% of altitude
Reliability: >5,000 h MTBF

Operational status

In production and service in transport aircraft and helicopters.

Contractor

Thomson-CSF Communications.

VERIFIED

AHV-17 digital radar altimeter

The AHV-17 is specifically designed for Rafale. It has very low probability of intercept due to an RF management system which adjusts output power as a function of altitude and enhanced receiver sensitivity. The AHV-17 has excellent jamming detection capability which enables it to have superior resistance even against advanced jammers.

The AHV-17's modular construction, based on a series of integrated tests, allows effective diagnosis of failures and simple replacement without the need to replace or adjust the defective module. Modules can be interchanged independently.

Specifications

Volume: <3 litres
Weight: <3 kg
Power supply: 28 V DC to MIL-STD-740D, 40 W
Frequency: 4.2-4.4 GHz
Altitude: up to 30,000 ft
Accuracy: 3 ft or 1%
Reliability: >5,000 h MTBF

Contractor

Thomson-CSF Communications.

VERIFIED

AHV-18 compact radio altimeter

The AHV-18 is the advanced military radio altimeter of the Thomson-CSF product line. It is multimode equipment, of modular design, which can be adapted for use with fixed- and rotary-wing aircraft, missiles, RPVs and flying weapons.

It has comprehensive ECCM capabilities using power management and jamming detection modes.

Specifications

Dimensions:
(indicator) 3 ATI
(transmitter/receiver unit) 124 × 81 × 81 mm
(antenna) 105 × 90 × 38 mm
Weight:
(indicator) 1.2 kg
(transmitter/receiver unit) 1.2 kg
(antenna) 0.13 kg
Altitude: up to 5,000 ft
Accuracy: 1 ft ±2%
Reliability: >5,000 h MTBF

Operational status

In production for the Agusta A 109 helicopter and Penguin Mk 2 missile.

Contractor

Thomson-CSF Communications.

VERIFIED

AHV-2100 digital radar altimeter

The AHV-2100 fully digital radio altimeter has been specifically designed for modern military helicopters. The simple hardware and up-to-date technology provide high reliability and performance at low altitude and in hovering.

The AHV-2100 has a dual-processing chain and DO-178 qualified software. In addition, power management of the RF output reduces the probability of interception at low altitude over water and the combination of a narrow receiver bandwidth with high-performance digital signal processing provides resistance to jamming.

Specifications

Dimensions: 110 × 90 × 190 mm
Weight: <2.2 kg
Power supply: 28 V DC
Altitude: up to 5,000 ft
Accuracy: 3 ft ±5%
Reliability: 5,000 h MTBF
Interfaces: ARINC 429, dual-redundant MIL-STD-1553B or analogue

Contractor

Thomson-CSF Communications.

UPDATED

AHV-2900 digital radar altimeter

The AHV-2900 is a fully digital radio altimeter designed for future combat aircraft and tactical fighters. It is optimised for low-altitude penetration. It includes dual-processing chains, and ARINC 429 or MIL-STD-1553B interfaces.

Contractor

Thomson-CSF Communications.

UPDATED

NRAI-7(·)/SC10(·) Identification Friend-or-Foe (IFF) transponder

The NRAI-7 is a solid-state Mk XII diversity transponder which inhibits replies to interrogator sidelobe transmissions, and automatically codes special replies to provide assistance in the position identification of particular aircraft and in emergencies. The diversity function is provided by a dual receiver with inputs connected to upper and lower antennas, a system for comparison of received signals and an antenna switch that directs the response to the antenna that has received the strongest interrogation signal. This allows more accurate identification, particularly during aircraft manoeuvres which can blanket or interrupt signals. The pilot may also insert codes such as radio failure alert and warning of hijackers aboard. It is available in one- or two-box housing (see NRAI-9A). The single-box version is claimed to be one of the smallest transponders in the world. A naval version is also available.

Operational status

NRAI-7 transponders are reported to have been installed on Mirage 2000, Mirage F-1, Mirage III, Mirage IV, AS 332 and AS 335, C130 Transall and other various aircraft as well as on board several ships of the French and other navies. It has also been integrated in various Polish aircraft, including MIG fighters. Over 2,000 systems have already been delivered to the French defence forces and those of other countries.

The Thomson-CSF NRAI-7(·)/SC10(·) transponder

Specifications

Peak power: 500 W
Sensitivity: −77 dBm
Frequency: 1,030 MHz (receive); 1,090 MHz (transmit)
Modes available: 1, 2, 3A/C and Mode 4 capability
No of codes: 32 (Mode 1); 4,096 (Modes 2 and 3/A); 2,048 (Mode C)
Dimensions: 130 × 127 × 145 mm
Weight: 3 kg

Contractor

Thomson-CSF Communications.

UPDATED

NRAI-9(·)/SC15(·) Identification Friend-or-Foe (IFF) transponder

The NRAI-9A is essentially a two-box version of the NRAI-7 which incorporates a number of improvements. These include the elimination of sidelobe response, and automatic special-code referral, with positive identification permitting a ground operator to locate a particular aircraft. Special emergency codes may also be employed, chosen by the pilot, such as radio failure. Dual-receiver channels connected to upper and lower antenna and comparison circuits provide a diversity function.

Operational status

As of this edition, NRAI-9 is thought to be fitted to ATL2 maritime patrol aircraft.

Specifications

Frequency: 1,030 MHz (receive); 1,090 MHz (transmit)
Power output: 500 W peak
Modes available: 1, 2, 3A/C and Mode 4 capability
Codes: 32 (Mode 1); 4,096 (Modes 2 and 3/A); 2,048 (Mode C)
Dimensions: 58 × 193 × 361 mm (transmitter/receiver); 127 × 130 × 80 mm (control unit)
Weight: 2.5 kg (transmitter/receiver); 1.4 kg (control unit)

Contractor

Thomson-CSF Communications.

UPDATED

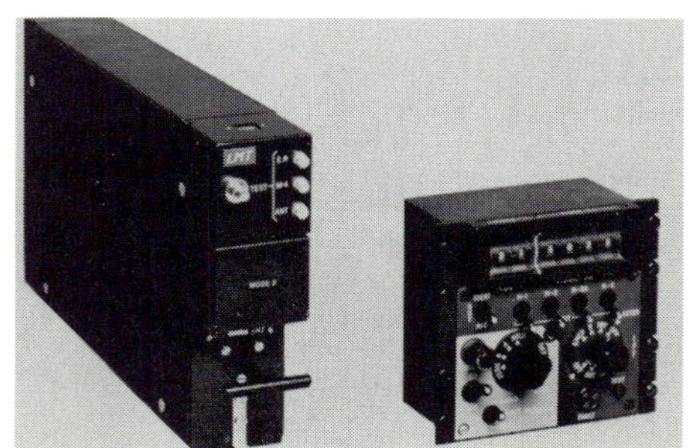
The Thomson-CSF NRAI-9(·)/SC15(·) IFF transponder

The Thomson-CSF NRAI-11/IDEE 1 Mk XII interrogator-decoder is in service in the Mirage 2000 **1995**

NRAI-11/IDEE 1 Mk XII interrogator-decoder

The airborne NRAI-11/IDEE 1 interrogator-decoder is used for identifying a friendly target and determining its range and azimuth. The interrogator-decoder and its control box are integrated in the IFF system of Dassault Mirage 2000 aircraft. The system includes all the functions needed for IFF air-to-air identification and includes an encoder, transmitter, RF switch, two receivers, analogue processing unit, defruiter, passive decoder, evaluator, extractor, staggering circuit and automatic self-test.

Volume and power consumption have been substantially reduced through the use of a solid-state transmitter, low-power integrated and monolithic high-density circuits, hybrid and custom LSI circuits and switching power supply.

Specifications

Dimensions: 124 × 80 × 194 mm
Weight: 12 kg
Power supply: 115 V AC, 400 Hz, 3 phase, 150 VA
Transmission power: 1 kW peak
Receiver sensitivity: −79 dBm between 1,087 and 1,093 MHz
Operating modes: 1, 2, 3/A and Mode 4 capability

Operational status

In service on the Dassault Mirage 2000.

Contractor

Thomson-CSF Communications.

UPDATED

TSB 2500 Combined IFF Interrogator and Transponder (CIT)

The TSB 2500 consists of two LRUs: the Combined Interrogator/Transponder (CIT) and the Antenna Control Unit (ACU) or the Antenna Adapter Unit (AAU). The system is compatible with various types of electronically (with ACU) or mechanically (with AAU) scanned antennas. The TSB 2500 is available in both interrogator and interrogator/transponder versions and is a highly flexible system capable of meeting the needs of many military platforms. It is modular in both design and operation.

It is a Mk XII and Mode S level 2/3 transponder and a Mk XII interrogator. It also incorporates provisions for the integration of future mode 5 new-generation IFF, for both transponder and interrogator functions.

The CIT integrates all transmission, reception, signal, and data processing functions required by an interrogator/transponder. It interfaces with the host platform via a MIL-STD-1553B databus.

The ACU controls electronically scanned antennas. The AAU functions as a booster and an RF front end. The packaging of ACU and AAU in separate boxes allows their installation close to the antenna system providing minimum RF losses in the cables.

The TSB 2500 can interface with any NSA Mode 4 crypto computer or secure crypto computer (for non NATO applications) that is interface compliant with STANAG 4193. It can also be fitted with a dual KIT/KIR appliqué crypto computer on the front panel.

Specifications

Interrogator
Power: >32 dBW
Frequency: 1.030 ±0.2 Mhz
Operating modes: 1,2,3/A,C,4 (Mode S upgradeable)

Transponder
Power: 500 W (±2 dB)
Freqency: 1.090 ±0.5 MHz
Operating modes: 1,2,3/A,4,S (Mode S upgradeable)
Dimensions
(CIT) 228.6 × 157.2 × 193.5 mm
(ACU) 230 × 115 × 105 mm
(AAU) 32 × 193 × 290 mm
Weights
(CIT) <10 kg
(ACU) 5.5 kg
(AAU) <4 kg

Operational status

Selected for Rafale (under flight test), Erieye, and NH 90.

Contractor

Thomson-CSF Communications.

NEW ENTRY

TSC 2050 IFF Mk XII/Mode S diversity transponder

The TSC 2050 IFF is a true diversity Mk XII and Mode S level 3, TCAS compatible transponder, which is fully compliant with ICAO Annex 10, STANAG 4193 and DOD-AIMS-65-100B standards. Moreover it includes all provisions for the integration of future (mode 5) for a new-generation IFF, in accordance with NATO recommendations.

The TSC 2050 has been designed to be shelf-mounted in non-pressurised zones, and is intended for both advanced aircraft or for retrofit purposes. It is provided with a MIL-STD-1553B bus interface, or can be operated through a control box.

The TSC 2050 transponder can interface with any NSA Mode 4 crypto computer or secure crypto computer (for non NATO applications) that is interface compliant with STANAG 4193. Optionally it can be provided with mechanical adapter to fit the crypto computer as an appliqué.

Specifications

Dimensions: 136 × 124 × 212 mm
Weight: <5.2 kg
Modes: 1,2,3/A,C,4 and S level 3

Operational status

The TSC 2050 has been selected for the UK Nimrod aircraft and for global retrofit of Romanian aircraft.

Contractor

Thomson-CSF Communications.

NEW ENTRY

Agrion maritime surveillance

Agrion is a member of the Iguane family of maritime surveillance radar systems and exists in several versions. It is designed primarily for use aboard helicopters or light aircraft forming a part of task forces, employed for support at sea or for coastal protection. Several types of antenna are available to meet the requirements of various aircraft. The Agrion 15 version allows the guidance of the AS 15TT Aerospatiale air-to-surface missile.

Agrion operates in the I/J-band, using pulse compression and frequency agility to ensure high performance on maritime targets in all combinations of weather, sea state and operating altitude. These same techniques also provide maximum protection against electronic countermeasures.

The system provides operational missions such as surface and anti-submarine warfare, over-the-horizon targeting for shipborne surface-to-surface missiles, search and rescue, marine environmental protection, navigation and weather avoidance.

Operational status

Agrion 15 radars are reported to be in service aboard Eurocopter AS 565SA Panther anti-submarine/anti-ship warfare helicopters operated by the Saudi Navy.

Contractor

Thomson-CSF Radars Contre-Mesures.

UPDATED

Chin-mounted Agrion 15 radar on an AS 565SA Panther anti-submarine/anti-ship warfare helicopter **1998**

The Thomson-CSF Cyrano IV-M multimode radar

Cyrano IV radars

The Cyrano IV is a development of the Cyrano series of airborne radars and is a multirole system. The Cyrano IVM is the newest version of this series. Basic functions performed by the radar are:
(a) air-to-air search
(b) automatic tracking
(c) interception and fire-control computations
(d) dogfight engagements
(e) home-on-jam mode
(f) ground-mapping
Additional options are:
(g) contour mapping
(h) terrain-avoidance
(i) blind let-down
(j) air-to-ground ranging.

The Cyrano IVM model incorporates track-while-scan facilities and is also suitable for air-to-sea search and tracking roles in addition to those listed for the Cyrano IV.

Data can be presented by means of a Type 196 gunsight or CRT HUD, and inputs to weapon systems are also available. Other aircraft systems providing inputs to the radar system include an inertial or gyro platform for altitude reference information, and an air data computer for aircraft performance and ambient parameters.

Operational missions include: interception, air superiority or interdiction using guns or missiles; all-weather penetration; air-to-ground attack with guns, bombs and rockets.

The Cyrano IVM, as well as being a multifunction radar, differs from its predecessor in the embodiment of new technology which confers improved reliability and maintainability on the later model. Cyrano IVMR is reported to have been developed for use on the Mirage F1-CR-200 tactical reconnaissance aircraft. It is said to incorporate ground mapping, contour mapping, air-to-surface ranging and blind let-down operating modes. Cyrano IVM3 is noted as having been designed for use on the Mirage 50 and as a retrofit equipment for the Mirage III and V. Incorporating technology from the RDM and RDI programmes, Cyrano IVM3 is reported to offer air-to-air, air-to-surface and maritime operating modes.

Operational status

Cyrano IV series radars are reported to be in service aboard Mirage F1 aircraft operated by the air forces of Ecuador, France, Greece, Iraq, Jordan, Kuwait, Libya, Morocco, South Africa and Spain. Cyrano IVM3 radars are noted as being used in Venezuela's eight aircraft Mirage 50EV/DV upgrade programme.

Contractor

Thomson-CSF Radars Contre-Mesures.

UPDATED

Iguane sea surveillance radar

Iguane is a maritime surveillance radar that has been produced to replace the DRAA2A sea surveillance radar fitted in the Breguet Alizé and for the Dassault Atlantique 2 long-range maritime patrol aircraft.

The system operates in the I/J-band, using pulse compression and frequency agility to ensure high performance against maritime targets in all combinations of weather, sea state and operating altitude, and to provide maximum protection against ECM. Operational missions performed by Iguane include surface and anti-submarine warfare, over-the-horizon targeting for shipborne surface-to-surface missiles, environmental protection, navigation and weather avoidance.

Operational status

In service on the Breguet Alizé update programme and for the Dassault Atlantique 2. The Italian Air Force has fitted 18 Iguane systems to its fleet of Atlantic 1 aircraft.

Contractor

Thomson-CSF Radars Contre-Mesures.

VERIFIED

The Thomson-CSF Radars Contre-Mesures ORB 32 radar is used for surveillance and fire-control applications in maritime aircraft

ORB 32 radar systems

ORB 32 is a range of I-band airborne radars. It features very low weight and power consumption and can be installed on a wide range of helicopters and aircraft. ORB 32 is built from modular subassemblies, enabling easy extension, and is designed for exclusive economic zone control, anti-surface warfare, anti-submarine warfare, active missile fire control, search and rescue, radar navigation and weather avoidance.

It features pitch and roll stabilisation, 360° azimuth and 30° elevation scanning and 60, 120, 180 and 240° sector scans, and azimuth, bearing or true motion stabilisation. The peak power is typically 70 kW.

The ORB 32 is available in a number of versions:

The ORB 3201 and ORB 3211 are simple compact and lightweight systems suitable for small aircraft or helicopters, specially designed for surface reconnaissance, exclusive economic zone control and search and rescue. They provide navigation information and weather avoidance.

The ORB 3202 and ORB 3212 are airborne reconnaissance and target designation radars. When integrated into a weapon system, their purpose is to detect, designate and accurately track two sea targets. Target co-ordinates may be automatically transmitted to active missiles carried by aircraft or helicopters or to a launch vessel for over-the-horizon targeting.

The ORB 3203 and ORB 3214 are one element of an anti-submarine warfare weapon system for helicopters or aircraft. They enable helicopter station holding in ASW, tactical situation information, guidance and aircraft attack on a designated target, navigation, weather and mapping. The ORB 3203 and 3214 perform both primary and secondary radar functions. Use of transponders makes identification of helicopters flying at low altitude possible, even if the primary echo is in sea clutter.

Operational status

In production for the French Navy and the armed forces of a number of other countries. Installed in Dauphin, Super Frelon, Nord 262, Super Puma and Boeing Vertol 107 helicopters.

Contractor

Thomson-CSF Radars Contre-Mesures.

VERIFIED

ORB 37 radar system

ORB 37 has been designed to meet the French Air Force navigation requirements for the C-160 Transall transport aircraft. It carries out weather avoidance and accurate ground-mapping functions, and is fitted with an interrogation facility for beacon homing.

The system consists of seven units; a slotted array flat-plate antenna, a transmitter/receiver, a power supply, a Plan Position Indicator (PPI) high definition circular display for ground-mapping at the navigator's station, a digital PPI display on the flight deck, and two control units, one for each station.

For maximum efficiency the antenna scans at low rate for the weather mode, and at high rate for the ground-mapping mode. The corresponding pulsewidths are 2.5 and 0.4 µs.

Specifications

Frequency: I-band (9,375 MHz)
Power output: 10 kW

Operational status

ORB 37 is reported to be in service with the French Air Force.

Contractor

Thomson-CSF Radars Contre-Mesures, Elancourt.

NEW ENTRY

Raphael SLAR 2000 surveillance radar

The SLAR 2000, also known as the Raphael TH, is an all-weather side-looking airborne radar employing synthetic aperture and pulse compression techniques to provide high-quality mapping. The airborne part of the system is pod-mounted on a combat aircraft. It can also be installed in the cargo bay of a commuter or transport aircraft. Radar information is transmitted via datalink to a ground station where it is displayed in real time.

The radar is highly directional and operates in the I/J-band. It has an effective beamwidth of a few mrads, providing a sharp and accurate radar map of the ground. The system also features an MTI capability.

The Thomson-CSF Radars Contre-Mesures SLAR 2000 is fitted on the Dassault Mirage F1 CR

Ground echoes are processed on the aircraft, then transmitted to a ground station which is air or ground transportable.

Operational status

In service with the French Air Force and the air forces of several other countries.

Contractor

Thomson-CSF Radars Contre-Mesures.

VERIFIED

RC 400 compact multimission multitarget radar

The RC 400 compact multimission multitarget radar has been designed for lightweight fighters and advanced training aircraft. RC 400 technology is derived from the RDY radar designed for the Mirage 2000-5. The baseline architecture comprises four line-replaceable units (LRUs): the antenna unit, the exciter/receiver, the transmitter, and the processor unit. This configuration can be reconfigured easily to fit the nose of most aircraft types and requires only non-filtered air cooling. Several antenna options are available, based on a low-inertia flat slotted plate design with very low sidelobes.

The RC 400 design features: track-while-scan, smart automatic management of scanning and target prioritisation to provide accurate, multiple, fire-and-forget capability, together with datalink requirements for missile control. The radar utilises high, medium and low PRFs to optimise performance in all combat situations. Within a given mode, the PRF is automatically managed from bat to bar, with respect to attack geometry.

It offers the following operational capabilities:

Air-to-air: all-aspect, look-up/look-down detection; automatic management of low, medium and high PRFs; automatic lock-on; simultaneous multitarget engagements; IFF; 55 n mile range

Air-to-ground: improved ground mapping (azimuth beam compression); Doppler beam sharpening; ranging

Air-to-sea: track-while-scan on two targets; target calibration.

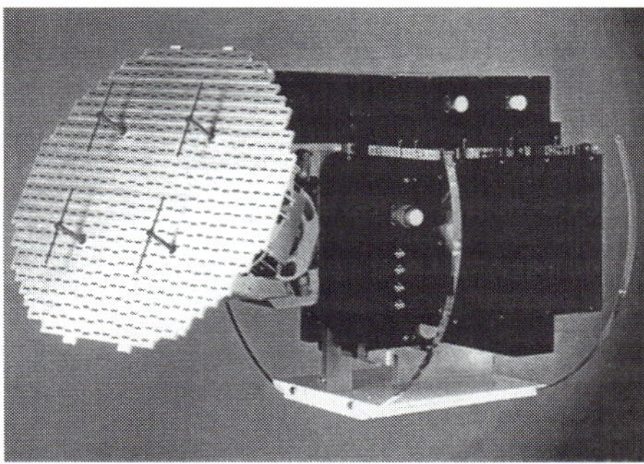

RC 400 compact multimission multitarget radar
***1998**/0011895*

Specifications

Transmitter unit: I/J-band; air cooled
Receiver unit: 2 channels; wide dynamic range
Processor unit: fully programmable signal and data processing; 1 Gflop
Antenna unit: monopulse flat slotted array; elliptical or circular; IFF interrogator
Weight: 130 kg
Power consumption: 3.6 kVA

Operational status

Flight testing in 1998. Initial deliveries planned for year 2000.

Contractor

Thomson-CSF Radars Contre-Mesures.

NEW ENTRY

The Thomson-CSF Radars Contre-Mesures RDI radar equips the Dassault Mirage 2000

ORB 37 weather and navigation radar ***1991***

RDI (Radar Doppler à Impulsions)

The RDI is one of two pulse Doppler radars (the other being the RDM) developed for France's Dassault Mirage 2000. It is intended for the all-altitude air superiority and interception version and is based on a travelling wave tube, I/J-band transmitter radiating from a flat, slotted plate antenna. The range is said to be around 90 km in look-down and the radar is designed to work in conjunction with the 40 km range Matra Super 530D semi-active homing air-to-air missile. The performance of the radar, and of other systems on the aircraft, benefits from the digital signal handling and transmission of information by databus. Considerable electronic countermeasures resistance is built into the equipment, which can operate in air-to-air search, long-range tracking and missile guidance, and automatic short-range tracking and identification modes. Although designed for air-to-air operation, the system incorporates ground-mapping, contour-mapping and air-to-ground ranging modes.

The high PRF available (100 kHz+) guarantees accurate target speed assessment with a Thomson CSF patented process to give range data at maximum range in search mode (as well as in tracking mode).

Air-to-air modes include: air-to-air search; long-range TWS or continuous target tracking and missile guidance; automatic short-range tracking for missiles or guns. Air-to-ground modes include: ground mapping; contour mapping; air-to-ground ranging. Total weight of the 11 line-replaceable units is 255 kg.

The system was developed by Thomson-CSF in collaboration with Dassault Electronique which undertook 30 per cent of the work.

Operational status

In production and in service with the French Air Force.

Contractor

Thomson-CSF Radars Contre-Mesures.

UPDATED

RDM (Radar Doppler Multifunction)

The RDM monopulse Doppler I/J-band radar is in production for the French Dassault Mirage 2000. Whereas the RDI is designed for interception and air combat, the coherent, multimode, all-digital, frequency-agile RDM is intended largely for the multirole export version. It operates in air defence/air superiority, strike and air-to-sea modes.

In the air-to-air role, the system can look up or down, range while searching, track-while-scan, provide continuous tracking, generate aiming signals for air combat and compute attack and firing envelopes. For the strike role it provides real-beam ground-mapping, navigation updating, contour-mapping, terrain-avoidance, blind let-down, air-to-ground ranging and GMTI (Ground Moving Target Indication). In the maritime role it provides long-range search, track-while-scan and continuous tracking, and can designate targets for active missiles.

For air-to-air combat, the RDM provides 120° cone of coverage, the antenna scanning at either 50°/s or 100°/s, with ±60°, ±30° or ±15° scan. A 5 m² target can be detected at up to 111 km range. For air-to-air gun attacks, the 3.5° beam can be locked to the target at up to 19 km range, with automatic tracking within the head-up display field of view, or in a 'super-search' area, or in a vertical search mode. In look-down, air-to-air, scenario, a 5 m² target can be detected at up to 46 km range.

Options include a continuous wave illuminator and Doppler beam-sharpening. Comprehensive electronic counter-countermeasures are incorporated.

Significant improvements to the radar, particularly to the look-down function, have been completed. Improvements include hardware and signal processing.

Operational status

In production. RDM radar equips the first Mirage 2000 squadrons in the French Air Force, as well as the air forces of Egypt, Greece, India, Peru and the United Arab Emirates.

Contractor

Thomson-CSF Radars Contre-Mesures.

UPDATED

RDY multifunction radar

RDY is the multifunction doppler radar for the Mirage 2000-5 aircraft.

In air-to-air mode, the RDY detects very low or high altitude targets at long range, irrespective of their angle of approach. It presents the pilot with tactical situation analysis, offers multitarget tracking and IFF interrogation. Three combat modes can be selected for close-in engagement. Air-to-ground modes are provided for all-weather low-altitude penetration through high-performance Doppler beam sharpening for navigation update, blind penetration, terrain avoidance, ground-mapping and contour-mapping. It provides the system with air-to-ground ranging for ground attack missions. In air-to-sea mode, the RDY is able to detect targets, even in high sea states up to 296 km, perform multitarget tracking and target allocation fo ASMs such as AM 39 or Kormoran II.

The radar uses high-, medium- and low-PRF waveforms. High PRF is used for long-range detection of low- flying targets; low PRF for high-flying targets and a medium PRF for all-aspect medium-range detection in ground clutter.

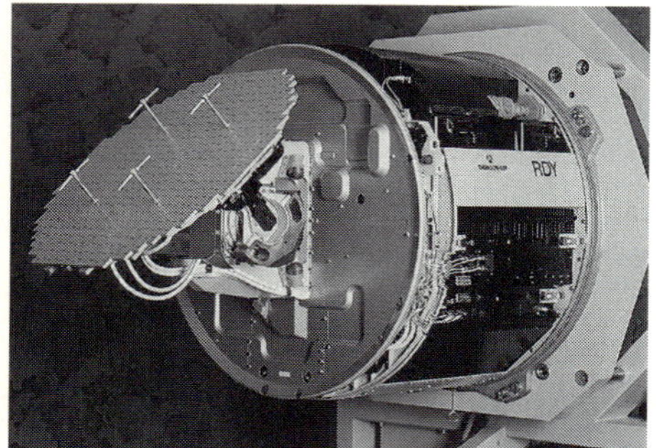

The Thomson-CSF RDY multitarget airborne fire-control radar **1996**

Signal processing is carried out by a programmable signal processor with a very large computational capability: a speed of 100 Mcops is claimed. This computational throughput enables additional options such as Doppler beam-sharpening, Ground Moving Target Indicator (GMTI) and efficient track-while-scan air-to-sea modes. Target handling capability is quoted as the ability to detect 24 targets and to track-while-scan the eight most threatening ones.

A 60 cm diameter flat-plate phased-array antenna is used with four integral high-frequency dipoles for IFF interrogation.

Operational status

In production since 1995. The RDY radar equips the Mirage 2000-5 for the French Air Force and export customers reported to be Qatar (12 aircraft) and Taiwan (48 single-seat and 12 two-seat aircraft). RDY is also being retrofitted into 37 French Air Force Mirage 2000C aircraft, and was reported to be in full operational service in 1997. RDY is the probable radar fit for the Mirage 2000-9 aircraft being negotiated by the UAE. At the Paris 1997 Air Show, Thomson-CSF unveiled a new SAR imaging function, which provides the RDY radar with an all-weather reconnaissance capability.

Contractor

Thomson-CSF Radars Contre-Mesures.

UPDATED

Romeo II obstacle avoidance radar

Romeo is an M-band obstacle avoidance radar intended primarily for helicopter use. It employs a solid-state transmitter/receiver, operating at a frequency of 94 GHz, with high density digital processing and a programmable computer. It has a very high detection capability on small targets such as power cables, pylons, antennas, bridges, trees and offshore rigs. The small size and weight of Romeo allows it to be fitted in a wide range of fixed-wing aircraft and helicopters.

Display presentation consists normally of a colour landscape image display, with different colours representing obstacles. Red is the danger colour and warns of close and dangerous objects. A head-up or helmet-mounted display can also be used for synthetic presentation of obstacle warning. Power output is in the region of 400 mW giving a range in the fully developed models of approximately 3,000 m.

Operational status

Romeo II is believed to be undergoing trials and a variant with a maximum detection range of 1 km is believed to have been developed. Romeo is also understood to have been tested by the US Army aboard an AH-1 attack helicopter.

The Thomson-CSF Romeo obstacle avoidance radar **1998**

Specifications

Frequency: M-band (94 GHz)
Range: up to 3,000 m in the fully developed model
Power output: 400 mW
Dimensions: 145 × 110 × 70 mm (control unit); 240 × 240 × 320 mm (antenna); 315 × 190 × 190 mm (processing unit)
Weight: 1 kg (control unit); 11 kg (antenna); 12 kg (processing unit)

Contractor

Thomson-CSF Radars Contre-Mesures, Elancourt.

NEW ENTRY

Varan sea surveillance radar

Varan is essentially an Iguane radar (see earlier item) with a smaller antenna which makes it suitable for virtually all the present and planned lightweight maritime patrol aircraft and helicopters. It has been fitted to the Dassault Falcon Gardian of the French Navy and selected for the naval version of the ATR 42 transport aircraft and Eurocopter SA 365F Dauphin 2 helicopter.

The system provides real-time pollution and ice detection. Key factors are I-band operation, pulse compression over several pulsewidths, frequency agility for electronic counter-countermeasures and beacon detection. The unspecified but low-peak power level, associated with high receiver sensitivity, increases the difficulty of detection by hostile radars. Typical detection ranges, in Sea State 3 to 4 are: snorkel 55 km, fast patrol boat 110 km and freighter 240 km. Total weight of the system's six units (including antenna) is 111 kg.

Operational status

In service on the French Navy's Falcon 20H Guardian aircraft. In service on Super Puma helicopters and Falcon 20 aircraft with foreign customers.

Contractor

Thomson-CSF Radars Contre-Mesures.

UPDATED

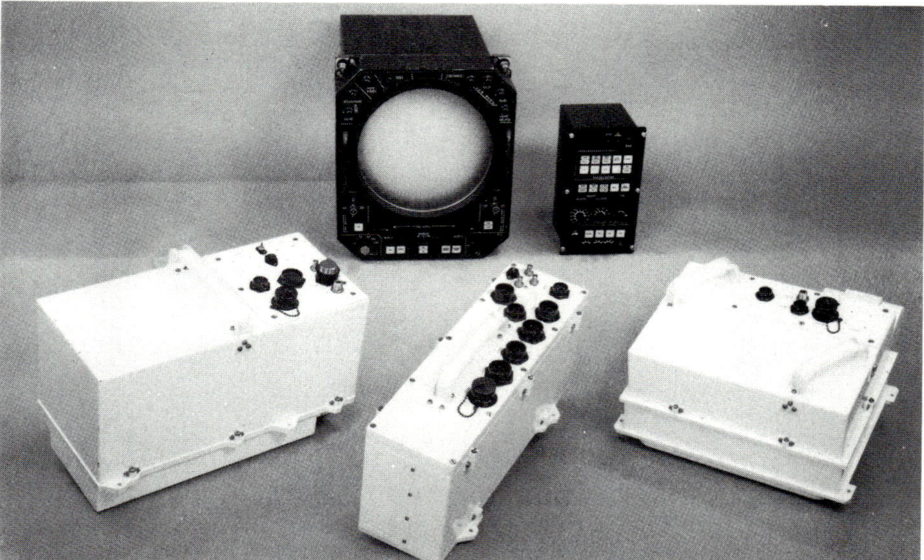

The Thomson-CSF Radars Contre-Mesures Varan surveillance radar

GERMANY

ATC 2000-(3)-R system

This small, lightweight air traffic control transponder is designed for installation where instrument panel space is limited. Its CU 5401 control unit is designed to be mounted into a standard 2¼ in (57 mm) round instrument panel cut-out, and is only 2½ in (63.5 mm) deep. The type 300 W mode A/C transmitter and receiver are housed in a separate unit, which can be installed remotely in the aircraft.

The unit uses a clear, high-contrast, double line LCD, which is readable under all lighting conditions including bright sunlight. When an altitude encoder is connected, the reported flight level is displayed below the transponder code, to verify correct operation of the entire system.

The standard 4096 identification codes are selected by the rotary selector switch.

Two preset identification codes, for VFR flight or other purposes, can be entered from the front panel and stored in non-volatile memory for instant recall by a single key stroke.

The ATC 2000-(3)-R transponder is suitable for all types of fixed- and rotary-wing aircraft. It operates from both 14 and 28 V input power, and is certified for both VFR and IFR operations. It complies with the requirements of JTSO C74c, class 1A, and can report altitudes up to 62,700 ft. Low-power consumption and small size and weight make it suitable for use as a standby transponder.

The ATC 2000-(3)-R transponder can be combined with other Becker CompactLine avionics systems, such as the COM 5200 or AR 4201 VHF transceivers or the NAV 5300 VOR/ILS navigation systems, which have similar control units. The transponder remote receiver/transmitter unit can be controlled by other types of CDU or FMS.

A special version of the ATC 2000-(3)-R, designated the ATC 2000-(3)-R62, is available for high-altitude operation up to 60,000 ft, fitted with a special pressure box.

Specifications

Dimensions: CU 5401 61.3 × 61.3 × 62 mm
ATC 3401-(1)-R 134 × 50 × 253 mm
Power supply: 10-32.2 V DC; 1.1 A at 14 V; 0.55 A at 28 V
Transmitter frequency: 1,090 MHz
Receiver frequency: 1,030 MHz
Modes: A and C
Control interface: RS-422

Contractor

Becker Avionic Systems.

UPDATED

ATC 2000-(3)-R system ***1997***/0001197

ATC 3401 Mode A/C transponder

The ATC 3401 is certified in accordance with FAA TSO C47c, Class 1A, for the highest level of unrestricted service and can report altitudes up to 62,700 ft.

The ATC 3401 transponder displays 4,096 identification codes in its left window, and can display its operating mode or flight level reporting altitude in its right window. A stored VFR code, such as 1200, can be quickly recalled with a VFR push-button. Provisions are included to enable IDENT to be controlled from a button on the control stick and for power to be turned on and off from an external control, if required.

The complete system is housed in a single compact unit, which complies with ARINC standards. It does not require remote boxes, interconnecting cables, or external forced-air cooling.

To facilitate dual-transponder installations, provisions are included for automatic transfer of one transponder to the standby mode, when the other transponder is selected for normal mode operation. This prevents both transponders transmitting at the same time.

The ATC 3401 Mode A/C transponder is a member of the Becker PrimeLine family of avionics equipments.

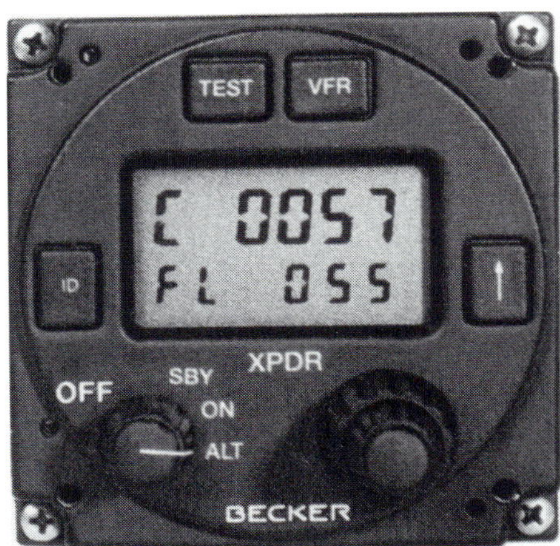

ATC 4401 solid-state transponder ***1997***/0001195

Specifications

Dimensions: 146 × 47.5 × 217 mm (not including clearance for connectors and cables)
Weight: 1.2 kg
Power supply: 10-32 V DC; at 28 V DC, 0.8 A normal, 0.3 A standby; at 14 V DC, 1.5 A normal, 0.5 A standby
Transmitter frequency: 1,090 MHz ± 0.3 MHz
Receiver frequency: 1,030 MHz ± 0.2 MHz
Modes: A, A + C

Operational status

In service.

Contractor

Becker Avionic Systems.

VERIFIED

ATC 4401 transponder

The ATC 4401 transponder is designed in a 57 mm round format for the restricted panel space in gliders and small single-engine aircraft. The transmitter and receiver, which operate on Modes A and C, can be installed separately in any convenient location in the aircraft.

Contractor

Becker Avionic Systems.

VERIFIED

DRA 100 digital radar altimeter

The DRA 100 is a fully digital multipurpose FMCW altimeter for all air vehicle types such as fighter aircraft, helicopters, missiles and UAVs, for both military and civil applications.

The system features extremely low probability of interception, high resistance to jamming and integrated self-test with periodic monitoring.

Specifications

Dimensions: 228 × 148 × 70 mm including antenna
Weight: 1.9 kg
Power: <70 mW
Accuracy: <2 ft ±2%
Altitude: up to 5,000 ft
Manoeuvrability: ±80° roll
Outputs: ARINC 429, MIL-STD-1553B
Reliability: >8,700 h MTBF

Operational status

In production for the JAS 39 Gripen, Eurocopter Tiger; selected for the RAH-66 Comanche.

Contractor

Daimler-Benz Aerospace AG, Defense and Civil Systems.

UPDATED

MIDAS Microwave Imaging DLR/ Dasa Airborne System

The Institute for Radio Frequency Technology of the German Aerospace Research Establishment DLR in Oberpfaffenhofen, with its experience of airborne Experimental SAR (E-SAR), and the Airborne Systems unit of Daimler-Benz Aerospace, Defense and Civil Systems in Ulm, with its knowledge of SAR and Inverse-SAR (I-SAR), are pooling their knowledge of Synthetic Aperture Radar (SAR) technology to develop MIDAS as an optimum solution to both reconnaissance and classification requirements of land and sea targets.

MIDAS missions are planned to be: all-weather sea and land monitoring; ground mapping and cartography; classification of agriculture, forestry and geological resources; disaster monitoring. Different configurations can be optimised to particular requirements.

Radar characteristics include: operation in D-, E/F-, or I-/J-band with either horizontal or vertical polarisation; high pulse compression ratios; single- and multiscan processing; real-time SAR processing; fully focused, motion compensated radar images, digital programmable processing; recording of raw and processed data (on EXABYTE).

The experimental DLR E-SAR mounted under a Do 228 aircraft ***1998***/0011894

Operational status

Experimental installation on Do 228 aircraft.

Contractors

Daimler-Benz Aerospace AG, Defense and Civil Systems.

Deutsche Forschungsanstalt für Luftund Raumfahrt e.V.

NEW ENTRY

STR 700 IFF transponder

The STR 700 IFF transponder is used by nearly all German armed forces aircraft. It features high reliability, small dimensions, low weight and diversity of operations.

The STR 700 meets AIMS specifications. It consists of two basic units: the receiver/transmitter and the control unit with logic section. The control unit has interfaces for the Mode 4 decoder/coder facility and for an altitude encoder. The STR 700 has been specifically designed for applications where diversity operation is required and is provided with two receiving channels. It can also, however, be equipped with a single receiving channel only. The receiver/transmitter of modular design is accommodated in a standard ½ ATR short case for both models. Usually the control unit consists of the actual control section, with switches and lamps, and of the logic section for decoding the interrogation signals and for coding the reply signals. If the available space in the cockpit is too narrow, the logic section can be accommodated separately from the control section.

Reliable identification of the target aircraft depends largely on the correct action of the transponder. Consequently the STR 700 transponder is provided with a large number of test circuits, to reveal the functional condition at any time automatically by internal interrogations. For system checking on the ground, the receiver/transmitter of the two-channel version is provided with 12 LEDs to indicate proper performance of the individual subassemblies.

Specifications

Dimensions:
(receiver/transmitter) 124 × 193 × 382 mm
(control unit/logic section) 146 × 134 × 155 mm
Weight:
(receiver/transmitter) 9.8 kg (single channel)
10.8 kg (two channel)
(control unit/logic section) 3.2 kg
Power supply: 16-32 V DC, up to 70 W
Frequency:
(receive) 1,030 ±1.5 MHz
(transmit) 1,090 ±3 MHz

Operational status

In production and service in aircraft of the German armed forces.

Contractor

Siemens AG Defence Electronics Group.

VERIFIED

INDIA

400AM IFF

The 400AM IFF Mk X operates in Modes 1, 2 and 3A/C, with 4,096 codes to full ICAO standards.

Specifications

Power supply: 27.5 V DC (nominal)
115 V AC, 400 Hz, 90 W (option)
Temperature range: −55 to +55°C
Altitude: (pressurised) up to 70,000 ft (unpressurised) up to 40,000 ft

Contractor

Hindustan Aeronautics Ltd.

VERIFIED

405A IFF transponder

The 405A IFF transponder is all solid-state and of modular construction. It provides automatic replies to appropriate ground or airborne interrogators operating on the Mk X IFF system, transmitting on 1,090 MHz and receiving on 1,030 MHz. It operates in Modes 1, 2, 3A and 3C on the full 4,096 codes.

Specifications

Dimensions:
(transponder) 122 × 201 × 201 mm
(control unit) 146 × 81.5 × 70 mm
Weight:
(transponder) 9 kg
(control unit) 0.6 kg
(mounting tray) 1.4 kg
Power supply: 115 V AC, 400 Hz, 150 VA (max)

Contractor

Hindustan Aeronautics Ltd.

VERIFIED

IFF 1410A transponder

The IFF 1410A transponder system has been designed for operation with a stand-alone control unit, or from a centralised controller through MIL-STD-1553B bus. The equipment uses all solid-state technology hybrid modules to achieve high reliability and ease of maintenance. Full operation in accordance with ICAO Annex.10 is provided. Extensive self-test diagnostic features help in identifying the faults to functional level. A secure mode of operation is available as an option.

Specifications

Operation modes: 1, 2, 3/A, C, Secure
Codes: 4,096 in Modes 1, 2, 3/A
2,048 in Mode C
Indentification facility
Military Emergency facility

IFF 1410A transponder ***1998***/0011893

Power supply:
(AC) 108-118 V, 400 Hz, single phase, 0.6 A (max)
(DC) 22-32 V, 0.5 A (max)
Dimensions: 310 × 100 × 194 mm (18 mm extra height with mounting tray)
Weight:
(without mounting tray) 7 kg
(with mounting tray) 8 kg

Contractor

Hindustan Aeronautics Ltd.

UPDATED

Indian ASWAC System

The Indian Airborne Surveillance Warning And Control (ASWAC) System is an indigenous programme to provide the Indian Air Force with an early warning aircraft. Development commenced in the mid-1980s.

The latest system is based on the HS-748 equipped with French and Indian surveillance equipment. Daimler-Benz has designed and built the radome; the Indian Electronics and Radar Establishment and Bharat Electronics have designed and manufactured the avionics. Most radar components are produced in India. No technical details on the system are available.

Operational status

Up to six ASWAC aircraft were originally envisaged, but the current requirement is unknown.

Contractor

Hindustan Aeronautics Ltd.

VERIFIED

RAM-700A radio altimeter

The RAM-700A radio altimeter features all solid-state modular construction. Operating on a frequency of 4.2 to 4.4 GHz, it provides indication of height over terrain up to 5,000 ft to an accuracy of ±2 ft up to 100 ft and four per cent above that.

Specifications

Dimensions: 124 × 97.2 × 366 mm
Weight: <4 kg
Power supply: 115 V AC, 400 Hz, single phase, 80 VA

Contractor

Hindustan Aeronautics Ltd.

VERIFIED

RAM 1701A radio altimeter

The RAM 1701A currently being developed by HAL, Hyderabad for the LCA programme, is an upgraded version of the existing RAM-700A, incorporating state-of-the-art technology. It can be interfaced to an onboard mission computer. The equipment is modular in design, highly reliable and small in size.

Specifications

Transmitter frequency: 4,200-4,400 MHz
Power: <35 W
Power supply: 22 V DC to 29 V DC
Altitude range: 0-1,500 m
Altitude accuracy: 0-30 m (±1 m +3%)
30 to 1,500 m (±4%)
Pitch limit: ±20°
Roll limit: ±25°
Dimensions:
(without mounting tray) 123 × 175 × 176 mm
(with mounting tray) 132 × 230 × 194 mm
Weight:
(without mounting tray) 3.5 kg
(with mounting tray) 4 kg

Operational status

Radio altimeters manufactured by Hindustan Aeronautics Ltd are fitted on An-32, MiG-21, Jaguar, Dornier aircraft and on Cheetah and ALH helicopters.

Contractor

Hindustan Aeronautics Ltd.

UPDATED

INTERNATIONAL

AMSAR (Airborne Multirole multifunction Solid-state Active array Radar)

GEC Thomson DASA Airborne Radar (GTDAR), is the joint venture set up by GEC-Marconi, Thomson-CSF (TCSF), and Daimler Benz Aerospace (DASA) to develop technology for a future Airborne Multirole Solid-state Active array Radar (AMSAR) under contract to the British, French and German ministries. The active array would succeed the Euroradar and RBE2 systems which are being developed for the Eurofighter and Rafale respectively. All three partners would participate in supplying complete production equipment to meet domestic requirements through GTDAR.

The project will culminate in a flyable technology demonstrator. The Defence Evaluation Research Agency handles British Government involvement, the Delegation Generale pour l'Armement looks after the French interests and Bundesant für Wehrtechnik und Beschaffung for Germany. Funding is divided between all three nations focusing on the development of Gallium Arsenide (GaAs) Monolithic Microwave Integrated Circuits (MMIC), active-array technologies.

This phase is nearing completion and the design of the flight demonstrator is underway.

AMSAR future active radar for combat aircraft ***1997***/0001198

Operational status

In development by GTDAR.

Contractor

GTDAR, Elancourt, France.

Subcontractors

GEC-Marconi Avionics Ltd, Radar Systems Division.
Thomson-CSF, Radars Contre-Mesures.
Daimler-Benz Aerospace AG, Defense and Civil Systems.

UPDATED

CLARA CO² Laser Radar

CLARA is an obstacle avoidance radar. Development of CLARA derives from an Anglo/French government-to-government initiative, resulting in a consortium comprising GEC-Marconi Avionics and Dassault Electronique. Under the work-share arrangements two identical demonstrator units are being produced and tested on a fixed-wing aircraft in the UK and on a helicopter in France.

CLARA is a self-contained CO² laser radar housed in an environmentally controlled pod and mounted on a helicopter or fixed-wing aircraft. It is designed for avoiding obstacles such as cables, pylons and so on, and will also provide other functions such as terrain-following, target ranging and designation, short-range true air-speed measurements, and moving target indication.

Operational status

In development under contract to the Service des Programmes Aeronautiques of France and the Defence Evaluation Research Agency in the UK, is understood that demonstration systems have been fitted to helicopters in both countries.

Flight trials were scheduled to start in 1996; present status uncertain.

Contractors

Dassault Electronique.
GEC-Marconi Avionics Ltd.

UPDATED

Euroradar ECR 90

The multimode radar for Eurofighter is currently in development by the Euroradar Consortium led by GEC-Marconi Avionics Ltd. The consortium was awarded the contract after a competitive evaluation with other systems.

ECR 90 is a third-generation coherent I/J-band multimode radar, based on the technology of Blue Vixen. It incorporates a significant increase in processing power which will exploit fully the high information content of the advanced transmission waveform, with wideband spread-spectrum, allowing operations to be maintained in a hostile EW environment. The radar comprises six line-replaceable units; the scanner; a waveguide unit; a two-module receiver; and a two-module transmitter.

Air-to-air features of the ECR 90 radar include: look-up and look-down capability; multitarget track-while-scan; target identification and prioritisation. Electronic CounterMeasures (ECM) resistance is incorporated to classify and counter jamming. Air-to-surface features include ground mapping/ranging, terrain avoidance, weapon release computation and sea surface search.

Operational status

ECR 90 radar is now in the latter stages of the development programme. Flight trials on a BAC 1-11 aircraft have shown that the radar performance is on schedule and the system is reliable. Prototype DA4 and DA5 Eurofighter aircraft both flew with operating ECR 90 radars on their maiden flights early in 1997; both aircraft are now engaged on a full evaluation of the ECR 90's radar properties for the detection and tracking of airborne targets. It is reported that in recent trials, the radar has detected and tracked fighter targets at over 160 km and large aircraft at up to 320 km.

GEC-Marconi has been awarded a contract by British Aerospace for the ECR 90 to equip Eurofighter. The contract covers the establishment of processes and facilities across the Euroradar Consortium to enable the production of radars through to the year 2015.

Contractors

GEC-Marconi Avionics Ltd, Radar Systems Division, (Prime Contractor).

Alenia Difesa, Avionic Systems and Equipment Division, FIAR.

Daimler-Benz Aerospace AG, Defense and Civil Systems.

ENOSA.

UPDATED

The Euroradar ECR 90 radar for the EF 2000

ELOISE CO^2 laser radar

ELOISE is a helicopter-borne laser radar modular Obstacle Warning System (OWS). It has been developed to meet the functional and technical requirements for a helicopter-borne sensor to improve safety during training and operational flying and to give aircrews improved confidence in poor weather conditions. ELOISE detects obstacles around the flight trajectory and provides a timely warning to allow an effective avoiding manoeuvre. In particular, it can detect wires and extended objects, critical during nap of the earth operations. In addition, despite being an active sensor, it is stealthy.

Specifications

Warning time for 5 mm cable: 10/s
Field of view: defined by the flight envelope
Optional interfaces: MIL-STD-1553-B, Video I/O
Power consumption: 280 W at 28 V DC
Overall weight: 20 kg

Operational status

The development of ELOISE has resulted in a low-cost modular approach to the OWS requirement. The consortium has the ability to offer a base level stand-alone system, or to integrate ELOISE with the rest of the sensor/avionic system to perform image motion uncoupling.

Contractors

GEC-Marconi Avionics Ltd.

Dassault Electronique.

Marconi Spa.

Zeiss-Eltro Optronic GmbH.

UPDATED

ENR European Navy Radar for the NH 90-NFH

Daimler-Benz Aerospace AG, FIAR, and Thomson-CSF have signed an agreement to co-operate in the design, manufacture and marketing of an airborne sea-surveillance radar — the ENR European Navy Radar — for the naval version of the NH 90 helicopter (NH 90-NFH).

The ENR is directly derived from the APS 784 produced by Eliradar (a consortium of FIAR and Officine Galileo), and the Ocean Master radar produced by the three ENR partners. It features state-of-the-art developments from these radars, including Inverse-Synthetic Aperture Radar (ISAR) processing.

Operational status

Based on the proposal submitted by the ENR partners, it is understood that NH Industries recommended the selection of this radar to the participating governments.

Contractors

Alenia Difesa, Avionic Systems and Equipment Division, FIAR.

Daimler-Benz Aerospace AG, Defense and Civil Systems.

Thomson-CSF Radars Contre-Mesures.

NEW ENTRY

IFF for Eurofighter 2000

Daimler-Benz Aerospace AG (Germany), Italtel (Italy), and Raytheon Systems Limited have teamed to design, develop and produce the IFF interrogator and IFF Mode S transponder for Eurofighter 2000. The system is compatible with IFF Mk XII, as defined by STANAG 4193.

Daimler-Benz Aerospace AG is also offering a derivative system for the NH 90-NFH helicopter.

Specifications

IFF Mode S transponder
Dimensions: ½ ATR intermediate length
Weight: 9 kg
Characteristics:
Mk XII (mode 1,2,3/A,C,4) compliant with STANAG 4193
Mode S
MIL-STD-1553B
Embedded cryptographic module
ADA software
Automatic Code Change (ACC)
Growth to Next Generation IFF (NGIFF)
BIT modes (P-BIT, C-BIT, I-BIT)

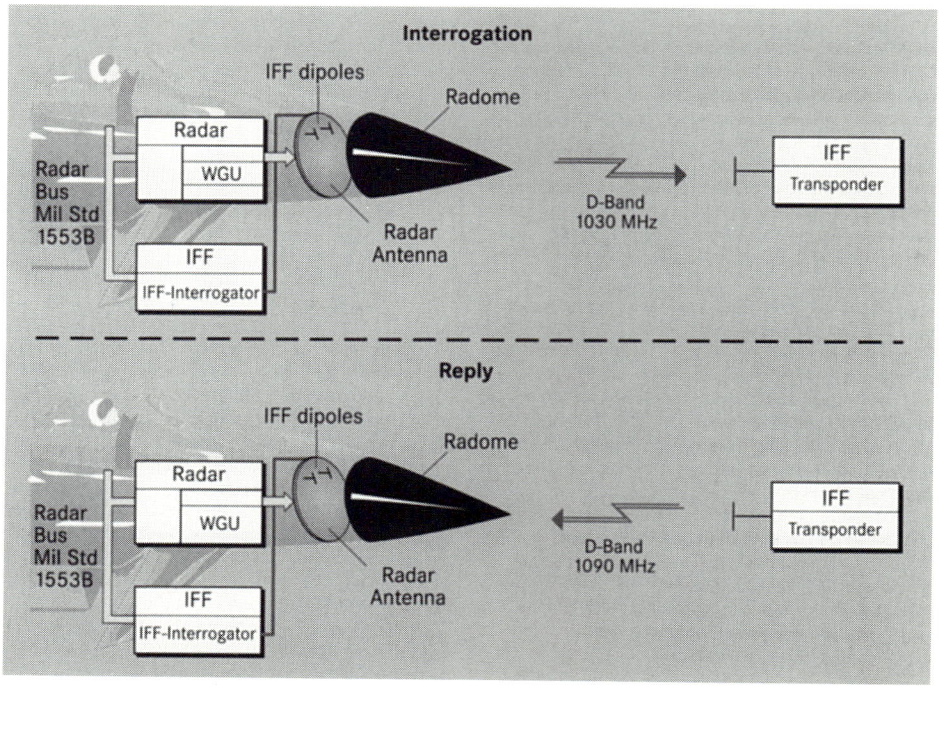

IFF for Eurofighter 2000
***1998**/0011892*

IFF interrogator
Dimensions: ¼ ATR intermediate length
Weight: 12.4 kg
Characteristics:
Mk XII (mode 1,2,3/A,C,4) compliant with STANAG 4193
MIL-STD-1553B
Embedded cryptographic module
Monopulse processing
ADA software
Automatic Code Change (ACC)
BIT modes (P-BIT, C-BIT, I-BIT)

Contractors

Daimler-Benz Aerospace AG, Defense and Civil Systems.
Italtel.
Raytheon Systems Limited, Electronic Systems Division.

NEW ENTRY

IFF for Eurofighter 2000
1998/0011891

Light Detection and Ranging Devices (LIDAR)

Optech Inc and Saab Dynamics are collaborating on the production of a family of airborne laser radars (LIDARs) for seabed mapping.

The system operates by emitting a series of short laser pulses. A programmable scanner directs the pulse to the desired location, across or ahead of the flight path. The backscattered light, including the echoes from beneath the surface, is collected using telescopic optics and the returned light is stored as a series of signals spread over time. The location of each echo is calculated in real time from onboard position and orientation instruments. The bottom is calculated in real time using special algorithms; the bottom depth and position are displayed for the operator. All data is also stored for post-flight processing.

Optech produced the first operational airborne laser bathymeter for the Canadian Hydrographic Service in 1985. This system, the Larsen 500, is still in use. In the late 1980s, Optech began developing a system with more advanced real-time detection and enhanced tactical capabilities.

At the same time, a more advanced version of this system was co-developed with Saab Dynamics and the Swedish Defence Research Institute. All systems are now jointly marketed by Optech and Saab as the Shoals Hawkeye family.

Shoals Hawkeye systems consist of an airborne portion for acquiring depth data and a ground-based portion for mission planning and data analysis. The airborne portion consists of a pod-mounted LIDAR transceiver and two electronics consoles. The system components are field transportable and the system can be installed in less than a day in a variety of aircraft, such as a Bell 212 helicopter or a Twin Otter.

The transceiver uses an Nd:YAG pulsed laser. A two-axis gyrostabilised programmable scanner, receiver optics and electronics for detection of the optical return signals, inertial reference instruments and a video camera are also included in the assembly. The transceiver is enclosed in a pod which is aerodynamically designed and is mounted under the fuselage or on the side of the aircraft.

The Shoals Hawkeye pod shown mounted on a Bell 212 helicopter

The two electronic consoles incorporate the computer system, operator display, data storage devices, inertial reference controllers and power distribution electronics.

The airborne portion of the system is managed by a single operator. Depths or objects are detected in real time and displayed to the operator, along with navigation information. The pilot uses an additional monitor to fly the aircraft along a predetermined flight path. The system can be used by day or night.

The ground-based system consists of a PC and digitising tablet for preflight mission planning and a more powerful workstation for post-flight data analysis.

Specifications

Dimension:
(pod) 3,000 × 500 mm cross-section
(electronics consoles ×2) 1,000 × 500 × 700 mm
Weight:
(pod including transceiver) 250 kg
(2 electronics consoles) 200 kg total
Altitude: 650-1,650 ft
Speed range: 10-200 kt

Contractors

Optech Inc.
Saab Dynamics AB.

VERIFIED

NGIFF (New Generation IFF)

The identification system used by NATO has long ceased to satisfy military requirements. Already in the mid-1980s a standard agreement, STANAG 4162, for a new identification system designated NIS (NATO Identification System) was produced. Although several system designs adapted to the changed threat-situation were evaluated, no final agreement has yet been achieved in NATO. Nevertheless, the studies conducted by the five nations (FR, GE, IT, UK and US) make it possible to define and develop equipment which is fully compliant with STANAG 4193 and ICAO Annex 10 for Mode S, and can be upgraded to NGIFF when it is defined in detail.

In the meantime, the request by civil aviation authorities for the Mode S function in military aircraft has changed the scenario. Independent from NGIFF, the Transponder has to perform Mode S functions from 1.1.1999 to satisfy Air Information Circular 13/92.

Operational status

France, Germany and Italy have decided to co-operate in this effort and to split the programme into two phases with the following content:
Phase 1: Development and procurement of MK X, XII, Mode S and NGIFF upgradeable transponders; definition and specifications of Mode S NGIFF upgradeable interrogators; studies for Crypto, STANAG and frequency supportability.
Phase 2: Development of NGIFF modules for the transponder; development and procurement of interrogators including Mode S; development and procurement of NGIFF-Crypto.

Italtel, Siemens and Thomson-CSF have agreed to create a combined development company (EURO-ID) for perfomance of the agreed programme.

Contractors

EURO-ID, a combined development company of:
Italtel, Italy
Siemens AG, Germany
Thomson-CSF Communications, France.

UPDATED

Ocean Master airborne maritime patrol radar

Thomson-CSF and Daimler-Benz Aerospace AG have teamed to develop the Ocean Master airborne maritime patrol radar, designed to meet the requirements of both fixed- and rotary-wing aircraft.

The two companies claim the following major features for the Ocean Master radar: outstanding radar performance in all situations; ultra-light design and simple installation to fixed- and rotary-wing aircraft; proven state-of-the-art technology.

Ocean Master is designed to fulfil both civil and military maritime missions including: Economic Exclusion Zone (EEZ) surveillance; Search And Rescue (SAR); Anti-Surface Vessel Warfare (ASVW) and Anti-Submarine Warfare (ASW); air-to-air detection.

Two basic versions of the radar are offered: the Ocean Master 100, and the Ocean Master 400. These designations relate to the use of either a 100 W fully coherent Travelling Wave Tube (TWT) amplifier, or a 400 W fully coherent TWT amplifier. Options and growth potential are available to meet specific user requirements, including Inverse Synthetic Aperture Radar (ISAR) processing to provide ship classification capability.

The companies claim reliable range detection and tracking of all types of targets, in all sea states, due to use of the following processing characteristics: high pulse compression; frequency agility; pulse-to-pulse and scan-to-scan integration; digitial programmable processing. Ocean Master also performs the following other modes: multitarget track-while-scan; ground mapping; target classification; weather detection and beacon mode.

Ocean Master in the basic version comprises only three units: antenna unit (360° rotation and/or sector scan); transmitter unit (100 or 400 W); exciter/receiver/processor. In addition, the man/machine interface includes display and controls.

Specifications

System:
Max range: 200 n/mile
Min range: 200 m
Min range resolution: 3 m
Power consumption: 115 V AC, 400 Hz; 2 kVA for 100 W model; 3.8 kVA for 400 W model
Interfaces: MIL-STD-1553B databus; ARINC 429; RS-422; video

Transmitter:
TWT type: fully coherent; 100 W or 400 W average power
Frequency: I/J-band; wideband frequency agility
PRF range: 300 Hz to 125 kHz
Dimensions: 415 × 275 × 320 mm or 566 × 208 × 260 mm
Weight: 29 kg or 39 kg

Exciter/Receiver/Processor:
Dimensions: 555 × 200 × 340 mm
Weight: 27 kg
Characterisitcs: pulse compression - high time-bandwidth product
automatic track-while-scan on 32 targets
pulse-to-pulse and scan-to-scan integration
digital radar map
target classification

Antenna:
Dimensions: 660 × 350 mm, or 940 × 350 mm, or 1,800 × 300 mm
Weight: 15 or 27 kg
Characteristics: stabilisation: 2-axis
tilt selectable; +4 to −29°
rotation rate; 6 to 30 rpm, automatically selected by mode
gain; 30.5 to 34 dB

Options:
Inverse Synthetic Aperture Radar (ISAR) for target classification
IFF compatibility
Interfaces with FLIR, IFF, ESM, sonics, datalink, video recorder
19 in diagonal display

Display/Control:
Dimensions: 375 × 300 × 415 mm
Weight: display 18 kg; controls 8 kg
Display unit: 14 in colour multifunction display; aircraft heading or North stabilised; 7.5 to 240 n/mile scales
Controls: touch-sensitive flat panel, plus tracker ball

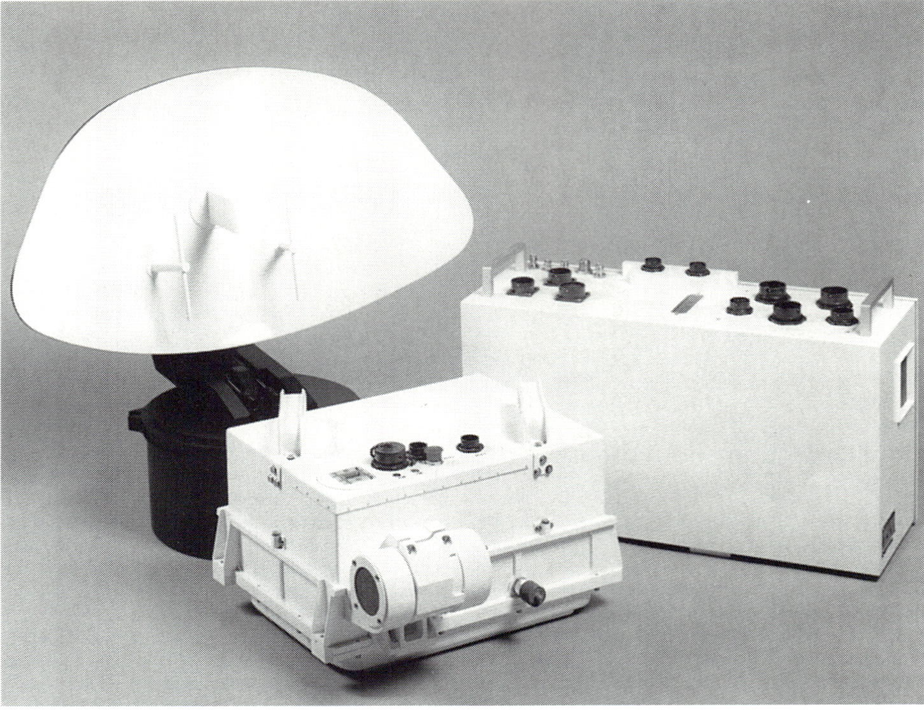

The Ocean Master airborne maritime patrol radar ***1998***/0011890

Operational status

In production, first deliveries were made at the end of 1994. More than 30 systems have been ordered by France, Indonesia, Japan, and Pakistan for use on Falcon 50, C212 maritime patrol aircraft/BO105 helicopters, US-1A helicopters, and Atlantique maritime patrol aircraft respectively.

Contractors

Daimler-Benz Aerospace AG, Defense and Civil Systems.
Thomson-CSF Radars Contre-Mesures.

UPDATED

RDR-1500B multimode surveillance radar

The RDR-1500B is a lightweight airborne digital colour display 360° multimode radar designed specifically for helicopters and fixed-wing aircraft in a multitude of low- and medium-altitude missions including anti-surface vessel operations, surveillance and patrol, search and rescue, customs and fishery protection. The system is available in single- and dual-display configurations. The dual-display configuration is provided when an operator console is available on the aircraft. It consists of eight units: receiver/transmitter, interface unit, two digital colour displays, cockpit control unit, cabin/console control unit, antenna assembly with antenna drive and flat plate array and switch unit.

The RDR-1500B is fitted to Agusta AB 412 helicopters ***1995***

The transmitter/receiver operates as a short-range pulse radar for high-resolution sea search and terrain-mapping, and also as a long-range pulse radar for long-range sea search, terrain-mapping and weather avoidance. Standard radar modes include weather detection, ground-mapping, search, beacon detection and identification. Ground-stabilised, aircraft heading and north-oriented display modes are available. The RDR-1500B has the facility to offset the sweep centre to any location on the display and provide target marker capability. Information from a variety of onboard navigation sensors such as INS, Omega, VOR, DME and FLIR can be displayed independently or as overlays (except for FLIR).

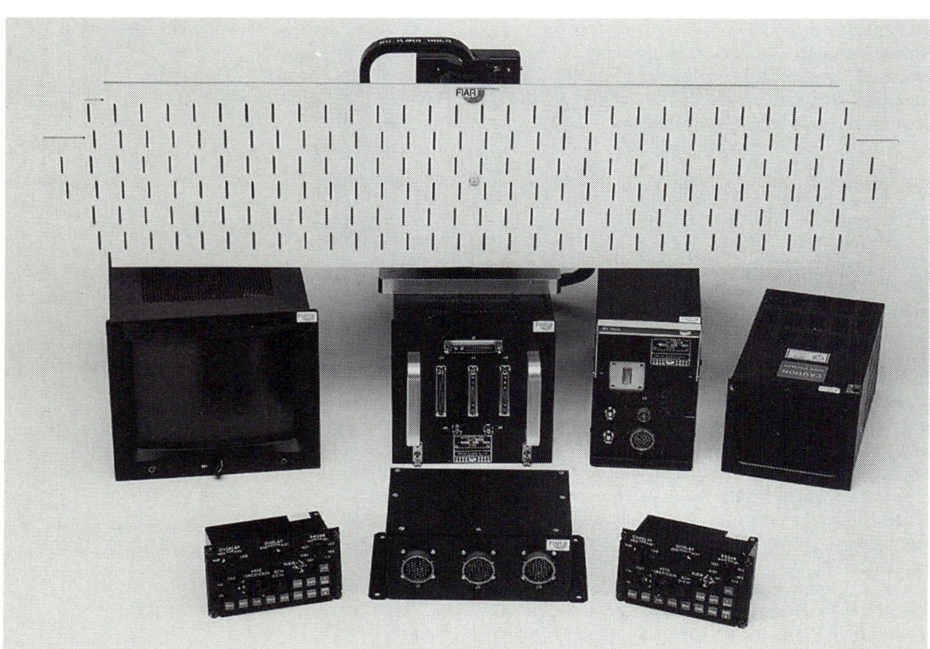

The dual configuration of the RDR-1500B multimode surveillance radar ***1995***

Other capabilities include target position transmission via datalink. An optional video processor allows use of CCP/PAL European standard colour display and videotape recording of the images on a VHS standard video recorder.

The video processor allows the radar operator to superimpose FLIR or TV colour images automatically over the radar picture. The video processor is also used to point the stabilised gimbal of FLIR/TV sensors to the selected target. Automatic target tracking is provided by the video processor unit. The modular design of the system allows for additional growth.

Specifications

Dimensions:
(transmitter/receiver) 194.1 × 123.2 × 320.5 mm
(colour indicator) 152.4, 228.6 or 254 mm
(control panel) 133.4 × 115.5 × 168.7 mm
(interface unit) 194.1 × 189.2 × 323.9 mm
Weight: 42.4 kg
Frequency:
(transmitter/receiver for weather and search) 9,375 ±5 MHz
(receiver for beacon) 9,310 ±5 MHz
Power output: 10 kW (nominal)

PRF: 1,600, 800 or 200 Hz
Pulsewidth: 0.1, 0.5 or 2.35 μs
Beamwidth:
(azimuth) 2.6°
(elevation) 10.5°
Range: 300 km

Operational status

In service. The RDR-1500B is installed on Agusta A 109 and AB 412 helicopters and the Westland Sea King and Lynx.

Contractors

AlliedSignal Commerical Avionics Systems.
Alenia Difesa, Avionic Systems and Equipment Division, FIAR.

UPDATED

SIT 421 (MM/UPX-709) transponder

The SIT 421 (MM/UPX-709) is a single-box airborne IFF transponder suitable for fitting in fixed-wing aircraft or helicopters. It operates in Modes 1, 2, 3/A, 4 and C. The receiver/transmitter includes a 500 W solid-state transmitter, dual-channel receiver and RF interface module. The first of these comprises a delay line oscillator, modulator, driver and power amplifier.

The controls for operation of the transponder, code and mode selection and so on, are mounted on the front of the equipment (which is designed for cockpit mounting) but versions are produced in which remote-control facilities are provided.

Specifications

Weight: 3.5 kg
Frequency:
(receiver) 1,030 MHz
(transmitter) 1,090 MHz
Sensitivity: −77 dBm (adjustable 69-77)
Dynamic range: 55 dB
Output power: 27 ±3 dBW at 1% duty cycle

Operational status

In production.

Contractors

GEC-Marconi Hazeltine Corporation.
Italtel.

VERIFIED

SIT 432 (AN/APX-104(V)) interrogator

The SIT 432 (AN/APX-104(V)) is a lightweight airborne IFF interrogator equipment suitable for installation on helicopters or fixed-wing aircraft to provide air-to-air and air-to-ship identification facilities.

The receiver/transmitter module contains a 1,200 W transmitter, a dual-channel receiver and an RF interface module. The receiver operates at 1,090 MHz and is of a dual-channel type which, in conjunction with a dual-channel antenna, provides for receiver sidelobe suppression.

The design employs surface acoustic wave technology in the local oscillator to obtain a reliable, simple design with good stability and no field alignment requirements. The transmitter is solid-state. It accepts coded video pulse trains from an external source and the internally generated Mode 4 ISLS pulse converts the coded video pulse trains into radio frequency pulse groups for transmission as IFF interrogation.

Specifications

Weight: 6.5 kg
Frequency:
(receiver) 1,090 ±0.2 MHz
(transmitter) 1,030 ±0.2 MHz
Sensitivity: −83 dBm
Output power: not less than 1,200 W
Duty cycle: 1% (max)
Dynamic range: 50 dB

Contractors

GEC-Marconi Hazeltine Corporation.
Italtel.

VERIFIED

SIT 434 IFF interrogator

The SIT 434 is a modular architecture IFF interrogator intended for both fixed-wing aircraft and helicopters. It has considerable growth potential. In the basic configuration, SIT 434 operates under the control of the SIT 905 control box.

SIT 434 is capable of interrogating on Modes 1, 2, 3A and 4, either separately or Modes 1, 2 or 3A interlaced with Mode 4. An associated crypto unit, with mounting and key-loading devices, is required for Mode 4.

Challenge control is possible by means of an enabling signal from the control box or radar system. Video output signals are generated by processing from the last interrogation cycle. Different symbols are generated for target, Mode 1/2/3A response and Mode 4 response.

Optional features on the SIT 434B include pulse-to-pulse defruiter to replace the internal decoding function and interface for an external active/passive decoder to replace the digital symbol generation.

For the SIT 434C, optional features include: control of the interrogator and target information interface by means of an embedded dual-redundant receiver/transmitter unit designed to MIL-STD-1553B, with the ability to separately interrogate azimuthal sectors, challenge management with interface standards based on expected target position; and an antenna synchro interface to associate azimuth IFF data with the radar plot.

Specifications

Dimensions: 1 ATR
Weight: 15 kg
Peak power: 1,200 W/300 W selectable
Frequency:
(transmit) 1,030 ±0.2 MHz
(receive) 1,090 ±0.2 MHz
Duty cycle: 1% (max)

Contractors

GEC-Marconi Hazeltine Corporation.
Italtel.

VERIFIED

SRT-5000 Mode S transponder

SRT-5000 is a product line of Mode S transponders that meets requirements of regional and business aircraft operators.

The Mode S transponder is an airborne system primarily designed for Air Traffic Control (ATC). It allows the identification and the location of the fitted aircraft. It is also capable of establishing a digital datalink between the ground and aircraft systems. It will be an integral part of the future Aeronautical Telecommunications Network (ATN).

Together with the level 4 capability and the ADS-B report transmission, SRT-5000 includes all features that will be required by future regulations, from the Mode S Enhanced Surveillance in Europe to the ACAS II mandatory in North America and in Europe. Moreover, it can fulfil the level 5 requirements by integration of an additional board which performs the Mode S Specific Services (MSSS) and the Airborne DataLink Processor (ADLP) functions.

Operational status

SRT-5000 is in development and meets ICAO and RTCA DO-181A/EUROCAE ED-73 standards. It will be certified in 1998.

Contractors

Dassault Electronique.
Becker Avionic Systems.

UPDATED

STR 2000 IFF transponder

The STR 2000 is a Mk X/Mk XII, Mode S Level 3 IFF transponder which is being developed in co-operation with Thomson-CSF Communications.

The equipment provides all IFF functions according to STANAG 4193 and for Mode S Level 3 according to ICAO Annex 10. Provisions to upgrade to NGIFF are incorporated. The transponder is designed so that it can easily replace the STR 700 without changes in the aircraft installation and has a dedicated control and display unit. An MIL-STD-1553 interface is provided.

Specifications

Dimensions:
(transponder) 124 × 193 × 382 mm
(control/display unit) 146 × 133 × 78 mm
Weight:
(transponder) 7.5 kg
(control/display unit) 1.5 kg
Frequency:
(receive) 1,030 ±0.5 MHz
(transmit) 1,090 ±0.5 MHz

Contractors

Siemens AG Defence Electronics Group.
Thomson-CSF Communications.

VERIFIED

ISRAEL

EL/M-2001B radar

The EL/M-2001B is a range-only I/J-band radar for single-seat tactical aircraft operating in air-to-air and air-to-ground modes. The target is detected visually while acquisition and tracking is accomplished automatically by the radar. The system can operate in heavy ground clutter. Information from the radar can be displayed on the head-up display or fed into the weapon control computer for weapon delivery computation. The six LRUs are based on solid-state technology, with the exception of the travelling wave tube, and have considerable reserves for future growth.

Specifications

Dimensions:
(diameter) 450 mm
(length) 790 mm
Antenna diameter: 195 mm
Weight: <50 kg
Power supply: 115 V AC, 400 Hz, 3 phase, 1 kVA, DC 30 W

Operational status

In service with IAI Kfir fighters of the Israeli Air Force.

Contractor

Elta Electronics Industries Ltd.

VERIFIED

The EL/M-2001B radar on an Israeli Air Force Kfir aircraft

EL/M-2032 radar

The EL/M-2032 is an advanced pulse Doppler multimode fire-control radar designed for multimission fighters, for both air-to-air and air-to-ground missions.

Modular hardware design, all-software control and flexible MIL-STD-1553B avionic interface ensure that the radar can be installed in various fighter aircraft such as the F-4, F-5, F-16, Mirage and MiG-21 and customised to meet specific requirements. Antenna size can be adapted to the space available in the aircraft nose.

The EL/M-2032 installed in MiG-21 ***1996***

In air-to-air operation the radar offers long-range target detection, automatic target acquisition in close combat situations, single-target track for weapon delivery and track-while-scan.

In air-to-ground missions the radar provides air-to-ground ranging, real-beam map, ground moving target indication, sea search, Doppler beam-sharpening for high-resolution mapping, terrain-avoidance and beacon modes, and a look-down/shoot-down capability.

The radar consists of the antenna, transmitter and receiver/processor.

The planar-array antenna features ultra-low sidelobes and two-axis monopulse operation.

The transmitter is a TWT coherent transmitter.

The receiver/processor includes a programmable signal processor which provides full software control of the system.

Specifications

Weight: 95-105 kg, depending on antenna
Power consumption: 2 kW
Range: 37-75 km on small fighter aircraft

Operational status

In development as part of the Northrop F-5 Plus package, and selected for retrofit to Chilean F-5E, Romanian MiG-21, Lancer, and Turkish F-4 Phantom 2000 aircraft.

Contractor

Elta Electronics Industries Ltd.

VERIFIED

EL/M-2075 Phalcon AEW radar

The EL/M-2075 Phalcon is a solid-state D-band conformal array radar system for use on a Boeing 707 and other aircraft. Phalcon is intended for airborne early warning, tactical surveillance of airborne and surface targets and intelligence gathering. It will also have the command and control capabilities needed to use this information.

The system uses six panels of phased-array elements: two on each side of the fuselage, one in an enlarged nosecone and one under the tail. Each array consists of 768 solid-state transmitting and receiving elements, each of which is weighted in phase and amplitude. These elements are driven by individual modules and every eight modules are connected to a transmit/receive group. Groups of 16 of these eight module batches are linked back to what is described as a prereceive/transmit unit, and a central six-way control is used to switch the pretransmit/receive units of the different arrays on a time division basis.

Each array scans a given azimuth sector, providing a total coverage of 360°. Scanning is carried out electronically in both azimuth and elevation. Radar modes include high PRF search and full track, track-while-scan, a slow scan detection mode for hovering and low-speed helicopters (using rotor blade returns) and a low PRF ship detection mode. These modes can be interleaved to provide multimode operation in any scanning sector. Typically, 2 to 4 second scan rates are used in high-priority sectors and 10 to 12 second rates in low-priority sectors.

Operational status

Development has been completed and two aircraft ordered by the Chilean Air Force.

Contractor

Elta Electronics Industries Ltd.

VERIFIED

The Elta Phalcon AEW radar is designed for installation in a Boeing 707

ITALY

APQ-706 radar

The APQ-706 search and attack radar is used in the Italian Navy Marte helicopter-launched anti-ship missile system. It consists of two I-band transmitter/receiver channels operating in frequency diversity, with frequency-agility facilities in one channel, and is provided with a data processing and extraction system and tactical display console. The APQ-706 can fulfil typical naval helicopter roles such as surface search, navigation, ASW and so on, in addition to providing target detection, acquisition and missile guidance facilities for the Marte system.

Operational status

Installed on Italian Navy SH-3D helicopters for the Marte anti-ship missile system. No longer in production.

Contractor

Alenia Difesa, Avionic Systems and Equipment Division, FIAR.

UPDATED

APS-705A search and rescue radar

The APS-705 I/J-band search and rescue radar was designed for naval helicopters, but is equally suitable for maritime patrol aircraft and land-based helicopters. The radar's main functions are navigation, search and detection, target localisation, target tracking and designation for weapon aiming and mapping. There are two multipurpose radar systems: APS-705 and APS-705A.

The antenna system is tailored to the space and location on the aircraft. For example, on the SH-3D the antenna is placed in the dorsal position on top of the fuselage. Line of sight stabilisation is provided and there are selectable antenna rotation rates of 20 or 40 rpm. Manually controlled antenna tilt provides ±20° of movement. Interfaces with sonar, Doppler navigation, IFF, ESM and so on, are available.

The display unit incorporates a 230 mm diameter CRT PPI with electronic and mechanical cursors and markers, complemented by a separate digital readout X-Y reference display.

The radar has two I-band transmitter/receivers for frequency diversity operation at 25 kW. There is an option for a 75 kW transmitter/receiver with frequency agility. Other facilities include sector transmissions and blanking, built-in test, datalink, track-while-scan and dense environment tracker. The radar can be

APS-705A search and rescue radar **1998**/0011886

integrated with the SMA UPX-719 beacon system. A pod-mounted version is also available for retrofit applications, where it can be used with anti-submarine missiles.

The APS-705A is an upgrade of the APS-705. The use of more modern techniques enables the system to offer additional features, such as digital processing, and to interface with airborne sensors such as FLIR, weapon systems, ESM, datalink and sonar. Other features comprise a freeze mode, colour raster scan display and an optional track-while-scan facility for multiple targets.

Specifications

Weight: 80 kg
Frequency: I/J-band
Pulsewidth: 0.05 and 1.5 μs
PRF: 1,600 and 650 Hz
Power output: 25 kW (75 kW option)
Range settings: 0.5, 1, 2, 5, 10, 20, 40, 80 n miles
Antenna rotation: 20 or 40 rpm

Operational status

In service with AB 212 and SH-3D helicopters of the Italian Navy and several other navies.

Contractor

Alenia Difesa, Avionic Systems and Equipment Division, FIAR.

UPDATED

APS-707 search and rescue radar

The APS-707 is a frequency-agile I/J-band radar for aircraft and helicopters designed for installation where low weight, power consumption and cost are important factors. It is a military qualified search radar featuring 360° scan for surface surveillance, small target detection for ASW operations, target designation for ASV attacks and radar mapping. The APS-707 can also display IFF, ESM, beacon and sonar data. An advanced video signal processor includes the presentation of data integrated with data from other sensors and so on, on a large TV monitor, with a cockpit repeater display being an option. Also available as an option is a MIL-STD-1553B interface, a sonar interface and a track-while-scan facility.

Operational status

In production for an unspecified country.

Contractor

Alenia Difesa, Avionic Systems and Equipment Division, FIAR.

UPDATED

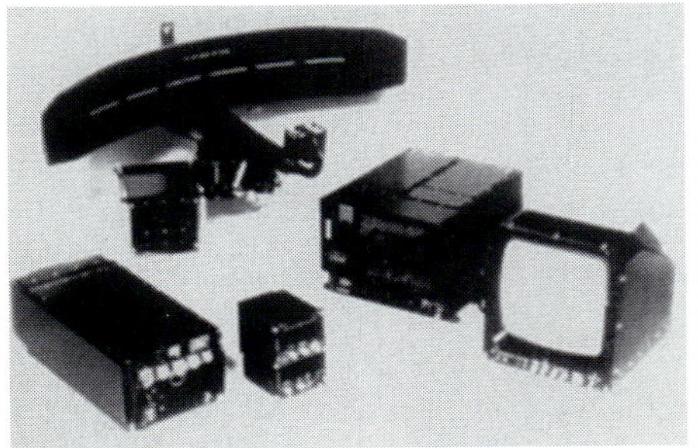

APS-707 search and rescue radar **1998**/0011885

APS-717(V)1 search and rescue radar

The APS-717 family consists of two search and navigation radar systems which are tailored to the requirements of individual customers. It is suitable for many roles, including search and rescue, surveillance, navigation and target designation.

The APS-717(V)1 is a lightweight radar which is suitable for both fixed-wing aircraft and helicopters. It operates in the I/J-band, providing detection over 180° in azimuth with automatic stabilisation. It can be integrated with the navigation system and a FLIR sensor. Other features include Constant False Alarm Rate (CFAR), scan-to-scan integration, pulse-to-pulse integration, a freeze mode and colour display with graphics.

The APS-717(V)2 is a high-performance upgrade of the APS-717(V)1 and offers a number of additional features. These include 360° azimuth coverage, integration and automatic initialisation of FLIR and LLTV, a video recorder output and an optional track-while-scan capability covering 32 targets.

Operational status

The APS-717(V)1 is in service on HH-3F helicopters of the Italian Air Force. The APS-717(V)2 is in service on Italian Harbour Authority AB 412 helicopters.

Contractor

Alenia Difesa, Avionic Systems and Equipment Division, FIAR.

UPDATED

APS-717(V)1 search and rescue radar **1998**/0011884

Creso airborne battlefield surveillance radar

Creso is one of the sensor systems under development as part of the Italian Army's Surveillance and Target

Creso airborne battlefield surveillance radar **1998**/0011889

Identification Subsystem (SORAO) CATRIN command, control, communications and intelligence system. It comprises both air and ground elements. The airborne element is installed aboard an Augusta AB 412 helicopter, and comprises the Creso battlefield surveillance radar, with ESM and FLIR sensors, together with a datalink system for air/ground data transfer.

Creso's operational roles are: detection of ground moving targets beyond the FEBA (Forward Edge of the Battle Area); production of a target count in battlefield areas designated to it; high-precision localisation of designated targets.

Creso is an I-band, pulse Doppler radar; it utilises a coherent TWT transmitter, and features: wideband frequency agility; high-resolution pulse compression; programmable FFT processor; zoom capability; high ECCM resistance; and growth capability for air-to-air surveillance.

Operational status

It is understood that there is one flight trials model fitted to an Augusta AB 412 helicopter. The Creso system is the Italian submission to the NATO multinational Alliance Ground Surveillance (AGS) programme.

Contractor

Alenia Difesa, Avionic Systems and Equipment Division, FIAR.

NEW ENTRY

Grifo multimode radar family

The Grifo multimode pulse Doppler I/J-band radar is a compact radar with a high degree of modularity and low life cycle costs designed for air superiority aircraft. It has look-up and look-down capabilities.

The radar features a monopulse flat plate antenna array, coherent TWT transmitter, pulse compression,

Augusta AB 412 helicopter showing Creso radar antenna beneath the nose; also shown the four port ESM system above the radar, and datalink below the tailboom **1998**

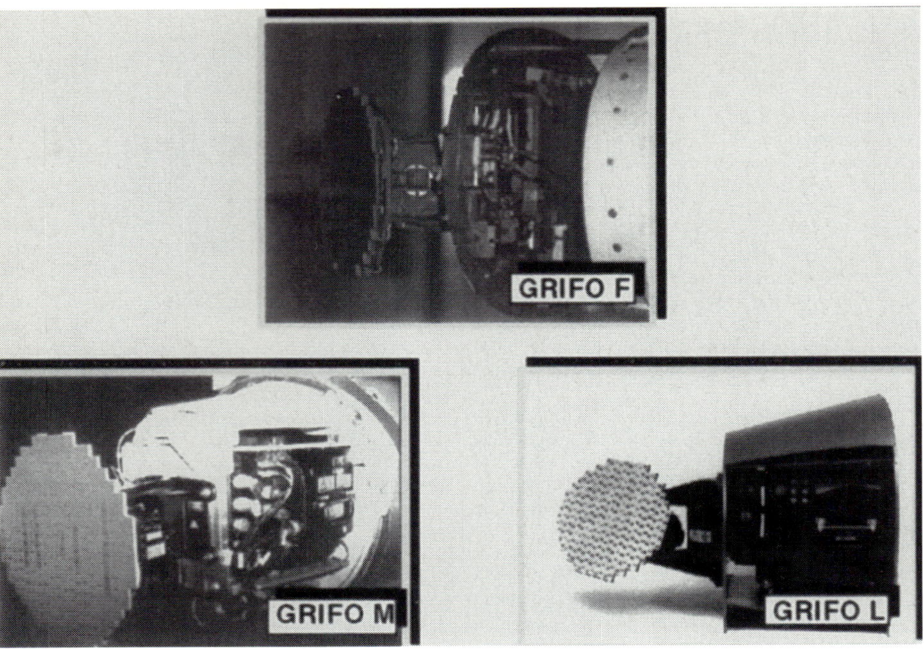

Grifo F for the F-5 E/F, Grifo L for the L-159, Grifo M for the Mirage III ***1998***/0011887

wideband frequency agility and a programmable waveform generator and FFT processor. It is fully compatible with semi-active and active missiles and has a high immunity to ECM.

The Grifo is suitable for fitting to a number of aircraft such as the Mirage, A-4, F-5, MiG-21, Super 7 and several trainer/light attack aircraft.

Grifo 7 radar

The Grifo 7 is designed to be installed in the nose of CAC F-7 aircraft. It has full look-up and look-down air-to-air capabilities through the use of Pulse Doppler and medium PRF waveform, plus an air-to-ground ranging mode to support CCIP/CCRP.

Two modes are selectable in air-to-air. Super-search is used for the acquisition and tracking of the highest priority target in the HUD field of view. The radar allows the missile seeker to be slewed to the target line-of-sight for offset delivery. In boresight, fixed antenna pointing is used for automatic acquisition and tracking of the nearest target.

Most of the hardware is common with the other versions of the Grifo family. Grifo 7 has compatibility with IR missiles, rockets, guns and free-fall bombs.

Specifications

Weight: 55 kg

Operational status

In production for the CAC F-7. 100 ordered by Pakistan for retrofit to the PAF F-7 fleet.

Grifo F radar

The Grifo F has been developed for retrofit in the Northrop F-5E/F. It is an I-band multimode pulse Doppler radar which offers eight air-to-air, four air combat and nine air-to-surface modes. It uses a TWT transmitter with both pulse compression and wideband frequency agility, a programmable FFT processor and a monopulse flat plate array antenna, of which two sizes are available.

Operational modes for air-to-air employment include: range-while-search, velocity search, track-while-scan, single-target track, situation awareness and air combat. For air-to-ground operation the modes are: real-beam mode, Doppler beam-sharpening, air-to-ground ranging, sea map, surface moving target indicator, surface moving target track and beacon.

Grifo 7 radar for F-7 aircraft ***1997***/0001199

Specifications

Weight: 80 kg

Operational status

In production for the Northrop F-5E/F. Selected by Republic of Singapore Air Force.

Grifo L radar

Grifo L is a variant of Grifo F selected by the Czech Air Force for integration into its Aero Vodochody L-159 fighter trainer.

Grifo M3 radar

Similar to the Grifo F, the Grifo M3 has been developed for the Mirage III. It makes use of a different antenna array to fit the nose of the Mirage III. Modes are as for the Grifo F.

The Grifo M3 is compatible with a variety of weapon systems including semi-active missiles. The Grifo M21 is similar to the M3 and is under development for the MiG-21.

Operational status

In production for the Mirage III, 35 ordered by Pakistan.

Grifo X Plus Radar

The Grifo X Plus is a variant of Grifo F for installation on the AM-X aircraft.

Contractor

Alenia Difesa, Avionic Systems and Equipment Division, FIAR.

UPDATED

Scipio radar family

The Scipio family is a series of coherent, lightweight, compact frequency agile radars. The range is designed to suit one-man operations on fighter aircraft with multi-role capabilities. It has a total weight of less than 75 kg, with a high MTBF and low MTTR. An extensive BIT system provides fault detection and location. Scipio offers an air-to-air mode with look-down capabilities, air combat mode with automatic detection, designation and tracking, sea mode for target detection and tracking of surface vessels and ground ranging and mapping.

The SCP-01 is the first member of the Scipio family. Its main features include I-band frequency, pulse compression and pulse Doppler techniques, frequency-agile TWT transmitter, high throughput, fully coherent, software reconfigurable signal processing, monopulse tracking, track-while-scan, MIL-STD-1553B interface and colour video output with graphics.

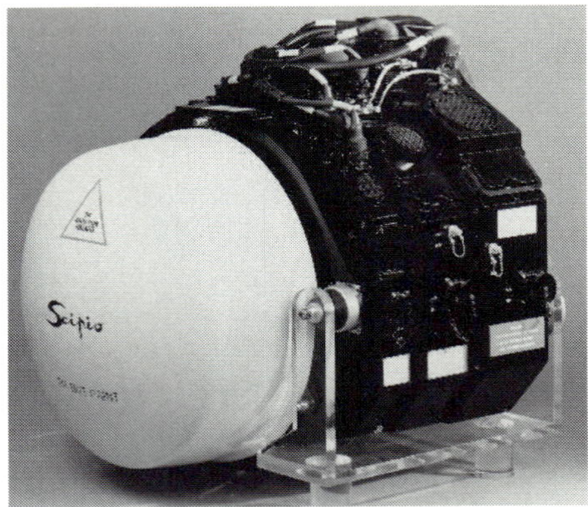

The SCP-01 airborne radar has been designed for Brazilian Air Force AMX aircraft

The SCP-02 is an upgrade of the SCP-01 with a larger antenna, higher transmitted power and better operational features such as Doppler beam-sharpening, terrain-avoidance and contour map.

Operational status

The SCP-01 has been designed for the Brazilian Air Force AMX and is in production. The SCP-02 is under development.

Contractor

Alenia Difesa, Avionic Systems and Equipment Division, FIAR

UPDATED

SIR SSR family

The SIR SSR family is a modular SSR system. By adding modules, SIR can be expanded from the SIR-R minimum version, through SIR-M to the most complex SIR-S version. The system is designed so that the cost of expansion is totally borne by the added unit.

The system consists of a transmitter, receiver, programmable processor/controller and power supply. The processor/controller is a firmware programmed unit which forms the extractor and provides all control signals and narrowband formatting. The processor/controller provides all the programmable functions for the transmitter and receiver.

SIR-R gives good performance and is especially well suited to areas where interference with other SSRs is not a problem. The system uses the programmable transmitter, standard SSR receiver and processor/controller. The antenna is a 14 or 28 element standard hog-trough antenna.

SIR-M has a full monopulse configuration and uses the same transmitter and processor/controller as the SIR-R. The receiver is configured in a sum and difference version and uses the same logarithmic amplifiers as the SIR-R. The antenna is an open array optimised for the monopulse capability.

SIR-S is an SIR-M system with additional modules to give a full Mode S capability. The receiver and processor/controller remain the same as for SIR-M.

Specifications

Frequency:
(transmit) 1,030 ±0.1 MHz
(receive) 1,090 MHz
Modes: 1, 2, 3A, B, C, D (Mode S optional)
Pulse duration: 0.8 μs
Duty cycle: 1% for SIR-R/SIR-M, 10% for SIR-S

Contractor

Alenia Difesa, Avionic Systems and Equipment Division, FIAR.

UPDATED

UPX-719 transponder

The UPX-719 transponder forms a part of the SMA Intra (interrogator/transponder) system which links ships with their co-operating helicopters. The UPX-719 is the airborne transponder; the UPX-718 is the shipborne interrogation part of the system. The shipborne transmission to the airborne transponder can be integrated with the main radar. Ships can identify up to 10 helicopters and vice versa, due to the characteristics of the coding system. Range is over 100 km for helicopters operating at 1,500 ft.

Operational status

In production for Italian and other navies.

Contractor

Alenia Difesa, Avionic Systems and Equipment Division, FIAR.

UPDATED

The UPX-719 transponder

APS-784 ASV/ASW radar

FIAR and Officine Galileo/SMA formed a consortium known as Eliradar to develop the APS-784 radar for the Italian Navy version of the EH 101 helicopter.

The radar provides high detection over a wide area and at long range, in adverse weather conditions. The APS-784 is able to detect and track even small targets such as liferafts, wooden boats and periscopes in rough sea states.

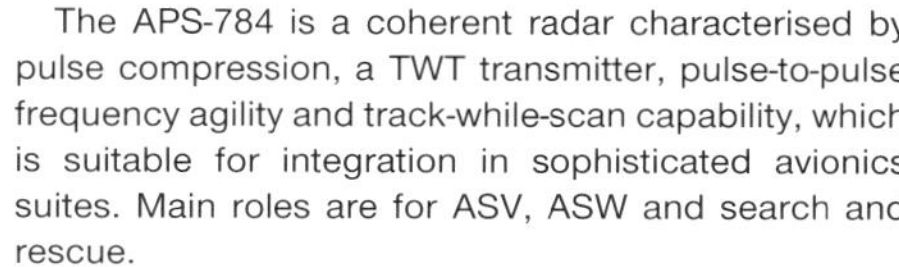

The APS-784 is a coherent radar characterised by pulse compression, a TWT transmitter, pulse-to-pulse frequency agility and track-while-scan capability, which is suitable for integration in sophisticated avionics suites. Main roles are for ASV, ASW and search and rescue.

The radar has 360° scan, missile launch assistance and weather detection modes.

The radar is packaged in four LRUs and transmits in the I/J-band. It has four operational modes featuring track-while-scan with adaptive strategies: anti-surface vessel, anti-submarine warfare, weather and short range. It offers 360° surveillance with linear and circular switchable polarisation and includes an IFF Mk XII antenna array. It has two independent scan converters and scan-to-scan integration.

Target classification (profiling) is an optional capability.

Operational status

In production for the Italian Navy EH 101 helicopter.

Contractor

Eliradar (FIAR and Officine Galileo/SMA).

UPDATED

APS-784 ASV/ASW radar for the Italian Navy EH 101 helicopter
1998/0011888

ANV-301 Doppler navigation system

The ANV-301 Doppler navigation system provides a self-contained worldwide precision navigation capability, utilising Doppler-derived measurement of aircraft velocity and external input of aircraft attitude and heading. The ANV-301 design approach allows system functions to be tailored to specific customer requirements at minimum cost and the electrical interface to be tailored for specific aircraft avionics and subsystems for additional enhancement of system capability.

The system comprises four LRUs: the receiver/transmitter radar, signal data converter, control display unit and pilot steering indicator.

Three pilot-selectable navigation co-ordinate systems with automatic co-ordinate conversion are available: latitude/longitude, worldwide alphanumeric UTM and arbitrary grid.

Contractor

Marconi SpA.

VERIFIED

ANV-351 Doppler velocity sensor

The ANV-351 Doppler velocity sensor provides precision measurement of helicopter velocity components. The ANV-351 is a single-unit sensor specifically designed for integrated avionic systems and weighing less than 7 kg. It comprises two main modules: the receiver/transmitter radar and the signal data converter.

The receiver/transmitter radar module is of fixed design and all-solid-state construction. It comprises separate four-beam transmit/receive antennas, utilising printed circuit planar-array technology. The module is available in two versions, optimised for either ASW or nap of the earth operation.

The signal data converter module comprises power supply, Doppler signature processing and MIL-STD-1553B interface.

Contractor

Marconi SpA.

VERIFIED

ANV-353 Doppler velocity sensor

The ANV-353 Doppler velocity sensor provides precision measurement of helicopter velocity components. It is a single-unit low-power sensor specifically designed for integrated avionic systems and weighing less than 5 kg. It comprises two modules: a receiver/transmitter radar and signal data converter.

The receiver/transmitter radar module is of fixed design and all solid-state construction. It comprises separate four-beam transmit/receive antennas utilising printed circuit planar-array technology. The redundant fourth beam provides high accuracy during extreme attitude manoeuvres. This module contains all RF signal generation and processing functions and receives its power input, modulation and timing signal from the signal data converter.

The signal data converter comprises power supply Doppler digital signal processing and discrete and ARINC 429 interfaces. An MIL-STD-1553B interface is an option. Individual beam frequency shifts are extracted from the spectrum received from the receiver/transmitter radar module, utilising a single time-shared IF channel and digital signal processing.

The extensive use of advanced components allows controllable RF transmitted power and automatic land/sea transition.

Contractor

Marconi SpA.

VERIFIED

JAPAN

Pulse radar altimeters

JAE pulse radar altimeters are used in military and civil fixed-wing aircraft and helicopters. The systems consist of a transmitter/receiver, an antenna and an indicator. The unit is provided with a low warning light or audio signal to provide a warning when the aircraft reaches a preset altitude.

Operational status

The APN-171 is fitted to the Shin Meiwa PS-1, Sikorsky HSS-2, Kawasaki/Vertol KV-107, Kawasaki P-2J and Kawasaki C-1. The APN-194 is fitted to the Lockheed P-3C and the JARN-P2 is fitted to the Boeing-Vertol CH-47, Bell AH-1S and Mitsubishi LR-1.

Contractor

Japan Aviation Electronics Industry Ltd.

VERIFIED

Laser radar

The Koito laser radar is a distance sensor with a high-speed response which may be used as an altitude sensor.

Since its light source uses a laser diode generating a near infrared ray, it is capable of detecting the position of an object at 100 m and detecting altitudes of about 50 ft above ground. The measurement cycle of 0.02 seconds provides a high-speed response and enables it to measure ever-changing distances between two objects. The use of the laser diode as the light source permits the circuitry to be simplified, leading to compact and lightweight construction.

Contractor

Koito Manufacturing Company Ltd.

VERIFIED

Radar for the FSX

A new airborne radar for the FSX close support aircraft has been developed by Mitsubishi Electronics Corporation under the management of the Japanese Defence Agency's Technical Research and Development Institute. The radar has a 66 cm diameter active phased-array antenna made up of 750 modules. It is reported to have track-while-scan facilities, with each module generating computer-controlled radar beams.

Operational status

In development.

Contractor

Mitsubishi Electronics Corporation.

VERIFIED

NETHERLANDS

Vesta transponder

Vesta is a landing and identification aid primarily intended for ship-based helicopters. It consists of two parts: the helicopter transponder and the ship receiver. Vesta enables accurate display and tracking of friendly helicopters on the radar display, even in heavy clutter environments. The transponder principle is based on a radar-triggered VHF reply. Operation is possible with any synchronised surveillance radar (either shipborne or shore-based) in the 1 to 10 GHz band.

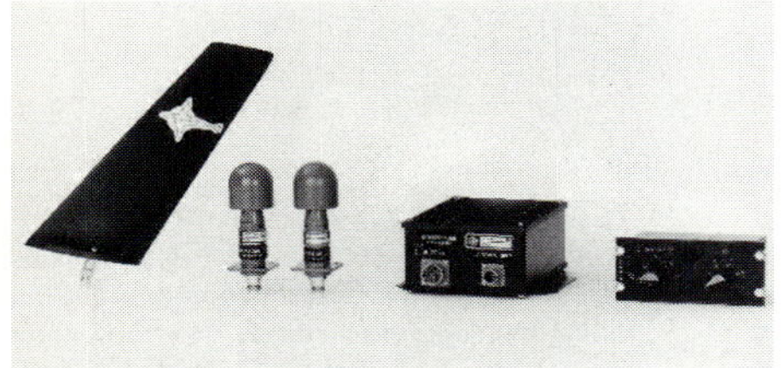

The Hollandse Signaalapparaten Vesta transponder system

For every radar pulse the transponder receives, a VHF reply pulse is transmitted, followed by a code pulse. Up to five helicopters can be identified by means of preselected codes (extension up to 64 is possible). The return signal is received and processed by the Vesta receiver in the ship, which identifies and decodes the transponder reply. Unwanted VHF reply pulses are rejected by digital filters controlled by the allocated radar on the ship.

The Vesta helicopter system consists of a fully solid-state transponder, a control unit, two radar pick-up antennas and a VHF transmitting antenna.

The two radar pick-up antennas are used to guarantee a combined sensitivity pattern which is virtually omnidirectional. The control unit has only two switches; one for sensitivity selection and power on/off and the other for code selection.

Specifications

Dimensions:
(transponder) 167 × 85 × 194 mm
(control unit) 146 × 66 × 68 mm
(radar pick-up antenna) 45 × 106 × 48 mm
(VHF transmitting antenna) 50 × 254 × 123 mm
Weight: 3 kg
Power supply: 28 V DC, 14 W
Frequency:
(transmitter) VHF A-band
(receiver) 1-10 GHz
Transmitter peak power: 10 W
Range: 0-230 km
Pulse duration: 2.2 μs nominal
Number of helicopter codes: 5 standard
(optional extension up to 64)

Operational status

Fitted to several types of helicopters. Over 55 Vesta airborne transponders have been delivered to several navies.

Contractor

Hollandse Signaalapparaten BV, Signaal Special Products.

VERIFIED

RUSSIAN FEDERATION AND ASSOCIATED STATES (CIS)

Russian radars

The table outlines Russian Federation radar systems and programmes. Full details are not available for all programmes, but data is provided in the subsequent equipment entries, where modern Russian source information is available.

Name	*Contractor*	*Aircraft*	*Type*
SPPZ	AeroPribor	Civil/military aircraft	Ground proximity warning system
IFF 60P	All-Russian JSC	Civil/military aircraft	IFF system
SRZO-KR	Kazan Scientific Research Institute	Military aircraft	IFF system
Berkut	Leninetz	IL-38 (May)	ASW radar
Kinzhal-V	Leninetz	Mi-28N, Ka-50	Attack radar. Possibly pod-mounted under stub-wing
Korshin	Leninetz	Tu-142 (Bear-F), Ka-25 (Hormone)	ASW radar
NIT	Leninetz		Side-looking airborne radar
Obzor	Leninetz	Tu-95MS (Bear-H)/Tu-160 (Blackjack)	
PN	Leninetz	Tu-22K, Tu-22M	
PNA-D	Leninetz	Tu-22M3 (Backfire-C)	

Name	Contractor	Aircraft	Type
Aisberg-Razrez	Scientific-technical centre Osnova, Leninetz Holding Co	Military/civil survey	Dual-band side-looking radar, ground/water surface surveying
Duet	Scientific-technical centre Osnova, Leninetz Holding Co	Civil aircraft	Dual-band weather/navigation radar
Neva	Scientific-technical centre Osnova, Leninetz holding Co	Civil aircraft mapping radar	Multifunction weather/obstacle/ground
VID-95	Scientific-technical centre Osnova, Leninetz Holding Co	Civil aircraft and landing radar	Short-range, high resolution, approach
Kvant	NIIP, Vega	An-71 (Madcap), YAK-44	Airborne early warning radar
Sabla	NIIP, Vega	MiG-25RB (Foxbat-D)	Side-looking airborne reconnaissance radar
Shmel	NIIP, Vega	A-50 (Mainstay)	Airborne early warning radar
Shomol	NIIP, Vega	MiG-25RB (Foxbat-D)	Side-looking airborne reconnaissance radar
Shtyk	NIIP, Vega	SU-24MR (Fencer-E)	Side-looking airborne reconnaissance radar
N011	NIIP, Zhukovsky	Su-27M/Su-35	
N011M	NIIP, Zhukovsky	Su-35/Su-37	Phased-array derivative of N011
N012	NIIP, Zhukovsky	Su-37	Tail radar
N014	NIIP, Zhukovsky	MiG-MFI (I-42)	Reportedly abandoned by NIIP
VEGA-M	NIIP, Zhukovsky	Tu-154M-ON	Airborne Surveillance: Open-skies
Zaslon	NIIP, Zhukovsky	MiG-31	Phased-array
Zaslon-M	NIIP, Zhukovsky	MiG-31M	Reportedly abandoned by NIIP
M002	NIIR, Moscow (now Phazotron)	Yak-41M	N010 Zhuk development. Multifunction air-to-air/ air-to-ground/map/terrain follow-avoid
N010 Zhuk	NIIR, Moscow (now Phazotron)	MiG-29M	Multifunction air-to-air/air-to-ground/map
Arbalet	Phazotron	Kamov helicopters: Ka-52 Aligator	Multifunctiion, air-to-surface, air-to-air
Gukol	Phazotron	Marketed for light strike/attack aircraft, transports, and so on	Weather/navigation radar
Kopyo (or Komar)	Phazotron	Marketed for MiG-21, MiG-23, F-5, Mirage F-1, Hawk 200	Multimode, multifunction, coherent PD air-to-air/ air-to-ground
Kopyo-25	Phazotron	Su-25TM (Frogfoot-B), Mi-28N	Multimode, multifunction, coherent PD air-to-ground
Kopyo-Ph	Phazotron		Phased-array version of Kopyo
Moskit	Phazotron	MiG-ATC advanced combat trainer aircraft	Multimode, multifunction, coherent PD air-to-air/ air-to-ground
Mosquito	Phazotron	Marketed for Jaguar upgrade	Multimode, multifunction, coherent PD air-to-air/ air-to-ground/air-to-sea maritime radar
Phathom	Phazotron/Thomson-CSF	Marketed for SU-22 upgrade	Co-operative derivative of Kopyo
RP-21 Sapfir	Phazotron	Many versions of MiG-21	Basic air-to-air radar
RP-22 Sapfir-21	Phazotron	Many later versions of MiG-21	Basic air-to-air radar
S-23 Sapfir-23	Phazotron	Many versions of MiG-23	PD air-to-air radar
RP-25 Sapfir-25	Phazotron	MiG-25 variants	PD look-down shoot-down air-to-air radar
N019 Sapfir-29	Phazotron	MiG-29	PD look-down shoot-down air-to-air radar
RP-35	Phazotron	Marketed for MiG-35	Multimode, multifunction, coherent, digital PD air-to-air/air-to-ground
Topaz N019M	Phazotron	MiG-29S, ME. Marketed for MiG-23, Mig-29 upgrades	Modified version of N019 Sapfir-29
Sokol	Phazotron	Marketed for latest Sukhoi aircraft	Multimode, multifunction, coherent, PD
Zhuk	Phazotron	Fitted to MiG-29M, and marketed for MiG-25, MiG-29, MiG-33 upgrades	Multimode, multifunction, coherent PD, AA/AG
Zhuk-27	Phazotron	Marketed for SU-27, SU-30MK	Derivative of Zhuk aimed at Su-27 upgrade market
Zhuk-Ph	Phazotron		Phased-array version of Zhuk
IFF 6201R/6202R	RadioPribor	Civil/military aircraft	IFF system

Korshin in the Ka-25 ASW helicopter

The Beriev/A-50 Mainstay AWACS carries Shmel radar

SPPZ ground proximity and warning systems

The SPPZ ground proximity warning systems can be used in all types of passenger and transport aircraft equipped with flight navigation systems that have digital information exchange. There are two models: the SPPZ-85 and SPPZ-2.

The SPPZ systems compute data obtained from the following systems to produce their warnings; the radio altimeter; air data computer system; ILS or MLS receiver; onboard inertial navigation system; landing gear and flap sensors. The SPPZ-2 system also utilises data from the flight management system and flight control system.

Warning data provided by the systems is as follows: excessive sink rate; excessive terrain closure rate; negative climb rate after take-off or missed approach; insufficient terrain clearance at landing with wrong configuration; inadvertent descent below glideslope; excessive difference in absolute altitude and pressure height; inadvertent flight into dangerous windshear (SPPZ-2 only).

Specifications

Outputs: 2 analogue; 40 voice; ARINC 429
Power: 115 V AC, 400 Hz, 20 VA (SPPZ-85), 25 VA (SPPZ-2)
Dimensions: 2 MCU
Weight: 3.5 kg

Operational status

Fitted to: An-70, IL-96, IL-114, Tu-204 and Tu-334 aircraft.

Contractor

AeroPribor-Voskhod Joint Stock Company.

NEW ENTRY

SPPZ ground proximity warning systems
***1998**/0011883*

NIT side-looking airborne radar

NIT is a side-looking radar designed for collecting detailed data on earth surface conditions. Intended roles include: survey and supervision of natural resources; ice survey; fire, flood, and disaster monitoring.

The radar provides coverage of a swathe up to 80 km wide, with different polarisations. Design capabilities include: on-line, real-time presentation of the radar picture on TV displays in the air; provision for digital or analogue recording; geographic referencing; a datalink to transmit information to ground and shipborne receiver/display systems.

Contractor

Leninetz Holding Company.

NEW ENTRY

VEGA-M Open-Skies airborne surveillance system

VEGA-M is an airborne surveillance system designed for Open-Skies procedures on the Tu-154M-ON aircraft, with a mission crew of five operators.

The complete system comprises: the airborne surveillance system installed on a Tu-154M-ON aircraft; an onboard digital recording system; a communications package; a ground-based data gathering and processing system.

The sensor package comprises: an aerial photography system; a RONSAR side-looking synthetic aperture radar, a RADUGA line-scan IR sensor; a TV-camera system. The sensor system is supported by an on-board computer system comprising five 486DX PCs, and specialised digital recording based on the VITYAZ'-ON recorder.

Specifications

Sensors	Height	Surface resolution
Aerial camera system		
panoramic cameras	8,000 m+	0.3 m
framing mapping cameras	1,500 m+	0.3 m
framing oblique cameras	1,500 m+ (slant range)	0.3 m
TV system		
mapping cameras and oblique	500 m+	0.3 m
IR system		
low altitude cameras	1,500 m+	0.5 m
high-altitude cameras	3,000 m+	0.5 m
Side-looking SAR	500-12,000 m	3 × 3 m

Contractor

Moscow Scientific Research Institute of Instrument Engineering MNIIP

NEW ENTRY

VEGA-M airborne surveillance system
***1998**/0011882*

IFF 60P system

The 60P is a new-generation Russian IFF system. The following elements of the system are described:

(a) the airborne radar interrogator (designated CP3-1P (facility 6231P))
(b) the special facility 61P which is described (verbatim) as: designed for automatic ciphering interrogation-reply correspondences in the general spoofproof identification mode (mode II) of the 60P IFF system' (designated 6110P-10 airborne configuration and 6110P-21 in land-/sea-based configuration).

CP3-1P (facility 6231P)

The airborne radar interrogator CP3-1P (facility 6231P) of the 60P IFF system is designed for identification of friendly air, land and sea platforms. The 6231P facility operates on a co-operative question and answer principle and provides the following capabilities:

(a) general identification 'friend or foe' of air, overwater and ground vehicles equipped with transponders and detected by radar
(b) selective identification of air, overwater and ground vehicles on the principle 'where are you?'
(c) position location of ground vehicles.

The 6231P facility comprises 8 LRUs: unit 551 - receiver-transmitter; unit 542 - reply evaluation unit; unit 541 - video processor; unit 591 - power supply unit; unit 581-1 - controls; unit 2410 - SLS modulator; unit 526 - RF switch; unit 531-5 - auxiliary controls unit.

Specifications

Maximum range of positive identification: not less than detection range of associated radars
Angular resolution: determined by the radiation pattern of the CP3-1P interrogator antenna (autonomous or built into the radar antenna)
Transmitter power output: 1.5 to 3.5 kW peak
Receiving/decoding sensitivity: −76 dBm
Total weight: not less than 26.3 kg
Total volume: not greater than 23.3 dm^3
Power consumption: 115 V AC (+/−5%), 380-1,050 Hz, <2.1A; +27 V DC ±10%, <0.7A; +10 V DC, <0.7A; −10 V DC, <0.2A

Operational status

Designed for both fixed- and rotary-wing aircraft.

Special facility 61P

The special facility 61P forms part of each element of the land, sea and airborne 60P system IFF installation. The description of its function is believed to represent the cryptographic function of the 60P IFF system.

The airborne equipment is designated 6110P-10. Code data is loaded using the 6110P-40 input device; data is loaded using a clockwork timer and time-set mechanism.

Special facility 61P cipher equipment of the 60P IFF system ***1998***/0011880

CP3-1P airborne radar interrogator of the 60P IFF system ***1998***/0011881

Special facility 61P comprises 5 units; unit 61C - Ciphering/Deciphering Unit (CDU); unit 61Y - rectifier, unit 61K - regulated power supply; unit 55 M - timer; unit 61E - interfacing unit.

Specifications

Weight: 8.95 kg
Volume: 7.2 dm^3
Power consumption: 115 V AC, 400 Hz; 27 V DC; <30 W

Contractor

All-Russian JSC Nizhegorodskaya Yarmarka

NEW ENTRY

SRZO-KR airborne interrogator-transponder

The SRZO-KR airborne interrogator-transponder is designed as an upgrade for earlier equipment of the Kremni-2 (2 m) IFF system.
(Editor note: believed to be the system known as SRO-2)

The SRZO-KR equipment is claimed to employ the newest circuit and technical design and component configurations, including microwave transistors, bodyless elements and thinfilm technology.

Four operating modes are described:
Mode I: general identification
Mode III: individual indentification
KO: checking identification
Distress: calling

Specifications

Transmitter power: >400 W
Receiver sensitivity: −72 dBm
Weight: <12 kg
Volume: <17 dm^3
Power consumption: 27 V DC, <195 W; 115 V AC, <115 VA
MTBF: 1,500 hrs

Contractors

Kazan Scientific Research Institute of Radio-Electronics (product designer).
RadioPribor (producer).

NEW ENTRY

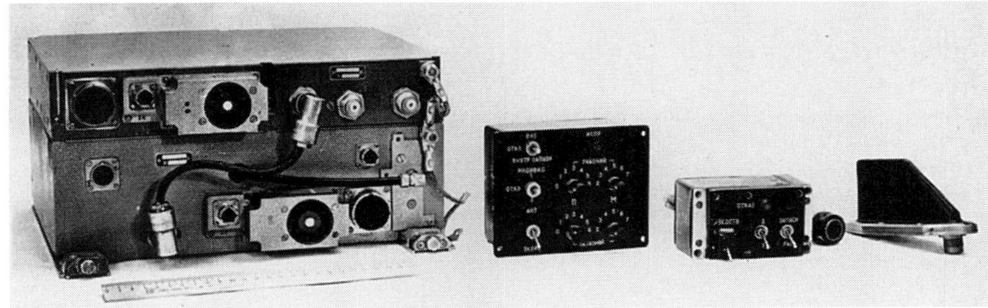

SRZO-KR airborne interrogator-transponder ***1998***/0011879

Aisberg-Razrez airborne dual-band side-looking radar

The Aisberg-Razrez radar is intended for observation of the Earth's surface. The radar is designed for detailed radar surveys of vast areas of ground and water surface. The system uses dual-band frequencies (centimetric and metric wavebands) to optimise efficiency regardless of weather conditions, time of day/night, and season of the year.

Radar data can be relayed over standard communication systems (including satellite datalinks) to specially equipped ground terminals to optimise operational efficiency. The system can also be integrated with other sensors (including the VIDS-95 radar) to increase operational flexibility.

Specifications

Aisberg Radar
Wavelength: 2 cm
Antenna beamwidth: 12 ang/min
Signal polarisation: HH, VV, HV, VH, circular
Range: 100 km
Scan strips: 64, 32, 16 km
Resolution:
(in range) 15 m
(along flight path) 12 ang/min
(in SAR mode) 5 m
Accuracy of map control: 30 m
Processing: non-coherent MTI, SAR
Display: TV, SVGA PC
Recording: Digital 2.5 Gbytes
Weight 320 kg
Power consumption: 2.1 kW

Razrez Radar
Wavelength: 3 m
Antenna beamwidth: 30°
Signal polarisation: HH, VV
Range: 50 km
Scan strips: 32, 16 km
Resolution:
(in range) 30 m
(along flight path) 30 m
Accuracy of map control: 30 m
Processing: SAR
Weight 120 kg
Power consumption: 1.4 kW

Contractor

Scientific-technical centre Osnova, Leninetz Holding Company.

VERIFIED

Duet dual-band weather/navigation radar

Duet is a dual-band (centimetric/millimetric) weather/navigation radar intended for civil aircraft. Basic operating modes include:

in centimetre-band operation: weather, turbulence, windshear, map
in millimetre-band operation: map, landing/takeoff, taxi

Specifications

Range scales:
(cm-band) 640, 320, 160, 80, 40, 20, 10 km
(mm-band) 20, 10, 4, 1 km
Range resolution:
(cm-band) 150 m
(mm-band) 7.5 m
Scan sector: ±90, ±45, ±25°
Scan rate: 45°/s
Azimuth resolution:
(cm-band) 3°
(mm-band) 0.7°
Antenna diameter:
(two-band) 760 mm
(one-band) 610, 560, 508, 457, 380 mm
Form factor of transceivers: 8MCU ARINC-600

Display: Duet display; aircraft multifunctional display; head-up display
Weight: two-band radar 50 kg

Contractor
Scientific-technical centre Osnova, Leninetz Holding Company.

VERIFIED

Neva multifunction civil weather/ navigation radar

The Neva radar is designed for heavy- and medium-sized civil/commercial aircraft. It provides the following three operating modes:
weather/turbulence indication: horizontal and vertical sections, four-colour display of rain, isolation of windshear and turbulence areas, automatic warnings
obstacle avoidance: high, and medium/low altitude modes, three-colour display, automatic warnings
ground mapping: panoramic and detailed ground images, beam sharpening up to 4 times, navigation interfaces.

Neva multifunction radar for civil aviation
1997/0002446

Specifications
Wavelength: 3.2 cm
Range scales: 600, 300, 150, 75, 40, 20, 10 km
Azimuth coverage: ±90, ±30°
Elevation coverage: ±15°
Weight: 50 kg
Power consumption: 115 V, 400 Hz, 300 V A
Antenna diameter: 762 or 559 mm, with polarisation selectivity and monopulse
Display: colour TV, 16 cm

Contractor
Scientific-technical centre Osnova, Leninetz Holding Company.

VERIFIED

VID-95 approach/landing radar

VID-95 is a short-range, high-resolution, approach and landing radar, intended for operation in ICAO Cat. II and IIIA conditions, at poorly equipped airfields. The radar operates at 8 mm wavelength, using a high-speed electronically scanning beam, and coherent received signal processing to control landings from 200 m altitude, provided the aircraft is positioned on the glide path using ILS, satellite or other navigation means.

Specifications
Runway detection range:
(VFR conditions) >5 km
(ICAO Categories 2 & 3A) >3 km
Frequency: 35.8 GHz
Pulse repetition frequency: 10,000 Hz
Scan sector: ±45°
Drift angle compensation:34°
Beam scan rate: 440°
Antenna beamwidth:
(horizontal) 0.5°
(vertical) 8°
Weight: 50 kg

Contractor
Scientific-technical centre Osnova, Leninetz Holding Company.

UPDATED

VID-95 antenna
1997/0002352

Arbalet combat helicopter multifunction radar for Kamov-52 Alligator

Arbalet is a multifunction air-to-surface and air-to-air radar proposed for the Kamov-52 Alligator attack helicopter. It provides day/night/all-weather combat capability.

Air-to-surface modes include the following over-land and over-water capabilities:
- target detection and localisation
- moving target indication and data track
- ground mapping
- terrain-following/avoidance
- air-to-surface missile and gun control

Air-to-air modes include:
- air target detection and tracking
- air-to-air missile and gun control

The Arbalet radar comprises two separate transmitter/antenna elements. The main antenna shown is located in the nose of the helicopter, and is said to operate in J-band, to provide the main air-to-ground capabilities. A separate antenna is located in

Arbalet combat helicopter multifunction radar
1997/0002353

Ka-52 Alligator helicopter showing nose mounted radar and mast-mounted radome
1998/0011878

the radome on top of the rotor shaft; this is said to operate in K-band, and to provide some elements of the surveillance and air-to-air modes.

Note that in some Phazotron literature, Arbalet is also spelled Arbalest.

Operational status
Arbalet appears to be more of a marketing proposal than a substantive system at present. It is noteworthy that the same Ka-52 Alligator airframe (061) is also (more often) shown with an electro-optic fit in the chin and no rotor mast radome.

Contractor
Phazotron Scientific & Production Company.

UPDATED

Gukol weather/navigation radars

Four variants of the Gukol radar have been marketed by Phazotron as I/J-band (8-12 GHz) weather/navigation radars for light strike/attack aircraft, helicopters and military/civil transports. Operating modes are claimed to include: weather and obstacle detection/avoidance; real beam and synthetic aperture mapping; navigation and beacon tracking modes.

The four variants have been offered with antenna sizes varying from 370 to 670 mm, and equipment weights from 15 to 65 kg.

The largest variant, marketed for larger transport/ tanker types is said to have an L/M-band blind landing capability.

Operational status
Not known.

Contractor
Phazotron Scientific & Production Company.

VERIFIED

Kopyo airborne radars

Four variants of the Kopyo radar are being marketed: Kopyo; Kopyo-25; Kopyo-Ph; and Phathom. The names 'Komar' and 'Super Komar' are sometimes used in the marketing material apparently interchangeably with Kopyo.

Kopyo airborne radar
The Kopyo radar is an all-weather, coherent, multimode, multiwaveform search-and-track radar that uses digital processing to provide the features and flexibility needed for both air-to-air and air-to-surface missions. It was derived from technology developed for the Zhuk radar.

Air-to-air modes include: range-while-search in look-up and look-down mode; single-target track; track-while-scan of eight targets and simultaneous engagement of two targets; air combat modes (vertical scan, HUD search, wide angle, boresight).

Air-to-surface modes include: real beam ground map; Doppler beam sharpening to 0.45° (1:10); synthetic aperture beam sharpening to give 30 × 30 m resolution; enlargement, freezing capability; track-while-scan four targets; ground moving targets indication/track; air-to-surface ranging.

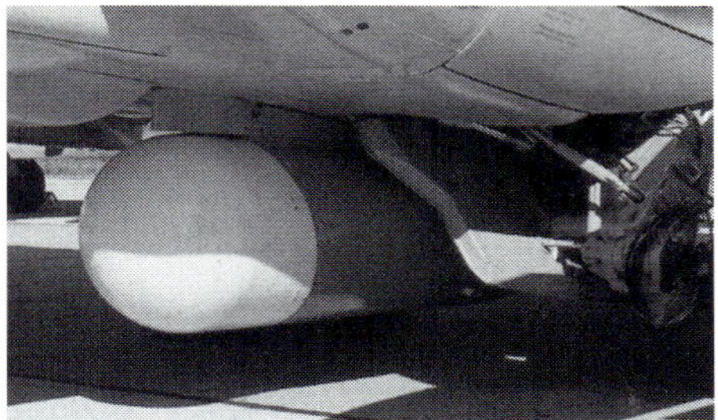

Kopyo-25 radar fitted to an SU-25TM to enable enhanced attack capability ***1996***

The Kopyo radar is compatible with many weapons including: Kh31A, R27R1, R27T1, R-73E, R60MK, RVV-AE, other precision weapons and iron bombs.

Specifications
Detection range:
57 km approaching targets
35 km receding targets
Angular coverage: ±10°, ±30° azimuth, 2 or 4 bars in elevation
Radar frequency: I/J-band
Peak power: 5 kW
Average power: 1 kW
Input power: 8.5 kVA, 400 Hz, 1 kW DC
Weight: 130 kg
Volume: 400 dm^3
Reliability: 120 hours MTBF
Cooling: air, liquid

Operational status
Fitment of Kopyo is part of the Indian Air Force upgrade of 125 MiG-21bis aircraft to MiG-21-93 configuration. Proposed to China for the F-7II and A-5 aircraft, as Komar, and for the new FC-1 aircraft as Super Komar. Being widely marketed for MiG-21, MiG-23, F-5, Mirage-1 Hawk 200 upgrades.

Kopyo-25 airborne radar
Kopyo-25 is a derivative version of the Kopyo radar, designed specifically for carriage on the Su-25TM aircraft. It is carried in an under-fuselage pod. Addition of the Kopyo-25 radar to the Su-25TM gives the aircraft day/night, all-weather capabilities as well as considerably enhanced efficiency in air-to-air combat.

Operational capabilities include: detect and track air targets in the automatic mode, (including targets flying at low altitude over land or sea); designate targets and engage them with radar- and IR-guided air-to-air missiles or guns; high-speed vertical search and automatic lock-on of visible targets in close combat, in association with the use of high-manoeuvrability dogfight missiles; ground mapping with real beam and Doppler sharpening and scaling-up of the chosen sector of the map.

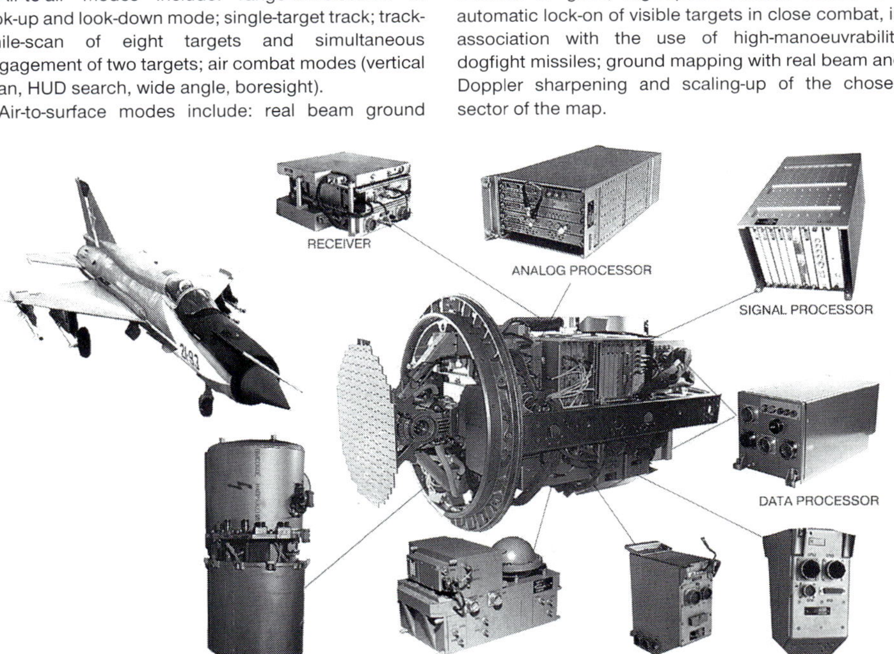

Kopyo-25 system units ***1997****/0001202*

Specifications
As for Kopyo.

Operational status
Designed for Su-25TM, also shown at Zhukovsky 97' Air Show on the Su-39 Strike Shield aircraft.

Kopyo-Ph
Kopyo-Ph is a derivative of the Kopyo radar that employs a phased-array antenna using technology derived from the N011M programme tested on Su-35.

Operational status
Development.

Phathom
Phathom is a collaborative venture between Phazotron and Thomson-CSF. From Phazotron come elements of the Kopyo radar including antenna, receiver, transmitter, and primary power supplies. Thomson provides the data and signal processing from its RDY radar. Analogue processors are a joint development.

Operational status
Development.

Contractor
Phazotron Scientific & Production Company.

UPDATED

Moskit/Mosquito radar

These two radars share a common name (Mosquito being the translation of Moskit), and appear to have identical specifications, despite the different roles marketed.

Moskit
The Moskit radar is a coherent, multi-mode, digital fire-control radar that provides weapon delivery and dogfight capabilities. It is smaller, lighter and less expensive than fighter radars of its class. Technology employed in the Moskit radar is reported to derive from the Kopyo programme. Moskit is claimed to detect and track targets at all aspects and altitudes, and to provide the following capabilities:
Air-to-air modes: range-while-search in look-up and look-down; eight targets track-while-scan and two target simultaneous engagement; air combat: HUD search, slewable scan, boresight, vertical scan.
Air-to-ground modes: real beam ground map; Doppler beam sharpening; synthetic aperture; enlargement; freeze; beacon; two targets track-while-scan; air-to-air ground ranging; ground moving targets track.

Moskit comprises five LRUs: a flat slot array antenna; air cooled TWT transmitter; monopulse four-channel coherent receiver; 280 Mflops programmable signal processor; 1 Mflops effective speed, 512K static RAM, 1.5M ROM data processor. The Moskit radar is compatible with such weapons as: Kh-29L, Kh-29TD, Kh-31A, Kh-31PE, Kh-38, RVV-AE, as well as KAB500KR iron bombs.

Specification
Detection range:
25 km approaching targets
15 km receding targets
Angular coverage: ±10°, ±30° azimuth, 2 or 4 bars in elevation
Radar frequency: I/J-band
Peak power: 4 kW
Average power: 0.3 kW
Input power: 2.1 kVA, 200 V, 400 Hz; 0.2 kW 27 V DC
Weight: 70 kg
Volume: 300 dm^3
Cooling: air

Operational status
Intended for MiG-AT.

Mosquito
Mosquito is being offered to the Indian Air Force as a maritime radar upgrade for its Jaguar aircraft. Mosquito is claimed to offer the following capabilities.
Air-to-sea modes: detection of sea targets to 100 km in Sea State 4 to 6, co-ordinates measurement accuracy of 300 m^2, and engagement using Sea Eagle air-to-surface missiles.
Air-to-surface modes: detection and co-ordinates

measurement of sea ports and fleet anchorage and engagement of them using unguided missiles.
Air combat modes: detection, lock-on and tracking of air targets and engagement of them using western and Russian guided air-to-air missiles; HUD screen; slewable scan; boresight; and vertical scan.

Operational status
Marketed for Indian Air Force Jaguar aircraft.

Contractor
Phazotron Scientific & Production Company.

UPDATED

RP-35 multimode airborne radar

The RP-35 is a coherent, multimode, digital fire-control radar that provides a comprehensive set of all weather air-to-air and air-to-surface modes, with superior dogfight and weapons delivery capabilities. The air-to-air modes provide the capability to detect, track and engage targets at all aspects, even in the presence of ground clutter. Air-to-surface modes provide extensive ground mapping, target detection, location and tracking capabilities, as well as navigation features.

The RP-35 radar is designed for use with the MiG-35 aircraft, and is compatible with a wide range of air-to-air and air-to-surface weapons, including: Kh-29T, Kh-31A, Kh-35U, Kh-38, R-27ER1, R-27ET1, R-27R1, R-27T1, R-73E, RVV-AE, KAB-500KR.

The RP-35 radar is designed for single-pilot operation. All combat-critical controls are integrated into the throttle grip and stick controller (HOTAS). Air-to-air and air-to-surface information is displayed on the Head-Up-Display as well as the Multifunction Cockpit display.

The antenna is a phased-array with electronic scanning, which provides high gain and low sidelobes at all scan angles. The transmitter is a liquid cooled TWT. The receiver is a three-channel system.

The RP-35 is claimed to have the following capabilities:

Air-to-air modes: 24 target track-while-scan simultaneously; range-while-search; air combat - vertical scan, HUD search, boresight, wide angle, velocity search; raid cluster resolution, automatic terrain avoidance.

Air-to-surface modes: four target track-while-scan; ground moving target indication/track; air-to-ground ranging; real beam ground map; Doppler beam sharpening; synthetic aperture; enlargement; freezing; beacon.

Specification
Detection range:
(approaching targets) 140 km
(receding targets) 65 km
Angular coverage: ±20°, ±60° azimuth, 2 or 4 bars in elevation
Radar frequency: I/J-band
Peak power: 8 kW
Average power: 2 kW
Input power: 12 kVA, 200 V, 400 Hz; 2 kW 27 V DC
Weight: 220 kg
Volume: 500 dm^3
Cooling: air/liquid
Reliability: >120 hours MTBF

Operational status
Designed to meet the requirements of the MiG-35.

Contractor
Phazotron Scientific & Production Company.

VERIFIED

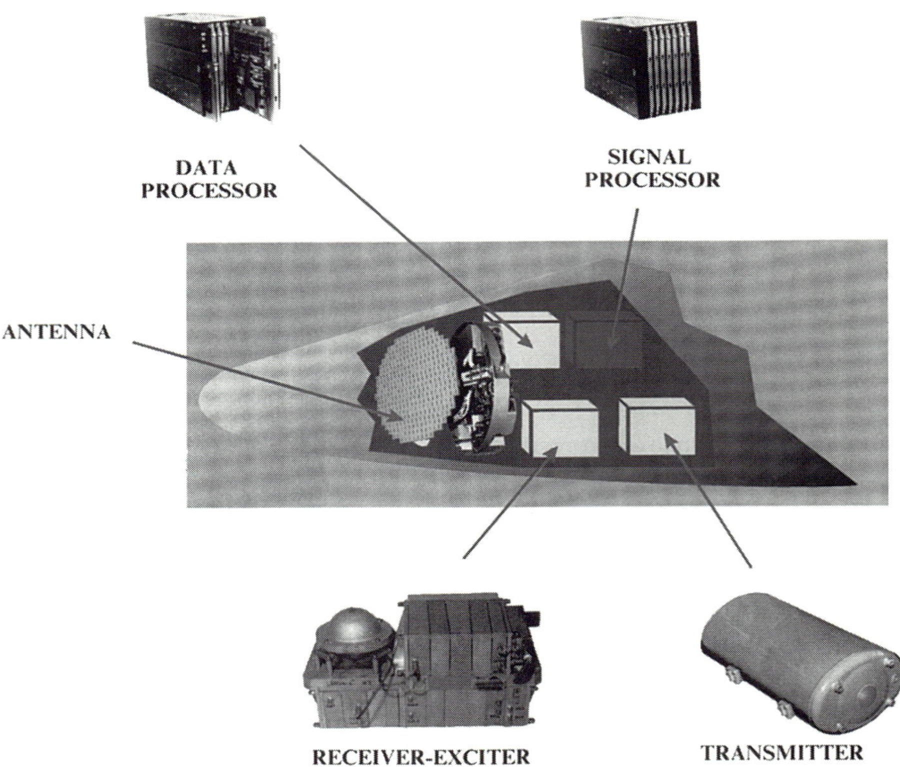

Moskit airborne radar for advanced combat trainers **1998**/0011877

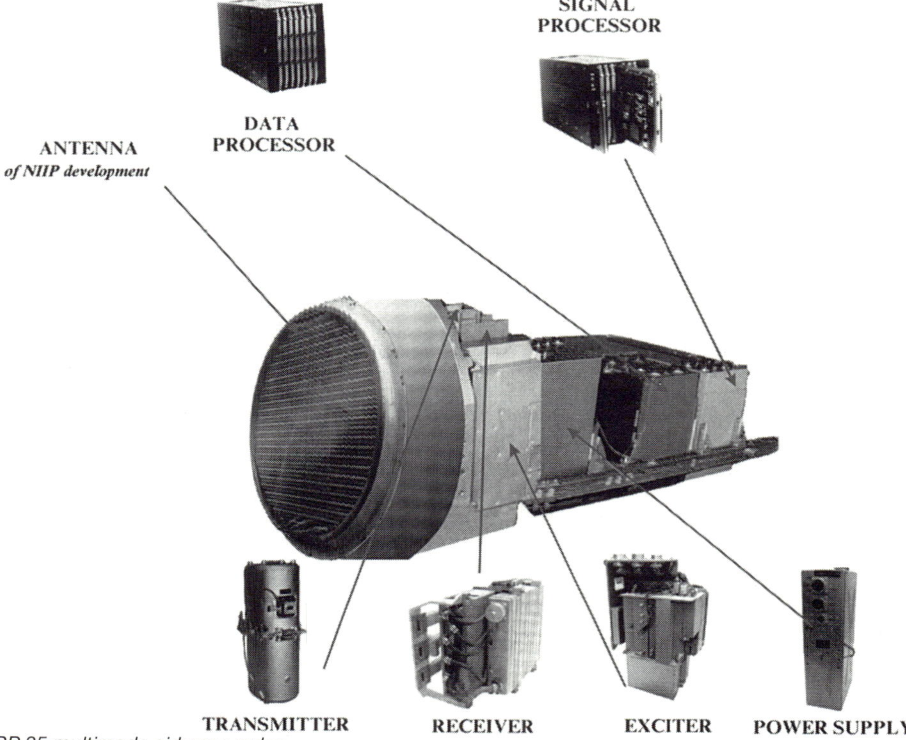

RP-35 multimode airborne radar **1998**/0011876

Sapfir airborne fire-control radars

The Sapfir series radars have been in use for up to 40 years on MiG-21, MiG-23, MiG-25, and MiG-29 aircraft. They were designed by predecessor organisations of Phazotron Scientific & Production Company.

RP-21 Sapfir
RP-21 Sapfir was widely fitted to early model MiG-21 variants. It is a basic I-band air-to-air fire-control radar, credited with the following operational capabilities: detection range 20 km; tracking range 10 km; azimuth cover 60°.

RP-22 Sapfir-21
RP-22 Sapfir-21 was widely fitted to later model MiG-21 variants. It is a basic low J-band air-to-air fire-control radar, credited with the following operational capabilities: detection range 30 km; tracking range 15 km; azimuth cover 60°; elevation cover ±20°.

S-23 Sapfir-23
S-23 Sapfir-23 was widely fitted to MiG-23 variants. It is a pulse Doppler low J-band air-to-air fire-control radar, credited with the following operational capabilities:

The MiG-25 carries the RP-25 Sapfir-25 interception radar

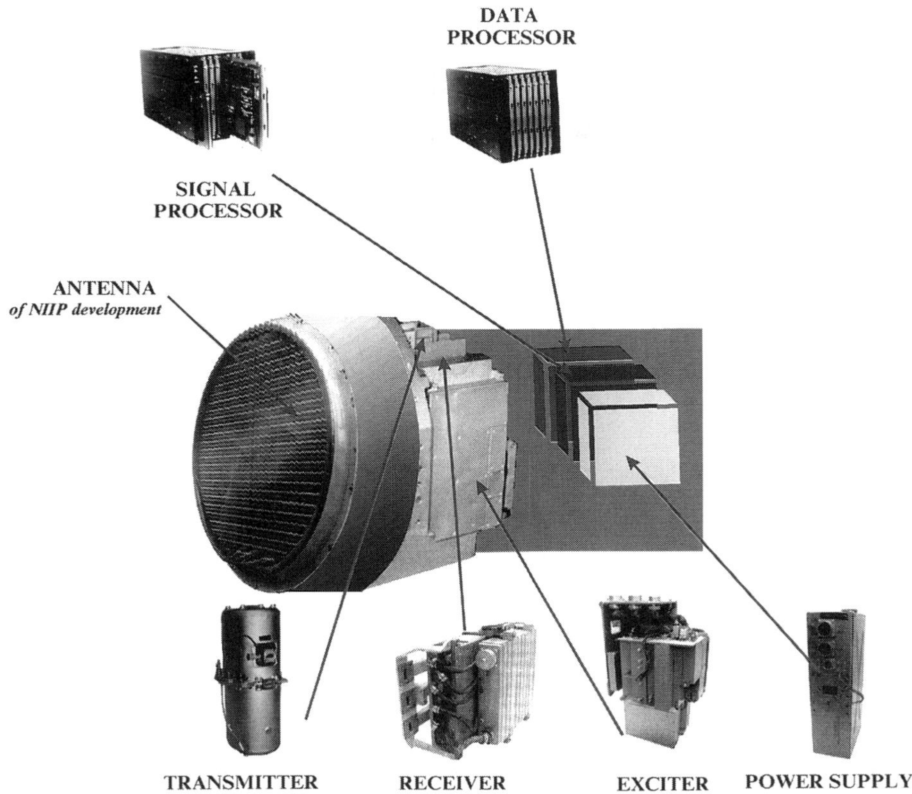

Sokol multimission airborne fire-control radar ***1998**/0011875*

detection range 70 km; tracking range 55 km; azimuth cover 60°; elevation cover ±60°.

RP-25 Sapfir-25

RP-25 Sapfir-25 was widely fitted to MiG-25 variants. It is a look-down/shoot-down I/J-band pulse Doppler radar, credited with the following operational capabilities: detection range 100 km; tracking range 75 km; azimuth cover 112°; elevation cover ±60°.

N019 Sapfir-29

N019 Sapfir-29 was installed on initial production MiG-29 aircraft. It is reported to be a look-down/shoot-down pulse Doppler fire-control radar, which operates in I/low-J band. Credited operational capabilities include: the ability to track 10 targets simultaneously; detection range 100 km; tracking range 70 km; azimuth cover 134°; elevation cover −38 to +60°.

N019M Topaz

N019M Topaz is reportedly installed on later model MiG-29 aircraft. It is a multimode, multifunction, coherent pulse Doppler radar, update of N019 Sapfir-29. Credited operational capabilities include: tracking range 80 km (RCS 3m^2) approaching; tracking range 40 km receding target; azimuth cover ±70°; elevation scan 4 or 6 bars; weight 350 kg.

Contractor

Phazotron Scientific & Production Company.

VERIFIED

Sokol multimission airborne fire-control radar

Sokol is a coherent, multimission, digital fire-control radar that provides a comprehensive set of all-weather air-to-air and air-to-surface modes. The diverse operating modes required to meet the multimission design concept are achieved by employing a variety of complex and flexible waveforms involving low, medium and high pulse repetition frequencies.

The Sokol radar is intended to be installed in next-generation multirole aircraft; it is compatible with the datalink requirements of a wide range of air-to-air and air-to-surface weapons, including: Kh-31A, R-27R1, R-27T1, R-73E, RVV-AE and other precision-guided weapons.

The antenna is a phased-array with electronic scanning, which provides high gain and low sidelobes at all scan angles. The transmitter is a liquid-cooled TWT. The receiver is a multiple-channel system, with low noise figure. The signal and data processors use flexible high order language programming.

Sokol is claimed to have the following capabilities:

Air-to-air modes: 24 target track-while-scan and simultaneous engagement of six targets; range-while-search look-up/look-down; air combat manoeuvring - vertical scan, HUD search, boresight, wide angle; velocity search; raid cluster resolution, automatic terrain avoidance.

Air-to-surface modes: four target track-while-scan; ground moving target indication/track; air-to-ground ranging; real beam ground map; Doppler beam sharpening; precision velocity update; synthetic aperture; enlargement; freezing; beacon.

Specification

Detection range:
(approaching targets) 180 km
(receding targets) 80 km
Angular coverage: ±20°, ±60° azimuth, 2 or 4 bars in elevation
Radar frequency: I/J-band
Peak power: 8 kW
Average power: 2 kW
Input power: 12 kVA, 200 V 400 Hz; 2 kW 27 V DC
Weight: 275 kg
Volume: 600 dm^3
Cooling: air/liquid
Reliability: >120 hours MTBF

Operational status

Designed to meet the requirements of next-generation multimission aircraft.

Contractor

Phazotron Scientific & Production Company.

VERIFIED

Topaz multifunction fire-control radar

Topaz is a multifunction, multimode, coherent, pulse Doppler, air-to-air, modernisation of the N019 radar, intended for installation in MiG-29 and MiG-23 aircraft. The modernisation is claimed to increase operational effectiveness, particulary in ECM conditions. Although designed as an air-to-air radar, Topaz can be modified to incorporate air-to-surface modes. It is compatible with a wide range of weapons, including: R27ER1, R27T1, R27ET1, RVV-AE, R73E.

Specifications

Detection range (against 3 m^2 target):
(approaching target) 80 km
(receding target) 40 km
Scan angle:
(azimuth) ±15°, ±25°, ±70°
(elevation): 4, 6 lines
MTBF: 100 h
Weight: 350 kg

Operational status

Installed in MiG-29ME and MiG-29S; suitable for retrofit to earlier MiG-29 and MiG-23 aircraft.

Contractor

Phazotron Scientific & Production Company.

NEW ENTRY

Topaz multifunction fire-control radar ***1998**/0011874*

Zhuk airborne radars

N010 Zhuk was designed by the NIIR Moscow (Scientific Research Institute for Radio Engineeering, Moscow), now part of Phazotron. N010 Zhuk (together with the N011 radar that equips Su-27M) was the first multifunction Russian radar capable of tracking air as well as surface targets. The radar uses a slotted flat-plate antenna and digital computer. It was introduced in

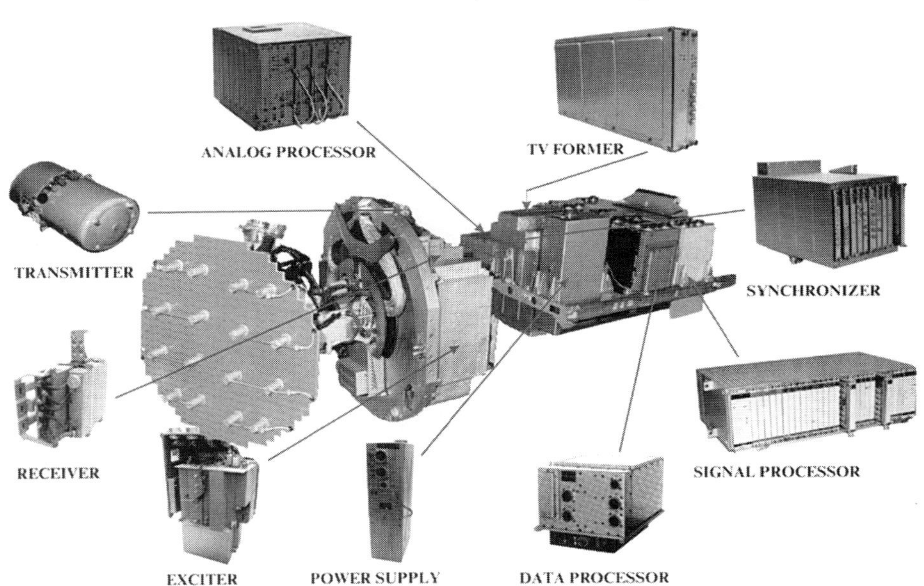

Zhuk airborne radar ***1998**/0011873*

1988 on the MiG-29M. Development has continued under Phazotron, and the following models are now being marketed:

Zhuk is the definitive type. It is a coherent, multimode, multimission, digital fire-control radar intended for MiG-29, MiG-33 and their upgrades.

Zhuk-27 is a variant marketed for Su-27 upgrades and Su-30 variants.

Zhuk-Ph is an advanced derivative of Zhuk that features a phased-array antenna system.

A key feature of the Zhuk radar is its programmability. The radar can be programmed to respond to new threats, to incorporate improved operating modes, to integrate new weapons, and to respond to new electronic countermeasures by software change.

Zhuk is compatible with the datalink requirements of a wide range of air-to-air and air-to-surface weapons, including: Kh-31A, R-27T1, R-73E, RVV-AE and other precision-guided weapons and iron bombs.

Zhuk is claimed to have the following capabilities:

Air-to-air modes: 10 target track-while-scan and simultaneous engagement of 2 to 4 targets; range-while-search look-up/look-down; air combat manoeuvring - vertical scan, HUD search, boresight, wide angle, automatic terrain avoidance.

Air-to-surface modes: four target track-while-scan; ground moving target indication/track; air-to-ground ranging; real beam ground map; Doppler beam sharpening; aircraft velocity measuring for navigation system updating; synthetic aperture; enlargement and freezing capability.

Specification

Detection range:
(approaching targets) 80 km
(receding targets) 40 km
Angular coverage: ±20°, ±60°, ±90° azimuth, 2 or 4 bars in elevation
Radar frequency: I/J-band
Peak power: 5 kW
Average power: 1 kW
Input power: 8.5 kVA, 200 V 400 Hz; 1.5 kW 27 V DC
Weight: 250 kg
Volume: 800 dm^3
Cooling: air/liquid
Reliability: 120 hours MTBF

Operational status

N010 Zhuk in service on MiG-29M; advanced versions being marketed for Russian Air Force and export aircraft. When fitted to the Chinese J-8IIM aircraft, it is called Zhuk-8II.

Contractor

Phazotron Scientific & Production Company.

UPDATED

IFF 6201R/6202R and 6231R/6232R systems

RadioPribor has manufactured airborne radio and navigation equipment for installation on all types of civilian and military aircraft and helicopters produced in Russia and the other CIS states.

The IFF responders 6201R and 6202R are fitted on all military and civil aircraft and helicopters. The model 6202R differs from the 6201R in that it incorporates additional signal amplifiers for use on heavy aircraft that have long SHF cable runs.

In aircraft equipped with radar, the 6231R or 6232R interrogators are installed to provide the interrogation function.

The combined system provides three modes of operation: general identification modes; individual identification modes providing 84 interrogation slots and 100,000 reply slots; identification of objects in distress (distress mode with interrogation/alarm modes without interrogation).

Operational status

Widely deployed on civil and military aircraft and helicopters manufactured in Russia and the RFAS countries.

Contractor

Production Association RadioPribor.

NEW ENTRY

RadioPribor's 6201R IFF responder **1998**/0010928

RadioPribor's 6231R IFF interrogator **1998**/0010929

Zaslon

Zaslon is an airborne fire-control radar fitted to MiG-31 'Foxhound' aircraft. The designators S-800 and SBI-16 have both been used in association with this radar. The NATO name is 'Flash Dance'.

Zaslon is an I-band pulse Doppler radar, with a look-down/shoot-down capability. It uses an antenna based on passive electronic scanning. Control of the beam is believed to be via ferrite phase shifters. Reported maximum range is 300 km.

The MiG-31M 'Foxhound' upgrade programme has an upgraded phased-array radar called Zaslon-M. Zaslon-M is a significantly improved radar which offers 150 to 200 per cent improvement. The working range is extended, whilst the radar also provides a genuine multifunction capability. The NIIP design bureau also confirms that integration work with the R-77 (AA-12) active medium-range AAM has been completed with Zaslon-M.

Zaslon on MiG-31 'Foxhound'

NIIP is also testing a phased-array radar (N-011M) for Su-35 (the advanced derivative of Su-27). The NIIP radar (N-011) previously associated with the programme uses a slotted flat plate antenna. This new radar builds on MiG-31 Zaslon phased-array experience, but it is not known whether the 'Zaslon' name is also applied to this development. NIIP says the N-011M will also be offered for Su-27, Su-30MK, and MiG 1-42.

Operational status

The Zaslon S-800 is in service with Russian Air Force MiG-31 'Foxhound'. Zaslon-M is associated with the MiG-31M programme.

Contractor

Scientific Research Institute of Instrument Engineering (NIIP), Zhukovsky.

UPDATED

SOUTH AFRICA

PA-5429 radar altimeter

The PA-5429 pulsed airborne radar altimeter provides the height between the altimeter and the underlying terrain/surface for heights from 0 to 5,000 ft. Control/data is via ARINC 429 or analogue/discrete interfaces. The unit operates in the mid J-band (~15 GHz) and features a single-LRU configuration, eliminating the need for separate RF feed cables and antennas. The altimeter has good ECCM performance with a Low Probablility of Intercept (LPI) and comprehensive anti-jamming features, making it suitable for a wide range of applications, including high-performance and transport aircraft, helicopters and missiles.

Specifications

Dimensions: 219 × 77 × 118 mm (direct mounting to aircraft skin)
Weight: <3 kg
Power supply: 28 V DC to MIL-STD-704
Consumption: 17 W nominal, 25 W max
Height range: 0 to 5,000 ft
Accuracy: ±3 ft for heights 0 to 100 ft
±3% for heights 100 to 5,000 ft

Contractor

Plessey South Africa Limited.

NEW ENTRY

PT-1000 IFF transponder

The PT-1000 provides Mk XII transponder capability and is suitable for airborne and marine applications. The transponder supports South Africa's national secure mode, which is available in a country-specific export version.

The transponder provides full diversity (dual-antenna) decoding and replies to modes 1,2,3/A,C and the secure mode, according to STANAG 4193.

The system consists of a tray-mounted transponder and an optional internal cryptographic module and panel-mounted Control/Display Unit (CDU).

The PT-1000 IFF transponder and control panel **1998**

Control and status interfacing to the transponder is via ARINC 429.

Specifications

Dimensions: 3/8 ATR (transponder)
5½ in panel, 127 × 113 × 43 mm (CDU)
Weight: <5 kg inclusive of the crypto module
Power supply: 28 V DC, to MIL-STD-704
Power consumption: 35 W nominal, 55 W max

Contractor

Plessey South Africa Limited.

UPDATED

XBT-2000 X-band radar transponder

The XBT-2000 is a radar transponder providing encoded replies to interrogations from airborne or shipborne X-band (NATO I-band) weather radars operating in the weather band (9,200 to 9,500 MHz). The transponder provides compatibility with the DO-172 radar beacon mode standard (encoded replies), as well as fully independent digital tuning of transmit and receive frequencies for alignment with specific radars. The transponder is suitable for man-pack deployment or may be mast-mounted. Applications include demarcation of remote runways or drop-zones as well as helicopter decks. Optional packaging is available for airborne use.

Specifications

Dimensions: 172 × 215 × 64 mm
Weight: <5 kg, including rechargeable battery pack and antenna
Power supply: 10.5 to 28 V external, or integral Ni/Cd battery
Power consumption: 2.5 W nominal, 5 W max
Frequency range: receiver and transmitter independently tunable between 8,500 and 9,500 MHz
beacon mode 9,375 MHz receive, 9,310 MHz transmit
Set-up: stored in non-volatile memory after configuration via PC RS-232 terminal interface

Contractor

Plessey South Africa Limited.

NEW ENTRY

SWEDEN

AESA Active Electronically Scanned Array antenna

Ericsson AESA (Active Electronically Scanned Array) is a new airborne radar project currently in development at Ericsson Microwave Systems.

It is intended for the next-generation Gripen aircraft as a multimode radar. The AESA technology will improve the radar's overall performance, especially its target detection and tracking capability. Beam direction can for instance change instantaneously, detection range will be considerably increased, and jamming suppression further improved.

The AESA radar will feature multibeam capability with all beams individually and simultaneously controlled. It can also operate simultaneously as a fire-control and obstacle warning radar, and be used both in intercept and ground attack missions. As a consequence of the very large number of transmitter and receiver modules, the radar will have a high system availability through graceful degradation.

Operational status

In November 1994 Ericsson received an order from the Swedish Defence Materiel Administration (FMV) for an airborne radar study utilising a new, active phased-array technology, utilising a thousand or more transmitter and receiver modules.

In mid-1997 Ericsson commenced the first test of the AESA system. Ground tests were performed with a model having some 100 transmitters and receivers. These tests are expected to continue until mid-1998 and the test results so far are promising.

The next development phase, which has already started, will lead to a full-scale active antenna system for airborne trials. It is expected to start flight trials in 2002, with potential application as a mid-life update for Gripen beyond the year 2010.

For this second phase Ericsson and FMV are looking for a partner and discussions are currently taking place with companies in the USA and Europe, notably with Northrop Grumman Electronic Sensors and Systems Division about the technology used in Agile Beam Radar being proposed for the F-16 C/D Block 60 configuration, and with the European GTDAR Consortium based on its AMSAR development work.

Contractor

Ericsson Microwave Systems AB.

UPDATED

AESA active electronically scanned array antenna ***1997***/0002218

The Swedish Air Force has selected the Saab 340B as the carrier for the Erieye mission system ***1998***/0010932

Erieye EMB 120 ***1997***/0002219

Ericsson Erieye AEW and control sensor in the Brazilian SIVAM system on the Embraer EMB 145 system ***1997***/0001210

Erieye AEW&C Airborne Early Warning & Control mission system radar

The Erieye AEW&C mission system radar features active, phased-array technology. The antenna is fixed, and the beam is electronically scanned, which Ericsson claims provides improved detection and significantly enhanced tracking performance compared with radar-dome antenna systems.

Erieye detects and tracks air and sea targets out to the horizon (and beyond due to anomalous propagation); instrumented range is 450 km. Typical detection range against fighter-sized targets is approximately 350 km, in a 150° broadside sector, both sides of the aircraft. Outside these sectors, performance is reduced in forward and aft directions.

Erieye is understood to operate as a medium- to high-PRF pulse Doppler, solid-state radar, in E/F-band (3 GHz), and to comprise 192 two-way transmit/receive modules, that produce a 1° pencil beam, steered as required within the operating 150° sector each side of the aircraft (one side at a time). It is understood that Erieye has some ability to detect aircraft in the 60° 'blind sectors' fore and aft of the aircraft heading, but has no track capability in this 'blind sector'. If this is so, the aircraft could be manoeuvred to permit tracking of targets in these sectors.

The electronically scanned antenna is controlled by an automatic intelligent energy management system, developed to utilise the phased-array technology implemented, pulse-by-pulse, to illuminate any desired azimuth. The ability instantaneously to direct the radar energy in any wanted direction is used by the operator to optimise power management for any particular scenario by assigning priorities to areas of interest, and thus optimising probabiities of detection and overall system performance.

Operational status

Erieye is in series production for the Swedish Air Force, where it is implemented in the Saab 340B aircraft. This system is understood to carry the system designation FSR 890, and the Erieye radar to be designated PS 890. Four aircraft have been handed over to the Swedish Air Force.

The Erieye system has been selected by the Brazilian Air Force as the airborne element of the SIVAM system for surveillance of the Amazonas. The contract for five Erieye systems is valued at SKr1.1 billion (US$145 million).

Delivery will begin in 1999 for installation in the Brazilian Embraer EMB-145 aircraft. Equipment includes the Erieye radar, with integrated IFF system, and command, control and communication system, that will be integrated into the SIVAM command and control system.

Marketing discussions are in progress with a number of other countries in Asia, Europe and the Americas.

Contractor

Ericsson Microwave Systems AB.

UPDATED

Microwave radiometer

The microwave radiometer has been developed for the Swedish Space Corporation Maritime Surveillance System. It produces real-time, colour-coded maps of oil spill thickness. The oil volume is calculated by the system processor. For easy interpretation of the data, coprocessing with infrared scanner data is used, but the radiometer can also be used as a stand-alone sensor. The microwave radiometer is contained in an aerodynamic pod that is easily installed even on small aircraft.

Specifications

Weight: 25 kg
Power supply: 28 V DC, 10 A
Frequency: 35 GHz
Altitude:
500 ft (325 ft swath width, 25 × 25 ft resolution)
1,000 ft (650 ft swath width, 50 × 50 ft resolution)
2,000 ft (1,300 ft swath width, 100 × 100 ft resolution)

Contractor

Ericsson Microwave Systems AB.

VERIFIED

PS-05/A multimode multimission radar for the JAS 39 Gripen

Ericsson Microwave Systems AB provides the multimode, multimission, I-band (9 to 10 GHz), pulse Doppler radar for the JAS 39 Gripen. Ericsson collaborated with GEC-Marconi in the 1980's on the original development of the radar. The PS-05/A radar of JAS 39 Gripen, and the Blue Vixen radar installed in the UK Royal Navy Sea Harrier FA 2 aircraft are understood to still share some elements (antenna, computer, radar exciter).

The PS-05/A radar provides the following multimode capabilities: air combat, ground attack, high-resolution mapping and reconnaissance to meet the multimission operational requirements of the JAS-39 Gripen aircraft. All principal modes are software-driven, and include:

(1) Air-to-air operation, using high-PRF and medium-PRF pusle Doppler waveforms, to provide: long-range search; auto-acquisition; multiple-target track-while-scan; multiple-priority target tracking; short-range, wide angle search and track for air combat; high-resolution single-target tracking; raid assessment; missile mid-course update

(2) Air-to-ground operation, using low-PRF pulse Doppler and frequency agility waveforms, to provide: search; tracking; high-resolution mapping; air-to-surface ranging.

The radar matches the datalink requirements of the AMRAAM and MICA air-to-air missiles, and is claimed to have excellent ECCM capabiities.

The main waveform modes are:

(1) HPD – a high-PRF, pulse Doppler mode for clutter rejection, primarily designed for use against airborne approaching targets

(2) MPD – a medium-PRF, pulse Doppler mode for clutter rejection, primarily designed for use against approaching and receding targets; a special high-resolution sub-mode is designed for target tracking

(3) LPD – a mode using Doppler processing for clutter rejection designed for use against surface targets

(4) LPRF – a low-PRF mode with pulse-to-pulse frequency agility for use against surface targets and for real-beam mapping

(5) AGR – a mode exclusively designed for ground target ranging

(6) DBS (Doppler Beam Sharpening) – a Synthetic Aperture Radar (SAR) mode utilising Doppler processing for high-resolution mapping, with high-angular coverage obtained by continuous antenna scanning

(7) SLM (SpotLight Mode) – an SAR mode utilising Doppler processing for very high resolution mapping.

The PS-05/A radar comprises five Line-Replaceable Units (LRU):

(1) Antenna Unit: a lightweight, slotted waveguide, planar array, featuring guard antenna, IFF dipoles and all-digital servo control, weight 25 kg

(2) Power Amplifier Unit, and Transmitter Auxiliary Unit: 1,000 W average power, flexible waveform, liquid cooled, TWT system, with two LRUs together weighing 73 kg

(3) High-Frequency Unit: multiple-channel receiver, pulse-to-pulse frequency agile, microwave integrated circuit design, internally software controlled, weight: 32 kg

(4) Signal Data Processor: D80 multiprocessor concept fully programmable unit, employing ASIC technology, with programming in High Order Language, Pascal, weight 23 kg.

Specifications

Weight: 156 kg
Power: 115/200 V AC, 400 Hz, 8.2 kW; 28 V DC, 250 W
Antenna: 600 mm planar-array
Interface: MIL-STD-1553B

Operational status

Some 50 units of the PS-05/A radar for the Gripen aircraft have been delivered.

Two significant improvements to radar performance are now being introduced: delivery of a new signal processor has started. It will be installed in all new production Gripen radars, and retrofitted into exisiting aircraft radars. This processor has higher capacity and improved functions in both air-to-air and air-to-ground modes. Radar system software has also been upgraded to facilitate future upgrades.

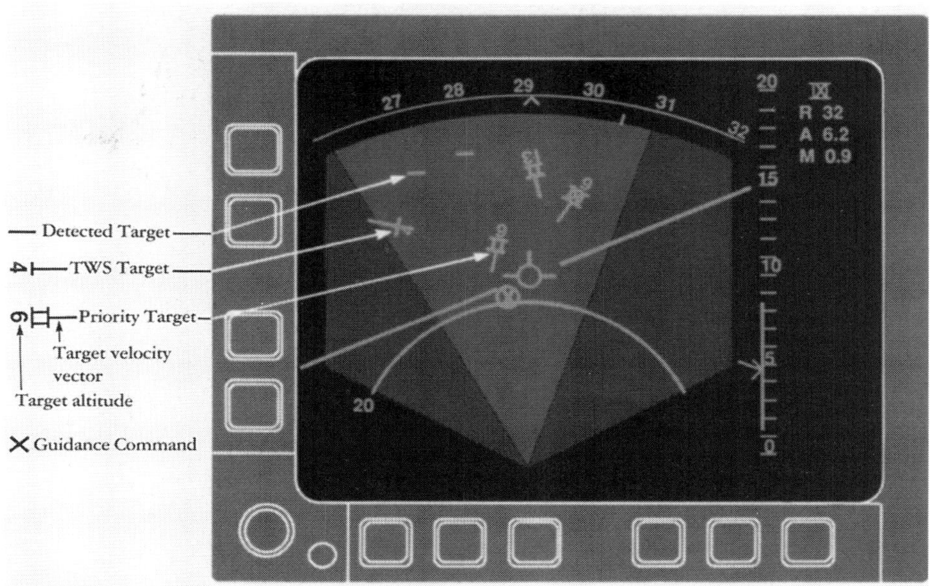

PS-05/A display with typical data format **1997**/0001209

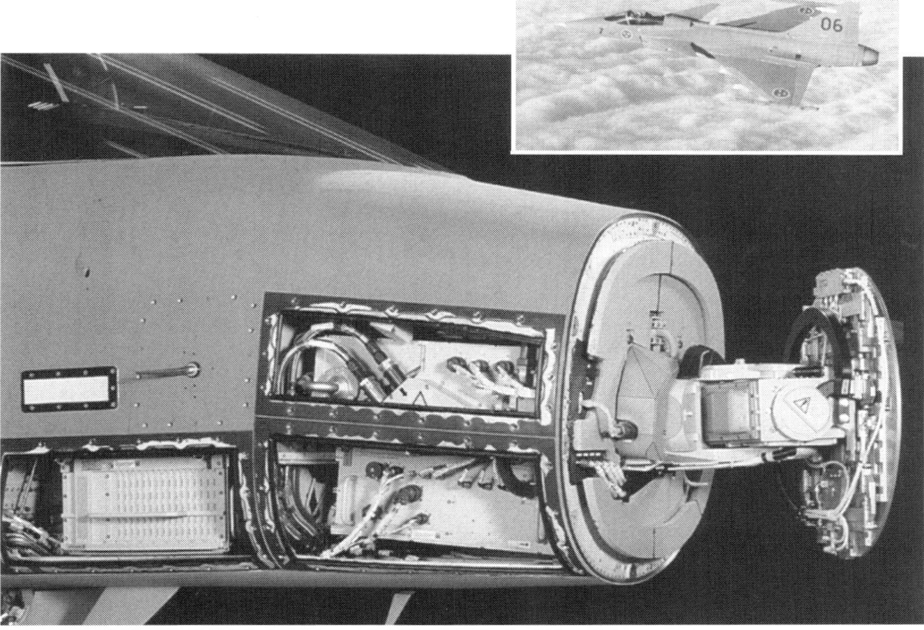

The Ericsson PS-05/A radar for the JAS 39 Gripen aircraft **1998**/0010930

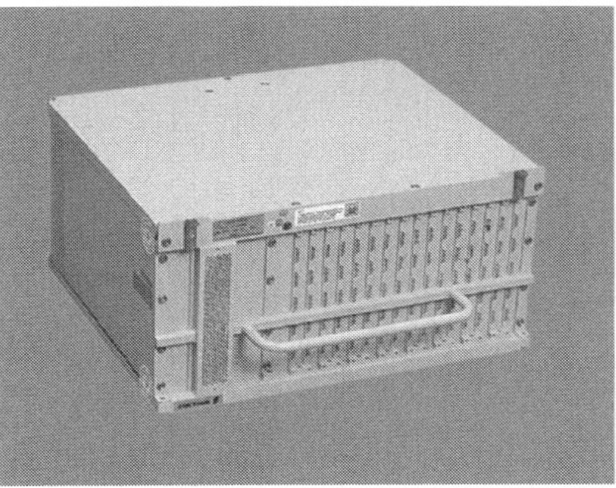

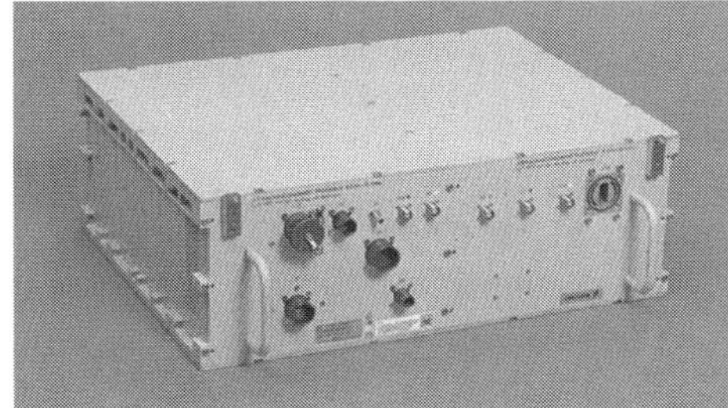

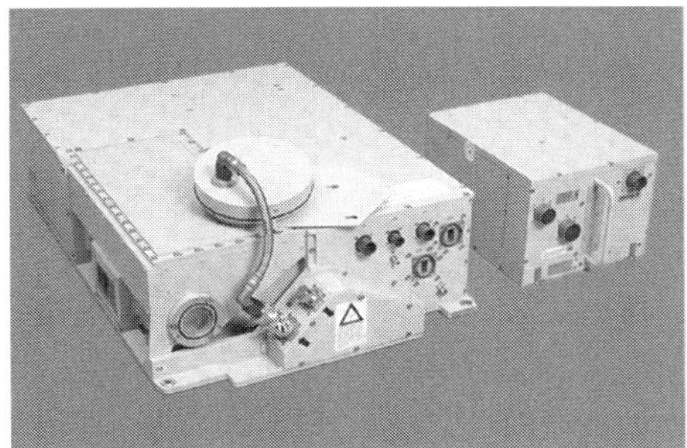

PS-05/A radar antenna unit, (bottom right) power amplifier and transmitter auxiliary units (bottom left), high-frequency unit (top right) and signal and data processor unit (top left) **1998**/0010931

Development of a third-generation signal processor, designated MACS, has also started in co-operation with Mecury Computer Systems Inc in the USA. MACS will replace the D80 computer as the aircraft's system computer in the radar, cockpit displays and the EW system from the third batch of aircraft. Ericsson is expecting an order for system computer retrofit for all Gripen aircraft in 1998, with deliveries to start in 1999. Introduction of Batch 3 aircraft will also include three 6 × 8 in multifunction colour displays to replace the three existing (EP17) 5 × 6 in monochrome displays.

Contractor

Ericsson Microwave Systems AB.

UPDATED

PS-37/A radar

The PS-37/A multimode I/J-band, monopulse radar was designed for the AJ 37 attack version of the Saab Viggen. It stems from mid-1960s' designs and is largely integrated with the navigation, display and digital computer-based data processing subsystems. It comprises two units: a scanner and a transmitter/receiver package, the latter being made up of 13 LRUs.

Except for some of the high-frequency components, the PS-37/A is of solid-state design. Elaborate signal processing provides a high degree of immunity from both natural interference and electronic countermeasures, and accuracy is improved by lobe-shaping to provide better resolution and reduce sidelobe effects. The radar is semi-automatic in operation to reduce the workload of the single-crew member, and information is presented on both head-up and head-down displays.

Operating modes are search, target acquisition, air-to-target ranging, obstacle warning, beacon homing and terrain-mapping. By adding a further unit a terrain-following capability can be provided, the pilot flying the aircraft in response to head-up display demands.

Three CASA 212 aircraft operated by the Swedish Coast Guard are fitted with the Ericsson SLAR **1995**

Operational status

In service. No longer in production.

Contractor

Ericsson Microwave Systems AB.

VERIFIED

PS-46/A radar

The software-controlled, multimode, pulse Doppler PS-46/A radar has been developed for the JA 37 fighter versions of the Swedish Air Force's Viggen. In view of the numerically small size of Sweden's defence force, great emphasis has been placed on operational availability and readiness; all-weather capability and effectiveness in an electronic countermeasures environment are also important requirements. Designed to cope with high-performance aircraft, transports and helicopters, the system has wide-angle coverage, look-down capability and can operate at all altitudes.

The multimode requirements of the PS-46/A are air-to-air and air-to-ground. The latter is met by using conventional non-coherent pulse waveforms, but the former calls for more sophisticated techniques. The standard radar functions are controlled by a data processor that extracts information from the raw radar and transfers it to other aircraft systems. A digital bus distributes all signals within the radar itself with minimum wiring. For the guidance of semi-active homing missiles an illuminator transmits a continuous wave RF signal through the radar antenna.

Control of the system through suitable software enables parameters to be changed or optimised according to the needs of flight development programmes without time-consuming equipment changes; similar modifications can be introduced during service according to changing military requirements, and radar signatures adopted for peacetime training and exercises can be easily changed during conflict to thwart enemy intelligence and countermeasures.

Specifications

Weight: 300 kg
Power:
(coherent transmitter) 500 W average
(continuous wave illuminator) 200 W
Frequency: I/J-band, bandwidth >10% of spectrum
Performance: detection range >50 km in look-down mode
Antenna: 900 mm diameter
Processor: high-speed 16-bit word length system with 32 k word programme memory
Modes: search, acquisition (automatic via HUD, semi-automatic via HDD), tracking (track-while-scan, continuous track), target illumination, air-to-ground ranging
High-resolution ground-mapping is optional
Reliability: 180 h MTBF

The Ericsson PS-37/A radar installed in the strike version of the Viggen

Ericsson PS-46/A radar installed in the Viggen fighter

Operational status

In service on the JA 37 Viggen. No longer in production.

Contractor

Ericsson Microwave Systems AB.

VERIFIED

Side-Looking Airborne Radar (SLAR)

This inexpensive, real aperture SLAR is designed exclusively for maritime patrol and surveillance (for fishery protection, oil pollution monitoring and search and rescue), with emphasis on simplicity and ease of operation. It consists of five LRUs: antenna, transceiver, digital signal processor, control unit and television display. The radar images can be recorded on standard video recorders. Digital radar video can also be recorded for subsequent computer processing. With a video link between aircraft and ground station or ship, images can be transmitted with no degradation of quality.

The digital signal processor is designed around a 1.6 Mbit television memory display. Up to 2,000 range cells can be processed to an accuracy of 6 bits (giving 64 grey-tone levels). The presentation and performance can be varied to suit customers' needs and other features include a reference grey scale, level mapping, positive or negative picture representation and automatic positioning of targets. Monochromatic television images recorded in the air can be converted to colour on the ground.

For double-sided coverage, a glass fibre pod containing two antennas is carried under the fuselage, or antennas are carried in individual pods on each side of the aircraft.

Specifications

Weight: 70 kg
Power supply: 28 V DC, 15 W
Frequency: 9.4 GHz
Pulsewidth: 0.5 μs
Beamwidth:
(horizontal) 0.5°
(vertical) 33°
Peak output power: 10 kW
PRF: 1 kHz

Operational status

In operation in one country outside Sweden. Three systems are also fitted to CASA 212 aircraft operated by the Swedish Coast Guard for oil pollution detection and ice mapping.

Contractor

Ericsson Microwave Systems AB.

VERIFIED

CARABAS surveillance radar

The Coherent All RAdio BAnd Sensing (CARABAS) is a prototype airborne low-frequency surveillance radar which can penetrate both foliage and the ground. It can be used by both military and civil authorities for tactical surveillance and remote sensing.

CARABAS is a wideband, synthetic aperture radar operating at 20 to 90 MHz frequencies necessary to penetrate the ground and foliage. The depth of ground penetration varies largely with the type of terrain, whereas foliage penetration is much less sensitive to the type of vegetation.

The CARABAS system has applications in detecting military targets such as underground bunkers and hardened shelters, as well as weapons and buildings screened by foliage.

The system uses a novel antenna design consisting of two inflatable booms which trail behind the aircraft in flight. Each boom is 5.5 m long by 0.3 m diameter, and are attached to a bracket on the aircraft tail. The two booms are inflated by an air compressor before take-off and during flight are kept rigid by the effect of ram air. Each boom consists of several thousand small components, making it act as a dipole antenna. Power is fed to each boom and the surface radiates a signal to the ground, alternately from left and right antennas, and receives the returns.

The data requires lengthy and highly sophisticated processing and, at the current stage of development, is recorded on tape and analysed on the ground. In the long term it is proposed that some of the processing is carried out on the aircraft.

Specifications

Frequency range: 20-90 MHz
Resolution: 2 × 1 m
Surveillance area: >1 km ^{2}/s

Operational status

CARABAS II has replaced the CARABAS I system as testing continues.

Contractors

Ericsson Microwave Systems AB.
Swedish National Defense Research Establishment (FOA).

UPDATED

UNITED KINGDOM

PA6150 airborne IFF/SSR transponder

The PA6150 MK XII IFF and Mode S SSR transponder is a remote terminal on a MIL-STD-1553B databus. A second, autonomous version uses a separate transponder control and display unit interfaced to the transponder LRI over RS-422A serial interfaces.

The transponder system is specified to meet the requirements of STANAG 4193, Part 1 for Mk XII IFF operation and ICAO Annex 10, Vol 1, Part 1, 4th Edition incorporating Amendment 68 for Mode S Level 3 operation. For the Mode S datalink, it will support Communications Capability Levels 1 to 3. Reply diversity operation is based on received signal level and timing at the two independent receiver channels. Provision is also being included for expansion to NG IFF at a later date.

The transponder system will accept the following Mk XII IFF interrogations and challenges: Modes 1, 2, 3A, 4 and C; automatic code changes for Modes 1 and 3A. Mode 2 will be set by switches mounted on the transponder LRI which are accessible when the LRI is installed in the aircraft.

For Mode S the transponder will accept the following interrogations/challenges: Intermode A/C/S all-call; intermode A/C only all-call; short air-to-air surveillance; altitude request surveillance; identity request surveillance; Mode S only all-call; comm-A, altitude request; comm-A, identity request and comm-C, ELM.

The transponder system will have comprehensive BIT facilities, including power-up BIT, continuous BIT and initiated BIT.

Specifications

Dimensions:
(transponder) 194 × 124 × 318 mm
(control unit) 95 × 146 × 100 mm
Weight:
(transponder) 9.5 kg
(control unit) 2.2 kg
Power supply: 28 V DC
Frequency:
(transmitter) 1,090 ±0.5 MHz
(receiver) 1,030 MHz

Contractor

GEC-Marconi Avionics Ltd, Defence Systems Division, Portsmouth.

VERIFIED

PTR283 Mk1/PVS1280 Mk II IFF interrogators

The PTR283 and PVS1280 interrogator systems have been designed to meet the requirements of in-flight secondary radar interrogators. The transmitter/receiver uses pulsed oscillator techniques employing automatic frequency control and a logarithmic receiver using silicon integrated circuits. The equipment interrogates on Modes 1, 2 and 3A, the pulses driving the modulator being generated by the encoder/decoder.

The PTR283 Mk1 equipment consists of a lightweight D-band transmitter/receiver unit, an associated encoder/decoder and a control unit. Other units associated with the system are an antenna switch, dual-antenna system and L-trace radar display.

The PVS1280 Mk II system consists of a D-band lightweight transmitter/receiver and encoder/decoder which offers the facility of active decoding and defruiting. This equipment is designed to integrate into an airborne primary radar system and the control of the PVS1280 is performed by the radar controller. It offers ISLS operation and an Interrogation Side-Lobe Suppression (ISLS) switch is available to enable the transmitted power to be distributed equally to the two antennas and the two antennas to be fed alternately in phase and anti-phase for ISLS operation.

Specifications

Frequency:
(transmit) 1,030 ±0.5 MHz
(receive) 1,090 ±0.2 MHz
Pulse length: 0.8 ±0.2 μs
Duty cycle: 0.11%
Decoder: 496 codes

Contractor

GEC-Marconi Avionics Ltd, Defence Systems Division, Portsmouth.

VERIFIED

PTR446A transponder

The PTR446A lightweight transponder identifies aircraft in response to secondary radar interrogation and covers civil and military modes. Emphasis in design was placed on reliability combined with small size and low weight and these qualities have been achieved by the use of specially designed micro-electronic circuits. A digital shift register replaces conventional delay lines in the decoder/encoder circuits, so providing time delays independent of temperature. Integrated circuits are used for the logic and video processing circuits and the logarithmic response intermediate frequency amplifier. Decoder, encoder and associated switches in the control unit reduce the number of interconnecting wires to five and substantially cut down the installation weight. The transmitter/receiver houses the pulse selection and power-supply modules. Three-pulse sidelobe suppression is incorporated.

Two control units are available for use with the transmitter/receiver. The smaller of the two is the PV447, of which there are six versions with the following capabilities:

PV447 - Mode 1 or 3A/B and Mode C or off
PV447A - Mode 1 or 2 and Mode C or off
PV447B - Mode 2 or 3A/B and Mode C or off
PV447C - Mode 1 or 3A and Mode 2 or off
PV447D - Mode A or B and Mode C or off
PV447E - Mode A/B or off and Mode C or off.

An alternative to the PV447 is the PV1447 control unit, which meets the requirements of NATO STANAG 4193 for IFF Mk XA and provides Modes 1, 2, 3/A and C. Automatic code changing is provided on Modes 1 and 3A with storage capacity for 48 codes in each mode. Manual code entry for these modes is via a front-panel keypad. Mode 2 codes are entered through screwdriver adjusted switches reached through the top cover of the unit.

The transponder can be used with either control unit without modification to the transmitter/receiver. Comprehensive self-test is incorporated in all units.

Specifications

Dimensions:
(transponder) 57 × 127 × 254 mm
(PV447 control unit) 146 × 57 × 102 mm
(PV1447 control unit) 146 × 95 × 165.1 mm
Weight:
(transponder) 1.7 kg
(PV447) 0.48 kg
(PV1447) 1.6 kg
Power output: 24.7 dBW
Pulse rate: 1,200 replies/s, each containing up to 14 reply pulses
Triggering sensitivity: −72 to −80 dBm

Operational status

In production and in service with helicopters of the Royal Navy, Army and Royal Air Force and on the British Aerospace Hawk.

Contractor

GEC-Marconi Avionics Ltd, Defence Systems Division, Portsmouth.

VERIFIED

Blue Fox interception radar

Blue Fox is a lightweight radar designed to fulfil the dual roles of airborne interception and air-to-surface search and strike. It was developed to form part of a fully integrated weapon system.

Blue Fox operates in the I-band and uses frequency agility to enhance the radar's immunity to ECM and improve its ability to detect small targets in bad weather or rough sea states.

For air-to-air interception, Blue Fox can be used for lead-pursuit or chase attacks and incorporates a transponder mode for identifying friendly aircraft or ships.

Blue Fox is built on the LRU principle. Each component part of the radar, such as the transmitter, receiver, processor and amplifier, can be checked easily or removed independently for servicing. The antenna is a flat aperture slotted array, stabilised in pitch and roll.

The radar display provides a bright digital scan-converted picture. Superimposed on the display are the flight symbols showing aircraft altitude, speed, heading and so on, so that the pilot can monitor and control the aircraft's manoeuvres while using the radar.

Operational status

In service in the British Aerospace Sea Harrier FRS.1 with the Royal Navy and the Indian Navy. Deliveries are complete.

Contractor

GEC-Marconi Avionics Ltd, Radar Systems Division, Edinburgh.

VERIFIED

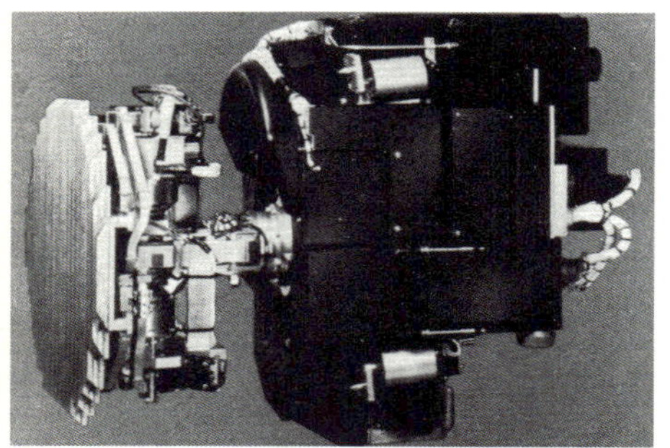

The Blue Fox scanner assembly showing the flat plate antenna

Blue Hawk radar

The Blue Hawk is a new I-band lightweight, coherent, multimode pulse Doppler radar designed for both new lightweight fighters and for the upgrade market. It employs low-, medium- and high-PRF waveforms and offers multiple functions for air interception, close air combat, air-to-ground and anti-ship operation. Aimed at today's typical threats, Blue Hawk has substantial processing capacity to allow it to be matched to a customer's particular requirements, and to adapt to future changes in the threat.

Blue Hawk weighs 107 kg and consists of four main elements; transmitter, receiver/exciter, combined display and data/signal processor, and the antenna. The modular design allows flexibility of installation, especially in platforms where space may be restricted. Blue Hawk is fully compatible with MIL-STD-1553B databus and may be integrated readily with a total avionic system. It is compatible with a wide range of current weapons and ordnance. A major feature of the design philosophy is the wide use of standard components and conventional manufacturing technologies. Blue Hawk employs a fully programmable digital unit comprising signal, data and display processing, with 50 per cent spare throughput and 50 per cent spare memory capacity for growth. It is optimised for HOTAS control. A number of inherent design features minimise the effects of ECM.

Specifications

Frequency: I-band (9.6-9.9 GHz) frequency-agile over multiple channels
Antenna: planar-array mechanically scanned +60° in azimuth and +60° in elevation. Antenna size can be varied to suit installation
Receiver/exciter: 2-channel receiver plus frequency generator
PRF: wide range from (low PRF) 800 Hz to (high PRF) 120 kHz
Power output: 8 kW peak; 160 W mean, a 400 W option is available
Detection performance: 51 n miles in look-up mode; 30 n miles in look-down mode; out to 80 n miles in ground-mapping mode
CW illuminator: can be offered as an integrated option for MRAAM control
Weight: 107 kg
Power requirements: 2.5 kW, 400 Hz, 3 phase

Operational status

Blue Hawk has completed the ground proving stage of development and has undergone a flight trials programme within the UK.

Contractor

GEC-Marconi Avionics Ltd, Radar Systems Division, Edinburgh.

UPDATED

Blue Kestrel maritime surveillance radars

The Blue Kestrel surveillance radar family consists of the Blue Kestrel 5000 and Blue Kestrel 6000. They are allied to the Seaspray radars using common modules throughout the range. This minimises cost and provides natural growth paths for the customer.

Blue Kestrel 5000 maritime surveillance radar

Blue Kestrel 5000 is a pulse compression radar sensor developed for the Royal Navy EH 101 Merlin helicopter. It consists of a large flat plate antenna and Travelling Wave Tube (TWT) transmitter to provide the Merlin with the optimum power aperture product, combined with a high-gain receiver and digital processor. The radar sensor is interfaced with and controlled by the platform's mission avionics management system via the MIL-STD-1553B databus.

This highly integrated sensor approach to the radar results in a high-performance pulse compression surveillance radar in a particularly compact and lightweight four-line replacement unit configuration. It is suitable for application in a range of sophisticated naval helicopters or maritime patrol aircraft.

Specifications

Scanner type: planar-array
Frequency: I-band
Transmitter: low-peak power, high-mean power TWT
Pulsewidths: selectable
PRF: selectable
Coverage: 360°
System weight: 102 kg
Features: pulse compression, CFAR, multiple TWS and operator-selectable scan-to-scan integration

Operational status

Blue Kestrel 5000 has completed development and production deliveries have commenced. The development phase included the building of 10 systems, all to full flying standard, and the completion of several thousand hours running time in ground-based rigs, two specially converted Sea King flying testbeds, and the EH 101 Merlin prototype. Lockheed Martin ASIC the Merlin programme prime contractor, placed the production contract for Blue Kestrel 5000 in October 1992.

Blue Kestrel 6000 maritime surveillance radar

Blue Kestrel 6000 is a coherent pulse Doppler radar sensor. It brings together GEC-Marconi Avionics' experience with the pulse compression Blue Kestrel 5000, the multimode pulse Doppler Blue Vixen, and the Inverse Synthetic Aperture Radar (ISAR) work which has been jointly funded by the UK Defence Evaluation

The main units of the GEC-Marconi Avionics Blue Kestrel 5000 radar

The EH 101 Merlin helicopter is being fitted with the GEC-Marconi Avionics Blue Kestrel 5000 radar

Research Agency and GEC-Marconi. Flight trials of a modified Blue Kestrel 5000 in a DERA Sea King testbed have been successfully completed under the latter programme, producing real-time images of co-operative and opportunity surface contacts from a helicopter. The resulting radar sensor provides enhanced air-to-surface and air-to-air detection, standoff classification capability, and a superior anti-submarine warfare capability with sub-clutter target detection. Additional options include moving target detection and high-resolution synthetic aperture radar ground mapping.

Specifications

Frequency: I-band
Type: coherent multimode
Transmitter: TWT fixed frequency/frequency agile
Scanner type: planar-array
Coverage: 360°
LRUs: 4
System weight: 125 kg
Features: CFAR, enhanced TWT, ISAR classification, ASW and air-to-air modes
Options: MTI and SAR ground mapping

Operational status

Blue Kestrel 6000 development is complete. It is conceived as a modular upgrade to the Blue Kestrel 5000 sensor.

Contractor

GEC-Marconi Avionics Ltd, Radar Systems Division, Edinburgh.

UPDATED

Blue Vixen radar

Blue Vixen is a lightweight multimode, coherent, pulse Doppler airborne interception radar operating in I-band. It is a true multimode radar, maintaining full power in all modes, with all-weather operation in look-down and look-up modes and over-sea and over-land detection of targets. It incorporates high-resolution ranging in air-to-air as well as its land and sea search modes.

Blue Vixen operates in low-, medium- and high-PRF modes. Selection of the appropriate PRF for optimum detection is automatic, and depends on background clutter and target density. Low PRF is used in the look-down mode to provide accurate range and velocity with all-aspect detection. High PRF provides for the look-down detection of targets approaching at high speed in a high-clutter environment.

Automatic track-while-scan and single-target track are available in air-to-air modes with electronic counter-countermeasures, and air combat modes. It is claimed to be the first AI radar in the world to be designed from the outset with full AMRAAM compatibility. It is also compatible with other MRAAM, Sea Eagle and Sidewinder.

Blue Vixen has been designed with a flexible LRU configuration. The antenna, receiver and transmitter are co-located in the nosebay while the radar power unit and the processor can be conveniently located nearby. Weight of the complete system is 145 kg.

Operational status

Currently in production for the Royal Navy Sea Harrier FA.2, a successful programme of trial firings of AMRAAM from a Sea Harrier fitted with Blue Vixen took place in 1993-94 in the USA. Blue Vixen enables up to four AMRAAM missiles to be ripple-fired, maintaining guidance datalink while simultaneously tracking the targets.

Contractor

GEC-Marconi Avionics Ltd, Radar Systems Division, Edinburgh.

UPDATED

Foxhunter airborne interception radar

Foxhunter is the airborne interception radar for Tornado F3. The design of Foxhunter provides a multimode system compatible with the size and weight limits of the RAF air defence version and the operational requirements of the next two decades. A substantial part of the signal processing is performed digitally in addition to digital radar data handling. The equipment anticipates trends in offensive tactics, such as low-level penetration and use of ECM, and the latest improvements give growth potential into supporting active missiles. Additionally, Foxhunter has the flexibility to operate as part of ground- or AEW-based control environments while retaining the ability to perform autonomously.

Foxhunter is designed to detect and track subsonic and supersonic targets at ranges in excess of 185 km at both low and high level.

Foxhunter operates in I-band (3 cm), and in its primary mode uses a pulse Doppler technique known as Frequency Modulated Interrupted Continuous Wave (FMICW). At the heart of the radar is a master timing and synchronising unit, employing phase-locked loop techniques, which generates the complex waveforms for amplification by the transmitter, and the accurately related reference signals and precise timing pulses employed within the receiver and signal processing circuits.

Compact and lightweight surface acoustic wave devices, provide signal waveforms for the pulse compression modes of the radar. The antenna uses the 'Elliot' twist-reflecting cassegrain principle which combines rigidity and light weight with extremely low levels of spurious radiation lobes, a key feature in the rejection of ground clutter and jamming signals. This type of antenna has been developed to a very advanced level of performance. The scanner employs a hydraulic drive mechanism and very high-grade servos to achieve the speed and precision of beam pointing and stabilisation demanded especially during manoeuvre.

The TWT transmitter has the inherent flexibility to handle the differing waveforms used in the various modes of operation. The heart of the signal processing system is a digital processor employing Fast Fourier Transform techniques of frequency analysis to filter the signal returns into narrow frequency channels. Target echoes are segregated from clutter returns and a particularly uniform detection threshold is achieved against all target velocities.

Foxhunter also provides target illumination for the Sky Flash medium-range semi-active radar air-to-air missiles and, since a significant ECM operating environment can be expected, the radar system incorporates many EPM features.

Product Improvement Programme

Since Foxhunter was introduced into service, the improvement programme has been defined in two main stages. Stage 1 introduced numerous modifications and permitted radar functions that were closely tied to HOTAS modes. The TWS and EPM algorithms were also refined. The Stage 2 standard introduces a number of major improvements, including a completely redesigned data processor and supporting software, as well as significant modification to both the transmitter and receiver paths. The resultant effect is a much higher quality of TWS performance in a high RF noise

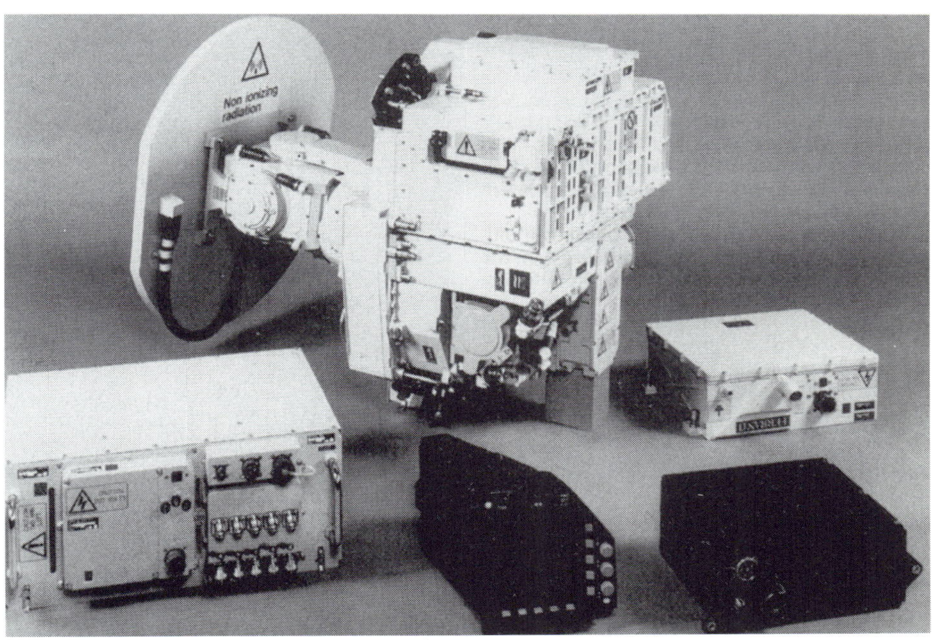

GEC-Marconi Blue Vixen B model components

environment. The new processor also enables sophisticated automatic and manual scan management techniques to be employed in TWS; it is also capable of handling a significantly greater number of tracks. Additionally, it provides scope for significant growth potential beyond Stage 2, including the ability to support active radar missiles.

Operational status

Foxhunter is in operational service with the Royal Air Force and the Royal Saudi Air Force and is presently being brought into service with the Italian Air Force. Stage 1 was introduced into service in 1989, and Stage 2 is currently being brought into service.

Contractor

GEC-Marconi Avionics Ltd, Radar Systems Division, Milton Keynes.

VERIFIED

The GEC-Marconi Avionics Foxhunter radar installed in the nose of a Royal Air Force Tornado F3 aircraft

The Seaspray Mk 1 in the nose of a Westland Lynx

Seaspray Mk 1 maritime surveillance radar

Seaspray is a family of maritime surveillance and targeting radar systems. Since the 1970's this family has been developed to include Seaspray 2000, Seaspray 3000 and Seaspray 4000, over 500 of these radars have been ordered. Allied with this family is the range of Blue Kestrel maritime surveillance radars. Commonality of modules is used throughout both the Seaspray and Blue Kestrel ranges. This minimises cost and provides natural growth paths for the customer. Details of the Seaspray range are given below. The Blue Kestrel range is described in a previous entry.

Seaspray Mk 1 is a high-performance I-band airborne maritime surveillance and targeting radar which was developed in the early 1970s as an integral element of the solution to a Royal Navy Staff Requirement. This requirement was raised as a result of the 1967 sinking of the Israeli destroyer Eilat by two Styx missile-armed fast patrol boats. The requirement demanded a light, agile shipborne helicopter which would be able to detect and neutralise long-range surface-to-surface missile-armed fast patrol boats while keeping them outside the engagement range of the helicopter's mother ship.

The solution was the Westland Lynx helicopter armed with British Aerospace Sea Skua missiles. To minimise the weight of the missiles and to enable four of them to be carried, semi-active radar homing was selected for their guidance. The link was the helicopters prime sensor and target illuminator, the Seaspray Mk 1. This lightweight radar provides the high-performance, frequency-agile detection of small targets in adverse conditions and tenacious monopulse lock-follow target illumination for the Sea Skua.

Specifications

Frequency: I-band
Peak power: 90 kW
Transmitter: high-speed, spin-tuned magnetron
Pulsewidths: 2 selectable
PRF: 3 selectable
Coverage: 180°
LRUs: 5
System weight: 75 kg
Target illumination: monopulse lock-follow

Operational status

Over 300 Seaspray Mk 1 radars have been delivered in GKN Westland Lynx helicopters to Brazil, Denmark, Germany, Netherlands, Norway and the UK. The Royal Navy also operates land-based coastal-sited versions of Seaspray Mk 1 as Checksite to monitor the effectiveness of shipborne EW equipment. Seaspray Mk 1 has also been delivered to the Republic of Korea for land-based trials, and to Pakistan in ex-Royal Navy Lynx helicopters. The Lynx, Seaspray Mk 1 and Sea Skua combination achieved success in the South Atlantic in 1982 and in the Gulf in 1991.

Contractor

GEC-Marconi Avionics Ltd, Radar Systems Division, Edinburgh.

UPDATED

Seaspray 2000 airborne radar

Seaspray 2000 is an I-band maritime surveillance radar which has been optimised for civil surveillance operations. A range of different sized, full-colour combined control and display units provides the man/machine interface. Control of the highly automated radar is via simple on-screen menus. Output ports are provided for hard-copy records, and video and one or more displays can be provided as control stations or repeaters. Superior contact detection in adverse

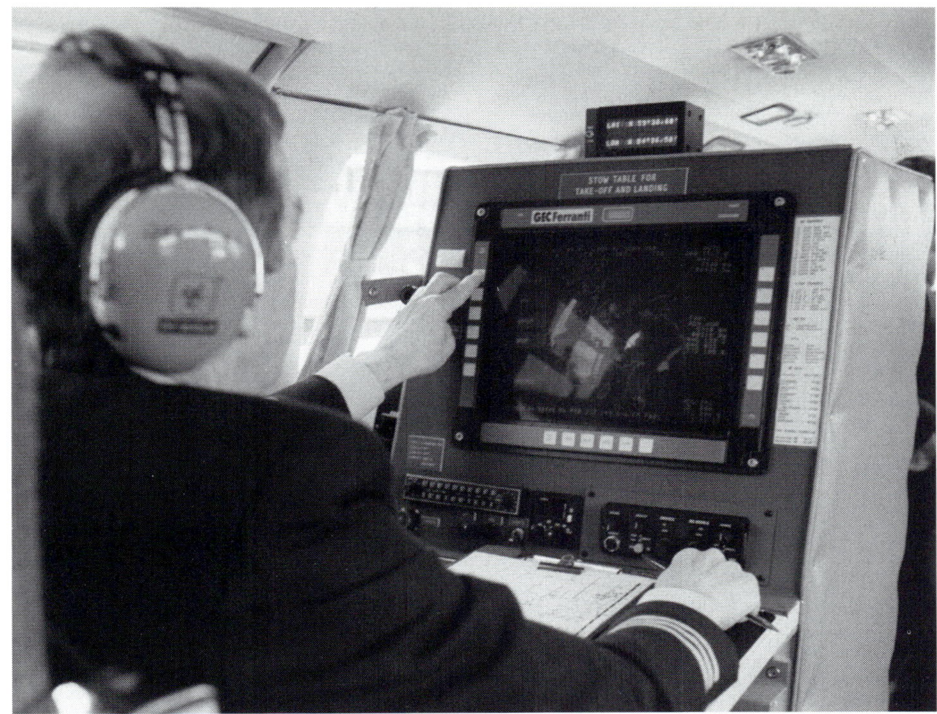

An operator's console for the Seaspray 2000 radar

conditions is provided by the high-peak power, frequency-agile transceiver, Constant False Alarm Rate (CFAR) processing, operator-selectable scan-to-scan integration, and a range of optional features. Options include an operator-selectable side-looking array radar for ground-mapping and pollution control, operator-selectable circular polarisation for enhanced performance in precipitation and multiple target track-while-scan. The display presentation is enhanced by a comprehensive set of synthetic overlay facilities, including digital coastlining, customer defined waypoints and operator defined variable waypoints.

Specifications

Frequency: I-band
Peak power: 90 kW
Transmitter: frequency-agile magnetron
Pulsewidths: 2 selectable
PRF: 4 selectable
Scanner type: front fed, elliptical section, double curved paraboloid
Azimuth scan: 360° continuous, 180 and 60° sector scan selectable
LRUs: 4
System weight: 80 kg
Features: CFAR and operator-selectable scan-to-scan integration
Options: operator-selectable SLAR, circular polarisation, multiple TWS and coastlining, a range of antenna sizes (up to 1 m diameter) and one or more of a range of display sizes (up to 0.35 m diagonal)

Operational status

Seaspray 2000 is in service in Reims F 406 Caravan II Vigilante aircraft with the UK government's Scottish Fisheries Protection Agency and in Dornier 228 maritime patrol aircraft with the Finnish Frontier Guard and in another UK government agency. In the Dornier 228 applications it is fully integrated with the GEC-Marconi Electro Optics multirole turret system FLIR.

Contractor

GEC-Marconi Avionics Ltd, Radar Systems Division, Edinburgh.

VERIFIED

Seaspray 3000 airborne radar

Seaspray 3000 (previously known as Seaspray Mk 3) introduces significant advances over Seaspray Mk 1, most notably through the provision of a new digital processor and full 360° scan. The system is designed primarily for operation in light, agile, shipborne naval helicopters to provide detection and tracking of small targets in adverse conditions for the Sea Skua missile. For this purpose it retains the combat proven monopulse lock-follow target illumination of its predecessor.

Consisting of six LRUs, Seaspray 3000 is configured to operate on a MIL-STD-1553B databus, with multiple additional standard interfaces being provided. The resultant lightweight, compact and flexible system is applicable to both the retrofit and new aircraft. Operating in the I-band, Seaspray 3000 uses a high transmitted power and very high speed agility to provide these platforms with high-detection performance in sea and weather clutter and in ECM conditions. The digital processor provides advanced features, including constant false alarm rate and track-while-scan facilities to optimise target detection and reduce operator work load.

The operator is provided with a comprehensive tactical situation display of scan-converted television format, in monochrome or colour. In addition, the output of other sensors (FLIR, ESM, Datalink and so on) may be displayed. The control unit employs a menu structure with soft key options.

Specifications

Frequency: I-band
Peak power: 90 kW
Transmitter: high-speed, spin-tuned magnetron
Pulsewidths: 2 selectable
PRF: 4 selectable
Coverage: 360°
LRUs: 6
System weight: 90 kg
Target illumination: monopulse lock-follow
Features: CFAR and operator-selectable scan-to-scan integration
Options: operator-selectable circular polarisation, a range of antenna sizes (up to 1 m diameter), and a range of display options

Westland Super Lynx helicopters of the South Korean Navy are equipped with the Seaspray 3000

GEC-Marconi Avionics Seaspray 3000 radar units

Operational status

Seaspray 3000 is in service with the Republic of Korea Navy and Brazilian Navy in GKN Westland Super Lynx, in the Turkish Navy in Agusta Bell 212, and in the German Navy in GKN Westland Sea King helicopters. The German Navy has also ordered Seaspray 3000 for its new Super Lynx helicopters and the Royal Navy are upgrading their Seaspray Mk 1 radars to Seaspray 3000 standard in their Lynx Mk 1 helicopters. Seaspray 3000 has also seen service in a Fokker F27 maritime patrol aircraft and has been proven with Sea Skua in land-based coastal battery and fast patrol boat applications.

Contractor

GEC-Marconi Avionics Ltd, Radar Systems Division, Edinburgh,

UPDATED

Seaspray 4000 airborne radar

Seaspray 4000 is a pulse compression airborne maritime surveillance radar capable of operating as a stand-alone radar or as the heart of a fully integrated avionics suite. It is the resultant system from combining the proven man/machine interface and processing of Seaspray 3000 with the company's pulse compression technology. Comprising six LRU's, Seaspray 4000 provides the optimum performance for medium-size maritime patrol aircraft and naval helicopters.

Specifications

Frequency: I-band
Transmitter: low peak power, high mean power TWT
Pulsewidths: selectable
PRF: selectable
Coverage: 360°
LRUs: 6
System weight: 80 kg
Features: pulse compression, CFAR, multiple TWS, operator selectable scan-to-scan integration
Options: operator selectable circular polarisation, range of antennas and displays

Operational status

Seaspray 4000 development is complete.

Contractor

GEC-Marconi Avionics Ltd, Radar Systems Division, Edinburgh.

UPDATED

Skyranger airborne radar

Skyranger is a lightweight airborne weapon control radar developed for light fighter and light attack aircraft in the retrofit market. It consists of three main units; antenna, transmitter/receiver and signal processor/power supply, and since the amount of space available for retrofit programmes can often be limited and irregular in shape, the modularity of Skyranger has been established at printed circuit card level. The individual cards can, therefore, be packaged into housings designed for the space available. The equipment has been designed as part of an integrated avionics suite, the other system being an air data computer, radar altimeter, head-up display, weapon aiming computer and secure communications.

Skyranger accepts discrete digital commands from a cockpit-mounted control panel and provides output data in the form of a digital serial link (ARINC 429) to a HUD and other weapon aiming systems. It has two main modes; guns and missiles; the former having a shorter range wide-angle beam. In the missile mode the radar energy is fed from the feed horn and reflected back from the parabolic antenna in a 6° beam with a maximum range of 15 km. For gun attacks, the radar energy is fed directly out from the antenna through a polarised window. This results in an 18° beamwidth and a range of 5 km. Minimum range is 300 m for guns and 150 m for missiles, with a ranging accuracy in the order of ±15 m below 3 km and ±30 m above. Target relative velocities from −500 to +1,000 m/s may be handled.

The equipment operates in I-band and has a 5 per cent pulse-to-pulse agility. MTBF is given as 200 hours and the equipment contains built-in test systems. The current version has a fixed antenna.

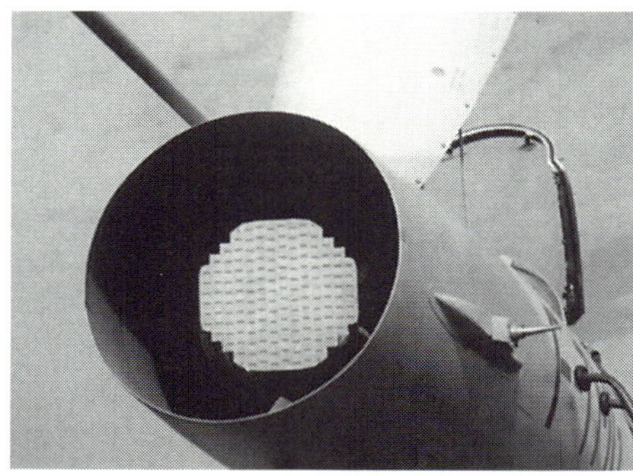

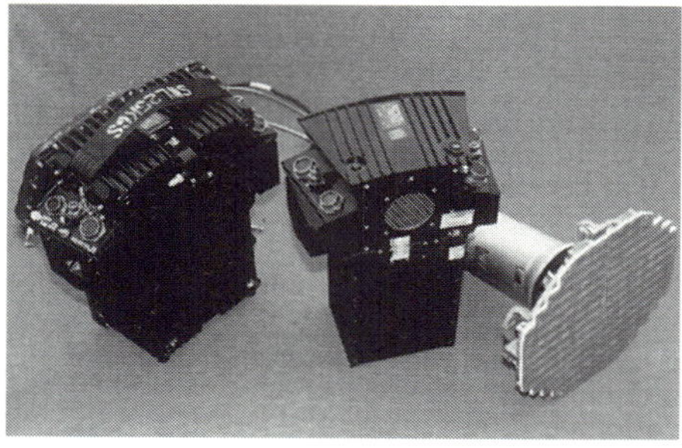

Super Skyranger airborne radar implemented on the Chinese F-7MG fighter ***1998***

Specifications

Frequency: I-band
Range: 15 km for missiles, 5 km for gun operation
Range resolution: 150 m
Pulse-to-pulse agility: 5%
Power supply: 27 V DC, <50 W; 115 V 400 Hz, single phase, <400 VA
Weight:
(antenna) 4 kg
(transmitter/receiver) 25 kg
(signal processor/power supply) 8 kg
(total installed weight) 40 kg

Operational status

In production for the F-7 fighter aircraft of the People's Republic of China, as part of the GEC integrated avionics system. More than 300 systems have been produced.

Contractor

GEC-Marconi Avionics Ltd, Radar Systems Division, Milton Keynes.

VERIFIED

Super Skyranger airborne radar

Super Skyranger is a low-cost multimode radar for light fighter and light attack aircraft. It is based on the Skyranger and is a direct replacement for Skyranger in the improved version of the Chinese F-7 fighter (designated F-7MG) and as an upgrade in the MiG-21 airframe.

Super Skyranger has a full look-down shoot-down capability using a planar-array antenna which can scan to ±30° dependent on the aircraft installation. It can provide target range, range rate and line of sight data to the aircraft's avionic system via ARINC 429 serial link and has retained the excellent ECCM features of the original Skyranger.

Contractor

GEC-Marconi Avionics Ltd, Radar Systems Division, Milton Keynes.

UPDATED

PA5000 series radar altimeters

The PA5000 is a software-controlled altimeter operating in the J-band using advanced microwave and signal processing surface mount VLSI integration techniques.

The transmit and receive antennas are both included within the unit outline and thus the PA5000 is fuselage-mounted, requiring only a single fuselage cutout. RF feeders are not required. Operating in the J-band, the PA5000 Series radar altimeter provides a covert system and gives precision accuracy and resolution for all high-performance fixed-wing, helicopter, UMA and RPV applications. In helicopter applications, excellent hover and nap of the earth performances are available.

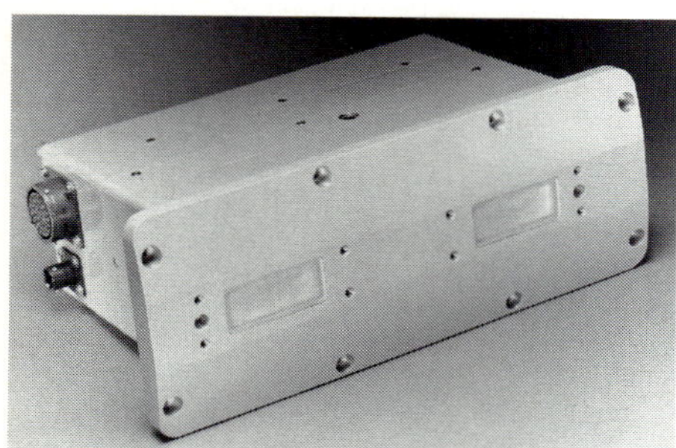

The PA5000 radar altimeter

PA5000 Series radar altimeters offer height ranges up to 5,000 ft with 2 per cent accuracy, are designed to be compatible with most types of aircraft digital and analogue interfaces and are qualified to MIL-STD-461C and MIL-STD-810D.

Specifications

Dimensions: 218 × 76 × 138.6 mm
Weight: 3.2 kg
Power Supply: 19-32 V DC, 25 W (max) at 28 V
Altitude: 0-5,000 ft options
Accuracy: ±2% height +2 ft
Reliability: 5,000 h MTBF

Operational status

Installed in Agusta/Sikorsky HH-3F and the NH500 helicopters.

Contractor

GEC-Marconi Electro Optics Ltd, Airadio Division, Portsmouth.

VERIFIED

PA5200 series radar altimeters

The PA5200 Series radar altimeter operates in the mid-J-band using microwave Field Effect Transistor (FET) technology. Software-controlled signal processing techniques are used to enable reliable performance to be achieved up to 5,000 ft with a transmitter power of only 0.5 W. Surface-mount technology is used to give a low-volume, high-reliability package which includes the antenna.

The PA5200 radar altimeter uses a dual leading-edge tracker to ensure tracking of the nearest object. Continuous automatic monitoring of the system ensures high reliability down to ground level.

Specifications

Dimensions: 225 × 76 × 117 mm
Weight: 3 kg
Power supply: 28 V DC, 26 W (max)
Altitude: up to 5,000 ft (can be extended)
Accuracy: ±(3 + 3% of height) ft
Temperature range: −40 to +50°C
Reliability: 5,000 h MTBF

Contractor

GEC-Marconi Electro Optics Ltd, Airadio Division, Portsmouth.

VERIFIED

PA5495 radar altimeter

The PA5495 radar altimeter operates in mid-J-band using microwave Field Effect Transistor (FET) technology. Software-controlled signal processing techniques are used to enable reliable performance to be achieved up to 5,000 ft with a transmitter power of only 1 W. Surface-mount technology is used to give low volume. Separate antennas are provided to be compatible with existing C-band installations.

The altimeter uses a dual leading-edge tracker to ensure tracking of the nearest object. Continuous automatic monitoring of the system ensures high reliability with acurate height indication right down to ground level.

Specifications

Dimensions: 140 × 220 × 85 mm
Weight: 4 kg
Power supply: 28 V DC, 26 W
Altitude: 0-5,000 ft (can be extended)
Accuracy: ±(3 + 3% of height) ft
Temperature range: −40 to +70°C
Reliability: 5,000 h MTBF

Contractor

GEC-Marconi Electro Optics Ltd, Airadio Division, Portsmouth.

VERIFIED

PVS1712 radar altimeter

The PVS1712 J-band pulse radar altimeter is designed for applications ranging from helicopters to high-performance fixed-wing aircraft. The use of a high operating frequency has enabled the design of a single-unit system, containing transmit, receive and electronic functions. The unit uses a pulse leading-edge tracking technique to measure precisely the time interval for radar pulse travel. It is an all-solid-state design and requires no warm-up. Very short 4 ns pulses are used. The system includes automatic error correcting circuitry which fully compensates for errors due to internal delays and their time/temperature drift.

The transmitter is a 5 W Gunn diode oscillator and a pseudo-homodyne receiver is used with a low-noise video preamplifier and main amplifier, the gain of the latter being geared to expected signal level at the range tracking point.

Specifications

Typical installation
Dimensions: 218 × 116 × 76 mm
Weight: 2 kg
Altitude: 0-1,000 ft
Accuracy: ±2% height + 2 ft at 500 ft
±5% height + 2 ft above 500 ft

Operational status

In service in British Army Gazelle helicopters and towed aerial targets.

Contractor

GEC-Marconi Electro Optics Ltd, Airadio Division, Portsmouth.

VERIFIED

AD1990 radar altimeter

GEC-Marconi Sensors is supplying a covert radar altimeter designed to meet the Royal Air Force's needs in the 1990s. The AD1990 radar altimeter directly replaces the existing altimeter in the Royal Air Force Tornado aircraft, using all existing fixtures and fittings.

The advanced digital signal processing techniques incorporated in the receiver allow the extraction and simultaneous tracking of height both above the ground and above obstacles such as trees. These two outputs enable the pilot to operate more safely when flying at low level and are also used by the Terrain Reference Navigation (TRN) system to enhance overall navigation performance. The AD1990's fast dynamic response time eliminates the need for groundspeed compensation of height data within the TRN system. Inherent in the signal processing technique is the ability to identify and reject unwanted signals from underslung stores and landing gear, a traditional problem for radar altimeters. Reliable operation is obtained from its maximum operating altitude of 5,000 ft down to ground level.

An important innovation is that the altimeter remains covert in operation, rendering it virtually undetectable by the enemy. Such Low Probability of Intercept (LPI) is achieved by spreading the transmitted signal over a very wide bandwidth through the application of pseudo-random phase modulation and adaptive power tailoring which, in addition, gives a high resistance to jamming. AD1990 can also be applied to other modern military aircraft where the ability to remain undetected is the key to mission success.

In addition to analogue height output, the system can be configured in either Panavia or MIL-STD-1553B interfaces.

Specifications

Dimensions: 109 × 154 × 318 mm
Weight: 5.25 kg
Power supply: 28 V DC, 55 W (max)
Frequency: 4.3 GHz
Range:
(height) 0-5,000 ft
(speed) 0-800 kt
(pitch) 0 to ±60°
(roll) 0 to ±60°
Accuracy: ±3 ft or ±3% whichever is greater
Temperature: −55 to +90°C

Contractor

GEC-Marconi Electro Optics Ltd, Sensors Division, Basildon.

UPDATED

ARI 5983 I-band transponder

The ARI 5983 I-band transponder provides a means of locating, identifying and providing navigational assistance to aircraft outside normal radar coverage and range. It is interrogated by a primary radar and gives an edge-of-band response. The response codes are selected on a simple control unit from which a comprehensive BIT routine can be initiated. The control unit also allows either manual or electronic switching between the transponder's two antennas to ensure optimum coverage.

The transponder receives interrogation signals, via the antenna, from pulse radars at any frequency in two bands 100 MHz wide. When interrogated, the transponder will respond with either a single-RF pulse, which provides enhancement of the radar return, or a coded group of up to six pulses, as selected on the control unit, which allows identification. There are 16 different reply codes available.

The transmitted power can be reduced by approximately 11 dB via another switch on the control unit. The transponder output will automatically be suppressed during the operation of other I-band equipment in the aircraft. Similarly, a pulse is supplied by the transponder to allow suppression of other equipment in the aircraft operating in the same frequency band when the transmitter is operating. The BIT self-test facility generates an interrogate signal which is fed into the transponder input. A green LED on the control unit indicates correct transponder operation.

Options include double- or multiple-pulse interrogation to minimise false triggering when several I-band radars are transmitting in the same area.

Both the transponder and control unit are fully NATO codified.

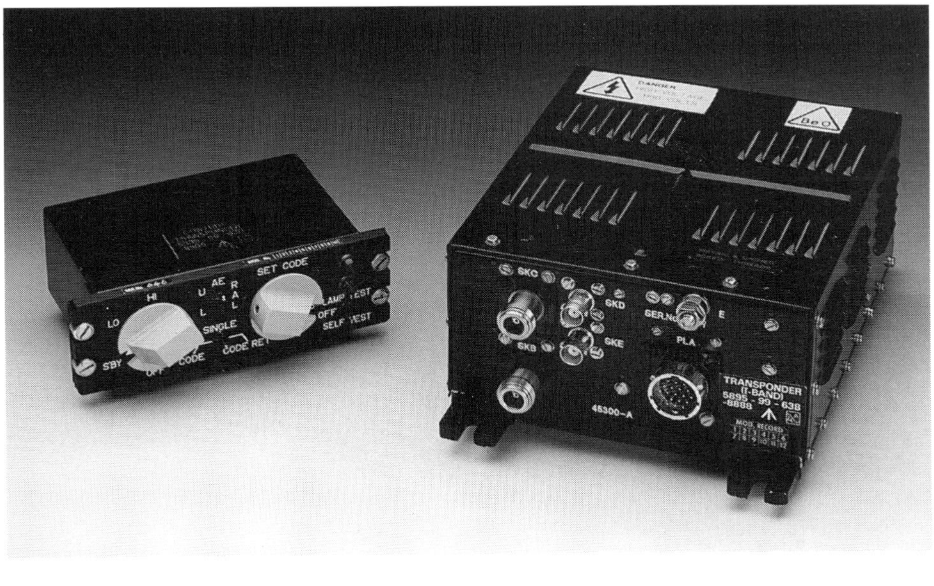

Specifications

Dimensions:
(transponder) 160 × 217 × 87 mm
(control unit) 147 × 117 × 48 mm
Weight:
(transponder) 2.7 kg
(control unit) 0.45 kg
Power supply: 28 V DC, 40 W (max)
Frequency:
(receive) 9,190-9,290 MHz and 9,360-9,460 MHz
(transmit) 9,310 ±7 MHz
Bandwidth: ±50 MHz
Sensitivity: −93 dBW
Output power: 135 W (min) to 300 W (max) peak
Pulse duration: 0.45 µs ±0.1 µs
Reply code: 6-pulse code, 16 settings
single-pulse reply capability
Pulse spacing: 2.9 µs nominal
Duty cycle: 0.005 (max)

The M/A COM ARI 5983 IFF transponder
1995

Operational status

In service with the Royal Navy on the Lynx, Sea King and Sea Harrier FRS.1 and F/A-2. Also exported for Dauphin and Lynx helicopters and the Do-228. Selected for the Royal Navy EH 101 Merlin.

Contractor

M/A COM Ltd.

VERIFIED

ML3500 radar transponder

To supplement the more sophisticated ARI5983 I-band transponder, M/A COM has introduced the ML3500 radar transponder. The ML3500 provides an edge-of-band response when interrogated by an I-band radar, and is ideal for enhancing the radar echoing area of small air and surface targets. Although in essence an active corner reflector, the transponder's response can be simply coded to aid target identification and, being edge-of-band, reduces primary plot clutter on radars with tunable receivers.

The ML3500 is lightweight and fully weatherproof.

Specifications

Dimensions: 120 × 170 × 55 mm
Weight: 1 kg
Power supply: 12 or 24 V DC, 10 W
Frequency:
(receiver) 9,000-9,600 MHz
(transmitter) 9,200-9,400 MHz (factory set)
Sensitivity: -43 dBm (min)
Interrogate pulsewidth: 0.15 to 1.5 μs
Stability: ±10 MHz
Pulse duration: 0.2 to 1 μs
Reply code: 5 output pulses - one for range mark, four customer settable identifiers
Duty cycle: 1% (max)
Temperature range: −20 to +50°C

Contractor

M/A COM Ltd.

VERIFIED

ASTOR airborne standoff radar

ASTOR is a UK MoD sponsored project designed to provide detailed over-the-border surveillance of the land battle and major hostile ground forces to fulfil UK MoD Staff Requirement (Land/Air) 925, valued at £750 million.

The objective is to provide high-resolution static imagery and the ability to detect moving targets to provide 24-hour observation and targeting intelligence of enemy first and second echelon forces and for peacekeeping operations. A secure datalink to mobile ground forces and ground-based interpretation facilities is also required.

It is envisaged that the primary sensor radar will be a synthetic aperture radar with moving target indication (SAR/MTI). The requirement is understood to be for several aircraft, plus supporting ground stations, to enter service from 2001.

Operational status

Two industrial teams have been competing to fulfil the ASTOR requirement, with a contract award due in 1998: Lockheed Martin UK Government Systems, teamed with Racal Radar Defence Systems (radar), Logica (communications), GEC Marconi Avionics (defensive aids), Lockheed Martin Ground Systems (ground station), Gulfstream (Gulfstream V aircraft), Marshall Aerospace (conversion).

Raytheon Systems Company teamed with Bombardier (Global Express aircraft), Raytheon Systems Limited (ASARS-2) GEC-Marconi Avionics, Thomson-CSF, UK Defence Evaluation Research Agency and Motorola.

Recently, however, Northrop Grumman has announced that it is submitting a bid for ASTOR based on the gulfstream V aircraft and JSTARS radar technology, notwithstanding that an earlier JSTARS bid had been rejected.

Contractors

See text for consortia data.

UPDATED

ARI 5955/5954 radar system

The ARI 5955/5954 radar system is designed for ASV, ASW and search and rescue roles, specifically for helicopters. ARI 5955 is the radar sensor and processing system and ARI 5954 the IFF transponder providing identification of friendly aircraft and surface craft.

The system operates in the I-band and has an antenna that can be gyrostabilised to compensate for aircraft motion. Antenna tilt is adjustable from the radar operator's position both above and below a horizontal datum.

In 1992 a contract was awarded to upgrade the ARI 5955 Royal Air Force Sea King search and rescue radars. This was achieved by incorporating the Super Searcher radar signal processing unit and displays, giving all the benefits of modern digital processing to a system which the Royal Air Force still regards as one of the best search and rescue radars. The upgraded system is designated ARI 5955/2.

Operational status

No longer in production, but still in service in a number of countries. However, the upgraded system using the ARI 5955 front end and the Super Searcher processor is still available as a low-cost high-performance option. Subsequent updates have been carried out on Royal Australian Navy Sea King Mk 50s. This system is a variant of the 5955 radar designated AW391(A).

Contractor

Racal Radar Defence Systems Ltd.

UPDATED

ARI 5980 searchwater radar

Searchwater is the commercial name for the ARI 5980 radar which is standard equipment on Royal Air Force Nimrod MR. Mk 2 maritime reconnaissance aircraft; it was designed to replace the ASV Mark 21 radar. The system formed part of a major mid-life refit and 31 Nimrod MR. Mk 1s have been converted to MR. Mk 2s by the addition of Searchwater and other improvements.

Searchwater is designed for all-weather, day or night operation outside the defensive range of potential targets.

The system comprises a frequency-agile radar which uses pulse compression techniques and a pitch and roll stabilised scanning antenna with controllable tilt and automatic sector scan. IFF equipment is included to interrogate surface vessels and aircraft.

The signal processor enhances the detection of surface targets (including submarine periscopes) in high sea states. An integrating digital scan converter permits plan-corrected presentation and classification of target and transponder returns. The single radar observer in the aircraft is presented with bright, flicker-free television-type PPI, B and A scope displays in a variety of interactive operating modes. Weather radar and navigation facilities are provided within the system.

A real-time dedicated digital computer relieves the radar operator of many routine tasks, while continuously and automatically tracking, storing and analysing data to provide position information and automatic classification for a number of ship targets at the same time. Built-in test facilities provide for automatic detection and diagnosis of faults.

The facilities offered by Searchwater reduce the vulnerability of the host aircraft by permitting operation in a standoff mode, avoiding the need to fly over the target for visual identification. Over-the-horizon targeting is also provided for such missiles as Sea Eagle and Harpoon.

The system is entirely modular, with the interfaces and mechanical construction designed for ease of fault location and replacement. Major units are functionally self-contained as far as possible with a minimum of interconnections. Extensive use is made of hybrid and integrated circuit techniques. The transmitter uses solid-state frequency generators and mixers, followed by two cascaded travelling wave tubes. A fluorocarbon liquid cooling system is employed. The scanner, which both transmits and receives radar and IFF signals, uses a reflector of lightweight construction based on resin-bonded carbonfibre.

The Racal Radar Defence Systems Searchwater antenna on a Royal Navy Sea King Mk 2 AEW helicopter. When in use the radome is inflated by engine bleed air. The system rotates clockwise through 90° to give clearance for landing

1997/0001211

The Racal Radar Defence Systems searchwater radar display in a Royal Navy Sea King Mk 2 AEW helicopter **1997**

Operational status
Searchwater entered service aboard Royal Air Force Nimrod MR. Mk 2s in 1979. The system has been evaluated on Lockheed Martin P-3B anti-submarine aircraft by the US Navy.

Searchwater in the Sea King
As a result of demands from the Royal Navy for more effective radar sensors for organic fleet defensive surveillance following operations in the Falklands, Searchwater was modified for use aboard converted Westland Sea King anti-submarine helicopters, redesignated Sea King AEW Mk 2. Ten Searchwater systems were acquired to permit 24 hour AEW coverage for the Royal Navy. Three Sea King AEW Mk 2s operate from each of the two active Royal Navy carriers and four helicopters are land-based.

The system was also supplied to the Spanish Navy for ASW helicopters under a £13 million deal announced in September 1984. Deliveries began in October 1986; the system entered service in mid-1987 and is operational on three Spanish Navy Sikorsky SH-3D helicopters.

Operational status
The first deliveries of Sea King AEW Mk 2s, fitted with modified Searchwater radars, were in April 1985 and were initially assigned to No 849 Naval Air Squadron. Deliveries to the Spanish Navy, began in October 1986. In 1989 the Royal Navy placed a further order for two systems for delivery in mid-1991.

Searchwater 2000AEW
In February 1997, the UK MoD appointed Racal Radar Defence Systems (RRDS) as prime contractor for the radar and mission system upgrade of the Royal Navy's Sea King AEW Mk 2 helicopters. Acting as prime contractor, RRDS will be responsible for all aspects of the upgrade, including equipment supply, installation, aircraft modification and certification, together with provision of logistic support.

GKN Westland Helicopters will be undertaking aircraft modifications and acting as aircraft design authority. Racal Avionics and Racal Instruments will provide cockpit upgrades, and Logica will develop the JTIDS Link 16 processing.

Racal will supply its new generation Searchwater 2000 AEW radar for the upgrade. This radar variant retains the high performance maritime surveillance features of the existing in-service Searchwater radar, but offers considerably improved AEW performance using new high power transmitters and advanced pulse Doppler signal processing techniques. The new radar was under development as a private venture for five years, and provides an advanced high technology hardware and software solution with considerable weight saving and improved reliability. Searchwater 2000MR (for Nimrod 2000) and Searchwater 2000 AEW (for Sea King AEW) have a high degree of commonality, and will provide operational and logistic standardisation benefits.

The updated helicopter is designated the Sea King Mk 7. The 10 in-service Sea King Mk 2 helicopters will be updated over the next seven years (1997-2003). The contract includes an extensive logistic support package covering mission and maintenance trainers and workshop support down to 2nd line, including the provision of support on-board aircraft carriers at sea.

Searchwater 2000MR
Racal Radar Defence Systems (RRDS) has been contracted by Boeing Space & Defense, as a result of their contract from British Aerospace for supply of the Nimrod MRA4 mission system, to supply 21 Searchwater 2000MR radars over the next eight years (1997-2004). The contract includes provision of integrated logistic support.

Searchwater 2000MR and Searchwater 2000AEW (see above) have a high degree of commonality.

Contractor
Racal Radar Defence Systems Ltd.

UPDATED

MAREC II Maritime Reconnaissance radar

The MAREC II (MAritime REConnaissance) radar is based on the Racal-Thorn helicopter radar system and was developed to meet the requirements for a small, efficient, low-cost system to conduct maritime patrol, including coastguard surveillance, search and rescue, fishery and oil rig protection and pollution control. Suitable for fixed-wing aircraft and helicopters, it has a true motion plotting table display covering 360° in azimuth and a range of 460 km, together with a pilot's display.

Specifications
Frequency: 9,345 MHz
Power output: 80 kW
Pulsewidth: 0.4 and 2.5 μs
PRF: 200 and 400 Hz

Operational status
The system has flown in the BAe Coastguarder demonstrator - a version of the HS748 - the prototype Airship Industries Skyship 500 airship and in Dornier 128 Skyservants for maritime patrol operations. Three MAREC II radars were supplied to the Indian Coast Guard; these have now been upgraded to Super MAREC standard.

Contractor
Racal Radar Defence Systems Ltd.

UPDATED

Super MAREC radar

Super MAREC (MAritime REConnaissance) is an upgrade of the MAREC II involving improved software and the replacement of the MAREC II's 430 mm plotting display with the 356 mm colour television-type display developed for Super Searcher. This saves weight and space and increases the tactical navigation facilities available to the operator. Additional facilities include multiple target track-while-scan and navigation overlay data.

Recent enhancements to Super MAREC include the addition of a full ISO contoured weather avoidance mode, a ground-mapping mode and an extra radar mode to give improved close-range target detection. Digital map overlays are also provided. A smaller display and joystick option is offered as a replacement for the large display and keyboard to permit radar operation from the cockpit by the second pilot.

Operational status
In production and service with the Indian Coast Guard, which has a requirement for 36 radars.

The latest version of the Super MAREC radar is the Super MAREC III.

Contractor
Racal Radar Defence Systems Ltd.

VERIFIED

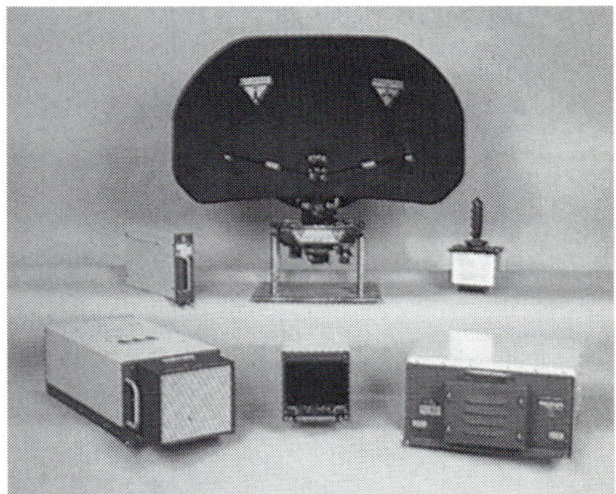

The latest version of the Super MAREC radar is Super MAREC III **1995**

Super Searcher airborne radar

Super Searcher is a development of the Sea Searcher which is fitted to Royal Navy Sea King helicopters. Designed for multithreat maritime operations, it is a lightweight I-band command and control radar with a horizontal aperture antenna providing a high-definition display on a range of colour raster scan CRT's that can show true motion or centre PPI with variable sector scan.

By comparison with Sea Searcher, Super Searcher has a greater detection, target tracking and guidance performance. The system has an inbuilt guidance capability which can be adapted for the fire-and-forget anti-surface vessel sea-skimming missiles.

The system incorporates three selectable pulsewidths, including a short one to give high definition of small targets in bad weather. Contact recognition is also improved by the use of the latest microprocessor techniques and signal processing algorithms.

The colour CRT facilities include freeze-frame and memory storage with graphical overlays of tactical and navigational symbology. The display can show either

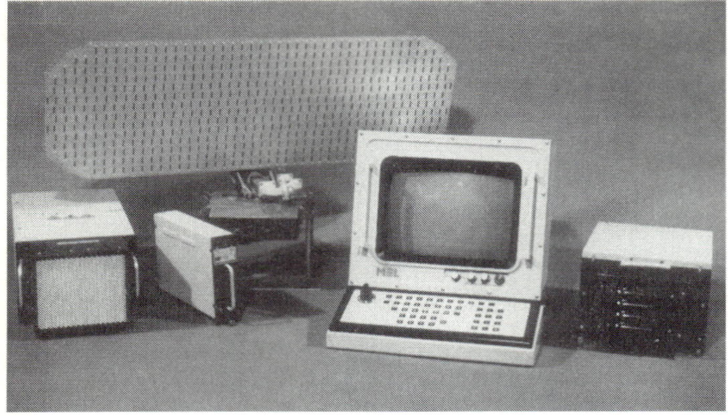

The Super Searcher GMR is installed in the Royal Air Force Dominie navigation trainer

The Super Searcher radar

true or relative motion with or without offsets. Data is displayed either in latitude and longitude or as a grid reference. The system has multiple track-while-scan capability and provides a range of navigational facilities including waypoint markers. As an option, digitised maps can be loaded prior to the mission, allowing the use of map overlays on the radar image.

The Super Searcher radar operates in primary and secondary modes, either separately or in combination. A comprehensive library of symbols that can be vectored as required eases the operator's task, particularly in intelligence storage and extraction. The system is compatible with IFF/SSR interrogators and can be interfaced with a range of other sensors such as FLIR, ESM and SONAR.

Super Searcher Ground-Mapping Radar (GMR) is a development of the Super Searcher radar and is a further addition to the Racal Radar Defence Systems Ltd family of lightweight airborne radars. The radar uses a three-axis cosecant^2 antenna combined with high-resolution displays and upgraded processing to enhance ground features. The low-cost system can be used for high-altitude radar fixing, ground-mapping and ground target attack training.

Operational status

In production. Super Searcher is the principal sensor on board the 20 Sea King Mk 42 ASW helicopters ordered by the Indian Navy. It has also been installed on the eight Sikorsky S-70B helicopters of the Royal Australian Navy. Brazil has been supplied 20 Super Searchers, for fitting to Embraer EMB-111 maritime patrol aircraft. Further systems have been fitted to Royal Air Force Dominie aircraft for navigation training. Over 100 Super Searchers have been sold.

Contractor

Racal Radar Defence Systems Ltd.

UPDATED

Cossor Interrogation and Reply Cryptographic Equipment (CIRCE)

Raytheon Systems Limited supplies a range of modules for all types of IFF applications to facilitate a nationally secure IFF operation in Secure mode which is similar to, but not interoperable with, NATO Mode 4. The modules can be added to any IFF system that is compatible with the NATO Mk XII standard (STANAG 4193).

Two key components of CIRCE are the programmer and the fill gun. The programmer is used to hold the key variable necessary to operate the CIRCE units. It is held in a secure location and the electronic fill gun is then used to transfer the key variable data from the programmer to the various IFF systems. The key variable data can be changed whenever required using the fill gun.

Specifications

Dimensions: 123 × 50 × 129 mm
Weight: 1.42 kg

Operational status

In service with two non-NATO countries.

Contractor

Raytheon Systems Limited, Electronic Systems Division.

UPDATED

CIRCE – IFF cryptographic unit **1998**/0011871

IFF 2720 transponder

IFF 2720 is a microminiature IFF Mk10A identification friend or foe/secondary surveillance radar transponder for use on all types of military aircraft and helicopters. On Modes 1, 2, 3/A and B the full 4096 codes are available and 2048 codes are available on Mode C for altitude reporting. In addition there are circuits for the identification facility (SPI or I/P) and military emergency. The system comes in two units: a transmitter/receiver and a controller, each of which features easily accessible circuit boards.

Early deliveries of Indonesia's British Aerospace Hawks have the IFF 2720. Later Hawks have the IFF 4720 transponder

Electronic warfare provisions include resistance to continuous wave, modulated continuous wave and pulse jamming, sidelobe rate limiting, short pulse and spurious interference protection, single-pulse rejection and long pulse discrimination.

The microprocessor-based control unit for use with this system is designated IFF 2743. It has a light-emitting diode display which can be fully dimmed.

Specifications

Dimensions:
(IFF 2720 transponder) ⅜ ATR short
90 × 194 × 314 mm
(IFF 2743 full facility control unit) 146 × 57 × 98 mm
Weight:
(IFF 2720 transponder) 4.6 kg
(IFF 2743 full facility control unit) 0.7 kg
Frequency:
(transmitter) 1,090 MHz
(receiver) 1,030 MHz
Power output: 27 dBW (500 W) (min)
Receiver sensitivity: −76 dBm
Dynamic range: >50 dB
Sidelobe suppression: 3-pulse
Qualification: STANAG 5017 Edition 2, ICAO Annex 10

Operational status

IFF 2720 is in production and equips many strike, transport, combat and trainer aircraft; it is fitted to early export versions of the British Aerospace Hawk. More than 1,000 systems have been delivered to over 20 countries.

Contractor

Raytheon Systems Limited, Electronic Systems Division.

UPDATED

IFF 3100 transponder

The IFF 3100 is a single-package transponder tailored to the Royal Air Force Panavia Tornado aircraft.

Claimed advantages over previous systems are small size, lower weight and simpler installation. Although the component density is high, reliability is ensured by the use of high-grade, close-tolerance circuits and a four-port circulator protects the output stages from the effects of any antenna mismatch. Open or short-circuit conditions at the antenna do not damage the transponder.

Extensive integrity monitoring is incorporated during operation and when the test button is pressed. Checks cover receiver sensitivity, receiver centre frequency, mode decoding, aircraft reply coding and transmitter power level.

Interrogation Modes are 1, 2, 3/A, B and C and the reply capability covers 4096 codes for Modes 1, 2, 3/A and B. Provision for the use of the X-pulse is included and there are 2048 codes for Mode C.

Specifications

Dimensions: 146 × 132 × 165 mm
Weight: 5.3 kg
Qualification: ICAO Annex 10, STANAG 5017 Edition 2

Operational status

In production. In service with Panavia Tornado aircraft with UK equipment fit including Saudi Arabian aircraft.

Contractor

Raytheon Systems Limited, Electronic Systems Division.

UPDATED

IFF 3500 interrogator

The IFF 3500 airborne interrogator employs monopulse techniques to achieve high accuracy in the measurement of target bearing. The IFF 3502 variant incorporates an automatic code-changing system to enhance security and eliminate the possibility of incorrect code setting.

The transmitter employs P2 emphasis to provide antenna beam-sharpening. P1 and P3 are transmitted on the antenna sum channel and P2 on the difference channel. Selectable 3 or 6 dB of P2 emphasis is available. Advanced video processing circuits for degarbling, defruiting, decoding and for echo and multipath suppression are contained within a single unit. Passive and active decoding are provided and two channels of passive decoding enable comparison during the overlap period between code changes. Active decoding provides serial readout of the 4096 reply codes.

Manual and continuous automatic built-in test circuitry checks transmitter power, interrogation coding, receiver sensitivity, defruiting/decoding, bearing accuracy and integrity of the transmission feeders.

Specifications

Dimensions: 1 ATR short case to ARINC 404A
Weight: 20.7 kg
Power supply: 115 V AC, 400 Hz, single phase
28 V DC
Frequency:
(transmitter) 1,030 MHz
(receiver) 1,090 MHz
Power output:
(P1, P3) 30.5 dBW
(P2) 0, +3 or +6 dBW above P1 power
Spurious outputs >76 dB below 1 W
Sensitivity (decoding): −80.5 dBm
Dynamic range: 60 dB
Spurious responses: 60 dB down outside pass-band
Bearing resolution: dependent on antenna configuration, but around 5% of angle between intersection points of control and interrogate patterns
Qualification: compatible with NATO STANAG 5017 Edition 3

Operational status

In service with the Royal Air Force Tornado F2 and F3 aircraft and Royal Navy Sea King AEW helicopters. Also fitted to Tornado, and Hawk aircraft delivered to Saudi Arabia.

The equipment will also be fitted during an extensive upgrade of Boeing 737 Surveiller aircraft which are used in a maritime patrol role by the Indonesian Air Force.

Contractor

Raytheon Systems Limited, Electronic Systems Division.

UPDATED

IFF 4500 interrogator

IFF 4500 has its origins in the Eurofighter 2000 where the UK input is supplied by Raytheon Systems Limited. However, it also incorporates some of the features of IFF 3500, its immediate predecessor in the Raytheon Systems Limited IFF inventory.

IFF 4500, which is a monopulse interrogator, is suitable for a wide variety of applications. Tactical fighter and maritime patrol aircraft surveillance platforms are already projected.

As well as operating in Modes 1, 2, 3A, C and 4, IFF 4500 has built in growth potential for upgrade to the Next-Generation IFF (NGIFF) system, as specified in STANAG 4193 Part V.ACC (Automatic Code Change) and Mode S can also be added as customer options.

Control of the interrogator from the host primary radar can be via a MIL-STD-1553B databus or via a discrete control alternative (IFF 4570). Target reports are fed again via a MIL-STD-1553B databus for display alongside the target information generated by the host primary radar.

Mode 4 or CIRCE cryptograhic units are built-in to the IFF 4500, but these 'add-ons' can be removed for IFF Mk10A only applications.

Specifications

Dimensions: 1 ATR short or ½ ATR medium (depending upon requirement)
Cooling: 1 ATR short-convection
½ ATR medium forced air

Operational status

Development is complete and IFF 4500 is just entering production.

Contractor

Raytheon Systems Limited, Electronic Systems Division.

UPDATED

IFF 4700 series transponders

There are three members of the Raytheon Systems Limited 4700/4800 Series of IFF transponders which are designed to meet the NATO IFF Mk XII specification (NATO STANAG 4193); all three versions use a common set of modules so they can be supported by a common spares inventory.

IFF 4720 transponder is designed for a wide range of applications, and is a plug-in replacement for the IFF 2720 system (see previously entry). The IFF 4720 has full dual-redundant decoders on each receiver channel to provide optimum anti-jamming performance and the system can be operated by a dedicated remote-control panel, or via a MIL-STD-1553 databus. Full standard coverage of Modes 1, 2, 3/A, 4 and C is offered and the IFF 4720 can operate up to 70,000 ft altitude.

Specifications

Dimensions: 90 × 194 × 314 mm
Weight: 4.5 kg

Operational status

In production. The IFF 4720 is standard equipment on the British Aerospace Hawk 100 and 200; also fitted to some MiG-29 aircraft and several Naval IFF applications.

IFF 4740 transponder is a single unit, for panel mounting, which is designed as a plug-in replacement for the IFF 3100 used on the Panavia Tornado. In performance it operates identically to the IFF 4720.

IFF 4720 transponder ***1998***/0011870

IFF 4770 control unit (transponder) ***1998***/0011869

Specifications

Dimensions: 127 × 132 × 132 mm
Weight: 4.5 kg

Operational status

In production.

IFF 4760 transponder is a NATO Mk XII compatible remotely controlled IFF designed also to act as a MIL-STD-1553 bus controller. The unit provides an alternative form factor to that of the IFF 4720. The transponder also offers tighter frequency tolerances, all-solid-state construction and VLSI processing to enhance performance; operation is again identical to the IFF 4720.

Specifications

Dimensions: 136 × 136 × 213 mm
Weight: 4.5 kg

Operational status

In production and in service. Over 250 sets have been purchased by the UK MoD.

IFF 4770 control unit is designed to operate with IFF 4720 and IFF 4760 transponders. It is an NVG-compatible package which is intended to occupy minimum cockpit space, while at the same time having good ergonomics.

Operational status

In service in CN-235, Hawk 100 and MiG-29.

Contractor

Raytheon Systems Limited, Electronic Systems Division.

UPDATED

IFF 4800 transponder

As well as operating in Modes 1, 2, 3A, C and 4, the IFF 4800 operates in Mode S which is the latest civil aviation selective address SSR system. By using Mode S, military aircraft will be able to continue to use civil airspace when conventional SSR modes are phased out in the early years of the next century.

Additionally, IFF 4800 incorporates the following facilities:

(a) Interface for Mode S Air Link Data processor
(b) Airborne Collision Avoidance System (ACAS) interface
(c) GPS position reporting interface
(d) Built-in Mode 4 or CIRCE cryptographic computer
(e) Provision for the incorporation for growth to the Next-Generation IFF (NGIFF) system (STANAG 4193 Part V)
(f) Automatic code change on Modes 1 and 3A

The IFF 4870 control unit is used together with the IFF 4800. For unified avionics control systems a MIL-STD-1553D databus version of IFF 4800 is available.

Specifications

Dimensions: ½ ATR short
Weight: 8.5 kg
Cooling: convection cooled

Operational status

Just entering production. Specified for Royal Australian Air Force Hawk, and IPTN CN 235 maritime patrol aircraft.

Contractor

Raytheon Systems Limited, Electronic Systems Division.

NEW ENTRY

AN/APN-194 radar altimeter

The AN/APN-194 is a low-profile, lightweight, all-solid-state radar altimeter which provides analogue and digital outputs to interface with an automatic flight control system. The unit conforms with the high environmental specifications of most military aircraft.

Specifications
Dimensions: 79.4 × 97.3 × 188.2 mm
Weight: 2 kg
Power supply: 115 V AC, 400 Hz
28 V DC
Frequency: 4.3 GHz
Pulse repetition frequency: 20 kHz
Radiated power: 5 W peak
Altitude: 0-5,000 ft standard
Output signals: digital and analogue
Accuracy: ±3 ft or 4%
Manoeuvrability:
(typical) pitch and roll ±45° within stated accuracy
Temperature range:
(operating) −55 to +95°C
(non-operating) −65 to +125°C

Operational status
In production.

Contractor
Smiths Industries Aerospace.

VERIFIED

HRA series radar altimeter

The HRA Series radar altimeter unit meets all requirements for a highly accurate measurement and flight-deck indicator on civil and military aircraft. The outputs are suitable for use with flight control systems including those used during automatic landing. Other applications include terrain-following and avoidance, reconnaissance and anti-submarine warfare.

Specifications
Dimensions: 203 × 150 × 96 mm
Weight: 2.9 kg
Power supply: 115 V AC, 400 Hz
28 V DC
Frequency: 4.3 GHz
Pulse repetition frequency: 10 kHz
Radiated power: 100 W peak (nominal)
Altitude: 0-5,000 ft, 0-2,500 ft, 0-1,000 ft
Output signals: analogue
Accuracy: ±3 ft or 3%
Manoeuvrability:
(typical) pitch and roll ±35° within stated accuracy
Temperature range:
(operating) −40 to +71°C
(non-operating) −54 to +95°C

Operational status
In service.

Contractor
Smiths Industries Aerospace.

VERIFIED

KTX series radar altimeter

The KTX Series is a lightweight pulse radar altimeter developed for applications requiring low-volume installation and is particularly suitable for executive aircraft, low-cost military aircraft and missiles. It incorporates a high proportion of integrated and solid-state circuitry, as well as the operational features of other Smiths Industries radar altimeters.

Specifications
Dimensions: 152 × 102 × 86 mm
Weight: 1.36 kg
Power supply: 28 V DC
Frequency: 4.3 GHz
Pulse repetition frequency: 8 kHz ±2 kHz
Radiated power: 50 W peak
Altitude: 0-5,000 ft, 0-2,500 ft, 0-1,000 ft
Output signals: analogue
Accuracy: ±3 ft or 3%
Manoeuvrability:
(typical) pitch and roll ±30° within stated accuracy
Temperature range:
(operating) −55 to +70°C
(non-operating) −55 to +90°C

Operational status
In service.

Contractor
Smiths Industries Aerospace.

VERIFIED

UNITED STATES OF AMERICA

AN/APS-144 airborne surveillance radar

The AN/APS-144 is a modular radar developed for over land and over water wide area surveillance for tactical, border and interdiction applications. Operating modes of the system are surface moving target indication, SAR imaging and airborne intercept. Its primary use is surface surveillance from manned aircraft, helicopters and UAVs. Typical targets are ground vehicles and personnel, watercraft and low-flying fixed-wing aircraft and helicopters. A cueing feature is included in the system for aiming electro-optical sensors for close examination and identification of moving targets. Target classification features are incorporated in the radar processor.

The modular configuration of the AN/APS-144 allows it to be configured for specific applications. For the airborne intercept mode, a pod-mounted configuration is utilised which is adaptable to light aircraft. An I-band version of the radar has been developed for long-range, wide area surveillance. The potential for D-, E-, F- and H-band versions is inherent in the modular design of the Series.

The AN/APS-144 consists of a receiver/exciter, power amplifier, antenna feed, pedestal control, digital control interface and control/display unit.

Specifications
Weight: 55 kg (manned aircraft)
Power supply: 28 V, 660 W
Frequency: J-band coherent, frequency agile, pulse Doppler
PRF: 3, 4, 5, 6.25 kHz
Coverage: 360° or sector (azimuth)
Range: 30 km
Antenna rotation: 11 and 18°/s

Operational status
A prototype was tested by the US Army in 1991 on a UH-60 helicopter and successfully demonstrated the capability to detect groups of personnel. Production is for use in drug interdiction operations.

Contractor
AIL Systems Inc.

VERIFIED

ALA-52A radio altimeter

The ALA-52A radio altimeter is a lightweight solid-state digital low-range unit which utilises a simplified microprocessor-based design.

The ALA-52's capabilities are achieved by an advanced microprocessor which handles all data computations, including: the application of correction factors for aircraft installation delay; the control of tracking filter gain bandwidth characteristics; collection and processing of the beat frequency count representing altitude information; the output of the altitude data for display via the ARINC 429 interface. Flag logic and monitor levels are also controlled by the microprocessor to reference criteria defined in the firmware.

As an added confidence factor, the ALA-52A utilises a second microprocessor of differing design architecture, to compute and verify altitude information by comparison independently.

One of the ALA-52A's other major advantages is its ability to perform continuous automatic self-calibration. By utilising a continuous feedback loop comprising the transmitter, quartz bulk-wave-delay device, a crystal reference and the modulator, the unit not only monitors the slope of the transmission but also maintains proper calibration. It complies with ICAO Annex 10.

Specifications
Weight: 4.54 kg
Altitude: up to 50,000 ft

Operational status
In production.

Contractor
AlliedSignal Commercial Avionics Systems.

VERIFIED

AN/APN-215(V) radar

The AN/APN-215(V) colour radar is a weather, surface search and precision terrain-mapping system derived from the successful and widely used RDR-1300 commercial system. It is designed for heavy twins, turboprops and transport helicopters. Low weight and a 445 km range suit it to utility and reconnaissance aircraft and it was chosen for the US Army versions of the Beech King Air, the U-21 and the RU-21.

In conjunction with other equipment the APN-215 can display navigation pictorial information overlaid on the weather map, together with pilot-programmable pages of checklist information such as en route navigation data and emergency procedures.

The system comprises three units: a 305 mm pitch and roll stabilised antenna, transmitter/receiver and colour control/display unit.

Operational status
In production and in service with the US Army and Coast Guard.

Contractor
AlliedSignal Commercial Avionics Systems.

VERIFIED

AN/APN-234 multimode radar

The AN/APN-234 is a lightweight airborne digital colour display multimode radar designed to provide sea search weather detection and terrain-mapping for a variety of military aircraft including rotary- and fixed-wing types ranging from light to heavy twins. The system consists of a receiver/transmitter, combined colour display/control unit, stabilised antenna and optional interface unit.

The AN/APN-234 is identical to the AN/APN-215 except for the addition of the sea search function.

Operational status

In service with the US Navy on the Northrop Grumman C-2A Greyhound and the Lockheed Martin EP-3E aircraft.

Contractor

AlliedSignal Commercial Avionics Systems.

VERIFIED

AN/APS-133 radar

The AN/APS-133 digital colour radar is a high-performance weather, beacon-homing and terrain-mapping system designed for large commercial and military transports. The RDR-1FB (Type 1) was originally launched for the retrofit of the US Air Force's C-141 Starlifter fleet and has since been installed on the C-5, E-3 and KC-10 aircraft among others.

The multicolour display can be used in conjunction with other equipment to show programmable checklists or to superimpose navigation or other information on the weather map. The system employs digital processing and microcomputer techniques, as well as a solid-state modulator.

In November 1984, the company delivered to the US Navy and Marine Corps the first units of the RDR-1FB (M) (Type 2) improved land-mapping version for its fleet of KC-130 tankers and C-130 transports; the unit now equips the entire fleet of 70 aircraft. It was specially suited to US Marine Corps requirements, with a high PRF, short pulsewidth, enhanced digital processor and selectable sector-scan antenna to improve radar navigation at low level.

The US Air Force has selected the Type 2 for its E-4 NEACP and VC-25A (Air Force 1 Boeing 747) aircraft and the US Navy uses the unit in its EA-6A Intruder aircraft fleet.

Significant landmarks and continental shorelines up to 555 km away can be portrayed in the ground-mapping mode by using the high-power output concentrated into a pencil beam. At the same time discrete details such as lakes, rivers, bridges, runways and runway approach reflectors readily show up on the colour display. To improve range resolution at short ranges the system operates with 0.4 μs (RDR-1FB(M)) or 0.5 μs (RDR-1FB) pulses in contrast to the 5 μs pulses used for long-range ground- and weather-mapping.

In the air-to-air mode, the APS-133(V) detects and tracks other aircraft during rendezvous, formation and air refuelling. Aircraft of C-130/C-141 size can be tracked to 56 km, but may still be resolvable at ranges as little as 550 m depending on relative bearing, aspect and altitude.

To provide long-range homing to remote ground destinations or tanker aircraft, the APS-133(V) operates at I-band frequencies — 9,375 MHz for beacon interrogation and 9,310 MHz for beacon reception. The identification of closely spaced pulse reply codes at long ranges is made possible by the marker and delay modes of the radar indicator. In the marker mode a variable marker is positioned on the screen just in front of the beacon reply. When switched to the delay mode the display presentation starts at the marker range. The range switch can then be moved to select a shorter range scale, yielding an expanded view of the area containing the beacon reply.

Derived from the Bendix/King RDR-1F used on many hundreds of airliners, the AN/APS-133(V) comprises five LRUs: a 762 mm fully stabilised split-axis parabolic antenna that provides specially shaped search or fan beams for terrain-mapping and skin painting, a transmitter/receiver, a colour display, a radar control unit and an antenna sector-scan control unit (RDR-1FB (M) only).

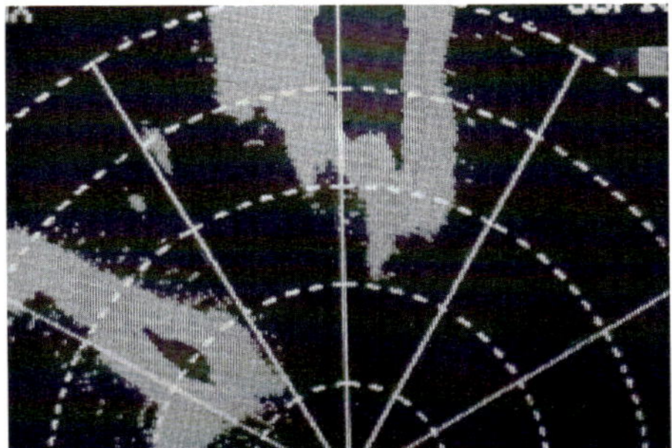

The moisture content of cloud is represented by three colours on the AN/APS-133 display

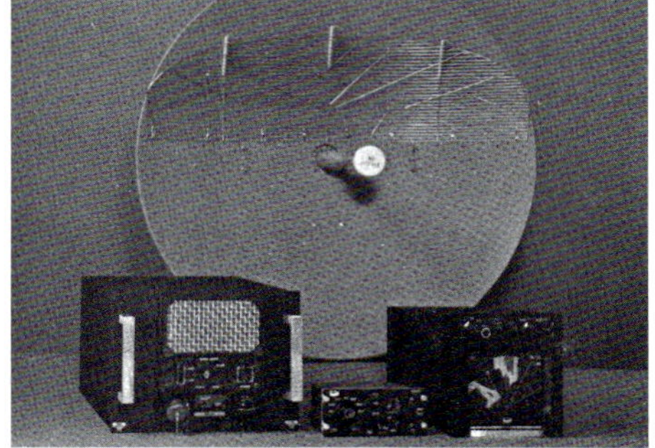

The AlliedSignal AN/APS-133 four-unit weather radar

Specifications

Weight:
(antenna) 15.8 kg
(transmitter/receiver) 22.2 kg
(sector-scan control unit) 1.2 kg
(colour indicator) 6.3 kg
Frequency: I-band (9,375 MHz transmit, 9,310 MHz receive)
Power output: 65 kW
PRF:
200 pps (Type 1 system)
200 and 800 pps (Type 2 system)
Pulsewidth:
(weather) 5 μs
(beacon) 2.35 μs
(mapping) 0.5 μs (or 0.4 μs Type 2, selectable)

Operational status

In service with US Air Force transport aircraft, notably C-5A, C-17 and C-141 and KC-10 Extenders, also E-4A, E-3A, VC-25A and in US Navy/USMC aircraft such as C/KC-130, EA-6A, E-6A, A-3 and YP-3C.

Contractor

AlliedSignal Commercial Avionics Systems.

VERIFIED

AN/APS-133(TTR-SS) multimode radar system

The AN/APS-133(TTR-SS) multimode radar is a coherent pulse Doppler system. It is designed for military tanker and transport aircraft.

The APS-133's advanced digital signal processing architecture provides operational advantages like frequency agility, pulse compression and Doppler beam-sharpening with monopulse resolution enhancement. The system has a precision ground-mapping capability and the capability to detect and display weather and turbulence. To aid in differentiating between mountain shadows and low reflectors such as lakes, and to allow detection of ridge lines, the system utilises a selectable fast time constant. When selected, this gives the appearance of a three-dimensional picture. Both the receiver/transmitter and the digital processor provide fault isolation to the LRM level, and fault isolation and storage to the LRU level.

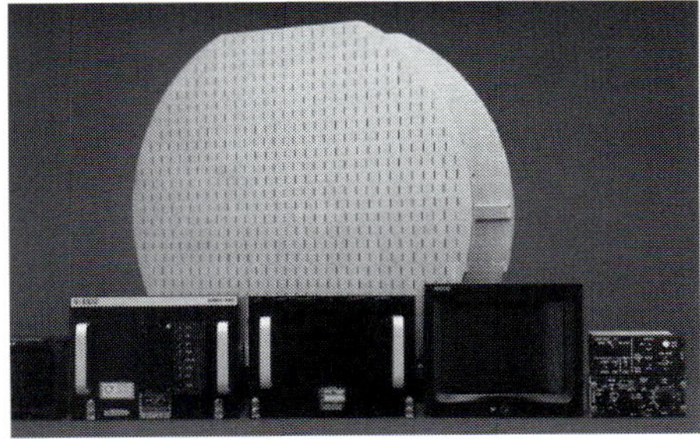

The AlliedSignal APS-133(TTR-SS) multimode radar is designed for military tanker and transport aircraft

The APS-133(TTR-SS) features calibrated turbulence detection and display on all range settings, signal attenuation compensation for more accurate weather display, solid-state digital design and independent roll axis stabilisation. There is automatic beacon decoding for APX-78 and APN-69 beacons, monopulse operation for separating closely spaced targets, and a freeze-frame facility. The system has a MIL-STD-1553B dual-digital bus for FMS and INS interface, antenna stabilisation amplifiers for pitch, roll and tilt and multishade monochrome display capability with colour enhancement. It can interface with and display station-keeping data from the AN/APN-169C and IFF data from the AN/APX-76.

Specifications

Weight:
(total system) 62 kg
Power supply: 115 V AC, 275 W
5 V AC, 10 W (panel lighting)
28 V DC, 100 W

Contractor

AlliedSignal Commercial Avionics Systems.

VERIFIED

Enhanced Ground Proximity Warning System (EGPWS)

The EGPWS includes all traditional GPWS functions, but also has a proprietary worldwide terrain database. Referencing aircraft location from a navigation system, the EGPWS can display nearby terrain and provides aural warnings approximately 60 seconds in advance of a terrain encounter, compared with 10 seconds for a traditional GPWS.

To present a terrain display and provide terrain warnings, the EGPWS computer is linked to other aircraft systems: the navigation system; the main computer, and the aircraft altimeter. Terrain from 2,000 ft below the aircraft altitude and higher can be shown on the weather radar display or electronic flight instrument system. It appears as a map-type presentation. Terrain is depicted green if below the aircraft's altitude, yellow if above, and red if well above. The display colour is denser as the height of the terrain increases. If the system issues an alert, the terrain that poses a threat is shown as a solid block of yellow or red.

The terrain database enables the EGPWS to provide audible alerts earlier than traditional GPWS. In the EGPWS, the computer constantly searches the database along the aircraft's projected flight path, giving the system a virtual look-ahead capability. The system searches down a sloping flight path, if the aircraft is banking, the computer searches along the projected turning path. If the EGPWS determines that the aircraft's flight path would take it too close to terrain in the database, it sounds the audible alert. This sound comes even earlier than a minute ahead if the terrain is particularly high above the aircraft's altitude.

The EGPWS currently interfaces with digital systems, although AlliedSignal plans a version for aircraft with analogue systems.

EGPWS is a form/fit replacement for the existing AlliedSignal GPWS, but some additional cable work will be required.

Operational status

In production. Over 1,000 units sold.

In December 1997, major US air carriers announced a voluntary programme to install EGPWS, in a six year programme extending until the year 2003 affecting about 4,500 aircraft operated by US major and national airlines. US regional airlines are not part of this voluntary programme, nor are charter carriers.

This initiative pre-empts a planned US FAA notice of proposed rule making for the fitment of EGPWS to US long-haul and most regional aircraft.

As of December 1997, it is understood that the AlliedSignal EGPWS is the only system currently certificated in the USA by the FAA. British Airways is the first airline in Europe to select the EGPWS. Following a successful two-year trial on a Boeing 747-400, British airways took delivery of its first production installation, on a Boeing 777, in late March 1998, and plans to equip its entire fleet.

AlliedSignal have integrated the EGPWS with the RDR-4B forward-looking wind shear radar; the combined system has been successfully flight tested for certification on the Boeing 777.

Contractor

AlliedSignal Commercial Avionics Systems.

UPDATED

E-TCAS Traffic alert and Collision Avoidance System

E-TCAS is a development of the AlliedSignal TCAS II system. It combines the ability for improved traffic alert and collision avoidance with a formation and rendezvous flying capability for military air transports. It provides a one mile range display that gives the pilot precise locations of nearby aircraft and enables him to fly safely in formation and refuel in flight.

The US Air Force and Lockheed Martin have let a contract to replace the current station-keeping equipment for providing tanker rescue/rendezvous mission and refuelling capability with a modified version of E-TCAS.

Operational status

E-TCAS has been designated as standard equipment on US Air Force C-130H aircraft and on the C-130J aircraft.

Contractor

AlliedSignal Commercial Avionics Systems.

UPDATED

GPWS: Ground Proximity Warning System, for helicopters

AlliedSignal Aerospace is seeking FAA certification of a ground proximity warning system (GPWS) for helicopter use.

The new system is a version of AlliedSignal's Mark VII GPWS that is optimised for unique rotary-wing requirements.

The system is similar to those engineered for fixed-wing aircraft, providing protection against excessive descent rate, descent after take-off, inadvertent descent below glide slope on instrument approaches and excessive bank angles. It also features helicopter-specific functions, which include: in the event of a power loss, altitude call outs are given during autorotation to assist the pilot in determining the appropriate time to apply control inputs; an annunciation of excessive pitch attitude to help prevent tail strike during landing; warnings of excessive terrain closure rate are applied at above-ground altitudes appropriate for helicopters and adjusted for forward speed; minimum terrain clearance settings are optimised for the lower operating altitudes of helicopters and are adjusted according to flight speed.

Operational status

First certification of the new system was planned for the Sikorsky S-76 series helicopter.

Contractor

AlliedSignal Commercial Avionics Systems.

VERIFIED

GPWS Mk II Ground-Proximity Warning System

The Mk II Ground-Proximity Warning System (GPWS) computer is designed for aircraft wired to ARINC 594 standard and is suitable for service in a wide cross-section of commercial, military or business aircraft. The Mk II model is claimed to be the first GPWS to use a Mach/airspeed input and therefore to have a much faster response time than previous GPWS computers. It was the first such system to offer voice alerts which specifically identified each warning mode, and the first to offer a warning mode for minimum approach conditions.

Warning modes for the Mk II unit are generally the same as for the Mk V digital GPWS although there are differences between the warning times and the warning envelopes themselves.

Specifications

Dimensions: ¼ ATR short
Weight: 3.63 kg (max)

Operational status

In production and service. Selected by the US Navy for the UC-12B.

Contractor

AlliedSignal Commercial Avionics Systems.

VERIFIED

GPWS Mk V Digital Ground-Proximity Warning System

The Mk V digital ground-proximity warning system computer is designed for service with aircraft equipped with ARINC 700 avionics. It provides the flight crew with back-up warning for seven potentially dangerous situations including windshear conditions.

Alerts and warnings are provided by steady or flashing visual indications and by audible warnings. Each audio warning is also annunciated to identify the particular situation such as excessive descent rate, excessive closure rate to terrain, significant altitude loss after take-off, insufficient terrain clearance, excessive descent below glide slope, altitude call-outs and windshear detection.

Windshear detection and annunciation are provided by the Mk V computer. When the computer detects an impending windshear situation, an optional amber light is turned on in the cockpit. If the aircraft experiences further windshear severity, a red warning light is displayed along with a voice message 'windshear' repeated three times. The windshear alert function takes priority over other GPWS alerts.

Other alerts are repeated twice. If the aircraft's performance continues to degrade, the message is repeated. A particular advantage of the variety of voice alerts is that its operationally orientated warnings permit confirmation by cross-checking of the panel instruments. Diagnosis of flight warnings can thus be quickly carried out and corrected. The speed of the ground-proximity warning system envelopes has been increased, providing longer warning times.

The system contains a number of features to assist in test maintenance and repair procedures. These include a non-volatile memory which stores both steady-state or intermittent faults occurring over the last 10 flight sectors and which can be erased only when the unit is removed from the aircraft for bench work. The accepted test procedure is programmed within the computer and a simple test fixture is all that is required to re-address computer output data back into the computer itself. An alphanumeric display on the front of the unit can be used to isolate faults and indicate specific LRUs which require replacement. Faults can be isolated on the bench to board level.

The GPWS complies with ARINC 600 standards and its subcomponents are grouped by circuit function on plug-in/fold-out removable printed circuit boards with easily removed captive hardware. Latitude and longitude are used to modify warning boundaries at certain locations to reduce nuisance probability, or increase available warning time.

Specifications

Dimensions: 2 MCU

Operational status

In production and service on Boeing 737-300.

Contractor

AlliedSignal Commercial Avionics Systems.

VERIFIED

GPWS Mk VI Ground-Proximity Warning System

The Mk VI ground-proximity warning system operates in six modes: excessive descent rate alert and warning; excessive closure rate to terrain; alert to descent after take-off; alert to insufficient terrain clearance; alert to inadvertent descent below glide slope; and altitude call-outs and bank angle alert. The Mk VI system is said to cost 40 per cent of the earlier Mk II GPWS.

The system has been refined to delete unwanted warnings by reducing glide slope and terrain clearance floor limits to trigger warnings at altitudes down to 750 ft above ground level on approach or to 925 ft above ground level with ILS acquired.

Manual functions have been added to minimise the chance of false warnings due to flapless landings or other operational modes, when landing at airports with unique terrain features or in the event of incompatible terrain clearance during approach and departure procedures.

Operational status

FAA TSO-C92b approval was given in September 1992.

Contractor

AlliedSignal Commercial Avionics Systems.

VERIFIED

GPWS Mk VII Ground-Proximity Warning System

The Mk VII warning computer is designed as a replacement for the Mk I and Mk II ARINC 594 ground proximity warning computers. The improved Ground Proximity Warning System (GPWS) dynamics provide the advantages of increased warning times, prioritisation of aural warnings and reduction in nuisance warnings in the cockpit, while implementing a cost-effective windshear warning system. The computer has a common part number used across a wide range of aircraft types, minimising the investment in spares.

The Mk VII warning computer meets the requirements of FAA AC 25-12 for windshear detection and alerting. It uses the existing ARINC GPWS interface, with additional signals provided through a second connector, for windshear detection, optional recovery guidance and custom altitude call-outs.

The GPWS features ground proximity warning and glide slope alerting; altitude call-out menus; bank angle alerting; reduced audio cockpit clutter; improved take-off monitoring for noise abatement procedures; automatic adjustment of warning modes for ILS approaches; windshear detection and annunciation; optional windshear recovery guidance; verbal annunciation of system faults and front replaceable software modules. It fits into existing GPWS rack space.

Specifications

Dimensions: ¼ ATR short
Weight: 2.72 kg (max)
Power supply: 115 V AC, 400 Hz, single phase, 15.7 W nominal
Environmental: DO-160B
Reliability: 15,000 h MTBF

Contractor

AlliedSignal Commercial Avionics Systems.

VERIFIED

KRA 10A radar altimeter

The KRA 10A radar altimeter is a low-cost system suitable for independent use or in combination with King Silver Crown avionics equipment and tailored to general aviation requirements. There is a standard facility for presetting decision height which produces a visual and aural warning on reaching the set altitude. Antennas suitable for flat and sloping skin installations are available. The KRA 10A is an all-solid-state system with short warm-up time and can be fitted with an auxiliary output to interface with flight director and autopilot installations.

Specifications

Dimensions:
(indicator) 100 × 83 × 83 mm
(transmitter/receiver unit) 79 × 89 × 203 mm
(antenna) approx 100 × 100 mm aperture
Weight:
(indicator) 0.4 kg
(transmitter/receiver unit) 0.9 kg
(antenna) 0.4 kg
Power supply: 28 V DC, 6 VA
Altitude: up to 2,500 ft

Accuracy:
(0-100 ft) 5 ft
(100-500 ft) 5%
(>500 ft) 7%

Operational status
In production.

Contractor
AlliedSignal Commercial Avionics Systems.

VERIFIED

KRA 405 radar altimeter

Part of the King Gold Crown avionics range, the KRA 405 is an all-solid-state radar altimeter suitable for twin-engine general aviation and regional airliner types. The KRA 405 interfaces with King KPI 553A HSI and KFC 300 autopilot to give smooth tracking of the glide slope beam. It can provide indications from 2,000 ft above ground level and a usable output is available from 2,500 ft above ground level for ground proximity warning system operation. Separate transmit and receive horn antennas are used.

Specifications KRA 405
Dimensions:
(indicator) 83 × 83 × 170 mm
(transmitter/receiver unit) 83 × 133 × 296 mm
(antennas) each 178 mm diameter
Weight:
(indicator) 0.8 kg
(transmitter/receiver unit) 2.9 kg
(antennas) 1.2 kg total
Power supply: 28 V DC, 24 VA
Frequency: 4,300 MHz
Altitude: up to 2,500 ft
Accuracy:
(0-500 ft) 5%
(>500 ft) 7%

The KRA 405B updates the KRA 405 by reducing the number of primary circuit boards from seven to two, reducing overall weight of the receiver/transmitter by 50 per cent, and adding updated software. System elements are: KRA 405B receiver/transmitter; KNI 415 (fixed-wing) or KNI 416 (rotary-wing) indicator; two KA 54A antennas; optional CM2000 configuration module.

Specifications KRA 405B
Dimensions:
KRA 405B: 279 × 76 × 90 mm
KNI 415/416: 170 × 83 × 83 mm
KA 54A: 93 × 89 × 19 mm
Weight:
KRA 405B: 1.36 kg
KNI 415/416: 0.77 kg
KA 54A: 0.09 kg (each)
Altitude: up to 2,500 ft
Accuracy:
(0-500 ft) ±5 ft
(>500 ft) ±7%

Operational status
In production.

Contractor
AlliedSignal Commercial Avionics Systems.

VERIFIED

KXP 756 Gold Crown III solid-state transponder

The KXP 756 is a third-generation system incorporating modern avionics techniques such as large-scale integrated circuitry, microprocessor data programming and all-solid-state transmitter design to provide reliability and simplicity of operation. It operates on Modes A, B and C up to 70,000 ft and can reply on any one of 4,096 preselected codes. Information is provided on one or two 2¼ in square gas discharge digital displays which are automatically adjusted in brightness by a photocell, for maximum visibility under all light conditions. The system is controlled by two concentric knobs — one for mode selection and the other for code selection — and incorporates identification, VFR code and self-test functions.

Specifications
Dimensions:
(control unit) 146.7 × 21.9 × 23.5 mm
(remote unit) 298.45 × 50.8 × 134 mm
Weight:
(control unit) 0.31 kg
(remote unit) 1.727 kg
Power supply: 11-13 V DC
Altitude: up to 60,000 ft
Frequency:
(receive) 1,030 MHz
(transmit) 1,090 MHz

Contractor
AlliedSignal Commercial Avionics Systems.

VERIFIED

MST 67A Mode S transponder

A third-generation Mode S transponder, the compact MST 67A offers all the capabilities of heavier airline-type units in a much smaller package. It incorporates a number of patent pending features, including such advances as a 16-bit microprocessor, programmable gate array digital signal processing and SAW technology. Fully TSO'd, the remote-mounted MST 67A is equipped with standard ARINC 400 series connectors.

A choice of control heads allows the MST 67A to fit virtually any corporate or regional airliner class cockpit. Featuring a photocell-equipped gas discharge display, the KFS 578A control unit supplies ARINC 429 data to all versions of the system. The KFS 578A can also serve as the aircraft's TCAS controller. For aircraft already equipped with a dzus-mount transponder control panel, the CTA 81A is available as a drop-in replacement. Fully compatible with ARINC 718 and ARINC 735 and providing many of the same interfacing and control functions as the KFS 578A, the CTA 81A features a high-contrast liquid crystal display.

In its non-diversity version, the transponder uses a bottom antenna only, for operators who do not anticipate installing TCAS in the aircraft but wish to ensure compliance with ATC reporting standards. A non-diversity MST 67A is fully compliant with air-to-ground/ground-to-air datalink applications.

The diversity version of the MST 67A uses inputs from two antennas, mounted top and bottom of the aircraft. Required for TCAS operations, the diversity option provides the aircraft with air-to-air datalink communications capability.

Enhanced BIT features constant monitor transponder status. A bidirectional interface between the transponder and the control unit also enhances diagnostic capabilities. With the test mode selected on the control panel, internally diagnosed problems can be viewed in real time and information stored for as many as the nine previous flights in non-volatile memory can be reviewed.

Specifications
Dimensions:
(MTS 67A transponder) 57.2 × 381 × 193.8 mm
(KFS 578A control unit) 53.1 × 57.2 × 187.5 mm
(CTA 81A control panel) 146.1 × 57.2 × 119 mm
Weight:
(MTS 67A transponder) 3.86 kg
(KFS 578A control unit) 0.45 kg
(CTA 81A control panel) 0.82 kg

Contractor
AlliedSignal Commercial Avionics Systems.

VERIFIED

RDR-4 radar

Chosen by Boeing as standard equipment for the 767 and 757 transports, the AlliedSignal RDR-4A is designed to meet the new ARINC 708 requirements. The I/J-band system features a solid-state transmitter and line of sight antenna with split-axis performance and is compatible with the EFIS flight decks of the Boeing 767, 757, MD-80, DC-10, Airbus A310, and Lockheed Martin L-1011 transports and other designs. The range is 592 km.

The RDR-4B incorporates forward-looking windshear detection and avoidance capabilities. The windshear detection capability is easily incorporated into existing RDR-4A radars, without form or fit changes to the installation.

The RDR-4B is a Doppler weather radar that measures actual horizontal windspeed using reflections from the moisture that is always present in the atmosphere and penetrates weather systems and detects microbursts embedded in rain. It provides specific windshear locations on a radar PPI presentation, giving 30 to 60 seconds of advanced warning, on a display free from interference such as ground clutter. The RDR-4B can display turbulence up to 46 miles ahead and 90° left or right. It can display windshear up to 5¾ miles ahead up to 40° left or right.

The system operates automatically any time the aircraft is below 2,300 ft AGL, although the mode is selectable at any time.

The RDR-4B radar display is a map-like presentation that shows areas of weather and their locations relative to the aircraft. Light rain is shown in green, moderate rain in yellow, and heavy rain in red. Turbulence is magenta and windshear is a symbol called a windshear icon — a pattern of red and white arc-shaped stripes.

The RDR-4B can generate audible windshear warnings and alerts whenever the aircraft is less than 1,500 ft AGL. During approach windshear ahead within 1.7 miles causes a 'go around' warning; in the landing phase this warning is reduced to 0.6 miles. During take-off windshear within 3.4 miles generates a warning.

Operational status
In production and in service with, among others, Airbus A300s of TOA Domestic Airlines of Japan, Southwest Airlines Boeing 737-300s and Singapore Airlines Boeing 747-300s. Also in service with Delta, American, Pakistan International, Northwest Airlines, Saudia, Varig, Finnair and Austrian Airlines.

First use of the RDR-4B in commercial service was on a Continental Airlines Boeing 737-300 in November 1994.

Contractor
AlliedSignal Commercial Avionics Systems.

UPDATED

RDR-1400 weather/multifunction radar

The RDR-1400 radar is designed for commercial helicopters, particularly those associated with the large international offshore oil and gas industry. It differs from almost all other AlliedSignal weather radars by having a beacon interrogator that exploits the increasing use of portable radar beacons in these industries. The system is suitable for search and rescue, surveillance, aerial survey work and law enforcement, as well as rig servicing.

The original monochromatic RDR-1400 has now been joined by a colour radar version with greater performance. The following operational modes are available:

Beacon navigation. The growing popularity of portable beacons is supported by several special RDR-1400 capabilities. A beacon signature is denoted by a short line or oblique on the display, the actual location of the device being determined by the middle of the line; the pilot can overlay beacon returns on the weather map. The beacon's discrete code can be displayed for positive identification, an important factor when the pilot is trying to locate a specific rig in a drilling farm where numerous rigs may be transponder-equipped. Beacon detection range is up to 296 km depending on altitude.

Beacon Trac. This mode, peculiar to AlliedSignal, generates and displays on the weather radar screen an inbound course to the discrete beacon. This course line can be rotated 360° about the beacon by rotating the horizontal situation indicator course selector, thus allowing the pilot to choose a convenient course to the beacon.

Obs Trac. This mode provides another course-following option. When in a weather or search mode, a track line or course-bearing cursor can be generated from the aircraft position and controlled by the horizontal situation indicator course selector to provide a course line to the chosen target. The Obs Trac heading is displayed digitally in the lower right-hand corner of the indicator. Left/right deviations can be determined by comparing heading information to this number and by observing the movement of the track line in relation to background targets.

Search. Three search modes are available. Search 1 has special sea clutter rejection circuitry to detect objects such as small boats or buoys down to the minimum range. Search 2 is for precision ground-mapping in situations where high target resolution is important. Search 3 includes maximum return clutter and can detect and track oil slicks.

Specifications

Dimensions:
(transmitter/receiver) 127 × 159 × 352 mm
(control/display unit) 159 × 159 × 276 mm
Weight:
(305 mm antenna) 15.47 kg
Frequency: I/J-band
Power output: 10 kW
Antenna size: 305 or 457 mm flat plate
Antenna scan angle: 120° or 60°
Display size: 110 × 85 mm
Qualification: TSO 63b

Operational status

In production and in service in civil and military helicopters.

Contractor

AlliedSignal Commercial Avionics Systems.

VERIFIED

RDR-1400C colour weather and search and rescue radar

The RDR-1400C is a weather detection system. It has a 445 km maximum display range, giving the user time to plan weather avoidance manoeuvres. For clear, detailed close-ups, two modes permit selection of full-scale ranges of either 1 n mile or 0.5 n mile, enhancing safety and precision of movement. The Target Alert feature flashes a warning whenever third-level (red) weather areas are detected up to 46 km beyond the selected range.

Different surveillance missions require different capabilities, so the RDR-1400C provides three specialised search modes. Search 1 incorporates special sea clutter rejection circuitry to help detect small boats or buoys down to a minimum range of 270 m.

Search 2 is designed for precision ground-mapping, where high target resolution is important. Search 3 mode, which includes normal ground-mapping, can also be used to detect and track sea-surface phenomena such as oil slicks.

The RDR-1400C complies with TSO C-102, enabling land or sea approaches in 200 ft ceiling, ½ mile visibility minimums. Its beacon tracking mode permits operation with either current beacon codes or the newer DO-172/16 format, changing with the push of a button.

In BeaconTrac® mode and equipped with a graphics interface unit, the RDR-1400C can automatically show target latitude and longitude along with NAV sensor data on a moving map display. A white course line, rotatable 360° around the centre of a target beacon, offers a digital flight path to the beacon from any direction. Left/right flight deviation cues are provided by a deviation control bar located on the indicator.

Delivering 10 kW of power, the system excels in the detection of small targets. For example, the RDR-1400C can locate a 9 m target in Sea State 2 from up to 18 km away, while operating at a typical helicopter altitude of 152 m.

Specifications

Dimensions:
(receiver/transmitter) 127 × 158.8 × 352.4 mm
(indicator) 158.8 × 158.8 × 276.2 mm
Weight:
(receiver/transmitter) 6.58 kg
(indicator) 5.22 kg
Power requirements:
4.2 A, 27 V AC
3 A, 115 V AC, 400 Hz
TSO compliance: C-102
Frequency: I/J-band
RF power output: 10 kW
Scan angle: 120 or 60°
Scan rate: 28°/s
Display range/marks: 0.5/0.125; 1/0.25; 2/0.5; 5/1.25; 10/2.5; 20/5; 40/10; 80/20; 160/40; 240/60 n miles
Min tracking range: 270 m
Beacon range: line of sight to 300 km

Contractor

AlliedSignal Commercial Avionics Systems.

VERIFIED

RDR-2000 vertical profile weather radar system

The RDR-2000 is the first of a new generation of vertical profile weather radar produced by AlliedSignal. In addition to normal weather radar features, the RDR-2000 adds a vertical display of weather, enabling the pilot to monitor storm development. Horizontal and vertical weather data can be presented simultaneously on the EHSI and MFD, coding four-colour display technology, with flashing 'Weather Alert' warnings.

Specifications

Dimensions:
254 or 309 mm antenna
Power output: 3.5 kW nominal (rated 4.0 kW)
Scan: 90/100 at 25°/s
Weather Avoidance: 400 km typical for 309 mm antenna system

Operational status

In production.

Contractor

AlliedSignal Commercial Avionics Systems.

VERIFIED

RDS-81 weather radar

The RDS-81 is a four-colour radar. The fourth colour, magenta, depicts the very heaviest areas of precipitation, enabling the system to provide a more clearly defined picture of where potentially severe weather is located. It also helps depict the steep rainfall gradients that are often associated with heavy turbulence or windshear. A selectable weather alert mode causes the magenta returns to flash on and off continuously as a further warning of intense storm cell activity. The 254 or 305 mm flat plate antenna is fully stabilised to keep it parallel to the horizon or at the selected tilt angle and automatic compensation is provided up to 25° of combined pitch and roll.

The digital control/display unit features a 127 mm display coated with a polarised optical filter to provide good colour, contrast and definition even in extreme lighting conditions. Range scales of 10, 20, 40, 80, 160 and 240 n miles (18.5, 37, 74, 148, 296 and 444 km) are available and a different combination of pulsewidth and PRF for each scale optimises radar transmission characteristics to give good long-range performance and high resolution at short ranges.

The RDS-81 can be combined with the GC 318A radar graphics unit and the KNS 81 integrated Nav/RNav system to provide a multifunction display with graphics, checklists and EFIS.

Specifications

Dimensions:
(control/display unit) 104 × 159 × 267 mm
Weight:
(control/display unit) 4.3 kg
(antenna/receiver/transmitter) 4.3 kg
Power supply:
(control/display unit) 28 V DC
Power output: 1 kW peak nominal
Frequency: 9,345 MHz
Pulsewidth: 0.6 to 11.5 μs depending on range
PRF: 128 to 1,026 Hz depending on range

Contractor

AlliedSignal Commercial Avionics Systems.

VERIFIED

RDS-82 radar

The RDS-82 is a digital, I/J-band radar. It has a four-colour display. Magenta indicates rainfall in excess of 51 mm/h and attenuation circuits automatically adjust the radar's sensitivity. The system is based on ARINC 429 digital data handling, so that all information and commands between the radar head and the control/display unit are carried on a twisted cable-pair. The antenna can be a 254 or 305 mm flat plate, scans 90° at the rate of 30°/s and radiates 1 kW. The system operates up to 55,000 ft without pressurisation.

Operational status

Launched in 1983. In service.

Contractor

AlliedSignal Commercial Avionics Systems.

VERIFIED

RDS-84 Series 3 radar

One of two digital colour radars announced by AlliedSignal in April 1984, the RDS-84 is an addition to the company's range of Series 3 third-generation digital avionics for general aviation. It is intended for heavy piston twins and light jet types. Providing a 120° antenna scan and weighing 9 kg, the RDS-84 features a combined transmitter/receiver and antenna like the RDS-82 introduced in 1983.

Operational status

In service.

Contractor

AlliedSignal Commercial Avionics Systems.

VERIFIED

RDS-84VP weather radar

The RDS-84VP (Vertical Profile) digital quadra colour radar takes advantage of the benefits of digital technology, to provide more precise readings of rainfall rates and a continuous bright picture of the weather and the navigation situation. Microprocessor technology allows the RDS-84VP to process more information than previous systems. All radar controls are multiplexed on an ARINC 429 control bus. Piezoelectric crystal oscillators provide a precise time reference to enable the RDS-84VP to track distant storms with great accuracy.

Another key advantage of the digital avionics design of the RDS-84VP is the system's built-in diagnostic capability, which allows it to serve as its own test set. The microprocessor-driven monitoring also alerts the pilot to any faults in the system with messages on the radar indicator.

The RDS-84VP employs autopulse, a technique that provides accurate displays of storm cells at all ranges. Autopulse enables the radar to optimise pulsewidth and PRF for each different radar range. This provides optimum long-range performance and high resolution at short range where it is most needed. Autopulse enables the RDS-84VP to operate with highly efficient peak output power.

Three features help the RDS-84VP to keep storms in perspective. The RDS-84VP uses Sensitivity Time Control (STC) logic to increase the amplifier gain for signals returning from greater distances. This means the storms that are further away will still have properly delineated areas of green, yellow, red and magenta. STC prevents storms from seeming to grow in intensity and size as they get closer. STC works only up to a specified range based on the point where amplifier gain has been increased to the maximum. The RDS-84VP also has an extended STC feature to compensate for weak signals returning from beyond the STC range. These weak signals are assigned display colours on a different schedule to ensure that areas of intense rainfall will be properly shown as red or yellow. The

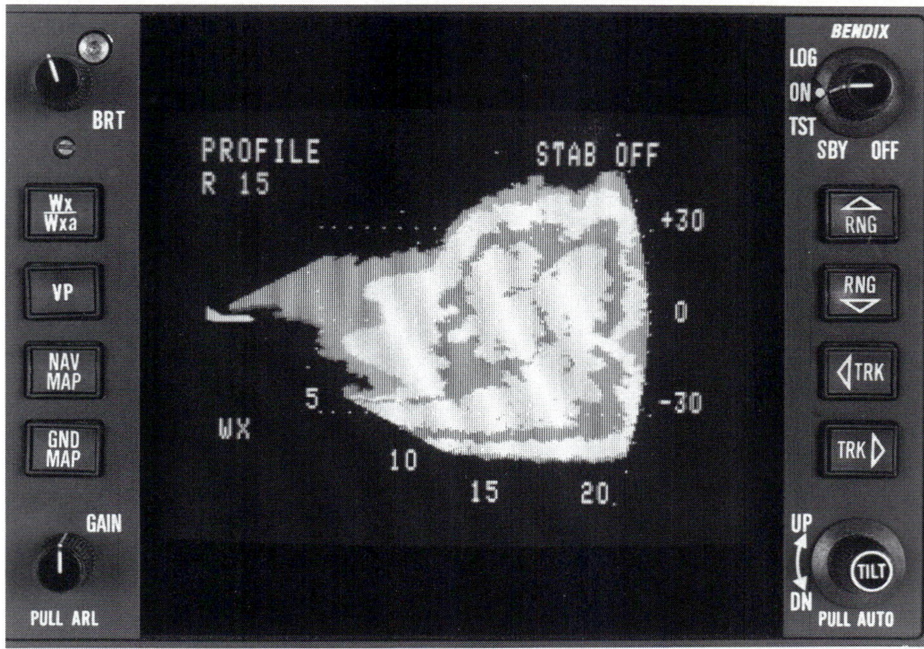

The AlliedSignal RDS-84VP weather radar scans both vertically and horizontally

RDS-84VP also compensates for radar signal attenuation. Radar energy sent out by the transmitter antenna may be partially absorbed by one storm before it reaches a second one. To help optimise the display of a second storm, special weather attenuation compensation circuitry increases receiver gain as needed to pick up storms hidden in the shadows of closer storms.

Specifications

Dimensions:
(IN 842A control/display unit) 114 × 163 × 290 mm
(GC 360A radar graphics unit) 163 × 34 × 342 mm
Weight:
(antenna/receiver/transmitter) 5.8 kg
(IN 842A control/display unit) 4.44 kg
(GC 360A radar graphics unit) 1.77 kg
Power supply: 28 V DC, 2.7 A
Power output: 1.3 kW peak
Frequency: 9,345 MHz
Pulsewidth: 0.8 to 15.3 μs depending on range
PRF: 128 to 1,086 Hz depending on range
Antenna size: 305 or 457 mm flat plate
Antenna scan: 120° at 30°/s
Antenna tilt: ±15°

Operational status

The RDS-84VP is designed for business jets and turboprops.

Contractor

AlliedSignal Commercial Avionics Systems.

VERIFIED

RDS-86 Series 3 quadra radar

The RDS-86 addition to the AlliedSignal range of Series 3 digital avionics for business aircraft was announced in April 1984. It is intended for heavy corporate jets such as the Dassault Falcon 900, Gulfstream IV and Canadair CL-600/601 Challenger. Like the RDS-84, the RDS-86 Series 3 radar has a combined transmitter/ receiver and antenna and displays four colours (as indicated by the description quadra). The system has automatic range limitation, whereby areas giving signal returns of an unreliably low level are painted blue as a warning; antenna stabilisation, to maintain a steady picture during climb and descent; and a long-range navigation mode in which data up to 1,850 km ahead can be displayed, in addition to eight selectable distance scales ranging from 9 to 592 km.

The system weighs 10 kg.

Operational status

In service.

Contractor

AlliedSignal Commercial Avionics Systems.

VERIFIED

TCAS Systems: CAS 66A/CAS 67A/CAS 81

CAS 66A TCAS I

The CAS 66A TCAS I features a computerised processor which interrogates the ATC transponders and then displays the positions of up to 30 surrounding aircraft. Two antennas are used, mounted on the top and bottom of the aircraft, to minimise blind spots.

Equipped with the power of a TCAS II, the CAS 66A display can be optimised for maximum situational awareness. It offers crew-selectable ranges of 3, 5, 10, 15, 20 and 40 n miles. A selectable altitude display window gives three perspectives on traffic: normal view; upward view (2,700-8,700 ft above); downward view (2,700-8,700 ft below). CAS 66A was designed to meet TCAS II standards, and features bearing accuracy of ±3°, rather than the ±30° TCAS I requirement.

If the transponders of surrounding aircraft are not able to report altitude, the CAS 66A provides position only. If they are able to provide altitude, CAS 66A reports this — either relative or absolute — and trend data.

CAS 66A provides coverage even in non-radar environments, and eliminates the need for Loran or GPS sensors. Eliminating only the capacity to show Resolution Advisories (RAs), the CAS 66A's display options include the choice of several weather radar indicators, any compatible Electronic Flight Instrument System (EFIS), or a dedicated Traffic Advisory (TA) display. CAS 66A interfaces through the ARINC 429 standard databus. CAS 66A can be upgraded to TCAS II standard.

CAS 67A TCAS II

CAS 67A system components include a processor which calibrates the antennas, choice of one or two directional antennas, a Mode S transponder, to provide datalink communications and a cockpit display. Display options include an EFIS or weather radar interface, a combination Resolution Advisory/Vertical Speed Indicator (RA/VSI) which provides RAs in TCAS mode, but otherwise functions as a conventional VSI, and a combination flat-panel electronic Traffic Advisory/ Vertical Speed Indicator (TA/VSI) which integrates TAs, RAs and VSI on one instrument. The CAS 67A system interrogates the transponders of surrounding aircraft, determining whether they are Mode A-, Mode C- or Mode S-equipped. If the intruder has Mode A, CAS 67A will provide position data. If the intruder has Mode C or Mode S the CAS 67A can show altitude and altitude trend data. If the intruder has Mode S the two transponders automatically establish communication and co-ordinate resolution manoeuvres for both aircraft.

CAS 81 TCAS II

Developed to meet the needs of airline operation, the CAS 81 TCAS II offers corporate aviation technology identical to the system in use today by most of the major airline carriers. The CAS 81 TCAS II system comprises a TCAS processor and a receiver/transmitter that supports all necessary surveillance and CAS logic functions. In addition, the CAS 81's processor serves as the central point of co-ordination for TCAS signal input and output, including aircraft systems interfaces, display drivers and audio outputs. Its modular design allows for flexibility in reconfiguring the processor I/O section to meet the needs of differing aircraft installations.

Patented interferometry processing provides for true omnidirectional reception of bearing data on the first reply from an intruder's transponder. It also provides bearing accuracy of ±3°, versus the ±15° FAA requirement. While especially important in crowded airspace, this increased accuracy helps minimise bearing errors due to an intruder's relative elevation, and reduces the effect of reflections from aircraft structures. Another feature of the CAS 81's processor is its ability to compensate for signal mismatches between antenna elements. This automatic antenna calibration also helps ensure increased bearing accuracy.

The central processor also co-ordinates beam steering, eliminating the need for a separate antenna steering unit. A typical CAS 81 installation features two four-element directional antennas, which provide extremely accurate signal reception from any direction and altitude within surveillance range.

Other key components include a Mode S transponder to co-ordinate air-to-air communication between approaching aircraft. This co-ordination ensures that the Resolution Advisories (RAs) issued by CAS 81 result in the proper complementary manoeuvres for the two aircraft. The Mode S transponder functions in the same manner as a Mode C ATCRBS transponder in providing altitude and identity information to ATC ground stations.

Contractor

AlliedSignal Commercial Avionics Systems.

VERIFIED

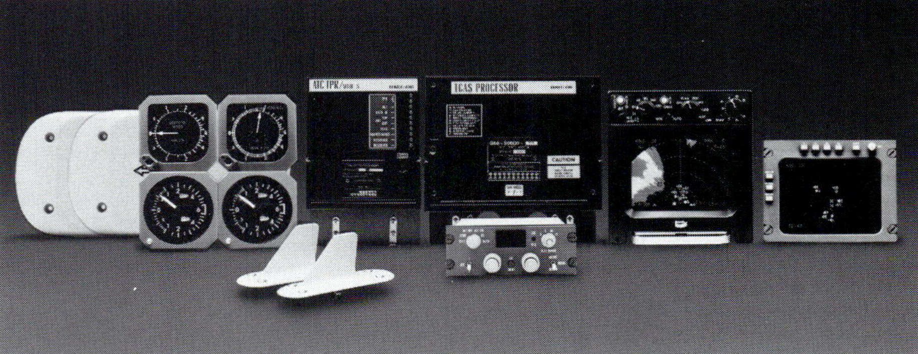

CAS 67A. Standard options include: TPU 67A processor; MST 67A transponder; IVA 81A display; KFS 578A controller; ANT 67A antennas; IVA 81B electromechanical display **1997**

TPR 2060 transponder

The TPR 2060 is a lightweight, compact air traffic control transponder designed for light aircraft and general aviation. It responds automatically to Mode A and Mode C interrogations and, with a suitable encoding altimeter input, will transmit aircraft altitude information with the normal reply pulses. A Mode B capability is optionally available for use in areas employing Mode B interrogation.

The TPR 2060 features special DME suppression circuitry to prevent interference between the transponder and DME installations when the antennas for the two systems are sited in close proximity. The system also permits transmission of a special identification pulse for a 20 second period by an ident button on the front panel. A reply lamp remains lit during this time to reassure the user that the transponder is identing.

Self-test facilities are incorporated. During self-test operation, the unit's coding and decoding circuits are exercised in the same manner as they would be during actual radar interrogation. The unit, which may be panel-, console- or roof-mounted, is in a single case and is of large-scale integrated circuit-type construction.

Specifications

Dimensions: 45 × 160 × 215 mm
Weight: 1.18 kg

Operational status

In service.

Contractor

AlliedSignal Commercial Avionics Systems.

VERIFIED

TRS-42 ATC transponder system

The TRS-42 ATC transponder system is a digital 325 W solid-state transponder for positive identification in the ATC environment. It consists of the TR-421 transmitter/receiver and the CD-422 control display unit.

The TR-421 transmitter/receiver solid-state design gives 4,096 codes of operation, plus Modes A, B and Mode C altitude reporting when connected to an encoding altimeter. The unit utilises a single chip microprocessor which ensures code data validity and display. To increase system reliability the TR-421 utilises a dual transmitter design. Under normal operating conditions the dual transmitters work together to provide a full 325 W of power. If one of the transmitters fails, the unit would continue to function, although at a reduced power capability. This feature is especially important in single transponder installations, where the loss of the transmitter would leave no identification capability.

The CD-422 control display unit has the capability to control a dual transponder installation via a single control head. The selection is made by simply pressing the selector button on the front panel. In a single transponder installation, this button is not provided. The CD-422 also provides an annunciation of the letters ID whenever the transponder replies to an interrogation. When the mode selector is in the VFR position, the active transponder is channelled to the VFR 1200 code. This code may be preprogrammed according to other international VFR codes.

The TRS-422 provides full-time self testing along with a pilot-selectable TEST mode. The self-testing monitors all key circuits such as the transmitter, receiver, encoder, decoder, video processor and central processor.

Specifications

Dimensions:
(control display unit) 63.5 × 79.38 × 63.5 mm
(front connector transmitter/receiver) 10.16 × 10.16 × 27.94 mm
(rear connector transmitter/receiver) 10.16 × 10.16 × 32.05 mm
Weight:
(control display unit) 0.27 kg
(front connector transmitter/receiver) 2.31 kg
(rear connector transmitter/receiver) 2.73 kg
Power supply: 18-33 V DC, 0.9 A nominal

Contractor

AlliedSignal Commercial Avionics Systems.

VERIFIED

AN/APX-100 IFF transponder

The AN/APX-100 IFF transponder is a panel-mounted IFF transponder using microminiature technology in both digital and RF circuitry. It is in production for a number of US military aircraft. The system is a completely solid-state, modular constructed equipment with a complete dual-channel diversity system, comprehensive BIT, digital coding and encoding and a high anti-jamming capability. Two antennas form part of the equipment and the diversity system receives signals from each and switches the transmitter output to the antenna which received the stronger signal. This is designed to cure the problems of poor coverage with a single antenna.

The transmitter is all-solid-state with a 500 W peak power output obtained from four parallel microwave transistors. Two additional transistors complete the transmitter oscillator and driver stages. The diversity system provides improved antenna coverage and allows improvement in performance in overloaded, jamming and multipath environments. Automatic overload control and anti-jamming features are also incorporated.

For aircraft with very limited cockpit space, an equipment bay-mounted configuration, the RT-1157/APX-100(V), is available. In addition, the RT-1471/APX-100(V), a databus version operating in accordance with MIL-STD-155B, is available.

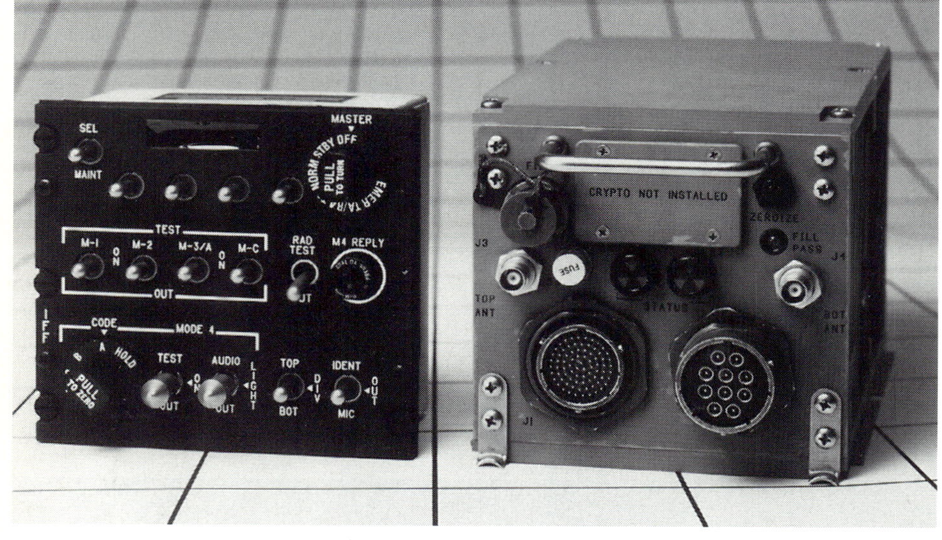

AlliedSignal's AN/APX-100 IFF transponder
1997/0001212

The latest in the APX-100 family includes Mk XII integrated crypto, full Mode S Level 3 and GPS position reporting for advanced air traffic control systems and TCAS systems. There are also NVG-compatible units available which operate at low light levels.

Specifications

Dimensions: 136.5 × 136.5 × 212.7 mm
Weight: 4.53 kg
Frequency:
(transmitter) 1,090 ± 0.5 MHz
(receiver) 1,030 ± 0.5 MHz
Peak power: 500 W ± 3 dB

Operational status

The APX-100 replaced the AN/APX-72 as the standard US military ATC transponder. Over 13,000 units have been built to date.

In production for all new US Navy, Army and Air Force aircraft including the AF-1, AH-1S/T, AH-64, AV-8B, C-5, C-12, C-17, C-20, C-21, C-23A, CH-47, C-130, EC-2C, EC-130, F-14D, F-18, F-22, HH-60, HH-65, LAMPS, MH-47, MH-60, OH-58, OV-10, RAH-66, SH-60, T-45A, UH-60, V-22 and VC-6.

The USA is delivering 113 APX-100 Mk 12 airborne transponders to Hungary under a US$12.7 million contract.

The equipment is also being produced under licence in Japan by Toyocom.

Contractor

AlliedSignal Electronic Systems.

VERIFIED

AV12X radar transponder

The Model AV12X radar transponder is a miniature precision radar augmentation device used to enhance the tracking capability of X-band (NATO I-band) radars. Utilised primarily for range safety functions, the AV12X is suitable for use in manned and unmanned vehicles, aircraft, missiles and target drones, both air and sea-borne.

The AV12X utilises solid-state circuitry. It has a tunable frequency range of 9.0 to 9.5 GHz. It has a long-life Gunn cavity transmitter and direct tuned radio frequency receiver. Frequency stability is ± 5.0 MHz under all conditions. Modulation type is AM pulse. Other features include: single- and double-pulse interrogations, single antenna connection, open and short-circuit antenna protection, reverse polarity protection and adjustable code and delay selection.

The AV12X operates on an input voltage of 22 to 32 V DC, with an operating temperature range of −35 to +60°C. It produces 12 W output power.

Contractor

AYDIN TELEMETRY.

UPDATED

AV400C radar transponder

The AV400C C-band (NATO G-band) radar transponder is a general purpose radar augmentation device used to enhance the tracking capability of C-band (NATO G-band) radars. Utilised primarily for range safety functions, the transponder is suitable for use in both manned and unmanned aircraft, missiles and target drones.

The AV400C features 400 W minimum peak power output and long-life beacon magnetron. It is tunable over 5.4 to 5.9 GHz and has a sensitive superheterodyne receiver, single and double-pulse interrogations and single antenna connection. The system has open and short-circuit antenna protection, reverse polarity protection and adjustable code and delay selection.

Operational status

In service.

Contractor

AYDIN TELEMETRY.

UPDATED

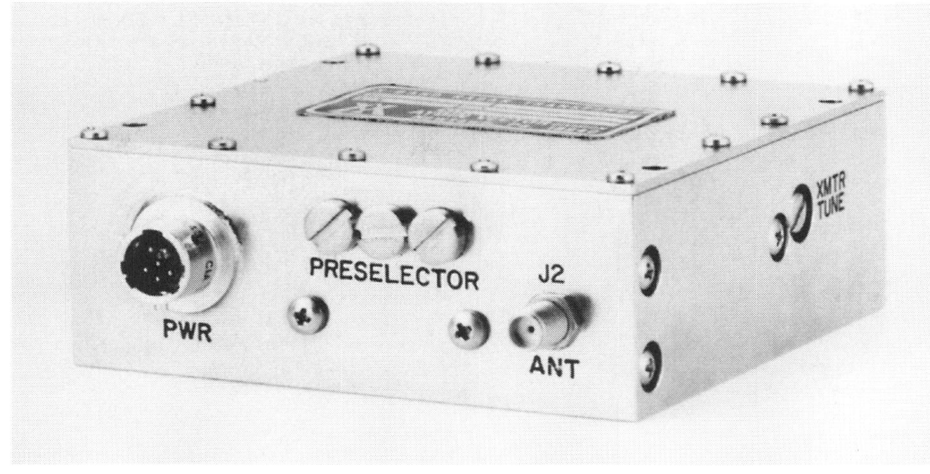

AV12X radar transponder
1997/0001213

SKYWATCH transmitter receiver computer, control display unit and low-profile directional antenna
1998/0018064

SKYWATCH™ traffic advisory system

SKYWATCH is an active surveillance system that operates as an air-to-air or ground-to-air interrogation device. It is derived from the TCAS 791 as a more affordable alternative to full TCAS systems for helicopter operators and general aviation aircraft.

When replies to SKYWATCH Mode-C type transponder interrogations are received, the responding aircraft's range, bearing, relative altitude and closure rate are computed to a fixed position and traffic conflicts are predicted. Visual targets are displayed using TCAS-like symbology, with aural traffic alerts.

The multifunction capability of the SKYWATCH system makes it possible to share a 3 ATI cathode ray tube display with late-model WX-1000 Stormscope weather mapping systems. SKYWATCH and Stormscope display functions are selected via a remote panel-mounted switching device. When operating in the Stormscope mode, the control display unit will temporarily switch to the SKYWATCH view if an intruder aircraft is detected which poses an immediate collision threat.

Specifications

Tracking capability: up to 30 targets
Display range: 2 and 6 n miles
Range accuracy: ±0.05 n miles (typical)
Bearing accuracy: 5° RMS (typical)
Altitude resolution: ±200 ft
TRC receiver/transmitter:
Dimensions: ARINC standard 404A ⅜ ATR short
Weight: 4.06 kg
Power required: 11-34 V DC
Multifunction control display unit:
(display) raster scan CRT
(resolution) 256 × 256 pixels
(dimensions) 3ATI × 209.3 mm
(weight) 1.03 kg
NY164 L-band (NATO D-band) directional antenna:
(dimensions) 279.4 × 158.8 × 35.6 mm
(weight) 1.04 kg

Operational status

In production.

Contractor

BFGoodrich Aerospace Avionics Systems.

UPDATED

Stormscope WX-900 weather mapping system

The Stormscope WX-900 weather mapping system maps electrical discharges and thunderstorm activity, clearly alerting crews to storms containing lightning.

Unlike radar, the WX-900 displays the electrical discharges associated with cumulus and mature and dissipating thunderstorms. It provides a full 360° view of weather or 180° during monitor modes, with pilot selectable ranges of 25, 50 and 100 n miles (46, 92 and 185 km). The Supertwist LCD is a self-contained panel-mount unit with electroluminescent back-lighting.

The WX-900 performs a self-test after turn-on, then performs tests continuously during system operation. The system features push-button selection of view, pilot initiated self-test programme, brightness adjustment, time mode which provides elapsed flight time and approach timer functions, battery monitor mode which monitors the aircraft electrical system and a noise analyser mode. An integral service menu includes strike test, spectrum analyser and board test modes.

Specifications

Dimensions:
(antenna) 254 × 87.6 × 167 mm
(display/processor) 86 × 86 × 192 mm
Weight:
(antenna) 0.42 kg
(display/processor) 0.71 kg
Power supply: 10.5-32 V DC, 8 W
Temperature range:
(display) 0 to +55°C
(antenna) −55 to +70°C
Altitude: up to 20,000 ft

Contractor

BFGoodrich Aerospace Avionics Systems.

VERIFIED

Stormscope® WX-950 weather mapping system

The Stormscope® WX-950 weather mapping system is the newest addition to the Stormscope® family. The WX-950 provides two modes of operation — cell mode and strike mode. It uses a high-resolution CRT within its 3 in ATI display.

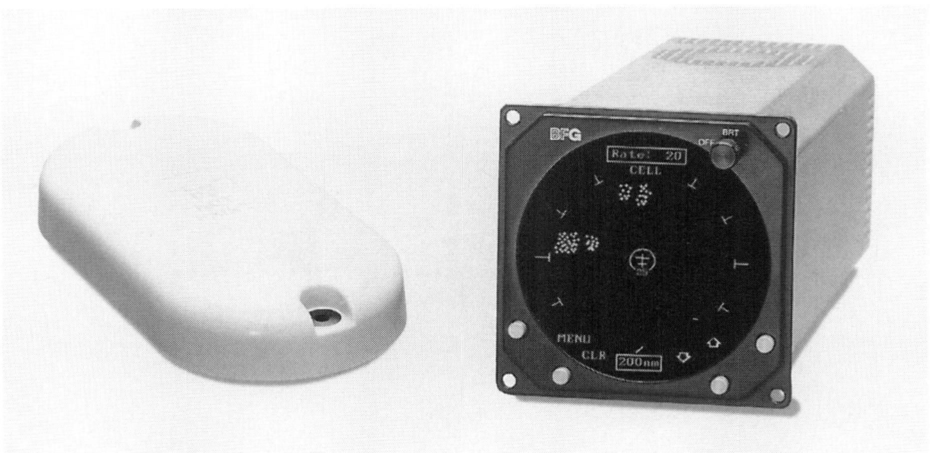

Stormscope WX-950 weather mapping system
1997/0001378

The WX-950 operates in 25, 50, 100 and 200 n mile ranges, displays a 360° view of electrical activity in all directions and a forward 120° view that doubles the resolution for analysing activity ahead.

When configured with a compatible heading system, the WX-950 provides heading stabilisation. A discharge rate indicator, integrity indicator and built-in self tests are other features of the unit.

Specifications

Dimensions:
(antenna) 25.4 × 87.6 × 174 mm
(display/processor) 3 ATI
Weight:
(antenna) 0.38 kg
(display/processor) 1.3 kg
Power supply: 11-32 V DC
Temperature range: -20 to +55°C
Altitude: 55,000 ft

Contractor

BFGoodrich Aerospace Avionics Systems.

VERIFIED

Stormscope WX-1000/WX-1000+/ WX-1000E weather mapping systems

The Stormscope WX-1000 and WX-1000+ series are weather mapping systems that map electrical discharges. Time and date information, stopwatch and timing information and checklists can be selected from the main menu. All units feature a continuous self-test with error messages that assure accurate weather detection.

The WX-1000 has pilot selectable ranges of 25, 50, 100 and 200 n miles (46, 92, 185 and 370 km), and provides either 360° or 120° viewing options. The high-resolution CRT display provides a sunlight-readable storm activity picture that can be used in tandem with conventional radar systems to determine the severity and stages of storms. Electrical discharge information is acquired and stored on all ranges simultaneously. Other features provide six programmable checklists, each containing a maximum of 30 lines with up to 20 characters per line. The WX-1000 can be upgraded to the WX-1000+.

The WX-1000+ is a heading stabilised WX-1000 which displays digital heading information in degrees when operating in the weather only modes and a flag advisory in the event of heading source malfunction. The WX-1000+ can be upgraded to the 'E' version with navaid option, to display navigational information from a Loran or GPS. The WX-1000+ is recognised by military services as the AN/AMS-2.

The WX-1000E with Navaid simultaneously displays thunderstorm information overlaid with a course line to 10 Loran or GPS-generated waypoints. Other features include a Loran/GPS-generated Course Deviation Indicator (CDI) and the display of six of 14 user-selectable Loran or GPS-generated flight parameters. Full flight plan within the selected range can be displayed.

The WX-1000E EFIS Interface allows thunderstorm activity to be displayed on a standard EFIS display. Thunderstorm data is transmitted via an ARINC 429 standard databus.

Stormscope Series II The WX-1000E with Navaid and WX-1000E EFIS Interface are also incorporated into the Stormscope Series II.

Specifications

Dimensions:
(processor) 86 × 124 × 322 mm
(display) 86 × 86 × 210 mm
(antenna) 29 × 114 × 256 mm
Weight:
(processor) 3.02 kg
(display) 1.03 kg
(antenna) 0.91 kg
Power supply: 10.5-32 V DC, 28 W
Temperature range:
(display) -20 to +70°C
(processor/antenna) -55 to +70°C

Contractor

BFGoodrich Aerospace Avionics Systems.

VERIFIED

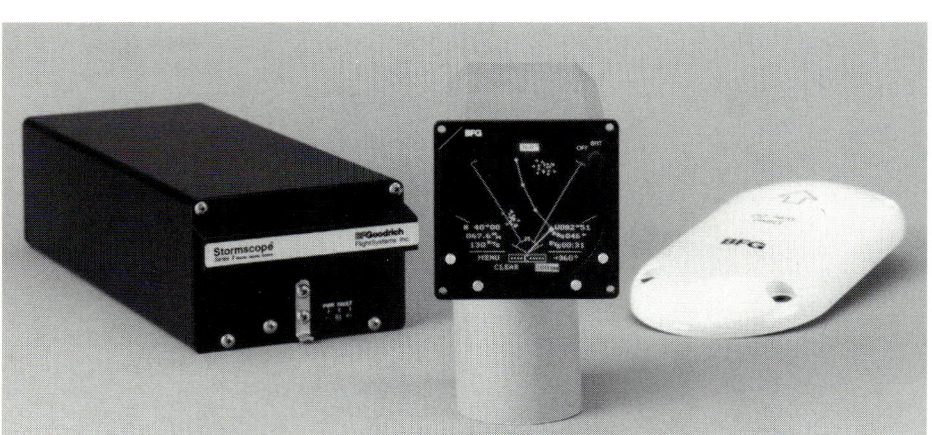

The Stormscope WX-1000 series consists of (left to right) the processor, display and antenna

Stormscope® WX-500 weather mapping system

The Stormscope® WX-500 weather mapping system is the newest addition to the Stormscope® family. The WX-500 has been designed to interface with the new generation of multifunction displays currently manufactured by Advanced Creations, ARNAV, Avidyne, Archangel, Eventide, and Skyforce Avionics. All sensor functions are controlled through the multifunction display.

The WX-500 with its advanced digital ranging algorithms combines the most popular features associated with the Series I and Series II systems to provide precision mapping of electrical discharges which are associated with thunderstorm activity.

There are two modes of operation – cell mode and strike mode. It operates in 25, 50, 100 and 200 n mile ranges, and displays a 360° view of electrical activity and a forward 120° view.

Heading stabilisation, a strike rate indicator, and continuous (and operator initiated) self-tests are other features of the unit.

Specifications

Dimensions:
antenna 25.4 × 87.6 × 174 mm
WX-500 processor w/o tray 130.8 × 44.6 × 228.6 mm

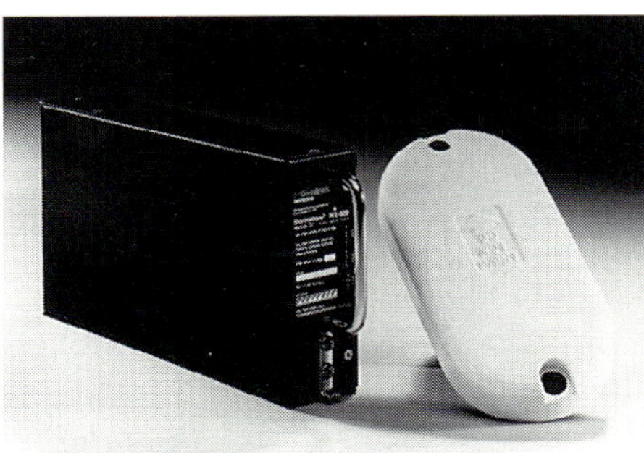

Stormscope® WX-500 weather mapping system
1998/0018063

Weight:
antenna 0.42 kg
WX-500 processor: 1.12 kg
Power supply: 11-32 V DC
Temperature range: -55 to +70°C
Altitude: 55,000 ft MSL

Contractor

BFGoodrich Aerospace Avionics Systems.

NEW ENTRY

TCAS 791 traffic alert and collision warning system

The TCAS 791 traffic alert and collision warning system consists of the TRC 791 receiver/transmitter, CD605 control display unit, NY156 L-band (NATO D-band) directional antenna and NY152 L-band (NATO D-band) omnidirectional antenna. It has been designed to operate as a TCAS 1 for regional airliners and other aircraft covered by the FAA mandate.

Working as an active air-to-air interrogation device, the TCAS 791 interrogates other airborne transponders in the surrounding airspace. It computes bearing, range, altitude and closure rates to plot traffic location and predict collision threats.

Utilising a combiner directional antenna, TCAS 791 determines bearing by a time difference across four poles. Distance and relative altitude is determined by comparing transponder Mode C to the intruder's transponder Mode C.

Using the dedicated CD605 3 in ATI enhanced visibility control/display unit (optional), traffic information is displayed utilising recognisable standard TCAS symbology. Traffic is displayed out to the horizontal range selected – 5, 10 or 20 n miles – and within ± 2,700 feet altitude relative to aircraft (normal mode). While the TCAS 791 tracks up to 35 intruder aircraft at a time, it automatically displays the eight highest-priority targets representing the greatest threats.

Should another aircraft present a collision threat, both visual and aural messages alert the pilot, allowing sufficient time to make visual contact with the threat aircraft and take appropriate action.

In addition to being able to select horizontal range displayed, TCAS 791 offers four relative altitude operational modes for "Look Up-Look Down" capability.

Ground Mode: Displays aircraft out to selectable ranges of 5 or 10 n miles and at altitudes up to 9,000 ft. This mode is particularly useful for checking potential traffic at uncontrolled airports.

Normal (NRM) Mode: Targets are displayed within ±2,700 ft altitude relative to the aircraft. This is the mode normally used in cruise.

Above (ABV) Mode: Targets are displayed up to 9,000 ft above and 2,700 ft below aircraft altitude. This mode would typically be set just prior to take-off to look for traffic during departure and climb-out.

Below (BLW) Mode: Targets are displayed up to 2,700 ft above and 9,000 ft below aircraft altitude. This mode is typically set prior to initiating a rapid descent from cruise altitude.

Specifications

Dimensions:
(TRC791 Receiver/transmitter) 193 × 157.5 × 345.4 mm
(CD605 control/display unit) 3 ATI short × 223.8 mm
(NY152 omnidirectional antenna) 57.1 × 47.2 × 68.1 mm
(NY156 directional antenna) 273.4 × 158.7 × 35.6 mm
Weight:
(TRC791 receiver/transmitter) 8.51 kg
(CD605 control/display unit) 1.36 kg
(NT152 omnidirectional antenna) 0.14 kg
(NT156 directional antenna) 1.04 kg
Power supply: 28 V DC, 95 W
Tracking: up to 35 targets
Range: 50 km nominal
Accuracy:
(range) ±0.1 km typical
(bearing) 5° RMS typical
(altitude) ±100 ft

Operational status

BFGoodrich claims that TCAS 791 was the first TCAS I to be TSO'd and the first to receive full unrestricted STC. STC approvals have been received for a large number of regional airline and corporate aircraft types.

Contractor

BFGoodrich Aerospace Avionics Systems.

VERIFIED

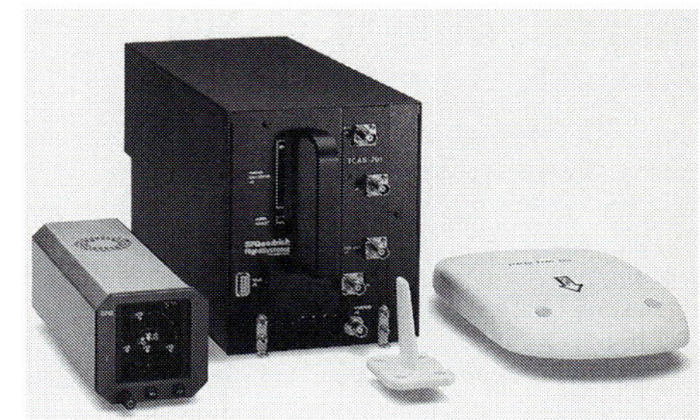

The TCAS 791 is designed for regional airliners
1995

Offensive avionics system for the B-1B

In October 1981, after a long evaluation of possible aircraft to carry cruise missiles, short-range attack missiles and nuclear and conventional weapons, the US government made the decision to resume development and production of the previously cancelled B-1 bomber. The B-1B programme called for the acquisition of 100 aircraft by the US Air Force Strategic Division Air Command (SAC) by June 1988.

Boeing was one of four associate contractors participating in the B-1B programme and was responsible for development and production or acquisition of the offensive avionics system and supporting elements of the defensive avionics system. Provision for growth to accept future mission requirements was also improved. The Conventional Mission Upgrade Programme (CMUP) is currently under way with the EMD phase authorised in 1995. This involves modifying a Weapons Interface Unit (WIU), developed originally for strategic applications, developing mission software and integrating new conventional weapons capability, including a MIL-STD-1760 interface for the incorporation of smart munitions in the weapon system.

The B-1B avionics are outgrowths of the original B-1 avionics developed, also by Boeing, for the update of the B-52. During the time between B-1 cancellation and initiation of the B-1B programme, extensive refinements were made in the systems, resulting in improved accuracy, reliability and overall performance. Provision for growth to accept future mission requirements was also improved.

The B-1B offensive avionics system can be divided into four interacting subsystems. These are: the computation complex subsystem; navigation subsystem; stores management subsystem; and controls and displays subsystem. These are interconnected with, and operate in conjunction with, aircraft systems originally provided by Rockwell International and the defensive system provided by AIL Systems. The Rockwell Aerospace and Defense segment was acquired by Boeing in December 1996. A brief description of each of the subsystems follows:

Computation complex subsystem

The computation complex is built around four dual-redundant MIL-STD-1553B databusses. These are at present connected to four avionics computer control units and a mass storage unit supplied by IBM and to two data transfer units from Sundstrand. The databusses also connect to various units of the other avionics subsystems. Together these units perform the functions of data/instruction loading, transfer, storage, distribution and processing. In 1997, Boeing was authorised to proceed with upgrading the computer complex. The computer units and mass storage device will be replaced with new systems developed by Lockheed Martin Federal Systems. The software will be converted from Jovial to Ada. Nearly all the avionics functions are performed either through avionics flight software and mission data loaded into the ACCs or by operator actions processed in the ACCs.

Navigation subsystem

The navigation subsystem enables the B-1B to navigate and penetrate to predesignated target areas, to align and deliver a variety of nuclear and non-nuclear weapons and to exit from the target area and return to a designated landing site.

The major components of the navigation subsystem are the inertial navigation units provided by Kearfott, a dual-channel radar system from Northrop Grumman ESSD and a Doppler velocity sensor from Litton Guidance and Control Systems. A radar altimeter manufactured by Honeywell provides height above the terrain at altitudes between 0 and 5,000 ft over both land and water.

Navigation data is processed in the computation complex computers and provided to other avionics subsystems for guidance, control and weapon delivery functions. Two IBM avionics computer controls, identical to those in the computation complex, are presently used to compute navigation and radar data for terrain-following functions. These computers will also be replaced with two Lockheed Martin Federal Systems computers during the computer upgrade.

Stores management subsystem

The B-1B is designed to carry and deliver both conventional and nuclear weapons of numerous types currently in the air force inventory or under development. It is capable of being adapted for new weapon types as they become available. Weapons are carried in either a two- or three-weapon bay configuration.

The major components of the stores management subsystem are the Weapon Interface Units (WIU). These are provided in several types and are employed according to the aircraft configuration and the weapon type to be carried. They provide the interface between the various weapons and aircraft and avionics systems. All power, control, status and alignment functions pass through the WIUs which are designed and manufactured by Boeing.

Controls and displays subsystem

The controls and displays subsystem provides the aircraft flight crew interface with the avionics system. Control and display components in the form of control panels and CRT displays are located at both the offensive and defensive system operator stations.

Control of both the offensive and defensive avionics is achieved via 12 Boeing-designed and manufactured control panel types and two tracking handles. Display is on three MultiFunction Display (MFD) units, two electronic display units and on radar target indicators. The MFDs are driven by a display electronics unit.

Operational status

In continuing development and in operational use. Systems for all 100 aircraft were delivered before the end of the first half of 1988.

Currently, Boeing is responsible for CMUP with the mission avionics being one component of the improvements to survivability, lethality and maintainability. This US Air Force programme adds improved weapons accuracy by integrating the Global Positioning System (GPS) and Joint Direct Attack Munition (JDAM), cluster bomb units, Wind-Corrected Munitions Dispenser (WCMD), and other weapons into the B-1B.

A modified BAC 1-11 continues to be used for additional flight development testing of offensive and defensive upgrades to the B-1B.

Contractor

The Boeing Company.

UPDATED

GCAS: Ground Collision Avoidance Systems

Cubic Defense Systems, Inc was awarded a US$1.34 million contract to manufacture 29 C-130 Ground Collision Avoidance Systems (GCAS) for the Canadian Forces – Air command C-130 Update Program.

Systems were delivered during 1996-97 to CAE Aviation Ltd of Edmonton, Canada for integration into the aircraft.

The Cubic GCAS system provides an aural and visual warning to aircrews of impending Controlled Flight Into Terrain (CFIT).

Cubic uses a dynamic, predictive algorithm which performs real-time calculations of the state of aircraft motion vectors. This algorithm is tailored to pilot reaction time, aircraft performance capabilities and avionics suite.

Operational status

Cubic recently completed testing of similar units on the CH-53 aircraft for the US Navy at the Patuxent River Naval Air Test Center. The follow-on for this developmental programme will be a production/installation programme for selected Navy/Marine helicopters.

Contractor

Cubic Corporation.

VERIFIED

GPWS: Ground Proximity Warning System

Adaptable to all classes of rotary-wing aircraft, GPWS can be installed as a ¼ ATR (short) avionics package interfaced with discrete sensors, or via the ARINC 429/MIL-STD-1553B databusses. It can also be embedded in computers already on board the aircraft.

The GPWS algorithm continually assesses the validity of input data and is optimised for highly dynamic tactical flight environments.

Contractor

Cubic Corporation.

VERIFIED

AN/AYD-23 GPWS: Ground Proximity Warning System

The AN/AYD-23 GPWS is designed specifically for rotary-wing and tactical aircraft. Packaged in a standard ¼ ATR avionics package, the AYD-23 has discrete, ARINC-429 and MIL-STD-1553B aircraft interfaces. The system uses the Intel i960, 32 bit processor and calculates aircraft flight parameters 10 times per second.

The key to the AYD-23 GPWS advanced capabilities is the dynamic predictive algorithm that provides highly accurate calculations of the aircraft's flight parameters with regard to flight performance and attitude with terrain. Based on these real-time calculations, the AYD-23 GPWS prevents inadvertent collisions with the ground.

The GPWS interfaces with existing sensors, including: radar altimeter, barometric altimeter, attitude gyros, accelerometers, and air data sensors, to predict amongst other calculations: altitude loss caused by pilot response, altitute loss caused by roll recovery, altitude loss during actual recovery, and altitude loss caused by terrain profile.

Contractor

Cubic Defense Systems.

UPDATED

AN/AYD-23 ground proximity warning system ***1998***/0018926

GTX 320 IFF transponder

The Garmin GTX 320 is a compact, panel-mounted, solid-state, 200 W, Class 1A transponder. It fits into existing installations, as a replacement upgrade for earlier-generation cavity tube transponders, to provide improved reliability and to eliminate warm-up time.

Specifications

Transmitter power: 200 W
Weight: 0.95 kg
Power: 11-33 V DC, 12 W
TSO compliance: C74c Class 1A
Mode A: 4,986 codes
Mode C: 100 ft increments −1,000 to +63,000 ft

Contractor

Garmin International Inc.

NEW ENTRY

Garmin GTX 320 IFF transponder ***1998***/0018067

AN/APX-111 Combined Interrogator/Transponder (CIT)

The AN/APX-111 CIT consists of an interrogator, a transponder and two associated cryptographic computers in a single, small, lightweight unit. For retrofit applications the CIT can replace six or seven boxes making up the existing interrogator and transponder, yielding reduced weight and freeing space for other avionics equipment.

The AN/APX-111's architecture enables it to support Mk X (SIF), Mk XII (Mode 4) or custom crypto systems. The crypto module is integral but removable from the front panel. Mode S transponder capability and growth to Mk XV (NIS) are also part of the system architecture.

The CIT meets international standards for IFF and ATC including US-DoD AIMS 65-1000B and NATO STANAG 4193. AN/APX-111 utilises miniature low-profile fuselage-mounted electronically scanned antenna arrays. It provides full interrogator range capability without the need for external amplifiers. The modular design is all solid-state with a unique approach to thermally efficient cooling. A MIL-STD-1750 processor and 1553 databus are included.

The latest IFF techniques are provided, including digital target reports for a clear operator display, monopulse processing for accurate target azimuth, a statistical reply evaluator for high-confidence identification, a defruiter for dense environments and cryptographic coding for security. Both continuous and operator-initiated built-in tests are provided, with reporting via the 1553 databus.

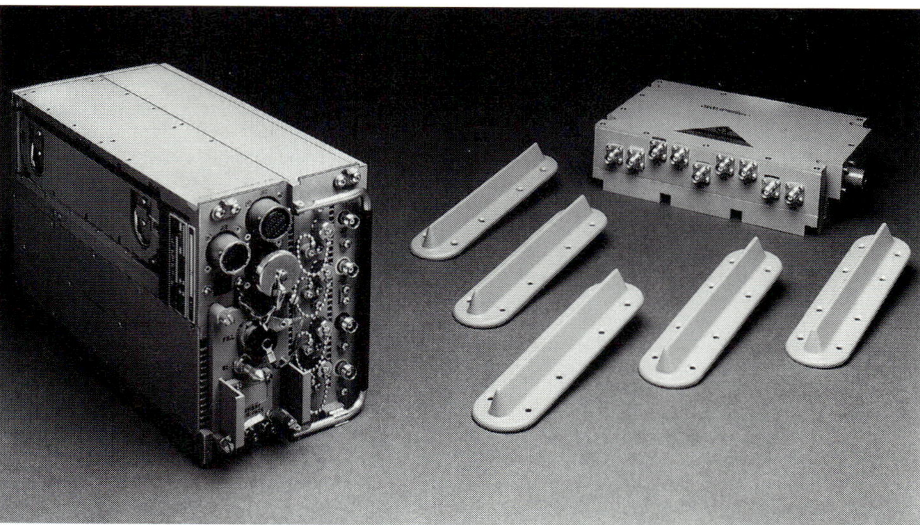

AN/APX-111 combined interrogator/transponder ***1997***/0002220

Specifications

Dimensions: 207.5 × 146 × 320 mm
Weight: 13.6 kg

Operational status

Fitted in F/A-18 FMS and F-16 MLU programmes, and Greek and Turkish air force F-16 aircraft.

Contractor

GEC-Marconi Hazeltine Corporation.

VERIFIED

The AN/APX-113(V) showing the combined interrogator/transponder (left), the lower interrogator antenna (centre, rear), the fuselage-mounted antenna elements (centre, front) and the beam-forming network (right) ***1995***

AN/APX-113(V) Combined Interrogator/Transponder (CIT)

The AN/APX-113(V) Combined Interrogator/Transponder (CIT) is a complete Mk XII identification system which includes crypto computers. It consists of one unit and incorporates growth for the next generation of IFF and combat aircraft identification equipment. The AN/APX-113(V) provides both interrogation and IFF responses on Modes 1, 2, 3A, C and 4, as well as incorporating Mode S Level 3.

The multiple antenna configurations feature electronic or mechanical scan. The system features Ada software and a MIL-STD-1553 bus interface.

Specifications

Dimensions:
(combined interrogator/transponder) 209.8 × 152.4 × 368.3 mm
(beam-forming network) 165.1 × 212.9 × 101.6 mm
(fuselage-mounted antenna elements) 39.4 × 82.6 × 332.7 mm
(lower interrogator antenna) 15.2 × 431.8 × 355.6 m
Weight:
(combined interrogator/transponder) 14.52 kg
(beam-forming network) 4.54 kg
(fuselage-mounted antenna elements) 0.23 kg

Power supply: 28 V DC, 200 W
Range: 185 km
Coverage:
(azimuth) ±60°
(elevation) ±60°
Accuracy:
(range) 500 ft
(azimuth) ±2°
In-beam targets: 32
Reliability: 1,600 h MTBF

Operational status

Developed specifically for F-16 Falcon. Fitted to F-16 Block 16 A/B MLU, Block 20 A/B, and Block 50 C/D aircraft. Also fitted to ASW/surveillance helicopters and the Japanese FS-X fighter.

Contractor

GEC-Marconi Hazeltine Corporation.

UPDATED

GRA-2000 Low Probability of Intercept (LPI) altimeter

The GRA-2000 LPI altimeter, has been selected by the US Joint Services Program Office to replace the AN/APN-194, -171, -209 and -232 series altimeters on the majority of tactical jet, helicopter and transport aircraft employed by the US Department of Defense.

The design is based on a very simple durable design employing a single I/F downconvert and specialised algorithms to provide exceptional stealth and jam resistance; it is an outgrowth of the AD-1990 SARA system being procured for Tornado aircraft in the UK.

Specifications

Dimensions: 185.4 × 97.3 × 77.5 mm
Weight: < 3.2 kg
Reliability: 8,000+ h
Accuracy: ±2 ft or 2% (1σ)
Performance: 0 to 35,000 ft AGL

Operational status

In development against the F-18E/F AN/APN-194 configuration. Options to provide additional units to the other configurations are expected within 18 months.

Contractor

GEC-Marconi Hazeltine Corporation.

VERIFIED

GRA-2000 LPI altimeter
1997/0002221

AA-300 radio altimeter

The AA-300 radio altimeter consists of RA-315 and RA-335 indicators, RT-300 transmitter/receiver and AT-220, -221 or -222 antenna.

The RT-300 transmitter/receiver is a solid-state unit offered in three optional configurations for different outputs.

The RA-315 indicator has a servo-controlled pointer display of radio altitude up to 2,500 ft. Below 500 ft the scale is expanded to enhance readability. There is an adjustable decision height bug and an amber decision height warning lamp. The RA-335 is similar to the RA-315 but is configured for helicopters, having a range of 0 to 1,500 ft. Below 200 ft the scale is expanded to improve readability.

Specifications

Dimensions:
(RA-315 and -335) 3 ATI × 1,143 mm
(RT-300) 104 × 116 × 281 mm
(AT-220) 63 × 159 × 142 mm
Weight:
(RA-315 and -335) 0.7 kg
(RT-300) 2 kg
(AT-220) 0.3 kg
Power supply: 21-32 V DC, 0.5 A
Accuracy:
(RT-300) 0-100 ft ±3 ft, 100-500 ft ±3%, 500-2,500 ft ±4%
(RA-315) 0-100 ft ±5 ft, 100-500 ft ±5%, 500-2,500 ft ±7%
(RA-335) 0-100 ft ±5 ft, 100-500 ft ±5%, 500-1,500 ft ±7%

Operational status

In production.

Contractor

Honeywell Inc Business & Commuter Aviation Systems.

VERIFIED

RCZ-852 diversity Mode-S transponder

The Honeywell RCZ-852 diversity Mode S transponder offers a small, light package that is optimised for corporate aircraft and regional airline applications. The RCZ-852 implements all currently defined Mode-S functions with provision for future growth. Current Mode-S transponders are used in conjunction with TCAS and ATCRBS to identify and track aircraft position, including altitude. This system transmits and receives digital messages between aircraft and air traffic control. The datalink provides positive and confirmed communications more efficiently than current voice systems.

The Honeywell design meets future needs by including growth capability to support the functions defined by CNS/ATM (Communications, Navigation, Surveillance/Air Traffic Management).

The Traffic Alert and Collision System (TCAS) is fully supported, including 'diversity' (top and bottom) antenna ports. Diversity provides reliable RF communication links between both ground-based and airborne interrogators. The transponder incorporates a TCAS II interface and is designed to be compatible with all TCAS II systems conforming to ARINC 718/735 characteristics.

The RCZ-852 transponder is an ICAO Level 3 system with growth to Level 4. Level 3 means that it will transmit and receive standard length (112 bit) datalink messages for 'COMM A' and 'COMM B' and receive 16-segment extended length datalink messages for 'COMM C'.

Honeywell has included full Built-in Test Equipment (BITE) and self-test capabilities to provide maximum reliability whilst minimising maintenance costs. The BITE system separates aircraft installation and aircraft system failures external to the transponder minimising the time required to return removed units to service.

Specifications

Dimensions: 107 × 84 × 318 mm
Weight: 2.3 kg
Power: 28 V DC

Contractor

Honeywell Inc Business & Commuter Aviation Systems.

NEW ENTRY

XS-950 Mode S ATDL (Air Transport DataLink) transponder

The Honeywell XS-950 transponder was designed for the air transport market and meets all ARINC 718 requirements. The XS-950 implements all currently defined Mode S functions with provision for future growth. Current Mode S transponders are used in conjunction with TCAS and ATCRBS to identify and track aircraft position, including altitude. This system transmits and receives digital messages between aircraft and air traffic control. The datalink provides positive and confirmed communications more efficiently than current voice systems.

The Honeywell design meets future needs by including growth capability to support the functions defined by CNS/ATM (Communications, Navigation, Surveillance/Air Traffic Management).

The Traffic Alert and Collision Avoidance System (TCAS) is fully supported, including 'diversity' (top and bottom) antenna ports. Diversity provides reliable RF communication links between both ground-based and airborne interrogators. The transponder incorporates a TCAS II interface and is designed to be compatible with all TCAS II systems conforming to ARINC 718/735 characteristics.

The Mode-S ATDL transponder is an ICAO Level 4 system with growth to Level 5. Level 4 means that it will transmit and receive standard length (112 bit) datalink messages for 'COMM A' and 'COMM B' and transmit and receive 16-segment extended length datalink messages for 'COMM C' and 'COMM D'.

Honeywell has included full Built-in Test Equipment (BITE) and self-test capabilities to provide maximum reliability whilst minimising maintenance costs. The XS-950 interfaces to all air transport OEM onboard maintenance systems.

Specifications

Dimensions: 124 × 194 × 325 mm
Weight: 5.7 kg
Power: 115 V AC or 28 V DC

Contractor

Honeywell Inc Business & Commuter Aviation Systems.

NEW ENTRY

Primus 700/701 Series surface mapping, beacon and colour weather radar

A three-box system including receiver/transmitter, indicator and antenna, the Primus 700/701 is compatible with certain Honeywell EFIS. Its powerful 10 kW magnetron transmitter has six pulsewidths, seven bandwidths and four PRFs for maximum performance on all ranges in all modes. Ten selectable range scales from 0.5 to 300 n miles (1 to 556 km) provide optimal range scales in every condition.

Five antenna sizes from 10 to 24 in make the Primus 700/701 system suitable for any airframe.

The Primus 700/701 dual EFIS interface capability and antenna sweep time-sharing in effect makes two radars available to the crew throughout the flight. Each crew member can select their own range, mode, gain and tilt display on their EHSI.

With the radar indicator, pilots will have the Honeywell features of a variable range mark and azimuth cursor with digital distance and bearing readouts. A new menu function allows the indicator switches to do double duty, controlling infrequently used features such as heading display on/off. Without a radar indicator installed, the system includes one or more WC-700 radar controllers.

Primus 700/701 weather radar features include a four-colour display of rainfall intensity. On ranges of 50 n miles (93 km) or less, turbulence detection shows areas where there is moderate or stronger levels of turbulence.

The Honeywell Rain Echo Attenuation Compensation Technique (REACT) safety feature performs three distinct functions. First, it maintains target calibration by compensating for attenuation caused by intervening rainfall. Returns remain properly calibrated for the storm behind the storm. Second, REACT advises pilots of areas where target calibration cannot be maintained even with maximum compensation. For those areas, REACT changes the screen background to blue, warning that calibration is no longer possible and attenuation may be hiding areas of severe weather. Third, any target displayed in the blue field will appear in magenta to alert the pilot of its probable severity.

The system also includes Honeywell weather radar features of ground clutter reduction and target alert.

The radar indicator interfaces with Honeywell's LSZ-860 lightning sensor system and Data Nav for complete severe weather avoidance, navigation and checklist capability.

High-resolution and high-sensitivity mapping modes include three display colours which differ from those used for the weather display and pilot-selectable sea clutter reduction. Range scales of 0.5, 1, 2.5, 5, 10, 25, 50, 100, 200 and 300 n miles are available and the shortest range scale provides a resolution of 55 ft.

The Primus 701 includes a beacon capability which makes low-visibility approaches possible where standard navaids may not be available. It also allows air-to-air rendezvous. The Primus 701 operates in radar only, beacon only or both beacon and radar modes. Beacon targets are shown in contrasting colours in weather- and ground-mapping modes.

Primus 700/701 BITE includes comprehensive and continuous internal fault monitoring and a menu function which allows access to monitor pages. A non-volatile memory records internal fault data for later retrieval by maintenance technicians.

Operational status

In production.

Contractor

Honeywell Inc Business & Commuter Aviation Systems.

UPDATED

Primus 880, 660 and 440 weather radars

The Primus 660 and 880 weather radars are high-power (10 kW) successors to the Primus 650 and 870 systems, respectively. The Primus 440 is designed as a powerful, reliable weather radar for light-class business aircraft.

All three systems have a stabilised antenna-up to 24 in for the Primus 880 and are packaged in a Transmitter/Receiver/Antenna (TRA) architecture that weighs 6.36 kg. Each system is compatible with Honeywell's LSZ-860 Lightning Sensor System and may be displayed on either the Electronic Flight Instrument System (EFIS) or on a dedicated weather radar indicator.

The Primus 880 features Doppler turbulence detection pulse pair processing that detects spectrum spreading caused by turbulence within any storm cell, regardless of rainfall rates. Once detected, turbulent areas are displayed in white on all ranges up to 50 n miles, allowing pilots to safely manoeuvre around potentially hazardous weather.

For the first time on a Primus radar system, Primus 880 also features Built-In Test Equipment (BITE) on two of the most important components of the system, the transmitter and receiver, providing a complete RF loop-back which continuously tests the transmitter power and receiver sensitivity and reports any faults to the pilot.

Other features of the Primus 880 include: Honeywell's exclusive Rain Echo Attenuation Compensation Technique (REACT) which alerts pilots to storms hidden behind other storms; Target Alert (TA) which notifies pilots of potentially hazardous weather directly in front of the aircraft; Altitude Compensated Tilt (ACT), which allows detection of weather that may affect the aircraft en route and reduces the amount of tilt management performed by the pilot; Ground Mapping (GM), which serves as a navigation aid by depicting terrain features not available in this clarity and detail with lower-power radars.

Honeywell's optional LSZ-860 Lightning Sensor System overlays lightning information onto the precipitation/turbulence display to provide a very powerful severe weather detection capability. The LSS accurately displays the position and lightning rate of up to 50 storm cells at the same time.

Operational status

In production.

Contractor

Honeywell Inc Business & Commuter Aviation Systems.

UPDATED

TCAS II (TCAS 2000) Traffic Alert and Collision Avoidance System

TCAS 2000 is Honeywell's latest TCAS II system. Compared with the earlier TCAS II, it is: smaller and lighter, offers double the range, and increases the computer capacity by 350 per cent. TCAS 2000 generates advisory information on targets up to 160 km away and can provide this information to other TCAS II-equipped aircraft to co-ordinate manoeuvres.

TCAS 2000 provides for standard TCAS II surveillance up to 32 km for ATCRBS-(Mode A/C) (Air Traffic Control Radar Beacon System) equipped aircraft, and up to 64 km for Mode S-equipped aircraft. As an option, TCAS 2000 can provide for extended range suveillance of up to 160 km for Mode S-equipped aircraft. TCAS 2000 is designed to handle closure rates of up to 1,200 kt and vertical rates of 10,000 ft/min. TCAS 2000 computes range, relative altitude, and bearing of nearby transponder-equipped aircraft and visually and aurally alerts pilots of potential collisions, recommending the least disruptive vertical manoeuvre for safe separation. Warning of potential collisions occur at least 20 to 30 seconds before predicted convergence, with more warning at higher altitudes.

The Honeywell TCAS 2000 consists of a computer unit, Mode S transponder, control panel, resolution and traffic advisory displays, and antennas.

The computer unit performs airspace suveillance, intruder tracking, traffic display, threat assessment, collision threat resolution and TCAS co-ordination. It uses data from airframe and other systems to change performance parameters for varying altitudes and aircraft configurations. Collision avoidance algorithms supplied by the FAA are used to determine whether a track aircraft is a threat and, if so, the best avoidance manoeuvre.

The Mode S transponder is specially designed for the air traffic control systems of the 1990s. It performs the functions of existing Mode A and Mode C transponders and provides data exchange between TCAS-equipped aircraft. It also communicates with ground-based Mode S sensors which set TCAS sensitivity levels based on traffic density. The transponder can transmit and receive on either the top or the bottom aerial to optimise signal strength and reduce interference.

The control panel selects and controls all TCAS elements including the computer, Mode S transponder, displays and conventional ATCRBS or second Mode S transponder. It includes a transponder failure lamp and four-character LED display for transponder codes which are set with concentric rotary switches. A variety of displays may be used for TCAS information. The Traffic Advisory (TA) and Resolution Advisory (RA) may be displayed on a colour flat panel display which integrates vertical speed indication (VSI/TRA). The display is packaged in a 3 ATI-sized indicator. The TCAS also interfaces with EFIS systems to display traffic and resolution advisory information in an integrated display format.

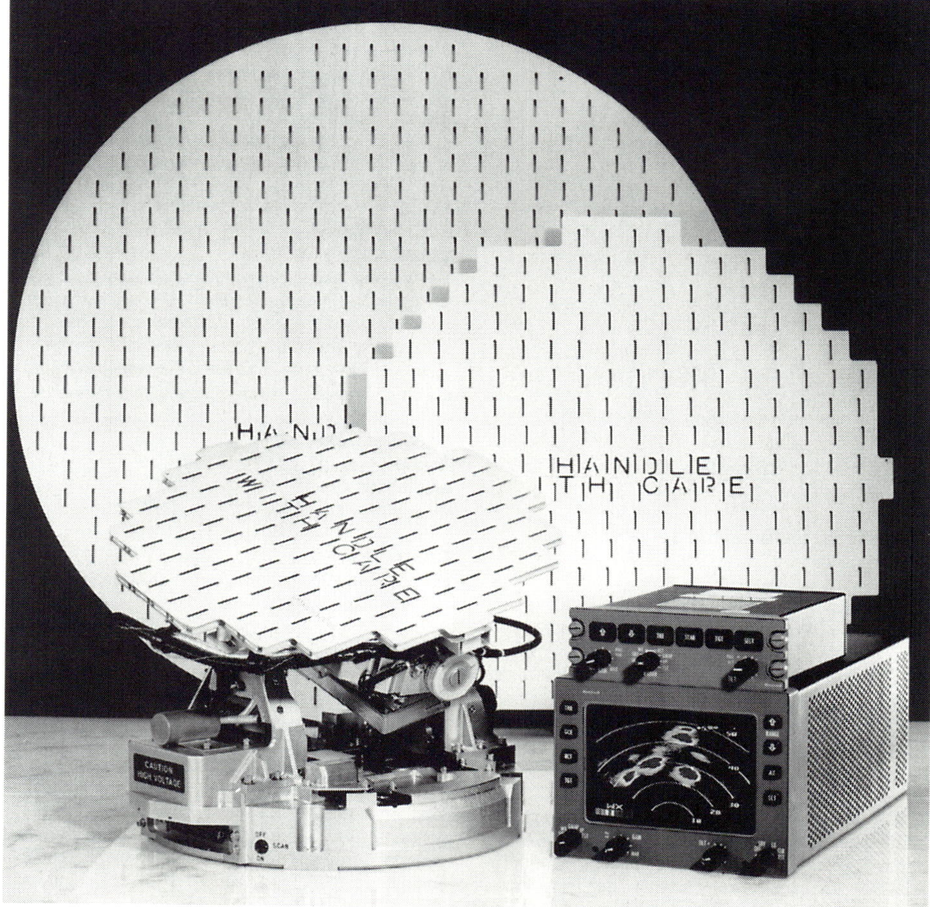

Primus 880 weather radar

1997/0001215

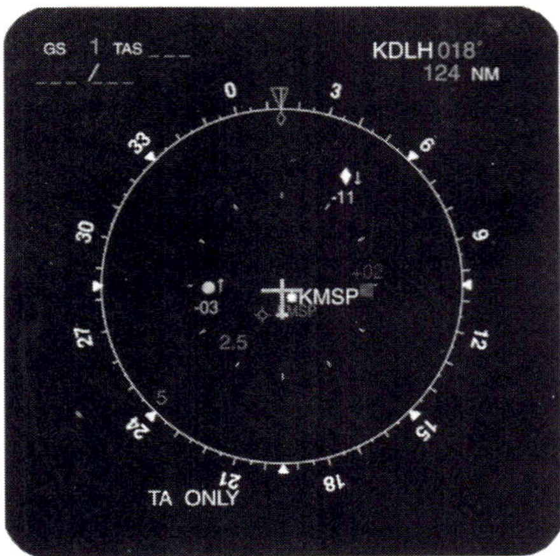

Honeywell TCAS 2000 displays: typical EFIS (above), and VSI/TRA (below) ***1998***/0018069

Honeywell RT-950 TCAS computer unit (6-MCU) (left) and RT-951 TCAS computer unit (4-MCU) (right) ***1998***/0018068

The colours used in TCAS are amber for alert, red for resolution advisory, and blue for non-hazardous traffic. The directional antenna features electronic sidelobe suppression and amplitude ratio tracking. The low-profile four-element antennas are mounted on the top and bottom of the fuselage and are capable of transmitting in four selectable directions and receiving omnidirectionally. The antenna transmits at 1,030 MHz and receives at 1,090 MHz.

TCAS 2000 is available in two sizes to meet most upgrade needs, as well as to forward fit a wide variety of aircraft. It is available in both 6-MCU (RT-950) and 4-MCU (RT-951) packages. Both versions offer 28 V DC power connections, whilst the 6-MCU version also offers a 115 V AC connection. The 6-MCU version is fully compatible in form, fit and function with current Honeywell TCAS II installations, while the 4-MCU version only requires the smaller tray for compatibility.

Specifications

Max range: 80 n miles to meet future Communications, Navigation Surveillance/Air Traffic Management (CNS/ATM) requirements
Display ranges: 5, 10, 20, 40 and 80 n miles
Tracks: 50 aircraft tracks (24 within 5 n miles)
Closing speed: 1,200 kt max
Vertical rate: 10,000 ft/min max
Normal escape manoeuvres: climb or descend rates; vertical speed limits
Enhanced escape manoeuvres: increased climb or descend; reversed direction of climb or descend

Operational status

Certified for aircraft such as the Gulfstream IV, Raytheon Hawker 800 and 1000, and Canadair CL-601-3A. Selected by British Airways for its European fleet of aircraft; also be Quantas and Saudia. Deliveries began in 1997.

Contractor

Honeywell Inc Business & Commuter Aviation Systems.

NEW ENTRY

AN/APN-59E(V) Search Radar

The APN-59E(V) was developed in the late 1970s to replace the earlier AN/APN-59B with improved performance and reliability for a variety of retrofits. The I/J-band radar was tailored to a closely defined mission profile with unusually stringent quality assurance demands. For example, component selection was made on the basis of a number of engineering and databank recommendations such as the Government/Industry Data Exchange Programme and the resulting system has been verified by rigorous testing to Advisory Group on Reliability of Electronic Equipment (AGREE) type testing. MTBF is given as 219 hours.

The principal modes are search, navigation, weather-mapping and beacon homing. To accommodate all these the operator can choose pencil or fan beam with a variety of pulse lengths and repetition rates; the system can be set up for angle sector or 360° scan.

All LRUs are interchangeable with those of the AN/APN-59B so that separate stocks of spares are not needed for flight line support and gradual upgrading of a system can be accomplished over a period of time and without aircraft stand-down.

Alternative configurations range from single azimuth/range displays, driven by the radar as an independent system, to more complex installations with up to three displays. Where requirements are particularly critical the system can be connected to a compass and a dead-reckoning computer for the most accurate navigation fixes. The weight of a typical configuration is about 84 kg.

Operational status

Retrofitted to C-130, C/KC-135 and RC-135 fleets, and C-130 fleets of various air forces.

This radar is now being replaced on US Air National Guard C/KC-135 aircraft by the Rockwell Collins FMR-200X radar as part of the PACER CRAG upgrade programme, due to be completed by the year 2000.

Contractor

Honeywell Inc Sensor and Guidance Products.

UPDATED

AN/APN-171 radar altimeter

For over 25 years, US Navy helicopters have been flying with the APN-171 (HG9000) as their standard altimeter. Many functions are available with this system including low-altitude warning, radar altitude warning set input, aircraft rate, aircraft altitude errors and landing gear warnings.

Three basic Honeywell AN/APN-171 radar altimeter systems are currently available: the HG9010 is a 0 to 1,000 ft system, the HG9025 is a 0 to 2,500 ft system and the HG9050 is a 0 to 5,000 ft system. All AN/APN-171 systems are available with standard or special output signals to represent a particular altitude range.

Specifications

Dimensions: 337 × 194 × 125 mm
Weight: 6.1 kg
Power supply: 115 V AC, 400 Hz, 10 VA
Altitude:
(HG9010) 0-1,000 ft
(HG9025) 0-2,500 ft
(HG9050) 0-5,000 ft
Accuracy: ±5 ft ±3%

Operational status

In production and service as standard on all US Navy helicopters. A modification kit is available to upgrade existing systems.

Contractor

Honeywell Inc Sensor and Guidance Products.

UPDATED

AN/APN-194 radar altimeter

The AN/APN-194 radar altimeter is standard on all navy fixed-wing and high-performance aircraft including the F-14 and F/A-18. Functions include low-altitude warning, radar altitude warning set input, aircraft rate, aircraft altitude errors, landing gear warnings and both analogue and digital outputs.

Several height indicators interface with the APN-194. The AN/APN-194 was introduced in the early 1970s as a form, fit and function replacement for the AN/APN-141 and Honeywell has now produced over 8,000 of these altimeters. Since then, the AN/APN-194 has been upgraded twice, incorporating a solid-state transmitter and producibility enhancements.

Options include transmitter power management. The AN/APN-194 can interface with up to four height indicators.

Specifications

Dimensions: 185 × 97 × 82 mm
Weight:
(radar altimeter) 2 kg
(height indicator) 0.7 kg
Power supply: 115 V AC, 400 Hz, 25 W
Frequency: 4,400 MHz
Transmit power: 5 W
PRF: 20 kHz
Pulsewidth: 0.02 or 0.20 ms
Altitude: 0-5,000 ft
Accuracy: ±3 ft ±4%

Operational status

In production and in service as standard on all US Navy fixed-wing high-performance aircraft.

Contractor

Honeywell Inc Sensor and Guidance Products.

UPDATED

AN/APN-209 radar altimeter

The AN/APN-209 radar altimeter is standard on all US Army helicopters. Functions include transmitter power management, low- and high-altitude warnings, NVG-compatibility, analogue and digital outputs and integration of indicator, receiver and transmitter. For more installation flexibility, a version of the AN/APN-209 with the transmitter/receiver separate from the indicator is available.

Honeywell has produced over 9,000 AN/APN-209 altimeters. It was introduced in the early 1970s and, since then, has been upgraded twice, incorporating a solid-state transmitter and producibility enhancements.

Options include MIL-STD-1553B databus, voice warning and LPI enhancements.

Specifications
Dimensions:
(receiver/transmitter) 145 × 83 × 83 mm
(indicator/receiver/transmitter) 199 × 83 × 83 mm
Weight:
(receiver/transmitter) 1.4 kg
(indicator/receiver/transmitter) 1.9 kg
Power supply: 28 V DC, 25 W
Altitude: 0-1,500 ft
Accuracy: ±3 ft ±3%

Operational status
In production and in service as standard on all US Army helicopters.

Contractor
Honeywell Inc Sensor and Guidance Products.

UPDATED

AN/APN-224 radar altimeter

The APN-224 was developed specifically for Strategic Air Command's Boeing B-52 and meets the nuclear hardening and high-reliability specifications of that aircraft. The system's performance and ability to withstand severe environments led to its selection by the US Air Force for the B-1B. The APN-224 has also been selected for the US Air Force's A-10 and the US Air National Guard's F-16.

Functions of the AN/APN-224 include nuclear hardness, aircraft rate, blanking pulse and both digital and analogue outputs.

A modified version of the AN/APN-224, which has a MIL-STD-1553B databus, has been developed for the F-16 LANTIRN aircraft.

Specifications
Dimensions:
(B-52) 213 × 127 × 86 mm
(F-16) 213 × 160 × 102 mm
Weight:
(B-52) 2.1 kg
(F-16) 3.6 kg
Power supply: 115 V AC, 400 Hz, 45 VA
Altitude:
(B-52) 0-5,000 ft
(F-16) 0-10,000 ft
Accuracy: ±5 ft ±4%

Operational status
In service on US Air Force/US Air National Guard B-52, B-1B and F-16 LANTIRN aircraft.

Contractor
Honeywell Inc Sensor and Guidance Products.

UPDATED

HG7170 radar altimeter

The HG7170 radar altimeter is a dual-redundant altimeter system. Functions of the HG7170 include MIL-STD-1553B databus, aircraft rate, transmitter power management, full BIT monitoring, continuous BIT, isolated grounding, nuclear event detector and nuclear hardness.

Specifications
Dimensions: 338 × 193 × 155 mm
Weight: 5.7 kg
Power supply: 28 V DC, 38 W
Altitude: 0-5,000 ft
Accuracy: ±2 ft ±4%

Contractor
Honeywell Inc Sensor and Guidance Products.

UPDATED

HG7500/HG8500 Series radar altimeters

HG7500 configurations available include analogue and/or digital altitude, altitude trips and ARINC 552A. The JG107X height indicator interfaces with the HG7500 and HG8500.

The HG8500 is a form, fit and function replacement for the HG7500. It has a solid-state transmitter and gallium arsenide receiver, and has been qualified to very stringent environmental and EMI requirements.

Options available on the HG8500 include transmitter power management, low-altitude performance in poor antenna installations and unique outputs to meet existing field applications.

Specifications
Dimensions: 137 × 83 × 83 mm
Weight: 1.3 kg
Power supply: 28 V DC, 16 W
Altitude:
(HG7502/HG8502) 0-2,500 ft
(HG7505/HG8505) 0-5,000 ft
(HG7508/HG8508) 0-8,000 ft
Accuracy: ±3 ft ±3% analogue altitude
±3 ft ±1% digital altitude

Operational status
In service on various helicopters and commercial airliners.

Contractor
Honeywell Inc Sensor and Guidance Products.

UPDATED

HG7700 Series radar altimeters

The HG7700 is specifically designed for the low-cost high-performance requirements of tactical manned and unmanned vehicles. The HG7700 features automatic test capability, noise immune tracker and low power consumption. Available options include digital output, transmitter power management and integrated antenna.

Specifications
Dimensions: 194 × 95 × 72 mm
Weight: 1.55 kg
Power supply: 28 V DC, 12 W
Altitude:
(HG7702) 0-2,500 ft
(HG7705) 0-5,000 ft
Accuracy: ±5 ft ±3%

Operational status
In production for fighter aircraft, as well as RPVs and tactical weapon systems. The HG7705 is fitted to the Saab JAS 39 Gripen.

Contractor
Honeywell Inc Sensor and Guidance Products.

UPDATED

HG9500 radar altimeter

The HG9500 is Honeywell's newest high-performance radar altimeter. With its modular design, the HG9500 can be configured to meet most conceivable form factor requirements, including US Air Force CARA requirements. A Honeywell patent design combines the high accuracy and programmable features of pulsed altimeter designs with the sensitivity and LPI advantages of coherent systems. Functions include MIL-STD-1553B databus, BIT monitoring, aircraft rate, LPI and immunity to jamming.

Specifications
Dimensions: 213 × 160 × 90 mm
Weight: 3.2 kg
Power supply: 28 V DC, 35 W
Altitude: 0-50,000 ft
Accuracy: ±2 ft ±2%

Contractor
Honeywell Inc Sensor and Guidance Products.

UPDATED

HG9550 LPI radar altimeter system

Honeywell claims that the HG9550 represents a quantum leap in altimeter capabilities over its earlier products in reliability, covertness, size, weight and cost.

The system was developed by Honeywell and jointly qualified with the US Air Force. The HG9550 was designed as a form, fit and function replacement for

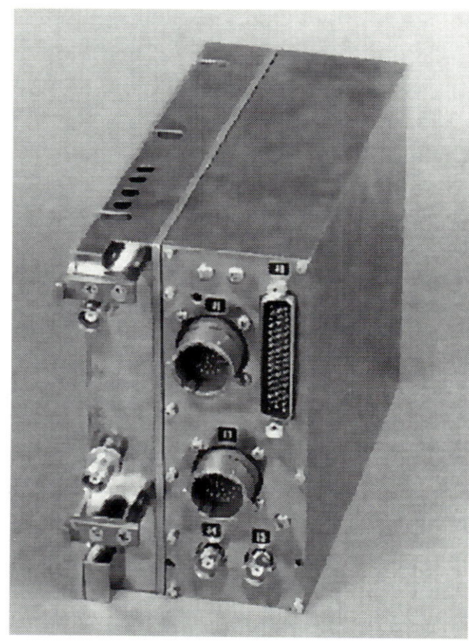

Honeywell HG9550 LPI radar altimeter system
1998/0018070

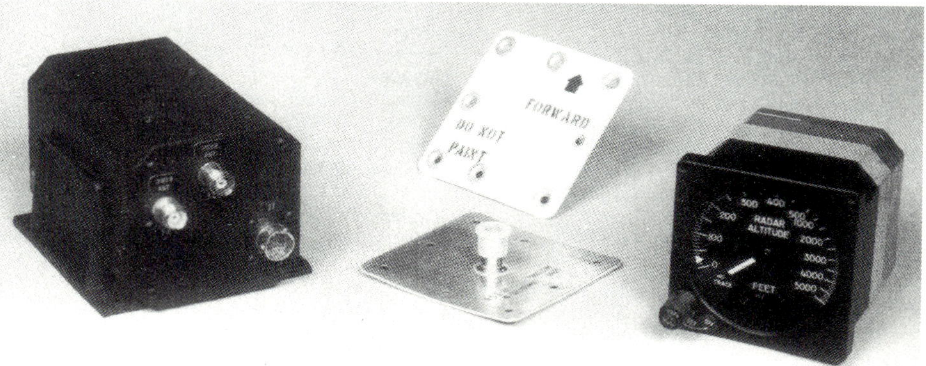

The Honeywell HG7500 radar altimeter

MIL-STD-1553B versions of the US Air Force CARA altimeter.

The HG9550 is designed to provide the high accuracy and programmable features of pulsed altimeter designs with the sensitivity and Low Probability of Intercept (LPI) advantages of coherent designs; with less than 1 W of power it is virtually undetectable. A microprocessor permits system characteristics such as track rate and ECCM response to be varied as a function of real-time inputs, or to be preprogrammed according to mission requirements. It is designed to be an element of GCAS systems.

Specifications

Dimensions: 90 × 60 × 222 mm
Weight: 4.43 kg
Altitude accuracy:
(analogue) ±4 ft or ±4% (whichever is greater)
(digital) ±2 ft (0 to 100 ft); ±2% (100 to 50,000 ft)
Manoeuvre angles: ±60° up to 3,000 ft; ±45° up to 5,000 ft; ±10° up to 50,000 ft
RF power: power managed: controlled at 10 dB above track threshold, with less than 1 W transmit power
Programmable features: track rate, ECCM response, sensitivity, altitude range, output formats
LPI features: frequency agility, power management, jittered code and PRF

Operational status

HG9550 is an off-the-shelf, fully qualified system, currently in production for: US Air Force C-130J, UK C-130J, Argentine A-4 upgrade, both Joint Strike Fighter configurations, and various other applications.

Contractor

Honeywell Inc Sensor and Guidance Products.

NEW ENTRY

Model 1044 altimeter

The Model 1044 radar altimeter was designed and developed to be a low-cost, lightweight, very accurate, high-rate production unit. The altimeter transmits and receives a very low RF signal that is Phase Shift-Key (PSK) modulated using a pseudo-random code. The PSK technique was selected because it provides the equipment with the capability to measure over a wide range of altitudes from 0 to 10,000 ft with low power and fewer parts than other types of radar altimeters. A commanded built-in test verifies that the unit is operational by testing 98 per cent of all the altimeter functions.

The altimeter comprises two subassemblies. The microwave subassembly contains a filter board assembly and four microwave hybrids: the transmitter, receiver, dielectric resonant oscillator and video amplifier. The microwave hybrids provide increased reliability with reduced parts count. The electronic assembly contains an analogue processor, digital processor and power supply.

Contractor

Litton Guidance & Control Systems.

VERIFIED

Model 2100 Doppler Velocimeter/ Altimeter (DV/A)

The DV/A combines both an altimeter and a Doppler velocity sensor into one compact package. The DV/A consists of two subassemblies: the antenna assembly and the Doppler processor board. The antenna assembly is mounted in the vehicle structure. The Doppler processor board is mounted in a navigation system box. The DV/A measures vehicle velocities and slant range. The output format to the navigation computer is in RS-422.

Specifications

Dimensions:
(antenna assembly) 170 × 386 × 52.1 mm
(Doppler processor board) 10 × 235 × 140 mm
Power: 27 W
Temperature range: −32 to +71°C

Contractor

Litton Guidance & Control Systems.

VERIFIED

Model 3044 radar altimeter

The Model 3044 radar altimeter was designed and developed for manned aircraft applications and operates from 0 to 10,000 ft. It employs hybrids, application specific integrated circuits and MIL-STD-1553B interface to provide a highly reliable lightweight production unit. The altimeter transmits and receives a very low RF signal that is phase shift-key modulated using a pseudo-random code. The phase shift-key technique provides the equipment with the capability to measure over the full altitude range using a very low-level RF output.

The altimeter comprises two subassemblies. The microwave subassembly contains a filter board assembly and four microwave hybrids: the transmitter, receiver, dielectric resonant oscillator and video amplifier. The microwave hybrids provide increased reliability at reduced part count. The electronic subassembly contains an MIL-STD-1553B interface, signal processor and power supply. The application specific integrated circuit incorporated into the signal processor enables digital signal processing at a greatly reduced part count.

Contractor

Litton Guidance & Control Systems.

VERIFIED

AN/APN-217 Doppler Radar Navigation Systems

The AN/APN-217 Radar Navigation System (RNS) is a lightweight, low-power, self-contained Doppler radar velocity sensor. The RNS unit detects and processes Doppler shifted frequency returns from continuous wave time multiplexed radar beams to determine three orthogonal velocities in aircraft heading (Vh), drift (Vd) and vertical (Vz) co-ordinates. The AN/ARN-217 is available in -217, -217(V)2, -217(V)3, -217(V)5 and -217(V)6 models and can accommodate output formats of ARINC or MIL-STD-1553 and provides DC analogue voltages for driving hover indicators and automated flight control systems. Applications encompass helicopters and medium-performance fixed-wing aircraft.

Specifications

Dimensions: 164 × 423 × 408 mm
Weight: 12.72 kg
Power supply: 28 V DC, 48 W
Velocity range:
(heading) −40 to 350 kt
(drift) −100 to 100 kt
(vertical) −4,500 to 4,500 ft/min
Accuracy:
(heading) <0.3% over land, <0.4% over water
(drift and vertical) <0.2%
Reliability: 15,257 h MTBF

Operational status

The AN/APN-217 is the standard Doppler for US Navy and US Marine Corps aircraft.

Contractor

Litton Guidance & Control Systems.

VERIFIED

AN/APN-218 Doppler Velocity Sensor (DVS)

The AN/APN-218 Doppler Velocity Sensor (DVS) is a reliable, high-performance, nuclear-hardened Doppler radar for fixed-wing aircraft. Usually referred to as the Common Strategic Doppler (CSD), it is combined with a GroundSpeed Drift Indicator (GSDI) to process digital velocity data from the radar and display the groundspeed and drift angle. The DVS is an LRU consisting of 10 Shop Replaceable Units (SRUs). The GSDI is also an LRU and consists of three SRUs.

The AN/APN-218 continuously measures three velocities: heading (Vh), drift (Vd) and vertical (Vz). These provide accurate in-flight data to navigation equipment. It can output data for all three co-ordinates from the microcontroller and process it into either MIL-STD-1553A or ARINC 575 formats.

Specifications

Dimensions:
(sensor) 636 × 708 × 168 mm
(GSDI) 146 × 76 × 155 mm
(CDU) 146 × 152 × 165 mm
Weight:
(sensor) 33.18 kg
(GSDI) 1.5 kg
(CDU) 3.9 kg
Power supply: 115 V AC, 400 Hz, 10 VA
Velocity range:
(heading) 96 to 1,800 kt
(drift) ±200 kt
(altitude) up to 70,000 ft
Reliability: >9,740 h MTBF

Operational status

In production. AN/APN-218 is in service in US Air Force B-52, C-130, KC-135 and MC-130 aircraft and on the C-130J.

Contractor

Litton Guidance & Control Systems.

VERIFIED

AN/APN-231 radar navigation system

Developed in 1984, the AN/APN-231 is the primary navigation system for the US Navy EA-6B aircraft. It also integrates with other equipment including the attitude/ heading reference system, air data computer, search radar, ECM sets and flight instruments.

The system comprises an AN/APN-200 Doppler velocity sensor, a CP-1573/APN-231 computer display unit and a CV-3780/APN-231 signal data converter.

Specifications

Dimensions:
(data converter) 226 × 259.1 × 426.7 mm
(velocity sensor) 149.9 × 632.5 × 652.8 mm
(computer display unit) 152 × 146 × 165 mm
Weight:
(data converter) 18.2 kg
(velocity sensor) 20 kg
(control display unit) 3.6 kg
Range: 50-999 kt forwards, 0-200 kt in drift, 0-5,000 ft/min vertically
Altitude range: up to 40,000 ft
Accuracy:
(velocity sensor) 0.13% +0.1 kt

Operational status

No longer in production. In service in the US Navy EA-6B.

Contractor

Litton Guidance & Control Systems.

VERIFIED

AN/APN-233 (220) Doppler velocity sensor

The AN/APN-233 can be used either as a single unit velocity sensor providing outputs to other aircraft systems or with a Control/Display Unit (CDU) and HSI as a self-contained navigation facility. It was designed for applications in which size, weight, performance and reliability are critical factors.

The APN-220 family evolved from a small, lightweight Doppler sensor originally designed for the US Army and was subsequently qualified by that service and by the US Air Force and the German Air Force.

An optimum velocity range and near-zone rejection are offered for each application.

The CDU combines the functions of navigation computer and control/display unit and contains an incandescent alphanumeric display panel and a keyboard for entering data and selecting operational modes. Up to 10 waypoints can be accommodated, and a non-volatile scratchpad memory holds critical information during power transients or interruptions.

Specifications

Typical fixed-wing applications
Dimensions:
(sensor) 426 × 291 × 113 mm
(CDU) 152 × 146 × 165 mm
Weight:
(sensor) 9.66 kg
(CDU) 3.86 kg
Power supply:
(sensor) 28 V DC, 28 W
(CDU) 28 V DC, 30 W
Output: heading, vertical velocity and groundspeed/drift to aircraft systems, for example AFCS, or to CP-1251 and HSI
Number of waypoints: 10 entered via front panel keyboard
Velocity range (typical system):
(speed) −40 to 600 kt
(drift) ±150 kt
(vertical) ±5,000 ft/min
Altitude: up to 50,000 ft
Accuracy:
(over land) 0.25% +0.2 kt
(over sea) 0.3% +0.2 kt
Self-test: BITE diagnostic programme locates faults at first line level to 95% confidence
Reliability: 2,600 h MTBF demonstrated in the Alpha Jet

Operational status

In production. The system has been produced for the US Navy C-2A Greyhound and US Marine Corps OV-10D Bronco observation post aircraft. Additional production for the S-2, DHC-5 and CH-47 aircraft has been completed.

The system has been chosen by the German Air Force for its Dassault/Dornier Alpha Jet strike/trainers as a velocity sensor to provide data to the navigation and weapons delivery systems. Versions of the equipment have also flown on RPVs and helicopters.

Contractor

Litton Guidance & Control Systems.

VERIFIED

AN/APX-92 IFF transponder

One of a family of IFF systems produced by Litton, the APX-92 is suitable for all types of aircraft, including helicopters. It offers response in Modes 1, 2, 3/C, 4 and C. In Mode 4 it operates in conjunction with a Mode 4 computer to give complete recognition and monitoring capability, while for Mode C it interfaces with an appropriate encoding altimeter. The 500 W transmitter and the receiver are both all-solid-state and the signal processor uses linear and digital circuitry.

Specifications

Dimensions: 133 × 146 × 254 mm
Weight: <4.54 kg
Power supply: 28 V DC, 55 W

Operational status

In service.

Contractor

Litton Guidance & Control Systems.

VERIFIED

AN/APX-101(V) IFF transponder

The AN/APX-101(V) diversity IFF transponder is all-solid-state and consists entirely of replaceable modules. The transponder is housed in a single LRU and is mounted in the airframe without the use of shock-mounts. Crystal-controlled pulsewidth, discrimination, decoding and encoding ensure accurate response to interrogations. BIT circuits monitor the critical parameters of the transponder and provide an immediate status indication.

The transponder operates in Mk XII Modes 1, 2, 3A, C and 4. It receives RF interrogations from two antenna systems, decodes the interrogation into the proper mode, encodes the selected reply and transmits the coded RF reply through the correct antenna.

Specifications

Dimensions: 152.4 × 127.3 × 278 mm
Weight: 6.53 kg
Power supply: 28 V DC, 65 W
Frequency:
(receiver) 1,030 MHz
(transmitter) 1,090 ±1.5 MHz
Temperature range: −54 to +70°C
Altitude: up to 100,000 ft

Operational status

In production and standard equipment on A-10, E-3A, F-5E, F-15 and F-16 aircraft. Over 3,000 units have been ordered.

Contractor

Litton Guidance & Control Systems.

VERIFIED

AN/APX-108 IFF transponder

The AN/APX-108 is a single unit, reduced size Mk XII diversity IFF transponder. Using advanced microwave packaging, high-speed analogue/digital converters, advanced digital signal processing and CMOS LSI gate array circuitry, the component count, volume, weight and power dissipation of the AN/APX-108 has been minimised. It operates in Modes 1, 2, 3A, C and 4.

Features include a COMSEC appliqué that uses electronic key fill to eliminate the requirement for a KIT-1A computer. For interface flexibility, the AN/APX-108 can be interfaced with either a MIL-STD-1553B multiplex bus or a C-6280 control box.

Specifications

Dimensions: 152.4 × 162.5 × 229 mm
Weight: 6.4 kg
Power supply: 28 V DC, 55 W nominal
Frequency:
(receiver) 1,030 ±0.5 MHz
(transmitter) 1,090 ±0.5 MHz
Output power: 500 W min at 1% duty cycle
Temperature range: −40 to +71°C
Altitude: up to 70,000 ft
Reliability: 2,500 h MTBF

Operational status

In production and in service.

Contractor

Litton Guidance & Control Systems.

VERIFIED

AN/APX-109(V)3 Mk XII IFF combined interrogator/transponder

The Litton AN/APX-109(V)3 represents a major advance in IFF system design. AIMS and STANAG requirements are both provided in hardware design and system implementation utilises a fully integrated system approach, making practical use of advanced technologies.

The AN/APX-109(V)3 is an efficient solid-state design that reduces the component count, weight and volume of the IFF system significantly. Employing advanced signal processing, the AN/APX-109(V)3 offers a prioritised four-channel operation; adaptive thresholding of the received video to improve performance in high-noise and jamming environments, and a monopulse processing of received target video for enhanced azimuth accuracy.

Specifications

Dimensions: 152 × 213 × 368 mm
Weight: 15.5 kg (including COMSEC appliqué)
Power supply: 28 V DC, 150 W
Frequency:
(receive) 1,030 ±0.2 MHz
(transmit) 1,090 ±0.2 MHz
Temperature range: −40 to +71°C
Altitude: up to 70,000 ft
Reliability: >2,000 h MTBF

Operational status

In production.

Contractor

Litton Guidance & Control Systems.

VERIFIED

AN/APX-( )MAT IFF transponder

The AN/APX-( )MAT contains, in a single unit, all Mk XII, Mode S and cryptographic IFF transponder functions with either a MIL-STD-1553B multiplex bus or a discrete control panel interface. It operates on Modes 1, 2, 3A, C, 4 and S. It is compatible with AIMS and STANAG requirements and uses advanced microwave packaging techniques to minimise space and weight.

Features include a KIV-2 COMSEC appliqué that uses an electronic key fill loader to eliminate requirements for a computer and mechanical code loader. Mode S Level 1 is provided as a standard feature, with optional growth to levels 2 and 3.

Specifications

Dimensions: 136.5 × 136.5 × 212.6 mm
Weight: 4.55 kg
Frequency:
(receive) 1,030 ±0.5 MHz
(transmit) 1,090 ±0.5 MHz
Power output: 27 dBW ±2 dB at 1.2% duty cycle

Contractor

Litton Guidance & Control Systems.

VERIFIED

OX-72/APX-109(V) combined interrogator/transponder

The OX-72/APX-109(V) combined interrogator/transponder is based on a fully integrated system approach and employs advanced technologies to provide a system which meets both AIMS and STANAG requirements. The solid-state design utilises advanced microwave packaging techniques, surface mount devices and CMOS LSI gate array cut-offs to effect significant reductions in component count, weight and volume.

The OX-72/APX-109(V) offers a prioritised four-channel AOC, adaptive threshold of the received video to improve performance in high-noise and jamming environments and monopulse processing of received target video to enhance azimuth accuracy.

The MIL-STD-1750A processor supports a built-in defruiter and statistical reply evaluator, to provide operational flexibility and allow optimisation of system parameters based on mission requirements. The 1750A processor also supports built-in test functions, allowing the OX-72/APX-109(V) to report up to 97 per cent of its critical failures over the MIL-STD-1553B bus.

The OX-72/APX-109(V) can be interfaced to either a mechanically or electronically scanned antenna.

Operational status

In production.

Contractor

Litton Guidance & Control Systems.

VERIFIED

TEC-60i combined interrogator/transponder

The airborne IFF TEC-60i combined interrogator/transponder, when functioning as an interrogator, challenges and identifies co-operative active targets. These targets are suitably equipped with compatible transponders and operate within effective IFF range. The unit also functions as a transponder responding to valid interrogations. The unit is a compact, lightweight system installed in the equipment bay, with a control panel mounted in the aircraft cockpit. Peak output power is 1 kW.

Operational status
In production and in service.

Contractor
Litton Guidance & Control Systems.

VERIFIED

AN/APN-59F(V) search radar

The AN/APN-59F(V) search radar is an upgrade of the AN/APN-59E(V). Principal modes are search, skin painting, ground-mapping/navigation, weather mapping and beacon homing. Pencil or fan beams with a variety of pulse lengths and PRFs are selectable. The antenna can scan a full 360° or in selected sectors. The LRUs are interchangeable with those of earlier AN/APN-59 systems and the system can be gradually upgraded without aircraft stand-down.

Many configurations are possible, ranging from single azimuth/range displays to more complex installations with up to three displays.

The APN-59(X) search and navigation radar is the ultimate upgrade of the AN/APN-59 series. It has been designed for simple direct backfit on a box-by-box basis for all C-130/135 type aircraft. All performance features of the older systems are retained or enhanced including the addition of multifunction colour displays. System MTBFs greater than 1,000 hours have been achieved by power reduction and the elimination of problem components. The system has been designated the AN/APN-242 under a USAF contract. See next entry.

Specifications
Weight: 84 kg (typical system)
Frequency:
(transmitter/receiver) 9,375 ±40 MHz
(beacon) 9,310 MHz
Peak power: 70 kW
Beamwidth: 3°
Pulsewidth: 0.35-4.5 μs variable

Contractor
Litton Marine Systems (Sperry Marine).

VERIFIED

AN/APN-242(V) search/navigation radar

The AN/APN-242, formerly the AN/APN-59(X), is a long-short-range colour weather and navigation radar with integrated navigation and air data overlays via the AN/ASN-165 Display Group. Principal features include weather detection and display in colour, black/white, or green for night vision goggle compatibility; enhanced terrain-mapping with navigation using a latitude/longitude-stabilised electronic cursor; aircraft detection/skin painting concurrent with other operating modes; beacon interrogation and display; and optionally, IFF interrogation and display.

The APN-242 was designed by Litton Marine Systems as a follow-on to the AN/APN-59 now in use worldwide on over 1,500 C-130 and C-135 aircraft. The radar is suitable for both new installations and as a form, fit and function replacement for the APN-59. Performance has been improved in every operating mode by increasing range, resolution and accuracy. By redesigning high failure rate components, the APN-242 achieves a system Mean Time Between Failure (MTBF) rate greater than 1,000 h.

The APN-242 antenna subsystem is a flat plate array which is stabilised to aircraft attitude reference systems and which, through the elimination of all gears, achieves reliabilities approaching 7,000 h. The array rotates 360° or can sector-scan. The antenna beam can be tilted vertically and the pattern can be instantaneously switched between pencil and fan to achieve the desired illumination. Pulsewidths and PRFs are operator-selectable. The low-noise receiver and lower-power transmitter improves APN-59 performance and the long life 10,000 h magnetron has the power to skin paint fighter aircraft at extended ranges through intervening rain showers. The AN/ASN-165 is the standard display group.

When replacing the APN-59, the APN-242 does not require an aircraft modification for installation and either the antenna or receiver/transmitter (R/T) subsystems can be dropped in as direct replacements for APN-59 subsystems. On these aircraft, the system can be installed on the flight line in a day by unit level maintenance using existing radar cabling, connections and mounting brackets.

Specifications
Frequencies:
(radar operation) I-band; 9,375 ±10 MHz
(beacon reception) 9,310 MHz
Transmitted power: 25 kW nominal peak (new high-reliability magnetron)
Ranges: 2.5 to 20 (2.5 n mile increments), 25, 30, 50, 100, and 240 n miles
Pulse length: multiple lengths (0.20, 0.8, 2.35, and 4.5 ms) automatically selected for different ranges and functions
Scanning features: 360°
Scan rates: 12 rpm on long-range functions; 45 rpm on short-range functions
Sector scan: (basic) approx 80°, centred about forward position, variable sector with alternate control
Antenna beam selection: pencil or equal energy (fan) beam; both with 3° azimuth beamwidth, instantaneous electronic switching
Antenna stabilisation: stabilised to existing aircraft reference throughout a range of ±15° pitch and ±30° roll
Controls: independent navigator and pilot controls
System components/weight:
(basic) 7 components/70 kg
Power:
(basic) 115 V ±5% (380 to 420 Hz), 800 VA average

Operational status
The AN/APN-242 is in production for a US Air Force customer.

Contractor
Litton Marine Systems (Sperry Marine).

VERIFIED

Longbow radar

A Lockheed Martin/Northrop Grumman joint venture company (Longbow Limited Liability Company LLLC) is working with The Boeing Company to supply the Longbow radar and RF Hellfire fire-and-forget missile to the AH-64D Apache helicopter for the US Army. The joint venture is also working with Westland Helicopter Ltd to supply the radar and missile for the WAH-64 Apache for the UK (contract selection in April 1996). A lightweight variant is expected to be integrated into the US Army's RAH-66 Comanche reconnaissance and light attack helicopter.

The AH-64D Longbow Apache helicopter with mast-mounted Longbow radar
1998/0018072

The Longbow system comprises a mast-mounted, millimetric-wavelength, fire-control radar, which allows the crew to search, detect, locate, classify, and prioritise mobile threats (tanks, air defence units, trucks and so on) in adverse weather, and an integrated, interferometer, radio frequency seeker to provide fire-and-forget capability for the Hellfire missile.

Operational status
The Longbow system has completed initial operational test and evaluation by the US government and entered production.

In December 1995, The Boeing company signed a five year contract valued at US$1.9 billion with the US Army for remanufacture of 232 AH-64A Apaches to AH-64D Longbow Apache standard by year 2002. Then, in November 1997, the joint venture team is reported to have been awarded US$107 million as part of a US$565 million firm fixed price multi-year contract for the fabrication, test and delivery of 207 Longbow fire-control radars and associated contractor services over a five-year period, by the US Army. However, some doubt is cast on the exact status of these contracts by continuing pressure from the US Government Accounting Office (US GAO) for further proving work to be completed before full multi-year production approval.

Contractors
Lockheed Martin Electronics & Missiles Company.
Northrop Grumman Corporation, Electronic Sensors and System Division.

UPDATED

AN/APG-67(F) attack radar

The AN/APG-67(F) has been developed for retrofitting into Northrop F-5 aircraft and other fighter aircraft.

A significant characteristic of the APG-67, stemming from its Modular Survivable Radar (MSR) ancestry, is the choice and layout of circuitry to enable the system to be updated for new technologies as they appear, such as very large-scale integration, without having to redesign the LRUs. Growth features can also be added in this way.

In air-to-air missions the system searches for, automatically acquires and tracks targets in both look-up and look-down situations. In its air-to-surface role the radar provides real-beam mapping, high-resolution Doppler beam-sharpened ground-mapping (40:1), and air-to-ground ranging. Variants of the radar can include synthetic aperture imaging.

The heart of the processor contains 16 Shark 2106D processors. Twelve perform signal processing functions and the other four perform Hardware I/O, radar mode control and aircraft mission computer interface functions, together with self-test radar processing associated with motion compensation, target tracking, antenna motion control and raster scan output to the cockpit display.

The processor integrates raw target data from the antenna so that reliable reports can be established. The unit incorporates a programmable signal processor and its use is considered to be the key to what is claimed to be this unit's exceptional performance. The processor performs a variety of functions, including fast Fourier transformation, moving target indication, pulse compression and motion compensation, and can operate with variable waveforms and bandwidth. The transmitter produces coherent I/J-band radiation from a travelling wave tube and can operate at low, medium and high PRFs, with variable power outputs and pulsewidths. The antenna is a flat plate slotted array with low sidelobe sensitivity and with ±60° scan in azimuth and elevation. Scan width is selectable and elevation search can be accomplished in 1, 2 or 4 bar modes. The receiver has a low-noise front-end to maximise detection range.

The AN/APG-67(F) version for the new aircraft and retrofit market offers a full range of air-to-air capabilities including range-while-search out to 148 km, three air combat modes and single target track. It also provides a complementary set of air-to-ground modes including map, expand, Doppler beam-sharpening, freeze and surface moving target indication and track over ground or rough seas.

It consists of three LRUs and incorporates a MIL-STD-1553B database interface and MIL-STD-1750A computer architecture. There is extensive built-in test.

Lockheed Martin has collaborated with Geophysical and Environmental Research Corp to develop GSAR, a variant of the AN/APG-67 radar to equip business aircraft for the SAR role. GSAR has a range of 150 km, with swath widths up to 40 km wide. Resolution is reported to be 3 m, improving to 1 m in spotlight mode.

Specifications - APG-67(F)

Volume: 0.049 ms
Weight: 74 kg
Frequency: I/J-band
Power: 2,600 VA
Transmit power: 350 W
Antenna size: 279 × 711 mm
Detection range (fighter-size targets):
(look-down) 39 km
(look-up) 72 km
(sea targets, 50 m²) 57 km Sea State 1
(max range, all modes) 148 km
Accuracy (air-to-ground): 15 m or 0.5% of range
Reliability: design 235 h MTBF

Operational status

In production. Taiwan has selected the APG-67 for its Indigenous Defence Fighter project.

Contractor

Lockheed Martin Ocean Radar and Sensor Systems.

UPDATED

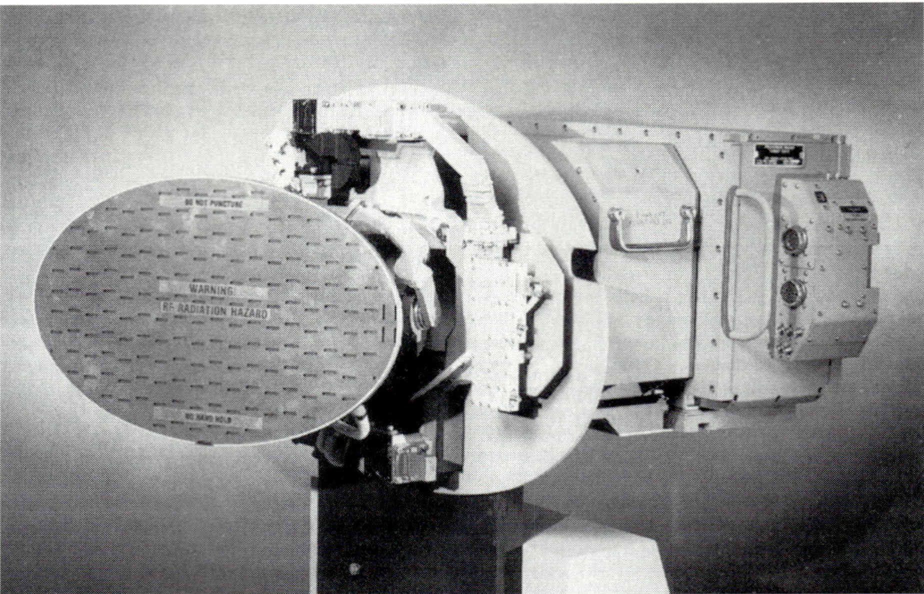

The AN/APG-67(F) radar for the F-5 retrofit market

AN/APS-125/-138/-139/-145 airborne early warning radars

The AN/APS-125/-138/-139/-145 series of airborne early warning radars have all been developed for the Northrop Grumman E-2C Hawkeye aircraft.

AN/APS-125

The AN/APS-125 radar was designed for the airborne early warning role, and operates in the UHF band. It was reported to have been the original fit in US Navy E-2C Group 0 aircraft, and in the E-2C Group 0 aircraft exported to Egypt, Israel and Japan, during the late 1970s and early 1980s.

AN/APS-138

The AN/APS-138 was an update of the AN/APS-125 during the mid-1980s. It reportedly included minor circuit changes and a low-sidelobe antenna, designated the Total Radiation Aperture Control Antenna (TRAC-A). AN/APS-138 radars were reportedly in production between 1983 and 1987, and the US Navy is understood to have updated AN/APS-125 radars to the AN/APS-138 standard. AN/APS-138 radars were also reportedly sold to Singapore on its E-2C aircraft.

AN/APS-139

The AN/APS-139 radar is a development of the AN/APS-138, that improves ECCM performance and surface surveillance capabilities, reportedly in production from 1987 to 1989. Reportedly no radars have been updated from AN/APS-138 to AN/APS-139 standard.

AN/APS-145

Latest in the series is the AN/APS-145. It is an update of the AN/APS-138 and AN/APS-139 radars, and has been retrofitted to all US Navy E-2C aircraft from 1990. The AN/APS-145 development reportedly specifically addresses the problem of overland clutter and provides fully automatic overland targeting and tracking capability, an improved IFF system, expanded processing and new colour displays. The present radar is said to perform very well over sea and desert, but to degrade rapidly when terrain becomes more rugged. To reduce the false alarm rate a feature known as 'environmental processing' is being developed. This adjusts the sensitivity of the radar cell by cell, according to the clutter and traffic in each cell. To enable large aircraft to be detected at long range (up to 400 n miles) a new lower PRF is used, and to match this development the E-2C's rotordome rotation rate is slowed from 6 to 5 rpm. A third PRF is also introduced allowing the radar to operate with different PRFs during scanning, to eliminate blind speed problems caused by single PRF operation. The AN/APS-145 entered service in 1992. It is fitted to US Navy E-2C Group II aircraft and to Taiwanese E-2C and to French E-2C aircraft.

E-2C Group II radar display **1998**/0018073

Lockheed Martin and Northrop Grumman have signed a MoU to collaborate in supplying airborne early warning and control (AEW&C) systems to overseas customers, based on Lockheed Martin supplying the AN/APS-145 radar and Northrop Grumman being responsible for integration of the mission system. This agreement applies to ADE&C variants of the Lockheed Martin C-130J and to other platforms, including the Northrop Grumman E-2C, but excluding the Boeing E-3 AWACS.

Operational status

Aircraft fits are reported to be as follows: E-2C Group 0 aircraft fitted with AN/APS-125/-138; E-2C Group I aircraft fitted with AN/APS-139; E-2C Group II aircraft fitted with AN/APS-145; all US Navy aircraft upgraded to AN/APS-145 configuration. Four US Coast Guard P-3 AEW&C aircraft are reportedly fitted with the AN/APS-138 radar.

Because the E-2C will serve the US Navy for many years, more capable Hawkeyes are on order. Northrop Grumman restarted its Hawkeye production line in 1994, after the US Navy ordered the first four of an expected 36 new Group II E-2Cs. In April 1996, the eighth and ninth new Hawkeyes were initiated under an advance procurement contract. In addition, Northrop Grumman is updating Group I Hawkeyes to Group II configuration for the US Navy. Modified aircraft are fitted with the AN/APS-145 radar system with fully automatic overland targeting and tracking capability, an improved IFF system, a 40 per cent increase in radar and IFF ranges, expanded processing capacity, new high-target-capacity colour displays, JTIDS for improved secure, anti-jam voice and data communications and GPS-based navigation capability.

Contractor

Lockheed Martin Ocean Radar and Sensor Systems.

UPDATED

The AN/APS-145 surveillance radar has been retrofitted to US Navy Grumman E-2C Hawkeye AEW aircraft

HOWLS experimental airborne radar

The HOstile Weapons Location System (HOWLS) radar is an extremely versatile coherent J-band phased-array radar, designed to investigate techniques from which solutions may be derived for detecting and locating fixed tactical targets in a clutter environment. It has been developed by Lockheed Martin under the sponsorship of the US Advanced Research Projects Agency and management of MIT's Lincoln Laboratory.

The key to the radar system is a lightweight, phased-array antenna. The antenna system utilises a low-cost, highly reproducible phased-array element that combines the radiating element and PIN diode phase shifter in a single integrated structure to minimise weight and cost.

HOWLS has six basic modes of operation. These are contextual ground map, fixed-target detection, ground moving target detection, low-level Doppler spectrum signature analysis, projectile detection and tracking, and Doppler beam-sharpening. Through microprocessor control, the radar has mode parameter flexibility and is capable of interleaved multimode operation. Selectable parameters include antenna scanning pattern and rate, PRF, data sampling window position and data sampling pattern. The radar is capable of switching among parameters in an orderly manner so that up to four operating modes can be interleaved.

Results of the experimental HOWLS programme will provide the technology base from which radars for airborne platforms can be configured to meet a variety of missions.

Contractor

Lockheed Martin Ocean Radar and Sensor Systems.

VERIFIED

Multimode radar for the F-5E

Lockheed Martin has developed a lightweight, advanced multimode radar for retrofitting into the F-5E aircraft. The equipment operates in the I/J-band, uses digital pulse compression and coherent signal integration techniques and has a high antenna gain as well as low sidelobes.

The equipment has two main modes of operation, air-to-ground and air-to-air, with the antenna scanning over ±40°, ±30° or ±10° in azimuth.

In the air-to-ground mode, facilities include ground-mapping up to 148 km range, ground track Moving Target Indication (MTI) up to 74 km, Doppler beam-sharpening to 36 km range, air-to-ground ranging, radar silent freeze-frame and scale expansion. Similar capabilities are possible over the sea, with detection of a moving or stationary patrol boat at 57 km range in seas of less than 1 m. For high sea states, a coherent MTI mode can be selected to detect targets moving at over 8 kt. Range resolution is 45 m and accuracy is 15 m.

In the air-to-air mode, facilities include look-up and look-down, search and track, air combat automatic acquisition in boresight, supersearch and vertical scan. In supersearch mode the radar scans the 20 × 20° field of view of the HUD; vertical scan is −10 × +40°. In this submode, and in boresight, the radar automatically locks on to a target and switches to single target track. Average detection range against a typical fighter aircraft is 56 km in the look-up mode and 35 km in look-down. In air-to-air search, angular accuracy is 3 mrad, range accuracy is 15 m or one per cent of range and velocity accuracy is 15 m/s or 0.4 per cent of range.

Specifications

Volume: 0.06 m^3
Antenna: 279 × 400 mm
Weight: 104 kg
Frequency: I/J-band
Power output: 350 W (average)
Reliability: 250 h MTBF

Operational status

In production.

Contractor

Lockheed Martin Ocean Radar and Sensor Systems.

UPDATED

AN/UPD-4 reconnaissance system

The AN/UPD-4 was developed for the US Air Force as a successor to the AN/APD-102A. It has since been enhanced and designated AN/UPD-8. The AN/UPD-4 side-looking radar is an all-weather, high-resolution reconnaissance sensor system for the airborne collection of tactical and strategic intelligence information. Utilising I/J-band radar energy to illuminate selected terrain swathes, it operates equally well in daylight, darkness and adverse weather.

The AN/UPD-4 consists of an AN/APD-10 synthetic aperture radar mounted in the aircraft, a ground-based correlator/processor set and test consoles for the maintenance of airborne and ground equipment. Airborne datalink transmitters and ground receiving and processing systems are optional additions to provide radar information to ground commanders more rapidly than the basic system.

The airborne system includes multi-element, phased, linear waveguide antennas individually gimballed for left- or right-side operation. Look-angle of the antennas is maintained in flight by a control system that receives error signals from the aircraft inertial navigation system and gimbal-mounted gyros and accelerometers.

The frequency converter/transmitter generates the stable radio frequency used for phase-locking the radar signals and contains the transmit/receive and antenna switching components. It also supplies the swept frequency for modulating the transmitted pulses, amplifies the reflected signals and converts the received RF to IF. The amplifier/modulator is a high-PRF pulse amplifier.

The signal data generator receives information relative to aircraft velocity and mode, and uses it to establish the radar PRF and generate the basic timing reference signals required throughout the system.

The radar mapping recorder contains the optical, electronic and electromechanical assemblies required to record the radar video data and associated coded data on photographic film. This information is displayed on two 127 mm CRTs as four intensity-modulated traces that are transferred by two mirror assemblies and four recording lenses to focus the images on the film. Mode strips that convey essential operational information and multi-element blocks containing information pertinent to the mission are recorded on the film. For datalink operation, an identical recorder is used at the correlator/processor on the ground to establish a continuous recording/correlation path. Film speed is carefully controlled in either application to maintain a fixed relationship to aircraft groundspeed.

The equipment has several modes of operation, providing a variety of standoff distances and altitudes and the option of recording only Fixed Target Indication (FTI) or both FTI and Moving Target Indication (MTI). Imaging of the terrain at either side can be obtained at the discretion of the operator. With datalink-equipped systems, the information may be transmitted to a ground station in real time for recording and processing. The final imagery is recorded on 241 mm wide film in four channels at a scale of 1:1 million for all modes. In the along track direction, targets and terrain features are imaged in terms of distance travelled and represent true ground separation. In the across track direction, the imagery is recorded in slant range or the distance from the aircraft to the target.

Specifications

Volume: 0.57 m^3
Weight: 282 kg
Frequency: I/J-band
Swath width: 18.5 km
Range: 55 km
Resolution: 3 m

Operational status

In service with RF-4EJs of the Japanese Air Self-Defence Force.

Contractor

Lockheed Martin Tactical Defense Systems, Goodyear, Arizona.

UPDATED

AN/UPD-8/-9 and AN/APD-14 side-looking reconnaissance systems

AN/UPD-8/-9

The AN/UPD-8 and AN/UPD-9 are updated and improved versions of the AN/UPD-4 (see previous entry), with much greater range. The AN/UPD-8 is a side-looking, synthetic aperture airborne reconnaissance radar system for the US Air Force. The system provides an all-weather, day/night, standoff, tactical and strategic reconnaissance capability and, with its extended range antenna pods and airborne datalink electronics, is the latest in a sequence of Lockheed Martin airborne reconnaissance radars.

Operating in the I/J-band, the system records with equal effectiveness during daylight and darkness and through periods of adverse weather that would render other types of reconnaissance systems ineffective. The airborne radar system's standoff capability, which is increased by 60 per cent in the new equipment, allows coverage of border regions, harbours and coastal areas without violating another nation's airspace or national waters. Maximum range, limited by power, is about 130 km. At such distances, resolution is 6.1 m in azimuth and 4.6 m in range but resolution increases at shorter ranges. At 50 to 90 km the radar resolution is 3 m, allowing large aircraft to be identified and showing clear images of vehicle formations.

Lockheed Martin has also developed a pod-mounted version of the UPD-8. Similar in size to a 1,250 litre fuel tank and weighing less than 682 kg, the pod includes the main antenna, support electronics and a datalink. A version of this pod has been flight tested on an RF-4B and is to be supplied for US Marine Corps F/A-18(RC) reconnaissance aircraft. The podded system can also be installed on an executive jet aircraft.

Another upgrade of the UPD-4 is the AN/UPD-9 developed for the US Marine Corps. This is similar to the UPD-8, with a datalink, but does not have the long-range features. Basic standoff range is about 80 km.

The airborne sensor part of the overall system, the AN/APD-12, operates in eight modes selectable by the operator according to aircraft altitude and the distance to the target area. The system can record to left or right of the aircraft track with the normal antennas or with the extended range antennas which are mounted in a centreline pod. Target information can be transmitted to the ground in real time through a datalink or can be recorded in flight for later processing on the ground. The former facility enables a ground commander, perhaps hundreds of kilometres away, to evaluate targets while the aircraft continues its mission.

Specifications

Volume:
(UPD-8) 1.305 m^3
(UPD-9) 0.58 m^3
Weight:
(UPD-8) 93.5 kg
(UPD-9) 321 kg
Frequency: I/J-band
Swath width: 18.5 km
Range: (UPD-8) 93 km, (UPD-9) 55 km
Resolution: 3 m

Operational status

In operational service with Israeli Air Force RF-4B aircraft.

AN/APD-14

Sandia National Laboratory has modified the AN/UPD-8 design to form the AN/APD-14 SAR for Open Skies (SAROS) set. The treaty on Open Skies specifies

SAR resolution no better than 3 m resolution. SAROS, installed aboard an OC-135 aircraft, has a centre frequency of 9.6 GHz and a mapping swath of 10 n miles.

Contractor

Lockheed Martin Tactical Defense Systems, Goodyear, Arizona.

UPDATED

LAIRS Lockheed Martin Advanced Imaging Radar System

The current LAIRS is the fourth-generation of upgraded Synthetic Aperture Radar (SAR) electronics based upon the original LAIRS that was developed from the Advanced Synthetic Aperture Radar System (ASARS-1). Although ASARS-1 was developed for the SR-71 Mach 3 environment, LAIRS has been specifically built and tailored to operate in the medium altitude, subsonic through trans-sonic speed regimes. LAIRS uses the latest in SAR algorithms and is composed of both an airborne segment and a ground segment.

The airborne segment is built in a modular design and can be easily modified to accommodate specific customer requirements. The system incorporates commercial-off-the-shelf (COTS) technology in many of the components, including the airborne or ground processing system. The collected data from the airborne sensor can be transmitted directly to the ground via a datalink as phase history data, recorded on the aircraft for later processing on the ground or processed in real time on the aircraft and transmitted to the ground as imagery data. The datalink also has a feature for both downlink and uplink of data.

LAIRS is capable of providing both fixed target imagery (FTI) and moving target imagery (MTI). The FTI modes are swath mode and spotlight modes. The swath mode provides a wide area (normally 19 km wide) image of the earth or sea for surveillance and provides medium-resolution imagery (normally 3 m resolution). The swath can be continuous in either a parallel path or an oblique path to the flight vector. Wider swaths and finer resolution swaths are available, and are easily incorporated into the modular design of LAIRS. The second FTI mode is spotlight mode. LAIRS has two spotlight modes, one for classification and a higher-resolution mode for identification of targets.

Specifications

	Swath mode	Spotlight mode	MTI mode
Min slant range	<19 km	<19 km	35 km
Max slant range	>185 km	>185 km	185 km
Resolution	<3 m	fine and ultra-fine	<3 m

Spotlight modes can image targets either forward or aft of the aircraft and within the squint limits of the antenna. LAIRS also has an MTI mode for imaging moving targets. The mode typically uses a sector scan image out to 185 km and can image moving targets within a 50 km sector that can be positioned anywhere from 35 to 185 km away from the aircraft.

The LAIRS ground station uses state of the art computer equipment and software to store, retrieve, combine, exploit and report. Typically LAIRS is displayed on soft copy so that the user can optimise data analysis from imagery produced by the LAIRS system. The number of workstations, the management of imagery software, and hard copy production and display equipment can be tailored to the specific requirements of the user.

Operational status

LAIRS radar equipment is reported to be operational aboard the US Customs Service P-3 aircraft, and is being built and will be fielded in an executive jet for export.

Contractor

Lockheed Martin Tactical Defense Systems, Goodyear, Arizona.

UPDATED

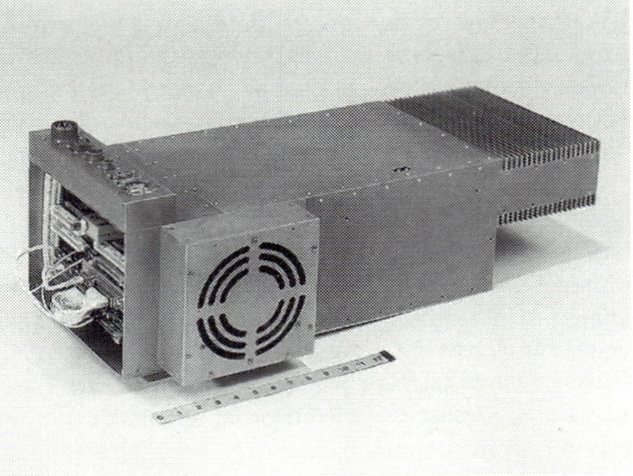

The radar electronics package used in STacSAR™ **1998**

STacSAR™ Small Tactical Synthetic Aperture Radar

The Small Tactical Synthetic Aperture Radar (STacSAR™) is a lightweight (less than 29.5 kg) J-band SAR system produced for use on unmanned aerial vehicles or light manned aircraft. It uses the latest radar technology and processing techniques. STacSAR™ provides fixed target imagery in both spotlight and search modes, as well as moving target indications. Onboard processing is included, allowing real-time exploitation. STacSAR™ is a modular design with inherent flexibility. System parameters are programmable under software control. SAR modes include both a strip map search mode and a higher resolution spotlight mode along with a moving target indicator mode. The user can specify mode parameters in selecting range and azimuth resolutions and swath widths. Transmitter pulsewidth and repetition frequency can also be varied to achieve maximum average power for selected modes. The system is designed for datalink operation where imagery can be downlinked in real time to user ground stations. STacSAR™ consists of two subsystems: an antenna/gimbal assembly and an electronics package.

Operational status

The system is performing fight demonstrations aboard manned light aircraft.

Specifications

Wavelength: J-band (17 GHz)
Resolution: 0.5-50 m programmable
Swath: 500-5,000 m programmable
Slant range: 10 km (with 12 W transmitter)
Weight: < 29.5 kg

Contractor

Lockheed Martin Tactical Defense Systems, Goodyear, Arizona.

UPDATED

The antenna/gimbal assembly used in STacSAR™ **1998**

The STacSAR™ radome configuration as used in a light aircraft application **1998**

AN/APQ-150A radar

The AN/APQ-150A is a pulsed I-band radar system for US Air Force AC-130A and AC-130H gunships. It is a beacon location and tracking system which operates in co-operation with a beacon on the ground. The system has been upgraded to remain in service for at least another 10 years.

The two major units of the radar are the receiver/transmit unit and the control indicator. The radar transmitter contains the antenna positioning functions and the antenna, transmitter, receiver and signal processor components of the system. The control indicator contains the system controls and the display.

Specifications

Frequency:
(transmit) 9,375 MHz
(receive) 9,310 MHz
Peak power: 5 kW min
Accuracy: 2 mils RMS

Operational status

In service on US Air Force AC-130A and AC-130H gunships.

Contractor

Motorola Inc Government & Systems Technology Group.

VERIFIED

AN/APS-94F side-looking radar

The AN/APS-94F is a side-looking radar carried by the US Army OV-1 Mohawk aircraft used for battlefield and forward area reconnaissance and surveillance. The radar also has survey and mapping applications. The main difference between the AN/APS-94F and earlier versions is the inclusion of an ECCM capability. The installation comprises six main subassemblies: the antenna, receiver/transmitter, power supply and mount, signal processor, interconnecting box and cockpit complex.

The antenna unit consists of a pod containing two slotted waveguide arrays mounted back-to-back on a gyrostabilised assembly which pivots at the centre. In flight, the antenna arrays are yaw-stabilised to preserve the quality of the radar picture.

The receiver/transmitter is tunable within a portion of the radar band. It is located in the Mohawk equipment bay with the signal processor and interconnecting box. The antenna and other units can be easily removed and replaced on the aircraft to permit change of role at short notice. Apart from high-power microwave devices and certain display components, the APS-94F is of solid-state design.

The cockpit complex consists of the units necessary for operator control of the radar, recording and presentation of the data gathered. Two radar area maps are available to the operator. One depicts the entire sensed area, including all detected ground targets as though they are stationary; this is the fixed target map. The other shows moving targets and displays them against a suppressed background map of the area to give a moving target map. The radar imagery is developed in the air using a dry silver film process.

An area of up to 100 km on each side of the aircraft can be mapped when both of the antenna arrays are in use. Either one can be selected if it is desired to map only one side. A range control determines the width of the swath to be mapped and presented on the photo-radar map. This control has three settings corresponding to 25, 50 and 100 km wide scans by each antenna. When used in conjunction with the antenna switch to select left, right or both arrays, maps corresponding to 1:250,000, 1:500,000, 1:1 million and 1:2 million can be presented on the display. Provisions are made for the insertion of markers and other data.

Operational status

The AN/APS-94 is no longer in production but remains in service on the US Army OH-1D Mohawk aircraft. It has also been fitted in Lockheed P-3 aircraft and the UH-1 helicopter.

Contractor

Motorola Inc Government & Systems Technology Group.

VERIFIED

AN/APS-131 side-looking radar

The AN/APS-131 is a Side-Looking Airborne Radar (SLAR) used for the detection of ships and boats, for search and rescue, and for the detection of oil pollution. The aircraft installation consists of six main subassemblies: antenna, receiver/transmitter, synchroniser, amplifier, recorder/processor/viewer and control unit.

The antenna unit consists of a pod containing two 2.44 m yaw-stabilised, vertically polarised, slotted waveguide arrays mounted back-to-back on a gyrostabilised assembly that pivots at the centre. The remaining equipment is mounted inside the cabin. The synchroniser provides radar timing and control functions and, based on inputs from the aircraft inertial navigation system, creates latitude and longitude lines for display on the radar imagery. The recorder/processor/viewer contains the components necessary to create the film imagery. Video data is impressed on the film by applying it to two CRTs as the film is pulled across the CRT faces. Typically, each CRT generates an image of the earth's surface on one side of the aircraft. The dry silver film is developed in near real time.

An area of up to 200 km on either side of the aircraft can be mapped when both arrays are in use and one can be selected if mapping of only one side is required. A range control determines the width of the target area to be mapped and presented on the photo-radar map. This control has four settings corresponding to 25, 50, 100 and 200 km wide scans by each antenna. When used in conjunction with the antenna switch to select either left, right or both arrays, maps corresponding to four standard scales can be presented on the display at 1:250,000, 1:500,000, 1:1 million and 1:2 million. Radar and aircraft operational data is annotated on the film. This data, together with the latitude and longitude printed on the film, helps the measurement of map co-ordinates for any feature observed on the radar image.

Operational status

The AN/APS-131 was developed under contract to the US Coast Guard and is in service on HU-25A aircraft. A similar radar, the AN/APS-135 is used on the US Coast Guard HC-130.

Contractor

Motorola Inc Government & Systems Technology Group.

VERIFIED

AN/APS-135 side-looking radar

The AN/APS-135 is a Side-Looking Airborne Radar (SLAR) used for the detection of ships and boats, and search and rescue and ice patrol in the North Atlantic. Apart from the position of the antennas, the system is virtually identical to the AN/APS-131. The aircraft installation consists of seven main subassemblies: antennas, antenna switching unit, receiver/transmitter, synchroniser, amplifier, recorder/processor/viewer and control unit.

The antenna unit consists of two pods mounted either side of the aircraft to provide an unobstructed view, each containing a 4.88 m horizontally polarised slotted waveguide array. In flight the arrays are yaw-stabilised to preserve the radar picture quality. The remaining equipment is mounted on a pallet inside the aircraft for easy removal. The synchroniser provides radar timing and control functions and, based on inputs from the aircraft inertial navigation system, creates latitude and longitude lines for display on the radar imagery. The receiver/transmitter contains the magnetron and low-noise receiver. The antenna switching unit directs the radiated power to the left or right antenna.

The process for creation of the film imagery and the mapping areas covered is identical to the AN/APS-131.

Operational status

The AN/APS-135 has been developed under contract to the US Coast Guard and is currently in service on HC-130 aircraft. The AN/APS-131 is used on the US Coast Guard HU-25A aircraft.

Contractor

Motorola Inc Government & Systems Technology Group

VERIFIED

Side-Looking Airborne Modular Multimission Radar (SLAMMR)

A version of the AN/APS-131/135, the SLAMMR, is offered internationally for a wide variety of maritime surveillance and border patrol applications. SLAMMR has been delivered, in a maritime surveillance configuration on the Boeing 737-200 aircraft, to Indonesia, and also to another country in a border patrol configuration on a C-130H aircraft.

The radar is designed for a variety of applications such as ice patrol, fishery protection, anti-drug smuggling, mapping and search and rescue. In particular, its performance over the sea is sufficiently good to reveal details of oil spills from tankers, and provide maritime reconnaissance data to a maximum range of 185 km on either or both sides of the aircraft. The sensor uses a planar array 2.5 or 5 m long, mounted below the aircraft or along the side of the fuselage, feeding an operator's control/display console. The two antennas are arranged to point perpendicular to the aircraft longitudinal axis. From this position the two antennas can survey the sea or ground on either side of the aircraft out to the horizon.

A notable application of SLAMMR is the patrol maintained by the US Coast Guard to provide iceberg warning to shipping. Before the advent of side-looking radar, iceberg observations were conducted visually and therefore could only be performed in good visibility. Now with SLAMMR, the US Coast Guard HC-130 Hercules can maintain a much more consistent monitoring.

Operational status

No longer in production. In service on Lockheed Martin C-130 and Boeing 737 aircraft.

Contractor

Motorola Inc Government & Systems Technology Group.

UPDATED

AT 150 ATC transponder

The Narco AT 150 transponder is a self-contained panel-mounted unit which meets the FAA's C74c Class 1A TSO specification for ATC transponders. Featuring full 4096 Mode A and A/C code capability up to 30,700 ft in 100 ft increments, the AT 150 is compatible with leading encoding altimeters and blind encoders such as the Narco AR 850 altitude reporter.

The AT 150 is both a transmitter and receiver which responds automatically to ground-based radar interrogation. The transponder reply is displayed on ATC radar displays as two short parallel lines which fill in when the system squawks ident.

Specifications

Dimensions: 159 × 45 × 286 mm
Weight: 1.68 kg
Power supply: 14 V DC, 1.6 A or 28 V DC, 1.6 A with adaptor
Power output: 250 W nominal
Frequency:
(transmit) 1,090 MHz ±3 MHz
(receive) 1,030 MHz

Operational status

In service.

Contractor

Narco Avionics.

VERIFIED

KWX 56 digital colour radar

The KWX 56 system is an inexpensive pitch/roll stabilised digital three-colour radar.

The system comprises two units: a 5 in (127 mm) diagonal KI 244 or KI 248 panel-mounted high-contrast black matrix display and a KA 126 or KI 128 combined antenna/transmitter/receiver. The latter can be stabilised using a flight director or vertical gyro; the addition of roll-stabilisation eliminates screen blanking caused by ground returns during medium or steep turns. The flat plate antenna has a diameter of either 10 or 12 in (254 or 305 mm), and can be supplied either unpressurised or pressurised for altitudes of up to 20,000 and 50,000 ft (6,100 and 15,200 m) respectively.

The KGR 356 radar graphics unit combines with the KWX 56 to give a self-contained panel-mounted area navigation system. With the KGR 356 in the Nav mode, the radar can display a weather plot with the superimposed location of the active VORTac and waypoints stored in the memory.

Specifications

Dimensions:
(antenna/receiver/transmitter) 254 mm diameter × 160 mm
(control/indicator) 314 × 159 × 121 mm
Weight:
(antenna/receiver/transmitter) 4.27-4.61 kg
(control/indicator) 3.9 kg
Power supply: 28 V DC ±10%, 3 A max
Power output: (peak) 7.5 kW nominal, 6 kW min
Frequency: 9,375 MHz
Pulsewidth: 3.75 μs nominal
PRF: 109 Hz nominal
Ranges: 10, 20, 40, 80, and 160 n miles (18.5, 37, 74, 148 and 296 km) by rotary switch
Display: conventional colours. In mapping mode red becomes magenta, green becomes cyan and yellow remains unchanged
Stabilisation: from KI 256 flight director and ARINC standard vertical gyros associated with most autopilot/flight director systems, the KWX 56 can be used with Century 41, Century IV, Cessna ARC 400, 800 and 1000 series autopilots

Operational status

In production and in service.

Contractor

Narco Avionics.

VERIFIED

KWX 58 digital colour radar

The KWX 58 is similar to the KWX 56 but includes additional features that are particularly suited to turbine-powered aircraft including 592 km range, and a 'target alert' feature to show significant weather beyond the display range. Weather plots are shown in green, yellow, red and magenta, according to rainfall density, on a high-contrast, non-fading black matrix screen. The system uses a 254 or 305 mm diameter planar-array roll/pitch stabilised antenna radiating 7.5 kW. Narco claims that the penetration compensation feature gives a more accurate detection of storms behind closer areas of rainfall. Display ranges are 5, 10, 20, 40, 80, 160 and 320 n miles (9, 18.5, 37, 74, 148, 296 and 592 km). The KWX 58 system weighs 8.5 kg.

Specifications

Dimensions:
(antenna/receiver/transmitter) 254 mm diameter × 160 mm
(control/indicator) 314 × 159 × 121 mm
Weight:
(antenna/receiver/transmitter) 4.27-4.49 kg
(control/indicator) 3.9 kg
Power supply: 28 V DC ±10%, 3 A max
Power output:
(peak) 7.5 kW nominal, 6 kW min
Frequency: 9,375 MHz ±30 MHz
Pulsewidth: 3.75 μs nominal
PRF: 109 Hz nominal

Operational status

In production and in service.

Contractor

Narco Avionics.

VERIFIED

AN/APN-232 Combined Altitude Radar Altimeter

The Combined Altitude Radar Altimeter (CARA) is an all-solid-state 0 to 50,000 ft FM/CW radar altimeter system operating at a nominal frequency of 4.3 GHz. It consists of a receiver/transmitter, antennas and indicators and offers inherent low probability of intercept and anti-jam capability as well as conventional analogue and digital outputs for aircraft avionic systems.

Low Probability of Intercept (LPI) performance is achieved by control of the power output so that the transmitted power is the least amount required for signal acquisition and tracking. The system mechanisation automatically adjusts the required transmitter power to maintain normal system operation over varying terrain, aircraft altitude and attitudes. The FM/CW technology, operating over a 100 MHz bandwidth, provides inherent spread-spectrum capability which further reduces detectability.

The AN/APN-232 is easily maintained and offers low life cycle cost. The equipment incorporates highly modular packaging. The assemblies, subassemblies and components are 100 per cent screened and tested to stringent military standards. Ease of maintenance is achieved by using proven microprocessor-based digital display fault monitoring. These elements combine to produce a predicted system MTBF of greater than 2,000 hours. Low life cycle cost is obtained from high system MTBF, automatic self-test with fault isolation to both the line- and shop-replaceable units and a two echelon maintenance concept which maximises the use of shop-replaceable units that may be discarded rather than repaired.

The Navcom AN/APN-232(V) CARA system **1995**

Specifications

Dimensions:
(receiver/transmitter) 88.9 × 160 × 222.2 mm
(indicator) 82.55 × 82.55 × 104.1 mm
(fixed-wing antenna) 101.6 × 101.6 × 5.3 mm
(rotary-wing antenna) 142.2 × 158.7 × 5.3 mm
Weight:
(receiver/transmitter) 4.76 kg
(indicator) 1.13 kg
(fixed-wing antenna) 0.23 kg
(rotary-wing antenna) 0.34 kg
Power supply: 28 V DC, 100 W max
Accuracy:
(analogue) ±2 ft ±2%
(digital) ±2 ft (0-100 ft), ±2% (100-5,000 ft), ±100 ft (5,000-10,000 ft), ±1% (>10,000 ft)

Operational status

In production, with over 10,000 APN-232s delivered. CARA is the US Air Force's standard altimeter.

Further technology enhancements are in development as kit module replacements to the basic system. Among these is a tested capability to track terrain and features simultaneously.

Contractor

NavCom Defense Electronics Incorporated.

UPDATED

AN/APG-66(V) Series radars

The AN/APG-66(V) Series radars are multimode, versatile systems installed on 16 different airborne platforms operating in 20 countries. The APG-66(V) Series provides improved performance, functionality, reliability and maintainability over its predecessor, the AN/APG-66 which was designed for the Lockheed F-16. Over 2,300 of the original version of the radar have been produced and deployed worldwide.

The AN/APG-66(V) Series radar is based on the multinational F-16 Mid-Life Update (MLU) APG-66(V)2 radar. Improvements made to incorporate the latest technology include a newly developed signal data processor, higher-power transmitter, low-power RF speed and sensitivity and faster antenna phase shifting.

Consolidating the functions of the radar computer and digital signal processor reduces system size, weight, power and cooling requirements, while providing seven times greater processing speed and 20 times greater non-volatile memory than the AN/APG-66 system. A Doppler correlator reduces the false alarm rate by classifying radar returns as either ground vehicles, weather, mutual interference or sidelobe returns. The processing and fully programmable graphics capability in the signal data processor provides numerous options for radar operational mode growth, colour display support and growth capabilities to perform combined multisensor processing functions such as for FLIR or missile warning.

Significant advantages over the original AN/APG-66 have been realised, including twice the operational range, 40 per cent higher reliability and 16 per cent reduction in system weight. A four to one improved ground-map resolution has also been achieved. Greater situation awareness is provided by Track-While-Scan (TWS), Situation Awareness Mode (SAM), multitarget track and improved protection against electromagnetic interference. Support for multiple employment of up to six missiles is provided and the radar is compatible with a variety of Beyond Visual Range (BVR) air-to-air missiles and anti-ship missiles.

The radar provides air-to-air and air-to-surface modes in all weathers against all-aspect targets in look-up/shoot-up and look-down/shoot-down engagements, including high-clutter environments. The flexible multimode fire-control sensor is designed to provide integrated target and navigation data. Air-to-air modes provide the capability to detect and track multiple targets in the presence of ground clutter and electromagnetic interference. Air-to-surface modes provide extensive mapping for navigation, target detection, tracking and ranging.

The upgrade of the air-to-air capability of both individual aircraft and the F-16 units provides a significant air superiority margin that will also include a degree of air space management, incorporating both control aircraft and ground station datalinks providing real-time target information. Operating modes include 10 target TWS, Range-While-Search (RWS) look-up and look-down, 6 shot AMRAAM against six targets, medium-resolution DBS (Doppler Beam Sharpening) and enhanced ECCM capabilities.

Key radar features in air-to-air are: Search and track mode incorporating TWS, RWS and SAM functions. SAM offers a better track quality than TWS on multitarget track, plus the capability to track one or two targets while maintaining RWS scan volume. The TWS mode can, in addition to providing information on up to 10 targets, provide search information on up to 64 other targets.

There is a multitargeting capability for up to 10 target tracks, two target SAM tracks, single target track and BVR engagements. Inclusion of a BVR capability is a key point in order to provide MLU modified aircraft with the ability to use medium-range radar-guided missiles such as Sparrow, AMRAAM and MICA.

Air Combat Manoeuvre (ACM) features automatic acquisition of targets at short range in four predetermined scan patterns. Other information available includes the search altitude display which provides a reference to the altitude of all radar search returns and an indication of target relative groundspeed as well as an indication of the target aspect.

In the air-to-surface mode the APG-66(V)2 radar provides a significant improvement in both ground attack and anti-ship roles, especially in conditions of extreme clutter. Functions exist for the provision of a radar map video for target detection and navigation as well as providing an improved capability against shadowed or hidden targets. Using the ground map as a base, the APG-66(V)2 is also able to provide an overlay of navigation and reference points especially for rendezvous locations, weapon release points and waypoints. Operating modes include real-beam map, enhanced ground map, medium and coarse Doppler beam-sharpening, fixed target track, ground moving target indication and SAM in ground map interleaved mode.

Continuous updating of the navigation system is also enhanced through air-to-ground ranging which, combined with mission data information, provides confirmation of reference points along a proposed mission route and information regarding terrain-avoidance for obstacles in either overfly or avoidance modes that can be preprogrammed into the mission data package.

Additional features include moving ground target tracking that can also be superimposed on an improved definition ground map and, in both ground and sea modes, the ability to detect and fix on targets in open country or at sea. The sea mode provides a capability for the accurate detection and tracking of targets in both low and high sea states. The long-range sea mode provides a detection range reported to be as much as 148 km.

The radar may be readily adapted to a platform through expeditious software modifications supported by the fast Express software development environment and by selecting one of the nine antennas available or by resizing the array as necessary.

Features providing for integration flexibility include NTSC and PAL/CCIR video formats, 480 × 480 colour video resolution with 64 grey shades or 256 colours, fully programmable multifunction displays, MIL-STD-1553B remote terminal or bus controller, discrete input/output for HOTAS, CW illumination LRU and RAM or ECS cooling air capability.

Possible growth capabilities include weather awareness, blind let-down, wide swath map at medium altitude, terrain-following/terrain-avoidance, synthetic aperture radar, internal FLIR, missile warning and ESM processing.

Specifications

Volume:
(APG-66(V)1) 0.102 m^3
(APG-66(V)2) 0.097 m^3
(APG-66H) 0.082 m^3
(APG-66T) 0.08 m^3
Weight:
(APG-66(V)1) 134.3 kg
(APG-66(V)2) 115.9 kg
(APG-66H) 107.7 kg
(APG-66T) 98.4 kg
Power supply: 115 V AC, 3 phase, 400 Hz, 3,209 VA
28 V DC, 115 W
Frequency:
(APG-66(V)1) 6.2-10.9 GHz
(APG-66(V)2) 8-10 GHz
Search angle: 120° (azimuth and elevation)
Azimuth scan: ±10°, ±30°, ±60°
Elevation coverage: 1, 2 or 4 bar
Range scale: 10, 20, 40, 80 n miles (19, 37, 74, 148 km)
Electronic protection: multiple features, EMI pulse editor, fast phase shifting, frequency agility.
Maintainability: 5 min MTTR
Reliability: >210 h MTBF (over 300 h MTBF demonstrated)

Operational status

By the end of 1997, more than 2,300 systems had been produced. The initial variants in the AN/APG-66 Series, the APG-66T and APG-66H, have been deployed in the AT-3 trainer and British Aerospace Hawk 200 respectively. The APG-66H is currently in production.

AN/APG-66(V)2 for European participating governments F-16A/B mid-life upgrade programme
***1997**/0001217*

The APG-66(V)1 was developed for the US Air Force. The contract for 270 systems was completed in 1991.

Development of the APG-66(V)2 for the MLU F-16 was initiated in 1992. The programme will update 48 Belgian, 61 Danish, 136 Netherlands and 56 Norwegian air force AN/APG-66 radars to AN/APG-66(V)2 standard. DT&E was conducted at Edwards AFB in May 1995. Test missions were flown by the Netherlands Air Force in Autumn 1996. Production radar software was completed in January 1997.

The APG-66(V)3 is in production for the Taiwanese Air Force F-16 and 157 systems are currently under contract. Deliveries began in November 1995. This variant has a CW illumination capability for the AIM-7 Sparrow missile.

Flight trials for both the APG-66(V)2 and (V)3 radars were conducted in 1994. These successfully demonstrated: doubled operational detection range; false alarms reduced by a factor of 10 even in the presence of multiradars and severe ECM; simulated six-shot AMRAAM capability and ground-mapping range improved to 148 km.

Northrop Grumman is also under contract to deliver 30 ARG-1 radars - a version of the APG-66(V) customised for the A-4.

Most recently, Northrop Grumman ESSD has proposed two advanced technology derivatives of the AN/APG-66(V) Series:

(1) the Agile Beam Radar and Integrated FLIR Targeting System (ABR/IFTS) proposed for F-16 Block 60 aircraft; this configuration would combine Active Electronically Scanned Array (AESA) technology with two FLIR/laser turrets (one optimised for air-to-air operations and one for air-to-ground laser target marking)

(2) a less advanced development of the AN/APG-66(V)2 and (V)3 configurations using new commercial-off-the-shelf (COTS) processing technology; marketing names used to refer to this proposed upgrade include AN/APG-66(V)X and APG-NU.

Contractor

Northrop Grumman Corporation, Electronic Sensors and Systems Division.

UPDATED

AN/APG-68 radar

The AN/APG-68 is a coherent, multimode, digital fire-control sensor that provides a complement of 25 all-weather air-to-air and air-to-surface modes with superior dogfight and weapons-delivery capabilities. The air-to-air modes provide the capability to detect, track and engage targets at all aspects and at all altitudes even in the presence of ground clutter. Target information in the air-to-air modes is presented by a synthetic video on a 'clean scope' display. Air-to-surface modes provide extensive mapping, target detection, location and tracking, as well as navigation capability.

The AN/APG-68 features a highly programmable modular architecture derived from the Northrop Grumman AN/APG-66. The major operational improvements incorporated into the AN/APG-68 include extended detection range and multiple target tracking, beyond visual range utilisation of AMRAAM (Advanced Medium-Range Air-to-Air Missile) and Sparrow (AIM-7F/M) missiles and enhanced ECCM (electronic counter-countermeasures).

Since its initial deployment, the APG-68 has undergone regular improvements to performance and reliability.

The AN/APG-68 features an electrically driven antenna, a Modular Low-Power Radar Frequency (MLPRF) unit that features plug-in modules for growth and ease of maintenance, and an air-cooled Dual Mode Transmitter (DMT) that permits the radar to operate using low-, medium- and/or high-pulse repetition frequencies (PRFs). This hardware capability and flexibility allow the radar to be optimised easily for any air-to-air or air-to-ground search, track or mapping scenario.

The Northrop Grumman AN/APG-68 multimode high-performance fire-control radar for the Block 30/40 F-16C/D aircraft has demonstrated a Mean Time Between Failures (MTBF) of more than 150 hours in the

The AN/APG-68 radar fitted to US Air Force F-16C/D aircraft ***1998**/0018074*

field. For Block 50, the AN/APG-68 has been upgraded with a more advanced airborne signal processor (APSP), which has increased the system reliability to over 300 hours MTBF while enhancing system performance and growth potential of the radar, at a reduced cost. It provides 25 separate modes of operation, including long-range, all-aspect detection and tracking, simultaneous multiple target tracking, and high-resolution ground mapping. The latest development of the AN/APG-68 is known as the AN/APG-68 ABR, where ABR describes the Agile Beam Radar capability associated with new active array antenna technology. It is being proposed by Northrop Grumman for the projected F-16C/D Block 60 configuration.

Specifications

Volume: 0.13 m^3
Weight: 172.3 kg
Frequency: 8-12 GHz
Transmitter: gridded, multiple peak power travelling wave tube
Antenna: planar array, 740 × 480 mm
Search range:
(air-to-air) 160 n miles (296 km)
(air-to-ground) 80 n miles (148 km)
Range scales: 10, 20, 40, 80, 160 n miles
Azimuth scan: ±10°, ±25°, ±30°, ±60 °
Elevation coverage: 1, 2, 3 or 4 bar
Number of units: 4 LRUs, all in the nose of the aircraft
Maintenance: 30 min MTTR
Reliability: 300 h MTBF

Operational status

Over 6,000 copies of the AN/APG-68 and its predecessor, the AN/APG-66, have been produced. Besides the US Air Force, the AN/APG-68 is operated by the air forces of Bahrain, Egypt, Greece, Israel, Korea and Turkey. Radars procured for FMS to Bahrain in April 1998 are noted as having the designation APG-68(V)5.

Contractor

Northrop Grumman Corporation, Electronic Sensors and Systems Division.

UPDATED

AN/APG-76 MultiMode Radar System (MMRS)

APG-76 MMRS is designed to provide the F-4 Phantom with enhanced air-to-air and air-to-ground capabilities. Air-to-air capabilities include look-up, look-down and beacon mode, as well as air track and air combat modes. Air-to-ground performance is enhanced through the inclusion of real-beam ground map, high-resolution Synthetic Aperture Radar (SAR) and Doppler beam sharpening, ground mapping with simultaneous ground moving target indication, and beacon modes. Software enables the multimode radar system to provide guidance for future standoff air-to-ground weapons. A powerful clutter suppression interferometer provides a clutter-free resolution SAR map, as well as multitarget tracking capability. The radar's air-to-ground capabilities have also been used in the Gray Wolf technology demonstration (see operational status). Here, the radar is mounted in a pod and is described as offering a number of operating modes including real beam, Doppler sharpening and spotlight. In real beam, the Gray Wolf application is noted as having a maximum detection range in excess of 185 km while the Doppler beam sharpening capability allows 26 × 26 km sectors to be scanned with high-resolution values. The equipment's spotlight mode is thought to incorporate three submodes, the most sensitive of which provides 0.3 m resolution on targets at ranges of up to 130 km. The Gray Wolf application is also noted as being able to track up to 75 moving surface targets simultaneously.

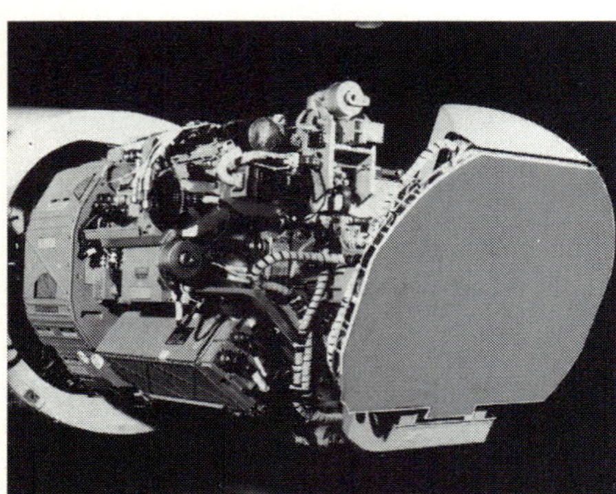

AN/APG-76 multimode radar ***1994***

Operational status

Northrop Grumman Norden Systems is reported to have supplied Israel with 50 APG-76 radars for use in Israel Aircraft Industries' F-4-2000 upgrade programme. A podded APG-76 codenamed Gray Wolf is also noted as having been tested as a theatre ballistic missile launcher detection and ground surveillance sensor aboard a USAF F-16 and a US Navy S-3B.

Contractor

Northrop Grumman Corporation, Electronic Sensors and Systems Division, Norden Systems, Norwalk, Connecticut.

UPDATED

AN/APG-77 multimode radar

A joint venture of Northrop Grumman's Electronic Sensors and Systems Division (ESSD) and The Raytheon Systems Company is developing the advanced AN/APG-77 radar for the US Air Force F-22. The radar's range will give an F-22 pilot unprecedented capability in air-to-air combat, allowing the pilot to track, target and shoot at multiple threat aircraft.

The first developmental F-22 radar has begun system level integration and testing at the ESSD facility in Linthicum, Maryland, US. The radar system is currently in a testing laboratory, where hardware and software integration is taking place as part of system level tests. The radar is the first of 11 systems to be delivered by the Northrop Grumman/Raytheon Systems Company team as part of the Engineering and Manufacturing Development (EMD) phase of the programme.

The radar employs an active electronically scanned antenna that features a separate transmitter and receiver for each of the antenna's radiating elements. This type of antenna provides the agility, low radar cross-section and wide bandwidth necessary to support the F-22's air superiority mission.

Systems integration and testing of the first F-22 radar is expected to extend over approximately 18 months. By the first quarter of 1998, the radar will be delivered to The Boeing Company's F-22 Avionics Integration Laboratory in Seattle, Washington, US, where engineers will integrate the radar with other F-22 avionics. Additional radar systems will be flight tested on a Boeing 757 testbed aircraft and on F-22 flight test aircraft. The radar is the first of 11 systems to be delivered as part of the EMD phase.

F-22 first flight took place in September 1997. Preproduction verification on three aircraft is due to occur between February 1998 and May 2002.

Full-scale production of the F-22 radar is scheduled to begin in 1999. IOC is planned for November 2004.

Specifications

Frequency: 8-12 GHz
Power per antenna module: 10 W
Reliability: 2,000 h MTBF estimated

Contractors

Northrop Grumman Corporation, Electronic Sensors and Systems Division.
Raytheon Systems Company.

UPDATED

AN/APN-241 airborne radar

The AN/APN-241 is a lightweight, fully coherent pulse Doppler radar. Based on the Northrop Grumman AN/APG-66/68 Series of fire-control radars, it has been developed to provide precision airdrop and navigation radar capabilities for military tanker and transport aircraft. The AN/APN-241 is operational on US and Royal Australian air forces' C-130H aircraft.

The AN/APN-241 has five radar modes: weather/turbulence, predictive windshear, high-resolution ground map, skin paint and beacon. The system has three display modes: station-keeping, flight plan and Traffic Collision Avoidance System (TCAS). The system can interleave radar modes, allowing the crew to view and control separate modes simultaneously while, at the same time, overlaying any of the three display modes. It is capable of accommodating a crew of two or three. The open systems architecture of the AN/APN-241 will allow advanced functions such as synthetic aperture radar, terrain-avoidance and maritime surveillance to be added without significant development costs.

Specifications

Antenna size: 610 × 890 mm
Weight:
(antenna) 31.8 kg
(electronics unit) 28 kg
Frequency: 9.3-9.4 GHz
Power output: 146 W peak, 11 W average
Range: 515 km
Coverage:
(azimuth) ±135°
(elevation) −15 to +10°

Operational status

The AN/APN-241 is in service on C-130H aircraft of the US Air Force, Royal Australian Air Force and Portuguese Air Force. It is also a candidate for upgrades to numerous tanker and transport aircraft worldwide. More than 230 radars of this type have been delivered worldwide.

Contractor

Northrop Grumman Corporation, Electronic Sensors and Systems Division.

UPDATED

AN/APQ-156/-148/APS-130 series multimode radars

AN/APQ-156/-148/APS-130 series radars are J-band airborne multimode equipments specially developed to combine the functions of two radars previously required by US Navy A-6A all-weather attack aircraft in a single radar. The APQ-156 system is a modification of the APQ-148 to accommodate the addition of a forward-looking infrared sensor/laser target recognition attack multisensor system to the A-6E aircraft. APS-130 is the system derivative fitted to some US Navy EA-6B electronic warfare aircraft. Functions performed by the APQ-156/-148/APS-130 include:
(a) search
(b) ground mapping
(c) tracking and ranging of fixed or moving targets
(d) terrain-avoidance or terrain-following
(e) beacon detection and tracking.

A track-while-scan capability provides simultaneous range, azimuth and elevation data for weapon delivery. As in other systems, range and azimuth markers must be placed on the target, but elevation data are available on a continuous basis and are derived from a separate phase interferometer array carried below the main scanner dish. The latter has a width of about 1 m and is illuminated by a conventional horn feed to produce a very narrow beam in azimuth.

The beam has a cosec^2 profile in elevation and this, with the interferometer elevation data provided, eliminates the need for mechanical scanning in the elevation plane. The interferometer array consists of two adjacent rows of 32 horns and moves with the main dish. Energy reflected from ground targets arrives at the upper and lower rows with a time difference which is measured by phase comparison techniques and translated into angular information.

There are two cockpit displays in the A-6E APQ-148 installation: a 13 cm storage tube unit for the pilot, and an 18 cm Direct View Radar Indicator (DVRI) for the bombardier/navigator. In the case of the APQ-156, only the DVRI is provided. Terrain data from the radar system are also presented on a vertical display for the pilot. The system incorporates comprehensive built-in test facilities. System weight is about 227 kg.

Contractor

Northrop Grumman Corporation, Electronic Sensors and Systems Division, Norden Systems, Norwalk, Connecticut.

UPDATED

AN/APQ-164 multimode radar

The AN/APQ-164 is the radar installed in the US Air Force B-1B aircraft. This radar combines technology from the F-16 AN/APG-68 radar and the Electronically Agile Radar (EAR) programme of the US Air Force.

The B-1B radar generates data for navigation, penetration, weapon delivery, and for certain other functions such as air refuelling. There are four modes in the AN/APQ-164 system that provide the navigation capability. The primary mode is a high-resolution synthetic aperture radar mapping mode, backed up by a monopulse enhanced real beam ground-mapping mode. The system also detects weather ahead and can display ground beacon returns over a real beam image. The penetration functions of the radar include automatic terrain-following and terrain-avoidance. For weapon delivery the radar provides four different functions. The first is a velocity update mode, similar to a Doppler navigator, which generates velocity information for the inertial navigation system. Coupled with an accurate Global Positioning System receiver in the avionics system, velocity update produces a dynamic, precision antenna calibration correction. Second, there is a ground moving-target detection and tracking capability for both fast and slow moving vehicles. Third is a high-altitude altimeter function that provides a very accurate measure of local height above the ground. Fourth is a monopulse targeting mode that provides accurate height to the on-scene selected fixed target.

The synthetic aperture mode provides the operator with a high-resolution image of an area of ground that can be chosen by the avionics system or the operator. Long-range maps can be made and five different map scales displayed. The synthetic aperture mapping mode accepts the co-ordinates of a waypoint from the avionics system and makes a map centered on that point. To make an image the antenna is electronically scanned to the waypoint location. The radar transmits a train of pulses, gathers data for the image, and then switches itself off. At the same time, the image is stored in the radar and presented on the display in a rectangular, ground co-ordinate display.

The radar provides the basic data required for automatic terrain-following. It scans the ground in front of the aircraft and measures the terrain in a range versus height profile out to 19 km and stores that data in the computer. The profile data is sent across the multiplex bus to the terrain-following control unit where the data is used to generate climb/dive commands. This flight profile is then automatically fed into the pilot's flight control system. Since the radar is not continuously scanning in terrain-following, a very low update is used, helping to reduce the risk of detection. This rate is variable and depends on aircraft altitude, manoeuvres, groundspeed and terrain roughness. Under normal conditions updates are made at 3 to 6 second intervals. However, if the terrain demands it, data can be gathered continuously.

The AN/APQ-164 in the B-1B is a dual-redundant system, with two complete and independent sets of Line-Replaceable Units (LRUs), except for the phased-array antenna. This was the first airborne application of this technology for combat aircraft. Only one set of LRUs is used at a time, the other being maintained on standby.

The phased-array is an outgrowth of the antenna developed on the EAR programme. It contains 1,526 phase control modules and allows virtually instantaneous beam movement to any point in the antenna field of regard. When the radar mission requires a forward, right or left region of regard, the antenna is physically movable to three different positions on a roll detent mount. The radar can, therefore, look off to either side of the aircraft or forward by rolling the antenna about an axis. The normal antenna position is looking forward. However, when the antenna is rolled to one side, the field of view extends from the aircraft nose back to about 115°, permitting a look off to the side of interest without having to change aircraft heading. Once physically moved to one of the three available positions, the antenna is locked into a detent. From the fixed spot, it can be scanned electronically ±60° in azimuth and elevation by means of a unit on the antenna called the beam-steering controller, which controls all 1,526 phase control modules.

Operational status

As of this edition, AN/APQ-164 is installed aboard US Air Force B-1B Lancer bomber aircraft.

Specifications

Frequency: I-band

Transmitter: gridded, multiple, peak-power TWT (similar to the AN/APG-68 transmitter)

Antenna: phased-array electronically scanned, 1,118 × 559 mm

Operating modes: (air-to-ground) high-resolution mapping, monopulse enhanced real beam mapping, automatic terrain-following, manual terrain-avoidance, velocity update, ground moving target detection and track, high-altitude calibrate, ground beacon; (air-to-air) weather mapping, air-to-air beacon, rendezvous mode. Growth for full conventional standoff capability and a full air-to-air mode complement is provided.

Weight: 570 kg

Contractor

Northrop Grumman Corporation, Electronic Sensors and Systems Division, Baltimore, Maryland.

UPDATED

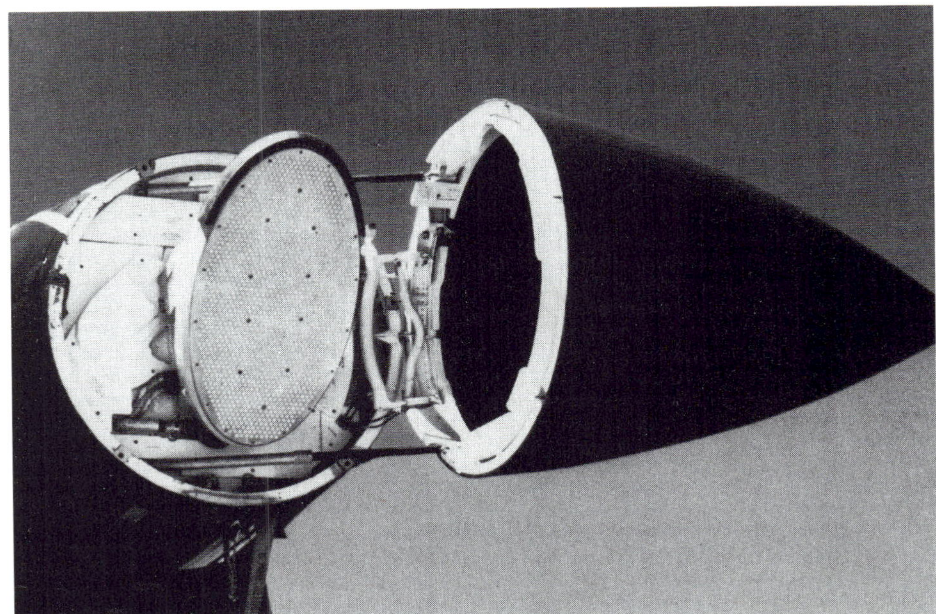

AN/APQ-164 airborne radar **1998**

AN/APY-2 radar for the E-3 AWACS

The Boeing Company is prime contractor to the US Air Force for the E-3 programme. Northrop Grumman's Electronic Sensors and Systems Division, based in Baltimore, Maryland, US, builds the production AN/APY-2 surveillance radar, which is installed in a Boeing 707-320 aircraft modified with a large radome for the radar and other antennas. The AN/APY-2 surveillance radar will also be installed on the Boeing 767 AWACS aircraft. The radar transmitter and receiver, communications gear, and other command and control systems are located inside the aircraft.

The E-3 radar provides full, long-range surveillance of high or low-flying aircraft. A maritime capability provides surveillance of moving or stationary ships. The E-3 aircraft operates during all kinds of weather and above all types of terrain.

The E-3 surveillance system detects and tracks both enemy and friendly aircraft in a large volume of air space. Low-flying aircraft, which can escape detection by ground-based radar, are detected by the E-3 aircraft.

The AN/APY-2 high-PRF pulse Doppler radar with its digital signal processing provides the downlook surveillance capability that is key to the E-3 airborne warning and control system.

The AN/APY-2 radar looks down at the ground, distinguishing between ground reflections (clutter) and radar returns from aircraft hugging the ground to escape detection. In addition to detecting either high or low-flying aircraft and ships out to the aircraft horizon, the E-3 radar also offers long-range aircraft surveillance above the horizon. For an E-3 aircraft altitude of 9,000 m (30,000 ft), the radar downlook range to the horizon is about 400 km (245 miles).

The E-3 aircraft provides a real-time assessment of both enemy action and friendly resources. With an E-3 aircraft the airborne commander has available the information that is needed to detect, assess and counter an enemy threat.

High-PRF pulse Doppler radar - concept: The basic advantage of high-PRF Doppler radar is that it provides better Doppler separation of moving targets from ground clutter. Additional radar techniques minimise ground clutter and maintain high sensitivity. Some of these are an extremely low-sidelobe antenna, ultrastable frequency generation, and digital processing techniques.

An inherent feature of the E-3 radar design is its flexibility to accommodate future growth through software control. Many signal processing functions are accomplished with programmed instructions rather than the specific hard-wired arrangements of circuitry so that functions or processes can be updated by altering programme instructions. These are programmable in flight; thus tactical programmes can be adjusted to respond to changes in the tactical situation.

Radar operating modes: The target-handling capability of the E-3 radar is enhanced by its operation in various modes, depending on the nature of the tactical situation. Each 360° radar scan can be divided into as many as 24 azimuth sectors and each sector can be operated with its own set of operation modes. Available radar modes are as follows:

(1) pulse Doppler non-elevation mode scans with high-PRF Doppler to provide downlook surveillance of aircraft out to the radar's horizon, but it does not measure target elevation.

(2) pulse Doppler elevation scan is similar to the previous mode, but includes an electronic vertical scan of the radar beam to provide target elevation.

(3) beyond the horizon mode of operation is used for long-range surveillance of medium- and high-altitude aircraft. Since the radar beam is above the horizon, there is no ground clutter and a low-PRF radar pulse is used to obtain the range and azimuth of target aircraft.

(4) passive scanning mode operates with the radar transmitter 'off' and the receiver 'on' to obtain ECM information, such as the locations of enemy jammers.

(5) maritime mode uses a very short radar pulse to provide the high resolution required to detect moving and anchored surface ships.

Multimode operation and sectoring permits maximum potential of the radar to be concentrated in sectors where the need is greatest. The pulse Doppler mode and beyond-the-horizon mode can be used simultaneously in an interleaved manner, as can the maritime and pulse Doppler non-elevation modes.

The slotted planar array antenna of the AN/APY-1/2 radar ***1998***/0018075

The rotodome of the Boeing E-3A Sentry houses the AN/APY-2 surveillance radar scanner

Technological features: To permit cancellation of the main-beam clutter in the radar receiver, extremely stable signal generation and advanced signal processing are used. The improvements in signal generation capability necessary for the E-3 radar were made through advances in circuit components, including oscillator crystals and transmitter tubes.

The signal-processing advances were made possible by the application of digital signal processing and by analogue-to-digital converter developments for the E-3 application. Radar control is accomplished by operating commands from the central computer.

Much of the high performance achieved by the E-3 pulse Doppler radar derives from the antenna and rotodome design. Low antenna sidelobes are necessary to minimise sidelobe clutter return and avoid reduction in detection performance for target returns that do fall within the sidelobe clutter region. The low sidelobe patterns that have been achieved in the E-3 radar antenna and rotodome represent a significant aspect in antenna design. The advance was made possible by the application of digital computing techniques to the design of the antenna, and by the use of high-precision digital techniques in manufacturing the antenna. The Boeing rotodome design enabled the low sidelobe characteristics of the antenna to be maintained when radiating through the rotodome.

In addition to minimising sidelobe clutter, the low-sidelobe antenna design is also a major contributor to the radar's resistance to jamming. The highly directional nature of the antenna when receiving rejects a jamming signal except when the antenna is pointed directly at the jamming sources. Hence, jamming signals can be easily identified and located.

E-3 radar Electronic Counter-CounterMeasures (ECCM) capabilities: The E-3 airborne surveillance radar system, with its inherent ECCM capabilities, can provide effective command and control under jamming conditions.

The E-3 ECCM features, such as low antenna sidelobes and inherent chaff rejection, are an integral part of the radar design, rather than add-on 'fixes'. This results from the pulse Doppler radar design to eliminate the effect of ground returns (clutter) and the requirement for the E-3 to operate effectively against a variety of electronic countermeasures.

Filtering is used in the pulse Doppler radar receiver to remove mainbeam clutter. This filter also eliminates the radar signals received from stationary or slowly moving targets, such as chaff.

Another form of ECCM available is the ability to switch to a radar frequency which is not being jammed. This negates the jammer effect even though the jammer remains within the radar line of sight.

The primary E-3 radar ECCM capability is derived from the very low antenna sidelobes. Flight tests of the E-3 radar, integrated with other E-3 subsystems such as a computer and tracking display consoles, have shown that the E-3 can operate successfully against powerful airborne and ground jammers.

The E-3 radar also has the ability to determine the relative bearing of a jammer with the radar in either an active or passive mode. The mobility of the E-3 enables it to perform self-triangulation. The system computer indicates the position data on an operator's console and computes the intercept path of the fighter aircraft designated to attack the jammer.

E-3 mobility also permits the use of jammer avoidance tactics to reduce or eliminate the jammer effects. The E-3 aircraft can drop below the radar horizon in relation to the jammers, thereby eliminating reception of the jamming energy. This tactic would result in some loss of low-altitude coverage but will enable target detection above the horizon line of sight to continue unaffected by the jammers.

Simultaneous operation of two or more E-3 aircraft enhances the overall system effectiveness against jammers and non-jamming targets. The E-3 has the ability to transmit information via a datalink system to other E-3 aircraft or to ground stations, producing a synergistic effect and improving the surveillance and battle management functions.

E-3 radar design: Radars for E-3 aircraft consist of three major subsystems: the slotted planar-array antenna located in the rotodome on top of the aircraft, the radar receivers and processors located in the centre of the aircraft cabin, and the radar transmitter located in the lower cargo bay. Total weight of the radar system is 3,742 kg.

The slotted planar-array antenna rotates with the aircraft rotodome at six revolutions/minute to provide horizontal radar scanning. The antenna face consists of 30 slotted waveguide sticks and measures 7.3 m by 1.5 m.

Vertical scanning and height-finding are performed by electronic scanning techniques using ferrite phase shifters. The phase shifters, phase-control electronics, receiver protectors and receiver paramplifiers are mounted on back of the antenna. The phase shifters are located on one side of the antenna and the electronics on the other side for weight balance and easy access during maintenance. Northrop Grumman-developed receiver protectors use a radioactive igniter power source for long life and fail-safe operation even during radar shutdown.

The radar transmitter consists of eight pressurised vessels located in the lower cargo bay. An overhead rail system permits easy removal of the transmitter units through a lower hatch.

The high-power transmitter chain is completely redundant with an inflight switchover capability should a malfunction occur. This transmitter redundancy, along with extensive redundancy in other parts of the E-3 system, assures high radar reliability and high probability of E-3 mission success.

Only two tubes are used in the transmitter chain: a high-power Klystron and a travelling wave tube driver. All other elements of the transmitter are solid state.

The radar receiver and digital signal processor are located in a single cabinet in the centre of the E-3 aircraft. Digital printed circuit boards and receiver electronics are accessed through hinged doors. Critical circuitry is backed up by redundant circuit boards.

A radar-dedicated digital computer and high-speed digital data processor control and monitor radar operation, reject ground clutter from radar returns, perform frequency analysis on signal returns, correlate radar returns to determine presence of legitimate targets, and digitally format the output target reports in range, velocity, azimuth and elevation.

Digitised radar information is passed in near-real time to the E-3 central processor and displays. The central processor correlates radar reports over successive scans to form target tracks. Navigation inputs are supplied to the radar computer to adjust for aircraft motion and altitude.

Digital radar signal processing provides significant advantages over conventional analogue processing in terms of cost, weight, complexity, reliability, maintainability and operability. Since the data processor is software programmable, system flexibility and room for potential system growth to meet changing environments also are advantages.

Use of high-reliability components throughout the radar, digital technology, use of integrated circuits and functional groupings of circuits, redundancy of critical circuitry with automatic switchover, built-in-test and fault isolation to an individual circuit board, and inflight maintenance capability all add together to make the E-3 radar highly reliable and easy to maintain.

Built-in test constantly monitors radar operation under control of radar computer software. Normal system operation is interspersed with fault detection tests. A 98.5 per cent probability of detecting online faults was demonstrated by the radar. In the event of a malfunction, the radar computer automatically reconfigures the system using available redundant circuitry. After correction, the results are displayed to the radar maintenance technician.

Spares for most non-redundant portions of the E-3 radar are carried on the aircraft and substituted while in the air. This attention to radar reliability and easy maintenance assures a high probability of mission success for the E-3 aircraft. Operational availability has consistently been greater than 95 per cent in the AWACS worldwide operation.

AWACS Radar System Improvement Programme (RSIP): Northrop Grumman's Electronic Sensors and Systems Division (ESSD), under a contract worth more than US$300 million from the US Air Force Materiel Command's Electronic Systems Center, is in the Engineering Manufacturing Development (EMD) phase of an E-3 AWACS Radar System Improvement Programme (RSIP). RSIP is a joint Air Force/NATO programme.

The RSIP contract includes the design, development and flight test of improvements to the AWACS AN/APY-1 and -2 radars to maintain operational capability against the growing threat from smaller radar cross-section targets, cruise missiles and electronic countermeasures. The contract also includes significant improvements in the man-machine interface and reliability and maintainability. RSIP represents the most significant upgrade to the E-3 radar since its development in the early 1970s.

The major portion of RSIP is devoted to increasing radar sensitivity against small targets through replacement of the digital Doppler processor and radar data correlator with a state-of-the-art Surveillance Radar Computer (SRC), and translation of the associated software into Ada language. In addition to handling the upgraded radar processing load, the SRC will contain adequate growth reserves to accommodate further radar upgrades.

Improvement in the man-machine interface will result from modification of the radar control and maintenance panel by incorporation of a spectrum analyser, special test equipment and new displays for monitoring the surveillance environment as well as the maintenance status of the radar system.

The RSIP EMD programme included fabrication of the first five modification kits for the US Air Force and one modification kit for NATO. In January 1997, the Boeing Company was authorised by the US Air Force, NATO and the UK to begin production of radar enhancement kits developed in the RSIP.

Under the initial production contract, Boeing and Northrop Grumman (ESSD) will build 18 modification kits for NATO, four for the US Air Force, eight for the UK. The first of three additional follow-on options for the US AWACS fleet has been exercised. The options ultimately will provide kits for 11 aircraft.

RSIP kits will be installed by: US Air Force at Tinker AFB for USAF; Daimler-Benz for NATO; British Aerospace for UK.

Operational status

The E-3 AWACS became operational with the US Air Force in early 1978. To date, 34 E-3s have been delivered to the US Air Force, 18 to NATO, and five to the Royal Saudi Air Force. Production of seven E-3 radars for the United Kingdom and four E-3 radars for the Republic of France has also been completed. Additionally, four 767 AWACS are being produced for Japan.

Contractor

Northrop Grumman Corporation, Electronic Sensors and Systems Division.

UPDATED

The JSTARS aircraft, showing the JSTARS radar under the forward fuselage ***1998***/0018076

Joint STARS AN/APY-3 Joint Surveillance Target Attack Radar System

The Joint Surveillance Target Attack Radar System (Joint STARS) designated AN/APY-3, is a long-range air-to-ground surveillance and battle management system. It is capable of locking deep behind hostile borders to detect and track ground movements, in both forward and rear echelon areas and to detect helicopter and fixed-wing aircraft. Joint STARS provides air and ground commanders with the intelligence and targeting data for management of their war-fighting assets.

Joint STARS is a complex of systems. It comprises an airborne platform, four major subsystems and a ground station module that receives, in near real time, radar data processed in the aircraft. The four major subsystems, all integrated on the airborne platform, consist of: an advanced radar; internal and external communications including UHF, VHF and HF voice links, the Joint Tactical Information Distribution System (JTIDS) and a newly developed surveillance and control datalink; operations and control including advanced computers and 18 operator display stations which perform the data processing and display functions for tens of thousands of targets and C^3I operation; and a self-defence suite which is in the process of being specified. Joint STARS is a joint US Air Force and US Army development, with the air force being responsible for the airborne segment and the army for the ground segment.

Joint STARS detects, locates, classifies, tracks and targets potentially hostile ground movements in virtually all weather. It operates in near real time and in constant communication through a secure datalink with army mobile ground stations that, in turn, can use TACFIRE and the advanced field artillery tactical data system to talk to artillery for fire support or to the all-source analysis system using US message format. Joint STARS will also maintain constant communication with air force tactical command posts via JTIDS. The platform for Joint STARS is a modified and militarised version of the Boeing 707-300 Series aircraft.

The technology employed in the radar was initially demonstrated as part of the Air Force/DARPA Pave Mower programme in the late 1970s. Major technological achievements of Joint STARS include: the software intensive displaced phased-centre slotted array antenna radar with several concurrent operating modes; the unique 8 m antenna mounted under the fuselage of the aircraft developed by Norden Division of United Technologies; very high-speed processors, each capable of over 600 Mops; high-resolution colour graphic and touchscreen tabular displays; the wideband surveillance and control datalink and over one million lines of integrated software code. The radar operates as either a side-looking synthetic aperture radar for the detection of fixed or stationary targets, or a Doppler radar to track slow-moving targets such as tanks or troop platforms.

In September 1987, the air force, following the August Joint Requirements Oversight Committee review, increased the number of E-8 aircraft from 10 to 22. In April 1988, the Conventional Systems Committee, after reviewing the Joint STARS airborne segment, concluded that the airborne platform should be changed from used aircraft to new 707s. However, in October 1989 the air force returned to the plan to continue the airborne portion of the programme with used 707 aircraft, on the grounds of the higher cost of new aircraft, to meet an initial operational capability in 1997.

In September and October 1990, operational field demonstrations in a dense electromagnetic environment and poor weather were completed in Europe. After six weeks of flying demonstrations, consisting of 25 missions and 110 flying hours, Joint STARS concluded the demonstration of the system's capabilities and data gathering for the continuing development of the programme. US Air Force and Army officials assessed the demonstration as a complete success. A further series of demonstrations to include operation of the associated ground terminal was planned for 1991.

The JSTARS radar antenna unit ***1998***/0018077

In November 1990, Northrop Grumman received a contract for the development of the third Joint STARS full-scale development aircraft and follow-on full-scale development work. Meanwhile, the two E-8A prototype aircraft (T-1 and T-2) achieved outstanding success in the 1991 Gulf War, and provided strong impetus for full development of the E-8C production configuration.

Specifications

Antenna: 7.3 m long, side-looking, phased-array housed in canoe-shaped radome under forward fuselage aft of nose landing gear; scanned electronically in azimuth, steered mechanically in elevation from either side of aircraft.

Operating modes: Wide area surveillance; fixed target indication; Synthetic Aperture Radar (SAR); moving target indicator; target classification.

Workstations: 17 identical workstations for system operators; one navigation/self-defence workstation; each operator can perform: flight path planning and monitoring; generation and display of cathographic and hyplographic map data; radar management, surveillance and threat analysis; radar; radar data review; time of arrival calculation; jammer location; pairing of weapons and targets, and other functions.

Communications: SCDL for transmission to ground stations; JTIDS; TADIL-J; Satcom; HF, UHF; VHF (SINCGARS)

Operational status

The two prototype E-8A aircraft are designated T-1 and T-2. The third aircraft - the first designated E-8C, is the preproduction Model (T-3). The first production E-8C (designated P-1), was handed over to Air Combat Command in June 1996. Joint STARS was approved in September 1996 for full-rate production for a planned buy of 19 production aircraft for US Air Force service (reportedly being reduced to 13 aircraft in recent budget cut proposals).

Initial Operational Capability (IOC) was achieved in December 1997, after delivery of the third production aircraft to the US Air Force in November 1997. Meanwhile, in June 1997, the US Air Force awarded Northrop Grumman two contracts for a computer replacement programme to install modern commercial-off-the-shelf (COTS) processor technology and fibre-optic links to the upgraded operator consoles.

Based on the success of the E-8A aircraft in the Gulf War, NATO expressed interest in procurement of the JSTARS system for its NATO Airborne Ground Surveillance (AGS) programme and Northrop Grumman has appointed Daimler-Benz Aerospace to be lead European contractor for its bid for the NATO AGS aircraft.

New higher-resolution SAR modes offered to NATO include: Enhanced (SAR) (ESAR); Inverse (SAR) (ISAR); and SAR Swath.

Contractor

Northrop Grumman Corporation, Electronic Sensors and Systems Division.

UPDATED

ASSR-1000 surveillance radar sensor

The ASSR-1000 is a radar sensor designed for the long-range detection of both air and maritime targets. It provides detection over rough terrain and high sea states. It is lightweight and adaptable for installation on a variety of platforms, including the Sentinel 1000 and 1200 airships.

The ASSR-1000 utilises technology from the Northrop Grumman Low-Altitude Surveillance System (LASS), with a solid-state transmitter, signal processor and post-processor. The lightweight radar has been carefully adapted for the airship airborne application and is planned to fly at altitudes up to 10,000 ft, providing a line of sight of 222 km. It is a NATO D-band fully solid-state coherent pulse compression system, featuring an ultra lightweight antenna, fully solid-state transmitter, frequency diversity and automatic target detection and tracking. It is capable of interfacing with operational displays, navigation equipment and datalinks. The ASSR-1000 includes an integrated AN/TPX-54 beacon interrogator.

It has a low probability of false alarm with a high probability of detection on small fluctuating targets, with a field-proven 60 dB moving target improvement factor over terrain and sea conditions. The antenna is a circular reflector, stabilised in both pitch and roll, with a wideband feed horn used for both the search and the beacon radars. The signal processor has special features that allow it to detect small targets in the surrounding environments. These features include full range I and Q processing, pulse compression, four pulse canceller with variable time periods between pulses, constant false alarm rate processing and dual-frequency transmission selectable from 21 frequencies.

The antenna for the ASSR-1000 surveillance radar is over 7 m wide and just under 4 m high

Specifications

Dimensions:
(antenna) 4,572 mm diameter
Weight: 850 kg nominal
Power: 12 kW
Frequency: 1,215-1,350 MHz
Pulsewidth: pulse pairs of 48 and 51 μs separated by 60 MHz
PRF: 375 pps average
Range:
(instrumented) 300 km
(2 m^2 air target) 215 km in clutter
(4 m^2 marine target) out to horizon in Sea State 3
Range resolution: 0.3 km
Azimuth coverage: 360°
Azimuth accuracy: 0.25°
Data refresh rate: 12 s
Reliability: >1,500 h MTBF

Operational status

In production and in service with the US Customs Service and several overseas customers.

Contractor

Northrop Grumman Corporation, Electronic Sensors and Systems Division.

UPDATED

Multifunction Integrated Radio Frequency System (MIRFS) for Joint Strike Fighter (JSF)

A team led by Northrop Grumman's Electronic Sensors and Systems Division (ESSD) is working under a $48.5 million contract from the Joint Strike Fighter (JSF) programme office to design, build and flight test the Multifunction Integrated Radio Frequency (RF) System (MIRFS).

Under the contract, the ESSD-led team is designing a new active electronically scanned array, a MultiFunction Array (MFA), which will help reduce the future JSF aircraft's avionics system cost by 30 per cent and its weight by 50 per cent. Other team members are Litton Amecon, Raytheon, GEC-Marconi Hazeltine and Harris.

The Northrop Grumman antenna will perform radar, high-gain electronic support measures, and Communication/Navigation/Identification (CNI) functions. The antenna will be flight-tested using existing radar support electronics from Northrop Grumman's APG-77 radar being developed for the F-22 aircraft, and a commercial-off-the-shelf emulator of the Integrated Core Processor (ICP). Software will be reused from the F-22 radar and from Northrop Grumman Norden Systems' AN/APG-76 radar. Northrop Grumman will flight-test the new MFA in the company's BAC 1-11 testbed aircraft against various airborne and ground-based targets, such as the SCUD missiles used during the Gulf War.

In parallel, Northrop Grumman is developing a Common Integrated RF (CIRF) concept that will combine radar and electronic warfare equipment into an advanced open architecture RF environment. The CIRF will replace the radar support electronics used in the initial flight tests and along with the advanced antenna and ICP will form a complete flying prototype of the MIRFS.

Operational status

Development.

Contractor

Northrop Grumman Corporation, Electronic Sensors and Systems Division.

VERIFIED

Multirole Electronically Scanning Aircraft system (MESA)

The MESA system is a low-cost, high-performance modular airborne surveillance system.

MESA supports a variety of air surveillance missions which include airborne, ground, maritime, environmental, drug interdiction and border patrol, and provides for the detection of low-flying aircraft, helicopters, tactical ballistic missiles and both stationary and moving targets in a clutter environment.

Long-range radar surveillance (460 km plus) missions are accomplished with a lightweight, electronically scanned array mounted either on top of the fuselage or in underwing pods, depending on the host aircraft chosen (Boeing 737, C-130, Fokker 50). The antenna is modular, allowing the radar system to be upgraded (with more powerful transmit/receive modules for greater range) as missions change.

Other sensor avionics can include IFF, ESM, FLIR and LLTV, all coupled to outboard workstations and radios.

Operational status

MESA continues to undergo ground integration and testing. A radar proof-of-concept flight demonstration was completed in 1994 and flight testing continued through 1996. A system flight demonstration was planned for summer 1997. Current status uncertain.

Contractor

Northrop Grumman Corporation, Electronic Sensors and Systems Division.

UPDATED

Airborne Shared Aperture Programme (ASAP)

The ASAP is sponsored by the US Naval Air Warfare Center and is intended to demonstrate a multifunction, broadband, shared aperture system for use on future air and sea systems. Wideband individual transmit and receive modules would make up the ASAP sensor aperture, providing the capability to transmit and receive radar, communications, ECM and ESM signals over a broad frequency band.

Operational status

A sole-source contract has been awarded to demonstrate ASAP technology.

Contractor

Raytheon Systems Company.

UPDATED

An F-15 Eagle fighter with nose radome swung aside to give access to the AN/APG-63 radar

AN/APG-63(V) fire-control radar

The AN/APG-63 is the principal sensor for the F-15 Eagle. The system was designed around three main objectives: capability, reliability and maintainability. Capability is aimed at providing one-man operation for the tracking of hostile aircraft at long range and close in, in a look-down situation; a clutter-free display with all appropriate information, including guidance and steering information, on a head-up display; simplified controls and co-ordination with weapons, and a secondary air-to-ground facility.

The requirement for the AN/APG-63 called for an MTBF of 60 hours, a level never before approached for this class of equipment. As a corollary, maintenance time would be reduced to about a quarter of that of previous systems. High reliability and relatively simple maintenance schedules reduce the turnround time. This is reflected in the smaller numbers of flight line and maintenance personnel needed. This is largely due to the use of hybrid and integrated circuits to reduce the number of components and to the use of only four basic module sizes in the entire system.

The system comprises nine LRUs: exciter, transmitter, antenna, receiver, analogue processor, digital processor, power supply, radar data processor and control unit. It has a wide look-angle and its antenna is gimballed in all three axes to hold target lock on during roll manoeuvres. The clutter-free head-down radar display gives a clear look-down view of target aircraft silhouetted against the ground, even in the presence of heavy clutter from ground returns. This look-down, shoot-down ability is achieved by using both high and medium PRFs, by digital data processing and by using Kalman filtering in the tracking loops. The system's gridded travelling wave tube permits variation of the waveform to suit the tactical situation.

False alarms are eliminated, regardless of aircraft altitude and antenna look-angle, by a low-sidelobe antenna and frequency rejection of both ground clutter and vehicles moving on the ground, so that only real targets are displayed.

Primary controls for the multimode, pulse Doppler I-/J-band radar are located on the control column, allowing the pilot to keep his head up during fast-moving situations. Three special modes are provided for close in combat: supersearch, vertical scan and boresight. These enable automatic acquisition of, and lock on to, targets within 18.5 km. In the supersearch mode the radar locks on to the first target entering the head-up display field of view. In the vertical scan mode the radar locks on to the first target that enters an elevation scan pattern at right-angles to the aircraft lateral axis. In the boresight mode the antenna is directed straight ahead and the radar locks on to the nearest target within the beam. In all tracking modes the target position is displayed on the HUD if the target is within the field of view. This greatly increases the range of visual detection.

The ability to handle equally effectively both closing and opening look-up and look-down situations is the result of combining high and medium PRFs. High PRFs are necessary to detect targets at long range but are not suitable for measuring range because there is insufficient time for a pulse to return and be correlated before the next one has been transmitted. However, medium PRFs that enable accurate range measurement and elimination of ground clutter do not have the power to give detection echoes at long range. The APG-63 interleaves, for the first time, both high- and medium-PRF waveforms. The key to this technology is the substitution of heavy and bulky Doppler filters with a digital signal processor. By this means, incoming signals are sampled and their frequency content analysed by performing Fourier transforms on individual samples.

Although designed specifically for the air-to-air role, the F-15 has emerged as a potent ground attack fighter, and the APG-63 has target ranging for automatic bomb release for a visual attack, a mapping mode for navigation and a velocity update for the inertial navigation system.

The APG-63 is compatible with AIM-12, AMRAAM, AIM-7F Sparrow and AIM-9L Sidewinder air-to-air missiles. Antenna search patterns and radar display presentations are selected automatically by means of a three-position switch on the throttle for the type of weapon to be used (medium-range radar homing, short-range infrared missiles or M61 cannon).

While the eventual reliability goal of 60 hours specified by MIL-STD-1781 was met consistently on bench tests, in the field, service equipment is currently giving about 30 to 35 hours MTBF.

All APG-63 radars produced since mid-1980 incorporate a programmable signal processor and a high-speed digital computer that enables the system to respond quickly to new tactics or weapons by changes to software rather than extensive hardware modifications. This change was introduced with the improved F-15C and F-15D models. Aircraft produced before the processor was introduced were scheduled to receive it as a retrofit item.

In December 1983, the US Air Force awarded a $274.4 million contract to upgrade the control computer and armament control system as part of the MultiStaged Improvement Programme (MSIP) launched in February of that year. MSIP arose out of a McDonnell Douglas study, beginning in June 1982, which showed the need to improve the avionics system, and specifically the radar, in order to maintain reasonable combat superiority over likely adversaries. Radar improvements include a memory increase to 1 M words, a radar data processor speed tripled to 1.4 Mops and improvements to both transmitter and receiver. The outcome of this is the AN/APG-70.

In March 1984, the F-15 was selected as the US Air Force's new dual-role air control and interdiction fighter. The upgraded radars for the two-seat F-15E version incorporate very high-speed integrated circuit technology to increase computational speed and improve ECM performance.

Specifications

Volume: 0.25 m^3
Weight: 221 kg total
Number of LRUs: 9
Frequency: I/J-band (selectable)
Transmit power: 12.975 kW
Reliability: 60 h MTBF

Operational status

In service with the F-15 Eagle. The last system was delivered in September 1986, by which time some 1,000 sets had been built, including co-production in Japan. Four P-3A Orions, on loan from the US Navy, have been modified to include a AN/APG-63 radar mounted in the aircraft nose, for anti-drug-smuggling operations with the US Customs Service.

The US Air Force plans to upgrade the AN/APG-63 radars in more than 350 F-15C/Ds at a cost of US$189.8 million. Flight testing began in 1997 with retrofitting starting in 1999 at the rate of 72 aircraft per year.

The upgraded radar, designated APG-63(V)1, utilised software from the later model APG-70 radar to provide a substantial increase in computing power and electronic counter-countermeasures (ECCM) capability, while improving the air-to-ground mode.

Contractor

Raytheon Systems Company.

UPDATED

AN/APG-65 multimission radar

In the air-to-air role the APG-65 radar incorporates the complete range of search, track and combat modes, including several previously unavailable in an operational radar. Specifically these modes are gun acquisition (the system scans the head-up display field of view and locks on to the first target it sees within a given range); vertical acquisition (the radar scans a vertical slice of airspace and automatically acquires the first target it sees, again in a given range); and boresight (the radar acquires the target after the pilot has pointed the aircraft at it). In the raid assessment mode the pilot can expand the region around a single target that is being tracked, giving increased resolution around it and permitting separation of closely spaced targets. As a gunsight, the radar operates as a short-range tracking and lead-computation device using frequency agility to reduce errors due to target scintillation. All the pilot has to do is to put the gunsight pipper on the target and press the firing button.

Other modes are long-range velocity search (using a high-PRF waveform to detect oncoming aircraft at high

The US Navy F/A-18 Hornet is equipped with the AN/APG-65 radar

The AV-8B Harrier II Plus is equipped with the AN/APG-65 radar

relative velocities); range-while-search (high- and medium-PRF waveforms interleaved to detect all-aspect targets, not only head-on but at any line of sight crossing angle), and track-while-scan which can track up to 10 targets simultaneously and display eight. When combined with autonomous missiles such as AMRAAM, this mode confers a launch-and-leave capability and the simultaneous engagement of multiple targets.

In the F/A-18's air-to-ground role, the APG-65 has six modes: terrain-avoidance, for low-level penetration of hostile airspace; precision velocity update, when the radar provides Doppler signals to update or align the aircraft's inertial navigation system; tracking of fixed and moving ground targets; surface vessel detection, in which the system suppresses sea clutter by a sampling technique; air-to-surface ranging on designated targets; and ground-mapping. Two Doppler beam-sharpening modes are provided for these air-to-ground modes.

During tests in early 1983, the system demonstrated (a year ahead of schedule) the 106 laboratory test hours MTBF required by contract. Two systems chosen at random for a reliability demonstration operated under test for a total of 149 hours without a failure.

All fault-finding is conducted with Built-In Test (BIT) which currently operates at module LRU level. The biggest improvement in BIT technology with the AN/APG-65 has been the reduction in false alarm rate. The US Navy requirement is to be able to detect and locate 98 per cent of all radar faults by BIT. The US Navy also specifies 12 minutes MTTR, which involves running a BIT test, locating the fault, removing and replacing the faulty line-replaceable unit and running a BIT check on the new unit.

In its first six months of operation the initial Marine Air Group II at El Toro in California had only six radar engineers to support a 12 aircraft unit flying 30 hours a month.

Under the terms of an Australian Industrial Participation (AIP) programme signed in December 1981, Philips Electronics Systems in Australia is engaged in final assembly and test of the radar and co-produces the data processor. In February 1984, it was announced that Marconi Española SA had been licensed to build low-voltage power supply modules for the radars being produced for Spain's F/A-18s.

German Air Force F-4F ICE programme

In May 1985, the German Ministry of Defence chose the AN/APG-65 as a major element in the Improved Combat Efficiency (ICE) programme for the German Air Force F-4F Phantom. This radar is also designated AN/APG-65GY.

Daimler-Benz built the radars under licence and ICE includes a Honeywell laser inertial navigation system, GEC Avionics air data computer, Daimler-Benz radar display and Raytheon Systems Company cockpit displays. DASA is also contracted to provide the ICE upgrade to the Hellenic Air Force as part of its F-4E upgrade.

Radar for AV-8B

A variant of the APG-65 entered service in the AV-8B Harrier II plus with the US Marine Corps and Italian and Spanish navies during 1993.

Specifications

(APG-65 in F/A-18)
Volume: <0.126 m^3 excluding antenna
Weight: 154.6 kg
Number of LRUs: 5
Frequency: 8-12 GHz
Antenna: low-sidelobe planar-array with fully balanced direct electric drive replacing hydraulics and mechanical locks of previous systems
Transmitter: liquid-cooled, contains software-programmable gridded travelling wave tube amplifier
Receiver: contains A-D converter
Radar-data processor: general purpose with 250 k 16-bit word bulk-storage disk memory
Signal processor: fully software-programmable, runs at 7.2 Mops
Reliability: 120 h MTBF. Built-in test equipment detects 98% of faults and isolates them to single replaceable assemblies that can be changed in 12 min without adjustments or setting up

Operational status

In service with US Navy, US Marine Corps, Canadian Forces, Royal Australian Air Force, Spanish Air Force and Kuwaiti Air Force F/A-18 aircraft; US Marine Corps, Spanish Navy and Italian Navy AV-8B Harrier II Plus aircraft and the German Air Force F-4F ICE programme.

In production for AV-8B Harrier II Plus aircraft of the US Marine Corps, Italian Navy and Spanish Navy. Contracted for the Hellenic Air Force F-4E upgrade in the ICE configuration.

Contractor

Raytheon Systems Company.

UPDATED

AN/APG-70 radar for the F-15E

The radar for the two-seat F-15E combat aircraft is a substantially improved version of the APG-63, designated AN/APG-70. It is also installed in a limited number of F-15C/D aircraft.

By comparison with the earlier system, the APG-70 has a far greater RF bandwidth, a larger look-down target detection range, a one-third increase in MTBF and is packaged in eight units instead of nine. Four of the LRUs (radar data processor, programmable signal processor, analogue signal converter and receiver/exciter) are completely new, while the transmitter and control unit have been modified; only the power supply and the antenna remain unchanged. The radar uses VLSI technology, making the programmable signal processor five times faster than previous designs while having three times the memory.

Expanded built-in test is provided, giving unambiguous fault detection and isolation and the ECCM capability is improved to combat new and more advanced threats. The system is compatible with existing and new missiles such as AIM-7F/M Sparrow, AIM-9 Sidewinder and AIM-120 Advanced Medium-Range Air-to-Air Missile (AMRAAM), and with 20 mm cannon.

In the air-to-air role the radar has five search modes: range-while-scan with high PRF, medium PRF or interleaved PRF, range-gated high PRF and velocity search. The radar also has single-target track, track-while-scan and raid assessment track modes and vertical search, super search, boresight and auto-guns target acquisition modes.

In the air-to-ground role the radar can produce a high resolution or a real-beam ground map; the specification calls for 2.6 m resolution at 75 km range. There is also a precision velocity update mode and air-to-ground ranging. For the future, terrain-following/terrain-avoidance, ground moving target track indicator and fixed-target track modes are planned.

In December 1987, Hughes (now Raytheon Systems Company) announced a US$58 million contract to supply a variant of the APG-70 radar as the fire-control system in the US Air Force Special Operations Force's AC-130U aircraft. This aircraft, of which 12 have been ordered, has a 105 mm howitzer and 25 mm and 40 mm cannon. Five new air-to-ground modes were developed for this application: fixed-target track, ground moving target indication and track, projectile impact point position, beacon track and a weather mode. The existing antenna and signal processor were modified and a digital scan converter added to the system. The first system was delivered at the end of 1988.

Specifications

Volume: 0.25 m^3
Weight: 251 kg
Frequency: 8-20 GHz selectable
PRF: multiple
Number of LRUs: 8

The AN/APG-70 in an F-15C aircraft

1996

Range:
(air-to-air) 185 km
(automatic acquisition) 500 ft-20 n miles
(ground map) 50 n miles+
Resolution:
(ground map) 2.6 m at 75 km
Reliability: 80 h MTBF

Operational status

In service in some F-15C/Ds and the F-15E.

Other variants of the F-15E equipped with a modified AN/APG-70 radar are being delivered to Saudi Arabia (F-15S) and Israel (F-15I).

Contractor

Raytheon Systems Company.

UPDATED

AN/APG-71 fire-control radar for the F-14D

AN/APG-71 is an enhancement of the AN/AWG-9 radar originally fitted to F-14 aircraft.

Compared with the AWG-9, the APG-71 offers better overland performance, expanded velocity search capability, a larger target engagement zone, a raid assessment mode and programmable electronic countermeasures and clutter control features.

The APG-71 is essentially a digital version of the radar section of the AWG-9, with greatly improved ECM performance acknowledging the new and vastly more sophisticated jamming technologies that have appeared since design of the F-14 was frozen. New modes include medium PRF all-aspect capability, monopulse angle tracking, digital scan control, target identification and raid assessment, but the number of boxes comes down from 26 to 14. The system also employs some of the elements developed by Hughes for the new F-15 radar, the AN/APG-70; for example the APG-71 has a signal processor which is 86 per cent common in modules with that in the APG-70. The AWG-9's transmitter, power supply and aft cockpit tactical information display are retained for the APG-71.

The APG-71 also incorporates non-co-operative target identification, by which radar contacts may be identified as friendly or hostile, at beyond visual ranges, through close examination at high resolution, of the returns; this technique obviates deficiencies and ambiguities in IFF equipment.

The AN/APG-73 radar for the F/A-18 **1995**

The APG-71's antenna retains the gimbal system of the AWG-9 and adds a new array with low sidelobes and a guard channel to eliminate sidelobe penetration of ground clutter and electronic warfare interference. The APG-71 also provides an improved radar master oscillator which significantly increases the number of radar channels and provides frequency-agile operation, with low sidelobes.

The signal processor has four processing elements as against the three found in the APG-70, giving an operating speed of 40 Mcops (million complex operations per second). The radar data processor also has a large degree of commonality with that in the APG-70, differing only in the interface cards, and operates at 3.2 Mips (million instructions per second).

The aft cockpit digital display for the APG-71 was originally developed for the AWG-9 but never put into production, while the tactical information display, originally designed for the F-106 aircraft, remains largely unchanged from the version in the F-14A. Significantly, the APG-71's software is written in Jovial.

The AN/APG-71 radar in an F-14D

Specifications

Volume: 0.78 m³
Weight: 590 kg

Operational status

The first system was delivered to Northrop Grumman in the second half of 1989.

Contractor

Raytheon Systems Company.

UPDATED

AN/APG-73 radar for the F/A-18E/F

The AN/APG-73 radar is based on the AN/APG-65, but uses new signal and data processors and a revised receiver/exciter. The Raytheon Systems Company claims this will give the new radar more than three times the speed and memory of the AN/APG-65 and will make it compatible with an electronically scanned antenna.

Specifications

Volume: (excluding antenna) 0.126 m³
Weight: 154 kg
Frequency: 8-12 GHz
Number of LRUs: 5 plus antenna

Operational status

In production for new F/A-18C/Ds and the F/A-18E/F for the US Navy and Marine Corps. First operational units of the AN/APG-73 were delivered in mid-1994. The AN/APG-73 was also specified for F/A-18C/Ds ordered by Finland and Switzerland.

Latest development models for the F/A-18E/F are believed to carry the designator AN/APG-73(V).

Contractor

Raytheon Systems Company.

UPDATED

AN/APQ-122(V) radar

The AN/APQ-122(V) is a dual-frequency nose radar developed for use in the US Air Force Adverse Weather Aerial Delivery System (AWADS) programme for installation in C-130E transport aircraft. This long-range navigation sensor is used for weather avoidance and navigation in supply dropping missions. The equipment provides ground-mapping out to more than 385 km, weather information up to 278 km and beacon interrogation up to 444 km when using the I-band frequency radar. J-band frequencies are used when short-range high-resolution performance and target location are required. In the J-band mode the radar provides a high-resolution ground map display to permit target identification and location for position fixing and aerial delivery missions. In this mode the radar will detect and display targets with a radar cross-section of 50 m^2 while operating in rainfall of 4 mm/h.

In addition to the dual-frequency system designed for AWADS, designated AN/APQ-122(V)1, three other configurations have been developed. The AN/APQ-122(V)5 is a single-frequency I-band radar which has been developed as a direct replacement for the AN/APQ-59 radar used in C-130 and E-4B aircraft. Facilities include long-range mapping, weather evaluation and avoidance and rendezvous. A navigation training version of the AN/APQ-122(V)5, the AN/APQ-122(V)7, has been designed for use in the T-43A aircraft. Another dual-frequency radar, the AN/APQ-122(V)8, incorporates a terrain-following capability and is used on Combat Talon 1 MC-130 aircraft.

Operational status

The AN/APQ-122(V) has been supplied to the US Air Force and the air forces of Argentina, Australia, Bolivia, Cameroon, Congo, Denmark, Ecuador, Egypt, Gabon, Greece, Indonesia, Iran, Israel, Italy, Jordan, Libya, Malaysia, Morocco, Nassau, New Zealand, Niger, Nigeria, Oman, Philippines, Portugal, Saudi Arabia, Singapore, Spain, Sudan, Thailand, Venezuela and Zaïre. The radar has been installed in C-130H, RC-130A, KC-135A, RC-135A, RC-135C and E-4B aircraft.

Contractor

Raytheon Systems Company.

UPDATED

AN/APQ-126(V) terrain-following radar

The AN/APQ-126(V) is a forward-looking variable configuration airborne navigation and attack radar which was produced for the US Navy A-7E and US Air Force A-7D aircraft. It operates in the J-band and its primary functions are ground-mapping, air-to-ground ranging and terrain-following/terrain-avoidance. The radar also features adverse weather look-through using selectable polarisation, slaved antenna pointing in air-to-ground ranging and variable tilt control which allows the pilot to optimise ground map displays and highlight points of interest.

Operational status

The AN/APQ-126 radar equips US Navy A-7E and US Air Force A-7D aircraft and was also produced for the A-7H aircraft of the Greek Air Force and for the A-7P and TA-7P aircraft of the Portuguese Air Force. The cost of 30 sets of APQ-126 for these latter aircraft was given as US$10.4 million. A variant, known as the AN/APQ-158, has been developed for the HH-53 helicopter. Production is now complete with over 1,000 units delivered.

Contractor

Raytheon Systems Company.

UPDATED

AN/APQ-158 radar

The AN/APQ-158 is a multimode forward-looking radar used primarily for terrain-following/terrain-avoidance at low altitudes in the Pave Low III night/adverse search and rescue helicopter, the Sikorsky MH-53J. The equipment is similar to the AN/APQ-126 but is modified for compatibility with the unique helicopter characteristics and the Pave Low III mission requirements. The radar contains 15 LRUs which provide the same basic modes of operation as the AN/APQ-126.

System upgrades have provided this radar with the ability to supply updates in all modes except terrain-following, and to perform terrain-following missions over very high clutter areas such as cities.

Operational status

In service in MH-53J helicopters.

Contractor

Raytheon Systems Company.

UPDATED

AN/APQ-168 multimode radar

The Sikorsky HH-60D Night Hawk helicopter is designed to penetrate hostile territory during darkness to rescue downed aircrews or deliver and retrieve special operations teams. The task calls for long-distance nap of the earth flying and accurate navigation.

The radar can operate in terrain-clearance, terrain-avoidance, air-to-air ranging and cross-scan modes, the latter combining ground-mapping or terrain-avoidance with terrain-following. A terrain storage facility permits the radar to have a reduced duty cycle thereby reducing the probability of detection by enemy ESM equipment.

The system has increased electronic countermeasures resistance, improved weather penetration, better guidance in turning flight, a power management function for semi-covert operation and low beam reflectivity. Extensive BITE provides a high degree of fault isolation and detection. The system is carried in a pod in the nose of the aircraft.

Specifications

Dimensions: 1,420 mm long × 330 mm diameter
Weight: 113 kg
Power supply: 115 V AC, 3 phase, 200 VA
28 V DC
Reliability: 144 h specified MTBF

Operational status

In service with the US Air Force Sikorsky HH-60D Night Hawk helicopter.

Contractor

Raytheon Systems Company.

UPDATED

AN/APQ-172(V) radar

The AN/APQ-172(V) forward-looking radar provides terrain-following and ground-mapping capabilities for the RF-4C aircraft, for all-weather day or night selective reconnaissance at high- or low-level flight. The AN/APQ-172(V) radar is a reliability and maintainability upgrade to the AN/APQ-99 radar, originally fielded in 1965. The upgrade is achieved through the incorporation of modification kits into the AN/APQ-99 terrain-following radar. Once upgraded, the system is redesignated as the AN/APQ-172(V) radar.

The radar provides three terrain-clearance modes, an air-to-air ranging mode for refuelling operations, two ground-mapping modes and two combined time-sharing modes. The terrain-following mode allows manual low-level flight in clear and adverse weather, utilising monopulse resolution improvement. Ground-mapping capability is provided up to 150 km range. The time-sharing modes include terrain-following/terrain-avoidance and terrain-following/ground-mapping. During the combined modes, terrain-following presentations are provided on the forward indicator and ground map or terrain-avoidance presentations are provided on the aft indicator. Operation is in the J-band.

Operational status

In operational service in aircraft of the German and Japanese air forces.

Contractor

Raytheon Systems Company.

UPDATED

AN/APQ-174 MultiMode Radar (MMR)

The AN/APQ-174 multimode radar has been developed for US Army, Navy and Air Force combat rescue and special operation missions, for use on aircraft such as the HH/MH-60, CH/MH-47, HH-53 and the V-22. The radar is a derivative of the LANTIRN terrain-following radar and the AN/APQ-168 multimode radar and maintains commonality with five of the six LANTIRN LRUs. The system will enable an aircraft to perform special operations and search and rescue missions at night, in adverse weather conditions and in a high threat environment.

AN/APQ-174 modes include normal, power management and weather, terrain-following, terrain-avoidance, ground-ranging, beacon and weather. Set clearances are 100, 150, 200, 300 and 500 ft. Weather performance is enhanced by the use of selectable circular polarisation and operation in 10 mm/h rain is claimed. The system includes extensive internal monitoring, periodic and manually initiated BIT and end-to-end test.

The AN/APQ-174 allows operations at low altitudes, down to 100 ft above the ground by day or night. MFR improvements include the addition of weather detection and beacon interrogation modes to help in navigation and rendezvous. Other upgrades to the radar include expansion of software memory, conversion to electrically erasable memory and addition of an obstacle warning signal to existing video displays. Another upgrade, adding a new set clearance altitude to the terrain-following mode, will allow the aircrew to train at a safer altitude.

Specifications

Dimensions:
(pod) 330 mm diameter × 1,090 mm
(radar interface unit) 760 × 330 × 480 mm
Weight: 114 kg
Reliability: 144 h MTBF specified

The AN/APQ-168 radar for the Sikorsky HH-60D Night Hawk helicopter

The AN/APQ-174 MultiMode Radar (MMR) provides the ability to operate at altitudes down to 100 ft in adverse weather by day or night

Operational status
The AN/APQ-174 MMR is deployed on the US Army Special Operations Aircraft (SOA) MH-60K and MH-47E.

Contractor
Raytheon Systems Company.

UPDATED

AN/APQ-181 radar for the B-2

The AN/APQ-181 for the Northrop Grumman B-2 operates in the J-band using 21 separate modes for terrain-following and terrain-avoidance; navigation system updates; target search, location, identification and acquisition; and weapon delivery.

The radar is a completely redundant modular system which employs two electronically scanned antennas, sophisticated software modes and advanced low probability of intercept techniques that match the aircraft's overall stealth qualities.

To meet reliability specifications, B-2 carries two complete radars consisting of 10 LRUs each, with all but the dual antennas able to function for either or both radar units. These LRUs and twin antennas weigh 953 kg and have a volume of 1.47 m^3. Six LRUs are symmetrically positioned on the sidewalls of the nose-wheel well and the two radar data processors are stacked vertically in the aft wall of the well.

Each 260 kg antenna is mounted in a cavity behind a large radome some 8 ft outboard of the aircraft centreline, just below the flying wing's leading edges. Antenna locations are marked by large, slightly darker, rectangular patches visible on the underside of the aircraft. The antennas look down and outward. They are electronically steered in two dimensions and feature a monopulse feed design that is claimed to enable fractional beamwidth angular resolution. A beam-steering computer establishes phase shifter settings based on pointing direction commands from the radar data processor. A Smiths Industries' motion sensor subsystem fitted to the antenna uses a modified strapdown inertial platform to measure antenna motion. This allows the radar to compensate for motion during synthetic aperture radar mode operation. The antenna design features its own power supplies, liquid cooling and line-replaceable modules.

The remaining LRUs in the radar are derived from other Raytheon Systems Company products used in aircraft such as the F-15C/D/E and F/A-18. All radar units communicate over a dual-redundant MIL-STD-1553 databus, and are hardened to withstand transient radiation and electromagnetic pulse effects. The radar was designed for very stringent environmental requirements exceeding those of other radars. This is because the vibration experienced by a B-2 operating at low altitude is considered to be especially severe for equipment because of the aircraft's stiffer, less flexible structure.

Operational status
In production.

Contractor
Raytheon Systems Company.

UPDATED

AN/APS-115 radar

The AN/APS-115 is one of the Raytheon Systems Company family of airborne search radars and is an I-band frequency-agile system of modular design. It is employed principally for ASW and maritime roles. The AN/APS-115 is a dual system to provide 360° coverage for the P-3C Orion land-based ASW aircraft. One antenna is mounted in the nose and the other in the rear. In addition to the underslung stabilised antenna assemblies, the equipment includes dual receiver/transmitters, an antenna position programmer, dual radar set controls and a common antenna control unit.

Operational status
No longer in production, but still in operational service.

Contractor
Raytheon Systems Company.

UPDATED

AN/APS-124 search radar

The AN/APS-124 search radar was specially designed to be part of the comprehensive avionics suite for the US Navy Sikorsky SH-60B Seahawk ASW helicopter built to satisfy the Light Airborne MultiPurpose System (LAMPS) Mk III requirement. One of the problems associated with the operation of these medium-size helicopters from the 'Spruance' class destroyers on which they serve is that of stowage, particularly the height limitation. The APS-124 is therefore designed around a low-profile antenna and radome and consists of six LRUs.

Optimum detection of surface targets in rough sea is accomplished by several unique features including a fast-scan antenna and an interface with the companion OU-103/A digital scan converter to achieve scan-to-scan integration. The system is associated with a multipurpose display and with the LAMPS datalink so that radar video signals generated aboard the aircraft can be displayed on LAMPS-equipped ships.

The system operates in three modes covering long- and medium-range search and navigation and fast scan surveillance. The display ranges are selectable up to 74 km and the false alarm rate is adjustable to suit conditions.

The system is designed around the MIL-STD-1553 digital databus to communicate with other aircraft equipment and the modular design facilitates installation on other aircraft.

Specifications
Weight: 95 kg
Coverage: 360° azimuth
Display range: out to 160 n miles (selectable)
Pulse length:
(long range) 2 μs
(medium range) 1 μs
(short range) 0.5 μs
PRF:
(long range) 470 pps
(medium range) 940 pps
(short range) 1,880 pps
Scan rate:
(long range) 6 rpm
(medium range) 12 rpm
(short range) 120 rpm

Operational status
Service deployment of the LAMPS III Seahawk helicopter began in 1983. More than 300 systems are in service.

Contractor
Raytheon Systems Company.

UPDATED

AN/APS-134(V) radar

The APS-134(V) anti-submarine warfare and maritime surveillance radar is the international successor to the US Navy's AN/APS-116 periscope detection radar. The APS-134(V) incorporates all the features of the former system while improving performance and adding capabilities, including a new surveillance mode.

The heart of the radar is a fast-scan antenna and associated digital signal processing which, says Raytheon Systems Company, form the only proven and effective means of eliminating sea clutter. This technique is used in two of the three operating modes, the third being a conventional slow scan for long-range mapping and navigation. The transmitter power is 500 kW.

In Mode 1, periscope detection in sea clutter, high-resolution pulse compression is employed with a high PRF and a fast-scan antenna, actual values being 0.46 m, 2,000 pps and 150 rpm. Display ranges are selectable to 59 km. There is an adjustable false alarm rate to set the prevailing sea conditions and scan-to-scan processing is employed.

Mode 2, long-range search and navigation, operates at medium resolution and with a low PRF, low scan and display ranges selectable to 278 km. Actual values are 500 pps and 6 rpm.

Mode 3 operates, again at high resolution, for maritime surveillance. A low PRF (500 pps) is used in conjunction with an intermediate scan speed of 40 rpm. Display ranges are selectable to 278 km and an adjustable false alarm rate is used together with scan-to-scan processing.

The system is also available in an offline configuration, with its own 10 × 10 in (254 × 254 mm) CRT control/display unit. Online operation linked in with other aircraft systems is accomplished via a MIL-STD-1553 digital databus, with the digital scan converter providing raster scan video for other aircraft displays. The weight of the entire APS-134(V), including the waveguide pressurisation unit, is 237 kg. The

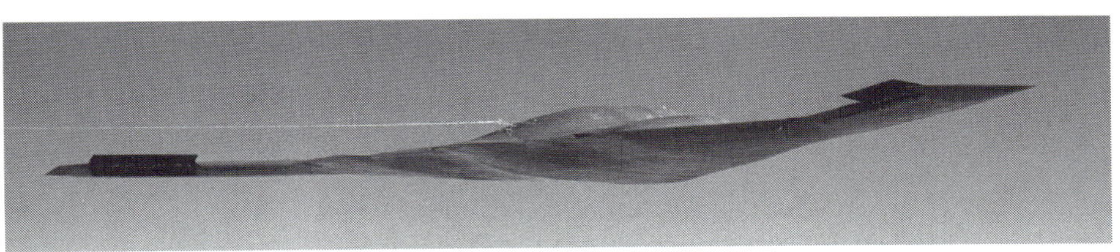

The Northrop Grumman B-2 bomber carries dual Raytheon Systems Company AN/APQ-181 radars

equipment is compatible with the inverse synthetic aperture radar techniques developed by Raytheon Systems Company for long-range ship classification.

Operational status

In service with Dornier/Dassault Aviation Atlantic ASW aircraft of the German Navy, Republic of Korea Navy and Pakistan Navy P-3C aircraft, Portuguese Air Force P-3P aircraft and as part of the Royal New Zealand Air Force Lockheed Martin P-3B Orion update programme.

The APS-134(V)6 radars on the Republic of Korea Navy P-3C aircraft are to be updated by the addition of ISAR (Inverse Synthetic Aperture Radar) capability. Some AN/APS-134(V) radars are also being upgraded to a so-called AN/APS-137(V)6 standard.

Contractor

Raytheon Systems Company.

UPDATED

AN/APS-134(V)7 radar

The AN/APS-134(V)7 offers technological improvements over its predecessor, the AN/APS-134(V). It features the developments of the US Navy's AN/APS-116 family of periscope detecting radar systems.

The AN/APS-134(V)7 radar system includes the periscope detection, long-range maritime surveillance and navigation modes from the AN/APS-134 radar. To that are added capabilities for improved periscope detection, advanced digital signal processing, multiple track-while-scan, dual-channel digital scan conversion, ESM countermeasures and 0-level built-in radar system diagnostics. There is also an optional capability to record radar and FLIR video.

The system is designed to detect small targets in high sea conditions at long range. Long-range performance is achieved by using a 500 kW high-power transmitter, 35 dB high-gain antenna and custom-developed low-noise preamplifier of less than 3 dB noise figure. These features provide the signal-to-noise ratio necessary to achieve long-range capability. Detecting small targets in the sea clutter environment is an inherent problem of maritime surveillance radar. To overcome this limitation, pulse compression and scan-to-scan processing are employed in the AN/APS-134(V)7.

The AN/APS-134(V)7 features simultaneous 32-target high-resolution tracking, 360° capable PPI coverage with sector scan, picture within picture PPI and B scan display format, dual-channel multilevel digital scan conversion and advanced digital signal processing. Multiple radar configurations are available for offline, MIL-STD-1553 and ANEW databusses. The system is adaptable to multiple radar video display configurations. The AN/APS-134(V)7 interfaces with other aircraft systems such as IFF, FLIR and ESM.

Operational status

In production for the Fokker Maritime Enforcer Mk 2 patrol aircraft.

Contractor

Raytheon Systems Company.

UPDATED

AN/APS-137(V) inverse synthetic aperture radar

The APS-137(V) is an improved version of the AN/APS-116 periscope detection radar which is standard on the US Navy's Lockheed S-3 aircraft. Over 200 units of the AN/APS-116 I/J-band radar have been produced and the APS-137(V) introduces an Inverse Synthetic Aperture Radar (ISAR) mode. Funding for this development, which increases radar processing and introduces a standard surveillance and automatic classification capability, started in 1982 and was scheduled to continue into 1994 as part of the avionics upgrade which denotes the S-3B version.

The APS-137(V) offers long-range detection and classification of ships, the radar producing a recognisable image of the target vessel. The image is derived from the Doppler shifts of the returns, compared with the reference level. Using pulse compression and fast scan processing to eliminate sea clutter, the APS-137(V) provides improved periscope detection and high-altitude maritime surveillance. Multiple target tracking (track-while-scan) is also available.

The APS-137(V) is compatible with the seekers in Harpoon, Tomahawk and other missiles, and can interface directly into weapons system computers. By comparing images before and after an attack, target battle damage can be assessed.

Operational status

Full production for the S-3B retrofit programme commenced in 1987. In January 1987, Raytheon Systems Company received a contract to supply AN/APS-137(V) radars to equip US Navy Lockheed S-3B and P-3C and US Coast Guard C-130 aircraft, with deliveries extending to 1992. This was extended at the end of 1987 to include a further nine sets and long lead items for 16, to equip the S-3B. A further contract was awarded in August 1991 for 24 AN/APS-137(V) systems. The AN/APS-137B(V)5 radar forms part of the P-3C Anti-surface warfare Improvement Programme (AIP) scheduled for 146 US Navy P-3C Update III-equipped aircraft.

Contractor

Raytheon Systems Company.

UPDATED

AN/AWG-9 weapon control system for the F-14A/B

The AN/AWG-9 is the aircraft element of the Phoenix AIM-54 air-to-air missile weapon control system in the US Navy Northrop Grumman F-14 aircraft. It comprises a fire-control radar and target illuminating radar, digital computer and displays. Provision is also made for the automatic exchange of datalink information between the AWG-9 and the Naval Tactical Data System (NTDS) or Airborne Tactical Data System (ATDS) for target designation and other functions. For onboard target acquisition, the AN/AWG-9 includes a long-range high-power pulse Doppler radar. This system has a look-down capability which enables it to pick up moving targets in the ground clutter that normally obscures targets from a conventional radar.

In addition to its long range, the AN/AWG-9 introduced, for the first time in an aircraft, the ability to track many targets at once with computer-aided selection of target priority. The system is designed for consecutive launch and simultaneous guidance of up to six AIM-54 missiles against separate targets, using time-sharing techniques for mid-course guidance.

In addition to the AIM-54, the AN/AWG-9 can be used with the AIM-7 Sparrow and AIM-9 Sidewinder, and for controlling the M-61 Vulcan 20 mm cannon.

The radar antenna is a planar slotted-plate array with a high-aperture efficiency for increased radar range and, when it was introduced, represented a significant step forward in design.

The pulse Doppler radar represents several years of development in transmitter tubes, crystal filters and planar-array techniques to enable the long-range requirements of the AIM-54 to be met. A low-noise parametric amplifier in the receiver section contributes to the long-range capability. Advanced Doppler techniques make look-down target acquisition possible. A flexible high-capacity computer permits simultaneous tracking of a large number of targets and aids the radar intercept officer in the assignment of priorities and in missile firing.

Target detection, tracking and ranging functions for all F-14 air-to-air weapon configurations are handled by the AN/AWG-9 radar and it can operate in either pulse Doppler or conventional pulse modes. A separate TWT provides CW illuminating energy for the semi-active homing AIM-7 Sparrow.

Data processing in the AN/AWG-9 is performed by a general purpose digital computer that features high-speed operation and a large memory capacity in an extremely compact package. The central computer keeps track of targets detected by the radar while the radar continues to search. Based on a preprogrammed logic, the computer evaluates threats, generates steering information for the pilot and paints a complete tactical situation for the radar intercept officer in standard symbology, based on data generated either internally or obtained through external links.

Information from the AN/AWG-9 is displayed on two CRTs. A 127 mm diameter unit is used as a multimode display for the presentation of raw radar target information and IFF returns. A larger 254 mm unit is used for displaying processed data. This includes target track information and alphanumeric and symbolic data obtained via datalink from other units. This display is also used as a computer readout device for the presentation of computer-generated missile and target assignments.

Specifications

Volume: 0.78 m^3
Weight: 590 kg

Operational status

The AN/AWG-9 radar is operational in US Navy F-14A/B aircraft. The last AN/AWG-9 weapon control system was delivered in August 1988; 695 were produced. Production of spares continued until mid-1989. The AN/AWG-9 has been replaced by the AN/APG-71 in the F-14D.

Contractor

Raytheon Systems Company.

UPDATED

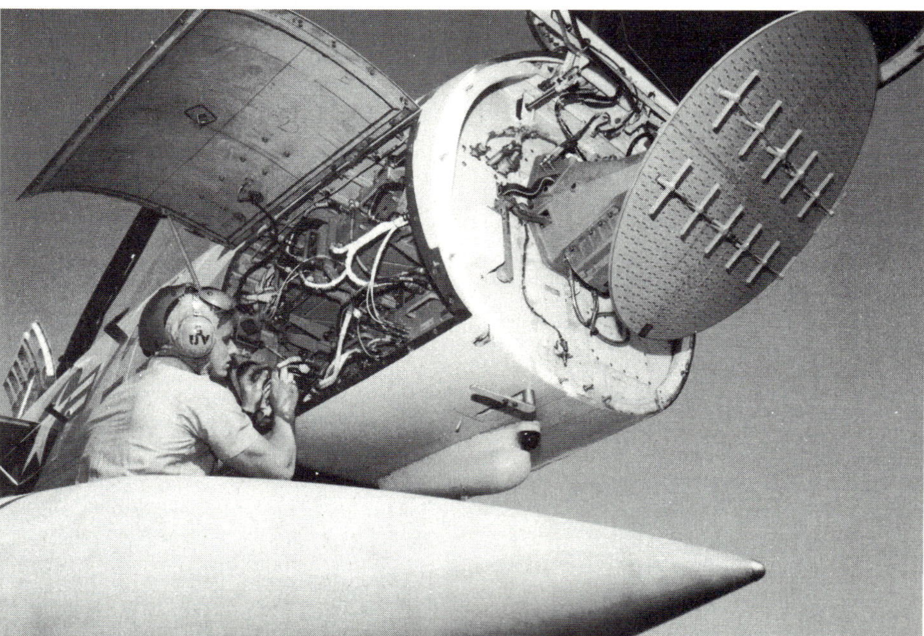

AN/AWG-9 weapon control system undergoing maintenance on the Northrop Grumman F-14. Note the IFF array on the front face of the flat-face antenna

The first production antenna of the ASARS-2 synthetic aperture radar being installed in the nose of a US Air Force U-2R aircraft

ASARS-2 advanced synthetic aperture radar system

The high-resolution ASARS-2 side-looking radar was designed for the US Air Force TR-1 high-altitude battlefield surveillance aircraft, and satisfies a US Air Force/Army requirement for a standoff intelligence gathering system. ASARS-2 was launched in 1977 under the designation AN/UPD-X and first test flown on board a U-2R reconnaissance aircraft in 1981. It is likely that the resolution approaches limits set by altitude, residual airframe vibration and atmospheric turbulence. The antenna is V-shaped to enable the ground on either side of the aircraft's track to be surveyed without the aircraft having to manoeuvre. Guidance for the ASARS-2 antennas is provided in part through several other sensors and information is transmitted via a datalink to a special ASARS-2 deployable processing station on the ground. Here the signals are converted into strip-maps and spotlights and are available within minutes for air force and army commanders.

ASARS-2 consists of a search mode for wide area ground coverage, featuring both a Moving Target Indicator (MTI) and a fixed target indicator and a spotlight mode for high-resolution coverage of smaller areas which can detect fixed targets only.

In 1988, Raytheon Systems Company was awarded US$20 million to develop an Enhanced Moving Target Indicator (EMTI) which, in part, is designed to add MTI to the spotlight mode. This entailed software modifications to the ground-based processing component and the addition of components in the airborne ASARS-2 receiver/exciter and processor control unit.

Operational status

In service with US Air Force TR-1 high-altitude reconnaissance aircraft. Details of the radar remain classified.

Contractor

Raytheon Systems Company.

UPDATED

Ground-mapping and terrain-following radar for the Tornado

The Tornado IDS radar has no official designation, but is often referred to as the Tornado Nose Radar (TNR). The TNR is also used in the ECR version of the Tornado.

The all-weather day or night radar comprises two essentially separate systems that share a common mounting, power supply and computer/processor. They are the Terrain-Following Radar (TFR) and the Ground-Mapping Radar (GMR). The first is used for automatic high-speed, low-level approach to the target and escape after an attack. The second is the primary attack sensor for the IDS Tornado and operates in air-to-ground and air-to-air modes to provide high-resolution mapping for navigation updating, target identification and fire control.

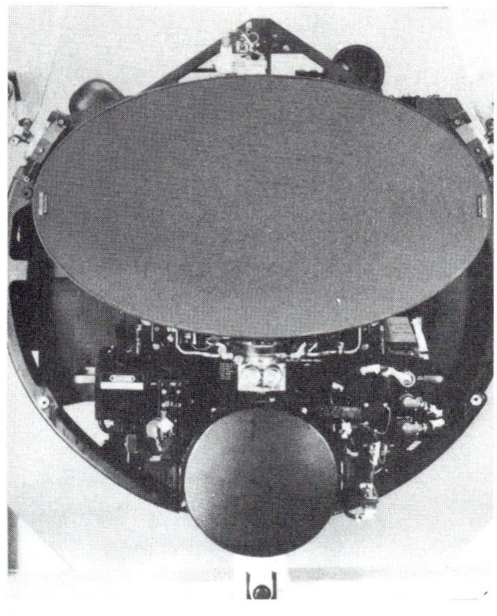

The attack radar for the Panavia Tornado IDS showing the ground-mapping antenna (above) and terrain-following antenna (below)

The radar enables the crew to fix the aircraft position by updating the Doppler-monitored inertial navigation system, provides range and tracking information for offensive or defensive weapon delivery, and commands, via the autopilot, a contour-hugging flight profile that reduces the chance of detection by hostile air-defence radars. ECCM, dramatically enhanced with the introduction of the Phase 1 upgrade improvement, is used to provide relative immunity from interference in severe ECM environments.

The three units comprising the system are the radar sensor (transmitter/receiver package for TFR and GMR), a digital scan converter and a radar display unit in which a moving-map image can be superimposed on a radar picture for navigation updating and target identification.

The GMR operates in the J-band frequency band with nine modes: readiness, test, ground-mapping, boresight contour mapping, height-finding, air-to-ground ranging, air-to-air tracking, land/sea target lock on and beacon homing.

The TFR operates at J-band frequencies and has four modes: standby, ground standby, test and terrain-following. In the latter mode the aircraft can be flown automatically or manually through head-up display steering information. The pilot can also select ride comfort (for a given speed, the closer the allowable ground clearance, the less comfortable the ride owing to the greater *g* levels needed to stay on the commanded flight profile).

Both systems have extensive built-in test features to ensure a high degree of fault isolation and comprehensive reversionary modes.

Operational status

In service on Tornado IDS and ECR aircraft.

Contractor

Raytheon Systems Company.

UPDATED

HISAR mapping & surveillance radar

Hughes Integrated Surveillance And Reconnaissance system (HISAR) is an I/J-band Synthetic Aperture Radar (SAR) mapping and surveillance system designed for border surveillance, remote sensing, and specialist monitoring of all types (oil pollution, deforestation, economic zones). It provides coverage to over 110 km with map-like presentation. It is assumed that the designation of this radar will be changed at sometime to reflect the merger of Hughes and Raytheon.

HISAR synthetic aperture radar installed aboard the Beech King Air 200T. ***1998***/0018071

HISAR Mapping & Surveillance Radar ***1997***/0001216

HISAR is based on experience with ASARS-2 and the AN/APQ-181 radar that equips the B-2 bomber.

Moving target detection capabilities include: a 10,000 km² area in a wedge from 30 to 120 km; a strip mode with 6 m resolution over a swath 37 km wide at ranges of 20 to 110 km; a spot mode with 1.8 m resolution over 10 km²; sea surveillance and air-to-air modes.

Specifications

Dimensions: 0.83 m³
Weight: 245 kg (full system)
Power consumption: 4,930 W

Operational status

Operational on US Army RC-7B special mission aircraft and the Beech King Air, known operationally as Airborne Reconnaissance Low-Multifunction (ARL-M). A derivative has been developed for the Tier II UAV.

HISAR is being marketed internationally for civil applications including environmental monitoring, border surveillance and maritime patrol. Optional, additional, complementary fits include the Raytheon Systems Company DB 110 electro-optical sensor and the AN/AAQ-16 infrared sensor.

Contractor

Raytheon Systems Company.

UPDATED

LANTIRN under F-16

Sea Vue (SV) surveillance radar

Raytheon Systems Company SV radars are modern, lightweight, high-performance systems. They employ proven long-range surface detection techniques that support the entire range of surveillance missions. Inverse Synthetic Aperture Radar (ISAR) and SAR modes provide real-time imaging of maritime, land-based and coastline targets. The APS-134 (LW) was the initial product of the Sea Vue (SV) radar family.

Radar features include: Sea Vizion™ Raytheon's unique combination of system parameters and target signal processing; fully coherent ISAR, SAR and MTI processing; pulse compression, pulse-to-pulse frequency agility and PRF jitter.

Radar modes available include: surface surveillance and small target detection; navigation and mapping; weather avoidance; ISAR and SAR processing. Available enhancements include: moving target discriminator; ISAR classification aids; Doppler Beam Sharpening (DBS); coherent look-down air target detection and tracking; control and display options.

Specifications

Detection performance:
(tanker) 230 n miles
(patrol boat) 95 n miles
(life raft) 30 n miles

Dimensions	**Width**	**Height**	**Depth**	**Weight**
antenna	1,240 mm	650 mm	n/a	23 kg
transmitter	330 mm	279 mm	498 mm	30 kg
RESP	391 mm	257 mm	498 mm	37 kg

Transmitter: coherent, solid-state, TWT; 8, 15 or 50 kW peak power options; 9.5-10 GHz; linear FM, fixed, frequency agile, or biphase coded waveforms
Antenna: parabolic or flat plate; stabilised, 360° scan; 6, 60, 120 rpm; sector and searchlight search, IFF capability
Receiver/Exciter/Synchroniser/Processor (RESP): linear FM pulse compression; digital pulse compression; Sensitivity Time Control (STC); Automatic Gain Contol (AGC)
MTBF: >500 h
Power: 28 V DC; 115 V AC, 400 Hz

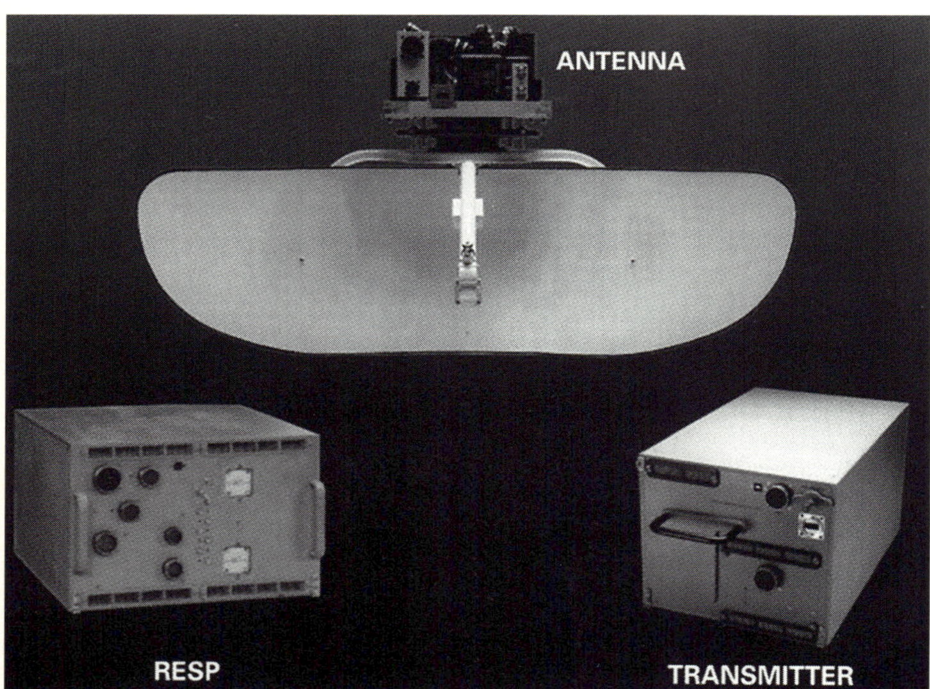

Sea Vue (SV) surveillance radar system, showing antenna, receiver/exciter/synchroniser/processor, and transmitter units ***1998***/0018078

Operational status

Demonstrated on PC-12 and IPTN CN-235 aircraft. Operational on Royal Australian Air Force Coastalwatch; Japanese Self Defense Force U-125 SAR; and Italian Customs ATR-42 aircraft.

Contractor

Raytheon Systems Company.

NEW ENTRY

Terrain-Following Radar (TFR) for LANTIRN

The LANTIRN TFR retains the basic facilities available in the Tornado, including the ability to operate under enemy jamming in bad weather and at very low level, but changes in system performance enable the aircraft to turn at a maximum rate of 5.5°/s compared with the Tornado's 2°/s and to bank up to 60° while still maintaining terrain-following capability. Whereas the Tornado and previous terrain-following radars employed magnetrons, the LANTIRN TFR has a travelling wave tube, giving considerably better performance, particularly in an ECM environment, and higher reliability. The pilot can control the aircraft's flight path by selecting terrain-clearance heights of from 100 to 1,000 ft.

The LANTIRN TFR subsystem began testing in a Lockheed Martin F-16 during 1983 and quickly demonstrated the ability to fly down to 100 ft above the terrain.

Operational status

In service on US Air Force F-15E and F-16D aircraft, and selected for F-15I and F-15S.

Contractor

Raytheon Systems Company.

UPDATED

860F-4 digital radio altimeter

The 860F-4 employs large-scale integrated technology and microprocessors but has an analogue output for driving existing non-digital instruments. It has been designed to ARINC 552/552A and can be used as a reference for Cat.IIIA automatic landings. Rockwell Collins says that the digital technology gives a 30 per cent reduction in parts and 40 per cent saving in weight compared with the earlier non-digital 860F-1 system. The equipment can be used from −20 to 2,500 ft. The outputs comply with ARINC 429 (digital) and ARINC 552/552A (DC analogue).

Specifications

Dimensions: ½ ATR short
Weight: 5.4 kg
Power supply: 115 V AC, 400 Hz, 50 VA
Frequency: 4,300 MHz

Contractor

Rockwell Collins.

UPDATED

900 series TCAS Traffic Alert and Collision Avoidance System

The 900 series TCAS combines advanced technology components, high-density circuit designs and comprehensive self-test circuitry to provide reliable operation. System installation time is minimised by the use of advanced test/diagnostic equipment and the availability of a variety of antenna baseplates to fit any aircraft fuselage.

The Rockwell Collins TVI-920 integrated display unit combines TCAS traffic and resolution advisories with a conventional vertical speed indicator **1998**/0018066

The Series 900 TCAS utilises a power supply requiring no forced air cooling or internal cooling fans. A full-colour active matrix liquid crystal display is employed to show TCAS advisories and commands.

The system consists of the TTR-921/920 TCAS receiver/transmitter, TPR-901/920 Mode S transponder, TTC-920 TCAS/Mode S control, TRE-920 TCAS antenna, TVI-920 vertical speed indicator and the WXI-711 TCAS traffic display.

Specifications

Dimensions:
(TRE-920 antenna) 25.4 × 159 × 280 mm
(TTR-921/920 receiver/transmitter) 193 × 196 × 389 mm
(TPR-901/920 Mode S transponder) 193 × 125 × 325 mm
(TTC-920 control) 57 × 146 × 185 mm
(TVI-920 vertical speed indicator) 3 ATI (191 mm diameter)
Weight:
(TRE-920 antenna) 0.7 kg
(TTR-921/920 receiver/transmitter) 8.16 kg
(TPR-901/920 Mode S transponder) 5.8 kg
(TTC-920 control) 0.9 kg
(TVI-920 vertical speed indicator) 2 kg
(WXI-711 traffic display) 5.89 kg
Power supply: 115 V AC, 400 Hz
Tracking capability: 60 targets
Range: 26 km
Accuracy: <5° RMS bearing

Contractor

Rockwell Collins.

UPDATED

ALT-50/55 Pro Line radio altimeters

The ALT-50 and ALT-55 Pro Line radio altimeters, the first with a range of 0 to 2,000 ft, the other 0 to 2,500 ft, have been designed for business aircraft. Both types provide decision height annunciation for Cat II landings. The decision height annunciators can be set at any desired altitude and both instruments can interface with high-performance flight directors and autopilots. The DRI-55 indicator is offered with a numeric readout of radar altitude and decision height.

Dual ANT-52 antennas are included.

Specifications

Dimensions:
(transmitter/receiver) ⅜ ATR short dwarf
(indicators) 77 × 77 × 152 mm
(antenna) 25 × 25 × 20 mm
Weight:
(transmitter/receiver) 2.54 kg
(indicator) 0.7 kg
(antenna) 0.1 kg
Power supply: 28 V DC

Operational status

In production and in service.

Contractor

Rockwell Collins.

UPDATED

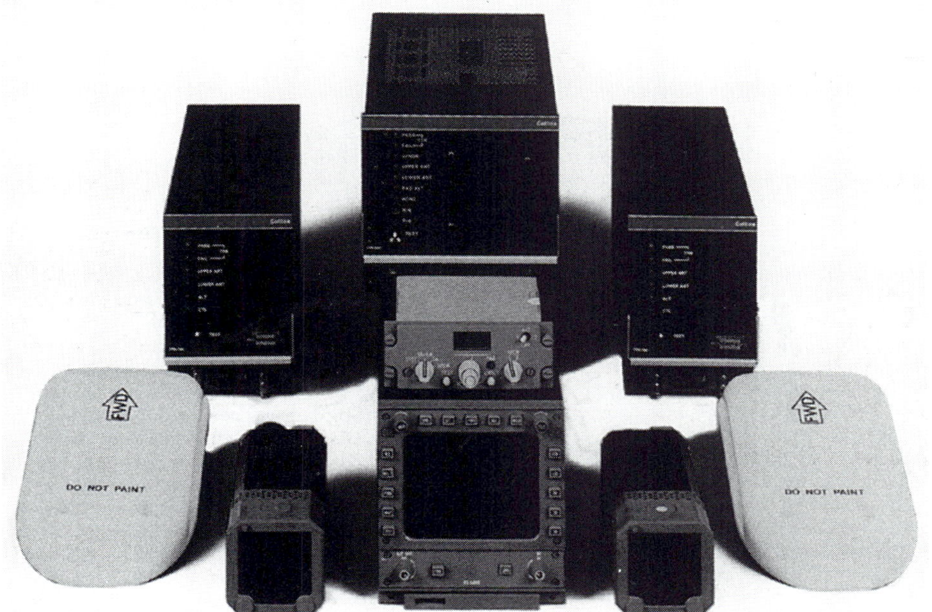

Units of the Rockwell Collins TCAS systems **1998**/0018065

LRA-900 low-range radio altimeter

The LRA-900 low-range radio altimeter includes enhancements to a digitally controlled variable bandwidth filter which provides improved noise rejection and leading-edge tracking. Signal microprocessors perform automatic calibration that continuously compares the received ground return signal frequency with the frequency produced by separate precise delay lines in each channel. This results in fine resolution, providing accuracy to support autoland flare and touchdown computations. Resolution at touchdown is quoted as better than 1.2 in.

Low-component count and low-stress level circuits give the LRA-900 high reliability and quick, easy and effective onboard fault isolation. A comprehensive self-test monitor in the system is capable of determining operational status to a 99 per cent confidence level.

The LRA-900 is designed to fulfil all new environmental requirements. The system is protected for Cat.K lightning and Cat.Y High-Intensity Radiated Fields (HIRF). In addition, it is fully functional over 200 ms of power interruptions. It meets ARINC 707, 429 and 600, TSO-C87, DO-160C, DO-155 and DO-178A.

Specifications

Dimensions: 3 MCU
Weight: 4.3 kg
Power supply: 115 V AC, 380-420 Hz, 24 W nominal
Frequency: 4,300 ±25 MHz
Altitude: −20 to 5,000 ft
Accuracy: ±1 ft or 2% of indicated altitude
Temperature range: −40 to +70°C

Contractor

Rockwell Collins.

UPDATED

TCAS II Traffic alert and Collision Avoidance System

The Rockwell Collins Traffic Alert and Collision Avoidance System (TCAS II) is a complete system which includes directional antennas, an advanced technology receiver/transmitter, Mode S transponder and a variety of display options. The Rockwell Collins TCAS features new functions as well as the FAA mandated Change 7.0 software for both the TTR-921/920 receiver/transmitter and TPR-720/900 Mode S transponder. Key features of these new functions include Automatic Dependent Surveillance-Broadcast (ADS-B) and enhanced surveillance (>100 n miles) range in the TTR-921 receiver/transmitter.

The TTR-921 receiver/transmitter incorporates advanced surface-mounted devices in high-density circuit designs for high reliability. It contains comprehensive self-test circuitry and is compatible with central maintenance computers. The receiver/transmitter includes new D-band RF concepts, the Rockwell advanced architecture microprocessor and high-speed signal/data processing. The TTR-921 is designed to fit any aircraft and interfaces with both analogue and digital systems. It weighs 8.2 kg, uses only 80 W of power and requires no cooling.

Operating with air traffic control radar beacon system Mode A and C interrogators, as well as Mode S, the Rockwell Collins TPR-900 transponder is also compatible with ARINC 735 TCAS systems. The TPR-900 has interfaces for dual Gilham, syncro, ARINC 429 and ARINC 575 input interface ports and is compatible with several types of barometric altimeters.

Rockwell Collins TCAS has been installed in a Russian International Airlines Ilyushin Il-86

The TPR-900 operates with a diversity selection function, using two receivers to improve air-to-air surveillance for TCAS. When a signal is received at the two antennas located on top and bottom of the aircraft, diversity selection determines which provides the stronger interrogation signal. The proper reply, depending on the type of interrogation, is then transmitted through the most efficient transmission antenna. Transponder operation is unaffected by aircraft position in relation to other aircraft or ground stations.

The TRE-920 TCAS directional antenna is an L-band array designed specifically to operate with the TTR-921/920 receiver/transmitter. The antenna has no integral electronics; a variety of baseplates is available to fit all aircraft fuselages.

Traffic Advisory (TA) and Resolution Advisory (RA) displays are required for TCAS. An electronic plan position indicator is required for TA. A variety of Collins displays is available for aircraft with limited panel space. These include vertical speed indicators, a combined display with weather radar or displays on electronic flight instruments.

The TTC-920 TCAS/Mode S control unit is a single panel-mounted unit which controls TCAS operational modes, altitude reporting, identification code, transponder operational modes and system test. Control of TCAS TA display range and mode is available as an option.

Operational status

Rockwell Collins TCAS II has been selected by over 128 major airlines which have ordered more than 4,500 systems.

Enhancements, planned for late 1998, will more than double en route range for surveillance to more than 100 n miles. Additionally, Automatic Dependent Surveillance-Broadcast (ADS-B) capability will be added to the TCAS/Mode-S surveillance systems through integration of GPS-based navigation and communication functions. Mode-S improvements will include Comm D/Level 4 communications external software data load capability and provisions for software growth to Level 5 message capability. Provisions are also included for enhanced surveillance as well as the new transponder ARINC 718 interface function.

Provisions will also be included for new GPS-based situational awareness capabilities which will be offered by service bulletin. Where size limitations are important, a smaller four MCU TCAS unit will be available. Offering maximum flexibility, this single unit will be capable of operating on either AC or DC power.

Contractor

Rockwell Collins.

UPDATED

TCAS-94 Traffic alert and Collision Avoidance System

The Rockwell Collins TCAS-94 meets the needs of regional and business aircraft operators for TCAS II capability in the new air traffic control environment. The TCAS-94 system operates as an airborne surveillance radar system in interrogating the ATC transponders of nearby aircraft within a 35 n miles range. Range, relative bearing and altitude of the nearby aircraft are measured by the TCAS-94 system for use in computing potential intruder aircraft. The TCAS-94 system provides aural warnings as well as visual indications. The visual warnings are provided in the form of Traffic Advisories (TAs) and Resolution Advisories (RAs). The TAs indicate the relative position of aircraft in close proximity, to assist the flight crew in visual acquisition. The RAs depict the commanded vertical manoeuvre necessary to achieve safe separation from nearby aircraft.

The TCAS-94 system comprises a TTR-920 TCAS II receiver/transmitter, TRE-920 directional antennas, TDR-94D diversity Mode S transponder, TVI-920D RA/TA/VSI liquid crystal display, TTC-920 TCAS/Mode S control, MFD-85C multifunction display and a CTL-92T TCAS control. A variety of combinations of this equipment will give operators full TCAS II capability necessary to operate in US airspace. The TCAS-94 system was designed for new aircraft as well as for the retrofit market. The Rockwell Collins TCAS-94 system is fully ARINC 735 compatible.

Specifications

Dimensions:
(TTR-920) 196 × 193 × 389 mm
(TRE-920) 159 × 25.4 × 280 mm
(TDR-94D) 124.5 × 84.6 × 353.2 mm
(TVI-920D) 76.2 × 76.2 × 190.5 mm
(TTC-920) 146.1 × 57.2 × 127 mm
(CTL-92T) 60.3 × 66.7 × 149.2 mm
Weight:
(TTR-920) 8.16 kg
(TRE-920) 0.7 kg
(TDR-94D) 3.18 kg
(TVI-920D) 1.6 kg
(TTC-920) 0.9 kg
(CTL-92T) 0.7 kg
Power supplies:
(TTR-920) 27.5 V DC, 80 W (115 V AC optional)
(TDR-94D) 27.5 V DC, 25 W
(TVI-920D) 27.5 V DC, 15 W (115 V AC optional)
(TTC-920) 115 V AC, 8 W
(CTL-92T) 27.5 V DC, 6.9 W

Operational status

In production and in service.

Contractor

Rockwell Collins.

UPDATED

TDR-90 transponder

The TDR-90 is an air traffic control Mode A and C transponder with 4,096 codes and an altitude reporting capability of up to 126,000 ft when used with an encoding altimeter. It is a remotely controlled system designed primarily for general aviation.

The system has a transmitter output power of 325 W nominal (250 W minimum) on a frequency of 1,090 MHz. Positive sidelobe suppression facilities are incorporated in order to provide a cleaner paint on the interrogator's trace. Two-way mutual suppression avoids interference with DME. Another feature is a stripline duplexer to control receiver front-end noise while retaining high sensitivity and frequency stability regardless of antenna matching.

A built-in test facility for both the transmitter and receiver functions is included. Test signals are injected at just above the minimum sensitivity level to ensure that receiver, decoder, encoder and transmitter are functioning correctly.

The system's CTL-90 control unit has two-knob code selection, ident, self-test, standby and altitude reporting on/off controls. An optional system selection switch can also be incorporated for use in dual installations. The display is of the gas-discharge type. The TDR-90 electronic unit can, however, interface with most conventional transponder controllers as well as the CTL-90 unit.

Specifications

Dimensions: ¼ ATR short
Weight: 1.59 kg

Operational status

In production and in service.

Contractor

Rockwell Collins.

UPDATED

TPR-900 ATCRBS/Mode S transponder

Operating with the Air Traffic Control Radar Beacon System (ATCRBS) Mode A and C interrogators, as well as Mode S, the TPR-900 transponder is also compatible with ARINC 735 TCAS systems. It has interfaces for dual Gilham, synchro, ARINC 429 and ARINC 575 input interface ports and is compatible with several types of barometric altimeters.

The solid-state TPR-900 transmitter meets FAA requirements for Class 4 (Comm D) datalink. Up to four segments of downlinked extended length messages can be sent by the transmitter at nominally 450 W power with expansion to 16 segments for Level 5 operation. Reported altitude, discrete address, maximum airspeed, sensitivity control, TCAS control data and Mode S ground station identification are provided as outputs from the system.

The TPR-900 operates with a diversity selection function using two receivers to improve air-to-air surveillance for TCAS. When a signal is received at the two antennas located on top and bottom of the aircraft, diversity selection determines which provides the stronger interrogation signal. The proper reply, depending on the type of interrogation, is then transmitted through the most efficient transmission antenna. Transponder operation is unaffected by aircraft position in relation to other aircraft or ground stations.

The system conforms to the requirements of DO-181A Change 1 and is designed to fulfil all new environmental requirements. It has protection for Cat.K lightning and Cat.U High-Intensity Radiated Fields (HIRF). In addition, the TPR-900 is fully functional for up to 200 ms of power interruptions.

Specifications

Dimensions: 125 × 193 × 325 mm
Weight: 5.6 kg
Power supply: 115 V AC, 400 Hz, 45 VA
Frequency: 1,090 MHz
Temperature range: −40 to +70°C

Contractor

Rockwell Collins.

UPDATED

TWR-850 Turbulence detecting Weather Radar

The TWR-850 uses solid-state electronics technology, has an integrated transmitter/receiver antenna and weighs less than 9 kg. The radar picture is displayed on a CRT which is part of the EFIS, using one or two WXP-850 control panels to select the various operating modes and ranges; two control panels enable pilot and co-pilot to select independent displays. Auto-tilt and ground clutter suppression are additional features which ease the important task of tilt management to discriminate weather from ground returns. A feature, offered to the general aviation and regional airline sectors for the first time, is picture rotation as the aircraft turns. Maximum weather detection range is 550 km, turbulence can be detected out to 85 km.

Specifications

Frequency: I-band, 9,345 MHz
Range: 550 km

Operational status
In production and in service.

Contractor
Rockwell Collins.

UPDATED

WXR-350 weather radar

Rockwell Collins introduced the WXR-350 weather radar in 1984. It is intended to work specifically in conjunction with electronic flight instrument systems and detects four levels of precipitation, the fourth being displayed in magenta. There is also a path attenuation correction alert function which informs pilots of possible weather cells which are hidden behind heavy rainfall areas. Antennas, of 12 and 18 in, are available to suit light turboprops to heavy corporate jets. The WXR-350 employs a proven magnetron with nominal peak power of 5 kW to give avoidance ranges of up to 300 n miles.

Operational status
In production and in service.

Contractor
Rockwell Collins.

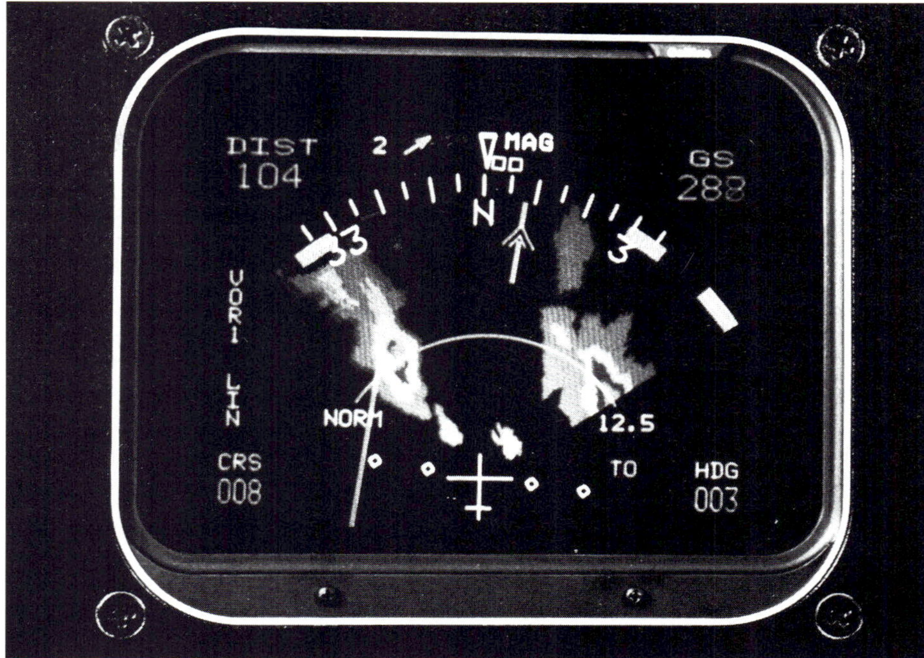

Rockwell Collins WXR-350 weather radar display

UPDATED

WXR-700C/WXR-700X radar

Rockwell Collins recently added Forward-Looking Windshear (FLW) detector capability to the WXR-700X radar system. This capability, per ARINC 708A characteristic, provides an alert to the pilot prior to entering a dangerous windshear event on landing or before takeoff. In addition to commercial airline applications, Rockwell Collins was selected to provide the US Air Force with the first Commercial-Off-The-Shelf (COTS) FLW radar with skin paint function in conjunction with the PACER CRAG programme.

The WXR-700 comprises four units: a slotted array flat-plate antenna (for good sidelobe reduction), a microprocessor-controlled transmitter/receiver, a display unit and a control unit. The microprocessor control system in the transmitter/receiver supervises all control and data transfers, programmes and controls the RF processes such as pulsewidth, bandwidth and PRF selection and directs antenna scan and stabilisation. The unit also contains circuits to reduce ground clutter suppression when operating in the weather mode. An optional feature is pulse-pair Doppler processing, whereby, with the addition of a single circuit board to the transmitter/receiver, the horizontal velocity of rainfall can be sampled. This technique is recommended by the NSSL as being particularly suitable for the analysis of storm cells.

The CRT indicator uses a high-resolution shadow-mask tube with a multicolour display scheme. The CRT provides alphanumeric identification of radar modes and incorporates annunciators and controls. The receiver has a sufficiently wide dynamic range to detect the Z-5 and Z-6 levels of rainfall that indicate a high probability of hail.

In December 1982, Rockwell Collins announced a new version of the WXR-700 series radar incorporating a facility to detect atmospheric turbulence. The facility permitted a considerable advance in weather interpretation, going beyond the conventional method of assessing turbulence by measuring areas of high rainfall. By means of a patented technique involving Doppler processing, the system can measure changes in rainfall velocity, which is now concluded to be a more reliable guide to the presence of turbulence than absolute rainfall rates. Weather radars formerly depended on the operator's ability to interpret radar echoes by their shape, intensity and gradient. The Rockwell Collins method shows, in magenta on the radar screen, areas of turbulence by a more direct (and therefore more reliable) method than has previously been possible.

Two factors are responsible for this capability. First, the use of a solid-state coherent transmitter/receiver that generates stable frequencies with very small dispersion, as opposed to the earlier magnetrons that produced a relatively broad band of frequencies. Secondly, Rockwell Collins' modification of the pulse-pair detection technique for measuring velocity variance permits much greater accuracy than with current radars. By receiving the in-phase and quadrature components of signals simultaneously, errors due to aircraft speed and antenna angles with respect to the weather under surveillance are effectively cancelled.

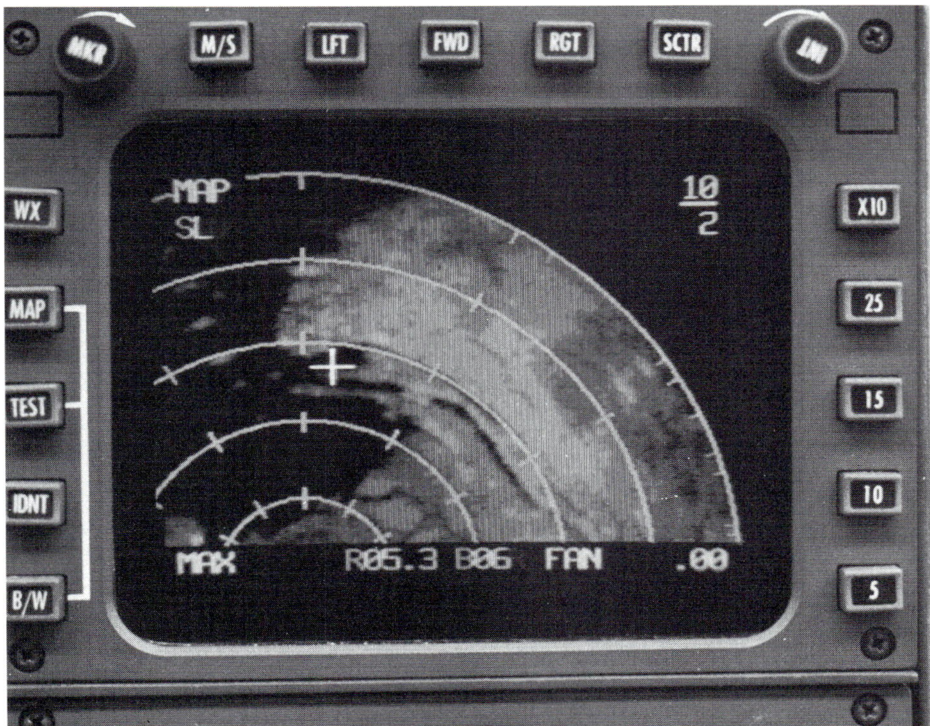

A Rockwell Collins WXR-700XM ground-mapping display shown offset to give maximum coverage to the right of the aircraft's track

Specifications
for C- and X-band systems (equivalent to NATO G- and I-band)

Dimensions:
(antenna) ARINC 708
(transmitter/receiver) 8 MCU
(indicator) ARINC 708 Mk II
(control unit) 146 × 67 × 152 mm

Weight:
(antenna) 12.25 kg
(transmitter/receiver) 12.25 kg
(indicator) 8.16 kg
(control unit) 1.04 kg

Frequency:
(C-band) 5,440 MHz
(X-band) 9,330 MHz

Power output:
(C-band) 200 W
(X-band) 100 W

PRF:
(C-band) 180-1,440
(X-band) 180-1,440

Pulsewidths:
(C-band) 2-20 μs
(X-band) 1-20 μs

Range:
(C-band) 445 km
(X-band) 593 km

Operational status
In production and service. The WXR-700X received certification in 1995. It provides both visual and aural alerts of windshear events occurring up to 90 seconds ahead of the aircraft flight path. The radar is automatically activated below 1,200 ft AGL and scans a detection path of 30° either side of the aircraft centreline. The WXR-700X is now flown on Airbus, Boeing and Fokker aircraft.

A special 'missionised' version of the commercial WXR-700X radar, designated the FMR-200X, is being fitted to US Air National Guard (ANG) C/KC-135 PACER CRAG upgrade aircraft as a replacement for the AN/APN-59 radar. The FMR-200X radar will provide a colour weather system incorporating forward-looking windshear, skin paint detection and limited ground map capability. Block 10 retrofit work covers the first 128 ANG aircraft starting in September 1996. Block 20 involves the remaining 602 C/KC-135 aircraft, with a planned completion date of 2000.

A derivative version of the WXR-700X, designated WXR-701X, upgraded with Doppler turbulence detection has been selected by the UK Ministry of Defence for the upgrade of its C-130K aircraft.

Contractor

Rockwell Collins.

UPDATED

Ground Collision Avoidance System (GCAS)

The GCAS is a generic sensor flexible unit which provides ground proximity warning to crews to prevent controlled flight into terrain, by the use of aircraft dynamics. The system is designed to meet or exceed the requirements of RTCA DO-161C and TSO C92b.

The state vector algorithm monitors aircraft and terrain position, anticipates aircraft manoeuvres and, considering various other available sensor data, matches the solution against warning criteria. The state vector approach predicts the aircraft's flight path, rather than simply detecting the aircraft's present position and issuing a warning based on the aircraft's entry into a predetermined flight envelope. The dynamic state vector method has no need for multiple modes of operation or system inhibits in flight, as false warnings are virtually eliminated.

Specifications

Dimensions: 57.1 × 158.7 × 320.5 mm
Weight: 3.72 kg
Power supply: 28 V DC, 1 A
Reliability: >15,000 h MTBF

Operational status

Originally developed for US military transport aircraft, the ground collision avoidance system is now available for use in commercial applications.

Contractor

SCI Systems Inc.

VERIFIED

AN/APQ-153 fire-control radar

The AN/APQ-153 airborne search target attack radar is a lightweight search and range tracking radar used on F-5E aircraft. This system operates in the I-band frequency range and provides stabilised search, automatic acquisition and target illumination, and automatic target ranging with boresight steering in the missile mode for head-down launch of Sidewinder AIM-9 missiles. In the gunnery modes, the radar automatically provides range and range rate outputs for targets within the sight lead-angle computation envelope.

The system consists of five main units: antenna assembly, receiver/transmitter, radar processor, indicator and set control. Combined weight is 50 kg. The antenna is a parabolic 300 × 400 mm dish with horizontal polarisation. The indicator uses a 127 mm direct view storage tube and provides a B-scope search display and a lock on missile mode display. Built-in test is provided for radar performance verification.

The APQ-153 has the following principal modes of operation:

Search mode

In search operation, a 7° elevation beamwidth is stepped up 3° at the right azimuth limit and down 3° at the left azimuth limit, providing a two-bar scan coverage of 10° in elevation. This two-bar 10° coverage can be adjusted ±45° in elevation by the pilot. Azimuth search coverage is ±45°. The search pattern is space-stabilised for aircraft pitch and roll motion to allow a given volume of space to be searched to prevent loss of the target and/or smearing of the display. Range of search coverage extends to 37 km.

Boresight missile mode

The missile mode enables the pilot to lock on to targets out to 19 km and provides aircraft steering information to align the acquisition envelope of the AIM-9 missile with the target. Once a target has been acquired a lock on occurs and azimuth and elevation steering data are provided on the indicator display in the form of a target steering bar. The aircraft is flown towards the bar to align it within the allowable AIM error circle scribed on the overlay. Missile acquisition occurs when the steering bar is within the allowable AIM error circle and the in-range indicator is illuminated.

Air-to-air gunnery modes

There are two air-to-air gunnery modes in the APQ-153: dogfight and AA1/AA2. Both are external commands to the radar. The gunnery modes are head-up and the radar automatically provides range and range rate information to the sight. Activation of a gunnery mode causes the antenna to align to the boresight in azimuth and 2.2° down in elevation. The range gate will automatically slew from 500 to 6,000 ft at 22,000 ft/s. Acquisition is automatic for the first target encountered.

Operational status

No longer in production. In operational use aboard Northrop F-5E and F-5F aircraft. Over 1,400 sets have been produced.

Contractor

Systems & Electronics Inc.

UPDATED

AN/APQ-157 fire-control radar

The AN/APQ-157 is a dual version of the AN/APQ-153 system. It consists of eight units: antenna assembly, receiver/transmitter, radar processor, front indicator, rear indicator, coupler power supply and two control units. Total weight is 64 kg. The indicators use a 127 mm direct view storage tube and provide a B-scope search display and a lock on missile display. Video trim is provided for the harmonisation of video displays. Each display can be adjusted to suit the individual operational preference and will correlate information within 1°. The control units are transferable for individual cockpit control.

All other details and modes of operation are as for the AN/APQ-153.

Operational status

No longer in production but still in service.

Contractor

Systems & Electronics Inc.

UPDATED

AN/APQ-159(V) radar

The AN/APQ-159(V) was designed in the mid-1970s as a technological upgrade of the AN/APQ-153/157 systems. It provides increased range, off-boresight lock on and angle tracking, frequency agility and a higher gain lower sidelobe planar-array. The AN/APQ-159(V)-1 and -2 incorporate a scan converter and television display compatible with the Maverick missile.

The latest derivatives are the AN/APQ-159(V)-5 and (V)-7. The AN/APQ-159(V)-5 includes off-boresight acquisition and angle track capability to improve search and acquisition performance, and offers 100 per cent increase in reliability over earlier variants. The AN/APQ-159(V)-7, in addition to improved reliability and performance, incorporates monopulse tracking and air-to-ground ranging to provide better angle tracking and optimum weapon delivery.

The AN/APQ-159(V)-5 is a lightweight forward-looking I/J-band radar designed for air-to-air detection and tracking and for installation in operational fighter aircraft. This version was developed with the intention of providing a cost-effective radar upgrade for the F-5E aircraft currently equipped with the AN/APQ-153 system. The criteria for the new radar were minimum aircraft modification and no impact on weight and balance, while providing improved performance, reliability and supportability.

To meet these criteria the reliability team analysed AN/APQ-153 and AN/APQ-159(V) data provided by users, and ascertained the major reliability concerns. The findings of this analysis resulted in the selection of higher-reliability components as appropriate, redesign of the processor power supply transformer to eliminate indicator noise, redesign of the servo amplifier to reduce power dissipation and incorporate protection against secondary failures, environmental stress screening on the receiver and modulator, and the incorporation of a gallium arsenide FET low-noise receiver.

The AN/APQ-159(V)-5 has been installed in the 17 US Air Force Aggressor F-5E aircraft at Nellis Air Force Base, and has demonstrated a constant detection of F-16 aircraft at 34 km and established lock on at 22 km in a realistic operational environment. This is a 30 per cent improvement compared to AN/APQ-153(V)-3 performance. The MTBF figure realised in the field will improve by 100 per cent from 62 hours for the AN/APQ-153 to 127 hours for the AN/APQ-159(V)-5.

Operational status

The AN/APQ-159(V)-5 is currently in production and is operational with the US Air Force Aggressor and US Navy Adversary units. It became operational in the US Navy Adversary unit in mid-1989 and is being considered by several countries for incorporation into the F-5E. Variations of the AN/APQ-159(V) are being considered for the F-5F, Mirage III and MiG-21 fire-control radar upgrades.

The AN/APQ-159(V)-1 to -4 versions remain in production and are installed in over 550 aircraft in 15 different countries.

The AN/APQ-159(V)7 has been selected by INDRA of Spain as the fire-control radar for the Spanish Air Force upgraded Mirage III. In addition, this version will be fitted on the 36 A-4M aircraft which are being delivered to the Argentine Air Force.

Contractor

Systems & Electronics Inc.

UPDATED

AN/APQ-170(V)1/AN/APQ-425 multimode radar

Systems and Electronics Inc developed the AN/APQ-170 radar to equip 28 MC-130H Combat Talon II aircraft procured for the US Air Force Special Operations Forces. The radar has terrain-follow/avoidance capabilities, as well as ground map, weather and beacon modes. Delivery of the first AN/APQ-170-equipped aircraft took place in 1989.

The Combat Talon II aircraft incorporate the avionics suite developed by Lockheed Martin, including the Raytheon Systems Company AN/AAQ-45 infrared system and a four-CRT cockpit display system.

Operational status

The current version of the radar in service on the 28 MC-130H Combat Talon II aircraft is the

AN/APQ-170(V)1. A programme to improve mission effectiveness and reliability and to reduce maintenance costs was awarded to Lockheed Martin Federal Systems (Owego) and Systems and Electronics Inc (SEI) as its subcontractor, in January 1997 for development, testing and installation for all 28 aircraft by November 1999. The upgraded configuration is designated AN/APQ-425.

Contractor

Systems & Electronics Inc.

UPDATED

AN/APQ-175 multimode radar

The AN/APQ-175 radar has been developed to replace the AN/APQ-122(V) for the US Air Force in the Adverse Weather Aerial Delivery System (AWADS) C-130E aircraft. It is designed to enable the C-130 to airdrop and air land personnel and equipment during poor weather conditions. Features of the equipment will include long- and short-range precision ground-mapping, weather detection and beacon integration and reception.

The AN/APQ-175(V)X is an X-band (NATO I-band) variant of the APQ-175. It provides the C-130 with a modern navigation and weather radar.

Operational status

Delivery of 50 production units began in 1990.

Contractor

Systems & Electronics Inc.

UPDATED

AN/APN-169/240/243/243A(V) Intraformation Positioning System (IFPS)

Sierra Technologies developed and manufactures the AN/APN 243 and 243A(V), a wideband datalink with multiple uses, including intraformation positioning, IFPS and stationkeeping (SKE). As an IFPS/SKE stationkeeping system it allows aircraft to fly fully instrumented formation flying, and to conduct aerial deliveries or instrumented approaches when using the ground-based zone marker sensor (AN/TPN 27B). The new version of Sierra's formation positioning system, AN/APN 243A(V) operates between 3,300 and 3,600 MHz. It is power programmable, uses direct sequence spread spectrum, and retains full interoperability with existing AN/APN-169 and AN/APN-240 systems already installed on over 750 aircraft.

The AN/APN 169, 240 and 243 SKE systems are limited to 18.5 km in-flight range and 36 participants on each of four channels. The AN/APN 243A(V) increases the internetted range to about 185 km. With the system's virtual channels 150 or more aircraft and/or surface contacts can be accommodated in the network.

System accuracies will exceed those of the current AN/APN-169/240 specifications and not be reliant upon global positioning systems for operation, thereby assuring continued system use should GPS be denied.

Operational status

In 1970 US Air Force C-130 tactical aircraft were equipped with the original AN/APN-169; over 750 C-1, C-17, C-130, and C-141 aircraft are now equipped with the SKE system. Sierra delivered its first AN/APN 243 (SKE-2000) with growth capability to the AN-APN 243A(V) for the C-17 and C-130J in 1996. The system's smaller size, lower weight, lower cost, higher MTBF, and use of military frequency band make it ideally suited for fixed-wing and rotary-wing aircraft, surface ships, vehicles, and fixed position sensors. The robust secure adaptive datalink enhanced the system for multiple uses.

Contractor

Sierra Technologies, Inc.

UPDATED

AN/APS-128A/B/C surveillance radar

The AN/APS-128A/B/C is one of the most widely used maritime patrol radars. It has come into its own largely as a result of the international 200 n miles (370 km) fishing boundaries which need constant patrolling, but by low-cost aircraft to be economically effective.

The system comprises a rectangular flat-plate antenna and pedestal, transmitter/receiver, radar control module, range and bearing control module and azimuth/range indicator.

Target detection ranges are:
Fishing vessel, in Sea State 3, 46 km
Trawler, Sea State 5, 93 km
Freighter, Sea State 5, 148 km
Tanker, Sea State 5, 185 km

The system also functions as a weather radar with a range of 370 km.

Specifications

Weight: 79.1 kg
Frequency: 9,375 MHz
Frequency agility: 85 MHz peak-to-peak
Power output: 100 kW peak
Pulsewidth: 2.4 and 0.5 μs
PRF: 267, 400, 1,200, and 1,600 Hz
Antenna rotation rate: 15 and 60 rpm
Antenna stabilisation: automatic compensation for pitch and roll up to ±20°, with tilt to ±15°
Azimuth/range indicator: PPI, P-7 phosphor, 178 mm CRT, north or aircraft heading orientated, range scales 46, 93, 231 km

Maritime patrol Beech Super King Air with the radome for the Telephonics AN/APS-128 surveillance radar on the underside of the fuselage

Operational status

In service. No longer in production. The AN/APS-128 is installed on all the Beech 200T maritime patrol aircraft operated by the Japanese Maritime Safety Agency and is operational with the Uruguayan Navy. The Brazilian Air Force bought 16 sets of equipment for its Embraer EMB-111 Bandeirante MR aircraft and Gabon and Chile also chose the system for EMB-111s. The Indonesian Air Force and the Royal Malaysian Air Force have installed the AN/APS-128 in their C-130 Hercules aircraft. It is also installed on the CASA 212 SAR aircraft of the Spanish Air Force.

Contractor

Telephonics Corporation.

UPDATED

AN/APS-128 Model D surveillance radar

An upgraded version of the AN/APS-128, the Model D is an all-digital radar with a scan converter to present information in television raster format. It contains target enhancement and clutter reduction circuits for frequency agility, sensitivity time control, constant false alarm rate and scan-to-scan integration.

The system comprises a flat-plate antenna and pedestal, transmitter/receiver, radar control unit, digital scan converter, a trackball or joystick cursor and bright display. A dual display for cockpit weather presentation is available, as well as cabin displays. A programmable microprocessor provides the alphanumerics and graphics to meet various mission requirements, while the antenna size can be altered to suit aircraft and operational needs. An alternative parabolic antenna with dual polarisation provides a pencil beam for sea search and a shaped beam for ground-mapping. The system has a 30 target track-while-scan facility and narrow/wideband datalinks. In addition, the APS-128D has successfully passed ground integration tests with the BAe Sea Skua missile.

Claimed target detection ranges are:
Snorkel or fishing vessel, Sea State 3, 55 km
Trawler, Sea State 5, 111 km
Freighter, Sea State 5, 185 km
Tanker, Sea State 5, 222 km

When functioning as a weather radar the range is 370 km.

Specifications

Weight: 91.8 kg
Frequency: 9.375 GHz
Frequency agility: 85 MHz peak-to-peak
Power output: 100 kW peak
Pulsewidth: 2.4 and 0.5 μs (0.1 μs pulse compression optional)
PRF: 1,400, 1,200 and 1,600 Hz

Operational status

In service but no longer in production. Over 90 systems have been sold worldwide. Versions of the APS-128D are in service with the Puerto Rican police and NASA, in the Falcon 900 for the Japanese Maritime Safety Agency, in CASA 212s for the Venezuelan Navy, Portuguese Air Force, Spanish Customs Service and Air Force, as well as the Argentine Coastguard and in the Skyvan for the Singaporean Air Force.

Contractor

Telephonics Corporation.

UPDATED

AN/APS-143(V)1 and (V)2 radars

The AN/APS-143 radar is designed to provide aircraft and aerostats with good detection capability against small targets in very high sea states. It is a lightweight, travelling wave tube radar with pulse compression. This provides better detection, resolution and clutter rejection, higher average power and greater receiver/transmitter reliability. The system includes a signal processor incorporating track-while-scan, scan conversion and databus interfaces.

Specifications

Weight:
(inc display) <110 kg
Frequency: 9.3 to 9.5 GHz
Agility: fixed, or can operate with 11 agility steps of around 20 MHz over the band
Peak power: 8 kW min, 10 kW nominal
Effective peak power: 700 kW up to 93 km range; 2.4 MW above
Compression ratio: 70:1 at ranges up to 93 km; 240:1 above
Pulsewidth: 5.0 and 17.0 μs uncompressed; 0.1 μs compressed (weighted)
PRF: 2,500, 1,510, 750 or 390 Hz (range dependent)

Operational status

In production. In operation with US government aerostats and US Air Force de Havilland DHC-8 aircraft used for test range surveillance. The Malaysian Air Force has taken delivery of the APS-143 for Beech 200T aircraft, as has the Japanese Maritime Safety Agency on the Saab 340B.

Contractor

Telephonics Corporation.

UPDATED

AN/APS-143(V)3 sea surveillance radar

The AN/APS-143(V)3 is a maritime surveillance and tracking radar designed for installation in a variety of fixed-wing aircraft and helicopters. The system uses frequency agility and pulse compression techniques and consists of three units: an antenna, receiver/transmitter and signal processor. Options available include track-while-scan, air search mode with MTI customised for the individual platform, air-to-surface missile guidance, integrated ESM interface, integrated IFF interface and a dedicated control unit.

The flat-plate antenna array can be fitted into any radome, or alternatively a parabolic type is available. It is stabilised for ±20° in pitch and roll. The transmitter is a TWT type with a peak power output of 8 kW. Options include the Array Systems, Canada TRISAR (ISAR, Spotlight SAR, and Strip Map), and colour displays.

The Boeing E-3 Sentry AWACS is equipped with the Telephonics AN/APX-103 IFF interrogator, with the antenna mounted back-to-back with the primary radar antenna in the rotating radome

Specifications

Weight: 81.8 kg
Frequency: 9.2-9.5 GHz
Frequency agility: 200 MHz (450 MHz for B(V)3)
Peak power: 8 kW
Pulsewidth: 5 or 17 μs, compressed width 100 ns (0.1 to 40 μs for B(V)3)
PRF: 2,500, 1,500, 800 or 400 Hz (2,500 Hz to 40 Hz for B(V)3)
Max range: 370 km
Compression ratio: 50:1/170:1 (1:1 to 3,600:1 for B(V)3)
Range resolution: 15 m
Azimuth accuracy: 0.5°

Operational status

In service with Republic of China Navy, the Hellenic Navy and the Royal Thai Navy S-70C helicopters, and ordered by the Israeli Navy, the New Zealand Navy, and the Turkish Navy. Also delivered to the Ecuadorean Navy for its Beech 200T aircraft.

An advanced version, the APS-143B(V)3 with ISAR, has been ordered by the US Navy for its SH-60R helicopters, and by Kaman Aerospace for the SH-2G helicopters for the Australian Navy. In addition, the Indian Navy has selected the APS-143B(V)3 for the DO-228 programme with ISAR, spotlight/stripmap SAR and MTI.

A variant of the radar, designated APS-143PC, forms part of the equipment kit of the four SH-2G(NZ) helicopters to be delivered to the New Zealand Navy, by Kaman Aerospace, in the year 2000.

Contractor

Telephonics Corporation.

UPDATED

AN/APS-147 multimode airborne radar

The AN/APS-147 multimode radar, designed for the US Navy SH-60R helicopter, is an inverse synthetic aperture radar which uses the latest in high throughput signal and data processing. Flexibility through programmability provides a product optimised for the maritime surveillance mission.

Advanced processing alllows the AN/APS-147 to use a variety of waveforms to perform its mission at an output power substantially lower than traditional counterparts in maritime surveillance radars. This results in a radar with an extremely Low Probability of Intercept (LPI). Using a low peak power waveform with frequency agility, the radar can detect medium- to long-range targets without the threat of ESM interception.

Radar modes include target imaging, small target and periscope detection, long-range surveillance, weather detection and avoidance, all-weather navigation, short-range search and rescue, enhanced LPI search and target designation.

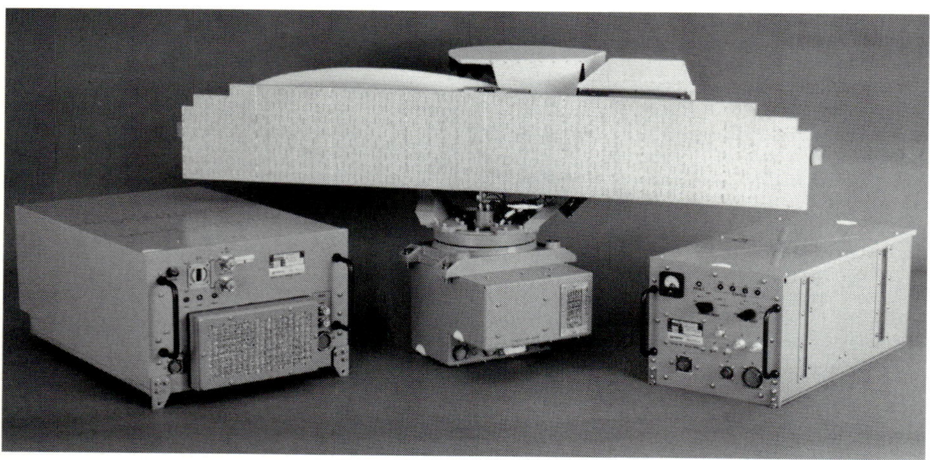
The AN/APS-143(V)3 is in service in S-70 helicopters

The AN/APS-147 features a flexible modular design which can be tailored to meet specific requirements and can be easily upgraded, LPI, high-resolution images for rapid classification, lightweight construction through the use of composite materials, low input power, simple design for high reliability and maintainability, fully programmable signal processor with multiple waveform exciter and high throughput rates and integrated IFF and SAR option.

Contractor

Telephonics Corporation.

UPDATED

AN/APX-103/AN/APX-103B IFF interrogator

The AN/APX-103 IFF interrogator system was developed for the Boeing E-3 Sentry AWACS which is operational with the US Air Force, NATO, UK, France and Saudi Arabia and will shortly be operational with the Japanese. The system utilises the features and functions of the co-operative beacon system commonly known as Mark XII Identification Friend or Foe (IFF). The AN/APX-103 provides state-of-the-art Air Traffic Control as well as Selective Identification and Friendly (SIF) identification of military aircraft for the airborne early warning mission. The output of the AN/APX-103 is digital target reports which are inputted to the E-3's central command and control computer for correlation of the IFF and primary radar information. The APX-103's digital processor operates accurately in dense target environments.

Operational status

The AN/APX-103 has completed a monopulse processing technology insertion upgrade and is now in service and available under the nomenclature AN/APX-103B. The AN/APX-103B technology improvements provide the operators with improved resolution and accuracy as well as improved target handling capacity. A further capability improvement, involving addition of the new civil Mode S feature to the equipment, is planned and expected to be available soon.

Contractor

Telephonics Corporation.

UPDATED

TRA 3000 radar altimeter

The TRA 3000 is a lightweight, panel-mounted radar altimeter for general use on executive and light aircraft and RPVs. The combined transmitter/receiver/antenna unit can be fuselage- or wing-mounted and used during all phases of flight. All indicators provide a decision height warning system. The top of the line TRI 40 indicator also includes a landing gear position indication.

Specifications

Dimensions:
(indicator) 35 × 88 × 189 mm
(transmitter/receiver/antenna) 192 × 126 × 25 mm
Weight:
(indicator) 0.27 kg
(transmitter/receiver/antenna) 0.68 kg
Frequency: 4,300 MHz
Altitude: 40-2,500 ft
Accuracy: 40-100 ft ±5 ft, 100-500 ft ±5%, 500-2,500 ft ±7%

Operational status

In service.

Contractor

Trimble Navigation Ltd, Avionics Products.

UPDATED

TRA 3500 radar altimeter

The TRA 3500 is a lightweight radar altimeter for general use on executive and light aircraft, helicopters and RPVs. It consists of a remote receiver/transmitter unit with two microstrip antennas which may be mounted on the fuselage or wings. The indicator provides visual and aural warnings.

Specifications

Dimensions:
(receiver/transmitter) 76 × 76 × 174 mm
(antennas) 11 × 106 × 174 mm
(indicator) 35 × 88 × 189 mm
Weights:
(receiver/transmitter) 1.48 kg
(indicator) 0.34 kg
Frequency: 4,300 MHz
Altitude: 0-2,500 ft
Accuracy: 0-100 ft ±5 ft, 100-500 ft ±5%, 500-2,500 ft ±7%

Operational status

In service.

Contractor

Trimble Navigation Ltd, Avionics Products.

UPDATED

TRT 250D transponder

The TRT 250D transponder is a lightweight solid-state transponder which meets FAA Modes A and C requirements. Operated by easy to use backlit push-buttons, the TRT comes in a flat-pack mounting. The TRT 250 is compatible with all current production height encoding altimeters.

Specifications

Dimensions: 41 × 81 × 254 mm
Weight: 0.77 kg
Frequency:
(receive) 1,030 MHz
(transmit) 1,090 MHz

Contractor

Trimble Navigation Ltd, Avionics Products.

UPDATED

Ballistic Winds programme

The Ballistic Winds programme is advancing airborne light detection and ranging (lidar) wind profiler technology to provide real-time 3-D wind information to improve delivery of various projectiles from high altitude. Wright Laboratory has been developing and field testing numerous lidar systems for several years. The first systems were ground based, and were used mainly for technology advancements, measuring and quantifying atmospheric backscatter, and supporting field tests. Field tests included measuring winds to correct for parachute drift, quantifying wind vortices from large aircraft, and base lining newer airborne lidar systems. Lidar technology advancements have permitted the transition from ground-based to airborne systems. The first airborne wind profiling system was demonstrated in 1995 on a C-141 aircraft. That system uses an eye-safe 2 μm laser source which shoots a laser pulse from the aircraft through the atmosphere. The returned laser light is Doppler shifted (wind velocity component) by the dust and water vapor particles that are traveling with the wind. Then, through a sequence of laser scans in a conical manner, the system can determine the directional component. The range resolved component is derived by the short pulse length of the laser and provides range slices of the atmosphere. As a result, the system can provide 3-D wind information from altitude to the ground with range slices of hundreds of meters and velocity accuracy of 0.5 m/s. This information can then be used to negate wind effects on projectile delivery.

The airborne laser wind profiler allows the navigator to incorporate real-time wind conditions from the aircraft to the ground and calculate a release point above 10,000 ft with high delivery accuracy. Wright Laboratory is planning to take the latest laser transceiver (smaller and more powerful than the original test transceiver) and install it in the forward section of a standard C-130 external fuel tank while keeping about 75 per cent of the fuel carrying capacity intact. The pod-mounted approach offers the ability to mount the lidar in any C-130 without permanent modifications to the airframe and do this in one day using standard flight line equipment.

Operational status

In development.

Contractor

US Air Force Materiel Command, Wright Laboratory.

VERIFIED

Moving and Stationary Target Acquisition and Recognition (MSTAR)

The objective of the MSTAR programme is to design, construct, and demonstrate in the laboratory an accurate and robust Automatic Target Recognition (ATR) system capable of locating and recognising time-critical targets in air-to-ground 2-D Synthetic Aperture Radar (SAR) imagery. The MSTAR programme emphasises an applied systems engineering approach that integrates advanced ATR algorithm modules to achieve accurate and robust performance in highly unconstrained ground-based image analysis scenarios. The scenarios focus on shallow hidden targets obscured by layover, partial masking, Cover, Camouflage and Deception (CC&D). The programme relaxes the constraints typically imposed by existing ATR development programmes on target, sensor, and background variants referred to as the MSTAR Extended Operating Conditions (EOCs).

The MSTAR EOCs include a minimum of 20 target classes at:

arbitrary viewing angles
radar squints off broadside at ±35°
radar depression angles up to 40°
modest levels of target camouflage and deception
30 per cent target obscuration levels in both flat and hilly terrains
multiple target configurations and articulations.

MSTAR algorithm design strategies employ a model-based or model-driven approach in which targets, backgrounds, target/background interactions, and the uncertainty with which they can be measured, extracted, and/or predicted, are each modelled to explicitly account for obscuration, clutter layover, camouflage, diffuse scattering, multipath reflections, and other operational conditions. A central tenet of this approach is that the large number of target, sensor, and background combinations found in realistic scenarios precludes the use of ATR algorithms that rely solely on precomputed templates, and which require large amounts of measured data. MSTAR will significantly advance the state of the art of existing SAR model-based algorithms such as the Automatic Radar Air-to-Ground Target Acquisition Programme (ARAGTAP) conceived of and developed by Wright Laboratory (WL) Avionics Directorate.

An associated team of module developers and a system integrator will develop the MSTAR system under a distributed, collaborative development strategy. The MSTAR modules are:

Focus of Attention - identifies potential targets and eliminates non-target areas in SAR imagery; uses contextual cues, map products, scene segmentation algorithms, map/image, and image/image registration algorithms, collateral information, image derived, and externally provided area delimitation techniques and advanced target detection algorithms.

Indexing - coarsely classifies target detections and hypothesises target orientation and background effects such as occlusion and layover.

Feature Prediction - uses phenomenon-based target and background models to predict the appearance of significant, discriminating SAR image features, along with appropriate probability distributions characterising feature attribute uncertainty.

Feature Extraction - excises and characterises stable, phenomenon-based target and scene features from image regions of interest.

Matching - compares corresponding predicted and measured target and/or background features to yield a probabilistic measure or score of their 'closeness'; refines estimates of key target parameters such as configuration, articulation, and orientation.

Search Management - formally accrues evidence and dynamically controls the overall hypothesis refinement process; controls the sequence of feature prediction and matching processes; assesses the outcome of prediction and matching; terminates the hypothesis refinement process.

Operational status

Early development.

Contractor

US Air Force Materiel Command, Wright Laboratory.

VERIFIED

Radar detection of concealed time critical targets (RADCON)

RADCON is a joint effort between Wright Laboratory (WL) and the Defense Advanced Research Projects Agency (DARPA). The objective is to extend and demonstrate foliage penetration (FOPEN) technology. RADCON will develop automatic target detection algorithms to enhance the performance of a radar sensor operating from an airborne platform to detect and classify time-critical targets concealed by foliage and/or camouflage. RADCON will: use real radar data to enhance the performance of target/clutter discrimination algorithms to achieve a useful probability of detection and false alarm rate at real-time operating speeds; develop a real-time signal processing capability; develop a concept of operation for a radar sensor capable of performing foliage penetration from an airborne platform and develop a foliage penetration radar system specification for an airborne platform consistent with the developed concept of operation and foliage penetration capability.

The RADCON Automatic Target Detection/ Classification (ATD/C) algorithm development began with the proven Concealed Target Detection (CTD) ATD/C algorithm as a baseline and will iteratively improve its performance. During the three years duration of the programme, radar data will be used to evaluate and enhance the performance of automatic concealed target detection algorithms. Target versus clutter discriminating techniques including, but not limited to: polarisation diversity, angle discrimination, wavelength, target groupings as well as other discriminants, are being developed, tested and evaluated to enhance the performance of the concealed target detection algorithms. These discriminants have been tested to ensure compatibility from the modeled data used in CTD. The performance of these algorithms will be evaluated by performing analysis of the real radar data containing targets concealed under foliage and/or camouflage. Once developed, RADCON's algorithms will be placed in the Khorus image processing environment and will be delivered as Khoros Glyphs as part of a Khoros toolbox to maximise development transportability and to minimise duplication of effort.

In order to properly evaluate the concealed target detection algorithms and validate the selection of target/clutter discriminants, RADCON has developed an advanced ground-based signal processing system. This demonstration system has been implemented from a Mercury i860 parallel processing system and will process data at real time rates. The system is a partially populated VME 9U chassis with 104 i860 processors with 16 megabytes of memory per node. These processors are on ruggedised 9U VME cards with 16 processors per card. The benchmark performance of this system ranges from 4 to 8 Gigaflops depending on instruction mix. The system has a total of 1.664 Gigabytes of memory. The system will process a single polarisation of radar data (IFP and RFI) at the real-time rate of 7.4 s/km^2 and 3.3 s for ATD/C. This system is available for further system trade studies and ATD/C algorithm development and performance analyses.

System, algorithm, and hardware trade studies have been performed to determine the ultimate system specification and concept of operation for real-time performance of IFP, RFI, and ATD/C processing of FOPEN data from an airborne platform.

Operational status

RADCON is a risk reduction programme and is in its last year of development at which time it will have automatic detection algorithms capable of integration into an airborne demonstration of a FOPEN radar capable of detecting time critical targets concealed under foliage and/or camouflage.

Contractor

US Air Force Materiel Command, Wright Laboratory.

VERIFIED

ELECTRO-OPTICS

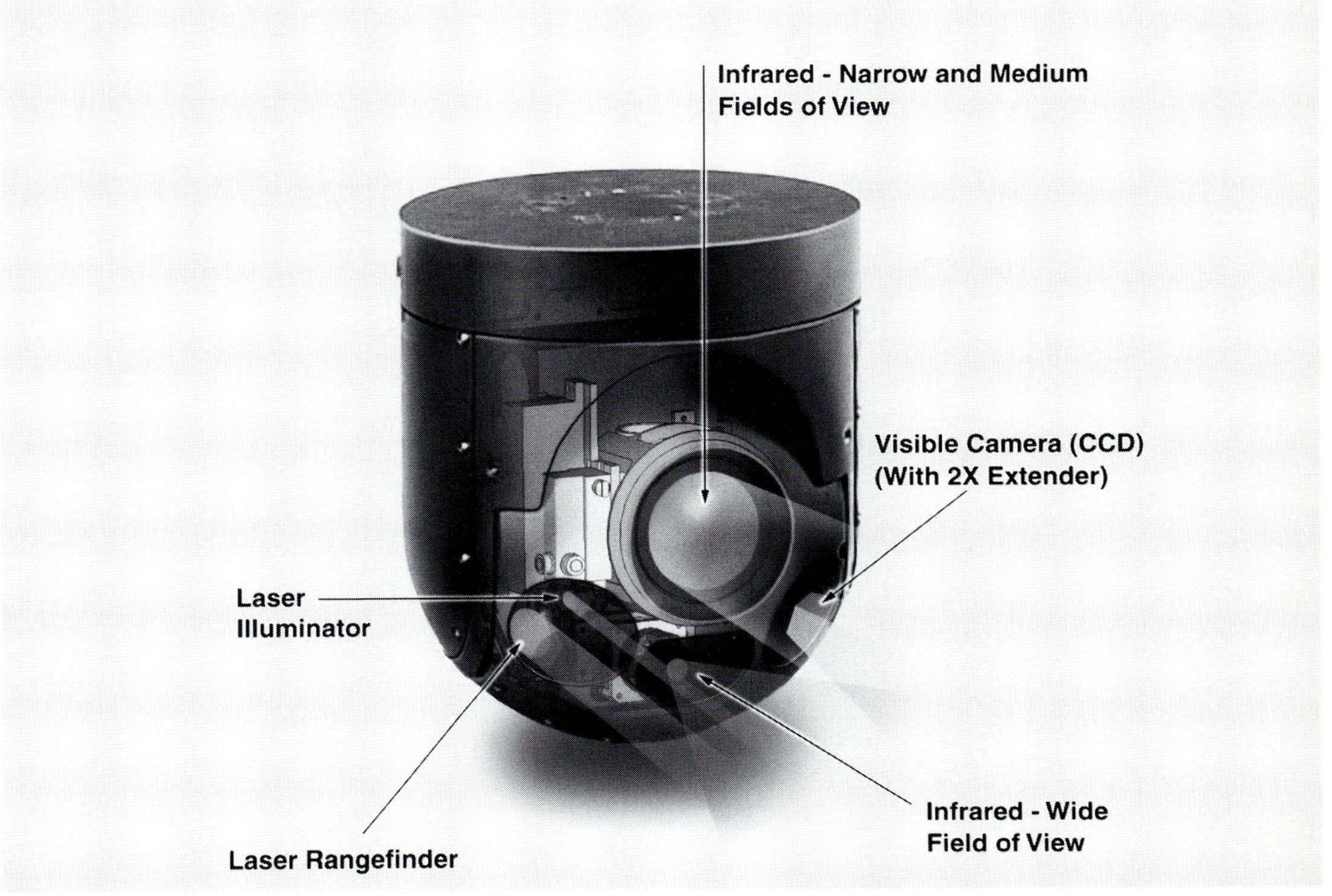

FLIR Systems Inc Star SAFIRE thermal imaging system AN/AAQ-22 ***1998***/0018313

AUSTRALIA

LRTS Long Range Tactical Surveillance sensor

British Aerospace Australia's LRTS thermal imaging sensor is part of a family of thermal imaging sensors for land, sea and air use. It is optimised for the maritime environment. Notable features are: a fourth-generation high-definition 640 × 486 pixel staring array detector, with high-performance optics and two or three fields of view; exceptional range performance in humid conditions.

Specifications

Detector: Platinum Silicide (PtSi): 640 × 486 pixels
Spectral response: 3-5 microns
Fields of view (degrees):
(narrow) 2.9 × 2.2
(wide) 11.3 × 28.5
Weight: 17 kg
Interfaces: RS422/232
Video format (frame rate): CCIR (25 Hz) or RS-170 (30 Hz); programmable 640 × 486 composite monochrome or RGB graphic overlay

Operational status

In service with the Royal Australian Navy.

Contractor

British Aerospace Australia Ltd.

NEW ENTRY

Ranger 600 eyesafe laser rangefinder

The Ranger 600 is one of a family of British Aerospace Australia's tactical sensor solutions for naval, land and airborne applications. The Ranger 600 eyesafe laser rangefinder is designed for a maritime environment and provides excellent sensitivity and range performance for a laser rangefinder that is eyesafe to the unprotected eye.

Ranger 600 uses the latest high-reliability, solid-state, diode-pumped laser technology that is OPO frequency shifted to eyesafe wavelengths. Notable features include: programmable internal false alarm and probability of detection settings; high performance transmit and receive optics, with transit alignment to within 0.2 mrad, and field of view less than 1.5 mrad; hardware-interlocked safety override and, to maximise tracking accuracy, the capability to synchronise laser ranging to an external imaging sensor.

Specifications

Detector: avalanche photodiode
Range: 30 km
Accuracy: ±2 m
Boresight: 0.2 mrad
Output power: 7.8 mJ per pulse (min)
PRF: up to 20 Hz
Fields of View (FoV):
(receiver) 1.5 mrad
(transmitter) 0.5 mrad
Weight: 10.9 kg
Power: 24-32 V DC
Interfaces: RS-422 (optional RS232)
Cooling: air cooled

Operational status

In production and in service with the Royal Australian Navy.

Contractor

British Aerospace Australia Ltd.

NEW ENTRY

SearchIR thermal imaging system: 3-5 μm multiple role

The SearchIR thermal imaging system is a small, lightweight unit suitable for air, land and sea use on military and civilian platforms, for new or retrofit installation.

The SearchIR system incorporates the following design features:
(a) long range and high resolution in warm, high-humidity environments, based on use of 3-5 μm staring array detector technology which simultaneously images the entire infrared (IR) scene with over 311,000 individual detectors to provide high-definition images;
(b) a gimbal system with precision four-axis stabilisation which eliminates image jitter, and two or three fields of view with which to optimise navigation, surveillance, and targeting/identification tasks;
(c) high resolution and a narrow field of view to optimise long-range surveillance capability;
(d) digital uniformity correction which improves infrared display imaging by eliminating the gain mismatch characteristic of scanned detector imaging systems, and automatic scan, track, and brightness functions to reduce operator workload;
(e) rugged construction, and flexible interfaces, to assist integration with other sensor/display systems and aid integration into a wide range of platforms.

British Aerospace Australia SearchIR thermal imaging system **1996**

Specifications

Fields of regard (degrees):
(azimuth) 360
(elevation) +35 to −120
Fields of view (degrees):
(narrow) 2.9 × 2.2
(intermediate) 11.3 × 8.6
(wide) 37.7 × 28.5
Spectral response: 3-5 μm
Detector: Platinum Silicide 486 × 640 pixels
Video format: RS-170 (30 Hz) or CCIR (25 Hz) available
Dimensions:
(turret) 406 × 550 mm (42 kg)
(control electronics) ½ ATR 350 × 124 × 194 mm (9 kg)
Power: 28 V DC, MIL-STD-704
Interfaces: dual redundant MIL-STD-1553B, RS-232 and RS-422

Operational status

The system is in use with the Royal Australian Navy, and it is under consideration by several other military forces.

Contractor

British Aerospace Australia Ltd.

VERIFIED

CANADA

Aviator's Night Vision Imaging System (ANVIS)

The Aviator's Night Vision Imaging System (ANVIS) is a helmet-mounted unity power image intensifier binocular which allows low or contour flying in fixed-wing aircraft or helicopters at night. The system is compatible with either second- or third-generation image intensifier tubes and can be used in conjunction with FLIR systems. The binocular is lightweight, can be fitted to any flying helmet and provides full peripheral vision. The complete ANVIS consists of the binocular assembly, image intensifier tubes, helmet visor interface and battery power pack.

Specifications

Weight:
(binocular) 0.463 kg
Field of view: 40°

Operational status

In service with the US Army.

Contractor

Hughes Elcan Optical Technologies Ltd.

VERIFIED

Gyro-stabilised video, infrared, laser and film surveillance systems

Wescam Inc produces a range of gyro-stabilised video, infrared, laser and film camera surveillance systems, together with associated datalink systems, controls and installation kits. They are suitable for fitment to a variety of air and ground platforms, including fixed-wing aircraft, helicopters, UAVs, and land/sea vehicles. Their primary roles include: civil/military surveillance, airborne law enforcement, environmental studies and sport/entertainment filming. A number of models are available, designated by their diameter and number of sensors (Model 12 DS having a diameter of 12 in and being fitted with a dual sensor).

An overview of capabilities is tabulated below for the Wescam Model 16/18/24/36 series. The overview is followed by details of particular systems.

Model 12DS
The Model 12DS is Wescam's smallest gyrostabilised camera, designed specifically to meet airborne law enforcement requirements. Weighing less than 23 kg, the 12 in diameter dual sensor camera features a high-resolution two field of view Indium Antimonide (InSb) staring array thermal imager and a colour CCD Daylight TV camera with ×14 zoom lens. The active gyrostabilisation and vibration isolation provide less than 35 micro-radians line of sight jitter. The Model 12DS can also be integrated with aircraft radar, navigation and map displays, GPS and with other sensor systems.

Model 14TS/14QS
The Model 14TS can accommodate three sensors, and the Model 14QS four sensors. The range of sensors available for the Model 14 include: high-resolution daylight TV with zoom; a 900 mm long-range spotter scope with camera; a 3-5 μm multi-field of view thermal

WESCAM Model 16/18/24/36 Series:

Characteristic	Model 16 Series	Model 18 Series	Model 24 Series	Model 36 Series
Key features	High performance day/night imaging with on-gimbal electronics	Four simultaneous sensor imaging (daylight TV, 900 mm spotter scope, 3-5 μm FLIR, laser rangefinder)	Ultra-high stability and long range zoom capability (to max 1,600 mm focal length)	Wescam's greatest stability and long range zoom capability (to max 2,200 mm focal length)
Performance	Identify a person at 130 m Detect a moving vehicle at 8 km	Identify a person at 400 m Detect a moving vehicle at 15 km	Identify a person at 600 m Detect a moving vehicle at 21 km	Identify a person at 1,000 m Detect a moving vehicle at 24 km
Stabilisation	Active, 4 axes inner & outer (pitch/yaw) gyroscopic line of sight, including downlook stabilisation. Third axis roll compensation available for Model 16DB	Active, 4 axes inner & outer (pitch/yaw) gyroscopic line of sight, including downlook stabilisation	Passive, 3 axes (pitch, roll, yaw) gyroscopic line of sight	Passive, 3 axes (pitch, roll, yaw) gyroscopic line of sight
Line of Sight – Jitter	<35 μrad rms of jitter*	<35 μrad rms of jitter with external isolator*	<5 μrad rms of jitter	<5 μrad rms of jitter
Vibration Isolation	6 axes isolation: (x,y,z, pitch, roll, yaw)	6 axes isolation: (x,y,z, pitch, roll, yaw)	6 axes isolation: (x,y,z, pitch, roll, yaw)	6 axes isolation: (x,y,z, pitch, roll, yaw)
Performance				
Slew rate: elevation	0-60°/s	0-90°/s	0-45°/s	0-45°/s
Slew rate: azimuth	0-60°/s	0-90°/s	0-60°/s	0-60°/s
Pan field of regard	Continuous 360°	Continuous 360°	Continuous 360°	Continuous 360°
Tilt field of regard	+30 to –120°	+30 to –120°	+30 to –90°	+30 to –90°
Physical				
Diameter (nominal)	0.40 m (16 in)	0.36 m (14 in)	0.61 m (24 in)	0.91 m (36 in)
Max height	0.55 m (20 in)/.61 m (24 in), (-W & -M versions only)	0.42 m (16.5 in)	0.81 m (32 in)	1.17 m (46 in)

* Note: Stable to the number specified for all airborne applications, including helicopters. Typical stability on a light aircraft is 2 × better (for example 15 μradians).

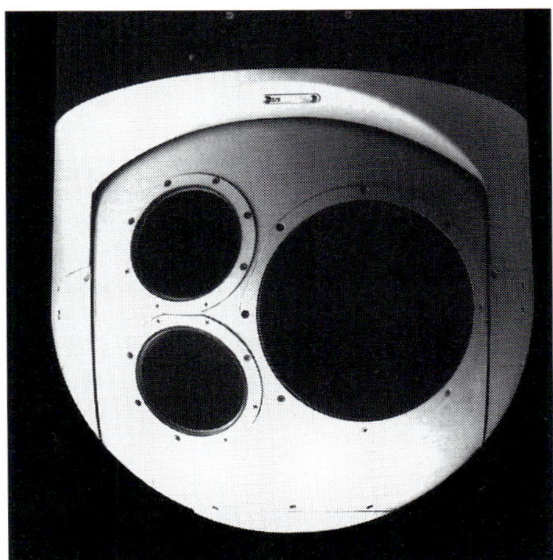

Model 14TS **1997**/0002205

imager, and an optional 'eye-safe' laser rangefinder. Stabilisation capacity is the same as Model 12DS.

Models 16DS-A/16DS-M/16DS-W

Models 16DS-A (AGEMA) and 16DS-W (Westinghouse) dual sensor systems provide high-resolution daylight television images together with 8-12 μm thermal imaging. The Model 16DS-A is optimised for environmental inspection, while the Model 16DS-W is configured to meet military specifications, with features such as higher temperature sensitivity for target detection. Model 16DS-M is configured for hot, humid and tropical climates and provides high-resolution daylight colour TV images together with 3-5 μm thermal imaging.

Model 16SS-A (AN/AAQ-501)

Model 16SS-A is a precision-stabilised infrared single sensor which detects, identifies and tracks distant objects in total darkness or poor weather. The sensor used is the AGEMA THV1000, 5-bar SPRITE focal plane, 8-12 μm system. It provides a narrow field of view (5.0 × 3.3°), and a wide field of view (20.0 × 13.0°). This 'Commercial-Off-The-Shelf' (COTS) hardware was designed to meet both military and commercial aviation requirements and has been given the military nomenclature AN/AAQ-501. Major operational roles are: surveillance, and search and rescue.

Model 16SS-B320/16SS-B750

Model 16SS-B single-sensor daylight broadcast TV system delivers high-resolution TV broadcast-quality images for electronic newsgathering applications. The Model 16SS-B750 single-sensor daylight broadcast TV system, complete with a three-CCD colour video camera and a ×36 lens, is used for sporting event

Model 16SS-B750 **1997**/0002207

reporting. A motorised extender provides double range performance (×72).

Model 24SS-550/24SS-1600

The Model 24SS-550 is a single sensor TV system used for long-range electronic newsgathering, giving magnification up to ×55 The Model 24SS-1600 provides yet greater magnification ×160 with gyrostabilisation and jitter control to less than 5 μrad. An optional low-light camera with spectral response from 550-900 nm is also available.

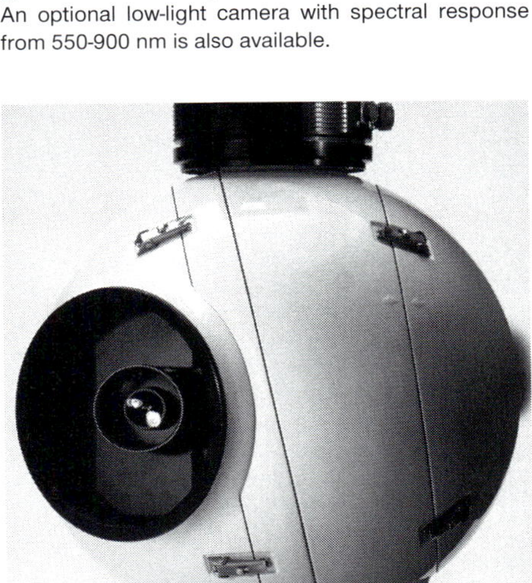

Model 24SS-1600 **1997**/0002208

AN/AAQ-501 **1997**/0002206

Model 36SS-2200

The Model 36SS-2200 provides ultra high-resolution broadcast TV images from very long standoff range. It provides ×220 magnification, with less than 5 μrad jitter. An optional interchangeable low-light GEN III camera is available for night-time use.

Contractor

Wescam Inc.

UPDATED

Model 36SS-2200 **1997**/0001219

P-3C Advanced Imaging Multi-spectral System (AIMS)

The Wescam Model 20 was developed specifically for long-range surveillance applications and combines high stabilisation with multiple, high magnification, day and night vision cameras. Based on the demonstrated performance of the Model 20, Wescam has been selected by Lockheed Martin Defense Systems, Eagan, to supply the Advanced Imaging Multi-spectral System (AIMS) for the US Navy P-3C Upgrade III Anti-Surface Warfare (ASuW) Improvement Program (AIP).

The AIMS replaces the capabilities of the current AN/AAS-36 InfraRed Detection Set (IRDS) and Electro-Optical Sensor (EOS) AN/AVX-1(V) in a turret system with full 360° azimuth field of regard. The AIMS will be located at the existing IRDS station.

Operational status

Wescam has been contracted to provide two systems, with options for follow-on orders of up to 160 systems. The AIMS variant of the Wescam Model 20 system is intended to be fielded on all AIP aircraft, and is compatible with non-AIP P-3C aircraft which currently incorporate the AN/AAS-36 IRDS. It can also be installed as a stand-alone installation, or integrated with other navigation and radar systems. Deliveries to AIP, Maritime Patrol Aircraft (MPA) and other customers with surveillance requirements are scheduled for early/mid-1998.

Contractor

Wescam Inc.

NEW ENTRY

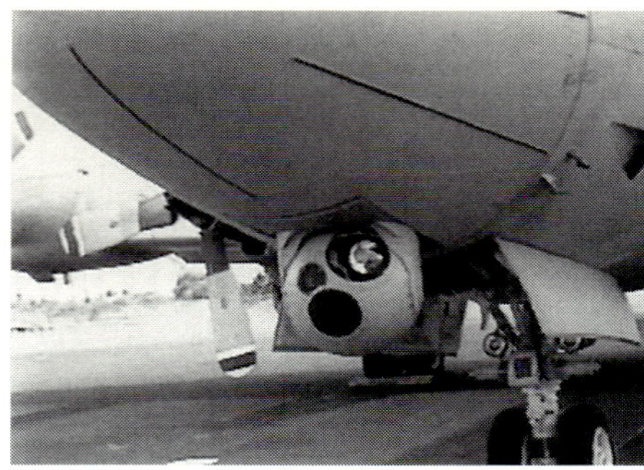

Wescam Model 20 Advanced Imaging Multi-spectral System (AIMS) US Navy P-3C Update III Anti-Surface Warfare (ASuW) Improvement Program (AIP) installation ***1998***/0018368

DENMARK

SIT/ISIT low-light level television system

JAI's intensified low-light television system is essentially a man-portable low-light level television system designed for use as electronic newsgathering equipment by broadcasting organisations. It has, however, been applied to a number of airborne military, paramilitary and civil roles, particularly aboard helicopters. These include reconnaissance, fisheries patrol and inspection, police night surveillance activities, mountain rescue and night electronic newsgathering. In such applications, the camera may be shoulder-mounted for use from a helicopter cabin footstep, or used with a Ronford 15S mounting head combined with a helicopter shockmount to counteract the effects of rotor-induced vibration.

The system operates over a wide range of lighting conditions, down to moonlight and starlight illumination levels. For helicopter work, the system is supplied with a 4.5 in (114 mm) viewfinder and a double pan bar for use with the 15S head. A full range of accessories is available and special equipment for outside broadcast applications includes: a video transmitter/receiver operating in the 2 GHz band, a real-time video image processing system and a man-portable videotape recording unit. The company also supplies a range of motorised zoom lens assemblies which meets the special optical requirements of low-light television systems. The manufacturer claims that, with these lenses, low-light television systems can operate over a range of high or low ambient lighting conditions.

Specifications

Weight: variable, according to lens system fitted, but typically approximately 6 kg (including camera, lens system and Ni/Cd belt-mounted battery power supply).

Operational status

In production and service.

Contractor

JAI A/S.

UPDATED

Modular Reconnaissance Pod (MRP)

The Royal Danish Air Force (RDAF) has contracted Per Udsen Company to develop an all-new reconnaissance pod to be sufficiently flexible to accommodate current and future sensor systems. The MRP comprises three parts: a common pod structure, a sensor-particular part, and a platform-particular part.

The common pod structure includes the podbody, strong-back, and electronic control system. The sensor-particular part is fitted with fixings for LRUs and sensors. The platform-particular part includes pylon attachments and electrical/mechanical interfaces.

The RDAF has selected the TERMA EWMS to control the MRP on its F-16 aircraft.

Payloads to be integrated are reported to include the: Recon/Optical CA-260, or most recently the CA-261; EO sensors; Vinten wet film and EO sensors; El-Op LAEO (Low-Altitude EO) and MAEO (Medium-Altitude EO) sensors; Lockheed Martin ATARS sensors.

Modular Reconnaissance Pod (MRP) ***1997***/0001220

Specifications

Dimensions: 4,496 × 762 × 610 mm
Weight:
(empty) 227 kg
(loaded) 544 kg
Flight envelope: +9 *g*
Data interfaces: tape recorders and datalink

Operational status

In development for RDAF F-16 aircraft. Initial flight trials on RDAF F-16 aircraft have been completed. More than 20 pods are reported to be on order by the following four air forces: Royal Danish Air Force, Belgian Air Force, Royal Netherlands Air Force, US Air Force. It is understood, that Per Udsen, TERMA Electronik and Recon/Optical may be teaming to fulfil these operational requirements.

Contractor

Per Udsen Company Aircraft Industry A/S.

VERIFIED

Airborne surveillance system

The TERMA surveillance system is designed for detection of oil spills, identification and documentation of fishing violations, performance of search and rescue missions and ice-mapping by day and night and during periods of poor visibility. It integrates a wide variety of surveillance sensors, navigation equipment and video systems.

A Side-Looking Airborne Radar (SLAR) has become the primary long-range sensor for oil pollution surveillance, typically covering a 37 km swath from preferred search altitudes. An oil slick is detected by variation in reflected radar signals between oil-covered water and normal seawater. In applications like ice-mapping and ship surveillance the SLAR covers a 74 km swath.

An Infrared/UltraViolet (IR/UV) scanner is provided for close-range imagery and allows a rough area estimation to be made, as the aircraft passes overhead, of the oil slick detected by the SLAR. The IR system can be operated by both day and night. It provides information on the spreading oil and indicates the relative thickness within the oil slick. The ultraviolet sensor is only used during daylight. It maps the complete area covered with oil irrespective of thickness.

A scanning radiometer system is provided for oil thickness measurements and quantification, enabling clean-up operations to attack the worst part of the spill first. The MicroWave Radiometer (MWR) measures microwaves originating from the sea surface at I/J- and K-band wavelengths.

Video cameras are used to secure evidence of oil pollution, fishery violation and other illegalities. Information can be recorded on videotape or stored as still photographs in the computer. Real-time navigation data is integrated into the picture. The video can be normal colour, highly sensitive low-light level TV or IR. A hand-held camera, with a real-time data annotation capability, can be integrated into the system.

Data downlink equipment is used for transmission of real-time or stored data to a ground- or ship-based station.

Information from microwave and optical sensors can be recorded either on a standard VTR or on a high-resolution digital tape recorder.

The 355 mm Sensor Image Display (SID) provides the operator with sensor information. The SID presents the current sensor image whether it is the SLAR image, the IR/UV scanner image or the radiometer image that is selected. Real-time navigational data is integrated into the bottom of the SID format. Information on aircraft position, heading, speed and altitude, as well as date and time is presented to the operator. By means of the trackerball, the operator can move a cursor on the SID and the target position is then annotated with real time.

The 254 mm colour map display provides the operator with an outlined map of the area under surveillance. The map display is integrated with the video system. The operator can select map information, video information or both simultaneously. Map information is available from customised map data. The map can be zoomed in close steps and the operator can insert symbols at any position.

The 254 mm control panel display facilitates the operation of all surveillance sensors, back-up stores and video systems. All possibilities in these systems are pre-arranged in the control panel display as logical menus, sectioned into two or three levels. Functions within the menu are accessed by use of a two-stroke keypress or the trackerball.

Specifications

Dimensions:
(operator console) 1,150 × 575 × 1,400 mm
(observer console) 1,150 × 775 × 1,190 mm
Weight:
(operator console) 145 kg
(observer console) 150 kg
Power supply: 28 V DC, 900 VA (max)

Contractor

TERMA Elektronik A/S.

VERIFIED

FRANCE

TIM laser rangefinders

TIM laser rangefinders are high-repetition rate eye-safe laser rangefinders designed to be integrated into an airborne fire control system to measure the distance to a ground-based, aerial or naval target.

They consist of a 1.54 μm transmitter integrating a 1.06 μm laser, a Raman conversion cell, a receiver, the power supplies, an interface chronometry/control and serial link RS-422 with the host system.

Specifications

Dimensions: 300 × 180 × 150 mm
Weight: 8.5 kg
Power supply: 220 V AC, 3 phase
or 115 V AC, 400 Hz

Operational status

TIM laser rangefinders have been selected for the Rafale aircraft and the Tiger helicopter.

Contractor

Compagnie Industrielle des Lasers (CILAS).

VERIFIED

TMS303 laser rangefinder

The TMS303 laser rangefinder is intended to be integrated into short- and medium-range weapon systems for range measurement of land and airborne targets, such as the SFIM Viviane sight for HOT missiles on the Gazelle helicopter.

The sight consists of a 1.54 μm transmitter integrating a 1.06 μm laser, a Raman conversion cell and a power supply; an interface/control/range processing card; a receiver, and a low-voltage converter.

To remain compatible with existing systems, the TMY303, a 1.06 μm version has been developed this could easily be upgraded to the 1.54 μm eye-safe configuration.

Specifications

Dimensions: 190 × 119 × 78.5 mm
Weight: 2.2 kg
Power supply: 28 V DC

Operational status

TMS303 and TMY303 are in series production.

Contractor

Compagnie Industrielle des Lasers (CILAS).

VERIFIED

TMS312 laser rangefinder

The TMS312 laser rangefinder is a version of the TMS303 (see previous item) which is designed for medium-range applications such as air-to-ground fire-control systems. It offers higher repetition rates of 3 Hz continuous wave or 6 to 8 Hz in short bursts.

Contractor

Compagnie Industrielle des Lasers (CILAS).

VERIFIED

IRIS new generation FLIR

IRIS is a high-sensitivity modular thermal imager with high resolution and image quality. IRIS's main features include: up to three switchable fields of view, automatic gain and offset control, polarity selection; ×2 zoom, athermalised focusing, extended BITE, boresight alignment; digital image enhancement, hot point detection and tracking, and low power consumption.

The latest enhancement given to SAGEM's nav/attack system (MAESTRO) has been the night operation capability, based on integration of the IRIS internal FLIR. The IRIS second-generation IRCCD camera provides a one-to-one infrared image, superimposed on the external world in the head-up display, in combination with the normal symbology. A narrow field of view image is also available to be displayed in the head-down display, for easy and precise target designation.

The IRIS thermal imager **1996**

This new configuration has recently been qualified in-flight onboard a Mirage III and Mirage 2000.

The compact design of IRIS allows for internal installation leaving all hardpoints free for weaponry.

Specifications

Wavelength: 8-12 μm
Detection module: integrated detector/dewar microcooler device
288 × 4 elements IRCCD focal plane array Cd Hg Te closed-cycle Stirling microcooler
NETP: T < 0.02°C
Video output: CCIR

Operational status

In production for Mirage 2000, Mirage III and UAV programmes. IRIS is also fitted into the main sight of several attack helicopters (French and export), including the Tiger HAP helicopter where it is part of the STRIX sight, and the Rooivalk sighting system. A variant called Condor 1 is used for the sight of the Tiger HAC helicopter, and Condor 2 is integrated in the Tiger HAC navigation FLIR.

Contractor

SAGEM SA, Defence and Security Division.

UPDATED

Cyclope 2000 infrared linescan sensor

The Cyclope 2000 infrared linescan sensor is designed for airborne applications including: reconnaissance, and battlefield observation and surveillance (on aircraft, helicopters or UAVs); as a navigation aid; the monitoring and surveillance of sensitive areas; forest fire detection, pollution detection (oil on the sea's surface) and mine detection. It provides day and night infrared images of the 8 to 12 μm spectral bandwidth to a thermal sensitivity of 0.1°C, with an angular resolution of 1 mrad.

The basic modular configuration can be adapted to various aircraft. Improved versions are offered with a different spectral bandwidth, multispectral detection or stereoscopy.

Specifications

Dimensions: 170 × 170 × 200 mm
Weight: <6 kg
Power supply: 115 V AC, 400 Hz, 3-phase, 70 W
or 28 V DC, 70 W

Contractor

SAGEM SA, Defence and Security Division.

UPDATED

CN2H Night Vision Goggles (NVGs)

The CN2H NVGs are part of a range of night vision systems produced for military applications; the goggles are specifically designed for use in helicopters and fixed-wing aircraft for night piloting in tactical situations. They are fixed to the helmet, with a power pack on the back, and a specially designed support bracket enables them to be immediately discarded in an emergency. Focusing and positional adjustments are available to suit the wearer. The goggles are compatible with a wide range of French, British and American helmets and they incorporate third-generation image intensifiers according to requirements.

Specifications

Weight: 0.95 kg incl battery pack
Power supply: 28 V DC or 3.5 V PS 31 battery
Battery life: 20 h
Field of view: 40°
Magnification: ×1

Operational status
In production. Adopted by the French Army Aviation (ALAT), Air Force and Navy, and by export customers.

Contractor
SFIM Industries.

UPDATED

CN2H-AA Night Vision Goggles (NVGs)

The CN2H-AA NVGs are a development of the CN2H. They are designed for aircrew in high-performance military aircraft and can be rapidly released with either hand before ejection. The NVGs can be equipped with Gen II and III image intensifiers, for high-resolution and high-sensitivity, and can be mounted on any type of aircrew helmet.

Specifications
Weight: 0.59 kg
Power supply: 28 V DC or PS31 battery

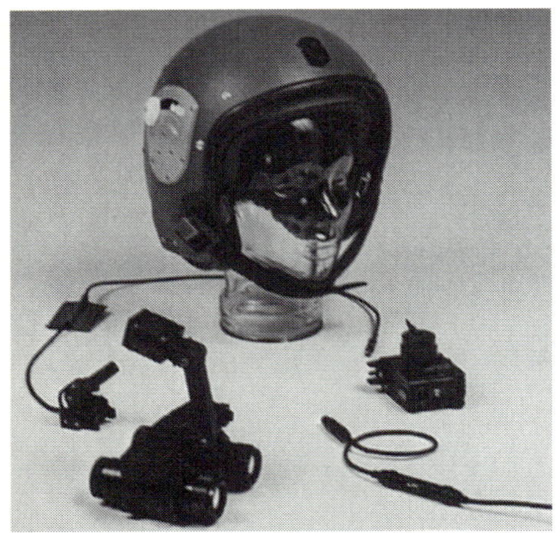

CN2H-AA NVGs have been successfully tested in a Mirage 2000 ***1995***

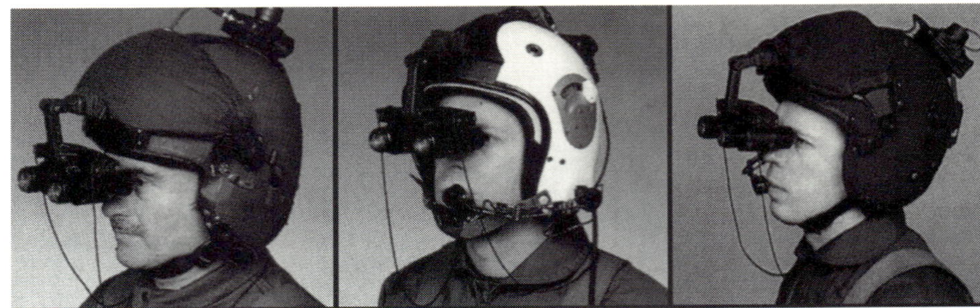

Some adaptations of the CN2H Night Vision Goggles already in service ***1998**/0018367*

Field of view: 40°
Magnification: ×1

Operational status
In production and in service with the French Army Aviation (ALAT).

Contractor
SFIM Industries.

VERIFIED

JADE night vision goggles

JADE is a projected-image NVIS, designed for pilots of combat aircraft and compatible with emergency ejection at speeds of up to 600 kt.

The goggles can be equipped with two third-generation, high-resolution tubes, or with 'Class IV' tubes. Monitoring of the Head-up Display and the cockpit instruments is facilitated by projected images of the relevant data, which can be superimposed over the night vision image into the pilot's field of view. JADE goggles are compatible with cockpit lighting designed in accordance with the STANAG 33800 standard.

Operational status
JADE has successfully completed 'windblast' qualification testing at the Centre d'Essais Aeronautiques de Toulouse (CEAT) and qualification testing at the Centre d'Essais en Vol (CEV).

JADE night vision goggles ***1998**/0018366*

Contractor
SFIM Industries.

NEW ENTRY

AA338,100 reconnaissance camera

The AA338,100 camera, with a very long focal length, is equipped with a high-definition lens. It offers lateral oblique photography of the ground, in daylight, from an aircraft flying at high and medium altitudes. The camera is installed in a pod manufactured by Avions Marcel Dassault. Aiming of the camera may be automatic for preplanned missions, or manual for targets of opportunity. The sight consists of a television camera displayed to the pilot on a radar scope and an inlay unit.

Specifications
Weight: 350 kg
Altitude: 13,000-49,000 ft
Range: 9-100 km
Resolution: 1 m at 100 km
Lens: 1,700 mm autofocus
Framing rate: 1 frame/0.6 s (max)
Film: 126 mm (114 × 111 mm format)

Contractor
Thomson-CSF Optronique.

VERIFIED

The Thomson-CSF Optronique AA338,100 reconnaissance camera

Airborne reconnaissance and surveillance system

The airborne reconnaissance and surveillance system provides photographic reconnaissance by day at low and medium altitudes for surveillance missions at sea through an assembly at the front or rear of the aircraft. The dropping of charges can also be photographed through the assembly at the rear of the aircraft.

The system consists of an AA3-35-100 camera, BC3-135-2 camera, BF3-135-2 computer and P11-1 vacuum

The airborne reconnaissance and surveillance photographic system for surveillance missions at sea showing (left to right) the three lenses, control unit and computer

pump. Three lens cones are available: the E3-150, E3-75 and E3-300.

Specifications

Dimensions (with E3-150 lens cone): 300 × 240 × 340 mm
Weight: 16 kg
Power supply: 200 V AC, 400 Hz, 3 phase
Format: 114 × 114 mm

Operational status

In production since 1988 for the French Navy Atlantique 2.

Contractor

Thomson-CSF Optronique.

UPDATED

AP 40 panoramic film camera

The AP 40 panoramic film camera is designed for medium-, low- and very low-altitude reconnaissance missions, at very high penetration speed.

Specifications

Lens focal length: 75 mm
Frame rate: 2 to 10 frames/s
Magazine: 75 m of 70 mm film; 300 exposures with standard film
Panoramic image: 180°
Dimensions: 386 × 316 × 197 mm
Weight: 19 kg
Optional fit:
(magazine) large capacity 150 m of 15 mm film; 600 exposures with standard film
(weight) 29 kg

Operational status

Current programme: upgraded Super Etendard

Contractor

Thomson-CSF Optronique

NEW ENTRY

AP 40 panoramic film camera under Super Etendard ***1998***/0018365

ATLIS II laser designator/ranger pod

ATLIS II is a pod-mounted laser targeting system with an automatic TV tracker, laser designator/ranger, tape recorder and interface electronics. The automatic TV tracker and laser stabilisation system reduce pilot workload, allowing tracking and designation in a single-seat aircraft. ATLIS II can be installed on a wide variety of strike or close support aircraft.

The system has an accuracy of 1 m on a target at an average firing range of 10 km and provides a high rate of success for low- and medium-level standoff attacks. It provides high image quality due to the dual-mode tracker which covers both the visible and near infrared portions of the spectrum. ATLIS has a long-range target acquisition capability even in the most adverse weather conditions. It features a high degree of magnification (up to ×20), damage assessment capability, video recording and a reconnaissance mode.

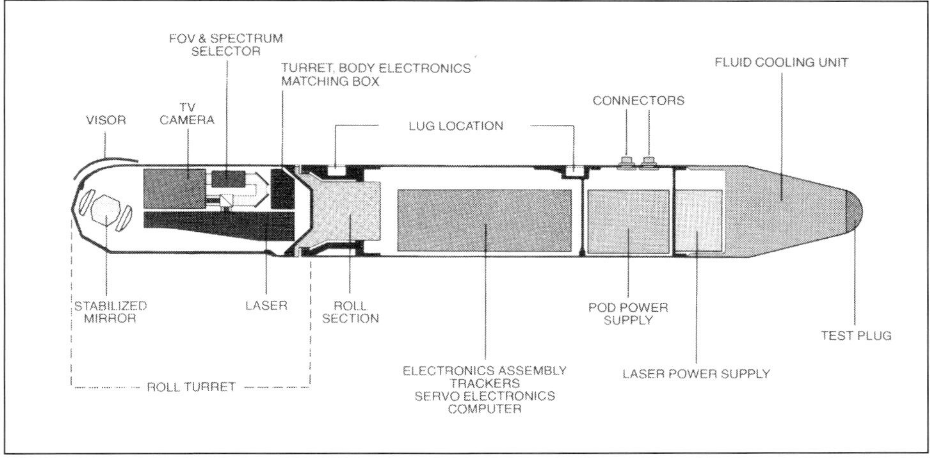

The Thomson-CSF Optronique ATLIS pod

Specifications

Dimensions: 2,520 (length) × 305 mm (diameter)
Weight: 170 kg
Power supply: 115 V AC, 400 Hz, 2.3 kW
Wavelength: 1.06 μm
Max operating range: dependent on the weapon selected
Angle of regard:
(roll) unlimited
(pitch) −160 to +15°
Attachments: standard 760 mm NATO bomb rack

Operational status

In production. The ATLIS 2 pod equips French Air Force Jaguars and has been ordered for export to equip Mirage 2000 and F-16 aircraft.

Contractor

Thomson-CSF Optronique.

VERIFIED

Chlio thermal imager

The Chlio long-range observation FLIR has been developed from the Tango thermal imager produced for the Atlantique 2 maritime reconnaissance aircraft. It is designed to be used for surveillance, as a flying aid in poor visibility, or as an aid for search and rescue operations.

The sensor is based upon the SMT common modules system and features focal length adjustment through the control panel. The thermal image is displayed on a CCIR 625-line 50 Hz television monitor. In CCIR format there are 780 pixels per line. It is fitted into a SERE BEZU lightweight gimbal with a simple mechanical interface, which may be fitted on the forward part on the side of the helicopter. Optional extras include line-to-line integration, auto-tracking, automatic video gain/offset control, target designation for radar.

The ATLIS II pod installed on an F-16 aircraft ***1998***/0018364

The Mirage 2000 with the Thomson-CSF Optronique ATLIS II laser designator pod mounted under the starboard engine

A typical image from the Chlio thermal imager ***1995***

The Chlio thermal imager on a Breguet-Alizé reconnaissance aircraft ***1996***

Operational status

In production for the French Air Force for installation on the Alouette III helicopter. Chlio has also been selected by the French Navy for Alizé reconnaissance aircraft and for five Dassault Falcon 50 aircraft for maritime surveillance. Chlio will equip 6 CN 212 maritime patrol aircraft of the Indonesian Navy as part of the Thomson-CSF AMASCOS mission system.

Specifications

Spectral band: 8-13 μm
Detector: 40-element CMT
Cooling: split-cycle Stirling
Fields of view
Zoom version: 24 × 16° to 10 × 6.7°, 12 × 8° to 5 × 3.3°
Bifocal version: 24 × 16°, 4 × 2.67°; 12 × 8°, 2 × 1.33°
Electronic magnification: × 2
Field of regard: +20 to −96°
Slew rate: 1 rad/s
Aiming accuracy: 0.5°
Turret dimensions: 422 × 405 × 562 mm
Sensor platform weight: 39 kg
Support electronics weight: 22.5 kg

Contractor

Thomson-CSF Optronique.

UPDATED

Chlio-S multisensor airborne FLIR

Thomson-CSF has developed a version of the Chlio thermal imaging system which incorporates a second-generation detector. Chlio-S is intended for search and rescue operations installed on helicopters.

The system uses the Synergi thermal imaging modules developed by Thomson-CSF, Pilkington Optronics and Zeiss-Eltro Optronic. The detector is a 288 × 4 cadmium mercury telluride focal plane array, operating in the spectral band 8-12 μm.

The system has four fields of view and is mounted on a gyrostabilised turret which is stabilised in two axes electro-mechanically and two axes electro-optically.

Operational status

Chlio-S is in production for C160, Alizé and French Navy Falcon 50 Surmar aircraft. In service on French Air Force Super Puma helicopters and with export customers.

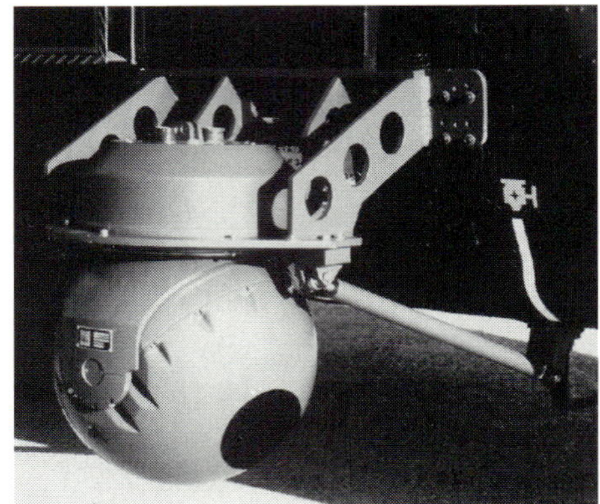

The Thomson-CSF Optronique Chlio ***1995***

Contractor

Thomson-CSF Optronique.

NEW ENTRY

Convertible Laser Designator Pod (CLDP)

The Convertible Laser Designator Pod (CLDP) has been developed from Thomson-CSF Optronique's ATLIS pod. Flight tests of the CLDP/TV began in mid-1986. Flight tests of the CLDP/CT version began in January 1988 using a Jaguar aircraft.

In September 1988, at the Landes flight test centre, a Jaguar successfully launched an AS 30 missile at a speed of 470 kt from an altitude of 213 m at a range of 8 km. The CLDP had acquired the target at a range of 13 km.

In CLDP the laser designator is supplemented by either a TV (CLDP/TV) or a thermal camera (CLDP/CT). The CLDP features a common body with laser transceiver, electronic assembly and environmental control system and two separate nose sections which can be changed in 2 hours.

A close-up of the Convertible Laser Designation Pod with thermal camera (CLDP/CT), mounted on a Mirage 2000 ***1998***/0018363

The TV head features a TV camera, gimballed mirror and a roll-stabilisation device. It also has a Field Of View (FOV) selector and a visible or near-infrared spectrum selector. This has four magnifications and corresponding fields of view.

The thermal imaging head features a gimballed optical head with a laser and thermal imager optics and roll-stabilisation device. There are four FOVs: 12/6° for navigation and 4/2° for target acquisition and tracking. The thermal imager is based upon the SMT modules.

Operational status

In production. CLDP has been ordered by Abu Dhabi for Mirage 2000 and Saudi Arabia for Tornado. It is being offered as part of the Sukhoi Su-22M5 upgrade.

Specifications

Weight: 290 kg
Length: 2.85 m
TV
Spectral band: 0.7-0.9 μm
Magnification: ×2.5, ×5, ×10, ×20
Fields of view: 0.75, 1.5, 3 and 6°
Laser
Wavelength: 1.06 μm

Contractor

Thomson-CSF Optronique.

UPDATED

CLDP/CT with thermal camera on a Mirage 2000 D ***1997***/0005517

Schematic of the CLDP/TV and CT Pods ***1998***/0005521

Damocles multimode multifunction laser designator pod

Damocles is a multimode, multifunction laser designator pod that incorporates a third-generation thermal imager. The thermal imager operates in the 3 to 5 μm waveband and uses a staring focal plane array. The pod also includes a CCD TV camera and laser spot tracker.

The modular pod is designed primarily for laser designation but can also be used for navigation, air-to-air identification and reconnaissance roles.

Operational status

Under development. First prototype is due to fly in 1998.

Specifications

Thermal imager
Spectral band: 3-5 μm
Laser
Wavelength: 1.06-1.54 μm
Electronic magnification: ×2
Dimensions (l × diameter): 2.5 m × 370 mm
Weight: 250 kg

Contractor

Thomson-CSF Optronique.

NEW ENTRY

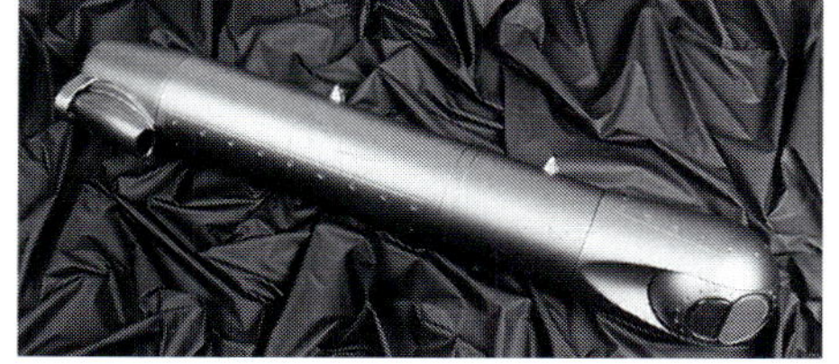

Damocles multimode multifunction laser designator pod **1998**/0018362

Irold reconnaissance camera

Irold is a long-range reconnaissance camera system carried in a pod under the fuselage. A video unit relays the camera picture to the pilot's radar screen and a symbology generator superimposes some flight data on the picture. The photograph can be taken with either left or right slant, there being a 90° mirror within the optical train. The slant angle can be varied between 65° and 86° from the vertical, giving a resolution of better than 0.3 m at 65° down to 1.5 m at 86° when the aircraft is at 40,000 ft altitude.

Specifications

Dimensions: 0.3 (length) × 0.53 m (diameter)
Weight: 340 kg
Power supply: 115 V AC, 400 Hz, single phase, 90 W 28 V DC, 417 W
Image format: 114 × 111 mm
Film capacity: up to 1,200 frames (according to film type)
Filming rate: up to 0.6 s/frame
Lens focal length: 1,700 mm
Field of view: 3.84° longitudinal × 3.74° lateral
Resolution: 1 m at 100 km

Operational status

In production and in service.

Contractor

Thomson-CSF Optronique.

VERIFIED

MDS 610 MultiDistance Sensor

A French Air Force Mirage F1 CR carrying the pod-mounted MDS 610 electro-optical sensor **1998**

The MDS 610 is a passive electro-optical airborne reconnaissance sensor designed for the following daytime intelligence gathering missions: low-, medium- and high-altitude tactical reconnaissance; standoff oblique and vertical reconnaissance; fixed- or moving-target localisation and identification.

MDS 610 is gyrostabilised in two axes and can be pod or fuselage-mounted on combat, reconnaissance and surveillance aircraft. The sensor has both pushbroom and panoramic scanning modes. It can be manually or automatically controlled with real-time image display in the cockpit and is compatible with real-time transmission via datalink. Automatic control is achieved by preprogramming according to the mission.

MDS 610 is of modular design with an electro-optical CCD detector (or film), giving unlimited mission duration, stereo viewing of small surface areas and in-flight recording and replay capability.

Operational status

MDS 610 has been fitted in the Thomson-CSF Desire reconnaissance pod demonstrator which has been tested on a Mirage F1 CR aircraft. It will be carried in the Presto pod, based on the Desire demonstrator. Five Presto pods have been ordered by the French Air Force and they are scheduled to equip Mirage F1 CR aircraft from 1999.

Specifications

Lens focal length: 610 mm — F/4
CCD detector: 10,000 pixels (0.4-1.1 μm)
Field of view: 11°
Lateral coverage: ±110°
Longitudinal coverage: ±20°
Weight: 120 kg
Dimensions: (d × l) 350 × 1,500 mm

Contractor

Thomson-CSF Optronique.

UPDATED

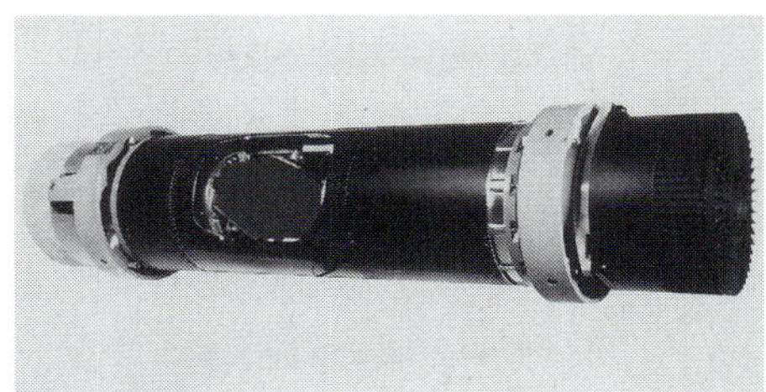

Thomson-CSF Optronique MultiDistance Sensor (MDS) 610 **1998**

NAVFLIR navigation and attack pod

Thomson-CSF has developed the NAVFLIR airborne forward-looking infrared navigation and attack pod. This provides assistance for low-altitude flight at night

Thomson-CSF Optronique NAVFLIR navigation and attack pod **1998**/0018360

and medium-range targeting in day or night-time conditions. Aircraft installation can be on the nose of a standard pylon or on a chin pod. The FLIR has a ×2 electronic zoom.

NAVFLIR can detect a target out to 20 km and perform reconnaissance at ranges of up to 10 km.

Operational status

Available.

Specifications

Spectral band: 3-5 μm
Fields of view
Wide: adapted to the HUD FOV
Narrow: 6°
Electronic zoom: ×2

Contractor

Thomson-CSF Optronique.

NEW ENTRY

Optronique Secteur Frontal (OSF) for the Rafale aircraft

Thomson-CSF Optronique and SAGEM SA are co-operating for the development and manufacture of the optronic, visual and infrared search and tracking system for the Dassault Aviation Rafale ACT and ACM.

The OSF (Optronique Secteur Frontal) is designed to aid covert missions, firing under jamming, visual identification, and damage assessment in air-to-air, air-to-ground and air-to-sea operations and to provide navigation/piloting assistance. Key features include infrared passive detection, very low false alarm rates, high-definition CCD imagery, an eye-safe laser rangefinder, very large field of regard and two optical heads to ensure simultaneous search/identification/telemetry functions.

Operational status

In development for the Rafale. First prototypes were available for flight test at the end of 1997.

Contractors
Thomson-CSF Optronique.
SAGEM SA.

VERIFIED

Presto/Desire reconnaissance pod

The Desire reconnaissance pod is a demonstrator built by Thomson-CSF Optronique under a French Ministry of Defence programme. This has led to the development of the Presto pod for standoff or high-speed penetration missions. The pod provides real-time imaging, digital recording and transmission to ground processing stations.

The system includes the Thomson-CSF Optronique MDS 610 multidistance sensor, a reconnaissance management system, a high-speed digital recorder and an associated ground station. Presto is capable of operating in both pushbroom and panoramic modes.

Operational status
The French Air Force has placed an order for five Presto pods for Mirage F1 CR aircraft, with deliveries scheduled to begin in 1999.

Contractor
Thomson-CSF Optronique.

NEW ENTRY

The Desire demonstrator reconnaissance pod **1997**/0005561

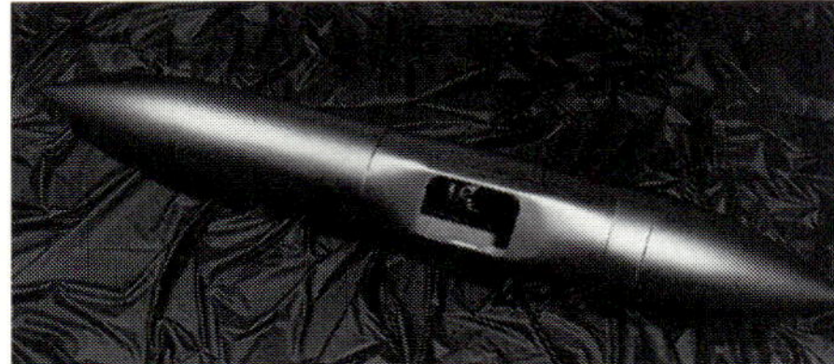

The Presto standoff electro-optical reconnaissance pod **1998**/0018359

Rubis navigation pod

The Rubis navigation FLIR pod is used as a flying aid for navigation and for air-to-ground and air-to-surface attack by day or night fighters. It is a multifunction pod with two fields of view. The wide field of view provides background overlay imagery in the pilot's HUD and mission specific symbology. It is used as a flying aid for navigation and target detection. The narrow field of view provides gyrostabilised lock for target identification.

The system consists of an infrared sensor operating in the 8 to 12 μm band, associated electronics and a processor. It has a line generation function, steerable line of sight, line of sight slaving, hotspot detection and automatic target tracking.

Specifications
Dimensions: 2,650 × 280 mm diameter
Weight: 110 kg
Fields of view:
(wide) 24 × 16°, 12 × 8° with zoom
(narrow) 6 × 4°, 3 × 2° with zoom

Operational status
In service with the French Air Force and foreign air forces on Mirage F1 and F-16 aircraft.

The Optronique Secteur Frontal (OSF) shown with the Rafale aircraft

1996

The Rubis navigation pod mounted on the centreline of an F-16

Rubis navigation FLIR pod on Mirage F1 aircraft **1998**/0018358

Contractor
Thomson-CSF Optronique.

VERIFIED

SDS 250 electro-optical reconnaissance sensor

The SDS 250 is a compact passive electro-optical airborne reconnaissance sensor designed to perform the following daytime intelligence gathering missions: low- and medium-altitude tactical reconnaissance, vertical reconnaissance and up to 20 km standoff oblique reconnaissance.

The sensor has pushbroom scanning mode, electronic roll stabilisation and selectable operating spectral band — either visible or near-infrared or both. The system provides real-time image display in the cockpit and line of sight position is selected from the cockpit. The system is compatible with real-time image transmission via datalink. The compactness of the sensor allows fuselage installation.

Thomson-CSF Optronique SDS 250 medium-/low-altitude electro-optic sensor ***1998***

Operational status

SDS 250 has been selected by the French Air Force for the Mirage F1 CR and by the French Aeronavale for upgraded Super Etendard aircraft.

Specifications

Lens focal length: 250 mm – f/5.6
CCD detector: 6,000 pixels (0.4-1.1 μm)
Field of view: 14°
Lateral coverage: ±85°
Weight: 30 kg
Dimensions: 410 × 400 × 230 mm

Contractor

Thomson-CSF Optronique.

UPDATED

Tactical Electro-Optical Reconnaissance System (TEORS)

The Tactical Electro-Optical Reconnaissance System (TEORS) provides a high degree of mission flexibility for low-level high-speed tactical aerial and maritime reconnaissance in standoff conditions, medium- and high-level vertical and oblique coverage surveillance missions, prestrike identification and post-strike reports and real-time or NRT data transmission.

TEORS features dual-mode operation in either film or electro-optic sensor and push-button, panoramic and zone tracking modes. It can be integrated in pods with other sensors for day and night missions to ensure compatibility with NATO operating systems.

The 24 in f4 focal-length lens has a gyrostabilised pointing mirror. The CCD sensor is a linear hybrid CCD array covering the 0.4 to 11 μm spectral bandwidth.

Specifications

Dimensions: 1,400 (length) × 300 mm (diameter)
Weight: 90 kg
Bandwidth: 0.4-11 μm
Field of view: 11°

Contractor

Thomson-CSF Optronique.

VERIFIED

Tango thermal imager

The Tango thermal imager is a modular thermal imaging system for long-range maritime patrol aircraft. It is an 8 to 12 μm CMT thermal imager with three fields of view. The system was developed for the Dassault Aviation Atlantique 2 maritime surveillance aircraft. Key features include: a large aperture for very long range imaging; high resolution; fine image stabilisation; automatic aiming towards designated targets; aircraft databus coupling; and line-to-line integration. It is incorporated in a Sere-Bezu gyrostabilised platform fixed under the nose of the Dassault Aviation Atlantique 2. Day and night missions include passive detection of ships and snorkels, long-range identification of surface vessels, reconnaissance, and search and rescue.

The Thomson-CSF Optronique Tango thermal imager mounted on the nose of the Dassault Aviation Atlantique 2 aircraft ***1996***

Tango 2G is a second-generation thermal imager based on the Synergi thermal imaging modules developed by Thomson-CSF, Pilkington Optronics and Zeiss-Eltro Optronic. Synergi uses a 288 × 4 IRCCD detector developed by Sofradir. Tango 2G has four fields of view. It is a multisensor integration and incorporates a CCD detector for the visible and near-infrared channel and has an auto-search mode. It is being developed for the Atlantique third-generation, the aircraft ATL3G.

Operational status

Tango is in production for French Navy Atlantique 2 aircraft.

Specifications

Tango modular thermal imaging system
Dimensions: 600 mm turret diameter
Turret weight: 85 kg
System weight: 120 kg
Gyrostabilised field of view
(Azimuth) ±110°
(Elevation) +15 to −60°
Tracking speed with speed/accuracy optimisation: 1 rad/s
Infrared channel
Detector: CMT
Cooling: Stirling engine
Spectral band: 8-12 μm
Fields of view
(Wide) 6.45 × 4.30°
(Medium) 2.15 × 1.43°
(Narrow) 1.07 × 0.7°
Tango 2G
Dimensions: 600 mm turret diameter
Turret weight: 75 kg
System weight: 98 kg
Gyrostabilised field of view
Azimuth: ±360°
Elevation: +15 to −93°
Tracking speed with speed/accuracy optimisation, auto-search mode: 1 rad/s
Infrared channel
Detector: 288 × 4 IRCCD CMT focal plane array
Spectral band: 8-12 μm

Tango 2G modular thermal imaging system for long range maritime patrol aircraft ***1998***

Fields of view: 15 × 11.2°
7.5 × 5.6°
1.5 × 1.1°
0.75 × 0.6°
Visible and near infrared channel
Detector: CCD
Field of view: 1.5 × 1.1°

Contractor

Thomson-CSF Optronique.

UPDATED

TMV 630 airborne laser rangefinder

The TMV 630 airborne laser rangefinder equipment has been designed to meet single unit, small installation requirements and can be fitted easily to a wide range of aircraft. It provides high-precision aircraft-to-target range measurement and is claimed to increase considerably the performance of conventional weapon-aiming systems. The large field of view provided is compatible with all head-up displays and the electrical interfaces are compatible with almost all aircraft types.

The Thomson-CSF Optronique TMV 630 airborne laser rangefinder

The high-speed and accurate laser beam-steering is specifically adapted for continuously computed impact point attacks irrespective of terrain or the nature of the weapons or aircraft altitude.

Specifications

Dimensions: 190 ×190 × 520 mm
Weight: 15 kg
Power supply: 28 V DC, 12 A
Wavelength: 1.06 μm
Range: up to 19 km
Accuracy: better than 1 mrad

Operational status

In production as part of nav/attack system on Dassault Mirage and Dassault Aviation/Dornier Alpha Jet aircraft.

Contractor

Thomson-CSF Optronique.

VERIFIED

The Thomson-CSF Optronique Victor camera on a French Army Gazelle helicopter ***1996***

TMV 632 airborne laser spot tracker and rangefinder

The TMV 632 ground attack laser rangefinder was developed at the request of the French DGA (Delégation Générale de l'Armament) and combines laser ranging and tracking functions into a single compact monobloc system, for use on tactical ground support and training aircraft. It provides the weapon systems with extremely accurate fire control data, for both laser-guided and conventional munitions. Several variants are available, with different digital interfaces. The dual-function TMV 632 offers the same level of performance and accuracy as single-function systems.

The TMV 632 is mounted in the airframe (embedded) or fitted in a mini-pod, or inside a store-carrying pylon. The TMV 632 detects and identifies the laser spot on illuminated ground-based targets. Acquisition and tracking are automatic. A sighting reticle in the head-up display allows the pilot to aim at the target. The laser beam of the rangefinder is locked to the position of the tracker, making accurate measurement of the aircraft-to-target distance possible. There are two independent safety interlocks in the TMV 632. If required, cooling air is supplied by the aircraft.

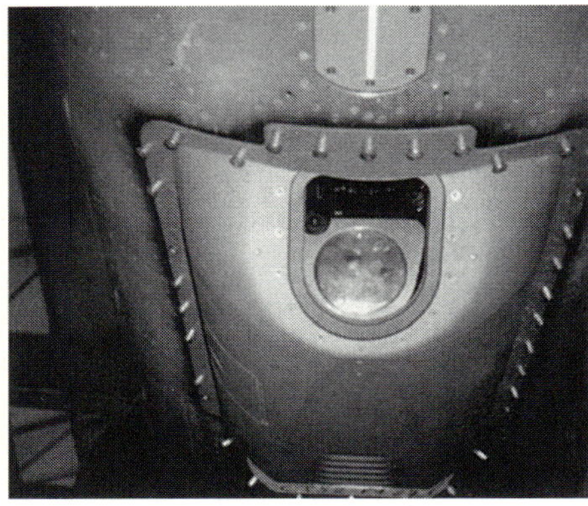

The TMV 632 laser spot tracker and rangefinder in underbelly housing ***1998***/0005531

Operational status

In contract production for the Mirage F1 CT.

Specifications

Overall Dimensions (L × W × H): 530 × 170 × 190 mm
Total weight: 18 kg
Wavelength
Laser: 1.54 μm
Spot tracker: 1.06 μm
Range: up to 20 km in rangefinding mode and 15 km in tracking mode
Field of regard: 40° azimuth, 20° elevation
Interfaces: TM632: digibus, TM632A: ARINC, TM632B: bus 1553
Power: 28 V DC, 8 A

Contractor

Thomson-CSF Optronique.

NEW ENTRY

The TMV 632 airborne laser spot tracker and rangefinder ***1998***/0018355

Victor thermal camera

The Victor thermal camera is designed to be connected to Viviane and Strix gyrostabilised sights, for SA342 Gazelle HOT and HAP Tiger escort helicopters. It can also be fitted on AS 365M Panther, BO 105 or other types of helicopters.

Victor converts the thermal radiation of landscape and objects into a visible image at TV standard, and is suitable for air-to-ground and air-to-air gun or missile firing, rocket firing and as a flying aid by day and night and under adverse weather conditions.

The system displays symbols in the eyepiece of the sight or on a TV monitor and has a magnifying function of ×2 to enlarge the image. There is a processing board option for improved performance. Initial sight stabilisation is by the platform and electronic fine stabilisation is by the imager. There is a specific video output for tracking.

Specifications

Weight:
(total) 25 kg
(on roof) 17 kg
Power supply: 20-32 V DC, 140 W
Wavelength: 8-13 μm
Trifocal lens: 30 × 20°, 6 × 4°, 2.4 × 1.6°
Range: up to 4,000 m

Operational status

Victor entered service in 1988.

Contractor

Thomson-CSF Optronique.

VERIFIED

GERMANY

Hellas helicopter obstacle warning system

Hellas is designed to provide warning of wires and other similar obstacles to helicopters flying nap-of-the-earth and other low-altitude missions, at up to 1,000 m range, both for military and civil operations.

Obstacle detection is performed with an imaging eye-safe ladar, which generates images of the scene in front of the helicopter, while range data processing is performed in the processing unit. Processing results can be configured for display as warning information for the pilot. The first processing action provides presentation of the colour-coded, real-time image on the HMI. Wire obstacles contrast with the background to become clearly visible and recognisable.

The imaging laser radar scanning architecture is designed to ensure the generation of range images which enhance the detection of wires and wire-like objects. The scanner, designed by Dornier, provides fast line scanning. Continuous column scanning is produced with an oscillating mirror. The range imaging system is capable of producing up to 100,000 pixels per second, allowing within the image generation for configuration of line and frame rates. The fibre optic scanner is based on two nutating mirrors on a single shaft. Laser pulses injected into a single fibre are imaged via the nutating mirror onto a circular fibre array. This array transforms into a linear array, positioned in the focal plane of the field optics.

Specifications

Sensor: range imaging laser radar
Wavelength: solid-state eye-safe at 1.54 micrometres
Laser pulse power: 4 kW
Receiver: InGaAs APD hybrid (Avalance Photo Diode)
Scanning: 2 axes. horizontal: fibre optic; vertical: oscillating mirror
Image repetition frequency: 2 Hz, 4 Hz possible
Field of view: 32 × 32°
Range: >1,000 m (extended area objects, in good visibility)
>400 m (extended area objects, poor weather)
>500 m (wires >10 mm, good visibility)
>300 m (wires >10 mm, poor weather, oblique incidence)
Range resolution: <1 m
Angular resolution: <0.35° horizontal
<0.2° vertical
Pixels: 95 horizontal
200 vertical
Volume: <36 litres
Weight: <30 kg

Operational status

Development contract awarded by the Federal Office of Defense Technology and Procurement. The system is understood to have been tested on a UH-1D helicopter, with a pre-production system being integrated onto a BK-117 aircraft.

Contractor
Daimler-Benz Aerospace, AG, Defense and Civil Systems, Airborne Systems.

NEW ENTRY

Hellas test unit on UH-1D helicopter
1998/0018354

IRLS InfraRed Linescanner Systems

The Honeywell Regelsysteme IRLS is designed to meet the German Air Force Tornado IDS reconnaissance aircraft requirement. It comprises the following line replaceable units:

Scanner Receiver Unit (SRU), supplied by Lockheed Martin IR Imaging Systems;

Reconnaissance Management Unit (RMU);

Digital Tape Recorder (DTR), supplied by Ampex Inc;

Reconnaissance Control Panel (RCP), supplied by Computing Devices UK;

Reconnaissance Power Supply (RPS).

The SRU, RMU, DTR, RPS and two aerial daylight cameras will be installed in the new modular pod designed by Daimler-Benz Aerospace.

The IRLS is optimised for operation between the altitudes of 200 and 2,000 ft, and at speeds ranging from 300 to 600 kt. Higher altitude operation can be achieved by software change.

The SRU senses infrared terrain radiation and converts it to 12 parallel video channels; an array of 34 detector elements is scanned across track. The SRU also contains patented scan compensation systems.

Operational status
The IRLS is in development for the German Air Force Tornado IDS reconnaissance aircraft. The Scanner Receiver Unit (SRU) is already in service as the Infrared Imaging System (IIS) on the Tornado ECR (Electronic Combat and Reconnaissance) aircraft.

Contractor
Honeywell Regelsysteme GmbH

NEW ENTRY

Thermal imagers

Multipurpose thermal imagers cover the 3 to 5 μm and 8 to 12 μm spectral wavebands. The thermal imagers have a high geometrical and thermal resolution and a high anti-blooming capability. A laser rangefinder or a CCD colour camera can be integrated to form a high-performance surveillance system. The system is based on a four-axis gyrostabilised platform and will operate with a correlation tracking system.

Operational status
An IR lens system with a dual field of view (of 8° and 2°) is currently under development.

Contractor
Steinheil Optronic GmbH.

VERIFIED

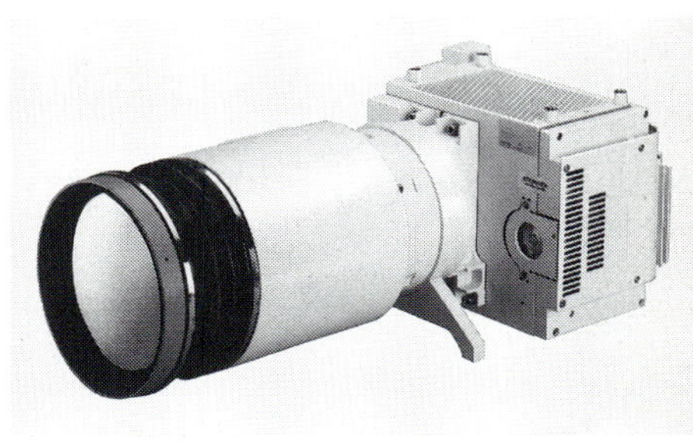

A Steinheil thermal imager
1995

KRb 8/24 F reconnaissance camera

The KRb 8/24 F camera achieves wide-angle panoramic coverage with undistorted framing camera geometry on a single 9.5 in (240 mm) wide film. The camera performs at highly survivable parameters for low- to medium-altitude reconnaissance missions. Each exposure covers 143° across track, with true angle forward motion compensation across the entire format. This format affords sequential along-track stereoscopic coverage. Images are without the cylindrical distortion inherent in panoramic cameras. Special Zeiss optics provide performance into the near infrared, allowing the use of all aerial film types, including colour and camouflage detection, without refocusing. Small size and low weight permit easy installation in RPVs, pods and aircraft.

Specifications
Dimensions: 356 × 374 × 311 mm
Weight: 22 kg
Power supply: 28 V DC, 250 VA
Focal length: 80 mm
Aperture range: f/2.56-f/16

Contractor
Zeiss-Eltro Optronic GmbH (ZEO).

UPDATED

LHM Laser Altimeter

The eye-safe Raman laser based on Nd:YAG laser technology has been developed to provide precise altitude information for film and electro-optical camera systems.

Specifications
Dimensions:
(laser altimeter) 180 × 125 × 112 mm
(control unit) 208 × 102 × 26 mm
Weight:
(laser altimeter) 3 kg approx
(control unit) 0.5 kg approx
Power supply: 28 V DC
Wavelength: 1.543 μm
Measuring frequency: 20 ppm continuously
Range: 100-12,000 m
Accuracy: ±5 m
Display: 16 lines of 24 characters
Temperature range: 0-40°C
Reliability: <100,000 measurements MTBF

Contractor
Zeiss-Eltro Optronic GmbH (ZEO).

UPDATED

Modular KS-153 camera system

The KS-153 is a modular camera system consisting of three different focal length configurations: The Pentalens 57, Trilens 80 and Telelens 80. The Pentalens configuration is the latest design development in the system. These three configurations have a common camera body, film cassettes and film cassette holder. The desired focal length lens and shutter assembly with format mask can easily be attached to the camera body to accommodate mission requirements. The high parts commonality provides a benefit to multiple-type camera users.

The modular KS-153 is a fully electronic,

microprocessor-controlled design and uses Carl Zeiss optics optimised for high resolution. The camera system is built and tested to meet and exceed reliability and maintainability criteria for use in modern military or commercial aircraft, pods and UAVs.

The KS-153 Telelens 610 is a pulse operated, sequential frame camera designed for medium- to high-altitude oblique photography from high-performance reconnaissance aircraft. The camera accommodates both 4 mm standard base and 2.5 mm thin base roll film in any panchromatic, infrared or colour emulsion. Each frame covers an angular field of view of 21.4° across track by 10.7° along track. The camera mounts in an integral ring bearing which allows in-flight rotation under electronic control to any desired oblique angle. The Telelens 610 has the same major camera features as the Trilens version of the KS-153 camera.

The KS-153 Trilens 80 is a high cycle rate, pulse operated, sequential frame camera designed for low- to medium-altitude wide-angle photography from high-performance reconnaissance aircraft. The camera accommodates both 4 mm standard base and 2.5 mm thin base roll film in any panchromatic, infrared or colour emulsion. Each frame covers an angular field of view of 143.5° across track and 48.5° along track. The wide lateral coverage is provided by the optical assembly of three S-TOPAR A1 2/80 lenses having an effective maximum aperture of f/2.56 and the field of view of the side lenses deflected by front-mounted prisms. Major camera features are true angle corrected Forward Motion Compensation (FMC), constant velocity focal plane shutter, integral intervalometer, Automatic Exposure Control (AEC), easily interchangeable interface card and continuously monitoring BITE. This BITE feature is further enhanced by an integral non-volatile BITE memory to enable post-flight analysis of in-flight transient failures.

Specifications

Dimensions:
(Pentalens 57) 439 × 467 × 502 mm
(Trilens 80) 439 × 467 × 506 mm
(Telelens 610) 423 × 809 × 470 mm
Weight:
(Pentalens 57) 59 kg
(Trilens 80) 57 kg with 500 ft of film
(Telelens 610) 110 kg with 200 ft of film
Power supply: 115 V AC, 400 Hz, 3 phase
Focal length:
(Trilens 80) 3.5 in (80 mm)
(Telelens 610) 24 in (610 mm)
Cycle rate:
(Trilens 80) 10/s (max)
(Telelens 610) 4/s (max)
Shutter speed: 1/2,000 to 1/150 s
Aperture range:
(Trilens 80) f/2.56 to f/16
(Telelens 610) f/4 to f/16
Format:
(Trilens 80) 72 × 222 mm
(Telelens 610) 230 × 115 mm
Film length:
(Trilens 80) 200 or 500 ft
(Telelens 610) 200 ft

Contractor

Zeiss-Eltro Optronic GmbH (ZEO).

UPDATED

Passive Airborne Modular InfraRed systems (PAMIR)

Utilising the IR portion of the spectrum, the PAMIR systems allow reconnaissance and navigation during night and day as well as under bad weather conditions. The thermal images of the PAMIR FLIR system are displayed on the HUD and/or HDD. Images can also be recorded on a VCR or datalinked to ground stations. The system consists of the sensor, power supply and cockpit control panel.

The modular design of PAMIR easily adapts to a variety of platforms such as fixed-wing aircraft or helicopters. It can be used for reconnaissance and surveillance, disarmament verification, law enforcement, search and rescue, environmental monitoring, navigation, border checks, drug enforcement, fishery monitoring or traffic control.

The modular design has made possible the development of the designs within the one concept: the navigation PAMIR-N, and the navigation/surveillance PAMIR-S system by simply replacing the fixed turret module with a stabilised rotatable turret on the sensor unit.

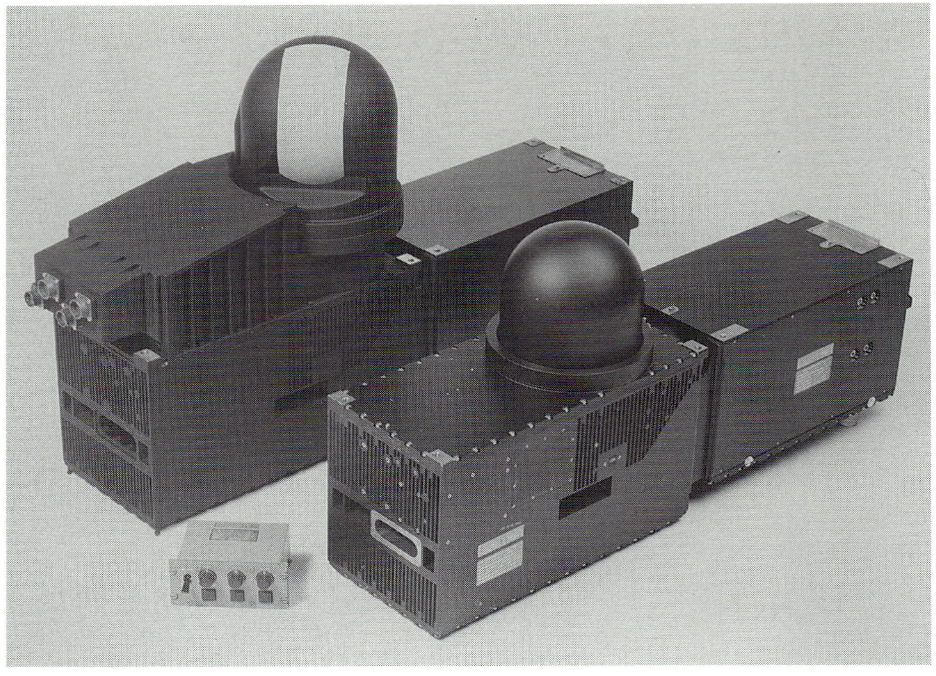

Sensors for the PAMIR-N (left) and PAMIR-S (right) FLIR systems ***1995***

Specifications

Dimensions:
(Pamir-N) 900 × 210 × 446 mm
(Pamir-S) 965 × 235 × 594 mm
(cockpit control panel) 146 × 67 × 135 mm
Weight:
(PAMIR-N) 48 kg
(PAMIR-S) 73.5 kg
(cockpit control panel) 0.8 kg
Power supply: 115/200 V AC, 400 Hz, 3 phase
(PAMIR-N) 350 W
(PAMIR-S) 500 W
Wavelength: 8-12 μm
Field of view:
(PAMIR-N) 16.8 × 22.8°
(PAMIR-S) 16.8 × 22.8°, 4.2 × 5.8°

Operational status

PAMIR-N is in service with German Air Force Tornado ECR (electronic Combat and Reconnaissance) aircraft.

Contractor

Zeiss-Eltro Optronic GmbH (ZEO).

UPDATED

RMK TOP aerial survey camera

The RMK TOP is an aerial photography system used for survey and cartographic purposes. It provides enhanced image quality with minimum distortion and

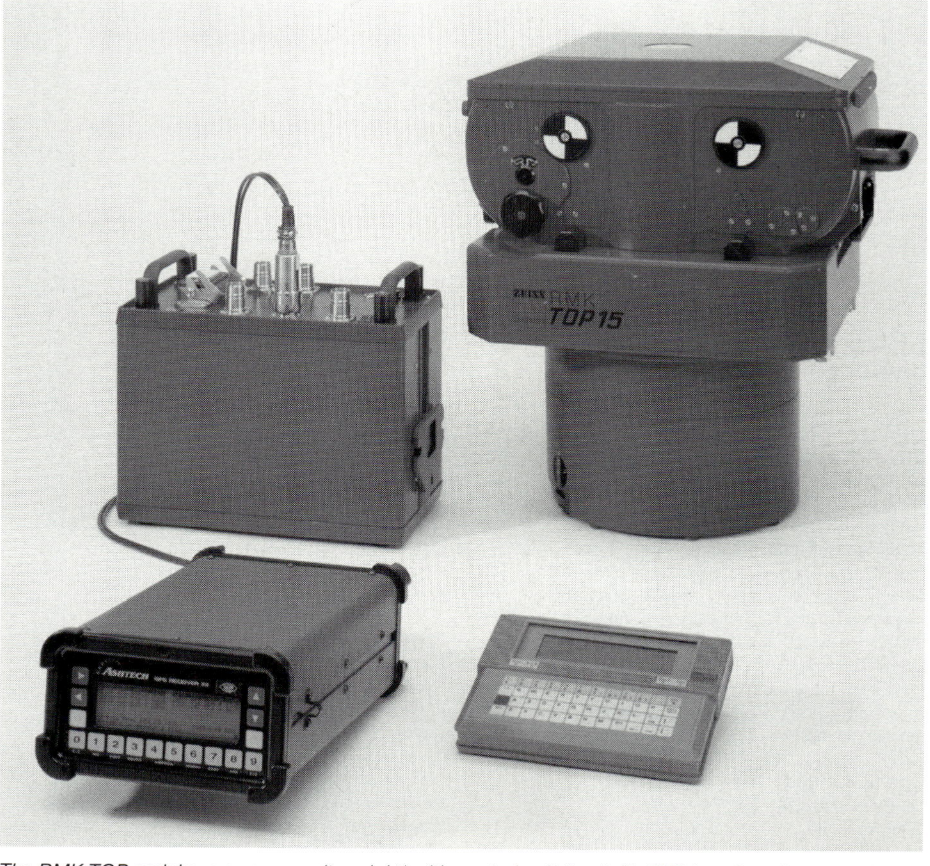

The RMK TOP aerial survey camera (top right) with control unit (top left), GPS interface (bottom left) and terminal (bottom right)

extensive image motion compensation by FMC and a gyrostabilised suspension mount. System control and function are monitored by a compact computer and microprocessor. It is particularly suitable for use with GPS-controlled navigation systems.

The basic components of an RMK TOP camera are the RMK TOP15 camera body with wide-angle PLEOGON A3 4/153 lens (or RMK TOP30 camera body with standard TOPAR A3 5.6/305 lens), T-TL compact computer for central operation control, T-MC film magazine with FMC and T-AS suspension mount gyrostabilised in three axes.

The following options are available for navigation and system control: T-FLIGHT GPS-supported photoflight management system for flight planning and mission documentation, T-NT visual navigation telescope, T-NA automatic navigation meter for automatic V/H measurement and interfaces for aircraft specific navigation systems.

Specifications

Weight:
(RMK TOP15) 176 kg total
Focal length:
(RMK TOP15) 6 in (153 mm)
(RMK TOP30) 12 in (305 mm)
Aperture:
(RMK TOP15) f/4 to f/22
(RMK TOP30) f/5.6 to f/22
Exposure time: 1/50 to 1/500 s

Contractor

Zeiss-Eltro Optronic GmbH (ZEO).

UPDATED

VOS 60 airborne electro-optical sensor

The VOS 60 sensor was designed for low- to medium-altitude reconnaissance missions. It is a high-performance electro-optical colour system and consists of the sensor head, electronic control unit and control panel. The image is acquired with a 3 × 6,000 pixel colour detector using the pushbroom technique. The digital output of the sensor can be recorded on a digital recorder or viewed on board on a high-resolution monitor. Additional mission-related data can be added. The system comprises a variety of high-performance lenses for different applications and image scale requirements.

The VOS 60 electro-optical sensor ***1995***

Specifications

Dimensions:
(sensor) 54 × 154 × 221 mm
(electronic control unit) 133 × 483 × 340 mm
(control panel) 213 × 129 × 107 mm
Weight:
(sensor) 6 kg
(electronic control unit) 11 kg
(control panel) 1.5 kg
Power supply: 28 V DC, 200 W
Field of view: 60° across flight path (with 60 mm lens)

Operational status

In production.

Contractor

Zeiss-Eltro Optronic GmbH (ZEO).

UPDATED

VOS 80C digital video colour camera

The VOS 80C is a high resolution digital video colour system designed for real-time collection of ground data, in support of environmental, control and surveillance functions.

VOS 80C sensor (front), and (behind left to right) sensor control unit, monitor, control panel, and a digital recorder ***1998***/0018353

The detector features three parallel CCD line arrays, each with individual colour filter (red, green and blue). In addition, a cut-off filter (700 nanometres) is attached to the lens front. The image data processed is fed to a monitor display and to a digital recorder.

The ground data can be displayed and viewed on the monitor either in continuous mode or with image freeze. Additional image annotation is possible, from mission or ground data.

Specifications

EO Sensor VOS 80C
Detector: Kodak, colour, 3 × 6,000 pixels (RGB)
Pixel size: 0.012 × 0.012 mm
Line space: 0.096 mm
Line rate: 1.6 kHz
Spectral response: 350 to 1,050 nanometres, colour with 700 nanometres filter
Data rate: 230 Mbits/s
Lens: Carl Zeiss Planar, fl 80 mm, f/2.8
Field of View (FoV): 48.5° across path
Weight: 3.5 kg
Dimensions: 154 × 154 × 210 mm
Power: 28 V DC, 28 W

Sensor Control Unit
Weight: approx 15 kg
Dimensions: 483 × 133 × 390 mm
Power: 28 V DC, 125 W

Sensor Control Panel
Weight: approx 1.5 kg
Dimensions: 150 × 200 × 105 mm
Power: from control unit

Contractor

Zeiss-Eltro Optronic GmbH (ZEO).

NEW ENTRY

INTERNATIONAL

Litening airborne laser target designator and navigation pod

Rafael Missile Division and Northrop Grumman Corporation are working under a teaming agreement involving sale and production of Rafael's Litening airborne laser target designator and navigation pod. On the International market Rafael is prime contractor, with Northrop Grumman as subcontractor. For the Litening system on US-built military aircraft sold to the US Government, or to allied nations through the US Foreign Military Sales (FMS) programme, Northrop Grumman is the prime contractor, with Rafael as its subcontractor.

Northrop Grumman is supplying the FLIR system for both US and International sales, except where other special arrangements are made.

Litening is a day/night precision laser targeting and navigation pod. It contains FLIR, TV, laser spot target tracker and laser marker sensors for use with either conventional or laser-guided bombs. An on-gimbal Inertial Navigation Sensor (INS) has a stabilised line of sight and automatic boresight capability. The INS and software design is optimised to permit easy integration in a wide variety of modern military aircraft such as the AV-8B, F-4, F-5, F-15 and F-16, as well as the Jaguar, Mirage 2000 and Tornado.

Rafael has supplied the Litening pod to several nations and integrated it in the following systems:

(1) Germany: IDS Tornado, as part of the Mid-Life Update (MLU) programme being carried out by Daimler-Benz Aerospace AG for the German Air Force and German Navy. In this instance, Litening will be fitted with electro-optical systems provided by Zeiss-Eltro-Optronic GmbH
(2) India: Jaguar and Mirage 2000 aircraft
(3) Israel: F-16C/D Block 30 and Block 40 aircraft
(4) Romania: MiG-21, as part of the Lancer upgrade
(5) Venezuela: F-16A/B Block 15 aircraft.

Specifications

Dimensions: 2,200 (length) × 406 mm (diameter)
Weight: 200 kg
FLIR
fields of view:
(narrow) 1.5 × 1.5°
(medium) 4.5 × 4.5°
(wide) 18.0 × 24.0°
spectral band: 8-12 micron
picture elements: 708 × 240
CCD camera
fields of view:
(narrow) 1.0 × 1.0°
(wide) 3.5 × 3.5°
picture elements: 768 × 494
Laser designator & rangefinder:
energy: 100 MJ per pulse
Trackers: advanced correlator/inertial laser spot search & track

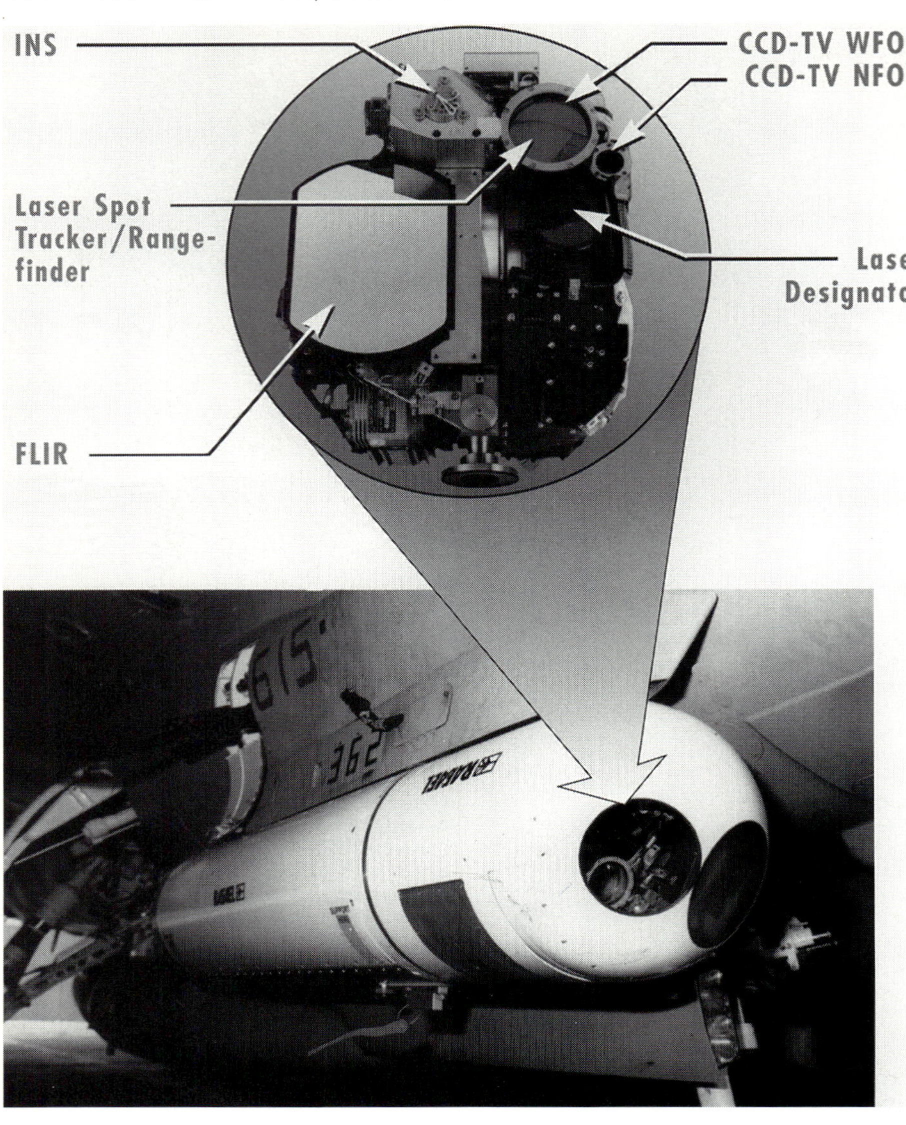

Gimbals:
(fields of regard) +45 to −150° pitch; ±400° roll
(stabilisation) 30 microradians
Flight envelope:
(at low altitude) 1.2M
(manoeuvre) 9G

Contractors
Rafael Missile Division.
Northrop Grumman, Electronics & Systems Integration Division.

NEW ENTRY

The Litening airborne laser target designator and navigation pod
1998/0018352

PIRATE InfraRed Search and Track (IRST) system

FIAR is the leader of the Eurofirst Consortium, also comprising Pilkington Optronics and Technobit, which is developing its Passive Infrared Airborne Track Equipment (PIRATE) for the EF 2000.

PIRATE will detect the infrared signature of aircraft at long range, over a wide field of view, under conditions of poor visibility. Being a passive sensor, it enables the aircraft to gather early intelligence of threats and to manoeuvre stealthily into an advantageous position without being detected by hostile EW systems.

The system will accurately track multiple high-speed targets, prioritise them and provide high-resolution images for visual identification. It provides highly reliable information for air-to-air and air-to-ground use. The system uses proven signal processing technology derived from the Pilkington Optronics Air Defence Alerting Device (ADAD) which demonstrates a very high suppression rate of potential false alarms.

PIRATE will be integrated with other onboard sensor systems for maximum sensor fusion effectiveness.

PIRATE will also locate low-level targets and provide cueing information. It provides data and imagery to head-up or multifunction head-down displays, facilitating navigation and terrain-avoidance in adverse weather conditions.

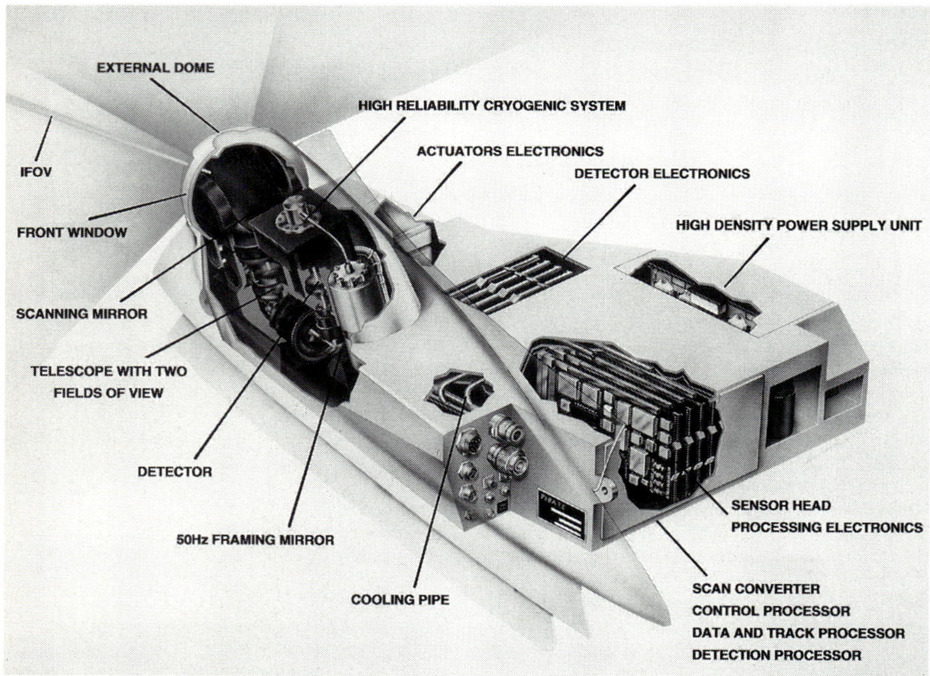

The PIRATE infrared search and track system ***1997***/0002209

Operational status
A development contract to supply equipment for EF 2000 aircraft was awarded in 1992. Flight trials will be conducted during 1998 using pre-production hardware.

Contractors
The Eurofirst Consortium, comprising:
Alenia Difesa, Avionic Systems and Equipment Division, FIAR company.
Pilkington Optronics, Glasgow
Tecnobit SA.

UPDATED

ISRAEL

DSP-1 Dual Sensor Payload

DSP-1 is a compact day/night observation system designed for use on light reconnaissance aircraft and helicopters (as well as UAVs and patrol boats). It is a four-gymbal system, gyrostabilised in azimuth and elevation. It uses two channels: a third-generation focal plane array InSb FLIR night sensor with a continuous (×22.5) zoom lens; and a high resolution colour CCD daylight channel equipped with a (×20) zoom lens.

Options include: ICCD; laser pointer; 8-12 micron FLIR (first or second generation); video tracker; radar designated pointing; MIL-STD-1553B databus; and GPS interface.

Specifications

FLIR sensor:
spectral range: 3.0-5.0 micron
detector: FPA 256 × 256 InSb
lens: × 22.5 zoom
fields of view:
(narrow) 0.98 × 0.92°
(wide) 21.7 × 20.6°
IFOV: 67 micro radians
NEDT: 0.02°K
video output: PAL or NTSC
cooler: closed cycle
Day sensor:
camera: high resolution colour CCD 768(H) × 494(V) pixels
lens: × 20 zoom
fields of view:
(narrow) 0.92 × 0.7°
(wide) 18.6 × 13.9°
Acquisition ranges:
truck detection: 25 km (daylight), 25 km (FLIR)
truck recognition: 10 km (daylight), 7.5 km (FLIR)
Electro-mechanical:
type: 4-gimbal
field of regard:
(elevation) +10° to −105°
(azimuth) 360° (xn) continuous
stabilisation: better than 25 micro radians RMS
pointing accuracy: 0.7°
Power: 28 V DC, 110 W
Dimensions: 500 (length) × 320 mm (diameter)
Weight: 26 kg

Contractor

Controp Precision Technologies Ltd.

NEW ENTRY

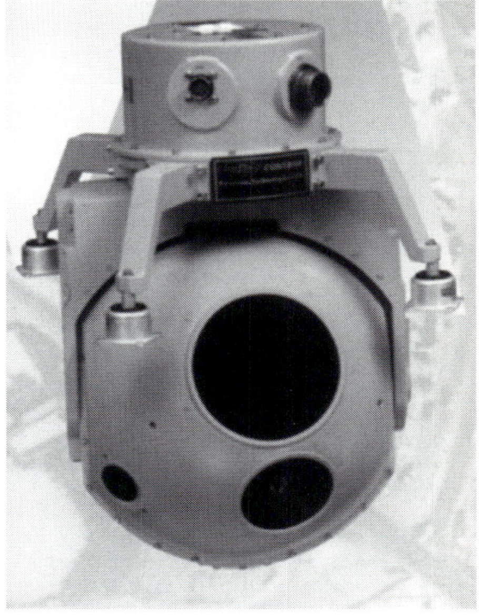

DSP-1 dual sensor payload ***1998***/0018350

All-Light Levels TV

The All-Light Levels TV (ALLTV) is based on an advanced high-resolution gated ICCD sensor with laser illumination for day and night surveillance. It is suitable for airborne, shipborne or land-based applications.

ALLTV comprises three subsystems: an electro-optical head, a motorised pan and tilt platform mount, and a portable control unit that includes joystick, operating panel and high-resolution display. Its features include a dynamic range of 10, very high sensitivity to allow usable pictures to be obtained at any light level, even during the darkest nights, and two simultaneous electro-optical channels: wide field of view for orientation and narrow field of view for long-range observation and identification.

ALLTV creates and transmits real-time TV pictures simultaneously to multiple users via wire or radio. The system's fast real-time image processor is designed for picture enhancement and reduction of quantum noise in night-time imagery. ALLTV also allows for recording of real-time TV pictures for analysis and debriefing purposes.

Contractor

Elbit Systems Ltd.

UPDATED

CTC-1/CTC-2 airborne colour TV camera

Elbit Systems' colour TV camera offers the user outstanding colour fidelity and full-band response for high-resolution colour applications. The camera is based on a 1.25 cm interline transfer CCD and micro-electronic assemblies.

The camera also incorporates a unique and specially designed Automatic Electronic Shutter (AES) function which adjusts the shutter speed automatically to maintain a wide dynamic range of light levels (0.5 to 20,000 foot candles) all day long. The AES function also ranges from a half hour before sunrise to a half hour after sunset. The AES feature eliminates the need for a moving mechanical iris, thus improving camera reliability.

The interline transfer technique significantly increases the camera performance when compared with the old frame transfer technique.

The CTC appears in two mechanical configurations: CTC-1 (housed) in one module, CTC-2 (housed) in two modules.

The electrical and mechanical interface is compatible with the GEC HUD. It features: high resolution (460 horizontal TV lines); high-sensitivity complementary colour filter; NTSC/PAL and Y/C (S-VHS) output; 1/60 to 1/31,000 second Automatic Electronic. Shutter (AES): wide dynamic range and excellent S/N ratio; AWB memory or auto-tracking (optional); low power consumption; military Qualification; single or double housing configuration; very fast response time.

Specifications

Sensor type: 1.25 cm interline CCD (6.4 × 4.8 mm)
Picture elements: 768 (H) × 494 (V)
Colour filter: Cy, Ye, Mg, G complementary filter
TV resolution: 460 (H) × 400 (V) lines
Spectral response: 460 to 630 nm
Field of view (FoV): 17 × 22° (optional other FoVs)
Response time: approx 1 s
Automatic Electronic Shutter (AES): 1/60 s to 1/31,000 s
Power source: 115 V AC, 400 Hz
Power consumption: <10 W

Contractor

Elbit Systems Ltd.

UPDATED

Nightwatch

Nightwatch is a modular concept designed for helicopter and transport aircraft. It assures reduced crew workload, increased accuracy, and avoidance of obstacles during night-time low-level flight. Elbit systems that can be integrated into the Nightwatch concept include: NVGs, NVD/HUD, HOCAS, IR filtered landing lights, IR imaging systems, moving map and cockpit mission management systems using Doppler/GPS/INS.

Operational status

Nightwatch is flying on CH-53, CH-47, CH-46, UH-60, HH-60, OH-58, UH-1, A 109, Gazelle, AS 350, AS 565, BO 105, AH-1, V-22 and C-130.

Contractor

Elbit Systems Ltd.

UPDATED

ACEM Aerial Camera Electro-optical Magazine (ACEM)

ELOP's ACEM replaces the film magazine with an electro-optical magazine featuring a linear array of butted CCD detectors in the focal plane.

The retrofitted camera employs the existing optics and structural mechanics of the original aerial film camera, the ACEM and modified scanning and control units.

Real-time collected image data can be recorded in-flight and/or transmitted in real-time to the ground station, providing rapid intelligence data.

Specifications

Output line rate: 2,000 lines/s
Focal plane array: 9,200 pixels
Spectral region: 0.5-0.9 μm
Output data dynamic range: 50 dB
Total weight: 15 kg
Power requirements: 28 V DC, MIL-STD-704C

Operational status

The systems have been integrated in IAF (Israeli Air Force) fighters.

Contractor

ELOP Electro-Optics Industries Ltd.

VERIFIED

ELOP's Aerial Camera Electro-optical Magazine (ACEM) ***1996***

Airborne Laser Rangefinder and Designator

The Airborne Laser Rangefinder And Designator (ALRAD) was developed to serve as the major component for a variety of high-performance airborne designation systems. The stringent requirements resulted in a compact, lightweight and easily maintainable and adaptable designator.

The airborne designation system increases the pilot's bombing effectiveness by marking the target with a laser beam and directing a laser-guided bomb on to the target. The system can be integrated into fighter aircraft as well as general utility aircraft. For this application it may be installed externally on any suitable aircraft station. The system is available in two basic configurations: either as a rangefinder, with a steering mirror for high-repetition precise ranging within a cone of 20 to 30° and slaved to the pipper like a radar ranger; or as a designator/ranger with steering mirrors for hemispherical coverage and slaved to the mission computer as well as to a manual control.

The ALRAD comprises the laser transceiver, a cooling unit, electronic unit and beam-steering unit. These four units can be combined modularly in accordance with system requirements. The laser transceiver and cooling unit are common to all systems while minor modifications to the electronic unit provide the flexibility to engineer the system adaptively for different requirements. Consequently, the system is well suited for upgrades, with rapid low-cost designs for the customer's applications.

Specifications

Weight: 7.5 kg nominal
Wavelength: 1.06 μm
Energy: 80 mJ
Pulse rate: up to 20 pps, codable
Beam divergence: 0.4 mrad
Interfaces: EIA-RS-429 or MIL-STD-1553B

Contractor

ELOP Electro-Optics Industries Ltd.

VERIFIED

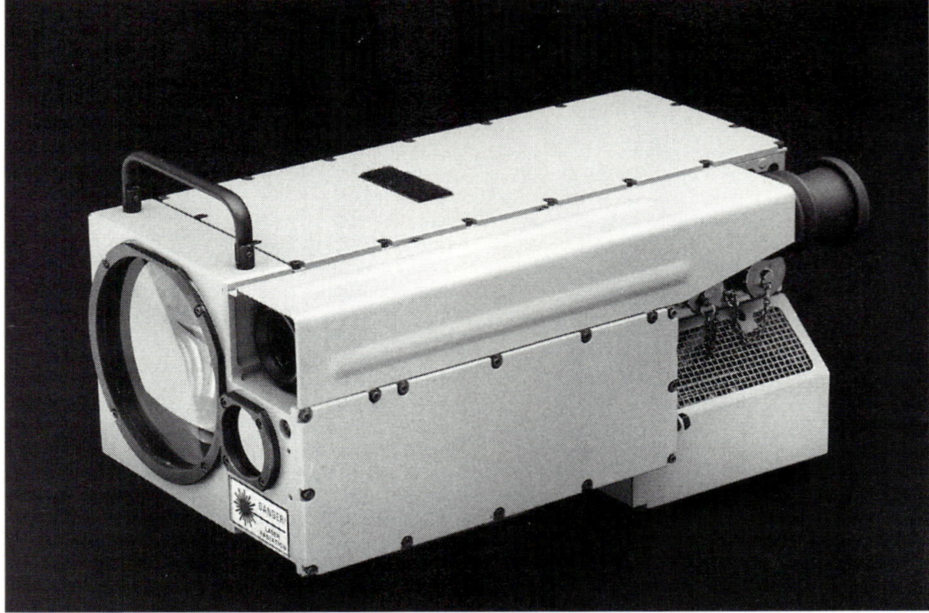

Airborne Laser Rangefinder and Designator ***1997***/0001223

COMPASS: COmpact MultiPurpose Advanced Stabilised System

COMPASS is designed to provide day/night search-and-track of land and sea targets, onboard weapon control, and day/night navigation. It is designed for land, sea, or airborne application. It is an adaptable, integrated concept, featuring modular components, add-on modules and interface/plug-in capability. It can be installed as a stand-alone system, or as part of a larger weapons system. It provides multi-operator control capability and, in a helicopter, it can be operated from the pilot and/or navigator and/or weapons operator positions.

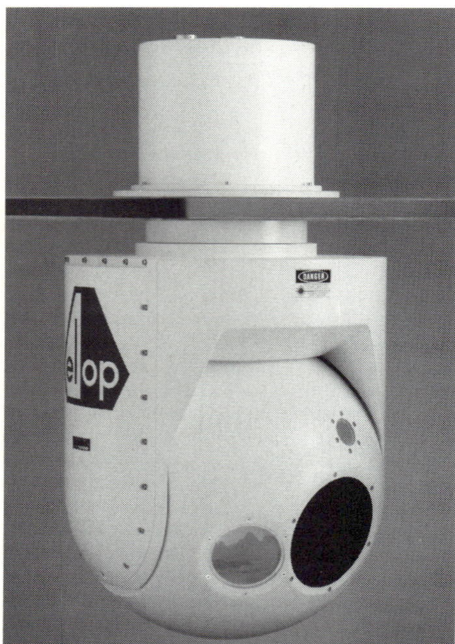

COMPASS: COmpact MultiPurpose Advanced Stabilised System for helicopters ***1997***/0001225

COMPASS incorporates three sensors that can all be brought to bear simultaneously. The unit combination of Forward-Looking InfraRed (FLIR), colour or black and white zoom CCD camera, and LaserRangeFinder/Designator (LRFD) enables crew-operated or remote search and track operations against land and sea targets, as well as weapons guidance and day/night navigation. All of these operations may be carried out at the same time, using the one combined sensor system.

Modular in design, the COMPASS concept system can be readily customised to individual user requirements. System integration is facilitated by provision of a MIL-STD-1553B databus interface and RS422 avionics interface.

Specifications

FLIR (interchangeable): 1st and 2nd generation FLIR, or mini-FLIR 8-12 micrometres, or matrix FLIR 3-5 micrometres
Angular coverage: azimuth: 360°; elevation: +35 to −85° (−100° optional)
Stabilisation: 20 microradians
Maximum flight speed: operating : 300 kt; endurance: 400 kt
Weight: 31 kg (without laser); 34 kg (with laser)

Contractor

ELOP Electro-Optics Industries Ltd.

UPDATED

COTIM Compact Thermal Imaging Module

COTIM is a very compact and lightweight thermal imaging sensor. It is designed for installation in miniature stabilised payloads for Remotely Piloted Vehicles (RPVs), light helicopters and small naval vessels.

It features: advanced focal plane detector technology with integral closed-cycle cooler; control via serial communication, with dual field of view (FoV). (Triple FoV or continuous zoom are optional.)

Specifications

Spectral region: 8-12 μm
Aperture: 100 mm
Fields of view:
(narrow) 2 × 1.5°
(wide) 7.1 × 5.3°
Instantaneous FoV:
(narrow) 0.11 mrad
Detector: MCT, more than 100 elements
Power: 40 W, 28 V DC
Weight: 2.9 kg

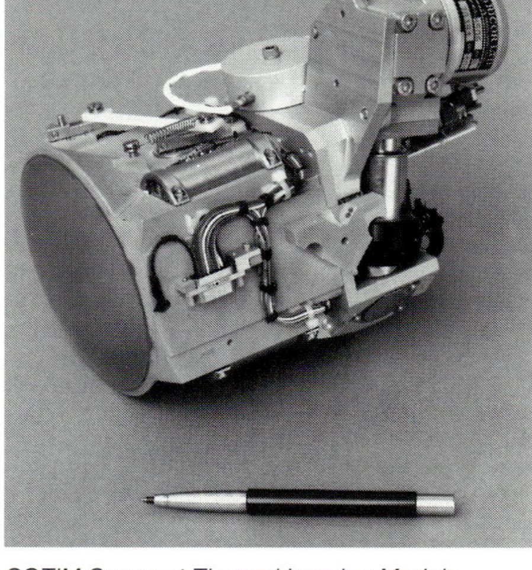

COTIM Compact Thermal Imaging Module ***1997***/0001224

Operational status

Development complete.

Contractor

ELOP Electro-Optics Industries Ltd.

VERIFIED

EO-LOROPS Electro-Optic Long Range Oblique Photography System

EO-LOROPS Electro-Optic Long Range Oblique Photography System outfits high-performance aircraft with an electro-optic, standoff reconnaissance system incorporating real-time data transmission or real-time data record/transmit options, in pod- or nose-mounted configurations.

The reconnaissance pod houses the camera, comprising the Cassegrain Ritchey-Chretien mirror telescope with a linear array of butted CCD detectors in the focal plane, the video processing unit and the scanning mirror.

The peripheral units, such as the datalink transmitter, digital VTR, air conditioning unit, power supply and reconnaissance management unit, also reside in the pod. Alternatively to the podded installation, El-Op offers an internal installation in which the reconnaissance equipment is mounted in the nose section of the aircraft.

System control is performed from the cockpit. Real-time collected image data can be recorded in flight and/or transmitted in real time to the ground station providing timely intelligence. The ground station incorporates the tracking antenna, datalink receiver, image enhancement and archiving capability and hard-copy and soft-copy displays.

Specifications

Camera type: E-O visible
Focal plane array: 10,000 pixels
Spectral region: 0.55-0.9 μm
Resolution:
57-70 cm/lp from 50 km, 40,000 ft
175/250 cm/lp from 90 km, 40,000 ft
Camera weight: 120 kg
Power requirements: 115 V AC, 400 Hz 3Ø, 28 V DC

Operational status

Under development.

Contractor

ELOP Electro-Optics Industries Ltd.

VERIFIED

Electro-Optic Long Range Oblique Photography System (EO-LOROPS) **1996**

Laser RangeFinder Designator (LRFD) systems

ELOP manufactures a range of Laser RangeFinder Designator (LRFD) systems, which have been selected for a number of modern aircraft and helicopter weapon systems, including:

(a) the Laser RangeFinder Designator (LRFD) for the RAH-66 Comanche reconnaissance attack helicopter of the US Army;
(b) the RangeFinder Target Designator Laser (RFTDL) of the Night Targeting System A (NTS A), integrated by TAMAM Division, Electronics Group, Israel Aircraft Industries and fitted to AH-1W Cobra helicopters of the US Marine Corps;
(c) the laser designator element of the F/A-18 NITE Hawk pod requirement for the US Navy, integrated by Lockheed Martin Electronics and Missiles Division where, following evaluation trails by the US Navy, ELOP has been selected as a second source supplier;
(d) the Laser Tracker Receiver (LTR) widely fitted to Apache attack helicopters;
(e) the Laser Designator element of the Litening pod, manufactured by Rafael Electronic Systems Division, for the Air Forces of Israel and Germany;
(f) the Laser RangeFinder Designator for the Kiowa (OH-58) helicopter of the US Army, where it will be incorporated into the mast-mounted sight for aiming laser-guided weapons, replacing the present laser system, with improvements in operational performance, reliability and training flexibility.

ELOP's Laser RangeFinder Designator (LRFD) for the RAH-66 Commanche helicopter Electro-Optic Sensor System (EOSS) **1998**/0018349

ELOP's Laser Designator for the F/A-18 NITE Hawk pod **1998**/0018348

Laser RangeFinder Designator (LRFD) for RAH-66 Comanche helicopter

A new LRFD developed by ELOP has been selected to form the LRFD element of the Electro-Optic Sensor System (EOSS), being integrated by Lockheed Martin Electronics & Missiles Division for the US Army's RAH-66 Comanche reconnaissance attack helicopter.

The new ELOP LRFD operates on two different spectral wavelengths, within the single system, to provide optimum performance for both the laser target designation/guidance function (1.06 micrometres), and the laser ranging fuction (1.54 micrometres) simultaneously.

The system is based on use of innovative diode-pumped technology, which improves both operation and reduces power consumption. A critical aspect of the design is that the basic wavelength is converted, through use of solid-state devices and advanced optics, to produce an eye-safe sytem, thus enhancing realistic crew training and training flexibility.

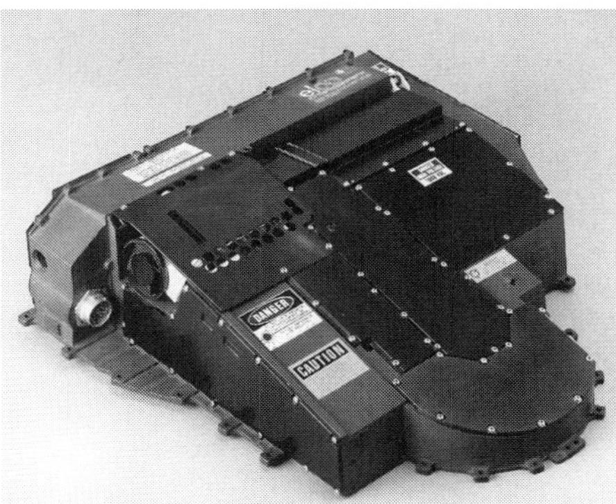

RFTDL for the Cobra helicopter **1997**/0001226

RangeFinder Target Designator Laser (RFTDL)

The RFTDL is a high-repetition rate laser designator which forms part of the Night Targeting system A (NTS A) of the US Marine Corps Cobra AH-1W helicopter. Integration of the RFTDL into the NTS A system is by TAMAM Division of Israel Aircraft Industries.

Contractor

ELOP Electro-Optics Industries Ltd.

NEW ENTRY

Multisensor Stabilised Integrated System (MSIS)

The Multisensor Stabilised Integrated System (MSIS) is a lightweight fully stabilised electro-optical system, for day and night passive surveillance and tracking of surface and airborne targets. The system is designed to detect, recognise and track the complete range of

The ELOP MSIS installed on a helicopter **1997**/0001227

manoeuvring targets from a rubber dinghy to a supertanker, as well as all types of helicopters, fixed-wing aircraft and sea-skimming missiles.

MSIS integrates three sensors: a CCD TV camera, a 6 in thermal imager and a laser rangefinder. It has a built-in automatic target-tracking computer. Optionally, it can also be equipped with a laser pointer. The stabilised ball turret provides a low wind resistance and the system obtains a picture with good resolution, high tracking accuracy and very good recognition ranges.

Operational status

In production.

Contractor

ELOP Electro-Optics Industries Ltd.

VERIFIED

Very Light Laser Rangefinder (VLLR-7)

The VLLR-7 is designed to respond directly to the need for greater accuracy, particularly when aiming at moving targets. It upgrades existing platforms and pods and enhances the capabilities of new systems. It is suitable for a variety of airborne systems, including RPVs, as well as seaborne and ground platforms. The VLLR-7 is compact, integrating easily into a variety of systems; modular, being adaptable to different systems and roles; high-speed, sending up to six laser pulses per second; ruggedised, to meet severe environmental conditions, and has trouble-free maintenance, with malfunction indicator, external test pins and plug-in PC boards. The housing is customised according to customer requirements.

The VLLR-7 features high efficiency, low heat dissipation to the pod and low divergence of the laser beam.

Specifications

Weight: <3.8 kg
Power consumption: 220 W (max)
Wavelength: 1.064 μm
Output energy: 80 mJ
PRF: single shot or up to 6 pps
Beam divergence: less than 0.4 mrad
Pulsewidth: 15 ns (max)
Range: 200-9,995 m
Range accuracy: ±5 m

Operational status

In production.

Contractor

ELOP Electro-Optics Industries Ltd.

VERIFIED

VLLR-7 Very Light Laser Rangefinder
1997/0001228

Cockpit Laser Designation System (CLDS)

The CLDS is a lightweight, small-size designator specially designed for light attack two-seater aircraft, mounted in the rear cockpit of the aircraft which commands the tactical operation. The CLDS can designate targets for other attacking aircraft equipped with laser-guided bombs and provide the ability to deliver the laser-guided bomb in a wide variety of modes, altitudes and ranges. It is designed for operation in close air support, battlefield air interdiction, deep strike or by a forward air controller using the magnifying sight for observation only and for sea surveillance.

The CLDS includes a TV camera and auto-tracker. This closed loop hands-off tracking system reduces the pilot's workload during the critical phase of the mission and significantly improves the tracking accuracy and manoeuvrability of the designating aircraft in the target area. The video camera, interfaced to an aircraft VTR, is also utilised for debriefing and intelligence purposes.

The CLDS can be interfaced to the aircraft Weapon Delivery and Navigation System (WDNS), enabling the WDNS to direct the CLDS line of sight towards a chosen target as a highly accurate sensor for marking targets and providing position updates to the WDNS.

The MBT Weapons Systems cockpit laser designation system

Specifications

Dimensions: 560 × 420 × 360 mm
Weight: 40 kg
Field of view:
(wide ×4 magnification) 12.5°
(narrow ×10 magnification) 5°
Field of regard:
(forward) 45°
(backward) 25°
(elevation) ±35°
Accuracy: 0.25 mrad
Range: 8-10 km typical

Operational status

In production and sold to several countries. The CLDS can be installed in the IAI Kfir, Dassault Aviation Mirage III and 5B, Northrop F-5F and Lockheed Martin F-16B aircraft.

Contractor

MBT Weapons Systems, Electronics Division IAI.

VERIFIED

High-performance FLIR

The high-performance FLIR is a compact parallel-scan thermal imaging system which operates in the 8-12 μm range. It is designed to provide maximum sensitivity and range performance with limited space and optical aperture, to give a serial scan-like image quality.

Combining a Newtonian-type telescope of relatively high magnification with a small element detector array with a high spatial resolution, the high-performance FLIR provides a fast compact optical design through a 1 W bidirectional scanner. The fast optics also guarantee high-efficiency cold shielding. Image quality is ensured by full DC restoration in conjunction with a very accurate channel-to-channel responsivity equalisation scheme. Other image enhancement features include dynamic range compression from 12 to 6 bits, electronic boost, interframe averaging, automatic gain and level and optional autofocus. An increase in range performance of approximately 20 per cent is achieved by a special electronic zoom underscan mode which also provides enhanced sensitivity and improved spatial resolution in both scan and cross-scan directions.

Utilisation of gate arrays and full custom VLSI designs have further reduced the size, weight and cost of the high-performance FLIR. All the preamplifiers, together with the electronic multiplexer, are housed in the Dewar package, while the rest of the electronics are included in the PC boards. Serially multiplexed detector outputs are provided for gimballed applications.

The combination of compactness and performance make the high-performance FLIR particularly suitable for long-range surveillance, target acquisition and tracking, and navigation. Typical applications include helicopter fire control sights, high-performance aircraft targeting and navigation sights, long-range detection of slow-moving aircraft and stationary concealed personnel, and observation of small fast-moving sea craft.

Contractor

Rafael Missile Division.

VERIFIED

Lilliput-1 observation and target acquisition system

Lilliput-1 is a miniature day observation and target acquisition system for helicopters, RPVs, ground and naval platforms. It consists of a gimballed module complete with CCD sensor and line of sight stabilisation, packaged in a composite shroud. The compact sensor has four fields of view. The narrowest, with 500 mm focal length and 20 μrad resolution, achieves exceptional performance in dynamic conditions. The CCD sensor can be replaced by an ICCD sensor for low-light conditions.

The system is lightweight, compact and self-contained and features very high resolution, line of sight stabilisation and low power consumption.

Specifications

Dimensions: 190 mm sphere
Weight: 1.45 kg (including laser rangefinder)
Power supply: 28 V DC, <15 W
Field of view: 20°, 7.5°, 2.7° and 1°
Coverage:
(azimuth) 360°,
(elevation) -85 to +10°

Contractor

Rafael Missile Division.

VERIFIED

Modular Thermal Imaging System (MTIS)

The MTIS is a high-performance, low-cost compact thermal imaging system in the 8-12 μm spectral range, consisting of a sensor and an electronic unit. It is suitable for night and adverse weather navigation, detection and recognition, search and rescue missions and surveillance.

MTIS can be adapted to a variety of platforms, such as helicopters, RPVs or land and sea platforms. It is available as an independent unit, or as part of a stabilised turret containing other sensors such as TV, laser rangefinder or designator and daytime optics. Integrated in such a multisensor, MTIS becomes the core of a 24 hour fire-control system.

Specifications

Weight:
(sensor) 6 kg
(electronic unit) 7 kg
Field of view:
(wide) 24.5 × 18.4°
(medium) 7 × 5.2°
(narrow) 2 × 1.5°

Contractor

Rafael Missile Division.

VERIFIED

TAWS-05 altitude warning sensor

TAWS-05 is an electro-optical sensor for monitoring the tail altitude of civilian and military helicopters. TAWS-05 uses a gallium-arsenide laser diode for high-frequency accurate measurement of the distance to the ground and warns the pilot when the limits of the safe range have been reached.

TAWS-05 represents a major improvement in safety measures for helicopters landing in poor visibility or under adverse atmospheric conditions. It interfaces with standard altitude warning systems and can be integrated into operational aircraft voice communication systems.

Specifications

Dimensions: 64 × 150 × 34 mm
Weight: 0.5 kg
Power supply: 22-32 V DC, <1.5 VA
PRF: up to 10 kHz
Range: 0.5-15 m
Temperature range: -40 to +70°C

Operational status

TAWS-05 has been successfully tested on an Israeli Air Force Sikorsky CH-53 in severe flight conditions.

Contractor

Rafael Missile Division.

VERIFIED

Topaz electro-optical targeting and surveillance system

Topaz is a targeting and surveillance system for day/night and adverse weather. Topaz incorporates FLIR, CCD, Laser RangeFinder (LRF) and automatic tracking capability. It is optimised for naval applications and adaptable to a wide variety of marine platforms and other platforms, such as UAVs and helicopters. Topaz meets all relevant MIL specs and is recommended for fire control, reconnaissance and search and rescue missions, coastal patrols and shore surveillance.

Operational Status

Rafael has supplied the Topaz targeting and surveillance system to several overseas customers.

Contractor

Rafael Missile Division.

NEW ENTRY

Toplite multisensor payload

Toplite is a compact multisensor payload which can be integrated into various weapon systems and is adaptable to a wide range of platforms. Toplite includes CCD, FLIR, LRF (Laser RangeFinder) and their respective electronics, all packaged in a turret. It can be used for observation, tracking, and pointing, slaved to either the platform's navigation, radar, or modular weapon sight, or to a helmet-mounted sight. Toplite can be integrated into helicopters, and marine platforms. The lightweight and sealed payload can be installed on boats and ships of all sizes. It can be upgraded to include a laser designator, laser spot tracker, laser illuminator and in-flight boresighting.

Operational status

The Toplite Multisensor Payload has been procured by several countries.

Rafael has entered a joint venture with Elbit Systems Ltd, for the integration of Toplite as part of an upgrading programme, signed with an overseas customer, for Puma helicopters.

Contractor

Rafael Missile Division.

NEW ENTRY

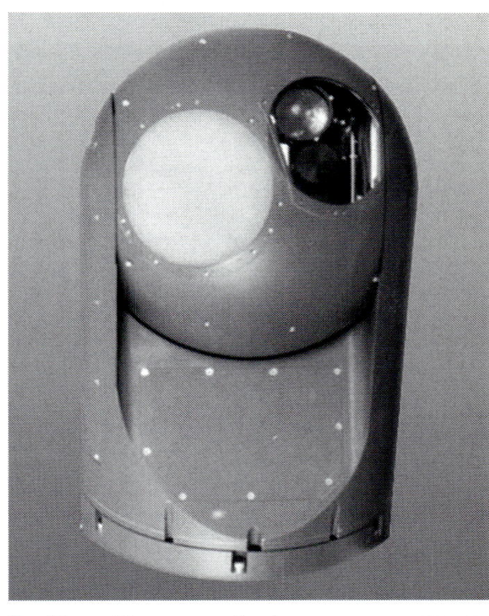

Toplite multisensor payload **1998**/0015291

Topaz electro-optical targeting and surveillance system **1998**/0015290

Multimission Optronic Stabilised Payload (MOSP)

Multimission Optronic Stabilised Payload (MOSP) is a lightweight dual or triple sensor, target acquisition and rangefinder/pointing payload.

MOSP has the capability to provide a stabilised view throughout the lower hemisphere, its coverage including the nadir point.

Sensor package options include: single monochrome (or colour CCD or triple colour CCD) day channel; 3 to 5 or 8 to 12 μm first, second or third generation Focal Plane Array (FPA) FLIR; laser rangefinder of up to 6 pulses per second, or laser target illuminator.

Specifications

Dimensions: 354 (diameter) × 548 mm (height)
Weight: 26-36 kg (varies with sensors carried)

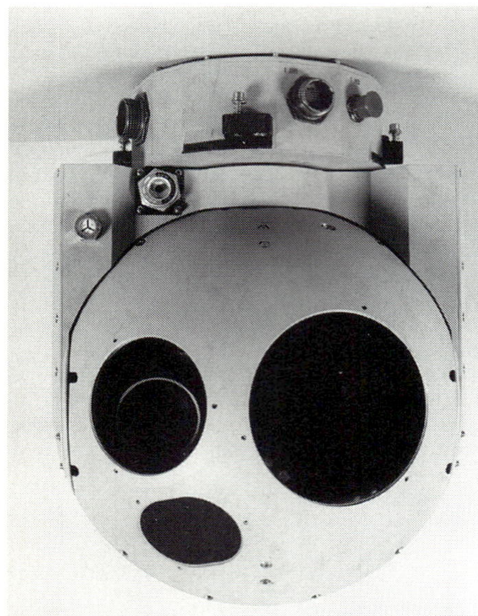

TAMAM HMOSP Helicopter Multimission Optronic Stabilised Payload **1998**/0018347

TAMAM HMOSP on Eurocopter SA 342 Gazelle helicopter **1998**/0018346

Spatial coverage:
(elevation) +15 to −105°
(azimuth) n × 360° (unlimited)
Fields of view:
TV channel: monochrome, or one- or three-colour CCD:
(narrow) 0.37° (for the long range MOSP)
(zoom) 1.3 to 18.2°
FLIR channel: 3-5 or 8-12 μm
(narrow) 2.4°
(medium) 8.2°
(wide) 29.2°
Power consumption: 280 Watt (day/night), 28 V DC

Operational status

Over 220 MOSP systems have been delivered for varying types of platform.

Contractor

TAMAM Division, Electronics Group, Israel Aircraft Industries.

UPDATED

Plug-in Optronic Payload (POP)

The Plug-in Optronic Payload (POP) is a modular, compact, lightweight electro-optical payload, designed for day/night surveillance, target acquisition, identification and location. POP is designed for light aircraft, helicopters and UAVs. It consists of a 260 mm diameter stabilised platform with a replaceable plug-in sensor unit, which may be configured to meet the customer's requirements.

The sensors include: a focal plane array infrared sensor (3-5 μm); a long range colour CCD TV; or a combination of both. The sensor unit can be replaced in the field within minutes. The payload includes an automatic tracker and is controlled via a parallel or serial communication channel.

POP installed on helicopter **1998**/0018344

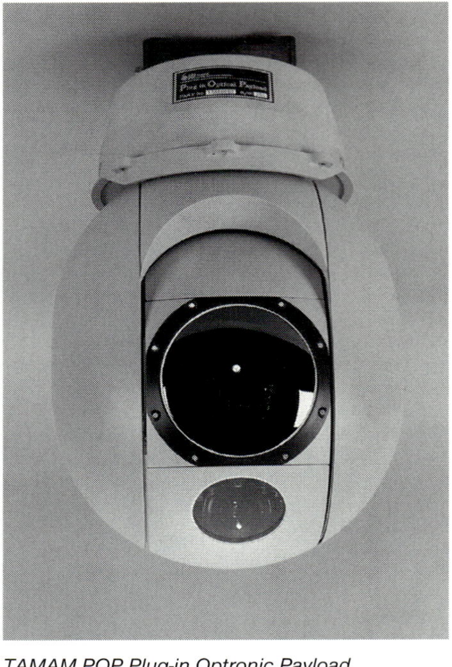

TAMAM POP Plug-in Optronic Payload **1998**/0018345

Specifications

Dimensions: 260 (diameter) × 380 mm (height)
Weight (typical): payload 15 kg; control unit 0.7 kg
Fields of regard:
(azimuth) 360°
(elevation) +40 to −110°
Fields of view:
focal plane array: 5 × 25° or 4 × 12°
TV: various zooms
TV and focal plane array: 4 × 3° to 16 × 12°

Operational status

In production.

Contractor

TAMAM Division, Electronics Group, Israel Aircraft Industries.

NEW ENTRY

ITALY

P0705 HELL laser rangefinder

P0705 HELL is an airborne Nd:YAG laser rangefinder designed for integration in (helicopter) stabilised electro-optic sight units with the functions of target ranging, gun and missile pointing and navigation fixing.

The fully qualified HELL system improves helicopter attack capabilities by providing extended reconnaissance and target detection ranges, increasing the probability of a kill, reducing flight times and enhancing flight safety.

Specifications

Wavelength: 1.064 μm
Pulse power: 4 MW
PRF: 2 Hz
Range: 300-10,000 m
Interface: RS-422 serial

Operational status

In production.

Contractor

Alenia Difesa, Avionic Systems and Equipment Division, FIAR company.

UPDATED

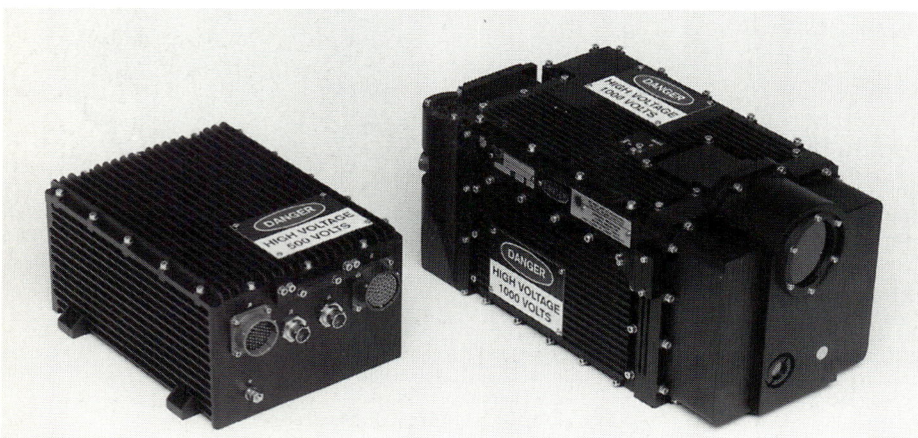

PULSE (Precision Up-shot Laser Steerable Equipment): steerable laser rangefinder for airborne applications **1997**/0001229

P0708 PULSE airborne steerable laser rangefinder

P0708 PULSE is an airborne steerable Nd:YAG laser rangefinder.

PULSE offers superior range performance, both in CCIP and CCRP attack modes and in navigation fixing, high-pulse power and repetition frequency, good precision and accuracy and an optimised configuration for easy installation on a wide variety of platforms. It is suitable for retrofitting to aircraft such as the A-4 Skyhawk, AMX, Mirage, F-5 and MiG-21.

Specifications

Wavelength: 1.064 μm
Steering angle: within 20° pointing cone
PRF: 10 Hz
Interface: MIL-STD-1553

Operational status
In production.

Contractor
Alenia Difesa, Avionic Systems and Equipment Division, FIAR company.

UPDATED

Galiflir avionic electro-optic multisensor system

Galiflir is a multisensor electro-optic system which is available in a number of configurations and is mainly intended for avionic applications such as navigation, surveillance, observation, reconnaissance, aiming and targeting.

Galiflir comprises a stabilised sensor platform and sensor pack. The sensor pack can include: FLIR, laser designator, day/night TV cameras to suit customer requirements. The FLIR has been designed on a modular basis to improve its flexibility. The modules for series parallel scanning have been designed for applications in fixed-wing aircraft, helicopters and RPVs.

The FLIR with dual field of view optics is mounted on a high-accuracy stabilised platform. A day and night TV camera can be mounted as an option on the same platform.

Galiflir undergoing operational testing on an Agusta A109 helicopter

Specifications
Volume: 10 litres
Weight: 32 kg
Power supply: 28 V DC, <100 W
Wavelength: 8-12 μm
Field of view:
(scanner) 40 × 27°
(wide) 16 × 10.8°
(narrow) 4 × 2.7°
Resolution: 0.15 mrad (narrow FOV)
Detector: Sprite, CMT 8 elements
IR lines: 512
Display: TV monitor (standard CCIR 625/50)

Operational status
In production.

Contractor
Alenia Difesa, Avionic Systems and Equipment Division, Officine Galileo SpA.

UPDATED

Pilot Aid and Close-In Surveillance (PACIS) FLIR

The PACIS FLIR is a thermal imaging system in the 8 to 12 μm range, designed to be installed on helicopters and fixed-wing aircraft in order to provide them with increased capability by day, night and in adverse weather operations. It creates a TV-compatible video signal for viewing in the cockpit on a standard display.

PACIS is provided with a telescope with two switchable fields of view and is steerable by a position control grip. It can be used for navigation, day/night surveillance, border patrol, search and rescue, remote sensing and monitoring or as a take-off and landing aid. PACIS can be interfaced with the aircraft avionic system.

The system is composed of a steerable platform, electronic unit and FLIR control grip connected by a cable to the electronic unit. The platform aims the FLIR optical axis in azimuth and elevation. The FLIR is equipped with a two field of view telescope: the wide field of view is used for navigation and surveillance, while the narrow is used to identify and track targets.

Specifications
Dimensions:
(steerable platform) 300 × 511 × 300 mm
(electronic control unit) 170 × 225 × 390 mm
(control panel) 146 × 124 × 165 mm
Weight:
(steerable platform) 23 kg
(electronic control unit) 8.5 kg
(control panel) 1.5 kg
Power supply: 28 V DC, 140 W (average), 450 W (peak)
Wavelength: 8-12 μm
Field of view:
(wide) 40 × 26.7°
(narrow ×4 magnification) 10 × 6.6°
Field of regard:
±170° azimuth,
+45 to −70° elevation
Video format: CCIR 625 lines at 50 Hz
Interface: RS-422

Contractor
Alenia Difesa, Avionic Systems and Equipment Division, Officine Galileo SpA.

UPDATED

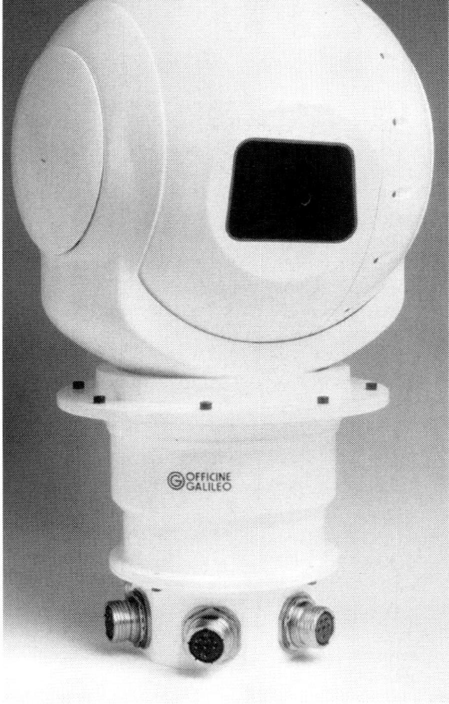

The steerable platform for the PACIS FLIR

Loam laser obstacle avoidance system

The Loam laser obstacle avoidance system is aimed at providing safe navigation capability to helicopters, particularly during low-altitude navigation. Based on laser technology, Loam will detect any obstacle such as wires, trees or masts. It is designed to be eye-safe.

Audio and visual warnings are given whenever an obstacle is detected along the aircraft flight path. The visual warning includes display of the shape of the obstacle, position, orientation and distance. Background information can be displayed as an option.

Loam comprises three units. The scan/detection unit provides full coverage of the aircraft trajectory. The processing unit provides obstacle detection and recognition. The display unit provides visual information on obstacle position, shape, orientation and distance.

Operational status
In prototyping phase.

Contractor
Marconi SpA.

VERIFIED

PILOT infrared integrated pod

The Marconi Pod-Integrated Localisation, Observation, Transmission (PILOT) is an electro-optical pod including a stabilised infrared sensor, automatic target detection/localisation/tracking unit and a bidirectional ECCM video and datalink.

The PILOT infrared sensor is a high-quality stabilised passive 24-hour all-weather infrared imager which can be steered in order to observe targets and background in front of and below the aircraft. The video output is a standard CCIR television signal which can be displayed on the aircraft head-up display or airborne monitors, or can be video recorded.

The Marconi PILOT infrared pod

By using image processing techniques PILOT allows automatic target detection, prioritisation and localisation with the presentation on the infrared image of the target position, priority and spatial co-ordinates.

PILOT provides the capability of automatic tracking of the target and can be controlled both by a local operator on board the aircraft and by a remote operator through a video and datalink. The video and datalink is frequency-hopping spread spectrum and offers robust ECCM performance. This link allows infrared image transmission to a remote station and bidirectional data communication between the pod and a remote station.

PILOT can be provided with the standard RS-422, ARINC 429 and MIL-STD-1553B avionics interfaces, which allow data communication with the airborne radar, the navigation system and the weapons management system.

Specifications

Dimensions: 2,062.9 (length) × 419.3 mm (diameter)
Weight: <128 kg
Power supply: 28 V DC, <750 W

Contractor

Marconi SpA.

VERIFIED

JAPAN

FLIR Pod for the FS-X

The Japan Defence Agency Technical Research and Development Institute is teamed with Mitsubishi Electric and Fujitsu in the development of a FLIR for the FS-X, to supplement the fire-control system for low-altitude night navigation.

The FLIR will have both air-to-air and air-to-ground applications and is specifically intended to enhance low-altitude night navigation. The sensor will have infrared CCD detectors operating in the 3-5 μm and 8-12 μm wavelengths.

Operational status

Full-scale development for installation on the FS-X. Development completion is planned for the FY2001.

Contractors

Japan Defence Agency Technical Research and Development Institute.
Mitsubishi Electric Corporation.
Fujitsu General Ltd.

UPDATED

RUSSIAN FEDERATION AND ASSOCIATED STATES (CIS)

GEO-NV-III-TV day/night tracking system

The GEO-NV-III-TV image-intensified, solid-state, charge-coupled device (CCD) camera is a versatile day/night tracking system, mounted on a gyrostabilised platform, that incorporates three separate channels: a CCD day sensor; an image intensifier; and a CCD night channel.

Specifications

Intensified CCD night channel:
scene illumination: 10^{-5} to 10^{-1} lux
sensor instantaneous field of view: 10°
combined sensor/platform field of view: +30 to −50°
focus range: 0.3 m to infinity
objective lens: f=75 mm; F/1.5
camera slew rate: 20°/s
system accuracy: 0.5°
dimensions: 200 (length) × 60 mm (diameter)
weight: 0.65 kg
CCD day channel:
pixels: 512 (H) × 582 (V)
pixel size: 7.6 × 6.3 mm
active imaging cell size: 4.6 (H) × 3.5 mm (V)
resolution: 480 TV lines
field of view: 13.5°
objective lens: f=25 mm; F/1.8
grey scales: 10
dimensions: 50 (height) × 60 mm (diameter)
weight: 0.1 kg
Image intensifier:
photocathode: GaAs

Operational status

Geophizika state that the GEO-NV-III-TV system is fitted to Kamov Ka-50 helicopters and to a wide range of Mil helicopters, including Mi-17, Mi-24, Mi-26 and Mi-28. Geophizika are also proposing it for the Ka-52 helicopter.

Contractor

Geophizika-NV.

NEW ENTRY

GEO-NVG-III NIGHT VISION GOGGLES

Geophizika claim that their GEO-NVG-III night vision goggles employ third-generation GaAs photocathode technology to provide high responsivity in starlight/overcast conditions. The goggles are ruggedised to meet Russian military requirements for low-altitude helicopter combat, reconnaissance, and search and rescue operations.

Features include: full 40° field of view, F/1.1 at 43 mm eye relief and 10 mm exit pupil; full binocular night vision; full peripheral vision; automatic brightness control; internal power supply; quick disconnect.

Specifications

Illumination: 10^{-5} to 10 lx
Magnification: 1
Field of view: 40°
Exit pupil & eye relief: 10 mm & 43 mm
Objective lens: fixed focus 25 mm, F/1.1
Focus range: 300 mm to infinity
Eyepiece lens: 25 mm
Weight: 0.78 kg
Voltage required: 3 V DC, 50 mA (2 × AA batteries)
Mechanical adjustment:
(vertical) 20 mm
(fore & aft) 24 mm
(tilt) 15°
(interpupillary) 56-73 mm
(eyepiece diopter) +4 to −4 diopters
Photocathode: GaAs
Sensitivity:
luminous 2,856K: 1,200 uA/1m
radiant (830 nm): 120 mA/W
equivalent brightness input (at 10^{-4}lx): 2.5 × 10^{-7}lx
S/N (at 10^{-4}lx): 15
centre resolution: 32 mm
useful cathode diameter: 17.5 mm

Operational status

Claimed to be in widespread use on Russian military helicopters.

Contractor

Geophizika-NV.

NEW ENTRY

Airborne Laser Radar Landing System (LRLS)

GosNIIAS has been involved with research in the area of laser radar technology for military aircraft surveillance and targeting. In 1994, work was started to convert this experience into production of a laser radar landing system for civil aircraft.

The LRLS is designed to enable general purpose aircraft and airliners to land safely in non-standard situations caused by adverse weather conditions, en-route obstacles, or the need to land on non-instrumented runways in emergency situations.

The LRLS provides day/night search, detection and auto-tracking of runways defined by optical reference marks in clear and adverse weather. The system determines aircraft angular and linear co-ordinate data, relative to the selected runway, and provides automatic steering commands to the flight control system, together with display of relevant information on the flight and navigation displays. The system is intended to support landings in Cat II and III conditions.

The system hardware comprises: the laser sighting unit; a computing unit; interfaces to the aircraft central computer and display systems.

Specifications

Runway detection range: over 5 km (in meteorological visibility range of not less than 1 km)
Detection range of objects on the runway and in the air: over 2 km
Weight (with two-axis stabilisation): 23 kg
Power consumption: 240 W

Operational status

Technical documentation and production models of all system components are available.

Contractor

State Research Institute of Aviation Systems (GosNIIAS).

NEW ENTRY

Airborne Multifunction Optical Radar System (AMORS)

Since 1982, GosNIIAS has been developing a generation of Airborne Multifunction Optical Radar Systems (AMORS), designed to provide automatic object and obstacle detection and recognition in poor visability conditions. The goals of this work were:

(a) to develop technologies, methods and algorithms as well as software and hardware for automatic scene analysis during complex processing using optical sensors with differing physical natures and data characteristics;
(b) the development of high performance image sensors, including Doppler laser radar, uncooled IIR systems, and pulse-illumination TV sensors;
(c) the operational development and full scale testing of optical radar prototypes.

AMORS is intended for air and ground monitoring, object detection, recognition and localisation against complex dynamic scenes. The system can detect and recognise various small-size objects, including vehicles and electrical power lines. Integration of AMORS into avionics sytems enhances navigation, safety in poor weather conditions, operation at poorly equipped airfields, and rescue and reconnaissance operations.

AMORS is based on range, velocity and image intensity movement, integrated image processing, and automatic object processing algorithms. The system features high jam resistance and is intended for day/night all weather operation. The system can comprise different laser radar, imaging infrared and TV sensor combinations, as well as signal processing software/hardware, object recognition modules, and a TV display control panel.

Examples of TV, laser radar and IIR images generated by the AMORS system ***1998***/0018343

Specifications

Range: up to 10 km
Resolution:
(angle) 0.4 min
(range) 1 m
(velocity) 0.3 m/s
Signal processing time for automatic object recognition: <1s
Weight: <50 kg

Operational status

Experimental prototypes have been developed and tested. Full scale testing is underway.

Contractor

State Research Institute of Aviation Systems (GosNIIAS).

NEW ENTRY

Laser rangefinder/target illuminators

Klyon

The Klyon laser rangefinder/target illuminator was developed for MiG-27M, Su-22, Su-25 and their variants.

Specifications

Ranging capability: up to 10 km; accuracy 5 m
Illumination range: up to 7 km
Weight: < 82 kg

Prichal

The Prichal laser rangefinder/target illuminator is employed on Su-25TK aircraft and on Ka-50 and Mi-28 helicopters. It is used for navigation, particularly cross-country over rough terrain, and for target ranging/illumination for laser-guided bombs and missiles.

Specifications

Ranging accuracy: 5 m
Weight: 46 kg

Technical specifications of laser emitters for Klyon and Prichal systems are as follows:

	Lasers				
	N1	**N2**	**N3**	**N4**	**N5**
Impulse energy (J)	0.07	0.18	0.38	0.4	0.25
Pulse repetition frequency (Hz)	up to 25	up to 25	25	10	25
Cooling	fluid cooling for all systems				
Weight (kg) (incl cooling system)	2.5	6.5	7.3	3.2	6
Overall dimensions(mm) (incl cooling system)	240× 188× 103	420× 180× 160	300× 155× 110	40× 390	400× 125× 120
Power supply	115 V; 400 Hz, 3-phase for all systems				
Power supply weight (kg)	6.5	6.5	10.5	20	9.8

Klyon laser rangefinder/target illuminator ***1997***/0003328

Operational status

In service.

Contractor

Production Association Urals Optical and Mechanical Plant (PA UOMZ).

UPDATED

Optical-electronic sight systems

OEPS-29

The OEPS-29 optical-electronic sight system is installed on MiG-29 aircraft. It provides: search, detection and tracking of targets at all altitudes, in free airspace, on the earth background, in daytime and at night and rangefinding for air and ground targets.

To increase battle effectiveness, the sighting system is integrated with the pilot's helmet-mounted display.

OEPS-27

The OEPS-27 optical-electronic sight system is installed in Su-27 variants. It provides the same functions as the OEPS-29, but it differs by providing a greater viewing angle and greater range for air target detection.

OEPS-29 optical-electronic sight system ***1997***/0003330

OEPS-27 optical-electronic sight system ***1997***/0003331

Specifications
Technical specifications of laser emitters for OEPS-27 and OEPS-29 sight systems are given in the previous entry, headed 'Laser rangefinder/target illuminator'.

Operational status
In service.

Contractor
Production Association Urals Optical and Mechanical Plant (PA UOMZ).

UPDATED

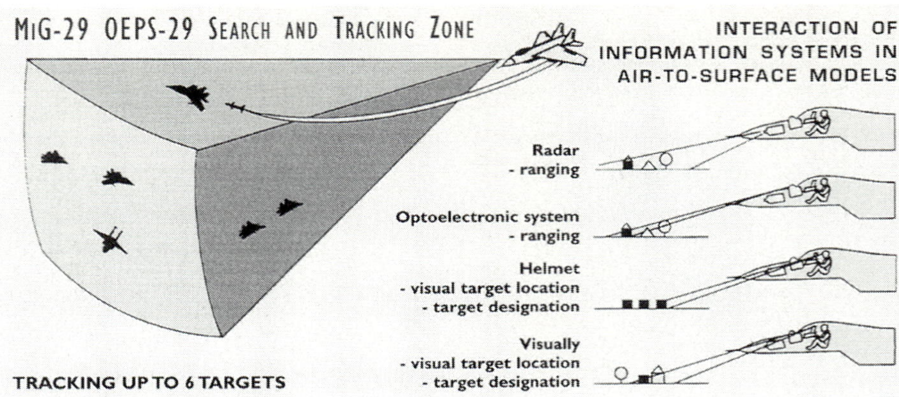

MiG-29 OEPS search and tracking zones ***1997***/0003329

Optical-TV 24-hour sight (OTV-124)

The optical-TV 24-hour sight is intended to provide detection, identification and tracking of ground and surface targets 24 hours per day. It has a high-resolution optical channel, with magnification, to aid identification of surface targets in the role of border control and maritime exclusion zone policing.

Operational status
The optical-TV 24-hour sight is produced for An-72P coastguard aircraft.

Contractors
Production Association Urals Optical and Mechanical Plant (PA UOMZ).

UPDATED

Optical-TV 24-hour sight ***1997***/0003332

YOM3 Gyrostabilised platform

The YOM3 Gyrostabilised platform is designed to carry thermal imagers, day and low-light-level TV cameras, cine- and video-cameras, laser rangefinders, and similar equipment. Four models are available (GOES-1/-2/-3/-4), designed to carry electro-optical payloads varying from 16 to 100 kg.

The system is suitable for both civil and military use. Payload specification is by customer choice. A fully defined variant of the GOES-3 system, designated GOES-320, carries the electro-optical payload specified below.

GOES-320
Thermal imager: AGEMA THV-1000
receiver: 5-bar SPRITE focal plane
spectral range: 8-12 micrometres
cooling: integral Stirling cooler
pixels: 580 × 386
field of view (H&V):
(narrow) 5.0 × 3.3°
(wide) 20.0 × 13.3°
NETD: 0.18°C
TV system: CCD Sony EVI-331 colour
pixels: 752 × 582
TV lines: 480
focal range: f=5.4 to 64.8 mm with 12 × optical zoom
field of view: 48.8 × 37.6° to 4.4 × 3.3°
Laser rangefinder (made by Production Association Urals Optical and Mechanical Plant (PA UOMZ):
wavelength: 1.54 micrometres
energy: 0.01 to 0.07 joules
beam divergence: 2 to 3 minutes of angle
PRF: 1 Hz

Specifications

Platform	GOES-1	GOES-2	GOES-3	GOES-4
Payload weight	100 kg	16 kg	30 kg	65 kg
Payload volume	85 dm³	13.5 dm³	20 dm³	55 dm³
Weight of optical turret	140 kg	25.5 kg	20 kg	85 kg
Dimensions of optical turret	∅720 × 980 mm	∅340 × 552 mm	∅460 × 613 mm	∅640 × 850 mm
Look angles				
(azimuth)	±170°	±170°	±235°	±135°
(elevation)	+80 to −40°	+10 to −30°	+45 to −115°	+10 to −30°
Stabilisation (micro radians):	50	70	50	50

Operational status
The company has stated that orders were placed for the GOES-320 system at the MAKS-97 air show. The GOES-320 specification is intended for civil application – the Kamov Ka-32A and Ka-226 helicopters have been indicated as appropriate.

Variant YOM3 systems are also fitted to the Kamov Ka-29, Ka-50N and Ka-52, and Mi-24 and -35 helicopters. The accompanying photographs taken at MAKS-97 show the YOM3 system mounted below Shkval on the Ka-50N Black Shark, although it has also been shown mounted above Shkval on the same

YOM3 mounted below Shkval on Ka-50N Black Shark helicopter (Paul Jackson) ***1998***/0018342

YOM3 mounted above the cockpit on Ka-52 Alligator helicopter (Paul Jackson) ***1998***/0018341

aircraft at other times. On the Ka-52 Alligator, it is mounted above the cockpit. The company stated that the production standard electro-optical sensor fit had not yet been finalised.

Contractor

Production Association Urals Optical and Mechanical Plant (PA UOMZ)

UPDATED

YOM3 on Ka-52 Alligator helicopter (Paul Jackson)
1998/0018340

A-84 panoramic aerial camera

The A-84 panoramic aerial camera is designed to provide wide-area photographic coverage of the earth's surface during daylight conditions, from medium and high altitudes. It can be set to operate automatically by onboard control system, or be controlled manually from its control panel.

The A-84 camera is equipped with image motion compensation and automatic exposure control. The camera film records navigation data from the aircraft navigation system.

Specifications

Focal length: 300 mm
Aperture: f/4.5
Frame size: 118 × 748 mm
Film size: 480 m (length) × 130 mm (width)
Nominal overlap in centre frame: 25%
Along track filming distance: 160 × aircraft altitude
Linear resolution: 0.4 m at range = 2 × altitude; 0.8 m at range = 6 × altitude
Power: 27 V DC, less than 300 W; 115 V AC, 400 Hz, less than 900 V A
Weight: 160 kg

Operational status

Fitted to Tu-22 medium bomber (presumably the Tu-22MR variant), and to the Tu-154 medium transport aircraft for 'Open Skies' operations. Also reportedly fitted to the M-17 high-altitude reconnaissance and research aircraft.

Contractor

Zenit Foreign Trade Firm, State Enterprise P/C S.A. Zverev Krasnogorsky Zavod.

NEW ENTRY

A-84 panoramic aerial camera ***1998***/0018339

AC-707 spectrozonal aerial camera

The AC-707 aerial camera has four separate photographic channels, four separate lenses, four filters and four automatic exposure control systems to provide photography in four pre-determined spectral bands. Each lens is corrected for focal length and distortion. The camera contains image motion compensation, vacuum back film platen, range focus and temperature focus compensation. It records navigation information obtained from the aircraft navigation system, and it can be fitted with gyrostabilisation. It can be controlled automatically or manually from its control panel.

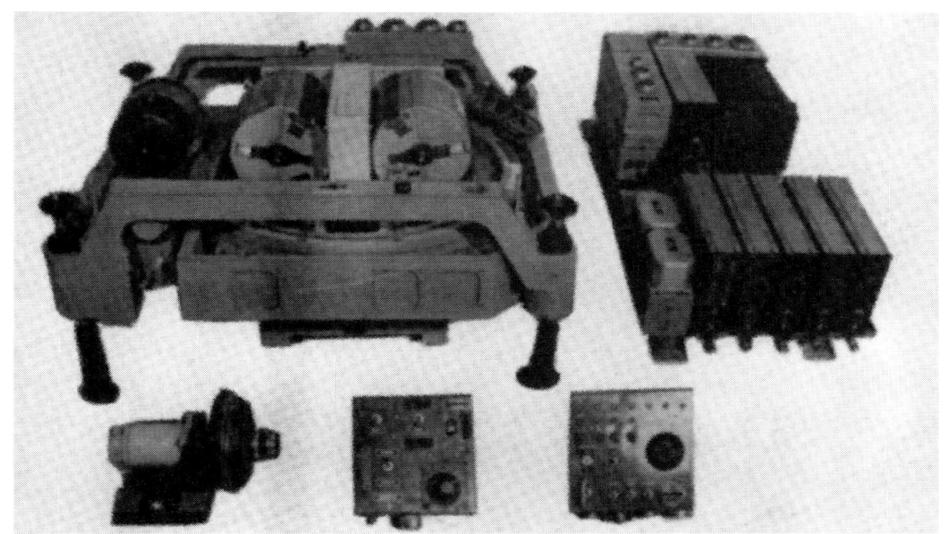

AC-707 spectrozonal aerial camera ***1998***/0018338

Specifications

Focal length: 140 mm
Aperture: f/2.8-f/22
Spectral zones:
Channel 1 (blue) 400-500 nm
Channel 2 (green) 480-600 nm
Channel 3 (red) 580-700 nm
Channel 4 (infrared) 700-860 nm
Frame size: 180 × 180 mm
Sub picture size: 70 × 70 mm
Film type: MLU-4, 240 m (length) × 190 mm (width)
Operating altitude: over 50 m
Exposure time: 1/20-1/300 s
Along track filming distance: 400 × altitude
Across track coverage: 0.5 × altitude
Power: 27 V DC, less than 30 W; 115 V AC, 400 Hz, less than 5 VA; 36 V AC, 400 Hz, less than 6 VA
Weight: 212.9 kg

Operational status

Installed on Mi-8 helicopters.

Contractor

Zenit Foreign Trade Firm, State Enterprise P/C S.A. Zverev Krasnogorsky Zavod

NEW ENTRY

AK-108Ph vertical and oblique aerial camera

The AK-108Ph aerial camera is designed for simultaneous vertical and oblique photography using three across-track width options (described as Routes 1, 2 and 3), together with three vertical/oblique angular options (defined as 0° (vertical), 75° and 80°). The camera head mirror and magazine can be rotated to provide the required coverage angle.

Camera features include: automatic exposure control; linear compensation of image shift during exposure; mirror stabilisation; temperature and pressure focus compensation and aircraft navigation recording.

Specifications

Focal length: 1.8 m
Aperture: f/5
Frame size: 180 × 180 mm
Angular field of view: 6°
Film size: 240 m (length) × 190 mm (width)
Route options: 1, 2 and 3
Photographic angle options: 0°, 75° and 80° from vertical

Angle of photography	Across track coverage, relative to height		
	Route 1	Route 2	Route 3
0°	0.1	0.18	0.26
75°	1.7	3.2	5.3
80°	3.8	8.6	21.4

Overlap, nominal: 20%
Power: 27 V DC, less than 300 W; 115 V AC, 400 Hz, less than 900 VA; 36 V AC, 400 Hz, less than 100 VA
Dimensions: 3.1 m × 600 mm diameter

AK-108Ph vertical and oblique aerial camera
1998/0018337

Weight: 600 kg

Operational status

Fitted to Su-24MR reconnaissance aircraft.

Contractor

Zenit Foreign Trade Firm, State Enterprise P/C S.A. Zverev Krasnogorsky Zavod.

NEW ENTRY

Shkval sighting system

The Shkval sighting system is designed as a comprehensive anti-tank electro-optical fire-control system, incorporating: TV sighting sensor, laser rangefinder and target designator, and laser beam-rider. It is also fitted with a three-axis field of view stabilisation system and an automatic image correlator to ensure tracking against ground, sea or sky backgrounds. 23 × magnification is provided to extend detection/tracking ranges. Tracking angles are quoted as +15° to −80° in elevation and ±35° in azimuth. Laser guidance system accuracy is quoted as 0.6 m.

Operational status

Shkval is part of the weapon system on the Ka-50 Black Shark and Ka-52 Alligator helicopters, and on the Su-24T and Su-39 multimission attack aircraft.

Contractor

Zenit Foreign Trade Firm, State Enterprise P/C S.A. Zverev Krasnogorsky Zavod.

UPDATED

Su-39 Strike Shield aircraft, showing: Shkval EO sighting system in the nose, Kopyo radar pod under belly, OMUL ECM pod under each wing, Pastil RWR on each wingtip ***1998***/0018336

Shkval window (above YOM3 'ball') on Ka-50N Black Shark helicopter (Paul Jackson)
1998/0018333

SOUTH AFRICA

LEO airborne observation systems

LEO is a range of high-performance gyrostabilised camera platform systems designed to provide steady vibration-free images for compact low-cost surveillance. Conforming to full aerospace standards, LEO can be quickly and easily installed on all helicopter and aircraft types. A variety of cameras is available, ranging from broadcast colour TV through low-light to a selection of infrared night sensors.

The LEO-400 Series is a range of compact 400 mm diameter two-axis stabilised platforms. The system is remotely controlled by an operator in the aircraft via a video monitor and laptop control unit. The video image can be recorded on tape and may be transmitted in real time by microwave downlink to fixed or mobile ground stations.

One of the new features making the AGEMA range of thermal imagers extremely useful in a SAR role is the use of colour isotherms, and/or colour thermal imagery. This enables the user to select a colour-enhanced thermal image, where a colour is assigned to a particular temperature band, thus giving a pseudo-colour image in 'rainbow' or 'hot iron' colour scale ranges.

Of even greater importance to the SAR user is the facility to retain a monochrome thermal image, but setting up to two thermal 'triggers', where a heat source at a temperature above or below these triggered temperature levels can be displayed in a pre-selected colour. Setting the trigger typically at just above the water temperature in the search area, would enable the subject to be highlighted on the operator's video screen, potentially cutting down dramatically on search time. Whereas this pseudo-colour imagery of the Agema Thermovision FLIR has always benefitted SAR operations, now longer focal lengths allow even greater detection ranges in the SAR role.

The LEO-400-SPIR/SPTV features image stability of better than 20 μrad and high-resolution colour TV with a ×32 zoom lens to give a wide field of view of 26° for panoramic surveillance and a narrow field of view of 0.8° for identification of people and vehicle licence plates from between 450 and 650 m. The compact single unit FLIR has full 12-bit dynamic range for image capture irrespective of the display setting, powerful 32-bit microprocessor software orientated design, ×8 continuously variable electronic zoom function, image freezing function for investigation of detail, temporal averaging function for low-contrast image enhancement and isotherm alarm levels for search and rescue detection. The FLIR has fields of view of 20° for panoramic surveillance and 5° for recognition and identification. FLIR detection ranges are 1 km for a swimmer in water, 3 km for a person on land and 10 km for a 10 m boat.

The operator's console and monitor for the LEO observation system mounted in a Eurocopter BO 105 helicopter ***1995***

The new LEO-400-LSPIR/SPTV, released at Eurocopter '96, is a LEO gyroscopically-stabilised camera turret, still in the 400 mm (16 in) category, but equipped with the new AGEMA Thermovision 1000 Compact Long Range (CLR) FLIR, plus the proven 3-CCD TV camera and ×32 zoom lens. The CLR thermal imager is now equipped with fields of view of 12° and 3°, rather than the 20° and 5° of the standard THV 1000. Additionally, specially selected high-grade military detectors ensure maximum resolution and sensitivity in the narrower fields of view. This enables SAR missions to be flown at higher altitudes, with greater detection ranges. FLIR detection ranges extend to 1.6 km for a swimmer in water, 5 km for a person on land and 18 km for a 10 m boat.

IRENCO offers full interface capability with onboard search radars, with dual-direction hand-over between the radar and LEO, as well as on-screen display of aircraft GPS position. To ease operator workload, a range of autotrackers is also optionally offered.

Specifications

Dimensions: 390 mm diameter
Weight:
(LEO-400-SPIR/SPTV platform) 35 kg
(monitor, control unit and electronics units) 23 kg
(LEO-400-LSPIR/SPTV platform with cameras) 36 kg
(total minimum system) 53 kg
Power supply: 22-32 V DC
Stabilisation (2 axis): <20 μrad RMS
Coverage:
(azimuth) 360°
(elevation) −100° to +20°

Contractor

Irene Commercial Enterprises (Pty) Ltd.

UPDATED

SWEDEN

Airborne laser rangefinder

The Ericsson Saab Avionics' airborne laser rangefinder is a very compact unit, which is easy to integrate into existing navigation and weapon delivery systems.

This high repetition-rate laser rangefinder uses a simple, modular design consisting of transmitter, receiver, range counter, and a deflection unit for the optical axis.

The laser is aimed at the target by slaving the deflection unit to the aircraft sighting system.

Specifications

Transmitter:
(laser type) Nd:YAG
(wavelength) 1.06 μm
(pulse energy) 20 mJ
(pulse length) approx 10 ns
(pulse repetition frequency) 1-10 Hz in bursts
(beam divergence) 0.7 mrad
Receiver:
(detector) Silicon avalanche diode
(field of view) 0.5 mrad
Range Counter:
(range, max) 20,000 m
(range, min) 200 m
(range resolution) 5 m
Range:
(typical range) ≥ 10 km at optical visibility > 20 km
Deflection Unit:
(azimuth travel) ±10°
(elevation travel) ±10°
(accuracy) < 1 mrad
(slow rate) > 60°/s
Interface:
(digital bus) ARINC 429
Weight:
(total weight) approx 14 kg

Ericsson Saab Avionics airborne laser rangefinder ***1997***/0002210

Dimensions: approx 500 × 160 × 160 mm
Power:
(power supply) 28 V DC
(power consumption) 225 W

Contractor

Ericsson Saab Avionics AB.

VERIFIED

Digital Reconnaissance Management System (RMS)

Ericsson Saab Avionics is developing a modular digital RMS for pod mounting or internal installation in tactical aircraft. The RMS is capable of processing information from several sensors, such as infrared linescan and CCD cameras, with a total data rate of more than 100 Mpixels/s. Information is handled and processed in a digital format and recorded on a digital MIL-STD-2179 VCR. A unique image compression facility allows data to be stored on suitable digital media.

Sensor imagery can, without having to stop the VCR, be frozen and analysed in flight with full resolution, using a quick-look memory. Data may be transmitted via datalink to a ground station.

Advanced automatic target recognition incorporates the latest developments in algorithms and other techniques which can be customised for specific applications.

The fully digital and modular design provides ready adaptation to new sensors and storage medias.

Operational status

Under development.

Contractor

Ericsson Saab Avionics AB.

VERIFIED

ARGUS 350

ARGUS 350 is a lightweight two-axis stabilised sensor platform, designed for: border surveillance (coast guard and customs, drug interdiction); general law enforcement and police operations; search and rescue; electrical powerline and pipeline inspections; environmental inspection (oilspill, ecological surveys, wildlife); and fire management.

Options available include: tracking systems and microwave downlink.

Specifications

Sensors:
(thermal imager) AGEMA Thermovision®1000
(daylight video) Sony EVI 311 Single CCD colour
Performance:
(elevation) +10° to −120°
(azimuth) ±140°
(max air speed) 140 kt
(voltage) 28 V DC
(power consumption) 50-200 W
(weight including sensors) 23 kg

Contractor

Polytech AB.

VERIFIED

IR-OTIS Infrared Optronic Tracking and Identification System

Saab Dynamics is currently working on design and test of a prototype of IR-OTIS, a multifunctional Infrared Search and Track (IRST) system, intended for the JAS 39 Gripen aircraft. The system can operate both as an IRST and a FLIR with a large scanning field. It provides silent situation awareness day and night and delivers excellent target acquisition data to the aircraft. It can also be used for ground attack and reconnaissance.

Since the sensor system is designed to be fitted on the Gripen and requires relatively little space, integration in other aircraft should be possible either in a new build arrangement or as retrofit equipment.

Operational status

In development for the Swedish Defence Material Administration for JAS 39 Gripen; the first development model has been delivered for flight testing on a JAS37 Viggen aircraft.

Contractor

Saab Dynamics AB.

UPDATED

Saab Dynamics IR-OTIS
1997/0002222

SEOS 200 helicopter observation system

The Saab SEOS, stabilised electro-optical multisensor system, is designed for a variety of helicopter applications.

The system is designed to give high probability of detection and it is stabilised to match weapon targeting requirements.

SEOS consists of a stabilised sensor platform, a conditioning unit for climate control, a processing unit for image enhancement and fusion of sensor signals, and a control unit with a high resolution colour monitor.

Mounted on the stabilised platform in the basic system are an 8 to 12 μm 2nd generation thermal imager, laser rangefinder and TV cameras. Optional sensors/designators include a laser designator, missile tracker/beam rider unit, low-light-level TV camera, 3 to 5 μm thermal imager.

Depending on sensor choice, SEOS can be integrated with various weapon interfaces, including optically controlled missiles, laser-guided munitions, turreted or fixed guns, rockets and artillery fire-control systems. The sensor head can be either mast or pedestal mounted.

The image processing system features automatic target tracking and rate-aided target detection. Image freeze/store, area tracking, thermal cueing, electronic magnification, sensor fusion, image integration, graphics generation and overlay are also included.

Specifications

Weight: 70 kg
Dimensions: 0.5 (height) × 0.6 m (diameter)
Azimuth coverage: ±200° (optional 360°)
Elevation coverage: −30 to +85°
Fields of view (horizontal):
(narrow) 1.5°
(medium) 4.5°
(wide) 18.0°
LOS jitter: <15 mrad rms
Interface: MIL-STD-1553B, RS 422 or others

Operational status

Under development.

Contractor

Saab Dynamics AB.

NEW ENTRY

TopEye survey system

Saab claims that its TopEye survey system utilises a unique laser rangefinder technique, integrated with the inertial navigation system of the host helicopter and digital GPS position data to produce an effective near-realtime topographical survey system.

The laser rangefinder can measure up to four different survey points in one and the same laser pulse, enabling TopEye to exactly define the structure of the area surveyed, and to produce a highly accurate digital terrain model, without recourse to photogrammetry. Accurate to within one centimetre, the laser rangefinder scans across the track of the helicopter by measuring the distance to the ground with up to 6,000 laser pulses per second. Four different distances can be identified and recorded by a single laser pulse, while CCD cameras simultaneously take pictures from the helicopter, both vertically and at a 45° oblique angle. The helicopter INS data and digital GPS data is used to support post-mission processing that produces the digital terrain product data.

Operational status

The airborne system can be installed on a Squirrel AS 350 helicopter within two hours, with no structural modifications. An underslung pod contains the complete data acquisition system (camera plus laser rangefinder).

Contractor

Saab Survey Systems.

NEW ENTRY

SWITZERLAND

RC30 aerial camera system

The RC30 aerial camera system is the newest system in the long series of WILD aerial survey cameras of LH Systems GmbH (formerly named LEICA AG, Heerbrugg) and Wild Heerbrugg AG, Switzerland. Wild RC10A was the first microprocessor-controlled aerial camera and was sold from 1981 to 1986 and the Wild RC20, which introduced Forward Motion Compensation (FMC) to the Wild aerial camera series, was sold between 1987 and 1992. Of these two models several hundred were sold. The current model RC30 was introduced in 1993 together with two new high-resolution lens cones and a serial connection to external computer and sensor systems. A further enhancement is the PAV30 gyrostabilised camera mount introduced in 1995.

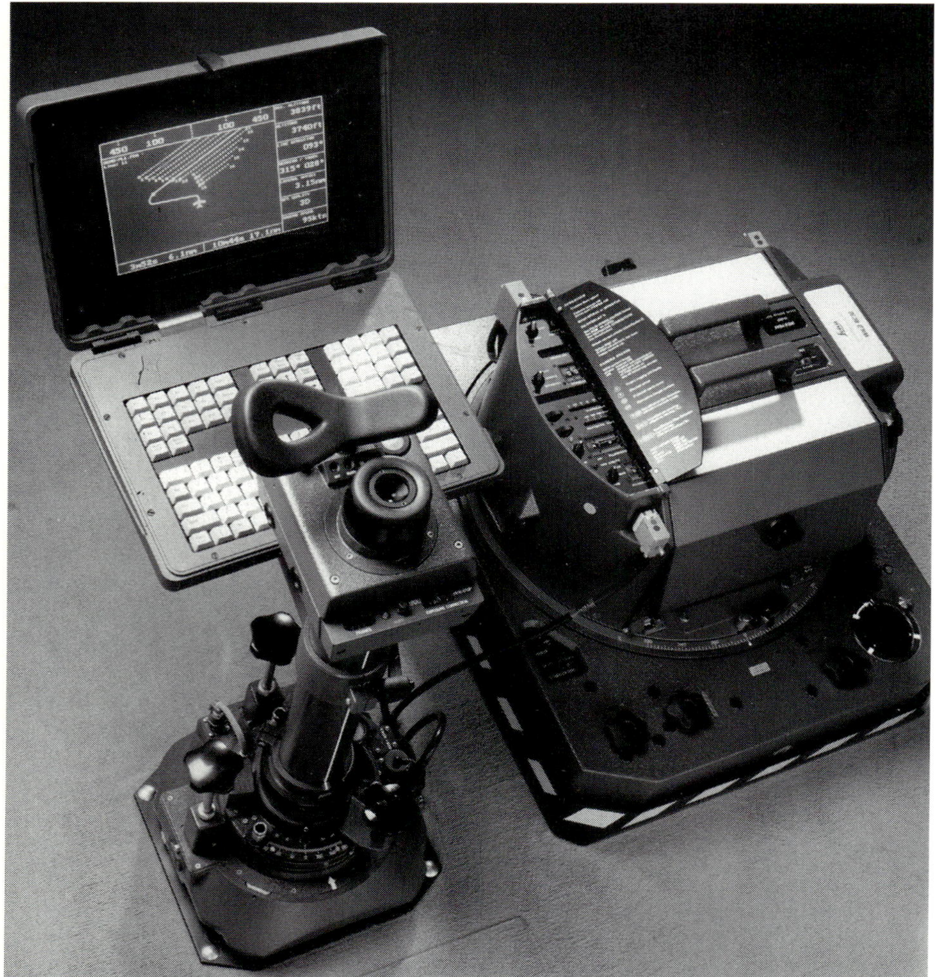

RC30 aerial camera system ***1998***/0018332

The RC30 camera system provides such features as:

FMC; the FMC counteracts forward image motion in a range from 1 mm/s to 64 mm/s;

PAV30 with AMC (angular motion compensation) is a camera mount with built-in gyros and sensors. It also functions as a sensor of aircraft attitude angles which are recorded in ASCOT (see navigation section);

S-type lens cones with AWAR resolutions of 107 Lp/mm (30/4 NAT-S) and 123 Lp/mm (15/4 UAG-s). The continuously adjustable shutter speeds go from 1/100s to 1/1000s and also the aperture between f/4 and f/22 is continuously adjustable;

PEM (automatic real-time exposure control). The PEM sensor is optimised for aerial photography and covers the full range of film types (B/W, B/W-Infrared, Color-negative and -positive, Falsecolor-Infrared);

EDI (external data interface), standard interface between the camera and the ASCOT (Aerial survey control tool) as well as the link to GPS receivers, aircraft navigation systems, or mission computers;

A complete array of interchangeable filters for all exposure conditions, absorption of unwanted spectral ranges, optimising light distribution in the image plane, correcting colour-balance of film emulsion.

ASCOT (Aerial Survey Control Tool) is a hardware/software system with an integrated 12-channel Leica GPS receiver to enable reliable survey flight navigation and provide the following additional functions:

interactive survey flight planning in the office or in flight;

guidance of the pilot and the operator/navigator on the flight line and during turns, as well as autosearch for nearest route;

automatic exposure release at pre-defined locations (pinpoint photography);

automatic in-flight data annotation on imagery (freely definable layout);

interface to camera mount PAV30 for drift signals provided by external gyros or other sensors and recording of residual angles of gyrostabilised mount;

logging of GPS raw data and camera data for post-processing, mission evaluation and reporting;

output to kinematic GPS post-processing software SKI-AERO.

SKI-AERO (kinematic GPS post-processing software) to determine camera projection centres and to transfer data to aerotriangulation packages such as, ORIMA, HATS PATB/GPS-Aero, ALBANY or BLUH.

RC30: microprocessor control, external connections and communication.

Specifications

Camera mount: PAV30 gyrostabilised mount
Stabilisation range in: Pitch and roll ±5°
Drift ±30°
Typical residual angular motion: < 0.3°/s
Typical residual deviation from vertical: < 0.2°/s
Navigation computer:
ACU30 with internal GPS
Internal Silicon Disk
up to 8 RS-232 interfaces
2 monitors (pilot/navigator)
Leica 9212 GPS, L1, 12 channels, DGPS, RTCM
Navigational accuracy: < 100 m
DGPS accuracy: < 5 m
Lens cones:
(15/4 UAG-S wide angle) 90°, f/4
AWAR: 123 Lp/mm
(30/4 NAT-S normal angle) 55°, f/4
(AWAR) 107 Lp/mm
(8.8/4 SAGA-F super wide angle) 120°, f/4
Film cassettes: single lightweight daylight cassettes

Operational status

In service.

Contractor

LH Systems GmbH.

UPDATED

TURKEY

ASELFLIR-200 second generation gyrostabilised airborne FLIR

The Aselsan ASELFLIR-200 forward-looking infrared (FLIR) system is a low-weight, multipurpose, thermal imaging sensor for pilotage/navigation, surveillance, search-and-rescue, automatic tracking, target classification and targeting. The ASELFLIR-200 incorporates a 4 × 240 focal-plane array detector which operates in the 8-12 micrometer band.

Key features of the ASELFLIR-200 include Electronic Image Stabilisation (EIS), Local Area Processing (LAP) for image enhancement, Multi Mode Tracking (MMT), analogue and digital video outputs for transmission and/or recording, MIL-STD-1553B/ARINC/other discrete databusses to interface with other on-board avionics such as rradar, navigation and weapon systems. The ASELFLIR-200 has three fields of view: Narrow Field of View (NFoV) for recognition and identification, Medium Field of View (MFoV) for detection and a unity Field of View (FoV) for navigation and pilotage.

The ASELFLIR-200 includes two weapon replaceable assemblies: turret unit (WRA-1) and electronics unit (WRA-2). The ASELFLIR-200 optionally incorporates a laser rangefinder and/or a CCD day TV camera.

Specifications

Field of View (FoV):
(wide) 22.5 × 30°
(medium) 5 × 6.67°
(narrow) 1.3 × 1.7°
Parallel detector channels: 240 × 4 FPA
Spectral Bandpass: 7.6 × 10.5 micrometers
Electronic zoom: 2:1 and 4:1
Gimbal angular coverage:
(azimuth) 360° continuous
(elevation) 40° up; 105° down
Gimbal acceleration: head steering compatible at aircraft speeds
Laser rangefinder: optional
Day TV: optional
Environmental: MIL-STD-E-5400
Video outputs: analogue and digital video outputs are provided
Cooling: self contained
Weight:
(turret unit) <31.8 kg
(electronics unit) <22.73 kg
Dimensions:
(turret unit) 323.85 (diameter) × 372.87 mm (height)
(electronics unit) 306.3 (width) × 413.5 (length) × 199.1 mm (height)
Power: standard aircraft power

Operational status

In production.

Contractor

Aselsan Inc, Microelectronics, Guidance and Electro-optics Division.

NEW ENTRY

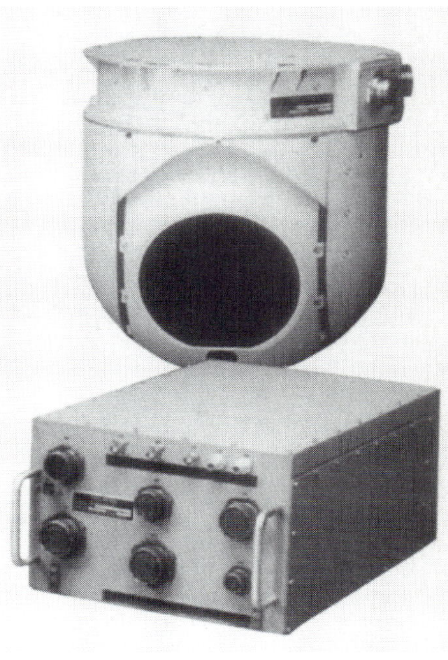

Aselsan ASELFLIR-200 turret unit (above) and electronics unit (below) ***1998***/0018331

UNITED KINGDOM

Thermovision 1000 series thermal imaging systems

AGEMA Infrared Systems manufactures a range of thermal imaging systems, optimised to different requirements, including: Thermovision 1000; Thermovision 1000LR (Long Range); Thermovision 1000CLR (Compact Long Range). High resolution real time images are obtained with excellent thermal sensitivity using the LK-4 scanning module, multi-element SPRITE detector and integral Stirling cooler, with 12-bit digital image processing for accurate image capture. Features include: automatic brightness and contrast adjustment; dual field of view lens for panoramic surveillance or close inspection (telescope option for optimum resolution); image freezing and electronic zoom; user defined alarm levels using isotherms or pre-set zone conditions; other user defined options available.

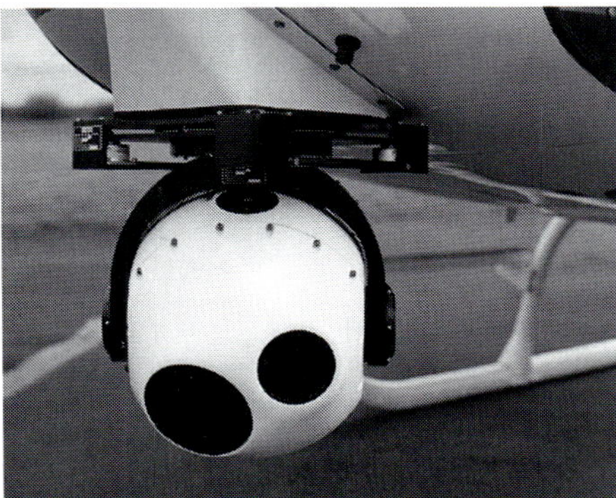

The Thermovision 1000 CLR fits inside the LEO-400-LSPIR/SPTV platform from Irenco ***1998***/0018330

The Thermovision systems can be integrated into the gyrostabilised four-axis platform from gimbal specialists, Irenco. Called the Leo-400-LSPIR/SPTV, this compact and rugged platform fits neatly underneath, or on the side, or on the nose of a wide variety of helicopters and fixed-wing aircraft. The

Thermovision 1000 thermal imager ***1997***/0001231

possibility to include a three-chip CCD broadcast quality daylight camera in the same gimbal provides the flexibility for day/night surveillance in the same mount.

Specifications

Dimensions: 310 × 164 × 221 m
Weight: 8 kg
Power supply: 28V DC, 55 W
Detector: Multi-element MCT SPRITE
Wavelength: 8-12 μm
Performance: <0.1°C
Optics:
Field of View (FoV) no lens 33° (H) × 21.6° (V)
Field of View (FoV) dual lens:
(wide FoV) 20° (H) × 12° (V); IFOV 0.6 × 0.6 mrad
(narrow FoV) 5° (H) × 3.0° (V); IFOV 0.15 × 0.15 mrad
Detection capability: man: 5 km; fast boat: 18 km; light aircraft: 20 km

Operational status

Widely used by the police in the UK and Europe.

Contractor

AGEMA Infrared Systems.

UPDATED

VITS automatic tracking systems

VITS is a family of high-performance products which employs an expanding architecture. They provide superior tracking capability, fast processing time and reduced operator workload.

VITS is fully compliant with the complete range of military specifications and provides both VME and RS-422/RS-232 serial interfaces. The family embodies Ada software, custom ASICs and extensive BIT facilities.

The VITS family consists of VITS1000, VITS2000, VITS3000 and VITS El-Stab. The basic VITS1000 features dual standard video input, high-precision centroid and correlation tracking, low data latency for high system bandwidth, automatic selection of optimum tracking mode, track quality measures to aid system performance, an aimpoint refinement and boresight alignment facility. Tracking performance is maintained in high-clutter backgrounds.

VITS2000 includes all these features plus multiple object tracking, target cueing and prioritisation, automatic target acquisition and robust tracking despite decoys and obscurants.

The top-of-the-range VITS3000 will include all the features of the other VITS models plus automatic target classification using neural networks and enhanced rejection of false alarms.

The VITS1000 tracking system has been integrated with the GEC-Marconi TIALD (Thermal Imaging Airborne Laser Designator) pod, and its performance proven through flight trials at DERA (Boscombe Down). This trials programme is now continuing with the integration of VITS2000 into the pod. VITS1000 is also being supplied to a variety of UK and offshore prime contractors and defence establishments.

VITS El-Stab is a new electronic image stabilisation package which has been introduced as part of the ongoing VITS development programme. The new stabilisation system offers a highly reliable, low-weight, low-cost alternative to gyrostabilisation in high-vibration environments and is adaptable for new-build or upgrade application. It uses VITS1000 tracking technology to reorganise and correct sensor vibration.

Used in conjunction with an electro-optical surveillance system, the electronic image stabilisation package provides a stable image to the operator, using advanced image processing techniques, and is particularly effective for both unstabilised systems and in high magnification systems, where mechanical solutions are inadequate and expensive. It can be integrated into the prime equipment enclosure, or separately boxed on the aircraft.

Operational status

VITS1000, VITS El-Stab in production.
VITS2000 under evaluation.

Contractor

British Aerospace Systems and Equipment.

VERIFIED

DAT 1000 video digital auto-tracker

The DAT 1000 video digital auto-tracker will acquire and track objects in a standard video field from normal light or thermal sensors and issue sensor platform correction data and/or target attributes to the sensor platform of the host system. It is available as a VME board pair for OEM applications or as a stand-alone black box unit with video and serial connections to sensor and host. Computing Devices can also offer a full system capability with tracker, sensor(s) and host system to suit individual customer requirements.

Unlike conventional centroid or correlation trackers, the DAT 1000 approach to target tracking is being used to unlock the potential for video processing applications. An example of this is the newly developed search and track facility as a complementary stage to the video tracking itself.

Operational status

In production.

Contractor

Computing Devices Company Ltd.

VERIFIED

RMS 3000 reconnaissance management system

The RMS 3000 Series combines scan conversion and image processing of electro-optic sensor imagery with in-flight display of the terrain overflown. The RMS 3000 Series system is fitted to Royal Air Force Tornado GR. Mk 1A reconnaissance aircraft and provides for the simultaneous recording and display of the imagery from multiple infrared sensors. Other system features include real-time and near real-time display of imagery, a wide range of rolling and updating display modes, display facilities including slow speed replay, magnification and processing to optimise the displayed image and aspect correction and rectilinearisation of displayed imagery to facilitate in-flight exploitation.

The system design is optimised to ensure that the ground exploitation time is minimised. Relevant features include combined video and aircraft data annotation recording on videotape, in-flight editing of evented imagery and rapid removal of video cassettes from the aircraft on return to base.

Operational status

In production and in service in Royal Air Force and Royal Saudi Air Force Tornado GR. Mk 1A aircraft.

Contractor

Computing Devices Company Ltd.

VERIFIED

Infrared images produced by a Computing Devices RMS 3000 being displayed in the rear cockpit of a Tornado GR. Mk 1A

RMS 4000 reconnaissance management system

The RMS 4000 Series is an all-digital system which is designed to be common across several platforms. Key features which have been used to enhance the system performance and increase the mission success rate are use of Ada high-order language, flexible modular design with high configuration commonality across platforms, high reliability and high level of functional redundancy to ensure no single defect can cause mission critical failure, design for two-level maintenance and growth potential to accommodate new sensors, image and target processes.

The RMS 4000 provides a selection of sensors with built-in growth potential, image data management, compression and decompression and routeing, multiple display formats for the selected sensor, recall of imagery for review and edit prior to data transmission, datalinking of key target imagery to provide real-time record and availability of archive data as soon as the aircraft lands.

Operational status

In production.

Contractor

Computing Devices Company Ltd.

VERIFIED

RMS 5000 reconnaissance management system

The RMS 5000 represents the latest modular design to accommodate multiple sensor types using standard interfaces. The function of the system can be chosen by selecting the configuration from a standard set of modules that provide for multiple sensor types including E-O, IR, SAR and MMWR. The system incorporates image enhancement to correct sensor and aircraft abnormalities, which is combined with sensor correlation and automatic target recognition. Imagery data undergoes high order data compression for in-flight display and datalinking for real-time exploitation. The system contains massive onboard data storage for in-flight and in-platform ground exploitation and a key feature is autonomous operation for single-pilot or unmanned aircraft. The RMS 5000 is available for built-in or podded use.

Operational status

In development.

Contractor

Computing Devices Company Ltd.

VERIFIED

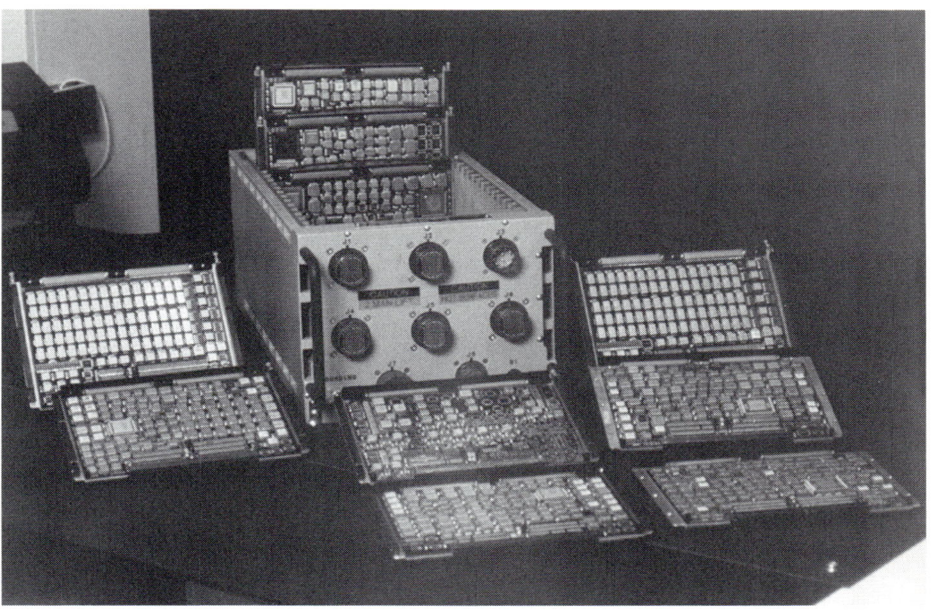

The Computing Devices RMS 5000 reconnaissance management system can be installed internally, or in a pod

---

Cats Eyes Night Vision Goggles (NVGs)

GEC Marconi's Cats Eyes NVGs system allows for the combination of both a direct visual and an intensified image to be presented to the pilot's eyes. The two images are combined in a 1:1 relationship and complement each other. The benefits of the system have been extensively proven in low-level night attack flying trials in a fully integrated NVG-compatible cockpit including Forward Looking Infrared (FLIR) generated head-up display imagery and a head-down multi-function display, as well as in the most demanding operational sorties such as the 1990/91 Gulf Conflict.

The system's advantages are many. Since the head-up display is seen through a direct visual path it is not degraded by unnecessary image intensification as it would be with conventional NVG systems. Additionally the direct vision path through the optical combiner arrangement makes monitoring of cockpit displays and instruments considerably easier whilst the ability to scan either side of the combiners enhances peripheral vision and ensures better spatial awareness. The direct vision path also removes problems normally associated with light to dark transitions as the intensified image becomes progressively more noticeable as the direct visual image fades. The system is compact and rugged and the restrictions on head mobility imposed by the depth of conventional NVG systems is avoided. Whilst the system incorporates a single handed quick release mechanism for the helmet interface, it can be configured to include an automatic separation system on ejection and designed growth will enable it to accept the latest image intensifier technology as it becomes available.

Specifications

Weight: (including mount and helmet bracket) 0.82 kg
Field of view: 30°
Magnification: Unity
Eye Relief: 1.0 in (25 mm)
Exit Pupil: 0.4 in (10 mm)

Operational status

In production. Selected as the standard NVG system for the US Navy and Marine Corps fixed-wing tactical aircraft. Over 800 systems now in service with the US Navy which are fully qualified to aircraft carrier EMC and environmental requirements.

Contractor

GEC-Marconi Avionics Ltd.

UPDATED

GEC-Marconi Avionics Cats Eyes night vision goggles ***1997**/0001235*

Nightbird Night Vision Goggles (NVG)

Designed specifically for fighter aircraft applications, the GEC-Marconi Nightbird NVG system has all the attributes of the NITE-OP system with the addition of full head-up display compatibility and the enhancement of pilot safety on ejection by means of automatic NVG detachment.

The NVG optics are designed to permit both HUD symbology and Forward Looking Infrared (FLIR) imagery to be viewed simultaneously. Additionally, the NVGs are designed to be employed with either Generation 2 or Generation 3 standard image intensifier tubes permitting growth if required. Finally, the ejection safe auto-detach mechanism has been fully qualified and avoids any risk of injury to the pilot from the goggles on ejection by safely releasing them approximately 4 ms after ejection initiation.

Specifications

Weight: 0.815 kg
Field of view (circular): 45°

Operational status

In service with the Royal Air Force Harrier GR. Mk 7, Tornado GR. Mk 1/1A and Jaguar.

Contractor

GEC-Marconi Avionics Ltd.

VERIFIED

GEC-Marconi Avionics Nightbird night vision goggles for fast-jet aircraft ***1997**/0001236*

NITE-OP Night Vision Goggles (NVG)

The GEC-Marconi Avionics NITE-OP NVG are specifically designed for helicopter aircrew, enabling them to fly visually at night.

NITE-OP NVGs have a fully circular 45° field of view.

GEC-Marconi Avionics Nite-Op Night Vision Goggles ***1997**/0001237*

The optical design provides large eye relief and exit pupil permitting the use of an eye protection visor or NBC protective mask. Manufactured using the latest composite materials, the goggles are lightweight and robust. Electrical configuration ensures high reliability and redundancy by powering each image intensifier tube separately by batteries integral to the NVGs. With no external wires or connectors required the system is completely self-contained and portable. Either Generation 2 or Generation 3 image intensifier tubes may be fitted.

Specifications
Weight: 0.8 kg
Power supply: 3.5 V batteries (independent for each channel)
15 h endurance at 0°C
Field of view (circular): 45°

Operational status
In service with UK armed forces and several overseas customers.

Contractor
GEC-Marconi Avionics Ltd.

VERIFIED

Type 105 laser ranger

The Type 105 is a high repetition rate, Nd:YAG steerable laser rangefinder developed privately by GEC-Marconi to provide compact low-cost accurate target ranging sensors for ground attack aircraft.

The 105 Series includes several variants to suit different aircraft installation and avionics requirements. The latest variant has been specifically designed as a two-box system for ease of installation in a wide range of aircraft. The 105 is now MIL-STD-1553B databus-compatible and is particularly suitable for aircraft updates such as for the F-5E, A-4, A-10, Hawk and AMX, as well as new ground attack aircraft.

Target range is measured to 3.5 m standard deviation to a range of 10 km, effectively removing the largest source of error in air-to-surface weapon delivery.

Low power consumption, ease of integration with aircraft avionic systems, flexible configuration, small size and frontal area are seen to be important factors in the suitability of the Type 105 for fit or retrofit in ground attack aircraft.

The GEC-Marconi Electro-Optics Type 105 laser ranger is fitted in the BAe Hawk 100

Specifications
Dimensions: 190 × 200 × 370 mm
Weight:
(laser) 12.5 kg
(electronics unit) 4.5 kg
Power supply: 28 V DC, 300 W
Wavelength: 1.06 μm
PRF: 10 pps
Angular coverage: within 10° semi-apex angle cone
Range: 10 km
Accuracy: 3.5 m standard deviation
Reliability: >1,000 h MTBF

Operational status
In service.

Contractor
GEC-Marconi Electro-Optics Ltd, Navigation and Electro-Optic Systems Division, Silverknowes.

VERIFIED

Type 118 lightweight laser designator/ranger

The Type 118 laser transceiver is suitable for integration with the visual optics of future or existing helicopter sights to designate targets for spot-tracker or laser-guided weapon applications. It is a lightweight Nd:YAG device, originally developed and produced for the mast-mounted sight on the US Army OH-58D Army Helicopter Improvement Programme (AHIP).

Specifications
Dimensions: 153 × 337 × 163 mm
Wavelength: 1.06 μm
Output energy: 110 mJ
PRF: 20 pps (max)
Range: 300 m-10 km
Resolution: 5 m

Operational status
In production.

Contractor
GEC-Marconi Electro-Optics Ltd, Navigation and Electro-Optic Systems Division, Silverknowes.

VERIFIED

Type 126 laser designator/ranger

The Type 126 is a high-energy Nd:YAG rangefinder and designator system developed for the UK MoD Thermal Imaging Airborne Laser Designation (TIALD) pod. The Type 126, comprising separate transmitter unit and power supply/control unit for ease of pod or aircraft installation, provides very high output energies at pulse repetition rates of up to 20 Hz in a temperate environment ranging from −54 to +71° C.

Operational status
In production and in service in the TIALD pod.

Contractor
GEC-Marconi Electro-Optics Ltd, Navigation and Electro-Optic Systems Division, Silverknowes.

VERIFIED

Type 221 thermal imaging surveillance system

The Type 221 thermal imaging surveillance system is designed for service with military helicopters. Developed by GEC-Marconi in conjunction with Pilkington Optronics, the system incorporates an IR18 thermal imager and telescope by the latter company, with sightline stabilisation steering provided by a GEC-Marconi stabilised mirror. The assembly is contained in a pod which either can be mounted beneath the nose of a helicopter or can project through an aperture in the aircraft floor.

The IR18 imager unit provides a normal field of view of 38° in azimuth and 25.5° in elevation. In the Type 221 application, users can choose a telescope magnification of either ×2.5 or ×9, with corresponding wide or narrow fields of view. The wider field of view (15.2° azimuth by 10.2° elevation) would be used for general surveillance, target acquisition or navigation; the narrow field of view (4.2° azimuth by 2.6° elevation) permits detailed observation for target identification or engagement of targets detected in the wide field of view mode. With the GEC-Marconi sighting mirror in the pod installation the system has fields of regard of +15 to −30° in elevation and ±178° in azimuth. The entire sensor system is vertically mounted above the mirror which is angled, periscope fashion, at 45° to the horizontal to provide views in the horizontal plane.

The sensor system employs SPRITE detector units cooled by a Joule-Thompson minicooler supplied with high-pressure compressed air. The air source is a bottle, mounted on the equipment and charged immediately before flight. This has a capacity of 1 litre and provides a system operation time of approximately 2½ hours. If greater endurance is required other cooling options, involving the use of mini-compressors permanently connected to the equipment, are available. The system operates in the 8 to 13 μm band and has a sensitivity of between 0.17 and 0.35° C to target background and surroundings.

The GEC-Marconi mirror has an aluminium reflective element which is diamond hined to a flatness of two fringes at 550 nm. This mirror has a reflectivity of greater than 97.5 per cent at 45° incidence, averaged over 8 to 12 μm. Its stabilisation system comprises a two-axis device with integrating rate gyros as rate sensors. The mechanism is driven in each axis by a direct-drive DC torque motor while steering is obtained by torquing the integrating rate gyro. Angular information is derived from a resolver fitted to each axis. The mirror sightline is controlled by signals from an electronics unit which also provides power and signals to the main turret azimuth drive on the pod. System control may be exercised through a digital computer or by a hand controller unit. According to GEC-Marconi, the use of the mirror system provides a higher degree of stabilisation than is normally attainable by other methods and the resultant image is blur-free under typical aircraft vibration conditions. Infrared spectrum vision is obtained via a germanium window on the front of the mirror turret assembly.

The display may be presented on either 525- or 625-line television monitors. The output is either in CCIR or EIA composite video formats, as required. This television-compatible output may be displayed on one or more monitors throughout the aircraft. Pilkington emphasises the desirability of having monitors at both the pilot and winch operator positions of a helicopter to co-ordinate crew members for better hover control during night and bad visibility winching operations.

The Type 221 system is intended for use in medium to large helicopters and roles envisaged include maritime reconnaissance, search and rescue and integration with onboard weapon systems to improve all-weather capability.

Specifications
Dimensions:
(pod unit) 865 long × 420 mm (max) diameter
(electronics unit) 127 × 432 × 330 mm (max)
Weight:
(pod unit) 75 kg
(electronics unit) 8 kg

Operational status
In service. The Type 221 thermal imager has been fitted to a number of Aerospatiale/Westland Puma helicopters operated by the Royal Air Force.

Contractors
GEC-Marconi Electro-Optics Ltd, Navigation and Electro-Optic Systems Division, Silverknowes.
Pilkington Optronics, Glasgow.

VERIFIED

Type 239 pilot's night vision system

The Type 239 is a compact steerable platform for the pilot's night vision system on the Agusta A129 helicopter.

The platform, on which is mounted a Loral thermal imager, has a two-axis movement, with ±130° range in azimuth and +75 to −60° in elevation. The platform's acceleration and slew rates for both axes are compatible with helmet tracking systems to enable the platform to function with a helmet display as a fully visual coupled system.

The platform is suitable for mounting on almost any helicopter, due to its small size and low weight, and can be fitted with a range of compact imaging systems to meet customer requirements.

Operational status

In service on the Agusta A129 and Italian HH3 helicopters.

Contractor

GEC-Marconi Electro-Optics Ltd, Navigation and Electro-Optic Systems Division, Silverknowes.

VERIFIED

Type 260 stabilised fast acting mirror

GEC-Marconi developed the Type 260 for an airborne Infrared Search & Track (IRST) demonstrator programme. The mirror is compact and rugged and requires a minimum volume external to the host airframe. It uses a novel strapdown stabilisation technique which permits the mirror sightline to be rapidly pointed, stabilised against 'own ship' motion and accurately registered with respect to inertial space.

In addition to IRST applications, the Type 260 mirror is suitable for other demanding optical tasks such as IRCM.

The modular nature of the system, mirror head, gyro package and compact electronics allows for greater flexibility in installation than usual. It is also possible to change the gyro type without major changes to the system.

Specifications

Mirror size: 130 × 182 mm
Field of regard (this can be expanded if required):
(azimuth) ±30°
(elevation) +25 to −20°
Acceleration: >150r/s²
Slew rate: >5r/s
Stabilisation: >100μrad
Electronics unit: 371 × 194 × 192 mm

Operational status

Prototype delivered.

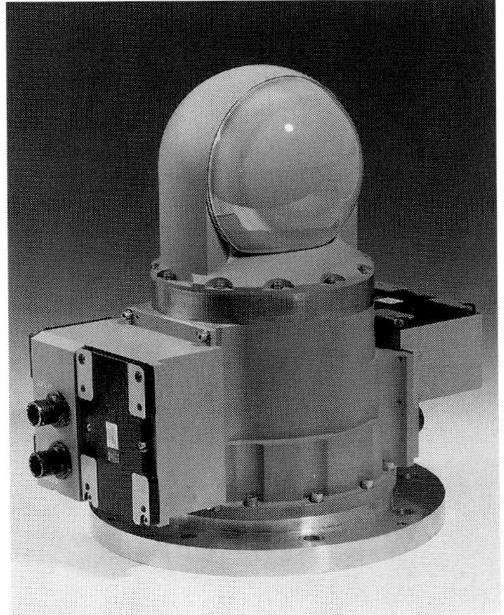

The GEC-Marconi Electro-Optics Type 260 stabilised fast-acting mirror **1996**

Contractor

GEC-Marconi Electro-Optics Ltd, Navigation and Electro-Optic Systems Division, Silverknowes.

VERIFIED

Type 629 modular rangefinder

The GEC-Marconi Type 629 Nd:YAG laser ranger is the standard transceiver in production for a number of military applications. It is based on an Nd:YAG laser to provide low-cost ranging over ranges of 300 m to 10 km with an accuracy of 2.5 m. Variants of the rangefinder are used in army and air force applications in a number of countries.

Specifications

Dimensions: 200 × 180 × 130 mm
Weight: 6 kg
Power supply: 22-30 V DC, 200 W
PRF: 10 pps nominal
Output energy: >60 mJ

Operational status

In production for a number of military applications.

Contractor

GEC-Marconi Electro-Optics Ltd, Navigation and Electro-Optic Systems Division, Silverknowes.

VERIFIED

Laser Ranger and Marked Target Seeker (LRMTS)

The LRMTS is a dual-purpose unit which can be used as a self-contained laser ranger or as a target seeker with simultaneous rangefinding. In the target seeking role it can be used to detect and attack any target designated by ground troops with a compatible laser, enhancing the effectiveness of battlefield close air support.

The LRMTS is an Nd:YAG laser mounted in a stabilised cage, which allows beam-pointing and stabilisation against aircraft movement. The seeker can detect marked targets outside the head movement limits. It operates at a relatively high pulse repetition frequency, thus allowing continuous updating of range information during ground attack. As range is a crucially important parameter for accurate weapon delivery, and yet virtually unobtainable on a non-laser equipped aircraft, the LRMTS is a vital additional sensor.

In a typical operation, a forward air controller with pulsed laser target-designation equipment directs the aircraft to a location within laser detection range before switching on the ground marker equipment. Radio communications between the forward air controller and aircraft crew are minimised and positive identification of even small, hidden or camouflaged targets is assured.

Once the LRMTS has detected the laser energy reflected from the target it provides steering commands to the pilot on the head-up display. Ranging data is also shown and fed directly into weapon aiming computations for the accurate and automatic release of weapons.

GEC-Marconi Electro-Optics LRMTS optics in the nose of the British Aerospace Harrier GR. Mk 3

The LRMTS is easy to install and harmonise with other aircraft systems and is said to be more effective than any alternative sensor during operations at the grazing angles used in low-level ground attack. By improving weapon delivery accuracy, the sensor ensures a high probability of success in single-pass, high-speed attacks.

Associated with the LRMTS head is an electronics unit which contains power supplies and ranging and seeker processing. The laser needs a transparent window and in the Jaguar is mounted behind a chisel-shaped nose, with two sloping panels. For the Harrier, where there is more chance of debris accumulation during VTOL operations, the optics are protected by retractable eyelid shutters.

A specially designed installation for the Royal Air Force Panavia Tornado has incorporated the LRMTS into an underbelly blister. In addition to the systems in service with the Royal Air Force, the equipment is also used by three overseas air forces.

An eye-safe version of LRMTS has been developed and successful trials have been completed. Eye-safe operation has been achieved by altering the output of the laser with a Raman cell from 1.06 μm to 1.54 μm which is in the eye-safe region of the spectrum. Existing LRMTS systems can be retrofitted with this modification which will significantly reduce safety restrictions for training.

Specifications

Dimensions:
(LRMTS head) 300 × 269 × 607 mm
(electronics unit) 330 × 127 × 432 mm
Weight:
(LRMTS head) 21.5 kg
(electronics unit) 14.5 kg
Power supply: 200 V AC, 400 Hz, 3 phase, 700 VA
28 V DC, 1 A
Wavelength: 1.06 or 1.54 μm
PRF: 10 pps
Angular coverage:
(elevation) +3 to −20°
(azimuth) ±12°
Roll stabilisation: ±90°
Detection angle: ±18° from aircraft heading
Max range: >9 km

The LRMTS mounted beside the nosewheel bay of the Royal Air Force Panavia Tornado GR. Mk 1

Operational status

In production. Development of the LRMTS began in 1968 under a government contract and prototype units were first flown in 1974. Deliveries to the Royal Air Force, accounting for over 200 units, for installation in nose housings of the British Aerospace Harrier and Sepecat Jaguar aircraft were completed in 1984. Deliveries continue and over 800 LRMTS have been delivered to the Tornado, Jaguar and Harrier programmes, in the UK and overseas.

Contractor

GEC-Marconi Electro-Optics Ltd, Navigation and Electro-Optic Systems Division, Silverknowes.

VERIFIED

TIALD day/night attack pod

The Thermal Imaging/Airborne Laser Designator (TIALD) provides military aircraft with laser designation and automatic tracking from both infrared and TV sensor imagery to maximise the effectiveness of both laser and conventional weapons on ground attack aircraft. It offers both thermal imaging and TV sensor data simultaneously during flight to provide a day, night and adverse weather capability. This allows the aircrew to choose the optimum sensor at the site of the intended attack.

TIALD is designed and manufactured by GEC-Marconi Electro-Optics Ltd, Navigation and Electro-Optic Systems Division, Silverknowes.

Apart from the cockpit display and controls the pod is self-contained, taking its power from the aircraft primary supplies. Interface with the avionics is via a MIL-STD-1553B databus.

The forward section of the pod contains the thermal imager and TV sensors, the telescope and the laser designator transceiver unit. The laser, TV and thermal imager optical paths are combined within the telescope and steered over a wide angle of regard by the pod roll and gimbal arrangement. The combined optical path is stabilised against aircraft movement and pod vibration by a stabilised mirror. The combination of optical paths is an important design feature which contributes to minimising the harmonisation errors between the three sightlines as well as enabling a narrow pod diameter with a large optical aperture.

The static rear sections of the pod contain the ram-air cooler, electronics units and power supplies. The automatic video tracker, produced by British Aerospace (Systems & Equipment), is included within the electronics units.

In operation the sightline would be directed on to the target area, either by commands from the aircraft avionics using data from the navigation system, radar and so on, or by manual control by the crew. When the target has been selected on the video display the tracker is engaged and locked to the target. Once this has been achieved the aircraft may be manoeuvred with the pod system automatically keeping the target on boresight and designated by the laser.

The equipment is used to designate the target for laser-guided weapons delivered from either the attacking or an accompanying aircraft. Alternatively, an attack using conventional weapons could proceed by updating the aircraft's weapon aiming computer with range and bearing information from the pod. TIALD also has a passive air-to-air and a reconnaissance/surveillance capability.

TIALD on USMC AV-8B ***1996***

Specifications

Dimensions: (length) 2,900 × (diameter) 305 mm
Weight: 210 kg
Power supply: 200 V AC, 400 Hz, 3 phase, 2 kW (max)
Field of view:
(wide) 10°
(narrow) 3.6°
Electronic zoom: ×2, ×4

Operational status

In production for Tornado GR. Mk 1 aircraft. GEC-Marconi Electro-Optics Ltd has been awarded a multimillion pound contract for the supply of TIALD pods for these aircraft.

TIALD was also selected for the Royal Air Force Jaguar and has been integrated and used successfully during operations in Bosnia. TIALD is being supplied as part of the Jaguar upgrade programme to the Royal Omani Air Force.

Contractor

GEC-Marconi Electro-Optics Ltd, Navigation and Electro-Optic Systems Division, Silverknowes.

UPDATED

All Light Level TV (ALLTV) system

The ALLTV system consists of low-light level television cameras and laser systems mounted on a steerable stabilised platform based on an existing family of platforms in production for civil and military applications. The ALLTV system enables an operator to detect, identify and track targets in daylight and at night, to direct aircraft guns automatically on to a target.

Operational status

The ALLTV system was selected for installation in US Air Force AC-130U Gunship aircraft.

Contractor

GEC-Marconi Electro-Optics Ltd, Sensors Division, Basildon.

VERIFIED

The GEC-Marconi ATLANTIC FLIR installed under the port wing of a Royal Air Force Jaguar T Mk 2A

ATLANTIC podded FLIR system

The Airborne Targeting Low Altitude Navigation Thermal Imaging and Cueing (ATLANTIC) pod-mounted FLIR system is designed to give ground attack aircraft night and poor weather capability on high-speed low-level missions.

The Atlantic pod employs the GEC-Marconi AN/AAR-51 modular FLIR system as used in the Harrier GR. Mk 7 and AV-8B and it can also include an advanced thermal cuer for early target detection and a laser spot tracker. The system, with multi-target compatibility, has a MIL-STD-1553B databus interface and can be integrated with existing weapons and avionics systems.

The telescope is selected to match the FLIR image to the aircraft HUD field of view. Pod cooling is by means of a self-contained environmental conditioning unit and the detector is cooled by a closed-cycle cooling engine. Time to readiness for the system is 3½ minutes typically from 20°C.

Specifications

Dimensions: 2,413 × 254 mm diameter
Weight: 100 kg nominal
Power supply: 28 V DC nominal, 210 W typical
115/200 V AC, 400 Hz, 146 W typical
Max speed: Mach 1.2 at 40,000 ft
Temperature range: −40 to +70°C
Video output: 525-line, 60 Hz or 625-line, 50 Hz

Operational status

In production for Royal Air Force Jaguar T Mk 2A aircraft. Integration with F-16 also demonstrated in trials with the Royal Netherlands Air Force.

Contractor

GEC-Marconi Electro-Optics Ltd, Sensors Division, Basildon.

VERIFIED

Heli-Tele television system for helicopters

The GEC-Marconi Heli-Tele is a broadcast standard television surveillance system designed for mounting on helicopters. The system provides long-range real-time airborne surveillance to meet the requirements of police, military, paramilitary, civil and other security forces.

Heli-Tele consists of a colour camera, with a high magnification zoom lens, mounted on a gyrostabilised steerable platform. The operator can steer the camera and adjust the magnification to display any selected area of the ground scene. Video information is transmitted to any number of ground stations via an air-to-ground microwave link with a range of over 90 km. For night or poor visibility surveillance, a standard modular thermal imaging sensor package may be installed in place of the TV camera payload, with a turnaround time of only 30 minutes.

Operational status

In service. The Heli-Tele has been certified by the UK's Civil Aviation Authority and is fitted to nine types of helicopter.

Contractor

GEC-Marconi Electro-Optics Ltd, Sensors Division, Basildon.

VERIFIED

Infrared Search and Track System (IRST)

The primary role of the IRST system is to detect and track multiple incoming aircraft at ranges compatible with those of modern air-to-air missiles and to display a thermal image of selected targets for identification at shorter range. The system is required to operate by day and at night and in adverse weather whilst the host aircraft is manoeuvring at high speed. Entirely passive target detection ensures covert operation, since there is no need to use the aircraft's search radar which could be detected at very long range by modern electronic warfare systems.

In the primary air-to-air role the IRST system consists of an infrared sensor whose field of view is progressively scanned across the likely target sector. It uses a highly agile mirror which is pointed and stabilised by a high-performance servo control system. The resulting IR data is electronically processed using advanced algorithms to extract target signals from unwanted clutter and background noise with a low false alarm rate.

In addition to the primary role, modern IRST systems are also used as a flying aid and for ground attack. In both these roles a high-definition wide field of view thermal image of the terrain ahead is presented on the pilot's HUD, superimposed precisely on his view of the outside world. He is thus able to fly fast at a low level, at night and in poor visibility, just as if in daylight, and to land at darkened airfields. In the ground attack role automatic thermal cueing highlights potential targets, alerting the pilot to their presence many seconds before they could be detected by his visual search alone.

GEC-Marconi is building an IRST demonstrator system which will make maximum use of infrared optical systems and thermal imaging equipment now in production. Much of this equipment is already in service with Royal Air Force Tornado GR. Mk 1 and Harrier GR. Mk 7, US Marine Corps AV-8Bs and other air forces worldwide. Current performance enhancements ensure the high capability of the demonstrator, paving the way for the production of IRST systems for next generation aircraft and possible retrofit to existing front-line military aircraft.

Operational status

GEC-Marconi has won a contract from the UK MoD for a demonstrator IRST system. The order is to supply flight-cleared equipment to the Defence Evaluation Research Agency for trials in the Tornado GR. Mk 1.

Contractor

GEC-Marconi Electro-Optics Ltd, Sensors Division, Basildon.

VERIFIED

Modular FLIR

The modular FLIR provides a passive solution to the demanding requirements of night navigation and target acquisition. The FLIR is installed within the aircraft's fuselage, with the optics looking forward through a small blister at the front of the aircraft. The system includes a lightweight miniaturised scanner and advanced signal processing to satisfy the demanding space and performance requirements of the airborne role. Hands-off fully automatic operation minimises aircrew workload and enhances combat survivability. The modular design of the FLIR permits simple reconfiguration to meet the space constraints of aircraft such as the Hawk 100. A variant of the production equipment for the Hawk 100 comprises two LRUs: the sensor head and the electronics unit. The electronics unit includes space provision for future growth in performance and capability. The configuration for the Harrier and AV-8B can be fitted within a 254 mm diameter pod.

By projecting a high-resolution image of the terrain ahead on the HUD, the FLIR permits the pilot to carry out aggressive manoeuvres at low altitude. In addition, an integrated thermal cuer detects hot objects within the scene, which may be potential targets, and marks them on the HUD. At night and in adverse weather conditions the FLIR significantly increases aircraft utilisation over a 24-hour period.

FLIR image on a pilot's head-up display

The telescope is selected to match the FLIR image to the aircraft HUD field of view. The detector consists of eight parallel CMT TEDs, cooled by a closed-cycle cooling engine. Time to readiness is typically 3½ minutes at 20°C.

Other versions are available for tactical transport aircraft, such as the C-130 Hercules, or as enhanced vision systems on civil aircraft.

Specifications

Power supply: 28 V DC nominal, 200 W typical
115/200 V AC, 400 Hz, 146 W typical
Temperature range: −40 to +70°C
Video output: 625-line, 50 Hz or 525-line, 60 Hz
Interface: MIL-STD-1553 and/or discrete hardwired

Operational status

In production for the Royal Air Force Tornado GR. Mk 4 and Harrier GR. Mk 7, AV-8B for the US Marine Corps, Spanish and Italian navies and the Hawk 100.

Contractor

GEC-Marconi Electro-Optics Ltd, Sensors Division, Basildon.

VERIFIED

MultiSensor Turret System (MST-S)

The MST-S is a versatile compact lightweight thermal imaging system suitable for helicopter and subsonic fixed-wing aircraft operations. It is NVG-compatible.

The gyrostabilised turret platform provides 360° steerable field of regard in both azimuth and elevation and is fully operational at airspeeds up to 300 kt.

The standard thermal imaging payload utilises UK TICM II modules, a closed-cycle cooling engine and a continuous zoom telescope with magnification of ×2.5 to ×10. Alternative payloads with a variety of IR and TV sensors and telescopes are available.

The basic system consists of two LRUs: the turret and a control switch joystick unit. This can be expanded by the addition of a third unit to cater for the range of options which provides full integration with

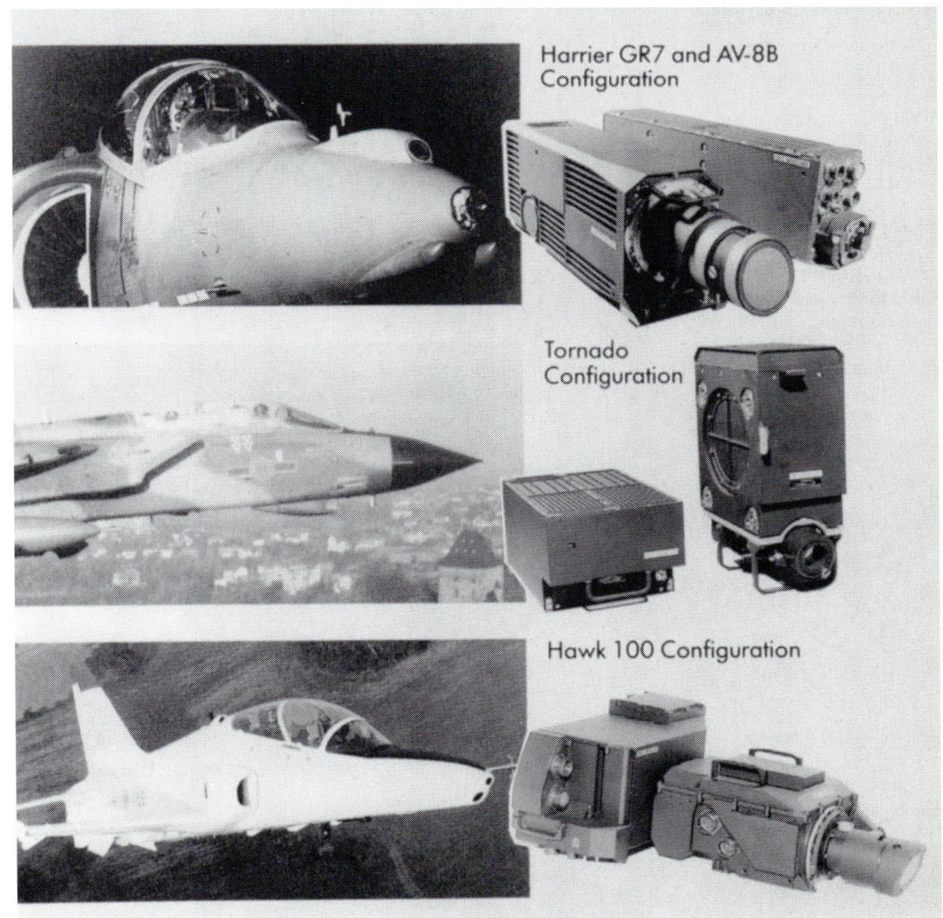

Configurations of the GEC-Marconi Sensors modular FLIR system

other aircraft systems. The options currently available include MIL-STD-1553B, ARINC 429 and RS-422 databus interfaces, video tracker, electronic magnification, radar designated target handover and a variety of control units.

Specifications

Weight:
(turret) 40 kg
(control switch joystick unit) 2 kg
(optional control electronics unit) 12-20 kg
Power supply: 28 V DC nominal, 340 W typical
Wavelength: 8-13 μm, 3-5 μm as an option
Video output: 625-line, 50 Hz or 525-line, 60 Hz

Operational status

In production for S-61, S-76, Lynx and Sea King helicopters and Fokker F50, Dornier 228 and CN-235 maritime patrol aircraft. A quantity of MSTs has also been fitted to Nimrod MR. Mk 2. Ordered for RAF EH101 Support Helicopters and as part of a sophisticated surveillance system being installed on Agusta-Bell 412 EP helicopters by an export customer.

Contractor

GEC-Marconi Electro-Optics Ltd, Sensors Division, Basildon.

VERIFIED

The multisensor turret fitted under the starboard wing of a Nimrod MR. Mk 2

Sea Owl passive identification device

The Sea Owl system provides a day and night long-range target detection and identification capability for helicopters. It uses a GEC-Marconi thermal imaging sensor with high-magnification optics mounted in a highly stabilised platform sited on the nose of the helicopter. The high-resolution thermal image, automatically optimised for maximum picture quality, is displayed in the cockpit. The employment of advanced signal processing provides tracking and target cueing automatically, further reducing aircrew workload.

The system is completely integrated with the helicopter central tactical system and other avionics systems and offers both automatic search and acquire operation and manual control via a joystick unit. The long-range standoff identification capability of this equipment has been proved around the world in a wide range of atmospheric conditions.

Specifications

Weight:
(turret) 64 kg
(signal processor) 17 kg
(tracking unit) 13 kg
(compressor) 10 kg
Power supply: 28 V DC nominal, 570 W typical
115/200 V AC, 400 Hz, 115 W typical
Wavelength: 8-13 μm
Detector: 8 parallel CMT TED
Magnification: ×5 to ×30 switched zoom
Field of regard:
(elevation) +20 to −30°
(azimuth) +120 to −120°
Video output: 625-line, 50 Hz or 525-line, 60 Hz

The GEC-Marconi Sea Owl passive identification device on a Lynx helicopter

Operational status

In production for the Royal Navy Lynx.

Contractor

GEC-Marconi Electro-Optics Ltd, Sensors Division, Basildon.

VERIFIED

Thermal Imaging common modules (TICM)

Thermal imaging common modules have been configured into a wide variety of thermal imaging systems. They are built to full military standard and use advanced optical and electronic components, including the high-performance GEC-Marconi TED detector, to give high resolution and sensitivity even at long range. Telescopes and displays can be selected by system designers to meet precise operational requirements.

These indirect view systems produce a high-quality video image which can be displayed on television monitors or head-up or head-down displays. They are fully automatic and minimise operator workload. The modular basis gives flexibility in system design, enabling application-specific sensor architectures to be configured by system integrators. The ease of maintenance and repair allows cost-effective use both in new weapons and for retrofit.

Operational status

In large-scale production for the UK MoD. Over 2,000 sets of modules have been supplied to the UK armed forces and to customers throughout the world.

Contractor

GEC-Marconi Electro-Optics Ltd, Sensors Division, Basildon.

VERIFIED

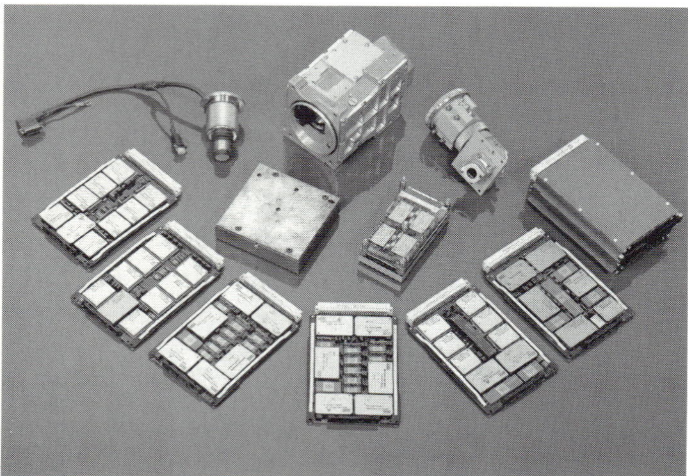

A set of thermal imaging common modules

A typical thermal imaging system configured from TICM II modules

ADEPT 30 automatic video tracker

The ADEPT 30 tracker is the latest in the range of Octec video trackers and is the core tracker used in both the CAATS and CAATS 2 systems. It comprises a well integrated software design on a single VME card designed for use in the Octec airborne CAATS systems, and easy integration into other manufacturers' avionics VME electronics.

The ADEPT 30 unit has a great deal of flexibility in its design, allowing closed loop control of a platform and easy integration to control panels.

A range of daughter boards gives even greater flexibility and performance, allowing higher input data rates, additional tracking channels, or further signal processing.

A range of enclosures is available from a commercial 19 in rack unit to fully ruggedised avionic enclosures. Build standards are available to meet different standards from commercial up to full MIL Spec.

Octec have integrated their trackers with over 50 different platforms, including most widely used helicopter sensor platforms.

Specifications

Dimensions: 233.4 × 160 mm Double Euro
Power: +5 V 3 A, +12 V 0.2 A, −12 V 0.2 A
Video input: Composite video 625/525 line CCIR or RS-170
No of video inputs: 2
Track modes: Centroid, Scenelock
Video output: 1 with symbology overlay
Interfaces available: VME, RS-232/422, Analogue, Discretes
Automatic video detection: variable from 2-90% of FoV

Operational status

Some 300 ADEPT 30 trackers have been shipped to customers all over the world, it continues in full production.

Contractor

Octec Ltd.

VERIFIED

ADEPT 30 Automatic Video Tracker ***1997***/0001238

CAATS 2 automatic video tracker ***1997***/0001239

CAATS Compact Airborne Automatic video Tracker

CAATS is a variant of the Octec ADEPT series video trackers, and has been configured specifically as commercial ruggedised equipment for use in aircraft and helicopters. It is extremely light in weight and of small dimensions, thus making retrofitting a straightforward task.

The interfaces and control techniques available in the standard unit allow for easy integration into existing airborne electro-optical surveillance systems.

The introduction of automatic tracking for airborne systems has proved to be a great advantage in reducing operator fatigue and increasing general tracking capability. Centroid and screen lock modes are provided, the latter particularly useful in locking onto a scene in the field of view.

Compact Airborne Automatic video Tracking System (CAATS) ***1996***

Specifications

Weight: 3 kg
Dimensions:
(tracking unit) 250 × 300 × 82 mm
Power: 28 V DC, 2A (max) or 115/230 V AC
Video input: 50-60 Hz, RS-170 or CCIR
No of video inputs: 2
Tracking modes: centroid and screenlock. Automatic target detection selectable over 90% of FOV.

Operational status

Selected by UK government agencies and US DoD.

Contractor

Octec Ltd.

VERIFIED

CAATS 2 - Compact Airborne Automatic video Tracker

This new unit offers significant advantages over the earlier CAATS system, which remains in production.

CAATS 2 provides the same or better tracking performance but is lighter, is less than half the size, and uses less power than the original design. It is a sealed unit using conduction cooling. Several versions are available to meet different environmental scenarios. CAATS 2 also has the capability, using daughterboards, to enhance performance including electronic image stabilisation and input of digital video.

The CAATS 'Scene Lock' tracking feature is retained and performance improved. This is a particularly important feature for the helicopter fit. Centroid tracking is also provided together with a range of submodes, including automatic cueing, which allows optimisation for particular target scenarios. A video symbology generator is integrated to allow on screen data to be shown and recorded.

Specifications

Weight: 2.8 kg
Dimensions: 250 (W) × 210 (D) × 50 mm (H)
Power: 28 V DC, 1.0A mean or 115/220V AC using adapter
Video input: RS-170 or CCIR video input 50/60 Hz
No of video inputs: 2
Tracking modes: Centroid, Scene Lock (other modes optional)
Video output: 1 with symbology overlay.
Automatic video detection: variable from 2-90% of FoV

Operational status

First shipments March 1997.

Contractor

Octec Ltd.

VERIFIED

EPIC identification thermal imaging

The EPIC thermal imaging system is constructed from the UK Ministry of Defence's ruggedised STAIRS C modules to provide thermal imaging for all land, sea and air platforms. EPIC offers greatly improved target recognition and identification ranges as a result of the extended spatial resolution, twice that of many first generation systems. The module interfaces have been defined to ease system integration, maintenance and testing.

Specifications

Field of view:
(single) 5 × 3°
(dual) 5 × 3° and 17.5 × 10.5°
Optical aperture: 110 mm (telescope)
Sightline stability: ≤0.1 mrad

Waveband: 8 to 9.4μm
Detector: 768 × 6 CMT diode array
Power (typical): 120 W
Weight: 20 kg

Dimensions:
(sensor head) 392 × 200 × 142 mm
(electronics) 220 × 200 × 142 mm
MTBF: 5,000 h

Contractors
Pilkington Thorn Optronics Limited, Hayes.
Defence Evaluation Research Agency, Malvern.

VERIFIED

Helicopter Infrared System

The Helicopter Infrared System (HIRS) is a high-resolution IR18 thermal imaging system, mounted in a steerable pod which provides real-time pictures to a display. The system is modular, making it adaptable to a number of applications and is normally gimbal-mounted, although a fixed mount can be provided. Objects as close as 1.5 m can be viewed and a ×6 infrared telescope is used for long-range detection. Pictures are presented on a 525/625-line monochrome display, which is also provided with the associated controls. More than one display can be linked to the sensors and the picture can be video recorded or datalinked to a ground station.

Specifications
Weight:
(total system including monitor and air bottle) 35 kg
(×6 telescope) 3.9 kg extra
Power supply: 24 V DC, 32 W
Field of view:
(normal) 38 × 25.5°
(×6 telescope) 6.3 × 4.25°
Coverage:
(elevation) +10 to −100°
(azimuth) ±100°

Operational status
No longer in production. In service.

Contractors
Simrad Optronics Ltd.
Pilkington Optronics Ltd.

VERIFIED

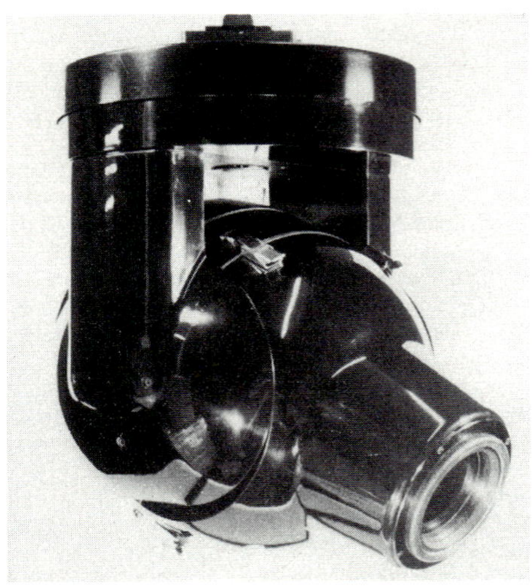

The Simrad Optronics Ltd/Pilkington Optronics Ltd IR18 helicopter infrared system

---

70 mm film format framing cameras

More than 20 variants of the W Vinten Ltd 70 mm film format framing camera have been produced over the past 30 years. Thousands of the cameras, which are simple to operate and require minimum maintenance, are in service worldwide. The compact cameras are suitable for low- and medium-altitude reconnaissance by day and may be fitted in a wide range of fixed-wing aircraft, helicopters, RPVs, drones and UMA internally or in a pod.

Principal features of the 70 mm film format framing camera are magazine loading, a range of interchangeable lenses from 38 mm up to 280 mm, high-speed focal plane shutters and optional data head.

The most common types in service are the F95 Mks 1-10 and the Types 360, 362, 512, 518, 547, 591, 618, 880 and 881.

Operational status
Most camera variants are available for production.

Contractor
W Vinten Ltd.

UPDATED

70 mm film format cassette-loaded panoramic cameras

Over 10 variants of the cassette-loaded 70 mm film format panoramic camera have been produced since the camera was first introduced into service in the early 1970s. The camera is fitted with a high-resolution 76 mm focal length lens with a cycling rate of 2 to 10 frames/s. It is normally operated in the autocycle mode but some variants may be fired at intervals in the pulse mode when interfaced with a suitable control. The cameras are inherently very reliable, due to the design concept.

The camera is suitable for low and medium reconnaissance by day or as a medium- to high-altitude tracking camera operated in the pulse mode in co-ordination with other sensors. It may be fitted internally or in a pod on a wide range of fixed-wing aircraft, helicopters, RPVs, drones or UMA.

Principal features of the cameras include a 76 mm high-resolution lens, autocycle or pulse control, optional remote control and optional data head.

The most common types in service are the 751, 751A-E, 752, 753, 755, 910 and 914 A/B/C.

Operational status
Large numbers of the camera are in service worldwide. Most variants are available for production. The Type 914A is in service on combat aircraft and drones.

Contractor
W Vinten Ltd.

VERIFIED

70 mm film format magazine-loaded panoramic cameras

The Type 900 magazine-loaded panoramic camera entered production in 1983. It is fitted with a high-resolution 76 mm focal length lens with a cycling rate of 2-17 frames/s. The Type 900A operates in the autocycle mode only, whereas the Type 900B, introduced into service in 1993, operates in either pulse or autocycle modes. The magazine accepts up to 750 ft of standard 0.004 in thick film or 1,000 ft of thin-based film.

Roles for the Type 900 panoramic camera include low- and medium-altitude reconnaissance by day, as a tracking camera for use with other sensor systems and area coverage. The camera is normally fitted inboard in aircraft or in a pod.

Principal features of the Type 900 include a 76 mm high-resolution lens, magazine loading, autocycle or pulse control (Type 900B) and a data head as standard.

Operational status
In production. Types 900A, 900B and F152 are in service.

Contractor
W Vinten Ltd.

UPDATED

Type 214 infrared linescan sensor

Linescan 214 is a high-performance day and night sensor operating in the 8-14 μm waveband. It has been designed for fixed-wing aircraft or helicopter airborne surveillance, with provision for in-flight imagery recording on film and, with additional equipment, real-time TV imagery viewing and recording.

The infrared signals from scanned terrain are recorded on 70 mm film contained within the unit. The sensor also produces a video output which is recorded on videotape in a linescan format video cassette recorder. The imagery is displayed on a TV monitor via a suitable scan converter such as the W Vinten Ltd Imagery Display Processor (IDP). Imagery is normally displayed on the IDP at a ground station. However, where a suitable installation can be provided on the aircraft, imagery may be displayed in real time, recorded and/or transmitted to the ground station via a datalink.

The Linescan 214 sensor may be installed in the aircraft fuselage or mounted in a pod under the fuselage or wing. The sensor and the cooling pack are low volume, lightweight and have low power consumption. A separate control panel is supplied for remote operation.

Specifications
Dimensions:
(linescan sensor) 258 × 437 × 317 mm
(cooling pack) 272 × 140 × 105 mm
(control unit) 113 × 146 × 113 mm
Weight:
(linescan sensor) 12 kg
(cooling pack) 1.7 kg
(control unit) 1.1 kg
Power supply: 28 V DC, 2.5 A
Resolution: 1.5 mrad
Sensitivity: 0.25° NET
V/H range: 0.2-0.7 rad/s
Field of view: 120° transverse

Operational status
No longer in production. Several hundred 201 systems have been delivered and are in service in a number of countries. The 201 system is, for example, the standard equipment aboard the Canadair CL-89 surveillance RPV. Another member of this series is the Linescan 212 system designed for use in light aircraft and helicopters. A number of these systems are also in service throughout the world.

Contractor
W Vinten Ltd.

VERIFIED

Type 401 infrared linescan sensor

Linescan 401 is a high-performance day and night sensor operating in the 8-14 μm waveband. Infrared signals from scanned terrain are recorded on 70 mm film contained within the unit.

The Linescan 401 sensor may be installed within the aircraft structure or in a pod under the fuselage or wing. The sensor is roll-stabilised and may be slewed 30° either side for scanning from the nadir out to the horizon.

Linescan 401 has been specifically designed for low-level day and night high-speed tactical airborne reconnaissance at speeds up to 600 kt and altitudes down to 200 ft.

Specifications
Dimensions:
(linescan) 604 × 320 × 280 mm
(roll swept diameter) 366 mm
(cooling pack) 390 × 230 × 120 mm

Weight:
(linescan) 34 kg
(cooling pack) 9 kg
Power supply: 200 V AC, 400 Hz, 3 phase, 600 W
28 V DC, 2 A
Resolution: >1 mrad
Sensitivity: >0.2° NET
V/H range: 0.025-5 rad/s
Field of view: 120° (30° left/right offset)
Reliability: >200 h MTBF

Operational status
In production and service. Seven countries have Linescan 401s in service and it is fitted to the Royal Air Force's Jaguar reconnaissance aircraft.

Contractor
W Vinten Ltd.

VERIFIED

Type 690 (126 mm) film format framing camera

The Type 690 (126 mm) film format framing camera entered production in 1976 and has been in continuous production with many enhancements since then. It is designed for day low-, medium- and high-altitude reconnaissance in the tactical standoff and LOng Range Oblique Photography (LOROP) roles and is fitted internally in fixed-wing aircraft and helicopters and in podded systems.

The camera was first introduced with 18 in (457 mm) and 36 in (914 mm) lenses, to be followed by 75 mm and 150 mm lenses in 1983. A high-resolution 900 mm lens replaced the 36 in lens in 1984 and a 300 mm lens was introduced in 1992. A 450 mm lens was added to the range in 1994. Both the 900 mm and 450 mm lenses are temperature and pressure controlled with a focus-for-range capability.

Type 690 cameras are most frequently fitted in pods, while others are fitted in fixed installations internally on the aircraft. In podded applications, the camera is fitted into a rotating nosecone assembly imaging through a mirror-box, enabling it to be rotated from vertical to port or starboard. On take-off and landing the camera window is rotated to point upwards to reduce contamination. The camera and rotating nose mechanism are controlled automatically through a dedicated sensor interface unit that forms part of the reconnaissance management system.

Some cameras are operated in helicopters on special anti-vibration mounts.

Operational status
In production. The Types 690, 690A and F144 are in service. A new ultra-high resolution 900 mm lens will shortly enter production.

Contractor
W Vinten Ltd.

VERIFIED

Type 950/955 cameras

The Type 950 and 955 panoramic cameras are designed to meet operational requirements for low-, medium- and high-altitude reconnaissance. The cameras can be fitted on a variety of airborne platforms including aircraft and UMA. Both the Type 950 and Type 955 are of compact design with a high degree of commonality which enables the cameras to be installed as role change units. The design enables the cameras to be fitted in small diameter reconnaissance pods in keeping with the dimensions of other sensors including IRLS, FLIR and EO sensors.

Both camera types are digitally controlled for operations on aircraft equipped with a MIL-STD-1553 databus or with conventional aircraft avionics systems. The camera control systems allow in-flight selectable across-track coverage angles from wide to narrow field of view.

The lenses have been designed specifically to meet the resolution requirements for a range of operational requirements at various altitudes. The lenses are very high resolution and are prefocused for optimum performance over the specified range of pressure, temperature and altitude without the need to refocus the lens whilst airborne. The cameras record on 5 in wide standard or thin-based film.

The Type 950 camera features a high-resolution 150 mm lens for low- and medium-altitude reconnaissance. The Type 955 camera features a 300 mm lens for medium- and high-altitude reconnaissance.

Specifications
Weight:
(Type 950A) 34 kg
(Type 950C) 41 kg
(Type 955B) 50 kg
Power supply: 28 V DC

Contractor
W Vinten Ltd.

VERIFIED

Type 8010 electro-optical sensor

The Type 8010 electro-optical sensor is a form and fit replacement for the F95 and other 70 mm film format framing cameras such as the Types 360, 518 and 544 manufactured by W Vinten Ltd, and can therefore be retrofitted into existing reconnaissance systems as well as integrated into new installations. The primary role for the sensor is day low- and medium-altitude tactical reconnaissance and surveillance. A secondary role is day low- and medium-altitude surveillance to aid various government agencies, including the police, coastguard, drug enforcement and fishery protection agencies.

The Type 8010 sensor may be fitted internally in a wide range of aircraft, RPVs, drones and UMA or in podded systems.

Principal features of the sensor include high spatial resolution, contrast stretch for low light and haze penetration, imagery recorded on S-VHS videotape, capability for onboard real-time display and in-flight data transmission and imagery exploitation on video monitors.

Specifications
Dimensions:
226 × 181 × 264 mm (1.5 and 3 in lenses)
226 × 181 × 258 mm (6 in lens)
Resolution:
(A version): 12μm, 4096 elements
(B version): 8μm, 6144 elements
Weight: 7.8 kg plus lens
Lenses: 152 mm, 76 mm, 38 mm
Field of view: 18.3°, 35.7°, 65.6°

Operational status
In production and in service on a variety of aircraft types.

Contractor
W Vinten Ltd.

UPDATED

Type 8040B electro-optical sensor

The Type 8040B utilises the same technology as the Type 8010, but provides a replacement for the 126 mm film format Type 690 camera.

Type 8010 E-O sensor compatible with 152 mm lens ***1997**/0001243*

Type 8040B E-O sensor ***1997**/0001244*

The Type 8040B sensor has an 8 μm 12,288 element charged coupled device mounted in the focal plane. The sensor electronics record the imagery onto either a digital or analogue output.

The pixel size within the focal plane array gives 'photographic quality' resolution in the imagery. When associated with the advanced technology 450 mm lens, which offers a corresponding high Modulation Transfer Function, the 8040B can offer long range resolution previously only associated with longer focal length systems.

The Type 8040B E-O sensor is designed to operate between 200 and 40,000 ft at slant ranges of 300 m to 40 km. Although the spatial resolution is comparable with that of film from the Type 690 film camera, the Type 8040B provides a significant advantage over film against low contrast targets, especially when imaged through a hazy atmosphere.

Type 8040B E-O sensor image **1997**/0001246

Specifications

Dimensions: depending on podded or internal installation
Weight: depending on podded or internal installation
Lenses: 450 mm
Fields of view: 12.4°, 6.2°

Contractor

W Vinten Ltd.

VERIFIED

IRLS 4000 infrared linescan sensor

The IRLS 4000 infrared linescan sensor has been designed to meet the exacting requirements for day and night reconnaissance by high-speed aircraft and may be installed internally in aircraft, RPVs, drones or UMA or in podded installations. The sensor is fitted on the Royal Air Force Tornado GR. Mk 1A reconnaissance aircraft and has been proved in combat. The IRLS 4000 is a high-performance day and night airborne sensor operating in the 8 to 14 μm waveband with horizon to horizon across-track coverage. The system comprises a scanning head and an electronics unit. The linescan is cooled by a continuous rated split-Stirling closed-cycle cooling engine.

Imagery is recorded on an S-VHS airborne video cassette recorder and displayed on an Imagery Display Processor (IDP). The IDP may be located on the aircraft, providing onboard real-time display, or on the ground for post-flight exploitation. The imagery may also be transmitted to the ground when suitable datalinking equipment is fitted to the aircraft.

Specifications

Dimensions:
(sensor) 309 × 254 × 247 mm
(electronics unit) 302 × 337 × 202 mm
Weight:
(sensor) 10.5 kg
(electronics unit) 12.5 kg
Power supply: 200 V AC, 400 Hz, 3 phase, 280 W
28 V DC, 115 W
Resolution: <1 mrad
Sensitivity: <0.2° NET
Field of view: 190° scanned, 180° displayed
Stabilisation: ±30° (electronically roll stabilised)
Outputs: Video, line sync, sample clock, BITE, system ready

Operational status

In service with the Royal Air Force Tornado GR. Mk1A and Royal Saudi Air Force Tornado IDS. Other export sales have also been achieved.

Contractor

W Vinten Ltd.

UPDATED

Tornado InfraRed Reconnaissance System (TIRRS)

The TIRRS for the GR1A incorporates two Side-Looking InfraRed (SLIR) sensors and an IRLS 4000 InfraRed Linescan manufactured by W Vinten Ltd. The SLIR sensors embody elements of the UK TICM SPRITE detector programme.

TIRRS is the first system of its type to enter service on a tactical reconnaissance aircraft anywhere in the world and was first employed operationally during the 1990-91 Gulf War. The system is mounted internally in the fuselage and provides horizon to horizon across-track coverage with roll stabilisation and gives a real-time display in the cockpit.

The output from the sensors is recorded on videotape and the operator in the rear cockpit can monitor the imagery while the sortie is under way, directly from the sensors or by replaying from the videotape recorders. The system offers a high-definition thermal picture which can be magnified or enhanced in flight.

Operational status

In service in Royal Air Force GR. Mk1A and Royal Saudi Air Force Tornado IDS reconnaissance aircraft.

Contractor

W Vinten Ltd.

UPDATED

Type 4000 IRLS sensor system including cassette recorder **1997**/0001245

VICON 18 Series 601 reconnaissance pod

The VICON 18 Series 601 pod is one of the wide range of VICON 18 reconnaissance pods designed for a variety of operational roles. Specifically, the Series 601 provides day time reconnaissance from a low-cost, lightweight pod for use at low, medium and high altitudes. Implicit in the pod design is the ability to integrate the pod with a variety of airframes. The pod is currently fully flight-cleared on the British Aerospace Tornado, Jaguar, Harrier and Hawk aircraft. VICON 18 Series pods are also operational on F5, MiG, Learjet and many other aircraft types.

The VICON 18 Series 601 pod contains two sensors: Type 690 (F144) 126 mm film format framing camera with a 450 mm focal length high-resolution lens which is mounted in a rotatable nose cone;

Type 900B 70 mm film format panoramic camera of 76 mm focal length which is mounted in the rear centre section.

The sensors are driven by a VICON 2000 Reconnaissance Management System under the control of the pilot or systems operator. The VICON 2000 System can be interfaced to a wide range of aircraft avionics systems including the MIL-STD-1533B databus. Navigational data can be imprinted on the imagery of both sensors.

The pod provides capabilities for the following operational roles: tactical stand-off photography; long-range oblique photography; low-level tactical reconnaissance; area coverage.

Other developments of the Vicon 18 Series reconnaissance pod include: the VICON 18-601 GP (1) fitted to RAF Jaguar aircraft; the VICON 18-601 Jaguar Replacement Reconnaissance Pod (JRRP), also known as the EO (GP) 1 or GP (1) EO; it carries a Type 8040B EO sensor in place of the F144 camera and records digital data directly. The JRRP also uses the Vigil IRLS in place of the earlier panoramic camera; the JRRP is to be fitted to RAF Jaguar aircraft as part of the continuous upgrade programme for this aircraft. A new Ground Imagery Exploitation System (GIES) will be required to support the upgraded pod both for imagery exploitation and for the mission planning activities needed to support automatic sensor operation.

Specifications

Dimensions:
(overall length) 2,250 mm
(standard diameter) 457.2 mm
(depth over saddle) 508 mm
(weight (estimated)) 254 kg
(centre of gravity) midway between the 14 in or 30 in attachments
Performance:
(sea level): up to 730 kt (M1.1) at 36,000 ft (11,000 m): up to 1,033 kt (M1.8)
Symmetrical normal (accelerations): +ve 9.0 *g*
−ve 2.0 *g*
Asymmetrical normal (acceleration): +ve 4.0 *g*
−ve 2.0 *g*
Rate of roll: 150°/s

Operational status

In service and in production.

Contractor

W Vinten Ltd.

UPDATED

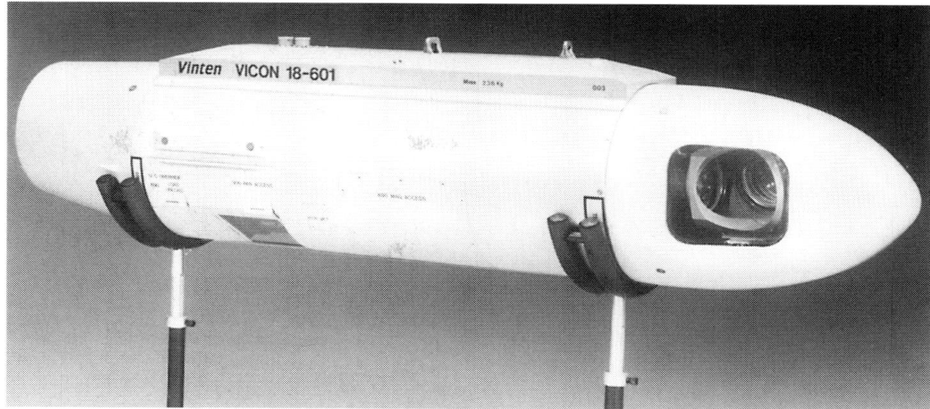

VICON 18 Series 601 Reconnaissance Pod ***1997***/0001247

Cessna Citation fitted with VICON 70 pod ***1997***/0001248

VICON 70 on Jaguar ***1997***/0001249

VICON 70 pod ***1997***/0002223

VICON 70 general purpose modular pod

Since the introduction of the VICON 70 Series of general purpose modular pods in the early 1980's over 50 different variants have been designed or built to meet a wide range of reconnaissance, surveillance, night illumination, Forward Looking InfraRed (FLIR) and Instrumentation requirements.

Variants have been supplied for operations on helicopters, light aircraft, civil aircraft and a range of fighter bombers flying at speeds up to M2.2.

Most variants have been designed to meet Vinten requirements but many have been built for other Original Equipment Manufacturers (OEM) particularly for research and development purposes.

Reconnaissance and Surveillance

The VICON 70 pod will accept a wide range of reconnaissance and surveillance sensors including film framing and panoramic cameras recording on 70 mm film, Infrared Linescan (IRLS) and Electro Optical (EO) sensors recording on either 70 mm film or on video tape and night illumination systems operating with either white light or IR flash.

Systems are designed to meet customers' operational requirements including the integration of CCD TV video sighting devices with on-board real-time display.

Combinations of day and night sensors have been supplied in VICON 70 pods from low V/H applications on helicopters to V/H 5 on fighter-bomber aircraft.

FLIR

A number of different FLIRs including GEC TICM II, HUGHES, and IR-18 have been integrated into VICON 70 pods for various OEMs and customers for a number of applications including navigation and attack.

Vinten-supplied FLIR pods are designed for reconnaissance and surveillance operations and are fitted with the GEC TICM II FLIR sensor, the associated processing electronics, and detector cooling system.

The imagery is recorded on video tape and in some applications the imagery is displayed in real time on board the aircraft on a video monitor.

Instrumentation pods

Variants of the VICON 70 pod have been designed to accept Photo-Sonics high-speed instrumentation cameras for recording release of weapons and other stores from combat aircraft. The pods will accept the 16 mm IVN or 16 mm IPL cameras operated from either aircraft or battery power.

Specifications

Performance: Variants cleared up to M 2.2
Attachment lugs:
Twin suspension. NATO standard 14 in or to customer specification
Connectors: Location and type to customer specification
Standard diameter: 355 mm (14 in)
Length: from 1447 mm (57 in) minimum up to 2,300 mm (100 in) maximum approx.

Operational status

In service and in production.

Contractor

W Vinten Ltd.

UPDATED

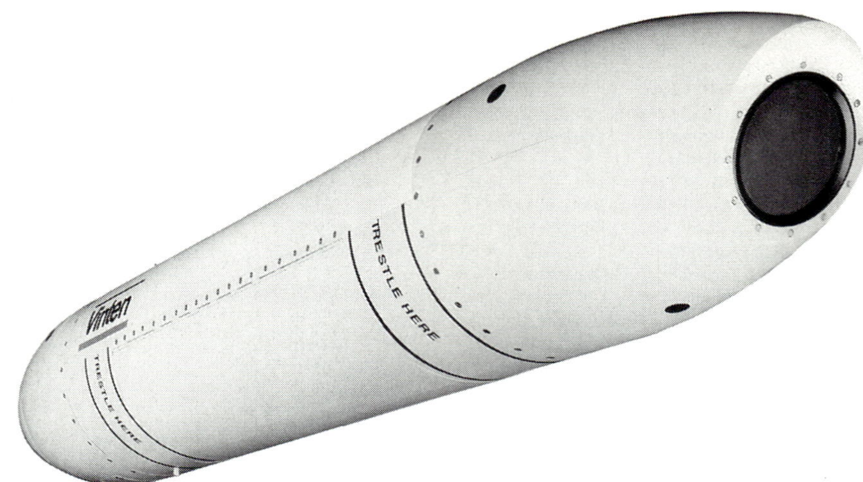

VICON 70 pod **1997**/0002224

VIGIL IRLS flown at 1,000 ft AGL-day **1997**/0001242

VIGIL IRLS flown at 1,000 ft AGL-night **1997**/0001241

VICON 2000 digital reconnaissance management system

The VICON 2000 modular digital reconnaissance management system entered service in 1986. It is now in service on a wide variety of combat aircraft types, interfacing with 64 K NAVHARS and MIL-STD-1553B databus and ARINC 407A, 429/10, 561 and 571. Some podded reconnaissance systems operate on multiple aircraft types, with minimal changes required to move the system from one aircraft type to another.

The VICON 2000 system is designed for use with all types of reconnaissance sensors fitted either internally or in podded systems. The system normally comprises a Navigational Interface Unit, Systems Management Unit and Sensor Interface Units (SIU). New sensors incorporate the SIU, reducing the number of LRUs in the system. The VICON 2000 may be controlled by a conventional cockpit control unit or via the aircraft computer system.

Components of the VICON 2000 are linked by RS-422A serial datalinks over two twisted pairs.

Operational status

In service in a wide variety of combat aircraft types.

Contractor

W Vinten Ltd.

VERIFIED

VIGIL InfraRed Linescan Sensor (IRLS)

The VIGIL IRLS is a single LRU sensor which incorporates the electronics and the control interface which may be either single discrete signals or RS-422 serial command link for interfacing to advanced avionic databusses.

VIGIL is designed for installation in RPVs, drones, UMA, helicopters and medium-performance fixed-wing aircraft up to a velocity/height ratio of 2.5.

It is a high-performance day and night airborne sensor with horizon to horizon coverage operating in the 8-14 μm waveband. The linescanner detector is a single element SPRITE CMT which is cooled by a continuous-rated Stirling closed-cycle cooling engine.

The infrared radiation from the terrain is scanned by the sensor, producing continuous along-track imagery. VIGIL produces an analogue signal containing both line sync and video. This signal is suitable for direct interfacing with either video cassette recorder for airborne or ground recording, datalink transmitter for transmission to a ground station or an imagery display processor for scan conversion and image manipulation for display on a conventional video monitor.

Specifications
Dimensions: 309 × 254 × 247 mm
Weight: 10.5 kg
Stabilisation: Mode + 30° roll (electronically roll stabilised)
Mode 2 locked to airframe
Environmental: MIL-STD-810 and DEF STAN 07-55

Operational status
In full production and operational service.

Contractor
W Vinten Ltd.

UPDATED

VIGIL IRLS including recorder
1997/0001240

UNITED STATES OF AMERICA

Airborne cameras

Ball's specialised airborne cameras give pilots a variety of exterior views, enhancing safety and security.

Ball designed the lightweight, charge-coupled device camera, both colour and monochrome, specifically for the flight environment. An electronic sensitivity control ensures clear images in both bright sunlight and low-light flight conditions. The environmentally sealed, self-contained camera is engineered to withstand extreme temperatures, shock and vibration.

The camera operates with electronic shutter control, improved low-light sensitivity and existing aircraft lighting to give flight crews optimal vision, day or night.

Camera features include: an environmentally sealed aluminum housing, adaptable for a variety of applications; lens sizes from 2.8 to 25 mm; electronic shutter dynamic range greater than 100,000 to 1; sapphire with ITO window coating for reliable de-icing/defogging; electronic shutter control for automatic low- to high-light level operation.

Operational status
Selected by Boeing Commercial Airplane Group for a 10-year contract to outfit its entire fleet of 777-300s with airborne cameras, as a ground manoeuvring system.

Contractor
Ball Aerospace and Technologies Corp.

VERIFIED

Day and night ruggedised CCD cameras

In developing a complete line of high-resolution, day and night, ruggedised CCD cameras, Ball altered its successful line of military programme Mil-Spec cameras, focusing on design flexibility to accommodate each user's unique requirements. Several standard package configurations are available, and custom packaging allows the cameras to fit into existing enclosures. Each camera comes in either the 'L' or 'square' housing.

The large format CCD provides high resolution and sensitivity, and it allows the cameras to be used as direct replacements for 1 in vidicon-tube cameras.

A number of options are available, including: 18 or 25 mm Gen-III image intensifiers to increase night camera resolution; modified intensifier photocathodes to vary the spectral reponse; high-resolution image intensifiers with resolution greater than 45 lp/mm; custom interface electronics; timing in accordance with STANAG-3350A video format; video conditioning (log amplifier, auto black); and full military qualification.
Cameras are available in three standard line rates:
525-line rate, 60 Hz, RS-170A;
625-line rate, 50 Hz, CCIR;
875-line rate, 60 Hz, RS-343A.

Contractor
Ball Aerospace and Technologies Corp.

VERIFIED

Intensified high-definition television camera

Ball Telecommunication Products Division has developed an intensified, high-definition television camera.

The camera uses a frame transfer, charge-coupled device paired with a high-resolution, 25 mm, Gen-III intensifier to obtain the absolute best in sensitivity and resolution.

The combination of an advanced automatic light control, based on intensifier gating, and an enhanced automatic gain control, ensures operation from starlight to full sunlight without the need for a mechanical iris or display brightness adjustments. The camera's 16:9 (H × V) aspect ratio, together with its outstanding hands-off performance, makes it an ideal sensor for helicopter operations including wide area, all-light-level surveillance or low-light-level microscopy.

Contractor
Ball Aerospace and Technologies Corp.

VERIFIED

Airborne Laser (ABL)

In November 1996, the US Air Force awarded Boeing Defense & Space Group a $1.1 billion Program Definition and Risk Reduction (PDRR) contract to develop the ABL system, also known as the YAL-1A Attack Laser. The Boeing team includes TRW Space & Electronics, Redondo Beach and Lockheed Martin Missiles & Space, Sunnyvale.

The ABL is a theatre ballistic missile defence system that will consist of a laser and a beam control apparatus fitted on a 747-400F freighter aircraft.

Under the terms of the PDRR contract, Boeing's ABL team over the next five years will demonstrate that relevant ABL technologies can be integrated onto an airborne platform to shoot down theatre ballistic missiles at ranges measured in hundreds of kilometres.

Following successful completion of the PDRR phase the US Air Force plans for an engineering and manufacturing development contract for a fully operational system, leading to a fleet of seven ABL aircraft.

Four adjunct missions will be studied during the PDRR: self-protection; protection of other high value assets; the role of the ABL in cruise missile defence; the role ABL can play in airborne surveillance.

Specifications
Laser: Chemical Oxygen Iodine Laser (COIL)
Laser power: Multimegawatt

Operational status
PDRR Phase I to 2001
PDRR Phase II to 2002, including lethal intercept demonstration
EMD to 2005
Production 2005 to 2008 for seven aircraft.

Contractor
Boeing Information and Communication Systems.

UPDATED

Airborne Surveillance Testbed (AST)

The AST project is a technology demonstration programme that supports development and evaluation of defensive systems to counter InterContinental and Theatre Ballistic Missiles (ICBMs and TBMs) and their warheads.

As an airborne platform, the AST also supports risk reduction testing and the evaluation of developing technologies.

Formerly known as the Airborne Optical Adjunct (AOA), AST is a key element of the US Department of Defense's Ballistic Missile Defense Organisation (BMDO).

Initial research was aimed at evaluating whether an airborne infrared (IR) sensor could reliably provide early warning using detection, tracking and target

Boeing Airborne Surveillance Testbed ***1996***

discrimination methods, as well as provide ICBM tracking information to ground radar. The programme was later renamed AST, and the original support mission was changed to data-gathering for ballistic missile defence development and resolving issues associated with target characteristics.

Under contract with the US Army's Space & Strategic Defense Command, Boeing has modified a 767 commercial jet aircraft to carry a large multicolour IR sensor housed in a cupola atop the 767's fuselage.

IR sensors detect the comparative heat of objects. The AST sensor, comprising more than 38,000 detector elements, is sensitive enough to detect the heat of a human body at a distance of more than 1,000 miles against the cold background of space. The AST sensor has demonstrated the capability to detect, track and discriminate warheads from missile components, debris and decoys, both from the ambient temperature of an object prior to launch and from the heat generated during re-entry into Earth's atmosphere.

Boeing was awarded the original AOA contract in July 1984. As the integration contractor, Boeing's responsibilities included procuring the data processor, as well as all recording, communications and support equipment. Hughes Aircraft Electro-Optical Data Systems Group designed, built and maintains the IR sensor.

The AST and its mission system equipment were initially used to gather data that confirmed the system's ability to acquire, track, discriminate warheads from decoys, and provide track information to ground units. Originally, the emphasis was on ICBMs. At the time of the 1990-91 Gulf War national emphasis shifted to defending military installations and troops from TBMs. To support the Army's studies on TMD, the AST system was used to gather data applicable to TBMs, as well as ICBMs.

AST has successfully participated in 55 missions, including eight operational exercises with real-time tactical communication links to Army, Navy and Air Force elements to demonstrate the utility of an airborne IR platform in a TMD role.

Information acquired during tests at national missile ranges is added to a database that is used in computer-simulation models, and to confirm the system's capability to discriminate between objects. Data acquired during flights has validated onboard software, computers, and communication equipment necessary for real-time processing of tracking data and transmission of that data to ground-based defensive units.

The AST frequently operates at the Western Test Range, off the coast of California, and observes missiles launched from: Vandenburg Air Force Base, California; the US Army's Kwajalein Missile Range in the central Pacific Ocean; Pacific Missile Range Facility, Hawaii; White Sands Missile Range, New Mexico; Eastern Test Range in Florida; and Wallops Flight Facility, Virginia.

During a typical strategic test mission, an ICBM launched from Vandenberg Air Force Base, for example, sends one or more ballistic missile re-entry vehicles into the Kwajalein Missile Range. Depending on the objectives assigned to AST, the aircraft will be positioned 150 to 300 miles to the side of the missile's trajectory or down range of the impact point to acquire data during boost, exoatmospheric flight, or re-entry.

For short-range TBMs, AST can observe the entire trajectory. The data acquired by the sensor is processed in real-time to generate a track that may be transmitted to the ground after each target-sighting update – about every 1.5 seconds. The computers on board AST extrapolate the track forward to predict an impact point, and can backtrack to estimate the launch point.

All sensor and tracking data generated by the computers is recorded on board AST for post-test analysis and system evaluation, which takes place at a laboratory in Kent, Washington. This information is also provided to the US Army's data facility in Huntsville, Alabama. It is available to all Department of Defense services and defence contractors engaged in ICBM and TBM defence projects to enhance their knowledge of missile threats and warheads.

The AST aircraft is based in Seattle, Washington, and can deploy to any national test range within one day. It carries a flight crew of 15 for typical missions, with room for observers from various government agencies and contractors. Missions last approximately six to eight hours, and the aircraft usually flies at an altitude above 42,000 feet, somewhat higher than commercial aircraft.

The technology developed for AST is applicable to many sectors of the defence community. AST personnel currently are supporting the US Navy Theater Wide Captive Carry (NTWCC) Standard Missile (SM)-3 and Airborne Laser (ABL) programmes.

For the NTWCC SM-3 programme, Boeing is working to integrate the last sensors onto the aircraft for risk reduction system tests.

Operational status

Successfully completed 55 data collection missions. Present development is funded by a 3-year, 10-month follow-on contract dated September 1995 at a value of $19.5 million a year. Studies are underway to assess the benefits of installing multiple sensors and experiments on the AST aircraft.

Contractor

Boeing Information & Communications Systems.

UPDATED

Gated Laser Illuminator for Night Television (GLINT)

The Gated Laser Illuminator for Night Television (GLINT) employs the same laser diode technology as the LTD/R (see next item) to produce a spotlight for night operations. GLINT is synchronised with a low-light-level camera to provide the AC-130U Spectre operators with clear pictures on a pitch black night and sends a virtually invisible beam of laser light to illuminate ground scenes.

Operational status

In production for the AC-130U Spectre all-light-level television system.

Contractor

The Boeing Company.

UPDATED

Laser Target Designator/ Rangefinder (LTD/R)

The Laser Target Designator/Rangefinder (LTD/R) provides range data for gunfire accuracy and gives a laser spot for smart weapons to use as a beacon. The laser is a diode-pumped solid-state laser which uses air-cooling and high-efficiency laser diodes. The LTD/R is smaller in size and weight and is more reliable and efficient than conventional flashlamp-pumped systems.

Operational status

In production for the US Air Force AC-130U Spectre all-light-level television system.

Contractor

The Boeing Company.

UPDATED

IR/UV maritime surveillance scanner

The AADS1221 IR/UV is a sensor subsystem of the Swedish Space Corporation Maritime Surveillance System. It is specifically designed for use in maritime surveillance applications.

The IR/UV scanner, operating in the 8.5 to 12.5 μm range for IR and the 0.32 to 0.38 μm range for UV, is used at low altitude to obtain high-resolution imagery of oil spills. IR data can be obtained by both day and night, providing information on the spreading of oil and also indicating the relative oil thickness within the oil slick. Usually 90 per cent of the oil is concentrated within less than 10 per cent of the visual slick. By using IR information, clean-up operations can be directed for maximum efficiency.

UV data is obtained during daylight and maps the entire extent of the slick, irrespective of thickness. The UV data adds confidence to the IR registration by distinguishing natural thermal phenomena, such as cold upwelling water, from suspected oil pollution.

In the Maritime Surveillance System, the IR/UV video outputs are presented in real time on a split screen format colour TV monitor, with IR on the left and UV on the right, with false colour coding for image enhancement.

Specifications

Dimensions:
(scan head) 380 × 380 × 380 mm
(operator console)180 × 490 × 340 mm
Weight:
(scan head) 17 kg
(operator console) 12 kg
Power supply: 28 V DC, 15 A
Field of view: 5 mrad instantaneous, 87° total
Scan rate: 160 scans/s

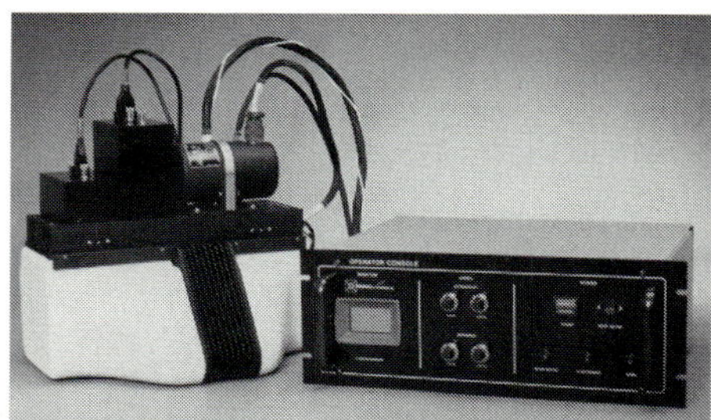

The IR/UV maritime surveillance scanner consists of the scan head (left) and an operator's console (right) ***1995***

Contractor

Daedalus Enterprises Inc.

VERIFIED

SAFIRE thermal imaging system AN/AAQ-22 **1998**/0015295

Series 2000 thermal imaging system AN/AAQ-21 **1998**/0015292

SAFIRE thermal imaging system AN/AAQ-22

Military-qualified as the AN/AAQ-22 thermal imaging system, the SAFIRE™ is a digital, high-resolution 8 to 12 micron wavelength system with three-axis gyrostabilisation and full 360° turret rotation. SAFIRE's wide field of view (28°) is designed for navigation and area searches. Its narrow field of view (5°) is supplemented by electronic zoom and freeze-frame features. The rugged turret has full performance at all pointing angles, including straight down. The system includes a hand-held controller and flight-hardened video display that supports recording for evidence and analysis. Options include compatibility with night vision goggles, autotracker, laser illuminator, laser rangefinder, searchlight slaving, and radar and navigation interfaces. The hermetically sealed SAFIRE turret is flight qualified for airspeeds up to 405 kt. SAFIRE sytems are certified on more than 20 fixed-wing and rotary aircraft.

Operational status

In service worldwide in 50 nations. Fielded by all branches of the US military. Other customers include: the Royal Danish Navy, the Japanese Maritime Safety Agency, the Royal Netherlands Navy, and the Royal Saudi Naval Forces.

Contractor

FLIR Systems Inc.

NEW ENTRY

Series 2000 thermal imaging system AN/AAQ-21

Military-qualified as the AN/AAQ-21 thermal imaging system, the Series 2000™ is used widely for surveillance, narcotics interdiction, police pursuit, search and rescue, and environmental patrols. The Series 2000 is a rate-stabilised, high-resolution analogue system that operates in the 8 to 12 micron wavelength range. The gimballed turret provides full hemispheric sensor pointing. A hand-held controller directs the imager. Thermal images are displayed on a flight-hardened display unit. Images can also be recorded to a VCR for evidence or analysis. Wide field of view is 28°; narrow field of view is 5°, and is available in either ×7.5 or ×10.5 magnification. Other options include a CCD colour camera and laser pointing system.

Operational status

In service worldwide. Customers include: law enforcement agencies throughout the United States, coastal patrols in Italy, and the US Air Force for use on UH-1N helicopters.

Contractor

FLIR Systems Inc.

UPDATED

Star SAFIRE airborne thermal imaging system AN/AAQ-22

Military-qualified as the AN/AAQ-22 thermal imaging system, the Star SAFIRE™ is a digital, high-resolution system enhanced for imaging over water and in humid atmospheric conditions. Star SAFIRE is a military-qualified COTS airborne infrared system equipped with a third-generation 3 to 5 micron indium antimonide (InSb) focal plane array detector (320 × 240). It offers three fields of view (1.4° narrow, 5.6° medium and 30° wide) for extended detection range capability and for surveillance, SAR, and flight safety. In addition to the thermal imager, Star SAFIRE has slots for three optional payloads (including CCD colour camera with ×2 extender, eye-safe laser rangefinder, or laser illuminator). Star SAFIRE is a three-axis gyrostabilised system with full 360° turret rotation, including straight-down. The system's footprint and quick-release mounting match that of the SAFIRE to give fleet users platform flexibility. The system includes a hand-held controller and flight-hardened video display that supports recording for evidence and analysis. System options include navigation and radar interfaces, autotracker, video downlinks, and video recorders. The hermitically sealed Star SAFIRE turret is flight qualified for airspeeds up to 405 kt and is designed for the marine environment.

Operational status

Star SAFIRE systems are certified on more than 20 fixed-wing and rotary aircraft.

Contractor

FLIR Systems Inc.

NEW ENTRY

UltraMedia 5-axis gyrostabilised camera system

The UltraMedia™ aerial camera system is designed to provide the highest quality long-distance surveillance capability for policing, Electronic News Gathering (ENG) and long-range sports coverage. UltraMedia systems have captured images from 'standoff' distances of more than two miles. The system's small size and weight make it ideal for nose mounting on many helicopters, including new fourth-generation models. The system uses a colour CCD camera and a continuous zoom lens with a ×2 extender for up to ×72 magnification. Intuitive controls can be adjusted to suit a range of user preferences. Five-axis gyrostabilisation ensures steady images, even in tightly-banked turns. The gyros cannot be unseated by extreme manoeuvres and require no daily balancing or set up. The system's

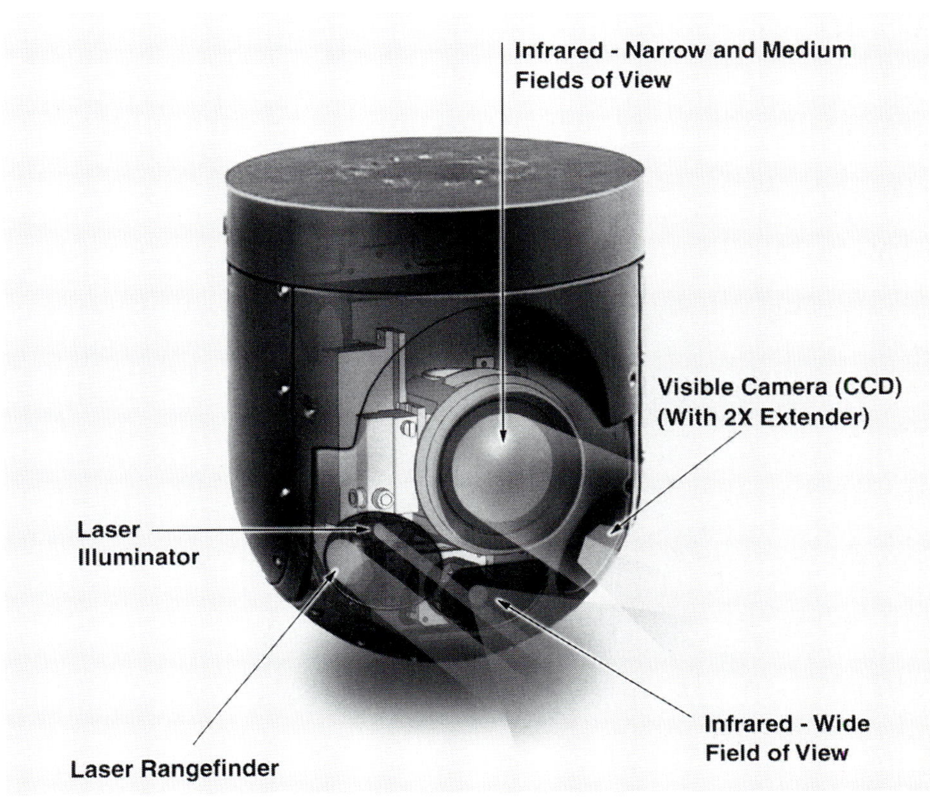

Star SAPHIRE airborne thermal imaging system AN/AAQ-22 **1998**/0018313

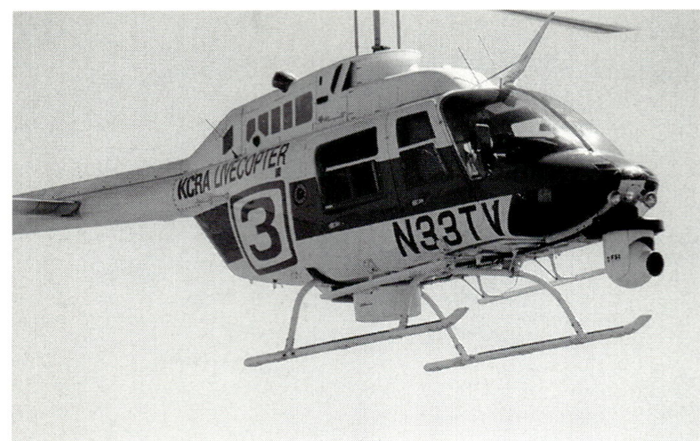

UltraMedia 5-axis gryostabilised camera system **1998**/0015297

ULTRA 4000 dual-camera 24 hour imaging system **1998**/0015293

low profile (diameter: 381 mm; height: 444.5 mm) minimises drag to improve fuel efficiency and increase flight time. Complete with camera and lens, the gimbal weighs 35.9 kg. Video images can be recorded to an onboard VCR for evidence gathering.

Operational status

Customers include: more than 50 television stations in the United States representing all four major networks (CBS, NBC, ABC, Fox) and sports and news coverage in Australia, Brazil, Japan and New Zealand. More than 50 UltraMedia systems are in use around the world.

Contractor

FLIR Systems Inc.

UPDATED

UltraMedia-RS 5-axis gyrostabilised compact system

Weighing only 15.9 kg, complete with camera and lens, the new UltraMedia-RS™ is designed for daylight surveillance and Electronic News Gathering (ENG) from small helicopters and light aircraft. The system uses a three-chip CCD colour camera and continuous zoom lens that provides up to ×40 magnification. The system's five-axis gyrostabilisation and advanced optics make it possible to capture details of unfolding events, even at considerable distance from the scene. The gimbal features 360° rotation and full look-down capability. The camera is operated by a laptop panel in the cockpit. A flight-hardened monitor provides instantaneous visual confirmation. The system's digital signal can be recorded to VCR for evidence or later analysis. An optional microwave downlink sends the signal live to the ground for real-time viewing or broadcast. The gimbal's small profile (diameter: 279.4 mm; height: 360.7 mm) minimises drag during flight. The UltraMedia-RS can be nose-mounted on many small aircraft.

Operational status

The UltraMedia-RS is used for news coverage by several television stations in the United States and for law enforcement surveillance.

Contractor

FLIR Systems Inc.

UPDATED

ULTRA 4000 dual-camera 24 hour imaging system

The ULTRA 4000™ combines a three-chip CCD colour camera for daylight surveillance and pursuit with the military-qualified SAFIRE thermal imaging system, for a complete, high-resolution package that is appropriate for policing and Electronic News Gathering (ENG), regardless of light levels. The system is gyrostabilised on three axes and offers 360° turret rotation, including full look-down capability. The SAFIRE is an 8 to 12 micron thermal imager with a 28° wide field of view for navigation and area searches. Its 5° narrow field of view is supplemented by electronic zoom and freeze-frame features. The colour camera has a 16:1 continuous zoom lens with a ×2 extender to capture close-up details. One hand-held controller operates both the visible light camera and the thermal imaging system. Video images are viewed on a flight-hardened display. Images can be fed live to ground units or a television station via an optional microwave downlink, or video can be recorded to an optional VCR. Systems can be configured for compatibility with night-vision goggles. Other options include autotracker, searchlight slaving, and radar and navigation interfaces.

Operational status

In service worldwide. Customers include: The Sussex Police, the Devon and Cornwall Police, the Central Counties Police (UK); the Indian Dept of Revenue through Pawan Hans Helicopters (India); and several television stations in the United States.

Contractor

FLIR Systems Inc.

UPDATED

ULTRA 6000 dual-camera 24 hour imaging system

The ULTRA 6000™ is a lightweight, four-axis gyrostabilised system designed to meet the operational demands of law enforcement. The system includes both daylight video and thermal imaging cameras to enhance an air support unit's search, surveillance and apprehension capabilities. The ULTRA 6000 is fitted with an advanced indium antimonide (InSb) 3 to 5 micron thermal imager, for sharp images over a range of operating temperatures and exceptional performance in the cooler temperatures encountered as a result of altitude, flight speed and seasonal weather. The CCD camera uses a ×15 continuous zoom lens for high-resolution colour video. The thermal imager is equipped with four Fields of View (FoV). For maximum imaging flexibility, the two optical FoVs (20° wide and 4° narrow) can be electronically extended for two additional FoVs (10° wide and 2° narrow). Loaded with both cameras, the gimbal (diameter: 279.4 mm; height: 381 mm) weighs less than 20 kg.

Operational status

Development complete.

Contractor

FLIR Systems Inc.

NEW ENTRY

UltraMedia-RS 5-axis gyrostabilised compact system **1998**/0015294

ULTRA 6000 dual-camera 24 hour imaging system **1998**/0015296

MERLIN night vision goggles

The Modular Ejection-Rated Low-profile Imaging for Night (MERLIN) night vision goggles have been developed for high-performance aircraft. The goggles incorporate a lower-cost Generation III tube and have a 35° field of view. Each of the two interchangeable monoculars weighs 0.25 kg and provides 20 mm of eye relief.

The company claims that MERLIN attached to a standard HGU-55 helmet extends the centre of gravity of the pilot's head forwards by only about 3 mm, allowing safe ejection. The identical halves of the goggles are attached to the helmet's visor mountings and a new visor is hung from a beak on the helmet. All optics, apart from the objective lenses, are placed behind the visor, further aiding ejection.

Specifications

Weight: 0.8 kg
Power supply: 3 V DC batteries (one per monocular)
Field of view: 35° circular
Focus range: 5 m to infinity
Magnification: ×1

Operational status

Under development.

Contractor

ITT Defense Electro-Optical Products Division.

VERIFIED

Night Targeting System (NTS)

The Night Targeting System (NTS) is an airborne electro-optical fire-control system designed to provide a night fighting capability for the AH-1 Cobra attack helicopter. It provides the capability to detect, acquire, track, designate and attack tactical targets at night and in limited visibility or adverse weather conditions. The system is operational in the US Marine Corps.

The NTS consists of a modified M-65 telescopic sight unit, a laser designator/rangefinder, a CCD TV sensor and a FLIR sensor to provide autonomous delivery of TOW and Hellfire missiles. A video cassette recorder is also included to provide intelligence gathering and training capability. An automatic target tracker is integrated to improve tracking accuracy for weapon delivery. An upgraded configuration, the NTS-Advanced (NTS-A), eliminates the Optical Relay Tube (ORT), incorporates reliability and maintainability improvements, and integrates a TOW 2A thermal tracker. The NTS-A is currently undergoing operational testing.

Operational status

In production for variants of the Bell AH-1 Cobra helicopter in service with the US Marine Corps, and friendly foreign militaries.

Contractor

Kollsman Inc.

VERIFIED

The Kollsman Night Targeting System Advanced (NTS-A)
1997/0001251

M-927/929 aviator's night vision goggles

The M-927/929 aviator's night vision goggles are based on the ANVIS design. The single point helmet-mounted system permits practically normal peripheral vision. A flip-up feature allows the pilot to pivot the assembly up out of the way whenever he wants to use unaided vision. The weight of the binocular is counterbalanced by a dual battery pack which mounts to the rear of the helmet.

The weight of the M-927/929 is 0.456 kg for the binocular assembly and 0.2 kg for the visor mount assembly, magnification is unity and the field of view is 40°. Brightness gains are 1,200 for the Generation II Plus M-927 and 2,000 for the Generation III M-929.

Operational status

In production to order.

Contractor

Litton Electron Devices.

VERIFIED

Litton M-927/929 aviator's night vision goggles

Dark Star laser designator

The Dark Star laser designator design is compatible with use in a broad range of airborne and ground targeting systems for laser-guided bombs, missiles, and artillery. The two-assembly system has been optimised for volume and weight, and includes transmitter and high energy converter line-replaceable modules. Incorporating a newly-developed high-brightness resonator, this air-cooled equipment provides state-of-the-art beam divergence at high efficiency.

Operational status

In production and in service.

Contractor

Litton Systems Inc, Laser Systems Division.

VERIFIED

Laser rangefinder/target designator for TADS

A Litton Laser Systems laser rangefinder/designator forms part of the Lockheed Martin Target Acquisition Designation Sight (TADS) which is fitted to the US Army's AH-64 helicopters. The Litton system comprises a laser transmitter and an electronics unit.

Operational status

In production and in service.

Contractor

Litton Systems Inc, Laser Systems Division.

VERIFIED

Mast-mounted sight laser rangefinder/designator

Litton Laser Systems produces the laser rangefinder and target designator for the Mast-Mounted Sight (MMS) which equips the US Army's OH-58D helicopters. Use of advanced packaging techniques enables the electronics, high-voltage power supplies, cooling system, range receiver and laser transmitter to be housed in a single LRU. The unit includes an asynchronous digital interface for precise communication with the MMS computer internally. All subassemblies interface with a common bus under control of a main processor which maintains system timing through software control. Built-in test circuitry periodically monitors system operation, allowing the processor to compensate automatically for component degradation.

Operational status
In production and in service. A suitable eye-safe tactical version has been developed. A modified unit has been integrated into the all-light-level television system in the Lockheed Martin AC-130U.

Contractor
Litton Systems Inc, Laser Systems Division.

VERIFIED

AN/AAS-35(V) Pave Penny laser tracker

The AN/AAS-35(V) is a miniaturised day and night laser-based target identification set. Used in conjunction with a laser designation system, either ground-based or in a co-operating aircraft, targets can be recognised and identified rapidly, and accurate steering data provided to ensure quick pilot reaction and accurate delivery of weapons.

A silicon pin diode detector head is used, with full lower forward hemisphere coverage plus some look-up capability. Pilot's controls permit selection of several seeker scanning patterns to improve early designator recognition. It can be used to improve the accuracy of conventional weapon delivery, or to lock up laser-guided munitions. The system is contained in a relatively small pod, which is either fuselage- or pylon-mounted to allow easy harmonisation with other onboard sensors. An aircraft adaptor module is used to integrate the sensor data with onboard processors and a pilot's control panel provides for easy use and built-in test operations.

Specifications
Dimensions: 833 (length) × 200 mm (diameter)
Weight: 14.5 kg
Power supply: 28 V DC, <10 A
Wavelength: 1.06 μm
Scan coverage:
(elevation) −90 to +15°
(azimuth) −90 to +90°
Selectable scan patterns: wide, narrow, depressed, offset
Output: direction cosines of line of sight

Operational status
In service. The first operational Pave Penny system for the A-10 was delivered in March 1977 and the US Air Force has equipped its entire fleet originally numbering 733 aircraft. Up to 380 US Air Force A-7D Corsairs also have Pave Penny installed. Starting in 1995, Pave Penny was also installed on US Air Force F-16 aircraft.

Contractor
Lockheed Martin Electronics & Missiles.

VERIFIED

AN/AAS-42 Infrared Search and Track System (IRST)

The Infrared Search and Track System is designed to permit the multiple tracking of thermal energy emitting targets at extremely long range to augment information supplied by conventional tactical radars. The system enhances performance against low radar cross-section targets while providing immunity to electronic detection and RF countermeasures. High-resolution IRST provides dramatically improved raid cell count at maximum declaration ranges - information that can stand alone or be fused with other sensor data to enhance situational awareness.

The IRSTS consists of a sensor head mounted beneath the nose of the F-14D and an electronics unit just aft of the cockpit. The system is integrated with the F-14's central computer system and complements the AN/APG-71 radar providing the aircrew both target track data and Infrared imagery displays. The AN/AAS-42 operates in 6 discrete modes, with selectable and individually controlled scan volumes in azimuth (±80°) and elevation (±70°). The system is suitable for installation on multiple tactical platforms as either an internal fit or in a self-contained podded configuration.

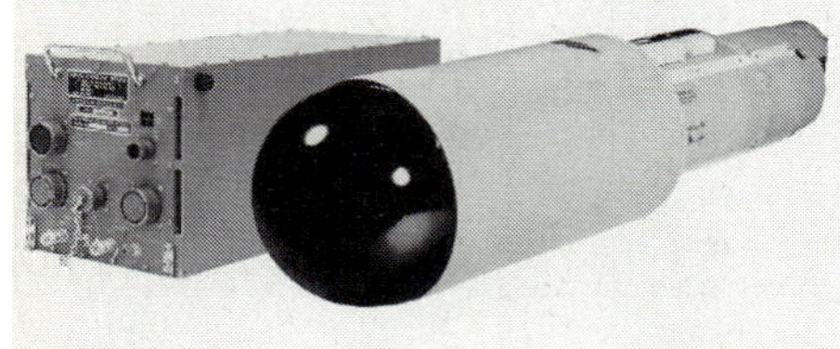

The AN/AAS-42 infrared search and track system for the US Navy F-14D

Specifications
Dimensions:
(sensor head) 914.4 × 228.6 mm diameter
(electronics unit) 190.5 × 190.5 × 482.6 mm
Weight:
(sensor head) 41.28 kg
(electronics unit) 16.78 kg

Operational status
The programme has completed production and integration for the US Navy F-14D aircraft, in which it has been operational since 1994.

Contractor
Lockheed Martin Electronics & Missiles.

UPDATED

AN/ASQ-145 low-light television system

The AN/ASQ-145 is a low-light television system designed to provide fire-control facilities for the US Air Force AC-130 Hercules gunship aircraft, on which it is a primary sensor. The system is unusual in that it uses a dual camera installation in order to provide both wide and narrow fields of view simultaneously.

Operational status
In production and service, currently being updated.

Contractor
Lockheed Martin Electronics & Missiles.

VERIFIED

AN/ASQ-173 Laser Spot Tracker/ Strike CAMera (LST/SCAM)

Developed specifically for the US Navy F/A-18 Hornet, the LST/SCAM permits crews to identify laser designated targets illuminated by either ground forces or co-operative aircraft, and to achieve more accurate tracking and weapon release performance than non-laser equipped aircraft. Target position data from the laser detector tracker is fed directly to the F/A-18 mission computer and used to provide weapon aiming and ordnance release information.

The laser spot-tracker optics are stabilised using attitude data passed to the unit from the aircraft inertial navigator. The detector is a four-quadrant photodiode type, mounted behind a hemispherical dome. Laser pulse decoder electronics are also contained in the detector.

Located in the same pod, for which Lockheed Martin is prime contractor, is a strike camera. This unit is in the aft section and has a wide field of view in the lower hemisphere. It can be slaved to the laser spot-tracker or independently controlled by the mission computer. Photographic data from this unit permits rapid assessment of damage after aircraft attacks.

For strike operations the laser spot-tracker/strike camera on the F/A-18 is mounted on the starboard fuselage side and an infrared target acquisition/tracker pod is carried on the port fuselage side.

Specifications
Dimensions: 2,290 (length) × 200 mm (diameter)
Weight: 73 kg
Wavelength: 1.06 μm
Scan patterns: preprogrammable/pilot selectable

Operational status
Production completed. In service in the US Navy F/A-18.

Contractor
Lockheed Martin Electronics & Missiles.

VERIFIED

EOSS Electro-Optic Sensor System for the RAH-66 Comanche

Lockheed Martin is developing a second-generation Electro-Optic Sensor System (EOSS) that will provide night navigation and targeting capabilities for the US Army RAH-66 Comanche helicopter. The EOSS is designed to permit the Comanche crew to fly safely at extremely low altitudes, even in darkness and poor weather, and detect targets faster and at greater distances than currently fielded EO systems.

The targeting system consists of a second-generation FLIR sensor, high-resolution TV camera and laser rangefinder/designator. The system uses a unique two bar scan technique to search for airborne

RAH-66 Comanche Helicopter

1997/0001253

and ground targets. Target imagery is prioritised according to threat potential and stored in computer memory, which permits the crew to remask while deciding combat tactics.

The night vision system will permit the pilot to manoeuvre safely at nap of the earth altitudes around the clock and in poor weather or limited visibility caused by smoke, dust or haze. The system is coupled to the pilot's line of sight and provides high-resolution day-like imagery at extended ranges using advanced focal plane array infrared technology.

The laser system has been developed by ELOP Electro-Optical Industries Ltd of Israel. It operates at two different wavelengths within a single unit. This enables target designation to be achieved using the appropriate weapon-compatible wavelength, whilst ranging is achieved using an eye-safe wavelength, with obvious benefits in training operations. A diode-pumped laser and wavelength translation technology is used to produce the eye-safe wavelength which minimise size and weight.

Specifications

Weight: 204.1 kg
Power supply: 270 V DC, 3.4 kW
Temperature range: −46 to +52°C
Altitude: up to 9,800 ft
Wavelength:
(solid-state TV) 0.65 to 1 μm
(FLIR) 8-12 μm
(laser) 1.06 μm (ranging/designation)
1.54 μm (ranging only)
Reliability:
(NVPS sensor) 270 h MTBF
(EOTADS sensor) 125 h MTBF

Operational status

In September 1991 Lockheed Martin received a contract from Boeing's Helicopter Division to build five EOSS systems for a demonstration/validation prototype phase. The first EOSS FLIR was completed and began testing in 1995. The first EOSS flight test is scheduled for the year 2000.

Contractor

Lockheed Martin Electronics & Missiles.

UPDATED

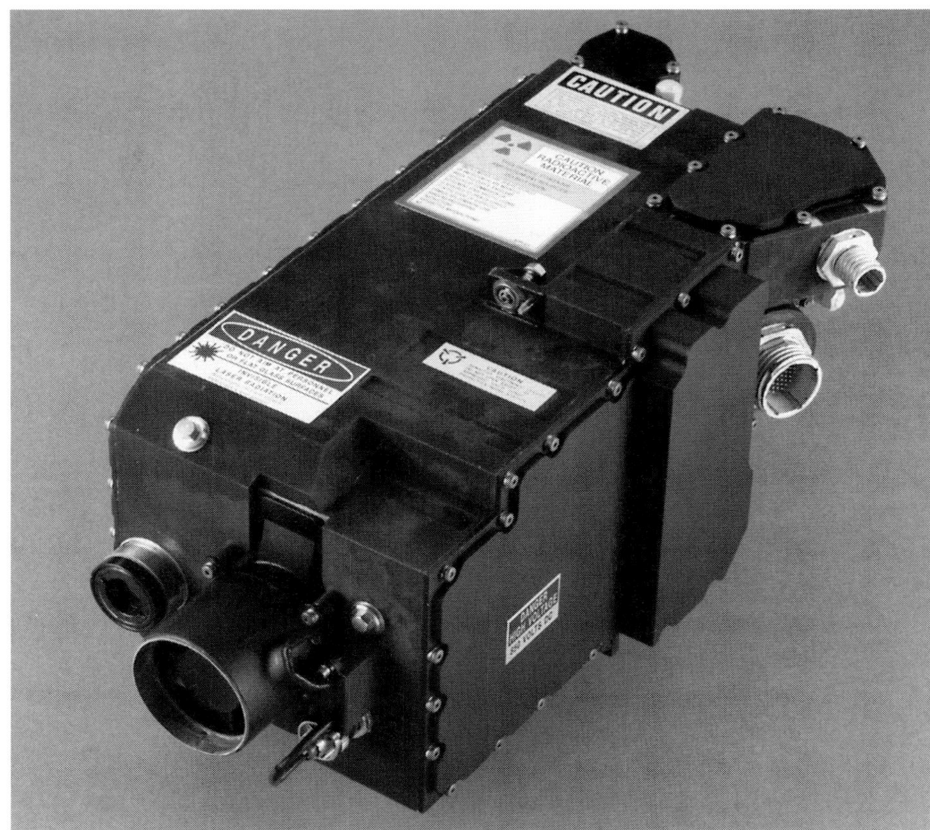

ELOP Laser Designator/Rangefinder element of EOSS ***1997**/0001254*

This Lockheed Martin F-16D carries both navigation and targeting pods of the Lockheed Martin LANTIRN system under its fuselage

LANTIRN system

The Low Altitude Navigation and Targeting Infrared for Night (LANTIRN) system consists of two pods: the AN/AAQ-13 navigation pod and the AN/AAQ-14 targeting pod.

LANTIRN provides the means by which F-15E and F-16C/D aircraft can penetrate hostile airspace at extremely low altitude and high speed, acquire their targets and deliver guided and unguided weapons around the clock. The initial application was the two-seat F-15E fielded by the US Air Force in 1989. LANTIRN also equips the F-16C/D. Compatibility of the LANTIRN system with single-seat navigation and targeting operations has been demonstrated on the F-16. In 1995 the US Navy selected the LANTIRN targeting pod for the F-14 Precision Strike programme.

The system

The LANTIRN system consists of two separate sets of equipment each contained in its own pod, suitable for underwing or underfuselage attachment. Either or both pods can be carried, depending on the particular mission requirement. This option enhances flexibility and diminishes support demands. The equipment within the pods is supplied by a number of manufacturers, but overall responsibility rests with Lockheed Martin as the prime contractor.

The AN/AAQ-13 navigation pod contains a wide field of view (21 × 28°) FLIR unit and a J-band terrain-following radar, together with the associated power supply, pod control computer and environmental system. FLIR imagery from the pod is displayed on a wide field of view holographic head-up display developed by GEC-Marconi Avionics. This provides the pilot with night vision for safe flight at low level. The Raytheon Systems Company Ku-band terrain-following radar permits operation at very low altitudes with en route weather penetration and blind let down capability.

Both the F-15E and F-16 have fully automatic terrain-following with inputs from the navigation pods to the digital flight control system in the aircraft. Additionally, terrain-following may be accomplished manually by means of directive symbology presented to the pilot on the HUD.

The AN/AAQ-14 targeting pod contains a stabilisation system, wide (6 × 6°) and narrow field (1.7 × 1.7°) FLIR, Litton Systems Inc, Laser Systems Division laser designator/ranger, automatic multimode tracker, automatic infrared Maverick missile hand-off system, environmental control unit, pod control computer, power supply and provision for an automatic target recogniser. The targeting pod interfaces with the aircraft controls and displays as well as the fire control system to permit low-level day and night manual target acquisition and semi-automatic weapon delivery of guided and unguided weapons. It may be configured as a laser designator-only pod, for use with laser-guided munitions and conventional weapons, by deleting the Maverick hand-off subsystem. This configuration of the LANTIRN targeting pod has been redesignated as Sharpshooter.

Both pods have environmental control units to ensure that their systems will function satisfactorily over a wide range of temperatures and flight conditions. Aircraft interfaces include a MIL-STD-1553 multiplex databus, video channels and power supplies. Service ground support employs typical three-level maintenance: organisational, where no special test equipment is required; intermediate, which employs automatic test equipment and depot servicing.

Specifications

AN/AAQ-13 Navigation Pod
Dimensions: 1985 × 546 × 355 mm
Weight: 195 kg
FLIR WFOV: 21 × 28°

AN/AAQ-14 Targeting Pod
Dimensions: 2500 × 381 mm
Weight: 245 kg
FLIR WFOV: 6 × 6°
FLIR NFOV: 1.7 × 1.7°

Operational status

In service on the US Air Force F-15E and F-16C/D and with a number of other air forces. A modified version of the AN/AAQ-14 targeting pod (as fitted to F-15E and F-16C/D), is being supplied to the US Navy for fitment to upgraded F-14 aircraft. It features a GPS/inertial navigation subsystem, eliminating the need for a dedicated navigation pod as used by the US Air Force.

The first production navigation pod was formally accepted by the US Air Force in April 1987, with the final item in the contract delivered in March 1992. US Air Force targeting pod deliveries were completed in April 1994. Total production is approximately 770 navigation pods and 724 targeting pods.

Lantirn is either operational or on order with the US Air Force, US Navy and nine other Air Forces: Bahrain, Egypt, Greece, Israel, Saudi Arabia, Singapore, South Korea, The Netherlands and Turkey.

Contractor

Lockheed Martin Electronics & Missiles.

UPDATED

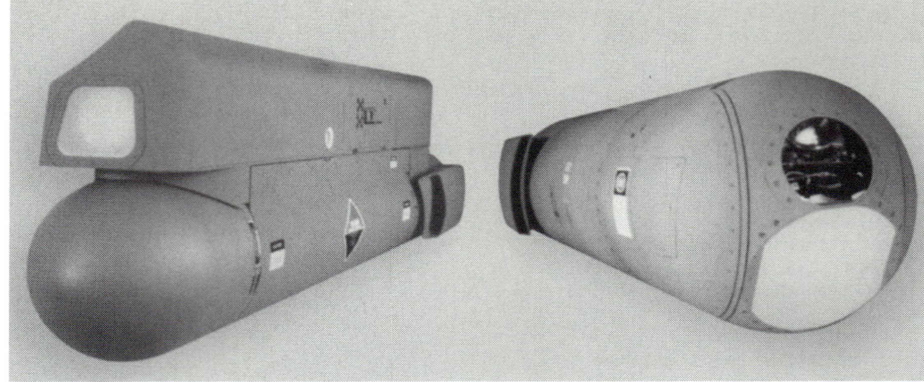

The LANTIRN navigation pod (left) and targeting pod (right) **1995**

LANTIRN 2000

Lockheed Martin has initiated work that it claims could advance the capabilities of its LANTIRN night navigation and targeting system to meet requirements of F-14, F-15 and F-16 jet fighter aircraft until 2020.

A baseline programme, called LANTIRN 2000, includes a third-generation Forward Looking Infrared (FLIR) sensor using Quantum Well technology; an Advanced Diode-Pumped Laser; and an enhanced computer system.

A collateral programme designated LANTIRN 2000+, offers US customers an Automatic Target Recognition (ATR) system; a Laser Spot Tracker (LST); and a digital disk recorder.

While significantly enhancing LANTIRN's traditional mission - day/night strike interdiction - LANTIRN 2000 will broaden aircraft mission capabilities to include air-to-air tracking, theatre missile defence and battle damage assessment. LANTIRN 2000+ refinements provide greater targeting flexibility and a reconnaissance capability.

LANTIRN 2000 technologies are being developed under Lockheed Martin's internal research and development programme, supported by several US Air Force programmes and the Air National Guard.

LANTIRN 2000 refinements are expected to produce dramatic results and savings. Use of Quantum Well FLIR technology means:

(a) Extremely dense detector arrays for sharp-edge imagery can be manufactured at a cost much lower than the costs of current detectors;

(b) Target detection, recognition and engagement ranges increase 50-60 per cent, improving an aircraft's first-pass kill capability, improving survivability and supporting near-term assessment of damage to high value targets;

(c) A level of FLIR sensitivity that allows air-to-air tracking of high-speed targets at difficult crossing angles;

(d) A 50 per cent increase in standoff ranges to perform assessment of near-term damage to high-value targets;

(e) A reduction in customer 'ownership costs' once the system is operational.

Insertion of a Diode-Pumped Laser for the critical task of spotting targets and guiding munitions will result in:

(a) A narrower laser beam that extends the Targeting Pod's operational range while permitting operations at 40,000 ft, in contrast to the current 25,000 ft;

(b) Fuller use of the operational envelope of the GBU-24 laser-guided weapon, ensuring more hits-per-sortie and enhancing air crew survivability;

(c) An increase in system reliability rates, owing to a more efficient power supply, fewer parts and cooler operating temperatures;

(d) A major range improvement in the eye-safe training laser, giving it the same range as the combat laser.

Refinements in LANTIRN 2000+ will be determined by customer requirements. The addition of a Laser Spot Tracker (LST) will provide: an ability to find and track targets that have been laser-designated by ground or other airborne sources; flexibility in tactical operations that improves target identification and limits collateral damage.

The Automatic Target Cuer (ATC) included in LANTIRN 2000+ will provide: a true automatic target recognition capability that supports the fighter reconnaissance mission; reduction in pilot workload because of its ability to automatically classify high priority targets.

Operational status

Many of the required technology upgrades will be demonstrated as part of US Air Force-funded flight tests to demonstrate FLIR performance, long-range staring array capabilities and automatic target recognition capabilities for the theatre defence mission.

Contractor

Lockheed Martin Electronics & Missiles.

VERIFIED

Low-Light-Level Television System (LLLTV)

Lockheed Martin has introduced the Low-Light-Level TeleVision (LLLTV), a general purpose multirole system. It is designed for maritime surveillance missions such as search and rescue, monitoring of territorial waters and policing offshore environmental laws. Principal features are high resolution and sensitivity at low-light levels and a small (16 mm diagonal) format. It is claimed to be capable of resolution densities in excess of 30 lines/mm and to have a wide dynamic range, which provides useful imagery around brightly lit parts of the area surveyed.

The camera head can be mounted in any attitude and is designed for hands-off operation. It uses a 16 mm vidicon tube and an 18 mm second-generation hybrid photocathode intensifier, which has extended sensitivity at the red end of the spectrum.

The system's electronic unit has a switchable line rate of either 525 or 875 lines a frame as standard but other line/frame rates are optionally available. Normal aspect ratio is 4:3 but a version with an aspect ratio of 1:1 is also available. Frame rate is 30 Hz. Resolution is 600 television lines horizontal and the dynamic range permits operation at face illumination levels from 0.1 ft candle down to starlight conditions.

The LLLTV has been chosen by the Spanish Navy for a shipboard fire-control application. The US Coast Guard has also selected an active gated version for possible use with its Dassault Aviation HU-25 Guardian aircraft, the application being to monitor and police territorial waters. In this configuration the equipment is designated AN/ASQ-174 Active Gated TeleVision (AGTV) and was flight-tested during 1984/85.

Specifications

Dimensions:
(camera head) 76 × 228 mm
(electronics unit) 241 × 184 × 165 mm
Weight:
(camera head) 1.36 kg
(electronics unit) 4.54 kg
Power supply: 28 V DC, 30 W

Operational status

In production and service.

Contractor

Lockheed Martin Electronics & Missiles.

VERIFIED

NITE Hawk targeting FLIR

Development of NITE Hawk began in March 1978 with the requirement to provide the F/A-18 Hornet with a day and night strike capability. The first systems, designated AN/AAS-38, were delivered to the US Navy in 1983.

The NITE Hawk presents the pilot with real-time passive thermal imagery in a television formatted display to assist in the location, identification and tracking of targets. The system provides the aircraft mission computer with accurate target line of sight pointing angles and angle rates. Automatic in-flight boresight compensation is used to correct for dynamic

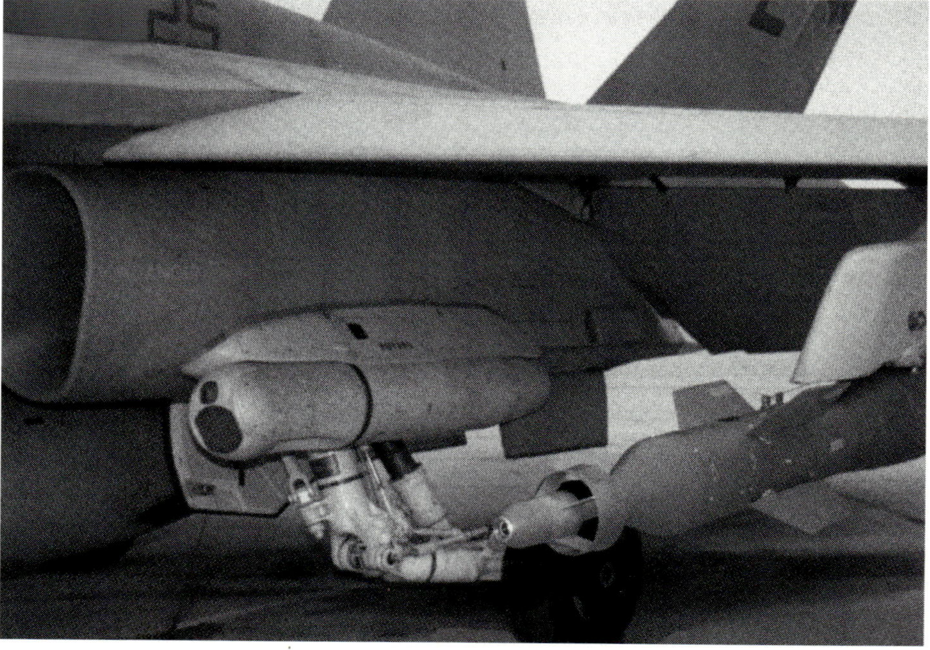

The NITE Hawk (AN/AAS-38B) carried by the F/A-18 aircraft **1997**/0001252

flex and thus maintain accurate pointing angles throughout the aircraft's full performance envelope. Increased weapon delivery accuracy is provided through a Laser Target Designator/Ranger (LTD/R) feature (AN/AAS-38A). The LTD/R provides precise target range information and designation capability for precision-guided munitions.

Additional operational capabilities were incorporated into the NITE Hawk AN/AAS-38B, which has been in production since 1992. A laser spot tracker has been included which allows the system to search for, acquire and track targets which have been laser designated by ground or other airborne sources. A multifunction autotracker also provides scene track, intensity centroid and geometric tracker algorithms for improved acquisition and maintenance of target lock on during attack of air-to-ground targets. Enhancements were also made to the air-to-air detection and tracking capabilities of the system.

Two further versions of the system have been developed: the AN/AAS-46 configuration for the F/A-18 E/F aircraft; and the NITE Hawk SC (self-cooled) system.

The NITE Hawk SC is designed to perform in a supersonic flight environment, providing single or multi-seat aircraft with 24-hour strike capability against land- or sea-based targets. NITE Hawk SC interfaces with other aircraft avionics over a MIL-STD-1553B multiplex databus.

Specifications

Dimensions:
(AN/AAS-38/38A/38B) 1,840 × 330 mm diameter
(NITE Hawk (SC)) 2,440 × 330 mm diameter
Weight:
(AN/AAS-38) 158 kg
(AN/AAS-38A/38B) 168 kg
(NITE Hawk (SC)) 195 kg
Field of view:
(wide) 12 × 12°
(narrow) 3 × 3°
Field of regard:
(pitch) +30 to −150°
(roll) ±540°
Stabilisation: 35 μrad
Tracking: 230 μrad
Pointing: 400 μrad

Operational status

The first production NITE Hawk system was delivered to the US Navy in December 1983. Systems have been delivered to, or are on order for: Australia, Canada, Kuwait, Malaysia, Spain, Switzerland and Thailand.

NITE Hawk SC has been flight tested on the F-14, F-15, F-16 and AV-8B. The NITE Hawk SC system is the baseline laser/designator pod for the Spanish EF2000.

Lockheed Martin completed development of the NITE Hawk AN/AAS-46 configuration for the F/A-18E/F at the end of 1996. Over 500 NITE Hawk systems have been delivered, and production into the 21st century is planned.

Contractor

Lockheed Martin Electronics & Missiles.

UPDATED

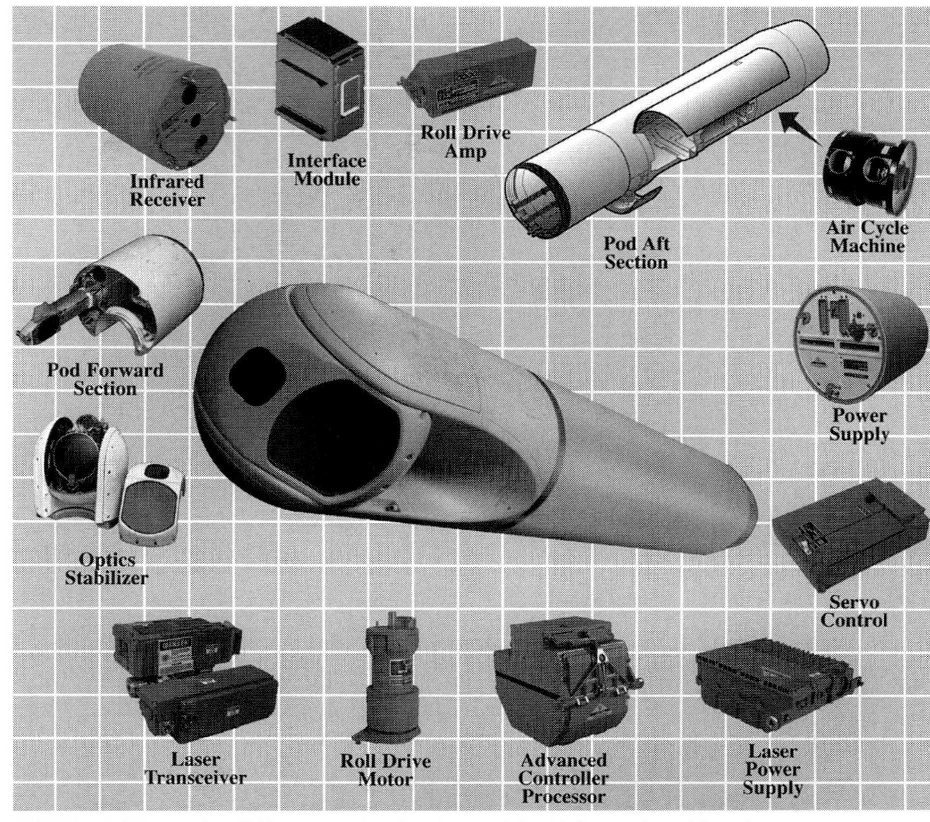

NITE Hawk SC targeting FLIR system showing the modular 12 line replaceable units **1998**/0018326

Pathfinder navigation/attack system

Pathfinder is an international derivative of the LANTIRN navigation system.

The Pathfinder navigation/attack system uses hardware derived from the LANTIRN night navigation system (see earlier item). It consists of three LRUs which can be integrated into a pod or embedded in the aircraft fuselage or pylon. These LRUs are an infrared sensor, power supply and environmental control unit. The first two of these LRUs are derived directly from LANTIRN but the environmental control unit is smaller and lighter, with reduced power requirements.

As with the LANTIRN navigation pod, Pathfinder imagery may be presented on a HUD or any other cockpit video display. Unlike the navigation pod, in addition to a ×1 wide field of view it also contains a ×3 magnified slewable field of view for standoff target acquisition.

Pathfinder has been demonstrated in a pod on the F-16, integrated into a pylon on the A-7 and embedded in the B-1B. It is suitable for installation on a wide variety of additional aircraft including the F-5, A-10, C-130, Dassault/Dornier Alpha Jet, Dassault Rafale, British Aerospace Hawk and Panavia Tornado.

Specifications

Dimensions: 1,950 (length) × 248 mm (diameter)
Weight: 90.5 kg
Power supply: 115 V AC, 400 Hz, 3 phase, 2.8 kW
28 V DC, 2 A
Field of regard: 77 × 84°
Field of view:
(wide) 21 × 28°
(narrow) 7 × 9°
Temperature range: −40 to +90°C
Reliability: 539 h MTBF

Operational status

Egypt has ordered 12 Pathfinder pods for the F-16.

Contractor

Lockheed Martin Electronics & Missiles.

UPDATED

Sharpshooter targeting pod

Sharpshooter is an international derivative of the targeting pod used in LANTIRN. Sharpshooter carries the designator AAQ-14(V1) to differentiate it from the AN/AAQ-14 pod of the LANTIRN system. It can be integrated with the Pathfinder navigation pod to permit pilots to fly at low altitudes in total darkness to the target area.

Sharpshooter provides round-the-clock targeting capabilities through the use of a 200 mm aperture infrared sensor that has a 1.7° narrow field of view and a 6° wide field of view. A stabilised line of sight includes a 150° look-back angle and continuous roll tracker, which can track either stationary or moving targets or track a scene using area correlation. The laser designator and ranger is boresighted to the centre of the field of view of the targeting pod and is programmable for coding. Sharpshooter includes an eye-safe laser for training.

Operational status

Orders include: three for Bahrain; 12 by Egypt for installation on F-16; Greece has ordered 16; Israel has ordered 30 systems for installation on F-16 and a further 10 for F-15I; and Saudi Arabia has 48 pods on order for F-15S.

Contractor

Lockheed Martin Electronics & Missiles.

UPDATED

Target Acquisition Designation Sight/Pilot Night Vision Sensor (TADS/PNVS)

Lockheed Martin's Target Acquisition Designation Sight/Pilot Night Vision Sensor (TADS/PNVS) is designed to provide day, night and limited adverse weather target information and navigation capability for the US Army AH-64D Apache attack helicopter. The TADS/PNVS system comprises two independently functioning subsystems known as TADS and PNVS.

The Lockheed Martin Pathfinder is a FLIR sensor pod derived from the LANTIRN navigation pod **1995**

An AH-64A Apache with the Lockheed Martin TADS/PNVS system mounted on the nose

TADS provides the co-pilot/gunner with search, detection and recognition capability by means of direct view optics, TV or FLIR sighting systems which may be used singly or in combinations according to tactical, weather or visibility conditions. The PNVS subsystem provides the pilot with flight information symbology which permits nap of the earth contour, on low-level flight to, from and in the combat area at altitudes low enough to prevent or delay detection by enemy forces.

TADS consists of a rotating turret, mounted on the nose of the helicopter and containing the sensor subsystems, an optical relay tube located at the co-pilot/gunner station, three electronic units in the avionics bay and cockpit controls and displays. TADS turret sensors have a field of regard covering ±120° in azimuth and from +30 to −60° in elevation.

By day, either direct vision or television viewing may be used. The direct vision system has a narrow field of view, 3.5° at ×18.2 magnification, and a wide field of view, 18° at ×3.5 magnification. The television system provides a narrow field of view of 0.9° and a wide field of view of 4°. For night operations the FLIR sensor has three fields of view: narrow (3.1°), medium (10.2°) and wide (50°).

Once acquired, targets can be tracked manually or automatically for autonomous attack with guns, rockets or Hellfire anti-tank missiles. A laser may also be used to designate targets for attack by other helicopters or by artillery units firing the laser-guided anti-armour Copperhead shell.

PNVS consists of a forward-looking infrared sensor system packaged in a rotating turret mounted above the TADS, an electronics unit located in the avionics bay and the pilot's display and controls. The system covers a field of ±90° in azimuth and from −45 to +20° in elevation. Field of view is 30 × 40°.

TADS is designed to provide a backup PNVS capability for the pilot in the event of the latter system failing. The pilot or the co-pilot/gunner can view, on his own display, the video output from either TADS or PNVS, raising the probability of mission success. Although designed primarily for combat helicopters flying nap of the earth missions, PNVS may also be used as a single entity in tactical transport and cargo helicopters.

Special attention has been paid to reliability, as has the ability to remove and replace units easily and rapidly on the flight line.

Operational status

In production for international customers, and for US Army AH-64D Apache helicopters.

Contractor

Lockheed Martin Electronics & Missiles.

UPDATED

ATARS Advanced Tactical Airborne Reconnaissance System

The Advanced Tactical Airborne Reconnaissance System (ATARS), under development by the US Navy and US Marine Corps, will provide high-resolution imagery for real-time or near-realtime reconnaissance. The Boeing Company is the prime contractor, and Lockheed Martin Fairchild Systems is the reconnaissance systems integrator.

F/A-18 (RC) aircraft equipped with the Low Altitude Electro-Optical (LAEO) and Medium Altitude Electro-Optical (MAEO) sensors for daylight operations and the InfraRed LineScanner (IRLS) for day and night operations will be capable of satisfying the requirement for deep penetration, under the weather, time-critical reconnaissance.

The MAEO sensor allows the aircrew to obtain high-resolution imagery from three to five miles standoff, without direct overflight of targets in high threat areas.

The Reconnaissance Management System (RMS) provides control of the sensors, recorders and datalink, manages the flow of data from the sensors to the digital recorder, and from the datalink to the ground station and to the cockpit displays. The system can store 12 preplanned point, strip or area targets and 20 targets of opportunity.

Specifications

Low Altitude Electro-Optical (LAEO) sensor:
200-3,000 ft above ground level
140° field of view
vertical or forward oblique
high resolution, dawn to dusk below the weather, high speed sensor
Medium Altitude Electro-Optical (MAEO) sensor
2,000-25,000 ft above ground level
22° field of view
horizon to horizon field of regard
high resolution, medium range standoff sensor for daylight operations in high threat environments
InfraRed LineScanner (IRLS) sensor:
200-10,000 ft above ground level
140° or 70° field of view
8-12 micron waveband
high resolution, low to medium altitude day/night reconnaissance capability

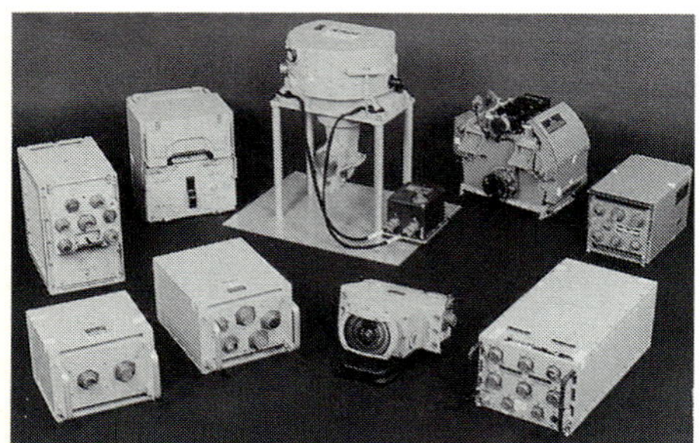

Lockheed Martin Fairchild Systems ATARS consists of three reconnaissance sensors and a number of electronic units ***1995***

Operational status

Originally in design as an internal fit for the F/A-18 (RC) and RF-4C aircraft, ATARS can be configured for internal or pod fit, and current production is for internal fit to US Marine Corps F/A-18D aircraft, and pod fit for US Air Force F-16 aircraft (when it is designated TARS – Theatre Airborne Reconnaissance System). ATARS is also being considered by the US Navy for the F/A-18F aircraft.

Contractor

Lockheed Martin Fairchild Systems.

UPDATED

AN/AAQ-23 Electro-Optical/Infrared Viewing System (EOIVS)

The AN/AAQ-23 Electro-Optical/Infrared Viewing System (EOIVS) provides high-performance day or night target acquisition and navigation imagery. The system consists of a mid-wave staring array Forward-Looking InfraRed (FLIR) sensor, sensor support electronics and video post-processor. Images are displayed on an 875-line display with TV compatibility in the cockpit. The EOIVS is based on an advanced development staring infrared focal plane array which uses a platinum silicide substrate. The focal plane array, measuring 640 × 480 pixels, has two fields of view, with the narrower field used for target acquisition.

Specifications

Weight: 43 kg without gimbal
Power supply: 115 V AC, 400 Hz, 3 phase, 208 W
Optics: Dual field of view f/1.6 fixed focus

Operational status

The EOIVS is designed to improve the reliability and maintainability of navigation systems for the B-52 fleet. The system could also be integrated on the B-1B. A derivative of the sensor is available for other military and commercial aircraft. Production deliveries began in March 1996.

Contractor

Lockheed Martin Fairchild Systems.

VERIFIED

AN/AVD-5 (LOROPS) electro-optical reconnaissance sensor system

The AN/AVD-5 is a high-resolution long-range reconnaissance system designed to provide identification of tactical targets at ranges beyond 75 km. The sensor has a 1,676 mm focal length, f/5.5 reflective lens coupled with a 12,000 pixel array which has sufficient sensitivity to capture near-realtime imagery and provide haze penetration. The AN/AVD-5 has five modes of operation which can be optimised for coverage, resolution or stereo performance, and used in both manual or automatic scan.

The AN/AVD-5 is modular in design, providing flexibility in installation in a variety of aircraft or in a reconnaissance pod on fighter aircraft. The system consists of an imaging LRU, sensor control unit, power amplifier unit, junction box and interconnecting cables. Additionally, it utilises a reconnaissance management system, a switchable main electronics unit and a digital recorder, all of which are common to the ATARS suite (see earlier item). The digital imagery from the sensor is capable of transmission for processing in a ground exploitation system.

A dual-band (EO/IR) sensor is in development.

Specifications

Dimensions:
(imaging unit) 130.1 × 50.8 × 50.8 mm
Weight: 310.7 kg
Power supply: 115 V AC, 400 Hz, 3 phase, 362 W
28 V DC, 799 W
Range: 4.8-80.5 km

Operational status
Production systems are being shipped under a US Air Force contract.

Contractor
Lockheed Martin Fairchild Systems.

UPDATED

AN/AXQ-16(V) cockpit television sensor

The AN/AXQ-16(V) cockpit television sensor allows recording of real-time gunsight, HUD symbology, instrument panels and audio in fighters, tactical strike aircraft and commercial aircraft. The camera can also drive a monitor to provide real-time displays for the second crew member in a two-seat aircraft. It is a small solid-state TV camera configured to meet customer requirements.

The camera uses a charge-coupled device for high sensitivity and dynamic range. The principal optical element is a 17 mm, f/1.2 lens with automatic iris control capable of operating at low-light levels.

The system can replace existing film cameras or interface with any HUD. It is a form, fit, function replacement for existing monochrome cameras. Event mark, BIT, configuration control and environmental requirements have been incorporated into the design.

Specifications
Dimensions: specific to requirements
Weight: <1.36 kg
Power supply: 115 V AC, 400 Hz, 3 phase, 20 W
Lens: various
Sensor: 768 × 494 pixels
Refresh rate: 30 frames/s
Line rate: 2:1 interface
Reliability: 32,000 h MTBF

Operational status
In production. The television sensor has been tested and fielded in several US Air Force, US Navy and foreign military tactical aircraft and is specified for the F-15 and F/A-18.

Contractor
Lockheed Martin Fairchild Systems.

VERIFIED

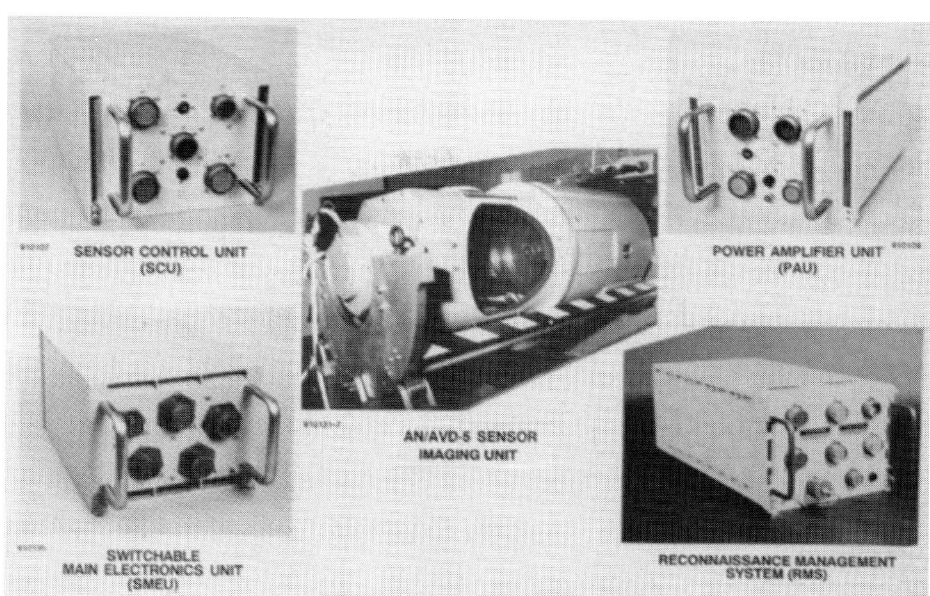

The AN/AVD-5 (LOROPS) electro-optical reconnaissance sensor system **1995**

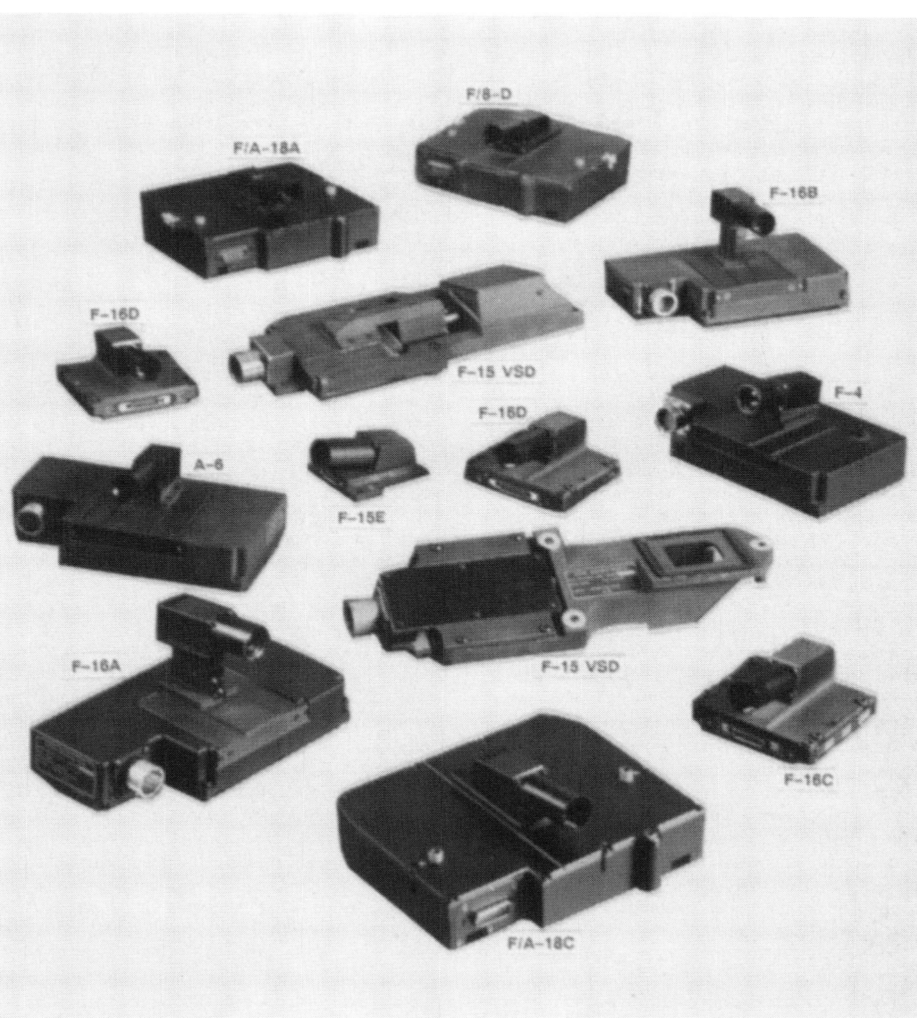

The Lockheed Martin Fairchild Systems AN/AXQ-16(V) cockpit television sensor is available in a large number of configurations **1995**

Colour cockpit TV system

The colour cockpit TV system is designed for airborne colour video recording, including HUD and gunsight recording, multifunction display recording, flight test instrumentation and mission strike recording. It can also be used as a UAV sensor.

It features a miniature camera, solid-state automatic electronic exposure control, a dynamic range of 12,000 to 1, optional boresight preset at the factory to within 0.56 mrad for various fields of view, event mark and multiplexing of alternative signals. It is capable of external synchronisation.

Operational status
Currently in production for the F-14D, F-15E, F/A-18 (3 per aircraft), MB-339 and T-45 aircraft.

Contractor
Lockheed Martin Fairchild Systems.

UPDATED

SU-172/ZSD-1(V) Medium-Altitude Electro-Optical sensor (MAEO)

The Medium-Altitude Electro-Optical sensor operates from 2,000 to 25,000 ft with a 22° narrow field of view, 12 in (304.8 mm) focal length, f/5.6 lens for daylight operations. It is designed for medium-altitude tactical airborne reconnaissance penetration and standoff missions. High-resolution imagery is captured on a 12,000, 10 × 10 μm pixel CCD.

The SU-172/ZSD-1(V) is designed with emphasis on an integrated system approach to reduce life cycle costs. The sensor is automated to reduce aircrew workload and for use in UAVs. Stabilisation allows for target tracking in the presence of aircraft manoeuvres.

Specifications
Dimensions:
(imaging unit) 511 × 381 × 330.2 mm
Weight: <62.14 kg
Power supply: 28 V DC, 440 W
Altitude: 2,000-25,000 ft

Operational status
In production for the US Marine Corps.

Contractor
Lockheed Martin Fairchild Systems.

VERIFIED

SU-173/ZSD-1(V) Low-Altitude Electro-Optical Sensor (LAEO)

The SU-173/ZSD-1(V) Low-Altitude Electro-Optical (LAEO) sensor operates from 200 to 3,000 ft with a low distortion 140° wide field of view and vertical or forward oblique fixed focus. The sensor provides high-resolution visible spectrum imagery for high-speed, low-altitude area coverage on tactical reconnaissance missions.

The SU-173/ZSD-1(V) consists of a sensor and electronics unit that collect imagery, provide roll correction of imagery, perform preliminary image processing and output digital image data. The low-altitude sensor system has been designed to be flexible and responsive to both mission and platform tailoring. The system may be internally mounted in manned aircraft such as the RF-4C/E or F/A-18(RC) or pod-mounted in the F-14, F-16, Tornado, Mirage, Jaguar and JAS 39, in addition to being configured for unmanned aerial vehicles.

Specifications

Dimensions:
(imaging unit) 284.5 × 121.9 × 363.2 mm
Weight: <24.5 kg
Power supply: 28 V DC, 231 W
Altitude: 200-3,000 ft

Operational status

In production for the US Navy and US Marine Corps.

Contractor

Lockheed Martin Fairchild Systems.

VERIFIED

AN/AAD-5 Infrared reconnaissance set

The AN/AAD-5 is a dual field of view high-performance infrared reconnaissance system which scans the terrain beneath an aircraft's flight path. The system consists of seven LRUs: the receiver, recorder, film magazine, control indicator, infrared performance analyser, cooler and power supply. These units may be as much as 6 m apart, although the film magazine must remain coupled to the recorder.

The receiver, which scans the traversed terrain and converts the received IR radiation into electrical signals, includes the scanning optics, cooler cryostat, two 12-element detector arrays, 24 preamplifiers and the associated buffer electronics. The detector arrays are Mercury Cadmium Telluride (MCT) photoconductors which are sensitive to infrared radiation in the 8 to 14 μm waveband. One array is used for the wide field of view and the other for narrow field of view. The receiver also contains the spin motor regulator and an encoder to provide system timing pulses to the recorder.

The recorder converts the video signals and time pulses from the receiver into the information required to produce a film record. It consists of a CRT, recording optical components, video and sweep electronic cards, digital timing circuits, a high-voltage power supply, film speed control circuits and an auxiliary data annotation set.

An improved version, the AN/AAD-5(RC), is derived from the AN/AAD-5. The AN/AAD-5(RC) imaging process is identical to that of the AN/AAD-5. The receiver is two-thirds the size of the AN/AAD-5 receiver and uses an integral 0.25 W split-Stirling miniature cryogenic refrigerator. This eliminates the separate cryogenic cooler LRU, as well as the need for the nitrogen supply tanks carried on board the aircraft.

In addition to improvements in size, weight, power consumption, reliability and supportability, the AN/AAD-5(RC) has been developed for rapid insertion into the tactical reconnaissance community by remaining highly compatible with the original AN/AAD-5. Imagery produced by the two systems is virtually identical and the AN/AAD-5(RC) operates in the same manner and meets all AN/AAD-5 performance specifications.

While the effective receiver aperture has been reduced, the imaging performance has been kept equal to that of the AN/AAD-5 receiver through improvements in the infrared detector performance. The same materials have been used in the scaled receiver. The spin mirror shaft encoder has not been scaled because of the need to maintain scan position accuracy. The scaling has resulted in a smaller, lighter and more reliable receiver.

The AN/AAD-5(RC) preamplifier assembly has also been upgraded. The role of the preamplifiers in the signal processing sequence is to boost the signal immediately after the detectors convert IR energy to analogue signals. Previously, preamplifiers occupied a significant portion of the receiver envelope. The AN/AAD-5(RC) hybrid monolithic preamplifier assembly occupies a smaller volume than the existing AN/AAD-5 set and provides higher reliability and maintainability.

Various electro-optical output options are available for both AN/AAD-5 and AN/AAD-5(RC) equipped systems. Upgrades using the Lockheed Martin IR linescanner real-time display system can add a cockpit real-time display with onboard video recording, with an optional datalink to ground stations, while maintaining the current film system operation.

Specifications

Dimensions:
(receiver) 460 × 380 × 330 mm (AN/AAD-5)
430 × 345 × 279 mm (AN/AAD-5(RC))
(recorder) 410 × 580 × 230 mm
(magazine) 380 × 180 × 250 mm
(control indicator) 150 × 100 × 80 mm
(cooler) 410 × 20 × 300 mm
(analyser) 150 × 50 × 300 mm
(power supply) 250 × 430 × 280 mm
Weight:
(AN/AAD-5) 130 kg
(AN/AAD-5(RC)) 63.2 kg
Power supply: 115 V AC, 400 Hz, 725 VA nominal
28 V DC, 118 W nominal
Magazine capacity:
106 m (conventional film)
213 m (thin-based film)

Operational status

In service with the US Air National Guard in the RF-4C and the US Navy in the F-14 TARPS. Also used by a number of foreign air forces including Australia for the RF-111 and Germany, Greece, Korea, Spain and Turkey for the RF-4. Over 600 AN/AAD-5 systems and 168 AN/AAD-5(RC) systems have been delivered. The AN/AAD-5(RC) is still in production.

Contractor

Lockheed Martin IR Imaging Systems.

UPDATED

D-500 InfraRed LineScanner (IRLS)

The D-500 InfraRed LineScanner (IRLS) is an advanced version of the AN/AAD-5 IRLS in service with the US military and several foreign armed forces. The D-500 is designed for pod, small aircraft and UAV applications and can fit within a 15 in diameter envelope so that it can be installed in all but the smallest pods. Its performance is the same as that of the AN/AAD-5. In addition to meeting the AN/AAD-5 film performance specification, the D-500 incorporates electronic oblique standoff viewing, BIT, videotape recording, real-time display and a datalink capability.

The D-500 consists of four basic LRUs: the receiver in which the IR image is formed at the Cadmium Mercury Telluride (CMT) detector array, the film recorder on which a visible image is formed through a CRT trace from the detector signal, the film magazine and a power supply. The intensity modulated CRT trace is coupled by a fibre optic faceplate to the recording film on which the image is composed on a line-by-line basis. The detector is composed of two linear arrays of 12 detectors, each aligned in the direction of flight. One to 12 detectors are used for recording, depending on the Velocity/Height (V/H) profile of the aircraft. The signal from the receiver is used to drive a 12-beam CRT which sweeps the signal on to film, mimicking the fashion in which the signals are received through the scan mirror at the detector. The scene sensed by the instrument can have a large intrinsic dynamic range which is first limited by the detectors to 12 bits. The range of levels displayed on the film is reduced below this by both the limited dynamic range of the CRT and by the film itself. This range is typically in the order of five bits or 32 grey levels in the film image.

The only major differences, other than size, between the AN/AAD-5 and the D-500 in the imaging process are the CRT/film coupling and the video electronics package.

Both optical efficiency and bulk have been improved by replacing the AN/AAD-5 CRT relay lens with the fibre optic faceplate CRT. The option for EO image recording was accommodated by integrating the video electronics with the receiver LRU. The D-500 fibre optic faceplate CRT eliminates the need for the relay lens.

In addition to these improvements, the D-500 has been developed for rapid insertion into tactical reconnaissance aircraft by remaining highly compatible with the AN/AAD-5(RC). Externally, the LRUs have been modified, yet internally some 80 per cent of components are identical. The D-500 receiver has been reduced in size and weight, compared to the AN/AAD-5, while the original image quality specifications have been maintained. The recorder and film magazine sizes have been reduced to complement the smaller receiver. The preamplifiers have monolithic hybrid circuitry.

While the effective receiver aperture has been reduced, the imaging performance has been kept equal to that of the AN/AAD-5 through improvements in the infrared detector. The D-500 can be operated as an EO videotape recording and/or film image generating system. The recorder and film magazine can also be remotely located in a ground station to provide the mission hard-copy database. Options for the D-500 include videotape recording, on aircraft real-time display and datalink capability.

The D-500 is electronically roll corrected. Up to ±20° of roll can be corrected in both the wide and narrow modes. The automatic roll correction is accomplished by shifting a display unblank pulse within the 180° timing window provided by each facet of the scan mirror. The display unblank pulse gates out the proper section of data from the number of degrees of video collected during a scan line.

Contractor

Lockheed Martin IR Imaging Systems.

UPDATED

Helicopter InfraRed Navigation System (HIRNS)

The Helicopter InfraRed Navigation System (HIRNS) is a low-weight television-compatible FLIR designed to aid navigation by day and night. The sensor head is integrated with a Ferranti steerable two-axis unstabilised platform.

A serial scanning thermal imager is used because balancing of amplifiers is not required to provide a high-resolution display. It also has the advantage of simpler processing electronics, while DC restoration ensures a bloom-resistant image and clear horizon definition. Energy from the scene is scanned horizontally with a continuously rotating eight-faceted mirror, while a flat nodding mirror carries out vertical scanning. The output signal from the SPRITE detectors is amplified and processed to provide gain and level control and is then displayed in a standard TV format.

The system may be linked with an Integrated Helmet And Display Sighting System (IHADSS) to allow the crew to perform nap of the earth missions.

Specifications

Weight:
(FLIR) 6.35 kg,
Field of regard:
(bearing) ±130°
(elevation) +20 to −60°

Operational status

In service on the Agusta A 129 helicopter.

Contractor

Lockheed Martin IR Imaging Systems.

UPDATED

Infrared Imaging System (IIS)

The Infrared Imaging System (IIS) is a near-realtime, 180° field of view, high-resolution IR linescanner system developed for the German Air Force Tornado ECR aircraft. The system consists of five interactive units. Two of these units, the Recorder/Film Processor Unit (R/FPU) and the Scanner Receiver Unit (SRU) are produced by Lockheed Martin. These two units provide the infrared detection and recording functions of the system. The other three units, the control, display and interface units, are produced by Honeywell Sondertechnik.

The SRU, used to detect horizon-to-horizon infrared radiation emanating from the terrain, senses medium band infrared radiation and converts it to 12 parallel video signal channels. SRU electronics provide a constant resolution footprint of the target, regardless of scan angle, throughout the widest practical ground coverage. These electrical signals are then sent to the R/FPU. The SRU is controlled by digital commands from the control unit, which are sent via the aircraft databus.

The SRU can be broken down into eight functional groups. The spin mirror group performs optical scanning of the IR scene. The IR sensor and amplifier group collect the IR energy of the scanned scene. The channel processor group performs the initial amplification and processing of the 34 preamplified detector signals. The mapping group processes signals from the channel processor group and combines them into 12 parallel video signals. The delay buffer and timing group provides proper timing for the 12 video channels and generates the file video output. The gyro and electronics group provides the roll compensation for maintaining the stability of the scanned imagery. The BIT and control group provide the token ring bus interface, central processing for the SRU and BIT circuitry. The video group provides the video output.

Only 12 infrared signal channels, video sync signals and roll sensor data are passed directly between the SRU and the R/FPU. The modular design concept is carried out within each unit in order to simplify maintenance and logistics.

The function of the R/FPU is to receive and process the video signals from the SRU, provide the aircraft crew with the ability to view the imagery in near real time and produce a permanent film record of the IR imagery scanned by the SRU.

A precision CRT converts the amplitude modulated video signals to intensity modulation of the CRT beam as it sweeps across the tube face in synchronism with the rotation of the spin mirror in the SRU. An autofocus detector and focusing circuit measure the intensity of the beam and automatically correct the focus. The film is exposed by the variable intensity CRT beam as it moves in front of the faceplate. Annotation data, obtained from the aircraft via the databus, is applied to the film at proper time intervals.

The dry process film is developed within the R/FPU for onboard viewing. The film moves, at constant speed, between a rotating drum and a heated shoe. A servo-controlled heater inside the shoe maintains the temperature at 130°, the level required for the dry process chemicals on the film to be activated by the heat, thus developing the film on board the aircraft.

Processed film is temporarily stored on the film manipulator. The film storage mechanism consists of two concentric powered drums on which the film is wound or unwound as required, to place the desired portion of the film in the viewing area. Each drum is controlled separately so that newly processed film can be wound on to the manipulator as it accumulates, while previously stored film is being wound back on to the manipulator as the operator commands viewing of earlier images.

As the film leaves the manipulator, it passes over a light table. This electroluminescent panel causes the image on the film directly above it to be projected on to the lens of a TV camera. The camera output is displayed on the operator's console. The operator can view any part of the exposed film. When previously exposed film is to be viewed, it is rerolled from its storage spool, moved back across the light table for viewing and temporarily stored on one of the two film manipulator drums. This action does not prevent newly exposed film from being stored on the other drum.

Selectable lenses give the TV camera three fields of view, effectively a zoom capability as the film is being viewed. Areas of interest on the film may be marked during viewing so that they may be quickly located at a later time.

A 107 mm reel of unexposed dry process film is available to the R/FPU. The film is provided at the correct time and speed for exposure by the CRT. The exposed film is stored on a separately controlled take-up reel.

Specifications
Dimensions:
(SRU) 520.7 × 381 × 509.8 mm
(R/FPU) 600 × 640 × 476 mm
Weight:
(SRU) 53 ±2 kg
(R/FPU) 87 kg

Operational status
In service on the German Air Force Tornado ECR.

Contractor
Lockheed Martin IR Imaging Systems.

UPDATED

InfraRed LineScanner (IRLS) Real-Time Display (RTD) system

The IRLS RTD system is a group of off-the-shelf LRUs that, when attached to a properly configured linescanner, will produce an instantly available video display of IR images. The RTD system is designed to accept the analogue outputs from the AN/AAD-5, AN/AAD-5(RC), D-500 and ATARS IRLS.

With the RTD system, detected IR energy may be viewed within milliseconds of occurrence. The real-time imagery may be viewed on board the aircraft or via data-link at a remote site, depending on the mission requirements. The RTD display video may be recorded by a VTR or normal film recording may be performed independently; the modified RS-170 EO output can easily be datalinked.

Switches on the joystick of the optional hand control LRU control all real-time display functions by controlling the microprocessor in the scan converter. Switch data is coded and transmitted to the scan converter as a serial data stream in RS-232 format. The analogue reticule control output is digitised and transmitted to the scan converter via the same RS-232 connection. If available, existing aircraft controllers can be used.

The optional CRT display is a video display terminal which has been ruggedised for use in the airborne environment. It accepts inputs with frame and line sync signals incorporated in the video format. From these inputs it produces an image display with a 4:3 aspect ratio. The portion of the IRLS scanned scene which is displayed can be varied by the operator using joystick controls. Because the output of the scan converter is RS-170, existing compatible aircraft displays can be used.

Specifications
Dimensions:
(scan converter) 570 × 258 × 194 mm
(CRT display LRU) 279 × 254 × 203 mm
(VTR LRU) 420 × 260 × 168 mm
(hand control unit LRU) 76 × 76 × 152 mm
Weight:
(scan converter) 18 kg
(CRT display LRU) 5.6 kg
(VTR LRU) 12.5 kg
(hand control unit LRU) 0.9 kg

Operational status
In service on Egyptian Air Force Beech 1900 aircraft.

Contractor
Lockheed Martin IR Imaging Systems.

UPDATED

Stabilised Thermal Imaging System (STIS)

The low-cost, high-performance Stabilised Thermal Imaging System (STIS) satisfies the surveillance and navigation requirements of fixed-wing aircraft, helicopters and UAVs.

STIS features the Lockheed Martin Mini-FLIR which, with two bar SPRITE detector, is a serial scanning IR sensor incorporating DC-restored electronics for a crisp, bloom-free display. STIS is designed in modular style. The standard modules consist of the miniature FLIR scanner, the split-Stirling cooler, the focal plane/preamplifier assembly and the monolithic electronics.

The Mini-FLIR provides a TV-compatible output, making it suitable for a wide variety of displays. The stabilised thermal imaging system incorporates a three field of view telescope and is mounted in a stabilised inner gimbal capable of providing smear-free imagery in a dynamic environment.

Available options for STIS include an autotracker, laser rangefinder, low-light-level TV camera and an alphanumeric display.

Specifications
Dimensions: 508 × 355 mm diameter
Weight: less than 34 kg
Waveband: 8-12 μm nominal
Field of view:
(wide) 20 × 30° (×1.2)
(medium) 7.2 × 10.8° (×3.3)
(narrow) 2 × 3° (×12)
Video format: RS-170, TV-compatible, CCIR option

Operational status
In service on Dornier 228 aircraft, Eurocopter Super Puma helicopters, Skyeye RPVs and various other helicopters and RPVs.

Contractor
Lockheed Martin IR Imaging Systems.

AN/AAS-40 Seehawk FLIR

The AN/AAS-40 Seehawk is a thermal imaging system using US Department of Defense common module FLIR components to provide high-resolution imagery. Designed for an aircraft or surface vessel, the system's current principal application is aboard a US Coast Guard Sikorsky HH-52A helicopter serving the primary role of search and rescue, law enforcement, maritime environmental control, marine and border control and navigational assistance. It is also installed on the Northrop Grumman S-2(T) ASW aircraft.

The Northrop Grumman AN/AAS-40 Seehawk lightweight FLIR system consists of a turret assembly, the control electronics unit and the power supply unit. FLIR imagery is fed to the cockpit-mounted display. An automatic scan capability provides constant search coverage in elevation and azimuth. The autosearch mode is enhanced by inclusion of automatic lock on which reacts to either large or small targets, as selected by the operator.

The Seehawk system has successfully completed two years' service aboard a Beech 200T aircraft. The system was fully operational in conjunction with the other onboard avionics systems, demonstrating maritime patrol, surveillance and reconnaissance applications. The latest generation AN/AAS-40 Seehawk was installed aboard a modified Sikorsky S-76 helicopter used as a demonstrator in support of the US Army RAH-66 Comanche programme.

Operational status
In production for the US Coast Guard HH-65A Dolphin helicopter and the Northrop Grumman S-2(T) ASW aircraft.

The AN/AAS-40 Seehawk FLIR on the US Coast Guard HH-52A helicopter

Contractor
Northrop Grumman Corporation.

VERIFIED

AN/ASX-1 Target Identification System Electro-Optical (TISEO)

The AN/ASX-1 TISEO, is a television-based, passive, daytime automatic target acquisition and tracking system. Comprising a high-resolution closed-circuit television sensor combined with a two fields of view telescope, TISEO is mounted on the port wing leading edge of many US Air Force F-4E Phantoms to enable the crew to recognise and identify targets at long range.

Specifications
Dimensions:
(television camera assembly) 813 (length) × 229 mm (diameter)
Weight: 26 kg
Field of view:
(wide) 1.4°
(narrow) 0.44°

Operational status
In service.

Contractor
Northrop Grumman Corporation.

UPDATED

AN/AVQ-27 laser target designator set

The AN/AVQ-27 designator has been developed for installation aboard two-seat F-5B and F-5F aircraft and can also be fitted in other two-seat aircraft. Several customers have selected the system, which permits target designation during operations using laser-guided munitions.

The system is installed in the rear cockpit, and can be removed when not required. The installation consists of a high-power laser, stabilised direct view optics with two fields of view and a 16 mm recording camera. The designator set package is only 100 mm wide and fits on the lower canopy rail, maintaining ejection-seat clearance and not interfering with other cockpit functions. A viewfinder/sight swings across into the pilot's field of view, enabling the pilot to track targets. This can be conducted manually, using a two-axis hand controller, and with rate or rate-aided tracking modes. The front cockpit is fitted with sight and canopy markings which allow the pilot to assist the rear-seat operator in initial target acquisition and then to maintain the target within the field of view. One-hand operation is claimed. Normal operations would call for designation by one aircraft, with weapon delivery conducted by an accompanying aircraft.

Operational status
In production and service.

Contractor
Northrop Grumman Corporation.

UPDATED

AN/AXX-1 Television Camera Set (TCS)

The AN/AXX-1 Television Camera Set (TCS) is the US Navy's version of the US Air Force's AN/ASX-1 TISEO but offers enhanced capabilities. It is a passive, daytime, automatic search and acquisition system which can be either manually operated or slaved to an air interception radar. The system not only acquires targets automatically, but presents multiple fields of view and is operational on US Navy Grumman F-14A aircraft.

Operational status
In service on F-14.

Contractor
Northrop Grumman Corporation.

VERIFIED

Helicopter Night Vision System (HNVS)

Northrop Grumman has designed and integrated a Helicopter Night Vision System (HNVS) to a Sikorsky detailed specification and is now the major subsystem integration contractor for the system being developed by Sikorsky Aircraft. This system is being installed in CH-53 Super Stallion transport helicopters for use on low-level tactical missions in adverse weather conditions. The system is based on equipment similar to that installed by Northrop Grumman on 10 AH-1S helicopters which were used as surrogate trainers for AH-64 pilots.

The system includes a data entry panel, master control assembly, the Honeywell Integrated Helmet And Display SubSystem (IHADSS) and control panel, video monitor/recorder and control panel, power

The AN/AXX-1 TCS system installed in a US Navy F-14

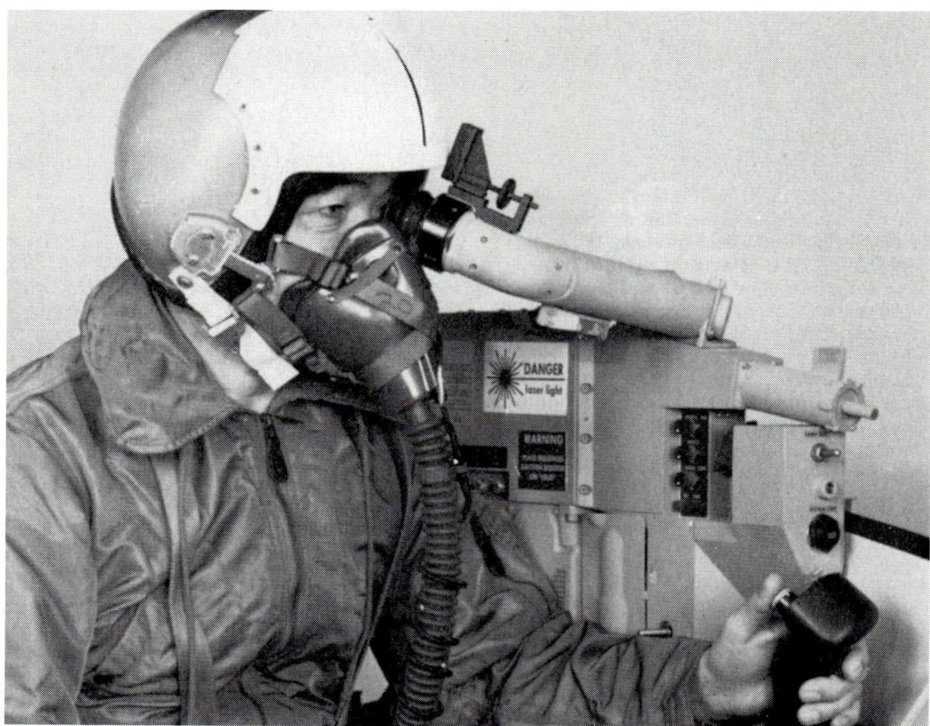

The AN/AVQ-27 laser target designator set

distribution unit, heater/filter assembly, system control electronics, vapour cycle unit, symbology generator, multiplexing remote terminal, control grips and the Lockheed Martin Pilot Night Vision System (PNVS).

The PNVS is a thermal imaging system which optically senses the heat emitted by objects and converts it into video image for display to the pilot and co-pilot. The PNVS, mounted as a turret on the chin of the CH-53E, is controlled by the IHADSS worn by the pilot or co-pilot. The IHADSS monitors head position and converts head movement into azimuth and elevation commands which are sent to the PNVS. Video from the turret is displayed and flight symbology is overlaid on the cockpit displays, allowing either the pilot or the co-pilot to fly the aircraft in a head-up attitude.

Operational status
In production and in service.

Contractor
Northrop Grumman Corporation.

VERIFIED

The Northrop Grumman sensor suite installed in the mast-mounted sight of a US Army OH-58D helicopter

Mast-mounted sight sensor suite

Northrop Grumman is producing the sensor suite for the mast-mounted sight of the US Army's Bell OH-58D helicopters. The suite consists of a FLIR, a day television sensor, laser rangefinder/designator, an advanced digital tracker and an automatic boresight device.

The mast-mounted sight has also been installed in US Navy ships.

Operational status
In production.

Contractor
Northrop Grumman Corporation.

UPDATED

Video Augmented Tracking System (VATS)

Northrop Grumman developed the Video Augmented Tracking System (VATS) for retrofit to the AN/AVQ-26 Pave Tack system aircraft. Pave Tack is an air-to-ground, laser designated weapons delivery system mounted in a pod containing a FLIR sensor. VATS automatically tracks ground targets, eliminating or reducing the weapons operator's need to follow the monitor by eye. It is microprocessor-controlled, with a self-contained power supply and a digital interface to the aircraft's flight computer.

Operational status
Ready for production; orders not yet placed.

Contractor
Northrop Grumman Corporation.

VERIFIED

AN/ASQ-153 Pave Spike laser designator/ranger

The Pave Spike development programme was initiated in 1971 and delivery of 156 pod sets to the US Air Force was completed by August 1977, by which time 327 F-4D Phantoms had been converted to accept the system. A further 82 sets for foreign use were delivered up to September 1979, including a substantial number for the Royal Air Force and some for the Turkish Air Force.

The system is contained within an externally mounted pod, the nose section of which revolves about the pod axis to provide roll stabilisation, and a cylindrical forward portion which rotates in pitch to provide elevation stabilisation. Virtually complete lower hemisphere coverage is thus provided in a relatively compact and light arrangement. The nose section is sealed and pressurised with nitrogen, maintained at a constant temperature for optimum sensor performance. The centre section provides umbilical connections between the nose and rotating sections, the aircraft and the aft electronics system. In the aft section is a cold plate on to which are mounted the electronic LRUs. These comprise a low-voltage power supply and pod control, servo drivers, laser control, laser power supply and interfaces. The pod contains a television tracking sensor and laser designator/ranger. The television sensor can be used for target acquisition and the designator permits accurate delivery of laser-guided munitions. Laser ranging can be used to improve the delivery accuracy of conventional weapons.

The overall AN/ASQ-153 system comprises the AN/AVQ-23 pod and several system components in the aircraft. These include a line of sight indicator, control panel, range indicator, modified radar control handle and weapon release computer. The system can be used with Paveway laser-guided bombs and several other laser-guided munitions.

Specifications
Dimensions: 3,660 (length) × 205 mm (diameter)
Weight: 193 kg
Wavelength: 1.06 μm

Operational status
In service. Initially the unit was procured for use only on US Air Force F-4D and F-4E Phantoms, but it is now also employed on several other types.

Contractor
Northrop Grumman Corporation, Electronic Sensors and Systems Division.

UPDATED

Electro-Optical Surveillance and Detection Systems (EOSDS)

EOSDS is a technology upgrade to an electro-optical system developed by Northrop Grumman called Night Giant, which comprises a forward-looking infrared sensor and television camera. The system will perform a variety of surveillance and detection missions as part of the Royal Air Force Nimrod MRA4 mission system.

Operational status
Northrop Grumman will deliver 21 EOSDS systems to The Boeing Company, as mission system integrators, with the first system to be delivered in October 1998.

Contractor
Northrop Grumman Electronic Sensors and Systems Division.

NEW ENTRY

Integrated MultiSensor System (IMSS)

The Integrated MultiSensor System (IMSS) was initially configured specifically to perform day and night all-weather air interdiction and maritime patrol as part of the anti-drug efforts of the US Customs Service, a role in which it has been exceptionally successful. By combining and integrating a high-performance multimode radar and an infrared imaging system, IMSS is also effective in performing reconnaissance, surveillance and search and rescue missions.

The IMSS consists of the AN/APG-66 radar fully integrated with an infrared detection set such as the Northrop Grumman WF-360, controls and displays and an inertial navigation system.

The AN/APG-66 is a digital, coherent, multimode radar system developed to serve both the strike and fighter demands of the F-16 aircraft. It provides long-range detection and acquisition of targets at all altitudes and aspect angles in the presence of heavy background clutter. The current version of the APG-66 incorporates a new signal data processor which replaces several radar units and eliminates the need for special interface hardware when used in IMSS. The result is significant improvement in system weight, cooling, power, capability and reliability.

The primary function of the infrared system is short-range tracking and observation of airborne, maritime and ground targets. As integrated in the IMSS, the infrared capability has, as a passive system, proved exceptionally effective in the covert tracking of aircraft. The infrared system can be applied independently to track a designated target by locking on to the heat differential generated by the target, it can be slaved to the radar so that the infrared line of sight follows radar tracked airborne targets and it can be directed to acquire and track ground targets automatically, using data from the INS. Additional capabilities include a TV camera, laser rangefinder and video recorder.

The IMSS controls and displays are integrated with the radar, infrared and INS through the radar signal data processor. A hand control unit provides the sensor operator with single-hand slew control of both sensors, including antenna elevation, radar cursor in azimuth and range, target designation and infrared line of sight. Separate displays are provided for the radar and the infrared, and these displays can be duplicated at several stations within the aircraft.

The INS provides the inertial references required by the radar and infrared systems, such as roll, pitch, yaw, heading and velocities in three axes.

The IMSS incorporates continuous self-testing and BIT to isolate a fault down to an easily replaceable unit.

Contractor
Northrop Grumman Corporation, Electronic Sensors and Systems Division.

VERIFIED

WF-360 surveillance and tracking infrared system

The WF-360 system features a Forward-Looking InfraRed (FLIR) sensor operating in the 8 to 12 μm long-wave spectrum, with options for adding a high-resolution day TV camera and an eye-safe laser rangefinder. These sensors are boresighted together on an optical bed housed in a stabilised gimbal designed for aircraft applications. The system provides high-resolution day and night detection and tracking of airborne, maritime and land-based targets. System capabilities include passive search and track in both air-to-air and air-to-surface modes. ARINC and MIL-STD-1553 bus interfaces are provided to communicate with other aircraft sensors such as the AN/APG-66/68 radar, flight management systems and inertial navigation systems. A high throughput processor provides cueing of the system to radar targets or navigation waypoints, real-time display of the track point co-ordinates and fire control solutions.

The Northrop Grumman WF-360 surveillance and tracking infrared system

The WF-360 infrared sensor marries US Army common modules with electronics providing DC restoration, automatic gain and level control and digital scan conversion for a standard US RS-170 525 line video output. These features, together with an advanced digital noise reducer and image enhancer, provide superior video imagery. Virtually all FLIR systems in the US DoD inventory use common module parallel scan long-wave technology because of its capability for widely varying operational scenarios throughout the world.

The aerodynamically streamlined turret utilises a two-axis stabilised platform to stabilise the optical lines of sight and point them throughout the entire lower hemisphere with look-up limited only by the aircraft structure. A broadband servo system featuring solid-state rate sensors and an inner acceleration loop provides stabilisation better than 35 μrad in typical aircraft environments ranging from helicopters to the US Air Force A-10. The turret is environmentally self-contained, employing an internal liquid-to-air heat exchanger to allow proper operation of the turret without the need for bleed or cabin air.

The WF-360's astronomical telescope provides two fields of view: a 4.5° narrow field of view and an 18.5° wide field of view. The TV sensor uses a ×6 zoom lens which can match either FLIR field of view or zoom continuously.

In addition to the TV and laser rangefinder, other optional plug-in modules provide capabilities including FLIR/TV image fusion, wide area correlation video tracking and covert laser illumination for night TV imagery. The computer-aided track mode, in conjunction with the automatic video trackers, provides an excellent coast mode in the event of short-term target obscuration.

Operational status

Used extensively for surveillance and drug interdiction by the US Coast Guard in the HU-25 Falcon jet and RU-38, the US Air Force in the C-26 and the US Army in the DH-7.

An improved system, featuring the second-generation common module 480 × 4 scanned focal plane array is available.

Contractor

Northrop Grumman Corporation, Electronic Sensors and Systems Division.

UPDATED

Colour video HUD cameras

Photo Sonics manufactures a family of colour video HUD (Head-Up Display) cameras for use on the A-4M, A-10 and F-16C/D aircraft; detail specification changes match the cameras to each aircraft type. All cameras utilise a mix of MIL-STD and COTS components; they incorporate extensive filtering to provide noise-free, high quality imagery, with automatic exposure control. The camera assembly, comprising periscope and colour camera, is mounted on the HUD to record what the pilot is viewing and the symbology on the HUD. Data is output to an onboard video cassette recorder.

Optical specifications

	Lens, focal length	FOV degrees		FOV milliradians	
		H	V	H	V
A-4M	16.0 mm, f1.4	22.6	17.1	395	298
A-10	25.0 mm, f2.5	14.6	11.0	255	191
	16.0 mm, f1.4	22.6	17.1	395	298
	15.0 mm, f1.4	24.1	18.2	420	317
F-16C/D	16.2 mm, f1.4	22.4	17.1	390	294.1

Specifications

Horizontal resolution: (NTSC) 470 TV lines
(PAL) 460 TV lines
Picture elements: (NTSC) 768 (H) × 494 (V)
(PAL) 752 (H) × 582 (V)
Power: 115 V AC 47-440 Hz
Weight: 1.9 kg

Contractor

Photo-Sonics Inc.

NEW ENTRY

Airborne Electro-Optical Special Operations Payload (AESOP)

AESOP, designated the AN/AAQ-16D, combines the AN/AAQ-16 Helicopter Night Vision System (HNVS) (see following item) with a three field of view telescope and a laser designator/rangefinder. The Raytheon Systems Company recently flight-tested an upgrade to the HNVS, called Hi-Mag, which provides both the high magnification capability required for target identification and the wide field of view needed for safe night and low-visibility pilotage.

The AESOP system is based on the AN/AAQ-16B Hi-Mag design, which upgraded the HNVS with a telescope featuring wide, medium and narrow fields of view, an enhanced autotracker and improved stabilisation. It also incorporates a lightweight laser target designator. Prototype systems will be used to demonstrate the ability to detect, recognise, track and direct Hellfire missiles to tactical targets. The system is designed for compatibility with the newer helicopters, in addition to test aircraft.

Operational status

In July 1994, a contract was awarded by the US Army for 15 systems, with options for a further 25 for installation on MH-60L Black Hawk and AH-6J Little Bird helicopters.

Contractor

Raytheon Systems Company.

UPDATED

AN/AAQ-15 infrared detection and tracking set

The AN/AAQ-15 is a small lightweight infrared tracking set that has been developed under contract to the USAF for fitting to the HH-60 helicopter and the MC-130H Combat Talon II aircraft.

The system uses the infrared receiver unit of the AN/AAS-36 with new electronics and lightweight gimbals. It has three fields of view, covers all of the lower hemisphere and has a 15° unobstructed look-up capability.

Operational status

In service.

Contractor

Raytheon Systems Company.

UPDATED

AN/AAQ-16 night vision system

The AN/AAQ-16 infrared imaging system entered initial production for the US Department of Defense in 1984. It has been selected by the US Army, Navy, Marine Corps and Air Force and international customers for a variety of helicopters.

The system provides 24-hour mission capability during night or degraded weather conditions in support of special operations, air assault, search and rescue, and anti-surface warfare.

An advanced thermal imaging system, AN/AAQ-16's high-resolution, TV-like imagery provides for low-level pilotage and navigation in the wide field of view, and in the narrow field of view, provides long-range target detection and identification. The AN/AAQ-16 combines automatic FLIR performance optimisation, and DC

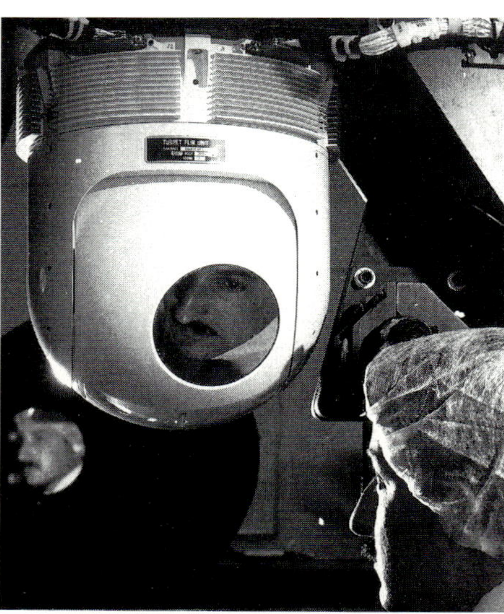
Close-up of the FLIR turret of the AN/AAQ-16 night vision system

restoration to respond to various environmental conditions, with hands-off features of automatic gain, level, and focus for a high-quality image and eased operator workload. Either a black-hot or white-hot image can be selected by the operator. Additional features which enhance mission success include an autotracker and operator-controlled autoscan.

Variants of the AN/AAQ-16 system include: the AN/AAQ-16B which utilises the 8 to 12 μm waveband and two fields of view (30 × 40° and 5 × 6.7°); the AN/AAQ-16C which also utilises the 8 to 12 μm waveband and three fields of view (30 × 40°, 5 × 6.7° and 1.9 × 2.5°) and a dual-mode tracker.

Specifications

Dimensions:
(FLIR turret) 356 (length) × 305 mm (diameter)
(electronics unit): 305 × 200 × 412 mm
(multifunction control unit) 76.2 × 76.2 × 157.5 mm
(system control unit) 38.1 × 146 × 115 mm
Weight:
(FLIR turret) 24.49 kg
(electronics unit) 20.87 kg
(multifunction control unit) 0.59 kg
(system control unit) 0.45 kg
Wavelength: 8-12 μm
Field of view:
(×1 magnification) 30 × 40°
(×6 magnification) 5 × 6.7°
(×16 magnification option) 1.9 × 2.5°
Reliability: >300 h MTBF

Operational status

More than 400 systems delivered or on order for the US Army UH-60L, CH-47D, MH-47E, MH-60K; US Air Force HH/MH-60G, US Marine Corps CH-53E, Australian S-70B. Co-produced with NEC for the Japanese Defence Force.

Contractor

Raytheon Systems Company.

UPDATED

AN/AAQ-16-27 MWIR staring sensor

The AN/AAQ-16-27 Mid-Wave Infrared (MWIR) staring sensor is an advanced infrared system that features substantial commonality with the combat-proven AN/AAQ-16B infrared system. It uses non-developmental, in-production, components to provide higher resolution imagery than current long-wave infrared systems, and incorporates an MWIR Indium-Antimonide (InSb) staring Focal Plane Array, with 480 × 640 detector elements.

The system features a turreted FLIR weighing 22.7 kg, with a total system weight under 42.3 kg. Options include a video autotracker, third field of view and eyesafe laser rangefinder.

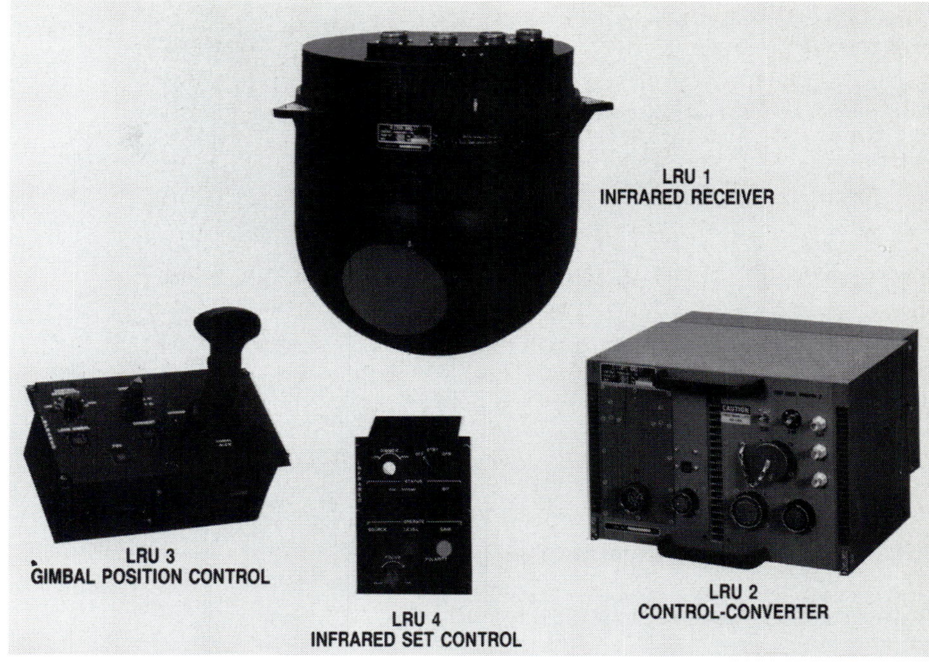

The AN/AAQ-17 infrared detecting set

Specifications

Field of view: 30 × 40° and 5 × 6.7°
Wavelength: 3-5 μm
Staring FPA: unprecedented imagery with less than half the aperture size
High performance staring array: 640 × 480 (InSb)
Options: third field of view; electronic zoom; dual-mode tracker; laser rangefinder

Operational status

The advanced generation system is in production for the US Marine Corps MV-22 Osprey, and on order for the Royal Australian Navy SH-2G helicopter.

Contractor

Raytheon Systems Company.

UPDATED

AN/AAQ-17 infrared detecting set

The AN/AAQ-17 infrared detecting set is a multipurpose thermal imaging system. Typical missions are navigation, search and rescue, surveillance and fire control. AN/AAQ-17 systems are fitted on US Air Force AC-130A, AC-130H and AC-130U gunships, HC-130P and HC-130 tankers and C-141 special operations transports.

In the US Air Force gunship programme the AN/AAQ-17 FLIR replaces the AN/AAD-7 FLIR on the AC-130H, providing improved performance and reliability at lower cost. The AN/AAQ-17 is a derivative of the AN/AAQ-15. Both of these systems are lightweight and easily adapted to new aircraft. Originally, the AN/AAQ-15 FLIR was developed for the US Air Force HH-60 Nighthawk helicopter and has also been produced for the MC-130H Combat Talon II aircraft.

The AN/AAQ-17 consists of four LRUs: the infrared receiver (LRU1), control-converter (LRU2), gimbal position control (LRU3) and infrared set control (LRU4). Environmental qualification, reliability demonstrations and maintainability demonstrations are complete. The AN/AAQ-17 uses standard DoD FLIR common modules to convert long wavelength radiation into a composite TV video.

A 13.7 × 18.3° wide field of view allows navigation, area search and detection of larger targets. The 3 × 4° narrow field of view allows small target detection and target recognition. A precision gimbal provides accurate line of sight angular measurement. An adaptive gate video tracker reduces operator workload by providing hands-off automatic line of sight control. Operation of the FLIR is through manual controls or over a MIL-STD-1553B databus. This produces a flexible system adaptable to many aircraft.

Specifications

Weight: (infrared receiver) 43.09 kg
(control converter) 21.77 kg
(gimbal position control) 2.27 kg
(infrared set control) 1.81 kg
Field of view: (wide) 13.7 × 18.3°
(narrow) 3 × 4°
Field of regard: (azimuth) ±200°
(elevation) +15 to −105°

Operational status

In service in Lockheed Martin AC-130A, AC-130H, AC-130U, HC-130N and HC-130P and C-141 aircraft.

Contractor

Raytheon Systems Company.

UPDATED

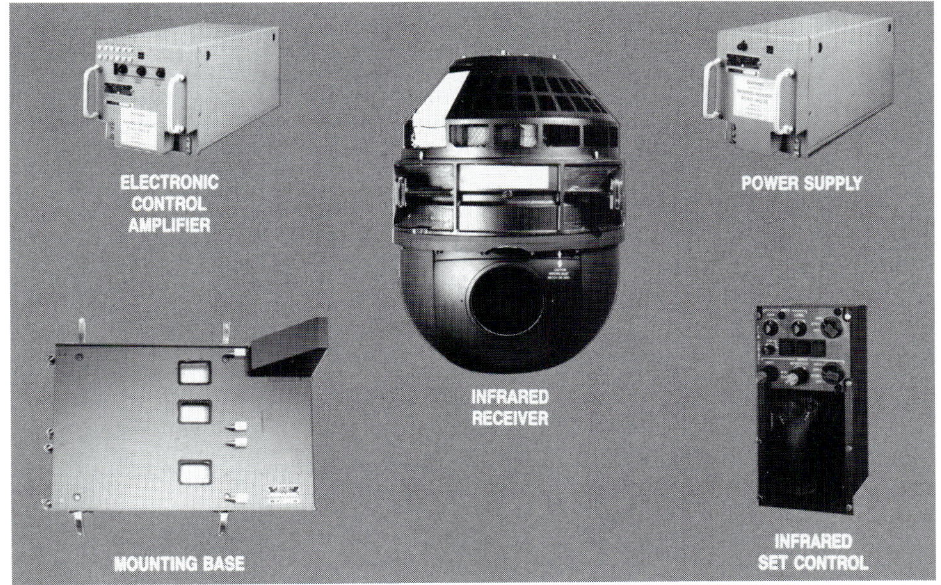

The AN/AAQ-18 FLIR system

AN/AAQ-18 forward-looking infrared system

The AN/AAQ-18 forward-looking infrared system is a common module update programme. The system improves reliability and readiness of the US Air Force Special Forces by replacing the AN/AAQ-10 system. System installations include the MH-53J helicopter and the MC-130E Combat Talon aircraft.

The primary mission of the AN/AAQ-18 is navigation.

It is also ideal for search and rescue missions. The AN/AAQ-18 consists of 5 LRUs.

Specifications

Power supply: 115 V AC, 400 Hz, 3 phase, 3kVA
28 V DC, 2 A
Field of view:
(wide) 18.2 × 13.7°
(narrow) 4.1 × 3.1°
Field of regard:
(azimuth) ±190°
(elevation) +15 to −105°

Operational status

Operational in the MH-53J helicopter and US Special Forces MC-130E Combat Talon aircraft.

Contractor

Raytheon Systems Company.

UPDATED

AN/AAQ-26 infrared detecting set

In August 1997, US Air Force Aeronautical Systems Command awarded Raytheon Systems Company a contract for the fabrication of 16 AN/AAQ-26 gunship infrared detecting sets and their installation on 9 AC-130U and 7 AC-130H aircraft. The FLIR provides the aircrew with the capability to operate safely outside the threat area, at greater standoff range.

The AN/AAQ-26 FLIR is based upon the the AN/AAS-44(V) product, which is the US Navy's Lamps Mk III FLIR set.

Operational status

In production for US Air Force AC-130H and AC-130U aircraft. Installation to be completed by 1999. The original development contract began in 1995.

Contractor

Raytheon Systems Company.

NEW ENTRY

AN/AAQ-27 (3 FOV) mid-wave infrared imaging system

AN/AAQ-27 (3 FOV) is a third-generation, 3 field of view, mid-wavelength infrared (MWIR) system for helicopter navigation, surveillance, and targeting applications. With three fields of view available, pilots can fly and navigate on low-level missions or detect and identify long-range targets from higher altitudes. Raytheon state that the system uses non-developmental production components, to provide good reliability, but that it provides higher-resolution imagery and better range performance than current long-wavelength infrared systems.

The AN/AAQ-27 (3 FOV) is a modified version of the AN/AAQ-27 infrared system which Raytheon (ex-Hughes) produces for the US Marine Corps MV-22 Osprey. It features substantial commonality with the combat proven AN/AAQ-16 infrared system. The AN/AAQ-27 (3 FOV) is compatible with existing AN/AAQ-16 system mountings and is easily interchangeable for upgrade programmes.

AN/AAQ-27 (3 FOV) mid-wave infrared imaging system showing the system electronic unit (above) and turret FLIR unit (right)

Specifications

Detector: staring array (640 × 480 InSb); spectral band 3-5 microns
3-FOV:
30 × 40° – 1×
5.0 × 6.7° – 6×
1.3 × 1.73° – 23×
Unit dimensions:
turret FLIR unit (TFU): 360.4 (height) × 303.5 mm (diameter)
systems electronics unit (SEU): 199.1 (height) × 413.5 (length) × 306.3 mm (width)
System weight: 42.27 kg
Tracker: dual-mode (centroid/correlation)
Options: eye-safe laser rangefinder; laser designator/rangefinder

Operational status

Selected by Kaman Aerospace International Corporation for the Royal Australian Navy Super Seasprite helicopter programme. The 3 FOV system provides over-water navigation and long-range target detection and classification. Deliveries are scheduled for April 1999 to February 2000.

Contractor

Raytheon Systems Company.

NEW ENTRY

AN/AAR-42 infrared detection set

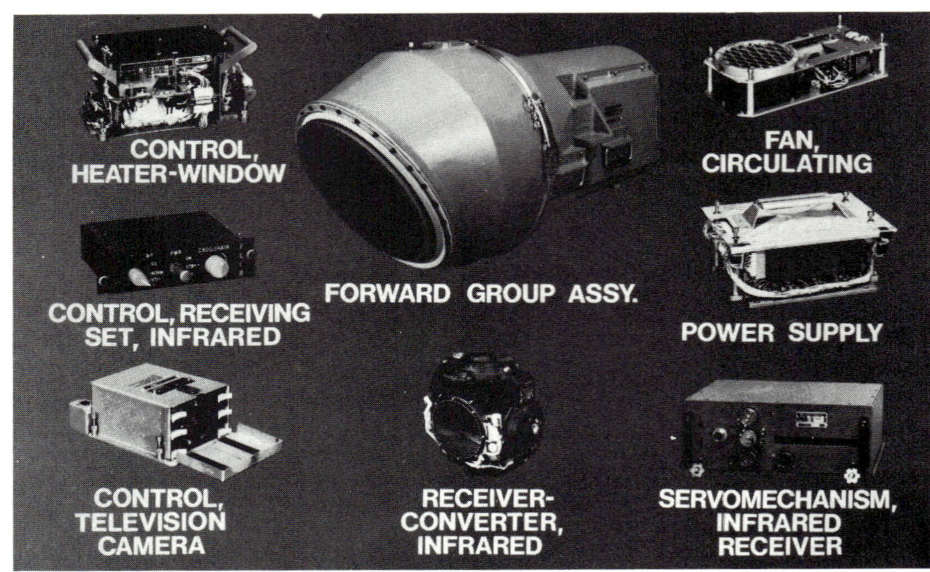

The AN/AAR-42 FLIR system

The AN/AAR-42 is a FLIR system for use on the US Navy A-7E strike aircraft. It is designed to provide a night window or bombsight which permits the pilot to perform single-seat close air support and reconnaissance missions by day or night and during poor weather conditions.

The AN/AAR-42 is installed in a pod with a gimballed FLIR unit which provides stabilised imagery on the pilot's head-up display system. It provides an azimuth coverage of ±20° and an elevation coverage from +5 to −35°. Both wide and narrow selectable fields of view are provided. The wide field of view, giving a ×1 magnification, is employed for pilot orientation, navigation update and target acquisition, while the narrow field of view, with a ×4 magnification, is used for target identification and weapon delivery. Features include automatic thermal focus compensation and sensor window de-icing.

Specifications

Weight:
(canister assembly) 95.45 kg
(servo electronics) 18.18 kg
Reliability: 390 h MTBF

Operational status

In service. The system is currently operational on the A-7E.

Contractor

Raytheon Systems Company.

UPDATED

AN/AAR-45 thermal imaging sight

The AN/AAR-45 forms a part of the Target Recognition and Attack Multisensor (TRAM) suite on the US Navy's A-6E aircraft, in which role it operates in conjunction with a GEC-Marconi Avionics head-up display.

The thermal imager is mounted on a gimbal system in a pod carried beneath the aircraft and the resultant picture is displayed with flight information on the HUD. This way the pilot can fly head-up during a low-level attack by night. The system is also linked to the aircraft inertial navigation system so that the imager is always kept pointing at the target, even during manoeuvres.

Operational status

Production of the AAR-45 is complete with 91 systems delivered.

Contractor

Raytheon Systems Company.

UPDATED

AN/AAR-50 Navigation FLIR (NavFLIR)

The AN/AAR-50 Navigation FLIR (NavFLIR) is a derivative of the AN/AAQ-16 night vision system installed in US Army helicopters. It uses a thermal imaging sensor to provide pilots of fixed-wing aircraft on low-level missions at night or in bad weather with a

TV-like image of the terrain ahead projected on to a head-up display. The system as configured for the F/A-18 is pod-mounted in a fixed forward-staring position but could be configured in different pods for a variety of aircraft.

NavFLIR consists of four major weapon-replaceable assemblies: FLIR sensor unit, pod electronics unit, thermal control unit and pod adaptor.

Specifications

Dimensions: 1,981 × 254 mm diameter
Weight:
(pod) 73.48 kg
(adaptor) 23.13 kg
Wavelength: 8-12μm
Field of view: 19.5 × 19.5° displayed
Reliability: >410 h MTBF

Contractor

Raytheon Systems Company.

UPDATED

The AN/AAR-50 NavFLIR mounted on the starboard intake of a US Navy F/A-18

AN/AAS-36 infrared detection set

The AN/AAS-36 infrared detection set is a FLIR system designed for US Navy P-3C maritime patrol aircraft to detect surface vessels, surfaced or snorkelling submarines and drifting survivors in darkness and limited visibility. The system was initially designed and developed to meet a P-3C update programme requirement, but the equipment has also been retrofitted to earlier P-3C and P-3B aircraft.

Production of the system commenced in 1977 following a testing, evaluation and demonstration programme which used 10 preproduction systems to assure the US Navy that design specifications were either met or exceeded. The Initial Operational Capability (IOC) was realised in 1979.

Based on US Department of Defense common modules which employ Cadmium Mercury Telluride (CMT) detectors, the AN/AAS-36 is a stand-alone system requiring only electrical power for operation. The common module infrared receiver is mounted in an azimuth-over elevation stabilised gimbal and provides lower hemisphere coverage of ±200° in azimuth and from +16 to −82° in elevation. Additional weapon-replaceable assemblies provide system power, servo control, FLIR system control, slew commands and a real-time video display. The display presentation is on an 875-line RS-343 composite television monitor which permits the operator to identify, as well as observe, vessels.

Features include automatic optical temperature compensation, gimbal pointing outputs for servo platform slaving, self-contained stabilisation and a two field of view optical system (15 × 20° or 5 × 6.7°). A digital computer interface is available for on-line gimbal control. The system contains built-in self-test facilities which permit checkout down to weapon-replaceable assembly level and these themselves are compatible with automatic test equipment.

The system is also being supplied to many non-US operators of the P-3 for upgrading to US Navy standards. The receiver-converter weapon-replaceable assembly of the AN/AAS-36 has been fitted to Sikorsky CH-53 helicopters and Cessna Citation, Beechcraft E-90, King Air 200 and other unspecified aircraft.

Specifications

Weight: 136.36 kg
Power supply: 115 V AC, 400 Hz, 3 phase, 2.5 kVA
28 V DC, 100 W

Operational status

In service.

Contractor

Raytheon Systems Company.

UPDATED

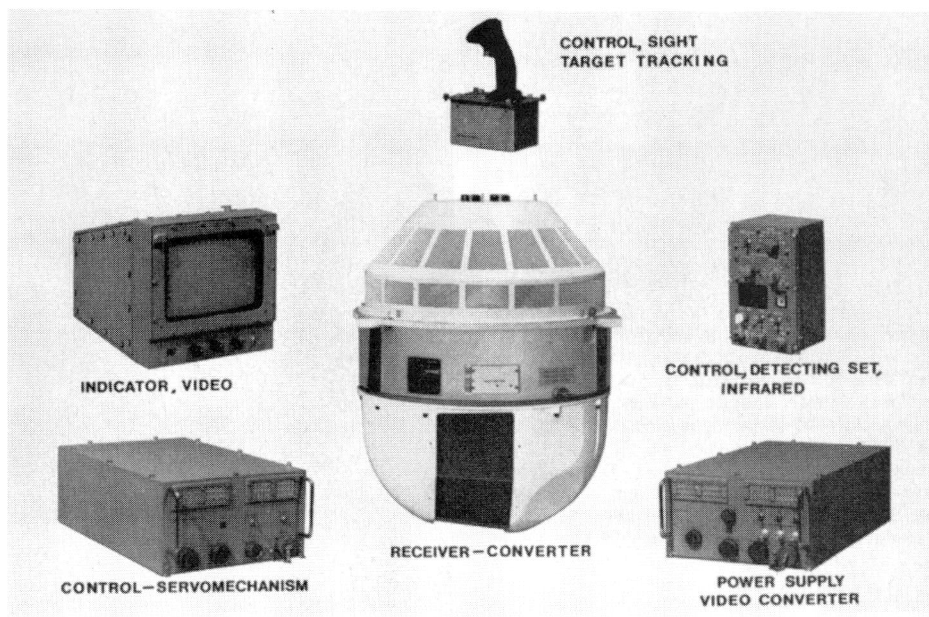

The AN/AAS-36 FLIR system is used in US Navy P-3C aircraft

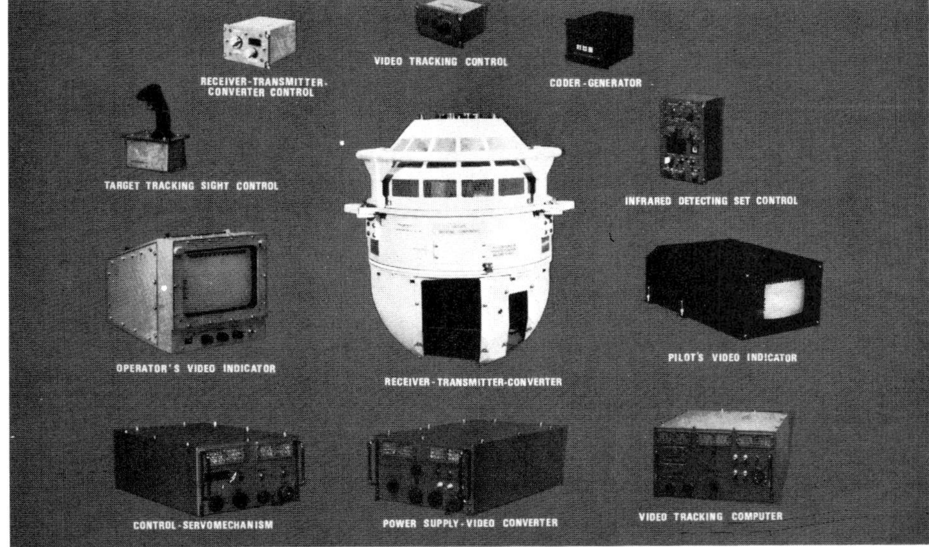

The AN/AAS-37 FLIR system was developed for US Marine Corps OV-10D aircraft

AN/AAS-37 infrared detection set

The AN/AAS-37 equipment is a variant of the AN/AAS-36 detection set but is a more sophisticated system, being combined with a laser designation and ranging capability. Developed for the US Marine Corps OV-10D forward air control aircraft, the AN/AAS-37 provides infrared vision for day or night operations under degraded environmental conditions, automatic target tracking and laser target-designation. The laser provides ranging and illumination of ground targets for laser-guided weapons such as the Paveway bomb or the Hellfire missile, either on an autonomous basis for aircraft equipped with the AN/AAS-37 or for a co-operating aircraft armed with these weapons.

Specifications of the FLIR sensor and associated equipment are virtually identical to those of the AN/AAS-36 system from which it was developed. It does, however, possess a number of additional features derived mainly from incorporation of the laser section. These include direct readout laser ranging and designation capability. The system can be used as a target sight aligned with the aircraft boresight by means of electronic adjustment and line of sight depression may be set by the operator for precision delivery of air-to-ground weapons. There are interfaces with aircraft systems for a radar altimeter and remote gyroscope and an accelerometer provides the system with rate-aided automatic target tracking capability using an adaptive gate centroid tracker. Offset tracking from a target or another landmark is also possible. The system's display is daylight visible.

Specifications

Laser designator/ranger
Coverage:
(azimuth) ±200°
(elevation) −82 to +16°
Weight: 189 kg
Power supply: 115 V AC, 400 Hz, 3 phase, <3 kVA
28 V DC, <1,800 W

Operational status

In service.

Contractor

Raytheon Systems Company.

UPDATED

AN/AAS-38A F/A-18 targeting pod

The AN/AAS-38A Forward-Looking Infrared (FLIR) targeting pod enables pilots of US Navy and Marine Corps F/A-18 Hornet aircraft to attack ground targets day or night with a precision strike capability. The system provides TV-like infrared imagery on a cockpit panel display and accommodates a laser rangefinder/designator that can pinpoint targets for both laser-guided and conventional weapon delivery.

The pod is integrated with the aircraft's avionics system through a MIL-STD-1553 databus, allowing the pod to receive command and cue signals from the onboard mission computer and provide status and targeting information to the cockpit display and weapon delivery system.

The AN/AAS-38A consists of 12 weapon replaceable assemblies (WRAs) that can be readily accessed and replaced without the need for calibration, alignment, special tools or handling equipment.

The AN/AAS-38A configuration accepts the two laser subsystem WRAs, a laser transceiver and laser power supply to provide the aircrew with the capability for laser target designation and ranging (LTD/R).

Design qualification and flight test were successfully completed in August 1994 thus providing the US Navy a second source for AAS-38 pods and spares (see also Lockheed Martin Electronics & Missiles entry entitled NITE Hawk).

The Targeting FLIR, when integrated with the AN/AAR-50 Navigation FLIR and Night Vision Goggles, provides the pilot/aircrew with the capability to maintain situational awareness, navigate/avoid terrain, acquire/designate targets and assess battle damage for deployment of the Pave Way/GBU-24 precision-guided weapon series.

Specifications:

Dimensions: 1,830 (length) × 330 mm (diameter)
Weight: 172.7 kg
NFOV: 3 × 3°
WFOV: 12 × 12°
Field of regard:
(pitch) +30° to −150°
(roll) ±540°
Video: RS-343 875 lines

Contractor

Raytheon Systems Company.

UPDATED

ATFLIR Advanced Targeting Forward-Looking InfraRed system

The ATFLIR pod includes both infrared targeting and navigation systems. The targeting FLIR uses the same third-generation Mid-Wave InfraRed (MWIR) staring focal plane technology that has been used in the US Marine Corps MV-22 Osprey (the AN/AAQ-16-27 system). The navigation FLIR will be provided by GEC-Marconi.

The design aim of the ATFLIR system is to achieve sufficiently accurate long-range performance for F/A-18E/F crews to be able to deliver their air-to-ground weapons from beyond the range of defensive anti-aircraft artillery and many surface-to-air missile systems.

Operational status

In development. Selected for the US Navy F/A-18E/F aircraft, and for retrofit to existing F/A-18 C/D models. The EMD contract is expected to be for 8 to 10 systems for qualification and flight testing in mid-1999, with initial low-rate production to follow, and an initial operational capability in mid-2002.

Contractor

Raytheon Systems Company.

NEW ENTRY

Cobra-Nite airborne TOW system upgrade

Cobra-Nite (C-Nite) is the designation given to the night targeting version of the airborne TOW missile system. This was developed to provide an enhanced night and adverse visibility capability for Cobra helicopters equipped with TOW missiles.

C-Nite also provides the Cobra with a target sight for use with unguided rockets and during conventional gun attacks. The system has been used effectively during exercises to monitor hostile forces and to direct ground troops from the helicopter, resulting in envelopment of the enemy by friendly forces.

The US Army Cobra attack helicopter is equipped with the Cobra-Nite night targeting system ***1995***

Operational status

In production. Deliveries began in 1989. Approximately 70 sets have been built for the US Army and 21 systems have been supplied to the US National Guard for installation in AH-1F helicopters. C-Nite systems are also being supplied to Bahrain and are being co-produced in Japan.

Contractor

Raytheon Systems Company.

UPDATED

Control By Light (CBL)

Raytheon's Control By Light (CBL) system provides significant reductions in aircraft weight, wiring complexity, and system integration cost. It is also fault tolerant giving increased safety and reliability, whilst at the same time providing immunity from lightning and other electromagnetic compatibility problems.

Design features include; independent bi-directional data flows; 1.25 Mbit/s data rate; message regeneration and other quality audit activities.

Operational status

Now being certified for FAA Part 25 operation by Raytheon.

Contractor

Raytheon Systems Company.

UPDATED

DB-110 Dual-Band reconnaissance system

The DB-110 Dual-Band reconnaissance system is a powerful day/night long-range sensor, combining electro-optical and infrared (IR) thermal imaging capabilities in a compact, lightweight, design. Modular design permits packaging as an internal fit, or within a pod (in a 275 gallon/1,500 litre fuel tank enclosure).

The high resolution achieved by the DB-110 satisfies intelligence requirements for active target detection and identification from long standoff range.

Multiple operational modes provide mission planning flexibility to interleave wide area search, spot collection, and target tracking/stereo modes. Both pre-programmed and manual control modes are provided, control being via a MIL-STD-1553B databus.

Specifications

Sensor
dimensions: 1,270 × 470 mm (diameter)
weight: 140 kg
power: 115 V AC, 400 Hz, 350 VA
Electronics
dimensions: 4 units: 257 × 206 × 422 mm; 356 × 343 × 89 mm; 257 × 175 × 66 mm; 107 × 287 × 81 mm
weight: 28 kg
power: 115 V AC, 400 Hz, 590 VA
Optical
type: Cassegrain reflector
focal length: to fit application
(visible) 2,800 mm nominal

DB-110 Dual-Band reconnaissance system ***1998***/0018327

(infrared) 1,400 mm nominal
aperture: 280 mm
(visible) f/10
(infrared) f/5
Focal planes
visible (0.5 to 1.0 micron): Silicon CCD array; 5,120 × 64 line array
infrared (2 to 5 micron): Indium antimonide (InSb) array; 512 × 484 area array; line arrays available
Data output: max data rate: 260 Mbits/s; data compression available to meet recorder or datalink budgets
Digital tape-recorder: data rate up to 240 Mbits/s continuously variable; 48 Gbytes on tape (equivalent to 20,000 nm^2)
Operation
standoff range: up to 45 km
altitude: 5,000-80,000 ft
ground speed: M0.1-M1.6
field of regard:
(across line of flight) 180°
(along line of flight) ±20°
type: panoramic/sector scan (4-28°)
overlap: variable from 0-100%
performance:
(visible) up to NIIRS 6
(infrared) up to NIIRS 5

Operational status
Selected by the Royal Air Force for Tornado GR. Mk 1A and GR. Mk 4/4A aircraft.

Contractor
Raytheon Systems Company.

NEW ENTRY

InfraRed Imaging Subsystem (IRIS)

The InfraRed Imaging Subsystem (IRIS) is part of the AN/AAS-38 FLIR imaging system. This is a self-contained pod designed for use on the US Navy F/A-18 aircraft for target acquisition and recognition and weapon delivery. It also allows reconnaissance under day and night and adverse weather conditions.

The IRIS includes dual fields of view of 12 × 12° and 3 × 3° with automatic thermal focus. Built-in image stabilisation provides a natural horizon display to the pilot.

A key feature of IRIS is the automatic video tracker contained in the controller-processor. This tracker provides automatic target acquisition, line of sight control and offset designation for accurate weapon delivery.

The controller-processor provides video processing necessary to produce 875-line RS-343 television video. The processor adds track and field of view reticles for cockpit displays. A MIL-STD-1553 digital databus provides communications with the aircraft AN/AYK-14 mission computer. The infrared receiver uses advanced switching regulator designs and high-density packaging to provide efficient primary to secondary power distribution, control and regulation for the IRIS and pod system.

Operational status
In service on the US Navy F/A-18 aircraft.

Contractor
Raytheon Systems Company.

UPDATED

RS-700 series infrared linescanner

The RS-700 airborne infrared linescanner operates in the 8 to 14 μm band, where absorption by carbon dioxide and water vapour is at a minimum. The Cadmium Mercury Telluride (CMT) detectors use the common module closed-cycle cooling subsystem. The optical system focuses radiation on to the detectors to produce video electrical signals that correspond with the picture formed by the radiation pattern scanned. The system converts the video signals to visible wavelengths by light-emitting diodes for recording on film.

The RS-700 is composed of three subassemblies mechanically mounted to form a single assembly for aircraft installation. Important operational features of the equipment include: manual or automatic gain selection, manual or automatic level control, video compression of unusually hot or cold objects; continuous scanning over whole velocity/height range, adjustable hot-spot marker, event marker and built-in test equipment.

Specifications
Weight:
(without roll stabilisation) 32 kg
(with roll stabilisation) 42 kg
Scan mirror facets: 4
Optical aperture: 38.4 cm^2
Detector cooling: closed-cycle, 77°K
Detector type: CMT
Recording light source: gallium-arsenide phosphor diodes
Film width: 70 mm
Film capacity: 46 m, 70 m
Velocity/height range: 0.2-5
Thermal resolution: 0.2°C
Spatial resolution: 0.5-1.5 mrad
Field of view: 120°

Operational status
Some 180 RS-700 series linescan systems have been supplied to the air forces of Denmark, Germany, Italy, Malaysia, Saudi Arabia, Singapore, Sweden, Switzerland and the USA.

Contractor
Raytheon Systems Company.

UPDATED

TIFLIR-49

TIFLIR-49 is a low weight, multiple purpose, thermal imaging sensor for navigation, surveillance, maritime search and rescue and troop-transport missions.

TIFLIR-49 includes two weapon replaceable assemblies: turret unit WRA-1 and electronics unit WRA-2. These units, together with off-the-shelf displays, controls and recorders provide a flexible, low-cost FLIR system for fixed- or rotary-wing aircraft.

TIFLIR-49 turret unit WRA-1 (below) and electronics unit WRA-2 (above) **1998**/0018314/5

Features include: a second generation focal plane array; electronic image stabilisation; dual-mode video tracker; multiple fields of view; local area processing; MIL-STD-1553B interface.

Specifications
Fields of view
(wide) 22.5 × 30°
(medium) 5 × 6.67°
(narrow) 1.3 × 1.7°
Electronic zoom: 2:1 and 4:1
Gimbal angular coverage
(azimuth) 360°
(elevation) +40 to −105°
Maximum airspeed: 300 kt

Operational status
Available. TIFLIR is understood to be essentially the same as the AN/AAS-44, but stripped of the laser, for international sales.

Contractor
Raytheon Systems Company.

UPDATED

CA-236 36 in E-O LOROP camera

Offering high performance in a mid-range focal length electro-optical camera, the CA-236 provides long-range oblique imaging in a highly compact configuration.

Key features of the CA-236 E-O LOROP camera include: panoramic scan, spot target acquisition and multi-aspect operation; medium- to high-altitude imagery collection; digital data captured to digital tape or transmitted to a ground station via a datalink; forward motion compensation (FMC); electronic time delay integration (TDI) for haze penetration and imaging under low-contrast conditions.

Specifications
Lens focal length and f/no: 36 in, f/8
Operating spectrum: 510-900 nm
Detector: 32 TDI silicon; 12,064 pixels
Field of view (FoV):
4,020 active pixels, 2.52°
6,024 active pixels, 3.77°
12,048 active pixels, 7.54°
Cross-track Field of Regard (FoR): horizon-to-horizon capable, window dependant
Cross-track scan size: variable 0.5-23°
Azimuth pointing angle: ±15°
Pointing accuracy: <0.2° (ref to inertial space)
Camera pupil diameter: 4.5 in
Camera resolution: 11 μrad/pixel
Camera output image data rate: 4,020 active pixels, 85.5 Mbit/s uncompressed

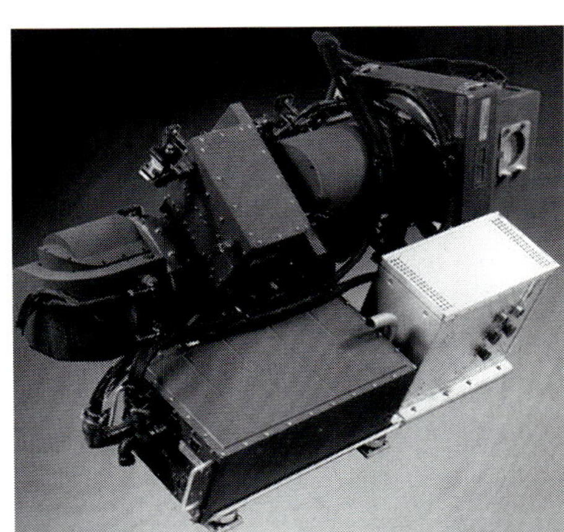

Recon/Optical Inc CA-236 E-O LOROP camera **1998**/0018324

An example of CA-236 E-O LOROP camera imagery at 2.5 × magnification **1998**/0018323

6,024 active pixels, 128 Mbit/s uncompressed
12,048 active pixels, 256 Mbit/s uncompressed
Compression: variable up to 14:1
Output image data: 8 bit ECL differential
No. of TDI stages: 1, 2, 4, 8, 16, 24, 32
Autofocus: continuous

Contractor

Recon/Optical, Inc.

NEW ENTRY

CA-260 and CA-260/25 E-O framing reconnaissance cameras

The CA-260 E-O framing camera is designed specifically to provide near-photographic quality images while enhancing the survivability of the tactical reconnaissance platform at low to medium altitudes. It is configured for external pod or internal aircraft mounting on a wide variety of reconnaissance platforms.

Advanced E-O framing technology reduces the amount of time required to cover a target area compared with conventional E-O linescan sensors. Wafer-scale processing has been used to develop the 4 M pixel or 25 M pixel array CCDs which are used in the CA-260 camera to give a wide field of view E-O image with continuous stereo coverage of targets at 56 per cent overlap. An on-chip motion compensation architecture eliminates the image blur normally associated with a wide field of view framing camera.

Because the mounting configuration and physical envelope are identical to that of the KS-87 film camera, the CA-260/4 or CA 260/25 can provide an interim E-O framing capability to users of existing KS-87 cameras.

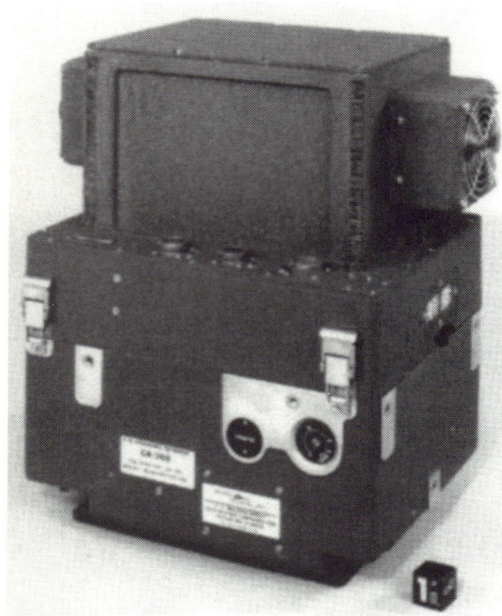

Recon/Optical Inc CA-260 E-O framing reconnaissance camera **1998**/0018322

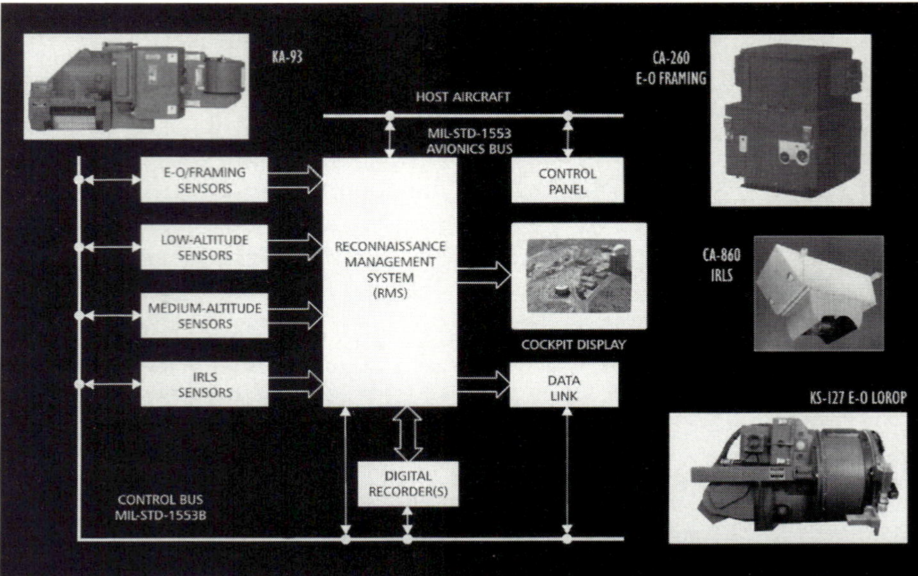

Recon/Optical Inc overall E-O reconnaissance concept **1998**/0018325

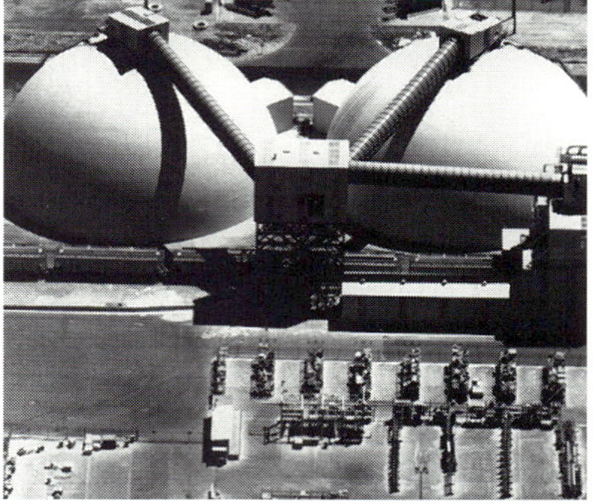

CA-260/25 25 Mega pixel imagery using 7:1 data compression and 16 × magnification **1998**/0018321

Specifications

Lens: 1.5 in fl, f/4.5; 3.0 in fl, f/4.5; 6.0 in fl, f/4.0; 12.0 in f1, f/4.0
Field of view (FoV):
(CA-260/4) 35.8° with 1.5 in fl lens; 18.3° with 3.0 in fl lens; 9.2° with 6.0 in fl lens; 4.6° with 12.0 in fl lens
(CA-260/25) 76.9° with 1.5 in fl lens; 43.3° with 3.0 in fl lens; 22.4° with 6.0 in fl lens; 11.3° with 12.0 in fl lens
Pixel size (pitch): 0.012 × 0.012 mm
Frame rate: 2.5 frames/s max
Dimensions:
(camera) 175.3 × 261.6 × 401.3 mm
(power supply) 134.6 × 274.3 × 464.8 mm
Weight:
(camera) 27.27 kg (without lens)
(power supply) 8.18 kg
Power: 115 V AC, 400 Hz, 210 VA; 28 V DC, 140 W

Operational status

Successful engineering tests were completed in July 1993. RF-4C and F-14 TARPS demonstrations were completed in 1994. The CA-260/4 is in service on US Air Force/Air National Guard F-16s. The CA-260/25 is now in production for the US Air Force Theater Airborne Reconnaissance System (TARS) Program.

Contractor

Recon/Optical, Inc.

UPDATED

CA-261 E-O step framing reconnaissance camera

The CA-261 E-O step framing camera is designed specifically to provide photographic quality images while enhancing the survivability of the tactical reconnaissance platform at medium altitudes. It is configured for external pod or internal aircraft mounting on a wide variety of reconnaissance platforms.

The CA-261 combines the proven performance of the CA-260 25-Mpixel E-O framing camera with the stepping capability of a proven two-axis stabilised step head with de-rotation prism. This combination produces a system that captures a series of 25-Mpixel images through a 12 in focal length E-O lens in the cross track direction. These images allow for wide area coverage of up to 180° in an E-O framing format.

The captured imagery can be displayed in mosaic form to provide a 'birds eye' view of the entire area of interest. This display retains the capability to manipulate individual images in the same manner as the current CA-260. De-rotation optics are employed to eliminate image rotation. The high resolution of the CA-260 is maintained by incorporating stabilisation electronics and software developed by Recon/Optical, Inc. This combination yields residual rates <.001 rad/s even at disturbance input rates in excess of 10°/s.

Advanced E-O step framing technology reduces the amount of time required to cover a target area compared to conventional E-O linescan sensors. Recon/Optical, Inc employed wafer-scale processing to develop the 25-Mpixel array CCDs used in the CA-260 to give a wide field of view E-O image. An on-chip motion compensation architecture eliminates the image blur normally associated with a wide field of view framing camera.

Specifications

Dimensions: 381 × 406.4 × 965.2 mm
Weight: 67.3 kg
Power supply: 300 W, 28 V DC; 630 VA, 115 V AC, 400 Hz, 3 Ø
Field of view (FoV): 11.3 × 11.3° each frame
Field of regard (FoR): 180 (in cross track) × 11.3° (in line of flight)
Frame rate: 2.5 frames/s
Angular resolution: 39.4 μrad/pixel
Max V/R: 0.44 rad/s with 10% overlap

Operational status

A successful flight test was conducted in July 1997 by the Royal Netherlands Air Force on an F-16 aircraft. Imagery was captured and presented in mosaic format, demonstrating the effectiveness of the wide area

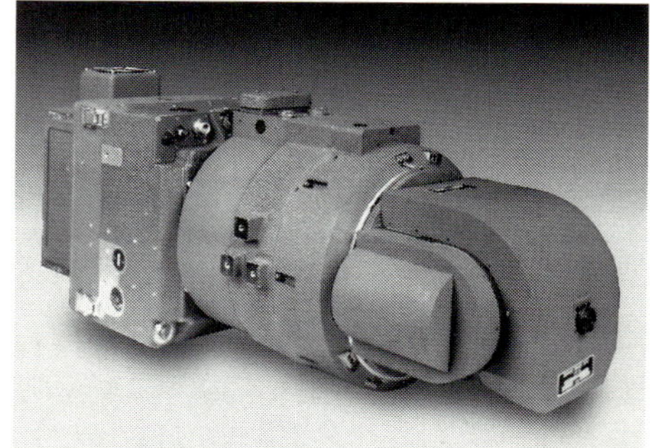

CA-261 E-O step framing camera **1997**/0001255

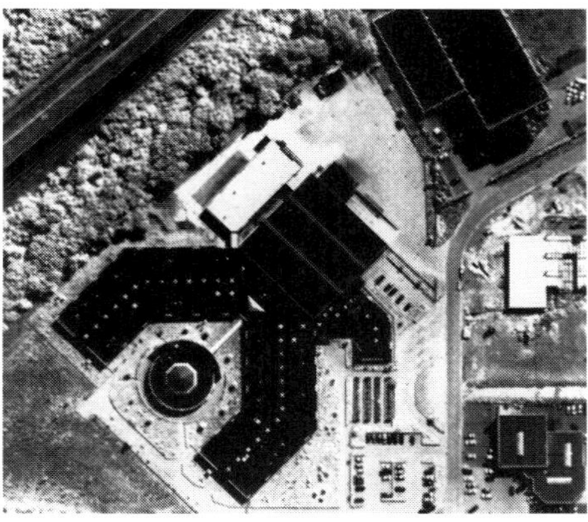

CA-261/25 E-O step framing imagery using 7:1 data compression and 16× magnification
1998/0018320

coverage approach using CA-260/25 framing technology. CA-261 development is complete and the camera is ready for production, with reported selection by the Royal Danish Air Force for its F-16 requirement (see Per Udsen entry).

Meanwhile, development is reported to be in progress on the CA-270, which will be a dual-band IR and daylight system with common aperture, aimed at the US Navy F/A-18E/F SHARP requirement.

Contractor
Recon/Optical, Inc.

UPDATED

CA-860 Infrared linescan system

The CA-860 infrared linescan system is designed for wide area surveillance. It operates in the 8 to 12 μm range and provides high-resolution images over a 120° field of view. Sensitivity exceeds 0.2°C.

The CA-860 is ideal for lightweight fixed- and rotary-wing aircraft installations. It outputs digital imagery that can be displayed in real time or stored on digital tape. The CA-860 is capable of resolving a 0.5 m target at a 3,000 ft altitude, while providing a field of view on the ground nearly 4 km wide.

Specifications
Dimensions: 457.2 × 558.8 × 406.4 mm
Weight: 38.56 kg
Power supply: 28 V DC, 600 W
Wavelength: 8-12 μm
Field of view: up to 120° continuously variable
Resolution: 0.25 mrad
Scan rate: 685.84 lines/s; 85.73 scans/s

Operational status
In production and in service with the US Army.

Contractor
Recon/Optical, Inc.

VERIFIED

KA-91 panoramic camera

The KA-91 is an 18 in focal length panoramic camera. It produces large-scale, wide-angle coverage photographs. The KA-91 offers a combination of features which make it ideal for medium-altitude reconnaissance. Using electronic synchronisation of all components, the camera is of modular design to simplify maintenance.

An important advantage of the KA-91 is sector scan. The camera can be set up for different across-track scan angles. The quality of the photography produced is enhanced by built-in roll stabilisation. To eliminate image blur arising from motion of the platform, the KA-91 uses translating lens forward motion compensation.

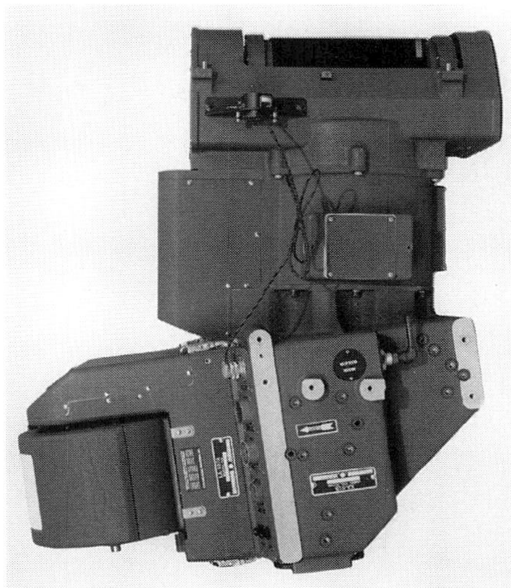

Recon/Optical Inc KA-91 panoramic camera
1998/0018319

Specifications
Weight: 76 kg including magazine, film and camera body
Power: 28 V DC, 115 V AC, 400 Hz, 3Ø
Lens: 18 in (450 mm)
Film:
5 in (127 mm) aerial roll film
1,000 ft of 2.5 mil film; 700 ft of 4 mil; 500 ft of 5.2 mil

Operational status
Currently in production and in use. An upgraded version of the KA-91 is being used in the US Open Skies treaty verification programme.

Contractor
Recon/Optical, Inc.

UPDATED

KA-93 panoramic camera

The KA-93 camera is a compact, prism scanning panoramic camera, which uses a 24 in f5.6 lens, for mounting internally in reconnaissance aircraft, reconnaissance pods or UAVs. It offers a combination of features which makes it ideal for medium-altitude reconnaissance.

An important advantage of the KA-93 is sector scan. The camera can be set up for six different across-track scan angles, each providing large-scale wide-angle, equivalent to the coverage provided by six framing cameras. The quality of the photography produced is enhanced by built-in roll stabilisation. To eliminate image blur arising from motion of the vehicle, the KA-93 uses translating lens forward motion compensation. Using electronic synchronisation of all components, the camera is of modular design to simplify maintenance.

Specifications
Weight: 95.26 kg
Power supply: 115 V AC, 400 Hz, 3 phase, 700 VA
28 V DC, 3 A
Lens: 23 in (609.6 mm) focal length, f/5.6
Film: 5 in (127 mm) aerial roll film, 1,000 ft (305 m)

Operational status
In production and in service.

Contractor
Recon/Optical, Inc.

UPDATED

CA-860 IR linescan system
1997/0001256

Recon/Optical Inc KA-93 panoramic camera
1998/0018318

KA-95 panoramic camera

The KA-95, Recon/Optical's most compact panoramic camera using 5 in (127 mm) film, is ideal for medium-altitude reconnaissance. Mounted internally in manned reconnaissance aircraft or externally in a pod, the camera utilises a 12 in lens which produces large-scale wide-angle photographs of up to 190° coverage, each being equivalent to the coverage provided by six framing cameras. Using electronic synchronisation of all components, the camera is of modular design to simplify maintenance.

An important advantage of the KA-95 is sector scan.

Recon/Optical Inc KA-95 panoramic camera
1998/0018317

The camera can be set up for six different across-track scan angles. The quality of the photography produced is enhanced by built-in roll stabilisation. To eliminate image blur arising from motion of the platform, the KA-95 uses translating lens forward motion compensation.

The KH-95B is designed for the RF-5E.

Specifications

Weight: 63.5 kg including film
Power supply: 115 V AC, 400 Hz, 3 phase; 28 V DC
Stabilisation: up to 10°/s roll
Lens: 12 in (304.8 mm) focal length, f/4
Film: 5 in (127 mm) aerial roll film, 1,000 ft 2.5 mil film

Operational status

In production and in service. The KA-95 is currently being used by a participant in the verification of the Open Skies treaty.

Contractor

Recon/Optical, Inc.

VERIFIED

KS-87 reconnaissance camera

The KS-87 has interchangeable lens cones of 3, 6, 12 and 18 in focal lengths. The camera incorporates automatic exposure control and forward motion compensation and is capable of 6 frames/s. It is the world's most widely used camera with over 4,000 cameras built to date.

Recon/Optical Inc KS-87 reconnaissance camera
1998/0018316

Operational status

Installed in the RF-4, RF-5 and many other reconnaissance aircraft. An upgraded version has been chosen for the Open Skies treaty verification programme.

Contractor

Recon/Optical, Inc.

VERIFIED

The KS-127B camera is carried by the RF-4

KS-146 (bottom) and KS-147 (top) Long-Range Oblique Photography cameras

KS-127 reconnaissance camera

The KS-127 is a reconnaissance camera mounted in the nose of an aircraft designed for Long-Range Oblique Photography (LOROP) missions. The film-only version is designated KS-127A, while the KS-127B is both film and E-O (electro-optic) capable. It has a 66 in (167 cm) focal length lens system, active and passive stabilisation, autofocus and thermal stabilisation. Typically, the KS-127B can take detailed photographs of sites over 35 km distant from an aircraft patrolling at 35,000 ft. This coverage permits missions to be conducted from safe, long-range, standoff distances. These missions eliminate the need for overflights and thereby avoid violation of foreign airspace. Aircraft survivability is also greatly improved.

Specifications

Power: 115/208 V rms, 400 Hz, 700 V A
28 V, 280 W
Lens: 1,676 mm fl, f/8.0
Film: 127 mm aerial roll film
1,000 ft of 2.5 mil film; 700 ft of 4 mil; 500 ft of 5.2 mil

Operational status

In production and service. It is currently operational in the RF-4, where a sight in the rear cockpit enables the crew to aim the camera directly at a specific target. The camera can also be upgraded to E-O operation.

Contractor

Recon/Optical, Inc.

VERIFIED

KS-146 reconnaissance camera

The KS-146 reconnaissance camera uses a 66 in focal length f5.6 lens to provide Long-Range Oblique Photography (LOROP). The KS-146 cameras currently in use are film cameras. Target coverage, on either side of the aircraft, can be selected from the cockpit. Typically, the KS-146 camera takes detailed pictures of sites to either the left or right side of the aircraft from 18.5 km to over 92.6 km distant from an aircraft flying at 30,000 ft or higher.

A programme is now under way to modify KS-146 film cameras to E-O cameras by using a CCD sensor array in place of the photographic film. This modification provides LOROP imagery in a digital form that can be transmitted to a ground station in real time and/or recorded on digital videotape on board the aircraft.

Specifications

Weight:
(camera) 317 kg
(total system) 385 kg
Power: 115/208 V rms, 400 Hz, 1,000 V A;
28 V, 300 W
Lens: 1,676 mm fl, f/5.6
Film: 127 mm roll film; 1,000 ft of 0.0025 in film

Operational status

In production and service. KS-146 cameras are in use by customers in Europe and Asia. They are also upgradeable to E-O operation.

Contractor

Recon/Optical, Inc.

VERIFIED

KS-147 reconnaissance camera

The KS-147 reconnaissance camera uses a 66 in focal length f5.6 lens to provide Long-Range Oblique Photography (LOROP). Target coverage, on either side of the aircraft, can be selected from the cockpit. The camera incorporates automatic exposure control, passive and active stabilisation, auto-focus and thermal stabilisation. Typically, the KS-147 takes detailed pictures of sites to either the left or right side of the aircraft from 18.5 km to over 92.6 km distant from an aircraft flying at 30,000 ft or higher.

A programme is now under way to modify the KS-147 to E-O camera configuration by using a CCD sensor array in place of photographic film. This modification provides LOROP imagery in a digital form that can be transmitted to a ground station in real time and/or recorded on digital tape on board the aircraft.

Specifications

Weight:
(camera) 245 kg
(total system) 270 kg
Power supply: 115/208 V AC, 400 Hz, 700 VA;
28 V DC, 280 W
Lens: 66 in (1,676.4 mm) focal length, f/5.6
Film: 5 in (127 mm) roll film, 1,000 ft of 0.0025 in film

Operational status

The KS-147 cameras presently in use are film cameras. They can also be upgraded to E-O operation

Contractor

Recon/Optical, Inc.

UPDATED

KS-157 reconnaissance camera

The KS-157 cameras presently in use are film cameras utilising a 66 in focal length f5.6 lens designed for a business jet. The system incorporates mounted sights that are utilised to point the camera. Target coverage on either side of the aircraft can be selected from the cabin. The camera has automatic exposure control, passive and active stabilisation, auto-focus and thermal stabilisation. Missions are flown at a standoff distance of 10 to 140 km at aircraft speeds of 480 kt and altitudes of up to 50,000 ft. Detection of 3 ft-size ground objects can be achieved at 60 n miles.

A programme is now under way to modify these cameras to E-O cameras by using a CCD sensor array in place of photographic film. This modification provides LOROP imagery in a digital form that can be transmitted to a ground station in real time and/or recorded on digital videotape on board the aircraft.

The Recon/Optical KS-157 reconnaissance camera for business jets

Specifications

Weight:
(camera) 317 kg
(total system) 385 kg
Power supply: 115/200 V AC, 400 Hz
28 V DC
Lens: 66 in focal length, f/5.6
Film: 5 in (127 mm), 1,000 ft (305 m) capacity
Format: 114 × 114 mm

Operational status

In production and in service. They can also be upgraded to E-O operation.

Contractor

Recon/Optical, Inc.

VERIFIED

AN/AAS-32 laser tracker

The AN/AAS-32 is a laser-seeker head, with associated components, which Rockwell produces for the Bell AH-1S Cobra light attack helicopter and, in derivative form, for the Target Acquisition Designation System/Pilot Night Viewing System (TADS/PNVS) in the US Army AH-64 armed helicopter.

The seeker is a wide field of view unit, sensitive to 1.06 μm radiation, that can detect target designations from ground troops or co-operative aircraft. Coded pulse data is used to minimise jamming and a four quadrant silicon detector head is used.

Production of AH-1S equipment was initiated in September 1981 and first delivery was in May 1984. TADS/PNVS production began in June 1981 and deliveries began in January 1983.

Specifications

Dimensions:
(receiver unit) 214 × 188 mm
(electronics unit) 152 × 152 × 188 mm
(control panel) 144 × 66 mm area
Weight:
(receiver unit) 9 kg
(electronics unit) 3.4 kg
(control panel) 0.6 kg
Power supply: 115 V AC, 400 Hz
28 V DC
Field of view (scanning):
(elevation) −60 to +30°
(azimuth) −90 to +90°
Field of view (instantaneous):
(elevation) 10°
(azimuth) 20°
Optical port diameter: 5 in (127 mm)
Focal length ratio: 0.3:1

Operational status

In service. Over 200 units have been produced.

Contractor

Rockwell Collins.

UPDATED

ATD-111 LIDAR

The ATD-111 is an airborne LIDAR (Light Detection and Ranging) system developed for the US Navy to detect and classify underwater threats, primarily submarines and mines. The system is a stabilised scanning laser unit mounted aboard a Navy SH-60 Seahawk helicopter.

The ATD-11 employs a solid-state laser self-contained in a pod. The onboard operator control unit provides detection information, images and maps in GPS coordinates on a flat panel colour display.

The ATD-111 is currently undergoing upgrades and testing and will compete for the US Navy's Airborne Laser Mine Detection System (ALMDS) award.

Operational status

Development.

Contractor

Sanders, a Lockheed Martin Company.

VERIFIED

Laser radar technology programmes

Laser radar ('ladar') offers benefits in ranging, imaging, and automatic target classification and recognition. The ladar functions best when used in conjunction with a wide area search sensor such as a radar or infrared detection system. Initial efforts exploited the CO_2 technology for imaging, Doppler measurement and homing applications.

More recently, advanced solid-state laser radar systems at 1,064 nm have been investigated for their compatibility with laser-guided bomb applications, while transmitters with wavelengths greater than 1,400 nm are of interest because of the increase in eye safety associated with those wavelengths. Methods to derive the so-called eye-safe wavelengths from the designator wavelength are subjects of active research and show considerable potential to be implemented in low cost, highly reliable multiple function systems. A multifunction system would be able to perform laser radar functions, such as imaging and ranging, and the designation function from a single laser source.

With the development of sources at the designator and eye-safe wavelengths comes the need to develop receivers with high sensitivity at these wavelengths. Recent advances in photoemissive detectors hold promise that image intensifier technology can be applied to imaging at wavelengths greater than 1,400 nm. Competing technologies include avalanche photodiode arrays based on InGaAs materials which are sensitive out to 2,000 nm and beyond. Such detector technologies have yet to reach the capabilities of detectors based on silicon, but they are being matured rapidly. Beyond just the detection process, there are investigations in readout electronics. Preliminary work to develop electronic readouts capable of supplying a three-dimensional image from a single laser pulse is under way.

Methods to shape and direct the transmitted and received laser energy are also of importance to laser radar systems. Typical methods include use of gimballed turrets and fast steering mirrors. More recently, methods of beam steering and shaping based on optical phased array technology have been investigated. Based on the same principles as phased array radar, optical phased arrays allow the steering and generation of multiple beams from an aperture with no moving parts. One method of phase modulation that has been investigated is based on liquid crystal technology. These electronically adaptable optical phased arrays are still in their infancy, but hold promise for low cost, lightweight apertures for multifunction systems.

Operational status

Solid-state laser and receiver component hardware is being fabricated and field tested, with plans for flight testing in the next few years. The US Air Force is continuing the development of systems which combine in a single aperture a ladar and a passive search sensor. In addition, work is being initiated to develop modular, multifunction, open architecture electro-optical systems which will combine, through a few conformal apertures distributed about an aircraft, all of the electro-optical functions required.

Over the next 3 years, the US Air Force plans to develop high power lasers suitable for long range tactical applications. In addition to applications where these 1 and 2 micron lasers can be directly applied, they will form the pump supplies to create 2 5 micron laser radiation through the use of optical parametric amplifiers for multi-spectral target identification and chemical remote sensing applications.

Contractor

US Air Force Materiel Command, Wright Laboratory.

VERIFIED

LIMAR laser imaging and ranging

LIMAR is an optical parallel processing system which utilises conventional video cameras as receivers. A short laser pulse is projected onto the object to be mapped and the returning reflected light is captured and tagged with a time-varying polarisation angle by LIMAR's electro-optical modulator subsystem. The polarisation angle, a known function of time, provides in turn the time of flight of the light and hence the distance to the reflecting surfaces. In contrast to slower scanning laser radar systems, LIMAR simultaneously captures a complete range image from a single laser pulse in the shortest possible time. The LIMAR range image is instantly registered with a corresponding video camera image of the scene. In other words, a LIMAR camera system provides a four-dimensional TV signal which encodes range data for each picture element or pixel along with three colour parameters.

The US Air Force Wright Laboratory has funded preliminary research and development efforts on LIMAR. It will provide the Air Force with a high speed ranging camera system for airborne target detection, identification, and tracking. Other possible aerospace applications include automation of mid-air refuelling, space docking operations, and the detection and mapping of space objects. Airborne LIMAR camera systems can facilitate topographic mapping of the earth's surface. Commercial applications include industrial inspection, robotic vision, and generation of computer models for virtual reality. For example, it is possible with LIMAR to generate three-dimensional movies or other dynamic displays of destructive testing of automobiles or aircraft. Another commercial concept is the use of a LIMAR camera for broadcasting four-dimensional signals that can be used by appropriate three-dimensional television display systems as well as by conventional TV sets.

Operational status

Current research and development efforts are focused on developing a low-cost, portable, eye-safe LIMAR prototype, which is necessary to the development, evaluation, and demonstration of the LIMAR technology and the transition of this technology to a wide class of effective military and civilian uses.

Contractor

US Air Force Materiel Command, Wright Laboratory.

VERIFIED

ELECTRONIC WARFARE

Vympel State Machine Building Design Bureau YB-3A expendable countermeasures dispenser system operating on MiG-29 **1998**/0018306

AUSTRALIA

ALR-2002 radar warning receiver

British Aerospace Australia is currently undertaking full scale engineering development of the ALR-2002 radar warning receiver. The contract, worth A$20 million, will result in the development of three systems which are scheduled for flight testing in 1999, on Royal Australian Air Force F-111C aircraft.

The ALR-2002 system architecture is designed specifically to meet the requirements of the F-111C aircraft, but its modular design will be suited to the Royal Australian Air Force F/A-18, air transport and Hawk Lead-in Fighter aircraft.

The system utilises a dual receiver architecture with parallel processing to maximise probability of intercept and situation awareness.

Key features of the ALR-2002 system include: near 100 per cent probability of intercept against specified emitters; full situation awareness to aircrew during high-density emitter environments; high angle of arrival accuracy over a wide RF bandwidth; the ability to detect a variety of Continuous Wave and Pulsed emitters; software developed in Ada.

The ALR-2002 RWR has been designed to operate in conjunction with other electronic warfare systems fitted to the host platform. The system can co-ordinate the responses to specific threats from a variety of sensors in addition to its own receivers. It has also been designed to act as the Bus Controller for an EW Mux Bus.

The system software is a key element of the design and will be 100 per cent developed by British Aerospace Australia. The signal processing algorithms are being produced using knowledge and experience gained through the ALR-2002 Concept Demonstrator program which was successfully demonstrated to the Royal Australian Air Force in 1993.

ALR-2002 is part of the Royal Australian Air Force's Project Echidna, which will provide an integrated EW upgrade to a range of Royal Australian Air Force, Army and Navy fixed- and rotary-wing aircraft, including F-111C, C-130J, S-70A-9, CH-47D and Sea King Mk 50A. As well as ALR-2002, Project Echidna aircraft will be equipped with ECM jammers, missile approach warners and chaff/flare dispensers. Later addition of towed radar decoys and a laser warning system is also possible.

Contractor

British Aerospace Australia Ltd.

UPDATED

MODIR Modulated InfraRed jamming system

Originally designed in conjunction with the Defence Science and Technology Organisation, British Aerospace Australia's MODIR jamming system is fitted to military and VIP aircraft to protect them against threats such as shoulder-launched surface-to-air missiles.

The MODIR system generates a high-power, modulated infrared signal which acts to disrupt the functioning of the signal processing circuits of a heat-seeking missile. This prevents the missile's guidance system from locking on to the target aircraft engine's infrared emissions and effectively means that the missile will not 'see' the aircraft.

Self-contained and operated from a small control panel in the cockpit, it provides for independent start-up, power and synchronisation monitoring and overheat shutdown override for each jammer unit. Fire suppression controls can be included on this panel or integrated with the aircraft fire extinguisher controls.

The MODIR system comprises the jammer units with applicable aircraft mounting, power conditioning and control units and the cockpit control panel. Normally, two jammer units only are required per aircraft.

MODIR operates from a standard aircraft three-phase 400 Hz 115 V supply. It can be mounted on the wing or fuselage structure, as a self-contained pod or as on the C-130 Hercules aircraft, attached to external fuel tanks.

The system has been developed to meet a Royal Australian Air Force requirement and has been fully environmentally qualified and certified by Royal Australia Air Force flight operations. British Aerospace Australia has also secured sales of MODIR to international customers.

Contractor

British Aerospace Australia Ltd.

VERIFIED

British Aerospace Australia's MODIR shown mounted on a Royal Australian Air Force C-130 Hercules

CANADA

Electronic warfare training system for CT-133 aircraft

Lockheed Martin Canada is to design, develop, manufacture, and integrate three electronic warfare training systems installed in Canadian Forces Challenger 600 aircraft, together with ground support facilities; the modified aircraft will be designated CT-133.

The airborne suite is designed to provide training in: threat evaluation; communications and signal detection; standoff jamming and chaff corridors; escort jamming; defensive ECM.

EW systems to be fitted to the aircraft include the following:

ESM: AN/ALR-504 (modified AN/ALR-76)

high power communications jammers: AN/ALQ-504 (Zeta ZS-1920)

high power C, D, E, F and G-band jammers: AN/ULQ-21

high power (coherent) I-band jammer: AN/ULQ-21(V)

chaff dispenser: AN/ALE-503(V)2

Operational status

Programme status uncertain.

Contractor

Lockheed Martin Canada Inc.

NEW ENTRY

CHILE

DM/A-104 radar warning receiver

The DM/A-104 is a wideband radar warning receiver for helicopters and combat aircraft that provides instantaneous detection of threat radar emitters. The system consists of four orthogonal spiral antennas, each connected to wideband crystal video receiving channels that provide 360° coverage in the 2-18 GHz frequency range for the detection of most search, acquisition and fire-control radars.

A powerful digital processor and advanced de-interleaving software ensure real-time automatic radar sorting and evaluation. A cockpit display presents information on the most dangerous threats. In addition, an audio warning system is sent to the intercom system. The threat itself is classified and the pilot is informed whether his aircraft is being targeted by a surveillance, acquisition or fire-control radar in lock on mode and whether the transmission is from a continuous wave radar.

The DM/A-104 can be interfaced with the DM/A-202 chaff/flare dispensing system, to provide an automatic self-protection capability, and onboard systems for blanking of self-initiated signals. It features a compact and modular design that does not demand much aircraft space and can be retrofitted easily on any combat aircraft. It is designed for a high level of reliability and maintainability, with a complete BIT capability for online diagnosis.

Specifications

Frequency: 2-18 GHz in 4 sub-bands
C/D-band 0.7-1.3 GHz
Sensitivity: −50 dBm
Accuracy: >10° RMS

Contractor

DTS Ltda.

VERIFIED

DM/A-202 chaff and flare dispensing system

The DM/A-202 is a self-protection system for helicopters and combat aircraft that provides chaff and/or infrared-flares to break the lock of radar and IR-guided missiles. It consists of five LRUs: the cockpit control unit and four launching units containing the chaff and flare cartridges. The control unit includes a bright display which shows the amount of chaff and flare cartridges remaining and a mode selector for selecting one of four different launching sequences.

Each launching unit comprises an easily removable magazine, allowing quick reloading of the cartridges, and the associated firing circuits. A typical configuration consists of 108 chaff cartridges and 54 flare cartridges contained in the four launching units. The system can be expanded to handle up to eight launching units.

Flexibility and ease of operation have been achieved

by the use of a fast reprogrammable microprocessor that ensures adaptability of the system to changing tactical situations. Simplicity of operation has also been enhanced by the installation of control switches on the aircraft throttle and/or stick, allowing the pilot to operate the system without removing his hands from the controls. Modes of operation are manual, semi-automatic and automatic. A safety switch is incorporated in the system so that all the stores can be jettisoned in an emergency.

To avoid reducing the aircraft operational load carrying capacity, the launching units are normally attached externally to the rear fuselage or to the sides of the ventral and wing pylons. Internal or semi-recessed installation can also be adopted, depending on the aircraft configuration and available space.

The DM/A-202 also provides an interface with the DM/A-104 radar warning receiver for full aircraft self-protection.

Operational status

The DM/A-202 is operational and being offered for export.

Contractor

DTS Ltda.

VERIFIED

EWPS-100 EW system

The EWPS-100 has been developed to protect helicopters and combat aircraft from present and future radar-controlled weapon systems. It operates over the 0.7 to 18 GHz frequency band and provides a low-cost and effective answer to operational requirements in the air-to-air and air-to-ground roles. An integrated EW system architecture has been developed using modular techniques, proven software and hardware building blocks. The EWPS-100 integrates the DM/A-104 radar warning receiver, DM/A-202 chaff/flare dispenser and DM/A-401 self-protection jammer.

The main features of the EWPS-100 are a high probability of threat interception, high sensitivity, initiation of countermeasures, power management in time and frequency and control of chaff/flare cartridge dispensing.

Specifications

Frequency: 0.7-18 GHz
Detection: pulse, pulse Doppler, CW
Sensitivity: −50 dBm
Maximum pulse density: 500,000 pps
Display: 16 threats simultaneously

Operational status

The EWPS-100 is operational and being offered for export.

Contractor

DTS Ltda.

VERIFIED

Caiquen II radar warning receiver

The Caiquen II radar warning receiver has four independent receivers, each with its own 2 to 18 GHz wideband crystal video receiver and antenna, offering coverage through a full 360° in azimuth and ±40° in elevation. PRFs between 200 Hz and 12 kHz can be processed.

There are eight LRUs: four antennas, two RF units, computer and a control and display unit. The cockpit display unit indicates the radar types being detected: 'A' for airborne and ground-based acquisition radars, 'V' for early warning radars, 'CW' for continuous wave fire-control radars and 'LKN' for locked on fire-control radars with pulse modulation. The display is divided into 12 sectors, indicating the bearing of the received signals.

Specifications

Volume: 0.011 m^3
Weight: 8 kg
Altitude: up to 40,000 ft

Operational status

In production and in service. The system equips the Chilean Air Force's British Aerospace Hunter FGA.71 aircraft and is understood to be under development for installation in the country's Mirage 50, among others.

Contractor

DTS Ltda.

VERIFIED

Caiquen III radar warning receiver

Caiquen III is a wideband radar warning receiver operating over the 2 to 18 GHz frequency range and giving full 360° coverage in azimuth and ±40° in elevation. The system is fully automatic and programmable, provides real-time analysis of the direction and type of threats and presents the data to the pilot on a bright dot matrix LED display divided into eight sectors. The threat is classified as a surveillance, acquisition or fire-control radar in lock on mode, or as a CW radar from a missile. Relative threat priorities can also be assessed.

The Caiquen III system comprises eight LRUs: four antennas, two RF units, a computer and a display and control unit. The four orthogonal spiral antennas are each connected to a wideband crystal video receiver, which ensures the detection of pulsed and CW signals. The receiver outputs are digitised and fed into a high-speed digital microcomputer, which processes the incoming signals in real time, de-interleaves individual pulse trains and establishes type, strength and direction of the emitters based on a comparison with a preprogrammed library. This information is presented to the crew on a dot matrix LED display, together with an audio warning signal. The display is capable of indicating radar type, received signal strength and octant of arrival for up to three threats simultaneously.

Caiquen III features a compact and modular design, which allows for very low demand on aircraft space and simple retrofit installation on existing aircraft. It also features a high level of reliability and maintainability through the use of MIL-qualified components and complete BITE for online fault diagnosis. It can be interfaced with the Eclipse chaff/flare dispensing system to provide an integrated self-protection capability, as well as onboard systems for blanking of self-emitted signals.

Specifications

Weight: 14 kg
Power supply: 28 V DC, 3.5 A

Operational status

Installed on Chilean Air Force Hunter and Mirage aircraft. Development is under way for installation on the F-5 aircraft.

Contractor

DTS Ltda.

VERIFIED

DTS Ltda Caiquen III radar warning receiver modules

Eclipse chaff/flare dispensing system

Eclipse is a self-protection system for helicopters and combat aircraft that provides chaff and/or infrared flares to break the lock of radar and infrared-guided missiles. It consists of three types of LRUs, including the cockpit control unit, and four launching units containing the chaff and flare cartridges. The control unit features a bright display, which indicates the amount of cartridges remaining, and a mode selector on which the pilot can select one of four different launching sequences.

Each launching unit comprises an easily removable magazine, allowing quick reloading of the cartridges, as well as the associated firing circuits. The normal payload of the system consists of 64 RR-170 chaff cartridges and 34 MJU-7B flare cartridges contained in the four launching units. The system can be expanded easily to handle up to eight launching units.

Flexibility and ease of operation have been achieved by the use of a fast and reprogrammable microprocessor that guarantees adaptability of the system to changing tactical situations. Simplicity of operation has also been enhanced by the installation of control switches on the aircraft throttle and stick, allowing the pilot to operate the system without taking his hands off the controls. A safety switch is incorporated in the system so that all the stores can be jettisoned in an emergency.

To avoid reducing the aircraft operational load-carrying capacity, the launching units are normally attached externally to the rear fuselage or to the sides of the ventral and wing pylons. Internal or semi-recessed installation can also be adopted, depending on the aircraft configuration. Eclipse can be interfaced with the Caiquen radar warning receiver to provide full aircraft integrated self-protection.

Operational status

Installed on Chilean Air Force Hunter aircraft. Installations for the F-5 and Mirage are under development.

Contractor

DTS Ltda.

VERIFIED

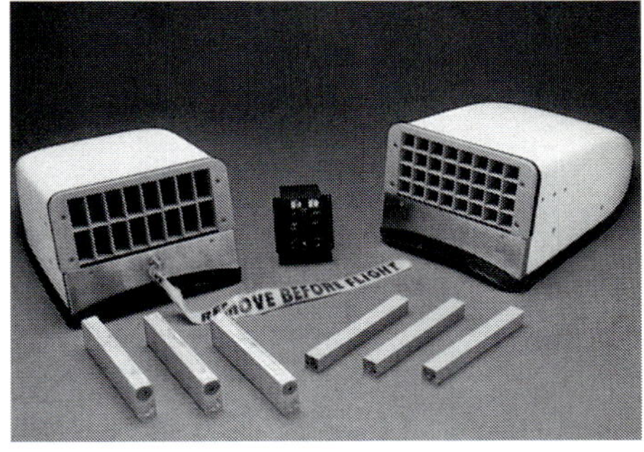

The DTS Ltda Eclipse chaff/flare dispensing system

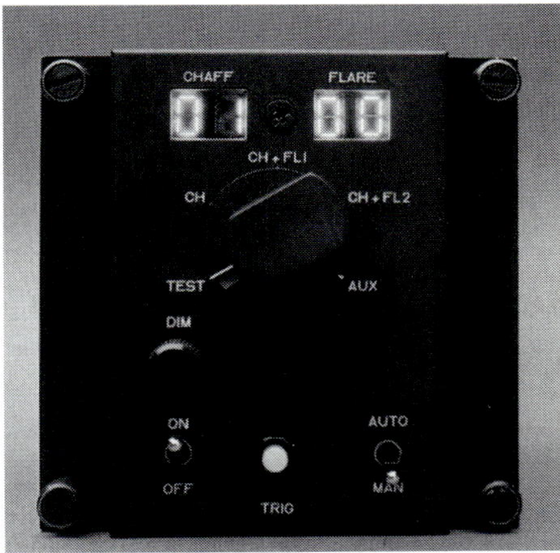

The DTS Ltda Eclipse control unit

Itata ELINT system

The Itata ELINT system is a high-sensitivity electronic intelligence-gathering system that can detect, locate and measure the parameters of emissions from search, acquisition and fire-control radars. Itata consists of a fully programmable superheterodyne receiver, a digital pulse analyser and a high-gain wideband rotating parabolic antenna which provides 360° coverage and bearing information to an accuracy of within a few degrees. Although intended primarily for light transport aircraft, Itata can also be installed in ships or ground vehicles.

The receiver operates over a frequency range of 3 MHz to 18 GHz in six bands. It can be used either in a wide open mode over the complete frequency range or in a selective mode over a single band. After detection of a transmission, the receiver locks on automatically and measures the frequency and other parameters. Digitised data of each intercepted signal can be recorded automatically for subsequent analysis.

Specifications
Frequency: 3 MHz-18 GHz in 6 bands
Azimuth coverage: 360°
Azimuth beamwidth: (E/F-band) 8°, (J-band) 1.8°
Polarisation: circular

Operational status
Operational with the Chilean Air Force on Beech 99A aircraft.

Contractor
DTS Ltda.

VERIFIED

CHINA, PEOPLE'S REPUBLIC

GT-1 chaff and infrared flare dispensing set

The GT-1 is a chaff and flare dispensing set which has been developed to provide self-protection to fixed-wing aircraft and helicopters. Chaff provides radar countermeasures in the 2 to 18 GHz frequency range and flares provide IR countermeasures in the 1 to 3 μm and 3 to 5 μm wavelengths. The complete system consists of a programme controller, an operations control, dispensers and cartridges and can be interfaced with a threat warning system to form a self-protection system. The standard configuration is 36 chaff and 18 flare cartridges. Dispensing can be manual or automatic and cartridges can be launched singly or in pairs.

Specifications
Weight: 40 kg (without cartridges and cables)
Power supply: 27 V DC, <0.7 A (static), <3 A (ignition)
380 V AC, 50 Hz, 3 phase, 2 kW
Coverage:
(azimuth) 360°
(elevation) 0 to +85°

Operational status
In production.

Contractor
China National Electronics Import and Export Corporation.

VERIFIED

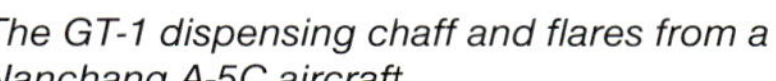

The GT-1 dispensing chaff and flares from a Nanchang A-5C aircraft

BM/KG 8601 repeater jammer

The BM/KG 8601 repeater jammer operates in the E/F- and G/H-bands and is available for installation in strike and fighter/bomber aircraft to counter airborne tracking, anti-aircraft fire control and SAM guidance radars. The jammer has a high power output, minimal repeater delay time, threat management through RF channelling and wide antenna coverage. It provides multijamming techniques.

Operational status
In production.

Contractor
Southwest China Research Institute of Electronic Equipment.

VERIFIED

BM/KG 8605/8606 smart noise jammers

The BM/KG-8605 operates in the I/J-band as a smart noise jammer performing a hybrid type of jamming which incorporates some of the features of both noise and deception jamming.

The BM/KG 8606 operates in the I-band and features both orthogonal and dual circularly polarised jamming techniques.

Both the BM/KG 8605 and the BM/KG 8606 operate in conjunction with chaff/flare dispensers providing cross polarisation jamming and fast and accurate set on. They are small, lightweight and have a low power consumption.

Operational status
In production.

Contractor
Southwest China Research Institute of Electronic Equipment.

VERIFIED

BM/KJ 8602 airborne radar warning system

The BM/KJ 8602 is a radar warning receiver designed for tactical and other combat aircraft. It consists of a digital signal analyser, a CRT display unit, control box, several receivers and a number of antenna units. It features wide frequency coverage in two bands, 0.7 to 1.4 GHz and 2 to 18 GHz, and is capable of dealing with multiple threats. Automatic sorting and identification of threat emissions are provided. The system can operate in conjunction with ECM units and chaff/flare dispensers.

The BM/KJ 8602 system is compact and lightweight, making it suitable for tactical aircraft where space and weight are strictly limited.

Specifications
Weight: 20 kg
Frequency: 0.7-14 GHz and 2-18 GHz
Response time: 1 s
Capacity: 16 threats simultaneously
Coverage:
(azimuth) 360°
(elevation) −30° to +30°
Accuracy: 15° RMS

The BM/KJ 8602 airborne radar warning system

Contractor
Southwest China Research Institute of Electronic Equipment.

VERIFIED

BM/KJ 8608 airborne ELINT system

The BM/KJ 8608 ELINT system detects, locates, identifies and analyses radar emitters deployed on the ground and at sea with high probability of intercept, high sensitivity and the accurate measurement of parameters. It features wide frequency coverage, high sensitivity and long operational range, automatic signal identification, emitter position fixing capability, operations in a dense RF environment and BITE.

Specifications
Power supply: 115 V AC, 400 Hz
28 V DC
Frequency: 1-18 GHz
Frequency accuracy: 5 MHz
Coverage: (azimuth) 360°
Accuracy:
(bearing) (1-8 GHz) 5°, (8-18 GHz) 3°

Contractor
Southwest China Research Institute of Electronic Equipment.

VERIFIED

DENMARK

Electronic Warfare Management System (EWMS) for the F-16

The Electronic Warfare Management System for the F-16 mid-life update, developed by Per Udsen and TERMA Elektronik in conjunction with the Royal Danish Air Force, consists of an Electronic Warfare Management System (EWMS), ALE-40 sequencer switch, Pylon Integrated Dispenser Station (PIDS), and Electronic Combat Integrated Pylon System (ECIPS).

The EWMS combines a single-point pilot control unit with an upfront indicator for the co-ordinated setting-up and operation of the F-16 EW suite. The two cockpit-mounted panels, the EW Management Unit (EWMU) and the EW Prime Indicator (EWPI), provide pilot/vehicle interface and enhance situational awareness.

The EWMU facilitates avionic and sensor input assessment, analysis and dynamic threat data processing to obtain the optimum countermeasures responses. The unit provides the prime control for all aircraft EW functions, including operation of the countermeasures systems with menu-driven set up of dispenser, programmes and techniques. The EWMU constitutes the main EW system computer, and provides co-ordinated control of the EW assets in manual, semi-automatic and automatic modes of operation.

The EWPI provides the pilot with head-up EW notifications, warnings and remaining payload counts, and furnishes dedicated prime controls for the RWR and ECM systems.

The hardware and software modular design of the EWMS features growth potential and flexibility to meet the need for further platform integration and databus expansion.

The enhanced sequencer switch assembly is an autonomous LRU compatible with the ALE-40 countermeasures dispensing system. The sequencer switch routes payload dispenser signals to the breech plate firing pins of the sequencer assemblies, executing the sequential firing of chaff, flares and other decoy countermeasures payloads. The sequencer can be set for single or double firing at one of eight preprogrammed intervals of 5 ms upwards. Fast firing is available down to the squib limit of 15 ms or less.

The PIDS is designed to increase the chaff/flare capacity of the F-16. All conventional weapon and ECM pod capabilities are retained. The PIDS is integrated with, and operated from, the onboard dispenser system. Existing weapon pylons are rebuilt to PIDS standard.

The ECIPS is a common denominator for several F-16 electronic combat systems as these systems are accommodated in a modified F-16 weapon pylon. The ECIPS features retention of all existing conventional weapon capabilities, the capability to carry alternative ECM pods on special missions, lack of impact on weapon separation or delivery and the use of existing aircraft wiring.

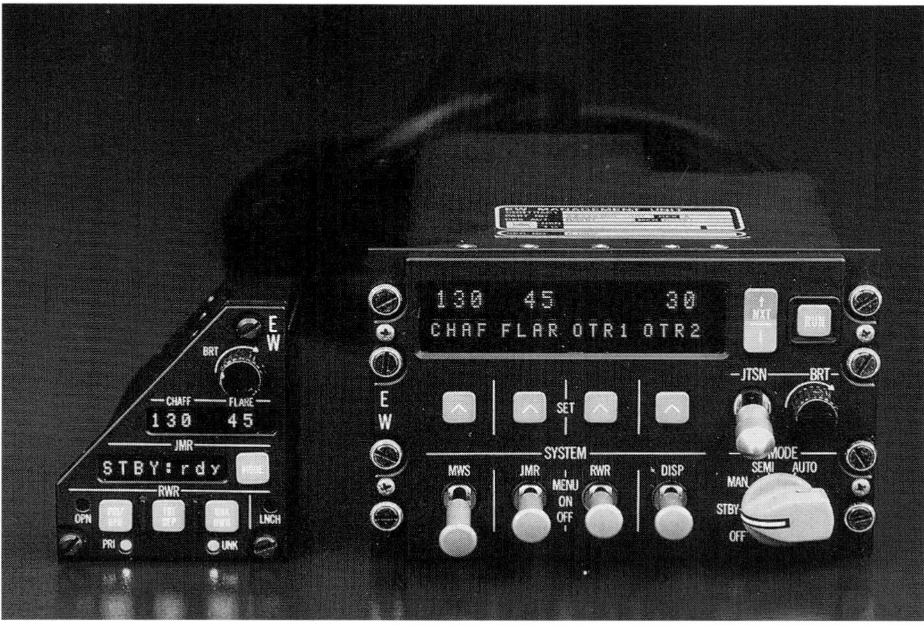

The EWPI (left) and EWMU (right) for the TERMA electronic warfare management system **1995**

Operational status

In service on Royal Danish Air Force C-130 and F-16 aircraft. Selected by the US Air Force under the Foreign Comparative Test program for use by Air National Guard and Air Force Reserve F-16 C/D and A-10 aircraft and some HH-60G helicopters of the US Air Force; in US service EWMS is designated AN/ALQ-213(V). Also part of the F-16 European MLU program.

Contractors

Per Udsen Company Aircraft Industry AS.
TERMA Elektronik AS.

UPDATED

Modular Countermeasures Pod (MCP-7)

The basic building block of Per Udsen's MCP-7 is a barrel module containing two ALE-40 or ALE-47 chaff and flare dispenser magazines and a sequencer switch. Two or four such barrels, which may be set at 15° intervals in order to dispense in the optimum direction, are attached to the strongback. The baseline system is completed by a nosecone containing a further dispenser magazine, and a tailcone accommodating a safety switch. The MCP is controlled from a TERMA EW Management System or similar installation. Options include the addition of RWRs, MAWs and towed decoys.

The MCP can be carried underwing, using standard NATO 350 mm lugs.

Specifications

Length: 2.27 m
Weight:
(empty) 65 kg
(with typical load) 132 kg

Operational status

Twin installation on Royal Netherlands Air Force Fokker 60 aircraft, typically carrying a load of 240 infrared flares and 360 chaff cartridges.

Contractor

Per Udsen Company Aircraft Industry AS.

VERIFIED

AN/ALR-DK Radar Warning Receivers (RWRs)

Royal Danish Air Force AN/ALR-69 radar warning receivers (RWRs) have been reconfigured to the AN/ALR-DK, the Mini-69, for use on lightweight helicopters of the Royal Danish Army Air Corps. The lightweight, miniaturised RWR has been developed by Danish industry in co-operation with the Royal Danish Air Force to decrease logistics and software support significantly while providing helicopters with an advanced RWR capability. The basic Mini-69 weighs 13.6 kg.

The AN/ALR-DK system consists of an azimuth indicator, prime control panel, auxiliary control panel, signal processor, frequency selective receiver, C/D-band receiver and four amplifier-detectors.

The Mini-69 has been modified from the AN/ALR-69(V) for Royal Danish Army Air Corps AS550 Fennec helicopters **1995**

The AN/ALR-69 has been modified to the AN/ALR-DK by a number of alterations to reduce size, weight and input power. The control panel has been rebuilt as a smaller LRU which optimises helicopter functions. In the CM-479, existing circuit cards have been repackaged in a smaller, lighter enclosure with the option for a second CPU/memory circuit card. The CM-479 power supply has been replaced by a new circuit card in the signal processor to match helicopter power requirements. The amplifier-detectors have been repackaged into two LRUs, each containing circuit cards from two of the AM-6639s. New antennas have been installed, but existing AN/ALR-69 antennas can be reused. The frequency selective receiver and C/D-band receiver are not used in the helicopter installation.

Operational status

The AN/ALR-DK system is installed in Royal Danish Army Air Corps AS 550 C2 Fennec helicopters.

Contractor

Radartronic A/S.

VERIFIED

Digital enhanced AN/ALE-40 sequencer switch

The fully digital sequencer switch is a direct replacement for the current AN/ALE-40 electromechanical sequencer switch. It operates in the existing aircraft installation without any requirement for modification.

The enhanced sequencer switch assembly is an autonomous LRU compatible with the AN/ALE-40 countermeasures dispenser system. It routes payload dispense signals to the breech plate firing pins in the dispenser assemblies, executing the sequential firing of chaff, flares and advanced decoy payloads.

Enhanced safety design is provided in accordance with MIL-E-5400. Extensive software and hardware monitoring of the squib firing current source and the complementary dual-level power switching circuits protect against inadvertent payload firing. To accommodate future dispenser system enhancements the sequencer switch is equipped with a databus interface, enabling intelligent, dynamic operation within an ACMDS advanced countermeasures dispenser system environment.

Operational status

Fitted to Danish AF C-130H aircraft, and Royal Netherlands AF C-130 and F-27 aircraft.

Contractor

TERMA Elektronik AS.

VERIFIED

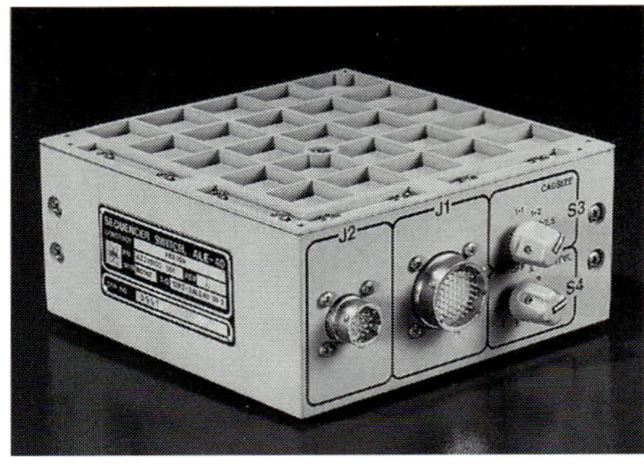

The digital sequencer switch is compatible with the AN/ALE-40 countermeasures dispenser system **1995**

FRANCE

5000 series chaff/flare countermeasure dispensers

Alkan has developed countermeasure dispensers that can accommodate either chaff or infrared flare cartridges or a combination of the two. The cartridges are arranged in interchangeable, easily handled magazines loaded in the modular dispenser. Typically, a single dispenser will accommodate five to seven modules. Each module houses a magazine containing, for example, eighteen 40 mm diameter chaff cartridges.

The dispenser electronic management system performs the firing sequences created by software whether it is connected to an RWR or not. It permanently manages the inventory of available cartridges and provides the necessary information to the cockpit control unit which displays the status of the complete equipment.

The CADMIR dispenser pylon using the same technologies is specifically designed for semi-conformal installation on the Mirage 2000 and 2000-5.

Operational status

The system is in service in French Air Force Jaguar aircraft. Two dispensers are fitted under the wing in a conformal installation near to the aircraft fuselage. Each dispenser contains seven modules (Alkan Type 5020).

The same system is also in service on the Dassault Mirage F1, on which it is installed on a special wing hard point.

The Alkan Type 5081 pod is designed to fit either to the JATO point of the MiG-21 or to any 14 in (356 mm) standard armament hard point. It is in service on the French Navy Super Etendard.

The same concept is used in an internal configuration on the Mirage 5. The Type 5013 dispenser which houses four standard magazines is in series production for a foreign customer.

Contractor

Alkan.

UPDATED

The Alkan 5020 decoy dispenser on a Jaguar aircraft

The ELIPS chaff/flare dispenser for helicopters and light aircraft **1995**

ELIPS helicopter self-protection system

The Electronic Integrated Protection Shield (ELIPS) has been designed for helicopters and light aircraft. It is a multidecoy launching system adaptable to any kind of chaff or flares, operating in automatic, semi-automatic and manual modes. Direct coupling to any type of Radar Warning Receiver (RWR) or Missile Approach Warner (MAW) for automatic sequence selection and launching is available.

The type and quantity of ammunition available, selected firing sequence and functioning status of the system are displayed either on the cockpit control unit or on a centralised multifunction ECM display. Loading of the various decoys is identified through codes allocated to magazines or through an integrated recognition system. Two, four or more magazines may be controlled without any modifications to the system.

Specifications

Weight:
(2 chassis configuration) <12 kg empty, <23 kg loaded with IR flares
Firing interval: 25 ms-1 min
System response: <50 ms

Contractor

Alkan.

VERIFIED

SPIRIT electronic warfare system

SPIRIT is an electronic warfare system for transport aircraft. It comprises a high-capacity dispenser providing up to 84 cartridges, containing up to 376 flares, on the aircraft. It consists of a Type 5160 decoy dispenser, scabbed on to each side of the fuselage, and an NVG-compatible control unit which offers automatic, semi-automatic or manual operation.

Operational status

Fitted to French Air Force C-160/C-160NG aircraft interfaced with the Thomson-CSF Sherloc radar warning receiver and an Elta missile approach warning system.

Contractor

Alkan.

UPDATED

ABD 2000 jammer

The ABD 2000 is a self-protection detector/jammer designed for the export version of the Mirage 2000. This multimission aircraft is intended to operate in a wide range of theatres and its self-protection system is able to counter all types of threats.

The ABD 2000 is installed internally, thereby avoiding using up ordnance pylon points and/or limiting the aircraft flight envelope. The complete system weighs 80 kg and consists of four LRUs: a main unit containing the receiver, jamming channels, transmitter and the aft antennas; a left-hand conformal unit containing the electrical power supplies for the system; a right-hand conformal unit containing the computers controlling the system operation; an antenna providing forward coverage and located on the vertical fin of the aircraft.

The ABD 2000 system is capable of detecting and identifying RF emitters selecting the more dangerous threats alerting the pilot on the countermeasures display unit; and automatically initiating jamming appropriate to the type of threat. It has a very wide frequency coverage, is entirely controlled by microprocessors, and can be programmed in accordance with user's operational scenarios. The system is operated by the countermeasures control panel fitted in the cockpit. Using this panel the pilot can select one of four positions: Off, SIL (the system detects a threat), Jam, Test.

The ABD 2000 can be integrated in a complete EW system, known as the Integrated CounterMeasures System (ICMS), which has been developed in collaboration with Dassault Aviation, Thomson-CSF and Matra Défense.

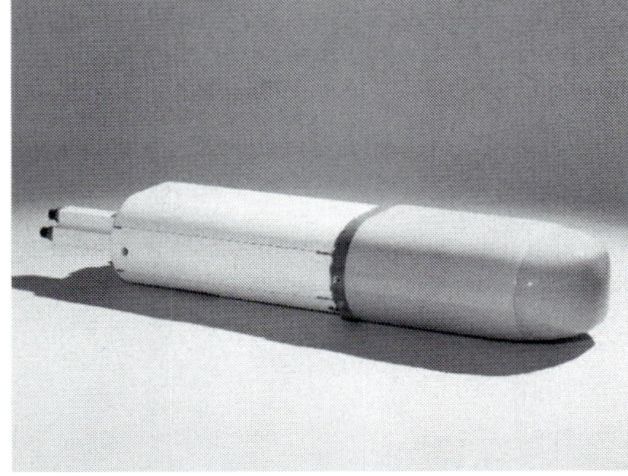

The ABD 2000 self-protection jammer

Specifications

Frequency: H-, I- and J-bands
Coverage: all sectors
Reaction time: 0.6 s
Detector sensitivity: up to −60 dBm
Measured parameters: RF frequency, frequency agility, pulsewidth, PRI, duty cycle, sector, polarisation, level, type
Multi-jamming modes: continuous and spot noise, barrage noise, cover pulse jamming, count down, velocity deception, blinking.

Operational status

In service on the Mirage 2000 aircraft of several air forces.

Contractor

Dassault Electronique.

UPDATED

ADELE ESM System

ADELE *(Alerte Détection et Localisation des Emetteurs)* is an ESM system for helicopters and maritime patrol aircraft based on interferometry techniques.

The essential features of ADELE are a high direction-finding accuracy, allowing discrimination between emitters and their localisation; very confident identification of all emitters; a powerful computer able to handle the densest electromagnetic scenarios.

Operational status

Fully developed.

Contractor

Dassault Electronique.

VERIFIED

BARRAGE radio jammer

BARRAGE is a tactical radio communication jammer that uses digital jamming mode generation techniques.

BARRAGE operates in the V/UHF bands and has a transmission power in excess of 100 W.

BARRAGE may be coupled with a direction-finding receiver to optimise jamming efficiency in dense environments.

Mirage 2000 aircraft fitted with the Dassault Electronique ABD 2000 self-protection jammer, showing the rear antenna above the jetpipe

Specifications

Weight: 10 kg
Power Consumption: 50 VA

Operational status

BARRAGE is suitable for helicopters, light multipurpose aircraft and UAVs.

Contractor

Dassault Electronique.

VERIFIED

Camel/BEL expendable jammer family

Camel/BEL is a miniaturised, expendable, electromagnetic, active, very wideband, coherent, decoy jammer for self-protection of aircraft against missiles and tracking radars. It presents a credible electromagnetic signature with regard to velocity, direction and range. Camel/BEL is, therefore, able to counter CW, pulsed and pulsed Doppler radar threats, in particular those radars used in active and semi-active missile seeker heads, either air-to-air or ground-to-air, and including radars with monopulse direction-finding or other counter-countermeasures.

Dimensions of the Camel/BEL cartridge are compatible with current-generation decoy dispensers.

Operational status

Limited-rate initial production.

Contractors

Dassault Electronique.
Etienne LACROIX.

VERIFIED

Carapace threat warning system

Carapace is a version of the passive part of the EWS-16 system for the F-16 aircraft (see later item). The main part of the system is a hybrid receiver, including IFM, crystal video and superheterodyne receivers, plus an interferometric direction-finding array.

Carapace is believed to cover the C to K frequency bands and is designed to detect, identify and localise all modern threats with great accuracy. It is able to analyse accurately, on a pulse-to-pulse basis, emitters at long range in severe ECM environments and offers an ESM capability. Accuracy of direction-finding is classified but is probably of the order of 1°.

Operational status

Carapace is in service with the Belgian Air Force F-16 aircraft.

Contractor

Dassault Electronique.

VERIFIED

DOPAGE/PHALANGER ESM/ELINT system

DOPAGE/PHALANGER is a new-generation ESM/ELINT payload for airborne platforms, including unmanned aerial vehicles, helicopters and light multipurpose aircraft. A pod version can be installed on combat aircraft.

Aimed at detecting, identifying, and localising ground-based radars, DOPAGE/PHALANGER delivers radar tracks for real-time display and analysis. The battlefield tactical situation and electromagnetic order of battle (EOB) can be displayed either on board the carrier platform or in a remote ground-based STRATEGIE family ESM/ELINT station collecting the data from the payload via datalink.

The key features of DOPAGE/PHALANGER are: high sensitivity and good direction-finding; small volume and weight (20 kg) for easy installation on UAVs and helicopters.

Operational status

DOPAGE/PHALANGER has been validated by the French MoD during ground and flight tests.

Contractor

Dassault Electronique.

VERIFIED

EWR-99/FRUIT radar warning receiver

The EWR-99 FRUIT radar warning receiver is designed for helicopters. It is a user programmable, database oriented system covering a very wide frequency range and making use of full-band Instantaneous Frequency Measurement (IFM) technology. The system is able to detect all RF signals including CW, pulse, and pulse Doppler emitters in high-density electromagnetic environments. More than a simple radar warning receiver, the EWR-99 offers smart management of onboard dispensers and can interface with a missile warning system.

Operational status

In service on French Army Aviation Forces Cougar, Gazelle and Puma helicopters; also selected by PZL - Swidnik to equip its Sokol multirole helicopters.

Contractor

Dassault Electronique.

UPDATED

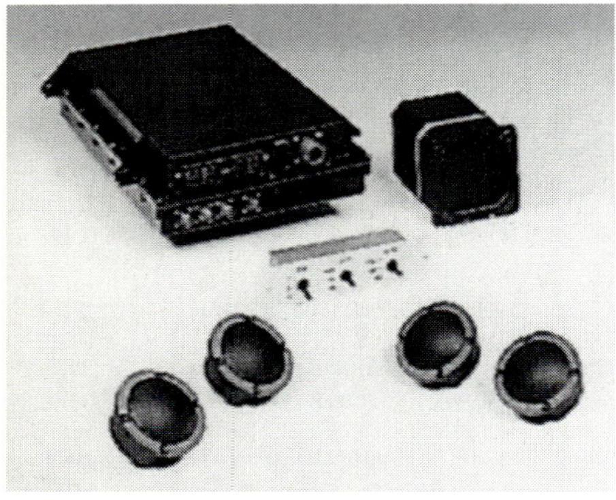

EWR-99 **1998**/0010927

EWS-16 self-protection system

The EWS-16 is an integrated system which consists of a threat warning system and an active jammer. The threat warning system can detect, identify and localise all modern threats with great accuracy. All threats can be identified in less than 1 second without any ambiguity, even in dense electromagnetic environments. The active jammer features high radiated power and can counter pulsed and CW radars. Analysis of threats is on a pulse-by-pulse basis, with very accurate identification and priority assessment, a clear pilot interface, and full integration with the weapon system. The EWS-16 is fully programmable and makes use of a separate threat library. The ESM subsystem can accurately locate ground-based radars and can record data during flight.

Features of the EWS-16 system include: a management and compatibility unit based on a 32-bit processor; a crystal video receiver and a high-speed wideband superheterodyne receiver; an instantaneous wideband direction-finding interferometer; an IFM receiver; real-time spectral analysis processing; a plug-in Emitter Identification (EID) and mission report module; a multiple threat jammer employing a high-power transmitter; and synergy between jammer and decoy dispenser.

Operational status

A version of the passive part of the EWS-16, known as Carapace, is in service in the Belgian Air Force for its entire fleet of F-16 aircraft.

Contractor

Dassault Electronique.

VERIFIED

EWS-21 radar warning system

The EWS-21 radar warning receiver is adapted from the EWS-A (see next item) for MiG-21 aircraft upgrade applications. It is effective against all types of radar emitters, offering full instantaneous band coverage. The system uses adaptive processing to deal with high-duty cycle radars and jammers and is capable of in-flight recording for post-flight analysis. In the passive target designation role the EWS-21 provides data for the MiG-21 Kopyo radar and modernised avionics. Integration within the aircraft has been studied thoroughly with Mikoyan.

Operational status

Ready for production, probably initially aimed at the update of India's MiG-21 fleet, as part of a new avionics suite.

Contractor

Dassault Electronique.

VERIFIED

EWS-A radar warning system (AIGLE)

The EWS-A (Electronic Warning System for Aircraft) is a compact radar warning system (10 kg) designed for modernisation for aircraft self-protection systems. The system is lightweight and can detect all radar threats including CW, pulsed and pulsed Doppler emitters, even in the densest environments. EWS-A is also sometimes designated as AIGLE.

The key features of the EWS-A are a very short reaction time, high confidence of emitter identification due to Instantaneous Frequency Measurement (IFM) and a 100 per cent probability of interception. The EWS-A will also control jammers and/or chaff launchers (if fitted to the aircraft), and can be integrated to the aircraft weapon system through serial link and/or multiplex bus interface. The EWS-A is fully user programmable.

Operational status

In full-scale production for the entire French Air Force Mirage F1 fleet of aircraft (110 units). In an associated programme, Dassault Electronique has been contracted to provide a mid-life update for the Barax self defence jamming pods, also fitted to the French Air Force Mirage F1 fleet.

Selected by Northrop Grumman as the RWR of choice for its Tiger IV upgrade proposals for F-5 aircraft.

Contractor

Dassault Electronique.

UPDATED

MWS 20/DAMIEN Missile approach Warning Systems

The MWS 20/DAMIEN family of active missile approach warning systems has been designed for self-protection of helicopters, and fixed-wing combat aircraft.

It detects approaching missiles and triggers the appropriate countermeasures at the optimum time. Its main functions are to: provide warning of all types of attacking missiles; calculate in real time the speed, range, direction of detected missiles, and time to impact; control a decoy-launcher in automatic mode in order to considerably reduce the decoy consumption.

Insensitive to weather conditions, MWS 20/DAMIEN is also not affected by infrared decoy (flares) illumination or other heat sources, and is designed to give a low false alarm rate.

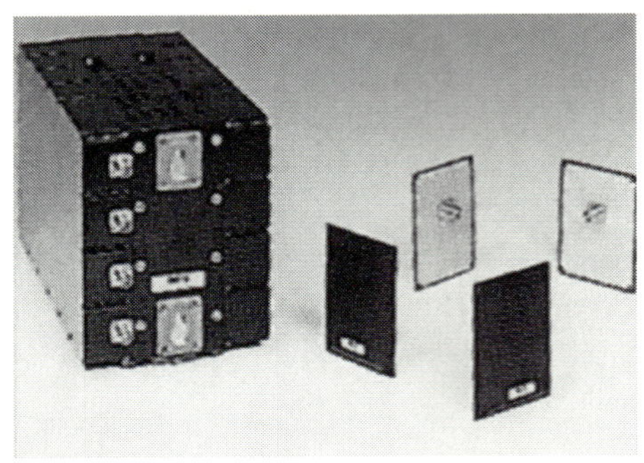

MWS20/DAMIEN **1998**/0010926

Due to the use of state-of-the-art technologies (ASIC components and Doppler solid-state transmitter), MWS 20/DAMIEN is extremely reliable. It has a BIT and modular architecture to simplify maintenance.

MWS 20/DAMIEN is highly miniaturised comprising a 10 kg airborne electronic unit and four conformal antennas.

Operational status

In production.

Contractor

Dassault Electronique.

VERIFIED

PAJ-95 detector/jammer

PAJ-95 is a pod-mounted airborne detector/jammer for self-protection of combat aircraft. It has a supersonic capability, with a diameter of 157 mm and a weight of 85 kg. Covering a very wide frequency band, and capable of multiple and independent jamming techniques, it can detect and counter all types of threat. It is fully automatic and is able to perform surveillance, alert and jamming functions over the complete frequency band. Threat and jamming techniques libraries can be reprogrammed by the user and loaded before take-off. Reaction time is very short and since it is entirely controlled by microprocessors, it can be adapted to future threat developments. PAJ-95 has the benefit of the most modern technology, through several updates.

Operational status

Versions of the PAJ-95 are in service for the French Mirage III, Mirage F1, and Jaguar. A particular version has been manufactured in co-operation with Spanish industry for the Spanish Air Force.

Contractor

Dassault Electronique.

VERIFIED

Spectra EW system

Dassault Electronique, Thomson-CSF and Matra BAe Dynamics are developing the Spectra self-protection system for the Rafale ACT/ACM aircraft. The system is the first ever in France to cover electromagnetic, laser and infrared domains. It makes use of sophisticated techniques, such as interferometry, digital frequency memory, electronic scanning, multispectral infrared detection, artificial intelligence, and substrate technologies (MMICS on GaAs substrates and VHS INS). Spectra includes an active phased-array transmitter which is currently in flight test.

The system is fitted internally in the Rafale, with over 10 locations distributed throughout the aircraft, and integrated through a specific EW databus and a central processor. Spectra may also be installed on the outside of the aircraft.

Matra BAe Dynamics will provide the DDM (Missile Launch Detector) (2-sensor type), the LCM (Modular Cartridge Dispenser) (with 4 cartridge dispenser modules and 2 chaff dispenser dual tubes).

Operational status

The programme was launched in 1990 under a French

MoD contract. The first prototype was delivered in 1993. First flights on Rafale occurred in 1994. Dassault Aviation is responsible for aircraft integration; Dassault Electronique for electromagnetic detection and jamming functions; Matra BAe Dynamics for IR detection and decoy functions; Thomson-CSF for systems integration and laser warning functions.

The Spectra design configuration has been sealed for production of the first operational system for the first French Navy Rafale aircraft delivery in 1999.

Contractors

Dassault Electronique.
Matra BAe Dynamics.
Thomson-CSF Radars Contre-Mesures.

UPDATED

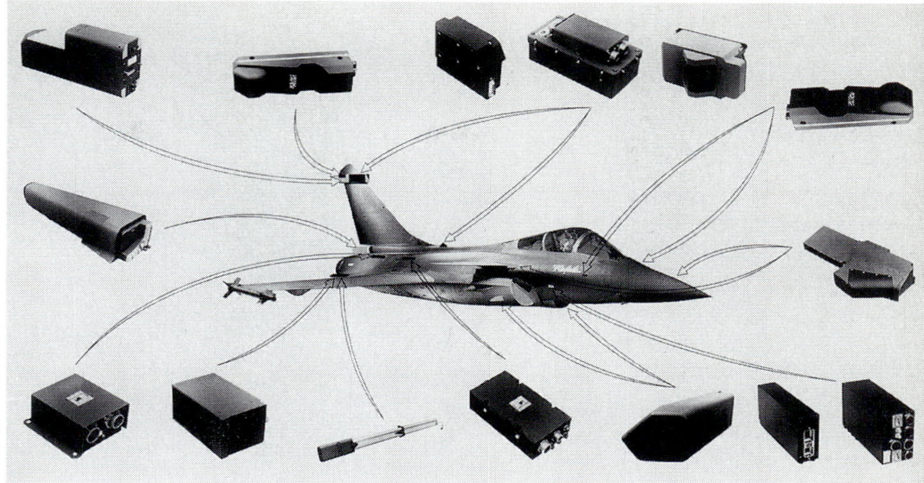

Rafale is equipped with the Spectra EW system **1996**

Corail countermeasures equipment

Corail is a radar and optronic countermeasures equipment designed for various versions of the Mirage fighter and can also be applied to transport aircraft. The system can be housed in a conformal or external pod; two pods can carry up to 256 decoy cartridges according to type - EM, IR and EO cartridges can be used.

Operational status

In production for Mirage F1 aircraft in service with the French Air Force.

Contractor

Matra BAe Dynamics.

VERIFIED

Corail self-protection countermeasures system under the French Air Force Mirage F1 CT wings **1996**

DDM missile launch detector

The DDM missile launch detector provides automatic detection of a missile launch plume, locates it in flight and instantaneously transfers threat data to the aircraft ECM system. It incorporates passive infrared detector techniques to ensure covert operation in severe ECM environments and is capable of locating missiles using any type of seeker. DDM features advanced signal processing to give extremely high probability of launch detection and a low false alarm rate. It includes a mosaic detector array to provide infrared signature discrimination and has the capability to handle 40 tracks simultaneously.

DDM can be configured for different applications by arrangement of its component modules to meet specific size, weight and coverage requirements.

The DDM-2000 has been developed for the Dassault Aviation Mirage 2000. In this, the electro-optical head and signal processing unit are integrated into a single unit mounted at the end of the Magic AAM launcher on either side of the aircraft.

In the DDM-Prime version, which forms part of the Rafale Spectra self-protection system, the electro-optical head and signal processing unit form separate modules.

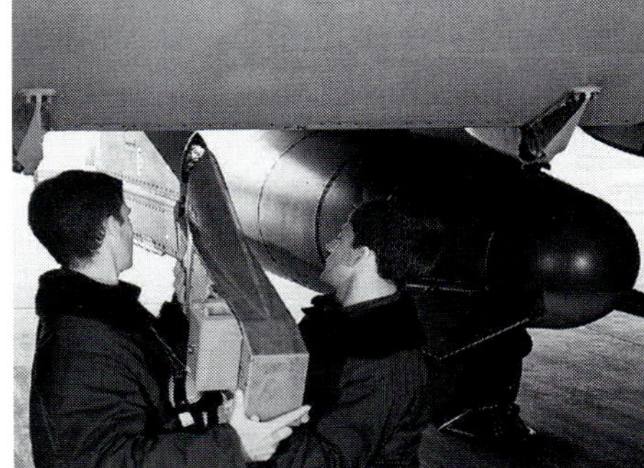

Missile launch detector (DDM) for the countermeasure equipment of a Mirage 2000 combat aircraft **1996**

Specifications

Weight:
(electro-optical head) 5.6 kg
(signal processing unit) 3.6 kg
Coverage: 180°
Accuracy: ±2°

Operational status

In final development for French Air Force Mirage 2000N and 2000D aircraft and Rafale.

Contractors

Matra BAe Dynamics.
SAGEM.

NEW ENTRY

LCM (Lance-Cartouches Modulaire) modular cartridge dispenser

Developed as part of the Rafale aircraft Spectra EW system, the LCM comprises a variable number of identical cartridge dispenser modules controlled by a single computer. LCM is the modular cartridge dispenser proposed by Matra BAe Dynamics for Spirale NG, Spectra, Myriad and Saphir M EW systems. The LCM system is compatible with the following standard cartridges: 1 × 1 in, 19 mm, 40 mm diameter, and 60 mm diameter. It is also claimed to be compatible with future smart ammunition rounds, digitally programmable in flight. Each module will provide a load capacity of between 6 (Lisca type) and 72 (Mucalir type) cartridges.

Contractor

Matra BAe Dynamics

NEW ENTRY

LEA (Leurre Electromagnétique Actif) active radar decoy

LEA is being developed by Matra BAe Dynamics and Thomson-CSF. It is an expendable pseudo-repeater mini jammer that is programmable in flight before firing.

Operational status

R+D trials complete. Production planned for year 2000.

Contractors

Matra BAe Dynamics.
Thomson-CSF Radars Contre-Mesures.

NEW ENTRY

LISCA (Leurre Infrarouge à Signature et Cinématique Adaptée) Smart IR decoy

LISCA is being developed by Matra BAe Dynamics and SNPE. Its propulsion and aerodynamics allow its separation path from the aircraft to be optimised to deceive IR-homing seeker heads. Its ejection system is common with that of LEA, and compatible with the LCM system.

Operational status

Development.

Contractors

Matra BAe Dynamics
SNPE Societé Nationale des Poudres et Explosifs.

NEW ENTRY

Phimat chaff dispenser

Phimat is a light low-drag radar decoying system designed for supersonic speeds and suitable for all operations in Jaguar, Harrier, Super Etendard and various types of Mirage aircraft. It can generally be adapted to all types of modern aircraft, particularly those with weapon stations capable of taking the R550 Magic or Sidewinder missiles.

The system consists of a cylindrical dispenser pod, chaff pack tubes, ejection mechanism, associated electronic block and a control box located in the cockpit.

Specifications

Dimensions: 3,600 mm long × 180 mm diameter
Weight (loaded): 105 kg

Operational status

Operational with the French Air Force and Navy and the UK Royal Air Force. Production ceased.

Contractor

Matra BAe Dynamics.

VERIFIED

Phimat chaff dispenser under Super Etendard aircraft ***1996***

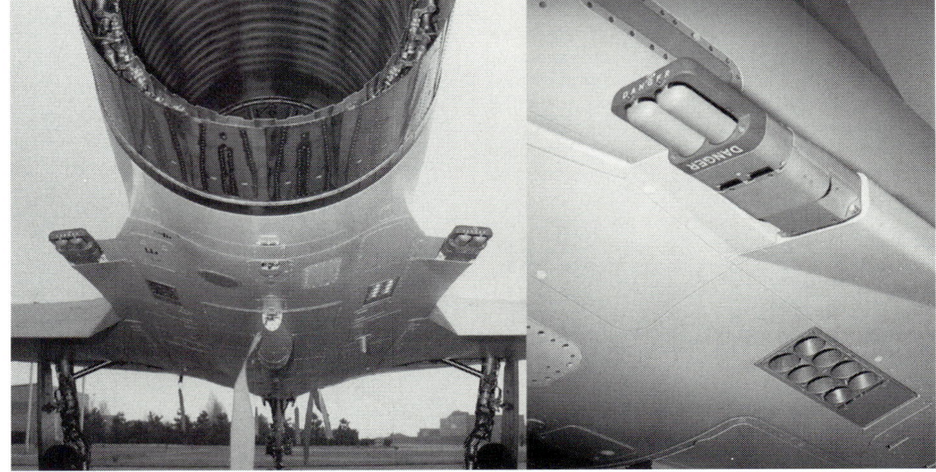

Spirale dispenser mounted on a Mirage 2000 aircraft

Saphir chaff and flare system

The Saphir chaff and flare system is optimised for use with helicopters. Four cartridge dispensers are used and the system can be deployed manually, fully or semi-automatically or in a survival mode.

Operational status

Saphir A was designed to equip the Horizon Super Puma helicopter. The system has been installed on Puma, Cougar, Lynx and Ecureuil helicopters. During the Gulf War, French Army Gazelles and Pumas and French Air Force Pumas were equipped with Saphir.

Saphir B was designed for light helicopters and is in service on the Gazelle and Lynx.

A version known as Saphir M, developed jointly by Buck Systems and Matra BAe Dynamics, has been selected for the NH 90 TTH (Tactical Transport Helicopter) and NH 90 NFH (NATO Frigate Helicopter) and Tiger helicopters. It uses Saphir technology and Buck expendables.

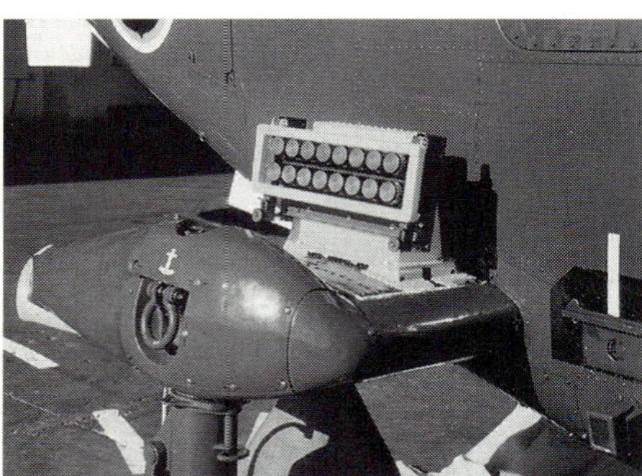

SAPHIR-B self-protection countermeasure system equipping a Lynx helicopter of the French Navy ***1996***

Contractor

Matra BAe Dynamics.

UPDATED

Spirale chaff and flare system

The Spirale chaff and flare system was developed for the Dassault Mirage 2000 fighter and entered service at the end of 1987. The system comprises two cartridge dispensers located under the rear fuselage, two chaff dispensers located at each wingroot, two electronics boxes and two fixed missile detectors in the Magic missile launchers. Spirale NG is a new version being developed for the Mirage 2000D. Spirale NG will have a new-generation computer as well as DDM (Missile Launch Detector) integration, with two outfits per aircraft and decoying capacity extensions – these will be carried on six ridge-mounted cartridge dispenser modules (LCM) and four cartridge dispenser modules under the fuselage, resulting in 12 cartridge dispenser modules being available on the Mirage 2000D.

The Sycomor ECM system on a Mirage F1 fighter

Operational status

Spirale is in service with several air forces on the Mirage 2000.

Contractor

Matra BAe Dynamics.

UPDATED

Sycomor chaff and flare system

The Sycomor chaff and flare system is intended for the various versions of the Dassault Mirage F1 fighter and can be packaged either in a 2.95 m long externally mounted pod or in a 2.5 m conformal pack. Each pack has three chaff dispensing tubes and seven cartridge magazines. Each pod has the capacity for two packs.

Operational status

Sycomor is operational with Mirage F1s flown by several air forces.

Contractor

Matra BAe Dynamics.

VERIFIED

SAMIR missile launch detector

The SAMIR (Systeme d'Alerte Missile Infra Rouge) missile launch detector provides automatic detection of a missile launch plume, locates it in flight and instantaneously feeds threat data to the aircraft ECM system. It incorporates passive infrared detector techniques to ensure covert operation in severe ECM environments and is capable of locating missiles using any type of seeker. SAMIR features advanced signal processing and a mosaic detector array to provide infrared signature discrimination. It can handle 40 tracks simultaneously.

SAMIR can be configured for different applications by arrangement of its component modules to meet specific size, weight and coverage requirements. It can be installed in a variety of aircraft types such as combat aircraft, helicopters and transport aircraft.

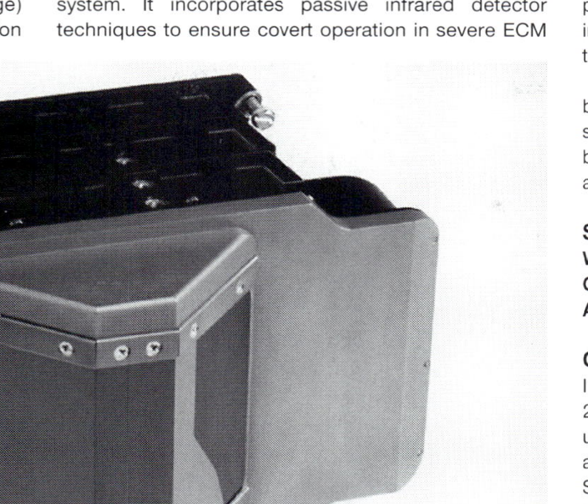

SAMIR missile launch detector **1997**/0001258

Specifications

Weight: 10 kg
Coverage: 180°
Accuracy: 2°±

Operational status

In production for French Air Force Mirage 2000N, 2000C, 2000D and in development for Rafale. It is understood that the version fitted to Mirage 2000 aircraft is configured into one LRU and designated SAMIR 2000, whilst the version being developed for Rafale is configured in two separate modules and designated SAMIR prime.

Contractor

SAGEM SA, Defence & Security Division.

UPDATED

ASTAC airborne ESM/ELINT system

ASTAC is an ESM/ELINT system intended primarily for detection and identification of enemy ground radar stations. It is suitable for medium- and high-altitude standoff reconnaissance missions, or at low altitude on the battlefield and in penetration where it can collect data for the avoidance or destruction of anti-air defences.

It consists of an airborne pod fitted with interferometers for high-precision direction-finding, two compressive receivers, a data recorder and a data downlink. One receiver is used to obtain a very precise measurement of radar frequency, and the two receivers working together can handle frequency-agile emitters. The pod is capable of handling up to 20 radars/s. ASTAC weighs about 400 kg and can be adapted easily to any combat or light transport aircraft equipped with accurate navigation equipment. Subsystems can also be installed on board the aircraft.

Information from the airborne sensors can be displayed on board and transmitted by secure datalink to a ground station where it is processed and the various radar parameters are displayed in alphanumeric format. These parameters include whether the radar is mobile or fixed, its type, frequency, PRF, pulsewidth, intra-pulse modulation and agility in the B- to K-bands. In addition to the alphanumeric display, a graphic display is provided which incorporates a map showing positions of detected emissions, together with known airfields, radar sites and so on. An airborne display provides real-time onboard information to the crew. Applications also exist for ground stations.

Specifications

Dimensions: 3,960 long × 406 mm diameter
Weight: 400 kg
Frequency: B- to K-bands 0.5-20 GHz (option to 40 GHz)
Accuracy: sub-degree

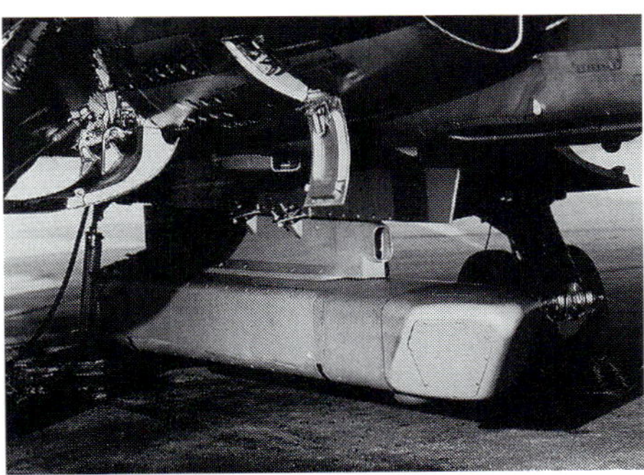

ASTAC airborne ESM/ELINT system **1996**

Operational status

In production for French Air Force Mirage F1CRs and for Japanese Air Self-Defence Force RF-4EJ aircraft. Demonstration flights have been carried out on F-16.

Contractor

Thomson-CSF Radars Contre-Mesures.

UPDATED

Barem jamming pod

The Barem self-protection jammer is designed to protect tactical aircraft and helicopters from radar-directed missiles and similar threats. Barem is able to jam pulse PD and CW threats. Its threat handling capability is reported to be two threats simultaneously fore or aft. Housed in a pod cleared for flight at speeds above M2, the system has front and rear receive and transmit antennas coupled to the superheterodyne receiver and TWT transmitter which work under automatic microprocessor control over the H-, I- and J-bands. Threats detected by the threat identification library and the jamming techniques generator are easily reprogrammable on the flight line. Received signals are recorded in flight for subsequent analysis and storage in the library.

Specifications

Dimensions: 3,450 (length) × 160 mm (diameter)
Weight: 85 kg
Power required: 700 VA

Operational status

In service on French and export Mirage, Jaguar and Super Etendard aircraft.

Contractor

Thomson-CSF Radars Contre-Mesures.

UPDATED

BF radar warning receiver

The Type BF radar warning receiver provides the crew with warning of most categories of airborne and surface radar threats and with an indication of their direction.

Four wideband antennas are used, linked to a video receiver and, when necessary, a synchronisation unit. In the Mirage the system control and display unit is integrated with other cockpit equipment, but it can be provided separately for other applications.

The receivers comprise photographically etched spiral antennas and microwave circuits for an RF test oscillator, limiter modulator diodes, high-pass filter,

The Thomson-CSF Barem jamming pod can be flown on tactical aircraft at speeds above M2

detector circuit, video modulation and pre-amplification. The two side-mounted antennas lie flush with the fin structure, while fore and aft antennas on the fin have conical radomes. An audio alarm is generated when threats are detected and approximate threat direction is indicated by one or more of four signal lamps. The threat is also categorised by one of three lamps which indicate conventional pulse radar, continuous wave or interrupted continuous wave radar or track-while-scan ground radar.

Specifications

Dimensions:
(flat antennas) 148 (diameter) × 53 mm (depth)
(conical antennas) 82 (diameter) × 360 mm (depth)
(synchronisation unit) 209 × 110 × 45 mm
(control box) 146 × 95 × 40 mm
(indicator unit) 68 × 61 × 61 mm
(video receiver) ¼ ATR short
Weight: 9.2 kg (total system)
Power supply: 200 V AC, 400 Hz, <500 VA

Operational status

No longer in production. In service in Mirage F-1A and F-1C, Mirage IIIZ, Super Etendard and Royal Netherlands Air Force F-5 aircraft. In widespread use in the aircraft of a number of countries in the Middle East. A version designated TMV 008H is produced for helicopters.

Contractor

Thomson-CSF Radars Contre-Mesures.

VERIFIED

Caiman noise/deception jamming pod

Probably developed from the earlier Alligator pod, Caiman is designed for the SEAD (Suppression of Enemy Air Defences) role. It can be installed underwing or on a fuselage pylon. The pod is self-contained, ram-air entering the unit through an annular intake to drive a power turbine and to provide cooling. Within the pod are fore and aft receiver antennas and two independent jammers each weighing 130 kg. The radiated power is 750 W in I-band and 3 kW in D-band. It can be operated in manual, semi-, or fully automatic mode. The system is combat proven.

Caiman Mk 2 is another podded Thomson-CSF jamming system which has been specifically designed for attack jamming against surveillance radars, including battlefield and AWACS type systems. It instantaneously covers two full octaves of hostile frequencies and can jam up to 20 threats simultaneously. The pod which contains TWT transmitters, superheterodyne receiver, autonomous power supplies and antennas, can be carried either under the wing or the fuselage of the aircraft. For tracking purposes, Caiman Mk 2 operates over a wider frequency range than Caiman and has an output jamming power in excess of 1,000 W. It incorporates an extensive memory capacity and modular design software for reprogramming to meet additional and changing threats. As with Caiman, the Mk 2 has manual and automatic operating modes.

Specifications

Dimensions: 5,950 (length) × 410 mm (diameter)
Weight: approx 500 kg

Operational status

In service, but no longer in production. Caiman is fitted to the Mirage F1, Mirage 2000, Jaguar and F-5 and has been supplied to export customers for several aircraft types.

The Thomson-CSF Remora jamming pod

Contractor

Thomson-CSF Radars Contre-Mesures.

UPDATED

DB-3141 noise jamming pod (low-band Remora)

The DB-3141 H- to I-band jamming pod has a single receiver, a travelling wave tube jammer and fore and aft transmitter antennas. It provides a simple active electronic warfare capability for Dassault Mirage fighters and possibly also for that company's Super Etendard. DB-3141 jams several threats simultaneously, including low-power Doppler radars. The system has a look-through capability, enabling it to discontinue jamming as soon as threat reception ceases.

Specifications

Dimensions: 3,520 (length) × 250 mm (diameter)
Weight: 175 kg
Power supply: 200 V AC, 400 Hz, 1.7 kVA

Operational status

In service on Dassault Mirage aircraft.

Contractor

Thomson-CSF Radars Contre-Mesures.

UPDATED

DB-3163 noise jamming pod (high-band Remora)

The DB-3163 noise jamming pod is designed to provide self-protection against both air and ground radar threats. Pulse and continuous wave emitters can be detected, identified and countered. A superheterodyne receiver performs a frequency scan search on emissions received by antennas at both ends of the pod. During preflight preparation, bands can be selected; up to three threats or groups of threats can be jammed simultaneously. An internal bootstrap air cooling system is employed and the system is energised from the aircraft's power supplies.

DB-3163 is combat proven.

Specifications

Dimensions: 3,520 (length) × 256 mm (diameter)
Weight: 175 kg
Power supply: 200 V AC, 400 Hz, 1.7 kVA
Frequency: I/J-band preprogrammed

Operational status

In service with a number of air forces.

Contractor

Thomson-CSF Radars Contre-Mesures.

UPDATED

DR 2000A ESM receiver

The DR 2000A is the airborne version of the DR 2000 Series which is used for land-based, shipborne and airborne applications. The basic system consists of intercept receiver, a Dalia 1000 Mk 1 or Mk 2 analyser, plus the necessary antenna units. The Dalia analyser provides alarm analysis and identification facilities through a programmable library of 1,000 radar modes and parameters.

The DR 2000A intercept receiver is built up of one omni-antenna and a number of DF antennas. It provides a virtual 100 per cent probability of intercept over the complete 360° of azimuth. The equipment carries out passive search and detection of all pulse and CW signals and gives an instantaneous visible and audio alert. The DR 2000A Mk 2 is an improved version with better sensitivity and coverage of the frequency bands from D to J and a DF accuracy of 5° RMS.

Operational status

Uncertain.

Contractor

Thomson-CSF Radars Contre-Mesures.

UPDATED

The Caiman jammer mounted under the starboard wing of a Mirage F1 fighter

DR 3000A ESM receiver

The DR 3000A is the airborne version of the DR 3000 Series designed for land, ship, submarine and aircraft operation. The 3000A is suitable for both fixed-wing aircraft and helicopters and consists of a processing unit, display unit and an antenna system and six DF and intercept aerials. It provides very high detection capability over the complete 360° combined with high

DR 3000 control and display unit 1996

sensitivity over the frequency bands from D to J. Reliable identification is based on efficient de-interleaving even in very dense electromagnetic environments, accurate parameter measurements and artificial intelligence techniques.

The DR 3000A is small, modular, flexible and can cope with the most commonly envisaged threats. It will meet all requirements for warning, surveillance, ELINT and target designation through ECM. The basic performance can be enhanced by various options to provide improved capabilities for ELINT and DF accuracy.

Total weight of the system, including processing unit, control and display console and the antennas, is below 80 kg.

Operational status

Procured by Pakistan for installation in Atlantique maritime patrol aircraft.

Contractor

Thomson-CSF Radars Contre-Mesures.

UPDATED

DR 4000A (TMV 202) ESM suite

The DR 4000A (TMV 202) is an ESM suite designed for high-sensitivity instantaneous intercept probability and automatic processing against electromagnetic threats in the D- to J-bands. It is intended for use in fixed-wing aircraft and helicopters for ELINT, surveillance, threat detection, identification and data handling system processing and is a version of the DR 4000 Series of equipments. The basic system consists of two sets of six DF antennas, an omnidirectional antenna, modules housing RF amplifiers, an IFM processing unit and an operator console containing controls and three-colour graphic and alphanumeric displays.

With reprogrammable logic and a three-colour display, the probability of interception in both direction-finding and frequency discrimination is claimed to be 100 per cent with only a single pulse, as a result of the crystal video amplifier techniques used. The sensitivity is sufficient to intercept pulse compression signals. The system can be interfaced with any data handling system and the chaff launcher or jammer components of the ECM suite through suitable databusses or point-to-point links.

Specifications

Weight:
(including antennas) 169 kg
Power: 1,800 VA

Operational status

Uncertain.

Contractor

Thomson-CSF Radars Contre-Mesures.

UPDATED

Gabriel SIGINT system

Thomson-CSF Radars Contre-Mesures has developed complete SIGINT electronic intelligence systems for integration on board aircraft such as the DC-8, Boeing 707, Transall and C-130. One of these, Gabriel, configures ASTAC technology for detection, analysis and localisation of radar emissions and a COMINT subsystem, provided by Thomson-CSF Communications for detection, interception, classification, listen-in, analysis and localisation of radio communications. The system offers a high degree of

French Air Force Transall aircraft are equipped with the Gabriel SIGINT system

automation to assist the operators to accomplish all types of missions.

Operational status

Two C-160 Transall aircraft equipped with the Gabriel system are in service with the French Air Force.

Contractor

Thomson-CSF Radars Contre-Mesures.

UPDATED

ICMS Integrated CounterMeasures Suite

Thomson-CSF Radars Contre-Mesures, Dassault Electronique and Matra, have developed an integrated internally mounted EW suite for the Mirage 2000 aircraft. This is a highly sophisticated system, known as the Integrated CounterMeasures Suite (ICMS), where all parts are linked to a central interface and management unit.

The system incorporates three warning receivers designed by Thomson-CSF. These receivers consist of a version of the Serval equipment (see later entry), a superheterodyne receiver to detect CW radar, pulse compression signals and low-power pulse Doppler signals, and a receiver/processor mounted in the aircraft nose to detect missile command links. The missile detector function can also incorporate an infrared warning receiver designed by Matra.

Two detector-jammers are included, each with its own receiver which allows it to operate should the basic radar warning receiver be out of action. These detector-jammers consist of a high frequency sub-unit (designed by Dassault Electronique) to counteract airborne and surface-to-air threats, and a low-frequency sub-unit (designed by Thomson-CSF Radars Contre-Mesures) to operate against surface-to-air threats in the lower part of the spectrum. The Matra Spirale chaff/IR flare dispenser is also included in the overall system to provide passive countermeasures. Spirale is an internally mounted equipment which dispenses stores through openings in the aircraft structure.

Operational status

In production for the Mirage 2000-5. In service on Greek Air Force aircraft and those of several other countries.

Contractors

Thomson-CSF Radars Contre-Mesures.
Dassault Electronique.
Matra BAe Dynamics.

VERIFIED

The Mirage 2000-5 is equipped with the ICMS integrated countermeasures suite

MSPS on Super Etendard 1996

MSPS EW system

The Modular Self-Protection System (MSPS) is a very lightweight, easy to install EW suite, mainly designed for retrofit in existing aircraft. It combines the Sherloc RWR with the Barem jamming pod to provide a combined crystal video/superheterodyne receiving system for accurate threat description and fast data transmission between the two equipments for speedy reaction.

The MSPS EW system can be installed on a large number of fixed-wing aircraft and helicopters. The jamming part can be installed internally. MSPS is fully automatic and reprogrammable, and can accept additions such as chaff and flare countermeasures, laser detection equipment and support jamming systems. The weight of the system is 100 kg when the jammer is pod-mounted and 80 kg when it is internally installed.

Operational status
In production for the French Navy Super Etendard upgrade.

Contractor
Thomson-CSF Radars Contre-Mesures.

VERIFIED

Myriad radar warning system

Myriad is a warning receiver in the millimetric-wave region. It is designed for the protection of helicopters, light aircraft or armoured vehicles from smart weapons using millimetre-wave seekers. It provides immediate warning over 360° coverage and weighs approximately 8 kg.

Operational status
In development.

Contractor
Thomson-CSF Radars Contre-Mesures.

UPDATED

SACRE non-comm ESM/ELINT system

The SACRE system is a lightweight fully automatic sensor and a programming/analysis workstation.

SACRE is designed for tactical situation assessment (ESM function) and EW intelligence data gathering (ELINT function).

Because of its small volume and light weight, SACRE may be integrated in light carriers such as UAVs or helicopters, as well as in highly mobile land platforms.

The data are automatically intercepted during the mission and can either be transmitted to the workstation for real-time processing by means of a datalink or stored in a mass memory.

Mission data are processed in the workstation to evaluate tactical situation assessment: radar interception, measurement, identification and localisation. Technical inter-pulse and intra-pulse analysis is also performed to refine technical data bases.

The SACRE sensor utilises interferometry techniques to provide very accurate direction-finding with a wide instantaneous bandwidth to precisely determine threat status and location.

Inter-pulse and intra-pulse (option) fine grain analysis allows precise identification of threats. A finger-print capability is a further option.

Specifications
Frequency: D to J band (C and K bands optional)
DF accuracy: 0.5° RMS
Instantaneous angular coverage: 2 × 120°
Weight: 25 kg

Operational status
In development.

Contractor
Thomson-CSF Radars Contre-Mesures.

VERIFIED

Sarigue SIGINT system

A French Air Force DC-8 aircraft has been fitted with a SIGINT suite, known as Sarigue. The size of the aircraft provides room for over 10 workstations.

In early 1993 Thomson-CSF Radars Contre-Mesures was awarded a contract for the Sarigue NG (Nouvelle Génération) in which a DC-8 will be fitted with advanced ELINT and COMINT suites. The output from these will be collected to provide the SIGINT picture.

Sarigue NG is based on ASTAC equipment (see earlier item) and the Thomson-CSF TRC 290/600 Series communications equipment. TRC 290 receivers operate in the VHF/UHF bands, while the TRC 600 Series is a range of receivers, analysers and DF equipment covering the frequency range 0.1 to 1,350 MHz.

Operational status
A DC-8 aircraft equipped with Sarigue is in service with the French Air Force. In May 1993, Thomson-CSF was awarded a contract to install Sarigue NG in a French Air Force DC-8, which is scheduled to enter service in 2000.

Contractor
Thomson-CSF Radars Contre-Mesures.

UPDATED

Serval radar warning receiver

Fitted to French Air Force and export versions of the Dassault Mirage 2000 fighter, Serval warns the pilot when the aircraft is being illuminated by surface or airborne radars of a hostile nature; friendly/hostile discrimination is done by comparing the characteristics of the illumination energy with those of emitters held in the reprogrammable system threat library. Frequency coverage is understood to be E-J band, utilising channelised, crystal video receiver technology.

Serval uses four detection antennas mounted on the wingtips and fin feeding a hybrid analogue/digital processor. The CRT display unit shows the strength and direction of the threat emitter and whether it is ground-based or airborne. Details of several emitters can be shown simultaneously. At the same time an audio alarm sounds in the pilot's headset.

Operational status
Fitted to French Air Force and export Mirage 2000 aircraft.

Thomson-CSF is understood to be developing an upgraded version of Serval, designated Serval NG-D (New Generation-Distance). The 'distance' descriptor is said to be added because Serval NG-D includes new software algorithms that better determine the distance of the emitter that is illuminating the RWR.

Contractor
Thomson-CSF Radars Contre-Mesures.

UPDATED

Sherloc TMV 011 radar warning receiver

The Sherloc TMV 011 RWR is a system designed for fixed-wing aircraft or helicopters. It incorporates a crystal video receiver, high-speed digital processor and a radar signals library which is easily reprogrammable on the flight line. It also delivers operational flight reports. Of modular design, there are many ways to build Sherloc units into a system to meet specific requirements for extensive self-protection. The latest version also includes an instantaneous frequency measurement receiver capability.

Threat data is presented to the pilot in the form of alphanumeric symbology on a colour display unit, the symbols denoting the identification of threat lethality and the position on the CRT being relative to the threat's bearing. Up to eight emitters can be presented simultaneously. Alternatively, a simple light-emitting diode display can be used, indicating threat classification and relative strength. Sherloc operates in the D- to J-bands. System weight is 11 kg for aircraft and 9.5 kg for helicopters.

The latest Sherloc version, Sherloc F, incorporates an instantaneous frequency measurement receiver.

The helicopter specific version is called Sherif.

Specifications
Weight: 13 kg
Frequency: D- to J-band; options for C- to K-band
Accuracy: >10°

Operational status
In production. Between 150 and 200 Sherloc RWRs have been sold. In service in the Mirage 50, Mirage F1, French Navy Crusader, French Air Force C-135F and C-160, and Super Puma and Dauphin helicopters. Sherloc also forms part of a proposed upgrade for Russian Su-22 aircraft.

Contractor
Thomson-CSF Radars Contre-Mesures.

UPDATED

Spider active expendable jammer

Thomson-CSF Radars Contre-Mesures, in association with Matra BAe Dynamics, is developing the Spider expendable self-protection jammer which can be fitted in current chaff and flare containers. This decoy is made of an electronic payload, using the latest technologies such as MMIC, a GaAs amplifier with a high degree of integration to fit in the limited volume, a vehicle housing the payload and a battery.

Spider is designed to counter most modern threats such as active coherent missile homing heads. It has the basic capability to defeat monopulse tracking.

Operational status
Status uncertain.

Contractors
Thomson-CSF Radars Contre-Mesures.
Matra BAe Dynamics.

UPDATED

Syrel ELINT pod

The Syrel pod is a fully automatic electronic reconnaissance system attached by a special centreline pylon on Dassault Mirage F1, Mirage 3 and Mirage 2000 aircraft. It can be used during medium- and high-altitude standoff missions, as well as low-altitude penetrations, automatically to acquire and record data relating to the identification and location of ground-based electronic systems. It is intended to provide reliable information on radars for early warning systems, search and acquisition, ground-control interception and fire control for anti-aircraft artillery or missiles.

The pod has two antenna sets at both front and rear,

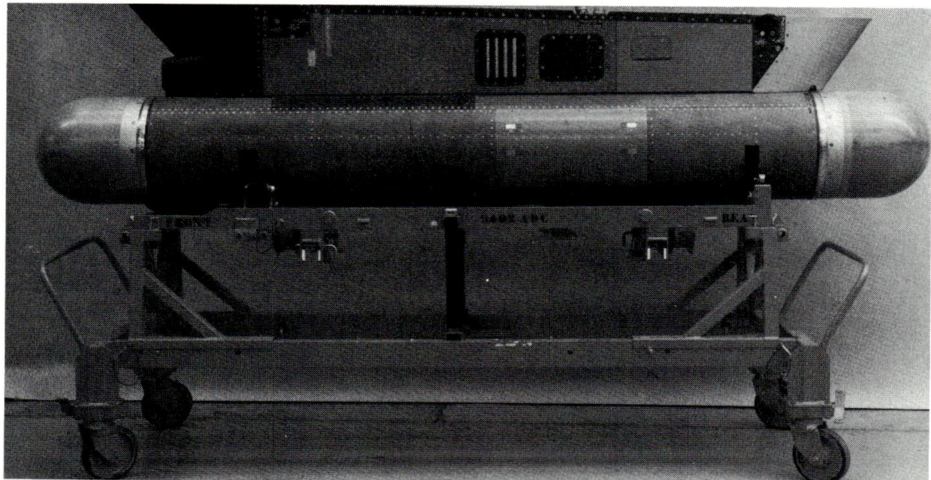

The Thomson-CSF Radars Contre-Mesures Syrel ELINT pod

receiver units, an amplifier and recorders in its centre section. The pylon houses a cooling system which has a ram-air intake in the pylon leading edge. High-speed operation is assisted by thick-film and microwave circuit assemblies on ceramic substrates. Thomson-CSF also produces first and second line maintenance equipment for use with the pod.

Syrel is combat proven.

Specifications

Dimensions: 3,570 length × 420 mm diameter
Weight: 265 kg

Operational status

No longer in production; believed to be in service on aircraft of the Spanish and other air forces, including Mirage III, Mirage F1 and Mirage 2000.

Contractor

Thomson-CSF Radars Contre-Mesures.

UPDATED

GERMANY

Helicopter Laser Warning Equipment (HLWE)

SEL has developed several types of laser warning sensors, one of which is the Helicopter Laser Warning Equipment (HLWE). The HLWE detects, identifies and locates the laser radiation source. Together with a threat warning system processor it can be used as a sensor for manual or automatic countermeasures.

The HLWE is designed to operate in helicopters such as the PAH-2 anti-tank helicopter but may also be modified for installation in fixed-wing aircraft such as EF 2000 or retrofit programmes.

Operational status

The HLWE has been successfully demonstrated in field trials.

Contractor

Alcatel SEL AG.

VERIFIED

Helicopter self-defence system

The Buck helicopter self-defence system is a modular, low-weight dispenser armed with chaff and flares that are tuned to the signature of the helicopter. It consists of a transport and firing unit installed on each side of the helicopter, with an operating and indication unit in the cockpit. A combination of firing sequences can be used, such as a single round of chaff or flares, a single round of chaff and flares, a multiple sequence of flares or a multiple sequence of chaff and flares.

The device is allied with, and activated by, a radar warning receiver or missile approach warner. Activation can be completely automatic, semi-automatic or manual.

Specifications

Weight: 25 kg including cartridges
Dispenser capacity: 8 rounds each
Ammunition type: 55 × 200 mm calibre chaff and flare cartridges

Operational status

Fully developed

Contractor

Buck System GmbH & Co.

VERIFIED

AN/ALQ-119GY/ALR-68

The German Air Force has decided to update its AN/ALQ-119 ECM pods to meet the modern airborne and groundbased threat environment. Daimler-Benz Aerospace AG has developed a modification kit consisting of RF modules, logic modules, wiring and software to fulfil the requirement. The main added features are: continuous frequency coverage; improved repeater techniques; improved receiver capability; new jamming techniques; introduction of an intelligent interface to the radar warning receiver; improved maintenance. The updated configuration is designated AN/ALQ 119GY and AN/ALQ 119GY/ALR-68, when it is integrated with the Litton Applied Technology AN/ALR-68 ARWS. The AN/ALR-68 ARWS Advanced Radar Warning System is described in the Litton Applied Technology section.

Operational status

The upgrade is designed for the German Air Force F-4F aircraft; the enhancement has been proven during formal testing, including tests against real threat systems.

Contractor

Daimler-Benz Aerospace AG, Defense and Civil Systems, Airborne Systems.

NEW ENTRY

AN/ALQ 119GY on German Air Force F-4F ***1998***/0018296

AN/ALQ 119GY ECM pod ***1998***/0018297

LWR Laser Warning Receiver

The Daimler-Benz Aerospace LWR is designed to be integrated with the Threat Warning Equipment (TWE) (see entry in International Section) to provide multiple threat warning capability for tactical helicopter systems.

The Daimler-Benz Aerospace LWR utilises high sensitivity near infrared detectors to provide precise measurement of the angle of arrival of laser pulses in azimuth, with single pulse detection close to 100 per cent, and high immunity against sunlight and explosion flashes.

The LWR provides detection, analysis, identification, bearing, measurement and transmission for visual display and audio alarm of threat laser illumination from laser range finders, illuminators and beam riders. Options include: an extension module for long wavelengths (far infrared); a stand alone version; Mil Bus interface; elevation angle measurement; tank and ship versions.

Specifications

Sensor coverage: 360° in azimuth, adjustable to customer requirements in elevation (up to 90°)
Azimuth accuracy: 10° rms
Reaction time: audio alarm within fractions of a second
False alarm rate (after processing): <1.5 per hour
Power consumption: 51 W (2 × 25.5 W)
Dimensions:
(height) 110 mm
(width) 129 mm
(length) 240 mm
Weight: 3.6 kg (2 × 1.8 kg)

Daimler-Benz Aerospace Laser Warning Receiver ***1998***/0018292

Operational status

Designed to be part of the TWE Threat Warning Equipment, selected for the NH90 TTH (Tactical Transport Helicopter).

Contractor

Daimler-Benz Aerospace AG, Defense and Civil Systems, Airborne Systems.

NEW ENTRY

Tornado Self-Protection Jammer (TSPJ)

The TSPJ is a third generation self-protection jamming pod, designed by Daimler-Benz Aerospace AG for German Air Force and German Navy Tornado aircraft.

Main design features are that it: is a generic, modular, jamming system with no threat-specific hardware; has an integrated receiver/processor; has a large repertoire of jamming techniques; all functions are user-programmable via Mil-Bus; there are autonomous and radar warning controlled modes; the modular design and intelligent interface permits adaptation to other platforms.

The system includes a Digital Radio Frequency Memory (DRFM) for the production of coherent jamming signals to defeat coherent threat radars; the DRFM developed by Daimler-Benz not only contains the RF and memory section to digitise and store signals precisely, but also all techniques generators needed for generation of the required jamming signals.

The TSPJ self-protection jammer is carried on the Tornado outer wing station

Operationally, the TSPJ provides a fully software programmable ECM system, capable of operation against both coherent and non-coherent threat systems. The pod can also be used for high resolution data collection, and as the basic building block of radar jammer simulator sytems.

Operational status

The first TSPJ systems have been delivered to the German Air Force. In production.

Contractor

Daimler-Benz Aerospace AG, Defense and Civil Systems, Airborne Systems.

UPDATED

Towed decoy

The Daimler-Benz Aerospace towed decoy is a high power (100 W) radar jamming and decoying system optimised for the protection of both fighter and transport/maritime patrol aircraft. It is effective against a wide range of aircraft including monopulse types.

It receives the emissions of radar threats and generates and transmits jamming signals to introduce angular errors in the tracking loops of airborne and ground-based target tracking radars and missile seeker heads.

Two versions of the towed decoy are available: towed decoy system with deployment system, allowing recovery of the towed body by parachute; towed decoy system with active winch, allowing retrieval of the towed body during flight.

Tornado pod installation of the towed decoy system **1998**/0018294

Daimler-Benz Aerospace towed decoy **1998**/0018295

The towed decoy has the following capabilities: high output power (100 W); frequency coverage for all relevant threats; ECM techniques against CW, pulse, and pulse Doppler radars; special techniques to confuse radar operators; techniques to protect the decoy body; redundant decoy resources; fast and reliable deployment and stable flight behaviour.

Operational status

The towed decoy system has been successfully flight tested on different aircraft, including Tornado and F-4; it will be commercially available in 1998.

The towed decoy has been evaluated on Tornado aircraft in single and multiship formations spanning most of the operational flight envelope of the aircraft.

Contractor

Daimler-Benz Aerospace AG, Defense and Civil Systems, Airborne Systems.

NEW ENTRY

Tornado ECR system

The Tornado ECR is based on the sixth Tornado production batch build standard and incorporates a MIL-STD-1553B databus, upgraded radar warning and active electronic countermeasures equipment and an improved missile control unit.

The Tornado ECR features: an Emitter Location System (ELS) to pinpoint, identify and display hostile radar emitters; an imaging infrared system for all-weather day and night reconnaissance; the Operational Data INterface (ODIN) for transmission of near real-time reconnaissance data to following aircraft and ground centres; FLIR to enable covert low-level flight in adverse weather conditions and at night; advanced displays and powerful computers to give the crew more time for tactical decision making; the HARM anti-radiation missile; an advanced interface concept to employ existing and future smart weapons and jammers and advanced avionic architecture that can grow and adapt to the demands of the threat in the 1990s and beyond.

The Raytheon Systems Company ELS allows the passive autonomous acquisition, identification and precise location in range and angle of radiating threats. The information is displayed to both members of the crew for target selection. After selection the co-ordinates are used for cueing the HARM missiles as well as for handover to follow-on forces or for onboard storage.

The ELS detects, identifies and locates radar emissions through the use of a high probability of intercept receiver system. It features multi-octave RF coverage, phase interferometric antenna arrays for precision direction-finding, passive ranging channelised receivers and a multiple MIL-STD-1750A digital processor. The system operates across the RF spectrum for all primary surface-to-air and airborne threats. Data acquired by the system is transferred to the tactical displays of both crew members for threat assessment. The ELS is interconnected with the aircraft avionic and defensive aids databusses, and the emitter library is loaded from the mission data transfer system.

The Tornado ECR is in service with the German and Italian air forces

The ELS can contribute to the ECR mission in other ways by assisting the crew during reconnaissance missions. It can be used to identify and locate mobile targets or targets of opportunity and also provide steering information for optimal sensor operation.

The fundamental Tornado ECR tasks are recce-attack or Pathfinder operations. For this the aircraft is equipped with the LITEF ODIN, a digital datalink. ODIN uses the UHF/VHF and HF frequency bands for transmitting near real-time reconnaissance information to following aircraft and to ground command posts. As a data interface ODIN converts signals from analogue to digital and vice versa. ODIN messages received by the communications system are automatically transferred for display to the crew via the avionic databus. Message formats on the weapon system operator's screens are used for preparing transmissions. Received and transmitted data can be recorded in the mission data transfer system. Voice communication is not affected by operating the datalink facility.

The internally mounted FLIR, developed by Zeiss, provides both aircrew with navigation, reconnaissance and attack information in adverse weather and in night conditions. FLIR enhances the covert penetration and attack capabilities of the Tornado ECR.

The ECR Infrared Imaging System (IIS) has a horizon-to-horizon capability which allows area and point reconnaissance. It provides a near real-time onboard display of the recorded image on the weapon system operator's screens. Evaluation of the displayed image by the crew results in a reconnaissance in-flight report which can be transmitted to ground stations and follow-on forces by the ODIN datalink. The recording medium is a high-resolution dry silver film which is fully developed seconds after the target images are recorded.

The IIS consists of the IR linescanner, electronic components for power distribution, formatting, processing and amplification, electromechanical components for film recording and developing and the control panel. It provides video images to the displays via the computer symbol generator. The IIS is also used to record ELS electronic intelligence data.

Operational status
In service with the German and Italian air forces.

Contractor
Daimler-Benz Aerospace AG Military Aircraft.

UPDATED

INTERNATIONAL

AN/AAQ-24(V) DIRCM Nemesis Directional InfraRed CounterMeasures suite

Nemesis is a Directional Infrared CounterMeasures (DIRCM) system that detects, acquires and tracks new IR threat missiles, and then defeats them by accurately focusing an intense beam of modulated IR energy onto the attacking missile. The system is intended to be updated later by the addition of a laser jammer.

Full fitting compatibility is designed into the system for a wide range of platforms from large fixed-wing transport aircraft to small helicopters.

Northrop Grumman is the team leader and overall system integrator; Northrop Grumman is also providing the passive warning element, based on its AN/AAR-54 system, and the jamming lamp and techniques generator; Rockwell provides the Fine Track Sensor (FTS) for the transmitter azimuth axis; British Aerospace (Systems and Equipment) Ltd is providing operator controls and threat assessment; GEC-Marconi is providing the transmitter unit.

The Northrop Grumman advanced, compact, high-performance, lightweight Missile Warning System utilised in the AAQ-24 DIRCM suite is based upon AN/AAR-54, also known as PMAWS (Passive Missile Approach Warning System). This passively detects missile plume energy, tracks multiple energy sources and classifies each source as a lethal missile, non-lethal missile (not intercepting the aircraft) or clutter. Its very fine Angle-Of-Arrival (AOA) capability delivers rapid and accurate hand-off to the IRCM pointing/tracking subsystem. Fine AOA processing provides detection ranges nearly double that of existing fielded passive systems and greatly reduces false alarm rates. This provides all-weather, all-altitude operation while protecting against multiple simultaneous engagements in dense clutter environments.

The system utilises a wide field of view sensor and compact processor. From one to six sensors can be employed, providing up to full spherical coverage.

The British Aerospace contribution to the AAQ-24 DIRCM program will include work in the development and assessment of system operating parameters for specific threats, the development, manufacture and qualification of prime system AC power source and the design and manufacture of operator/system interface control panels. BASE will manufacture various circuit card assemblies, provide operational system support and participate in the design verification and testing of AAQ-24.

GEC-Marconi Electro-Optics Ltd, Navigation and Electro-Optic Systems, Edinburgh, has developed the high-performance transmitter unit for both the fixed- and rotary-wing AAQ-24 DIRCM suites. The transmitter unit consists of a pointing turret with a payload of a Fine Track Sensor (FTS) and an IR jammer. The transmitter unit, when cued by the Missile Warning System, acquires the incoming missile, tracks it and projects a high-intensity IR beam at the target.

The agile pointing system ensures that the AAQ-24 DIRCM transmitter unit can rapidly acquire the approaching missile and the four-axis tracking system ensures that the maximum jamming energy is maintained on the target throughout the engagement. Each AAQ-24 DIRCM transmitter is equipped with a laser path to enable the system to be upgraded at a future date.

Rockwell's Fine Track Sensor is the eye of the high-performance transmitter, located on the azimuth axis. During a threat situation, the image of the incoming missile is electronically processed by the FTS. The electronic image is used by the AAQ-24 DIRCM system to 'close the track loop' by locking the transmitter onto and maintaining an IR beam on the incoming missile until it is defeated. The capability of the FTS to carry out the system requirements is based on proven Mercury Cadmium Telluride technology.

AAQ-24 takes advantage of the high-resolution, high-sensitivity, large area focal plane array detector based imaging system. The FTS features fast cool-down time and high sensitivity, which affords post-burnout tracking over an extended temperature range. The FTS operates in the mid-wave sector of the IR spectrum to take advantage of the higher contrast and increased discrimination capabilities within that region.

Specifications — fixed-wing
Total weight:
(without laser) 219.3 kg
(with laser) 228.8 kg
Power (simmer):
(without laser) 3.2 kVA, 0.6 kW
(with laser) 3.2 kVA, 0.75 kW
Power (transmit):
(without laser) 19.7 kVA, 0.6 kW
(with laser) 19.7 kVA, 0.96 kW

Specifications — rotary-wing
Total weight:
(without laser) 55.7 kg
(with laser) 60.5 kg
Power (simmer):
(without laser) 0.98 kW
(with laser) 0.98 kW
Power (transmit):
(without laser) 2.85 kW
(with laser) 3.00 kW

Operational status
The Preliminary Design Review (PDR) and Critical Design Review (CDR) have been completed. Effectiveness testing and aircraft integration are under way for a range of selected fixed- and rotary-wing platforms. The system is ready for integration of the laser countermeasures element.

Initial flight trials began aboard a Sea King helicopter in October 1997 at GKN Westland Helicopters facility in Yeovil, UK. Formal UK MoD trials are expected to begin in 1998, together with flight tests in the USA. Service introduction in the UK is expected in 1998/9.

Present plans call for the UK to instal Nemesis in 186 aircraft, including airlifter, tanker and VIP aircraft and six types of helicopter. The US Air Force SOCOM plan is for fitment to 60 C-130 AC/MC aircraft.

Contractors
Northrop Electronics Systems International Inc.
British Aerospace Systems and Equipment Ltd.
GEC-Marconi Electro-Optics Ltd, Navigation and Electro-Optic Systems.
Rockwell International.

UPDATED

The Nemesis fixed-wing directional IRCM suite **1996**

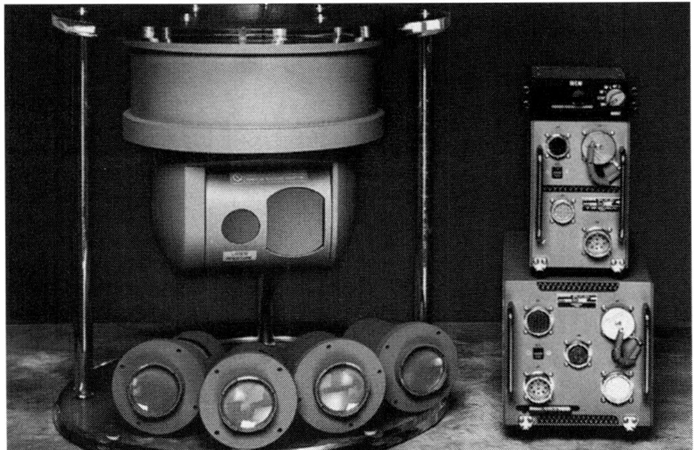

The Nemesis rotary-wing directional IRCM suite **1996**

Advanced Self-Protection Integrated Suite (ASPIS)

Litton Applied Technology, Raytheon Systems Company and Tracor Aerospace have teamed to produce the family of Advanced Self-Protection Integrated Suites (ASPIS) which automatically detect and counter hostile threats.

ASPIS uses interchangeable subsystems, so the configuration is flexible. The suite can grow by upgrading existing subsystems, by adding subsystems or by reprogramming to counter the changing threat environment. This flexibility also facilitates installation in a wide variety of aircraft.

ASPIS consists of Litton threat warning systems, Raytheon active jammers and Tracor countermeasures dispensers.

The threat warning system is the heart of the ASPIS suite. The AN/ALR-93(V)I threat warning system is incorporated in ASPIS.

Interfaced with the above threat warning system, the Raytheon AN/ALQ-187H is a fully integrated, power-managed electronic countermeasures system. It can counter multiple pulse, pulse Doppler and CW threats, including SAM, AAA and air-to-air weapons. The ALQ-187 features user-programmable threat data and ECM techniques files for automatic detection of single or multiple threat radars.

The Tracor countermeasures dispensers contribution to ASPIS is the AN/ALE-47 TACDS (Threat Adaptive Countermeasures Dispenser System). The AN/ALE-47 is a threat adaptive dispensing system that can prioritise and automatically dispense the correct type and amount of chaff and/or flares to counter single or multiple threats. Also available as part of the ASPIS system is the Daimler-Benz Aerospace AG/Litton ATD MILDS III Missile Launch Detection System, which utilises UV technology detection to minimise false alarms and provide accurate bearing data on approaching missile threats, and the Grinaker Avitronics LWS-250 Laser Warning System which can detect and identify laser rangefinders, target designators and beamriders operating in Ider bands I, II and III.

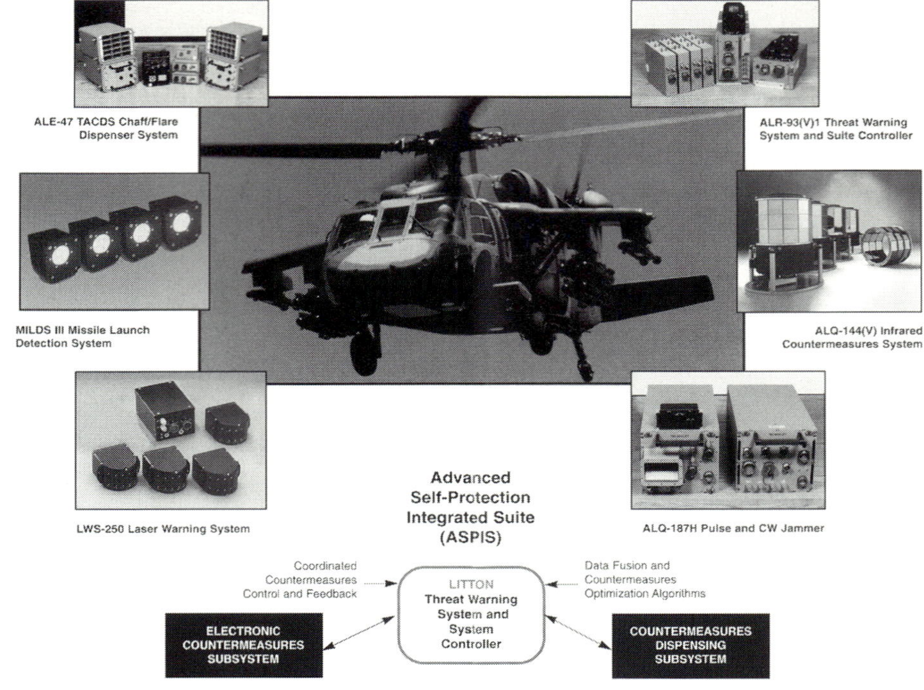

Advanced Self-Protection Integrated Suite (ASPIS) ***1998***/0018268

Operational status

ASPIS suites are available for a wide range of attack aircraft and helicopters.

In February 1993, it was announced that the ASPIS team had been selected by Greece for the upgrade of the F-16 defensive aids system. Deliveries of the ASPIS system to Greece began in mid-1994.

Contractors

Litton Applied Technology.
Raytheon Systems Company.
Tracor Aerospace, Inc.
Daimler-Benz Aerospace AG, Defense and Civil Systems.
Grinaker Avitronics.

UPDATED

EF 2000 Defensive Aids SubSystem (DASS)

The EF 2000 Defensive Aids SubSystem (DASS) is an integrated electronic warfare suite comprising an ESM system, ECM, towed offboard decoy, computer, chaff/flare dispenser, laser warner and missile approach warner. The development of this system by a consortium, consisting of GEC-Marconi Defence Systems in the UK and Elettronica in Italy, is expected to take about five years. Although Germany, one of the four nations in the EF 2000 project, opted out of DASS, Spain has opted into the project, and Germany is now reconsidering purchase of DASS for its EF 2000 aircraft and workshares are being re-negotiated.

Operational status

The Eurodass consortium was awarded a £200 million contract in March 1992 for the development of DASS. The work-share is split approximately 60 per cent to the UK and 40 per cent to Italy. In March 1993, a contract was awarded to GEC-Marconi, Electro-Optics Ltd, Navigation and Electro-Optics Systems, Edinburgh for the laser warning system for the EF 2000.

Contractors

GEC-Marconi Avionics Ltd, Defence Systems Division, Stanmore.
Elettronica SpA.

VERIFIED

EF2000 Eurofighter carries DASS ***1998***/0018290

Erijammers A100 (ALQ-503)/A110

The Erijammer A100 jammer system is a manually or automatically computer-controlled jammer pod for tactical use and ECCM training of air defence fighters and AAA operators. The system provides Responsive Electronic Warfare Training and Support (REWTS) by giving the operator situational awareness with built-in RWR, look-through capability and a set on receiver. Over 50 smart noise, advanced range, velocity and angle deception modes and combinations of these modes are available. The pod is also capable of providing simulation of missile seeker radars. Single or multi-threat capability is provided by an advanced frequency memory loop, a set on receiver and selectable bandwidths.

The pod is entirely self-contained and requires only power from the carrier aircraft. The system is controlled by an ECM operator, through a cockpit control box, and

The Erijammer A100 ECM training pod carried under a Pilatus PC-9

The Ericsson Saab/Rodale Erijammer A100/AN/ALQ-503 pod

programmable EEPROM; it is 100 per cent reprogrammable in the air. The analysis and subsequent jamming of incoming signals over 360° with coverage for the selectable high- and low-gain antennas give the system high flexibility in tactical flying and training.

The Erijammer A110 represents a new generation of training jammers with combined tactical capability.

Specifications

Dimensions: 3,235 (length) × 426 mm (diameter)
Weight: 210 kg
Power supply: 115 V AC, 400 Hz, 3 phase, 3 kVA
28 V DC, 5 A
Frequency: H- and I/J-bands
Output power: 350 W, ERP 1-2 or 10 kW
Coverage:
(horizontal) 360°
(vertical) ±30°
Speed: M0.2 to M1 +

Operational status

In service in the Canadian, Swedish and Swiss air forces.

Contractors

Ericsson Saab Avionics AB.
Rodale Electronics Inc.

UPDATED

ERWE II Enhanced Radar Warning Equipment

Daimler-Benz Aerospace and Litton Applied Technology Division have teamed to develop and produce the Enhanced Radar Warning Equipment and its updates. The ERWE II detects and analyses hostile radars illuminating the aircraft. It provides the crew with display and audio information, and provides other elements of the EW system with data on which to effect appropriate responses.

The main features of the ERWE II are that it is: a multi-receiver and processor architecture, with wide instantaneous bandwidth, monopulse, E-/J-band and receiver, high sensitivity narrow band receiver; special C-/D-band receiver for detecting missile guidance signals. It includes two Mil Bus interfaces; discrete interfaces for look-through management; embedded maintenance facilities; provision for the user to programme threat libraries, identification and recording requirements.

The operator is provided with alphanumeric threat displays, audio and two selectable modes of operation: terminal mode: providing smart analysis algorithms and identification/classification of threats; bypass mode: providing high probability of intercept of scanning emitters and real-time indication of signals.

Contractors

Daimler-Benz Aerospace AG, Defense and Civil Systems, Airborne Systems.
Litton Applied Technology Division.

NEW ENTRY

Enhanced Radar Warning Equipment ERWE II units
1998/0018293

Long Star jamming system

Long Star is a major SEAD (Suppression of Enemy Air Defences) sytem intended to meet US Army needs for an organic Army Support Jamming (ASJ) capability. It provides support jamming of enemy radars to enable close support aircraft to carry out attack missions. Long Star employs the MultiBeam Array Transmitter (MBAT) used in many Rafael systems and provides detection, identification, direction-finding and jamming of threat emitters. It can also be used for ELINT.

The system consists of the MBAT, identification unit, ESM/IFM units, ESM/IDF unit, power management system and operator console. The system's effective radiated power gives radar burn through and its modularity provides for rapid installation and removal from a variety of platforms. The modular transmitters provide frequency coverage that can be tailored for specific missions. System weight is approximately 900 kg, depending on configuration, and it consumes 35 kVA of electrical power.

Operational status

Status uncertain. It is understood that Long Star is being evaluated by the US Army and has been offered as a support jammer for the UH-60 Black Hawk helicopter.

Contractors

Northrop Grumman Corporation.
Rafael Armament Development Authority.

UPDATED

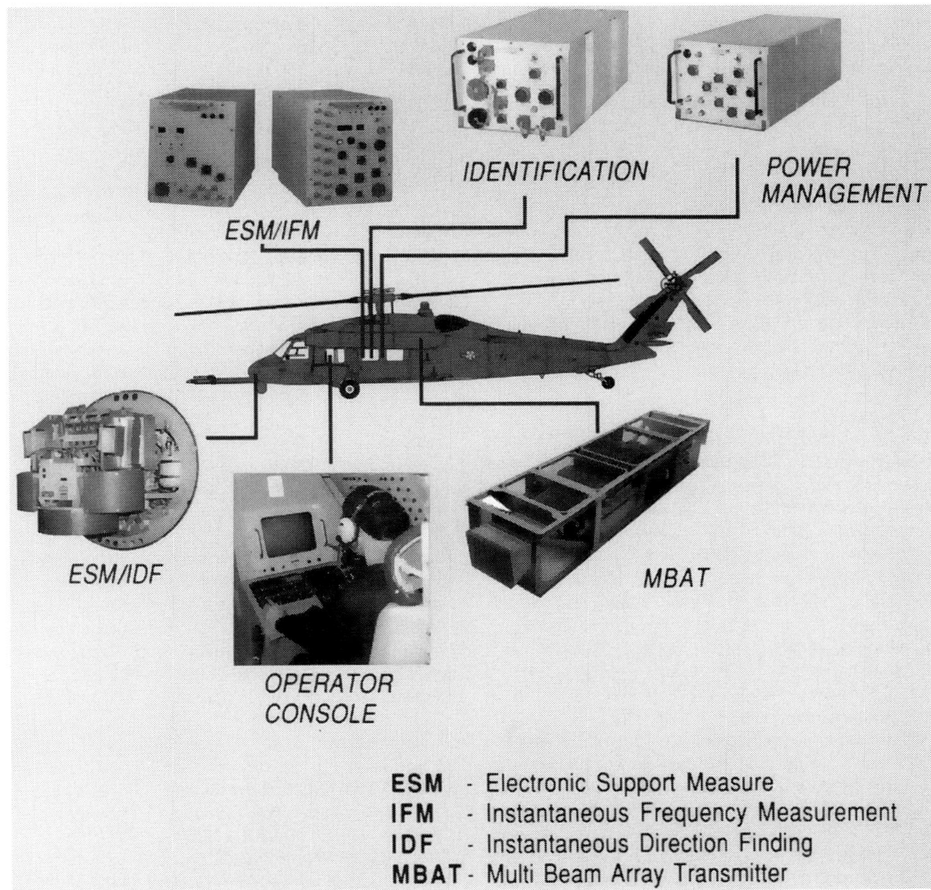

Long Star ASJ installation in UH-60A helicopter
1998/0018289

MILDS III missile launch detection system AN/AAR-60

The MILDS III missile launch detection system detects and declares potentia incoming missiles and indicates direction of arrival and time to go. MILDS III also automatically triggers self-protection actions such as release of flares, IR jamming and flight manoeuvres. It senses and images the ultraviolet emission of missiles' plumes and uses advanced software algorithms to declare and track a threat in real time.

MILDS III is designed for six sensor LRUs functioning in a master/slave configuration. Optional sensor LRUs can be supplied for greater field of view coverage. Each sensor contains preprocessing, signal processing and a communications processor. The master contains an additional software function for sensor fusion. Inertial navigation system data are used for threat declaration. Threat data are passed through standard datalinks to the EW display or EW suite.

Specifications

Dimensions: 120 × 120 × 120 mm
Weight: 12 kg total
Field of view: 360° azimuth, 95° elevation (95 × 95° for each sensor head)
Response time: <0.5 s
Resolution: <1°
Detection range: >5 km typical
Sensor: passive (imaging UV)
Interfaces: RS422 or MIL-STD-1553B

Operational status

Development complete. MILDS III has been designated AN/AAR-60. Recommended by Eurocopter for the NH 90 TTH transport helicopter

Contractors

Daimler-Benz Aerospace AG Defense and Civil Systems, Airborne Systems.
Litton Applied Technology.

UPDATED

MILDS six sensor LRUs functioning in master-slave configuration ***1998***/0018288

NRWE and German Air Force Tornado ***1998***/0018249

NRWE Tornado New Radar Warning Equipment

Daimler-Benz Aerospace and Litton Applied Technology Division (ATD) have teamed to design and develop a new generation of radar warning equipment for Tornado aircraft of the German Air Force by introducing new and proven elements into the basic RWR.

The main features of the NRWE architecture include: a highly integrated multichannel dual receiver front-end; multi-processor architecture; parallel processing; hardware pre-processing. Specific features include:

(1) dual E-/J-band receiver front end providing: a monopulse E-/J-wideband multi-channel receiver system, with full instantaneous bandwidth, and improved sensitivity throughout; a monopulse E/J band intermediate bandwidth receiver for detection of high PRF missile guidance and low power (pulse and CW) signals; detection of slow scanning threats such as modern airborne interceptors;
(2) detection and identification within the K-band frequency range;
(3) a digital receiver providing special signal analysis measurement and analysis functions;
(4) a special receiver for detecting C/D-band missile guidance signals;
(5) pulse on pulse, and pulse on CW performance provided by advanced detector log video amplifiers;
(6) hardware pre-processing and data reduction provided by spatial screens and multiple temporal filters;
(7) parallel data processing and adaptive analysis algorithms for identification and classification;
(8) processing architecture with growth to a full defensive aids sub-system capability and multifuction displays;
(9) scenario adaptive programming by the user of threat libraries, identification and recording features;
(10) Mil Bus for system integration and software loading/down-loading;
(11) discrete interfaces for look-through;
(12) embedded maintenance functions.

Operational status

Designed for German Air Force Tornado aircraft.

Contractors

Daimler-Benz Aerospace AG, Defense and Civil Systems, Airborne Systems.
Litton Applied Technology Division.

NEW ENTRY

REWTS Turbo Crow

The Turbo Crow Responsive Electronic Warfare Training System (REWTS) consists of a high-power broadband radar jammer and a high-power radar simulator combined with a sophisticated monitoring and control system. Turbo Crow is typically installed in a commercial jet (Learjet, for example) or similar military aircraft and is a dual-role system - used in both training and tactical applications.

Jamming system

The jammer provides high-power broadband jamming in user-selected bands. Jamming techniques include noise, deception, cover pulses, false targets, Doppler and combinations thereof. The system is capable of simultaneous operation in multiple bands. The jammer is modular and can be configured to suit a particular aircraft's physical characteristics. Modularity also results in significant maintenance and repair advantages. The system's operating architecture and mechanical design readily accept changes and enhancements.

Radar simulator

The Radar Simulator provides the Turbo Crow with the capabilities of simulating a wide variety of airborne and ground-based radars including search and tracking, missile guidance and fire control. The installation provides for simultaneous operation of jammers and radar simulator. The programmable aspects of the Digital Airborne Radar Threat Simulator (DARTS) brings a new level of sophistication to EW training and test of shipboard/ground-based ESM systems.

Monitoring and control system

The Monitoring and Control System is the heart of the REWTS system providing the operator with full control over the engagement scenario - whether it be training or tactical. He or she knows where the target is; can monitor the target's response to the jamming - both ECCM and manoeuvres; is able to adjust the jamming to suit the situation; and can also record any aspect of the engagement.

Optionally the Turbo Crow can be enhanced for ELINT/ESM missions with capabilities to detect, analyse and classify radar data as well as passive ranging/positioning of radars.

The system can either be carried on board or in wing-mounted pods using a hybrid configuration with the high-power amplifiers and antennas housed in pods carried on wing hard points. This configuration results in significant savings in installation costs, better aerodynamics, less on board high-power equipment to cool, and more flexibility in aircraft utilisation.

Specifications

RF range:
(jammer) 0.85-18 GHz (3 bands)
or 0.2-18 GHz (4 bands)
(DARTS) 7.8-8.5, 8.6-9.6, 12-13.2, 14-15.2 GHz
(ELINT) 0.4-18 GHz (optional 40 GHz)
Jammer output power: 200 or 400 W typical
DARTS output power: 100 KW (min)
ELINT sensitivity: −70 dBm
DF accuracy: 3-5°
Jamming modes: noise, smart noise, deception, cover pulses, multiple false targets, Doppler, co-ordinated and combined modes
DARTS modes:
(scan) circular, steady, sector centered, sector off centre
(PRF) stable, jitter, stagger
(adjustable) pulsewidth, PRF, delay, frequency
Monitoring and control: situational awareness, measurement, control, recording and databank system
Pods: ALQ-167 or ALQ-503

Contractors

Ericsson Saab Avionics AB.
Rodale Electronics Inc.

UPDATED

S200 self-defense jammer

S200 is a small cost-effective modular pod-mounted countermeasure system capable of generating ECM techniques with radiated power of 1 KW ERP typical. The system uses flight proven hardware from both the ERIJAMMER A100 and the Turbo Crow system.

The S200 is housed in a standard ALQ-167 pod shell and uses a standard ALQ-167 tray and proven radomes and antennas.

Specifications

Frequency range: 7.5-18 GHz
(optional) 5.3-10.5 GHz; 2.5-6 GHz; 0.85-1.4 GHz
Sensitivity:
(low duty signals) −42 dBm
(high duty signals) −42 dBm
(ERP) 1 KW typical
(antenna coverage) 2 × 110° azimuth, 40° elevation
Range deception:
(incoming pulse length) (min) 150 ns
(min delay) (max) 150 ns
(memory time) up to 10μs or more (automatic longer memory time for pulse compression radars)
Velocity deception:
(frequency translation) up to ±127.5 KHz
(gain (including antennas)) up to 70 dB (installation dependent)
Amplitude modulation:
(swept square wave frequency) 0.01-1,000 Hz
(duty cycle) 5-95%
(sweep time) 0, 5-10 s
(modulation depth) (min) 50 dB
(input power) 200/115 V, 400 Hz, 3 phase, 28 V DC, 10 A, <3 KW
Physical characteristics:
(pod length) (min) 3,200 mm
(pod diameter) 254 mm
(pod weight) (max) 180 kg

Contractors

Rodale Electronics Inc.
Ericsson Saab Avionics AB.

VERIFIED

Silent Sentinel

Silent Sentinel is a passive payload aimed at detecting, identifying and locating ground-based radars.

Designed to be integrated on platforms such as UAV, helicopters or light aircraft, Silent Sentinel either delivers radar tracks for real-time display and analysis or uses its high-capacity recording feature for post-flight analysis.

The battlefield tactical situation can be displayed either on a compact laptop computer or on a SPARC-based ESM/ELINT workstation.

Silent Sentinel uses interferometry to give accurate direction-finding. A compressive receiver offers long-range detection, superior parameters measurement and very short acquisition times.

Specifications

Frequency range: 2-20 GHz (0.5-40 GHz optional)
Field of view: 180°
Measurement accuracy:
(frequency) 2 MHz class
(bearing) 1° class
Payload physical characteristics (typical):
(weight) 20 kg
(diameter) 200 mm
(length) 900 mm
(power consumption) <300 W

Contractors

Electronics & Space Corp.
Dassault Electronique.

VERIFIED

TWE Threat Warning Equipment

The TWE includes a laser warner, made by Daimler-Benz Aerospace, and an Instantaneous Frequency Measurement (IFM) radar warner made by Thomson-CSF Radars Contre-Mesures.

The TWE was designed initially for the Eurocopter Tiger helicopter. It is intended to provide warning of hostile surveillance and fire-control radars, tracking radars, CW illuminators, and missile seekers, as well as laser rangefinders, laser illuminators and laser missile guidance systems.

The TWE is designed to be the core of a defensive aids suite, being linked to a chaff and flare dispenser, missile launch detector or missile approach warner, with growth to include ECM capability via MIL-STD-1553B interface.

The TWE system can be easily programmed to meet differing system and scenario requirements.

Specifications

Weight: <12 kg
Frequency:
(radar) E- to K-bands
(laser) 0.4-1.1 μm (option: 0.4-1.7 μm)
Coverage: 360° azimuth, ±45° elevation
Accuracy:
(angular) better than 10° rms
(frequency) less than 20 MHz
Resolution:
(laser) 4°
Library size: >1,000 threats
Dimensions:
(CPU) 158 × 194 × 321 mm
(antenna) ∅100 × 88 mm
Power: <170 W

Operational status

TWE has been developed for installation aboard the Tigre HAC (Hélicoptère Anti-Char)/ HAP (Hélicoptère d'Appui et de Protection) and UHU (Unterstützungs Hubschrauber) battlefield escort / anti-tank helicopters which are on order for the French and German armies respectively.

TWE threat warning equipment **1997**/0001259

TWE also forms the basis of a French – German helicopter applicable Electronic Warning System (EWS) which has been selected for the NH90 TTH (Tactical Transport Helicopter) version. In this application the TWE radar and laser warning system is teamed with a MILDS passive missile approach warner.

In the NH90 NFH (NATO Frigate Helicopter) version, TWE is replaced by the Elettronica DETE 90 ESM, which interfaces with the Saphir-M system.

Contractors

Thomson-CSF Radar Contre-Mesures.
Daimler Benz Aerospace AG, Defense and Civil Systems, Airborne Systems.

UPDATED

ISRAEL

AES-210/E ESM/ELINT RWR system

The AES-210/E is a sophisticated system which performs ESM, ELINT and radar warning system tasks in very dense RF environments. It can be installed in small- or medium-size aircraft or ships for applications such as maritime and overland surveillance, ELINT information gathering and platform self-protection. As a programmable system designed with a modular and flexible architecture for future growth, the AES-210/E is optimised to receive all known and foreseeable radar signals, identifying highly complex signals and threats and employing an advanced combination of receiving techniques.

The AES-210/E offers very high intercept probability, sensitivity and reliability, providing very accurate direction-finding. A number of different modern DF techniques can be provided to meet operational requirements, including advanced interferometric methods to achieve the most accurate data. The system has five fast CPUs and can be operated automatically or by a single operator.

Specifications

Weight: 45 kg typical
Frequency: 0.5-18 GHz
Coverage: 360° azimuth

Accuracy: 7° RMS coarse, 3° RMS fine
Library: >1,000 emitters
Interface: RS-232, RS-422, MIL-STD-1553B, ARINC
Environmental: MIL-STD-5400T

Contractor

Elisra Electronic Systems Ltd.

UPDATED

Units of the Elisra AES-210/E ESM/ELINT/RWR system **1995**

LWS-20 laser warning system

Four laser sensors receive and detect laser pulsed signals and send the detected pulses to the Laser Warning Analyser (LWA). The LWA characterises each laser pulse for its angle of arrival, relative time of arrival and amplitude. It then processes the single pulses or pulse trains and identifies the threatening emitter, based on preprogrammed library information. Identified threats are displayed on the screen of the self-protection system as alphanumeric symbols. These symbols enable the crew to establish the type, azimuth and lethality of the threats. In addition, audio warnings are sent to the crew through the aircraft intercom system.

The LWA incorporates a powerful military 16-bit microprocessor with EEPROM for combat software, programmable tables and in-flight recording. The LWS-20 may include a MIL-STD-1553B RWR interface or other ECM interfaces.

Specifications

Dimensions:
(analyser) 146 × 106 × 78.7 mm
(control unit) 54 × 146 × 19 mm
(display unit) 172 × 82.5 × 82.5 mm
(sensors) 150 × 640 × 860 mm
Weight:
(analyser) 1.5 kg
(control unit) 0.2 kg
(display unit) 1.2 kg
(sensors) 0.6 kg each
Power supply: 28 V DC, 50 W

Operational status

In production and in service.

Contractor

Elisra Electronic Systems Ltd.

VERIFIED

LWS-20 laser warning receiver **1997**/0002225

PAWS Passive Airborne Warning System

The Passive Airborne Warning System (PAWS) is a lightweight missile launch and approach warning system designed for fighter aircraft and attack helicopters. It consists of one or two infrared sensors, each with an associated processor based on T805 32-bit transputers. The system may operate independently or may be integrated with a radar warning receiver.

PAWS detects the missile exhaust plume radiation and tracks it even in high-clutter environments. It determines when a missile threatens the aircraft, provides an accurate readout of approach direction and an estimate of the time to intercept. It can also select the appropriate narrowbeam countermeasures and activate them automatically. It is said to have a very low false alarm rate even in violent manoeuvres and when operating in a high-clutter environment.

PAWS consists of a high computation power parallel processor and an IR sensor built by El-Op Electronic Industries. It features missile launch and approach warning, discrimination between threatening and non-threatening missiles, multithreat warning capability, accurate approach direction and time-to-go estimation and automatic activation of countermeasures.

Specifications

Power source: 28 V DC (MIL-STD-704D)
115 V, 400 Hz, 1Ø
Power consumption: 260 W
Dimensions:
(sensor) 132 × 187 × 365 mm
(processor) short ½ ATR (203 × 389 × 127 mm)
Weight:
(sensor) 6.5 kg
(processor) 9 kg
MIL specification: MIL-STD-5400, Class I B
Communication interfaces: RS 422, MIL-STD-1553B MUX BUS, transputerlink

Operational status

Fully developed.

Contractor

Elisra Electronic Systems Ltd.

VERIFIED

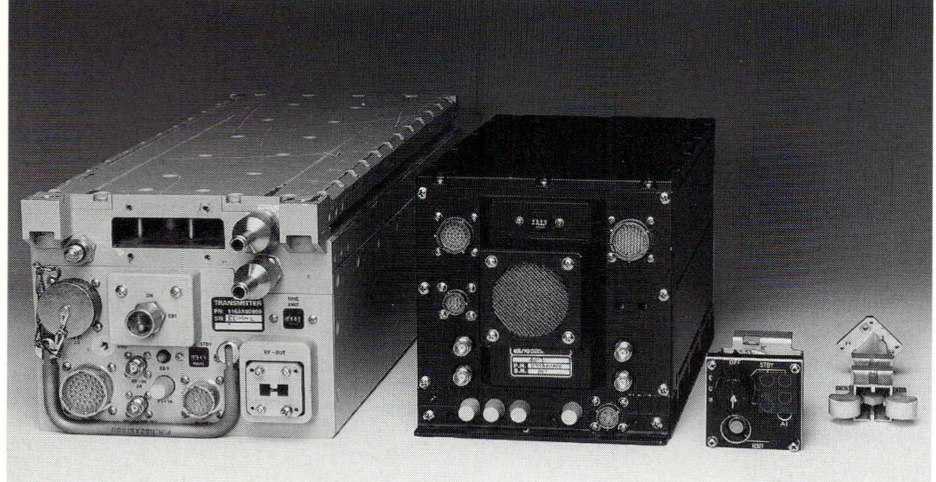

The SPJ-20 self-protection jammer **1995**

SPJ-20 self-protection jammer

The SPJ-20 self-protection jammer is a lightweight jammer. The system provides attack aircraft and helicopters with comprehensive active self-protection capability for combat sorties on the modern battlefield.

The SPJ-20 features a powerful software programmable jammer, wideband frequency range, ECM for pulse and Doppler threats and an interface to RWR systems. ECM includes a CW repeater, pulse repeater and two RF sources. ECM software programmable techniques include range gate pull-off, velocity gate pull-off, amplitude modulation, false targets and noise. It is easy to install and maintain.

Specifications
Volume: 15 litres
Weight: 18 kg
Power supply: 28 V DC, 500 W
115 V AC optional
Frequency: 6-18 GHz
Environmental: MIL-STD-5400

Contractor
Elisra Electronic Systems Ltd.

VERIFIED

SPS-20(V) airborne self-protection system

The SPS-20(V) is a low-cost, low-volume and lightweight radar warning system designed to fit existing helicopters and aircraft. It detects and displays pulsed radar threats operating within the 0.7 to 18 GHz frequency range. A 3 in (76 mm) display unit provides an alphanumeric and special graphic symbols representation of the type, course, angle of arrival, relative lethality and status of the analysed radar threats. The SPS-20(V) is equipped with a fast microprocessor which executes data processing and interfacing tasks.

The system features wideband acquisition including C/D-band, optional high-sensitivity CW detection and optional recording of flight events and emitter parameters for playback and evaluation.

The system is fully programmable and there is an optional provision for a tie-in with the flare/chaff dispensing system.

Specifications
Dimensions:
(analyser) 146 × 106 × 78.7 mm
(dual-channel receiver) 148 × 65 × 95 mm
(control unit) 54 × 146 × 19 mm
(display unit) 172 × 82.5 × 82.5 mm
Weight:
(analyser) 1.4 kg
(dual-channel receiver) 1.1 kg each (2)
(control unit) 0.2 kg
(display unit) 1.2 kg
Power supply: 28 V DC (MIL-STD-704)
Environmental: MIL-STD-5400

Operational status
In production and in service.

Contractor
Elisra Electronic Systems Ltd.

VERIFIED

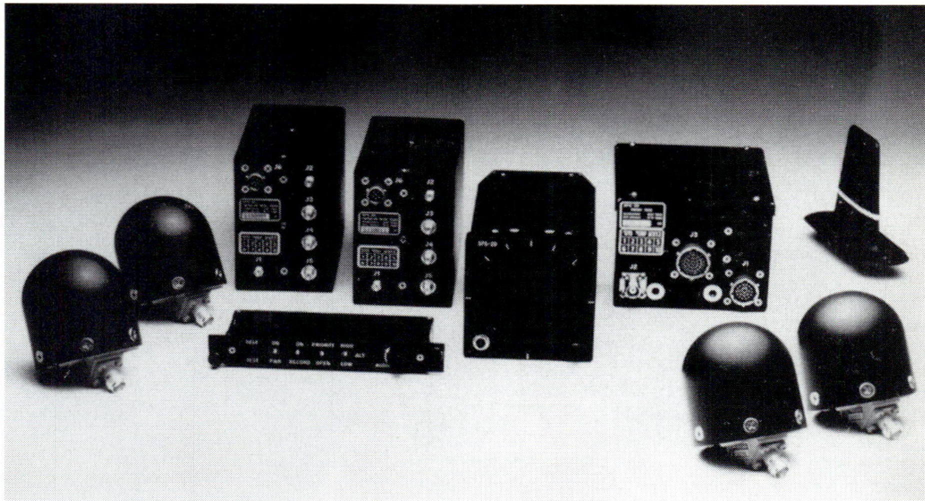
The Elisra SPS-20(V) warning system

SPS-45(V) integrated airborne Self-Protection System

The SPS-45(V) is an airborne self-protection system which integrates two subsystems: the SPS-20(V) which detects pulse radars from the low band to 18 GHz and the SRS-25 which is a superheterodyne receiver detecting CW, high PRF and low ERP radars. The LWS-20 may also be incorporated as an option to detect laser pulses. Identified threats detected by the subsystem are displayed to the pilot on a 3 in unit which provides an alphanumeric representation of type, angle of arrival, relative lethality and status.

The SPS-45(V) offers complete threat reprogrammability for new and changing environments through an extensive, easily updated emitter library. A portable memory loader/verifier enables the loading or unloading of the operational software and emitter tables in the field.

Four orthogonally dispersed cavity-backed spiral antennas receive pulsed signals in the 2-18 GHz frequency range and send them to the two dual receivers. The video pulse output from both receivers is fed to the analyser which tags every pulse with its angle of arrival, amplitude, pulsewidth and time of arrival.

The analyser CPU processes these characterised pulses, de-interleaves individual pulse trains and identifies the threatening emitter by comparing the pulse parameters with preprogrammed library information. Pulse signals in the low band are also detected and characterised to provide additional information and threat status identification.

The superheterodyne receiver is connected to an additional set of four cavity-backed spiral antennas to enable signal DF. As an option, the superheterodyne receiver can be fed via couplers from the dual receiver antennas.

The remote extension enables selection and amplification of the signals detected by the aft antennas. The input circuitry enables selection of one of the four antennas. The signal detected from the selected antenna passes a preliminary filter, mixes with the output signal of a voltage-controlled oscillator and goes to an IF section. The signal is then amplified, filtered and detected. The video signal is transferred to the video circuit for conversion to digital values. The final analysis and identification of the threat is done by an 80386 microprocessor and the results are transferred to the analyser via a serial datalink.

The analyser is responsible for integrating the data from the subsystems into one common threat file.

The analysed threats are displayed on the display unit screen. In addition, audio warnings are dispatched to the pilot through the aircraft intercom system. The threat representations on the display screen allow the pilot to establish the azimuth, lethality and emitter type for up to 16 of the most lethal threats. Up to 250 different symbols can be programmed to represent specific threat types. Additional alert capability for missile launch status is provided.

The system operating mode is controlled by front panel push-buttons on a single control unit. This unit enables the pilot to power the system, activate the automatic self-test procedure, delete preprogrammed threats from the display and define modes of operation.

Each of the two subsystems incorporates its own microprocessor with EEPROM and RAM memory. Each microprocessor works autonomously using its own emitter and mission table.

The SPS-45(V) consists of SPS-20(V) and SRS-25 subsystems

Specifications
Weight:
(SPS-20) 7.5 kg
(SRS-25) 3.8 kg
Power supply: 28 V DC (MIL-STD-704)
Frequency: low band to 18 GHz
Environmental: MIL-STD-5400, Class II

Contractor
Elisra Electronic Systems Ltd.

VERIFIED

SPS-65(V) and SPS-65V Self-Protection Systems

The SPS-65(V) is a sophisticated lightweight system with outstanding sensitivity which is capable of intercepting and analysing a wide range of known and anticipated emissions from the battlefield, including laser radiation sources. The design of the SPS-65(V) has been enhanced by technically implementing combat experience.

The SPS-65(V) comprises three subsystems. The SPS-20(V) low-volume lightweight radar warning system detects, processes and displays radar threats operating within the 0.7 to 18 GHz frequency range. The SRS-25 which is a superheterodyne receiver detecting modern CW, high PRF and low ERP radars. The LWS-20 laser warning receiver detects, identifies and locates hostile laser-guided weapons.

A 3 in display unit provides an alphanumeric representation of the type, relative bearing, relative lethality and status of up to 16 analysed threats. In addition, audio warnings are dispatched to the pilot through the intercom system. A flare/chaff dispensing system, which can be automatic or crew-activated, is available for countermeasures capabilities.

Software and emitter tables can be loaded in the field into the emitter library in a matter of seconds. This provides complete and immediate adaptability to changing threat environments. The analyser records flight events and unloads them into a tape cassette for playback after flight.

The SPS-65V is an upgraded new generation system based on the SPS 65(V). It is a lightweight, small volume, low cost self-protection system dedicated to helicopters. Already in use by major defence forces worldwide, the SPS-65V has recently been fully qualified by the German military authorities and is currently installed in a CH-53G helicopter.

Specifications

Weight:
(SPS-20(V)) 7.5 kg
(SRS-25) 3.8 kg
(LWS-20) 2.5 kg
Power supply: 28 V DC
Frequency: 0.7-18 GHz
Environmental: MIL-STD-5400 Class II

Contractor

Elisra Electronic Systems Ltd.

The Elisra SPS-65(V) self-protection system for helicopters **1997**/0002226

UPDATED

SPS-1000V-5 self-protection system

The SPS-1000V-5 is an airborne self-protection system which integrates two subsystems. The basic SPS detects radars within the low to Ku-bands, coverage of the millimetre waveband is an option. The multichannel Superheterodyne IFM Receiver Subsystem (SRS) provides detection and unambiguous identification of CW, high PRF, low ERP and pulse Doppler radars.

Threats are presented to the pilot on a 3 in display which provides a graphic representation of threat type, angle of arrival, relative lethality and status.

The SPS-1000 offers complete threat reprogrammability for new and changing environments through an extensive, easily updated emitter library. A portable field loader unit enables updated emitter tables to be loaded in the field and recorded flight data to be unloaded for pilot debriefing.

Analysed threats are displayed on the display unit screen and audio warnings are sent through the aircraft intercom system. The displayed threat representations allow the pilot to establish the azimuth and emitter type for up to 16 of the top lethal threats.

Up to 128 different symbols can be programmed to represent specific threat types. Alert capability for missile launch status is provided as well.

The operating mode is controlled by front panel switches on a single control unit. The unit enables the pilot to power the system, activate the automatic self-test procedure, delete preprogrammed threats from the display, define high- or low-altitude emitter priorities and activate recording of threats.

The SPS-1000-V-5 self-protection system **1997**/0002227

Specifications

Weight:
(central receiver processor) 14 kg
(pilot display and control) 2.4 kg
(antennas and front end receivers) 6 kg
(RF switch) 0.8 kg
Power supply: 115 V AC, 400 Hz
28 V DC (MIL-STD-704)
Environmental: MIL-STD-5400, Class II

Contractor

Elisra Electronic Systems Ltd.

UPDATED

SRS-25 airborne receiver

The SRS-25 detects and analyses CW, high-PRF and medium-ERP radar signals and determines their direction of arrival. It forms part of the SPS-45(V) and SPS-65(V) integrated helicopter self-protection systems. Continuous wave or high-PRF signals in the 6.5 to 18 GHz frequency range are analysed and signals can be recorded for later playback, analysis or training. An extensive threat library, which can be updated easily between flights, is incorporated in the unit. Four cavity-backed spiral antennas are used for accurate direction-finding.

Specifications

Dimensions: 101 × 122 × 250 mm
Weight: 4 kg
Frequency: 6.5-18 GHz
Accuracy: ±10 MHz
Resolution: 2 MHz
Coverage: 360° azimuth, ±30° elevation

Operational status

Available.

Contractor

Elisra Electronic Systems Ltd.

VERIFIED

ALR-2001 ESM

A sophisticated airborne ESM able to measure radar operating parameters with sufficient precision to 'fingerprint' individual radars from others of the same type.

ALR-2001 also has COMINT capabilities from V/UHF to microwave datalink.

It provides direction-finding capability, reportedly to better than 1°.

Operational status

Selected by the Royal Australian Air Force for its P-3C aircraft.

Contractor

Elta Electronics Industries Ltd.

VERIFIED

COMINT/ELINT system for the Arava

The COMINT/ELINT system for the Arava twin-engined light transport can be rolled on or off a multipurpose aircraft or be permanently installed. The COMINT operator uses sensors such as the Elta EL/K-1250 with omnidirectional antennas, direction-finding and signal recording. The ELINT operator has 0.5 to 18 GHz bandwidth coverage with omnidirectional antennas, signal analyser, data processor and recorder facilities. Jamming can be accomplished using antenna arrays in the aircraft tail, giving approximately 240° azimuth coverage in the lower hemisphere. Spot and band jammers of 20 to 400 W power are used.

Specifications

In production.

Contractor

Elta Electronics Industries Ltd.

VERIFIED

EL/K-1250T VHF/UHF COMINT receiver

The compact EL/K-1250T synthesised receiver operating in the 20 to 510 MHz band is used as a building block for larger COMINT or EW systems such as the Arava EW. The unit has four selectable intermediate frequency filters which demodulate AM, FM, continuous wave and single sideband signals. Intermodulation protection is claimed from RF preselection by voltage tracking filters and the fast tuning synthesiser settles within 500 ms between channel changes. Remote digital control operation is possible and the compact dimensions, low weight and power consumption are achieved by extensive use of advanced microcircuit technology.

Specifications

Dimensions: 114 × 193 × 356 mm
Weight: 7 kg
Frequency: 20-510 MHz
Frequency accuracy/stability: ±1 ppm
Synthesiser settling time: 500 μs
IF bandwidth: select 4 of 10, 20, 50, 100, 300, 600, 1,000 kHz
Noise figure:
(20-180 MHz) 12 dB
(180-510 MHz) 11 dB

Operational status

In production. The unit is in service with Israeli and other armed forces.

Contractor

Elta Electronics Industries Ltd.

VERIFIED

EL/K-7010 tactical communications jammer system

The EL/K-7010 is a modular family of communication jamming systems designed for stand-alone operation and providing for signal search and acquisition, preset channel monitoring and automatic jamming modes. It can also be integrated within a larger electronic warfare system. Computer control of power management, fast reaction jamming and multikilowatt effective radiated power provide simultaneous multiple target capability. New interception tasks are accomplished rapidly by a fast scanning receiver and automatic signal sorting in the system computer. In addition to airborne installations, ground-based systems suitable for armoured personnel carriers and air conditioned shelters are available.

Operational status

In production.

Contractor

Elta Electronics Industries Ltd.

VERIFIED

EL/K-7032 airborne COMINT system

The EL/K-7032 is designed for the surveillance and interception of radio signals in the 20 to 500 MHz frequency range, operation being largely computer-controlled.

A typical airborne system would include a supervisor's station having a computer controller, two VHF/UHF radios, a display and a data recorder. This station can work with up to four traffic collection stations, each having up to four radios and data recorders.

The EL/K-7032 has a frequency resolution of 1 kHz (with 10 Hz an option) and can intercept AM, FM, CW and HF-SSB transmissions as required.

Specifications

Frequency: 20-500 MHz
Resolution: ±1 kHz (10 Hz optional)
Modes: AM and FM (CW and SSB optional)

The Elta EL/L-8202 advanced self-protection jammer pod mounted under the wing of an Israeli Air Force Kfir

Operational status

In production and in service. Fitted to the Phalcon early warning aircraft.

Contractor

Elta Electronics Industries Ltd.

VERIFIED

EL/K-7035 all-platform COMINT system

The EL/K-7035 COMINT system intercepts, monitors, locates, analyses and reports on radio communications in the 20 to 500 MHz range. It is suitable for air, sea or ground applications.

The standard EL/K-7035 system includes a supervisor console, 2 to 5 operator consoles, a system controller and mass storage, RF distribution and antenna, a DF system, plotter position and datalink. The supervisor and operator consoles would each be equipped with two to four COMINT receivers, two to four controllers, two to four dual-channel tape recorders and a ruggedised computer. As an option these consoles could also have an IF panoramic display and time code generator/reader.

Contractor

Elta Electronics Industries Ltd.

VERIFIED

EL/L-8202 advanced self-protection jamming pod

Specifically designed to counter Soviet radars operated by the Syrian forces during the 1982 Lebanon war, the EL/L-8202 combined receiver/jammer system is compatible with the IAI Kfir fighter and any similar aircraft able to accommodate the pod, which resembles that of the US Air Force AN/ALQ-131. F- to J-band (3 to 20 GHz) coverage is provided with high-power broad-band jammer outputs through either fore or aft antennas. Beam-shaping features are incorporated and the threat library plus jamming techniques and stored mission data can be reprogrammed on the flight line. There is an integral ram-air/liquid cooling system. The logic unit within the pod interfaces the jammer with remote radar warning receivers and cockpit displays. Other systems in the family are the smaller pod-mounted EL-8200 system which weighs 113 kg, but offers 80 per cent of the capability of Cerberus. Two versions are available, the EL-8222 and EL-8224, as well as the EL-8223 internal fit system.

Specifications

Dimensions: 2,900 × 260 × 390 mm
Weight: 200 kg
Power: 1.7-2.3 kVA
Frequency: F- to J-bands

Operational status

In production and in service with the Israeli Air Force. A version known as Cerberus was developed with

The Elta EL/K-7032 airborne COMINT system

Daimler-Benz Aercspace for German Air Force Tornado IDS aircraft.

Contractor
Elta Electronics Industries Ltd.

VERIFIED

EL/L-8230 internal self-protection jammer

Suitable for IAI Kfir and Lockheed Martin F-16 sized aircraft, the EL/L-8230 self-protection system is designed to combat both surface and airborne threats. It operates across the G- to J-bands (4 to 20 GHz) with separate receiver and transmitter antenna groups and can generate noise or repeater jamming signals. It is designed to use standard avionics bay cooling air and, being fully contained within the aircraft, does not increase drag. The radio frequency unit, radio frequency power amplifier and logic unit are combined in one box.

Specifications
Dimensions: 250 × 240 × 600 mm
Weight: 42 kg
Power consumption: 1.8 kVA

Operational status
In production and in service with the Israeli Air Force.

Contractor
Elta Electronics Industries Ltd.

VERIFIED

EL/L-8231 internal self-protection system

The EL/L-8231 is an internally mounted ECM set, operating in the H-, I- and J-bands, designed to protect combat aircraft or helicopters from attack by missiles with CW radar guidance. Incoming signals are automatically analysed and the appropriate jamming signal transmitted. The set also interfaces with the aircraft's RWR.

Specifications
Volume: 15 litres
Weight: 18 kg
Power consumption:
(transmit) 600 VA

Operational status
In production and in service.

Contractor
Elta Electronics Industries Ltd.

VERIFIED

EL/L-8233 Integrated Self-Defence System (ISDS)

Elta's EL/L-8233 ISDS was specifically designed to meet the operational requirements of small, light, fighter aircraft. Featuring a modular architecture, advanced technology, proven software and hardware, ISDS as developed for F-5E and F-5F aircraft includes: a Radar Warning Receiver (RWR); mini CW repeater jammer; and chaff/flare dispenser, all integrated through a MIL-STD-1553B databus.

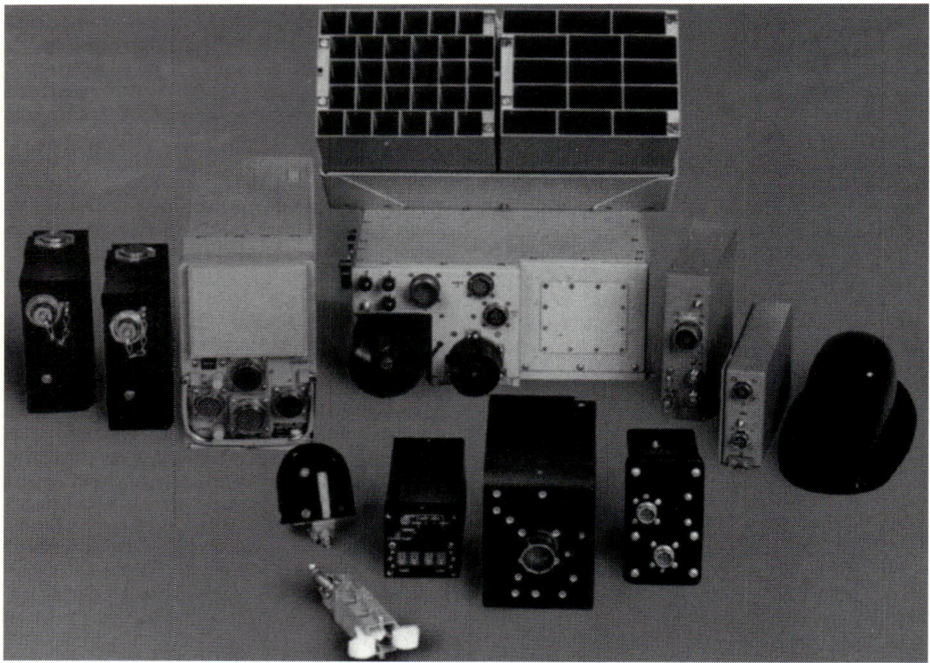

EL/L-8233 Integrated Self-Defence System (ISDS) ***1998***/0018251

The RWR provides warning of: pulse, pulse Doppler, CW and other exotic radar threats. It provides high probability of detection beyond lethal threat envelopes, in a very short cycle time, with positive identification based on threat frequency and time domain characteristics. It offers high integration with the active jamming system and automatic release of chaff and flares, together with recording of unidentified threats, and field programmability.

The CW repeater jammer provides self defence deception against enemy weapons that use pulse Doppler radar and semi-active radar missile seekers in conjunction with CW illumination. Both Velocity Gate Pull Off (VGPO), and Angle Deception (SSW) techniques are exploited in the CW repeater. The system is programmable using ELINT data and Pre-Flight Message (PFM).

The chaff and flare dispenser is a smart, threat-adaptive design, that can be tailored in capacity to meet customer requirements.

Operational status
Available, possibly ordered for Chilean Air Force F-5 aircraft.

Contractor
Elta Electronics Industries Ltd.

NEW ENTRY

EL/L-8240 self-protection system

The EL/L-8240 is an internally mounted self-protection system suitable for modern combat aircraft. It offers radar warning, jamming and deception over a large frequency range and with wide angular coverage it can be integrated with other avionics systems. A distributed processing system is used and the unit can be readily reprogrammed to meet changing threats.

Specifications
Weight: 136 kg
Power required: 7 kVA

Contractor
Elta Electronics Industries Ltd.

VERIFIED

EL/L-8300 airborne SIGINT system

In operation in large aircraft such as the Israeli Air Force's converted Boeing 707s, the EL/L-8300 is a long-range highly sophisticated SIGINT system said to be capable of detecting communications and other electronic signals at ranges up to 450 km. Received and processed data can be transmitted to the EL/L-8353 ground command and control centre for further processing, evaluation and dissemination of data.

The EL/L-8300 incorporates the EL/K-7032 COMINT system (see earlier entry), the EL/L-8312A ELINT system (see later entry) and the EL/L-8350 command and analysis station on the aircraft. Training for operators of the EL/L-8300 can be undertaken on the

The Elta EL/L-8300 SIGINT system in operation in an Israeli Air Force Boeing 707 aircraft. Also installed on Boeing 707 ECM aircraft of the Spanish Air Force

L-8351 simulator and post mission analysis is done on the EL/L-8352 system.

Specifications

Frequency:
(ELINT) 0.5-18 GHz (0.03-40 GHz optional);
(COMINT) 20-1,000 MHz (2-1,500 MHz optional)
Coverage: 360° azimuth
Accuracy: ±3°
Library: 2,000 emitters (5,000 optional)

Operational status

In service in Israeli Air Force Boeing 707 aircraft and incorporated in the Santiago Boeing 707 ECM aircraft of the Spanish Air Force. It is also reported that the South African Air Force operates at least one Boeing 707 with a version of the EL/L-8300 SIGINT system.

Contractor

Elta Electronics Industries Ltd.

VERIFIED

EL/L-8303 ESM system

The EL/L-8303 is a computerised ESM system for shipborne, airborne and ground-based applications. It is designed to intercept, digitise, identify, display and record radar signals. The processed data can be examined by an operator or used to activate automatic ECM or chaff systems. The system incorporates computer control and processing features. It has an omnidirectional instantaneous direction-finding capability which is claimed to be highly sensitive and accurate with a high intercept probability across a wide bandwidth. Alphanumeric keyboard inputs and graphical display output of data are standard features and the operator can access part of the emitter library. The system also includes built-in test circuits.

Intercepted radar signals are measured and classified according to pulsewidth, frequency, azimuth, power and time of arrival and are then fed to the processor. This conducts signal sorting, recognition functions and emitter identification and produces graphical data outputs. A modular system configuration permits optimisation of the system to meet customer cost and performance requirements.

Specifications

Power supply: 115 V AC, 400 Hz, 4 kVA
Frequency: 2-18 GHz (optional down to 0.5 GHz)
Azimuth accuracy: 5° RMS (3° option)
Max pulse rate: 500,000/s
Sensitivity:
(wideband) −60 dBm
(narrowband) −80 dBm
Dynamic range: 60 dB
Reaction time: 1 μs (typical)
Installation: two 19 in (487 mm) racks
Environmental: MIL-E-5400 and MIL-E-16400

Operational status

In production.

Contractor

Elta Electronics Industries Ltd.

VERIFIED

EL/L-8312A ELINT/ESM system

The EL/L-8312A system forms a part of the EL/L-8300 SIGINT set (see earlier entry) or can operate alone, covering the 0.5 to 18 GHz frequency band to detect, analyse and identify signals out to the radar horizon. Direction-finding is fast and accurate and bearings are stored in the associated computer and correlated with aircraft navigational data to give a display of the actual location of selected transmissions on a colour graphic console. The EL/L-8312A incorporates the EL/L-8312R receiver, EL/L-8320 signal parameters measurement equipment and the EL/L-8610 computer.

Operational status

In production and service with the Phalcon and a variety of other aircraft. In June 1991, it was announced that Singapore had awarded a $20 million contract for EL/L-8312A systems for four Enforcer Mk 2 maritime patrol aircraft. The EL/L-8312A has also been supplied for retrofit to Royal Australian Air Force P-3C Orions.

Contractor

Elta Electronics Industries Ltd.

VERIFIED

EL/M-2160 missile approach warning system

The EL/M-2160 missile approach warning system provides warnings of missile attack and automatically activates chaff and flare dispensers to protect the aircraft. The EL/M-2160 is an all-solid-state pulse Doppler radar. It provides time-to-impact and direction information to enable timely and effective response.

The equipment consists of the Transceiver Processing Unit (TPU), RF head, antennas and cockpit control unit.

Specifications

Dimensions:
(TPU) 508 × 254 × 216 mm
Weight:
34 kg (TPU and RF head)
Power consumption: 400 W

Operational status

The system has been designed for F-16 aircraft and adapted for transport aircraft and helicopters. The system is currently in service with several European and Asian air forces on a variety of aircraft and helicopters including C-130, C-160, Fokker, Antonov-32, Bell-212 and Mi-17. The system has been successfully tested against live missiles firing and has been proven effective in protecting the platforms in several missile firing engagements.

Development of a low weight (15 kg) modern pulse Doppler radar missile approach warning system is in progress.

Contractor

Elta Electronics Industries Ltd.

UPDATED

Guitar-300 passive missile warning system

Guitar-300 is a lightweight totally passive, autonomous warning system which detects missiles by sensing the electro-optical emissions from their engines. The system provides audio and visual alarms, allowing time for the aircraft to activate its defensive systems. It has been developed for the protection of helicopters and C-130 aircraft.

Guitar-300 is based on a patented sensor and automatically controls the infrared countermeasures subsystem. It emits no electromagnetic or electro-optical signals which might be detected. It is designed to detect ground-to-air and air-to-air missiles and has an extremely low false alarm rate.

Guitar-300 features coverage of all possible attack angles and high discrimination against background interference. It withstands adverse environmental conditions and is compatible with cockpit displays.

Guitar-300 passive missile warning system ***1998***/0018287

Specifications

Azimuth coverage: 360°
Elevation coverage: 90°
Warning time: 4-6 s
Weight: less than 15 kg
Power: less than 200 W
Interface: MIL-STD-1553B

Operational status

Test flown. Suitable for helicopters, fighter and transport aircraft.

Contractor

Rafael Electronic Systems Division.

UPDATED

Kingfisher ESM system

Kingfisher is an ESM system which is designed to provide fast and accurate surveillance of the electronic battlefield over the 2 to 18 GHz frequency range. It may be installed in fixed-wing aircraft or helicopters. Using broadband, fast response Instantaneous Frequency Measurement (IFM) and Instantaneous Direction-Finding (IDF) receivers, Kingfisher provides almost 100 per cent probability of radar pulse intercept. It is designed with modularity, allowing for customised hardware and retrofit into existing systems, and has considerable growth potential.

Kingfisher is able to operate successfully in dense electromagnetic environments. It receives signals over a wide range and amplitude simultaneously, evaluating parameters such as frequency, direction, pulsewidth, pulse type and time of arrival. The system then activates filtering and correlation algorithms, after which the data are further processed for additional information such as PRI, frequency range and scan and emitter antenna scan.

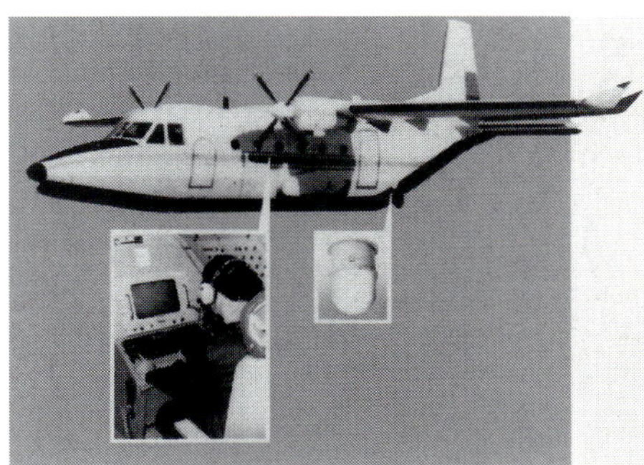

Kingfisher ECM system ***1998***/0018286

If required, Kingfisher can carry out automatic activation of passive and active countermeasures to form a fully power-managed EW suite. In addition, it will support the targeting of optically or thermally guided or unguided weapons.

Specifications

Frequency: 2-18 GHz (0.5-40 GHz optional)
Coverage: 360° azimuth, 40° elevation
Accuracy:
(frequency) ±1.5 MHz RMS
(bearing) 1-2°
Sensitivity: −60 dBm

Operational status

Fully developed and tested in real scenarios, in simulators and on EW ranges.

Contractor

Rafael Electronic Systems Division.

UPDATED

Rattler power management radar jammer

The Rattler power management radar jammer is designed to operate as a standoff and direct support jammer in a dense EW environment. It is suitable for operation in aircraft, ships and land vehicles and is optimised against search, tracking and surveillance radars. The wide frequency coverage can be set according to customer requirements.

Rattler enables the simultaneous jamming of up to three threats by using time-sharing techniques and concentrates transmitted power in spots or barrage to achieve a high jam/signal ratio. Up to 16 systems can be connected to the same bus for simultaneous computer-controlled operation, or a single system can be operated manually through the control unit.

Specifications

Dimensions:
(HP amplifier) 620 × 254 × 36 mm
(LP source) 605 × 254 × 271 mm
(control unit) 180 × 127 × 228 mm
(power supply) 232 × 125 × 180 mm
Weight:
(HP amplifier) 38 kg
(LP source) 28 kg
(control unit) 2 kg
(power supply) 5.5 kg
Power supply: 105-125 V AC, 400 Hz ±5%, 3 phase

Operational status

In service.

Contractor

Rafael Electronic Systems Division.

UPDATED

Advanced Digital Dispensing System (ADDS)

The Advanced Digital Dispensing System (ADDS) is a computer-controlled, threat adaptive countermeasures dispensing system. It is designed to protect aircraft from both ground and air threats by dispensing decoy payloads. The system is configured for high-performance aircraft, helicopters, transport aircraft and maritime patrol aircraft.

ADDS can dispense chaff, flares and RF expendables in single shot or multiple simultaneous dispense programmes to provide multispectral responses that best counter the threat.

ADDS incorporates improvements derived from combat experience and hundreds of installations on a wide variety of aircraft, including the A-4, AH-64, F-4, F-5, F-15 and F-16.

Operational status

In service, with several ADDS delivered worldwide.

Contractor

Rokar International Ltd.

UPDATED

CDF-3001 airborne V/UHF COMINT/DF system

CDF-3001 is an airborne self-contained COMINT/DF system designed to acquire, monitor and measure the Direction Of Arrival (DOA) of tactical radio transmissions. The operational system is driven by a Windows-based man machine interface, and advanced hardware implementation provides fast scan and DF capability. The system is linked to the host aircraft navigation and command and control system via an RS232 interface. The CDF-3001 system is suitable for both fixed- and rotary-wing application.

Specifications

Frequency coverage: 20-500 MHz (optionally 1.5-1,000 MHz)
Monitored signals: AM and FM

Contractor

Tadiran Electronic Systems Ltd.

NEW ENTRY

CDF-3001 airborne V/UHF COMINT/DF system installed under helicopter **1998**/0018285

TACDES SIGINT system

TACDES is an airborne SIGINT system consisting of the RAS-1B or RAS-2A ELINT system, the ACS-500 automatic VHF/UHF COMINT system and the TDF-500 automatic direction-finding unit integrated with a computerised C³I system. TACDES is installed on large transport aircraft such as the Boeing 707 or Lockheed Martin C-130 and performs real-time acquisition, location, processing and reporting of communications and radar signals between 20 MHz and 18 GHz.

Accurate instantaneous wide-angle coverage is provided by interferometric COMINT and ELINT DF systems. The system features an enhanced command and control capability and real-time secure voice and data communications.

ACS-500 Automatic VHF/UHF COMINT System

The ACS-500 is an airborne self-contained system designed for acquisition and monitoring of tactical radio nets which operate in the 20 to 500 MHz frequency band. The system features spectrum scanning in programmable frequency sectors and provides active display of up to 100 preprogrammed frequencies with identification, direction, signal level and time of reception data, and fully automatic direction-finding. It has an operator oriented hands-off monitoring capability and built-in test equipment. Operation has been simplified by adopting a user-friendly man/machine interface.

Main functions are to provide: spectrum scanning from 20 to 500 MHz, with reporting of new active, previously unknown or unrecognised stations; scanning of known frequencies and updating operators' activity displays; and a monitor, analysis and recording facility for any active stations.

Specifications

Frequency: 20-500 MHz (1.5-1,000 MHz optional)
Preprogrammed frequencies: 100
Accuracy: 2.5° RMS typical
Modes: AM, FM

RAS-1B ELINT and ESM system

The RAS-1B airborne interferometric ELINT system, installed on large transport aircraft, covers the 0.7 to 18 GHz frequency band (0.5 to 40 GHz optional) to obtain the enemy Electronic Order of Battle (EOB). It can measure frequency, pulse repetition frequency, amplitude and Direction Of Arrival (DOA) data on an average of two emitters per second, with exceptionally high accuracy. Operation has been optimised for weak emitter recognition in high-density electromagnetic environments and an activity file is maintained which contains identification, classification and updated data on recognised threats. A combination of wide bandwidth acquisition IFM receiver and a narrowband superheterodyne analysis receiver is used to ensure 100 per cent probability of detection and high system throughput. Operation can be fully automatic or man/

machine oriented in simple manual mode to accommodate up to two operators.

Specifications

Frequency: 0.5-18 GHz (0.5-40 GHz optional)
Frequency resolution: 1.25 MHz
Frequency accuracy: 2.5 MHz
DOA accuracy:
(from 2-18 GHz) 0.7° (typical), 0.3° (optional)
(from 0.5-2 GHz) 1.3° (typical), 0.5° (optional)
PRI resolution: 100 ns
PRI accuracy: 200 ns
Instantaneous bandwidth:
(analysis) 40 MHz
(acquisition) 8 GHz
Sensitivity: −70 dBm typical
De-interleaving capability: Up to 8 pulse trains in the same frequency/DOA cell

RAS-2A ELINT and ESM system

The RAS-2A airborne, wingtip-mounted, interferometric tactical ESM system is deployed on small- and medium-size transport aircraft. In the 0.5 to 18 GHz frequency band it can measure frequency, PRI, amplitude and direction of arrival data with excellent accuracy, with 360° azimuth coverage. Operation has been optimised for a single operator who can both display an EOB scene and passively target an anti-radiation or anti-ship missile. The high sensitivity of the system ensures weak signal processing even at long ranges. Automatic file activity is maintained for quick identification and classification of threats.

The RAS-2MT is the configuration for maritime surveillance and targeting.

Specifications

Frequency: 0.5-18 GHz
Frequency resolution: 0.2 MHz typical
Frequency accuracy: 0.4 MHz typical
DOA accuracy:
(left/right) 0.8°
(fore/aft) 5°
Coverage:
(azimuth) 360°
(elevation) ±10°
Sensitivity: −75 dBm
Instantaneous bandwidth: 1, 5, 15, 40 MHz

TDF-500 VHF/UHF automatic direction-finding system

The TDF-500 is an airborne automatic direction-finding system using a powerful central computer, high-performance twin-channel receiver and antenna array mounted on the aircraft wings and under-surfaces. It covers the 20 to 500 MHz band and can determine direction of arrival to 1.5°. The antenna ports connect to an RF switch matrix which selects antenna pairs and feeds them to the twin-channel receivers. In each sampling period a phase measurement is performed between one antenna pair and successive phase measurements are processed to obtain direction of arrival information.

Specifications

Frequency: 20-500 MHz
Power consumption: 400 W
Response time: less than 300 ms
Modulation: AM, FSK, PSK SSB
Accuracy: 2° RMS between 30 and 500 MHz
4° RMS between 20 and 30 MHz
Coverage:
(azimuth) 360°
(elevation) −2 to −7°
Polarisation: vertical

Operational status

In production for use on large transport aircraft that are required for special surveillance tasks.

Contractor

Tadiran Systems Ltd.

UPDATED

The RAS-1B ELINT and ESM system operators' consoles ***1998***

ITALY

SL/ALQ-34 ECM pod

The SL/ALQ-34 ECM pod is available as a self-defence aid for high-performance aircraft. The SL/ALQ-34 provides a radar warning facility and can jam anti-aircraft artillery or surface-to-air missile radars. Current versions operate across a 6 GHz bandwidth in the H- to J-band (6 to 20 GHz) but lower frequency systems at C-, D-, E- and F-bands, effective against air defence search radars, are in prospect.

The receiver processor conducts threat assessment and ranking using a stored library of threat data, and performs jamming power management tasks. Probably two travelling wave tube transmitters are used and cooled by a closed-loop liquid cooling system. Supersonic high-altitude clearance has been obtained.

Operational status

In production.

Contractor

Alenia Difesa, Avionic Systems and Equipment Division.

UPDATED

SL/ALQ-234 self-defence pod

The SL/ALQ-234 self-defence pod fulfils the requirements for aircraft self-defence with a minimum of penalty to flight and combat capabilities. The SL/ALQ-234 is designed for standard fuselage or wing pylon installation. A small cockpit control panel provides the pilot with threat warning and jamming reaction information. Self-contained ram-air turbine power generation and cooling, as well as the modular pylon attachment technique, make the SL/ALQ-234 suitable for immediate installation on a variety of supersonic aircraft.

Real-time processing and power management are the main factors leading to the multiple threat self-defence capability of the SL/ALQ-234. For pulse threats the system covers the I- and J-bands and provides noise and deception jamming. CW coverage is provided in the H- to J-bands and the threat is countered by deception jamming.

The system can operate even in the complete absence of prestored threat data, through software generated self-adaptive modes. Against known threats

SL/ALQ-234 self-defence pod

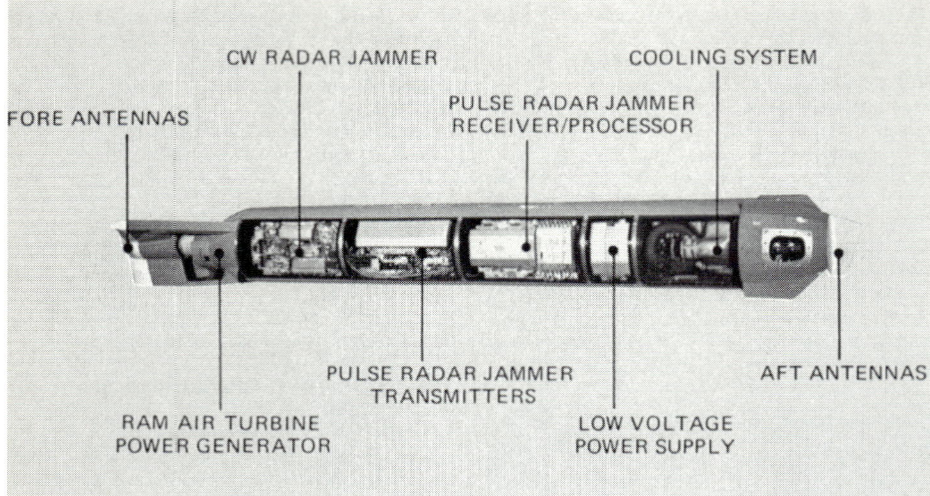

The internal arrangement of the SL/ALQ-234 self-defence pod

prestored in the system memory it can perform more complex and specific reactions. Positive self and mutual screening of the aircraft is achieved against the most advanced ground-based air defence systems.

Specifications

Dimensions: 3,825 (length) × 414 mm (diameter)
Weight: 270 kg
Frequency:
(pulse) I/J-bands
(CW) H/J-bands
Power output: 7.5 kVA
Altitude: sea level to 30,000 ft
Speed: M1.1 at sea level, M1.5 at 30,000 ft

Operational status

In service with various air forces.

Contractor

Alenia Difesa, Avionic Systems and Equipment Division.

UPDATED

Apex jamming pod

The Apex deception jamming pod uses the existing ELT 555 pod shell and is claimed to be effective against all types of fire-control radars. Frequency coverage against both pulse and CW emitters is quoted as being extended. The pod weighs 140 kg.

Apex uses fast digital RF memory technology to facilitate coherent response techniques and the system is front-line programmable.

Operational status

Apex is suitable for aircraft such as the Hawk 100/200, MB-339, F-5 and Mirage F1.

Contractor

Elettronica SpA.

VERIFIED

Aries tactical support and training EW system

Aries can be used either for training EW specialists or for tactical ELINT/COMINT and surveillance and support ECM missions. It is designed to be housed in a small transport aircraft and typically has a supervisor's position and two operator stations. Data collected during tactical ELINT/COMINT and surveillance missions can be transmitted over a secure datalink to the ground for near-realtime evaluation.

The Aries system consists of two subsystems: Aries-A which is dedicated to all activities in the radar frequency band and Smart Guard (see later entry) which covers the communication bands.

Aries-A detects, analyses, identifies and locates emissions in the C- to J-bands and, during ECM support missions, drives a set of jammers covering the D- to J-bands.

Smart Guard carries out similar functions in the VHF and UHF frequency ranges, with HF available as an option.

Aries is fully compatible with navigation systems such as inertial, Omega and GPS and is provided with a look-through facility to permit signal reception during jamming.

The Elettronica Aries modular airborne EW system for tactical support and training

Operational status

In service since 1980. More than 60 Aries systems have been sold to a number of air forces.

Contractor

Elettronica SpA.

VERIFIED

ARWE Advanced Radar Warning Equipment

ARWE was designed for the Italian Tornado IDS aircraft. It is based on a multiprocessor architecture, with MIL-STD-1553B interface connections to other elements of the Defensive Aids Sub-System (DASS).

The ARWE system architecture comprises:
(1) 4 DF channels with 360° azimuth coverage;
(2) E-/J-band frequency capability;
(3) frequency selective receiver with: multiple IF bandwidths (wide bandwidth to optimise probability of detection and response time; medium bandwidth as a compromise to meet general processing requirements; narrow bandwidth to provide discrimination of CW signals);
(4) adaptive bandwidth selection controlled by the signal environment and the active electronic countermeasures requirements;
(5) instantaneous frequency measurement within the IF band to ensure fast emitter identification and designation of active countermeasures.

Operational capabilities include:
(1) 3 in cathode ray tube display;
(2) E-/J-band frequency coverage;
(3) detection of Low Probability of Intercept (LPI) and

ARWE advanced radar warning equipment
1998/0018284

Multi-Function Radar (MFR) emitters;
(4) exotic emission detection capability;
(5) compatiblity with on-board high duty factor radars and jammers;
(6) DASS management capability;
(7) flightline reprogrammable.

Operational status
Operational on Italian Air Force Tornado IDS aircraft.

Contractor
Elettronica SpA.

NEW ENTRY

Colibri integrated ESM/ECM system

The Colibri system is an integrated ESM/ECM equipment designed for installation on any type of medium-sized helicopter operating in the naval surveillance, anti-submarine and anti-ship roles, to extend the ESM/ECM range of the parent ship. The system provides a helicopter with the facilities for passive surveillance, DF analysis, identification and location of radar emitters, even when the system is operating in a dense electromagnetic environment. It provides for effective self or mutual protection against radar-guided missiles.

Colibri can be configured to meet particular operational roles by combining a variety of basic modules. The basic configuration is represented by the ELT/161 ESM and the radar warning system. A helicopter's radar warning function can then be supplemented by the ELT/562 deception jammer to obtain effective self-protection and the configuration can be further expanded by the addition of the ELT/261 for ESM capability and the ELINT role. A further unit, the ELT/361, can be added to achieve even greater protection capability.

In the basic configuration, Colibri comprises the ELT/161, consisting of receiver, antenna system, warning processor and radar warning display. The ELT/161, operating in conjunction with the ELT/261 system, gives overall capability in that field and is designed to fulfil the maritime surveillance and EW roles to extend the electronic warfare range of the parent ship. In this role, the system supplies instantaneous DF and display of intercepted emissions, frequency measurement of pulse and CW emissions, automatic analysis, identification of source and automatic tracking of selected emissions. The addition of the ELT/562 deception jammer provides the helicopter with self-protection against sea-to-air and air-to-air missiles. The ELT/361 noise jammer is added to suppress enemy surveillance radars. The system is capable of the following instantaneous surveillance coverage:

Altitude	Effective radius against radar type (in miles)	
	I-Band (8-10 GHz)	F-Band (3-4 GHz)
300 ft	26	34
1,000 ft	41	50
2,000 ft	47	63

Operational status
No longer in production. In service with helicopters of several navies.

Contractor
Elettronica SpA.

VERIFIED

ELT/156(V) radar warning receiver

The Elettronica ELT/156(V) radar warning receiver

Suitable for fighter, light strike/attack aircraft and helicopters, the ELT/156 series of lightweight passive detection systems provides broadband 360° azimuth coverage and can distinguish anti-aircraft artillery, surface-to-air missile and air-to-air missile illuminations in bearing and range. Miniature crystal video receiver technology and advanced signal processing, including built-in test equipment, contribute to minimise system weight. A dual-mode CRT display, on which synthetic or raw video data can be shown, is standard. An audio output is available to complement the display. Latest versions of the equipment are lighter than earlier production systems.

Specifications
Weight:
(4 antennas) 0.15 kg each
(2 RF heads) 1 kg each
(signal processor) 5.2 kg
(display unit) 1.9 kg
(control panel) 0.6 kg
(total weight) 10.3 kg

Operational status
In service.

Contractor
Elettronica SpA.

VERIFIED

ELT/156X radar warning receiver

The Elettronica ELT/156X radar warning receiver features very wideband miniature crystal video receiving systems and provides full azimuth and frequency coverage up to millimetric-waves and including distributed multiprocessing. The use of bit-slice microprocessing and EEPROM technology allows flexible and high-speed processing, giving fast reaction times and adaptability to threat evolution. Threat parameter software is reprogrammable at flight line level. The dual cockpit display incorporates both a raw mode, to give an immediate picture of the surrounding environment, and a synthetic mode, to give a clear indication of threat identity. The system also includes an aural alarm.

The ELT/156X may be integrated with other onboard EW systems and sensors via two RS-422A serial links.

The system consists of two antennas/RF heads, signal processor, display and control panel. The version for helicopters, the ELT/156X(V2), employs four single-channel RF heads.

Specifications
Weight:
(antenna/RF head × 2) 2.2 kg
(signal processor) 10.5 kg
(display) 2.1 kg
(control panel) 0.7 kg
(total weight aircraft version) 17.7 kg
(total weight helicopter version) 20.1 kg

Operational status
Now in production for a number of air forces.

Selected by the Italian Navy for its AS/ASVW and AEW EH 101 helicopters, also selected for the Brazilian AMX light fighter aircraft.

Contractor
Elettronica SpA.

UPDATED

ELT/158 radar warning receiver

The ELT/158 is designed for combat aircraft and provides high-speed detection and indication of threats through a full 360° coverage. The dual-mode cockpit display can show raw or synthesised data. The system can be integrated with a laser warning receiver, has fully automatic built-in test facilities and can be reprogrammed on the flight line.

Contractor
Elettronica SpA.

VERIFIED

The ELT/156X radar warning receiver showing (from left to right) antennas, processor and (front) the control panel, (rear) the display

ELT/263 ESM system

The ELT/263 airborne ESM is designed principally for use in maritime aircraft such as the Beech 200T but has been configured for other types such as the Guardian, CASA C212, Learjet 35A, F27 Maritime, Bandeirante, Piaggio P166 and Britten-Norman Maritime Defender. The surveillance of coastal waters is its main function. Detection, analysis and identification of emissions in the E- to J-bands is performed, in addition to bearing measurement; this permits emitter location by using a series of successive bearing measurements.

The ELT/263 equipment, in addition to its surveillance tasks, provides for emitter location by the triangulation method. Emission analysis, identification and bearing information can be transferred to external users such as ground stations, naval units and the aircraft's radar operator and tactical navigator and can also be recorded on an optional printer.

The main items of the ELT/263 are a set of four DF antennas, an omnidirectional antenna set, DF receiver, IFM receiver, ESM display and control console, radar warning processor and radar warning display unit.

Operational status

In service. The equipment is in current production for several customers.

Contractor

Elettronica SpA.

VERIFIED

ELT/457-460 supersonic noise jammer pods

A set of four related noise jammer pods is available for use on any high-performance strike/fighter or light attack aircraft, mounted on an underwing pylon. All units are self-contained with a ram-air turbine generator in the nose section and each is dedicated to a particular waveband. A heat-exchanger is situated behind the turbine and there are fore and aft antennas in all pods. The ELT/459 and ELT/460 versions have additional antennas beneath the body of the pod. Processing within the system allocates threat jamming power on a proportional basis against any type of pulsed radar threat and includes built-in test equipment and control of blanking with other aircraft systems. All units are claimed to be highly resistant to ECCM, including any type of frequency agility, and incorporate a large threat library. Maximum operating speeds are M1.1 at sea level and M1.5 at 40,000 ft.

Specifications

Dimensions: 3,120 (length) × 340 mm (diameter)
Weight: 145 kg
Interface: STANAG 3726 and MIL-A-8591D

Operational status

In service.

Contractor

Elettronica SpA.

VERIFIED

ELT/553(V)-2 airborne pulse and CW jammer

The ELT/553(V)-2 airborne deception jammer features fully automatic operation and is programmable at flight line level. It is designed to ensure maximum flexibility, operating either as a stand-alone system, or in co-operation with on board RWRs.

ELT/553(V)-2 comprises:
(1) a low band pulse module (E- to H-band)
(2) a high band pulse module (H- to J-band)

ELT/553(V)-2 airborne pulse and CW jammer
1998/0018283

(3) a CW module dedicated to counter semi-active missile threats.

ELT/553(V)-2 utilises:
(1) a channelised receiving section architecture for quick and reliable lock-on and recognition, even in very dense environments;
(2) repeater jammer philosophy to increase effective jamming power and hence jamming effectiveness in the presence of ECCM techniques;
(3) dedicated CW mode to counter semi-active missile threats;
(4) effective operation against staggered and jittered PRI emitters, as well as against frequency diversity and frequency agility threats;
(5) automatic deception of locked-on radars;
(6) multi-threat capability;
(7) RS422 interfaces.

Optional capabilities include:
(1) automatic deception of pulse and CW radars, TWS included;
(2) pulse by pulse power management and optimisation across 10 channels of operation;
(3) very high ERP;
(4) flight line programmability;
(5) complete BITE.

Specifications

LRUs
High-band pulse module:
(quantity) 1
(weight kg) 40
Low-band pulse module:
(quantity) 1
(weight kg) 37
CW module:
(quantity) 1
(weight kg) 20.4
Control panel:
(quantity) 1
(weight kg) 1.4
High-band antenna:
(quantity) 2
(weight kg) 0.2 × 2
Low-band antenna:
(quantity) 2
(weight kg) 0.4 × 2
Power consumption: 2.5 kVA

Operational status

Understood to be selected for the Brazilian AMX light fighter aircraft and possibly Italian Tornado aircraft.

Contractor

Elettronica SpA.

UPDATED

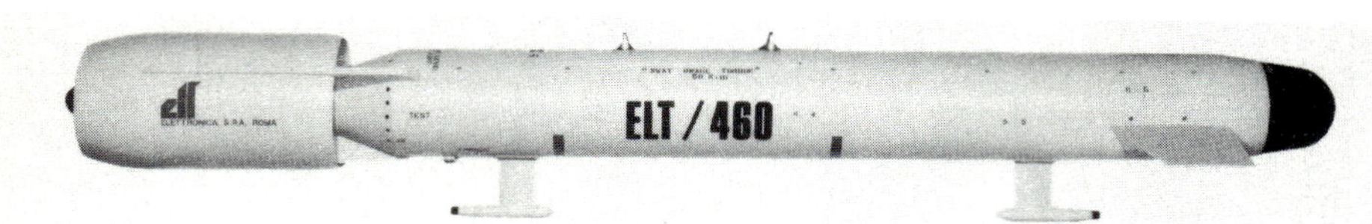

The Elettronica ELT/460 supersonic pod

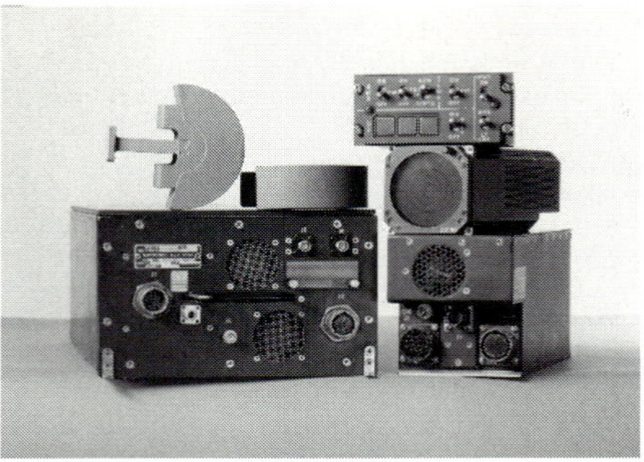

A self-protection suite for modern attack helicopters comprising the ELT/554 lightweight deception jammer (left) and the ELT/156 radar warning receiver

ELT/554 deception jammer

The ELT/554 is specifically designed for installation in attack helicopters and operates in the J-band. It is lightweight and compact and reacts rapidly to incoming signals. The jamming programmes are intelligent and several threats can be jammed simultaneously.

Contractor

Elettronica SpA.

VERIFIED

ELT/555 supersonic self-protection pod

The ELT/555 has a similar configuration to the ELT/457 system, with the same ram-air turbine on the nose of the pod. Internal equipment includes fore and aft facing antennas which receive pulsed and continuous wave signals and transmit pulsed responses, plus separate fore and aft facing continuous wave transmitter antennas on the undersurface. The system is sensitive to H- to J-band (6 to 20 GHz) threats and is designed to operate in dense electromagnetic environments. It has multiple target contrast capability and features BITE. Operating envelope and mechanical interface specifications are the same as those of the ELT/457. The cockpit display can be tailored to customers' requirements.

Specifications

Dimensions: 3,000 (length) × 340 mm (diameter)
Weight: 150 kg

Operational status

In service.

Contractor

Elettronica SpA.

VERIFIED

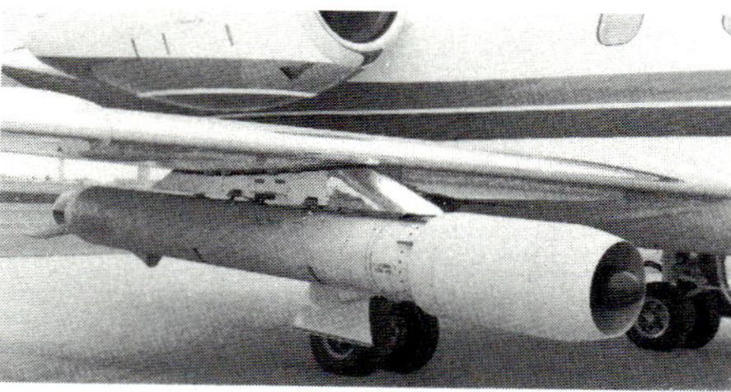

The Elettronica ELT/555 deception jammer pod on a Learjet 35A multimission trainer

ELT/558 self-protection jammer

The ELT/558 self-protection jammer operates at the lower end of the frequency spectrum and features wide-angle coverage, fully automatic responsive jamming operation, high ERP, low false alarm rate, high immunity to ECCM, very short reaction time and effective reaction against many threats. The system can be integrated with the aircraft's avionics.

The ELT/558 receiver unit (left) and ELT/558 transmitter unit (right) ***1998***/0018282

Operational status

In service on the Mirage 2000.

Contractor

Elettronica SpA.

VERIFIED

ELT/562 and ELT/566 deception jammers

The ELT/562 and ELT/566 are repeater jammer systems for internal installation on aircraft and helicopters. ELT/562 is designed to counter pulse threats, while ELT/566 combats continuous wave threats. One or both jammers may be associated with an airborne ESM system such as the Elettronica Colibri. The jammers are effective against H- to J-band threats and are intended to protect aircraft and helicopters in battlefield environments.

Operational status

In production.

Contractor

Elettronica SpA.

VERIFIED

Fast Jam ECM system

Fast Jam is designed for operations over the VHF and UHF communications frequencies. It is intended mainly for airborne applications, where maximum advantage is obtained by the platform elevation and speed, although the compact and lightweight construction of the system also allows installation on land vehicles and ships. The main features of the system include modularity, coverage of the VHF and UHF bands with optional coverage of the HF band, high speed and intercept capability and multiple frequency jamming capability.

In the Fast Jam configuration the system consists of a search and analysis receiver subsystem, monitoring subsystem and jammer subsystem. The first two subsystems perform the same functions as the basic ESM system described in the later entry for Smart Guard. The addition of the jammer subsystem allows the jamming of intercepted communications channels, to disrupt enemy communications.

The jammer subsystem operates under the control of the system computer and is fully automated. By time-sharing, it is able to jam up to six channels simultaneously without performance degradation. The channels to be jammed are either preset or can be selected by the operator according to the specific operational requirements. A look-through function allows an adaptive jam/receive time management so that simultaneous search and jamming functions can be performed.

Manual control of the jammer subsystem is available as a back-up to fully automatic operation. A high-power wideband all-solid-state amplifier is used in the jammer subsystem and VSWR and thermal automatic switch off circuits are employed to avoid permanent damage to the amplifier in case of overheat or antenna failure. Various types of modulation are available from external modulation sources.

Operational status

In service. The system has been installed in a number of different types of aircraft.

Contractor

Elettronica SpA.

VERIFIED

Sea Petrel RQH-5(V) airborne ESM/ELINT systems

The Sea Petrel RQH-5(V) is a family of systems which can meet EW requirements ranging from threat detection and analysis to electronic intelligence. The various configurations and options allow the system to be tailored to a specific requirement. All the RQH-5(V) Series of equipments have capabilities for integration with other systems and provide target parameters and direction of arrival information for weapon systems. The small size and low weight of RQH-5(V) components make the system readily adaptable to current airborne, naval and ground installations with a minimum of effort. The system can be operated after minimal instruction and is designed for maximum operating time with minimum servicing.

The system covers frequencies from 0.65 to 18 GHz, with an optional extension to 40 GHz, and is entirely automatic. It provides real-time automatic extraction, analysis and tracking of all incoming radar signals. It also provides pulse, intrapulse and fine analysis for ELINT, including frequency fine measurement, measurement of jitter and stagger, frequency or PRI agility analysis, histogram preparation, detection and recording of antenna pattern and related amplitude histogram. The operator has only to view the system display which provides data on up to 200 emitters. The RQH-5(V) can operate without any prior knowledge of the electromagnetic scenario with no significant reduction in performance. On the other hand, the ELINT capability in terms of emitter parameter statistical analysis and pulse and intrapulse analysis, allows complete characterisation and analysis of the radar signals and keeps records of them for post-flight data collection.

The basic components of the RQH-5(V) are the antennas, direction-finder receiver, IFM receiver and data extractor.

The antenna group includes one omnidirectional antenna and four or eight DF antennas. Various DF and fine DF antenna types are available to cover different frequency ranges. The modular approach of each antenna module allows different installation configurations, to cater for any platform constraints.

Each antenna unit incorporates the associated electronic circuitry. Direction-finding is performed by a wide open omnidirectional and instantaneous receiver using amplitude comparison monopulse techniques. An eight port configuration is normally adopted, but it can also use a four element DF antenna subsystem for radar warning receiver applications.

The IFM receiver features high sensitivity and high probability of detection, wide open operations, fast response in order to operate in a multimillion pps environment and very high instantaneous dynamic range. Two versions are available — the FR-6 and FR-7 — providing different RF measurement accuracy.

The data extractor consists of a very powerful multiprocessor structure, specially adapted for real-time applications as a derivative of the AYK-204 airborne computer. The resulting automatic data extraction process has a very high acquisition speed of up to 60 new emitters every 20 ms, even in a completely unknown environment. This unit includes some standard I/O interfaces, such as serial lines, video graphic, memory expansion and MIL-STD-1553 bus developed for the Alenia Difesa AYK-204 airborne computer and its derivative versions.

Several graphic and alphanumeric display modes are available to the operator. This includes frequency and direction of arrival of the signal, tabular lists, emitter characteristics, true or relative bearing, frequency-agile deviation, PRF, jitter and stagger values, emitter name and threat level and emitter scan period and type. Tactical, panoramic and geographic modes for ESM, radar and navigation are also available.

The identification library handles up to 3,000 modes of emitter parameters. The operator can store previous known data together with pre-assigned threat level and confidence level. The system compares extracted emitter data with the library and provides an immediate alert on high-interest emitters.

Several optional components are available. A high-accuracy direction-finder uses a multiple beam antenna system and a crystal video amplitude sectoral monopulse receiver. Several multibeam flat arrays, covering different frequency ranges, are available to provide different instantaneous fields of view and very high DF accuracy. A K-band high-gain steerable antenna and a formatter unit provide high-sensitivity detection and direction of the emitter signals. A fine analysis ELINT receiver gives supplementary information on the active extracted emitter, including the presence of simultaneous multiple RF or intrapulse modulations.

The following configurations are available:

The SL/ALR-730 Series offers electronic support measures and ELINT for all types of platforms including large and medium maritime patrol aircraft and large, medium and small helicopters. It is based on superheterodyne receiver technology.

The SL/ALR-740 Series offers RWR functions combined with automatic signal analysis for post-flight intelligence, and is designed for installation on small aircraft or helicopters. Average DF accuracy of 10° RMS is provided, with automatic warning and emitter parameter measurements. The ALR-741-R uses multiple-IFM receivers.

The SL/ALR-780 Series allows the integration of ECM modules in the above ESM equipment.

Specifications

Frequency: 0.6-18 GHz
Sensitivity: −60 dBm
Accuracy: ±2.5° RMS

Operational status

In service. Current and future applications include the EH 101 and AB-412 helicopters and Nimrod, P-3 Orion and Atlantique maritime patrol aircraft. A version of the SR/ALR-730, the ALR-735(V)3, has been selected by the Italian Navy for the EH 101 helicopter in its AS/ASVW (Anti-Submarine/Anti-Surface Vessel Warfare) and AEW (Airborne Early Warning) roles.

Contractor

Elettronica SpA.

VERIFIED

Smart Guard COMINT system

The Smart Guard COMINT system is designed for intelligence monitoring of the VHF/UHF communications band. Although it is mainly intended for airborne applications, where maximum advantage is obtained by the platform height and speed, the

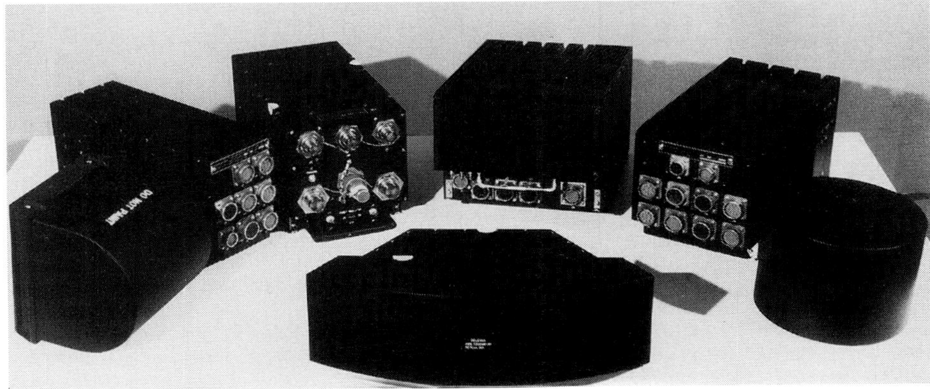

The Sea Petrel RQH-5(V) ESM/ELINT system ***1995***

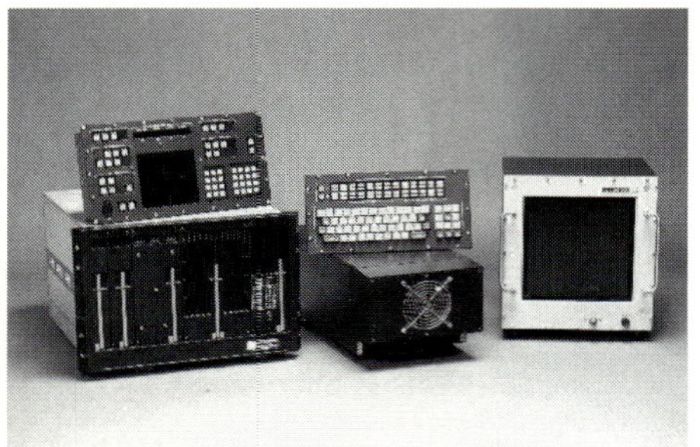
The Smart Guard basic ESM subsystem

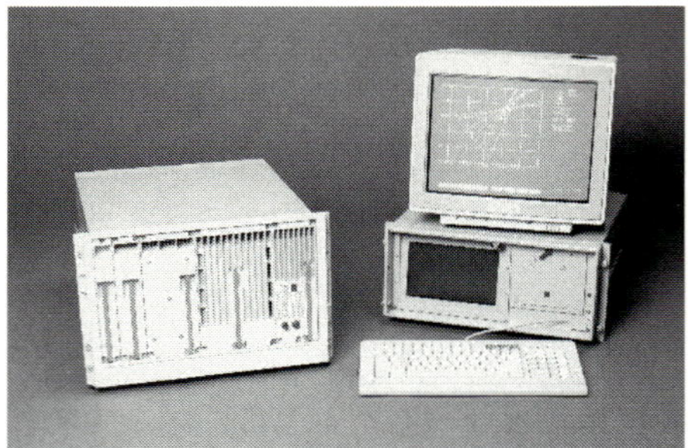
The Smart Guard DF and emitter fixing subsystem

The Smart Guard monitoring and recording subsystem

compact and lightweight construction of the system allows installation on ground vehicles and in ships.

The basic ESM system consists of a search and analysis receiver and a monitoring subsystem. In this configuration the system allows the search and intercept of communications through a computer-controlled operation. The search is carried out over the complete bandwidth, or on specified sub-bands or channels. On operator control the receiver stops on a specific channel, so allowing the demodulation and analysis of the particular channel. A continuous monitoring of up to eight channels is provided by up to eight remotely controlled receivers and associated recorders. The frequency tuning of each monitoring receiver is set automatically by the system computer. Voice and associated data are recorded for subsequent analysis. In this configuration the system is manned by a single operator.

In the Smart Guard configuration, the basic ESM system is augmented by a DF and fixing subsystem. The latter aids the capability to measure the DOA of the communications transmissions, and to determine their location. The DOA measurements are performed by a specific dual-channel superheterodyne receiver controlled by the system computer, and are obtained through an interferometric measurement. In addition to the DOA, the frequency value and signal strength are measured for each specific channel.

An interface with the navigation system provides the actual position of the platform in such a way that for each specific emission a set of data is stored. This contains DOA, frequency value, signal strength and platform position. The fixing of specific communication emission is performed by a dedicated computer and associated software algorithms by using the DOA and platform position data of a specific channel. At least two significant DOA measurements are required for a fixing computation.

In the Smart Guard configuration, two operators are required: one to control the basic ESM system and one dedicated to the fixing operation. A ground-based retrieval and analysis system is available as an option.

Integration of the Smart Guard COMINT system with the ELT/888 ELINT system makes it possible to monitor effectively the full electromagnetic environment and to identify and locate all enemy weapon systems.

Operational status
In service.

Contractor
Elettronica SpA.

VERIFIED

RALM-01 laser warning receiver

The RALM-01 laser warning receiver provides laser threat detection and classification capability. Specially designed for airborne platforms, the RALM-01 can either be integrated with other EW equipment or work as a stand-alone equipment and directly drive countermeasures systems.

Based on two side-mounted head sensor units, the RALM-01 standard version provides full 360° azimuth coverage and 90° vertical coverage in the visible and near infrared bands, while the RALM-01/1 version makes use of two additional side-mounted head sensor units to provide far infrared band coverage.

The threat direction is shown, to an accuracy of 45° in azimuth, on the display together with the threat type. An audio alarm is also generated. Either an RS-232C or MIL-STD-1553 interface is provided for integration with other EW equipment. A blanking input signal and a countermeasures start output signal are also available.

Specifications
Azimuth coverage: 360°
Elevation coverage: 90°
Sensor band: 0.5 to 1.8 μm
Extended sensor: 8 to 12 μm
Weight: 4.5 kg
Dimensions:
(sensors (× 2)) 90 × 50 × 40 mm
(processor) 3/8 ATR short

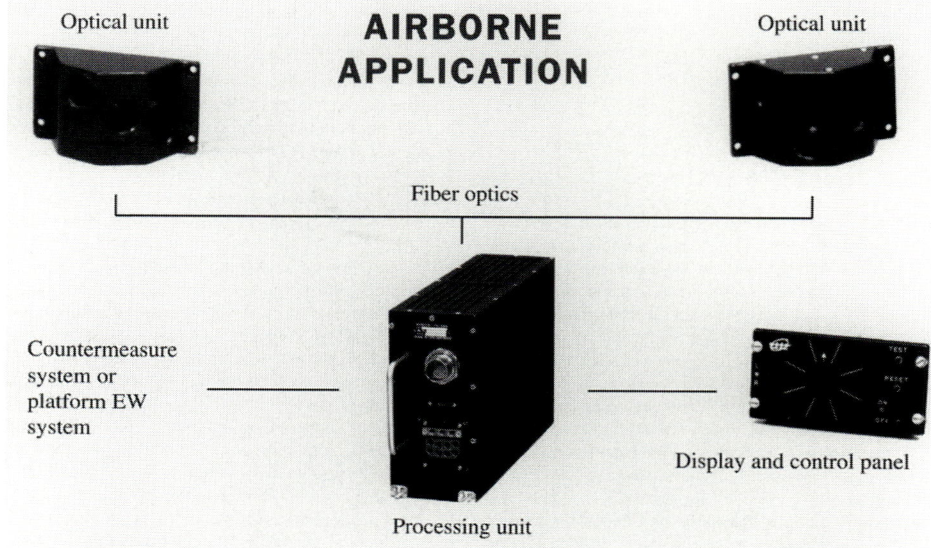

RALM-01 laser warning receiver **1998**/0018252

Operational status
In production for Mangusta A129 and HH-3F SAR helicopters, and C-130J aircraft.

Contractor
Marconi SpA.

UPDATED

JAPAN

J/ALQ-5 ESM system

Used on ESM variants of the Kawasaki C-1 medium transport, the J/ALQ-5 ESM system receives and jams surface-to-air missile radars.

Operational status
In service on the Kawasaki C-1 ESM variant.

Contractors
Mitsubishi Electric Corporation.
NEC Corporation.

VERIFIED

J/ALQ-6 jamming system

The J/ALQ-6 jamming system is in development to provide an airborne radar jamming capability for F-4EJ and Mitsubishi F1 fighters.

Operational status
In development for F-4EJ and Mitsubishi F1 aircraft.

Contractors
Mitsubishi Electric Corporation.
NEC Corporation.

UPDATED

J/ALQ-8 jamming system

The J/ALQ-8 is in development for the F-15J; it may be a variant of the J/ALQ-6 set. It is understood to provide countermeasures in the 1 to 4, 4 to 8 and 7.5 to 18 GHz bands. ALQ-8 jammers integrate with the J/APR-4 radar warning system.

Operational status
In development for the F-15J.

Contractors
Mitsubishi Electric Corporation.
NEC Corporation.

VERIFIED

J/APQ-1 rear warning receiver

The J/APQ-1 rear warning receiver has been developed for the F-15J aircraft. It is designed to cope with both radar and infrared threats and will automatically activate chaff and flare countermeasures. The starboard side of the tail of the F-15J will be modified to allow the fitting of a radar antenna with a diameter of 200 mm. A visual indicator and audio alert are positioned in the cockpit.

Operational status

Developed for the F-15J. Believed to have entered service in 1992.

Contractors

Mitsubishi Electric Corporation.
NEC Corporation.

VERIFIED

J/APR-4/4A radar warning system

The J/APR-4 was designed for the F-15J/DJ aircraft. It is able to process multiple inputs simultaneously in a dense electromagnetic environment, and has a digital computer with a reprogrammable software package to allow reconfiguration for future requirements. The indicator provides for daylight viewing and multithreat data presentation in alphanumeric and graphic format. The system is also designed to interface with other EW equipment such as the J/ALQ-8.

The J/APR-4A is the advanced model of the J/APR-4 radar warning receiver. Its specification calls for the ability to process multiple inputs simultaneously in a dense electromagnetic environment. The system incorporates a digital processor with reprogrammable software which permits reconfiguration to meet developing threats. A tactical situation CRT display presents multiple-threat data in alphanumeric and graphic form. Interfaces with other onboard electronic countermeasures systems, such as J/ALQ-8 jammers, can be accommodated.

Operational status

In production for the F-15J/DJ.

Contractor

Tokimec Inc.

VERIFIED

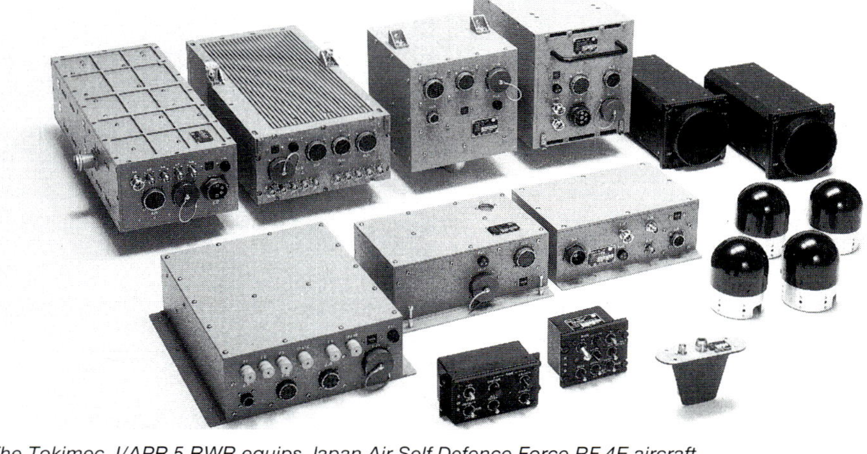

The Tokimec J/APR-5 RWR equips Japan Air Self-Defence Force RF-4E aircraft

J/APR-5 and J/APR-6 radar warning systems

The J/APR-5 and J/APR-6 are developments of the J/APR-4 system, with additional capability designed to cope with current threats. Actual sizes and weights vary according to application.

Operational status

All RF-4E reconnaissance aircraft are fitted with the J/APR-5. The J/APR-6 equips the F-4EJ Kai aircraft.

Contractor

Tokimec Inc.

VERIFIED

The J/APR-6 is installed in Japan Air Self-Defence Force F-4EJ Kai aircraft

NORWAY

Samover jammer

Samover is a pod-mounted programmable responsive noise jammer operating in the 8 to 16 GHz band. It is fitted to F-5 aircraft to provide standoff jamming for F-16s launching Penguin Mk 3 anti-ship missiles.

Samover automatically identifies and responds to threats via a multiprocessor control system and provides CW, ICW and pulse signals. It is a VCO-base, liquid cooled pod which can be integrated with a radar warning receiver and chaff and flares, if required for self-protection.

Specifications

Dimensions: 2,750 long × 254 mm diameter
Weight: 150 kg

Operational status

In service with Royal Norwegian Air Force F-5 aircraft.

Contractor

Kongsberg Gruppen AS.

VERIFIED

RUSSIAN FEDERATION AND ASSOCIATED STATES (CIS)

20 SP M-01 airborne flare dispenser system

The 20 SP M-01 flare dispenser system provides protection against infrared homing missiles. It has the following modes of operation:

(1) automatic: the system automatically selects the optimum flare dispense programme, responds to the commands of the EW suite and allows the aircrew to concentrate on their primary mission;

(2) manual: pre-programmed dispense programmes are available for manual selection and activation by the aircrew;

20 SP M-01 airborne flare dispenser system
1998/0018280

(3) accelerated: employment takes place at the maximum dispense rate, on aircrew command – emergency ejection.

The system provides multiple firing pulses to enable single, double, triple and quadruple payload dispensing when required. Emergency ejection is also available.

System architecture comprises: a program loading unit; a control unit; two switching units; and four dispenser magazines.

Specifications

Maximum payload: 120 flare cartridges
Salvo length: 1-4 flares
Salvo spacing: 0.01-10.0 s
Power: 27 V DC, 100 W

Opertional status

Fitted to MiG-29 and Su-27 aircraft.

Contractor

Aviaavtomatika.
Joint Stock Company PRIBOR.

NEW ENTRY

Irtysh EW system on Su-39 Strike Shield

The Central Scientific Institute for Radiotechnical Measurements TSNIITI, Omsk is the system designer for the complete EW system - known as Irtysh - on the Su-39 Strike Shield aircraft (also known as Su-25TM). Data from the following elements of the system is displayed on both the Head-Down Displays (HDDs) and the Head-Up Display (HUD).

Pastil radar warning receiver

The Pastil RWR covers radio frequencies from 1.2 to 18 GHz, with the ability to intercept pulse, pulse Doppler, and continuous wave signals. It can operate in a stand-alone mode or be integrated with the electronic countermeasures system. It has antennas in the front and side of both wing tip fairings, and in the tail-sting of the aircraft.

Su-39 starboard wing view, showing the Omul ECM pod on the outer weapon station, and the Pastil RWR antennas on the forward end and side of the wing tip fairing. The port wing carries an identical installation
1998/0018334

Omul MSP-25 Electronic CounterMeasures ECM pod

The ECM system is located in the two pods on the outer weapon stations of both wings. It is said to 'cover the necessary radio frequency bands to counter expected threats', and to provide essentially 360° angular coverage, except for a 15° half angle cone each side of the normal to the aircraft centreline. The configuration comprises two identical pods, both of which have receive and transmit capability fore and aft; the operational configuration being to receive on one pod and to transmit on the other to overcome isolation problems. The system is reported to provide both noise and deception countermeasures, including range, angle and velocity gate pull-off.

Shokogruz InfraRed CounterMeasures IRCM

The main IRCM system, known as Shokogruz, is an active jammer mounted in the tail-sting. It is a modulated IR power source, which is claimed to protect the engines at all thrust levels up to 95 per cent. Above 95 per cent thrust, flares are used to augment the system; however, a new active jammer called SNOP is being developed to cater for the higher thrust level requirement. The IR flare dispenser system, known as UV-26, holds a total of 192 flares.

Operational status

Installed in the Su-39.

Contractor

Central Scientific Institute for Radiotechnical Measurements TSNIITI, Omsk (possibly also known as Omskavtomatiki).

NEW ENTRY

Su-39 tail view, showing the Shokogruz active infrared countermeasures system, and above it the rear-facing antennas of the Pastil RWR
1998/0018335

Su-39 Strike Shield aircraft, showing: Shkval EO sighting system in the nose, Kopyo radar pod under belly, Omnl ECM pod under each wing, Pastil RWR on each wingtip
1998/0018336

Airborne ARM control/ESM systems

Avtomatiki is offering the ATsU-1 and ATsU-2 pods, which detect emitter activity, and generate target designation data for Anti-Radiation Missile (ARM) launch.

The ATsU-1 pod is described as being capable of processing pulse and CW data in one system-specific frequency band, whilst the ATsU-2 pod is reported to handle pulse radars in one specific band, and 'quasi-CW' emitters in a second frequency band.

Both systems are quoted as being able to respond to 10 separate emitters and to cue two ARMs at a time.

The ATsU-2 pod has two antennas (one on each side of the pod), whilst the ATsU-1 pod has only a single antenna array.

Avtomatiki, also produces an associated wide-angle SRR system (presumably an ESM) to cue the ATsU pods. The SRR equipment is reported to cover 2 to 18 GHz and ±50°.

Specifications

Weight: each pod weighs 216 kg
Angular coverage:
(azimuth) ±32°
(elevation) 0 to −16°
Azimuth accuracy: ±1°
Ranging accuracy (in co-ordination with the aircraft navigation system): ±10%

Contractor

CKB Avtomatiki, Omsk.

VERIFIED

SPO series radar warning receivers

The SPO-10 is an analogue radar warning receiver, which covers H/J-bands, and is fitted to MiG-21SM 'Fishbed J' and MiG-29.

The SPO-15 RWR covers the frequency band 5 to 10 GHz.

The SPO-23 is the latest development RWR for the CKB Avtomatiki, Omsk design bureau. SPO-23 is applicable to MiG-21, -23, -29 and Su-22 aircraft, and to the Ka-50 and Ka-52 attack helicopters.

SPO-23 can be configured to cover 4 to 11 GHz or 4 to 18 GHz. The antenna system is matched to the application. Ultra-broadband cones are used for helicopter applications, whilst a Broadband Beam Length (BBL) array, comprising conformal radome units, each with four or five outputs, to provide adequate angular resolution to match targeting requirements for the Kh-25P and Kh-31P anti-radiation missiles. A threat library of 128 emitters is incorporated. Weight is 18 to 30 kg depending on configuration.

Operational status

In service.

Contractor

CKB Avtomatiki, Omsk.

VERIFIED

L-166B1A airborne fixed source IR jammer

The L-166B1A fixed-source infrared jammer is designed for helicopter application on Mi-8MT/Mi-17, Mi-24, Mi-25 and similar foreign helicopter types. The system uses an IR radiation source and mechanical modulator. It is claimed to offer protection against IR guided missiles such as Sidewinder, Mica, Strela, Redeye, and Chaparral. System weight is said to be no more than 20 kg and power consumption 2.8 kW.

Contractor

Elers-Electron Ltd.

VERIFIED

Gorizont chaff/flare dispensers

UP-P1, UV-P2 and L-028K airborne chaff dispensers fitted to aircraft such as Tu-22, Il-76, Su-24 and Su-27.

APP-50MR and APP-50MA chaff/flare decoy launchers are fitted to a variety of aircraft including: An-22, Tu-160, Su-24 and Su-27.

Operational status

In service.

Contractor

Gorizont.

VERIFIED

Airborne radar jammers — Gardeniya, Schmalta and Sorbtsiya

Gardeniya, Schmalta and Sorbtsiya are all airborne jammers, that are believed to belong to the same family.

The Gardeniya family includes both pod-mounted and internal radar jammers, that can operate autonomously and automatically. Family capabilities are reported to include: self-protection; noise jamming in the 10, 20 and 70 cm wavebands; and communications jamming (Gardeniya IFUE — fitted to Mi-17P). Internally mounted versions of the Gardeniya family are reported to be fitted to MiG-29 aircraft.

Sorbtsiya (also reported as Sorbtsiya-S) wide spectrum jamming pod has been shown fitted in wing tip pods on Su-271B, naval Su-27 'Flanker', Su-34 and the Su-35 advanced 'Flanker' derivative.

Operational status

Many systems are in service; it is probable that development continues.

Design Bureau

Pleshakov Scientific & Industrial Corp (GosCNRTI), Moscow.

VERIFIED

YB-3A flare dispenser 32 × 26 mm cartridge dispenser unit (Paul Jackson) **1998**/0018301

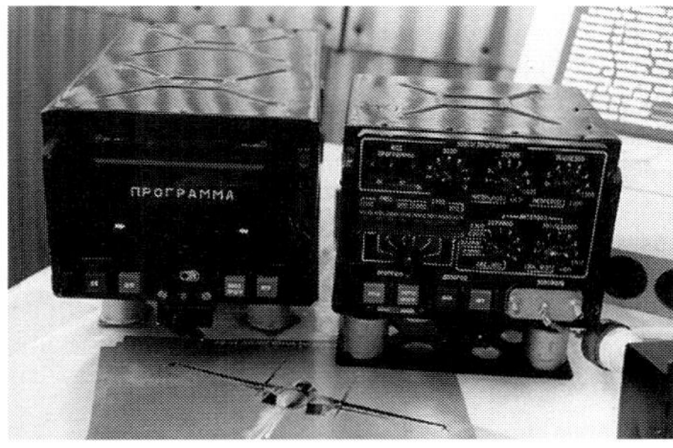

YB-3A flare dispenser double program selector unit (Paul Jackson) **1998**/0018303

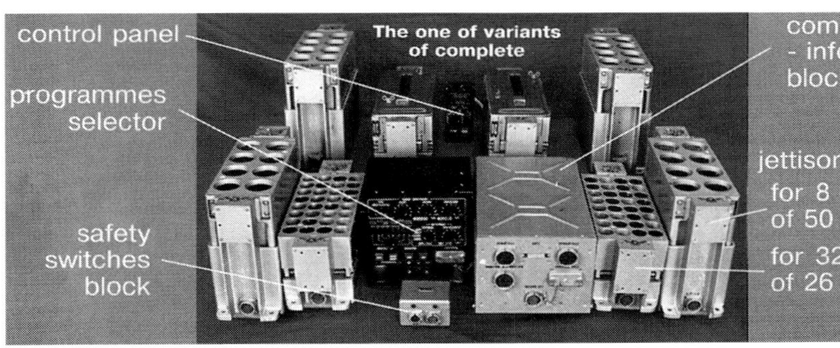

YB-3A flare dispenser - typical configuration **1998**/0018305

YB-3A flare dispenser system

The YB-3A flare dispenser is a modular design, that offers a large number of hardware configuration options, and considerable programming flexibility.

The basic configuration comprises four electronic line replaceable units: the control panel, a program selector (double unit), a computer unit, and a safety unit. Two types of dispenser units are available, capable of dispensing 8 × 50 mm calibre cartridges and 32 × 26 mm calibre cartridges respectively. The control system is capable of addressing up to 512 flare locations, contained in multiple dispenser unit configurations.

Control capabilities include:

(1) 50,000+ flexible random interval programming options
(2) 8 preliminary installed programs, with flexible selection in flight
(3) 5 variable program parameters
(4) 1-8 cartridges in salvo
(5) 0.025-16 seconds interval between salvos
(6) 18 minutes maximum program duration, or until 'all gone'

Sorbtsiya on an Su-34

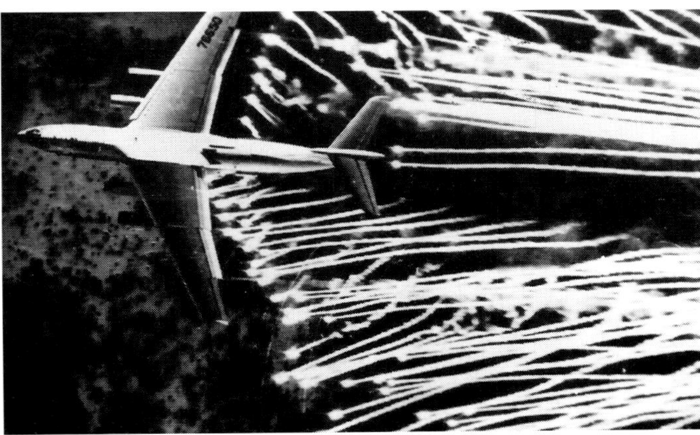

YB-3A emergency ejection of all flares **1998**/0018299

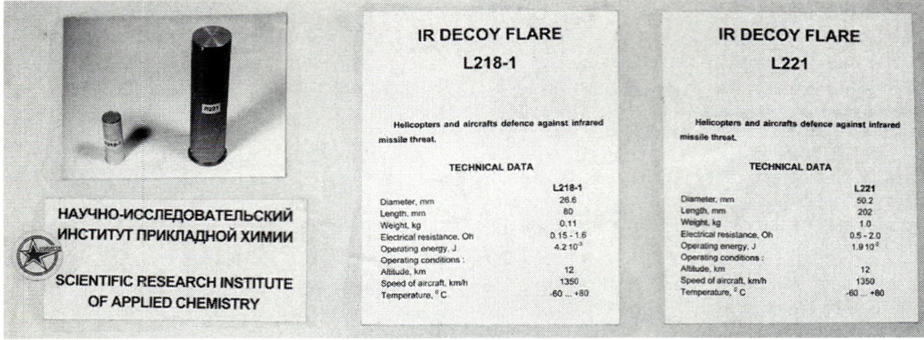

L-218-1 and L221 flare cartridges for YB-3A flare dispenser (Paul Jackson) ***1998***/0018300

YB-3A flare dispenser control panel (Paul Jackson) ***1998***/0018304

(7) 8 - 512 flare payload
(8) one year programmed life
(9) emergency ejection of all flares

The control panel provides the following capabilities: program loading, fire/stop control, payload remaining indication, rapid fire and built-in-test control, variable brightness display.

The program selector provides the following features: automatic operation, random intervalometer capability, one year programmed life.

Specifications

Dimensions and weights:
(cockpit control panel) 164 × 64 × 104 mm; 0.5 kg
(program selector) 186 × 221 × 212 mm; 5.5 kg
(computer unit) 172 × 373 × 214 mm; 9.8 kg
(safety unit) 94 × 90 × 66 mm; 0.35 kg
(dispenser units) 130 × 384 × 172 mm for the 32 × 26 mm cartridge dispenser; 7.4 kg
130 × 384 × 285 mm for the 8 × 50 mm cartridge dispenser; 8.5 kg
Power: 24-30 V DC, 150 W; 115 V AC, 400 Hz, 200 VA
Cartridges: Developed by the Scientific Research Institute of Applied Chemistry.

	L218-1	L-221
Calibre	26.6 mm	50.2 mm
Length	80 mm	202 mm
Weight	0.11 kg	1.0 kg
Aircraft speed (max)	1,350 km/h	1,350 km/h
Dispenser barrel life	700 bursts	250 bursts

Operational status

Designed for both aircraft and helicopter application; widely deployed.

Contractor

Vympel State Machine Building Design Bureau.

NEW ENTRY

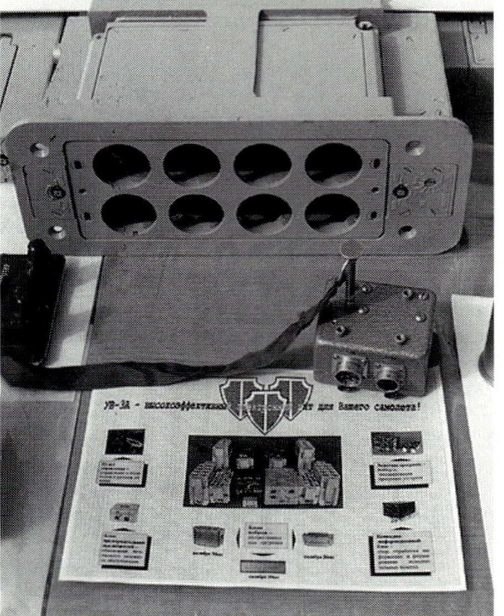

YB-3A flare dispenser 8 × 50 mm cartridge dispenser unit (Paul Jackson) ***1998***/0018302

Zenit L166VIAZ infrared jammer

Fitted to Mi-24 helicopters since 1983. Two basic variants exist, UZV-1 and UZV-2 reported to be similar to Sanders ALQ-144

Operational status

In service.

Contractor

Zenit Foreign Trade Firm State Enterprises p/c SA Zverev Krasnogorsky Zavad.

UPDATED

ASO series expendables

The ASO-2, developed from the ASO-21 fitted to the Yak-28PP 'Brewer-E' and MiG-21, was originally fitted to the Mi-24 'Hind' helicopter but proved unsatisfactory since the helicopter's slow speed and the range of the flares was insufficient to provide a reasonable separation between aircraft and flares. An improved version, believed to be the ASO-2bis, was fitted to helicopter fuselages, firing forward of the centreline in two banks of three launchers. This version has been seen on Mi-24/Mi-25 helicopters.

A further development, the ASO-3, has been seen on the Mi-8MT Hip helicopter, as well as on other combat helicopters. It features a neater installation and offers much better protection to the electronic components. The ASO-3 has a more powerful rocket discharge system and has been fitted extensively to a number of combat aircraft, including the MiG-27 and MiG-29, where it may also have the designation BVP-30-26M.

Operational status

In service.

VERIFIED

LIP missile approach warner

LIP is a microwave Doppler missile approach warning system. In its original fit, on board Mi-24 helicopters serving in Afghanistan, a single set was installed to cover the rear and underneath of the helicopter. This coverage was inadequate since it provided no warning of missiles approaching from the forward arc or fired from high ground down on to the helicopter. The most recent installation, on the Ka-29 marine assault helicopter, has a second antenna covering forward arcs.

LIP antennas have also been observed on board the Su-25 Frogfoot ground attack aircraft. In this case, a single antenna is fitted in the tailcone directly underneath the Sirena 3 radar warning system.

Operational status

In service.

VERIFIED

R-949 communications jammer

The R-949 is fitted to Mi-8SMV specialist EW 'Hip' helicopters. It is reported to have four transmitters covering the communications band.

Operational status

In service.

VERIFIED

Sirena-3 radar warning receiver

Sirena-3 has been the standard radar warning receiver known to NATO fitted to aircraft of the erstwhile Soviet bloc. The relationship of Sirena to the SG-1 RWR, or SPO series RWRs described in separate entries is not clear.

Operational status

In service.

VERIFIED

SPS series radar jammers

The SPS series of jammers includes a large number of systems, fitted to a wide variety of aircraft, in both dedicated EW roles and self-protection fits.

SPS-5-28, known as Fasol, is fitted to the specialist EW aircraft: Su-24MP 'Fencer-F' and Tu-22P 'Blinder-E'.

SPS-22-28 and SPS-44-29 are obsolescent/obsolete systems fitted to Yak-28PP 'Brewer-E' and Tu-16PP 'Badger-J' EW aircraft.

SPS-63, -66, -68 form part of the Mi-8PPA 'Hip-K' battlefield EW helicopter.

SPS-130, -140 Gvozdika jammers were first-generation systems fitted to MiG-27 and MiG-25 aircraft.

SPS-150 Lyutik self-protection jammers fitted to MiG-25RB 'Foxbat-B'.

SPS-160 Geran family of self-protection jammers fitted to Su-24 and Su-27.

SPS-170 family of self-protection jammers scheduled for installation in Su-34, -35 aircraft (may include Sorbtsiya-S).

Operational status

In service.

VERIFIED

SRO/SRZO series IFF transponder/interrogator

The SRO/SRZO series IFF transponder/interrogator is standard fit in RFAS aircraft. A large number of versions exist.

Operational status
In service.

VERIFIED

SRS series airborne electronic intelligence equipment

Little information is available on the SRS series electronic intelligence equipment, which is fitted to MiG-25 specialist aircraft.

Operational status
In service.

VERIFIED

SOUTH AFRICA

CFD-300 Chaff and Flare Dispensing (CFD) systems

The CFD-300 series of Chaff and Flare Dispensing (CFD) systems provides self-protection against radar and infrared guided weapons by ejecting chaff or flare cartridges. The series is available in integrated (CFD-300) or stand-alone (CFD-310) configurations. It uses NATO standard 25 × 25 mm chaff and 25 × 25 mm or 50 × 25 mm flare cartridges. Chaff and flares can be fired in combination or simultaneously.

Firing sequence parameters, such as delay time, burst time and numbers, subsalvo time and numbers, salvo time with numbers and mode, are user programmable. Firing sequence programmes can be downloaded into the system non-volatile memory on the flight line using the preflight data downloader. The system has a feedback loop that verifies that a cartridge has been fired and performs a 'remaining stores' count for both chaff and flares that is separately available for selection on display. Built-in safety features prevent inadvertent firing while the aircraft is on the ground and the CFD-300 has extensive BIT features.

The integrated CFD-300 configuration interfaces with a radar warning system or multisensor warning system for control, display and automatic response. A separate cockpit control unit is not required. Firing may be completely automatic, semi-automatic or manual. Additional aircraft parameters, such as speed and altitude, may also be processed, allowing real-time modification of dispensing sequences for changing flight profiles.

The stand-alone CFD-310 utilises a cockpit control unit for the control, display and selection of preprogrammed firing sequences.

Specifications

Dimensions:
(cockpit control panel) 74 × 132 × 157 mm
(control/dispenser electronic unit) 136 × 101 × 252 mm
(single dispenser block housing) 130 × 228 × 71 mm
(chaff dispenser block) 121 × 178 × 228 mm
(flare dispenser block) 121 × 178 × 226 mm
(cassette) 35 × 288 × 45 mm
(chaff or flare magazine) 31 × 302 × 210 mm

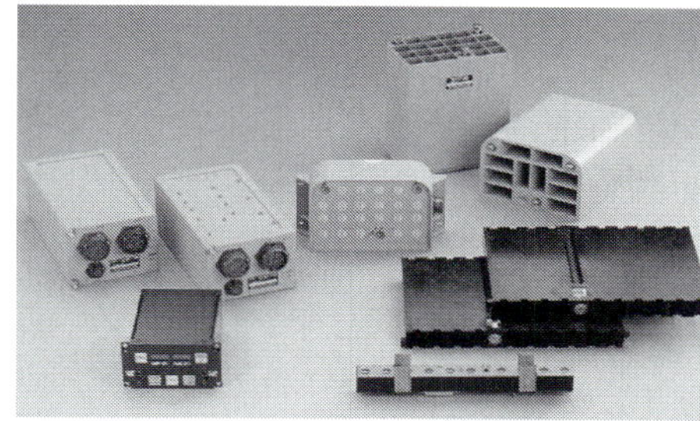

The CFD-300 Chaff and Flare Dispensing system

Weight:
(cockpit control unit) 0.7 kg
(control/dispenser electronics unit) 4 kg
(single dispenser block housing) 1.4 kg
(chaff dispenser block) 4.9 kg
(flare dispenser block) 6.3 kg
(cassette) 0.4 kg
(chaff magazine) 2.3 kg
(flare magazine) 3 kg

Contractor

Grintek Avitronics, Grintek Electronics Limited.

UPDATED

ELS Emitter Location System

Accurate direction-finding is important to ESM systems for the geo-location of emitters and for directing jammers and weapons. Integrated with the MSWS multi-sensor warning system, the ELS provides the high accuracy DF required for these tasks.

The ELS consists of an integrated receiver and controller and a number of antenna arrays, dependant on the system requirements. The main features of the system are: low mass and volume by using a single channel switched receiver; intra-pulse channel switching for a single pulse DF capability; high DF accuracy using a combination of phase and amplitude comparison technique; pulse Doppler handling capacity; high sensitivity.

The ELS is designed as an integral part of MSWS to enhance the DF capability of the radar warning function. All display and control is via the host system. The ELS functions are: accurate bearings for emitters designated from the host system; frequency measurement for designated emitters; detection of low probability of intercept emitters in an autonomous mode; gathering of emitter data for ESM/ELINT analysis.

ELS emitter system **1997**/0001261

Specifications

Dimensions:
(controller) 193 × 127 × 343 mm
(hi band array) 110 × 90 × 250 mm
(lo band array) 180 × 80 × 500 mm
Mass:
(controller) 15 kg
(hi band array) 5 kg
(lo band array) 7 kg
Frequency coverage: 0.5-2 Ghz (lo band); 2-18 Ghz (hi band)
Frequency resolution: 2 MHz in narrowband mode
Instantaneous bandwidth: 1 GHz/80 MHz
Direction finding: 1° RMS
Field of view: ±35° in azimuth per 2 antenna array (±45° with reduced DF accuracy); ±70° in azimuth per 3 antenna array

Contractor

Grintek Avitronics, Grintek Electronics Limited.

UPDATED

LWS-200 laser warning system

The LWS-200 consists of four or six advanced laser warning sensors, an analyser and a multifunction display unit. The laser sensors detect direct incidence on the detectors, providing directional information, and scattered laser energy when only omnidirectional information will be available. These sensors pass the detected information to the analyser. The processed data is compared to the preprogrammed threat library, for threat identification and subsequent display using defined symbols, on a multi-function colour display.

Specifications

Dimensions:
(analyser) 79 × 105 × 146 mm
(laser warning sensor) 64 × 86 × 103 mm
(threat display/control unit) 130 × 146.5 × 113.5 mm

The LWS-200 showing (rear) the laser warning analyser (left) and threat display/control unit (right) and (front) four laser warning sensors

Weight:
(analyser) 1.2 kg
(laser warning sensor) 0.7 kg
(threat display /control unit) 1.1 kg
Wavelength: 0.6-1.8 μm - extended range also available
Threats: Ruby, GaAs, Nd:YAG, Raman shifted lasers
Coverage:
(azimuth) 360° (180° per sensor)
(elevation) −40 to +20°
Sensitivity: 20 W/m
Azimuth AOA accuracy: ±6° RMS

Contractor

Grintek Avitronics, Grintek Electronics Limited.

UPDATED

MAW-200 Missile Approach Warning system

MAW-200 Missile Approach Warning system is an Integrated or stand-alone Missile Approach Warning system. It is a totally passive system which provides detection and timely warning of the approach of suface-to-air and air-to-air missiles. Upon positive detection of the approaching missile, a priority interface to the chaff and flare dispensing system is activated for immediate and automatic dispensing of countermeasures against the threat. A visual Direction-Finding (DF) warning is provided to the aircrew via a EW suite's display unit (or in the case of a stand-alone system the MAW's display unit) accompanied by the appropriate audio alarms.

The MAW subsystem consists of four sensors and a processing card which resides in the EW controller of the EW suite. On the stand-alone version, processing is done in a dedicated processor unit. Each sensor is responsible for processing its detection algorithms. The processing card inside the EW contraller of the EW suite is responsible for the further processing of data received from the various sensors and for the built-in test control and management of the MAW sensors.

Main features include: totally passive operation; low false alarm rate; no in-flight recalibration required; no cooling requirec; instantly on-line, no cooling time required; comprehensive self-test routines.

MAW-200 stand-alone system **1997**/0001260

The MSWS multisensor warning system covers a wide range of threats **1997**/0001262

Specifications

Sensor dimensions: 130 × 164 × 190 mm
Sensor weight: 2 kg
Detection method: passive ultra-violet
Detection range: >5 km for shoulder-launched missiles
Spatial FOV per sensor: 90° azimuth, 60° elevation
DF resolution (azimuth): 10°

Contractor

Grintek Avitronics, Grintek Electronics Limited.

UPDATED

MSWS MultiSensor Warning System

The MultiSensor Warning System (MSWS) provides tactical aircraft with a complete warning capability for self-protection. The capability includes radar warning, laser warning, and missile approach warning. The architecture provides for a variety of sensors to be integrated into and managed by the system, allowing the user to upgrade the system.

Generic design and low unit count allow easy installation in aircraft ranging from helicopters to fighters. Complete spherical coverage is available and the system provides full threat identification. Threat identification parameters are user definable. The MSWS-240 is flight line programmable and includes extensive BIT facilities.

The system includes a radar warning function for pulse Doppler and CW radars in high pulse density environments, a man/machine interface via a multifunction display and interface to and control of automatic chaff and flare dispensing systems.

The RWR features an Instantaneous Frequency Measurement (IFM) receiver, covering the 2 to 18 GHz band in 4 GHz steps. It is reported to be able to cope with pulse densities up to 2 Mpps and to display worst situation threats within 500ms, using 32-bit parallel processors.

The LWR is reported to cover 0.5 to 12 μm wavelengths, offering detection capability against laser rangefinders, designators and missile guidance lasers, providing both threat classification and bearing.

The MAW uses ultraviolet detection techniques and is said to typically provide 5,000 m warning of shoulder-launched missiles.

The standard configuration comprises four sensor heads for each of the RWR, LWR and MAW functions. The display shows the nature and status of received signals, together with relative bearing, lethality and tabulated parametric date. Audio warning is provided to alert the crew to display data.

The system can be integrated with recording facilities for use in the Intelligence gathering role.

Growth options include an interface with an active ECM system activated automatically on threat detection, avionic system interface via a MIL-STD-1553B bus.

Specifications

Dimensions:
(multisensor warning analyser) 193 × 127 × 374 mm
(threat display and control unit) 144 × 165 × 140 mm
(front end receiver ×4) 158 × 45 × 176 mm
(spiral antenna ×4) 110 × 110 × 67.5 mm
(laser warning sensor ×4) 64 × 86 × 103 mm
Weight:
(multisensor warning analyser) 15 kg
(threat display and control unit) 1.7 kg
(front end receiver ×4) 2 kg
(spiral antenna ×4) 0.7 kg
(laser warning sensor ×4) 0.7 kg
Frequency (radar):
(pulse) 0.7-40 GHz
(CW) 0.7-18 GHz
Wavelength:
(laser) 0.5-12 μm
Threats: ruby, GaAs, NdYAG, Raman shifted lasers
Coverage:
(radar) 360° (azimuth), ±45° (elevation)
(laser) 360° (azimuth), +20 to −40° (elevation)

Contractor

Grintek Avitronics, Grintek Electronics Limited.

UPDATED

The RWS-50 radar warning system **1995**

RWS-50 radar warning system

The RWS-50 system provides tactical aircraft with a comprehensive radar warning capability for self-protection. This capability can be extended to include laser and missile approach warning.

In its basic configuration, the RWS-50 consists of four 2 to 18 GHz spiral antennas, two-dual detector amplifiers, an analyser unit and a colour multifunction display and control unit. This configuration includes an interface for automatic or manual control of a chaff and flare dispensing system.

The RWS-50 features a versatile threat library, flexible architecture, parallel processing, high sensitivity, high probability of intercept/low cycle time and low power consumption.

Upgrade options include; 0.7-1.4 GHz detection capability (omni or full DF); CW detection capability (omni or full DF); extended RF range 0.7-40 GHz (in one antenna); frequency measurements (via external superheterodyne/IFM subsystem); increased sensitivity/dynamic range; MIL-STD-1553B interface; LWR capability; MAW interface; spherical coverage.

Specifications

Dimensions:
(analyser) 193 × 127 × 374 mm
(threat display and control unit) 144 × 165 × 140 mm
(dual-detector amplifier) 158 × 45 × 176 mm
(spiral antenna) 70 × 70 × 110 mm
Weight:
(analyser) 15 kg
(threat display and control unit) 1.7 kg
(dual-detector amplifier ×2) 1.5 kg
(spiral antenna ×4) 0.4 kg
Power supply: 28 V DC, 50 W
Frequency: 2-18 GHz
Coverage:
(azimuth) 360°
(elevation) ±45°
Accuracy: 15° RMS

Contractor

Grintek Avitronics, Grintek Electronics Limited.

UPDATED

GSY1500 VHF/UHF communications jamming system

The main purpose of the GSY1500 VHF/UHF jamming system is the disruption and/or jamming of communications channels and emissions in the 20-500 MHz frequency band. The GSY1500 is modular and can be mounted in airborne platforms, such as helicopters and transport aircraft, or land-based or shipborne platfoms.

Jamming features include look-through capability to monitor continued target presence, selectable optimised counter modulation types for various target signals, time-division multiplex jamming or frequency division multiplex jamming, jamming of up to 20 prioritised target frequencies, effectiveness against fixed-frequency voice and data communication links, selection of different output power settings and suppression of unwanted harmonics and spurious signals.

GSY1500 V/UHF communications jamming system mounted on a helicopter with 100-500 MHz log-periodic dipole array antenna ***1998***/0018278

The GSY1501 airborne communications EW system can be expanded to accommodate up to 15 operators ***1995***

Intercept features include rapid scanning of the entire frequency band, noise-riding signal detection with digital signal processing techniques, high sensitivity with ranges of up to 450 km, demodulation facilities for both AM and FM modulation and rapid look-through monitoring and detection during jamming cycles.

Control features include computer control for high system flexibility, display of system status and detected activities, real-time operator interaction and short system reaction time, prediction of jamming effectiveness based on an analysis of ground and air communication links, pretasking of the system via a floppy disk, programmable safeguards for own force signals, mission log and analysis on hard disk, comprehensive BIT and software flexibility to accommodate specific user requirements.

The GSY1500 consists of a wideband fast setting receiver, graphical display and computer unit with keyboard, countermeasure generator unit, fast setting RF synthesiser, power amplifier with transmit/receive switches and harmonic filters, audio intercom panel and RF filter unit.

Specifications

Dimensions:
(control console) 700 × 820 × 1,700 mm
(amplifier console) 580 × 650 × 1,110 mm
Weight: <500 kg, depending on options
Power supply: 115 V AC, 400 Hz, 3 phase
or 220 V AC, 50 Hz, single phase
28 V DC
Frequency: 20-100 MHz or 100-500 MHz or both
Output power:
(20-400 MHz) 500 W or 1,000 W options
(400-500 MHz) 400 W or 800 W options
Scanning speed: 80 channels/s
System reaction time: ≤10 ms
Number of pretasked channels: 20

Operational status

In service with aircraft of the Air Force of the Republic of South Africa.

Contractor

Grintek System Technologies, Grintek Electronics Limited.

UPDATED

GSY1501 airborne communications EW system

The GSY1501 is a comprehensive airborne communications EW system ideally suited for installation on passenger aircraft, cargo carriers, business jets or similar types. The system provides a complete capability to enable the detection, interception, direction-finding, recording and disruption by jamming or deception of enemy command and control communication networks. It provides coverage of the frequency spectrum from 20 to 1,000 MHz and can be extended to operate from 1.5 MHz. Because of its modular approach in design, the GSY1501 can be adapted to accommodate up to 15 operators and additional capabilities can be added as required.

The GSY1501 consists of a number of subsystems as detailed on following page.

RF reception and distribution
A 20 to 500 MHz and 500 to 1,000 MHz blade antenna with omnidirectional response in the azimuth plane provides reception over the full frequency range. An antenna distribution unit feeds received RF signals to the receivers in the system and a blanking unit protects the antennas during active transmission.

Spectrum scanning
Scanning receivers perform scanning of the spectrum in the band selected by the operators.

Direction-finding
A seven-channel interferometer direction-finder determines the bearing of a signal in the 20 to 500 MHz frequency band or, optionally, up to 1,000 MHz. The DF is fed by an antenna array with wide aperture to provide high accuracy.

System control
ESM and ECM system control units consist of a number of processors which control the scanning receivers and direction-finder, power amplifiers, synthesisers and counter modulation generators.

Operator workstations
The number of workstations is configurable, depending on the available space and platform limitations. A workstation consists of a monitoring receiver, digital voice recorder, spectral display unit, operator interface unit and intercom control unit. The operator interface is an environmentally hardened personal computer with a colour flat panel display and conventional PC keyboard.

Jamming
The system consists of 30 to 100 MHz and 100 to 500 MHz high-power antennas capable of transmitting more than 1 kW of RF power, RF jamming synthesisers each having three independent RF channels capable of AM or FM modulation, counter modulation generators which generate the baseband signals from a digital source and are used as modulation sources for the jamming signal and power amplifiers covering the 20 to 100 MHz and 100 to 500 MHz bands which are capable of generating up to 1 kW of RF power.

Specifications
Frequency:
(ESM) 20-1,000 MHz (1.5-30 MHz option)
(ECM) 30-500 MHz
Scan rate:
(1.5-30 MHz) 0.3 MHz/s
(20-1,000 MHz) 4-16 GHz/s
Accuracy: ±1.5° RMS
DF agility: 40 bearings/s

Operational status
In service with aircraft of the Air Force of the Republic of South Africa.

Contractor
Grintek System Technologies, Grintek Electronics Limited.

UPDATED

SPAIN

ELIOS ELINT system

ELIOS (ELINT Identification and Operating System) is a powerful tool for processing and utilising an ELINT database and is complementary to other ELINT systems. It is capable of a number of functions, ranging from gathering and processing data to mapping processes. It provides fast identification, electronic database management and library loading. The main characteristics of ELIOS are a library of threats with a high-powered capacity for response and reaction, tracking and repetition of missions on the ground, and the mapping and position-fixing of threats.

ELIOS simplifies mission preparation, records collected data and allows detected emissions to be compared to memory files. It increases the speed of management, allows the system operator to access all databases and allows tailoring of the electronic database format. High security of data is provided by double protection against loss, software inviolability and cryptographing of files.

The system is supplied in rugged hardware and can be interfaced and installed in a variety of land, sea and air platforms.

Operational status
Fully developed.

Contractor
ELT SA.

VERIFIED

NIDJAM jammer

The Navigation/Identification Deception Jammer (NIDJAM) operates over the frequency range 950-1,250 MHz in the band used mainly for Tacan, DME and IFF systems.

The NIDJAM superheterodyne receiver detects emissions over the band in continuous and discrete scan modes. It controls the activity in a set of previously selected frequencies by means of the discrete scan and identifies nav/ident systems and their modes of operation. Deception signals are selected from a series of preset signals or fixed patterns, although random patterns may be employed.

Scanning sensitivity is better than −88 dB and scan speed is up to 1,000 channels/s. The radiated power in the jammer assembly exceeds 300 W CW per channel. Blade type monopole antennas are used for scanning and jamming in airborne applications.

Specifications
Frequency: 950-1,250 MHz
Sensitivity: −88 dBm
Scan speed: up to 1,000 channels/s
Radiated power: 300 W CW per channel

Operational status
Fully developed.

Contractor
ELT SA.

VERIFIED

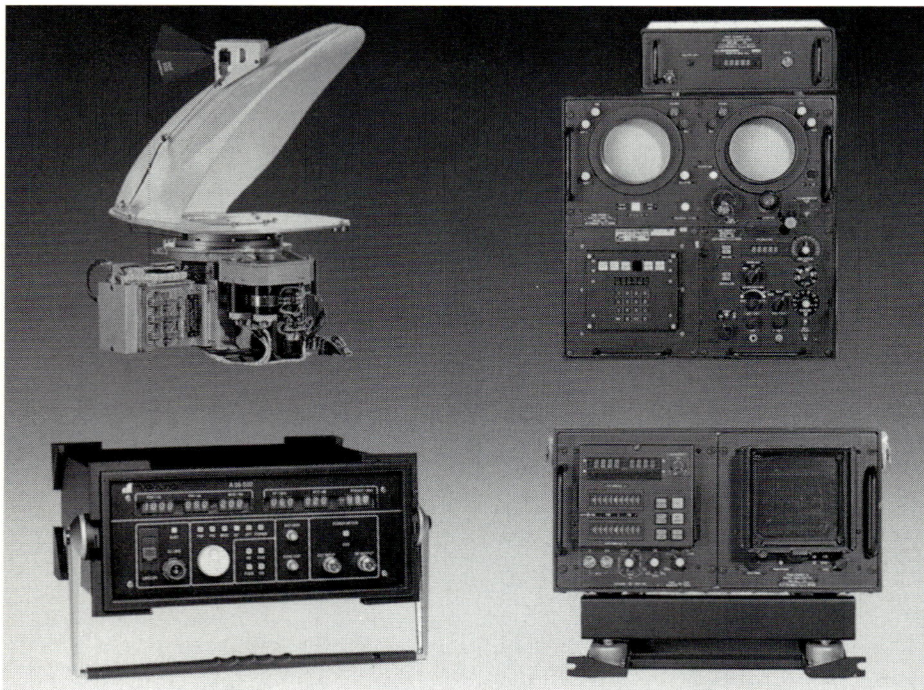

The Signal Identification Mobile System

Signal Identification Mobile System (SIMS)

The Signal Identification Mobile System (SIMS) is configured for airborne, shipborne and land-based applications. It allows detection, direction-finding, analysis and library storage over the 1-18 GHz frequency range and transfers all data to remotely located sites through a built-in datalink.

SIMS consists of an antenna unit, RF unit, direction-finding and panoramic signal displays, a computerised system controller, video analyser and peripherals.

Specifications
Power supply: 115 V AC, 400 Hz
Frequency: 1-18 GHz pulse or CW signals
Accuracy:
±4° (1-2 GHz)
±3° (2-4 GHz)
±2° (4-18 GHz)

Contractor
ELT SA.

VERIFIED

SOCCAM COMINT system

The modular communication observation and control system (SOCCAM) is configured for airborne, shipborne and land-based applications. It is a COMINT system for tactical and strategic missions and spectrum control over the frequency range 20-500 MHz.

SOCCAM provides functions for scanning, searching and detecting active transmissions. The system detects activity in a series of discrete bands and analyses intercepted signals to allow the operator to determine transmission characteristics and store them for subsequent analysis.

SOCCAM has two different operating modes: operation in an unknown scenario when the system searches for active transmissions in the mission area, and operation in a known scenario, where the operator has prior knowledge of threats in the area and seeks to locate and monitor them.

Specifications

Frequency: 20-500 MHz
Sensitivity: −100 dBm
Modes: AM, FM, CW and PLS
Accuracy: 4° RMS
Interfaces: RS-422, IEEE-488

Operational status

In service.

Contractor

ELT SA.

VERIFIED

Taran airborne ESM/ECM system

Taran is a modular computer-controlled airborne EW system for both ESM and ECM in the communications and tactical navigation/identification bands. The system is intended primarily for airborne applications and can be installed on a wide range of fixed-wing aircraft and helicopters.

Two different bands can be operated: low band for communication systems and high band for navigation/identification systems. The low-band subsystem consists of a search and analysis receiver, monitoring console and jamming set. The high-band subsystem comprises a search and analysis receiver, monitoring console and a deception jammer.

The main features of Taran include the ability to operate in a dense electromagnetic environment, and to resist multiple threats simultaneously. There is a high degree of automation in order to reduce reaction time, and growth potential to cater for the evolving threat. Taran has total communication capacity with the command and control system and a simple man/machine interface and ergonomic design. The system is modular, able to adapt to the needs of different platforms and has high reliability and ease of maintenance.

The system is installed with a suitable set of consoles with interconnection routeing depending on the type of airborne platform.

Specifications

Power supply:
115 V AC, 400 Hz, 1.36 kVA (2.6 kVA with jammer operating)
28 V DC, 1.5 kW (15.2 kW with jammer operating)

Operational status

In production for the Spanish Air Force and in service on the Falcon 20.

Contractor

ELT SA.

VERIFIED

EN/ALR-300(V)1 Radar Warning Receiver

The EN/ALR-300(V)1 is a fully programmable radar warning receiver for combat aircraft and helicopters. The ALR-300 system provides detection, identification, direction finding, and location of emitters in the C to J bands.

The equipment consists of: four DF spiral antennas; four channelised crystal video receivers; a processor control unit and an azimuth indicator display with synthetic audio warning.

The system provides identification of all detected pulse and CW radar emitters and warns the pilot of threats by means of both an alphanumeric graphical CRT display and voice synthesised messages in accordance with the lethality of the detected emitters. The system includes the ability to record up to 100 emitters during flight.

The ALR-300 system includes full mission and maintenance hardware and softwre support facilities. Reprogramming on the flight line can be accomplished by use of EEPROM mission loading equipment that forms part of this support capability.

Specifications

Weight: 20 kg
Power consumption: 225 W
Frequency: C- to J-bands in five sub-bands
Coverage: 360° azimuth
Accuracy: <12° RMS

Operational status

In operation on Mirage F-1 fighter aircraft, CASA C-101 trainer aircraft, and on Chinook and Super Puma helicopters. In production for Beech aircraft.

Contractor

INDRA.

UPDATED

Receiver

D/F antenna

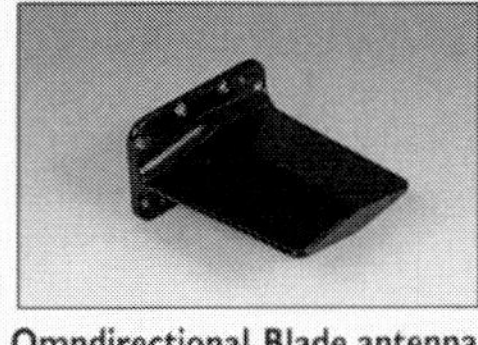

Omndirectional Blade antenna

Control unit

Display

Processor

The EN/ALR-310 is fitted in Spanish Army helicopters

EN/ALR-300(V)2 Radar Warning Receiver

The EN/ALR-300(V)2 radar warning receiver upgrades the EN/ALR-300(V)1 by the addition of a superheterodyne/DFIM receiver, a power supply, and new software to improve the equipment performance.

Radar data are presented to the operator through alphanumeric symbols on a high-brightness display. The position of the symbols on the display indicates lethality and bearing of the detected signal and identification of the radar. Symbol blinks and an audio alarm indicate high-priority radars.

Specifications

Weight: 38 kg
Power consumption: 400 W
Frequency: C- to J-bands in five sub-bands
Coverage: 360° azimuth
Accuracy: >12° RMS

Operational status

The EN/ALR-300(V)2 is the standard equipment for Spanish combat aircraft. It is in operation on the Mirage F1 fighter aircraft, and in production for Super Puma helicopters.

Contractor

INDRA.

UPDATED

EN/ALR-310 Radar Warning Receiver

The EN/ALR-310 is a fully programmable crystal video radar warning receiver which incorporates a digital computer to provide emitter identification in complex signal environments. The system incorporates a readily reprogrammable emitter library and provides unique identification of all detected pulsed and CW emitters by alphanumeric symbology on the CRT display. Up to 15 radars may be displayed simultaneously.

Specifications

Weight: 20 kg
Power consumption: 180 W
Frequency: 0.7-1.5 GHz and E/J-band, in 2 bands
Coverage: 360° azimuth
Accuracy: >15° RMS

Operational status

Selected for Spanish Army helicopters.

Contractor

INDRA.

UPDATED

SWEDEN

AR-830 threat warning system

The AR-830 is a fully programmable threat warning system incorporating a special purpose high-speed signal processor, a general purpose computer and broadband receivers for radar and laser signals. It is designed for multirole aircraft and its performance is optimised for attack, surveillance and fighter missions. The system incorporates an easily programmable signal library and is capable of rapid and unambiguous threat identification, even in complex signal environments.

Threat information is displayed on a CRT or integrated with the aircraft display systems.

The AR-830 can be interfaced to jammers, chaff and flare dispensers, recorders and other systems. BITE provides fault isolation down to LRU level.

Operational status

In production.

Contractor

CelsiusTech Electronics AB.

VERIFIED

The AR-830 threat warning system

AR 961 RWR/ESM

The AR 961 system is an advanced airborne RWR/ESM system offering excellent performance and flexibility at low cost. The AR 961 system is composed of: an antenna unit with four or six antennas for the basic 2 to 18 GHz band protected by a circular radome (additional antennas for the 28 to 40 GHz band can optionally be included); a Receiver and Signal Processing Unit (RSPU) with receivers, pulse processor and interface to cockpit display, operator's console or Data Handling System (DHS).

The receiver subsystem is based on wide open amplitude monopulse IFM receivers giving high sensitivity as well as accurate DF and pulse parameter measurements. For improved performance, superheterodyne receivers can optionally be integrated in the RSPU.

The pulse processor, the core of the AR 961, is of unique and proven design. All processing is done in real time to ensure a rapid response and that no pulses are lost. CelsiusTech's deinterleaving and emitter identification algorithms have been developed, extensively tested and refined over the past ten years.

The AR 961 system can be operated stand-alone or be tied into an integrated EW/DHS system in the aircraft. In a typical RWR application the identified threats are presented on a polar cockpit display. In the ESM application an operator's console can be provided. Graphics under Windows™ give the operator extensive opportunities to analyse real time as well as stored data.

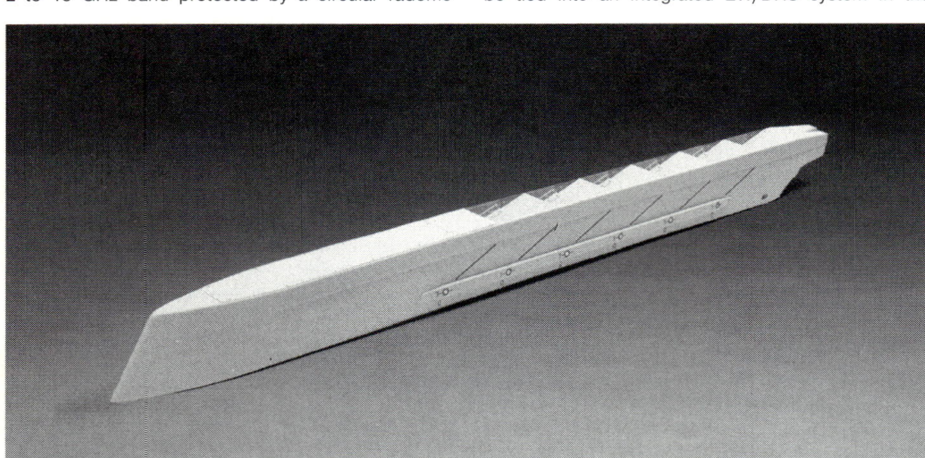

BOP/A pyrotechnical dispenser ***1997***/0001265

BOP chaff dispenser on Gripen aircraft ***1998***/0018277

Specifications

Frequency range:
2-18 GHz; 28-40 GHz (optional)
Azimuth coverage: 360°
DF accuracy:
(4 port) 7° RMS
(6 port) 3.5° RMS
RF accuracy:
(wide band) 5 MHz RMS
(narrow band) 1 MHz RMS
Dynamic range: 60 dB
Tracked emitters: 500 max
Emitter modes in library: 10,000 max
Reaction time: 1 s max
Installation:
Dimensions and weights
(Antenna Unit) Ø × H 250 × 180 mm
(Weight) 3 kg
RSPU:
(W × D × H) 256 × 387 × 194 mm (1-ATR Short)
(Weight) 16 kg
Interface: Ethernet RS 422
MIL-STD 1853 optional
Power: 3 × 116 V
400 Hz
300 W (approx)

Operational status

In service.

Contractor

CelsiusTech Electronics AB.

VERIFIED

BO 300 passive countermeasures system

The BO 300 passive countermeasures system consists of BOL chaff dispensers, BOP flare dispensers and a BOC controller module.

The BOL dispenser module is mounted in the rear of an existing missile launcher. It consists of a chaff compartment, electromechanical feed mechanism and electronics unit. Each dispenser holds 160 chaff packages. In the chaff compartment, chaff is stored in a large number of RR-183 NATO standard flat cassettes that interlock with each other.

The BOL 300 is designed to integrate with a variety of missile launchers. The LAU-7 launcher has been redesigned, to incorporate a BOL module in the rear half and a new gas bottle for cooling the missile seeker in the nose, this configuration is known as the LAU-138. The BOL 300 has also been integrated with the common rail launcher. Yet other versions have been integrated into the LAU-127, LAU-128 and LAU-129 launchers. There is also a scab-on version of BOL for use with aircraft that lack a suitable launcher. The chaff is stored in plastic frames and each BOL module can take 160 packs {US name RR-184) which are ejected from the rear and spread in the turbulent airstream. Chaff ejection is electromechanical, which is claimed to be more precise than pyrotechnic ejection as well as being safer and easier to maintain. Chaff cloud dispersion is synchronised by an onboard countermeasures computer.

The photo shows the BOL chaff dispenser installed on the US Navy's F-14 aircraft. BOL is installed inside the missile launcher which means no reduction in either weapon payload capacity or flight performance.

The BOP pyrotechnic dispenser comes in A, B and C versions for compatibility with installation configurations on most types of aircraft. BOP dispensers have been developed with an emphasis on high reliability and safety. BOP/A is mounted as an extension of a flap-jack. The dispenser is loaded with twelve 55 mm diameter standard NATO flare cartridges giving a capacity of up to 36 flares. BOP/B is scabbed-on or mounted as an extension of a pylon. It has capacity for 15 MJU-7, six MJU-10 or six standard 55 mm diameter cartridges. BOP/C is intended for internal mounting flush with the fuselage. The dispenser is loaded with 20 MJU-7 or 8 MJU-10 flares. BOP/B and C can also be loaded with 40 RR170 chaff cartridges. All BOP dispensers can be loaded with expendable jammers.

The BOC is a combined cockpit control unit and system programmer. The controller can easily be connected to different types of warners and aircraft systems by means of databusses and discrete signals. Reprogrammable non-volatile memories are used for storage of mission-specific data. It can control up to eight BOL or BOP dispensers simultaneously and can also control the most commonly occurring pyrotechnic dispensers. The BOC controller can provide adaptive processing, to optimise response in a specific threat situation and stores management, to keep an accurate record of the remaining load and give full flexibility in load configuration and optimum use of expendables.

The BO 300 system operates in automatic, semi-automatic and manual modes, depending on warning equipment, tactical situation and operator workload.

Specifications

Dimensions:
(BOP dispenser) 154 × 210 × 715 mm
(BOC control unit) 76 × 147 × 165 mm
Weight:
(4 BOP dispensers, empty) 32 kg
(4 BOP dispensers, loaded) up to 66 kg
Power supply: 115 V AC, 400 Hz, 100 VA

Operational status

BOL is in service on the following aircraft: RAF Harrier GR7 and Tornado F3; Swedish Air Force JA-37 Viggen, and US Navy F-14. BOL will also be integrated in a pylon on the EF2000 Eurofighter, and a solution is being adopted for the JAS-39 Gripen. Installations are being studied for A-4, AV-8B, Hawk, Jaguar, L-159, P-3 and Sea Harrier. Meanwhile US DoD Foreign Comparative Test (FCT) trials of BOL are being conducted using the LAU-128 MRL launcher on US Air Force F-15 and for a scab-on version on the US Navy S-3 aircraft. The US Navy has allocated funds for certification of BOL on F/A-18C/D Hornet aircraft with BOL in the Sidewinder launcher on the wing-tips of the aircraft.

Contractor

CelsiusTech Electronics AB.

UPDATED

BOH 300 ECM dispenser

The BOH 300 is a chaff/flare dispenser designed to provide protection for helicopters in hovering and forward flight. It has processor control, reprogrammable operational features and databus interface capability. The system is suitable for deployment of radar, optical and infrared decoys from attack, anti-tank or scout helicopters. Dispensers can be mounted on landing gear, fuselage structure or weapon hardpoints.

Operational status

Under development.

Contractor

CelsiusTech Electronics AB.

VERIFIED

BOZ 3 training chaff dispenser

BOZ 3 is a high-capacity unit originally developed to a Swedish Air Force requirement and since adapted for training use in view of its electromechanical design. It can be manually or automatically initiated and has both break-lock and corridor chaff modes.

Operational status

In service with several countries for training purposes.

Contractor

CelsiusTech Electronics AB.

VERIFIED

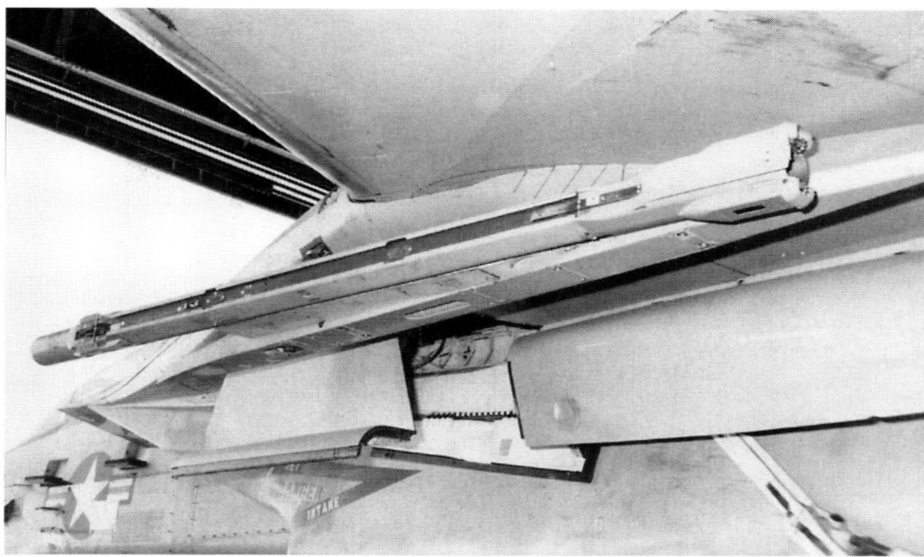

The BOL chaff dispenser installed on the US Navy's F-14 aircraft

1997/0001263

BOL chaff dispenser ***1998***/0018276

The BOZ 3 training chaff dispenser

The BOZ 100 chaff/flare dispenser for high-performance aircraft ***1997***/0001266

BOZ 100 ECM dispenser

The BOZ 100 is an advanced ECM chaff and IR flare dispenser originally developed for the Swedish Air Force, but also supplied to overseas customers. It is capable of sophisticated break-lock and corridor operation at subsonic and supersonic speeds. The unit has a comprehensive EW system interface and is suitable for use in deep penetration, strike, reconnaissance and electronic warfare roles. It is microprocessor-controlled and has a reprogrammable program memory.

Specifications

Dimensions: 4,000 (length) × 380 mm (diameter)
Weight: 325 kg
Attachments: 14 or 30 in (356 or 762 mm) bomb locks

Operational status

In service with the Swedish Air Force (under the designation BOX 9) and with Tornado and Jaguar aircraft. About 700 pods have been supplied for the Tornado programme. The German Air Force and Navy use the BOZ 101, the Italian Air Force uses the BOZ 102 and the Royal Air Force uses the BOZ 107.

Contractor

CelsiusTech Electronics AB.

VERIFIED

EWS 39 - Gripen integrated EW suite

Ericsson Saab Avionics is responsible for the EW suite in the Gripen fighter, designated EWS 39. EWS 39 integrates the RWR/ESM and passive missile warning system, with the radar, infrared search and track system, and the ground-air data link, to provide an overall integrated defensive system.

Sensor data is compared with library data for all the sensors and systems to evaluate threat level, alert the pilot, and initiate appropriate countermeasures responses.

EWS 39 is designed to handle all types of off-board, integrated and pod-mounted jammers, chaff and flare dispensers.

CelsiusTech Electronics AB supplies the RWR/ESM, and the BOP 300 flare dispensers; it is understood that CelsiusTech Electronics AB is also developing a towed radar decoy for the system, to be integrated with the BOP 300 dispenser.

Contractor

Ericsson Saab Avionics AB.

NEW ENTRY

TURKEY

AN/ALQ-178(V)3 and (V)5 integrated self-protection systems

The AN/ALQ-178(V)3 and (V)5 integrated self-protection systems are manufactured by MiKES Microwave Electronic Systems Inc. for installation in Turkish Air Force aircraft. The AN/ALQ-178 is an advanced, internally mounted, self-protection system specifically designed for high performance fighter aircraft, including the F-16 and F/A-18. It has been fully operational since 1986. See also entry on AN/ALQ-178 Rapport III ECM system under Lockheed Martin Electronic Defense Systems.

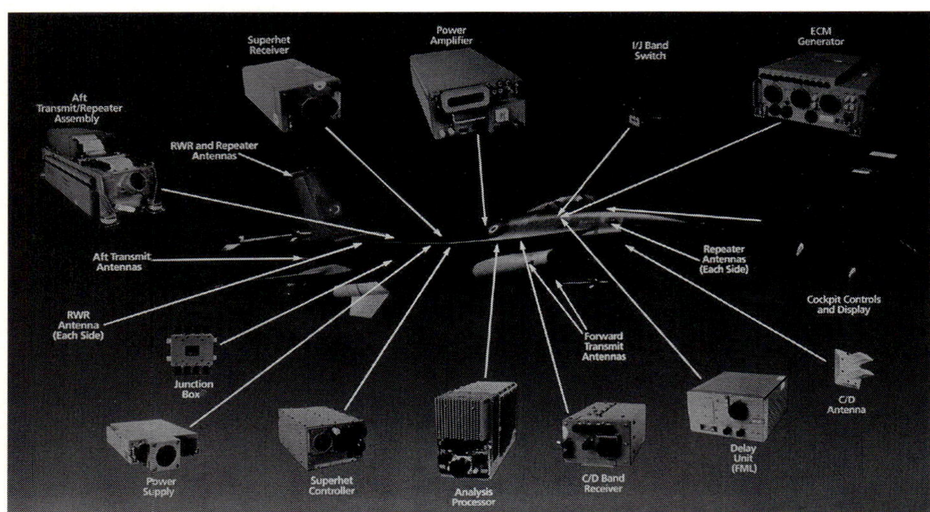

The AN/ALQ-178 receiver utilises a superheterodyne receiver to provide precision frequency measurement and required sensitivity. The integrated architecture of the AN/ALQ-178, utilising a common receiver system, results in rapid threat identification and counter responses (both RF ECM and chaff and flare) for an exceptional level of aircraft survivability in dense environments.

Operational status

The AN/ALQ-178(V)3 is currently operational on 160 F-16s of the Turkish Air Force, and is now being considered for the Turkish Air Force F-4 modernisation programme, which covers 54 aircraft. The AN/ALQ-178(V)5 is being offered for the second batch of F-16s for the Turkish Air Force, which is 80 aircraft.

Contractor

MiKES Microwave Electronic Systems Inc.

VERIFIED

MiKES Microwave Electronics Systems Inc AN/ALQ-178 integrated self-protection system
1997/0002228

UNITED KINGDOM

Infrared jammer

The British Aerospace infrared jammer is designed to provide protection against missiles using an infrared seeker head. It is intended to be mounted externally on a helicopter and operates by radiating modulated energy from an infrared source. The output from the jammer is modulated mechanically by means of shutters. The output signal enters the missile seeker head, impairing its ability to track. The IR source is a graphite radiating element which is hermetically sealed within a sapphire envelope.

The optical assembly which surrounds the lamp, rotating at high speed, passes between the lamp and slots in the outer drum. The drive to the optical assembly comes directly from a brushless DC motor integrally contained within the jammer mounting structure.

Surrounding the optical system is a cylinder with 16 longitudinal slots spaced around its circumference. The slots are covered by a window which provides environmental protection to the working parts and also serves to transmit IR wavelength radiation and to block visible emissions. The jammer is therefore covert in the visible spectrum.

The modulated IR signal emitted by the jammer has the effect of degrading the tracking ability of IR-guided missiles beyond the point at which they can be effective. The jammer radiates adequate power in the spectral band used by the seeker head to protect most helicopter types.

Field of view of 360° in azimuth and ±30° in elevation can be provided. These apply to the jammer prior to its installation on a helicopter. The pilot requires only a simple on/off control; two signals are provided to monitor lamp and motor operation. Weight of the complete system is 18 kg.

Operational status

In service.

Contractor

British Aerospace Systems & Equipment.

VERIFIED

AMIDS Advanced Missile Detection System

GEC-Marconi is developing a family of advanced performance missile approach warners based on the pulsed Doppler radar detection techniques of the PVS2000 (see later item). The Advanced MIssile Detection System (AMIDS) will provide extended azimuth detection coverage, detection range and warning time in comparison to present generation missile approach warning systems, and be capable of being configured for any airborne platform.

A contract for the first derivative of AMIDS has been awarded to GEC-Marconi to develop an advanced missile approach warner for the EF 2000. This will be fully integrated with the Eurodass Defensive Aids SubSystem. A major share of the development is being undertaken by Elettronica SpA.

The EF 2000 missile approach warner will be configured into transmitter and receiver/processor main LRUs together with two forward and one rear antenna assemblies. This configuration allows for the later integration of a passive missile launch detector, giving expanded detection performance.

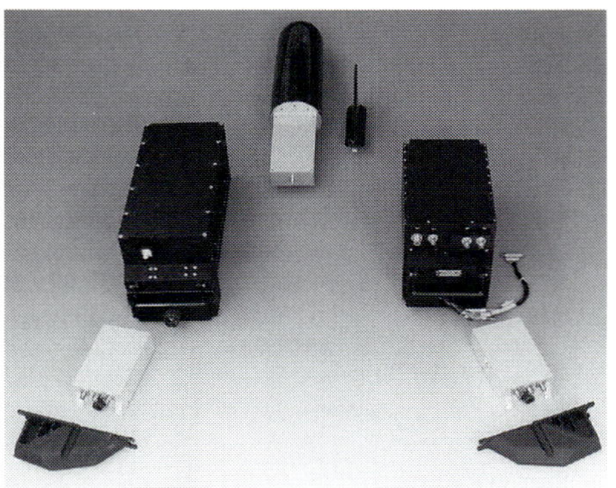

The Advanced Missile Detection System ***1996***

Specifications

Dimensions:
(processor) 480 × 160 × 190 mm
(transmitter) 380 × 160 × 190 mm
Weight:
(processor) 18 kg
(antennas) 2 kg each
(transmitter) 13 kg
(low-noise amplifiers) 2 kg each

Operational status

In development for the EF 2000.

Contractor

GEC-Marconi Avionics Ltd, Defence Systems Division, Portsmouth

VERIFIED

PA7010 Expendable System Programmer (ESP)

The PA7010 Expendable System Programmer (ESP) is designed to interface with radar warning receivers, jamming systems and chaff and flare dispensers, and to integrate the various EW sensors for the maximum operational effectiveness. The interface is produced via digital datalinks and discrete signals; crew monitoring and intervention is embodied in the design.

On detection of a threat, ESP is alerted to command the release of expendables. An easily reconfigurable stores library determines the appropriate release pattern and sequence. ESP then commands the response from one or more dispensers.

ESP has been designed for military combat aircraft and has a predicted reliability in excess of 1,000 hours MTBF. Modular construction and comprehensive BIT simplify maintenance.

Specifications

Dimensions: 127 × 127 × 190.5 mm
Weight: 2.7 kg
Power supply: 28 V DC, 10 W

Operational status

In production. The system is fitted in the Royal Air Force Harrier GR. Mk 7, in which it interfaces with the Zeus ECM system and onboard countermeasures dispensers. It has also been selected for the EF 2000.

Contractor

GEC-Marconi Avionics Ltd, Defence System Division, Portsmouth.

VERIFIED

PVS2000 Missile Approach Warner (MAW)

GEC-Marconi has developed a radar-based Missile Approach Warning (MAW) equipment designed to equip the Royal Air Force's Harrier aircraft. The MAW uses a low-powered pulse Doppler to detect approach of missiles, and automatically initiates countermeasures. The system is optimised to provide protection against air-to-air and ground-to-air heat-seeking missiles.

The MAW comprises: a transmitter/receiver unit, lightweight antenna, signal processing unit and cockpit control unit.

PVS2000 family of equipment ***1996***

The Transmitter/Receiver Unit (TRU) contains a low-power solid-state transmitter and receiver, analogue processing and digital conversion circuitry. It is designed for installation in an unconditioned area of the aircraft tail within 1 m of the antenna.

The remote lightweight antenna is connected to the TRU, using low-loss Gore cable, or may be attached directly to the TRU. The antenna must have a clear view aft.

The Signal Processing Unit (SPU) houses all the digital signal processing necessary for detection and assessment of signal returns. All interfaces with other aircraft avionics systems, BITE monitoring and module level diagnostics are handled by this unit. The SPU is installed in a conditioned avionics bay.

The optional Cockpit Control Unit (CCU) provides on/off control and incorporates a BITE status indicator lamp. A small auxiliary blade antenna is mounted on any suitable site with a relatively clear view forward.

The radar is mounted in the tail of the aircraft and should be linked to an expendables dispenser for the automatic deployment of flares and chaff.

Specifications

Dimensions:
(transmitter/receiver unit) 156 (length) × 250 mm (diameter)
(signal processing unit) ½ ATR short
(optional cockpit control unit) 38 × 146 × 75 mm
(main antenna/radome) 100 (length) × 260 mm (diameter)
(auxiliary antenna) 150 × 100 × 260 mm
Weight:
(transmitter/receiver unit) 5.5 kg
(signal processing unit) 7 kg
(cockpit control unit) 1 kg
(main antenna/radome) 0.7 kg
(auxiliary antenna) 0.25 kg

Operational status

In production for Harrier GR. Mk 7 aircraft. Airborne live fire missile detection trials were carried out in 1992, successfully demonstrating the MAW's specified performance.

Contractor

GEC-Marconi Avionics Ltd, Defence Systems Division, Portsmouth.

VERIFIED

1220/1223 Series laser warning receiver

The 1220 Series laser warning receiver is a modular expandable system with centralised processing and miniaturised remote sensor heads. It may be operated as a stand-alone LWR with its own display or as part of an integrated system with, for example, the Marconi Sky Guardian 200 RWR.

A premission programmable threat library provides the capability for selective threat identification. The panel-mounted display is NVG-compatible and provides sector indication and other threat data.

An optional module provides a record of up to several thousand date and time tagged laser detections.

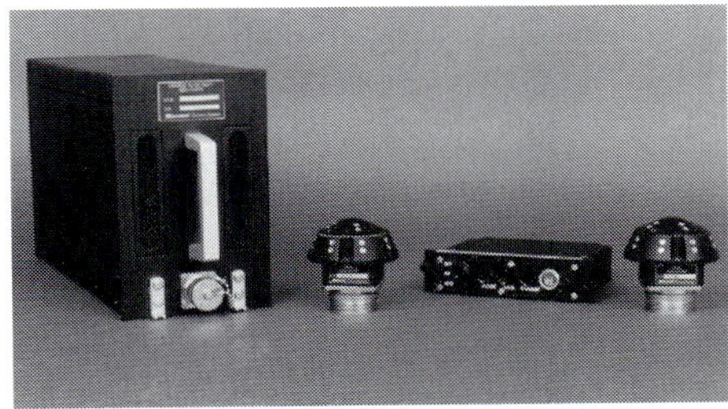

The 1220 Series laser warning receiver consists of electronic unit, control unit and sensors

Specifications

Dimensions:
(sensor) 25 (length) × 50 mm (diameter)
(electronics unit) 127 × 194 × 323 mm
(display) 80 × 80 mm
Weight:
(sensor) 0.4 kg
(electronics unit) 5 kg
(control panel) 0.4 kg
(display panel) 0.4 kg
Power supply: 24-28 V DC or 115 V AC, 400 Hz
Field of view: 360° azimuth × 45° elevation
Angular resolution:
(azimuth) ±22.5° (15 or 10° option)
Spectral range:
(basic system) 0.35-1.1 μm
1-1.8 μm and 8-11 μm options
Spectral resolution: 15-100 nm
Interfaces: RS-422 or MIL-STD-1553B

Operational status

Testing complete and undergoing delivery for high-performance aircraft applications. The system has been sold to unspecified customers in Europe and the Far East.

An updated variant, the 1223 LWR, forms part of the GEC-Marconi HIDAS system selected for the British Army WAH 64 Apache helicopter. The model 1223 incorporates very high senstivity and extended wavelength coverage to ensure timely detection of laser beamriding threats at the point of launch, as well as rangefinders and target designators.

Contractor

GEC-Marconi Avionics Ltd, Defence Systems Division, Stanmore.

UPDATED

Apollo radar jammer

Apollo is an advanced radar jammer for aircraft self-protection. It is designed to ensure survival in hostile environments by countering a wide range of threats and is available as a pod or installed within the airframe.

Apollo comprises modules developed from in-service GEC-Marconi systems. These include warning receivers, jammers and digital processors combined in a lightweight compact system. Apollo operates automatically, so that jamming management does not add to the pilot's workload.

The podded Apollo fits on a standard hardpoint and can be quickly installed and removed, giving commanders the freedom to redeploy it quickly on other aircraft.

Different performance specifications are available: from a simple repeater jammer, to a more capable system that incorporates a frequency set on receiver and noise deception capability, through to advanced single or dual DRFM options.

Apollo's jamming responses are software driven. As well as producing noise, it generates carefully modulated response patterns to confuse and deceive fire-control radars. By using various jamming

The GEC-Marconi Avionics Ltd Defence Systems Apollo pod ***1995***

techniques Apollo is effective against pulse and CW radar signals. It is capable of dealing with multiple threats simultaneously.

Where the aircraft is fitted with a radar warning receiver such as the GEC-Marconi Sky Guardian (see later entry), data can be passed between Apollo and the RWR to optimise intelligent interaction. This combination offers the advantages of an integrated defensive aids system and controls the decoy dispensers to fire chaff and flares at the best possible moment to co-ordinate with the jamming.

Apollo is designed to be as self-sufficient as possible. If the aircraft's own electrical system cannot generate the power needed, a ram-air turbine is fitted to the podded variant to supply the unit.

Apollo's digital processors identify radars by comparing their parameters with the built-in EW library. The jamming technique most likely to defeat that radar is selected and used. The control system organises jamming responses in order of priority by assessing various factors. These include whether a signal is known to be a threat, whether it is locked on and its direction relative to the aircraft. Apollo can be programmed on the flight line with the latest threat intelligence.

In an aircraft fitted with a radar warning receiver, Apollo momentarily blanks out the jamming signal so that the RWR can sample the threat environment and this updated threat information is then passed back to the jammer. The operation is synchronised to maintain peak jamming performance.

Specifications

Dimensions: 2,650 (length) × 280 mm (diameter) (length with RAT) 2,985 mm
Weight:
130 kg, 160 kg (with RAT)
40 kg (inboard)
Power supply: 115 V AC, 400 Hz, 3 phase
28 V DC
Frequency: H- to J-bands (other bands optional)

Operational status

Fully developed and flight tested.

Contractor

GEC-Marconi Avionics Ltd, Defence Systems Division, Stanmore.

UPDATED

ARI 18223 radar warning receiver

The ARI 18223 is a self-contained radar warning receiver system suitable for single-seat aircraft and is installed in Royal Air Force Jaguar aircraft. Systems have also been supplied to some overseas operators of these two aircraft. Frequency coverage is 2 to 18 GHz and a lamp display is used to show the quadrant of the highest priority threat. Antennas are on the vertical fins of the aircraft and an audio alarm is also triggered when a threat is displayed.

Operational status

No longer in production. In service on Royal Air Force Jaguar aircraft.

A Royal Air Force Jaguar equipped with the GEC-Marconi ARI 18223. Receiver units are installed near the fin tip

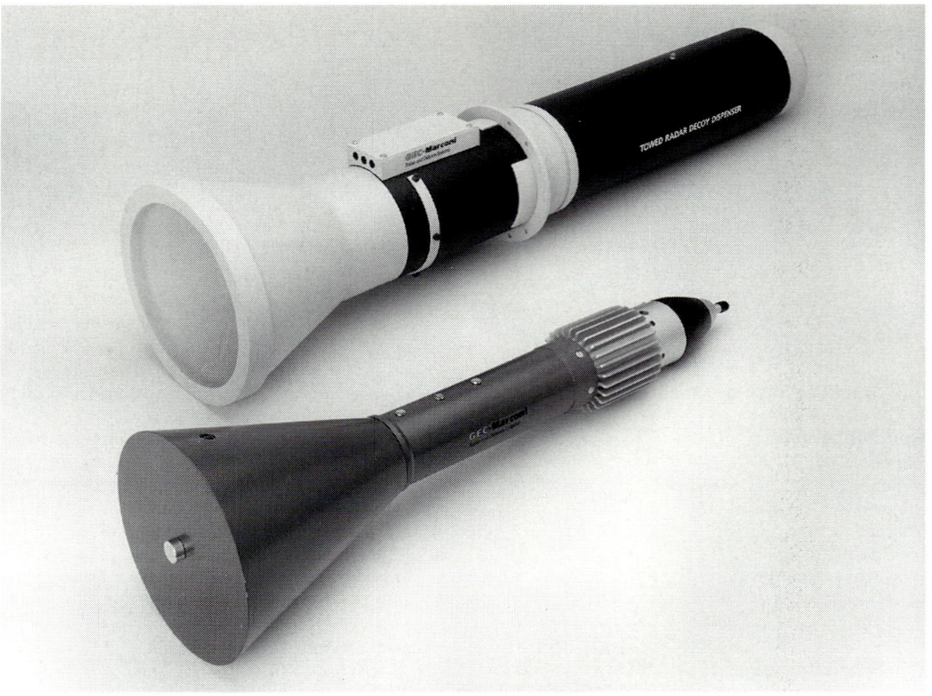

Ariel airborne towed radar decoy (below) and dispenser (above) ***1998**/0018275*

Contractor

GEC-Marconi Avionics Ltd, Defence Systems Division, Stanmore.

VERIFIED

Ariel airborne Towed Radar Decoy (TRD)

Ariel is a second generation compact, lightweight and cost-effective radar jamming system which is recoverable during or after flight for repeated operational employment on subsequent missions.

Ariel may be mounted within the airframe of the host aircraft, in a pod on a standard underwing pylon or as a scabbed-on fit to the fuselage skin. Two different types of decoy have been developed. The early first-generation device was a relatively high-power RF linked repeater decoy which is equivalent to towing a slaved transmitter. The current second-generation version, which is now the standard operational variant in RAF service is a high-power datalinked decoy which can function as both a towed radar repeater and deception jammer. Both variants depend on equipment aboard the host aircraft, including the host aircraft's RWR, the EHT power supply unit and the winch control unit. The tow cable is of multi-element fibre optic construction and includes electrical conductors and strength members. The towed body includes the preamplifiers and final TWT amplifiers, the transmit antennas and the power conditioning unit. The RF-linked decoy uses the aircraft radar warners and an onboard ECM techniques generator, with the decoy interface module passing signals down an optical fibre to a photo detector in the decoy. The decoy transmits all types of ECM signals, as generated by the onboard techniques generator or the existing onboard ECM systems.

For large aircraft, the decoy is deployed and recovered using an associated onboard winch mechanism. After deployment on smaller combat aircraft, it is jettisoned before landing and may be recovered by optional parachute for reuse.

Regardless of its deployment method, the Ariel fibre optic tow cable is constructed to allow the decoy to be deployed throughout a full sortie, rather than just when a threat is detected.

Ariel is effective against monopulse, semi-active radars and home-on-jam systems. It has fully programmable countermeasures techniques and can operate either in a stand-alone mode or be integrated with an existing onboard EW suite. It has been flight proven between 150 kt and M1.2 and will be progressively cleared throughout the EF 2000 flight envelope, as the basis of the TRD fit for EF2000 which is understood to include a double installation in a pod on the aircraft's starboard wingtip.

Specifications

Frequency: 2-18 GHz
Coverage: spherical
Techniques: noise, repeater and advanced countermeasures

Operational status

First fitted on Royal Air Force Nimrod MR2 aircraft in the 1991 Gulf War, Ariel has been installed on Tornado F3 aircraft operating over Bosnia, where it is fitted in a modified BOZ pod. Flight testing on a C-130 aircraft, fitted with an ALQ-131 pod as host is also reported from the US Air Force.

Contractor

GEC-Marconi Avionics Ltd, Defence Systems Division, Stanmore.

UPDATED

AWARE ESM system

AWARE is a product inherited by GEC-Marconi as part of the break-up of the Ferranti Company. It is not, at present, actively marketed.

The Advanced Warning of Active Radar Emissions (AWARE) range of equipments provides advanced warning of radar emissions for use in light strike and transport aircraft and helicopters or in marine craft and ground vehicles.

The system is modular in concept and design, providing variants of the basic system by simple changes of subassembly or LRUs. The expansion facilities offered provide an ESM suite and integration with ECM systems and will cover from C-band up to the millimetric region.

The flexibility of design of AWARE allows systems to be configured to match user requirements, from a basic RWR to a full ESM suite, and it is suitable for operation in a wide range of airframes, from light battlefield helicopters to high-performance aircraft. There are currently four variants in the AWARE range: AWARE-3, AWARE-4, AWARE-5 and AWARE-6. All have the same basic modules, the only difference being in the man/machine interface.

AWARE-3

AWARE-3 is the basic ESM system, covering the E- to J-bands and intended for battlefield helicopters, light tactical and logistic aircraft. The system detects and identifies pulse, pulse Doppler, CW and ICW signals. An immediate identification of the threat is provided and the particular threat type is identified.

Identification is achieved by comparison with a stored threat library. The library, threat priorities and ECM interfaces are under software control which is loaded prior to a mission, allowing rapid updating of

The AWARE ESM system is in service in Lynx and Gazelle helicopters ***1995***

threats and priorities on a mission by mission basis. Additionally, security of system and intelligence data is enhanced by the ability to clear the library and software. The system operates in peak pulse densities of several hundred thousand pulses/s, radiated by many simultaneous emitters.

The AWARE display is heading compensated with computer-controlled symbology. An extensive symbol library can be programmed with symbols of the user's choice.

AWARE-3 consists of four planar spiral antennas, a hand-portable programme loading unit, two dual-channel crystal video receivers, a fast Instantaneous Frequency Measurement (IFM) receiver and a powerful signal processor. Control of the system can be achieved either as an autonomous system, where the control unit and a 3 in display are employed, or as an integral part of the mission management system, with control being exercised via a MIL-STD-1553B standard interface. All LRUs are convection cooled.

The four antennas are mounted in mutually orthogonal azimuth directions. The signals received are passed to the two dual-channel crystal video receivers. These receivers provide the preamplification, detection and compression of the video signals for input to the signal processor. They also provide controlled RF outputs to the IFM receiver. The output of the IFM receiver is presented to the signal processor as a digital number.

The signal processor determines the direction of arrival and time of arrival of the signals. The results of these operations, when combined with output of the IFM receiver, allows de-interleaving of the incoming signals to be performed. Once de-interleaved and characterised, the signals are classified by comparing them with a stored library of known emitters. Following classification, the source of emission and its range and bearing may be displayed in simplified plan form on the monitor or passed to the aircraft mission management system via a monitor or via a MIL-STD-1553 databus.

AWARE-4
AWARE-4 is an ESM system for maritime helicopter applications. The right-hand screen retains the PPI presentation of emitters to maintain the radar warning role, while the left-hand screen is a tote for the display of emitter parameters.

AWARE-5
AWARE-5 retains the AWARE-3 control unit but has no display. In this variant, the data are displayed on either the aircraft HUD or the AI radar display, via the MIL-STD-1553B or ARINC 429 databusses. AWARE-5 is for use in light strike aircraft.

AWARE-6
AWARE-6 is similar to AWARE-4, but the display unit is replaced by two 10 in colour monitors. AWARE-6 is for use in maritime patrol aircraft operations where more space is available and a larger tote that can display the whole track table is required.

Specifications
System weight:
(AWARE-3) 13 kg
(AWARE-4) 15 kg
Power supply: 28 V DC
150 W (AWARE-3)
180 W (AWARE-4)
Frequency: 2-18 GHz
Accuracy: better than 10° RMS

Operational status
In production for customers in the UK and overseas. AWARE-3, under the designation ARI23491 (Rewarder), equips Lynx and Gazelle helicopters of the British Army. AWARE-3 also equips Lynx helicopters of the Royal Netherlands Navy. Systems are being used for both airborne and shipborne applications.

Contractor
GEC-Marconi Avionics Ltd, Defence Systems Division, Stanmore.

UPDATED

Hermes ESM system

Hermes is an EW surveillance system for the interception and analysis of radar signals, providing data for planning operations and for defence. It can form part of an air defence network, both in the maritime surveillance role and from land-based sites.

In the airborne mode, Hermes can be configured from a basic radar warning receiver to a complete ESM system for the AEW role and reconnaissance. It can also be fitted into UAVs.

Hermes intercepts radar signals at long range using wideband superheterodyne receivers that give unrivalled handling of continuous wave and high-duty cycle signals. The modular design means that Hermes can be installed in a wide variety of platforms and can be fitted inboard on helicopters and medium-size aircraft or in a pod on fast jets.

Hermes is fitted with interfaces for navigation, data exchange and the automatic triggering of electronic countermeasures. These interfaces can be tailored to specific platform requirements.

Specifications
Weight: 110 kg approx
Frequency: C- to J-bands
Signal types: pulse, CW, ICW, RF agile, PRI agile, jammers
Coverage:
(azimuth) 360°
(elevation) ±45°
Emitter library: >1,000 modes

Operational status
The system is in service with the Indian Navy on its Westland Mk 42B Sea King helicopters.

Contractor
GEC-Marconi Avionics Ltd, Defence Systems Division, Stanmore.

VERIFIED

HIDAS Helicopter Integrated Defensive Aids System

Founded on the needs of the demanding attack helicopter role – where short engagement ranges demand rapid reaction times – HIDAS components are designed to operate together to provide radar, laser and missile threat warning, together with advanced radar and IR countermeasures and off-board decoys.

Component systems of HIDAS are as follows:

(1) RWR radar warning receiver: Sky Guardian 2000 RWR from GEC-Marconi Avionics Ltd; Sky Guardian 2000 covers the E-J band, with optional frequency extensions downwards to C/D bands, and upwards to K band.
(2) LWR laser warning receiver: Series 1223 LWR from GEC-Marconi Avionics; with full beamrider detection capability.
(3) CMWS common missile warning system: AN/AAR-57(V) from Sanders, a Lockheed Martin Company; the AN/AAR-57(V) is the Common Missile Warning System (CMWS) of the ATIRCM/CMWS; each CMWS set provided for the HIDAS system will include an electronic control unit (ECU), which processes threat data, and four passive electro-optic missile sensors, being produced by Lockheed Martin Infrared Imaging Systems.
(4) RF radar frequency jammer: Apollo RF jammer from GEC-Marconi Avionics Ltd; covering the H-J band, with optional extensions down to E-G bands, and upwards to K band.

HIDAS helicopter integrated defensive aids system ***1998***/0018274

(5) CMDS countermeasures dispensing system: Vicon 78 series 455 from W Vinten Ltd; comprising three dispensers, one for chaff and two for flares. The chaff will be fired upwards towards the tail rotor. Flares will be fired from dispensers on the rear fuselage. GEC-Marconi envisages that HIDAS will employ 'smart flares' emerging from current R&D programmes to improve IRCM capability.
(6) DIRCM directional infrared countermeasures from the current NEMESIS international programme to provide bands I-IV IR band jamming.

To achieve the high degree of integration essential to timely reaction, the DAS Management module is incorporated within the Sky Guardian 2000 RWR. This enables data from all the DAS sensors (RWR, LWR, CMWS) to be correlated against the large threat library. The DAS Manager then determines and automatically responds with the optimum mix of radar, IR and off-board countermeasures. The DAS communicates with the aircraft mission system over a MIL-STD-1553B databus and can display data or system status reports on the aircraft multifunction displays or on dedicated EW displays.

By combining and correlating data from all its sensors, HIDAS is able to present to the crew a comprehensive picture of the surrounding environment. Thus, crews have the option to take pre-emptive action; for instance, the HIDAS detection can be used to cue the on-board weapon system.

HIDAS is configured to ensure radar, laser and missile threats are detected, identified, declared to the crew and, where appropriate, countermeasures automatically instigated, in under 2 s—typically 1 s from threat detection.

HIDAS automatically selects the appropriate form or combination of radar jamming, IR jamming, chaff or flare dispensing. Where the crew require it, the countermeasures can be restricted to semi-automatic operations and a manual override is available at all times.

HIDAS is fully user reprogrammable at the flight line and provision is made for a mission library of up to 4,000 threat modes. Individual modes can be linked readily to enable the positive identification of weapon systems.

A 'smartcard' PCM port can be provided on the control unit to facilitate the loading of pre-flight messages. The large capacity of these cards allows the same device to be used for recording in-flight mission data. Software loading and recording functions can also be carried out by aircraft systems over the MIL-STD-1553B Bus.

The GEC-Marconi MERLIN is a PC-based software support system that enables the user to quickly create their own mission libraries, specify the required countermeasures and carry out post flight analysis of mission data.

Control of HIDAS is exercised either through a simple dedicated control unit with associated display, or through 'soft' controls on multi function displays via the aircraft mission computer.

HIDAS supports both MIL-STD-1553B and RS422 interfaces, and provides a combined situation picture and status report for display to the crew on the aircraft multi-purpose display.

Operational status
Selected for the British Army WAH-64 Apache attack helicopter programme.

Contractor
GEC-Marconi Avionics Ltd, Defence Systems Division, Stanmore.

UPDATED

PA7030 laser warning equipment

GEC-Marconi laser warning equipment offers both instantaneous warning and bearing information on radiation by ruby and neodymium lasers. The system consists of a direct detector head unit and up to four indirect detectors mounted externally on the aircraft. A compact display and control unit is mounted in the cockpit.

The indirect detectors are sensitive to pulsed laser radiation scattered by the aircraft and so can detect designators and rangefinders even when these do not directly illuminate the direct detector. Electronic filtering discriminates against glints and flashes to give an extremely low false alarm rate. The direct sensor gives instantaneous bearing on a radiation source to within 15°.

Specifications
Dimensions:
(PA7031 direct detector) 152 × 48 mm
(PA7032 scatter detector) 105 × 80 × 130 mm
(PA7034 signal conditioner) 260 × 160 × 87 mm
(PA7033 display) ¼ ATR short
Weight:
(direct detector) 2 kg
(scatter detector) 1.5 kg
(signal conditioner) 2 kg
(display) 2 kg
Power supply: 28 V DC, <12 W
Coverage:
(direct detector) 360° (azimuth), −45 to +10° (elevation)
(scatter detector) as required

Operational status
Development equipment has had trials in helicopters and fixed-wing aircraft in the UK, USA and Canada.

Contractor
GEC-Marconi Avionics Ltd, Defence Systems Division, Stanmore.

VERIFIED

Radar Homing and Warning Receiver (RHWR)

The Radar Homing and Warning Receiver (RHWR) is produced in two versions: an amplitude comparison version for strike aircraft and a high-accuracy version for air defence aircraft. Both versions are effective against radar threats likely to be encountered and are combat proven. Frequency coverage is available from C- to J-bands. RHWR can detect and identify threats up to and beyond the radar horizon. The information obtained is correlated with the aircraft interception radar tracks to allow for the classification and allocation of target priorities. High-accuracy DF is obtained by using interferometers. RHWR uses a powerful digital processor which accepts the parametric data from the receivers, including frequency, PRI, pulsewidth and scan characteristics. The information is compared with a comprehensive emitter library. The identified emitters are passed to the cockpit displays which provide the crew with both tabular and graphical presentation of threats, as well as the overall radar scenario.

The RHWR has recently been upgraded by GEC-Marconi to meet the requirements of UK MoD Air Staff Requirement 907. Increased processing power and additional memory has been added and the capability for the RHWR to exchange data with the Sky Shadow ECM pod has been provided. Sky Shadow has also been upgraded under the same programme. Whilst each system retains the ability to operate independently, under normal conditions the RHWR acts as the defensive aids system controller of an integrated system.

Operational status
In service in the Tornado F3 of the Royal Air Force and other air forces.

Contractor
GEC-Marconi Avionics Ltd, Defence Systems Division, Stanmore.

VERIFIED

Sky Guardian 200 radar warning receiver

Sky Guardian 200 uses crystal video receivers and the latest digital processors to give analysis of threats and high probability of intercept of signals. Information is presented to the crew by alphanumeric presentation on either an indicator bearing unit, a multifunction display or a HUD. Sky Guardian 200 will control a range of countermeasures, including jamming and chaff and infrared flare dispersion.

Specifications
Weight: 24 kg
Frequency: E-J in 3 bands (C/D- and K-bands optional)
Frequency measurement: pulse, pulse Doppler, CW, ICW
Coverage: 360° azimuth instantaneous
Accuracy: better than 10°
Response time: <1 s
Emitter library: 400 emitters with 2,500 modes

Operational status
In production. Sky Guardian 200 is in service on many fixed-wing aircraft and helicopters of the Royal Air Force, such as the Jaguar and VC10 tanker, Royal Navy Sea Harriers, Matadors of the Spanish Navy, Omani Air Force Hawk 200 aircraft and aircraft of the Austrian Air Force and other overseas air forces. It has recently been selected for the new Czech L-159 aircraft.

Contractor
GEC-Marconi Avionics Ltd, Defence Systems Division, Stanmore.

UPDATED

Sky Guardian 300 ESM system

Sky Guardian 300 is an advanced airborne electronic support measures system which detects radar emitter signals from outside their own detection range. It can recognise the bearing and mode of operation of an emitter, provide fine DF and prioritise threats. The

The Sky Guardian 200 radar warning receiver

1995

system includes a library capable of recognising up to 4,000 different signals, a central processor providing identification times of less than a second and programmable display with a variety of graphic and tabular formats on full colour raster scan monitor.

Sky Guardian 300 allows the operator to prioritise and select relevant data and also has the ability to compile and update emitter libraries at both first line and national level. The system will integrate with other onboard sensors to provide a comprehensive picture. It will provide automatic warnings and initiate countermeasures.

A variant of the Sky Guardian 300, the Sky Guardian 350, includes a rotating antenna and special receiver facility to provide long-range detections and raid assessment. It offers improved signal measurement and high accuracy DF.

Specifications

Weight: 28 kg plus display
Frequency: E- to J-bands (C/D-band optional)
Frequency measurement: DIFM
Coverage: 360° instantaneous
Accuracy: better than 2.5°
Response time: <1 s
Library: over 4,000 emitters

Operational status

Fully developed and on order for specified customers.

Contractor

GEC-Marconi Avionics Ltd, Defence Systems Division, Stanmore.

VERIFIED

Sky Guardian 2000 radar warning receiver

Sky Guardian 2000 is configured for rapid response in a high-density electromagnetic environment and has the processing capability to display multiple threat signals in less than a second. The system operates either as a stand-alone radar warner or integrated with an ECM system such as the Apollo radar jammer or other countermeasures. Sky Guardian 2000 will accept data from the Type 1220 laser warning receiver and display the threats on a common indicator bearing unit.

Sky Guardian will provide long-range detection of airborne and surface threat radars, emitter identification and built-in recording for post-flight analysis. Other key features are display persistence for fleeting intercepts, low-band targeting options and the ability to integrate into a multifunction cockpit control and display. The system is extendable to meet future threats.

Sky Guardian 2000 employs advanced signal processing algorithms and provides high sensitivity as well as accurate RF measurement. An emitter library of 4,000 emitter descriptions can be loaded by a PCMCIA smart card incorporated into the control unit. Sky Guardian 2000 has been designed from the outset as the core of a Defensive Aids System (DAS) and includes DAS control functions.

Specifications

Weight: 13 kg
Frequency: E- to J-bands (C/D- and K-bands optional)
Frequency measurement: pulse, pulse Doppler, CW and ICW
Coverage: 360° instantaneous
Accuracy: better than 10°

Operational status

Fully developed and on order for specified customers. It is understood to have been ordered by the Austrian Air Force for J350E Draken fighters, and it has been selected for RAF EH 101 support helicopters.

Contractor

GEC-Marconi Avionics Ltd, Defence Systems Division, Stanmore.

VERIFIED

The Sky Guardian 2000 radar warning receiver ***1995***

Sky Shadow ECM pod

GEC-Marconi Avionics Ltd, Defence Systems is the prime contractor in the production of the ARI 23246/1 Sky Shadow ECM pod developed for use with the Royal Air Force Panavia Tornado. Other components are provided by Racal, British Aerospace and GEC-Marconi Avionics. GEC-Marconi is responsible for overall design, development and subsequent testing. Sky Shadow flight tests were completed in 1980.

A high-power travelling wave tube amplifier is used with a dual-mode capability for deceptive and continuous wave jamming. A voltage-controlled oscillator uses a varactor-tuned Gunn diode to cover the full frequency band. The set-on receiver is of Racal design and signal processing uses GEC-Marconi Avionics hardware.

The pod incorporates both active and passive electronic warfare systems and includes an integral transmitter/receiver, processor and cooling system. It has radomes at both ends and is stated to be capable of countering multiple ground and air threats, including surveillance, missile and airborne radars. Power management is automatic and modular construction allows the system to be adapted to differing operational missions.

A Royal Air Force Tornado carrying the ARI 23246/1 Sky Shadow ECM pod on each outer pylon

An enhanced version of Sky Shadow has been developed. This features a high-performance data processor running on Ada software. The range of jamming activities has been extended, with greater flexibility for front line reprogramming.

Sky Shadow has been upgraded, with the Tornado RHWR, as part of the UK MoD Air Staff Requirement 907. The improved Sky Shadow is able to exchange data with the RHWR. When operating in the linked mode, RHWR acts as a Defensive Aids System (DAS) controller and both equipments offer the synergistic benefits of an integrated system.

Specifications

Dimensions: 3,350 (length) × 380 mm (diameter)

Operational status

In production. GEC-Marconi Avionics Ltd, Defence Systems was awarded a contract to upgrade the Sky Shadow ECM pod for the Tornado mid-life update. Deliveries of modules and modification kits are nearing completion.

Contractor

GEC-Marconi Avionics Ltd, Defence Systems Division, Stanmore.

VERIFIED

Zeus ECM system

Zeus is a compact multipurpose EW integrated defensive aids system for combat aircraft. It provides a complete defensive system consisting of a radar warning receiver and a jammer. The RWR is the latest in a series of passive intercept systems and is based on IFM and fast superheterodyne techniques. It is able to intercept and measure the characteristics of all radar-controlled systems which may threaten the aircraft. The RWR will measure parameters including direction of arrival, time of arrival, frequency, PRI, pulsewidth, amplitude, scan interval and scan rate. Signals are passed to the Zeus digital processor which identifies radar type, displays it and gives audio warning. The processor also controls the jammer and other means of countermeasures such as chaff, flares and decoys. The transmitter will jam both pulse and CW radars, while control features ensure that home-on-jam radars are not given sufficient time to lock on to the aircraft. Different jamming modes are available with Zeus for the operational requirements of individual customers.

Specifications

Weight: 118 kg
Frequency: C- to J-bands
Coverage:
(azimuth) 360°
(elevation) ±45°
Response: 1 s typically
Jamming: noise, VGPO, deception, co-operative

The GEC-Marconi Zeus internal ECM system
1995

Operational status

Zeus is in service with the Royal Air Force on the Harrier GR. Mk 7 aircraft. Installations have been designed for other aircraft, including the JAS 39 Gripen, AV-8B, F-16 and F/A-18. Zeus has successfully completed technical and operational evaluations in France, the UK and USA.

Contractor

GEC-Marconi Avionics Ltd, Defence Systems Division, Stanmore.

VERIFIED

Type 453 laser warning receiver

The Type 453 laser warning receiver is designed for use in fixed-wing aircraft, helicopters and armoured fighting vehicles. It provides a countermeasure to laser-guided weapon systems and provides an audible alarm and visual display showing the direction from which the threat has originated. Output from the receiver can be combined with radar warning equipment to give an integrated battlefield threat warning system or with a defensive aids system.

The system uses a number of dispersed sensors to detect incident laser radiation. This counters the problem of detecting a very narrow beam which at any instant would be illuminating only a small part of the airframe. The sensor head configuration can be tailored to provide spherical or hemispherical cover and presents the laser threat bearing to the crew as sectoral information. The number of sensors required is tailored to the aircraft type. The sensor protrudes through the aircraft skin as a 25 mm hemisphere and occupies a space of 50 mm depth behind the skin.

The sensor heads are completely passive. Laser radiation is routed to the central processing unit by fibre optic cables. This feature eliminates risks of false alarms being generated by radio frequency interference.

Operational status

In advanced development. In March 1993, Eurofighter GmbH awarded a subcontract for a derivative of the Type 453 laser warning receiver for the EF 2000.

Contractor

GEC-Marconi Avionics Ltd, Radar Systems Division, Silverknowes.

UPDATED

Hostile Fire Indicator (HOFIN)

The Hostile Fire Indicator (HOFIN) uses shockwave sensors to detect gunfire aimed at the host aircraft, and is a passive warning system suitable for helicopter use. It uses a five-armed sensor (four arranged in one plane at 90° to each other and one perpendicularly) which is mounted beneath the helicopter. Shockwaves generated by projectiles are converted to electrical signals and then processed in an electronic unit which occupies a ⅜ ATR short box. This produces outputs which can be used to generate audio or visual warning of nearby arms fire. An audio warning lasts for approximately 1 second after a shockwave has been detected and a visual warning is presented on a 4 ATI size unit which has eight 45° wide segments. Four segments which indicate the approximate direction from which the shock is detected will illuminate for about five seconds.

Specifications

Dimensions:
(sensor array) 305 × 305 × 195 mm
(computer) 94 × 418 × 228 mm
(indicator) 106 × 106 × 125 mm
Weight:
(sensor array) 1.93 kg
(computer) 2.72 kg
(indicator) 1 kg
Power supply: 22-28.5 V DC, 60 W
Sensitivity: responsive to supersonic projectiles at 20 m
Temperature range: −20 to +50° C
Humidity limit: 95% non-condensing

Operational status

In service with British, Italian and Canadian forces.

Contractor

MS Instruments plc.

VERIFIED

AN/ULQ-19(V)3 ECM system (RACJAM AIR)

The AN/ULQ-19(V)3 (RACJAM AIR) system is a helicopter-mounted VHF responsive jammer which provides an airborne communications ECM capability without the need for the aircraft to be assigned permanently to the jamming role. In US service it is known as the AN/ULQ-19(V)3; in service with UK forces it is known as RACJAM AIR.

The system consists of an ESM subsystem used to find targets for the jammer, the RJS3100 100 W VHF responsive jammer and a blade antenna fitted to the aircraft.

The jamming signal is radiated through an aerodynamically efficient blade antenna mounted on a suitable inspection panel on the underside of the aircraft, with a coaxial feeder routed into the cargo bay. There is no requirement to remove this antenna when the aircraft is not being used for jamming. The antenna is tuned rapidly to each target frequency as it is attacked, thus ensuring maximum transmitted power and effective range. The antenna tuning rate is compatible with the rapid retuning capability of the jammer system.

The rest of the system is contained within two easily deployed, shockmounted containers, each weighing less than 85 kg, secured to the aircraft standard cargo lashing points. It requires no special wiring or modifications to the existing power or RF distribution systems and can be installed in less than 15 minutes. Once installed, the system is connected to the aircraft power supply and intercom system and is then ready for operation.

Operational status

In service. It has been certified on the UH-1H and UH-1N helicopters.

Contractors

Racal Communications Inc.
Racal Radio Ltd.

VERIFIED

DRFM TG Digital Radio Frequency Memory Techniques Generator

Racal Radar Defence Systems (RRDS) has been contracted by Lockheed Martin Fairchild Systems to provide its DRFM TG for integration into the Nimrod MRA4 Defensive Aids Sub-System (DASS).

System architecture provides for the Lockheed Martin Fairchild Systems AN/ALR-56M(V) RWR to cue the DRFM TG, which is then used to drive the Fibre-Optic Towed Decoy (FOTD) version of the Raytheon Systems Company AN/ALE-50 towed decoy system.

The DRFM TG has been developed by RRDS in its work on advanced jamming system technology in association with the UK Defence Evaluation Research Agency (DERA). It is understood that a twin-DRFM installation is to be used to meet the high data throughput specification for the Nimrod MRA4 system.

Operational status

RRDS are to deliver the first integration model to Lockheed Martin Fairchild Systems in late 1998. All 21 Nimrod MRA4 are due to be fitted with the system.

Contractor

Racal Radar Defence Systems Ltd.

NEW ENTRY

Griffin Radar Warning Receiver (RWR/ESM)

Griffin is designed for fixed- and rotary-wing aircraft. It provides full surveillance, identification and threat warning in dense signal environments. It uses advanced SADIE customised very large scale integration (VLSI) ESM processors to give continuous radar identification within 1 second from a mission library of 3,500 radar modes. It provides automatic data recording of intercepts for post-mission analysis using the programmer/loader unit (PLU). The same PLU is used for loading pre-flight generated ESM library and mission data. Griffin uses advanced instantaneous frequency measurement (IFM) receivers, with improved fidelity of measurement.

Griffin detects pulse, pulse Doppler and continuous wave radars beyond lethal range, and provides prioritised audio and visual cues in the primary threat bands in multi-emitter and high pulse density areas.

System elements include:

(a) four circularly polarised, planar cavity backed spiral antennas, designed for surface mounting;

(b) Dual Receiver Units (DRU); each pair of antennas feeds into a DRU; the DRU supplies both RF and video to the signal processor unit;

(c) a Signal Processor Unit (SPU); which contains the IFM that accurately measures the RF on a pulse-by-pulse basis, or repetitively sampled CW; and the SADIE processor

(d) Display and Control Unit (DCU): a dual control CRT unit that provides threat warning and tote separately; the DCU is NVG compatible; display options include a single CRT indicator with separate control, and remote display via MIL-STD-1553B databus.

(e) programme loader unit (PRU).

Specifications

RF range: 2-18 GHz; options for C/D, K and M (millimetric) bands
Azimuth coverage: 360°
Elevation coverage: ±35°
Sensitivity:
(pulse) 45 dBmi (typical)
(CW) 55 dBmi (typical)
POI/Pulse data rate: 100%; >500 kpps
Bearing accuracy: better than 8° RMS
Frequency agiles; simple/agile/hopper
PRI: fixed/stagger/jitter/grouped
Scan types:
circular/sector/conscan/lock-on/unknown
Pulses to form track: 6
Number of tracks: 200
Threat report time: 1 s

Contractor

Racal Radar Defence Systems Ltd.

VERIFIED

Griffin radar warning receiver on Hawk aircraft ***1997***/0001267

A Royal Navy Westland Lynx is equipped with the Racal MIR-2 warning receiver

Kestrel ESM system

An ESM system in production for ELINT, surveillance and reconnaissance tasks, Kestrel has a near 100 per cent probability of intercept of any radar emission within its frequency range. Kestrel also utilises the latest processing techniques to give the operator virtually real-time information. In service with UK forces, it is known as Orange Reaper.

The system receives and processes radar emissions over 0.6 to 18 GHz. A six-port amplitude comparison system measures the bearing of the emissions and provides instantaneous digital information over 360° in azimuth. At the same time, a frequency measurement receiver provides an instantaneous frequency indication.

Digitised information on all threats, as accumulated pulse by pulse, is passed to the processor. In this equipment overlapping pulse trains from different radars are derived and their pulse repetition frequency and frequency agility characteristics determined. The digitised data is then passed into the main processor where long-term information is extracted. Radar identification is made by comparing measured and derived radiation parameters with those stored in a library of known emitters. Full information about the radar signal environment is then presented to the system operator on an ordered tabular or tactical display or, alternatively, the data can be transferred to remote displays using a standard data highway.

The Racal Kestrel ESM system is fitted to the Royal Navy EH 101 Merlin helicopter

Orange Reaper has been selected for the Westland/Agusta Royal Navy EH 101 Merlin helicopter.

Specifications

Weight: 55 kg (max)
Frequency: 0.6-18 GHz (optionally up to 40 GHz)
Frequency accuracy: 5 MHz RMS
Accuracy: ±3.5° RMS
Radar storage capacity: 2,000 modes (min)
Processing time: 1 s (max)

Operational status

In production and operational on Lynx helicopters of the Royal Danish Navy, with Orange Reaper versions ordered for 44 EH 101 for the Royal Navy under a £30 million contract.

An export version of Kestrel, known as Kestrel II, is being evaluated by a number of countries.

Contractor

Racal Radar Defence Systems Ltd.

UPDATED

MIR-2 ESM system

The MIR-2 airborne ESM system, also known as Orange Crop, combines airborne warning system and search receiver functions using an advanced digital receiver served by a fully solid-state wideband antenna system covering the 0.6 to 18 GHz radar frequencies. Six antennas monitor a full 360° in azimuth and provide a very high intercept probability, improving the crew's location and identification of friendly and enemy radars. The control-indicator unit, on the cockpit coaming, is a compact light-emitting diode display which indicates frequency band, amplitude and relative bearing on the main display. On an associated fine bearing unit the true bearing and radar PRF are indicated to a high degree of accuracy. The complete system of eight units comprises six antennas, a processor and the control/indicator unit.

Specifications

Weight: 47.7 kg
Frequency: 0.5-18 GHz
Pulsewidth: 0.15-10 μs
PRF: 0.1-10 kHz
Coverage: 360°

Operational status

In production and in service with UK Royal Navy Lynx and Sea King helicopters and Royal Air Force C-130 Hercules aircraft. in August 1994, Racal announced a $10 million order for MIR-2 systems for Brazilian Navy Lynx helicopters. Sales now exceed 300 sets.

Contractor

Racal Radar Defence Systems Ltd.

VERIFIED

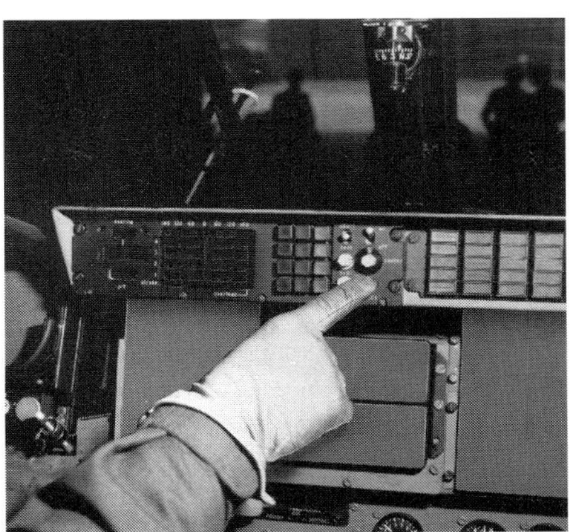

The Racal MIR-2 ESM cockpit control indicator unit

Prophet radar warning receiver

Prophet is an operationally proven radar warning system designed to reduce combat aircraft vulnerability to radar associated threats on the battlefield. It provides the crew with timely and unambiguous threat warning, permitting appropriate countermeasures, such as manoeuvres or chaff discharge, to be employed. System design has taken account of the likely need for frequency extensions in order to increase the survivability of the battlefield helicopter.

The receiver detects signals over 2 to 18 GHz and, by means of a processor with a programmable threat library, alerts the pilot to imminent hostile action. The system is designed to operate in a dense RF environment. A four-port antenna system provides direction-finding, using signal amplitude comparison to derive bearing. The system has an extremely low false alarm rate and threats are recognisable with a high degree of confidence. An audible warning is provided by a tone generator and fed to the aircraft internal communications system.

The Prophet display uses light-emitting diodes and has three lines, each dedicated to a threat. Each line can show an arrow indicating threat direction and a three-character alphanumeric identifier. Display colour and brightness is such that it can be viewed in bright sunlight and through night vision goggles. Also provided, as an option for fixed-wing aircraft, is a full threat display on the HUD and multifunction displays.

The system can be readily reprogrammed and has high in-service reliability.

Specifications

Weight: 10 kg
Power supply: 28 V DC, 80 W
Frequency: 2-18 GHz (pulse and CW); options: millimetric wave and laser threat warning
Coverage: more than 100 radar modes
(azimuth) 360°
(elevation) ±30°
DF accuracy: 10° RMS
Detection time: <1 s

Operational status

In service with Royal Navy Sea King HC4s and British Army Gazelles and other air forces. Prophet is in production for the British Aerospace Hawk 100 and 200.

Contractor

Racal Radar Defence Systems Ltd.

VERIFIED

The Racal Prophet radar warning receiver

Corvus ELINT system

Corvus is a self-contained ELINT receiving system which includes signals analysis and magnetic recording. The equipment provides a complete turnkey system requiring only an antenna and turntable system.

A major feature of the system is its ability to record and analyse intrapulse modulation on wideband frequency-agile emitters. This modulation is measured at up to 200 MHz sampling rate and the measurements are correlated with the standard pulse descriptors.

Analysis of the live or recorded data can be performed by the integral pulse analyser. Alternatively, the EP4240 separate off-line analysis system is available. This allows a more detailed analysis to be made and ELINT libraries to be compiled, with rapid retrieval of data.

The system covers the frequency range of 0.5 to 18 GHz and options are available for wider ranges. The instantaneous bandwidth of the channelised receiver is 1 GHz.

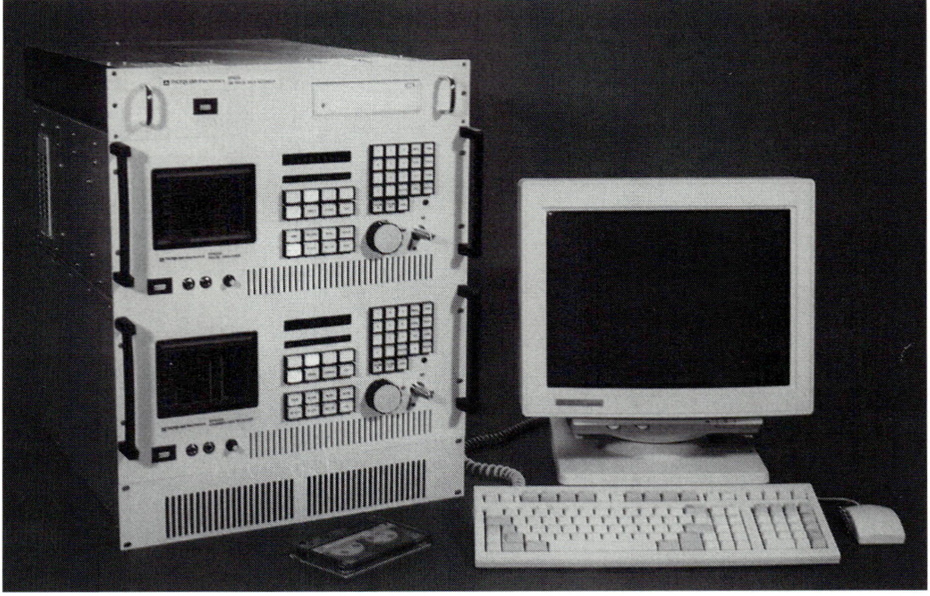

The Corvus ELINT system ***1995***

Modern radars employ modulation during each pulse in order to improve range resolution and produce profiles of targets. Some radars have unintentional modulation during each pulse and this can be used to provide unique identification.

Corvus can record every pulse of a radar over many antenna scans, complete with intrapulse data. This is measured by sampling every 10 ns and measuring both instantaneous amplitude and instantaneous frequency. These samples are then digitised and recorded directly on to a magnetic tape recorder.

The sampling and digitising technique produces a recording bandwidth of up to 50 MHz and this is essential for observing the intrapulse modulation of modern complex military radars without distortion.

Corvus links the recording technique to the channelised receiver to enable every pulse of a frequency-hopping radar to be reproduced, even if the hop range is as wide as 1 GHz.

Contractors

Racal Radar Defence Systems Ltd.

UPDATED

EP4000 Series of EW modules

A range of advanced modules for electronic warfare has been developed by Racal. The range includes the EP4410 pulse analyser and the EP4220 channelised receiver.

The new EP4410 is designed to interface with the EP4220 or conventional swept superheterodyne receivers. It is an advanced pulse analyser and uses qualification on descriptor parameters to enable signals to be selected for analysis and displayed on the high resolution colour graphics display. In addition to pulse descriptor information, the VME processor-based analyser measures intrapulse frequency and amplitude information by frequent sampling of the signals through each pulse. The EP4410 has the capacity to perform an analysis on up to 64k pulses, with intrapulse information automatically correlated with pulse descriptor data.

The EP4220 is a wideband channelised receiver and has been specifically designed for ELINT, ESM and signal monitoring applications which require high sensitivity and wide instantaneous bandwidth. Operating in the 0.4 to 18 GHz frequency band, it offers very high performance, detecting complex signals in a dense electromagnetic environment. Its wide 1 GHz bandwidth enables full examination of frequency agile transmitters. Within each channel, the detection bandwidth is sufficiently narrow to give measurements comparable with narrowband ELINT receivers.

Specifications

EP4410 pulse analyser
Dimensions: 483 × 178 × 575 mm
Weight: 20 kg
Power supply: 115/240 V AC, 40-60 Hz, 300 W

EP4220 channelised receiver
Dimensions: 483 × 310 × 580 mm
Weight: 50 kg
Power supply: 115/240 V AC, 40-60 Hz, 400 W
Frequency: 0.5-18 GHz
Temperature range: +10 to +30°C

Contractor

Racal Radar Defence Systems Ltd.

UPDATED

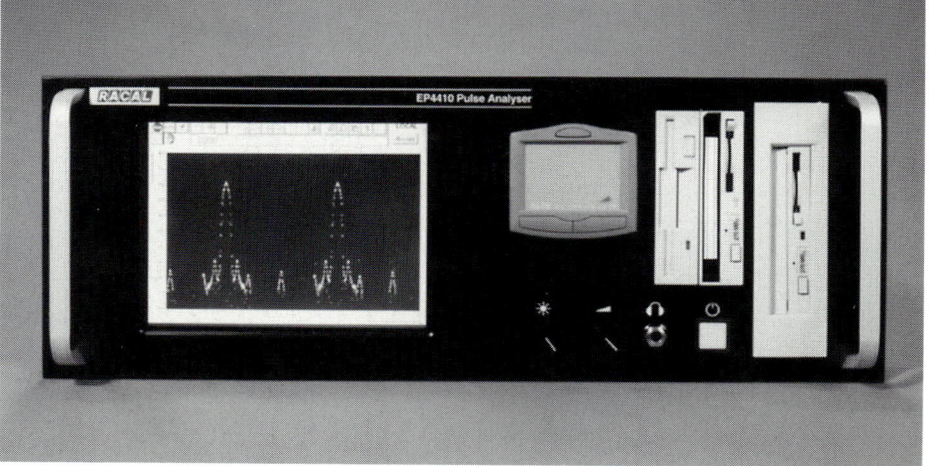

The EP4410 Pulse Analyser ***1997***/0001269

Vicon 70 countermeasures pod

Vicon 70 is a modular, lightweight dispensing pod which uses components of the Vicon 78 Series 455 system to provide self-protection against radar threats and heat-seeking missiles. It is designed for helicopters and fixed-wing aircraft where structural or other constraints preclude the use of airframe-mounted dispensers.

Vicon 70 can be configured for two dispensers, or for up to eight dispensers by adding modules to the pod. Modules can also be used to house radar warning receiver and missile approach warner LRUs. It can be fitted to any fixed-wing aircraft or helicopter type equipped with weapons racks or fuselage or wing pylons with 14 in NATO attachments. The firing trajectory of the dispenser modules can be preset to optimise the effectiveness of the chaff and flares. Manual or fully automatic versions of Vicon 70 are available. The automatic version interfaces directly with the radar warning receiver via a serial datalink or databus.

Unlike the internal CMDS installation, the podded solution requires no major airframe modification and provides a greatly enhanced chaff dispensing capability. The additional benefit of the pod is that countermeasures dispensing may be carried as role equipment and can be downloaded should self-protection not be required.

Specifications

Dimensions: 2,705 length × 356 mm diameter
Weight: 210 kg fully loaded with 8 dispensers

Contractor

W Vinten Ltd.

VERIFIED

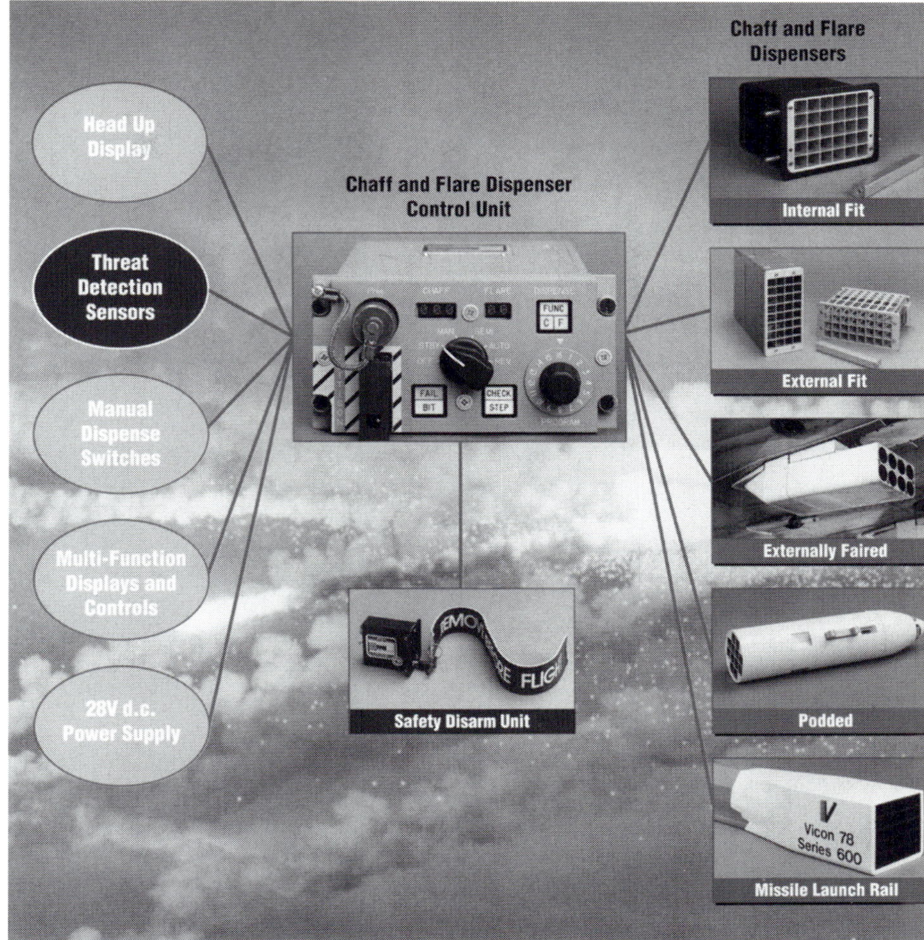

Vinten countermeasures dispensing systems ***1997***/0001268

Vicon 78 airborne decoy dispensing systems

Vicon 78 is a family of advanced lightweight chaff and IR decoy dispensing equipment in production for high-performance combat, tactical transport and maritime patrol aircraft, helicopters and RPVs. Vicon 78 systems can be fitted as original equipment or upgrade existing systems. It is a defensive aids system which provides self-protection by passive ECM against radar-guided and IR-seeking air- and ground-launched missiles and radar-directed anti-aircraft artillery fire.

Vicon 78 systems provide manual or fully automatic threat adaptive capability. Fully automatic operation, greatly reducing aircrew workload, is achieved by interfacing the dispensing system to a threat detection sensor. In automatic mode the programme to be dispensed is downloaded from the Radar Warning Receiver (RWR) or Missile Approach Warner (MAW). In semi-automatic mode the programme to be dispensed is downloaded from the RWR and initiated by the crew. In manual mode the crew can dispense preset chaff or IR flare programmes.

Systems can be offered with multidispenser configurations to cater for large transport and maritime patrol aircraft down to a single lightweight dispenser for smaller aircraft, each of which can accept interchangeable chaff or flare magazines containing up to 64 cartridges. Dispensers can also be configured to dispense expendable jammers, towed decoys and other format expendables as required. The dispensers may be internally mounted, semi-recessed with fairings or carried in a lightweight pod.

Vicon 78 Series 200

Vicon 78 Series 200 equipment has been supplied for the Sea Harrier programme. The Series 200 CounterMeasures Dispensing System (CMDS) is a two-dispenser system, controlled by a salvo control processor unit and manually operated by the pilot. Vicon 78 Series 200 equipment has successfully undergone EMC testing at Boscombe Down and has been selected by the UK MoD. The Series 205 CMDS has the ability to dispense ALE-39 and ALE-40 format rounds from a common dispenser.

Vicon 78 Series 203 equipment was developed and supplied to CASA of Spain for an export CASA 101 programme. It is a four-dispenser system controlled by a two-box processor with a fully automatic mode of operation via an interface to the aircraft Litton AN/ALR-80(V) RWR. The Series 203 system also equips the CASA CN-235 tactical transport aircraft, interfacing to the Litton AN/ALR-85 RWR.

The Vicon 78 Series 210 was selected by the Royal Air Force for operations by the Tornado F3 in the Gulf War. It utilises a new format dispenser that fires a 55 mm two shot infrared decoy flare. After the 1990-91 Gulf War, the Series 210 was incorporated as a fully approved modification on the Tornado F3 aircraft.

Vicon 78 Series 300

The Vicon 78 Series 300 equipment was developed for the BAe Hawk 100 and 200 Series aircraft. In the Hawk, the Series 300 CMDS is a digitally controlled, fully automatic two-dispenser system that interfaces via an RS-422 link to the Racal Prophet or Marconi Sky Guardian RWRs. The Series 300 system has been designed with the capability to control up to 16 dispensers, with each dispenser having the capability to be configured to hold up to 64 chaff or flare cartridges. Each dispenser can accept interchangeable chaff and flare magazines. When power is applied, the system will identify the number and type of expendables in each dispenser and display the count of chaff and flare cartridges on the cockpit control unit.

A continuous BIT facility is executed by the system; in the event of failure, fail indications are displayed on the cockpit control unit. The BIT is also designed as a first line maintenance facility to identify faults and so speed rectification.

The Vicon 78 Series 300 CMDS has been designed to interface with any RWR or ESM system that has a serial datalink interface capability. The RWR/CMDS interface has been successfully achieved with Litton, Marconi and Racal. Agreements are also in place with other RWR manufacturers. Vicon 78 CMDSs have the ability to interface to MAWs and missile launch detectors.

Vicon 78 Series 400

The Vicon Series 400 incorporates the advanced electronics of the Series 300, which have been improved and miniaturised, with the eight cartridge 16 shot 55 mm dispenser format of the Series 210. The

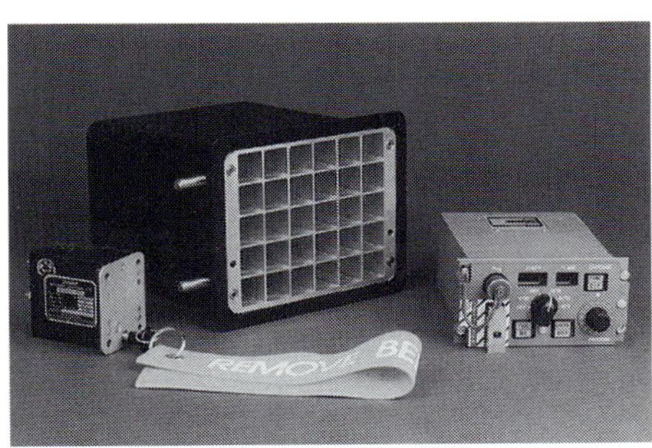

Vicon 78 Series 455 ***1996***

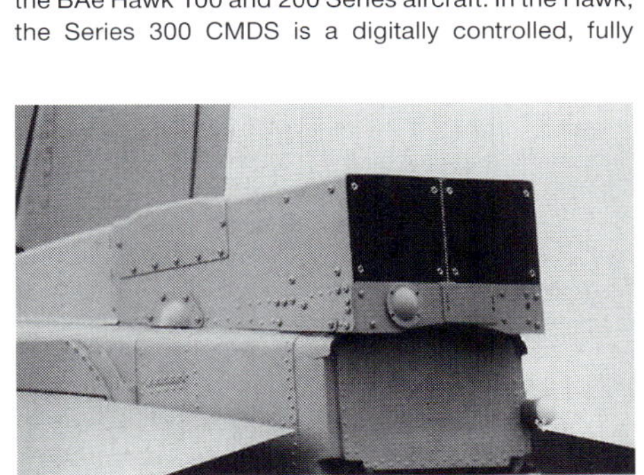

Rear view of a Hawk showing the Vicon 78 Series 300 countermeasures dispenser ***1995***

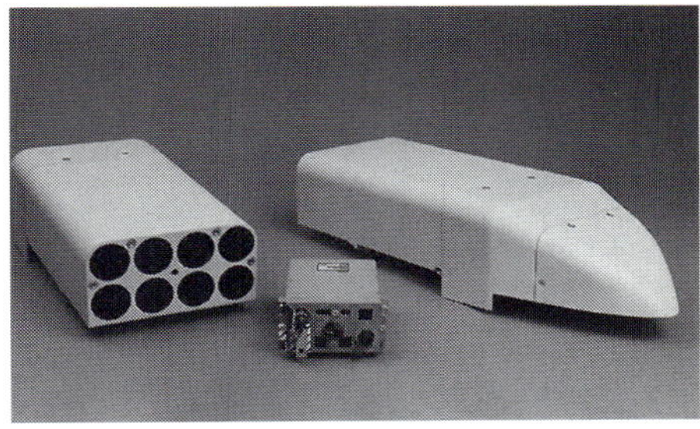

The Vicon 78 Series 400 is fitted to UK and Italian Tornado F Mk3 aircraft ***1995***

system utilises a single miniature LRU Chaff and Flare Dispenser Control Unit (CFDCU), which can be either cockpit- or remote-mounted, and provides the system control, processing and programme loading functions.

The Vicon 78 Series 400 has been selected for a fleetwide upgrade of Royal Air Force Tornado F3 aircraft.

Vicon 78 Series 420

The Series 420 is based on the 400 system electronics, with the Countermeasures Dispensers (CMD) being reconfigured to a 1 in × 1 in payload, 8 × format suitable for helicopter platforms. Each 8 × 4 CMD can carry 32 expendables and the system can be configured for up to four dispensers. The 420 system is currently equipping British Army Air Corps Lynx and Gazelle helicopters.

Vicon 78 Series 455

The Series 455 CMDS is the latest advanced lightweight system suitable for helicopters, fast jet, MPA and transport aircraft. Dispensers are mechanically interchangeable with ALE-40/47 LRUs. Due to its low mass and reduced number of LRU's, the 455 CMDS is ideal for retrofit and upgrade programmes whilst allowing ease of integration into new-build aircraft. The system can control up to 24 Countermeasures Dispensers and two BOL chaff dispensers. The Dispensers may be in different sized formats offering a wide selection of payload choice including Vintens new 6 × 5 sized dispenser. The system is NVG-compatible, has a number of different threat sensor interfaces and may also be integrated via a MIL-STD-1553B databus. For upgrade purposes the system may be grown into a large multidispenser equipment by the addition of further Dispensers without the need for sequencer LRU's. The Vicon 78 Series 455 is configured for Hawk 100/200 upgrades. It has been selected as the CounterMeasures Dispensing System element of the UK WAH-64 Apache attack Helicopter Integrated Defensive Aids System (HIDAS).

The Vicon 78 Series 300 countermeasures dispenser has been developed for Hawk 100 and 200 aircraft

The Vicon 78 Series 600 countermeasures dispenser is designed for the common rail launcher

The Vicon 78 Series 420 is currently equipping British Army Corps Lynx and Gazelle helicopters

1997/0001270

Vicon 78 Series 500

The Vicon 78 Series 500 lightweight stand-alone dispenser utilises the electronics of the Vicon 78 Series 300 system reconfigured to provide a simple and cost-effective dispensing solution for light aircraft, helicopters and RPVs. The 12-shot dispenser has a simple mechanical and electrical interface and can be operated from the cockpit or, in the case of an RPV, via an RF link. The system can be configured to interface to an MAW.

Vicon 78 Series 600

The Vicon 78 Series 600 is a chaff and flare dispenser which can be mounted on the end of a missile launch rail. The system allows additional countermeasures to be carried without requiring major airframe modifications and with only a small weight and drag penalty.

Operational status

Vicon 78 dispenser applications are as follows: Vicon 78 Series 200 Sea Harrier FRS Mk 51 aircraft; Vicon 78 Series 203 export-CASA 101 (interfacing with AN/ALR-80(V) RWR) and CN235 (interfacing with the AN/ALR-085 RWR) programmes; Vicon 78 Series 210 fully approved modification for Royal Air Force Tornado interceptors post 1990/1991; Vicon 78 Series 300 all export Hawk training/light strike aircraft; Vicon 78 Series 400 UK and Italian Tornado F Mk3 aircraft; Vicon 78 Series 420 British Army Air Corps Gazelle and Lynx helicopters; Vicon 78 Series 455 Royal Air Force Nimrod MR2 Mk2 aircraft, export Hawk training/light strike aircraft, Aerovodochody L159 light multirole combat aircraft, export C-130 aircraft, and several rotary-wing aircraft and UK WAH-64 Apache HIDAS.

Contractor

W Vinten Ltd.

UPDATED

UNITED STATES OF AMERICA

AN/ALQ-161 ECM system for the B-1B

AIL Systems is prime contractor for the AN/ALQ-161 suite for the US Air Force Rockwell B-1B. This was launched in 1972 under AIL Systems leadership for the Rockwell B-1A bomber, but the effort came to a halt when the aircraft was cancelled in June 1977. The programme was, however, reinstated in October 1981. AIL Systems accordingly resumed its task as ALQ-161 project leader, having participated in limited flight trials with one aircraft since 1979. While the system has undergone some redesign to incorporate new requirements and technology that have arisen during the dormant period, a basis of confidence was built up on the results of more than 400 hours of operation during 95 flights with a B-1A prototype. The principal goal of these early trials was to demonstrate that the system would operate in harmony with the B-1A's flight control system and offensive avionics system. The current equipment was developed under restart and initial production contracts, worth more than US$1,700 million, which were awarded to AIL Systems and its subcontractors. Each aircraft set is understood to cost around US$20 million, approximately one-tenth the cost of a complete B-1B, and comprises no fewer than 108 LRUs, with a total weight of 2,300 kg exclusive of cabling, displays and controls. Total cost of the ALQ-161 programme is estimated at more than US$2,000 million.

The operating frequency range is approximately 0.5 to 10 GHz, to cover early warning, ground-controlled interception, surface-to-air missile and interceptor radar frequencies. Jamming signals in the higher regions of the ECM spectrum are emitted from three electronically steerable, phased-array antennas: one in each wing-glove leading-edge and the third in the fuselage tailcone. Each antenna provides 120° azimuth and 90° elevation coverage. Lower frequency signals are emitted from quadrantal horn antennas mounted alongside the high-frequency equipment.

Major subcontractors in the AN/ALQ-161 programme are Northrop Grumman, Litton Industries and Sedco Systems. All companies serve as subsystem managers and have responsibilities for receivers, data processing and jamming techniques. Northrop Grumman provides the low-frequency jamming antennas and Sedco is supplying the phased-array antennas.

AIL Systems is responsible for system integration and for the LRUs, including 51 unique designs. Most of the LRUs are about 0.03 to 0.06 m^3 in volume and weigh between 18 and 36 kg. The majority are readily accessible and can be removed easily or installed by one or two people. Total power required is about 120 kW.

Key improvements introduced since the programme restarted include: a frequency extension into the K-band to improve the warning and jamming performance; extension into the low-frequency domain below 200 MHz to improve the response of the warning system; introduction of a digital radio frequency memory to permit the deception jamming of more advanced radars, notably those employing pulse Doppler techniques; and introduction of a new tail warning system. The last was originally to have been the Westinghouse AN/ALQ-153 radar, chosen in competition with AIL Systems' own AN/ALQ-154. Both systems had been flight tested in 1978 and the ALQ-153 was selected to update the Boeing B-52G. Subsequent investigation by AIL Systems, however, showed the tail warning function could be integrated with the ALQ-161 to a very large extent with considerable savings in weight and cost. AIL Systems was accordingly awarded a US$9.1 million contract in September 1983 to extend the ALQ-161 system to include this function. In another change, an IBM 101D computer replaced the earlier Litton system as the central processor, making for commonality with a similar machine elsewhere in the aircraft. The computer uses software based on Jovial to identify the function of every hostile radar, assess its potential threat and assign a jamming priority. Three sets of phased-array antennas mounted in the wing leading-edges and tail provide full 360° coverage in Bands 6, 7 and 8, antennas for lower frequencies being located fore and aft in the airframe.

Some indication of the magnitude of the system is provided by the observation during early design that although the preferred subsystem configuration involved an increase in uninstalled weight of 57 kg, it eliminated 304 kg of cabling.

Two complete systems were delivered for continuing trials with a modified Rockwell B-1A in the Summer of 1984 and for the first production B-1B; the latter made its initial flight in October 1984. Production rate was built up to four systems a month by May 1986. Continued testing and evaluation was supported by a comprehensive integration test-rig at AIL Systems Deer Park, New York plant. In August 1985 AIL Systems was awarded an US$1,800 million contract for the final 92 ALQ-161 shipsets, believed to be the largest ever EW contract.

Problems in development were reported in mid-1986 leading to the first 22 B-1Bs not being equipped with the tail warning system which had been scheduled to enter service in mid-1987. Problems with the tail warning system and with repeatability of results from the system as a whole were a part of a list of deficiencies in the B-1B's performance made public at the end of 1986. Of a US$600 million request made by the US Department of Defense at that time to extend the capabilities of the B-1B, almost US$100 million was allocated to bring the ALQ-161 up to its original specification and solve other problems (according to the DoD), while the bulk of a US$131 million element within the US$600 million total requested for expenditure in 1988/89 was to upgrade the defensive avionics suite to meet capabilities which have emerged since the original baseline was set in 1982.

The AIL AN/ALQ-161 system for the B-1B

Operational status

Production of 100 systems is complete and the system is installed in the B-1B. In February 1991 AIL was awarded a US$16.5 million contract for AN/ALQ-161A systems for the Northrop B-2 and an additional US$5.491 million to upgrade systems for the B-1B.

Contractor

AIL Systems Inc.

VERIFIED

AN/ALQ-176(V) ECM pod

The AN/ALQ-176(V) is a pod-mounted jamming system designed, developed and tested to fulfil aircraft mission requirements for ECM support, standoff jamming and combat evaluation and training. The pod system comprises transmitter modules and a complementary antenna housed in canisters to form a slim pod with a 10 in (25 cm) diameter. The AN/ALQ-176(V) is mounted on standard aircraft wing or fuselage stores/ munitions stations. The pod offers the option of using aircraft internal power or a ram-air turbine.

The modular design of the AN/ALQ-176(V) pod allows flexibility in ECM configurations. A two- or three-canister arrangement is available. The AN/ ALQ-176(V)1 two-canister configuration houses up to three voltage-tuned magnetron transmitters. The AN/ ALQ-176(V)2 three-canister version provides housing for five transmitters. Dependent on frequency band and tube selection, each transmitter produces 150 to 400 W CW (up to 30 per cent bandwidth) at efficiencies of greater than 50 per cent. The transmitters provide ECM across the specified threat frequencies. Transmitter protection circuits safeguard the system and provide fault status to the cockpit control panel.

The AN/ALQ-176(V) has been specifically engineered to allow rapid turnround in meeting the demands of new threat frequencies. It can accommodate new demands through selection of different stored parameters or through easy replacement of the transmitter tube or antenna. Transmitter spares, maintenance costs, times and training skills required are greatly reduced by the use of standard transmitter design. In addition the design incorporates built-in test and provides for parameter/ mode selection at the cockpit control panel.

The flexibility in ECM pod configuration and standard aircraft stores/munitions stations installation enables the AN/ALQ-176(V) ECM pod to be used on a wide variety of tactical and support aircraft.

Specifications

Dimensions: 250 mm diameter
(AN/ALQ-176(V)1) 1,990 mm length
(AN/ALQ-176(V)2) 2,590 mm length
Weight:
(transmitter module) 17 kg
(AN/ALQ-176(V)1 pod) 112 kg
(AN/ALQ-176(V)2 pod) 196 kg
Input/output power:
(AN/ALQ-176(V)1) 2.6 kVA/1.2 kW
(AN/ALQ-176(V)2) 4.5 kVA/2 kW
Frequency: 0.8-10.5 GHz
Transmitter power output: 150-400 W per tube CW (dependent on frequency and tube selection)
Modulation: noise (various)
Control: ALQ-10725 and C-9492 aircraft ECM cockpit control panel

Operational status

The US Air Force, the Royal Norwegian Air Force, the Royal Thai Air Force, and the Canadian Ministry of Defence have the AN/ALQ-176(V) in their inventories.

Contractor

Alliant Defense Electronics Systems Inc, a wholly owned subsidiary of Alliant Techsystems Inc.

VERIFIED

The AN/ALQ-176(V)1 in two-canister configuration with ram-air turbine electrical supply and internal electronics tray showing the modular design

AN/ALE-29A/29B system

The AN/ALE-29A countermeasures dispensing equipment is used on tactical aircraft for the controlled ejection of chaff and IR decoy flares as a means of self-defence against radar-directed and IR homing missiles. The equipment comprises a programmer, two sequencer switches, two dispensers each consisting of a block and printed circuit board, two dispenser housings and interconnecting cable. The ALE-29A system carries 30 chaff cartridges or IR flares per dispenser. RR-129 and RR-144 chaff cartridges and Mk 46 and Mk 47 IR flares are used.

The programmer is normally located in the aircraft cockpit and carries the necessary information for selection of dispensing patterns, timing sequences and so on. An optional feature is a dual-channel programmer which allows the simultaneous dispensing of chaff and flares via separate programmes.

The AN/ALE-29B is an improved version allowing greater payload flexibility.

The AN/ALE-29 system can be interfaced with threat warning devices and forms part of the tactical jamming system on the US Navy EA-6B Prowler aircraft.

Operational status

Operational on US Navy EA-6B Prowler aircraft.

Contractor

Alliant Defense Electronics Systems Inc, Lundy Products Group.

VERIFIED

AN/ALE-43 chaff cutter/dispenser pod

The AN/ALE-43 is a high-capacity chaff system which holds rolls of chaff material and cuts it to the appropriate dipole length during operation. Lundy claims to have overcome the difficulties associated with past chaff cutters by developing a unique chaff roving supply system.

The system has a chaff roving hopper behind which is the chaff cutter. A remote processor controls operation of the cutter.

Chaff is drawn simultaneously from up to nine chaff roving packages in the hopper. Each roving package passes through a guide tube which terminates at draw rollers and a cutting roller. As dipole lengths are cut, they are discharged into a turbulent airflow and distributed efficiently behind the pod. Each cutter assembly consists of a drive motor, clutch/brake unit and three indexing cutting rollers, the latter embodying blades which yield specific combinations of dipole lengths.

The system can be podded or mounted internally. It is frequently used as a training aid, but its primary wartime functions would be for anti-ship missile defence, area saturation operations over battlefields, corridor seeding and aircraft self-protection.

Specifications

Podded version
Dimensions: 3,370 (length) × 480 mm (diameter)
Weight:
(empty) 139 kg
(loaded) 284 kg
Chaff payload: 8 × RR-179 roving packages
Max dispensing rate: 7.2 × 10^6 dipole in/s

Internal version
Weight:
(basic) 37 kg
(max loaded) 191 kg
Chaff payload: 9 × RR-179 roving packages
Max dispensing rate: 8.1 × 10^6 dipole in/s

Common characteristics
Power supply: 115 V AC, 400 Hz, 1.7 kVA
28 V DC, 2.5 A
Max continuous dispense time: 11 min
On-time: select 1-9 s in 1 s steps or continuous
Off-time: select 1-9 s in 1 s steps
Altitude: 0-50,000 ft

Operational status

In service. The podded version can fit a variety of standard pylon attachments and is used on the B-52, F-4 Phantom, EA-4A Skyhawk and EA-6B Prowler. Internally mounted versions have been used on the ERA-3B Skywarrior and NKC-135.

Contractor

Alliant Defense Electronics Systems Inc, Lundy Products Group.

VERIFIED

AN/ALE-43(V)1/(V)3 chaff countermeasures dispenser set

The AN/ALE-43(V)1/(V)3 chaff countermeasures dispensing set is designed for two types of installation: as a pod (mounting per MIL-A-8591 both 14 inch and 30 inch mounting), or internal mount. In the pod installation, the chaff cutter assembly is mounted on a structural bulkhead at the rear of the pod centre section. This assembly consists of a drive motor, clutch/brake unit, cutting mechanism and supports. The cutting mechanism comprises a rubber platen roller and three cutter rollers.

Specifications

AN/ALE-43(V)1 Pod Version
Dimensions: 3,370 (length) × 482 mm (diameter)
Weight (empty): 154 kg
Power requirements: 115/208 V, 400 Hz, 3-phase 1,700 VA and 28 V DC, 2.5 A
Chaff supply: standard RR-179/AL roving bundle 18 kg
Chaff payload: 144 kg (8 bundles)
Dispensing rate: 216 g/s per 8 roving bundles
Modes (2):
(pulse): 1-9 s in 1 s intervals
(continuous): 660 s

AN/ALE-43(V)3 Internal Version
Dimensions: 408 × 338 × 310 mm
Weight (empty): 36 kg
Power requirements: 115/208 V, 400 Hz, 3-phase 1700 VA and 28 V DC, 2.5 A
Chaff supply: standard RR-179/AL or special roving bundle
Chaff payload: up to 225 kg (up to 9 roving bundles)
Dispensing rate: 27 g/s per roving bundle
Modes (2):
(pulse): 1-9 s in 1 s intervals
(continuous): 880 s

Operational status

The AN/ALE-43(V) pod version has been proven on SH-3 helicopter, EA-6A, EA-7A, F-4, F-16, F-18, P-3, Learjet 36 and other aircraft. Internal installation versions are operational on such aircraft as: EW Falcon, EW Challenger, ERA-3B, NKC-135 and Learjet 35.

Contractor

Alliant Defense Electronics Systems Inc, Lundy Products Group.

VERIFIED

AN/ALE-44 dispenser pod

The AN/ALE-44 is a chaff or flare dispenser system suitable for supersonic aircraft which is usually installed as a two-pod system with a control unit in the cockpit. Each pod houses two dispenser modules and a sequencer. The pods are lightweight units which can be installed on wingtip, underwing or underfuselage store locations. They have dual-channel dispensing capability and can be quickly reloaded with RR-129 chaff or Mk 46 infrared flare cartridges. The cockpit unit permits selection of burst rates, burst interval and units per burst. Flares and chaff can be dispensed simultaneously.

Specifications

Dimensions:
2,060 (length) × 110 × 180 mm (section)
Weight:
(empty) 13.6 kg
(with 32 chaff cartridges) 19.9 kg
(with 32 flare cartridges) 22.6 kg
(control unit) 1.1 kg
Modes: 1 or 2 units per burst
Programmes: 1, 2, 4, 8 or continuous bursts
Rate: 4, 2, 1 or ½ bursts/s

Operational status

In production.

Contractor

Alliant Defense Electronics Systems Inc, Lundy Products Group.

VERIFIED

Countermeasure system

A recent development has been a 14-cartridge chaff/flare dispenser which can be accommodated on virtually any surface by using a contoured mounting plate. Aircraft countermeasure loads can be tailored to missions by using different numbers of dispensers, and chaff cartridge capacity is claimed to be adequate to protect aircraft such as Lockheed F-16 and McDonnell Douglas F-4 Phantom against radars in the E- to J-bands within 2 seconds of ejection.

Specifications

Dispenser
Dimensions: 710 × 130 × 150 mm
Weight: 14.2 kg with flares

Cartridge
Dimensions: 127 (length) × 40 mm (diameter)
Weight:
(I/R) 0.32 kg
(chaff) 0.3 kg
Flare: 15 kW output, 2-5 m range
Chaff: radar cross-section of 20 m² within 1-2 s of ejection

Operational status

Under development.

Contractor

Alliant Defense Electronics Systems Inc, Lundy Products Group.

VERIFIED

FAC dispenser

The Lundy FAC countermeasures dispenser is a compact lightweight system which fires both chaff and flares. Designed for the US Air Force to dispense the percussion-initiated Mk 50 flare, the FAC dispenser has been successfully flight-tested on Cessna O-2 aircraft and Bell UH-1 helicopters.

To augment its operational usefulness, Lundy has developed and flight-tested a percussion-initiated chaff unit which is compatible with the FAC dispensing mechanism. The chaff unit provides effective protection against known radar-controlled threats operating in the 3 to 16 GHz frequency range.

The basic FAC system contains a cockpit control/display unit, two payload modules and two firing mechanisms. A cascade feature in the control circuitry allows any number of dispenser assemblies to be added to the basic system. The payload module holds 20 chaff units or flares interchangeably and is removed for loading from the aircraft during ground operations. The FAC dispenser is configured for both internal and external mounting.

Although primarily developed for use with forward air control aircraft types, the FAC dispenser can be used to fire any 40 mm munitions from air or land vehicles.

Contractor

Alliant Defense Electronics Systems Inc, Lundy Products Group.

VERIFIED

AN/ALE-54 mini-chaff cutter

The AN/ALE-54 mini-chaff cutter cuts chaff dipoles in flight to provide flexible chaff size. It can be used on a variety of platforms and configurations, such as small and large fixed-wing aircraft and helicopters or UAVs, with internal or pod mounting, and on ground vehicles. Based on AN/ALE-43(V) technology, the mini-chaff cutter is a low-cost replacement for AN/ALE-39/40/47 chaff dispensers. It is claimed that the cutter can generate, in a very short burst, chaff equivalent to the dispensing of three RR-129 or RR-170 cartridges.

The mini-chaff cutter features: threat adaptable cutter rollers, used to determine the length of chaff dipoles, to cover frequencies from 1 to 100 GHz; is easily replaceable to counter changing threats; low dispenser and payload cost; enhanced break lock capability, with up to six times increase in effective chaff use compared to standard pyrotechnic self-protection dispensers and a reduced weight of 11 kg including a self-contained payload package sufficient for 60 break lock events. The system has flexible payloads including economic self-contained RR-194 mini-roving bundles in standard throwaway flight line replaceable packages, a standard, separately located 18 kg RR-179/AL roving bundle for high-volume chaff requirements and variable chaff fibre count for specific mission requirements. The mini-chaff cutter enhances survivability by increasing chaff capacity to allow for greater flare carriage. There is no ejection of plastic parts and the device emits no pyrotechnic flash to disclose position during night operations.

Contractor

Alliant Defense Electronics Systems Inc, Lundy Products Group.

VERIFIED

Supersonic countermeasure pod

The supersonic countermeasure pod is suitable for a wide range of strike aircraft. It comprises a control unit and two dispensing pods, each pod holding two dispensing modules and a sequencer. Total system capacity is 64 RR-129 chaff dispensers or Mk 46 infrared flare cartridges. Flight qualified for supersonic flight, and with a low frontal area, the system is marketed as being suitable for installation in a pod or on a pylon. The control unit permits selection of burst rate, burst interval and units per burst. Specification is identical to the AN/ALE-44 unit.

Operational status

In development.

Contractor

Alliant Defense Electronics Systems Inc, Lundy Products Group.

VERIFIED

AN/ALR-75(V) surveillance receiver

The AN/ALR-75(V) is a surveillance receiver providing coverage over the frequency range 0.1 to 18 GHz. It consists of an aircraft style set of control panels and a number of remote-control units in ATR configurations. The system includes complete digital control, computer compatibility, scratchpad memory and digital refreshed displays. The AN/ALR-75(V), also known as the SCR-2100, is one of a number of equipments in the SCR series of receiving equipments.

Operational status

In operational service in the ERA-3B, NKC-135A and EC-24A aircraft, and also as a ground-based equipment in AN/ULQ 13(V)1 vans.

Contractor

Andrew SciComm Inc.

VERIFIED

AN/ALQ-133 ELINT system

The AN/ALQ-133 Quick Look II is a tactical ELINT system deployed in the OV-1D Mohawk aircraft for emitter detection and passive surveillance by the US Army. It has also been proposed for use in a number of A-10 tactical aircraft.

The AN/ALQ-133 consists of two pods: one contains ELINT receivers covering the frequency range 400 MHz to 18 GHz and equipped with the appropriate broadband antennas, the other contains data processing equipment. The pod antennas are phased interferometer elements capable of providing direction of arrival measurements over a 90° sector abeam of the aircraft to a typical accuracy of 0.5°. If some degradation of bearing accuracy is acceptable, coverage can be enlarged to 120°.

The data processor is responsible for control of the search receivers, analysis of intercepted signals and comparison against a file of known hostile emitters. It also records the signal parameters for future use. Information can be relayed by datalink to a ground station.

Operational status

In operational service with the US Army. Quick Look II is being replaced by the improved Guardrail sensor.

Contractor

ARGOSystems Inc.

VERIFIED

AN/ALR-52 ECM receiver

The AN/ALR-52 is a multiband Instantaneous Frequency Measuring (IFM) receiver for airborne or land applications. A naval variant is also produced, as the AN/WLR-11. Typical systems cover the microwave frequency band from 0.5 to 18 GHz, using receiver modules that cover octave bandwidths. Additional capabilities include: provision for selection of a particular signal or frequency band; blanking of signals which are of no further interest; separation of interleaved pulse trains; measurement of radar parameters; analysis of CW in addition to pulse trains; and direction of arrival measurement. The complete equipment provides for two operator positions, each equipped with a control unit and a display. The two IFM control units are capable of complete control over any band selected. If the same band is selected by both units, the video processing unit gives priority control to one position only. All intercepted signals, however, are presented on the displays at both positions.

A modified version of the ALR-52 has been developed and deployed, with one of the operator positions replaced by an interface unit to link the IFM receiver with a digital computer. This system employs a 1 to 18 GHz DF antenna system. The computer uses the data to tag every received pulse with its frequency, pulsewidth, time of arrival and azimuth. This information is stored and used to develop a library of emitter parameters.

Operational status

In operational service. The AN/ALR-52 is fitted to the EP-3E ELINT aircraft operated by the US Navy.

Contractor

ARGOSystems Inc.

VERIFIED

AN/AYR-1 ESM system

The AN/AYR-1 is an advanced ESM system which is part of the upgrade of US Air Force and NATO E-3 AWACS aircraft. It will carry out passive interception, identification and analysis of radar and radio signals. The antennas are believed to be installed in a blister on either side of the aircraft, just aft of the cockpit, and in nose and tail arrays, to provide 360° azimuth coverage. The extra processing units required to analyse signals quickly and classify radar returns, plus associated equipment, are expected to add some 855 kg to the overall aircraft weight.

Operational status

In production for the US Air Force and NATO. A January 1997 firm fixed-price contract award to Boeing Defense and Space Group provided for integration of four ESM systems to E-3F AWACS aircraft at a value of US$32.4 million. Contracting authority is Electronic Systems Center, Hanscom AFB.

Contractor

ARGOSystems Inc.

VERIFIED

AR-700 ESM/DF system

The AR-700 ESM/DF system is a passive electronic warfare system designed for tactical operation in a dense signal environment. It automatically receives and processes signals in the frequency range 2 to 10 GHz to identify and determine threat potential and the bearing of intercepted signals. The information is presented to the operator on alphanumeric and tactical graphical displays. Alarms warn the operator if specific threats are detected. The processed information can be logged on a printer and can be routed to the datalink processing system.

A cassette recorder is used to load emitter and platform libraries into the signal processor. The emitter library can be created, expanded or changed by keyboard entry at the operator console. The keyboard is also used to operate the system and select various display formats through the use of special function keys and alphanumeric inputs. Additional special function keys are located on the operator control unit.

A special interface in the advanced signal processor sends emitter reports to the datalink interface. The system also receives aircraft heading from the inertial navigation system in the bearing processor unit.

The AR-700A is a higher performance version of the AR-700, with missile configurations for use on a variety of platforms. It has wide open 0.5 to 18 GHz frequency coverage, a near 100 per cent probability of intercept, a one second reaction time and excellent DF accuracy.

Specifications

Frequency: 2-8 GHz, 8-18 GHz
Frequency accuracy:
3 MHz (2-8 GHz)
5 MHz (8-18 GHz)
PRF: 123 to 20,000 pps
PRF resolution: 0.1 Hz
Accuracy:
(coarse) 5° (70% angular coverage),
8° (95% angular coverage)
(fine) 2.3° (70% angular coverage),
4° (95% angular coverage)
Library: 990 emitters
Emitters tracked: 200

Operational status

In production.

Contractor

ARGOSystems Inc.

VERIFIED

AR-730 ESM/DF system

The AR-730 ESM/DF system is a modular equipment designed for maritime patrol aircraft and anti-submarine warfare applications. It provides surveillance and threat warning over the frequency range 0.5 to 18 GHz with automatic signal and data processing and precision DF. The system consists of antenna assemblies, an RF processing assembly and a control/processing assembly.

The antenna assemblies are designed for simple installation on any aircraft and consist of an omnidirectional antenna channel for signal acquisition, rotating DF antenna for direction-finding and optional coarse DF antennas and crystal video receivers. For the AR-730, maximum weight savings are realised by mounting the high-gain DF antenna on the back of the radar antenna. Alternative configurations include separate antenna/radome mountings on the underside of the aircraft.

The RF processing assembly contains the receivers. A DIFM receiver provides high probability of intercept. A superhet provides added sensitivity for finding low-power emitters, allowing searching of dense portions of the frequency spectrum and the 0.5 to 18 GHz band.

The control/processor assembly includes a signal processor and optional cassette recorder, printer and operator display/keyboard. Automated acquisition and identification of intercept is provided, with prompt notification to the avionics computer system. Information provided to the avionic system includes intercept reports, intercept updates, threat alerts and systems alarms. The only information required for processing is a threat/identification database.

Contractor
ARGOSystems Inc.

VERIFIED

AR-900 ESM system

The AR-900 is a high-performance, fully automatic, passive surveillance system that has been designed for both Electronic Warfare (EW) and ELINT applications. It provides both 100 per cent Probability Of Intercept (POI) and full 2 to 18 GHz frequency coverage. The system comprises three major assemblies: the antenna assembly, the receiver/processor unit, and the operator's workstation. The antenna assembly uses a wideband omnidirectional antenna and a wide open amplitude monopulse DF subsystem that provides better than 3° RMS bearing accuracy.

The receiver/processor unit comprises: two Digital Instantaneous Frequency Measurement (DIFM) receivers, an amplitude-monopulse bearing processor, and a signal processor.

This equipment can process 1,000,000 pps. The signal processor receives digitised high-resolution frequency measurements for each received Radio Frequency (RF) pulse and Continuous-Wave (CW) signal from the DIFM receivers, while the amplitude-monopulse bearing processor provides direction of arrival and amplitude of each pulse. The signal processor processes this information in parallel, compares the pattern with those in a signal library to identify the emitter and its platform, and alerts the operator to a high-threat signal within 1 second of acquiring the signal. Emitter libraries can be programmed with up to 10,000 emitter modes. The operator's workstation is made up of a keyboard with integrated trackball; a high-resolution colour display; a 3.5 in floppy disk drive; CD-ROM; printer; and an embedded computer.

The AR-900's operator is provided tactical and intelligence information through simple, comprehensive displays. The activity, tactical graphics, and tactical summary pages give the operator threat-warning information, while the frequency × azimuth, frequency × PRI, and frequency × amplitude pages are provided for analysis. Various displays aid in system setup, and an intercept report generator is provided to prepare intelligence reports.

AR-900 ESM system **1998**/0018250

Specifications

Frequency range	**2 to 6 GHz**	**6 to 18 GHz**
System sensitivity	−65 dBm	−65 dBm
Azimuth coverage, instantaneous	360°	360°
DF measurement accuracy, rms	3.5°	2°
Displayed resolution	1.0°	1.0°
Dynamic range	>70 dB	
Signal types received	Conventional pulse trains, agile frequency (±10%), staggered PRI (2-16 positions), jittered PRI (±10%), frequency/phase/amplitude or pulse, CW, FMCW, pulse Doppler	
Frequency measurement		
(displayed resolution)	1 MHz	1 MHz
(accuracy, rms)	2 MHz	3 MHz
Pulsewidth measurement		
(range)	0.1-230 μs (0.05-230 μs with 2 dB reduction in sensitivity)	
(resolution)	0.05 μs	
Amplitude measurement		
(range)	>60 dB	
(resolution)	1 dB	
PRI measurement		
(range)	2-20,000 μs	
(resolution)	0.1 μs	
(display)	PRI (μs) or PRF (Hz)	
Scan types	Circular, conical, bidirectional, unidirectional	
Polarisations received	Horizontal, vertical, slant linear, circular	
System alarms	Threat, steady illumination, CW, amplitude	
Threat library capacity	10,000 emitter modes	
Number of signals tracked	500	
System reaction time	1 s max	
Pulse density capacity	1,000,000 pps	
Options		
(RF extensions)	0.5-2.0 and 18-40 GHz	
(superheterodyne receiver)	0.5-18 GHz	
(fine DF)	1.0°	

Contractor
ARGOSystems Inc.

NEW ENTRY

AR-7000 airborne SIGINT system

The AR-7000 is designed to provide airborne reconnaissance, direction-finding and geolocation of emitters in all bands from 20 MHz to 18 GHz. Three subsystems are used to provide this full band coverage of ELINT, COMINT and ESM.

The ELINT subsystem provides for operator controlled signal search, analysis and recording of all pulsed signals from 0.1 to 18 GHz. Three superheterodyne tuners are provided for the 0.5 to 18 GHz tuning range; a fourth heterodyne tuner covers the 0.1 to 1 GHz range. A pulse analyser performs detailed numeric emitter parameter measurements.

The ESM subsystem provides instantaneous wideband intercept, automatic detection, identification and DF on all signals from 0.5 to 18 GHz. The wideband instantaneous coverage in a dense complex environment is provided by three DIFM receivers and by crystal video receivers which furnish monopulse DF data. Additional DF capability is given by a fine DF subsystem which uses phase interferometers for high-accuracy DF measurement on a tasked basis. An emitter library of up to 5,000 emitters may be tracked in the active emitter file.

The COMINT subsystem provides up to 20 receivers from as many as five operators to use in monitoring the 20 MHz to 1 GHz band. In addition, a DF subsystem furnishes DF on communication signals from 20 to 500 MHz. The DF subsystem uses two phase-matched receivers and two antenna arrays to perform highly accurate multiple differential phase measurements on signals from 20 to 500 MHz. The multiple differential phase measurements are converted to DF estimates by the subsystem processor.

Operational status
The AR-7000 forms the basis of the Fokker 50 SIGINT Black Crow aircraft proposal.

Contractor
ARGOSystems Inc.

VERIFIED

F-15 Precision Direction-Finding (PDF) system for SEAD missions

Designed to detect and accurately locate enemy radar and other RF transmitters, the new PDF system consists of a receiver and digital processor that is connected to conformal antenna arrays developed by Boeing. The system was jointly developed with Litton Amecom and TRW.

For demonstration, the PDF receiver processor has been installed in the F-15's 20 mm cannon ammunition bay. The conformal antennas are positioned around the nose of the aircraft, aft of the radome.

By providing aircrews with precise targeting, situation awareness and threat information, the PDF will add a key sensor to the existing suite of integrated avionics.

When coupled with the JTIDS/Link-16 system, the PDF could be used to help direct attacks against SAM sites and other defence weapon systems, or to warn other aircraft of their presence.

Operational status
Boeing began a series of flight tests in August 1996 to demonstrate the use of a new Precision Direction-Finding (PDF) system on the F-15 Eagle for conducting air defence suppression missions.

Contractors
The Boeing Company.
Litton Amecom.
TRW.

UPDATED

AN/AAR-44 missile warning receiver

The AAR-44 infrared passive airborne warning receiver continually searches a hemisphere while tracking and verifying missile launches. It warns the crew of missile position and automatically controls countermeasures to neutralise threats and enhance survival.

Design features of the equipment include: automatic pilot warning and countermeasures command, continuous track-while-search processing, multiple missile threat capability and countermeasures discrimination capability. Multidiscrimination modes against solar radiation and terrain and water reflection, minimise false alarms. MIL-STD-1553B bus interface capability is incorporated. A number of sensor unit configurations are available to fulfil specific field of view requirements.

Specifications
Control and display unit
Dimensions: 102 × 142 × 150 mm

Weight: 1.12 kg
Processor
Dimensions: 200 × 212 × 252 mm

Weight: 8.82 kg
Sensor/Dimensions: 400 × 369 mm (diameter)
Weight: 17.91 kg
Displays: sector threat indicators; external countermeasures command and audio tone

Operational status

The AAR-44 is thought to be in service aboard US Air Force C-130 aircraft. AAR-44 equipments operated by the US Air Force's Special Operations Command are understood to have been the subject of a processor upgrade programme.

Contractor

Cincinnati Electronics Corporation.

UPDATED

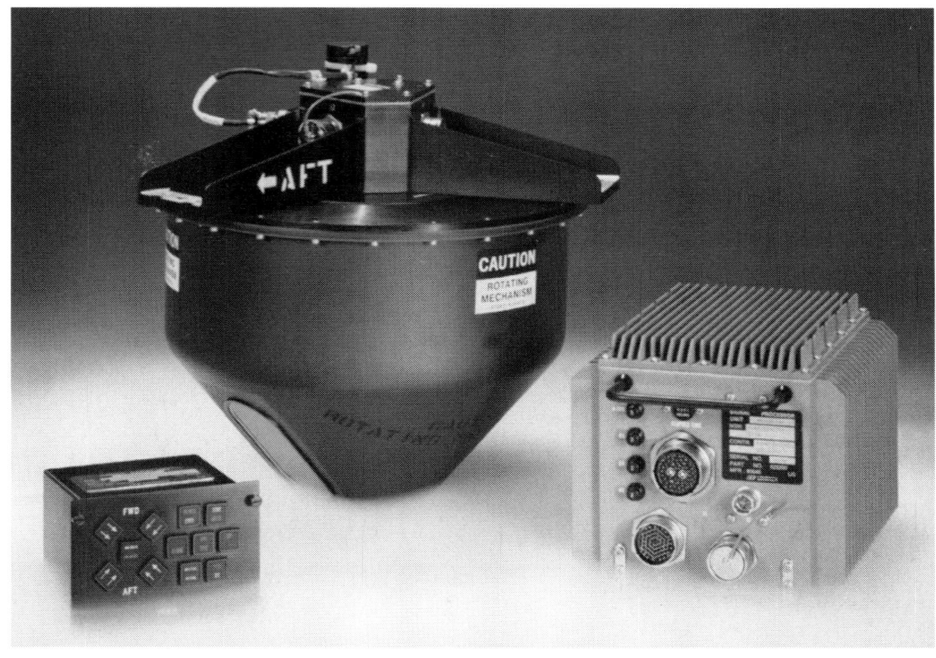

AN/AAR-44 missile warning system with (left to right) control display unit, conical detector head and processor ***1991***

AN/AAR-44(V) advanced missile warning system

The AN/AAR-44(V) advanced missile warning system is a collaborative development of the earlier AN/AAR-44 equipment by Cincinnati Electronics Corporation and Raytheon Systems Company (Goleta).

The AAR-44(V) features include: passive multicolour infrared detection for positive missile warning, with minimum false alarm rate, using colour, size, intensity, and trajectory discriminants; multiple simultaneous threat detection, not confused by flares or multiple objects; wide volume coverage (±135 × 360°); accurate angular detection (better than 1°), with laser-pointing growth potential.

Designed for use on virtually any airborne platform, the AAR-44(V) can be mounted internally, via a 4 in diameter opening in the aircraft skin, externally into a weapon-carrying pylon, or into the underside of the lower gondola of the ALQ-184 ECM pod.

Specifications

Dimensions:
(baseline) 101.6 × 152.4 × 330.2 mm + 152.4 mm dome
(alternate) 101.6 × 152.4 × 406.4 mm
Weight: 9.09 kg
Detection range: beyond lethal range
Altitude: to 45,000 ft
False alarm rate: claimed to be insignificant (classified)
Coverage (per sensor head): ±135 × 360°
Cueing accuracy (AOA): better than 1°
Multiple threats: virtually unlimited
Warning time: classified
Reliability: over 1,000 h
Interfaces: MIL-STD-1553B, RS-422, RS-232
Power: 28 V DC, 1.1 A; 115 V AC, 400 Hz, 1.0 A

Operational status

Extensively proven in field tests; not yet in operation.

Contractors

Cincinnati Electronics Corporation.
Raytheon Systems Company.

UPDATED

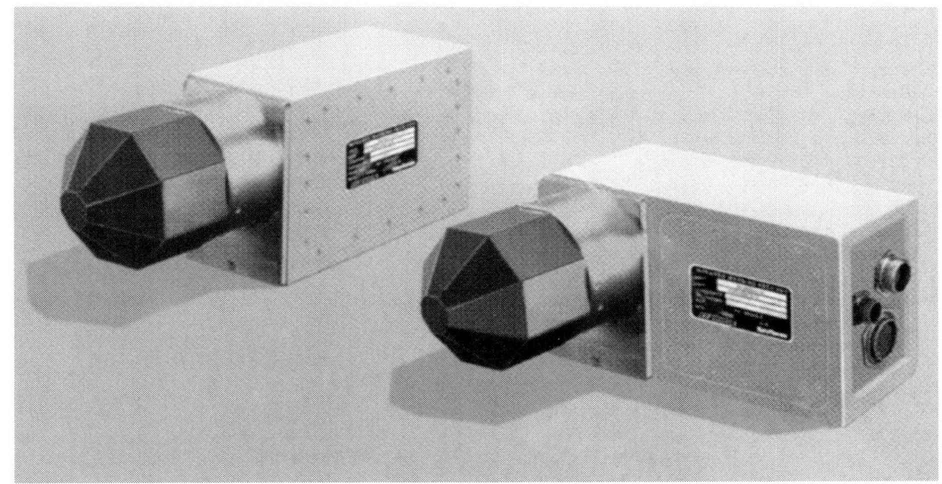
AAR-44(V) Baseline Configuration (right) and Alternate Configuration (left) ***1997***/0002111

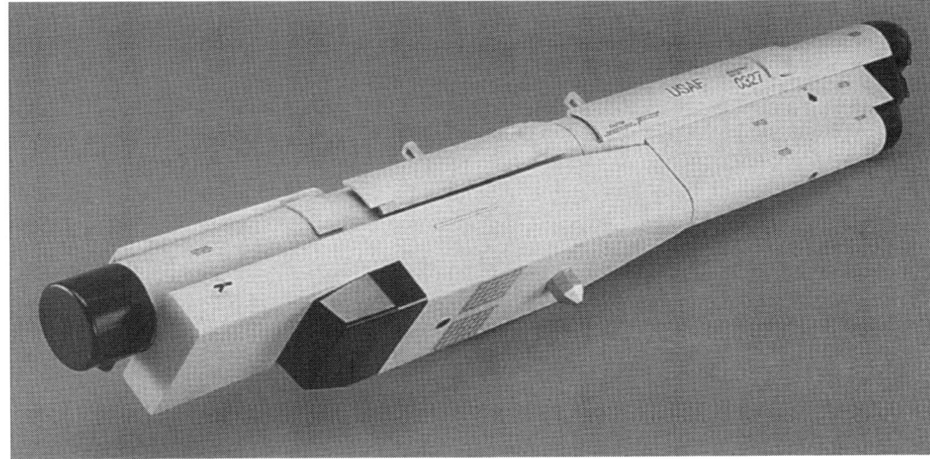

AAR-44(V) integrated into the underside of the lower gondola of the ALQ-184 ECM pod ***1997***/0001271

AN/ALQ-167(V) Yellow Veil Jamming Pod

The AN/ALQ-167(V) Yellow Veil dual-band airborne jammer provides jamming against surface and airborne sensors. Antenna coverage is both forward and aft. The pod is suitable for both helicopters and fixed-wing aircraft and can be used for tactical self-protection or standoff escort jamming roles or for training.

Cockpit selectable frequency and jamming mode provisions allow manual updates in flight or the jammer can be interfaced with ESM equipment. A broad range of jamming modes is available in the standard configuration and special jamming requirements can be accommodated.

Pod-mounting is the standard configuration, but the electronics can also be configured for internal carriage or for drone or RPV requirements.

Specifications

Dimensions: 3,700 (length) × 23 mm (diameter)
Weight: 100-150 kg
Output power: 100-200 W, ERP 1 kW (min)
Frequency: 0.1-18 GHz in customer specified bands

Operational status

Yellow Veil is pod-mounted on UK Royal Navy Lynx and Sea King helicopters.

Contractor

Condor Systems Inc.

UPDATED

AN/ALR-81(V)1 and ALR-81(V)3 ESM/ELINT systems

The AN/ALR-81(V)1 is a fully synthesised microprocessor-controlled receiving system designed for general search, collection, signal analysis and recording. It consists of two units: a tuner and a control/display/demodulator. The tuner is a fully synthesised microwave tuner able to cover the 0.5 to 40 GHz frequency range. It features low-phase noise combined with instantaneous IF bandwidth. The second unit includes a multitrace digitally refreshed spectrum/IF pan display, plus a complete analysis demodulator. It is a single compact unit and includes MIL-STD-1553B remote-control interface.

The AN/ALR-81(V)1 employs high-resolution, digitally refreshed RF and IF pan displays. Spectrum information is provided over the entire frequency display in the form of combined multitrace band scan and IF pan displays with alphanumeric annotation. In the manual mode of operation, a split screen display is provided to allow the operator an overview of the frequency band and a high-resolution IF pan output.

The AN/ALR-81(V)3 is an integrated ELINT system. It features a broadband spinning DF antenna, multiple low-phase noise, wide instantaneous bandwidth superheterodyne receivers and an operator control and display station. The system includes a millimetre-wave antenna and an integral down converter, packaged on the DF antenna.

The operator control unit allows control of both antenna and receiver parameters. It features a high-resolution vector graphics display with alphanumerics. Additional features include multitrace, digitally refreshed spectrum/IF pan displays, multimode DF displays and a complete analysis demodulator. The millimetre-wave downconverter features automatic threshold detection. This allows the tuner to scan bands of interest and then switch to millimetre-wave bands when activity is detected. The AN/ALR-81(V)3 receiver/DF system can be remotely controlled via a MIL-STD-1553B interface bus.

The Condor AN/ALR-81(V)3 ELINT system **1996**

Operational status

In production and in service.

Contractor

Condor Systems Inc.

UPDATED

AN/ALR-801 ESM/ELINT system

The Condor ALR-801 ESM/ELINT system **1996**

The ALR-801 is a multipurpose maritime patrol ESM/ELINT system for airborne applications, even in smaller aircraft. It features NDI subsystems and can receive over the frequency range 0.1 to 18 GHz and provide automatic DF from 0.4 to 18 GHz. Both the receive and auto-DF frequency range can be extended to 40 GHz with NDI hardware.

Using a high-gain shaped beam-spinning DF antenna and associated pedestal control unit, a preamplifier, TN-618 superheterodyne tuner (also noted as using IFM receiver technology) and the SP-103 signal processor, the ALR-801 searches the environment, de-interleaves multiple pulse trains, identifies emitters using a user-supplied library and automatically performs direction-finding on each signal.

The combination of a high-gain antenna, tuner with 500 MHz instantaneous bandwidth and the intelligent search strategies resident in the signal processor results in near 100 per cent probability of interception. Radar parameter measurements are made with extreme precision to eliminate ambiguities in signal identification.

The ALR-801 weighs less than 100 kg and consumes less than 1 kW of prime power. Growth potential includes interactive controls and displays to place a human operator in the loop, the CS-D-250 high-speed digitiser for broadband digital recording of radar signals and the Signal Processing Tools (SPT) software for post-flight analysis of signals, and digital receivers for FMCW and LPI radar detection.

Operational status

Designed for maritime patrol aircraft; available.

Contractor

Condor Systems Inc.

UPDATED

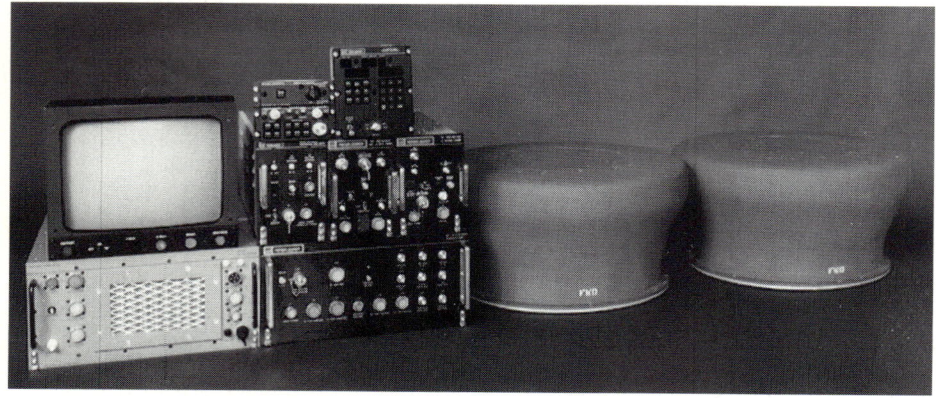

The Condor AN/APR-46 receiving system **1996**

AN/APR-46 receiving system

The AN/APR-46 receiving system, also known under the company designation WJ-1840, is a rapid scanning highly sensitive heterodyne receiver, able to detect threat warning radars at very long range. The receiver is also equipped for emitter bearing determination, as well as signal parameter extraction and identification. It can be interfaced to onboard avionics through a dual-redundant MIL-STD-1553B interface.

The intercept portion of the AN/APR-46 provides frequency coverage between 30 MHz and 18 GHz. Direction-finding capability is available between 0.5 and 18 GHz. Antenna installations containing the omnidirectional and DF antennas can be mounted on both the top and bottom of the aircraft. A single operator can select either or both and configure the receiver scan strategy to optimise coverage against expected threats. Fully automated search, identification and reporting functions are possible for stand-alone systems.

Threat identification is provided to assist the operator by analysing the radio frequency, pulse repetition frequency and pulsewidth of the intercepted signal. These values are then compared with parameters contained in an emitter library.

Specifications

Frequency: 0.5-18 GHz
Emitter PRF: CW to 100 kHz
Emitter pulsewidth: 200 ns to CW
Probability of intercept: 100% within bandwidth of receiver
Accuracy:
10° RMS (0.5-2 GHz)
5° RMS (2-18 GHz)

Operational status

In production for a US Air Force airborne application. Conversion to monopulse DF capability is in progress.

Contractor

Condor Systems Inc.

VERIFIED

AN/APR-46A and AN/APR-46A (V)1 ESM systems

Both the AN/APR-46A and AN/APR-46A(V)1 systems are designed for the panoramic threat avoidance role. They are fitted with highly sensitive and selective superheterodyne receivers covering the frequency band 30 MHz to 18 GHz, utilising one or two channels of microwave tuners.

Digital scan control with multimode operation and an eight-trace digitally refreshed pan display is implemented to provide band scan, sector scan, and manual modes. AN/APR-46A(V)1 adds monopulse direction-finding to the AN/APR-46A capability, with DF accuracy of 5° rms over the 0.5 to 18 GHz band.

Contractor

Condor Systems Inc.

NEW ENTRY

CS-2010 HAWK Receiver/DF system

The CS-2010 is a multiple operator, pooled resource ELINT/ESM system covering the frequency range 20 MHz to 40 GHz, designed for airborne, shipborne or land-based applications. The family of equipments includes microwave tuners, demodulators, VHF/UHF up converters, 18 to 40 GHz down converters and spinning DF antenna subsystems. System elements, or resources, are maintained by the system as a pool of resources and are shared by the operators. Operators acquire resources from the pool as required and release them back to the pool when no longer needed. Operators interface to the CS-2010 through a resources control unit, a keyboard and a pair of high-resolution raster scan RGB display units. The RCC provides the operator with control of the pooled resource elements and provides the operator with real-time displays of both antenna and receiver functional outputs.

The CS-2010 features a variety of operational modes, including common band scan, analysis and DF mode. Common band scan is a display synthesised from the RF scan data produced by all unused tuners in the system resource pool - that is, the display is produced by parallel scanning all tuners which have not been acquired by an operator for use in analysis of DF operations. The normal and expanded analysis displays provide the operator with RF, IF pan displays and LIFM for receiver channels acquired by the operator from the resource pool. The operator has control over five independent traces representing up to five independent digitally generated polar, inverse polar and rising raster rectilinear, and supports the manual DEF capabilities of the system.

The system is built around VersaNet, a time domain multiple access bus. VersaNet distributes control information between system elements and carries digitised display and audio information gathered in the system elements to the RCC. VersaNet enables additional resources such as tuners, demodulators, resource control units and pedestal electronic units to be added to the system as required. The principle tuner in the family is the TN-618 microwave tuner which features very low phase, 500 MHz wide IF and 10 kHz step size over 0.5 to 18 GHz. Block up converters and down converters extend the frequency coverage. The tuner also includes a built-in single bandwidth demodulator. An external full-function demodulator with eight IF bandwidths, IF pan and Log/Lin/FM video inputs is also available.

The Condor Systems CS-2010 HAWK receiver/DF system ***1995***

Operational status

In production.

Contractor

Condor Systems Inc.

UPDATED

CS-3360 Lightweight ESM system

The CS-3360 is a complete ESM system with antennas, receivers, pulse processor and multifunctional display. It combines the functions of radar warning, tactical surveillance and ELINT collection. It is designed for installation on aircraft, ships and land vehicles, where space is limited but high performance is required. Emitters are detected, identified and displayed with direction of arrival.

The SP-2300 processor can process normal, stagger, jitter, multipulse, CW and complex radar signals. De-interleaving software is able to separate pulse trains in dense signal environments and measure the parameters of individual signals. Precision pulse interval measurements, with sub-nanosecond accuracy, are achieved with adaptive pulse tracking algorithms and an internal rubidium timing standard.

The SP-2300 can quickly identify signals using a built-in signal identification library with space for over 2,000 emitter modes. An internal editor allows field

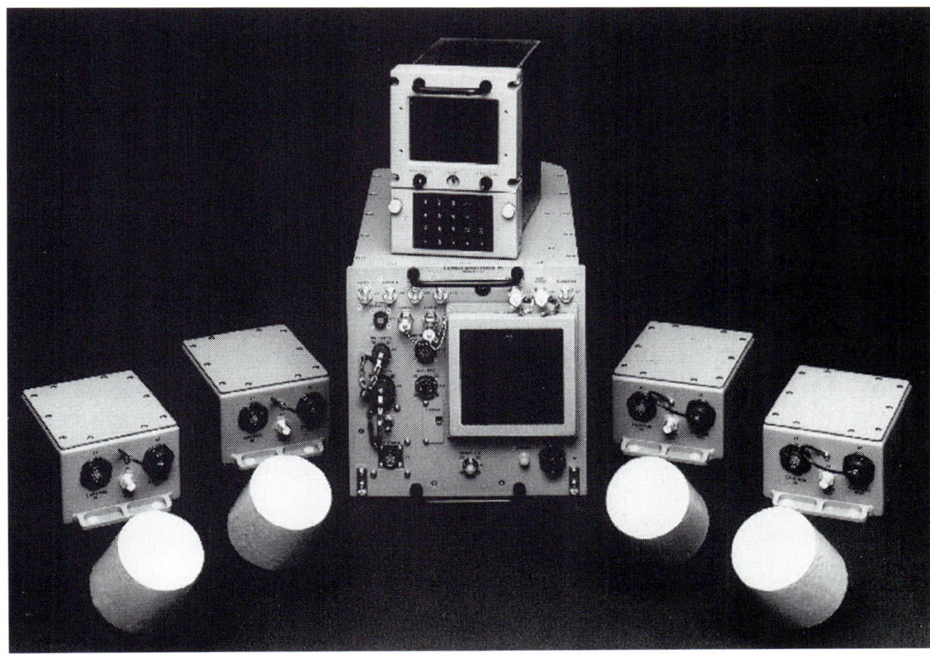

The Condor CS-3360 ESM system ***1996***

updates to library entries. Each library entry allows for information on signal priority, high/low frequency, high/low PRF/PRI, high/low pulsewidth, library entry number, signal identification, scan type, scan period, illumination time, function of emitter, emitter mode for a set of parameters, up to two lines of text information, power in kW and antenna gain in dB.

Features of the CS-3360 also include 0.5 to 18 GHz frequency coverage, IFM receiver, high sensitivity and probability of intercept, high-speed pulse DF processing, instantaneous azimuth coverage, library loading from floppy disk and interfaces to time and navigation, external computer and SIGINT equipment. It weighs less than 30.4 kg.

Other systems in the series include: CS-3300 rapid deployment systems designed for applications requiring maximum capability, minimum weight and volume, and rapid set-up time. This includes the CS-3350 and CS-3360 systems. The latter is a lightweight, multiplatform equipment. Both systems operate over the 0.5 to 18 GHz band; CS-3500 tactical signal detection and classification systems; CS-3700 wideband ELINT intercept systems, latest version listed as CS-3701; CS-3900 remotely controlled ELINT receiver systems which can be used in remote fixed sites or unmanned aerial vehicles. They operate automatically or manually allowing remote control of frequency and pulse processing modes.

Contractor
Condor Systems Inc.

VERIFIED

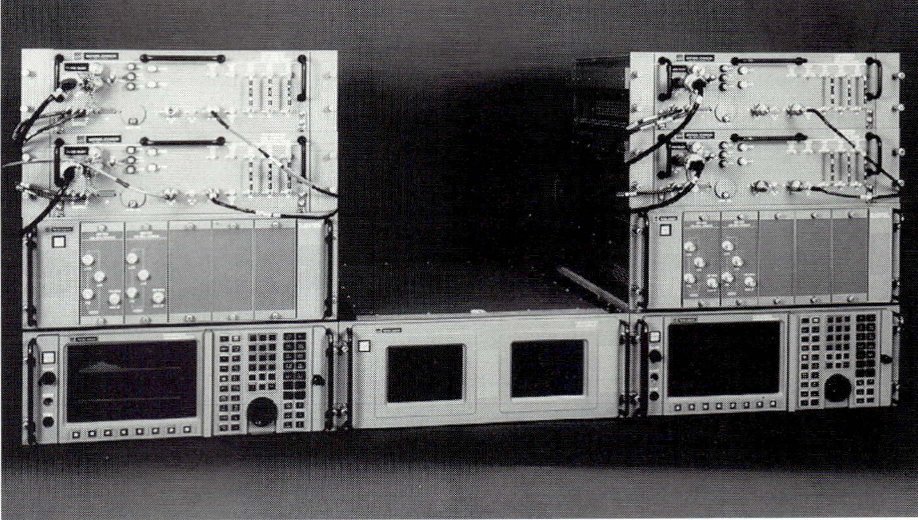

The Condor CS-36500 signal intelligence receiving system ***1996***

CS-6700 ACES automatic processing ESM systems

CS-6700 ACES is a high-performance ESM/ELINT system for use on airborne, ground-mobile, shipborne or fixed-site installations. Frequency coverage is 0.5 to 18 GHz (expandable to 40 GHz). High-sensitivity, wide band superheterodyne and IFM receivers are implemented, together with automatic and manual radar signal analysis facilities. A Windows™-based HMI includes: scan control, library editing, and graphical signal analysis.

Contractor
Condor Systems Inc.

NEW ENTRY

CS-36500 Signal Intelligence Receiving System (SIRS)

The CS-36500 Signals Intelligence Receiving System (SIRS) is a 0.03 to 40 GHz ESM/ELINT/COMINT receiver which maximises current capability and future adaptability. It can be configured as a single/independent or as multiple/interactive operator positions. This modularly designed superheterodyne receiver consists of a C-100 control/display unit; an optional C-200 scan display; a CY-100 equipment frame with one or more demodulator plug-ins, depending upon desired IF bandwidth and video performance; and one or more octave or multioctave tuners, or the FXT-1XX millimetre extensions. Narrowband and wideband (to 500 MHz) tuners and demodulators are available.

The C-100 controls SIRS operations; displays operating parameters, status, IF Pan and analysis mode video; and accepts configuration programming. This unit has internal removable mass memory and supports multiple standard interfaces.

The C-100 panel includes eight soft keys for maximum versatility. Controller panels are backlit and may be NVG-compatible. Parameter values are entered via three methods; numeric keypad, cursor increment (arrow keys) and slew knob (shaft encoder). Operation is user-friendly with automatic prompting. An optional external keyboard is available for touch-typist execution.

The C-100 ElectroLuminescent (EL) display provides operation, configuration, and diagnostic reports; BITE status; flexible RF, IF and time spectrum display with a selectable refresh/decay rate; as well as an AM/FM display mode for accurate measurement of broadband emitter's frequency excursions. SIRS may also include an audio alarm which alerts the operator to important but infrequent events such as detection of energy above 18 GHz.

The C-200 scan display is capable of RF panoramic display for up to eight additional tuners.

The SIRS controller design incorporates features which facilitate acquisition of difficult-to-detect emitters, such as those with a low duty cycle. For example, scan coverages and sweep rates may be optimised to enhance probability of intercept for specific signals. In addition, a scan priority feature allows the operator to choose the number of scan repetitions over each frequency sector relative to the other sectors. Mission profiles, called Receiver Instruction Sets (RISs), can be created online (or offline on a personal computer) and stored in the C-100. They include receiver control parameters such as mode, frequency scan limits, frequency markers, frequency lock-out sectors, priority, dwell times, attenuation, sweep rate, threshold levels, video select, IF pan settings, and IF bandwidths.

The CY-100 equipment frame, containing internal power supplies and video switching, accepts plug-ins including five demodulator families, various switching matrices, and custom units.

The CS-36500's architecture provides the user with the flexibility necessary to meet varying mission requirements and facilitate future system expansion. Its modular design permits adjustment of the receiver's frequency coverage and configuration. Single- and multiple-operator configurations with narrowband and/or wideband analysis ability are standard. Future expansion is accomplished by connecting additional units. The system executive will automatically recognise the additions. Controllers in multioperator systems are interactive and may hand-off tasking to each other, or borrow currently unused assets.

The C-100 uses multiple microprocessors and incorporates MIL-STD-1553B digital bus architecture. The system can include any combination of octave and multioctave-band tuners. Retrofits may utilise existing assets, either analogue or digitally tuned. The basic system is able to control 15 tuners and demodulators. Tuners offering special characteristics, such as phase-locked scan, are exploited by the system software.

Contractor
Condor Systems Inc.

UPDATED

CS-6700 ACES automatic processing ESM systems ***1998***/0018273

SP-2060 pulse processor

The SP-2060 pulse processor measures radar signal parameters, providing fast, accurate results. Operation is automatic, giving the operator a display of signal activity with identification of specific radar types. A continuous log of activity is stored in memory for post-mission analysis. Requiring only a single coaxial connection, it is compatible with any type of radar receiver. Two modes of control/display units are available. The EI-5100 control/display unit has a 9 in (178.6 mm) display and mounts in a standard 19 in rack. The EI-1400 control/display with a 5 in (127 mm) display is available for aircraft installations where space and weight are limited. Both models have built-in floppy disk drives.

The SP-2060 processes normal, stagger, jitter, CW and complex radar signals. De-interleaving software is able to separate pulse trains in dense signal environments and measure the parameters of individual signals. A high-speed digitiser and a maths co-processor allow processing of high pulse rate radars. Precision pulse interval measurements with sub-nanosecond accuracy are achieved with an internal rubidium timing standard and adaptive pulse tracking algorithms.

A time of arrival phase tracking algorithm is able to separate and measure two emitters with identical pulse repetition intervals. The algorithm remembers the time of arrival phase from scan to scan and does not require continuous signal reception.

The SP-2060 can quickly identify specific signal types using a built-in signal library with space for over 2,000 emitter modes. An internal text editor allows field updates to library entries. Entries are loaded from, and stored on, floppy disks.

The disk recording capability enhances the utility of the SP-2060. A single disk can store multiple libraries, enabling quick updates to match the local threat environment. Digitised pulse-by-pulse descriptors can be recorded for post-mission analysis. Recording of the intercept file on disk ensures a continuous log of mission signal activity.

Contractor

Condor Systems Inc.

VERIFIED

ASPJ AN/ALQ-165 Airborne Self-Protection Jammer

The AN/ALQ-165 Airborne Self-Protection Jammer (ASPJ) is designed for F-14D, F-16C, F-18C/D, A-6E and AV-8B aircraft and can be installed internally or in a pod. The equipment incorporates the latest technology in TWTs, microwave components and packaging. It can be electrically reprogrammed on the flight line and the BIT and modular plug-in design permits rapid replacement of assemblies at the operational level without external test equipment. Weight of the system is between 91 and 150 kg, depending on configuration; it occupies a space of 0.6443 m^3.

The AN/ALQ-165 has the ability to select automatically the best jamming techniques to use against any given threat, based on the jamming system's own computer data and real-time data of the threat signal from the receiver/processor. The computer software can be modified to accommodate new threats as they arise. The equipment covers the frequency range in two bands and is technically expansible to cover a greater frequency range if required.

The transmitters within the system can jam a large number of threats simultaneously over various ranges and in different modes. The computer selects the power and duty cycle criteria based on threat parameters detected and processed by an advanced receiver subsystem. An augmented version with an additional transmitter power booster is also available.

AN/ALQ-165(V)

Following renewed enthusiasm for the ASPJ system in the US, the US Naval Air Systems Command in December 1997 issued a US$45 million firm fixed price contract to procure 36 AN/ALQ-165(V) ASPJ systems, data support services and 15 AN/ALQ-165(V) ASPJ systems as spares for the US Navy; and 10 AN/ALQ-165(V) ASPS spares for the US Air Force. Contract fulfilment to be by December 1999. The nature of the upgrade represented by the addition of (V) to the nomenclature is not known.

Specifications

Dimensions:
(processor) 141.7 × 199.4 × 403.9 mm
(low-band receiver) 141.7 × 199.4 × 400.6 mm
(high-band receiver) 141.7 × 199.4 × 400.6 mm
(low-band transmitter) 121.4 × 208 × 644.7 mm
(high-band transmitter) 121.4 × 208 × 644.7 mm
Weight:
(processor) 17.69 kg
(low-band receiver) 16.78 kg
(high-band receiver) 16.78 kg
(low-band transmitter) 30.84 kg
(high-band transmitter) 29.48 kg

Operational status

The system is now in production for the Finnish and Swiss F/A-18 Hornet programmes. ALQ-165 (ASPJ) was installed on US Marine Corps F/A-18C/D aircraft for Bosnian operations.

In May 1996, the US Navy approved 'unrestricted fleet use' of ASPJ on F-14D, and in August 1996, funding was approved for 36 more ASPJ systems for US Navy F/A-18C/D aircraft.

In January 1997, the Republic of Korea Air Force (RoKAF) placed an order, valued at more than US$100 million, for AN/ALQ-165 ASPJ for its F-16 fighter programme.

Contractor

Consolidated Electronics Countermeasures.
(An ITT Defense and Electronics Avionics Division/Northrop Grumman Corporation, Electronic Sensors and Systems Division joint venture).

UPDATED

Internal Installation

F/A-18

F-16

WRAs/LRUs

Hi-Band Transmitter | Hi-Band Receiver | Lo-Band Receiver | Computer Processor | Lo-Band Transmitter

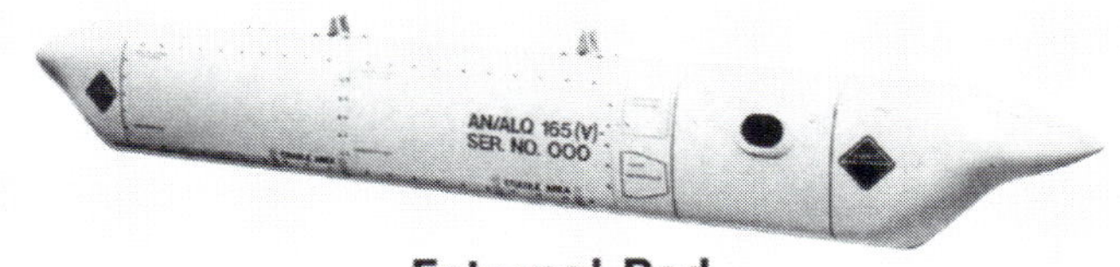

External Pod

ASPJ AN/ALQ-165 pod installation (below) and internal fit configuration (above) **1998**/0018272

ASPJ AN/ALQ-165 internal fit on F-14D **1998**/0018271

AN/ALQ-150 ESM receiver

The AN/ALQ-150 is a tactical airborne ESM system used as part of the Cefire Tiger electronic warfare programme. It is employed against multichannel communication transmitters and has an electronic support measures capability for integration with jamming systems. The complete system is deployed in four modular units in three US Army RU-21 aircraft and can be used against Frequency Division Multiplexed (FDM), Frequency Modulated (FM) and Time Division Multiplexed (TDM) emitters.

The Cefire Tiger airborne ESM system interfaces with the Le Fox Grey control processing centre and the forward control and analysis centre to provide overall control of all ECM missions on a real-time basis. The control processing centre has wideband datalinks which communicate with up to three Cefire Tiger systems simultaneously. The effective radiated power is from 3 to 10 kW, depending on frequency. Frequency ranges in each system are:

Aircraft 1: 60 to 115 and 1,500 to 9,000 MHz;
Aircraft 2: 115 to 480 MHz;
Aircraft 3: 450 to 1,500 MHz.

Operational status

The first units were deployed during mid-1984.

Contractor

GTE Government Systems Corporation, Electronic Defense Sector.

VERIFIED

AN/ALQ-136 radar jamming system

The AN/ALQ-136 family of airborne jamming systems is designed to protect rotary- and fixed-wing aircraft from radar-guided weapons. The AN/ALQ-136 enables the host aircraft to carry out a mission from a low-level approach, pop up and complete its mission before the hostile radar can initiate a fire sequence.

The system consists of three main units: a control unit, two spiral antennas - one each for transmit and receive - and a transmitter/receiver. When the aircraft is illuminated by a hostile radar, the jammer automatically analyses the received pulses, compares them with its threat library, assigns a priority and then provides the most appropriate countermeasure response.

The AN/ALQ-136 is software reprogrammable to allow adaptation to the swiftly changing threat scenarios and can simultaneously handle multiple threats. The system is internally mounted and integrated on numerous US Army helicopters.

The AN/ALQ-136(V)1/5 jamming system is carried by US Army AH-64 Apache attack helicopters

Operational status

In service. More than 1,400 systems have been delivered to the US Army and are deployed on AH-64 Apache and AH-1 Cobra attack helicopters. Other platform applications include the MH-53J Pave Low, MH-60 Black Hawk, EH-60 Quick Fix, MH-47E Chinook and AH-1W SuperCobra helicopters and the A-10 and RC-12/RU-21 fixed-wing aircraft.

AN/ALQ-136(V)2

The AN/ALQ-136(V)2 is an enhanced version with increased receiver sensitivity, frequency range and threat-handling capability. The system design employs extensive use of Thick Film Hybrid technology (TFH) and multilayered PCBs. A MIL-STD-1553B databus is used to facilitate system integration.

Specifications

Dimensions:
(receiver/transmitter) 177.8 × 330.2 × 408 mm
Weight: 31.8 kg

Operational status

200 systems have been delivered to the US Army.

Contractor

ITT Defense and Electronics Avionics Division.

UPDATED

AN/ALQ-172 Electronic CounterMeasures system

The AN/ALQ-172 is an advanced ECM system which was combat proven during Operation Desert Storm. It is installed in US Air Force B-52 strategic bombers and Special Operations MC-130E/H Combat Talon I and II and AC-130U Gunships.

System design features of the AN/ALQ-172 include full automation, multiband coverage, simultaneous multiple threat recognition and jamming, digital computer control, advanced jamming techniques, high effective radiated power, threat reprogrammability, high-gain array antenna, threat warning display, dual MIL-STD-1553B databus interface and extensive BIT.

The Combat Talon II MC-130H is being upgraded with the AN/ALQ-172(V)3 jamming system

There are three versions of the system: the AN/ALQ-172(V)1, AN/ALQ-172(V)2 (which incorporates a phased-array antenna) and the AN/ALQ-172(V)3.

Operational status

In service in US Air Force B-52 and Special Operations MC-130E/H Combat Talon I and II, AC-130H and AC-130U gunship aircraft. System upgrades under way include expansion of AN/ALQ-172 system memory and extended frequency coverage designated AN/ALQ-172(V)3 for the AC-130H upgrade programme.

Contractor

ITT Defense and Electronics Avionics Division.

UPDATED

AN/ALQ-211 Suite of Integrated RF Countermeasures (SIRFC)

In June 1996, the AN/ALQ-211 Suite of Integrated RF Countermeasures (SIRFC) programme successfully completed the Critical Design Review (CDR). (Note: The AN/ALQ-211 SIRFC system was formerly called the Advanced Threat Radar Jammer (ATRJ). In June 1996, the system received its official nomenclature of AN/ALQ-211 SIRFC.)

The AN/ALQ-211 SIRFC system is a fully integrated airborne electronic combat system that provides real-time battlefield situational awareness, radar warning and self-protection capabilities to a multitude of aircraft.

SIRFC intercepts, analyses, passively ranges and geolocates multiple air- and ground-based RF threat signals and co-ordinates appropriate countermeasures or weapons cueing.

The AN/ALQ-211 SIRFC system provides an embedded capability which directly supports the battlefield digitisation initiatives of the US Army. SIRFC's real-time situational awareness reports and

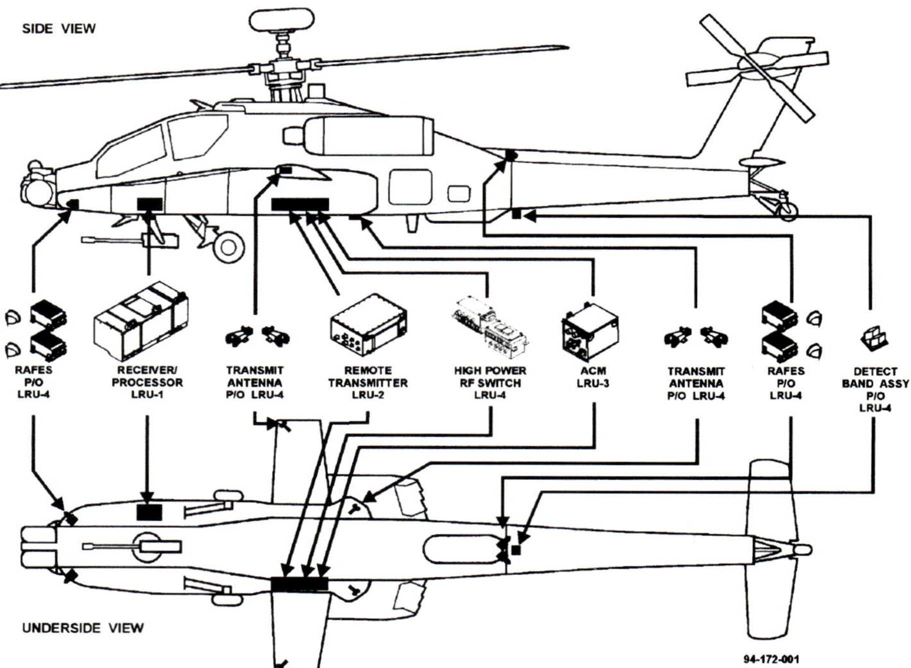

SIRFC installation in the AH-64D Apache Helicopter ***1998***/0018270

updates can be disseminated horizontally and vertically to aircrews and battlefield commanders. The location of radar-directed threats can be transmitted from the aircraft to the ground commander over the tactical datalink and displayed on appliqué hardware. The situational awareness and onboard countermeasures capabilities of the AN/ALQ-211 SIRFC system allows aircrews to evade or defeat a diversity of airborne and ground-based threats, significantly enhancing survivability.

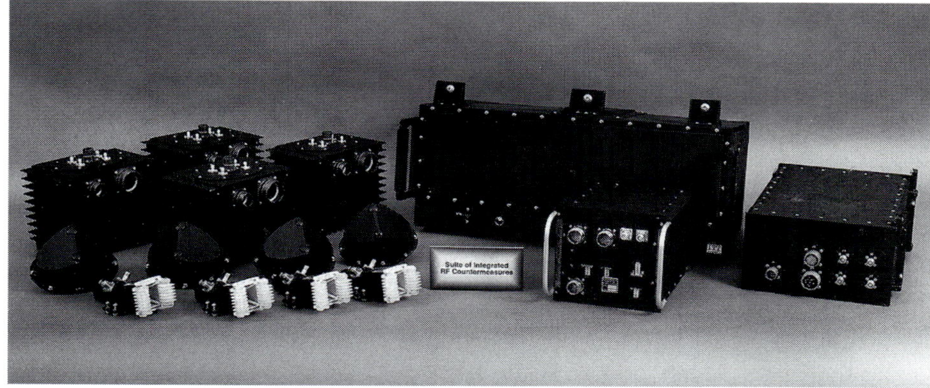

Suite of Integrated RF Countermeasures (SIRFC) ***1996***

Operational status

ITT Avionics, and its team members Tracor Aerospace Electronic Systems and EMS Technologies, are under contract with the United States Army for the Engineering & Manufacturing Development (EMD) of the ALQ-211. The AN/ALQ-211 SIRFC programme is managed by the US Army Program Executive Office (PEO) for Aviation, St Louis, Missouri. In a complementary programme, the ITT Avionics team is under contract by Bell Textron to integrate the AN/ALQ-211 SIRFC EMD system on the CV-22 Osprey.

The AN/ALQ-211 is slated for deployment on the Longbow Apache helicopter and CV-22 Osprey aircraft in addition to a variety of other platforms including the CH-47 and UH-60 transport helicopters and Special Operations' MH-47 and MH-60 helicopters.

Initial deliveries of AN/ALQ-211 SIRFC EMD units to the US Army were scheduled for late 1997, followed by system deliveries for the CV-22 beginning in mid-1998.

Contractor

ITT Defense and Electronics Avionics Division.

UPDATED

AN/ALR-73 detection system

The AN/ALR-73 passive detection system is an airborne ESM equipment developed for the US Navy's E-2C airborne early warning aircraft. It is an improved version and successor to the AN/ALR-59 and, because of the extensive update, has been given its own designation. The ALR-73 is intended to augment the AEW, surface, subsurface and command and control functions of the E-2C by enhancing the threat detection and identification capabilities of the aircraft. It is a completely automatic, computer-controlled, superheterodyne receiver processing system that communicates directly with the E-2C command and control central processor. The design of the system was motivated by four major considerations: very high probability of intercept in dense environments; automatic system operation; high reliability; and ease of maintenance.

Features of the ALR-73 which are related to its intercept probability performance include 360° antenna coverage, four independently controlled receivers, dual-processor channels and a digital closed-loop rapid tuned antenna. Features concerned with automatic operation are low false alarm rate, automatic overload logic, an AYK-14 computer which adaptively controls hardware and degraded mode operation.

The system uses 52 grouped antennas in four sets, one for each of several wavebands. Each complete set is positioned to look at a 90° sector. The forward and aft antennas are in the fuselage extremities and the sideways-looking aerials are in the tailplane tips. All receiver sets are under separate control, so wavebands are scanned independently and simultaneously in all sectors. Each antenna has dual-processing channels and uses digital closed-loop rapid tuned local oscillators. The latter provide instantaneous frequency measurement with fast time response and high-accuracy frequency determination.

Receiver outputs are collected at a signal preprocessor unit which performs pulse train separation, direction-finding correlation, band tuning and timing and built-in test equipment tasks. Data is then in a form suitable for the general purpose digital computer, which has overall control of electronic surveillance measures operations and will vary frequency coverage, dwell time and processing time according to prescribed procedures. Control of these parameters is aimed at maximising the probability of intercepting signals on particular missions. Other onboard sensor data and crew inputs will determine the technique adopted. Data such as signal direction of arrival, frequency, pulsewidth, pulse repetition frequency, pulse amplitude and special tags are sorted by the computer and transmitted to the E-2C central processor.

The ALR-73 can measure direction of arrival, frequency, pulsewidth and amplitude and PRI simultaneously. Scan rate information is also available if called for by the central processor. Special emitter tags can be provided. The ALR-73 detects and analyses electromagnetic radiation within the microwave portion of the spectrum and sends emitter reports of pulsewidth, PRI, direction of arrival, frequency, pulse amplitude and special tag to the E-2C's central processor via its own data processor. The ALR-73 immediately reports new emitters to the central processor which performs the identification function. It eliminates redundant data on emitters for a programmable period of time, thus significantly reducing the data rate to the central processor. The ALR-73, a multiband, parallel scan, mission programmable system covers the frequency range in four bands through step sweeping. Programmable frequency bands and dwell time permit very rapid surveillance of priority threat bands. Non-priority bands are also monitored, but at a reduced rate. Probability of intercept is increased without sacrificing sensitivity through the detection of both real and image sidebands.

The US Navy E-2C Hawkeye carries the Litton AN/ALR-73 ESM system

Operational status

In production and service on the US Navy Northrop Grumman E-2C and with the forces of France, Japan, Singapore and Taiwan.

The AN/ALR-73 has also been successfully installed on C-130 aircraft.

Contractor

Litton Amecom.

UPDATED

LR-100 Warning and Surveillance Receiver

The LR-100 is a lightweight radar signal receiver covering 2 to 18 GHz. Options are available for 70 to 200 MHz and 18 to 40 GHz. This receiver has been built to provide precision RWR, ESM and ELINT measurements with a total installed weight of less than 23 kg, including receiver, antennas, cables and brackets. The LR-100 system is a complete, two channel interferometer receiver system with a 500 MHz bandwidth and VME-based processor.

Phase and amplitude calibration signals are injected at the antenna to achieve precision angle and location measurements.

Emitter identification and location data are passed digitally for warning, display and analysis and recording. The only support equipment needed for the LR-100 is a Windows™-compatible PC. Help and maintenance manuals are built into Windows™-based software. User defined receiver function and identification parameters are programmed with the same software tool. The Litton software tool can graphically display this information on the vehicle or via datalink. The LR-100 can be modified in real time for special receiver modes or directed tuning.

Specifications

Dimensions:
(receiver) 254 × 177.8 × 127 mm
(array) 304.8 × 304.8 × 90 mm
Weight:
(receiver) 11 kg
(array) 1.8 kg
Field of view: ±65° (azimuth and elevation)
Accuracy: 0.78° RMS

Operational status

Built as a CoTS product. Proposed to the United Arab Emirates as part of both the Lockheed Martin F-16 bid. It has been selected for the Royal New Zealand Navy Kaman Aerospace SH-2G Super Seasprite multimission helicopters.

Contractor

Litton Amecom.

UPDATED

LR-4500 microwave collection system

The LR-4500 microwave collection system was designed to perform ESM and ELINT functions over the frequency range of 0.5 to 18 GHz. Information is gathered by onboard antenna and receiver/processor equipment, which may be installed on aircraft, on ships or at ground sites. Aircraft installations may be either internal or external in a pod. Collected data may be processed locally or transmitted to an operations centre by datalink. Information may also be recorded on magnetic tape for post-mission analysis. The airborne system can be installed in either manned or unmanned vehicles.

The complete system includes airborne, ground

processing and maintenance support equipment. The airborne component of the system weighs 102 kg.

Operational status

Equipment has been delivered to an unnamed customer.

Contractor

Litton Amecom.

VERIFIED

LT-500 Emitter Targeting System

LT-500 is a passive precision interferometric RF Emitter Targeting System (ETS). The fully militarised LT-500 ETS is a recent joint development of Litton Amecom and TRW Avionics. Designed to achieve the goal of balancing performance, cost, and reliability in a single receiver/processor electronics package, its primary application is passive precision RF targeting of modern air-ground weapons via retrofit and upgrade of existing tactical fighter aircraft.

This cost-effective implementation consists of a totally modular 'building block' architecture (SEM-E size electronics modules weighing 34 kg total without antenna) that may be easily tailored in capabilites to match specific emitter tracking/targeting mission requirements. The LT-500 ETS may be installed with a variety of interferometer antenna array configurations and is operated via a MIL-STD-1553 control/display interfaces. An embedded, 32-bit JIAWG compliant, RISC processor hosts the Ada language 'Operational Flight Program'.

Employing Litton-patented interferometric direction-finding and ranging techniques, the LT-500 achieves very rapid situation awareness, emitter identification and precision geolocation in dense electromagnetic environments. The LT-500 ETS is compatible with either internal or pod installation on board the F-15, F-16, F/A-18 or similar tactical aircraft. Because of its modularity, it may be easily adapted to a wide variety of other airborne, shipboard, or ground-based platforms.

Contractor

Litton Amecom.

VERIFIED

Airborne Digital Automatic Collection System (ADACS)

The Airborne Digital Automatic Collection System (ADACS), an ESM and data collection suite, is an advanced configuration of the AN/ALR-66B(V)3 ESM system. It encompasses integrated ESM capabilities in conjunction with precision emitter parameter collection and measurement. The system handles modern threats including CW, interrupted CW, pulse Doppler, low probability of intercept, jitter/stagger/agile, pulse compression and frequency-agile radars, and performs precision measurements on all intercepted emitters. The design is flexible for easy installation in both fixed- and rotary-wing aircraft.

The system provides full sensitivity and over-the-horizon detection, covering a full 360° in azimuth and ±45° in elevation in the C- to J-bands in high-density EW scenarios. ADACS has automatic operation of both ESM and data collection with precision parameter measurement and recording of all intercepted emitters. The virtually infinite memory capacity permits full mission recording. A removable data record module and library module ensure complete data security.

Operational status

Lockheed Martin ordered eight systems for the P-3C for a foreign customer in March 1991.

Contractor

Litton Applied Technology.

VERIFIED

AN/ALR-66B(V)3 surveillance and targeting system

The AN/ALR-66B(V)3 was developed for the US Navy as an enhancement in capability for its multimission aircraft. The ALR-66B(V)3 surveillance and targeting system is the successor to the US Navy's AN/ALR-66A (V)3 system.

Integrated with the aircraft's radar antenna, the system provides ultra-high system sensitivity and precision DF accuracy. Interfaced with other aircraft sensors, the AN/ALR-66B(V)3 provides operation on a non-interfering basis, as well as interfacing with aircraft navigation systems, central computer and display.

Radar antenna modification techniques provide C- to J-band signal reception and precision direction-finding capabilities required for over-the-horizon targeting. Simultaneous operation of the radar and ESM surveillance and targeting functions is allowed. Alternatively, a dedicated spinning antenna may be used for 360° C- to J-band precision measurements.

Advanced signal processing techniques for instantaneous, positive emitter identification in high-density environments are used. A multimode emitter library, which uses EEPROM technology, permits rapid reprogramming of total library scenarios and individual emitters. Emitters not found in the library are displayed by their generic characteristics with precise parameter measurements. The use of EEPROM technology for the AN/ALR-66B(V)3 eliminates the need for hardware modification when changes to the memory are made.

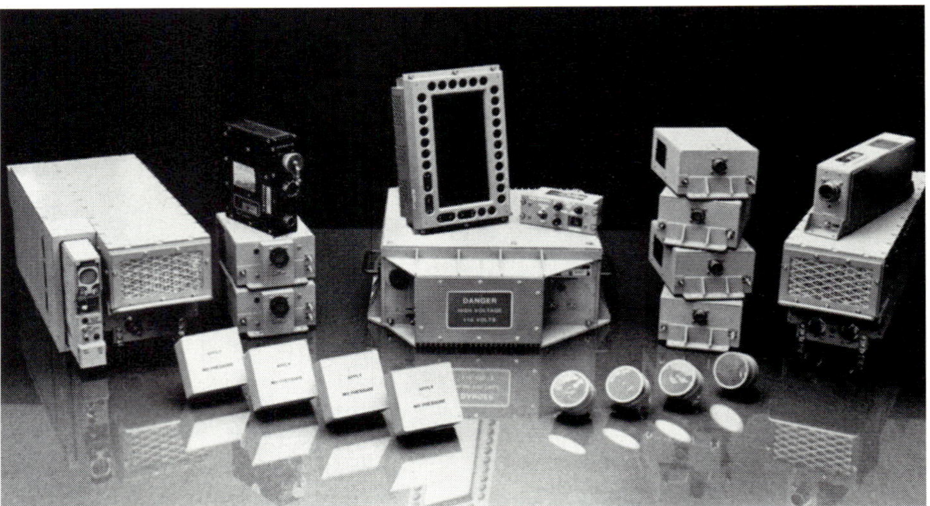

The AN/ALR-66B(V)3 surveillance and targeting system

The basic modes of operation are:

Surveillance mode: the highest priority emitters are presented on the plasma display in positions corresponding to their range and bearing, using unique symbology which is either emitter or platform related.

Targeting mode: precise targeting data on any emitter is presented on the plasma display. This operating mode provides the OTH targeting data required for weapon activation.

Emitter waveform analysis: video is accepted from the computer-converter and data is presented directly on the plasma display, allowing the operator to analyse emitter waveforms.

Specifications

Weight: 87.3 kg
Receiver: crystal video
Frequency: contiguous over the C- to J-bands
Warning and identification: all pulsed radars including pulse Doppler, CW, ICW, LPI, 3-D, jitter/stagger, pulse compression and agile
Radar storage: >2,000 emitter modes in removable library storage module
Symbology: 1, 2 or 3 symbols per emitter as desired, programmable

Operational status

The AN/ALR-66(V)3 has been procured by the US Navy and other international customers for Lockheed Martin P-3B and P-3C aircraft. The AN/ALR-66(V)3 has been upgraded to the AN/ALR-66B(V)3 for the US Navy.

Contractor

Litton Applied Technology.

VERIFIED

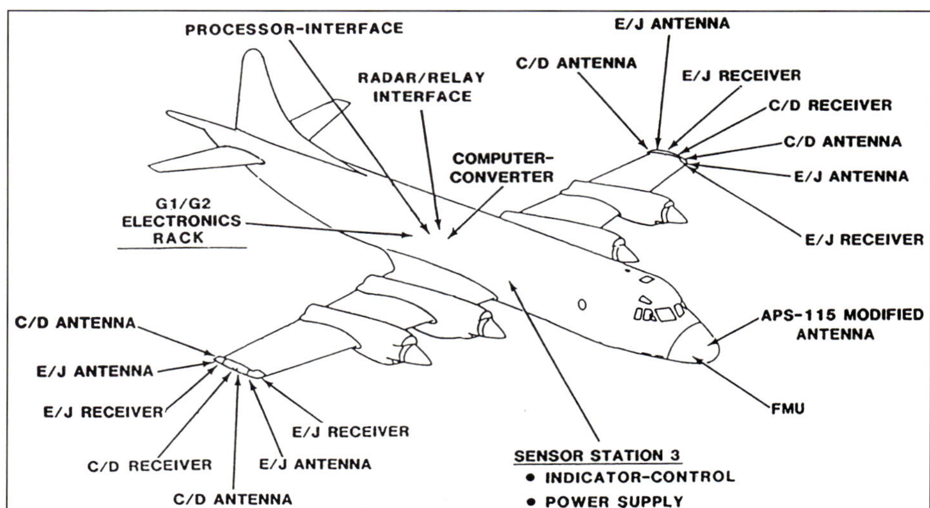

The AN/ALR-66B(V)3 layout in the Lockheed P-3C Orion ASW aircraft

AN/ALR-66(V)4 high-sensitivity ESM system

The AN/ALR-66(V)4 countermeasures receiving set is an airborne electronic system which detects and identifies radars in the C- to J-bands. It is currently in operation on US Navy Boeing E-6A aircraft.

Key features of the system include ultra-high RWR sensitivity, excellent DF accuracy, positive emitter identification in high-density environments, self-protection for the aircraft, automatic operation and complete BIT capability. The system is designed with EEPROM data memory so that the library can be reprogrammed to recognise the identifying characteristics of radar emitters other than those contained in the original threat library. Utilising the computer memory loader, the system can be totally reprogrammed on the flight line. As the system is software intensive, the life-cycle costs are extremely low.

Operational status

In service on US Navy Boeing E-6A aircraft.

Contractor

Litton Applied Technology.

VERIFIED

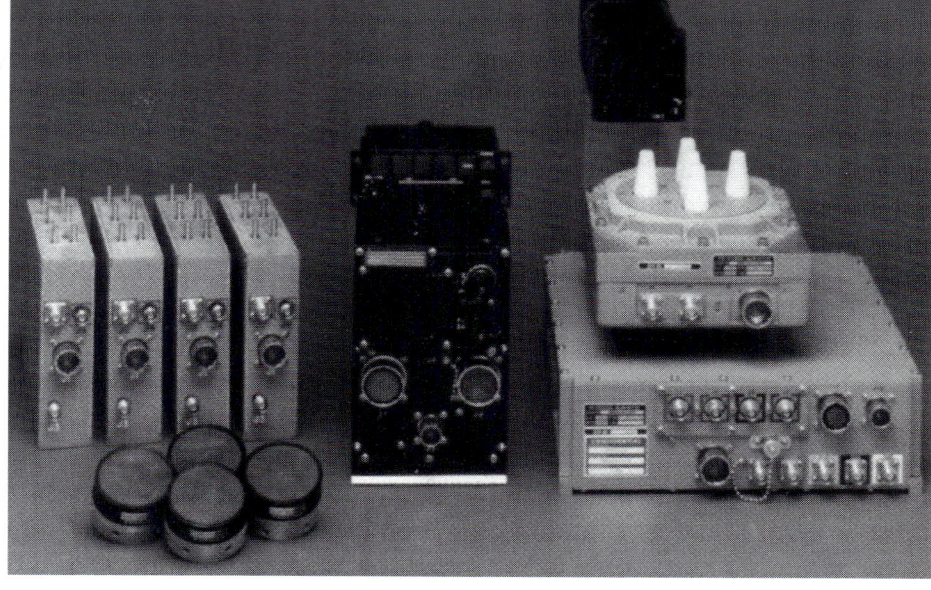

Litton Applied Technology AN/ALR-67(V)2 countermeasures warning and control system ***1998***/0018269

AN/ALR-67(V) countermeasures warning and control system

The AN/ALR-67 countermeasures warning and control system is the successor to the US Navy's AN/ALR-45 system installed in A-6E and F-14A tactical aircraft. It is the standard threat warning system for tactical aircraft and was specifically designed for the A-6E/SWIP, AV-8B, F-14B, F-14D and F/A-18. The system detects, identifies and displays radars and radar-guided weapon systems in the C to J frequency range. The system also co-ordinates its operation with onboard fire-control radars, datalinks, jammers, missile detection systems and anti-radiation missiles. The dispensing of expendables, such as chaff, flares and decoys, is controlled by the AN/ALR-67.

The AN/ALR-67(V) consists of broadband crystal video receivers, a superheterodyne receiver, an integrated low-band receiver, an antenna array and an alphanumeric azimuth indicator, all under the control of a dual ATAC 16M CP-1293 threat processor. The system is field-programmable and includes provision for software updates at squadron level. The CP-1293 and IP-1276 used in the ALR-67 and ALR-45F configurations are identical and interchangeable. The system is fully compatible with MIL-STD-1553A databus requirements and features interface control of such systems as HARM, ALQ-126A/B, ALE-39, ALQ-162 and ALQ-165.

AN/ALR-67(V)2

The AN/ALR-67(V)2 was the first upgrade of the AN/ALR-67(V) system; it comprises the following units (see photo):

(1) four small spiral high-band antennas to provide 360° azimuth RF coverage
(2) four wideband, high-band quadrant receivers
(3) a low-band array plus receiver to provide 360° azimuth low-band coverage
(4) a narrowband superheterodyne receiver for signal analysis functions
(5) twin CPU
(6) threat display
(7) control unit.

The AN/ALR-67(V)2 in turn has been given a significant enhancement in capability, through Engineering Change Procedure ECP-510 to the AN/ALR-67E(V)2 standard.

AN/ALR-67E(V)2

ECP-510 provided a card-for-card upgrade of the AN/ALR-67(V) to the AN/ALR-67E(V)2 standard. It provides a significant increase in system sensitivity in the presence of strong signals and offers a significant increase in computer pulse processing capability. The upgrade also features the capability to detect and exploit unique signals for improved tactical awareness. To better manage high density signal environments, the AN/ALR-67E(V)2 incorporates Application-Specific Integrated Circuit (ASIC) chips that increase the system's processing power five-fold. The AN/ALR-67E (V)2 provides additional enhancements including a 10-fold improvement in detection ranges when in the presence of a wingman's radar signals; it also incorporates INS stabilisation for accurate display in high 'G' manoeuvres.

The AN/ALR-67E(V)2 modular architecture permits further enhancement, without change to the aircraft system, to provide yet more computer power, increased detection range, and improved target discrimination.

AN/ALR-67(V)3/4

For clarity, note that the AN/ALR-67(V)3/4 developments of the AN/ALR-67(V) are made not by Litton, but by the Raytheon Systems Company.

Operational status

In production. Over 1,600 AN/ALR-67(V) and AN/ALR-67(V)2 systems have been sold.

The AN/ALR-67(V) has been supplied for the F/A-18 to the US Navy and US Marine Corps, Australia, Canada, Finland, Kuwait, Malaysia, Spain and Switzerland. Orders, for delivery in 1999, have been awarded by Boeing on behalf of the Spanish Air Force and Royal Thai Air Force for their F/A-18 aircraft. The configuration is understood to be designated AN/ALR-67B(V)2.

The AN/ALR-67(V) system has also been supplied for US Marine Corps, Italian and Spanish AV-8Bs. An updated version, designated AN/ALR-67(V)2E is being requested by the US Marine Corps for a proposed FY2000 upgrade of its AV-8B aircraft. The AN/ALR-67(V)2E is understood to include all-new software and a range of improvements originally developed for the Super Hornet's ALR-67.

Contractor

Litton Applied Technology.

UPDATED

AN/ALR-68 Advanced Radar Warning System (ARWS)

The Advanced Radar Warning System (ARWS) is now deployed operationally by the German Air Force. It is a digital threat warning receiver system based on improvements to the ALR-46. Conversion to digital operation is achieved by the addition of the Applied Technology ATAC computer.

The ALR-68 is a wide open crystal video, field-programmable system which provides in-cockpit threat parameter programming and hand-off to tactical ECM systems. It uses a digital threat processor containing software provisions unique to the German threat scenario and is the first wide open digital, general purpose computer-based RWR to become operational in Europe.

Operational status

Manufactured in association with Daimler-Benz Aerospace, the AN/ALR-68A(V)2 is installed in F-4F and RF-4E aircraft of the German Air Force.

Contractor

Litton Applied Technology.

VERIFIED

AN/ALR-69 radar warning receiver

An outgrowth of the earlier models of the ALR-46 warning receiver, the AN/ALR-69 employs a Frequency Selective Receiver System (FSRS) and a low-band launch alert receiver (Compass Sail) added to the basic ALR-46. Compass Sail detects and analyses SAM guidance beams to warn the pilot of missiles tracking toward the aircraft. The system is designed to activate ECM resources automatically. The signal processor provides executive control for the FSRS. It accepts video inputs from five receivers and processed information from the FSRS, sorts and analyses the data, identifies and labels the received radar signals, tracks the status of these radar signals and generates signals to provide threat warning to the operator. Functions performed by the FSRS include warning and direction-finding on CW signals, accurate frequency measurements on pulse signals for ambiguity resolution, threat antenna scan type and rate analysis and jammer frequency set on for jammer power management and blanking functions.

The AN/ALR-69 RWR upgraded with Litton's CM518 digital processor, AM-9101 Single Channel Amplifier Detectors (SCADs) and R-2094M Frequency Measurement Receiver (FMR) is known as the AN/ALR-91(V)3M threat warning system.

Operational status

More than 3,500 systems have been delivered to the US Air Force and to NATO countries. Examples of the AN/ALR-69 system are installed in F-16s supplied to Denmark, Norway, Netherlands, and US Air Force. The AN/ALR-69 is also used on the A-10, F-4 and C-130 aircraft. The AN/ALR-69(V) is also fitted in Danish Air Force C-130 aircraft.

Contractor

Litton Applied Technology.

UPDATED

AN/ALR-87 radar warning system

The AN/ALR-87 advanced threat warning system is designed with all the necessary performance features to meet identified and postulated threat environments. It is a replacement for the AN/ALR-46 and AN/ALR-69

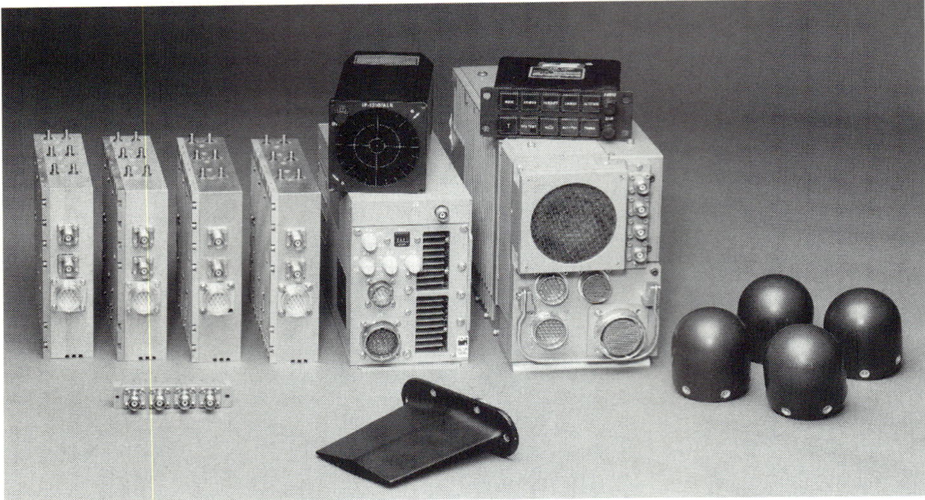

The Litton AN/ALR-87 threat warning system

family of radar warning receivers. The AN/ALR-87 is based on a digital IFM receiver and is controlled by MIL-STD-1750A microprocessors. The system is fully programmable with in-flight recording capability, operates in the C- to J-bands to provide rapid emitter identification in complex and dense threat environments and interfaces with active and passive onboard systems.

It consists of four high-band antennas, four quadrant receivers, a low-band antenna, low-band receiver and power supply, processor/receiver, azimuth indicator and indicator control unit.

Specifications

Dimensions:
(high-band antenna × 4) 50.8 × 70.1 mm diameter
(quadrant receiver × 4) 175.3 × 154.9 × 40.6 mm
(low-band antenna) 101.6 × 127 × 76.2 mm
(low-band receiver/power supply) 254 × 152.4 × 101.6 mm
(processor/receiver) 127 × 193 × 370.8 mm
(azimuth indicator) 81.5 × 81.5 × 222.1 mm
(indicator control unit) 45.7 × 129.5 × 109.2 mm
Weight:
(high-band antenna × 4) 0.45 kg
(quadrant receiver × 4) 10 kg
(low-band antenna) 0.13 kg
(low-band receiver/power supply) 5.45 kg
(processor/receiver) 11.82 kg
(azimuth indicator) 1.36 kg
(indicator control unit) 0.68 kg
Frequency: C- to J-bands
Power: 580 W
Processors: three MIL-STD-1750A microprocessors
Memory: 128 k EEPROM, 162 k RAM
Mission data recorder: 1 M × 17 dynamic RAM
Interfaces: MIL-STD-1553B AV and EW bus, RS-422, RS-232

Operational status

200 systems have been procured by the Swiss Air Force for Northrop Grumman F-5E and F and Dassault Mirage III aircraft.

Contractor

Litton Applied Technology.

VERIFIED

AN/ALR-91(V)3 series threat warning systems

The AN/ALR-91(V)3 series of threat warning systems are upgrades of earlier Litton radar warning receivers. The AN/ALR-91(V)3 is the designation given to the AN/ALR-46 RWR upgraded with Litton's CM-518 digital processor replacing the CM-442A processor, and the AM-9101 Single Channel Amplifier Detectors (SCADs) replacing the AM-6639 high-band receivers. Replacement of the R-1854A low-band receiver with the R-1854M Frequency Measurement Receiver (FMR) is a further option. The updated system offers the following improvements, relative to the ALR-46 RWR: up to 10 times the detection range; operation in pulse densities five times greater; detection and DF of pulse, PD and CW emitters.

AN/ALR-91(V)3 radar warning receiver
1998/0018267

AN/ALR-91(V)3M radar warning receiver
1998/0018266

The AN/ALR-91(V)3M is the designation given to the AN/ALR-69 RWR upgraded with Litton's CM-518 digital processor replacing the CM-479 processor; the SCADs replacing the AM-6639A high-band receivers; the R-2094M FMR replacing the R-2094 FSRS receiver/controller. The updated configuration offers the following improvements, relative to the AN/ALR-69 RWR: greater than four times the detecton range; operation in five times the pulse density; detection and DF of pulse, PD and CW signals.

Specifications

RF coverage: 5-18 GHz
Spatial coverage: 360° azimuth; ±60° elevation
DF: 12° RMS

Contractor

Litton Applied Technology.

NEW ENTRY

AN/ALR-93(V)1/2 radar warning receiver

The AN/ALR-93(V)1/2 radar warning receiver is a lightweight high-sensitivity RWR covering the C- to J-bands, designed to operate in dense complex emitter environments with 100 per cent probability of intercept. Its unique architecture, combined with sophisticated packaging concepts, results in a high-performance, small footprint, radar warning system weighing less than 20 kg. It is specifically designed for weight critical platforms desiring a high-performance RWR system. It provides threat warning and situation awareness, allowing the crew to execute missions with significantly increased effectiveness, and meets the critical requirements for today's fighter aircraft applications.

The ALR-93(V)1/2 is designed to provide superior performance in an extremely lightweight modular low-power package. By combining the wideband acquisition of an Instantaneous Frequency Measurement (IFM) receiver with the selectivity of a SuperHeterodyne Receiver (SHR), the ALR-93(V)1/2 is able to provide the best possible system for mission success. The ALR-93(V)1/2 utilises an IFM receiver whose wide acquisition bandwidth processes a 4 GHz instantaneous portion of the RF spectrum, while maintaining 100 per cent probability of intercept on signals as narrow as 100 ns. The wide instantaneous bandwidth guarantees pulse-to-pulse measurement on even the most frequency-agile radars, while minimising the time required to cover the entire RF spectrum.

The SHR provides high sensitivity and high selectivity with its two narrowband frequency search modes. The SHR microscanning capability includes resolving and/or looking around multiple pulse Doppler, CW or interrupted continuous wave emissions.

Specifications

Weight: <20 kg
Frequency: C- to J-band
Radar types: all pulsed radars including pulse Doppler, CW, ICW, LPI, jitter and stagger, pulse compression, frequency-agile and agile-agile
Azimuth: 360° coverage
Radar storage: more than 2,000 emitters
Reliability: >1,000 h MTBF

Operational status

In July 1993, a contract was signed for 80 systems for F-16C/D aircraft for a NATO customer.

Contractor

Litton Applied Technology.

VERIFIED

AN/ALR-93(V)2,3,4 ESM system

The AN/ALR-93(V)2,3,4 ESM is designed for over-the-horizon detecting and targeting. It is an automatic system with continuous 360° coverage for use in the C- to J- frequency bands. It provides instantaneous emitter and platform classification, passive targeting for precise weapon delivery, generic threat processing of modern threats such as CW, pulse Doppler, agile-agile and low probability of intercept radars.

The AN/ALR-93(V)2,3,4 features high accuracy, interferometer DF emitter data display with pulse analyser, modular architecture and high reliability and

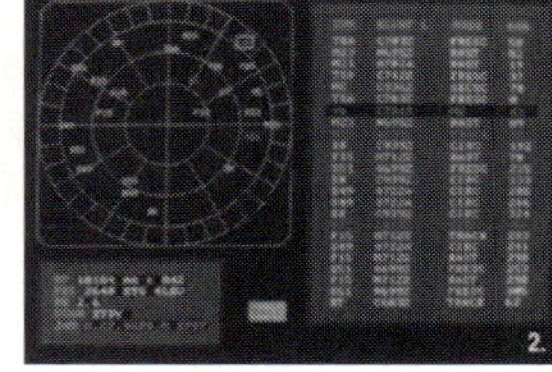

AN/ALR-93(V)2,3,4 ESM system showing: 1. System hardware 2. ESM display 3. Butler Matrix antenna option 4. Interferometer antenna option
1998/0018265

maintenance. Over 1,000 threats can be tracked simultaneously.

The AN/ALR-93(V)2,3,4 employs three RF intercept receiving techniques — IFM, superheterodyne and amplified crystal video — gaining the benefits derived from each of these technologies. Its multiple 32-bit architecture provides the preprocessing and processing power to make it fully capable of handling future threat environments. The modular approach allows multiple missions to be readily accomplished. The system maximises the use of pattern recognition and calibration techniques to ensure accurate high-speed contact identification and platform classification.

Two antenna options are available, a Butler Matrix precision DF phase-amplitude antenna, which gives 3° RMS DF accuracy, or an interferometer precision DF antenna, which gives 1° RMS DF accuracy.

Specifications

Dimensions:
(C-J antenna - 4 per system) 73.4 × 97 mm
(C-J receiver - 4 per system) 35.6 × 139.7 × 133.3 mm
(directional antenna) 127 × 381 mm
(directional receiver) 35.6 × 139.7 × 133.3 mm
(ESM processor) 193.5 × 123.7 × 319 mm
Weight:
(C-J antenna) 0.9 kg each
(C-J receiver) 5.44 kg each
(directional antenna) 2.72-4.54 kg
(directional receiver) 1.36 kg
(ESM processor) 10.89 kg

Contractor

Litton Applied Technology.

UPDATED

AN/ALR-606(V)1 series radar warning receivers

The AN/ALR-606(V)1 ESM system is a fully digital radar warning receiver. It is based on the AN/ALR-66A(V)1, with which it shares the same hardware configuration.

The ALR-606(V)1 is designed especially for the export market by offering item for item replacement of older analogue and digital RWR systems in previous generation maritime patrol aircraft such as the S-2 Tracker, TA-7 and C-130. It also provides full digital signal processing, high-intensity CRT, E- to J-band coverage and alphanumeric symbology.

Use of EEPROM allows on-site reprogramming capability for display symbology and program parameters through the use of a separate Computer Memory Loader (CML).

The ALR-606(V)1 has recently been upgraded to an ALR-606A(V)1 configuration which incorporates new higher-sensitivity receivers, a liquid crystal display, increased EEPROM memory, enhanced processing and allows complete OFP and emitter library reprogramming through the use of a CML.

Specifications

Weight: 28 kg
Frequency: E- to J-band continuous

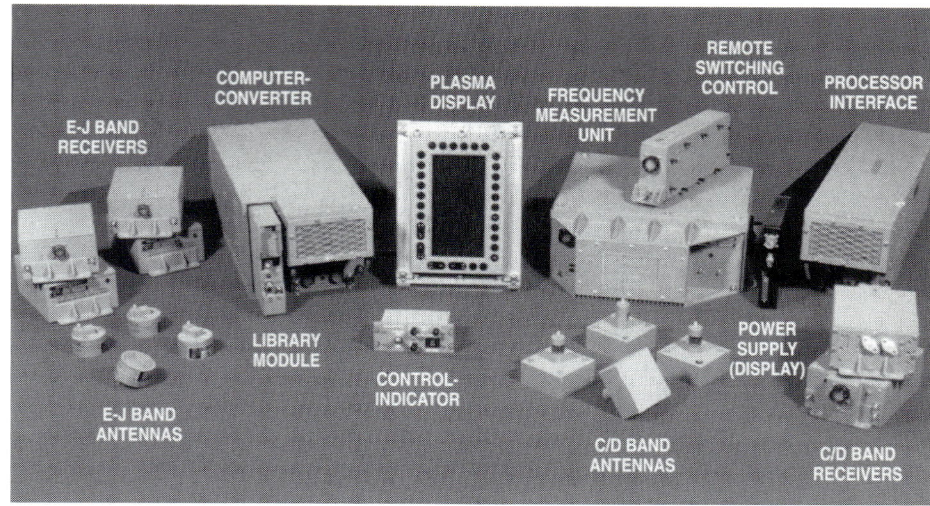

Units of the AN/ALR-606(V)2 surveillance and direction-finding system

Warning I/D: all pulse types; CW, ICW and pulse Doppler, LPI, jitter/stagger, pulse compression, and frequency-agile radars
Receiver type: crystal video
Azimuth: 360° coverage
Symbology: up to 3 symbols per emitter, programmable by customer
Interface: RS-232 or MIL-STD-1553
Reliability: >630 h MTBF demonstrated (MIL-STD-781)

Operational status

No longer in regular production, having been replaced by the newer ALR-606A(V)1, the ALR-606(V)1 is in worldwide use in numerous types of aircraft.

Contractor

Litton Applied Technology.

UPDATED

AN/ALR-606(V)2 series surveillance and direction-finding systems

The AN/ALR-606(V)2 surveillance and direction-finding system is specifically designed for use in maritime patrol aircraft, offering coverage in the C- to J-bands and over-the-horizon emitter location. It was derived as an export version of the AN/ALR-66A(V)3 system. The equipment provides advanced capabilities in such areas as precision DF accuracy, high sensitivity for over-the-horizon detection, precise frequency measurement, advanced signal processing coupled with expanded data memory, multimode operator interactive display and controls, precision emitter parameter measurements, integration with other aircraft primary sensors and EEPROM flight line reprogramming. Upgrades to the ALR-606A(V)2 can be accomplished with minimal or no aircraft wiring change. Upgrade includes higher sensitivity E- to J-band receivers, expanded emitter library capacity, improved processing and in-flight signal-of-interest programming. Additional upgrades to ALR-606B(V)2 have been developed which include incorporation of a dedicated, controllable DF antenna and the addition of a data recorder to support ground analysis of ESM intercepts.

Operational status

The AN/ALR-606(V)2 has been provided for the Northrop Grumman S-2 maritime patrol aircraft, Sikorsky S-70 helicopter and for numerous international customers. The ALR-606(V) as well as the upgraded ALR-606A(V)2 and ALR-606B(V)2 systems are in production.

Contractor

Litton Applied Technology.

VERIFIED

AN/APR-39A(V)2 and 'AN/APR-39A(V)3 with CW' threat warning systems

AN/APR-39A(V)2

The AN/APR-39A(V)2 is designed specifically for use on the US Army Special Electronic Mission Aircraft (SEMA), US Navy HH-60 helicopters and for all US Marine Corps helicopters and non-high-performance fixed-wing aircraft such as the C-130, OV-10 and CV-22. In addition to threat warning, the system acts as the controller for the Marine Corps' integrated electronic warfare survivability suite. The AN/APR-39A(V)2 is integrated with laser warning, missile warning systems and countermeasures dispenser systems.

'AN/APR-39A(V)3 with CW'

The AN/APR-39A(V)3 utilises APR-39A(V)2 technology and includes dual-channel crystal video receivers, plus a Tuned RF (TRF) receiver, to provide enhanced direction-finding of CW and pulse Doppler radars, enhanced frequency measurement for improved ambiguity resolution, and better high-pulse density performance. The 'AN/APR-39A(V)3 with CW' is fully programmable using a memory/loader verifier unit. The system is deployed by the US Army, US Air Force and NATO aircraft and can be exported to countries meeting US Department of State licensing requirements.

AN/APR-39A(VE)/(V)3 threat warning system

Litton Applied Technology is to supply its 'Viking CW Upgrade' version of the AN/APR-39A system, designated AN/APR-39A(VE)/(V)3, to the Royal Norwegian Air Force for its BA-412 helicopters (25 units).

Specifications

Weight:
(AN/APR-39A(V)2) 15.9 kg
(AN/APR-39A(V)3) 8 kg
Power supply: 28 V DC
((V)1/(V)3) 58 W
((V)2) 200 W

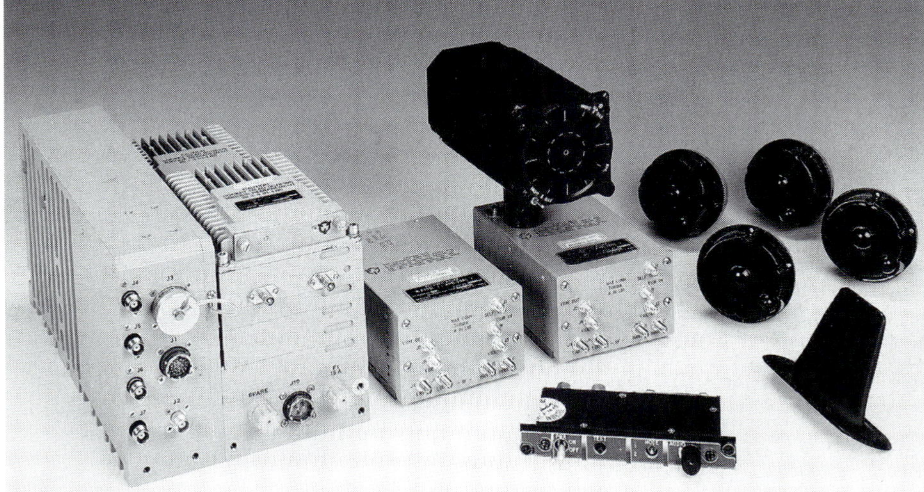

The Litton AN/APR-39A(V)2 threat warning system **1996**

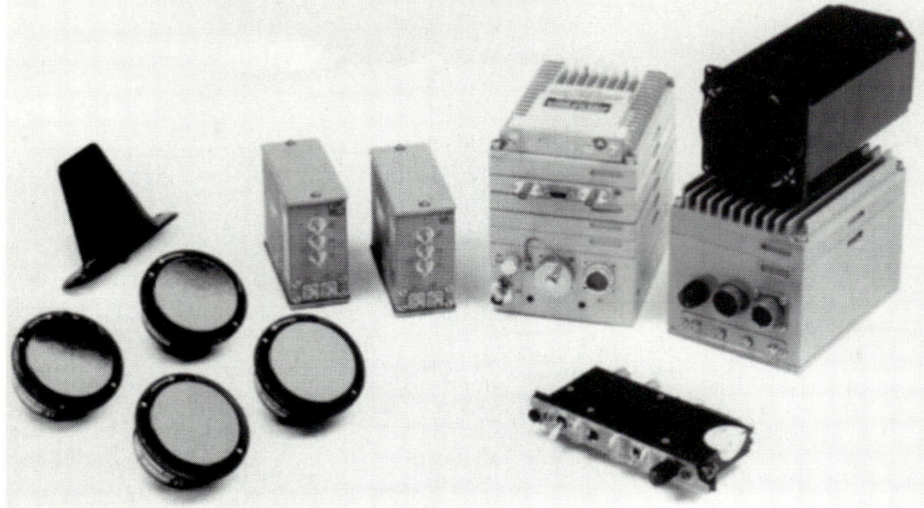

The Litton 'AN/APR-39A(V)3 with CW' threat warning system (shown with optional MIL-STD-1553B databus) ***1998***/0018264

Frequency: E-K and C/D bands
Interfaces: RS-422 databus, MIL-STD-1553B databus optional, E-O warning systems, missile launch detectors, radar jammers, CW warning receivers

Operational status

The AN/APR-39(V)2 completed flight evaluation and entered production during 1996. It has been selected for the upgrade of the Bell SuperCobra AH1(4B)W and Huey UH-1(4B)N.

Contractor

Litton Applied Technology.

UPDATED

ERWE Enhanced Radar Warning Equipment

The Enhanced Radar Warning Equipment (ERWE) consists of a fully programmable digital threat processor and three separate receiver elements which provide C- and D-band coverage and both wideband and narrowband coverage for the E- to J-bands. A crystal video receiver is employed to provide wideband coverage for rapid threat response and a superheterodyne receiver provides high sensitivity and selectivity. Both receiver sections employ the Litton Applied Technology ATAC computer. The equipment provides a prioritised alphanumeric display and a high-intensity azimuth indicator and is capable of rapid change of threat parameters and priorities from mission to mission. The ERWE was developed for the European threat scenario and is now operational with German Air Force and Navy Tornado IDS aircraft.

Operational status

In service with German Air Force and German Navy Tornado IDS aircraft. ERWE I and ERWE II improvements are now in production.

Contractor

Litton Applied Technology.

UPDATED

Litton Enhanced Radar Warning Equipment

Apollo

Apollo is the name for a US Air Force foreign military sales programme that provides a Defensive Aids Suite (DAS) for C-130 aircraft.

Apollo comprises the AN/ALQ-156 pulse Doppler MAW, the AN/AAR-44 and AN/AAR-47 electro-optic MAWs and an AN/ALE-40 chaff/flare dispenser. Alternative versions may also include the AN/ALR-56M RWR and the AN/ALE-47 smart chaff/IRCM dispenser.

Operational status

The US government initiated a 'quick-fix' programme to support aircraft involved in UN peacekeeping duties, and Apollo installations have now been fitted to C-130 aircraft from a number of foreign air forces, including the Australian, Norwegian and Swedish air forces.

Contractor

Lockheed Martin Aircraft Services.

VERIFIED

Compass Call electronic warfare aircraft

The EC-130H Compass Call version of the Hercules transport is designed to identify and disrupt enemy radars and communications.

The Compass Call aircraft was a response to an urgent request from Tactical Air Command for a communication jammer. Development of equipment and operational tactics are the responsibility of the Tactical Air Warfare Center. The aircraft entered service in mid-1982 with the 41st Electronic Combat Squadron. It is distinguished from all other Hercules C-130 variants by the large vertical aerial forward of the vertical fin, and by two more underwing antennas.

Compass Call has been modified and upgraded over the past decade, mainly to improve its jamming power and characteristics, its capability in multithreat situations and the overall software and system management. It is understood that a US Air Force programme is under way to enable the system to jam more signals simultaneously and at longer ranges.

Systems on the aircraft include the Raytheon Systems Company ALQ-173, ALQ-175 and ALQ-198 high-band systems with digitally tuned receivers.

Operational status

The standard configuration in operational service is designated Block 20. A Block 30 upgrade, aimed at improving reprogrammability, signal acquisition, threat identification and threat location, is under way; three aircraft were delivered in 1997; the remainder will be delivered by mid-1999. US Air Force plans call for six Block 20 aircraft to be updated to Block 35 standard starting in FY2000, and subsequently for the rest of the force to be upgraded to Block 40 standard. It is reported that Block 35 modifications will replace remaining analogue technology systems, and instal the Tactical RAdio Countermeasures System (TRACS) and digital compressor omitted from Block 30. Block 40 requirements have yet to be finalised.

Contractor

Lockheed Martin Aircraft Services.

UPDATED

Data Collection and Processing System

The Data Collection and Processing System (DCPS) is an airborne SIGINT collection system housed in a shelter which can be installed in any C-130E/H aircraft. The shelter is 11.89 m long (39 ft), houses 23 racks of electronic equipment and utilises 11 operators plus one airborne maintenance technician. The DCPS concept includes a modular, distributed computer-managed COMINT subsystem and a fully automated ELINT subsystem. An extensive voice and data communications suite provides real-time reporting and mission command and control. The DCPS antennas are mounted on removable aircraft components such as the paratroop and wheel-well doors and escape hatches. The system can be loaded into a non-dedicated aircraft within 10 hours, providing a quick response capability.

The communications intercept subsystem features manual or computer-directed search, intercept, direction-finding, geolocation, analysis and recording of communications signals in the HF to SHF frequency bands. Each operator is supported with computer-assisted processing for mission tasking, signal interpretation, signal data management, graphical situation display and textual display and management.

The ELINT subsystem provides automated acquisition, identification, direction-finding, geolocation, analysis and reporting of radar signals. The search spectrum and threat profiles are downloaded and easily tailored to specific environments, missions or tasking. This subsystem is supported by a single operator position for system control and reporting management. In addition, a manual SHF collection capability is provided to support both high-capacity communication signals intercept and manual technical analysis of radar signals.

In terms of platform performance, the shelter and mission equipment weigh approximately 12,250 kg, giving a maximum endurance of 10 hours. Service ceiling is 25,000 ft, giving a collection range of well over 370 km from the aircraft.

Contractor

Lockheed Martin Aircraft Services.

VERIFIED

The Lockheed Martin data collection and processing system shelter being loaded into a C-130 Hercules

SAMSON system

SAMSON (Special Avionics Mission Strap On Now) is a quick turn around kit facilitating reconfiguration of C-130 aircraft for special missions such as search and rescue, photo reconnaissance, drug interdiction, electronic surveillance/countermeasures, air sampling and radio relay. The kit comprises an equipment pod which replaces the C-130E/H external fuel tank, a cargo pallet-mounted operators' console, interface wiring, and any necessary additional external components. The kit may be installed or removed within one work shift by a crew of four to six persons and is accomplished with no permanent modification to the aircraft.

Operational status

The system prototype has undergone numerous field tests by US and foreign C-130 users. A production version of the system is currently in service in a reconnaissance role.

Contractor

Lockheed Martin Aeronautical Systems.

VERIFIED

SATIN EW system

SATIN (Survivability Augmentation for Transport Installation) is a minimum modification approach to equip C-130 aircraft with self-protection sensors and countermeasures. The installations are designed to be installed by a small team in a few days. All external or cockpit components are provided with replacement cover panels, so that the systems can be removed when not required, or used on other aircraft.

SATIN components are selected from the standard military inventories of the particular user to minimise logistics problems. Systems which have been installed for various users include the AN/ALR-69 Radar Warning Receiver, APR-39A(V)1 Radar Signals Receiving Set, AN/ALQ-156 ESM system, AN/AAR-47 Missile Warning Set, AN/ALQ-157 Self-Protection System and the AN/ALE-39, ALE-40 and ALE-47 Countermeasures Dispenser Systems. Other systems can be incorporated, as dictated by user requirements.

Operational status

SATIN has been installed on various versions of C-130 aircraft for several users and the basic installation approach has been incorporated into production C-130 aircraft.

Contractor

Lockheed Martin Aeronautical Systems.

VERIFIED

Targeting Avionics System (TAS)

Targeting Avionics System (TAS), previously known as the Multiple Emitter Targeting Receiver (METR), is a lightweight system for targeting airborne and ground-based receivers. It can be installed in an aircraft pylon and was designed to provide targeting against multiple threats. Its accuracy in azimuth and elevation enables it to be used to cue other sensors or seekers such as FLIR and IR missiles.

Operational status

A pylon-mounted version of TAS was successfully flight tested on an F/A-18 in October 1996, as part of the US Navy Targeting Avionics System Demonstration programme. During this flight test, a HARM was launched in the 'range known' mode using cueing data provided by TAS. Additional development is continuing as part of this programme.

Contractor

Lockheed Martin Electronics & Missiles.

UPDATED

AN/ALQ-157 infrared countermeasures system

The AN/ALQ-157 provides multiple simultaneous protection for large heavy-lift helicopters and medium-size fixed-wing aircraft against SAM and AAM threats. The system employs advanced components and microprocessor technology to allow operator jamming code selection and reprogrammability for future threats. The two fuselage-mounted synchronised jammer assemblies provide continuous 360° protection against threats launched from any direction. The power module, line filter and pilot control indicator can be placed anywhere within the aircraft.

Specifications

Weight: 99.8 kg
Power supply: 115 V AC, 3 phase, 400 Hz
28 V DC

Operational status

In production. Lockheed Martin received the initial

production contract in December 1983 for systems for the US Marine Corps' CH-46 helicopter. US and international aircraft equipped with the AN/ALQ-157 include the SH-3, CH-46, CH-47D, H-53, Lynx, C-130 and P-3C. Nearly 700 systems have been sold worldwide.

Contractor
Lockheed Martin Electro-Optical Systems.

VERIFIED

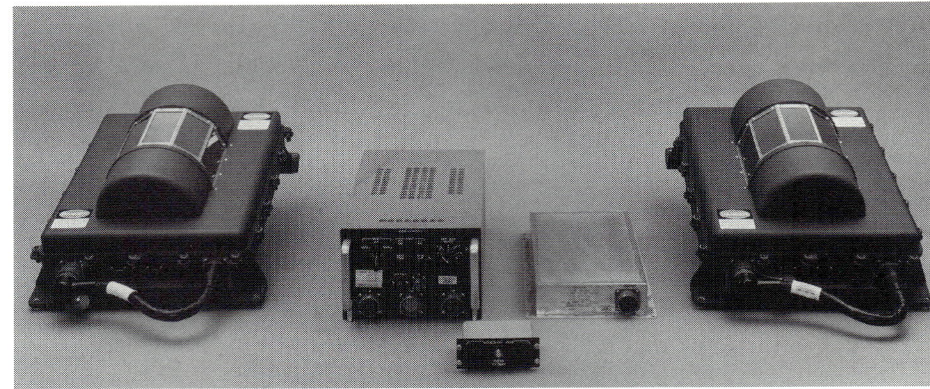

The AN/ALQ-157 showing the two fuselage-mounted jammer assemblies (left and right) and, in the centre, the power module/pilot control indicator, line filter

AN/ALQ-204 Matador IR countermeasures system

The AN/ALQ-204 Matador is a family of infrared countermeasures systems which has 11 different configurations. The AN/ALQ-204 is also known as Matador. The system is suitable for all types of large transport aircraft with unsuppressed engines, one transmitter per engine being recommended for maximum protection. The basic system consists of transmitters, a Controller Unit (CU), Power Supply Unit (PSU) and an operator's controller. Transmitters are electronically synchronised by the CU which controls and monitors one or two transmitters. The operator's controller is common to all configurations, controls from one to seven transmitters and incorporates a system status display.

Each transmitter contains an IR source which emits pulsed radiation to combat an IR missile. Preprogrammed multithreat jamming codes are selectable on the operator's control unit and all new codes can be entered as required to cope with new threats. Each transmitter weighs 22 kg nominal and can be pod mounted for aircraft attachment.

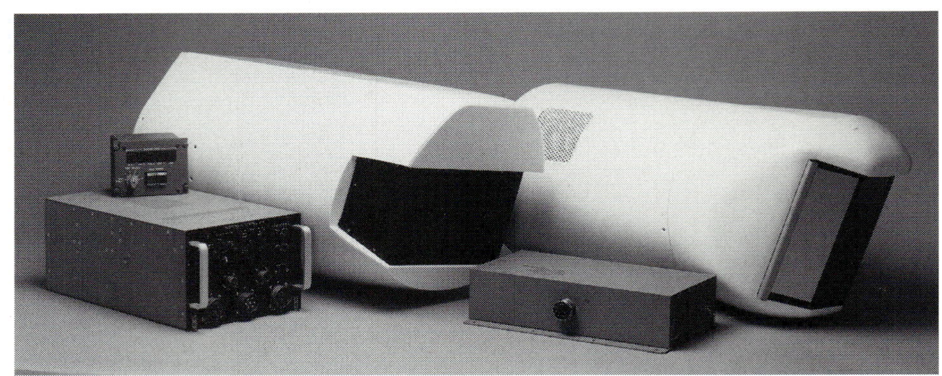

The AN/ALQ-204 infrared countermeasures system is installed in a number of large fixed-wing aircraft

Operational status
In production. In service on the Boeing 747 and 707, Lockheed Martin L-1011 and British Aerospace VC10, BAe 146 and Andover aircraft.

Future applications include the Boeing E-3A, 737 and 767, Airbus A300 and A340.

Contractor
Lockheed Martin Electro-Optical Systems.

VERIFIED

Challenger IR jammer

A small omnidirectional IR jammer, especially suitable for helicopters and light aircraft, Challenger can be installed in fixed aircraft apertures or deployed on retractable platforms. The jammer head weighs about 9 kg and power consumption is within the range 1 to 4 kW. The jammer assembly is driven by a compact controller in the aircraft. Single or dual transmitter configuration can be selected to provide 360° azimuth protection. Challenger, an upgraded AN/ALQ-157, was designed to be flexible and to offer a wide variety of installation options.

Operational status
In production and in service on Lynx and UH-1 helicopters.

Contractor
Lockheed Martin Electro-Optical Systems.

VERIFIED

Defendir IR jammer

Defendir is a family of systems which can be used for small and large helicopters, as well as turbojet transport fixed-wing aircraft. It can be supplied in various configurations, such as an aft-mounted jammer for aft protection, an aft-mounted jammer with side windows for aft and side protection, a nose-mounted jammer for forward protection or aft and nose jammers for all-round protection.

Operational status
In development.

Contractor
Lockheed Martin Electro-Optical Systems.

VERIFIED

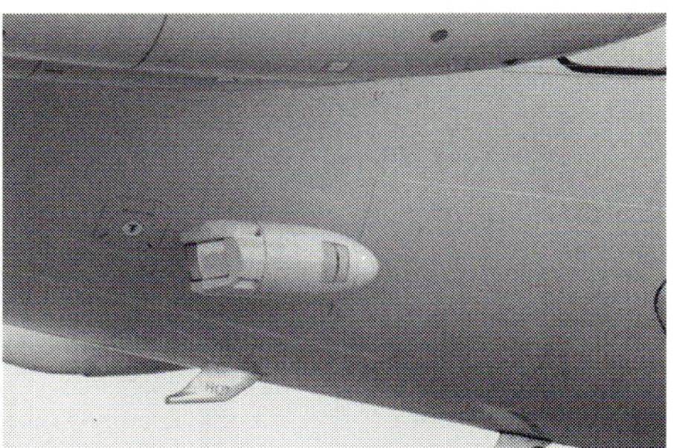

The Lockheed Martin AN/ALQ-204 mounted on the fuselage of a Royal Air Force VC10 (Paul Jackson)

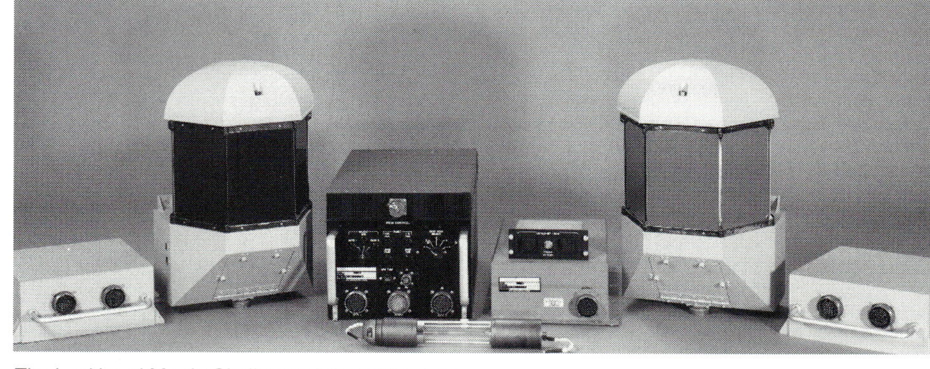

The Lockheed Martin Challenger infrared jammer

AN/ALQ-78 ESM system

The AN/ALQ-78 ESM system is used aboard the Lockheed Martin P-3C Orion maritime patrol aircraft as its electronic support measures sensor. The antenna is carried on a pylon under the inner wing. It automatically detects and measures the characteristics and bearings of intercepted radar signals of anti-submarine and electronic warfare interest. The measured parameters and bearing of the intercepted signals are supplied to the aircraft central data processing system for evaluation, recording and presentation on the aircraft displays.

The system uses a high-speed rotating antenna and a scanning, superheterodyne receiver for acquisition of signals in specific frequency bands of particular interest to the P-3C. Operation is mostly automatic, based on parametric data. The countermeasures set normally operates in an omnidirectional search mode. When a radar signal of interest is acquired and analysed, ALQ-78 automatically initiates a direction-finding routine. The signal data is processed by the central data computer and formatted for readout on a multipurpose display.

Operational status
ALQ-78 is understood to have been in service aboard Lockheed Martin P-3C Orion patrol aircraft since 1969. As part of the US Navy's Orion update effort, ALQ-78 is noted as being scheduled for replacement by the AN/

ALR-66(V)5 system. ALQ-78 is also known to have been built under licence by Mitsubishi for use on board Japanese P-3C aircraft. In US service, ALQ-78 has been upgraded to ALQ-78A standard.

Contractor

Lockheed Martin Fairchild Systems.

UPDATED

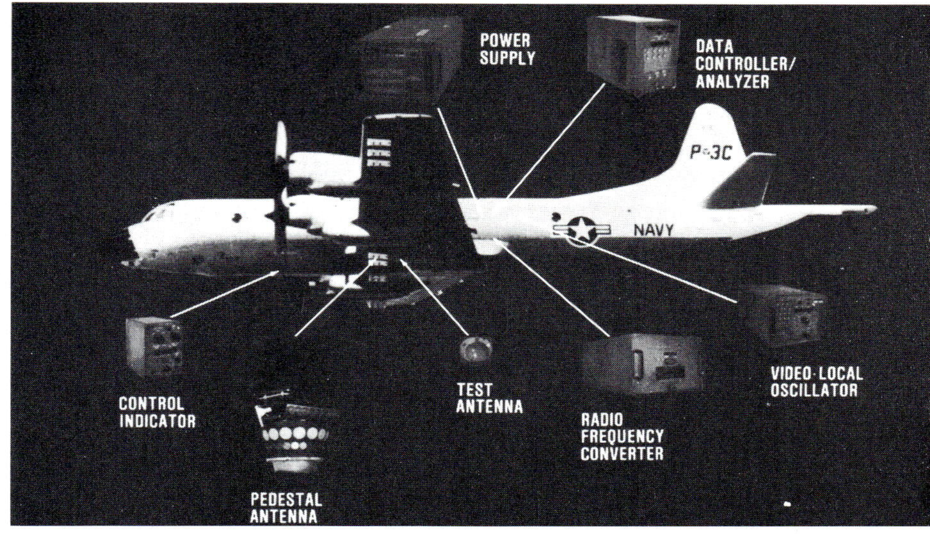

Lockheed Martin AN/ALQ-78 ESM equipment on the US Navy Lockheed Martin P-3C Orion

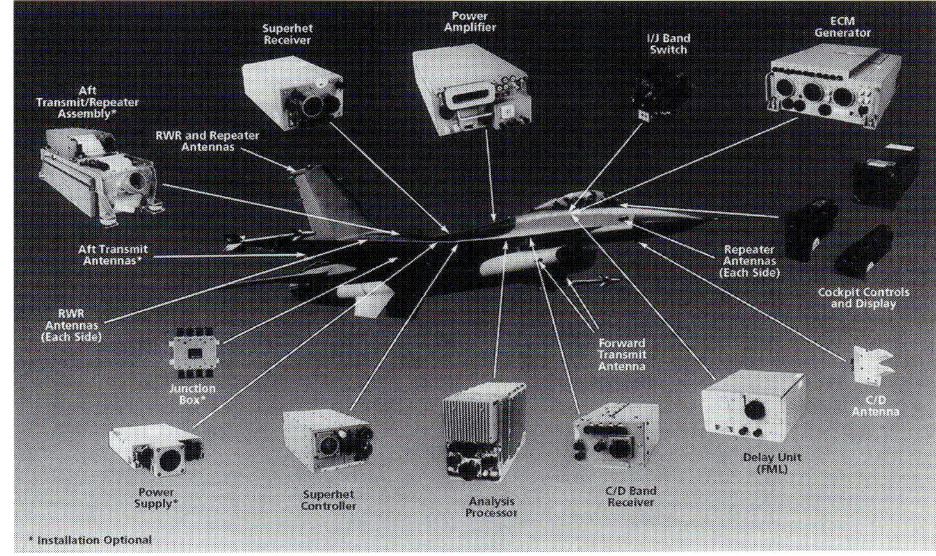

The elements which make up Lockheed Martin's ALQ-178(V) ECM system **1998**

AN/ALQ-178(V) radar warning and Electronic CounterMeasures (ECM) suite

AN/ALQ-178(V) is an integrated radar warning and active countermeasures suite for tactical fighter aircraft, such as the F-16. An earlier version was known as RAPPORT III. The system utilises a central programmable computer for data analysis and system control, with independent microprocessors to direct Radar Warning Receiver (RWR), display, jamming and countermeasures dispensing functions. The wideband RWR continuously scans the threat radar environment. Detected signals are de-interleaved and identified by radar type and displayed to the pilot on a cathode ray tube in legible, unambiguous format. The power management algorithm optionally matches the countermeasures to the RWR's constantly changing threat picture. Separate forward and aft jammers are used for maximum spatial coverage. The jammers are accurately set for maximum power output at each victim radar's frequency to maximise effective radiated power and jammer effectiveness. The jammer frequency range provides coverage of the entire threat band while the jammer power is sufficient to counter these radars throughout their lethal area. The countermeasures dispenser is automatically controlled, including the pilot's display and cues, to provide optimum response to the threat between active jamming and chaff dispensing.

Although the ALQ-178(V) consists of an RWR, an active jammer and countermeasures control, the RWR can be installed as an independent threat warning system (controlling a countermeasures dispensing system), with the jammer added later.

The latest derivative of the ALQ-178(V) family incorporates technological advancements and packaging to increase capabilities significantly. Enhanced performance is provided by Digital RF Memory (DRFM) technology, agile jamming channels, distributed high speed/capacity processors and precision direction-finding capabilities.

A follow-on to ALQ-178(V), combined with the AN/ALR-56M or other RWR, is known as the AN/ALQ-202(V) autonomous jammer.

Operational status

ALQ-178(V)3 was selected by Turkey in January 1989 for its F-16 aircraft and, in October 1989, it was announced that Turkey had awarded a contract worth $325 million for supply of systems, test and support, and ancillary equipment. Deliveries ran from October 1991 to early 1996. Lockheed Martin and the Turkish company Kavala have formed a joint venture known as MIKES to produce the ALQ-178(V)3 in Turkey after initial deliveries from the USA. ALQ-178(V)1 is fitted to F-16s of the Israeli Air Force with the threat warning subsystem. A portion of the system is built under licence in Israel by Elisra. The Israeli application may cover aircraft up to batch 3 F-16C/Ds but excludes the country's F-16D-30 aircraft. ALQ-178(V)1 incorporates a set of crystal video (DF) receivers which have been eliminated from the ALQ-178(V)3.

Contractor

Lockheed Martin Fairchild Systems.

UPDATED

AN/ALQ-202 autonomous jammer

The AN/ALQ-202 autonomous jammer is an advanced capability ECM jamming system for internal installation on F-16, F/A-18, and other aircraft. The ALQ-202 provides automatic and prioritised ECM responses to the full spectrum of radar threats, including advanced airborne interceptors and surface-to-air missiles. The ALQ-202 is fully integrated with onboard avionic systems including Radar Warning Receiver (RWR), Fire-Control Radar (FCR), chaff/flare dispenser and mission computer. System interfaces are designed for non-interference, allowing fully compatible operational performance of all onboard avionic systems. The ALQ-202 integration with the CounterMeasures Dispenser (CMD) system enables fully automated/co-ordinated electronic and dispensed countermeasures operations. The ALQ-202 is designed to operate independently, or interfaced with any RWR, including the ALR-56M Advanced Radar Warning Receiver (ARWR), and earlier ALR-67 and ALR-69 systems.

The ALQ-202 incorporates an internal Jammer Support Receiver (JSR). The JSR automatically scans through a programmed frequency range of interest, providing a rapid and independent threat detection capability against both pulsed and CW threat signals. The ALQ-202 further determines the direction of arrival

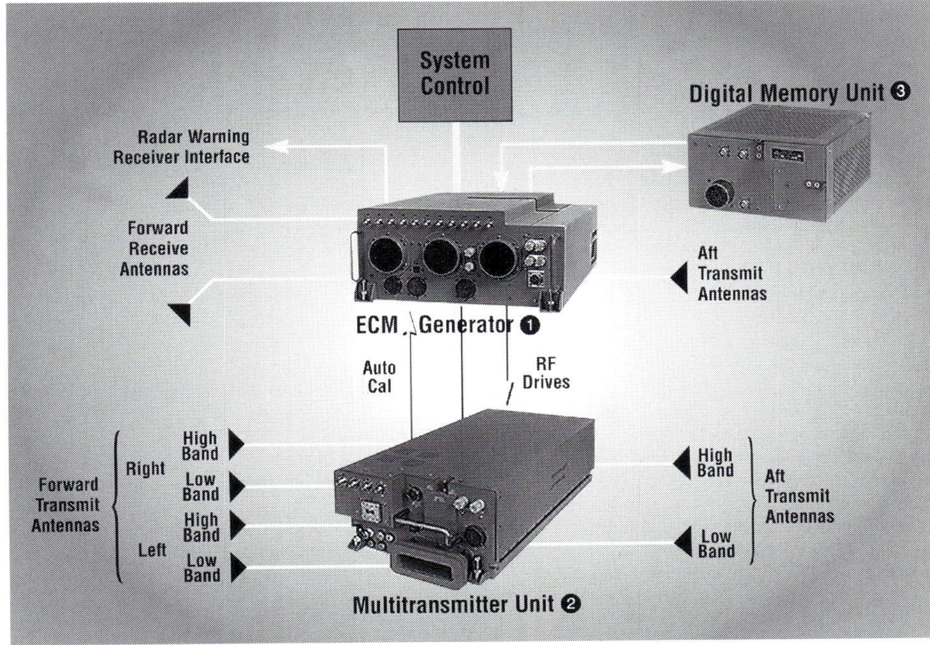

Lockheed Martin AN/ALQ-202 autonomous jammer units **1998**/0009915

of each identified threat radar, and selects from a prioritised multilevel list of ECM techniques to automatically counter threat radars with an optimised directional response.

The ALQ-202 continually searches for new threats and updates system information on those previously detected. Jamming is both transponder and repeater type, and includes continuous and time-gated responses. The system utilises an extensive library of complex deception and denial jamming techniques, with proven effectiveness.

The ALQ-202 system comprises an ECM generator, multitransmitter unit(s), Digital RF Memory (DRFM) and system control unit. The ECM generator contains the jammer support receiver and ECM technique generating functions. The DRFM is provided for additional jamming techniques capability. Multitransmitter units utilise new technology travelling wave tubes for improved efficiency and reliability, with critical low (non-interfering) thermal noise level, to assure compatible operation with other aircraft systems.

Contractor

Lockheed Martin Fairchild Systems.

UPDATED

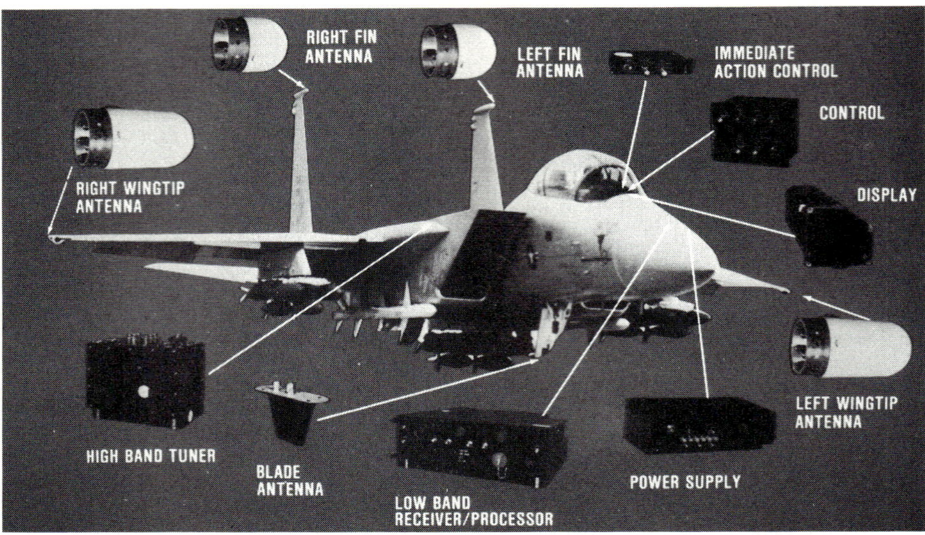

Lockheed Martin AN/ALR-56C radar warning equipment on the US Air Force F-15 Eagle

AN/ALR-56A/C Radar Warning Receiver (RWR)

The ALR-56 RWR is used with the AN/ALQ-135 Internal Countermeasures Set jamming equipment to form the Tactical Electronic Warfare System (TEWS) of the F-15 fighter aircraft.

The basic ALR-56 system incorporates the R-1867 processor/low-band receiver, the R-1866 high-band receiver, the IP-1164 display, the C-9429 immediate action control unit, the PP-6968 power supply, a TEWS controller and an antenna array. In more detail, the processor and low-band receiver unit contains three major sections: a single-channel low-band superheterodyne receiver, a dual-channel Intermediate Frequency (IF) section, and a processor.

The low-band receiver is electronically tuned under control of the processor. The dual-channel IF section operates with either the high-band dual-channel receiver or the low-band single-channel receiver. Selection of the receiver to operate with the dual-channel IF section is under the control of the processor.

The processor contains a preprocessor and a general purpose digital computer. The preprocessor contains all the video circuits for analysing intercepted signals. It also provides digital outputs to the computer which represent the measured signal parameters. On the basis of these measurements, a digital output is provided to the display and an audio signal generated. The computer also controls all ALR-56 system functions and is software programmable for the expected threat environment.

The solid-state, digitally controlled, dual-channel high-band receiver is capable of scanning an extremely large portion of the electromagnetic spectrum. Its single conversion superheterodyne design provides high sensitivity and the use of dual yttrium indium garnet Radio Frequency (RF) preselection is claimed to afford excellent selectivity and spurious rejection. High frequency accuracy is obtained through the use of a frequency synthesised local oscillator.

The RF input ports are configured to accept two main antenna inputs to each channel and two additional RF inputs which may be designated to either channel. All switching functions are provided internally.

The receiver is capable of receiving signals over a large dynamic range for analysis and Direction-Finding (DF) measurements. A self-contained precision signal source is provided for calibration and/or built-in test over the entire tuning range. Provision is also made for multiplexing out important receiver functions.

All receiver functions are controlled by serially generated NRZ Manchester coded data. The precise data rate also serves as the reference clock for the receiver frequency synthesiser.

On the countermeasures display, rapid threat evaluation for aircraft defensive manoeuvres is accomplished through automatic intensity control utilising ambient light sensing techniques. Sharp, high-contrast, unambiguous alphanumerics and special symbology allow immediate assessment of the overall tactical threat, without requiring pilot display adjustments as external ambient light fluctuates. Phosphor screen-optical filter matching provides contrast enhancement to overcome direct sunlight contrast degradation.

A range-bearing data presentation is provided with special clutter elimination programmes as well as special threat status. Built-in test circuits determine display system malfunction and automatically indicate GO conditions. An integral lighting panel provides control illumination for night-time viewing.

The antenna system consists of four circularly polarised spiral antenna assemblies (each within its own radome) and a blade antenna. Collectively, this antenna system provides omnidirectional acquisition and direction-finding over the operating frequency range of the RWR. The four spiral antennas, which cover the high-band frequency range, are mounted in a manner that provides 360° azimuth coverage. DF operation is accomplished under computer control. The blade antenna provides omnidirectional coverage of the low-band frequency range.

The TEWS control unit provides for control of the RWR as well as other countermeasures subsystems. A single switch controls the entire RWR on/off function, and separate switches and relays are provided for the subsystems under RWR control. Audio tone cues are controlled from this unit also.

The other control panel is for activating in-flight functions that require instant pilot reaction. In addition to countermeasures subsystem mode control, interrogated data readout on this unit can be called up from the subsystems by means of push-button operation.

A major update known as the ALR-56C has now been completed with improvements in the processor to handle new threats, greater signal densities and other changes. ALR-56C is a digitally controlled, dual-channel superheterodyne receiver covering the E- to J-bands, and is capable of sorting and identifying all current and projected threats to the aircraft. RF input ports are configured to accept two main antenna inputs to each channel and two additional RF inputs that may be designated to either channel, with all switching functions provided internally.

ALR-56C was designed to improve on the ALR-56A signal acquisition capability to include the latest threat parameters and to improve on situational awareness and terminal radar direction. The system maintains the ALR-56A capabilities in the areas of analysis, establishment of priorities, jammer management, passive countermeasures management, and visible and audio displays and warnings.

Operational status

ALR-56 is in service aboard US Air Force and Saudi Arabian F-15 aircraft. ALR-56C equipments are reported to be installed on F-15E strike aircraft and as being retrofitted to F-15Cs and Ds as part of the two types' Multinational Staged Improvement Programme. ALR-56C is reported to be replacing ALR-56A throughout the F-15 inventory as well as being the standard installation on all new build aircraft.

Contractor

Lockheed Martin Fairchild Systems.

UPDATED

AN/ALR-56M(V) Radar Warning Receiver (RWR)

The AN/ALR-56M(V) has become the US Air Force's standard RWR and is designed to meet the 21st century threat environment and operational requirements. ALR-56M(V) was selected as the replacement for the AN/ALR-69(V) RWR in a US Air Force competitive fly-off programme in 1988. ALR-56M (V) has successfully completed all required US government testing, including US DoD operational test and evaluation and has been in production since 1989. Over 900 ALR-56M(V) systems are known to be in service with the US Air Force or export customers.

ALR-56M(V) is stated to have an internal jammer hardware interface; sensitivity greater than maximum threat engagement ranges; adaptive selectivity options to provide threat sorting in dense environments; frequency agility to allow rapid adaptive RF scanning; measurement ability to sort and identify time of arrival, pulse repetition frequency, signal strength, pulsewidth, angle of arrival and RF frequency on a monopulse basis; throughput to handle high pulse density within the response time requirement; and an interface capability to all aircraft avionics via dual redundant MIL-STD-1553B busses - one for aircraft avionics (inertial navigation system, fire-control radar and control/ display systems) and one for electronic warfare system integration (AN/ALE-47 countermeasures dispenser, internal or external jammer, flight test instrumentation and memory loader/verifier unit). ALR-56M(V) also provides the ability to store multiple, in-flight selectable mission data tables. The operational flight programme and the mission data tables are completely separate entities, allowing for easy reprogramming of parametric data. ALR-56M(V) supports two-level maintenance, enhanced by extensive built-in test.

The system consists of four direction-finding receivers, each connected to one of the four high-band quadrant antennas, a C/D-band receiver/power supply, connected to a low-band antenna assembly, a superheterodyne receiver, a superheterodyne controller and an analysis processor which communicates with the control panel and azimuth indicator and provides two MIL-STD-1553B busses. ALR-56M(V) is effectively a time-shared, rapid, agile single-channel superheterodyne receiver for processing the signals intercepted by the four high-band antennas and the low-band antenna assembly. Wideband, fast, frequency-agile, sensitive receivers provide intra-pulse parametric measurement capability. A computer-controlled tunable radio frequency notch filter, adaptive digital signal processing and parametric screening provide the ability to control the input to the high-speed control processor unit resulting in an unambiguous rapid threat warning, even when operating in the densest, most sophisticated threat environment.

Operational status

ALR-56M(V) is reported to be mandated for all US Air Force AC-130U, CV-22 and Block 50/52 F-16C/D aircraft. The system is also understood to be a retrofit item for US Air Force Block 40/42 F-16 aircraft and to be under consideration for use in B-1B strategic bombers. In addition, it is the recommended replacement for the AN/ALR-69(V) on the F-4, MH-53J, C-130, Joint STARS, A-10, MC-130E, EC-130 and C-141.

As of 1997, ALR-56M(V) is reported to be in full rate production for the US Air Force, and a January 1998 contract provided for delivery of 490 replacement computer kits. More than 1,000 units have been ordered. It entered worldwide operation in the US Air Force in 1992. ALR-56M(V) has been selected for overseas application on F-16 and C-130 aircraft, including Foreign Military Sales contracts for South Korea and Taiwan.

Contractor

Lockheed Martin Fairchild Systems.

UPDATED

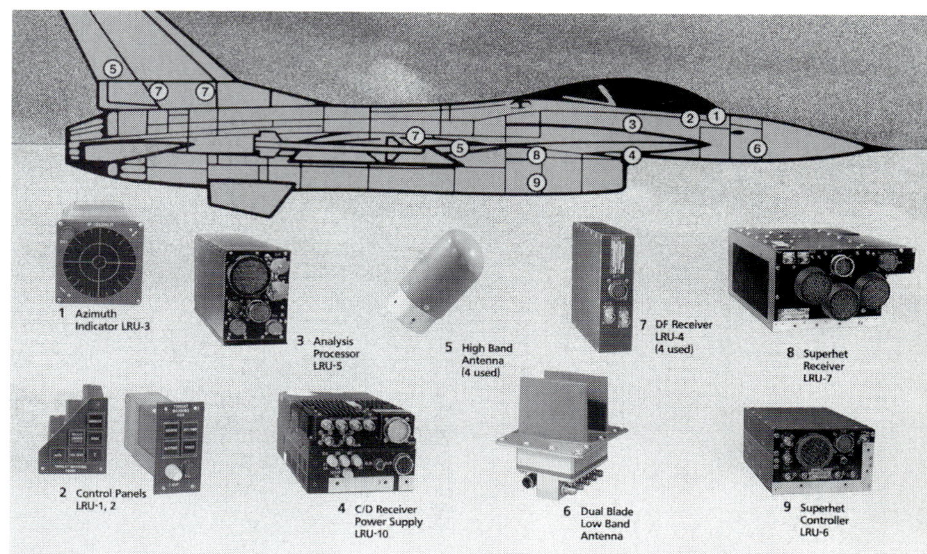

The elements making up ALR-56M(V) as applied to the F-16

AN/APR-39(V) Radar Warning Receiver (RWR)

The AN/APR-39(V)1 was designed and developed by E-Systems (now Raytheon Systems Company). The APR-39(V)2 processor was designed and developed by Lockheed Martin under contract to the US Army Electronic Command. A third version, the AN/APR-39A (V)1 is also a Lockheed Martin product. This version incorporates a number of updates, including the provision of millimetric-wave threat warning.

Description

The APR-39(V)1 RWR equipment provides automatic warning of emitters in the E, F, G, H, I and most of J radar bands, as well as the appropriate portions of C- and D-bands. It is intended for use on either fixed-wing aircraft or helicopters. The equipment provides indications of bearing, identity and the mode of operation of detected signals with the acquired data being displayed on a cockpit indicator. Proportional pulse repetition frequency of displayed signals and alarm tones are presented to the crew via an integral audio warning subsystem.

The basic APR-39(V)1 system comprises two dual-video receivers, four spiral cavity-backed antennas, one blade antenna, an indicator unit, a comparator and a control unit. Without cables and brackets, this weighs 3.63 kg.

An updated version, APR-39(V)2 has been produced by Lockheed Martin. In this variant, the comparator is replaced by a Lockheed Martin CM-480/APR-39(V) digital processor. This performs signal sorting, identification of emitters, bearing computation and character generation for the presentation of threat details in alphanumeric form on the cockpit display. The unit weighs 6.5 kg and incorporates an adaptive noise threshold and angle-gate, programmable pulse repetition interval filters, and coded emitter outputs. A 19 k word programmable read-only memory/random access memory is provided.

Units of AN/APR-39(V)1 RWR system **1998**

Operational status

As of this edition, APR-39(V) series equipments are understood to be in service. They are noted as having been installed in 15 types of aircraft and patrol/fast attack craft. Several thousand APR-39(V) systems are reported to have been produced, a figure which includes 230 for installation aboard German PAH-1 anti-tank helicopters.

Contractors

Lockheed Martin Fairchild Systems.

NEW ENTRY

AN/APR-39A(V) threat warning systems

AN/APR-39A(V)1

The AN/APR-39A(V)1 is an upgrade of the earlier analogue APR-39(V)1 radar warning system. It is a lightweight system designed for helicopters and light fixed-wing aircraft. It provides visual and aural warning of hostile radar energy incident on the host aircraft over a very wide frequency range. The system displays multiple threats, (with the highest priority indications highlighted) while keeping track of other emitters in the environment.

A digital display clearly identifies the threat type and azimuth from the aircraft. It also indicates if the threat is searching, locked and tracking, or when the radar lock is broken. The alphanumeric display will automatically separate symbols that are grouped too closely together. The aural warning is by synthetic voice.

AN/APR-39A(V)1 threat warning system **1998**

AN/APR-39A(V)3

AN/APR-39A(V)3 uses APR-39A(V)1 technology and provides continuous coverage through dual-channel crystal video receivers. This system is deployed on US Army and NATO aircraft. It is specifically designed for helicopters and other light aircraft operating at very low levels. It consists of 10 line-replaceable units which are form-fit compatible with the analogue APR-39(V)1. Frequency range covered is E/J and C/D.

Operational status

As of this edition, more than 4,000 APR-39A series equipments are noted as having been delivered.

Contractor

Lockheed Martin Fairchild Systems.

NEW ENTRY

AN/TLQ-17A countermeasures set

The modular AN/TLQ-17A is an advanced tactical communications countermeasures set developed for the US Army. It is configurable to meet user requirements for a number of host platforms including helicopters.

The AN/TLQ-17A incorporates computer-controlled search, surveillance and jamming in the HF and VHF ranges, microprocessor technology for advanced EW applications, BITE and modularity for ease of maintenance.

Specifications

Frequency:
(band 1) 1.5-20 MHz
(band 2) 20-80 MHz
Operating modes: search/lockout, priority search/lock on, monitor/automatic and scan band sectors
Effective radiated power: 10-550 W
Receiver tuning time: 1 ms
Preselected frequencies: 255
Reliability: 400 h MTBF

Contractor

Lockheed Martin Fairchild Systems.

VERIFIED

DASS 2000 Electronic Warfare (EW) suite

The Lockheed Martin DASS 2000 electronic warfare suite is a turnkey, fully integrated defensive aids suite which can be configured to meet specific customer requirements and is applicable to a wide range of platform types. The suite's long-range, multispectral threat detection and identification capabilities provide the aircrew with full situation awareness through sensor fusion, and ensure the application of optimum self-protection countermeasures. These co-ordinated capabilities ensure platform survivability in all phases of mission prosecution.

DASS 2000 resources typically include a radar warning receiver, missile warning system, laser warning system, chaff/flare dispensers, electronic countermeasures jammer, towed radar decoy and infrared countermeasures. Lockheed Martin will adapt the configuration to user requirements, integrate it into the platform ensuring full compatibility, and provide a full range of training and logistics support.

DASS 2000 features interactive functions and a full colour display that provides the EW operator with tactical awareness. The display enables threat assessment as well as automatic countermeasures response by using the defensive aids resources.

Operational status

No details of any operational deployment are known; the system has been marketed in the maritime patrol aircraft scenario.

Contractor

Lockheed Martin Fairchild Systems.

UPDATED

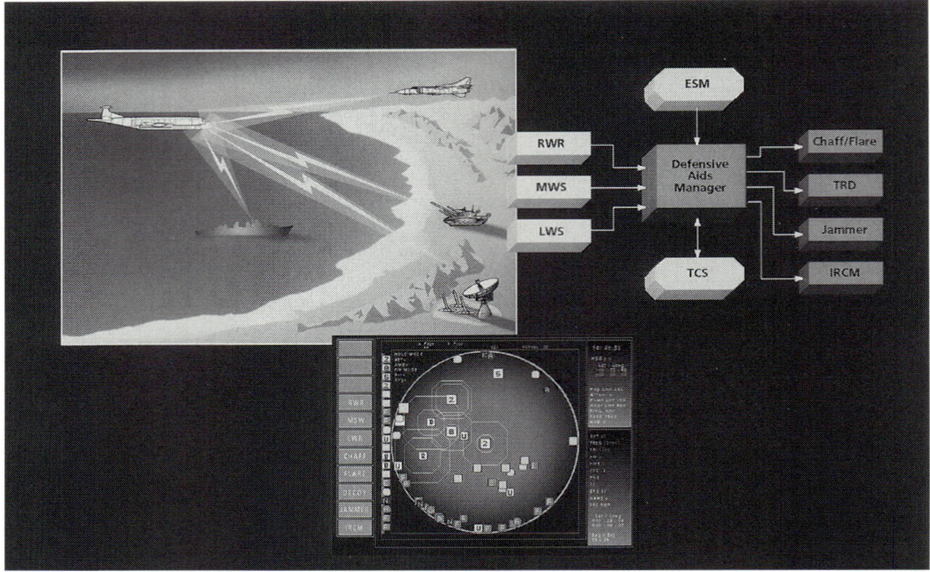

DASS 2000 provides maritime patrol with an effective, integrated EW capability ***1998***

EW-1017 surveillance system

As an airborne electronic surveillance system, EW-1017 automatically acquires and identifies emissions within the C- to J-bands. The system is also designed to receive and identify all those emissions illuminating the aircraft (including short bursts) in particular when it is operating in very dense signal environments. The warning of possible danger is given both visually and aurally on a display unit while preferential scan is used to ensure immediate recognition of possible lethal threats.

EW-1017 consists of antenna arrays, receiver, a processor system, a pilot display, and ESM operator interactive display subsystem. Broadband spiral antennas are used to provide omnidirectional coverage which, together with their separate multiband receivers, are mounted in pods on each wingtip. This location drastically limits aircraft 'shadowing' and the proximity of the receiver cuts signal losses to a minimum. Angular bearing of the emissions is determined by using selected pairs of antennas.

The hybrid superheterodyne receiver combines high acquisition probability, high sensitivity, frequency accuracy and a high degree of frequency selection and selectivity. A broad bandwidth is used in the acquisition mode to obtain the initial intercept with a narrow bandwidth then being used for accurate bearing measurement and analysis. To ensure processing capability in highly dense signal conditions a high-speed digital computer performs the data processing functions, supplemented by microprocessors. This enables the receivers to scan the frequency band continuously on a reprogrammable basis, so that conventional, continuous wave and agile signals are processed for identification. For special applications, the receiver can be interfaced via a smart post processor unit to a centralised tactical display and control system.

An interactive display subsystem provides a full range of operator facilities to manage and optimise collection. It also provides a readily accessible real-time emitter and platform library storage and analysis capability, facilitated by a modern data management system. A control/display unit allows the operator to monitor and control the automatic surveillance function to resolve possible ambiguities and evaluate and use to the best advantage the data displayed.

Operational status

EW-1017 is reported to be in service with the German Naval Air Arm (Atlantique maritime patrol aircraft) and the Royal Air Force (Nimrod MR Mk 2 maritime patrol and Sentry AEW Mk 1 airborne warning and control system aircraft). In UK service, the system is designated as ARI-18240 Yellow Gate.

Contractor

Lockheed Martin Fairchild Systems.

UPDATED

Nimrod MRA.4 Defensive-Aids SubSystem (DASS)

The Nimrod MRA.4 DASS is not yet fully defined, but the following elements are understood to be included: the Lockheed Martin Fairchild Systems AN/ALE-56M radar warning receiver; the Raytheon Systems Company ALE-50 towed radar decoy; the Racal Radar Defence Systems Digital Radar Frequency Memory Techniques Generator (DRFMTG).

System architecture is based on receipt of threat warning via the AN/ALE-56M and/or an as yet unspecified missile approach warning receiver; production of a suitable threat countermeasure in the techniques generator, and its transfer via EO cable to the AN/ALE-50 towed radar decoy.

The Racal techniques generator is based on advanced twin Digital Radio Frequency Memory (DRFM) units, which are used to classify the threat and customise the most coherent jamming response.

Operational status

Ground integration tests are due to begin in autumn 1998.

Contractor

Lockheed Martin Fairchild Systems.

NEW ENTRY

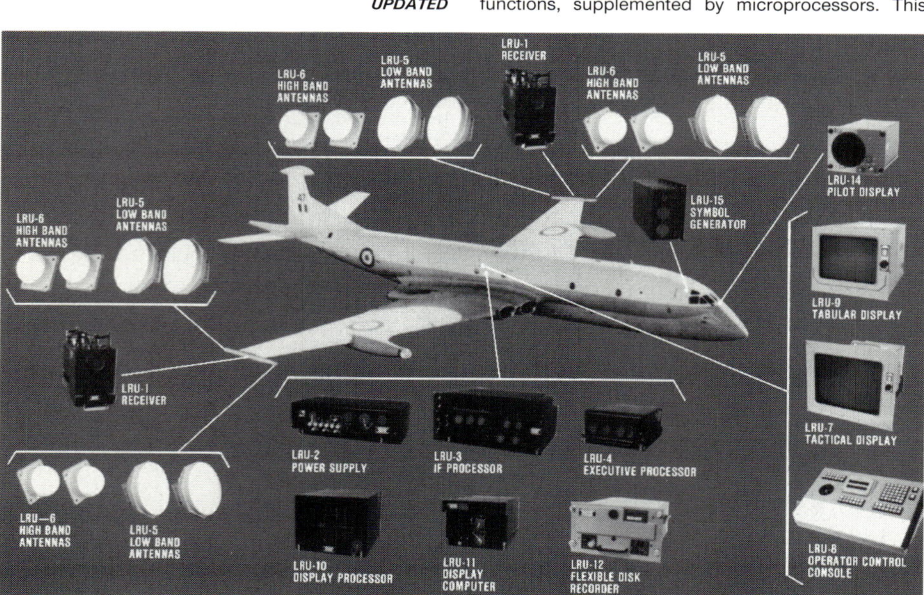

The Lockheed Martin EW-1017 ESM system showing the location of units on the Nimrod MR. Mk2 aircraft

RAPPORT II ECM system

The Rapid Alert and Programmed Power management Of Radar Targets (RAPPORT) is unusual, perhaps unique, in being an American system procured initially by a European customer after competitive evaluation and development, while not already being available off-the-shelf for US applications.

RAPPORT is designed to protect aircraft against airborne and ground-based radar-directed weapons. The system continuously analyses, identifies and measures the bearings of threat radars and can rank up to 14 such threats in order of priority according to reprogrammable data held in the system's memory. An alphanumeric display presents data to the pilot on the threat and its bearing from the aircraft, with audio alert tones. Automatic noise or continuous wave pulse countermeasures action can then be initiated.

RAPPORT II consists of seven LRUs and associated receiver and jamming antennas. The LRUs comprise a receiver/processor, techniques generator, control and display unit, two dual-channel direction-finding receivers and two dual-band power amplifiers. The system brought to ECM technology for the first time such features as solid-state power amplifiers and distributed processing, but most importantly it incorporated, in the same suite of equipment, both passive radar warning and active radar jamming functions. Such internal integration of tasks within the system obviated the need for space-consuming and expensive interface boxes.

Specifications

Volume: 0.14 m^3
Weight: 128 kg
Digital processor: TI 2520
Memory: 32 kbytes
Growth flexibility: MIL-STD-1553

Operational status

In service with the Chilean Air Force Mirage V aircraft. No longer in production.

Contractor

Lockheed Martin Fairchild Systems.

UPDATED

AN/ALR-47 radar homing and warning system

Developed for the US Navy S-3A Viking ASW carrier-borne aircraft, the AN/ALR-47 is a comprehensive passive electronic warfare system which uses four cavity-backed planar spiral antennas in each wingtip. The aerials are orthogonally directed to enhance monopulse direction-finding, ensuring that threat direction is measured very accurately. Associated with them are twin, highly sensitive, narrowband receivers and a comprehensive processor. Manual or automatic system operation is possible, control being exercised over frequency band limits, speed of tuning and signal selection. The processor indicates the frequency scanning limits, scan speed, pulse length, pulse repetition frequency and bearing limits of any detected radar transmissions.

Operational status

No longer in production. In service on Canadian Forces CP-140 Auroras, in which application it is designated ALR-502, and formerly in the S-3A. In the S-3A it has been replaced by the AN/ALR-76 as part of the avionics upgrade which denotes the S-3B version.

Contractor

Lockheed Martin Federal Systems.

VERIFIED

AN/ALR-76 ESM system

The AN/ALR-76 is a form, fit and function replacement for the ALR-47 and is part of the Weapons Systems Improvement Program (WSIP) for the S-3B ASW aircraft. Compared with the ALR-47 the ALR-76 has an extended frequency range and improved capabilities which include the automatic classification and location of emitters. The ALR-76 utilises VLSI technology and is of modular construction to facilitate future growth.

Specifications

Weight: 61 kg

Operational status

The AN/ALR-76 is in full production for the US Navy S-3B weapons system improvement programme, as well as the EP-3E and ES-3A electronic reconnaissance aircraft. It is also installed on the Canadian EST Challenger aircraft.

Contractor

Lockheed Martin Federal Systems.

VERIFIED

AN/APR-48A Radar Frequency Interferometer (RFI)

The AN/APR-48A RFI is designed to minimise target acquisition and platform exposure times and operates in conjunction with radar and electro-optical sensors to extend the useful range, decrease acquisition time and provide positive identification. It provides high sensitivity and precision sensor cueing in a lightweight and modular configuration suitable for both airborne and ground platforms. Weight of the system on the AH-64D Longbow Apache is 14.5 kg, including the antenna and receiver. The system can be packaged as a single LRU or multiple LRUs to meet platform requirements. Expansion modules will allow growth to millimetric-wave capabilities.

The RFI system design incorporates a four-element interferometer coupled with a three-channel phase receiver. A four-element coarse DF array is used for initial signal acquisition. When a fine DF measurement is required, a four-element, long baseline interferometer is used which is said to provide a DF accuracy of better than 1°. The receiver is a full four-channel amplitude and three-channel phase measuring unit.

The signal parameter measurement function includes analogue video and IF circuits for measuring the signal amplitudes, radio frequency and interferometer phases. The digital processing is carried out in the processor which sorts the incoming pulses, computes DF, characterises and identifies emitters and determines threat priority.

Operational status

The AN/APR-48A has been developed for the US Army AH-64D Longbow Apache and OH-58D Kiowa helicopters. Some 222 systems are currently in production for the AH-64D.

Contractor

Lockheed Martin Federal Systems.

VERIFIED

Intelligence and Electronic Warfare Common Sensor (IEWCS)

Lockheed Martin Federal Systems will provide the US Army's IEWCS system under a 5 year contract worth $276.5 million. Lockheed Martin Federal Systems, with its major subcontractors including Condor Systems and Sanders, will initially integrate the system aboard HMMWV trucks and EH-60L helicopters.

Operational status

Development.

Contractor

Lockheed Martin Federal Systems.

VERIFIED

AN/AAR-47 missile warning set

The AN/AAR-47 missile warning set detects the plume of approaching surface-to-air missiles and gives the pilot an indication of range and bearing. Decoy systems, such as the AN/ALE-39 flare/chaff dispenser, can be automatically triggered and the system is also compatible with the AN/APR-39A radar warning receiver.

The feasibility of such a warning system was first demonstrated by Loral in December 1977 and, following development of the AAR-46 system in 1979, full-scale development of the AAR-47 started in March 1983. The AAR-47 is small, needs no cryogenic cooling and has an MTBF of 1,500 hours.

The system consists of six sensors, a central processor and a control indicator. The sensors are based on ultraviolet technology. They are hard-mounted on the skin of the aircraft and provide full coverage, with overlap to protect against blanking. The complete system weighs only 18.2 kg. Within 30 seconds of electrical power a self-initiated BIT programme is completed and the system is operational

The Lockheed Martin AN/AAR-47 missile warning set

with no in-flight down time for recalibration. The processor analyses the data from each sensor independently and as a group, and automatically deploys the appropriate countermeasures. In the case of a flare failure, the AAR-47 automatically commands a second flare.

Multidirectional threats are automatically analysed and prioritised for countermeasures sequencing. The control indicator displays the incoming direction of the highest priority threat for tactical manoeuvres.

A shorter sensor has been developed for use in locations, such as the F-16 fuselage, where depth is limited.

AN/AAR-47 technology is used as the basis of the Common Missile Warning System (CMWS) element of the Advanced Threat InfraRed CounterMeasures (ATIRCM) and Common Missile Warning System (CMWS) being developed by Sanders and Lockheed Martin IR Imaging Systems.

Specifications

Dimensions:
(sensor) 120 × 200 mm
(processor) 203 × 257 × 204 mm
Weight:
(4 sensor system) 14 kg
(sensor) each 1.5 kg
(processor) 7.9 kg
Power supply:
(4 sensor system) 28 V DC, 75 W
(4 W per sensor, 59 W for processor)
Coverage: 360° azimuth (given by 6 sensors)

Operational status

AN/AAR-47 systems are installed on transport aircraft and helicopters operated by Australia, Canada, UK and USA.

Contractor

Lockheed Martin IR Imaging Systems.

UPDATED

Passive Missile Warning Set

The Passive Missile Warning Set is designed for the C-130 to detect potential missile threats. It discriminates against false targets, declares the approaching missile threat when this is within the optimum countermeasure interval and sends a signal to the countermeasures dispenser. The system is totally passive, requiring only the E-O signature from the missile plume. Non-threat false alarms are eliminated by E-O spectral selection and signal processing algorithms.

The four sensors are hard-mounted on the aircraft skin to provide 360° horizontal coverage. These are wide field of view staring E-O receivers that collect in-band radiation and convert it to electrical signals. Background clutter is rejected spectrally. The signal processor receives the signals from the four sensors and uses temporal algorithms to distinguish threats from non-threat or false signals. There are no moving parts in the sensors or the signal processor and no cooling air is required. The pilot control indicator contains BIT initiation plus a quadrant threat indicator.

Contractor

Lockheed Martin IR Imaging Systems.

VERIFIED

AN/ALQ-122 ECM system

The AN/ALQ-122 is the power-managed radar jamming system in use on the B-52 bomber. The equipment searches for, acquires and tracks radar threat signals, and generates narrowband, low-duty cycle ECM signals to deny range and azimuth to enemy radars. Signals are linear amplified by solid-state amplifiers in the aircraft.

Operational status

In operational use on B-52 aircraft. An advanced mixed-mode deception jammer is reported to be in development by Motorola for the US Air Force based on work on this equipment and the Northrop Grumman ESSD AN/ALQ-131(V) ECM pod.

Contractor

Motorola Government & Systems Technology Division.

VERIFIED

AN/AAR-54(V) Passive Missile Approach Warning System (PMAWS)

Northrop Grumman and the US Department of Defense (DoD) have developed and tested a PMAWS which can be used on a wide variety of aircraft and ground vehicles. The system is available to provide internal advanced missile warning for tactical and transport aircraft, helicopters and armoured fighting vehicles. It is also used to provide identification, missile tracking information and target cueing to Directed InfraRed CounterMeasures (DIRCM) systems. Because of the adaptive design of AAR-54(V), all applications can use common hardware and software.

The fine 1° angle of arrival discrimination capability of AAR-54(V) provides greatly reduced false alarm rates as well as detection ranges nearly double that of existing ultraviolet systems. Passive time to intercept is another system feature providing optimum cueing to countermeasures dispensers. AAR-54(V) provides all-weather, all-altitude operation while protecting against multiple simultaneous engagements in dense clutter environments.

The system consists of wide field of view, high-resolution ultraviolet sensors and a modular electronics unit. From one to six sensors can be utilised, providing up to full spherical coverage. Full in-flight built-in test and fault isolation to a single sensor or electronics line-replaceable unit provides level-2 maintenance.

Operational status

Northrop Grumman, the US DoD and the UK Ministry of Defence are believed to have completed AAR-54(V) design verification testing. This trials programme is understood to have included several hundred live-fire demonstrations with the warner installed on QF-4 drones, a cable car test rig and ground vehicles. The system has also been integrated into and demonstrated with a DIRCM system and the AN/ALQ-131 electronic countermeasures system. In April 1995, the UK's Ministry of Defence and the US Special Operations Command awarded Northrop Grumman a US$35 million engineering, manufacturing and development/production contract for the AN/AAQ-24(V) DIRCM system which includes AAR-54(V). In this application, the equipment provides missile detection, lethal missile declarations and fine angle of arrival hand-off to the DIRCM. AAQ-24(V) is scheduled to be installed on 14 types of UK and US fixed-wing aircraft and helicopters.

Outside the AAQ-24(V) application, AAR-54(V) has been selected by the Portuguese Air Force for use on a percentage of its C-130H fleet. Here, a six sensor AAR-54(V) configuration is teamed with Tracor's AN/ALE-40 dispenser system and TERMA's Electronic Warfare Management Unit to create a defensive aids suite. Northrop Grumman is also understood to be working with Danish contractor Per Udsen to integrate AAR-54(V) into an F-16 pylon application. A flight demonstration is scheduled in the Netherlands in an F-16 MLU aircraft in spring 1998.

Contractor

Northrop Grumman Corporation, Electronic Sensors and Systems Division.

UPDATED

The AN/AAR-54(V) Passive Missile Approach Warning System (PMAWS) consists of up to six sensors and an electronic unit ***1996***

AN/ALQ-119 noise/deception jamming pod

Initiated as project QRC-522 in 1970 and one of the most numerous jamming pods in service with the US Air Force, the AN/ALQ-119 was one of the first dual-mode noise and deception jammers to appear. It was used initially on the F-4 Phantom, but was subsequently adopted for the Fairchild A-10 and Lockheed Martin F-16 as well as the F-111 and F-15. The system has a three-band frequency range transmitter which covers the terminal threat range. Both noise and deception jamming modes can be employed. Each pod has dual-mode travelling wave tube emitter elements. The pod has the gondola cross-section introduced by Northrop Grumman on the AN/ALQ-101 pod.

The AN/ALQ-119(V)15 model is recognisable alongside earlier versions of the pod by the addition of a radome below the front end of the gondola portion. Many earlier versions of the pod with the US Air Force and other air forces were upgraded to this standard. Features of the (V)15 version include automatic control of power radiated, frequency selection and signal type. A shorter body version, the AN/ALQ-119(V)17, is also operated by the US Air Force.

With the appearance of new air-to-air and ground-to-air threats, the US Air Force, in the early 1980s, instituted a major improvement programme, designated AN/ALQ-119A (Seek Ice). Raytheon was appointed to strip out and replace much of the existing electronics and incorporate current Rotman lens technology. The changes facilitate reprogramming and make for greater reliability and ease of servicing. The upgraded pod is designated AN/ALQ-184.

Daimler-Benz Aerospace AG has recently developed the AN/ALQ-119 pod for use by German Air Force F-4F Phantom aircraft; this variant is designated AN/ALQ-119GY (see Daimler-Benz Aerospace AG entry).

Operational status

In service with F-4, F-16 and A-10 aircraft. The total production run exceeds 1,600 units. Sets are operated by Germany and Israel. Superseded in US Air Force service by the AN/ALQ-131 system.

Contractor

Northrop Grumman Corporation, Electronic Sensors and Systems Division.

UPDATED

AN/ALQ-131 noise/deception jamming pod

The AN/ALQ-131 is an automatic, highly reliable modular self-protection system. It was designed to provide advanced broadband coverage against all types of modern radar-guided weapons. The ALQ-131 is carried externally on a variety of front-line, high-performance aircraft. It is certified on front-line aircraft such as the A-7, A-10, C-130, F-4, F-15, F-16, F-111 and Harrier.

The modular design of the pod structure and electronic assemblies, plus its central computer software architecture, enable the ALQ-131 system to adapt quickly to a broad spectrum of EW applications. This feature proved valuable in the Gulf War, where

ALQ-131s provided over 48 per cent of the US Air Force tactical aircraft EW self-protection. More impressive, the latest Block II version experienced no combat losses in over 12,000 combat sorties. It is also the pod chosen to protect the aircraft in Bosnian operations on USAF F-16, A-10, C-130 and Royal Netherlands Air Force F-16.

Increased effectiveness can be achieved by incorporating various mission modules, including missile warning systems, offboard and towed countermeasures dispensers and advanced technique generators. This capability allows maximum flexibility in countering all threat types.

The basic structure elements of the pod are modular canisters that provide structural support, cooling and environmental protection. Each canister is an I-beam structure which also serves as a cold plate. Both sides of the I-beam form equipment bays into which functional equipment modules are mounted. The modules can also be mounted in lower equipment bays on the bottom surface of the I-beam. Using common mounting techniques, each canister can accommodate several equipment modules which can be removed directly from the bays without disassembly of the pod. The system is 2.83 m long and its weight ranges from 260 to 324 kg.

The functional organisation of the system is centred around the Interface and Control (I/C) module which contains a programmable digital computer as the system controller. The modules required for a given configuration are connected to the I/C by a digibus that carries all sensor and control data. A memory loader/verifier allows operational flight and mission specific program software to be loaded into the pod on the flight line in less than 15 minutes.

The I/C module also contains a digital waveform generator that can permit up to 48 simultaneous waveforms for deception modulation. When any ECM technique requires a deception waveform, the latter's values are transmitted to the onboard equipment via a waveform distribution bus.

Maintenance of the ALQ-131 is based on the pod's Centrally Integrated Test System (CITS) which provides a comprehensive functional check of system operation, both in flight and on the ground. During flight, the CITS continuously monitors the operational status of the equipment, including repeater channel modulation, high voltage of the TWTs, noise power output, primary bus voltage and the integrity of the computer memory.

The pod is a software reprogrammable system which allows a tactical commander to tailor the ALQ-131's responses for the mission requirements. Utilising the ALQ-131's ability to be flight line reprogrammed, mission specific data can be created in response to threat changes and, by using a memory loader verifier, changes loaded into the pod's digital computer.

Another capability of the system is its self-contained power management feature. This is included in a receiver/processor module which detects radar threats, measures their key parameters and performs weapon type and operational mode identification. This information is then used by the ALQ-131's I/C module to select optimum jamming techniques automatically and tailor their parameters for countering all detected threats. Numerous threats can be countered simultaneously, each with independent techniques, using the receiver/processor's PRI tracking capability.

Specifications

Dimensions:
(one band shallow) 2,210 × 297.2 × 533.4 mm
(two band shallow) 2,819.4 × 297.2 × 533.4 mm
(three band deep) 2,819.4 × 297.2 × 635 mm
Weight:
(one band shallow) 175.54 kg
(two band shallow) 263.1 kg
(three band deep) 298.92 kg
Power supply: 115 V AC, 400 Hz
Reliability: 125 h MTBF

Operational status

More than 560 Block I and 460 Block II pods have been delivered to the US Air Force. A production programme to update the original Block I configuration to the latest Block II version is under way. Over 345 pods have been procured by the air forces of Bahrain, Belgium, Egypt, Israel, Japan, Netherlands, Pakistan, Portugal and Thailand. International deliveries are continuing and are expected to continue until 2000.

In October 1993, the US Air Force completed EMI/EMC testing of the ALQ-131 modified with an AN/ALQ-153 active missile warning system and AN/ALE-47 countermeasures dispenser on an F-16 aircraft. In April 1994, the ALQ-131 was demonstrated with a fully integrated AN/AAR-54 passive missile warning system and AN/ALE-47. A self-contained external ECM suite configuration of ECM pod, missile warning system and countermeasures dispenser is being considered, to provide tactical aircraft with protection against IR-guided missiles while eliminating Group A aircraft modifications.

The pod has been combat proven in Bosnian operations on US Air Force A-10, C-130 and F-16 aircraft and Royal Netherlands Air Force F-16s.

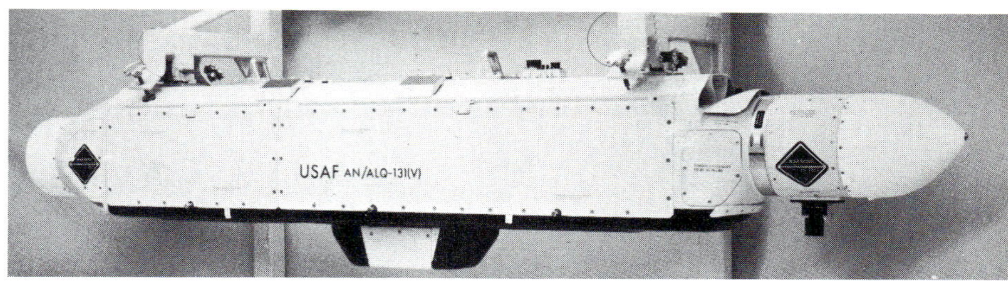

The Northrop Grumman AN/ALQ-131(V) ECM pod

The F-16 carries the AN/ALQ-131 pod under the fuselage **1995**

Contractor

Northrop Grumman Corporation, Electronic Sensors and Systems Division.

UPDATED

AN/ALQ-153 threat warning system

The AN/ALQ-153 threat warning system is a frequency/PRI-agile pulse Doppler radar that detects and discriminates between approaching missiles and aircraft. It is a range-gated Doppler system capable of detecting the approach of the entire spectrum of hostile aircraft and land-, sea- and air-launched missiles. Range and time-to-impact are computed automatically, updated continuously and compared with in-flight selectable timing criteria so that automatic countermeasures dispensing and aircraft manoeuvres can be generated. The system continuously displays the most imminent threat while tracking multiple missile engagements. Both 360° all-aspect and tail-only coverage versions are available.

Specifications

AN/ALQ-153
Dimensions:
(radar receiver transmitter) 408.9 × 251.5 × 342.9 mm
(analogue data signal processor) 457.2 × 254 × 160 mm
(digital data signal processor) 457.2 × 279.4 × 121.9 mm
(antennas) 177.8 length × 152.4 mm diameter
Weight:
(radar receiver transmitter) 10.34 kg
(analogue data signal processor) 21.55 kg
(digital data signal processor) 16.78 kg
(antennas) 1.59 kg

AN/ALQ-153(V)
Dimensions: 660.4 × 203.2 × 152.4 mm
Weight: 31.75 kg

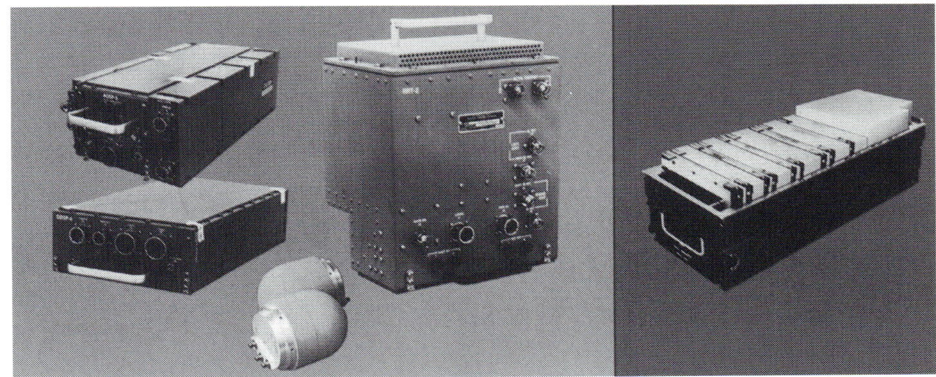
The four LRUs which comprise the AN/ALQ-153 threat warning system (left) and the single unit of the AN/ALQ-153(V) (right)

Operational status

Production is complete with the delivery of 321 systems for US Air Force B-52G/H aircraft. Spare LRU production was completed in 1991, bringing the total number of systems up to 400.

The 360° all-aspect AN/ALQ-153(V) has recently undergone extensive flight testing on tactical aircraft, while the tail-only system has successfully demonstrated in-flight missile detection, deception and low false alarm performance on the B-1B and B-52 in low-level, high-clutter environments. Current efforts are being focused on repacking the system for installation in both an ECM pod and internally for tactical aircraft. Using modern components, the volume is being reduced to approximately 0.023 m³ and weight to less than 32 kg.

The ALQ-153 is one of a number of systems being evaluated by the US Air Force as the basis for a future missile attack warning equipment for internal carriage on the B-1B and fighters, as well as in ECM pods. In October 1993, the US Air Force completed EMI/EMC testing of the ALQ-153 installed in an ALQ-131 ECM pod on an F-16 aircraft.

Contractor

Northrop Grumman Corporation, Electronic Sensors and Systems Division.

VERIFIED

AN/AAQ-8(V) (QRC 84-02) infrared countermeasures pod

The AN/AAQ-8(V) is a multithreat infrared countermeasures system capable of operating in a supersonic environment. This pod is a second-generation system updated to meet new and continuing threats, and has been extensively deployed. The system is mounted in an aerodynamically faired pod and can be configured with a ram-air turbine, allowing protection independent of aircraft power and cooling resources.

Specifications

Dimensions: 2,290 (length) × 254 mm (diameter)
Weight:
(AAQ-8(V1)) 107 kg
(AAQ-8(V2)) 120 kg
Power supply: 115 V AC, 400 Hz, 3 phase, 4 kVA
28 V DC, 20 W

Operational status

In service on fixed-wing combat and transport aircraft including A-7, C-130, F-4, F-5E and Mirage F-1C.

Contractor

Northrop Grumman Corporation, Electronic Systems, Electronics & Systems Integration Division.

UPDATED

The US Navy Northrop Grumman EA-6B Prowler is scheduled to remain in service until 2020

AN/ALQ-99(V) Tactical Jamming System (TJS)

The AN/ALQ-99 is a large and sophisticated ECM jammer designed for the SEAD Suppression of Enemy Air Defences role for the US Navy EA-6B and US Air Force EF-111A aircraft.

It has been operational for over 20 years, and has undergone numerous upgrades to meet changing operational requirements, and to take advantage of technical advances. During this time, it has progressed through nomenclature changes up to the current configuration, which is believed to be known as AN/ALQ-99F(V).

It was originally designed in two configurations: as a full internal-fit for the EF-111A; and as a five external underwing/fuselage pod-fit for the EA-6B.

Since the EF-111A is now being withdrawn from US Air Force service, the EA-6B aircraft, and with it the AN/ALQ-99 ECM system is being further upgraded to meet both the US Navy and the US Air Force SEAD requirements, and many more years operational service is expected from the EA-6B/ALQ-99 system.

The EA-6B configuration comprises either two or three operator positions, from which specialist crew members exercise automatic/semi-automatic/manual control over the System Integrated Receiver (SIR) situated in the aircraft tail fin, and the five underwing/fuselage pods. Each pod includes transmitter elements and high-gain electronically steerable antennas (in both nose and tail), together with ram-air turbines that provide the requisite electrical power.

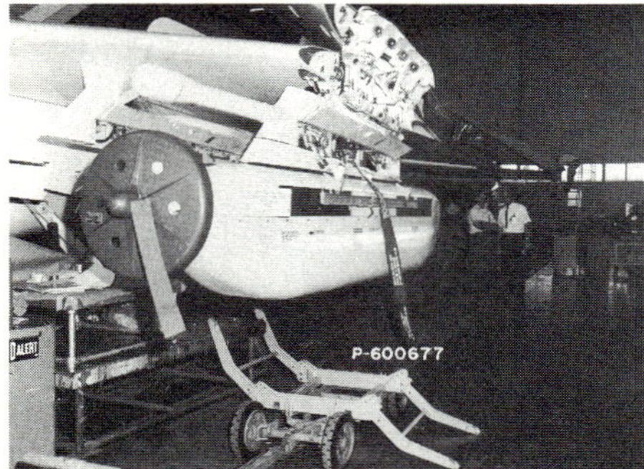

The AN/ALQ-99 jammer pod installed on a US Navy Grumman EA-6B Prowler

The ALQ-99 was designed from the outset to meet specific threat requirements with very high effective radiated power. To meet this design aim the pod has always been fitted with high-gain, directional antennas, and specific limited bandwidth transmitters. The original version only covered four bands, but additional bands have been added through the years to meet new operational requirements, and coverage is now understood to be from VHF up to I/J-band in separate frequency-specific bands: designated from band 1 up to band 9/10.

Recent upgrades to the ALQ-99 pod have been integrated with EA-6B aircraft upgrades. In 1976, ICAP 1 (Increased CAPability) improvements were introduced to reduce processing time, add new displays, and improve communication, navigation and IFF functions. Then, in 1982, ADVCAP (ADVanced CAPability) was implemented to update the OR-262 receiver/processor and ALR-73 passive detection system.

In the 1980s, ICAP 2 configurations, designated -82, -86 and -89 for the years of their introduction were implemented. Current ICAP 2 Prowlers carry five integrally powered pods with 10 very high-power jamming transmitters. Each pod contains an exciter that optimises the jamming signals for transmission by two powerful transmitters. The ICAP 2 Prowler carries exciters that generate signals in any of seven frequency bands. Each pod can jam in different bands simultaneously. In addition, the Prowler can carry any mix of pods, depending on mission requirements.

The latest defined configuration is designated ICAP 2 Block 89A. The ICAP 2 Block 89A configuration was flown for the first time in June 1997. Following testing at the US Navy's test centre at Patuxent River, four validation/verification kits are to be tested, before fitment to all 125 Prowler aircraft commences.

ICAP 2 Block 89A is understood to include the following upgrades to the ALQ-99 system: an upgrade to the universal exciter, low- and high-band transmitters; and the AN/USR-113 communications jammer. It is also understood to include the following avionics upgrades: new radios (AN/ARC-210); an embedded inertial navigation system; a global positioning system, as well as hardware and software enhancements to the navy's standard mission computer AN/AYK-14; new instrument landing system and COTS electronic flight instrumentation system.

In January 1998, Tracor Aerospace Electronic Systems Inc was awarded a US$60 million contract to manufacture 120 AN/ALQ-99 Band 9/10 transmitters for the US Navy Air Systems Command (with options for 61 additional transmitters and spares). Tracor Aerospace Electronic Systems Inc is also involved with production of AN/ALQ-99 Bands 1 and 2, as well as development and production of the Low-Band Transmitter.

After a common configuration is achieved, ICAP 3 will be launched; it will add a reactive jamming capability, which is understood to include a receiver upgrade, upgrades to the HARM targeting system, and improved narrowband jamming capabilities against modern SAM systems. ICAP 3 will also include a variety of communication datalinks and displays for enhanced situational awareness.

Contractors

Many contractors have been involved with the AN/ALQ-99/EA-6B system through the years. It is understood that the current ICAP 2 Block 89A team is led by the Northrop Grumman Corporation, with Sanders, Litton Industries Amecom, and PRB Associates.

At the time of going to press, it was announced that the ICAP 3 programme valued at US$150 million had been awarded to Northrop Grumman, Bethpage. Initial operating capability is planned for early 2004, and the aircraft are expected to remain in operational service until at least 2015. Subcontractors involved in the project include:

(1) Litton Amecom for the passive receiver technology
(2) Sanders for upgrades to the AN/USQ-113 tactical communications jamming system, and for new tactical displays
(3) PRB Associates for an updated mission computer and new display software
(4) Comptek Federal Systems for advanced jamming software.

UPDATED

AN/ALQ-135 jamming system

The AN/ALQ-135 Internal Countermeasures Set (ICS) is a component of the Tactical Electronic Warfare System (TEWS) for the US Air Force F-15; it is installed in various configurations in the F-15A, C, D and E variants and operates with the AN/ALR-56 radar warning system and the AN/ALQ-45 countermeasures dispenser.

The AN/ALQ-135 is an advanced jamming system which uses high-powered transmitters. All equipment is mounted internally and jamming system management is self-contained in the F-15C, D and E.

The basic ALQ-135 consists of seven LRUs plus appropriate waveguides and antennas. Three LRUs are control oscillators and four are amplifiers. Over the years the system has continued to evolve along with the capabilities of the aircraft and changes in the threat. While maintaining commonality with the original system and support electronics, the AN/ALQ-135 has been updated to include full band coverage and extremely effective technique flexibility.

For installation in the two-seat F-15E the frequency range of the system has been expanded by the addition of a Band 3 transmitter/receiver/processor and power amplifier for the aft radiating antenna. Bands 1 and 2 have individual jammers in older F-15s, but in the F-15E variant they are being combined into a 'Band 1.5' low-band jammer that is half the size of the bands 1 and 2 jammer. The band 1.5 subsystem is scheduled to begin US Air Force operational test and evaluation in 1998. Four extra LRUs are added to the system.

Receiver: channelised, multimode for fast response and activity detection throughout the band, variable bandwidth superheterodyne with IFM and sophisticated blanking/lookthrough.

Control: 20 microprocessors in a federated architecture with parallel processing for fast system response and flexibility/reprogrammability.

Techniques: demonstrated capability for emerging/developing radar threats with advanced techniques and coherent response.

Operational status

In service with F-15. Northrop Grumman has produced more than 1,400 AN/ALR-135 systems.

Contractor

Northrop Grumman Corporation, Electronic Systems, Electronics & Systems Integration Division.

UPDATED

The Northrop Grumman AN/ALQ-135 jamming system is carried internally in the F-15

AN/ALQ-155(V) B-52 power management system

The AN/ALQ-155(V) countermeasures power management system forms part of the upgrading programme to improve the defensive avionics of the B-52 aircraft. It provides integral set on receivers for each jamming transmitter, plus increased effective radiated power density through accurate frequency set on. The system is a power management evolution for the ALT-28(V) active ECM set providing automated hand-off from the ALR-46 radar warning receiver with near-instantaneous jammer response. It is computer managed and field programmable. The system contains automatic frequency control in all modes and a wide variety of ECM techniques that are automated, semi-automated or manual. A 12-transmitter upload capability is provided.

A variety of improvements is incorporated in the ALQ-155(V) including frequency agility against multiple threats, pulse repetition interval trackers, cover pulse jamming techniques, false target generation through pseudo-random noise, coherent and incoherent jamming and downlink jamming. It has a hybrid IFM receiver and central receiver capability, programmable noise optimisation, increased pulse-up power for CW to pulse operations, electronically steerable antenna system compatibility and compatibility with the AN/ALQ-117 deception I/J-band jammer.

Operational status

Production complete. About 300 ALQ-155s were produced.

Contractor

Northrop Grumman Corporation, Electronic Systems, Electronics & Systems Integration Division.

VERIFIED

AN/ALQ-162 countermeasures set

Development of the ALQ-162 was started in 1979 under contract to the US Navy; the US Army later joined the programme. It is a small, reprogrammable radar jamming system which can be supplied with its own receiver/ESM management processor, or made compatible with many existing types of radar warning receiver processor systems. The system is software-programmable to meet new threats, and can be installed in pod, pylon or internal fit configurations.

ALQ-162 is fully integrated with the following systems: radar warning receivers (AN/APR-39, -43, -45, -46, -66, -67 and -69); pulse jammers (AN/ALQ-126 (A and B) and AN/ALQ-136); CM dispensers (AN/ALE-36, -40 and M-130).

The AN/ALQ-162 Pulse Doppler (PD) upgrade, also known as Shadowbox II, has advanced pulse Doppler capabilities to counter emerging threats and provide increased aircraft survivability. No changes to form, fit, or aircraft wiring is required to implement this upgrade.

Specifications

Dimensions: 161 × 184 × 420 mm
Weight: 18 kg
Power supply: 115 V AC, 400 Hz, 3 phase, 650 W
Heat dissipated: 480 W
Reliability: >300 h MTBF demonstrated

Operational status

Current installations include: US Navy — A-4M, A-7E, AV-8B, F-4S and RF-4B; US Army — EH-1, EH-60, OV-1D, RC-12D, RU-21, RV-1D; NATO — C-130, F-16 and Saab F-35 Draken, more than 650 installations. Enhancements planned by Northrop Grumman for the ALQ-162 include:

(1) Power Plus: utilising advanced microwave power technology, which will be available in conjunction with the Shadowbox II PD upgrade, the effective radiated power will more than double
(2) Increased Threat Handling: the current ALQ-162 defeats multiple threats in the same band simultaneously, incorporating new technology will allow the system to defeat multiple threats in different bands simultaneously
(3) Additional Pulse Capability: the current Shadowbox II defeats CW and PD threats. When combined with available technologies currently used in other Northrop Grumman EW products — Tactical Radar Jammer (TRJ) and AN/ALQ-135 — the ALQ-162 will be a more capable system able to defeat CW, PD and pulse threats
(4) Potential Growth: other growth areas include integration with towed radar decoys and integration with DIRCM.

Contractor

Northrop Grumman Corporation, Electronic Systems, Electronics & Systems Integration Division.

UPDATED

AN/APR-50 defensive management suite

AN/APR-50 is the US Air Force designation for the ZSR-63 ESM suite and radar warning receiver for the B-2 aircraft. The system is classified and very few technical details have been released. It would appear, however, that the complete installation consists of a number of antennas feeding nine radio frequency front ends to detect and analyse a wide variety of signals. It is assumed that each of the front ends is tuned to a different part of the frequency spectrum. Five receivers receive the outputs of the front ends and pass these to the processor. The APR-50 is believed to be linked to the ZSR-62 system, a classified part of the B-2 defensive management suite in development by Northrop Grumman which employs advanced concepts in EW technology.

AN/ALQ-162 installed in the outboard pylon on F/A-18 **1998**/0018248

Operational status

In January 1993, Northrop Grumman was awarded a US$117 million contract to continue development of the AN/APR-50. Northrop Grumman was also awarded US$53.9 million to carry on with ESM development, including extension of the frequency range. It is believed that this originally was for Band 2 and the extension was to cover Band 4 from 500 MHz to 1 GHz.

Contractor

Northrop Grumman Corporation, Electronic Systems, Electronics & Systems Integration Division.

VERIFIED

Modularised InfraRed Transmitting System (MIRTS)

The Modularised InfraRed Transmitting System (MIRTS) is a derivative of the AN/AAQ-8(V). It is an advanced subsonic, infrared countermeasures system for deployment in a wide range of aircraft, including helicopters, which can be carried internally or pod mounted. MIRTS utilises advanced jammer technologies including a variable optics/reflector design to provide optimum aircraft infrared signature coverage, combined with advanced digital electronics and mode-switching power supplies to enhance system reliability, maintainability and versatility.

The MIRTS installed on a Royal Air Force Raytheon Hawker 125 (Paul Jackson)

Specifications

Transmitter/receiver
Dimensions: 228 × 240 × 635 mm
Weight: 23.6 kg
Power supply: 115 V AC, 400 Hz, 3 phase, 2.7 kVA
28 V DC, 5.6 W

Control unit
Dimensions: 190 × 259 × 318 mm
Weight: 10.3 kg
Power supply: 115 V AC, 400 Hz, 3 phase, 340 VA
28 V DC, 5.6 W

Operational status

Under development. The system has been tested on the rear fuselage of a Royal Air Force VC10. Committed aircraft installations include the Boeing 707, 747 and DC-8, Eurocopter Puma; Falcon 20, Fokker 27, and Raytheon Hawker 125.

Contractor

Northrop Grumman Corporation, Electronic Systems, Electronics & Systems Integration Division.

VERIFIED

Starfire self-protection suite

The Starfire laser-based InfraRed CounterMeasures (IRCM) self-protection suite has been developed as the ATIRCM solution. Its jamming effectiveness is increased by the use of superior sightline stabilisation in the most demanding flight environment. The heart of the system is an advanced and combat proven pointing and tracking system, keeping Starfire's laser accurately on the target. Starfire represents an integrated and reliable IR defensive self-protection suite that can be fitted on military and commercial aircraft.

The Starfire suite provides fast accurate threat missile location in a combined missile approach and warning surveillance system. The high-power jammer acts only on demand from an alert from an approaching missile. The suite tracks missiles in all modes of operation. All-aspect self-protection is provided with a power-managed architecture. Full fitting compatibility is designed into the system for a wide range of helicopters and fixed-wing aircraft.

Contractor

Northrop Grumman Corporation, Electronic Systems, Electronics & Systems Integration Division.

VERIFIED

AN/ALE-50 Towed Decoy System (TDS)

The ALE-50 TDS provides protection against RF threats. When deployed, the decoy seduces RF guided missiles away from the host aircraft. This stand-alone system requires no threat specific software and communicates health and status to its host aircraft over a standard databus.

The ALE-50 TDS consists of a launch controller, launcher and towed decoy. The decoy control/monitor electronics and power supply are contained in the launch controller. The launcher, which holds the decoy magazine, can be customised to fit any candidate aircraft. The decoy has a 10 year shelf life and is packaged in a sealed canister which also contains the pay-out reel.

Operational status

The ALE-50 programme was a joint US Navy/US Air Force/Raytheon Systems Company development. The MultiPlatform Launch Controller (MPLC) is the standard launch controller for all installations. US Air Force F-16 aircraft are the first operational users, for which decoy production began in early 1997. Other users, for which launcher-specific installations have been developed, are the US Air Force B-1B and US Navy F/A-18E/F. An installation is also being developed for integration with the AN/ALQ-184 ECM pod, designated AN/ALQ-184(V)9 (see later entry).

Raytheon Systems Company, using IR&D funds, is currently developing an infrared towed decoy to expand the capability of the ALE-50 system to protect against both RF and IR threats.

A derivative variant of the ALE-50 is also being developed for the Royal Air Force Nimrod MRA.4 maritime patrol aircraft, where it will be linked to the AN/ALR-56M RWR for threat cueing, and to a techniques generator being developed by Racal Radar Defence Systems based on DRFM technology.

In December 1997, Raytheon Systems Company was awarded a US$35.5 million contract to provide 1,522 towed decoy rounds (1,372 for the US Air Force and 150 for the US Navy) applicable to the AN/ALE-50(V) countermeasures system on the F-16 and F/A-18 aircraft.

Contractor

Raytheon Systems Company.

UPDATED

AN/ALQ-108 IFF jamming pod

The AN/ALQ-108 jamming pod is used in the US Navy's Northrop Grumman E-2C Hawkeye, Lockheed Martin EP-3A Aries and S-3A Viking types, and some German Air Force F-4 Phantoms, to improve survivability in ASW and ELINT operations by jamming IFF transmissions.

Operational status

Production of about 300 sets is reported.

Contractor

Raytheon Systems Company.

UPDATED

AN/ALQ-128 threat warning receiver

The AN/ALQ-128 is the standard threat warning receiver on the US Air Force's F-15 aircraft, forming part of the Tactical Electronic Warning System (TEWS) with the ALR-56, ALE-45 and ALQ-135 systems. The ALQ-128 has been in production since 1980. Little is known about this system's performance; it possibly provides coverage of the higher-frequency bands above the J-band limit of the ALR-56.

Operational status

In service on the US Air Force F-15.

Contractor

Raytheon Systems Company.

UPDATED

AN/ALQ-184(V) self-protection jamming pod

The AN/ALQ-184(V) ECM pod is designed to provide effective countermeasures against surface-to-air missiles, radar-directed gun systems and airborne interceptors by selectively directing high-power jamming against multiple emitters. An upgrade of the AN/ALQ-119(V) pod, the AN/ALQ-184(V) functions as a repeater, transponder or noise jammer. Multibeam architecture based on proven Rotman lens antenna technology provides substantially reduced countermeasures response time with a tenfold increase in ERP at 100 per cent duty factor. High reliability is inherent in the architecture due to redundant mini-TWTs, reduced operating voltages and lower operating temperatures. Maintainability has been improved by the use of digital circuitry, automatic gain optimisation, elimination of manual adjustments and extensive computer-driven built-in test.

AN/ALQ-184(V) self-protection jamming pod
1998/0018258

Operational status

Raytheon Systems Company has delivered more than 880 pods to the US Air Force since 1989. The ALQ-184 operates on US Air Force A-10, F-4, F-15 and F-16 aircraft. AN/ALQ-184(V) has also been selected by the Taiwanese government for its F-16 aircraft, and deliveries began in 1997.

In 1996, the US Air Force awarded Raytheon Systems Company a contract to upgrade the ALQ-184. Under the contract, Raytheon Systems Company will build 225 reprogrammable low-band kits, which will provide the pods with rapid flight line programming, improved life cycle costs and additional modulation jamming techniques. The contract also includes options for upgrade kits to all ALQ-184 ECM pods produced to date.

Contractor

Raytheon Systems Company.

UPDATED

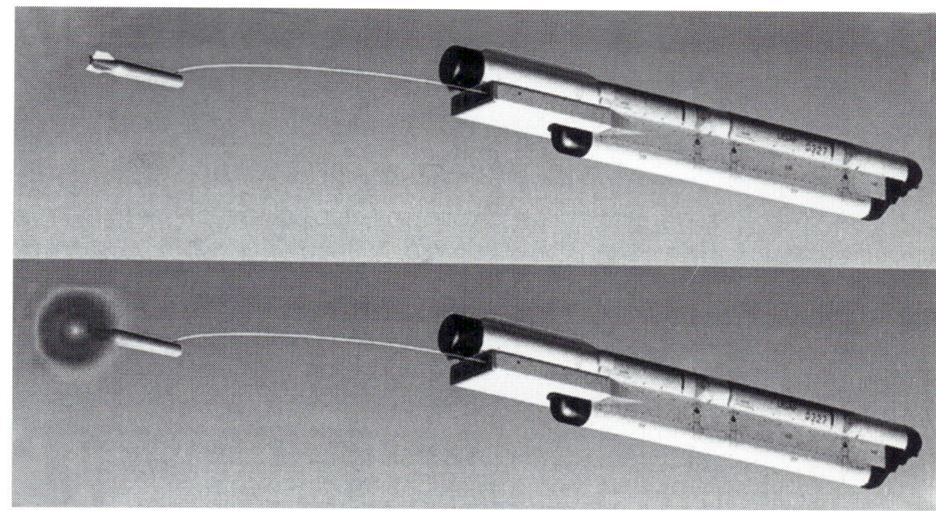

AN/ALQ-184(V)9 combined ALQ-184 ECM pod and ALE-50 towed decoy system, indicating the design concept of radar and infrared decoy operation ***1998***/0018256

AN/ALQ-184(V)9 combined ALQ-184 ECM pod and ALE-50 towed decoy system

The AN/ALQ-184(V)9 is a development of the AN/ALQ-184 ECM pod, in which a scab-fit unit of four ALE-50 towed decoy system units is added to the aft end of the pod.

The ALE-50 system consists of a launch controller, launcher and towed decoy. The decoy protects the host aircraft against radar-guided missiles by providing a more attractive target and seducing them away from the aircraft.

Technique co-ordination between the two systems will be managed by an Advanced Correlation Processor (ACP). The ACP in the (V)9 pod has been tested by the US Air Force Air Warfare Center.

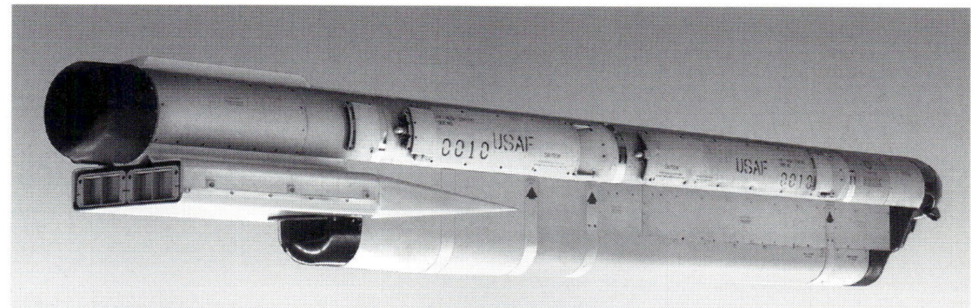

AN/ALQ-184(V)9 configuration, showing 1 × 4 ALE-50 launcher unit at centre rear ***1998***/0018257

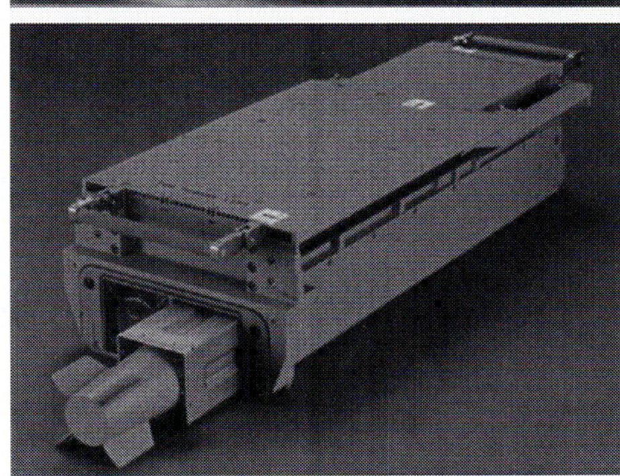

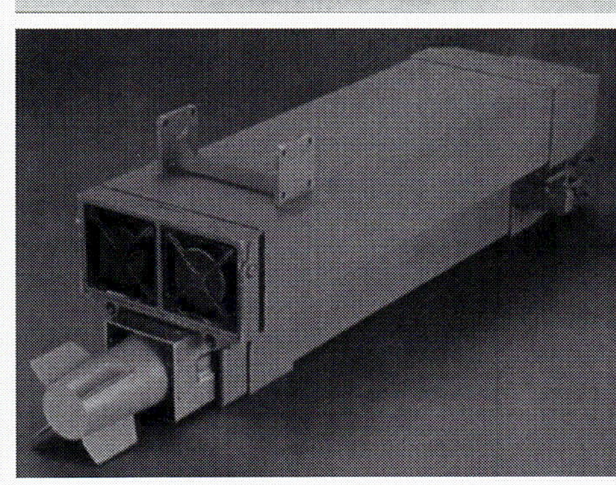

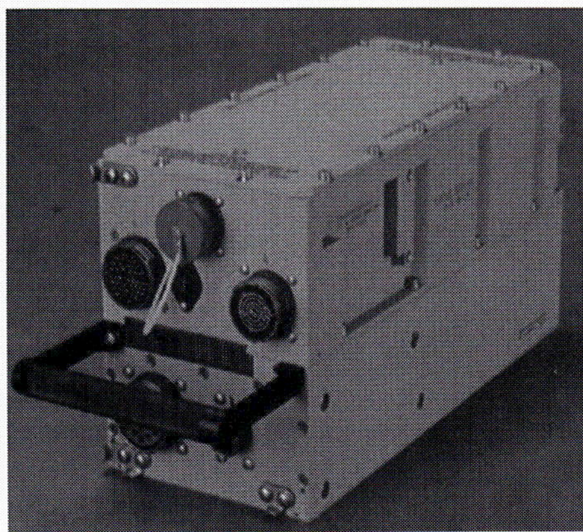

Composite 9-view photo, showing: F-16 Launch/Launch Controller; (left); F/A-18E/F T-3 Launcher (centre); B-1B 1 × 4 Launcher (right); MultiPlatform Launch Controller (bottom centre); ALQ-184(V)9 pod with 1 × 4 launcher (bottom right); and ALE-50 production (bottom left) ***1998***/0018259

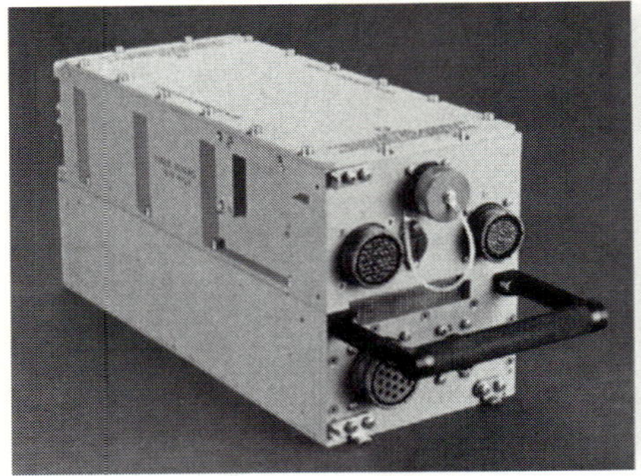

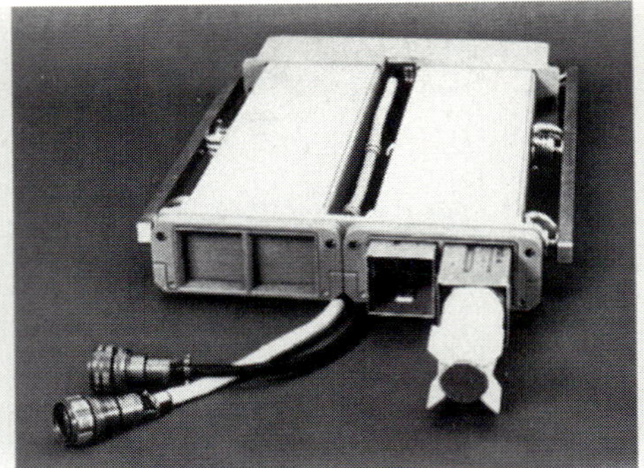

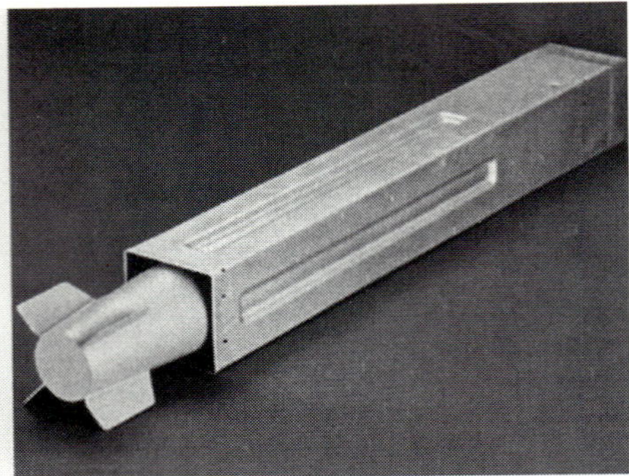

AN/ALQ-184(V) 9 units: (left) multiplatform launch controller, (centre) four-decoy launcher, (right) decoy and canister ***1998**/0018255*

The ACP will make the decision between the ALQ-184 and ALE-50 threat responses in order to employ the most effective counter to the threat.

Raytheon Systems Company is currently developing an infrared towed decoy to expand the capability of both ALE-50 and ALQ-184(V)9 systems to provide equally effective protection against both RF and IR threats.

Integration of ALE-50 into ALQ-184 also results in significant improvement of basic ALQ-184 capability. To provide space for the ALE-50 launch controller and four decoy launcher, the current low-band controller is modernised and made field programmable by conversion of 12 1970s vintage circuit cards into two 1990s technology circuit cards. The two-card low-band modification improves MTBF to the degree that the addition of ALE-50 LRUs are completely offset and the resultant ALQ-184(V)9 MTBF is better than the basic pod.

ALQ-184(V)9 will have an improved ability to communicate with its host aircraft and other onboard systems, such as missile warning receivers and radar warning systems, via newly installed dual redundant MIL-STD-1553B interfaces.

As with the basic ALE-50, ALQ-184(V)9 will offer growth to fibre optic decoy capability using proven technique generation in the ALQ-184 pod.

Operational status

Development.

Contractor

Raytheon Systems Company.

NEW ENTRY

AN/ALQ-187 internal countermeasures system

The AN/ALQ-187 is an internally housed, fully automatic jammer system integrated with radar warning and flare/chaff systems for tactical aircraft self-protection. Primarily intended for F-4, F-16, A-7 and Mirage 2000 aircraft, it detects and defends against surface-to-air missiles, anti-aircraft artillery and air-to-air interceptor weapon systems. It can interface with the AN/ALE-39 or AN/ALE-40 chaff/flare dispensers and the AN/ALR-66, AN/ALR-69 or AN/ALR-74 radar warning receivers.

The fully software programmable system automatically detects single- and multiple-threat radars of the same or different technologies and selects ECM programmes to counter pulse, pulse Doppler or CW ground-based, shipborne or airborne emitters. The system is flight line reprogrammable, allowing immediate incorporation of the latest intelligence threat data and EM techniques. Advanced power management techniques ensure maximum jamming effectiveness. A variant of the AN/ALQ-187 jammer, designated the AN/ALQ-187H is used as part of the Litton Applied Technology ASPIS Advanced Self-Protection Integrated Suites system.

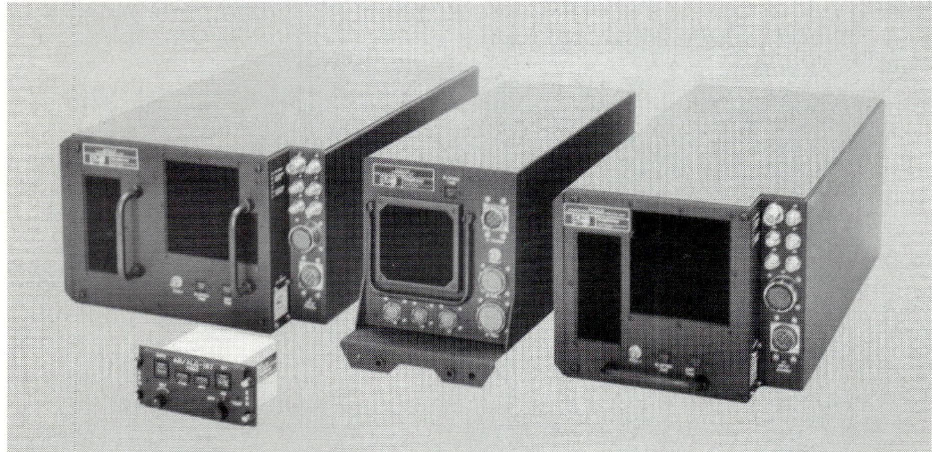

The AN/ALQ-187 internal jamming system consists of four LRUs

Specifications

Dimensions:
(control unit) 86 × 146 × 83 mm
(forward transmitter unit) 222 × 336 × 533 mm
(aft transmitter) 256 × 222 × 533 mm
(processor) 203 × 188 × 458 mm
Weight:
(control unit) 1.35 kg
(forward transmitter) 43.1 kg
(aft transmitter) 29.5 kg
(processor) 14.1 kg
Frequency: 6.5-18 GHz

Operational status

The AN/ALQ-187 is operational with the A-7, F-4 and RF-4 aircraft of a NATO country.

Contractor

Raytheon Systems Company.

UPDATED

US Navy F-14 Tomcats are fitted with AN/ALR-50 radar warning receivers

AN/ALR-50 radar warning receiver

The AN/ALR-50 was part of a very substantial US Navy programme throughout the early 1970s.

Operational status

At least 1,300 sets were delivered to the US Navy and used on A-4 Skyhawk, A-7 Corsair, EA-6A Intruder and EA-6B Prowler, F-8J/RF-8G Crusader, F-14 Tomcat, RA-5G Vigilante and RF-4B/F-4N Phantom. The AN/ALR-50 is no longer in production.

Contractor

Raytheon Systems Company.

UPDATED

AN/ALR-67(V)3/4 countermeasures receiving set

The AN/ALR-67(V)3/4 is commonly known as the Advanced Special Receiver (ASR) set. It will be the standard US Navy Radar Warning Receiver (RWR) for carrier-based tactical aircraft such as the F/A-18, F-14, A-6E and AV-8B.

The system comprises seven Weapon Replaceable Assembly (WRA) types to give a total of 13 WRAs. These consist of four integrated antenna detectors, four quadrant receivers, countermeasures receiver, countermeasures computer, low-band integrated antenna, control status unit and azimuth display indicator.

Each Integrated Antenna Detector (IAD) includes a dual-polarised microwave antenna, MMW antenna and supporting electronics. For ease of installation in various aircraft types, the IAD is available in three different housings.

Each IAD feeds signals into an associated Quadrant Receiver (QR) which conditions the received energy for processing. Conditioning involves filtering, amplification and frequency conversion to IF. Band structure has been optimised to minimise interference from onboard jammers.

The Countermeasures Receiver (CR) accepts preconditioned signals from the QR and low-band WRAs and generates digital words describing the parameters of the pulsed and CW radar waveforms detected. Parameters include amplitude, angle of arrival, time of arrival, frequency, pulsewidth and modulation.

The Countermeasures Computer (CC) incorporates an Ada programmable 32-bit JIAWG-compatible computer. Software in the CC processes the pulse and CW data to characterise, identify and prioritise intercepted threats based on their potential lethality. The CC manages all external interfaces, including two MIL-STD-1553B busses and special interfaces to various jammers and missile launch computers.

The low-band WRA receives and conditions signals in the lower frequency region.

The Control Status Unit (CSU) is a push-button WRA in the cockpit that enables the pilot or maintenance technician to issue commands to the system.

The Azimuth Display Indicator (ADI) is a 3 in (76.2 mm) diameter CRT cockpit display used to show intercepted threats.

The AN/ALR-67(V)3/4 is understood to form part of the IDECM programme.

Operational status

Production for the AN/ALR-67(V)3 is scheduled for the F/A-18 from 1996-2008, the F-14 from 1998-2000 and the AV-8B from 1998-2002.

Contractor

Raytheon Systems Company.

UPDATED

AN/ALR-89(V) integrated self-protection system

The AN/ALR-89(V) radar and laser warning system incorporates three subsystems: the AN/ALR-90(V) RWR which detects pulsed radars in the C- to J-bands, the AN/APR-49(V) RWR which is a superheterodyne receiver for detecting modern pulse and non-pulse radars and the AN/AVR-3(V) laser system which detects laser emissions.

Identified threats which are detected by the three subsystems are presented to the pilot on a 3 in (76.2 mm) display that provides an alphanumeric representation of the type, angle of arrival, relative lethality and status of the threats.

The ALR-89 offers complete threat reprogrammability for new and changing environments through an extensive, easily updated emitter library file. A portable loader unit enables loading of operational software and emitter tables and downloading of recorded data in the field. The integrated ALR-89 or one of its subsystems is suitable for the self-protection of helicopters or fixed-wing aircraft regardless of their size. Electrical design and mechanical configuration both adapt to new installations and provide an upgrade of AN/APR-39(V)1 installations. This upgrade can be implemented in various levels of performance using the basic APR-39 system and selected LRUs from the ALR-89 system.

Specifications

Weight:
(AN/ALR-90(V)) 5.76 kg
(AN/APR-49(V)) 4.04 kg
(AN/AVR-3(V)) 3.86 kg
Power supply: 28 V DC, 210 W (MIL-STD-704)
Frequency: C- to J-bands, 2 laser bands
Coverage: 360° azimuth
Environmental: MIL-E-5400, Class II

Operational status

In production.

Contractor

Raytheon Systems Company.

UPDATED

AN/ALR-90(V) self-protection system

The AN/ALR-90(V) is a microprocessor-controlled airborne self-protection system which employs a crystal video detector system to detect pulsed radars in the C- to J-bands. Detected radars, which are analysed and identified as threats, are presented on a 3 in (76.2 mm) display that provides an alphanumeric representation of the type, angle of arrival, lethality and status of the threats.

The AN/ALR-90(V) offers complete threat programmability for new and changing environments.

The AN/ALR-67(V)3/4 ***1996***

This is accomplished with an extensive, readily updated emitter library file. A portable loading unit enables field loading of the operational software and emitter tables and downloading of recorded threat data.

The system has provisions for integration with various avionics systems, such as radar, for blanking, and other digital data and ECM sensors. It can also be interfaced with the AN/APR-49(V) radar warning receiver and the AN/AVR-3(V) laser warning system.

Specifications

Weight: 5.77 kg
Frequency: C- to J-bands in three bands
Signal types: pulsed radars
Coverage: 360° azimuth

Contractor

Raytheon Systems Company.

UPDATED

AN/APR-49(V) self-protection system

The AN/APR-49(V) is a microprocessor-controlled airborne self-protection system which uses a superheterodyne receiver to detect pulsed and CW signals in the H- to J-bands. An integral high-speed microprocessor controls the receiver, executes data processing and performs the required interfacing tasks. The receiver scans the programmed frequency ranges, measures frequency of detected signals, calculates the angle of arrival, analyses emitter parameters, identifies threat radars and provides data for display. The processed threat data are presented on a 3 in (76.2 mm) display that provides an alphanumeric representation of the type, angle of arrival, relative lethality and status of threats.

Other features of the AN/APR-49(V) include high-resolution emitter separation and identification; the ability to record emitter parameters in flight for post-flight playback, signal analysis and training; simple and effective interface with laser and other pulse radar warning receivers; field loading of the emitter library and downloading of the system recorder; and provisions for interface with chaff/flare dispensing, ECM and missile warning systems.

The AN/APR-49(V) superhet receiver is connected to an array of four spiral antennas. The receiver input enables selection of any four antennas in a controlled sequence to enable angle of arrival determination. The optional remote RF preamplifier enables selection and amplification of signals detected by the remote antennas. An integral signal processor not only determines DF but also characterises each signal, including frequency, and identifies a threat emitter by comparing the signal parameters with preprogrammed library information.

The digital signal analyser in the receiver incorporates a computer featuring a 32-bit microprocessor with EEPROM for combat software and emitter tables and RAM for temporary storage. When the APR-49(V) is configured with a wideband radar warning receiver it is capable of effective and accurate determination of radar warning for self-protection.

Specifications

Weight: 6 kg with indicator, remote preamplifier and 4 antennas
Frequency: H- to J-bands
Signal types: high-PRF and low-ERP pulsed signals and CW signals
Coverage: 360° azimuth or selected coverage

Contractor

Raytheon Systems Company.

UPDATED

AN/ASQ-213 HARM Targeting System (HTS)

The AN/ASQ-213 HTS is fitted to US Air Force Block 50/52 F-16 aircraft to provide targeting information for the AN/AGM-88A HARM missile in the manned SEAD role. The HTS detects, identifies, and locates hostile radars and provides data needed by HARM for calculation of targeting and launch parameters. It is carried on the starboard chin station of the F-16. It weighs 40 kg. In the manned SEAD role, HTS/HARM-equipped F-16 aircraft often operate with RC-135 Rivet Joint or EA-6B Prowler aircraft to optimise effectiveness.

Operational status

In service. Raytheon is said to be discussing an HTS integration on F/A-18C/D with the US Navy.

Contractor

Raytheon Systems Company.

UPDATED

AN/AVR-2A(V) laser detecting set

The AN/AVR-2A(V) laser detecting set detects, identifies and characterises optical signals and provides audible and visible warning of laser threats to the aircrew. It consists of four staring array SU-130A(V) sensor units and a CM-493A interface unit comparator. The AN/AVR-2A(V) interfaces with all variants of the AN/APR-39(V) Series radar signal detecting set to

US Air Force Block 50 F-16 carrying HTS on the starboard chin station ***1996***

function as an integrated radar and laser warning receiver system.

The AN/AVR-2A(V) system provides 360° coverage and can identify laser rangefinders, designators and beamriders. It contains a reprogrammable, removable, user data module. P³I expansion includes an MIL-STD-1553B interface.

Operational status

In production for the US Army, Navy, Marine Corps and Special Forces for the AH-1F Cobra, AH-64 Apache, AH-64D Longbow Apache, AH-1W SuperCobra, MH-60K Black Hawk, MH-47E Chinook, OH-58D Kiowa Warrior, HH-60H Combat Rescue, UH-1N Huey, and V-22 Osprey helicopters. Also selected for the UK EH 101SH and WAH-64D helicopters. Under development for: RAH-66 Comanche, C-130 Hercules, SH-60B LAMPS, and SH-60F CV helo aircraft.

Contractor

Raytheon Systems Company.

UPDATED

The AN/AVR-2A laser detecting set consists of four sensor units and an interface unit comparator **1995**

AN/AVR-3(V) airborne laser warning system

The AN/AVR-3(V) is an airborne microprocessor-controlled warning system for detecting laser emissions in two threat bands, with provision for a third. Identified threats, which are detected and analysed, are presented to the pilot on a 3 in (76.2 mm) CRT display that provides an alphanumeric representation of the type of laser and the angle of arrival.

The AVR-3 offers complete threat reprogrammability for new and changing environments as laser weapons increase and mature. This is accomplished by an extensive and easily updated emitter library file. A portable loader unit enables the loading of operational software and emitter tables and the downloading of recorded signal data in the field.

The AVR-3 is suitable for the self-protection of helicopters and fixed-wing aircraft regardless of size. Electrical design and mechanical configuration provide for the use of up to eight laser sensors to ensure complete spatial coverage for all aircraft.

The AVR-3 features detection of modern pulsed lasers, direction-finding of received signals, high-resolution emitter separation, sophisticated signal processing, alphanumeric azimuth threat display and full system BITE. The system is able to record emitter parameters during flight for post-flight playback and signal analysis. It has provision for a tie-in with flare and chaff dispensers, ECM and missile warning systems.

Specifications

Dimensions:
(laser sensor) 63.5 × 85.1 × 149.9 mm
(laser analyser) 146 × 106.2 × 78.7 mm
(control unit) 54.1 × 146 × 19 mm
(display unit) 158 × 80.5 × 80.5 mm
Weight:
(laser sensor) 0.59 kg
(laser analyser) 1.5 kg
(control unit) 0.2 kg
(display unit) 1.2 kg
Power supply: 28 V DC, 90 W (max) (MIL-STD-704)
Frequency: 2 laser bands, provision for a third
Coverage: 360° azimuth
Environmental: MIL-E-5400

Contractor

Raytheon Systems Company.

UPDATED

Airborne remote-controlled ESM system

The airborne remote-controlled ESM system consists of a number of radio receivers and antennas operated by remote radio control from a ground facility which is equipped with the necessary processing and display equipment. It is intended to be carried in small, relatively inexpensive aircraft and the system configuration can be tailored to the particular aircraft type. Onboard equipment would normally consist of radar detection units covering the frequency range 0.5 to 18 GHz, radio monitoring receivers covering 1.5 MHz to 2 GHz and radio direction-finding systems covering 20 to 1,200 MHz. The datalink is a full-duplex highly directional system.

Operational status

Fully developed as a private venture.

Contractor

Raytheon Systems Company.

UPDATED

Direction-finding system

Received RF signals between 20-1,500 MHz route to distribution and switching circuitry contained in a modular RF unit. This unit provides 14 reconfigurable HF/VHF/UHF RF inputs and is designed to be cascadable, allowing for additional antenna elements or arrays, simply by adding other RF units. The RF unit also includes a switchable dual-channel HF upconverter for coverage of the 2-20 MHz frequency range.

The IF outputs of the two Nanomin receivers are each routed into coherently clocked A/D converters and then into a two-channel DDC. The DDC digitally fine tunes to the specified frequency, generates in-phase and quadrature components of each channel, down-converts the desired portion of the IF bandwidth to baseband, and digitally filters the signal with linear phase Finite Impulse Response (FIR) filters. The DDC also performs basic preprocessing and controls data flow over the VME bus to the Motorola 68040 processor.

Switchable IF filters operate digitally within the DDC. The set of IF bandwidths can be changed easily by downloading a different set of coefficients. The digital decimation filter uses FIR filter designs that exhibit ideal phase characteristics as well as excellent roll-off features. This digital demodulation concept offers unique flexibility for tailoring the DF receiver to specific user needs.

Accurate angle of arrival measurements can be obtained for both data and voice signals using virtually any type of modulation, including AM, FM, continuous wave, single sideband, and independent sideband. Precision results on digital signals can also be achieved using frequency-shift keying or pulse-shift keying modulation.

The DF processor is a single board computer using the Motorola 68040 processor, with its floating point math co-processor running at 25 MHz. Along with 4 MB of random access memory (RAM), a 16 MB electrically erasable programmable read-only memory (EEPROM) board is included for non-volatile storage of calibration data. Interface boards are included for Ethernet, IEEE-488 and navigation systems (ARINC and Synchros).

Angle-of-arrival and position information feed into the DF processor for line-of-bearing (LOB) calculation.

The system handles pulse-type signals with pulse-widths down to 0.4 microsecond and with duty cycles as low as 0.0005.

The AN/AVR-3(V) airborne laser warning system consists of a number of sensors (left and right), a laser analyser (centre left), a display unit (centre right) and a control unit (front centre)

Specifications

Frequency range: 2-1,500 MHz
DF Accuracy: <1° RMS estimated for fine DF using typical airborne VHF/UHF arrays
DF Sensitivity: −116 dBm (10 dB S/N in 6.4 kHz bandwidth)
DF Output: LOB, quality factor, frequency, time, in/out of geosort limits, emitter location (calculated in workstation)
Options:
(search): 100 freq/s (software upgrade only)
(classification): AM, FM, FDM, FSK, PSK
(copy): digital demod-AM, FM, SSB

Contractor

Raytheon Systems Company.

UPDATED

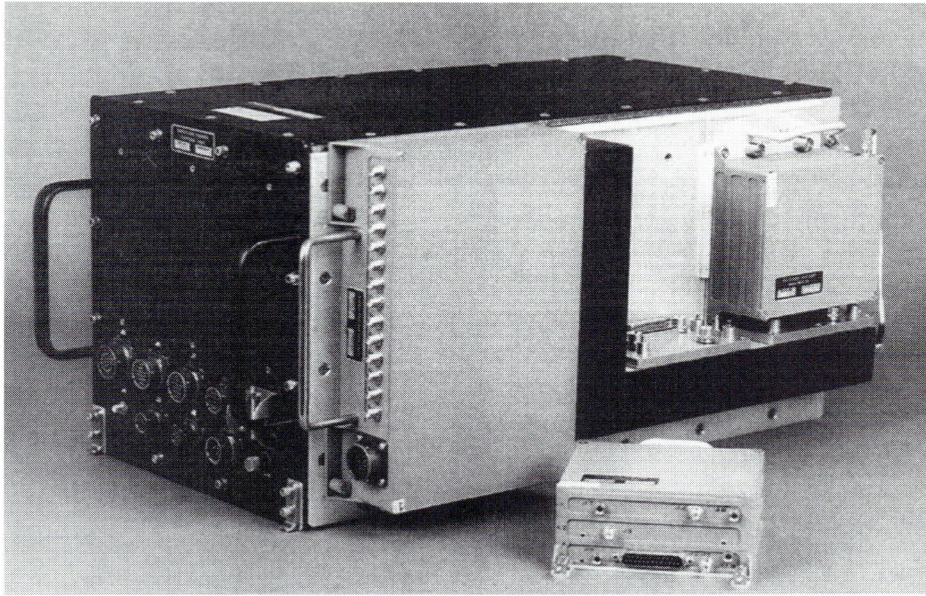

Direction-finding system  ***1997**/0001273*

EC-24A electronic warfare training aircraft

The former Chrysler Technologies Airborne Systems (CTAS) converted a DC-8-54 as an electronic warfare aircraft for use by the US Navy's Fleet Tactical Readiness Group (FTRG) which acts as the Orange Force in exercises. The EC-24A, which is the platform for the Orange Force Commander, is based at Waco and operated and maintained by the civilian crews of CTAS.

The EC-24A has a crew of 10, including three flight crew, six systems operators and the Orange Force commander. There are seats for a further 20 people on the aircraft.

The aircraft is equipped with: two AN/ALT-40 radar jamming systems with steerable antennas; two Collins AN/USQ-113 communications systems jammers; two AN/ALE-43 chaff dispensers effective over the A/J-bands; two Scientific Communications AN/ALR-75 ESM receiver systems with pulse analysers to give onboard signal identification capability; six Collins AN/ARC-159 UHF; four Collins AN/ARC-190 HF and two Collins AN/ARC-186 VHF transceivers; two OE-320 direction-finding systems; two Hewlett-Packard HP-322 computers; one KY-58 secure communications system and one AN/PSC-3(V)1 Satcom radio.

Raytheon also operates and maintains two US Navy NKC-135A aircraft in support of FTRG, in a similar role to the EC-24A.

Contractor

Raytheon Systems Company.

UPDATED

The interior of the EC-24A electronic warfare training aircraft

EH-60L Advanced Quick Fix

The EH-60L Advanced Quick Fix system will provide the US Army with a significant improvement in EW capability over the existing EH-60A Quick Fix helicopter. It will operate as the jamming component of the Division's Intelligence Electronic Warfare Common Sensor (IEWCS) assets to provide expanded signals intelligence, increased jamming performance and datalink interoperability with other Common Sensor platforms.

Operational status

Full-scale engineering development models have been delivered for the test programme.

Contractor

Raytheon Systems Company.

UPDATED

The EH-60 L Advanced Quick Fix ***1995***

Emitter Location System

Raytheon Systems Company, under contract to Daimler-Benz on behalf of Panavia, developed an Emitter Location System (ELS) for use on the Tornado ECR aircraft. The ELS detects, identifies and locates radar emissions through the use of a high probability of intercept system. It features multi-octave frequency coverage, phase interferometric antenna arrays for precision direction-finding and passive ranging, channelised receivers and multiple 1750A digital processors for operation in dense signal environments.

The ELS operates across the frequency spectrum for all primary surface-to-air and air-to-air threats. Data acquired by the system is transmitted, via a MIL-STD-1553B databus, to the tactical displays of both crew members. Threat assessments can then be made by the crew and the appropriate countermeasures activated. Threats which have been located by the ELS can also be communicated to follow-on forces through the Operational Data Interface (ODIN). This data can then be used for threat avoidance or suppression operations.

The design uses a SAW channeliser in a cued analysis receiver configuration. In this configuration the delay lines allow the channeliser to measure the frequency of the incoming signal and then, using a fast-settling local oscillator, tune and cue up the narrowband receivers for subsequent analysis and pulse report generation. The channeliser provides the cued analysis contiguous high signal-selectivity across a wide instantaneous bandwidth. The basic channeliser

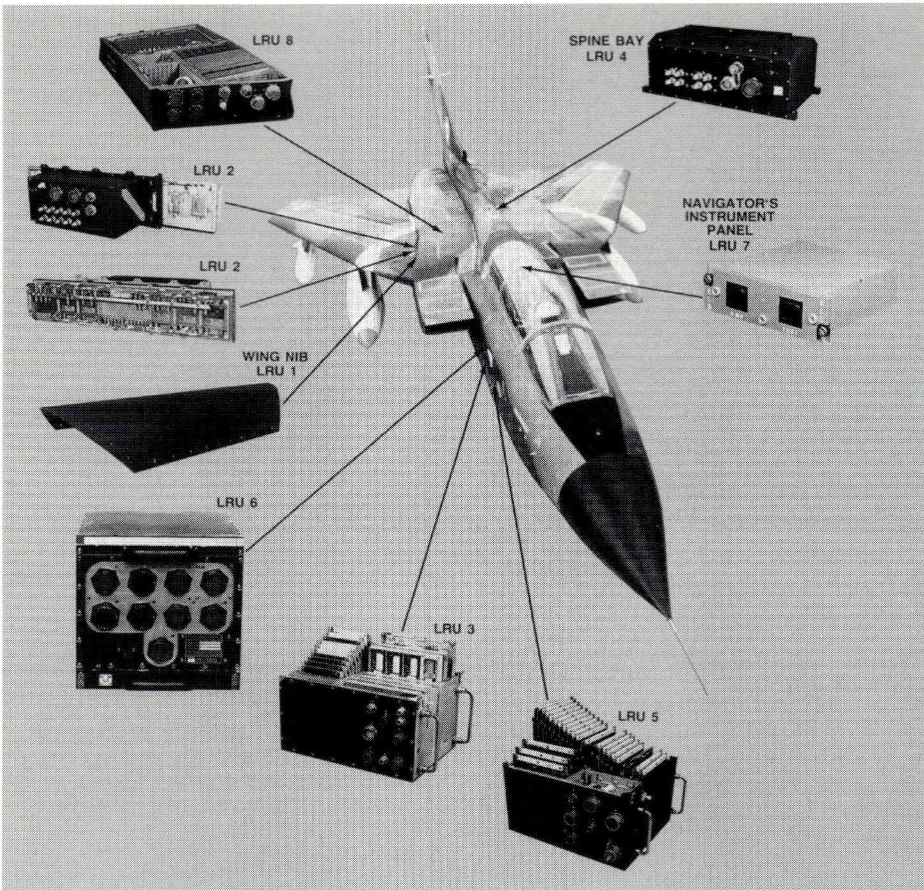

Layout of the emitter location system in the German Air Force ECR Tornado

design consists of multiple parallel SAW filter banks, each bank having multiple signal channels.

High-speed real-time emitter analysis is accurately processed for positive threat identification. In addition to this, accurate DF and passive ranging is computed within the processor, correctly locating and displaying large numbers of emitters. The processed information can then be used to cue weapons, and could be configured to steer jamming pods.

The configuration of the ELS in the Tornado ECR consists of 10 individual LRUs. The two antenna/RF converter units are mounted conformally with the aircraft's left and right wing nibs near the base. The remaining units are distributed within the wing shoulder and gun bays. Operator control is provided by the ELS control panel and the Tornado TV tab displays.

Operational status

In service in German Air Force Tornado ECR aircraft.

Contractor

Raytheon Systems Company.

UPDATED

Escort ESM system

Escort is a family of electronic support measures for maritime patrol aircraft, small ships and ground-based vehicles. It is generally applicable to adapted versions of business aircraft. Escort provides signal detection, sorting, analysis, identification and reporting of radar frequency emitters. Wide instantaneous spatial and frequency coverage, combined with sophisticated processing techniques, provides high probability of intercept. All mission data can be recorded on tape cartridge for post-mission analysis. The ELINT configuration provides precision direction-finding by using phase interferometer antennas in conjunction with the basic receiver processor and display system.

The basic system operates over the 2 to 18 GHz frequency band with growth options available. It uses two antenna units, each consisting of a four-element antenna array, a high-speed RF switch and various control and amplification circuitry. The receiver unit accepts RF from port and starboard antenna units and performs frequency, time of arrival, amplitude and angle of arrival measurements on each received pulse. The processor consists of a 16-bit general purpose computer with 128,000 16-bit words of memory and a variety of interface circuits. The search and dwell process used by the receiver as it steps between sub-bands to intercept and collect emitter data is controlled by the processor software. Because of the wide bandwidth of the receiver, the search cycle over the entire frequency/azimuth domains is typically accomplished within one main beam illumination of a radar, ensuring high probability of intercept and a very low intercept time for new emitters.

The ESM operational software program maintains an extensive emitter data file, which includes the measured parameters as well as time of initial and most recent intercepts. Formats of CRT displayed outputs vary according to user requirements.

Specifications

Volume: 0.112 m^3
Weight: 91 kg
Frequency: 2-18 GHz
Processing: 100 emitters
Accuracy: 8° DF, 5 MHz frequency, 1 μs PRI

Operational status

Fully developed. The system probably forms part of the RC-350 Guardian airborne reconnaissance system.

Contractor

Raytheon Systems Company.

UPDATED

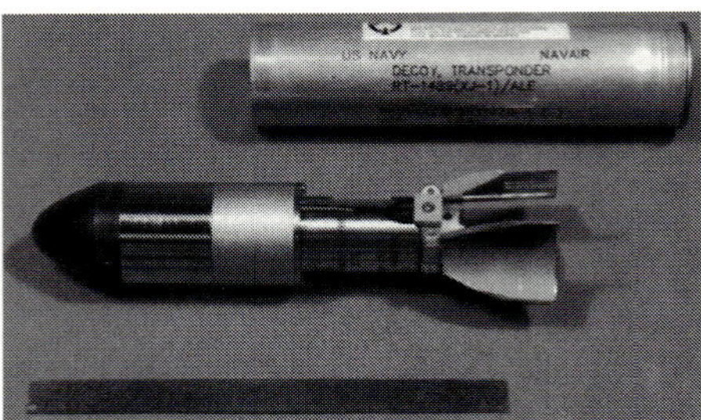

The GEN-X expendable decoy

GEN-X expendable decoy

Raytheon Systems Company has developed the generic expendable, GEN-X, as a follow-on to the AM-6988A POET system. GEN-X is an active transponder which emits a signal, after being ejected from its aircraft-mounted AN/ALR-39 or AN/ALE-47 dispenser, to decoy radar-guided missiles. GEN-X has a much greater frequency band than POET and can be programmed with its operating frequency before flight or in flight by information from the RWR.

Internally the system is made up of four primary electronic modules which make extensive use of monolithic, hybrid and silicon integrated circuitry. Individual rounds are fired from their sleeve by a gas squib and deploy three spring-loaded fins for flight stability.

An ignition primer activates an onboard battery and, once powered up, the round begins a sequential search of its predefined frequency bands until a high-priority threat signal is identified. The unit's transmitter is then activated, to provide a highly attractive radar target for the incoming missile.

Operational status

Advanced development was completed in 1986 and full-scale development in 1990. The first production contract, worth US$67.8 million, for 7,000 GEN-X decoys was awarded in May 1992. Production deliveries for US Navy and Marine Corps aircraft began in 1994.

Contractor

Raytheon Systems Company.

UPDATED

NKC-135A electronic warfare training aircraft

A Boeing KC-135A aircraft has been converted into an NKC-135A electronic warfare aircraft for use by the US Navy, NAVSEA and other research and development organisations.

The aircraft is fitted with a comprehensive suite of EW equipment which includes two ALT-40 radar jamming systems covering the A/B- to I/J-bands, an ALR-75 ESM system, an OE-320 DF system interfaced to the ALR-75, a Fleet Tactical Readiness Group Airborne EW System equipped with HP 9826 computers and a colour graphic display, two ALE-43 chaff systems and a USQ-113 comms jamming system covering VHF and UHF bands. Two pylons have been fitted, each capable of carrying up to 1,000 kg of external stores in pods. Additional power for the EW

The NKC-135A US Navy electronic warfare aircraft

systems is provided by the installation of four 90 kVA generators.

Operational status
In service with the US Navy.

Contractor
Raytheon Systems Company.

UPDATED

RC-135 Rivet Joint

Rivet Joint is the programme name given to the 14 US Air Force RC-135V and RC-135W SIGINT aircraft. Details of the fit are classified, but they are reported to have up to 17 operator positions, with one automated ELINT, two manual ELINT, and up to 13 COMINT stations.

A further two RC-135U aircraft (Combat Sent) are employed on technical SIGINT tasks.

All these aircraft are derived from the C-135, itself a derivative of the Boeing 707.

Operational status
All aircraft are in active service. They are reported to have played an important role in the 1990-91 Gulf War, and to be in use in Bosnia.

Contractor
Raytheon Systems Company.

UPDATED

Threat Deception System (TDS) electronic countermeasures system **1997**/0002229

Towed Angular Deception System (TADS)

The Towed Angular Deception System (TADS) comprises the Threat Deception System (TDS) and a fibre modified towed decoy system. Compatible with standard aircraft power, TADS uses complex digital signal processing hardware, software and firmware, combining onboard and offboard radio frequency (RF) countermeasure techniques to defeat emerging radar threats. Using the TDS in conjunction with the Towed Decoy System provides additional jamming techniques beyond standard towed decoy effects to increase threat miss distance and enhance platform survival. The system's MIL-STD-1553B interface bus provides hardware and software status messages and allows external user system control, if desired. The TDS controls the Towed Decoy System via a separate RS-422 interface bus to achieve the highest system throughput.

The Threat Deception System (TDS) is a low-cost jammer which uses the cueing of available radar warning receivers to automatically provide effective electronic countermeasures (ECM) and enhanced miss distance for the platform. The TDS combines off-the-shelf RF, and microwave and digital components including multiple C40 DSP microprocessors in a VME open architecture environment with embedded C code, offering a flexible and powerful solution to requirements for low-cost ECM.

The standard towed decoy system is designed to 'pull missiles' off protected platforms using stand-alone ECM techniques. The system consists of a mission programmable launch controller, decoy launcher and decoy magazine. The system can be programmed to suit a variety of platform requirements.

The fibre-optic modified option downlinks additional techniques generated on board the platform to the decoy for offboard transmission, enhancing system performance. The decoy system can be used alone or in conjunction with the onboard transmitter.

The use of the TDS in conjunction with the towed decoy system results in increased miss distance, especially against newer monopulse radars.

Contractor
Raytheon Systems Company.

UPDATED

AN/ASQ-191 system

The AN/ASQ-191 has been developed for use in airborne, fixed or mobile ground applications and covers the frequency range 225 to 400 MHz. It has two basic modes of operation: a communications mode for voice comms or deception, and a command, control and communications countermeasures mode for monitoring and jamming. It can operate against fixed-frequency or slow frequency-hopping radios. The AN/ASQ-191 incorporates a number of automatic scanning, monitoring and jamming functions which cover the complete frequency range, parts of it or selected frequencies. The jammer has an output of about 100 W and the jamming sequence includes a 6 ms look-through. A 400 W jamming variant is available.

Operational status
Fitted to the EC-24A US Navy EW training aircraft.

Contractor
Rockwell Collins.

UPDATED

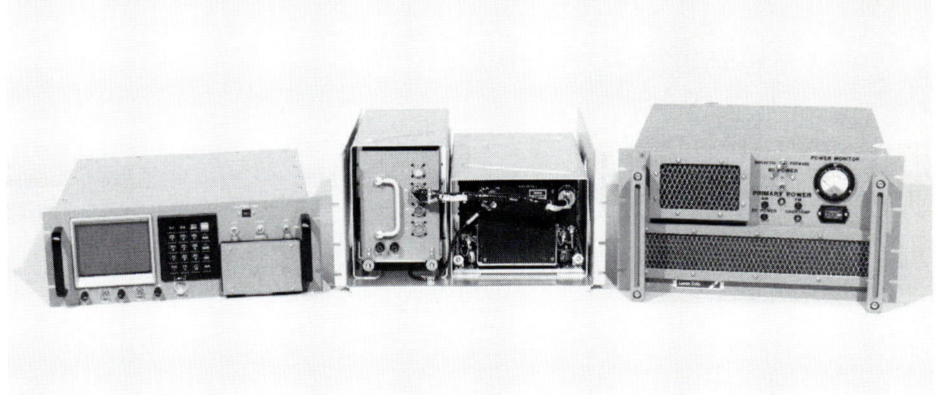
The Rockwell Collins UST-104 radio countermeasures set

UST radio countermeasures sets

The Rockwell Collins UST radio countermeasures equipment provides reliable HF, VHF and UHF ESM and ECM over the 2 to 500 MHz frequency range. The system is suitable for installation on airborne and ground-based platforms. Equipment operation and control is accomplished with the operator control unit which can be mounted remotely. Hard-copy capability is available through a front panel printer. All parameters, system commands and system operation are shown on either a high-resolution CRT in the operator display in a large aircraft or an electroluminescent display in a smaller tactical aircraft.

UST capabilities include scanning, display and jamming of target activity throughout the entire range, a selected range or selected frequencies.

The range of UST equipment includes:

UST-103, for large aircraft, covering the frequency range 225 to 400 MHz with a power output of 100 W.

UST-104A, for large aircraft and ships, covering the frequency range 20 to 500 MHz with a power output of 400 W.

UST-104B, for tactical aircraft, covering the frequency range 100 to 500 MHz with a power output of 400 W.

UST-105A, for large aircraft and ships, covering the frequency range 20 to 30 MHz with a power output of 400 W.

UST-105B, for large aircraft and ships, covering the frequency range 2 to 500 MHz with power outputs of 1.5 kW for VHF and UHF and 2.5 kW for HF.

UST-106, a lightweight airborne HF system covering the frequency range 2 to 30 MHz with a power output of 100 W. Frequency extension is available up to 2.5 GHz for jamming and monitoring functions.

Operational status
The UST-103, UST-104A, UST-104B and UST-105A are in service. The UST-105B and UST-106 are in development.

Contractor
Rockwell Collins.

UPDATED

AN/ALQ-167 ECCM/ECM jammer pod

The AN/ALQ-167 pod generates a wide selection of noise and deception jamming. The jamming system is modular and is based on building blocks; as a result, frequency range and performance characteristics can be tailored to a particular need. The system is designed for use on manned aircraft or aerial targets, or for laboratory applications for radar and missile system evaluation.

The AN/ALQ-167 is digitally controlled using a cockpit-mounted control box. The parameters and features that can be selected include operating bandwidth, jamming techniques, set on receiver, forward and aft radiation and power level. Up to 24 noise and deception modes are currently available.

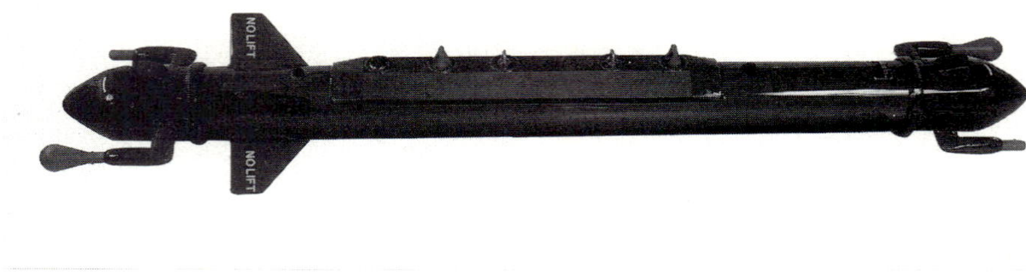

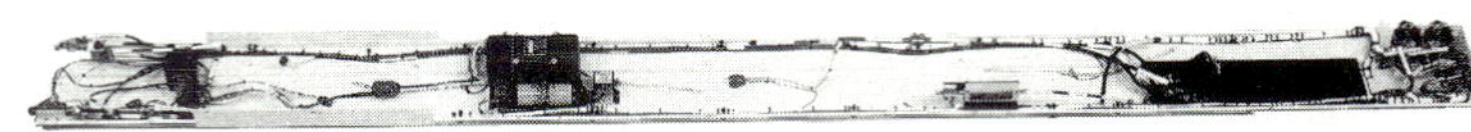

The AN/ALQ-167 ECCM/ECM training pod

Specifications

Dimensions: 2,300 × 3,700 mm
Weight: 140-180 kg
Output power: 100-200 W, ERP 1 kW min
Frequency: 0.85-18 GHz in customer-specified bands

Operational status

In service.

Contractors

Rodale Electronics Inc.

VERIFIED

Digital Airborne Radar Threat Simulator (DARTS)

The Digital Airborne Radar Threat Simulator (DARTS) is capable of providing radar signal emission simulation of airborne search radars, targeting radars, terrain-following radars and missile seeker radars. DARTS provides for in-flight selection of emitter frequency, pulse repetition frequency, pulsewidth, stable, stagger or jitter PRF mode selection and antenna scan simulation.

DARTS is a band reconfigurable magnetron-based emitter system employing an FET modulator. This enhancement permits the selection of transmitted pulsewidth by the operator over the specified 0.1 to 2 μs range with a 0.1 μs resolution. Remote magnetron frequency tuning from the control indicator is also provided.

DARTS is housed in a standard AN/ALQ-167 pod.

Specifications

Dimensions: 3,100 × 250 mm
Weight: 147 kg
Power supply: 115 V AC, 400 Hz, 3 phase
Output power: 12.5 MW typical
Frequency: H-, I- and J-bands

Operational status

In service.

Contractors

Rodale Electronics Inc.

VERIFIED

REWTS Turbo Crow

The Turbo Crow Responsive Electronic Warfare Training System (REWTS) consists of a high-power broadband radar jammer and a high-power radar simulator combined with a sophisticated monitoring and control system. Turbo Crow is typically installed in a commercial jet (Learjet, for example) or similar military aircraft and is a dual-role system - used in both training and tactical applications.

Jamming system

The jammer provides high-power broadband jamming in user-selected bands. Jamming techniques include noise, deception, cover pulses, false targets, Doppler and combinations thereof. The system is capable of simultaneous operation in multiple bands. The jammer is modular and can be configured to suit a particular aircraft's physical characteristics. Modularity also results in significant maintenance and repair advantages. The system's operating architecture and mechanical design readily accept changes and enhancements.

Radar simulator

The Radar Simulator provides the Turbo Crow with the capabilities of simulating a wide variety of airborne and ground-based radars including search and tracking, missile guidance and fire control. The installation provides for simultaneous operation of jammers and radar simulator. The programmable aspects of the Digital Airborne Radar Threat Simulator (DARTS) brings a new level of sophistication to EW training and test of shipboard/ground-based ESM systems.

Monitoring and control system

The Monitoring and Control System is the heart of the REWTS system providing the operator with full control over the engagement scenario - whether it be training or tactical. He or she knows where the target is; can monitor the target's response to the jamming - both ECCM and manoeuvres; is able to adjust the jamming to suit the situation; and can also record any aspect of the engagement.

Optionally the Turbo Crow can be enhanced for ELINT/ESM missions with capabilities to detect, analyse and classify radar data as well as passive ranging/positioning of radars.

The system can either be carried on board or in wing-mounted pods using a hybrid configuration with the high-power amplifiers and antennas housed in pods carried on wing hard points. This configuration results in significant savings in installation costs, better aerodynamics, less onboard high-power equipment to cool, and more flexibility in aircraft utilisation.

Specifications

RF range:
(jammer) 0.85-18 GHz (3 bands)
or 0.2-18 GHz (4 bands)
(DARTS) 7.8-8.5, 8.6-9.6, 12-13.2, 14-15.2 GHz
(ELINT) 0.4-18 GHz (optional 40 GHz)
Jammer output power: 200 or 400 W typical
DARTS output power: 100 kW (min)
ELINT sensitivity: −70 dBm
DF accuracy: 3-5°
Jamming modes: noise, smart noise, deception, cover pulses, multiple false targets, Doppler, co-ordinated and combined modes
DARTS modes:
(scan) circular, steady, sector centered, sector off centre
(PRF) stable, jitter, stagger
(adjustable) pulsewidth, PRF, delay, frequency
Monitoring and control: situational awareness, measurement, control, recording and databank system
Pods: ALQ-167 or ALQ-503

Contractor

Rodale Electronics Inc.

UPDATED

Rodale 200 self-defence pod

The Rodale 200 is a pod-mounted fully automatic self-defence system effective against pulse and CW radar transmissions. Velocity, range and angular deception modes are available and the equipment can communicate with the aircraft avionics via a standard databus. The system consists of two units: the pod-mounted jammer with built-in electronics and antennas and a cockpit-mounted control panel.

Specifications

Dimensions: 3,020 (length) × 254 mm (diameter)
Weight: 180 kg
Frequency: G- to J-band
Output power: 200 W min
Speed: Mach 1+

Operational status

Under development.

Contractors

Rodale Electronics Inc.

VERIFIED

Smart Crow ECCM/training jammer

The Smart Crow jammer is digitally controlled from a control box and a computer and is designed for an onboard EW station with jammers mounted inside the aircraft. Its modularity enables a distributed packaging configuration to fit particular limitations of the specific aircraft. Digital spot, barrage and swept noise bandwidth and velocity and range deception parameters are all digitally controlled. Frequency can be set digitally or by using the set on receiver. The system has built-in test capabilities that provide both fault isolation and calibration.

The system provides up to 25 jamming modes with all standard noise and deception techniques. A built-in set on receiver connected to antennas featuring integral amplification is used against any frequency-agile radar in the threat band. Different antenna spatial

The Rodale Smart Crow ECCM/ECM jammer is carried in the Learjet 35/36

coverages can be selected. The system is modular, allowing easy installation and maintenance.

Specifications

Power supply: 115 V AC, 400 Hz, 3 phase, 5 kVA 28 V DC, 1 A
Output power: 200 W typical, 1 kW ERP (min)
Frequency: 0.5-18 GHz in customer-specified bands

Operational status

In service but no longer in production being replaced by REWTS Turbo Crow.

Contractor

Rodale Electronics Inc.

UPDATED

AN/ALQ-126B Deception Electronic CounterMeasures (DECM) system

AN/ALQ-126 is a DECM system developed by the US Naval Air Systems Command for its Tactical Air Electronic Warfare Program. It provides wider frequency coverage than the preceding AN/ALQ-100 system and was initiated in response to new threats from anti-aircraft, surface-to-air missile and airborne interception systems. The initial production model (ALQ-126A) was widely used by the US Navy aboard aircraft such as the A-4, A-6, A-7 and F-4 aircraft.

An improved version, the ALQ-126B, offers increased frequency coverage, and incorporates a digital instantaneous frequency measurement receiver, improved deception techniques and more modem construction, packaging and cooling arrangements. ALQ-126B includes a distributed microprocessor control system to enable the system to be reprogrammed on the flight line and improve signal processing. Covering the 2 to 18 GHz band frequency band, the system is capable of generating a variety of jamming modulations including inverse conical scanning, range gate pull off, swept square wave and main lobe blanking.

ALQ-126B operates either autonomously or as part of an integrated weapon system comprising the AN/ALR-45F, APR-43 or ALR-67 radar warning equipment, the AN/ALQ-162 continuous wave radar jamming system, and the HARM, Sparrow, Phoenix and AIM-120 missiles. ALQ-126B is compatible with the A-4, A-6, A-7, F-4, F-14 and F/A-18 aircraft. ALQ-126B also forms part of the ALQ-164 ECM pod (see later entry).

Specifications

ALQ-126B
Dimensions: 411 × 270 × 609 mm
Weight: 86.3 kg

Operational status

The US Navy has awarded Sanders contracts to the value of nearly US$500 million for the production of the ALQ-126B with production and deliveries continuing. The equipment is also installed on Australian F/A-18, Kuwaiti F/A-18, Malaysian F/A-18, Canadian CF-18 and Spanish EF-18 aircraft. Deliveries to these countries are complete.

Contractor

Sanders, a Lockheed Martin Company.

UPDATED

AN/ALQ-144 and -144A InfraRed CounterMeasures (IRCM) system

ALQ-144 is an electrically powered IRCM set which provides medium-sized helicopters and small fixed-wing aircraft with protection against heat-seeking missiles. It is an omnidirectional system consisting of a cylindrical source surrounded by a modulation system to confuse the seeker of the incoming missile. A number of variants available; ALQ-144(V)1, ALQ-144(V)3, ALQ-144A and ALQ-144(VP) phase lock systems. The electrically heated graphite source has extremely long life and the complete set weighs less than 14 kg.

Specifications

Weight: 12.7 kg (transmitter); 0.5 kg (control unit)
Dimensions: 241 × 241 × 336 mm (transmitter)
Field of view: 360° azimuth

Operational status

The ALQ-144(V)1/3 and ALQ-144A(V)1/3 variants of the basic system are reported to be in service with the US Army, Air Force and Marine Corps. ALQ-144(VP) has been used on the SH-60, H-3 and SH-2 aircraft of the US Navy.

The Sanders AN/ALQ-144 infrared countermeasures set mounted on the upper fuselage of a helicopter

Sanders is understood to have received contracts worth nearly US$100 million covering upgrade kits to bring ALQ-144(V)1 and (V)3 systems up to ALQ-144A (V)1 and (V)3 standard. All current US Army ALQ-144 helicopter applications are noted as being of the A(V)1/3 standard. Currently, over 5,000 ALQ-144 systems are thought to be in service with the US military and a number of foreign military services.

Contractor

Sanders, a Lockheed Martin Company.

UPDATED

AN/ALQ-147A IR countermeasures system

The AN/ALQ-147A airborne IR countermeasures system was developed specifically for the OV/RV-1D aircraft and was fitted to the aft end of the fuel tank. The system uses a ceramic IR source that is heated by burning JP fuel mixed with ram air. The output of the ceramic IR source is then mechanically modulated to provide an output signal designed to confuse heat-seeking missiles. It is one of three IR systems referred to as the Hot Brick.

Operational status

Fitted to OV/RV-1D aircraft of the US Army.

Contractor

Sanders, a Lockheed Martin Company.

UPDATED

AN/ALQ-156 Missile Approach Warner (MAW)

The AN/ALQ-156 is an airborne internally mounted pulse Doppler missile warning system, originally developed for slow- and low-flying helicopters and fixed-wing aircraft, but later modified for use in other types of aircraft. The system is solid-state, incorporating four doughnut-shaped antennas, a transceiver and a control unit. The system recognises missile threats by comparing closure rates and other ballistic parameters, warns the pilot aurally and triggers the countermeasures command to dispense flares, chaff or other expendables. When coupled with radar and laser warning sensors, the ALQ-156 can initiate

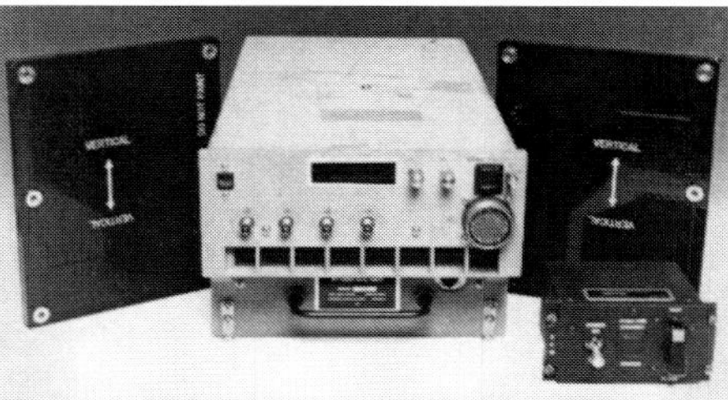

The AN/ALQ-156(V) detects the approach of IR missiles **1995**

automatic selection of the appropriate countermeasure.

A later model, the AN/ALQ-156A, was developed for the US Navy and was designed for use in tactical aircraft. It integrates an RWR and deceptive ECM data to allow smart control of decoys and flares and has a much longer detection range.

Operational status

The AN/ALQ-156 is fitted to a variety of aircraft, including the US Army CH-47, EH-1H, EH-60B, BV-10, REC-12D/H/K, RU-21A/B/C/D and RV-1D platforms. The US Air Force uses the system on its C-130 transports and the Royal Air Force on its Chinook helicopters. Several other nations use the system on C-130s. Nearly 1,000 systems have been delivered. An external pod-mounted configuration has been flight tested successfully.

Contractor

Sanders, a Lockheed Martin Company.

UPDATED

AN/ALQ-164 ECM pod

The AN/ALQ-164 is an airborne pod-mounted jammer for both pulse and CW modes. The pod incorporates unmodified ALQ-126B and ALQ-162 jammers. It can operate fully autonomously or integrated with an RWR and other avionics on board the aircraft. It is reprogrammable to provide for threat changes. The ALQ-164 is intended for fitting on the centreline pylon of the AV-8B aircraft and has applications for a wide variety of tactical and transport aircraft.

Specifications

Dimensions: 2,160 × 445 × 406 mm
Weight: 188 kg
Power supply: 115 V AC, 400 Hz, 3 kVA
Frequency: G- to J-bands

Operational status

In service with the US Marine Corps AV-8B. A contract for 74 pods was awarded in June 1988. The Spanish Navy has ordered the ALQ-164 for the EAV-8 Harrier. It is also reported that ALQ-164 has been ordered by the Italian Navy for its AV-8B+ aircraft.

Contractor

Sanders, a Lockheed Martin Company.

VERIFIED

AN/USQ-113 communications jammer

The AN/USQ-113 provides the EA-6B Prowler aircraft with its communications jamming capability. To meet the expanded role of the aircraft now that it has subsumed the US Air Force support jamming tasks (after retirement of the EF-111A) as well as those of the US Navy, an upgrade programme has been initiated. Three preproduction systems will be developed during a non-recurring Engineering and Manufacturing

Development (EMD) phase. The contract also calls for the upgrade of 30 existing USQ-113 systems, beginning in July 1997, scheduled for completion in May 1998.

The upgrade will consist of new receivers, power amplifiers and transmitters developed to extend the frequency range of the system and to replace the operator display. The new display has an improved Liquid Crystal Display (LCD) that runs Windows-based software. Additionally, a signal recognition device is being developed for insertion in a new system controller. With the device, aircrews will be able to analyse the signals they receive.

Contractor
Sanders, a Lockheed Martin Company.

UPDATED

ATIRCM/CMWS AN/ALQ-212(V)

The AN/ALQ-212(V) Advanced Threat IR CounterMeasures (ATIRCM) system, combined with the AN/AAR-57 Common Missile Warning System (CMWS), is in development by Sanders and Lockheed Martin Infrared Imaging Systems. The system combines the missile warning (CMWS), with an advanced jammer that integrates laser and xenon lamp technology, and an enhanced dispenser to counter current threats. The design also includes pre-planned product improvement plans for countering future threat systems. When the missile warning sensors detect a threat, the ATIRCM system will slew one or two jamming heads in the direction of the missile. A missile tracker on the jammer head will further refine the direction of the jamming transmission, ensuring that the main beam remains on the target. The system will also cue the release of flares from the dispenser.

AN/AAR-57(V), CMWS
The AN/AAR-57(V) CMWS is a Tri-Service programme to develop a passive missile warning system. The system consists of four or six missile warning sensors, depending on coverage required around the aircraft, and an Electronic Control Unit (ECU). The sensors detect the launch and motor burn of an approaching missile, provide warning to the crew and automatically dispense IR flares.

The missile warning sensor has been sized to fit on a wide variety of aircraft, from helicopters to tactical jet fighters, without modification. When installed, a crew member inserts a data card, called the User Data Module (UDM) into the ECU during flight preparation which initialises the system and provides threat

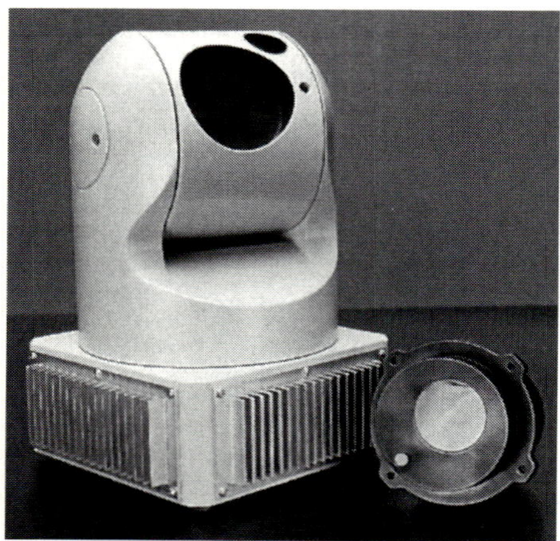

ATIRCM/CMWS counters advanced threats using the Common Missile Warning System to cue directable jammers and dispensers **1998**/0018254

Specifications

	AN/AAR-57(V)		AN/ALQ-212(V)	
	power (W)	weight (kg)	power (W)	weight (kg)
Missile warning sensor (with anti-icing)	166.1	8.8	110.8	5.86
Electronics control unit	381.0	8.5	347.0	8.50
Jam head control unit			240.7	6.05
Jam head (2)			1,456	27.05
Laser jam source (1)			410.0	8.87
Totals	547.1	17.3	2,564.5	56.33
	6 sensor suite		4 sensor, 2 head suite	

information to the system processor. When removed at the end of a mission, the overall CMWS system becomes unclassified.

Operational status
The US Army is the main customer for the AN/ALQ-212(V), and it has issued Sanders with a contract to integrate the system onto the AH-64D Longbow Apache. Other services and several foreign countries are considering application to their platforms, and it is expected that the system will be fitted to about 25 different types of US Army, Navy, Marine Corps and Air Force rotary-wing, transport and fighter aircraft.

The CMWS system will be deployed on US Navy and US Air Force tactical aircraft and on a number of foreign aircraft. The AN/AAR-57(V) CMWS has been selected for the HIDAS system on the UK WAH-64 Apache attack helicopter programme.

Contractors
Sanders, a Lockheed Martin Company.
Lockheed Martin IR Imaging Systems.

UPDATED

Improved Self-Defense System (ISDS) InfraRed CounterMeasures (IRCM)

Developed as an improvement on Sanders' Self-Defense System, ISDS is a fuselage- or pylon-mounted IR countermeasures system which is designed to provide protection against IR homing anti-aircraft missiles. The system comprises a new multiband IRCM transmitter, an electronic control unit and an operational control unit. System configuration varies according to aircraft type with a typical installation incorporating an electronic control unit/transmitter package for each engine. A single, cockpit-mounted operation controller can control up to four transmitters. ISDS has also been flight tested in a ram-air turbine-powered pod configuration. Within the system as a whole, the individual transmitters weigh 29.5 kg; the electronic control unit, 2.3 kg and the 146 × 125 × 57 mm operator control unit, 1.56 kg.

Operational status
ISDS is in production and in service on both turboprop and jet-powered aircraft. More than 150 ISDS systems are reported to have been delivered.

Contractor
Sanders, a Lockheed Martin Company.

UPDATED

Integrated Defensive Electronic CounterMeasures (IDECM) Radio Frequency CounterMeasures (RFCM) system

The AN/ALQ-214(V) IDECM RFCM system is a US Navy lead, joint US Navy/US Air Force programme which is designed to provide a range of aircraft types with next-generation protection from RF threats. The system's primary application is the F/A-18E/F carrier-borne, multirole combat aircraft where it is integrated with a radar warning receiver, the AN/AAR-57 Common Missile Warning System (CMWS) and the Advanced Strategic/Tactical Expendable (ASTE) infrared decoy flare to create a total defensive aids package. System design objectives include improved situational awareness through the use of the CMWS and the fusion real-time information; improved IR threat countering through the use of the kinetic ASTE round and 'smart' dispensing routines, and a counter RF capability which can defeat radars which incorporate features such as monopulse angle tracking, signal coherency, and manual tracking.

The IDECM RFCM system comprises a Techniques Generator (TG), an Independent WideBand Repeater (IWBR) and the ALE-55 Fibre Optic Towed Decoys (FOTD). Both the TG and the IWBR interface with the FOTDs through the decoy dispensing system. Electronic CounterMeasures (ECM) techniques are synthesised in the TG and transduced to optical frequencies for transmission to the FOTD via its tow line. Within the FOTD, the optical data is converted back to RF format for amplification and transmission. The IWBR provides an alternate source of ECM by passing the threat signal (as received on the host aircraft) to the FOTD.

The IDECM RFCM system is mandated for the F/A-18E/F and USAF B-1B and is under consideration for application to F-15 aircraft. Additionally, the IDECM RFCM system is understood to be under study for possible application to a wider range of types including the US Air Force's AC/MC-130, F-16 and U-2 aircraft. As installed in the F/A-18E/F, the IDECM RFCM system makes maximum use of the existing AN/ALE-50 and AN/ALQ-165 Group A provisioning with the described TG and IWBR functions being implemented in four weapon replaceable assemblies; a receiver, modulator, processor and signal conditioning unit. The onboard transmitters are an option for the F/A-18E/F application.

Operational status
In November 1995, the contractor team of Sanders and ITT Avionics announced that Sanders (as prime) had been awarded a US$26.8 million, five year duration, IDECM RFCM system Engineering and Manufacturing Development contract. The contractor team is required to deliver and integrate five TGs, 150 FOTDs and 100 FOTD mass models for use in development flight testing. Under a contract option, work is also being done on a common, high-powered Towed Decoy System design, development and test effort for the US Air Force F-15 and B-1B aircraft, together with a B-1B architecture study to determine how the IDECM RFCM system can support the B-1B's Defensive System Upgrade Program (DSUP). An IDECM RFCM system operational assessment is planned for autumn 1998 followed by low-rate initial production during 1999.

Contractors
Sanders, a Lockheed Martin Company.
ITT Defense and Electronics Avionics.

UPDATED

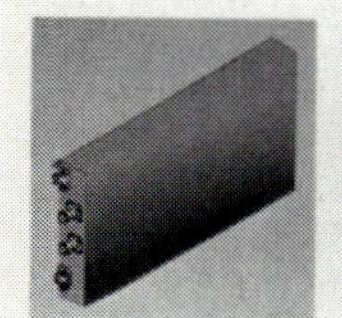

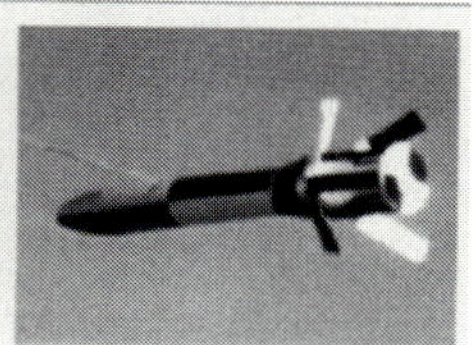

IDECM RFCM LRUs, showing, from left: the signal conditioning assembly; the receiver/processor/techniques generator; the onboard transmitters (optional); the fibre optic towed decoy **1997**/0001274

Joint SIGINT Avionics Family (JSAF) Low-Band SubSystem (LBSS)

In February 1997 Sanders, a Lockheed Martin Company, was selected by the US DoD Joint Airborne Signals Intelligence (SIGINT) Program Office for development and demonstration of the Joint SIGINT Avionics Family (JSAF) Low-Band SubSystem (LBSS).

The LBSS is a key element in the planned upgrade of US Army, Navy and Air Force airborne SIGINT capabilities.

Sanders will serve as prime contractor and system integrator and will also develop signal processing hardware and software for the system. Major subcontractors include: Radix Technologies, Inc of Mountain View, California; Applied Signal Technologies (APSG) of Sunnyvale, California; and TRW System Integration Group, also of Sunnyvale. Radix will provide radio frequency (RF) and digital signal processing subsystems; APSG will develop special signal processing subsystems; and TRW will be responsible for high-speed networking and computing subsystems.

The JSAF low-band subsystem is a platform-independent, modular, reconfigurable suite of hardware and software that can address multiple mission scenarios aboard a variety of aircraft. It will significantly enhance the ability of reconnaissance platforms to detect and locate modern enemy communications systems and provide real-time intelligence on enemy intentions and capabilities to the warfighter.

Operational status

Under the development contract, the consortium will design, develop, integrate and flight test four demonstration units. Delivery of the first development unit will be completed by June 1999. The programme is sponsored by the Pentagon's Defense Airborne Reconnaissance Office and will be administered by the US Air Force's Aeronautical Systems Center, Wright-Patterson AFB, Ohio.

Initially, JSAF LBSS will be deployed on US Air Force RC-135 Rivet Joint aircraft and other special Air Force platforms as well as the US Army's RC-7 (Airborne Reconnaissance Low) and the US Navy's EP-3 aircraft. JSAF LBSS will also be capable of deployment on unmanned aerial vehicles (UAVs) in the future.

Contractor

Sanders, a Lockheed Martin Company.

UPDATED

TACJAM-A ESM/ECM Equipment

TACJAM-A will provide the US Army with its next generation of communications intercept, DF and countermeasures equipment. It is designed to provide the communications ESM and ECM subsystem building blocks of the US Army's IEW Common Sensor System and will be integrated into fixed-wing aircraft and helicopters as well as ground vehicles. The programme is currently under contract to Sanders and AEL.

The most important features of the TACJAM-A system are improved speed and accuracy in producing ESM information for the tactical commander and the advanced ECM capabilities this affords. The ESM subsystem uses modular building blocks and can be tailored to a number of platforms. The ECM subsystem is also modular and can be configured as a single- or dual-channel system.

TACJAM-A includes many automated functions previously performed by the operator or maintainer. The extensive BIT, down to card level, means that field repair is quick and can be accomplished without special test equipment. If a system component fails in operation, the system automatically reconfigures itself to a degraded mode and alerts the operator to the system capabilities that remain.

Contractors

Sanders, a Lockheed Martin Company.
Tracor Aerospace Inc.

VERIFIED

AN/APD-13 Advanced Quick Look ESM Subsystem

The AN/APD-13 is the ELINT subsystem for the US Army's Guardrail/Common Sensor RC-12 aircraft. This system, with the addition of Advanced Quick Look (AQL) and communications intercept improvements, has become the Guardrail/Common Sensor system. AQL provides up-to-date non-communications intercept capabilities to the intelligence processing facility on the ground.

The system provides non-communication threat signal intercept direction-finding and identification, 360° field of view all-weather operations, real-time sensor system control via datalink and electronic order of battle assessment.

Operational status

In service in the US Army RC-12 aircraft.

Contractor

Systems & Electronics Inc.

UPDATED

The AN/APD-13 Advance Quick Look ESM subsystem is carried on the US Army RC-12 **1995**

TC-510/AC Series HF/VHF/UHF direction-finders

The TC-510/AC Series of airborne HF/VHF/UHF direction-finder systems is designed for installation on small and large military or commercial aircraft. The system utilises a single small DF antennna, flush-mounted or externally mounted on the fuselage. Single or simultaneous independent multiple direction-of-arrival information is provided to the operator for received signals in the 1.5 to 1,300 MHz frequency range.

Variants of the system provide acquisition and direction-finder information on signals up to 2.6 GHz, including INMARSAT and new communication frequency allocations.

Operational status

In production and in service with various military and civilian agencies. TC-510/AC systems have been installed on both fixed-wing aircraft and helicopters.

Contractor

Tech-Comm Inc.

VERIFIED

Advanced Digital Dispenser System

The Advanced Digital Dispenser System (ADDS) is a microprocessor, computer-controlled, threat adaptive countermeasures dispensing system providing aircraft self-protection against current and future ground and airborne threats. ADDS utilises user programmable dispensing programmes, adapted for the specific threat and optimum engagement parameters, to control and dispense chaff, including dual chaff cartridges, flares, RF and future types of offboard expendable decoys.

With simple software modifications, aircraft configurations can be tailored for high-performance aircraft, helicopters and transport and maritime patrol aircraft. ADDS incorporates improvements derived from combat experience and hundreds of installations on a wide variety of aircraft including the A-4, F-4, F-5, F-15, F-16 and others.

ADDS features include automatic, semi-automatic and manual operational modes; payload inventory; discrete inputs from aircraft MIL-STD-1553B and RS-422 databusses to interface with onboard RWR, MLD and TWS warning sensors; solid-state sequencer switch providing multiple firing pulses with low burst intervals and automatic and crew initiated BIT.

Operational status

In production for many air forces. Installed as standard equipment on a number of aircraft types.

Contractor

Tracor Aerospace Inc.

VERIFIED

ALE-47(V) Threat Adaptive Countermeasures Dispenser System (TACDS)

The ALE-47(V) TACDS provides an integrated, threat-adaptive, reprogrammable, computer-controlled capability for dispensing expendable decoys (chaff, flares and others) to enhance aircraft survivability in sophisticated threat environments. The threat-adaptive features of the TACDS are provided by the customer-defined Table-Look-Up Mission Data File (MDF). The TACDS Operational Flight Program (OFP) derives all (fixed and adaptive) dispensing programmes based entirely upon the customer-developed MDF. The system interfaces with other onboard EW and avionics systems through two MIL-STD-1553 databusses, two RS-422 busses or discrete lines.

The following operating modes are available: manual, semi-automatic and fully automatic. Mode selection is made by the crew on a Cockpit Display Unit (CDU). An emergency jettison mode is available to dispense all expendables on command. The manual mode requires pilot preselection and automatic mode selects the appropriate expendables as well as the optimum dispensing rate. In semi-automatic mode the system requires the pilot to initiate an automatically optimised routine.

Other features include: Built-In Test (BIT) to the Shop

Repairable Unit (SRU); a Shop Replaceable Assembly (SRA); hardware/software growth provisions; reprogrammable manual dispense routines; ALE-40 form-fit compatibility and ALE-39 Sequencer form-fit compatibility. Support equipment available for the TACDS includes the Tracor Mission Planning System/ Threat Matrix Generator (TRACOMPS/TMG) that permits efficient development of MDFs customised to the customer's threat environments; Memory Loader Verifier (MLV) for loading MDFs into the system; and a Maintenance Test Station (MTS) to provide depot level repair capability.

Operational status

Currently in service for fighter, cargo and helicopter aircraft for several international users. Contracts for the US Air Force continue until 2002.

Contractor

Tracor Aerospace Inc.

UPDATED

AN/ALE-36 dispenser pod

The AN/ALE-36 (QRC-490) is a modified AN/ ALE-38/41 tactical fighter countermeasures dispenser pod adapted to carry RR-136 and RR-137 chaff units. The QRC/TBC-600 version is a modified AN/ALE-38 pod carrying 600 RR-170 chaff units for saturation or corridor chaff applications. Payloads are carried in 10 dispenser blocks of 60 payloads each and include an integral dispenser. Each dispenser can be loaded with chaff units or IR flares. The fully loaded weight is 309 kg.

Operational status

No longer in production but still in service.

Contractor

Tracor Aerospace Inc.

VERIFIED

AN/ALE-38/41 dispenser system

The AN/ALE-38 (US Air Force) and AN/ALE-41 (US Navy) equipments are high-capacity bulk chaff dispensers. They employ dispensing techniques which provide continuous dipole dispersal and instantaneous bloom for laying chaff corridors. They can also be used for aircraft self-protection and can be turned on automatically by the radar warning receiver. Precut dipoles are sandwiched between two wraps of mylar film with six 22.7 kg rolls of this composite being carried in each pod.

Operational status

ALE-38 and -41 systems are understood to be in US Air Force and US Navy service respectively.

Contractor

Tracor Aerospace Inc.

NEW ENTRY

AN/ALE-39 series ECM dispensers

The AN/ALE-39 is a first-generation system for the protection of tactical aircraft from ground- and air-launched missiles and radar-directed anti-aircraft guns. It represents the culmination of over 10 years of dispenser system refinement by the US Navy and is designed with flexibility of response to meet present and future needs in a changing threat environment.

The system is capable of accommodating up to three types of expendable payloads loaded in any combination of multiples of 11. Chaff, IR flares or expendable jammers can all be dispensed manually or automatically in accordance with preset programmes. The dispensing function can be initiated by the pilot, or crew member in the case of the F-14. The system is also capable of accepting dispense commands from aircraft warning receivers.

The programme flexibility, payload loading versatility, operational modes and wide selection of payload dispensing options and sequences, including simultaneous multiple flare ejection, random chaff ejection and rapid dump of flares, combine to make the AN/ALE-39 one of the better first-generation analogue-programmable countermeasures-dispensing systems in the US inventory.

ALE-39B is an upgrade of ALE-39 and is designed to protect tactical aircraft from missile and radar-directed anti-aircraft gun threats. It is capable of accommodating up to three types of expendable payloads (chaff, InfraRed (IR) flares and expendable jammers), loaded in any combination of multiples of 10. All three types of payload can be dispensed manually (single payload) or automatically in accordance with preset programmes. The dispensing function can be initiated by the pilot (or the weapon systems operator in the case of the F-14). The system is also capable of accepting dispense commands from aircraft warning receivers.

Operational status

ALE-39 series dispensers are reported to have been installed on A-6, AV-8B, EA-6B, ES-3A, F-14, F/A-18, S-3B and SH-2F/G aircraft of the US Navy, US Marine Corps and other air forces/naval air arms. The system is also compatible with POET and GEN-X active offboard expendables. More than 1,000 systems have been delivered since the initial production release in October 1973. ALE-39 is no longer in production where it has been replaced by AN/ALE-47.

Contractor

Tracor Aerospace Inc.

UPDATED

AN/ALE-40(V) dispenser system

The AN/ALE-40 offboard expendable countermeasures dispenser system was the precursor to the AN/ALE-47. This intervalometer and stepper-switch-based system was developed in the 1960s and has been widely installed. AN/ALE-40 is being upgraded or replaced with the form-fit-compatible AN/ ALE-47 LRUs.

Operational status

Obsolete. Widely deployed in F-16 export aircraft. Upgrade or replacement required to support operational readiness.

Contractor

Tracor Aerospace Inc.

UPDATED

AN/ALE-45 countermeasures dispenser

The AN/ALE-45 is a microprocessor-controlled countermeasures dispenser system developed for the F-15 aircraft. It responds automatically to threat notification from the radar warning receiver, tail warning set or pilot.

The system consists of a programmer assembly and four dispensing switch assemblies. Eight magazines containing expendable stores are attached to the switch assemblies. The programmer assembly houses the central processor and the input/output circuitry. An operational flight program which contains the dispenser programs, and other preprogrammed data defining the functional operation of the system, is changed easily through a front panel connector. The software is designed to provide automatic priority selection of dispensing program based on computations of threat sources from the radar warning receiver or the pilot and aircraft flight data. The dispensing program parameters are selectable, including payload class, burst count and interval and payload count and interval.

The four dispensing switch assemblies are identical and interchangeable. Their primary function is the firing of payloads.

The AN/ALE-45 has the capability to dispense RR-170 and RR-180 chaff cartridges, MJU-7 and MJU-10 flare cartridges and other payload types of similar form factor.

Operational status

In service on F-15.

Contractor

Tracor Aerospace Inc.

VERIFIED

AN/ALE-47 countermeasures dispenser system

The AN/ALE-47 is an offboard expendable countermeasures dispenser system that is intended to replace the AN/ALE-40 and AN/ALE-39 on currently equipped aircraft and to be fully integrated into numerous new aircraft.

The system interfaces with other onboard EW and avionics systems through two MIL-STD-1553 databusses, two RS-422 busses or discrete lines. The AN/ALE-47 uses threat data received over these interfaces to assess the threat situation and determine the appropriate countermeasures response. Dispense routines tailored to the immediate aircraft and threat environment may be automatically, semi-automatically or manually dispensed, depending on the mode selected by the user.

Threat response determination is embedded in the system operational software. The software consists of the Operational Flight Program (OFP) and the Mission Data File (MDF). The OFP processes input threat and aircraft position data to compute the optimum offboard response. The MDF contains the aircraft and mission specific parameters used by the OFP and is fully flight line reprogrammable.

Other features include modular automated test equipment; a Consolidated Automatic Support System (CASS) compatibility; Built-In Test (BIT) to the Shop Repairable Unit (SRU); a Shop Replaceable Assembly (SRA); hardware/software growth provisions; reprogrammable manual dispense routines; ALE-40 form-fit compatibility and ALE-39 Sequencer form-fit compatibility.

Operational status

In service on the F-16 and F/A-18, C-5, C-17, C-130, C-141, HH-60, MH-47, MH-60 and the P-3 and slated for service on the E-8C, V-22, VH-60 and the VH-3. A

The AN/ALE-40 chaff/flare dispenser for the US Air Force Lockheed F-16 fighter

combined ALE-47/AAR-47 system, called the Airlifter Defense System, (ADS), was due to be fitted to the entire C-17 fleet by the end of 1997.

Contracts totalling more than US$100 million have been awarded to Tracor for several production lots in a programme with an estimated value of about US$150 million. Contract awards continue.

Contractor
Tracor Aerospace Inc.

UPDATED

M-130 Dispenser

The M-130 aircraft general purpose dispenser is a lightweight countermeasures dispenser system developed for the US Army, using AN/ALE-40 technology. The system is designed for employment on tactical helicopters or fixed-wing aircraft, such as the AH-1, CH-47, OH-58, UH-1, RU-21 and OV-1D, to provide self-protection from radar-directed weapons by dispensing chaff and from infrared homing weapons by dispensing flares.

The M-130 system is of modular design to allow flexibility of operational configuration. The modules consist of a cockpit control unit, electronics module, one or more dispensers, cabling and aircraft adaptors. The cabling and adaptors are unique to the aircraft type, whereas the other modules are common among all aircraft. The M-130 uses the M-1 countermeasures chaff and the M-206 countermeasures flare as payloads. Each payload module accommodates 30 chaff units or flares of a single type. The system configuration may be either a single dispenser of 30 capacity or a double dispenser of 60 capacity. The system weight, with 60 chaff units or flares, is 21.8 kg.

The M-130 will function with input from the infrared threat warning device directly to the electronic module, obviating the need for pilot action.

The chaff units and flares have a nominal 25 by 25 mm square configuration and are 210 mm in length.

The Tracor Inc M-130 (with the cover removed) on a Royal Air Force Chinook helicopter (Paul Jackson)

The form factor is identical to the RR-170A/AL chaff used in the AN/ALE-40 and is functionally interchangeable.

Operational status
In service on a number of aircraft types.

Contractor
Tracor Aerospace Inc.

VERIFIED

M-130A Threat Adaptive Countermeasures Dispenser System (TACDS)

M-130A TACDS is a lightweight form and fit replacement for the M-130 helicopter dispensing system using solid-state microprocessor technology to perform threat adaptive, automatic programming for dispensing chaff, flare and advanced expendables as they become available. The system interfaces with other onboard EW and avionics systems such as radar warning receivers, missile warning systems, laser warning systems and the aircraft avionics, compares these inputs against the threat response files and selects the optimum dispense program.

M-130A TACDS provides three in-flight selectable modes of operation: manual whereby four preprogrammed dispense programs can be activated from the cockpit control unit; semi-automatic where the system prompts the aircrew that a threat exists and selects the optimum expendable/sequence for aircrew activation; automatic where the threat analysis (based on sensor and avionics inputs), dispense programs and activation performed by TACDS are fully automatic without aircrew action.

M-130A TACDS can be easily retrofitted into any aircraft equipped with the M-130 system or as new installations. M-130A TACDS has been designed to communicate over several communications busses such as the MIL-STD-1553A/B and RS-422. Built-In Test (BIT) is included with both continuous automatic and aircrew-initiated BIT providing fault isolation to the system replaceable unit.

Capabilities designed into M-130A TACDS provide offboard countermeasures to counter existing IR and RF threats with growth provisions for future threats. Flexible load/mission mixes are available with five payload types per magazine, 15 magazine mixes, multiple simultaneous firing pulses and extremely short burst intervals provided by the solid-state sequencer switch. Automatic misfire detection and correction eliminates the M-130's unreliable flare detector to ensure that decoys are dispensed at the critical moment. Threat response files and mission-specific dispense programs can be developed by the user and uploaded into the system on the flight line.

Operational status
M-130A TACDS has been developed and is in production for international users.

Contractor
Tracor Aerospace Inc.

VERIFIED

AN/APR-44(V) radar warning receiver

The AN/APR-44(V) has been adopted by the US Army and Navy.

The system consists of three units. The Model AS-3266 antenna is a monopole unit which provides omnidirectional, vertically polarised coverage within a 50° elevation sector and is designed to be mounted on a flat horizontal surface. The second unit, the receiver, is a radio frequency chopped, crystal video type with a bandpass filter followed by a radio frequency switch/detector module, a linear video amplifier and processing circuitry. The video amplifier output is routed to a comparator before being compared with a radio frequency chopping signal in the processing logic. This procedure detects any continuous wave signal and triggers a 3 kHz audio output, a logic output, an alert lamp drive and a 2 Hz logic blink signal. The third unit is a control panel which includes the alert light indicator. Three versions of the system are available. The 44(V)1 incorporates the R-2097 receiver for coverage of the H/I-bands, the 44(V)2 uses the R-2098 J-band receiver and the 44(V)3 covers H-, I- and J-bands.

Specifications
Dimensions: 165 × 94 × 229 mm
Weight: 0.85 kg
Frequency: selectable
Bandwidth: selectable
Sensitivity:
(min) −45 dBm
Max RF input level:
(integral limiter) 1 W CW

Operational status
In service in different versions with various US Army and Marine Corps helicopters.

Contractor
Tracor Aerospace Electronics Systems.

VERIFIED

CrossJam 2000 training and testing jammers

CrossJam 2000 is a family of modern jamming systems designed to be versatile in meeting the requirements for ECCM training or radar testing. It is designed to simulate numerous self-protection and escort jamming systems and can be configured from a simple noise jammer to a sophisticated deception system capable of providing ECCM training for the most advanced fighters or AEW and GCI radars.

CrossJam 2000 is available in bands covering 0.5 to 18 GHz. It can be configured with various high-power amplifiers ranging from 100 to 500 W and is available in an AN/ALQ-167 pod or custom mounted internally in a business jet, such as Lear, Falcon or Challenger, or a fighter such as the F-5.

The full system is capable of generating all significant ECM techniques in use. Available in the CrossJam 2000 are noise, range, set-on and Doppler techniques.

Operational status
CrossJam 2000 is used to train NATO fighter pilots. The Canadian Department of National Defence has selected the system as an integral component of the Electronic Support and Training Challenger aircraft.

Contractor
Tracor Aerospace Electronics Systems.

VERIFIED

Threat Emitter Simulator System (TESS)

The Threat Emitter Simulator System (TESS) is a programmable threat simulator that is tunable in flight. It can be used to simulate threat aircraft radar and air-to-surface emissions and is used for training weapon systems operators.

TESS can be housed in either a standard AN/ALQ-167 pod shell for external carriage or EIA standard 19 in racks for internal carriage. The major assemblies are the transmitter assembly, magnetrons, antennas, associated RF components, interface assembly and low-voltage power supply. Both configurations have identical assemblies. For the pod configuration, the assemblies are mounted on an electronics tray assembly which slides into the pod.

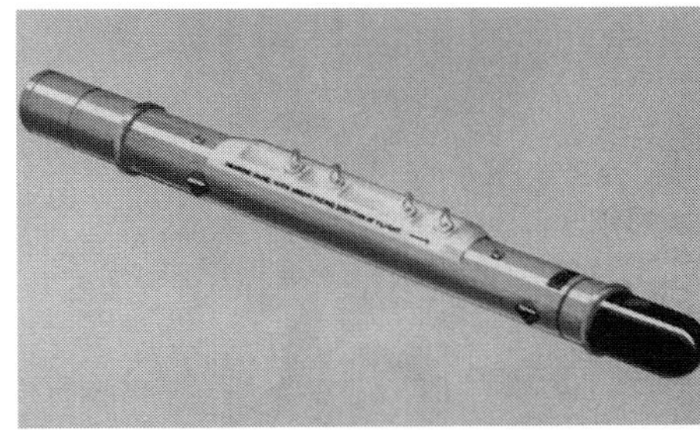

The Threat Emitter Simulator System pod **1995**

CrossJam 2000 is an integral component in the Canadian Forces Challenger EW training aircraft **1995**

In flight, the operator can choose from 15 programmed scenarios which vary frequency and scan patterns. Closed-loop remote magnetron tuning is aided by a frequency measurement device which prevents excessive drift.

TESS has been designed for flexibility, while requiring a minimum of component replacement between bands. For the lower bands of 7.8 to 8.5 GHz and 8.5 to 9.6 GHz, only a change of magnetron is required. For the upper bands of 12 to 13.2 GHz and 14 to 15.5 GHz, the magnetron and associated RF components are replaced. Mechanical packaging of TESS allows rapid replacement of components for band change or maintenance functions.

The TESS pod has been designed and tested to meet military environmental requirements and is qualified for carriage on commercial and military aircraft at speeds of up to 750 KCAS or M1.3.

Specifications
Dimensions:
(pod) 3,284 length × 254 mm diameter
(internal) 482.6 × 488.9 × 609.6 mm
Weight:
(pod)170.1 kg
(internal) 79.38 kg
Power supply: 115 V AC, 400 Hz, 3 phase
Frequency:
7.8-8.5 GHz, 8.5-9.6 GHz
12-13.2 GHz, 14-15.2 GHz
Environmental: MIL-E-5400 Class 1A

Operational status
Reported in service on the CT-133 aircraft, with the Canadian Forces.

Contractor
Tracor Aerospace Electronics Systems.

VERIFIED

ULQ-23 FutureJam countermeasures

The ULQ-23 electronic countermeasures set, known as FutureJam, was developed by the US Navy and has been manufactured by Tracor Aerospace Electronic Systems. Future applications are being investigated, including airborne rack-mounted systems and increased technique capabilities.

FutureJam provides programmable ECM at a lower power level for jamming a radar system. Coupled with an external amplifier and antenna, the system can be used as a complete airborne jammer. An enhanced jamming system can be configured with the addition of a real-time controller and an external set on receiver.

FutureJam is capable of various noise jamming techniques, including barrage, narrowband, spot and swept. Various deception techniques are also available, either alone or in combinations, including AM, swept AM, RGPO, random Doppler, multiple frequency repeater, velocity gate stealing and narrowband repeater noise. With the addition of a real-time computer, multiple technique scenarios can be generated.

Contractor
Tracor Aerospace Electronics Systems.

VERIFIED

Star receiver system

The Special Threat Analysis and Recognition (STAR) receiver system is intended for augmentation of ESM or radar warning receivers. The main part of the system is a dual-path, self-calibrating superhet receiver using a fast tuning phase-locked synthesiser and encoders for front end sorting of radar parameters. An encoder chip digitises the input signal and controls phase interferometry measurements for determination of angle of arrival of the signal. Features include a programmable screening pulse environment and adaptable threshold. Also included in the system is a data processor based on an Intel 80186 machine which executes Pulse Recognition Interval De-interleaving (PRIDE) algorithms and formats contact reports for transmission to the host system.

The processing and control architecture of the STAR system allow it to accept high-level commands or to operate in a slave mode in the structure of a complex ESM system. Demonstrations of the system have shown an angular accuracy of 1°. It is claimed that pulses as narrow as 100 ns can be detected and between 30 and 100 targets can be tracked simultaneously.

In operation, the STAR system executes a search and tracking strategy defined by a host computer or system operator. The system scans the frequency rapidly, detects signals of interest, analyses them and calculates the angle of arrival, pulse amplitude and pulsewidth on a monopulse basis.

Specifications
Volume: 0.0141 m³
Weight: <8.2 kg
Frequency: 0.5-18 GHz
Pulsewidth detection: <100 ns without DF, 600 ns with DF
Target tracks: 30-100 simultaneously
Dynamic range: >50 dB
Accuracy: 1° RMS

Operational status
Flight tests have been carried out.

Contractor
TRW Military Electronics & Avionics Division.

VERIFIED

Target Locating Radar

The Target Locating Radar (TLR) is a self-contained radar detection, location, classification and tracking system. It can be used in aircraft or missiles or UAVs and is totally reconfigurable externally. TLR is available with a variety of wideband antennas. It searches for targets of interest and detected pulses of interest are processed to determine signal characteristics and angle of arrival. These data are reported over the bus interface for multiple targets. TLR can provide data for use by other sensors, the crew or as a direct input to a guidance system or can be used as a mid-course or terminal sensor. Other features of the TLR include light weight, small size and high accuracy.

Specifications
Volume: 0.0142 m³
Weight: 8.2 kg
Frequency: 2-18 GHz
Pulsewidth detection: 100 ns without DF, 600 ns with DF
Target tracks: 30 simultaneously
Accuracy: 1° RMS

Contractor
TRW Military Electronics & Avionics Division.

VERIFIED

AN/ALQ-151 airborne jammer

The AN/ALQ-151 is the major EW sensor in the US Army's Bell EH-1H and Sikorsky EH-60A ECM helicopters designated Quick Fix I and Quick Fix II respectively. Design commenced in 1975 and the system entered production in 1984. The ALQ-151 is used to deny the enemy the use of his radios for battlefield communications in helicopters.

Operational status
Quick Fix I in service on US Army Bell EH-1H helicopters.

Contractor
TRW Systems Integration Group.

VERIFIED

ES4000 SIGINT system

The ES4000 is a complete airborne reconnaissance and surveillance system that consists of a mobile ground control and processing centre and modular surveillance and electronic warfare payloads. The system is intended primarily for light aircraft. It has 360° azimuth field of view and provides broad area surveillance with a communications signal intercept payload. Frequency range is 20 to 500 MHz, which can be extended to 1,200 MHz. Sensors are remotely controlled from a small, mobile shelter on the ground and SIGINT information is transmitted to the ground station in real time.

In addition to the signal intercept package, the aircraft can be fitted with a video and infrared camera.

Specifications
Frequency: 20-500 MHz (1,200 MHz optional)
Frequency resolution: 100 Hz
Signal types: AM, FM, CW, USB, LSB
Accuracy: 2° RMS
Bearing files: 2,000

Operational status
The ES4000 is installed in the Dornier Speed Canard SCO1-B optionally piloted vehicle.

Contractor
TRW Systems Integration Group.

VERIFIED

ES5000 COMINT/ELINT system

The ES5000 is a combined communication and electronic intelligence system which can be installed in most types of aircraft. It performs signal interception, direction-finding and analysis, data collection and location of communication emitters from 2 to 1,200 MHz and non-communication emitters from 0.02 to 18 GHz. Information is transmitted by microwave link to a ground station. The airborne SIGINT equipment can be controlled either by operators on board the aircraft or remotely from the ground analysis centre.

On a typical mission, one or two airborne collection centres communicate by microwave link with the ground centre. One ES5000 can perform signal intercept, direction-finding and emitter location, producing emitter bearing data. A two aircraft configuration can perform instantaneous emitter location on short burst transmissions. Both systems pass data to the ground analysis centre.

Specifications
COMINT
Frequency: 2-1,200 MHz
Instantaneous coverage: 500 MHz
DF rate: 12/s coarse, 4/s fine
Field of view:
120° each side (primary)
±30° front and rear (secondary)
Elevation: −10 to +2°
Accuracy:
<2° RMS (primary)
<3° RMS (secondary)

ELINT
Frequency: 0.02-18 GHz
Coverage: 360° azimuth
Accuracy: 1-3° RMS
Measurements: PRI/PRF, pulsewidth, amplitude, scan time, DF, time of arrival, intrapulse
PRI range: 2-10,000 µs
Radar types: pulsed, 16 level stagger, CW, interval pulse, complex
Emitter library: 10,000 modes
Active emitters: 1,024

Microwave link
Frequency: E/F-band
Range: 350 km at 10,000 ft altitude

Operational status
In production and in service.

Contractor
TRW Systems Integration Group.

VERIFIED

Guardrail system

Guardrail is an airborne SIGINT system that intercepts, locates and classifies target systems and transmits data to ground processors to provide real-time intelligence information. The Guardrail suite consists of a transportable ground-based system and RC-12 aircraft carrying remotely controlled mission equipment.

The Guardrail system is primarily for COMINT, but an ELINT capability has been added to later aircraft. The concept behind Guardrail is that data from multiple receivers are microwave downlinked to a ground-based Integrated Processing Facility (IPF).

The airborne relay facilities part of the Guardrail system is the airborne communications intelligence payload installed in electronic mission RC-12K, N and P aircraft. They contain intelligence intercept, signal classification and direction-finding equipment, and include provisions and interfaces for other signal intelligence subsystems, mission datalinks and aircraft navigation and communications systems. Collected data is datalinked directly to the IPF, or via an RC-12Q mothership.

Through a series of demonstrations, known as Precision SIGINT Targeting System (PSTS), Guardrail has proven its ability to work with national resources to provide targeting data for cross-programs operations. Guardrail has participated in the US Office of Naval Research (ONR)-sponsored Advanced Concept Technology Demonstration (ACTD), and is claimed to be already compliant with the US Joint Avionics SIGINT Architecture (JASA), with Guardrail/Common Sensor System 2 being the first operational implementation.

For the 21st century, Guardrail/Common Sensor will be part of the US Army's Aerial Common Sensor system, which will also use the Direct Air-to-Satellite Relay (DASR) operational with Guardrail.

Operational status
Guardrail has been in operational service with the US Army for a number of years on RU-21H and RC-12D aircraft. The original Guardrail II was updated to IIA and was later updated to Guardrail IV. Further updating provided the Guardrail V under the nomenclature AN/USD-9(V)2. The US Army has developed a modification to Guardrail V that includes the Advanced Quick Look ELINT system produced by Electronics & Space Corporation and the Communications High-Accuracy Airborne Location System (CHAALS) provided by Lockheed Martin. This integrated system, known as GuardRail/Common Sensor (GRCS), was deployed in the RC-12K.

Latest versions RC-12K, N, P and Q are fitted with upgraded CHAALS equipment and have access to the Defense Satellite Communications System (DSCS).

Contractor
TRW Systems Integration Group.

UPDATED

WTSS Wideband Tactical Surveillance System

The TRW WTSS is designed to be fitted in the Gulfstream IV-SP aircraft by means of an optional cargo door fit. The WTSS is based on COTS surveillance equipment to minimise risk.

Specifications
RF coverage: 20-2,000 MHz; 2 MHz and 40 GHz extension options
Signal search rate: 15-45 GHz/s
DF accuracy: <1° RMS
SIGINT sensors: fast signal search; D/F-emitter location; pool of fast set-on receivers; digital and audio playback; co-channel mitigation; ELINT option
Communications: air to ground datalink - G.703 compatible - 2 Mbps
Operator stations: 4-6 positions; Sun workstations; C and Ada software

Operational status
Installation proven.

Contractor
TRW Systems Integration Group.

NEW ENTRY

TRW wideband tactical surveillance system
1998/0018253

Advanced laser electro-optic/infrared countermeasures

The US Air Force Wright Laboratory's Avionics Directorate is sponsoring the research and development of advanced technology for laser-based infrared and electro-optic IR/EO countermeasures to fill operational requirements to defeat IR missiles and EO/IR tracking systems. This technology is being structured to meet both near-term and far-term demands driven by the continuing advancement and proliferation of EO/IR missiles and tracking systems.

Near-term programme
The near-term aim is to demonstrate and evolve the technology necessary for a threat adaptable IR jamming suite capable of protecting large and high-value aircraft against the proliferated threat of surface-launched IR missiles. This IRCM suite would include provision for missile warning, fine tracking, missile identification or classification and laser jamming. The technology and performance requirement derived from this Advanced Technology Demonstration (ATD) programme would feed an offshoot development that could upgrade the specifications of current system developments, for example, the Army Advanced Tactical Infrared Countermeasures (ATIRCM) or USSOCOM DIRCM to meet US Air Force requirements. Initial Operational Capability for such a system is planned for just after the year 2000 requiring the programme to develop the technology in Fiscal Year 1999 (FY99). Functional testing at White Sands Missile Range in FY97/98 will include numerous live fire tests. One major aim is to satisfy user needs for data on technology needed to meet overall requirements at minimal cost. The programme will address the capability of basic open-loop jammers as well as more complex threat adaptive closed-loop techniques, signature suppression and missile warning system requirements. Wright Laboratory is currently under contract with Lockheed Martin Defense Systems at Akron, Ohio to develop the demonstration IRCM testbed to be used in field testing. The development plans for the IRCM testbed will require interface for multiple candidate missile warners and several candidate fine tracking/CM laser systems. Current plans call for using mature developmental systems such as the ATIRCM and the SOCOM/UK DIRCM as the baseline for upgrading to a higher level of capability. The programme will use IRCM lasers being developed by ARPA for future IRCM systems but is also seeking other laser source technologies, for example, the diode pumped approaches from Massachusetts Institute of Technology.

Long-term programme
In the longer term, the emerging focal plane array imaging seekers and laser-guided threats are advancing countermeasure technology. Also driving technology is the need to improve high-performance combat aircraft survivability against a threat base that is turning towards EO/IR to overcome the limitation caused by application of stealth radio frequency (RF) signature suppression techniques. The longer term programme is researching multiple approaches to defeat robust threats while meeting the integration and operating requirements of high-performance combat aircraft. The philosophy of the programme is to pursue an integrated modular avionics capability in which the multiple functionalities offered by solid-state optical systems, high-power multiband lasers and advanced tracking and imaging recognition algorithms would provide a cost-effective compact system. Such a system may be able to provide the warning, tracking and identification functions against missiles, aircraft and ground EO/IR threats, and apply multiband countermeasures techniques against these threats. The countermeasures techniques may include both jamming and damaging options offered through application of directed energy technologies.

IR Decoys
Wright Laboratory is continuing to research expendable decoys which may be the only IRCM option for some aircraft in the future. As the US Air Force's Advanced Strategic and Tactical Infrared Expendables (ASTE) programme/effort has entered into Engineering and Manufacturing Development and will be fielding the next-generation expendables, Wright Laboratory is funding technology programmes to support product improvements needed to meet third-generation IR threats that are included in the intelligence order of battle estimates. The Special Material Decoy (SMD) project will search for answers to two unique decoy requirements while the more mature mixed expendables programme will move toward injecting advanced decoy technology into operational use.

Contractor
US Air Force Materiel Command, Wright Laboratory.

VERIFIED

AN/ALQ-504 airborne VHF/UHF communication signals intercept/DF/jamming/deception system

The AN/ALQ-504 (formerly ZS-1920) integrated airborne intercept, Direction-Finding (DF), jamming and deception system is designed to be installed in a range of aircraft, from small twin turboprop or jet aircraft to large aircraft such as the C-130 and Boeing 707 and 757. Its role is primarily as an EW training system. The system performs search and detection of communication signals in the VHF and UHF bands between 30 and 400 MHz. Frequency extensions down to 20 MHz and up to 1,000 MHz are available. The system measures DF lines of bearing and computes the location of these signals. It also allows the operator to jam selected signals or to transmit false (spoofing) messages for deception or battle training purposes. Signal jamming can also be linked to DF bearing, providing jamming of signals within a selected azimuth sector.

All functions of the AN/ALQ-504 system are controlled by a single operator. The AN/ALQ-504 system comprises a single ruggedised PC-compatible computer workstation, a precision high-speed multichannel DF receiver/exciter, an RF processor and a high-power amplifier and receive and transmit antennas.

The AN/ALQ-504 system operates in a number of passive and passive/active modes. Signal search and detection are available in both continuous scan (band-search) and Signal Of Interest (SOI) scan/jam modes. In continuous scan mode, operator-defined frequency bands are continuously scanned with all signal activity displayed on the screen. Selected signals can be transferred by the operator to any of the SOI lists. A hop plot mode is also provided for operations against frequency agile signals.

In monitor/jam mode, the operator is able to DF at a single frequency and locate signals of interest, and can listen to and record signals using the built-in Digital Audio Recording and Playback (DARP) module. The DARP is also used in monitor/jam mode to develop 'sound bullets' built-up from individual words or fragments of recorded transmission as false messages for rebroadcasting.

Contractor

Zeta.

NEW ENTRY

ZS-1910 ECM system

The ZS-1910 is a communications intercept, direction-finding and jamming system. It covers the 20 MHz to 1 GHz frequency range and is suitable for installation on small and large fixed-wing aircraft and helicopters.

The system is software based, using a ruggedised PC-based microcomputer, dual-channel receiver, fast switching synthesised exciter and solid-state power amplifier. It uses an advanced correlative interferometer DF technique combined with software-based signal processing to provide a high probability of intercept with a low false alarm rate during passive operations.

In the active mode of operation the ZS-1910 provides for the disruption and denial of enemy communication links on several different frequencies simultaneously. The system effective radiated power is sufficient to permit aircraft operation at considerable standoff distances.

System operation is automatic under the menu-driven tasking and control of the single system operator. It allows the operator to search and establish a qualified target list according to the frequency and direction of arrival, and to direct the ECM asset in accordance with the target list on a prioritised basis.

Operational status

In production.

Contractor

Zeta.

UPDATED

DATA PROCESSING, MANAGEMENT AND DISPLAYS

Data handling
Data recording
Navigation and nav/attack
Flight management and control
Cockpit displays, instruments and indicators
Head-up displays, helmet-mounted displays and weapon aiming sights
Stores management

Smiths Industries Aerospace Displays and Mission Computer is standard equipment on the next-generation Hawk aircraft for the Royal Australian Air Force lead-in fighter programme
1998/0018925

DATA HANDLING

CANADA

CMA-2060 Data Loader System (DLS)

The CMA-2060 DLS provides a general purpose solution to data storage and transfer problems in military avionics applications. The system consists of a removable 1 to 2 Mbyte non-volatile Data Transfer Cartridge (DTC) and a Data Transfer Unit (DTU). The DLS provides real-time data exchanges with an RS-422 databus for functions such as recording single or multiple data and/or maintenance information for subsequent downloading to ground support equipment, and interfacing with ground support equipment for the downloading of flight data and/or the uploading of flight configuration, subsystem configuration and mission profile data.

The CMA-2060 data loader system ***1995***

Specifications

Dimensions :
(DTU) 44 × 146 × 168 mm
(DTC) 83 × 18 × 131 mm
Weight:
(DTU) 1.36 kg
(DTC) 0.45 kg
Power supply: 28 V DC, 1 A (nominal)

Operational status

In production and service. Selected by the Canadian DND and installed on the CH-146 Griffon and CC-130.

Contractor

Canadian Marconi Company.

UPDATED

CMA-2074CDG Colour Display Generator

The CMA-2074CDG is a digital computer-based system that combines any two of the incoming eight subsystem video Red, Green, Blue (RGB) channels with colour graphics and text to produce three independent RGB video output channels. The CDG comprises three independent graphics engines, controlled by an application processor, that are able to generate complex, high-performance displays for aircraft. Typical applications include: tactical display generation; digital map presentation; radar and FLIR symbology overlay; video management; mission processing; and cockpit display upgrades.

Specifications

Dimensions: 5 MCU (160 × 196 × 321 mm)
Weight: 7.5 kg
Power supply: MIL-STD-704B, 70 W (max), 115 V AC, 400 Hz
Interfaces: Up to 8 i/p video channels, monochrome/colour; programmable video, RGB RS-170 or RGB CCIR 472-3; up to 3 independent o/p video channels-RGB; double buffered o/ps to drive up to 6 colour displays; up to 4 cursor control devices - RS-232; master sync o/p based on external or internal video sync generator; MIL-STD-1553B notice 2 — Bus A and B; RT address and parity; ground support interface - RS-422

Operational status

In production. Selected by GKN Westland Helicopters for the UK MoD EH 101 Merlin.

Contractor

Canadian Marconi Company.

UPDATED

CMA-2074 Data Interface Unit (DIU)

The CMA-2074 DIU is a small, rugged, highly adaptable system for interfacing discrete, digital and analogue sensors to MIL-STD-1553B and ARINC 429 avionic buses for a variety of fixed-wing aircraft and helicopters. The DIU provides Ada software-controlled acquisition, processing and multiplexing of avionics data from multiple interfaces in either stand-alone or dual-redundant applications and provides 1553B bus control, remote terminal and non-1553B avionic subsystem control. The unit can be software-configured to specific customer requirements and expanded to include functions such as maintenance recording and exceedance detection.

A typical application alllows for interfacing, filtering, signal processing and data acquisition of up to 256 inputs. These are digitised, scaled, validated and multiplexed on to one or more serial buses. This data is then available to all avionics systems connected to the same serial bus. Data received on the same MIL-STD-1553B bus or another can be reformatted as required by the DIU. Commands received via the avionics buses can be used to control DIU analogue and digital outputs. Up to 256 inputs and outputs can be accommodated, exclusive of avionic buses.

The DIU is contained in a ½ ATR short (4 MCU) enclosure; other enclosure options are available. Ada operational software allows the unit to be adapted for use in many aircraft types and configurations. The DIU provides complete closed-loop BIT and performance monitoring of all system components. Programme software can be modified by data upload via either the RS-422A GSE interface or the 1553B interface. By multiplexing many signals on to a MIL-STD-1553B or ARINC 429 bus, aircraft wiring and installation costs, as well as weight, are greatly reduced.

Specifications

Dimensions: 127 × 185 × 305 mm
Weight: 4.3 kg (max)
Power supply: 16-36 V DC, 30 W (max) (MIL-STD-704A)
Interfaces:
(digital) MIL-STD-1553B, 3 RS-422A/232, 4 ARINC 429 inputs/2 outputs per card
(analogue) 2-wire differential/28 channels per card, 4 synchro and 4 LVDT channnels per card, 12 channel synchros per card, 6 strain gauge channels per card. Frequency, pulse, ratiometric, variable reluctance, piezoelectric, capacitance and thermocouple available
(discrete) single-wire input/56 channels per card, single-wire outputs
Environmental: MIL-E-5400 Class 1A, MIL-STD-461 Parts1/2 EMI/EMC
Reliability: >5,000 h MTBF

Operational status

In production and service. Variants of the CMA-2074 are currently in service and installed on the US Coast Guard HC-130/KC-130 and US Air Force MH-53J Pave Low III and KC-135 Speckled Trout aircraft.

Contractor

Canadian Marconi Company.

UPDATED

AN/AYK-23(V) airborne military computer

The AN/AYK-23(V) is a 32-bit general purpose computer capable of being configured to suit many real-time applications. Its militarised construction, to MIL-E-5400 Class 1A, makes it suitable for severe airborne environments.

Using Motorola's 68030 technology and militarised VME architecture, the AN/AYK-23(V) is a functional superset, and a form, fit replacement for the AN/AYK-10(V) computer used on the S-3B. Additional acoustics system WRAs captured by the AN/AYK-23(V) include one PDP, two drums, and one drum power supply. The original CMS-2 and Ultra-32 Tactical Mission Programme was translated and recompiled to the 68030, while the original acoustics software was rewritten in Ada and complied to 68030.

The AN/AYK-23(V) consists of a set of multilayer printed circuit cards including central processors, graphics processors, monitor drive cards, global memory cards, maintenance interface cards, input video cards, and 6 input/output interface card types. Up to 72 of these circuit cards can be configured in the system. The AN/AYK-23(V) provides user-friendly and efficient maintenance diagnostics and debug capabilities through a basic operator panel and remote terminal interface. All WRA's are front-removable. With its open architecture, the AN/AYK-23(V) is capable of significant future performance and input/output growth.

The AN/AYK-23(V) is designed to control time-critical real-time systems such as command and control, fire control, electronic warfare and countermeasures, communications, navigation, acoustics and radar.

Specifications

Dimensions: 906 × 1,513 × 463 mm
(module) 152 × 229 mm
Weight:
(baseline) 125 kg
(max configuration) 193 kg
Power supply: 115 V AC, 3 phase, 400 Hz
(baseline) 635 W
(max configuration) 3,000 W

Contractor

Lockheed Martin, Canada.

VERIFIED

AN/UYK-507(V) high-performance computer

The AN/UYK-507(V) computer is a 16/32-bit general purpose computer capable of being configured to suit many real-time applications. Its militarised construction, to MIL-E-16400, makes it suitable for severe environments.

Using ASIC technology and high-performance architecture, the AN/UYK-507(V) is functionally compatible with the AN/UYK-502(V), AN/UYK-505(V) and AN/UYK-44 standard military computers.

The Lockheed Martin, Canada AN/UYK-507(V) high-performance computer

While emulating these earlier processors, the AN/UYK-507(V) computer provides additional features for object code optimisation and system performance improvements. Expanded memory reach, included as part of the processor design, allows absolute addressing to four billion memory locations. Five Mips performance is achieved with 90 per cent cache hits out of the 64 kbytes of cache memory. Higher performance rates are achievable with increased cache hit rates.

The AN/UYK-507(V) consists of a set of multilayer printed circuit cards including a central processor, input/output processor, maintenance processor, semiconductor memory, 13 input/output interface types and a power supply. The AN/UYK-505(V) provides user-friendly and efficient maintenance diagnostics and debug capabilities through a basic operator panel with a remote terminal interface.

The AN/UYK-507(V) features an open architecture and can be easily incorporated into existing applications since it continues to use the AN/UYK-502(V) cabinet, power distribution system and input/output interface modules. This allows present systems to achieve a significant performance improvement by a simple substitution of the present AN/UYK-502(V) processor/memory nine-card set with the new AN/UYK-507(V) processor/memory three-card set.

Either as a stand-alone unit or embedded in the system, the AN/UYK-507(V) is designed to control time-critical real-time systems such as command and control, fire control, electronic warfare, communications, navigation, logistics and naval tactical data systems. Other uses are radar and signal processing, message handling, information management and air traffic control.

Operational status

Dimensions: 444.5 × 482.6 × 469.9 mm
(module) 171 × 228 mm
Weight: 52.6 kg (max)
Power supply: 115 V AC, single phase, 575 W

Contractor

Lockheed Martin, Canada.

VERIFIED

CHINA, PEOPLE'S REPUBLIC

MD-90 Central Air Data Computer (CADC)

The MD-90 CADC is a product of codevelopment by Chengdu Aero-Instrument Corporation and Honeywell Inc of the USA and features a 32-bit Motorola MC68332 microprocessor.

Operational status

First flight test was in 1993 and FAA certification in October 1994.

Contractor

Chengdu Aero-Instrument Corporation (CAIC).

VERIFIED

MD-90 CADC and MD-90 aircraft **1997**/0001275

SS/SC-1, -1A, -1B, -1G, -2, -4, -5, -10, -11 air data computers

This series of air data computers are all configured for use in fighter aircraft of the J-7 type and trainers such as the K-8 Karakorum. They take data from pressure, temperature and attitude sensors, process it using INTEL 8086 series processors and pass the data to the navigation, weapons management and flight control systems, and to other Chengdu flight displays.

Operational status

In production and in service.

Contractor

Chengdu Aero-Instrument Corporation (CAIC).

VERIFIED

Y7-200B Air Data System (ADS)

The Y7-200B ADS is designed for the Y7-200B aircraft, and it features the following components: 8903 air data computer; 8904 altimeter; 8905 airspeed indicator; 8908 vertical speed indicator; 89008 altitude preselector/altitude alerter; total air temperature/static temperature/true airspeed indicator.

The Y7-200B ADS provides outputs to the FMS, EFIS and navigation systems, and aural warnings/ground proximity warnings to the pilot.

Operational status

In production and in service.

Contractor

Chengdu Aero-Instrument Corporation (CAIC).

VERIFIED

SS/SC-5 air data computer and K-8 Karakorum trainer **1997**/0002361

Y7-200B aircraft and Air Data System **1997**/0002345

9416 air data computer

The 9416 air data computer is used on the K-8 Karakorum trainer. It features an INTEL 8086 processor and combines with CAIC's 9414 preselector and 9415 altimeter into an air data system.

Operational status

In production and in service.

Contractor

Chengdu Aero-Instrument Corporation (CAIC).

VERIFIED

FRANCE

2084 mission computer

Like the M182-84 (see further item), the 2084 belongs to a series of general purpose digital computers designed to operate with high reliability under extremely severe environmental conditions. The arithmetic unit uses microprocessors assembled in hybrid modules, offering an extremely high-performance/volume ratio.

Within a ½ ATR case, the 2084 computer essentially comprises: an arithmetic unit; a working memory of up to 512 k 18-bit words; a fully comprehensive input/output system (including the simultaneous management of two multiplexed buses); and a power supply.

Very comprehensive basic software, such as LTR and Ada compilers, real-time monitor, assemblers, micro-assemblers and programmes to assist in the development and debugging of application programmes and microprogrammes, has been developed for these computers.

Operational status

No longer in production. The 2084 computer is in service in all versions of the Dassault Mirage 2000 aircraft including the 2000DA, 2000N and 2000-5 (two computers per aircraft).

Contractor

Dassault Electronique.

VERIFIED

2084-XR mission computer

The 2084-XR computer belongs to the 84 series of digital computers especially designed to meet the requirements of combat aircraft such as the Mirage 2000, and to operate with a high degree of reliability in harsh environmental conditions.

The 2084-XR is an enhanced version of the 2084 computer. Using enhanced technology, such as VHSIC, ASICs and CMOS battery back-up RAMs, it provides 1.2 Mips throughput.

The 2084-XR is compatible with other computers in this range and has the same software available.

Specifications

Dimensions: ½ ATR
Addressing capacity: 4 M words
Computing speed: 1.2 Mips

Operational status

In production for the Mirage 2000D and 2000-5.

Contractor

Dassault Electronique.

VERIFIED

CICS 68040 computers

The CICS 68040 computers are intended for helicopter and aircraft upgrade applications. These computers interface with the aircraft systems via two redundant 1553 buses, ARINC 429 links and other specific links.

Their modular architecture is based on a VME backplane bus and can include, other than the bus input/output and acquisition functions, several real-time 68040 central processing units programmable in Ada or in C.

Contractor

Dassault Electronique.

VERIFIED

M182-84 mission computer

Using hybrid micro-electronic technologies, the M182-84 computer has been designed for combat aircraft. It features light weight and small volume, modular construction for easy extension and maintenance, a standard digital input/output unit and a specific input/output unit for adaptation to all weapon system requirements.

Operational status

No longer in production. The M182-84 is in service in French and export versions of the Mirage F1 aircraft.

Contractor

Dassault Electronique.

VERIFIED

M182-XR mission computer

The M182-XR is an enhanced version of the M182-84 computer (see previous item). Using advanced technology such as VHSIC and CMOS battery back-up high-speed RAMs, it provides 1 Mips throughput, 512 k words of memory and an extensive set of input/output capabilities. The M182-XR represents a three-fold increase in processing power over the basic M182-84, while retaining compatibility with its predecessor to enable previously developed application software to be used.

The M182-XR is small in size and weight, with improved MTBF and power consumption, modularity and integrated maintenance. It offers significant life cycle cost savings and is ideally suited for equipping or retrofitting weapon delivery and navigation systems.

Operational status

Production started in mid-1991 for retrofitting French Air Force Mirage F1-CT and F1-CR aircraft. Production is now complete.

Contractor

Dassault Electronique.

VERIFIED

Military real-time LANS

The DIGIBUS GAM-T-101 became a French triservice standard multiplex databus in 1982. It is used in many military systems (Mirage F1, Mirage 2000, ATLANTIQUE 2, missiles, submarines, navy ships and land-based applications).

STANAG 3910 is a dual-speed version of the MIL-STD-1553B which is widely used in numerous onboard military systems. This new databus has been adopted for the EFA and the Rafale aircraft, it can also be used for the modernisation of any MIL-STD-1553-based system.

Operational status

Over 35,000 remote terminal bus interface units and 3,500 bus controllers have been produced.

Contractor

Dassault Electronique.

VERIFIED

Rafale Mission Computer

Based on a VME backplane, the Rafale mission computer is composed of a dual-SPARC processor CPU board, one 8 Mbyte memory board and three NATO STANAG 3910 bus controller boards. It delivers 24 Mips computing throughput while achieving outstanding gateway performance in terms of data transfer rate and low latency time. A high level of integration and real-time capabilities has been achieved by the use of dedicated ASICs.

Dassault Electronique has developed a full basic software package with real-time executive, Ada compiler and libraries. A complete software development workshop and validation tools are also available.

Specifications

Dimensions: ARINC 600 - 4 MCU

Operational status

In production for the Rafale aircraft.

Contractor

Dassault Electronique.

VERIFIED

15M/125X general purpose processor

The 15M/125X has been selected as the main computer in the Dassault Aviation Atlantique 2 aircraft to process the data received by the various sensors; it is the most recent addition to the Sextant Avionique range of general purpose units and employs many internal operating procedures common to previous units. It is a 32-bit machine with up to one Mops capability. A 1 Mbyte memory can be addressed and versions of the processor are typically installed in a 1 ATR case. LTR real-time software support is provided and numerous applications for high-performance military equipment are anticipated.

Operational status

No longer in production. In service in French Navy Atlantique aircraft.

Contractor

Sextant Avionique.

VERIFIED

Type 130 Air Data Computer (ADC)

The modular design Type 130 air data computer is intended for use on new-generation aircraft and for the retrofit of fighter aircraft, and has already been chosen for various Mirage retrofits.

The ADC comprises two pressure sensors for measuring static and total pressures; a CPU board including arithmetic unit; memories and sensor measurement circuits; an input/output board for all analogue and digital multiplexed bus interfacing; and a power supply board. A large number of built-in tests is available and permanent monitoring of all functional circuits takes place during flight, the results being expressed as maintenance words stored in a protected memory or dispatched on the digital lines.

Pressure sensors are the Sextant Avionique vibrating quartz blade Type 51 pressure sensors.

Specifications
Dimensions: 2 MCU
Weight: 4.6 kg
Power supply: 115 V AC, 400 Hz, single phase, 30 VA
Reliability: 10,000 h MTBF

Operational status
In production for retrofit of Mirage aircraft.

Contractor
Sextant Avionique.

VERIFIED

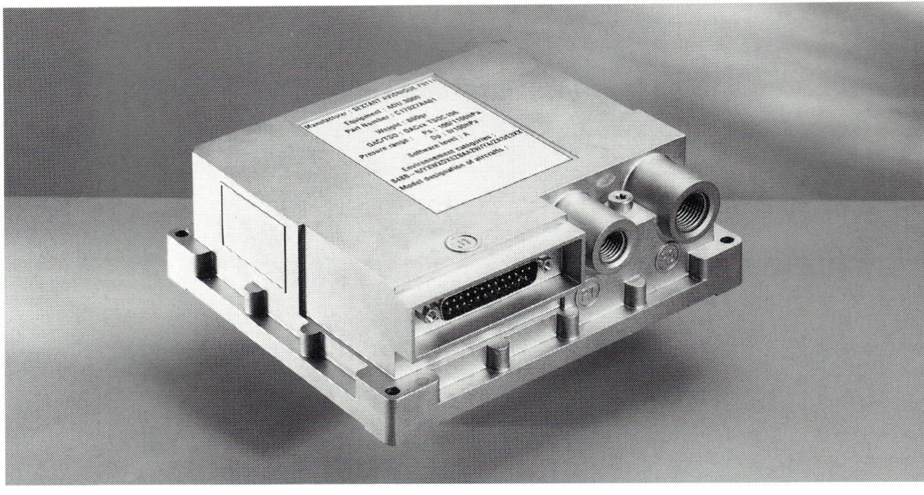

Type 3000 ADU **1997**/0002321

Type 300 Air Data Unit (ADU)

The Type 300 ADU is designed to measure static and differential (or pitot) pressures and impact temperature. From these data, the ADU computes the true airspeed and provides air data parameters through an ARINC 429 bus. Analogue outputs are available as an option.

The unit is ideally suited to provide the primary reference in helicopters and for retrofits in military aircraft.

Specifications
Weight: 1.2 kg

Operational status
In production and in service for Super Puma, Tiger and Rooivalk helicopters.

Contractor
Sextant Avionique.

VERIFIED

Type 300X Air Data Units (ADUs)

The Type 300X family of ADUs is designed to measure static and pitot pressures, static and total temperatures, and to compute and send over to the digital databus all the air data parameters.

These air data units have been miniaturised and are ideally suited as primary or secondary air data references on civil or military aircraft, helicopters and missiles.

Specifications
Dimensions: 200 × 130 × 50 mm
Weight: 0.8 kg
Power supply: 28 V DC, 5 W
Reliability:
(military) 30,000 h MTBF
(civil) 80,000 h MTBF

Operational status
ADU 3000, helicopter version, selected on Dauphin N4 and EC-135. ADU 3008 TSO C106 certified, civil aircraft version, selected on DASH 8-400.

Contractor
Sextant Avionique.

UPDATED

AC 68 multipurpose disk drive unit/airborne data loader

The multipurpose disk drive unit/airborne data loader enables the uploading and downloading of onboard computers and allows remote loading of the computer programme database or programme initialisation for ACMS, FMC, ACARS and TCAS, memory computer downloading and recording data reports, QAR function and other functions, from aircraft computers. The AC 68 data loader is also compatible with other data loaders in accordance with the ARINC 603 standard.

The storage medium used for transfer is a 3.5 in (89 mm) double-side high-density magnetic diskette. The formatted capacity is 1.44 Mbytes.

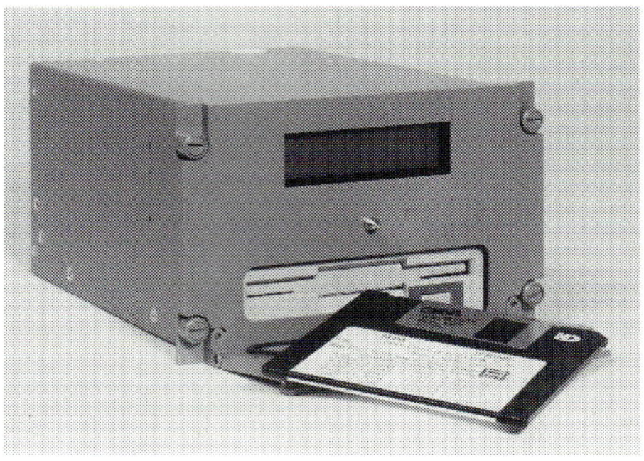

The SFIM Industries AC 68 multipurpose disk drive unit/airborne data loader

After insertion of the floppy disk, data transfer starts immediately. A configuration file contained on the floppy disk defines the transfer characteristics: input/output links used; ARINC 429 transmission rates (Hi or Lo); and uploading/downloading actions. Exchanges are driven by the data loader of the computer, and checks of the data transfer are automatic.

The front panel provides a one line, 16 character LCD.

Operational status
In production for Airbus and Boeing aircraft.

Contractor
SFIM Industries.

UPDATED

Airborne data loader ARINC 615/603

The SFIM Industries airborne data loader is compliant with ARINC 615/603 and designed to meet the requirements of all commercial aircraft.

Specifications
(cockpit installation)
Dimensions: 146 × 95.2 × 171.4 mm
Weight: 3 kg
Power: 115 V, 400 Hz, <10 VA

Contractor
SFIM Industries.

NEW ENTRY

SFIM Industries airborne data loader ARINC 615/603 (left) and ground support portable data loader (right) **1998**/0015267

DFDAU-ACMS Digital Flight Data Acquisition Unit — Aircraft Condition Monitoring System

SFIM Industries produces a range of DFDAU-ACMS units tailored to the requirements of the following

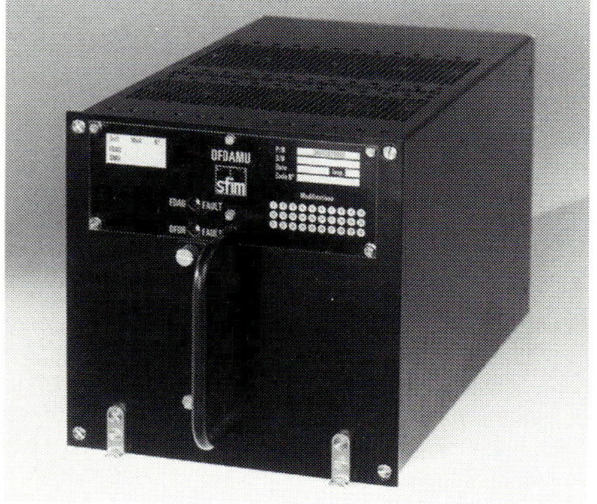

A300-600/A310 common unit DFDAU-ACMS **1998**/0015268

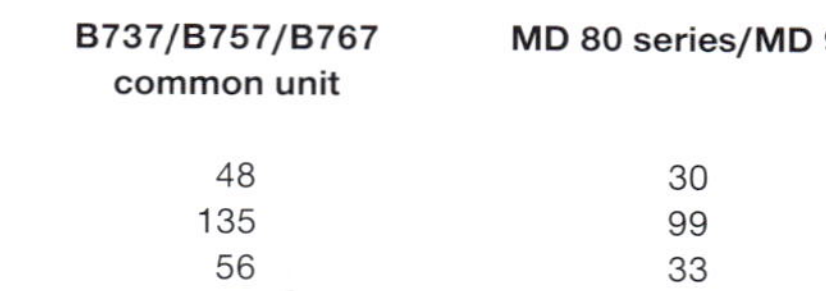

Acquisition	A300-600/A310 common unit	B737/B757/B767 common unit	MD 80 series/MD 90
(analogues)	48	48	30
(discretes)	100	135	99
(ARINC 429 buses)	56	56	33

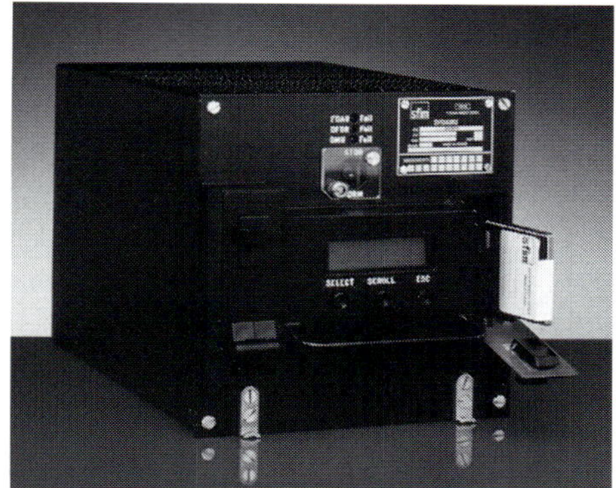

B737/B757/B767 common unit DFDAU-ACMS
1998/0015269

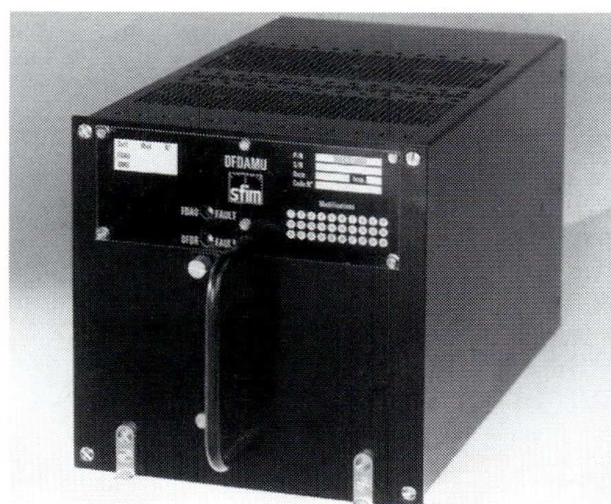

MD 80 series/MD 90 DFDAU-ACMS
1998/0015270

aircraft types: A300-600/A310; B737/B757/B767; MD 80 series/MD 90. All units are designed to acquire engine and aircraft data reports, including aircraft life data, weather, and other parameters by user programmable requests.

Specifications

Interfaces:
(DFDR/SSFDR recorder) ARINC 717
(MCDU) ARINC 739
(On-board printer) ARINC 740 or 744
(QAR/DAR recorder) ARINC 591
(ACARS MU) ARINC 724 or 724B
(ADL/PDL) ARINC 615
Dimensions: 6 MCU
Weight: 8.5 kg
Power: 115 V AC, 400 Hz, 90 VA
Further specifications can be found in the table above.

Contractor

SFIM Industries.

NEW ENTRY

DMU-ACMS Data Management Unit — Aircraft Condition Monitoring System

The SFIM Industries DMU-ACMS is a common unit for use on the A319/A320/A321 series of commercial aircraft. The unit is designed to acquire engine and aircraft data reports, and data required for airline operational management, as may be programmed by the users.

Specifications

Interfaces:
(MCDU) ARINC 739
(on-board printer) ARINC 740 or 744
(DAR recorder) ARINC 591 (fully programmable frame)
(ACARS MU) ARINC 724B
(MDDU/PDL) ARINC 615
(CFDS (Centralised Fault Data System)) ABD 0048

A319/A320/A321 DMU-ACMS common unit
1998/0015271

Acquisition:
(discretes) 20
(ARINC 429 buses) 55
Dimensions: 3 MCU
Weight: 4.5 kg
Power: 115 V AC, 400 Hz, 40 VA

Contractor

SFIM Industries.

NEW ENTRY

FDAU-ACMS Flight Data Acquisition Unit — Aircraft Condition Monitoring System

The SFIM Industries FDAU-ACMS is designed as a complete package to monitor the mandatory aircraft data parameters, together with data required for engine maintenance, on the ATR42/ATR72 commuter aircraft. The complete system comprises: the DFAU (Flight Data Acquisition Unit); the FDEP (Flight Data Entry Panel); and a 3-axis accelerometer (TAA).

The FDEP displays: data, time and flight number; together with engine maintenance data.

The FDAU performs: acquisition of mandatory parameters; generation of a data frame to a DFDR/SSFDR; generation of the same parameters to the QAR and generation of time code. The system also incorporates BITE.

The ACMS function includes all reporting functions via the display memory terminal, disk and ACARS.

ATR42/ATR72 FDAU-ACMS ***1998***/0015272

Specifications

Interfaces:
(DFDR/SSFDR recorder) ARINC 573
(QAR recorder) ARINC 591
objective torque output
(DMT (Display Memory Terminal)) via RS-232
(ACARS MU) ARINC 724
Dimensions:
(FDAU) ½ short ATR (127.5 × 199 × 318 mm)
(FDEP) 146 × 66.6 × 114.3 mm
(TAA) 101.6 × 91.7 × 63.5 mm
Weight:
(FDAU) 5 kg
(FDEP) 0.7 kg
(TAA) 0.4 kg
Power:
(FDAU) 28 V DC, 60 W (with FDEP)
(TAA) 28 V DC, 3.5 W

Contractor

SFIM Industries.

NEW ENTRY

FDIU Flight Data Interface Unit

The FDIU is designed as a common unit for the A319/A320/A321/A330/A340 aircraft. Its function is to interface with a flight data recorder to ensure recording of all mandatory aircraft parameters. It acquires both discretes and ARINC 429 bus data, and interfaces to: the DFDR/SSFDR — ARINC 717; the QAR — ARINC 591; the CFDS/OMS — ABD 0018 or 0048.

Specifications

Dimensions: 2 MCU
Weight: 2 kg
Power supply: 115 V AC/400 Hz, 25 VA

Contractor

SFIM Industries.

NEW ENTRY

A319/A320/A321/A330/A340 FDIU ***1998***/0015273

NH 90 Flight Control Computer (FCC)

SFIM Industries has been selected by Eurocopter France to design and build a flight control computer for the NH 90 helicopter.

The FCC is an essential component of the electrical flight control unit and comprises the following two systems: the primary flight control system which provides basic control capabilities to the helicopter and its stabilisation; and the automatic flight control system which provides the 'hands off' high-level control modes that are required by the helicopter to fulfil its mission.

Operational status

Flight tests in progress.

Contractor

SFIM Industries.

UPDATED

GERMANY

Data transfer system

The Dornier Division of Daimler-Benz Aerospace AG, Defense and Civil Systems data transfer systems comprise portable solid-state data carriers as key elements with easily expandable memory capacity and associated data carrier adaptors for various ground and onboard installations. The data carrier adaptors have a similar design and are easily interchangeable by replacing the interface card which interfaces the equipment to ground computers or onboard systems. Onboard data carrier adaptors are available for MIL-STD-1553B and RS-422 interfaces. The portable and removable pocket-sized solid-state data carriers are used like a floppy disk to store any kind of data in an intelligent way.

Dornier is currently producing several types of CMOS EEPROM-based portable solid-state data carriers with storage capacities ranging from 64 kbytes to 8 Mbytes. The new generation of data carriers is equipped with a powerful built-in RISC processor combined with a specific controller ASIC providing autonomous data storage management and a high-speed serial datalink with 10 Mbits/s. The firmware routines implemented in the RISC processor provide application independent data management procedures which are directly comparable to the hard disk file system of standard DOS PCs. Programmable applications enable a large number of files of different sizes and a flexible structure of hierarchical directories limited only by the storage size of the data carrier. Up to 25 files can be opened simultaneously. For multi-user operation, a file can be opened for read access by up to 25 users simultaneously.

Specifications

Dimensions:
(Alpha Jet) 180 × 127 × 36 mm
(Tornado, C-160, EF 2000) 180 × 127 × 73 mm
(PAH-2 Tiger) 180 × 127 × 53.9 mm
(data carrier) 80 × 55 × 25 mm
Weight:
(Tornado) 1.7 kg
(Alpha Jet) 1.2 kg
(PAH-2 Tiger) 1.3 kg
(data carrier) 0.1-0.2 kg
Power supply: 28 V DC, 25 W

Operational status

Qualified for carriage on the Alpha Jet, Tornado, EF 2000 and PAH-2 Tiger.

Contractor

Daimler-Benz Aerospace AG, Defense and Civil Systems.

UPDATED

LAC-L LITEF avionic computer landing aid

LAC-L is used as the central processing unit of the AJAX landing aid made by ADS, St Petersburg. In the AJAX system, LAC-L processes TV-picture data to autonomously determine relevant navigation parameters including distance to touch down, velocity and height in real time and provides it for display to the pilot.

Specifications

Dimensions: 320 × 57 × 194 mm ARINC-600-6 2MCU
Weight: 2.3 kg
Power consumption: 28 V DC/25 W
Data input: Video from 2 CCD cameras
Data output: 2 video outputs (PAL, RGB)

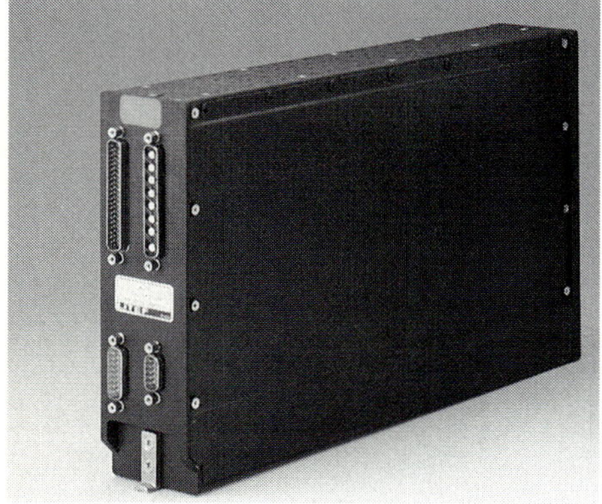

LAC-L LITEF avionic computer landing aid
***1997**/0005426*

Contractor

LITEF GmbH.

VERIFIED

Main computer for the Tornado

The in-service main computer for navigation and weapon delivery in the Tornado has been upgraded to fulfil increased computing requirements. It is now based on a multiprocessor architecture which still uses the original LR-1432F digital airborne computer but adds one or more high-performance 68040 processor modules. This new structure enables execution of all existing software, as well as recently introduced Ada software running on the 68040 processors.

The computer is equipped with various input/output interfaces including a high number of Panavia interfaces, one or two MIL-STD-1553 bus systems, discretes and special type interfaces.

Powerful graphics capabilities have been added.

The upgraded main computer of the Panavia Tornado
1995

Operational status

The upgraded main computer is in production and in service.

Contractor

LITEF GmbH.

VERIFIED

Mission computer/weapon computer for the F-4F

The LITEF mission computer/weapon computer for the German Air Force Phantom F-4F features microprocessor technology based on the Motorola 68020 with 32-bit architecture, fast CMOS-SRAM/EPROM, CMOS, ASICs to reduce power, space and weight, high-density packaging with melted core multilayer and surface-mounted technology, high reliability and extensive BIT. The computer contains two spare slots for future expansion.

Specifications

Dimensions: ½ ATR
Weight: 6.5 kg
Power supply: 28 V DC, 50 W (max)

Operational status

German Air Force F-4F Improved Combat Efficiency (ICE) programme.

Contractor

LITEF GmbH.

VERIFIED

Air data computer/air data transducer for the JAS 39 Gripen

The air data computer/air data transducer for the Saab JAS 39 Gripen combine the primary air data computer with an air data transducer channel as back-up in a single electronics box. Both functions are fully independent of one another and are electrically decoupled. The air data computer contains two high-precision pressure sensors which convert the static pressure and the total pressure into an electronic frequency signal. The signal is digitised and fed to a 16-bit microprocessor. Using software algorithms, the classic air data are computed, taking into account further parameters such as barometric altitude correction and total air temperature. Air data such as speed, altitude and Mach number are passed to aircraft systems via MIL-STD-1553B interfaces.

The other part of the unit contains two additional identical pressure sensors which operate in the same way as in the air data computer. In contrast to the computer, however, the final data are not computed. Instead the static and total pressures are passed to the flight control computer via an ARINC 429 interface. Due to the complete separation of the functions of the air data computer and air data transducer, the air data transducer has its own power supply unit, a separate processor and a separate software programme. In the flight control computer, the pressure values supplied by the air data transducer are used to calculate the air data, taking into consideration the barometric altitude correction and total air temperature, and the result compared with the values from the air data computer.

Both channels of the Nord-Micro air data computer and air data transducer have a highly developed self-testing capability and are able to check their own correct function or to provide information on any faults which may have occurred. In the event of an error, the faulty function is detected. The use of hybrids, gate arrays and LCC circuits enable the units to be housed in a 4 MCU box.

Operational status

In service on the Saab JAS 39 Gripen.

Contractor

Nord-Micro Elektronik Feinmechanik AG.

UPDATED

JAS 39 Gripen air data computer/air data transducer
***1998**/0015274*

Modular avionic computers

A range of modular avionic computers are designed by Teldix for a variety of airborne roles.

The Missile Control Unit (MCU) is designed to control intelligent missiles fitted to aircraft such as the Panavia Tornado. The MCU serves as a computing and interface system between the aircraft's main digital computer and the relevant input/output circuits of the missiles.

The MCU is a 16-bit multimicroprocessor containing the interface circuits to convert, process and pass internally fed data to the missiles via the MIL-STD-1553B databus. The Launcher Decoder Unit (LDU) provides the interface between the MCU and individual missiles. The MCU and LDU are in production.

Operational status

In production.

Contractor

Teldix GmbH.

VERIFIED

INTERNATIONAL

Avionic computers for EF 2000

The four avionic computers for EF 2000 provide processed data for attack, navigation, data recording and defensive aids functions. These computers are based on a modular multiprocessor architecture and are equipped with common MC 68020 and MC 68882 processing modules, with common housing and power supplies. A special purpose coprocessor was developed at LITEF, based on the high-performance Reduced Instruction Set Computer (RISC) technology.

The equipment software is developed mainly in Ada. It includes BIT capabilities, an adapted Ada run-time system and a set of Ada packages to provide a defined interface between target specific input/output interfaces and the application software.

Each computer is equipped with various interfaces. The attack/navigation computers are equipped with two STANAG 3910 high-speed fibre optic databusses, the interface processor unit has two STANAG 39210, analogue, audio and discrete interfaces and the defensive aids computer has one STANAG 3910 interface, with discretes for chaff and flares and for blanking.

Operational status

In development for the EF 2000. Formal qualification is in progress for the attack/navigation computer and the interface processor unit. The defensive aids computer is still in development.

Contractors

Alenia Difesa, Avionic Systems and Equipment Division, GF-Sistemi Avionici.
Computing Devices Company Ltd.
INDRA.
LITEF GmbH.

UPDATED

EF 2000 air data transducer

The EF 2000 air data transducer, currently being developed by an international consortium consisting of GEC-Marconi Avionics (Rochester), Bavaria Avionik Technologie and Tecnobit, provides high integrity and fast dynamic response air data information.

The air data transducer features a combined multifunction pitot-static and flow angle mobile vane developed by Sextant Avionique. It provides local air data parameters, such as pitot, static and differential pressures, via a dedicated MIL-STD-1553B databus to the flight control computer. Calculations of airspeed, altitude, Mach number, angle of attack and angle of sideslip are generated to support the EF 2000's artificial aerodynamic stabilisation.

Specifications

Dimensions: 138 × 125 × 155 mm
Weight: 3.2 kg
Power supply: (heater) 115 V AC, 450 W
±20 V DC, 15 W

Operational status

In advanced development. Prototypes are available.

Contractors

GEC-Marconi Avionics, Rochester.
Bavaria Avionik Technologie GmbH.
Tecnobit SA.

VERIFIED

EF 2000 Cockpit Interface Unit (CIU)

The EF 2000 CIU was initially designed for the British Aerospace EAP programme and has been selected for the EF 2000. The system is being produced jointly by Teldix, Alenia Difesa, INDRA and GEC-Marconi Avionics.

The CIU collects discrete signals from various buttons and switches in the cockpit and data from other avionic systems being passed along the data highway, for onward transmission to other cockpit systems.

Operational status

Under development.

Contractors

GEC-Marconi Avionics, Rochester.
Alenia Difesa, Avionic Systems and Equipment Division, GF-Sistemi Avionici.
INDRA.
Teldix GmbH.

UPDATED

EF 2000 front computer

ENOSA, leading a consortium with GFSA, Smiths Industries and VDO, has been awarded the development contract for the front computer for the EF 2000. The front computer forms part of the Utilities Control System (UCS) of the aircraft; it is of compact design and is fully integrated with other primary aircraft systems.

The front computer provides all the functions for control and monitoring of: the environmental and temperature control system; the life support system; the crew escape system.

Design features of the front computer are that it is fully compliant with EF 2000 requirements and that it is databus MIL-STD-1553B compatible, allowing it to communicate with the rest of the UCS. It is based on a Motorola 68020 microprocessor.

Operational status

Now in full-scale development under contract from Eurofighter. Prototypes are currently fitted in Eurofighter prototype aircraft, undergoing flight evaluation.

The unit has completed 6,000 hours of qualification testing.

Contractors

ENOSA.
Alenia Difesa, Avionic Systems and Equipment Division, GF-Sistemi Avionici.
Smiths Industries Aerospace.
VDO.

VERIFIED

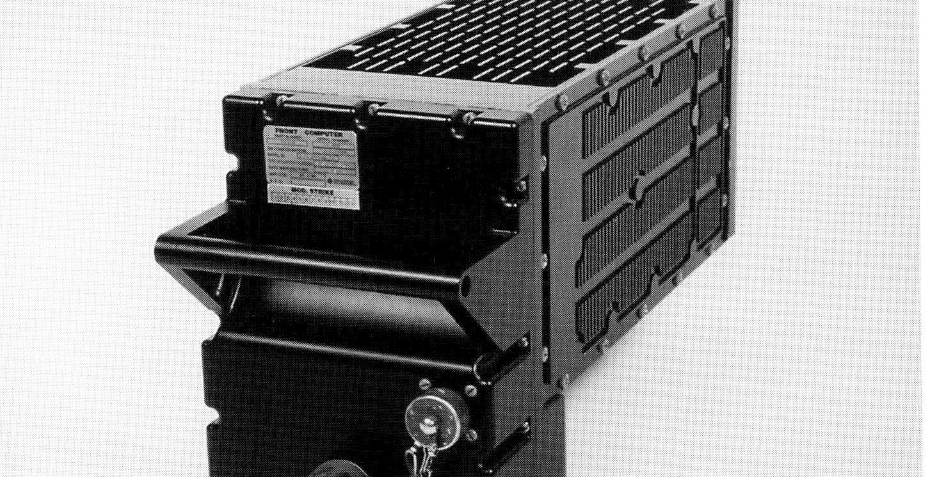

EF 2000 front computer
1997/0001276

EF 2000 Maintenance Data Panel (MDP)

ENOSA, leading a consortium with Dornier GmbH, ESD-Plessey and Alenia Difesa, has been awarded the development contract for the MDP and Portable Maintenance Data Store (PMDS) for the EF 2000. The MDP and PMDS apply the most advanced technology for monitoring and recording information from the aircraft for maintenance operations.

EF 2000 maintenance data panel
1997/0001277

The MDP and PMDS display the following types of data: refuelling/defuelling; weapon stores data loading; mission data loading; failure data (actual and last flight status); limit exceedances data; status of consumables; life usage data; and aircraft tail number.

Limit exceedances and fatigue data generated during the last five flights, together with data from the engine and other systems connected to the MDP (via databus) are available and recorded through the PMDS.

Operational status
Now in full-scale development under contract from Eurofighter. The delivered prototypes are in flight trials.

Contractors
ENOSA.
Dornier GmbH.
ESD-Plessey.
Alenia Difesa, Avionic Systems and Equipment Division, GF-Sistemi Avionici

UPDATED

Harpoon Airborne Command, Launch and Control System II

GEC-Marconi Avionics (Rochester) and The Boeing Company have developed a range of command, launch and control systems compatible with all standards of Harpoon and SLAM missiles.

The Harpoon Airborne Command, Launch and Control System II (HACLCS II) adds new capability at greatly reduced cost and increases the potential for Harpoon carriage on smaller aircraft, such as the IPTN-235 and Fokker 50.

The GEC-Marconi Avionics HACLCS II computer, as the main subsystem of HACLCS II-2, provides all the control and power signals necessary to interface with the aircraft systems, including a MIL-STD-1553B databus, and to control four or six missile interface configurations on the P-3 aircraft and on fast-jet applications.

Specifications
Dimensions: 191 × 194 × 309 mm
Weight: 9.5 kg
Power supply: 115 V AC, 2 A (max)
28 V DC, 1 A (max)

Operational status
In preproduction.

Contractor
GEC-Marconi Avionics.
The Boeing Company.

UPDATED

Interface unit for the Tiger helicopter

The interface unit is a data acquisition, management and preprocessing computer for the basic avionics of the French/German Tiger helicopter. It is based on a modular multiprocessor architecture and includes the MC 68020 processor. It is equipped with input/output modules for data transfers via interfaces of different types, such as MIL-STD-1553B, ARINC 429, analogue, frequency measurement and discrete.

The equipment and the application software is developed in Ada. It includes BIT capabilities, an adapted Ada run-time kernel and flexible data transmission, handling, formatting and processing so that interface parameters are configurable via a parameter table which allows changes in acquisition rate or selection of preprocessing without any modification of the software itself.

Operational status
Qualified.

Contractors
TELDIX GmbH.
Sextant Avionique (Co-op partner).
VDO Luftfahrtgeräte Werk (Co-op partner).

VERIFIED

ISRAEL

ACE-5 computer

The ACE-5 is a modular high-performance airborne general purpose computer which is fully compatible with MIL-STD-1750A and MIL-STD-1553A and B. It is designed for multirole fighter mission management.

Computation performance is 1.7 Mips, with growth potential to 2.5 Mips. Some 256 k words of memory with battery back-up are programmable at flight line level, with growth capacity to 1 M words. BITE provides continuous monitoring of functional integrity.

Specifications
Dimensions: 137 × 198 × 472 mm
Weight: 10 kg
Power supply: 115 V AC, 400 Hz, 3 phase
28 V DC
Environmental: MIL-E-5400 Class II
Reliability: >500 h MTBF

Operational status
In service on Israeli F-16C/D aircraft.

Contractor
Elbit Systems Ltd.

UPDATED

Enhanced airborne Communication, Navigation and Identification (ECNI) system

The Enhanced airborne Communication, Navigation and Identification (ECNI) system enables integrated, centralised and computerised control and resource management of communication, radio navigation and identification devices, using avionics controls and displays as the man/machine interface. It also enables back-up control and management of vital CNI functions during failures, using a dedicated control and display. Management and control of all audio sources such as the crew, radios, alarms, weapons and EW is provided using a digital audio matrix which also provides intercom and alarm generation as by-products. The ECNI manages and controls synthetic voice and tonal alarms.

ECNI consists of the main control unit which contains all the electronics required for system implementation, an integrated back-up control box which is used as a redundant unit for control and audio during failures and additional control and audio panels for the use of other crew members. In the normal mode the ECNI is designed to be operated using the avionics controls and displays, and in the back-up mode the system is operated via the dedicated control box. The main control unit includes a digital/audio switching matrix which enables intercommunication between system subscribers according to predefined communication maps or dynamic real-time resource selection. Alarm control and management are principal features of ECNI. There are two types of alarms: synthetic voice and regular tone. The system is capable of translating existing tonal alarms into synthetic voice alarms, according to the latest human engineering requirements.

Contractor
Elbit Systems Ltd.

UPDATED

Modular MultiRole Computer (MMRC)

The MMRC performs tasks for a full avionic suite. The MMRC is a single-multimodule computer designed to handle a broad spectrum of tasks previously performed by several individual units.

It is easily upgradeable to accept new systems and capabilities, and has built-in growth potential in processing capability, memory and interfaces.

Each module of the MMRC performs a different function: fire control; stores management; display processor (MFDs, MFCDs, HUD, HMD); display and sight helmet; integrated communication radio navigation and identification system; digital image processing and communication system; and video and VTR controller.

The MMRC is also the central element of the HALO Advanced Helicopter Avionics Suite offered as a flexible system upgrade for a wide range of helicopters.

Operational status

In production.

Contractor

Elbit Systems Ltd.

UPDATED

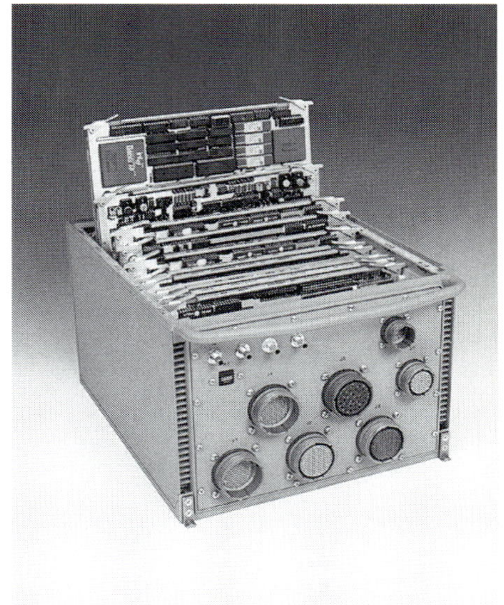

The Modular MultiRole Computer (MMRC)
1996

EL/P-8930 Advanced Programmable Signal Processor (APSP)

The APSP is a signal processor for computation intensive airborne, naval and other military and civil applications. It is based on the new generation of off-the-shelf DSP chips that ease programme development and maintenance. A high-speed parallel Data Transfer Network (DTN) provides flexible communications between modules.

The APSP consists of Programmable Processing Modules (PPMs), special processing modules, data processors, memory modules, a master controller, fast input/output channels and a high throughput DTN. For ease of programming, the PPMs comprise pairs of tightly coupled modules sharing a common memory. One five-card PPM is capable of 700 Mflops and has up to 22 Mbytes of memory, expandable by adding more memory modules.

The DTN is a flexible parallel message passing unit providing 32-bit wide data transfer from sensors and among processors. Up to eight stations can be connected simultaneously with a rate of up to 22 Mbytes/s between each pair of stations.

The APSP is packaged in various size chassis and cooled either by forced air or fans. The processor is available in military or commercial versions.

The main features of the EL/P-8930 are high-level C and assembler language programming with no micro-coding, a distributed real-time operating system, highly optimised 32-bit floating-point calculations, fault tolerant design for high mission availability, modular open architecture for easy adaptability to specific applications, computing from 165 Mflops to 3,500 Mflops depending on configuration and special processors including FFT and convolver modules.

Typical applications for the EL/P-8930 include multimode fire-control radars, phased-array radars, sonar, electronic warfare, COMINT, multisensor fusion, target recognition, cryptography, image processing, multimedia, video compression/decompression and communications.

Contractor

Elta Electronics Industries Ltd.

VERIFIED

EL/S-8600 computer

There are several versions of the basic EL/S-8600 main computer, all of which use similar central processing units and architectures and are customised for various military applications. The EL/S-8610 has 15 modules (each with two printed circuit boards) in a 1 ATR box; the EL/S-8611 has only four modules in a customised box. In addition to their airborne applications, other computers in the range are used for land and sea operations. Typical applications are for inertial navigation and weapon delivery systems, artillery tactical fire-control systems, message switching centres and mobile communications control. Software support is available for C, Fortran-77, Iso-Pascal, APL and Cobol.

Specifications

EL/S-8610
Dimensions: 394 x 257 x 194 mm
Weight: 21.5 kg
Computer type: binary, two's complement, micro-instruction (15 register), fixed point
Word length:
(32-bit instruction words) 16-bit
Max address range: 64 kbyte (expansible to 2 Mbyte with mapping unit)
Instruction set: 90
Micro-instruction cycle time: 225 ns
Micro-instruction depth: 1,024 words (expansible to 2,048 words)
Typical execution time:
(add/subtract) 0.66 μs
(load) 1.34 μs
(multiply) 6.56 μs
(divide) 9.26 μs
Input/output options: wide variety, customised to use

Operational status

In service.

Contractor

Elta Electronics Industries Ltd.

VERIFIED

EL/S-9005 military AVIION computers

The EL/S-9005 is a high-performance military computer family based on the Motorola 88000 32-bit RISC processor. It is software-compatible with the Data General AVIION 400/4000 systems.

The EL/S-9005's advanced architecture supports a wide variety of modules using the system's fast GM bus and industry standard VME bus. The systems are supported by Data General's DG/UX and Ready System's VRTX32 operating systems.

The EL/S-9005 features a multiprocessor/parallel processing configuration and supports Ethernet, SCSI, RS-232, RS-422 and Graphic Controller SPSP hardware interfaces. It is housed in a 19 in (482.6 mm) rack.

Contractor

Elta Electronics Industries Ltd.

VERIFIED

Data Transfer Equipment (DTE)

Rada has developed a family of data transfer systems of different memory types, sizes and interfaces. These systems are used in advanced aircraft (fixed- and rotary-wing) for data exchange between the squadron mission planning stations and the aircraft's onboard computers and systems. The system comprises a portable solid state memory cartridge on which the mission parameters are loaded in the squadron and an interface unit through which the cartridge is downloaded into the aircraft computers for system initialisation and set up.

During the flight there is a flow of information (maintenance, intelligence and so on) to and from the aircraft systems. After the mission the recorded data is downloaded into a ground station.

Rada's DTE are available in the following configurations:

IDTE Specifications

Dimensions:
(DTU) 177.8 × 146 × 114.5 mm
(Cartridge) 190.5 × 120 × 41.3 mm
Weight:
(DTU) 2.5 kg
(Cartridge) 0.78-1.0 kg
Power supply: 115 V AC, 400 Hz, single phase, 17 W
Memory type: RAM/EEPROM/Flash
Memory size: 0.5 M-80 MB
Interface: MIL-STD-1553A/B or any tailored interface

DRE Specifications

Dimensions:
(DRU) 230 × 145 × 32 mm
(Cartridge) 97 × 110 × 20 mm
Weight: (DRU + Cartridge) 1.6 kg
Power supply: 28 V DC

Data Transfer Equipment (DTE) **1997**/0001279

Memory type: Flash
Memory size: 4 -32 MB
Interface: RS-232/422

Rada Data Recorder Equipment (DRE)
1997/0001278

Operational status

Rada's DTE are in service on F-16A/B/C/D, F-5E, F-15 and Mirage types of aircraft.

Contractor

Rada Electronic Industries Ltd.

VERIFIED

Expanded Data Transfer and processing System (EDTS)

The EDTS provides avionics integrators with a powerful tool for aircraft data acquisition, to handle both operational and maintenance data. The EDTS is functionally divided into the Data Transfer and processing Unit (DTU), Data Transfer Cartridge (DTC), Maintenance Monitoring Panel (MMP) and the Electronic Storage Unit (ESU). These modules are connected through RS-422 serial communication lines. The DTU is the main controller which interfaces with the avionic MIL-STD-1553A/B dual-redundant multiplexed bus.

The DTU, the central unit of the EDTS, is a computer which communicates with the other DTS units via digital communication links. The DTU also performs the interfacing of the EDTS subsystem with the avionics system via the two avionics multiplex buses. The DTU manages the major DTS functions: preflight data transfer, in-flight data transfer and processing storage, and BIT management.

The DTC consists of a removable data transfer cartridge and a housing located in the aircraft. The cartridge is a 21 × 94 mm rugged solid-state portable memory. It contains a microcontroller and data storage of 208 kbytes to 512 kbytes. The memory has RAM and EEPROM CMOS chips packaged in leadless carrier technology.

Mission planning data is loaded into the cartridge at the ground mission planning station. It is then inserted into the rackmount housing in the aircraft. The DTC provides the medium for data storage. After the flight it is removed and brought to the ground mission planning station where its contents can be read.

The MMP is a smart unit containing a microcontroller and a data storage memory. A dot matrix display and five push-buttons are located on its front panel. The MMP needs to be installed in a location with easy access and good visibility for the ground crew. It receives and stores a list of failed LRUs. The MMP provides, without DTU intervention, display of one failed LRU code at a time when called up by the ground crew, clearance of the LRU code after the LRU has been replaced, display of the code for the next failed LRU after depression of the control button and communication with other avionics for display of preflight test information and maintenance data.

The ESU is a small unit with a volume of less than 0.2 litres. In order to increase its operational efficiency it is energised only as called for by the DTU. Its purpose is to store parameters related to the aircraft structure such as harmonisation parameters and the aircraft identification number. It is fixed to the aircraft structure as part of the aircraft wiring.

Specifications

Data transfer and processing unit
Dimensions: 122 × 193.5 × 233.7 mm
Weight: 6.35 kg
Power supply: 115 V AC, 400 Hz, 3 phase
28 V DC

Data transfer cartridge
Dimensions:
(DTC) 21 × 165 × 94 mm
(housing) 47 × 165 × 146 mm
Weight:
(DTC) 0.47 kg
(housing) 0.59 kg
Power supply: 7 V DC, 15 V DC

Maintenance monitoring panel
Dimensions: 190 × 124 × 69.8 mm
Weight: 1.27 kg
Power supply: 28 V DC

Electronic storage unit
Dimensions: 89.9 × 70.1 × 30 mm
Weight: 0.27 kg
Power supply: 7 V DC and 15 V DC from the DTU

Contractor

Rada Electronic Industries Ltd.

VERIFIED

ITALY

MARA general purpose computers

The Modular Architecture for Real-time Applications (MARA) family of high-power computers is based on Intel microprocessors and is intended for real-time military applications. The modular architecture permits processing power, memory type and size and input/output modules to be tailored to a wide variety of requirements. They are highly fault-tolerant, and comprehensive fault detection circuits are incorporated. MARA has been used by Alenia as an embedded computer in its stores management and electronic warfare systems, in data acquisition units and as a symbol generator in head-up and multifunction displays. As a stand-alone computer, MARA has been used for onboard data handling for both aircraft and mission systems management in the EH 101 helicopter.

Specifications

Typical physical characteristics:
Dimensions:
(6 modules) ½ ATR short
(10 modules) ¾ ATR short
(13 modules) 1 ATR short
Weight:
(6 modules) 9.5 kg
(10 modules) 13 kg
(13 modules) 16 kg
Power supply: 115 V AC, 400 Hz
80 VA (6 modules)
130 VA (10 modules)
160 VA (13 modules)
or 28 V DC
Input/output module options: MIL-STD-1553B/ARINC 429 parallel and serial channels/tape recorder driver/display drivers, raster and stroke/analogue, digital, discrete, synchro, frequency multichannel, A/D and D/A converter multichannel

Contractor

Alenia Difesa, Avionic Systems and Equipment Division.

UPDATED

	MARA 203	**MARA 204**	**MARA 205**	**MARA 206**	**MARA 207**
Microprocessor type and instruction set	8086	286	386	960	960
Number of microprocessors	1	1-4	1-4	1	1-4
Word length	16-bit data	16-bit data	32-bit data	32-bit data	32-bit data
Max addressable memory in modules of various RAM/EPROM/PROM combinations	1 Mbyte	16-24 Mbytes	16 Mbytes	16 Mbytes	16 Mbytes
Typical speed	0.3 Mips	0.8-3.2 Mips	1.2-4.8 Mips	3.2 Mips	3.2-12.8 Mips
Numeric co-processor	optional	optional	optional	internal	internal

ANV-801 computer display unit

The ANV-801 computer display unit replaces multiple control layouts with a single, easy to read unit, allowing the concentration of flight management information and data presentation. Versatile design, modular hardware and software architecture make it easy to reconfigure the ANV-801 to meet customer requirements.

A 16-bit Intel 80C186 Series CPU with 1 Mbyte memory or a 32-bit Mips R3000 Series RISC CPU with up to 8 Mbytes memory provide the ANV-801 with the computational power for solving complex tasks in real time.

The ANV-801 integrates external navigation sensors into sophisticated navigation packages providing a high-accuracy computation capability, including navigation information for the crew and steering output to flight director and autopilot. It is also available with a built-in GPS receiver sensor, providing self-contained navigation capabilities and a built-in modem for input/

output datalink. The ANV-801 interfaces with a wide range of analogue and digital equipments using ARINC 429, MIL-STD-1553B, serial and modem digital interfaces and analogue interfaces. These include communications, landing and IFF equipments.

Specifications

Dimensions: 185 × 146 × 181 mm
Weight: <4.5 kg
Power supply: 28 V DC or 115 V AC, 400 Hz

Contractor

Marconi SpA.

VERIFIED

ANV-803 computer and interface unit

The ANV-803 was designed to handle real-time environments in modern avionic systems where powerful computation capability, small size and low weight are required. The powerful RISC engine and modular design of the ANV-803 meet a wide range of applications while keeping a simple monoprocessor architecture, resulting in a good price/performance ratio and maintenance cost.

System options include a complete range of input/output support modules and a comprehensive set of software development tools for Ada programming. Different interface cards are available, including MIL-STD-1553B BC/RT/BM, ARINC 429, discrete, serial and analogue.

Specifications

Dimensions: ⅜ ATR short
Weight: <7.5 kg
Power supply: 28 V DC or 115 V AC, 400 Hz

Contractor

Marconi SpA.

VERIFIED

The Marconi ANV-803 computer and interface unit
1995

JAPAN

Altitude computer

The altitude computer receives pitot and static pressure and total temperature inputs for the calculation of aircraft altitude. It has a microprocessor and performs digital calculations. The altitude computer includes BIT and performance monitoring.

Operational status

In production for the T-2 trainer.

Contractor

Shimadzu Corporation.

VERIFIED

The altitude computer is in production for the T-2 trainer
1995

Compact airdata transducer

Tokyo Aircraft Instrument Co Ltd has started to produce a compact airdata transducer which can be used on board as a compact airdata computer or as an airdata measurement device. The transducer has a micro-vibration pressure sensor and a semiconductor pressure sensor both of which make it possible to provide accurate airdata information at DC voltage. The transducer can be expanded to provide the interface with ARINC 429, ARINC 629, MIL-STD-1553B and other serial or parallel buses as may be required by customers.

Operational status

In production for commercial and military aircraft.

Contractor

Tokyo Aircraft Instrument Co Ltd.

UPDATED

Truncated pyramid-shape multihole pitot probe system

This system is composed of truncated pyramid-shape multihole pitot probe tube and processing control computer (air data computer), both of which have been developed and patented by Tokyo Aircraft Instrument Co Ltd in conjunction with the National Aerospace Laboratory of Science and Technology Agency, Japan (NAL). The system measures alpha-angle, beta-angle, airspeed and altitude precisely and quickly with airflows from truncated pyramid-shape multihole pitot probe tube and provides air data information to other avionic systems via serial databusses. The system contributes to flight performance and the enhancement of STOL capability.

Operational status

In production and service in the Dornier 228 of NAL, the aerospace experimental vehicle 'ALFLEX' of NAL/NASDA.

Contractor

Tokyo Aircraft Instrument Co Ltd.

UPDATED

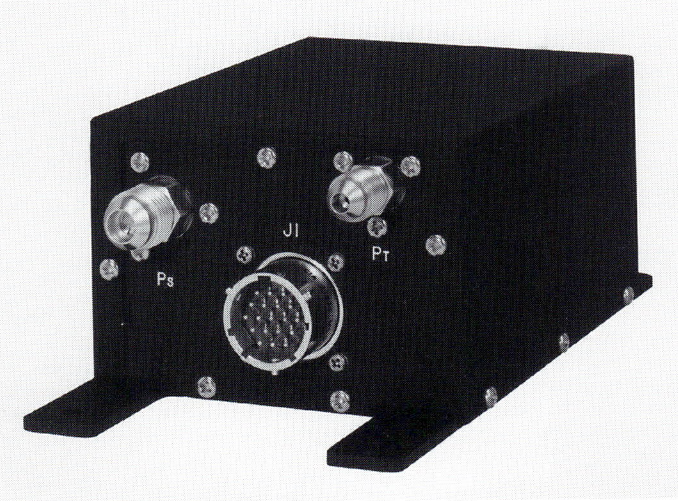

Compact airdata transducer **1997**/0001283

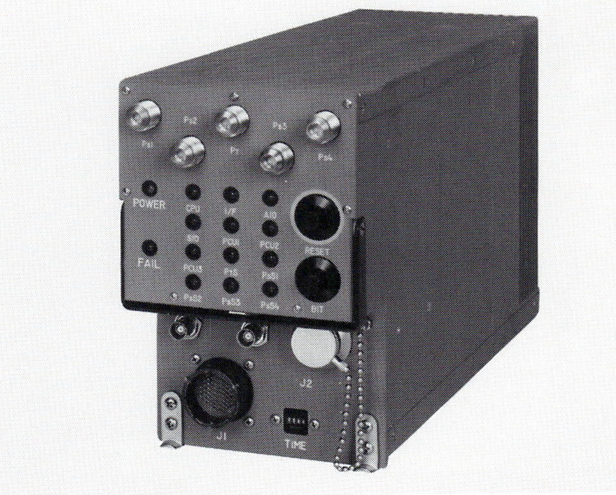

Processing control computer **1997**/0001284

RUSSIAN FEDERATION AND ASSOCIATED STATES (CIS)

SVS series of digital air data computers

The SVS series of digital air data computers provide measured and computed data on: true altitude; vertical speed; Mach number; stagnation temperature; outside temperature; angle of attack; and angle of slip. Inputs include: static pressure; total pressure; stagnation temperature; local angle of attack; local angle of slip; QFE and QNH.

Operational status
AeroPribor state that they are the only producer in Russia, and that these systems are fitted to all military and civil aircraft.

Contractor
AeroPribor Voskhod Joint Stock Company.

NEW ENTRY

SVS series of digital air data computers
1998/0018928/0015275

Specifications

	SVS-85	SVS 2Ts-00	SVS 2Ts-002	SVS PKR-1
Inputs				
(static pressure)	11.555-107.6 kPA	1.07-133.3 kPa	1.07-133.3 kPa	17.3-107.5 kPa
(total pressure)	11.55-135.4 kPa	6.66-279.9 kPa	6.66-279.9 kPa	11.5-115 kPa
(stagnation temperature)	−60 to +99°C	−60 to +350°C	−75 to +400°C	−70 to +150°C
(local angle of attack)	−60 to +60°	−60 to +60°	−60 to +60°	−60 to +60°
(QFE and QNH)	57.7-107.4 kPa	70.1-107.4 kPa	70.1-107.4 kPa	57.7-107.4 kPa
Outputs				
(true altitude)	−500 to 15,000 m	−500 to 30,000 m	−500 to 30,000 m	−500 to 13,000 m
(vertical speed)	−100 to 100 m/s	−500 to 500 m/s	−500 to 500 m/s	−60 to 60 m/s
(Mach No)	0.1-1.0	0.2-3.0	0.2-3.0	0.1-1.0
(stagnation temperature)	−60 to 99°C	−60 to 350°C	−75 to 400°C	−60 to to 100°C
(outside temperature)	−99 to 60°C	−60 to 60°C	−75 to 60°C	−90 to 60°C
(angle of attack)	−60 to 60°	−60 to 60°	−60 to 60°	−60 to 60°
(angle of slip)		−30 to 30°	−30 to 30°	−30 to 30°
Power	115 V AC, 400 Hz, 50 VA	115 V AC, 400 Hz, 50 VA	115 V AC, 400 Hz, 50 VA	5 V DC, 6.5 W 15 V DC, 2 W
Weight	6.5 kg	7.0 kg	4.5 kg	0.8 kg
Dimensions	4 MCU	4 MCU	3 MCU	240 × 190 × 40 mm

C300 programmed signal processor

The C300 programmed signal processor is designed for real-time processing of radar information from the latest generation multifunction airborne radars. It comprises the following modules; signal input and buffering module; data processor and control module; computing module; video processor; non-volatile memory; input/output interfaces module; specialised communications control modules and co-processors are further options.

Specifications
Max throughput at signal processing: 3,200 Mflops
Max throughput of data processing: 64 Mips
Main memory: 48 Mbytes
Non-volatile memory: 10 Mbytes
Signal speed: 4 × 12 Mbytes/s (4 channels)
Video image resolution: 1,024 × 768 pixels
Power supply: 200 V, 400 Hz, 1,000 W
Weight: 39 kg
Software: assembler; C; parallel processing system

Contractor
Leninetz Holding Company.

NEW ENTRY

BTsVM-386 airborne digital computer

The BTsVM-386 airborne digital computer is a 32-bit, PC-compatible, multitask, modular, open architecture system, which can be extended by peripheral and graphics coprocessors. Three processor options are available. Program languages: C++; Modula-2; Assembler.

Specifications
Memory:
(ROM) 1.5 Mbytes (extended to 65 Mbytes)
(static RAM) 0.5 Mbytes (extended to 64 Mbytes)
(flash memory) extended to 256 Mbytes

	i386/387-20	i486DX-50	i860-25
Word length	32-bit	32-bit	64-bit
Speed			
(fixed point)	10 Mops	50 Mops	50 Mops
(floating point)	0.7 Mops	3.2 Mops	50 Mops
(graphics)			25 k polygons/s

Input/output:
(ARINC-429) 32 independent input; 16 independent output
(multiplex MIL-STD-1553B) 3 channels
(event signals) 32 input; 16 output
MTBF: 10,000 hours
Power: 115 V AC; 26 V DC, 90 W
Weight: 8 kg

Contractor
Ramenskoye Design Company AO RPKB.

NEW ENTRY

BTsVM-386 airborne digital computer **1998**/0015276

SOUTH AFRICA

Mission computer

The ATE mission computer is designed to meet the requirements of modern integrated avionics systems. Advanced RISC architecture and modular design provides for a high-performance generic mission computer.

A distributed processor architecture with intelligent Input/Output (I/O) peripherals allows the application software to run independently from the I/O system software. Real-time colour symbology generation/video mixing, as well as HUD stroke symbology generation, is managed by high-performance intelligent symbol generation modules. I/O interfaces include MIL-STD-1553B BC/RT, synchro, analogue, serial and discrete. A user-friendly PC-based applications software development environment is fully implemented and validated. An offline PC-based graphics generation tool allows pilot evaluation of symbology before downloading to the mission computer. Extensive BIT ensures failure detection of between 95 and 98 per cent.

Specifications
Dimensions: 330 × 129 × 197 mm
Weight: 7 kg
Power consumption: 70 W
Reliability: 4,000 h MTBF

Operational status
Flight tested.

Contractor
Advanced Technologies & Engineering Co (ATE).

VERIFIED

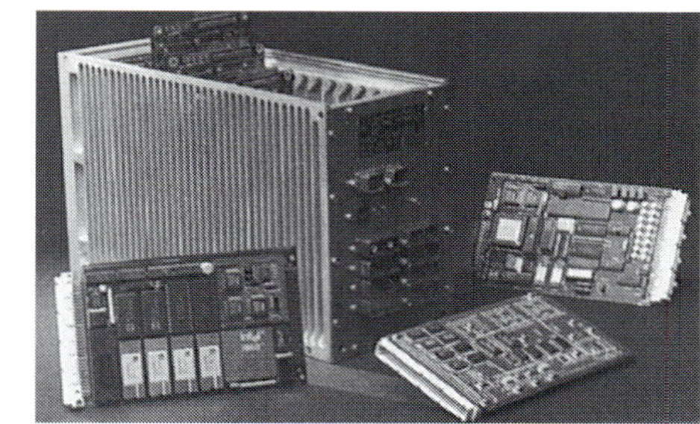

The ATE mission computer
1995

AM1000 distributed architecture data acquisition and processing system

The AM1000 distributed architecture data acquistion and processing system has been designed as an expandable modular COTS avionic architecture to service Health and Usage Monitoring (HUMS), Flight Data Recorder (FDR), and Cockpit Voice Recorder (CVR) systems.

Individual modules are combined, from two to six, to provide a required capability in an integrated unit. Communication between units takes place using industry standard protocols. Primary computing takes place within each decentralised functional unit. Commercially available software is utilised.

The COTS hardware is ruggedised to meet user requirements, using convection cooling and hardware enclosures to match the application concerned. Functional units include: Data Concentrator Unit (DCU) AM1-1000; Vibration Monitoring Unit (VBU) AM1-2000; Engine Monitoring Unit (EMU) AM1-3000; Engine Debris Monitoring Unit (EDMU) AM1-4000; together with supporting processor; power supply and interface units.

Contractor
Analysis, Management & Systems (Pty) Ltd.

NEW ENTRY

AM1000 distributed architecture data acquisition and processing system **1998**/0015277

AM3000 harsh environment computer

The AM3000 range of high-performance, high-reliability electronic modules is based on the Multibus Manufacturers Group standard for implementation of the IEEE 1296 Multibus II on MIL-STD-1389D SEM-E. This allows the performance, flexibility and reliability of Multibus II to be combined with a rugged and compact module, providing an ideal computer for harsh environments. AM3000 modules can be supplied in either military or industrial screening levels.

The military screened modules are suitable for airborne use, as well as for land and naval applications, where custom input/output modules may be added to the standard range in order to meet the requirements of new systems or mid-life upgrades to existing systems. Modules can be packaged in standard or customised enclosures, using either forced air or convection cooling. Interface standards include Multibus II, MIL-STD-1553B, RS-232, RS-422, RS-485, FDDI, ARINC 429, ARINC 404, MIL-STD-704D and MIL-STD-1275A.

Standard modules include the AM3010 32-bit RISC CPU module, AM3080 32-bit CISC CPU module, AM3030 input/output processor module, AM3020 MIL-STD-1553B communications module and the AM3060 power supply module.

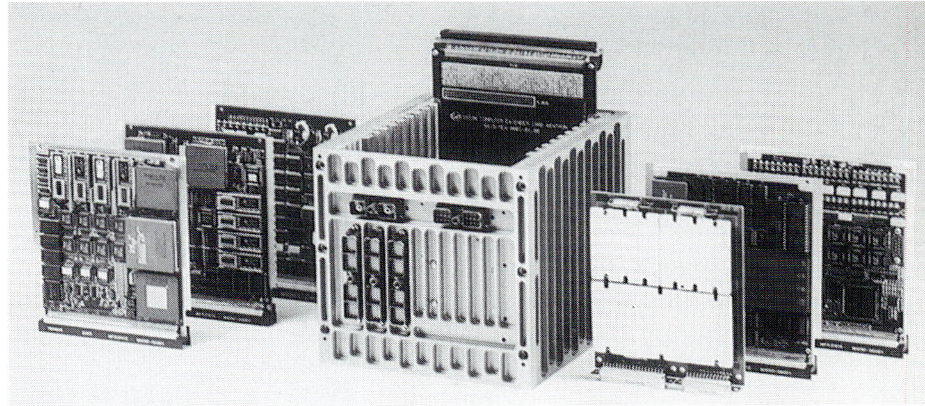

Modules from the AM3000 harsh environment computer ***1995***

Specifications
Dimensions: ¾ ATR or ½ ATR short (6-12 modules)
Weight: 7.5 kg (9 modules)
Power consumption: 80 W

Contractor
Analysis Management & Systems (Pty) Ltd.

VERIFIED

SWEDEN

SDS80 standardised computing system

Designed for airborne, land and naval applications, the SDS80 is intended as a general purpose processor with cost savings extending over software development as well as hardware. The current design consists of three modules: the D80 computer, PUS80 programme development system and Pascal/D80 high-order language. By adding D80 modules the system can be expanded from a relatively simple unit to an extremely high-performance embedded computer system. Using an intermodule bus, up to 15 processing units, input/output modules and bulk storage facilities can be connected. A low-cost microprocessor complement, the D80M, has been developed, as have Ada software compilers for the D80 computer and microprocessor.

The D80 computer is the system computer for the radar, display and other systems in the Saab JAS 39 Gripen combat aircraft and for the mid-life update programme for the Royal Navy's Sea Harriers.

The company is producing an Ada environment, based on the D80 computer, under contract to the Swedish Government. Designated SDS80A, the system will include the Ada language and an Ada integrated development environment.

The latest, improved version of the D80 computer, retrofitted into the first batch of Gripen aircraft
1998/0015278

Specifications

Dimensions: 125 × 193 × 318 mm typical
Weight: 6.5 kg
Power supply: 56 W AC or DC
Word length: 32-bit
Typical speed: 1 Mips
Instruction time: from 125 μs
Interfaces: include MIL-STD-1553B
Cooling: forced air or fan

Operational status

In production. In service on the JAS 39 Gripen. Also selected for the British Aerospace Sea Harrier F/A Mk 2. The latest, improved version of the D80 computer has been retrofitted into the first batch of Gripen aircraft, which originally carried an earlier version of the system.

Contractor

Ericsson Microwave Systems AB.

UPDATED

CD 207 mission computer

The CD 207 mission computer is based on three 486 computer module systems configured for multiprocessing. This unit replaces the predecessor CD 107 in the JA 37 Viggen Fighter as a form and fit replaceable LRU.

The CD 207 includes a MIL-STD-1553B RT/BC interface, a mass storage of 40/80 Mbyte and a master controller for the aircraft data communication to all major avionics.

Operational status

Prototypes delivered to customer.

Contractor

Ericsson Saab Avionics AB.

VERIFIED

JAS 39 Gripen aircraft interface unit

The JAS 39 aircraft interface unit, which is based on a 486 computer module system, is used to interface aircraft sensors in the JAS 39 Gripen to the main electronic system. The unit contains the logic to handle warning information and operate the control panel.

Power supply control of the electronic system, normal control of the undercarriage and operation of a crash recorder are other functions of the unit. The main part of the logic in the unit is provided by software. Critical functions are implemented in dedicated logic.

The aircraft interface unit features modular design, LRU concept to facilitate maintenance, chassis integrated power supply for optimal heat dissipation, SMT for increased packing density and built-in redundancy.

Operational status

In production for the JAS 39 Gripen.

Contractor

Ericsson Saab Avionics AB.

VERIFIED

JAS 39 Gripen environmental control system controller

The environmental control system controller is a computer which controls cockpit temperature, cockpit air flow, cooling pack temperature and avionics air flow. The controller receives input signals for the cockpit control panel, various sensors and trip switches, and produces output signals for the control valves and indicator lamps. Communication to and from the aircraft avionics systems is made via a redundant interface MIL-STD-1553B serial bus.

The aircraft interface unit features modular design, LRU concept to facilitate maintenance, chassis integrated power supply for optimal heat dissipation, SMT for increased packing density and built-in redundancy.

Operational status

In production for the JAS 39 Gripen.

Contractor

Ericsson Saab Avionics AB.

VERIFIED

PP1, PP2 AND PP12 display processors

First-batch JAS 39 Gripens are fitted with PP1 and PP2 display processors. Second batch aircraft will be fitted with the smaller lighter PP12 processor.

Operational status

PP1 and PP2 in service. PP12 on order.

Batch one Gripen aircraft will also be fitted with PP12, as an upgrade at a later stage.

Contractor

Ericsson Saab Avionics AB.

VERIFIED

UNITED KINGDOM

ACCS 2000 and 2500 general purpose airborne computers

The ACCS 2000 (Airborne Computing and Communications System) is a family of microprogrammed computers designed to provide standard low-cost data processing for a wide range of vehicles and missions; it is a derivative of the AN/AYK-14 which is the standard airborne computer for the US Navy (see entry under Computing Devices International in the USA section). The computers can be used as multiprocessor units with an extensive interface capability. There is a high degree of functional and mechanical modularity for flexible growth. The system architecture is not affected by modular configuration changes and is the feature most responsible for providing low cost and versatility.

The Single Card Processing (SCP) module, the heart of the family of computers, contains the microprogrammed control, arithmetic unit, registers, micro-memory, real-time clocks, interrupt logic, bootstrap memory and bus interface.

The ACCS 2500, an upgrade of the earlier ACCS 2000, is fitted as the mission computer to all Royal Air Force Harrier GR. Mk 7 aircraft and will be fitted to the T10. The greater processing power of the ACCS 2500 was designed to cope with the greater demands of the night attack and poor visibility missions of the GR7. The ACCS 2500 features power conversion, SCP, memory control, two 64 k × 18-bit word core memories, input/output unit and three MIL-STD-1553B bus controllers.

Further modular upgrades are available by supplanting the SCP with up to four VHSIC processor modules, each of which increases the processing power by between two and four times that of a single SCP. Each VHSIC processor module carries 1 Mword onboard memory and additional semiconductor memory modules are also available.

The Computing Devices ACCS 2500 is the mission computer for the Royal Air Force Harrier GR. Mk 7

Specifications

Word length: 16-bit
Typical speed:
up to 1.2 Mips (one SCP)
up to 6 Mips (one VPM)
up to 12 Mips (two VPMs)
Input/output: 3 MIL-STD-1553B ports, up to 16 other channels
Memory:
32 k words 18-bit core modules (900 ns access time)
32 k words 18-bit semiconductor modules (400 ns access time)

Operational status

In service on Royal Air Force Harrier GR. Mk 7 aircraft. The 88 ACCS 2000 computers have been modified to ACCS 2500 standard.

Contractor

Computing Devices Company Ltd.

UPDATED

ACCS 3000 mission computer

The ACCS 3000 series is a family of computers designed to meet the requirements of the next generation of military aircraft such as the EF 2000, or for major avionic upgrade programmes. It uses a processor-independent multiprocessor architecture

EF 2000 processor modules

with an integrated Ada environment. Modules are available using any standard 16-bit or 32-bit CISC or RISC microprocessor family or with non-standard families. The processors are augmented by modules performing signal processing, display graphics processing and offering databus and standard signal interfaces, to give an integrated computing and communications system.

Specifications

Dimensions: 240 × 150 mm
Weight: 0.7 kg
Power consumption: 15 W
Processing: 2 × 68020 32-bit 20 MHz processor and floating-point coprocessor, 2 Mbyte SRAM and 2 Mbyte EPROM giving 2.5 Mips performance. Software in Ada.

Operational status

The processing subsystem has been developed and selected for four main computers for the EF 2000: the navigation computer, attack computer, interface processing (utilities) unit and defensive aids computer.

Contractor

Computing Devices Company Ltd.

VERIFIED

Helicopter Air Data System (HADS)

In 1979, GEC-Marconi Avionics designed and developed a low airspeed Helicopter Air Data System (HADS) for the AH-1S Cobra helicopter.

The system provides full three-axis, prime accuracy, air data information by utilising the GEC-Marconi Avionics Airspeed And Direction Sensor (AADS). Recently, GEC-Marconi Avionics has created variants of the HIADC that can interface to the AADS probe. This now means that three-axis air data information is available on modern digital busses such as MIL-STD-1553B. The major parts of the system are shown below:

Airspeed And Direction Sensor (AADS)

The AADS probe is mounted externally to the aircraft below the rotor and swivels to align with the local airflow. The AADS contains pitot and static pressure ports for measuring the magnitude of the local airspeed vector and a pair of resolvers for determining its direction. This enables the system to provide data for: rotor downwash velocity; ground effect; forward, rearward and lateral airspeed (to zero kt); vertical airspeed; wind direction, drift and lift margin (when integrated into an avionics suite); enhanced pilot awareness.

Installation of the system has demonstrated major improvements for both operational and flight test environments in: fire control; low airspeed, low altitude manoeuvres.

HIADC Interface

The HIADC interfaces with the AADS by measuring its probe angle, air temperature and pitot and static pressures. Air data calculations are performed and the resultant parameters are made available on digital databusses.

Specifications

Airspeed and direction sensor
Dimensions: 97 × 317 × 246 mm
Weight: 1.1 kg

High integration air data computer
Dimensions: 140 × 102 × 83 mm
Weight: 1.2 kg

Operational status

System proven in flight evaluation on 16 different types of helicopter including UH-1, UH-60, CH-47, BO 105, EH 101, Puma and Sea King. Over 1,500 systems supplied for attack helicopters including Bell AH-1S Cobra, Agusta A-129 and Super Puma. A major order was received in 1997 from McDonnell Douglas Helicopter Systems (now The Boeing Company) for HADS in support of US Army AH-64D Longbow and British Army WAH-64 Apache helicopters.

Contractor

GEC-Marconi Avionics Ltd, Rochester.

UPDATED

1285

GEC-Marconi Avionics Helicopter Air Data system (HADS)

High-Integration Air Data Computer (HIADC)

The High-Integration Air Data Computer (HIADC) utilises advanced production techniques and miniaturised air data transducers. It features low individual component count and power consumption to provide an accurate, highly reliable, compact air data computer to meet the growing market demand for distributed pressure sensing devices.

The flexible configuration concept of HIADC offers the ability to satisfy applications for both fixed-wing aircraft and helicopters.

HIADC typically interfaces to the aircraft total air temperature and angle of attack sensors; measures pitot, static and differential pressures and, having corrected for systematic error characteristics, computes a full range of accurate air data parameters. Data is digitally distributed to various aircraft systems in either ARINC 429, RS-422 or MIL-STD-1553B output formats.

HIADC has been designed as a fit-and-forget air data system, which virtually maintenance and pr benefits.

Specifications

Dimensions: 1
Weight: 1.2 k
Power supp
Reliability

1997/0002346

GEC-Marconi Avionics High-Integration Air Data

s for cost

Operational status

In production for numerous fixed- and rotary-wing programmes worldwide.

Contractor

GEC-Marconi Avionics Ltd, Rochester.

UPDATED

Miniature Standard Central Air Data Computer (MSCADC)

The Miniature Standard Central Air Data Computer (MSCADC) is a modular digital air data computer which provides a lightweight, high-reliability system. The MSCADC can easily be configured for retrofit applications, such as for the A-4 or F-5, or new aircraft where centralised air data functions are required. MSCADC outputs include digital ARINC 429 and MIL-STD-1553B formats, analogue, synchro and discrete.

Specifications

Dimensions: 115 × 140 × 229 mm
Weight: 4.77 kg
Power: 35 W (max)
Reliability: >7,800 h MTBF

Operational status

MSCADC is in full-scale production for F-5 applications worldwide.

Contractor

GEC-Marconi Avionics Ltd, Rochester.

VERIFIED

Standard Central Air Data Computers (SCADC)

At the end of 1981 GEC-Marconi Avionics was contracted to develop a range of Standard Central Air Data Computers (SCADCs) to upgrade US Air Force and Navy aircraft. The common standard replaces the great variety of air data systems installed in some 40 variants of 15 different types in US military service.

To satisfy these requirements SCADC has been ...oped in six configurations, each having more than ... cent core commonality with the others. One or ...ecial-to-type modules are included in each ...ation to accommodate unique aircraft .... The high commonality design has produced ... life cycle cost benefits. The equipment is ...o the latest software, computing and data ... specifications used in US military aircraft. ... cent failure detection and isolation is ... the built-in test systems, and ground ...enable 98 per cent of all LRU failures to ... isolated.

...urations are:

...**140/A.** This unit is fitted on six US Air ... aircraft types and includes a MIL-... capability.

LRUs for the range of GEC-Marconi Avionics SCADC units ***1997**/0001286*

Specifications

Dimensions: 191 × 191 × 302 mm
Weight: 11.1 kg
Power: 72 W

SCADC unit CPU 141/A. This is fitted on the US Air Force C-5A, C-5B, C-141A, C-141B military transports. The system also includes a MIL-STD-1553B interface capability.

Specifications

Dimensions: 191 × 124 × 545 mm
Weight: 15.3 kg
Power: 102 W

SCADC unit CPU 142/A. This equiped six versions of the US Air Force General Dynamics F-111, and was designed to include MIL-STD-1553B interface capability.

Specifications

Dimensions: 191 × 356 × 483 mm
Weight: 38.1 kg
Power: 98 W

SCADC unit CPU 143/A. This computer has been designed to include MIL-STD-1553B for eight versions of the F-4 Phantom.

Specifications

Dimensions: 191 × 420 × 305 mm
Weight: 20.68 kg
Power: 80 W

SCADC unit CPU 152/A. This two-channel unit is designed for all variants of the US Navy S-3 aircraft and includes dual MIL-STD-1553B navigation system and cockpit instrument interfaces.

Specifications

Dimensions: 222 × 142 × 485 mm
Weight: 15 kg
Power: 30 W

SCADC unit CPU 175/A. This computer has been designed to be fitted to the US Navy Northrop Grumman F-14A, A+ and D aircraft and includes MIL-STD-1553B interface. In addition to performing normal CADC functions, this unit also provides dual-channel control of the wing sweep.

Specifications

Dimensions: 160 × 218 × 510 mm
Weight: 31 kg
Power: 26 W

Operational status

SCADC has recently completed quantity production with over 5,500 units delivered to the US and other users worldwide.

Contractor

GEC-Marconi Avionics Ltd, Rochester.

UPDATED

...nd Transfer Set

...and transfer set is currently ...emory capacity in a fully ...n in military aircraft.

...as a quick-release handle ...tion and handling. The ...ompatible with aircraft ...d includes the power ... for control and data ...read/write capability

...used for upload of ...tional data or for ...sion data. Higher capacity variants, up to 400 Mbytes are in development and flight evaluation.

Operational status

In production for the Royal Air Force Harrier GR. Mk 7 and T. Mk 10.

Contractor

Normalair-Garrett Ltd.

UPDATED

The solid-state data transfer system has been selected for Royal Air Force Harrier GR. Mk 7 and T. Mk 10 aircraft
1995

Miniature primary Air Data Computer GADU D60350

The GADU miniature ADC is a compact, lightweight unit with integrated pitot/static solid state sensors in a rugged enclosure. The GADU is available with either ARINC 429 or MIL-STD-1553 bus interfaces. A 'height-lock' function is provided on the GADU, which makes it particularly suitable for helicopter application.

Specifications

Dimensions: 89 × 132 × 172 mm
Weight: 1.3 kg
Power supply: 28 V DC, 10 W peak

Operational status

In service on a wide range of helicopters; also on Boeing 727/737 and Lockheed Martin C-130 aircraft.

Contractor

Penny & Giles Avionic Systems Ltd.

VERIFIED

Primary Air Data Computers (ADC)

Penny & Giles Aerospace produces primary accuracy ADCs in both miniature and ARINC 565 standard form factors. The Generic Air Data Unit (GADU) miniature ADC is a compact lightweight unit with integrated pitot/static solid-state sensors in a rugged enclosure. The GADU is available with either ARINC 429 or MIL-STD-1553 databus interfaces. A height lock function is provided on the GADU and is therefore particularly suited to rotary-wing applications.

Specifications

Miniature primary ADC GADU D60350
Dimensions: 89 × 132 × 172 mm
Weight: 1.3 kg
Power: 28 V DC, 10 W peak
Standards: RTCA DO-160C/178B
Interfaces: ARINC 429/575, airspeed and height lock, MIL-STD-1553 (optional)
MTBF: 15,000 flight hours

Generic Air Data Unit (GADU) ADC **1997**/0005416

Operational status

Fitted to a wide range of aircraft and helicopters including: Sikorsky S70/S76, Westland Lynx/Sea King, Eurocopter Dauphin, MBB BK117 helicopters; and Boeing 727/737, Lockheed Martin C-130 aircraft. Applications include: primary air data system; rotary- and fixed-wing versions for RSVM retrofit; FMS; TCAS; GPWS; windshear detection.

Contractor

Penny & Giles Aerospace Ltd.

VERIFIED

Penny & Giles Aerospace Digitas ADC **1997**/0005417

Secondary Air Data Computers (ADC)

The Digitas and TAS/Plus ADCs provide a wide range of outputs including: altitude; airspeed; airspeed rate; baro corrected altitude; static air temperature; true airspeed, Mach and total air temperature. Depending on the model, these outputs are available on various avionic interfaces, including ARINC 429/575, MIL-STD-1553, RS 232/422 and analogue.

Specifications

Miniature primary ADC GADU D60350
Dimensions: 89 × 132 × 160 mm
Weight: 1.1 kg
Power: 28 V DC, 10 W peak
Standards: RTCA DO-160B/178B
Interfaces: ARINC 429/575, airspeed
Optional interfaces: MIL-STD-1553
MTBF: 15,000 flight hours
Applications: Secondary air data source, rotary- and fixed-wing flight data recording, HUMS

Secondary ADC TAS/Plus 90004
Dimensions: ⅜ ATR short
Weight: 2.04 kg
Power: 28 V DC, 10 W peak
Standards: RTCA DO-160C/178B, TSO-C106
Interfaces: ARINC 429/575, RS-232/-422, altitude rate/airspeed
Optional interfaces: Analogue altitude rate and Mach
MTBF: 10,000 flight hours
Applications: Business and commuter aircraft GPS, FMS, TCAS, GPWS, flight data recording data display

Operational status

Fitted to a wide range of rotary- and fixed-wing aircraft.

Contractor

Penny & Giles Aerospace Ltd.

VERIFIED

Secondary air data sources

The analogue Air Data Module (ADM) has been especially designed for GPWS (Ground Proximity Warning System) applications requiring a low-cost source of secondary air data information via an analogue channel. The ADM can be mounted in a wide variety of locations to aid retrofit of a GPWS capability.

The TP91 range of altitude and airspeed transducers provides a passive, lightweight, low-cost solution for applications such as flight data recording where a secondary or independent source of altitude or airspeed data is required.

Specifications

Air Data Module D60286
Dimensions: 89 × 132 × 159 mm
Weight: 1 kg
Power: 28 V DC, 5 W peak
Standards: RTCA DO-160C
Applications: GPWS

Altitude and Airspeed Transducer TP91
Dimensions: 83 × 90 × 82 mm
Weight: 0.8 kg
Power: 12 V DC, 4W peak
Standards: TSO C51a
Applications: Flight data recording

Contractor

Penny & Giles Aerospace Ltd.

VERIFIED

Advanced Memory Unit AMU

The Smiths Industries AMU is a sophisticated data transfer and recording system which is designed to operate on both a standard MIL-STD-1553A/B communications bus and the high-speed interface bus over which the AMU and the associated new US Navy Tactical Aircraft Moving MAp Capabiity (TAMMAC), for which it is designed, will communicate. Each AMU can accept two PCMIA cards.

The AMU will replace earlier US Navy data storage and mission data loader equipment.

Operational status

The TAMMAC system is scheduled to be fitted initially to five baseline US Navy TAMMAC aircraft: F/A-18, AV-8B, AH-1W, UH-1N, and V-22, with the following aircraft likely to be fitted later: F-14, S-3, CH-53, CH-60, SH-60, and P-3.

Contractor

Smiths Industries Aerospace.

NEW ENTRY

Digital air data computers for civil aircraft

A range of digital air data computers has been produced for both turboprop and civil jet transports. These computers meet the requirements of ARINC 706 specification, with outputs in both ARINC 429 and analogue format. They incorporate high-performance vibrating cylinder sensors and CMOS technology to minimise weight, space and power requirements.

Extensive BITE facilities are available, including a 10-flight memory to facilitate onboard checks and bench diagnostic routines.

This range of computers is demonstrating extremely high reliability on the Boeing 737 with a MTBF in excess of 30,000 flying hours.

Specifications

Dimensions: ⅜ or ½ ATR
Weight: 4-5.9 kg
Power: 20 VA

Operational status

In production and service on the Boeing 737-300, 400 and 500, the BAe ATP and Boeing 727 upgrade.

Contractor

Smiths Industries Aerospace.

UPDATED

Digital air data computers for military aircraft

A range of digital air data computers is available for both conventional and VTOL aircraft applications. Additional channels are incorporated where increased integrity is required as in the case of VTOL aircraft.

Depending on specification, high- or low-range pressure vibrating cylinder sensors are used and outputs are provided in either analogue or MIL-STD-1553 format. Comprehensive BITE facilities are incorporated, including continuous in-flight monitoring and fault location down to module level.

Specifications

Dimensions: ⅜ ATR
Weight: 5 kg
Power: 35 VA

Operational status

In service on the BAe Sea Harrier.

Contractor

Smiths Industries Aerospace.

UPDATED

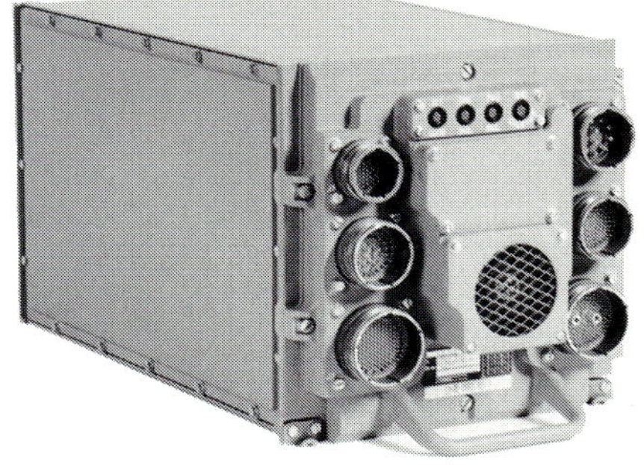

Displays and Mission Computer DMC

1998/0015266

Displays and Mission Computer DMC

Smiths Industries Aerospace has developed a modular Displays and Mission Computer DMC, which can be configured to match the exact system requirements of each application by selecting modules from an extensive library of standard electronic cards, power supplies and ATR short cases.

The standardised internal databus allows the system to be updated or expanded by inserting appropriate cards as technology evolves.

The size of the DMC depends on the number of modules needed to meet the operational requirements; ½ ATR, ¾ ATR, and 1 ATR sizes are available.

Electrical connections are via a rear panel which can be adapted to different connector configurations including: DPX and ARINC 600. The panel is detached to provide access to the backplane. Cases are cooled to ARINC 404A/600 specifications.

Operational status

The DMC is standard equipment on the next-generation BAe Hawk for the Royal Australian Air Force lead-in fighter programme, and for a number of other programmes including: the BAe Harrier F/A-2, Boeing AV-8B Harrier II and T-45 Goshawk.

Contractor

Smiths Industries Aerospace, Defence Systems.

NEW ENTRY

UNITED STATES OF AMERICA

Air data computer for the A-10

Designed to full military environment and MIL-STD-1553 digital databus standards, the air data computer for the US Air Force Fairchild A-10A features a digital pressure transducer and a ferro-resonant power supply. Modular programming is provided and the memory has a 6 k word read-only memory and a 256 word random access memory.

Specifications

Dimensions: ½ ATR
Weight: 6.3 kg
Power: 36 W

Operational status

In service on US Air Force A-10 aircraft.

Contractor

AlliedSignal Engine Systems & Accessories.

VERIFIED

Digital air data computer for the AV-8B

In common with all AlliedSignal's air data computers, this model uses a quartz pressure transducer. The unit is similar in construction to that used on the US Air Force A-10 aircraft. It features a non-volatile memory for the retention of fault messages, even if power is turned off.

Specifications

Dimensions: 129 × 193 × 355 mm
Weight: 7.2 kg
Inputs: 10
Outputs: 78

Operational status

In service on the US Marine Corps AV-8B V/STOL close support aircraft.

Contractor

AlliedSignal Engine Systems & Accessories.

VERIFIED

Digital air data computer for the B-1B

This computer was developed for the US Air Force B-1B and features a digital MOS large-scale integrated circuit processor and digital pressure transducers. It has an altitude reporting facility and the built-in self-test system provides continuous failure monitoring.

Specifications

Dimensions: 158 × 195 × 500 mm
Weight: 12.6 kg
Inputs: 19
Outputs: 92

Operational status

Developed for the US Air Force B-1A and in service in the B-1B.

Contractor

AlliedSignal Engine Systems & Accessories.

VERIFIED

Secondary air data sources

The analogue Air Data Module (ADM) has been especially designed for GPWS (Ground Proximity Warning System) applications requiring a low-cost source of secondary air data information via an analogue channel. The ADM can be mounted in a wide variety of locations to aid retrofit of a GPWS capability.

The TP91 range of altitude and airspeed transducers provides a passive, lightweight, low-cost solution for applications such as flight data recording where a secondary or independent source of altitude or airspeed data is required.

Specifications

Air Data Module D60286
Dimensions: 89 × 132 × 159 mm
Weight: 1 kg
Power: 28 V DC, 5 W peak
Standards: RTCA DO-160C
Applications: GPWS

Altitude and Airspeed Transducer TP91
Dimensions: 83 × 90 × 82 mm
Weight: 0.8 kg
Power: 12 V DC, 4W peak
Standards: TSO C51a
Applications: Flight data recording

Contractor

Penny & Giles Aerospace Ltd.

VERIFIED

Advanced Memory Unit AMU

The Smiths Industries AMU is a sophisticated data transfer and recording system which is designed to operate on both a standard MIL-STD-1553A/B communications bus and the high-speed interface bus over which the AMU and the associated new US Navy Tactical Aircraft Moving MAp Capabiity (TAMMAC), for which it is designed, will communicate. Each AMU can accept two PCMIA cards.

The AMU will replace earlier US Navy data storage and mission data loader equipment.

Operational status

The TAMMAC system is scheduled to be fitted initially to five baseline US Navy TAMMAC aircraft: F/A-18, AV-8B, AH-1W, UH-1N, and V-22, with the following aircraft likely to be fitted later: F-14, S-3, CH-53, CH-60, SH-60, and P-3.

Contractor

Smiths Industries Aerospace.

NEW ENTRY

Digital air data computers for civil aircraft

A range of digital air data computers has been produced for both turboprop and civil jet transports. These computers meet the requirements of ARINC 706 specification, with outputs in both ARINC 429 and analogue format. They incorporate high-performance vibrating cylinder sensors and CMOS technology to minimise weight, space and power requirements.

Extensive BITE facilities are available, including a 10-flight memory to facilitate onboard checks and bench diagnostic routines.

This range of computers is demonstrating extremely high reliability on the Boeing 737 with a MTBF in excess of 30,000 flying hours.

Specifications

Dimensions: 3/8 or 1/2 ATR
Weight: 4-5.9 kg
Power: 20 VA

Operational status

In production and service on the Boeing 737-300, 400 and 500, the BAe ATP and Boeing 727 upgrade.

Contractor

Smiths Industries Aerospace.

UPDATED

Digital air data computers for military aircraft

A range of digital air data computers is available for both conventional and VTOL aircraft applications. Additional channels are incorporated where increased integrity is required as in the case of VTOL aircraft.

Depending on specification, high- or low-range pressure vibrating cylinder sensors are used and outputs are provided in either analogue or MIL-STD-1553 format. Comprehensive BITE facilities are incorporated, including continuous in-flight monitoring and fault location down to module level.

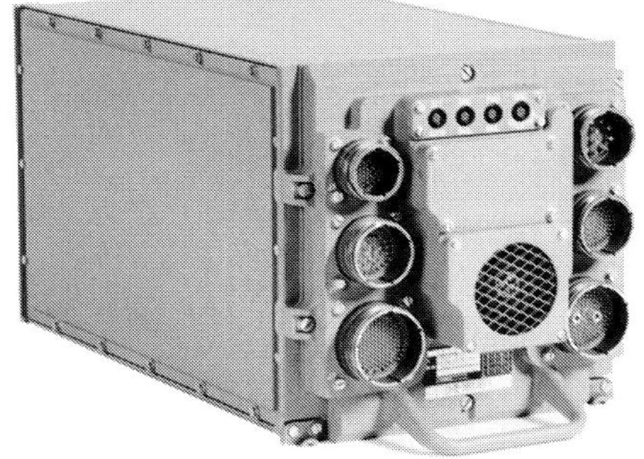

Displays and Mission Computer DMC
1998/0015266

Specifications

Dimensions: 3/8 ATR
Weight: 5 kg
Power: 35 VA

Operational status

In service on the BAe Sea Harrier.

Contractor

Smiths Industries Aerospace.

UPDATED

Displays and Mission Computer DMC

Smiths Industries Aerospace has developed a modular Displays and Mission Computer DMC, which can be configured to match the exact system requirements of each application by selecting modules from an extensive library of standard electronic cards, power supplies and ATR short cases.

The standardised internal databus allows the system to be updated or expanded by inserting appropriate cards as technology evolves.

The size of the DMC depends on the number of modules needed to meet the operational requirements; 1/2 ATR, 3/4 ATR, and 1 ATR sizes are available.

Electrical connections are via a rear panel which can be adapted to different connector configurations including: DPX and ARINC 600. The panel is detached to provide access to the backplane. Cases are cooled to ARINC 404A/600 specifications.

Operational status

The DMC is standard equipment on the next-generation BAe Hawk for the Royal Australian Air Force lead-in fighter programme, and for a number of other programmes including: the BAe Harrier F/A-2, Boeing AV-8B Harrier II and T-45 Goshawk.

Contractor

Smiths Industries Aerospace, Defence Systems.

NEW ENTRY

UNITED STATES OF AMERICA

Air data computer for the A-10

Designed to full military environment and MIL-STD-1553 digital databus standards, the air data computer for the US Air Force Fairchild A-10A features a digital pressure transducer and a ferro-resonant power supply. Modular programming is provided and the memory has a 6 k word read-only memory and a 256 word random access memory.

Specifications

Dimensions: 1/2 ATR
Weight: 6.3 kg
Power: 36 W

Operational status

In service on US Air Force A-10 aircraft.

Contractor

AlliedSignal Engine Systems & Accessories.

VERIFIED

Digital air data computer for the AV-8B

In common with all AlliedSignal's air data computers, this model uses a quartz pressure transducer. The unit is similar in construction to that used on the US Air Force A-10 aircraft. It features a non-volatile memory for the retention of fault messages, even if power is turned off.

Specifications

Dimensions: 129 × 193 × 355 mm
Weight: 7.2 kg
Inputs: 10
Outputs: 78

Operational status

In service on the US Marine Corps AV-8B V/STOL close support aircraft.

Contractor

AlliedSignal Engine Systems & Accessories.

VERIFIED

Digital air data computer for the B-1B

This computer was developed for the US Air Force B-1B and features a digital MOS large-scale integrated circuit processor and digital pressure transducers. It has an altitude reporting facility and the built-in self-test system provides continuous failure monitoring.

Specifications

Dimensions: 158 × 195 × 500 mm
Weight: 12.6 kg
Inputs: 19
Outputs: 92

Operational status

Developed for the US Air Force B-1A and in service in the B-1B.

Contractor

AlliedSignal Engine Systems & Accessories.

VERIFIED

Miniature primary Air Data Computer GADU D60350

The GADU miniature ADC is a compact, lightweight unit with integrated pitot/static solid state sensors in a rugged enclosure. The GADU is available with either ARINC 429 or MIL-STD-1553 bus interfaces. A 'height-lock' function is provided on the GADU, which makes it particularly suitable for helicopter application.

Specifications

Dimensions: 89 × 132 × 172 mm
Weight: 1.3 kg
Power supply: 28 V DC, 10 W peak

Operational status

In service on a wide range of helicopters; also on Boeing 727/737 and Lockheed Martin C-130 aircraft.

Contractor

Penny & Giles Avionic Systems Ltd.

VERIFIED

Primary Air Data Computers (ADC)

Penny & Giles Aerospace produces primary accuracy ADCs in both miniature and ARINC 565 standard form factors. The Generic Air Data Unit (GADU) miniature ADC is a compact lightweight unit with integrated pitot/static solid-state sensors in a rugged enclosure. The GADU is available with either ARINC 429 or MIL-STD-1553 databus interfaces. A height lock function is provided on the GADU and is therefore particularly suited to rotary-wing applications.

Specifications

Miniature primary ADC GADU D60350
Dimensions: 89 × 132 × 172 mm
Weight: 1.3 kg
Power: 28 V DC, 10 W peak
Standards: RTCA DO-160C/178B
Interfaces: ARINC 429/575, airspeed and height lock, MIL-STD-1553 (optional)
MTBF: 15,000 flight hours

Generic Air Data Unit (GADU) ADC ***1997***/0005416

Operational status

Fitted to a wide range of aircraft and helicopters including: Sikorsky S70/S76, Westland Lynx/Sea King, Eurocopter Dauphin, MBB BK117 helicopters; and Boeing 727/737, Lockheed Martin C-130 aircraft. Applications include: primary air data system; rotary- and fixed-wing versions for RSVM retrofit; FMS; TCAS; GPWS; windshear detection.

Contractor

Penny & Giles Aerospace Ltd.

VERIFIED

Penny & Giles Aerospace Digitas ADC ***1997***/0005417

Secondary Air Data Computers (ADC)

The Digitas and TAS/Plus ADCs provide a wide range of outputs including: altitude; airspeed; airspeed rate; baro corrected altitude; static air temperature; true airspeed, Mach and total air temperature. Depending on the model, these outputs are available on various avionic interfaces, including ARINC 429/575, MIL-STD-1553, RS 232/422 and analogue.

Specifications

Miniature primary ADC GADU D60350
Dimensions: 89 × 132 × 160 mm
Weight: 1.1 kg
Power: 28 V DC, 10 W peak
Standards: RTCA DO-160B/178B
Interfaces: ARINC 429/575, airspeed
Optional interfaces: MIL-STD-1553
MTBF: 15,000 flight hours
Applications: Secondary air data source, rotary- and fixed-wing flight data recording, HUMS

Secondary ADC TAS/Plus 90004
Dimensions: 3/8 ATR short
Weight: 2.04 kg
Power: 28 V DC, 10 W peak
Standards: RTCA DO-160C/178B, TSO-C106
Interfaces: ARINC 429/575, RS-232/-422, altitude rate/airspeed
Optional interfaces: Analogue altitude rate and Mach
MTBF: 10,000 flight hours
Applications: Business and commuter aircraft GPS, FMS, TCAS, GPWS, flight data recording data display

Operational status

Fitted to a wide range of rotary- and fixed-wing aircraft.

Contractor

Penny & Giles Aerospace Ltd.

VERIFIED

Operational status
In production for numerous fixed- and rotary-wing programmes worldwide.

Contractor
GEC-Marconi Avionics Ltd, Rochester.

UPDATED

Miniature Standard Central Air Data Computer (MSCADC)

The Miniature Standard Central Air Data Computer (MSCADC) is a modular digital air data computer which provides a lightweight, high-reliability system. The MSCADC can easily be configured for retrofit applications, such as for the A-4 or F-5, or new aircraft where centralised air data functions are required. MSCADC outputs include digital ARINC 429 and MIL-STD-1553B formats, analogue, synchro and discrete.

Specifications
Dimensions: 115 × 140 × 229 mm
Weight: 4.77 kg
Power: 35 W (max)
Reliability: >7,800 h MTBF

Operational status
MSCADC is in full-scale production for F-5 applications worldwide.

Contractor
GEC-Marconi Avionics Ltd, Rochester.

VERIFIED

Standard Central Air Data Computers (SCADC)

At the end of 1981 GEC-Marconi Avionics was contracted to develop a range of Standard Central Air Data Computers (SCADCs) to upgrade US Air Force and Navy aircraft. The common standard replaces the great variety of air data systems installed in some 40 variants of 15 different types in US military service.

To satisfy these requirements SCADC has been developed in six configurations, each having more than 80 per cent core commonality with the others. One or two special-to-type modules are included in each configuration to accommodate unique aircraft interfaces. The high commonality design has produced significant life cycle cost benefits. The equipment is designed to the latest software, computing and data transmission specifications used in US military aircraft. Over 95 per cent failure detection and isolation is achieved with the built-in test systems, and ground crew tests will enable 98 per cent of all LRU failures to be detected and isolated.

The six configurations are:

SCADC unit CPU 140/A. This unit is fitted on six US Air Force and US Navy aircraft types and includes a MIL-STD-1553B interface capability.

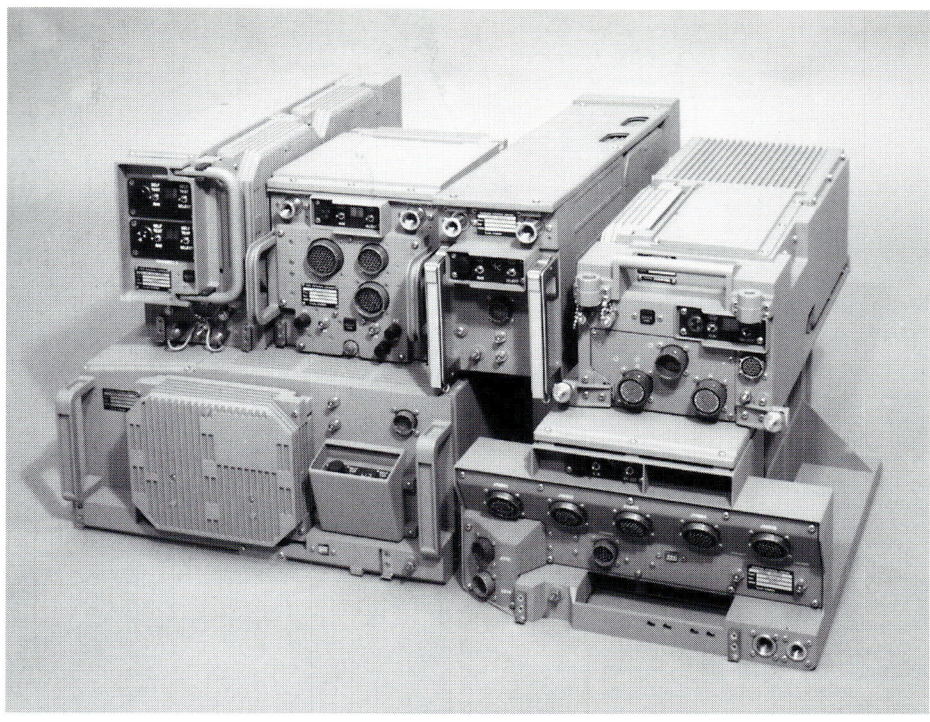

LRUs for the range of GEC-Marconi Avionics SCADC units ***1997***/0001286

Specifications
Dimensions: 191 × 191 × 302 mm
Weight: 11.1 kg
Power: 72 W

SCADC unit CPU 141/A. This is fitted on the US Air Force C-5A, C-5B, C-141A, C-141B military transports. The system also includes a MIL-STD-1553B interface capability.

Specifications
Dimensions: 191 × 124 × 545 mm
Weight: 15.3 kg
Power: 102 W

SCADC unit CPU 142/A. This equiped six versions of the US Air Force General Dynamics F-111, and was designed to include MIL-STD-1553B interface capability.

Specifications
Dimensions: 191 × 356 × 483 mm
Weight: 38.1 kg
Power: 98 W

SCADC unit CPU 143/A. This computer has been designed to include MIL-STD-1553B for eight versions of the F-4 Phantom.

Specifications
Dimensions: 191 × 420 × 305 mm
Weight: 20.68 kg
Power: 80 W

SCADC unit CPU 152/A. This two-channel unit is designed for all variants of the US Navy S-3 aircraft and includes dual MIL-STD-1553B navigation system and cockpit instrument interfaces.

Specifications
Dimensions: 222 × 142 × 485 mm
Weight: 15 kg
Power: 30 W

SCADC unit CPU 175/A. This computer has been designed to be fitted to the US Navy Northrop Grumman F-14A, A+ and D aircraft and includes MIL-STD-1553B interface. In addition to performing normal CADC functions, this unit also provides dual-channel control of the wing sweep.

Specifications
Dimensions: 160 × 218 × 510 mm
Weight: 31 kg
Power: 26 W

Operational status
SCADC has recently completed quantity production with over 5,500 units delivered to the US and other users worldwide.

Contractor
GEC-Marconi Avionics Ltd, Rochester.

UPDATED

Data Storage and Transfer Set (DSTS)

The solid-state data storage and transfer set is currently available with 1 Mbyte of memory capacity in a fully qualified format for installation in military aircraft.

The data transfer module has a quick-release handle mechanism for ease of operation and handling. The system is configured to be compatible with aircraft standard console mountings and includes the power supply and a databus interface for control and data transmission. The system has full read/write capability for upload and download of data.

The data transfer system can be used for upload of mission, navigation and other operational data or for the download of maintenance and mission data. Higher capacity variants, up to 400 Mbytes are in development and flight evaluation.

Operational status
In production for the Royal Air Force Harrier GR. Mk 7 and T. Mk 10.

Contractor
Normalair-Garrett Ltd.

UPDATED

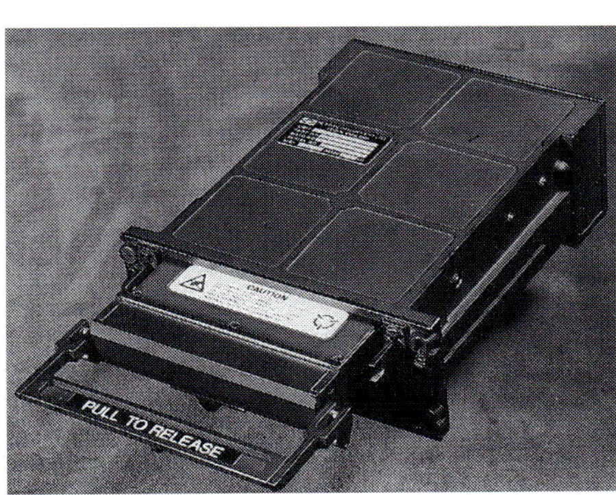

The solid-state data transfer system has been selected for Royal Air Force Harrier GR. Mk 7 and T. Mk 10 aircraft
1995

with an integrated Ada environment. Modules are available using any standard 16-bit or 32-bit CISC or RISC microprocessor family or with non-standard families. The processors are augmented by modules performing signal processing, display graphics processing and offering databus and standard signal interfaces, to give an integrated computing and communications system.

Specifications

Dimensions: 240 × 150 mm
Weight: 0.7 kg
Power consumption: 15 W
Processing: 2 × 68020 32-bit 20 MHz processor and floating-point coprocessor, 2 Mbyte SRAM and 2 Mbyte EPROM giving 2.5 Mips performance. Software in Ada.

Operational status

The processing subsystem has been developed and selected for four main computers for the EF 2000: the navigation computer, attack computer, interface processing (utilities) unit and defensive aids computer.

Contractor

Computing Devices Company Ltd.

VERIFIED

Helicopter Air Data System (HADS)

In 1979, GEC-Marconi Avionics designed and developed a low airspeed Helicopter Air Data System (HADS) for the AH-1S Cobra helicopter.

The system provides full three-axis, prime accuracy, air data information by utilising the GEC-Marconi Avionics Airspeed And Direction Sensor (AADS). Recently, GEC-Marconi Avionics has created variants of the HIADC that can interface to the AADS probe. This now means that three-axis air data information is available on modern digital busses such as MIL-STD-1553B. The major parts of the system are shown below:

Airspeed And Direction Sensor (AADS)

The AADS probe is mounted externally to the aircraft below the rotor and swivels to align with the local airflow. The AADS contains pitot and static pressure ports for measuring the magnitude of the local airspeed vector and a pair of resolvers for determining its direction. This enables the system to provide data for: rotor downwash velocity; ground effect; forward, rearward and lateral airspeed (to zero kt); vertical airspeed; wind direction, drift and lift margin (when integrated into an avionics suite); enhanced pilot awareness.

Installation of the system has demonstrated major improvements for both operational and flight test environments in: fire control; low airspeed, low altitude manoeuvres.

HIADC Interface

The HIADC interfaces with the AADS by measuring its probe angle, air temperature and pitot and static pressures. Air data calculations are performed and the resultant parameters are made available on digital databusses.

Specifications

Airspeed and direction sensor
Dimensions: 97 × 317 × 246 mm
Weight: 1.1 kg

High integration air data computer
Dimensions: 140 × 102 × 83 mm
Weight: 1.2 kg

Operational status

System proven in flight evaluation on 16 different types of helicopter including UH-1, UH-60, CH-47, BO 105, EH 101, Puma and Sea King. Over 1,500 systems supplied for attack helicopters including Bell AH-1S Cobra, Agusta A-129 and Super Puma. A major order was received in 1997 from McDonnell Douglas Helicopter Systems (now The Boeing Company) for HADS in support of US Army AH-64D Longbow and British Army WAH-64 Apache helicopters.

Contractor

GEC-Marconi Avionics Ltd, Rochester.

UPDATED

GEC-Marconi Avionics Helicopter Air Data system (HADS) **1997**/0001285

High-Integration Air Data Computer (HIADC)

The High-Integration Air Data Computer (HIADC) utilises advanced production techniques and miniaturised air data transducers. It features low individual component count and power consumption to provide an accurate, highly reliable, compact air data computer to meet the growing market demand for distributed pressure sensing devices.

The flexible configuration concept of HIADC offers the ability to satisfy applications for both fixed-wing aircraft and helicopters.

HIADC typically interfaces to the aircraft total air temperature and angle of attack sensors; measures pitot, static and differential pressures and, having corrected for systematic error characteristics, computes a full range of accurate air data parameters. Data is digitally distributed to various aircraft systems in either ARINC 429, RS-422 or MIL-STD-1553B output formats.

HIADC has been designed as a fit-and-forget air data system, which virtually eliminates the requirements for maintenance and provides significant life cycle cost benefits.

Specifications

Dimensions: 140 × 82 × 102 mm (max)
Weight: 1.2 kg
Power supply: 28 V DC, 6 W
Reliability: >25,000 h MTBF

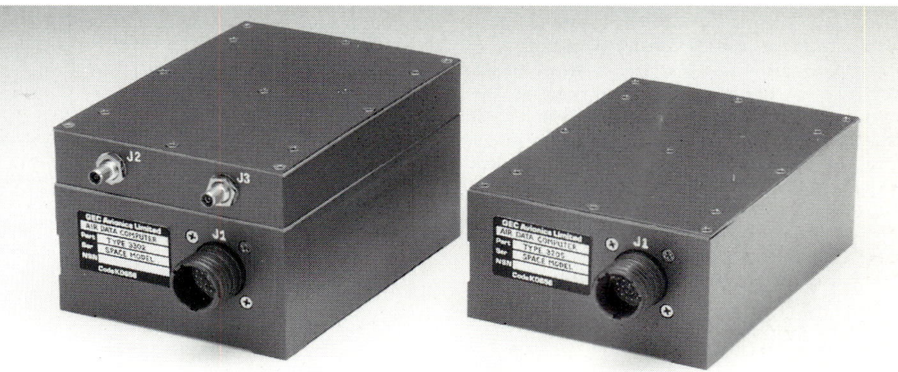

GEC-Marconi Avionics High-Integration Air Data Computer (HIADC) **1997**/0002346

Digital air data computer for the JA 37

This is an all-digital computer with similar features to those of the B-1 air data computer and has bidirectional digital communication with the aircraft subsystems. It contains both static and dynamic self-test facilities.

Specifications

Dimensions: 157 × 213 × 439 mm
Weight: 11.4 kg
Inputs: 16
Outputs: 37

Operational status

In service in the Saab JA 37 interceptor.

Contractor

AlliedSignal Engine Systems & Accessories.

VERIFIED

Miniature Air Data Computer (MADC)

The simplification of the interface to MIL-STD-1553B with only a few analogue inputs has made possible the Miniature Air Data Computer (MADC).

Inputs to the baseline MADC are indicated static and total pressure, total temperature (50 or 500 ohm probe), self-test and identification discrete signals with commands via the 1553 databus. Analogue inputs of barometric correction and angle of attack may also be provided.

In addition to the digital databus and altitude reporting code outputs, the MADC has provision for a dual-synchro drive compatible with AAU-19/A, AAU-34/A and AAU-37/A altimeters. These synchro outputs can also be software-programmed to provide other air data functions such as Mach number or true airspeed. Other miscellaneous analogue and discrete outputs tailored to a particular application can be provided. The pressure transducers are self-contained plug-in modules using fused quartz RC sensors.

Alternative mounting arrangements enhance the flexibility offered by this small package. A front panel Go/No-Go fault annunciator signals the result of the built-in test. Test points are furnished in the front panel input/output connector for fault isolation at the intermediate and depot levels of maintenance.

Specifications

Dimensions: 127 × 108 × 255 mm
Weight: 3 kg
Power: 15 W (max)

Operational status

In production.

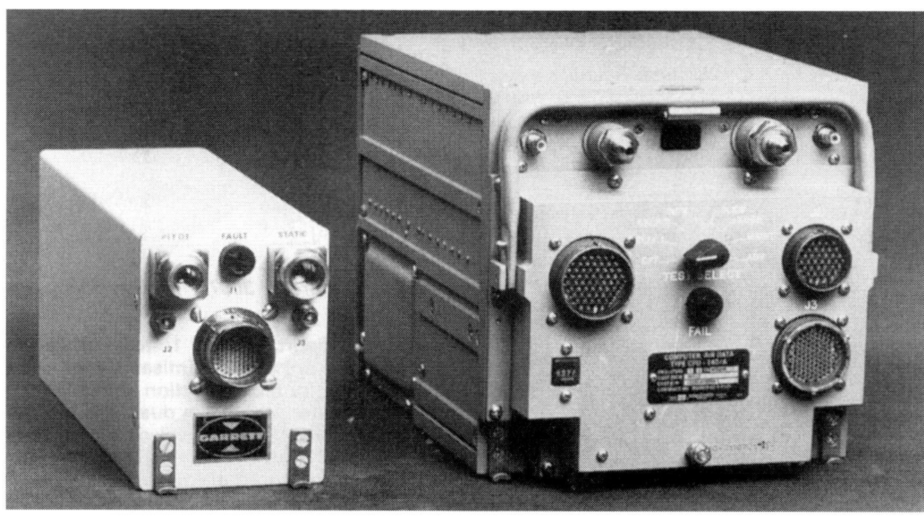

The miniature air data computer (left) is half the size and weight of a standard central air data computer (right)

Contractor

AlliedSignal Engine Systems & Accessories.

VERIFIED

Standard Central Air Data Computers (SCADC)

AlliedSignal has completed contracts to supply standard central air data computers to Lockheed Martin for the C-5B Galaxy and to Northrop Grumman for the C-2A Greyhound carrier onboard delivery aircraft.

The AlliedSignal SCADC configuration uses a non-volatile memory for storing air data and subsystem fault information. There is a high standard of fault detection and high reliability ensures compatibility with a two-level maintenance strategy.

Specifications

CPU-140/A for the C-2A
Dimensions: 143 × 140 × 242 mm
Weight: 9.4 kg
Power supply: 115 V AC, 400 Hz, 60 W

CPU-141/A for the C-5A and C-5B
Dimensions: ½ ATR
Weight: 14.4 kg
Power supply: 115 V AC, 400 Hz, 72 W

Operational status

In service on US Air Force C-5B Galaxy and US Navy C-2A aircraft.

Contractor

AlliedSignal Engine Systems & Accessories.

VERIFIED

True airspeed computer

The true airspeed computer provides outputs of true airspeed for RNav, INS and display applications. It features a vector pressure ratio transducer which facilitates computation of true airspeed using only one device. This servoed force-balance unit maintains a high level of performance under all aircraft operating conditions.

An accuracy of ±4 kt is obtained over a true airspeed range of 95 to 500 kt. A static air temperature output is available as an option.

Specifications

Dimensions: ⅜ ATR
Weight: 4.2 kg
Power supply: 115 V AC, 400 Hz, 30 VA

Operational status

In production.

Contractor

AlliedSignal Engine Systems & Accessories.

VERIFIED

Airborne computer for the B-1B

Ametek supplies airborne computer-based systems for both military and aerospace markets. The first application of a microprocessor to an airborne product built by Ametek was the signal conditioning and distribution unit for the B-1A in the early 1970s. With the revival of the aircraft as the B-1B a redesign of the system took advantage of newer technologies which include a 16-bit I^2L radiation-hardened design microprocessor and large-scale integrated components. The system interfaces with 64 engine sensors plus 12 more from the aircraft and electrical multiplex subsystem.

The B-1B system includes: a dual-redundant signal processing and power supply to enhance single-point failures; signal conditioning; sensor excitation; data conversion; linear computations; thrust computations; and periodic and initiated self-test parameter compensations for adverse temperatures to generate warning messages. Output information is transmitted via a dual-redundant digital Emux word-generated interface with aircraft avionics. The computer can perform 7 Mips.

Operational status

In service; production for the B-1B is complete.

Contractor

Ametek Aerospace Products.

VERIFIED

Data Acquisition Unit (DAU)

The Data Acquisition Unit (DAU) is designed for analogue or digital processing applications where large amounts of data need to be processed or consolidated. The system can be configured as a single- or dual-channel unit. A dual-channel configuration provides complete hardware redundancy for all parameters.

The DAU is designed using a modular approach with motherboard and plug-in function boards, which can be added as required to accommodate any aircraft sensor or communication bus. Custom boards can be designed to handle specialised sensors or communications. The system utilises advanced technology to process analogue, digital and discrete engine and airframe signals. Typical signals are filtered, converted to digital data, scaled and formatted for cockpit display or use by other aircraft signals.

In addition to analogue inputs, each DAU channel includes two ARINC 429 inputs and an RS-232 interface for communications with Sentinel displays or other aircraft systems such as the flight management system and FADECs.

All processed or stored information is transmitted via the standard ARINC 429 output bus. Optional databusses or buffered analogue and discrete outputs can be incorporated into the DAU. The built-in RS-422 port can be used to output real-time data or as an access point for maintenance interrogation.

In addition to the normal tasks of conditioning and processing signals, the DAU can perform a variety of maintenance functions. Health monitoring, exceedance recording and trending algorithms can be run in a background mode while the processor would normally be idling.

Extensive BIT isolates problems to the faulty sensor or circuit. Self-test is run continuously, with results stored in non-volatile memory for evaluation. Cross-channel communications are used to verify channel integrity.

Specifications

Weight: 4.08 kg
Power supply: 10-32 V DC or 115 V AC, 400 Hz
Temperature: −40 to +70°C

Operational status

Fitted to Bombardier Global Express aircraft.

Contractor

Ametek Aerospace Products.

VERIFIED

Engine data converter

The engine data converter is used for reduction of large amounts of analogue, digital and discrete data to ARINC 429 format. Some typical sources of data are the engine and transmission sensors, prop speed sensors and all the various aircraft system discretes.

The engine data converter features extremely flexible architecture, permitting maximum cost effectiveness in any application. Included on individual, easily serviced plug-in cards are all the signal conditioning and data conversion circuitry needed to interface the engine sensors with the digital databus. For use in systems where reliability is of the utmost importance, two totally independent channels are provided in each engine data converter.

For use with aircraft systems, the engine data converter offers additional optional features such as internal storage and periodic reporting of engine limit parameters, engine serial number encoding and non-volatile storage of limit exceedances, durations and engine cycle. It can also be structured to interface with other databusses, such as MIL-STD-1553B.

The design is sufficiently flexible so that it can be used with any type of engine. The software, as well as the analogue to digital circuit cards, is modular, so that changes to the engine sensors and engine performance characteristics can be readily accommodated.

One engine data converter is assigned to each engine. It is a dual-redundant device. Each half is a totally independent functioning unit having its own power supply, input converter cards, discrete units, microprocessor and digital output circuits. It shares only a common interconnect between its two halves. Isolation between the two halves is such that a failure of one half will not affect the other half's operation.

Specifications

Weight: 5.22 kg
Power consumption: 7 W per channel typical, 10 W per channel (max)
Reliability: 5,000 h MTBF

Contractor

Ametek Aerospace Products.

VERIFIED

Engine Monitoring System Computer (EMSC)

Ametek manufactures the Engine Monitoring System Computer (EMSC) for the General Electric F110 engine which equips the US Air Force F-15 and F-16C/D aircraft. The computer provides in-flight monitoring of engine exceedance, faults or trends. Engine-related signals are acquired from the engine monitoring system processor via the MIL-STD-1553B engine signal databus. Diagnostic data is retained in non-volatile memory which annunciates this data visually to the cockpit.

A secondary function downloads the data via the RS-232C series communication link to the data display and transfer unit for ground support evaluation. This link also uploads information to the EMSC such as the aircraft engine diagnostic information, time data, aircraft serialisation and life usage data.

Operational status

In production for the General Electric F110 engine on the F-15 and F-16C/D. Most of this system's modules are also used in the engine analyser unit for the Northrop Grumman E-2C's Allison T56-A-427 engine.

Contractor

Ametek Aerospace Products.

VERIFIED

Digital Air Data Computer (DADC)

The digital air data computer meets full US military standards. Present aircraft applications are the A-4, F-5, F-16, L-159, Mirage and a number of additional military aircraft. It is microprocessor-based, using precision solid-state vibrating quartz pressure transducers, with analogue potentiometers and synchro outputs as well as dual-redundant serial digital databusses.

Built-in test equipment allows a high degree of self-diagnosis and the computer has considerable growth potential. It has been designed as a standard air data computer both for retrofit and new aircraft programmes.

Operational status

In production and operational in the A-4, F-5, F-16, L-159 and Dassault Aviation Mirage.

Contractor

Astronautics Corporation of America.

UPDATED

The Astronautics digital air data computer

Modular Mission and Display Processor MDP

The Astronautics modular MDP is a high-performance, advanced technology computer, using CMOS technology. It provides the basic functions required by military display and control applications, including complete Weapon Delivery Navigation Systems. The modular MDP is the heart of each one of the display, navigation and weapon control units.

The modular MDP comprises five main items: the controller, which includes the CPU and the main memory; the graphics engine, a high-speed micro-programmed controller capable of driving stroke and raster displays in various combinations and formats; the video front end which mixes selected inputs with synthetic symbols; the input/output modules; power supply.

Specifications

CPU: 32-bit, 12 Mips at 25 MHz (up to 20 Mips for R-3081 CPU)
Memory: 2 Mbyte standard (4 Mbyte by upgrade)
Graphics engine: 5 million pixels/s/channel (three graphics engines)
Graphics capability: up to 6 independent graphics channels; up to 512 graphics characters/symbols; 8 colours
ARINC buses: 429 and 629 available
Data interface: RS-232, RS-422, discretes, analogues, synchro
Number of card slots: 12
Power: 115 V AC, 400 Hz at 220 VA
Dimensions: 279.4 × 261.6 × 193 mm
Weight: 15.9 kg
MTBF: 1,200 h

Contractor

Astronautics Corporation of America.

NEW ENTRY

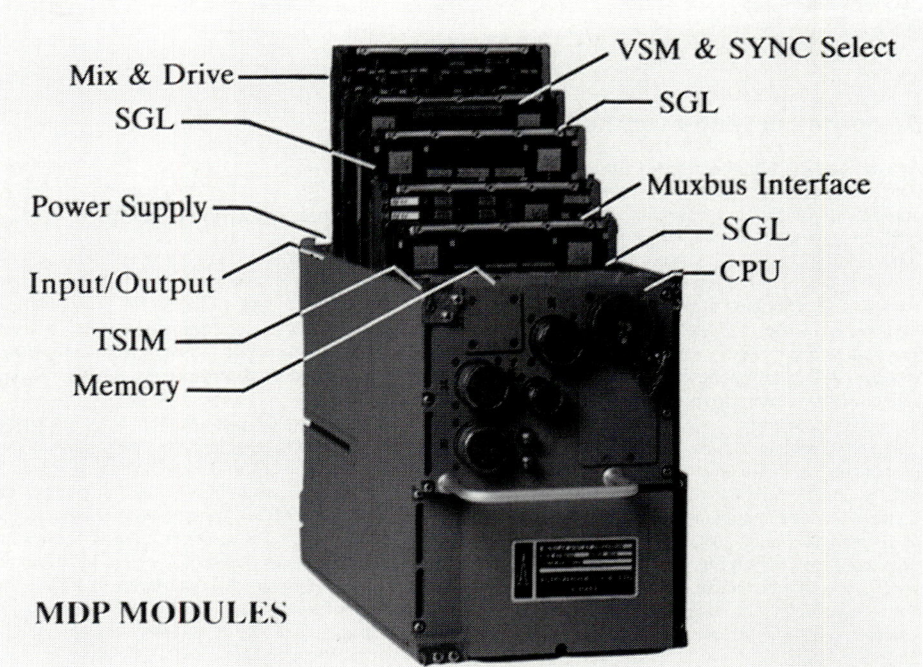

Astronautics modular mission and display processor
***1998**/0015279*

Albus-1553-8 bus monitor

The Albus-1553-8 is an advanced MIL-STD-1553 bus monitor. The Albus is designed to monitor all traffic on up to eight MIL-STD-1553A/B buses and provide PCM and tracksplit output streams for each bus channel, a composite stream consisting of bus traffic from all eight channels and digitised voice and parallel time code in the PCM outputs. The Albus is ideally suited for aircraft systems which require IRIG-106-93 Chapter 8 all-bus monitoring.

The Albus uses the standard functional elements of AYDIN TELEMETRY'S programmable bus monitor product line combined with new architecture and flight-proven high-density mechanical packaging. A fully configured package measures 127 × 190.5 × 251.4 mm and weighs less than 6.8 kg. The Albus can monitor bus traffic and from it extract messages, merge bus words with time code and parallel user input data to create IRIG-106-93 compatible PCM data streams. It also provides tracksplit PCM output streams for recording, composite PCM output consisting of bus traffic from the eight channels and hardware selectability of BIT rate and number of output tracks and format.

The Albus has been designed to accommodate various connector configurations, power input options, channel quantity options and bus protocols.

Contractor

AYDIN TELEMETRY.

UPDATED

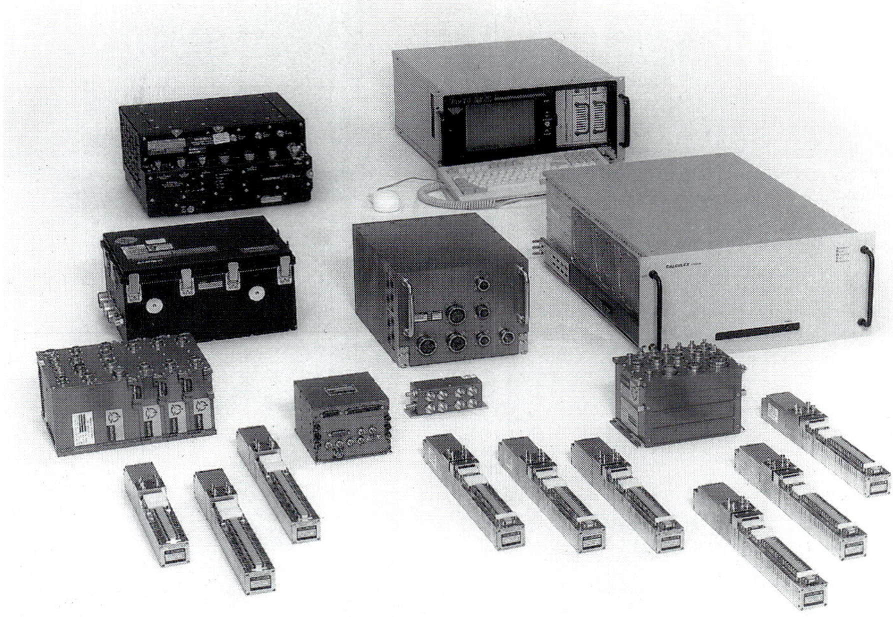

AYDIN TELEMETRY representative CAIS system ***1997**/0002349*

Common Airborne Instrumentation System (CAIS)

The CAIS was developed under the auspices of the US DoD to promote standardisation, commonality, and interoperability for flight testing.

The central characteristic of CAIS is a common suite of equipment used across service boundaries and in any airframe or weapon system testing.

CAIS products comprise airborne equipment items, that can be configured to meet project requirements, and comprehensive ground facilities to support the airborne effort:

MDAUs: Miniature Data Acquisition Units to support both analogue and digital data acquisition
PMU-700-C5: Programmable Master Controller Unit
PBC-800: Programmable Bus Controller
PCU-800C: Programmable Conditioning Unit
MPC-800C: Miniature Programmable Conditioner
MiniARMOR-700: High-Speed Multiplexer
ATD-800-II: Airborne Tape Deck
CCU-800: Cockpit Control Unit
GSU-800: Ground Support Unit
CBE-850: CAIS Bus Emulator
Lab ARMOR-715: High-Speed Demultiplexer

Operational status

CAIS is used in the F-18E/F and F-22 aircraft programmes. AYDIN TELEMETRY is the instrumentation system integrator for the F-22 programme, and is responsible for delivering a complete turnkey system.

AYDIN TELEMETRY has been selected as sub-contractor by Boeing Aircraft, Missile Systems Division to provide flight test instrumentation for its Joint Strike Fighter concept demonstration programme. AYDIN TELEMETRY will also manufacture much of the flight test instrumentation equipment and integrate the AYDIN Data Acquisition System for the two next-generation X-32 fighter aircraft.

Contractor

AYDIN TELEMETRY.

UPDATED

DPM-800E PCM encoder

The DPM-800E PCM encoder is a fully programmable high-performance 8- to 12-bit resolution data system for acquiring conditioned signals in severe airborne applications where flexibility and reliability are the primary requirements.

Features of the DPM-800E include EEPROM programmable via parallel IF port, 2 Mbit operation, eight programmable bit rates, user programmable for format, bits per word and gain/offset, filtered data outputs, single-ended or differential analogue inputs, single-ended discrete bilevel inputs and subcommutation and supercommutation.

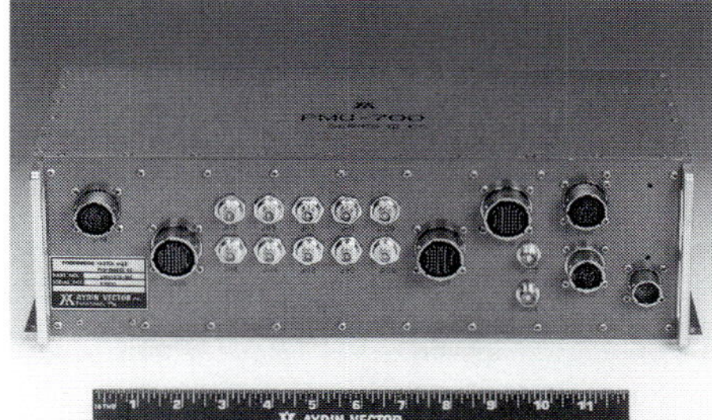

PMU-700-C5 programmable master controller unit ***1997**/0002347*

The DPM-800E can be configured as a stand-alone, master or remote unit. Up to seven DPM-800Es can be configured in a master/slave or cluster configuration via the AYDIN TELEMETRY standard 10-wire differential interface. This 10-wire interface can be utilised with other AYDIN TELEMETRY data acquisition products such as MIL-STD-1553 bus monitors, signal conditioning units and master controllers for large distributed systems.

Specifications

Dimensions: 88.9 × 82.5 × 118.1 mm
Weight: 0.79 kg
Temperature range: −35 to +85°C

Contractor

AYDIN TELEMETRY.

UPDATED

MDM-700 modem

The MDM-700 modem is ruggedised for airborne use and provides a relatively inexpensive means of interfacing a computer to a duplex RF link which utilises conventional FM transmitters and receivers. The MDM-700 is a full-duplex interface. However, the modem can be used in the simplex mode without degradation and without special considerations.

The MDM-700 transmit section accepts the RS-232 signal and converts each transition into a half-sine shaped pulse. A positive transition creates a positive pulse and a negative transition creates a negative pulse. Should there be no transition activity, the modem continues to create pulses of the same polarity as the last but at a low frequency. A result of this is that a signal can be AC-coupled; DC response is not required.

The modem receive portion accepts the half-sine pulse train and recreates the original RS-232 data. Since the conversion is one of edge coding, the modem is insensitive to data rate, except for the maximum of 9,600 bps. Clock is not used nor is it required.

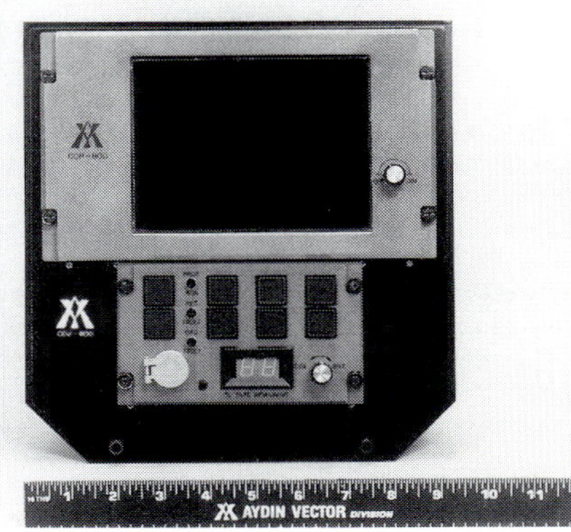

CCU-800 cockpit control unit ***1997**/0002348*

Specifications

Weight: 0.23 kg
Power supply: 28 V DC

Contractor

AYDIN TELEMETRY.

UPDATED

MME-64 Miniature Multiplexing Encoder

The MME-64 miniature multiplexing PCM encoder is a low-cost data system designed for airborne data acquisition and telemetry applications to accommodate up to 64 single-ended analogue inputs or up to 24 discrete bilevel inputs. Unfiltered PCM output is NRZ-L 0 to 5 V TTL, CMOS-compatible. Filtered PCM output is six pole Bessel response, factory-adjustable from 0±50 mV to 0±2.5 V. Bit rate can be programmed up to 1 Mbit. The frame pattern is controlled by EPROM and is factory-programmed.

Specifications

Dimensions: 82.6 × 76.2 × 30.4 mm
Temperature range: −30 to +70°C

Contractor

AYDIN TELEMETRY.

UPDATED

MMSC-800 MicroMiniature Signal Conditioner

The MMSC-800 microminiature signal conditioner/PCM encoder combines AYDIN TELEMETRY's experience of analogue and digital conditioning into a 12-bit PCM encoder. A complete family of modules provides signal conditioning for various sensors. EEPROM programmable gain/offset and sample rates provide the user with hands-off control of measurement characteristics and output data formatting. The MMSC-800 utilises AYDIN TELEMETRY thick film hybrid modular assemblies and has been qualified for missile and aerospace environments. It is available in stand-alone or master remote configurations.

The MMSC-800 features 12-bit PCM encoder with integral signal conditioning and multiplexing, EEPROM programmable gain, offset, sample rate and PCM format, ruggedised modular construction, extremely small size and RS-232 programming capability. It is military and airborne qualified and has high reliability. The system accuracy is quoted as 0.5 per cent.

Contractor
AYDIN TELEMETRY.

UPDATED

MMSC-800-RPM/E remote pressure multiplexer/encoder

The MMSC-800-RPM/E remote pressure multiplexer/encoder accepts pressure inputs directly, electronically scans the pressure inputs and conditions and encodes the resulting information. It is based on the MMSC-800 signal conditioner and PCM encoder, which has been used on many commercial test programmes and applications. Operating as a remote unit, the RPM/E can be placed in areas of the aircraft well away from the Model PCU-700 Series III central controller. Communication between the master and remote is via a 10-wire command response interface. The channel sequence and gain/offset combinations are fully programmable through the central controller. The temperature reference of the PSI scanner is electronically monitored and transmitted to the central controller to allow the user to correct for errors due to temperature changes in the scanner. Features are controlled EEPROM, programmable via an RS-232 from any personal computer terminal.

Specifications
Dimensions: 135.9 × 45.7 × 96.5 mm
Weight: 3.63 kg
Input channels: 32
Temperature range: −20 to +80°C

Contractor
AYDIN TELEMETRY.

UPDATED

MPC-800 miniature signal conditioner and encoder

AYDIN TELEMETRY's new miniature signal conditioner and encoder (MPC-800) is a data acquisition system component that provides all the most common required signal conditioning and encoding functions in a small modular package. The design minimises size, weight and cost, and improves survivability in extremely hostile environments. The unit is constructed from stackable conditioning and overhead modules which contain discrete, thick film hybrid and custom ASIC circuit technologies.

By utilising modular architecture, the MPC-800 can be configured to meet the exact needs of any flight test program and still allow the user the ability to reconfigure the unit in the field to adapt to changing mission requirements. Full programmability of gains, offsets, filter cutoffs, and channel sampling rates is offered through a Graphical User Interface (GUI). The MPC-800 may be operated as a master, remote, or stand-alone system.

A wide variety of standard analogue and digital signal conditioning modules are available for virtually any sensor. Custom modules can be created for unique measurement requirements.

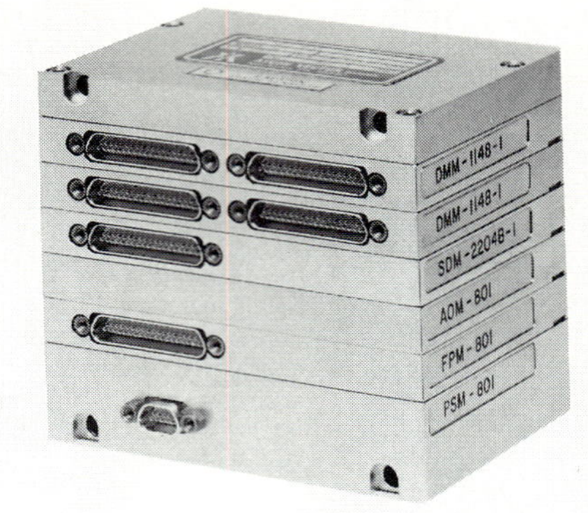

MPC-800 miniature signal conditioner and encoder ***1997***/0002350

Contractor
AYDIN TELEMETRY.

UPDATED

PCU-800 series signal conditioner and PCM encoder

The PCU-800 series signal conditioner and PCM encoder provides a ±0.5 per cent system accuracy over its operating temperature range. Available in 2, 4, 8 or 16 card slot configuration, the PCU-800 features include: EEPROM programmable format and channel parameters via RS-232; plug-in system overhead and signal conditioning cards, both digital and analogue types; universal bridge conditioning with completion, calibration and substitution; programmable balance via software control and 12-bit resolution.

The PCU-800 will operate as a stand-alone or master controller. Channel capacities vary depending on the type of signal inputs, but they can range up to more than several hundred channels per unit. Up to 40 PCUs can be clustered in a distributed data acquisition system to provide a total channel capacity of more than several thousand.

The PCU-800 is directly compatible with many other ADAS-7000 and ADAS-8000 airborne products such as the PMU-700 Series III programmable master controller, Albus-1553 all-bus monitor, MPBM-1553 microbus monitor and ATD-800 Series II ruggedised tape deck.

Contractor
AYDIN TELEMETRY.

UPDATED

SBS-500 single-rate bit synchroniser

Processing PCM data in the SBS-500 single-rate bit synchroniser suppresses receiver noise that may accompany the digital input signal and converts the input to a clean digital output. Available as a stand-alone 275 cm³ package, the SBS-500 is suitable for missiles, aircraft and space applications.

The SBS-500 single-rate bit synchroniser

It has a clock synthesis oscillator and a data detection match filter determined by the specific bit rate. The SBS-500 can also perform code conversion from digital transmission codes and provide biphase or NRZ outputs. The SBS-500 performs well under low-input signal-to-noise conditions and is typically within 2 dB of theoretical bit detection accuracy. Acquisition time is 10 clock cycles typical and the SBS-500 has output options of TTL, HC, RS-232 and RS-422.

Specifications
Power supply: 28 V DC
Bit rate: 1-250 kHz
Temperature range: −40 to +85°C

Contractor
AYDIN TELEMETRY.

UPDATED

SSC-2008 super signal conditioner

The SSC-2008 super signal conditioner is a complete ultra precision signal conditioning module that contains its own 12-bit 6 µs digitiser, excitation sources and bridge balancing circuits. It is intended to excite, balance, condition and digitise a wide range of transducers such as strain gauges, RTDs, pressure transducers, potentiometers, microphones with frequency outputs up to 20 kHz and accelerometers with the addition of external capacitors.

The SSC-2008 incorporates a technologically advanced digitally programmable instrumentation amplifier. Gain range for this amplifier is 0.125 to 2,048 in 15 steps, with a gain accuracy of better than 0.08 per cent. With an input impedance of 1,000 Megohms, the input bias current is only 15 nA and linearity is better than 0.005 per cent. It contains a programmable excitation voltage from 0 to 10.24 V in 4,096 steps. This excitation voltage is capable of supplying 5 mA and is foldback protected. In order to eliminate error produced by the system wiring and to allow current amplification, a remote sense input and drive outputs are provided for the excitation output.

The SSC-2008 contains a sample and hold and a 12-bit, 6 µs digitiser. Incorporating the analogue/digital eliminates most system noise problems, substantially increases effective system speed and allows simultaneous sampling capability. The SSC-2008 also provides two tracking constant current sources with a maximum output compliance of 12 V to enable extremely accurate measurements.

Contractor
AYDIN TELEMETRY.

UPDATED

The SSC-2008 super signal conditioner ***1995***

DB-682/DB-683 data storage units

The DB-682 is a data storage unit designed for use with the Global GNS500A Series 3 and Series 4 systems. The DB-682 features a pilot-replaceable database cartridge that includes position information for all VORs, VORTacs, NDBs, outer markers, high- and low-altitude intersections and all public use airports with runways of 3,000 ft or longer in the US and 4,000 ft or longer in the remainder of North, Central and South America. There is also an international database cartridge available which includes position information for all VORs, VORTacs, VOR/DMEs, Tacans, high and low intersections and all public use airports with runways of 4,000 ft or longer for virtually the entire free world.

The DB-682 offers several mounting options each allowing easy access to the database cartridge, including a dzus rail cockpit mount which provides quick-change capability for the flight crew.

The DB-683 is a data storage unit similar to the DB-682. It features a dual-port ARINC 429-compatible interface format and offers a choice for VLF and inertial operators looking for an affordable database storage system.

Specifications

Dimensions: 57.1 × 101.6 × 177.8 mm
Weight: 0.69 kg
Power supply: 28 V DC, 250 mA

Operational status

In production.

Contractor

BFGoodrich Aerospace Avionics Systems.

VERIFIED

AN/ASQ-195 signal data converter set

The AN/ASQ-195 signal data converter set is a multifunctional digital processor that provides control and data interchange for radios, instrument landing systems, Tacan, automatic direction-finders, sensors such as LANTIRN and radar altimeters, air-to-air interrogation, IFF and avionics BIT. The system has been designed for the F-15E but can be adapted to other aircraft. Each of the two units serves as back-up or redundant data processor for the other unit for many of the functions.

The AN/ASQ-195 features three dual-redundant 1553B databus interfaces, three RS-422 databus interfaces, three MIL-STD-1750A microprocessors, six serial microcontrollers, 120 k EEPROM memory, 240 programmable discrete output interfaces, 355 programmable discrete input interfaces, four synchro interfaces and 28 analogue interfaces.

Display of the INS parameters, frequencies and channel selection is provided by the up-front control panel, communicating with the AN/ASQ-195 via the RS-422 databus.

Contractor

The Boeing Company.

UPDATED

Integrated Sensor System (ISS)

The four-year Integrated Sensor System (ISS) programme is sponsored by the US Air Force Wright Laboratory. The Boeing Company heads an industrial team which includes the Raytheon Systems Company and TRW. The goal of the programme is to integrate the Radio Frequency (RF) functions of Communications/ Navigation/Identification (CNI), radar and electronic warfare systems in order to cut cost, size and weight of aircraft avionic systems and improve system reliability.

ISS will build on the US Air Force Pave Pillar/Pave Pace programmes which were aimed at a more integrated approach to advanced sensor processing. The keys to achieving cost and reliability improvements are the concepts of common modules, resource sharing and reconfiguration of the RF sensor systems.

The modular, open architecture design of ISS has the potential for a 50 per cent improvement in overall avionics systems costs and a 200 per cent improvement in reliability, at half the size and weight of contemporary systems. In conceptual studies, the ISS approach reduced the number of RF modules in a notional fighter/attack aircraft from 345 to 169 and the different types of modules from 70 to 33.

The team will design an ISS system architecture and then build key modules and integrate them into a system capable of demonstrating the required RF functions across a broad range of frequencies and waveforms. The team will focus on the integration of what is traditionally referred to as analogue RF support electronics units such as RF switches, converters, receivers, transmitters and other equipment.

The goal is to validate the architectural feasibility of an integrated RF system that can provide the performance and operational needs of future weapon system platforms and also be capable of insertion into current aircraft systems.

Operational status

The Boeing Company has been awarded a US$16.5 million contract by the US Air Force to develop and demonstrate an integrated sensor system.

Contractor

The Boeing Company.

UPDATED

Rotorcraft Pilot's Associate

The Boeing Company is teamed with Lockheed Martin Federal Systems to develop a Rotorcraft Pilot's Associate (RPA) for the US Army. The RPA programme aims to establish revolutionary improvements in combat helicopter effectiveness through the application of knowledge-based systems for cognitive decision-aiding and the integration of advanced pilotage, acquisition, armament and fire control, communications, controls and displays, navigation, survivability and flight control equipment.

RPA builds on advanced avionics technologies being developed by the US Army for the updated AH-64 Apache and the RAH-66 Comanche. These systems enhance the automation of such functions as flight control, information processing and weapons management. In addition, RPA technology has applications in the civilian marketplace and offers an opportunity to aid human performance in a variety of fields.

RPA moves into the cognitive realm of data interpretation, hypothesis formulation, planning and decision making. The result is an intelligent associate that will assist the pilot in understanding the vast array of battlefield information, planning the mission and managing the complex systems in modern military aircraft. Development and evaluation will occur in three stages. Initial design and assessment will be accomplished in a rapid prototyping laboratory environment. As the design matures, it will move into full mission simulation, where more rigorous and in-depth evaluations will be conducted. The third stage will install RPA in an AH-64D Apache attack helicopter, incorporating the Boeing Company's advanced digital flight control system as well as the US Army's advanced helicopter pilotage sensor system.

The Boeing Company will be responsible for system integration, architecture design, prototype development, full mission simulation, offensive systems, vehicle management and flight test. Lockheed Martin Federal Systems will provide computing technology, including the Massively Parallel Processor, and has responsibility for data distribution, mission planning, defensive systems management and external situation awareness.

Operational status

The Boeing Company and Lockheed Martin Federal Systems have received a US$70 million contract for development and demonstration of the RPA. Flight and operational demonstrations were planned for 1997; current status not known.

Contractors

The Boeing Company.
Lockheed Martin Federal Systems.

UPDATED

AN/AYK-14(V) standard airborne computer

The AN/AYK-14(V) is a high-performance general purpose computer with both 16-bit and 32-bit processing elements. It consists of a family of interchangeable processor, memory, power, enclosure and input/output modules that can be configured to meet specific price, performance and functionality needs for a wide range of applications. The AN/ AYK-14(V) computer offers high performance and high reliability with a low life cycle cost, while meeting airborne MIL-E-5400, shipboard MIL-E-16400 and land MIL-E-4158 environments. It is the US Navy's standard airborne computer and is currently being used on a wide variety of military platforms by the US and its allies.

The 16-bit version of the AN/AYK-14(V) is currently in its third generation, which is known as VHSIC AN/ AYK-14(V). It provides performance of up to 20 Mips or more within a single enclosure and has a memory addressing capacity of 16 Mbytes. The VHSIC computer is fully compatible with software written for either of the prior generations. The instruction set is compatible with that of the AN/UYK-44 and AN/UYK-20 and is supported by the MTASS software development environment. Other 16-bit modules include 128 kbyte core memory, MIL-STD-1553A/B, NTDS, RS-232, Proteus, discrete Input/Output (I/O) and application specific I/O modules.

The 32-bit version of the AN/AYK-14(V), also known as the Advanced AYK-14, is based on commercial RISC technology. Up to 225 Mips of processor performance and 180 Mbytes of memory capacity are currently available within a single enclosure. The use of commercial open system backplane and processor standards provide built-in performance and capability growth potential that can increase the performance of commercial technologies. Like the 16-bit AN/ AYK-14(V) processing elements, the Advanced AYK-14 also provides modular processing and I/O components that can be configured to meet specific price and performance requirements of a range of applications. Currently, a MIPS R4400 RISC computing module processor, other processor modules, and SCSI, MIL-STD-1553A/B, RS-422, Proteus and discrete I/O modules are available.

The Advanced AYK-14 is based on the IEEE Futurebus+ backplane interface standard, which allows the integration of other supplier modules and functions in AN/AYK-14(V) systems. The Advanced AYK-14 supports all three Futurebus+ primary module formats: 10 SU; 12 SU and SEM-E. A complete, commercially supported, Ada development environment is available for commercial workstation Sun and RS-6000

The F/A-18 Hornet avionics system incorporates two AN/AYK-14(V) mission computers

networks, providing full development and real-time debugging support. A real-time operating system with POSIX-compatible services, priority management capabilities and Rate-Monatonic scheduling support completes the Advanced AYK-14's capabilities. Ada compilation, debugging and runtime can be obtained from multiple commercial suppliers. Current implementation utilises the Rational VADS-Advanced commercial product. Device drivers and hardware support software are available from Computing Devices International.

To ease the transition from 16-bit to 32-bit processing, the AYK-14 supports systems configured with both 16-bit and 32-bit processing elements. This allows platforms such as the EA-6B Prowler that currently use the 16-bit AN/AYK-14(V) to take advantage of the 32-bit Ada processing elements, without discarding the investment that has been made in 16-bit software development and testing. The 16-bit and 32-bit AYK-14 processing elements communicate over the standard 16-bit backplane bus, sharing memory and messages to co-ordinate the system processing.

The AYK-14 is adaptable to a variety of enclosures as a general purpose processor, emulator, controller, dedicated processor or algorithm unit. Modules have also been embedded in applications where size, weight and power are critical, such as the AN/ALQ-149 ECM system. Representative applications include weapon delivery/fire control, guidance, communications, navigation, avionics mission computing, sensor data processing, display subsystem control, radar or sonar processing system control, electronic countermeasures, electronic surveillance measures management and digital flight control.

On the F/A-18, two 16-bit AN/AYK-14(V) computers are installed in a dual-redundant system configuration. One computer serves as the navigation and engine controller, while the other acts in a mission management role, handling and processing weapons and target information. Each computer can accommodate up to 16 Mbytes of memory and provides up to 20 Mips of processing throughput.

The Advanced AYK-14 is currently being integrated into the LAMPS III Block II, providing up to 225 Mips of throughput and 180 Mbytes of memory; it takes advantage of the AYK-14 open systems architecture and integrates video and graphics modules developed by Loral Federal Systems along with the processing and I/O modules from Computing Devices International. The Advanced AYK-14 is also being used on the V-22 Osprey in a lower cost configuration that provides 45 Mips throughput and 12 Mbytes of memory on a single board computer. The EA-6B programme is adding the 32-bit capability to its 16-bit machine through the use of a split backplane.

Specifications

Dimensions: 194 × 257 × 356 mm typical
Weight: 11-16 kg
Power: 50-300 W typical
Computer type: binary, fixed or floating point
Word length: 16-bit or 32-bit with double precision and floating point
Typical speed: 500 Kips to 225 Mips
Max addressing: 16 Mbytes (16-bit) to 4 Gbytes (32-bit)
Input/output options: discretes, MIL-STD-1553A or B (dual-redundant buses), NTDS fast, slow, ANEW and serial, 6 MHz Manchester, RS-232C, 85323 Proteus, RS-422, SCSI, RS-485, TM-bus

Operational status

In production and selected for the Boeing F/A-18 (which has two AYK-14 mission computers), Sikorsky SH-60B Seahawk LAMPS Mk III helicopter, Northrop Grumman E-2C Hawkeye, Boeing/British Aerospace AV-8B, Northrop Grumman EA-6B Prowler, F-14D, EP-3E, ES-3A, Bell/Boeing V-22 Osprey, Joint STARS, Lockheed Martin P-3C Orion and embedded modules in the AN/ALQ-149, US Air Force TAOM/MCE system.

A modified version, designated ACCS 2500, is used on the RAF's Harrier GR. Mk 7 aircraft (see entry under Computing Devices Company in the UK part of this section).

Contractor

Computing Devices International.

UPDATED

MAXION/ATR multiprocessor system

The MAXION/ATR multiprocessor system is based on the Concurrent MAXION real-time standards-based system. It is a ruggedised system using the standard commercial off-the-shelf MAXION board set and satisfies the requirement for a real-time multiprocessor system which is capable of being deployed in operational aircraft. It conforms to the industry standard ATR form factor. MAXION/ATR provides more than 600 Mips of power within a VME-6u double height Eurocard.

The system is based on the Mips Technologies R4400 RISC, an industry leading microprocessor. With a system bandwidth of 1.2 Gbytes/s, the system architecture is flexible and extensible. It incorporates the Scalable Multiprocessor Architecture for Real-Time (UltraSMART). This provides maximum total system performance, both in terms of computation and handling I/O data, and delivers virtually the maximum processing power possible from each multiprocessor. As requirements grow, MAXION/ATR users will be able to add new processors.

Operational status

Available in configurations of one to four processors with up to eight processor models available.

Contractor

Concurrent Computer Corporation.

VERIFIED

M362F general purpose processor

The M362F is a general purpose unit which uses parallel, binary, floating point, two's complement 16-bit processing. A typical instruction mix yields an operating speed of about 340 k instructions per second.

The M362F can be tailored to particular operations by specifying from a wide choice of memory types and standard input/output circuit modules. It is mechanised on two operating modules. The instruction repertoire can be varied by adding microprogramme memory or changing the existing microprogramme memory which consists of 10 integrated circuits. The M362F will operate with core and/or semiconductor memory. The processor provides 72 basic machine instructions, including nine special purpose types that are mechanised using the basic input/output instructions. Micro-instructions are held in a 512-word control memory, but this can be expanded to 2,048 words, providing for such operations as byte (8-bit) control jumps, skips and transfers, register variable shifts, register/register floating point arithmetic, logical immediate and register/register, and macro instructions. The last provides square root and trigonometric functions much faster than by using subroutine executions.

Development of M362F software can be accomplished on support equipment such as mini-computer directed systems, IBM 360/370 or similar facilities. Mini-computer equipment enables software development on a stand-alone basis in the laboratory.

The M362F is used as the fire-control computer in the Lockheed Martin F-16 and has a comprehensive set of Jovial or assembly coded software facilities.

Specifications

Dimensions: ½ ATR case
Weight: 6.5 kg
Power: 28 V DC
Computer type: parallel, binary, fixed and floating point, two's complement
Word length: 16-bit
Typical speed: 340 Kips
Max address range: 65,536 (64 k) memory locations
Instruction set: 72 (expansible)
Input/output options
analogue/digital/analogue converter
analogue input/output multiplexer
discrete input/output (28 V)
MIL-STD-1553 bus
multipurpose serial digital processor

Operational status

In service as the fire-control computer in the F-16.

Contractor

Delco Electronics.

VERIFIED

M362S general purpose processor

Related to the M362F, the M362S is a 32-bit, high-speed general purpose processor. It uses microprogrammed, parallel, binary, fixed and floating point, two's complement operations and has a typical operating speed of 750 Kips.

The processor provides 96 basic machine instructions, microprogrammed in a 512-word control memory, with possible expansion to 1,024 words. A RAM system is available and comprises semiconductor CMOS memory modules, memory controller and an error detection and correction unit. A total of 65,536 (64 k) 16-bit memory locations can be accommodated.

Packaging is either for forced-air cooling, as in aircraft bays, or a radiant cooling frame for space applications. The system has been selected for NASA's Inertial Upper Stage programme. A comprehensive set of Jovial and assembly coded software facilities is available.

Specifications

Dimensions: 365 × 365 × 152 mm
Weight: 24.5 kg
Power supply: 28 V DC, 220 W
Computer type: binary, fixed and floating point, two's complement
Word length: 32-bit
Typical speed: 750 Kips
Max address range: 65,536 (64 k) memory locations
Instruction set: 96 (expansible)
Input/output options
analogue/digital/analogue converter
analogue input/output multiplexer
discrete input/output (5 or 28 V DC)
MIL-STD-1553 bus
multipurpose serial digital processors

Operational status

In production.

Contractor

Delco Electronics.

VERIFIED

M372 general purpose processor

The M372 is a development of the M362F used as the fire-control computer for the Lockheed Martin F-16. It has been designed for applications requiring extended performance in memory, input/output, throughput and computational speed. Operating speed is typically between 570 and 720 Kips, and non-volatile, electrically alterable memory of 32 k, 64 k, 128 k or 256 k words capacity can be accommodated and addressed. In addition to analogue and discrete input/output channels, up to three MIL-STD-1553 dual-redundant digital databus channels can be accommodated and the computer executes the US Air Force MIL-STD-1750 standard instruction set architecture.

The system is supported by Jovial software, and packaging dimensions can be between ½ and 1 ATR, depending on the features incorporated.

Specifications

Dimensions: 112 × 170 × 190 mm
Weight: 7.3 kg
Power supply: 28 V DC, 100 W
Computer type: binary, parallel, microprogrammed, fixed and floating point, two's complement
Word length: 16-, 32-, or 48-bit
Typical speed: 570-720 Kips
Max address range: 1,576,058 (1,024 k) words
Instruction set: MIL-STD-1750A, Notice 1, 262 instructions

Operational status

In production. The M372 computer is used as the F-16's enhanced fire-control computer, as the Lockheed Martin LANTIRN pod control computer and as the C-5B MADAR multiplexer/processor.

Contractor

Delco Electronics.

VERIFIED

Magic IV general purpose processor

The hardware used in this processor is unrelated to earlier Delco processors. Magic IV is a high-performance, all-large-scale integrated microcomputer system, which promises to provide lower cost, reduced power, weight and size and greater modularity and reliability advantages over existing machines. Typical operating speed is 250 Kops. There are three basic machines: the M4116, M4124 and M4132 available with 16-, 24- and 32-bit word architectures respectively. They are seen as suitable for remote terminal or sensor orientated processor applications.

Almost exclusive use of large-scale integrated circuits has resulted in a great reduction in the number of components compared to a conventional processor. The large-scale integrated units are:

	M4116 (16-bit)	**M4124** (24-bit)	**M4132** (32-bit)
CPU control unit	1	1	1
CPU arithmetic unit	4	6	8
Input/output control unit	2	3	4
Memory controller	2	3	4
	—	—	—
Total	9	13	17

Additional large-scale integrated circuits may be needed to meet programmable communication interface and digital input requirements. MTBF estimates range from 27,000 hours for a simplex configuration to 150,000 hours for a dual-redundant system. NMOS large-scale integrated circuits are used with typical component densities of 1,000 gates and 5,000 transistors per 250 mil chip, with pair gate delays below 5 ns. Delco claims better nuclear radiation tolerance than contemporary dynamic NMOS circuits due to the static logic design, substrate bias design and exclusive use of NOR gate. Digital inputs compatible with ARINC 561, 575 and 583 can be provided.

Specifications

Dimensions: 167 × 69 × 127 mm
Weight: 1.4 kg
Power: 25.3 W
Computer type: binary, parallel, fixed point, two's complement
Word length: 16-, 24- or 32-bit
Typical speed: 250 Kops
Max address range: 32,768 (32 k) words
Instruction set: 89 (expansible)

Operational status

In production. The computer is used as the Fuel Savings Advisory and Cockpit Avionics System (FSA/CAS) computer for the US Air Force's C-135 and KC-135 aircraft. It is also used in the performance management system in the Boeing 747 and DC-10 and MD-80 series.

Contractor

Delco Electronics.

VERIFIED

Magic V general purpose processor

The Magic V processor is an all VLSI implementation of a CPU and memory system that executes the MIL-STD-1750A, Notice 1 instruction set architecture.

The Magic V processor utilises 3 μm bulk CMOS technology to configure a complete MIL-STD-1750A central processor in just 10 VLSI chips. An eleventh VLSI chip provides extended memory management to address up to one million words and a twelfth chip provides an IEEE-488 bus interface which enables external communication with the CPU and memory for monitoring performance and for developing software. These 12 VLSI chips are mounted on one side of a single ½ ATR size circuit card assembly. The back side of the single circuit card assembly mounts the VLSI memory controller that interfaces the CPU with the memory devices mounted on this side. Typical configurations of the M572 include up to 192 k words of CMOS RAM plus a start-up ROM, or various combinations of RAM, EEPROM and UV PROM devices, all capable of being addressed by the programmable VLSI memory controller.

The M572 typically operates at 850,000 to 1 million operations running the DAIS instruction mix. A built-in feature of the M572 is the ability to couple multiple CPU/memory cards in a multiprocessor configuration.

Specifications

Dimensions:
(CPU and memory) 109 × 163 × 13 mm
Weight: 0.45 kg with memory side fully populated
Power:
(CPU) 2.2 W
(CPU and memory) 7.3 W
Computer type: general purpose, microprogrammed, fixed and floating point, custom CMOS VLSI, TTL compatible interfaces
Word length: 16-, 32- and 48-bit; 48-bit logic unit
Instruction set: MIL-STD-1750A, Notice 1; all specified options
Memory size: up to 192 k words CMOS RAM on ½ ATR size circuit card; up to 256 k words CMOS RAM on ¾ ATR size circuit card
Max address range: 1 million words
Addressing modes: direct, indirect, immediate, indexed, non-indexed, relative, base relative, BIT, byte
Test interface: built-in IEEE-488 interface for DMA operations, test communications, software development, and real-time performance monitoring
Input/output: all mandatory features and non-application dependent MIL-STD-1750A options implemented, discrete and digital outputs/inputs, analogue outputs/inputs, and MIL-STD-1553B bus interfaces implemented in various system level applications

Operational status

In service on C-17A and F-14D in the display systems. Selected for engine monitor and control systems on the RAH-66 helicopter.

Contractor

Delco Electronics.

UPDATED

Data Transfer Equipment (DTE)

The Data Transfer Equipment (DTE), also known as the Data Insertion Device (DID), fulfils five main functions. These considerably reduce the workload imposed on crews and eliminate the risk of human error. The aircraft is automatically initialised and mission start procedures are launched simply by inserting a cartridge programmed during the preflight preparation. The DTE takes full control of navigation during the automatic ground-hugging phases of the flight and operates the weapons system. It also has electronic mapping and mission playback and debriefing on the various phases. The DTE reduces ground maintenance since it records and analyses the main aircraft parameters throughout the mission.

The DTE is capable of interfacing with all ground mission preparation and debriefing systems. It is operable from any allied or friendly logistic platform. The DTE also allows training in an intensive electronic warfare environment without requiring any special infrastructure.

Specifications

Dimensions:
(data transfer unit) 178 × 127 × 113 mm
(data transfer cartridge) 191 × 119 × 41 mm
Weight:
(data transfer unit) 3 kg
(data transfer cartridge) 0.7 kg
Power supply: 115 V AC, 400 Hz, single phase
Cartridge capacity: 8 k words expandable to 2 M

Operational status

In service with F-14, F-16, A-6, SH-60 and C-17.

Contractor

Fairchild Defense OSC.

VERIFIED

Upgraded Data Transfer Equipment (UDTE)

Fairchild's Upgraded Data Transfer Equipment (UDTE) comprises: an aircraft resident receptacle called an Upgraded Data Transfer Unit (UDTU) and a removable data transfer cartridge called a Mega Data Transfer Cartridge with Processor (MDTC/P); UDTE is a form, fit, and function replacement for Fairchild's Data Transfer Equipment. In addition, UDTE provides an expanded array of functions.

Data transfer capabilities include: a Digital Terrain System (DTS); an Embedded Data Modem (EDM); an Embedded GPS Receiver; an Embedded Display Processor; an Embedded Training System; an Integrated Electronic Combat System; a high-performance processor to back up flight critical systems; up to 2 Gbytes of solid-state mass storage.

UDTU

Using state-of-the-art circuit design and packaging techniques, the three circuit cards in the original DTU have been reduced to a single-circuit card. This card uses an industry standard 32-bit microprocessor which has enough processing throughput to retain all of the original DTU's functionality while also hosting advanced software algorithms such as those required for Embedded Training and an Integrated Electronic Combat System. In addition, new capabilities may be added to the UDTU via the two unused card slots. The list of circuit cards which can populate these spare slots is continually growing and includes a ruggedised GPS receiver, a Fairchild Defense designed EDM and a Display Processor.

MDTC/P

The MDTC/P is a form, fit, and function replacement for Fairchild's standard Data Transfer Cartridges and remains fully AFMSS compatible. The MDTC/P contains up to 280 Mbytes of non-volatile, solid-state mass memory and a high-performance Digital Signal Processor (DSP) ideally suited for mathematically complex algorithms. Typical uses for the MDTC/P include mission planning, storage of Digital Map Data, storage of avionics data gathered during flight, in-flight execution of British Aerospace's TERrain PROfile Matching (TERPROM) algorithm which is used as part of the Digital Terrain System and post-mission playback associated with training.

Ground Support Peripherals

The MDTC/P is fully AFMSS compatible. For those applications where an AFMSS is not available, Fairchild's SCSI Cartridge Interface Device (ScsiCID™) allows any personal computer or workstation to communicate with the MDTC/P over a standard SCSI-2 interface. Data rates in excess of 5 Mbytes/s allow data to be rapidly loaded into the MDTC/P's mass memory.

Specifications

Dimensions:
UDTU: 178 × 127 × 113 mm
MDTC/P: 191 × 119 × 41 mm
Weight:
UDTU: 3.0 kg
MDTC/P: 1.6 kg
Input power: 115 V AC, 400 Hz, single phase, 32 W max
Software: Ada

Contractor

Fairchild Defense OSC.

VERIFIED

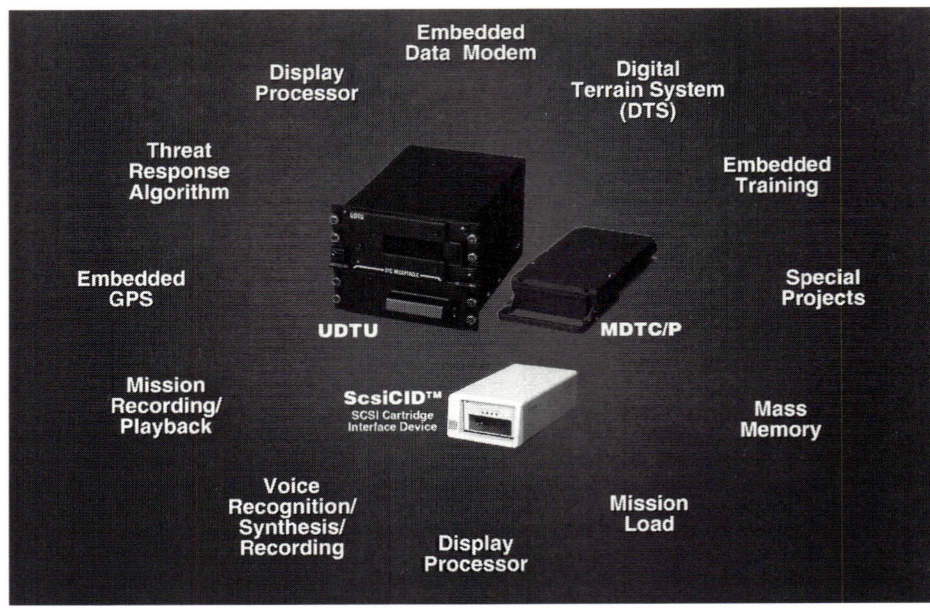

Fairchild's Upgraded Data Transfer Equipment (UDTE) ***1997***/0002320

Mega Data Transfer Cartridge with Processor (MDTC/P)

The Mega Data Transfer Cartridge with Processor (MDTC/P) stores Digital Terrain Elevation Data (DTED) and the Digital Vertical Obstruction File (DVOF) used in terrain referenced navigation. The first application of the MDTC/P is for a digital terrain system provided by British Aerospace Systems & Equipment TERPROM software for the F-16.

INS, barometric and radar altimeter data are transmitted over the aircraft databus to the MDTC/P and INS update and pilot cueing and warnings are transmitted by the MDTC/P over the bus to the INS, multifunction display, HUD and voice warning system.

Other applications for the MDTC/P include the digital moving map database for the F-22, EW simulation and the Tactical Air Combat Mission Analysis and Training System.

Operational status

Ordered by the US Air Force Reserves for digital terrain systems in the F-16.

Contractor

Fairchild Defense OSC.

VERIFIED

AN/ASQ-215 Digital Data Set (DDS)

Fairchild's Digital Data Set (DDS) is a high capacity solid-state military airborne data storage and retrieval system. It provides rapid mission initialisation of the aircraft's avionics suite, as well as recording pertinent mission and maintenance data during flight; the DDS also supports in-flight processing. The AN/ASQ-215 has been adopted as the US Navy Standard Digital Data Set (DDS). The DDS will support Global Positioning System data entry requirements for all MIL-STD-1553 multiplex databus-equipped Navy aircraft. However, because many added features have been incorporated into this data management system, the DDS is suitable for a variety of tri-service military applications.

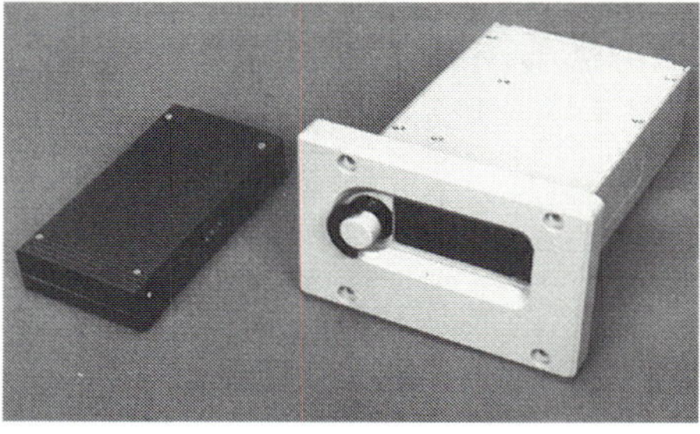

The mission data loader consists of (left) the data transfer module and (right) the interface receptacle unit ***1995***

The Digital Data Set consists of two elements, a hand-carried, non-volatile solid-state data storage device known as the Data Transfer Module (DTM), and the cockpit-mounted, intelligent receptacle for the DTM, the Interface Receptacle Unit (IRU). The IRU manages and controls the data exchange between the DTM and other avionic subsystems via the aircraft's serial digital MIL-STD-1553 multiplex databus. The DTM serves as the transportable storage medium for both pre- and post-mission information exchange between ground computer stations and the airborne system.

Operating primarily as a remote terminal on the MIL-STD-1553 multiplex bus, the IRU also has the provisions to operate as a bus controller. The IRU provides an exhaustive Self-Test/Built-In-Test function for all circuitry, including the DTM while performing real-time bulk memory error detection and correction.

On a single circuit board, the DTM provides 2 Mbytes, expandable to 40 Mbytes with an additional card, of high-speed random access non-volatile data storage. The DTM contains only bulk memory and address decode logic to reduce overall operating expense and simplify future growth. A high degree of data integrity is assured through the use of a proven error detection and correction method.

Specifications

Dimensions:
(IRU) 203 × 127 × 53 mm
(DTM) 152 × 81 × 32 mm
Weight:
(IRU) 1.50 kg
(DTM) 0.31 kg
Input power: 28 V DC at 15 W (max)

Contractor

Fairchild Defense OSC.

VERIFIED

CP-2108A (3007A) data controller/Mission Computer (MC)

The GEC-Marconi Hazeltine CP-2108A (3007A) data controller/Mission Computer (MC) was designed to satisfy the combined requirements of the MC-130E Combat Talon 1, the AC-130H gunship and other programmes. Using a single part number MC suitable for multiple missions greatly reduces spares and logistics support. Level 1 test equipment requirements are also minimised by the use of extensive BIT for fault detection and isolation.

The computer employs a dual-CPU architecture. Each processor consists of a high-speed MIL-STD-1750A CPU and a dedicated local memory. The amount of local memory can be varied to suit the application.

The MC has 512 k words of installed memory, 2.6 Mips of processing capacity and extensive I/O capability high accuracy. There is built-in memory growth to the million word limit of MIL-STD-1750A, multiplexer growth to permit a third dual-channel of MIL-STD-1553B and additional analogue and digital I/O. The unit is mechanised with several blank circuit cards that can be populated with either currently available card designs or new I/O. The 3007A MC is therefore suitable for a variety of other applications.

A plug-in CPU replacement has been developed by GEC-Marconi which allows one or both of the MIL-STD-1750A processors to be replaced by a 25 Mips 32-bit RISC processor. This increases total available throughput to over 50 Mips for applications such as digital terrain and digital map database systems, knowledge-based system health monitoring, threat correlation and so on. A corresponding increase in memory to 68 Mbytes can also be supported. A ½ ATR variant of the 3007A has also been developed which uses the same dual-CPU configuration.

To assist in developing and debugging software on the 3007A, a Computer Support System (CSS) has been developed. The CSS includes mainframe-based flight software development tools and utilises a micro-VAX host computer for real-time programme debug and validation. An IEEE interface is provided for memory loading and verification.

Specifications

Dimensions: 241 × 330 × 478 mm
Weight: 35.5 kg
Power supply: MIL-STD-704, 356 W (+130 W blower)
CPU: dual MIL-STD-1750A
Throughput: 2.6 Mips (DAIS), growth to 6 Mips
Memory: 512 k × 16 words with growth to 2 million words
Input/output: two MIL-STD-1553B dual-mux channels, approximately 100 discrete, various digital/DC, digital/AC, synchro/digital, digital/synchro, digital resolver and serial digital channels
Environmental: MIL-E-5400 Class 1A Category III

Operational status

In service in US Air Force AC-130H gunship and MC-130E Combat Talon I aircraft.

Contractor

GEC-Marconi Hazeltine Corp.

VERIFIED

SKC-3140 bus controller and main computer for the AMX

Two GEC-Marconi Hazeltine SKC-3140 bus controllers and computers equip each Alenia/Embraer AMX aircraft, providing the main and back-up computations and data control.

A single ¾ ATR LRU, containing computer, bus controller and power supplies, controls data from navigation and attitude sensors, radar, head-up display, weapon aiming and communications controllers and the various control and display panels in the aircraft, and data is passed around the aircraft on a MIL-STD-1553B dual-channel databus.

Specifications

Weight: 9 kg
Power supply: 115 V AC, 400 Hz, <200 W
Reliability: > 3,600 h MTBF

Operational status

In production for the Alenia/Embraer AMX.

Contractor

GEC-Marconi Hazeltine Corp.

VERIFIED

The SKC-3140 bus controller and main computer for the AMX

Air Vehicle Interface and Control System (AVICS)

The Air Vehicle Interface and Control System (AVICS) provides the primary interface between a number of aircraft subsystems and mission equipment packages.

AVICS is implemented as the AVIC 1 and AVIC 2 independent units which provide analogue, discrete and serial digital interfaces, signal conditioning, analogue/digital conversion, and data processing to support aircraft subsystem monitoring and control. Each AVIC channel is composed of several plug-in SEM-E Line Replaceable Modules (LRMs) mounted within a backplane/enclosure LRM. The AVIC channels are electronically identical except for the inductive debris monitoring/electrostatic engine monitoring LRM and the vibration monitoring LRM which are both installed in AVIC 2 and the two environmental control system LRMs installed in AVIC 1. Both AVIC units also contain a power supply LRM, process/input/output controller LRM and analogue LRM which are common to each channel.

Specifications

Dimensions: 203.2 × 264.2 × 292.1 mm per unit
Weight: 11 kg total

Operational status

Currently being flight tested.

Contractor

Hamilton Standard Division of UTC.

VERIFIED

Multi-Application Control Computer (MACC)

The Multi-Application Control Computer (MACC) is a flexible computer that can be applied to a wide range of flight and subsystem control fixed-wing aircraft and helicopter applications. A key feature of the MACC architecture is that any number of MACCs can be interconnected to form a simplex, dual, triplex or quad configuration without hardware or software changes. For example, the quad configuration can be reconfigured as two separate, but linked, dual-channel computers performing different tasks. This flexibility is made possible by the Input Output Computer (IOC) which performs all Input/Output (I/O) processing in firmware, leaving the application specific processing to the general purpose processors.

The hard separation between I/O management and application processing is key to the success of the MACC architecture. Both the IOC firmware and application software can be loaded and monitored over an RS-422 link from a portable PC. The same interface can also be used for system testing and integration.

The IOC contains redundancy management, BIT and fault detection and isolation algorithms which are performed at significantly faster speeds compared to general purpose processors. Many of these algorithms are embedded in silicon microcircuits, giving the MACC system significant I/O processing bandwidth. Specifically, the IOC can perform all the I/O processing needed for 50,000 sensor sets/s. The result is a system that is robust, fault tolerant and highly flexible for meeting the requirements of most embedded control systems.

Specifications

Dimensions: 257.8 × 193.8 × 254 mm
Weight: 7.76 kg
Power supply: 115 V AC, 28 V DC, 55.5 W

Operational status

In production. Present applications include flight control, vehicle management system control, actuator and subsystem controllers.

Contractor

Hamilton Standard Division of UTC.

VERIFIED

Boeing 777 Air Data Module (ADM)

The Boeing 777 Air Data Module (ADM) is a pressure transducer with a single pressure input port and an ARINC 629 bus output that transmits linearised, digital serial pressure data. Each Boeing 777 aircraft is configured with six ADM units: three with pitot pressure sensors and three with static pressure sensors. The ADM transmits ARINC 629 digital serial air data to the Air Data Inertial Reference System (ADIRS) and the Secondary Attitude Air data Reference Unit (SAARU).

Specifications

Dimensions: 88.9 × 63.3 × 171.4 mm
Weight: 1 kg (max)
Power supply: 28 V DC, 10.6 W (max)
Temperature range: −15 to +70°C
Accuracy:
(100-147.4 mb) ±0.25 mb
(147.4-175.3 mb) <±0.3 mb
(175.3-1,400 mb) ±0.3 mb
Reliability: 50,00 h MTBF predicted

Operational status

In production and service in the Boeing 777.

Contractor

Honeywell Inc Air Transport Systems.

VERIFIED

DC-9 Digital Air Data Computer (DADC)

The Honeywell DC-9 Digital Air Data Computer (DADC) replaces the existing DC-9 electromechanical air data computer. The new DADC is a form, fit and functional replacement unit featuring solid-state components, including silicon chip pressure sensors that are more accurate, stable and reliable than the vibrating diaphragm predecessors.

Specifications

Dimensions: ⅜ ATR
Weight: 5.44 kg
Power supply: 115 V AC, 400 Hz, 20 W
Temperature range: −55 to +70°C
Accuracy:
(ARINC 429 altitude) ±15 ft at sea level, ±80 ft at 50,000 ft
(synchro altitude) ±20 ft at sea level, ±55 ft at 45,000 ft
(ARINC airspeed) ±4 kt at 60 kt, ±1 kt at 450 kt
Reliability: 14,000 h MTBF predicted

Operational status

In service in the DC-9.

Contractor

Honeywell Inc Air Transport Systems.

VERIFIED

Digital air data computer

Honeywell produces an all-digital ARINC 706 standard air data computer for Airbus Industrie A310 and Boeing 757 and 767 airliners. These all-solid-state systems use large- or medium-scale integrated technology, CMOS master transmitter/receiver chips and a Z8000 microprocessor. Built-in test sequences run continuously in flight to provide a 95 per cent fault-detection capability, with a 99 per cent probability of correct fault identification. Test failure information is stored in the non-volatile memory for post-flight analysis. External sensor failures can also be detected and signalled visually. The system can accommodate a 20 per cent increase in output parameters, more than 50 per cent computational growth and more than 100 per cent memory growth. The unit incorporates Honeywell patented transducers and has ARINC 429 transmitter and receiver interfaces.

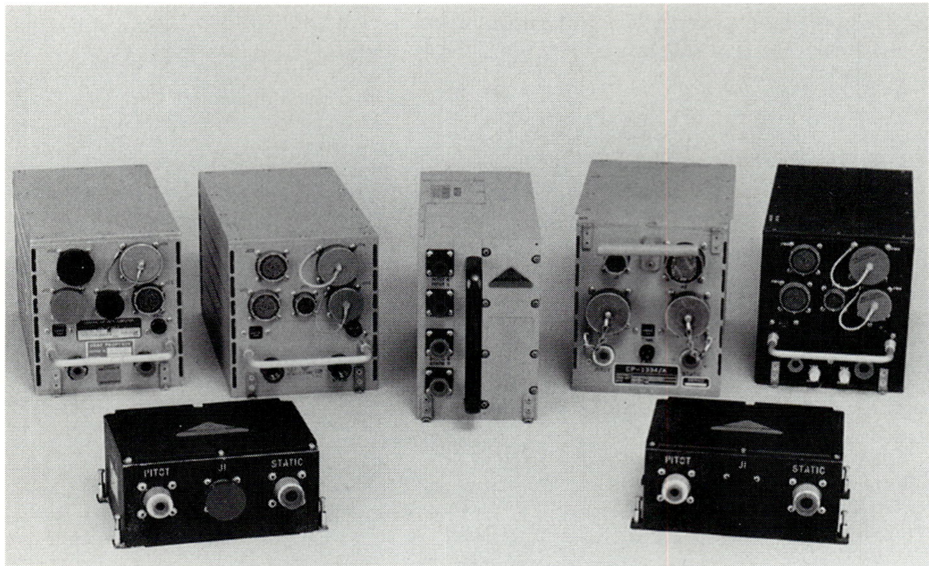

Pictured (left to right, rear) are the central air data computers for the F-15, F-16, C-17, F/A-18 and KC-135R, and (front) the B-52

Specifications

Dimensions: 4 MCU
Weight: 5.7 kg
Power: 25 W
Accuracy:
(airspeed) ±2 kt at 100 kt, ±1 kt at 450 kt
(altitude) ±15 ft at sea level, ±80 ft at 50,000 ft
(Mach) ±0.003 at 0.8-0.9 in the range 25,000-45,000 ft
Failure memory: 6 failures/10 flights
Reliability: 15,000 h MTBF

Operational status

In service.

Contractor

Honeywell Inc Air Transport Systems.

VERIFIED

HG280D80 air data computer

Standard for the DC-10 and MD-80 transport aircraft, the HG280D80 microprocessor-based ARINC 576-type digital system replaced the earlier HG280D5 computer. The transducers are new dual sensors, with high accuracy and a fast warm-up time.

The central processor memory and input/output functions have been reduced from the 360 integrated circuits on nine boards of the earlier D5 to 22 circuits on a single card. Built-in test includes both continuous monitoring and manually activated self-test. The D80's non-volatile memory automatically stores information for five flights and data may be retrieved by selecting a single switch on the front panel. If a failure has occurred either of two warning lights will illuminate.

The D80 can also accommodate limited power interruptions. A capacitatively maintained supply voltage and a CMOS memory can retain critical data during power losses of several milliseconds. When power resumes the device automatically restores all outputs based on this stored data.

Specifications

Dimensions: ½ ATR long
Weight: 5.9 kg
Power supply: 115 V AC, 400 Hz, 30 W
Temperature range: −45 to +55°C
Accuracy:
(altitude) ±15 ft at sea level, ±80 ft at 50,000 ft
(airspeed) ±5 kt at 60 kt, ±3 kt at 325 kt
Reliability: 13,300 h MTBF

Operational status

Over 1,500 units are in service. The HG280D80 is standard equipment on DC-10, KC-10 and MD-80 aircraft.

Contractor

Honeywell Inc Air Transport Systems.

VERIFIED

HG480B and HG480C digital air data computers

The HG480B and HG480C digital air data computers meet ARINC 545 and replace the analogue HG180 on Boeing 727 and 737 aircraft. The unit incorporates ARINC 429 buses.

A single-board microcomputer controls the functions and built-in test (which includes both continuous monitoring and manually activated self-tests) reduces maintenance. Ground built-in test has been expanded and includes q-pot stimulation and transmission of fixed values on all signal output lines.

The HG480 is available in several versions. The basic configuration allows for differences in aircraft wiring but is functionally identical for what are designated the B1, B2 and B3 versions. Designed to operate with conventional analogue equipment, the system also provides outputs for digital instruments and digital autopilot. The B4 version is designed primarily to interface with an all-digital flight control system but can still operate with synchro-driven instruments.

Specifications

Dimensions: ½ ATR long
Weight: 7.48 kg
Power supply: 115 V AC, 26 V AC, 400 Hz, 44 W
Temperature range: −45 to +55°C
Accuracy:
(altitude) ±15 ft at sea level, ±80 ft at 50,000 ft
(airspeed) ±2 kt at 60 kt, ±1 kt at 300 kt
Reliability: 17,600 h MTBF

Operational status

In production and in service in the Boeing 707, 727-200/300, 737-200/300 and 747-200/300. Over 2,000 units are in service.

Contractor

Honeywell Inc Air Transport Systems.

VERIFIED

HG480E1 digital air data computer

The baseline for the HG480E1 is the Sperry ARINC 706, repackaged into a standard ARINC 404 ½ ATR case. The basic circuitry of the ARINC 706 digital air data computer remains unchanged. Two new cards were incorporated for the additional analogue outputs required for Boeing 737-300 and -400 applications.

Specifications

Dimensions: ½ ATR long
Weight: 7.48 kg
Power supply: 115 V AC, 26 V AC, 400 Hz, 35 W
Temperature range: −45 to +55°C
Accuracy:
(altitude) ±15 ft at sea level, ±80 ft at 50,000 ft
(airspeed) ±2 kt at 100 kt, ±1 kt at 300 kt
Reliability: 12,000 h MTBF

Operational status

Designed for the Boeing 737-300/400. In service.

Contractor

Honeywell Inc Air Transport Systems.

VERIFIED

MD-90 Central Air Data Computer (CADC)

The MD-90 Central Air Data Computer (CADC) is a derivative of the HG280D80. The CADC provides both ARINC 429 digital serial air data outputs and ARINC digital serial and DC analogue air data outputs. Two unique features of the CADC - Trigger-On-Failure (TOF) BIT and an internal temperature probe - operate in unison to record BIT failures along with corresponding CADC internal temperature data. BIT failure records and other CADC features can be accessed through the front panel connector via the RS-232 bus. Similar to its predecessor, the CADC uses two vibrating cylinder transducers for sensing pitot and static pressure inputs. The vibrating cylinder transducers can be substituted with silicon pressure sensors. Inputs are also provided from a total air temperature probe, baroset signals and programme discretes. Jointly developed by Honeywell and the Chengdu Aero Instruments Corporation.

Specifications

Dimensions: ½ ATR long
Weight: 5.35 kg
Power supply: 115 V AC, 400 Hz
Temperature range: −55 to +75°C
Accuracy:
(altitude) ±15 ft at sea level, ±80 ft at 50,000 ft
(airspeed) ±5 kt at 60 kt, ±1 kt at 450 kt
Reliability: 13,000 h MTBF predicted

Contractor

Honeywell Inc Air Transport Systems.

UPDATED

ADZ air data system

The ADZ air data system uses an AZ-241 or -242 computer, depending on whether the aircraft is a turboprop or a jet. These are now complemented in production with the AZ-600, for turboprops, and AZ-800, for jets, digital air data systems. All systems use the Honeywell patented vibrating diaphragm pressure sensor to provide outputs for altitude, airspeed, vertical speed, true airspeed, true air temperature and total air temperature information.

If required, a single computer can handle a dual-flight director installation. Consequently, no matter which director is driving the autopilot the full complement of modes is available.

Specifications

AZ-241 (turboprop)
Dimensions: 193 × 79 × 361 mm
Weight: 3.9 kg

AZ-242 (jet)
Dimensions: 193 × 124 × 361 mm
Weight: 5.2 kg

AZ-600 (turboprop)
Dimensions: 193 × 92 × 362 mm
Weight: 3.7 kg

AZ-800 (jet)
Dimensions: 193 × 92 × 362 mm
Weight: 4.08 kg

Operational status

In production for a wide range of executive turboprop and jet aircraft; one of the latest is the Raytheon Hawker 1000.

Contractor

Honeywell Inc Business & Commuter Aviation Systems.

VERIFIED

IC-800 Integrated Avionics Computer (IAC)

Honeywell has developed an Integrated Avionics Computer (IAC) which incorporates the functions of four major avionics subsystems into a compact lightweight package that is 75 per cent smaller than the units it replaces.

The IC-800 is a key component of Honeywell's Primus 2000 advanced avionics system. The units incorporated in the IAC include a digital automatic flight control computer which has all the features and functions of the SPZ-8000 digital autopilot, dual-display processors, a fault warning computer and a full Flight Management System (FMS) computer. The IAC, which is a single LRU, has hardware expansion capability for incorporating additional functions.

The IAC provides at least 50 per cent more processor capacity and 100 per cent more memory for each function than was previously available. Its FMS computer has a 1.9 Mbyte database. This additional memory and processor power will allow substantial growth as each function is enhanced in the future.

The IAC's reduced size and weight result from the extensive use of advanced circuit and packaging technologies. These technologies include Very Large Scale Integration (VLSI), Surface Mount Technology (SMT), Application Specific Integrated Circuits (ASIC), high-density memories, rigid flex motherboards, multilayer circuit boards and advanced thermal management. The IAC is housed in an ARINC ½ ATR short package with an integral cooling fan.

Operational status

In production for the Dornier 328 and the Bombardier Global Express business jet.

Contractor

Honeywell Inc Business & Commuter Aviation Systems.

VERIFIED

Micro Air Data Computer (MADC)

The AZ-840 Micro Air Data Computer (MADC) provides both ARINC 429 and bidirectional Avionics Standard Communications Bus (ASCB) digital serial interfaces for the Honeywell Primus 2000, flight control and navigation systems. The MADC uses extensive surface-mount technology for minimising unit size and weight. Full air data outputs and miscellaneous inputs/outputs are provided, including ATC digitiser and input/output discretes for switching functions and aircraft identification selection. Extensive BITE monitors MADC operation and records faults on non-volatile memory for fault analysis on the ground.

The MADC supports Reduced Vertical Separation Minimum (RVSM) requirements.

Specifications

Dimensions: 106.7 × 149.9 × 157.5 mm
Weight: 2.04 kg
Power supply: 28 V DC, 16 W (max)
Accuracy:
(altitude) ±20 ft at sea level, ±150 ft at 60,000 ft
(airspeed) ±2 kt at 100 kt, ±4 kt at 400 kt
Reliability: 6,000 h MTBF predicted

Operational status

Versions of the MADC are used on the Citation III, Dornier 328 and Raytheon Hawker 1000.

Contractor

Honeywell Inc Business & Commuter Aviation Systems.

UPDATED

HG1140 multirole air data computer

Honeywell's HG1140 multirole air data computer features high accuracy (0.019%f.s.) absolute pressure sensors and the associated electronics for accepting analogue inputs from the aircraft's Total Air Temperature (TAT) sensor and Angle of Attack (AOA) sensor. The result is a full complement of air data parameters for display, flight control, and other onboard system functions.

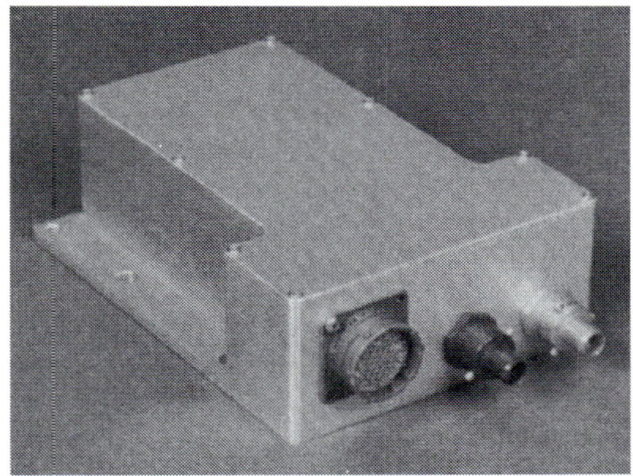

HG1140 multirole air data computer ***1998***/0015280

Specifications

Dimensions: 152.4 × 111.76 × 50.8 mm
Weight: 0.91 kg
Power: 28 V DC, 6 W
MTBF: 100,000+ h (AIC); 50,000 h (AUF)

Operational status

Fully qualified to both RTCA DO-160 and MIL-STD-810E requirements. The multirole air data computer and its air data module variant have been selected for the T-38C Talon Avionics Upgrade Program and for the Lockheed Martin Joint Strike Fighter. The HG1140 is an extension of the company's current air data computer fitted to F-16, F-18, C-17 and other aircraft.

Contractor

Honeywell Inc, Sensor and Guidance Products.

NEW ENTRY

Military air data computers

Honeywell military digital air data computers are in production and in service in the following aircraft: B-52, C-17, C-130, F-16, F-117, F/A-18, FSX and KC-135R.

The latest air data computers use patented solid-state pressure sensors and are said to be so reliable that they will never require recalibration.

Operational status

In production and service.

Contractor

Honeywell Inc Sensor and Guidance Products.

UPDATED

C-17 Warning and Caution Computer System (WACCS)

Developed for use on the US Air Force C-17 transport aircraft, the Litton Warning and Caution Computer System (WACCS) accepts discrete aircraft data and data from the MIL-STD-1553B bus connecting specialised peripheral LRUs. It contains dual-redundant MIL-STD-1553B bus communication ports and can function as the bus controller or a remote terminal. WACCS processes signals from all sources and manipulates the inputs via Boolean logic to provide warning messages on the warning panel or by lighting annunciator lamps.

Specifications

Dimensions: 190 × 188 × 315 mm
Weight: 8.09 kg
Power supply: 28 V DC, 24 W
Reliability: 20,611 h MTBF

Operational status

In production for US Air Force C-17 aircraft.

Contractor

Litton Guidance & Control Systems.

VERIFIED

F-16 general avionics computer

The F-16 general avionics computer acts as the mission computer for the F-16C/D aircraft, providing avionics and weapons control solutions, IFF processing and navigation functions. Designed for MIL-STD-1750 ISA performance, the computer offers as standard features throughput of up to 6 Mips DIAS, 512 k or greater RAM or EPROM, battery back-up MIL-STD-1553B input/output with up to four channels per module, with bus controller or remote terminal. Options include single module and custom packaging, custom input/output and BIT functions.

Specifications

Dimensions: 135 × 312 × 333 mm
Weight: 9.09 kg
Power supply: 115 V AC, 400 Hz, 70 W
Reliability: >4,000 h MTBF

Operational status

In production for the F-16C/D in service with the US Air Force and other air forces.

Contractor

Litton Guidance & Control Systems.

VERIFIED

Pilot's Associate

Current funded studies include examining the use of distributed associate systems to increase safety and expand capacity of the National Airspace System. This effort considers air traffic management procedures, airline operations, pilot and controller workload, and flight deck avionics. Live technology and systems demonstrations are being developed in conjunction with National Aeronautics and Space Administration (NASA) and Federal Aviation Administration (FAA) efforts towards air traffic management modernisation and free flight.

Operational status

Commercial applications of Pilot's Associate technology are being pursued.

Contractor

Lockheed Martin Aeronautical Systems.

VERIFIED

Core Integrated Processor (CIP)

Lockheed Martin's Core Integrated Processor (CIP) for the C-17 Globemaster III is the central computer that controls all of the aircraft's avionics systems. The computer provides all existing mission computer functions and has growth capability to interface with, and provide data management for aircraft controls, displays and sensor data.

The CIP is a high-speed general purpose computer designed to meet the real-time processing requirements of the C-17 application. The baseline CIP is housed in a single Line Replaceable Unit (LRU) containing four subassemblies. The LRU is a full ATR cross-section, 362 mm in length, providing 10 Versa Module Eurocard (VME) module slots. The MIPS R4400 was selected as the Central Processing Unit (CPU) based on off-the-shelf multisource availability, performance and the error detection and correction capability. The VME backplane is used as the intermodule communication channel providing up to 80 Mbytes/s bus bandwidth. A fundamental requirement for the CIP design is open systems such as VME, POSIX, UNIX and VxWorks for both hardware and software. The entire software approach promotes open architectures which feature industry standards and off-the-shelf solutions for operating systems, kernels, programming languages and software development environments.

Two units will be installed on each C-17, replacing the existing three unit mission computer. The CIP will be retrofitted on existing C-17s and will be installed on all future C-17 aircraft.

Operational status

In production and in service on C-17.

Contractor

Lockheed Martin Control Systems.

VERIFIED

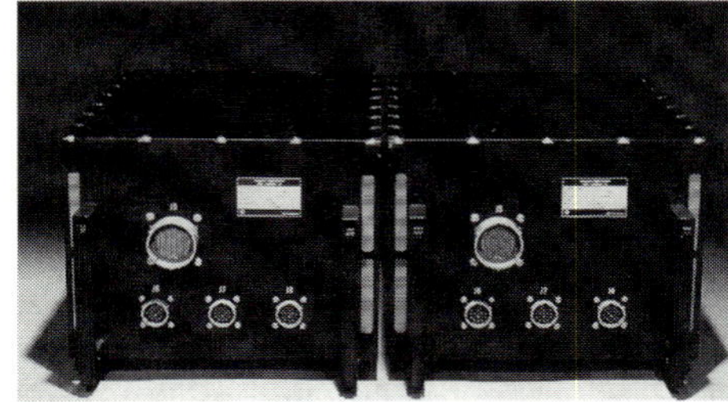

C-17 and CIP **1997**/0001287

CC-2E Data Processing System (DPS)

The CC-2E Data Processing System (DPS) is the central processor for the Boeing E-3 AWACS aircraft which provides early warning of threats and airborne control of friendly aircraft. It receives and processes data from onboard sensors, sends the completed airborne view to mission operators and links critical messages to ground commanders. The CC-2E DPS is the fourth-generation multipurpose airborne processor for the E-3.

A recent value engineering change proposal effort resulted in the development of increased memory in the CC-2E DPS, reducing the number of monolithic memory units from five to three per system.

Operational status

In service.

Contractor

Lockheed Martin Federal Systems.

UPDATED

Airborne Battlefield Command Control Center (ABCCC III)

The Airborne Battlefield Command and Control Center (ABCCC III) is an airborne node in the US Air Force Tactical Air Control System. It maximises the efficient use of fighter forces and other air resources by gathering real-time battlefield data from forward areas and by managing aircraft in air-to-ground operations.

Designed for the EC-130E, the ABCCC III self-contained capsule houses 15 automated workstations, allowing the battle staff to manage the tactical air assets conducting over 150 sorties/h effectively. During each 10 hour mission of the EC-130E, the ABCCC III provides a communications link to higher headquarters and co-ordinates forces in the battle area, updating aircraft on their way to the target and collecting their in-flight reports on the way out. Automated capabilities allow the battle-staff to analyse ongoing combat quickly and direct offensive air support toward fast developing targets.

The ABCCC III consists of four major airborne subsystems: The Communications Subsystem (CS), the Tactical Battle Management Subsystem (TBMS), the Airborne Maintenance Subsystem (AMS) and the Capsule Subsystem. There is also a major subsystem on the ground. The Mission Planning Subsystem (MPS) provides the tactical database used by the airborne maintenance technician for system initialisation. On mission completion the MPS can also be used for post-flight playback and analysis of mission data.

Once airborne, the CS and its Automated Communications and Intercom Distribution System (ACIDS) controls and secures all communications between the capsule and forward units, other aircraft and rear bases. Communications within the capsule and between the capsule and the flight crew are also handled by ACIDS.

The TBMS provides comprehensive battlefield management capabilities for up to 12 operators stationed at individual battle-staff consoles. Fast accurate access to communications, as well as tactical and map databases, enhances operator effectiveness.

The AMS provides diagnostic and fault isolation to detect hardware or software malfunctions during a mission. System initialisation and control are also integrated into AMS functions.

The Capsule subsystem consists of the facilities providing physical, environmental and life support for operating the airborne subsystems.

Contractor

Lockheed Martin Tactical Defense Systems.

VERIFIED

AN/ASQ-212 mission processing system

The AN/ASQ-212 system consists of the CP-2044 computer and several interconnection devices which comprise a form, fit and function replacement for the AN/ASQ-114 computer, data analysis logic units and the signal data converter. The extended memory upgrade of the AN/ASQ-114 computer is transferred to the CP-2044 and used both for global and secondary memory. The CP-2044 incorporates Motorola 68030 processors to provide a throughput ranging from 10 to 25 Mips, which is 30 times greater than the current system in the P-3C Update I/III aircraft at a fraction of the current size, weight and power requirements. In the full Update III Ada implementation, less than 50 per cent of the CP-2044 minimum throughput and memory capacity is utilised.

The CP-2044 VME bus open architecture can be configured with additional processing, memory and input/output modules to meet the requirements of new subsystems such as GPS and Satcom, and of processing intensive functions such as sensor post-processing and data fusion.

Initially designed for retrofit into P-3C Update I/III aircraft, the AN/ASQ-212 can be easily tailored to the requirements of other P-3C configurations as well as new aircraft.

Operational status

The AN/ASQ-212 system was developed for the US Navy P-3C aircraft under a two-phase programme that began in September 1989. The first production systems were installed in Navy test aircraft and training facilities beginning in May 1993, at a rate of four systems per month. Also in service with P-3C export customers.

Contractor

Lockheed Martin Tactical Defense Systems.

UPDATED

ASPRO ASsociative PROcessor

The ASPRO Associative Processor is a small but powerful fully military-qualified 0.44 cu ft parallel processor embedded in the Northrop Grumman E-2C Hawkeye's avionics suite. It performs millions of operations per second to track potentially thousands of targets. It also performs additional display processing functions on the aircraft.

Since it was originally developed, ASPRO's processing throughput has been increased and its memory doubled, with no increase in volume. In addition, a system to meet MIL-E-16400 has been built, for use in 'Los Angeles' class submarines. Lockheed Martin has developed the ASPRO-VME, which features increased memory, system throughput, and expansion capability to accommodate new applications.

Specifications

Dimensions: 203 × 229 × 254 mm
Weight: 14.5 kg
Power: 200 W
Throughput: 50 Mops
Reliability: 4,500 h MTBF

Operational status

In service in the E-2C Hawkeye aircraft.

Contractor

Lockheed Martin Tactical Defense Systems.

UPDATED

ASPRO-VME parallel/associative computer

The ASPRO-VME, Lockheed Martin's fourth-generation parallel processing computer, is a modular open architecture VME-compatible card set. It is capable of performing between 150 Mflops and 2.4 Gflops. The basic three-module 512 processor configuration can be expanded from 512 to 8192 processor elements. Each of the parallel processors, called Processing Elements (PEs), contains a full 32-bit IEEE floating point processor and a bit-serial processor.

A module occupies a single-VME card slot and consists of two printed circuit boards attached to a cold plate. The entire ASPRO-VME, in its basic configuration, requires three VME slots.

Programmable in Ada, ASPRO-VME is supported by a powerful software development tool set which allows application programmes to be easily developed. Because of ASPRO's single instruction multiple data stream architecture, modular expansion from 512 PEs to 8,192 PEs can be accomplished without rewriting software.

The parallel architecture and associative search capability produce significant gains over conventional processors for sophisticated tracking, correlation, data fusion, and situational awareness algorithms.

ASPRO-VME is available in a rugged and full MIL-SPEC configuration. It can operate in a stand-alone mode or be directly embedded in commercial and rugged workstations with 6U VME slots. Applications for ASPRO-VME include command and control, correlation and tracking, data fusion, database management, signal processing, expert systems, neural networks and image processing.

Specifications

Weight: 2.59 kg
Power: 80 W

Operational status

Production units available.

Contractor

Lockheed Martin Tactical Defense Systems.

UPDATED

Avionic common module systems

Lockheed Martin Tactical Defense Systems is the technology leader for development, application, implementation and integration of JIAWG/MASA/SHARP avionics common module processing clusters and systems. These capabilities and resources are aimed at allowing module users complete integration control from subsystem development to system platform implementation.

The foundation of common module information processing systems is extensive work in VHSIC design. A limited number of these standard VHSIC chips enables creation of a family of modules that can be used in different avionics systems. Designed in the Standard Electronic Module (SEM) E format, each module measures 149.3 × 162.6 × 14.7 mm.

The Lockheed Martin Tactical Defense Systems family of avionics common modules consists of seven processing and Input/Output (I/O) SEM-E module types, a power supply SEM-E module, backpanels, liquid or air-cooled racks and active and passive star coupler technology. Reusability and standardisation have been stressed within the design philosophy for the avionic equipment family. Lockheed Martin Tactical Defense Systems has created a generic modular, module functional design to promote reuse of ASIC devices and to support integrated diagnostics.

The Advanced General Purpose Processor Element (AGPPE) is a 32-bit RISC-based data and signal processor module ideally suited for avionic systems applications. Equipped with a Mips R3000 RISC processor, its pipelined architecture yields very high throughput while its standard SEM-E format and three input/output interfaces allow easy integration in various systems.

Advanced ceramic circuit boards and surface-mount components contribute to AGPPE's light weight, low power consumption and dense packaging. Full 32-bit operation is enhanced by a five-stage pipeline, on-chip cache control and an on-chip memory management unit. Block refilling of both instruction and data caches is supported.

The Mips R3000 processor executes instructions up to 20 times faster than the VAX 11/780. In addition the R3010 floating point co-processor chip handles floating point arithmetic compliant with the ANSI/IEEE standard.

Onboard memory resources include a 1 Mbyte SRAM which can be accessed synchronously in two CPU clock cycles. Bootstrap and debugging code can be stored in a 128 kbyte EEPROM. Separate 16 kbyte data and instruction caches, which effectively double the available cache memory bandwidth, provide instructions and operands at the CPU clock rate. Other features that enhance performance include a four-word buffer for block refill of each cache and a one-word write buffer for writes to main memory.

The module features TM bus and a Lockheed Martin designed maintenance controller ASIC with an IEEE-488 channel for console operations. A network interface controller provides a one-chip interface from the AGPPE to a data flow network, a 32-bit parallel bus.

MIL-STD-1750A processor

The MIL-STD-1750A processor module is a 3.85+ Mips VHSIC processor with 512 k words of local SRAM, plus 8 k words of start-up ROM and PI-bus, TM-bus and IEEE-488 standard I/O interfaces on a single-width double-sided ¾ ATR size SEM-E module. The module's ceramic printed circuit boards and surface-mount components provide a dense, lightweight, low-power, general purpose data processor module. The highly parallel pipelined architecture enables high throughput. Advanced built-in test techniques reduce life cycle costs by supporting two-level maintenance. In addition, the standard SEM-E size and three standard I/O interfaces allow easy integration into a variety of systems.

The processor module has an onboard MIL-STD-1750A maintenance controller to automate built-in test on the module, communicate with the other modules via the TM-bus and handle console operations via the IEEE-488 bus.

The MIL-STD-1750A data processor's three CMOS gate arrays, the CMOS semi-custom maintenance controller common to the Unisys common module family and the ECL clock chip are equivalent to over 160,000 gates.

High-speed databus interface

The High-Speed DataBus (HSDB) interface module integrates the high-performance processor with a linear token-passing high-speed fibre optic databus on a single-width double-sided ¾ ATR size SEM-E module.

The HSDB module provides a dual-redundant 50 Mbit/s fibre optic system interface, processor, 256 k words of local SRAM, a subsystem interface, maintenance via the TM-bus and IEEE-488 bus I/O interfaces.

The module functions are highly integrated, using surface-mount components. The HSDB can be configured with processor/bus combinations consisting of either the high-performance 1750A processor or the MIPS-based RISC processor and with either a PI or N bus subsystem interface. The HSDB interface module has a CMOS gate array, an ECL gate array and hybridised fibre optic transmitters and receivers for the HSDB interface.

DC-DC converter

The Lockheed Martin Tactical Defense Systems DC-DC converter is a single-width ¾ ATR SEM-E size power supply that provides over 200 W of regulated power from an unregulated 270 V DC bus. The converter uses a standard power supply topology with hybrids and high-frequency techniques to attain high efficiency, high reliability and small size.

Input power to the converter is +135 V DC and −135 V DC (270 V DC line-to-line) MIL-STD-704D. Output voltages are +5 V DC, −15 V DC, −5.2 V DC and +80 V DC. The efficiency of the DC-DC converter is 80 per cent at full load. Up to six modules may be paralleled to increase current capacity and/or provide redundancy.

The converter has a patented digital controller to enhance stability, improve testability and provide fault detection. The module has a slave TM-bus interface to communicate with other modules in the subsystem.

Contractor

Lockheed Martin Tactical Defense Systems.

VERIFIED

Modular Airborne Processor (MAP)

The MIL-STD-1750A Modular Airborne Processor (MAP) is a 1.5 Mips VHSIC, CMOS processor for use in UAVs such as smart weapons, RPVs, targets, and land, sea and air autonomous vehicles.

It has up to 256 k words each of SRAM and EEPROM with 8 k start-up memory. The basic MAP consists of two 6 × 9 in (152.4 × 228.6 mm) PCs: a processor board and an Input/Output (I/O) board. The processor board contains the 1750A chip set and resource controller ASIC along with system SRAM. The I/O board contains a dual-redundant MIL-STD-1553B bus controller interface, a MIL-STD-1553B remote terminal interface compliant with the requirements of MIL-STD-1760A, MIL-STD-1760A discretes and EEPROM memory. Two additional cards can be provided to accommodate user-defined I/O requirements.

All MAP power is provided by a pluggable PC card DC/DC converter requiring MIL-STD-704 28 V DC input. A spare slot is reserved for the optional Programme Development Card/Performance Monitor Module (PDC/PMM).

The PDC provides a standard IEEE-488 control interface for applications development and test. The PDC also provides a plug-in PMM interface to a standard VAX DRQ3B I/O channel providing enhanced Ada software development capability, including real-time interactive data collection and debugging.

The MAP is housed in a lightweight EMI compliant enclosure utilising MIL-C-38999 connectors. It is designed for conduction cooling requiring no fans or environmental controls. The MAP utilises low-cost glass-epoxy boards with pin-grid array ASICs and DIP integrated circuits.

The MAP design utilises a 3.8 Mip 1750A chip set with the RC coupled to a modified system information transfer and execution bus open architecture. This permits modular expansion on the I/O bus as required by the application. The MAP contains BIT coverage of 98 per cent as per MIL-STD-2084.

Contractor

Lockheed Martin Tactical Defense Systems.

VERIFIED

U1638A MIL-STD-1750A computer

The U1638A computer is a militarised radiation-hardened general purpose avionic computer that meets all the requirements of MIL-STD-1750A Notice 1. The U1638A computer uses the U1635 high-speed MIL-STD-1750A central processing unit. The U1638A combines this high-density high-speed microprogrammed CPU with memory, input/output interfaces and power conditioning to provide the high-speed data computation, data storage and data transfers required for avionic systems.

The U1638A computer is packaged as a MIL-STD-1788 type 10 chassis which includes 12 shop-replaceable units. One is a power supply/power conditioner and 11 are processor and Input/Output (I/O) printed circuit cards.

Using the U1635 CPU in conjunction with a high-performance cache memory, the U1638A achieves a DAIS mix performance in excess of 1.5 Mips. The main memory consists of one million words of non-volatile core memory organised as two 512 k modules. The primary I/O interfaces are the four MIL-STD-1553B dual-redundant and multiplex databusses.

The U1638A includes a programmed I/O channel and 16 input and 16 output discretes which are programme-controlled and can be defined to meet the user's needs. Four external interrupts and five identification bits are also provided. The unit is radiation-hardened with a radiation detector, power dump circuitry, a microcontrolled main memory radiation event recovery system independent of macro software and EMP protection.

Specifications
Dimensions: 193.5 × 322.3 × 319 mm
Weight: <23 kg

Power supply:
(dual AC input) 115 V AC, 400 Hz, 3 phase, 350 W

Operational status
In service on the US Air Force B-2.

Contractor
Lockheed Martin Tactical Defense Systems.

UPDATED

Hawk/32 MIL-SPEC computers

The MIL-SPEC commercially developed Hawk/32 super mini-computer has been selected for a variety of air, land and sea military applications. In production since 1984, this computer offers compatibility with military hardware, filling a niche between rugged and full MIL-SPEC computers.

The Hawk/32 is a single-processor 32-bit computer. CPU performance is 1.1 Mips. Modules are available with memory capacities of 2 Mbytes, 32 Mbytes (Models 1901 or 1904) or 8 Mbytes (Model 1906). Radiation hardening is available for the Model 1900. TEMPEST hardening is also available.

Specifications
Dimensions: 195 × 320.5 × 581.7 mm
Weight: 40.82 kg
Altitude: up to 50,000 ft
Temperature range: −55 to +54°C
Vibration: 10 *g* with isolation, 2 *g* hard-mounted
Reliability: >MIL-HDBK-217E

Operational status
Over 350 units have been delivered since 1986 for programmes such as the US Army/Air Force Joint STARS and the US Navy TACAMO.

Contractor
Lockheed Martin Western Development Labs.

VERIFIED

AN/AYK-42(V) processors

The most significant machine in the AN/AYK-42(V) range is the Norden Systems PDP-11/34M processor. All machines in the series are airborne processing units which are software and interface compatible with Digital's PDP-11 system. Fully militarised, they are suitable for applications ranging from tactical avionics to complex command and control.

The PDP-11/34M unit includes a complete processor, memory, peripheral interfaces and power supply on a single chassis. Modular unit construction ensures quick and easy replacements.

Features include: a complete PDP-11 instruction set (over 400 instructions); 1 k word cache memory option; memory expansible up to 128 k words; 16 k or 32 k word memory modules; 900 ns cycle time in core memory; memory management and protection; hardware multiply and divide; floating point processor option; hardware stack processing; and an input/output rate up to 1.1 M words/s.

Another processor in this range, the PDP-11/70M, can comprise up to four 1 ATR boxes dedicated to processing, power supply, 256 k words memory and expander (extra input/output options) facilities respectively. This unit can perform up to 850 Kips. The LSI-11M is a single-card version of the same processor design which operates at approximately 200 Kips. It has 4 k words of random access memory and can be associated with 16 k or 32 k word core storage modules.

Specifications
Norden PDP-11/34M
Dimensions: 498 × 257 × 194 mm
Weight: 22.7 kg
Power: 410 W (max)
Computer type: binary, fixed and floating point
Word length: 32- or 64-bit
Instruction set: > 400 instructions (as PDP/11)
Typical speeds:
(without cache) 275 Kips
(with cache) 400 Kips
Input/output: 5-12 slots. Up to 1.1 M words/s
Memory:
(core) up to 128 k words
Environmental: MIL-E-5400

Operational status
In production.

Contractor
Northrop Grumman, Electronic Sensors and Systems Division, Norden Systems.

UPDATED

Omnidirectional Air Data System (OADS)

Developed for helicopter applications, the Omnidirectional Air Data System (OADS) consists of a mast-mounted sensor, connected cable and air data computer, and provides an airspeed output over the range 0 to 200 kt irrespective of direction. Additional outputs include forward, rearward and sideways speed components, air density, altitude, air temperature and pressure. The original Pacer OADS was used in the Bell X-22 ducted propeller tilt engine V/STOL research aircraft built in the early 1970s. The current system is in production for the US Army AH-64A Apache anti-tank and US Coast Guard Eurocopter HH-65A Dolphin search and rescue helicopters. In July 1987 a technology transfer contract was signed with the China National Aero-Technology Import and Export Corporation (CATIC). This will involve production of helicopter air data systems configured for the Z-9 helicopter, which is built in China under licence from Aerospatiale.

Specifications
Dimensions:
(sensor) 254 × 76 × 76 mm
(computer) 184 × 127 × 241 mm
Weight:
(sensor) 1 kg
(computer) 2.7 kg

Operational status
In production for the Boeing Company AH-64, Eurocopter HH-65A and Chinese Z-9 helicopters.

Contractor
Pacer Systems Inc.

VERIFIED

F-16 modular mission computer

The computer system for the F-16 provides multiple processing functions in a single chassis. It replaces three present computers in the earlier configuration and provides processor power to support the addition of capabilities such as FLIR and digital terrain functions. SEM-E format modules provide data processing, avionic display, power supply and bus interface for the computer. Various aircraft system functions are defined by the application software operating on the computer. The primary computing module processors are based on the R3000 32-bit RISC instruction set architecture. Each processor module provides 16 VAX/Mips throughput and has 1 Mbyte of on-module memory. The computer, as configured for the F-16 with 16 digital modules installed, weighs 17.78 kg.

Operational status
Initial deliveries began in 1993.

Contractor
Raytheon Systems Company.

UPDATED

Integrated Core Processor (ICP)

The Raytheon Systems Company has been selected by Lockheed Martin Tactical Aircraft Systems to provide the ICP for its two JSF Joint Strike Fighter demonstrator aircraft.

The ICP is the central computer system for the JSF, including all the embedded computing elements for multiple subsystems. It provides the digital processing resources for sensors, communications, electronic warfare, guidance and control and cockpit displays. The ICP implements an open system architecture that maximises the use of commercially supported products and standards, to define, implement and support an affordable ICP that uses open-system concepts to enable seamless incorporation of new technologies.

Contractor
Raytheon Systems Company.

NEW ENTRY

Matchwell integrated data fusion system

Matchwell is an advanced development project funded by the US Air Force Wright Laboratory. The goal of Matchwell is to provide early threat warning to aircrew. As air defence systems proceed through the targeting process, they emit sufficient information to allow an observer armed with the appropriate sensors and intelligence data to determine their location and intentions. Matchwell accomplishes this by performing fusion at Levels 1, 2 and 3 to provide data fusion and situation assessment. Use of artificial intelligence techniques allow all this to be executed in real time.

Matchwell is implemented on two DEC 3100 workstations. One hosts the environment simulator that generates an integrated set of signal reports; the other hosts the threat warning software. The workstations provide high-resolution graphic displays of the actual and inferred situation. The software is written in a combination of Ada and C to facilitate ease of transition to an embedded real-time environment.

Contractor
Raytheon Systems Company.

UPDATED

Processor Interface Controller and Communication (PICC) module for the F-22

Raytheon Systems Company produces the MIL-STD-1750A instruction set architecture-based Processor Interface Controller and Communication

(PICC) module for the F-22 aircraft. Each module provides data processing, MIL-STD-1553B bus communication, intermode communications via serial interchannel datalink and analogue and digital input/output for the vehicle management system.

Contractor
Raytheon Systems Company.

UPDATED

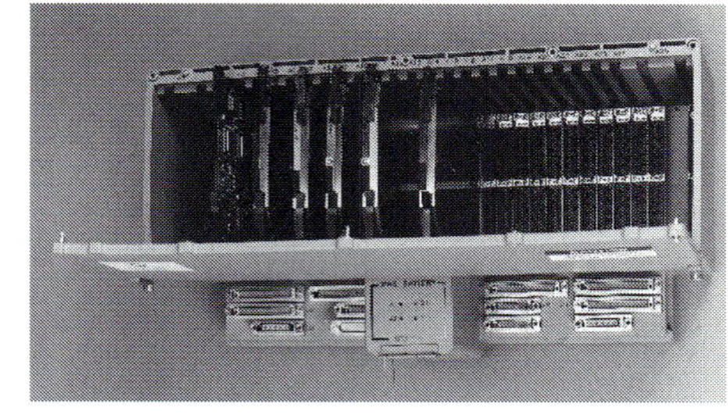

The processor interface controller and communication module for the F-22

ADS-82 digital air data system

Rockwell Collins developed the ADS-82 air data system as part of the Pro Line range of avionics equipment. The system processes data derived from pitot and static pressures and interfaces with the aircraft flight control and flight director systems.

To adapt the system to a variety of aircraft a single plug-in module is mounted at the rear of the computer. Operating parameters are coded in the module according to aircraft type.

Optional output instruments include: a true airspeed/temperature indicator; Mach/airspeed indicator; altitude preselector/alerter; encoding altimeter; and vertical speed indicator. These indicators are in 3 ATI format. An ARINC 575 true airspeed output is available for use in inertial or Omega/VLF navigation systems.

Specifications
Dimensions: ½ ATR
Weight: 2.99 kg
Power: 23 W
Altitude: −1,000 to +50,000 ft
TAS range: 50-600 kt

Operational status
In production and in service.

Contractor
Rockwell Collins.

UPDATED

ADS-85/86/850 digital air data systems

The ADS-85/86/850 Series is similar to the ADS-82 except that it uses a solid-state piezo-resistive sensor. The ADS-85 interfaces with the Collins electromechanical air data instruments and has an ARINC 429 bus for auxiliary systems. The ADS-86 interfaces with Rockwell Collins CRT air data instruments. The ADS-850 contains an ARINC 429 bus for flight control and attitude and heading communications.

Specifications
Dimensions: ½ ATR
Weight: 2.52 kg
Power: 18 W

Operational status
In production and service.

Contractor
Rockwell Collins.

UPDATED

Data Management System (DMS)

The Data Management System (DMS) is a step towards achieving paperless operations, with electronic control of information. In the cockpit, approach and departure information, taxi and ground diagrams are electronically maintained and can be narrowed to the definition required by the pilot with the pan and zoom feature. Maintenance, cabin and operations data can be accessed at a single terminal, which also serves as the data load and retrieval centre. The Collins DMS is packaged in a single cabinet, using line-replaceable modules. These allow significant reduction in space and weight requirements and need no forced-air cooling.

The DMS consists of a multichannel computer, with information contained in mass storage modules for quick and efficient data access. Information is displayed on high-resolution colour flat-panel displays. The entire system is networked with a fibre optic distributed data interface.

The Rockwell Collins DMS is adaptable to virtually all aircraft types.

Operational status
Available as an option for the Boeing 777. First deliveries were made in 1995.

Contractor
Rockwell Collins.

UPDATED

TIMIS Tactical Information Management Integrated System

The Rockwell Collins TIMIS provides a fully integrated Command, Control, Communications Computer and Intelligence, Surveillance/Reconnaissance (C4IS/R) System that provides the user with the capability to incorporate existing/new sensors, tactical data communications, satellite data, Link-11, Link-16 and Link-22 air traffic management datalinks (FAA, ICAO) or user-specific links (Link-Y). The TIMIS exploits existing LOS, BLOS and satellite communications and is growth-oriented to incorporate emerging communications systems and media (GBS, EHF and SHF). It consists of high-quality, field-proven, COTS components. It can be implemented in airborne (fixed- or rotary-wing), sea-based (surface and subsurface) and ground (fixed and mobile) configurations specifically tailored to the needs of the user. TIMIS is a data management system that contains interfaces for organic and external platform sensors, mission processors, displays, communications and tactical navigation. It is composed of two major core components - a Tactical Data Processor (TDP) and a DataLink Processor (DLP). The TDP with its attendant five processors control the display, image, generic-front-end communications processor and communications preprocessor as well as the DLP. The TDP and the DLP are mounted in a ruggedised VME chassis. The TDP processes and routes all the tactical message traffic to the tactical system controller operator displays and the flight deck. The TDP also controls all the communications to/from the platform, including satellite sensor/imagery data. The TDP performs correlation using attribute functions to aid in target definition including identification (friend or foe).

The DLP provides automatic interface with datalinks (L-11 and L-16) using M and J series messages and attendant data terminal sets and provides simultaneous datalink operation. The DLP provides Link-11 and Link-16 message generation and translation. The Data Handling System (DHS) is a high capacity and high-speed central data store that provides the capability for data centric operations. The DHS is used for loading of mission programs, recording of mission data (including video and voice) and is the data store for mission historical tactical data, simulation, onboard training and maintainability aids. It may also be implemented as the central data store for other platform sensor or processor functions.

The high-resolution colour displays are universal workstations interfacing with the TDP via EIA-485 (data) and EIA-170/343 (video). The TDP also provides interfaces via RS 232, RS 422, ARINC 429, ANEW, Manchester and MIL-STD-1553. The TIMIS offers upgrades to existing control display units or control display navigation units as an option.

The TIMIS interfaces directly to a sensor or to a mission processor without impacting the existing interfaces or processor software, thus reducing integration risk and cost. TIMIS real-time interfaces with flight management and onboard sensor systems, Electronic Support Measures (ESM), Infrared Detection Systems (IRDS), digital cameras (optics), conventional radar and Inverse Synthetic Aperture Radar (ISAR) provide the operator the means to overlay previous missions' sensor and imagery data with on-station sensor information to assess changes in tactical situations, sensor and emitter variations and acoustic anomalies.

The TIMIS provides the controller/operator with a complete picture of the tactical situation via high-resolution colour displays with access to real-time sensor videos, geographic sensor and identity filters, digital maps and datalink information.

In addition to having a large track database (5000 track capacity) the TIMIS provides assistance with the managing and correlation of track data. The mission operator is able to archive/save data and record video to magneto-optical disk, recall selected tracks desired for display and filter undesired tracks by geographic identity or sector.

A Defense Mapping Agency digital map with GPS/INU inputs provides accurate position and heading for a detailed display of track data. Waypoints for navigation aid the flight personnel in track correlation and are accessible to the tactical operator as a manual entry in flight or as pre-flight inputs. TIMIS options include expanded sensor integration, mission system interfacing and digital mapping for flight station display.

Contractor
Rockwell Collins.

UPDATED

Aircraft systems processor

SCI is currently producing two distinct versions of the Aircraft Systems Processor (ASP) for the AH-64D Longbow Apache. These are the systems processor and the weapons processor, which vary in input/output and content. These processors provide all mission data, housekeeping and weapons processing and have been successfully integrated and flight tested in the first Longbow Apache prototype aircraft. Extensive use of ASICs, SMT, double-sided circuit card assemblies and controlled impedance motherboards has resulted in a full MIL-E-5400 processor that has 330 input/output channels, dual-1750A CPUs, dual-MIL-STD-1553B interfaces and extensive BIT implementation, including built-in logic analyser and console debugging functions.

The ASP features high-density input/output combined with high-computational throughput and flexibility and low weight and size. The ASP chassis is an eight-slot air-cooled aluminium housing. Five of the slots are general purpose, one slot is dedicated to the power supply module and one slot is double width to accommodate a piggyback module or other oversize assembly. Each ASP housing can contain up to 14

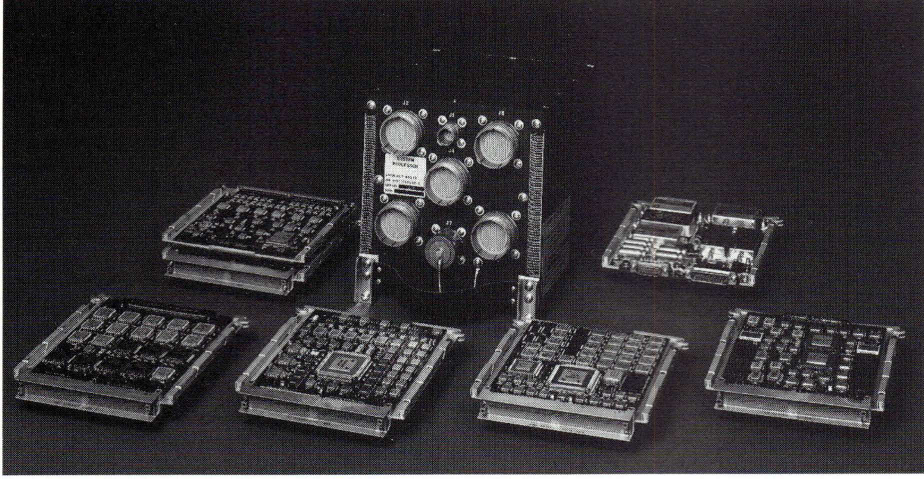

The SCI aircraft systems processor can house up to six modules

CCAs on six modules, plus a power supply. All ASP modules are double-sided SEM-E size with an aluminium core to provide efficient heat transfer.

Available as an option is an R3000 CPU to replace the 1750 CPU for those applications that need significant processing power above that offered by the 1750 CPU. The R3000 CPU is based around the LSI Logic LR33000, the first integrated Mips architecture microprocessor. Within a single chip, the LR33000 combines the scalar processor and cache/memory manager portions of the Mips architecture, as well as 1 kbyte of high-speed data cache memory and 8 kbytes of high-speed instruction cache memory. Operating at 30 MHz, the R3000 CPU offers performance figures of 15 Mips or greater.

Specifications

Dimensions: 190.5 × 193 × 203.2 mm
Weight: 6.35 kg
Reliability: 6,000 h MTBF

Contractor

SCI Systems Inc.

VERIFIED

Two different versions of the ASP are being produced for the AH-64D Longbow Apache

Pave Pace integrated avionics architecture

The Pave Pace programme continues the further development of integrated system avionics architectures begun by its predecessor, the Pave Pillar programme. Pave Pace will both enhance and extend the Pave Pillar architecture.

Several enhancements are planned. One will employ advanced digital multichip packages, using silicon-on-silicon technology. This approach will allow SEM-E-sized line replaceable modules, roughly 150 × 150 × 15 mm, to operate at speeds approximately one billion Floating point Operations Per Second (Flops). Substantial weight and volume reduction and concomitant reliability increases result from the need for three to four times fewer modules, the need for dramatically fewer module-to-backplane connectors and more than a three-fold decrease in solder joints. However, the densely packaged circuitry and clock rates of around 100 MHz will generate heat removal requirements in excess of 150 W per module. Pave Pace solves this problem with another Pave Pillar enhancement by using liquid flow-through modules, where the coolant, normally confined within the rack enclosure, is brought into the centre of a hollow module frame that supports circuitry on either side. Within this frame, the fluid interacts with a metallic heat exchanger, allowing a shorter thermal path between the heat source and coolant. Data shows 200 W can be removed at circuit temperatures of 83°C. Projections show that a 10,000 to 20,000 hours MTBF reliability for such modules is possible at an estimated cost of US$25,000 to US$30,000.

Another enhancement will be the use of modular high-speed optical network crossbar switches, operating at 1 to 2 Gbits/s, that will allow fault-tolerant interconnections between sensor-based signals and signal processing centres located in the racks. These same switches will be used for rack-to-rack communication and fault-tolerant rack display routeing for digitised video signals.

Extensions to the Pave Pillar architecture are the most dramatic, however. An integrated sensor system will be implemented. Here, the same Pave Pillar concepts of building a small family of modular digital building blocks to create system processing functions is used, except the modular building blocks are multifunction apertures and the family is a small number of analogue-based SEM-E-sized synthesisers, receivers, transmitters and switches. Projections show that an integrated RF system has the potential to reduce cost and weight of RF support electronics by approximately 50 per cent and improve sensor system

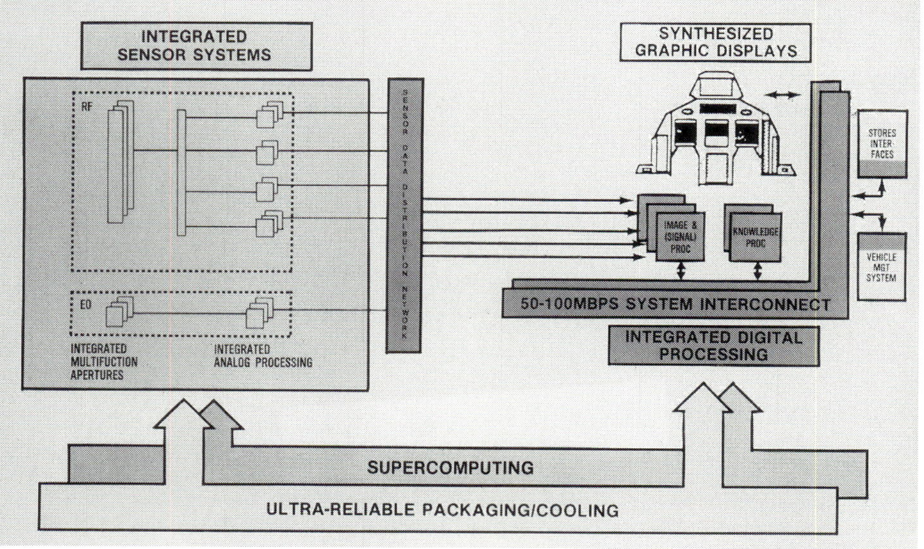

The Pave Pace programme continues the development of integrated system avionics architecture started under the Pave Pillar programme

availability by over a factor of three to four by using new circuitry and fault-tolerance techniques. Key technologies to be used include millimetre and microwave-integrated circuits, low temperature co-fired ceramic packaging, multi-arm spiral antennas for multifunction use spanning 200 MHz to 6 GHz, and a broadband active array capable of performing radar and EW functions. The modular family concept is projected to reduce the number of modules required, relative to federated sensors, from about 186 to 105, the number of module types from 40 to 22 and the number of RF apertures from 40 to 13.

The next stage in the Pave Pace programme is to develop an integrated RF sensor system. In this, the common pools of resources will be capable of performing the different functions that are needed in the RF area.

Common frequency converters, which convert the analogue signal from the antennas via an RF interconnection network to digital common IF signal, process the signal first. The objective of the frequency converter is to filter, amplify and convert the analogue signal to a common IF signal to be sent over an IF network to the pool of receiver modules.

A small number of types of receiver modules would process the signal and convert it to its digital I and Q data form. Critical technologies include programmable digital filters and wideband analogue to digital converters.

For some signals, it is still anticipated that there will be a need for digital signal preprocessing before they are sent to the common avionics system data and signal processors. However, because of the speed and throughput performance of current digital processing technology, the requirement for a dedicated preprocessing function might become obsolete.

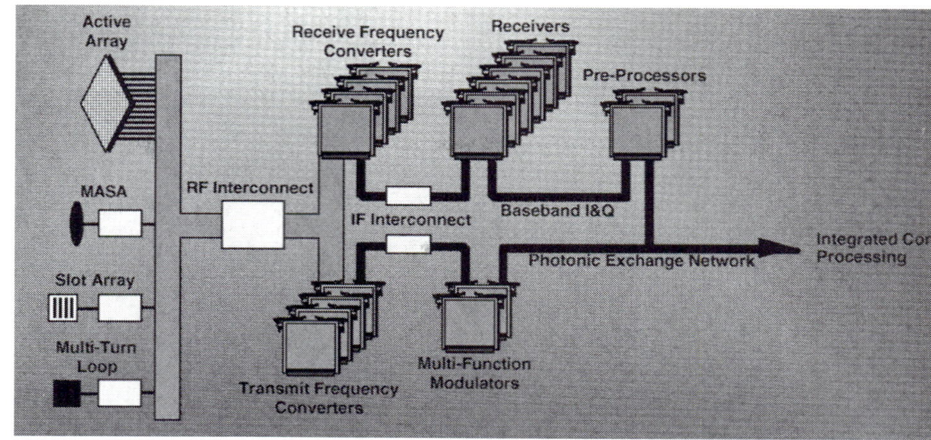

The Integrated RF System is the next stage in the Pave Pace programme

During the Integrated Sensor System programme, the architectural features and concepts of this type of common modular design for RF resources will be demonstrated.

Operational status

Three functional design contracts for the Pave Pace system have been completed by McDonnell Douglas (now Boeing), Boeing and Lockheed. A flow-through cooling module and rack have been built and successfully tested by AT & T. RF packaging using LTCC and MMIC circuits has been demonstrated by a project with Westinghouse (now Northrop Gumman); a 2 Gbits/s optical switch and a SEM-E co-fired digital processor operating at 800-1,000 Mflops were started in 1992. Further system-level design work will result in both hardware and software specifications which will finalise the Pave Pace configuration. In 1994, a programme to build and integrate the Integrated Sensor System began, with real-time laboratory-based demonstrations to be conducted during 1998-99.

The two teams currently designing the integrated RF system are Boeing/Raytheon Systems Company/TRW and Lockheed Martin/Boeing/Harris/Northrop Grumman.

Contractor

US Air Force Materiel Command.

UPDATED

Emergency Avionics System (EAS), showing the EAS 3000 (left), Cockpit Control Unit (CCU) (centre) and Data Acquisition Unit (DAU) (right), produced by DRS Flight Safety and Communications, a DRS Technologies Canada Company ***1998***/0015298

AUSTRALIA

Aircraft Fatigue Data Analysis System (AFDAS)

British Aerospace Australia's Aircraft Fatigue Data Analysis System (AFDAS), directly monitors and permanently records all relevant fatigue load spectra occurring at selected locations on aircraft structures.

AFDAS provides information for the planning of aircraft maintenance repairs and spares, fatigue predictions, assessment of fatigue damage or crack growth damage or crack status and a basis for correlation of those features with operational usage patterns.

Central to the AFDAS concept is a computer database of actual flight loads. Each aircraft is fitted with strain gauges permanently installed at critical locations in its structure. An airborne unit, the Strain Range Pair Counter (SRPC) automatically processes and stores the information from the gauges. During aircraft servicing, this data is transferred to a floppy disk by means of a portable data readout computer.

The AFDAS comprises up to 12 transducers within the aircraft, including strain gauge bridges and accelerometers, the SRPC, the ground-based portable data readout computer and data analysis software subsystem.

The strain gauge transducers are mounted in the aircraft at the critical strain points together with the SRPC. The SRPC monitors the output from each transducer and extracts range pairs which provide the basis required to compute the fraction of the fatigue life of a structure that has been used and the extent of potential crack growth. The conditioned amplified transducer signals are repetitively sampled and digitised, under the control of the Central Processor Unit (CPU). The CPU then performs additional processing to extract range pairs which are counted and stored in a non-volatile memory. Data from individual flights can be stored separately in memory before performing a data readout operation.

The ground-based portable readout computer is connected to the SRPC when a memory full indication is provided by the SRPC. This computer interrogates the SRPC and transfers fatigue data gathered since the last readout operation on to a mini-floppy disk. The SRPC memory is then cleared in readiness for continued data acquisition.

A portable terminal is also used to initialise the SRPC with information such as aircraft tail number and time and date. This terminal also allows a wide range of diagnostic routines built into the SRPC to be initiated. The SRPC processing algorithms can be tailored to suit the requirements of individual users.

Specifications

Dimensions: 139 × 275 × 154 mm
Weight: 4.5 kg
Power supply: 28 V DC, 21 W
Data storage: 64 kbytes
Interface: RS-232
Environmental: MIL-STD-810C

Operational status

AFDAS is currently in service with Royal Australian Air Force aircraft including the PC-9, F-111C and F/A-18.

Contractor

British Aerospace Australia Ltd.

VERIFIED

CANADA

CMA-2071 Structural Usage Monitor (SUM)

The CMA-2071 Structural Usage Monitor (SUM) acquires and processes structural usage data from rotary- and fixed-wing aircraft. The system comprises a single LRU mounted on the aircraft which monitors aircraft sensors in real time and stores the acquired data digitally in a large solid-state non-volatile memory. The recorded data is classified according to type and flight regime and then compressed and stored in the form of time history, histograms and aircraft header information.

When interfaced with a crash survivable memory unit, the CMA-2071 can provide flight data and cockpit voice recording in accordance with ED-55/56. Associated with the SUM is the CMA-2081 Ground Support Equipment (GSE), the purpose of which is to retrieve the acquired data from the onboard SUM, process it, and provide display readouts to facilitate maintenance functions such as predictive requirements and the determination of aircraft structural integrity. The GSE also allows for single or two-point SUM sensor calibration, updating of SUM system and parameter configuration data, display of SUM BIT results and the transfer of flight data from the GSE to remote post-processing facilities.

Specifications

Dimensions: ½ ATR
Weight: 2.3 kg
Power supply: 28 V DC, 40 W
Processor: CMOS MIL-STD-1750A or i80846 CPU with memory management
Interfaces: DC analogue, AC analogue, discretes, thermocouple, resistance bulb, pressure, bridge strain gauge, synchro, pulse, vibration, MIL-STD-1553B, RS-422, RS-232C
Inputs: up to 250

Operational status

In production. Selected for the Canadian DND for installation on the CC-130.

Contractor

Canadian Marconi Company.

UPDATED

Emergency Avionics Systems (EAS)

DRS Flight Safety and Communications' family of products integrates the functionality of a Flight Data Acquisition Unit (FDAU), Flight Data Recorder (FDR), Cockpit Voice Recorder (CVR) and Emergency Locator Transmitter (ELT), into a deployable beacon offering highly versatile survival and recovery capability. In a deployable system, the beacon increases the survivability of flight and voice data by avoiding the intense destructive forces which occur during a crash by automatically deploying from the airframe away from the accident site.

The beacon, which contains the solid-state crash-protected data recorder memory and ELT, immediately initiates transmission of search and rescue distress signals, providing immediate location of the accident site, thereby assisting early recovery of survivors, protection of valuable flight/voice data and reduced search and rescue and recovery costs. The beacon floats on water indefinitely in the event of an over-water incident. The beacon transmits at a frequency of 121.5 MHz and 243 MHz. As an option, a 406 MHz beacon can be chosen. This allows identification and messaging information to be added to the distress signal to significantly improve localisation of the downed aircraft via the COSPAS/SARSAT system.

The EAS3000 can be configured as a Combi-Cockpit Voice/Flight Data Recorder (CVR/FDR) or optionally a dedicated FDR or CVR. In each configuration the beacon contains the Emergency Locator Transmitter (ELT). An ELT-only configuration (ELB3000) is available for customers who do not require CVR/FDR capability. The ELB3000 can be upgraded with CVR/FDR capability when required. A Cockpit Control Unit (CCU) enables preflight checks to be performed on EAS3000 integrity. The CCU also allows manual deployment of the Beacon Airfoil Unit (BAU).

By combining the recorder and locator functions, installation, operation and maintenance costs are reduced. In addition, the EAS3000 offers significant weight advantages over conventional installations of multiple, fixed onboard systems.

The EAS3000 provides full data acquisition capability enclosed in a ½ ATR short Data Acquisition Unit (DAU) accommodating various inputs, including: ARINC-429; MIL-STD-1553B; analogue; discrete; thermocouple; synchro; and RVDT. The DAU contains all sensor interfaces and processor electronics for data inputs plus audio input for the CVR. Provision is made for receiving HUMS data.

Specifications

Cockpit voice recorder
Functionally compliant to ED-56A
4 audio channels, 1 hour storage per channel
Bulk erase option, ARINC 757 compatible

Flight data recorder
Compliant to ED-55
Selection of data acquisition options: MIL-STD-1553B bus; ARINC-429; direct to sensor interface suite: analogue, synchro, discrete, pulse, thermocouple; 25 hour storage; multiple data storage configurations; ARINC 747 compatible

Recorder system
Solid-state
ED-56A compliant record controls
Download and playback maintenance system allows: FDR data analysis; audo reply: BITE system

Emergency locator transmitter
DO-183 compliant
Operates on 121.5 MHz and 243 MHz
Option for 406 MHz
COSPAS/SARSAT compatible
Automatically deployable
BITE system

Compliance:
Transport Canada/Civil Aviation Authority (CAA) appliance approval, including TSOs: C123 (CVR); C124 (FDR); and C91a (ELT)

Operational status

Deployable emergency avionics systems are currently installed on the following fixed-wing aircraft and helicopters: C-5, C-9B, C-135, CC-115, CC-130, CP-140, E-3A AWACS, E-4A, E-6A, P-3 (A, B and C variants), T-43A, and Tornado aircraft; 212/412, CH-46, CH-47, Dauphin, EH 101, H-3, HH-1H, HH-1N, Lynx, Puma, S-61, Super Lynx and Super Puma helicopters.

Contractor
DRS Flight Safety and Communications.
A DRS Technologies Canada Company.

UPDATED

Deployable Flight Incident Recorder Set (DFIRS)

DRS Flight Safety and Communications' (DRS FS&C) Deployable Flight Incident Recorder Set (DFIRS) includes a DFIRS bus interface unit and a deployable flight incident recorder unit. This product, which is qualified for fighter aircraft application, provides the rapid location of downed aircraft and protection for the survivability of recorded flight data memory. The DFIRS is also adaptable to subsonic speed fixed-wing aircraft.

The recording electronics, and the transmitter system, are encased in a deployable aerodynamic airfoil that separates clear of the aircraft and is designed to land safely on land, or in water, away from the point of the aircraft impact. The transmitter, which automatically activates on release, provides the important homing signal for search and rescue crews.

System features include solid-state recording technology capable of expansion of up to 2 hours recording time and a fully qualified bus interface which interfaces to the aircraft mission computer via a MIL-STD-1553B serial bus. The bus interface can also effectively operate with Crash Survivable Memory Units (CSMU).

Specifications
Operating frequency: 243 MHz
Recording time: 30 minutes, expandable to 2 hours

Operational status
Installed on F/A-18s C&D and E&F variants

Contractor
DRS Flight Safety and Communications.
A DRS Technologies Canada Company.

UPDATED

EAS3000 (left), cockpit control unit (centre), and data acquisition unit (right)
1998/0015298

CZECH REPUBLIC

AMOS Aircraft Monitoring System

AMOS is designed to provide acquisition, processing and storage of information for the L159 light attack aircraft; it monitors airframe fatigue data; flight data; incident data; pilot response data. There is an optional facility for in-flight transmission of the acquired data.

AMOS comprises a Flight Data Acquisition Unit (FDAU); crash-protected Flight Data Recorder (FDR-159); Function Signalling Panel (FSP); Flight Data Entry Panel (FDEP); Ground Evaluation Equipment (GEE); and Ground Support Unit (GSU).

Using the GEE/GSU, it is possible to replay the mission in 3-D, as well as to analyse and evaluate aircraft system data.

Contractor
VZLU-SPEEL Ltd, Aeronautical Research and Test Institute – Special Electronics.

UPDATED

Specifications (AMOS Aircraft Monitoring System):

	FDR-159	FDAU	FSP
Dimensions	185 × 134 × 330 mm	240 × 260 × 390 mm	145 × 32 × 100 mm
Weight	<8 kg	<10 kg	<0.35 kg
Power	28 V DC, <8 W	28 V DC, <55 W	28 V DC, <10 W
Memory capacity	16 MB (custom)	4 MB (custom)	
MTBF	>5,000 flight hours	>3,000 flight hours	>20,000 flight hours
Communications interfaces	RS-232; RS-422; ARINC 429; MIL-STD-1553B		
Compliances	MIL-STD-810E; EUROCAE ED-55; TSO-C124		

Flight Data Recorders (FDRs)

VZLU-SPEEL has developed a family of solid-state FDRs that are designed to be direct replacements for the mechanical crash recorders in Russian aircraft, where SARPP-12, TESTER U3 (2T-3M), TESTER U3L

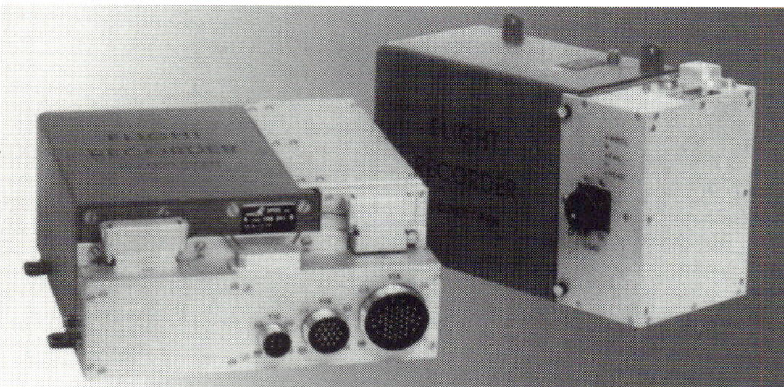

FDR-39 (left) and FDR-59 (right)
1998/0015300

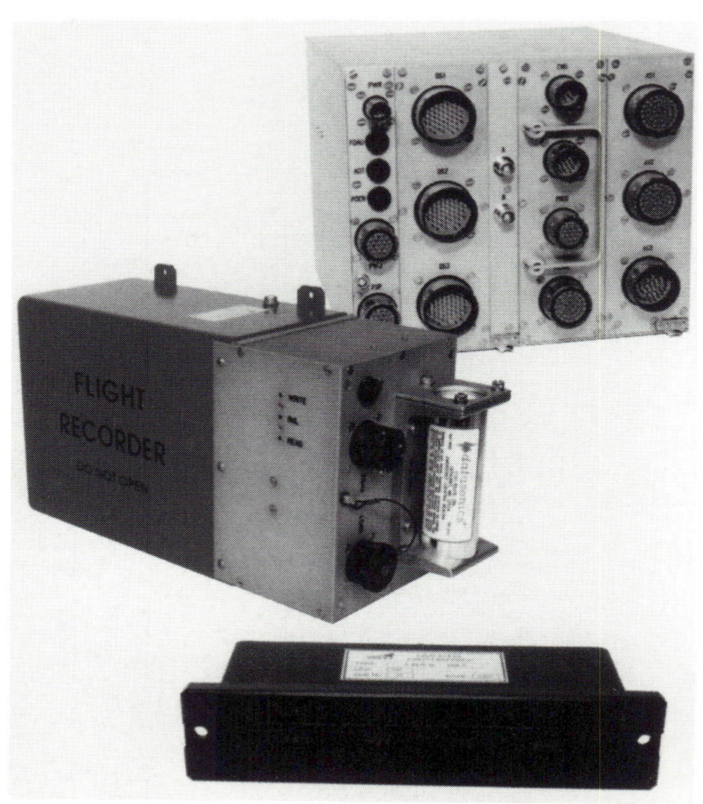

AMOS: FDAU, FDR-159, FSP
1998/0015299

(M2T-3), BUR-1 (ZBN1) and PARES systems are used.

VZLU-SPEEL's FDR recorders include the following models:

FDR-39H for Mi-17 and Mi-24 helicopters
FDR-39T for L39, L410 and MiG-21 aircraft
FDR-59A for L59MS aircraft
FDR-59B for SOKOL helicopters and L410 UVP aircraft
FDR-59T for MiG-29, Su-22, and Su-25 aircraft

The above FDRs are supported by the PMU Portable Memory Unit for downloading and by the Ground Evaluation Equipment (GEE) for evaluating data.

FDR-39H and FDR-39T solid-state flight data recorders

The FDR-39H and FDR-39T solid-state FDRs are designed as replacements for the SARPP-12DM (IM) and SARPP-12 mechanical crash recorders respectively. They comprise three modules: power supply; measuring unit; and solid-state memory. Download can be accomplished via RS-232C interface to a portable Notebook, or via RS-422 interface to Speel's Portable Unit (PMU-F).

FDR-59A and FDR-59B solid-state flight data recorders

The FDR-59A and FDR-59B solid-state FDRs are designed as replacements for the PARES 59E and the ZBN-1 from the BUR-1 system mechanical tape crash recorders respectively. They comprise two modules, connection and communication module, and solid-state memory unit. Both have 4 MB recording capacity, corresponding to approximately eight flight hours.

Contractor

VZLU-SPEEL Ltd, Aeronautical Research and Test Institute – Special Electronics.

UPDATED

Specifications

	FDR-39H and FDR-39T	FDR-59A and FDR-59B
Dimensions	258 × 241 × 117 mm	185 × 125 × 264 mm
Weight	8 kg max	7 kg max
Power	28 V DC, <9 W	28 V DC, <15 W
Interfaces	RS 232C; RS-422	RS-232C; RS-422
MTBF	10,000 hours	10,000 hours

FRANCE

BSDM mission data storage unit

The BSDM mission data storage unit is fitted to Rafale in a twin extractable cassette configuration.

Contractor

Dassault Electronique.

VERIFIED

ESPAR static-memory flight-data recorder

ESPAR is a flight data recorder jointly manufactured by Dassault Electronique and SFIM Industries. It is the latest product in a new generation of static-memory crash recorders intended for both civilian and military aircraft. Its modular design allows integration into standard cases (½ATR long or short) or cases adapted to the aircraft structure, and ensures interchangeability with magnetic-tape recorders.

The flight-data recorder has a modulable memory capacity, whilst appropriate differential encoding allows the memory capacity needed for flight data recording to be reduced.

A specific memory storage method provides for recovery of recorded data even in the event of a memory failure.

Data can be transmitted from the recorder unit in different modes: Harvard biphase; NRZ with clocks; RS-232C or RS-422.

ESPAR has withstood the tests imposed by TSO-C51a and TSO-C111 certifications: shock; crush force; perforating force; exposure to flames; immersions; electrical protection; EMI protection.

Specifications

Dimensions: 118 × 120 × 220 mm
Volume: 3.10 dm^3
Weight: <8 kg
Power consumption: <20 W
Power supply: aircraft DC or AC system
Option: underwater localisation beacon

Operational status

In service on Mirage F1 and 2000, and Rafale. Also proposed for civil aviation; chosen by Collins for the Ilyushin Il-96M 'Westernisation' and by IPTN for the N-250 aircraft.

Contractors

Dassault Electronique.
SFIM industies.

VERIFIED

ESPAR static-memory flight-data recorder **1997**/0001291

Extended Storage Quick Access Recorder (ESQAR)

Designed and manufactured by Dassault Electronique, the Extended Storage Quick Access Recorder (ESQAR) is jointly marketed and supported worldwide by Dassault Electronique and AlliedSignal Aerospace.

ESQAR is used as a quick access recorder and/or digital ACMS recorder. Based on rewritable optical disk technology, the ESQAR, which is fully interchangeable with existing magnetic-tape recorders, provides increased storage capacity, enabling more parameters to be recorded at a higher rate. Recorded data can be accessed directly from DOS files read through a PC.

Designed to be fitted on civil aircraft, ESQAR benefits from Dassault Electronique's experience in developing ruggedised systems for operating in severe environments. The system provides full-time recording throughout the flight, with maximum data integrity.

Specifications

Dimensions: ARINC 404/600 formats
Input data rate: 64-512 words/s
Storage capacity: 128 Mbytes extendable to 256 Mbytes
Environmental: DO160C

Operational status

Certified and operated on most commercial aircraft including those manufactured by Airbus and Boeing. Selected by more than 20 airlines worldwide.

ESQAR is available and has been in commercial service since December 1993. It has been selected by several airlines and is also operated by Aerospatiale aircraft.

SABENA Group has placed an order with Dassault Electronique for 51 ESQAR units. This optical disk-based recorder will be fitted on the following SABENA Group fleets: Boeing B737 and B747, Airbus A310 and A340, Avro RJ85. Egyptair has placed an order for its incoming A340 and B777 fleets, Air Jet, for its BAe 146 fleet and Air Mauritius, for its A340 fleet (retrofit of magnetic-based QAR and forward-fit).

Contractor

Dassault Electronique.

VERIFIED

Solid-State Flight Data Recorder (SSFDR)

Designed in collaboration with SFIM Industries, the Solid-State Flight Data Recorder (SSFDR) has been

Dassault Electronique/SFIM Industries SSFDR solid-state flight data recorder ***1998***/0015304

developed to replace conventional magnetic tape airborne recorders.

Fitted with an entirely electronic memory, this new-generation recorder features considerably improved reliability, elimination of scheduled servicing because of the absence of moving parts, and appreciable savings in weight and volume. These savings are achieved by the use of a small-size static memory made possible by the application of a specific data storage algorithm adapted to all types of aircraft and capable of recording more than 25 hours of flight in commercial aircraft. It also has an improved environmental resistance due to the protection of data provided by the case, better thermal protection, protection against crushing, resistance to pressure and protection against corrosion.

In the event of an accident, the recorder memory is extracted from its protective case and then installed on a device linked to a data analysis workstation. During data retrieval, a recovery algorithm supplies the right value of each parameter according to the required resolution. A further advantage is the extremely fast rereading of flight data without the removal of the recorder. In this case, recorded data is transmitted over a high-speed line to a static data retrieval set which is then connected to the workstation.

The SSFDR packaging may be easily adapted to any military or civilian aircraft and versions have been produced for the Mirage F1 and 2000.

SSFDR complies with TSO C124, EUROCAE ED-55 and ARINC 545/747.

Specifications

Dimensions:
(basic version) 118 × 120 × 220 mm
(Mirage F1) 319 × 124 × 193 mm
(Mirage 2000) 294 × 133 × 245 mm
Weight:
(basic version) <8 kg
(Mirage F1) 7 kg
(Mirage 2000) 10.5 kg
Power supply: 28 V DC
Options:
voice: 1 or 2 channels
data: radio frequencies; synchro data; pressure sensors; discretes; databus A429 and 1553 data

Operational status

SSFDR is available and in commercial operation. It is certified on all Airbus types and the Boeing 737, 757, MD-90, and on the IPTN N-250. It has been selected by a number of airlines in Europe, the Near East, the Middle East and Asia.

Contractors

Dassault Electronique.
SFIM Industries.

VERIFIED

Test and measurement system

For more than 30 years Dassault Electronique has studied test and measurement systems for airborne applications. Dassault Electronique now provides a new generation of systems built around an innovative architecture. Implementing the most recent technology, this set of equipment uses very high-performance components to allow more high-density circuit integration and significantly reduce size, weight and costs. It results in a very flexible system that can acquire several thousand parameters.

Core of the system is SERPAN, a digital acquisition unit. When used as a stand-alone unit, this device receives most airborne busses in full or filtered mode, including MIL-STD-1553 and ARINC 429 protocols. When equipped with PCM interface boards, it becomes a PCM concentrator, allowing connection of other SERPANs or devices with PCM outputs (Irig or Daniel).

Analogue signals are acquired by means of CCE, SAMPLE and/or CANAR units. CCE is a miniaturised container located near the sensors. Up to 15 CCEs can be coupled to the system by a high-speed bus to allow acquisition of parameters from thermocouples, strain gauges, low-level sensors, temperature and piezoelectric probes. SAMPLE is an analogue acquisition unit. Designed for vibration acquisition, this fully automatic unit outputs a PCM stream for coupling to a SERPAN core unit. CANAR is a special design of SAMPLE for analogue acquisition on helicopter rotors.

ELEFAN is a static flash-memory recorder that is utilised instead of tape devices in severe environment. The high data storage capacity of this unit (up to 2 Gbytes) allows several hours of recording. Both ELEFAN and SERPAN functions can be mixed in the same ruggedised unit.

A standard PC workstation runs ground software for configuration, ELEFAN downloading and data analysis.

The SAMPLE and CANAR units may also be used as stand-alone systems for analogue acquisition.

SAMPLE

Although SAMPLE is a part of Dassault Electronique's test and measurement system, this unit may also be used as a stand-alone unit for analogue acquisition. Fully programmable, SAMPLE is designed for vibration acquisition. Main applications are measurement during load release, dummy ejection, stress analysis and other tests in very severe environments. SAMPLE is capable of absorbing shock at 200 *g* for 11 ms. When housed in MAGISS, a patented technology from Dassault Electronique for device protection, this unit is shock absorbent at 500 *g* for 6 ms.

The SAMPLE unit is equipped with analogue acquisition and memory boards. Each acquisition board can receive up to four channels (differential inputs) with programmable excitation, gain, offset, filtering and sampling rate. Storage capacity is from 64 to 384 Mbytes. SAMPLE can also acquire up to 12 digital pulses through differential inputs with opto-electronic isolation. Each input may be used to trigger the recording process with a programmable timer. Coding of signals is made on 12 bits with 0.1 per cent accuracy.

For particular applications, SAMPLE can acquire both analogue information and ARINC 429 busses. Merged data are stored into memory, when demultiplexing is made by ground workstation at the time of downloading.

SAMPLE is available in two widths, according to the customer needs. The smaller unit can support 16 channels and can be equipped with a battery for autonomous powering. The larger unit has 32-channel capability. A special design known as the CANAR is also available for helicopter rotor installation.

A standard PC workstation runs ground software for configuration, downloading and data analysis.

SAMPLE is currently in use for testing parachutes, ejection seats and released loads. Also in use for stress measurement on military carrier aircraft.

CANAR

CANAR is a special design of the SAMPLE unit for helicopters; it is installed on top of main or tail rotors. Although CANAR is a part of Dassault Electronique's test and measurement system, this unit can also be used as a stand-alone unit for analogue acquisition. Because it uses the same boards and similar embedded software, the CANAR unit has the same functions as the SAMPLE unit.

CANAR is a hexagonal box available in two heights according to rotor type. The smaller unit has 24 channels for the tail rotor. The larger unit has 64 channels for the main rotor.

A standard PC workstation runs ground software for configuration, downloading and data analysis. CANAR is in development.

Operational status

Currently in use for flight testing of Mirage 2000, Rafale, and several foreign customers' retrofit programmes.

Contractor

Dassault Electronique.

VERIFIED

CP 1654 airborne life monitor

The CP 1654 calculates the residual engine service life of the main engine components. It provides definition of a short- and medium-dated schedule for engine components' removal before shipment to the second-level maintenance shop as well as the planning applicable to the fourth-level maintenance of the modules.

Operational status

In production.

Contractor

ELECMA, the Electronics Division of SNECMA.

VERIFIED

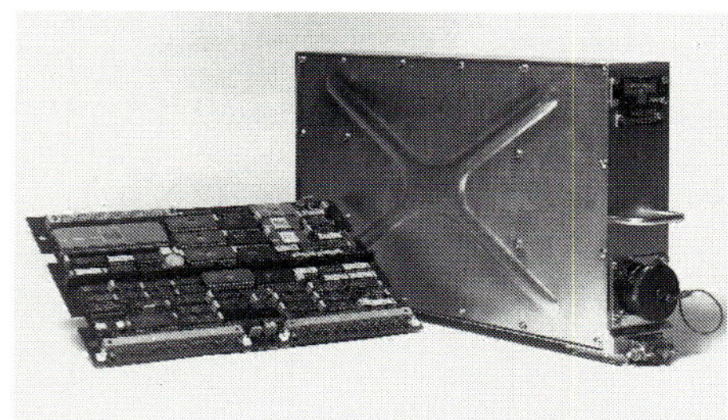

The ELECMA CP 1654 airborne life monitor for the M 53-P2 engine

DV 6410 series high-speed helical scan recorders

The DV 6410 series of airborne recorders includes the DV 6410 and the DV 6210. Data rates are 10 to 240 Mbits/s for the DV 6410 and 5 to 120 Mbits/s for the DV 6210. Recording time varies between 25 hours at 5 Mbits/s and 24 minutes at 240 Mbits/s.

The DV 6410 Series has a storage capacity of 350 Gbits on a 19 mm D1-M cassette in a format compatible with MIL-STD-2179 or ANSI IDI. It is a compact, fighter- and space-proof unit. The DV 6410 also provides for two auxiliary channels and a full remote-control interface.

Specifications

Dimensions:
(record-reproduce unit) 314 × 240 × 520 mm
(power supply unit) 314 × 140 × 302 mm
Weight:
(record-reproduce unit) 35 kg
(power supply unit) 12 kg
Power supply: 28 V DC
115/220 V AC, 50/400 Hz, single phase
115 V AC, 60/400 Hz, 3 phase, <400 W
Temperature range: −55 to +85°C
Altitude: up to 50,000 ft

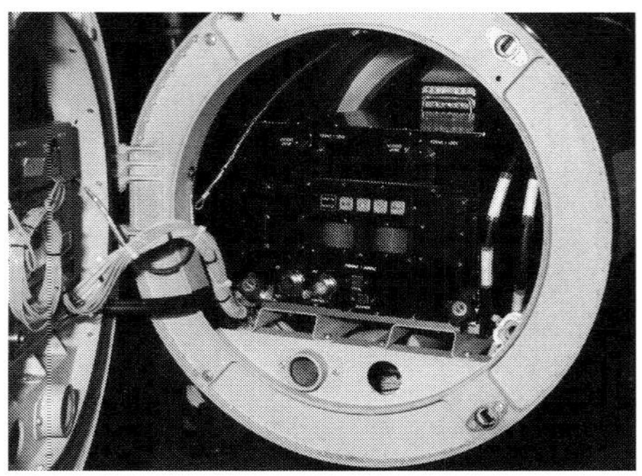

The DV 6410 helical scan recorder used in the central pod of a Mirage 2000

Operational status

In production and in service for various airborne, shipborne and space applications. The DV 6411 recorder has been selected for the Royal Air Force Project RAPTOR (Reconnaissance Airborne Pod for Tornado). RAPTOR is to integrate a day-night medium level stand-off imaging capability in Tornado GR. Mk 1A by December 1998, and also in Tornado GR. Mk 4 aircraft.

Contractor

Enertec, the Data Acquisition and Recording Division of Schlumberger Industries.

UPDATED

DV 6420 series high-speed helical scan recorders

The DV 6420 series of airborne recorders suitable for large aircraft consists of the DV 6420 and DV 6220. Data rates are 10 to 240 Mbits/s for the DV 6420 and 5 to 120 Mbits/s for the DV 6220. Recording time varies between 44 hours at 5 Mbits/s and 55 minutes at 240 Mbits/s.

The DV 6420 is the rack-mounted version of the DV 6410. The basic characteristics are common to both versions, but the rack-mounted version accepts D1 medium or large cassettes and has a storage capacity of 790 Gbits on a D1-L cassette.

A number of standard computer interfaces are available such as HIPPI, SCSI2 and UME64. A higher rate version, the DV 6820, provides 480 Mbits/s I/O rate capability.

Specifications

Dimensions: 483 × 311 × 593 mm
Weight: 65 kg
Power supply: 115/220 V AC, 50/400 Hz, single phase 115 V AC, 400 Hz, 3 phase, <450 W 28 V DC
Temperature range: −55 to +85°C
Altitude: up to 10,000 ft

Operational status

In production and operating in a number of airborne, shipborne and ground platforms.

Contractor

Enertec, the Data Acquisition and Recording Division of Schlumberger Industries.

UPDATED

ME 4110 airborne instrumentation recorder

The ME 4110 is a lightweight and compact recorder designed for use during the flight testing of military fighters and other aircraft. It uses 10 in reels of 1 in wide magnetic tape.

Specifications

Dimensions: 400 × 285 × 160 mm
Weight: 20 kg
Power supply: 22-32 V DC, 200 W
Recording time: 7 min - 16 h
Number of tracks: 14 or 28
Recording modes:
FM (up to 500 kHz)
direct (up to 2 MHz)
PCM (up to 4 Mb/s)

Operational status

In production and in service.

Contractor

Enertec, the Data Acquisition and Recording Division of Schlumberger Industries.

UPDATED

ME 4115 airborne recorder/ reproducer

The ME 4115 is an advanced, microprocessor-controlled recorder designed for use during the testing of aircraft, ships or vehicles or as an ELINT, ASW or reconnaissance mission recorder and can be configured for any application. It uses 15 in reels of 1 in wide magnetic tape and has built-in error correction electronics and read/write heads to give 14 or 28 tracks IB, WB or DD magnetic heads.

Depending on the disposition of the electronics chassis relative to the tape deck, several standard versions are proposed: 14 or 28 track record/ reproduce either for anti-vibration mount or 19 in bay installation, tape deck separated from the electronics for integration on board small fighters and 28 track record-only and monitoring.

This latest configuration is fully compliant with MIL-STD-1610 and STANAG 4238 Annex B analogue acoustic recording standards.

The ME 4115 accepts a wide range of high data rate digital formats up to 200 Mbits/s including standard interfaces for MIL-STD-1553 bus monitoring. It has been selected as a flight test recorder by most airframe manufacturers for programmes such as the Alenia/ Embraer AMX, Northrop Grumman F-14 and Airbus A340.

Specifications

Dimensions: 581 × 394 × 176 mm
Weight: 35 to 45 kg
Power supply: 22-32 V DC or 115 V AC, 300 W
Recording speed: 4.75-394 cm/s
Number of tracks: 14 or 28
Recording modes:
FM (up to 1 MHz)
direct (up to 4 MHz)
PCM (up to 8 Mb/s)
Cumulative data rate: over 200 Mbits/s

Operational status

In service on French and Italian Atlantique maritime patrol aircraft. The ME 4115 is operated by the US forces in several airborne and shipborne programmes as the AN/USH-33(V)2. More than 900 units have been produced.

Contractor

Enertec, the Data Acquisition and Recording Division of Schlumberger Industries.

UPDATED

PC 6033 general purpose digital cassette recorder

The PC 6033 records serial data on a continuous basis over a long period of time. Typical applications are performance and maintenance recording, engine health monitoring, aircraft testing and reconnaissance.

The PC 6033 performance/maintenance recorder
1995

The system complies with ARINC 591. Cassettes can be replayed at 80 times the recording speed.

Specifications

Power supply: 19-32 V DC or 115 V AC, 400 Hz
Recording capacity, data rate:
50 h, 138 Mbits
50 h at 768 bits/s
25 h at 1,536 bits/s
12.5 h at 3,072 bits/s
Number of tracks: 12
Recording code: biphase L or M
Error rate: <1 bit in 10^5

Operational status

In production for and in service with several airlines; selected for the A320.

Contractor

Enertec, the Data Acquisition and Recording Division of Schlumberger Industries.

UPDATED

PE 6010 and PE 6011 digital flight data accident recorders

The PE 6010 is a lightweight digital flight data accident recorder, in which weight saving is accomplished by a relaxation of the dynamic penetration requirement. In terms of accident protection, however, it meets all other requirements of FAA TSO C51a.

The PE 6011 is more compact, meets TSO C51a completely with only a small weight increase and has 50 per cent more capacity than the PE 6010. The new design of tape deck is simpler, with attendant gains in reliability and maintainability.

Specifications

Dimensions:
(PE 6010) ½ ATR short (length 319 mm)
(PE 6011) ½ ATR (length 296 mm)
Weight:
(PE 6010) 7.7 kg
(PE 6011) 10.2 kg
Power supply:
(PE 6010) 18-30.5 V DC
(PE 6011) 12-32 V DC
Recording capacity, data rate:
(PE 6010) 16 h, 768 bits/s
(PE 6011) 8 h, 2,308 bits/s
Replay data rate:
(PE 6010) 4,608 bits/s
(PE 6011) 18,464 bits/s
Accident survival:
(PE 6010) TSO C51a (except penetration is static)
(PE 6011) TSO C51a

Operational status

The PE 6010 is no longer in production. In service with Dassault Mirage F1 fighters flown by the air forces of Egypt, France, Greece, Iraq, Kuwait, Morocco and Spain. It has been supplied for the Indian HAL Ajeet trainer and French Navy Super Frelon helicopters.

Contractor

Enertec, the Data Acquisition and Recording Division of Schlumberger Industries.

UPDATED

PS 6024 cassette memory system

The PS 6024 is a digital magnetic cassette tape recorder/reproducer designed for operation under severe environmental conditions. It comprises a cassette drive unit, microprocessor-controlled central processing unit and electronic interface circuits. Its chief function is as an easily integrated mass memory unit for military computers with standardised interfaces.

Specifications

Dimensions: ⅜ ATR short
Interfaces: V24/RS-232, MIL-STD-1553B, Digibus HDLC as required
Capacity: 6 Mbytes formatted (1 k blocks)
Data density on tape: 1,600 bits/in

Max recording rate: 2.2 kbytes/s usable data
Recording format: forward and backward serial recording in biphase code
Recording speed: 12 in/s (300 mm/s)
Rewind speed: 50 in/s (1,270 mm/s)
Max length of data blocks: 2,048 (2 k) bytes
Interblock gap: 20 mm

Operational status

In service in German and Italian Panavia Tornado aircraft.

Contractor

Enertec, the Data Acquisition and Recording Division of Schlumberger Industries.

UPDATED

VLDS-BR digital cassette recorder

The VLDS-BR responds to digital acquisition needs at rates or recording time ranges situated below those covered by large format recorders such as MIL-STD-2179 or ID-1. It allies the helical scan recording technique to an efficient on-tape format, allowing storage of more than 83 Gbits of data on a single VHS cassette, the equivalent of 43 minutes of recording time at maximum data rate. The performance of the VLDS-BR, benefiting from an input/output buffer memory, also accommodates any data rate within its range, with no need for manual adjustments or hardware configuration.

The VLDS-BR digital cassette recorder **1995**

The VLDS-BR also provides an auxiliary channel for recording housekeeping signals, such as time code or annotation data. This data is converted into digital form, then embedded in the main data on the helical tracks. It can be controlled manually with front panel push-buttons in a similar way to a home VCR or remotely via an RS-422 serial line. An SCSI interface, offered as an option, permits both data transfer and control by computer. The VLDS-BR has been optimised for use in severe environments, with rugged mechanical and electronic elements, internal isolators and military interface. All the relevant environmental parameters have been taken into account.

Specifications

Dimensions: 222 × 483 × 559 mm
Weight: 34 kg
Power supply: 220 V AC, 47/440 Hz
115 V AC, 47/400 Hz 28 V DC
Data rates:
(continuous) 800 kbits/s to 32 Mbits/s
(burst mode) 0 to 160 Mbits/s
Altitude: up to 15,000 ft

Contractor

Enertec, the Data Acquisition and Recording Division of Schlumberger Industries.

UPDATED

MONITAIR flight data recorder

The MONITAIR flight data recorder is designed to record the airframe and engine parameters of turbine engined helicopters.

MONITAIR memorises: the date, time and duration of each flight (with partial flights) and makes daily reports with cumulated flight time; it records maximum values of engine pitch/torque, turbine temperature, turbine speed and rotor rpm during start sequence and throughout flight; it also records the value and duration of all turbine temperature exceedances.

MONITAIR is tamperproof, and all disconnections are indicated. It is supported by a WINDOWS-based support system.

Specifications

Storage capacity:
(single-engined helicopter) 9 h
(twin-engined helicopter) 6.30 h
Flight memory:
(single-engined helicopter) 1,400 flights
(twin-engined helicopter) 1,000 flights
Autonomous storage: 75 days
Dimensions: 260 × 155 × 90 mm
Weight: 3.2 kg
Power: 15-35 V DC, 250 mA

Operational status

The system can be installed on every type of turbine helicopter, and it has been integrated on to all Eurocopter helicopters, and a wide range of Bell and Sikorsky types.

Contractor

MONIT'AIR.

NEW ENTRY

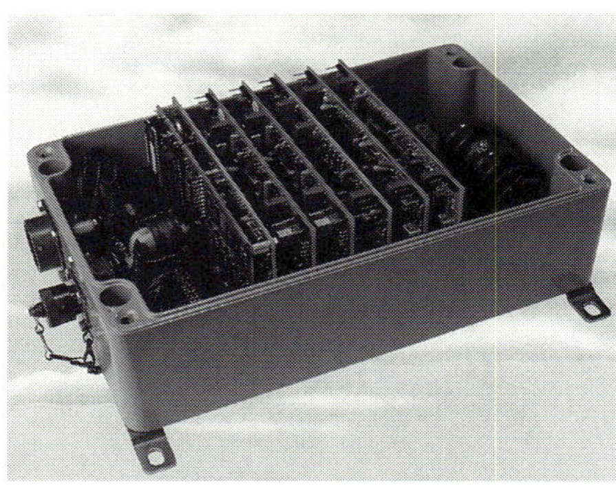

MONITAIR flight data recorder **1998**/0015301

EBS 2501 magnetic bubble memory equipment

EBS 2501 magnetic bubble memory equipment consists of a controller, one to four holders and 16 or 32 Mbit memory cartridges. It has been selected for the Dassault Mirage 2000N, 2000D and 2000-5 to store mission data including terrain altitudes. A version is used in the Dassault Atlantique 2 to store mission data as well as main computer programs for restarting after inadvertent shutdown. The equipment has also been selected for missile telecommunications systems and nuclear submarines.

Specifications

Dimensions:
(controller) 58 × 194 × 240 mm
(holder) 95 × 127 × 165 mm
(cartridge) 27 × 104 × 164 mm
Power supply: 28 V DC, 70 W
Capacity: up to 4 Mbytes
Average access time: 20 ms
Transfer frequency: 600 kbits/s
Interface: HDLC, Digibus
Interface options: MIL-STD-1553B and SCSI

Operational status

In production for the Dassault Mirage 2000N, 2000D, 2000-5 and Atlantique 2.

Contractor

SAGEM SA, Defence and Security Division, Paris.

UPDATED

Jumbo memory equipment

Jumbo is a compact mass memory suitable for demanding onboard data processing or exchange applications, such as: mission planning interface, digital terrain storage, mission data recording/replay and flight test instrumentation. Designed for use on airborne, naval or ground systems, Jumbo is composed of a data transfer unit and a family of data transfer cartridges.

The Jumbo data transfer unit can house different types of cartridges implementing solid-state Flash EEpRoM chips or ruggedised 1.8 in magnetic disk drives, thus fulfilling a wide spectrum of environmental specifications.

A dedicated version of Jumbo for ground applications is available for connection to any standard data processing unit (PC or workstation).

As an option, an add-on board and dedicated software give Jumbo the capability of a high-performance digital map generator (Mercator equipment).

Specifications

Dimensions:
(data transfer unit) 95 × 127 × 222 mm
(cartridge) 27 × 105 × 165 mm
Weight:
(data transfer unit) 2.5 kg
Power supply: 28 V DC, 30 W
Temperature range: −35 to +71°C
Cartridge capacity: 20-1,000 Mbytes

Operational status

In production.

Contractor

SAGEM SA, Defence and Security Division, Paris.

UPDATED

Flight Data Acquisition Unit (FDAU)

The Sextant Avionique Flight Data Acquisition Unit (FDAU) and its associated data entry panel are compatible with the digital flight data recorder requirements of ARINC 573 and 717.

The FDAU codes aircraft parameters into digital format and transmits them to the data recorder. The data entry panel allows the crew to enter flight number, time and events into the system. It also acts as a fault monitor.

Operational status

In production.

Contractor

Sextant Avionique.

VERIFIED

Full format printer for the Airbus A330 and A340

The full format printer, as envisioned for use on A330 and A340 commercial transport aircraft, has to comply with a high-performance standard level. Primarily located in the cockpit, the full format printer provides for high-speed and high-quality printing capacities, while achieving a low-noise level and a high operational reliability.

The full format printer communicates with multiple onboard systems such as the flight management system, maintenance system, ACARS and aircraft condition and monitoring systems. It has to be able to process large quantities and various types of data transmitted through up to 12 ARINC 429 databusses.

Operational status

In production for the Airbus A330 and A340.

Contractor

Sextant Avionique.

VERIFIED

Full format printer for B747-400/ MD-11/MD-90

The full format printer for the Boeing 747-400 and MD-11/MD-90 is based on ARINC 744 recommendations and uses thermal printing technology. It features 300 DPI resolution and high buffer capacity, to interface with major onboard systems including the maintenance computer. To meet future requirements including FANS and electronic library systems, the full format printer features a graphic interface capability associated with a high-speed datalink. It can handle a wide variety of texts and complex diagrams, making it an ideal link for the crew to provide quality hard-copy reports. The full format printer is installed in the cockpit and may also be installed in the cabin (centronics interface).

Operational status

In production.

Contractor

Sextant Avionique.

VERIFIED

Full format printer for the Boeing 777

The full format printer for the Boeing 777 is based on thermal printing technology. It features full graphics capability and interfaces with major onboard systems, including the Aircraft Information Management System (AIMS) and the Electronic Library System (ELS).

Featuring 300 DPI resolution for quality equal to a laser printer, the full format printer handles a wide variety of texts and complex diagrams, making it an important link in the onboard data management chain.

The full format printer is installed on the central pedestal of the Boeing 777, behind the throttles.

To meet future requirements for FANS or electronic library services, the full format printer features graphic hardware interface capabilities with a high-speed datalink. A cabin installation is optional.

Operational status

In production for the Boeing 777.

Contractor

Sextant Avionique.

VERIFIED

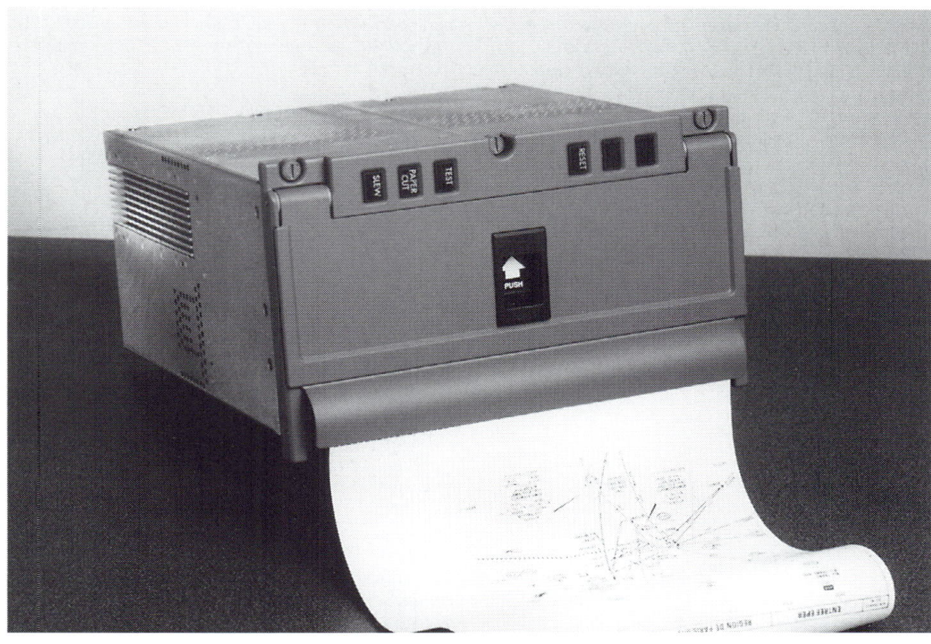

Full format printer for B747-400/MD-11/MD-90
1997/0001292

Aircraft Condition Monitoring System (ACMS)

The Aircraft Condition Monitoring System (ACMS) has been designed for the acquisition of aircraft parameters in accordance with the latest airworthiness regulatory agency requirements. It processes engine, APU and aircraft monitoring, airline incident and specific investigation report data and transmits this data to a digital flight data recorder or quick access recorder. Provision is also made for customised reports. The customisation is realised through Ground Support Equipment (GSE); software is PC-compatible. The system interfaces with ACARS, multifunction Control Display Unit or CDU, onboard printer, data loader and centralised fault display system.

A new option that utilises a removable PCMCIA drive offers flight data storage equivalent to use of external recorders.

Operational status

In aircraft such as the Airbus A320, A330 and A340 and Boeing 747-400, the ACMS is a two-box configuration with separate flight data acquisition unit and data management unit. These two boxes are combined into a single unit in the Airbus A300-600 and A310, Boeing 737, 757 and 767, MD-87, -88 and -90 and the ATR 42 and 72.

Contractor

SFIM Industries.

UPDATED

CF 368C CCD colour camera

The CF 368C onboard colour video camera is designed for use with an HUD. It is equipped with a high-resolution and high-sensitivity CCD HAD matrix sensor. The camera operates to the PAL standard.

Specifications

Dimensions: 125 × 92.5 × 81.5 mm
Weight: 0.88 kg
Power supply: 28 V DC, <10 W
Field of view: 24° (horizontal) × 18° (vertical)
Pixels: 752 (horizontal) × 582 (vertical)
Definition: >400 points per line, >350 line vertically

Contractor

SFIM Industries.

VERIFIED

CF 369C CCD colour camera

The CF 369C onboard colour video camera is designed for use with an HUD. It is equipped with a high-resolution and high-sensitivity CCD HAD matrix sensor and operates on PAL or Y/C standards.

Specifications

Dimensions:
(camera head) 40 × 60 × 65 mm
(electronic control unit) 125 × 66 × 81.5 mm
Weight:
(camera head) 0.2 kg
(electronic control unit) 0.88 kg
Power supply: 28 V DC, <10 W
Field of view: 24° (horizontal) × 18° (vertical)
Pixels: 752 (horizontal) × 582 (vertical)
Definition: >400 points per line, >350 lines vertically

Contractor

SFIM Industries.

VERIFIED

CF 370 CCD colour camera

The CF 370 CCD colour camera is a dual unit equipped with a high-resolution and high-sensitivity CCD HAD matrix sensor. It is available in PAL or Y/C standards.

Specifications

Dimensions:
(electronics unit) 164 × 92 × 78 mm
(video sensor head) 50 mm length × 20 mm diameter
Weight:
(electronics unit) 0.98 kg
(video sensor head) 0.1 kg
Power supply: 28 V DC, <10 W
Temperature range: −10 to +60°C
Pixels: 752 × 582

Contractor

SFIM Industries.

VERIFIED

CF 371 CCD colour camera

The CF 371 onboard colour video camera is designed for use with an HUD. It is equipped with a high-resolution and high-sensitivity CCD HAD matrix sensor and operates on PAL or Y/I standard specifications.

Specifications

Dimensions:
(camera head) 31 × 53 × 60 mm
(electronic control unit) 150 × 80 × 100 mm

Contractor

SFIM Industries.

VERIFIED

CF 372 CCD colour camera

The CF 372 camera is a CF 371 camera with a multiplexing board allowing a temporal multiplexing of four colour or black and white videos.

Contractor

SFIM Industries.

UPDATED

Damien 6-UAM modular mixed acquisition unit

The Damien 6-UAM system is designed for the acquisition of all types of analogue and digital parameters and may be used in combat aircraft, civil aircraft, helicopters, ships and land vehicles. It is available in several configurations either as a centralised stand-alone acquisition unit, a decentralised system comprising several units and managed by one of these units, a decentralised acquisition system comprising several units managed by a central control unit. This versatility and modularity permits the system to acquire anything from 50 to 1,600 parameters.

Damien 6-UAM is a modular system with a growth capability which is suitable for both large and small systems. It features a wide selection of inputs and outputs, with the analogue parameters grouped and segregated from digital parameters. Programming is by means of a plug-in module.

A Damien 6-UAM unit consists of an assembly of mechanical sections, into which printed circuit boards are plugged. Input/output data is fed through a single connector located on the front of each section. Sections are interconnected by a variable length flex background circuit fixed to the rear of each section. The 28 V DC power supply is fed to the central part of the unit.

Contractor
SFIM Industries.

UPDATED

ED 3333 data acquisition unit

For military aircraft applications, the ED 3333 is intended to acquire flight parameters; it automatically increases the flight number recorded at the beginning of each flight. The system monitors the performance of sensors and transducers and signals a warning in the event of a failure and transmits data to the crash recorder.

Specifications
Dimensions: ¼ ATR short
Weight: 3 kg
Power: 28 VA

Operational status
In production.

Contractor
SFIM Industries.

VERIFIED

ED 34XX data acquisition and processing unit

The ED 34XX has been designed for installations with weight or space constraints. It accepts the mandatory crash parameters, meets the requirements of ARINC 573 and can also take additional digital inputs from ARINC 429 databusses. After processing, information is transferred to a crash-protected recorder. The system can be expanded by adding a microprocessor and associated electronics to the same box to monitor engine and flight parameters. A printer and quick access recorder may be connected to store maintenance data.

The system can store data for between 50 and 100 flights. Information can be replayed on the ground through a low-cost commercial microcomputer system such as an IBM or Apple PC.

A new option that utilises a removable PCMCIA drive offers flight data storage equivalent to use of external recorders.

Specifications
Dimensions: ½ ATR short
Weight: 4.2 kg
Power: 30 VA

Operational status
In production for the ATR 42 and 72.

Contractor
SFIM Industries.

VERIFIED

ED 41XX, ED 44XX, ED 45XX, ED 47XX data acquisition units/data management units

The ED 41XX, ED 44XX, ED 45XX, ED 47XX are designed for the new-generation transports with digital avionics and electronic flight deck displays and are contained in ARINC 600 housing. They are designed as an expandable system by adding electronic boards. Several configurations are available: the basic FDAU for DFDR/QAR data supply; the expanded Data Management Unit (DMU) version for engine monitoring with basic engine trend monitoring reports; the expanded DMU with enhanced report monitoring for engine/APU monitoring, aircraft performance, environment reports, autoland performance as well as additional programmable reports; the expanded version with either solid-state memory for crew proficiency and flight data storage or with integrated and removable PCMCIA drive for flight data storage equivalent to external recorder.

All those units can be connected to a quick access recorder, cockpit printer, portable or onboard data loader, flight deck display (MCDU/CDU) and ACARS Management Unit for transmission of data to and from the ground.

Specifications
Dimensions:
3 MCU for DMU
6 MCU for DFDAMU
Weight: 4.5-8.5 kg (max)
Power: 40-85 W (max)

Operational status
In production for Airbus and Boeing commercial aircraft.

Contractor
SFIM Industries.

VERIFIED

ED 43XXXX flight data interface unit

The ED 43XXXX unit is intended to acquire the aircraft parameters and deliver a PCM message to the DFDR and QAR on A320, A330 and A340 aircraft.

It also interfaces with the onboard centralised maintenance system.

Specifications
Dimensions: 2 MCU
Weight: 3.7 kg
Power: 40 VA

Operational status
In production for Airbus A319, A320, A321, A330 and A340 aircraft.

Contractor
SFIM Industries.

UPDATED

EVS 901 R videotape recorder

The EVS 901 R videotape recorder is used for storage and video restoration in harsh environments. It is equipped with a Hi-8 deck and features both CCIR-PAL and Y/C standards.

Specifications
Dimensions: 162.5 × 109 × 192.5 mm
Weight: 2.8 kg
Power supply: 28 V DC, 20 W
Temperature range: −40 to +55°C

Contractor
SFIM Industries.

VERIFIED

EVS 906 videotape recorder

The EVS 906 videotape recorder is a EVS 901 R with an RGB video input.

Contractor
SFIM Industries.

VERIFIED

EVS 925 videotape recorder

The EVS 925 videotape recorder is used for storage and video recovery in harsh environments. It is equipped with a Hi-8 mm deck and features both CCIR-PAL and RGB standards. The EVS 925 is controlled by a serial RS-422 link.

Operational status
In production.

Contractor
SFIM Industries.

NEW ENTRY

EVS 1001 R videotape recorder

The EVS 1001 R videotape recorder is used for storage and video recording in harsh environments. It is equipped with a Hi-8 mm deck and complies with EIA/NTSC standards. The recording capability is 90 minutes. The EVS 1001 R provides the following functions: video and audio recording (two channels), video and audio playback, rewind, forward wind, pause, battery-driven autonomous cassette ejection function.

Specifications
Dimensions: 162.5 × 109 × 192.5 mm
Weight: 2.8 kg
Power supply: 28 V DC, 20 W
Temperature range: −40 to +55°C

Contractor
SFIM Industries.

VERIFIED

Mini-ESPAR 2

Designed by SFIM Industries, the Mini-ESPAR 2 has been developed to be fitted as a crash recorder on helicopters or military aircraft. The Mini ESPAR 2 can acquire the basic information to cover the crash requirement.

Fitted with an entirely electronic memory, this new-generation recorder features considerably improved reliability, elimination of scheduled servicing because of the absence of moving parts and appreciable savings in weight and volume. These savings are achieved by the use of a small-size static memory, made possible by the application of a specific data storage algorithm adapted to all types of aircraft and capable of recording up to 25 hours of flight data, and voice recording of up to 1 hour, in commercial aircraft. It has improved environmental protection against crush, pressure, temperature and corrosion.

During data retrieval, a recovery algorithm supplies the expected value of each parameter for comparison,

SFIM Industries Mini-ESPAR 2 **1998**/0015302

to speed data analysis. A further advantage is the ability to re-read flight data without removal of the recorder. In this case, recorded data is transmitted over a high-speed line to a static data retrieval set which is then connected to the workstation.

The Mini-ESPAR 2 can be adapted easily to any military or small civilian aircraft. It complies with TSO C124 and EUROCAE ED-55.

Specifications

Acquisition capacity:
(basic) 3 synchros, 15 analogues, 12 discretes, 3 frequencies
(options) 1 or 2 mixed audio channels, serial bus ARINC 429, or serial bus MIL-STD-1553B, or 2 pressure transducers
Recording duration:
(parameters) up to 24 h
(audio) up to 1 h
Frame rate: 64 or 128 words/s
Average parameter sample rate: 0.25-16 Hz programmable
Dimensions: 183 × 148 × 273 mm
Weight: 9 kg
Power supply: 28 V DC or 115 V AC
Options: voice recording, Arinc 429 recording

Contractor

SFIM Industries.

UPDATED

Socrate 2/Saturne data acquisition systems

Socrate 2 is an acquisition subsystem of the Damien 6 family. It was designed to be installed on the top of a helicopter main rotor or in the centre of a helicopter tail rotor in order to acquire parameters from rotor blades, especially strain gauges and temperatures, and to transmit up to 144 parameters on databusses through a slip ring to the cabin. The equipment is powered through the slip ring.

Saturne has the same function as Socrate 2, with a ring shape to allow the use of a mast-mounted sight on the helicopter. The acquisition capability is up to 128 parameters and the slip ring system is included in the equipment.

The equipment can also deliver the angular position of the rotor in order to correlate the position of blades with the values of parameters.

Specifications

Dimensions:
(Socrate 2) 75 to 275 mm diameter
(Saturne) 220 × 440 mm external diameter
260 mm internal diameter
Weight:
(Socrate 2) 2.6-12.5 kg
(Saturne) 30 kg

Operational status

The equipment is in production for Eurocopter Tiger helicopters.

Contractor

SFIM Industries.

VERIFIED

SSCVR Solid-State Cockpit Voice Recorder

The SSCVR has been designed in accordance with ARINC 757 – TSO C 123 – ED 56(A) by SFIM Industries, Dassault Electronique and TEAM, to meet the requirements of all commercial aircraft for cockpit voice recording systems. The system has no moving parts, and uses compressed voice storage to provide up to two hours recording time.

Specifications

Recording attributes:
(method) non-sequential adaptive differential pulse code modulation and code exciter linear parameter modelling
(medium) flash memory modules
(inputs) 3 crew microphones; one area microphone; GMT (ARINC 429 format); rotor speed; flight data recorder time marker
(time) 30, 60 or 120 minutes
Dimensions: ½ ATR short – ARINC 404
Weight: 8.5 kg
Power:
(DC version) 28 V DC, 0.8 A
(AC version) 115 V AC, 400 Hz, 0.26A

Contractors

SFIM Industries.
Dassault Electronique.
TEAM (Télecommunications, Electronique Aéronautique et Maritime).

NEW ENTRY

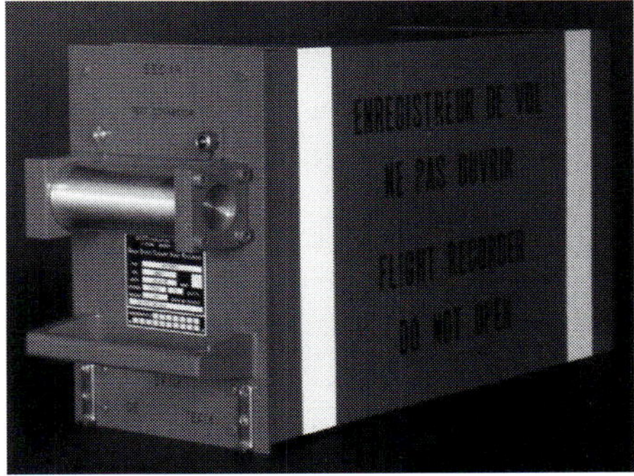

SFIM/Dassault Electronique/TEAM SSCVR solid-state cockpit voice recorder **1998**/0015303

AA8-123 head-up display camera

The AA8-123 is a single module head-up display camera which is used in conjunction with the VE-120 CRT display on the Super Etendard.

Specifications

Dimensions: 146.5 × 110 × 65 mm
Weight: 1.13 kg
Capacity: 9 m of standard film
Frame format: 10.4 × 7.5 mm
Framing rate: 16 frames/s
Running time: 70 s

Operational status

No longer in production. In service with the Super Etendard.

Contractor

Thomson-CSF Optronique.

VERIFIED

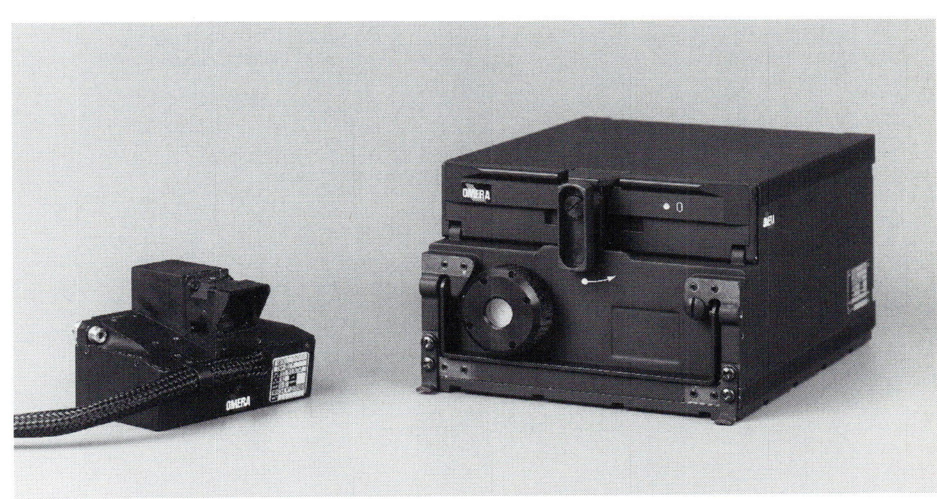
The Thomson-CSF Optronique video recording system showing (right) the OEV 301 magnetic tape recorder

OEV 301 magnetic tape recorder

The OEV 301 is a standard 8 mm magnetic-tape recorder. Its compact design enables it to be installed on fixed-wing aircraft and helicopters and allows recording and distribution of monochrome and colour visual or audio information.

Specifications

Weight: 2.7 kg
Power supply: 28 V
Consumption: 10 W
Standard: CCIR PAL 625-lines
EIA NTSC 525-750 lines
Signal/noise ratio: >47 dB

Operational status

In service.

Contractor

Thomson-CSF Optronique.

VERIFIED

Video recording systems

Mission sight recording is accomplished by a system consisting of, an OTA 204 monochrome or OTA 300 colour video camera and an OEV 301 onboard magnetic-tape recorder. The CCD video camera replaces the earlier AA8-400 film camera. A two-unit version of the video recording system has been developed for the holographic HUDs of future combat aircraft, sights used on present and new-generation helicopters and sights on the gyrostabilised turrets of future tanks.

Operational status

The OTA 208 video camera and the OEV 301 magnetic-tape recorder are fitted on upgrade French Navy Super Etendard aircraft.

Contractor

Thomson-CSF Optronique.

VERIFIED

GERMANY

Airborne Video Data Acquisition System (AVDAS)

The Airborne Video Data Acquisition System (AVDAS) is designed for observation and surveillance missions, as well as to support training and debriefing. All aspects of critical situations can be recorded, stored and evaluated. Data recorded on the crash-protected airborne video recorder include navigation, radar FLIR, inside and outside scene and audio information. All information is converted into a video signal, so that standard tapes and ground replay equipment can be used.

Operational status

AVDAS is on Tornado aircraft and is used in helicopters for test and evaluation purposes.

Contractor

Bavaria Keytronic Technologie GmbH.

VERIFIED

Display Video Recording System (DVRS)

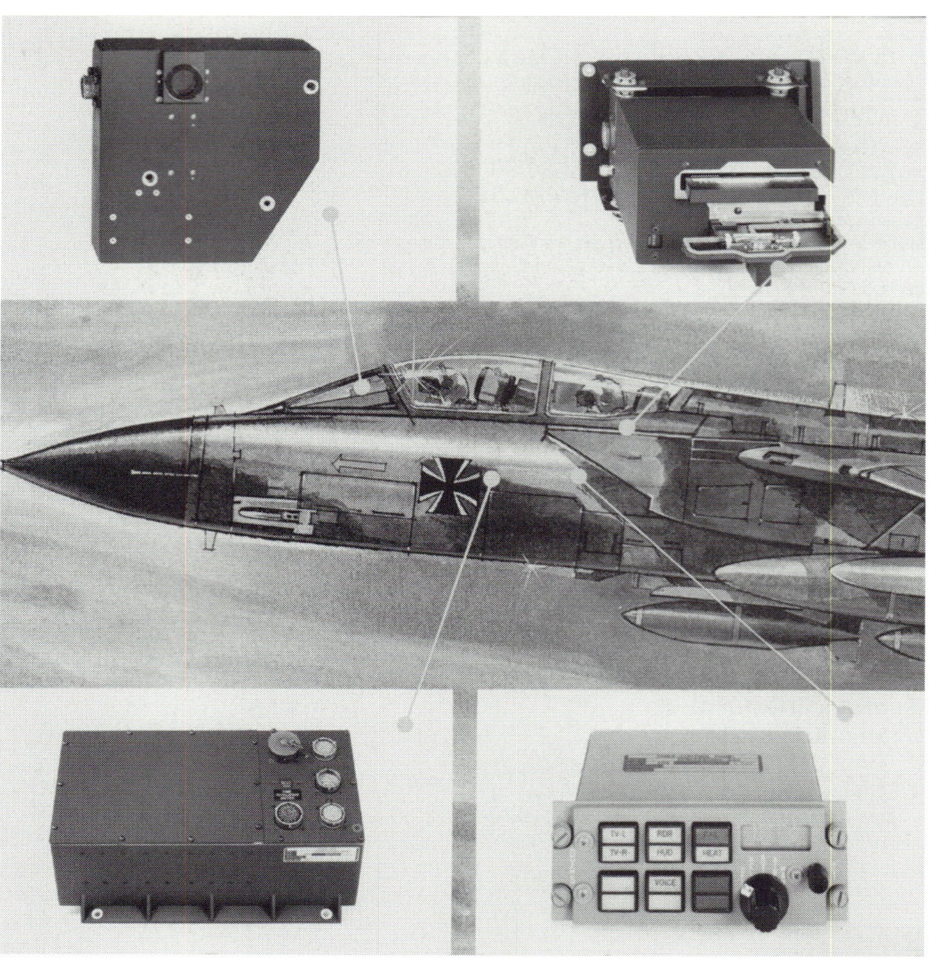

The display video recording system for the Tornado ECR showing the HUD camera (top left), videotape recorder (top right), electronics unit (bottom left) and control panel (bottom right)

The Display Video Recording System (DVRS) has been designed by Bavaria Keytronic Technologie for the German Air Force Tornado ECR and is pin-to-pin and mounting compatible with all versions of the Tornado IDS. It is also suitable for army, navy and commercial applications. The system can handle up to seven video inputs with a bandwidth of 9 MHz each, plus audio channels, event and weapon release markers, databusses, radar and FLIR.

The system consists of a HUD camera, electronics unit, videotape recorder with 90 minutes' recording tape and a control panel.

The HUD camera consists of the video sensor head with optical components and CCD video colour sensor module, and an electronics assembly consisting of power supplies, control logic, BIT circuitry and video amplifiers. HUD symbology is combined with the picture of the outside world and transmitted in colour PAL format via the multiplexing unit to the VTR.

The electronics unit consists of an analogue-to-digital converter unit, multiplex circuitry, microprocessor, control BIT circuitry and power supplies.

The AR 250 MM-R videotape recorder, based on the TEAC V 250 AB-R, is a compact airborne video cassette recorder which uses miniature ¼ in tape video cassettes. This recorder can be replaced by a TEAC V 80 recorder. The microprocessor-controlled panel gives the operator complete up-to-date information on the system, including remaining recording time. The control panel consists of the mode selection switch, video input, selection control, alphanumeric display and amplifiers.

The system provides 90 minutes' recording time on each cassette which can be changed easily during flight.

Specifications

Dimensions:
(HUD camera) 190 × 45 × 210 mm
(electronics unit) 120 × 325 × 170 mm
(videotape recorder) 107 × 139 × 156 mm
(control panel) 55 × 145 × 127 mm

Operational status

In service in German and Italian Air Force Tornado ECR aircraft.

Contractor

Bavaria Keytronic Technologie GmbH.

VERIFIED

Flight Safety Recording System (FSRS)

The Flight Safety Recording System (FSRS) provides information for analysis of accidents and severe crashes, training of pilots, flight attendants and ground crew and surveillance of passenger and cargo compartments and exterior equipment. It consists of cameras, an electronic control unit and a Hi 8 format crash-protected video recorder.

The cameras may be black and white or colour cameras installed inside or outside the aircraft. The cameras may be installed in the cockpit to record instruments and actions, in the passenger/cargo compartment to observe and record the situation and outside the aircraft to observe and record equipment and the outside world. They may be designed for low-light level operation, with infrared illumination. The electronic control unit multiplexes all video signals, controls the recording, and has an optional interface to GPS and aircraft busses to record navigation and other aircraft data. The crash-protected video recorder is an ARINC 404A ½ ATR short unit which provides recording of the last 2 hours and is qualified for MIL-STD-461 and TSO C51A environmental conditions.

Operational status

FSRS is designed for installation in helicopters and fixed-wing aircraft and can be tailored to different applications or customer requirements.

Contractor

Bavaria Keytronic Technologie GmbH.

VERIFIED

Multiple Dislocated Flight Data Recorder System (MDFDR)

The new-generation ultra lightweight Multiple Dislocated Flight Data Recorder (MDFDR) is part of an error tolerant modular solid-state crash and cockpit voice recording system which comprises up to four MDFDR recorders providing sufficient recording capacity. Due to the modularity, small dimensions and standardised interfaces, this recording system is suited for installation into helicopters and general aviation aircraft.

The MDRDR system consists of a Multiple Recorder Data Acquisition Unit (MRDAU), one to four MDFDR, Hand-Held Terminal (HHT) and ground station, designed to collect onboard data (audio, analogue, frequency, mil-bus, HDU and discrete inputs). It filters data to avoid exceedance and logistic relevant data and stores the resulting different data groups in the FDR and SSDC.

After flight, the data can be checked on board by the HHT or can be transferred via the HHT or SSDC to the ground station for detailed checkout.

Specifications

Dimensions:
(steel ball) 55 mm diameter
Data interface: RS-485 up to 1 Mbit/s (asynchronous datalink)
Power: 8-12 V DC, 1 W

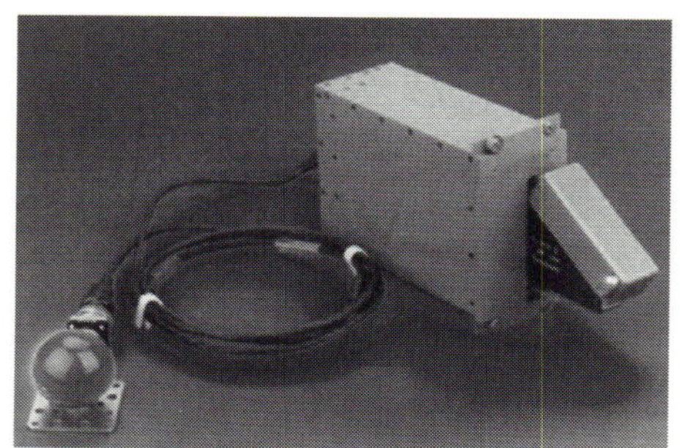

MRDAU Multiple Recorder Data Acquisition Unit
1998/0015306

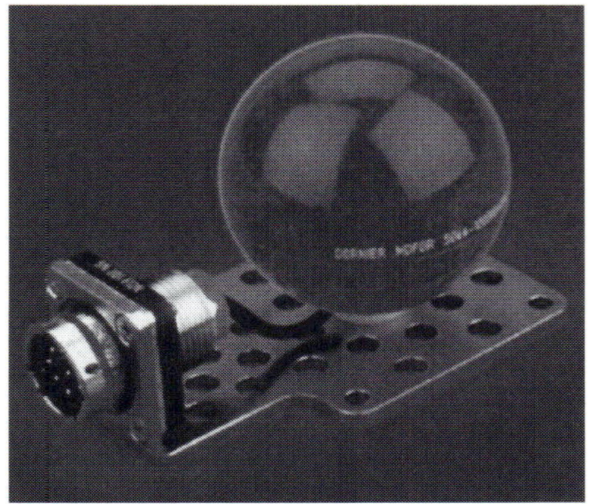

Multiple Dislocated Flight Data Recorder ***1998***/0015305

Temperature range:
(operating) −55 to +85°C
(storage) −55 to +85°C
Weight:
(one MDFDR) 0.35 kg
(MRDAU) 1.5 kg
Memory capacity: 4-32 Mbyte, 1 to 8 tracks, min track size 2 Mbyte

Multiple Recorder Data Acquisition Unit (MRDAU)

The MRDAU controls and supplies up to four MDFDR. The application specific programmable data interfaces of the MRDAU provide flexibility to adapt the MDFDR system to various helicopters and general aviation aircraft. For maintenance recording functions, the MRDAU has an optional data carrier interface which can store up to 170 Mbyte of flight data.

Optional data carrier ***1998***/0015307

Specifications

Dimensions:
(front panel) 146 mm × 76.2 mm × 6 mm
(case) 180 mm × 127 mm × 73 mm
Weight: 1.5 kg (excluding data carrier)
Power: 28 V DC, 25 W
Temperature range:
(onboard equipment, operating) −55 to + 71°C

Hand-Held Terminal (HHT) (readout unit)

The Dornier HHT is a powerful battery-powered product to read out the MDFDR and the MRDAU. It is capable of 7 to 10 hours of continuous use. It provides a large 640 × 480 pixels, easy to read backlit display, with VGA graphics. The HHT is IBM PC/AT compatible, it is possible to connect an external keyboard, mouse, printer, monitor and floppy disk drive. Optional plug-in modules are available for SCSI, PCMCIA and other ISA bus interfaces.

Hand-held terminal ***1998***/0015308

Contractor

Daimler-Benz Aerospace, Defense and Civil Systems.
Dornier GmbH.

NEW ENTRY

Flight Control Data Concentrator (FCDC)

The Flight Control Data Concentrator (FCDC) is part of the flight control system of the Airbus Industrie A319, A320 and A321 aircraft. Its main functions are to validate, concentrate and store in-flight status and failure data of the flight control system. The electrical interfaces are a number of ARINC 429 and discrete inputs and outputs. Outline and mounting are in accordance with ARINC 600.

Specifications

Dimensions: 2 MCU
Weight: 3.4 kg
Power supply: 28 V DC, 16 W
Interfaces:
(input/output) ARINC 429, SGS and SAV discrete
(input) DC analogue
Reliability: >18,000 h MTBF

Operational status

In production. The FCDC entered operational service in 1988 on board the A320, which is equipped with a dual FCDC installation. It is also in service on the A319 and A321 aircraft.

Contractor

LITEF GmbH.

VERIFIED

DATaRec-A4 acoustic recording system

The DATaRec-A4 is a complete portable four-channel acoustic/analogue recording system. The system provides connectors, preamplifiers and power supplies for Bruel and Kjäer microphones. With 24-bit digital signal processing and filtering, exceptional phase error specification allows accurate correlation of data between channels. Weighting filters can be selected and, with pre-emphasis, the dynamic range will be extended to 130 dB.

A front panel LCD provides bar graph and alphanumeric displays of input/output voltage, power and spectra. The IRIG generator/decoder allows synchronisation to the internal or external IRIG source, accurately time-stamping all data. Additional data that may be recorded includes voice, RS-232 and opto-coupled rpm and speed data.

With eight different channel configurations, a patented self-calibration system, rugged but lightweight chassis and battery operation, the A4 may be used wherever portability is of importance.

Specifications

Dimensions: 270 × 88 × 265 mm
Weight: 6 kg (incl battery)
Power supply: battery operated for up to 2 h
10-36 V DC or 100-240 V AC (with adaptor)

Contractor

Racal Heim GmbH.

VERIFIED

DATaRec-A16 analogue recording system

The DATaRec-A16 is a comprehensive analogue recording system, expandable from 16 to 80 channels. Internal signal conditioning modules allow direct connection of common transducers. With all channels sampled simultaneously, 24-bit digital signal processing and filtering, exceptional phase error and dynamic specifications are achieved. The large dynamic range leads to extremely low harmonic distortion and noise levels.

A quick-look facility for all input and output signals can display minimum, maximum and mean levels.

The IRIG generator/decoder allows accurate time-stamping of all data, using internal or external time sources. Additional data that may be recorded includes voice, RS-232 and opto-coupled rpm and speed data.

With 51 different channel configurations, patented self-calibration system and a rugged but lightweight chassis, the A16 can be used in a wide variety of applications.

Specifications

Dimensions: 325 × 133 × 320 mm
Weight: 12 kg
Power supply: 10-36 V DC or 100-240 V AC, 50 W

Contractor

Racal Heim GmbH.

VERIFIED

DATaRec-D3 digital recording system

The DATaRec-D3 is a versatile digital recording system for PCM, MIL-STD-1553 and RS-422 data systems. An optional PCM merger card allows four independent PCM streams to be recorded. The IRIG generator/decoder provides accurate time-stamping of all data using internal or external time sources.

The MIL-STD-1553 interface card permits a dual-redundant 1553 bus to be recorded. The clock, either internal or external with 1 μs accuracy, allows the RT response and intermessage gap times to be measured and the 1553 stream to be accurately replayed.

Additional data that may be recorded includes voice, RS-232 and opto-coupled rpm and speed. Memory buffers allow bursts of higher rate data to be recorded or the 3 hour record time to be significantly extended.

The use of rugged but lightweight chassis and universal auto-sensing power supplies allow the D3 to be used in a wide variety of applications.

Specifications

Dimensions: 270 × 88 × 230 mm
Weight: 4.2 kg
Power supply: 10-36 V DC or 100-240 V AC (via adaptor), 25 W

Contractor

Racal Heim GmbH.

VERIFIED

DATaRec-D4 digital recording system

The DATaRec-D4 digital recording system is able to record at the rate of 4 Mbits/s for 2 hours, with 1μs resolution time stamping. The DATaRec-D4 contains an IRIG-B time code generator and decoder. This locks to an external source or can be used as the source for equipment external to the D4. The time is maintained by battery back-up. The IRIG-B time code is reconstructed during replay.

The D4 will store different configurations in non-volatile memory. One of these, the default configuration, is loaded when power is applied to the D4. This configuration includes the mechanical status of the tape drive. The recorder may therefore be installed in inaccessible locations and recording is made and ended simply by operating a 28 V power switch. There are simple remote-control function switches for play, record, stop, rewind and fast forward.

There are many useful accessories for the D4, such as a hand-held remote-control unit and a cockpit mounting remote control with a sunlight readable LED display and simple control for the pilot's use. The D4 has a large amount of recording capacity. This allows the recording and playback of event markers, voice commentary, RS-232 data and rpm pulse input. Up to 32 thermocouple, or low-frequency analogue, signals may be recorded at the same time as the main PCM stream or MIL-STD-1553 data, via the RS-232 input, by connecting an optional T32 system.

The D4 is of modular construction, which provides for easy upgrade or modification. For example, an additional MIL-STD-1553 recording interface or replay module may be incorporated. There are five free slots and all optional interface boards can be mixed up to the sum bit rate of 4 Mbits/s.

The D4 operates from power supplies in the range 20 to 36 V DC and will operate without errors even when subjected to intermittent power failures in the region of 50 ms. There is also an accumulator option which saves all data during a power interruption of more than 50 ms. It is easily installed in an interface slot, like an interface card.

The construction of the D4 is extremely robust but lightweight and is designed for aerospace and automotive applications. The tape drive is mounted on built-in vibration isolators and these allow error-free operation under vibration up to MIL-STD-810C of 5 *g* and 10 *g* acceleration.

Specifications

Dimensions: 124 × 180 × 260 mm (½ ATR)
(cassette) 73 × 54 × 10.5 mm
Weight: 6 kg
Power supply: 20-36 V DC, 70 W (max)
Temperature range: −10 to +60°C

Operational status

Selected for the Royal Navy Westland Lynx Mk 8 helicopter.

The DATaRec-D4 high-speed digital data recording system **1997**/0001293

Contractor

Racal Heim GmbH.

UPDATED

DATaRec-E8 digital cassette tape recording system

The DATaRec-E8 is an eight-channel PCM recorder. Data is recorded on a 3.8 mm wide magnetic tape using the helical scan recording system. Digital Audio Tape (DAT) cartridges and drives are used for the storage of data. An efficient error-correction process provides data security for the system.

Incorporating the input/output filters and the cassette drive, the eight-channel PCM electronics comprise a compact unit. The inputs and outputs are connected to BNC jacks without intermediate adaptor. The number of channels can be selected to one, two, four or eight. Signal bandwidth is 5 kHz for each channel with eight-channel operation. Input and output filters can be dispensed with in many applications.

All analogue channels are sampled at the same time. The unit incorporates eight analogue/digital and eight digital/analogue converters with a resolution of 14 bits.

In additon to data, date and time are recorded continuously. A voice channel for recording comments is also provided. A built-in loudspeaker allows the voice recording to be replayed. A headphone connection and line output are provided. A standard IRIG-B time code can be recorded instead of voice recording. Index marks which were set automatically or manually during recording can be searched at high speed during replay.

The DATaRec-E8 magnetic tape system is especially suitable for mobile use. Small dimensions and low power consumption are important features. The units operate correctly even in vibration of up to 5 *g*. Three alternative power supply sources allow universal use. The 6 V, 1.8 Ah storage battery permits the PCM to operate for 2 hours. A DC voltage adaptor or a mains adaptor can be used instead of the storage battery. Voltage range is from 10 to 30 V.

Specifications

Dimensions: 235 × 92 × 176 mm
Weight: 2.8 kg
Power supply: 6 V, 1.8 Ah battery
10-30 V DC

Contractor

Racal Heim GmbH.

VERIFIED

DATaRec-D10/D12/D40 digital cassette tape recording system

The D10/12/40 have been designed for aerospace and defence applications and are 10, 12 and 40 Mbits/s nominal (with compression turned off) digital tape recorders. Their construction is extremely robust, yet lightweight. The heavy-duty 0.5 in Digital Linear Tape (DLT) tape drives are mounted on built-in vibration isolators which provide error-free operation in harsh environments to MIL-STD-810C (5 *g* vibration and 10 *g* acceleration).

Modular construction is provided with six free slots, which allow users to configure the recorder for particular applications and to modify it for future programs. All optional interface boards can be mixed without limitation (for example, 8 + channel analogue input and output combined with multiple MIL-STD-1553B bus recording and reconstruction). The recorders can merge and record multiple digital and analogue data streams up to combined maxima of 10, 12 or 40 Mbits/s (actual data rates due to low overheads) without the use of compression and can achieve data rates up to 150 per cent using built-in compression. Analogue bandwidths from 300 kHz (D10) to 1.2 MHz (D40) make the systems ideal for use as mission recorders particularly in sonobuoy recording applications. RS-232 serial, SCSI, and power-on remote control is available and remote contact mission configuration. The D10/12/40 have a

DATaRec-D12 **1997**/0001294

built-in IRIG-A/B/G/J time code generator and decoder, and a serial GPS time code decoder. A high-stability oscillator (0.44 ppm) is available to improve time code accuracy.

Specifications

Recording time:
D10: 2 h, at full bit rate, 10 Mbits/s
D12: 3.6 h, at full bit rate, 12 Mbits/s
D40: 1.8 h at full bit rate, 40 Mbits/s
Recording time can be increased with lower bit rate, and with use of built-in data compression (data dependent)
Recording capacity:
10-15 Gbyte, 20 Gbyte and 35 Gbyte (uncompressed)
Dimensions: ¾ ATR wide

Operational status

Selected by Sikorsky for the flight test certification programme for the S-92 Helibus.

Contractor

Racal Heim GmbH.

UPDATED

OptoRec Q1 and Q2 quick access recorders

The OptoRec Q1 and Q2 quick access recorders are based on Motorola 56001 DSPs. The Q1 includes a VGA module and only a keyboard and monitor need to be connected to make the recorder act as a PC. The Q2 uses a 96001 CPU, running at 33 MHz, and can be connected to a PC via the RS-232 port. Both recorders provide a standard Centronics port for a printer.

The Q1 and Q2 will record ARINC 573 and 575, with options that allow multiple ARINC 429 and analogue data streams to be recorded simultaneously. An intelligent data decoder unit automatically detects and decodes the different data types. A clock with battery back-up provides accurate date and time annotation on the recordings. An ASCII character stream can be recorded at up to 9,600 baud for report writing. Status and error conditions are indicated via a 12 digit front-panel LED display.

The standard recording medium is 3.5 in optical disk conforming to ISO/CD 10090, with an option for a DAT streamer. The disk allows up to 43 hours record time at 512 words/s and all data is recorded in DOS format files. The disk may be removed and read by any standard 3.5 in DOS-compatible optical drive. The Q1 also has an option for a hard disk drive.

Data to be recorded is buffered via a 128 kbyte non-volatile RAM and reliable recording is achieved even when subjected to power failures up to several seconds. Up to 16 Mbytes of memory are allocated separately for retaining application, or mission specific programs.

Contractor

Racal Heim GmbH.

VERIFIED

INTERNATIONAL

AH-64D Apache Longbow HUMS

Created jointly by Stewart Hughes, Base 10, and the Boeing Company, the AH-64D Apache Longbow HUMS provides comprehensive health and usage monitoring. The diagnostics for avionics, rotors, transmission, engines and airframe are integrated with the helicopter's data management and display systems. It includes exceedance monitoring and cockpit voice and flight data recording.

The system comprises the Expanded Maintenance Data Recorder (EMDR); MultiPurpose Display (MPD); PCMCIA card receptacle and portable ground station.

System features include displays, warnings and pilot initiated functions fully integrated with the Longbow data management system covering 1,800 fault codes; data download via PCMCIA card or MIL-STD-1553B databus to portable PC ground station; ground station for HUMS data management and maintenance. EMDR features include data recording based on the Longbow Integrated Maintenance Support System (LIMSS); Stewart Hughes HUMS; 28 vibration inputs; eight tachometer inputs; blade tracker (day/night); voice recorder; 80 MB crash-survivable memory; up to 160 MB non-crash survivable memory; MIL-STD-1553B databus terminal; RS-422, RS-232, RS-485 serial communications.

Contractors

Stewart Hughes Ltd.
Base 10.
The Boeing Company.

NEW ENTRY

AH-64D helicopter, which utilises the Stewart Hughes/Base 10/Boeing HUMS
1998/0015309

Helicopter Flight Data Recording/ Health and Usage Monitoring System (FDR/HUMS)

The HUMS part of the FDR/HUMS system combines vibration and usage monitoring of critical dynamic power-train components with techniques such as chip detection, rotor track and balance, engine power assurance, cycle counting, exceedance monitoring and oil analysis. Integrated with the FDR, the FDR/HUMS supports helicopter operations, safety and maintenance — both civil and military.

The system comprises a data retrieval unit, sensors and data sources (to customer requirements), crash survivable FDR, and a ground station for fleet data storage.

Operational status

FDR/HUMS in the forms of North Sea HUMS, EuroHUMS™ and AHUMS™ are in service in North America, Europe, Australia and South East Asia. The HUMS element of the system is being supplied by Stewart Hughes Ltd. FDR/HUMS installations have been produced for Bell 412/212 and Sikorsky S-76 helicopters. Applications for these systems include North Sea HUMS: S61N, AS332 Mk 1, BV234 and S76; EuroHUMS: AS332 Mk 1, AS332 Mk 2, AS532 Mk 1 and AS532 Mk 2; AHUMS: Bell 412, CH47D and S-76.

Contractor

Teledyne Controls.
Stewart Hughes Ltd.

NEW ENTRY

Helicopter FDR/HUMS equipment
1998/0015310

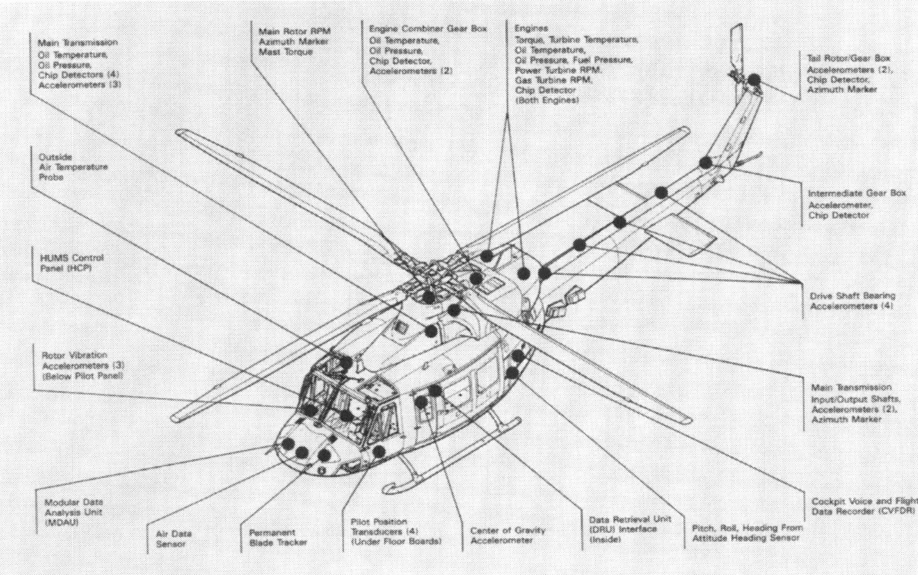

Bell 412 helicopter installation FDR/HUMS
1998/0015311

ISRAEL

Colour HUD camera

The colour HUD camera is a high-sensitivity, wide dynamic range, high-resolution camera which is mounted on an HUD to record head-up display symbology and views of the real world. It incorporates a ½ in CCD. Output is in NTSC, PAL or S-VHS formats.

Specifications

Weight: 0.25 kg
Power supply: 12 V DC, 500 mA
Resolution:
460 × 400 (NTSC)
450 × 450 (PAL)
Pixels:
768 × 494 (NTSC)
752 × 582 (PAL)
Shutter speeds: 1/60 to 1/10,000 s

Operational status

Fitted to the ELOP 982/3 HUD.

Contractor

ELOP Electro-Optics Industries Ltd.

VERIFIED

Autonomous Combat manoeuvres Evaluation (ACE) system

The Autonomous Combat manoeuvres Evaluation (ACE) system is a flight training debriefing system aimed mainly at multiparticipant flight training. The development of ACE is the direct outcome of an operational requirement for a squadron-level replacement for the current Air Combat Manoeuvring Instrumentation (ACMI) ranges.

ACE, unlike ACMI, is not ground range dependent and this enables debriefing of any kind of flight executed anywhere, without the need for carriage of external pods or telemetry. It is installed internally on each aircraft. Installation is simple and no software or cockpit changes are required for aircraft equipped with an MIL-STD-1553 databus. The ground debriefing station is based on commercial hardware and is designed to be used at squadron level on a daily basis.

Aircraft flight data required to reconstruct manoeuvres and the operation of the avionics system is recorded independently on board each aircraft by the ACE flight monitor unit on the aircraft's VTR. The use of a C/A or P code GPS as an integral part of the ACE airborne system provides position accuracy and synchronisation, without being limited to a specific antenna range and telemetry. After landing, the cassettes of all the participants, containing the digital data and video and audio playbacks, are processed on a ground debriefing station from which a unified display file is produced and displayed graphically. Synchronised HUD video playback and audio complement and complete the debriefing material.

Rada Autonomous Combat manoeuvres Evaluation system onboard unit ***1997***/0001295

Specifications

Dimensions: 70 × 165 × 320 mm
Weight: 4 kg
Power supply: 28 V DC

Operational status

In service on Israeli Air Force F-16 aircraft with an option for F-15 and F-4 2000 aircraft, to become the standard fit on all Israeli Air Force fighters.

In service on Chilean F-5E aircraft.

Contractor

Rada Electronic Industries Ltd.

VERIFIED

Flight Monitoring System (FMS)

The Flight Monitoring System (FMS) is an airborne data acquisition system, designed to perform the two major tasks of fatigue data processing and recording and Air Combat Manoeuvring Instrumentation (ACMI). In addition, the open architecture and large storage capacity of the FMS enable it to be used as an onboard integral flight test instrumentation kit.

The fatigue recorder consists of airframe fatigue recording and engine fatigue recording. In airframe fatigue recording, the FMS monitors and records a set of strain gauges and multiplexer bus flight data parameters that are stored on a solid-state memory cassette. The data is downloaded into a ground station in which the aircraft fatigue index is calculated. In engine fatigue recording, the FMS monitors engine parameters via the F-100 Digital Electronic Engine Controller (DEEC) and counts and records engine cycles from which engine fatigue data is extracted. Monitoring the DEEC enables the FMS to detect and record engine parameters during engine fault occurrences. This optional feature already exists in the DEEC. However, centralising this information in the FMS cassette may be an advantage.

In Autonomous Combat Evaluation (ACE), the FMS monitors selected multiplexer bus and DEEC parameters and records them on a video cassette recorder. The cassettes are replayed on the ACE ground station, providing an ACMI debriefing for multi-aircraft formations.

The FMS open architecture makes it suitable for use as an onboard instrumentation kit. Programmable selected flight test data may be recorded on the system's solid-state memory cassette and on the VTR cassette, which makes it an efficient and flexible tool for flight test programmes.

Monitoring the multiplexer bus enables the FMS to record maintenance fault lists when they appear on the bus.

The FMS consists of a Flight Monitoring Unit (FMU), Data Recording Cartridge (DRC) and Video Tape Recorder (VTR). Ground equipment for debriefing purposes is also associated with the FMS.

The FMU performs processing of the aircraft video signal, originally connected to the VCR, by integrating it with the ACE and flight test data; transfer of the combined video signal to the VCR, retaining its original visual data in addition to new information; monitoring of the aircraft strains by continuous measurement of the strain gauge sensors; acquisition of engine data through a serial measurement channel connected to the DEEC; acquisition of multiplexer bus data through its interface, and monitoring of the fuel quantity by continuous measurement of the applicable signals.

The DRC stores all the data sent from the FMU. It is based on non-volatile solid-state memory modules. The DRC is used for the transfer of all the data acquired during flights to the ground station for further processing and analysis.

The VTR cassette is used to store selected flight data and engine parameters for post-mission debriefing or flight test analysis.

Contractor

Rada Electronic Industries Ltd.

VERIFIED

Helicopter Power Plant Recording and Monitoring system (PPRM)

The Power Plant Recording and Monitoring system (PPRM) is an add-on system for recording and monitoring critical helicopter power plant parameters during flight. It enables analysis of power plant performance and flight event data and provides for direct diagnostics and fault detection capability. The PPRM improves mission availability and reliability and saves time and cost in flight-line and depot maintenance.

System components are Engine History Recorder (EHR) – this airframe-mounted unit automatically records routine maintenance data and out of range events of critical parameters of aircraft power plants such as the twin/triple engine power plant of the CH-53, the twin engine power plant of the Bell 212 and the single engine power plant of the Bell Cobra AH-1; Performance Display Panel (PDP) – this cockpit-mounted panel unit continuously displays measured and calculated power plant performance parameters to the aircrew; Test and Data Extractor Unit (TDEU) – this ground crew support equipment is a hand-held computer which interfaces with the EHR and enables rapid power plant data extraction for flight line and depot-level maintenance activities.

PPRM display panel ***1997***/0001408

Main features are identification and recording of specific power plant subsystem failures and fatigue cycles accumulation for 'on condition' maintenance; automatic or flight crew initiated data recording; immediate detection of power plant failures or performance degradation; continuous calculation and display of flight-critical information to the pilot including engine power and maximum hovering weight; real-time pilot warning and recording of over-limit conditions such as engine speed, temperature and torque; calculation and accumulation of Low Cycle Fatigue (LCF) and blade life.

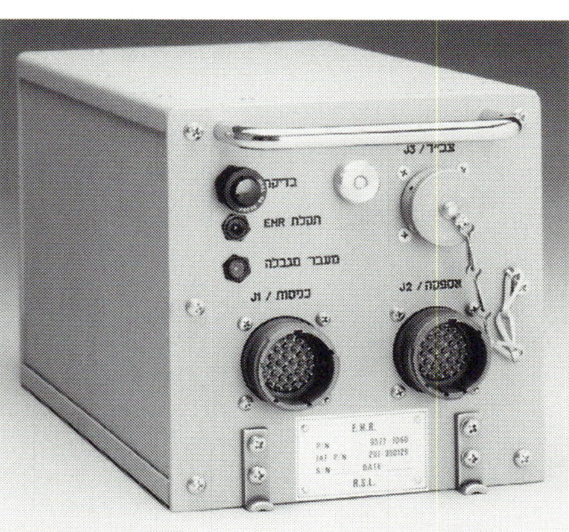

PPRM EHR unit ***1997***/0001407

Specifications:
EHR
Dimensions: 150 × 150 × 250 mm
Weight: 3.2 kg
Supply Voltage: 18-33 V DC IAW MIL-STD-704B
Power: 10 W
Qualification:
(environmental) MIL-STD-810C
(electromagnetic) MIL-STD-461B
Communication: RS-232

PDP
Display: 2 row, 16 characters NVG-compatible
Dimensions: 127 × 76 × 98 mm
Weight: 0.8 kg
Power: 6 W
Qualification:
(environmental) MIL-STD-810C
(electromagnetic) MIL-STD-461B

Operational status
In service with the Israeli Air Force.

Contractor
RSL Electronics Ltd.

VERIFIED

ITALY

Crash/Maintenance Recorder System (CMRS) for the Tornado

Alenia has developed the Crash/Maintenance Recorder System (CMRS) selected for the Tornado. The Tornado CMRS includes onboard and ground-based equipment for the collection, management and recording of flight data and subsequent computation and analysis.

The Tornado CMRS is composed of an Extended Data Acquisition Unit (EDAU), Maintenance Recorder Unit (MRU), Accident Data Recorder (ADR), Mobile Quick-Look Facility (MQLF) and automatic ground station and semi-automatic test equipment.

The EDAU is based on an Alenia advanced MARA avionics computer and is used for the collection, conversion and compression of a wide range of flight signals.

The MRU records on an easily removable cassette all the EDAU processed data.

The ADR, supplied by British Aerospace, records on a crash-protected endless loop tape all the critical data for analysis in the event of an accident.

The MQLF is employed as first line ground support to perform fast analysis of the flight recorded data and to assess the aircraft serviceability status.

The CMRS has crash recording time of up to 2 hours and a maintenance recording time of 3 to 9 hours depending on the mission profile. The system records up to 125 maintenance parameters, of which up to 37 are for structural fatigue analysis and life computation, up to 32 are for engine low-cycle fatigue and up to 20 are for aircraft system monitoring. The CMRS has VDU, printer and plotter output devices and presents information in tabular or plotted form.

Operational status
In service on the Italian Air Force Tornado.

Contractor
Alenia Difesa, Avionic Systems and Equipment Division.

VERIFIED

Airborne Strain Counter (ASC)

The Airborne Strain Counter (ASC) unit, is designed to determine the different load spectra and/or local stresses that arise during operation in all aircraft structures submitted to variable loads. The unit is suitable for monitoring the fatigue life of complex structural elements, where the local stresses cannot be easily correlated to the standard usage parameters.

The ASC automatically accomplishes acquisition of a maximum of eight parameters, received from strain gauges or accelerometers positioned on the most significant points of the aircraft structure; conversion of the acquired data into digital form; processing of the data and outputting of a stress matrix; storage of the data in a non-volatile memory; interfacing with the dedicated ground support unit to allow data transfer and subsequent processing; supply of power to the sensors; and performance of self-test to check serviceability of the equipment.

The ASC features eight fully independent channels for acquisition, preprocessing and storage of data. Each channel is capable of managing the interface signals of the associated strain gauge or accelerometer. Each channel amplifies the analogue signal received from the sensor and converts it into a digital signal through an A/D converter, providing information on number of stress cycles experienced by the structure, mean values of the stress for every cycle and peak-to-peak value of the stress for every cycle.

Information is stored in a non-volatile memory on aircraft serial number, card serial number, ASC serial number, total number of stress cycles, status word, stress matrix, total number of minutes of aircraft motion and total number of minutes of operation in the presence of failures.

An RS-422 databus is used to transfer the data stored from the measurement channels to the ground support unit.

At system power on, each channel performs a self-test to check hardware and software integrity.

Operational status
In service on the Aermacchi MB-339 and the Aermacchi/Alenia/Embraer AMX.

Contractor
Logic SpA.

VERIFIED

JAPAN

Crash Protected Video Recorder (CPVR)

The Crash Protected Video Recorder (CPVR) is designed to provide a record of the glass cockpit data presented to the flight crew. This is a natural follow-on from the standard flight data recorder. Knowledge gained over many years in the design and manufacture of crash recorders has been utilised in the design of a prototype CPVR.

The Hi-8 mm video standard has been chosen as it is lightweight, reliable and has a bandwidth approaching 4 MHz. The CPVR provides recording times of 4 hours for NTSC and 3 hours for PAL. These two video standards, plus RS-170, provide 400 TV lines resolution. The system also has a single audio channel with a frequency response of 100 Hz to 10 kHz.

The CPVR will form part of a system utilising closed-circuit television cameras located in suitable positions both within and outside the aircraft. Camera positions will be chosen to provide the aircrew with real-time viewing of internal and external features on a suitable display. This follows recommendations from the UK Aircraft Accident Investigation Branch. The number of cameras will depend upon specific requirements, with the chosen outputs being recorded on the CPVR to enable in-depth analysis to be carried out.

Specifications
Dimensions: ¾ ATR
Power supply: 28 V DC, 15 W
Environmental: MIL-STD-810D, MIL-STD-461B, MIL-STD-462

Operational status
Under development.

Contractor
TEAC Corporation.

VERIFIED

V-83AB-F airborne video cassette recorder/reproducer

The V-83AB-F video cassette recorder/reproducer is a three-deck version of the TEAC V-80. It records three signals for 2 hours each or one signal for 6 hours and provides 2 hours' playback time on each cassette. It has applications in a variety of markets including high-performance and commercial aircraft, RPVs, helicopters, test monitoring and surveillance. Use of the Hi-8 mm format offers the user the highest possible resolution, plus access to inexpensive commercially available playback units and video cassettes.

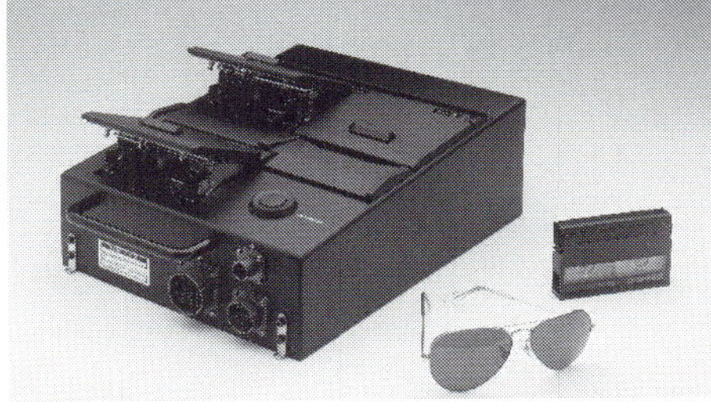

The V-83AB-F video cassette recorder/reproducer

A total of 400 lines of horizontal resolution and a signal-to-noise ratio approaching broadcast quality are available. Life expectancy exceeds 10,000 hours. All three decks are controllable from an RS-422 full-duplex interface or by discrete switch-closure signals which render the V-83AB-F compatible with control panels for earlier TEAC video recorders. An internal 28 V power supply simplifies power wiring connections to the unit. The recorder is fully ruggedised, equipped with an internal heater and designed to withstand high vibration levels and temperature extremes. Total weight is 8.16 kg.

Contractor
TEAC Corporation.

VERIFIED

V-250 series videotape recorders

The V-250 series of airborne videotape recorders are high-performance units designed for colour or monochrome recording on ¼ in video cassettes. It is ruggedised to withstand the effects of shock, vibration, high *g* loading, altitude, explosive decompression and wide temperature variations.

The V-250 AB-R has been designed as a record-only model for minimum size and weight. The V-250 AB-F both records and allows playback for in-flight and post-mission analysis. Both are available in 525- or 875-line EIA/NTSC or 625-line CCIR/PAL video formats and provide 60 minutes' (EIA/NTSC) or 51 minutes' (CCIR/PAL) recording.

Specifications

Dimensions:
(V-250 AB-R) 107 × 139 × 156.5 mm
(V-250 AB-F) 192 × 107 × 156.5 mm
Weight:
(V-250 AB-R) <3 kg
(V-250 AB-F) <4 kg
Power supply: 20-32 V DC unregulated

Contractor

TEAC Corporation.

VERIFIED

V-1000 AB-F videotape recorder

The V-1000 AB-F videotape recorder and reproducer uses high-capacity ¾ in U-matic 'S' video cassettes and is designed to record FLIR and IR linescan displays and for other high-resolution recording applications. The rugged construction provides protection against shock, vibration and low pressure.

The system is available in both 525- or 875-line EIA and 625-line CCIR video formats, providing recording times of up to 72 minutes in 525-line EIA, 60 minutes in 875-line EIA and 69 minutes in CCIR.

Specifications

Dimensions: 271.8 × 166 × 367.5 mm
Weight: <12.5 kg
Power supply: 115 V AC, 400 Hz, single phase

Contractor

TEAC Corporation.

VERIFIED

NETHERLANDS

OTA series cockpit cameras

The OTA series is a range of black and white, or colour, cameras designed for the HUDs of helicopters and fixed-wing aircraft as part of a mission recording system. Used with an airborne magnetic-tape recorder, they superimpose and record HUD symbology on real-time video pictures together with flight data. The OTA cameras enable in-flight permanent recording of all the visual information received by the pilot from take-off to landing. The high-sensitivity cameras feature an automatic exposure control to provide fast adjustment to sudden changes in brightness levels. An in-flight replay capability is provided together with display options for the co-pilot.

Current cameras in the OTA range include the OTA-222 black/white HUD camera; and the OTA-1320 red/green/blue colour HUD camera, selected for Rafale.

Specifications

Dimensions: 75 × 65 × 95 mm
Weight: 0.65 kg
Field of view: 30° horizontal × 22.5° vertical
Line of sight: <1 mrad
CCD sensor: 756 horizontal × 575 vertical pixels
Depth of focus: 3 m to infinity

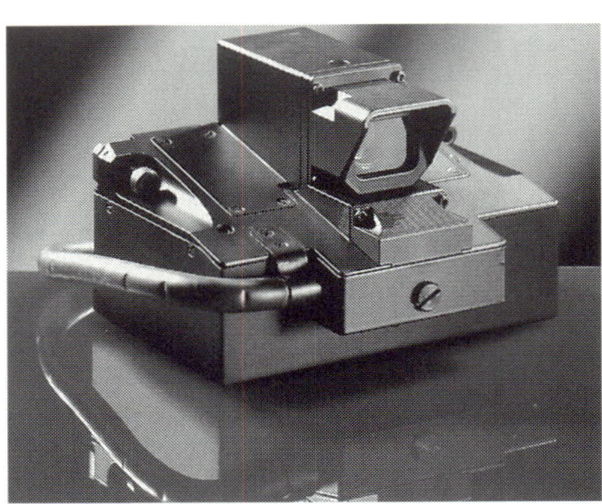

OTA 222 B/W HUD camera **1997**/0001296

Spectral range: 0.4-0.7 μm
Bandwidth: 5.5 MHz
Dynamic range: 3-100,000 lux
Video signal: CCIR/STANAG 3350B EIA/RS-70
Synchronisation: int/ext input (RS-422)
Power: 12-60 V DC, 4 W
Temperature range: −15 to +55°C
Reliability:
(operating flight hours) 5,000 h MTBF

OTA-1320 RGB HUD camera selected for Rafale **1998**/0015312

Operational status

OTA cameras are in production for the Mirage 2000-5 and Rafale fighter aircraft.

Contractor

Signaal USFA.

UPDATED

NORWAY

EE 235 solid-state recorder

The EIDEL Eidsvoll Electronics AS EE 235 solid-state recorder is designed for use by the aerospace, military and industrial markets, and is qualified for both aircraft and missile use. Its prime purpose is telemetry data storage, and features include 32 to 224 Mbyte data storage in non-volatile memory; serial data recording at up to 10 Mbits/s; an input data buffer for burst data; extended storage by interconnecting any number of recorders; configurable PCM encoder module for analogue and digital data collection; direct, or control box operation. The system can be configured to user requirements by adding memory and encoder modules as required.

Variants include the 32 Mbyte, EE 235-M2, basic unit: EE 236 power module + EE 237 solid-state controller module
128 Mbyte, EE 235-M3, basic unit + 1 × EE 238 solid-state memory unit
224 Mbyte, EE 235-M4, basic unit + 2 × EE 238 solid-state memory units

Supporting modules include: the EE 240 PCM encoder module, and the EE 241 BiØ-L Bitsynchroniser module.

Specifications

Dimensions:
(EE 235-M2) 105 × 143 × 49 mm
(EE 235-M3) 105 × 143 × 74 mm
(EE-235-M4) 105 × 143 × 99 mm
Weight:
(EE 235-M2) 0.8 kg
(EE 235-M3) 1.2 kg
(EE-235-M4) 1.6 kg
Input bit rate: 0-5 MHz
(option) 0-10 MHz
(low power mode) 0-1 MHz
Playback rate: 375 kHz; 1,500 kHz, 3 MHz or external
Input: NRZ-L with CP. TTL and RS-422 level
Output: NRZ-L, CP, BiØ-L. TTL and RS-422 level

Operational status

Widely used in the aerospace, military and industrial markets.

Contractor

EIDEL Eidsvoll Electronics AS.

NEW ENTRY

EIDEL EE 235-M3 solid-state recorder, showing EE 236 power module, EE 237 controller module, and EE 238 memory module **1998**/0015313

POLAND

Solid-State Quick Access Recorder (SSQAR) family

The ATM company has designed the Solid-State Quick Access Recorder (SSQAR) with contactless data transmission between recorder and cartridge. This solution with a RAM-based removable cartridge has been developed to give the user high-quality data recording and extremely fast data replay with no pins to damage, nor tape to be mangled. The ATM-SSQAR family is also easy and flexible to operate. Its flexibility allows a wide range of different applications with most major recording systems installed on board civil and military aircraft as well as helicopters and gliders.

Programmable and able to make online analyses, the ATM-SSQAR family can display and export warning signals indicating previously programmed parameter exceedences.

To be able to meet all customer requirements ATM has designed a family of recorders consisting of: ATM-QR3 (modular, most sophisticated type); and ATM-QR4 (modular, advanced type).

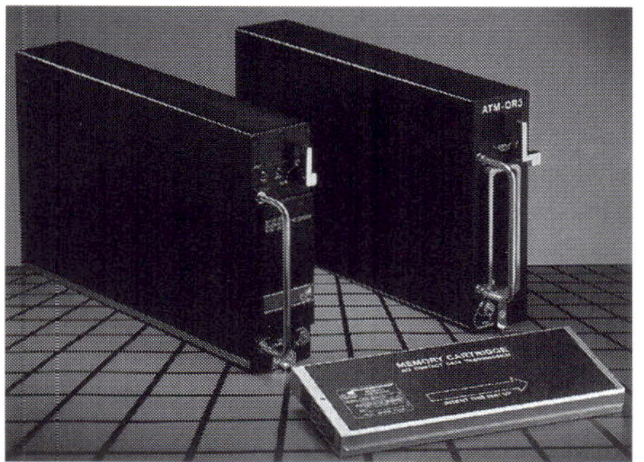

ATM's solid-state quick access recorder ***1998***/0015314

Versatility is the main feature of the ATM-SSQAR family. In a single unit it combines the functions of quick access recorder with built-in real-time clock; data acquisition unit; and data management unit (programmable reports depending on installed options).

Recording time depends on the cartridge size and data format. The table below shows examples of recording time for different size ATM cartridges. Values in the table are estimated for the most popular recording standards.

Examples of recording time (hours)

Type of cartridge	Recording system			
	ARINC 717			**MSRP-64**
	64 words/s	128 words/s	256 words/s	
ATM-MC5/30	34	17	8	60
ATM-MC5/70	80	40	20	140
ATM-MC5/70	120	60	30	220

Data replay from the cartridge is performed by the ATM-RD3 reader, and complete operation using a PC-compatible computer, takes up to 3 minutes, for 30 flight hours (for 64 words/s transmission speed).

The basic components include built-in real-time clock; display for recorder status and online programmable flight data analysis (ACMS); data output for external equipment; interface for entry panel; test connector for data monitoring; and data dump.

The optional components: Depending on SSQAR type, the following optional modules are available:

ATM-QR3

Optional internal full-size modules: QR3DC (interface operating two ARINC 429 busses); QR3FT (programmable vibration spectrum analyser); QR3PE (module operating as a DAU).

Optional small size module: QR3AR (ARINC 573/717 bus interface).

ATM-QR4

Optional internal full-size modules: QR4DC (interface operating two ARINC 429 busses); QR4PE (module operating as a DAU).

Optional small size module: QR4AR (ARINC 573/717 bus interface).

Optional external modules

ATM-RT allows connection to any recorder of the ATM-SSQAR family to MSRP-64 and 256 data recording systems. ATM-DP analogue multiplexer provides the ability to process additional input signals (not available for QR2).

Specifications

Dimensions: 317 × 57 × 193 mm
Power supply: 115 V AC, 40 Hz, 25 W
or 27 V DC, 20 W

Operational status

In service on the Boeing 737 and 767, Airbus A310, Antonov An-28PT, ATR 72, Ilyushin Il-62M and -76 and Tupolev Tu-134/154M/204 and Swift S-1 glider. ATM has won a US$250,000 contract to fit SSQARs to all LOT Polish Airlines Boeing 737 and 767 and ATR 72 aircraft. Also fitted to military aircraft: I-22; PZL-130TC; Su-22M4; TS-11; W-3 helicopter.

Contractor

ATM Inc.

UPDATED

RUSSIAN FEDERATION AND ASSOCIATED STATES (CIS)

KARAT integrated monitoring and flight data recording system

The KARAT system comprises KARAT-B the airborne monitoring and recording system and KARAT-N a portable protected computer ground data processing and analysis system.

KARAT-B comprises two line replaceable units FDAU Flight Data Acquisition and processing Unit and MPFDR MultiPurpose Flight Data Recorder.

KARAT-B performs the following functions in the FDAU, acquisition, recording and processing of flight information from sensors and onboard systems; in-flight monitoring of onboard equipment and display on aircraft displays of relevant information; determination of maintenance requirements in the MPFDR, crash-protected storage and protection of recorded data; readout to KARAT-N of recorded data via high-speed datalink using RS-232 or RS-422 protocols.

KARAT-B employs an open architecture design that makes it suitable for all types of military and civil aircraft and helicopters. Data is protected in accordance with TSO C 124 requirements.

KARAT-N provides readout, processing and analysis of KARAT-B data; preparation of KARAT-B software and data files.

Specifications

MPFDR

Dimensions: 120 × 143 × 320 mm
Weight: 10 kg
Power: 27 V DC
Memory: up to 64 Mb
Inputs: ARINC 717 (1 channel); ARINC 429 (1 channel)
Net interface: IOLA
Crash protection (100% at):
(impact shock) 3,400 g for 6.5 ms
(fire) 1,100°C for 30 min
(deep sea pressure) 20,000 ft for 1 day; 10 ft for 30 days
(pierce) 500 lb from 10 ft with 6.35 mm steel penetration pin
(static crush) 23 kN

Operational status

KARAT is installed in the upgraded MiG-21 being supplied to the Indian Air Force. KARAT replaces the BASK and Tester U3L systems of the original MiG-21.

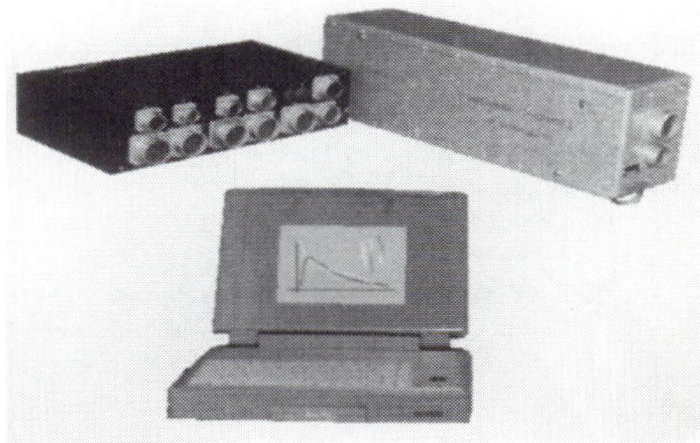

KARAT-B MPFDR (top right), KARAT-B FDAU (top left), and KARAT-N (bottom) ***1998***/0015315

Contractors

GosNIIAS State Research Institute of Aviation Systems.
JSC Pribor Design Bureau Aviaavtomatika.

NEW ENTRY

SOUTH AFRICA

Data Transfer System (DTS)

The Data Transfer System (DTS) is a modern ruggedised solid-state non-volatile device that provides a centralised storage and archiving medium for avionics systems. It consists of the Data Transfer Unit (DTU), which is fixed in the cockpit, and a Portable Data Store (PDS). The PDS is a removable cartridge which provides for the transfer of data between a mission planning centre and the avionics host computer and records flight and maintenance data for flight evaluation and analysis.

The PDS electrical interface to the DTU takes place via the dual spring-loaded ejection mechanism, conveying both data and power. The communication between the DTU and PDS is serial, with a data transfer rate of 150 kBaud. The PDS is inserted through a slot into the DTU and is not orientation sensitive.

Specifications

Dimensions: 165 × 126 × 52 mm
Weight: <1.5 kg
Interface: RS-422, RS-485, RS-232
PDS memory: up to 4 Mbytes non-volatile RAM
Reliability: 5,000 h MTBF

Operational status

In production.

Contractor

Advanced Technologies & Engineering Co (ATE).

VERIFIED

Health and Usage Monitoring Systems (HUMS)

AMS Health and Usage Monitoring Systems (HUMS) is a modular system comprising data acquisition processing unit; control and display; Flight Data Recorder (FDR) and Cockpit Voice Recorder (CVR); data transfer device; and ground replay station.

The system covers the full functional spectrum, including flight data recording, cockpit voice recording, engine health and usage monitoring, performance trending, structural fatigue monitoring, limits and exceedance monitoring, vibration monitoring and data storage and transfer.

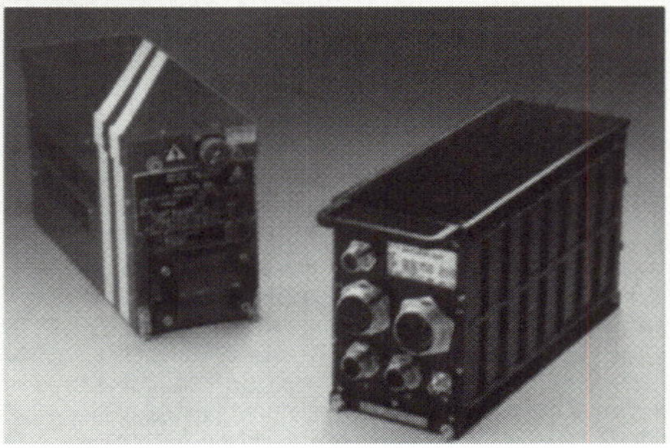

BAe Hawk Mk 100 HUMS ***1998***/0015316

The data acquisition unit is based on the AM3000 harsh environment avionics computer. This features IEEE 1296 Multibus II, MIL-STD-1398D SEM-E modules, Intel 80960 MC 25 MHz CPU, MIL-STD-704D power supply, multiples of 16 analogue and 32 discrete inputs, extensive BIT capability, ½ ATR short ARINC mounting or customised packaging and convection cooling.

The digital flight data recorder has a solid-state memory array as the recording medium, crash protection in accordance with EUROCAE ED-55/56 and is a ½ ATR short ARINC mounting unit.

The data extraction computer is a PC-based hardware platform that can be interfaced with the Global Logistics Information System. It operates in a Microsoft Windows-based environment and can function as a data transfer device as well as a powerful analysis station.

Operational status

Variants of the system have been engineered for the Rooivalk attack helicopter HUMS, and the SMR 95 engine HUMS on Cheetah fighter aircraft.

AMS is also developing a version for the BAe Hawk Mk 100 series jet trainer aircraft that will feature airframe fatigue monitoring; FDR/CVR in accordance with ED55/ED56A (BASE SCR500-660); avionics equipment maintenance recording; flight line maintenance support station; fleet airframe usage management system.

Contractor

Analysis Management & Systems (Pty) Ltd.

UPDATED

GAE1200 digital audio recorder

The GAE1200 is a unit for recording and replaying audio signals, based on digital signal processing techniques. It uses a hard disk as its storage medium. This improves access time over conventional tape recorders and allows the user to move from one position in a recording to another almost instantaneously. It is thus ideally suited to communications, EW and radio monitoring applications. The GAE1200 operates either from its front panel or via a remote control.

The GAE1200 may be used with a range of SCSI hard disk sizes, yielding differing recording times. A 180 Mbyte hard drive gives a recording time of some 7 hours.

The GAE1200 was designed to take advantage of the ability to access information almost instantaneously and in virtually any order. It is also a robust replacement for a cassette recorder.

Specifications

Dimensions: 132 × 360 × 122 mm
Weight: <6 kg
Power supply: 18-36 V DC, 36 W
110/240 V AC, 50-40 Hz, 50 VA
Temperature range: 0 to +55°C
Interfaces: RS-232, parallel, transputer link

Contractor

Grintek System Technologies GST,
Grintek Electronics Limited.

VERIFIED

The Grintek System Technologies GAE1200 digital audio recorder
1995

UNITED KINGDOM

AE6100HW casette data recorder

Avalon Electronics has launched a new series of high-capacity data recorders based on Digital Linear Tape (DLT) cassette technology.

Developed originally as a low-cost mass storage peripheral for PC and workstations, the multitrack DLT linear format offers very high-rate data capture.

The AE6100HW range records between one and eight separate serial bit streams in NRZ TTL, ECL or MIL-STD-1553B format at 40 Mbits/s with error rates better than 1 in 10^{14}. It can store 280 Gigabits of data on a single DLT tape.

Data can be output in the same format as it was recorded or via an optional SCSI-2 replay interface for immediate computer analysis.

The complete system including record/replay electronics, power supplies and environmental control is enclosed in a 19 in rack unit measuring 176 × 220 × 500 mm. MTBF is specified >12,500 hours. The system architecture allows for a number of recorders to operate in parallel without the need for sophisticated resynchronisation techniques, for example five 40 Mbits/s units can operate logically as a single 200 Mbits/s data acquisition system.

For the future, a range of multichannel analogue/digital front-ends are planned allowing users of open-reel IRIG recorders an easy migration path to the new DLT technology.

Contractor

Avalon Electronics Ltd.

NEW ENTRY

Avalon's AE6100HW cassette data recorder
1998/0015317

FlightVU aircraft video flight recorder

British Aerospace Systems & Equipment and DM Aerospace, a division of Dedicated Microcomputers Group Limited, are to develop an aircraft video flight recorder. The FlightVU recorder system will provide a crash-protected environment for the digital recording of video data from the aircraft's flight deck or from external cameras.

The Memorandum of Understanding signed by the two companies brings together the latest solid-state, crash-protected memory technology from BASE and the compression techniques and miniature ruggedised camera technology from DM Aerospace.

Full colour video images will be digitised, compressed and then recorded into a solid-state memory module, fully crash protected to the equivalent of ED56A plus 1 hour fire test. The standard memory module will contain up to 30 minutes of video at three frames/s and will be capable of upgrading to a higher capacity if required. Data can be made available immediately for in-flight monitoring, or downloaded for post-flight analysis.

Operational status
In development.

Contractor
British Aerospace Systems & Equipment.

VERIFIED

SCR 200 flight data/voice recorder

The SCR 200 consists of a crash-protected accident data recorder and a data acquisition unit. It has been designed for the Tornado. The accident data recorder comprises a continuous single-spool tape transport with four recording tracks. Recording rate is 128 12-bit data words/s. Total recording time is 111 minutes data and 37 minutes voice. Audio information is recorded on one track simultaneously with data on the other three.

Specifications
Dimensions:
(ADR) 425 × 172 × 81 mm
(DAU) 383 × 128 × 200 mm
Weight:
(ADR) 7.5 kg
(DAU) 7 kg
Power supply: 115 V AC, 400 Hz, 25 W
28 V DC

Operational status
In production. Over 750 systems are in service on Royal Air Force, Royal Saudi Air Force and Italian Air Force Tornado aircraft.

Contractor
British Aerospace Systems & Equipment.

VERIFIED

SCR 300 data/voice recorder

The SCR 300 flight data recording system is designed for use on military aircraft, where space is at a premium. It is believed to be the smallest and lightest tape combined CVR/ADR system currently available.

The main function of the SCR 300 is the preservation of a flight audio and data record in the event of an accident, but it can also be used for maintenance monitoring. The system consists of a crash-protected Accident Data Recorder (ADR) and a Data Acquisition Unit (DAU).

The ADR is mechanically and thermally protected to recent FAA and CAA standards and has provided 100 per cent data and voice retrieval. The system comprises a continuous single-spool tape transport with six 30 minute tracks of information, with tracks assigned to audio and data as required. The recording rate is 128 12-bit data words/s, giving a total recording time of 3 hours. The ADR is fitted with a sonar locating beacon.

The DAU collects information from signal sources in a variety of digital, analogue and discrete forms, samples each at an appropriate rate and converts them into a stream of serial digital data which it then transmits to the ADR. The number of parameters monitored depends on the application. A typical system has more than 40 digital, analogue, audio and discrete signal inputs. Uniquely, the DAU stores up to 64 different aircraft frame formats within the processor. This feature greatly eases logistics in mixed aircraft fleets, where the single standard DAU is totally interchangeable between numerous aircraft types.

Specifications
Dimensions:
(ADR) 115 × 150 × 250 mm
(DAU) 89 × 125 × 200 mm
Weight:
(ADR) 6 kg
(DAU) 4 kg
Power supply: 28 V DC, 40 W

Operational status
Available against special order and in service worldwide on over 200 Jaguar aircraft, Harrier, Sea Harrier, Hawk 60/100/200, Andover, BAC 111 and Raytheon Hawker 125 fixed-wing aircraft and Chinook, Lynx, Sea King, Gazelle, Scout and Wessex helicopters.

Contractor
British Aerospace Systems & Equipment.

UPDATED

SCR 500 solid-state cockpit voice and flight data recorders

The SCR 500 combines the latest digital recording and flash memory techniques, to provide a lightweight cost-effective range of Cockpit Voice Recorder/Digital Flight Data Recorders (CVR/DFDRs). The SCR 500 range complies with the latest requirements of EUROCAE -55/56A, the latest TSO C123A/C124A for increased fire protection and ARINC 757/747, with recording duration to meet both current and future requirements.

The SCR 500-030 and SCR 500-120 CVRs respectively have 30 and 120 minutes' audio recording duration. The combined CVR/FDRs have a range recording duration from 30 minutes audio/10 hours data to 2 hours audio/25 hours data at 128 words/s or 50 hours at 64 words/s data rates. All SCR 500 recorders have instant real-time audio and data playback.

The SCR 500-1530 and SCR 500-1560 solid-state CVR/FDR combined recorders provide 30 minutes or 1 hour, four-channel cockpit voice and 25 hour digital data. They are designed to provide a long duration audio/digital recorder in a single box, for large helicopters.

The SCR 500-1620 solid-state CVR/FDR combined recorder provides a full 2 hours, four-channel cockpit voice and 25 hour digital data. It is designed to replace both CVRs and FDRs in large passenger aircraft giving the safety benefits of dual redundancy. Utilising an SCR 500 Combi in place of a CVR enables regional aircraft operators to implement cost-effective FOQA procedures in conjunction with the latest analysis software package.

All models can be supplied with control panels and an underwater beacon. The audio provision includes four analogue channels recorded digitally and one digital data channel plus timebase. Test equipment, designed around common PC architecture, provides Downloading/Testing and Replay of all the SCR 500 family of recorders. A hand-held data downloader is also available.

Specifications
Dimensions: ½ ATR short
Weight: <7 kg for all versions
Power supply: 115 V AC, 400 Hz
28 V DC, 12 W (nominal)

Operational status
Certified to CAA ED 55/56A, and FAA TSO C123/C124, and compliant with TSO C123A/C124A. In production and in service.

The recorders are being fitted into several major fixed-wing and helicopter programmes: prime fit for Sikorsky S92 and S76C+ (IHUMS) programmes; prime fit for AVRO RJ series; prime fit for RAAF Hawk LIF; prime fit for ATLAS Rooivalk attack helicopter; prime fit for UK MoD DHFS Bell 412s; selected for RAF BAe 146 and Raytheon Hawker 800 fleets.

Contractor
British Aerospace Systems & Equipment.

UPDATED

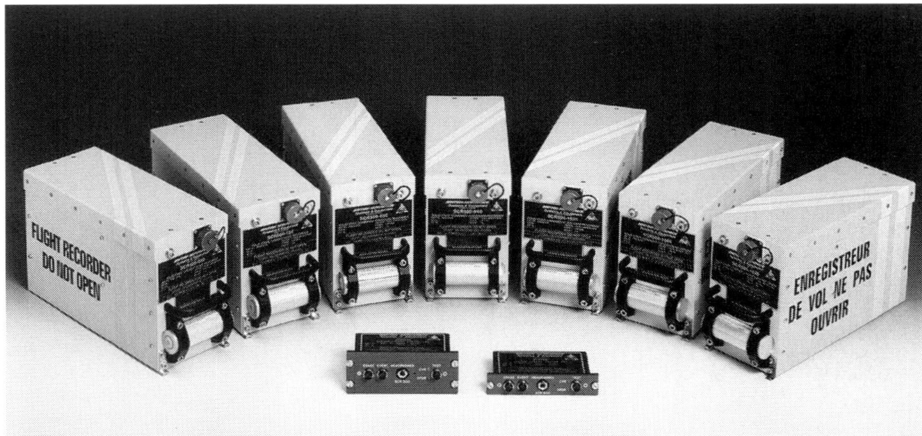

Seven variants of the SCR 500 solid-state cockpit voice and flight data recorders **1996**

ACCS 3200 fatigue monitoring and computing system

The fatigue monitoring and computing system provides real-time fatigue data gathering and analysis, via parallel processing of 16 strain gauge sensor data channels. Real-time analysis isolates structural fatigue to individual aircraft structures. A comprehensive ground analysis facility allows analysis of full sortie raw sensor data and a complex load history log by airframe, together with fleetwide data correlation.

Specifications
Dimensions:
(fatigue monitoring computer) ½ ATR rack or custom housing
(raw data recorder) 83.8 × 158 × 221.2 mm
Weight:
(fatigue monitoring computer) 9 kg
(programme and data transfer unit) <1 kg
(raw data recorder) 3 kg
Processing: 68020 20 MHz processor, 640 kbyte EEPROM, 128 k SRAM, software in Ada

Operational status
Deliveries for Royal Air Force Harrier GR. Mk 7 aircraft complete.

Contractor
Computing Devices Company Ltd.

VERIFIED

DCR 1055/1080 airborne digital cassette recorders

The DCR 1055 Super VHS airborne recorder and the DCR 1080 Hi-8 mm airborne recorder are both available with a digital encoder module to facilitate the recording of digital data in a variety of formats. Formats currently available are MIL-STD-1553 and RS-422 with ARINC 429, SAVA, Biphase L and customer-definable interface available shortly.

The single LRU has a maximum data rate of up to 2.2 Mbits/s with a bit error rate of better than 1 in 10^{10}. An internal card set is also soon to be available for both the DCR 1055 and the DCR 1080.

Specifications

Dimensions: 117 × 140 × 153 mm
Weight: <3 kg
Power supply: 28 V DC, <20 W

Operational status

Stand-alone encoder/decoder in production. The internal card set is in development.

Contractor

Computing Devices Company Ltd.

VERIFIED

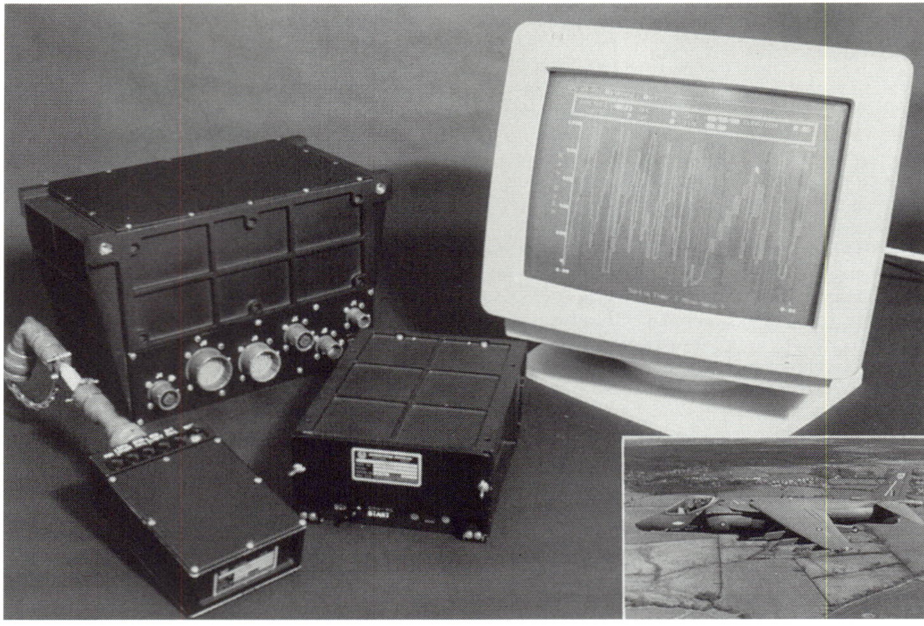

The Computing Devices fatigue monitoring system

DCR 2056 airborne digital cassette recorder

The DCR 2056 is a full Mil-Spec airborne digital rotary recorder/reproducer. It is able to record for almost 9 hours at a data rate of 2.5 Mbits/s using standard E300 VHS cassettes. The recorder incorporates Reed Solomon error correction systems and has an error rate of better than 1 in 10^8 with zero bit slip. Utilising standard E300 VHS cassettes, the DCR 2056 is able to store 8.75 Gbytes of user data.

The DCR 2056 has RS-422 or MIL-STD-1553 control ports with a 12-bit parallel data port. BITE aids rapid diagnosis to improve first line maintenance.

Although currently configured for 2.5 Mbits, the DCR 2056 can be configured for other specific data rates. For enhanced ground-based analysis, the DCR 2056 has a ground replay unit for greater manipulation of the data.

Specifications

Dimensions: 200 × 257 × 324 mm
Weight: 12 kg
Power supply: 115 V AC, 400 Hz,
150 W (max) (with heaters on), 80 W nominal
Environmental: MIL-E-5400
EMC: MIL-STD-461

Operational status

In development.

Contractor

Computing Devices Company Ltd.

VERIFIED

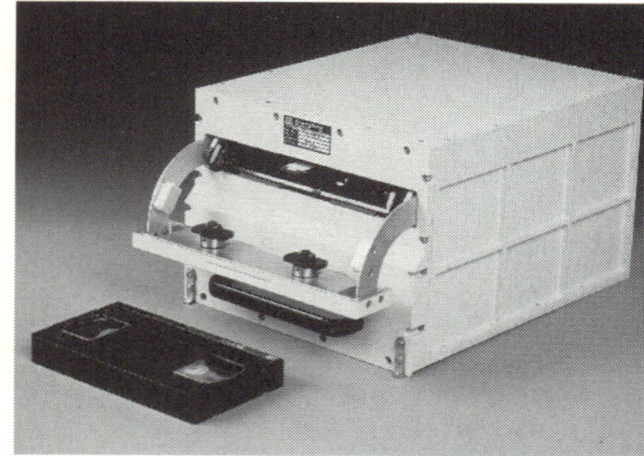

The Computing Devices DCR 2056 airborne digital cassette recorder

DCR 3000 series airborne digital cassette recorders

The DCR 3080 is the first unit in the DCR 3000 series of airborne digital cassette recorder/reproducers. It is an 8 Mbits/s recorder that uses standard Super VHS cassettes.

Using a Reed Solomon error correcting system, an error rate of better than 1 in 10^{10} TWX can be achieved. A variable rate buffer allows the DCR to record and reply to data rates from 0 to 8 Mbits/s via an RS-422 level data I/O. Control of the DCR can be achieved either via an RS-422 or RS-232 control port or the integral control and status panel mounted on the front of the unit.

Specifications

Dimensions: 200 × 257 × 324 mm
Weight: 12 kg
Power supply: 28 V DC,
150 W (heaters on)
80 W (nominal)
Temperature range: −20 to +55°C
Environmental: MIL-E-5400
Vibration: MIL-STD-810D

Operational status

In production. The DCR 3080 has been selected as the flight instrumentation recorder for Airbus Industrie A330 and A340 wide-body passenger aircraft.

Contractor

Computing Devices Company Ltd.

VERIFIED

DCR 9000 airborne digital cassette recorder

The DCR 9000 is an airborne digital data recorder specifically designed for use in the full spectrum of fixed-wing and rotary platforms, including fast jets. With a current rate of up to 16 Mbits/s, soon to be expanded to 32 Mbits/s, the DCR 9000 is compatible with the Metrum VLDS ground replay recorder which provides replay data rates of up to 32 Mbits/s.

Bit error rates are better than 1 in 10^{10} and commercial standard Super VHS cassettes are utilised so as to provide approximately 10 Gbytes of data storage on an E180/T120 standard cartridge. Operating times are dependent on the selected data rate, but at 8 Mbits/s a record duration of approximately 90 minutes can be achieved.

Features of the DCR 9000 include full remote control via RS-232 or MIL-STD-1553B bus. Data interfacing is either via SCSI or native interface. Customer specific interfaces are also available.

Specifications

Dimensions: 168 × 258 × 324 mm
Weight: 12 kg
Power supply: 28 V DC, 90 W

Operational status

In development.

Contractor

Computing Devices Company Ltd.

VERIFIED

Fatigue Monitoring and Computing System (FMCS)

The Fatigue Monitoring and Computing System (FMCS) is designed to provide real-time onboard analysis of fatigue as it occurs in the aircraft structure, for the lifetime of the aircraft. The system provides the benefits of increased safety, reduced maintenance and support costs, improved availability, increased operational life and timely embodiment of structural modifications.

The FMCS onboard computer unit consists of a data processor, memory and optimally located structural sensors to capture fatigue and aircraft performance data. This data is processed in real time to produce cumulative damage profiles for each critical structural component. In addition, as a system option, the raw sensor data may be recorded to provide further detailed analysis.

The fatigue and aircraft performance data can be fully analysed on the supporting ground station to give accurate assessment of the severity and the time that damage occurred. The ground station also has database support for individual aircraft debriefing and across-fleet correlation.

Operational status

First development equipments delivered. Original equipment and its derivatives are now finding application for both fixed- and rotary-wing operational load monitoring and health and usage monitoring programmes for aircraft.

Contractor

Computing Devices Company Ltd.

VERIFIED

Programmable aircraft warnings panel

The programmable legend warnings panel offers a high-intensity LED display which is NVG-compatible. The display has programmable legends which are controllable over a dual-redundant MIL-STD-1553B avionics databus. Internal redundancy is incorporated to ensure continued safe operation in the event of defects and dedicated discrete inputs are also incorporated to ensure correct warning displays should the databus sources fail.

Aircraft status, health and fault data received over the databus is categorised by the internal Ada software for display on a simple menu; amber and red characters are available, with a dedicated display area for catastrophic warnings. An interface is provided to the audio management unit to give the aircrew co-ordinated visual and audible or digitised voice warnings of any new conditions.

Specifications

Dimensions:
(display unit) 118 × 156 × 70 mm
(electronics unit) 100 × 220 × 237 mm
Weight:
(display unit) 1.7 kg
(electronics unit) 2.95 kg
Power supply: 28 V DC, 270 W (max)

Operational status

Deliveries for EF 2000 integration work began in 1992.

Contractor

Computing Devices Company Ltd.

VERIFIED

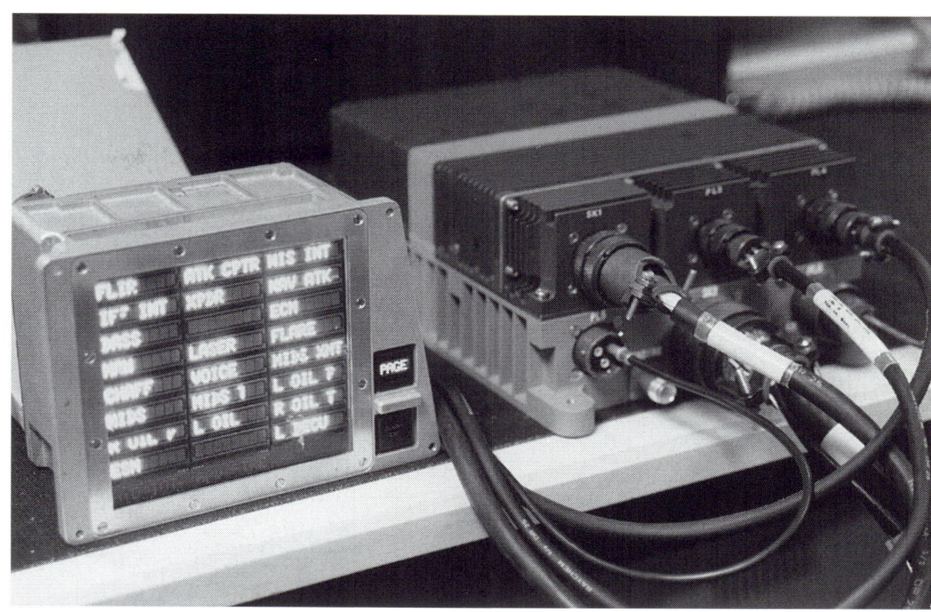

The programmable aircraft warnings panel has been selected for the EF 2000

VCR 1053 video cassette recorder

The VCR 1053 is a conventional VHS transport ruggedised by Computing Devices. It offers record and reproduce compatibility with existing VHS PAL and NTSC systems.

A record time of up to 5 hours can be achieved using conventional E300 commercial VHS cassettes. A full Mil-Spec power supply is fitted to the 1053 to protect it from the spikes and surges which are commonplace on modern airborne systems. The recorder is fully remotely controllable via RS-232, RS-422 and MIL-STD-1553 databusses and is internally, anti-vibration mounted.

Specifications

Dimensions: 145 × 245 × 330 mm
Weight: 4.5 kg
Power supply: 28 V DC or 115 V AC, 40-450 Hz option, 15 W nominal to MIL-STD-704
Temperature range: 0 to +45°C
Environmental: MIL-E-5400

Operational status

In production. The recorder is currently in service on a variety of UK and French helicopters.

Contractor

Computing Devices Company Ltd.

VERIFIED

VCR 1055 video cassette recorder

The VCR 1055 is an S-VHS transport ruggedised by Computing Devices. It offers recording and reproduction of high-resolution imagery to either PAL or NTSC standards. Fully cross-compatible with existing S-VHS commercial systems, the 1055 is optimised for recording high-definition image data.

The recorder is able to record up to 5 hours of video using either VHS or S-VHS tapes, has internal anti-vibration mounts and is remotely controllable via RS-232, RS-422 or MIL-STD-1553 databusses. The 1055 also offers up to four auxiliary tracks for voice annotation and so on.

Specifications

Dimensions: 185 × 245 × 330 mm
Weight: 6 kg
Power supply: 28 V DC or 115 V AC, 40-450 Hz option, 15 W nominal to MIL-STD-704
Temperature range: 0 to +45°C
Environmental: MIL-E-5400

Operational status

In production.

Contractor

Computing Devices Company Ltd.

VERIFIED

VCR 1066 video cassette recorder

The VCR 1066 is a full Mil-Spec video cassette recorder specifically designed for high-speed military aircraft and, in particular, is fitted to Tornado GR. Mk 1A reconnaissance aircraft of the Royal Air Force and the Royal Saudi Air Force. It is used to record and replay high-resolution infrared sensor information in flight, as part of the Cockpit Display of InfraRed Reconnaissance System (CDIRRS).

A record time of 1 hour is achieved with a standard VHS video cassette, using a two-head helical scan system for recording. A modified VHS format is employed to give a 4 MHz video bandwidth, with a signal-to-noise ratio of better than 43 dB.

In the Tornado application, non-standard TV signals are recorded but the system can record 625-line 50 Hz and 525-line 60 Hz signals. Two linear record/replay tracks are also available.

Specifications

Dimensions: 168 × 257 × 324 mm
Weight: 12 kg
Power supply: 115 V AC, 400 Hz, single phase, 120 W
Environmental: MIL-E-5400
EMC: MIL-STD-461

Operational status

In production for Tornado GR. Mk 1A reconnaissance aircraft, with orders approaching 500 systems.

Contractor

Computing Devices Company Ltd.

VERIFIED

VCR 1076 airborne analogue cassette recorder

The VCR 1076 is a full military specification airborne recorder capable of recording analogue signals in a variety of format configurations. In its current configuration the VCR 1076 is capable of recording two simultaneous and separate channels of analogue information at up to 6 MHz per channel. The use of a revised formatter can reconfigure the VCR 1076 so as to record a single channel up to 12 MHz bandwidth.

Record time is up to 1 hour in either mode using standard 0.5 in media, in a VHS pattern cassette.

Specifications

Dimensions: 168 × 257 × 324 mm
Weight: 12 kg
Power supply: 115 V AC, 90 W
Environmental: MIL-E-5400
EMC: MIL-STD-461

Operational status

In development and selected for the Northrop Grumman A-6E common mission recorder under the US designation AN/USH-42.

Contractor

Computing Devices Company Ltd.

VERIFIED

VCR 1080/1036 video cassette recorders

The VCR 1080 is an airborne video cassette recorder/replay unit, specially configured for use on a wide variety of platforms including fast jets and helicopters. A format fully compliant with the Hi-8 mm tape footprint allows ease of use in ground replay and duplication machines. Both PAL and NTSC standards are supported, allowing for the widest possible use of these products throughout the world.

Principal features of the VCR 1080 are that it records luminance and chrominance channels separately so as to provide the highest possible resolution on replay or duplication. Local controls or bus controls allow for the greatest possible systems integration and flexibility. A feature of the VCR 1080 is the provision of additional card slots in the recorder housing, to allow incorporation of either systems controller cards or customer-specific interface or digital encoder cards.

The VCR 1036 is a high-specification version of the VCR 1080. It has been selected as the cockpit video, voice and data recorder for all EF 2000 variants. Specific differences from the VCR 1080 include full MIL-STD-1553B control interfaces, RGB-to-video converters, digital encoders to allow simultaneous data to be recorded with the video channel, and two separate audio channels for two-man crew operations.

Specifications

Dimensions: 186 × 200 × 218 mm
Weight: <4.5 kg
Power supply: 115 V AC, 400 Hz
28 V DC, 30 W
Environmental: MIL-STD-810B
EMC: MIL-STD-461

Operational status

In production. The VCR 1036 recorder has been selected for all variants of the EF 2000.

Contractor

Computing Devices Company Ltd.

VERIFIED

8 mm and Hi-8 mm sealed video recorders

Various configurations of Sealed Video Recorder (SVR) have been produced to suit different cockpit and bay mount installations. All have integral anti-vibration mounts configured for bulkhead or 5 in ARINC rack mounting, with fixed or flying connectors and local or remote controls and indicators.

Operational status

In production.

Contractor

GEC-Marconi Avionics Ltd, Mission Avionics Division, Edinburgh.

VERIFIED

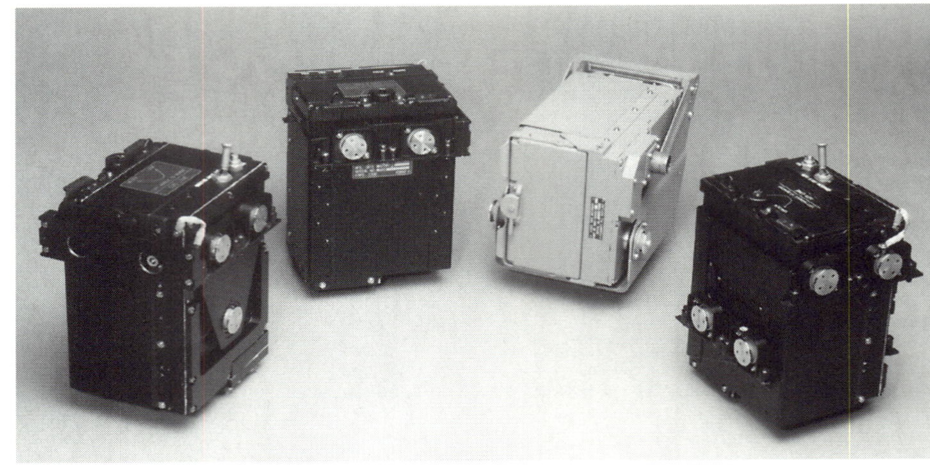

GEC-Marconi Avionics 8 mm and Hi-8 mm sealed video recorders **1997**/0001298

F-16 HUD video camera

This single unit high-resolution colour camera was developed to fit the F-16 C/D HUD. It replaces the existing two-unit monochrome camera.

The latest technology solid-state sensor provides high resolution, excellent sensitivity and accurate colour. Alignment accuracy is high for mounting the camera forward of the HUD. Either 525-line/60 Hz NTSC or 625-line/50 Hz PAL configurations are available, and the output can be Y/C or composite video.

Operational status

In production.

Contractor

GEC-Marconi Avionics Ltd, Mission Avionics Division, Edinburgh.

VERIFIED

Harrier GR. Mk 7 video recording system **1997**/0001303

Harrier GR. Mk 7 video recording system

The video recording system for the Night Attack Harrier GR. Mk 7 was developed as an upgrade to the GR. Mk 5 system, to incorporate recording of the FLIR video. Using two sealed video recorders, one for 'head-up' recording and one for 'head-down' recording, the system records colour HUD camera, FLIR and dual-mode tracker. Growth is built in to enable recording of all head-down displays at a later date.

Operational status

In production.

Contractor

GEC-Marconi Avionics Ltd, Mission Avionics Division, Edinburgh.

VERIFIED

GEC-Marconi Avionics F-16 HUD video camera (left) and camera montage **1997**/0001299/0001300

MOdular Data Acquisition System (MODAS)

The Modular Data Acquisition System (MODAS) has been designed to gather and record any type of signal to be found on an aircraft during flight trials. A large range of signal input types are supported by both the Mk I and Mk II systems. Inputs include strain gauges, voltages, synchros, thermocouples, tachos, ARINC 429, MIL-STD-1553B and RS-232.

MODAS Mk I

MODAS Mk I has up to 4,096 input channels and is expandable to meet specific requirements at sampling rates of up to 128 k samples/s. Eight separate sampling programs can be selected in flight. A pulse coded modulation digital technique is employed for recording data in either simultaneous multitrack, serial streams or IRIG-106 format. A comprehensive ground replay facility is also available and MODAS can interface with a telemetry system.

MODAS Mk II

The Mk II version of MODAS is less than half the size of the earlier model and is four times as fast. The PA 3101 processor and recorder interface unit can accept data from eight acquisition units, or up to 64 with an expansion unit. The PA3120 general purpose acquisition unit houses up to eight interfaces and two MIL-STD-1553B databusses can be monitored with the PA 3130 databus acquisition unit. Data acquisition is up to 512 k parameter samples/s.

Specifications

MODAS Mk I

Dimensions:

(acquisition/processing unit) ½ ATR
(control unit) 146 × 191 × 151 mm
(monitor unit) 146 × 191 × 166 mm
(small recorder) ¾ ATR short
(large recorder) 533 × 360 × 200 mm

Weight:

(acquisition/processing unit) 10 kg
(control unit) 3 kg
(monitor unit) 3 kg

(small recorder) 13 kg
(large recorder) 39 kg
Power supply: 115 V AC, 400 Hz, 3 phase or 28 V DC

MODAS Mk II
Dimensions:
(PA 3101, 3110 and 3130) 135 × 230 × 210 mm
(PA 3120 control and monitor unit) 66 × 146 × 132 mm
Weight:
(PA 3101, 3110, 3130) 5 kg
(PA 3120) 0.9 kg
Power supply: 28 V DC, 60 W (max)

Operational status
In service in a wide range of UK and European aircraft.

Contractor
GEC-Marconi Avionics Ltd, Flight Systems Division, Data Systems, Portsmouth.

VERIFIED

PA3520 Aircraft Integrated Monitoring System (AIMS)

The PA3520 Aircraft Integrated Monitoring System (AIMS) records accident data, engine life information and structural data. It comprises the GEC-Marconi Avionics Ltd PA3521A data acquisition and processing unit and the Penny & Giles D50330 Mk 2 accident data recorder.

Accident-related data can be downloaded directly into a portable recording unit for later analysis, thus avoiding the need to remove the data recorder itself. Engine and structural data can be downloaded into a unit with a display unit, allowing instant viewing of critical data, or the information can also be stored for subsequent analysis. The PA3521A can accept discrete, shaft rotation and analogue inputs, or can interface with an MIL-STD-1553B databus.

Specifications
Dimensions: ½ ATR short
Weight: 4.55 kg
Power supply: 28 V DC, 28 W

Operational status
In production.

Contractor
GEC-Marconi Avionics Ltd, Flight Systems Division, Data Systems, Portsmouth.

VERIFIED

PA3584 solid-state acquisition and recorder unit

The PA3584 solid-state acquisition and recorder unit is designed for applications where space and weight are at a premium. It offers solid-state flight data recording, cockpit video recording and data acquisition in a single ½ ATR long unit.

Suitable for helicopters and transport, business, regional and commuter aircraft, the combined unit simplifies installation and support requirements.

It features an integral data acquisition unit for more than 33 parameters, four 2 hour voice channels and integral solid-state 25 hour duration data recorder, direct recording with no data compression or decompression techniques needed, full compliancy with TSOC124, EUROCAE ED-55, ED-56A and CAA Specifications 10A and 18 and fast data retrieval using a 386-based PC and 64, 128 and 256 words/s operating options.

Specifications
Dimensions: ½ ATR long
Weight: 13 kg typical
Power supply: 115 V AC or 28 V DC, 25 W (max)

Contractor
GEC-Marconi Avionics Ltd, Flight Systems Division, Data Systems, Portsmouth.

VERIFIED

PA3700 Integrated Health and Usage Monitoring System (IHUMS)

GEC-Marconi Avionics Ltd has developed an Integrated flight data recording Health and Usage Monitoring System (IHUMS) for use in many aircraft applications.

The IHUMS has been designed to fulfil the functions of a Digital Flight Data Recorder (DFDR), a Cockpit Voice Recorder (CVR) and a Health and Usage Monitoring System (HUMS). The system architecture has been optimised to provide a minimum hardware solution by the combination of these functions. Each airborne system comprises a Data Acquisition and Processing Unit (DAPU), Cockpit Voice and Flight Data Recorder (CVFDR), Card Maintenance Data Recorder (CMDR) and Control and Monitor Unit (CMU). In addition to this core system, other options are available, including a control and display unit, pilot interface panel, cockpit warning panel and quick access recorder.

The PA3701 DAPU contains all the conditioning circuitry necessary to sample and accurately monitor a wide range of different types of electrical inputs for subsequent recording, measurement or processing. The mandatory data output interfaces to a standard ARINC 573/747 flight data recorder and a standard ARINC quick access recorder. Selected mandatory data, together with raw and partially processed HUM data, is also fed to a ruggedised maintenance data recorder. Data can be displayed on the optional control and display unit as required.

There are several versions of the crash-protected recorder available to customers with differing requirements. For instance, one CVFDR provides 5 hours of continuous digital recording at a data rate of 128 12-bit words/s and 1 hour voice on each of three separate tracks. Another FDR provides 8 to 10 hours of continuous digital data recording at a data rate of 128 12-bit words/s and is used in conjunction with a separate CVR.

The CMDR was designed to meet ARINC 615 as a high-speed data loader. The recording medium is a 2 Mbyte SRAM card. The data interface is bidirectional using RS-232C protocol. The unit is rugged, compact and the recording medium easily transportable, which makes it ideal for use as a card maintenance data recorder. The bidirectional interface to the DAPU enables upload of documentary data from a card as well as download of raw and preprocessed data for maintenance purposes.

The downloaded airborne DAPU data is supported by a comprehensive ground replay and analysis computer system.

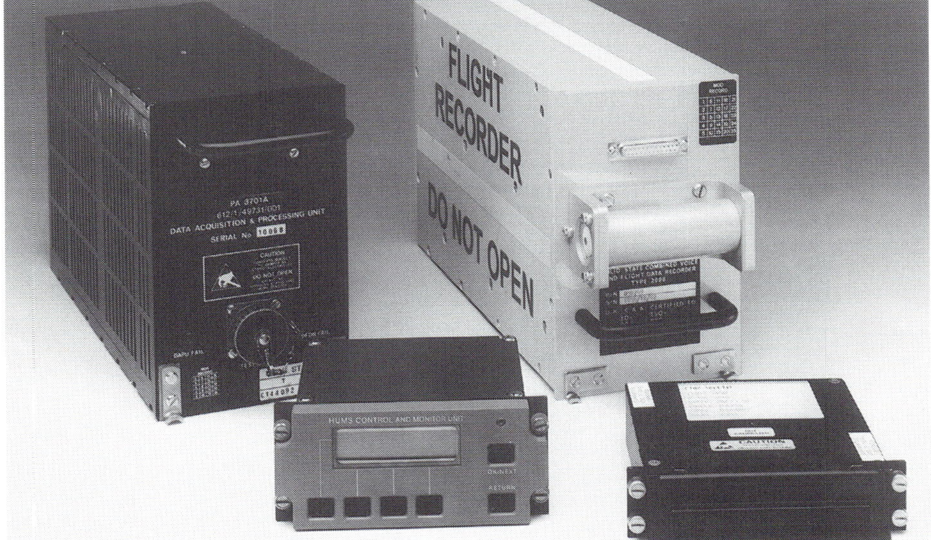

Units of the IHUMS II equipment showing (left to right) the DAPU, CMU, CVFDR and CMDR **1995**

Specifications
Dimensions:
(PA3701 DAPU) ½ ATR short
(CVFDR) 115.6 × 173.5 × 440.5 mm
(CMDR) 38.1 × 133.35 × 132.05 mm
Weight:
(PA3701) 6.2 kg
(CVFDR) 8.5 kg
(CMDR) 0.5 kg
Power supply: 28 V DC, 35 W (max)

Operational status
The system is fitted to over 100 helicopters operating over the North Sea. Customers include Bristow and Bond Helicopters, Morefly and Braathens of Norway and Shell Brunei Petroleum. The system has undergone trials on a Royal Navy Sea King.

The Mk II IHUMS has been selected by Sikorsky for the S-76C and S-92 production helicopters. The system architecture has been optimised to provide a minimum hardware solution and comprises a DAPU, CVFDR, CMDR and CMU.

Contractors
GEC-Marconi Avionics Ltd, Flight Systems Division, Data Systems, Portsmouth.

VERIFIED

PA3800 series flight data acquisition units

The PA3800 series, of flight data acquisition units, was developed by GEC-Marconi Avionics Ltd to meet the new FAA and CAA requirements for flight data recording applicable from 1991. The unit samples data from a variety of input signals which may be analogue, digital or discrete. The information is sampled in a programmable predetermined sequence and assembled into a digital data stream in a format compatible with any standard ARINC 573/717/747 digital flight data recorder. There are four or five PCB positions available for expansion of the system, perhaps taking the form of extra signal conditioners, or the unit may be expanded into an integrated microprocessor-based monitoring system to provide engine or airframe usage monitoring.

The PA3810 is for accident data acquisition, with four-card expansion available. The PA3820 is for helicopter health and usage monitoring, including accident data. The PA3830 is for fixed-wing aircraft for the Aircraft Integrated Monitoring System (AIMS), including accident data.

Specifications
Dimensions:
(PA3810) ⅜ ATR short case to ARINC 404A
Weight: 4 kg typical
Power supply: 115 V AC or 28 V DC, 15 W (max)

Operational status
In full production for both military and civil applications on fixed-wing aircraft such as the de Havilland Dash 8, Cessna Citation and Canadair Challenger and on Bell and Eurocopter helicopters.

Contractor
GEC-Marconi Avionics Ltd, Flight Systems Division, Data Systems, Portsmouth.

VERIFIED

PRS2020 engine monitoring system

Developed from the PVS1820 system, the PRS2020 is lighter, smaller, uses less power and is compatible with MIL-STD-1553 databus interfaces. It also includes a 32-bit high-speed microprocessor, plus hybrid and uncommitted logic array-based electronics.

The PRS2020 consists of several component units including:

PRS2021 engine monitoring unit: a single box data acquisition and processing unit for analogue signal conditioning, discrete digital conditioning and/or digital databus parameter recovery. The heart of this unit is a single, very large-scale integrated hybrid circuit comprising a 32-bit 19-register central processor with 4 kbytes of random access memory and 8 kbytes of read-only memory reserved for applications programs.

PRS2023 display: available for taking a quick look at data, the display can be mounted on the engine monitoring unit or remotely. The format of the display can be defined by the customer.

PRS2026 data retrieval unit: a rugged battery-powered unit for extracting data from the engine monitoring unit for subsequent analysis and for system test and calibration.

GEC-Marconi Avionics Ltd also offers the options of a bulk storage recorder, a portable ground replay unit and comprehensive systems software packages.

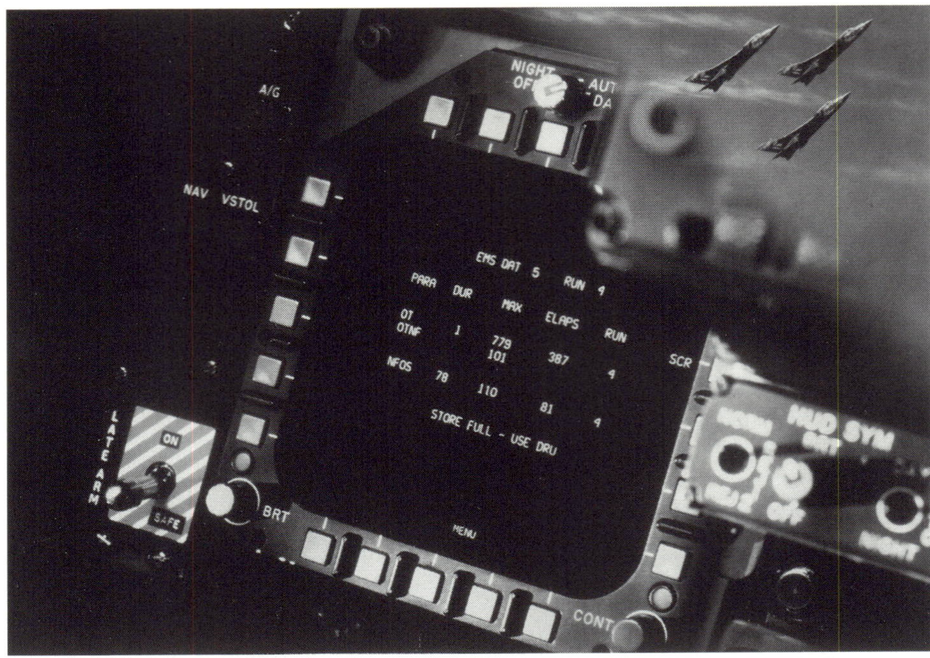

The cockpit display for the PRS2020 engine monitoring system

Specifications

Dimensions:
(PRS2021) 200 × 194 × 124 mm
(PRS2026) 280 × 260 × 80 mm

Operational status

Under development.

Contractor

GEC-Marconi Avionics Ltd, Flight Systems Division, Data Systems, Portsmouth.

VERIFIED

PRS3500A data/voice accident recorder

The PRS3500A consists of the GEC-Marconi Avionics Ltd PRS3501A data acquisition unit and the Penny & Giles D50330 accident data recorder.

The system provides 2 hours' continuous recording time, taking in up to 50 parameters at 240 words/s. The small size of the equipment makes it suitable for aircraft with limited cockpit space and digital transmission reduces the size and weight of wiring looms while increasing data integrity. Recorded data can be extracted in 6 minutes using a portable transfer device. The system comprises two units: the flight data recorder and a data acquisition unit.

Specifications

Dimensions:
(data acquisition unit) ½ ATR short
(flight data recorder) 115 × 172 × 440 mm
Weight:
(data acquisition unit) 4.5 kg
(flight data recorder) 8.3 kg

Operational status

In production and in service in Royal Air Force Harrier GR7 aircraft.

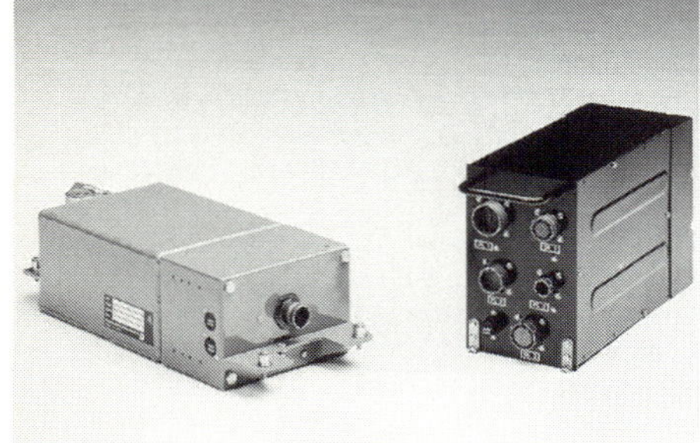

GEC-Marconi PRS 3501A data acquisition unit, and Penny & Giles D50330 accident data recorder
1996

Contractor

GEC-Marconi Avionics Ltd, Flight Systems Division, Data Systems, Portsmouth.

VERIFIED

PV1584 data recorder

A compact unit combining recorder and acquisition electronics within a single ½ ATR case, the PV1584 is designed to replace ARINC 542 electromechanical recorders. GEC-Marconi Avionics Ltd claims that the system is unique in meeting the full requirements of the FAA and CAA in a single unit of this size. The system is approved by all major certification authorities, including those of Australia, Canada, Germany, UK and the USA. Its signal handling capability meets all current requirements, including those of the CAA's Specification 10 and the FAA's Part 121 section 343, with 10 per cent excess capacity.

The system can be used in conjunction with a copy recorder to extract data for analysis *in situ* without disturbing the equipment. A PV1591 flight data entry panel can be used for manual insertion of documentary information into the recording.

The PV1584 comprises a protected flight recorder at the front of the unit and a data acquisition electronics section at the rear. Design baseline is ARINC 573, input parameters being sampled at the rate of 64/s, each sample being converted into a 12-bit binary word. The resultant digital data stream is routed to the recorder section and placed on a Mylar tape loop giving 25 hours' continuous recording. The sampling rate for given parameters can be varied as required. The entire capacity of the tape can be dumped in 16 minutes by means of a Lockheed 235 data dump recorder. The unit contains an EEpRoM, giving eight programs selected by changing wire links in the aircraft. The device can, therefore, be reprogrammed to meet requirements so that interchangeability could be maintained between eight aircraft types in an operator's fleet, for example. Signal types comprise 26 V synchro, DC ratio and absolute, AC ratio, potentiometer, variable resistance and frequency (tacho or pulse probe). Discretes comprise DC series and shunt and AC series and shunt.

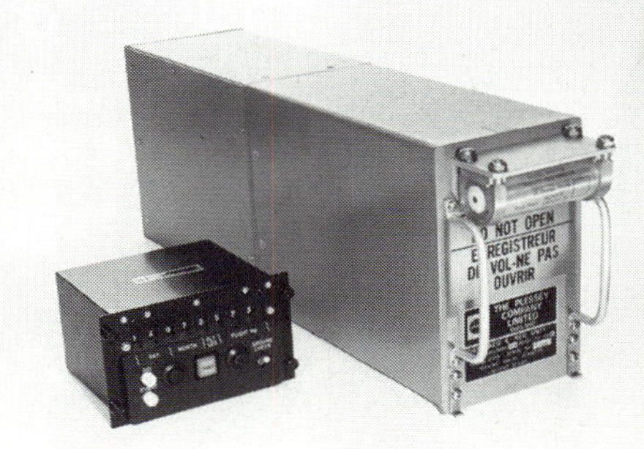

The PV1584 data recorder ***1996***

Specifications

Dimensions: ½ ATR long
Weight: 11.34 kg
Power supply: 115 V AC, 400 Hz, 25 W

Operational status

No longer in production. During 1985, the PV1584 was selected to equip the Royal Air Force's Nimrod aircraft and the British Aerospace ATP. It is standard equipment on the BAe 146 and 748, the Raytheon Hawker 125 and on the Shorts 360.

Contractor

GEC-Marconi Avionics Ltd, Flight Systems Division, Data Systems, Portsmouth.

VERIFIED

PV1591 Flight Data Entry Panel (FDEP)

The PV1591 Flight Data Entry Panel (FDEP) provides a display of the flight data recorder system status and a means of manually inserting documentary data into the recording.

Aircraft trip and date data is inserted into the system by use of eight thumbwheel switches on the front panel of the unit. The FDEP signal output represents the number indicated on one of these switches. Each of the eight switches is sampled in turn, for one data frame of the associated Flight Data Acquisition Unit/Data Acquisition and Recording Unit (FDAU/DARU). At the end of this cycle an invalid frame is output for eight frames to provide a synchronisation signal. The 16-frame sequence repeats continuously.

There are two push-buttons provided, one marked EVENT and the other marked DDI for documentary data insert, to insert markers into the recording to assist post-flight analysis of the data.

Two alternative status lights are available. One is a single indicator for use with a single box DARU. The other is a split indicator, one half marked FDAU and the other Digital Flight Data Recorder (DFDR), for use with a two-box system. Several alternative logic combinations are available, dependent on the type of FDR and the aircraft installation.

Specifications

Dimensions: 85.5 × 146 × 117 mm
Weight: 0.9 kg

Operational status

FDEPs are available for the whole range of FDAU, DARU and DFDR units.

Contractor

GEC-Marconi Avionics Ltd, Flight Systems Division, Data Systems, Portsmouth.

VERIFIED

PV1820C structural usage monitoring system

Designed to monitor airframe fatigue by continuously measuring structural loads during flight, the PV1820C is based on the company's earlier PVS1820 engine usage life monitoring system. Data and the computed results are recorded on a cassette-loaded quick access recorder for ground replay and analysis. The system comprises a PV1820C structural monitoring unit, a PV1819C control unit, quick access recorder Type 1207-003 and a transient suppression unit Type PV1845. Microprocessor operation simplifies the measurement of stress by enabling the output from a number of strain gauges (typically 16) to be monitored along with other flight parameters on a cassette recorder or solid-state data card.

Specifications

Dimensions:
(PV1820C) 94 × 194 × 394 mm
(PV1819C) 146 × 76 × 194 mm
(1207-003) 146 × 51.6 × 169 mm
(PV1845) 95 × 53 × 190 mm
Weight:
(PV1820C) 5.44 kg
(PV1819C) 1.59 kg
(1207-003) 1.77 kg
(PV1845) 1.4 kg
Sampling rate: Each parameter is defined by 10 bits of a 12-bit word, the 11th bit being a compression flag and the 12th being reserved for parity. The sampling rate for each quantity is programmable in binary steps from 1 to 128 times/s
Data compression: programmable 16:1, 8:1 or 4:1
Self-test: comprehensive automatic self-test and fault diagnosis is built in

Operational status

In production and in service.

Contractor

GEC-Marconi Avionics Ltd, Flight Systems Division, Data Systems, Portsmouth

VERIFIED

PV1954 flight data acquisition unit

The PV1954 fulfils all of the requirements for a 32 parameter Flight Data Recorder (FDR) and provides expansion capability for additional maintenance monitoring.

Conditioning circuits enable the PV1954 to sample data from a wide variety of input signals. These inputs may be analogue, digital or discrete. The information is sampled in a predetermined sequence and assembled into a digital data stream in a format compatible with any standard ARINC 573/717/747 Digital FDR (DFDR).

A programmable read-only memory controls the input signal sampling sequence. This method of control permits the user to define the content of all the words in the frame, except synchronisation words.

The PV1954 provides an output of 64 12-bit data words/s in Harvard bi-phase format to the DFDR. As an option, this data rate may be increased to 128 or 256 words/s.

An auxillary data output in RZ format is provided for use with an optional quick access recorder at a rate of 64 words/s, with an option to increase the data rate to 128 or 256 words/s.

A time synchronisation output in frequency shift key format is provided to synchronise the DFDR to the cockpit voice recorder.

Specifications

Power input: 28 V DC nominal, 15 W (max)
Dimensions: ½ ATR short case to ARINC 404A
Weight: 5 kg typical

Operational status

In service on Gulfstream IV, British Aerospace 146 Series 2 and ATP aircraft.

Contractor

GEC-Marconi Avionics Ltd, Flight Systems Division, Data Systems, Portsmouth.

VERIFIED

V3500 miniature camera

The V3500 is a high-resolution solid-state sensor camera for monochrome, low-light and colour recording. It is fully automatic and has a range of auto-iris lenses available which make it well suited for head-up display recording applications as well as for helicopter sights and air vehicle stabilised platforms. The V3500 can be supplied in either a single- or two-unit configuration; the camera head can be separated from the electronics unit if required.

Specifications

Dimensions:
(camera head) 90 × 28 × 35 mm
Weight:
(camera head) 0.15 kg

Contractor

GEC-Marconi Electro Optics Ltd, Sensors Division, Basildon.

VERIFIED

High-capacity digital data recorder

The Normalair-Garrett high-capacity digital data recorder utilises the Ampex DCRSi technology, and is ruggedised for operation in severe environment applications such as helicopters.

The Normalair-Garrett recorder, including interfaces to a typical ASW mission system, is contained in a 16 MCU volume envelope and has a total weight of 30 kg. The recorder is ideally suited to mission recording, as data can be recorded at any rate up to 107 Mbits/s and, because of the incremental tape motion, uses the minimum tape consistent with recording the output of the mission system. One cassette holds 385 Gbits of data, which gives over 8 hours' recording of a typical helicopter ASW mission.

High-capacity digital data recorder showing the Tape Transport Unit (TTU) (right) and the Data Acquisition Unit (DAU) (left) ***1997***/0001304

The associated replay system can replay the data at up to 240 Mbits/s, and will reconstitute with high integrity an exact replica of the input signals.

Specifications

Max input data rate: 107 Mbits/s
Record time per cassette: 1 h at 107 Mbits/s; 8 h at 13 Mbits/s
Tape speed: incremental at 5.31 ips
Bit error rate (corrected): better than 1 in 10^7
Total storage per cassette: 380 Gbits
Dimensions:
(Tape Transport Unit (TTU)) 11 MCU
(Data Acquisition Unit (DAU) 5 MCU
Weight: 30 kg
Power: 115 V AC, 400 Hz, less than 350 VA

Operational status

Fully qualified and in production for the Royal Navy EH 101 Merlin helicopter.

Contractor

Normalair-Garrett Ltd.

UPDATED

Central Maintenance Panel (CMP)

Incorporating a dual-microprocessor system, the Central Maintenance Panel (CMP) is designed to record aircraft fault data obtained from LRUs whilst the aircraft is either in the air or on the ground. It comprises a standard ATR enclosure constructed from light aluminium alloy.

The CMP monitors up to 192 bipolar discrete outputs from equipment or systems through remote sensors. Recorded fault data can be interrogated by ground crew for diagnostic purposes via either the operator interface or an RS-422A interface. The operator interface comprises a 32-character LCD which displays the CMP status and fault data. Interrogation push-buttons are mounted on the front face of the panel. The RS-422A interface is a serial data transmission interface enabling the operator to download fault data stored in the CMP to customer automatic test equipment. This enables data to be output to a printer to provide hard copy of the faults that have occurred in the LRU.

The CMP has six modes of operation: normal; self-test; standby; fault indication; data transmission and reprogramme. The fault indication and data transmission modes can be manually selected; all the other modes are automatically selected according to the access door position.

Specifications

Dimensions: ½ ATR
Weight: 2.91 kg
Power supply: 28 V DC, 30 W (max)
Temperature range: −54 to +71°C
Altitude: up to 45,000 ft

Operational status

Fitted to the Embraer AMX aircraft.

Contractor

Page Aerospace Ltd.

VERIFIED

---

900 series combined voice and flight data recorders

The 900 series of flight data recorders is derived from military recorders fitted to aircraft such as the Harrier GR. Mk 7, Tucano and EH 101 Merlin. It records 5 hours of data at 128 words/s plus three channels of audio, each of 1 hour duration. An enhanced version gives 8 hours of data.

Data can be transferred at high speed for rapid extraction. The system includes an ultrasonic locator beacon with a six year battery life. The system is CAA approved and the audio section meets requirements of CAA Specification 11 and is compatible with ARINC 557; crash protection meets or exceeds TSO C51a. The 900 series has comprehensive BITE and integrity of recorded data can be confirmed on a read-after-write basis.

Specifications

Dimensions: 420 × 175 × 118 mm
Weight: 8.5 kg
Power supply: 28 V DC, 14 W

Operational status

900 series recorders have been delivered to Bristow helicopters as part of the IHUMS system on the Sikorsky S-61 helicopter. No longer in regular production, but available on special order.

Contractor

Penny & Giles Aerospace Ltd.

VERIFIED

2000 series Solid-State Combined Voice and Flight Data Recorder (SSCVFDR)

The 2000 series Solid-State Combined Voice and Flight Data Recorder (SSCVFDR), integrates both cockpit voice and flight data recording functions into a single unit. The recorder complies with the latest EUROCAE ED-55 and ED-56A and the FAA TSO-C123/C124 requirements for airborne recording equipment. Several options are available on the unit, such as extended audio duration, onboard maintenance system interface or integral area microphone preamplifier. The baseline unit records a minimum of 25 hours of flight data, at 128 words/s, with 30 minutes' voice on four channels and can interface to any ARINC 573 or ARINC 717 digital flight data acquisition unit or form part of a health and usage monitoring system. Continuous self-test, front panel interface, for *in situ* replay and diagnostics checks, and an underwater locator beacon are standard features.

The SSCVFDR can be hard mounted in the aircraft without shockmounts, as there are no moving parts. Because both data and audio recording is achieved in the combined crash-protected unit, the SSCVFDR is particularly suitable for airframes where weight and space are restricted. Advanced design techniques and manufacturing processes, coupled with rigorous environmental testing, mean high reliability and no periodic maintenance.

Specifications

Dimensions: 124 × 320 × 194 mm, ½ ATR Short
Weight: <9 kg
Power supply: 115 V AC, 400 Hz or 28 V DC, 20 W (max)
Temperature range: −55 to +70°C
Altitude: −15,000 to 55,000 ft
Environmental: DO-160C
Reliability: 15,000 h MTBF

Operational status

In service with military and civil operators. Current production item.

Contractor

Penny & Giles Aerospace Ltd.

VERIFIED

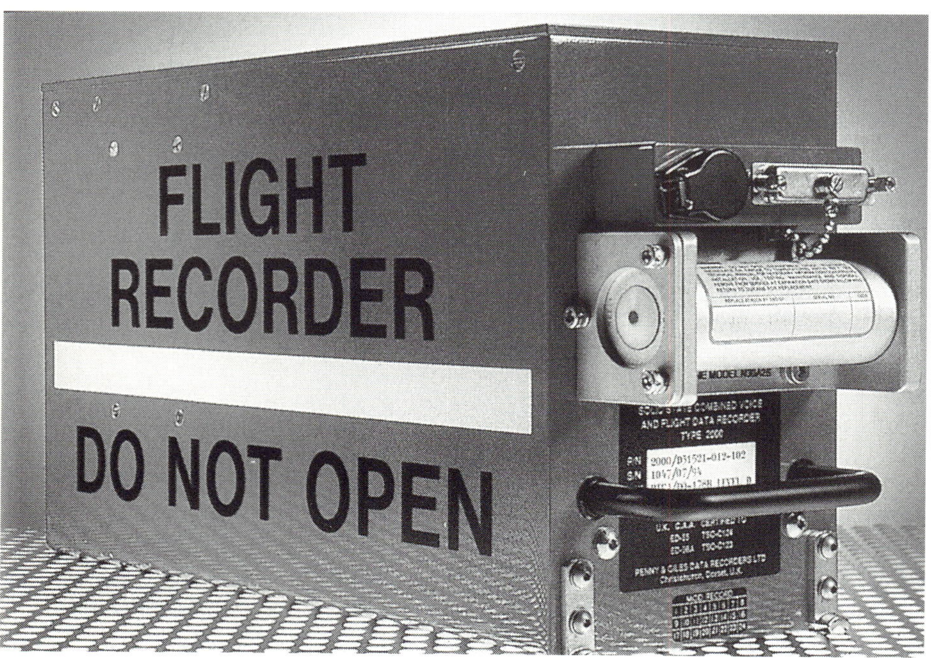

2000 series solid-state combined voice and flight data recorder ***1997***/0005418

Accident Data Recorders (ADRs)

The family of accident data recorders has been developed by Penny & Giles around a recycling mechanism using plastic-based magnetic recording tape. The track configuration is variable to meet customer needs, with up to eight tracks being currently available in various combinations of voice and data. These ADRs are fully protected according to all the major civil and military airworthiness requirements. Versions are currently in production for the Harrier GR. Mk 7, Tucano and the EH 101 Merlin.

Harrier GR. Mk 7 This unit has four data and two voice tracks arranged in two channels, each having two data and one voice track. Recording duration is 2 hours. Penny & Giles, in conjunction with GEC-Marconi Avionics Ltd as prime contractor, was chosen in December 1982 to provide the combined accident data and cockpit voice recorder for the Royal Air Force Harrier GR. Mk 7. The system has also been proposed as a private venture development for the US Marines' AV-8B Harrier. GEC-Marconi Avionics Ltd will work on the data handling and normalising unit and Penny & Giles on the tape transport system. Penny & Giles' activity grew out of work undertaken for the ADR 7000 accident data recorder, but with much more emphasis on the voice recording aspect.

Tucano This has two data channels of two tracks each and three voice tracks which are recorded all the time the recorder is operating. Recording duration is 2 hours for data and 1 hour for voice.

EH 101 helicopter ADR This ADR is under development and will be broadly similar to the Harrier GR. Mk 7 unit noted above.

Operational status

In service on the Harrier GR. Mk 7, Tucano and in production for the EH 101 helicopter.

Contractor

Penny & Giles Aerospace Ltd.

UPDATED

ADR 800 accident data recorder

The ADR 800 family of accident recorders, designed for civil and military fixed-wing aircraft and also used for helicopter flight trials, continues in service. It has a capacity of 138 Mbits and can interface either directly or via ARINC 573. Packaged into a ½ ATR short unit, with or without vibration isolators, the system is protected according to the requirements of TSO C51a. Operating temperature is between −65 and +70°C.

Operational status

In service on Concorde aircraft and some older Boeing B707 aircraft; no longer in production.

Contractor

Penny & Giles Aerospace Ltd.

VERIFIED

Cockpit Voice Recorder CVR-90

The CVR-90 is compliant with both FAA TSO C-123 and EUROCAE ED-56A survivability and environmental requirements. Voice recording on four channels is provided with a capacity of 30 minutes per channel. The CVR supports ARINC 557 and 757 compatible avionics.

Specifications

Dimensions: ½ ATR short
Colour: International orange
Operating temperature: −55 to + 70°C
Non-operating temperature: −55 to +85°C
Power: 28 V DC or 115 V AC, 400 Hz less than 15 W

Weight: 8.2 kg
MTBF: 20,000 h
Features: built-in test; fail safe erase through double electrical interlock

Contractor

Penny & Giles Aerospace Ltd.

VERIFIED

Flight Data Recorder FDR-91

FDR-91 is compliant with FAA TSO C-124 and EUROCAE ED-55 survivability and environmental requirements. A data recording capacity of 25 hours is provided. The FDR supports ARINC 573, 717 and 747 compatible avionics.

Specifications

Dimensions: ½ ATR long
Colour: International orange
Operating temperature: −55 to + 70°C
Non-operating temperature: −55 to +85°C
Power: 28 V DC or 115 V AC, 400 Hz less than 15 W
Weight: 8.4 kg
MTBF: 20,000 h
Features: built-in test; Harvard bi-phase input 64/128 bits/s; high-speed download

Contractor

Penny & Giles Aerospace Ltd.

VERIFIED

Optical Quick Access Recorder (OQAR)

Penny and Giles new high-capacity Optical Quick Access Recorder (OQAR) enables airlines to transfer data quickly from the flight line to the Flight Operations Quality Assurance (FOQA) system. The compact, ruggedised unit features a fault tolerant design with extensive internal diagnostics and provides up to 40 hours of recorded data at 1,024 words/s and 640 hours of data at 64 words/s. This data allows an airline to conduct the following types of analysis: Engine Condition Trend Monitoring (ECTM); flight operation analysis; aircraft and autopilot performance; engine exceedance reporting; automatic landing analysis and engine derate measurement.

The Optical Quick Access Recorder is designed to provide all the necessary data for effective Flight Operations Quality Assurance (FOQA) programmes.

Penny & Giles airborne and ground program loaders

The OQAR can be programmed to record over 2,000 separate parameters, including data from the APU, brake system, and navigation computer which is typically not recorded by other flight data recorders. Airlines that implement FOQA programmes can select which parameters to record, in order to increase safety while reducing maintenance costs, to identify problems and trends more easily and to optimise crew training.

These new high-capacity data recorders document operational characteristics of the aircraft's entire flight, including take-offs and landings. Fully ruggedised, they record even when subjected to vibration or shock, so transient or abnormal events such as hard landings or fast rotations will not be missed. The unit is designed so the rewritable magneto-optical disk can be quickly removed from the aircraft during a level A check, at weekly intervals or even during a normal gate turn. As part of the FOQA programme, the flight data is then downloaded to a ground station where after de-identification, analysis for significant operational patterns can be conducted. Additional ARINC 429 inputs are provided for supplementary TCAS (Traffic Collision and Avoidance System) and GPS (Global Positioning System) recording, or for use as part of a FANS (Future Air Navigation System).

Operational status

USAir, United Airlines and Continental are the first three US airlines to order the OQAR for use in their FOQA programmes. The OQAR units are available in configurations for all new Airbus and Boeing aircraft; and are fully compatible with all digital bus aircraft including ARINC 573/717, and ACMS. They feature embedded control software, which is field upgradable, with a front panel RS-232 port allowing for self-diagnostic checks, relay of the stored data, or configuration changes. The recorders use industry standard, 3.5 in rewritable magneto-optical disks with MS-DOS-compatible format, that are also ISO 10090 compliant.

Contractor

Penny & Giles Aerospace Ltd.

VERIFIED

PLU 2000 program loader

Penny & Giles' computer program loaders are based on industry standard DC300A data cartridges and expanded capacity DC300XL units. They provide a range of features appropriate to military or commercial fixed-wing aircraft and helicopters, and land- and sea-based vehicles and vessels, in severe environments where low error rates are essential.

The PLU 2000 program loader can accommodate up to three optional modular interfaces, has a capacity of 3.2 Mbits expansible to 6.4 Mbits and can load up to 10 preselected programs. It is operated by a remote-control unit.

Operational status

In service on Royal Air Force Nimrod aircraft and Royal Navy Sea King helicopters.

Contractor

Penny & Giles Aerospace Ltd.

VERIFIED

Quick access recorder

The quick access recorder is an avionic data logger of modular construction designed to provide rapid access to aircraft performance data recorded in flight. The media used is an industry standard ¼ in magnetic tape cartridge, giving low-cost data storage.

The unit fulfils the quick access recorder requirement in a civil aircraft integrated monitoring system; it also provides automatic read-after-write error detection and correction and an onboard replay facility with GMT search. The quick access recorder is microprocessor controlled and conforms to ARINC 591. It has a built-in power supply, with battery back-up for power drop-out immunity.

Specifications

Dimensions: 124 × 320 × 194 mm
Weight: <7 kg
Power supply: 115 V AC, 400 Hz, 70 VA peak
Reliability: 4,500 h MTBF

Operational status

Selected by a number of major airlines for use on a wide variety of aircraft.

Contractor

Penny & Giles Aerospace Ltd.

VERIFIED

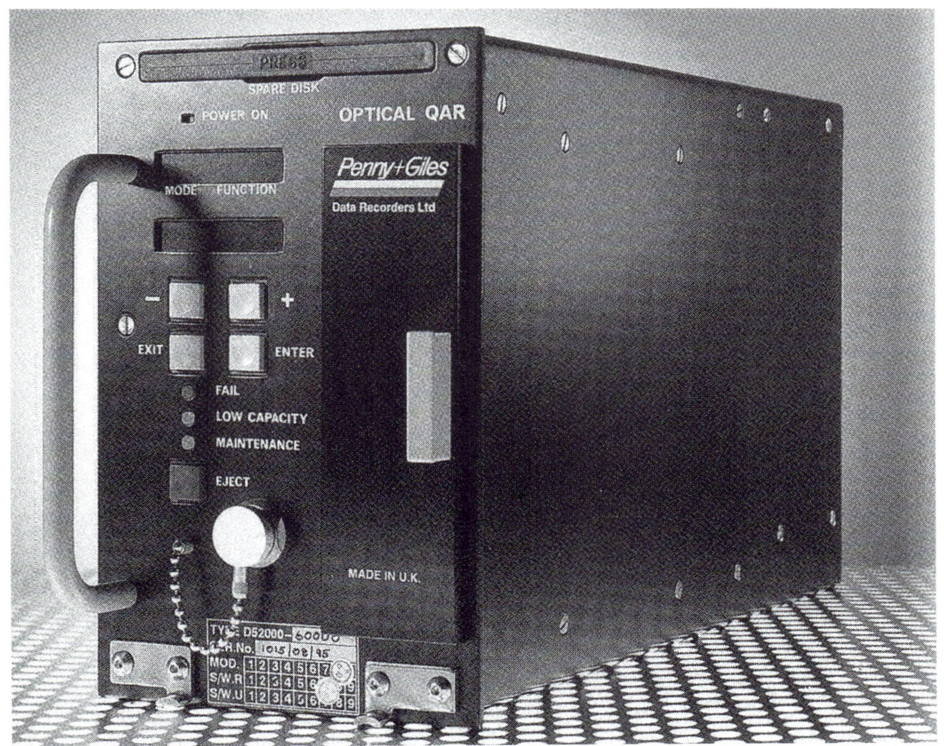

Optical Quick Access Recorder (OQAR) ***1997***/0005419

Solid-State Quick Access Recorder (SSQAR) D51555

The Solid-State Quick Access Recorder (SSQAR) D51555 has been developed for light turbine-engined helicopters and turbofan corporate aircraft for offline trend analysis, aircraft performance monitoring and maintenance recording. It is a ruggedised, lightweight and environmentally sealed data recorder which uses industry standard solid-state memory cards. The unit acquires navigational data, typically from a GPS receiver, as well as engine and airframe parameters from various ARINC 429/573/717, RS-422, analogue and discrete sources. These inputs are sampled and decoded as necessary and the acquired data is time-stamped with a GMT value before being written to flash memory cards. Three card slots are provided, each capable of addressing up to 32 Mbytes of memory, subject to card availability of higher-capacity cards in the future.

Specifications

Power supply: 28 V DC nominal
Interfaces: ARINC 573, ARINC 429 (2), RS-422 (2), analogue (4), discrete inputs (16), run control, discrete outputs
Reliability: 15,000 h MTBF

Operational status

In late stage of development.

Contractor

Penny & Giles Aerospace Ltd.

VERIFIED

The solid-state quick access recorder D51555

1995

Pegasus high-capacity digital cassette data recorder

Pegasus is designed for airborne, naval and ground-based data capture and storage applications. It incorporates fast start/stop buffering to write and read any digital data rate up to 100 Mbits/s without adjustment. The standard D-1 tape cassette provides a rate independent storage capacity of 240 Gbits. High-rate serial ECL and 8-, 16- or 32-bit parallel data can be accommodated and a range of resident control/status interfaces eases the problems of system integration.

The Pegasus 64 records 8 Mbytes/s for 1 hour, or longer at lower rates, and fills a gap in the market between the S-VHS products and high-rate large format recorders.

The Pegasus 100 can record continuous data rates up to 12.5 Mbytes/s or burst data up to 16 Mbytes/s.

Specifications

Dimensions: 575 × 482 × 550 mm
Weight: 52 kg
Power supply: 120/240 V AC, 48-63 Hz, 650 W (max) (400 Hz optional)

Contractor

Penny & Giles Data Systems Ltd.

VERIFIED

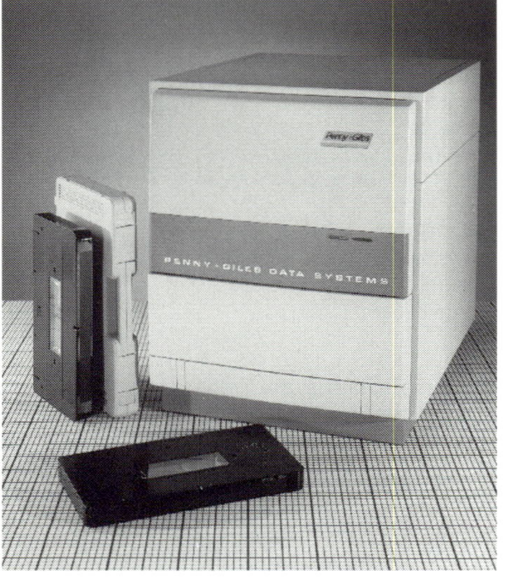

The Pegasus high-capacity digital cassette data recorder

HER 200 series tape transport

The HER 200 series uses 73 m of 0.5 in tape, with five recording speeds from $^{15}/_{16}$ to 15 in/s. The transport is supported on anti-vibration mounts within a cylindrical or rectangular case, so no added shock insulation is needed.

The Racal Radar Defence Systems Ltd Type HER 601 ruggedised tape transport

Specifications

Dimensions:
(HER 200 cylindrical case) 96 × 215 mm
(HER 202 rectangular case) 245 × 107 × 108 mm
Weight:
(HER 200) 2.4 kg
(HER 202) 3.1 kg
Power supply: 22-32 V DC, 25 W
Tape speeds: $^{15}/_{16}$, 1, 3¾, 7½ and 15 in/s
Tape running time: 640 s at 3¾ in/s
Altitude: up to 70,000 ft
Max linear acceleration: 50 *g*

Contractor

Racal Radar Defence Systems Ltd.

VERIFIED

HER 402, 600 and 601 tape transports

The HER 402, 600 and 601 ruggedised tape transports operate to full military specifications and are packaged in conventional rectangular boxes. Each offers six tape speeds up to 30 in/s, with up to 28 record tracks being possible using a pair of multitrack heads. The Type 400 takes 650 ft of 1 in wide tape, the Types 600 and 601 1,400 ft.

Specifications

Dimensions:
(402) 279 × 159 × 152 mm
(600) 381 × 221 × 178 mm
(601) 381 × 221 × 199 mm
Weight:
(402) 8 kg
(600) 15.5 kg
(601) 16.3 kg
Power supply:
(402) 28 V DC, 54 W
(600) 28 V DC, 54 W nominal
(601) 28 V DC, 63 W
Tape running time: (at 3¾ in/s)
(402) 34.6 min
(600/601) 64.6 min
Altitude: up to 70,000 ft
Max linear acceleration: 30 *g*

Contractor

Racal Radar Defence Systems Ltd.

VERIFIED

Storehorse and Storeplex range of instrumentation recorders

The Storehorse range provides record and replay in 7, 14, 28 and 42 track configurations. Heads meeting IRIG intermediate band, wideband and double density standards are available for operation in the speed range $^{15}/_{32}$ to 120 in/s. A wide selection of interchangeable data channels is available to provide DR, FM, dual-mode FM/DR, digital and voice recording modes up to a maximum of 4 MHz bandwidth at 120 in/s tape speed. The Storehorse features automatic calibration and equalisation, microprocessor control and other advanced techniques to ensure efficient and accurate recording. The extended bandwidth capability of Storehorse DD-4 is of particular benefit in predetection and telemetry recording.

The Storeplex range offers a new modular concept in helical scan digital recording. The tape transport module uses economical S-VHS tape, a reliable high-capacity easy-to-handle medium. Up to 64 analogue and/or digital channels are housed in a compact signals module. Channels can be recorded using any combination of bandwidths and clock rates up to the maximum system aggregate data rate of 51.2 Mbits/s. Storeplex analogue channels feature fully automatic calibration to provide a dynamic range of 96 dB and channel-to-channel phase accuracy of better than 0.5° at 45.5 kHz. Digital channels offer from 1 to 16-bit parallel input at rates between 2 kbits/s and 51.2 Mbits/s. An integrated SCSI interface allows rapid

The Racal Storeplex Delta instrumentation recorder offers 51.2 Mbits/s recording capacity and variable tape speed ***1995***

28 Storehorse instrumentation recorders are installed on the Advanced Range Instrumentation Aircraft (ARIA) ***1995***

direct-to-computer download of signal data. The system offers a flexible accurate solution for acoustic, vibration and telemetry data acquisition.

Operational status

The Storehorse range is established worldwide for use in radio, radar and sonar monitoring, with many applications in aerospace development and range telemetry recording.

The Storeplex range has become a standard solution for acoustic, sonar and vibration measurement applications.

Contractor

Racal Recorders Ltd.

VERIFIED

0730 KEL series low-cycle fatigue counter

The 0730 KEL series calculates cumulative low-cycle fatigue and records the number of engine starts and engine hours. It also provides information on exceedances, banding of speeds, voltages and thermocouples, spool-up and spool-down times and snapshots of input data. The 0730 Series records a maximum of 18 aircraft or engine input analogue parameters including up to six speed, six voltage, two thermocouple and four discrete parameters. These are all configurable in range and sensitivity to customer requirements. It provides outputs via RS-232 link to a data transfer device, printer or maintenance computer, and current and voltage sources for aircraft sensor excitation. It also provides an output to an integral LED display for manual interrogation of life usage exceedance fault codes.

Specifications

Dimensions: ¼ ATR dwarf
Weight: 2.5 kg
Speeds: pulse probe or tachogenerator 10 Hz to 25 kHz
Voltage: range selectable 0-0.5 V, 0-5 V, 0-40 V FSD
Thermocouple: 0-1,000°C Ch/Al
Discretes: 0-5 V, 0-28 V nominal

Operational status

In service with Swiss Air Force, the Royal Air Force Red Arrows aerobatic team and the Sultan of Oman's Air Force British Aerospace Hawks and Swiss Air Force Eurocopter Super Pumas.

Contractor

Smiths Industries Aerospace.

UPDATED

0826 KEL health and usage monitor

The 0826 KEL health and usage monitor provides snapshots of input data, exceedance monitoring and incident monitoring. It also calculates cumulative low-cycle fatigue. It records up to eight speed, 32 voltage, 16 discrete and four vibration aircraft and engine parameters. Outputs are provided via RS-422 link to a data transfer device, printer or maintenance computer and the system has an integral LED display for manual interrogation of input parameter values.

Specifications

Dimensions: ⅜ ATR
Weight: 5 kg
Speed: pulse probe 10 Hz to 4 kHz
Voltage: 0-5 V FSD
Discretes: 0-28 V nominal
Vibration: 20-750 Hz buffered

Operational status

In production for and in service in the BAe 146.

Contractor

Smiths Industries Aerospace.

UPDATED

0829 KEL health and usage monitor

The 0829 KEL health and usage monitor provides snapshots of input data, exceedance monitoring and incident monitoring. It has capacity available to incorporate LCF counting. There are 30 aircraft and engine input parameters available, including up to eight speed, 15 voltage and seven discrete parameters, and the system incorporates an ARINC 429 databus. Outputs are via RS-422 link to a data transfer device, printer or maintenance computer and there is an integral LED for manual interrogation of input parameter values.

Specifications

Dimensions: ⅜ ATR
Weight: 5 kg
Speed: pulse probe 10 Hz to 4 kHz
Voltage: 0-5 V FSD
Discretes: 0-28 V nominal
Vibration: 20-750 Hz buffered

Operational status

In service with the British Aerospace ATP.

Contractor

Smiths Industries Aerospace.

UPDATED

Onboard Maintenance System (OMS)

The Smiths Industries' engine life computer and standard flight data recorder have been combined into a single Onboard Maintenance System (OMS). The engine life computer records running hours and engine starts, and provides real-time calculations of low-cycle fatigue and snapshots of data to highlight deviations, as well as recording excursions outside defined limits. The standard flight data recorder records structural, engine, tracking, maintenance, mishap and training or mission replay data.

The benefits of the OMS are crash survivability, crash position location and water recovery, fast accurate data transfer, management of major engine components, improved flight safety and longer life and lower cost of ownership. The system is suitable for both new build aircraft and retrofit. It consists of a Signal Acquisition Unit (SAU), Crash Survivable Memory Unit (CSMU) and engine life computer. The SAU serves as the interface to aircraft data signals and power inputs. It acquires, computes, compresses and stores data. Choosing which data to record is controlled by software and can be tailored to any application. The SAU is entirely solid state.

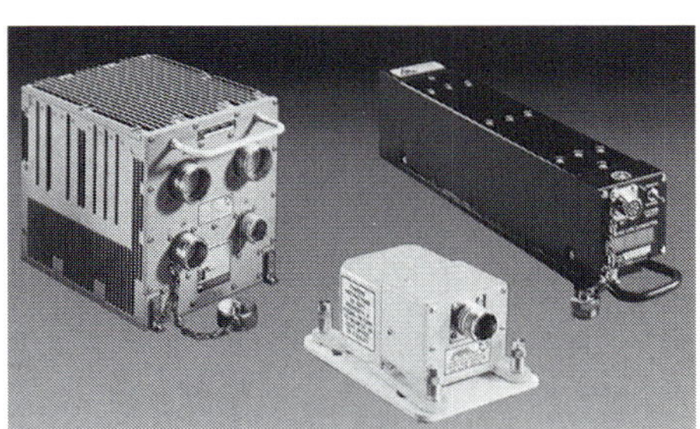

The Onboard Maintenance System consists of (left to right) the signal acquisition unit, crash survivable memory unit and engine life computer ***1995***

The CSMU can be adapted to a variety of requirements. Retention of the last 15 minutes or a full 25 hours of mishap data is available and the capability to store the last two hours of voice is in development.

The engine life computer processes data to compute engine life usage, thermal creep, exceedances and performance information. This is transferred into internal non-volatile memory for retrieval either manually via the integral LED display or semi-automatically via a data transfer device. The computer uses standard hardware modules and the software architecture is structured for multiple applications. External programming enables general engine algorithms, such as low-cycle fatigue and exceedance levels, to be factory programmed to suit specific applications and customer requirements. The computer includes 64 kbytes of non-volatile memory to store cumulative life usage totals, performance data and incident records. Extensive BIT is incorporated to monitor the system for correct functionality. The computer can be configured to drive annunciators for predefined input conditions, exceedances and BIT. A Deployable Flotation Unit (DFU) houses the CSMU and is equipped with a radio beacon and strobe. The buoyant DFU is released at impact or at a specified depth. The beacon and strobe provide visual and radio position location and will operate for at least 72 hours continuously.

Contractor

Smiths Industries Aerospace.

VERIFIED

Voice And Data Recorder (VADR)

The VADR provides survivable voice and data recording in a compact form-factor. Flexible configurations of voice and data recording offered by the Smiths Industries' VADR bring modern recorder technology to platforms previously limited by the weight and bulk of traditional recording systems.

The compact, lightweight VADR is built on Smiths Industries' solid foundation of military data recording expertise. Over 4,000 solid-state recorders were supplied worldwide in the last 10 years.

Smith Industries' VADR is available in a variety of configurations including CVR/FDR combination; CVR only; and FDR only.

The recorded voice and data is available for electronic download directly through the front connector. Ground servicing, replay, and data analysis capabilities, hosted on readily available PC computers, are also available.

Features include EUROCAE ED-56 CVR; EUROCAE ED-55 FDR; TSO compliant; light weight (3 kg); mounts in virtually any location or orientation; and options of ARINC 404 adaptor, area microphone, and ARINC 757 control panel.

Specifications

Dimensions:
(3253A) 86 × 108 ×190 mm
(3253C) 86 × 127 × 190 mm
Weight:
(3253A without beacon) 3 kg
(3253C without beacon) 3 kg
(3253A with beacon) 3.3 kg
Power:
(input) 28 V DC (115 V AC option - 3255)
(consumption) <9 W (3253A)
<12 W (3253C)

Contractor

Smiths Industries Aerospace.

UPDATED

Health and Usage Monitoring System (HUMS)

HUMS combines vibration and usage monitoring of critical dynamic power-train components with techniques such as chip detection, rotor track and balance, engine power assurance, cycle counting, exceedance monitoring and oil analysis. Integrated with a Flight Data Recorder, HUMS helps the aviation industry meet rising standards demanded by regulatory bodies, and provides a complete 'picture' of aircraft health and usage.

Stewart Hughes Ltd has teamed with various avionics manufacturers to produce HUMS. The first installation was flown on an S61N in November 1991. More than 200 systems have now been sold. Established systems exist for the following helicopter types: Bell 412/414, Boeing AH-64D and CH-47, Eurocopter Congar and Super Puma AS332 Mk 1 and 2, Sikorsky S-61 and S-76. Customers include US, North Sea operators, Republic of Singapore Air Force, Royal Netherlands Air Force.

Operational status

In production and in service.

Contractor

Stewart Hughes Ltd.

UPDATED

Series 2768 Super VHS video cassette recorder

The small lightweight Series 2768 helical scan video recorders are suitable for recording both video and data signals in hostile environments and alternative case designs enable the recorder to be located in either cockpit or equipment bay.

The recorder employs a Super VHS-C/VHS-C cassette to achieve a minimal size. Cassettes may be replayed on the ground by means of a mechanical adaptor in standard Super VHS-C and VHS video equipment, resulting in low-cost and readily available playback systems. Vinten video recorders are capable of producing recordings through high *g* aircraft manoeuvres, high vibration or gunfire.

In operation the Series 2768 recorder is sealed against sand, dust and water and contains a conditioning heater for low-temperature operation.

A significant feature of the Series 2768 is the incorporation of anti-vibration/shockmountings within the case of the recorder. This enables the recorder to be hard-mounted to the aircraft in almost any orientation and requires no sway space. The Series 2768 recorder may be used for unusually formatted signals such as those from linescan sensors.

The Airborne Recorder for IRLS and EO Sensors (ARIES) Video Cassette Recorder (VCR) is one of the Vinten Series 2768 recorders which are designed specifically for airborne environments. The recorder is available in both cockpit and bay-mounting versions.

The VCR is designed for recording linescan formatted video imagery derived from IRLS or EO sensors. It can be fitted internally or in a pod in manned aircraft or in RPVs, drones and UAVs.

Specifications

Dimensions: 125 × 152 × 226 mm
Weight: 3.5 kg
Recording system: rotary helical scan

Series 2768 video cassette recorder **1998**/0015318

Video signal system: PAL colour/CCIR monochrome 625-lines or NTSC colour/EIA monochrome 525-lines, plus other formats for non-standard signals
Recording time: up to 2 hours
Event marker: visual and audio
Recording bandwidth: up to 4.8 MHz

Operational status

In production and in service.

Contractor

W Vinten Ltd.

VERIFIED

Series 3150 colour video camera

The Series 3150 colour video camera is a modular CCD camera specifically designed for airborne recording applications.

The camera is designed to provide high-resolution colour images and is readily installed as either original equipment or as an aircraft upgrading. There are two versions of the camera: the low-profile and the universal. The latter has the ability to separate the lens and sensor from the remainder of the camera by a flying lead. This new concept for an airborne colour camera enables the ideal positioning of the camera lens in a location where there may otherwise be limited room for the complete camera. Previously the only alternatives were either to produce a special camera of dedicated design or to provide an optical periscope and suffer the resulting reduction of light to the sensor.

The camera is able to produce television pictures over a very wide illumination range and operate continuously in very demanding environmental conditions.

For cockpit installation the Series 3150 camera's standard modules are flexible enough in configuration to be orientated to suit most HUD and gunsight mounting requirements. This approach offers both an optical and cost-effective solution. The equipment can be used in association with Vinten airborne video recorders such as the Series 2768.

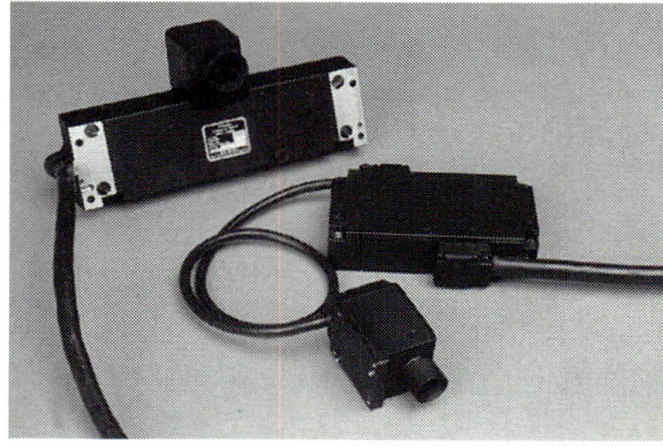

Alternative configurations of the Series 3150 colour video cameras **1998**/0015319

Specifications

Dimensions:
(camera head) 105 × 56 × 24 mm
Weight:
(camera head) 0.75 kg
Power supply: 28 V DC, 18 W nominal when used with Series 2768 VCR
Camera head sensor: solid-state imager
Field of view: 25° × 19° or 20° × 15°
Refresh rate: 50 or 60 Hz
Dynamic range: 5-170,000 lux

Operational status

In service and in production.

Contractor

W Vinten Ltd.

VERIFIED

Type 1192 recycling wire recorder

The Type 1192, based on a recording medium of stainless steel wire, preserves the last 25 hours of flight data for crash analysis. Wire has long been used as a recording medium and gives the maximum data storage for a given volume and weight. The Type 1192 mechanical transport can be used in conjunction with any data processing and normalising system providing analogue or digital information. The system is designed to withstand 100 *g*, 5 m/s shocks, the impact of a 227 kg steel bar dropped from 3 m, a static crush force of 2,270 kg, exposure to flame at 1,100°C for 5 minutes over half the outside area, and a combination of fuel, hydraulic fluid, lubricating oil, fire extinguisher fluid and salt water for 36 hours.

Specifications

Dimensions: 203 mm sphere reducing to 178 mm cylinder over central region
Weight: 7.27 kg
Power supply: 115 V AC, 400 Hz, 3 phase
28 V DC
Recording medium: 0.058 mm diameter stainless steel wire, breaking force >0.34 kg
Data transfer rate: 1,536 flux reversals/s

Operational status

In service. No longer in production.

Contractor

W Vinten Ltd.

VERIFIED

Type 1200 cockpit voice recorder/reproducer

The Type 1200 voice recorder/reproducer is designed for mounting in the confined areas typical of fighter cockpits. When connected to the communications control system, it provides intelligible speech and record/replay facilities. The system has only two controls, a selector switch for record/replay and a toggle switch for on/off, rewind and fast forward. When a microphone in the recorder detects speech it switches on automatically and continues to operate until speech stops. The recorder then continues to run for a preset period and switches itself off until the next speech sequence causes it to repeat the cycle. Cassettes may be replayed through the aircraft system or through a ground-based playback unit. The system employs standard Philips C60 and C90 audio cassettes. A version designated the Type 1200-004 can record and reproduce data as well as voice.

Specifications

Dimensions: 147 × 43 × 168 mm
Weight: 1.36 kg
Tape cassette: Philips Compact C60 (30 min audio/track) or C90 (45 min audio/track)
Tape erase: automatic prior to recording

Operational status

In service. No longer in production.

Contractor

W Vinten Ltd.

VERIFIED

Type 1300 pilot's display recorder

The Type 1300 camera photographically records head-up display symbology superimposed on the pilot's view of the outside world. The record can be employed for the training of combat tactics, for debriefing or for assessment of weapon aiming and target damage. Mounted on the side of the head-up display it gives the least visual obstruction to the pilot and preserves a clear rear surface on the head-up display combiner glass for control purposes. The system comprises a camera unit, film magazine and interchangeable lenses. Automatic exposure control and variable shutter speed permit the use of film emulsion speeds within a wide brightness range. The exposure is controlled by a built-in electronic sensing unit and emulsion speed is adjusted automatically by a coding feature in the film magazine that presets the appropriate controls.

Specifications

Dimensions: 210 × 190 × 45 mm
Weight: 3 kg
Power supply: 28 V DC
2 A (starting)
1 A (running)

Type 6051 video conversion unit ***1998***/0015320

Film type: 22 m × 16 mm × 0.1 mm double perforated cine
Framing rate: 16 or 32 frames/s
Film movement indicator: visual indication
Exposure control: integral with camera, automatic over brightness range 78-5,028 ft-lambert
Running time: 3 min at 16 frames/s
Shutter speed: 1.25-10 ms
Event marker: remote operation
Lens (with periscope):
f/2.8-22, F=25 mm, FoV 16 × 22°
f/2.8-22, F=50 mm, FoV 8 × 11°

Operational status

In service. No longer in production.

Contractor

W Vinten Ltd.

VERIFIED

Type 6051 video conversion unit

The Vinten Type 6051 video conversion unit is designed as a 525-line/60 Hz to 625-line/50 Hz video converter for use in combat aircraft. It enables 525-line/60 Hz video sources to be interconnected to 625-line/50 Hz video switching, display and recording equipment and is appropriate for use where multiple video standards exist on an aircraft. The converted video output can be synchronised to another onboard 625-line/50 Hz source.

Operational status

In production.

Contractor

W Vinten Ltd.

VERIFIED

UNITED STATES OF AMERICA

Aircraft condition monitoring system

The aircraft condition monitoring system is an integrated data management system. It receives data from aircraft monitoring systems and processes the data to facilitate various aircraft functions including maintenance, aircraft and flight crew performance and communications.

Operational status

In service. First deliveries (for Boeing 747-400 aircraft) in 1998.

Contractor

AlliedSignal Commercial Avionics Systems.

UPDATED

AN/ASK-7 data transfer unit

The AN/ASK-7 data transfer device, originally designed for use with the Boeing AGM-86 air-launched cruise missile, has now been adapted for the US Air Force B-52 and B-1B aircraft.

The unit transfers large quantities of digital information from ground-based master systems to aircraft equipment. It will also monitor performance parameters as part of the central integrated test system on the B-1B. In the B-52, it forms a small part of the current Offensive Avionics System (OAS) update programme to improve the effectiveness of the aircraft.

Major features of the AN/ASK-7 include full record/replay capability, MIL-STD-1553B interface, error correction code, remote electronics, extensive built-in test and nuclear hardening. The system comprises a control unit, containing most of the recorder electronics, and an instrument-mounted cassette drive unit accommodating two cassettes at a time.

Specifications

Weight:
(cassettes) 3.1 kg
(cassette drive unit) 3.6 kg
(control unit) 6.8 kg
Power supply: 115 V AC, 400 Hz, 20 W

Operational status

In service.

Contractor

AlliedSignal Commercial Avionics Systems.

UPDATED

AV-557C Cockpit Voice Recorder (CVR)

The AV-557C CVR provides a crash survivable record of the last 30 minutes of flight crew conversation by recording it simultaneously on four separate tracks. The new device succeeds the earlier AV-557A and AV-557B CVRs, offering improvements in reliability and maintenance cost by incorporating integrated circuits.

The specially coated polymide-base magnetic tape recording medium is less sensitive to heat than conventional polyester film, so that data can be recovered after exposure to 240°C, compared with 110°C in conventional equipment. Recorder electronics are mounted on functional plug-in modules for ease of test and replacement. The tape has a clear window at each end so that the beginning and end of tape can be detected, capstan motor rotation electronically reversed and the record, erase and monitor heads switched for bidirectional recording.

The recorder is used in conjunction with a microphone monitor that can be installed on the flight deck. It is housed within a unit that also contains an amplifier, filter, monitor indicator, push-test and bulk-erase buttons and a headphone jack for auditing recorded information. An external microphone can also be plugged into the unit.

Specifications

Dimensions: ½ ATR
Weight: 10.5 kg
Tape type, drive and speed: 0.25 in polymide-base, AC hysteresis synchronous motors, 2.75 in/s (70 mm/s)
Bandwidth: 300-5,000 Hz ±3 dB
Maintenance period: 6,000 h
Environmental:
(fire) meets requirements of TSO C84 and TSO C51a; temperature of 1,100°C over 50% of equipment surface for 30 min
(salt water immersion) 30 days
(explosion) meets or exceeds DO 160 requirements

Operational status

In production and in service. Over 2,000 units have been built.

Contractor

AlliedSignal Commercial Avionics Systems.

VERIFIED

ED-55 Solid-State Flight Data Recorder (SSFDR)

The ED-55 Solid-State Flight Data Recorder (SSFDR) is an all-solid-state implementation of a crash survivable flight data recorder. It conforms to ARINC 573, ARINC 717 and ARINC 747. The unit eliminates all moving parts and uses solid-state flash memory as the recording medium. The SSFDR comprises three shop replaceable units: the crash survivable memory, power supply and interface and control assembly.

Memory capacity allows for the last 25 hours of flight data to be stored at an input rate of 64 words/s or 128 words/s. Data is stored without the use of any data compression algorithm, which provides for maximum reliability and integrity of critical flight information. An underwater locator beacon is available as an option.

Specifications

Dimensions: ½ ATR long or short
Weight: 8.2 kg
Temperature range: −55 to +70°C
Reliability: >15,000 h MTBF

Contractor

AlliedSignal Commercial Avionics Systems.

VERIFIED

ED-56A Solid-State Cockpit Voice Recorder (SSCVR)

The ED-56A Solid-State Cockpit Voice Recorder (SSCVR) is ARINC 757 compliant. It uses industry standard audio compression in both 30 minute and 2 hour models. Interface provisions include: one wideband and three narrowband audio channels; dedicated rotor tachometer input; FSK and GMT time recording; two ARINC 429 interfaces reserved for future ATC datalink messages; and maintenance data.

The ED-56A utilises FLASH EPROM memory and a patent Crash-Survival Memory Unit (CSMU). It is designed to replace older tape-based units.

Specifications

Dimensions: ARINC 404 ½ ATR short
Weight: <7.3 kg
Power: 35 W maximum

Contractor

AlliedSignal Commercial Avionics Systems.

VERIFIED

Flight Data Acquisition Management System (FDAMS)

The Flight Data Acquisition Management System (FDAMS), combines the functionality of an ARINC 717 Digital Flight Data Acquisition Unit (DFDAU) with the real-time monitoring and troubleshooting capability of a Data Management Module (DMM) in a single 6 MCU box. It is a powerful computing machine with 2 Mbytes of solid-state mass memory.

The FDAMS provides a standard set of 64 maintenance, operational and special algorithms, four of which can be customised for troubleshooting. The airline can also enable or disable any of the 64 algorithms directly by using ground support software.

The ARINC 717 DFDAU provides the mandatory ARINC 573 flight data to the flight recorder. With its seven databases, the DFDAU meets the FAA 88-1, CAA and ICAO requirements for all newly built aircraft. To accommodate any future aircraft modifications easily, Sundstrand's databases can be reloaded from a floppy disk using an ARINC 615 data loader.

The DMM monitors a predefined set of parameters in the aircraft. It then collects and downlinks both exceedance and routine data via a cockpit printer, ACARS or data loader. The data is processed to provide both maintenance and flight operations reports necessary for monitoring the engines and other aircraft systems. Using an IBM-compatible PC, the readout programs provide information in numerical or graphical format.

Contractor

AlliedSignal Commercial Avionics Systems.

VERIFIED

Flight Data Recorder/Fault Analyser

The three components comprising the basic solid-state Flight Data Recorder/Fault Analyser (FDR/FA) are the crash survivable memory unit, the signal acquisition unit and the auxiliary memory unit with standard electronic memory cartridge.

The Crash Survivable Memory Unit (CSMU) is a miniaturised solid-state recorder packaged in a protective enclosure, designed to ensure retention and recovery of critical aircraft flight data in the event of a crash.

The protective enclosure comprises a hardened steel case with thermal protection features including a high-temperature stable insulation material and a synthetic thermal mass designed to ensure maximum protection of the memory element. The CSMU has been successfully tested to FAA and triservice survivability limits.

The CSMU contains a 32 k word EEPROM memory, a memory management and communications controller function and an RS-422 serial interface to allow remote or local installation.

The Signal Acquisition Unit (SAU) is the heart of the FDR/FA system. It programs the acquisition process, performs the data compression algorithms and directs the recording process. The SAU's capability to accept up to 52 analogue parameters, 119 discrete parameters and 10 special purpose parameters while using MIL-STD-1750A processor architecture gives it the flexibility for a variety of applications.

Aircraft sensor information may be collected from a remote data acquisition unit, a MIL-STD-1553B multiplex bus or from sensors connected directly to the SAU. The SAU collects this sensor information and produces aircraft data which is then stored in either the CSMU or the Auxiliary Memory Unit (AMU). Power to both the CSMU and the AMU is distributed by the SAU. Also, access is provided by the SAU for ground readout equipment.

The auxiliary memory unit/data transfer system includes a Standard Electronic Memory Cartridge (SEMC). The SEMC offers high environment data transportability in a compact shirt-pocket-sized cartridge. The SEMC employs a family of high-density end-stacking memory slices to allow full freedom to configure a cartridge to the exact amount and type of memory desired. These memory slices also enable the user to change the memory configuration at the unit level, eliminating the need to ship the SEMC back to the manufacturer.

The modular family includes EEPROM and CMOS static RAM slices. UVPROM can be accommodated. Memory capacity to 4 Mbytes of 8-bit words can be achieved.

Specifications

Dimensions:
(CSMU) 177.8 × 152.4 × 114.3 mm
(SAU) 171.4 × 215.9 × 273 mm
(AMU/DTS) 130.2 × 146 × 152.4 mm
(SEMC) 34.9 × 82.6 × 139.7 mm
Weight:
(CSMU) 2.95 kg
(SAU) 6.89 kg
(AMU/DTS) 2.04 kg
(SEMC) 0.68 kg

Contractor

AlliedSignal Commercial Avionics Systems.

VERIFIED

Micro-Aircraft Integrated Data System (Micro-AIDS)

The Micro-Aircraft Integrated Data System (Micro-AIDS) has been developed to automate the various data logging functions on an aircraft, including the gathering and processing of data concerning engine condition and operational performance monitoring, exceedances and fuel usage and so on. A single unit is used to gather the data which is stored in solid-state memory.

The Micro-AIDS operates on the ARINC 573 data stream and recorded data can be recovered by ground crew using a portable data retrieval unit.

Specifications

Dimensions: ARINC 600 2 MCU
(3 MCU in expanded version)
Weight: 2.3 kg (3.6 kg expanded)
Power supply: 115 V AC, 400 Hz, 15 W
Capacity: 25 flights

Operational status

In production.

Contractor

AlliedSignal Commercial Avionics Systems.

VERIFIED

Mini-Flight Data Acquisition Unit (Mini-FDAU)

The Mini-Flight Data Acquisition Unit (Mini-FDAU) satisfies the mandatory acquisition requirements of the various regulatory agencies and is expandable to the currently proposed ICAO requirements. The Mini-FDAU is available in either the ARINC 404 or 600 form factor.

Output to any ARINC digital flight data recorder is generated in standard ARINC 573 format at 768 bits/s, Harvard biphase. Provisions are included to receive the DFDR BITE signal and individually display both the FDAU and DFDR status. Provisions are also included for interfacing with a Sundstrand flight data entry panel.

Input flexibility is available to permit interfacing with any aircraft configuration and to comply with diverse regulatory requirements. Data acquisition modules are available to accommodate use with analogue or digital ARINC 429 systems or with a combination of both analogue and serial databus inputs.

Input acquisition capacity is provided for recording of the FAA 17 mandatory parameters requirement. Capacity can be expanded to comply with the proposed ICAO requirements by the addition of one module.

The analogue acquisition design achieves a universal capability to acquire synchro, AC ratio, low-level DC and potentiometer signals with a single input circuit. The unit also allows automatic determination of the input signal based only on the aircraft wiring connections. This allows the Mini-FDAU to be interchanged among different aircraft without the need for reprogramming of a configuration PROM.

The Mini-FDAU is expandable to accommodate future additional input to include parameters such as those required for engine condition monitoring, a separate CPU and memory capable of screening data for storage in onboard solid-state memory. An ACARS interface expansion allows access to stored information and downlinking automatically or on demand.

Specifications

Dimensions: ARINC 600, 3 MCU
Weight: 3.63 kg (max)

Contractor

AlliedSignal Commercial Avionics Systems.

VERIFIED

PTA-45B airborne data printer

The AlliedSignal PTA-45B data printer uses a high-efficiency switching power supply which provides sufficient power to continuously print 'all-black' (all 600+ dots printing at the same time). Since normal operation runs at only 15 per cent of the machine capacity, the printer is claimed to be capable of providing up to 10 times the reliability of previously available models.

The printer permits flight crew access to hard copies of uplinked flight plans, messages and graphics, and can also print a copy of datalink information displayed on the AlliedSignal weather radar display. The printer may be driven by a datalink system or a maintenance computer. User features include paper guides to help prevent paper jamming, a hinged spool to ease paper

Units of the AlliedSignal PTA-45B data printer family

changing and a paper level indicator to tell how much paper is left. The unit has a MTBF in excess of 5,000 hours.

Operational status

In service on British Aerospace 146 aircraft. The PTA-45B is standard equipment on the Airbus A320 and has been selected for the Boeing 737, 747-400, 757 and 767, MD-11, MD-80 and MD-88 and Airbus A310-300, -400 and -500 airliners.

Contractor

AlliedSignal Commercial Avionics Systems.

UPDATED

Quick Access Recorder (QAR)

The QAR is an Aircraft Integrated Data Systems (AIDS) recorder.

Twelve-track sequential recording provides up to 50 hours of operation per cassette at an input data rate of 768 bits/s. Total storage up to 138 million bits is available at a conservative packing density of 2,400 bits/in. The QAR can accept either Harvard biphase or bipolar RZ data input.

It features remotely selectable tape speeds to accommodate four input data rates, a ruggedised metal cassette to protect the tape in or out of the recorder, precision tape guidance within the cassette to ensure interchangeability between recorders and remote equipment status indication as a function of the BITE capability.

Specifications

Dimensions: ARINC 404 ½ ATR long
Weight: 8.62 kg
Power supply: 115 V AC, 400 Hz, 60 W (max)

Contractor

AlliedSignal Commercial Avionics Systems.

VERIFIED

Ruggedised Optical Disk System (RODS)

The RODS is a digital data storage and retrieval device utilising a fully erasable and rewritable 5.25 in diameter optical disk as the storage medium. RODS is designed for use in special mission aircraft and for other military applications requiring highly reliable data storage and transfer in harsh operating environments.

Typical areas of application include digital map terrain data storage, electronic document storage, signal and mission data recording, maintenance and structural data recording and general purpose data loading and data transfer.

Data capacity is 300 Mbytes user data per side and sustained transfer rate is 518 kbytes/s. Access is 100 ms maximum.

Specifications

Dimensions: 127 × 190.5 × 317.5 mm
Weight: 6.8 kg
Power supply: 115 V AC, 400 Hz or 28 V DC
Reliability: 5,000 h MTBF

Contractor

AlliedSignal Commercial Avionics Systems.

VERIFIED

SETS-II Severe Environment Tape System

The Severe Environment Tape System (SETS-II) is claimed to be the lightest and most compact digital data recording system currently available. It emphasises a very low error rate during recording and playback in the most harsh military air, sea and land environments, and comprises two modules: a tape drive system, with the recorder electronics and mechanical elements, and a tape unit containing the recording medium, magnetic recording head tape, tension device and tape-end sensor.

The system accommodates 300 ft of 0.25 in tape, with a four-track single gap read/write head, a recording density of 1,600 bits/in, transfer rate of 48,000 bits/s (phase encoded) and total capacity of 23 M bits. The nominal error rate is 1 in 10^7 bits. Tape speed is 30 in/s write and 120 in/s search and rewind. Total system weight is 2.02 kg.

Operational status

In development.

Contractor

AlliedSignal Commercial Avionics Systems.

VERIFIED

Solid-State Cockpit Voice Recorder (SSCVR)

The Solid-State Cockpit Voice Recorder (SSCVR) is designed for the commercial airline, business and commuter and general aviation markets. It complements the flight data recorder in providing a solution for crash survivable recorder requirements.

The SSCVR is based on the existing Sundstrand design providing 90 per cent commonality between the two products. A 30-minute four-channel version is offered for existing ARINC AV557 applications and a 2 hour four-channel variant will be available for future JAR operational rules.

Contractor

AlliedSignal Commercial Avionics Systems.

VERIFIED

Solid-State Flight Data Recorder (SSFDR)

The Solid-State Flight Data Recorder (SSFDR) is an all-solid-state crash survivable flight data recorder. The unit eliminates all moving parts and uses solid-state flash memory as the recording medium. The SSFDR comprises the crash survivable memory, power supply and interface and control assembly. The memory capacity allows the last 25 hours of flight data to be stored at an input data rate of 64 words/s or optionally 128 words/s. An underwater locator beacon is available as an option.

The SSFDR has been designed to minimise the life cycle cost through high reliability, low unit weight and field upgradability.

Specifications

Dimensions: ½ ATR long or short
Weight: 8.16 kg
Power supply: 115 V AC, 400 Hz or 28 V DC, 35 W (max)
Reliability: >15,000 h MTBF

Contractor

AlliedSignal Commercial Avionics Systems.

VERIFIED

Supplemental Flight Data Acquisition Unit (FDAU)

The FDAU has been designed to provide a low-cost convenient means for increasing the number of flight parameters recorded by AlliedSignal's standard five-parameter Universal Flight Data Recorder (UFDR). Without replacing the UFDR, the Supplemental FDAU and recorder can together accommodate up to 11 parameters. This is accomplished by merging new parameters into the ARINC 573 data stream generated by the UFDR's internal FDAU and returning the data to the recorder section of the UFDR. The Supplemental FDAU provides a status output that is compatible with the ARINC 542 status-reporting system, causing minimal impact to existing wiring. If the Supplemental FDAU is not installed or its status is invalid, the UFDR automatically reverts to its normal five-parameter operation.

The Supplemental FDAU can accept a variety of sensor and signal types. The programmable analogue inputs accept synchro, AC ratio No 1, AC ratio No 2, 0-5 V low-level DC and potentiometer signals. Dedicated inputs include shunt and series discretes, ARINC 573 serial data from the UFDR and UFDR validity status.

The BIT capabilities of the Supplemental FDAU determine whether the recording is functioning and enunciate system status. Additionally, the Supplemental FDAU will provide validity indication that will allow faults to be isolated to the LRU.

Specifications

Dimensions: ARINC 600, 2 MCU
Weight: 2.27 kg

Contractor

AlliedSignal Commercial Avionics Systems.

VERIFIED

Tactical Optical Disk System (TODS)

The Tactical Optical Disk System (TODS) is a severe environment digital data storage and retrieval device utilising a fully erasable and rewritable 5.25 in diameter optical disk as the storage medium. TODS is designed for unrestricted operation in high-performance military aircraft and other platforms requiring highly reliable data storage and transfer in very severe operating environments.

Areas of application include digital map terrain data storage, signal and mission data recording, maintenance and structural data recording and electronic document storage.

Data capacity is 300 Mbytes user data and sustained transfer rate is 518 kbytes/s. Access time is 120 ms maximum.

Specifications

Dimensions: 127 × 165.1 × 254 mm
Weight: 8.16 kg
Power supply: 115 V AC, 400 Hz or 28 V DC
Reliability: 5,000 h MTBF

Contractor

AlliedSignal Commercial Avionics Systems.

VERIFIED

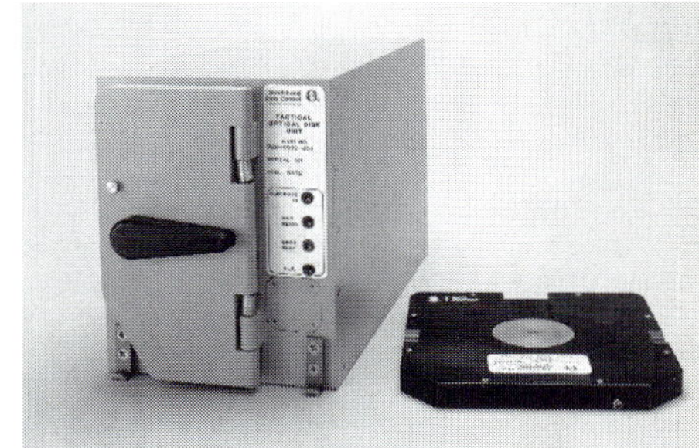

The tactical optical disk system

Universal Flight Data Recorder (UFDR)

The 25-hour-capacity, crash-protected digital UFDR has a variety of configurations. Together with optional accessory equipment, the system can be made completely interchangeable with existing flight data recorders to ARINC 542, with digital recorders to ARINC 573/717, or expanded parameter systems intended for CAA-regulated applications.

The recorder uses Kapton tape for durability, a single motor and co-planar reel-to-reel tape transport. It can be hard mounted in the aircraft, and considerable attention has been given to reducing the number of moving (and therefore wearable) parts. High-speed data retrieval permits the entire tape content to be transferred to a copy recorder in less than 30 minutes, so that the operation may be conducted during normal turnround.

The system uses a patented feature called Checkstroke which monitors the data recorded each second and verifies its accuracy against the continuous stream of information received in the recorder's electronics. This built-in test feature ensures the validity of the recorded data.

With the release of the FAA NPRM 85-1, new versions of the UFDR have been designed. The new H model is considered a six-parameter unit (meeting that portion of the NPRM) and the new G Mod 10 model meets the expanded 11-parameter portion of the NPRM. The H Model is expansible to the G Mod 10 simply by adding one circuit card and a 5 V DC power supply.

A new feature of these models is the addition of programmable inputs which allows the user to tailor the UFDR to his particular transducer types. Another plus is the increased accuracy of the system.

The following inputs are available for the H Model: vertical acceleration, heading, pneumatic airspeed, pneumatic attitude, marker beacon, time, trip and date and two programmable inputs for altitude and airspeed.

The following inputs are available for the G Mod 10 Model: the same as for the H Model plus pitch attitude, pitch trim position, roll, flap, engines 1 to 4, total air temperature, one spare programmable and two spare 0-5 V DC.

Specifications

Dimensions: ½ ATR long per ARINC 404
Weight: 13.6 kg
Power supply: 115 V AC, 400 Hz, single phase
Compatible transducers: synchros, 0-5 V DC 3 wire, 0-18 V DC 3 wire, 0-5 V DC 2 wire, 0-15 V 2 wire, AC ratio type 1, resolvers and selsyns.
Recording capacity: 25 h at 768 bits/s
Input interface:
(digital flight data system) ARINC 573/717 data stream from remote flight data acquisition unit
(flight data recorder) vertical acceleration, altitude, airspeed, heading, marker beacon and three discrete signals (on/off, to/from and so on). Alternatively can be ARINC 573/717 input from a remote flight data acquisition unit
(expanded channel flight data recorder) same as for flight data recorder plus seven additional synchro inputs
Tape format: blocked incremental, biphase encoded, with 768 data bits, preamble, postamble and IRG
Tape type: Kapton, 0.25 in magnetic
Tape speed: recording, incremental 5 in/s = 0.414 in/s net; playback 5 in/s
Tape capacity: 450 ft for 25 h recording
Number of tracks: 8 bidirectional sequential
Playback time: 30 min with copy recorder, 2 h with sequential track transcription
Crash survivability: TSO C51a
Environmental: RTCA DO 160, TSO C51a

Operational status

In production and service for over 20 years. In US Air Force service on KC-10A Extender tankers and US Army service on Black Hawk helicopters. Also in widespread airline service.

Contractor

AlliedSignal Commercial Avionics Systems.

VERIFIED

Engine Analyser and Synchrophase Unit (EASU)

The EASU provides continuous, real-time monitoring of engine operating parameters. It monitors engine vibration and converts the information to spectral data points, comparing it against defined limits. Exceedances are reported to the cockpit and time-stamped records are stored in non-volatile memory for downloading and analysis by maintenance staff.

For turboprop applications, the EASU functions as a propeller speed/phase synchroniser to reduce aircraft cabin noise and vibration.

Contractor

Ametek Aerospace Products.

VERIFIED

Sentinel Data Acquisition Unit (DAU)

The DAU is used to acquire and transmit critical information from a variety of aircraft systems. It is a dual-redundant 3 MCU unit for monitoring systems such as the electronic engine controller, electrical system, air data computer, hydraulic system and a variety of other aircraft subsystems.

The DAU transmits data through both ASCB and ARINC 429 busses. Information will then be processed by the IC-800 integrated avionics computer and may be shown on the EICAS.

Operational status

The DAU has been integrated with the Honeywell Primus 2000 XP integrated avionics system for the Bombardier Global Express business jet.

Contractor

Ametek Aerospace Products.

VERIFIED

DCRsi Clip-On 1 Gbit/s airborne imagery recorder

The Ampex 1 Gbit/s airborne ultra-high rate recorder, known as the DCRsi Clip-On™, offers a 1 Gbit/s snap-shot imaging capability with instant access to cached data.

Clip-On is designed to be used in conjunction with an Ampex DCRsi™ digital cartridge recorder in an airborne image gathering role. Typically, data acquired at 240 Mbits/s or lower will be recorded on cache and tape simultaneously using the high-rate DCRsi 240. Data streams faster than 240 Mbit/s will be stored in cache and then automatically backed up to tape when the cache is nearly full or on a command from the operator. The baseline solid-state storage capacity is 5 Gbytes although larger memory sizes can be specified.

The system combines the benefits of extremely fast solid-state memory and permanent non-volatile tape storage in one inexpensive, fully integrated package. It has major operational advantages for airborne tactical reconnaissance since up to 5 Gbytes of cached imagery are always available for immediate access, giving the operator the ability to prioritise targets and downlink images almost instantaneously. The Clip-On system has already been selected for a number of classified programs in the USA and is compatible with the whole range of Ampex ruggedised recorders including DCRsi 75, DCRsi 107 and DCRsi 240.

Operational status

Launched June 1997.

Contractor

Ampex Corporation Data Systems Division.

VERIFIED

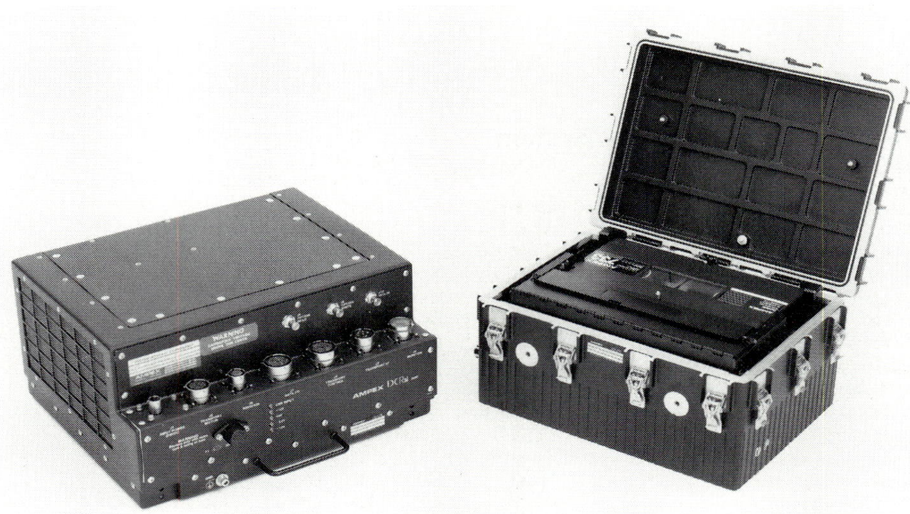

The Ampex Clip-On system offers a 1 Gbit/s snap-shot image recording and instant replay capability for airborne reconnaissance applications **1997**/0002443

DCRsi 75R digital cartridge recording system

The DCRsi 75R is an inexpensive 75 Mbit/s DCRsi variant designed to fill an important niche between 'top-end' S-VHS units and Ampex's existing 107 and 240 Mbit/s DCRsi models. This recording system is targeted particularly at the growing need, particularly within the anti-submarine warfare, airborne instrumentation and telemetry areas, for a severe-environment recorder with a lower rate capability and proportionately lower cost. DCRsi 75 retains the field-proven DCRsi transverse scan recording footprint for full crossplay compatibility with other models in the range.

Contractor

Ampex Corporation Data Systems Division.

VERIFIED

DCRsi 107/107R Digital Cartridge Recording system

The DCRsi 107/107R rack-mount and modular ruggedised systems are 1 in transverse scan, rotary digital recorders capable of recording and reproducing at any user data rate from 0 to 13.4 Mbytes/s (0-107 Mbits/s). This capability can be sustained for over 1 hour in one tape cartridge with a total storage capacity of 48 Gbytes which is equal to four 14 in tape reels recorded on a conventional 28 track HDDR.

Capitalising on improvements in integrated circuit density and the use of ASICs has resulted in a DCRsi 107 system that is smaller, lighter and has 50 per cent less power consumption than the DCRsi system it replaces. It provides a computer-friendly mass storage data peripheral to any air, sea or land platform.

The DCRsi 107 features a format and interface that is compatible with the DCRsi and the DCRsi 240, constant packing density, 96 Mbit internal I/O buffer and data block, time code addressing and searching and automatic playback alignment. Data transfer can be continuous, in bursts or changing. A ruggedised version, the DCRsi 107R, is designed for hostile environments. The system has RS-232 and RS-422 control interfaces.

The DCRsi 107 is available in both single module, 19 in rack-mount and modular ruggedised configurations. The ruggedised DCRsi 107R is available either as a two module record only or as a three module record/reproduce configuration.

Specifications

Dimensions:
(tape transport module) 373.4 × 274.3 × 175.3 mm
(rec/rep electronics module) 360.7 × 485.1 × 152.4 mm
(optional AC input power module) 254 × 152 × 76.2 mm
Weight:
(tape transport module) 14.75 kg
(record electronics module) 15.88 kg
(reproduce electronics module) 15.42 kg
(optional AC power module) 4.54 kg
Power supply: 28 V DC
210 W (record only)
430 W (record/reproduce)
Temperature:
−30 to +50°C (operating)
−54 to +70°C (non-operating without tape)
Altitude: up to 50,000 ft operating

Operational status

In production for German Air Force Tornado IDS aircraft reconnaissance pods.

Contractor

Ampex Corporation Data Systems Division.

VERIFIED

DCRsi 120 Digital Cartridge Recording system

The DCRsi 120 recorder is equipped with a fully integrated internal memory buffer which provides total isolation between the user interface and the instantaneous timing demands of the tape transport. Via this front-end buffer, the DCRsi recorder will unconditionally follow the user's data clock – any data rate from zero up to 120 Mbits/s can be recorded or played back either continuously, in bursts, or while fluctuating at any slew rate. Instant-on record capability provides for event capture without time lag associated with tape speed lock-up. No operator adjustments are required as the data rate changes. The DCRsi recorder completely emulates a solid-state FIFO memory. The system is a slave to user interface equipment, thus simplifying the task facing the data and control interface designer.

The DCRsi 120 recorder utilises a field-proven transverse-scanning tape transport design, featuring mechanical simplicity, compact size and a short, co-planar tape path. The front loading/unloading of the tape cartridge is done instantly, simply and reliably because there are no elevators or other complex mechanisms used to transport the cartridge. The compact transverse scanner assembly contains six azimuth record/reproduce heads. The scanner assembly is self-contained and is easily replaceable in the field with a typical head life exceeding 3,000 head-to-tape hours.

DCRsi 120 digital cartridge recording system
1998/0015321

The DCRsi 120 recorder is available in both a two-module and a single-unit rack-mount configuration.

The rugged 50 Gbyte-capacity DCRsi tape cartridge is made of fibre-reinforced polycarbonate. It is flame retardant, UV resistant, non-toxic, with high impact strength. Tape durability is rated at more than 200 passes.

Contractor

Ampex Corporation, Data Systems Division.

NEW ENTRY

DCRsi 240 Digital Cartridge Recording system

The DCRsi 240 rack-mount and modular system is a 1 in transverse scan, rotary digital recorder capable of recording and reproducing at any user rate from 0 to 30 Mbytes/s (0-240 Mbits/s). The byte parallel data interface used in the DCRsi, DCRsi 107 and DCRsi 240 consists of eight parallel data lines, one common clock and one enable signal. This interface compatibility, along with the common format on tape, results in full tape interchange among all three DCRsi models. Tapes recorded on the DCRsi 240 can be played on DCRsi and DCRsi 107, and vice-versa. The 240 Mbits/s transfer rate capability is accomplished by increasing the number of record/playback heads from 6 to 12 and adding a second data channel, while retaining the single channel I/O architecture intact. Data is now recorded or played back on two heads simultaneously. The data from each of the two heads is processed by separate data channels at 120 Mbits/s each and the two data streams are combined in the single 72 Mbyte data buffer which yields a sustained total throughput of 240 Mbits/s with a peak rate of up to 300 Mbits/s. This capability can be sustained over the entire tape cartridge, resulting in a storage capacity of 48 Gbytes which is equal to four 14 in tape reels recorded on conventional 28 track HDDRs.

Capitalising on improvements in integrated circuitry density and the use of ASICs has resulted in a DCRsi 240 system that is smaller, lighter and consumes less power than the previous DCRsi systems. The DCRsi 240 provides the power of a computer-friendly mass storage data peripheral to any air, sea or land platform.

The DCRsi 240 features constant packing density providing 48 Gbytes user storage per cartridge regardless of data rate, data block and time code addressing and searching and automatic playback alignment. Data transfer can be continuous, in bursts or changing. The two-module ruggedised DCRsi 240 is designed for hostile environments. The system is configured for RS-232 and RS-422 control interfaces.

Specifications

Dimensions:
(tape transport module) 373.4 × 274.3 × 175.3 mm
(rec/rep electronics module) 388.6 × 317.5 × 195.6 mm
(cartridge) 266.7 × 165.1 × 41.9 mm
(optional AC power modules) 254 × 152.4 × 76.2 mm
Weight:
(tape transport module) 14.74 kg
(rec/rep electronics module) 14.06 kg
(cartridge) 1.13 kg
(optional AC power module) 4.54 kg
Power supply: 28 V DC, 450 W typical
Temperature range:
−30 to +50°C (operating)
−54 to +70°C (non-operating without tape)
Altitude: up to 50,000 ft operating

Contractor

Ampex Corporation Data Systems Division.

VERIFIED

PAR 1000 Mini-HUMS

PAR 1000 is a compact, lightweight, Mini-HUMS system that provides automatic recording of turbine engine parameters for exceedance monitoring, health trending, and maintenance diagnostic purposes. It also monitors and records engine and airframe hours and cycle data. It can be fitted to both fixed- and rotary-wing turbine-powered aircraft.

The standard PT6 system comprises a computer pilot's display, warning lights, associated transducers, and installation kit. Associated ATS (DOS)-based or GBS (Windows)-based ground support systems are used to analyse the data.

Operational status

The approved application list includes a large number of Bell, Bolkow, Eurocopter and Hughes helicopter types, together with Cessna, Shorts and Raytheon fixed-wing aircraft.

Contractor

Avionics Specialties Inc.

NEW ENTRY

ADAS-7000 Airborne Data Acquisition System

AYDIN TELEMETRY manufactures a wide range of telemetry and data recording systems for missile and air vehicle flight testing including the ADAS-7000, originally designed for the IAI Lavi flight test programme. This provides a master/slave recording system for aircraft use, using advanced signal conditioning and encoding hardware. A PMU-700 Series III program master unit is at the heart of the ADAS-7000; this is used as a central encoder/controller of the whole system and communicates on a MIL-STD-1553 digital data highway with a number of slave units. Over 5,000 channels can be monitored and recorded.

Contractor

AYDIN TELEMETRY.

UPDATED

ATD-800-II airborne tape system

The ATD-800-II ruggedised tape deck is a low-cost, high-performance digital data record/reproduce tape media subsystem suitable for use in the harsh environmental conditions usually associated with flight test applications.

All components of the ATD-800-II are contained in a ruggedised chassis with internal shockmounts to physically isolate all devices from externally imposed vibration. The chassis is sealed from the potential contamination of the surrounding atmosphere and an internal temperature control system is provided to maintain temperature within operating limits. The recording system is the industry recognised DLT-4000 cartridge tape mechanism by Quantam. The front panel of the ATD-800-II contains a comprehensive set of operating status indicators.

The ATD-800-II is designed to interface directly to a standard SCSI-2 controller such as is readily available on computer systems as well as on the AYDIN TELEMETRY MiniARMOR-700 Data Multiplexer/Demultiplexer. When used in conjunction with the MiniARMOR-700, the ATD-800-II supports recording of multiple combinations of serial and parallel data

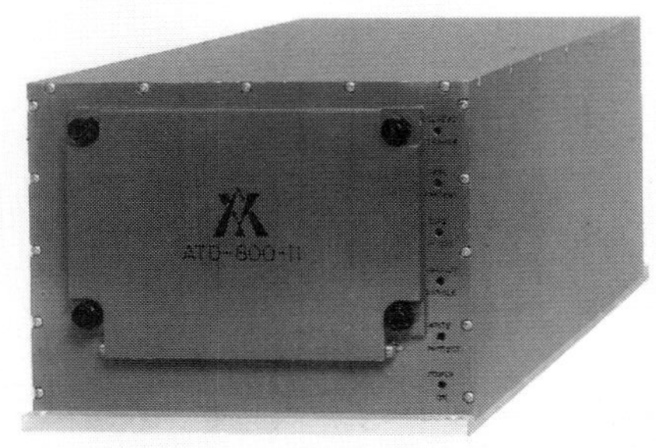

Series II ATD-800 digital data tape recorder ***1997***

sources. For example, several channels of PCM may be combined with time, voice, MIL-STD-1553, parallel and digital and analogue inputs. Playback of the data may be accomplished by direct connection of the SCSI-2 interface to a computer system or by using a playback configuration of the MiniARMOR-700. The latter approach permits coherent reconstruction of the original data streams.

Key parameters and features include record/reproduce rates of 1.5 Mbytes/s sustained, 5.0 Mbytes/s burst, 20 Gbytes data storage capacity per cartridge, non-compressed, BER less than one error in 10E17 bits, record time of 3.6 hours at maximum rate, single-ended or differential SCSI-2 interface, tape dubbing software (requires record and playback unit) high-speed access to stored data, remote control option. Records PCM, 1553, voice, time.

Specifications

Dimensions: 387 × 173 × 136 mm
Weight: 6.8 kg
Operating temperature range: –20 to +50°C

Contractor

AYDIN TELEMETRY.

UPDATED

MiniARMOR-700 multiplex/demultiplex system for digital data recording

The AYDIN TELEMETRY MiniARMOR-700 multiplex/demultiplex system for digital data recording is based on the design concepts of the Asynchronous Real-time Multiplex and Output Reconstructor (ARMOR) developed by Calculex, Inc.

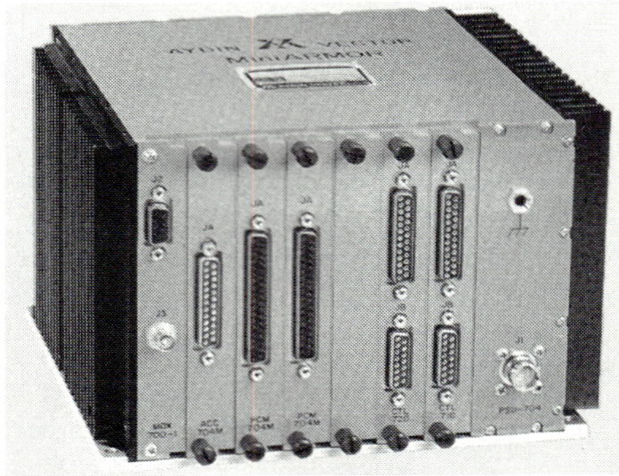

MiniARMOR-700 miniature asynchronous real-time multiplexer and output reconstructor **1997**

MiniARMOR-700 systems combine a wide variety of analogue and digital signals into a single high-speed (up to 240 Mbps) composite digital data stream for digital recording. A system configured to reconstruct the composite signals does so while maintaining inter-channel data coherency.

The MiniARMOR-700 supports two operational formats to maintain full backward compatibility with Calculex ARMOR 1 products while affording the full flexibility and performance of the advanced MiniARMOR-700 design.

The MiniARMOR-700 is a modular system with a full compliment of modules and options to support virtually any mix of analogue and digital signal types to support specific user applications. Modules may be changed in the field for user ease.

The MiniARMOR-7000 is available in a 5, 7, 9 and 11 slot configuration, with a maximum measurement of 136 × 190.5 × 261.6 mm and a maximum weight of 6.16 kg. It operates from 28 V DC, 115/220 V AC, 46-63 Hz and has an operating temperature range of –30 to +70°C.

Contractor

AYDIN TELEMETRY.

UPDATED

Health and Usage Management System (HUMS) for helicopters

The BFGoodrich HUMS acquires data from 150 input channels at a high sample rate, storing the complete data only if an anomalous condition is detected, displaying it to the pilot as a caution using a 3 in indicator. The stored data is written on to an onboard memory card for post-flight insertion into the ground station Windows-based computer to produce a series of operations, maintenance and engineering reports. The onboard system comprises the Main Processor Unit (MPU) a ½ ATR short package; the Cockpit Display Unit (CDU), a 3 ATI five line display area multifunction key interface; and a Data Transfer Unit (DTU), containing 20 Mbytes transfer storage, typically representing 10 hours of flight data retention, and a PCMCIA Flash card.

Operational status

BFGoodrich Aerospace has received FAA STC certification for its HUMS system installed on the Agusta 109K2 flown by Rega International Air Ambulance, and has been awarded a contract from the US Navy to fit the system on its Sikorsky SH60 and CH53 fleets.

Contractor

BFGoodrich Aerospace, Aircraft Integrated Systems.

NEW ENTRY

The BFGoodrich health and usage management system showing left to right: CDU/DTU/Flash card, MPU, ground station **1998**/0015322

Embedded and Special Application mass storage (E/SA)

The E/SA is intended for those applications where embedded or form factor restrictions require a compact militarised disk. E/SA is based on a field-proven high-performance 3.5 in magnetic disk. It consists of a sealed disk cartridge with 426 or 1,054 Mbytes of formatted data storage, an optional cartridge interconnect module and an optional mounting frame.

Specifications

Dimensions:
(disk cartridge) 58.4 × 124.5 × 177.8 mm
(interconnect module) 61 × 125.7 × 57.1 mm
(mounting frame) 292.1 × 94 × 262.6 mm
Weight:
(disk cartridge) 2.27 kg
(interconnect module) 0.18 kg
(mounting frame) 1.81 kg
Power supply:
(disk cartridge) 5 V DC, 0.75 A and 12 V DC, 0.8 A
(interconnect module) 5 V DC and 12 V DC, 1.3 W

Contractor

Computing Devices International.

VERIFIED

Hard Disk Subsystem (HDS)

The HDS provides a highly reliable form and fit replacement for Miltope and Ampex magnetic tape transports in an airborne environment. Based on field-proven 3.5 in magnetic disk technology, the HDS provides a ×65 increase in storage capacity and reductions in weight and power. Each HDS enclosure emulates up to three tape transports with electrically isolated and redundant hardware for each magnetic tape transport equivalent set. An operator control panel provides a user-friendly interface, while extensive BIT capability allows for easy field maintenance and eliminates the need for special test equipment. The disk media is in a removable cartridge which currently uses a drive with 1 Gbyte of formatted storage.

The HDS is designed to support incorporation of future higher-capacity disk technology by simply replacing the drive in the removable cartridge with new 3.5 in disks. Capacities of 8 Gbytes are planned.

Specifications

Dimensions:
(enclosure) 469.9 × 787.4 × 314.2 mm
(1,054 Mbyte cartridge) 58.4 × 124.5 × 177.8 mm
Weight:
(enclosure) 43.1 kg
(1,054 Mbyte cartridge) 2.27 kg
Power supply: 115/200 V AC, 400 Hz, 3 phase, 165 W
Reliability: 14,500 h MTBF

Contractor

Computing Devices International.

VERIFIED

Model R Mobile Mass Storage System (MMSS)

The MMSS/R provides solutions to meet a broad range of applications. It is based on field-proven high-capacity high-performance 5.25 in (133 mm) Winchester disk products.

Through the use of common module assemblies, many configuration options are available to meet specific system and application requirements. The system is designed for easy operator use and maintenance.

The MMSS/R is packaged in a convection-cooled ATR enclosure. Computer system interface is SCSI or ANSI 3.131-1986, with single-ended or optional differential input/output. It is configured with one disk cartridge of 1,054 Mbytes capacity. The disk cartridges are designed to incorporate higher-capacity disk storage technology.

Specifications

Dimensions:
(enclosure) 194 × 257 × 395 mm
(1,054 Mbyte cartridge) 93.5 × 180.3 × 277.1 mm
Weight:
(enclosure) 18.6 kg
(1,054 Mbyte cartridge) 4.8 kg
Power supply: 115 or 220 V AC, 50/60/400 Hz and 28 V DC
Reliability: 10,000 h MTBF (one cartridge)

Contractor

Computing Devices International.

VERIFIED

Operator workstation Embedded Disk (OED)

The OED provides a high-performance, low-cost solution to meet a broad range of applications, including operating systems and data storage. The OED is based on high-capacity 3.5 in Winchester disk products currently used on several airborne and other programmes.

The OED consists of two 1 Gbyte Removable Transportable Media Modules (RTMM) and a Power Converter Module (PCM). The OED is packaged in an air-cooled 19 in rack-mounted enclosure. The system is designed for easy operational use and maintenance. System architecture allows daisy chaining of up to three OEDs over a single SCSI channel.

The OED is designed to support incorporation of future higher-capacity disk technology by simply replacing the drive in the removable cartridge with new 3.5 in disks. Capacities of 8 Gbytes are planned.

Specifications

Dimensions:
(OED) 525.8 × 180.3 × 381 mm
(RTMM) 58.4 × 124.5 × 177.8 mm
Weight:
(OED) 18.14 kg
(RTMM) 3.63 kg
Power supply: 28 V DC, 75 W
115/200 V AC, 400 Hz, 3 phase and 115/200 V AC, 60 Hz, single phase options
Temperature range:
(OED) −40 to +55°C
(RTMM) −20 to +55°C
Altitude: up to 42,000 ft
Reliability: 5,000 h MTBF

Contractor

Computing Devices International.

VERIFIED

Versatile Mass Media Memory (VM3)

The VM3 is a 19 in rack-mounted, very large capacity mass storage system for applications where high performance is required. Designed to meet MIL-E-5400, VM3 can be used for airborne and other applications.

Each VM3 suite consists of up to four environmentally sealed SCSI-2 Removable Transportable Media Modules (RTMM), up to two disk controllers and up to four Input/Output (I/O) interfaces. A standard VM3 with four TRMMs provides a total system storage capacity of 4 Gbytes, with a 5 Mbyte/s burst transfer on each SCSI I/O channel.

The fan-cooled VM3 offers an internal architecture that provides a very high data throughput of up to 64 Mbytes/s and multitasking. The internal bus structure permits burst multiplexing of data as well as overlapped disk access by up to four host computers.

VM3 may be configured with up to two disk controllers, each with a 256 kbyte data buffer and up to four I/O interfaces. Available I/O interfaces include SCSI-3, NTDS with AN/UYH-3 emulation and NTDS with ANEW. The SCSI I/O adaptors and disk controller architecture support an internal sustained data transfer rate of 5 Mbytes/s.

Each 1 Gbyte RTMM has a 240 kbyte buffer for read and write cacheing. The RTMMs support a burst data transfer rate of 10 Mbytes/s with sustained data transfer rates ranging from 1.3 to 2.1 Mbytes/s. The RTMM has an average latency time of 6.8 ms and an average seek time of 12 ms. It is mounted in a shock/vibration isolation frame. RTMMs are interchangeable within the VM3 unit and with other VM3 systems.

The VM3 features comprehensive BITE, providing fault detection and executing independent diagnostics with greater than 85 per cent accuracy at the replaceable module level. BITE can be executed via the operator keyboard/display in the front cover or remotely via an RS-232 interface.

The VM3 is designed to support incorporation of future higher-capacity disk technology by simply replacing the drive in the removable cartridge with new 3.5 in disks. Capacities of 8 Gbytes are planned.

Specifications

Dimensions: 482.6 × 660.4 × 355.6 mm
Weight: 63.5 kg
Power supply: 115/200 V AC, 400 Hz, 3 phase, 450 W
115/200 V AC, 60 Hz, single phase and 28 V DC options
Temperature range: −40 to +55°C
Altitude: up to 42,000 ft
Reliability: 5,000 h MTBF

Contractor

Computing Devices International.

VERIFIED

8 mm Colour airborne Video Tape Recorder (CVTR)

The 8 mm CVTR incorporates the latest technology in a line of 8 mm video recorder products developed by DRS Technologies Inc. Lightweight, cost-effective, low-powered and ruggedised, the system provides high-resolution video recording of the head-up display and other cockpit data for the combat missions of the A/OA-10A aircraft. It is specifically designed to operate under the harsh environments experienced by aircraft supporting active ground operations.

The Commercial-Off-The-Shelf (COTS)-based design of DRS Technologies' 8 mm recorder product line has resulted in a substantial cost saving for its customers, while still meeting or exceeding critical performance, reliability and long-term support requirements.

Operational status

DRS Technologies has successfully delivered over 1,200 recorders in this product line for air, sea and land platforms spanning the US Navy, Army and Air Force. Applications include the US Navy's F/A-18 aircraft, the US Army's OH-58D Kiowa Warrior helicopter and the US Air Force's A/OA-10A aircraft.

Contractor

DRS Technologies Inc
DRS Precision Echo Inc.

VERIFIED

AN/AQH-9 mission recorder system

AN/AQH-9 mission data recorder system showing the interface unit (left), remote control unit (centre) and rotary-head data recorder (right) **1998**/0015323

The AN/AQH-9 mission recorder system was designed as the ASW system recorder for the US Navy SH-60F CV inner zone helicopter; it consists of four weapon replaceable assemblies: a Mission Tape Recorder Interface Unit (MTRIU), a VHS video cassette rotary-head data recorder, a Remote-Control Unit (RCU), and a video cassette. To minimise weight, the AN/AQH-9 was configured for record-only; other systems are available that offer in-flight playback.

The rotary-head data recorder accepts a modified digital data stream for recording black-and-white video signals The recorder incorporates two side tracks (time code and auxiliary voice) for expanded capabilities. Standard VHS tape cassettes are used, with a recording rate of 3.3 Mbits using a modified RS-170 format.

Specifications

Tape cassette: standard VHS (½ in) tape format
Playing time: 2 h with T-120 tape; 2 h 40 min on T-160 tape
Dimensions:
(recorder) 210.8 × 289.5 × 93.9 mm
(interface unit) 256.5 × 213.3 × 193 mm
(remote-control unt) 146 × 152.4 × 66 mm
Weight: 16.9 kg
Power: 28 V DC, 25 W; 115 V AC, 400 Hz, 550 W

Operational status

Developed as the ASW mission data recorder for the US Navy's SH-60F CV inner zone helicopter, and already fitted to approximately 140 US Navy SH-60F and 14 Taiwanese Navy SH-60F CV aircraft; it could be tailored for similar helicopter and aircraft ASW applications.

Contractor

DRS Technologies Inc
DRS Precision Echo Inc.

NEW ENTRY

AN/AQH-11 high-density digital mission recorder system

The AN/AQH-11 high-density digital mission recorder system is a high-performance, compact, lightweight, half-inch VHS system, designed specifically for military helicopters.

When configured in a single-channel mode, the system captures 2 hours of data at a user rate of 8.33 Mbps (T-120 S-VHS cassette). The system comprises three weapon replaceable assemblies interface unit, remote control unit and tape transport.

Specifications
Signal acquisition interface:
RS-343 raster video
AN/ARR-84 sonobuoy receiver (4 channels)
ICS/UHF voice audio (4 channels)
IRIG-B time code
sensor audio (2 channels)
MIL-STD-1553B databus
Dimensions:
(interface unit) 256.5 × 213.4 × 193 mm
(tape transport) 210.8 × 290 × 95 mm
(remote-control unit) 147.3 × 152.4 × 66 mm
Weight: 17.7 kg
Power:
(mission recorder) 28 V DC, 25 W; 115 V AC, 400 Hz, 100 W
(interface unit) 115 V AC, 400 Hz, 30 W

Operational status
Derived from the AN/AQH-9 system used on the US Navy SH-60F CV inner zone helicopter.

Contractor
DRS Technologies Inc
DRS Precision Echo Inc.

NEW ENTRY

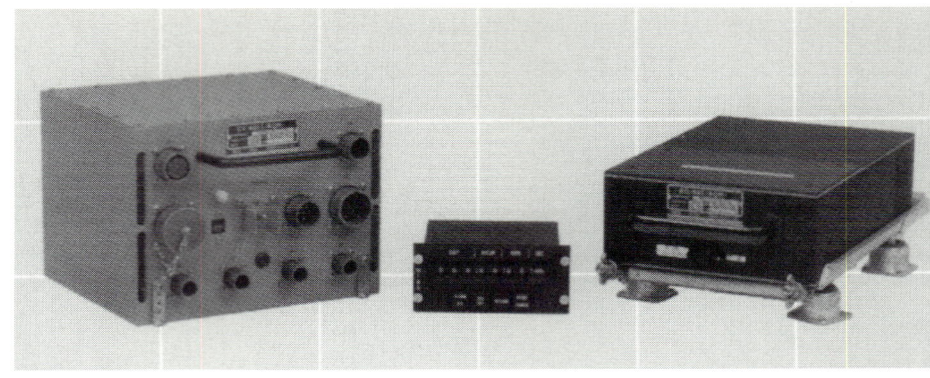

AN/AQH-11 high-density digital mission recorder system showing the interface unit (left), remote control unit (centre) and tape transport (right) ***1998***/0015324

AN/USH 42() mission recorder/reproducer set

The AN/USH-42() is a high-performance compact, lightweight, Mission Recorder/Reproducer Set (MR/RS) comprising two weapon replaceable assemblies: the Recorder Reproducer Unit (R/REU) and the Remote Control and Advisory Panel (RCAP).

The MR/RS is capable of recording two channels of video data, two audio channels and digital annotation data. The two video channels to be recorded are selected from up to four video inputs. Dual self-contained digital scan converters provide input processing of radar, FLIR, missile video and other data for recording on dual Hi-8 mm tape transports. Annotation data is extracted from the MIL-STD-1553B databus. The AN/USH-42() is capable of airborne and ground playback.

Specifications
Video resolution: 400 horizontal lines nominal
Video recording time: 2 h/channel
Output: RS-170 video (2 channels)
Number of channels: 2 video, 2 audio
Dimensions:
(R/REU) 330.2 × 431.8 × 222.25 mm
(RCAP) 76.2 × 254 × 146.05 mm
Weight: 18.18 kg
Power:
(R/REU) 28 V DC, 200 W
(RCAP) 5 V AC, 400 Hz

Operational status
The AN/USH-42() is part of the US Navy S-3B Viking avionic system; it was developed initially for use in the US Navy's A-6E Intruder aircraft.

Contractor
DRS Technologies Inc
DRS Precision Echo Inc.

NEW ENTRY

AN/USH-42() mission recorder/reproducer set ***1998***/0015325

DCMR-24 Digital Cassette Mission Recorder

The DCMR-24 is a new addition to DRS Precision Echo's line of products for recording solutions. The DCMR-24 is a multifunctional digital recorder/reproducer designed around a digital 8 mm transport, utilising standard VME (6 μ) circuit cards and backplane. The digitial cassette is housed in a rugged package designed for severe environmental conditions.

A three slot, industry-standard, VME backplane provides the DCMR-24 with a wide variety of recording configurations to meet changes in mission recording requirements. The DCMR-24 supports multiple analogue, digital (serial), video, and MIL-STD-1553B bus monitoring channels.

The DCMR-24 utilises AIT (Advanced Intelligent Tape) technology developed by Sony for the computer industry. With Ad-Me (Advanced Metal Evaporated) tape, the AIT provides storage capacity of up to 25 Gbytes (non-compressed) and recording speeds of up to 24 Mbps. The Digital Cassette Mission Recorder utilises the AIT tape cassette, which incorporates a 64 k memory chip for file, and header information providing fast data retrieval. This recorder can be controlled/operated with front panel switches, remotely over the MIL-STD-1553B bus, or remotely via an RS-232 port running a terminal utility programme using an optional remote control panel.

Specifications
Recorder type: 8 mm rotary, head helical scan with AIT cassette
Interface options: digital (serial), analogue, video and MIL-STD-1553B bus
Sustained data rate: 3 MB/s (up to 9 MB/s/w/compression
Burst data rate: ~12 MB/s (asynchronous mode enabled); ~20 MB/s (synchronous mode enabled)
Buffer size: 4 MB
Media format: AIT-1
Tape capacity: 25 GB (up to 75 GB w/compression)
Interface options:
Serial data:
up to 24 I/O channels
~ (max aggregate of 24 Mbps = 2 h record time)
impedance 120 ohm or selectable
single-ended or differential
aggregate data rate 24 Mbps
RS-422, RS-232, HDLC, asynchronous, and synchronous

DCMR-24 Digital Cassette Mission Recorder ***1998***/0015326

PCM NRZ, biphase, Manchester and others
Video channels:
1 or 2 channels
~ colour (560 × 480 × ¸24 bit near full motion = 2 h record time
~ monochrome (560 × 480 × 8 bit full motion = 2 h record time
SCSI-2 fast/wide:
3 MB/s (up to 6-9 MB/s w/compression)
~ (24 Mbps = 2 h record time)
12 MB/s (asynchronous mode enabled)
20 MB/s (synchronous mode enabled)
Analogue channels:
up to 30 I/O channels
MIL-STD-1553B:
up to 8 dual-redundant channels
Bus interface:
single/multimode (remote terminal and/or bus monitor)
Power: 28 V DC, 120 W
Dimensions: 152 × 247.65 × 368.3 mm
Weight: 9.09 kg

Contractor
DRS Technologies Inc
DRS Precision Echo Inc.

NEW ENTRY

DCMR-100 Digital Cassette Mission Recorder

The DCMR-100, DRS Precision Echo's new digital instrumentation recorder/reproducer, is a user-configurable multifunctional recorder. Designed around the Sony Betacam® transport, the DCMR-100 incorporates industry-standard VME circuit cards in a five slot VME backplane. By embedding the digital (serial), analogue, video, and/or MIL-STD-1553B interfaces in the DCMR-100, the need for a separate interface box and cabling is eliminated, therefore reducing the space, weight and power requirements for installation.

The DCMR-100 can record/reproduce up to 64 channels (128 channels record only) of instrumentation data. Sampling rate and resolution can be configured to meet requirements. Eight channels of video can be recorded, or 32 channels of digital (serial) data at maximum aggregate of 12 MB/s sustained rate. With the optional MIL-STD-1553B bus interface, the DCMR-100 can record up to 16 dual-redundant channels.

The DCMR-100 also provides a standard SCSI-2 Fast/Wide interface for raw data input/output (data dump). At the maximum data rate of 12 MBs, the DCMR-100 writes 42 GB of data to a single commercially available DTF™ cassette in less than 1 hour, with a bit error rate of 1×10^{17}. The DCMR-100 operates off 28 V DC power, at 250 W. The DCMR-100 is controlled/operated from front panel controls, or remotely with the MIL-STD-1553B bus, or remotely via an RS-232 port running a terminal utility programme (an optional remote-control panel will be available spring, 1998).

Specifications
Recorder type: ½ in rotary helical scan, digital cassette
Interface options: digital (serial), analogue, video
Sustained data rate: 12 MB/s

Burst data rate:
(asynchronous mode enabled) ~12 MB/s
(synchronous mode enabled) ~20 MB/s
Media format: DTF
Tape capacity: 42 GB
Interface options:
Digital (serial) data:
Up to 32 I/O channels
impedance 120 ohm or selectable
single-ended or differential
aggregate data rate 12 Mbps
RS-422, RS-232, HDLC, asynchronous and synchronous
PCM NRZ, biPhase, Manchester and others
Video channels:
up to 8 I/O channels
~ colour (560 × 480 × 24 bit near full motion = 2 h record time)
~ monochrome (560 × 480 × 8 bit full motion = 2 h record time)
SCSI-2 fast/wide:
up to 12 MB/s
12 MB/s (asynchronous mode enabled)
20 MB/s (synchronous mode enabled)
Search speed: 300 MB
Data buffer: 32 MB
Analogue channels:
up to 64 I/O (128 channels record only)
MIL-STD-1553B:
up to 16 dual-redundant channels
Bus interface:
single/multimode (remote terminal and/or bus monitor)
Power: 28 V DC, 250 W
Dimensions: 482.6 × 393.7 × 292.1 mm
Weight: 25 kg

Contractor
DRS Technologies Inc
DRS Precision Echo Inc.

??????

DCMR-100 Digital Cassette Mission Recorder
1998/0015327

Replacement Data Storage System (RDSS)

DRS Technologies has been awarded the contract to provide the Replacement Data Storage System (RDSS) for the Royal Norwegian Air Force P-3 Upgrade Improvement Programme (UIP) by Lockheed Martin Tactical Defense Systems, Eagan. The RDSS employs Commercial-Off-The-Shelf (COTS) magneto-optical laser technology disk drive to provide 2.6 Gbyte of storage. As a result, the RDSS will allow data file storage on a single disk instead of multiple tape cartridges.

Designed for compatibility with the existing NTDS (Navy Tactical Data System) interfaces and future signal processors, the RDSS will provide a file management capability for supporting multiple clients on an Ethernet interface. No change to existing software is required, but mission performance and reliability will be considerably enhanced.

Contractor
DRS Technologies Inc
DRS Precision Echo Inc.

VERIFIED

WRR-812 airborne video tape recorder

The WRR-812 airborne video tape recorder/reproducer is a small lightweight 8 mm recording set for ruggedised aviation or ground mobile applications. The system conforms to the high band 8 mm commercial standard for recording RS-170 or NTSC colour video and accompanying audio. Replay is also possible on any commercial desktop VCR conforming to the same commercial standards. Typical installations would be for capturing HUD camera video or TV seeker output. Flight test applications for instrumentation or monitoring purposes are common.

A microprocessor-controlled command translator interface provides adaptive, user selectable options for functional control. Serial digital command links such as RS-232 and RS-422 are used. A compatible remote-control unit is available. Recording medium is the commercial standard P-6120 MP tape cassette which is compatible with commercial 8 mm VCRs. Tape cassette loading is from the front, with mechanical unloading without host platform power.

With an expanded command set, studio-type editing functions are available. Features such as frame by frame advance/reverse, slow motion, full frame stills and tear-free high-speed modes are provided.

Specifications
Dimensions: 185 × 131 × 216 mm
Weight: 2.5 kg
Power supply: 28 V DC, 15 W
Record time: 2 h of video formatted data (RS-170, colour or monochrome)
Video inputs: 1 channel EIA RS-170
Audio inputs: 1 channel
Playback resolution: 400 horizontal lines nominal (4.5 MHz bandwidth)
S/N ratio: >-45 dB
Environmental: MIL-E-5400 Class 1b

Operational status
In service on A-10 and F/A-18 aircraft for tactical data recording.

Contractor
DRS Technologies Inc
DRS Precision Echo Inc.

UPDATED

WRR-812 airborne video tape recorder
1998/0015328

WRR-818 airborne video tape recorder

The WRR-818 airborne video tape recorder, with its internal transport isolator, is designed to support airborne and ground mission applications while recording up to 2 hours of video. The WRR-818 accepts EIA RS-170, NTSC colour, or monochrome. It is also available for recording PAL or CCIR format. The Hi-8 mm recording system delivers more than 400 lines of horizontal resolution and improved signal-to-noise ratio. The Audio Frequency Modulation (AFM) provides excellent fidelity and a wide dynamic range. Also in the same line of Hi-8 mm tape format products, DRS makes the WRR-812 airborne video recorder.

Applications include airborne and ground imagery recording; recording of cockpit Heads-Up Display (HUD) and MultiFunction Display (MFD) video on all modern fixed-, or rotary-wing aircraft; recording/playback of tactical Forward Looking InfraRed (FLIR), Low-Light-Level TV (L3TV) and radar video of tactical identification, targeting, and assessment; recording and in-air remotely commanded playback of Unmanned Air Vehicle (UAV) FLIR and TV for Over-The-Horizon (OTH) missions; applications requiring lightweight, low power, and ruggedised performance. A variant with front panel control selection is also available.

Specifications
Dimensions: 119.6 × 147.8 × 160.5 mm
Weight: <2.7 kg
Power supply: 28 V DC, 12 W
Record format: Hi-8 or standard 8 mm format; 2 hours record time per cassette
Control interface: RS-422A user selection

Operational status
In service on F/A-18 aircraft and OH-58D helicopters for recording tactical data on combat missions. Selected for US Navy F/A-18C/D and F/A-18E/F aircraft.

Contractor
DRS Technologies Inc
DRS Precision Echo Inc.

UPDATED

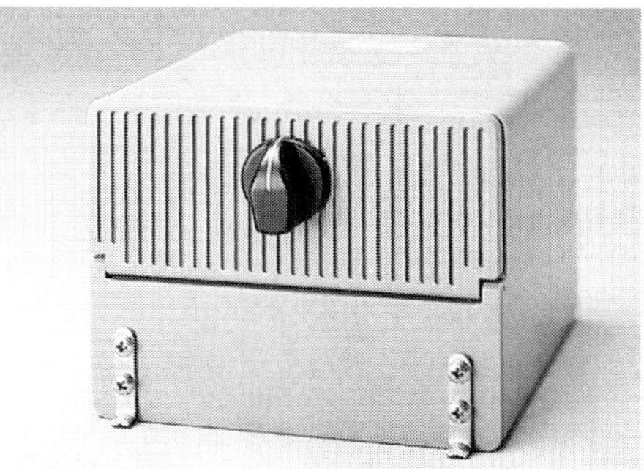

WRR-818 airborne video tape recorder
1998/0015329

WRR-833 tri-deck cassette video recorder/reproducer

The WRR-833 tri-deck cassette video recorder/reproducer is a high-performance, compact, lightweight, highly versatile video recorder and playback unit. It is designed specifically for rugged applications and the physical constraints of military fixed- or rotary-wing environments.

A simple switch allows three channels to record for 2 hours or one channel for up to 6 hours. In case of a transport failure, the system is designed for graceful degradation that allows the operator to designate priorities. Each Hi-8 mm recorder system delivers more than 400 lines of horizontal resolution. An optional video/digital (internal) multiplexes system permits multichannel video recording. The optional digital interface is ideal for capturing one or more of the dual MIL-STD-1553B, or asynchronous digital data streams.

Applications include: airborne and ground mobile imagery recording; simulated recording of cockpit Heads-Up Display (HUD); mission analysis.

Specifications
Number of channels: 3 NTSC or RS-170
Video interface: monochrome or NTSC colour, EIA standards, 1 V peak-to-peak composite video, sync negative
Max record time: 2 h (Sony P6-120 cassette) per channel; 4-6 h with sequential recording
Playback resolution:
(Hi-8 mm) 400 horizontal lines nominal
(Std-8 mm) 230 horizontal lines nominal
Dimensions: 243.8 × 149.9 × 355.6 mm
Weight: <9.1 kg
Power: 28 V DC, 120 W
Control interface: RS-232C, RS-422A or discrete

Contractor
DRS Technologies Inc
DRS Precision Echo Inc.

UPDATED

High-speed Solid-State Recorder (HSSR)

Fairchild Defense has completed design and development of its new High-speed Solid-State Recorder (HSSR) for initial application in airborne reconnaissance systems. This COTS product's I/O emulates two existing standard tape recorder interfaces: Ampex DCRsi 240™ and MIL-STD-2179. HSSR features two removable, non-volatile memory cartridges with total capacity variable up to 54 Gbytes using currently available FLASH memory technology, compete solid-state design, 240 Mbytes/s input rate, direct access with virtually zero 'search' time, and read-while-write functions.

Operational status

Development complete. The HSSR has been demonstrated in a US Air National Guard F-16, when it successfully recorded and reproduced digital reconnaissance imagery. The HSSR is now available for customer use in airborne reconnaissance programmes.

Contractor

Fairchild Defense OSC.

VERIFIED

AN/UYH-15 recorder-reproducer set, sound

The AN/UYH-15 is a voice recording and reproduction system that uses modern digital speech processing technology. It is a compact system that allows operators to monitor, record and instantly recall any recorded message. Designed primarily for use with signal acquisition systems as the standard replacement for the US Army AN/UNH-17A analogue cassette recorder, the AN/UYH-15 is ideally suited for all real-time voice transcription and analysis applications. The AN/UYH-15 may be controlled by either a host computer or by one or two control display panels.

The AN/UYH-15 can record six analogue input channels simultaneously. During recording, each operator can either monitor input or play back recorded files. Operator commentary can be recorded and time-correlated to a given signal. The signal and its related commentary can be combined for output to aid in analysis.

Each recorded signal is digitally sampled and compressed before being stored on the AN/UYH-15's hard disk, which makes it possible to store six hours of voice input. The voice compression algorithm offers proven performance in the noisy military environment, as well as high-quality reproduction independent of the signal being reproduced.

Specifications

Dimensions:
(chassis) 133.4 × 482.6 × 412.8 mm
(control/display panel) 50.8 × 228.6 × 152.4 mm
Weight:
21.32 kg (with 2 control/display panels)
16.78 kg (without panels)
Power supply: 105-130 V AC or 208-240 V AC, single phase, 47-400 Hz or 22-30 V DC, 80 W typical
Audio channels: 6 input, 2 output
Bandwidth: 300 Hz - 4.4 kHz
Capacity: 6 h
Recording media: 170 Mb formatted hard disk
Interfaces: RS-232C, IEEE-488
Environmental: MIL-STD-810D, MIL-STD-461/462, TEMPEST

Contractor

GTE Government Systems.

VERIFIED

Aircraft Propulsion Data Management Computer (APDMC)

The APDMC is primarily an engine monitoring, aircraft warning and display and data transfer unit. It acquires engine data from electronic engine controls and provides engine health, status, maintenance and limit functions.

Avionics systems are connected to the warning and caution systems MIL-STD-1553B bus and ARINC 429 bus through the APDMC with 37 other aircraft LRU analogue and discrete input/output parameters. The APDMC provides aircraft stall, overspeed, take-off and horizontal stabiliser warning and display; records engine maintenance data and system flight data; and provides cockpit display data for system LRUs, fault history, propulsion and exhaust gas temperature.

The signal input/output complement includes dual-redundant MIL-STD-1553B multiplex databusses; eight receive and seven transmit ARINC 429 serial databusses; two ARINC 573-7 serial databus outputs of which one is used for an AIMS recorder output; 35 analogue inputs including a mix of RVDTs and LVDTs plus excitation; a mix of AC and DC absolute and ratiometric signals; 26 discrete inputs including series and shunt; and six discrete outputs.

The MIL-STD-1750 microprocessor and memory management unit controls an expandable 128 k onboard ROM, 32 k RAM and 32 k NV fault storage. System BIT allows fault isolation to the offending LRU with 97.5 per cent confidence. Internal BIT provides onboard unit status. Non-volatile fault storage allows ground-based interrogation of LRU faults.

Specifications

Dimensions: 123.9 × 193 × 320 mm
Weight: 5.55 kg
Power supply: 28 V DC, 40 W
115 V AC

Operational status

In production for the US Air Force C-17.

Contractor

Hamilton Standard Division of UTC.

VERIFIED

DFDAU 120 Digital Flight Data Acquisition Unit

Complying with ARINC 717 and performing the same functions as the FDAU for the mandatory flight recording systems on the new-generation transports built to ARINC 700, the DFDAU 120 Digital Flight Data Acquisition Unit contains a microprocessor permitting it to record some AIDS information in addition to its crash recorder functions.

Specifications

Dimensions: 6 MCU
Weight: 6.8 kg
Power: 40 W

Operational status

The DFDAU 120 is standard equipment on the Boeing 767 and 757 and is a basic option on the Airbus A310. It is also used as a building block for an expanded ACMS on the A300 and A310.

Contractor

Hamilton Standard Division of UTC.

VERIFIED

DMU 100 and 101 Data Management Units

The DMU 100/101 Data Management Units are the brain for the ACMS system, analysing the real-time information from the FDAU, the ADAU or the DFDAU. They control the digital ACMS recorder, having program logic that determines what information to record and when to do so. They also provide data to displays, including the FDEP and airborne printer, and contain extensive built-in test equipment.

Specifications

	DMU 100	DMU 101
Dimensions	1 ATR long	6 MCU
Weight	16.36 kg	5.7 kg
Power	180 W	55 W

Operational status

The DMU is part of the company's Mk II and expanded Mk III ACMS system in the following aircraft:
DMU 100: 747, DC-10 and A300
DMU 101: A310 and 767.

Contractor

Hamilton Standard Division of UTC.

VERIFIED

DMU 120 Data Management Unit

The DMU 120 is a powerful microprocessor-based data acquisition and processing tool utilised on the A320 aircraft for aircraft, engine and APU condition monitoring and for intensive aircraft systems troubleshooting. It collects comprehensive aircraft and engine data via ARINC 429 serial input ports. Real-time data processing and analysis is performed. Information in the cockpit is made available in convenient format via the MCDU and printer. Automatic, cockpit and uplink requested information supplied by the DMU is transmitted to the ground by ACARS. The DMU provides for data reduction recording to the DAR which permits economical ground-processing of airborne data.

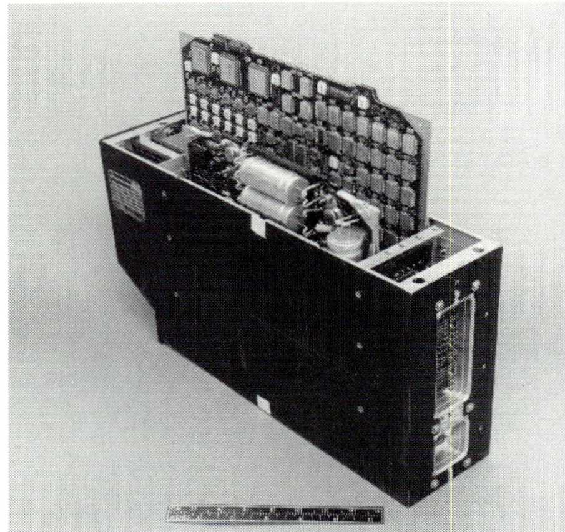

The DMU 120 is installed in the Airbus A320

Specifications

Dimensions: 195 × 95.5 × 287.6 mm
Weight: 4.4 kg
Power: 30 VA

Operational status

In production for the Airbus A320.

Contractor

Hamilton Standard Division of UTC.

VERIFIED

DSS-100 Data Storage Set

The DSS is a major element of the US Navy Flight Incident Recorder and Aircraft Monitoring System (FIRAMS) on the F/A-18 aircraft. It also performs data storage functions on the A-6, F-14, AV-8B and V-22 aircraft. The DSS consists of a Data Storage Unit (DSU) and a Data Storage Unit Receptacle (DSUR). The DSUR is mounted in the cockpit. The DSU is a modular electronic memory unit which slides into the DSUR for flight and is removable for analysis off the aircraft. The DSU employs a microprocessor to extract data from a MIL-STD-1553B bus for storage in memory and vice versa. The microprocessor also expands the effective amount of internal memory by executing a data compression algorithm. The memory medium is non-volatile EEPROM installed in blocks of from 2 to 8 Mbytes. The DSUR contains no electronics or active components.

Specifications

Dimensions:
(DSU) 43 × 119 × 205 mm
(DSUR) 53 × 127 × 229 mm
(faceplate) 102 × 165 mm
Weight: 1.9 kg
Power: 12 W

Operational status

In production and service fitted to the A6, AV-8B, F-14, F/A-18.

Contractor

Hamilton Standard Division of UTC.

VERIFIED

Engine Diagnostic Unit (EDU)

Pratt and Whitney's F100-PW-220 and 229 high-performance engines come with Hamilton Standard's advanced engine monitoring system which includes both engine-mounted equipment and the associated ground-based diagnostic units.

An EDU is the on-engine module which acquires engine data from controls and sensors, records operating time and cycles, detects critical events and stores selected event parameters. The EDU also performs extensive self-health and data validity checks and stores this data, along with Digital Electronic Engine Control (DEEC) diagnostic data, for further analysis.

The hand-held Data Collection Unit (DCU) is used to gather data from the EDU, clear the EDU's memory and perform EDU/DEEC system diagnostics. Data from multiple aircraft may be stored in the DCU's removable memory module, providing easy data transfer between the flight line and a ground-based computer for logistics data recording.

The ground-based portable Engine Analyser Unit (EAU) interfaces with the EDU or the DEEC for data acquisition, memory examination or modification and data monitoring of the EDU/DEEC output data streams during engine operation. The EAU is also capable of performing diagnostic testing of the DEEC or EDU and exercising combined EDU/DEEC fault logic and event codes, providing a fast, accurate determination of engine problems.

Specifications

Dimensions:
(EDU) 241 × 279 × 127 mm
(DCU) 475 × 424 × 412 mm
Weight:
(EDU) 4.13 kg
(DCU) 23.6 kg

Operational status

In production.

Contractor

Hamilton Standard Division of UTC.

VERIFIED

FDAU 100 Flight Data Acquisition Unit

Essentially a data gatherer, the FDAU 100 flight data acquisition unit contains the signal conditioning needed to rationalise the many types of signals from engine, airframe and systems sensors. Signals are multiplexed and digitised so that they can be recorded on the DFDR for accident investigation purposes. The unit complies fully with ARINC 573 and meets regulatory agency flight data acquisition requirements.

Specifications

Dimensions: ½ ATR long
Weight: 1.8 kg
Power: 70 W

Operational status

The unit is used on Boeing 727, 737, 747, DC-9 and DC-10, and Airbus A300 airliners.

Contractor

Hamilton Standard Division of UTC.

VERIFIED

Data storage and retrieval unit

The radiation-hardened data storage and retrieval unit has evolved from the optical disk digital memory unit developed for night attack aircraft. The 300 Mbyte unit is small and lightweight, offering a significant size advantage over larger magnetic storage systems. The unit reduces the risk of loss or compromise of vital information and makes storage of information on board more efficient.

In the Northrop B-2, two rewritable data storage and retrieval units are located in the cockpit console, providing the crew with immediate access to the mission database and the ability to record mission data and performance information in flight.

Operational status

In service on US Air force B-2 aircraft.

Contractor

Honeywell Inc Sensor and Guidance Products.

UPDATED

Model A100S Solid-State Cockpit Voice Recorder (SSCVR)

The Model A100S Solid-State Cockpit Voice Recorder (SSCVR) is available for new installations or as a direct replacement for existing ARINC 557 CVRs without wiring changes to the aircraft. Existing Fairchild users installing the Model A100S SSCVR do not require new control units, microphones, or changes to existing operating procedures.

While mounted on the aircraft, recorded data cannot be extracted from the SSCVR. This protection is necessary to protect the rights of pilots and other flight crew members. The model 100S SSCVR meets the following specifications in full compliance with worldwide regulatory requirements, including EUROCAE (ED56) and CAA specifications together with FAA ISO-C123.

Specifications

Dimensions: ½ ATR short ARINC 404
Weight: 7.3 kg
Power supply: 115 V AC, 12 W

Contractor

L-3 Communications Corp, Aviation Recorders.

UPDATED

SSCVR open **1997**/0001306

Model A200S Solid-State Cockpit Voice Recorder (SSCVR)

The Model A200S solid-state cockpit voice recorder provides 2 hours of recording, with the last 30 minutes being redundant in a very high-quality format. This unit meets EUROCAE ED-56A and TSO C123a. Readout and copying of data is available instantly by using a hand-held downloading device (digital audio playback unit) without the need to go through a long computer conversion process. The solid-state characteristics of the Model A200S eliminate periodic maintenance requirements, and greatly increase reliability. It meets the severe vibration and environmental requirements of DO-160C without the need for a vibration-mounted tray. A built-in test function includes a tone generator that superimposes the signal on the audio channel to ensure the unit is functioning correctly. In addition the unit has a continuous BITE that monitors the entire memory and power supply to ensure that the unit is operating to the highest standards. The A200S accepts four channels of audio including pilot, co-pilot, area microphone and third crew member, PA system and data/timebase source via 429. The A200S includes a front-mounted underwater locator beacon.

With the A200S, audio retrieval and playback is instantly available by using a simple hand-held device without requiring complex computer conversion. However, when mounted on the aircraft, recorded audio data cannot be extracted from the recorder, to protect the rights of pilots and other flight crew.

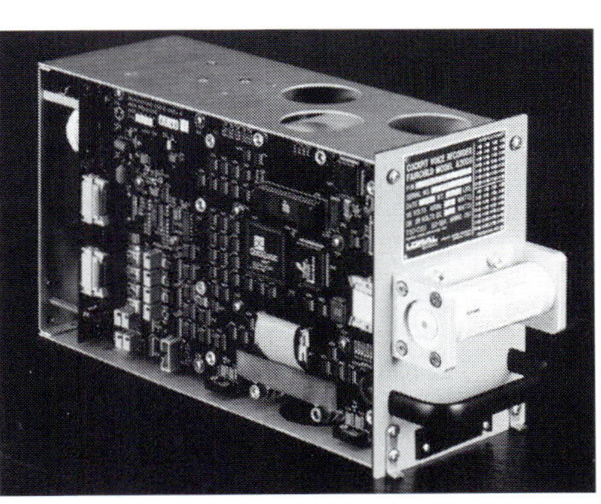

The Model A200S solid-state cockpit voice recorder **1996**

Specifications

Dimensions: ARINC 404 ½ ATR short
Weight: 8.44 kg
Power supply: 115 V AC, 400 Hz
or 28 V DC, 12 W (max)
Frequency response:
150-3,500 Hz (three inputs)
150-6,000 Hz (one input)

Contractor

L-3 Communications Corp, Aviation Recorders.

UPDATED

Model F1000 Solid-State Flight Data Recorder SSFDR

The Model F1000 SSFDR was the first production flight data recorder to provide solid-state operation. An onboard data retrieval capability eliminates the need to remove the unit from the aircraft to recover flight data. The recorder meets or exceeds all requirements and characteristics of ARINC 542A, ARINC 573/717, ARINC 747, EUROCAE ED-55, TSO C124a and RTCA DO-160C. The F1000's solid-state design also eliminates the need for a vibration-mounted tray.

With its microprocessor-based architecture, the Model F1000 incorporates the intelligence to perform sophisticated self-tests, as well as provide fault-tolerant features. The F1000 can receive data from an ARINC 573/717 flight data acquisition unit or process raw data in the ARINC 542A mode by accepting analogue signals in synchro, DC, pneumatic, frequency and discrete formats. The recorder can accommodate parameter storage requirements of 32, 64 or 128 words/s with a recording time of 25 hours. A high-speed data dump can be accomplished in 1 minute. The Model F1000 includes a front-mounted underwater locator beacon.

Specifications
Dimensions: ARINC 404 ½ ATR long
Weight: 10.2 kg
Power supply: 115 V AC, 400 Hz
or 28 V DC, 26 W (max)
Reliability: 15,000 h MTBF

Contractor
L-3 Communications Corp, Aviation Recorders.

UPDATED

Model FA2100 recorder family

Basic family
Model FA2100 was originally designed by Fairchild to meet industry demands for lighter weight and higher reliability.

Solid-State Cockpit Voice Recorder (SSCVR)
Model FA2100 SSCVR was introduced to meet the needs of commercial, regional, business and government fleets. The FA2100 meets the requirements of EUROCAE ED-56A, FAA TSO-C123a RTCA DO-160C and ARINC 757. Readout and copying (both in analogue and digital forms) data is available instantly by using a hand-held downloading device (Portable Interface – PI) without the need to go through a long computer conversion process. The solid-state characteristics of the Model FA2100 eliminate periodic maintenance requirements and greatly increase reliability. It meets the requirements of DO-160C without the need for a vibration-mounted tray. A built-in test function includes a tone generator that superimposes the signal on the audio channel to ensure the unit is functioning correctly. In addition the unit has a continuous BITE that monitors the entire memory and power supply to ensure that the unit is operating to the highest standards. The FA2100 accepts four channels of audio including pilot, co-pilot, area microphone and third crew member, PA system, data/timebase source via RS-429 interface; rotor speed encoding is available for helicopter installations. The FA2100 includes a front-mounted underwater locator beacon.

With the FA2100, audio retrieval and playback is instantly available in real time using a simple hand-held device without requiring complex computer conversion. However, when mounted on the aircraft, recorded audio data cannot be extracted from the recorder, to protect the rights of the pilots and other flight crew.

Specifications
Dimensions: ARINC 404 ½ ATR short – footprint – height is 139.7 mm
Weight: <4.36 kg
Power supply: 115 V AC, 400 Hz or 28 V DC, 9 W (max)
Frequency response:
(three inputs) 150-3,500 Hz
(one input) 150-6,000 Hz

Solid-State Flight Data Recorder (SSFDR)
Model FA2100 SSFDR was the second generation of solid-state flight data recorders designed by Fairchild. An onboard data retrieval capability eliminates the need to remove the unit from the aircraft to play back flight data. The recorder meets or exceeds all requirements and characteristics of ARINC 573/717/747, EUROCAE ED-55, TSO-C124a and RTCA DO-160C. The FA2100 solid-state design also eliminates the need for a vibration-mounted tray.

With its microprocessor-based architecture, the Model FA2100 incorporates the intelligence to perform sophisticated self-test, as well as provide fault-tolerant features. The recorder can accommodate parameter storage requirements of 64, 128 or 256 words/s with a minimum recording time of 25 hours. Future expansion of memory to allow 512 words/s is planned. A high-speed data dump can be accomplished in approximately 2 minutes. The Model FA2100 includes a front-mounted underwater locator beacon.

Specifications
Dimensions: ARINC 404, ½ ATR – footprint – height is 139.7 mm. This flight recorder is available in both long and short versions
Weight: <4.36 kg
Power supply: 115 V AC, 400 Hz or 28 V DC, 9 W (max)

Model FA2100 combination recorder
The Model FA2100 combination recorder combines the features of the SSCVR with the SSFDR to provide one combined recorder with both features. The recorder meets the characteristics of TSO-C123a/C124a, EUROCAE ED-56A/ED55, ARINC 757 and RTCA DO-160C. Readout and copying of data is available instantly by using a hand-held downloading device (Portable Interface – PI) for both flight and audio data. The solid-state characteristics of the FA2100 eliminate periodic maintenance requirements and greatly increase reliability. It meets the requirements of DA-160C without the need for a vibration-mounted tray. A built-in test function includes a tone generator that superimposes the signal on the audio channel to ensure the unit is functioning correctly. In addition, the unit has a continuous BITE that monitors the entire memory and power supply to ensure that the unit is operating to the highest standards. The FA2100 accepts four channels of audio including pilot, co-pilot, area microphone and third crew member, and PA system. The flight data is accepted via an ARINC 573/717 data stream from an onboard acquisition unit.

With the FA2100 combination recorder, the audio and data retrieval and playback is instantly available by using a simple hand-held device. However, when mounted on the aircraft, recorded audio data cannot be extracted from the recorder, to protect the rights of pilots and other flight crew.

Specifications
Dimensions: ARINC 404 ½ ATR Short – footprint – height is 139.7 mm
Weight: <4.36 kg
Power supply: 115 V AC, 400 Hz or 28 V DC, 9 W (max)
Frequency response:
(three inputs) 150-3,500 Hz
(one input) 150-6,000 Hz
Recording duration:
30 min of voice/25 h of data
60 min of voice/10 h of data

Contractor
L-3 Communications Corp, Aviation Recorders.

NEW ENTRY

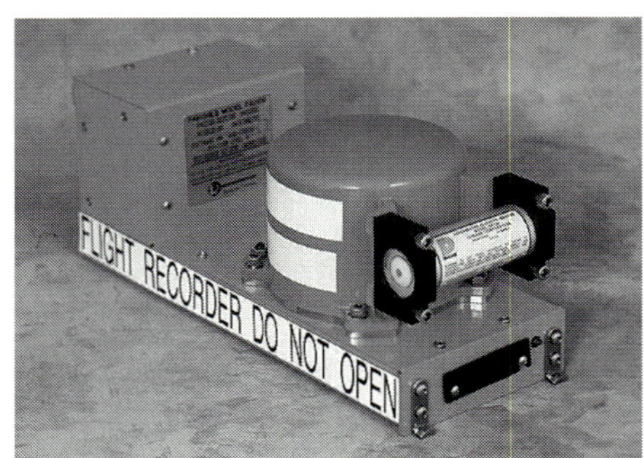

L-3 Communications Corp, Aviation Recorders Model FA2100 recorder **1998**/0015330

Colour Cockpit TV Sensor (CCTVS)

Lockheed Martin Fairchild Systems CCTVS updates the company's Cockpit TV System (CTVS), of which over 16,000 have already been sold to 28 countries and over 30 different aircraft types.

The CCTVS is an exact form, fit and function replacement for the current Fairchild Systems' monochrome Heads-Up Display (HUD) camera.

CCTVS cameras provide higher-resolution colour imaging sensor and recording capability, with expanded low light night performance.

Specifications
Pointing accuracy: ±0.56 mrad (at factory)
Light levels: 0.05-16,000 fL (both ALC and AEEC)
Resolution: >470 TVL/PH(); >350 TVL/PH(v)
MTBF: 32,000 h

Contractor
Lockheed Martin Fairchild Systems.

NEW ENTRY

F/A-18 low-light colour camera configuration
1998/0015331

MARS-II data recording system

MARS-II provides a systems approach to data recording. An SCSI computer enables an electronic module to utilise tape, disk or solid-state memory, with one electronic module supporting up to seven storage modules. Combined in this system is the capability to record or reproduce eight completely asynchronous, electronically configurable channels of PCM or dual-redundant MIL-STD-1553B data in addition to one channel of IRIG time code data and one channel for voice annotation. In addition, MARS-II accepts NRZL, R-NRZL, biphase and analogue data formats.

Specifications

Dimensions:
(electronic module) 127 × 304.8 × 342.9 mm
(storage module) 127 × 238.1 × 342.9 mm
Weight:
(electronic module) 15.88 kg
(storage module) 6.8 kg
Power supply: 28 V DC
115/220 V AC, 47-400 Hz
Temperature range: -54 to +50°C
Altitude: up to 50,000 ft

Contractor

Metrum-Datatape Inc.

UPDATED

The Metrum Datatape MARS-II consists of (left) the storage module and (right) the electronics module
1995

Model 32HE variable speed digital recorder

The first in series of harsh environment recorders by Metrum, the Model 32HE is designed, built and qualified specifically for harsh conditions. The Model 32HE is a sealed unit; once the tape door is closed, the media, electronics and transport mechanism are protected from the environment. Suitable for use in fixed- and rotary-wing aircraft operations, including ASW, flight testing and other situations demanding large storage volumes.

Using the latest 100 kbpi version of Metrum technology, the Model 32HE captures data at variable streaming rates from 0 to 32 Mbps and burst rates from 0 to 160 Mbpi. It records data at a density of 100,000 bpi onto broadcast-quality, high-energy, Metrum-certified ST-160 S-VHS cassettes, that provide 13.8 Gbytes of data storage, giving 57 minutes minimum (linearly increasing as data rate reduces).

The Model 32HE includes three interfaces: a single-board mux - 8 channel PCM and 2 channel analogue; series RS-422, TTL, ECL; parallel RS-422, TTL, ECL.

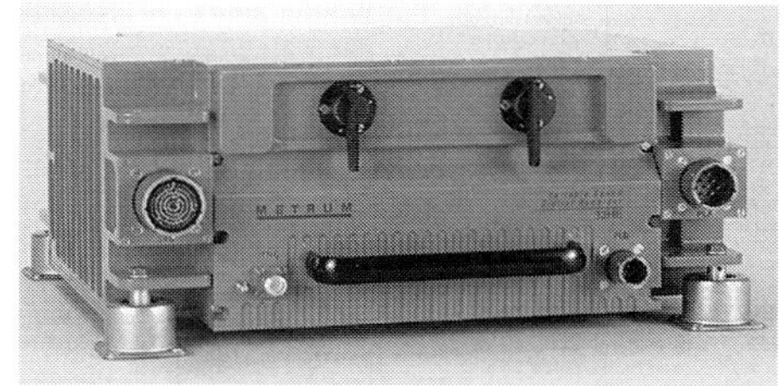

Model 32HE variable speed digital recorder
***1998**/0015332*

Specifications

Dimensions; 381 × 165.1 × 386.1 mm
Weight: 15.9 kg
Power: 28 V DC, 100 W (300 W with heaters operating)

Operational status

Introduced in 1997. A new 64 Mbps version of the Model 32 HE, designated Model 64HE is scheduled for shipment in the first quarter of 1999. It will provide a variable data rate up to 64 Mbits/s and will be able to store up to 27.5 Gbytes on a single, low cost, S-VHS tape. Model 32HE recorder/reproducers will be upgradeable to 64HE capability also in the first quarter of 1999.

Contractor

Metrum-Datatape Inc.

UPDATED

Series 2000 camera

The Series 2000 16 mm camera is capable of time-lapse, normal speed and up to 500 fps operation. It features interchangeable 200, 400 or 1,200 ft daylight-loading film magazines and has a synchronous phaselock option allowing synchronisation of several cameras.

The camera is extremely small when using the 200 ft magazine. Overall length and height are increased as the magazines containing greater lengths of film are installed. Magazines can be changed in a few seconds.

The Series 2000 camera is similar to the KB-21C camera system used by the US Air Force.

Specifications

Dimensions:
(with 200 ft magazine) 139.7 × 114.3 × 203.2 mm
Weight: 2.72 kg
Power supply: 28 V DC,12 A
115 V AC, 50-400 Hz, 3.5 A optional
Temperature range: -54 to +70°C
Acceleration: up to 25 *g*

Contractor

Photo-Sonics Inc.

VERIFIED

The Photo-Sonics Series 2000 16 mm camera

Super SVCR-V301 high-resolution airborne video recorder

The high-resolution Super SVCR-V301 recorder is lightweight and compact. It is designed to record video camera, infrared sensor and multifunction displays in the stringent environment of fighter aircraft. It is designed and tested to meet MIL-STD-810C/D, including rain, sand and dust, and EMI-tested to MIL-STD-461C and -462. It is designed specifically to meet the stringent electrical, mechanical and environmental requirements encountered in modern flight test applications.

The V301 incorporates Super VHS format, rewind and playback, over two hours of recording, high-speed forward and reverse search, a visual event marker, comprehensive BIT, electronic frame indexing, serial and parallel interface, three audio channels and 525-, 875- and 1,023-line scan rates.

The V301's Super VHS format is not just an improvement to standard VHS. It is a distinct new format providing significantly higher picture clarity with a full 400 lines of horizontal resolution in both colour and black and white recording, providing significant improvement in line picture detail. The Super SVCR-V301 provides higher luminance signal frequency and wider frequency deviation and separates luminance and chrominance signals to minimise the degradation of image quality from cross colour and dot interference. The signal-to-noise ratio in the V301 has been significantly increased by broadening the frequency deviation from 1 to 1.6 MHz. Raising the carrier frequency also reduces interference with chrominance signal and substantially increases contrast range.

The V301 will record in both standard and Super VHS formats. This allows the use of existing VHS ground playback equipment until it is replaced with higher resolution Super VHS equipment. A host of full-function commercial ground playback equipment is currently available, all capable of playing cassettes recorded on the V301. Super VHS format tapes cannot be played back on standard VHS systems. However, they can be transferred or edited down to ¾ in Umatic or the standard VHS format.

Specifications

Dimensions: 111 × 212 × 290 mm
Weight: 7.2 kg
Power supply:
115 V AC, 400 Hz, 120 W (heater only)
22-30 V DC, 33 W

Operational status

Flight test aircraft on which the V301 has been installed include the F-15, E-2C, F/A-18, P-3 and RF-4C. Operational programmes include the Tornado GR. Mk 4, AH-1W Upgrade and F/A-18 foreign military sales. The V301 is under consideration for a number of other advanced operational programmes.

Contractor

Photo-Sonics Inc.

VERIFIED

SVCR-120R series airborne video cassette recorder

The SVCR-120R series recorder is engineered and manufactured exclusively for airborne video and recording. Low bit rate PCM recording is also possible with appropriate encoding and decoding hardware. The SVCR-120R uses only proven components that are tested to withstand the humidity, vibration and temperature common to the flight test or battlefield environment. The SVCR-120R has been environmentally qualified and complies with MIL-STD-810C with tests performed by the US government and independent laboratories.

The use of standard VHS cassettes allows recording of head-up display cameras, infrared sensors and multifunction displays for 2 hours in colour or black and white.

The SVCR-120R is available for colour recording when it is designated SVCR-120RC-A. This capability is available for existing black and white recorders which can be converted to colour by exchanging plug-in circuit cards. Configuration is controlled in accordance with MIL-STD-483, to ensure maintainability and consistent environmental capabilities.

Specifications

Dimensions: 93.8 × 210 × 290 mm
Weight: 6.12 kg
Power supply: 115 V AC, 400 Hz, single phase, 100 W (heater only)
20-32 V DC, 25 W at 28 V

Operational status

Over 1,400 SVCRs have been delivered for use at flight test facilities and in operational aircraft. It was used on flight test programmes for the B-1, F-15, F-16, F/A-18, F-111 and AHIP, among others, and is standard equipment on the OV-10D Bronco, SH-60 ASW helicopter, Special Operations MH-47 and MK-60K helicopters and the US Coast Guard Nightstalker.

Contractor

Photo-Sonics Inc.

VERIFIED

AN/ASH-28 signal data recorder

The AN/ASH-28 records digital data on a 25 hours capacity quickly removable cassette. This system is designed for the F-15 and is housed within a single box containing twolinear accelerometers, one angular accelerometer and three gyros and interfacing with the F-15's MIL-STD-1553 digital databus.

Operational status

In service in US Air Force F-15 aircraft but no longer in production.

Contractor

Smiths Industries Aerospace.

VERIFIED

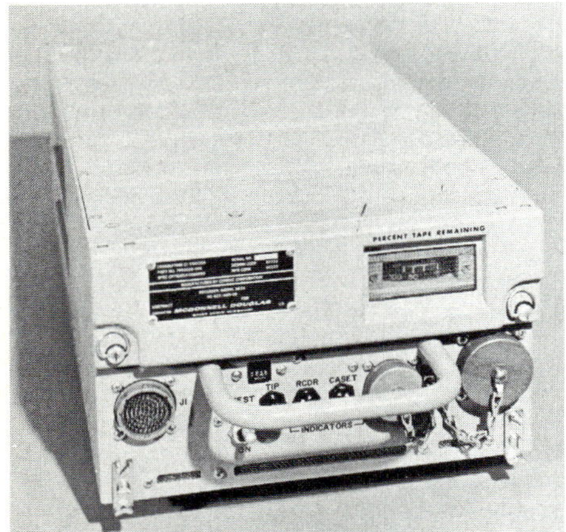

The Smiths Industries AN/ASH-28 signal data recording set **1996**

Data Transfer System (DTS)

The Data Transfer System (DTS) is a selectable family of automated digital data loading, recording and post-mission downloading subsystems for aircraft and surface vehicles using digital avionics. The first solid-state DTS was initiated by Smiths Industries in the 1970s to obviate digital information transfer problems encountered during the preflight loading of mission-related information. Errors and delays in aircraft readiness were being caused by the manual insertion of preflight information such as target co-ordinates and mission waypoints via aircraft system keyboards. The introduction of the DTS has basically eliminated preflight data entry errors while, at the same time, reducing cockpit of flight deck initialisation time from about 30 minutes to a few seconds.

In addition to the loading and retrieval of normal mission data, DTS equipment is being used to initialise Joint Tactical Information Distribution System (JTIDS), Navstar GPS (Global Positioning System), missile guidance control units, digital mapping system and voice control interactive devices. DTS equipment can also provide immediate post-mission printout and analysis of operational and flight test information. It is also used within Flight Data Recorder systems as an information recording and download mechanism.

A typical DTS includes a small portable solid-state data transfer memory media, a receptacle into which the media is inserted and a ground-based mission data computer terminal with a software database.

Available memory media includes a small shirt pocket-sized Data Transfer Module (DTM, or cartridge) which is offered with SRAM, Flash EPROM and combined SRAM/Flash EPROM, depending on customer preference/need. DTM memory capacities range from 16 kbytes to 150 Mbytes with planned growth to 500 Mbytes by 2000. Smiths Industries has introduced seven backward-compatible DTM improvement configurations since its initial DTM design in the 1970s. Optional PCMCIA (PC Card) portable memory media is also offered for selected Smiths Industries DTS applications.

The data transfer module receptacle, mounted in a convenient location within the aircraft, accepts the DTM and/or PC Card(s) via a safety spring-loaded door. Depending on user needs, Smiths Industries receptacles are modularly expandable to accommodate standard electronic interfaces, such as RS-422, MIL-STD-1553A/B, SCSI-2, Fibre Channel and combinations of these interfaces. Newer DTM/PC Card receptacles also offer a media data management microprocessor, additional memory and/or other electronics to increase system functionality. The receptacle electronics can be provided either within the receptacle or packaged remotely to conserve cockpit space. Smiths Industries cockpit receptacles can be configured to accommodate the DTM, PC Cards (PCMCIA) or a combination of DTM and PC Card memory media.

DTM/PC Card ground interface and mission planning computer terminals incorporate user-friendly software to accommodate efficient digital data loading and retrieval functions. Mission planning terminals have progressed from large expensive machines to smaller, more capable and less expensive ground computer systems. Smiths Industries offers its own tactical Mission Data Ground Terminal (MDGT) based upon personal computer (PC) technology. Also offered is a family of DTM ground interface and test equipment for use within customer designated mission planning systems. Selectable DTM ground interface and test unit/kit solutions include DTM to PC-ISA/EISA, RS-422, IEE-488 and SCSI-2 interfaces. Included within these standard mission planning systems are the Computer Aided Mission Planning System (CAMPS), Mini-CAMPS, Mission Support System-1 (MSS-1), MSS-2, MSS-2+, Air Force Mission Support System (AFMSS), Portable Mission Planning System (PMPS) and Army Aviation Mission Planning System (AMPS).

Operational status

In production. Over 70 types of fighters, multi-engine aircraft, helicopters and surface vehicles have been equipped with the Smiths Industries DTS, including over 7,500 systems to date.

Contractor

Smiths Industries Aerospace.

VERIFIED

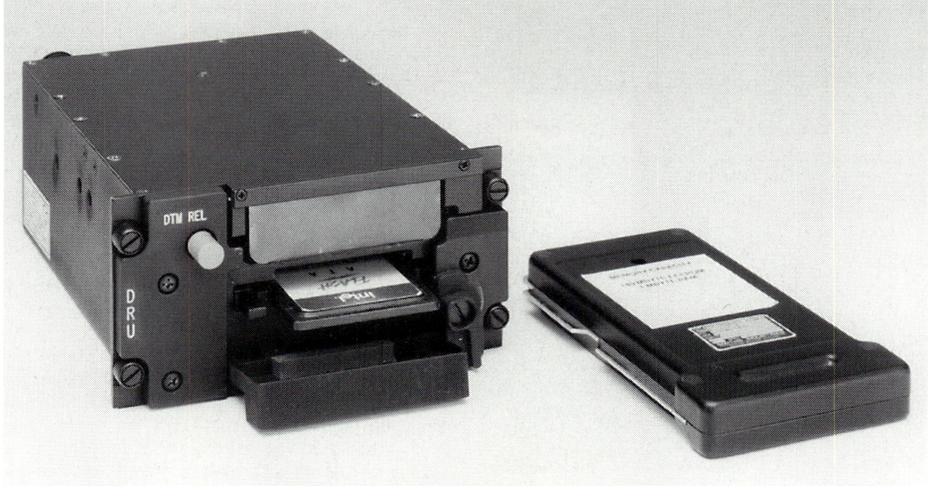

Data Transfer System **1997**/0001307

GenHUMS

GenHUMS offers a multi-aircraft capable generic Health and Usage Monitoring System capability, using proven airborne and ground technology, based on µHUMS. GenHUMS comprises the airborne GenHUMS equipment; and the HUMS Ground Station (HGS). The airborne GenHUMS components are the Data Acquisition and Processing Unit (DAPU); Cockpit Control Unit (CCU); Cockpit Interface Panel (CIP); Data Transfer System (DTS); and Optical Blade Tracker (OBT).

GenHUMS functionality implements the following health and usage monitoring functions: transmission health monitoring; engine health and usage monitoring; rotor track and balance; rotor and airframe health monitoring; aircraft usage monitoring. Acquired data is stored in crash-protected memory. Cockpit voice recording is also provided.

Operational status

Selected by the UK MoD for RAF Chinook helicopters, with options for other UK MoD helicopters including Sea King, Puma and Lynx. On the Chinook nearly 200 parameters vital to operations are monitored. Future developments will include an upgrade with the capability to predict fatigue in real time.

Contractor

Smiths Industries Aerospace.

NEW ENTRY

Smiths Industries GenHUMS **1998**/0015333

Health and Usage Monitoring System (μHUMS)

Smiths Industries Aerospace combines the proven Cockpit Voice Recorder/Flight Data Recorder aircraft monitoring technology with proprietary Health and Usage Monitoring System (HUMS) technology to provide an advanced multi-aircraft capable, 'single box' system—the μHUMS. The μHUMS has been developed using off-the-shelf technology and a proven ground support system.

Integrating the capabilities of Smiths Industries' divisions DSNA, DSC, BAS, and MJA Dynamics with a team of leading companies in key HUMS technologies, the μHUMS provides critical flexibility for cross platform and future growth requirements. The μHUMS represents the logical evolution of multibox systems to a single box utilising the latest monitoring, recording, and diagnostic systems technology.

Design features include: ED55/56A automatic rotor balancing; onboard rotor track and balance; aircraft usage monitoring; rotor, engine, transmission, and gearbox health monitoring; powertrain, gearbox, and bearing diagnostics; engine and structures life usage monitoring; flight regime recognition; crash-protected voice and flight data recording, 10 to 150 Mbyte PCMCIA or ruggedised data transfer system; wide interface capability MIL-STD 1553B/RS-429/ARINC 422/717; extensive recording/monitoring/playback/analysis capability; optional onboard oil and lubrication system monitoring (Inductive Debris Monitoring System—IDM); optional engine exhaust debris monitoring (Electrostatic Engine Monitoring System—EEMS).

Contractor

Smiths Industries Aerospace.

VERIFIED

Inductive Debris Monitor (IDM)

The IDM provides real-time damage detection and is claimed to give earlier warning than any other monitoring technique for fluid wetted systems, without false alarms.

The IDM is guaranteed to detect magnetic and non-magnetic metal chips, is independent of fluid velocity and viscosity and is insensitive to foam, bubbles and dielectric properties. Metal chips are classified by type, size and rate of occurrence, detecting damage long before performance degradation and without periodic inspections, to provide up to 10 hours early warning of impending failure.

Operational status

In testing for the RAH-66 Comanche.

Contractor

Smiths Industries Aerospace.

VERIFIED

Removable Auxiliary Memory Set (RAMS)

RAMS includes a Data Transfer Interface Unit (DTIU), for managing data and providing system communications, and a receptacle for the associated memory module, the Data Transfer Module (DTM) or programmable cartridge. It is about the size of a cassette tape and reduces the need for large ground readout equipment. RAMS performs the functions of downloading memory, uploading operational flight plans and documentary data, formatting memory and resetting usage accumulators, flight counters and BIT history.

Data for the Standard Flight Data Recorder (SFDR) is recorded on the DTM during flight. The receptacle displays percentage of memory used and system faults. The module and receptacle perform the following SFDR functions, eliminating the need for a digital computer: load Signal Acquisition Unit (SAU) operational flight program; load documentary data into SAU non-volatile memory; format standard flight data recorder memories; clear usage accumulators and BIT records; download CSMU data; download SAU non-volatile memory; and download SAU auxiliary memory unit.

DTMs are available in 1 to 8 Mbytes battery-backed RAM capacities and 10 to 86 Mbytes EEPROM. Near-term growth to over 500 Mbytes is planned.

Specifications

Dimensions:
(module) 82.55 × 20.57 × 150.1 mm
(DTIU receptacle) 128 × 77 × 190 mm
(DTMR receptacle) 128 × 32 × 133 mm
Weight:
(module) 0.34 kg
(DTIU receptacle) 1.81 kg
(DTMR receptacle) 5 kg
Power supply: 28 V DC, 20 W

Operational status

RAMS has been installed on US Air Force B-52, F-15, KC-135 and T-38 aircraft.

Contractor

Smiths Industries Aerospace.

VERIFIED

Standard Flight Data Recorder (SFDR) system

The Smiths Industries solid-state Standard Flight Data Recorder (SFDR) system has been developed under a US tri-service specification. The solid-state SFDR can replace older oscillograph or tape recorders as well as add more comprehensive aircraft monitoring functions. It was initially introduced by the US Air Force on the Lockheed F-16 fighter.

The Series SFDR consists of the Signal Acquisition Unit (SAU) with auxiliary memory unit, the Crash Survivable Memory Unit (CSMU), the optional Flight Data Panel (FDP) and the Removable Auxiliary Memory Set (RAMS) consisting of an intelligent Data Transfer Interface Unit (DTIU) or a simple Data Transfer Module Receptacle (DTMR) and Data Transfer Module (DTM). The CSMU is also being used on Sweden's Saab JAS 39 fighter.

The SAU is capable of receiving a combination of over 600 discrete, analogue and digital multiplex bus parameters which are converted into digital information, data compressed and stored within the SAU's auxiliary memory unit, RAMS data transfer module and/or CSMU. Daily monitoring and recording of general airframe, avionics and engine health structural loads and engine low-cycle fatigue is typical. All conversion and data management functions are managed witihn the SAU as well as data monitoring and customer designated alerts. The SAU also supports rapid download to ground data logging systems and to graphical replay and analysis systems.

Selected flight data parameters are sent to a compact armoured and insulated CSMU to ensure data recovery following a flight incident. The CSMU is designed to withstand stringent mishap conditions including temperatures of up to 1,100°C, mechanical shock of 3,400 g and penetration. The CSMU non-volatile memory has a life expectancy of 60,000 hours which, together with the built-in test facility, permits it to be mounted in inaccessible regions of the airframe.

Routine maintenance data is retrieved from the aircraft by means of flight line Ground Readout Equipment (GRE) or an installed RAMS. Retrieval data is then transferred to a Data Recovery and Playback Evaluation System for data decompression, analysis and implementation. The STS is developing and loading Operational Flight Program (OFP) data and providing maintenance support for the SFDR.

Specifications

SAU
Dimensions: 157 × 178 × 184 mm
Weight: 6.51 kg
Memory:
(program) 572 kbytes
(scratchpad) 512 kbytes
(non-volatile) 16 kbytes
(auxiliary) up to 3 Mbytes
Environmental: MIL-E-5400 Class II
Power: <50 W

CSMU
Dimensions: 76 × 76 × 117 mm
Weight: 1.58 kg
Memory: 64 k or 256 kbytes
Min recording time:
(attack fighter/trainer aircraft) 15 min active flight, 60 min normal flight
(transport aircraft) 25 h
Environmental:
(impact) 3,400 g for 6 ms
(penetration) 500 lb for 15 ms
(static crush) 5,000 lb for 5 min
(fire) 1,100°C flame for 30 min, or equivalent oven test
(fluid immersion) 48 h at 1,500 ft
Power: <4 W

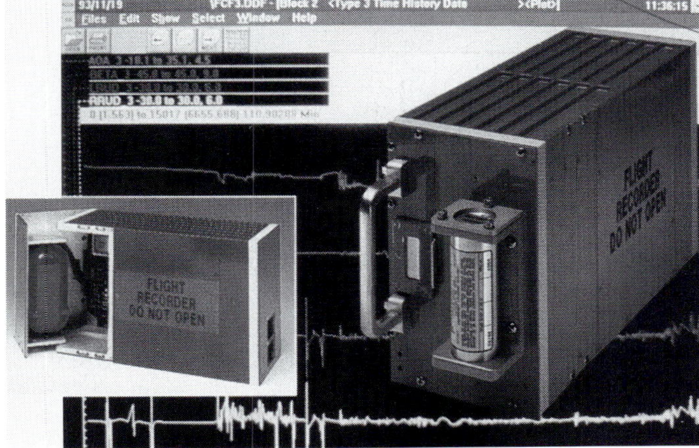

Smith Industries μHUMS ***1997***/0001446

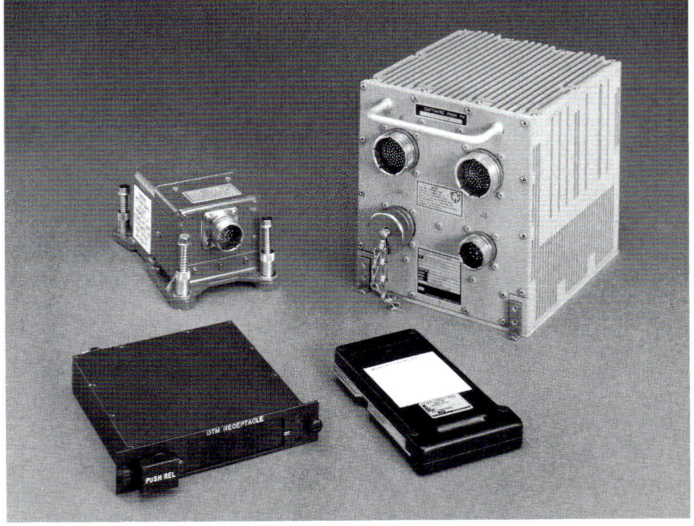

The Smiths Industries standard flight data recorder

Locator: acoustic beacon or crash position indicator available

Operational status

In service on both military and commercial fixed- and rotary-wing aircraft.

Contractor

Smiths Industries Aerospace.

VERIFIED

Voice And Data Recorder (VADR)

The dual-use Voice And Data Recorder (VADR) combines the functions of a Cockpit Voice Recorder (CVR) and Flight Data Recorder (FDR) into a highly reliable, lightweight package. The compact recorder system can be hard mounted in virtually any location or orientation, affording the original equipment manufacturer or avionics integrator increased installation flexibility. The VADR utilises solid-state memory technology, offering increased reliability and low power consumption. The crash-protected memory packaging techniques are based on the US Air Force Standard Flight Data Recorder (SFDR) Crash Survivable Memory Unit (CSMU).

The VADR provides data collection and mishap recording of audio data and aircraft flight and system parameters to support post-incident analysis. It can be configured as a TSO-C124 Flight Data Recorder (FDR), a TSO-C123 Cockpit Voice Recorder (CVR) or an ARINC 757 combined FDR/CVR. This family of recorders meets the survivability requirements of EUROCAE ED-55 for FDRs and ED-56/56A for CVRs for both ejectable and non-ejectable recorders. The VADR offers mounting provisions for an underwater locator beacon.

Additional configurations of the VADR include an ARINC 404 adaptor tray for form, fit and function replacement of existing recorder systems, a CVR/FDR combining voice and MIL-STD-1553 or ARINC 429 serial bus interfaces and a high-capacity data only FDR for MIL-STD-1553, ARINC 429, or RS-422 serial data recording.

The VADR is available in a number of models offering audio capacities from 1 channel × 30 minutes to 4 channels × 120 minutes. Data capacities from 2 to 30 hours are available depending on application.

Specifications

Dimensions vary with model
Weight varies from: 3-3.5 kg
Audio frequency response:
3 channels: 150-3,500 Hz
1 channel: 150-6,000 Hz

Operational status

In production and in service on a wide variety of fixed- and rotary-wing aircraft, including: US presidential helicopters, AV-8B, B1-B, ALX Super Tucano, AH-64A, C-2, C-130J, CH-47D, EC-135, F/A-18E/D, F-111, HH-60J, HH-65A, JPATS, MH-47E, MH-60K, OH-58D, OH-X, RAH-66, SH-60J, UH-60, UP-3, US-1A, VH-3, VP-3, VH-60 and WAH-64 aircraft.

Contractor

Smiths Industries Aerospace.

UPDATED

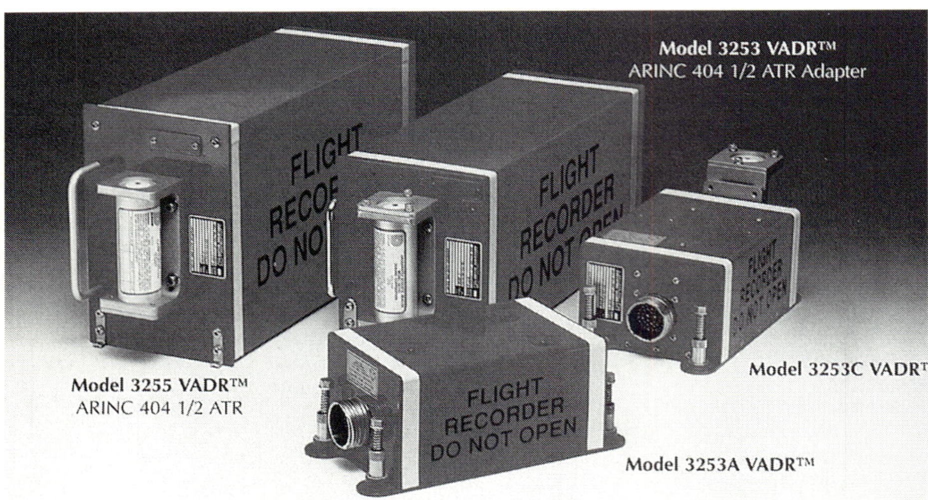

The Smiths Industries family of voice and flight data recorders
1998/0015334

NuHums™ health and usage monitoring system

The NuHums is a modular, lightweight design that performs all signal processing in a single ½ ATR unit. The NuHums can be upgraded to add an optional flight data/cockpit voice recorder or third party diagnostic system.

The NuHums performs real-time onboard diagnostic processing for in-flight analysis of helicopter rotors, drivetrains, gearboxes and engines. It utilises the RADs-AT rotor track and balance system, and provides immediate feedback on any parameter exceedance.

NuHums modular architecture permits the customer to select features to match the requirement, and to add new features when required. The only essential element is the Data Acquisition Unit (DAU); sensors, cockpit displays, recorders, and other elements are options.

Specifications (DAU)

Channels: 48 analogue and 48 digital channels
Internal memory: 64 Mbytes
Dimensions: ½ ATR
Power: 28 V DC, 30 W
Weight: 4.55 kg

Operational status

In production and in service on a number of helicopter types.

Contractor

SPS Signal Processing Systems.

NEW ENTRY

NuHums health and usage monitoring system
1998/0015335

V-80AB-F 8 mm recorder

TEAC has been contracted by the US Navy to supply 209 airborne video tape recorders for US Marine Corps AV-8B Harrier II aircraft.

Contractor

TEAC America Airborne Video Products.

VERIFIED

T'AIMS I (Teledyne Aircraft Integrated Monitoring System)

T'AIMS I

T'AIMS I evolved from the Teledyne Modular Data Analysis Unit (MDAU) and Ground Support Equipment (GSE). T'AIMS I performs exceedance monitoring based on cockpit displays, and it automatically detects any limit exceedance in the engine, rotor or transmission. It also monitors flight hours, engine hours and engine cycles, performs health checks and displays results to the flight crew. Exceedances and performance data are saved in non-crash-protected memory.

There are three system elements: the Modular Data Analysis Unit MDAU; T'AIMS I control panel; and the ground station computer – a laptop PC to which flight data can be downloaded on the flight line. The MDAU interfaces to the engine, gearbox and rotor sensors; the air data system; cockpit displays and the control panel.

T'AIMS II

T'AIMS II expands the capabilities of T'AIMS I by adding a crash-protected solid-state voice and data recorder; increased capability for aircraft performance monitoring; and increased capability for data analysis in the ground station system.

Contractor

Teledyne Controls.

NEW ENTRY

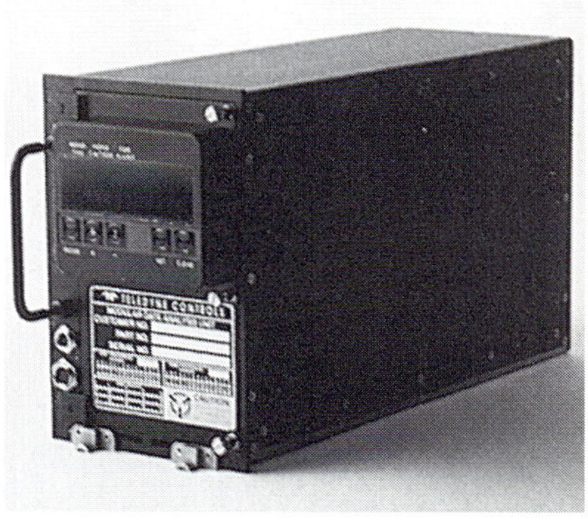

T'AIMS modular data analysis unit

1998/0015336

Solid-State CVR-30A Cockpit Voice Recorder

The SSCVR-30A solid-state cockpit voice recorder provides 30 minutes of history. It accepts four channels of cockpit audio, converts the audio to digital format and stores the data in solid-state non-volatile flash memory. A fifth channel digitally records helicopter rotor speed. The recorder is fully ARINC 557 compatible.

Two control units are available. One conforms to ARINC 557 specifications, the other is a slimline unit designed for installation where cockpit panel space is at a premium and ARINC 557 specifications are not required. Both utilise an LED display for signal level.

Specifications

Dimensions:
(recorder) ½ ATR short
(ARINC control unit) 57.2 × 148 × 92.1 mm
(slimline control unit) 38.1 × 146 × 98.6 mm
Weight:
(recorder) 10.5 kg
(ARINC control unit) 0.5 kg
(slimline control unit) 0.34 kg
Power supply: 27.5 V DC, 1 A
115 V AC, 400 Hz, 0.2 A option
Reliability: 18,000 h MTBF

Operational status

In production.

Contractor

Universal Avionics Systems Corporation.

VERIFIED

Solid-State CVR-30B/120 Cockpit Voice Recorders

The SSCVR-30B/120 solid-state cockpit voice recorders are the latest Universal Avionics Systems Corporation solid-state CVRs. They feature the same capabilities as the SSCVR-30A and are fully ARINC 757/557 compatible. The SSCVR-30B records 30 minutes of data; the SSCVR-120 records 120 minutes. The recorders accept four channels of cockpit audio, convert the audio to digital format and store the data in solid-state, non-volatile flash memory, together with helicopter rotor speed and time.

Specifications

Dimensions: ½ ATR short
Weight: 5.85 kg
Power supply: 27.5 V DC, 1 A
115 V AC, 400 Hz, 0.2 A optional
Reliability: 30,000 h MTBF

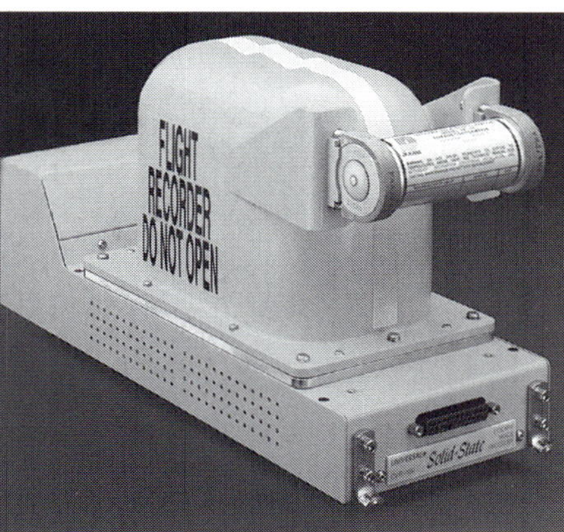

Solid-state CVR-30B/120 cockpit voice recorder

1998/0015337

Operational status

FAA certified TSO 123a and ED56A.

Contractor

Universal Avionics Systems Corporation.

UPDATED

The Ramenskoye Design Company SINUS integrated navigation, flight management and display system ***1998***/0015370

CANADA

AN/APN-208 and AN/APN-221 Doppler navigation systems

The AN/APN-221 Doppler set is supplied for night/adverse weather search and rescue helicopters. This system was derived directly from the AN/APN-208(V) helicopter Doppler navigation system developed to meet the requirements of ASW and search and rescue helicopters, where versatility in operation and interface options are essential. A recent addition has been the integration of GPS inputs into the navigation solution to provide enhanced performance.

A specific application of the APN-221 Doppler navigation set, which was developed under the sponsorship of the US Air Force Systems Command, Aeronautical Systems Division, is the Pave Low III/Sikorsky HH-53 medium-lift helicopter. The operational requirements of this aircraft called for an extension of the AN/APN-208(V)'s capability to include guidance in poor weather or darkness through the use of advanced computer techniques and cockpit displays. Improvements in the AN/APN-221 have since been incorporated into the AN/APN-208.

The four-unit APN-221 comprises a four-beam lightweight antenna, ¾ ATR short signal data converter with 16-bit microcomputer, 6-line by 12-character control and display unit and a steering/hover indicator. The displays are also available in a form that is compatible with Generation III ANVIS night vision goggles. It provides three pilot-selectable navigation co-ordinate systems, with automatic conversion from one to another. These are latitude/longitude, worldwide alphanumeric UTM and arbitrary grid. Data can be stored for up to 75 mission waypoints, 10 targets of opportunity or 25 library waypoints (including Tacan beacon locations). There are also three pilot-selectable search patterns with automatic turning point computation and navigation/guidance outputs: creeping line, expanding square and sector. The system provides aircraft velocity outputs to the flight control system, enabling coupled hover manoeuvres and automatic approaches to the hover to be conducted.

The transmitter/receiver/antenna uses a Gunn diode RF source to generate four Janus configuration beams, and FMCW modulation gives high accuracy and immunity from carrier noise, precipitation, surface spray and reflections from nearby objects such as airframe structure and sling loads.

Specifications

Dimensions:
(antenna) 439 × 439 × 113 mm
(signal data converter) 194 × 198 × 319 mm
(control/display unit) 146 × 114 × 165 mm
(steering/hover indicator) 83 × 83 × 127 mm
Weight:
(antenna) 5.59 kg
(signal data converter) 9.45 kg
(control/display unit) 3.54 kg
(steering/hover indicator) 1.14 kg
Transmission: 4 beam Janus, time-shared (200 ms/cycle), 3 × 6.7° beamwidth from Gunn diode, 13.325 GHz with FMCW modulation optimised for flight envelope
Velocity range:
(forward) −50 to 300 kt
(lateral) 100 kt
(vertical) ±5,000 ft/min
Accuracy:
(forward) 0.3% speed along velocity vector ±0.2 kt

The Canadian Marconi AN/APN-208(V) Doppler system

(lateral) 0.32% speed along velocity vector ±0.2 kt
(vertical) 0.2% speed along velocity vector ±20 ft/min
Inputs: pitch/roll attitude and heading from AHARS or INS. Various optional interfaces with map displays, Tacan, sonar, radar and air data computer systems.
Microcomputer:
(architecture) word addressed, 16-bit parallel microprogrammed, single/multiprecision
(instruction set) 70, Hewlett-Packard compatible
(addressing) immediate, direct, indirect up to 32 k words
(memory) RAM/ROM up to 32 k words
(software) Assembler/Fortran mixed, complete Hewlett-Packard support software library

Operational status

The AN/APN-208 is in service with the armed forces of many NATO and other countries. The AN/APN-221 is in service for the US Air Force Sikorsky Pave Low III HH-53 helicopter.

Contractor

Canadian Marconi Company.

UPDATED

CMA-900 GPS navigation/flight management system

The CMA-900 provides full-performance GPS navigation and extensive flight management features, including company routes and worldwide ARINC 424 navigation databases, SID/STAR navigation, GPS instrument approaches, offset tracks, and autopilot-coupled holding patterns and procedure turns. As a multisensor navigator, the CMA-900 integrates other approved navigation sensors, and offers several navigation modes including GPS, DME/DME, DME/VOR, Omega/VLF and optional INS/IRS. This enables the system to provide seamless navigation for all phases of flight with maximum accuracy, integrity and availability. Other features include RNP/ANP, RTA and fuel management functions, and auto-tuning of navigation and communications radios. Growth to vertical navigation and other FMS features is planned for late 1998.

The system is authorised to TSO-C129, Class A1 and fully compliant with both the required and desired performance requirements of the FAA Notice N8110.57. Designed for operation in the CNS/ATM environment, the CMA-900 incorporates growth capacity for future datalink, and ATN compatibilities, as well as local and wide-area augmentation differential GPS (LAAS/WAAS), and GIC for precision approaches and landings. This makes the CMA-900 compatible with any of Eurocontrol's possible P-RNAV requirements.

The CMA-900 incorporates a 12-channel GPS Sensor Module (GSM), which is directly derived from the GPS sensor unit CMA-3012. This system has been selected for the Boeing 777, 737 and Airbus 330/340 series, and has provisions to meet all RTCA requirements for SCAT-I landing equipment. The DGNS capability includes pseudo-ILS approaches and hardware provisions for ARINC 429 or RS-232 interfaces with the local and wide-area datalink receivers and a CMA-2014 multipurpose display. This is a 14-line 24 character fully ARINC 739-compatible colour AMLCD display.

The technology used in the GSM includes integrity-related features such as proprietary algorithm for high-performance, real time, satellite Fault-Detection and Exclusion (FDE) including Receiver Autonomous Integrity Monitoring (RAIM), and software written in Ada to critical category standards. Performance is provided by high-dynamic, all-in-view reception of up to 12 channels (including Inmarsat integrity overlay) and carrier phase tracking. The all-in-view tracking and very fast acquisition capabilities of the CMA-900 will be particularly important, in the future, to offset any satellite blanking by the aircraft during terminal-area

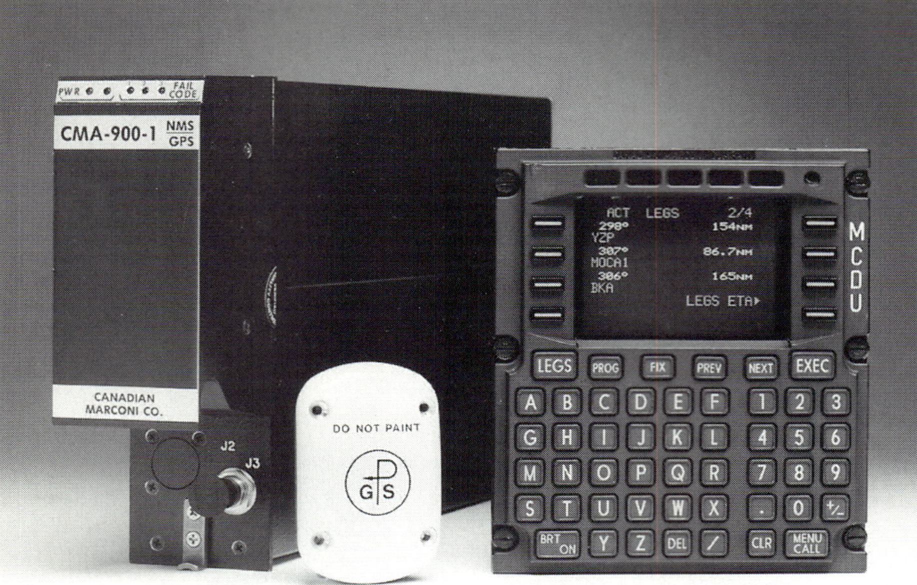

The CMA-900 GPS/FMS components (left to right) FMU, antenna, MCDU ***1996***

and precision approach manoeuvres. For aircraft currently equipped with older DME receivers, the DME/DME mode can be implemented using either an ARINC 709 scanning DME, or a digital DME sensor dedicated to the CMA-900.

The CMA-900 operating procedures conform to airline practices, including keyboard and page layouts, full scratchpad, line-select and menu functions. This approach results in fleet-wide commonality in crew interface and a very flexible system with the ability to meet the FANS CNS/ATM environment without keyboard redesign. It also ensures excellent transfer of training as pilots progress through a mixed fleet of aircraft.

The CMA-900 is equipped with the optimum suite of analogue and digital interfaces for retrofit applications, including outputs to conventional and Electronic Flight Instruments (EFIS), suitably modified weather radars, autopilot and flight director systems, and inputs from other navigation sensors, and both analogue and digital air data and heading systems. Data loader interfaces are in accordance with ARINC 615.

Other capabilities include full ARINC 739 compatibility and ARINC 429 file transfer protocols, which allow the FMU to interface with a number of aircraft systems, including ACARS and other ARINC 739-compatible CDUs. The CMA-900 also supports ARINC 429 fuel-computer inputs for fuel management functions in conjunction with operator inputs of fuel on board, and so on.

Specifications

Dimensions:
(FMU) 200 × 57 × 324 mm
(MCDU) 171 × 146 × 214 mm
(GA/AEU) 19 × 74 × 119 mm
Weight:
(FMU) 3.63 kg
(MCDU) 4 kg
(GA/AEU) 0.5 kg
Power supply: 28 V DC, 73 W (max)
Accuracy:
(horizontal) 33 ft
(vertical) 130 ft
(velocity) 1 kt
(time) 2 μs
Reliability:
(FMU) 10,000 h MTBF
(CDU) 8,000 h MTBF

Operational status

In production. The CMA-900 has been delivered for Boeing 727, 737, DC-9, and MD-80, and Airbus A-300 aircraft. Orders have also been placed for Boeing 747 and DC-10s.

Contractor

Canadian Marconi Company.

UPDATED

CMA-2012 Doppler navigation sensor (AN/ASN-507)

The CMA-2012 is a newly developed, single LRU Doppler navigation sensor designed for both rotary- and fixed-wing aircraft applications where navigation aiding, back-up navigation and hover are of primary importance. Based on Canadian Marconi's experience in the design and development of Doppler radar systems and related technologies, the design parameters of the CMA-2012 were set to achieve significant reductions in size, weight and cost compared to current systems, while enhancing performance capabilities and improving reliability.

With the CMA-2012, Canadian Marconi has achieved these improvements by implementing innovative design features, including: digital signal processing for real-time signal analysis; optimised hover-hold mode for precision hover with drift rates significantly less than 1 m/minute; tactical modes include silent, horizontal beam cutoff and EW equipment compatible intermittent track.

Specifications

Velocities (land):
Vx (forward) −50 to 250 kt ±0.3% Vt
Vy (lateral) −100 to 100 kt ±0.3% Vt
Vz (vertical) ±5,000 ft/min ±0.3% Vt
$Vt^2 = Vx^2 + Vy^2 + Vz^2$
Altitude: 2 to 15,000 ft
Power: 28 VDC/45 W max
Cooling: convection
Weight: 5.5 kg
Environment: MIL-E-5400 Class 1A/DO-160C
Reliability: 6,700 h MTBF
Dimensions: 372.6 × 345.3 × 49.5 mm
Standard interfaces: MIL-STD-1553B, ARINC-429

Operational status

The CMA-2012 has been selected for five models of scout/anti-tank helicopters, as well as a variety of other tactical helicopters in Canada, Europe, Asia and Africa. 150 systems delivered to date (December 97).

CMA-2012 Doppler navigation sensor ***1998***/0015338

Contractor

Canadian Marconi Company.

UPDATED

CMA-3000 Single Unit Navigator (SUN)

The CMA-3000 Single Unit Navigator (SUN) is an evolutionary step in the Canadian Marconi Company's navigation and GPS products. The CMA-3000 is a one component hybrid made up of the development of CMA-900 GPS/FMS and CMA-2014 Mk III Multi Control Display Unit. The system is integrated into a single control display unit with the CMA-900s multi-sensor navigation and optional external GPS sensor functions. The system is designed to provide helicopters, regional and commuter aircraft with a versatile GPS-based flight management system housed in one line replaceable unit, featuring a multipurpose control and display function; a full-function embedded navigator and a radio management system.

The basic CMA-3000 features eight ARINC 429 inputs and three ARINC 429 outputs, eight discrete inputs, four discrete outputs and one RS-422 I/O port which makes it suitable for most helicopter applications.

For retrofit into aircraft with analogue interfaces, the CMA-3000 can be complemented with an external analogue adapter unit. This configuration provides a cost effective solution while still preserving a common pilot interface.

The CMA-3000 is designed for multi-sensor RNAV operation during the worldwide transition to GPS primary means navigation. Since, certification approvals have already obtained for the CMA-900, the CMA-3000 GPS/FMS will support installation approval for GPS primary means oceanic/remote operations. In addition, the CMA-3000 will be authorised for GPS-based instrument approaches under TSO-C129 Class A1. This full-function navigation system can also be used with a wide variety of navigation sensors.

The CMA-3000 uses state-of-the-art colour Active-Matrix Liquid Crystal Display (AMLCD) technology. It features a 5 inch diagonal display screen with 14 lines of 24 characters each. The system includes a full alphanumeric keyboard; 12 line-select keys 15 function keys; and nine dedicated annunciators.

The CMA-3000 display and back-lighting are sunlight readable, and compliance with MIL-L-85762A for Class B Night Vision Goggle (NVG) operation is available as an option.

The system's operating procedures conform to current airline practices of 'glass cockpit' aircraft, including keyboard and page layouts, and full scratchpad, line-select and menu functions. The CMA-3000 conformity allows fleet-wide commonality in the pilot interface. All navigation and radio tuning functions utilise the same scratchpad/line-select crew interface philosophy.

Waypoint, navigation and guidance information is generated in both geographic and track-related reference frames. This information is clearly visible to the pilot on the unit's display. In addition, the system will output the information to flight instruments and autopilot/flight director systems. Its capabilities include:

(1) Flight planning and route creation, selection and modification;
(2) Complete oceanic, en-route, terminal and non-precision approach navigation and guidance;
(3) GPS instrument approaches;
(4) Continuous and manually-initiated predictive FDE;
(5) Outputs to Electronic Flight Instrument Systems (EFISs), and digital autopilot and flight director systems;
(6) Direct-to and leg/course intercept navigation, holding patterns, DME arcs, procedure turns, and offset tracks;
(7) Automatic leg change with fly-by and fly-over leg transitions;
(8) Time and fuel management, including Required Time of Arrival (RTA) computation and display;
(9) Required and Actual Navigation Performance (RNP/ANP) computation and display;
(10) Search pattern navigation;
(11) ARINC 615-3 data loading capability for software maintenance and database update;

(12) Sensor status information display;
(13) Navigation and communication radio tuning;
(14) Digital map display interface to support route exchange and positioning;
(15) Kalman Filter Integration of GPS/AHRS (INS) (option);
(16) Compliance with all relevant RTCA and TSO requirements;
(17) Operation in severe (100 V/m) HIRF environments.

Specifications

Dimensions: 172 (h) × 146 (w) × 160 (d) mm
Mass: 3.0 kg
Power requirements: 28 V DC, 70 W (max)

Operational status

In service by end of 1998.

Contractor

Canadian Marconi Company.

NEW ENTRY

CMA-3012 Global Navigation Satellite Sensor Unit (GNSSU)

The CMA-3012 GNSSU meets the requirements of primary means oceanic/remote area operation as specified by FAA order 8110.60. The Sureflight software package is available to complement all CMC GPS receivers for primary means navigation flight planning and dispatch. Key characteristics include:

(1) Twelve simultaneous channels, all of which can be used for continuous satellite tracking, and any two of which are assignable as GPS/WAAS Integrity Channels (GIC)
(2) Comprehensive end-to-end receiver Built-In Test (BIT)
(3) Carrier phase tracking
(4) Differential GPS (SCAT 1) functionality
(5) Full compliance with TSO-129A B1/C1 and RTCA DO-208 sensor requirements
(6) Growth provisions for WAAS

Specifications

Receiver: 12 parallel channels
Frequency: L1, 1,575.42 MHz, C/A code
Time to first fix: 95% confidence 75 s max
Time to reacquisition: 5 s max
Accuracy:
(horizontal position) 22.5 m, 95%, S/A off
(differential) <2.4 m, 95% (optional)
(altitude) 30 m, 95% S/A off
(velocity) 0.1 kt, 95%, S/A off
Position update: 5 s (optional)
Dimensions: 66 × 216 × 241 mm
Weight: 2.55 kg
Input power: 18 to 36 V DC, 20 W max

CMA-3012 GPS sensor unit ***1998***/0015339

MTBF: 45,000 h
Inputs: 8 ARINC 429, 1 RS-232
Outputs: 3 ARINC 429, 1 RS-232; 1 28 V valid discrete; three 1 Hz time marks

Operational status

FAA approved as primary means of oceanic/remote area navigation. Installed with Racal RNav 2 system on helicopters operating over the North sea.

Since 1993, over 2,500 units have been installed in numerous air transport and general aviation aircraft types.

Contractor

Canadian Marconi Company.

NEW ENTRY

Inertial Referenced Flight Inspection System (IRFIS)

The Litton Inertial Referenced Flight Inspection System (IRFIS) is a totally self-contained en route and terminal navigation aid calibration system. The IRFIS is suitable for calibration of Cat I, II and III ILS and MLS facilities, as well as VOR, Tacan, DME, NDB and other navigation aids, in accordance with ICAO Doc 8071, Volumes I and II.

Unlike current systems, the IRFIS does not require ground personnel operating tracking equipment, nor is its operation impeded by inclement weather conditions. Automatic inspection capability is provided using updates from the Litton aircraft position sensor, multiple DME reception or pilot fixes to refine the positional data supplied by the inertial navigation system.

The IRFIS performs automatic real-time calculations, following completion of each inspection run, and instantly displays the summary of results to the operator. A large plasma monitor provides an unambiguous display in a clear format. Interactive software dialogue validates all operator inputs and, through the use of programmed function keys, requires minimal keyboard entry.

The Rolm 1866 militarised computer performs reliable and consistent data processing functions, including automatic calculation of ILS and MLS parameters and analysis of VOR/Tacan error curves, bends, scallops and roughness. A Rolm 2150 interface unit handles the receiver signal conditioning and automatic frequency tuning functions.

Inertial guidance is provided by the LTN-92 ring laser gyro INS which performs all required navigation and steering computations, provides autopilot inputs for waypoint to waypoint navigation and orbital steering about any pilot-designated point.

The Aircraft Position Sensor (APS) is a self-contained position sensor used to perform fully automatic inspection of ILS, MLS and PAR facilities. Mounted on the underside of the aircraft fuselage, the APS updates at each end of the runway are used by the software resident in the Rolm computer to improve the INS position data. The aircraft's position determining accuracy is further enhanced through the use of an 18-state Kalman filter and a Bryson-Frazier smoother error estimator.

Non-fading hard-copy calibration data is presented via an RMS GR 33-1 printer/plotter which allows the transfer of screen data to paper. A Targa data storage unit records inspection data on bubble memory for long-term storage.

The IRFIS components may be packaged in a standard side-facing console or in a compact forward-facing console. With the forward-facing style, a separate equipment rack is normally mounted opposite the console, facing the operator. The rack contains the dedicated IRFIS computer, receivers and test equipment, while all controls required for flight inspection are available at the console.

The inertial referenced flight inspection system ***1995***

Operational status

In service with Transport Canada, the Civil Aviation Administration of China, the Japanese Air Self-Defence Force, the Netherlands Department of Civil Aviation, the Royal Air Force, the South African Department of Transport, the Republic of China (Taiwan) Air Force and the Thailand Department of Aviation.

Contractor

Litton Systems Canada Ltd.

VERIFIED

Next-generation Litton flight inspection system

Litton's next-generation flight inspection system combines the latest digital technology with carefully developed software to provide high performance and reliability with excellent operational characteristics. The common core and building block design allows each system to be customised to meet the requirements of individual users. Possible system configurations range from the basic Semi-Automatic Flight Inspection System (SAFIS) to the fully Automatic Flight Inspection System (AFIS) with a range of options. The console can be easily installed and removed to allow the aircraft to be used for other roles. Next-generation flight inspection systems are designed to accommodate all current and planned flight inspection functions in a format that is flexible, versatile, user-friendly and cost-effective. The system consists of the flight inspection processing subsystem, the position reference subsystem and optional equipment.

The flight inspection processing subsystem includes the operator's console, with computer, recording equipment and radio receivers. It features a high-resolution colour display terminal with keyboard and trackerball, a data storage unit which uses removable magneto-optical diskettes, a printer, a graphic recorder and flight inspection receivers. The VMEbus computer houses dual processors and interface modules. A 68040 processor provides high-speed data processing while an 80486 processor supports the operator

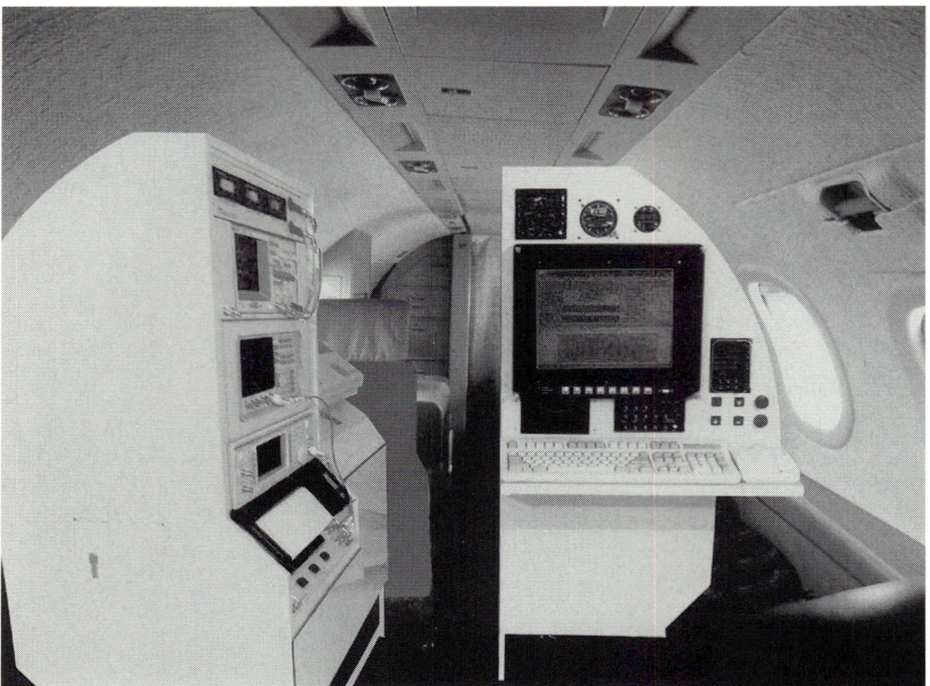

The next-generation flight inspection system ***1995***

interface. Software in C++ runs on DOS with Windows, providing a very user-friendly environment. Navigation information is displayed on an EHSI, while an optional EADI is also available.

The position reference subsystem provides the reference against which navaid facilities are calibrated. GPS is used for checking en route facilities but calibration of landing aids requires greater precision. For ILS, MLS and PAR, the AFIS uses a combination of differential GPS, INS and over the runway position updates to achieve a positional accuracy of better than 30 cm in three axes at the runway threshold. For the SAFIS, a telemetry tracking system is used to track the aircraft during approaches.

Optional accessories are available for the flight inspection system. These include an oscilloscope which allows selected parameters to be examined and a spectrum analyser which allows RF signals to be analysed for interference. A signal generator will allow onboard calibration of receivers under control of the flight inspection computer. Although tuning of the receivers is normally performed automatically by the computer, an optional radio tuning unit provides conventional manual tuning capability from a single universal tuning head.

Specifications

Dimensions:
(basic system) 0.4 m^3
(average system) 0.56 m^3
(fully loaded system) 0.75 m^3
Weight:
(basic system) 136 kg
(average system) 295 kg
(fully loaded system) 360 kg

Operational status

In production. In service with the Directorate General of Air Communications Indonesia installed in Learjet 31A aircraft. A further system has been delivered to the Korean Ministry of Transportation for installation in a Canadair Challenger CL 603-3R.

Contractor

Litton Systems Canada Ltd.

VERIFIED

CZECH REPUBLIC

SENAP-signal emulator of aeronavigation and landing

The SENAP system was developed as a substitution system to replace the RSBN navigation system. Replacement of RSBN by SENAP releases the 700 to 1,000 MHz band for other civil purposes. The system utilises a GPS receiver, and can integrate VOR/DME, Loran, ILS and other navigation data sources.

SENAP retains the RSBN pilot control, and the original RSBN operational capabilities and accuracy: range 0 to 500 km; azimuth 0 to 360°. SENAP complies with ICAO standards. Operation in either SENAP or RSBN modes can be selected by the pilot.

Specifications

Dimensions/Weight:
Basic Unit 319 × 194 × 157 mm/5.8 kg
Adapter 319 × 194 × 90.5 mm/5.8 kg
Power supply: 27 V DC.

Operational status

A version is available for fitment to Su-22Mk4 aircraft.

Contractor

Elektrotechnika-Tesla Kolin, a.s.

UPDATED

SENAP
1997/0001309

DENMARK

Aeronautical CAPSAT

Aeronautical CAPSAT uses British Telecom's Standard Positioning Service (SPS), a worldwide fax, telex and data satellite service. It provides automated Global Positioning System (GPS) position reporting and two-way messaging capability between aircraft and ground controllers located anywhere in the world within Inmarsat coverage.

Aeronautical CAPSAT enables ground controllers to monitor the position of helicopters from a central location. Since helicopters typically operate at low altitudes, the use of satellites provides a more reliable two-way communications link than other methods of transmission.

The system can be programmed to send regular, automatic GPS position reports to a PC connected to the worldwide telephone and data network. The system's data format is compatible with those of the international Future Air Navigation System (FANS) and Automatic Dependence Surveillance (ADS).

The ground controller also has a satellite connection, making the system ideal for operations in areas where the communications infrastructure is unreliable or non-existent. The accuracy of the satellite GPS is integrated into Aeronautical CAPSAT and the system is geared towards safety and is very user-friendly.

Able to report the aircraft's position automatically every 2 minutes, or on direct request by ground operators, no physical aircrew input is needed. In the event of a forced landing, ground operators will have rapid access to accurate position data that is not limited by geographical position.

Fully two-way, the system can operate certain modes in near real time and lets aircrew alter flight plans on the basis of the latest meteorological information. They can also make immediate changes to forward maintenance and refuelling plans and report aircraft position, regardless of where it is in the world.

The system comprises a small aerodynamic fin antenna; an internal transceiver and amplifier subsystem; a compact message terminal with a full QWERTY keyboard and integrated LCD (Liquid Crystal Display); and a mini printer unit for receiving hard copies of messages.

Ground operators connect to the telephone or data network via a modem or when in remote regions, to British Telecom's Goonhilly ground earth station using a portable satcoms unit. The CAPSAT Manager PC software is a full featured fleet and solo aircraft tracking program that lets ground operators monitor flight paths on a moving map display and communicate directly with aircrew, over the satellite link, via their onboard system.

Operational status

Supplied to the Royal Air Force, the UK Customs and Excise, and to the UK Meteorolgical Research Flight.

Contractor

Thrane & Thrane A/S.

UPDATED

FRANCE

COSPAS-SARSAT emergency locator transmitters - A06 range

The A06 range of COSPAS-SARSAT emergency locator transmitters includes the following types:

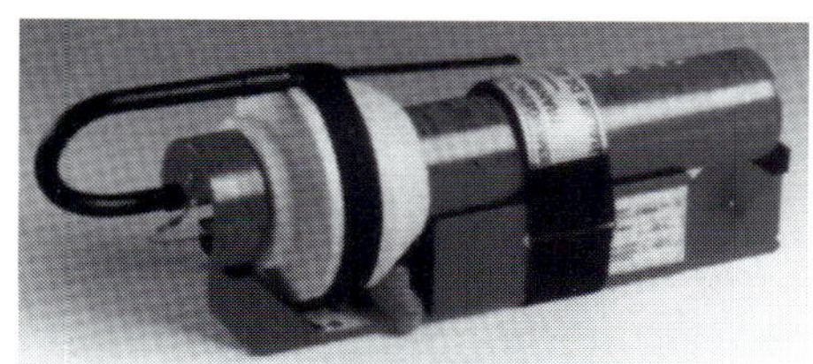

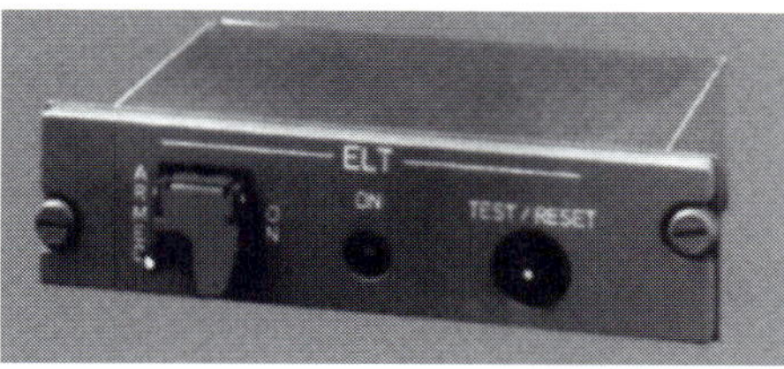

COSPAS-SARSAT emergency locator transmitters - A06 range **1998**/0015341

for ED-62 requirements; A06-A06V1 and the A06-A06V2 models;

for TSO requirements; A06T model.

The A06 range is packaged in an emergency distress orange container. The transmitter includes a 3-postion switch; red high-intensity light and buzzer; antenna connector; 3-frequency antenna; remote control (whip, or aircraft skin antenna). Activation is automatic by G-switch.

Transmissions are on the 121.5 and 243 MHz international distress frequencies, at 100 MW power; and on 406 MHz COSPAS-SARSAT frequency at 5 W power.

A remote control unit option is available for Airbus cockpit fit, and as a custom option for other aircraft types. Weight is 1.3 kg.

Contractor

CEIS TM — LCD Division

NEW ENTRY

51RV-4/5DF VOR/ILS receivers

The 51RV-4DF VOR/ILS receiver forms one part of a flight inspection system, providing test signals and measurement data for flight inspection of VOR localiser and glide slope installations. It is a direct retrofit for ARINC 547 VOR/ILS receivers.

All parameters that need to be accessed are available on the front panel and no wiring harness modification is required. The 51RV-4DF features digital bearing output filtering of the basic bearing information, remote control by two out of five binary control, remote control by serial information in ARINC 429 binary broadcast form, dual conversion of 200 VOR/Loc channels and 40 glide slope channels, dual-instrumentation circuits with internal comparison and level monitoring and BITE.

The 51RV-5DF is identical to the 51RV-4DF except that it meets ICAO FM immunity requirement.

Operational status

In production for numerous airport and military calibration authorities worldwide.

Contractor

Rockwell-Collins France, Blagnac.

UPDATED

The Rockwell Collins France 51RV-4/5 DF VOR/ILS receiver **1998**/0015342

DF-206NF ADF radio navigation system

The DF-206NF ADF radio navigation system is the replacement system for the DF-206, of which thousands of units are in service on many types of aircraft. The DF-206NF features improved reliability, extended frequency range, ARINC 429 compatibility, light weight/small size, easy maintenance and direct retrofit of the DF-206.

The DF-206NF ADF radio navigation system with (left to right) the antenna, receiver and control unit

The heart of the DF-206NF system is the 51Y-7NF receiver. The receiver is associated with the ANT-206NF combined sense/loop antenna, BCD control unit and an operational ARINC 429 control unit. The 51Y-7NF is a preplan product improvement to allow incorporation of additional features such as scanning on two channels and slow hop rate for ECCM compatibility.

The BC-206NF control unit has been designed to provide frequency and mode control of any version of the receiver. The control unit exists in several versions, depending on voltage and panel lighting, and is also NVG compatible.

Specifications

Dimensions:
(receiver) 335 × 57 × 112 mm
(control unit) 146 × 57 × 110 mm
(antenna) 216 × 149 × 43 mm
Weight:
(receiver) 1.8 kg
(control unit) 0.8 kg
(antenna) 1.5 kg
Power supply: 27.5 V DC, 0.4 A
26 V AC, 400 Hz, 0.3 A
Frequency: 190-2,999.5 kHz
Channel spacing: 500 Hz
Preset channels: 20

Operational status

Operational on French Army Aviation Puma and Fennec helicopters, and French Air Force Tucano trainer aircraft. Selected for the Tiger European helicopter programme.

Contractor

Rockwell-Collins France, Blagnac.

UPDATED

DF-301E direction-finder

The DF-301E (military designation OA-8697 or OA-8697A) is a solid-state direction-finder which utilises digital electronic circuitry to achieve improvements in bearing accuracy, acquisition speed and stability. Operating in the UHF and/or VHF frequency range of 100-400 MHz, automatic direction-finding capability is provided within one unit, giving cost, space and weight savings.

Used in conjunction with associated receiver and bearing indicator, the electronically commutated

The Rockwell-Collins France DF-301 EF direction finder **1998**/0015343

antenna provides relative bearing information to the UHF/VHF signal source transmitter. In the airborne environment, the DF-301E is used for course navigation, or to determine the relative direction of another transmitting aircraft. The unit meets MIL-E-5400 Class 2 requirements for environmental conditions. The unit weighs 3.4 kg and measures 88 × 134 mm.

Operational status

In production. Over 8,000 units are in operation.

Contractor

Rockwell-Collins France, Blagnac.

UPDATED

DF-430F tactical direction finder

The DF-430F is Rockwell-Collins France's new generation of automatic tactical V/UHF direction-finder.

The DF-430F features an embedded synthesised receiver that covers the 30 MHz to 400 MHz frequency band, thus enabling the DF-430F to function as a full stand-alone tactical direction finder.

The DF-430F has been designed for simple installation on any type of aircraft (fixed- or rotary-wing). It provides rapid and accurate bearing acquisition.

The DF-430F is a three LRUs system including:
one DF Antenna (ANT-430F);
one Receiver and Processing Unit (RPU-430F);
one Control and Display Unit (BC-430F), as an option.

DF-430F main characteristics are, 30 to 400 MHz frequency range; AM or PM antenna modulation according to frequency band; fully solid-state antenna; fast bearing acquisition (50 ms burst); high bearing accuracy; dead reckoning capabilities; remote control capability via ARINC 429 or MIL-STD-1553B interface; easy DF-301E mechanical retrofit; watertight package; and flush-mount installation.

Specifications

Typical range (up to the line of sight):
30-88 MHz: 100 nm (20 W)
100-400 MHz: 100 nm (5 W)
Bearing Accuracy
<3° Forward/Backward axis
<5° for other bearings
Dimensions;
ANT-430F: diameter: 278.5 mm; height: 107 mm
RPU-430F: ARINC 600 ¼ ATR Short
BC-430F: width: 146 mm; depth: 150 mm; height: 95.5 mm
Weight:
ANT-430F: 2.6 kg
RPU-430F: 3.2 kg, ARINC 600 ¼ ATR short
BC-430F: 0.8 kg

Operational status

Selected for NH-90 multinational helicopter programme.

Contractor

Rockwell-Collins France, Blagnac.

UPDATED

IPG-100F GPS receiver

The IPG-100F is a five-channel C/A and P/Y code single-frequency GPS receiver with Selective Availability and anti-spoofing capabilities.

The IPG-100F houses the receiver and its associated control panel in a single dzus-mounted box. The display includes two lines of 16 LED alphanumeric characters, each providing direct access to all GPS data, waypoints and navigational information. A keyboard with 12 keys provides access to system mode selection and data entry. Both the display and the keyboard are NVG-compatible.

The receiver is built with five independent channels providing continuous tracking of satellites in view, with permanent selection of the best four to calculate a navigation solution.

The IPG-100F includes a circular connector on the front panel to allow the operator to load crypto keys into

The Rockwell-Collins France DF-430F tactical direction finder showing (left to right) the BC-430F control display unit, the RPU-430F receiver processing unit, and the ANT-430F DF antenna **1998**/0015344

The display for the Rockwell-Collins France IPG-100F GPS

the PPS-SM modules. The unit is still unclassified even when the crypto keys are loaded. Two installation configurations are available: Stand-Alone as an autonomous RNav or Remote Sensor connected with the aircraft navigation computer through standard RS-422 or ARINC 429 ports.

The positions of up to 999 waypoints, airports or navaids can be loaded from a Jeppesen database on an RS-232 line either from a personal computer or magnetic card reader.

Specifications

Dimensions: 146 × 85 × 150 mm
Weight: 1.7 kg
Power supply: 10-32 V DC, 25 W
Accuracy:
3D position error (SEP) 16 m, full dynamics
3D velocity (RMS) 0.3 m/s, full dynamics
(time) 100 ns RMS

Operational status

In production for the French Army Aviation, the French Air Force, Royal air Force AWACS, Royal Navy Sea Harrier and Polish Air Force MiG-29 aircraft.

Contractor

Rockwell-Collins France, Blagnac.

UPDATED

IPG-120F GPS receiver

The IPG-120F GPS receiver is derived from the earlier IPG-100F system; it embeds Rockwell's latest GPS core engine. The IPG-120F is a 12-channel P/Y code L1/L2 GPS receiver. It is a form/fit replacement for the IPG-100F unit.

The IPG-120F GPS core engine provides the following features: RAIM (Receiver Autonomous Integrity Monitoring); WAGE (Wide Area GPS Enhancement); secure DGPS; fast initial acquisitions/direct Y code acquistion; high jam-resistance; all-in-view tracking and navigation, using Jeppesen database information and user waypoint data; NVG compatibility.

Specifications

Dimensions: 146 × 85 × 155 mm
Weight: 1.7 kg
Power: 10-32 V DC, 25 W
Position accuracy:
PPS: <15 m SEP
SPS: <100 m horizontal

Operational status

Upgrade in progress from IPG-100F.

Contractor

Rockwell-Collins France, Blagnac.

NEW ENTRY

MDF-124F direction-finder

Developed for Search and Rescue (SAR) operators, the function of the MDF-124F direction-finder is to localise the signal emitted by a distress beacon in order to rescue survivors on the ground or at sea.

The MDF-124F, with its embedded receiver, constantly scans and monitors 121.5, 243 and 406 MHz frequencies and meets the needs of SAR operators for an airborne system that reduces mission time by homing in on a localised beacon. It can also home in on a non-localised beacon because of its long-range detection capability (better than 100 nm). By combining the scanning of both 406 MHz, for long-range signal reception, and 121.5 and 243 MHz, for short range reception, rescue time is reduced. Information provided by the COSPAS-SARSAT satellite detection system concerning position of a beacon is updated every 90 minutes.

The MDF-124F consists of the MDF-124F antenna and distress receiver unit, packaged in a single watertight unit, and the BC-124F control unit.

The antenna is a fully static rotating antenna controlled by a unique patented driver. This antenna creates an AM modulation of incoming signals in the 100-400 MHz frequency range and a PM modulation on a 406 MHz incoming SARSAT signal.

The MDF-124F distress receiver is made of a dual-frequency AM receiver, on 121.5 and 243 MHz, and of a 406 MHz PM receiver. The two receivers are fully compatible with all types of modulation produced by International distress beacons. The MDF-124F also includes a VHF/UHF low-noise preamplifier for use with an external receiver in the 100-400 MHz frequency range.

The signal processor receives the demodulated audio signal from the internal distress receiver or from the external VHF/UHF receiver. The MDF-124 is based on phase comparison performed by a microprocessor. The signal processor delivers relative bearing in ARINC 407 (three-wire synchro) and ARINC 429 formats. When the MDF-124F is on COSPAS-SARSAT frequency, the bearing is estimated during beacon silence if the aircraft heading is available.

The BC-124F control unit allows the selection of the desired distress frequency or of one external receiver out of four. The control unit also indicates the reception of a distress signal on a non-selected distress frequency. If several COSPAS-SARSAT signals are received simultaneously, the BC-124F allows the selection of the desired one.

Specifications

Dimensions:
(MDF-124F) 90 × 315 mm diameter
(BC-124F) 66.6 × 146 × 150 mm (dzus mounted)
Weight:
(MDF-124F) 3.5 kg
(BC-124F) 0.7 kg

Power supply: 27.5 V DC, 700 mA
Interfaces: bearing output and heading input, ARINC 429 and ARINC 407

Operational status
In production for the French MoD, Australia, Canada, Japan, Spain, Poland and China.

Contractor
Rockwell-Collins France, Blagnac.

UPDATED

MDF-124F(V2) direction-finder

Recent miniaturisation of technology together with the experience gained, especially in maritime surveillance operation, have led Rockwell-Collins France to develop and release a new version of MDF-124F called MDF-124F(V2).

MDF-124(V2) is a full stand-alone SAR and tactical direction-finder; its embedded synthesised receiver allows it to operate the V/UHF 100-407 MHz frequency range without any external receiver.

MDF-124F(V2) comprises: MDF-124F(V2): antenna and receiver unit;
BC-124F(V2): control and display unit.

MDF-124F(V2) features the following direction-finding capabilities: direct access of any of the three international distress frequencies or auto-scanning/ alert on these frequencies; direct access to ARGOS channel; direct access to VHF-FM maritime channels 16 and 70; direct access to manual/preset user's frequencies in DF Tactical mode; combined tactical/ auto alert SAR modes allowing DF operation on one user's tactical manual/preset channel while still monitoring the three international distress frequencies. Other characteristics are identical to the MDF-124F.

Specifications
Dimension, weight, interfaces and power supply are the same as for the MDF-124F.

Operational status
In production for Australia, Belgium, China and Japan.

Contractor
Rockwell-Collins France, Blagnac.

UPDATED

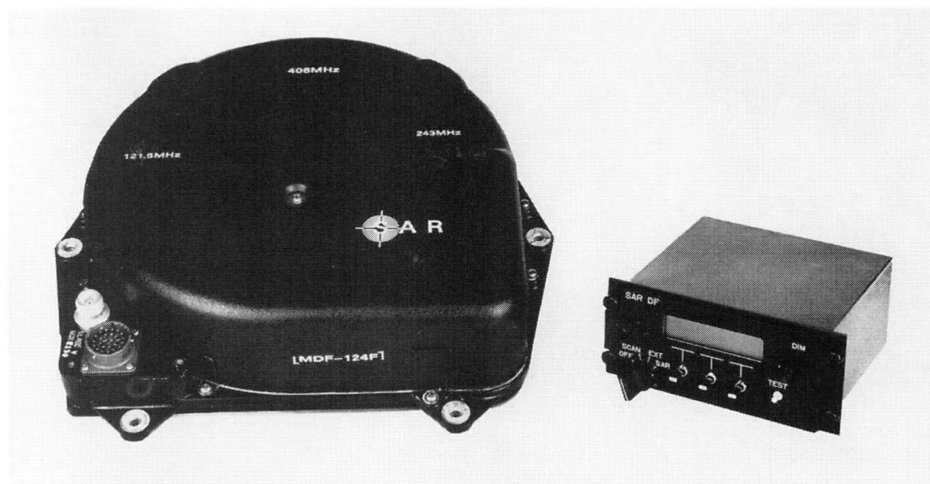

MDF-124F/MDF-124F(V2) direction-finder, showing (left) the antenna/receiver unit and (right) the control unit
1997/0001311

RSC-125F Personnel locator system, showing BC-125F (left) LPT-125F (centre) and ANT-430F (right)
1997/0001310

RSC-125F personnel locator system

RSC-125F is a personnel locator system which provides combat SAR platforms with the advanced capabilities required for the very demanding combat SAR mission.

This equipment has been developed to recover downed pilots or aircrew members using personal locator beacons.

RCS-125F is a full stand-alone system, with embedded receiver, which can be installed on both fixed- and rotary-wing platforms. Different interfaces give RSC-125F great flexibility and allow it to be integrated within ARINC-429 or MIL-STD-1553B architectures.

RSC-125F also includes tactical DF capabilities in the 30 to 400 MHz band and civil SAR frequencies, when operating all International distress frequencies.

RSC-125F is a 3 LRUs system including one DF antenna (ANT-430F); one Localisation Processor and Transmitter unit (LPT-125F); one control and display unit (BC-125F) (optional).

The main features of RSC-125F are compatibility with existing AN/PRC-112 and AN/PRC-434 Personal Locator Beacons (PLB); civil SAR (121.5-243 MHz-406.025 MHz COSPAS-SARSAT, VHF/FM maritime channels 16 and 70) and Argos capabilities; tactical DF in the 30 to 400 MHz band (25 kHz step); automatic scanning function on distress frequencies; remote control capabilities through ARINC 429 or MIL-STD-1553B interfaces; NVG compatible; high accuracy bearing and distance information on beacon; computed bearing during gap between the interrogations; easy flush-mount installation.

Specifications
Dimensions:
ANT-430F: diameter: 278.5 mm, height: 107 mm
LPT-125F: ARINC-600 ¼ ATR Short
BC-125F: 146 × 66.6 × 150 mm
Weight:
ANT-430F: 2.6 kg
LPT-125F: 4.5 kg (ARINC 600 ¼ ATR short)
BC-125F: 0.8 kg

Operational status
In production for French military aircraft.

Contractor
Rockwell-Collins France, Blagnac.

UPDATED

Embedded GPS receiver

The SAGEM embedded GPS receiver consists of a radio frequency and a digital function. The digital processing function implements 12 parallel channels to acquire and track up to 12 satellites at the same time. One single processor is dedicated to signal processing. Internal data transfer between the GPS and the INS processor is accomplished through a direct interface which minimises data latency and allows accurate time tagging, which is difficult to achieve in the case of separate GPS and INS units.

The receiver has the capability to track all visible satellites. This capability maximises the continuity and smoothness of navigation outputs, the precision of navigation outputs (which increases with the number of satellites tracked) and the reliability and integrity of navigation outputs as it reduces the detection time of a faulty satellite and allows for hardware redundancy.

Most of the GPS signal processing is achieved digitally. From the down conversion in the RF module, digitising is carried out using ASICs and highly integrated circuits. This has the advantages of high miniaturisation of the GPS receiver, signal-to-noise optimisation, operation independent of temperature conditions and time, and high reliability and flexibility.

Specifications
Accuracy (absolute positioning):
(standard positioning service) 100 m horizontal
(precise positioning service) 18 m horizontal
Accuracy (differential positioning):
(standard positioning service) 5 m
(precise positioning service) 1 m

Operational status
In production for French Air Force Mirage F1 and export customers.

Contractor
SAGEM SA, Defence and Security Division, Paris.

VERIFIED

MAESTRO nav/attack system

SAGEM SA has designed the Modular Avionics Enhancement System Targeted for Retrofit Operations (MAESTRO) with flexibility so it can be easily tailored to any customer's specific requirements and offer capabilities at par with those of current front line fighters.

MAESTRO features full inertial and GPS performance provided by SAGEM navigation and mission computer systems implementing the Embedded GPS-Inertia (EGI) concept, Terrain Contour Matching (TERCOR) for stealth navigation and covert attack in combination with the SAGEM high-capacity data transfer system offering digital moving map display capability, wide field-of-view FLIR-compatible HUD, glass cockpit, Hands On Throttle And Stick (HOTAS) interface, air-to-ground and air-to-air fire control, including multimode pulse Doppler radar and/or laser rangefinder, full EW suite comprising radar warning, missile launch detection, chaff and flare self-protection and/or jamming systems

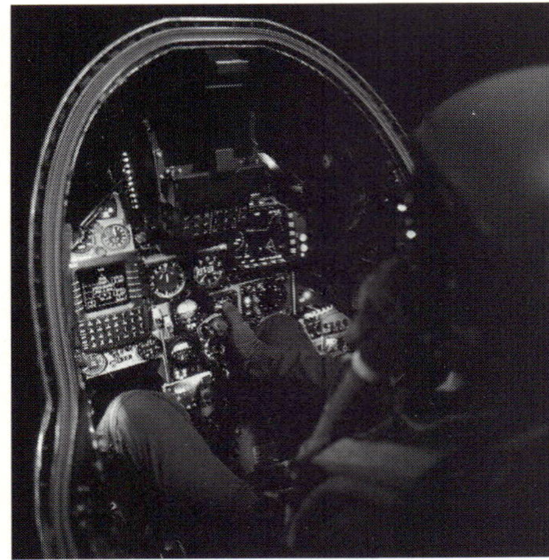

The SAGEM MAESTRO is designed for retrofit applications

and the all digital CIRCE 2001 mission planning system.

MAESTRO can be adapted to new aircraft of western or eastern origin.

Operational status

In production. Applications include the upgrade of Belgian Air Force Mirage 5, Chilean Air Force Mirage 5 and C-101 combat/trainer aircraft, Indian Air Force Jaguar, PZL Irdya Polish trainer/attack aircraft and Pakistan Air Force Mirage III and 5.

Contractor

SAGEM SA, Defence and Security Division, Paris.

UPDATED

Mercator digital map generator

The Mercator digital map generator provides pilots of modern aircraft and helicopters with presentation of a colour moving map display.

To increase flight safety and operational effectiveness, Mercator displays colour terrain information, related to the aircraft altitude. Mercator is derived from the Jumbo mass memory equipment and offers up to 1 Gbyte capacity, corresponding to a very large geographical area at various scales.

SAGEM Mercator, onboard digital map generator
***1998**/0015345*

Key features of the Mercator digital map generator include a north-up or track-up map/DTED display; real-time smooth rotation, pan/scroll and zoom; the ability to drive up to two multifunction colour displays; up to 1 Gbyte mass storage cartridge, allowing 1,000,000 km^2 (1:100k) on line map data; standard MIL-STD-1553B/ RS422 databus interface; simultaneous functions of mass memory server on MIL-STD-1553B (mission planning data) and digital map generation.

Specifications

Dimensions: 95.2 × 127 × 222 mm
Weight: 2.7 kg
Power: 28 V DC, 35 W

Operational status

In production.

Contractor

SAGEM SA, Defence & Security Division, Paris.

NEW ENTRY

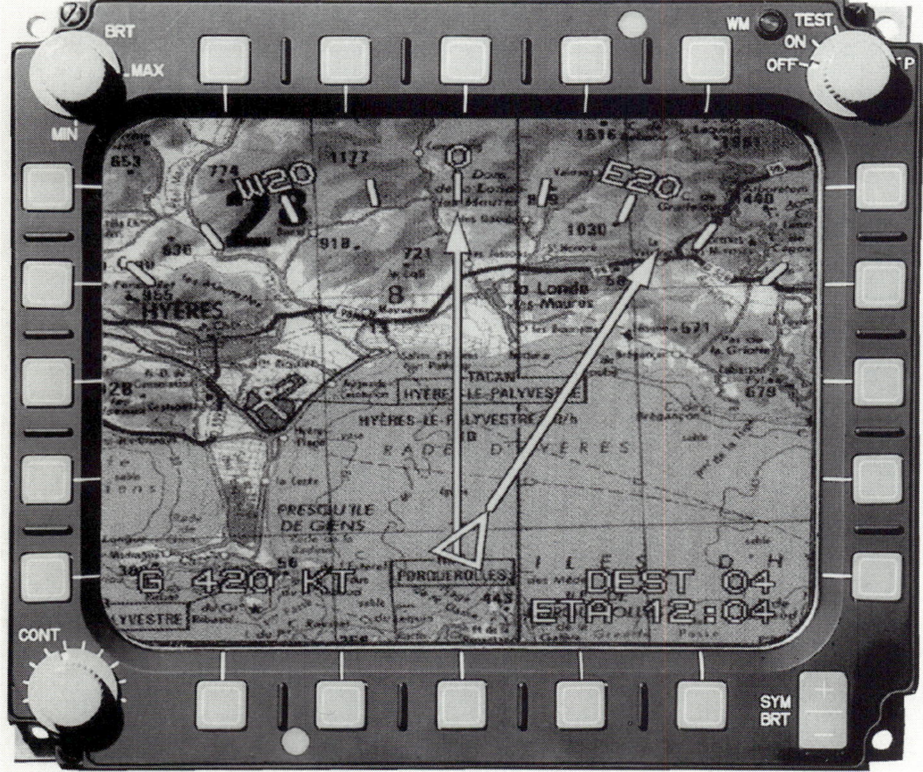

SAGEM Mercator map display in moving map mode, North oriented ***1998**/0015346*

Sigma ring laser gyro inertial navigation systems

The Sigma family of inertial navigation systems implements a combination of high-performance ring laser gyro sensors, accelerometers and a multichannel GPS receiver. Such systems are intended for use in aircraft equipped with a multiplexed databus.

Sigma systems offer all the benefits of tight hybridisation between ring laser gyros, accelerometers and a GPS receiver by a multi-sensor Kalman filter for both alignment and navigation. The synergy between these three elements brings the following advantages: reduction in size, weight and power consumption through the integration of inertial and GPS functions; short alignment time on the ground, in the air or at sea; monitoring of sensor performance and integrity for GPS and INS; automatic in-flight calibration of inertial sensors; long-term stability of inertial performance and higher resistance to jamming and improved dynamic behaviour of the GPS.

Some versions combine the navigation and fire-control functions. All versions provide aircraft position, velocity and attitude information, computation of navigation and steering information to waypoints; position updating by navigation fixes; terrain reference updating; weapon delivery computations, consisting of ballistics, determination of release point, ripple spacing of weapons, safety pull-up information; head-up display information for target acquistion and commands for the blind release of weapons; attack modes; air data computations and multiplex bus control.

Sigma 95MF inertial nav/attack system

The Sigma 95MF (MultiFunction) uses three high-accuracy ring laser gyros and three accelerometers and is fitted with an embedded GPS receiver. Sigma 95MF is the heart of MAESTRO, SAGEM's avionics system. Sigma 95MF is a highly integrated system which provides a combination of high-performance hybrid inertial navigation, attack computations and mission management tailored to advanced multirole combat or tactical transport aircraft.

Specifications

Dimensions: 209 × 400 × 380 mm
Weight: 17 kg
Power supply: 28 V DC, <90 W (115 V AC/400Hz optional)
Accuracy: 0.6 nm/h (inertial mode)

Operational status

In production for the Pakistan Air Force Mirage III upgrade.

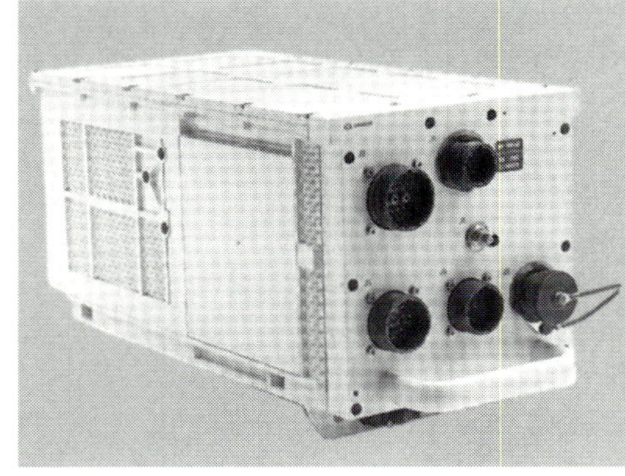

Sigma 95MF (MultiFunction) inertial nav/attack system ***1998**/0015347*

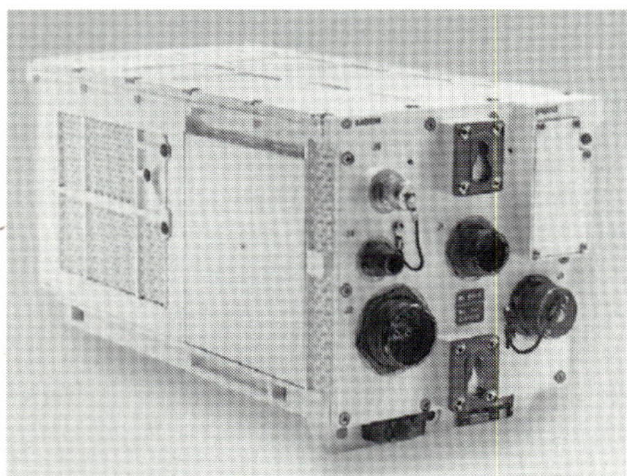

Sigma 95N (Navigation) inertial navigation system ***1998**/0015348*

Sigma 95N inertial navigation system

The Sigma 95N (Navigation) uses three high-accuracy ring laser gyros and three quartz accelerometers. It performs hybrid inertial/GPS navigation. It interfaces with the aircraft avionics systems through a MIL-STD-1553B multiplexed databus. The versatility of its interfaces enables the Sigma 95N system to be integrated easily with a wide range of carriers, avionic configurations and therefore operational situations.

Specifications

Dimensions: 209 × 200 × 385 mm
Weight: <15 kg
Power supply: 28 V DC, <45 W
Accuracy: 0.6 nm/h (inertial mode)

Operational status

In production for the French Air Force/French Navy Rafale aircraft (under programme name RL-90) and for the Cougar helicopter.

Sigma 95L inertial navigation system

The Sigma 95L (Light) is a compact inertial reference system fitted with a SAGEM embedded GPS. It has been designed for applications on helicopters and fixed-wing aircraft requiring lightweight, small size, high-performance navigation systems.

Sigma 95L can function as a self-contained attitude and heading reference system or as a full inertial navigation system. It performs hybrid inertial/GPS navigation. It interfaces with the aircraft avionics system through a MIL-STD-1553B multiplexed databus and ARINC 429.

Specifications

Dimensions: 180 × 125 × 280 mm
Weight: <8.5 kg
Power supply: 28 V DC, <35 W
Accuracy: 1nm/h (inertial mode)

Operational status

Off-the-shelf production. A (SAPHIR) version of Sigma 95L which includes an integrated GPS receiver, has been selected for the NH 90 helicopter programme, where it will provide the main navigation data and inertial references for the fly-by-wire system.

Contractor

SAGEM SA, Defence and Security Division, Paris.

UPDATED

Uliss inertial navigation and nav/attack systems

Uliss modular systems all employ high-accuracy inertial components consisting of two dynamically tuned gyros and three dry accelerometers, a microprocessor-controlled computer working at 1 Mops with EPROM memory and highly integrated and hybrid circuits. As an option, Uliss can be equipped with an embedded GPS receiver for high-performance INS/GPS coupling and a higher-performance RISC processor using Ada language.

Sigma 95L (Light) inertial navigation system
***1998**/0015349*

Uliss systems fall into two categories, in both of which the main inertial navigation unit is contained within a ¾ ATR short case. In the first category are navigation versions with a position accuracy of better than 1 n mile/h. In the second category the navigation function is combined with the computation necessary for weapon delivery. Alignment time is 90 seconds for stored heading and 5 to 10 minutes for self-contained gyrocompassing. Standard interfaces permit the systems to be linked to other equipment via MIL-STD-1553B databusses or ARINC serial data lines. A failure detection system can detect faults at module level with 93 per cent confidence, isolate them and signal their presence on a magnetic annunciator without external test equipment.

More than 80 per cent of the components and subassemblies are common to all members of the Uliss family, the principal differences being in specific interfaces and computation functions.

In all, close to 2,000 Uliss units are in operation on 25 different types of aircraft.

Uliss 45 Inertial Navigation System

The Uliss 45 system has been optimised for high accuracy in long-range navigation and certain other special applications. Its interfaces comply with ARINC 561 and embody significant flexibility, for example, in order to communicate with two DME or Tacan receivers, with Kalman filtering for better accuracy. The equipment is used on long-range transport aircraft and as an accurate position and velocity reference for flight development purposes.

Specifications

Dimensions:
(navigator) 420 × 194 × 191 mm
(control/display unit) 209 × 114 × 127 mm
Weight:
(navigator) 16 kg
(control/display unit) 3 kg
Power supply: 115 V AC, 400 Hz, 250 VA
Accuracy: 1 n mile/h CEP

Operational status

In service in Mirage F1-CTs, and Boeing KC-135 tankers of the French Air Force.

Uliss 52 inertial navigation system

The Uliss 52 navigation system is designed for high-performance combat aircraft. The program is based on Ada. Avionics information and commands are distributed by a digital multiplexed databus. It comprises three units: a UNI 52 inertial navigator, a PCN 52 control/display box and a PSM 52 mode selector fitted with an automatic insertion module to allow information such as flight plan, system data and maintenance information to be fed in.

Specifications

Dimensions:
(navigator) 386 × 194 × 191 mm
(control/display unit) 208 × 114 × 127 mm
Weight:
(navigator) 15 kg
(control/display unit) 3 kg
Power supply: 115 V AC, 400 Hz, 3 phase, 250 VA
Accuracy: 1 n mile/h CEP

Operational status

In production for French Air Force Dassault Mirage 2000DA, N and D aircraft and the export version Mirage 2000-5.

Uliss 92 Inertial Nav/Attack System

The Uliss 92 combines in a single box all the functions of an inertial navigation and fire-control system. It is based on Ada programming and comprises three units: UNA 92 inertial/attack box, PCN 92 control/display unit and PSM 92 mode selector.

Specifications

Dimensions:
(navigator) 386 × 194 × 191 mm
(control/display unit) 216 × 116 × 153 mm
(mode selector) 151 × 41 × 135 mm
Weight:
(navigator) 16 kg
(control/display unit) 3.5 kg
(mode selector) 1 kg
Power supply: 115 V AC, 400 Hz, 3 phase, 220 VA
Accuracy: 1 n mile CEP position, 5 mrad weapon delivery

Operational status

In production for the Belgian Air Force Mirage 5 and the Pakistan Air Force Mirage III upgrades.

Contractor

SAGEM SA, Defence & Security Division.

UPDATED

ELT 90 series Emergency Locator Transmitters

Satori makes three ELTs:

ELT 90-ELT 92 a bi-frequency (121.5/243 MHz) automatic fixed/portable equipment conforming with EUROCAE ED-62/RTCA DO-182

ELT 96-406 a tri-frequency (121.5/243/406 MHz) automatic fixed/portable equipment conforming with EUROCAE ED-62/RTCA DO-182

ELT 96S-406 a tri-frequency (121.5/243/406 MHz) automatic fixed/portable equipment conforming with EUROCAE ED-62/RTCA DO-204.

Specifications

Dimensions: 83 × 103 × 215 mm
Weight: 1.52 kg
Output power: ELT 90-ELT 92: 100 mW PERP 48 hours
ELT 96 (both models): 5 W PERP 48 hours.

Contractor

Satori.

VERIFIED

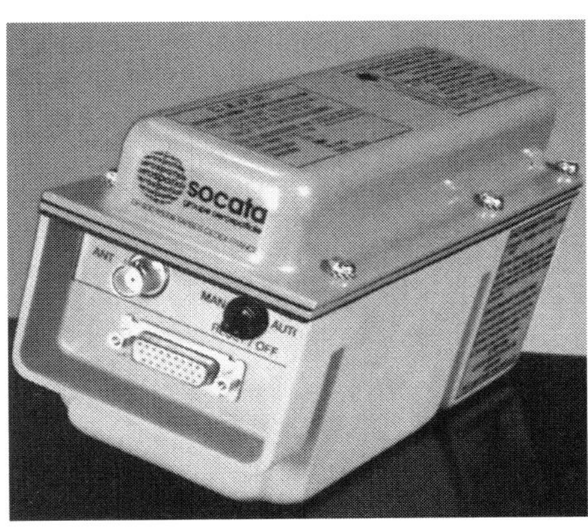

ELT 96-406 emergency locater transmitter
***1997**/0001194*

Automatic Dependent Surveillance (ADS) unit

The Sextant Avionique Automatic Dependent Surveillance (ADS) unit was introduced in 1993. The reduction of separation distance, and thus the creation of more airspace, depends on establishing the exact position of an aircraft in space and time. To do this, the flight data used by the pilot must be processed and transmitted to a receiving station. The ADS system features the automatic transmission of this information and immediate reporting of data to the airline.

ADS will extend surveillance capability to areas not covered by radar. By adding a Secondary Surveillance Radar (SSR), the service can be extended to terminal areas and high-density continental airspace.

ADS is expected to provide a number of benefits to both pilots and airlines, including significantly enhanced flight safety, extension of ATC services to oceanic regions and areas not covered by radar, reduced separation minima, and highly accurate surveillance protected against interference in high-density airspace.

The Sextant Avionique approach is two-fold: refining the ADS concept through a stand-alone computer designed for current aircraft and developing an embedded ADS function for the avionics suites on future aircraft.

Contractor

Sextant Avionique.

VERIFIED

Cirus attitude and heading reference system

Cirus is a hybrid inertial attitude and heading reference system. Its basic design involves the use of inertial/magnetic heading and air data hybridisation techniques. The system generates heading, attitude, angular velocities, accelerations, true airspeed, pressure altitude and temperature.

As an option, it can also deliver position and groundspeed data in conjunction with a Doppler navigation radar or a GPS receiver. The groundspeed vector supplied by the GPS system, with its medium-term accuracy, can be combined with the angular and linear speed data generated by Cirus, with its excellent short-term accuracy, to form a versatile high-performance AHRS/GPS coupling system.

Specifications

Dimensions: 260 × 150 × 150 mm
Weight: 5 kg
Power supply: 28 V DC, 50 W
Alignment time: <1 min

Operational status

In production. The Cirus system is fitted to the Eurocopter Super Puma Mk 2 helicopter. It has been selected for the French Air Force C-160 Transall retrofit and maritime patrol Alizé aircraft.

Contractor

Sextant Avionique.

VERIFIED

Meghas™ new-generation avionics suite for helicopters

Sextant Avionique and SFIM Industries have designed a new-generation of avionics systems for helicopters, called Meghas. Based on an integrated family of equipment, this new avionics system is designed to satisfy the specific requirements of helicopter operators. Meghas has been chosen for Eurocopter's new civil helicopters, up to the six-ton class.

Meghas provides the following functions needed for helicopter operations, autopilot, guidance and navigation, engine and vehicle control, onboard maintenance, radio communications and radio navigation.

By minimising the crew's workload, Meghas allows helicopter crews to accomplish all missions — EMS, SAR, surveillance and offshore transport, quickly and efficiently. The system's modular design facilitates addition of new mission functionality, as well as providing weight savings, lower cost of ownership and higher reliability.

Meghas new-generation avionics suite for helicopters ***1998***/0015350

The major components in the Meghas avionics system include:

(1) A central control unit with a display system comprising two active-matrix liquid crystal displays. This is the Vehicle and Engine Management Display (VEMD), which replaces conventional indicators, and allows immediate verification of both vehicle and engine parameters. The design used complies with all High-Intensity Radiated Field (HIRF) and lightning requirements. The display system version for twin-engine helicopters has been selected as basic equipment on the IFR version of the EC 135, with certification scheduled for the end of 1997. It will also be standard on the Dauphin N4, scheduled for certification in mid-1998.
(2) Flight Display SubSystem (FDSS), with very high-resolution multifunction LCDs. For this function, Sextant Avionique developed a new 'smart' multifunction display, the SMD45H, a colour, high-resolution, rectangular unit. The FDSS has been chosen for the IFR version of the EC 135 and the Dauphin N4. The SMD45H can display inputs from various mission-specific equipment, including FLIR, map generator, and weather radar. A new display, the SMD68H, was recently selected for the first EC 135 IFR programme, for delivery and certification in early 1998.
(3) Automatic Flight Control System (AFCS), designed and built by SFIM Industries, includes an attitude and heading reference system (APIRS), by SFIM and Air Data System (ADS 3000). The AFCS can be upgraded, changing it from a simple stability augmentation system to a complete four-axis autopilot system that can be coupled to a GPS receiver for auto-approach.
(4) Health and Usage Monitoring System (HUMS), by SFIM Industries, which increases crew safety and enhances maintenance automation processes.
(5) Navigation equipment, including a GPS receiver and Search and Rescue (SAR) capability.
(6) Centralised Radio Control (CRC), comprising radio communications, radio-navigation and identification control systems.

Operational status

Selected for the IFR version of the EC 135 helicopter, with certification was scheduled for the end of 1997, and for the Dauphin N4 helicopter, with certification scheduled for mid-1998.

Contractors

Sextant Avionique.
SFIM Industries.

NEW ENTRY

MultiMode Receiver (MMR)

The Sextant MultiMode Receiver (MMR) is a new landing and precision approach sensor that provides ILS, MLS and GLS functions in one LRU. In addition the MMR provides position, time and velocity in a permanent GPS navigation mode.

The receiver has been designed according to ARINC 755 specifications for dual or triplex installations providing Cat III operations. The MMR is immune to FM radio signals in accordance with ICAO annex 10.

The MMR offers a high degree of integrity and reliability due to the digital technology used in the design.

Specifications

Dimensions: 3 MCU
Weight: 5.4 kg
Power supply: 115 V AC, 400 Hz single phase
Temperature range: −40 to +70°C
Compliances:
ILS ARINC 710-9
MLS ARINC 727
GPS ARINC 743A
MMR ARINC 755

Operational status

In production September 1997.

Contractor

Sextant Avionique.

VERIFIED

The Sextant Avionique MMR multimode receiver ***1997***/0001314

Nadir Mk 2 navigation/mission management system

The Nadir Mk 2 is a multipurpose processing and display system that can store details of up to 100 waypoints and the characteristics of up to 100 VOR/DME stations. It is not limited to Doppler but can operate with many other navigation sensors. Nadir Mk 2 comprises a 4 MCU central processing unit and a general purpose control/display unit. It can provide the following functions: navigation management based on Doppler, VOR/DME, Tacan, Omega/VLF, GPS, heading sensor, inertial sensor and air data inputs; flight management, with guidance for optimum cruise conditions, fuel and weight management and engine monitoring; air data computations involving speed, altitude and outside air temperature and interface with autopilot, radar and navigation indicators.

Nadir Mk 2 may be used in a dual-system configuration, in which one computer is responsible for navigation and flight management, while the second deals with weapons and aircraft management.

The Nadir 1000 navigation and mission management system for helicopters **1998**/0015351

Operational status

In production. Nadir Mk 2 has been selected for 30 Eurocopter helicopter programmes, and for fixed-wing aircraft operations.

Contractor

Sextant Avionique.

UPDATED

Nadir 1000 integrated navigation and mission management system

The Nadir 1000 is an integrated navigation and mission management computer developed to provide mission assistance for military and civil light helicopters over both land and sea. The Nadir 1000 is designed to provide multisensor navigation from Doppler, radio navigation and GPS, for flight management, navigation and mission management, and system links for SAR, hover and ASM roles.

The Nadir 1000 weighs 3 kg in the basic version and consists of two modules: a front panel module and a processing/power supply module. Optionally, analogue/digital and input/output modules can be added.

Operational status

In production. Selected for several Eurocopter export programmes. More than 100 systems delivered for Fennec and Super Puma.

Contractor

Sextant Avionique.

NEW ENTRY

NASH-Night Attack System for Helicopters

NASH – Night Attack System for Helicopters – is designed to provide the capability required for combat helicopters to carry out attack and counter-insurgency missions at night and in all weather conditions. In addition to flight management and navigation functions, this system displays information on observation, target selection and fire control.

Observation, target detection and fire-control functions are handled by the following systems:

(1) The Victor thermal imager, integrated either in the gyrostabilised sighting system, or in a Chlio gyrostabilised platform, coupled with the existing sighting system. The FLIR image is displayed on the SMD66 smart multifunction liquid crystal display

(2) Two Topowl® binocular helmet-mounted sight/displays, which can display images from an infrared sensor or integrated light intensifiers. Topowl® offers a very wide field-of-view, with images projected directly on to the visor.

Navigation and mission management functions are provided by a Nadir 1000 navigation system, working in conjunction with a GPS receiver, Stratus attitude and heading reference system and a Doppler radar.

All flight information for pilot and co-pilot are shown on the SMD66 head-down display and the Topowl HMS/D.

Operational status

Sextant Avionique's night navigation system has been selected for the Rostvertol Mi-35 and the Kamov Ka-52.

Contractor

Sextant Avionique

NEW ENTRY

STRATUS altitude and heading reference system **1997**/0001315

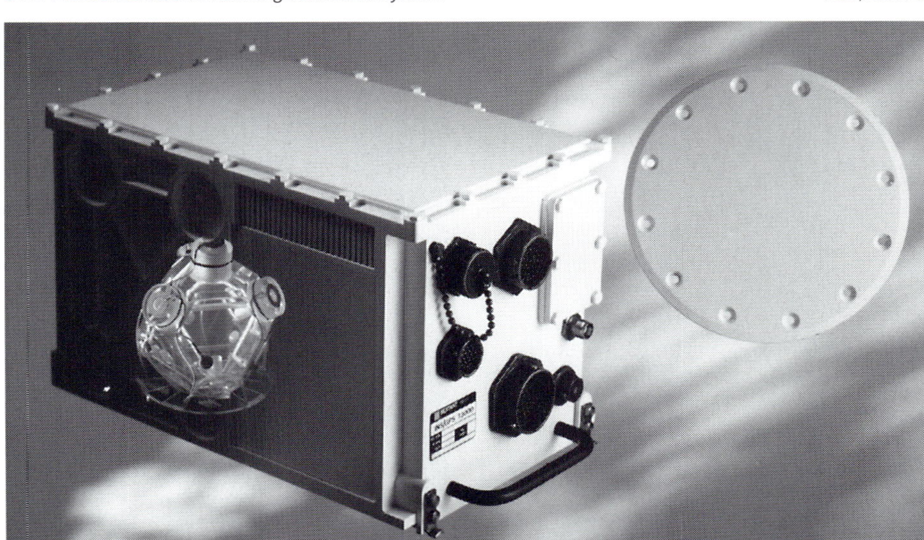

TOTEM 3000 navigation and flight management system **1997**/0001317

Stratus and Totem 3000 flight systems

Sextant Avionique currently produces three types of Ring Laser Gyros (RLG):

a three axis monolithic PIXYZ®, 14 cm path length, 0.1°/h in run stability;

a three axis monolithic PIXYZ®, 22 cm path length, 0.001°/h in run stability;

a single axis, 33 cm path length, 0.001°/h in run stability.

With this family of RLG, Sextant Avionique produces several inertial reference units for a large range of applications.

STRATUS:

Stratus is an attitude and heading reference system that uses the RLG PIXYZ® 14 cm path length. It is used as a basic reference system for attitude and hybrid navigation (with doppler radar and/or air data) on military helicopter, advanced trainer and missile systems.

Specifications

Attitude: 0.2°
Magnetic heading: 0.3°
Position: 1°/° of distance (with Doppler)
Weight: 5.2 kg
Power supply: 115 V AC, 45 VA
Interfaces: ARINC 429 and MIL-STD-1553B
Environment: as per MIL-STD-810D

Operational status

Selected for the Tiger and Rooivalk helicopters.

TOTEM 3000:

Totem 3000 is an inertial/GPS navigation system for high-performance military aircraft. It utilises the PIXYZ® 22 cm path length RLG and the Sextant Avionique

Topstar® 1000 GPS receiver board. The use of these two highly integrated and high-performance subassemblies leads to significant improvement of cost, size, and reliability compared with conventional single axis RLG.

All the functions are in compliance with SNU 84-1 and STANAG 4294

Specifications

Attitudes/heading: 0.05°
Position: 0.5 n mile/h CEP (inertial mode)
21 m (95%) (inertial/GPS PPS mode)
Velocity: 0.7 m/s RMS (inertial mode)
0.1 m/s (95%) (inertial/GPS PPS mode)
Dimensions: 177.8 × 177.8 × 279.4 mm
Weight: 8 kg including GPS
Power consumption: 40 W
Interfaces: standard MIL 1553 B
Environment: as per MIL-STD 810 D

Operational status

Selected on MIG AT, I22 and retrofit programmes including MiG-21, Mirage F-1 and Lockheed Martin C-130 aircraft.

Contractor

Sextant Avionique.

NEW ENTRY

Rooivalk helicopter avionics system

In August 1994, Sextant Avionique signed a contract with the South African avionics integrator ATE Pty Ltd to supply key avionics equipment for the Rooivalk attack helicopter being developed and manufactured by South African aircraft manufacturer Atlas Aviation of the Denel Group.

Equipment supplied by Sextant Avionique comprises the liquid crystal displays, helmet-mounted displays, laser gyro navigation system with GPS, standby instruments and pilot handgrips.

Sextant Avionique has now completed delivery of all basic avionics, and has started delivery of equipment for an expanded basic avionics package.

Basic avionics

The basic avionics of the Rooivalk are off-the-shelf equipment which were all qualified on the Tiger combat helicopter.

Sextant Avionique is providing a complete, redundant navigation system, including two STRATUS inertial navigation units (each based on a PIXYZ® three-axis ring laser gyro), air data equipment, two magnetometers and an NSS 100-1 GPS receiver.

The navigation system is built around the STRATUS unit, which functions as inertial sensor, navigation computer and controller of other peripherals.

Three complete systems have been delivered. Initial flight tests were on a Puma helicopter; Rooivalk integration started in mid-1996.

The Rooivalk will be equipped with Sextant Avionique's MFD 66 liquid crystal displays. They are part of a complete family of LCDs developed by Sextant Avionique for both civil and military applications, including rotor-wing aircraft, and already chosen for a number of programmes, including Rafale, Mirage, MIG-AT, LCA and MB-339CD.

Expanded basic avionics package

The helmet-mounted display supplied by Sextant Avionique was developed within the scope of development contracts for the Rafale and Tiger programmes.

The TOPOWL® HMD selected for the Rooivalk is one of a new generation of binocular, wide-field-of-view, day/night capable, head-up display/sights developed by Sextant Avionique. It provides a significant improvement in pilot comfort and safety, based on the projection of images on the helmet visor - either video or synthetic images - or from and integrated image intensifier.

The first HMD was delivered for tests in April 1996.

Contractor

Sextant Avionique.

VERIFIED

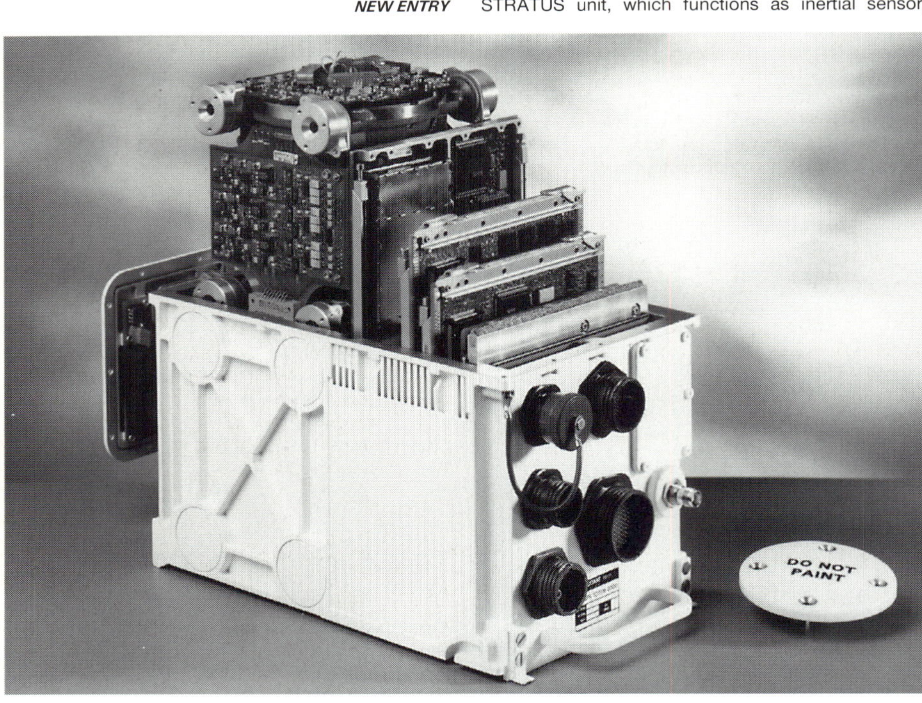

TOTEM 3000 navigation and flight management system. **1997**/0001316

The Rooivalk helicopter carries the Sextant Avionique avionics system **1997**/0001318

Topflight avionics suite

The Topflight avionics suite has been designed for basic or advanced training aircraft, as well as for combat aircraft such as the Mirage F1 or Su-22, either as original equipment or for retrofit. It enhances aircraft operational capabilities in a complex environment by supporting the pilot in all navigation tasks and in both air-to-air and air-to-surface missions.

Topflight incorporates advanced technologies, such as holography, liquid crystal displays, laser gyros, head position direction, new-generation computers, GPS, digital mapping and voice command. These are employed in a helmet-mounted display, a key part of the man-machine interface tailored to night mission capability; multimode smart head-up display; multifunction liquid crystal displays; laser gyro navigation system; digital autopilot, and the Precise Position Service (PPS) version of GPS. Topflight is built around a modular mission computer and symbol generator.

The compact design, incorporating smart liquid crystal displays, gives Topflight a multiple mounting capability and the modular computer provides the performance levels needed for multirole aircraft.

Operational status

In production for the MiG-AT upgraded avionics programme.

Contractor

Sextant Avionique.

VERIFIED

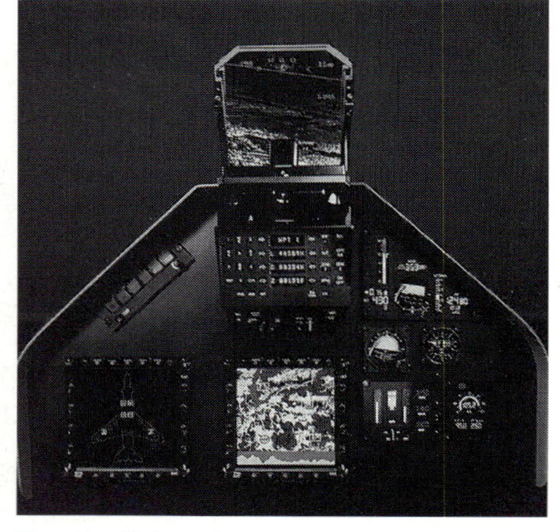

The Topflight avionics suite for training and combat aircraft **1995**

Topstar family of GPS receivers

Sextant Avionique is producing a complete family of GPS sensors for aeronautical, missile, and space applications, military and civil.

Topstar sensors are multichannel continuous tracking sensors. They provide time and three-dimensional position and speed at a high data rate (up to 10 Hz).

They are characterised by the capability to provide data for speeds up to 800 kt and for high acceleration, with accuracy in compliance with international standards. Additional functions such as DGPS, relative navigation, combined GPS/GLONAS processing, and RAIM are offered as options.

Four sub-families of products are available: (three for aircraft applications, detailed below, and one Topstar 300 for space use).

Topstar 100: stand-alone sensors for high dynamic military applications in aircraft, helicopters, missiles.

Specifications

SPS or PPS operation
8 parallel channels
Dynamic operation: 1500 m/s - 10 *g*
Accuracy: as per STANAG 4294
Dimensions: 130 × 240 × 80 mm
Weight: 2 kg
Power supply: 28 V DC, 20 W
Interfaces: MIL-STD-1553B, ARINC 429, RS-422
Environment: as per MIL-STD 810-E

Operational status

In production for Mirage 2000, Rafale, Rooivalk, Cougar, Apache, Mirage F1.

Topstar 1000 is a new version of Topstar 100, designed as a PCB module to be integrated in host equipment (IRS). Topstar 1000 has the GLONASS processing capability and interfaces with host computer through a dual-port RAM and RS-422 bus.

Specifications

Dimensions: 150 × 145 × 150 mm
Weight: 0.5 kg
Power construction: <10 W

Operational status

In development, embedded in TOTEM 3000 IRS, selected in several retrofit programmes, including MiG-21.

Topstar 200: ARINC 743A stand-alone sensors for civil aviation.

Specifications

C/A (L1) + W.A.A.S. capability
15 parallel channels
Dynamic operation: 2000 kt, 3 *g*
Accuracy: as per STANAG 4294, RAIM GIC and differential RAIM
Dimensions: ARINC 743A 2MCU format or ARINC 743A alternate format (190 × 240 × 63 mm)
Weight: 2 kg
Power supply: 28 V DC, 15 W
Interfaces: as per ARINC 743A
Environment: as per RTCA DO 169 C
TSO C 129 C1 label

This product is also available as a PCB to be integrated in host equipment (IRS or multimode receiver MMR)

Operational status

Selected for the Boeing MD 82 and for the Airbus family.

Contractor

Sextant Avionique.

VERIFIED

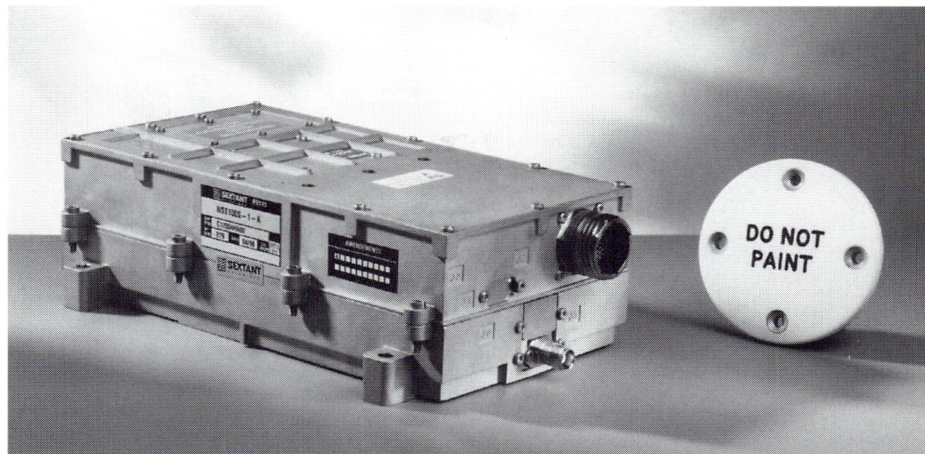

Topstar 100 ***1997***/0001320

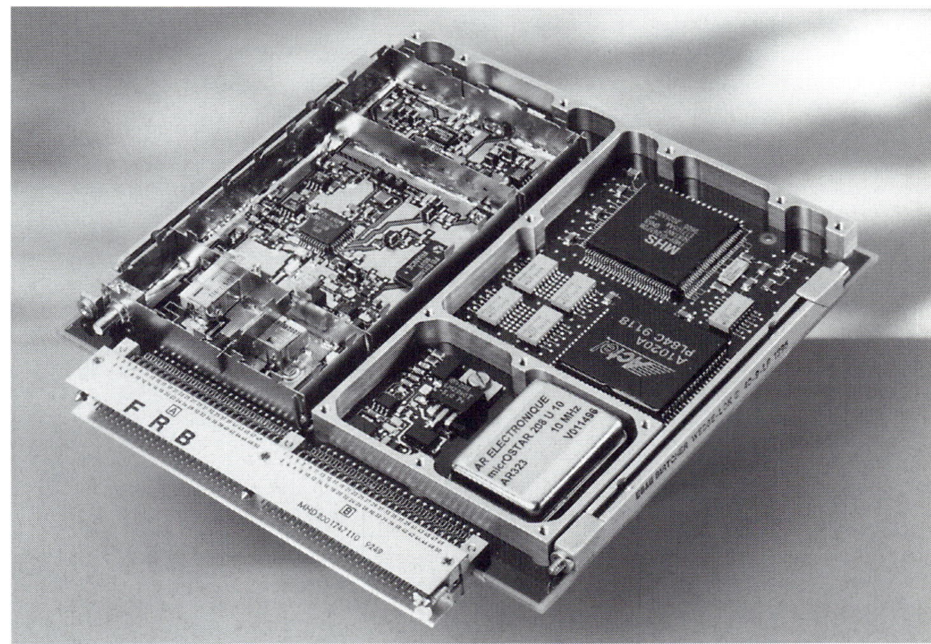

Topstar 1000 ***1997***/0001321

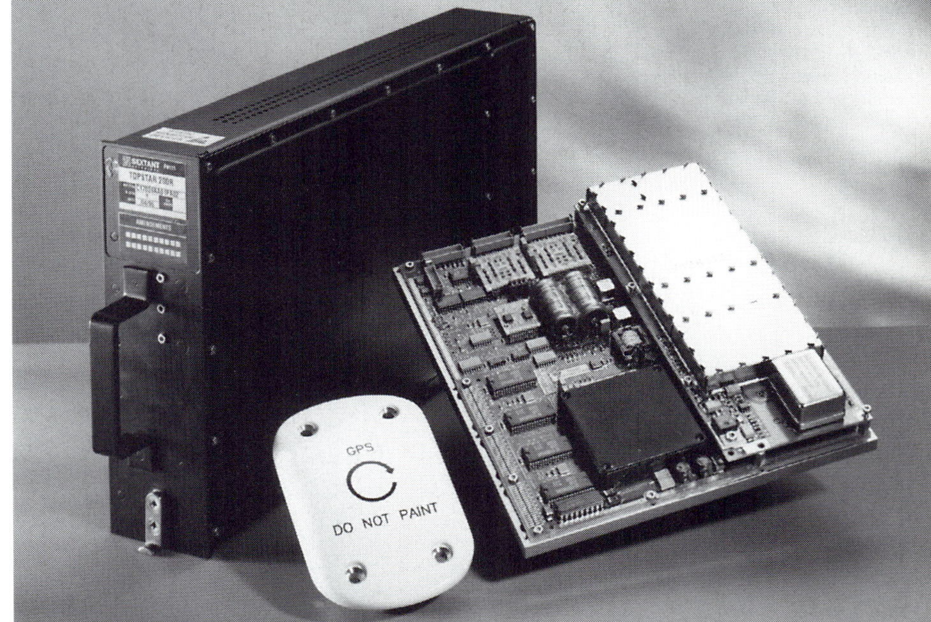

Topstar 200 ***1997***/0001319

APIRS Aircraft Piloting Inertial Reference Strapdown Sensor

APIRS is the latest fibre optic AHRS designed by SFIM Industries for commercial and military aircraft. This AHRS replaces conventional vertical gyro, directional gyro, rate gyros and accelerometer packages with one single highly reliable low-size lightweight box. Used in conjunction with a static magnetometer, APIRS provides, through an ARINC 429 databus, three-axis attitude, magnetic heading, angular rates and acceleration data about the body axes of the aircraft for AFCS and EFIS. When interfaced with an air data system, it also furnishes inertial baro-altitude and inertial vertical speed data.

APIRS is of modular construction and consists of an inertial measurement unit which contains the primary sensors and associated electronics, a power supply

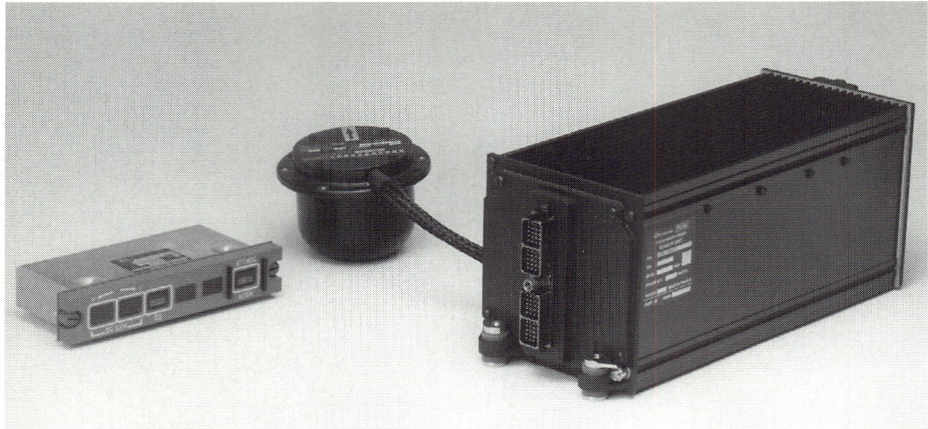

The SFIM Industries APIRS AHRS (Attitude Heading Reference System) ***1997***/0001322

module, an EMI/lightning protection and connector module and a harmonisation unit for safeguarding the installation and magnetometer calibration data. Augmentation by GPS is an option.

Specifications

Dimensions: 289 × 105 × 114 mm
Weight: <3.5 kg
Power supply: 28 V DC, <30 W

Operational status

In production. APIRS has been selected by Eurocopter for the EC-135, NH 90 and Dauphin N4 helicopters. APIRS has also been selected for the DASH 8-400 aircraft.

Contractor

SFIM Industries.

UPDATED

NC 12 airborne Tacan interrogator

The NC 12 is a very small and lightweight Tacan interrogator which fully complies with STANAG 5034, the improved MIL-STD-291. It is an all-digital system fitted with ARINC 429, ARINC 582 or MIL-STD-1553B standard outputs. It provides the pilot with digital distance and bearing relative to Tacan beacons on the NR 13 control box indicator. The NC 12 interrogator can be used either on new aircraft or for retrofit applications.

In air-to-air mode the pilot is supplied with distance to another aircraft fitted with an airborne beacon. A warning light indicates the presence of any jamming of the bearing and distance information.

The NC 12 interrogator is also fully adapted to advanced radio navigation through the provision of W and Z channels for DME-P compatibility.

Specifications

Dimensions:
(NR 13 control box) 144 × 63 × 57 mm
(NC 12 transmitter/receiver) 318 × 91 × 193.5 mm
Weight:
(NR 13 control box) 0.7 kg
(NC 12 transmitter/receiver) 5.5 kg
Power supply: 28 V DC, 1.4 A, 40 W
Transmission power: 350 W
Frequency:
(transmit) 1,025-1,150 MHz
(receive) 962-1,213 MHz
Number of channels: 126 X channels, 126 Y channels
Accuracy: 0.1 n miles and 1°
Altitude: up to 100,000 ft

Operational status

In production for the Mirage 2000 and F1-CT, Super Puma, Dauphin, UH-60, CH-47 and Tucano. Selected for the Rafale.

Contractor

Thomson-CSF Communications.

UPDATED

The Thomson-CSF Communications NC 12 airborne Tacan interrogator ***1998***/0015352

Over-The-Horizon Target (OTHT) designation system

The Over-The-Horizon Target (OTHT) designation system is designed to detect and locate targets beyond the horizon and to transmit the co-ordinates to a coastal battery or ship. The complete system consists of equipment for acquisition and transmission.

The ORB-32 panoramic radar detects, locates, identifies through the use of IFF data and automatically tracks the targets. The ORB-32-03 version for helicopters contains an antenna with IFF capability in the radome and, in the fuselage, a frequency-agile transceiver, junction box, radar control unit, scan converter, 9 in TV scope and a track-while-scan processing unit. The IFF option includes an IFF adaptation unit and IFF control unit.

The transmission of data takes place by conventional or jam-protected UHF radio over the frequency range 225-400 MHz. The jam-resistant version consists of a TDP-500 Series unit which allows data transmission in the frequency-hopping mode at 64 hops/s and generates system synchronisation, and an ERM-9000 transceiver equipped with a frequency-hopping unit. The conventional version consists of an MSA-300 modulator which converts the message into two LF tones for the UHF transceiver, and a CDM-6000 which receives information in the form of LF tones and converts them back to digital form.

Contractor

Thomson-CSF Communications.

VERIFIED

TLS-2020 Multi-Mode Receiver (MMR)

The TLS-2020 Multi-Mode Receiver is a new VOR/ILS receiver designed for fighters and helicopters. It fully complies with new ICAO - Annex 10 regulations (FM immunity requirements). It also includes a VHF receiver for the DGPS datalink.

The TLS-2020 receiver can optionally be configured with Marker, MLS, DGPS and GPS functions by simply adding modules.

The TLS-2020 MMR utilises the latest technology developed both for the commercial aviation and French military programmes. The TLS-2020 belongs to the TSL-2000 product line.

The Thomson-CSF Communications, TLS-2020 Multi-Mode Receiver (MMR) ***1998***/0015353

Specifications

Basic functions: VOR/ILS/VHF receiver for DGPS
Options: Marker, MLS, DGPS and GPS
Dimensions: ¼ ATR short
Weight: <4 kg
Power: 28 V DC
Interfaces: MIL-STD-1553B, ARINC 429, analogue interfaces

Operational status

Selected for the RAFALE and the NH 90.

Contractor

Thomson-CSF Communications

NEW ENTRY

TLS-2030 Multi-Mode Receiver (MMR)

The TLS-2030 Multi-Mode Receiver is a new VOR/ILS receiver which complies with ARINC 755.

The TLS-2030 is the military version of the TLS-755 MMR that has been developed for commercial aircraft. The VOR function has been added to the functions of the TLS-755.

The Thomson-CSF Communications, TLS-2030 Multi-Mode Receiver (MMR) ***1998***/0015354

It offers VOR/ILS capability and a VHF receiver for DGPS functions.

Optionally it can be configured with MLS, DGPS and GPS. The TLS-2030 belongs to the TLS-2000 family.

Specifications

Basic functions: VOR/ILS/VHF receiver for DGPS
Options: MLS, DGPS and GPS
Dimensions: 3 MCU
Weight: 4 kg
Power: 115 V, 400 Hz
Interfaces: ARINC 429

Contractor

Thomson-CSF Communications

NEW ENTRY

TLS-2040 Multi-Mode Receiver (MMR)

The TLS-2040 Multi-Mode Receiver is a new VOR/ILS receiver designed to replace the existing ARINC 547 VOR/ILS receivers. It fully complies with ICAO Annex 10 FM immunity requirements and includes a VHF receiver for the DGPS datalink. It is form-and-fit exchangeable with existing ARINC 547 receivers.

The TLS-2040 receiver can optionally be configured with Marker, MLS, DGPS and GPS functions by simply adding modules.

The TLS-2040 utilises the latest technology developed for commercial aviation and military programmes. The TLS-2040 belongs to the TLS-2000 product line.

Specifications

Basic functions: VOR/ILS/VHF receiver for DGPS
Options: Marker, MLS, DGPS and GPS
Dimensions: ½ ATR short
Weight: <4 kg
Power: 28 V DC
Interfaces: As defined by ARINC 547, MIL-STD-1553B, ARINC 429

Contractor

Thomson-CSF Communications

NEW ENTRY

The Thomson-CSF Communications, TLS-2040 Multi-Mode Receiver (MMR) **1998**/0015355

GERMANY

AeroNav Integrated Navigation System (INS)

AeroNav is a TSO C-129 certified GSP navigation system. It provides a 12-channel GPS receiver, integrated with an Inertial Measurement Unit (IMU) for improved navigation performance and Aircraft Autonomous Integrity Monitoring (AAIM). Flight management software with Jeppesen database supports the crew during mission planning and in flight. AeroNav's open interface architecture provides links to moving maps like the Dornier DKG-3 and special mission equipment, for example FLIR, or communications equipment (Satcom or GSM).

Specifications

Dimensions: NCU (Navigation and Communication Unit): 198 × 126 × 360 mm
CDU (Control and Display Unit): 76 × 146 × 185 mm
Weight: NCU 5.7 kg; CDU 1.4 kg

Operational status

In production and in service.

Contractor

Aerodata Flugmesstechnik GmbH.

UPDATED

AeroNav Integrated Navigation System (INS)  **1997**/0005427

AD-FIS Flight Inspection System

The AD-FIS is a highly integrated, digital flight inspection system, which utilises advanced methods of real-time data processing and satellite navigation technologies. AD-FIS is compliant with ICAO recommendations, and performs calibration of all ground-based navigation aids including, ILS, MLS, TACAN, NDB, DME, PAR, SSR, VHF/UHF Comms, VDF and UDF.

The system consists of the following subsystems: flight inspection receivers; position reference sensors; data processing equipment; operator console and aircraft interface.

Data from the system's sensors, which include airborne receivers and ground-based position reference equipment, is interfaced to the computer via an ARINC 429 databus. Data from ground-based sensors is transmitted to the aircraft via a UHF datalink. Sensor information received by the computer is stored in raw, unprocessed format on a magneto optical disk, allowing future access to all sensor data, and is simultaneously processed by the computer. Processed data is displayed, in real time, to the operator in graphical format via the operator's LCD screens, and finally printed on the system's colour printer.

The system hardware is housed in a compact rack design. The rack houses the AD-VC6 computer, flight inspection receivers, colour printer, operator station and additional items such as a spectrum analyser. Ground-based reference equipment includes a highly accurate laser tracking system and DGPS station.

Specifications

Dimensions: vary according to customer specifications
Weight: (approximate) 150 kg. Varies according to system design

Operational status

In production and in service.

Contractor

Aerodata Flugmesstechnik GmbH.

VERIFIED

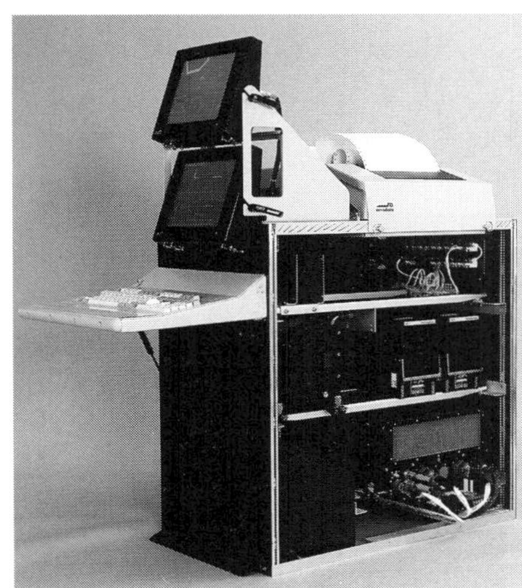

AD-FIS Flight-Inspection System **1997**/0005428

DME/P 400 dual-mode airborne interrogator

The DME/P 400 airborne unit was developed in 1985 as a precision distance-measuring equipment for use with the Microwave Landing System (MLS). The DME/P 400 is based on the Initial Approach Mode (IAM) and Final Approach Mode (FAM) of the ground element of MLS and combines the conventional DME and DME/P functions in one unit. The IAM is fully compatible with DME, which is important as there may not always be a DME/P ground system to complement the airborne precision equipment.

The airborne interrogator switches automatically from IAM to FAM and has the same linear broadband and narrowband receiver as the MLS ground system.

The DME/P 400 offers high reliability and continuity of service through the use of new technology micro-units and solid-state RF power devices. Continuous microprocessor fault monitoring and BITE capability allow fault detection down to module level. As with the MLS ground equipment, the DME/P 400 has high-speed signal processing with internal calibration loops to eliminate drift effects.

Contractor

Alcatel SEL AG.

VERIFIED

Globos NRP-6 GPS satellite navigation receiver

The six-channel Globos NRP-6 satellite navigation receiver is currently available in two versions for receiving the C/A code. Globos M is for ships and Globos A for aviation use. Globos A is based on a highly dynamic six-channel receiver. The integrated navigation computer permits storage of up to 264 waypoints and computes the required waypoint data in three dimensions, enabling the GPS receiver to provide glideslope data. An 11-step Kalman filter provides supplementary data for the receiver, safeguarding the integrity of the system in the event of satellite shadowing.

The system consists of the core module, antenna preamplifier and antenna.

The core module comprises the power supply, RF, signal processing and navigation computer modules on two printed circuit boards.

The GPS antenna is a stripline antenna for surface mounting on cylindrical or flat bodies. It is optimised to receive the L1 frequency with a bandwidth of 2 MHz. The pattern enables satellite acquisition 5° above the horizon at any pitch and roll angle below 5°.

The external preamplifier compensates for antenna cable loss, provides additional gain and filters the RF signal.

During periods of signal loss from one or more satellites the NRP-6 will accept navigation information from sensors such as INS, AHRS, ADC, heading or rate gyros, barometric altimeter or pitot tube.

The Globos system is characterised by low weight and volume and is suitable for integration with complex navigation and radio systems.

Specifications

Dimensions:
(core module) 150 × 104 × 61 mm or 86 × 130 × 76 mm
(antenna) 120 × 120 × 45 mm
(preamplifier) 95 × 116 × 35 mm
Weight: 0.95 kg ± 50 g
Power supply: 10-32 V DC, 15 W
Accuracy (SEP):
(3D position) <30 m
(velocity) <0.1 m/s
(3D position differential) <4 m

Operational status

Entered production during 1990.

Contractor

Alcatel SEL AG.

VERIFIED

3300 series VHF navigation systems

The series 3300 navigation systems have been designed for use in fixed- or rotary-wing aircraft, and have been certified for high-altitude operation. The series includes a variety of VOR/ILS receivers, indicators, special application converters and a glide slope receiver, which can be combined to tailor systems to meet the exact requirements of specific installations. Interfaces are provided for most types of electromechanical or electronic HSI, RMI and CDI dsplays.

Complete VOR/LOC receiver, VOR/LOC converter and glide slope receiver systems are housed in a single compact unit, which complies with ARINC standards. They do not require the individual control boxes and remote receiver units with interconnecting cables, or external forced-air cooling, which are typical of older navigation receiver designs in this performance class.

Clear LCD displays, which are easily read in the brightest sunlight show both active and standby frequencies. Remote tuning via databus can be provided, and tandem tuning heads are available for dual-cockpit installations. Built-in test functions for the displays and for the proper operation of the receivers and converters are standard features.

The NR 3300 navigation receivers are certified to the stringent requirements of all applicable FAA TSO, LBA, ICAO and RTCA specifications.

Specifications

Dimensions: 146 × 47.5 × 225 mm (excluding clearance for connectors and cables)
Weight:
(NR 3300-(1)/(3)) 1 kg
(NR 330-(2)/(4)) 1.2 kg

Operational status

In service and production.

Contractor

Becker Avionic Systems.

UPDATED

NR3300 and Indicator ***1996***

ADF 3500 system

In addition to the standard frequency range of 190 to 1,799.5 kHz, the ADF 3500 also receives the international maritime distress frequency of 2,182 ±5 kHz, making it suitable for search and rescue, and other offshore operations. It is certified for high-altitude operation in turbine-powered aircraft.

The complete system is housed in a single compact unit, which complies with ARINC standards. It does not require remote boxes, interconnecting cables, or external forced-air cooling.

Clear LCD displays, which are easily read in the brightest sunlight show both active and standby frequencies. Preselection of a standby frequency enables instant switch over to a second station for position fixing by cross bearings. A low-profile combined sense and loop antenna is suitable for mounting on high-speed aircraft. Becker RMI converters are available to enable the system to drive most types of indicators.

The ADF 3500 systems are certified to the requirements of applicable FAA TSO, JTSO, RTCA, EUROCEA and FTZ specifications.

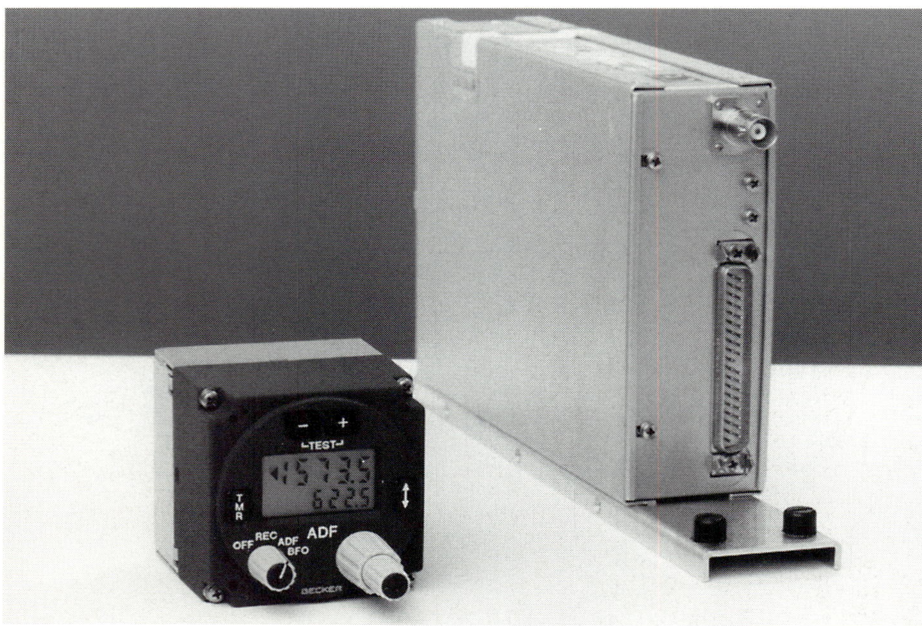

RA 3502 ADF receiver system ***1997***/0001323

Specifications

System model numbers	System components
ADF 3502-(1)	Standard version AD 3502 receiver AN 3500 antenna ID 3502 indicator
ADF 3503-(1)	For RMI with standard synchro input AD 3502 receiver AN 3500 antenna AC 3503-(1) converter
ADF 3504-(1)	For RMI with standard 2.5 V DC Sin/Cos input AD 3502 receiver AN 3500 antenna AC 3504-(1) converter
ADF 3504-(2)	For RMI with standard 5-10 V DC Sin/Cos input AD 3502 receiver AN 3500 antenna AC 3504-(2) converter
CU 3502-(01)	Tandem Control Unit can be added to the system to enable dual control operation

Dimensions
AD 3502 receiver: 146 × 47.5 × 245 mm
Weight: 1.0 kg
ID 3502 indicator: 82.55 × 82.55 × 135 mm
Weight: 0.5 kg
AN 3500 antenna: 190 × 54 × 330 mm

Weight: (1.7 kg)
Supply voltage: 25 - 30 V DC
(20 V DC in emergency)
26 V AC, 400 Hz, for AC 3503-(1) converter

Power consumption: At 27.5 V DC, without panel lights
ADF 3502 650 mA
ADF 3503 1.15 A
ADF 3504 0.6 A

Frequency Ranges: 190-1,799.5 kHz and 2,182 kHz ± 5 kHz

Channels Spacing: 500 Hz

Bearing accuracy: <3° at 70 μV 190-850 kHz
<8° at 70 μV >850 kHz

Contractor
Becker Avionic Systems.

VERIFIED

AirScout moving map system

The Becker AirScout moving map display is designed for use on aircraft and helicopters. It uses GPS data to indicate position, velocity and current track information on a full-colour electronic moving map.

The core of the system is the Jeppesen database which includes data on all airports, navaids and communication frequencies.

Scanned images can be read into the system from any available source, including maps, satellite imagery and aerial photography.

The AirScout display consists of one panel-mounted full-colour LCD with flexible graphic capabilities, a dzus-mounted Control Unit (CU), and ¼-ATR short LRU, which houses the GPS sensor, processor and mass memory. The display can be mounted in the panel, or specially configured for use in helicopters on a side panel where it can be folded away when not in use.

The display is swivel mounted to provide a wider viewing angle and to reduce glare. The display is readable in bright sunlight.

Becker Avionic Systems dzus-mounted AirScout display ***1997***/0001327

Specifications
Dimensions:
Display: 165 × 200 × 72 mm
Control unit: 47.5 × 145 × 127 mm
Main LRU: ¼ ATR
Weights:
Display: 1.45 kg
Control unit: 0.48 kg
Main LRU: 1.95 kg
Power supply: 28 V DC,1 A
GPS sensor: 6 channels, 8 satellites tracked; WAAS and P-code GPS sensor optional; DGPS receiver optional

Operational status
In service with military, police and medical services in Europe and the USA.

VERIFIED

AirScout GA

AirScout GA is a new-generation system, based on the existing AirScout system, intended for general aviation. With a width of 159 mm, it is ideal for general aviation installations. It weighs less than 3.7 kg total, and includes a sunlight readable control and display unit, with remote main unit and antenna. It matches Becker's PrimeLine II communication and navigation system. Navigation data is based on the Jeppesen database, which is available with worldwide coverage, loadable via PC or CD-ROM. With an integrated 12-channel GPS receiver (5Hz update rate, WAAS supported) Becker is is the process of obtaining system certification for enroute and approach operations.

Additional features including HSI, altimeter and direction-finder capabilities are planned during 1998.

Contractor
Becker Avionic Systems.

UPDATED

AirScout new modular moving map system

AirScout new modular moving map system allows step by step update. Together with the Becker PrimeLine system, the new modular AirScout moving map system

Airscout military system pilot's display ***1997***/0005420

Airscout military system observer's display ***1997***/0005421

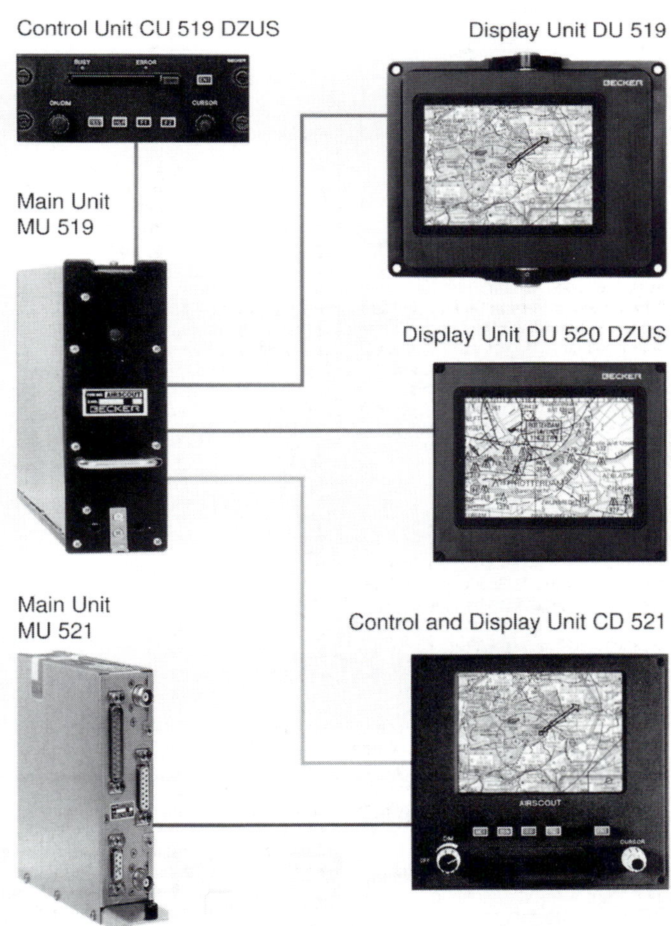

Some of the configurations available with the new Airscout components ***1997***/0005422

combines the benefits of compact packaging and installation flexibility to meet future mission requirements.

With ARINC 429 and RS-232 interfaces, AirScout is also able to communicate with other systems such as FLIR, EHAS, air data computers and radio altimeters. NVG compatibility is offered as an option.

Contractor

Becker Avionic Systems.

UPDATED

Automatic Direction-Finder system

The Automatic Direction-Finder (ADF) system operates with a sensitive combined sense-loop antenna. As well as the normal operating frequency range of 190 to 1,799.5 kHz, the system also provides the international maritime distress frequency of 2,182 kHz. A tandem version of the equipment is available and special converters have been designed so that other indicators can be used in place of the standard manual one.

Contractor

Becker Avionic Systems.

VERIFIED

PrimeLine II IFR Packages

Becker Avionic Systems produces a family of communications and navigation systems, designated PrimeLine II, that it has now integrated into a single unit to further reduce weight, balance and volume problems.

Designated 2. Block and 4. Block, they utilise small controls, coupled via databus to remote transmitters and receivers, to make best use of cockpit panel space.

For IFR-use, the 2. Block configuration includes one navigation display, together with integrated marker lights and an ADF system.

The 4. Block system (see photograph) includes the ATC 2000 (3)R transponder system; COM 5200 VHF communication system; NAV 5300 VOR/ILS navigation system; and RA 3502 ADF system.

Status

Becker is planning to introduce the PrimeLine II products in the first quarter of 1998.

Contractor

Becker Avionic Systems.

UPDATED

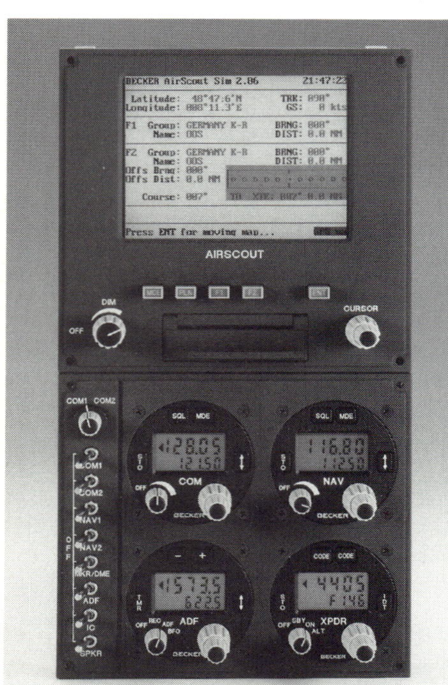

4. Block system and AirScout **1997**/0001324

PrimeLine II communication and navigation system

The PrimeLine II is a remotely-controlled system consisting of COM, NAV, ADF and ATC systems with an integrated audio panel. The control panel has a standard width of 159 mm and fits in modern general aviation cockpits.

The connection between the control units and the transceivers is achieved by two prefabricated cable harnesses. The interfaces to all COM and NAV modules is standard and universal. This minimises installatiion time and eases addition and replacement of individual modules as desired, without the need of additional cables or major changes.

Operational status

Certification for all components planned for December 1997, with deliveries starting in Spring 1998.

A completely digital control panel will replace the existing conventional control units to ensure upgrade potential for the future.

Contractor

Becker Avionic Systems.

UPDATED

IN 3300 series VOR/LOC/GS indicators

The IN 3300 series indicators are precision engineered instruments which display LOR, localiser and glide slope deviation information for en route navigation and approaches.

Either 5, 14 or 28 V lighting is provided and non-reflective glass is used to ensure reliable readability under all operating conditions.

The model IN 3300-(10) contains a course selector and display, rectilinear VOR/LOC and glide slope cross-pointers and warning flags, and a TO-FROM indicator. The IN-3300-(3) also contains a built-in marker beacon receiver and automatically photocell-dimmed indicator lamps for airway marker, outer marker and inner marker beacon indications.

IN 3300 series indicators can be combined with Becker's NR 3320/30 navigation receivers to form a dependable, lightweight, easily installed VHF navigation system.

The IN 3300 series indicators are approved for operation to 50,000 ft, and are certified to the rigorous requirements of all applicable FAA and RTCA specifications.

Specifications

Dimensions: 82.55 × 82.55 × 130 mm (excluding clearance for connectors and cables)
Weight:
(IN 3300-(3)) 850 g
(IN 330-(10)) 820 g

Operational status

In production and in service.

Contractor

Becker Avionic Systems.

UPDATED

IN 3300-(3)/-(5)/-(6) **1997**/0001328

IN 3360-(2)-B compact VOR/LOC/GS indicator

The IN 3360-(2)-B indicator is a precision engineered instrument which displays VOR, localiser and glide slope deviation information for en route navigation and approaches.

The indicator contains a course selector and display, VOR/LOC and glide slope cross-pointers, warning flags, and a TO-FROM indicator. It provides all steering information needed for IFR flying, and is ideal for use in aircraft with limited panel space, or for use as part of an 'emergency bus' avionics package for large aircraft.

The IN 3360-(2)-B indicators are approved for operation to 40,000 ft, and are certified to the requirements of TSO C52a.

Specifications

Dimensions: 60 × 60 × 110 mm
Weight: 0.40 kg

Operational status

In production and in service.

Contractor

Becker Avionic Systems.

VERIFIED

IN 3360-(2)-B VOR/LOC/GS indicator **1997**/0001329

NAV 5300 VOR/ILS navigation systems

The NAV 5300 systems are part of the Becker Compact Line of equipments that can be customised at installation to customer requirements.

The system is ideally suited to installations where minimum panel space is to be used. They utilise a small, lightweight CU 5301 control unit, which fits into a standard 57 mm round instrument panel cut-out, and is only 63.5 mm deep. Lightweight remote receivers, with VOR/LOC or VOR/ILS beacon capabilities are mated with this control unit, to complete the systems.

The receivers can be installed at any convenient place in the aircraft. The CU 5301 control unit uses a clear, high-contrast, double line LCD display, which is readable under all lighting conditions, even bright sunlight.

Both active and standby frequencies are displayed, and can be transferred by a single stroke of the 'flip-flop' button. Up to 99 preset frequencies can be entered from the front panel, and stored in non-volatile memory. Parallel outputs are provided for automatic DME channelling. The steering signals and flag drive outputs are compatible with most commonly used CDI, HSI, flight director, and autopilot systems.

All systems will be JTSO certified for either VFR or IFR use in all types of fixed-wing and rotary-wing aircraft, and comply with ICAO requirements for VHF radios.

The NAV 5300 systems can be combined with other Becker Compact Line avionics systems, such as VHF transceivers and ATC transponders which have similar control units. The remote receiver units can be controlled by other types of CDU or FMS.

Specifications

RN 3330 -(1): VOR/LOC receiver and converter
RN 3320 -(1): VOR/LOC receiver and converter and GS receiver
CU 5301: control unit
RM 3300- ( ): converters to drive RMI

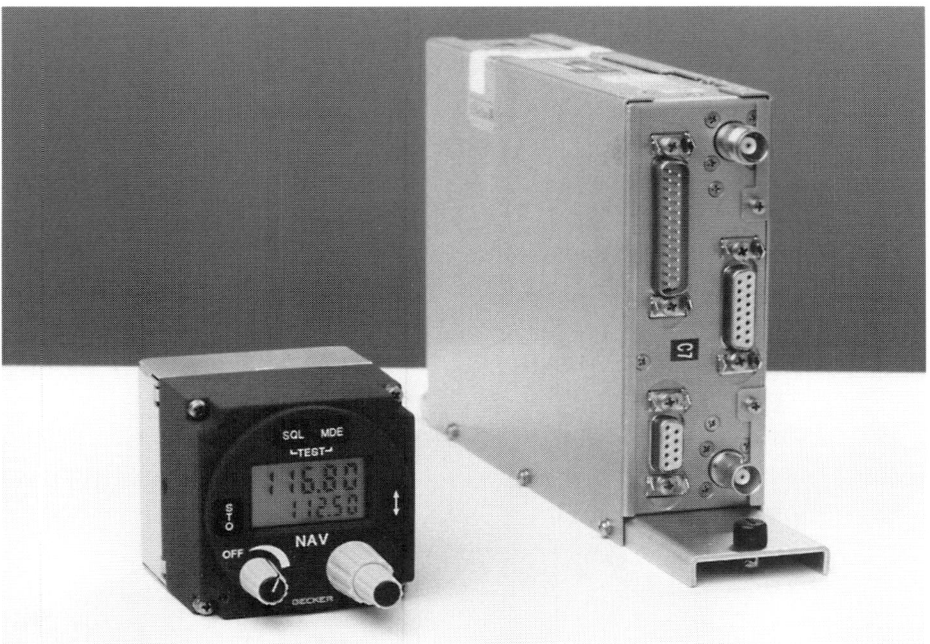

NAV 5300 VOR/ILS navigation system ***1997***/0001325

Dimensions:
CU 5301 61.3 × 61.3 × 62 mm
RN 3320/30 134 × 50 × 243 mm
RM 3300- ( ) 134 × 50 × 214 mm
Operation voltage:
10 to 32 V DC 26 V AC, 400 Hz, 10 mA (RMI converter)
Power consumption:
At 28 V DC.
without panel lights
RN 3330-(1) 250 mA
RN 3329-(1) 330 mA
Frequency ranges:
VOR/LOC 108.00-117.95 MHz
200 channels
Glide slope 329.15-335.00 MHz
40 channels
Channel spacing
VOF/LOC 50 kHz
Glide slope 150 kHz

Contractor
Becker Avionic Systems.

UPDATED

NR 3300 series VHF navigation receivers

The NR 3320-(02)-(01) and NR 3330-(02)-(01) are new navigation receivers designed as retrofit systems for the Becker NAV 2000 series receivers with which they are pin compatible without mechanical modification.

The NR 3320-(02)-XXX provides a composite navigation signal with glide slope receiver capability, while the NR 3330-(02)-XXX provides its composite navigation output without the glide slope.

Active and standby (preset) frequency read out is given on two sunlight readable LCDs.

Specifications
Dimensions: 126 × 46 × 186 mm
Weight: 1.2 kg
Power supply: 13.5 V/27.5 V DC
NAV Receiver:
Frequency range: 108-117.950 MHz
Channels: 200
Channel spacing: 50 kHz
Memory channels: 20
Glide slope Receiver:
Frequency range: 329.150-335 MHz
Channels: 40
Channel spacing: 150 kHz

Contractor
Becker Avionic Systems.

UPDATED

PrimeLine communications and navigation system for business aviation

The Becker PrimeLine communications and navigation system fulfils the need of the business aviation market for a complete dzus rail-mounted family of Cat. I avionic products. The basic PrimeLine products are completely housed in single units, requiring no remote boxes. Products in the range include:

AR 3202 20W airborne VHF transceiver, or the compact AR 4201 7W airborne VHF transceiver;
NR 3300 VOR/LOC/GS navigation receivers;
IN 3300 series VOR/ILS navigation indicators, including a 3 in VOR/LOC/GS CDI, a 3 in VOR/LOC CDI, and a 2.25 in VOR/LOC/GS CDI;
ADF 3202 automatic direction-finder systems;
RMI 3337 radio magnetic indicator;
HSI 421 (4 in) and HSI 8131 (3 in) horizontal situation indicators;
Audio selector and indicator systems;
ATC 3401 transponder;
AirScout Moving map system.

Operational status
In service.

Contractor
Becker Avionic Systems.

UPDATED

ProfiLine communication and navigation system

The ProfiLine communications and navigation system comprises:

CU900	dzus width controller with ARINC 429 and ARINC 410 interface for COM transceivers and NAV receivers (optionally NVG-compatible).
NAV900	NAV receiver with built-in glide slope, pin compatible to the COLLINS 51RV-1C, based on a modern design with integrated BITE (test also the HF section). It complies with the new ICAO annex 10 requirements. The NAV900 has an ARINC 410, an ARINC 429 control interface, and a built-in marker receiver. A DGPS receiver and a MIL-STD-1553 interface will be available optionally in the future.

Contractor
Becker Avionic Systems.

UPDATED

RM 3300 series RMI converters

The RM 3300 series converters enable Sine/Cosine or composite video signals from navigation receivers to be used to drive RMI indicators and other display devices which require XYZ synchro inputs.

By varying the circuitry of the internal modules in the RM 3300 family, the converters can be used to provide inputs for XYZ synchros per ARINC 407, Sin/Cos DC or Sin/Cos AC indicators. There is even a model which simply functions as a synchro amplifier, to enable low-power synchro signals to drive corresponding high-power synchros.

These converters are approved for installation in unpressurised areas of aircraft operating up to 50,000 ft. There are no altitude restrictions for installations in pressurised areas.

Specifications
Dimensions: 54 × 139 × 214 mm (excluding mating connector)
Weight: 0.75 kg

Operational status
In production and in service.

Contractor
Becker Avionic Systems.

VERIFIED

Becker ProfiLine NAV900 receiver and control unit
1997/0005424

NFS-3000 series navigation management system

Using differential technology (DGPS), NFS-3000 products achieve maximum accuracy, thus permitting precision approaches and landings. The combined GPS/GLONASS receiver offers maximum integrity and permits PRAIM (Predictive RAIM).

The NFS-3000, uses an open architecture that ensures unlimited access to enhanced functionalities, under development, including, the Wide Area Augmentation System (WAAS), the European Geostationary Navigation Overlay System (EGNOS), the Ground Collision Avoidance System (GCAS) and the Surface Movement Guidance and Control System (SMGCS).

Contractor

Daimler-Benz Aerospace AG, Defense and Civil Systems.

NEW ENTRY

Specifications

	NFS-3001	NFS-3002	NFS-3003
Monochrome display	•	•	
Colour display with map presentation			•
12-channel GPS receiver with RAIM and PRAIM	•	•	•
RNP5 requirements	•	•	•
Combined 18-channel GPS/GLONASS receiver with RAIM and PRAIM	optional	optional	optional
Dimensions (mm)	66.7 × 146 × 254	52.1 × 160 × 254	104.2 × 160 × 254
GP receiver module			
Channels	12	12	12
Accuracy			
(horizontal) (DGPS)	1 m (95%)	1 m (95%)	1 m (95%)
(vertical) (DGPS)	3 m (95%)	3 m (95%)	3 m (95%)
Time-To-First-Fix (TTFF)	<75 s	<75 s	<75 s
Reacquisition time	<5 s	<5 s	<5 s
GPS/GLONASS receiver module			
Channels (GPS/GLONASS)	12/6	12/6	12/6
Accuracy			
(horizontal) (DGPS)	4.4 m (95%)	4.4 m (95%)	4.4 m (95%)
(vertical) (DGPS)	6 m (95%)	6 m (95%)	6 m (95%)
Time-To-First-Fix (TTFF)	<90 s	<90 s	<90 s
Reacquisition time	<5 s	<5 s	<5 s
VHF receiver module			
Frequency range	108.00-117.95 MHz	108.00-117.95 MHz	108.00-117.95 MHz
Channel separation	25 kHz	25 kHz	25 kHz
Modulation	D8PSK	D8PSK	D8PSK
ADS frequency range	118.00-136.00 MHz	118.00-136.00 MHz	118.00-136.00 MHz

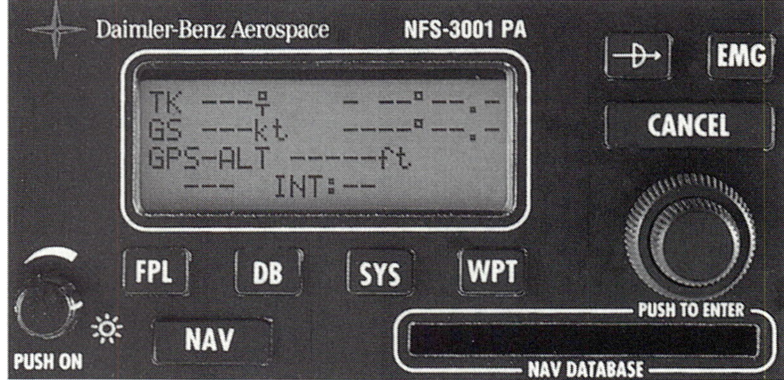

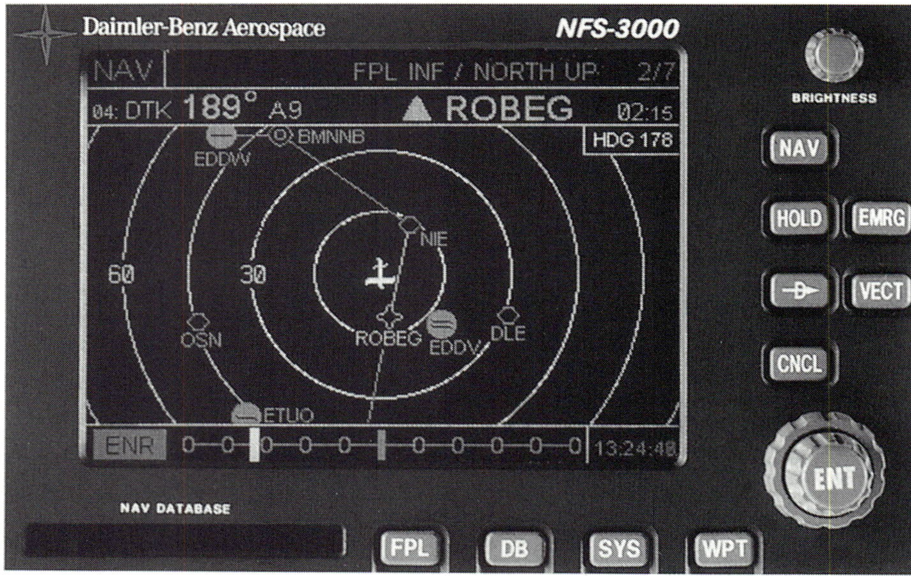

NFS-3003 panel-mount size, colour graphic display with integrated CDI and map presentation
1998/0018927

EuroNav III task management system

The EuroNav III task management system is part of the overall EuroNav AH Guidance System. EuroNav AH is a mission control management system for up to 500 helicopters, consisting of the fixed EuroNav z Central Control Station and up to 500 helicopter EuroNav III task management systems. The system utilises the GPS navigation system for determining helicopter positions, and all participants are linked via appropriate communications facilities.

The EuroNav z Central Control Station collects, evaluates, and stores data on all aspects of the operation, maintenance, management, and financing of the helicopter fleet concerned.

Each helicopter is fitted with the EuroNav III task management system, which comprises a RN4-2.9 navigation computer, with its database and navigation program; a DYNT51-1.7 display 5.1 inch; a CLN14-1.8 control unit; a SBN-2.8 switch box, and a GPS antenna.

The EuroNav III task management system is operated by the air observer, who can use the system to perform all operational actions required for the following types of task: flight planning; flight execution; flight data recording; emergency situation recovery.

The navigation computer is based on the high end

industrial standard and includes a 12-channel GPS receiver with capability to accept DGPS correction data. It is housed in a ½ ATR box, and includes 2 × HDD, up to 1.3Gbit each.

The standard monitor provides a full colour 5.1 in display, with 320 × 240 pixel resolution. As options, 6.4 in and 10.4 in displays with 640 × 480 pixels resolution and colour depth from 4 K to 256 K are available. The control unit and switch box are used to control and interconnect all system components.

Operational status

Widely used by police forces and search-and-rescue organisations in many countries of Western Europe, and also in Asia.

Contractor

EuroAvionics Navigationssysteme GmbH & Co.

NEW ENTRY

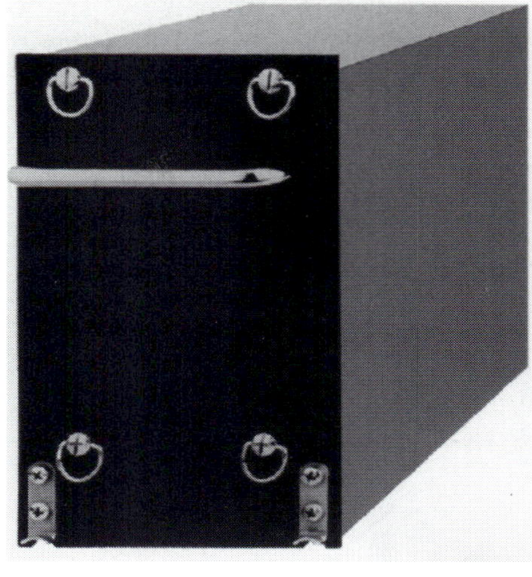

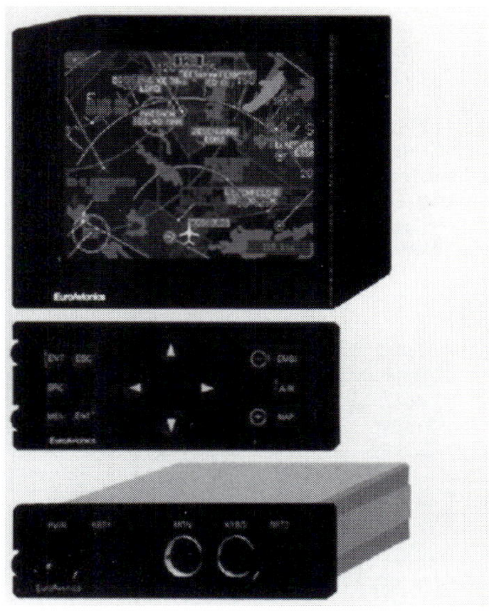

EuroNav III task management system, showing the navigation computer (above), and on the right (top to bottom), the monitor, control unit, and switch box ***1998***/0015357

H-764G Embedded GPS/INS (EGI)

The H-764G EGI is a small, low weight, navigation system with embedded GPS. It is based on a 32-bit Intel 80960 microprocessor, MIL-STD-1553B databus, and four RS-422 buses on one card. It is programmed in Ada and provides a triple navigation solution (pure inertial, GPS-only, and blended GPS/INS), using a Collins GEM™ GPS receiver module.

Specifications

Dimensions: 177.8 × 177.8 × 248.9 mm
Weight: 8.4 kg
Accuracy:
(Position) <1.0 n mile/h CEP
(Velocity) <1.0 m/s
Blended GPS/INS Performance:
Position accuracy (SEP) <16 m
Velocity accuracy (rms) <0.03 m/s
Align time:
(Gyrocompass) 4 min
(Stored heading) 30 s
(in air align) 4 min
Power: 28 V DC, 40 W
Interfaces: MIL-STD-1553B (RT or BC) and four RS-422 serial databusses (optional)
Synchro, analogue, discrete

Contractor

Honeywell Regelsysteme GmbH

NEW ENTRY

µ-INS/GPS

Flight control, guidance, integrated navigation, and mission management are integrated in the µ-INS/GPS navigation system HD991A1.

The core µ-INS/GPS contains the inertial measurement module HG1700, a system/processor module, a Global Positioning System (GPS) and a power supply module. Several additional modules are available to meet application specific requirements.

These enhancement modules include radar altimeter, air data transducer and terrain correlation modules.

Flexible system architecture permits use of specific application software, using Honeywell's Embedded Computer Toolbox and Operating System (ECTOS).

The IMU integrates three Honeywell GG1308 miniature Ring Laser Gyros (RLG) and three digital accelerometers as well as electronics in a strapdown configuration. It provides fully compensated inertial data on a digital, serial output.

The µ-INS/GPS system processor module is based on the Intel 80960MC microprocessor and is the processing centre for navigation processing. It also contains digital inputs/outputs and three user-configurable RS-422 serial interfaces with selectable baud rates up to 1 Mbaud.

The embedded 12-channel L1 C/A code GPSCard™ represents NOVATEL's newest generation of GPS receivers with DGPS capability. It provides excellent position and velocity performance with a very short reacquisition time even under high dynamic conditions.

Specifications

Dimensions: 210 × 120 × 125 mm
Weight: <3.0 kg
Power: 20 W, 11 to 36 V DC
Interfaces: 3 serial RS-422, GPS receiver with coarse acquisition (SA on)
Accuracy:
(position) <34 m (1σ)
(velocity) (N,E,D) <0.5 m/s (1σ)
(pitch/roll) <0.1° (1σ)
Heading:
(stationary align) <6.0°
(in-flight align) <1.0° (C/A code)

Contractor

Honeywell Regelsysteme GmbH

NEW ENTRY

Fibre Optic Rate Sensors

LITEF's family of miniature Fibre Optic Rate Sensors (µFORS) is designed to meet the most demanding requirements in a wide range of air, land and sea applications. µFORS provide error compensated angular or turn rate outputs via a standard serial interface. Free from the effects of gravity-induced errors and with no moving parts, the sensors are virtually insensitive to shock and vibration. They offer extremely high reliability, without the need for periodic maintenance.

The key features of the µFORS family are: selectable digital output data with flexible interface characteristics such as 16-bit or 24-bit resolution, choice of baud rate and measurement range. The µFORS offers short initialisation time, comprehensive BIT and continuous status reporting, high reliability and small volume, low weight and low power consumption.

Specifications

Dimensions: 100 cm³
Weight: 0.15 kg
Power supply: 5 V DC, 2 W
Measurement range: ±1,000°/s
Reliability: 100,000 h MTBF

Operational status

In series production.

Contractor

LITEF GmbH.

VERIFIED

Global Positioning Laser Inertial Navigation (GPIN) equipment for German Tornado

The Global Positioning Laser Inertial Navigation (GPIN) is part of a German Tornado upgrade programme. GPIN is an advanced technology lightweight ring laser system with a fully embedded GPS (INS/GPS) based on the Litton LN-100G (EGI) product family.

Operational status

Modification/development for German Tornado.

Contractor

LITEF GmbH.

VERIFIED

Inertial Measurement Unit (IMU) for EF 2000

The IMU is part of the EF 2000 quadruplex fly-by-wire flight control system. It has a strapdown design and is housed in a single box which is internally separated into four channels.

Measurement of aircraft body angular rates and linear accelerations and calculation of attitude and heading angles are accomplished by the IMU. It performs autonomous heading alignment at system start up. To provide damping of air data disturbances and to provide a back-up source of air data. The IMU computes angle of attack, sideslip, TAS and altitude data based on inertial measurement and augmented by signals from the air data sensors, when available.

The IMU contains four dual-axis dynamically tuned LITEF gyroscopes and eight single-axis LITEF pendulum accelerometers mounted in a skewed-axis orientation. This combination of accelerometers and gyroscopes is about half the number of sensors used in conventional systems architecture and therefore represents a considerable weight saving.

Operational status

Under development for the EF 2000 and substantially flight tested.

Contractor

LITEF GmbH.

VERIFIED

LCR-88 Attitude and Heading Reference System

The LCR-88 is a strapdown Attitude and Heading Reference System (AHRS) specifically designed to meet the requirements of the commuter and general aviation market. A version with increased angular rate capability, designated LCR-88A, is available for applications in trainer aircraft. It contains a pair of two degrees-of-freedom dry-tuned gyros and three linear accelerometers. The design is based on LITEF's reliability proven LTR-81 ARINC 705 AHRS. The system operates with 28 V DC power. Several configurations are available to meet the interface requirements for integration into the available avionics packages on the market.

Specifications

Dimensions: 388 × 124 × 194 mm
Weight: 6.3 kg
Power supply: 28 V DC, 85 W
Reliability: >7,000 h MTBF

Operational status

In service in business jets, turboprops, commuter aircraft and helicopters. The LCR-88M has been selected for the German Air Force C-160 Transall.

Contractor

LITEF GmbH.

VERIFIED

LCR-92 μAHRS Attitude and Heading Reference System

The LCR-92 is an extremely small and light strapdown reference system using LITEF fibre optic gyros. It provides pitch, roll, magnetic heading information and angular rates around the aircraft body axes, replacing conventional vertical/directional gyro installations in one single box.

The system features full compatibility with modern digital cockpit instruments with an ARINC 429 databus. Analogue outputs for use with conventional mechanical indicators are available as an option.

Specifications

Dimensions: 278 × 102 × 128 mm
Weight: 2.1 kg
Power supply: 28 V DC, <25 W
Accuracy (2σ): heading 1°
Reliability: >6,000 h MTBF

Operational status

FAA certification obtained. In production for business jets, turboprops and helicopters.

Contractor

LITEF GmbH.

VERIFIED

LCR-92 μAHRS Attitude and Heading Reference System **1997**/0001330

LCR-93 μAHRS Attitude and Heading Reference System

The LCR-93 is a new member of the μAHRS family based on the LCR-92. It has the same housing and dimensions and offers the same interface configurations. The mechanical layout and connectors ensure mechanical plug-in interchangeability with the LCR-92.

In addition to the LCR-92 capabilities, the LCR-93 standard version provides body axis referenced accelerations as well as inertial altitude and inertial vertical speed. The LCR-93 performance is increased by using air data augmentation.

The advanced version LCR-93V combines GPS and air data inputs to provide the necessary outputs to Head Up Display (HUD) with a speed vector indication.

Operational status

FAA certification in process. Has been selected by Cessna as standard AHRS for the Excel and replaces the LCR-88 in the Pilatus PC-9.

Contractor

LITEF GmbH.

VERIFIED

LCR-93 μAHRS Attitude and Heading Reference System **1997**/0001331

LKS-91 gyro stabilisation system

The LKS-91 gyro stabilisation system is an accurate, robust and reliable stabilisation system. A two degrees-of-freedom dry-tuned gyro, miniaturised digital electronics and mechanical mounting are combined into a compact unit. Typical applications are for use as an angle or turn rate sensor for the stabilisation of antennas, periscopes and other items fitted to vehicles and other platforms.

Specifications

Dimensions: 65 × 78 mm diameter
Weight: 0.42 kg
Power supply: ±15 V AC, 5 V DC
Reliability: >15,000 h MTBF

Operational status

In series production.

Contractor

LITEF GmbH.

VERIFIED

INDIA

ARC-610A automatic direction-finder

The all-solid-state modular construction ARC-610A automatic direction-finder covers the frequency range 190 to 1,700 kHz in steps of 0.5 kHz to indicate on an LED display the direction of a ground beacon. It has 10 preset channels and includes self-test. The system consists of a receiver and a controller.

Specifications

Dimensions:
(receiver) 288 × 60.5 × 200 mm
(controller) 146 × 66 × 146 mm
Weight:
(receiver) 5 kg
(controller) 1 kg
Power supply: 27.5 V DC, 2 A
Bearing accuracy: ±2° excluding quadrantal error
Temperature range: −55 to +55°C
Altitude: up to 70,000 ft

Contractor

Hindustan Aeronautics Ltd.

VERIFIED

ARC 1610A automatic direction-finder

The ARC 1610A automatic direction-finder provides bearing information of the known ground beacons operating in medium frequency along with code reception to identify the selected ground beacons.

The ADF system ARC 1610A is a compact lightweight system which provides accurate, stable bearing information with a high degree of reliability and maintainability using hybrid technology. The system is compatible with ARINC 570.

Specifications

Frequency range: 190-1,700 KHz
Frequency indicator: LED display
Modes of operation: ADF and ANT
Bearing accuracy: ±2°
Hunting: ±2°
Bearing resolution: 6 s
Power input: 27.5 V DC, 2 A

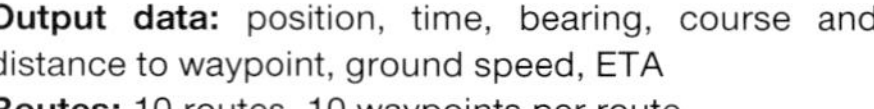

Preset channels: 10
Audio output: 100 mW across 600 Ω
Dimensions:
(receiver) 250 × 195 × 90 mm
(controller) 163 × 14.6 × 66 mm
Weight:
(receiver) 4 kg
(controller) 1 kg

Contractor

Hindustan Aeronautics Ltd.

VERIFIED

GNS-642 Global Navigation System

The GNS-642 system tracks all satellites in view to provide complete positional and navigation data. It is able to handle up to 100 waypoints and 10 routes.

Specifications

Receiver: 5 channel, L1 frequency
Sensitivity: −160 dBW, SNR 10 dB
Horizontal position accuracy: 25 m (RMS)
Vertical position accuracy: 50 m (RMS)
Velocity accuracy: 0.15 kt (RMS)
Output data: position, time, bearing, course and distance to waypoint, ground speed, ETA
Routes: 10 routes, 10 waypoints per route
Output display: 3 lines, 16 characters LED
Dimensions and weight:
(receiver) 124 × 120 × 120 mm; weight <1.5 kg
(display unit) 146 × 56.5 × 90 mm; weight <0.5 kg
Power supply: 27.5 V DC, 20 W
Interfaces: RS-232 with NMEA-0183

Contractor

Hindustan Aeronautics Ltd.

UPDATED

Hindustan Aeronautics ARC 1610A automatic direction-finder ***1996***

GNS-642 Global Navigation System ***1997**/0002441*

SSMR II marker beacon receiver

The SSMR II solid-state marker beacon receiver is a fixed-frequency radio receiver which processes amplitude-modulated 75 MHz marker beacon signals into voltages for operating the indicator lamp provided in the cockpit and providing an aural warning. The system does not differentiate between the three beacon frequencies 400 Hz, 1,300 Hz and 3,000 Hz and the visual and aural indications remain the same for all three frequencies. The system consists of a receiver and the antenna which is a part of the aircraft.

Specifications

Power supply: 22-31 V DC, 10 mA
Frequency: 75 MHz ±70 kHz

Operational status

In service on Indian Air Force MiG aircraft.

Contractor

Hindustan Aeronautics Ltd.

VERIFIED

UHS 190A UHF homing system

The UHS 190A UHF homing system for fixed-wing aircraft and helicopters is used in conjunction with the normal communications transceiver for locating ground transmitters and personal rescue beacons. It operates in the frequency range 225 to 399.975 MHz with two UHF antennas. The system provides an accuracy of ±5° for homing. The system has the capability to drive two homing indicators.

The UHS 190A consists of a UHF homing adaptor for receiving input signals from the two UHF antennas, homing controller for processing the ADF audio signal from the V/UHF communication set and a homing indicator for indicating the relative bearing of the ground station.

Specifications

Dimensions:
(adaptor) 146 × 98 × 32 mm
(controller) 146 × 51 × 105 mm
(indicator) 10 × 100 × 50 mm
Weight:
(adaptor) 0.3 kg
(controller) 0.55 kg
(indicator) 0.45 kg
Power supply: 22-31 V DC

Operational status

The UHS 190A is fitted on Indian Navy Sea King, Chetak, Il-38 and Dornier aircraft.

Contractor

Hindustan Aeronautics Ltd.

VERIFIED

INTERNATIONAL

ADS - Automatic Dependent Surveillance

Led by the UK CAA National Air Traffic Services (NATS), the ADS Europe consortium is investigating the applications and benefits of using satellite communications systems to meet the anticipated doubling of air traffic by the turn of the century. Through more accurate monitoring and control of aircraft position, ADS will enhance safety standards while increasing airspace capacity by reducing the separation needed between aircraft.

Using the Racal Avionics ADS (Automatic Dependent Surveillance) whilst operating in an ICAO (International Civil Aviation Organisation) Aeronautical Telecommunications Network (ATN) FANS (Future Air Navigation System) standard environment, the UK CAA plans to introduce an ADS system to cover North Atlantic airspace in the late 1990s.

Operational status

In the latest trial, Racal's ADS equipment supported the full complement of ARINC 735 ADS message sets using ATN communication protocols and responded to 'contracts' established. During the flight the aircraft were required to supply different sets of data at differing rates independently to the two ground stations. Data intervals ranged from 5 minutes en route to 1½ minutes in the approach to landing.

The ADS Europe consortium, partly funded by the EU, includes Racal Avionics and other British, French and Dutch companies from the regulatory, avionics and telecommunications sectors. The ADS Europe project is the first ADS trial using live aircraft and using protocols compatible with the future ICAO ATN. It is also the first use of Data-3, which is fundamental to ATN, as the satellite communications medium.

Contractor

ADS Europe Consortium.

VERIFIED

CMA-3012/3212 GPS sensor units

The CMA-3012 and CMA-3212 are designed for certification under anticipated new rules which will allow the use of GPS as a sole means of navigation. Key characteristics for this application include 12 receive channels, all of which can be used for continuous satellite tracking and any two of which are assignable as GPS integrity channels; comprehensive end-to-end receiver BITE; carrier phase tracking; differential GPS input and growth provisions for GPS/GLONASS.

The CMA-3012 and CMA-3212 conform to ARINC 743 and 429-12, DO-160C, 178B and 208, TSO C129 and MIL-STD-810. Inputs consist of eight ARINC 429 and an RS-232. Outputs consist of three ARINC 429, three 1 kHhz time marks and RS-232 and 28 V valid discrete.

Specifications

Dimensions:
(CMA-3012) 66 × 220 × 240 mm
(CMA-3212) 199.6 × 57.1 × 388.6 mm
Weight:
(CMA-3012) 3.2 kg
(CMA-3212) 3.6 kg
Power supply: 18-36 V DC, 20 W (max)
Frequency: 1,575.42 MHz, C/A code
Accuracy (95%):
24 m (horizontal)
30 m (vertical)
0.3 kt (velocity)
Reliability: 40,000 h MTBF

Operational status

TSO-C129 B1/C1 was received in mid-1995, and SCAT I implementation was completed in mid-1996. The system has been certified for Primary Means operation to FAA N8110.57. Ground preflight software (SureFlight) is also available.

Contractors

Canadian Marconi Company.
Honeywell Business & Commuter Aviation Systems.

VERIFIED

Global Navigation Satellite Sensor Unit (GNSSU)

Honeywell and Canadian Marconi Company are collaborating in the production of the Global Navigation Satellite Sensor Unit (GNSSU). The 12-channel GNSSU tracks all GPS satellites in view to provide better than 25 m position accuracies. It offers sensor computational errors of no more than 1.5 m, receiver autonomous integrity monitor and ARINC 743 design. The GNSSU provides accurate worldwide oceanic, en route and approach navigation, simplified pilot interface, time and position for automatic dependent surveillance and inertial reference system integration.

Increased capabilities result from combining the GNSS data in an 18-state Kalman filter inside the laser IRU. With this, the inertial error of 2 n miles/h is bounded by the accurate GNSS satellite measurements. There is continued system accuracy during periods of less than four satellites and continued integrity with less than five satellites. The high-frequency inertial sensor measurements integrated with the high-accuracy GNSS measurements provide the optimum navigation solution. The GNSS integration in the IRS eliminates the possibility of an added 25 m track error that could occur with blending of GNSS and IRS in the FMS. The GNS/IRS is designed to enable calibration of the inertial sensors after sole source GNSS certification.

Specifications

Dimensions: 63.5 × 215.9 × 241.3 mm
Weight: 3.18 kg
Accuracy:
(position) 25 m
(velocity) 1 kt
(time) 2 ms
Reliability: 55,000 h MTBF predicted

Operational status

The GNSSU has been selected as the standard option on the Boeing 777. It was TSO'd by the FAA in January 1994. Southwest Airlines has ordered the GNSSU for its Boeing 737-700 aircraft.

Contractors

Honeywell Inc Business & Commuter Aviation Systems.
Canadian Marconi Company.

UPDATED

ILS-85 Instrument Landing System

The ILS-85 is designed to integrate with the EFIS-85, FCS-85 and other systems in the -85 series utilised by the Il-96-300 and Tu-204 aircraft.

Specifications

Frequency range:
(localiser) 108.00-111.95 MHz
(glideslope) 329.15-335.00 MHz
Channels:
(ILS mode) 40
(LS-50 mode) 20
Precision:
(ILS mode) ICAO Cat III
(LS-50 mode) Cat II
Receiver dimensions: 3 MCU
Power: 115 V AC, 400 Hz, 30 VA
Weight:
(receiver) 5.5 kg
(antenna) 0.53 kg
MTBF: 5,000 h

Contractor

NIIAO Institute of Aircraft Equipment, Moscow.
Ukraine Research Institute of Radio Equipment, Kiev.

NEW ENTRY

LANE: Low-Altitude Navigation Equipment

LANE is a full low-altitude navigation system. Design is modular and flexible to meet operational requirements. It includes the following features:
database: a high-density, high-compression, large area coverage, rapid reload digital database;
digital map: selectable colour palette (day/night), variable orientation, multiple scales display, on which radar/threat data can be superimposed;
Terrain Referenced Navigation (TRN): giving automatic continuous navigation, with accuracy better than 50 m CEP, waypoint cueing and target ranging;
Terrain Following (TF): covert and passive TF, automatic and manual modes, obstacle and threat avoidance; perspective display: real-time display, with zoom, and tactical information overlays.

Specifications

Dimensions: ¾ ATR
Weight: 17 kg
Power: 150 W

Contractors

Dassault Electronique, St-Cloud.
GEC-Marconi Avionics Ltd, Rochester.

VERIFIED

LANE: Low-Altitude Navigation Equipment **1997**/0001332

Satellite-based navigation and landing systems

Daimler-Benz Aerospace AG and Rockwell Collins are developing a family of navigation and landing systems for civil aviation, based primarily on satellite navigation. It comprises airborne components and differential GPS ground stations.

The aircraft equipment contains a GPS or a combined GPS/GLONASS receiver module. The functionality of an ARINC 743A receiver is expanded by the addition of navigation integration through Kalman filtering and navigation management functions. Several types of equipment comprise the family, depending on the required implementation capability. This includes area navigation, non-precision approach, precision approach down to Cat III limits, flight management functions and the capability to use wide and local area augmentation systems. Depending on the functional configuration, the housing will vary between two and three MCU. All software is written in Ada according to DO-178B.

Application areas are in the air transport and regional aircraft markets, as well as for helicopters and general aviation. Existing navigation equipment, such as VOR/DME, ADC and AHRS can be interfaced via ARINC 429 or analogue inputs. The system will be operated via an external CDU or a digital map display.

Operational status

In production. First installations in Jeddah (Saudi Arabia), Cottbus-Drewitz (Germany) and Cedar Rapids (USA); operational test at Munich airport (Germany).

Contractors

Daimler-Benz Aerospace AG.
Navigation and Flight Guidance Systems GmbH.
Rockwell Collins.

UPDATED

TERPROM®

TERPROM® is a digital terrain system, which provides highly accurate navigation, plus predictive ground proximity warning, passive terrain following, and weapon aiming benefits with no external aids, and no forward electronic emissions.
TERPROM® is a software-based product which uses stored digital terrain elevation data, together with inputs from the aircraft's inertial navigation system to predict aircraft height above the ground. This prediction, measured against the true height from the radar altimeter, enables the Kalman filter to provide corrections to navigation systems inputs.

Ground proximity warning
TERPROM® generates 'predictive' ground proximity warnings which do not rely on current radar altimeter inputs. These warnings will be given if, for example the aircraft banks towards a cliff face or approaches the ground (even in inverted flight). While TERPROM® modes are reliant on radar altimeter information in the long term, in the short term (seconds rather than minutes), full performance is maintained without radar altimeter information.

Terrain following
TERPROM® generates terrain-following signals that can be displayed to the pilot. Unlike conventional terrain-following radars, TERPROM® achieves total freedom of manoeuvre by the use of its in-built map. Through its constant awareness of the shape of the terrain beyond the immediate horizon, TERPROM® enables an aircraft to follow ground contours more closely, and thus further reduce exposure to enemy attack.

Obstacle warning
TERPROM® will provide advanced warnings to advise the pilot that there are obstacles in the predicted flight path, in addition to 'pull up' warnings to prevent collision with terrain or obstacles.

Passive target ranging
Successful ground attack requires undetected approach and accurate weapon aiming. As a passive system TERPROM® gives no forewarnings.

With knowledge of target elevation (from the database) and aircraft altitude, height above targets is continuously available for target ranging calculations.

Thus, weapon aiming is significantly improved.

Navigation

Through sophisticated Kalman filtering tecniques, TERPROM® builds an error-corrected model of the aircraft's inertial navigation system, achieving a drift-free navigation accuracy in the order of 30 m CEP over terrain with slopes as low as 2 per cent. The robustness of the system model is such that, once established, accurate navigation can be maintained even on long over-water passages or during total radio silence.

Navigation performance

The performance assumes terrain data to DLMS level 1 DTED, valid radar altimeter information for 75 per cent of the time and flight below 5000 ft. agl.

Horizontal <30 m CEP

Vertical <4 m RMS

Inputs	**Outputs**
Systems navigational data	Navigation corrections
Barometric altitude	Ground proximity warning
Radar altimeter	Terrain following command
Selected TF and GPW	Obstacles warning
Height	Ranging data

Options

TERPROM® is available:

In the Fairchild Defense OSC Mega Data Transfer Cartridge with Processor.

Embedded in the Honeywell inertial navigation system H764G.

Embedded in the AYK14 family of mission computers from Computing Devices.

Operational status

TERPROM® has been selected by the air forces of Belgium, Denmark, Netherlands, Norway, Oman, UK, USA, (plus 2 others) for fitment to F-16, Harrier GR. Mk 7, Jaguar, Mirage 2000 and Tornado aircraft.

Contractors

British Aerospace Systems & Equipment.

Computing Devices Company Limited.

Honeywell Inc, Sensor and Guidance Products.

Fairchild Defense OSC.

UPDATED

Tiger combat helicopter avionics system

The Eurocopter Tiger combat helicopter's navigation system was developed by Sextant Avionique, Teldix GmbH and Daimler-Benz Aerospace AG; it features a hybrid design, including Doppler radar, radio altimeter, Pixyz® 3-axis ring laser gyros, and air data modules, which together provide the data required by the flight control (developed by Nord-Micro Electronik Feinmechanik AG and Sextant Avionique), navigation, guidance, head-down display systems (Sextant Avionique and VDO-Luftfahrtgerate Werk), and control and display unit (Rhode & Schwarz GmbH and Sextant Avionique).

LITEF GmbH, Sextant Avionique and VDO-Luftfahrtgerate Werk have together developed the five onboard computers: the ACSG (Armament Computer Symbol Generator); MCSG (Mission Computer Symbol Generator); BCSG (Bus Computer Symbol Generator); RTU (Remote Terminal Unit); and the CDD mission data concentrator.

The Helmet-Mounted Sight/Display (HMS/D) has been developed by Sextant Avionique and VDO-Luftfahrtgerate Werk; it features a 40° field of view, with a thermal imager slaved to the pilot's head position, allowing the crew to fulfil day/night all-weather missions.

Operational status

Equipment prototypes have been delivered and are installed on the PT2, PT3, PT4 and PT5 prototypes.

Contractors

Sextant Avionique.

Daimler-Benz Aerospace AG.

LITEF GmbH.

Nord-Micro Electronik Feinmechanik AG.

Rohde & Schwarz GmbH & Co KG.

VDO-Luftfahrtgerate Werk GmbH.

NEW ENTRY

Cockpit of the Eurocopter Tiger combat helicopter
1998/0015358

Timearc 6 GPS navigation management system

The Timearc 6 GPS navigation management system is a joint venture by Flight Components and Alcatel SEL. It provides accurate position information from an advanced six-channel GPS receiver designed with the latest technology for military and commercial applications. The Jeppesen NavData card contains information on all airports, VORs, NDBs and intersections in Europe or North America. The three-line 16-character LED alphanumeric display, which is

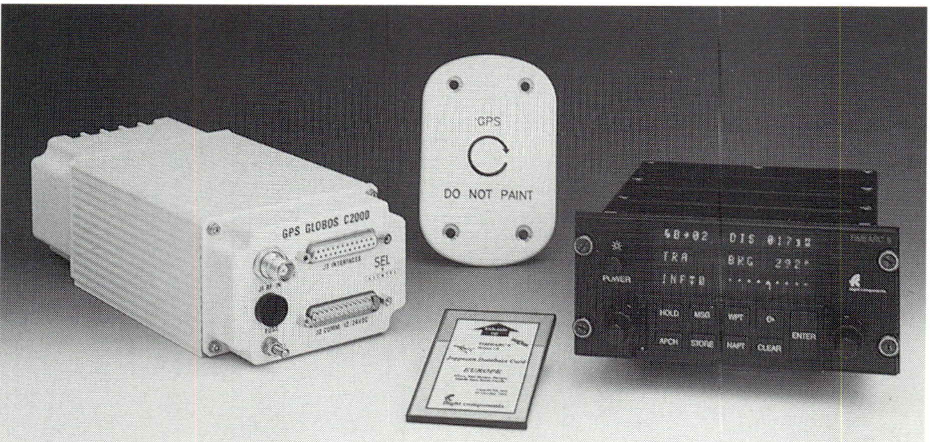

The Timearc 6 GPS navigation management system showing (left to right) the core module, (rear) the antenna with preamplifier, (front) the Jeppesen NavData card and the control and display unit
1995

extra bright and adjustable in intensity, is easy to read and is equipped with a left/right cursor indexer.

The Timearc 6 uses GPS satellites for position calculation and navigation and provides worldwide continuous and precise navigation data, offering operation under dynamic conditions up to 10 *g* , time to first fix of less than one minute and optional precise position data with differential GPS.

The associated mission planning system provides the ability to augment the Jeppesen NavData card with user defined waypoints and individual mission data. In addition, besides Universal Transversal Mercator (UTM), positions can be displayed in various local grids, including British National Grid and Swiss Grid to compare directly with local maps. Optionally, the Control and Display Unit (CDU) is equipped with an ARINC 429 interface, and the CDU features SID, STAR, and approaches capability.

Specifications
Dimensions: 66 × 146 × 177 mm
Weight: 1.37 kg
Power supply: 10-40 V DC, 0.26 A at 28 V
Temperature range: −20 to +55°C
Altitude: up to 35,000 ft
Accuracy - 3D position (95%):
(SPS not degraded) 51 m
(SPS with selective availability) 174 m
(DGPS without selective availability) 10 m
(DGPS with selective availability) 20 m

Contractors
Alcatel SEL AG.
Flight Components AG.

UPDATED

ISRAEL

D-Map digital moving map system

Designed for a two-seater cockpit, Elbit's digital moving map system (D-Map) single- or twin-map generators are capable of producing numerous types of images presented on a MultiFunction Colour Display (MFCD). Various map modes and features are controlled by keys integrated as part of the display. The system also features a Mass Storage Device (MSD) for storing databases such as a 300 × 300 km map in three different scales, obstacle overlay and library. The powerful graphic unit can draw raster and vector maps simultaneously.

Using a pointing device switch, the pilot can prepare and display his emergency flight plan in seconds. Working at a ground station, it is possible to plan an entire mission in advance, incorporating tactical mission data, flight plan, obstacles and communication data. The accumulated data is subsequently downloaded to the MSD or directly to the Map Display Generator (MDG) via serial communication channels. The end result is a significant increase in flight safety and reduction in pilot workload.

The system is capable of operating in two major display modes: as a digitised map display or aerial photographs, in which a digitised paper map in a range of linear scales from 1:50,000 to 1:2 million is displayed; or as a digital map display in which the display is a pseudo two-dimensional terrain model consisting of coloured altitude planes and vector type ground features calculated from a digital terrain elevation database.

The image presented on the MFCD is clear, crisp and easily interpreted. The system also offers the user tools such as zoom, heading up, scale and freeze for map manipulation. The online computing power of the MDG can calculate the line of sight between any given positions or point altitudes for threat coverage or safety corridors for routes and approaches.

Specifications
Dimensions:
(MMDG) 137 × 198 × 472 mm
(MFCD) 216 × 216 × 400 mm
(MSD) 225 × 140 × 303 mm
Weight:
(MMDG) 10 kg
(MFCD) 11 kg
(MSD) 6 kg
Transmits: flight plans prepared on maps
Video:
(RS-170) R, G, B
(RS-422) serial channel
(RS-232) obstacles database update
Options: MIL-STD-1553B mux bus, NVG-HUD

Operational status
D-Map has been selected by Rega, a prominent air ambulance service, for the A 109 EMS helicopter and has already been selected by the Israeli Air Force for the CH-53 helicopter modernisation programme. Also selected for the V-22 Osprey. EFW Inc, an Elbit subsidiary in the USA, will act as prime contractor in the contract.

Contractor
Elbit Systems Ltd.

UPDATED

The Elbit Systems Flightsight rangeless AACMI
1996

Flightsight rangeless AACMI

Flightsight is a rangeless Autonomous Air Combat Manoeuvring Instrumentation (AACMI) system that provides in-flight recording for highly effective comprehensive post-mission debriefing and analysis. It can be installed in a wide range of fighter aircraft without any expensive modifications.

Flightsight is an autonomous integral system connected to the aircraft avionics that gathers information into standard storage devices. The system offers debriefing capability with no geographical limitations. Navigation data from the aircraft arrives from the INS and GPS, giving exact position and performance information. The digital data is downloaded to the debriefing computer, together with other aircraft data. The computer can give an online display from its disks.

Flightsight provides a comprehensive review of the mission. It can be installed permanently either separately or as part of an aircraft upgrade. It does not use a missile pod and thus can be used in actual operational missions, in addition to training exercises. Voice and video are synchronised with the situation display. Flightsight interfaces with any existing avionics system based on a MIL-STD-1553 databus, and uses existing data from the navigation radar, EW and stores management systems.

Contractor
Elbit Systems Ltd.

UPDATED

Halo advanced helicopter avionics

Halo is an integrated suite of avionics tailored for attack helicopters. The suite largely comprises Elbit's own systems being developed for the Israeli Air Force. Included are a head-up display projected on to night vision goggles, precision navigation system and an extensive database which permit the pilot to fly at high speed and low level by night.

Halo integrates vision and perception, accurate navigation and geographic intelligence and obstacle database. Vision and perception are improved by night vision goggles and a head-up display which provide flight symbology collimated with the outside view. This includes a built-in fire-control FLIR capability for fighter attack helicopters. For navigation, Halo integrates a Doppler radar with GPS to provide position accuracy better than 50 ft, regardless of flight duration and terrain. Symbolic display of the flight path and helicopter position along with other data enables the pilot to fly on track and at high speed. The intelligence and obstacle database is programmed into a data transfer unit and loaded into the computer before flight. Display of data to the crew is via an easy to read symbolic map which eliminates the need for expensive digital maps.

Specifications
Dimensions:
(central computer) 134.6 × 203.2 × 431.8 mm
(multifunction display) 134.6 × 203.2 × 254 mm
(control display unit) 228.6 × 203.2 × 304.8 mm
(NVG and HUD control unit) 134.6 × 203.2 × 203.2 mm
Weight:
(central computer) 15 kg
(multifunction display) 8 kg
(control display unit) 4 kg
(NVG and HUD control unit) 8 kg

Operational status
Deliveries to the Israeli Air Force began in 1988. In October 1992 a contract was signed to supply Halo for upgrading Romanian Puma 330 helicopters.

Contractor
Elbit Systems Ltd.

UPDATED

EL/K-7200 DME/P-N airborne interrogator

Elta has been awarded a contract by the US Department of Transportation to develop DME/P-N airborne interrogators for MLS. Elta's EL/K-7200 Precision (P) and Normal (N) DME operates as a subsystem of the MLS, providing precision range and range rate information to the aircraft.

DME/P ground transponders will soon be available for installation in MLS equipment at selected major airports. In its N mode, the DME is compatible with existing transponders. Each DME/P-N includes an interrogator and controller unit for installation in the cockpit.

The EL/K-7200 utilises large-scale integration components and advanced mass production technologies. It rapidly acquires normal and precision ground transponders and remains locked-on during all flight manoeuvres including initial approach, final approach and missed approach through the implementation of proprietary algorithms and adaptive digital filters.

Operational status

FAA qualified.

Contractor

Elta Electronics Industries Ltd.

VERIFIED

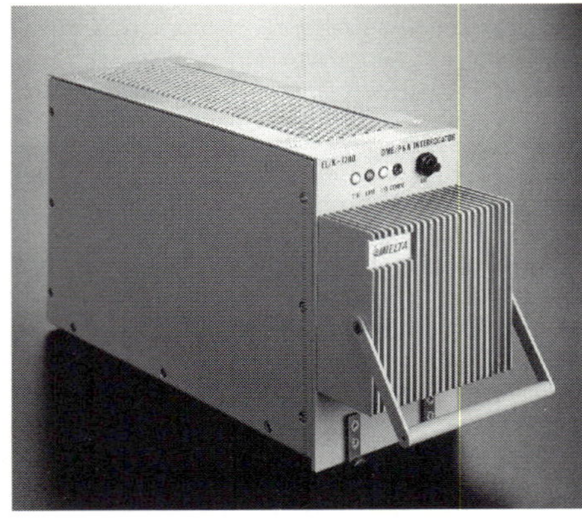

The Elta EL/K-7200 DME/P-N interrogator

3-D GPS attitude determination receiver

The Rokar 3-D GPS attitude determination receiver is a 6 to 24 channel receiver with 1 to 4 antenna inputs, designed to provide 360° coverage in high-dynamic airborne scenarios.

Each antenna ia connected to 6 to 12-channels depending upon the configuration and application.

This line of receiver products provides ready-to-use solutions for attitude determination and for antenna coverage, eliminating the need for specialised GPS antennas.

Key features are fully spherical antenna coverage with no SNR degradation; attitude determination; off-the-shelf, low-cost antennas to replace wraparound antennas; differential operation; high-dynamic scenarios operation; INS integration; multipath and spoofing rejection.

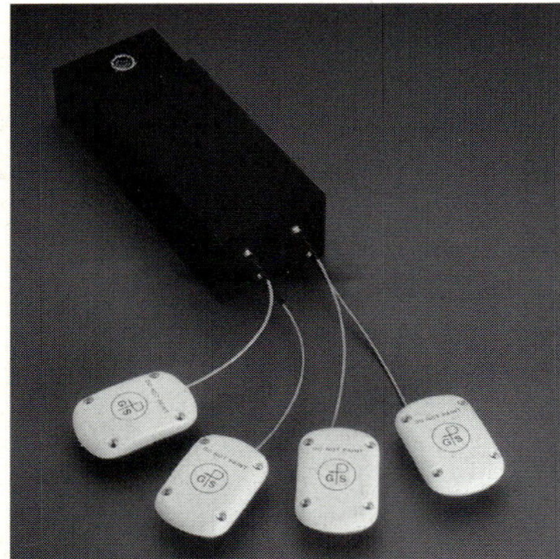

The Rokar 3-D GPS attitude determination, 360° coverage, high-dynamics scenario GPS receiver, four antenna configuration **1998**/0015359

Contractor

Rokar International Ltd

NEW ENTRY

GPS NAVPOD rugged receiver

The Rokar GPS NAVPOD is 'an all-in-view', parallel tracking, 12-channel GPS receiver, designed for demanding pod and combat aircraft applications.

It is a fast reacquisition C/A code receiver, which processes navigation data signals transmitted from all satellites in view. In addition to regular navigation data, the NAVPOD also outputs pseudo ranges, delta ranges, system time, measurement quality and other data required for coupling with Inertial Navigation Systems (INS).

The system is capable of operating in differential navigation mode, and of receiving/producing differential GPS corrections.

The receiver incorporates the latest in digital ASIC design and Receiver-On-Chip (ROC) technology.

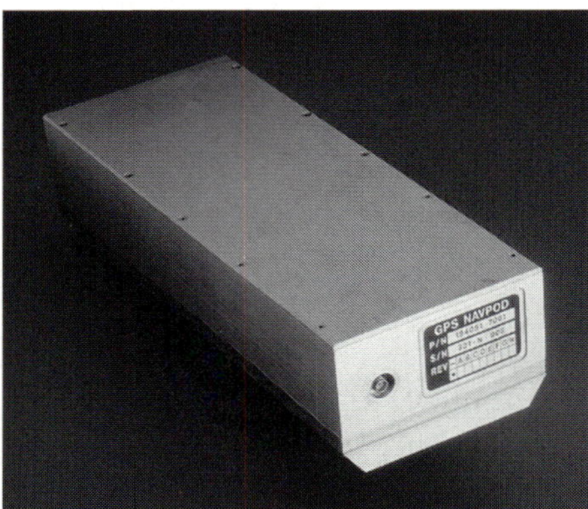

The Rokar GPS NAVPOD high-dynamics, rugged GPS receiver **1998**/0015360

Contractor

Rokar International Ltd.

NEW ENTRY

GPS SWIFT high-velocity, high-acceleration receiver

The Rokar GPS SWIFT is an 'all-in-view', parallel tracking, 12-channel GPS receiver. The GPS SWIFT is intended for use in platforms that travel at high velocity and high accelerations.

The GPS SWIFT provides accurate position, velocity, acceleration and acceleration-rate (jerk), as well as time and other GPS data. The system is capable of operating in differential navigation mode, and of receiving/producing differential GPS corrections. It is also designed for tight-integration with other sensors such as Inertial Navigating Systems (INS).

Key features are high-velocity (2000 m/s) and high-acceleration (15 g) operation; provision of navigation data including acceleration and jerk; INS integration.

Contractor

Rokar International Ltd.

NEW ENTRY

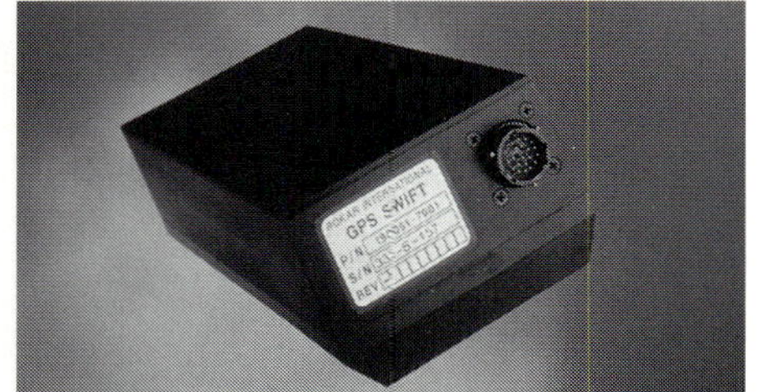

The Rokar GPS SWIFT high-velocity, high-acceleration, GPS receiver **1998**/0015361

NTS/NTS-A Night Targeting System

Since 1982, Tamam has been involved in upgrade programmes to provide Cobra attack helicopters with laser ranging, designation and night attack capabilities. Developed by Tamam under sole-source contract from the US Marine Corps and Israeli Air Force, NTS has accumulated considerable operational use.

Night Targeting System (NTS-A) is a new version of the NTS, with enhanced performance suitable for a wide range of different attack helicopters (including the AH-1 W/S/P Cobra). In NTS-A, the original optical tube is removed, freeing space in the helicopter cockpit, for the use of flat-panel displays, and allowing the system to be compatible with different types of attack helicopter, and with modern glass cockpit architectures, such as the US Marine Corps AH-1W 4BW upgrade programme.

NTS-A mounted on Cobra AH-IS helicopter **1998**/0015362

NTS-A night targeting system-A ***1998***/0015363

NTS-A has the following capabilities:
Installation flexibility with TOW I, TOW II, Hellfire, and other weapons;
Target acquisition during day, night and limited visibility conditions;
Laser Designator and Rangefinder System (LDRS) for laser-guided weapons and for measuring target ranges;
TV Tracker (TVT), providing target auto-tracking during day and night;
Guidance for all types of TOW missiles;
Fully-automatic in-flight boresight capability;
Navigation and tactical data display;
Display of NTS-A standard video signal (both the FLIR and TV camera pictures) on a multifunction display in both the gunner and pilot cockpits;
Built-in growth potential for future integration with other systems onboard, via (2)MIL-STD-1553B databusses.

Specifications

FLIR - 2nd generation
FLIR Fields of View (FOV) and magnification (relative to display viewing angle of 40 × 25°):

Wide FOV (H × V)	18.0 × 24.0°	×2.0
	30.0 × 40.0°	an option for pilotage
Medium FOV (H × V)	5.2 × 3.9°	×9.0
Narrow FOV (H × V)	1.47 × 1.1°	×32.1
Zoom NFOV (H × V)	0.73 × 0.55°	×64.3

Automatic TV Tracker (TVT) - with prediction, adjust and offset modes:
Field of regard:
(azimuth) 90° to the left, 95° to the right
(elevation) up 30°, down 60°
Turret slew rate:
(low-angular velocity) 2°/s
(high-angular velocity) 90°/s
(acceleration) 60°/s³
TVC features:
(frame transfer Charged Couple Device (CCD)), ½ in
(number of pixels): 780 × 576
Missile interfaces available:
Hellfire missile; Rafael NT-D missile
all types of TOW missiles (TOW, 1-TOW, TOW 2A, TOW 2B)
Weight: 129 kg (excluding aircraft installation kit)
Power:
(average power) 400 W
(max power) (during laser operation) 650 W

Operational status

In production. Over 300 systems have been ordered.

Contractors

TAMAM Division, Electronics Group, Israel Aircraft Industries.

NEW ENTRY

TN-90 compact inertial navigation system

The TN-90 is a small lightweight low-cost strapdown inertial system customised according to mechanical and electrical requirements.

Typical applications are for guided and smart missiles and bombs, unmanned air vehicles, helicopters, electro-optic and targeting systems, instrumentation pods, flight control and as a back-up for the INS.

Specifications

Dimensions: 128 × 127 × 119 mm typical
Weight: 2.4 kg
Outputs: linear acceleration, velocity and angular rates, position, azimuth and Euler angles
Interface: multiple RS-422 serial, MIL-STD-1553B mux-bus electrical
Accuracy (with CA code GPS):
(position) 100 m SEP
(velocity) 0.1 m/s
(attitude) 0.2°
(azimuth) 0.2-0.4°

Operational status

In production.

Contractor

TAMAM Division, Electronics Group, Israel Aircraft Industries.

UPDATED

TR-90 compact inertial navigation system

TAMAM has developed and produced a family of small lightweight low-cost strapdown inertial reference units, customised according to mechanical and electrical requirements.

Typical applications are for guided and smart missiles and bombs, unmanned air vehicles, optronic and targeting systems, instrumentation pods, antenna stabilisation, flight control and as a back-up navigation system.

Specifications

Dimensions: 128 × 127 × 82 mm typical
Weight: 1.8 kg typical
Outputs: velocity and angular increments, linear acceleration, angular rates
Interface: multiple RS-422 serial

Operational status

In production.

Contractor

TAMAM Division, Electronics Group, Israel Aircraft Industries.

UPDATED

TN-90Q/G compact inertial navigation system

TAMAM's TN-90Q/G is a compact lightweight unit that integrates strapdown, inertial and GPS navigation. The system incorporates into the package, a proven off-the-shelf GPS receiver tightly coupled with an inertial measurement unit.

The system provides positioning, velocity, time, attitude, heading, angular rate and acceleration.

Specifications

Dimensions: 209.4 × 114.2 × 93.9 mm
Weight: 3.5 kg
Power: 45 W typical, 28 V DC
Interfaces: RS-422 or MIL-STD-1553B mux-bus
Accuracies:

	C/A code GPS	with P(Y) code
Position	120 m	20 m
(after 5 min)	350 m	250 m
Velocity	0.6 m/s	0.1 m/s
Attitude	0.1°	0.03°
Azimuth	0.2°	0.1°

Operational status

In production.

Contractor

TAMAM Division, Electronics Group, Israel Aircraft Industries.

NEW ENTRY

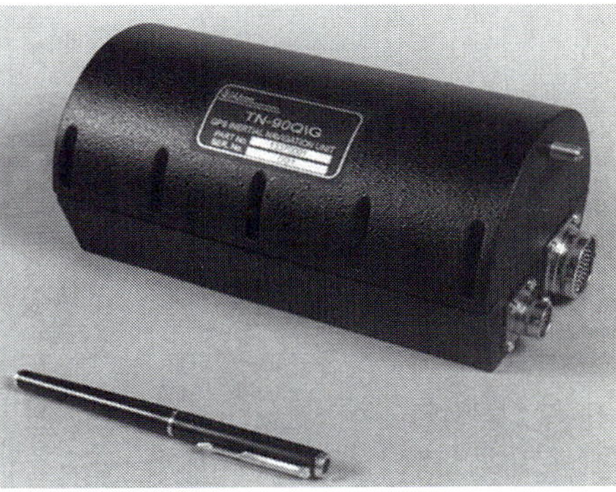

TAMAM's TN-90Q/G compact inertial navigation system ***1998***/0015364

ITALY

Lisa-4000 strapdown inertial reference unit

The Lisa-4000 performs the basic functions of flight reference and navigation. It provides heading and attitude for display and autopilot and high-speed low-latency anti-aliasing filtered body angular rate and linear acceleration data for autopilot inner loop stability augmentation.

Navigation is performed by coupling the Lisa-4000 with GPS, Doppler radar and air data systems. The Lisa-4000 also outputs high-speed inertial velocity data required for accurate weapon release. It therefore provides all the inertial parameters required for autopilot, navigation, display and weapon delivery.

The Lisa-4000H version is a full-aided inertial navigator with the option of a MIL-STD-1553B or ARINC 429 interface. It can also be supplied with additional synchro outputs to drive back-up cockpit instruments.

Specifications

Dimensions: ½ ATR short/short
Weight: 6.2 kg
Power supply: 28 V DC, 90 W unregulated
Cooling: none to +71°C
Accuracy (RMS):
(heading) 0.2° + compass
(pitch and roll) 0.2°
(body rates-PQR) 0.15°/s, 200 Hz, 2.5 ms latency
(body rates-XYZ) 5 m*g*, 200 Hz, 2.5 ms latency
(horizontal velocity) 2 m/s
(vertical velocity) 1 m/s
(position) 1% of distance travelled
(altitude) 100 ft + 1%
Inputs: GPS, Doppler velocities, barometric altitude, true airspeed, compass, up to 100 waypoints
Outputs: bearing to destination, distance to go, time to go, wind speed and direction, groundspeed

Operational status

The Lisa-4000B is in production for the A 129 attack helicopter. The Lisa-4000D is operating on the Royal Navy EH 101 helicopter. The civil qualified Lisa-4000E is for the EH 101 civil variant. The GPS coupled Lisa-4000G is in production for an RPV. The Lisa-4000H is in production for Italian Navy AB-212, AB-412, SH-30 helicopters, and the EB version on Romanian MiG-21.

Contractor

Litton Italia SpA.

VERIFIED

LN-93EF ring laser gyro inertial navigation system

The LN-93EF is based on the LN-93 ring laser gyro standard navigator built by Litton to SNU-84 standards. However, the EF version, modified to EF 2000 requirements, includes upgraded electronics, a fibre optic databus, and Ada software. The result is a system with the same accuracy and maturity but with a 30 per cent reduction in size and weight than the SNU-84 system.

Specifications

Dimensions: ¾ ATR (length 360 mm)
Weight: 16 kg
Power supply: 115 V AC, 400 Hz, 85 W
28 V DC back-up
Temperature range: −40 to +71°C
Align time:
(gyrocompass) 4 min
(memorised) 30 s
Accuracy:
(position) 1 n mile/h CEP
(velocity) 0.8 m/s RMS
(attitude) 0.05° RMS
(heading) 0.1° RMS

Operational status

Delivered to the EF 2000 development programme.

Contractor

Litton Italia SpA.

VERIFIED

ANV-141 VOR/ILS airborne navigation set

The Marconi ANV-141 is a VHF navigation system for the reception and processing of VOR and ILS signals. The receiver is packaged in a short/dwarf ½ ATR case which is designed to be mounted on either a hard or shockmount.

The ANV-141 is a receiver which provides 200 VOR/Loc channels from 108 to 117.95 MHz with manual and automatic instrumentation outputs and 20 paired glide slope channels from 329.5 to 335 MHz.

The ANV-141 is a compact all-solid-state avionic equipment that exceeds TSO specifications and meets military requirements.

Specifications

Dimensions: ½ ATR short/dwarf
Weight: <5.5 kg
Power supply: 27.5 V DC, 42 W (max)
26 V AC, 20 VA (max)

Contractor

Marconi SpA.

VERIFIED

ANV-201 airborne microwave landing system

The ANV-201 airborne receiver is a precision angular position sensor which is capable of operating with any Microwave Landing System (MLS) ground equipment facility that transmits the standard ICAO signal format.

The ANV-201 receives C-band signals radiated by the ground stations and processes the angle and rate functions to yield offset corrected two-dimensional angular position information. This information is sent to external peripherals in the form of both analogue and digital signals. The difference between the actual aircraft angular position and the desired azimuth and approach angle can be sent in analogue form to the Automatic Flight Control System (AFCS) and Course Deviation Indicator (CDI). Digital information exchange with aircraft avionics is through ARINC 429 or MIL-STD-1553 interfaces.

In conjunction with precision measuring equipment receivers, the ANV-201 is a highly accurate three-dimensional positioning system for terminal area navigation.

The control panel allows selection of azimuth radials up to 60° either side of the centreline of the runway in 1° increments and glide slope angles from 0 to 19.9° in 0.1° increments. It also allows selection of any one of the 200 available C-band channels.

The ANV-201 uses a powerful 16-bit microprocessor and has a comprehensive self-test capability. It is designed to meet the EMC and environmental requirements of MIL-STD-461 and MIL-STD-810.

Specifications

Dimensions: ⅜ ATR short
Weight:
(receiver) 5.5 kg

Contractor

Marconi SpA.

VERIFIED

ANV-211 airborne DME/P interrogator

Marconi's DME/P ANV-211 airborne equipment is a precision range sensor and processor capable of operating with any Precision (P) or Narrow (N) ground transponder. The ANV-211 DME/P together with the angle receiver ANV-201 comprise the microwave landing system for terminal area navigation and landing.

The DME/P interrogator is composed of two LRUs: the interrogator and the control panel. The control panel is available either as a dedicated DME/P or as a combined control panel that manages both MLS and DME/P.

The ANV-211 conforms to MIL-STD-461 and MIL-STD-810D for EMC and environmental conditions.

Specifications

Dimensions: ⅜ ATR short
Weight: 5.5 kg

Contractor

Marconi SpA.

VERIFIED

ARG-80 automatic direction-finder

The ARG-80 automatic direction-finder is a completely solid-state instant digital tuning system suitable for updating ADF capability on new fixed- and rotary-wing aircraft.

A single crystal digital stabilised frequency synthesiser covers the frequency range from 190 to 1,794.5 kHz at 0.5 kHz spacing. Circuitry features eliminate all electromechanical moving parts, optimising bearing performance under high-noise conditions.

Easy access plug-in modules are assembled on a rugged frame which provides both mechanical integrity and heat dissipation. Any suitable RMI may be used with the ARG-80. Frequency channels can be selected by the remote-control unit having BCD code. Two different types of controllers are available for providing receiver switching, selection of ADF or antenna operation, self-test, tone identification and audio frequency gain control. The first type of controller is a standard single frequency version; the second offers a four-channel preselected facility, enabling instantaneous selection of frequencies previously stored in the memory.

Specifications

Dimensions:
(receiver) 124 × 87 × 319 mm
(single-channel controller) 146 × 67 × 76 mm
(four-channel controller) 146 × 67 × 123 mm
(antenna) 450 × 240 × 40 mm
Weight:
(receiver) 3.5 kg
(single-channel controller) 0.5 kg
(antenna) 1.5 kg

Contractor

Marconi SpA.

VERIFIED

JAPAN

JSN-8 strapdown laser Attitude and Heading Reference System (AHRS)

The JSN-8 high-performance laser AHRS is suitable for military and civil applications. The sensors are three Honeywell laser gyroscopes with three JAE JA-5 accelerometers. All the sensor equipment and associated processing is enclosed in a single unit and there is a separate control panel. In its current form the equipment operates solely as an attitude and heading reference system, but it can be developed to include inertial navigation functions and so has inertial navigation type interfaces.

It can operate in high angular rate conditions of up to 400°/s roll rate and does not suffer any performance degradation due to acceleration. It has a short reaction time and does not require a flux valve device. Reliability is quoted as in excess of 2,300 hours. In-flight alignment can be accomplished, aided by such systems as Doppler, Tacan or Navstar, and magnetic heading can be provided as an output.

Specifications

Dimensions:
(control panel) 100 × 146 × 76 mm
(attitude/heading reference unit) 250 × 330 × 180 mm
Weight:
(control panel) 0.8 kg
(AHRU) 19.8 kg
Power supply: 115 V AC, 400 Hz, 30 W
28 V DC, 115 W
Input/output:
(digital) ARINC 429 or MIL-STD-1553B or special
(analogue) synchro attitude signal outputs, and analogue} turning rate signal
(discrete) valid signal and bit command output
Alignment time:
(normal) 2.5 min
(stored heading) 1 min
Environmental: MIL-E-5400T Class 2X
Electromagnetic interference: MIL-STD-461A, -462

Operational status

In production.

Contractor

Japan Aviation Electronics Industry Ltd.

VERIFIED

Laser gyro inertial navigation system

The strapdown ring laser gyro inertial navigation system is installed on the CH-47 transport helicopter. The use of ring laser gyros in the system permits increased reliability, longer life and so on, than a conventional platform-type inertial navigation system, considerably reducing life cycle costs, including maintenance and overhaul costs.

Operational status

In service in Kawasaki/Boeing CH-47 helicopters of the Japanese Air Self-Defence Force.

Contractor

Japan Aviation Electronics Industry Ltd.

VERIFIED

Strapdown Attitude and Heading Reference System (AHRS) for the SH-60J

The strapdown attitude and heading reference system installed on the Mitsubishi/Sikorsky SH-60J anti-submarine helicopter detects pitch, roll and azimuth angles and turn rates with reference to body axes, and supplies this data to other airborne equipment. Combined with a Doppler navigation system, the strapdown attitude and heading reference system can also function as an inertial navigation system. The data is supplied to other equipment through a MIL-STD-1553B databus. The strapdown system uses a small ring laser gyro, is compact and has higher performance, reliability and life than a conventional attitude and heading reference system.

Operational status

In service in Mitsubishi/Sikorsky SH-60J helicopters of the Japanese Maritime Self-Defence Force.

Contractor

Japan Aviation Electronics Industry Ltd.

VERIFIED

Strapdown Attitude and Heading Reference System (AHRS) for the T-4 aircraft

The strapdown attitude and heading reference system for the Kawasaki T-4 is an integrated sensing system. It provides aircraft acceleration signals, angular velocities and attitude in three axes to other airborne equipment. Data transmission is via ARINC 429 databusses. Laser gyros are incorporated in the system, leading to high reliability and service life.

Operational status

In service in Kawasaki T-4 aircraft of the Japanese Air Self-Defence Force.

Contractor

Japan Aviation Electronics Industry Ltd.

VERIFIED

Hybrid Inertial Sensor Unit (HISU)

The Hybrid Inertial Sensor Unit (HISU) is a core sensor of the Attitude and Heading Reference System (AHRS) now planned. The core sensor, the development of which has been completed, incorporates three Fibre Optic Gyros (FOGs) and three accelerometers, which provide three-axis angular rates and three-axis accelerations.

The sensor is to be integrated with a strapdown system (also under evaluation), into one component.

Operational status

Under evaluation.

Contractor

Tokyo Aircraft Instrument Co, Ltd.

VERIFIED

Hybrid Inertial Sensor Unit (HISU)
1997/0001333

NORWAY

Navaids flight inspection system

The Navia Aviation navaids flight inspection system is a digitally computerised system for the checking of radio navigation aids. The basic system is configured with a powerful 32-bit computer and very high-capacity data storage for flight data. The computer performs all the routine operations such as sensor tuning, recorder parameter set-up, pilot guidance, final report printout and sensor parameter calibration. In addition, the computer performs real-time processing, storage and colour presentation, which greatly improves the speed of inspection procedures.

The system consists of three components: the digitally computerised airborne console, the ground-based position reference system and the maintenance and ground support system.

Optimised integration of data presentation for the airborne controller leads to a compact panel design. Due to the console design, operators are able to face forward even for installations in the smallest aircraft.

The basic system may be used for checking VOR, DME, NDB, VDF, PAR, ILS, SSR, markers and VHF comms. Optionally, UDF, Tacan, MLS, SLS and UHF comms may be checked.

Specifications

Dimensions: 500 × 610 × height (mm) dependent on configuration
Weight:
(console) 95 kg
(rack) 75-95 kg
Power supply: 28 V DC, 15-25 A
26 V AC, 400 Hz
115 V AC, 400 Hz, 40-350 VA
Crash certification: FAR 25

Contractor

Navia Aviation AS.

UPDATED

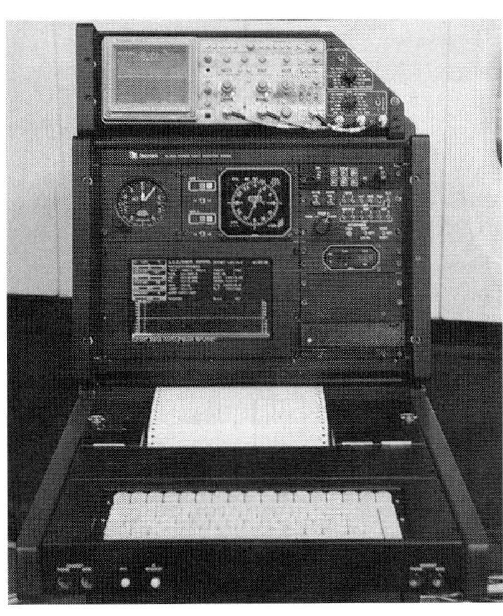

The operator console for the navaids flight inspection system ***1995***

RUSSIAN FEDERATION AND ASSOCIATED STATES (CIS)

DME/P-85 navigation system

The DME/P-85 navigation system provides slant range, in kilometres and nautical miles, from DME/N, TACAN and DME/P ground stations.

Specifications

Weight:
(interrogator) 9.5 kg
(protuding antenna) 0.33 kg
(suppressed antenna) 0.55 kg
Measurement range: 300 nm; 555 km
Frequency range: 960-1,213 MHz
Channels: 352
Dimensions: 4 MCU
Power: 115 V AC, 400 Hz, 70 VA

Operational status

Development contract award in 1985. Standard fit in Russian airliners.

Contractors

All-Russia Research Institute of Radio Equipment.
NIIAO Institute of Aircraft Equipment.

NEW ENTRY

VOR-85 VHF Omni Range system

VOR-85 provides identification of ground VOR and marker beacons, and magnetic azimuth relative to VOR ground beacons.

Specifications:

Dimensions: 3 MCU
Weight:
(receiver) 5 kg
(navigation antenna) 1.7 kg
(marker antenna) 0.9 kg
Frequency range:
(VOR) 108.00-117.95 MHz
(marker) 75 MHz
Channels (VOR): 160
Accuracy (95%): 0.5°
Power: 115 V AC, 400 Hz, 30 VA

Operational status

Development contract award in 1985. Standard fit in Russian airliners.

Contractors

All-Russia Research Institute of Radio Equipment.
NIIAO Institute of Aircraft Equipment.

NEW ENTRY

BINS-85 Inertial Navigation System

Designed for the An-70 transport aircraft, the BINS-85 inertial navigation system provides the performance specified below.

Specifications

Error performance:
(position coordinates (latitude and longitude)) in 10 h of operations) 37 km (max)
(ground speed) 14.4 km/h
(true heading) 0.4° (max)
(roll and pitch angles, in navigation mode) 0.1° (max)
Status ready time:
(at 0 to +55°C) 10 min (max)
(at 0 to −20°C) 18 min (max)
Temperature range: −20 to +55°C
Digital outputs: 32 parameters
Power: 115 V AC, 400 Hz, 200 VA (max) 27 V DC, 180 W
Mean time between failures: 5,000 h
Dimensions: 319 × 322 × 194 mm
Weight: 20 kg (max)

Operational status

Fitted to An-70 transport aircraft.

Contractor

Aviapribor

NEW ENTRY

GINS-3 Gravimetric Inertial Navigation System

GINS-3 is designed for two applications: stand-alone navigation of aircraft; and aerogravimetry. System components are the gravimetric inertial unit, radio altimeter, air data system, satellite navigation system, electronics unit, and computer.

BINS-85 inertial navigation system **1998**/0015365

Specifications

Navigation errors:
(position coordinates, independent of flight time and distance) 0.2 to 10 km
(ground and vertical speed components) 0.1 to 1 m/s
(altitude above sea level) 1 to 5 cm
Aerogravimetry error of gravitational anomaly in free air: up to 0.7 mGal
Scales of maps produced: 1/200,000 and more
Power: 27 V DC, 1,000 VA (max)
Weight: up to 120 kg

Contractors

Aviapribor

NEW ENTRY

GINS-3 gravimetric inertial navigation unit **1998**/0015366

I-21 inertial navigation system

The I-21 inertial navigation system is designed to autonomously determine flight/navigation data, and provide it to the aircraft system for generation of steering commands between defined waypoints. It conforms to ARINC 61.

Specifications

Navigational errors:
(geographical coordinates (in 10 h)) 37 km (max)
(ground and speed components) 12.6 km/h
(true heading) 0.2 + 0.025°/°C
(angles of roll/pitch, gyro heading) 0.1°
Readiness time: <15 min at 20°C
Mean time between failures: 1,200 h
Power: 500 VA (1,500 VA (max)
Weight: <50 kg

Operational status

Fitted to An-124, An-224, Il-62, Il-76, Tu-154, Tu-160 aircraft

Contractor

Aviapribor

NEW ENTRY

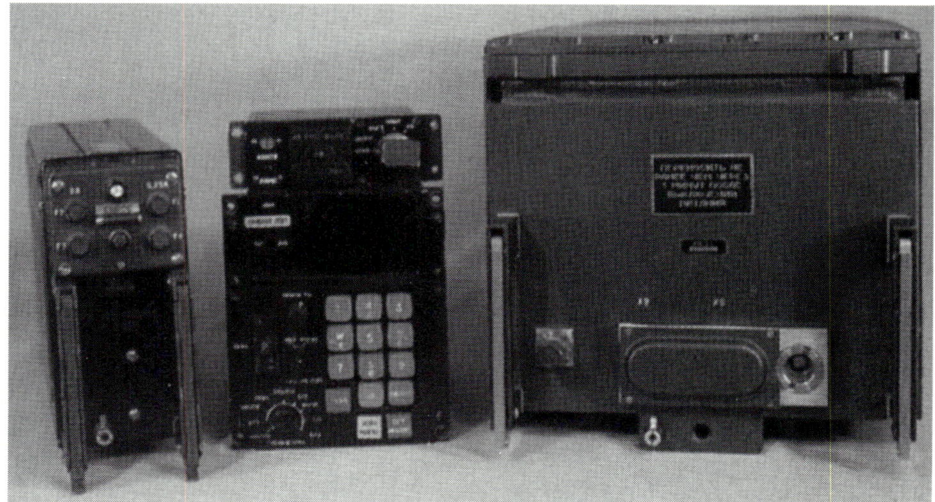

I-21 inertial navigation system **1998**/0015367

CH-3301 GLONASS/GPS airborne receiver

NAVIS receivers are able to process signals from both GLONASS and GPS satellites, thereby increasing accuracy and reliability by comparison with receivers that process only one source of data.

Specifications

Memory: 500 waypoints, 50 routes
Display: alphanumeric LCD display; 20 symbols on each of 2 lines
I/O: 2 input/output ports RS-232C; 1 output channel in ARINC 429 format; 1 analogue barometric input
Antenna: to ARINC 743A
Dimensions: 384 × 159 × 51 mm
Weight: 2.4 kg
Power: 27 V DC; 115 V AC, 400 Hz
Temperature:
(receiver) −20 to +50°C
(antenna) −55 to +55°C

Contractor

NAVIS.

NEW ENTRY

INS-80 Inertial Navigation System

The INS-80 system is made as a single unit, comprising gyrostabilised platform, platform electronic units, a computer, and I/O devices.

Two free dynamically-tuned gyroscopes and three accelerometers are mounted on the gyrostabilised platform. The electronic units are mounted on separate PCBs.

Specifications

Position error: 3.7 km/h
Velocity error: 2.0 m/s
Roll, pitch and heading error: 0.1°
Readiness time: 1 to 10 min
Volume: 18 dm^3
Weight: 19 kg

INS-80 inertial navigation system **1998**/0015368

Contractor

Ramenskoye Design Company AO RPKB

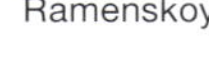

NEW ENTRY

NS BKV-95 integrated Navigation System

The NS BKV-95 integrated navigation system performs three functions: strapdown attitude and heading reference system, NAVSTAR receiver, navigation computer. The system is provided with control displays to the customer specification.

Specifications

Dimensions: 388 × 124 ×194 mm (4 MCU)
Weight: 8 kg

NS BKV-95 integrated navigation system **1998**/0015369

Power: 27 V DC
Outputs: GOST 18977-79 Sup 3; (ARINC 429)
Accuracy (2 Sigma):
(position) 100 m
(ground speed) 2 km/h
(roll/pitch angles) 0.25°
(magnetic heading) 1°

Operational status

Suitable for all types of civil aircraft

Contractor

Ramenskoye Design Company AO RPKB

NEW ENTRY

SINUS integrated navigation, flight management and display system **1998**/0015370

SINUS integrated navigation, flight management and display system

SINUS is an integrated navigation and flight management system, designed for the aircraft and helicopter retrofit market. It comprises systems that perform the normal functions of an integrated navigation and flight management system, including inertial navigation, satellite navigation (using both GLONASS and NAVSTAR), air data and altimeter integration. The inertial flight computer integrates these functions with radar and FLIR data, via an airborne graphics system to provide navigation, radar and flight management data to the pilot by any combination of the following methods: multifunction AMLCD, head-up display, helmet-mounted display, and aural warnings. The SINUS system also provides appropriate outputs for automatic control of aircraft flight controls and engine systems.

Specifications

Navigation errors:
(inertial mode) 1 nm/h
(satellite mode) 50 m
(FLIR mode) 20-100 m
Velocity error: 0.3 m/s
Angle errors:
(roll/pitch) 0.5°
(yaw) 1°
Weight: 28 kg

Operational status

Designed for full internal, or partial podded, fit for upgrade of military aircraft and helicopters such as MiG-21, Su-24, Mi-8, Mi-24.

Contractor

Ramenskoye Design Company AO RPKB

NEW ENTRY

A-744 radio-navigation receiver

The A-744 is an integrated GLONASS/GPS navigation receiver that determines position, time, velocity (horizontal and vertical components) in real time. It is designed to be used either independently, or as part of an integrated aircraft system. It can be commanded to GPS only mode, or to GLONASS-GLONASS/GPS mode.

Specifications

Dimensions: 281 × 191 × 64 mm
Weight: 2.7 kg
Power supply: 27 V DC, 18 W
Antenna: active, non-protruding
Interfaces: GOST 18977-79 PTM 1495-75 (ARINC 743A, 429) multiplex channel GOST.26765.52-87 (MIL-STD-1553B)
Number of channels: 6 (universal)
Operating frequency:
GLONASS, EA code: 1598-1616 (F1)
GPS, C/A code: 1575.42 (L1)
Output data: position, velocity, time, quasi-range and Doppler shift measurement
Position error (2e) m:
GLONASS: 45 horizontal/70 vertical
GPS (SA including GLONASS/GPS: 100 horizontal/160 vertical

DGPS/Diff.GLONASS: 6 horizontal/10 vertical
Time to first fix: less than 150 s
Dynamic characteristics: (not more than)
(velocity) 500 m/s
(acceleration) 40 m/s^2
(acceleration increment) 30 $m/s^2/s$
Temperature limits:
(operation) −50 to +50°C
(storage) −55 to +85°C
Antenna dimensions/weight:
active antenna: 110 mm diameter, 41 mm height, 0.31 kg
passive antenna: 80 × 80 mm, 33 mm height, 0.27 kg

Contractor
Scientific-technical centre Osnova, Leninetz Holding Company.

VERIFIED

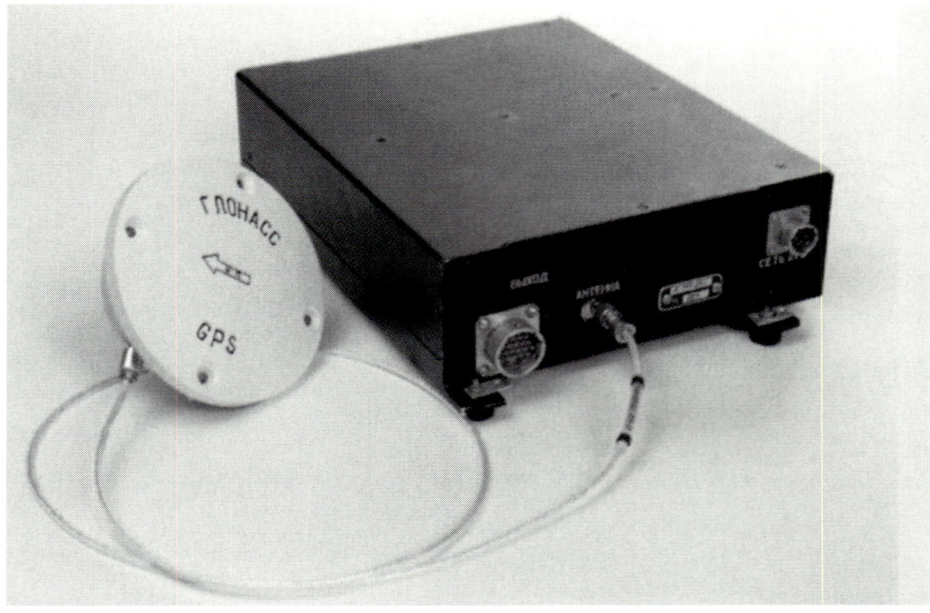

A-744 radio-navigation receiver
1997/0005425

SOUTH AFRICA

PT-730 airborne TACAN

The PT-730 airborne TACAN interrogator provides distance, relative bearing, range rate, morse-decoded identification and time to a selected TACAN ground station in T/R mode or airborne beacon in A/A mode. In A/A mode, the equipment also provides distance to the nearest TACAN-equipped responding aircraft, as well as replies to a maximum of five TACAN-equipped interrogating aircraft.

The equipment consists of a receiver/transmitter unit and an optional control panel and DME display. Control and data interfacing is via either ARINC 429, RS-422 or MIL-STD-1553B. It features microprocessor based design, dual antenna with manual or automatic selection, solid-state 700 W transmitter, continuous, pilot-initiated and power-up self-test.

Specifications
Dimensions: ¼ ATR
Weight: <5 kg
Power supply: 28 V DC to MIL-STD-704
Consumption: 50 W nominal, 65 W max
Channels: 126-channels, with X or Y coding. May be linked to VOR/ILS or VHF COMM channel tuning
Range: limited by line of sight, instrumented to 399 n miles
Accuracy: ±0.1 n miles

Contractor
Plessey South Africa Ltd.

UPDATED

ANM-90 GPS navigation system

The ANM-90 GPS navigation system has been designed to provide precise three-dimensional position, velocity and time for air, land and sea military forces. The equipment is aimed at a variety of military users, therefore, particular attention has been paid to the logistics elements, environmental conditions, volume, weight and life cycle cost.

The ANM-90 is available in four configurations: ANM-90 manpack, ANM-90 vehicle-mounted manpack, ANV-90 integrated vehicle system and ANA-90 airborne GPS sensor. The ANA-90 operates on the ARINC 429 databus and adheres to ARINC 743 which defines standards for interfacing GPS with airborne navigation computers. The ANA-90 provides position, velocity, heading and time to a range of navigation computers such as the RNS 252 and RNav 2 Racal computers.

Specifications
Dimensions: 63.5 × 215.9 × 241.3 mm
Weight: <3.5 kg
Power supply: 10-32 V DC, <0.4 A
Temperature range: −15 to +65°C
Accuracy (with selective availability) (95%):
(2-D position) 100 m
(altitude) 160 m
(velocity) 1 kt
(differential) (2D position) <25 m
(altitude) <25 m
Reliability: >4,000 h MTBF

Contractor
RDI (Pty) Ltd.

VERIFIED

ENAV 100 - Doppler velocity sensor

The ENAV 100 Doppler velocity sensor is a single unit comprising a solid-state transmitter, receiver and associated microprocessor circuitry to provide complete velocity component information.

When used in conjunction with a navigation computer, such as the ENAV 150, an autonomous navigation system can be provided. The ENAV 100 has been designed specifically for helicopters and low dynamic fixed-wing aircraft.

The standard output format conforms to ARINC 429 although other outputs, for example MIL-STD-1553B or conventioanl pulsed output, can be provided. Low-range DC analogue outputs are also available to drive conventional hovermeters.

The single transmitter is switched sequentially into each of three beams in a Janus configuration. This method allows a reduction in the number of components, weight and cost and leads to improved reliability.

Continuous wave transmission is normally used and gives better performance and freedom from height aberrations, compared to modulated systems. Other modes include interrupted CW for reduction of rain returns when flying in precipitation, and a low-power (or Stealth) mode.

Adaptor rings can be provided to retrofit older Doppler installations with the modern, lightweight ENAV 100.

Specifications
Dimensions:
(overall including flange) 117 × 403.2 × 436.4 mm
(aircraft cut-out size) 357.2 × 390.3 mm
Weight: 11 kg (max)
Microwave transmitter: J-band varactor multiplier 13.325 GHz ±20 MHz continuous wave; transmitter power 200 mW nominal (2 mW low power mode)
Power supply: 28 V DC, 90 W (max)
Velocity range:
−50 to +300 kt, along heading, Vx
±100 kt, across heading, Vy

Operational status
Currently in service in aircraft of the South African Air Force.

Contractor
Reutech Systems.

UPDATED

ENAV 150 - airborne navigation computer

The ENAV 150 airborne navigation computer is a low-cost, lightweight solution designed to meet the demand for an advanced navigation system that can be fitted to both fixed-wing aircraft and helicopters.

The ENAV 150 is compatible with the ENAV range of Doppler velocity sensors, to provide an autonomous Doppler navigation system. The ENAV 150 can also accept inputs from a range of inertial attitude and heading reference systems, to give a fully integrated hybrid navigation system with accuracies superior to a pure inertial navigation system. The computer will also accept position updates from radio navigation aids such as GPS, Loran C and Omega/VLF.

The variable intensity display can be clearly viewed in bright sunlight.

Up to 100 waypoints can be stored. They may be loaded manually via the keyboard, or from an optional Data Transfer Device.

The system can provide guidance to the autopilot for route and tactical steering as well as search and rescue patterns.

The ENAV 150 has synchro, analogue, discrete and ARINC 429 inputs to cope with a variety of attitude and air data signal formats. The computer's ARINC 429 input/output capability is fully compatible with a wide range of modern avionic architectures.

Specifications

Dimensions: 124 × 146 × 201 mm
Weight: 3.5 kg
Power supply: +28 V DC, 40 W
Waypoint capacity: up to 100 waypoints
Route capacity: single route of up to 20 waypoints

Operational status

Currently in service in aircraft of the South African Air Force.

Contractor

Reutech Systems.

UPDATED

ENAV 200 - hovermeter

The primary function of the ENAV 200 hovermeter in any helicopter installation is to provide indication of low speed velocities either during or when entering and leaving the hover flight configuration. The indicator is designed to operate directly from the ENAV 100 Doppler velocity sensor.

The ENAV 200 contains three moving coil movements which show helicopter velocities over the following ranges:

Along heading	10 kt backward to 20 kt forward
Across heading	15 kt left to 15 kt right
Vertical	500 ft/min downwards to 500 ft/min upwards

The meter also includes warning flags for both 'memory' and for 'power off' and is internally lit.

Overall dimensions meet ARINC Specification Number 408 for 3 ATI size.

Specifications

Weight: 0.9 kg
Power: 28 V DC
Accuracy: the error in displayed velocity will not exceed 1 kt in the along heading and across heading directions and 40 ft/min in the vertical direction
Flight envelope:
(speed) −50 to +350 kt
(altitude) 0 to 20,000 ft

Operational status

Currently in service in aircraft of the South African Air Force.

Contractor

Reutech Systems.

UPDATED

ENAV 300 VOR/ILS

The ENAV 300 system includes a number of VOR/ILS units which can be tailored to suit most fixed-wing and rotary-wing aircraft installations. The ENAV 300 offers interfaces to most types of RMI, HSI or PNI indicators. Microprocessor-controlled frequency generation, non-volatile memory, double frequency LC-displays for active and present channels and optoelectronic friction-free frequency selection are the most innovative characteristics of the VOR/ILS/GS receiver ENAV 300 (Series).

The built-in VOR/LOC converter can drive either a standard cross pointer OBS indicator or an RMI indicator via its DC Sin/Cos output (via a TMI amplifier).

Various options are available to drive various types of indicators.

Contractor

Reutech Systems.

VERIFIED

ENAV 400 - Distance Measuring Equipment (DME)

The ENAV 400 (DME) is a modern state-of-the-art system which is built and tested to exact standards. This system offers a cost effective solution for distance measuring equipment replacement for the upper general aviation market.

The ENAV 400 offers the following features as standard: channels; BDC or ARINC two out of five interface; Nav 1 or Nav 2 system selection; high-visibility plasma display with automatic dimming; range display up to 200 n miles; ground speed display up to 999 kt; time to station display up to 99 minutes; AF station ident; and standard 11 to 33 DC voltage operation.

As an option the system may be configured with a slave unit or alternatively with two slave units and a remote switching unit.

The ENAV 400 (Series) consists of the DME Interrogator Type ENAV 420; Remote Master Indicator Type ENAV 430; and Remote Slave Indicator Type ENAV 435.

Contractor

Reutech Systems.

VERIFIED

SPAIN

NAT-5 tactical navigation system

The NAT-5 is a computer-based navigation system for airborne ASW tactics. It comprises a control unit, central processor unit and display unit, and accepts information from any or all of Doppler, true airspeed, AHRS, GPS, sonar and radar.

The display unit shows updated positional information plus seven predefined contacts in latitude and longitude, X-Y co-ordinates or any of four different polar co-ordinate systems, with graphic scales from 4.5 × 3 up to 288 × 192 n miles.

The control unit is a microprocessor-based control panel with a keyboard including function, numeric and control keys and a joystick for quick data entry. The control unit allows easy system operation by means of up-down menus on the screen.

Graphics operation is provided by means of a powerful graphics processor with up to 100 kbytes of stored standard search patterns and a cartridge, located in the control unit, with up to 512 kbytes for predefined maps, tactics and tactical scenarios. The cartridge can also store system airborne data for debriefing purposes.

The system provides a serial output for a sensor operator control keyboard and interfaces for cockpit instruments such as the BDHI.

The operational software and the interface electronics are designed in a modular form to accommodate each future change with a minimum impact on the configuration of the system.

Optionally, the system includes ground support equipment and a navigation sensors simulator for training, testing and maintenance facilities. The ground support equipment has a special software utility which allows the user to create his own tactics and maps for loading into the cartridge.

Specifications

Dimensions:
(control unit) 135 × 146 × 181 mm
(central processor unit) 194 × 124 × 320 mm
(display unit) 265 × 325 × 333 mm
Weight:
(control unit) 2.5 kg
(central processor unit) 10 kg
(display unit) 17 kg

Operational status

In production and in service in Spanish Navy Sikorsky SH-3D/G Sea King and SH-3D Sea King AEW helicopters.

Contractor

INDRA.

UPDATED

SWEDEN

Global Positioning and Communication (GP&C) system

The Global Positioning and Communication system (GP&C) is a system designed for radio transmission of position data among a large community of users. Each user broadcasts positioning data on the radio network and thus every user is able to keep track of all other users within radio range. The basic GP&C system comprises the GP&C transponder, antennas and a presentation unit.

The core of the system is the GP&C transponder which within itself comprises a satellite receiver, a radio transceiver and a communications computer. The satellite receiver is a commercially available GPS receiver, both standard and military receivers can be integrated in the transponder. The radio transceiver and the communications computer with special software are unique units specially manufactured for integration in the transponder. The special software, executed in the communications computer, implements the actual Self-Organising Time Division Multiple Access (SOTDMA) transmissions on the radio network. To achieve global synchronisation in the SOTDMA network, the time signal from the GPS receiver is used by each user to determine when data shall be transmitted on the radio network.

In the basic GP&C system position data from its own GPS receiver is both passed to the local presentation unit as well as broadcasted on to the radio network where it is received by other users. In a similar fashion, GPS position from other users is received by the radio processed by the communications computer and passed to the local presentation unit.

For efficient use of the radio channel, a specially designed version of Time Division Multiple Access (TDMA) is used, SOTDMA. During operation every unit participating in the radio network is assigned or selects a free slot which is a fixed duration of time in the SOTDMA cycle. One cycle equals 1 minute and is currently divided into 2,250 slots. A unique feature in the SOTDMA concept is the ability to handle units entering and leaving the radio network without two or more units transmitting in the same slot. An enhanced version of the software is currently under test where capacity is extended to 4,500 slots per 1 minute cycle.

Primarily designed for exchange of position information, the system is capable of transmitting all types of information embedded in GP&C messages. The system can handle 256 different types of messages of which the majority is available for the user to define in his specific application. Information that does not fit in to one slot are composed in a special long message by chaining several slots together.

In applications where enhanced position accuracy is required, differential corrections can be transmitted over the radio network, thus providing all users with a position accuracy well under 5 m.

Operational status

At present, all systems in operation are operating on the VHF frequency band, but radio units for UHF and SHF are currently being evaluated.

Contractor

CelsiusTech Electronics AB.

VERIFIED

The GP&C system gives an overview of the air traffic situation. The GP&C display is shown here in a Swedish CAA aircraft
1995

R3 GP&C (Global Positioning & Communication) transponder

The R3 GP&C transponder is designed for air-to-air and air-to-ground navigation. The transponder consists of a GPS receiver for navigation and a VHF radio which communicates with other transponders. The VHF radio link is a self-organising robust Time Division Multiple Access (TDMA) datalink synchronised by the very accurate time pulses derived from the GPS satellite system.

The datalink is capable of transmitting more than 2,000 position reports per minute. The limited radio range for VHF dictates a cellular system. This means that time slots in the TDMA datalink can be used freely in each cell. The total capacity of the system is therefore very large.

The GP&C system operates in two modes: autonomous and polled. In autonomous mode, the slot allocation process is controlled in each transponder by utilising a sophisticated algorithm to ensure that interference between slots is avoided. In polled mode, the slot allocation process is controlled by a base transponder and the mobile transponders are commanded to use a certain slot in the TDMA frame and reporting rate. Messages consist of position messages and text messages. The position messages include identity, latitude and longitude, altitude, velocity, course, navigation mode and GPS time. Text messages are 40 characters long and may be broadcast, without acknowledgement, or addressed with acknowledgement. The transponder is connected to display systems onboard aircraft and to a network on the ground for ATC services.

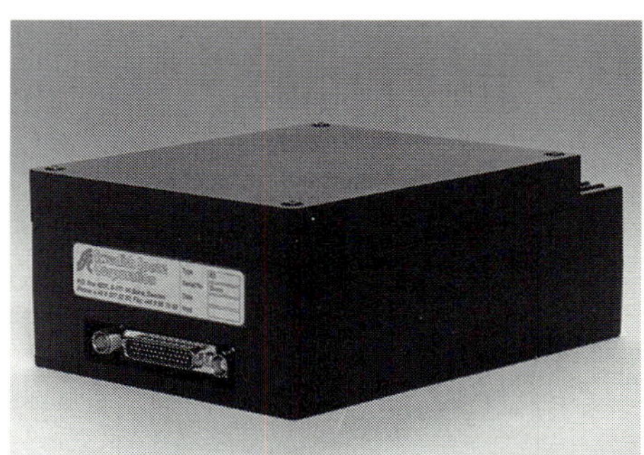

R3 GP&C global positioning & communication transponder demonstration model ***1998***/0015371

Differential corrections (DGPS) are also transmitted on this network.

Specifications

Dimensions: 152 × 88 × 210 mm
Weight: 2.8 kg
Power supply: 14 or 28 V DC
Frequency: 136-174 MHz
Channel spacing: 25 kHz

Operational status

Presently under test in several European countries for ADS-B and fleet management applications.

Contractor

GP&C Sweden AB.

VERIFIED

SWITZERLAND

ASCOT Aerial Survey Control Tool

The Aerial Survey Control Tool (ASCOT) is a GPS-supported navigation system for use in aerial survey. It provides interface to airborne equipment such as the GPS receiver, aircraft navigation system, cameras and mission recording. It offers interactive mission planning, programming of photo annotation, better survey flight navigation, automatic triggering of the camera, accurate overlaps and sidelaps, reductions in aircrew from, typically, three to two, the capability to repeat a specific portion of a flight, in-flight annotation of photos and availability of all key mission data in digital form for graphical processing, documentation and further processing.

ASCOT consists of a control computer, control station and pilot display. In addition, there are software modules for mission planning, flight execution, post flight analysis and utility programs.

The control computer is based on a powerful industrial computer designed for use under extremely demanding conditions and suited to the airborne environment. The choice of control computer takes into account aspects such as system and function reliability, safety and electromagnetic tolerance. A silicon disk has been selected as the standard data storage device, mainly because of its reliability under turbulent air conditions and in unpressurised survey aircraft at high altitudes. The architecture of the computer is designed around an Intel processor for fast real-time computations. The operating system is MS-DOS. Depending on the type of application, alternative RAM memory sizes and clock frequencies are available.

The control station is used for general management and aircraft guidance during survey missions. It can be mounted on the NSF3 navigation sight of a Wild camera system and allows supervision and control of the survey flight from the usual camera operator's seat. It can also be mounted in any other suitable location. The navigation station combines a monitor and alphanumeric control keyboard. The standard display consists of a large, flat-panel electroluminescent display with a high resolution. The bright flicker-free image is especially contrasted and clear. A liquid crystal colour display is also available. The alphanumeric keyboard is used for the control of the system, as well as for mission comments. With this set-up, the navigator can access all the functions of the system during flight.

The graphic guidance display for the pilot consists of either a liquid crystal display or an electroluminescent monitor. For aircraft with a pressurised cabin, a CRT display is available.

Contractor

LH Systems GmbH.

UPDATED

TURKEY

LN100GT EGI embedded GPS inertial navigation system

The LN100GT EGI system is an advanced technology unit that includes a sensor assembly with three Zero-Lock™ Laser Gyros (ZLGs), an A-4 accelerometer triad, and five electronics assemblies together with two spare carc slots, all installed in a small, lightweight package.

The core LN100GT INS/GPS is optimised for individual applications by appropriate additions of I/O cards and other modules (radar altimeters, dedicated mission processors, and other LRUs) installed in the spare card slots; by modification of the software for different I/O messages and formats, modified mode control, and tuning of the Kalman filter. The LN100GT is a hardware/software flexible unit which can be adapted to various air platforms without the need to change vehicle architecture.

The LN100GT provides three simultaneous navigation solutions; hybrid GPS/INS, free inertial, and GPS only. The LN100GT optimally combines GPS and INS features to provide accurate position, velocity, attitude and pointing performance as well as excellent acquistion and anti-jam capabilities. The embedded GPS module is an L1/L2 CA/P(Y) code unit capable of accepting RF (or IF) inputs from the GPS antenna system. Stand-alone GPS PVT data and stand-alone INS data are provided for integrity and fault monitoring purposes.

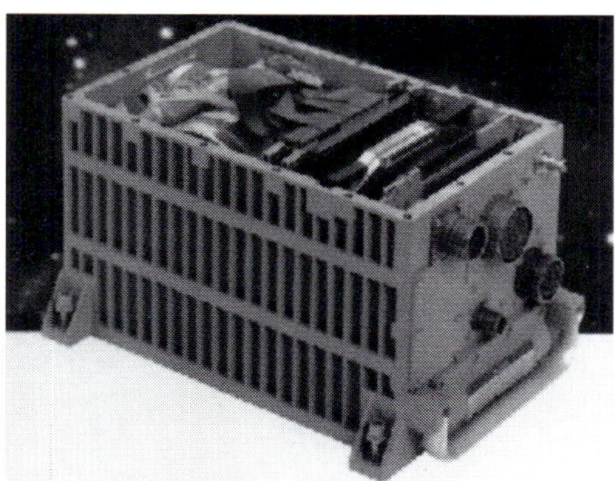

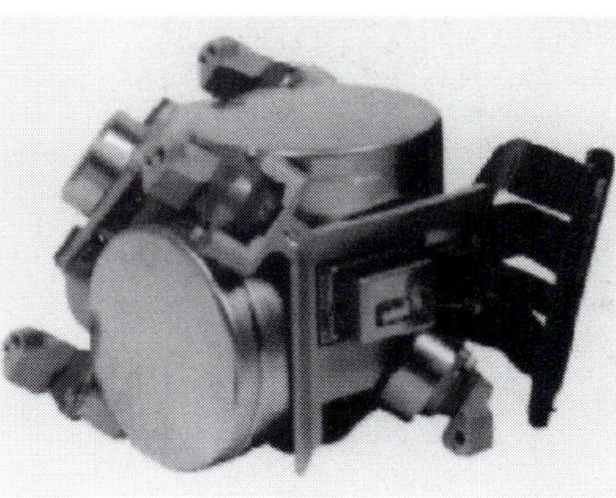

Aselsan LN100GT EGI (left) and ZLG/miniature accelerometer (right) **1998**/0015372

Specifications

Laser gyro features: (18 cm nondithered, Zero-Lock™ Laser Gyro and miniature accelerometer technology)
Accuracy:
(position) < 10 m CEP
(velocity) <0.01 m/s
Dimensions: 280 × 180 × 180 cm
Weight: 9.8 kg
Interfaces: MIL-STD-1553B/ARINC/discrete databusses

Status

In production.

Contractor

Aselsan Inc, Microelectronics, Guidance and Electro-optics Division.

NEW ENTRY

UNITED KINGDOM

LINS 300 Laser Inertial Navigation System

British Aerospace Systems and Equipment manufactures a family of LINS 300 SNU 84-1 inertial navigation equipment which meets the requirements of many different types of aircraft. LINS 300 provides vehicle acceleration, velocity, position, heading and attitude. Outputs are available in MIL-STD-1553B, ARINC 429 and synchro formats. For the EH 101 Merlin, LINS can align when airborne using a Kalman filter to interface with a Doppler velocity sensor or GPS; this also enables shipborne alignments. An ARINC 429 interface and Doppler/GPS air alignment mode are unique among SNU 84-1 format systems.

LINS 300-10 is the baseline SNU-84 system used on fixed-wing aircraft and helicopters. LINS 300-20 is the baseline ARINC 429 system with air alignment.

Specifications

Dimensions: 460 × 191 × 194 mm
Weight: 20 kg
Alignment time:
(gyrocompass) 8 min
(stored heading) 1.5 min
Accuracy:
(position) 0.8 n mile/h CEP
(velocity) (horizontal) 2.5 ft/s
(vertical) 2 ft/s
(rates) 400°/s

Operational status

In 1990 LINS 300-11 was selected as baseline equipment for the BAe Hawk 100 and 200 Series aircraft, and is in service with three air forces. In 1995 LINS 200-21 was also selected for the RAF utility variant of the EH 101 helicopter.

Contractor

British Aerospace Systems & Equipment.

VERIFIED

ADELT CPT-600/609 Automatically Deployed Emergency Locator Transmitter

The basic ADELT CPT-600 system comprises the beacon, carrier and ejection mechanism, deployment battery and control panel. The beacon is contained within the carrier and mounted externally at a suitable location. The control panel provides for testing of system integrity, system arming, crew activation, and deployment confirmation. Various remote activation devices may be incorporated in the installation design, such as frangible, float, inertia, saline, or hydrostatic switches. At least one sensor should couple with the deployment battery to avoid dependency on the aircraft electrical supply. When operated, whether by crew or remote sensor activation, a small cartridge triggers the release of two coiled springs, which eject the beacon safely away from the aircraft. Once upright and in the water, the beacon will automatically transmit homing signals on 121.5/243 MHz, while the transponder reacts to aircraft or ship's radar, pinpointing its position on the search vessel's radar screen.

A derivative of the ADELT CPT-600, the CPT-609 is comprised of a homing transmitter on 121.5/243 MHz, radar transponder on 9 GHz compatible with aircraft or ship's radar, and satellite transmitter operating on 406 MHz.

The beacon is programmed with the country code and the aircraft registration marking or radio call sign. When the beacon is activated by deployment into water, the satellite transmitter transmits its programmed information as a burst of coded signals to the orbiting COSPAS/SARSAT satellites receiving on 406 MHz. The message is stored by the satellite and downloaded to the nearest local user terminal ground station. Here the signal is processed to obtain latitude and longitude of the aircraft in distress and its identity. This in turn is passed to a mission control centre which routes the information to the rescue co-ordination centre nearest the incident, from where search and

The ADELT CPT-600 **1996**

rescue forces will be sent. Accuracy of the 406 MHz signal is to within 1 to 2 km, although testing has proved accuracies of 0.2 km can be achieved.

Once in the general vicinity of the beacon, SAR teams can locate the scene with the aid of the 121.5/243 MHz homing transmitter and the 9 GHz radar transponder which will guide SAR forces to within 10 m of the beacon.

The ADELT CPT-609 can be field programmable, greatly reducing the cost and impact of changing aircraft identity. The CPT-609 can be reprogrammed at remote locations without the need for costly and inconvenient disassembly for replacement of EPROMS.

Contractor

Caledonian Airborne Systems Ltd.

VERIFIED

770-13 panel-mounted homer

The 770-13 panel-mounted homer is intended to provide comprehensive homing facilities to an existing installation with the minimum of additional items and simple modifications to the aircraft wiring. The panel space occupied is small, the fascia measuring only 83 by 45 mm.

The instrument is designed to be fitted with a Chelton 7-25-26 antenna feed unit and a suitable pair of homing antennas. It interfaces with the existing UHF comms radio and its antenna and contains all the necessary circuitry to allow the user to home on signals received by either its own internal guard receiver or by the comms radio.

The Chelton 770-13 panel-mounted homer

The 770-13 combines the simplicity of the single box approach with the flexibility of the Chelton modular system, giving an easy-to-fit, yet sophisticated, installation.

Specifications

Dimensions: 83 × 45 × 210 mm
Weight: 0.625 kg
Frequencies:
(Channel 1) 243 MHz
(Channel 2) ±1.5% of Channel 1

Contractor

Chelton (Electrostatics) Ltd.

VERIFIED

Direction-finding system 930

The System 930 V/UHF direction-finding antennas enable a standard AM communications receiver to be converted into a 360° direction-finding system. These antennas are suitable for installation on fixed- or rotary-wing aircraft as well as maritime, land-mobile and ground installations. With the adaptor plate, P/N 23468, the 931-1 antenna is essentially a drop-in replacement for the DF301E direction finder. The 930-1 is a replacement for the AN18 direction finder.

A typical system interconnection utilises an existing transmitter/receiver. The RF changeover relay enables the transmitter/receiver to be switched between the direction-finding antenna and the normal communications antenna. The action of the direction-finding antenna is to modulate the received RF signal, the sense and depth of this modulation being related to the direction of arrival of the received signal. The receiver demodulates the modulation and passes the resulting audio back to the direction-finding antenna for processing. After processing in the antenna unit, the resulting bearing is available on an ARINC 407 output to drive a synchro indicator. An optional ARINC 429 or RS-422 databus output capable of driving an electronic display can also be specified.

The System 930 has been specifically designed to be compatible with direction-finding on pulsed tone personal locator beams (PLBs), including SARSAT beacons, when used with a SARSAT compatible receiver such as Chelton 7-28-406.

Operational status

System 930 series of direction-finding antennas have already been installed on the following platforms: Augusta AB412; BAe Nimrod; Bell 212, 214ST, 412SP and CH146; Cessna 416 and Caravan; DHC Dash 8; ECD UH1D; ECF Puma and Super Puma; Learjet; PBN Islander; PZL Mielec AN28; Shrike Commander; and Westland Lynx.

Contractor

Chelton (Electrostatics) Ltd.

VERIFIED

ACCS 3300 digital map generator

Computing Devices has designed rugged map display equipment intended for installation in harsh airborne environments. Most recently the ACCS 3300 digital map generator has been selected for the Royal Navy Merlin helicopter, under contract from Loral. This design has a scaleable architecture in order to meet a range of performance criteria, from static north-up maps to full rotating tracking maps with overlays. This will allow presentation of both raster and true vector data with efficient man/machine interface to facilitate zoom and a list of additional functions. The growth capability enables additional functions such as intervisibility, two- and three-dimensional displays. Map data is optimally compressed and stored to provide the operator with control of the map, to enable display of key map and overlay information.

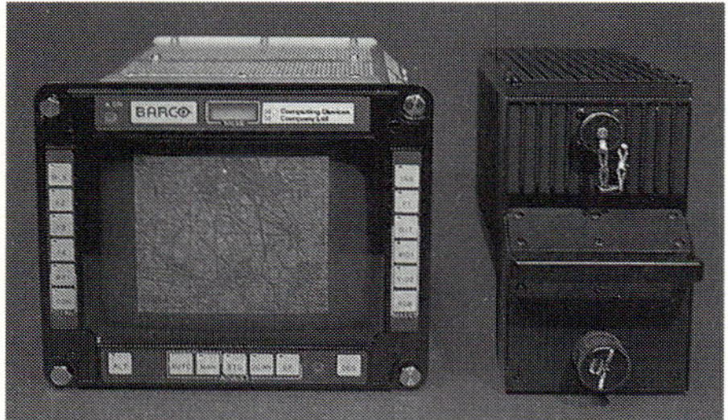

The ACCS 3300 digital map generator for the EH 101 Merlin helicopter with a Barco display
1995

The complexity of overlays can include text with sensor data, providing the ability to perform sensor and map fusion and object correlation presented on a range of low- to high-resolution displays.

Map database expertise has been achieved in compressing raster data and, more significantly where a structured dataset is required, in automatically generating vector data from a raster map using its own Unix-compatible knowledge system. The advantage of utilising vector map data is its easy interpretation, enhancement and attribution. The vector dataset and its associated overlays are more easily maintained and updated than raster.

Specifications

Dimensions: ½ ATR
Power supply: 115 V AC, 400 Hz, 110 W
28 V DC optional
Outputs: RS-170, RGB 625-line CCIRR
Environmental: MIL-STD-810E
Reliability: >2,000 h MTBF

Operational status

Selected for the Royal Navy EH 101 Merlin helicopter.

Contractor

Computing Devices Company Ltd.

VERIFIED

Terrain referenced mission systems

Research is continuing into the exploitation of terrain referencing principles and database technology used in conjunction with Electro-Optic (E-O) sensor systems. Given precise knowledge at all times of aircraft spatial position and attitude and access to a digital topographical database, it is possible to synthesise real-time perspective imagery of the terrain scene around the aircraft. This image, when properly fused with information from an E-O sensor such as a FLIR, can form the basis of a low-flying aid in situations where information from the FLIR is less than adequate. The concept is readily extended to portraying other database information such as targets, flight path obstructions and power cables. This last application is particularly attractive and is perceived as offering a major contribution to flight safety during fast jet and helicopter operations.

Terrain and obstacle scene generation combined with FLIR has recently been shown to be a viable concept. The basic philosophy is that of enhancing the FLIR image by cueing features that are not otherwise clearly visible due to the prevailing thermal conditions. Development of the concept is continuing using ground rig facilities together with flight trials on Andover and Hunter laboratory aircraft equipped with the necessary enabling technologies. On the Hunter the system is implemented as part of the PENETRATE demonstrator programme.

PENETRATE (Passive Enhanced Navigation with Terrain Referenced Avionics) is a venture to develop an integrated covert mission system exploiting terrain reference technology. The system, mounted in a small pod, comprises a mass data store and a computing suite undertaking such functions as terrain referenced navigation, terrain/obstacle scene generation, tactical routeing and mission management. The mass data store is a 400 Mbyte capacity optical disk system developed for use in a combat aircraft environment.

Operational status

Continuing research and development.

Contractor

Defence Evaluation Research Agency, Farnborough.

VERIFIED

Visually Coupled System (VCS)

A Visually Coupled System (VCS), currently under development at the Defence Evaluation Research Agency, Farnborough, will enable research into Helmet-Mounted Display (HMD) flight symbology, platform characteristics and the human factors of VCS operation at night and in adverse weather conditions. Navigation accuracies of better than 20 m are anticipated from an integrated INS/GPS/terrain referenced navigation suite.

A Lynx helicopter has been extensively modified to

accept a chin-mounted pan and tilt platform containing a thermal imaging sensor. The platform moves in synchronisation with head movements which are detected by DC head trackers, and a stabilised 1:1 thermal image of the outside world is displayed on a 53° field of view HMD, together with essential flight symbology. The symbology can be rapidly reprogrammed in order that optimised displays for specific flight conditions can be designed and proven.

The scope of the research will shortly be expanded to investigate the potential of obstacle warning devices, enabling operation at low level or close to obstructions with increased confidence.

The benefits of this research may be enjoyed by military and civil helicopter operators alike. Nap of the earth flight at night or in adverse weather allows military helicopters to operate covertly on the battlefield, increasing survivability and mission effectiveness. There are advantages for search and rescue and offshore rig helicopter operators where knowledge of the presence of sometimes highly dynamic obstructions, such as ship superstructures, is essential.

Contractor

Defence Evaluation Research Agency, Farnborough.

VERIFIED

Terrain Reference Navigation (TRN)

Terrain elevation data can be exploited to achieve automatic micro-navigation and covert terrain-following of the necessary high integrity. The micro-navigation employs modern radar altimeters incorporating variable output power and spread spectrum techniques to provide low probability of detection. In Spartan, the radar altimeter is used to map the vertical cross-section of the terrain beneath the aircraft for matching within the database. This Terrain Referenced Navigation (TRN) produces fixes every 2 seconds. The fix matching algorithm is robust, recovering quickly from the larger inertial errors likely to be met after a prolonged water crossing or from any local errors in the terrain data.

The TRN fixes are used to Kalman filter the INS outputs and thus, even with medium-grade INS, the aircraft position is known with extremely good accuracy and high confidence. With this knowledge, the database can be scanned over the projected flight path to predict the ground profile ahead. The TF algorithm is therefore able to define and demand a totally safe kinematic flight path that ensures the closest maintenance of clearance level, even in manoeuvring flight. The sudden unmasking of close terrain and ballooning over hill crests that can occur with conventional TF radars is avoided. The system also monitors achieved and forecast ground clearances against those required and gives ground proximity warning where necessary. Additionally, obstacles are included in the database and their location and indicated height projected on to the pilot's HUD or helmet-mounted display for obstacle cueing.

Specifications

Dimensions: ¾ ATR short
Weight: 15 kg
Power: 200 W
Coverage: 230,000 sq miles
Database programming: 13 min
Environmental: MIL-STD-810, MIL-STD-461 and 462
Reliability: 3,000 h MTBF

Digital colour map

Advances in techniques for lossless compression, storage and recovery of digital data, combined with developments in cockpit colour displays, have been exploited by GEC-Marconi Avionics to produce two- and three-dimensional digital map presentations with wide area coverage. Topographical and cultural data can be stored in pixel or vector formats and recalled for colour video display as realistic reproductions of paper maps or as terrain representations. Additional switchable colour palettes are available to suit different applications, for example to suit cockpit lighting conditions for night operations. Depending upon how the database is compiled, the resulting display can be decluttered of unwanted detail. The viewing area can be rotated in orientation, swiftly and smoothly manoeuvred about the stored area and viewed at any one of the standard aeronautical map scales or with a zoom capability. Additionally, on a sortie-to-sortie basis a mission routeing and intelligence overlay can be generated and loaded into the database for use in flight.

The addition of terrain elevation information to the database generates a wide range of enhanced tactical head-down display options to overlay the basic map. For example, dynamic relative height shading provides terrain-avoidance assistance, whilst ground-to-air intervisibility displays, with ground threat positions and effective ranges, aid threat-avoidance manoeuvring. Conversely, dynamic air-to-ground intervisibility displays demonstrate the achievement of terrain-masking and provide a pseudo radar display to ensure that any mapping radar transmissions are initiated only when the target or fix point is in radar view. Head-up display enhancement is achieved by the generation and precisely placed display of ridge lines to give added pilot confidence in poor visibility or when limited to flat contrast FLIR pictures.

Operational status

In service in Harrier II, Jaguar, Tornado and C-130J aircraft.

Contractor

GEC-Marconi Avionics, Rochester.

VERIFIED

PA9052SM GPS receiver

The PA9052SM GPS receiver is GEC-Marconi Electro Optics Ltd standard airborne user equipment. It provides full Precise Positioning Service (PPS) capabilities and includes all the interfaces expected of a military GPS receiver including MIL-STD-1553B, ARINC 429/575, PTTI and RS-422. Standard Positioning Service (SPS) variants are also available. As well as generating a basic navigation solution, the PA9052SM can provide area navigation facilities and is suitable for use in an integrated navigation system.

GEC-Marconi is also providing a five/six-channel GPS receiver module for the EF 2000 project. This module also provides full PPS capability and is suitable for use in embedded applications.

PA9000 Series of GPS equipment showing the airborne GPS receiver, antenna, preamplifier and control/display unit module, together with the naval receiver, control/display unit and antenna ***1995***

Specifications

Dimensions:
(PA9052 receiver) 216 × 194 × 90 mm
(PA9915 antenna) 89 mm diameter
Weight: 4.3 kg

Operational status

Deliveries of GPS receivers to customers worldwide started in 1989. They were selected for the development phase of the AH-64D Apache Longbow helicopter and are currently in service on Royal Air Force Tornado, Jaguar, Harrier and Nimrod aircraft.

Contractor

GEC-Marconi Electro Optics Ltd, Airadio Division, Portsmouth.

VERIFIED

PA9360 GPS modules

The PA9360 family of GPS modules provides a flexible solution for applications requiring an embedded GPS receiver capability. The PA9361 module is the first in the family and is designed for applications in the airborne, naval and land environments. Advanced ECL and VLSI ASIC technology has been used to achieve full six-channel, dual-frequency capability within a single module.

The PA9361 module uses software from the PA9000 Series of military GPS receivers and is suitable for embedded use in a range of navigation equipments. Installation and support are simplified by the incorporation of a PPS-SM device which ensures that the module is unclassified, even when loaded with encryption keys, and the ability to power feed a remote preamplifier via the single-cable RF input. Performance is maximised by the availability of 10 Hz GPS measurements and full navigation capability in high dynamic environments, even in the unaided mode. Extensive BITE and a high inherent reliability combine to lower the customer's support tests.

Specifications

Dimensions: 150 × 150 × 25 mm
Weight: 1.4 kg
Power supply: 5 V DC, 15 V DC or battery
Temperature range: –54 to +71°C
Accuracy:
(position) 16 m (SEP)
(velocity) 0.2 m/s (95%)
(time) 100 ns (1σ)
Reliability: >10,000 h MTBF calculated

Contractor

GEC-Marconi Electro Optics Ltd, Airadio Division, Portsmouth.

VERIFIED

FIN 1000 series inertial navigation systems

FIN 1000 is the designation of a family of inertial navigation systems based on the GEC-Marconi gimballed inertial platform using floated rate integrating gyros and precision force feedback accelerometers. The group includes particular systems optimised for long-term high accuracy and for rapid reaction alignment. Versions include:

FIN 1010 Developed for the Panavia Tornado, the FIN 1010 has all-digital interfaces and an accuracy of better than 1 n mile/h. Another version of the Tornado system, with analogue interfaces and comprehensive route navigation, is fitted to the Mitsubishi F-1 advanced trainer.

FIN 1012 Fitted to Royal Air Force British Aerospace Nimrod MR. Mk 2 aircraft, the FIN 1012 is optimised for long-term accuracy.

FIN 1031 The FIN 1031 Navigation Heading and Attitude Reference System (NavHARS) is fitted in Royal Navy British Aerospace Sea Harrier F/A-2s. The inertial platform is stabilised by two two-axis ruggedised oscillogyros developed by GEC-Marconi. NavHARS utilises other aircraft sensors which provide Doppler radar velocities, true airspeed and flux valve magnetic heading.

A two-minute alignment can be achieved on land or sea with further refinements when the aircraft is airborne.

The GEC-Marconi FIN 1031B NavHARS has been fitted as part of the mid-life update installation in the Royal Navy Sea Harrier F/A-2. It is similar to the FIN 1031 system but has two independent dual-redundant MIL-STD-1553B databusses. The addition of the MIL-STD-1553B databusses is primarily to facilitate interfacing with the Blue Vixen radar, AMRAAM and revised avionics such as the new bus control interface unit.

Specifications

Inertial platform unit
Dimensions: 212 × 215 × 332 mm
Weight: 11.9 kg

Processor unit
Dimensions: 261 × 199 × 381 mm
Weight: 13.64 kg

Control/display unit
Dimensions: 147 × 152 × 139 mm
Weight: 2.5 kg
Power supply: 200 V AC, 400 Hz, 3 phase
28 V DC for switching and lighting

FIN 1064 FIN 1064 is an integrated navigation and attack system, providing inertial navigation and a wide range of weapon delivery modes. The system has been in service since the early 1980s in the Royal Air Force Jaguar ground attack/reconnaissance aircraft, and is also fitted to Jaguars of the Oman and Ecuador Air Forces. Navigation data is provided by a FIN 1000 series gimballed inertial platform. The system was fitted in a mid-term upgrade of the aircraft systems and provides a suite of analogue and digital interfaces to the aircraft sensors and control/display systems; it also provides the capability to upgrade the system software at LRU level using a portable, solid-state, programme loader. Weapon aiming computation is provided for both air-to-air and air-to-ground modes. More recently, FIN 1064 has been updated to incorporate a MIL-STD-1553B databus, integration of inertial and GPS navigation data, and integration with the GEC-Marconi TIALD (Thermal Imaging and Laser Designator) Pod. The mission and weapon-delivery capability of the system has been significantly enhanced by in-service software upgrades, in line with the expanding role of the Jaguar aircraft.

FIN 1075 The GEC-Marconi FIN 1075 inertial navigation system was selected for the Harrier GR. Mk 5 and GR. Mk 7 and is currently in service with the Royal Air Force in both the UK and Germany. The system is form, fit and function interchangeable with the AN/ASN-130 navigation system in the US Marine Corps AV-8B. The inertial platform uses gyros and accelerometers manufactured by GEC-Marconi and interfaces with the Harrier GR. Mk 7 databus, avoiding the need for a dedicated control/display unit.

The inertial platform can be aligned on land, at sea or in flight. The ground alignment mode has a wander azimuth and does not require an initial heading input.

Specifications

Dimensions: 193 × 286 × 356 mm
Weight: 20 kg
Reaction time: <3 min
Gyrocompass align: 7 min for 0.8 n mile CEP
Accuracy:
(navigation) 0.8 n mile CEP
(heading) ±0.1°
(attitude) ±0.1°
Reliability: 1,500 h MTBF

FIN 1075G The FIN 1075G is a variant of the FIN 1075 which is designed to operate in conjunction with a stand-alone GPS receiver to provide a continuous, precision navigation solution under high dynamic conditions with significant periods of GPS outage.

FIN 1075G has been evaluated at Boscombe Down and has been used extensively in recent overseas Harrier GR. Mk 7 operations.

The same system is also capable of a GPS aided moving base alignment and is currently being assessed to provide the Harrier GR. Mk 7 with an 'at sea' alignment capability.

Contractor

GEC-Marconi Electro Optics Ltd, Navigation and Electro Optics Division, Silverknowes.

VERIFIED

FIN 1110 two gimbal inertial navigation system

The FIN 1110 is a lightweight and compact inertial navigation system for helicopters and maritime and transport aircraft. The system is contained in a ½ ATR short box weighing less than 10 kg and costing about one third the price of a full inertial navigation system. Complexity is avoided by the use of only two gimbals (hence the designation two gimbal INS) in place of the four needed in aerobatic aircraft. At the same time the system has a strapdown azimuth sensor gyro. The adoption of a gimballed platform rather than a strapdown system for this purpose has several advantages. The most significant is the rapid and independent alignment capability, the system finding true north to an accuracy of 0.4° within 4 minutes of powering up. The system is thus freed from the inherent errors caused by variations in the earth's magnetic field. Continuing alignment in the air permits operation from moving platforms such as ships.

The original concept was to incorporate sufficient computing power to permit integration with an external sensor. Since Doppler is widely used in military helicopters, this was chosen for initial trials. Working closely with UK Doppler manufacturers, GEC-Marconi modelled the potential errors in both inertial navigation and Doppler, the outcome being a 15-state Kalman filter providing a much greater degree of accuracy than could be obtained by either Doppler or INS in isolation.

In October 1986, the MoD ordered 16 FIN 1110s to operate as vertical and heading reference systems in conjunction with the Racal-Thorn Defence Searchwater radar which equips the Sea King AEW helicopters of the Royal Navy.

Specifications

Dimensions: 334 × 125 × 194 mm
Weight: 9.8 kg
Alignment time: 2-5 min

Operational status

No longer in production. In service in Royal Navy Sea King AEW helicopters to stabilise the radar antenna and improve navigation performance and in the Swedish Air Force Super Puma.

Contractor

GEC-Marconi Electro Optics Ltd, Navigation and Electro Optics Division, Silverknowes.

VERIFIED

FIN 3110G ring laser gyro INS/GPS

GEC-Marconi manufactures the FIN 3110 INS/GPS integrated system to meet the requirements of military aircraft, helicopter, self-propelled howitzers, artillery, land vehicles and marine craft.

The GEC-Marconi FIN 3110 is an Integrated Navigation System (INS) consisting of a ring laser gyro inertial sensor and an embedded Global Positioning System (GPS) receiver module. This system is capable of providing precise and continuous outputs of navigation heading and attitude data to the weapon and flight control systems.

The FIN 3110 is a small (177.8 × 177.8 × 279.4 mm), lightweight (10.5 kg) unit consuming just 55 W from a 28 V DC power source. The INS and GPS are closely integrated in a Kalman filter, which combines the high position accuracy of the GPS receiver with the angular rates and linear acceleration of the INS on a 1553B databus. Other interfaces could be available if required. The applications software is written in Ada and is executed on a Motorola 68040 processor.

The FIN 3110 is the culmination of 15 years of research and development of ring laser gyro technology. The advent of the single module GPS receiver and processor in 1993 allowed the integrated navigation system to be manufactured for the UK MoD and export programmes.

Specifications

Alignment times:
(gyrocompass) 4 min
(rapid reaction) 30 s
Inertial performance:
(position) <0.8 n mile/h CEP
(velocity) (N,E) <2.5 ft/s RMS
(velocity) (vertical) <2 ft/s RMS
GPS performance:
(position) (spherical error) <16 m SEP
(velocity) (per axis) <0.1 m/s RMS
MTBF: >5,000 h
Power: 28 V DC, <55 W
Dimensions: 177.8 × 177.8 × 279.4 mm
Weight: 10 kg
Environmental requirements:
MIL-E-5400T;
Temperature altitude operation to Class 2 (optionally Class 2X for forced cooling);
Tested to MIL-STD-810E (temp, altitude, vibration), MIL-STD-461C (EMC), MIL-STD-704E (power supply)
Interfaces:
1 or 2 dual-redundant MIL-STD-1553B (RT or bus control); PTTI/Have Quick; RF for GPS antenna; Dual RS-422 instrumentation; ARINC 429; Synchro/analogue; Panlink

Operational status

In production.

Contractor

GEC-Marconi Electro Optics Ltd, Navigation and Electro Optics Division, Silverknowes.

VERIFIED

FIN 3110 GTI

The FIN 3110 GTI (GPS, Terrain, Inertial) is a variant of the FIN 3110 navigation system designed for standoff missiles and aircraft.

The system is based on a Ring Laser Gyroscope (RLG) inertial system capable of autonomous operation to 0.8 n mile/h and includes two sophisticated Kalman filters for the provision of horizontal and vertical integrated navigation solutions using additional sensor data. In addition to the basic inertial system the unit has the capacity to be equipped simultaneously with an embedded military GPS receiver, a Digital Terrain System (DTS) and special-to-type analogue and/or digital interfaces.

The GPS receiver can be either military code or civil C/A code.

The DTS module provides both terrain referenced navigation and terrain following functions and is comprised of: terrain elevation data storage memory (EEPROM); mission specific digital terrain elevation data; data processing hardware; terrain-referenced navigation update algorithms; and terrain-following algorithms providing steering commands.

A feature of the FIN 3110 GTI is the direct use of GPS and TRN measurement data in the same integrated navigation Kalman filter to give improved performance and increased robustness through improved sensor cross-mounting.

The overall size of the FIN 3110 GTI is the same as the basic FIN 3110.

Contractor

GEC-Marconi Electro Optics Ltd, Navigation and Electro Optics Division, Silverknowes.

VERIFIED

AD380 and AD380S Automatic Direction-Finders (ADF)

Latest in a number of automatic direction-finding systems developed and produced by GEC-Marconi, the AD380 is designed to ARINC 570 and covers the frequency range 190 to 1,799.5 kHz in 0.5 kHz steps while its variant, the AD380S, covers 190 to 1,599.5 kHz. The latter additionally covers the international maritime distress frequency of 2,182 kHz with the ability to tune to ±0.5 kHz on either side of the nominal frequency, rendering it particularly suitable for search and rescue. This difference apart, both systems are designed to the same standard.

AD380 systems have automatic, crystal-controlled frequency selection and are of all-solid-state construction with instantaneous electronic tuning. Built-in test facilities are incorporated.

A range of controller options is available, permitting single, dual or programmable operation. Standard controllers come in three versions, all of which provide selection of frequency and mode, ADF, antenna, test or beat frequency oscillator, together with volume control. The decade frequency selectors display the operational frequency in 1 in (25 mm) high numerals. Frequency controls comprise three concentric knobs which operate the logic frequency circuitry.

The first type of controller is the G4032E, a single frequency version, and the second type, the G4033E, is used for tuning two receivers. Mode facilities in each case are selected by toggle switches and each controller has a test button. The third controller type is the G4034E, designed for rapid retuning of a single receiver to one of two frequencies which can then be selected by operation of a transfer switch when a white bar is displayed across the figures of the frequency not in use. Mode facilities are selected by a rotary switch.

A programmable controller, the AA-3809, conforms to the dimensions, form factor and electrical requirements of ARINC 570. It provides a four-channel preselect facility and can control both versions of the AD380. Six push-buttons permit instantaneous selection of frequencies previously entered into the memory store, which retains information when the equipment is unpowered.

The displayed frequency is always that to which the receiver is tuned and the operator can change stored information while in flight or select the 'N' button which allows normal operation of the controller. A brightness control, which is independent of the main aircraft panel lighting system, is incorporated. Normal mode facilities are also available for pilot operation.

Specifications

Dimensions:
(receiver) ¼ ATR short
Weight:
(receiver) 4.5 kg

Operational status

Over 700 ADF 380 units have been manufactured.

Contractor

GEC-Marconi Electro Optics Ltd, Sensors Division, Basildon.

VERIFIED

AD620C navigation system

The AD620C is a microprocessor-governed integrated navigation and control system based on the earlier AD620 navigation computer. It provides not only navigation facilities but also control of sensors such as Tacan and VOR/DME and, in addition, can compute true airspeed from basic air data information. TAS provides smoothing for Tacan and VOR/DME signals and dead-reckoning navigation in the absence of Tacan and VOR/DME. It can operate in polar or cartesian co-ordinates and will automatically tune a Tacan or VOR/DME when waypoints are selected. Other facilities include full slant range correction and outputs to an automatic flight control system or weapon aiming system. Fix and mark facilities are also available. Data, entered by a control and display unit keyboard, is retained when power is cut-off.

The system comprises three units: a Navigation Computer Unit (NCU), a Control and Display Unit (CDU) and a Remote Readout Unit (RRU). Each controls a microprocessor to allow simple interfacing with its companion units to form an ARINC 429 digital data highway. The inclusion of a microprocessor in the control and display unit also permits a rapid display response to keyboard input while at the same time allowing display formatting to take place within the control and display unit itself. The three units have considerable spare computer capacity for future development and growth.

Features of the navigation computer unit include an expanded memory capacity for both programme and in-flight data acceptance, improved interface circuit design for better performance and higher integrity and a new form of mechanical construction which makes each module a self-contained entity and allows testing down to component level with simple equipment. A built-in test programme monitors all input and output signals as well as the computer software.

Specifications

Dimensions:
(NCU) 194 × 127 × 318 mm
(CDU) 95 × 146 × 165 mm
(RRU) 31 × 80 × 144 mm
Weight:
(NCU) 3.5 kg
(CDU) 1.8 kg
(RRU) 0.34 kg

Operational status

No longer in production. In service in the Aermacchi MB-339A trainer.

Contractor

GEC-Marconi Electro Optics Ltd, Sensors Division, Basildon.

VERIFIED

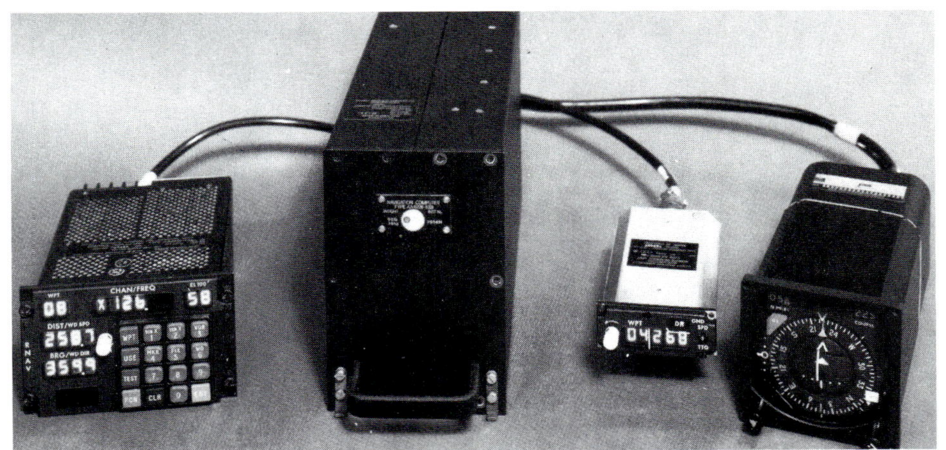

The AD620 integrated navigation and control system

AD620K integrated navigation system

Designed for the single-seat Aermacchi MB-339C Veltro light strike aircraft, the AD620K is a development of the earlier AD620C navigation system provided for the two-seat Aermacchi MB-339A trainer. The AD620K has an improved performance compared with earlier versions owing to the addition of an AD660 Doppler velocity sensor and associated processing and displays. The system operates in conjunction with an inertial platform, nav/attack information and guidance being presented on a wide-angle head-up display. The navigation computer can accept data from a wide variety of navigation sensors together with an attitude and heading reference system, to compute present position.

Operational status

The system was selected for the MB-339C in October 1984 and is in service with the Royal New Zealand Air Force.

Contractor

GEC-Marconi Electro Optics Ltd, Sensors Division, Basildon.

VERIFIED

AD660 Doppler velocity sensor

Tha AD660 Doppler sensor is packed into a single unit containing the antenna, transmitter/receiver, tracker, and digital and analogue inputs and outputs, all under microprocessor control. The transmitter is a Gunn diode generating about 200 mW of power at 13.325 GHz. The signal is frequency modulated, the actual frequency being selected by the microprocessors to give the best return signal in any set of conditions, eliminating for example height hole effects. The system employs four beams, any three of which provide satisfactory performance, the fourth being used to enhance operation at extremes of altitude and to provide a self-check function. The standard output for groundspeed and drift angle is an ARINC 429 serial digital data highway, but ARINC 582 and 561 digital and ARINC 407 synchro outputs can also be provided.

Specifications

Dimensions: 379 × 237 × 132 mm

Operational status

No longer in production. The AD660 is a standard option on Boeing 727 and 737 transports and has been adopted by British Airways and Lufthansa among other operators. It has been selected by CASA for its C-101DD Aviojet advanced trainer/ground attack aircraft.

Contractor

GEC-Marconi Electro Optics Ltd, Sensors Division, Basildon.

VERIFIED

AD2770 Tacan

The AD2770 Tacan navigation and homing aid is suitable for all types of aircraft. This system is now used on the majority of front-line aircraft in service with UK forces. It provides range and bearing information from any selected ground Tacan station or from any suitably equipped aircraft and is available in a number of forms offering outputs in digital ARINC 429 or analogue form or a combination of the two. Output signals may be provided to drive range/bearing or deviation indicators on a pilot's panel or to interface directly with a computer.

The system has full 252 channel X and Y mode capability and operates up to 300 n miles range with a range accuracy of better than 0.1 n mile. Bearing accuracy is said to be better than 0.7° on normally strong input signals. It comprises two units: a transmitter/receiver hard-mounted in the avionics bay and a panel-mounted remote-control unit. A switching unit, also installed in the avionics bay is required when two antennas are fitted. A mounting tray with a cooling air blower is necessary if a cooling air supply is not available from the aircraft's own air-conditioning system.

The transmitter/receiver section with a digital interface only is contained in a ¾ ATR short case with a front doghouse. Versions with analogue outputs are accommodated in a case of longer dimensions to house the additional circuitry. Signal processing circuitry is largely digital in the interests of system reliability, and continuous integrity monitoring techniques eliminate the risk of erroneous outputs.

Range and bearing analysis and output formats are prepared in a general purpose computer module called the analyser. This allows both range and bearing signal processing to use the same circuitry, with a consequent reduction in the number of components. The range system uses a parallel search method, said to be unique, which by making use of all signal returns achieves a very rapid lock on.

The AD2770 system operates in three modes: receive (giving bearing information only); transmit and receive; and air-to-air (providing range information only). Transmitter frequency range is from 1,025 to 1,150 MHz with an output of 2.5 kW peak pulse power. Receiver frequency coverage is from 962 to 1,213 MHz. The tracking speed range is from 0 to 2,500 kt.

The design of the AD2770 system is flexible both electrically and mechanically and alternative configurations with appropriate form factors, output characteristics and mechanical and electrical interfaces can be provided for new aircraft or for retrofit.

Specifications

Dimensions:
(control unit) 57 × 146 × 83 mm
(transmitter/receiver standard unit) 194 × 191 × 380 mm
(antenna switch) 69 × 130 × 56 mm
Weight:
(control unit) 0.45 kg
(transmitter/receiver standard unit) 14 kg
(antenna switch) 0.25 kg

Operational status

In service in the Tornado GR. Mk 1 and F3, Nimrod MR. Mk 2 and Sea Harrier. Currently in production for BAe and for Boeing Helicopters Division.

Contractor

GEC-Marconi Electro Optics Ltd, Sensors Division, Basildon.

VERIFIED

AD2780 Tacan

A follow-on from the AD2770 series supplied for the Panavia Tornado and other front-line types, the AD2780 is also proposed for military applications and is, says GEC-Marconi, lighter, smaller and less expensive than its predecessors, with extensive use of large-scale integration and microprocessor technology. The system provides slant range and relative bearing to a standard Tacan station, range rate (which approximates to groundspeed when not used in conjunction with the company's area navigation system), time to go to waypoint or station, ARINC 429 serial data output and 252 channels in X and Y modes. Outputs of range are available in digital format to ARINC 429 and in analogue to dial and pointer displays.

Specifications

Dimensions: 127 × 153 × 318 mm
Weight: 3.5 kg
Frequency:
(transmitter) 1,025-1,150 MHz
(receiver) 962-1,213 MHz
Range rate output: 0-999 kt with accuracies ±15 kt for 0-300 kt, and ±5% for 300-999 kt
Time to station output: 0-99 min
Tracking speed: 0-1,900 kt, 0-20°/s
Memory:
(range) 10 s
(bearing) 4 s

Operational status

In production and service in the British Army Gazelle and Royal Netherlands Air Force Eurocopter BO 105 helicopters and Royal Air Force Shorts S312 Tucano trainers.

Contractor

GEC-Marconi Electro Optics Ltd, Sensors Division, Basildon.

VERIFIED

NavSymm Sharpe XR6 12-channel GPS receiver

The NavSymm Sharpe XR6, 12-channel receiver, is a totally redesigned unit based on the earlier NavSymm XR5 series; it offers much improved levels of accuracy and performance. The NavSymm Sharpe XR6 receiver provides 2-way communication on all three ports for transfer of information at up to 20 Hz. Satellite tracking reacquisition is achieved in under one second, at acceleration rates of up to 4 g. It also incorporates an event marker allowing other equipment to demand position information. The event marker is activated in one of two ways, either by the arrival of a pulse or an ASCII string. In the latter case, the position information is incorporated with the ASCII string and either output on a data port or stored in the internal memory. Uplink of Ephemeris data is also possible via a datalink using RTCM messages thus ensuring optimum performance in highly-dynamic situations.

The NavSymm Sharpe XR6 has been designed with open system architecture, offering access to many types of raw data for system integration purposes. It can also be used as a base station for the purpose of RTCM messages.

Specifications

Receiver: 12-channel C/A code, L1 frequency
Update rate: 10 Hz
Max speed: 1,000 kt
Acceleration: 4 g
Time to first fix: 20 s (with current Ephemeris)
Accuracy RMS (PDOP<3):
(position) <15 m stand alone
(with DGPS) <2 m beacon
Velocity: 0.03 m/s (differential mode)
Time output: 1 pps +/−100 ns
Latency, navigation mode: 80 ms
Dimensions: 175 × 80 × 57 mm
Weight: <1 kg
Power: 11-32 V DC; 8 W

Operational status

Available.

Contractor

Navstar Systems Ltd.

NEW ENTRY

NavSymm Sharpe XR6 12-channel GPS receiver
1998/0015373

AMS 2000 multifunction Control Display Navigation Unit (CDNU)

The AMS 2000 multifunction CDNU is a flexible navigation and management system with embedded P(Y) code GPS that is easily configurable to suit the customer's requirements and the host airframe. Typically Search and Rescue (SAR) specific functions are added for aircraft/helicopters, having a SAR role.

This CDNU is configured as a navigation computer using GPS and, if available, a combination of sensors including IN, Doppler and Air Data. From this baseline, the CDNU capability may be increased to meet additional customer requirements - thanks to the modular approach adopted for hardware and software. The CDNU can provide a full mission management facility in single or dual configuration using MIL-STD-1553B and ARINC 429 databus.

The AMS 2000 CDNU provides centralised control and display of the chosen avionics suite. Data and instructions are inserted manually via the keyboard or Data Transfer Device (DTD). The DTD is used for flight planning and for post-flight data retrieval. A customised mission planning station is available.

The CDNU software is written in Ada. The non-volatile Flash memory is reprogrammable via the DTD. The CDNU's large memory capability allows a wide range of equipment interfaces to be supported - in addition to navigation sensors.

The AMS 2000 multifunction CDNU is compatible with GNSS and provides a control and display function for the Racal Avionics Satellite TRansceiver (STR) system. The CDNU may be used as a cockpit mission system or a tactical system, without any additional interfacing hardware being required.

Specifications

Size: 161.43 × 145.5 × 216.5 mm
Weight: 5 kg (max)
Power: 28 V DC, 45 W (max - for full module complement), typically 36 W
Cooling: convection - no forced air required
Environmental: MIL-STD-810E; MIL-STD-461C/D RTCA/DO-160C
Memory: 4 Mbyte Flash PROM
I/O: MIL-STD-1553B; ARINC 419/429 (Hi/Lo) RS-232, -422; DC, AC, synchro, discretes
GPS: embedded C/A or P(Y) 5 channel to STANAG 4294. Includes Have Quick, BID 250 and KYK 13 interface

Operational status

Under contract to UK MoD for Sea King Mk 3, 4, 5 and 6; Lynx MHA8; Merlin Mk 1 (RN); and Nimrod.

Contractor

Racal Avionics Ltd.

UPDATED

CDU/IN/GPS

Racal Avionics is responsible for full integration of an Inertial Navigation and Global Positioning System (IN/GPS) and a Control Display Unit (CDU) to interface with the Central Tactical System of the UK Nimrod MRA4 aircraft. The CDU concerned is a derivative of the AMS 2000 Control Display and Navigation Unit, made by Racal Avionics, and the combined unit is to be known as the CDU/IN/GPS.

Operational status

In development for Nimrod MRA4.

Contractor

Racal Avionics Ltd

NEW ENTRY

Doppler velocity sensors

There are five models in the Racal Doppler velocity sensor family: the Doppler 71 and 72 antenna units for helicopters and fixed-wing aircraft respectively; the Doppler 80 for helicopters and light aircraft; and Doppler 91 and 92 antenna units for helicopters and fixed-wing aircraft respectively.

The Doppler 91 and 92 units use the best features of the previous models and incorporates microprocessor technology and new manufacturing techniques. These units use waveguide antennas and varactor multiplier transmitters for high accuracy at greater altitudes.

The Doppler 80, with printed antennas Gunn diode RF source and switched beams, is for low-level helicopters operations where low weight is particularly important.

Antenna units for helicopters have a speed range of −50 to +300 kt forward and 100 kt laterally. The fixed-wing sensors have a corresponding speed range of −50 to +1,000 kt and 200 kt laterally. Velocity data can be provided in either analogue or ARINC 429 digital format and MIL-STD-1553 databus may be specified. The microwave signals produced by the Doppler 90 series are specially tailored to reduce errors created by heavy rain, snow and hail.

The transmission characteristics of the Doppler 91 and 92 units can be remotely controlled. A low-power stealth mode can be selected for minimum detectability, or the transmitter switched off if the beam goes above the horizon, for example when the aircraft is banking. They transmit information to other aircraft systems in ARINC 429 serial digital data form, but the MIL-STD-1553B remote terminal format can also be supplied.

The Doppler 91 can be supplied in a configuration that is optimised for rotary wing over-water operations, where transition to hover and auto-hover are autopilot functions dependent upon the maximum amount of continuously available velocity data.

Specifications

Dimensions:
(Doppler 71/72) 406 × 406 × 127 mm
(Doppler 80) 356 × 381 × 80 mm
(Doppler 91/92) 358 × 391 × 118 mm

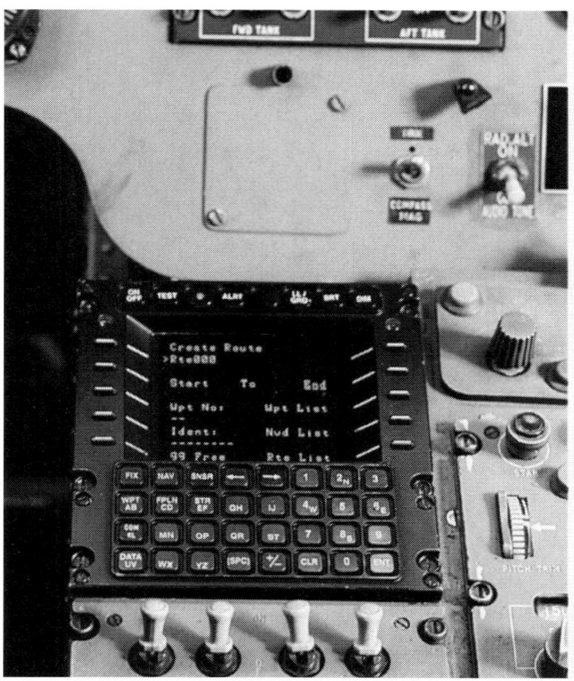

AMS 2000 CDNU

1997/0002442

Weight:
(Doppler 71/72) 16.5 kg
(Doppler 80) 8.6 kg
(Doppler 91/92) 11 kg
Power supply:
(Doppler 71/72) 115 V AC, 400 Hz
(Doppler 80 and 91/92) 28 V DC

Operational status

In production. The Doppler 91 has been selected for the EH 101, Sea King, and Lynx helicopters. The Doppler 71 is installed on Royal Air Force Aerospatiale/Westland Puma, Sikorsky/Agusta SH-3D, Westland Sea King and Royal Navy and British Army Lynx helicopters and Belgian Air Force Sea King Mk 48s. Over 3,500 Doppler 71s are in service.

Contractor

Racal Avionics Ltd.

UPDATED

LWCCU-LightWeight Common Control Unit

The LWCCU is based on the AMS 2000 CDNU (Control Display and Navigation Unit). The LWCCU is the main pilot/co-pilot interface with the aircraft's mission, tactical, navigation and communications systems. It provides significant airframe weight saving (9.8 kg), as well as increased reliability and maintainability over earlier systems.

Racal Avionics will install its LightWeight Common Control Unit into Royal Navy Merlin and Royal Air Force EH 101 support helicopters **1998**/0015374

Operational status

Selected by the UK MoD for the Royal Navy Merlin Mk 1 helicopters and the Royal Air Force EH 101 support helicopters. Four LWCCUs will be fitted in each of the Royal Navy's 44 Merlin Mk 1 helicopters: two for the pilot/co-pilot and one each for the mission equipment operators in the cabin. Each of the Royal Air Force's 22 Merlin Mk 3 helicopters will carry only one LWCCU, in the cabin, for use in a maintenance role.

Contractor

Racal Avionics Ltd.

NEW ENTRY

RNav 2 navigation management system

Certificated by the UK Civil Aviation Authority in 1984, and now TSO C129 A/C115 B compliant, the RNav 2 navigation management system is in widespread use by helicopter operators supporting the North Sea oil industry and search and rescue, corporate and special mission operators worldwide. It is also in use in military rotary- and fixed-wing aircraft where the operating environment does not justify the expense of extended military environmental specifications.

The equipment comprises a Control and Display Unit (CDU) and a Navigation Computer Unit (NCU), and can accept inputs from GPS, Decca, VOR/DME, Loran C and Doppler. Four separate navigation plots are maintained within the system, any of which may be selected for guidance. In the latest systems any navigation plot with temporarily invalid sensor input is able to revert to Doppler updating. With certain sensor combinations a total of five input sensors is therefore possible.

The computer provides a variety of display and guidance outputs. Analogue outputs are available for the retrofit installation whilst digital databusses permit interfacing with EFIS and digital AFCS. Inputs of fuel flow can be accepted from a variety of flowmeter types. Fuel computations include range, endurance, fuel remaining over each waypoint and at destination and estimates of these values when a helicopter is in the hover.

A choice of CDU types is available with letters first or numbers first alphanumeric keys, Gen II or Gen III NVG capability and compatibility with Health and Usage Monitoring Systems (HUMS).

NCU hardware variants provide for interfacing with a variety of VOR/DME types, HSIs and RMIs. Software variants offer options such as grid navigation, vertical navigation, transition down and compatibility with the Eurocopter Super Puma Mk II integrated flight data system.

RNav 2 is operating worldwide using GPS as one of its input sensors. It will interface with GPS receivers which conform to ARINC 743. It forms the core of search and rescue systems certificated by a growing number of airworthiness authorities for IFR use in helicopters such as the Eurocopter Super Puma and the Sikorsky S-76. It is also in demand by fishery and environmental protection agencies in a variety of fixed-wing aircraft.

Specifications

Dimensions:
(control/display unit) 146 × 114 × 208 mm
(navigation computer unit) 124 × 184 × 324 mm
Weight:
(control/display unit) 2.7 kg
(navigation computer unit) 5.5 kg

Operational status

In production. Used in large numbers for offshore oil support. Military users include the UK MoD, United Arab Emirates AB-412 SAR and Royal Norwegian Air Force Sea King Mk 43 update.

Also fitted to fixed-wing aircraft such as Dornier 228s of the UK Ministry of Agriculture, Food and Fisheries, Cessna Caravan IIs of the Scottish Department of Agriculture, Food and Fisheries, Dutch police Turbine Islanders and British Army Defenders.

Contractor

Racal Avionics Ltd.

UPDATED

RNS 252 navigation system

RNS 252 is a single-unit panel-mounted navigation computer which can accept inputs from Doppler and one additional sensor such as GPS or Loran C. Sensor control is exercised through the computer keyboard and sensor data is accessed on the computer's dot matrix display.

The system accommodates 200 waypoints which are numbered but may also carry a five-character ident. Any

The display for the Racal Avionics RNav 2 area navigation system

or all waypoints may be vectored if required. Waypoints may be loaded manually through the keyboard or automatically through a data transfer device. A variety of steering modes is available. Steering guidance is available in the basic form of steering arrows on the computer display, but outputs are available for driving instrumentation and both AC and DC analogue autopilots. A weather radar output provides a navigational overlay for use with digital colour radars.

RNS 252 is compatible with Racal Avionics Type 70, 80 and 90 Doppler sensors and with GPS sensors which conform to ARINC 743.

Specifications

Dimensions:
(standard version) 124 × 146 × 201 mm
(Supertans version) 161 × 146 × 240 mm
Weight:
(standard version) 3.5 kg
(Supertans version) 4.2 kg
Power supply: 28 V DC, 40 W

Operational status

In production. In service in fixed- and rotary-wing aircraft worldwide including British Antarctic Survey Twin Otters, Indonesian Army BO 105s, Norsk Luftanbulanse BK 117s and Royal Moroccan Air Force Puma and Gazelle helicopters. Versions with GPS are in service in Royal Air Force Chinook and Puma, Royal Navy Sea King and British Army Lynx helicopters.

Contractor

Racal Avionics Ltd.

UPDATED

Supertans integrated Doppler/GPS navigation system

The Supertans integrated Doppler/GPS navigation system is a replacement for the TANS series of equipments. Based on the RNS 252 (see previous item), it combines a six-channel GPS receiver with any of the Racal family of Dopplers. The package has been designed as a drop-in replacement for current TANS equipments.

Supertans uses all existing connectors and cables, with only minimal additional cabling required for the GPS installation. The system maintains and enhances all the present capabilities of the TANS/Doppler system while adding the precise navigational accuracy of GPS.

Specifications

Dimensions:
(Supertans) 240 × 161 × 146 mm
(GPS receiver) 368 × 197 × 57 mm
(antenna) 102 × 10 × 95 mm
Weight:
(Supertans) 4.2 kg
(GPS receiver) 2.75 kg
(antenna) 0.5 kg
Power supply: 28 V DC, 40 W

Operational status

In series production and in operational use in Royal Air Force Chinook, Royal Navy Sea King and British Army Lynx AH. Mk 7 helicopters.

Contractor

Racal Avionics Ltd.

UPDATED

LandStar GPS

The Racal Survey LandStar satellite-based differential GPS system is available in Europe, Middle East, Southern Africa, North and South America, Australia, New Zealand and Indonesia. LandStar's precise positioning capability may now be used in a wide range of applications in these regions including surveying, mineral exploration, forestry and agriculture and also for transport and emergency services.

The new European service follows the launch of Italsat F2 communications satellite which transmits differential corrections to users throughout an area from the Western Atlantic to East of Moscow and from the Straits of Gibraltar to Scandinavia.

Users of the new LandStar America system receive differential GPS corrections via the AMSC (American Mobile Satellite Corporation) L-band (NATO D-band) spot beam geostationary satellite. Differential corrections are obtained from a network of 11 precisely co-ordinated reference stations which ensure effective accuracy and coverage throughout the region. The data is gathered at Racal's LandStar hub in Houston where it is processed and monitored for quality control. It is then uplinked to the satellite via the AMSC facility in Reston, Virginia.

The Australian LandStar DGPS service uses the Optus B1 geostationary satellite to broadcast the differential corrections generated at nine reference stations over the Australian continent. By increasing the transmitted signal strength Racal has extended the coverage to include the additional countries of Indonesia, New Zealand and Papua New Guinea.

The LandStar differential GPS service was developed by Racal from experience gained on SkyFix, which is the world's most extensive DGPS network. The power of the LandStar satellite spot beam transmissions have made it possible to incorporate the LandStar receiver and antenna within a small, lightweight package suitable for backpack, vehicle and light aircraft applications. The unit processes and receives the differential messages, including wide area multireference signals, together with proprietary quality control and other messaging. Data is provided in industry standard RTCM Sc 104 format so that LandStar can be used with virtually every differential capable GPS receiver. Accuracies achieved are typically better than 1 m.

LandStar differential GPS supporting geomagnetic surveying **1996**

LandStar receivers are available in an OEM or a stand-alone form which may have a high-quality GPS unit fitted as an option.

Operational status

Available.

Contractor

Racal Survey Ltd.

UPDATED

Microwave Aircraft Digital Guidance Equipment (MADGE)

The Microwave Aircraft Digital Guidance Equipment (MADGE) is a military landing system, which can be used to provide night and all-weather landing guidance and terminal area navigation information, for all types of fixed-wing aircraft and helicopters under the strictest emission control conditions, to reduce detection by the enemy. The system consists of ground and airborne equipment. The ground equipment may be installed on board a ship, in addition to its use for tactical operations in the field.

MADGE can be used for multiple simultaneous approaches and is passive unless correctly interrogated. It includes a two-way datalink between aircraft and ship or ground and selective aircraft identification. Navigation information is provided to an aircraft out to a range of 30 n miles (55.6 km) and landing guidance is provided out to 15 n miles (27.8 km) on ILS or Automatic Carrier Landing System (ACLS) indicators.

Aircraft equipment consists of the pilot's controller, logic unit, transmitter/receiver, two antennas and an antenna switching unit, plus the use of the aircraft normal instrument display. All equipment operates from 28 V DC.

Specifications

Dimensions:
(controller) 65.5 × 146 × 84 mm
(logic unit) 194 × 91.9 × 384 mm
(transmitter/receiver) 198.9 × 91.4 × 383.5 mm
Weight:
(controller) 0.8 kg
(logic unit) 4.9 kg
(transmitter/receiver) 5.7 kg

Operational status

In service in Royal Navy British Aerospace Sea Harrier aircraft.

Contractor

Racal-Thorn Defence.

VERIFIED

GAS-1 GPS adaptive antenna system

The GAS-1 system protects GPS systems from deliberate countermeasures and radiated interference. Using a Controlled Reception Pattern Antenna (CRPA) and the associated Antenna Electronics (AE), the system employs a null steering technique to provide protection against six jamming signal sources. The speed of null steering is greater than the dynamics of any known production or developmental aircraft platform, ensuring that the tracking of jamming sources is not disrupted by aircraft manoeuvres.

The system provides protection for multiple GPS based navigational/targeting aids including inboard stores and weapons. All up system weight is less than 9.1 kg.

Operational status

The GAS-1 system is a derivative of equipment already in service with air forces in the UK and Australia, and GAS-1 has been selected to enter service with the US Air Force from early 1998.

The GAS-1 system replaces the earlier STR-2200 CRPA and STR-2400 ACU models, and is compatible with both RF and IF variant GPS receivers.

Contractor

Raytheon Systems Limited, Electronic Systems Division.

UPDATED

STR 2515 series receiver processor unit

The STR 2515 series receiver processor units are high-dynamic, five- or six-channel, ¼ ATR short units. The RPU provides Standard Position Service (SPS) C/A code and, to US DoD authorised users, the very accurate Precise Position Service (PPS) P/Y code. A major advantage of the STR 2515 Series is that the C/A code version of the RPU can be simply upgraded at the factory to P/Y code once the customer has an agreement with the US DoD.

The STR 2515 Series features separate L1 and L2 channels, for higher resistance to jamming, and five or six dedicated receiver channels. It is capable of handling all selective availability and anti-spoofing aspects required by US DoD authorised users and is the first European GPS RPU to be unclassified when keyed. It has parallel operation of MIL-STD-1553B, ARINC 429, ARINC 572 Gray coded baro altitude, RS-422 and PTTI/Have Quick interfaces and operates with either FRPA or CRPA systems without modification.

A Kalman filter is available which, when embedded in the host GPS, helps to overcome the problems associated with close coupled navigation solutions and the security implications associated with the export of GPS corrected pseudo-range and range rate data. Kalman filter solutions provide a flexible method of calibrating inertial, altitude and heading reference systems and Doppler, or any combination of all three.

Specifications

Dimensions: 320 × 60 × 194 mm
Weight: 4 kg
Power supply: 115 V AC, 400 Hz or 28 V DC, 40 W
Reliability: >7,000 h MTBF

Operational status

More than 2,100 units have been sold worldwide to date.

Contractor

Raytheon Systems Limited, Electronic Systems Division.

UPDATED

SAR homing systems

Series 406

Fully compatible with the latest 406.025 MHz COSPAS/SARSAT emergency locator beacons, the 406 system is a complete self-contained unit interfacing with a single pair of antennas to provide 'left/right' steering information against a transmission source.

Designed to monitor four distress frequencies: 121.5, 156.8, 243.0 and 406.025 MHz, the unit processes the information and displays it on an analogue indicator. Adjacent-frequency test modes are provided.

Specifications

Radio frequencies: 121.5, 156.8, 243.0, 406.025 MHz
Dimensions: 146 × 66.6 × 155 mm
Weight: 0.9 kg

Series 406 derivatives

406-1 Homer: The initial system covers 121.5, 156.8, 243.0 MHz. Fitted to both rotary- and fixed-wing aircraft.

406-2 Homer: Identical in function to the 406-1 Homer, with the addition of 406.025 MHz capability. Both standard and night vision compatible; additional remote indicators optional.

406-3 Homer with extended frequency range capability: With a remote controller can be extended over the whole 100-400 MHz band, whilst retaining the 406-2 Homer features. Can be interfaced to ARINC 429 or 1553 databus.

406-053 Full NVIS compatible homing system: Four frequency capability: 121.5, 156.8, 243.0,

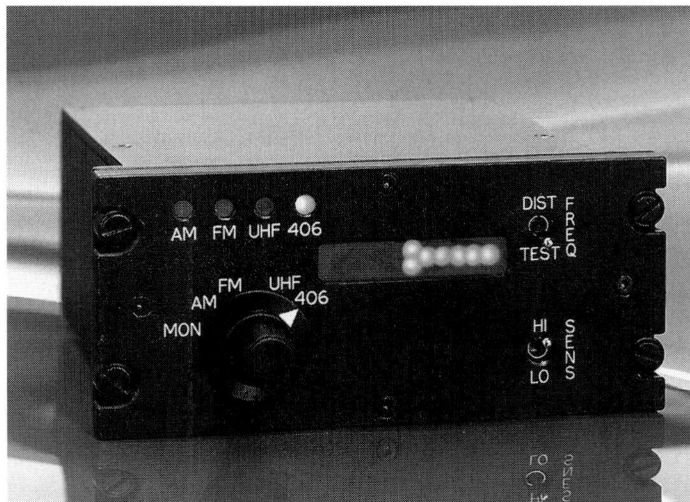

Series 406-053 NVIS-compatible homing system **1997**/0001181

RAF Rescue helicopter with Series 406 Homer and Series 500 personal locator beacon
1997/0001180

406.025 MHz, plus additional frequencies at ±1 MHz of each distress frequency. 5 V or 28 V Green NVIS; choice of white or red lighting for non-NVIS units.

Series 407
The series 407 remote indicators are for use where operators need a lightweight indicator either as a secondary instrument for navigators' use or where panel space precludes use of the 406 series. The 407 provides a visual indication of the 'left/right' steering information to track a transmission source, the directional data is indicated only when a valid 'homing' signal is received from the 406 homing unit.

Specifications
Dimensions: 86 × 48 × 123 mm
Weight: 0.3 kg

Contractor
Techtest Limited.

VERIFIED

UNITED STATES OF AMERICA

Apollo 360 round GPS

The Apollo 360 round GPS moving map display slides into a standard 3⅛ in instrument hole. The back-lit LCD display can be configured to display standard numeric information or a moving map which shows position relative to flight plan and track, airspace boundaries, and nearby waypoints. Fitted as standard is an extensive Jeppesen database that includes public use airports, VORs, NDBs, intersections, and all special use airspaces. Satellite tracking is provided by a six-channel parallel sensor. Both the database and the operating software can be updated by means of a serial data port without removing the unit from the cockpit panel. Coupled with the optional Apollo six-channel GPS receiver the 360 map can be transformed into a GPS navigator.

Specifications
Dimensions: 3⅛ in (79 mm) (diameter), 200.4 mm (deep)
Weight: 1.36 kg
Power supply: 10-40 V DC, 3 W nominal
Temperature range: −10 to +55°C
Altitude: up to 55,000 ft
Accuracy: 15 m RMS (100 m, DRMS with S/A)

Contractor
II Morrow Inc.

VERIFIED

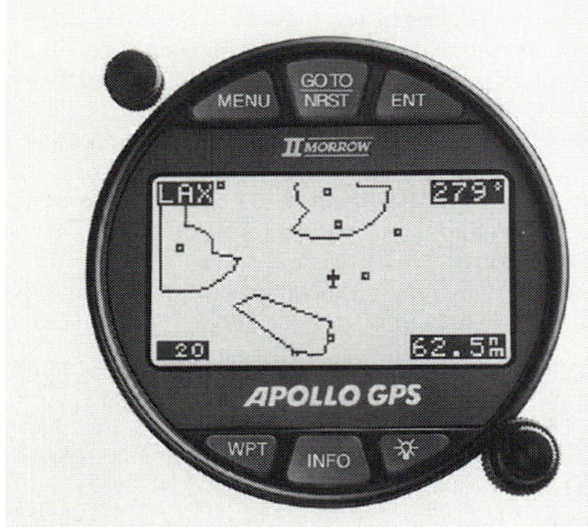

The Apollo 360 round GPS **1996**

Apollo 2001 GPS Navigation Management System (NMS)

The Apollo 2001 GPS NMS is a modular system which the pilot customises to his particular requirements. The current configuration allows for input from Loran C and eight-channel built-in GPS position sensors and a fuel flow/air data sensor which provides desired heading, winds aloft, TAS and fuel management. All navigation information is displayed on an LCD panel mount navigation management computer. Sensors are mounted remotely. Features include direct-to navigation, datacards with airports, VORs, NDBs, intersections, waypoint information, MSA/MESA information and airspace for the designated region, alerts for altitude deviation and VNav, 10 20-leg flight plans and emergency search. Apollo 2001 has built-in eight-channel parallel GPS capability and RAIM. It is certified TSO C129 for en route, terminal and non-precision approaches.

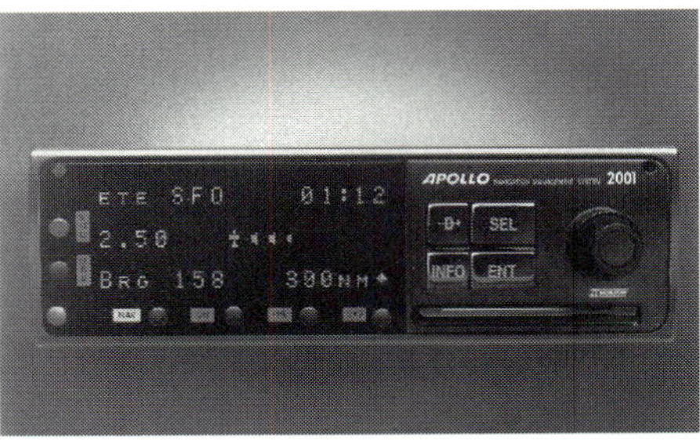

The II Morrow Apollo 2001 GPS navigation management system

Specifications
Dimensions:
(navigation management computer) 159 × 51 × 267 mm
(multichain Loran sensor) 38 × 160 × 295 mm
(global positioning sensor) 57 × 160 × 295 mm
(fuel flow/air data sensor) 57 × 160 × 295 mm
Weight:
(navigation management computer) 1.5 kg
(multichain Loran sensor) 1.6 kg
(global positioning sensor) 1 kg
(fuel flow/air data sensor) 1.7 kg
Waypoint capacity: selection of user-replaceable datacards plus 200 user definable

Operational status
VFR/IFR approved by Transport Canada.

Contractor
II Morrow Inc.

VERIFIED

APOLLO 2101 GPS Navigation Management System (NMS)

The Apollo 2101 GPS NMS is certified TSO C129 for IFR en route and non-precision approach. Designed for fixed- or rotary-wing aircraft, commercial, corporate and military. It features a full alphanumeric keyboard and high-contrast LED, with NVG option.

The integrated database provides instant data on airports, VORs, NDBs and intersections. GPS approach data is automatically displayed for destination during flight planning.

The Apollo 2101 fits any dzus rail and requires no external OBS wiring. The NAVNET interface, high-speed serial bus, and ARINC-429/-561 support provides compatibility with CDI, MSI and moving map displays as well as air-data, ACARS and EFIS systems.

APOLLO 2101 GPS NMS **1997**/0001334

C115A multisensors certification allows use of additional positioning and data systems to improve safety. The Apollo 2101 features eight-channel parallel GPS and RAIM.

Specifications
Dimensions:
(2101 NMS computer) 146 × 76 × 135 mm
(2102 keypad) 146 × 38 × 139.7 mm
Weight: 1.7 kg
Accuracy:
(horizontal) 15 m RMS (100 m at 2 Drms with S/A)
(vertical) 156 m at 2 Drms with S/A (velocity) 0.5 m/s

Operational status
In production.

Contractor
II Morrow Inc.

VERIFIED

Apollo Loran C receivers

II Morrow manufactures a range of Loran C receivers including the model 604, model 612 and model 614. Each provides Loran C navigation with outputs on an LED display, with bearing and range, groundspeed, estimated time en route, cross-track error, ground track angle and latitude and longitude being selectable. Most systems can be loaded with a library of navigation aids and airport data within the USA and Canada, via a solid-state cartridge memory device. The model 614R is a two-box system with separate control display and receive/control units.

The 618 series receivers have LED displays and are available in three configurations: 618 standard, 618C round and 618R dzus-rail mount. Features include a built-in database of airports, VORs, NDBs and intersections in Canada, the Caribbean, Central America, Mexico and the USA; airspace alert; emergency search; a 20-leg flight plan; and three-dimensional navigation features such as VNav, altitude alert and three-dimensional airspace when the Loran receiver is coupled with the Apollo altitude encoder or altitude converter.

The Flybuddy and Flybuddy Plus are both low-cost standard panel mount receivers with LCD displays. Flybuddy features include direct-to navigation, ten 10-leg flight plans, emergency search, alerts for waypoint arrival and countdown timer, and a built-in database of public use airports and VORs in the USA and Canada. Flybuddy Plus adds airport city search and a user-replaceable datacard with extensive waypoint information such as runways, frequencies and fuel.

Specifications
Model 602
Dimensions: 50 × 159 × 279 mm
Weight: 1.7 kg
Waypoint capacity: 200
Range resolution: 180 m

The Apollo 618R Loran C receiver

The Apollo Flybuddy Loran C receiver

Model 604FB
Dimensions: 50 × 159 × 284 mm
Weight: 1.5 kg
Waypoint capacity: library plus 100 user definable
Range resolution: 180 m

Model 612 (IFR approved) and **612B** (VFR)
Dimensions: 50 × 159 × 279 mm
Weight: 1.7 kg
Waypoint capacity: library plus 100 user definable
Range resolution: 180 m

Model 612C
Dimensions: 256 × 82 mm diameter
Weight: 1.6 kg
Waypoint capacity: library plus 100 user definable
Range resolution: 180 m

Model 614R
Dimensions:
(CDU) 57 × 146 × 122 mm
(RCU) 152 × 48 × 284 mm
Weight:
(CDU) 0.7 kg
(RCU) 1.25 kg
Waypoint capacity: library plus 100 user definable
Range resolution: 180 m

618 standard panel mount
Dimensions: 159 × 51 × 265 mm
Weight: 1.7 kg
Waypoint capacity: built-in database plus 500 user definable

618C round panel mount
Dimensions: 257 × 826 mm diameter
Weight: 1.6 kg
Waypoint capacity: built-in database plus 500 user definable

618R dzus-rail mount
Dimensions:
(CDU) 57 × 127 × 117 mm
(RCU) 160 × 46 × 286 mm
Weight:
(CDU) 0.7 kg
(RCU) 1.3 kg
Waypoint capacity: built-in database plus 500 user definable

Flybuddy and Flybuddy Plus
Dimensions: 159 × 51 × 285 mm
Weight: 1.25 kg
Waypoint capacity: built-in database plus 100 user definable

Contractor
II Morrow Inc.

VERIFIED

Apollo SL40, SL50 & SL60 slimline nav/comm series

The first three models of the slimline series, the SL40 comm transceiver, SL50 GPS navigator and SL60 GPS/comm, share the 33 × 159 mm form factor and sunlight-readable LED display technology. They utilise the advanced technology of II Morrow's high-end GPS receivers.

The SL40 comm transceiver, (8 W carrier transmit power) is TSO certified and features direct access to National Weather Service information and automatic standby frequency monitoring. The SL50 GPS navigator is TSO C129 approved for en route and terminal operations. The SL60 GPS/comm carries these features, plus a navigation database to automatically select communication frequencies throughout the flight plan.

Specifications
Comm radio: 118-136.975 MHz; 760 channels; 10 W o/p; 2 × 8 RF memory.
GPS receiver: 30 reversible flight plans with 30 legs; 200 user defined waypoints; 8-channel parallel GPS; certified TSO C129 class A2 for en route and terminal approach.

Operational status
In production.

Contractor
II Morrow Inc.

VERIFIED

ADF-2070 Automatic Direction-Finder (ADF)

The ADF-2070 is a panel-mounted unit designed for the general aviation sector. It is claimed to be able to receive signals from exceptionally long ranges and has two sensitivity settings: extended range reception and conventional ADF which is used primarily on approach. This extended range facility is provided by a coherent detection feature which results in good reception characteristics with high immunity from thunderstorms and other static interference. Continuous digital tuning ensures lock on to the desired frequency.

The system also features a blade antenna which serves both the communications radio and the ADF systems. The ADF sensor is installed in the base of the blade and feeds signals to the receiver via a small amplifier which is mounted adjacent to the antenna blade but within the aircraft skin. It is claimed that this configuration provides nearly twice the gain of other combination antenna units, results in a reduction in cable length and has no impedance matching requirement.

Provision is made for correction of quadrantal error either on the ground or while airborne. The receiver output can drive either a standard ADF indicator with rotatable azimuth card or an HSD-800 horizontal situation indicator.

Specifications
Dimensions: 45 × 159 × 234 mm
Weight: 2.63 kg

Operational status
In service.

Contractor
AlliedSignal Commercial Avionics Systems.

VERIFIED

DFA-75A ADF receiver

The AlliedSignal DFA-75A utilises advanced LSI and microprocessor designs to achieve a greater level of dependability, accuracy and performance. While it uses advanced techniques to deliver performance advantages over its predecessors, its design is based

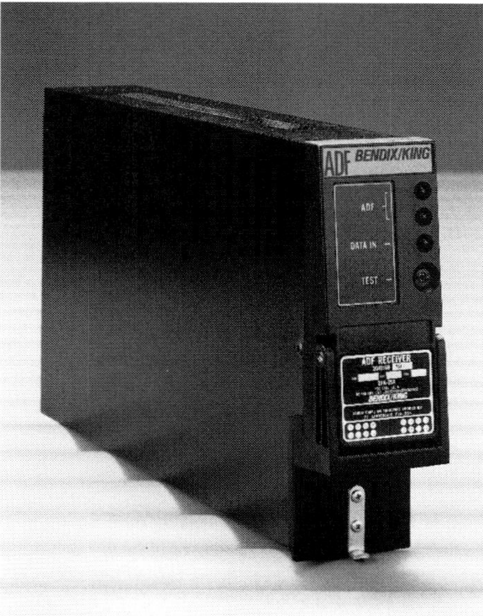

The AlliedSignal DFA-75A ADF receiver

on proven and efficient components which increase reliability and reduce weight and complexity.

The DFA-75A meets or exceeds all ARINC 712 characteristics for form, fit and function. One of the ways it meets the requirement for improved installations is by using a combined loop/sense antenna and digital interface with the receiver. Another major improvement is its ability to deliver more stable bearings in the presence of thunderstorms and during severe ionospheric conditions. This is achieved through a scheme of signal modulation, coherent demodulation and adaptive digital filtering from the antenna. The DFA-75A also provides automatic bearing adjustments to compensate for circular and quadrantal errors. Adjusted bearing data is smoothed by a second order digital filter.

Microprocessor monitoring permits a higher level of fault isolation diagnostics to be performed within the unit during BIT. The flight fault memory provides a non-volatile memory with the capacity to store 10 faults in each of 64 individual flight segments. It complies with ICAO Annex 10.

Specifications
Weight: 4.175 kg
Frequency: 190-1,750 kHz
Channel spacing: 0.5 kHz

Operational status
In production.

Contractor
AlliedSignal Commercial Avionics Systems.

VERIFIED

DFS-43 direction-finder system

The DFS-43 is an all-digital automatic direction-finding system. It is a lightweight system consisting of the DF-431 receiver, CD-432 control display unit and AT-434 combined loop and sense antenna.

The DF-431 receiver provides accurate reception of en route NDBs, locator outer markers and commercial AM broadcast stations. It is available with front- or rear-mounted connectors.

The CD-432 control display unit provides for a dual-frequency readout, with one active and one standby. Frequencies are alternated via the frequency transfer button. The unit provides frequency control from 190 to 1,860 kHz, with half or whole kHz incremental spacing. It also features a non-volatile frequency memory which retains the last frequency used, eliminating the possibility of frequency loss due to power interruptions.

The DFS-43 automatic direction-finder employs full time monitoring and self-testing of key functions such as the power supply, synthesiser lock, receiver lock and signal processing.

The AT-434 combined loop/sense antenna system is designed specifically for use with the DFS-43 system.

Specifications
Dimensions:
(antenna) 146 × 152.4 × 332.7 mm
(control display unit) 63.5 × 79.4 × 63.5 mm
(front mount receiver) 101.6 × 101.6 × 279.4 mm
(rear mount receiver) 101.6 × 101.6 × 320.5 mm
Weight:
(antenna) 1.72 kg
(control display unit) 0.27 kg

(front mount receiver) 2.36 kg
(rear mount receiver) 2.36 kg
Power supply: 18-33 V DC, 0.6 A at 28 V

Contractor
AlliedSignal Commercial Avionics Systems.

VERIFIED

DMA-37A DME interrogator

The DMA-37A is a fast-scan DME interrogator which meets all ARINC 709 characteristics. The design utilises advanced all-digital technologies to provide pilots with accurate and reliable DME and Tacan slant range information which can be transmitted via the ARINC 429 bus for use by both the visual display instruments and AFCS.

The DMA-37A features a centralised microprocessor system which offers superior signal processing capabilities and handles all unit control monitoring functions. Channel and control information is received through one of two selectable ARINC 429 frequency/function data input ports.

The primary processor automatically decodes the Morse code signal received from the channel selected for Ident both for single channel timing and when in the multiscan mode. A dedicated second microprocessor translates the logic signal representing the dots and dashes into alphanumeric characters before being formatted into ARINC 429 for transmission on the output ports by the main CPU.

The unit's microprocessor also helps to simplify maintenance through a rigorous self-monitoring and self-diagnostic routine. To further reduce down time, the fault memory and BITE are interfaced with the central fault display system and access to strategic points within the unit is provided via the automatic test equipment connector. It complies with ICAO Annex 10.

Specifications
Weight: 5.85 kg
Frequency:
(transmitter) 1,025-1,150 MHz
(receiver) 962-1,213 MHz
Channel spacing: 1 MHz

Operational status
In production.

Contractor
AlliedSignal Commercial Avionics Systems.

VERIFIED

The DMA-37A DME interrogator

DMS-44 Distance Measuring System

The DMS-44 is a digital solid-state dual transmitter distance scanning measuring system which can receive three stations simultaneously. It is a lightweight all-digital system consisting of the DM-441 transmitter/receiver and the SD-442 sector display.

The all-solid-state transmitter provides the capability simultaneously to scan stations for Nav 1 and Nav 2 and a third station which is transparent to the pilot for computation. It utilises two separate microprocessors and a video processor. The master processor provides signal processing, control computations and analogue range information. The slave processor performs the digital input/output generation, including the input of frequency tuning interfaces. The range processor can lock on to data in less than 200 ms and provides an LSB accuracy of better than 0.01 n miles. The system is available with either front or rear connector mounts.

The SD-442A panel-mounted selector/display unit provides full-time display of distance and groundspeed to the selected VORTac. The active Nav is annunciated below the distance readout.

The DMS-44 employs full-time self-monitoring of key circuits such as the synthesiser, receiver, transmitter, power supply and master processor.

Specifications
Dimensions:
(selector/display) 82.6 × 39.4 × 63.5 mm
(front mount transmitter/receiver) 127 × 101.6 × 279.4 mm
(rear mount transmitter/receiver) 127 × 101.6 × 320.6 mm
Weight:
(selector/display) 0.204 kg
(front mount transmitter/receiver) 2.77 kg
(rear mount transmitter/receiver) 3.22 kg
Power supply: 18-33 V DC, 1.2 A

Contractor
AlliedSignal Commercial Avionics Systems.

VERIFIED

Global positioning unit

The global positioning unit is a ¼ ATR short unit which is a repackaging of the GPS module described above. Developed to meet the market demand for a GPS sensor for the GNS-X without having to add VLF/Omega capability, the unit will initially be targeted at continental rather than transoceanic aircraft. The global positioning unit retains all the features of earlier Global Wulfsberg systems.

Operational status
In service.

Contractor
AlliedSignal Commercial Avionics Systems.

VERIFIED

KLN 35A GPS/KLX 135A GPS/COMM systems

The panel-mounted KLN 35A GPS incorporates moving map graphics, a high-visibility display and a choice of customised Jeppesen NavData databases - including coverage for America, Atlantic and Pacific areas.

The moving map, useful for providing situational awareness, displays Special-Use Airspace (SUA) boundaries and provides SUA alerting. All three databases contain appropriate Flight Service Station (FSS) and Air Route Traffic Control Center (ARTCC) frequencies, and airport runway data.

The KLN 35A presents this information via an advanced double super-twist nematic LCD. Offering improved viewing in direct sunlight and extended side-to-side visibility.

The KLX 135A offers greater capability. Incorporating all the same GPS performance and features, the KLX 135A GPS/COMM also integrates a TSOd, 760-channel, Very High-Frequency (VHF) communications radio with its navigation functions.

A new capability developed specifically for the KLX 135A, QuickTune, allows the pilot to enter the standby COMM frequency directly from the GPS database, saving effort and reducing the chances of making an entry error.

Specifications
KLN 35A GPS
Dimensions: 158.7 × 50.8 × 289.1 mm
Weight: 0.94 kg
Power requirements: 11-33 V DC

KLX 135A GPS/COMM
Dimensions: 158.7 × 50.8 × 289.1 mm
Weight: 2 kg
Power requirements: 14 V DC (28 V DC with available KA 39 voltage converter)
VHF communications transceiver transmitter power: 5 W (min) (7 W nominal)

Contractor
AlliedSignal Commercial Avionics Systems.

VERIFIED

KLN 88 Loran navigation system

The KLN 88 is a multichain Loran navigation system certified to TSO C60b standards. It tracks up to eight Loran stations in up to four different chains simultaneously.

The database, covering the United States, Canada, Mexico, Central America and the Caribbean, contains over 40,000 elements and includes all public use and military airports with runways of 1,000 ft or longer, VORs, NDBs, published intersections and outer markers, plus up to 250 user-defined waypoints. In addition, ARTCC and special use airspace boundaries are outlined. Virtually every airport communications, flight service station, ATIS and navaid frequency in North America is included. The information, provided by Jeppesen, is contained in a cartridge which plugs directly into the back of the KLN 88. Information is updated by direct replacement of the cartridge every 28 days. The 3.3 in (83.8 mm) split-screen CRT display allows two whole pages of data to be viewed simultaneously and includes a built-in moving map graphics facility.

Specifications
Dimensions: 160.3 × 50.8 × 334 mm
Weight: 2.82 kg
Power supply: 11-33 V DC, 2.5 A (max)

Contractor
AlliedSignal Commercial Avionics Systems.

VERIFIED

KLN 89/KLN 89B GPS navigation systems

The KLN 89 is a panel-mounted, eight-channel, GPS-based navigation system with a database that can be updated by the pilot. The KLN 89B adds IFR-certifiable en route, terminal, and approach capability. A basic system comprises: panel-mounted unit, altitude input, and KA 92 antenna. Among additional components that may be added to increase capabilities are an external Course Deviation Indicator (CDI) or HSI; RMI, some Shadin or ARNAV fuel management systems; several external moving maps, and certain Shadin air data systems.

Specifications
Dimensions: 160.3 × 50.8 × 272.3 mm
Weight: 1.16 kg
Power: 11-33 V DC at 2.5 A
TSO (KLN 89B only): C129 Class A1

Contractor
AlliedSignal Commercial Avionics Systems.

VERIFIED

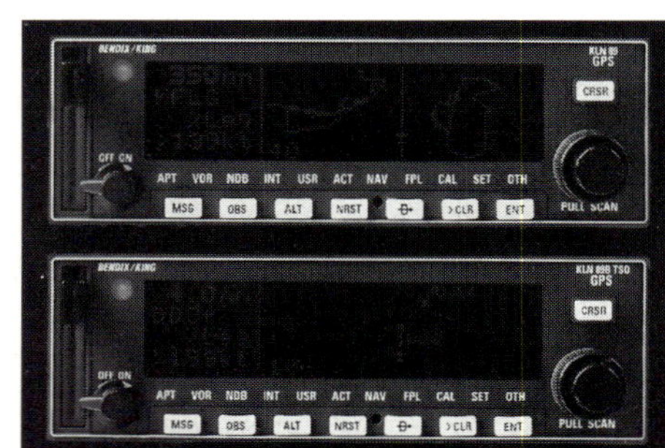

KLN 89/89B GPS navigation systems **1997**/0001335

KLN 90B approach-certified GPS navigation system

Together with the capability to perform non-precision GPS approaches, the KLN 90B navigation system offers an easy-to-read CRT map display and a comprehensive Jeppesen database. Designed to meet the FAA's C129 A1 specifications, the KLN 90B features an improved eight-channel parallel GPS receiver for even more reliable satellite tracking. Other enhancements include a more pilot-friendly interface and an expanded database, complete with SID and STAR waypoints and approaches.

Providing all the benefits of satellite-derived input - worldwide coverage, a high degree of accuracy and immunity to atmospheric disturbance - the KLN 90B harnesses the power of GPS to make non-precision approaches easier for the pilot.

Navigation pages were designed specifically for use during approaches, to provide the greatest amount of information with the least amount of effort. In fact, most necessary and en route information is presented concisely and logically on one large, easy-to-read screen. This page also incorporates a moving map, to show progress throughout the approach. Additional approach-specific pages are readily accessed, significantly reducing pilot workload during this critical phase of flight.

KLN 90B database information includes airport data (identifier, name, runway length/surface/lighting, instrument approach availability, customs, types of fuel, oxygen, landing fees, and other data comparable to that found in an airport services guide, and a runway diagram for airports providing published runway threshold co-ordinates); communications frequencies (ATIS, clearance, ground control, tower, CTAF, advisory and various VFR frequencies); VOR information; NDB information; intersection information (data about low altitude, high altitude, approach and SID/STAR intersections, with outer markers and compass locators; ARTCC and FIR boundaries and frequencies); flight service station frequencies and locations; minimum safe altitudes; special-use airspace.

Specifications

Dimensions: 160.3 × 50.8 × 334 mm
Weight: 2.66 kg
Temperature range: −40 to +70°C
Altitude range: up to 50,000 ft
Power inputs: 11-33 V DC at 2.5 A (max)
TSO: C129 A1

Contractor

AlliedSignal Commercial Avionics Systems.

VERIFIED

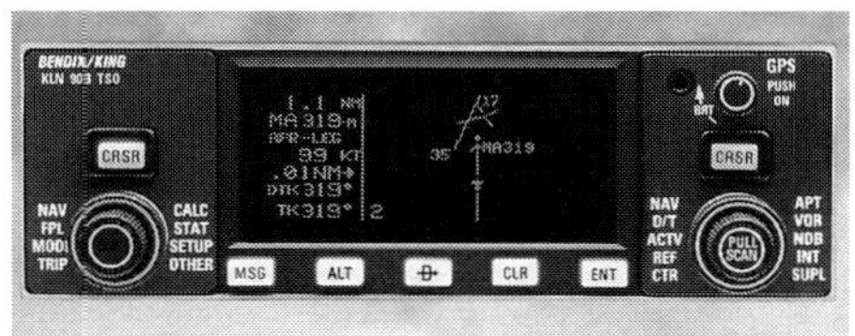

KLN 90B approach-certified GPS navigation system ***1997***/0001336

KN 62A/63/64 digital DMEs

The compact KN 62A digital DME is a fully self-contained 200-channel system in the Silver Crown range. It features an all-solid-state transmitter and four LSI chips and requires only 33 mm of panel height.

The KN 62A can be channelled remotely through almost any navigation equipment receiver, or tuned directly with its own frequency selection knobs. Dual channelling makes two DME frequencies available at all times. The self-dimming digital display provides information on simultaneous DME distance, groundspeed and time-to-station, or DME distance and internally selected frequency.

The KN 62A meets the US FAA's TSO performance and environmental standards; the KN 64 offers similar performance but is not TSOd.

The KN 63 is also TSOd, and can be integrated with the KNS 81 integrated Nav/RNav to increase sophistication.

Specifications

Weight: 1.18 kg
Power supply: 11-33 V DC, 15 W

Contractor

AlliedSignal Commercial Avionics Systems.

VERIFIED

KNR 665 Gold Crown integrated area navigation system

A member of the Gold Crown family of avionics for the business, corporate turboprop and jet upper end of the general aviation spectrum, the KNR 665 can operate in conjunction with the KFC 300 flight director and autopilot system.

The KNR 665 can memorise and display up to 10 waypoints and/or VORTac stations. Waypoint information in Jeppesen format includes VORTac frequency, outbound and reciprocal courses and waypoint bearing and distance. All waypoint information is displayed simultaneously for ease of reference. Individual parameters for any waypoint can be changed without resetting other data; the information is inserted by push-button and checked on a scratchpad before entering it into the system. An autocourse facility provides air traffic control clearance to a particular waypoint by pressing the 'autocourse' button, so that the route to that particular waypoint is instantly computed and displayed. The new course is automatically displayed on the horizontal situation indicator. The system can be checked by a push-button.

Operational status

In service.

Contractor

AlliedSignal Commercial Avionics Systems.

VERIFIED

KNS 80 Silver Crown integrated navigation system

The KNS 80 comprises a 200-channel VOR/Loc receiver, 200-channel digital DME, 40-channel glide slope receiver and digital RNav computer that can store up to four VOR/Loc frequencies and waypoints. Information is displayed on a full-width light-emitting diode display, and the system can operate in conjunction with an ARINC horizontal situation indicator or course deviation indicator. Extensive use is made of Large Scale Integration (LSI) technology, the single LSI chip accomplishing the work that would have required 40 conventional integrated circuits. As with other members of the Silver Crown family, the KNS 80 is aimed at the lower end of the general aviation sector.

Specifications

Dimensions: 160 × 76 × 305 mm
Weight: 2.7 kg
Power supply: 11-33 V DC, 25 W
Distance to next waypoint: 199.9 n miles in 0.1 n mile increments, 0.1° angle increments selectable on display

Operational status

In production.

Contractor

AlliedSignal Commercial Avionics Systems.

VERIFIED

KNS 81 Silver Crown integrated navigation system

A development of the KNS 80, the KNS 81 integrated navigation system can accommodate up to nine waypoints and embodies a number of new features. A remotely mounted DME enables a KDI 572 indicator to be positioned directly in front of the pilot, providing simultaneous digital readouts of distance, groundspeed and time to either a VORTac station or an RNav waypoint. There is simultaneous display of bearing, distance and frequency waypoint details on the KNS 81 panel for easy programming and updating of navigation information. A radial push-button permits a rapid bearing check to a chosen VORTac or RNav waypoint, displayed on the DME panel indicator in place of groundspeed and time to next waypoint. A check push-button permits a rapid cross-check of bearing and distance from the VORTac without disturbing other navigation instrument settings. A radio magnetic indicator output to a KI 229 or KNI 582 indicator gives an accurate bearing to a selected RNav waypoint or VORTac station.

Specifications

Dimensions: 160.3 × 5.1 × 291.2 mm
Weight: 2 kg
Power supply: 11-33 V DC, 15 W
Number of waypoints: 9
Distance to next waypoint: 199.9 n miles in 0.1 n mile increments; 0.1° angle increments

Operational status

In production.

Contractor

AlliedSignal Commercial Avionics Systems.

VERIFIED

KNS 660 flight management system

Aimed at the turbine-powered corporate and commuter sector of general aviation, KNS 660 is the designation for a family of great circle navigation management systems to work in conjunction with Gold Crown III avionics and other compatible units.

The KNS 660 has its own database, ARINC 429 input/output formats, analogue processing, air data function, vertical navigation computation, frequency management and optional Omega/VLF sensor. It has a choice of two control/display units, the 6 in (152 mm) KCU 568 or the 4.5 in (114 mm) KCU 567 for installations with limited space. The KNS 660 will handle signals from VOR/DME, Omega/VLF, Loran C, INS, GPS, AHRS (Attitude Heading and Reference System) and compass. AlliedSignal's KTU 709 Tacan system can replace the DME if required.

An optional receiver/processor providing a choice between Omega/VLF and Loran C is now available. A potential growth area is MLS.

Operational status

Production began in July 1984. AlliedSignal has developed the KLN 670 GPS Navstar sensor as part of the KNS 660. The system was certified in 1987.

Contractor

AlliedSignal Commercial Avionics Systems.

VERIFIED

KR 86 digital ADF

The all-solid-state KR 86 ADF offers positive and precise crystal digital tuning between 200 and 1,750 kHz and includes BFO for LF stations which broadcast an unmodulated signal. Constant audio output over wide variations of signal strength is assured by automatic gain control. The KR 86 is compact and self-contained and is easily fitted in a standard panel cutout. Self-test is provided.

Contractor

AlliedSignal Commercial Avionics Systems.

VERIFIED

KR 87 ADF system

The basic KR 87 system includes the KR 87 receiver, KI 227 indicator, KA 44B combined loop and sense antenna, and mounting racks and connectors. The all-solid-state receiver operates on DC voltages between 11 and 33 V and draws only 12 W of power. Electronic timers in the KR 87 provide aids to flight management as a flight or elapsed timer.

The KR 87 features a crystal filter for better long-range reception, coherent detection circuitry to lock on to weak stations, electronic tuning using a microprocessor and an LSI single-crystal digital frequency synthesiser circuit, EAROM non-volatile storage of frequencies during shut down or power interruptions and a fold-out modular construction for easy access to all circuits and components. The display is self-dimming.

Contractor
AlliedSignal Commercial Avionics Systems.

VERIFIED

KTU 709 Tacan

Particularly suited to corporate aircraft, where weight, cost and power requirements are notably important, the KTU 709 Tacan transmitter/receiver is based on AlliedSignal's extensive DME experience and the extensive use of Large Scale Integration (LSI) technology; the system specifically is based on the new-generation KDM 706 distance measuring equipment. This Tacan provides bearing, slant range, range rate and time to station or waypoint information to a KDI 572 control/indicator unit. Transistors provide a 250 W peak-to-peak output for a typical range of 250 n miles and the system covers 252 channels; all tuning is done electronically, using a digital frequency synthesiser designed around an AlliedSignal LSI chip.

Specifications
Dimensions: 76.2 × 127.0 × 260.4 mm
Weight: 2.6 kg
Frequency:
(transmit) 1,025-1,150 MHz
(receive) 962-1,213 MHz
Number of channels: 250
Reliability: 2,000 h MTBF design

Operational status
In production.

Contractor
AlliedSignal Commercial Avionics Systems.

VERIFIED

KX 125 Nav/Com system

The self-contained KX 125 nav/com system is very compact and utilises a back-lit LCD for its display. It operates on all 760 comm frequencies, from 118 to 136.975 MHz in 25 kHz steps. These are displayed on the left display for both active and standby frequencies. The nav receiver offers 200 VOR/Loc frequencies in 50 kHz steps. These are displayed on the right display for both active and standby frequencies. The KX 125's middle display shows the course deviation when the CDI mode is selected. In the bearing mode, the middle window displays three-digit bearing-to-station information. Selecting the radial mode allows radial information to be shown.

Whenever a transmitter has been activated continuously for more than 35 seconds, the unit reverts to receive mode and the comm frequency displays flash to alert the pilot to a stuck microphone.

A built-in audio amplifier is standard. Other features include autopilot interfacing and channelling of separate DME and glide slope receivers. The KX 125 can also drive an external CDI.

Contractor
AlliedSignal Commercial Avionics Systems.

VERIFIED

MLS-20A Microwave Landing System receiver

The MLS-20A is a 200-channel receiver with associated control/display unit and omnidirectional antenna and is based on custom LSI and microprocessor technology. It is compatible with conventional analogue flight instruments, is approved to FAA TSO 104 and meets the requirements of the new international standards. The system has a coverage of 20 n miles and registers from 0 to 20° in elevation and up to 60° either side of the extended runway centreline. It is designed to work with all MLS ground stations transmitting the FAA-ER-700-08C signal format.

Specifications
Dimensions:
(receiver) 128 × 100 × 301.2 mm
(control/display unit) 147.4 × 67.3 × 166.7 mm
Weight:
(receiver) 4.1 kg
(control/display unit) 1.82 kg
Power supply: 28 V DC, 1.5 A
Frequency: 5,031.00-5,090.70 MHz
Number of channels: 200

Operational status
No longer in production. The system is installed on Panavia Tornado aircraft of the Italian Air Force.

Contractor
AlliedSignal Commercial Avionics Systems.

VERIFIED

MLS-21 Microwave Landing System receiver

AlliedSignal announced the MLS-21 microwave landing system receiver in September 1985. The receiver is compatible with ARINC 727, can automatically tune DMEs and can accept analogue or digital inputs and outputs; it is also compatible with the EFS-10 and other electronic flight instrument systems.

The MLS-21 can be used in either automatic or manual modes. In the latter the pilot can select ±60° in azimuth and up to +20° in elevation for the approach. The maximum selectable angles can be limited to suit the capabilities of the aircraft.

The MLS-21 can receive 200 MLS channels in the frequency band 5,031 to 5,090.7 MHz (channels 500 to 699).

Specifications
Dimensions:
(receiver) 102 × 124 × 330 mm
(control/display unit) 63 × 80 × 60 mm
Weight:
(receiver) 2.9 kg
(control/display unit) 0.86 kg
Power supply: 28 V DC, 0.75 A
Frequency: 5,031.00-5,090.70 MHz
Number of channels: 200

Operational status
In production.

Contractor
AlliedSignal Commercial Avionics Systems.

VERIFIED

Quantum nav/com avionics family

Quantum is a line of navigation and communications equipments. Its features will extend reliability over previous designs, with a 30,000 hour MTBF and a guarantee on Mean Time Between Unscheduled Removals. Quantum also offers commonality of numerous parts, assemblies and modules.

A common main processor with digital signal processor provides industry analogue processing and repeatability. Quantum uses software that is 80 per cent common, provides High-Intensity Radiation Frequency (HIRF) and lightning protection and uses power supplies of common design. A common front LCD panel permits easier maintenance with easily understood fault message displays. A flash card is provided for onboard data loading, rapid evaluation of system performance and data recording. A PC-compatible maintenance access port allows for diagnostics and fault isolation on the aircraft. Upgraded mechanical packaging improves protection for electromagnetic interference, reduces heat rise and ensures structural integrity.

Performance has been improved by providing 8.33 kHz channel spacing on the voice data radio and raising the radio altimeter range to 5,000 ft.

The Quantum line consists of:
ALA-52B radio altimeter
DFA-75B automatic direction-finder
DMA-37B distance measuring equipment
RIA-35B instrument landing system
RTA-44D voice data radio
RVA-36B VOR

Contractor
AlliedSignal Commercial Avionics Systems.

VERIFIED

Quantum™ Line Communication and Navigation Systems

Quantum is a line of navigation and communications equipments. The Quantum line consists of:
ALA-52B radio altimeter
DFA-75B automatic direction-finder
DMA-37B distance measuring equipment
MMR multimode receiver
RIA-35B instrument landing system
RTA-44D voice data radio
RVA-36B VOR

Quantum features will extend reliability over previous designs, with a 30,000 hour MTBF and a guarantee on Mean Time Between Unscheduled Removals. It also offers commonality of numerous parts, assemblies and modules.

Quantum line equipment utilises more than 80 per cent of components and software that are common across the line. Quantum line features include: 486 main processor; common monitor processor in the ILS and radio altimeter; liquid crystal display; packaging to ARINC 650; HIRF/lighting protection; ICAO FM immunity; ETOPS. A common front LCD panel permits easier maintenance with easily understood fault message displays. A flash card is provided for onboard data loading, rapid evaluation of system performance and data recording. A PC-compatible maintenance access port allows for diagnostics and fault isolation on the aircraft. Upgraded mechanical packaging improves protection for electromagnetic interference, reduces heat rise and ensures structural integrity.

Performance has been improved by providing 8.33 kHz channel spacing on the voice data radio and raising the radio altimeter range to 5,000 ft.

Operational status
Widely employed on civil aircraft.

Contractor
AlliedSignal Commercial Avionics Systems.

UPDATED

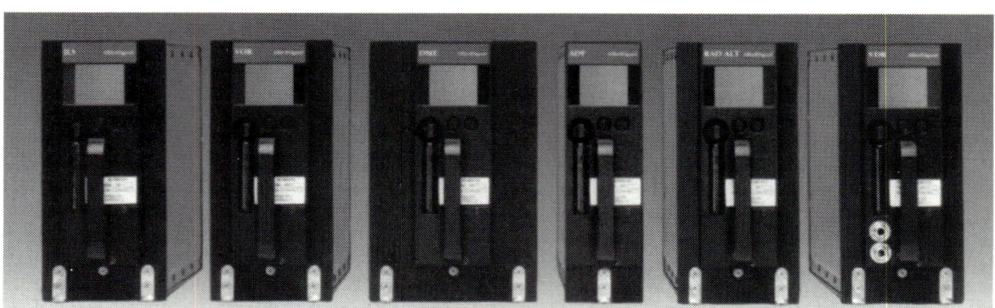

The Quantum™ line is composed of the RIA-35B ILS receiver, DMA-37B DME interogator, DFA-75B ADF receiver, ALA-52B radio altimeter, RTA-44D VHF Data Radio (VDR), and MMR ***1998***/0018183

RIA-35A ILS receiver

The RIA-35A is an ARINC 710-compatible ILS receiver, part of the CNI700 digital nav/com family. It uses dual microprocessors and custom large-scale integration design, with continuous monitoring of all unit subassemblies. It complies with ICAO Annex 10.

Specifications

Dimensions: 102 × 124 × 330 mm
Weight: 2.9 kg

Operational status

In service and production. In September 1985, AlliedSignal announced that the RIA-35A had been selected as standard on the Gulfstream G-IV aircraft and in June 1986 the RIA-35A was named as standard equipment on the Airbus A320. It is in service with numerous airlines worldwide.

Contractor

AlliedSignal Commercial Avionics Systems.

VERIFIED

RN-262B (AN/ARN-127) VOR/ILS system

The RN-262B VOR/ILS system comprises a compact, solid-state 200-channel VOR/Loc remote-mounted receiver, a 40-channel glide slope receiver and marker beacon receiver, and a panel-mounted control/display unit. All three receivers operate independently.

Specifications

Dimensions:
(receiver) 183 × 130 × 319 mm
(control unit) 146.1 × 66.7 × 114.3 mm
Weight:
(receiver) 4.5 kg
(control unit) 1 kg
Power supply: 26 V AC, 400 Hz, 25 VA
28 V DC, 1.5 A

Operational status

In production and in service. The system has been adopted by the US Air Force, Army and Coast Guard.

Contractor

AlliedSignal Commercial Avionics Systems.

VERIFIED

RVA-36A VOR/marker receiver

The RVA-36A airborne VOR/marker receiver is designed in accordance with ARINC 711. The design provides a highly accurate digital omni-bearing output in ARINC 429 format.

With the advent of pure digital interface and reliable microcomputers, the RVA-36A was designed to make optimum use of a microprocessor, or microcomputer, which permits features and operational flexibility that could not have been achieved in earlier generation equipment. Signal processing, control, monitoring and detailed fault analysis are all under the control of a programmable microprocessor.

Some of the features offered in the RVA-36A which are attributable to the microprocessor are improved accuracy, comprehensive monitoring, detailed fault isolation and precise predictable and more selective receiver design. In addition, BITE monitors the performance of all assemblies and if a fault occurs it is stored in EEPROM and can be sent to the central fault display system. It complies with ICAO Annex 10.

Specifications

Dimensions: 324.1 × 90.93 × 194.06 mm
Weight: 4.6 kg
Frequency: 108-117.95 MHz
Channel spacing: 50 kHz

Operational status

In production.

Contractor

AlliedSignal Commercial Avionics Systems.

VERIFIED

VNS-41 VHF navigation system

The VNS-41 navigation system is a digital VOR/ILS receiver and processor system providing VOR, localiser, glide slope and marker beacon reception. It is a lightweight system consisting of the VN-411 receiver and the CD-412 panel-mounted control display unit.

The VN-411 receiver contains the VOR/Loc, glide slope and marker beacon receiver and processors. It employs full-time self-test monitoring of the key internal circuits such as the power supply, synthesisers and automatic gain control and all receivers and converter circuits. Advanced signal processing provides VOR accuracy within 1°, as well as steady navigation signals.

The CD-412 control display unit provides for a dual-frequency readout, one active and one standby. The frequencies are alternated via the frequency transfer button. The CD-412 displays 200 channels with 50 kHz spacing from 108 to 117.95 MHz. It also features a non-volatile frequency memory which retains the last frequency used, eliminating the possibility of frequency loss due to power interruptions. The CD-412 can display digitally either the bearing or radial to the selected VOR in the standby window.

Specifications

Dimensions:
(control display unit) 63.5 × 79.4 × 63.5 mm
(front connector receiver) 101.6 × 101.6 × 279.4 mm
(rear mount receiver) 101.6 × 101.6 × 320.5 mm
Weight:
(control display unit) 0.27 kg
(front connector receiver) 2.04 kg
(rear mount receiver) 2.81 kg
Power supply: 18-33 V DC, 0.8 A
Frequency:
(navigation receiver) 108-117.95 MHz
(glide slope receiver) 329.15-334 MHz
Channel spacing:
(navigation receiver) 50 kHz
(glide slope receiver) 150 kHz
Number of channels:
(navigation receiver) (VOR) 160, (Loc) 40
(glide slope receiver) 40

Contractor

AlliedSignal Commercial Avionics Systems.

VERIFIED

The RN-262B (AN/ARN-127) VOR/ILS system

FMS 5000 Flight Management System

The FMS 5000 consists of a multichain master independent Loran capable of tracking two Loran chains and up to 12 ground stations simultaneously. All world Loran chains are available, including the US mid-continent NOCUS/SOCUS chains. The FMS 5000 automatically selects the strongest master and secondary Loran stations, providing hands-off operation.

Three remote GPS receiver options are available for the FMS 5000. A five-, six- or 12-channel tracking receiver can be supplied with the FMS 5000, or added in the future. Together, the Loran and GPS receivers continuously scan up to 12 Loran ground stations and all satellites in view. The FMS 5000 displays the navigation solution with instant exchange of sensor positioning, providing hands-off Loran/GPS operation.

The FMS 5000 interfaces with analogue instruments, fuel computers, air data computers, moving maps, autopilots, CDI/HSI and annunciators. An optional ARINC 429 interface allows coupling to EFIS systems.

A two-line 40-character sunlight-readable LED display shows bearing, distance, groundspeed, CDI, waypoint identity, altitude and track. Waypoints can be located by identity, city, local proximity or adjacent waypoints.

The FMS 5000 uses a Jeppesen NavData card containing 40,000 waypoints. The North America coverage extends from Alaska through Central America. The International card contains all areas outside North America. Worldwide navigation, combining North America and International data, is available on a single World card. Airports, VORs, NDBs, terminal and en route intersections are stored on the crew updatable NavData card. Special use airspace includes floors and ceilings, TCAs, ARSAs, ATAs, MOAs and restricted, prohibited and alert penetration warnings.

Coupled to an optional Mode C encoder interface, pressure altitude with a pilot input barometric correction is displayed. VNav becomes automatic with known present altitude and known waypoint elevations. An altitude hold advisory will notify the pilot of altitude deviation.

Contractor

Arnav Systems Inc.

VERIFIED

FMS 7000 Flight Management System

The FMS 7000 is a small dzus-mounted Flight Management System (FMS) for business and commuter aircraft. It is designed to provide the pilot with a comprehensive primary or separate secondary navigation tool and database facility. It can be configured with a Loran sensor and a high-precision GPS sensor. The FMS 7000 uses sensor assessments of signal quality to provide the best position information.

A standard Jeppesen North American or International database with over 60,000 aviation facilities is contained on a high-capacity NavData card that can be easily updated. Each card provides all worldwide port-of-entry airports with runway lengths in excess of 6,000 ft, plus all hard-surfaced airports, facilities, frequencies and navigation information for the North American or International geographic areas. An optional Jeppesen World NavData card provides comprehensive worldwide navigation on a single card.

The FMS 7000 automatically builds SID and STAR route waypoints, transition routes, crossing altitudes, arrival routes and arrival frequencies. Jeppesen terminal navigation is also included on the World NavData card.

The Arnav advanced communication network, Arnet, and the ARINC Standards Interface (ASI) have been developed specifically for the FMS 7000. Arnet is a high-speed serial communication device that facilitates integration of other avionics such as autopilot, air data and fuel computers. ASI allows the FMS 7000 to communicate with virtually all cockpit systems, including analogue and digital format DME, EFIS, HSI and flight directors. ARINC 429, ARINC 419, six-wire ARINC 568, synchro, MIL-STD-1553 and several DME controls or data formats are supported.

Cross-talk capability and system redundancy is optional on the FMS 7000. Tandem control/display units allow the co-pilot to view all flight, fuel and air data information and perform waypoint search routines.

The Arnav GridNav mission management system program is a software option for Loran, GPS and combination FMS. The GridNav option allows the pilot to plot and fly a grid pattern of user specified dimensions using a minimum number of waypoints and with simplified navigation programming. An event trigger provides a precise timed pulse to a camera shutter or target drop release.

Sensor options for the FMS 7000 include Loran only, GPS only or both Loran and GPS. The Loran-based system automatically selects the proper Loran chains and stations based on aircraft location. The FMS tracks up to 12 Loran stations and automatically selects the signal geometry that provides the best fix. Loran is approved for IFR flight and permits the pilot to file IFR direct when within the boundaries of Loran coverage.

The GPS-based receiver is differential ready and conforms to TSO C-129 certification. The FMS 7000 may be populated with either a five-channel or 12-channel GPS receiver. Both work in any weather and are not subject to precipitation static, low-frequency thunderstorm emissions or any other weather interference. Through a powerful microprocessor, the FMS 7000 tracks all satellites in view and then selects the satellite geometry to acquire the most precise fix. Signal acquisition and tracking are continuous throughout all dynamics of flight. Initial time to fix a reliable position is less than 1 minute and position is updated every second.

Both Loran and GPS sensor options can be installed in the FMS 7000 to provide multisensor blended mode navigation. The combined sensor output can be used for IFR direct navigation when operating within Loran coverage areas and when TSO is achieved. The FMS has TSO C-60b for IFR en route navigation and terminal navigation, TSO C-115a multisensor certifications and Loran approach certification.

Specifications
Dimensions:
(Loran/GPS LRU) ¼ ATR short
(CDU) 57.1 × 139.7 × 127 mm
Weight:
(Loran/GPS LRU) 1.82 kg
(CDU) 0.68 kg
(Loran/GPS antennas) 0.64 kg
Power supply: 11-35 V DC, 15 W (max)
Temperature range:
(LRU) −55 to +70°
(CDU) −20 to +70°C
Altitude: up to 55,000 ft
Certification: TSOs C-60b, C-115a, C-44a, C-106
Environmental: Do-160c cats A1/D2, MIL-STD-167

Operational status
The FMS 7000 is installed in a wide range of civil and military aircraft and helicopters.

Contractor
Arnav Systems Inc.

UPDATED

GPS-506 Global Positioning System receiver

The GPS-506 is a six-channel GPS continuous tracking remote sensor which interfaces with the R-50i Loran C receiver (see later item). The R-50i has a European database. The R-50i Loran/GPS is approved by the FAA for IFR operations.

Operational status
Selected for the Citation V aircraft by the Great Lakes Chemical Corporation.

Contractor
Arnav Systems Inc.

VERIFIED

Navision 50 moving map

The Navision 50 is a general aviation version of the Arnav airborne moving map. It interfaces with the R-50 and R-50i Loran C and is compatible with most general aviation Lorans. With an adjustable zoom feature ranging from 1 n mile to 1,000 n miles, the pilot can view airport diagrams including runway layout and runway numbers, or look up to 1,000 n miles ahead on the flight path. The display can be viewed north up or it can be aligned with the aircraft ground track.

About the size of a 2 in thick approach plate, the Navision 50 is portable and can be used anywhere in the cockpit. Also included is the Jeppesen North American NavData card and flight planning mode for developing flight plan legs shown in columnar form or graphically. The user waypoint feature allows pilot entry of mountain peaks or other high terrain and actual elevations can be entered to be shown on the map.

The 3 × 5 in (76 × 127 mm) electroluminescent display is easily read in bright sunlight and may be dimmed for night flight. The Navision 50 includes many of the features found on the Navision 1000 which was designed for larger aircraft.

Contractor
Arnav Systems Inc.

VERIFIED

R-50 Loran C receiver

The R-50 IFR Loran C is a development of the successful AVA-1000 and the performance has been improved to permit operation in areas of poor Loran coverage, such as the North Slope of Alaska, Bermuda, the Caribbean and parts of the mid-West. The R-50 includes the Jeppesen NavData slip-in database card which provides information on airports, VORs, NDB intersections and special use airspace alerts. The panel-mounted system offers extended range, automatic chain nomination, flight planning and 100 flight plan waypoints and can accommodate up to 1,500 waypoints. Present position is shown on a two-line 40-character dot matrix display and outputs are provided to an autopilot, a CDI and to equipment using RS-232 signal format.

Operational status
In production.

Contractor
Arnav Systems Inc.

VERIFIED

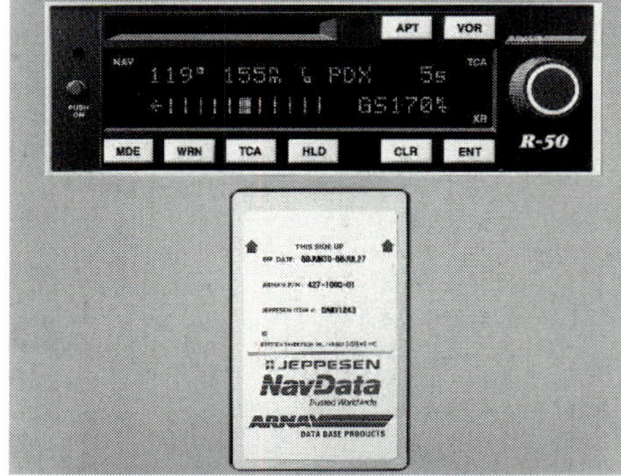

The Arnav R-50 Loran

R-50i Loran C receiver

The IFR and TSO C60b approved multichain R-50i Loran C simultaneously tracks up to 12 Loran stations and can be used as a primary means of en route and terminal IFR navigation. Up to 120 waypoints can be stored and flown as flight plans. A sunlight-readable two-line 40-character LED display provides bearing, distance, waypoint identifier, course deviation indication, track guidance, groundspeed and suggested vertical navigation.

The Jeppesen NavData card database contains North American airports, VORs, NDBs and intersections and can store 150 custom waypoints. The database can be accessed by identification or partial city name or identifier.

Navigation management features include vertical navigation, extended range, true airspeed/winds aloft calculations, fuel range and minimum safe altitudes.

The R-50i can instantly display the 15 closest airports, VORs or intersections at the touch of a button. Incorporated in the system is a software feature called Pilot Prompt which guides the pilot through every operation with simple messages. Included with each R-50i is a home power supply and simulation mode for flight planning and training.

Specifications
Dimensions: 50.8 × 158.75 × 271 mm
Weight: 1.13 kg
Power supply: 10-35 V DC, 1.5 A

Contractor
Arnav Systems Inc.

VERIFIED

STAR 5000 panel mount GPS

The STAR 5000 blends GPS technology with the Arnav R-50i IFR-certified Loran. The panel mount GPS receiver delivers five-channel continuous and parallel tracking of all satellites in view. Its onboard network interfaces with an array of avionics including analogue instruments, fuel computers, air data computers, moving maps, autopilots, CDI/HSI and annunciators.

A two-line 40-character sunlight-readable LED display shows bearing, distance, groundspeed, course deviation, waypoint identification, altitude and track. Waypoints can be located by identification, city, local proximity or proximity to other waypoints. The system uses a 40,000 waypoint Jeppesen NavData card, known as the Gold Card because of the wealth of information. Jeppesen NavData is offered for three geographical areas. The North America card coverage extends from Alaska to Central America. The International card covers all areas outside North America. Worldwide navigation combining the data on the other two cards is available on a single World card.

Airports, VORs, NDBs, terminal and en route intersections are stored on the pilot updatable NavData card. Special use airspace includes floors and ceilings and the card provides TCA, ARSA, ATA, MOA, restricted, prohibited and alert area penetration warnings. Coupled to an optional Mode C encoder interface, pressure altitude with a pilot input barometric correction is displayed. VNav becomes automatic with known present altitude and known waypoint elevations. An altitude hold advisory will notify the pilot of an altitude deviation.

Standard features of the STAR 5000 include nearest airport, VOR, intersection search, true airspeed, wind and fuel calculations, MSA/MESA, 150 waypoint flight planning and 300 user waypoints and the pilot's choice of the 40,000 waypoint North America or International NavData cards. The STAR 5000 has UTM and military grid reference co-ordinates as a standard feature. Options include the altitude encoder interface, GridNav mission management software, ARINC 429 interface and an air data computer.

Contractor
Arnav Systems Inc.

VERIFIED

ELT Emergency Locator Transmitters

Artex Inc manufactures a series of FAA approved emergency locator transmitters that meet TSO C91a and TSO C126. The ELTs are designed for installation in fixed- or rotary-wing aircraft. The ELTs for fixed-wing aircraft activate via a single-axis 'G' switch and the ELTs for helicopters activate via a six-axes 'G' switch module. Models available include the ELT 110-4 basic model and ELT 100HM (helicopter version), approved to TSO C91a (121.5/243 MHz) the ELT 110-406 and ELT 110-406HM (approved to TSO C126), and ED-62 (121.5/243/406.025 MHz).

Artex also manufactures an ELT-to-navigation interface for use with the ELT 110-406 systems. This interface allows the ELT to couple via ARINC 429 to GPS, Loran or FMC thus broadcasting latitude/longitude as part of the 406 MHz digital message.

Specifications

All models provide 50 mW output at 121.5/243 MHz for 50 hours. In addition the ELT 110-406 systems operate at 5 W on 406.025 MHz for 24 hours.

Operational status

Widely fitted by fixed- and rotary-wing aircraft manufacturers.

Contractor

Artex Aircraft Inc.

UPDATED

AN/ARN-154(V) Tacan

The AN/ARN-154(V) lightweight airborne Tacan is a remotely controlled multistation tracking system utilising large-scale integrated circuits and CMOS technology. The system consists of the RT-1634 receiver/transmitter, ID-2472 indicator unit, MT-6734 Tacan control unit mounting base and antenna.

The AN/ARN-154(V) has the capability to track up to four ground stations simultaneously in range and two in bearing. Tracking velocity is up to 1,800 kt. It incorporates a sine/cosine bearing output, along with an ARINC 547 CDI interface and ARINC 547/579 low- and high-level flags. Bearing information is pilot selectable from either of the two tracking channels. The system has both X and Y mode MIL-STD-291B air-to-air ranging, plus antenna switching to allow the aircraft to be configured for dual antennas.

The AN/ARN-154(V) provides an ARINC 568 digital range output from either of the two tracking channels for display on remote EFIS or HSI displays.

The RT-1643 is offered in a number of configurations to interface to most navigation flight management systems. The AN/ARN-154(V) can be used as a pilot-controlled positioning system or as a blind navigation sensor controlled by one or more long-range navigation systems.

An ARINC 429 version is available as an option. This version will output range on three separate channels and bearing on two. The input tuning is standard ARINC 429 bus. This unit is ideal for updating VLF/Omega or inertial navigation systems in helicopters and fixed-wing aircraft, or as a stand-alone Tacan system.

The AN/ARN-154(V) system can utilise an ID-2472 indicator unit which displays distance, Tacan radial or bearing, decoded station ident, groundspeed and time to station. Either or both Tacan stations may be displayed simultaneously. A second ID-2472 can be installed to provide independent cockpit instrumentation.

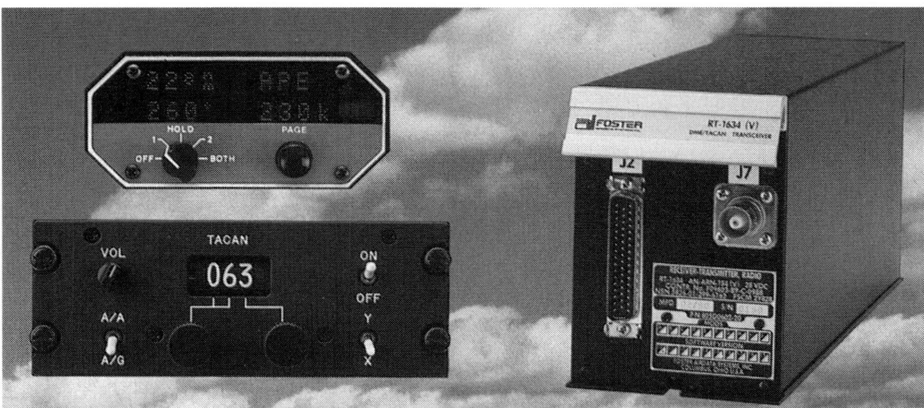

The AN/ARN-154(V) Tacan showing (top left) the ID-2472 display unit, (bottom left) a control unit and (right) the RT-1634 receiver/transmitter ***1995***

Specifications

Dimensions:
(display unit) 3ATI × 175.3 mm
(receiver/transmitter) 290.1 × 87.4 × 127.8 mm
(control units) 50.8-56.9 × 71.1-146 mm
Weight:
(display unit) 0.48 kg
(receiver/transmitter) 2.81 kg
(mounting base) 0.2 kg
(control units) 0.57-0.91 kg
Power supply: 18-32 V DC,1.25 A (max)
Frequency:
(receive) 962-1,213 MHz
(transmit) 1,025-1,150 MHz
Range: up to 400 n miles
Accuracy:
(range) ±0.1 n mile
(bearing) ±0.5°
Temperature range: −54 to +71°C
Altitude: up to 70,000 ft

Contractor

BFGoodrich Aerospace Avionics Systems.

VERIFIED

Reuseable navigation software module

Boeing demonstrated its new reuseable navigation software module in an F-15 Eagle equipped with a commercial MIPS R4400 SC processor. The single processor card exceeds the processing power of the IO cards in the computer. The same navigation software module was successfully flown in an AV-8B Harrier II equipped with a commercial power PC processor. The reuseable software has also been trialled in an F/A-18 again equipped with a power PC processor.

The reuseable software project is part of a McDonnell Douglas affordability initiative to develop aircraft with more flexible, open architectures. The approach is based on a modular object-orientated design, as well as the commercial POSIX standard and hardware independent high-order languages. This can cut the costs of software development and maintenance in half. The use of commercial processors can also significantly reduce recurring and non-recurring hardware costs.

Contractor

The Boeing Company.

UPDATED

Reusable navigation software module demonstrated in an F-15 Eagle equipped with a MIPS R4400 SC processor ***1997***/0001339

Global Positioning System (GPS)

The Wulfsberg Electronics Division GPS is designed to provide a high-accuracy satellite position fix to complement the existing navigation sources utilised by the GNS-500A Series 4 and Series 5 VLF/Omega navigation system, GNS-1000 flight management system and GNS-X flight management system.

The GPS comprises an antenna, a down converter, an RF/IF module and a GPS main processor. No additional boxes or racks are required. With the exception of the antenna system, all components are housed within the major assemblies of existing systems.

The GPS receiver features five independent channels offering signal acquisition and tracking throughout the dynamics of a flight and is connected to the antenna by a single coaxial cable. Drag is minimised by the low-profile antenna which protrudes only 6.4 mm above the skin of the aircraft.

The Wulfsberg Electronics Division GNS-500A Series 5 control/display unit

Specifications

Dimensions:
(antenna) (diameter) 88.9 mm × 6.4 mm
(down converter) 139.7 × 139.7 × 20.32 mm
Weight:
(RF/IF module) 2.04 kg
(main processor) 0.18 kg
(down converter) 0.79 kg
Power supply: 28 V DC, 0.5 A

Contractor

Chelton Avionics, Inc, Wulfsberg Electronics Division.

VERIFIED

GNS-500A series Omega/VLF receivers

The GNS-500A was approved by the FAA for IFR en route navigation in US national airspace shortly after Omega was introduced in 1976. The GNS-500A can receive both Omega and VLF transmitters and has undergone a number of improvements since it was originally introduced.

GNS-500A Series 2
The Series 2 was introduced to offer non-volatile retention of waypoints and present position, automatic computation of magnetic variation and a continuous clock to maintain Greenwich Mean Time and date.

GNS-500A Series 3
The Series 3 was introduced in 1980 to answer the need for more waypoints and a faster and clearer data presentation from any combination of up to 5 letters and numbers on 9 flight plans of up to 20 waypoints each. Information is presented on a sunlight-readable CRT providing a 14-character by 8-line matrix. In the Series 3 system all data is organised into one of three sections: navigation, data and flight plan. Each section in turn contains information laid out on a number of pages.

GNS-500A Series 4
The Series 4 offers further operational features; it can directly replace earlier models. The Series 3 and 4 operate with the AlliedSignal Aerospace Global Wulfsberg NDB-2 navigation data bank and can implement 127 waypoints and nine flight plans with up to 20 waypoints each. The system, which comprises three units, is fully compatible with ARINC 561 and 571. The Series 4 features a compact full alphanumeric control display unit, a 'Direct-To' key and AFIS compatibility.

GNS-500A Series 5
The GNS-500A Series 5 offers both GPS and Loran C receivers housed inside the receiver computer unit. A new navigation filter has been implemented which blends these two position inputs with the VLF/Omega position to arrive at a best computed position. All older GNS-500A models can be upgraded to the Series 5; the impact on aircraft wiring is minimal and is confined to the antenna system.

Specifications

Weight:
(total system) 17.5 kg
Power supply: 28 V DC, 7.5 A

Contractor

Chelton Avionics Inc, Wulfsberg Electronics Division.

VERIFIED

AN/ARS-6(V) personnel locator

The personnel locator system consists of the AN/ARS-6(V) guidance system manufactured by Cubic Defense Systems and the Motorola AN/PRC-112 survival radio. The survival radio remains silent until interrogated by the airborne guidance system, when it responds with a coded 300 ms pseudo-noise burst transmission. The AN/ARS-6(V) guidance system then displays range and steering directions in the rescue aircraft.

The AN/ARS-6(V) includes a compact radio transceiver, high-gain antenna set, control unit and instrument panel display.

Light, compact and modular, the system is easily transportable. It can be installed in a UH-1H or UH-60 in less than 30 minutes, and the system's simple controls and logical sequence make it extremely easy to operate.

The AN/ARS-6(V) is operated from the Control/Display Unit (CDU). The operator programmes Channel A and B frequencies from the 3,000 channels between 225 and 300 MHz or selects 243 and 282.8 MHz on preset channels. Steering commands and ranging information are displayed on a small liquid crystal Remote Display Unit (RDU) which is easily mounted on the aircraft instrument panel. Range to target is measured first in miles, then in feet. Steering commands are presented on a bar graph display that indicates the position of the target relative to the heading of the aircraft. Both the RDU and the CDU have sunlight-readability and full NVG compatibility.

The Receiver/Transmitter (RT) is a fully synthesised UHF transceiver optimised to operate with the AN/PRC-112 survival radio. It weighs less than 10 kg and is packaged in a ¾ ATR case. The shock-isolated RT mounting base is adaptable to either temporary installation on the aircraft floor or permanent mounting. The RT's 1553B bus interface can configure the AN/ARS-6(V) as a remote terminal on the databus, integrating control and display with existing aircraft functions and eliminating the need for the RDU and CDU.

The antenna group consists of twin blade antennas on the aircraft ventral surface. By mounting the antennas on the aircraft centreline, signal scattering effects from aircraft structures are minimised to achieve the specified accuracy.

Specifications

Dimensions:
(antenna set) 228.6 × 116.8 × 50.8 mm
(receiver/transmitter) 317.5 × 193 × 198.1 mm
(antenna switching unit) 144.8 × 193 × 33 mm
(control display unit) 101.6 × 146 × 76.2 mm
(remote display unit) 38.1 × 76.2 × 48.3 mm
Weight:
(antenna set) 0.59 kg
(receiver/transmitter) 9.89 kg
(antenna switching unit) 1.04 kg
(control display unit) 1.3 kg
(remote display unit) 0.27 kg
Power supply: 28 V DC, 4.5 A (max)
Power output: 10 W
Frequency: 225-300 MHz
Available channels: 3,000 (25 kHz steps)
Reliability: 2,650 h MTBF

Operational status

In production since 1988 for the US Army, Navy and Air Force and other customers. Over 1,000 units have been ordered.

Contractor

Cubic Defense Systems.

VERIFIED

Carousel IV Inertial Navigation System (INS) (AN/ASN-119)

During the late 1960s the Carousel IV inertial navigation system was the subject of the largest ever single military procurement of such equipment, when it was chosen by the US Air Force for its fleet of C-5A Galaxy and C-141 StarLifter transports and KC-135 tankers. It had earlier been chosen as standard fit for the Boeing 747, when it was fitted as a three-system installation to become the first certificated commercial inertial navigation system. A guaranteed MTBF of greater than 1,250 hours was an important factor in Boeing's choice of the Carousel IV. Some 7,000 sets have now been delivered to support 30 military programmes and more than 60 airlines. The Carousel IV has a demonstrated MTBF of more than 3,000 hours on commercial aircraft.

Improved versions of the Carousel IV have been built for the Boeing E-3A Sentry AWACS under the designation AN/ASN-119, the Titan III ICBM missile fleet, numerous helicopter applications and the airborne element of the US Army's Guardrail intelligence programme. The AWACS installation comprises two ASN-119 platforms operating in conjunction with a single Northrop Grumman AN/ARN-129 Omega system and a Litton AN/ADN-213 Doppler velocity sensor. The combination provides an accuracy of better than 1 n mile on a 10-hour mission.

Each system comprises three elements: an inertial navigation unit with gyros, accelerometers and computing functions; a control/display unit; and a mode selector unit. A battery unit to maintain operation during power transients is optional.

Specifications

Dimensions:
(inertial navigation unit) 215.9 × 259.1 × 510.5 mm
(control/display unit) 114.3 × 146.1 × 152.4 mm
(mode selector unit) 146.1 × 38.1 × 50.8 mm
Weight:
(inertial navigation unit) 25 kg
(control/display unit) 2.1 kg
(mode selector unit) 0.45 kg

Operational status

In production and in service.

Contractor

Delco Electronics.

VERIFIED

Carousel 400 series inertial systems

The Carousel 400 series of inertial reference systems is designed to meet the requirements of advanced aircraft and provide an upgrade for older aircraft. It consists of the Carousel 411, 424, 444 and 448 systems. These systems are based on the hemispherical resonator gyro which has no moving parts and no known wear-out mechanism.

Carousel 411. The Carousel 411 inertial navigation system is a direct replacement for ARINC 561 systems and no operational, electrical or mechanical changes are required. It provides the advantages of a new system, while maintaining all the current interfaces. Any combination of Carousel IV and 411 systems can be installed on the aircraft. The Carousel 411 provides a completely self-contained worldwide navigation capability. In addition, it provides digital outputs in both ARINC 561 and 704 formats. The Carousel 411 can make use of the existing Carousel IV CDU or an optional RNav CDU with no wiring changes. The optional RNav CDU has full alphanumeric capability and, along with an optional navigation database and autotune circuit card, makes the system a lateral flight management system. Also available as an option is an embedded GNSS receiver/processor.

Carousel 424. The Carousel 424 MG small inertial navigation system consists of a strapdown platform, digital computer and associated input/output hardware. It is designed for the military environment and has provision for an embedded GPS receiver. The Carousel 424 also provides an extra card slot for unique aircraft functions or interfaces such as flight control, radar or air data. It can make use of existing CDUs for independent display of navigation data. An optional CDU is available.

Carousel 444. The Carousel 444 is a 4 MCU ARINC 704 inertial reference system. It is designed for use in new production aircraft. The reference unit consists of a strapdown platform, digital computer and associated input/output hardware. Use of this solid-state instrument, combined with supporting electronics, makes accuracy and reliability an inherent feature of the reference unit. An inertial system display unit is available to provide mode selection and insertion of initial position. The Carousel 444 can be used in 10 MCU configurations, with the addition of an adaptor tray. An embedded GNSS receiver/processor is available as an option.

Carousel 448. The Carousel 448 is a 4 MCU inertial reference system and air data computer which meets the ARINC 738 ADIRS specification. In addition to all the information available from the Carousel 444, the Carousel 448 provides full air data outputs including true airspeed, Mach number, air temperature and angle of attack. An embedded GNSS receiver/processor is also available for the Carousel 448.

Contractor
Delco Electronics.

VERIFIED

Low-Cost Inertial Navigation System (LCINS)

The Low-Cost Inertial Navigation System (LCINS) development programme extends Delco's Carousel class systems' activity into applications that require somewhat lower performance at substantially lower cost. Typical navigation accuracy rating for the LCINS is 2 to 4 n miles/h. The LCINS is a strapdown configuration, utilising Incosym Inc two-degree-of-freedom gyros. The entire inertial reference assembly is substantially reduced in size. A digital microprocessor performs all the measurement data processing, instrument torquing computation, scaling, attitude and navigation functions. Steering commands and other autopilot interfaces are provided.

Specifications
Dimensions: 152 × 152 × 215 mm
Weight: 3.0 kg
Power supply: ±15 and ±5 V DC, 35 W

Operational status
The LCINS has been selected as the Three Axis Inertial Measurement System (TAIMS) for a classified military aircraft programme and as the inertial sensor assembly for a second classified military programme.

Contractor
Delco Electronics.

VERIFIED

Argus moving map displays

Argus moving map displays place all vital navigation and position information on an instantly readable display in front of the pilot. By minimising head-down time and giving the pilot more time to manage the aircraft, the Argus makes flying safer. The display can interface with, and receive its navigation data from most Loran C, GPS and other long-range navigation systems.

The Argus 3000 is designed for VFR use in light single- and twin-engined aircraft. Like the Argus 5000, it interfaces with most popular Loran C and GPS receivers. Its database contains over 11,000 landing facilities, 6,500 navaids and every special use airspace, including TCAs and ARSAs.

The ARGUS 5000's comprehensive versatile moving map display shows all TCAs, ARSAs, navaids and landing facilities. At the touch of a button, the information submode presents detailed information about any on-screen facility from its own field-replaceable database. The Argus 5000 also provides a convenient digital readout of bearing or radial and distance to any selected facility.

The larger Argus 7000 provides all the information contained in the Argus 5000 on a bigger display. The Argus 7000 fits in a similar tray to the Argus 3000 and 5000, but has 2.3 times the screen area of the standard size Argus models. It can be updated without the need to remove it from the instrument panel.

An international database, containing navigation information for airports, navaids and special use airspace worldwide is available for use with the Argus 3000, 5000 and 7000.

The Argus range of moving map displays also includes an integrated flight planning feature as an option for new units or an upgrade for units already in service. This enables pilots to create and activate flight plans directly on the moving map. It allows up to 20 waypoints to be displayed at one time on one screen

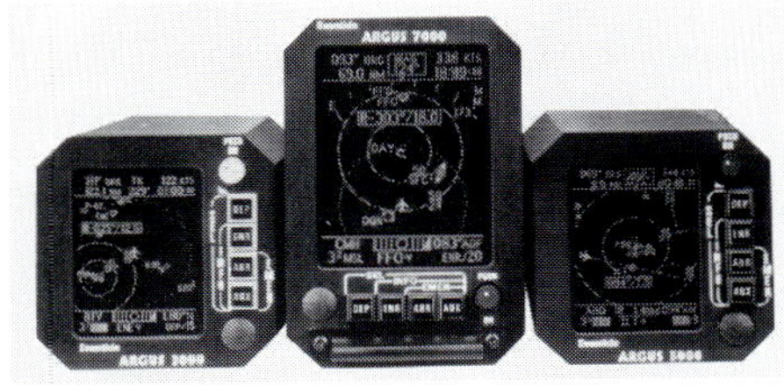

Argus moving map displays showing (left to right) the 3000, 7000 and 5000 models

and up to 150 user-defined waypoints can be stored in non-volatile memory. With the Flight-Plan-plus software the Argus can now store 30 flight plans with up to 20 waypoints per flight plan - 600 user waypoints in total.

With the RMI adaptor, the Argus moving map display can indicate up to two ADF or VOR pointers as with a traditional RMI display and also provides a digital bearing. This mode is approved for ADF or VOR approaches.

The RMI adaptor plugs directly in to the back of the Argus or can be mounted remotely. An RMI/ARINC adaptor incorporates a data converter that allows Argus to accept ARINC 419 and ARINC 429 data from VLF/Omega, inertial and flight management systems.

The Argus 5000/CE and 7000/CE units display colour screen graphics and add several new hardware and software enhancements including flight recording, an internal barometric pressure sensor, a rotary encoder for easy data entry and convenient database updates.

Advanced display technology allows the Argus 5000/CE and 7000/CE to present graphics in red, green and yellow, while retaining the sharp, bright, sunlight-readable qualities of the monochrome Argus models (which continue to be available). With these new colour Argus models, the pilot can select a colour scheme to display a variety of graphic data - including the ability to 'colour code' Class B and Class C airspace.

The built-in barometric pressure sensor gives these Argus models new capabilities which provide the pilot greater situational awareness; the aircraft's current altitude is taken into account when displaying restricted areas.

Flight recording is another new feature. The Argus will record current latitude, longitude, time, date, altitude, ground speed and track and heading in its non-volatile memory for up to 10 hours of flight, which can be 'played back' in real or compressed time on the Argus screen.

With the Argus 5000/CE and 7000/CE, updating the database is easier and less expensive. Updates are available over the Internet, on floppy disk and on PC (PCMCIA) Cards.

Argus can interface with, and receive its navigation data from, most GPS, Loran C and other long-range navigation systems. The units are available with a US or international (worldwide) database. Both databases contain navigation information for airports, navaids and special use airspace. The Argus 5000 and 7000 units are TSOd and IFR-approvable. The Argus 3000 is TSOd as well, when equipped with the international database.

Specifications
Dimensions:
(Argus 3000/5000) 81.3 × 81.3 × 269.4 mm
(Argus 7000) 81.3 × 121.9 × 273 mm
Weight:
(Argus 3000/5000) 1.6 kg
(Argus 7000) 2 kg
Power supply: 11-33 V DC, 15 W

(RMI Adaptor)
Dimensions: 79.4 × 79.4 × 69.34 mm
Weight:
(RMI) 0.37 kg
(RMI/ARINC) 0.41 kg
Power supply: 11-33 V DC, 2-3 W
Environmental: RTCA-DO-160B

Contractor
Eventide Avionics Division.

VERIFIED

Weather Display Adaptor (WDA)

With the Weather Display Adaptor (WDA), the Argus moving map displays weather information from BFGoodrich Stormscope equipment. A single, high-resolution display gives the benefits of Stormscope plus the advantages of the Argus moving map.

The Eventide 5005 and 7005 WDAs are compatible with WX-10A and WX-11 Series II Stormscopes. By replacing the Stormscope display with an Argus moving map display and tieing the WDA into the Stormscope processor, the Argus screen can display information from the WX-10A or WX-11 separate WD mode. All Stormscope functions are executed through the Argus function keys, essentially allowing for two instruments in one panel space.

The 5007-05 and 7007-05 WDAs interface with the WX-1000E weather mapping Stormscope system and emulate the Stormscope functions on the Argus display. Weather mapping information can be shown either on both the Argus and the Stormscope displays or solely on the Argus screen, if panel space is a problem.

Contractor
Eventide Avionics Division.

VERIFIED

IFR/VFR avionics stacks

Garmin have integrated their products into IFR and VFR avionics stacks comprising the following units:

IFR avionics stack
GMA 340 TSOd audio panel
Dual GNC 300 TSO IFR certified GPS/Comm
GTX 320 Class 1A transponder
MD 41 annunciator
GI 102 course deviation indicator
GI 106 course deviation indicator with glide slope

VFR avionics stack
GMA 340 TSOd audio panel
GNC 250XL GPS/Comm with moving map
GPS 150XL GPS/Comm with moving map
GTX 320 Class 1A transponder
GI 102 course deviation indicator

Contractor
Garmin International.

NEW ENTRY

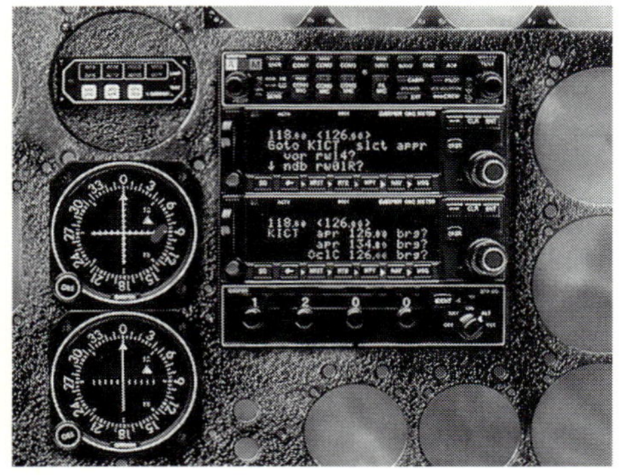

Garmin IFR stack ***1998***/0015379

GNC 250 and GNC 300 TSO GPS/Comm

GNC 250: VFR GPS/Comm
The panel-mounted GNC 250 comprises a GPS sensor and a fully-TSOd, 760-channel VHF communications radio. Supplemented with a transponder, a GNC 250 serves as a complete avionics package for VFR operation. The GNC 250 includes Garmin's MultiTrac8™ receiver, which provides clear, accurate navigation information linked to a choice of three navigation databases (Americas, International or Worldwide) and Jeppesen® information.

The GNC 250 incorporates a TSOd, full-function digital transceiver. Standard features include flip-flop frequency selection - with frequency prompts from take-off to touchdown - automatic squelch and a 'cursor select' function, which enables selection of a standby frequency from the GPS database.

GNC 300: IFR-Certified GPS/Comm
The GNC 300 adds IFR certification to the nav capabilities of the GNC 250. Complying with the FAAs TSO C129 A1, the GNC 300 offers non-precision approach, en route and terminal GPS capability under Instrument Flight Rules (IFR). Like the GNC 250, the GNC 300 incorporates a VHF communications radio. It presents its information on the same 80-character vacuum-fluorescent display. Also, like the GNC 250, the GNC 300 is ideal for either new aircraft installations or as an update to the current panel.

Incorporating the same MultiTrac8™ GPS receiver, the GNC 300 streamlines 'direct-to' navigation. Its comprehensive Jeppesen® database offers airport, runway, VOR, NDB, intersection and communication frequency information for either the western hemisphere (Americas option) or the rest of the world (International option). Ample room for trip planning, waypoint storage and a 'nearest airports' list provides additional capability. Sectorised, altitude-sensitive SUA warnings are also included.

The GNC 300 also features a NAV/Com page that automatically provides frequencies from pre-loaded flight plans.

Fully-TSOd, the GNC 300 also incorporates a

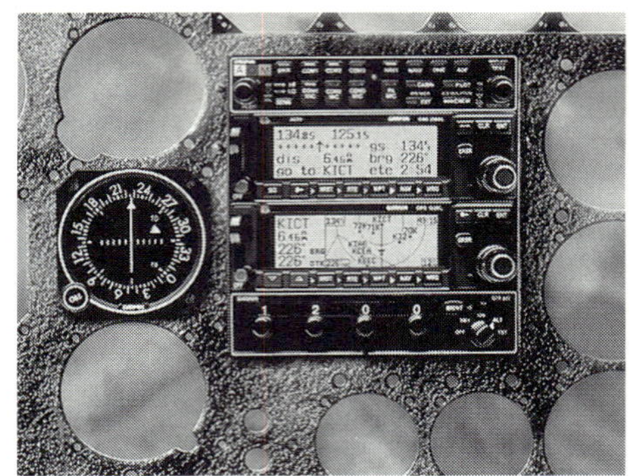

Garmin VFR stack ***1998***/0015380

powerful 760-channel digital transceiver for audio communications. As with the GNC 250, the GNC 300s Comm section includes such standard features as flip-flop frequency selection, automatic squelch and 'cursor select' function.

Both the GNC 250 and 300 interface with most flight control, EFIS, HSI and other cockpit systems.

Contractor
Garmin International.

VERIFIED

GNC 250XL GPS/Comm system

The GNC 250XL navigation system is a GPS 150XL 12-channel global positioning system, to which a 760 channel VHF transmitter has been added.

Operational status
The GNC 250XL has been available since May 1997.

Contractor
Garmin International.

NEW ENTRY

Gamin GNC 250XL GPS/Comm system ***1998***/0015378

GPS 150/GPS 150XL Global Positioning System

The GPS 150 has an extensive Jeppesen database, 1,000 waypoint storage and 60 character screen. It features a fully dimmable display which provides three full 20 character lines of information in a 2 in (50.8 mm) panel space. The vacuum fluorescent dot matrix display and an advanced optical filter provide readability from almost any angle, even in daylight.

The screen displays information which includes airports, VORs, NDBs, all intersections and up to 1,000 user-defined waypoints. Additional information includes all comms frequencies, runway lengths, FSS frequencies, minimum safe altitudes, Class B/C frequency information and city, state, facility name and country. The data is stored on a front-loading miniature data card.

The GPS 150 gives one button access to instantaneous and continuously updated lists of the nearest waypoints, including airports, VORs, NDBs, intersections, FSS frequencies and user-defined positions. Some 20 reversible flight plans with up to 31 waypoints each can be stored.

Additional features include a Direct To function to set an instantaneous course to any waypoint, an AutoStore function to build routes easily while en route, a

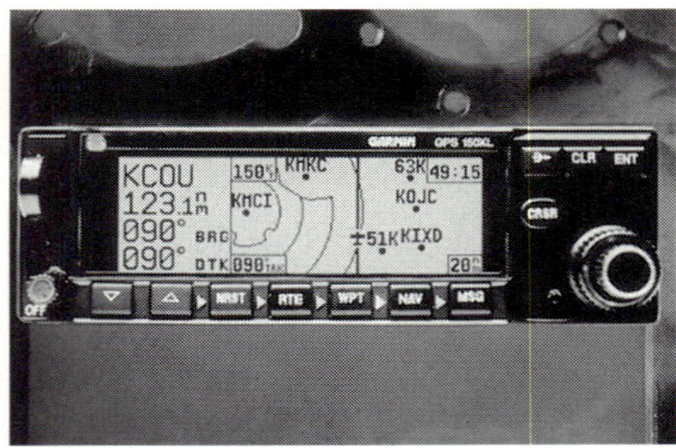

Garmin GPS 150XL global positioning system ***1998***/0015377

message function to provide informative, warning messages or time-scheduled messages for trip and maintenance planning, Offset Navigation for parallel tracking and AutoSearch for quick search and rescue grid activation.

The GPS 150 continuously tracks and uses up to eight satellites with one second updates and horizon to horizon coverage. The unit has a built-in back-up battery that can provide up to four hours of operation, even if aircraft electrical power is lost.

The GPS 150XL updates the GPS 150 by adding very high-resolution moving map graphics; and a 12 parallel-channel GPS receiver.

The display is a DSTN (Double Super Twist Nematic) system, which doubles the number of pixels available to improve the quality of the display.

The GPS 150XL has been available since May 1997.

Specifications
Dimensions: 159 × 51 × 148 mm
Weight:
(unit with battery pack) 0.95 kg
(rack) 0.32 kg
Power supply: 10-33 V DC
115 V AC or 230 V AC with battery charger
Rechargeable battery pack

Contractor
Garmin International.

UPDATED

GPS 155 TSO Global Positioning System

The GPS TSO is a panel-mounted receiver similar in design and features to the GPS 150 (see previous item) which can be upgraded to TSO standards. The database in the GPS 155 TSO includes non-precision approaches for US airports, all published SIDs and STARs, 1,000 user-defined waypoints and 20 reversible routes with up to 31 waypoints in each. It contains an extensive, updatable Jeppesen database providing detailed information on airports, VORs, NDBs, intersections, comms frequencies, runways, FSS and MSAs. It also lists the nine nearest airports, VORs, NDBs, waypoints and FSS frequency. The GPS 155 TSO has a built-in Ni/Cd battery back-up, capable of providing emergency power in the event of aircraft electrical power failure.

The GPS 155/165 TSO can fully interface with flight controls, EFIS, HSI, moving map, altitude encoder, fuel management and other aircraft systems.

In February 1995, the GPS 155 TSO received FAA A1 certification, clearing it for en route, terminal and non-precision approach IFR operations.

Specifications
Dimensions: 158.7 × 50.8 × 133.3 mm
Weight: 0.97 kg
Power supply: 115-230 V AC with battery charger
10-33 V DC, rechargeable battery
Battery life: up to 2 h
Temperature range: −30 to +70°C
Accuracy (RMS):
(position) 15 m
(velocity) 0.1 kt

Contractor
Garmin International.

VERIFIED

GPS 165 TSO Global Positioning System

The GPS 165 TSO is the dzus-rail configuration of the GPS 155 (see earlier item). It is currently being considered by the FAA for Class A1 certification under TSO C129. The GPS 165 TSO tracks and uses up to eight satellites continuously, providing 1 second updates with horizon to horizon coverage.

GPS 165 TSO features include 1,000 user-defined waypoints, 20 reversible routes with up to 31 waypoints each, single key Direct To operation, Jeppesen NavData stored on an updatable removable data card, Auto Search search and rescue grid activation function, optional remote battery pack back-up providing two hours of operation in the event of aircraft power failure and interfaces for ARINC 429, GAMA 429, RS-232 and RS-422.

In April 1995, the GPS 165 TSO achieved FAA C129 A1 certification for IFR approach operations.

Specifications

Dimensions: 146 × 57.1 × 133.3 mm
Weight: 0.63 kg
Power supply: 115-230 V AC with battery charger
10-33 V DC, rechargeable battery
Battery life: up to 2 h
Temperature range: −30 to +70°C
Accuracy (RMS):
(position) 15 m
(velocity) 0.1 kt

Contractor

Garmin International.

VERIFIED

ARN-155 Precision Landing System Receiver (PLSR)

The PLSR is a multimode precision landing system receiver designed to replace existing DoD navaid avionics. PLSR is interoperable with both civilian (ICAO/FAA compliant) and military Cat.I and II landing aids, including Instrument Landing System (ILS), Microwave Landing System (MLS) and Differential Global Positioning System (DGPS). PLSR meets the new international FM interference requirements (a critical requirement in Europe for ILS operation) and also incorporates the latest GPS JPO GRAM compliant, all-in-view, P/Y GPS receiver technology. In addition to the embedded VHF datalink for local area DGPS augmentation, PLSR also contains VOR capability. A growth version adds the J-band Pulse Coded Scanning Beam (PSCB) capability for US Navy carrier and marine operations.

The Receiver Processor Unit (RPU) is capable of being controlled via a MIL-STD-1553B digital databus or ARINC 429 interface, or with an optional Control Display Unit (CDU). The RPU also contains discrete inputs and outputs for compatibility with older aircraft.

The AN/ARN-155 comprises two units: the R-2556/ARN-155 Receiver Processor Unit (RPU) and the C-12377/ARN-155 radio and Control and Display Unit (CDU). It accepts DME, TACAN and GPS ranging information from external avionics for computed mode MSL approaches and accommodates both split site offset and collocated landing site scenarios.

Specifications

Dimensions:
(RPU) 258 × 127 × 91 mm
(CDU) 146 × 121 × 38 mm
Weight:
(RPU) 4.49 kg
(CDU) 1.22 kg
Power supply: 28 V DC, <30 W
Interfaces: MIL-STD-1553, ARINC 429, discretes
Environmental: MIL-E-5400 Class 2
Reliability: >10,500 h MTBF per MIL-HDBK-217E

Operational status

GEC-Marconi Hazeltine Corporation was selected by the US Air Force in June 1993 to build and test 30 preproduction ILS/MLS 2-Band PLSR systems, including a full qualification programme. In August 1996 the Air Force awarded GEC-Marconi Hazeltine Corporation an Engineering Change Proposal (ECP) to add DGPS with a 12-channel, P/Y code, all-in-view GPS receiver, thereby providing an ILS/MLS/DGPS three-band PLSR system. In December 1996 the USAF selected PLSR for the C-17 fleet, adding VOR as a new mode and initiating a production programme with deliveries in late 1997. Follow-on full-scale production is scheduled to start in 1998.

The AN/ARN-155 PLSR is a leading candidate for future landing avionics in the tri-service US DoD's Joint Precision Approach and Landing System (JPALS) study.

Precision landing system receiver **1997**/0003107

Contractor

GEC-Marconi Hazeltine Corporation.

VERIFIED

AN/ASN-128B Doppler navigation system

The AN/ASN-128B is the US Army's standard lightweight helicopter airborne Doppler navigator and comprises three units: a receiver/transmitter/antenna, signal data converter and computer/display unit. A steering hover indicator can also be included as an option. With inputs from heading and vertical references, the system provides aircraft velocity, present position and steering information from ground level to above 10,000 ft.

GEC-Marconi Hazeltine is under contract to the US Army to insert GPS capability into the AN/ASN-128. Identified as the AN/ASN-128B, the contract is to add a P/Y code GPS receiver, in the AN/ASN-128B signal data converter, and the modification of the computer display unit software to enable display of both Doppler navigation and GPS data. The contract is to modify 1,693 AN/ASN-128 systems currently installed in UH-60A/L Black Hawks and CH-47D Chinooks. These kits would also be applicable to most other AN/ASN-128 systems.

GEC-Marconi Hazeltine has also developed a field kit enabling the AN/ASN-128 Doppler navigator to interface with a stand-alone Trimble GPS receiver, for continuous update of Doppler-derived present position with valid GPS data.

Specifications

Volume: 20,724 cm^3
Weight: 13.61 kg
(hover indicator) 0.9 kg
Propagation: 4-beam configuration operating FM/continuous wave transmissions in K-band. Beam shaping eliminates the need for a land/sea switch. The single transmit/receive antenna uses full aperture in both modes to minimise beamwidth and reduce fluctuation noise
Number of waypoints: >100
Self-test: localisation of faults at LRU level by BITE
Reliability:
(complete system) >2,100 h MTBF

Operational status

In production. Approximately 4,500 sets AN/ASN-128 have been manufactured, of which approximately 1,700 are AN/ASN-128B sets.

In service with US Army Sikorsky UH-60A and UH-60L, Bell AH-1F, Boeing AH-64A and CH-47D helicopters; also in Royal Australian Air Force Bell UH-1H, UH-60 and CH-47D, Hellenic Air Force UH-1H, Jordanian Army AH-1S, Pakistan AH-1S, South Korean CH-47, UH-60 and AH-1S, Spanish Army Eurocopter BO 105 and CH-47B, Taiwan Army CH-47B and Turkish UH-60. The system, with hover indicator, has been provided for the German Army Eurocopter BO 105, PAH-1, UH-1D and VBH helicopters. Also built under licence in Japan for the JASDF AH-1S and CH-47D. In use in Austria, Bahrain, Brunei, China, Denmark, Dubai, Egypt, Greece, Netherlands, Singapore, Spain, Taiwan and Thailand. Version kit deliveries started in the second quarter of 1996.

Contractor

GEC-Marconi Hazeltine Corporation.

UPDATED

AN/ASN-137 Doppler navigation system

The AN/ASN-137 is the multiplexed version of the AN/ASN-128. It is compatible with the MIL-STD-1553B databus and has ARINC 575 or 429 outputs.

The AN/ASN-137 uses the same receiver/transmitter/antenna as the ASN-128 and has an optional control/display unit of the same size, with the

The GEC-Marconi Hazeltine AN/ASN-128 Doppler navigation system: from left, velocity sensor (transmitter/receiver), control/display unit and signal data converter

same front panel as the ASN-128. Additional features in the AN/ASN-137 include hover bias correction for precision hovering, 12-point magnetic deviation entry and the addition of latitude and longitude and UTM grid zone outputs in MIL-STD-1553 output.

Specifications

Weight: 11.7 kg
Power supply: 28 V DC, 87 W
Reliability: 2,800 h MTBF calculated

Operational status

In production and in service. Approximately 1,600 sets have been manufactured.

The AN/ASN-137 is installed on the US Army's Bell OH-58D and the US Navy's Sikorsky VH-3D and VH-60 helicopters. It is also currently installed as a break-in change on AH-64A helicopters, on the US Army MH-47E and MH-60K helicopters and the US Air Force MH-60G Pave Hawk helicopter.

Contractor

GEC-Marconi Hazeltine Corporation.

VERIFIED

AN/ASN-157 Doppler navigation set

The AN/ASN-157 is a Doppler navigation set in a single LRU, emulating the two-unit AN/ASN-137 and providing all ASN-137 and ASN-128 functions.

Specifications

Dimensions: 370 × 342 × 57 mm
Weight: 5.7 kg
Reliability: 7,300 h MTBF

Operational status

In production and in service. Installed on the Agusta A 109 helicopter for the Belgian Army, the CH-47D glass cockpit Chinook cargo helicopter for the Netherlands Air Force and the AH-64D Longbow Apache attack helicopter for the US Army. Selected for the Westland/Boeing WAH-64 Apache attack helicopter for the British Army.

Contractor

GEC-Marconi Hazeltine Corporation.

UPDATED

GRD-2116 overwater Doppler navigation system

The GRD-2116 Doppler velocity sensor/navigation system is designed to provide optimum performance during both overwater and overland operations.

Improvement in received signal-to-noise is accomplished by increasing transmitter power, using a low-noise receiver amplifier and transmitting and receiving RF energy via separate electrical apertures to improve isolation. Low sidelobes and power management are used to achieve low probability of intercept.

Mechanically, the unit is the same size and weight as the AN/ASN-157, has the same mounting configuration and the same ARINC 575 and MIL-STD-1553 interfaces. A digital frequency tracker provides rapid signal acquisition and accurate hover.

Contractor

GEC-Marconi Hazeltine Corporation.

VERIFIED

GA/1000 VHF nav/com system

The Genave GA/1000 is a VHF/AM nav/com system for light aircraft. It can be panel-mounted as a single unit or, alternatively, installed so that the VOR/Loc indicator section can be retained as a panel instrument with the control head mounted elsewhere in the cabin.

The system's communications section covers 720 channels in the VHF band 118 to 135.975 MHz at 25 kHz spacing. The independent navigation receiver covers the band 108 to 117.95 MHz with 50 kHz separation and covers 200 navigation channels (160 VOR and 40 localiser). The use of separate receivers for communications and navigation functions permits radio operation without disrupting reception of navigation signals. Both sections employ hot filament digital readout displays for confirmation of the frequency selection. This is carried out in each case through use of dual-frequency selector knobs on the front case of the control unit.

Transmitter output power is a nominal 4 W carrier signal and a radio frequency actuated LED is incorporated as a transmit indicator. Communications receiver gain is automatically controlled and automatic squelch with manual disable is also provided. Audio output may be either through a 4 ohm speaker for which a 3 W output is available, or into 600 ohm earphones from a 100 mW output. The same values are applicable to the audio outputs from the navigation receiver section. In VOR mode, transmissions from the system's transmitter section cause no visible deflection of the course deviation indicator needle. Both VOR and localiser have ARINC standard autopilot outputs.

The GA/1000 uses solid-state integrated circuitry and a single crystal digital synthesiser is employed for frequency generation. Receiver demodulation circuitry is of the single conversion type. A range of antenna systems and associated antenna coupler units is available to match differing aircraft installation requirements and a speaker muting relay is also obtainable.

Operational status

In production and in service.

Contractor

Genave Inc.

VERIFIED

Digital Map Generator (DMG)

The Digital Map Generator (DMG) uses stored terrain data to generate moving terrain map displays, to eliminate the need for paper maps in the cockpit and to improve the pilot's situational awareness. These map displays can either be digitised representations of standard aeronautical charts or topographical displays created from digitised elevation and cultural feature data. Symbology overlays are added to show mission features such as waypoints, flight paths and threat and target positions, and assist the pilot with terrain-aided navigation and ground collision avoidance.

Harris is also providing the high-speed databus interface for the data transfer unit, a mass memory storage device which stores digital map and mission planning data.

Operational status

In development for the RAH-66 Comanche helicopter.

Contractor

Harris Corporation.

VERIFIED

Digital Terrain Management and Display (DTM/D)

The Digital Terrain Management and Display (DTM/D) has been test flown on the US Air Force Advanced Fighter Technology Integration (AFTI)/F-16 aircraft. It combines a Harris digital map generation system with a colour multifunction display from AlliedSignal, and is expected to provide a navigation system for combat aircraft 10 times more accurate than conventional systems but with only minimal electromagnetic emissions, at the same time leaving the aircraft's radar free for other purposes.

The DTM/D correlates radio altimeter information with a digitally stored map, providing for covert navigation which will, according to the USAF, enhance terrain-masking tactics and give the pilot the ability to look beyond hills in front of and to the side of the aircraft with a passive ranging capability. This will allow the pilot to work out the best route well in advance and be able constantly to relate his position to threats and targets.

Operational status

Flight testing in the AFTI/F-16 has been completed. Harris worked under a US$5.9 million contract jointly funded by the US Army and Air Force.

Contractor

Harris Corporation.

VERIFIED

Advanced GNS/IRS integrated navigation system

Honeywell has integrated laser inertial functions with the Global Navigation Satellite System (GNSS). The advanced GNS/IRS offers high reliability, long life, low power consumption, small size, light weight, fast alignment, full performance for alignment up to 78° latitude and improved BITE.

The advanced 4 MCU IRS is 60 per cent smaller, 40 per cent lighter and uses 50 per cent of the power of Honeywell's 10 MCU systems, while meeting the same performance specifications. The heart of the system is the Ring Laser Gyro (RLG). The advanced IRS is fully provisioned for GNSS integration. The blending of these two systems into one navigation solution offers precise navigation accuracies.

The GNSS unit will continuously track all satellites in view to give accuracies of 25 m or better with selective availability switched off. The GNSS unit offers growth potential for Wide Area Augmentation System (WAAS) and Local Area Augmentation System (LAAS) as well as the capability for differential GPS. Further enhancements under development include capabilities to support precision approaches to unimproved runways, automatic dependent surveillance and sole-means navigation.

Specifications

Dimensions: 124.5 × 317.5 × 193 mm
Weight: 12.25 kg
Accuracy:
(navigation) (IRS) 2 n miles/h,
(GNS/IRS hybrid) 25 m
(velocity) (IRS) 12 kt
(GNS/IRS hybrid) 0.3 kt
(attitude) 0.1°
(heading) 0.4°

Contractor

Honeywell Inc Air Transport Systems.

UPDATED

Air Data Inertial Reference System (ADIRS)

The ARINC 738 Air Data Inertial Reference System (ADIRS) is an air data computer combined with an Inertial Reference System (IRS).

The ADIRS provides complete ARINC inertial reference system outputs including primary attitude and heading, body rates, acceleration, groundspeed, velocity and aircraft position and ARINC 706 air data outputs, which include altitude, true airspeed, Mach number, air temperature and angle of attack.

Each ADIRS is equipped with three air data inertial

reference units, one control display unit and eight air data modules mounted remotely, adjacent to the pitot and static pressure sensors.

The air data reference electronics and the laser gyro inertial reference units are packaged in a 10 MCU box or 4 MCU box and require a nominal power of 109 W. The unit meets the functional requirements of ARINC 738 and the environmental requirements of DO-160b. On the A320, aircraft maintenance is simplified by extensive reporting of ADIRS LRU Operational Status to the centralised fault data system.

The air data module requires a nominal power of 1.8 W. The unit meets the environmental requirements of DO-160b and features a solid-state pressure transducer.

The control display unit is packaged in accordance with ARINC 738 and requires a nominal power of 5 W exclusive of warning lights. The unit meets the functional requirements of ARINC 738 and the environmental requirements of DO-160b. It features a liquid crystal display.

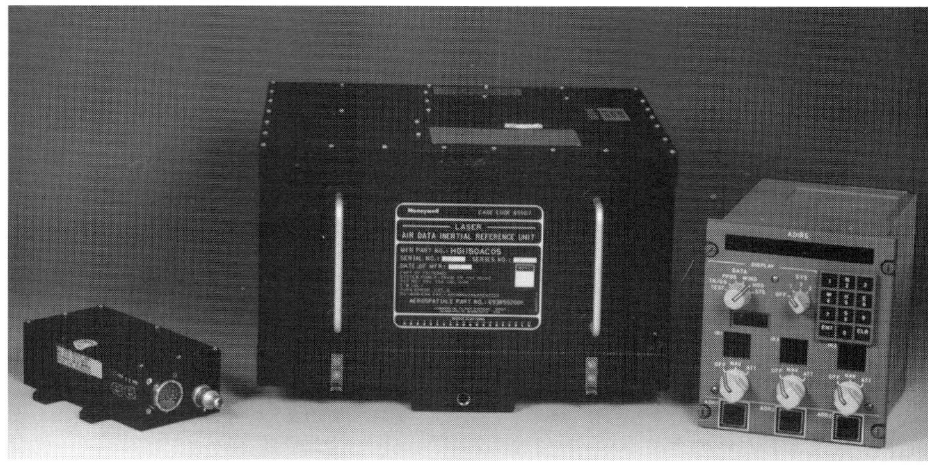

The Honeywell ADIRS air data inertial reference system showing (left) the air data module, (centre) the integrated air data/inertial reference unit and (right) the integrated control display unit

Specifications

Dimensions:
(ADIRU) 322.6 × 322.6 × 193 mm
(ADM) 50.8 × 76.2 × 152.4 mm
(CDU) 170.2 × 146 × 152.4 mm
Weight:
(ADIRU) 19.5 kg (10 MCU box)
or 10 kg (4 MCU box)
(ADM) 0.63 kg
(CDU) 2.27 kg

Operational status

ADIRS equips Airbus A320, A330 and A340 aircraft.

Contractor

Honeywell Inc Air Transport Systems.

VERIFIED

Fault Tolerant Air Data Inertial Reference System (FT-ADIRS)

The Fault Tolerant Air Data Inertial Reference System (FT-ADIRS) consists of a Fault Tolerant Air Data Inertial Reference Unit (FT-ADIRU), a Secondary Attitude and Air data Reference Unit (SAARU) and six Air Data Modules (ADMs). The FT-ADIRU provides attitude and heading data for inertial navigation as well as air data computations. The SAARU provides a back-up source of attitude and air data computations. The ADMs provide both the FT-ADIRU and SAARU with three redundant sources of static and pitot pressure data.

Operational status

The FT-ADIRS is standard on the Boeing 777 aircraft.

Contractor

Honeywell Inc Air Transport Systems.

VERIFIED

HT 9100 GNSS navigation management system

The HT 9100 system is a full flight regime lateral navigation system, designed to provide state-of-the-art navigation performance. The system combines the benefits of Global Navigation Satellite Sensor (GNSS) technologies and innovations with established airline Flight Management System (FMS) operational procedures.

The HT 9100 is capable and certified for RNP/ANP (Required Navigation Performance/Actual Navigation Performance), similar to the Honeywell 747-400 FANS 1 software. (RNP is the required performance accuracy of a particular segment of airspace. ANP is a measure of the uncertainty in the position estimate of the system.)

The HT 9100 is designed to permit GPS navigation under Instrument Flight Rules (IFR) conditions for en route, terminal area and approach operations. In addition, the system complies with FAA Notice N8110.60, permitting GPS as the sole navigation system for oceanic and remote operations.

The system shipset consists of a Navigation Processor Unit (NPU), Multifunction Control and Display Unit (MCDU) and an Antenna Coupler Unit (ACU). The NPU is the primary processing and control unit of the HT 9100. It contains a 12-channel Global Positioning System (GPS) receiver, central processing units (CPUs) and navigation database for flight management functions. The MCDU controls the user input and display function. Data entry and display to the pilots are accomplished through the use of a 66-key alphanumeric keyboard with 12 line-select keys and 5.5 in diagonal multicolour active-matrix liquid crystal flat panel dislay. The ACU receives, amplifies, conditions and sends GPS signals to the receiver in the NPU.

Operational status

HT 9100 GNSS Navigation Management System has received Technical Standard Order (TSO C129 A1) approval from the US Federal Aviation Administration. This approval encompasses Standard Instrument Departures (SIDs) Standard Terminal Arrival Routes (STARs), GPS overlay and GPS approaches as well as Company routes, J-routes and V-route capability.

In addition, the HT 9100 has also received Supplemental Type Certificate (STC) approval from the FAA on 727, DC-10, L-1011 and MD-80 aircraft. The certification gives operators the ability to fly IFR (Instrument Flight Rules), supplemental en route, terminal, non-precision approaches and primary oceanic/remote navigation with the system.

The Honeywell/Trimble team forged a strategic alliance in June 1996 and announced in December a launch order for more than 500 HT 9100s from American Airlines. Many other orders have followed.

Contractors

Honeywell Inc Air Transport Systems.
Trimble Navigation Limited.

UPDATED

Integrated Global Positioning/ Inertial Reference System (GPIRS)

The integrated Global Positioning/Inertial Reference System (GPIRS) combines the best of global positioning and inertial reference systems to provide very accurate worldwide navigation.

The inertial reference system is upgraded to an integrated GPIRS by adding GPS processing software in the inertial reference unit and coupling it to an ARINC 743 GPS sensor unit.

The GNSSU (Global Navigation Satellite Sensor Unit) is a remote-mounted unit that provides all the functions necessary for either integrated or stand-alone configurations. All existing Honeywell IRUs can be modified into GPIRUs which are one-way interchangeable with existing IRUs.

The GPS sensor unit receives satellite data using a 12-channel design. Data are received from all satellites in view, with updates once per second. The all-digital multiple correlator design allows for satellite tracking during periods of low signal-to-noise. This enables the receiver to track satellites to a zero degree elevation angle.

Specifications

Dimensions:
(IRS) 317.5 × 320 × 198.1 mm
(GPSSU) 190.5 × 215.9 × 55.9 mm
Weight:
(IRS) 19 kg
(GPSSU) 2.27 kg
Time to first solution:
(IRS) 10 min
(GPSSU) 4 min typical
Accuracy:
(position) (IRS) 2 n miles/h 95%
(GPSSU) 25 m

Honeywell/Trimble HT 9100 global positioning/inertial reference system **1997**/0003108

(velocity) (IRS) 8 kt
(GPSSU) 1.8 kt
(time) (GPSSU) 350 ns
Reliability:
(IRS) 5,000 h MTBF
(GPSSU) 20,000 h MTBF

Operational status

Selected as standard equipment by British Aerospace for its smaller air transport aircraft, and in a dual fit for the Boeing MD 90-30.

Contractor

Honeywell Inc Air Transport Systems.

UPDATED

Laser Inertial Reference System (IRS)

The world's first production ring laser gyro Inertial Reference System (IRS) was chosen by Boeing as part of the avionics package common to both the 767 and 757. It was also selected for the Boeing 737-300/400/500, MD-80 and MD-11, and the Airbus A320, A330 and A340, Fokker 100, and BAe 146-300.

The strapdown configuration is so called because the gyrostabilised platform of current conventional inertial navigation and attitude reference systems is replaced by three ring laser gyro units mounted rigidly to the aircraft and at right angles to one another. The laser gyro detects and measures angular rates of motion by measuring the frequency difference between two contrarotating laser beams made to circulate (hence the term ring) in a triangular cavity by mirrors. When the units are at rest the distances travelled by each beam are the same, as are the frequencies. When the unit rotates, one path lengthens while the other shortens and so a frequency difference is established proportional to the rate of rotation of the unit. The difference is measured and processed digitally in ARINC 704 format as aircraft attitude in pitch, roll and yaw.

Since the accelerometers are mounted rigidly in the box, their signals are related to aircraft axes and have to be processed to convert them to the external inertial reference frame necessary to provide navigation and flight control information and guidance.

A strapdown system has no moving parts to wear, fail or become misaligned; no gimbals, torque motors, spin-motors, slip-rings, or resolvers, and no scheduled maintenance, realignment or recalibration requirements are anticipated. A typical installation comprises three inertial reference units (containing the sensing and computing elements) and a display unit. The Honeywell laser device is contained within a low expansion, triangular glass block, with a 34 cm path length. It has demonstrated a MTBF of 20,000 hours during more than 50 million flight hours.

Specifications

Dimensions: 4 MCU or 10 MCU
Weight:
(4 MCU) 12.24 kg
(10 MCU) 19.5 kg
Power:
(4 MCU) 44 W
(10 MCU) 86 W
Outputs: primary attitude information to displays and Automatic Flight Control Systems (AFCS), linear accelerations, velocity vector and angular rates to AFCS, wind shear detection and energy management, magnetic heading for displays and AFCS and long-range navigation data

ARINC 704
Accuracy (10 h flight - 95% probability):
(position) 2 n miles/h
(velocity) 12 kt
Self-test: BIT (initiated and continuous) detects 95% of failures with 95% confidence level
Reliability: >5,000 h MTBF predicted

Operational status

In production for Boeing 767, 757 and 737-300/400/500, MD-11 and MD-80 Series aircraft; and Airbus A300-600, A310, A320, A330 and A340.

Contractor

Honeywell Inc Air Transport Systems.

UPDATED

Secondary Attitude and Air data Reference Unit (SAARU)

The SAARU, a fail-safe and highly reliable device, operates as a secondary system to the fault-tolerant air data inertial reference unit (see earlier item).

SAARU measures the aircraft's linear and rotational motions and computes air data measurements to provide fail-safe secondary attitude and air data reference information. The 10 MCU device also provides digital attitude and air data reference information to the cockpit LCD standby displays.

Operational status

Selected for the Boeing 777 aircraft.

Contractor

Honeywell Inc Air Transport Systems.

UPDATED

AHZ-800 Attitude Heading Reference System (AHRS)

The AHZ-800 is the next generation Attitude Heading Reference System (AHRS) designed for high performance and high reliability while attaining lower power dissipation and reduced size and weight. This is accomplished through the use of advanced manufacturing techniques, such as very large-scale integration and application specific integrated circuits, and fibre optic rate-sensing advanced sensor technology.

Honeywell has developed a practical interferometric fibre optic gyro sensor that replaces the heavier less reliable spinning iron rate-sensors used in the conventional AHRS. The advent of the interferometric fibre optic gyro sensor makes the AHZ-800 a truly solid-state device. The AHZ-800 is a 4 MCU package that outputs attitude, heading and rate data on ARINC 429 and ASCB digital buses. The attitude and heading source approaches the performance of an inertial reference system, but at a significantly lower cost.

Operational status

The AHZ-800 is in service with the Dornier 328 regional airliner.

Contractor

Honeywell Inc Business & Commuter Aviation Systems.

VERIFIED

GPS sensor unit

The Honeywell two-channel fast sequencing receiver uses data from all satellites in view. This makes it capable of FAA certified supplemental navigation. The GPS Sensor Unit (GPSSU) is an improved version of the Honeywell system that has been delivered to special mission operators over the past five years. The advanced receiver design has demonstrated superior acquisition and tracking of satellites right down to the horizon.

The Honeywell GPSSU, which meets the ARINC 743 requirements for GPS, provides autonomous GPS data such as position and velocities directly to a Honeywell flight management system or Lasernav II navigation management system. It also provides range, range rate and time, for blending with inertial data in the Laseref III inertial reference system and the GPS/INU configurations of the Laseref and Laseref II.

Specifications

Dimensions: 215.9 × 190.5 × 55.9 mm
Weight: 2.27 kg

Operational status

In production.

Contractor

Honeywell Inc Business & Commuter Aviation Systems.

VERIFIED

HT1000 GNSS navigation management system

The HT1000 system consists of three Line-Replaceable Units (LRUs): Navigation Processor Unit (NPU); Multifunction Control and Display Unit (MCDU); Antenna Coupler Unit (ACU);

The NPU is the primary processing and control unit of the HT1000. It contains a 12-channel GPS receiver, multiple microprocessors, a navigation database and external system interfaces.

The GPS receiver receives and processes GNSS signal data from the antenna coupler unit to compute aircraft position and velocity. Using its 8 megabyte navigation database and GNSS position, the NPU performs all the functions necessary to provide aircraft guidance through the waypoints of a selected flight plan.

Built for adaptability in all applications, the NPU features analogue, digital and discrete interfaces to support all aircraft types. The NPU functionality is packaged in a small, lightweight 2-MCU size unit.

Data entry and display to the pilot are accomplished through the use of a colour ARINC 739 compatible MCDU. The MCDU controls the user input and display function of the HT1000, as well as other ARINC 739 subsystems. The MCDU offers high reliability and a flexible presentation. The keyboard includes a 66-key full alpha and full numeric keyboard, dedicated function keys and 12 line-select keys. The display is a 5.5 inch diagonal, multicolour active-matrix liquid crystal flat panel.

The ACU receives, amplifies, conditions and sends GPS signals to the GPS receiver in the NPU. The ACU contains an omnidirectional flat microstrip antenna with integral preamplifier and is referred to as 'active' since it performs the first stage of signal amplification. The ACU permits ACU to NPU cable runs in excess of 100 feet, which enhances the ease and flexibility of installation design. Moreover, the ACU is designed to optimise the GPS receiver performance and has improved filtering to prevent SATCOM interference.

The HT1000 is designed to achieve Federal Aviation Administration TSO C129, Class 1A certification, permitting GPS navigational use for Instrument Flight Rules (IFR) en route, terminal and approach operations. The HT1000 includes Receiver Autonomous Integrity Monitoring (RAIM).

The HT1000 can accept inputs from auto-tuned Distance Measuring Equipment (DME) and Inertial Navigation System (INS) to provide multisensor navigation per TSO C115a.

The HT1000 will implement Required Navigation Performance/Actual Navigation Performance (RNP/ANP) similar to the Honeywell FANS 1 software. RNP is the required performance accuracy of a particular segment of airspace. ANP is a measure of the uncertainty in the position estimate of the system.

On aircraft with enhanced flight guidance equipment, dual HT1000s will allow the airlines to apply to operate under the 'G' flight plan category.

The HT1000 is a full-flight regime lateral and vertical navigation system with interfaces to the aircraft flight instruments, flight director/autopilot and ARINC 739 compatible systems through the MCDU. Based on the programmed flight plan, the NPU provides coupled lateral guidance. Using the flight plan and fuel flow inputs, the NPU provides advisory climb, en route and descent vertical guidance.

The system is designed to take advantage of a pilot's knowledge of current air transport FMS operation, thereby providing 12 line-select keys, standard ARINC 739 keys, dedicated function keys and a scratchpad field for data input and messages.

A flight planning function allows pilots to create new

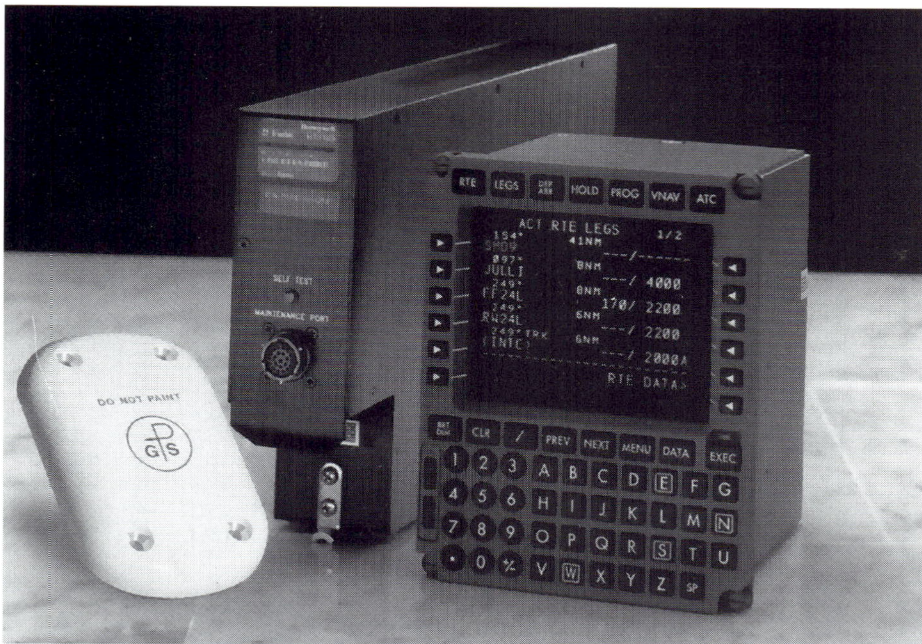

HT1000 GNSS navigation management system showing antenna coupler unit (left) navigation processor unit (centre) and multifunction control and display unit (right) ***1998***/0015382

flight plans via traditional air transport methods or through company route entry.

The HT1000 system's 12-channel GPS receiver provides continuous all-in-view satellite tracking. The digital receiver computes position updates five times per second, measures position to 15 m and measures velocity to 0.1 kt*. The GPS receiver is capable of receiving and utilising digital differential corrections for applications requiring increased accuracy and integrity. Slant range information received from DME-DME is used for position calculation. DME-DME position is automatically utilised as the primary source in the event of a GPS failure.

A configuration module, installed on the mounting tray, reduces maintenance actions by eliminating connector strapping. The module is programmable via the MCDU (or separate connector port) enabling individual aircraft type definitions to be easily accomplished.

An installation checkout and maintenance selection uses the MCDU display to show system installation status as well as inputs. Maintenance and historical BITE information is also available for downloading.

A single NPU has the capability to interface with two MCDUs and provide each crew member with input/output capability. Dual HT1000 system hardware consists of two each of the LRUs listed in the system description. With dual units, each system can be interfaced with its onside instrumentation and by request, data entry from offside units can be transferred between units. In the event of sensor failure, the offside data can be used in both systems.

Operational status

The HT1000 is in production. Growth options include ARINC 724B ACARS datalink. ARINC 610A SimSoft, Wide Area Augmentation System (WAAS) en route, WAAS Cat I approach, Automatic Dependent Surveillance (ADS), Local Area Differential GPS (LADGPS) (special) Cat I, II approach, and Global Orbiting Navigation Satellite System (GLONASS) capability.

Contractors

Honeywell Inc, Business and Communication Aviation Systems.

Trimble Navigation Ltd, Avionics Products.

NEW ENTRY

HT 9000 GPS navigation system

The Honeywell/Trimble HT 9000 remote-mounted GPS navigation system is designed to meet the demanding certification, operational and maintenance requirements of today's commercial, air transport and corporate operators. The HT 9000 is FAA TSO C-129

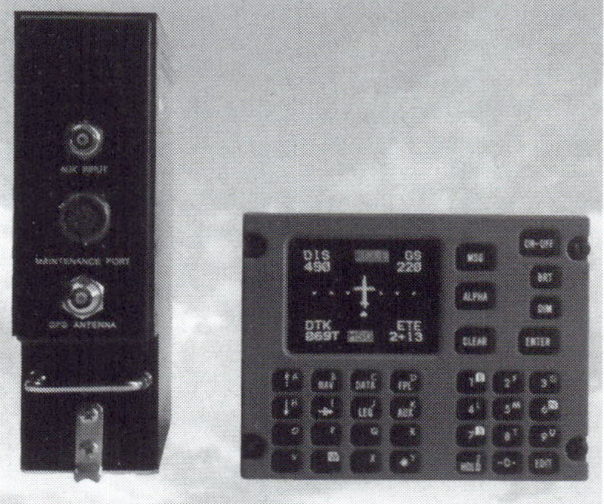

HT 9000 GPS navigation system showing the navigation processor unit (left) and control/display unit (right) ***1998***/0015381

(A1) approved for IFR en route, terminal and non-precision approach procedures as well as primary oceanic and remote means of navigation. The nine-channel all-in-view Navigation Processor Unit (NPU) interfaces to most digital and analogue aircraft systems. It also provides the user with the ability to fly all 24 ARINC leg types with the accuracy and precision of GPS. The HT 9000 provides the option of a compact, ARINC, or an FMS-like multifunction Control Display Unit (CDU).

Contractors

Honeywell Inc Business and Commuter Aviation Systems.

Trimble Navigation Limited.

UPDATED

Laseref inertial reference system

A derivative of the Lasernav inertial navigation system, Laseref is intended as an attitude and heading reference or primary sensor, for flight and navigation management equipment. Laseref shares many of the Lasernav modules and assemblies.

The solid-state, strapdown sensor generates present position, groundspeed, heading and windspeed and direction for flight management systems and navigation equipment, attitude and heading for flight instruments, weather radar stabilisation and autopilots. It replaces vertical and directional gyros, compass systems, fluxgate sensors and other independent navigation equipment with self-contained sensors and computing circuits to provide digital outputs, together with the ARINC 407 synchro outputs still needed by current generation avionics in ARINC 429 format. As with Lasernav II, alignment time is greatly reduced in comparison with gimballed inertial systems; typically 2½ to 10 minutes depending on latitude.

GPIRU configurations are available which contain additional electronics that blend inertial data with satellite information obtained from the Honeywell GPS sensor unit.

The Laseref system comprises two components: an inertial reference unit containing the sensing and computing elements, and a mode select unit that provides power to the former and governs its operation. Data insertion is through the control/display unit of the flight management system used in conjunction with the Honeywell system. The company offers an optional device, the inertial sensor display unit, as an alternative means of initialisation. This unit has a small display for reading out inertially computed present position, groundspeed, wind data and heading. Comprehensive built-in test equipment is provided; the system performs a rapid preflight self-test and monitors its operation throughout flight. Like Lasernav II, Laseref can be installed in an unpressurised bay.

Specifications

Dimensions:
(inertial reference unit) 322 × 324 × 193 mm
(mode select unit) 146 × 38 × 63.5 mm
(inertial sensor display unit) 146 × 114 × 167 mm

Weight:
(inertial reference unit) 21.1 kg
(mode select unit) 0.45 kg
(inertial sensor display unit) 2.3 kg

Power supply:
(total) 115 V AC, 400 Hz, 137 W
or 28 V DC

Operational status

In production and in service on the Gulfstream II, Raytheon Hawker 800 and other business aircraft, also the Boeing 727-200 and DC-10.

Contractor

Honeywell Inc Business & Commuter Aviation Systems.

VERIFIED

Laseref II inertial reference system

Laseref II performs the same functions as Laseref and uses the same inertial sensor assembly, but in addition is designed to interface with new-generation digital avionics, such as flight management systems, using the Aircraft Standard Communications Bus (ASCB). The Laseref II may also interface with aircraft having the ARINC 429 digital databus. Laseref II has been selected as the standard factory IRS installation on the Dassault Falcon 900, Gulfstream IV and Canadair CL-601-3A.

GPIRU configurations are available which contain additional electronics that blend inertial data with satellite information obtained from the Honeywell GPS sensor unit.

Specifications

Dimensions:
(inertial reference unit) 322 × 324 × 193 mm
(mode select unit) 146 × 38 × 62.5 mm

Weight:
(inertial reference unit) 21.1 kg
(mode select unit) 0.45 kg
(inertial sensor display unit) 2.3 kg

Power supply:
(total) 115 V AC, 400 Hz, 137 W
or 28 V DC

Operational status

In production. Laseref II is standard in a dual configuration on the Gulfstream IV, Dassault Falcon 900 and Canadair Challenger 601. Other applications include the Raytheon Hawker 800, Cessna Citation III and de Havilland Dash 8.

Contractor

Honeywell Inc Business & Commuter Aviation Systems.

VERIFIED

Laseref III inertial reference system

The Laseref III all-digital laser system is 60 per cent smaller, 45 per cent lighter and uses 50 per cent less power than its predecessor, the Laseref II. The heart of the Laseref III IRU is a smaller ring laser gyro sensor. In addition to the new sensor, Honeywell is utilising surface mount technology, very large-scale integration, application specific integrated circuits, more powerful and faster processing and enhanced software in the new system. The Laseref III IRU includes integrated GPS processing which further enhances position and velocity data with a hybrid blending of raw inertial and satellite data.

The Laseref III IRU is pin for pin compatible with the Laseref II IRU and can be installed in the latter's 10 MCU tray using a mechanical adaptor. It operates with the same mode-select unit and optional Lasertrak navigation display unit as existing systems.

Specifications

Dimensions: 124.5 × 320 × 198.1 mm
Weight: 11.79 kg
Align time: 2.5-10 min
Accuracy:
(navigation) 2 n miles/h
(velocity) 12 kt
(attitude) 0.1°
(heading) 0.4°

Operational status

In production since September 1991. The Laseref III has been selected for the Dornier 328 regional airliner and the upgrade of the Dassault Falcon 2000. Certified on the Raytheon Hawker 1000 in a dual configuration.

Contractor

Honeywell Inc Business & Commuter Aviation Systems.

VERIFIED

Laseref SM inertial reference system

The Laseref SM inertial reference system comprises the Mode Select Unit (MSU) and the Inertial Reference Unit (IRU). The MSU selects the IRU mode of operation and displays Operational Status messages on the annunciator panel. The IRU contains the laser inertial components, a processor and associated electronics and BIT. The INU is designated to add a GPS navigation processor card to incorporate an integrated GPS receiver.

By configuring the Laseref SM with an appropriate flight management system, a comprehensive special mission system is created. The combination of these units results in the ability to perform missions requiring special patterns, automatic camera control and computed air release point, in addition to standard flight management functions.

Laseref SM interfaces provide required data to a variety of special mission equipment. Standard ARINC interfaces output essential flight data to digital and analogue flight instruments, autopilots, radars, sensors and other special devices. Additional outputs are provided for special interface requirements.

The Laseref SM IRU is designed to add, as a growth option, an integrated GPS receiver. By adding one card to the IRU, a GPS PreProcessor Module (PPM) and an antenna, the standard Laseref SM becomes a fully integrated inertial/GPS system. The PPM receives satellite data using a two-channel fast sequencing design. Data is received from all satellites in view, up to a maximum of eight. The low signal-to-noise/fast sequencing design allows useful satellite tracking to zero degree elevation angle with rapid acquisition of satellite data. Pseudo range and pseudo range rate data is transmitted from the PPM to the IRU where the added GPS navigation processor card processes the pure GPS solution. This card also contains the Kalman filter that blends the GPS and inertial data to provide the GPS hybrid solution.

Specifications

Dimensions:
(inertial) 317.5 × 320 × 198.1 mm
(GPS) 152.4 × 177.8 × 50.8 mm
Weight:
(inertial) 21.32 kg
(GPS) 1.36 kg
Reaction time: 2.5-10 min
Accuracy:
(position) (inertial) 0.8 n miles/h CEP
(GPS) 25 m SEP
(velocity) (inertial) 10 ft/s
(GPS) 0.1 m/s
(time) (GPS) 350 ns
Reliability:
(inertial) 5,000 h MTBF
(GPS) 20,000 h MTBF

Contractor

Honeywell Inc Business & Commuter Aviation Systems.

VERIFIED

Lasernav laser inertial navigation system

Introduced during early 1983, Lasernav (a Honeywell trademark) exploits the strapdown inertial reference system developed for the Boeing 767 and 757, but also has facilities that make it suitable for long-range business and corporate jet aircraft.

In addition to normal navigation functions, Lasernav can replace all customary attitude and heading sensors, including compass system components such as the flux detector, resulting in a reduction of up to 14 separate boxes. This is said to result in a weight saving of up to 60.3 kg, and a volume reduction of up to 50 per cent by comparison with equivalent dual installations in other systems.

Lasernav comprises two units; an inertial navigation unit and a control/display unit. The memory can store the co-ordinates of up to 255 waypoints in 20 routes and up to 20 waypoints per flight plan, for immediate recall. In dual installations each system can store a different flight plan, but can share it with the other system if required. For international routes or long overwater sectors, the system can use a Global Wulfsberg NDB-2 database and only the departure point and destination co-ordinates need to be inserted. Lasernav then computes a great circle route, selects and identifies air traffic reporting points and lists bearing and distance to the nearest VOR/DME on the eight-line by 14-character control/display unit.

A notable advantage is the sharp reduction in alignment time; the interval from switch-on to ready is as little as 2½ minutes at the equator and 10 minutes at 60° latitude.

Specifications

Dimensions:
(inertial navigation unit) 322 × 324 × 193 mm
(control/display unit) ARINC 561
Weight:
(inertial navigation unit) 21.1 kg
(control/display unit) 3.2 kg
Power supply:
(total) 115 V AC, 400 Hz, 146 W
28 V DC, 28 W

Operational status

In service.

Contractor

Honeywell Inc Business & Commuter Aviation Systems.

VERIFIED

Lasernav II navigation management system

Introduced in 1984, Lasernav II is an inertial system for airlines and general aviation which combines self-contained, strapdown laser inertial position and aircraft motion sensors with externally sensed radio signals to provide a single efficient integrated guidance package. The notable advantages of previous Honeywell laser inertial systems are maintained in Lasernav II: 2½ to 10 minutes alignment time dependent upon latitude, three to four times the reliability of conventional systems and reduced size, weight and power consumption. The system can also be mounted in an unpressurised environment.

Position data obtained from the Honeywell GPS sensor unit and VOR/DME and Omega/VLF stations is blended with inertial position information using high-speed digital computing techniques. DME/DME updating is obtained by way of an auto-tuning function that requires no pilot inputs. VOR/DME updating is obtained by manually tuning the radios. Triple inertial navigation system mixing combines inertial data from two other Lasernav II systems to calculate a composite inertial position.

In addition to normal navigation functions, Lasernav II can replace all conventional attitude and heading sensors including vertical and directional gyros, flux valves and compass controllers. In a typical dual installation, the total box count can be reduced by up to 14 separate boxes, resulting in weight savings of up to 60 kg and a greatly simplified installation.

Lasernav II comprises a navigation management unit and a control/display unit. An internal non-volatile memory can store 20 flight plans of up to 20 waypoints each. A Global Wulfsberg NDB-2 worldwide database (required for DME auto-tuning) facilitates the automatic flight planning feature that requires the pilot to input only the departure and destination points. A great circle route is computed, intermediate waypoints are selected and range and bearing to nearest VOR/DME is displayed, all automatically.

Specifications

Dimensions:
(navigation management unit) 322 × 324 × 193 mm
(control/display unit) ARINC 562
Weight:
(navigation management unit) 22.1 kg
(control/display unit) 3.2 kg
Power supply:
(total) 115 V AC, 400 Hz, 160 W
28 V DC, 19 W
Accuracy (95% probability):
(position) 2 n miles/h
(velocity) 8 kt
(heading) 0.4°
(pitch and roll) 0.1°

Operational status

In service. Installations approved include Cessna Citation III, Gulfstream II and III, Dassault Falcon 50 and Canadair Challenger CL-600 and CL-601 business jets, and Boeing 737-200 airliners.

Contractor

Honeywell Inc Business & Commuter Aviation Systems.

VERIFIED

Lasertrak navigation display unit

A companion device to the Honeywell Laseref, Laseref II and Laseref III inertial reference systems, the Lasertrak navigation display unit provides back-up waypoint navigation using positions from up to three IRS. In the event of a flight management system failure the Lasertrak allows continued navigation along the planned route. Accepting up to 10 waypoints, Lasertrak computes and displays desired track and cross-track error for the intended course.

The Lasertrak can be used in flight management and inertial reference systems architecture when flight dispatch with a failed FMS is desired.

Specifications

Dimensions: 146 × 114 × 152 mm
Weight: 2.3 kg
Power supply: 27 V DC, 10 W

Operational status

Production deliveries started in March 1987. Installation certified in the Dassault Falcon 50 and 900, Gulfstream IV and Canadair Challenger CL-601-3A.

Contractor

Honeywell Inc Business & Commuter Aviation Systems.

VERIFIED

MLZ-850 Microwave Landing System (MLS) receiver

The stand-alone MLZ-850 consists of an ML-850 digital MLS receiver, the CM-850 controller and one or two AT-851/852 antennas. The small AT-851/852 antenna is normally located on the nose section of the aircraft centreline. A second antenna is mounted on the aft lower surface of the fuselage and is required on fixed-wing aircraft to be able to receive the MLS on all headings. All connections between the antennas and the receiver are with conventional coaxial cable in common use. An optional antenna-mounted preamplifier is used for aircraft requiring long cable runs.

The pilot selects azimuth and elevation angles on the CM-850 cockpit controller which is also used to tune one of the 200 MLS channels and auto-tune the DME.

The MLZ-850 is compatible with conventional cockpit displays, providing analogue deviation outputs to ADIs, HSIs and compatible flight control systems. An ARINC 429 formatted digital output provides interfaces with other digital technology systems such as EFIS. BITE provides a preflight test of the digital receiver system.

Specifications

Dimensions:
(receiver) 354 × 84 × 99 mm
(controller) 139.7 × 60.4 × 66.7 mm
(blade antenna) 63.5 × 38.1 × 19 mm
(flush mount antenna) 6.6 × 76.3 mm diameter
Weight:
(receiver) 2.09 kg
(controller) 0.454 kg
(blade antenna) 0.1 kg
(flush mount antenna) 0.18 kg
Power supply: 28 V DC, 15 W
Frequency: 5,031-5,090.7 MHz
Channels: 200

Operational status

In production.

Contractor

Honeywell Inc Business & Commuter Aviation Systems.

UPDATED

AN/ASN-131 SPN/GEANS precision inertial system

The Standard Precision Navigator/Gimballed Electrostatic Aircraft Navigation System (SPN/GEANS), designated AN/ASN-131, was developed primarily under sponsorship of the US Air Force Avionics Laboratory at Wright-Patterson Air Force Base, Ohio.

The basic SPN/GEANS system consists of an Inertial Measurement Unit (IMU), an Interface Electronics Unit (IEU) and a complete software library. For a stand-alone system capability, these two units are supplemented with a Digital Computer Unit (DCU). The IMU contains, in addition to the Velocity Measuring Unit (VMU) and two ESGs, temperature control electronics, accelerometer pulse rebalance and V readout electronics, precision timing reference, gimbal control electronics and Built-In Test Equipment (BITE) functions, as well as serial digital databus communication electronics. The IEU provides power conversion, control and sequencing electronics, additional BITE circuits and a common serial databus interface with other units of the inertial system and other subsystems.

The ESG has only one moving part, a suspended hollow beryllium ball, which is combined with two optical pick-offs to give error and timing signals. These in turn are used to drive the IMU platform gimbals to maintain a stable reference base for the accelerometers. Three highly accurate, single-axis accelerometers (contrasted with the two-degree-of-freedom ESGs) are used within the VMU which is mounted on the stable platform inner element. These accelerometers, oriented in an orthogonal triad configuration, measure accelerations directly and provide the incremental velocity pulses to the computer which uses them in its software algorithms to calculate velocity and position parameters.

The ESG system requires little or no reliance on other navigation aids for most aircraft applications and thus can be described as self-contained.

In addition to the traditional function of position determination or basic navigation, the higher accuracy outputs of velocity and attitude data have opened up a new realm of possibilities for stabilisation and/or motion compensation for other non-inertial sensors. These include high-precision radars, sonars, lasers, optical and electro-optical devices.

SPN/GEANS is being deployed throughout the entire US Air Force B-52 strategic aircraft and F-117A stealth fighter fleets. Widespread use is also expected for long-range reconnaissance and patrol missions, specialised cargo and transport usage in both military and civil applications and tactical military aircraft.

Operational status

Selected by the US Air Force for the B-52 bomber and F-117A stealth fighter fleets.

Contractor

Honeywell Inc Sensor and Guidance Products.

UPDATED

Digital map system

Using the technology and experience gained from the US Navy's AV-8B and F/A-18 night attack programme as a baseline, Honeywell has developed an advanced digital imaging system.

The capabilities of the base system have been expanded to include terrain reference navigation, ground collision avoidance, perspective view, terrain-following computations, threat intervisibility and sensor blending for advanced avionics systems.

Digitised aeronautical charts and Defense Mapping Agency Digital LandMass System (DLMS) Level 1 and 2 terrain and feature data are stored and recalled from the digital memory unit to generate a full-colour moving map which provides precise navigation and route selection to and from the point of target acquisition and weapons delivery.

The digital memory unit is a stand-alone mass memory unit designed to provide increased storage to military systems requiring high reliability and quick access times. The unit contains a flight-proven fully militarised optical disk capable of storing 520 Mbytes of data. Information is written to the disk by a laser diode in the head assembly. A beam of light is focused on the medium to produce a hole. During the read operation, the hole is recognised as a bit of data. Unlike magnetic media, which can be scratched by a head crash, data stored on the optical disk is written to a thin alloy encased in a protective substrate and a clear protective cover. Digitised reconnaissance photos, radar data, Landsat data, emergency procedures, let-down and approach plates, mission data and flight plans can be stored with digitised aeronautical charts and DLMS data for a variety of missions.

The digital map computer performs all the airborne map generation. The configuration on board the US Navy AV-8B night attack aircraft contains 11 circuit cards, a motherboard, a power supply and additional space to accommodate functions such as terrain reference navigation, terrain-following, ground collision avoidance, target hand-off, in-flight route planning, ridgelines, perspective view, sensor blending and threat intervisibility. The computer receives operational parameters, which include aircraft state vector, map scale, zoom factor, north up/track up, database type, declutter select/deselect and video output mode, from the mission computer via a MIL-STD-1553B multiplex bus. Data requests are then sent to the digital memory unit via a fibre optic link.

DLMS elevation data is compared to absolute elevation limits or limits relative to the aircraft altitude, resulting in a clearance band display. Colour is assigned, based on the band and sun angle shading. Variably spaced contour lines can be added. The overlay selection includes linear, area and point feature data such as roads, rivers and towers in the DLMS mode, and flight path, waypoint symbols, text and threat symbols in the DLMS and digitised chart modes. The video generator combines the background scene memory with the overlay memory to provide a composite video output. The system supports two red/green/blue video outputs and two monochrome video outputs at either 525/60 Hz or 625/50 Hz line rates.

Processing is distributed across several high-performance microprocessors rather than a single unit. Data flow is pipelined through the digital map computer and parallel processing is performed wherever possible. The use of parallel processing provides a level of fault tolerance, since a failure in the map generation does not prohibit the graphics overlay operation.

Operational status

Production deliveries for the US Marine Corps AV-8B began in June 1989. Honeywell is also under contract to provide the digital video mapping system for the V-22 Osprey tiltrotor aircraft.

Contractor

Honeywell Inc Sensor and Guidance Products.

UPDATED

H-423 ring Laser Inertial Navigation System (LINS)

The H-423 Laser Inertial Navigation System (LINS) was developed according to the US Air Force SNU-84-1 specification which was the RLG version, an update to ENAC77-1. The system is a self-contained unit comprising three Honeywell GG1342 RLGs, three solid-state Sundstrand QA2000 accelerometers and associated electronics along with a dual-redundant MIL-STD-1553B databus and a complete MIL-STD-1750A navigation processing package. It also contains built-in test circuitry to achieve a greater than 95 per cent fault detection.

Under the US Air Force contract, Honeywell is guaranteeing the H-423 will achieve 2,000 hours MTBF in a fighter/helicopter environment and 4,000 hours MTBF in a transport environment. Honeywell claims the H-423 has the highest reliability and maintainability and lowest life-cycle costs ever achieved by a military aircraft inertial navigation system, allowing the US Air Force to go from a three-level to two-level maintenance system. These performance and reliability advantages have been the primary drivers for the US Defense Services now adopting RLG technology for all future aircraft inertial navigation systems.

Specifications

Dimensions: 459.7 × 193 × 200 mm
Weight: 22 kg
Power supply: 140 VA AC, 125 W DC
Interface: dual 1553B digital databus
Accuracy: <0.8 n miles/h with full performance from 22 s stored heading alignment
Specification: SNU-84-1, FNU-85-1

Operational status

In August 1985, the US Air Force selected the H-423 as the standard inertial navigation system for its C-130, a number of its fixed- and rotary-wing aircraft. Up to the end of 1989 more than 1,000 military systems had been delivered.

In November 1990 the system was selected for Royal

Australian Air Force F-111 aircraft in a US$5 million contract and for US Air Force F-16 upgrade programmes. Since then, H-423 systems have been selected for upgrades of Belgian and Danish F-16s and for the Swedish JAS 39, Taiwanese IDF and the Indian Light Combat Aircraft.

In September 1992 the H-423E, a high-accuracy version of the H-423, was selected to upgrade the F-117A fleet. The high standard of accuracy is achieved through the use of an enhanced specialised version of the H-423 standard software.

Contractor

Honeywell Inc Sensor and Guidance Products.

UPDATED

H-764/H-764G small common inertial navigation system

In October 1987 the H-764 was selected for the AH-64 Advanced Apache helicopter programme. The self-contained unit consists of three Honeywell GG1320 RLGs, three solid-state Sundstrand QA2000 accelerometers and associated electronics. The unit also includes dual MIL-STD-1553B databusses and a MIL-STD-1750A microprocessor. It has provisions for synchro outputs, embedded GPS and dedicated high-speed databus for flight control interface.

The INS portion of the H-764G consists of three GG1320 RLGs, three solid-state Sundstrand QA2000 accelerometers and associated electronics. The unit includes dual MIL-STD-1553 databusses and a MIL-STD-1750A microprocessor. Embedded in the H-764G is a tightly coupled Texas Instruments GPS module with six channels and P(Y) code capability which utilises pseudo range and pseudo range rate satellite data. The INS/GPS has provisions for synchro outputs and a dedicated high-speed databus for flight control interface.

Honeywell also offers a single-card radar altimeter module (HG7805) using MMIC, advanced RF packaging, digital signal processing, ASIC, surface mount technology and IF hybrid circuits.

The design of the H-764G offers improved resistance to jamming and faster reacquisition after jamming occurs. It also allows error bounding with as few as one satellite in view. The system reduces the workload on the mission computer of the platform in which it is installed since it contains its own Kalman filter. Life cycle costs are very low compared to platforms using separate INS and GPS units due to Honeywell's design of only one unit with a significantly lower parts count. The H-764 INS only model can easily be retrofitted with GPS at any time.

Specifications

Dimensions: 177.8 × 177.8 × 279.4 mm
Weight: 9.07 kg
Power supply: 28 V DC, <70 W
Alignment time:
(gyrocompass) 3-8 min
(stored heading) 30 s
Accuracy (H-764):
(position) 1-3 n miles/h CEP
(velocity) 3-8 ft/s RMS
(heading) 0.1°
(pitch and roll) 0.05° RMS
Accuracy (H-764G):
(position) <1 n miles/h CEP pure inertial
<16 m SEP GPS/INS blended
(velocity) <3 ft/s RMS pure inertial
<0.1 ft/s RMS GPS/INS blended
(heading) 0.02° RMS
(pitch and roll) 0.01° RMS
Reliability:
(H-764) 2,000 h MTBF
(H-764G) 4,000 h MTBF predicted

Operational status

The H-764 is in service in the AH-64 Apache helicopter.

Contractor

Honeywell Inc Sensor and Guidance Products.

UPDATED

H-770 ring laser gyro inertial navigation system

Honeywell developed the H-770 ring laser gyro inertial navigation system for the F-15.

The H-770 uses the same ring laser gyro and accelerometers as the H-423 (see earlier entry), but contains two separate databusses for the F-15 applications. The system can determine for itself in which variant of the F-15 it has been installed, and perform accordingly. The H-770 has a reliability 20 times better than the AN/ASN-109 system that it is replacing in the F-15.

Specifications

Dimensions: 381 × 213 × 330 mm
Weight: 26.8 kg
Power required: AC 140 VA, DC 130 W
Accuracy: 0.4 n mile/h CEP
Alignment:
(gyrocompassed) 4 min
(stored heading) 30 s
Reliability: 2,000 h MTBF

Operational status

Honeywell had been contracted to retrofit the US Air Force fleet of F-15A, B, C, D and E variants.

Contractor

Honeywell Inc Sensor and Guidance Products.

UPDATED

SRS 1000 Attitude and Heading Reference System (AHRS)

The SRS 1000 strapdown Attitude and Heading Reference System (AHRS) uses advanced technology to meet modern AHRS requirements with very low life-cycle cost. A 15-second reaction time, independent of ambient temperature or vibration, is claimed.

There are two units: the attitude heading reference unit and compass controller unit. The attitude heading reference unit accepts magnetic compass, air data and Doppler inputs to provide attitude, heading, body rates and accelerations, groundspeed and drift angle information to other aircraft systems. The compass controller provides systems information and control functions to the crew.

An inertial measurement unit within the reference unit contains two flexure suspended gyros and two toroidal accelerometers for the X and Y axes, plus a standard force-feedback accelerometer for the Z axis.

Operational status

In production. The system is standard equipment on the Airbus Industrie A300 and A310 wide-body airliners.

Contractor

Honeywell Inc Sensor and Guidance Products.

UPDATED

KN-4060 series ring laser gyro inertial navigation systems

The KN-4060 family of navigation systems offers a low-cost solution for a wide variety of fixed-wing aircraft and helicopter and UAV applications. The systems feature the Kearfott ring laser gyro currently under contract to the US Navy and US Army.

The KN-4060 is designed to operate in conjunction with embedded or federated GPS receivers for enhanced navigation performance and advanced satellite reacquisition. The model KN-4068GC includes an Embedded GPS Receiver (EGR). The system's modular architecture allows for alternative navigation aiding inputs, including Doppler radar, baro-altimeter and airspeed or remote located GPS receivers via MIL-STD-1553 or RS-422 data protocols. The KN-4060 family provides navigation functions in digital formats including MIL-STD-1553B, RS-422 and ARINC 429. As a full featured navigation system, the KN-4060 provides precision position and velocity over extended periods, as well as the attitude and heading output of a Standard Attitude Heading Reference System (SAHRS). The small, lightweight, rugged chassis can accommodate field replaceable sensor assemblies of three levels of performance to meet customer requirements.

Specifications

Dimensions: 177.8 × 177.8 × 279.4 mm
Weight: 9.5 kg
Power supply: 28 V DC, 35 W or 115 V AC, 400 Hz back-up battery capability
Outputs:
(digital) MIL-STD-1553B, RS-422, ARINC 429
(analogue) synchro, discrete, MAD
GPS receiver:
(operating frequencies) L1/L2, L1
(antispoof/enhanced antijam) P(Y), C/A code
Channels: 5
Performance:

	GPS/Inertial			Pure Inertial		
	KN-4062G	KN-4065G	KN-4068G	KN-4062	KN-4065	KN-4068
Accuracy						
(heading) (°RMS)	0.1	0.05	0.04	0.40	0.12	0.08
(pitch/roll) (° RMS)	0.01	0.01	0.01	0.03	0.01	0.01
Navigation position						
INS/GPS (ft)	50	50	50			
(inertial) (n/mile h)				3.5	1.5	0.8
Velocity						
(ft/s)	0.5	0.5	0.5	17	5	2.5
Weight (kg)	7.27	8.5	9.09			

Operational status

The system has been flight-tested by A & AEE Boscombe Down, UK; Holloman AFB, New Mexico; and Patuxent River, Maryland.

Contractor

Kearfott Guidance and Navigation Corporation.

UPDATED

KN-4060 series navigation system **1998**/0015383

KN-4065 Improved Standard Attitude Heading Reference System (ISAHRS)

The KN-4065 Improved Standard Attitude Heading Reference System (ISAHRS) provides both attitude heading reference and full navigation outputs. ISAHRS features Monolithic Ring Laser Gyro (MRLG) technology for high reliability, low cost and weight and less power. The ISAHRS provides pure inertial navigation accuracies of better than 2 n miles/h as well as AHRS data. Hybrid inertial performance of better than 1 n mile/h is achieved by using an optical embedded GPS receiver or user-furnished velocity/position aids. Outputs are provided via MIL-STD-1553B multiplexer bus, RS-422 serial interface and synchro. The size of the unit makes it suitable for most aircraft.

Items in this family of Kearfott ring laser gyro systems include:
The Improved Standard Heading Reference System (ISAHRS);
The Modular Azimuth Position System (MAPS);
The Low Cost Attitude Heading Reference System (LAHRS);
The Attitude Motion Sensor Set (AMSS);

Specifications
Dimensions: 279.4 × 177.8 × 177.8 mm
Accuracy (1σ RMS):
(heading) 0.5°
(pitch/roll) <0.1°
Outputs: MIL-STD-1553B, synchro
Performance:
(heading) 0.5° (1σ) RMS (8 min alignment time)
(pitch/roll) <0.1° (1σ) RMS
(dynamic range) 400°/s max (all axes) ± 10 *g* acceleration
(backup navigation) 2 n miles/h

Operational status
The KN-4065 has been certified by the US Department of Defense for use in military aircraft and is in production for several types of US Navy aircraft.

Contractor
Kearfott Guidance and Navigation Corporation.

UPDATED

KN-4071 Attitude Heading Reference System (AHRS)

Kearfott's KN-4071 AHRS uses Ring Laser Gyro (RLG) technology in a strapdown configuration. Based on off-the-shelf hardware and software, the system provides digital as well as analogue information for cockpit display.

The system performs in the following modes: compass mode, directional gyro mode, slaved mode, enhanced mode, and initiated bit mode.

In the enhanced mode, the embedded GPS supports navigation and in-flight alignment with an accuracy of ±0.25° heading and ±0.15° attitude. The KN-4071 comprises line replaceable units and a mode select panel. It weighs 9 kg.

Contractor
Kearfott Guidance & Navigation Corporation.

NEW ENTRY

SKH-4210 series of ring laser gyro Standard Attitude Heading Reference Systems (SAHRS)

The Kearfott SKH-4210 series of ring laser gyro Attitude Heading Reference Systems (AHRS) are strapdown systems in use on a number of rotary- and fixed-wing aircraft applications in all three US services. They are used as an AHRS as well as a back-up navigator providing 2 n miles/h navigation performance.

AHRS consists of modular hardware, MIL-STD-1553B digital interface bus and self-diagnostic built-in capability. It provides outputs of aircraft pitch, roll, heading and angular rates for cockpit instrumentation, weapons delivery and autopilot flight control. It also provides acceleration, velocity and position information for medium accuracy navigation. Modes of operation include MAD slaved, directional gyro, compass emergency, operate or navigate and in-flight restart and calibrate.

Items in this family of Kearfott ring laser gyro systems include:
The Carrier Aircraft Inertial Navigation System (CAINS II);
The Standard Attitude Heading Reference System (SAHRS);
The Modular Azimuth Position System (MAPS);
The Milstar Antenna Reference Unit (ARU);
The Attitude Motion Sensor Set (AMSS).

Specifications
Dimensions: 177.8 × 177.8 × 279.4 mm
Weight: 12.7 kg
Interface: MIL-STD-1553B
Performance:
(heading accuracy) 0.5° (1σ) RMS (8 min alignment time)
(pitch/roll accuracy) <0.1° (1σ) RMS
(dynamic range) 400°/s max (all axes); ± 10 *g* acceleration
(back up navigator) 2 n miles/h

Operational status
No longer in production. The AN/USN-2(V) is fitted to the EA-6B, F-14A(Mod), F-14D, OV-1D, T-45A and V-22. Other candidates are the SH-2B, SH-3G/H, SH-60B and SH-60F.

Contractor
Kearfott Guidance and Navigation Corporation.

UPDATED

AN/ARC-73A VHF nav/com radio

Lapointe's AN/ARC-73A is an AM VHF combined navigation and communications transmitter/receiver system for light military aircraft.

The transmitter section covers the frequency band 116 to 149.95 MHz in which it provides 680 channels. The receiver, which can drive ILS and VOR indicators, covers a wider band to receive ground-based navigation aid signals. This band extends from 108 to 151.95 MHz and provides 880 channels. Channel spacing is at 50 kHz increments in both transmitter and receiver sections. Transmitter power output is 20 W. The system is remotely controlled.

Specifications
Dimensions:
(transmitter) 401 × 89 × 193 mm
(receiver) 318 × 89 × 191 mm
(controller) 160 × 145 × 56 mm
Weight:
(transmitter) 6.7 kg
(receiver) 4.7 kg
(controller) 0.76 kg

Operational status
In service.

Contractor
Lapointe Industries.

VERIFIED

LTN-72 inertial navigation system

The LTN-72, introduced in 1972, is a self-contained, all-weather, worldwide navigation system for commercial aircraft that is independent of any ground-based aids. Designed to incorporate area navigation facilities, the INS provides continuous position, navigation and guidance data.

The system comprises three units: Mode Selector Unit (MSU), Control/Display Unit (CDU) and Inertial Navigation Unit (INU). The MSU is used to energise and align the system prior to flight and to select navigation or attitude reference modes of operation. The CDU permits the crew to enter present position and waypoint co-ordinates, select track steering and display information generated by the system. The INU houses the gimbal structure with its gyros and accelerometers, associated electronics, power supply and data converter.

Ease of maintenance has been emphasised; for example, the principal mechanical elements - gyros and accelerometers - can be removed and replaced in 20 minutes using only screwdrivers. The gimbals are cantilevered, permitting the servo electronics to be mounted directly on the platform. This permits the use of flexible leads instead of slip-rings in some cases, improving reliability. The platform has only two slip-rings compared with four on a conventional platform.

The LTN-72 has very extensive self-test and failure detection facilities. It complies with ARINC 561 in that the probability of an undetected failure in attitude during the last 30 seconds before touchdown is less than 1 in 10^6. Again, an analogue output test feature permits tests not only of the INS but also of the flight instruments by driving them to various test readings.

The LTN-72R inertial navigation system is a development of the LTN-72, with automatic radio position update, automatic Omega position update and triple system mixing capability. It may be operated in the area navigation mode, using range and bearing information from selected VORTac stations or from a combination of Omega transmitters, providing very high accuracy independent of time.

The system uses newly developed gyros, platform and accelerometers and a new expansible C-4000 digital computer.

The LTN-72RL is an advanced, worldwide inertial navigation system able to automatically update itself by radio navigation fixes. It has a control/display unit which functions as an intelligent data terminal and incorporates a five-line by 16-character light-emitting diode display for presentation of operator-entered or computer-processed data. Waypoint data and VOR/DME locations can be prestored in the computer and the system contains an algorithm of magnetic variation that can be used to compute magnetic heading, track and desired track independently of the aircraft compass system; this algorithm is limited to latitudes between 60°N and 60°S. The prestored database contains information specified by the operator on selected VHF navaids, airports and some high-altitude waypoints. This bulk data is programmed in read-only memory.

A section of electrically alterable memory is allocated for particular waypoints or fixes not contained in the standard databases; up to 160 routes with an average of 20 waypoints per route can be stored in this way and recalled for use at any time. The total number of waypoints in all routes is limited to 3,200.

Specifications
Dimensions:
(inertial navigation unit) 267 × 219 × 507 mm
(control/display unit) 146 × 114 × 157 mm
(mode selector unit) 146 × 38 × 51 mm
Weight:
(inertial navigation unit) 26.8 kg
(control/display unit) 2.3 kg
(mode selector unit) 0.5 kg
Power supply: 115 V AC, 400 Hz
Number of waypoints: 9 to 99 plus remote entry capability
Waypoint offset capability: up to 399 n miles worldwide
Display: 7-segment incandescent numerals for all ARINC terms, together with display of INS parameters on flight director, horizontal situation or remote indicators
Inputs: self-contained inertial guidance. The avionics interface is designed to ARINC 561 and 575 and compatible with all flight directors and autopilots
Outputs: actual track, track angle error, cross-track and desired track, plus access to computer during flight for great circle distance computations

Operational status

In service. The Anglo-French Concorde is one application of the LTN-72. The LTN-72RL was initially certificated in July 1981 for Saudia Airlines Boeing 747s.

Contractor

Litton Aero Products.

VERIFIED

LTN-90 ring laser gyro inertial reference system

The LTN-90 comprises an inertial reference unit, a mode selector unit and an inertial sensor display unit. At the heart of the system, and contained within the inertial reference unit, are the Ring Laser Gyros (RLGs) which measure rotation accelerations and rates about the three aircraft axes, and the three single-axis accelerometers that measure accelerations and rates along the aircraft axes. The RLG system and accelerometers are mounted at right angles to each other and are rigidly secured to the case.

Unlike RLGs which are based on a triangular light path, the Litton LG-8028 units use a square path configuration, with a 28 cm path length. The reason for this, says Litton, is that for the same scale factor a square is smaller than a triangular gyro, resulting in a more compact overall sensor assembly. The square gyro is said to produce less backscatter at each reflection because of the improved 45° angle of incidence, versus 30° for a triangle, which makes for lower random noise.

The substantially greater data processing power required by strapdown systems over conventional gyro-based inertial navigation equipment is provided, in the case of the LTN-90, by three Zilog 8000 microprocessors, each of which is assigned a particular task. This computing power is also employed to correct temperature variations. Traditionally, gyroscopes are susceptible to changes in temperature, causing inaccuracies, and the normal approach to dealing with this effect is to maintain a constant temperature by trickle heating. In the LTN-90 the effect is compensated by mathematically modelling the way in which the characteristics of the system and accelerometers vary with changing temperature and then by applying a correction to their output signals.

Easy maintenance is a key advantage claimed by Litton. It is achieved by functional partitioning, so that each circuit board has all the components needed to support a major activity, more effectively isolating faults and simplifying troubleshooting. This partitioning is achieved by reducing the number of electronic plug-in modules from between 15 and 20 to 7.

Litton describes the LTN-90 as a sensor rather than (in the case of navigation) as a complete system and this is the way in which it is used in the A310. With inputs only from an air data computer, the LTN-90 can provide accurate outputs of attitude, heading, present position, drift angle, ground track, flight path angle, groundspeed, vertical velocity and windspeed, as well as aircraft angular rates and linear accelerations. Alternative control/display and mode selector unit configurations can be supplied to optimise specific aircraft requirements.

Specifications

Inertial reference unit
Dimensions: 194 × 322 × 318 mm
Weight: 19.9 kg
Power: 110 W
Cooling: ARINC 600
MTBR: 2,500 h

Mode selector unit
Dimensions: 38 × 146 × 51 mm
Weight: 0.45 kg
Power: negligible
Cooling: none
MTBR: 50,000 h

Inertial sensor display unit
Dimensions: 114 × 146 × 152 mm
Weight: 2.27 kg
Power: 15 W
Cooling: none
MTBR: 15,000 h

The A310 has a triple-redundant Litton LTN-90 installation

Accuracy (95%):
(heading) 0.4°
(pitch and roll) 0.1°
(position) 2 n miles/h
(groundspeed) 8 kt
(flight path angle) 0.4°
(body rates) 0.1%
(body accelerations) 0.01 *g*
Reaction time: 10 min

Operational status

In production for Airbus A310 and A300-600 and US Navy Boeing E-6A TACAMO.

Contractor

Litton Aero Products.

VERIFIED

LTN-92/LTN-99 Inertial Navigation Systems (INS)

The LTN-92 provides proven technology in a standard ARINC 561 1 ATR box. The inertial sensors and much of the electronics are the same as in the LTN-90-100 inertial reference unit which is currently flying on the Airbus A310, Gulfstream III, Canadair Challenger and other high-technology aircraft. The LTN-92 has been chosen for installation on the Cathay Pacific Boeing 747-300s and -200s as well as for retrofit on its Lockheed L-1011 aircraft. Other customers include Hawaiian, Continental, Western and Federal Express.

The LTN-92 INS comprises three separate units: the Inertial Navigation Unit (INU), the Control/Display Unit (CDU) and the Mode Select Unit (MSU). Three 28 cm ring laser gyros and a triad of force rebalanced accelerometers comprise the instrument cluster of the INU. The instrument electronics digitise the instrument control signals and the instrument outputs for easy microprocessor interface. There are three microprocessors used to process the instrument data, perform input and output interface functions and perform navigation calculations. As heat is not required and the latest technology parts are used, power consumption is reduced to a maximum of 175 W total operating power. The CDU has a five-line light-emitting diode dot matrix display and keyboard, providing the INU interface with the crew. The MSU controls the operational mode of the INS.

The LTN-92 is pin-for-pin compatible with existing ARINC 561 INS systems and ARINC 571 IRS systems. In addition there are three 429 high-speed digital output buses, nine 575/429 low-speed input buses, 12 programmable synchro outputs, four synchro inputs, four analogue DC or AC two-wire outputs, DME pulse pair input and 2 × 5 radio tuning line interfaces. This allows the LTN-92 to interface with existing avionics equipment installations as well as being a part of new installations.

Other capabilities are programmed into the LTN-92 to make it a versatile navigation unit. The system will accept RNav, Tacan, Omega, triple INS mixing or manual position updates for improved performance. With an external database or using the internal 16 k × 16 EEPROM data storage the crew may programme a flight plan and the INS will steer the aircraft along the entered flight plan. An interface to a weather radar system is provided for modification of the flight plan during flight. Intersystem communication is utilised to check system performance and send flight plan and initialisation data between systems. An extensive software built-in test monitor program of more than 150 performance checks is continually run to ensure the validity of the computed data.

The LTN-92 has been designed for future growth capability. Space has been provided for incorporation of an air data computer, GPS or Omega dual-system capability.

Litton Aero Products has contracted the Interstate Electronics Corporation (IEC) to add GPS and flight management system (FMS) capability to the LTN-92. IEC will supply their GPS navigation unit and a multifunction control display unit.

Specifications (LTN-92)

Dimensions:
(inertial navigation unit) 257 × 218.9 × 507.5 mm
(control/display unit) 146 × 114.3 × 157.5 mm
(mode selector unit) 146 × 38.1 × 50.8 mm
(battery unit) 129 × 193.8 × 365.5 mm
Weight:
(inertial navigation unit) 25.85 kg
(control/display unit) 2.27 kg
(mode selector unit) 0.45 kg
(battery unit) 12.24 kg
Power supply: 115 V AC, 400 Hz, single phase
Accuracy:
(position) 2 n miles/h (95%)
(pitch, roll and attitude) 0.05°
(heading) 0.4°
(groundspeed) ±8 kt
(vertical velocity) 30 ft/min
(body angular rates) 0.1°/s
(body accelerations) 0.01 *g*

The Litton LTN-92 inertial navigation system showing (rear left) the inertial navigation unit, (right) the control/display unit and (front left) the mode selector unit
1995

LTN-92E GPIFMS ***1998***/0015384

Operational status
In production and service. Applications include Boeing 747, DC-8 and DC-10, Gulfstream II and III, Lear 35, Lockheed Martin L-1011, US Army Boeing Vertol CH-47 and the US Presidential Boeing 747-200. The French Air Force has selected the LTN-92 in a dual configuration with the Litton LTN-211 Omega navigation system for installation in 10 C-130 aircraft and this has now been adopted as the baseline configuration for the C-130/L-100. Over 1,000 systems have been installed on 25 aircraft types.

LTN-92E GPIFMS
LTN-92E GPIFMS updates the LTN-92 as an economic solution to CNS/ATM (Communication Navigation Surveillance/Air Traffic Management) requirements for classic aircraft; it meets RNP R-Nav requirements, and includes a WAAS-compatible 12-channel GPS and Autonomous Integrity Monitoring Extrapolation (AIME) integrity; it also provides FMS data in ARINC 739 format and an ARINC 739 compatible Multifunction Control Display Unit (MCDU).

Specifications (LTN-92E GPIFMS)

Parameter	Hybrid Nav Mode	Altitude Mode
Position	0.3 nm, 99%, TSO-C11513	
Pitch/roll attitude	0.05°	TSO C4c
True heading	0.05°	
Ground speed	±4 kt	
Vertical velocity	0.6 ft/s	30 ft/min
Body angular rates	0.1°/s	0.1°/s

Operational status
TSO activity starts in July 1998. First deliveries September 1998. Two launch orders for 75 systems.

LTN-99
Capabilities of the LTN-99 will include: ACARS, GPS approach, advanced flight planning and CNS/ATM, as well as Autonomous Integrity Monitored Extrapolation (AIME), a patented software algorithm developed by Litton that provides 'sole-means-of-navigation' GPS performance. The LTN-99 will allow all current INU users to upgrade their equipment to CNS/ATM functionality. Target markets for the LTN-99 include all aircraft using existing LTN-92, LTN-72, LTN-58, LTN-51, Carousel IV and VI INS.

Contractor
Litton Aero Products.

UPDATED

LTN-101 FLAGSHIP global positioning, air data, inertial reference system

Four-mode Laser Gyro (FLAG) Software/Hardware Implemented Partitioning (SHIP) is designed for a wide range of applications. It can be used in single, dual or triple installations as an Inertial Reference System (IRS), combined IRS and Global Positioning System (GPS), Air Data Inertial Reference System (ADIRS) or ADIRS and GPS. At switch on, the system automatically recognises aircraft type and configures itself for either an ARINC 704 or 738 installation. An adaptor tray allows FLAGSHIPS's 4 MCU ADIRU to fit directly into a 10 MCU rack without system or rack modification.

FLAGSHIP integrates navigational functions and offers reductions in size, weight and power by eliminating the need for external air data computers and their interconnections. This also leads to savings in spares and maintenance and yields significant increases in reliability.

The GPS-IRS integration is an ideal combination because the two functions are highly complementary. The self-contained IRS contributes to GPS dynamic performance by facilitating satellite acquisition and tracking. The GPS in turn supplies ultra-precise position and velocity to the IRS, whose inertially derived position and velocity accuracies degrade with time. The GPS also makes possible inertial alignment during taxi or flight and furnishes long-term correction data which is used to calibrate the IRS inertial sensors.

The FLAGSHIP system consists of the LTN-101 Air Data Inertial Reference Unit (ADIRU), Sextant Avionique Air Data Module (ADM), Global Positioning System Sensor Unit (GPSSU), Mode Select Unit (MSU) and Control Display Unit (CDU). The ADIRU contains the inertial instrument package and performs all system computations with the exception of GPS sensor calculations. Critical air data and inertial reference functions are hardware partitioned to facilitate fault containment. The four-mode laser gyro requires no dithering and is free of conventional ring laser gyro lock-in and other dithering associated errors. The Litton A-4 accelerometer triad completes the sensor package. Surface mount devices and ASIC contribute to an ADIRU 60 per cent smaller than other ARINC 738 systems. Designed for a variety of environments, the ADIRU will operate up to 18 hours without cooling air.

The ADMs interface air data sensors with the ADIRU. Using an aneroid capsule and resonating quartz blade sensor, the ADM converts static and dynamic pressure into electrical signals. These signals are temperature corrected, converted to ARINC 429 format and transmitted to the ADIRU on a digital databus.

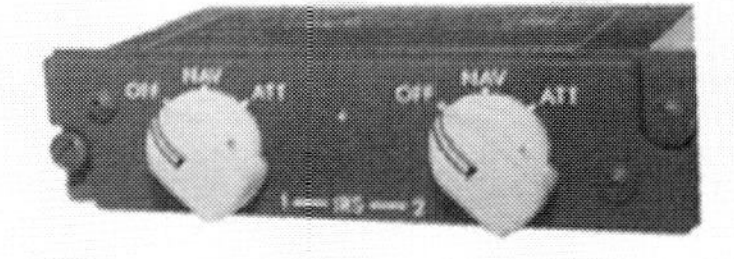

LTN-101 FLAGSHIP showing system units;
Air Data Inertial Reference Unit (ADIRU) - ARINC 738
Air Data Module (ADM) - ARINC 738
Global Navigation System Sensor Unit (GNSSU) - ARINC 743
Mode Select Unit (MSU)
Control Display Unit (CDU) ***1998***/0015385

The stand-alone GPSSU is a third-generation Litton design. The eight-channel continuous tracking receiver features enhanced integrity monitoring and rapid time to first fix. GPS usability is maximised by early acquisition of low-elevation satellites and minimal loss of satellite reception during aircraft manoeuvres. GPS outputs of position, velocity, time and raw satellite data are supplied to the ADIRU and other avionics. Both ARINC 743 configurations are offered: a 2 MCU avionic bay-located unit using an antenna with an internal preamplifier or a remote GPS sensor designed for installation near a passive antenna.

The MSU is a switching device used to apply power, annunciate system operating modes and indicate when the system is running on battery power.

The optional CDU, offered as a flight management computer back-up, provides a keyboard for initialisation data and a data display for auxiliary readout. Rotary switches select individual system modes. Push-buttons and annunciators allow inertial and air data output databusses to be turned off by the operator and indicate faults.

Detailed module BIT history is stored on each module. This includes the identity of the aircraft, ADIRU and other modules and LRUs in the system. System BIT history is stored on the computer module, along with its own history. If this module is replaced, system history is transferred to the new module.

Specifications

Dimensions:
(ADIRU) 4 MCU
(GPSSU) (rack mount) 2 MCU
(remote) 64 × 216 × 241 mm
(ADM) 145 × 97 × 53 mm
(MSU) 89 × 146 × 76 mm
(CDU) 171 × 146 × 152 mm
Weight:
(ADIRU) 12.3 kg
(GPSSU) 3.6 kg
(ADM) 2.4 kg
(MSU) 2.7 kg
(CDU) 10 kg

Operational status

Selected by many airlines for some 130 aircraft, including Airbus A321, A330 and A340 and Canadair RJ85 and RJ100 aircraft. Also selected for the Ilyushin Il-96M, Tu-204-200, Saab 2000, CC604, C-130 and An-28.

Contractor

Litton Aero Products.

VERIFIED

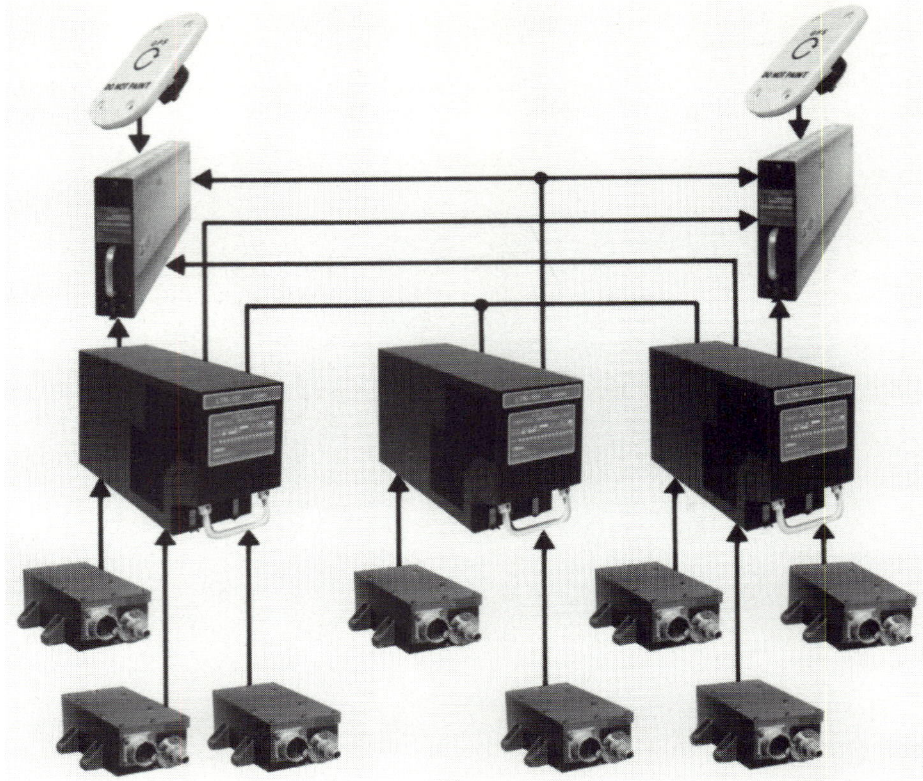

LTN-101 FLAGSHIP in a triplex configuration **1998**/0015386

LTN-211 Omega/VLF system

The LTN-211 airborne equipment, in conjunction with the Omega worldwide network of very low frequency stations or the US Navy's chain of VLF transmitters, provides a limited error display of guidance information for long-range great circle navigation. It is programmed to adopt, under certain circumstances, a back-up dead-reckoning mode of operation based on available aircraft velocity and heading. Switchover to the dead-reckoning mode is accomplished automatically when the number and quality of signals received falls below the requirement needed for acceptable position tracking and navigation.

When the initial position, time and date have been entered, the system automatically chooses the stations to be used on the basis of measured signal/noise ratios of the transmissions being received at the time. This method yields the best position accuracy since it considers both the quality of the stations selected as well as their propagation stability, and uses those stations least likely to be affected by diurnal signal changes. The basic LTN-211 uses signals of the three standard Omega frequencies, 10.2, 11.33, and 13.6 kHz. Capability to process the fourth Omega frequency, 11.05 kHz, is an available option. In the Omega/VLF system, this fourth frequency is used in conjunction with a VLF converter to process signals from the US Navy transmitters.

The system comprises three units: receiver/processor, control/display and antenna coupler.

Specifications

Dimensions:
(receiver/processor) 191 × 194 × 498 mm
(control/display unit) 146 × 114 × 158 mm
(antenna) 457 × 105 × 45 mm
Weight:
(receiver/processor) 11.8 kg
(control/display unit) 1.72 kg
(antenna) 3.63 kg
Power: 87 W total
Number of waypoints: 9
Inputs: signals from all 8 Omega stations and the US Navy VLF chain in ARINC 599 format plus true airspeed input for computation of windspeed and direction
Computer: TMS 9900 second-generation microprocessor, with 16 bits a word, and directly addressing 24,576 words of ultraviolet reprogrammable read-only memory
Outputs: groundspeed, track, heading, drift angle, cross-track, track angle error, present position in latitude/longitude, waypoints, distance and time to go, wind, desired track, station Operational Status and signal quality, GMT, date, self-test and malfunction codes. All outputs are to ARINC standard. Digital outputs, autopilot command steering signals and synchro outputs to the horizontal situation indicator can be provided
Failure detection: C9000 computer/processor has extensive BIT facilities within itself, as well as controlling the BIT section of the LTN-211 system. Probability of the BIT system detecting a fault anywhere in the navigation system is more than 99 per cent.

Operational status

In service. LTN-211 equipment has been installed aboard Boeing 707, 727, 737, 747, DC-8, DC-10, Gulfstream II, III, Lockheed Martin C-130, P-3, L-1011, Jetstar, Dassault Falcon, Raytheon Hawker 800, Cessna Citation and Fokker 27 aircraft.

The LTN-211 and LTN-72 INS form a part of the AN/ARN-99 Omega navigation system developed by Northrop Grumman for US Navy P-3C aircraft and CH-53D/E and RH-53D helicopters. For these and other non-airborne applications, over 150 ARN-99s have been produced.

Contractor

Litton Aero Products.

UPDATED

The control/display unit, receiver and aerial for the Litton LTN-311 Omega/VLF system

LTN-311 Omega/VLF/GPS system

The LTN-311 is a pin-for-pin replacement unit for the LTN-211 Omega system. It provides over 300 waypoints and adapts to special user software.

The compact control/display unit provides an active display screen 15 in (381 mm) high formatted into eight lines of 16 LED characters. The display incorporates an alphanumeric keyboard, four soft keys and a full menu-driven scenario. The top line of the screen displays the page function and the bottom line the soft key legends. Red LEDs may be substituted for marine applications. The enhanced colour and user-friendly menu-driven scenario combine with the other features of the CDU to lessen operator workload and the possibility of operator error. The CDU is self-leading, needing no pilot's guide to operate the system.

The LTN-311 Receiver Processor Unit (RPU) is an advanced development based on Litton's extensive Omega design experience. The power supply is an AC/DC non-interruptible power supply module; the computer and memory are combined on a single card which fulfils both the memory and processing roles. The RPU has a spare card slot which can incorporate the Litton single card Micro Nav GPS. The Micro Nav provides increased navigation capability as well as dual-system operation at minimal cost. The RPU can communicate with ACARS via an ARINC 429 port. Alternatively, this port could be used for access to a databank or to a recording device.

A significant feature of the LTN-311 is the incorporation of extensive BITE. The alphanumeric LED CDU allows plain language reporting of all detected malfunctions along with recommended maintenance codes. All multibox radio-based products of the past have had major non-verified removal problems. The LTN-311 communicates with the pilot and maintenance engineer in English to break through this barrier. This improved BITE system is designed to maximise the MTBR/MTBF ratio, minimise spares requirements, increase line maintenance efficiency and reduce training requirements.

Operational status

In production. Early in 1987 the Peruvian Air Force ordered five LTN-311 systems for installation in Antonov An-32 aircraft. In September 1988 the Civil Aviation Administration of China chose the LTN-311 for the China Eastern Airlines fleet of MD-80 and MD-82 aircraft.

American Airlines is the major LTN-311 user with over 600 systems on MD-80 and DC-10 aircraft.

Contractor

Litton Aero Products.

VERIFIED

LTN-450 navigation management system

The LTN-450 is a full capability navigation management system which incorporates a powerful navigation computer, GPS receiver and control/display functions in one compact, lightweight, low-cost unit.

The LTN-450 features an active matrix LCD flat-panel display which provides a large eight line by 14 character display. The full alpha and numeric keyboards, along with 10 function keys, provide easy, quick and efficient data entry and display control. Data displayed on the LTN-450 is organised in selectable display pages under 10 functions: Data, FPL, Nav, Vnav, DTO, Fuel, List, Menu, MSG and Hold. Within each function, pages are placed sequentially in order of usefulness or priority. The Menu and List functions both augment the user-friendly operation and increase data input accuracy.

The fully integrated LTN-450 navigation management system provides the pilot with centralised control of the aircraft navigation sensors, computer-based flight planning using an extensive navigation database, lateral and vertical navigation, three-dimensional approach mode and fuel management.

The navigation database is contained on a flash memory card which inserts into a slot in the faceplate. The Jeppesen database card can hold up to 80,000 waypoints and nav aids and includes airports, VORs, DMEs, VOR/DMEs, ILS, VORTacs, NDBs, as well as en route and terminal waypoints. In addition to the flash card navigation database, internal memory capacity exists for extensive user-defined data. Up to 200 routes with up to 98 waypoints in a single route for a total of up to 3,000 waypoints, 100 user-defined locations or waypoints, 100 approaches and 25 temporary radar waypoints can be included.

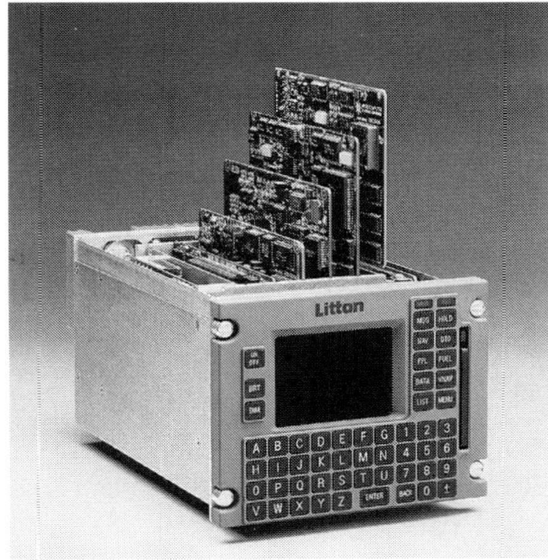

The LTN-450 navigation management system

1995

The LTN-450's internal multichannel GPS receiver continuously processes pseudo-range and range rate measurements from four or more satellites. The navigation solution includes three-dimensional position and velocity, time and Operational Status. in degraded constellation geometrics, the GPS receiver will continue to provide two-dimensional navigation using measurements from three satellites, along with pressure altitude.

In addition to the internal GPS receiver, the LTN-450 has the capability to use up to three external long-range navigation sensors, a scanning DME and, during approaches, VOR to determine aircraft position. The type and number of sensors can be tailored to optimise specific operational requirements, from combinations of GPS, inertial, Loran C and Omega/VLF sensors. In addition to these inputs, the system receives true airspeed and altitude from the air data computer, heading from selectable sources and fuel flow information. A best computed position is derived using a Kalman filtering technique. This is used for navigating the aircraft along the programmed flight plan and defined approaches. Course to waypoint, wind velocity and groundspeed are all computed and displayed.

The LTN-450 provides desired track, bearing, crosstrack, lateral deviation and related data to the flight guidance system for mechanical HSI and EFIS displays. Roll steering command is output to the autopilot or flight director system and bank angles are scaled commensurate with altitude.

The LTN-450 accepts two fuel flow sensor inputs and/or manual entries integrating real-time fuel management information with the navigation functions. During flight, it automatically updates the figures for the fuel on board and gross weight, and provides continuous estimates of the fuel requirements for the programmed flight plan based on fuel flow, true airspeed and groundspeed. Specific range is presented in both air and ground n miles/lb.

The vertical navigation function allows a desired vertical flight profile to be defined. Multiple VNav waypoints are available and are automatically prefilled with those on the flight plan. The LTN-450 computes and displays aircraft deviation from the programmed descent profile.

The LTN-450's three-dimensional approach mode provides both lateral and vertical guidance while flying non-precision approaches. Guidance can be generated for VOR, VOR/DME and RNav approaches. Flight director/autopilot coupling is available, complete with scaling, sensitivity and displays similar to an ILS DME approach.

Specifications

Dimensions: 114.3 × 146 × 241 mm
Weight: 2.9 kg
Power supply: 19-32 V DC, 35 W (max)
26 V AC, 400 Hz

Contractor

Litton Aero Products.

VERIFIED

LTN-2001 global positioning system

The LTN-2001 C/A code GPS provides continuous worldwide precision three-dimensional navigation data and offers a means to upgrade and enhance the performance of ARINC 561, 599, 704 and 738 navigation systems. The eight-channel continuous tracking receiver is a mature third-generation Litton design. It maximises GPS usability and features advanced integrity monitoring to deliver performance superior both to sequencing type receivers and those with fewer channels.

The system tracks low-elevation satellites as they rise above the horizon and continues to track them until they disappear, minimises loss of satellite reception during aircraft manoeuvres, exceeds the ARINC 743 external interference specification and resists multipath reception caused by terrain features and aircraft surfaces. The LTN-2001 integrates with upgradeable navigation systems and provides in-flight alignment, continuous sensor calibration and error bounding capability for inertial navigation systems, plus GPS positional accuracy and worldwide capability.

The LTN-2001 provides GPS outputs of position, velocity, altitude, time, pseudo range and delta range to inertial navigation systems, flight management systems, smart CDUs and other avionics via ARINC 429 high- and low-speed databusses. An RS-422 serial output is available as an option. Both ARINC 743 configurations are available: a 2 MCU avionic bay-located unit and antenna with internal preamplifier or a remote sensor unit designed for installation within 10 ft of a passive antenna.

Specifications

Dimensions:
(rack) 2 MCU
(remote) 63.3 × 215.9 × 241.3 mm
Weight: 3.63 kg
Power: 15 W
Accuracy (2 DRMS): 100 m
Antenna: active or passive conformal
Reliability:
(receiver) 40,000 h MTBF
(antenna) 100,000 h MTBF

Contractor

Litton Aero Products.

VERIFIED

LTR-81-01 and LTR-81-02 Attitude and Heading Reference Systems (AHRS)

The LTR-81-01 is an advanced design LTR-81 AHRS providing the full set of ARINC 705 digital outputs plus a complete set of analogue outputs capable of interfacing with analogue aircraft systems. Greater accuracy, higher reliability and a significant weight reduction are achieved without the need for vertical, directional or rate gyros, body-mounted accelerometers and compass couplers. The LTR-81-01 and LTR-81-02 are identical, except that the LTR-81-01 has two additional analogue output cards.

The LTR-81-01/02 accepts inputs from the magnetic compass, air data computer and VOR/DME. The system outputs consist of attitude, heading, groundspeed, vertical speed, drift angle, flight path angle and linear accelerations and angular rates. The LTR-81-01/02 mechanisation ensures no degradation of accuracy during aircraft manoeuvres. Two advanced K-273 two-degree-of-freedom tuned rotor gyroscopes are incorporated in the design. Three high-accuracy accelerometers complete the inertial instruments.

The LTR-81-01/02 computer module dual-processor mechanisation which has been implemented in the system offers numerous advantages including fewer components, improved monitoring, lower power consumption and commercially available support hardware and test equipment.

The LTR-81-01/02 features modular construction, permitting ready access, and simple removal for ease of maintenance. Single layer printed circuit plug-in modules permit simpler repair and result in reduced repair time. BITE technology has been implemented, including redundant comparison checks of all gyro loops.

The system will continue to perform accurately after brief power interruptions and is not adversely affected during absence of air data or VOR/DME inputs for periods of several minutes.

The LTR-81-01 has been certified as a critical system in accordance with FAR 25.1309 and its software is certified to RTCA DO-178.

Designed with future growth as a primary consideration, the system is capable of interfacing with all analogue and digital equipment envisioned to the turn of the century, including GPS.

Specifications

Dimensions: 8 MCU
Weight:
(LTR-81-01) 13.15 kg
(LTR-81-02) 10.89 kg

Power supply: 115 V AC, 400 Hz, single phase
(LTR-81-01) 100 W
(LTR-81-02) 85 W
Alignment time: 45 s
Accuracy (95%):
(heading) 2°
(pitch and roll) 0.5°
(groundspeed) 8 kt
(flight path angle) 1°
(body rates) 0.1°/s or 1%
(body accelerations) 0.01 *g*
Reliability: >7,000 h MTBF

Operational status

Selected by over 50 airlines with over 2,000 units delivered. Applications include the Airbus A300, Boeing MD-82, MD-83, MD-87 and MD-88, Fokker 50 and 100, British Aerospace ATP, and business jets and helicopters.

Contractor

Litton Aero Products.

VERIFIED

AN/ASN-130A (LN-38A) carrier aircraft inertial navigation system

The AN/ASA-130A, designated LN-38A by Litton, is a complete system contained within a single weapons replaceable assembly, the principal elements being a P-1000C platform, G-1200 gyros, A-1000 accelerometers and LC-4516C digital computer. The P-1000C provides attitude, heading and velocity information, while the LC-4516C computer is a 16-bit general purpose device with 30 k memory operating at 238 Kops. It has a 10 MHz serial digital databus in MIL-STD-1553 format and synchro, analogue and discrete signal input/output capabilities.

Specifications

Dimensions: 427 × 287 × 190 mm
Weight: 15.9 kg
Power supply: 115 V AC, 400 Hz, 3 phase
(start) 800 W
(run) 185 W
Accuracy: 1 n mile/h CEP
Reliability: 1,950 h predicted MTBF

Operational status

No longer in production. About 2,000 of these systems have been produced. The system is used on US Navy and US Marine Corps AV-8B, A-6, EA-6B, F/A-18A and F-14D aircraft. Also in service in Australia, Canada and Spain in the F/A-18 and in the UK in the Harrier GR. Mk 7.

Contractor

Litton Guidance & Control Systems.

VERIFIED

AN/ASN-139 (LN-92) carrier-based aircraft inertial navigation system

The Litton LN-92 is being produced as the US Navy AN/ASN-139 Carrier Aircraft Inertial Navigation System (CAINS II) ring laser gyro INS as a form, fit and function replacement for the AN/ASN-130A. It is designed for operation by high-performance carrier-based aircraft. The accuracy specifications are better than 1 n mile/h CEP and 3 ft/s velocity with a 4 minute reaction time.

The gyro in the LN-92 is the Litton LG-9028 28 cm ring laser gyro used in conjunction with the A-4 accelerometer triad. This system incorporates MIL-STD-1750 processors for navigation and signal data processing. The reliability of the unit permits the Navy to employ a two-level maintenance concept.

Specifications

Dimensions: 427 × 287 × 190 mm
Weight: 21.46 kg
Reliability: 3,992 h MTBF

Operational status

In production and in service with US Navy F/A-18, F-14D, EA-6B and AV-8B aircraft. The system has been selected by Finland, Kuwait, Malaysia and Switzerland for the F/A-18 and has been flight-tested in the Royal Air Force Harrier GR. Mk 7.

Contractor

Litton Guidance & Control Systems.

VERIFIED

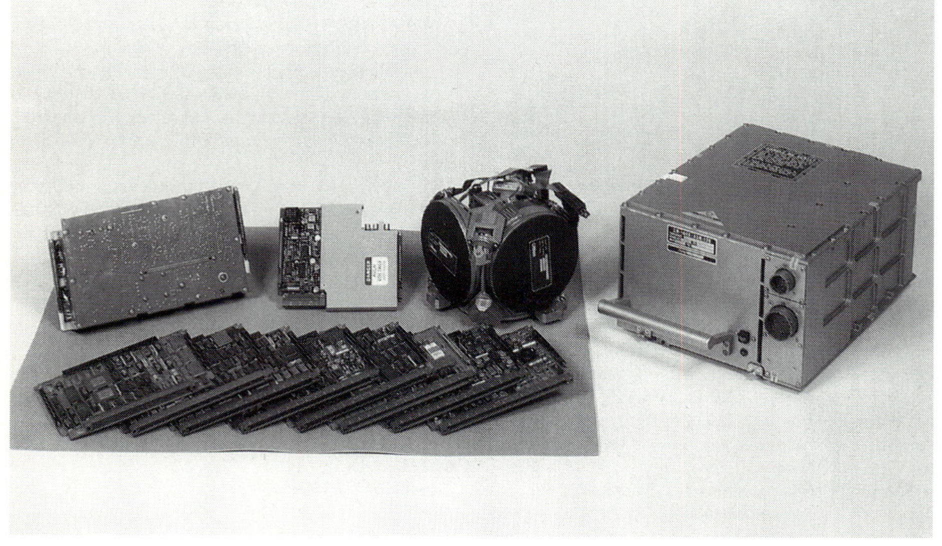

The Litton AN/ASN-139 INS is installed in the Northrop Grumman F-14D **1995**

AN/ASN-141 (LN-39) inertial navigation unit

The AN/ASN-141, designated LN-39 by Litton, is the sensing and data processing device that was chosen by the US Air Force as the basis for the standard inertial navigation system on the A-10 and F-16.

The unit contains a P-1000 platform, stabilised by two G-1200 gyros and mounting three A-1000 accelerometers, and an LC-4516C general-purpose computer.

Particular emphasis has been placed on reliability. This is accomplished partly by the use of large/medium-scale integrated circuits and hybrid components, allowing a significant parts count reduction.

Specifications

Dimensions:
(inertial navigation unit) 191 × 193 × 460 mm
(control/display unit) 146 × 152 × 185 mm
Weight:
(inertial navigation unit) 17.36 kg
(control/display unit) 3.73 kg
Power:
(starting) 550 W
(running) 180 W
Accuracy (position, velocity):
(gyrocompass) 0.8 n mile/h CEP, 2.5 ft/s
(stored heading) 3 n miles/h CEP, 3 ft/s
Align time:
(gyrocompass) 8 min at 21°C
(stored heading) 1.5 min at 21°C
LC-4516C computer: 16-bit single or 32-bit double precision, 65,536 words, semiconductor RAM/ROM or EPROM 24 k words total
Reliability: 740 h MTBF specified goal over full military environment

Operational status

In service in A-10 and F-16 aircraft. The system is also in service in the Brazilian/Italian AMX, Japan Air Self-Defence Force F-4EJ, Thai Air Force F-4 and F-16s in Bahrain, Egypt, Korea and Turkey. Over 3,000 AN/ASN-141s have been ordered.

Contractor

Litton Guidance & Control Systems.

VERIFIED

AN/ASN-142/143/145 (LR-80) strapdown Attitude and Heading Reference System (AHRS)

The AN/ASN-142/143/145, designated LR-80 by Litton, was developed as a flexible system to meet the growing operational requirement for an all-attitude dynamically accurate attitude reference for air, land and sea applications. It is a lightweight self-contained LRU that provides heading, pitch and roll and angular rate reference information for a wide variety of military applications including cockpit display, sensor stabilisation, fire control and autopilot reference.

The equipment senses vehicle angular rates and translational accelerations by means of two Litton G-7 gyroscopes and three Litton A-4 accelerometers.

Specifications

Dimensions: 194 × 192 × 259 mm
Weight: 8.1 kg
Power supply: 28 V DC, 60 W nominal
115 V AC, 60 W (620 W warm-up)
Accuracy:
(heading) 0.5° RMS
(pitch and roll) <0.25° RMS
(angular rate) 0.25°/s RMS

Operational status

In production. Over 1,000 systems have been delivered to several programmes for US and overseas services, primarily for the Boeing AH-64 Apache, Bell OH-58D Kiowa Warrior and Sikorsky UH-60.

Contractor

Litton Guidance & Control Systems.

UPDATED

CN-1655/ASN (LN-94) ring laser gyro INU

The CN-1655/ASN, designated LN-94 by Litton, is a variant of the LN-93 (see next item) developed as a ring laser gyro form, fit and function replacement for the Litton LN-31 conventional gyro INS previously fitted in the F-15. The system uses the same inertial assembly and most of the electronic circuits of the LN-93, but is packaged in the LN-31 chassis and has input/output circuitry that is specific to the F-15.

Operational status

Orders have been placed for retrofit into all US Air Force F-15A, B, C and D aircraft, as well as for Israel and Japan. The LN-94 is also qualified for the F-15E.

Contractor

Litton Guidance & Control Systems.

VERIFIED

CN-1656/ASN (LN-93) ring laser gyro INU

The CN-1656/ASN, designated LN-93 by Litton, is the US Air Force standard ring laser gyro INU. This unit is now in production, employing 28 cm path length ring laser gyros. Following a contract awarded in 1985, units have been widely fitted to US Air Force aircraft and to aircraft in many other countries.

The LN-93G is form, fit and functionally interchangeable with the LN-39 and LN-93 standard INUs, as well as any other systems that are in compliance with SNU-84-1 used in numerous types of US and foreign tactical aircraft. The LN-93G ring laser gyro INU/GPS is in a single box which contains the US Air Force standard navigation unit integrated with a Collins militarised six-channel P code GPS receiver module.

Specifications

Dimensions: 521 × 204 × 193 mm
Weight: 22 kg

Operational status

In production. The LN-93G is in development and a large number of units has been ordered for US fighter aircraft.

Contractor

Litton Guidance & Control Systems.

VERIFIED

GPS Guidance Package (GGP)

The GPS Guidance Package (GGP) is an ARPA sponsored programme to develop a family of low-cost, modular, miniature GPS-based guidance systems that can easily be configured to support a broad spectrum of US military platforms. The GGP wil consist of the Litton Miniature Inertial Measurement Unit (MIMU) and the Rockwell Collins Miniature GPS Receiver (MGR).

Specifications

Dimensions: <1,630 cm³
Weight: 3.19 kg
Accuracy: <10 m SEP with GPS
Output: position, velocity, attitude, heading, angular rate and acceleration

Operational status

In development.

Contractor

Litton Guidance & Control Systems.

VERIFIED

LN-100 advanced navigation system

In October 1989, Litton was awarded a contract to develop a low-cost inertial system featuring low power, weight and volume. The LN-100 employs non-dithered Litton zero-lock laser gyro and A-4 accelerometer instrument technologies, together with a 22-state Kalman filter which integrates Doppler velocity and GPS position and velocity inputs, to serve as a dual-flight control reference and high-accuracy navigation system for the AH-64D Longbow Apache programme.

In April 1991 the LN-100 INU was selected by Boeing to provide the Navigation Quality Inertial Sensor (NQIS) for the RAH-66 Comanche helicopter inertial navigation system. The LN-100 is also used as the inertial reference system for the Lockheed/Boeing F-22 advanced tactical fighter.

Specifications

Dimensions: 241.3 × 177.8 × 177.8 mm
Weight: 8.6 kg
Power supply: 28 V DC, 25 W
Accuracy: 0.8 n mile/h
Reliability: 5,000 h MTBF

Operational status

Selected for the AH-64D Longbow Apache and RAH-66 Comanche helicopters and the F-22A Raptor.

Contractor

Litton Guidance & Control Systems.

VERIFIED

LN-100G navigation system

The LN-100G has evolved from the proven LN-100 product line. All LN-100 systems use common hardware and software elements, affording economies of scale from high-rate production. Major US DoD programmes that require lightweight inertial navigation systems have contracted for the LN-100. These include the RAH-66 Comanche helicopter and the F-22. The LN-100G, with an embedded GPS receiver, has been selected by the US EGI tri-service programme office for the F-18 and EA-6B upgrades, by the US Navy for the T-45A Cockpit 21 programme and by Boeing for the Japanese 767 AWACS.

The LN-100G provides three simultaneous navigation solutions; hybrid GPS/INS, free-inertial and GPS only. The processing heart of the LN-100 product line is the 32-bit PowerPC Motorola microprocessor; the software is Ada.

The LN-100G optimally combines GPS and INS features to provide enhanced position, velocity, attitude and pointing performance as well as improved acquisition and anti-jam capabilities.

The addition of a GPS receiver package on a single circuit card provides a complete hybrid navigation unit with very low size and volume. Two spare card slots allow for the addition of analogue I/O modules, ARINC interfaces, Low Probability of Interception (LPI) radar altimeter, air data and other expansion modules.

The embedded GPS module is an L1/L2 CA/P(Y) code unit capable of accepting RF (or IF) inputs from the GPS antenna system. Stand-alone GPS PVT data and stand-alone INS data are provided for integrity and fault monitoring purposes.

Contractor

Litton Guidance & Control Systems.

VERIFIED

LN-200 inertial measurement unit

The LN-200 is a three-axis strapdown inertial measurement unit designed for tactical missile and unmanned vehicle guidance, flight control and sensor stabilisation applications. It is very small and light and employs fibre optic gyros and silicon chip accelerometers for extreme ruggedness and high reliability.

A version of the LTN-200, identified as the LN-201, has been selected as the flight control quality inertial system for the US Army Boeing/Sikorsky RAH-66 Comanche helicopter.

Litton, teamed with Rockwell Collins, has a contract with the US Advanced Research Projects Agency for an integrated inertial/GPS system. The system, known as GPS Guidance Package and identified as the LN-250, has a goal of a navigation grade inertial system with full GPS capability in a 0.0016 m³ package.

Operational status

In development.

Contractor

Litton Guidance & Control Systems.

VERIFIED

RINU(G) navigation system

Litton's Guidance & Control Systems division has been awarded a contract by the US Navy to provide replacement navigation systems for all of the US Navy's P-3 maritime and C-130 patrol and cargo-type aircraft.

Production options for up to 784 systems to be delivered up to 2001 could bring the total contract value to Litton to more than US$50 million.

The systems will be direct replacements for existing Litton units aboard the aircraft since the late 1970s. Two systems are installed in each aircraft.

The new systems, called RINU(G), combine in a single unit the continuous, long-term accuracy of the latest laser gyroscope navigation technology with the geographical precision of a global positioning system satellite signal receiver.

The P-3/C-130 navigation units will be based on Litton Guidance & Control Systems division's production LN-100G system.

Operational status

In development.

Contractor

Litton Guidance & Control Systems.

VERIFIED

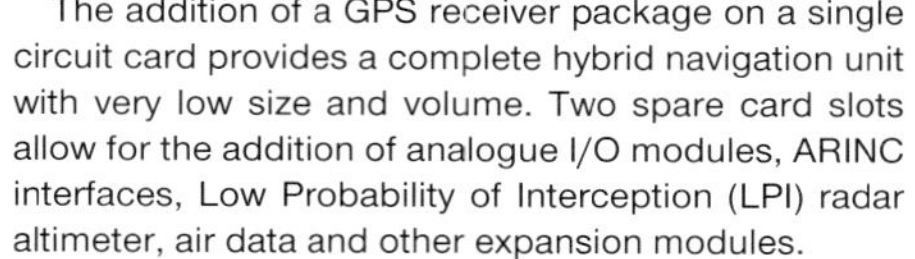

The LN-100 has been selected for the AH-64D Longbow Apache, RAH-66 Comanche and F-22
1995

Low-Altitude Safety and Targeting Enhancement (LASTE)

The Low-Altitude Safety and Targeting Enhancement (LASTE) system for the Fairchild A-10 improves flight safety, survivability and lethality. Increased flight safety is achieved through the Ground Collision Avoidance System (GCAS) and Low-Altitude AutoPilot (LAAP). Survivability and lethality are improved through a blend of enhancements to bombing, air-to-ground gunnery and air-to-air gunnery.

GCAS provides both aural and visual warnings in potential ground collision situations. The two types of warning are a hard warning when radar altitude is less than 90 ft and a predictive warning when a minimum safe altitude will be penetrated as a consequence of continuing the present flight path.

LAAP provides a path hold mode, an altitude/heading hold mode and an altitude hold mode. This reduces pilot workload and thus fatigue, particularly during long-range missions or periods of low-altitude operation.

Bombing and gunnery are improved by the integrated flight and fire-control system called the Precision Attitude Control Augmentation/Improved Air-to-Ground Sight System (PACA/IAGSS).

Air-to-ground gunnery achieves greater accuracy

due to a Continuously Computed Impact Point (CCIP) which computes fire control estimates based on ballistics, range, inertial velocity, angle of attack, sideslip, wind and bank angle for firing ranges out to 12,000 ft.

Precision Attitude Control (PAC) provides accurate gun aiming during the long firing bursts that are necessary to achieve the required hit density. PAC also removes the disturbing effects of wind gusts on gun aim.

CCIP for bombing is supplemented with a Projected Bomb Impact Line (PBIL) to provide the pilot with precision control of his flight path in a curvilinear approach and an accurate bomb release in a banked turn. This permits much greater flexibility in the approach to a target and minimises the time to deliver a bomb accurately, while the manoeuvre complicates the prediction problem for defensive fire-control systems.

An air-to-air gunsight provides a defensive capability for the A-10. The excellent ballistics of the GAU-8 and the large overnose vision of the A-10, when combined with an effective gunsight, provide a formidable capability against armed helicopters. An especially important feature of the sight is the firing evaluation display system which is used to evaluate simulated shots against manned aircraft in realistic training exercises.

This system uses a computer originally designed for flight control applications expanded to include GCAS and weapon delivery algorithms and a display symbol generator. A coupling to the existing stability augmentation system helps to achieve precision control during gunfire as well as autopilot control.

Specifications

Dimensions:
(computer) 177.8 × 359.7 × 144.8 mm
Weight:
(computer) 9.19 kg

Operational status

In service in the A-10.

Contractor

Lockheed Martin Control Systems.

VERIFIED

CNS-12 ACARS/GPS/ADS communication, navigation and surveillance system

The Magellan CNS-12 features a WAAS-capable 12-channel GPS receiver with Receiver Autonomous Integrity Monitoring (RAIM), an optional D8PSK receiver card for differential GPS and aircraft navigation technology to meet the ICAO VDL requirements. The GPS, together with the ARINC VHF Aircraft Communications Addressing and Reporting System (ACARS) two-way datalink, provides timely and accurate Automatic Dependent Surveillance (ADS) position reports from any aircraft via the ACARS network.

The CNS-12 employs a PCMCIA card capable of storing information on nearly 21,000 airports, navaids and waypoints. The Jeppesen database card contains detailed information on airports, VORs, DMEs, ILS/DMEs, VORTacs, Tacans and NDBs, as well as en route and terminal intersections. GPS overlay approaches, SIDs and STARs are included on the expandable database card. Additionally, a pilot-defined waypoint database stored in RAM can accommodate up to 500 waypoints. The information is used to create up to 50 flight plans with up to 40 waypoints in each. User flight plans and waypoints can be stored on a programmed PCMCIA card. The active flight plan and stored waypoints can be selected from the internal database, user datacard or Jeppesen NavData card and displayed in a variety of ways. En route, approach and vertical navigation is displayed on the CNS-12 display.

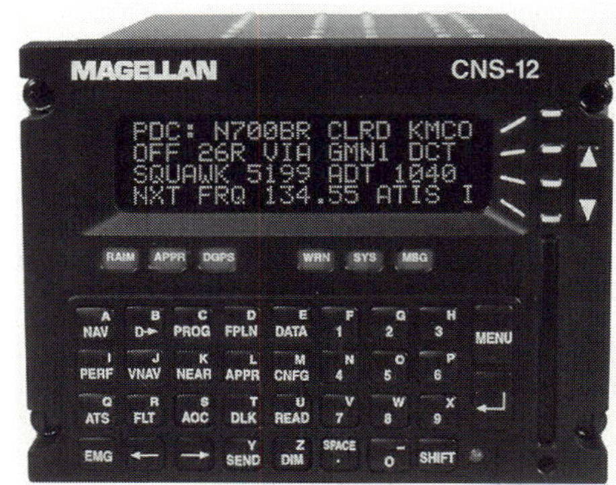

CNS-12 ACARS/GPS/ADS communication, navigation and surveillance system ***1998***/0015387

A second, dedicated D8PSK VHF receiver for differential GPS precision approach information is optional. The receive only module will make precision differential GPS approaches possible as the ground infrastructure becomes available.

The CNS-12 enables two-way communication for sending and receiving clearances, text messages, system essential messages, weather reports, Out/Off/On/In (OOOI) reports, engine data and other operational messages over the GLOBALink/CNS network. It also automates predeparture clearance and delivery of flight papers. The CNS-12 communications element relays digital Automatic Terminal Information Services (ATIS) arrival and departure messages for any airport during flight.

The two-way VHF datalink complies with ARINC 745-2 for ADS. The CNS-12 will automatically transmit data derived from its integrated GPS to air traffic service centres, providing surveillance beyond the radar horizon.

Specifications

Dimensions: 146 × 107.9 × 203.2 mm
Weight: 3.08 kg
Power supply: 14 or 28 V DC, 7 A (max)
Temperature range: −15 to +55°C
Altitude: up to 15,000 ft
Frequency: 108-136.975 MHz
Accuracy:
(position) 15 m RMS
(with DGPS) 8 m RMS
(velocity) 0.1 kt (with DGPS)
(time) UTC to nearest µs

Operational status

In service on a variety of commercial aircraft, CNS-12 links via ARINCs GLOBALink/CNS datalink service and can be integrated with the ADS system.

Contractors

Magellan Systems Corporation.
ARINC.

UPDATED

ADF 841 Automatic Direction-Finder

The ADF 841 combines the functionality of frequency storage, transfer and timing with simple operation. It features frequency storage, frequency transfer, crystal clear audio, frequency memory, a gas discharge display and automatic dimming. Since both the mode and timer functions are selected through the same switch, pilot workload is minimised, especially in IFR conditions.

Supplied with a combined loop sense antenna and indicator with rotatable azimuth card, the ADF 841 employs coherent detection which both reduces interference and increases the reception range.

Specifications

Dimensions:
(receiver) 158.7 × 38.1 × 279.4 mm
(indicator) 82.5 × 82.5 × 14.3 mm
(antenna) 152.4 × 292.1 × 10.2 mm
Weight:
(receiver with tray) 1.5 kg
(indicator) 0.34 kg
(antenna) 1.6 kg
Power supply: 11-32 V DC, 1 A at 13.73 V

Operational status

In production.

Contractor

Narco Avionics Inc.

VERIFIED

The Narco Avionics ADF 841 automatic direction-finder

DME 890 Distance-Measuring Equipment

DME 890 distance-measuring equipment features instant lock on, remote channelling capability, high-visibility gas discharge displays, microprocessor-based electronics and automatic display dimming. It is RNav compatible and has a two-frequency storage capability. Simultaneous display of distance, groundspeed and time to station are provided.

A flexible unit, the DME 890 can be remotely channelled by the Mk 12D and Mk 12E nav/com, Nav 824/825 or other navigation receivers. The DME 890 is ideally suited for use with the Narco NS 801 RNav system.

Specifications

Dimensions: 158.1 × 38.1 × 279.4 mm
Weight: 1.7 kg
Power supply: 11-32 V DC, 15 W
Number of channels: 200
Display range: 160 n miles in 0.1 n mile increments
Accuracy:
(range) ±0.1 n mile
(groundspeed) ±3 kt

Operational status

In service.

Contractor

Narco Avionics Inc.

VERIFIED

Mk 12 series NAV/COM receivers

The Narco Mk 12 is a VHF communications/navigation radio system designed principally for light and general aviation aircraft. The communications section is a transmitter/receiver covering the VHF band from 118 to 136.975 MHz and encompassing 760 channels. Two channels are preselectable, one for active and the other for standby use. Transmitter power output is nominally 8 W.

The navigation section consists of a VOR/Loc receiver covering the VHF band from 108 to 117.95 MHz providing 200 channels; again, two of these are preselectable for active and standby use.

In each section, new frequencies may be entered into the standby mode at any time and transfer buttons are activated to exchange the selected frequency between active and standby positions.

Navigation section output will drive compatible horizontal situation indicators, area navigation systems, VOR radio magnetic indicators or the system's companion ID 824 VOR/Loc indicator. It will also drive the ID 825 VOR/ILS indicator and, for full ILS capability, a combined 40-channel glide slope receiver, covering the band 329.15 to 335.0 MHz, is available.

Mark 12D+

The TSO'd Mark 12D+ significantly reduces cockpit workloads with its 10 COM frequency storage, digital radial readout in NAV, and full760 COM channels.

The Mark 12D+ is outwardly similar to the Mark 12D with the exception of a mode selector knob which is located beneath the COM display windows. The new radio retains the standby and active frequency display and flip-flop transfer features, but allows the user to program each of the 10 channels. The frequency selector also doubles as the channel selector.

Active and standby navigation frequencies are displayed in the NAV window. Except during entry of a standby NAV frequency, the right position of the NAV window always displays a digital radial readout from the VORTAC (or dashes if the signal is too weak).

Specifications

Dimensions: 159 × 64 × 279 mm
Weight:
(nav/com with glide slope receiver) 2.0 kg
(without glide slope receiver) 1.9 kg
(mounting tray) 0.34 kg

Operational status

The Mark 12D+ updates the earlier Mark 12 A to D models.

Contractor

Narco Avionics Inc.

UPDATED

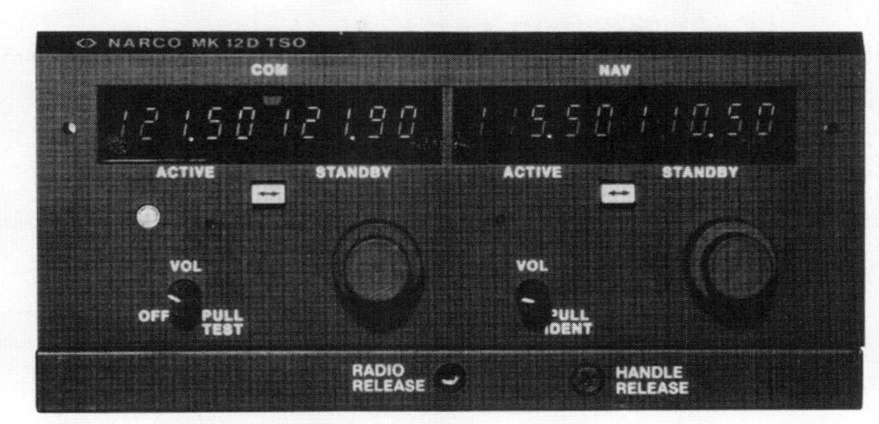

Narco Avionics Mark 12D/2 NAV/COM receiver **1997**/0001342

Narco Avionics Mark 12D+ NAV/COM receiver **1998**/0015388

NAV 122D and NAV 122D/GPS self-contained NAV receiver indicators

The NAV 122D and NAV 122D/GPS self-contained NAV receiver indicators are being relaunched as technology updates of the earlier discontinued 3 in NAV instruments carrying the NAV 122 nomenclature. The new NAV 122D and NAV 122D/GPS instruments will be form/fit replacements for all earlier models from the NAV 122 back to the NAV 12 (except that replacement of the NAV 12 model will require an interconnect cable).

The new NAV 122D includes self-contained VOR/Localiser/Glideslope receiver and converter, with VOR/Glideslope indicator and Marker Beacon lights; DME channelling and autopilot interface. It utilises, as a self-contained item, the following elements: 200 channel NAV receiver, VOR, Localiser and Glideslope Indicator, with Marker Beacon Lights.

Narco Avionics NAV 122D self-contained NAV receiver indicator **1998**/0015389

The new NAV 122D/GPS optional version additionally includes the resolver required to interface with GPS, where VOR-style, 'Left/Right' indication is required for GPS IFR approval.

Specifications

VOR/Loc receiver: 108.00-117.95 MHz
Glideslope receiver: 329.150-335.00 MHz
External interfaces:
(DME channelling) 2 out of 5
(autopilot) left/right, to/from, up/down, glideslope flag
(Marker lamps) inputs for amber, white, blue lamps

Operational status

Availability planned for the first quarter of 1998, subject to FCC approval.

Contractor

Narco Avionics Inc.

NEW ENTRY

NS9000 multisensor navigation system

The NS9000 provides multisensor functionality by blending signals from its integral VOR/DME and GPS receivers, and interfaces with conventional VOR/LOC/GS indicators.

The NS9000 overcomes the inherent weakness of GPS and VOR/DME receivers (scalloping, P-static and satellite unavailability).

To derive latitude/longitude positions, the NS9000 selects the 10 closest VORs and 10 closest DMEs from its Jeppesen database. Based on geometry and distance, the NS9000 then selects the two VORs and two DMEs with the highest accuracy potential. It then automatically scans these stations and computes up to four fixes. The NS9000 then blends those resulting fixes with the position data derived from its five-channel GPS sensor (and Loran if it is RS-232 equipped). Three microcomputers reject any fix which varies significantly from others.

It integrates information from GPS, VOR/DME, and Loran to provide the most accurate navigation. Operation is entirely automatic after selection of the waypoint. Frequency only has to be entered when switching from multisensor mode to standard mode, to enter an ILS frequency for example. This enables use of the 40-channel localiser and glide slope receiver for precision approach.

The Narco NS9000 is FAA TSO'd and STC'd for non-precision and full-precision approach. It is also compliant with ICAO FM immunity requirements and received full Basic RNAV (BRNAV) approval in November 1997.

Contractor

Narco Avionics Inc.

UPDATED

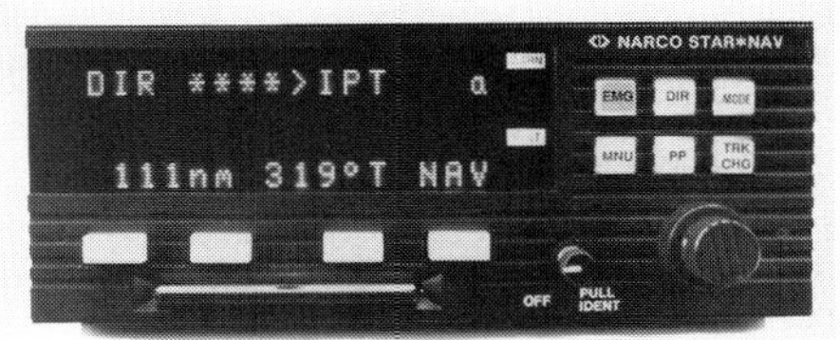

*The Narco Avionics NS9000 STAR*NAV multisensor navigation system* **1997**/0001343

AN/ARN-84 Tacan

Operational features of the ARN-84 Tacan include X-Y position fixing, air-to-ground link with any standard Tacan station, air-to-air bilateral ranging, receive only for bearing and beacon identification, radio frequency datalink, air-to-air bearing and inverse mode. The system provides automatic antenna switching together with digital and analogue range and bearing outputs to aircraft instruments and computers. The solid-state design gives a level of reliability previously unmatched according to NavCom. The range is 300 n miles.

Operational status

In service. The system is fitted to 15 types of US military aircraft including the A-4M, A-7E, C-130, E-2C, F-14A, S-3A, AH-1T and SH-3H. Over 3,200 ARN-84s have been produced. The AN/ARN-84 is no longer in production.

Contractor

NavCom Defense, Electronics Inc.

VERIFIED

Northstar M3 GPS Approach

The Northstar M3 GPS Approach uses 12-channel, parallel tracking, with user-replaceable FliteCards to provide an IFR (Instrument Flight Rules) navigation capability. When used in conjunction with the Northstar Smart-Comm system it automatically organises required frequencies to calculated position.

The M3 has received TSO C-129, Class A1 approval for en route, terminal and non-precision approach operations.

The M3 is designed so that the pilot only has to select the approach and follow the indicator. The entire approach, complete with each leg of procedure turns and holds, is programmed into the database. If radar vectors are received to the final approach course, the pilot is able to select 'Vectors to Final' and automatically pass intermediate waypoints, whilst retaining situational awareness relative to final approach course and final approach fix.

Specifications

Channels: 12 continuous, parallel-tracking
Navigation accuracy:
15 m RMS (30 m 2DRMS)
100 m 2DRMS with S/A activated
Navigation update rate: 1 s
Time to fix: 1 min (typical)
Operating modes:
2D Nav, 4 satellites visible
3D, 5 or more satellites
Automatic cold start: neither time nor position input required
Voltage: 10-35 V DC negative ground
Power: 35 W nominal
Size: 159 × 299 × 51 mm
Weight:
(with mounting tray) 1.9 kg

Operational status

Listed as recommended equipment in the American Champion Aircraft Corporation list of avionics options.

Contractor

Northstar Avionics.

UPDATED

Northstar VFR GPS-60

The 12-channel, parallel-tracking GPS-60 provides all-weather, worldwide navigation. The VFR-only GPS-60 offers many of the navigation features found in Northstar's M3 IFR GPS (except GPS approaches). The GPS-60 provides immediate access to flight information, such as distance and bearing to destination; ground speed; NDBs, VORs and intersections; winds aloft, and the Class B and Class C airspaces.

The Northstar GPS-60 will interface directly with today's CDIs, HSIs, Autopilots and Moving Maps.

Every GPS-60 comes complete with a user-updatable FliteCard containing a comprehensive Jeppesen database of over 8,000 US public airports. With Northstar's exclusive SmartComm intelligent frequency software, the GPS-60 organises all database frequencies. Alternative databases available for the GPS-60 include North America; International; and helicopter, all of which include private airports in the United States in addition to the public-use airports.

Specifications

Type: L1 frequency, C/A code (SPS)
Multichannel: continuous tracking
Navigation accuracy:
15 m RMS (30 m 2DRMS)
100 m 2DRMS with S/A activated
Navigation update rate: 1 s
Time to first fix: 1 min (typical)
Operating modes:
2D Nav, 3 satellites visible
3D Nav, 4 or more satellites
Automatic cold start: neither time nor position input required
Annunciator output: warn, parallel offset, waypoint alert, VFR
Serial ports: RS-422/RS-485
Power: 10-36 V DC at 14 W
Size: 159 × 298 × 51 mm
Weight: 2 kg

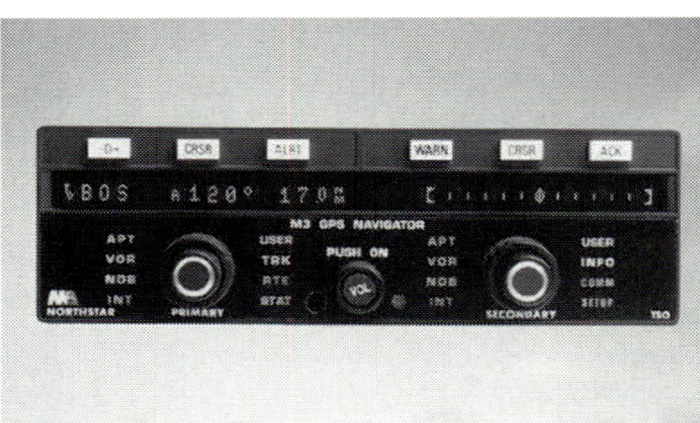

Northstar SmartComm **1996**

GPS antenna
Type: low-profile patch with integral L1 preamplifier
Length: 87 mm
Width: 56 mm
Weight: 0.14 kg
Mounting: surface-mounts to top of aircraft; requires 12.5 mm cut out in aircraft skin

Contractor

Northstar Avionics.

UPDATED

SmartComm intelligent frequency management

SmartComm is an intelligent position-based communications management system. SmartComm organises all database frequencies relative to present position. A standby stack keeps track of the last five frequencies used. The local category list allows the selection of frequencies by type with up to eight frequencies in each list.

All new Northstar navigators feature SmartComm and include fully operational SmartComm software. Optional 760-channel remote-mounted transceivers instantly tune the frequency selection directly from the Northstar navigator, tuning it into a GPS/comm unit.

Specifications

Nominal voltage: 13.75 V DC
Current (transmit): <2.5 A
Frequency range: 118 to 136.975 MHz
Channels: 760
Weight: 0.55 kg
Dimensions: 70 × 61 × 210 mm

Operational status

In production.

Contractor

Northstar Avionics.

UPDATED

AN/ASB-19(V) Angle Rate Bombing Set (ARBS)

The AN/ASB-19(V) Angle Rate Bombing Set (ARBS) was designed for US Marine Corps aircraft to improve day and night bombing accuracy when operating in the close support role using unguided weapons. The system provides accurate delivery, irrespective of target velocity, wind velocity, target elevation or dive angle. It is compatible with guided ordnance and can be used to direct gunfire and air-to-ground rockets. ARBS was originally designed for application to the US Marine Corps A-4M and is also the primary weapons delivery system for the US Marine Corps and Spanish Navy AV-8B and the Royal Air Force Harrier GR.Mk 7 aircraft.

ARBS comprises three main subsystems: a dual-mode tracker, weapon delivery computer and control unit. The tracker includes a laser and pilot-controlled television tracking equipment, both using a common optical system. The dual-mode tracker automatically tracks targets designated by a laser (either ground-based or by an accompanying aircraft) or targets which are television-designated by the pilot of the ARBS-equipped aircraft itself. The tracking system's common optics enable transition from laser to television tracking mode to be accomplished without losing the target.

Tracking information is passed from the dual-mode tracker to the weapon delivery computer. The weapon delivery computer performs computations for weapon trajectory and fire control, position control of the dual-mode tracker during target acquisition, digital filtering of the dual-mode tracker angular rate signal outputs and automatic fire or weapon release signals to the armament and other systems.

The weapon delivery computer receives aircraft to target line of sight angle and angle rate data from the tracker when that unit has achieved target lock-on. This data, combined with true airspeed and altitude information from the air data computer, when processed by the weapon delivery computer, provides the weapon delivery solution. Target position, weapon release and azimuth steering information are generated and presented to the pilot via a head-up display.

The function of the ARBS control unit is to interface with the weapon delivery computer and to provide the pilot with control of the tracker modes and entry of display, target, navigation and maintenance information.

When operating in the laser mode, the sensor automatically acquires the target, which is illuminated from an external laser source. The sensor, via the weapon delivery computer, presents steering signals to the pilot on the head-up display. In television mode, visible light from the common optical system is directed to form an image on the television display. Tracking of small low-contrast poorly defined moving targets is said to be possible even with changes in aspect ratio, in the presence of competing clutter or when the target is partially lost to view behind ground obstructions.

Following head-up acquisition in the television mode, the pilot is shown a ×7 magnified image of the target on the cockpit television monitor display. He may at this stage use a hand control to slew the tracker gate on to a new track point or on to a nearby alternative target.

Weapon release in either laser or television tracking mode may be made automatically or manually, with a weapon release time and a continuously computed impact point being generated by the weapon delivery computer. A weapons data insert panel provides entry into the weapon delivery computer of weapon characteristics and rack type for those stores being carried for a specific mission.

ARBS permits delivery of any type of weapon at any attitude and airspeed combination at any dive angle or in level flight. The system accuracy is said to be sufficient for first-pass precision delivery in close air support or interdiction missions against land or seaborne targets, but navigation steering commands are also generated to provide second passes against hardened targets. This re-attack facility provides the pilot with head-up steering information for return to a designated target, the location of which is retained in the system memory until a new target is designated by the pilot.

The system's television tracking element also provides a limited air-to-air capability for daylight operations, its ×7 magnification being particularly useful for visual identification and tracking of airborne targets.

Angular coverage of ARBS extends from ±35° in azimuth and from +15 to −70° in elevation at roll angles of ±450°. These limits are sufficiently wide to allow extensive flexibility in weapon selection and delivery profiles and permit multiple roll manoeuvres to take place on the attack approach run without loss of target lock on. The tracking equipment's laser is NATO coded, but is compatible with other current laser designators, such as Hughes' TRAM, at ranges adequate to ensure the success of first-pass attacks.

The ARBS weapon delivery computer processor on the A-4M is a 16 k word capacity system extensible to 32 k words. With 16 k words, the system has far more capability than is required for weapon release and steering command computation and may be used for additional purposes such as the pilot-initiated self-test of the ARBS system which performs operational readiness testing and fault isolation functions. By modification of the computer input/output scaling factor, the ARBS can be configured to interface with aircraft avionics other than those of the A-4M and the Harrier AV-8B. Three basic aircraft interfaces are required: a vertical reference to provide pitch and roll data, true airspeed and a servoed optical sight or head-up display on which to present azimuth steering commands, continuously computed impact point and bomb release information.

ARBS systems can be configured, in either internally or pod-mounted versions, to suit the requirements of other aircraft. Several pod configurations have been formulated for candidate aircraft with attachment hooks at 356 mm and 762 mm centres, compatible with standard pylon or fuselage centreline mountings. Where existing cockpit controls cannot be used, two small ARBS control units may be fitted at any accessible position. The A-4M control units are configured with standard 146 mm widths adaptable to most cockpit consoles.

Specifications

Volume: 0.07 m^3
Weight: 58.18 kg

Operational status

In service in US Marine Corps and Spanish Navy AV-8B aircraft and Royal Air Force Harrier GR. Mk 7s.

Contractor

Raytheon Systems Company.

UPDATED

The ARBS in the nose of a US Marine Corps AV-8B

GPS-Aided Targeting System (GATS)

The US Air Force B-2 bomber is equipped with the GPS-Aided Targeting System to reduce target location error. GATS uses the B-2s GPS navigation system and the Hughes-built synthetic aperture radar to provide accurate target location data, that is then passed to the GPS-Aided Munitions (GAM); together they form the GATS/GAM targeting system of the B-2.

Operational status

In service.

Contractor

Raytheon Systems Company.

UPDATED

MX4200 GPS receiver

The MX4200 GPS receiver uses six parallel channels to provide continuous code and carrier tracking of signals from six satellites simultaneously. The unit provides latitude, longitude, altitude, groundspeed and course, precise time and GPS satellite almanac and ephemeris data.

The receiver uses carrier-aided smoothing to reduce pseudo range measurement noise, averages the effects of multipath signals and has an eight-state Kalman filter that solves for three position co-ordinates, three velocity co-ordinates and clock and frequency offsets.

Specifications

Dimensions: 139.7 × 171.4 × 45.7 mm
Weight: <0.9 kg

Contractor

Raytheon Systems Company.

UPDATED

MX8000 series military GPS receivers

The MX8000 Series are single card, high-performance military GPS receivers designed for embedded applications. The MX8100 provides six parallel continuous tracking, switchable channels, each independently tracking code and carrier measurements for smooth navigation results, high accuracy and greater data availability. Designed to the requirements of GPS-EGR-600, both hardware and software interfaces provide quick and easy mating to military inertial instruments so that initial communication is quickly established.

MX8000 receivers feature a flexible modular architecture, intended for growth and adaptation to changing military requirements. High computer throughput, memory margins and software modularity allow new requirements to be added with minimum impact on the weapon system. The field reprogrammability feature allows mission software to be changed as required to meet specific tactical requirements. The receivers can be cold keyed using standard techniques. Receiver default condition is the Y-code, allowing direct P/Y code acquisition in time critical applications.

Specifications

Dimensions: 149.4 × 148.6 × 16.5 mm
Weight: 0.27 kg
Power supply: 5 V DC, 7.5 W
Accuracy:
(position) 15 m SEP, 2-3 m with DGPS
(velocity) 0.1 m/s
(time) 100 ns
Temperature range: −40 to +85°C

Contractor

Raytheon Systems Company.

UPDATED

51Z-4 marker beacon receiver

The Rockwell Collins 51Z-4 marker beacon receiver automatically provides aural and visual indication of passage over airway and instrument landing system marker beacons. The system is approved for Category II approaches in a number of Rockwell Collins' all-weather avionic system certifications. Operating at a frequency of 75 MHz, the receiver sensitivity can be varied between two preadjusted levels through a cockpit Hi-Lo switch. The Hi position is used to gain early indication of a marker beacon, the Lo position is then subsequently used closer to the beacon for a sharper position fix. Alternatively, the receiver sensitivity is continuously variable from the cabin or flight-deck by means of a potentiometer.

The system is of all-solid-state construction and contains triple-tuned circuitry for the rejection of spurious signals generated by television and FM broadcast transmitters. It is designed for three-lamp indication but may be easily modified for single-lamp operation by removal of a resistor and wiring the three lamp outputs together. In either type of operation, outputs can operate two sets of indicator lamps in parallel.

An optional self-test facility causes the internal generation of 3,000, 1,300 and 400 Hz marker signals which are detected in sequence by the receiver. The indicators light in order and corresponding aural tones are also generated.

The unit is suited to retrofit installation since it is mechanically and electrically interchangeable with a number of other marker beacon receivers, including the 51Z-2 and 51Z-3 units. An associated marker beacon antenna, the 37X-2 system, is also available. Designed for operation with the 51Z-4 and other compatible receivers, the antenna is plastic-filled and sealed to reduce the effects of precipitation static. This unit, which weighs less than 0.45 kg, can be mounted without cutting into the airframe structure and has negligible drag.

Specifications

Dimensions: ¼ ATR short low
Weight:
(without self-test option) 1.36 kg
(with self-test option) 1.47 kg

Operational status

In production and in service.

Contractor

Rockwell Collins.

UPDATED

ADF-60A Pro Line Automatic Direction-Finder (ADF)

The ADF-60A is a lightweight and rugged ADF operating from 190 to 1,749.5 kHz in 500 Hz steps.

Two combined loop and sense antennas are available: the ANT-60A for single system installations or the ANT-60B for dual system installations.

Specifications

Dimensions:
(receiver) ⅜ ATR short/dwarf
(ANT-60A) 419 × 216 × 41 mm
(ANT-60B) 604 × 269 × 25 mm
Weight:
(receiver) 1.9 kg
(ANT-60A) 1.5 kg
(ANT-60B) 2.3 kg

Operational status

In production and in service.

Contractor

Rockwell Collins.

UPDATED

ADF-462 Pro Line II Automatic Direction-Finder (ADF)

The ADF-462 is an all-digital ADF operating from 190 to 1,799 kHz together with the 2,182 kHz distress frequency. Two antennas are available: the ANT-462A for single system installations or the ANT-462B for dual system installations. The ADF-462 is compatible only with CSDB or ARINC 429 controls. Outputs are also serial digital and are provided in both CSDB or ARINC 429 characteristics. In addition, a DC sine/cosine output is available for interfacing with conventional RMIs.

Specifications

Dimensions:
(receiver) ⅜ ATR short/dwarf
(ANT-462A) 452 × 218 × 41 mm
(ANT-462B) 604 × 269 × 25 mm
Weight:
(receiver) 1.7 kg
(ANT-462A) 1.5 kg
(ANT-462B) 2.3 kg

Operational status

In production and service.

Contractor

Rockwell Collins.

UPDATED

ADF-700 Automatic Direction-Finder (ADF)

Designed in accordance with ARINC 712, the ADF-700 automatic direction-finder was introduced in 1980. It is based on experience gained with the earlier DF-203 and DF-206 receivers. These technical advances were pioneered by Rockwell Collins and incorporated into draft ARINC characteristic 712, which also provided for ARINC 429 interfaces and an integral loop/sense antenna. Centralised fault monitoring is now incorporated into the design consistent with ARINC 604 characteristics.

In addition to mechanical and reliability improvements introduced by the new characteristic, a significant performance improvement came with the change from analogue to digital technology. In the ADF-700 all bearing signal baseband processing is performed digitally. The reference and bearing signals are converted into digital form using a 12-bit CMOS analogue-to-digital converter and these are subsequently handled by an Intel 8086 16-bit microprocessor. The ARINC 429 input/output functions are performed by an Intel 8049 processor in conjunction with a Rockwell Collins universal asynchronous transmitter/receiver.

The advent of third-generation microprocessors permits the introduction of digital ADF signal processing to improve accuracy and reliability, a significant advantage being the elimination of mechanical adjustments. The system incorporates improved self-test capabilities as a result of using the 8086 microprocessor to control a test sequence incorporating a digital test signal synthesis. The same power supply subassembly is also used in the VOR-700 and ILS-700, leading to reductions in spares inventories and maintenance costs.

Specifications

Dimensions: 2 MCU per ARINC 600, 604
Weight: 2.9 kg
Power supply: 115 V AC, 400 Hz, 26 VA
Frequency: 190-750 kHz
Channel spacing: 0.5 kHz
Modes: ANT-aural receiver, ADF navigation, CW/MCW
Tuning: ARINC 429 dual serial bus
Accuracy: better than 0.9° with ARINC 712 antenna in 35 V/m field, exclusive of antenna error

Operational status

In production and in service.

Contractor

Rockwell Collins.

UPDATED

AHS-85 strapdown Attitude and Heading reference System

Designed for business aircraft and regional airlines, the AHS-85 attitude and heading reference system is aimed at a market largely ignored before 1983. The emphasis is on simplicity and all sensing and processing functions are grouped in one unit. The system uses a sensor package which eliminates all common high failure rate components and uses rotors that spin at one-seventh the speed of conventional gyros. The primary sensors are two piezoelectric multisensors, each with four independent pick-offs, which determine two axes of angular rate and linear acceleration data, thus providing one axis of redundant information for system monitoring. These devices are complemented by a flux detector mounted independently in a portion of the airframe without magnetic disturbances. The replacement of any sensing element does not require adjustment to the installation and compass compensation remains unchanged if the AHS-85 is replaced.

Processed data is compatible with flight instruments, autopilot systems, weather radar stabilisation and flight data recording. True airspeed can be accepted from a remote air data system. The system provides digital and analogue outputs and three-axis body rate and acceleration data and does not suffer directional gyro gimbal errors.

Specifications

Dimensions: 124.5 × 193 × 320 mm
Weight: 6.9 kg
Power supply: 28 V DC, 50 W
Accuracy:
(pitch and roll) 0.5° nominal,
1° during manoeuvres
(heading) 1° nominal,
2° during manoeuvres

Operational status

In production and in service.

Contractor

Rockwell Collins.

UPDATED

AHS-3000 Attitude Heading System

The AHS-3000 Attitude Heading System is Rockwell Collins' first avionics application of Digital Quartz Gyro (DQG) technology developed by Systron Donner. This DQG's small, lightweight, low-power, solid-state micromachined gyroscope is fabricated chemically from material similar to that used in quartz watches. It has been proven reliable in demanding automotive, railroad and maritime applications.

Because the construction has no moving mechanical parts, the DQG used in the AHS-3000 has an estimated mean time between failures of more than 10,000 h — up to 10 times improvement over most existing spinning-wheel and fibre optic systems.

The attitude heading computer used in the new AHS-3000 is one-third the weight and size of the current AHRS computer.

Operational status

The AHS-3000 was developed for integration with the Rockwell Collins' Pro Line 21 avionics system. First application will be the Raytheon Premier I business jet.

Contractor

Rockwell Collins.

NEW ENTRY

Airborne PC card receptacle DR-200

The Rockwell Collins airborne (PC Card) receptacle is an intelligent microprocessor-based unit that permits access to an aeronautical worldwide database and mission-specific data from two PC card sockets and communications via the MIL-STD-1553B databus. The unit accepts standardised memory-type PCMCIA PC cards of type-I, II or III sizes. The unit provides connector, pin-out and software compatibility with the Collins DR-902(v) data receptacle and cartridges. When used in conjunction with a Digital Aeronautical Flight Information File (DAFIF) or Jeppesen formattted PC card, the unit will provide instant retrieval of all stored airport, heliport, communication and navigation information. Multiple data files and subdirectories can be accessed using DOS-like commands via the databus. The PC cards can be programmed with the off-the-shelf desktop and laptop computers using DOS file protocol and plugged directly into the DR-200 receptacle.

Airborne PC Card receptacle DR-200 and mission planning station **1997**/0001338

All DR-200 operations are controlled by the aircraft bus-controller using MIL-STD-1553B messages. The unit provides status, type and directory information for each inserted card. Data file access and control flexibility is provided through DOS-like commands. When given an ICAO lookup command, the unit performs data retrieval and display formatting for the requested worldwide airport, heliport, navaid or intersection. Companion ground station software, designed to operate on a desktop or laptop computer, provides transfer and filtering of DAFIF information from CD-ROM to PCMCIA cards. The unit has the capability to continously search for and monitor the worldwide database for the 12 nearest airports, navaids or waypoints. The processor module is designed around a high-performance 80C186 microprocessor operating at 12 MHz. This module contains 64 k RAM, 64 k programme memory, 64 k non-volatile memory. Interfaces include: MIL-STD-1553B, discrete I/O, RS-422 and ARINC 429.

Specifications

Size: 57 × 146 × 153 mm
Weight: (max) 1 kg
Power: (max) 5 W

Contractor

Rockwell Collins.

UPDATED

AN/ARN-118(V) Tacan

The AN/ARN-118(V) Tacan provides the crew and flight control systems with reliable navigation data. Advanced digital circuitry is employed to achieve space and weight savings. All-solid-state, with the exception of the transmitter tubes, the AN/ARN-118(V) unit is packaged in a ¾ ATR short configuration. Both X and Y channels are standard on the system and a self-contained automatic antenna switch enables the unit to be plugged into aircraft with either single or dual Tacan aerial installations. With full digital readout, distance and bearing can be acquired in less than 1 and 3 seconds respectively. Input and output hardware is designed to accommodate complete digital serial-controlled Tacan navigation functions, including automatic tuning by RNav or air navigation systems. In addition, the system's output is compatible with INS, RNav and automatic navigation systems to provide full Rho-Theta real-time update information to these systems. Range rate information is available for area navigation.

Specifications

Dimensions:
(control) 127 × 76.2 × 81.3 mm
(receiver/transmitter) 190.5 × 172.7 × 317.5 mm
(adaptor) 43.2 × 172.7 × 317.5 mm
Weight:
(control) 0.77 kg
(receiver/transmitter) 12.23 kg
(adaptor) 1.81 kg
Power supply: 115 V AC, 400 Hz, 125 W

Operational status

In production. A recent application is the US Air Force KC-10 Extender. The original US Air Force production contract for AN/ARN-118(V) units was placed in 1975 and the system is standard equipment with that service and with the US Coast Guard. It is also becoming standard for the US Army and Navy and has been chosen by over 35 countries, some 32,000 sets having been built. The ARN-118(V) will remain in production for some years.

Contractor

Rockwell Collins.

UPDATED

AN/ARN-139(V) Tacan

The AN/ARN-139(V) transforms an aircraft into a flying Tacan station by providing air-to-air bearing and distance information. This capability is added to all the standard features of the AN/ARN-118(V) Tacan which provides standard air-to-ground and air-to-air Tacan information. Rendezvous with an ARN-139(V)-equipped aircraft or ship simplifies critical missions by providing a reliable readout of bearing and distance during the approach.

The ARN-139(V) is derived from the ARN-118(V), of which more than 32,000 units have been built. The ARN-139(V) provides distance and bearing transmission and inverse Tacan operation. The latter function allows a tanker aircraft, for example, to read the bearing and distance to a Tacan-equipped aircraft in need of air refuelling. Inverse Tacan also enables a pilot to determine the bearing to a DME ground station. The selectable range ratio capability permits the pilot, by means of a switch, to limit all replies to within four times the range of the nearest aircraft, or to concentrate on aircraft more than 30 times the distance of the nearest aircraft.

Specifications

Dimensions:
(AN/ARN-139(V)) 254 × 196 × 494 mm
(C-10059/A and C-10994 control units) 127 × 76 × 81 mm
Weight:
(AN/ARN-139(V)) 31.3 kg
(control unit) 0.9 kg
Frequency:
(transmitter) 1,025-1,150 MHz
(receiver) 962-1,213 MHz
Transmitter power:
(min) 500 W
(typical) 750 W
Receiver sensitivity: −92 dBm
Modes: X/Y channels, inverse beacon, inverse air-to-air, inverse transmit/receive, inverse receive
Range: 390 n miles
Track rate: 3,600 kt, 20°/s
Accuracy:
(distance) (digital) 0.1 n mile,
(analogue) 0.2 n miles
(bearing) (digital) 1°
(analogue) 1.5°
Reliability: 1,000 h design MTBF

Operational status

In service.

Contractor

Rockwell Collins.

UPDATED

AN/ARN-144(V) VOR/ILS receiver

The ARN-144(V) VOR/ILS receiver is said to be the first such system to be compatible with the MIL-STD-1553B digital databus. All the standard VOR, localiser, glide slope and ILS beacon facilities are available with 160 VOR channels and 40 localiser/glide slope channels being selectable at 50 kHz spacing.

A number of configurations is produced to meet specific military applications. The R-5094/ARN-514, a version of the ARN-144, is standard on F/A-18. A number of different control panels is also available.

Specifications

Dimensions: 104 × 127 × 304 mm
Weight: 3.6 kg
Power supply: 28 V DC, 25 W

Operational status

In production and in service.

Contractor

Rockwell Collins.

UPDATED

AN/ARN-147(V) VOR/ILS receiver

The AN/ARN-147(V) is fully compatible with the MIL-STD-1553B databus. This receiver combines all VOR/ILS functions such as VOR and ILS localiser, glide slope and marker beacon in one compact system.

In late 1997, Rockwell Collins released both a new model of the AN/ARN-147(V) and a modification kit for existing receivers that provides FM interference immunity in accordance with ICAO Annex 10.

All-solid-state modular construction makes the ARN-147 a reliable receiver for either new or retrofit applications on fixed- and rotary-wing aircraft.

Rotor modulation suppression circuitry is pin-selectable in the ARN-147 for reliable operation in rotary-wing aircraft. FAA split-channel requirements are met by providing 50 kHz spacing for 160 VOR and 40 localiser/glide slope channels.

Digital and analogue outputs are compatible with the latest high-performance flight control systems, digital indicators and analogue instruments. In addition, high- and low-level deviation and flag outputs for VOR, Loc and glide slope are provided.

The ARN-147 meets the following US military standards and specifications: MIL-E-5400 Class II environment; MIL-STD-810 vibration, including gunfire vibration; MIL-STD-461/462 electromagnetic interference, and MIL-STD-704 power characteristics. An FM immunity upgrade is now available.

Specifications

Dimensions: 104 × 127 × 304 mm
Weight:
(with MIL-STD-1553B) 3.6 kg
(without MIL-STD-1553B) 3.4 kg
Power supply:
28 V DC, 25 W with MIL-STD-1553B
28 V DC, 20 W without MIL-STD-1553B

Operational status

In production and service as the standard VOR/ILS receiver for the US Air Force and currently installed on aircraft such as the C-130, C-5, C-141, CH-53, UH-1H and UH-60. In mid-1985, Rockwell Collins received a contract to supply the US Air Force with over 25,000 AN/ARN-147(V) sets for use on several types of transport aircraft.

The system has also been selected for the V-22 Osprey, US Navy T-45 and US Air Force C-17. It is being retrofitted to the US Air Force T-38 aircraft.

Contractor

Rockwell Collins.

UPDATED

AN/ARN-149(V) (DF-206A) Automatic Direction-Finder (ADF)

The AN/ARN-149(V) was the first low-frequency automatic direction-finder to provide an internal field upgradable MIL-STD-1553B digital multiplex bus capability. This system, consisting of a receiver, control, antenna and mount, provides a low-frequency automatic direction-finding function in a lightweight easily installed set. The all-solid-state receiver eliminates all moving parts such as goniometers, synchros and mechanical tuners. Quadrantal Error Correction (QEC) is set by aircraft connector strapping, eliminating corrector modules and airframe specific internal adjustments. The antenna combines the loop and sense antennas and preamplifiers in one compact housing, eliminating expensive sense panels, couplers and special prefabricated cable assemblies. A dual version of the same antenna is also available in one aerodynamic package in either white or black colours. The receiver is controlled by a four-wire serial bus from the control. The system meets MIL-E-5400 and has an MTBF of 4,000 hours.

Specifications

Dimensions:
(receiver) 79 × 127 × 279 mm
(control) 146 × 57 × 96 mm
(antenna) 216 × 43 × 419 mm
Weight:
(receiver) 2.5 kg with MIL-STD-1553B capability
(control) 0.7 kg
(antenna) 1.4 kg
Power supply:
26 V AC, 7.8 W
28 V DC, 18 W

Operational status

In production.

Contractor

Rockwell Collins.

UPDATED

AN/ARN-153(V) Tacan

The AN/ARN-153(V) consists of one or two antennas, a cockpit control, the receiver/transmitter and a mounting tray.

Key features of the system are digital outputs for both distance and bearing (with optional analogue outputs), input and output 1553 capability, microprocessor-based design, dual-antenna ports, inverse bearing capability in certain applications with the 938Y-1 antenna, pilot-selectable air-to-air range ratio capability, 390 n miles range, signal-controlled search, solid-state 500 W transmitter, compatibility with ARINC 568, 582 and 429, enhanced BIT and W and Z channels for MLS compatibility.

The AN/ARN-153(V) has four basic modes of operation: receive, transmit/receive, air-to-air receive and air-to-air transmit/receive. When used in conjunction with the optional 938Y-1 rotating antenna and an optional control unit, the AN/ARN-153(V) can also provide bearing in certain applications to an air-to-air Tacan if it can transmit unmodulated squitter, as well as bearing to any DME-only ground station.

Specifications

Dimensions:
(control) 144.8 × 76.2 × 99.1 mm
(receiver/transmitter) 104.6 × 172.2 × 304.8 mm
(mounting tray) 114.3 × 50.8 × 312.4 mm
Weight:
(control) 0.9 kg
(receiver/transmitter) 6.49 kg
(mounting tray) 0.54 kg
Power supply: 28 V DC, 1.5 A nominal
Frequency:
(transmitter) 1,025-1,150 MHz
(receiver) 962-1,213 MHz
Number of channels: 126 X and 126 Y
Provision for W and Z
Range: up to 390 n miles
Accuracy:
(distance) (digital) ±0.1 n mile,
(analogue) ±0.2 n miles,
(bearing) (digital) ±0.5°,
(analogue) ±1.5°

Contractor

Rockwell Collins.

UPDATED

APR-4000 GPS approach sensor

The APR-4000 combines essential navigation functions with critical landing functions into a single, integrated GPS sensor that supports en route, terminal area, precision approach and non-precision approach operations. The sensor uses uplinked final approach waypoints and differential corrections of raw GPS satellite signals to determine the aircraft's precise lateral and vertical position in relation to the associated approach path. The information presented to the flight crew is ILS look-alike deviation data.

The APR-4000 GPS approach sensor utilises a 12-channel GPS receiver and incorporates Receiver Autonomous Integrity Monitoring (RAIM) and predictive RAIM that improves fault tolerance and the integrity of the computed solution by verifying and anticipating the availability of GPS signals throughout the flight plan. The sensor is engineered to provide growth to local area augmentation system specifications for Cat I and Cat II GPS precision approaches as well as growth to regional area augmentation systems such as WAAS, EGNOS and MT-SAT.

Operational status

Flight testing of the APR-4000 approach sensor is under way. Flight tests, involving two different types of special Cat I ground stations, are being used to measure system accuracy and performance parameters for production sensors, which are expected to be available in 1998.

Contractor

Rockwell Collins.

NEW ENTRY

AVSAT navigation systems

The Rockwell Collins AVSAT is a family of satellite-based precision navigation systems designed for all phases of flight, including take-off, en route, approach and landing. It is grouped into series, to match the mission of the aircraft and caters for long-range corporate jets, medium to light jets and turboprops and 30- to 100-seat regional airline passenger aircraft.

At the core of each AVSAT series is a Collins GPS sensor. The capabilities of a flight management system are incorporated within the navigation system. The avionics GPS engine features 12-channel satellite tracking, including those comprising the Wide Area Augmentation System, providing satellite coverage sufficient to meet the FAA-required navigation performance and precision approach criteria; a new software code to ensure certification to DO-178B Level A; Receiver Autonomous Integrity Monitoring (RAIM) and predictive RAIM, and differential correction computation to qualify the receiver for Cats. I and II.

The AVSAT Series provides the capability to calculate position and navigation information accurately anywhere in the world; provide VNav guidance and comprehensive pilot annunciations; fly complete airways, SIDs, STARs and approach legs, and predict fuel consumption. AVSAT also significantly reduces pilot workload by automating many calculations and functions. Information is presented on multifunction display pages, allowing easy, efficient movement through the system. Pilots can remain eyes-up during all phases of flight, resulting in enhanced situational awareness and greater control. Flight planning is simplified to the point where even the most complicated procedures can be generated with a few keystrokes. AVSAT is a component system of the technology required for the planned CCA 'free flight' regime in which aircraft en route separations can be reduced at the turn of the century.

AVSAT 3000

The AVSAT 3000 is designed for light jet and turboprop aircraft. It provides the capability for en route, terminal and non-precision approach and lateral/vertical navigation, and includes the growth potential to support precision approach computations.

The 3000 Series installation consists of the CDU-3000 control/display unit, FMC-3000 flight management computer (FMC) and AVSAT GPS sensor. Updated navigation information may be uploaded in the field and 100 flight plans, with 100 waypoints per plan, can be stored in the navigation database.

AVSAT 3000 also provides access to ground-based messaging, weather data and flight plan transmission through the Airborne Flight Information System (AFIS). Free form messages can be sent to and from the CDU. Frequently used messages can be formatted and stored.

AVSAT 4000

The AVSAT 4000 meets the needs of regional airline aircraft, providing worldwide GPS navigation, simplifying cockpit management and enhancing flight crew and aircraft performance. In addition to GPS, long-range navigational inputs such as IRS/AHRS, VLF/Omega, VOR and DME can be used to provide accurate determination of aircraft position. Coupled with flight management technology, the system provides the integrity, flexibility and operational capabilities required by regional airline operators.

AVSAT 4000 provides the capability for preflight and general flight planning. Complete lateral and vertical navigation is available. Radio sensor management is also available. Two flight plans are available: the primary flight plan is used for active guidance, while a secondary flight plan can be stored as an alternative and activated if needed. Full time and fuel management performance are provided, with fuel prediction available. Up to 1,000 standard company routes, each with up to 100 waypoints, can be stored in the system's database. The database contains navaids, waypoints, non-directional beacons, airports, airport reference points and runway thresholds. Flight planning includes the capability to execute SIDs, STARs, airways and holding patterns.

Once a flight plan is activated, it is presented graphically on a MultiFunction Display (MFD). The MFD also displays navaids, intersections, airports, terminal waypoints and non-directional beacons. In addition, the

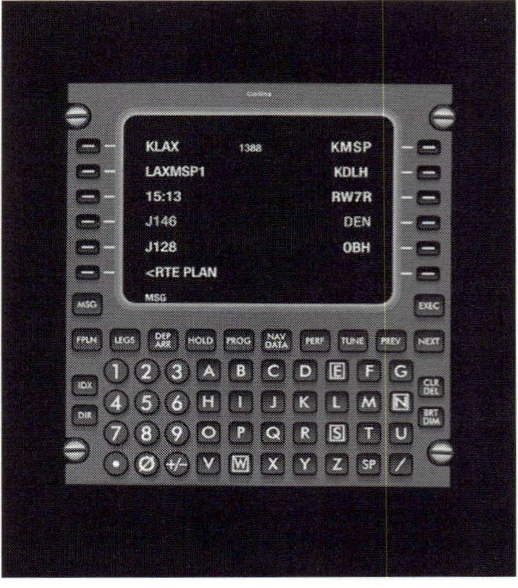

The AVSAT 3000 CDU **1996**

MFD can present complete operational status, progress and flight plan summary text.

AVSAT 4000 comprises three LRUs: the CDU-4100 Control/Display Unit, FMC-4200 Flight Management Computer (FMC) and GPS-4000 GPS sensor. The FMC is packaged as a line-replaceable module, housed in the Collins Integrated Avionics Processing System (IAPS). Pro Line 4 integration allows the AVSAT 4000 to perform self-diagnostic functions and informs the Pro Line 4 maintenance diagnostic computer of internal faults.

AVSAT 4000 has been certified for use on the Saab 2000 aircraft.

AVSAT 5000

Designed specifically around the needs of medium to light jet and turboprop aircraft, the AVSAT 5000 centralises control of EFIS, navigation, weather radar, TCAS and radio management in one convenient location to maximise cockpit space. It includes features such as the ability to calculate accurately position and navigation information anywhere in the world, provide VNav guidance and comprehensive pilot annunciations, fly complete airways, SIDs, STARs and approach legs, predict fuel consumption, and offer the right combination of performance, efficiency and ease of use required to perform in complex and demanding operating environments.

AVSAT 5000 combines GPS technology with flight management capability to provide satellite-based precision navigation systems for take-off, en route, terminal area, approach and landing. Other long-range navigational inputs include IRS/AHRS, VLF/Omega, VOR and DME. These may be used to provide the most accurate determination of aircraft position. Vertical navigation is available in each phase of flight. Included is the ability to execute a vertical direct-to, generating a vertical path from the aircraft's current position to a designated altitude.

AVSAT 5000 also provides access to ground-based messaging, weather data and flight plan transmission through the Airborne Flight Information System (AFIS). Free-form messages can be sent to and from the CDU. Frequently used messages can be formatted and stored.

AVSAT 5000 comprises three LRUs: the CDU-5000 control/display unit, FMC-5000 Flight Management Computer (FMC) and AVSAT GPS sensor. The FMC is packaged as a line-replaceable module, housed in the Collins Integrated Avionics Processing System (IAPS). Pro Line 4 integration allows the AVSAT 5000 to perform self-diagnostic functions and informs the Pro Line 4 maintenance diagnostic computer of internal faults.

AVSAT 6000

The AVSAT 6000 is designed for long-range business aircraft. It features the ability to calculate position and navigation information accurately anywhere in the world, provide VNav guidance and comprehensive pilot annunciations, fly complete airways, SIDs, STARs and approach legs and manage fuel consumption.

AVSAT 6000 comprises three LRUs: the CDU-6000 or CDU-6100 control/display unit, FMC-6000 flight management computer and GPS-4000 sensor. It can be configured for single, dual or triple operation by installing the desired number of FMC modules and the same number of control/display units. It is designed to be integrated with the Pro Line 4 avionics system.

AVSAT 6000 is designed for worldwide navigation, including polar navigation. Long-range inputs in addition to GPS can be used, including IRS/AHRS, VLF/Omega, VOR and DME to provide the most accurate determination of position. It contains a global database of current information used for both flight planning and navigation.

AVSAT GNLU

To meet the requirements of the Air Traffic Management (ATM) system, Rockwell Collins is offering the AVSAT GNLU system. The system defined by AVSAT GNLU is of a single navigation unit integrating all the GNSS-based (Global Navigation Satellite Systems) en route, terminal and landing system capabilities required to operate in a global CNS/ATM environment (Communications Navigation Surveillance/Air Traffic Management).

The AVSAT GNLU contains an advanced GPS receiver (FAA order 8110.60 compliant) with the required integrity to support precision GNSS-based Landing System (GLS) approach certification to Cat I and II levels.

Because the role of the GPS sensor is so important, Rockwell Collins has designed and developed a specialised GPS engine for use in air transport system applications. Optional MLS modules, developed in conjunction with Daimler-Benz Aerospace, and optional ILS modules can be integrated within the GNLU to provide classic aircraft with analogue interfaces. This is an efficient update path for compliance with new landing sensor requirements, such as FM immunity.

As part of the total Rockwell Collins AVSAT FMS package, the GNLU adds substantially more than GPS receiver inputs to existing flight management capabilities; GNLU is the primary processing and control unit of the AVSAT GPS/FMS system. The AVSAT FMS supports complete lateral navigation and provides vertical guidance on approach.

The complete system comprises the cockpit-mounted Multifunction Control/Display Unit (MCDU), the remote 4-MCU sized GNLU, and a GPS antenna.

The AVSAT GNLU is structured to add the interfaces and processing needed for precision GNSS approaches, using an integral uplink data receiver to simplify installation. The GNLU navigation computer has the volume and throughput provisions to support the Wide Area Augmentation System (WAAS) and Local Area Differential GNSS (LADGNSS) receiver and modem capability; all essential WAAS hardware is provided in the baseline configuration. The system also provides ACARS and ARINC 622 datalink interfaces and is provisioned to interface with ATN compliant architectures.

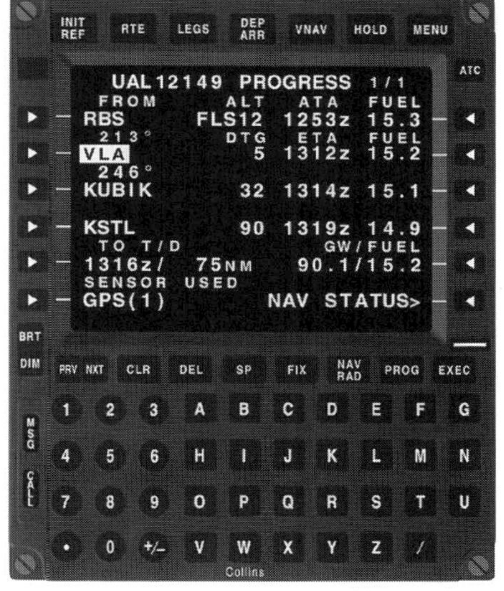

AVSAT GNLU multifunction control/display unit
***1998**/0015375*

GNLU-900 GNSS ***1998**/0015376*

Specifications (GNLU-900 GNSS)

GPS performance:
(channels) 12-channel, all-in-view tracking
(frequency) 1575.42 MHz (L1) C/A code
Accuracy:
(horizontal) 23 m (4.5 m with DGPS); 1.5 m/s (velocity)
(altitude) 30 m (6.0 m with DGPS); 1.2 m/s (velocity)
Dimensions: 4 MCU per ARINC 600
Weight: 7.3 kg
Power: 28 V DC

Contractor

Rockwell Collins.

UPDATED

Communications, Navigation and Identification (CNI) modules for the F-22

Under the CNI concept for the F-22, SEM-E modules are placed in an integrated avionics rack. CNI represents a modular avionics approach, as opposed to LRUs. These CNI modules perform avionics tasks that are accomplished today by individual LRUs. The programme is designed to provide reliable advanced avionics in a much smaller space than that occupied by current units.

Modules developed by Rockwell Collins include the VHF/UHF receiver, VHF/UHF transmitter, GPS receiver, GPS antenna electronics, single and five-channel L-band receiver and antenna interface unit. In addition to hardware modules, Rockwell Collins is also supplying the software for GPS, VHF/UHF and Have Quick II.

Operational status

Under a US$102 million contract, initial module deliveries for the F-22 were due to continue until the end of 1997.

Contractor

Rockwell Collins.

UPDATED

Rockwell Collins is supplying communications, navigation and identification modules for the F-22

DME-42/442 Pro Line II DME systems

The DME-42/442 are all-digital DME systems which can provide complete information on up to three DME stations using a single receiver. They were designed for business or commuter aircraft where only a single DME facility is needed, but where additional information is useful and DME-42 can directly replace the earlier DME-40 system. Station information is presented on the associated IND-42 display.

The DME-42/442 systems can provide the display with data to show distance to station up to 300 n miles, time to station up to 120 minutes and groundspeed up to 999 kt.

The DME-442 is compatible only with CSDB or ARINC 429 controls and displays.

Specifications

Dimensions:
(R/T) ½ ATR short/dwarf
(IND-42) 42 × 86 mm
Weight:
(R/T) 2.4 kg
(IND-42) 0.41 kg
Power supply: 28 V DC, 0.7 A plus 0.3 A for display
Altitude: up to 70,000 ft

Operational status

In production and in service.

Contractor

Rockwell Collins.

UPDATED

DME-900 Distance Measuring Equipment

The DME-900 distance measuring equipment has been designed to provide highly accurate distance data outputs in all modes of operation. Error sources in the interrogator are identified and reduced to the lowest level. A highly efficient parallel ranging technique allows the DME-900 to perform less than 15 interrogations, half the allowable maximum average. Only two output devices are required to achieve a 700 W nominal power output. The synthesiser/driver operates at L-band transmit channel frequencies, significantly reducing the parts count and circuit complexity. In addition, optimum filtering methods reduce naturally occurring noise contained in the DME/Tacan signal.

Specifications

Dimensions: 4 MCU
Weight: 6 kg
Power supply: 115 V AC, 380-420 Hz, 35 W (max)
Frequency:
(transmit) 1,025-1,150 MHz
(receive) 962-1,213 MHz

Channels: 252
Range: −1 to 320 n miles
Accuracy: ±0.1 n mile

Contractor

Rockwell Collins.

UPDATED

GEM II/III/IV standard positioning service embedded GPS receiver

The GEM (GPS Embedded Module) series are five-channel P/Y Code receivers on a single circuit card, with both 12 and 24-channel SEM-E single circuit card receivers available in 1998.

GEM can operate in stand-alone GPS mode (unaided) or aided through integration with INS or Doppler. The primary data interface is a Dual Port Random Access Memory (DPRAM), which provides data according to ICD-GPS-059 and ICD-GPS-104 formats as well as satellite position and measurement data.

Units compliant with the US DoD GRAM performance specification are also being made available in 1998.

Specifications

	C/A code without SA	C/A code with SA
Position accuracy	10 m horizontal	100 m horizontal
Velocity accuracy 3D	0.2 m/s	0.7 m/s
Time accuracy	70 ns	270 ns

Dynamic conditions: 1,200 m/s, (min) satellite signal power level, and 31 dB jammer signal level. SA degradation is assumed at the current nominal SPS level of 100 m
Number of channels:
(GEM II/III) 5, (GEM IV) 12/24, GRAM 12/24
Frequency:
(GEM II/III/IV) L1/L2
Dimensions: 150 × 145 × 15 mm
Weight: (GEM II/III/IV and GRAM) 0.55 kg

Operational status

The system has been selected for the US DoD's first embedded GPS and inertial programme (GINA) for the Navy T-45 trainer aircraft, as well as for the tri-service Embedded GPS/IVS (EGI) programme.

Contractor

Rockwell Collins.

UPDATED

GLU-900/GLU-920 Multimode Receivers

The Rockwell Collins GLU-900 MMR provides two or more landing system standards. Both the ILS and GNSS functions are basic to the landing receiver, while the MLS and GLS functions are categorised as options. The optional MLS modules were developed in conjunction with Daimler-Benz Aerospace as part of its strategic relationship with Rockwell Collins Avionics and Communications.

The design provides ILS lookalike interfaces to the existing aircraft autopilot and display systems and accommodates requirements for both dual-dual and triplex auto-land architectures. It also offers the accuracy, reliability and integrity required for critical performance in Cat III landings.

The MMR will continue to support existing all-weather landing capability and will expand capabilities as the industry moves toward the future GLS.

The pilot interface to the system is mechanised through a multipurpose control display unit which uses standard ARINC operational philosophy. This allows interoperability training benefits to mixed fleet operators, and future integration of the datalink control function.

The GPS flight management processor and navigation database software can be loaded on the flight line, simplifying system growth.

There are two versions of the series 900 MMR. The GLU-900 is a digital system, whilst the GNLU-900 interfaces analogue systems.

The GLU-920/GNLU-920 system indicates a colour liquid crystal multifunction Control Display Unit (CDU), a remote-mounted navigation computer and a GPS receiver.

Specifications

GNLU-900 specifications
Dimensions: 4 MCU per ARINC 600
Weight: 7.3 kg approx
Power: 29 V DC
TSO: C129 B1
Qualification: RTCA DO-160 C, DO-178 B

GLU-900
Dimensions: 3 MCU
Weight: 3.86 kg
Power: 115 V AC, 400 Hz
TSO: -C36e, -C34e, (A2D2 YBA (BCL) EIXXXXXZEAEZYZLXX
General: DO-192; DO-195; DO-160C; FCC part 15, EUROCAE ED-46; ED-47A

GPS Performance
Channels: 12-channel, all-in-view tracking
Frequency: 1575.42 MHz (L1) C/A code transmissions
Sensitivity:
(acquisition) −121 dBm
(tracking) −125 dBm
Accuracy (95%):
(horizontal position) 23 m (4.5 DGPS)
(altitude) 30 m (6.0 m DGPS)
(horizontal velocity) 1.5 m/s
(vertical velocity) 1.2 m/s
(time) 100 ns
Time to fix first (95%): 75 s max with valid initialisation; 10 min, max without valid initialisation

MLS Performance
Channels: 200, per ICAO Annex 10
Antenna connections: 3 (2 passive, 1 active)
Datalink frequency (optional): C-band

ILS Performance
Frequency range: LOC 108.10 to 11.95 MHz; GS 329.15 to 335.0 MHz
Channel spacing: LOC 50 kHz; GS 150 kHz
Navigation outputs: ARINC 429

Operational status

Certified for Airbus Industrie forward production aircraft, initially in ILS/GPS configuration. Certification is also planned for Boeing 737-600/-700/-800, 747-400, 757, 767, and 777 aircraft. United Airlines has selected the series 900 MMRs for its fleet; the order could be for up to 770 GLU-900 MMRs and 330 GLNU-900 MMRs.

Contractor

Rockwell Collins.

UPDATED

GPS-4000

The Collins GPS-4000 enables aircraft equipped with flight management capability to perform GPS-based en route, terminal area and non-precision approach navigation, as well as primary means of oceanic/remote operations.

The GPS-4000 is part of the continued development of the Collins AVSAT satellite-based communication and navigation system, providing operators with the advanced capabilities required to operate within the evolving CNS/ATM environment.

The 2 MCU GPS-4000 sensor processes the transmissions of up to 12 GPS satellites simultaneously, calculating navigation solutions based on information from all satellites in view. A minimum of four satellites with acceptable geometry, or three satellites plus calibrated barometric altitude, are required to calculate navigation solutions.

The heart of the GPS-4000 is the Collins Avioincs GPS Engine. The self-contained 12-channel GPS receiver outputs Earth-Centered Earth-Fixed (ECEF) position, velocity and GPS time once a second. This information is used by the flight management system to calculate a flight plan-based navigation solution. In addition, Predictive Receiver Autonomous Integrity Monitoring (PRAIM) allows crews to determine whether the satellite geometry at the destination airport will support approach at the planned time of arrival.

The GPS-4000 is designed to support sole means of navigation for en route, terminal and non-precision approach in today's Air Traffic Control environment. It is also capable of being upgraded via service bulletin to provide precision approach capability as the infrastructure evolves. The resulting APR-4000 approach sensor accommodates use of the Wide Area Augmentation System (WAAS) and Local Area Augmentation System (LAAS) in support of precision approach operations.

Operational status

The Rockwell Collins GPS-4000 Global Positioning System sensor has been certified for use on the Beechjet 400A, Canadair RJ, Learjet 60 and Saab 2000.

Contractor

Rockwell Collins.

UPDATED

ILS-900 Instrument Landing System receiver

The ILS-900 instrument landing system is a high-integrity ILS receiver with software verified to DO-178A Level 1. This system performs to Cat III fly-by-wire failure and monitor requirements in dual and triplex autopilot installations.

The receiver has been designed for Cat III dual-autopilot operations which place heavy reliance on internal monitoring within the ILS receiver. A high degree of monitoring integrity has been included in the ILS-900 and system reliability has been improved with digital technology. In addition, the system features audio mute in cruise mode and automatic Morse decoding. Absolute partitioning between BIT and the primary monitoring deviation circuitry has been achieved by using four separate processors.

Specifications

Dimensions: 3 MCU
Weight: 3.86 kg
Power supply: 115 V AC, 400 Hz, single phase
Frequency:
(localiser) 108.1-111.95 MHz
(glide slope) 329.15-335 MHz
Channel spacing:
(localiser) 50 kHz
(glide slope) 150 kHz
Temperature range: −55 to +71°C

Contractor

Rockwell Collins.

UPDATED

LRN-85 Omega/VLF navigator

Designed for the worldwide navigation of commercial, corporate and military aircraft, the LRN-85 is claimed to be one of the most comprehensive Omega/VLF systems available. By comparing all available signals from Omega and VLF ground stations against a precise rubidium frequency standard, the system can navigate a direct route between departure and arrival points, eliminating dependence on VORTacs and providing continuous position information to the crew.

As well as maintaining 100 per cent duty cycle on the Omega transmitters, the system receives both data sidebands from the seven worldwide VLF stations.

A notable feature of the system is the mass memory, which can store the co-ordinates of all the world's VORTacs and major airports, and the waypoint memory, with capacity to hold the co-ordinates of up to 100 waypoints. For the majority of airports, therefore, the crew need only enter the identifying codes of the start and destination airports and the LRN-85 computes a direct course between them. Should this still not be adequate to define a route or series of regularly used routes, the crew can insert the appropriate waypoints and destinations and designate them by alphanumeric codes.

Two dedicated microcomputers provide the LRN-85 with power and flexibility to drive or command many functions, for example provide desired track, cross-track and distance to go to a horizontal situation indicator; automatic and manual waypoint advance; roll command with turn anticipation; databus feed for automatic transmission of waypoint information to a second LRN-85 or other navigation system; internally computed magnetic variation; true and magnetic display of bearing, desired track and windspeed and direction; computed offset waypoint; automatic computation of diurnal shift and test mode for station deselect and for computer diagnostics.

The system comprises five units: an E-field VLF blade antenna or H-field loop, antenna coupler, optional equipment unit (containing the atomic frequency standard, a battery and additional interfaces), receiver/processor and control/display unit. In its ARINC 599 version, the optional equipment unit is incorporated into the receiver/processor.

Specifications

Dimensions:
(control/display unit) 146 × 114 × 159 mm
(receiver/processor) 190 × 193 × 498 mm
Accuracy: 2 n miles CEP with a minimum of 2 usable stations
Qualification: ARINC 599

Operational status

In service, no longer in production.

Contractor

Rockwell Collins.

UPDATED

MAGR/E-MAGR receivers

The MAGR/E-MAGR military GPS receivers provide the user with a wide variety of input, selective availability/anti-spoofing and power supply options. All of these receivers provide MIL-STD-1553, RS-422, ARINC 429 and Have Quick interfaces.

The ability to process selective availability/anti-spoofing is controlled by the National Security Agency and the approved Rockwell Collins precise position service-security module allows selected users to process GPS crypto-variables. These crypto-variables give the user the maximum accuracy possible at all times.

The Miniaturised Airborne GPS Receiver (MAGR) is a five-channel receiver that fits within a ⅜ ATR form factor. It provides high levels of GPS performance and excellent anti-jamming capabilities on high-dynamic aircraft, including fighters and bombers. MAGR is designed to be a stand-alone GPS receiver which can operate with a MIL-STD-1553 or ARINC 429 interface or with a stand-alone control display unit. It is offered in versions that operate on 115 V AC or 28 V DC. The receiver can also be ordered with an intermediate frequency or radio frequency interface.

MAGR provides the maximum levels of navigation accuracy in the highest levels of jamming and dynamics. Operating in the Precise Positioning Service (PPS) with velocity to within 0.1 m/s, MAGR also supports accurate time information for radio system co-ordination and synchronisation.

The Enhanced MAGR (E-MAGR), with 24 channels, precision real-time kinematic operation and Wide Area Augmentation System (WAAS) and Local Area Augmentation System (LAAS) capability was produced in 1997.

An MAGR with embedded GRAM receiver compliant with US DoD GRAM performance specification will be available in 1999.

Specifications

Dimensions: 82 × 174 × 305 mm
Weight: 5.9 kg
Number of channels: MAGR 5; E-MAGR 24
Interface: MIL-STD-1553B, ARINC 429, RS-422, Have Quick/1 PPS, US Standard Crypto
Accuracy:
(position) 16 m SEP
(velocity) <0.1 m/s
(time) <100 ns
Reliability: >5,000 h MTBF

Operational status

The MAGR is a standard US DoD airborne GPS system. It has been chosen by all four US military services.

The receiver provides full accuracy, P/Y-code GPS capability to a wide variety of platforms, including F/A-18, AV-8B, F-16, F-117A, B-2 and S-3B. MAGR is fitted to several international aircraft, as well as selected special use platforms. It can accommodate many interfaces, including SA/AS with PPS-SM and can be integrated with existing INS and Doppler systems.

Contractor

Rockwell Collins.

UPDATED

VIR-32/432 Pro Line II navigation receivers

The VIR-32/432 are digital VOR/ILS navigation receivers designed for business or commuter aircraft. The VIR-32 can directly replace the VIR-30A or be installed in an all-digital aircraft. Several versions of the VIR-32 are available, but all are identical in format, the variations being selected by making the appropriate wiring connections. The system can receive 200 VOR/localisers and the associated 40 glide slope channels.

The VIR-432 is compatible only with CSDB or ARINC 429 controls.

Specifications

Dimensions: ⅜ ATR short/dwarf
Weight: 2 kg
Power supply: 28 V DC, 1.4 A
Altitude: up to 70,000 ft

Operational status

In production and in service.

Contractor

Rockwell Collins.

UPDATED

VOR-700 VOR/marker beacon receiver

The VOR-700 VHF omnidirectional range/marker beacon receiver incorporates digital technology combined with the background derived from previous industry standards such as the Collins 51RV-2, 51RV-4 and 51Z-4 VORs. The system was designed in accordance with ARINC 711.

All bearing signal baseband processing in the VOR-700 is accomplished digitally. The 30 Hz reference and variable signals are converted into digital form using a 12-bit CMOS analogue-to-digital converter and are thereafter handled by an Intel 8086 16-bit microprocessor. The ARINC 429 input/output functions are performed by an Intel 8048 microprocessor in conjunction with a Collins universal asynchronous transmitter/receiver large-scale integration-based circuit. Additional functions, such as self-test, auto-calibration and monitoring, are also conducted digitally.

Digital processing improves the accuracy of measuring bearings by reducing the effects of temperature variation and ageing. Implementation of 30 Hz bandpass filters in firmware, compared with previous analogue methods, permits improved tracking of ground station modulation frequency variations, increased navigation sensitivity and better rejection of undesired components in the modulation of received signals.

VOR-700 parts count has been reduced by 40 per cent compared with the most recent analogue technology VOR systems such as the 51Z-4 marker beacon receiver and the VOR portion of the 51RV-4. The system is fully compliant with ARINC 711 and extends many parameters of previous generation equipment. In particular, bearing measurement accuracy is five times better and there are 40 per cent fewer adjustments. Centralised fault monitoring capability is offered.

Specifications

Dimensions: 3 MCU per ARINC 600
Weight: 3.9 kg
Power supply: 115 V AC, 400 Hz, 30 VA
Frequency:
(VOR) 108-117.95 MHz
(marker beacon) 75 MHz
Channel spacing:
(VOR) 50 kHz

Operational status

In production and in service.

Contractor

Rockwell Collins.

UPDATED

VOR-900 VOR/marker beacon receiver

The VOR-900 maximises bearing accuracy by minimising the effects of temperature variation and ageing by using digital techniques. Implementation of 30 Hz bandpass filters in firmware, instead of analogue techniques, provides improved tracking of ground station modulation frequency variations. It also contributes to increased navigation sensitivity and better rejection of undesired components of received signal modulation.

Low parts count, quality components and straightforward design all contribute to reliable operation of the VOR-900. The system accommodates ground station and environmental anomalies. Functional test is provided by employing microprocessors to control a test sequence using digital test signal synthesis. LRU and card testing are simplified by an organised modular design which results in easily tested functional modes.

Specifications

Dimensions: 3 MCU
Weight: 3.86 kg
Power supply: 115 V AC, 400 Hz
Frequency:
(marker beacon) 75 MHz
(VOR) 108-117.95 MHz
Channel spacing:
(VOR) 50 kHz
Temperature range: −55 to +71°C

Contractor

Rockwell Collins.

UPDATED

GPS Navstar navigation systems

SCI produces two- and five-channel GPS receivers in multiple configurations; the two-channel UH receiver for helicopters, the five-channel 3A receiver for aircraft and the five-channel 3S for shipboard use. The receivers are fully qualified for operation on highly dynamic platforms in a high-jamming environment. As multichannel receivers, they are capable of processing signals from multiple satellites simultaneously, expediting initial acquisition times. The receivers are capable of providing accuracies of better than 16 m SEP in position, 0.1 m/s in velocity and 100 ns in time, even in the presence of the DoD selective availability and anti-spoofing environment.

Specifications

Dimensions:
(3A receiver) 193 × 191 × 484.8 mm
(UH receiver) 193.6 × 191.3 × 382.6 mm
Weight:
(3A receiver) 18.2 kg
(UH receiver) 11.5 kg

Power supply: 120 V AC, 400 Hz
160 V DC
Interfaces:
(3A receiver) ARINC 429 and 575, MIL-STD-1553, PTTI, KYK-13
(UH receiver) ARINC 561, 572, 575, 582, Have Quick, RS-422, KYK-13

Reliability:
(3A receiver) >1,350 h MTBF
(UH receiver) >160 h MTBF

Operational status
In service.

Contractor
SCI Systems Inc.

VERIFIED

AN/ARN-136A(V) Tacan

The AN/ARN-136A(V) lightweight airborne Tacan is a remotely controlled system utilising large-scale integrated circuits and CMOS technology. The system consists of the RT-1321A/ARN-136A radio receiver/transmitter with a range unit and the CP-1398/ARN-136 azimuth computer with bearing unit. The units contain no moving parts and through exclusive use of CMOS circuits, diodes and LSI chips are rated for continuous operation at +70°.

The units meet or exceed all FAA TSO C66a requirements for high-altitude operation up to 70,000 ft. The Tacan system has been designed to work with all Tacan or VORTac stations meeting MIL-STD-291B and FAA selection order for the US National Aviation Standard 1010.55 and ICAO Annex 10.

The AN/ARN-136(V) (ruggedised/EMI lightweight airborne TACAN) has been designed to meet ElectroMagnetic Compatibility (EMC) requirements of MIL-STD-461B Methods CE03, CS01, CS02, CS03, CS04, CS05, CS06, RE02, RS02 and RS03. In addition, RTCA Do-160B A2E1/A/MNO/XXXXXX2BABA and induced signal suspectibility category A and Z have been met, together with Shock and Random Vibration per MIL-STD-810C Method 514.2 procedure 1A and Method 516.2 procedure I30Gs. This enhancement is in use by the US Navy.

The AN/ARN-136A system utilises an ID-2218/ARN-136 range indicator which displays distances up to 399 n miles, groundspeed up to 999 kt and time to Tacan station up to 99 minutes. To minimise the number of wires between the cockpit and avionics bay where the remote boxes are installed, the AN/ARN-136A Tacan incorporates a three-wire serial databus that is used for both range and tuning.

The radio receiver/transmitter utilises two LSI circuits to provide digital computations of distance, groundspeed and time to station. All tuning is performed electronically, using a digital frequency synthesiser. The unit provides 63 channels of X mode air-to-air ranging.

The azimuth computer accepts receiver video, pulse-pair decoding information and DC power from the range unit. The bearing unit derives aircraft bearing angle, with respect to local magnetic north, to or from the VORTac or Tacan ground beacon. The derived bearing information is then converted into several bearing data formats to enable proper interfacing with a variety of aircraft instruments. The unit is designed with two multilayer printed circuit cards, divided so that one card derives the bearing signal digitally and the second interfaces with the aircraft instruments. This provides a bearing accuracy of ±0.5°.

Specifications
Dimensions:
(receiver/transmitter) 63.5 × 133.3 × 298.4 mm
(azimuth computer) 63.5 × 133.3 × 298.4 mm
Weight: 3.13 kg
Power supply: 11-33 V DC, 15 W

Contractor
Sierra Technologies.

UPDATED

PFIS-II Portable Flight Inspection System

PFIS-II Model 9106 is the second generation of the PFIS models 7802 and 8009 which were introduced in 1978. The PFIS-II, developed for both civilian and military markets, can be used as the primary flight inspection system or to augment current in-service systems.

Designed for small aircraft, the PFIS-II is installed on the aircraft seat rails and has a 30-minute installation or removal time to reconfigure the aircraft. Modification of the aircraft is kept to a minimum by adding six antennas, two standard indicators, five switches and a 28 V DC, 26 V AC feed from the aircraft power bus. The size and weight are approximately the same as the size and weight of a single aircraft seat and a 77 kg person.

Configuration of the basic PFIS-II is delivered with a single VOR/ILS and marker beacon receiver, VHF communications, radio telemetering theodolite receiver and DME. In addition, the PFIS-II may be enhanced with an ADF, Tacan system, a second VOR/ILS receiver and a 115 V AC, 400 Hz inverter for test equipment use. MLS capability is under consideration as a future expansion element.

Contractor
Sierra Technologies.

VERIFIED

Model 6000 Attitude and Heading Reference System (AHRS)

The Model 6000 represents a series of AHRS designed and manufactured by Smiths Industries for a large variety of aircraft applications. Licence production is also undertaken by Smiths Industries Newmark (UK) and KKK (Japan). Over 3,200 examples were produced between 1974 and 1990.

The AHRS provides multiple synchro outputs of roll, pitch and heading through all aircraft attitudes, including loops and spins. Heading rate and validity outputs are also provided. The system is tolerant of power interruptions of up to 30 seconds and can accommodate the body rates attainable in any aircraft type.

The Model 6000 AHRS consists of three units: an all-attitude twin gyro platform Displacement Gyroscope Assembly (DGA), an Electronic Control Amplifier (ECA) and a Compass System Controller (CSC).

Specifications
Dimensions:
(DGA) 150 × 140 × 260 mm
(ECA) 155 × 188 × 231 mm
(CSC) 146 × 38 × 102 mm
Weight: 14.1 kg incl mounting bases
Power supply: 115 V AC, 400 Hz, single phase, 90 W
Gyro drift:
(vertical) 3°/h
(directional) 0.5°/h
Environmental: MIL-E-5400 Class II

Operational status
No longer in production. Aircraft applications include the A-10A, T-45A, T-46A, Alpha Jet, Hawk, S.211, AT-3, S-70 and the IA-63. Retrofit applications include the F-106, C-141, B-52G/H, J-35 Draken, HSS-2B, PS-1, A-4 and F-5A/B.

Contractor
Smiths Industries Aerospace.

UPDATED

AN/ARN-101(V) nav/attack system

The AN/ARN-101(V) was produced for the avionic update programmes on US Air Force RF-4C, F-4E and F-4G aircraft. It is a digital, integrated system, offering improved reliability and facilities compared with the former system employed in those aircraft.

A central digital computer with a memory of 64 k words interfaces with aircraft sensors to provide positional, reconnaissance and weapon delivery facilities. An inertial navigation system, fire-control radar, Doppler navigator, Pave Tack, TISEO and optical sight systems all interface with the computer. A multisensor implementation of offset target location and co-ordinate computation is used to update the navigation position and to provide information to resolve the tactical problem.

In the rear cockpit of the aircraft two displays are used in conjunction with the system. One is a digital display with three lines of 24 characters each of alphanumeric information. The other unit is a full 36-key control unit. These units are software-controlled to give a comprehensive and easily followed procedure to the weapons systems officer. Mission parameters can be prepared at a ground station and transferred to the aircraft using a data transfer unit.

The ARN-101(V) is a complete navigation and weapon delivery system. It can be used for navigation to targets, target attacks, pull-up cueing, target of opportunity data freeze, danger area avoidance, autopilot coupling, flight director steering, return to base, approach to landing and many other functions; the inertial measurement unit can be restarted and realigned while airborne.

Operational status
In service. Over 400 systems were delivered. Some 20 systems have been installed in export F-4E aircraft.

Contractor
Smiths Industries Aerospace.

VERIFIED

AN/ASN-162 Attitude and Heading Reference System (AHRS)

The AN/ASN-162 AHRS is currently in development as the US Air Force and Navy standard compass/attitude and heading reference system. Smiths Industries interferometric fibre optic gyros are incorporated into the system which is the first military AHRS to use these as inertial sensors.

The AN/ASN-162 will provide multiple synchro outputs of roll, pitch and heading for all aircraft attitudes. Body axis rate and acceleration data will be available via a MIL-STD-1553B databus. The inclusion of an improved Magnetic Azimuth Detector (MAD) will provide magnetic compass information to the system. The MAD and AHRS will be calibrated automatically for ease of maintenance.

The AN/ASN-162 consists of four units: the inertial reference unit, control indicator, flight data memory unit and the MAD. It is being designed for a wide range of aircraft applications. Specific aircraft interface parameters are stored in the flight data memory unit. This allows a single inertial reference unit configuration to be used in all aircraft configurations.

Operational status
In engineering development.

Contractor
Smiths Industries Aerospace.

VERIFIED

Navigation Attack System (NAS)

The core of the Navigation Attack System (NAS) is the Computer Interface Unit (CIU), a specially developed computer based on dual MIL-STD-1750A processors which combines mission and navigation management with stores management and weapon delivery and which contains the main controller for the MIL-STD-1553 databus. It also generates symbols for the

HUD and radar display. Some of the CIU functions, such as bus control and navigation management, are backed up by the display processor. The CIU and display processor are designed to permit easy expansion to meet future needs.

The NAS includes two MultiFunction Displays (MFDs). Normally, the right-hand MFD displays radar data, with navigation, stores or other information on the left screen. As with other modern avionics, a primary mode switch on the throttle puts the HOTAS controls in navigation, air-to-air or air-to-ground mode.

The SNU-84-1 ring laser gyro inertial navigation unit is a standard component. NAS offers the same choice of radar modes as the standard F-16 radar, selectable through the HOTAS controls.

The NAS includes a Sidewinder control system for enhanced AIM-9 off-boresight capability. Another feature of the NAS is a Smiths Industries Data Transfer Module (DTM), a read-write device using removable memory cartridges, located on the right side of the cockpit. The DTM is used in conjunction with a Mission Data Ground Terminal (MDGT). The MDGT can also be linked to a map and digitiser pad to allow the direct entry of navigation data or used to download post-flight and maintenance data.

Operational status
In service in New Zealand Air Force A-4 Skyhawk aircraft.

Contractor
Smiths Industries Aerospace.

VERIFIED

The Smiths Industries Self-Contained Navigation System for the US Air Force C-130

Self-Contained Navigation System (SCNS) for the C-130

The Self-Contained Navigation System (SCNS) is a fully integrated navigation and communications management system designed to enhance mission performance. Independent navigation capability is achieved through integration with the ring laser gyro INS, GPS and a Doppler velocity sensor. SCNS hardware is interchangeable on C-130 models and is easily adapted to other military aircraft.

The pilot, co-pilot and navigator each have independent control through the use of an Integrated Control and Display Unit (ICDU) which contains a 16-bit MIL-STD-1750A processor; it can control up to 12 radios. The ICDU communicates with the other boxes in the system via a MIL-STD-1553B databus and features a 10-row by 24-character display which is compatible with night vision goggles.

The main functions of the SCNS include navigation, flight planning, aircraft guidance and steering and control of radio and navigation systems. As well as conventional navigation facilities such as time and distance to waypoint, the SCNS can also display information regarding eight different types of airdrop.

The system is standard equipment on US Air Force C-130E and H aircraft.

Operational status
In service in C-130 aircraft.

Contractor
Smiths Industries Aerospace.

VERIFIED

VIR-351 VHF-nav receiver

The S-TEC VIR-351 VHF navigation receiver is a solid-state, 200-channel unit which provides VOR/Loc deviation. VOR/Loc flag output, VOR to/from information to the 350 series indicators, and is capable of tuning the GLS-350 glide slope receiver and DME-451 Distance Measuring Equipment system.

Specifications
FAA TSO: C40ac, C36c class D, DO-138
Dimensions: 79 × 66 × 316 mm
Weight: 1.4 kg

Operational status
In production.

Contractor
S-TEC Corporation.

UPDATED

AN/APN-217 Doppler velocity sensor

The AN/APN-217 is a microprocessor-controlled solid-state single-unit radar with a unique continuous wave space-duplex design that avoids modulation losses, eliminates altitude holes and bypasses the limitations associated with modulated systems. At low altitudes the APN-217 is claimed to be 100 times more sensitive than modulated Doppler navigation radars. A combination of six special features prevents acquisition or tracking of antenna vertical sidelobe returns, making the system suitable for coupled transitions from forward flight to hover and vice versa.

Specifications
Dimensions: 420 × 414 × 175 mm
Weight: 15.5 kg
Power supply: 28 V DC, 55 W
Outputs: DC signals to AFCS and flight instruments such as hover indicator, digital outputs to MIL-STD-1553 provide velocities to navigation computer and one version of APN-217 has digital outputs to ARINC 575
Accuracy (digital):
(groundspeed) 0.2%
(heading/drift) 0.3 kt
(vertically) 35 ft/min
Display: outputs are compatible with a number of displays, including JD-2440/APN, multipurpose indicators and with groundspeed/drift and hover instruments
Conversions: the APN-217 Radio Navigation Set (RNS) can interface through the CV-4010/APN-217(V) RNS converter to AFCS systems previously using the AN/APN-182 Doppler velocity sensor
Controls: land/sea select. The system is controlled via MIL-STD-1553A or B databus
Speed range:
(along track) −40 to +350 kt
(drift) ±100 kt
(vertically) ±5,000 ft/min
Self-test: BITE detects 98% of predictable failure modes
Reliability: 11,000 h MTBF demonstrated

Operational status
Production deliveries began in 1982 and are expected to continue throughout the 1990s.

The AN/APN-217 is the primary navigation sensor for US Navy Sikorsky SH-60B/F Seahawk helicopters. It also equips the Navy's Sikorsky MH-53E and RH-53D minesweeper helicopters and US Marine Corps AH-1, UH-1N and CH-46 helicopters. Deliveries of a slightly modified version, the AN/APN-217(J), have been made to Japan for Sikorsky HSS-2 Sea Kings and SH-60Js. In December 1992, Teledyne Ryan was awarded a US$12 million contract for 140 AN/APN-217(V) sets for a variety of US Navy helicopters.

Contractor
Teledyne Electronic Tech.

UPDATED

AN/APN-218 Doppler velocity sensor

The AN/APN-218 Doppler velocity transmitter/receiver is a high-performance, nuclear-hardened radar chosen by the US Air Force as its Common Strategic Doppler (CSD) navigation system for fixed-wing aircraft. Competitive development of the APN-218 began in 1976. Production awards followed in 1978 and deliveries began the following year.

The performance and reliability of the system stems from two characteristics that it shares with the parent APN-213 Doppler; use of a continuous wave space-duplex transmit/receive system that is typically 7 to 10 dB more efficient in its use of transmitter power than a modulated system and a high-gain narrow-beamwidth planar-array antenna for high accuracy and sensitivity.

The features that commended it to the US Air Force as its CSD are nuclear hardness, high performance, choice of ARINC 575 or dual-redundant MIL-STD-1553B databus interface and integral radome.

The system can be used as a source of velocity information to other equipment or in conjunction with a groundspeed and drift indicator.

Specifications
Dimensions:
(sensor) 716 × 645 × 170 mm
(GDSI) 146 × 76 × 155 mm
(CDU) 146 × 152 × 165 mm
Weight:
(sensor) 31.9 kg

(GDSI) 1.5 kg
(CDU) 3.9 kg
Power supply: 115 V AC, 400 Hz, 170 VA
Frequency: 13.325 GHz
Speed: 96-1,800 kt
Altitude range: 0-70,000 ft
Beam geometry: 4 beams time-shared
Accuracy: 0.14% RMS
Self-test: BIT both continuous and commanded
Reliability: >3,000 h MTBF

Operational status

AN/APN-218 has been retrofitted to US Air Force B-52 and KC-135 aircraft. The system is also fitted to the US Air Force C-130 aircraft. The modified version of AN/APN-218, designated AN/APN-230, is fitted to the US Air Force B-1B bomber.

Contractor

Teledyne Electronic Tech.

UPDATED

AN/APN-231 radar navigation system

Developed in 1984, the AN/APN-231 is the primary navigation system for the US Navy EA-6B aircraft. It also integrates with other equipment including the attitude/heading reference system, air data computer, search radar, ECM sets and flight instruments.

The system comprises an AN/APN-200 Doppler velocity sensor, a CP-1573/APN-231 computer display unit and a CV-3780/APN-231 signal data converter.

Specifications

Dimensions:
(data converter) 226 × 259.1 × 426.7 mm
(velocity sensor) 149.9 × 632.5 × 652.8 mm
(computer display unit) 152 × 146 × 165 mm
Weight:
(data converter) 18.2 kg
(velocity sensor) 20 kg
(control display unit) 3.6 kg
Range:
(forwards) 50-999 kt
(drift) 0-200 kt
(vertically) 0-5,000 ft/min
Altitude range: up to 40,000 ft
Accuracy:
(velocity sensor) 0.13% +0.1 kt

Operational status

Production deliveries for the US Navy EA-6B were made in 1984.

Contractor

Teledyne Electronic Tech.

UPDATED

AN/APN-233 (220) Doppler velocity sensor

The AN/APN-233 can be used either as a single-unit velocity sensor providing outputs to other aircraft systems or with a Control/Display Unit (CDU) and HSI as a self-contained navigation facility. It was designed for applications in which size, weight, performance and reliability are critical factors.

The APN-220 family evolved from a small, lightweight Doppler sensor originally designed for the US Army and was subsequently qualified by the service and by the US Air Force and the German Air Force.

An optimum velocity range and near-zone rejection are offered for each application.

The CDU combines the functions of navigation computer and control/display unit and contains an incandescent alphanumeric display panel and a keyboard for entering data and selecting operational modes. Up to 10 waypoints can be accommodated, and a non-volatile scratch-pad memory holds critical information during power transients or interruptions.

Specifications

Typical fixed-wing applications
Dimensions:
(sensor) 426 × 291 × 113 mm
(CDU) 152 × 146 × 165 mm
Weight:
(sensor) 9.66 kg
(CDU) 3.86 kg
Power supply:
(sensor) 28 V DC, 28 W
(CDU) 28 V DC, 30 W
Output: heading, vertical velocity and groundspeed/drift to aircraft systems, for example AFCS, or to CP-1251 and HSI
Number of waypoints: 10 entered via front panel keyboard
Range (typical system):
(forward) −40 to 600 kt
(drift) 150 kt
(vertically) 5,000 ft/min
Altitude range: up to 50,000 ft
Accuracy:
(over land) 0.25% +0.2 kt
(over sea) 0.3% +0.2 kt
Self-test: BITE diagnostic program locates faults at first line level to 95% confidence
Reliability: 2,600 h MTBF demonstrated in the Alpha Jet

Operational status

In production. The system has been produced for the US Navy C-2A Greyhound and US Marine Corps OV-10D Bronco observation post aircraft. Additional production for the S-2, DHC-5 and CH-47 aircraft has been completed.

The system is used by the German Air Force for its Dassault/Dornier Alpha Jet strike/trainers as a velocity sensor to provide data to the navigation and weapons delivery systems. Versions of the equipment have also flown on RPVs, such as Teledyne's BGM-34C, and on helicopters.

Contractor

Teledyne Electronic Tech.

UPDATED

AN/ARA-63 microwave landing system

The ARA-63 is the airborne portion of the US Navy's standard aircraft approach control system for landing on aircraft carriers and equips aircraft such as the A-6E, F-14A, S-3A and F/A-18. Known as a radio receiving decoder set, it works in conjunction with the AN/SPN-41 on board ships and the AN/TRN-28 transmitters at naval air stations. Pulse-coded microwave transmissions are received by the ARA-63, decoded and displayed on a standard cross-bars indicator in the cockpit. There are 20 channels in the range 14.688 to 15.512 GHz. The system has three LRUs: receiver, pulse decoder and control unit.

Operational status

In service in US Navy aircraft such as the A-6E, F-14A, S-3A and F/A-18 and also overseas. Over 3,000 units have been produced. Development and early production was by Telephonics. Follow-on production by Stewart-Warner.

Contractor

Telephonics Corporation.

VERIFIED

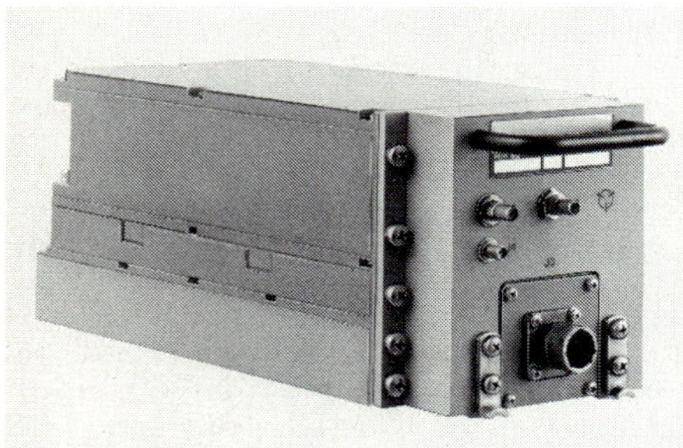

The TILS II receiver for the JAS 39 Gripen

TILS II Landing System receiver for the JAS 39 Gripen

Telephonics is producing the Tactical Instrument Landing System receiver (TILS II) for the JAS 39 Gripen. The airborne unit weighs less than 5 kg and is approximately 120 × 280 × 125 mm in size. It provides outputs to a MIL-STD-1553B databus.

The system works in conjunction with the TILS ground equipment developed by Telephonics and currently deployed throughout Sweden and Finland. This receiver and ground system are currently in production.

A variation of the receiver with DME has also been supplied for the Space Shuttle programme.

A multimode version incorporating ILS, GPS and J-band MLS is in development. This version will provide users with the flexibility to land at airfields equipped with ILS as well as Tactical Instrument Landing System (TILS) equipped roadways. The GPS feature can be used for en route navigation. All three functions will be contained in the same form factor as the current TILS II version.

Operational status

In service on JAS 39 Gripen. The predecessor TILS I receiver remains operational on Saab Viggen and Drakken aircraft.

Contractor

Telephonics Corporation.

VERIFIED

Aerial surveyor

The Aerial Surveyor is a complete survey system which precisely identifies the position of the aircraft at the time an aerial photograph is taken. Measurements are made with respect to a referenced station, such as another Aerial Surveyor or geodetic surveyor, and then post-processed. Knowing the precise position of the aircraft reduces the need for expensive ground control.

Measurement rates as fast as every half second minimise the effect of the aircraft dynamics and are ideal for efficient aerial photogrammetry. Interface capability is provided through dual RS-232 ports. Synchronisation of cameras is possible by utilising the 1 pps output and/or the event marker which allows camera shutter operation to be directly linked to precise GPS time and positioning.

On board the aircraft the receiver/datalogger will operate at aircraft speeds as high as 700 kt and maintain measurement integrity at accelerations up to 2 *g*. A preamplifier which allows connection to an aircraft antenna is provided.

Optional features, like the external frequency input or RTCM input, provide the flexibility to design a system which meets exact requirements.

Specifications

Dimensions: 300 × 350 × 130 mm
Weight: 7.2 kg
Power supply: 10.5-35 V DC, 8.5 W
Temperature range: −20 to +55°C

Contractor

Trimble Navigation Ltd, Avionics Products.

VERIFIED

TDF 100D Automatic Direction-Finder (ADF)

The Trimble TDF 100D automatic direction-finding system comprises three units: the TDF 100D panel-mounted receiver/control unit, the TA 10 combination sense and loop antenna and the TDI 10 ADF indicator.

The TDF 100D receiver/control unit tunes from 200 kHz to 1,800 kHz in 1 kHz increments and also provides capability for receiving the marine distress frequency on 2,182 kHz. The front panel controls provide on/off volume control, a three-position toggle switch for selection of ADF, ANT and BFO modes, a push-push filter switch to clarify identification signals and digital frequency selector switches. The unit is designed to mount beside the Terra transponder, comms or navigation units.

The TA 10 ADF antenna contains a loop antenna, sense antenna and amplifier circuits for each antenna. It operates between temperatures of −55 and +70°C at altitudes up to 50,000 ft. The antenna may be top- or bottom-mounted and provides for quadrantal error adjustments.

The TDI 10 indicator mounts in any 3 in instrument housing. The indicator unit contains the power supply for the system, requiring 11 to 33 V DC at a nominal 8.5 W. A heading knob positions the compass card manually, while the pointer is driven by a stepper motor. Bearing accuracy is ±3° and the pointer will move from 175° off bearing to the correct bearing in less than 5 seconds.

Specifications

Dimensions:
(indicator) 139.7 × 83.8 × 83.8 mm
(receiver/control unit) 79.4 × 40.6 × 261.6 mm
(antenna) 203.2 × 134.6 × 66 mm
Weight:
(indicator) 0.59 kg
(receiver/control unit) 0.68 kg
(antenna) 0.64 kg

Contractor

Trimble Navigation Ltd, Avionics Products.

VERIFIED

TN 200D navigation receiver

The TN 200D is a companion to the TX 760 nav/com receiver. It provides navigation and audio signals to operate Tri-Nav, Tri-Nav-C and most other VOR/ILS indicators. An optional 40-channel glide slope receiver can be added at any time and becomes a part of the TN 200D. Provision for remote channelling of DME systems is provided when the glide slope option is installed. A digital frequency synthesiser generates all 200 navigation channels from 108.00 to 117.95 MHz with digital channel selection.

Specifications

Dimensions:
264.6 × 159 × 84 mm (with mounting tray)
Weight: 2.09 kg
Power supply:
(transmit) 13.75 V DC, up to 2.5 A
Frequency:
(nav) 108-117.95 MHz at 50 kHz spacing
(glide slope) 329.15-335 MHz (40 channels)

Operational status

In service.

Contractor

Trimble Navigation Ltd, Avionics Products.

UPDATED

Trimble 2000 Approach Plus airborne IFR GPS navigation system

The Trimble 2000 Approach Plus, certified to C-129 (A1), may be used for supplemental en route, terminal and approach IFR operations.

The Trimble 2000 Approach Plus has been designed to minimise pilot workload during execution of approach procedures. The pilot is automatically prompted to enable the approach at 30 n miles from the destination airport, an action that requires a single press of the Enter Key. Procedure turns are fully automatic, no holding or suspension of procedures is required. All system functions required during the approach are automatically selected; all messages are automatically presented and removed. If a missed approach is required, the pilot selects the Direct Key when located at a position anywhere from inside the Final Approach Fix (FAF) to beyond the Missed Approach Point (MAP). Flight path presentation is accurate and only database waypoints are presented.

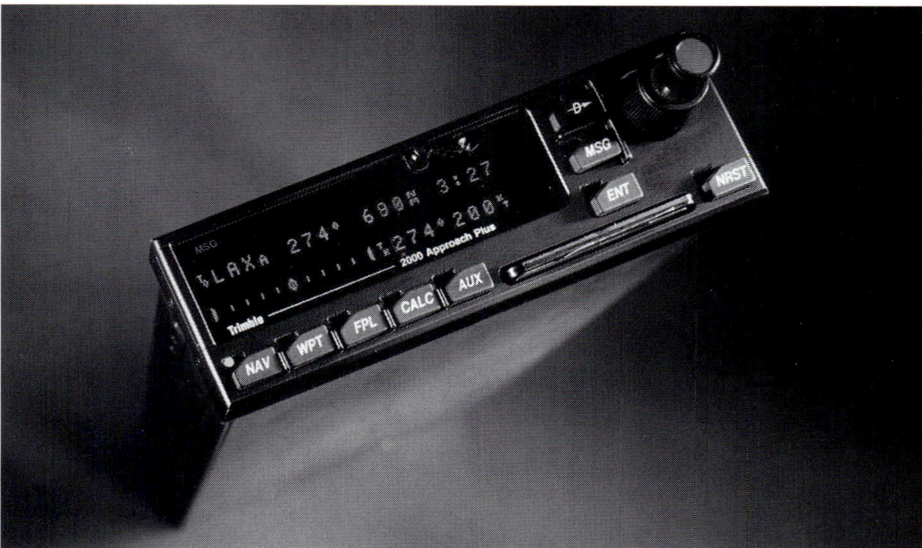

Trimble 2000 Approach Plus airborne IFR GPS navigator ***1998***/0015390

The Trimble 2000 Approach Plus incorporates an advanced 12-channel GPS receiver. It continuously tracks all satellites in view, determining and displaying a new position five times every second and measuring speed to better than a tenth of a knot. It calculates and displays position, bearing, distance, ground speed and ground track and the estimated time en route on a high-contrast LED display. An electronic Course Deviation Indicator (CDI) and a patented track angle error display (TKE) are included. The Trimble 2000 Approach continuously checks accuracy using Receiver Autonomous Integrity Monitoring (RAIM) and predicts RAIM conditions for the approach arrival time.

Installation of the Trimble 2000 Approach Plus is simplified. Only four external annunciators are required; Message (MSG), Waypoint (WPT), Approach (APR), and Hold (HLD). Connection to an external CDI/HSI for navigation and autopilot interface is available. External switching of resolvers is not required, resulting in substantially reduced installation costs. When integrated to an optional air data computer, Trimble 2000 Approach Plus will display true air speed, density altitude, pressure altitude and calculate and display winds aloft. The system can automatically sequence through up to 40 flight plans with up to 40 waypoints each (maximum waypoint total is 400), show the nearest airport, plan vertical descents, display minimum safe altitudes and provide a variety of other functions.

The Trimble 2000 Approach Plus incorporates Jeppesen's NavData card, including airports, approaches, SIDS, STARS, VORs, intersections, NDBs, airspace boundaries and MEAs.

Specifications

Type: 12-channel receiver, L1 frequency, C/A code, continuous all-in-view tracking
Acquisition time: 1.5-3.5 min
Position update rate: 5 times/s
Dynamics: 800 kt (4 *g* tracking)
Accuracy:
(position) 15 m RMS
(velocity) 0.1 kt steady-state
(altitude) 35 m RMS (msl)
(time) UTC to nearest μs
Computation range: Great Circle: 0-999 n miles
Distance resolution: 0-9.99 n miles: in 0.01 n mile increments; 10-99.9 n miles in 0.01 n mile increments; 100 n miles + in 1.0 n mile increments
Interfaces:
EAI Standards, RS-422
CDI, flags and external annunciators, altitude input
GPS antenna: Omnidirectional flat microstrip with integral preamp. SMA connector
Display:
LED: 2 lines of 20 characters each
High-intensity, orange alphanumeric
Dimensions:
(receiver) 158.75 × 274.32 × 50.8 mm
(antenna) 95.25 × 101.6 × 19.05 mm
Weight:
(receiver) 0.92 kg
(antenna) 200 g
Power: 10-32 V DC, negative ground, 0.9 A at 14 V DC, 0.5 A at 28 V DC, 12 W
Compliance: FAA TSO C-129 (A1); FAA AC 20-138

Contractor

Trimble Navigation Ltd, Avionics Products.

UPDATED

Trimble 2101 I/O Approach Plus airborne IFR GPS navigation system

The Trimble 2101 I/O Approach Plus is the full input/output version of Trimble's GPS Dzus rail mount navigator series. Certified to TSO C-129(A1), it may be used for supplemental en route, terminal and approach IFR operations.

The 2101 I/O Approach Plus is also approved for 8110.60 GPS as Primary Means of Navigation for Oceanic Operations and B.RNAV European Airspace Operations.

The Trimble 2101 I/O Approach Plus minimises pilot workload during approaches. It automatically selects all the required navigation functions and automatically presents and removes all messages.

At the heart of the Trimble 2101 I/O Approach Plus is an advanced 12-channel GPS receiver with a defined upgrade path to WAAS capability. It continuously tracks all satellites in view, determining and displaying a new position four times per second and measuring speed to better than a tenth of a knot. It calculates and displays position, bearing and distance to waypoint, ground speed and ground track and estimated time en route. It includes an electronic Course Deviation Indicator (CDI) and a patented track angle error (TKE) display. The 2101 I/O Approach Plus presents the data on a high-contrast LED display.

The 2101 I/O Approach Plus continuously monitors its accuracy via Receiver Autonomous Integrity Monitoring (RAIM). It also predicts RAIM conditions for the approach arrival time.

The Trimble 2101 I/O Approach Plus is easy to install. Installation requires only four external annunciators: Message (MSG), Waypoint (WPT), Approach (APR) and Hold (HLD). Expensive resolvers are not required.

The 2101 I/O Approach Plus supports CDIs, HSIs, altimeters, RMIs, fuel flow gauges and many other flight instruments.

When integrated with an air data computer, the Trimble 2101 I/O Approach Plus displays true air speed, density altitude and pressure altitude and calculates and displays winds aloft and applies current wind ETE and ETA calculations. It also automatically sequences through up to 40 waypoints in a flight plan, shows the nearest airport, plans vertical descent profiles and performs a variety of other functions.

The Trimble 2101 I/O Approach Plus incorporates Jeppesen's NavData card–an avionics database that includes airports, approaches, SIDS, STARS, VORs, NDBs, intersections, airspace boundaries and more.

Specifications

Type: 12-channel receiver, L1 frequency, C/A code, continuous all-in-view tracking
Acquisition time: 1.5-3.5 min
Position update rate: 4 times/s
Dynamics: 800 kt (4 *g* tracking)
Accuracy:
Position: 15 m RMS
Velocity: 0.1 kt steady-state
Altitude: 35 m RMS (msl)
Computation range: Great Circle: 0-999 n miles
Distance resolution: 0-9.99 n miles: in 0.01 n mile increments; 10-99.9 n miles in 1.0 n mile increments
GPS antenna: Omnidirectional flat microstrip with integral preamp. TNC connector
Display:
LED: 2 lines of 20 characters each
High-intensity, orange alphanumeric
Dimensions: 146.05 × 196.85 × 76.2 mm
Weight: 1.26 kg
Power: 10-32 V DC, negative ground, 1.0 A at 14 V DC, 0.5 A at 28 V DC, 14 W

Contractor

Trimble Navigation Ltd, Avionics Products.

UPDATED

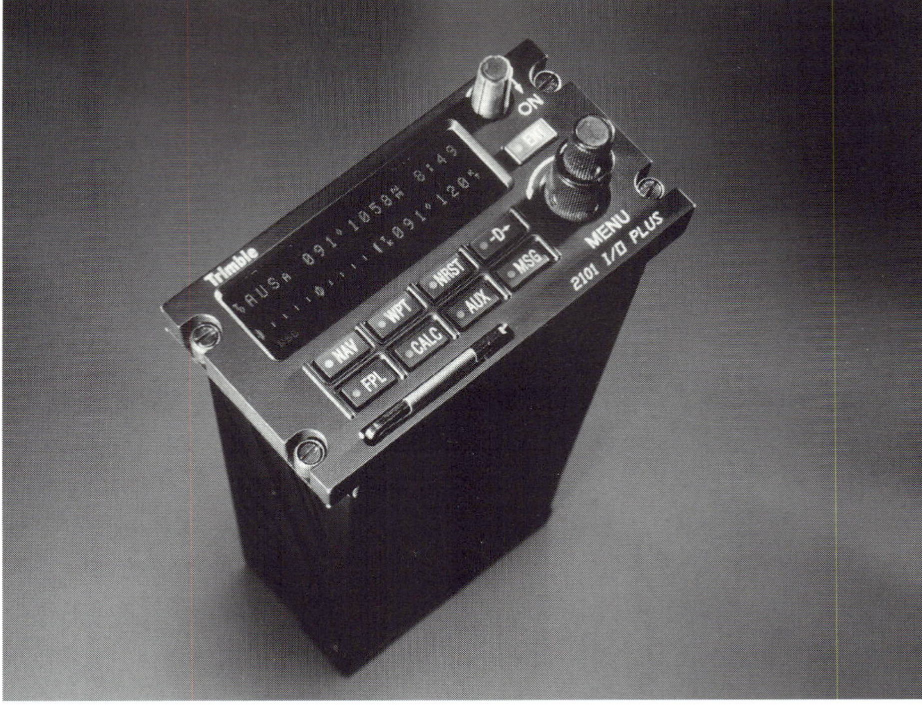

Trimble 2101 I/O Approach Plus airborne IFR GPS navigation system **1998**/0015391

Trimble 8100 Airborne GPS Navigation System

The Trimble 8100 advanced GPS navigation system is designed to meet the certification, operational, and maintenance standards required for commercial, air transport and corporate operations.

The Trimble 8100 is an advanced nine-channel GPS receiver with a defined upgrade path to 12-channels and WAAS capability. This system provides four-dimensional aircraft guidance for timed great circle and vertical navigation through selected flight plan waypoints. Trimble's 8100 provides transitions between all phases of flight navigation, including oceanic, en route, terminal, and non-precision approach.

Certified to FAA TSO-C129 and RTCA DO-208, Trimble's 8100 provides flight planning information using Company Routes, J-Routes, SIDS, STARS and approaches, and supports all 23 ARINC 424 leg types. The system is also approved as a primary means of navigation for remote and oceanic operations.

Trimble's 8100 input/output capability allows a wide range of analogue and digital interface options. It can also be used as a stand-alone ARINC 743 sensor, allowing a one-box GPS solution for an entire aircraft fleet. The Trimble 8100 is capable of interfacing to a DME transceiver to support multisensor navigation (TSO-C115a) and European P-RNAV.

Trimble 8100 airborne GPS navigation system **1997**/0001345

Flight planning and navigation information is entered on one of two different CDUs (Compact and ARINC). Each CDU features a seven-colour display on which alphanumeric and symbolic data, including a graphical CDI presentation, are displayed.

Trimble's 8100 is Differential GPS (DGPS) capable, in anticipation of future Cat. I, II and III approach and landing certifications.

Specifications

Navigation Processor Unit:
Dimensions: ARINC 2 MCU (¼ ATR Short)
368.3 × 196.6 × 572 mm
Weight: 2.73 kg
Power: 28 V DC input
Control Display Unit: (compact)
Dimensions: 203.2 × 95.3 × 146.1 mm
Weight: 2.07 kg
General: Nine-channel, L1 frequency (1,575 MHz), C/A code, digital GPS receiver, tracks all satellites in view
Inputs: Synchros: TAS, HDG, ALT, pitch, roll ARINC 429:10 (high or low speed)
Fuel flow
Discretes: 32
Outputs: Synchros: 3 programmable, (synchro or SIN/COS)
Roll steering; left-right; glide slope
ARINC 561/568
ARINC 429: 5 (high or low speed), GAMMA compatible
Discretes: 16
Database: Expanded Jeppesen Data Base (available via ARINC 615, or separate data loader)
Compliant:
TSO-C129, RTCA DO-208
TSO-C115a
Acquisition time: 1 min, nominal
Update rate: 5 Hz
Dynamics: 0-800 kt (3 *g* acceleration)
Time: Universal co-ordinated time to the nearest microsecond
Accuracy typical:
(Position) 15 m RMS)
(Altitude) 35 m RMS
(Velocity) 0.1 kt RMS steady rate

Contractor

Trimble Navigation Ltd, Avionics Products.

UPDATED

AN/ASN-175 Cargo Utility GPS Receiver (CUGR)

Trimble's CUGR receiver is a dzus-mount P(Y) GPS navigation system for worldwide military aviation operations. It is a PPS version of Trimble's 2101 I/O Approach and meets the performance standards for Instrument Flight Rules (IFR) for en route, terminal and non-precision approach phases of flight as specified in the US Army Aircraft GPS Integration Guide (GIG). The unit continuously checks accuracy using Receiver Autonomous Integrity Monitoring (RAIM) and predicts RAIM conditions for approach. Trimble's CUGR can automatically sequence up to 20 flight plans with 20 waypoints each, showing nearest airport, planning vertical descents, displaying minimum safe altitudes as well as many other features. The receiver incorporates Jeppesen's NavData card including airports, approaches, SIDS, STARS, VORs, intersections, NDBs, airspace boundaries and MEAs. The unit can be interfaced with HSI, CDI, autopilot or air data computers to maximise the flexibility of the system, all within a single unit. The receiver is also AN/AVS-6 compatible.

Operational status
In October 1996, Trimble was awarded a contract from the US Army Communications-Electronics Command (CECOM) to provide Cargo Utility GPS Receivers (CUGRs) totalling over US$12 million. The contract provides for an initial procurement with two, one year options; over the life-time of the contract, CECOM may procure as many as 3,600 receivers. The CUGR system will be installed on US military helicopters including selected UH-1, UH-60, OH-58 and CH-47 aircraft.

Specifications
Receiver: 6-channel P(Y) code, continuous tracking, digital receiver; L1, L2 capable; 12-channel upgrade planned.
GPS antenna: L1/L2 flat panel
Dimensions/weight:
(receiver) 146 × 197 × 76 mm; 1.38 kg
(amplifier) 115 ×89 × 31 mm; 0.27 kg
(antenna) 96 × 102 ×26 mm; 0.32 kg
Performance:
Autonomous: PPS position accuracy: 3-D 16 m SEP; PPS velocity accuracy: 0.2 m/s
DGPS: position accuracy: 5 m; velocity accuracy: 0.2 m/s RMS

Contractor
Trimble Navigation Ltd, Military Systems.

UPDATED

AN/PSN-10(V) Trimpack GPS receiver

The AN/PSN-10(V) Trimpack small lightweight GPS receiver is a completely self-contained navigation system. It can pinpoint position to within a few metres anywhere on earth and is built to operate under battlefield conditions. Its rugged electronics are sealed in a metal impregnated high-impact case that also houses a built-in unbreakable antenna. It is suitable for helicopters and medium-performance fixed-wing aircraft, as well as land applications.

Trimpack can operate in temperatures ranging from −30 to +65°C. Solid construction, coupled with electronics tested to MIL-STD-810D, results in a calculated MTBF of over 15,000 hours.

All controls are designed for easy operation. A rotary switch selects a wide variety of navigation functions including position and altitude, range and bearing, cross-track error, velocity and time to go. Up to 26 waypoints can be stored and then transferred from set to set over a datalink interface. No initialisation for position or time is ever required.

Trimpack automatically selects the optimal mix of satellites and when only three are visible it automatically operates in an altitude-hold mode.

With three independent channels of GPS, Trimpack can navigate under a wide range of dynamic conditions, including speeds up to 600 kt. Its multichannel design also allows precise velocity measurement and guarantees quicker acquisition of satellites. It can go from a cold start to a three-dimensional fix in 2½ minutes and can calculate a fresh position every second thereafter.

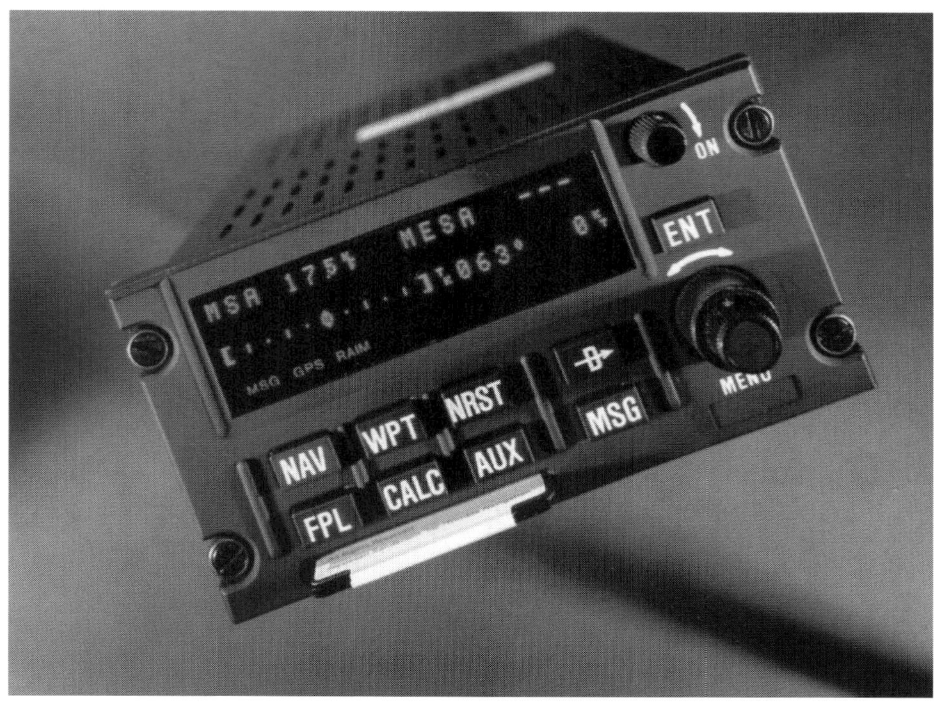

AN/ASN-175 Cargo Utility GPS Receiver (CUGR) ***1998***/0015392

Specifications
Dimensions: 198 × 223 × 63 mm
Weight: 1.9 kg
Power supply: 2 lithium batteries, rechargeable Ni/Cd battery pack or 8 alkaline AA cells
9-32 V DC, 5 W external power option

Contractor
Trimble Navigation Ltd, Military Systems.

VERIFIED

Mobile Reporting Unit (MRU)

The Mobile Reporting Unit (MRU) is the mobile segment of the Trimble Mobile Automated Command and Control System and its Advanced Range Tracking System. The MRU is a fully integrated, stand-alone unit designed to withstand MIL-STD shock, vibration and temperature extremes and to maximise reliability. It may be installed on fixed-wing aircraft and helicopters, as well as on land vehicles and naval vessels.

The MRU provides real-time unit identification, precise time-tag, course, speed and autonomous or differential GPS positioning with regular telemetry updates at the rate of one per second. System Operational Status information and two-way messaging are also available when appropriate system options are specified. When equipped with the standard low-noise C/A code GPS receiver, the MRU delivers differential GPS position and velocity.

While the standard low-noise receiver is adequate for most high-performance applications, the MRU accommodates other Trimble GPS receivers for special applications. Options include full military P(Y), standard navigation and other special purpose Trimble receivers.

Specifications
Dimensions:
(1 ATR) 318 × 257 × 191 mm
(½ ATR) 318 × 129 × 191 mm
Weight:
(1 ATR) 7.5 kg
(½ ATR) 6 kg
Frequency: 403-430 MHz, 450-470 MHz
Temperature range: −30 to +60°C
Altitude: −100-50,000 ft
Accuracy - autonomous GPS:
(position) 15 m
(velocity) 1 ft/s
(time) 0.1 μs
Accuracy - differential GPS:
(position) 2.5 m
(velocity) 0.3 ft/s
(time) 0.1 μs

Contractor
Trimble Navigation Ltd, Military Systems.

VERIFIED

TANS GPS navigation sensor

The TANS is a rugged two-channel sequencing GPS navigation sensor which receives the L-band C/A code signals broadcast by GPS satellites. It uses the GPS signals to provide three-dimensional positioning, velocity and time over dual RS-422 digital interfaces.

TANS II is a six-channel continuous tracking GPS sensor that provides a three-dimensional position within 25 m anywhere on earth. With its continuous tracking design, TANS II provides data at speeds up to 800 kt. Initialisation is not required and a three-dimensional fix is calculated from a cold start nominally within 1½ minutes. For increased accuracy, differential positioning capability is built in.

The TANS is intended for applications where size, weight and power are at a premium. Features include no requirement for initialisation, automatic satellite selection algorithms, high reliability and low maintenance costs and easy integration, interface and operation.

Specifications
Dimensions:
(receiver) 127 × 207 × 56 mm
(antenna) 96 × 102 × 19 mm
Weight:
(receiver) 1.27 kg
(antenna) 0.17 kg
Power supply: 9-32 V DC, <3 W
Temperature range: −40 to +71°C
Accuracy:
(position) 25 m SEP
(velocity) 0.2 m/s RMS
(time) UTC to 1 ms

Contractor
Trimble Navigation Ltd, Military Systems.

VERIFIED

TANS Vector GPS attitude determination system

The TANS Vector is a solid-state attitude determination and position location system. The compact and rugged receiver offers a comprehensive range of capabilities. It is ideally suited for airborne navigation applications that currently use gyro-based systems to determine three-axis attitude and position. The TANS Vector provides azimuth, pitch and roll angles, position, velocity, time, differential positioning and waypoint navigation using a continuous tracking L1 C/A code receiver.

The system comprises two major subassemblies: the Receiver Processor Unit (RPU), with bidirectional RS-422 interface, and a three- or four-element antenna array. An optional control display unit is available or the TANS Vector can be integrated into the user's system. The attitude receiver consists of two circuit cards, which make it ideal for RPU embedded applications. The system is designed to withstand extreme shock and vibration and a broad range of operating temperatures.

Specifications

Dimensions:
(RPU) 127 × 207 × 56 mm
(antenna) 96 × 102 × 19 mm

The Trimble TANS Vector receiver system consists of a receiver processor unit and antenna array
1995

Weight:
(RPU) 1.4 kg
(antenna) 0.18 kg
Power supply: 9-32 V DC, <4.5 W
Temperature range: −40 to +71°C
Altitude: up to 60,000 ft
Accuracy (RMS):
(attitude) 0.15-0.3°
(autonomous GPS position) 50 m horizontal, 80 m vertical
(differential GPS position) 2-5 m horizontal, 5-8 m vertical
(velocity) 0.2 m/s
Reliability: >10,000 h MTBF

Contractor

Trimble Navigation Ltd, Military Systems.

VERIFIED

Tasman GPS sensor

Tasman is a six-channel, Precise Positioning Service (PPS) GPS receiver that uses both the L1 and L2 frequencies. It corrects for selective availability, and utilising Receiver Autonomous Integrity Monitoring (RAIM), protects against spoofing. It provides positioning accuracies of better than 16 m and also supplies velocity and time for navigation and tracking applications.

With its continuous tracking design, Tasman provides position data over a wide range of dynamic conditions and velocities. Initialisation is not necessary and a three-dimensional fix is calculated within 1½ minutes. Position and velocity updates occur every second thereafter.

Data is transferred through two serial input/output ports, allowing Tasman to interface with vehicle sensors, control systems and a control/display unit. The low-profile antenna provides excellent satellite visibility. Tasman also reports GPS system health and receiver operational Operational Status. A differential GPS capability is built in.

Housed in a lightweight, sealed aluminium case, Tasman is built to withstand extreme temperatures and severe vibration.

Specifications

Dimensions:
(receiver) 127 × 207 × 56 mm
(antenna) 96 × 102 × 19 mm
Weight:
(receiver) 1.3 kg
(antenna) 0.28 kg
Power supply: 9-32 V DC, <3.5 W
Temperature range: −40 to +71°C
Altitude: −1,300 to 66,000 ft
Accuracy - autonomous GPS:
(position) 16 m SEP
(velocity) 0.2 m/s RMS
(time) 100 ns RMS
Accuracy - differential GPS:
(position) 5 m SEP
(velocity) 0.2 m/s RMS
(time) 100 ns RMS

Contractor

Trimble Navigation Ltd, Military Systems.

VERIFIED

GPS-950 sensor

The GPS-950 GPS sensor has been specifically designed to interface with UNS flight management and navigation management systems.

The GPS-950 utilises a five-channel receiver designed to process continuously pseudo range and range rate measurements from the four or more satellites having the best geometry and to output a navigation solution which includes three-dimensional position and velocity, time and Operational Status. When necessary, the fifth channel is used to gather ephemeris data from all the satellites in view, allowing rapid adaptation to obscuration. Otherwise, it is used to gather navigation data. In degraded constellation geometries, the GPS-950 will continue to provide two-dimensional navigation, using measurements from three satellites along with either GPS-corrected pressure or baro-corrected altitude.

Operation of the GPS-950 is controlled through and by the UNS systems with no additional control or display required. The GPS-950 provides outputs for up to three, and can accept inputs from two, UNS systems. Digital communications via low-speed ARINC 429 format are utilised between the sensor and UNS systems. These can integrate position data from the GPS-950 sensor with other long- and short-range navigation sensors to derive a highly accurate best computed position.

Integrity of the GPS position solution is monitored by both the GPS-950 sensor and the UNS systems with appropriate messages provided to the pilot.

Specifications

Dimensions:
(sensor) 194 × 25.1 × 386.8 mm
(antenna) 119.4 × 73.7 × 57.2 mm
Weight:
(sensor) 1.5 kg
Power supply: 18-32 V DC, 20 W

Operational status

Qualified in March 1994 to TSOs C-115a and C-129 B/C3.

Contractor

Universal Avionics Systems Corporation.

VERIFIED

GPS-1000 sensor

The GPS-1000 long-range navigation sensor features a 12-channel engine which tracks all satellites in view simultaneously and includes real-time and predictive Receiver Autonomous Integrity Monitoring (RAIM) capabilities. It features high acquisition and tracking sensitivity, carrier phase smoothing and position updates every second and will be able to accept pseudo-range correction signals from local area differential GPS in the future.

The GPS-1000 can be certified for en route and terminal operations, and non-precision approaches.

Specifications

Dimensions: 194.1 × 25.1 × 386.8 mm
Weight: 1.6 kg
Power supply: 18-32 V DC, 0.255 A at 28 V
Certification: TSO C-129 Class B1/C1

Operational status

In production.

Contractor

Universal Avionics Systems Corporation.

VERIFIED

GPS-1200 sensor

The GPS-1200 long-range navigation sensor utilises a more highly advanced 12-channel GPS engine than the GPS-1000. It features real-time and predictive Receiver Autonomous Integrity Monitoring (RAIM), satellite fault detection and exclusion capability and carrier phase tracking, and is Wide Area Augmentation System (WAAS) ready. It also includes local area differential GPS signal integration capability. Besides en route and terminal operations, the GPS-1200 can be certified as one of the two required long-range sensors for NAT MNPS navigation across the North Atlantic. It can also be certified for non-precision approaches and incorporates DO-178B critical level software which will be required for precision approach certification in the future.

Specifications

Dimensions: 194.1 × 25.1 × 386.6 mm
Weight: 1.6 kg
Power supply: 18-32 V DC, 0.255 A at 28 V
Certification: TSO C-129 Class B1/C1

Operational status

In production.

Contractor

Universal Avionics Systems Corporation.

VERIFIED

LCS-850 Loran C sensor

The LCS 850 Loran C sensor has been specifically designed to interface with UNS flight management and navigation management systems with no additional control or display required. Using a true multiple station solution, the LCS-850 provides accurate latitude/longitude position information to the UNS management systems. It tracks up to eight stations in a multi-GRI position solution, automatically selecting the best signals from multiple chains. Position is calculated using a minimum of three stations on one GRI or two stations on each of two GRIs. Notch filters reduce interference and provide increased performance in weak Loran signal areas.

The LCS-850 meets the rigorous standards of TSO C-60b and also meets both system and accuracy requirements for RNav operations in the US National Airspace System and has NAT MNPS airspace approval.

Specifications

Dimensions: 194.1 × 25.1 × 386.6 mm
Weight: 1.6 kg
Power supply: 18-32 V DC, 0.225 A at 28 V

Operational status

In production.

Contractor

Universal Avionics Systems Corporation.

VERIFIED

UNS-764 Omega/VLF sensor

The UNS-764 Omega/VLF sensor was designed to operate in conjunction with the Universal's UNS-1 flight management and navigation management systems, working either alone or in combination with inertial, GPS or Loran C sensors. The system comprises an antenna and receiver, the control/display unit of the UNS-1 system providing the necessary management functions. Gate-array and surface-mount technologies combine to reduce the size and weight of the UNS-764 system over previous Omega/VLF sensors. Operation is automatic, control being exercised via the UNS-1. Five independent receive channels enable all available Omega, and up to eight VLF, stations to be monitored simultaneously.

Specifications

Dimensions: 194.1 × 56.9 × 387.6 mm
Weight: 2.9 kg
Power supply: 28 V DC, 20 W (max)

Operational status

In production.

Contractor

Universal Avionics Systems Corporation.

VERIFIED

UNS 764-2 GPS/Omega/VLF sensor

The UNS 764-2 is a combination GPS/Omega/VLF long-range navigation sensor. It incorporates a 12-channel GPS receiver that features real-time and predictive Receiver Autonomous Integrity Monitoring (RAIM), satellite fault detection and exclusion capability, carrier phase tracking. It is Wide Area Augmentation System (WAAS) ready.

The Omega/VLF receiver features five independent channels which track all available Omega and up to eight VLF stations simultaneously.

Specifications

Dimensions: 194.1 × 56.9 × 387.6 mm
Weight: 3.6 kg
Power supply: 28 V DC, 30 W nominal
Certification: TSO C-129 Class B1/C1, C-94a

Operational status

In production.

Contractor

Universal Avionics Systems Corporation.

VERIFIED

UNS-RRS radio reference sensor

The UNS-RRS radio reference sensor has been designed to provide DME, VOR and Tacan radio data for the UNS-1 flight and navigation management system. It is remotely tuned through the UNS-1 CDU via an ARINC 429 databus and each RRS can support two UNS-1 systems.

Specifications

Dimensions: 194.1 × 56.9 × 387.6 mm
Weight:
(receiver) 352 kg
Power supply: 28 V DC, 1 A (max)

Operational status

In production.

Contractor

Universal Avionics Systems Corporation.

VERIFIED

Real-Time Information into the Cockpit (RTIC)

The Avionics Directorate, Wright Laboratory, Wright-Patterson AFB and US Air Force have demonstrated Real-Time Information into the Cockpit (RTIC) technologies through multiple programmes and flight demonstrations. A building block approach has been used to develop and demonstrate successively more advanced RTIC technologies and capabilities. The US Air Force and US Navy have co-ordinated efforts; the US Navy is concentrating on the F/A-18, AV-8 and AH-1 platforms, while the US Air Force is concentrating on the F-15E and the F-117.

The RTIC technology has primarily been developed as a part of the Off Board Targeting Experiments (OBTEX) programme. OBTEX has been designed to provide the warfighter with the latest information for precise targeting and situation awareness. This RTIC information has proved effective for aiding aircrew members in the acquisition and targeting of time critical mobile targets.

OBTEX is part of a larger overall sensor-to-shooter concept which includes imaging sensors, ground processing and dissemination systems, advanced datalinks and various deep strike aircraft. The primary components of the OBTEX system are onboard image processing, the associated datalink to receive the off-board data and a ground processing element called a Joint Targeting Workstation (JTW). The JTW prepares the data by annotating the imagery with a transparent triangle over the selected target and attaching any necessary textual information. The imagery along with threat and weather information are put into a targeting package that is sent to the F-15E. The onboard processing consists of a 1/2 ATR box with five cards to decompress incoming data from the Joint Tactical Information Distribution System (JTIDS) and process it such that it can be displayed on the MultiPurpose Display (MPD) for execution by the aircrew. The JTIDS (LINK-16) datalink is crucial to the functionality of the OBTEX system.

Operational status

Recently, OBTEX was demonstrated as part of the Air Combat Command sponsored Project Strike II demonstration. This demonstration showed the ability to perform targeting against time critical mobile targets by providing real-time intelligence data to an F-15E and an F-117A. The data consisted of imagery, threat updates and textual information used for targeting. The OBTEX technology included the ground node used to package and disseminate the targeting folder and the on board processor to decompress the data and display it to the aircrew. The result of this demonstration was a much reduced targeting timeline for locating and prosecuting time critical mobile targets. There will be a follow on programme to OBTEX entitled Integrated RTIC/RTOC for Combat Aircraft (IRRCA) which will continue to develop advanced RTIC technologies in addition to developing Real Time Information out of the Cockpit technology. Also, IRRCA will develop a system solution which will marry emerging technologies offering increased capabilities for offensive targeting and defensive situation awareness with applications to varying mission areas.

Contractor

US Air Force Materiel Command, Wright Laboratory.

VERIFIED

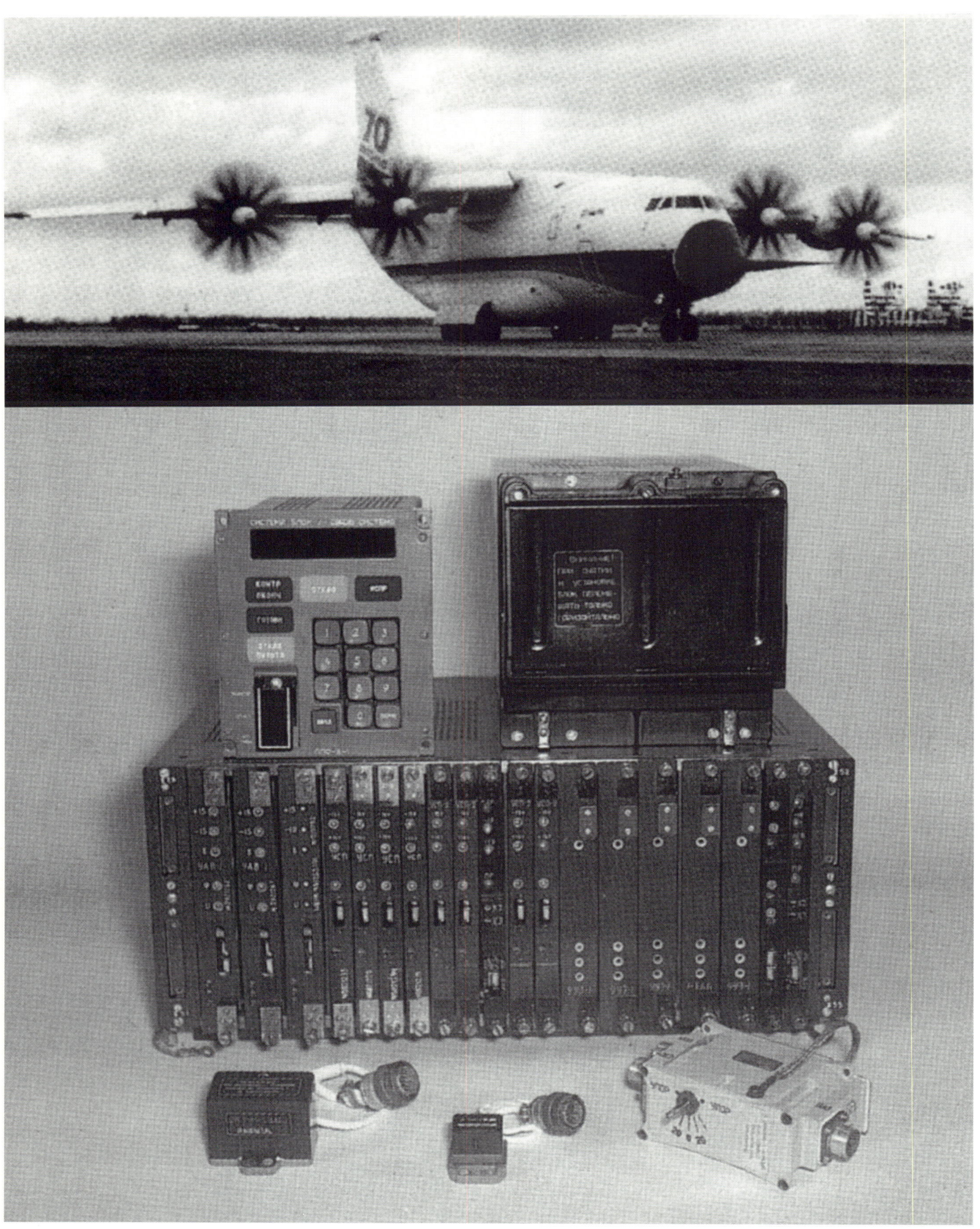

AviaPribor EDSU-77 fly-by-wire flight control system for the An-70 four prop-fan, medium-range, wide-bodied, tactical transport aircraft 1998/0018226

CANADA

Contaminant and Fluid Integrity Measuring System (C/FMIS)

The Contaminant and Fluid Integrity Measuring System (C/FMIS™) comprises a Control and Display Unit (CDU), a Central Control Unit (CCU) and up to eight flush-mounted sensors, depending on aircraft configuration. The electrical properties of the substance detected by each sensor is evaluated by the CCU and a corresponding output advisory signal is displayed on the CDU.

C/FIMS™ is designed to operate independently as a

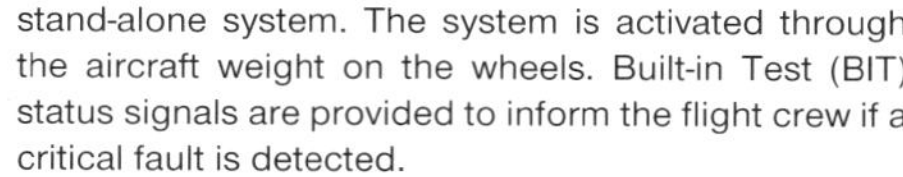

stand-alone system. The system is activated through the aircraft weight on the wheels. Built-in Test (BIT) status signals are provided to inform the flight crew if a critical fault is detected.

C/FIMS™, is an enhanced software version of the previously developed Clean Wing Detection System (CWDS™). The sensing technology detects, discriminates and measures the thickness of the contaminant and layers of contaminants on the surface and specifically frost, ice, snow and de/anti-icing fluids; measures and indicates the integrity of the de/anti-icing fluid; predicts the onset of fluid failure and the imminence of ice adhesion and accurately measures aircraft skin temperature.

Operational status

Operationally evaluated.

Contractor

AlliedSignal Aerospace Canada.

NEW ENTRY

Contaminant and Fluid Integrity Measuring System (C/FMIS)
1998/0018229

CMA-2082 Avionics Management Systems (AMS)

CMA-2082 Avionics Management Systems (AMS) are microprocessor-controlled management systems that can interface with a wide range of avionics equipment via a MIL-STD-1553B or ARINC 429 databus. A discrete interface assembly can be added to integrate other avionics systems.

The AMS uses a Thin Film ElectroLuminescent (TFEL) flat panel display with contrast enhancement/ bandpass filters to achieve both sunlight-readability and night vision goggles compatibility. The display is made up of active picture elements (pixels) arranged in an active display area of approximately 76 × 76 mm. A larger display, known as the CMA-2082-5, with an active display area of 76 × 127 mm, is also available. A comprehensive keyboard with 31 keys forms an integral part of the front panel and includes 10 soft keys whose functions are assigned by the system software and displayed on the screen. Extensive processing capability is based on an Intel 80486.

The CMA-2082D avionics management system
1996

Typical applications for the CMA-2082 include Communications, Navigation and Identification (CNI) functions; navigation control and display; integrated weapons controller and digital map controller. Other capabilities can be added to meet customer specifications with built-in maths co-processor and cache, and an Intel 82786 graphics processor.

Variants include the CMA-2082A, recently fitted to UH-60Q and CH-146 UTTH helicopters, and the CMA-2082D.

The CMA-2082D integrates navigation sensors, including optional built-in GPS, communications and the control of weapons and FLIR or radar systems. A dual-redundant MIL-STD-1553B interface, functioning as either a bus controller or remote terminal, is standard. Dual ARINC 429 interfaces are available, as are other customer specified input/output options that permit a direct interface with subsystems. An optional speech recogniser data input and control, designed for use in high-noise environments, is also available.

Specifications

Dimensions:
(CMA-2082) 146 × 181 × 170 mm
(CMA-2082-A) 146 × 239 × 170 mm
(CMA0-2082D) 146 × 239 × 170 mm
Weight:
(CMA-2082) 4.54 kg
(CMA-2082-A) 5.44 kg
(CMA-2082D) 5.9 kg
Power supply: 28 V DC, 75 W (max)
Display size:
(CMA-2082) 76 × 76 mm, 192 × 192 pixels
(CMA-2082A) 76 × 127 mm, 192 × 320 pixels
(CMA-2082D) 102 × 102 mm, 512 × 512 pixels
Display capacity:
(CMA-2082) 12 lines, 21 characters per line
(CMA-2082A) 20 lines, 21 characters per line
Visibility: readable in 10,000 ft candles of incident light
Memory: 256 kbytes UVPROM expandable to 512 kbytes; 16 kbytes RAM; 8 kbytes EEPROM; 512 to 2,048 kbytes FLASH, programmable via RS-232 or MIL-STD-1553 interface
Environmental: MIL-E-5400T, Class 1

Operational status

In production and service. Selected by the US Army/ Sikorsky Aircraft for the UH-60Q Black Hawk. Variants of the CMA-2082 are currently in service, installed on the US Air Force MH-60G Pave Hawk, MH-53J Pave Low III, AC-130 Gunship and MC-130E Combat Talon. The CMA-2082 is in service and installed on the US Navy ES-3A Viking and E-2C Hawkeye. It is in service with the Canadian DND, installed on the CH-146 Griffon.

Contractor

Canadian Marconi Company.

UPDATED

FRANCE

Mirage Fly-by-Wire (FBW) systems

In common with most high-performance military aircraft of its era, the Dassault Mirage III was developed with electrically signalled flight control information and some analogue electronic processing for improved capability. There is also a mechanical back-up. Powered control operation without fly-by-wire signalling allows the aircraft to be recovered from most flight conditions although its handling qualities are degraded. The Mirage III system was retained in the Mirage 5 and the Mirage 50.

The Dassault Mirage IV is aerodynamically similar to the Mirage III but is considerably larger. Having been designed during the same decade it incorporated a similar fly-by-wire system. Philosophy and operational modes are understood to be approximately the same as the Mirage III.

The Dassault Mirage F1 first flew in 1966 and uses a flight control system which is essentially like that of the Mirage III in that it uses electrical signalling, some electronic processing and powered controls with a mechanical back-up. The fly-by-wire system configuration was altered to match the flight control surface layout of a conventional aircraft planform.

A fully fly-by-wire analogue system was specified for the Dassault Mirage 2000, which has a similar configuration to that of the Mirage III but is conventionally unstable. The fully FBW flight control system provides an appropriate degree of stability and mixes control inputs from the pilot and systems so that airframe stresses and aerodynamic limits, including those during supersonic flight, are not exceeded. The aircraft's available performance can therefore be fully exploited with better manoeuvrability than with a conventional FBW system.

The wing trailing-edge control surfaces (elevons) are used for control in the pitch and roll axes and are signalled electronically without any mechanical back-up. There are two surfaces on each wing, each served by a twin body servo-jack and each body taking demands from two of the four available electronic channels. The single rudder is signalled via a triple fully FBW arrangement. A fifth channel emergency system, which uses battery supported signalling to provide the pilot with a degree of actuator control, is also employed, suggesting that in some portions of the flight envelope the aircraft has a reasonable degree of natural stability.

The system has quadruplex sensors associated with the pilot's controls and multiple gyros, accelerometers and air data sensors, which feed in to a quadruplex processor arrangement using hybrid computing techniques. Each processor is largely analogue, with digital computation of control gain parameters. Dassault is responsible for all flight control system development and provides interfaces which enable automatic flight control system demands to reach the flying controls. The Mirage 2000 is in production, having entered service in July 1984.

Operational status
All types are still in service.

Contractor
Dassault Aviation.

VERIFIED

AFDS 95-1 flight control system

The AFDS 95-1, designed for IFR operation with light and medium helicopters, is a modular system consisting of a basic two- to four-axis autopilot computer, coupler flight director computer and associated equipment. In the event of a coupler flight director computer failure the system is reconfigured in its basic mode, which allows the pilot to complete the mission.

By using inputs from vertical and directional gyros, the digital autopilot computer provides references for automatic control relative to attitude and heading. The basic autopilot functions are cyclic and yaw damping, long-term pitch and roll attitude hold and long-term heading hold in cruise and hover. The fourth axis facility is intended for operators needing transition and hover capability. The autopilot computer drives the electromechanical actuators in the control circuit and, when coupled to the coupler flight director computer, the trim and artificial feel actuators. The coupler flight director, digital panel-mounted computer, interfaces the navigation equipment to the attitude director indicator and autopilot computer.

With up to 11 push-buttons, the coupler flight director computer permits the selection of operating modes which include HDG, NAV, IAS, VS, GS, B ALT, APP 1, APP 2, HOV, R ALT and CLB allowing en route and approach applications at cruise speeds, or search and rescue, or anti-submarine warfare operations. The coupler flight director computer includes co-ordination and collective, yaw and cyclic coupling, together with longitudinal and lateral speed control when using Doppler signals.

The AFDS 95-1 flight control system is transparent, therefore the pilot's control inputs are detected and the attitude hold terms removed to keep the autopilot from resisting pilot control in manoeuvring flight. The system is FAA certified in VFR and single-pilot IFR configurations.

Operational status
In service on the B 222, AB 212 and A 109 K2 helicopter and in production. Selected by Bell Helicopters for its Model 427 light twin-turbine helicopter. Certification on the PZL Swidnik Sokol SW-3A is planned for 1998.

Contractor
Sextant Avionique.

VERIFIED

AP 205 autopilot

The AP 205 autopilot was designed for the French version of the Sepecat Jaguar. It is a simple and rugged system allowing pilots to fly hands-off while conducting other operational tasks.

Autopilot facilities include pitch attitude and altitude hold in the pitch axis and heading hold in the roll axis. Trim functions provide automatic pitch trim and manual trim assist in roll. Further modes for navigation or guidance purposes can be added. Modular construction is claimed to minimise maintenance and the system is designed to be retrofitted into aircraft not equipped with autopilot. The first autopilot was delivered to the French Air Force in November 1981.

Specifications
Dimensions: 124 × 210 × 253 mm
Weight: 5.4 kg
Power supply:
200 V AC, 400 Hz, 3 phase, <40 VA
26 V AC, 400 Hz, 3 phase, <1 VA
28 V DC, 15 W

Operational status
In service in the French Air Force Sepecat Jaguar.

Contractor
Sextant Avionique.

VERIFIED

AP 305 autopilot/flight director system

The AP 305 autopilot and flight director system is derived from an earlier Sextant Avionique system, the Tapir, which equipped civil Fokker F27 and French Air Force Nord 262 light transports. For French Navy Nord 262s a radio altimeter low-altitude mode has been added for low-flying operations over the sea. It is an analogue system with sideslip suppression, turn co-ordination and Cat. I auto-approach capability. Flight director/autopilot facilities include altitude capture and hold, selected heading capture and hold, VOR/ILS capture and tracking, attitude hold and appropriate operation indications and alerts.

There is also a sideslip detector and four servo actuators.

Specifications
Dimensions:
(system computer) 421 × 125 × 194 mm
(control panel) 180 × 146 × 57 mm
(mode selector) 135 × 146 × 27 mm
Weight:
(excl actuators) 8.6 kg
(incl actuators) 21.6 kg
Power supply: 115 V AC, 400 Hz, single phase, <80 VA
28 V DC, <150 W

Operational status
In service in French Navy and Air Force Nord 262 aircraft.

Contractor
Sextant Avionique.

VERIFIED

AP 405 autopilot and autothrottle system

The AP 405 has been specially designed to monitor height above sea level, to meet the requirements of the French Navy for the low-altitude and high-speed operations of its carrierborne Dassault Super Etendard. Precise angle of attack holding is provided by the autothrottle, including during carrier approaches when precise control is a deciding factor in achieving consistency and safety of operations.

Autopilot functions are pitch hold, altitude capture and hold, heading hold (with control wheel steering override), semi-automatic pitch trim and the provision for instinctive disconnect to make rapid flight path changes. At very low altitudes there is also a radio altitude hold mode with continuous flight path monitoring. The autothrottle system can capture and hold a selected angle of attack.

Specifications
Dimensions:
(system computer) 409 × 198 × 94 mm
(control/indicator panel) 115 × 26 × 150 mm
(autothrottle actuator) 120 × 120 × 200 mm
(angle of attack selector panel) 37 × 45 × 50 mm
Weight: 7.9 kg
Power supply:
115 V AC, 400 Hz, single phase, <60 VA
26 V AC, 400 Hz, single phase, <1 VA
28 V DC, <60 W

Operational status
In service in French Navy Super Etendard aircraft.

Contractor
Sextant Avionique.

VERIFIED

AP 505 autopilot

The AP 505 has been designed for M2.0+ combat aircraft. Pilots of the Dassault Mirage F1, which is equipped with the system, claim to be satisfied with its reliability and ease of use. The system can be switched on before take off, and engaged or disengaged by a handgrip trigger on the control column. In basic mode it maintains the longitudinal attitude, held when the pilot releases the stick trigger, and either the heading or bank angle, depending on whether re-engagement is effected at a bank angle of less or more than 10°. Autopilot modes provide automatic flight at a preselected altitude, heading or VOR/Tacan/ILS bearing. Limits on attitude hold facilities are ±40° in pitch and ±60° in roll.

Specifications
Dimensions:
(system computer) 190 × 202 × 522 mm
(function selector unit) 132 × 142 × 27 mm
(heading selector unit) 99 × 35 × 80 mm
Weight: 14.9 kg
Power supply:
200 V AC, 400 Hz, 3 phase, <100 VA
26 V AC, 400 Hz, single phase, <6 VA
28 V DC, <15 W

Operational status
In service in Dassault Mirage F1 aircraft. More than 600 sets have been built and supplied to eight countries.

Contractor
Sextant Avionique.

VERIFIED

AP 605 autopilot

The AP 605 is a digital flight control system developed for the Dassault Mirage 2000 supersonic combat aircraft. The high-capacity Sextant Avionique Series 7000 computer allows future extensions to the basic auto-pilot, to provide new modes suited to the particular missions flown by the aircraft. Flight control signals which pass between the autopilot and the Mirage 2000 fly-by-wire flight control system use a digibus serial data transmission system.

The AP 605 provides semi-transparent control, meaning that it is engaged or disengaged simply by releasing or taking hold of the control stick, which has a trigger switch in the handgrip. There is a high degree of internal monitoring, and computer design and organisation have been configured to reduce onboard maintenance. In basic mode the autopilot will hold pitch angle to any value within the range ±40°, or bank angle in the range ±60°. Additionally there are altitude capture and hold modes, preset altitude acquisition and fully automatic approach capability down to 200 ft. The Mirage 2000N version autopilot includes a terrain-following mode.

The Mirage 2000D version includes coupling with the air-to-ground fire-control system and the capability of very low-altitude capture and hold over the sea.

Specifications

Dimensions:
(system computer) 194 × 124 × 496 mm
(control unit) 24 × 146 × 115 mm
Weight: 12.7 kg
Power supply:
200 V AC, 400 Hz, 3 phase, <100 VA
26 V AC, 400 Hz, single phase, <1 VA
28 V DC, <30 W

Operational status

In service in Dassault Mirage 2000 aircraft.

Contractor

Sextant Avionique.

VERIFIED

AP 705 autopilot

The AP 705, equipping the Atlantique 2 ASW aircraft, is designed to provide precise flight path control and a high level of safety at very low altitudes above the sea. The system can hold a course while maintaining altitude. These levels of performance are made possible by the quality of inertial data available on the aircraft and by the versatility of the microprocessor-based computer. Built-in automatic testing enables the operation of the system and its safety devices to be checked before take off, and also facilitates onboard maintenance.

The three-axis autopilot includes pitch trim and can drive a flight director system. Autopilot modes permit pressure altitude hold, glide slope beam tracking (in Category I weather minima) and radio altitude hold over the sea at very low levels in reduced visibility. Lateral modes provide for holding the heading or course at the time of engagement, heading hold and homing or tracking on radio navaids, or navigation waypoints.

Specifications

Dimensions:
(system computer) 384 × 256 × 194 mm
(control unit) 200 × 164 × 67 mm
(servo-actuator) 185 × 183 × 101 mm
Weight: 23 kg
Power supply:
200 V AC, 400 Hz, 3 phase, <50 VA
28 V AC, <150 VA

Operational status

In production for the Dassault Atlantique 2 maritime patrol aircraft.

Contractor

Sextant Avionique.

VERIFIED

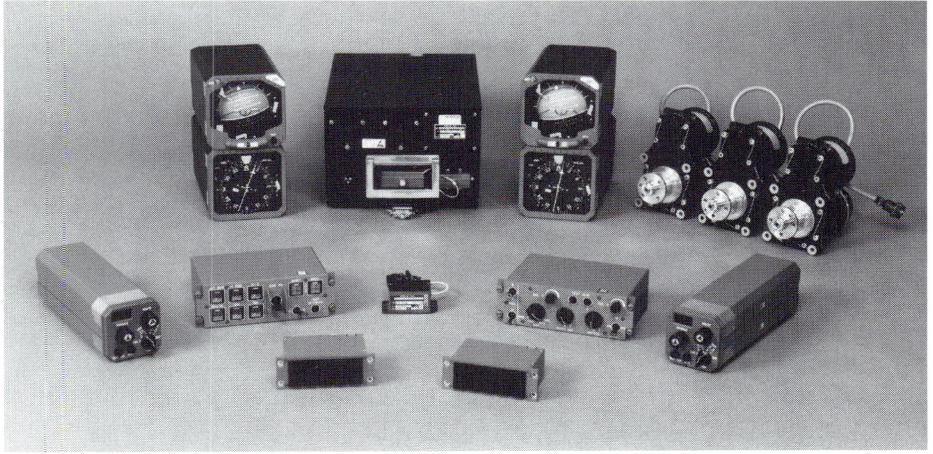

AP 705 autopilot

The APFD 800 autopilot/flight director has been selected in the upgrade of French Navy Alizé aircraft

AP 2000 autopilot system

The AP 2000 autopilot system is designed specifically for VFR or IFR operation with light and medium helicopters. The system performs attitude stabilisation, basic cruise and approach modes for two or three axes of the helicopter. Built-in automatic testing enables the operation of the system and its safety devices to be checked before take off and also facilitates onboard maintenance.

The AP 2000 autopilot system consists of one fully digital computer, one controller/programmer and serial and trim actuators. The AP 2000 autopilot computer drives serial actuators on the cyclic and yaw axes to perform stabilisation, attitude hold with beep trim control and heading hold. The parallel actuators are driven by the computer to achieve an automatic trim function. With the AP 2000 system, cruise and approach can be flown using standard modes such as altitude hold, heading select, IAS hold, VOR navigation, glide slope and localiser approach or back course localiser approach. Altitude and vertical speed select modes are optional.

One of the main features of the AP 2000 system is its adaptability towards the various possible helicopter configurations in terms of certification, operational or flight characteristics requirements.

The AP 2000 system also incorporates numerous features for simplified automatic line and shop maintenance.

Operational status

In development.

Contractor

Sextant Avionique.

VERIFIED

APFD 800 Autopilot/Flight Director

The APFD 800 all-digital autopilot/flight director has been developed for a wide range of aircraft, from light jet or prop fighters to medium transport, ASW or SAR aircraft.

It has a modular architecture. All the subassemblies are designed for easy installation on existing aircraft as part of the upgrade operation. It offers basic and higher modes, with many specific functions in the horizontal and vertical planes. Safeties are designed for use down to very low altitude, with or without autothrottle. It can be coupled with radio navigation, INS or the flight management system and is proposed with brushless DC motor actuators.

The APFD 800 provides semi-transparent control of the aircraft, as it can be instinctively engaged or disengaged hands-on-stick.

Specifications

Dimensions:
(computer) 4 MCU
(control unit) 146 × 57 × 140 mm
(mode selector) 38 × 140 × 146 mm
Weight: 7 kg total
Power supply: 28 V DC, 73 W

Operational status

Selected by the French Navy for upgrade of its Alizé maritime patrol aircraft.

Contractor

Sextant Avionique.

VERIFIED

Autoflight system for the A330 and A340

The autoflight system for the Airbus A330 and A340 is composed of two flight management guidance and envelope computers, a flight control unit and three multipurpose control and display units.

The flight control unit is used for short-term control of the aircraft autopilot and to select the display modes.

The multipurpose control and display units are installed on the centre pedestal in the cockpit. They are used for long-term control of the aircraft, initialisation of the fuel management system and to provide the interface between the aircrew and the maintenance system, ACARS, IRS and GPS.

The autoflight system can operate under autocontrol, using references computed by the flight management guidance and envelope computers on the basis of data selected by the aircrew through the multipurpose control and display units, or under manual control.

Operational status
In production for the Airbus A330 and A340.

Contractor
Sextant Avionique.

VERIFIED

B 39 autocommand autopilot

Introduced into French Air Force Dassault Mirage III interceptors since 1975 to replace older autocommand systems, the B39 is available for retrofit to other versions of the same aircraft type. It is a relatively simple system, but uses modern integrated circuit techniques to confer benefits in terms of performance, safety and maintenance standards. The new autocommand computer is physically interchangeable with the original equipment.

Autopilot functions are reduced to attitude and altitude hold modes. Attitude hold includes short-term capability, stability augmentation and uniform artificial feel load against load factor irrespective of flight conditions.

Specifications
Dimensions: 264 × 200 × 140 mm
Weight: 6.5 kg
Power supply: 200 V AC, 400 Hz, 3 phase, <25 VA
28 V DC, <1 A

Operational status
In service in Dassault Mirage III aircraft.

Contractor
Sextant Avionique.

VERIFIED

Centralised Fault Display Interface Unit (CFDIU) for the A320

The concept of a Centralised Fault Display Interface Unit (CFDIU), pioneered for the A320, suits the need for effective maintenance tools that enable troubleshooting down to the faulty LRU. The CFDIU collects and processes all built-in test equipment messages from the main aircraft avionics subsystems. Maintenance information can be displayed for diagnosis either by the flight crew or ground personnel.

The CFDIU receives all failure message output and stores it for subsequent use. It also performs a correlation check between messages separated by a short time interval that are assumed to have been generated by the same fault. It is possible for the CFDIU to transmit messages to the ground during flight by a datalink, or printout all the in-flight messages after the aircraft has landed.

The CFDIU is packaged in a 4 MCU ARINC 600 box and consumes 25 W at 28 V DC.

Operational status
In service in the Airbus A320.

Contractor
Sextant Avionique.

VERIFIED

Digital Automatic Flight Control System (AFCS) for the A300, A310 and A300-600

Digital processors replace the largely analogue elements of the original automatic flight control system in the Airbus Industrie A300 wide-body airliner, bringing the standard of this AFCS up to that of an almost wholly digital system. This is available in all A300 production with the forward-facing crew compartment. The system was first flown on the A300 in December 1980 and entered service with Garuda Indonesian Airways in January 1982. It is certificated for Cat IIIB operation. The newer A310, the first of which entered service in April 1983, has the new digital automatic control system as standard.

The digital automatic control system provides the flight augmentation functions of pitch trim in all modes of flight, yaw damping, including automatic engine failure compensation when the autopilot is engaged, and flight envelope protection. It has a comprehensive complement of autopilot and flight director modes that permit automatic operations from take off to landing and roll out, a thrust control system which operates throughout the flight envelope, and a derate capability. It contains protection features against excessive angle of attack and has a fault isolation and detection system for line maintenance.

The Airbus A310 has the Sextant Avionique digital automatic flight control system

The Sextant Avionique AFCS also integrates windshear detection as a combination of vertical and horizontal shear components, with annunciation and flight guidance functions.

Design has been in accordance with ARINC 600 and 700 characteristics and has led to the adoption of ARINC 429 databusses between the automatic control system processors and sensors. Four to six processors are used, comprising two flight augmentation computers, one or two flight control computers (the second unit being necessary only if Cat III automatic landing capability is required), together with a thrust control computer. A second thrust control option is available, and the system also includes a flight control unit providing pilot interface with the autopilot/flight director and autothrottle functions, and a thrust rating panel which allows crew access to thrust limit computations.

A further new item of equipment is an engagement unit, with pitch trim and yaw damper engage levers, autothrottle arm and engine trim controls, which is mounted in the flight deck roof panel. Two pitch and roll dynamometric rods are also used as control wheel steering sensors. There are two trim actuators, an autothrottle actuator, a coupling unit on each engine and two further dynamometric rods connected to the throttle control linkage.

Flight deck controller equipment has been revised and a new autopilot/flight director and autothrottle mode selector is installed in the centre glareshield. Variable data can be entered by rotating selector knobs and shown by liquid crystal display readouts. The various modes are engaged by push-buttons. Modes available are altitude capture and hold, heading select, profile to capture and maintain vertical profiles and thrust commands from the flight management system, localiser, landing and speed reference. Autothrottle modes include delayed flap approach, speed/Mach number select, and engine N1 or engine pressure ratio selection.

The thrust rating panel is mounted above the centre pedestal. This has comprehensive controls permitting thrust levels to be selected, depending on operating mode and providing for selection of such facilities as derated thrust take off.

The fault isolation and detection system has a dedicated maintenance/test panel which, on a two-line by 16-character display, provides written alert messages based on fault information from automatic testing activities. This is conducted in all LRUs and includes fault isolation, tests to check for correct operation after maintenance action and on the ground automatic landing availability checks. Up to 30 faults from six flights can be stored and retrieved.

Production of the system is jointly conducted by Sextant Avionique (as prime contractor) and Smiths Industries (UK) with Bodenseewerk (Germany).

Computer units
1 or 2 flight control computer(s) (10 MCU size)
1 or 2 thrust control computer(s) (8 MCU size)
2 flight augmentation computers (8 MCU size)

Control units
flight control unit (glareshield)
thrust rating panel (centre panel)
FAC/ATS engagement unit (roof panel)
Maintenance/test panel

Other units
2 pitch dynamometric rods
2 roll dynamometric rods
2 trim actuators
autothrottle actuator
2 engine coupling units
2 engine dynamometric rods

Operational status
In service in the Airbus A300-600ST and A310.

Contractor
Sextant Avionique.

VERIFIED

FDS-90 Flight Director System

The FDS-90 system comprises an attitude director indicator and a navigation coupler/computer unit.

The H140 attitude director indicator can operate autonomously using self-contained gyros and power inverters. It uses a ball-type real world display and has a three-cue command capability. There are annunciators for go-around, decision height and flight director mode monitor. The gyro can be caged.

The B152 nav coupler/computer is a small panel-mounted unit which interfaces the attitude director indicator to navigation equipment. It may have up to 11 push-buttons that permit selection of different flight director operating modes. These are:

HDG: captures and tracks the heading selected on the horizontal situation indicator

NAV or V/L: captures and tracks VOR and ILS localiser beam for short-range navigation or FMS data for long-range navigation. It has the capability to interface with Tacan, Loran or Doppler sensors, depending upon the mission

BC: tracks the back course localiser

BALT: maintains the baro-altitude existing at the time of selection

GS: captures and tracks an ILS glide slope beam

VS: maintains the vertical speed that exists at the time of engagement

IAS: maintains the airspeed that exists at the time of engagement

In addition to the various functions incorporated in the FDS-90, that are more particularly concerned with en route and approach applications at cruise speeds, the search and rescue and anti-submarine warfare coupler functions are specific to hover mode and to the various transition phases. These include:

APP 1: from the initial conditions of height above 200 ft and IAS higher than 60 kt it provides the simultaneous commands for a descent to 150 ft min and deceleration to 60 kt

APP 2: from the final APP 1 conditions it provides the commands required to control the helicopter's descent down to the height selected by the pilot and deceleration to hover

HOV: holds the zero lateral and longitudinal groundspeeds provided by the Doppler radar or the zero sonar cable angles, depending on the submode selected by the pilot

CLB: from the initial conditions of height above 200 ft and IAS higher than 60 kt it provides the commands for a climb at 300 ft/min and acceleration up to 60 kt.

The SAR and ASW coupler is connected to the ADI for manual use in the FD mode or to the basic stabilisation system (Ministab, AFDS 95-1 or any other AFCS) for automatic operation.

Specifications

Dimensions:
(attitude director indicator) 4 ATI
(nav coupler/computer) 3 ATI standard case
Weight:
(attitude director indicator) 2.5 kg
(nav coupler/computer) 1.3 kg
Power supply:
28 V DC, <20 W (attitude director indicator)
28 V DC, <15 W (nav coupler/computer)
Environmental: DOO 160 C, TSO'd

Operational status

In production and in service, on S-61 (HH3-F) and B206.

Contractor

Sextant Avionique.

VERIFIED

Flight Management and Guidance System (FMGS) for the A320

In the A320 the autopilot, flight director and flight management functions are integrated in a single system, the Flight Management and Guidance System (FMGS), bringing a 60 per cent reduction in the number of LRUs and a 50 per cent saving in volume and weight over earlier systems such as that in the A310.

Two flight augmentation computers in 8 MCU boxes provide dual-channel command signals for yaw damping, rudder trim, rudder travel limit system and flight envelope protection. The entire system is monitored by fault detection and isolation software, with warning indications and appropriate vital actions and responses being automatically displayed on the EFIS CRT displays. Airbus is sufficiently confident about the integrity of the system to believe that the reversionary mode will never be employed operationally.

The computer has about 2.5 Mbytes of non-volatile memory and takes in data from the ADIRS, radio navigation aids, radio altimeter and the FADEC. Through interfaces with the crew via the flight control unit, multifunction control and display unit, it provides autoland, autothrust, autopilot cruise, four-dimensional navigation control, flight management and performance management in the vertical profile.

Operational status

In production for the Airbus A320.

Contractor

Sextant Avionique.

VERIFIED

The A320 flight-deck showing flight management and guidance system controls on the glare shield (British Airways)

Fly-by-wire spoiler and elevator control for the A320

The Airbus A320 is the first commercial transport aircraft to employ digital fly-by-wire technology in the primary flying controls (although the A310 uses FBW to command the flap, slat and airbrake secondary control movement). In the aircraft, miniature sidesticks command pitch and roll attitude. The sticks are centred by simple springs providing 'return to neutral' forces independent of speed or altitude. The rudder continues to be mechanically operated via conventional rudder pedals.

Elevators and ailerons are controlled by four independent channels, contained in two 8 MCU boxes, the quadruplex system commencing with eight electrical movement linear variable differential transformer pick-offs (four per axis) in the sidestick controllers. Each pilot has his own sidestick mounted on the side console, the captain operating his with the left hand, the first officer with the right.

While two of the three primary flying controls are activated by fully FBW equipment and thus have no manual reversion, total failure of the aileron and elevator channels can be compensated by way of the mechanically operated tailplane trim and rudder; tailplane trim replaces elevators and rudder stands in for ailerons. As with the A310, FBW is also used for flaps, slats and speedbrakes and in the form of Full Authority Digital Engine Control (FADEC) is also used for engine management.

Sextant Avionique is responsible for supplying the three Spoiler and Elevator Computers (SEC) providing primary control of the five wing spoilers and back-up elevator control.

The multiplexing of the control signals in the SEC provides roll and speedbrake control and also gust load alleviation. In addition, the SEC provides back-up main and trim pitch control in case of failure of the associated computer.

Each A320 has three SECs to meet safety, availability and system configuration requirements. The SEC is self-monitored and its internal architecture is designed to meet the highest safety standards. It is based on two processors (one functional and one for monitoring) which are totally segregated in terms of hardware and have dissimilar software. Each SEC is contained in an 8 MCU box.

Operational status

In production for the Airbus A320.

Contractor

Sextant Avionique.

VERIFIED

Fly-by-wire systems for the A330 and A340

As on the A320, the A330 and A340 flight controls are hydraulically actuated and electrically or mechanically controlled. Pilot controls in the cockpit consist of two sidesticks, conventional rudder pedals, mechanical pitch trim and electrical rudder trim.

Electrical flight control is achieved by seven computers of three different types: three Flight Control Primary Computers (FCPC) are in charge of generating control laws and controlling surfaces, two Flight Control Secondary Computers (FCSC) are also in charge of controlling surfaces and two Flight Control Data Concentrators (FCDC) interface the flight control system with other aircraft systems to provide an isolation function.

Each of the two FCSCs can control the power elements used to activate the aircraft control surfaces. In normal operation, the FCSC achieves spoiler control, rudder trim control and rudder travel limiting. In back-up mode, as in the case of a failure of the FCPC, the FCSC achieves aileron control, elevator control and yaw damping.

The FCDC performs data concentration, warning and maintenance functions. In data concentration, the FCDC transmits information, such as control surface positions, to the display management computers and to the flight data interface unit. In warning, the FCDC indicates flight control failure status to the flight warning computers and to the display management computer. For maintenance the FCDC isolates and memorises the flight control system failures and interfaces with the centralised maintenance computer.

Operational status

In production for the Airbus A330 and A340.

Contractor

Sextant Avionique.

VERIFIED

Fuel Control and Monitoring Computer (FCMC) for the A330 and A340

The A330 and A340 have two fuel tanks in each wing, one in the tailplane and one in the fuselage.

The A330 and A340 fuel control and monitoring system is composed of about 96 fuel height probes and other sensors for temperature, densitometers and level detectors installed in the fuel tanks, one to three refuelling panels and two identical Fuel Control and Monitoring Computers (FCMC).

The FCMC in the A330 and A340 concentrates functions previously scattered in distinct computers in the A310 and A320 such as fuel quantity measurements, Hi-Lo measurement and centre of gravity control.

The role of the FCMC is to measure the fuel quantity available on the aircraft and indicate it to the aircrew, control automatic refuelling of tanks to preselected levels, monitor associated pumps and valves, monitor and control the fuel temperature in the tanks, control the fuel tank utilisation sequence, detect too low or too high levels in tanks and manage the associated warnings, control fuel distribution for wing load alleviation and control the centre of gravity of the aircraft by transferring fuel from one tank to another.

Operational status

In production for the Airbus A330 and A340.

Contractor

Sextant Avionique.

VERIFIED

Gemini 10 navigation and mission management computer

The Gemini 10 computer contains a Jeppesen database and, in addition to carrying out conventional FMS functions, also performs tactical military functions. Connected to the VH100-T HUD, it supports missions such as assault landings, airdrops and tactical low-altitude flights.

The computer's compact design is due to its monobloc concept, with the navigation computer and control and display unit integrated in a single item of equipment. This layout offers advantages for installation of lower weight and smaller volume and higher reliability, easier maintenance and lower power consumption in operation. The Gemini 10's dual architecture helps increase the probability of mission success by allowing reconfiguration on the validated FMS, in case of failure of one of the two units.

The software comprises 150,000 lines in Ada language, conforming to the requirements of DO178, and allows the display of approximately 150 pages of information. The display offers 12 lines of 20 characters each and five variable label keys with functions depending on the mission.

Operational status

Production of the Gemini 10 started in 1993 for the C-160 Transall upgrade.

Contractor

Sextant Avionique.

VERIFIED

Helistab stability augmentation system

Helistab is a modular system consisting of a basic three-axis Ministab system, plus an autopilot computer and associated equipment. In normal operation sufficient Ministab functions are in use for this to constitute an adequate reversionary mode in the event of an autopilot failure.

The autopilot replaces the rate gyro integrated terms by vertical and directional gyro inputs, providing references for automatic control relative to attitude or heading. Basic autopilot functions are long-term pitch and roll attitude hold, long-term heading hold in cruise and hover and auto trim in pitch. Functions that can be added include turn co-ordination and heading capture in cruise, automatic hands-off recovery from unusual attitudes, barometric altitude or airspeed hold in cruise, Doppler-based longitudinal and lateral speed control in hover and collective/yaw coupling.

Helistab is transparent, therefore the pilot's control inputs are detected and the attitude hold terms removed to keep the autopilot from resisting pilot control in manoeuvring flight.

Operational status

In service on the Agusta-Bell AB 212 helicopter.

Contractor

Sextant Avionique.

VERIFIED

Integrated Modular System (IMS) avionics suite

IMS is an avionics suite designed for 70- to 130-passenger regional and business aircraft. It provides primary air data and inertial reference, long-range navigation using GPS, flat panel LCD standby instruments, automatic flight control, warning management, flight management, data acquisition for other systems and built-in maintenance of all Migrator subassemblies.

IMS also offers three optional functions: a Central Maintenance Function (CMF) which centralises all failures recorded by the other systems such as radio, weather radar and servo-systems connected to IMS; an Aircraft Condition Monitoring System (ACMS) which monitors and records a number of parameters, in particular engine data, and an Aircraft Communications and Automatic Reporting System (ACARS) for VHF transmission of data from the ACMS and other systems to the ground. The IMS is available in various configurations that can be tailored to the needs of particular regional and business jets. The basic configuration is for regional airlines flying mainly scheduled flights between cities. This configuration offers an autopilot allowing Cat II approaches and an RNav flight management system which fully meets the minimum operational procedure objectives that European and North American ATC authorities will implement in the next few years. Adding spare modules to the integrated flight computer also allows aircraft with this configuration to stay in service for several days after an initial failure, before returning to their maintenance station. Higher configuration standards will include a fail-passive autopilot incorporated in the basic IFC volume, providing Cat IIIA autoland capability which can be extended, if needed, to Cat IIIB. In addition, a flight management system handles automatic operation of the flight plan.

Contractor

Sextant Avionique.

UPDATED

Warning and maintenance system for the A330 and A340

The A330 and A340 warning and maintenance system is composed of two major subsystems: the flight warning system and the onboard maintenance system.

The flight warning system presents the aircrew with visual and warning messages related to failures. It also indicates the seriousness of each failure and the corrective actions to be taken. The system consists of two flight warning computers, two data acquisition concentrators and an ECAM control panel.

The onboard maintenance system generates and displays maintenance data on the multipurpose control and display units for use by aircrew and ground maintenance personnel. It consists of two centralised maintenance computers.

Development and production is conducted jointly by Sextant Avionique and Aerospatiale.

Operational status

In production for the Airbus A330 and A340.

Contractors

Sextant Avionique.
Aerospatiale.

VERIFIED

AFCS 85 autopilot/flight director system

The basic AFCS 85 system is a full-time two-axis series autopilot which is the first step in a building block concept to provide automatic stabilisation for light helicopters.

The basic version provides long-term hands-off VFR capability, including heading select mode, baro-altitude and airspeed hold.

Adding the FDC 85 flight director coupler and pitch and roll autopilot monitor gives single pilot IFR capability by providing facilities such as automatic lateral and vertical guidance and by driving command bars on the attitude director indicator.

The basic system weighs 6.8 kg.

Operational status

In production. The system has been chosen for the Alouette III, Gazelle and single- and twin-engine Ecureuil helicopters.

Contractor

SFIM Industries.

VERIFIED

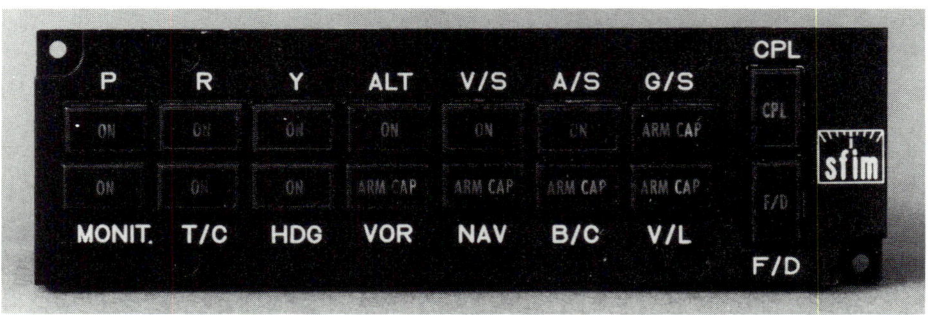

Controller for the SFIM Industries 85T31 automatic flight control system

AFCS 155 autopilot/flight director system

The AFCS 155 is a series duplex, fail-passive autopilot. Optimised for both single and dual pilot IFR operations, covering the whole flight envelope from hover to VNE, it is IFR-certificated on the Eurocopter Super Puma and Dauphin helicopters and can be integrated with SFIM couplers. These include the FDC 85 three- or four-axis and FDC 155 four-axis digital coupler. In the duplex configuration each channel has its own power supplies, sensors and interconnections.

Basic functions include long-term attitude and heading hold, turbulence compensation, collective link mode and autotrim. All autopilot configurations satisfy single pilot IFR requirements and there are three upper modes: heading select, altitude hold and airspeed hold.

Fly-through or transparent handling characteristics allow the pilot to make quick attitude and heading changes while benefiting from dynamic damping. On releasing the controls the autopilot returns to long-term stabilised flight at the previously set attitudes. If new attitude settings are required the pilot can either use the stick release button on the cyclic pitch grip or can change the settings at a slow rate by using the stick-top four-way 'beep' trim button. The 'beep' trim button is also used to alter the reference airspeed slowly when the automatic airspeed hold mode has been selected.

All versions of the system include automatic trim which keeps the series actuators centred so that the autopilot has full authority.

Additional flight director coupler facilities provide IFR automatic navigation, radio navigation and approach capabilities, including steep approach MLS beam capture and track. There are also additional modes for anti-submarine warfare and offshore or search and rescue operations. The latter includes automatic pattern following, automatic up and down transitions to and from selected radio altimeter height, Doppler/radio height or sonar cable hover and low Doppler speed automatic hold. Compatible couplers include FDC 85 three- and four-axis navigation and approach couplers (the latter is Cat II certified), and the FDC 155 all mission digital coupler.

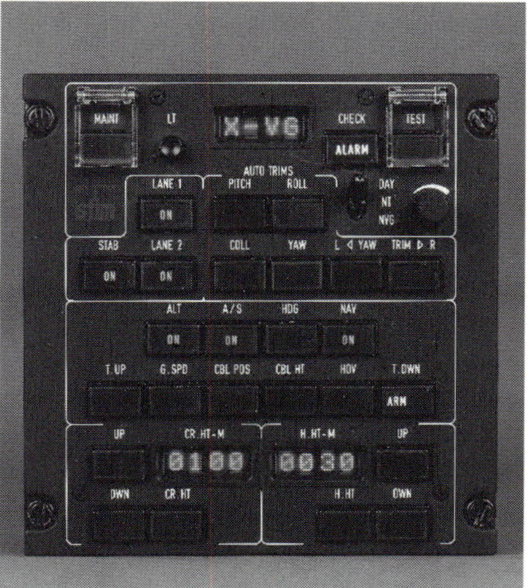

AFCS 166 control unit as fitted in the naval version of the Hindustan Advanced Light Helicopter (ALH)

Specifications

Weight:
(autopilot computer) 8 kg
(servo amplifier) 4.2 kg
(autopilot control box) 1.6 kg
(4 actuators) 1 kg each
(3 trim servos) 1.3 kg each
(barosensor) 1.5 kg
(FDC 85) 2 kg
(FDC 155) 8 kg
(flight director coupler box) 2 kg
(collective pitch motor) 2 kg

Operational status

In production and installed on Eurocopter AS 332 Super Puma and AS 365N Dauphin helicopters.

Contractor

SFIM Industries.

VERIFIED

AFCS 165/166 flight director systems

The AFCS 165/166 and variants are four-axis flight control and flight director systems intended for medium and heavy-capacity helicopters. The power components are either electromechanical or electrohydraulic, the sensors being digital or analogue. A dual/dual microprocessor system provides fail-passive and fail-operational capabilities without performance loss.

The basic AFCS 165/166 comprises Cat. II navigation and approach modes and allows single pilot IFR operations. A modular system approach is also possible. Modes relating to specific missions such as ASW, SAR or anti-tank are offered as an option, software integrated.

In addition to the monitoring and safety functions, the

AFCS is fitted with preflight test, a computer self-test and built-in first and second line maintenance test facilities.

Operational status

In production. The AFCS 165 has been selected by Eurocopter for the Super Puma Mk II and the AFCS 166 for the Advanced Light Helicopter developed by Hindustan Aeronautics.

Contractor

SFIM Industries.

VERIFIED

AP 146 autopilot

The AP 146 autopilot is designed for anti-submarine helicopters equipped with MAD systems. It provides accurate radio height hold with filtering to compensate for swell up to Sea State 5. Safety circuits inhibit active or slow failures in the pitch channel, for extended low-altitude operation.

Operational status

In service.

Contractor

SFIM Industries.

VERIFIED

APM 2000 autopilot module

The APM 2000 is the core of the Automatic Flight Control Subsystem of the new generation of 2 to 3 tonne helicopters.

The functions are: stability and control augmentation, long term attitude hold, and control of flight. Depending on the configuration, the system can operate either in simplex APM operation or in duplex APMs. The APM 2000 is a fail passive digital computer that uses highly integrated technology in the modular concept known as 'Avionique Nouvelle'. The APM 2000 is the basis of a computer family that can meet requirements from 2-axis to 4-axis autopilot system implementation. Two identical APM 2000 modules can be integrated to build a fail operational configuration, to improve the mission reliability.

The APM 2000 can interface either with the EHA through a SAS 2000 computer or directly with SEMA.

A typical APM 2000 installation could include one SAS 2000, two trim actuators, one to four SEMA actuators, one APMS, two AHRS (digital), two ADU, two FCDM/EFIS systems, the cyclic grip, the collective grip. According to version, the APM 2000 could be connected to one Doppler system, one VEMD, one HUMS, one MFDAU, one APM 2000 (duplex configuration).

Specifications

Dimensions: 337 × 183 × 38 mm
Power supply: 16 to 32.2 V, 25 W
Weight: 2 kg
Environmental: DO 160 C, TSO

Operational status

In production and installed on Eurocopter EC135, Dauphin N4.

Contractor

SFIM Industries.

VERIFIED

FCS 60B automatic Flight Control System

The FCS 60B is designed for commuter aircraft to provide ease of operation for flight profiles ranging from initial climb out to Cat II ILS or MLS approach.

In the BAe Advanced Turboprop, the system consists of a pair of three-axis autopilot/flight director fully digital computers which drive dual primary and trim servos. The FCS 60B also acts as a back-up fly-by-wire system for the three axes. This mechanisation allows independent and redundant en route operation of the pilot and co-pilot flight directors, as well as cross-monitoring for Cat II approaches.

Operational status

Certified on the British Aerospace Advanced Turboprop aircraft.

Contractor

SFIM Industries.

VERIFIED

Units of the SFIM Industries FCS 60B autopilot and automatic flight control system for commuter aircraft

SAS 2000 Stability Augmentation System

The SAS 2000 computer (Stability Augmentation System) is used to control the stability of helicopters (2-3 tonnes). The SAS 2000 is configured as two separate optional equipment packages in order to offer the maximum customer flexibility and choice: yaw SAS and/or pitch and roll SAS.

The yaw SAS consists of an integrated yaw rate gyro and control law computer, which provides command signals to an electromechanical series actuator driving the input lever of the tail rotor hydraulic boost.

The equipment consists of: a Fibre Optic Gyro (FOG) and Smart Electro-Mechanical Actuator (SEMA). The FOG measures the yaw angular rate and is used to provide damping about the vertical axis. In addition to the rate signal, the unit is also used to implement the control laws and output signals to feed the SEMA.

The SEMA is a high performance electromechanical actuator using modern brushless motors with rear earth permanent magnets. It incorporates a complete position servo feedback loop using Hall effect sensors within the same housing.

Pitch and roll SAS computer: Designed to be interfaced with a conventional attitude sensor, the SAS analogue pitch and roll computer controls the slaving loop of an electro hydraulic actuator. The pitch and roll SAS computer interconnects a full digital auto pilot module to the EHAs. In the VFR configuration the pitch and roll SAS computer is connected to helicopter generation, EHAs, trim actuator units (pitch and roll), cyclic stick grip and instrument panel switches, a remote vertical gyroscope or panel-mounted gyro horizon and pitch/roll axes solenoids.

Specifications: FOG

Mass: 0.75 kg
Software: RTCA DO 178B
Power supply: 16 V DC to 32.2 V DC
Environmental: DO 160 C

Specifications: SEMA

Input: Arinc 429
Adjustable working stroke (± 2.5 and 8 mm)
Working speed up to 18 mm/s
Stall load: > 50 N
Limit load: > 4000 N
Mass: < 0.8 kg

Specifications: pitch and roll SAS computer

Dimensions: 200 × 122 × 112 mm
Weight: 1.3 kg
Power supply: 28 V, 20 W
Environmental: DO 160 C

Operational status

In production and installed on Eurocopter EC 135.

Contractor

SFIM Industries.

VERIFIED

SFIM Industries APM 2000 autopilot module **1997**/0001398

SFIM Industries pitch and roll SAS computer **1997**/0001396

GERMANY

Flight Control Unit (FCU) for the Airbus A319/A320/A321

The FCU for the Airbus A319, A320 and A321 represents a smart, multipurpose control and display unit, interfacing between the pilot and the autoflight and electronic flight instrumentation systems. Installation in the glareshield and separate controls for pilot and co-pilot for communicating with the primary instrumentation system enable the crew to work head-up. The FCU allows the pilot to engage autopilot, flight director and autothrust systems, and to set flight altitude, speed and course.

The FCU consists of two independent computers with automatic switchover to ensure redundant signal processing. All data exchange between FCU and external systems is done via a discrete interface and ARINC 429 serial datalink. Contrast and illumination of the displays and panels are adaptable to extreme environmental conditions, from direct exposure to sunlight at high altitude to the special requirements of night approaches.

Application of specifically designed LCDs to display set values allows wide viewing angles with high contrast and sunlight readability, even with the pilots wearing polarised sunglasses. Set values are introduced into the FCU using optical encoders. These are incremental opto-electronic devices with a notched input and push-pull capability.

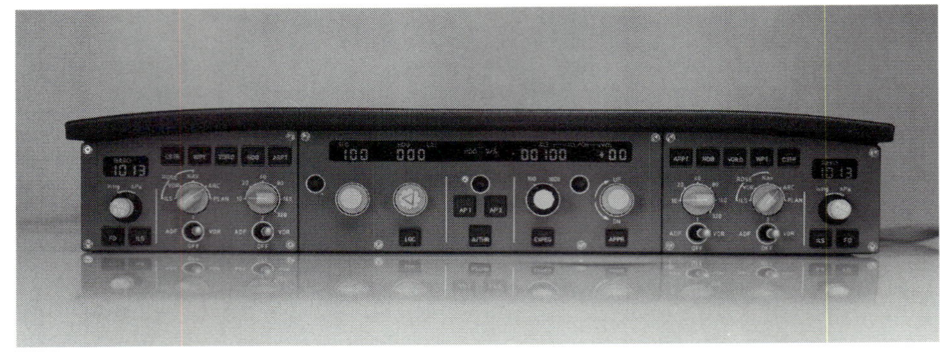

Flight control unit for the Airbus A319, A320 and A321

LED keys for the selection of operating modes are based on push-buttons with redundant lighting. Each push-button shows a green confirmation bar and a white illuminated legend. Misreading of non-illuminated bars in full sunlight is prevented by integrated layers of optical filter coating. The push-buttons have double poles to ensure precise operation and repeatable tactile properties.

Operational status

In production for the Airbus A319, A320 and A321.

Contractor

Bodenseewerk Gerätetechnik GmbH/BGT.

VERIFIED

Air data systems

Nord-Micro Electronik Feinmechanik AG manufactures a range of air data systems including Air Data Computers (ADC) and engine air intake control computers for both civil and military subsonic and supersonic aircraft. Modular design concepts are used, together with ASICS and hybrid circuits to produce lightweight, compact, low power consumption systems. For critical data, redundant dissimilar algorithms are used to avoid single point failure and improve fault tolerance. Extensive built-in test and health monitoring functions are also included.

Tornado air intake control system

The Tornado air intake control system comprises two digital air intake control units and one pilot's panel. It controls the engine intake ramp position via an electrohydraulic actuator.

Airbus A300 and A310 ADCs

Two identical ADCs are used in each type of aircraft; they comply with ARINC 706 and include extensive built-in-test capabilities.

JAS39 Gripen air data computer/air data transducers (ADC/ADT)

The ADC/ADT is configured as two separate computers in one common housing. They communicate with the fly-by-wire flight control computer via ARINC 429 as well as MIL-STD-1553B databus interfaces.

EF 2000 ADCs

Two ADCs supply pressure, temperature, altitude rate and air stream direction information to the fly-by-wire flight control computers.

JAS39 Gripen air data computer/air data transducers (ADC/ADT) **1998**/0018244

Airbus A300 and A310 ADCs **1998**/0018245

Operational status

In service.

Contractor

Nord-Micro Elektronik Feinmechanik AG.

NEW ENTRY

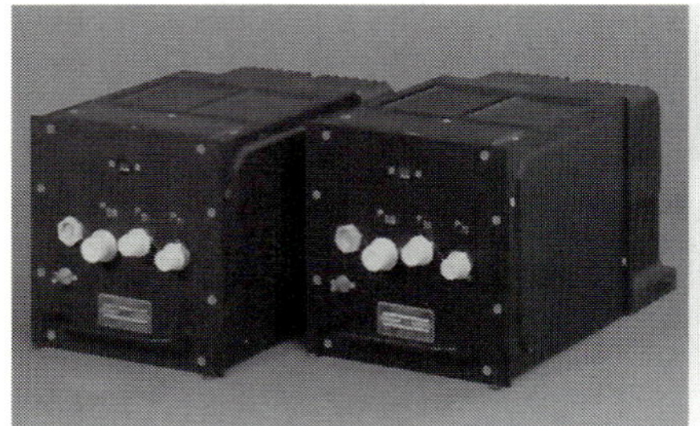

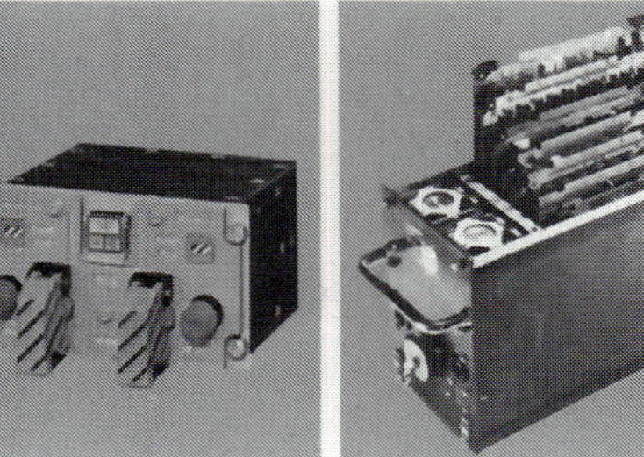

Tornado air intake control system **1998**/0018246

EF 2000 air data computers **1998**/0018243

INTERNATIONAL

Aria — EFIS-95 — airborne integrated avionics system

Aria — EFIS-95 is a joint venture development between NIIAO Institute of Aircraft Equipment, Moscow and AlliedSignal Aerospace of the USA to develop an airborne integrated avionics system for the Be-200 medium capacity amphibian aircraft.

The modular nature of the system is such that it could also be applied to a large number of other advanced civil aircraft, including: Il-96-300, Il-114, Tu-204/-214/-324/-330/-334, and YAK-142M; it could also be applied as an upgrade to Il-86, Tu-154/-154M and YAK-42.

NIIAO is primarily responsible for software development and aircraft integration management, whilst AlliedSignal (and its subsidiaries) is the supplier of most of the system hardware (except the air data system), including: FMS, EFIS, GPWS, FDE, AHRS (LITEF LCR-88), radar (Bendix King RDR-4B). Provision is being made for the system datalinks to accommodate CNS/ATM requirements.

Specifications

System weight:
(Be-200 class aircraft) 400 kg
Power:
(Be-200 class aircraft) 6 kW
MTBF: 20,000 h

Contractors

NIIAO Institute of Aircraft Equipment, Moscow.
AlliedSignal Aerospace, USA.

NEW ENTRY

Automatic Flight Control System (AFCS) for Concorde

GEC-Marconi Avionics jointly developed with Sextant Avionique the Concorde Automatic Flight Control System (AFCS), which comprises eight separate subsystems with no fewer than 38 individual units, and provides fully automatic control of the aircraft from take-off, through the entire flight envelope and down to approach and landing phases. The system is duplex-monitored and fail-operational to meet the operational requirements of ICAO Cat IIIA and is cleared for automatic landing with 15 ft decision height and 200 m runway visual range. It was the first AFCS to use a push-button controller fitted into the flight deck glareshield.

Operational status

In service in British Airways and Air France Concorde aircraft.

Contractors

GEC-Marconi Avionics Ltd.
Sextant Avionique.

VERIFIED

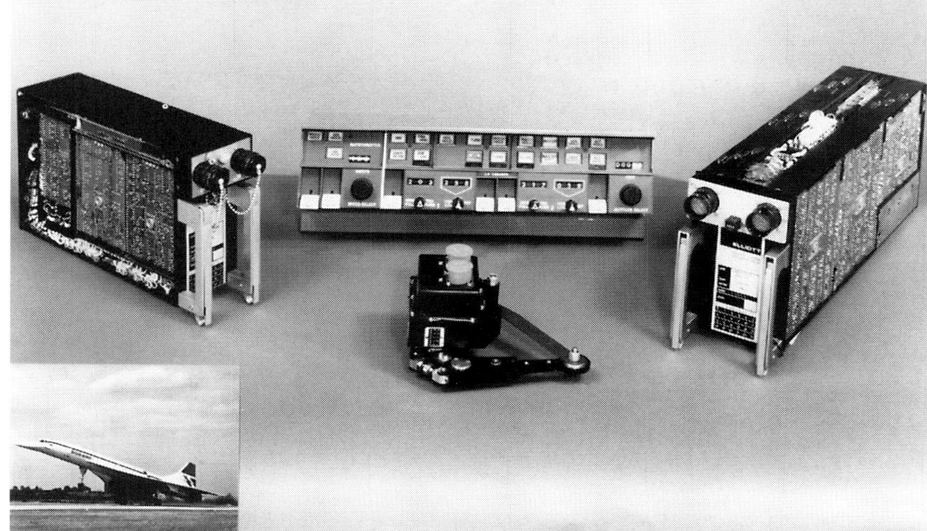

Automatic flight control system for Concorde **1997**/0001403

Automatic Flight Control System (AFCS) For Eurocopter Tiger helicopter

The Automatic Flight Control System (AFCS) is a digital autopilot, which is superimposed on the mechanical flight control system of the helicopter. The duplex digital AFCS provides the following functions: command and stability augmentation/attitude hold, to stabilise the aircraft as a weapon platform and reduce gust sensitivity; tactical modes, including three axis hover, a line-of-sight mode for weapon aiming/weapon delivery and gunfire compensation; standard cruise modes providing four axis control.

Each AFCS consists of two identical Flight Control Computers (FCC) and two identical Control Panels (CPs). The FCC contains five modules with the following elements/ functions: CPU Board with two microprocessors, one for system management functions and the other one for running the control laws; Input/Output Controller (IOC) for input/output control, analogue I/O, cross-channel datalink, communication and hardware synchronisation; input/output board for discrete I/O requirements; Power Supply Unit (PSU) for the generation of internal voltages and LVDT/RVDT excitation; rear module for EMC filtering and lightning strike protection.

The AFCS can be controlled from either of the two control panels, which are installed in the front and rear cockpits. Apart from the ARINC link, communication between avionics equipment and the AFCS is established with a MIL-STD-1553B bus.

The AFCS software is programmed in Ada using state-of-the-art software development techniques.

The Tiger AFCS is a joint development between Sextant Avionique and Nord-Micro (acting as prime contractor).

Operational status

First flight of an AFCS equipped Tiger was in 1993. Flight testing on Tiger prototype helicopters continues.

Contractors

Nord-Micro Elektronik Feinmechanik AG.
Sextant Avionique.

VERIFIED

Tiger PAH-2 automatic flight control system
1997/0001399

Cockpit Weather INformation system (CWIN)

An advanced Cockpit Weather INformation system (CWIN) project sponsored by NASA-Langley Research Center and developed by a consortium led by The Boeing Company is designed to give pilots real-time information on weather conditions anywhere in the contiguous US.

The CWIN, an integrated in-flight communication and navigation system, blends data from communication and Global Positioning System satellites, ground weather radar images and lightning strikes and airport observations and forecasts. Flight crews are provided with complete, up-to-date weather information, including trends, enabling them to make early decisions to avoid adverse conditions.

The automated CWIN provides graphic information on a colour 10.4 in diagonal liquid crystal display and updates the information every 15 minutes by satellite. Conveniently located near the cockpit weather radar display, the CWIN provides a touchscreen menu.

Commercial off-the-shelf hardware and software is used. Hardware units include a 486-based data processor, several gigabytes of mass memory storage and an input-output server to interface with the SATCOM, Global Positioning System and existing avionics systems.

Other applications being developed include windshear displays, three-dimensional graphics, an airport runway taxi map, flight operations information and a terrain and weather overlay.

Additional CWIN team members include Computing Devices International; Inmarsat; COMSAT Aeronautical Services; Astronautics Corp of America; Pulse Embedded Computer Systems Inc; Sextant Avionique; Trimble Navigation; Honeywell Inc; Canadian Marconi Corp; and WSI Corp.

Operational status

In development.

Contractor

The Boeing Company (consortium leader).

UPDATED

Digital fly-by-wire fight control system for the EF 2000

GEC-Marconi Avionics leads a consortium comprising BGT of Germany, Alenia Difesa, Avionic Systems and Equipment Division of Italy, and Empresa Nacional Optica Sociedad Anonima (ENOSA) of Spain. This consortium is under contract to Daimler Benz Aerospace (DASA) to develop the Flight Control Computer (FCC) and Stick Sensor and Interface Control Assembly (SSICA) for the European fighter aircraft EF 2000.

The Flight Control System (FCS) provides the aircraft with a full-time fly-by-wire control system which controls the aircraft via its 11 primary and secondary flying control surfaces to give the aircraft carefree handling characteristics together with outstanding agility and manoeuvrability throughout the flight envelope. The EF 2000 has no mechanical back-up system and is therefore totally dependent on this digital system. Redundancy of the system is ensured through replication of both the hardware and software, and should a failure occur in any one computer, this can be isolated and the aircraft will continue to fly normally.

To provide the necessary integrity and safety, the system uses a quadruplex design with each of the four FCCs containing identical hardware and software. Each computer contains Motorola 68020 32-bit microprocessors and software compiled in Ada. Comprehensive built-in test provides a continuous system monitoring for both ease of system maintenance and assurance of the continued high integrity performance. ASICs have been widely used to ensure a low component count to maximise performance and reduce system size and weight. The FCC also contains an interface to the aircraft's utility bus to provide integrated vehicle management and to a STANAG 3910 fibre optic avionics bus to allow integration with the mission avionics.

Operational status

Currently in development.

Contractors

GEC-Marconi Avionics Ltd.
Bodenseewerk Gerätetechnik GmbH/BGT.
Alenia Difesa, Avionic Systems and Equipment Division, GF-Sistemi Avionici SpA.
ENOSA.

UPDATED

Flight control system for the AMX

GEC-Marconi Avionics is co-operating with Alenia Difesa in the Fly-By-Wire (FBW) flight control system for the Italian/Brazilian AMX strike aircraft. The system comprises two dual redundant flight control computers each based on 16-bit microprocessor hardware and incorporating fail-safe software. The system commands the movement of seven control surfaces and incorporates a recently developed autopilot facility and automatic pitch, roll and yaw stabilisation. Analogue computing is used for the actuator control loops, the pilot command paths and rate damping computation. Digital computing handles gain scheduling, electronic trim and integration of the airbrake. Redundant microprocessors in the flight control computer units monitor system performance and are associated with built-in test facilities designed to provide a high confidence and rapid comprehensive system check. Testing is initiated by the pilot before flight and is conducted automatically thereafter. In addition to the flight control computers, the FBW system also includes pilot control position sensors, three-axes rate gyros and air data components.

Operational status

In production.

Contractors

GEC-Marconi Avionics Ltd.
Alenia Difesa, Avionic Systems and Equipment Division, GF-Sistemi Avionici SpA.

UPDATED

AMX flight control computer
1997/0001411

Flight monitoring system for Ilyushin Il-96M/T aircraft

Rockwell Collins has signed contracts with the Russian State Research Institute for Aviation Systems (GosNIIAS) and the Elara Joint Stock Company in Cheboksari for co-production of the Il-96M/T's Aircraft Systems Central System (ASCS). The system will perform flight engineering functions and will provide a two-member flight crew capability for the long-range, four-engine airliner produced by the Ilyushin Aircraft Association.

Operational status

Development.

Contractors

Rockwell Collins.
Elara Joint Stock Company, Cheboksari.
Russian State Research Institute for Aviation Systems (GosNIIAS).

UPDATED

Fly-by-wire system for the Tornado

The Panavia Tornado flight control system uses triplex electronic signalling and processing, with quadruplex actuation, to provide a high degree of manoeuvrability throughout the entire flight envelope and has a mechanical back-up system for emergency control though with degraded handling qualities. The fly-by-wire processing is incorporated in the Command and Stability Augmentation System (CSAS) and the automatic flight functions, which include safety critical capability such as automatic terrain-following, are integrated with the automatic flight control system. These are separately configured but very closely allied systems. The triplex CSAS components are associated with a duplex Spin Prevention and Incidence Limiting System (SPILS) which ensures that the crew can fly the aircraft to its structural and aerodynamic limits without the risk of loss of control.

Design, development and production of the systems have been a trinational venture between GEC-Marconi Avionics in the UK, BGT in Germany and Alenia in Italy.

Command stability and augmentation system

The CSAS is an analogue FBW manoeuvre demand system. It provides electrically signalled pitch, roll and yaw control and automatic stabilisation of aircraft response to pilot command or turbulence. Gain scheduling improves handling qualities and control stability over the entire flight envelope and operates in conjunction with the spin prevention and incidence limiting system. 'Carefree' manoeuvring allows the exploitation of the aircraft's full lift capability under all flight conditions without the risk of structural damage resulting from a pilot's control demand for control surface movement that would exceed the design strength of the airframe.

Individual LRUs are the CSAS pitch computer, CSAS lateral computer, CSAS control unit, pitch, roll and yaw rate gyros and pitch, roll and yaw position transmitters.

Spin prevention and incidence limiting system

SPILS is a duplex analogue system which operates in conjunction with the CSAS to achieve maximum aircraft lift in low-level flight. It limits aircraft incidence, irrespective of the pilot's demands, when maximum safe angles of attack are reached. Individual LRUs are the SPILS computer and SPILS control unit.

Automatic flight director system

The digital Autopilot and Flight Director System (AFDS) automatically controls the flight path in all modes, including terrain-following, and sends signals to the director instruments enabling the crew to monitor autopilot performance or to fly the aircraft manually. Pitch autotrim is also incorporated. The duplex self-monitoring processor configuration provides high-integrity automatic control, permitting low-altitude cruise with appropriate safety margins. The flight director remains available after most single failures. Autopilot manoeuvre demand signals are routed to the control actuators through the command and stability augmentation system. A 12-bit processor is used with 6 kbits words of stored program and 1 kbits words of data store. Typical computing speed is around 160 Kops and program cycle time is 32 ms. Individual LRUs are the two AFDS computers, AFDS control unit, autothrottle actuator and pitch and roll stick-force sensors.

Operational status

In service.

Contractors

Alenia Difesa, Avionic Systems and Equipment Division, GF-Sistemi Avionici SpA.

Bodenseewerk Gerätetechnik GmbH/BGT.

GEC-Marconi Avionics Ltd.

UPDATED

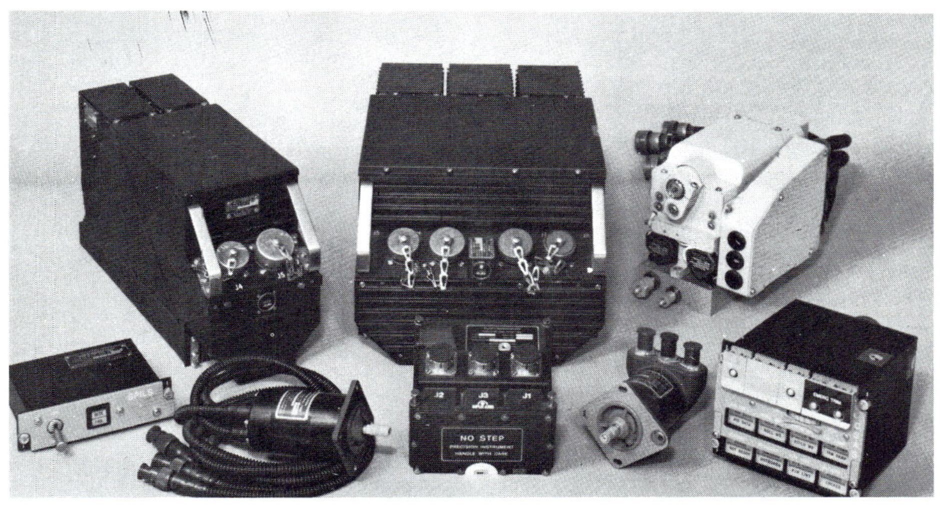

Panavia Tornado CSAS equipment includes computer units, control panels, triplex position transmitters, triple gyro packs and quadruplex first stage actuators

Panavia Tornado SPILS computer opened up to show the four-board front connector layout

---

SLS Satellite Landing System

In January 1995, Honeywell and Pelorus Navigation Systems Inc of Calgary, Canada, teamed to develop and manufacture a differential GPS (DGPS) ground reference station called the Satellite Landing System (SLS). The SLS allows for improved all-weather operations that reduce delays and operating costs while maintaining high integrity and safety. It can provide Special Cat I (SCAT I) capability to all runway ends within a 30 n mile radius, making it more cost effective than traditional landing aids that are limited to one runway end. It ensures local airport control and enhances satellite coverage and it can provide the flexibility to design approaches that minimise flight time and meet noise abatement objectives.

SLS will be Cat I and II capable with growth to Cat III, and SLS will enable variable geometry precision approaches and departures.

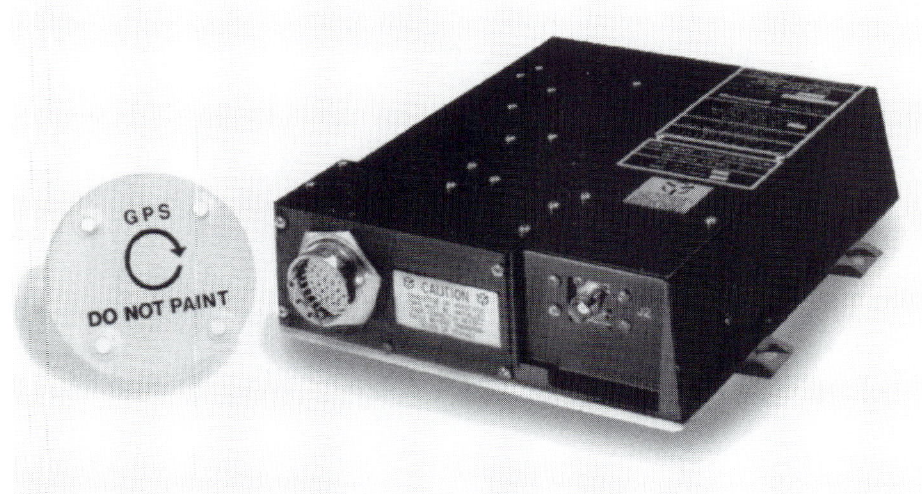

Antenna Differential Global Navigation Satellite Sensor Unit (DGNSSU) ***1997***/0001405

SLS-1000 and SLS-2000

The SLS ground station is available in two configurations, SLS-1000 and SLS-2000. Both systems comprise three major subsystems: ground reference station, Remote Satellite Measurement Units (RSMUs) and VHF (Very High Frequency) datalink transmitter.

The SLS-1000 unit is a fail-safe system that is designed to continuously perform self tests to determine its 'health.' If it detects a problem, it will notify the operator and any aircraft in the area that it is not capable of sending accurate data.

The SLS-2000 unit is a fail-operational system that is not effected by single component failures. The system operator is notified of a component failure and can call for service while the unit continues to operate.

The SLS-1000 and SLS-2000 are self-calibrating. The systems are designed with performance monitors, which eliminate the need for periodic flight checks. Both systems come with a fault-tolerant power supply and battery back up to ensure continuity of service in the event of a power outage.

Airborne equipment complement

For the forward fit market, Honeywell and Pelorus are working with industry and aircraft manufacturers to design the optimum solution for future aircraft.

In the retrofit market, most aircraft will need some combination of the following systems: Flight Management System (FMS) or SLS Controller; VHF datalink receiver; Differential Global Navigation Satellite Sensor Unit (DGNSSU); Analog Interface Unit (AIU).

Maintaining the high integrity of the system is critical to certification and safety. To do this, the Honeywell

DGNSSU, designed to ARINC 743 standards, calculates and directs the flight controls. The DO 178B level B, critical level software, ensures that the system will perform the calculations correctly.

Annunications and specific interface requirements are resolved on an aircraft-by-aircraft basis.

The FMS or SLS Controller tunes the VDL-500 to the SLS ground station datalink frequency at a given airport. The range error corrections and path points received by the VDL-500 are sent to the DGNSSU.

The DGNSSU makes the necessary corrections and calculates the approach path.

The approach path, based on SLS position information, is transmitted to the flight controls as an ILS look-alike signal. The approach is flown by the flight controls using this input.

Contractors

Honeywell Inc Business and Commuter Aviation Systems.

Pelorus Navigation Systems Inc.

VERIFIED

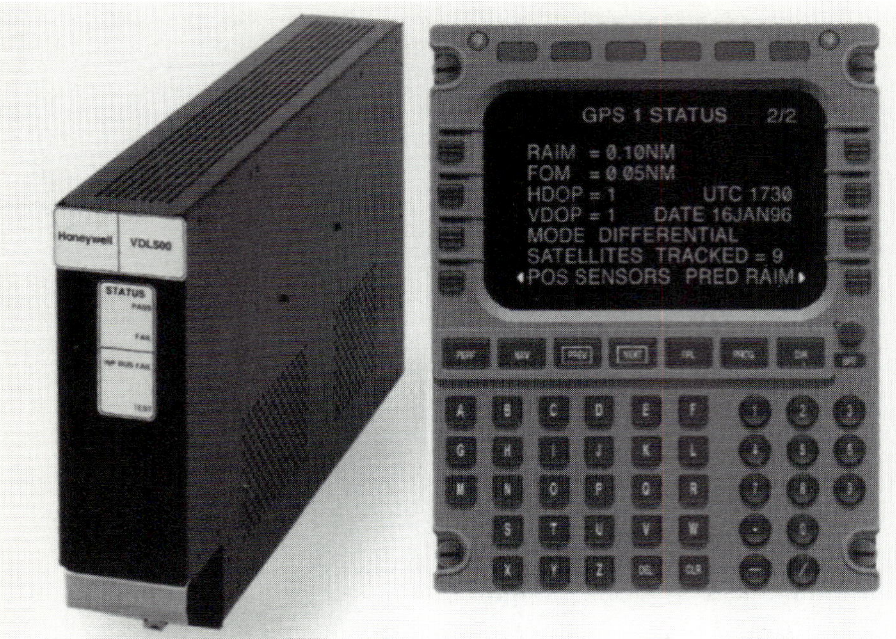

VDL-500 Fllight Management System (FMS)
1997/0001404

Utility control system for the EF 2000

The EF 2000 utility control system is based on the utilities management system which was successfully demonstrated on the Experimental Aircraft Programme (EAP). The system integrates, via an MIL-STD-1553B databus, the control and monitoring of a number of aircraft functions, including environment, cabin temperature, life support, crew escape, fuel management, fuel gauging, secondary power, hydraulics and other miscellaneous systems. The main benefits derived from the system are in weight saving and reliability. In addition, there is a greater fault tolerance and better damage resistance, system health is monitored and trends can be identified, aiding planned maintenance.

Operational status

In development for the EF 2000.

Contractors

Alenia Difesa, Avionics Systems and Equipment Division.

ENOSA.

Smiths Industries Aerospace.

VDO Luftfahrgeräte Werk Adolf Schindling GmbH.

VERIFIED

ISRAEL

Up-Front Control Panel (UFCP)

The ELOP Up-Front Control Panel (UFCP) is the pilot's input and control terminal for activating and operating the various avionic systems such as the Weapon Delivery and Navigation System (WDNS), stores management system, navigation equipment, communications equipment, IFF and so on, in advanced aircraft. It constitutes the main man/machine interface between the pilot and the aircraft computers.

The UFCP displays pilot-initiated or automatic alphanumeric information received from the aircraft avionics. Communication with the avionics is through the UFCP's digital push-buttons and information is displayed either on the UFCP or on the HUD. The UFCP provides the pilot's control terminal for the aircraft avionics systems, the pilot's input terminal for digital information and a display for alphanumeric information from the avionics systems.

The UFCP is located on the front of the pilot display unit. Electrical interface with the WDNS is via MIL-STD-1553A or B, RS-422A or as specified by the customer. A microcontroller manages the UFCP functional display, illumination, validation, analogue to digital converter and communication with the WDNS. Management is performed in real time.

Contractor

ELOP Electro-Optics Industries Ltd.

VERIFIED

ELOP Up-Front Control Panel shown fitted on an HUD

Autocommand Mabat flight control system

Autocommand Mabat is an analogue flight control system used in the IAI Kfir fighter and developed from a Sextant Avionique unit used in the Dassault Mirage III fighter. Autocommand Mabat is a pitch-only stabilisation system which alleviates undesirable transient handling characteristics (including during the transonic region of the flight envelope); improves dynamic stability with consequent benefits in target pursuit, aiming, firing and weapon release accuracy; provides stick-force/*g* scheduling; and provides attitude hold for a limited period after stick release and automatic barometric height hold (except during transonic flight).

Autocommand Mabat is a single LRU comprising three interconnected modules: an analogue processor,

Israel's two-seat Kfir TC2 has the IAI autocommand Mabat analogue flight control system

air data system and power supplies. The main unit is the processor, which uses DC operational amplifier technology. Integrated circuit AC/DC and DC/AC conversion units and self-test circuits are incorporated and the unit uses multilayer printed circuit boards and a motherboard to reduce internal wiring to a minimum.

Pitch auto-stabilisation is provided to requirements set out in MIL-F-8785, with a stick force/*g* characteristic set by the manufacturer, and pitch attitude hold in the ±20° pitch range. After releasing, the stick maximum pitch angle change is less than 5°/min. The maximum change of altitude due to a malfunction cannot exceed 33 ft during the first second following the failure and any normal acceleration produced is within a specified envelope. Self-test can be initiated by the pilot and if the test-button lamp does not light within 10 seconds this is taken as an indication of failure. System engagement/disconnection can take place anywhere within the flight envelope. Automatic disconnection will occur only if a malfunction occurs in the compensator system, the autocommand safety system, or if the number one hydraulic system disconnects and the preliminary servo changes over to the number two. Manual disconnect is achieved via several separate override or disconnect buttons, or by exerting a high stick-force.

Altitude hold is available only if pitch stabilisation has been engaged. It is usable throughout the flight envelope, except in the M0.95-1.15 transonic region. In level flight altitude errors are within ±30 ft or 0.1 per cent of altitude, or double these values at 30° roll angle and not outside a linear extension to 150 ft error at 60° roll angle. Aircraft oscillations are limited to periods in excess of 20 seconds and maximum normal acceleration is less than 0.1 *g*. The hold mode can only be engaged if normal acceleration is less than 0.5 *g*. Automatic disconnect will occur if rate of height change exceeds 2,000 ft/min, when the speed is between M0.95-1.15 or if the undercarriage is extended.

Sensors and output devices associated with the unit include a rate gyro unit, accelerometer unit, left- and right-hand compensation servos, pitch pre-servo, stick dynameter and trim motor. A range of dedicated cockpit controls and annunciators is also installed.

System performance conforms to MIL-18224C and MIL-F-8785 and electrical characteristics are to MIL-STD-704A.

Specifications

Dimensions: 306 × 198 × 195 mm
Weight: 10 kg
Power supply: 115 V AC, 400 Hz, single phase, 100 mA, 28 V DC, 1 A
Reliability: >1,400 h MTBF

Operational status

In service.

Contractor

Israel Aircraft Industries Ltd.

VERIFIED

JAPAN

AFMS controller for the SH-60J

The Automatic Flight Management System (AFMS) controller is used in the automatic flight management system installed in the Mitsubishi/Sikorsky SH-60J anti-submarine helicopter. The AFMS provides a flight management integrated display, alarms and flight management. The AFMS controller is a digital computer and provides the data processing both for the automatic flight management system and for navigation equipment management.

Operational status

In service in the Mitsubishi/Sikorsky SH-60J helicopter.

Contractor

Japan Aviation Electronics Industry Ltd.

VERIFIED

Automatic flight control system for the F-1

The automatic flight control system installed on the Mitsubishi F-1 fighter uses a digital computer in a single system. The system monitors attitude, azimuth and altitude hold, reducing the pilot's workload and contributing to flight safety and mission performance. It includes fail-safe control.

Operational status

In service in the Mitsubishi F-1 aircraft.

Contractor

Japan Aviation Electronics Industry Ltd.

VERIFIED

AFCS for helicopters

Production of the Automatic Flight Control System (AFCS) for helicopters has started, as well as for fixed-wing aircraft. The system consists of the FCC-20 Digital Flight Control Computer (FCC), DCP-20 Flight Control Panel and ADC-20 Air Data Computer. The AFCS is a single channel system, but the FCC in the system features not only digital auto-pilot and digital SCAS functions, but also analogue backup SCAS functions and a dual PSU system, which contribute to flight safety.

Operational status

In production.

Contractor

Tokyo Aircraft Instrument Co, Ltd.

VERIFIED

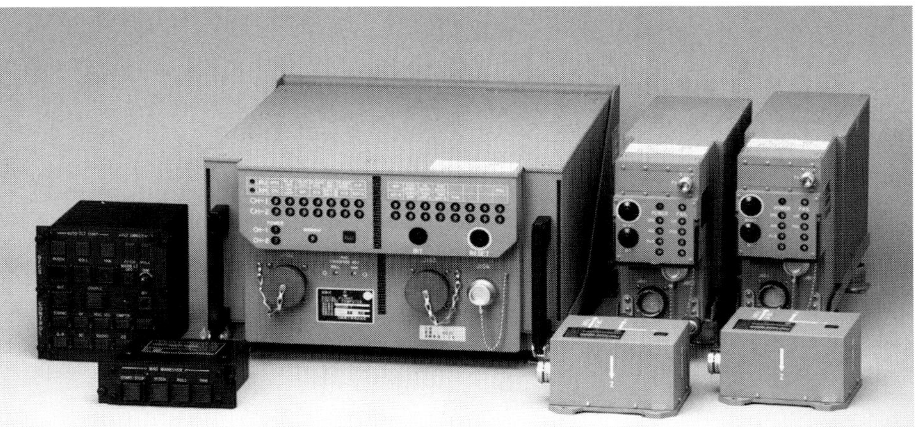

The Tokyo Aircraft Instrument Co AP-120 DFCS

AP-120 DFCS

AP-120 Digital Flight Control System (DFCS) is a high-grade comprehensive flight control system which consists of Digital Flight Control Computer (DFCC), two altitude controllers, two accelerometers, main and sub control panels. The DFCC, as a core unit, has both autopilot and flight director functions. It features dual-redundant fail-operative/fail-passive capability and is inertially smoothed to give high-accuracy stable altitude-hold capability. High maintainability is ensured through extensive use of an In-Flight Performance Monitor (IFPM), Built-In Tester (BIT) and non-volatile maintenance memories, and the use of selected high-reliability parts and advanced thermal design techniques. The system can be expanded to provide a MIL-STD-1553B databus interface and other enhancements.

Operational status

In production and service in the Lockheed/Kawasaki P-3C aircraft and its derivatives.

Contractor

Tokyo Aircraft Instrument Co, Ltd.

VERIFIED

Four-axis sidestick controller

The four-axis sidestick controller has been developed by Tokyo Aircraft Instrument Company in association with Kawasaki Heavy Industries. It consists of a sensor unit and a control box. The system has quadruple redundancy in each axis.

The sensor is a force deflection right-side stick, having four-bridge solid-state strain gauges, which provide electric signals for pitch, roll, yaw and CP to a flight control computer.

Operational status

Under evaluation in the Kawasaki BK-117 fly-by-wire experimental helicopter.

Contractor

Tokyo Aircraft Instrument Co, Ltd.

VERIFIED

The four-axis sidestick controller is under evaluation in a BK-117 fly-by-wire helicopter **1995**

POLAND

AP-2 digital autopilot

The AP-2 digital autopilot was designed and produced according to FAR23 regulations and the AS402A document. It is a two-axis flight control and flight director system for light aircraft.

The autopilot provides pitch and roll attitude stabilisation and control, altitude stabilisation, heading stabilisation, interception and tracking of VOR radials and localiser and glide slope beam interception and tracking and localiser back course mode for ILS approaches. Two basic modes of operation are available: automatic flight control for basic autopilot functions and flight director mode.

The AP-2 includes the main processing unit, selector panel, servo amplifiers, optional diagnostic console and optional optical flight director indicator.

The device may be equipped with two different operators' consoles. The basic console, the autopilot selector panel, enables the pilot to engage or disengage the autopilot, select any mode of operation and receive signals about correct or incorrect autopilot operation. The optional console provides additional diagnostic functions.

The AP-2 autopilot generates analogue signals for the flight director indicator and digital signals for the external mode annunciator and other onboard instruments such as the flight recorder.

Specifications

Dimensions:
(flight computer) 160 × 75 × 280 mm
(selector panel) 160 × 75 × 40 mm
Weight: 9.7 kg total typical
Power supply: 27.7 V DC
(flight director) 0.8 A
(flight control) 3.5 A

Operational status

The present version of the AP-2 is adjusted for Piper Seneca II aircraft and Edo-Aire flight instruments.

Contractor

ATM Inc.

VERIFIED

RUSSIAN FEDERATION AND ASSOCIATED STATES (CIS)

ASShU-334 fly-by-wire flight control system

The ASShU-334 fly-by-wire flight control system is a digital and analogue system designed to provide stability and control for regional aircraft during manual, automatic and override manual flight conditions. System components are as follows:

stability and control computers	4 units
analogue units	2 units
pre-flight maintenance panel	1 unit
linear acceleration sensors	6 units
angular rate sensors	18 units
position sensors	7 units
transformer units	3 units

Specifications

No of redundant channels: 6
Weight: 90 kg
Power: 500 W

Operational status

Fitted to the An-334 twin turbofan medium-range airlifter aircraft.

Contractor

AviaPribor.

NEW ENTRY

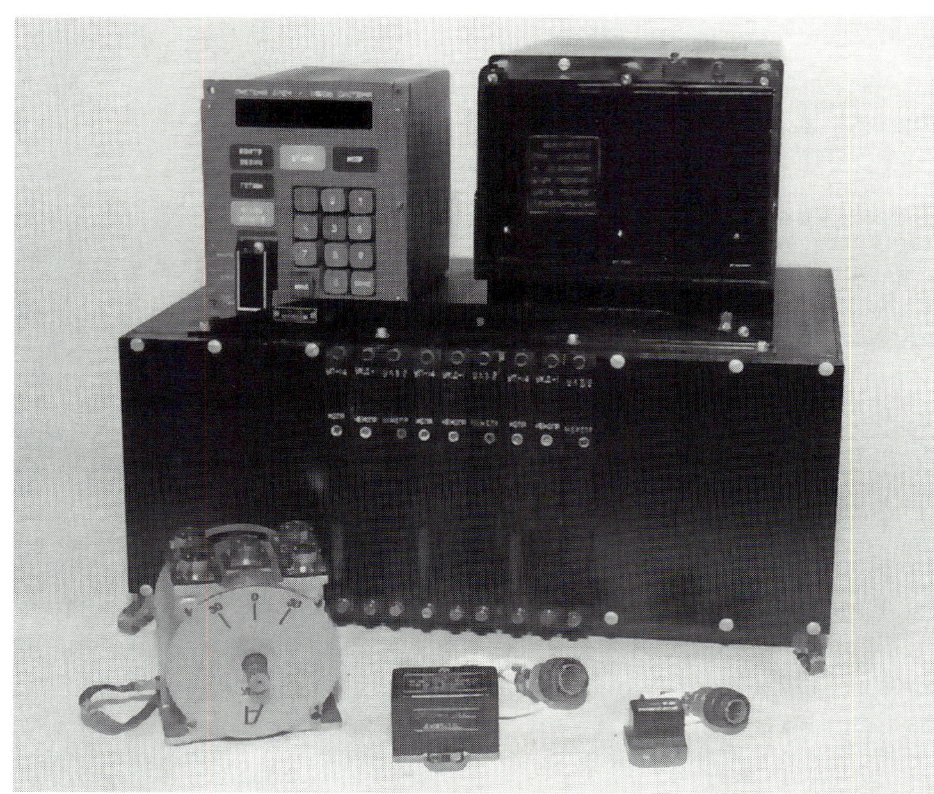

ASShU-334 fly-by-wire flight control system **1998**/0018242

ASUU-96 automatic control and stability augmentation system

The ASUU-96 control and stability augmentation system is a digital-analogue system, designed for the Il-96 aircraft; it comprises the following units:

control and stability computer unit	5 units
control monitor unit	4 units
actuator unit	1 unit
pre-flight maintenance panel	1 unit
linear accelerator transducer	4 units
triplex position pickoff	8 units
angular rate sensor	12 units

Specifications

No of redundant channels: 4
Weight: 200 kg
Power: 115 V AC, 400 Hz, 2,000 VA

Operational status

Fitted to Il-96 wide-bodied airliner.

Contractor

AviaPribor.

NEW ENTRY

ASUU-96 automatic control and stability augmentation system **1998**/0018241

EDSU-77 fly-by-wire flight control system

The EDSU-77 is designed to provide aircraft stability and control in pitch, yaw and roll for the An-70 tactical transport aircraft in manual, automatic and override control modes of flight operation; it comprises the following units:

stability and control computers	4 units
analogue interfaces	2 units
maintenance panel	1 unit
linear acceleration sensors	12 units
angular rate sensors (unified)	18 units
position sensor (triple)	20 units
emergency trim control	1 unit
electromechanical actuators and sensors as required	

Specifications
No of redundant channels 6
Weight: 250 kg
Power: 115 V AC, 400 Hz, 1,000 VA

Operational status
Fitted to An-70 propfan medium range, wide bodied, tactical transport aircraft.

Contractor
AviaPribor.

NEW ENTRY

EDSU-77 fly-by-wire control system for the An-70 aircraft **1998**/0018228

Contractor
AviaPribor.

NEW ENTRY

SVS-V1 helicopter air data computer system

The SVS-V1 air data computer is designed for measurement of altitude and speed data on helicopters; it comprises two line-replaceable units: computer, airspeed vector transmitter.

Specifications
Airspeed longitudinal component: –90 to +450 km/h, (±3.5-5 km/h)
Airspeed lateral component: 0-90 km/h, (±3.5 km/h)
Indicated airspeed: 30-450 km/h (±2-6 km/h)
Vertical baroinertial speed: –30 to +30 m/s, ±0.3 ms
Pressure equivalent altitude: –500 to +7,000 m, ±6 m
Outside air temperature: –60 to +60°C, ±2°C
Weight: 6 kg
Power: 30 VA

SVS-V1 helicopter air data computer system
1998/0018240

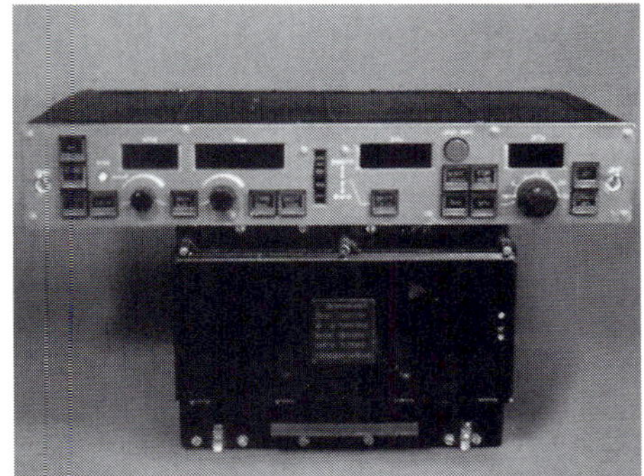

VSUP-85 flight control computer system
1998/0018239

VSUP-85 flight control computer system

The VSUP-85 flight control computer system is designed to generate control data for all phases of flight of modern transport aircraft types; it comprises two types of line-replaceable units: flight control computer (3 units) and cockpit control unit (1 unit).

Specifications
Compliance: NLGS, DO-160
Weight:
(Tu-204) 51.5 kg
(Il-96) 36 kg
Power: 115 V AC, 400 Hz, 525 VA
27 V DC, 30 W
36 V AC, 400 Hz, 4.5 VA
6 V AC, 400 Hz, 25 VA
MTBF:
(computer) 5,000 h
(control unit) 6,000 h

Operational status
Fitted to the Tu-204 twin turbofan medium-range airliner, and the Il-96 wide-bodied airliner. The -85 element of the designator indicates that the original development contract was awarded in 1985.

Contractor
AviaPribor.

NEW ENTRY

VSUPT-334 flight and thrust control computer system for Tu-334

The VSUPT-334 flight and thrust control computer system performs both flight control and thrust control functions for the Tu-334 medium-range airlifter and its twin turbofan ZMKB Progress D-436T1 engines; it comprises the following line-replaceable units: flight/thrust computer (3 units), thrust unit, cockpit control unit, thrust actuator.

Specifications
Compliance: NLGS, DO-160
Weight: 67 kg
Power: 115 V AC, 400 Hz, 525 VA
27 V DC, 400 W
36 V AC, 400 Hz 4.5 VA
6 V AC, 25 VA

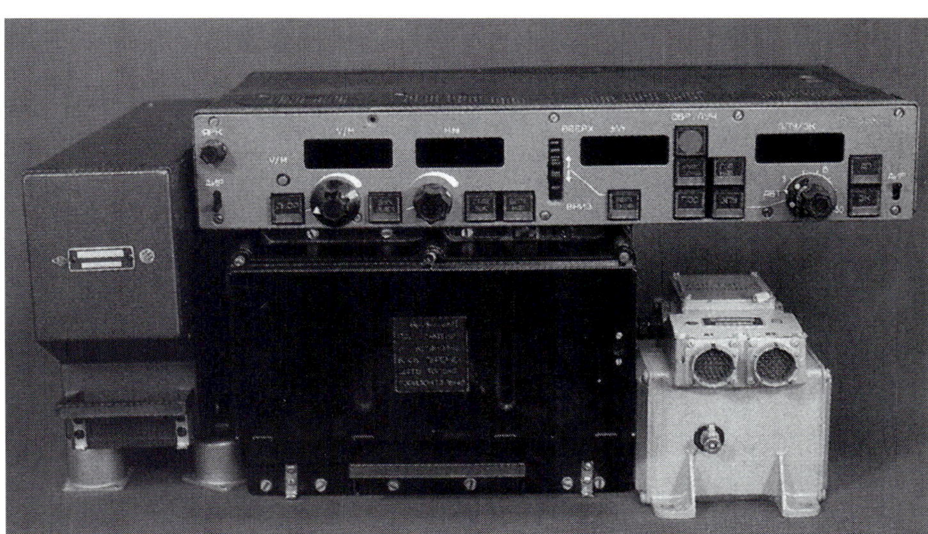

VSUPT-334 flight and thrust control computer system for Tu-334 **1998**/0018238

MTBF:
(computer) 5,000 h
(cockpit control unit) 6,000 h
(thrust actuator) 5,000 h

Operational status

Fitted to the Tu-334 twin turbofan medium-range airlifter.

Contractor

AviaPribor.

NEW ENTRY

VSUT-85 thrust control computer system

The VSUT-85 thrust control computer system is designed to provide automatic control/hold of indicated airspeed or Mach number by controlling power plant thrust on Tu-204 and Il-96 aircraft; it comprises the following line-replaceable units: thrust control computer (2 units), thrust control, thrust actuator.

Specifications

Compliance: NLGS, DO-160
Weight:
(Tu-204) 34 kg
(Il-96) 38 kg
Power: 115 V AC, 400 Hz, 325 VA
(Tu-204) 27 V DC, 432 W
(Il-96) 486 W
MTBF:
(computer BVUT) 5,000 h
(thrust actuator PRT-204) 5,000 h
(thrust control panel PUT-3) 8,000 h

Operational status

Fitted to the Tu-204 twin turbofan medium-range airlifter, and Il-96 four turbofan, wide-bodied, airliner.

VSUT-85 thrust control computer system
1998/0018237

The -85 element of the designator indicates that initial development contract award was in 1985.

Contractor

AviaPribor.

NEW ENTRY

Aircraft Systems Control System (ASCS) — Electronic Flight Engineer

GosNIIAS has developed the Electronic Flight Engineer: Aircraft Systems Control System (ASCS), as an outgrowth of its work with Collins and Smith Industries on the following elements of the Il-96 M/T avionics development programme: the Integrated Display System (IDS), Flight Management Computer System (FMCS), Automatic Flight Control System (AFCS), Thrust Management Computer (TMC) and Central Maintenance Computer System (CMCS); and on experience gained in production of the Collins Traffic alert and Collision Avoidance System (TCAS).

The Electronic Flight Engineer is intended for aircraft systems automation and serviceability/maintenance enhancement. The automation element permits reduction of the flight crew to two.

The Electronic Flight Engineer automates control of over 20 aircraft systems, including: hydraulics, electrical power, air conditioning, pressure regulation and other systems. The system comprises three computer units (ASCC) and eight MFUs.

Operational status

Certification by the RF Aviaregister and the US FAA of all GosNIIAS-developed Il-96T systems was scheduled for March 1998.

Contractor

GosNIIAS State Research Institute of Aviation Systems.

NEW ENTRY

Antonov-70 airborne information system

The airborne information system designed for the Antonov-70 medium-range transport aircraft comprises: centralised firmware, unified information acquisition, processing and output to the flight crew and ground-based maintenance personnel of data on the status of the aircraft functional systems.

The airborne information system is designed to reduce the cost for maintenance and to permit aircraft operation at poorly equipped airfields for 30 days or 200 flight hours.

The airborne information system comprises two subsystems: the Information and Warning Subsystem (IWS), and the Monitoring and Maintenance Subsystem (MMS).

The IWS provides:

(1) reception of parametric and warning information from functional systems via digital data links and equipment sensors;
(2) preprocessing of received information;
(3) continuous monitoring of aircraft functional systems' technical status and crew actions;
(4) generation and transfer of parametric information and monitoring results to the electronic display system, monitoring and maintenance subsystem and emergency warning system.

The MMS provides:

(1) reception of parametric and warning information, generated by IWS and functional systems;
(2) information processing;
(3) presentation of parametric information and processing results on the displays;
(4) recording of information in operation recorder in automatic and manual modes;
(5) output of information to the printer in automatic and manual modes;
(6) manual input of changed-over constant and other auxiliary information from the control panel;
(7) generation and record of monitoring results in non-volatile memory;
(8) presentation of information stored in non-volatile memory and operation recorder on the MMS display;
(9) initiation of aircraft systems' test through the MMS control panel.

'MONITORING' is the basic operating mode of the system, in which the monitoring of the aircraft systems' technical status and crew action, as well as the record of parametric information and monitoring results are provided.

Flight crew and maintenance personnel receive the following information:

(1) operating information on the electronic display system, emergency warning system's panel and MMS screen (it is used by the crew in-flight and ground-based services when controlling the monitored systems and for evaluating their status);
(2) report information, recorded on paper tape (it is used on the ground for post-flight evaluation of the monitored systems' technical status and the crew actions, as well as for detected failure unit removal);
(3) report information, recorded in the operation recorder, with pre- and post-histories of errors in the aircraft systems, and information about the aircraft systems failures, stored in non-volatile memory.

'TESTS' is a mode, in which the IMS and MMS technical status monitoring and their failure location up to a line-replaceable unit are performed.

'OR OUTPUT' is a mode, in which the readout and analysis of information, recorded in the Operation Recorder (OR), are performed directly on board the aircraft. The information processing and its subsequent displaying and printing in the form of digital information and plots are envisaged in this mode.

'FAILURES' is a mode, in which the displaying and printing of the reference information, stored in non-volatile memory, are produced on the ground.

Specifications

Information and warning subsystem:
(input analogue signals) 560
(input digital signals) 800
(ARINC-429 input channels) 20
(No of types of channels) (analogue) 12; (digital) 2
(multiplex exchange channels) 8
(total no of input parameters) 8,000
(output digital signals, +27 V) 32
(No of generated messages) 2,000
Presentation of parametric and warning information on 5 displays of the electronic display system
(ROM capacity) 300 KByte
(main memory capacity) 80 KByte
Monitoring and maintenance subsystem:
Ground-based display:
(screen diagonal) 23 cm
(No of lines) 28
(No of displayed character types) 121
Operation recorder:
(recorded data capacity) 300 Mbyte
(frame capacity, 16-bit word) 1,024
(continuous recording time) 24 h
accelerated playback at readout
(capacity of information, recorded in non-volatile memory) 100 Kbyte
(No of input parameters) 8,000
(No of generated messages) 8,000
(ROM capacity) 600 KByte
(Main memory capacity) 40 KByte

Contractor

Leninetz Holding Company.

NEW ENTRY

Antonov-70 aircraft
1998/0018227

EFIS-85 electronic flight instrument system

EFIS-85 is an integrated navigation, flight information and cathode ray tube display system. It comprises four displays, three symbol generators and two control panels. The panels are interchangeable and may function as Pilot Flight Displays (PFD) or Navigation Displays (ND).

PFDs display attitude, altitude and speed information, flight director and heading information; it also displays glide-slope and localiser data, and information from automated systems.

NDs display navigation data, radio navigation data, weather radar and TCAS data.

Each symbol generator can support the operation of both ND and PFD for one pilot or both pilots' displays can be fed from the same symbol generator.

All units have built-in-test facilities, together with redundancy of image generation and display options.

Specifications

Display
Dimensions: 203 × 230 × 356 mm
Weight: 17.3 kg
Power: 250 W

Symbol generator
Dimensions: 194 × 127 × 324 mm
Weight: 7.0 kg
Power: 120 W

Control panel
Dimensions: 146 × 176 × 200 mm
Weight: 2.5 kg
Power: 25 W

Operational status

Development contract award in 1985. Fitted to Il-96-300 and Tu-204 airliners.

Contractors

NIIAO Institute of Aircraft Equipment.
Ulyanovsk Instrument Design Office.

NEW ENTRY

FCS-85 flight control system

FCS-85 is part of a standard digital avionics system designed for aircraft automatic control during flight and for director control during take-off and approach.

In conjunction with other avionics control systems FCS-85 provides control functions including:

(1) in conjunction with TCS-85 thrust control system, automatic control in altitude and heading, automatic hold of barometric altitude;
(2) with the control column, control of pitch, heading and roll;
(3) automatic approach and automatic landing in compliance with ICAO Cat IIIA requirements using ILS and MLS radio beacons;
(4) automatic approach to decision height in compliance with ICAO Cat II requirements using ILS, MLS, Landing System-50 (LS-50) radio beacons;
(5) automatic director approach to decision height in compliance with ICAO Cat I requirements using ILS, MLS, LS-50, SRNS radio beacons;
(6) automatic monitoring of FCS operation;
(7) prevention of exceedance conditions.

FCS-85 comprises the following LRUs: flight control computers (FCC-1) 3 linked units, each 8 MCU, ARINC 429 compatible; control panel (CP-56) 1 unit.

Specifications

FCC-1
Power: 115 V AC, 400 Hz, 150 VA; 36 V AC, 400 Hz, 1.5 VA; 27 V DC, 30 W
Weight: 12 kg
MTBF: 5,000 h

CP-56
Power: 115 V AC, 400 Hz, 75 VA; 6 V AC, 400 Hz, 12 VA; 27 V DC, 1 W
Weight: 8 kg
MTBF: 6,000 h

Operational status

Development contract award in 1985. Fitted to Il-96-300 and Tu-204 airliners.

Contractors

NIIAO Institute of Aircraft Equipment.
Moscow Institute of Electromechanics and Automatics.

NEW ENTRY

FILS Fault Isolation and Localisation System

FILS is the top level, of three levels, of onboard BITE and fault isolation system. It integrates the BITE activities and data management and storage of all major flight management systems, such as EFIS, FCS, TCS.

FILS is designed to: process fault diagnosis and isolation; output fault data on the databus and displays and ground test facilities; store data on faults for up to 10 flights and 30 faults per flight in non-volatile memory; input data on faults into aircraft utility systems.

FILS is a single LRU, with alphanumeric display, mounted in the flight engineer position, that complies with ARINC 604 and 734.

Specifications

ROM capacity: 40 kwords
Non-volatile memory: 16 kwords
Inputs; up to 25 channels (ARINC 429)
Outputs: up to 6 channels (ARINC 429); up to 12 discrete instructions
Weight: 10 kg
Power: 115 V DC, 400 Hz, 60 VA
MTBF: 5,000 h

Operational status

Development contract award in 1985. Fitted to Il-96-300 and Tu-204 airliners.

Contrctors

NIIAO Institute of Aircraft Equipment.

NEW ENTRY

FMS-85 flight management system

As part of the standard avionics system for the Il-96-300 and Tu-204 aircraft, FMS-85 provides the following function: generation of information and control signals for 4-D navigation and control of en route flight, SID/STAR procedures, data on navigation aids, optimisation of fuel use.

The FMS-85 comprises the following LRUs: TsVM 80-40001 flight management computer; POOI-85M multifunction control and display unit (MCDU). The MCDU is used to input navigation and flight control data to the central computer and their output to the EFIS; the control of the FMS-85 modes of operation, and control of the I42-1S long-range navigation and satellite navigation systems.

Specifications

RAM capacity: 19 kwords
Non-volatile memory capacity: 236 kwords
ROM capacity: 128 kwords
Inputs; 72 channels; 32 discretes
Outputs: 10 channels; 8 discretes
Power: 115 V AC, 400 Hz, 190 VA; 27 V DC, 2.5 W
Weight: 15 kg

Operational status

Development contract award in 1985. Fitted to Il-96-300 and Tu-204 airliners.

Contractors

NIIAO Institute of Aircraft Equipment.

NEW ENTRY

705-6 attitude heading reference system

The 705-6 attitude heading reference system is part of the inertial navigation system of the MiG-29. It provides the following data outputs: gyroscopic heading, pitch and roll angles, and absolute orthogonal accelerations.

Specifications

Normal erection: 15 min
Fast erection: 3 min
Errors:
(normal erection) 0.15°
(fast erection) 0.7°

Operational status

Fitted to MiG-29 aircraft.

Contractor

Ramensky Instrument Engineering Plant.

NEW ENTRY

705-6 attitude reference system
***1998**/0018236*

STR7-4 fuel quantity and flow metering system

The STR7-4 fuel quantity and flow metering system fulfils the following purposes: computing aircraft total fuel mass during refuelling and in flight, computing and indication of fuel mass transferred in flight, indication of fuel mass in the aircraft tanks, and low fuel-remaining warnings, indication and fuel mass flow, control of fuel usage, fuel system BITE.

System line-replaceable units include: nine capacitance-type fuel quantity sensors; three fuel flow transmitters; 13 level switches; six temperature sensors; two electromechanical indicators; two control and monitoring panels; four electronic converter units.

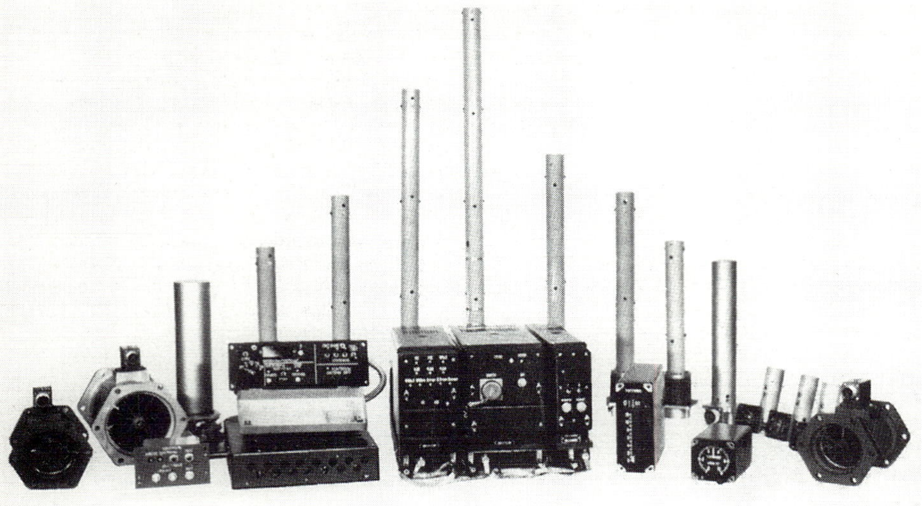

STR7-4 fuel quantity and flow metering system ***1998***/0018235

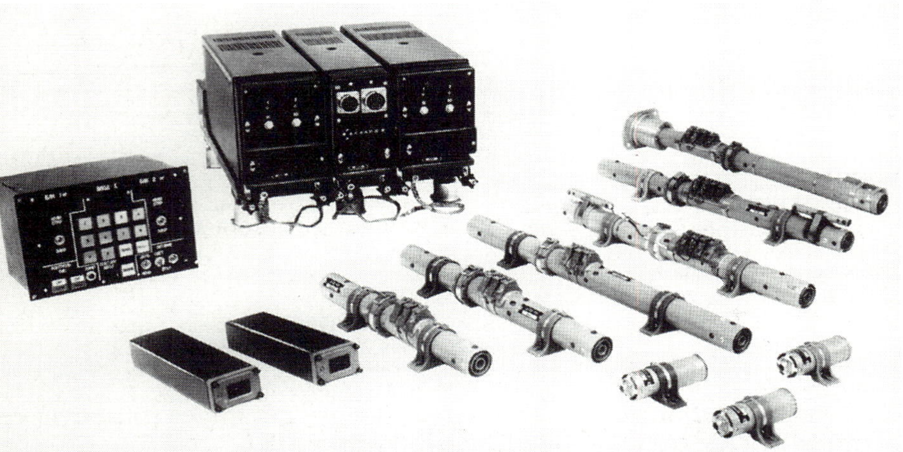

SUIT8-10 fuel management and indicating system ***1998***/0018234

Specifications

Fuel mass error: ±2%
Fuel level error: ±5 mm
Data exchange: ARINC 429
Unit design: ARINC 404
Mass:
(excl transmitters) 25 kg
(incl transmitters) 38 kg
Power: 27 V DC; 115 V AC, 400 Hz

Contractor

TekhPribor State Enterprise.

NEW ENTRY

SUIT8-10 fuel management and indicating system

The SUIT8-10 fuel management and indicating system provides the following: measurement and indication of fuel mass in aircraft tanks, and total aircraft fuel mass, control of fuelling and fuel use, warning of critical fuel reserves and unbalance.

Composition of the system for the Il-114 aircraft is as follows: six capacitance-type fuel quantity transmitters; six fuel quantity compensator transmitters; six level switches; three electronic converter units; two indicators; one control panel.

Specifications

Fuel mass error: ±1.5%
Fuel level error: ±5 mm
Data exchange: ARINC 429
Unit design: ARINC 404
Mass:
(excl transmitters) 20 kg
(incl transmitters) 26 kg
Power: 27 V DC; 115 V AC, 400 Hz

Contractor

TekhPribor State Enterprise.

NEW ENTRY

UNITED KINGDOM

Active noise control

GEC-Marconi Avionics has developed an active noise control system which lowers the noise within the aircraft cabin to provide a reduction in the fatigue level for the cabin staff and an improvement in the comfort level for the passengers. Primarily targeted at turboprop aircraft, it can be fitted to any aircraft where tonal noise is considered to be intrusive.

All major tonal noise sources present within a particular spectrum are identified and monitored via an array of microphones discretely placed within the aircraft trim and connected to a central controller. Antiphase signals are then calculated using powerful Digital Signal Processors which then control the tonal noise present in the cabin via a set of loudspeakers also discretely placed in the aircraft trim.

GEC-Marconi Avionics active noise control ***1997***/0001409

Flight trials have taken place on a number of aircraft including the Jetstream 41, for which the system has now been selected and is being actively marketed by both Jetstream and GEC-Marconi Avionics. The system can either be specified as an initial fit or retrofitted with equal simplicity.

Trials have also been conducted on the Lockheed C-130 military transport where significant reductions in noise level were demonstrated in both the cockpit and cargo bay areas.

Operational status

System developed to production standard and selected for Jetstream 41.

Contractor

GEC-Marconi Avionics Ltd.

VERIFIED

Automatic Flight Control System (AFCS) for the Lynx helicopter

GEC-Marconi Avionics has produced several variants of the AFCS for the Lynx helicopter. The auto-stabiliser provides attitude stabilisation in the pitch and roll axes with yaw stabilisation in heading hold. These stabilisation demands are fed to limited authority series actuators which drive the main powered flying controls. The autostabiliser is active throughout the flight envelope and operational during autopilot flying.

All autostabilisation channels are fully duplicated with monitoring facilities to indicate discrepancy between lanes. Additional pitch axis stability is provided by the inclusion of a Collective Acceleration Control (CAC) which senses normal aircraft acceleration and applies a collective pitch demand to compensate for the pitch rate divergence and instability at high forward speed and aft Centre of Gravity.

The autopilot modes of Barometric Altitude Hold, Radio Altitude, Radio Altitude Acquire and Heading Hold operate in their respective axes individually or in combination. Automatic transition and cable hold modes operate simultaneously in pitch, roll and collective axes. The autopilot is simplex in the pitch, roll and yaw axes but is duplicated In the collective axis. Control signals are fed into the limited authority actuators and also the full authority parallel actuators to prevent series actuator saturation during autopilot manoeuvres.

Operational status

In service with the British Army since 1975, the Automatic Flight Control System for the Lynx helicopter has been delivered to 10 different users and in all, over 600 systems have been sold.

Contractor

GEC-Marconi Avionics Ltd.

VERIFIED

The GEC-Marconi Avionics flight control system for the Lynx helicopter includes computer, controller and drive units ***1997***/0001412

Digital Flight Control System (DFCS) for the F-14

A digital version of the existing Flight Control System (FCS) has been developed by GEC-Marconi Avionics for the F-14 Tomcat under the US Foreign Comparative Test Program.

This new digital system will replace the analogue version in form, fit and function but with an improved set of high angle of attack and power approach configuration flight control laws combined with a redundancy management scheme which provides a fail operational/fail safe capability.

The control laws for the power approach configuration are designed so that there is virtually no requirement for the pilot to provide co-ordinating rudder pedal inputs in response to lateral stick deflections and the redundancy management scheme has the aim of providing a near fail operational architecture.

Each of the three DFCCs are of similar design but each contains specific aircraft interface and software functions appropriate to its function. The DFCCs fulfil the existing FCS functions and contain growth provision to accommodate additional specified functions, if required later.

Operational status

On March 28, 1996, the US Navy and GEC-Marconi Avionics signed a not to exceed price for the first phase of the contract to complete the development and manufacture of 93 ship-sets of digital flight control computers and control panels. Further phases of the contract are expected to cover the manufacture of an additional 151 ship-sets to cover other variants of the F-14 aircraft and also provide US support for the programme.

Evaluation and flight test programme continues.

Contractor

GEC-Marconi Avionics Ltd.

VERIFIED

Digital flight computer for the F-14 ***1997***/0001410

Fly-by-wire control stick assemblies

GEC-Marconi Avionics has produced high-integrity position sensors for a wide range of aircraft including the pilot's stick sensors and the rudder pedal position sensors for the Tornado aircraft; pilot position sensors for the AM-X; and 'relais' jack sensors for the Concorde AFCS. This range of sensors has now been extended to include fly-by-wire passive control sticks for the EF 2000 and F-22 in centre and side console configurations respectively.

F-22 side-stick controller ***1997***/0001413

EF 2000 centre stick controller
The 'passive' unit utilises springs and dampers to provide feel to the pilot and contains high integrity electrical position sensors. They are configured as two axis displacement units utilising multi-redundant position sensing to provide high integrity input demands to the flight control system. No mechanical outputs are provided from these passive units and the force and damping characteristics are fixed. Mass balancing can also be specified which minimises stick movement caused by aircraft acceleration (see also International Section for details of consortium partners).

F-22 side stick controller
The modular construction provides easy maintenance and also enables the force/displacement characteristics of the units to be tailored to meet the particular aircraft requirements. All of the units are high integrity, flight safety critical items manufactured to the highest standards using class 1 components and manufacturing techniques. Careful choice of materials ensures the strength-to-weight ratio is optimised to meet aircraft weight constraints, provide a robustness capable of withstanding an applied pilot load in excess of 200 lb and still give reliable precision manoeuvring throughout the flight envelope and life of the aircraft.

Further developments
A design currently under development is an electrically back-driven active pilot controller which will allow adjustable spring rates, breakout forces, detents and damping ratios to provide situational awareness to the pilots at all times. With feel characteristics capable of being tailored to suit individual applications, it will also permit electrical connection between two sticks such that if one stick is moved, the other will follow.

Active throttle units using the same technology and possessing a high degree of commonality with the active stick unit are also under development.

In aircraft with a requirement for dual sticks, such as a civil aircraft, helicopter or military trainer, and fitted with a full-time fly-by-wire system, linked active control sticks can be provided, packaged to suit the individual requirements.

Operational status
Stick Assemblies are being manufactured and supplied as part of the EF 2000 and the F-22 programmes.

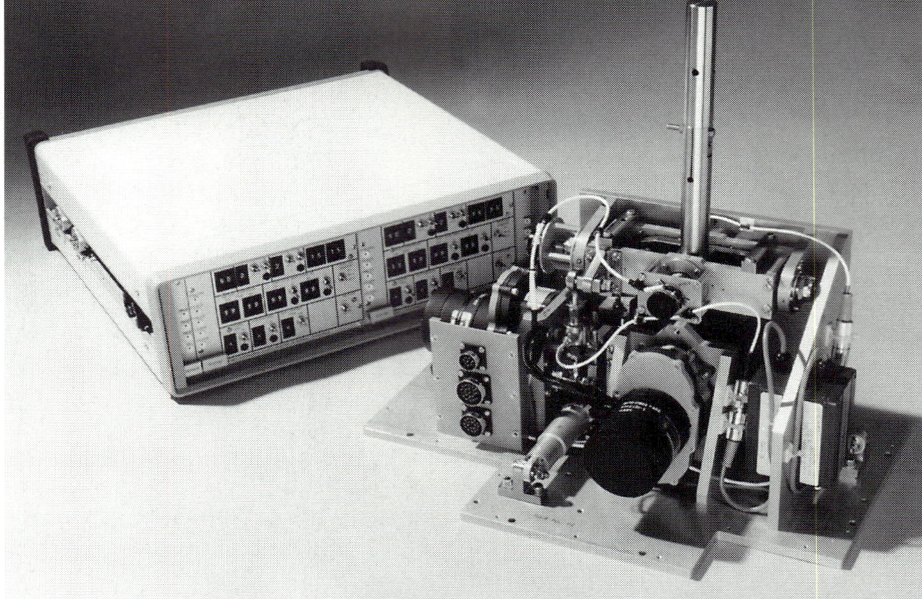

GEC-Marconi Avionics 2-axis active stick demonstrator ***1997***/0001414

The most recent development is for the supply of active stick units for the US Joint Strike Fighter programmes for both Lockheed Martin and Boeing.

Contractor
GEC-Marconi Avionics Ltd.

UPDATED

Primary Flight Computers (PFCs) for the Boeing 777

The three Primary Flight Computers (PFCs) form the core element of the Fly-By-Wire (FBW) system for the Boeing 777. The complete Primary Flight Control System (PFCS) provides control in all three axes together with protection and compensation facilities.

Primary flight computer for Boeing 777 ***1997***/0001415

The FBW system provides full time control in pitch, roll and yaw axes with pilot commands being input to the PFCs via a triplex ARINC 629 bus and Actuator Control Electronics (ACE) units. Pilot commands are provided from conventional control column and pedal inputs via multiple position transducers to the ACEs which convert the signals into digital form for transmission to the PFCs.

The PFCs compute the pilot commands for control of a single tabbed rudder surface, two elevators, a single all moving tailplane, an outboard aileron, a flaperon and seven spoiler surfaces on each wing. These computed commands are then routed via the ACEs to the appropriate electrohydraulic actuator. The computation also implements the C* control laws, together with the protection and compensation facilities for ease of crew workload.

All three primary flight computers are identical and contain three dissimilar computing lanes. The secondary redundancy management of these lanes allows fault tolerant operation in the event of failures, Within each primary flight computer, each lane uses a different 32-bit microprocessor. Three different compilers are used to convert the Ada high order software design into the microprocessor's object code. This level of dissimilarity provides protection against residual compiler and processor errors.

Extensive use has been made of ASICs ensuring a low component count and high reliability. This high reliability, coupled with the fault tolerant architecture, provides benefits to the operator through reduced life cycle costs, improved maintainability and increased system availability.

Operational status
In service, with over 400 computers delivered for Boeing 777 aircraft.

Contractor
GEC-Marconi Avionics Ltd.

UPDATED

N-250 Flight Management System (FMS)

The Flight Management System (FMS) for the IPTN N-250 aircraft provides all the control and monitoring functions for the hydraulically driven flight control system for the N-250 aircraft. It provides full authority fly-by-wire system control for the elevators, ailerons, rudder, spoilers and flaps. Mechanical back-up, by way of a standby actuation system, is also provided to control the critical elevator and aileron surfaces.

A control/monitor function is used in the N-250 FMS. Each lane uses a different processor board: an Intel 80C196 processor and a Motorola 6802 processor, and associated dissimilar software. This encourages improved system integrity and graceful degradation, without reducing system performance, and reduces common mode failures.

Nine identical Electronic Control Units (ECUs) control the five different types of aircraft surfaces. A master pin coding arrangement allows the ECU automatically to lock into the appropriate software program for the system to which it is connected. Each ECU transmits signals such as surface position, system status and fault indication to the EICAS system via an ARINC 429 databus. These compact modules are totally interchangeable between the different actuator subsystems, encouraging a low ECU spares level, low maintenance and improved aircraft despatch rates.

The N-250 FMS is a lightweight, easy to install, expandable system. It can be used on aircraft with medium to large technology content.

Specifications
Dimensions: ARINC 600 1 MCU
Weight: 1.8 kg (max)

Operational status
In development.

Contractor
Lucas Varity Aerospace.

VERIFIED

Racal Avionics Management System (RAMS)

The RAMS is a family of avionics management systems that can be configured to meet the operational requirements of military and commercial operators for helicopters and fixed-wing aircraft. The purpose of the equipment is to reduce the aircrew workload and so enhance flight safety and the chance of mission success.

The RAMS family starts with the basic RAMS 1000 navigation management system. The RAMS 3000 is similar, but configured around a MIL-STD-1553B databus. The RAMS 2000 is the non-MIL-STD-1553B equivalent of the RAMS 4000.

The RAMS family has a flexible hardware and software architecture for the user-friendly control and display of avionics systems, as well as accomplishing a variety of interfacing and processing tasks. Simplex and duplex system configurations can be provided, depending on the level of redundancy, interfacing and processing required. Management of the Communication, Navigation and Identification (CNI) subsystems can be provided, as well as mission management for the sensors and subsystems specific to the operational role of the aircraft. Other functions provided are performance management, including fuel management and Health and Usage Monitoring and Sensing (HUMS) of engine and transmission parameters.

The interface between the aircrew and the avionics systems is one or more Control and Display Units (CDUs) with alphanumeric keyboard and dedicated function keys; the CDUs are night vision goggle compatible. Associated with the CDU are one or more Processor Interface Units (PIU), commonly called the mission computer, with a hardware and software configuration specific to the operational requirement.

MIL-STD-1553B is the primary interface between the PIU, CDU and other compatible sensors and subsystems, with the bus control provided by the PIU; other interfacing can be analogue or digital, with a variety of industry standards such as ARINC 429, RS-232 and RS-422. A pocket book-sized Data Transfer Device (DTD) with solid-state memory is used to enter information and instructions from a DTD ground loader into the various battery-supported RAMS databases maintained in the PIU.

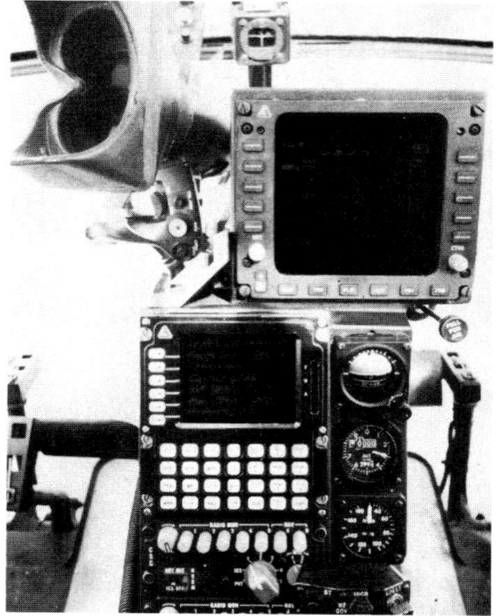

The integrated crew station on the Boeing 530MG helicopter with the RAMS multifunction display (top) and control/display unit

A variety of liquid crystal display remote indicators, monochrome and colour raster and stroke displays with associated symbol generators, stiff stick controllers and auxiliary control panels are available to make up any required configuration.

A Racal Avionics RAMS 4000 forms the heart of the Royal Navy Lynx Central Tactical System (CTS), with two specially extended control and display units and a central tactical situation display. Additional equipment includes a dual-stores management unit for weapons control and two processor interface units to link RAMS to existing navigation, communications and other systems. A data transfer device allows mission data to be loaded quickly into the system prior to flight. The high flexibility and wide range of enhancements of RAMS allows it to be tailored for present day and future requirements.

Specifications

Type 5401 processor interface unit
Dimensions: 129 × 194 × 414 mm
Weight: 5.9 kg (depending on number of interface modules)

Type 5407 control and display unit
Dimensions: 146 × 229 × 248 mm
Weight: 4.5 kg

Type 5404/5 series data transfer device with receptacle
Dimensions: 146 × 38 × 171 mm
Weight: 0.6 kg

Operational status

In production although superceded by the AMU 2000 CDNU. Notable applications of RAMS are the Central Tactical System (CTS) for the Royal Navy Lynx helicopter; the tactical data system in the Royal Danish Navy Lynx; and the Boeing 530MG Defender multirole helicopter.

RAMS has also been fitted in Royal Swedish Air Force Super Puma rescue helicopters and a RAMS version with multifunction displays has undergone trials on German Army PAH-1 anti-tank helicopters. It has been configured for use in maritime patrol fixed-wing aircraft and helicopters.

Contractor

Racal Avionics Ltd.

UPDATED

Electrical Load Management System (ELMS)

ELMS provides a comprehensive range of functions including distribution and protection of primary and secondary electrical power, and control of aircraft utilities sub-systems. The system also interfaces with the cockpit controls and displays via an ARINC 629 databus for control switching and system status reporting.

Smiths Industries ELMS for the Boeing 777 ***1997***/0003120

The Boeing 777 ELMS has three primary panels and four secondary panels: left, right and auxiliary power and ground handling/service distribution, left power, right power, and standby power.

The design of the ELMS features compact primary panels with line replaceable contactors. Three of the secondary power management panels incorporate dual ARINC 629 bus interfacing, dual processing elements and line replaceable electronic and switching relay modules. The design makes the maximum use of multiple redundancy techniques and immunisation against high intensity radio frequencies. Contributing to the integrity of the system is the incorporation of technology already proved on the Eurofighter and Longbow Apache load management systems.

Operational status

In production for, and in service on, Boeing 777.

Contractor

Smiths Industries Aerospace.

VERIFIED

Electrical Power Management System (EPMS) for the AH-64C and D Apache

The EPMS for the AH-64C and D Apache helicopters provides switching, monitoring and protection of aircraft primary and secondary distribution of AC, DC and battery power. The main functions of the system are to provide automatic load switching, power source and load distribution protection, as well as electrical system diagnosis. Key benefits are reduced aircraft weight, build time reduction and improved BIT and maintainability.

Operational status

Six AH-64Ds have been equipped for trials of the system.

Contractor

Smiths Industries Aerospace.

VERIFIED

Flight Management Computer System (FMCS)

Conforming to the full ARINC 702 specification, and a standard option on the Airbus A310 and A300-600 aircraft, this Flight Management Computer System (FMCS) is the prime interface between crew and aircraft and enables optimum performance to be achieved from take off to final approach. Main functions include flight planning, navigation, performance optimisation, flight guidance (with coupling to autopilot and autothrottle) and display processing. The operational procedures create a working routine which is easy to implement and is similar for all phases of flight, optimising the factors affecting flight profile to give greater economy of fuel consumption, flight time and aircrew workload.

The system design is based on a parallel multiprocessing arrangement of microprocessors within the flight management computer unit. This technique permits high-processing capability and gives the flexibility to accommodate future expansion of functions and procedures. Two sets of dual 16-bit microprocessors – one dedicated to navigation, the other to performance functions – provide overall throughput of over 1 Mops. Additional microprocessors are dedicated to input/output and database control functions. A bubble memory provides 256 k words of memory for navigation and performance database storage. There is provision for up to 56 discrete inputs and 16 discrete outputs, plus 32 input and 12 output ARINC 429 channels. The system contains its own built-in test routines which constantly monitor system operations and fault detection.

The crew interface is with the control/display unit which has a 14 lines by 24 character CRT format. The bottom line can be used for scratchpad entries. A full alphanumeric keyboard is provided, together with function keys and 12-line select keys adjacent the CRT. Self-contained built-in test provides a cued step-by-step test of all push-buttons, annunciators and the CRT display. For routine operations, most of the information is defaulted from the navigation database, requiring a minimum of manually entered data.

The Enhanced Flight Management Computer System (EFMCS) supersedes the FMCS. This provides one million words of EEPROM memory for navigation and performance database storage, replacing the 256 k words of bubble memory in the FMCS. Further reliability and functional improvement are provided, including the facility for interfacing to an ARINC 615 high-speed data loader. The EFMCS is in the preproduction stage.

Control/display unit for the Smiths Industries flight management system

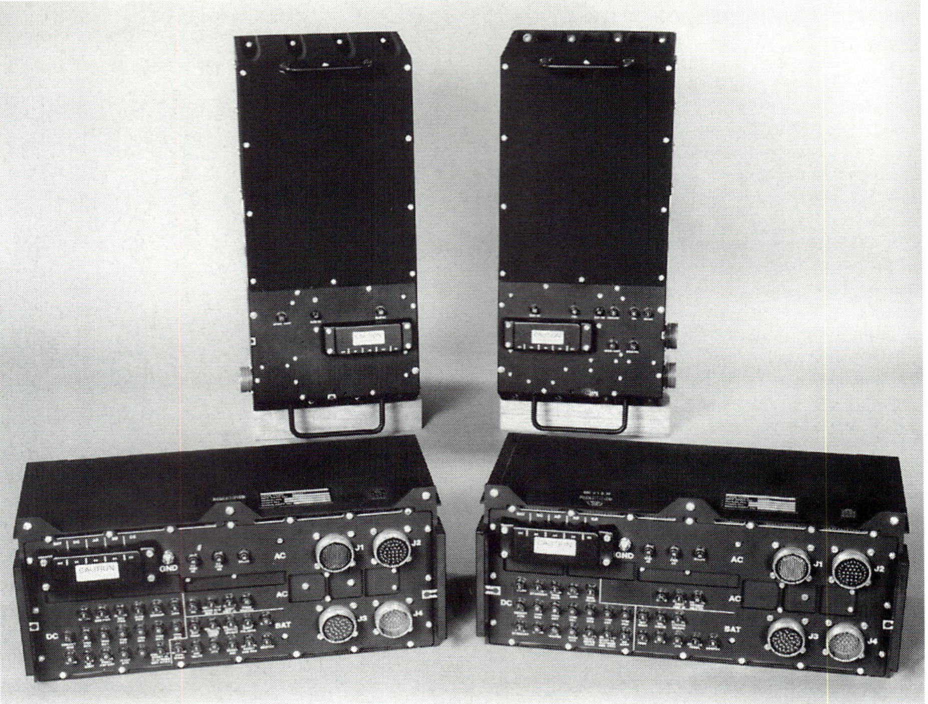

Electrical Power Management System for the AH-64D Longbow Apache **1997**/0001395

Singapore Airlines was the launch customer for the second-generation enhanced management computer for six new build Airbus A310 and retrofit of the existing fleet of 15 A310s. The enhanced system offers four times the current database capacity; an essential feature for extended route operations and improved performance and reliability.

Specifications

Dimensions:
(computer unit) 8 MCU
(control/display unit) 267 × 229 × 146 mm
Weight:
(computer unit) 12.7 kg
(control/display unit) 6.3 kg
Power:
(computer unit) 200 W
(control/display unit) 87 W

Operational status

In production. Airline customers include Kuwait Airways, Saudia, Air France, Sabena, Nigeria Airways, Air India, Cyprus Airways, Air Nuigini, Singapore Airlines, Air Algerie and Monarch Airlines. The Smiths Industries FMCS has also been chosen for the Boeing E-6A of the US Navy.

Contractor

Smiths Industries Aerospace.

VERIFIED

Fuel quantity gauging and indication

Smiths Industries designs and manufactures electronic equipment and systems for the measurement, management and indication of fuel in civil and military aircraft. Current fuel system applications include large civil transport aircraft such as the Boeing 777 and Airbus family; commuter aircraft and business jets such as the Raytheon Hawker 800/1000, Avro RJ-146 and Jetstream 41 and 61; and military aircraft such as the BAe Hawk, AMX and EF 2000. Smiths Industries is also a partner in collaborative programmes, including for the Airbus A300, A310, A319 and A320.

Analogue and digital displays

Fuel contents indicators range from simple moving coil analogue types to servo-pointer and digital multitank displays driven from an ARINC 429 databus. Internal illumination is optional and dial presentation and colour are displayed to specification. Solid-state LED or LCD indicators can be used to provide numeric, analogue or graphical presentations of fuel quantity.

Analogue and digital signal processors

The analogue output from capacitance probes can be processed by entirely analogue means and used to drive either analogue or digital indicators. Where higher accuracy or additional facilities are required, probe signals can be digitised and processed digitally to provide outputs of fuel mass in ARINC 429 or MIL-STD-1553 formats. All analogue and digital processors incorporate BITE, which performs levels of self-test varying from basic confidence checks to comprehensive system testing and calibration.

Fuel Quantity Indicating System (FQIS)

The FQIS developed for the Airbus A300, A310, A319 and A320, incorporates advanced digital computing technology which brings improved accuracy and reliability to the system compared with analogue equipment installed on earlier aircraft. BITE, failure recording and a recall facility are included.

Fuel level sensing systems

Fuel level sensing systems provide an accurate and safe means to detect fuel levels. They feature a low-cost, solid-state fluid level sensor connected to a separate switch unit which can either stand alone or be incorporated into another unit within the fuel system. The lightweight sensor unit is small enough to be mounted on a tank wall or fuel gauging probe and is immune to temperature effects over a wide operating range. Applications include high- and low-level warning indication or control, automatic shut-off switching for refuelling, automatic control of liquid transfer and sequential draining and filling of tanks. Fuel level sensors may also be applied to other fluids such as oil or hydraulic fluid.

Ultrasonic and capacitance tank probes

Fuel height within tanks can be measured using either capacitance or ultrasonic probes. Capacitance probes can be used in analogue or digital fuel systems. Systems using ultrasonic probes are digital throughout. It is usual to install several probes in each tank so that fuel levels can be gauged accurately over a wide range of aircraft attitudes and fuel contents. Non-linear height and volume characteristics of fuel tanks can be accommodated either by mechanically profiling the linear electrodes of the probes or, in digital systems, by incorporating appropriate software in the processor.

Ultrasonic fuel quantity gauging system

Fifty-four ultrasonic quantity sensors are distributed among the Boeing 777's wing and fuselage tanks. A central processor addresses each sensor individually and computes fuel volume and mass in each tank. Each fuel tank has a densitometer and a water detector, and one of the wing tanks is fitted with a fuel temperature

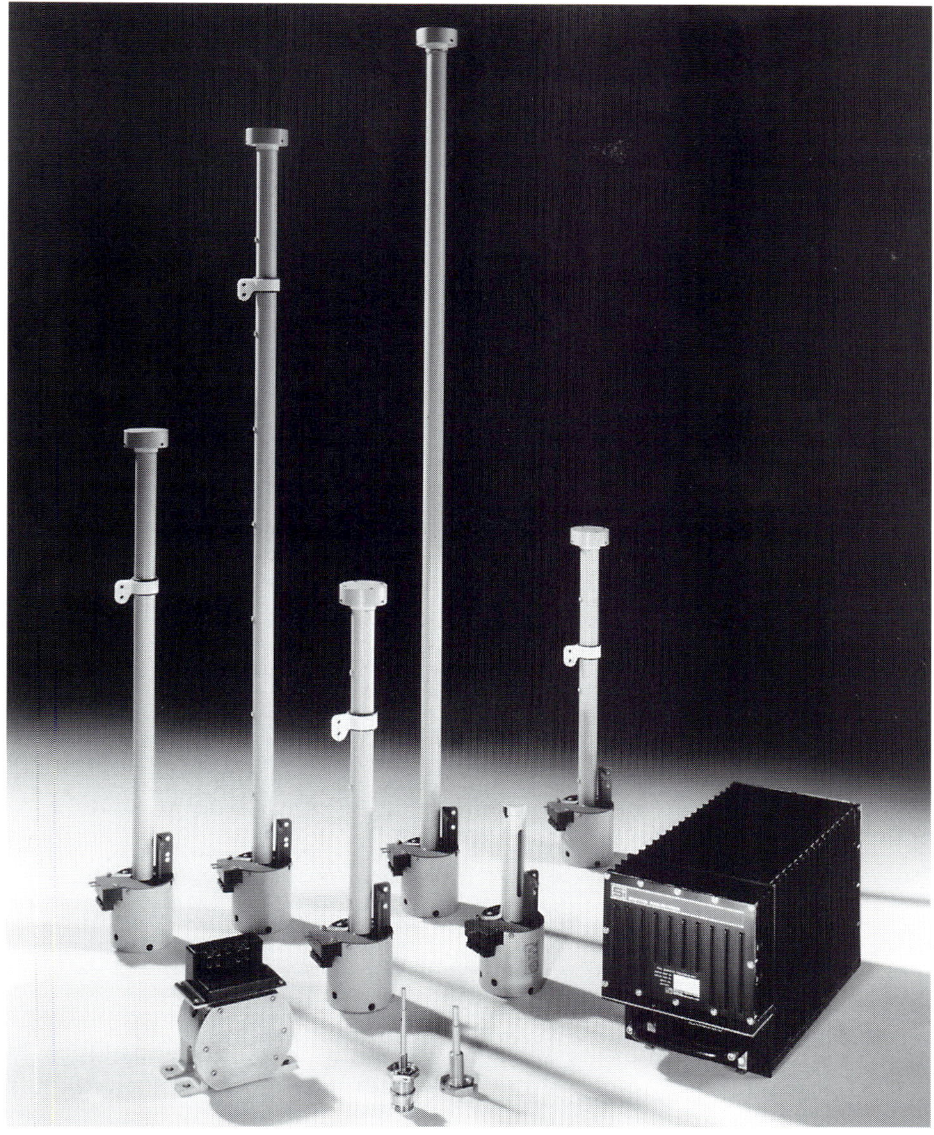

Ultrasonic fuel quantity gauging system ***1997***/0001423

sensor. Fuel management data is provided to the crew via an ARINC 629 databus linked to the cockpit displays. The data includes: quantity in each tank; total quantity; fuel imbalance and low-level warnings. There is a complete built-in test facility and the total system is designed to have a high level of tolerance to faults. In addition to the cockpit display of information, there is a refuel control and display panel linked to the central processor via an ARINC 429 databus.

Operational status

In production for, and in service in, the Boeing 777.

Contractor

Smiths Industries Aerospace.

VERIFIED

Ground roll director system

The Smiths Industries ground roll director system provides pilots with head-free guidance information. It meets the requirements for ground roll guidance in Cat IIIB weather conditions. It has been designed so that pilots can easily revert to para-visual guidance when forward visual references disappear in conditions of deteriorating visibility such as drifting fog, RVR reporting failure, or differing fog densities.

The system is based on the Para-Visual Director (PVD) concept developed during the early 1960s. A display unit is positioned in the glareshield directly in front of each pilot. During head-free operation, while concentrating on external visual cues, the pilot immediately registers any movement of the black and white bands in his peripheral vision and makes instinctive corrections to the azimuth steering controls without looking directly at the display unit. For the Boeing 777, an Active Matrix Liquid Crystal Display (AMLCD) replaces the original electromechanical 'barber's pole' indicator, providing a 100 × 25 mm usable area that offers the additional use as a multifunction display for messages such as those associated with an ATC datalink or FMC.

Specifications

Dimensions:
(display) 204 × 63.6 × 33.8 mm
(computer) 321 × 193.5 × 61.5 mm
Weight:
(display) 0.55 kg
(computer) 2 kg

Operational status

In production for the Lockheed Martin TriStar L-1011 and Boeing 747-400, 757, 767, 777 and MD-11.

Contractor

Smiths Industries Aerospace.

VERIFIED

Nimrod MRA. 4 - Avionics Systems

Smiths Industries Aerospace will supply and support four major elements of the new avionics suite for Nimrod MRA. 4 - the UK's updated maritime patrol aircraft. They comprise the Navigation and Flight Management System, based on the latest FMS for the Boeing 737-600/700/800, with added military capabilities; the Utility Systems Management System controlling utilities such as hydraulics and environmental systems; the Stores Management System handling weapons delivery; and the Full Authority Digital Engine Control System for the four BR710 engines (developed by RoSEC, a joint venture between Rolls-Royce and Smiths Industries).

Operational status

In development. Equipment for the first two development aircraft will be supplied in July 1998. Production deliveries commence in April 2001.

Contractor

Smiths Industries Aerospace.

UPDATED

Project ICHTHUS

A consortium of UK companies, led by Smiths Industries Aerospace, has launched a three year research and technology demonstrator project in the developing field of integrated health and usage monitoring systems for aircraft.

Supported by the UK Department of Trade and Industry (DTI) Aerospace and Defence Industries Directorate, Project ICHTHUS (Integrated and Coherent Technology for aircraft Health and Usage Support Systems) will examine a number of emerging technologies for monitoring engines, and fixed- and rotary-wing aircraft systems and structures. In particular it will address the integration issues surrounding such technologies.

The project's partners are British Aerospace Airbus and GKN Westland Helicopters as aircraft constructors; Rolls-Royce as engine manufacturer; and Smiths Industries Aerospace, GEC-Marconi Avionics, Lucas Varity Aerospace and Ultra Electronics as equipment manufacturers. The consortium will consult major operators in assessing the economic implications of the technologies investigated.

The project will be supported by the DTI, under its Civil Aircraft Research and Demonstration Programme (CARAD).

Contractor

Smiths Industries Aerospace (leading a consortium of UK companies).

VERIFIED

SEP10 Automatic Flight Control System (AFCS)

The SEP10 automatic flight control system is fitted to the British Aerospace 146. It provides three-axis control or stabilisation and incorporates a pitch and roll two-axis autopilot, elevator trim, flight director and yaw damping facilities. It uses simple, well-proved control laws and the minimum of sensors. There is also a synchronise control facility which allows the pilot to disengage the autopilot clutches and sensor chasers temporarily and to manoeuvre the aircraft manually, so adjusting the data of the basic and manometric autopilot modes.

The autopilot is based on rate control laws. Pitch and roll rate signals are derived from ARINC three-wire attitude references, thus eliminating the need for rate gyros. Other ARINC standard interfaces accept a wide range of sensor inputs, including those from barometric and radio navaid sensors, and allow systems to be tailored to suit operators' needs. The autopilot computer uses digital computing techniques to provide outer loop control and to organise the mode logic, and has capacity to accommodate optional facilities. Analogue computing is used for the inner loop stabilisation computing, servo-drive amplifiers and safety monitors.

The system can be supplied with either a parallel acting yaw damper, which uses a rotary servomotor to drive the rudder and rudder pedals, or a duplicated series yaw damper which drives linear actuators in series with the rudder control run. In each case, the yaw damping system is self-contained and consists of an analogue yaw computer, sensor and the relevant actuator or servo motor.

Flight director computations are performed within the digital section of the autopilot computer, which can supply commands to V-bar or split-axis flight directors. The flight director and autopilot share common mode selection and outer loop guidance but, if desired, they can be operated independently.

Emphasis has been placed on maintainability and ease of testing, both for the installed system and for individual units in the workshop. Routine testing is designed to confirm correct functioning of safety devices, the tests being performed by operating a test-button in conjunction with buttons on the mode selector. Modular construction has been used extensively to ensure that faulty equipment can be corrected and recertified easily and quickly.

The following descriptions of individual LRUs outline the operation of a full SEP10 system.

Autopilot controller The autopilot controller, in addition to providing autopilot and yaw damper engage or disengage controls, also includes pitch rate and roll angle selectors, and the elevator and rudder trim indicators. Engagement of the autopilot and yaw damper is confirmed by the illumination of a legend within each selector. Pitch control uses a spring-centred lever which has a non-linear feel so that minor adjustments can be made instinctively. Roll control is accomplished by rotation of the control knob, which remains offset by a displacement proportional to the roll angle demanded in the basic mode, but returns automatically to the central position on selection of an alternative mode.

Mode selector There are 11 push-button switches, each illuminating as mode indicators for the selection of both autopilot and flight director functions; control mode engagement is confirmed by the illumination of a white triangle on the appropriate button.

A turbulence facility is included to soften flight disturbances in turbulent air. This reverts the autopilot to the basic stabilisation mode and at the same time reduces the overall gain of the system. Autopilot and engagement lights are provided so that the engagement state of the system can be seen on the mode selector. There is also provision for remote mode indication.

Autopilot computer The autopilot computer receives both analogue and logic information from sensors, controllers and selectors and processes it to formulate the pitch and roll axis demands and the flight director commands. The majority of autopilot computing is performed digitally, although analogue techniques are used to provide pitch and roll stabilisation and authority limitations. Correct functioning of the computer safety circuits is verified by a test facility at a convenient remote station.

Yaw computer This unit takes short-term damping information from a yaw rate gyro, a lateral accelerometer and the roll VRU; it drives the series rudder actuator to provide yaw damping and turn co-ordination. For aircraft types requiring a parallel damper, such a system is available.

Altitude selector The altitude selector provides facilities for altitude preselect mode as well as the normal altitude alerting functions. Altitude information is obtained from either a servo altimeter or an air data source. Selected altitude is presented on a counter display. A warning flag obscures this display in the event of a power failure or absence of altitude valid signal, and a test facility allows checking of the associated audio-visual signals and altitude preselect function.

Air data unit Where there is a requirement for a Mach hold facility to secure better fuel economy, the basic airspeed sensor can be replaced with an air data unit providing the necessary extra outputs.

Monitor computer For operation to Cat II weather minima, this unit computes the performance monitor functions necessary to provide a fail-safe pitch channel. It independently monitors autopilot pitch rate, localiser and glide slope deviation and provides outputs that can be used to disconnect the autopilot and provide warnings to the pilot. The computer is completely independent of the autopilot, and a self-test facility allows a check to be made on the correct operation of all the monitoring functions.

Operational status

Selected for the BAe 146.

Contractor

Smiths Industries Aerospace.

VERIFIED

SEP20 Automatic Flight Control System (AFCS)

Both the autostabiliser and the autopilot in the SEP20 AFCS are fully digital and achieve levels of reliability and repeatability higher than was possible with earlier analogue systems. In addition, digital technology allows the mode of operation to be modified according to varying flight conditions or aircraft configurations. If required, the pilot can uncouple the autopilot and fly the aircraft using flight director commands provided by the system.

A comprehensive built-in test facility is incorporated, providing output data on the status of the system. Maintenance and release test functions are included, together with preflight safety checks, in-flight monitoring and the ability to identify a faulty LRU quickly.

The full AFCS comprises two identical digital Flight Control Computers (FCC), a Pilot's Control Unit (PCU), a Dynamic Sensor Unit (DSU) and a Hover Trim Control (HTC) unit for helicopter SAR applications. Extensive use is made of ARINC 429 both for external communications and for communications within the system. The equipment is designed to suit widely differing primary roles, such as anti-submarine operations or civil passenger transport, necessitating exacting safety standards.

The two identical FCCs are packaged in a 6 MCU configuration. Each comprises nine printed circuit cards incorporating four microprocessors of two widely used types - the Intel 80286 and Motorola 68000. Each microprocessor has been programmed independently to minimise the possibility of common mode faults.

The PCU enables the pilot to select the required AFCS control mode and displays the state of mode engagement. The unit is divided into two segregated sections in order to maintain integrity and fault survivability.

To achieve the level of sensor signal redundancy for failure survival in the yaw axis, the DSU incorporates a yaw rate gyro, lateral accelerometer and normal accelerometer. Output signals are provided to ARINC 429 digital format.

For helicopter applications the HTC is integral with the winch controller and provides the winchman with limited authority control of the aircraft through the AFCS hover trim mode.

Operational status

Selected for the Westland/Agusta EH 101 helicopter.

Contractor

Smiths Industries Aerospace.

VERIFIED

'Smart' Throttle Actuators

The 'Smart' Throttle Actuators for the Boeing 777 represent a new generation design using 'Application Specific Integrated Circuit' (ASIC) technology. ASIC reduces volume, weight and the number of wiring connectors by integrating the avionics within the electromechanical actuator of each throttle lever.

The smart actuator of each throttle lever receives digital commands from the 777s Airplane Information Management System (AIMS). The actuators drive the throttles to the commanded position as well as feeding back throttle positional data to the AIMS.

Operational status

In production for, and in service on, Boeing 777 and 737.

Contractor

Smiths Industries Aerospace.

VERIFIED

STS 10 full flight regime autothrottle

Designed for the Boeing 737-300 and now installed as standard equipment, the Smiths Industries autothrottle has been developed from the highly successful system supplied to Boeing for the 727-200 and 737-200 aircraft. In 1990 the STS 10 was fitted to the 737-500 and is capable of operating with intermixed engine situations.

A single unit will fit either 737-300, 737-400 or 737-500. It interfaces with flight management systems, digital air data systems, inertial reference systems and digital autopilots and uses advanced digital techniques for higher reliability, easier maintenance and lower cost of ownership.

The system comprises a digital computer with independent electromechanical drive to each throttle lever. The computer, which is housed in a single ½ ATR long box, accepts analogue and digital information from sensors and systems on board the aircraft. After processing this data the computer generates outputs to drive servo-actuators which adjust the position of each throttle lever independently, so achieving optimum engine performance. A further output from the same computer drives the fast/slow indicators on the ADIs or EFIS displays.

The autothrottle includes a number of unique features designed to enhance performance and promote flight safety. A particular feature of the system is the ability to override the actuator drive and adjust the throttle levers manually, without the pilot applying more force than he would normally use in manual operation.

To achieve precise control throughout the full flight regime, the autothrottle computer continuously monitors all the necessary engine and aircraft parameters and adjusts the thrust in accordance with the prevailing flight conditions. Protection is included to prevent exceeding predetermined N1 engine limits and maximum aircraft incidence.

The system includes damping controls which are designed to minimise throttle activity during normal flight conditions.

If a large change in vertical windspeed occurs during the approach a command is inserted which enables the system to achieve the required level of thrust more quickly.

With the launch of the Boeing 737-600/700/800 the Autothrottle Computer is being repackaged into a smaller lighter case. By the use of ASIC technology the Autothrottle Computer is being re-engineered into a ⅜ ATR Short case. The software embodied in the new unit will remain largely unchanged but is adapted for the differences in airframe performance and changed engine characteristics. The Full Flight Regime features of the current B737-300/400/500 Autothrottle Computer will be embodied in the new Autothrottle Computer for the 737-600/700/800.

The single largest change is that the 737-600/700/800 aircraft will use FADEC controlled engines in the same way as the 777. This similarity has lead to a change in the Servo Drive for the autothrottle system. In the 737-300/400/500 it was necessary to drive the cables which routed the throttle level commands to the engine. In FADEC controlled aircraft it is only necessary to drive the pilots' throttle levers and therefore the technology used in Autothrottle Servo Motor (ASM) developed for the 777 has been adapted for use in the 737-600/700/800. The main change for Smiths Industries was the need to adapt the 777 unit into a smaller envelope for the new 737 variants.

Operational status

In production. The unit is standard fit on 737-300, 737-400, 737-500 and 737-700 for which the new ⅜ ATR unit will be in production.

Contractor

Smiths Industries Aerospace.

UPDATED

Versatile electronic engine controller

Developed for the Pratt & Whitney PW305 turbofan engine, this is a versatile digital engine control unit which can be integrated with minimal redesign on a wide range of aero engines and airframes. Reduced unit cost from volume production, together with minimal non-recurring costs and benefits from reliability growth in related applications, combine to make this an attractive and cost-effective unit for both military and civil applications.

Operational status

In service in the Raytheon Hawker 1000, Learjet 60 and Astra Galaxy aircraft.

Contractor

Smiths Industries Aerospace.

VERIFIED

Self-contained attitude indicators

Smiths Industries - Newmark manufacture a number of self-contained attitude indicators which feature electrical erection and lockable mechanical caging. These high quality units can be supplied with features such as synchro outputs, pitch trim, various sphere markings, case mountings and lighting configuration including NVG-compatibility.

Specifications

Accuracy: ±0.5°
Range:
(roll) ±360°
(pitch) ±85°
Power supply: 28 V DC
Installation: Case ARINC 3ATI, 0-20° panel
Weight: 1.7 kg

Operational status

In production.

Contractor

Smiths Industries - Newmark.

UPDATED

SN100 strapdown Attitude Heading Reference System (AHRS)

The SN100 AHRS provides high performance and flexibility at a competitive price.

Internal solid-state sensors generate angular rate and linear acceleration signals, and a three-axis Fluxgate Magnetometer corrects long term drift. A microprocessor computes the vehicle heading, attitudes and rates and provides outputs in analogue or digital form. The system features a continuous built-in test facility.

This unit can be interfaced with an electronic display to provide an aircraft standby system.

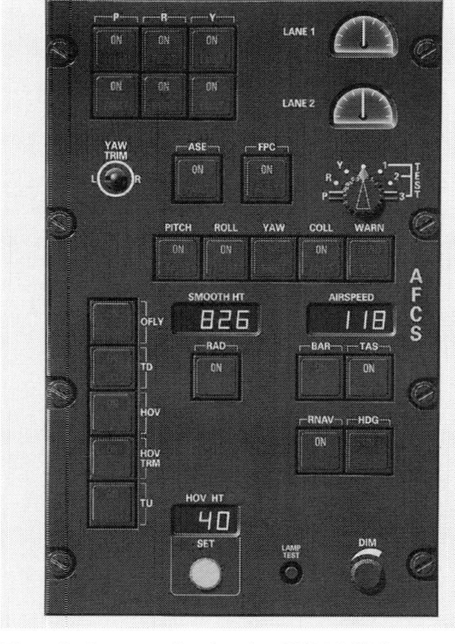

The pilot's controller for the SN500 flight control system has engagement buttons for the duplex autostabiliser and autopilot modes **1995**

Specifications

Performance:
(heading) 360° ±1°
(roll attitude) ±180° ±1°
(pitch attitude) ±90° ±1°
(pitch and yaw rates) 100°/s ±1%
(roll rate) 300°/s ±1%
(acceleration) 25 *g* ±1%
Weight: 2 kg
Power supply: 28 V DC

Operational status

Prototype stage.

Contractor

Smiths Industries - Newmark.

UPDATED

SN500 Automatic Flight Control System (AFCS)

The SN500 is a high-integrity AFCS designed for the most demanding applications. The system can be integrated both with existing and future facilities, such as MLS, and is suited for all Sea King roles including ASW, OTHT, SAR, AEW and utility. It features duplex stabilisation, attitude and heading hold, fly-through manoeuvring, automatic trimming, barometric height hold, airspeed hold, heading acquire, RNav, radar altitude hold, transition up and down, hover with hover trim and overfly.

The analogue autostabiliser provides tight hands-off control combined with flexible hands-on manoeuvring. The duplex lanes of computing are integrated with existing simplex series actuators and, in the cyclic axes, with existing hydraulic beepers. The system is designed to provide a level of safety suited to demanding environmental conditions. The computing circuits are modular and use separate plug-in boards in each lane. Each board is self-contained with its own stabilised power supply. Lane 1 and Lane 2 circuits are separated by a partition and use separate wiring. BITE facilitates ground testing and preflight check-out provides continuous interlane monitoring.

The digital autopilot uses the latest technology to provide coupled control of a wide range of modes. The use of digital techniques allows the specific requirements of each mode, both standard and special to role, to be taken into account. The use of extensive self-testing facilities and performance monitoring within the simplex microprocessor give unambiguous indication to the crew of any detected failures and, where safety requires it, initiates a fly-up procedure. In all axes but collective, autopilot outputs are summed into the duplex lanes of the analogue autostabiliser. In the collective axis, functions are monitored using analogue techniques.

Specifications

Dimensions:
(sensor unit) 65 × 130 × 212 mm
(autostabiliser computer unit) 194 × 125 × 398 mm
(pilot's controller) 245 × 146 × 213 mm
(autopilot computer unit) 194 × 125 × 398 mm
(hover trim controller) 86 × 146 × 277 mm
Weight:
(sensor unit) 1.9 kg
(autostabiliser computer unit) 5 kg
(pilot's controller) 5 kg
(autopilot computer unit) 5 kg
(hover trim controller) 1 kg
Power supply: 115 V AC, 400 Hz
28 V DC

Operational status

Developed as a retrofit for the SH-3 Sea King helicopter. Now in production for Royal Air Force Sea Kings.

Contractor

Smiths Industries - Newmark.

VERIFIED

SN501 Automatic Flight Control System (AFCS)

This version of the SN500 AFCS has been certified for civil use in Sikorsky S61 helicopters and similar types.

Specifications

As per SN500.

Operational status

In production.

Contractor

Smith Industries - Newmark.

VERIFIED

SN510 Automatic Flight Control System (AFCS)

This version of the SN500 system incorporates a much higher level of monitoring in the ASE which automatically deselects a failed part of the system without reducing the overall performance. In this form it is suitable for helicopters capable of high speeds and levels of agility.

Specifications

As per SN500.

Operational status

Under development.

Contractor

Smith Industries - Newmark.

VERIFIED

UltraQuiet active noise control systems

The UltraQuiet active noise control systems for passenger aircraft cabins reduce noise by duplicating the primary sound field with an additional secondary sound field in anti-phase. The UltraQuiet system can operate over the complete flight envelope including take-off and landing, reduce the harmonics of the blade passing frequency, track the variation in the blade passing frequency which occurs during air turbulence and when the aircraft banks, and reduce beat caused by poor propeller synchrophasing.

A typical system comprises 24 loudspeakers, 48 microphones and an electronic controller. The loudspeakers are lightweight devices specially designed for optimum performance at low frequencies. They are positioned in the trim beside the seats and in

UltraQuiet Active Noise Control System

the roof. The exact positions are determined by measurement and practicalities of the trim design. The loudspeakers each have a low-power amplifier which is mounted on, or adjacent to, the loudspeaker. This gives maximum flexibility when deciding the number of loudspeakers required. The microphones are positioned in the rear of the trim. The electronic controller consists of printed circuit boards inside a ½ ATR short ARINC 600 case. The unit has a comprehensive internal built-in test system.

The performance of the system is optimised for cabin wide noise reduction, with no areas where noise is unacceptable. The system compensates automatically for the changes in sound field caused by staff and passengers moving around the cabin. The system typically achieves attenuations of 8 to 12 dBA throughout the flight envelope.

Operational status

UltraQuiet Active Tuned Vibration Attenuators (ATVAs) are the latest addition to Ultra Electronics' portfolio of active noise control systems. The ATVAs are used in place of loudspeakers, and produce the anti-phase sound-field by vibrating the fuselage of the aircraft. This reduces both cabin noise and vibration. The launch customer for the ATVA system is de Havilland; the Dash-8Q Series 100, 200, 300 and the new 70-seat Series 400 are all fitted with the UltraQuiet ANVC system as standard.

Contractor

Ultra Electronics Ltd, Noise & Vibration Systems.

UPDATED

UNITED STATES OF AMERICA

LoFLYTE

LoFLYTE™ was developed under funding from the US Air Force and the National Aeronautics and Space Administration, using technology previously developed under contracts with the US Navy, the National Science Foundation, and the National Aerospace Plane Joint Program Office. The 'waverider' shape, so named because it is designed to ride on the shockwave created at hypersonic speeds, was proved to be aerodynamically sound through extensive wind tunnel testing at NASA Langley Research Center. In August 1996, LoFLYTE™ was introduced at the Experimental Aircraft Association Fly-in Convention in Oshkosh, Wisconsin. LoFLYTE™ is currently undergoing flight testing at Edwards Air Force Base in California.

LoFLYTE™ incorporates several important innovations in its design, including the M5 Waverider design, neural air data 'smart skin' technology, neural network inner and outer loop flight controls, neural network engine controls, advanced real-time data acquisition and control system, adaptive 'fly-by-light' communications, neural network fault diagnosis, hypersonic flowpath demonstration and nozzle concepts, tiperons, 'pilot induced oscillation' detection and compensation, and small-scale jet-powered aircraft prototyping. Once flight testing of the current 100 in model is completed, the next phase of the project is to build a 23 ft version of the design.

At the heart of LoFLYTE™'s innovative control system is Accurate Automation's Neural Network Processor (NNP®). The NNP®, developed under SBIR funding from the US Navy, is the only multiple instruction/multiple data neural processor currently on the market. Artificial neural networks are a class of important computational methods that loosely emulate biological neural assemblages of neurons and their dendritic connections that form the brain. Artificial neural networks work like the brain to recognise patterns, make decisions, and perform many other common tasks. They exploit massively parallel processing and distributed memory in ways similar to the brain.

The LoFLYTE™ Waverider flight model ***1997***/0003301

A chief goal of the LoFLYTE™ programme is to demonstrate that neural networks can safely and efficiently control an aircraft during flight. This neural network design adapts and learns during flight. As the aircraft is flown, the controller adapts its actions to varying flight conditions. When it later encounters the same conditions, it recalls the best control strategy. Adaptation and learning are achieved within the scope of a carefully constructed proof of stability. The learning consistently results in improved performance.

Operational status

Development.

Contractor

Accurate Automation Corporation.

VERIFIED

LoFLYTE™ neutral network processor
1997/0003302

Airborne Flight Information System (AFIS)

AlliedSignal has expanded the capabilities of its GNS-1000 flight management system and GNS-500A Series 4 and Series 5 navigation systems into a full airborne flight information system, with comprehensive facilities for flight plan creation before flight and amendment in flight. The Airborne Flight Information System (AFIS) is also available for AlliedSignal's GNS-X flight management system.

Before flight the pilot can access the Global Data Center via a computer, to obtain flight planning, wind and en route weather information. This information is recorded on a mini-computer disk, which is later loaded into the AFIS system via an onboard data transfer unit. During flight the comparison of planned and actual flight plan information can be viewed at any time on a control and display unit.

During flight, the air-to-ground and ground-to-air datalink can be used to obtain additional wind and weather information, send and receive flight related messages and update flight plans.

Operational status

Fully operational.

Contractor

AlliedSignal Commercial Avionics Systems.

VERIFIED

FCS-60 series 3 digital Flight Control System

The designation FCS-60 embraces a family of four flight control systems with performance to suit different categories of regional commuter and business aircraft.

The FCS-60 is basically a three-axis system under microprocessor control. The simplest system available has one channel per axis, weighs 14 kg and is driven by a single air data sensor. Navigation and sensor units are controlled by a single fail-passive flight controller, the outputs from which operate pilot and co-pilot flight director instruments. The system can be expanded to dual-simplex or duplex configuration, the second of which may be certificated for Cat II operations. Both have dual air data sensors and the second version has dual-drive motors and three microprocessors. The system provides pitch stabilisation with automatic trim, roll stabilisation, heading hold, yaw damping and turn co-ordination. The yaw damper can be used independently of the other channels. Additional modes provide lift compensation during turns, compensation for turbulence and pilot-commanded inputs.

Operational status

In production.

Contractor

AlliedSignal Commercial Avionics Systems.

VERIFIED

FCS-870 automatic Flight Control System

AlliedSignal has combined integrated circuit technology with several new automatic flight control features to produce a system which offers optimum performance over a wide range of general aviation aircraft. The FCS-870 is an autopilot with flight director and independent yaw damper options which form the basis of many options. It is designed for installation in a broad range of aircraft types, from heavy singles to most turboprop-powered types. The system meets or exceeds the TSOs for these classes of aircraft.

The complete FCS-870 consists of the cockpit instruments (including flight controller, attitude director indicator, horizontal situation indicator and mode annunciator) and a remote-mounted computer amplifier and servoes. The yaw damper option adds a side-slip sensor, a panel-mounted turn and slip indicator and a remote yaw servo.

Flight controller This is a small panel-mounted unit used to select the desired operational modes of the system. All nomenclature on the panel is back-lit for easy night viewing.

Mode annunciator This can be mounted in any convenient head-up panel location so that the pilot can monitor the autopilot or flight director functions in use. The unit also alerts the pilot to some fault and armed system conditions.

Computer amplifier This is the main flight control system unit and is an all-solid-state device which houses all the lateral and longitudinal computational circuitry, power supplies and altitude transducer. It also contains the calibration circuits which ensure compatibility with specific aircraft sensor and output requirements, plus an additional circuit board providing input signals for flight director operations. Lateral and longitudinal data circuits are segregated on opposite sides of the unit, so reducing the amount of inter-wiring and augmenting reliability and serviceability. Relays have been eliminated and all heat generating components are near the outside of the unit to improve cooling.

Servos Advanced design servos provide the greatest torque required for the highest-performance aircraft likely to use the system. Three similar units are used for pitch, roll and trim control.

Yaw damper The independent yaw damper provides positive turn co-ordination and rudder control in all flight conditions. In twin-engined types the yaw damper is claimed to provide substantial assistance in maintaining directional control during an engine-out sequence.

Instrumentation The company recommends integration with the following flight instruments: DH-886A 4 in director horizon indicator or DH-841V 3 in director horizon indicator, plus HSD-880 4 in or HSD-830 3 in horizontal situation indicators.

Specific features of the flight control system are: command turn (half- or full-rate co-ordinated turns can be initiated by rotating a knob); full pitch integration (provides smooth capture of desired altitude and eliminates standoff errors); automatic altitude preselect; pitch synchronisation (keeps elevator surfaces aligned with trim to eliminate disengagement disturbances); control wheel steering (manoeuvre to desired attitude with button depressed and then release to leave flight control system maintaining pilot's demand); coupled go-around, automatic crosswind correction; all angle intercepts and pitch rate command and manual or automatic glide slope capture.

Specifications

Weight:
(basic autopilot) 8.94 kg
(with 3 in FD/HSI) 14.11 kg
(with 4 in FD/HSI) 17.08 kg
(optional yaw damper) 3.33 kg
TSO compliance: C9C, C52A, DO160.

Operational status

In service.

Contractor

AlliedSignal Commercial Avionics Systems.

VERIFIED

Global Star 2100 Flight Management System (FMS)

AlliedSignal's new Global Star 2100 system, now under development, will be standard in the Cessna Citation Excel, de Havilland DHC-8 Series 400, and Avro International Aerospace 146/RJ. It will be an option on Bombardier's Learjet 45.

Fully compatible with FANS, the Global Star 2100 will: comply with standards that will facilitate denser air route structures; interface with digital communication devices such as Airline Communication and Reporting Service (ACARS) and Airborne Flight Information System (AFIS); fly non-precision GPS instrument approaches and offer an upgrade to fly precision instrument approaches; couple to an autopilot to automatically fly entire procedures requiring vertical as well as lateral manoeuvres; have a performance calculator (as a future upgrade) to determine such items as required runway length, operating speeds, and take-off thrust; offer the power to accommodate FANS requirements, made possible by greater processing speed and database capacity.

Operational status

First deliveries to aircraft manufacturers and retrofit customers are expected in 1998.

Contractor

AlliedSignal Commercial Avionics Systems.

VERIFIED

GNS-1000 Flight Management System (FMS)

The GNS-1000 flight management system offers a choice of flight management capabilities including GPS, VLF/Omega, DME/DME, VOR/DME and IRS. It houses both a five-channel GPS receiver that continuously tracks up to five satellites simultaneously and a VLF/Omega sensor. A single module formats the system to existing interface requirements.

Specifications

Dimensions: 384.6 × 195.1 × 124.5 mm
Weight: 9.58 kg

Contractor

AlliedSignal Commercial Avionics Systems.

VERIFIED

GNS-X Flight Management System (FMS)

The GNS-X flight management system combines a Navigation Management Unit (NMU) and a CDU to provide comprehensive navigational capability. The internal database has 4 Mbits of memory and the GNS-X has interfaces with aircraft navigation systems (GPS, Omega, INS and so on) and aircraft systems (autopilot, air data system, AFIS and so on). Internal DME/DME and DME/VOR processing is standard. Also featured is an internal Loran C, frequency management and automatic fuel flow.

Specifications

Dimensions:
(NMU) ¼ ATR short
Weight:
(NMU) 3.2 kg
(CDU) 2.9 kg

Operational status

In production from early 1988. The GNS-X is certified on the King Air 200 and Falcon 900.

Contractor

AlliedSignal Commercial Avionics Systems.

VERIFIED

GNS-XL/GNS-XLS Flight Management Systems (FMS)

The GNS-XL and GNS-XLS systems are derived from the earlier GNS-X series of Global Wulfsberg systems. Both are compact, single-box, flight management systems that incorporate eight-channel GPS, and feature flat-panel liquid crystal displays. Both provide control of aircraft navigation sensors, communication, radio and fuel management. With full analogue/digital interfaces, they are well suited for both new programmes, and upgrade/retrofit requirements.

The receiver incorporates RAIM for enhanced reliability, and to meet FAA TSO C129 Class A1/B1/C1 requirements enabling the operator to make GPS-derived IFR approaches as well as en route and terminal navigation. Both systems include Fault Detection and Exclusion (FDE), enabling them to be used as primary means of navigation during transoceanic and remote area operations. Position and velocity data is accepted from internal and external sensors, using a special navigation filter to generate a composite system position. As well as the built-in GPS receiver, the system includes a VORTAC Position Unit (VTU) processor, which automatically selects the best available DME/DME and VOR/DME measurements from VOR/DME, VORTAC, Tacan and ILS DME units. External interfaces include those for VLF/Omega, Inertial Reference Systems (IRS), or Inertial Navigation System (INS).

Data communication interfaces include a full AFIS (Airborne Flight Information System), and optional Satellite Data Communications (SDC) connection.

These systems have growth potential for Wide Area Augmentation System (WAAS) and Differential GPS.

Specifications

GNS-XL
Display: Full-colour 5½ in diagonal display
Dimensions: 181 × 146 × 200 mm
Weight: 3.64 kg
Inputs:
(analogue) fuel flow, air data, heading, VOR/DME
(digital) air data, heading, VOR/DME, weather radar, EFIS, radio frequencies
Outputs:
(analogue) HSI course and bearing, XTK and vertical deviation, To.From, autopilot steering, annunciators
(digital) EFIS/flight director, autopilot, radio tuning

GNS-XLS
Display: Full-colour 4 in diagonal
Dimensions: 114 × 146 × 165 mm
Weight: 3.18 kg
Inputs: analogue and digital: fuel flow, air data, heading, weather radar, EFIS, VOR/DME/DME
Outputs: analogue and digital: EFIS/flight director, autopilot, radio tuning

Operational status

GNS-XL: Certified by FAA to TSO C-129 Class A1; can be upgraded for compatibility with FANS. Standard equipment on Citation Ultra business jets from January 1997.

GNS-XLS: FDE given FAA approval November 1996. All future deliveries with FDE. Selected by British Aerospace Asset Management as a navigation upgrade for BAe 146 fleet.

Contractor
AlliedSignal Commercial Avionics Systems.

VERIFIED

KAP 100 Silver Crown autopilot

The KAP 100 is a panel-mounted single-axis, wings level, digital flight control system which includes a KG258 horizontal reference indicator and KG107 directional compass. An optional slaved compass system can be substituted for the latter item. Options include manual electric trim, control wheel steering and a yaw damper. Lateral modes include heading select, navigation tracking, approach and localiser back course modes. The system was announced in June 1982 and is certificated on several single-engined aircraft types.

Specifications
KAP 100 (KG 107 directional gyro, no yaw damper)
Weight: 4.9 kg
Power supply: 14 V DC, 3.1 A or 28 V DC, 1.6 A

Operational status
In production.

Contractor
AlliedSignal Commercial Avionics Systems.

VERIFIED

KAP 150 Silver Crown autopilot

The KAP 150 is a panel-mounted two-axis digital autopilot providing pitch and lateral control facilities. It is integrated with standard flight instrument packages which are available from the company's range of products. Slaved compass and remote mode annunciator options are provided. Autopilot modes include pitch hold, heading select, altitude hold, navigation tracking, approach, glide slope and localiser back course, vertical trim and control wheel steering. A yaw damper is available as an optional extra. The system is certified on several single-engined aircraft types.

Specifications
KAP 150 (KG 107 directional gyro, no yaw damper)
Weight: 8.2 kg
Power supply: 14 V DC, 5.1 A
28 V DC, 2.5 A

Operational status
In production.

Contractor
AlliedSignal Commercial Avionics Systems.

VERIFIED

KAP 150H helicopter digital flight control system

The KAP 150H two- or three-axis autopilot is designed to meet the needs of single-engine turbine-powered helicopter operators in emergency medical transport, law enforcement, pipeline patrol or a variety of other uses. A compact panel-mounted system, the KAP 150H is engineered for easy console installation.

A derivative of the KAP 150, the KAP 150H autopilot operates in heading select, altitude hold, vertical trim and control wheel steering modes, providing workload reduction for a single pilot during VFR cruise operation. The KAP 150H also reduces pilot fatigue, enhancing safety and delivering a smooth ride for passengers.

System options include a yaw axis and either a standard directional gyro or the AlliedSignal KCS 55A compass system, an electrically slaved 3 in (76 mm) horizontal situation indicator.

Specifications
Weight: 9.07 kg
Power supply: 6.5 V AC, 3.5 V DC

Operational status
Certified for the Bell 206B JetRanger and LongRanger series of helicopters.

Contractor
AlliedSignal Commercial Avionics Systems.

VERIFIED

KFC 150 Silver Crown autopilot/ flight director system

The KFC 150 is a KAP 150 autopilot with a (76 mm) air-driven gyro, flight instrument package. Single-cue V-bar presentation is used on the flight director. Also included in the package is a KCS 55A slaved compass system with KI 525A pictorial navigation indicator. The KFC 150 provides pitch attitude hold, altitude hold, flight director, heading select, navigation, approach, glide slope, back course, vertical trim and control wheel steering.

Specifications
KFC 150 (without yaw damper)
Weight: 11.5 kg
Power supply: 14 V DC, 8.7 A or 28 V DC, 4.4 A

Operational status
In production.

Contractor
AlliedSignal Commercial Avionics Systems.

VERIFIED

KFC 200 Silver Crown autopilot/ flight director system

The KFC 200 automatic flight control system is suitable for a wide range of single- and twin-engined light aircraft. It comprises a two-axis autopilot with flight director instruments and can be configured as the full KFC 200 or the lower-cost KAP 200 system. Both variants have a two-axis autopilot, the low-cost option providing wings-level and pitch attitude hold with altitude, navigation, approach, localiser, back course and heading select modes. The larger variant additionally includes go-around and control wheel steering modes.

Flight instrumentation options include the KG 258 flight command indicator and KI 525A horizontal situation indicator for the low-cost KAP 200, or the same horizontal situation indicator with a KI 256 flight command indicator in the more comprehensive KFC 200 system. Manual electric pitch trim facilities are included and options include slaved gyro and yaw damper installations.

Specifications
KFC 200 (without yaw damper)
Weight: 12.8 kg
Power supply: 14 V DC, 15.5 A or 28 V DC, 9.5 A

Operational status
In production.

Contractor
AlliedSignal Commercial Avionics Systems.

VERIFIED

KFC 250 Gold Crown autopilot/ flight director

The KFC 250 system is effectively the computation and control function of the Silver Crown KFC 200 in conjunction with the (108 mm) KFC 300 flight director. A solid-state computer generates flight director commands in parallel with three-axis autopilot control signals. The system comprises a KAP 315 mode annunciator, KCL 310 flight director, KPI 552 pictorial navigation indicator (essentially a horizontal situation indicator), KAS 297 altitude selector, KC 290 mode selector and KC 291 yaw mode controller.

Specifications
Weight: 20 kg
Power supply: 115 V AC, 400 Hz, 80 VA
28 V DC, 11 A
26 V AC, 400 Hz, 42 VA

Operational status
In production for high-performance piston twins and medium-size business turboprops.

Contractor
AlliedSignal Commercial Avionics Systems.

VERIFIED

KFC 275 flight control system

The KFC 275 digital flight control system is designed primarily for piston twins. This system uses the same KCP 220 autopilot computer with four microprocessors and the same KDC 222 air data sensor as the KFC 325. However, the KFC 275 uses a different flight instrument system and (76 mm) electromechanical instruments, including the KI 256 V-bar flight command indicator and the KCS 55A slaved pictorial navigation indicator system.

The KMC 221 mode controller provides mode selection and annunciation for most modes, along with the KAP 185A mode annunciator. Annunciations are provided for both armed and coupled modes when appropriate.

Like the KFC 325, the KFC 275 can be configured with optional altitude/vertical speed preselect. There is, however, a choice of either the KEA 130A three pointer encoding altimeter or the KEA 346 counter-drum pointer servoed altimeter to go with two different versions of the KAS 297 altitude/vertical speed preselector.

Operational status
In production.

Contractor
AlliedSignal Commercial Avionics Systems.

VERIFIED

KFC 325 digital flight control system

The KFC 325 is a three-axis digital flight control system designed to suit high-performance turboprop aircraft. An extensive preflight test of full-time monitors ensures system integrity while airspeed compensated control maintains proper response to changing aircraft configurations. Manual electric trim speed is also adjusted to aircraft speed.

The remote-mounted flight computer contains four microprocessors, with one dedicated to each of the following functions: roll, pitch, yaw damp and logic. In addition to providing these computations the microprocessors provide extensive preflight test of pitch, roll and accelerometer monitors which ensures system integrity during flight operations.

The KFC 325 is configured with a 4 in (102 mm) electromechanical ADI and HSI with growth provisions for interface with the EFS-10, EFS-40 and EFS-50 electronic flight instrument system EADI and EHSI displays. The electromechanical instruments include the KCI 310A rotating sphere flight command indicator and the KPI 553A pictorial navigation indicator with radiate of DME distance, groundspeed and TTS plus radar altitude and distance from 1,000 ft AGL to touchdown.

In addition to such standard modes as altitude hold (ALT), heading select (HDG), nav (VOR/RNav), approach (APR), glide slope (GS), reverse localiser (BC), control wheel steering (CWS), indicated airspeed (IAS), hold and yaw damp (YD), the KFC 325 also has the standard comfort modes of soft ride and half bank. Altitude/vertical speed preselect is optional. The optional KAS 297C altitude and vertical speed selector is a panel-mounted unit which can interface with a KEA 346 counter-drum pointer servoed altimeter.

The KFC 325 also includes as standard equipment

the KDC 222 air data sensor which provides altitude and airspeed as well as normal and sideslip inputs to the flight control system.

The servos use capstan assemblies which may be left in the aircraft should servo repair be required, thus allowing the aircraft to remain rigged.

Operational status

In production.

Contractor

AlliedSignal Commercial Avionics Systems.

VERIFIED

KFC 400 digital flight control system

Intended for turboprop and turbine-powered general aviation aircraft, the KFC 400 is a dual-channel system designed to fail-passive. The system is microprocessor driven and when used in conjunction with AlliedSignal's KNS 660 navigation and frequency management system can provide full three-dimensional flight guidance with automatically scheduled climb and descent profiles.

The system includes a digital air data computer providing altitude, airspeed, vertical speed and Mach number via an ARINC 429 digital databus, and it can also drive a range of electromechanical flight instruments. Continuous fault monitoring is incorporated.

Operational status

In production. It has been certified as part of the avionics suite on the Beechjet 400.

Contractor

AlliedSignal Commercial Avionics Systems.

VERIFIED

Warning Computer Mk VII

The warning computer Mk VII is a VLSI technology single ¼ ATR computer that can provide windshear detection and alerting, recovery guidance and advanced ground proximity in a single integrated economical flight safety system. The Mk VII is designed for all existing and new aircraft with ARINC 500 analogue avionics. It offers many benefits to owners of analogue aircraft.

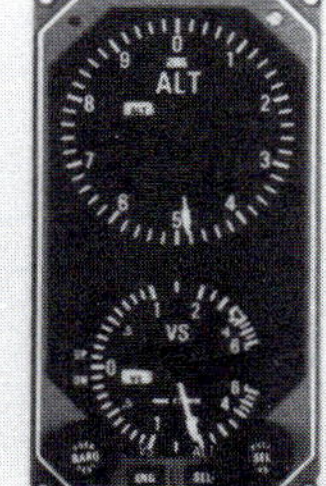
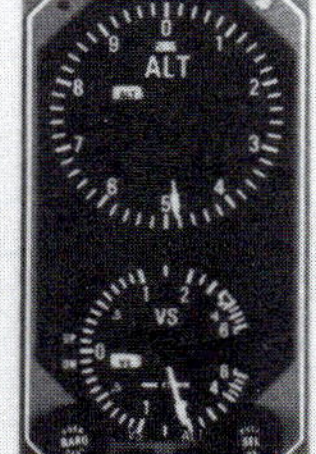

KFC 400 digital flight control system

KFC 325 digital flight system showing (top left) the EFS 40 EADI, (bottom left) the EFS 40 EHSI and (right) the KMC 321 digital mode selector

Specifications

Dimensions: ¼ ATR short
Weight: 2.63 kg
Power supply: 115 V AC, 400 Hz, single phase, 11.5 W (max)
Reliability: 30,000 h MTBF

Operational status

In production and in service on Boeing and Airbus aircraft. In service on US Navy P-3C, E-2, C-2 and S-3 aircraft.

Contractor

AlliedSignal Commercial Avionics Systems.

VERIFIED

Stall warning computer

The AlliedSignal stall warning computer provides output signals to such devices as stick-shaker and stall warning horn, based on the aircraft's angle of attack, flap/slat settings and rate of increase of angle of attack. Self-monitoring circuits minimise false alarms and failures. Front panel LEDs allow isolation of misaligned and faulty sensor inputs.

Specifications

Dimensions: ⅜ ATR short
Weight: 4 kg

Operational status

In production and service.

Contractor

AlliedSignal Commercial Avionics Systems.

VERIFIED

Digital fully fly-by-wire system for the F-16C/D

The system has evolved from the equipment developed by AlliedSignal for the US Air Force Advanced Fighter Technology Integration (AFTI) programme based on an F-16 airframe. A fundamental change from the flight control system in the AFTI aircraft is, however, the progression from three digital plus one analogue channels per axis to four digital channels, the configuration favoured for operational reliability (the system has to remain operational after any two failures). The four digital channels use MIL-STD-1750A architecture utilising Jovial high-order language and each has a processor, memory, input/output functions, discrete failure logic, MIL-STD-1553B digital databus and serial link to other channels. The computer is designed to remain fully operational following any two consecutive failures within the quadruplex part of the system. The single unit, weighing 22.7 kg and housed in a 1 ATR long box, has the equivalent function of the four separate units on the AFTI aircraft and occupies about half the space. In all other respects the system meets the form, fit and function requirements that permit it to replace directly the earlier analogue system.

Computing and processing technology is based on large-scale and very large-scale integration and gate arrays. It is designed to accommodate further growth, incorporating VHSIC components, without the need for new software. Flight critical functions are hard-wired into the system for the highest integrity, but less critical signals communicate with other equipment via the digital databus. The system has 48 k words of PROM memory, 2 k words of RAM scratchpad memory, 8 k words of input/output scratchpad memory and a 2 k word memory to record faults.

Benefits of the all-digital system over the previous analogue one are given as better reliability, lower power density, smaller number of components, greater ease of tuning to meet changes and lower life cycle costs. Cost per aircraft set is said to be less than $100,000.

AlliedSignal considers this new flight control system to be an important step forward, since it can be seen as the basis for future integrated flight-critical full authority flight control systems incorporating thrust/side-force vectoring, terrain-following and fire control.

Specifications

Dimensions: 514 × 273 × 222 mm
Weight: 22.7 kg
Power: 200 W
Reliability: >2,150 h demonstrated MTBF

Operational status

Installed in later F-16C and F-16D single-seat and two-seat fighters.

Contractor

AlliedSignal Electronic Systems.

UPDATED

Airborne vibration monitoring system

The airborne vibration monitoring system is designed for engine rotor imbalance monitoring and onboard two-plane engine trim balancing. The system consists of two major components: a remote charge converter and a signal conditioner.

Vibration data is transmitted to the cockpit for display via an ARINC 429 or 629 databus. Advanced digital signal processing techniques provide accurate vibration magnitude and phase management. Utilising in-flight data provides superior trim balance information, allowing accurate one shot engine balancing and eliminating requirements for engine ground runs for balance verification. The system simplifies engine balance operations by providing specific maintenance instructions. Extensive BIT capabilities isolate faults between accelerometers, the two LRUs and the interconnection cabling.

For system retrofit applications, an integrated signal conditioner, incorporating the functions of a remote charge converter, is available in the ARINC 429 configuration.

Specifications

Dimensions:
(signal conditioner) ARINC 600 3 MCU
(remote charge converter) 56 × 61 × 114 mm
Weight:
(signal conditioner) 2.8 kg
(remote charge converter) 0.45 kg

Operational status

Selected for the Boeing 777.

Contractor

Ametek Aerospace Products.

VERIFIED

Airborne vibration monitoring system for the Boeing 777

MFD 5000 cockpit management system

The MFD 5000 cockpit management system is an interactive graphics map, terrain and obstruction proximity system, air data system and EICAS, all presented on a single sunlight-readable multifunction display. It displays the aircraft position on a comprehensive 200,000 waypoint plan view chart. Jeppesen NavData, including airports, runways, frequencies, TCA/ARSA boundaries, VORs, NDBs, airways, intersections, SIDs/STARs and approaches combines with digital elevation mapping, geography, hydrography, manmade obstructions, highways and more, presenting VFR and IFR charts in scale levels ranging from the airport to hemisphere. Custom databases for EMS and vehicle tracking are also available.

The MFD 5000 monitors up to 35 engine and airframe conditions. By interfacing with a host of aircraft systems, the EICAS constantly scans for out of range conditions, decreasing pilot workload while increasing safety. All caution advisories are cross-referenced, providing audio and visual alerts to the MFD 5000 and external annunciators. The EICAS and map are fully interactive, to deliver crew advisories on position, environment, navigation, fuel management, air data and engine and airframe conditions.

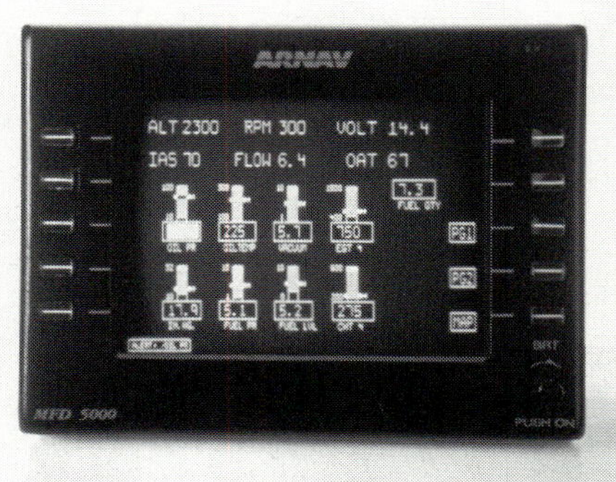

Arnav MFD 5000 displays both navigation and aircraft systems data

The EICAS couples to many existing aircraft systems including analogue instruments, rpm, voltage, oil temperature, encoding altimeters, fuel computers, Loran/GPS receivers, other engine monitoring systems or a host of Arnav transducers. The EICAS contains several pages of information with three assignable priority levels. For mixture purposes, the hottest EGT and CHT are automatically presented on the main display.

The remote LRU 5020 EICAS computer measures all channels 10 times/s to ensure constant accuracy. Bright columns display operating ranges with a digital value written below. Pilot programmable alarms advise on out of range conditions. When an alert condition occurs, the display switches to the failed system, followed by the associated checklist.

Through altitude encoders and Loran/GPS receivers, the MFD 5000 compares the aircraft altitude with the digital terrain-mapping database. When below minimum safe altitude, the terrain/obstruction proximity system advises the pilot of ground proximity, taking account of both terrain and manmade obstructions.

The Arnav FMS 5000/7000 and STAR 5000 navigation management systems are interactive with the MFD 5000. All flight planning and search routines can be programmed through the FMS 5000/7000 for display on the MFD 5000. The remote Arnav GPS-505, GPS-506 and GPS-512 receivers also interface with the MFD 5000, eliminating the need for a panel mount navigator. Other Loran and GPS receivers with RS-232 protocol are compatible.

Contractor

Arnav Systems Inc.

VERIFIED

Flight director autopilot systems

The three-cue flight director system for helicopters combines a versatile flight director computer with a three-command bar ADI, dual-bearing pointer HSI and a multimode computer controller. The system provides ILS, VOR and ADF approach capability.

The three-cue system adds a collective command steering bar to the pitch and roll command steering bars used in two-cue systems. Continuous altitude, airspeed, vertical speed, VOR/ILS and optional Doppler and altitude alert inputs permit the flight director computer to respond to pilot-selected flight modes. A pilot can execute VOR/ILS intercepts, glide slope intercepts, vertical and airspeed holds, deceleration rates and altitude hold. The computer provides automatic intercept and tracking of VOR, glide slope and localiser, and initiation of deceleration for ILS approaches. Pilot and co-pilot displays and controls are effected by optional slaved horizontal situation indicators and transfer controls.

The indicators are 5 in (127 mm) units, hermetically sealed with a dry nitrogen/helium atmosphere. Direct current servos used in these units are claimed to result in considerably less heat dissipation, less power drain, higher torque and greater reliability than typical AC servoed units.

The mode controller has four switches for mode, nav select, vertical speed and airspeed selections. An optional remote dual-course selector can be added to supplement the basic controller.

The flight director computer accepts inputs from the mode controller and an array of sensors. It computes the pitch, lateral and collective commands required to adhere to selected and/or scheduled flight parameters and displays the commands on the attitude director indicator.

Specifications

Dimensions:
(ADI) 127 × 133 × 194 mm
(HSI) 127 × 108 × 174 mm
(flight director computer) 59 × 194 × 319 mm
(mode controller) 146 × 105 × 127 mm
(remote course selector) 146 × 32 × 165 mm
Weight:
(ADI) 3.2 kg
(HSI) 2.7 kg
(flight director computer) 2.9 kg
(mode controller) 1.6 kg
(remote course selector) 0.7 kg
Power supply:
115 V AC, 400 Hz, 45 VA
28 V DC, 0.5 A
5 V AC, 8 VA (for lighting)

Operational status

In production.

Contractor

Astronautics Corporation of America.

VERIFIED

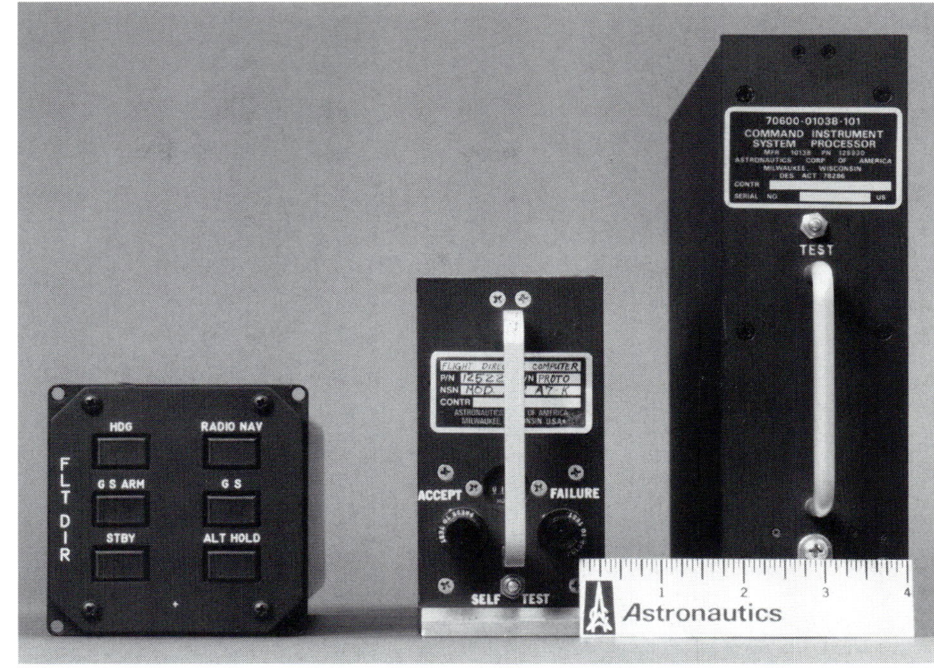

The Astronautics Macchi MB-339 flight director computer (left), A-10 flight director computer (centre) and Black Hawk command instrument system processor (right)

Flight director computers

Astronautics manufactures both analogue and digital flight director computers for most military fixed-wing aircraft and helicopters. The flight director computer provides both two- and three-cue steering information to the pilot which may be presented on either cross-pointers or a combined single indicator.

A flight director computer which is compatible with GPS is currently being manufactured for the US Army Sikorsky UH-60A helicopter. The flight director computer accepts digital signals directly from the GPS receiver and other radio navigation aids and converts them to analogue signals for display on the flight instrument. The navigation information received from the GPS receiver provides horizontal deviation, vertical deviation and glide slope angle.

Operational status

In production.

Contractor

Astronautics Corporation of America.

VERIFIED

The Astronautics three-axis helicopter autopilot system

Three-axis autopilot for the 500/530 helicopter

Astronautics has developed an autopilot for light helicopters which has been FAA certified and is available for Boeing 500D, 500E and 530 helicopters. The autopilot provides full three-axis control to reduce pilot fatigue. There are seven basic operating modes, plus hands-off stabilisation. In basic attitude retention mode, the helicopter can be flown hands-off not only in straight and level flight but also during climbs, descents and turns. The desired attitude will be held even in autorotation and all turns are automatically co-ordinated. In altitude hold mode the desired altitude is maintained to within ±20 ft. The altitude can be captured with vertical velocities as high as 1,000 ft/min. Following high-vertical velocity climb or descent the helicopter will smoothly change pitch attitude and capture the selected altitude. In heading hold mode the pilot may select the desired heading either before or after the mode is engaged. All turns to the new heading are automatically co-ordinated at a bank angle of 20°. Hands-off hover capability is provided either in or out of ground effect. Heading hold may be engaged at any time during hover. Any new desired heading is entered by moving the 'bug' on the heading gyro, and yaw damping is provided throughout the autopilot flight envelope.

Operational status

In production and in service in Boeing 500/530 helicopters.

Contractor

Astronautics Corporation of America.

UPDATED

PAR Power Analyser and Recorder

The PAR Power Analyser and Recorder is an intelligent turbine health monitor designed to analyse and record aircraft operation and display to the pilot the right information at the right time.

PAR automatically displays the DC bus voltage which remains displayed until the engine is started. Once the engine has been started the outside air temperature and computed density altitude are displayed to provide the pilot with performance guidelines. The PAR measures temperatures, N_1 and N_2 turbine speeds, torque, rotor speed or propeller rpm, outside air temperature, pressure altitude, airspeed and fuel flow.

Should an exceedance of any parameter occur, the display will show 'Exceedance in progress' and the memory will record the exceedance and the value of all other parameters at the time the exceeding parameter is at its peak. Time, date and engine S/N will be included for identification. Pressure altitude and outside air temperature will also be recorded for every exceedance.

The PAR system comprises a computer and pilot's display panel in a quarter-dwarf ATR-style case with an RS-232 communication port for downloading recorded data to a hand-held printer or personal computer. Information may be transmitted via telephone modem. The system also includes a torque pressure transducer and independent sensors for outside air temperature and pressure altitude. The pilot's display will indicate when memory is 80 per cent full or greater. If the signal is ignored, recording will continue over the oldest information.

Specifications

Dimensions: 57.1 × 152.4 × 203.2 mm
Weight: 2.27 kg
Power supply: 12-32 V DC
Temperature range: −40 to +75°C
Memory capacity: 175 events

Contractor

Avionics Specialities.

UPDATED

Steep approach monitor

The steep approach monitor was specially designed for use on the Saab 340 commuter aircraft, but can also be used on other aircraft. It cuts out unwanted ground proximity warnings during a planned high-angle approach.

The steep approach monitor fits between the Collins ADS 81 air data computer and the Sundstrand Mk II ground proximity warning computer. When selected by the pilot, it intercepts and modifies the signals sent from the ground proximity warning computer to the air data computer. During a deliberate steep approach, the pilot can descend towards the runway without unwanted, potentially distracting proximity warnings.

Operational status

No longer in production. In service in the Saab 340.

Contractor

AVTECH Corporation.

VERIFIED

Digital Fuel Quantity System upgrade FQS/(DFQS) for P-3 aircraft

The technology utilised in this capacitance-type DFQS offers P-3 users performance benefits over the original equipment system. Solid-state, micro-processor-based DFQS architecture incorporates robust Built-In-Test (BIT) and pre-calibrated components which guarantee system mean time between failure rates of 5,000 h minimum. The DFQS consists of six flight station indicators, five repeater indicators, 23 DC full-height compensated fuel tank units, four attitude compensation units, one low-level warning control unit, and all associated wiring. The DFQS provides accuracy which meets MIL-G-26988C Class II, and will operate with a range of fuel including JP-4, JP-5, JP-8 and NATO F-35.

Operational status

Selected by the US Navy in 1994 for the P-3C sustained readiness program.

Contractor

BFGoodrich Aerospace Aircraft Integrated Systems.

UPDATED

Digital fuel quantity system for P-3 aircraft upgrade **1997**/0001431

Fire Detection Suppression system (FiDS)

The BFGoodrich Aerospace FiDS is an integrated system which includes smoke detectors, flight deck panel, maintenance bay panel, wiring, plumbing, and halon bottles. It is designed to satisfy the requirements of FAR Pt 28.855.

The system provides 60 second detection and 60 minute suppression. Should smoke be detected the existing master caution warning light illuminates both on the glare shield and on the flight deck panel, and an audible warning is provided to the flight deck crew. Despatch reliability is assured using redundant dual-loop smoke detectors operating independently in each zone and redundant electronic circuitry.

FiDS is adaptable to several models of aircraft including: Boeing 727, 737 and DC-9, and is supplied as a turnkey installation kit.

Contractor

BFGoodrich Aerospace Aircraft Integrated Systems.

NEW ENTRY

Fire/Overheat Detection System (F/ODS)

The Fire and Overheat Detection System (F/ODS) provides continuous fire and overheat detection in engine nacelles and the auxiliary power unit compartment. The system also detects hot air leakage from engine bleed air ducting and identifies the location of the overheat. On detection of an overheat, the system sends commands to the bleed air isolation valves. The F/ODS reports this information to the mission computer via the MIL-STD-1553 databus.

The system consists of a dual channel controller that monitors 44 sensor loops and operates as a master and slave with cross-channel fault checking to achieve high-integrity fire detection.

Based on this technology, a system can be designed and customised to meet the fire detection needs of most aircraft. The system may also be incorporated into an integrated utility system where several aircraft subsystems are combined in a single package.

Specifications

Dimensions: 318 × 191 × 194 mm, ARINC 404 ¾ ATR Short
Weight: 6.9 kg

Contractor

BFGoodrich Aerospace Aircraft Integrated Systems.

UPDATED

Fuel Quantity Indicating Systems (FQIS)

BFGoodrich Aerospace provides fuel quantity indicating systems for both commercial and military aircraft, fixed- and rotary-wing including: Boeing, Airbus, Saab, Embraer, de Havilland, Lockheed Martin, Sikorsky, Bell, and other OEMs. These fuel quantity indicating systems provide high reliability, ease of maintenance and guaranteed 1 per cent accuracy.

Retrofit FQIS systems similar to the production configurations have been chosen by numerous airlines for their Boeing 747-200, -300, and 757/767 cargo and passenger aircraft.

The in-tank hardware for the fuel quantity indicating systems on these aircraft include all new harnesses, fuel quantity sensors, compensators for providing signal variation in dielectric constant of the fuel, plus densitometers which provide direct fuel density measurement information to the processor. Flight deck and refuelling panel indicators provide visual verification of total fuel and individual tank quantities. The processor unit controls the entire fuel system by calculating fuel quantity, controlling the refuelling sequence and performing extensive health monitoring on both wiring and electronics. The processor provides messages on operational health and history.

Other programmes for which BFGoodrich produces specific retrofit fuel quantity indicating systems are: Boeing 727, Lockheed Martin C-130 and P-3.

BFGoodrich provides fuel management and measurement for the Boeing B-1B, Northrop Grumman B-2 and the Airbus A330/340. In addition to measuring the fuel quantity, these systems control pumps and valves to transfer the fuel between tanks thus maintaining centre of gravity for optimum performance.

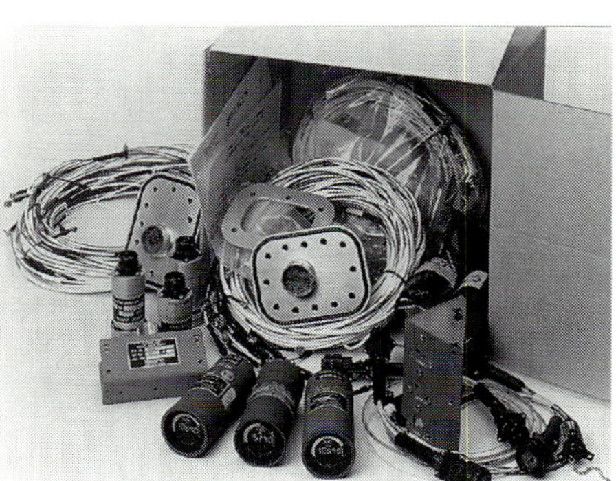

Retrofit fuel quantity indicating system for Lockheed Martin C-130 aircraft **1998**/0018232

Boeing 737-700 FQIS tank unit hardware **1998**/0018231

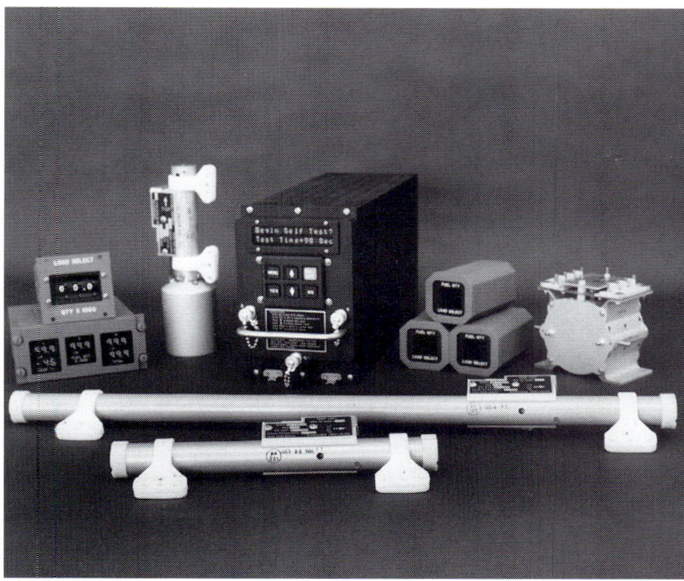

Boeing 757/767 fuel quantity indicating system **1998**/0018233

The BFGoodrich Integrated Utilities Management System is being fitted to the Lockheed/Boeing F-22

Contractor
BFGoodrich Aerospace Aircraft Integrated Systems.

NEW ENTRY

Integrated Utilities Management System

The Integrated Utilities Management System (IUMS) is a system that allows for the functional and physical integration of utilities subsystems, and the effective sharing of resources. What traditionally required 20 to 30 stand-alone black boxes can now be configured in a flexible lightweight, small network of enclosures, each sharing common electronics, processors and power supplies. For logistics support, the departure from traditional subsystems will result in greater system fault isolation, allow a two-level maintenance structure and reduce maintenance time and cost.

Operational status
The system is being provided for the Lockheed/Boeing F-22 aircraft.

Contractor
BFGoodrich Aerospace Aircraft Integrated Systems.

VERIFIED

Vibration, Structural Life and Engine Diagnostic system (VSLED)

The Vibration, Structural Life and Engine Diagnostic System (VSLED) is a health monitoring and diagnostic system designed for the V-22 engines and transmission units.

The airborne unit performs data acquisition and signal processing in real time on the aircraft and also stores data for later analysis on the ground. This system may be adapted for application on other aerospace platforms.

The design of the VSLED airborne unit is modular in both hardware and software so that the system can be easily modified for a different interface complement.

Specifications
Dimensions: 229 × 203 × 203 mm
Weight: 9.1 kg

Contractor
BFGoodrich Aerospace Aircraft Integrated Systems.

UPDATED

ACTIVE - Advanced Control Technology for Integrated Vehicles

ACTIVE - Advanced Control Technology for Integrated Vehicles is a joint NASA, US Air Force, Boeing and Pratt & Whitney programme. ACTIVE is assessing the use of engine thrust vectoring for improving the range, fuel economy, manoeuvrability and survivability of advanced aircraft.

The test results from the 60 planned ACTIVE flights over the next few years will provide useful information for improving the performance of a variety of advanced civil and military aircraft.

The use of thrust vectoring for pitch and yaw control reduced the need for aerosurface controls, which add drag and weight to an aircraft and work well only within limited flight regimes. Plans for future ACTIVE flights include testing the F-15's manoeuvrability with about half of its vertical tail area removed.

The F-15 demonstrator is equipped with Pratt & Whitney axisymmetric thrust vectoring nozzles attached to two high performance F100-PW-229 engines. As prime contractor for the ACTIVE programme, The Boeing Company is responsible for integrating the nozzles and engines with the F-15 demonstrator.

Contractor
The Boeing Company.

UPDATED

F-15 ACTIVE test aircraft
1997/0001444

Altitude Preselect/Alerter 1D960

The 1D960 Altitude Preselect/Alerter provides precise altitude preselecting and/or alerting capabilities, in addition to ATC (100 ft resolution) or RS-232 (1 ft resolution) altitude encoding. The display is a sunlight-readable dot matrix format.

During climb or descent, the desired capture altitude is entered on the 1D960 with the rotary switch. The 1D960 is armed by pressing the 'AP Arm' switch on the face of the unit. The Altitude Preselect/Alerter compares the selected altitude with the information provided by the encoding altimeter. As the altitude approaches to within 1,000 ft of selected altitude the display will flash and give two audible tones. Upon closure of 200 ft from selected altitude the display stops flashing and gives one audible tone. At selected altitude the autopilot will smoothly transition to altitude mode and the 1D960 will change from Arm Mode to display current encoder altitude. If the aircraft deviates more than 200 ft from selected altitude the display will flash and give one audible tone.

The 1D960 Altitude/Preselect/Alerter can be installed with the two-axis Century 2000 or Century 41 Flight Control Systems, or used as a stand-alone altitude alerter.

Specifications

Dimensions: 37 × 81 × 229 mm
Panel cutout: 3 in ATI
Weight: 0.61 kg
Power: 14 or 28 V DC

Contractor

Century Flight Systems Inc.

VERIFIED

Century I autopilot

The Century I autopilot is an all-electric, rate-based lightweight single-axis wings level/heading system. An electric actuator in the aileron circuit provides the control power for attitude stabilisation and pilot commanded, knob controlled turn rates of up to 200° a minute. A tilted rate gyro inside a standard 3 in (76 mm) case senses roll rate and rate of turn, for both instruments and servo, and the system can be slaved to VOR/Loc. Century I can also be used as a back-up to the Century IIB, III or IV vacuum/electric systems (see later items), sharing the same roll servo.

Specifications

Dimensions: 89 × 89 × 194 mm
Weight: 3.17 kg
Power supply: 14 or 28 V DC, 1.25 A

1D960 altitude preselect/alerter **1997**/0001429

Operational status

In production.

Contractor

Century Flight Systems Inc.

VERIFIED

Century IIB autopilot

A single-axis development of the Century I, the Century IIB has two panel-mounted instruments, a directional gyro with a heading bug and an attitude gyro as sensors, and additional navigation options. The system includes a solid-state computer, control panel and roll servo providing lateral stabilisation, roll command and heading select. An optional radio coupler permits heading select, VOR capture and tracking, increased sensitivity for VOR approaches and localiser capture. An HSI may be substituted for the directional gyro.

Specifications

Dimensions:
(directional gyro) 85.6 × 82.6 × 161.8 mm
(control unit) 92.3 × 50.8 × 108 mm
Weight: 4 kg
Power supply: 14 V DC, 2 A or 28 V DC, 1.5 A

Operational status

In production.

Contractor

Century Flight Systems Inc.

VERIFIED

Century 2000 autopilot/flight director **1997**/0001432

Century III autopilot

The Century III two-axis autopilot comprises directional and attitude gyro panel-mounted instruments, control unit, computer/amplifier and electric servos. It is claimed to be approved for more makes and models of aircraft than all of its competitors combined, and to be the preferred altitude hold system for dealer installed retrofit installations in a wide range of singles and twins. The standard features include separate roll and pitch engagement, altitude hold, pitch and roll command and automatic or manual electric pitch trim. The optional radio navigation coupler is identical to that of the Century IIB.

Specifications

Dimensions:
(control unit) 127 × 57.2 × 63.5 mm
(panel instruments) as for Century IIB
Weight: 8.9-10.9 kg depending on options
Power supply: 14 or 28 V DC, 4.5 A

Operational status

In production.

Contractor

Century Flight Systems Inc.

VERIFIED

Century 41 autopilot/flight director

The first Century autopilot to feature digital processing, the Century 41 is a two-axis system with built-in VOR/Loc/GS couplers and outputs to drive a single cue V-bar flight director. It employs both position and rate signals to command the flight control servos. Other features include synchronised pitch attitude and altitude hold modes, a pitch modifier, automatic and manual electric trim and automatic preflight check schedule. VOR/Loc/GS capture intercepts are tailored to groundspeed, intercept angle, wind direction and distance from the ground station. The VOR tracking circuitry incorporates gain reduction to ensure 'soft' passage around the station.

When using the go-around mode, the pilot merely has to press a single button, clean up the aircraft and add power; the autopilot flies to a calibrated pitch-up attitude appropriate to the single engine safety speed for that particular type of aircraft, and turns on to a new heading preset by the pilot. The NSD-360A slaved or unslaved HSI may be substituted for the standard directional gyro. Both vacuum and electric variants are available.

Specifications

Dimensions: 3 ATI panel displays
Weight (installed): 16.2-17.5 kg

Operational status

In production.

Contractor

Century Flight Systems Inc.

VERIFIED

Century 2000 autopilot/flight director

The Century 2000 is the company's newest model. The panel-mounted one-, two- or three-axis modular design autopilot provides economical upgrading from any of several levels of capability.

Any Century 2000 can be expanded to include fully coupled roll and pitch plus yaw damper three-axis autopilot with flight director and altitude preselect by installing expansion kits. The basic system is roll axis with heading and built-in VOR/Loc coupler. Automatic proportional pitch trim and glide slope coupler are standard with the pitch axis.

The Century 2000 has a number of automatic performance and safety features such as soft track and internal circuit checks. The system will interface with the NSD-360A HSI and any other ARINC HSI. The flight director steering horizon is the single-cue delta/vee presentation. Non-flight director and heading only directional gyros are also compatible.

Specifications

Dimensions:
(panel space) 158.3 × 57 mm
Weight: 8.3 kg
Power supply: 14 or 28 V DC, 4.5 A

Operational status

In production.

Contractor

Century Flight Systems Inc.

VERIFIED

Fuel Savings Advisory and Cockpit Avionics System (FSA/CAS)

Delco provides the Fuel Savings Advisory and Cockpit Avionics System (FSA/CAS) for the upgrading of the current Boeing C-135 and KC-135 fleet of about 700 aircraft.

In this system, a CRT control and display unit replaces one of the standard inertial navigation system units to simplify crew management tasks; in particular, all basic data for flight use can be entered at this one point. In addition to the usual engine and navigation information, a new fuel management panel and centre of gravity display have been installed for inputs of fuel state, disposition and usage. This latter function is invaluable in tanker operations and ensures efficient fuel allocation and usage. The panel is linked to other avionic systems by a MIL-STD-1553B databus.

The FSA provides commands to the pilots during climb, cruise and descent using flight path optimisation algorithms and flight manual data for lift, drag and thrust. An additional mode computes all required take-off and landing parameters based on crew inputs for present conditions.

Many KC-135s are being refitted with GE/SNECMA CFM56 engines, which provide extra thrust over earlier types. FSA/CAS is integrated with this engine, offering more economic operation.

Specifications

Dimensions:
(FSA computer) ½ ATR long
(control/display unit) 146 × 181 mm
(bus controller) 207 × 131 mm
(fuel panel) 457 × 123 mm
(fuel management computer) 235 × 249 mm
(remote display unit) 83 × 83 mm
Weight:
(FSA computer) 11.9 kg
(control/display unit) 4.5 kg
(bus controller) 4.1 kg
(fuel panel) 12.5 kg
(fuel management computer) 21.8 kg
(remote display unit) 0.95 kg
Power:
(FSA computer) 95 W
(control/display unit) 44 W
(bus controller) 34 W
(fuel panel) 92.8 W
(fuel management computer) 139.5 W
(remote display unit) 10 W

Operational status

In service on US Air Force Boeing C-135 and KC-135 aircraft.

Contractor

Delco Electronics.

VERIFIED

Mission computer and display system for the C-17

In September 1986, Delco Electronics was awarded a contract for the development of a prototype mission computer and electronic display system for the C-17.

Although the computer is similar to a flight management system, it features a number of other functions including parachute cargo extraction guidance and automatic flight control system coupling for station-keeping in formation flying.

Operational status

In service on US Air Force C-17 transport

UPDATED

Performance management system

The Delco performance management system was initially developed for use in the Boeing 747 as a retrofit system and as an option on the 747-200 and 747-300.

It is a two-box system with the addition of an engine interface unit and a switching unit. The first device processes engine discrete signals, such as fuel flow, into a format usable by the computer. The switching unit handles all switching of autopilot, autothrottle and instrument signals from standard aircraft configuration to performance management control.

The performance management system achieves closed-loop control of the aircraft by providing commands to the autopilot for vertical steering, and either directly or via the inertial navigation system to the autopilot for lateral steering and the autothrottle for engine thrust control.

Delco has also adapted the performance management system for all models of the DC-9, although the engine interface unit is not required, and in the case of the MD-80 the switching unit is not needed. The system has also been adopted for the DC-10 and Boeing 727 airliners.

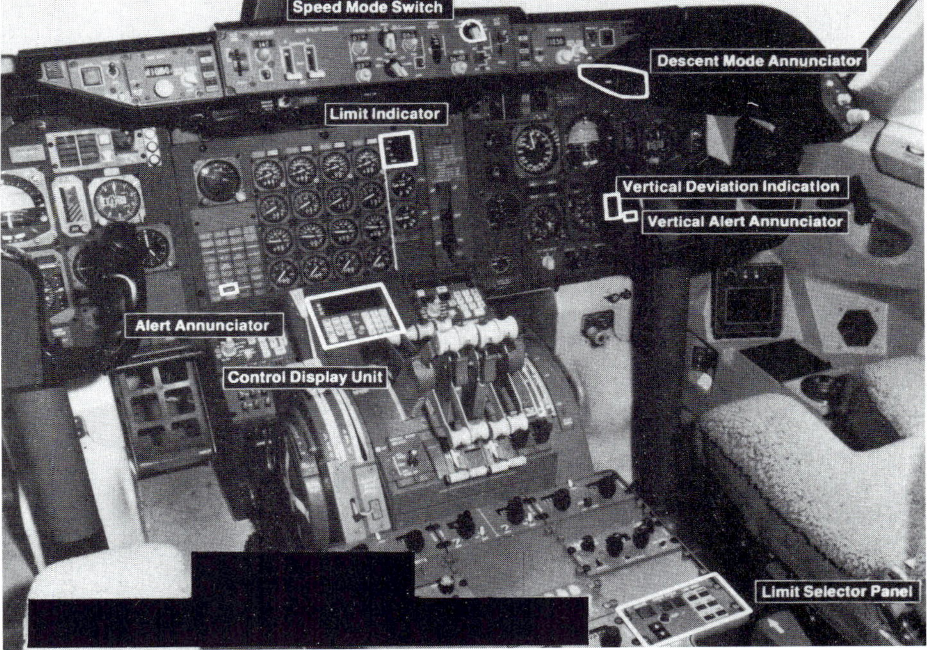

Installation of the Delco performance management system in a Boeing 747

Specifications

Dimensions:
(computer unit) ½ ATR long
(control/display unit) 146 × 114 mm high
(engine interface unit) ¼ ATR long
(switching unit) ½ ATR long
Weight:
(computer unit) 13.2 kg
(control/display unit) 2.7 kg
(engine interface unit) 10 kg
(switching unit) 10 kg
Power:
(computer unit) 200 W
(control/display unit) 38 W
(engine interface unit) 4.2 W
(switching unit) 110 W

Operational status

In service on Boeing 727, 747, and DC-9, DC-10 and MD-80 aircraft.

Contractor

Delco Electronics.

UPDATED

Microtrac II engine vibration monitor

Microtrac II is a microprocessor-based engine vibration signal conditioner. Vibration is detected by piezoelectric accelerometers mounted at sensitive regions in the engines. The electrical signals proportional to vibration are transmitted by the accelerometers to the signal conditioner in the avionics bay.

The Microtrac II employs digital signal processing and high-performance digital finite impulse response filters. The unit provides data which is processed by both programmable digital broadband and digital narrowband filters. The narrowband tracking filter is controlled by the output from a tachometer on the engine to isolate the vibration frequency. The narrow bandwidth means that the signal-to-noise ratio of the final vibration indication is high, and the digital filter's transient response is fast so that it can track the fundamental vibration frequencies of each of the engine's rotors during rapid accelerations.

Digital data is input and output via an ARINC 429 compatible serial digital databus. Analogue data is output as voltage or current for flight deck meter display and flight recording systems. A variety of engine vibration data can be recorded in flight and stored in non-volatile memory for later use in determining engine health. This flight history is output over the ARINC 429 bus.

All Microtrac II models provide in-flight phase angle data which can be used by maintenance personnel for engine trim balancing. The phase and vibration amplitude data can be stored in the unit or in separate data acquisition systems. The in-flight balancing measurement can be accomplished more accurately than on the ground and repeated ground run-ups are avoided, saving time and fuel.

The system uses a generic approach to produce a

standard hardware unit which will satisfy the specifications for most engine/aircraft configurations. Software is then developed to satisfy the specific needs of each individual engine/aircraft configuration. This software is incorporated into a personality module which is fitted to the front of the Microtrac II.

The design utilises ASICs and digital processing circuitry to reduce the number of components required and simplify the mechanical packaging. The easily accessible front panel connector provides ARINC 429, conditioned wideband velocity and tachometer outputs, and analogue discrete inputs to control operation of the unit. The front panel LED display is used to display generated BITE faults and stored flight history information.

Specifications

Dimensions: 93.2 × 193.5 × 320 mm
Weight: 2.8 kg
Reliability: 30,000 h MTBF

Operational status

Standard equipment on Boeing aircraft, including the MD-11. Microtrac II has also been selected as the EVM signal conditioner on the MD-90 and as the vibration monitor for the Tay-engined Boeing 727.

Contractor

Endevco.

UPDATED

FCC100 automatic flight control system

The FCC100 system was designed specifically for the US Army Sikorsky UH-60 Black Hawk helicopter. It meets requirements calling for quick and easy field maintenance and repair and is claimed to have been the first application of a dual-digital processor-based stability augmentation system for a production helicopter. Features include three-axis stability augmentation with turn co-ordination, pitch, roll and yaw attitude and altitude/airspeed hold modes.

Specifications

Dimensions: 356 × 305 × 217 mm
Weight: 8 kg
Power: 75 W

Operational status

In production for the Sikorsky UH-60 Black Hawk helicopter.

Contractor

Hamilton Standard Division of UTC.

VERIFIED

FCC105 automatic flight control system

The FCC105 automatic flight control system introduced modern digital technology features in Sikorsky HH-60, CH-53E, SH-60 and MH-53E helicopters. It is a dual-redundant system which provides three-axis stability augmentation, turn co-ordination and hands-off flight. It can also include a stick-force feel system and has four degrees of freedom facilities and hover augmentation. The most recent version, on the Sikorsky MH-53E helicopter, includes MIL-STD-1553B digital data interfaces. A non-redundant derivative is utilised on the SH-60B. A further derivative of this, with the nomenclature 'general computer', has been developed for use on H-60 helicopters other than the SH-60B.

The baseline 53E computer used in the Sikorsky SH-60B includes automatic approach to hover, three-axis Doppler and radio altitude coupling, with provision for tactical navigation computer coupling.

Specifications

Dimensions: 190 × 432 × 266 mm
Weight: 11.5-13.5 kg
Power: up to 83 W

Operational status

In production for Sikorsky CH-53E, MH-53E, HH-60 and SH-60 helicopters.

Contractor

Hamilton Standard Division of UTC.

VERIFIED

Advanced Common Flightdeck (ACF) for DC-10

Honeywell Air Transport Systems is working with Boeing and Federal Express to install a new 'Advanced Common Flightdeck' (ACF) on the Federal Express company's fleet of DC-10 trijets.

The cockpit upgrade is part of a long-term, two-phase project that could eventually convert many DC-10 transports to advanced technology freighter aircraft. The upgraded aircraft will be redesignated MD-10s; the first are expected to join the Federal Express fleet in mid-1999.

The 'glass cockpit' programme will give the upgraded aircraft the latest in cockpit and control systems technology, including MD-11-derived Category IIIb autoland capability and navigation systems/sensors, as well as commonality with new aircraft coming off assembly lines.

The ACF is based on Honeywell's integrated cockpit design for the two-crew MD-11, with six across 8 x 8 in displays showing all flight and systems information.

The electronic flight instruments, based on Honeywell's liquid crystal 'flat panel' displays, will actually take the MD-10 cockpit a generation beyond the MD-11.

The ACF derives from the aircraft system controllers originally designed for the MD-11. This system manages the functions of all major aircraft systems including the hydraulic and fuel systems, eliminating the need for a flight engineer as a crew member.

The ACF also uses a new generation of high speed, high capacity computers built around Honeywell's Versatile Integrated Avionics (VIA) architecture.

The VIA architecture is based on shared computing resources and robust partitioning developed by Honeywell for its Integrated Modular Avionics (IMA) concept.

The VIA system design for the DC-10 upgrade will integrate multiple functions in the VIA computers, including: displays; flight management; central aural warning.

Three of the VIA computers, coupled with a pair of new aircraft interface units linking them to the aircraft systems, replace 22 separate computers in the existing DC-10 design. With other changes in the avionics bay, an overall saving of approximately 1,000 lb in removed equipment is predicted as a result of the ACF modification.

The VIA-based ACF concept is also being designed into the new MD-95 100-passenger twin jet.

Operational status

An initial ACF installation in a flight test DC-10 will begin in early 1998, with first flight of the ACF-configured aircraft expected in the third quarter of 1998.

Contractor

Honeywell Inc Air Transport Systems.

UPDATED

ADIRS Air Data/Inertial Reference System

Honeywell has integrated the air data and inertial functions for the 777 into an innovative fault-tolerant configuration replacing triple-redundant inertial and dual-redundant air data systems with a single, self-monitored system.

This system, which includes a fault-tolerant integrated Air Data/Inertial Reference Unit (ADIRU), a fiber optic-based secondary Air Data and Attitude Reference Unit (SAARU) and six Air Data Modules (ADMs), offers significant airline benefits in the form of lower cost of ownership and high dispatch reliability.

The ADIRU's fault-tolerant design allows the airline to defer maintenance to scheduled intervals without compromising either the system's functionality or integrity.

At the heart of the Fault Tolerant (FT)/ADIRS subsystems are the latest advancements in sensor technology — Honeywell's GG1320 ring laser gyro and interferometric fiber optic gyro.

The fault-tolerant architecture uses a combination of six gyros in a 'hexad' configuration, six accelerometers, four independent processors, three power supplies and three fail-operational ARINC 629 digital databusses for high performance and reliability. This six-gyro/accelerometer design provides the functional equivalent of four conventional triple-sensor inertial systems — replacing a total of 12 gyros and 12 accelerometers.

The SAARU operates as a fail-safe secondary system to the FT/ADIRS. The unique 'tetrad' sensor geometry uses four fiber optic gyros and four linear accelerometers to provide the functional equivalent of

Air Data/Inertial Reference System ADIRS **1997**/0001439

six gyros and accelerometers arranged in a conventional triad.

Operational status

In service on Boeing 777 and Airbus A320, A330 and A340 aircraft.

Contractor

Honeywell Inc, Air Transport Systems.

UPDATED

AIMS Airplane Information Management System

Honeywell's AIMS architecture, for the B-777 is an advanced avionics system designed to meet very high requirements for functionality, maintainability and dispatch reliability.

The AIMS consists of dual integrated cabinets that contain all of the central processing and input/output hardware to perform the following functions: flight management; displays; navigation; central maintenance; airplane condition monitoring; flight deck communications; thrust management; digital flight data; engine data interface; data conversion gateway.

By eliminating the need for separate Line Replaceable Units (LRUs) for each subsystem – each with its own power supply, processor, chassis, operating system, utility software, input/output ports and built-in test – the AIMS concept saves significant weight, space and power consumption on board the airplane while improving overall system reliability and maintainability.

Displays

The 777's flight deck features six 'D' size (8 × 8 in) Active Matrix Liquid Crystal Displays (AMLCDs) in a horizontal layout similar to the 747-400. These advanced technology screens display primary flight, navigation and engine information with automatic reversion capability.

The flight management function for the Boeing 777 takes advantage of Honeywell's mature systems developed for the 757, 767 and 747-400 airplanes while at the same time providing greatly improved performance, functional capability and growth potential for the Future Air Navigation System (FANS).

The system includes three passively cooled Multifunction Control Display Units (MCDUs) featuring full-colour flat panel LCDs. The MCDU controls the flight management function and is capable of controlling other ARINC subsystems.

The Honeywell CMF consolidates the maintenance activity of virtually all systems on the B-777. A primary objective is to reduce maintenance by improving fault isolation and detection capability, thus reducing the number of spares needed as well as minimising dispatch delays and cancellations.

The B-777's ACMF enables the flight crew to make informed decisions about the state of the aircraft. The system continuously monitors engines and aircraft systems and generates reports that can be customised to individual airline needs.

The DCMF/FDCF is the intermediate system of the airborne datalink infrastructure. It serves as the router between various airborne applications, the printer, the

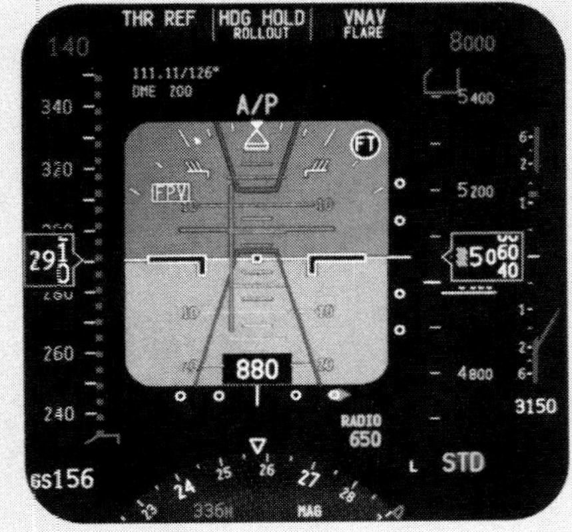

AIMS Primary flight display ***1997***/0001436

AIMS cabinet ***1997***/0001435

onboard Local Area Network devices and ground-based datalink applications via various air-ground subnetworks (such as VHF radio and SATCOM).

It also functions as the crew interface to the datalink system by means of cockpit displays, the CDU and cursor control device. It supports airline programmable ACARS applications.

The Boeing 777 AIMS provides capability for the airlines to customise various aircraft systems such as the Airplane Condition Monitoring Function (ACMF) as well as for flight management, flight deck and data communications, central maintenance and the electronic checklists. This capability is enabled by the AIMS philosophy of fault containment and software partitioning.

Operational status

In service on Boeing 777 aircraft.

Contractor

Honeywell Inc, Air Transport Systems.

UPDATED

Flight Management Systems

Flight Management Systems (FMSs) provide real-time, full flight regime navigation, guidance and performance management. Lateral, vertical and speed coupling to autoflight systems reduces crew workload, while providing enhanced safety and fuel efficiency. Primary crew interface to the Flight Management Computer (FMC) is through the Control Display Unit (CDU) or Multifunction CDU (MCDU). The FMC also provides key navigation and guidance data to the Electronic Flight Instrument System (EFIS) for crew situational awareness.

The FMS receives a number of sensor inputs such as from air data computers, inertial reference systems/inertial navigation systems, GPS, DME and ILS. Using a loadable navigation database, the FMS can fly preplanned routes or allow the crew to access stored airport, navaid and waypoint data. The database also

AIMS Multifunction Control/Display Unit ***1997***/0001437

contains Standard Instrument Departures (SIDs) and Standard Terminal Arrival Routes (STARs) compatible with published procedures and charts. New FMS-based procedures are being defined to allow for more efficient arrivals and to avoid noise sensitive areas. Navigation database sizes range from 400 kbytes to 4 Mbytes. Newer systems will support growth to 16 Mbytes.

Many of the FMS have two-way digital communications capability over the ACARS network to provide uplink and downlink of winds, flight plans, position and performance data. Introduced on the Boeing 747-400 (FANS 1 package), the FMS will also interface to Air Traffic Control (ATC) to allow direct pilot-controller datalink clearance delivery and request. Using Automatic Dependent Surveillance (ADS) over Satcom, ATC centres will continue to track aircraft outside radar coverage. These features, compliant with the ICAO Future Air Navigation System (FANS) for improved airspace management, will provide significant reductions in fuel burn and time savings through improved operational procedures.

Current systems use Honeywell-designed SDP-175/185 CMOS processors. Due to the tremendous growth in functionality, many of the systems have dual SDP-185 processor designs. New systems will use high-speed RISC processors and advanced partitioning architectures.

On the Boeing 777, the FMS functions described above are hosted on the Aircraft Information Management System (AIMS) cabinet. On the Airbus A320, A330 and A340 family of aircraft, the FMS is hosted within the flight guidance computer. All other systems are 8 MCU LRUs.

Specifications

Dimensions: 8 MCU
Weight: 12.73 kg (varies with aircraft)
Power: 145 W (varies with aircraft)

Operational status

The Honeywell FMS is standard on the Boeing 747-400, 757, 767 and 777; the MD-11, MD-87/88 and MD-90; the Airbus A300-600, A310, A320, A330 and A340; and the Fokker 100/70. In addition, British Airways has retrofitted dual FMS on the 747-100/200 fleet and SAS has retrofitted it on the MD-80 series aircraft.

Pegasus FMS

The latest derivative system, designated Pegasus, was scheduled to receive its first FAA certification in Spring 1998. Airlines will get FANS-1/FANS-A capability, together with significantly enhanced system response times, greater operational capability and increased navigation database capacity.

Pegasus architecture is designed to be used across multiple aircraft hardware platforms: Airbus A320 family, A330, A340; Boeing 717, 757, 767, MD-11 and MD-90. It has been selected by more than 20 airlines and the US Air Force.

Contractor

Honeywell Inc Air Transport Systems.

VERIFIED

SFS-980 digital flight guidance system for the MD-80

Used exclusively on the MD-80, the SFS-980 is an automatic flight guidance system based on relatively few LRUs, especially in respect of the range of functions performed. In addition to a comprehensive selection of conventional autopilot/flight director operating modes, the system is FAA cleared for Category IIIA automatic landings of 50 ft decision height and 700 ft runway visual range and has a full-time autothrottle. The large-scale use of digital computing has led to the installation of a comprehensive built-in self-monitoring and maintenance diagnosis capability. Major LRUs and functions are:

Digital Flight Guidance Computer (DFGC) There are two identical 1 ATR long (13.15 kg) DFGC units individually capable of handling all system functions including fail-passive automatic landing. Within each unit analogue/digital conversions take place and the complete system occupies 31 boards against a maximum capacity of 51. Each digital processor has

30 k words of read-only memory and 4 k words of RAM. In addition to all autopilot/flight director processing, each unit also generates thrust rating indication signals, plus maintenance data storage and status test panel data. The aircraft can have a head-up display for take-off and go-around data presentation and guidance signals for this unit are also generated in each DFGC.

Flight Guidance Control Panel (FGCP) This unit fits in the centre glareshield and contains mode selection and control functions for full-time autothrottle, both flight directors, the autopilot and altitude alerting. Pilots may select which DFGC controls all functions. Autothrottle speed/Mach number, selected heading, vertical speed and selected altitude readouts are shown on seven-segment incandescent lamps which may be dimmed by a control knob on the bottom of the panel. The speed/Mach number knob is a three-position control, while the heading selection is one of two concentric knobs. The outer knob provides selection of maximum bank angle for all autopilot/flight director lateral modes except Loc. The inner knob is a four-position device providing heading and autopilot/flight director heading mode selection. The three-position Alt knob provides altitude selection and autopilot/flight director altitude preselect mode arming.

VHF/nav control panels Two panels on either side of the FGCP allow selection of VORTac station frequencies and courses. Displays are incandescent lamp readouts.

Flight Mode Annunciator (FMA) One FMA on each pilot's panel provides instrument failure warning for ILS, attitude, heading, automatic landing, auto-trim or instrument monitor functions. These indicate which autopilot or flight director system is engaged and warn of autothrottle or autopilot disconnects. The units also annunciate which autothrottle, flight director and/or autopilot modes are armed and in which mode the system is currently controlling.

Autopilot duplex servo drive Three duplex servos drive the ailerons, elevators and rudder. The duplex servo functions only during ILS, land or go-around mode. A linear actuator provides normal yaw damping functions in other flight regimes.

Each servo has two separate DC electric motors whose outputs are summed in a differential gear train with a single output. Each servo sends position and rate feedback signals to the DFGC, where servo models monitor actual servo operation. The duplex servo design provides fail-passive protection against hard over manoeuvres. A fault in one servo channel is cancelled by the mechanical velocity summing conducted in the dual-servo differential.

Autothrottle/speed control system Conducted in the DFGC, this function provides fast/slow attitude director indicator commands throughout the entire flight regime. During take-off and go-around it provides pitch guidance for the flight director, autopilot and optional head-up display. The speed control system provides a speed margin above stall (alpha speed) for all autothrottle modes, in addition to the autopilot stall protection feature.

The autothrottle system automatically prevents excursions beyond maximum operating airspeed and Mach values, slat and flap placard speeds and engine EPR limits. It also keeps the aircraft at or above alpha speed for prevailing flap/slat position and angle of attack, using limit data stored in the DFGC solid-state memory.

Automatic reserve thrust enhances safety, operational economy and noise reduction. In the event of an engine failure, as indicated by a difference of more than 30 per cent between engine N_1 (fan) speeds simultaneously with slats extended indication, the good engine thrust is automatically boosted by about four per cent. Automatic reserve thrust also allows use of less than maximum certified thrust for take-offs and go-arounds without corresponding reduction in gross take-off weight, with benefit to fuel and maintenance costs.

Windshear computer Initial production units were delivered for use on Delta Air Lines MD-88 aircraft in 1988.

Operational status
In service in MD-80 series aircraft.

Contractor
Honeywell Inc Air Transport Systems.

VERIFIED

SP-150 automatic flight control system for the 727

The SP-150 automatic flight control system embodies an automatic landing function and is used extensively in Boeing 727 airliners. It superseded the earlier SP-50 automatic landing system and employs more solid-state components for rate sensing and integration. The system has a dual-channel configuration which meets FAA Category IIIA requirements of 50 ft decision height and 700 ft runway visual range.

A wide range of stabilisation, attitude hold and external sensor steering modes is available in the basic autopilot. Additionally there is provision for area navigation steering and for a radio altimeter input, so control law gains can be varied during the approach. Fail-operational performance is assured by the dual-computer configuration and a dual-channel yaw damper is also part of the overall flight control system.

Computational equipment used in the system includes two ½ ATR pitch computers, two ⅜ ATR yaw damper couplers and a single ¾ ATR roll control computer.

Operational status
In service in the Boeing 727.

Contractor
Honeywell Inc Air Transport Systems.

VERIFIED

SP-177 automatic flight control system for the 737

The SP-177 is an integrated digital/analogue automatic flight control system for the Boeing 737 twinjet airliner providing Cat. IIIA automatic landing capability. Main elements of each installation are two flight control computers and a glareshield-mounted controller. These are associated with an AD-300C ADI and an RD-800J HSI for each pilot. The system can be integrated with a performance management system and autothrottle.

Each computer performs pitch and roll computation for autopilot and flight director functions. The system has been configured so that pilot involvement is minimised and to ensure the flexibility of the system is available with simple and logical crew control operations. The pitch axis uses pitch attitude and rate, altitude rate, vertical acceleration and longitudinal acceleration for stabilisation. Vertical acceleration is blended with the altitude rate signal from the air data system to provide filtered altitude rate for altitude acquisition, altitude hold, vertical speed and glide slope control. Radio altitude and radio altitude rate are used to command automatic landing flare. Additionally, there is a flight director take-off mode which, based on flap setting and an angle of attack submode, keeps speed above the stall.

The design aims to provide high maintainability and reliability, has built-in self-test features and incorporates the minimum number of LRUs. Operation of the built-in test facilities can be selected from the flight deck and provides a readout on a performance data computer system display.

Specifications
Dimensions:
(flight control computer × 2) 1 ATR long
(mode control panel) 440 × 74 × 348 mm
System weight: 38 kg
Digital interfaces: ARINC 429
Power: 160 W

Operational status
In service. The SP-177 system is installed in Boeing 737-200s.

Contractor
Honeywell Inc Air Transport Systems.

VERIFIED

SP-300 digital automatic flight control system for the 737-300

The SP-300 is essentially an all-digital version of the hybrid analogue/digital SP-177 automatic flight control system designed for earlier versions of the Boeing 737-200 (see previous entry). The previous system employed analogue circuitry to compute the safety critical Cat IIIA automatic landing functions and digital circuits for en route flight control. The configuration of the SP-300 meets the requirements for fail-passive Cat IIIA automatic landing and independent computation for Cat II flight director approach. To meet the safety criteria for approach and landing, the system has a dual/dual configuration; each of the two control channels having two different processors with different software. In this way the possibility of common mode failures and software errors is reduced to a very low level.

The mode control panel on the glareshield provides centralised control for all autopilot, flight director and autothrottle functions.

Operational status
In service in the Boeing 737-300.

Contractor
Honeywell Inc Air Transport Systems.

VERIFIED

SPZ-1 autopilot/flight director for the 747-100/200/300

The SPZ-1 system, with minor changes, is fitted to 100, 200 and 300 versions of the Boeing 747 airliner. It comprises the pitch and roll functions and associated electromechanical ADI and HSI flight directors. Boeing is the design authority, Honeywell making the equipment to Boeing's specifications and drawings. The current SPZ-1 is a three-channel system, pitch and roll computations being accommodated in three separate boxes. The two sets of flight director instruments are normally driven by different computers but can be switched as necessary in the event of a failure. Each of the three channels has its own set of attitude, air data and other sensors and the entire system is designated as fail-operational (defined as the situation whereby no single failure will cause the performance of the system to fall below the limits required by the autoland manoeuvre).

Three SPZ-1 configurations have been developed. As originally planned, the system incorporated two channels in pitch and roll and was designated as a fail-passive (no single failure will cause a hardover control demand) Cat II system. The two channels are compared one with another by means of a single box Monitor and Logic Unit (MLU). This system also incorporates a one-box automatic stabiliser trim. From this was developed the basic three-channel fail-operational system, and with it the 747 became the first US transport aircraft to be certified for Cat IIIA autoland. The upgrading was accomplished partly by adding the third autopilot, partly by changes in the mode select

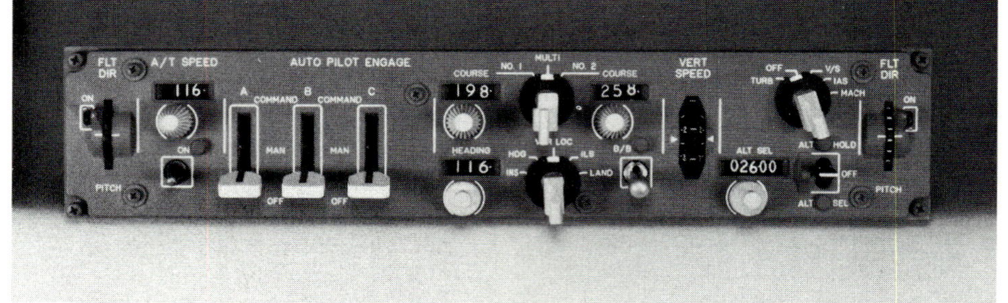

Glareshield controller for the Honeywell SPZ-1 autopilot/flight director system

panel and by the addition of three Landing Control Logic Units (LCLUs) which replace the MLU. In 1976, the autoland capability was extended to include automatic roll-out control along the runway, representing Cat IIIB. This capability was gained by replacing the LCLU with a Landing Roll-out Control Unit (LRCU) and by adding a microprocessor-controlled built-in test schedule.

The performance of the SPZ-1 has since been enhanced by the Analogue Autoland Improvement Programme (AAIP). This was launched in 1977 with the initial aim of minimising the number of disconnects being experienced at ILS capture. It was expanded in 1980 to include optimisation of the ILS tracking capability and touchdown footprint, go-around and landing flare performance.

Specifications

Dimensions:
(roll computer × 3) 3/8 ATR each
(pitch computer × 3) 3/8 ATR each
(autostabiliser) 3/8 ATR
(LRCU × 2) 3/4 ATR each
(LCLU × 2) 3/4 ATR each
Weight:
(roll computer × 3) 7.7 kg each
(pitch computer × 3) 8.1 kg each
(autostabiliser) 4.5 kg
(LRCU × 2) 14.1 kg each
(LCLU × 2) 10.4 kg each

Operational status

All three configurations in service in Boeing 747-100, -200 and -300 aircraft.

Contractor

Honeywell Inc Air Transport Systems.

VERIFIED

Windshear systems

Honeywell windshear systems provide detection, alert and guidance in a single unit with two levels of detection - 'caution' and 'warning'. During take-off and approach, the most critical phases of flight, the systems offer an angle of attack reference on the flight director. The ADI also gives an immediate pitch cue to help in exiting a shear. With extensive filtering and an automatic compensation for aircraft manoeuvres and configuration changes, the windshear systems integrate safety with reliability.

Key windshear features enhance the value of the system. The low installation cost is complemented by nearly universal compatibility. Modification of existing aircraft systems is not required and self-contained sensors reduce the proliferation of system configurations. Pin-programmable for multi-aircraft application, the windshear system meets FAA reliability requirements with a 99.9 per cent availability rating and an undetected failure ratio of .00001.

Honeywell windshear systems are available in two configurations: as a stand-alone unit installed in a 3/8 ATR short box or as a system integrated with the advanced flight management computer system or flight control computer for new airliners.

Specifications

Dimensions: 3/8 ATR short
Weight: 6.8 kg
Power: 22 W
Reliability: 20,000 h MTBF

Operational status

In production and standard fit on Avro 146/RJ, Boeing 727, 737 and 747, Fokker 100 and F28 aircraft, Lockheed Martin L-1011, Boeing DC-8, DC-9, MD-11, MD-80 and MD-90 aircraft.

Contractor

Honeywell Inc Air Transport Systems.

UPDATED

Automatic flight control system for the A 109

The Agusta A 109 flight control system has been designed to reduce pilot workload and improve reliability and safety at a realistic cost. A typical installation includes two autopilot (helipilot) systems for redundancy, which may be used with or without a flight director system.

This duplex system consists of one directional and two vertical gyros, controller panel, two computer units and five series actuators, two each for pitch and roll and one for yaw control. A trim computer and two trim actuators for pitch and roll adjustments are also part of the system.

A flight director facility can be incorporated but requires the addition of a mode controller and navigation receiver inputs, plus ADI and HSI. Flight director functions include heading, vertical speed, barometric altitude and airspeed select, go-around procedure selection, and VOR, ILS, MLS, Omega/VLF, RNav system and Loran or Tacan coupling.

Operational status

In service in the Agusta A 109 helicopter.

Contractor

Honeywell Inc Business & Commuter Aviation Systems.

VERIFIED

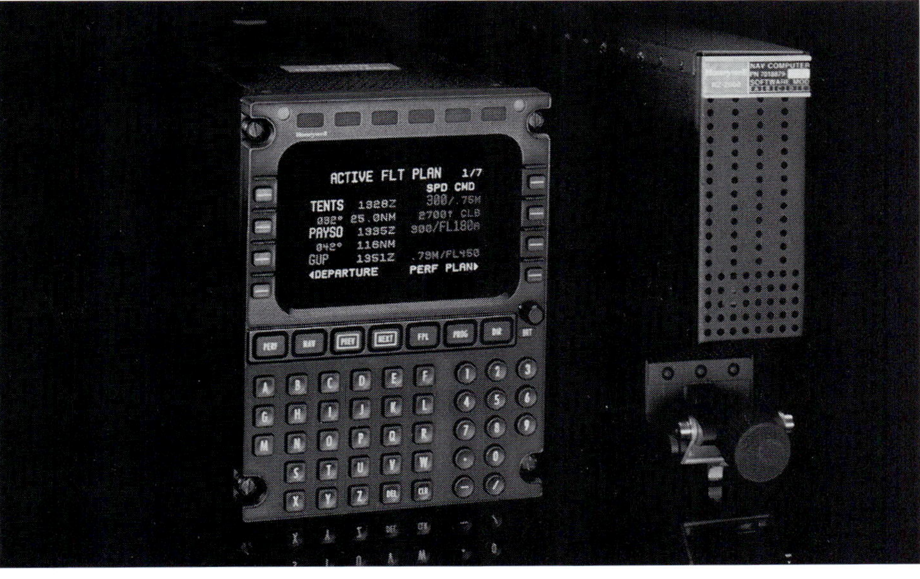

This configuration of the FMZ-2000 flight management system features a control/display unit and a stand-alone navigation computer **1995**

FMZ-2000 Flight Management System (FMS)

The FMZ-2000 is Honeywell's current generation Flight Management System, designed for applications in corporate and regional airliners. The system is available in two configurations: a stand-alone navigation computer or a circuit card in the Honeywell IC-800 integrated avionics computer. Both configurations are compatible with the Avionics Standard Communication Bus (ASCB) and ARINC 429 system architecture. They include similar operational benefits and feature a full colour Control and Display Unit (CDU) and associated data loader.

Both FMZ-2000 configurations provide a premium user interface including an easy-to-read keypad with select keys and a colour display. In addition, the system provides a worldwide navigation database including Standard Instrument Departures (SIDs), Standard Terminal Arrival Routes (STARs) and approach procedures. Multiple sensor inputs, including GPS, IRS, and DME/DME positioning are accommodated. The system provides vertical navigation in climb and descent, multiple holding patterns, and a non-precision approach capability with high precision waypoints.

The system also includes, Honeywell's patented algorithm, SmartPerf™. SmartPerf™ effectively 'learns' aircraft-specific performance. The SmartPerf™ function enables the navigation computer to learn the performance of a specific aircraft, thus affording operators access to performance calculations previously available only to aircraft equipped with a dedicated performance management computer. Using the aircraft's performance database, and both entered and sensed atmospheric data, SmartPerf™ provides performance calculations of time, fuel and predicted altitude at all waypoints; time, fuel, and distance to top of climb and top of descent; maximum endurance targets; optimum cruise altitude; time and distance to bottom of step climb; figure of merit indicating accuracy of fuel calculations and predictions for stored flight plans. Pilots have the option of manually entering altitude and calibrated airspeed or Mach speed constraints.

The FMZ-2000 also performs coupled vertical guidance to multiple, three-dimensional waypoints. Using multiple waypoint vertical navigation allows operators to enter altitudes for each waypoint, both climbing and descending. From this data, the navigation computer displays precise altitude crossings of all predefined waypoints and a complete vertical profile, including top of climb and top of descent on compatible Electronic Flight Instrument Systems (EFIS).

When equipped with Global Positioning System (GPS) sensors and receiving valid GPS data, the FMZ-2000 will navigate entirely by GPS and still maintain input from all available sensors. This affords operators the accuracy of GPS with the integrity and safety provided by a multiple sensor system.

The latest upgrade, known as FMZ-2000 Version 5.0, has completed its TSO flight requirements for the military C-20 aircraft and approval was imminent at the time of going to press. TSO approvals were anticipated for the Citation X, ATR 42 and Gulfstream V by April 1998. The Version 5.X upgrade features a Pentium processor together with a larger 16 Mbyte database memory providing additional capacity for future expansion to meet CNS/ATM requirements. Software improvements to be offered include: parallel database loading, terminal area speed targets, SLS/LAAS compatibility, multiple flight plans on one data loader disk and support of 8.33 kHz communications tuning for European airspace. Take-Off and Landing Data (TOLD) software enhancements, which reduce pilot workload by automating the computation of take-off and landing data, can also be selected by the customer for specific aircraft types.

To make training even easier, Honeywell has optional Personal Computer (PC)-based training for the FMZ-2000. This product will allow pilots and maintenance personnel to accomplish self-paced, interactive training using a CD-ROM-based instruction set. Training is designed in modules and can now be accomplished anywhere the customer has access to a suitably equipped personal computer.

Operational status

TSO'd in a number of aircraft versions, the stand-alone

system is currently standard equipment on the Gulfstream IV, Dassault Falcon 900B, and Raytheon Hawker 800XP. It is optional on the Dassault Falcon 2000 and the Falcon 50EX, Cessna Citation VII, and Embraer 145. it is also STC'd in a number of retrofit applications including the Challenger 600/601, Hawker 800/1000, Boeing 727, Cessna Citation V and Beechjet 400. The integrated system is standard equipment on the Gulfstream V, Cessna Citation X, Dassault Falcon 900EX and Dornier 328. The system will also be standard equipment on the Bombardier Global Express when the aircraft is certified in 1998.

Contractor

Honeywell Inc Business & Commuter Aviation Systems.

UPDATED

SPZ-200A autopilot/flight director system

The SPZ-200A autopilot/flight director system is available for corporate business aircraft such as the Beech King Air and British Aerospace Jetstream 31. It is available with either 4 or 5 in (102 or 127 mm) instruments and includes an air data system. Options include vertical navigation control.

Operational status

In service in Jetstream 31. No longer in production.

Contractor

Honeywell Inc Business & Commuter Aviation Systems.

UPDATED

SPZ-500 automatic flight control system

The SPZ-500 is an integrated autopilot/flight director system suitable for corporate aircraft that provides all the necessary pilot interface controls, air data computation and control servos to fly a selected flight profile automatically. It integrates with companion flight director and air data systems.

The autopilot is a full-time system which provides continuous control through all phases of climb, cruise and descent, and with a full complement of lateral and vertical modes. These may be flown automatically by engaging the SPZ-500 or manually by following computed steering commands presented on the flight director instruments. The latter also enables the pilot to monitor autopilot performance.

The SPZ-500 contains an air data system to provide information over a wide range of flight profiles.

The SPZ-500 is a full three-axis autopilot with a yaw damper. It has acceleration and rate limiting circuits to provide smooth autopilot performance without compromising positive control action. Turn entry and exit is smooth and by programming the roll rate limit as a function of selected mode, rates are matched to the required manoeuvres.

Control of the autopilot for basic stabilisation and attitude command is provided through the autopilot controller. Engaging the system with no flight director mode selected causes the aircraft to maintain the existing pitch attitude, roll to wings level and then hold the existing heading. With a navigation or vertical path mode selected, engaging the autopilot automatically couples the selected mode. When the autopilot is engaged the yaw damper is automatically engaged to provide yaw stabilisation through control of the rudder. When the autopilot is not engaged, the yaw damper may be used separately to assist the pilot during manual flight.

The autopilot controller includes the turn knob and pitch wheel, allowing the pilot to insert pitch and roll commands manually. The amount of bank or pitch change is proportional to the command selected. The soft ride engage button reduces autopilot gains for smoother operation in turbulence, and the automatic elevator trim annunciators show any out of trim condition. The autopilot may be preflight checked with the test button.

Touch control steering allows the pilot to take control of the aircraft momentarily without disengaging the system. The pilot can push the button on the control wheel and manually change the aircraft flight path. While the touch control steering button is pressed, the autopilot synchronises on the existing aircraft attitude. On releasing the button the system holds the new attitude and resumes the coupled flight mode.

Several 4 or 5 in (102 or 127 mm) flight director instrumentation sets can be integrated with the flight control system. If the Honeywell ADZ-241/242 air data system is installed this includes Honeywell air data instrumentation.

Specifications

Autopilot
Dimensions:
(controller) 67 × 146 × 114 mm
(computer) 194 × 71 × 321 mm
(3 servos and mounts) 100 × 129 × 224 mm
(normal accelerometer) 51 × 25 × 61 mm
(trim servo) 56 × 84 × 175 mm
Weight: 13.39 kg total

Air data system
Dimensions:
(computer) 193 × 124 × 361 mm
(altimeter) 83 × 83 × 159 mm
(VNav computer/controller) 38 × 83 × 272 mm
(vertical speed indicator) 83 × 83 × 140 mm
(Mach/airspeed indicator) 83 × 83 × 186 mm
Weight: 9.44 kg total

Flight director system
Dimensions:
(ADI (AD-650B unit)) 129 × 129 × 223 mm
(HSI (RD-650B unit)) 103 × 129 × 208 mm
(remote controller) 38 × 146 × 66 mm
(computer) 194 × 71 × 321 mm
(mode selector) 48 × 146 × 114 mm
(rate gyro) 46 × 52 × 95 mm
Weight: 10.16 kg total

Operational status

In service. SPZ-500 variants are available for the Dassault Falcon 10 and 20, Citation I, II, III, V and VI, Cheyenne II, Cessna Conquest, Mitsubishi Marquis and Solitaire business jets.

Contractor

Honeywell Inc Business & Commuter Aviation Systems.

VERIFIED

SPZ-600 automatic flight control system

The SPZ-600 automatic flight control system is designed specifically for long-range high-performance business jets and is a complete dual-channel system from the sensors to the servos. All roll, pitch and (optional) yaw channels are fully operational at all times and the performance of each is continuously monitored and compared. If a failure that resulted in a hardover manoeuvre occurs in any channel, it is immediately shut down, resulting in single channel operation in that axis only. Performance status is continuously displayed on the autopilot status panel, as well as on the autopilot master warning annunciator. Honeywell says that the unique monitoring system and duplex servo design have been thoroughly tested in transport aircraft. Each of the duplex servos incorporates two independent servo motors that operate from signals applied by their own autopilot channels. The common tie to a single control surface is accomplished through a mechanical differential gear mechanism.

In the event of a system fault, the master warning annunciator flashes amber and may be cancelled by pressing the annunciator. The status panel then indicates the system has automatically disconnected the malfunctioning channel and is a single channel in that axis only. The status panel also enables the manual selection of single channel operation in any axis and has provisions for testing the dual-autopilot channels and monitoring circuits before flight.

Control of the autopilot for basic stabilisation and attitude command is provided through the autopilot controller. Engaging the system with no flight director mode selected causes the aircraft to maintain the existing pitch attitude, roll to wings level and hold the existing heading. With a navigation or vertical path mode selected, engaging the autopilot automatically couples the selected mode. When the autopilot is engaged the yaw damper is automatically in use, when not engaged the yaw damper may be selected separately to assist the pilot during manual flight. The autopilot controller also has a turn knob and pitch wheel, which allow the pilot to insert pitch and roll commands manually. The amount of bank or pitch change is proportional to the command selected. A soft ride engage button reduces autopilot gains for smoother operation in turbulence and a couple button selects which flight director is driving the autopilot.

Touch control steering allows the pilot to take control of the aircraft momentarily without disengaging the system. He pushes the touch control steering button on the control wheel and manually changes the flight path as desired. While the button is pressed, the autopilot synchronises with the existing aircraft attitude and on releasing the button the system holds the new attitude or resumes the coupled flight mode.

The flight director system uses standard 5 in (127 mm) instruments and there is a choice of displays, with either split cue or V-bar directors and vertical scale differences. The ADZ-E 242 air data system includes a full set of appropriate instruments, including altitude alert controller and true airspeed/temperature indicator.

Specifications

Autopilot
Dimensions:
(controller) 69 × 146 × 114 mm
(status/switching panel) 48 × 146 × 114 mm
(2 duplex servos and brackets) 106 × 205 × 298 mm
(computer) 194 × 71 × 321 mm
(yaw actuator) 54 diameter × 232 mm
(normal accelerometer) 31 × 25 × 61 mm
Weight: 14.17 kg total

Flight director system
Dimensions:
(ADI (AD-650B unit)) 129 × 129 × 224 mm
(HSI (RD-650B unit)) 103 × 129 × 203 mm
(remote controller) 38 × 146 × 66 mm
(computer) 194 × 71 × 321 mm
(mode select) 47 × 146 × 114 mm
(rate gyro) 46 × 52 × 65 mm
Weight: 10.16 kg total

Air data system
Dimensions:
(computer) 193 × 125 × 361 mm
(altimeter) 83 × 83 × 159 mm
(vertical speed indicator) 83 × 83 × 146 mm

HSI and ADI flight director displays, computers and controllers are part of the Honeywell SPZ-600 automatic flight control system

(Mach/airspeed indicator) 83 × 83 × 186 mm
(VNav computer/controller) 38 × 87 × 500 mm
Weight: 9.44 kg total

Gyro references
Dimensions:
(vertical gyro) 157 × 165 × 238 mm
(directional gyro) 191 × 154 × 229 mm
(flux valve and compensator) 121 diameter × 73 mm
Weight: 6.08 kg total

Operational status

In service. SPZ-600 variants are available for the Fokker F27, Raytheon Hawker 125-700, Canadair Challenger, Gulfstream III and Cessna Citation aircraft.

Contractor

Honeywell Inc Business & Commuter Aviation Systems.

VERIFIED

SPZ-700 autopilot/flight director for the Dash 7

Chosen as the standard autopilot/flight director combination for the de Havilland Dash 7 transport, the SPZ-700 is a full-time three-axis system for Cat II approaches. It provides all usual vertical and lateral flight director modes, together with RNav and MLS guidance. These vertical modes are altitude hold, altitude select, vertical speed hold and IAS hold, while the MLS mode provides for the steep approach path appropriate to STOL aircraft. The lateral modes are standard heading, navigation, ILS, back course, VOR approach and RNav.

The system is controlled via a single control panel that annunciates functions engaged. Manual demands can be fed in through the pitch trim wheel and a proportional roll knob. For manual flight a yaw damper can be chosen independently of the autopilot.

Operational status

In service in the de Havilland Dash 7 aircraft.

Contractor

Honeywell Inc Business & Commuter Aviation Systems.

UPDATED

SPZ-4000/4500 automatic flight control system

Specifically designed for turboprop aircraft, the SPZ-4000/4500 system provides the accuracy, smoothness and reduced weight and volume advantages of a digital control system. It also offers more control authority than an analogue system and has comprehensive self-test capability. Autopilot, flight director and air data functions are integrated into a single system.

The functions of the three-axis autopilot and flight director are combined in a single flight control computer, and lightweight flight control servos are used. Engaging the autopilot without selection of flight director mode causes the aircraft to maintain the existing pitch attitude and to roll wings level on the existing heading. With a navigation or vertical path mode selected, engaging the autopilot automatically couples the selected mode. The yaw damper is automatically engaged in autopilot operations and can be manually engaged to assist in hands-on flying.

Control functions include pitch wheel and turn knob inputs for pitch and roll commands. There is a preflight autopilot check facility and an annunciator to draw attention to out of trim conditions.

Operational status

The SPZ-4500 is in service on the British Aerospace Jetstream 41, C-160 Transall, Convair 5800, and Beech King Air 200.

Contractor

Honeywell Inc Business & Commuter Aviation Systems.

UPDATED

SPZ-5000 integrated avionics system

The SPZ-5000 integrated avionics system is designed for light business jets and regional airline or business turboprops. Advanced processing technologies and the integration of key functions result in increased capability and flexibility while achieving significant reductions in size, weight, power requirements and installation costs.

The SPZ-5000 is based on the IC-500 display/guidance computer which integrates the EFIS symbol generator, the flight director and the autopilot into a ½ ATR size box. The SPZ-5000 utilises an ARINC 429-based architecture and is composed of an EFIS electronic display offering a choice of either 5 × 5 in (127 × 127 mm) or 5 × 6 in (127 × 152.4 mm) size displays (the latter will incorporate integrated air data displays as an option), single or dual flight director and an optional single or dual fail-passive autopilot. The SPZ-5000 utilises the full range of Honeywell's advanced sensors including the AZ-840 micro air data computer or the AZ-429 air data sensor, separate vertical and directional gyros or the AH-800 fibre optic AHRS, Primus II radio system and the Primus 650 weather radar system. Standard options are Honeywell's TCAS, MLS, lightning sensor, Primus 870 turbulence detection weather radar, Primus 700 or Primus 450 weather radar and the Laseref III inertial reference system.

Operational status

In production. First certificated on the Cessna CitationJet 525 in October 1992. The Loadmaster is to be certified in 1999.

Contractor

Honeywell Inc Business & Commuter Aviation Systems.

UPDATED

SPZ-7000 series digital flight control system

The SPZ-7000 is Honeywell's fourth-generation helicopter autopilot, and is claimed to be the first digital pitch, roll, yaw and collective four-axis system to receive civil certification. The system is microprocessor-based and has two flight computers providing full autopilot and stability augmentation, dual-flight director mode selectors, automatic preflight and en route testing features and comprehensive diagnostic circuits.

A version of the system designated SPZ-7300 was chosen by Agusta for search and rescue versions of the A 109, AB 212 (a licence-built version of the Bell 212) and AB 412. The SPZ-7300 couples a helipilot system with a digital flight path computer added to provide important new functions for the search and rescue mission. In addition to the standard flight director modes, it generates a two-stage decelerating approach to the hover, hover augmentation and the ability to hover while coupled to a Doppler navigation reference for better hover performance.

Operational status

The system was certified aboard the Sikorsky S-76 Mk II in November 1983 (now also certified on the S-76B model) and in December 1984 was specified by the airframe company as the factory standard option for this upgraded version of the corporate helicopter; in the S-76 Mk II the dual-channel system permits Instrument Meteorological Conditions (IMC) operation with one pilot. The system has also been certified on the Eurocopter AS 365N helicopter and for the Bell 222UT utility transport helicopter. All versions in production.

Contractor

Honeywell Inc Business & Commuter Aviation Systems.

UPDATED

SPZ-7600 integrated SAR avionics

The SPZ-7600 is a system designed to fill the civil search and rescue role for operations in all weather conditions worldwide. It includes dual FZ-706 flight control computers, helicopter optimised EDZ-705 EFIS displays with special SAR symbology, the Primus 700 Series surface mapping/weather/beacon radar and solid-state AA-300 radio altimeter systems with circuitry and displays specifically for rotary-wing applications.

The SPZ-7600 is built around the single pilot IFR technology of the SPZ-7000 (see item above), proven in service on Sikorsky S-76 helicopters in global conditions.

During critical operations, the SPZ-7600 can be programmed to execute an automatic hands-off approach to hover and auto-hover with velocity hold at an electronically pinpointed datum.

Operational status

In production. Certified for Bell 412 and Sikorsky S-76 helicopters.

Contractor

Honeywell Inc Business & Commuter Aviation Systems.

VERIFIED

The Honeywell SPZ-7000 digital automatic flight control system is installed in the Eurocopter AS-365N helicopter

SPZ-8000 digital flight control system

The SPZ-8000 digital flight control system was designed with emphasis on corporate aircraft such as the Raytheon Hawker 800. The system is unusual among its kind in having a bidirectional databus - the Honeywell ASCB avionics standard communications bus - and includes a flight management system accommodating both lateral and vertical guidance. Honeywell believes that its ASCB system provides both the higher refresh rates needed for flight control applications and a greater flexibility in use than the one way ARINC 429 standard. The system is built around two FZ-800 flight computers and is fail-operational.

SPZ-8000 all-digital autopilot for fixed-wing aircraft

Operational status

In production. The first aircraft to be certified was the de Havilland Canada Dash 8. This was followed by the Raytheon Hawker 800 and 1000, Aerospatiale/Alenia ATR 42, Cessna Citation III and the Dassault Falcon 900. The Gulfstream G-IVSP and Canadair Challenger CL-601-3A followed in 1987. Latest standard installations include the ATR-72 and Citation VII and Gulfstream IV.

In the Gulfstream IV installation, the SPZ-8000 system integrates: the Electronic Flight Instrument System (EFIS); the Engine Instrument and Crew Alerting System (EICAS); Digital Flight Control System (DFCS); Flight Management System (FMS); Laseref II Inertial Reference System (IRS); and Primus 870 Weather Radio System.

Contractor

Honeywell Inc Business & Commuter Aviation Systems.

UPDATED

SPZ-8500 integrated avionics system

Honeywell's SPZ-8500 integrated avionics system for the Gulfstream V aircraft includes a six-tube EFIS/EICAS with 8 × 8 in CRT displays, dual IC-800 integrated avionics computers, dual FMZ-2000 flight management systems and dual fail-operational autopilot/flight directors. Other system features include triple micro air data computers, triple Laseref III inertial reference system and Primus 870 turbulence detecting weather radar.

Operational status

Certified on Gulfstream V in April 1997.

Contractor

Honeywell Inc Business & Commuter Aviation Systems.

VERIFIED

Honeywell SPZ-8500 integrated avionics system on Gulfstream V flight deck
1997/0003303

AMS 2000 Avionics Management System

Honeywell's AMS 2000 is a fully integrated modular avionics concept/system designed to meet the needs of military tanker/transport/maritime patrol aircraft. It integrates flight management systems, flight controls, cockpit instruments and cockpit controls. Modular architecture is built around the system processor with growth to TCAS, MLS, HUD, digital map, NVIS-compatibility, DGPS and CNS/ATM.

Operational status

In service with C-130, C-141 and P-3 aircraft of various countries.

Contractor

Honeywell Inc Defense Avionics Systems.

VERIFIED

Royal Netherlands Air Force (RNLAF) CH-47 upgraded cockpit ***1996***

Avionics Control and Management System for the CH-47D (ACMS)

The Avionics Control and Management System (ACMS) for the CH-47D helicopter provides the primary interface for the pilot and co-pilot to aircraft pilotage, mission and air vehicle systems. Although some components are essential to more than one function, the ACMS can be divided into the following general functional areas: pilotage subsystem, mission subsystem and air vehicle subsystem.

The **pilotage subsystem** provides control and display of aircraft sensor data, selection of navigation sources, and other functions essential to flying the helicopter. The EFIS, a primary element of the pilotage subsystem, replaces traditional mechanical gauges, indicators and control heads with electronically generated displays, allowing the operator to select and tailor presentations to provide the optimum level of information during each mission phase.

The **mission subsystem** provides mission planning utilities, storage and retrieval of navigation reference points, and control of communication and navigation radio lists. Pre-mission planning is normally accomplished on a ground-based workstation and downloaded to the aircraft via a data transfer cartridge. The air vehicle subsystem allows continuous performance monitoring of all major aircraft systems including warning/caution/advisory annunciations and the capability to interface to the health and usage monitoring system.

The **air vehicle subsystem** monitors engines and transmissions, fuel and hydraulic systems, flight controls and actuators, the electrical system, and other aircraft systems and equipment. Critical information about the engines and rotors is continuously available on torque/cruise/RPM displays.

System design incorporates redundancy to minimise vulnerability to failures and damage and to enhance the accuracy of navigation equipment. The ACMS interfaces with other aircraft equipment via two dual-redundant MIL-STD-1553B multiplex databusses, special purpose high- and low-speed ARINC 429 buses, and aircraft discrete, synchro and proportional analogue I/O interfaces. The ACMS also receives video signals from the weather radar system for display in the cockpit. A digital map function provides real-time tracking of helicopter position overlaid on paper chart or digital terrain-elevation data displays. Processing and subsystem interfaces are managed by four integrated system processors (two mission, two air vehicle). The primary crew interface is supported by four multifunction displays (two mission, two air vehicle) and two control/display units. Additional information is displayed on two electronic flight instrumentation systems, each comprising of an electronic attitude/direction indicator and an electronic horizontal situation indicator. Two optical display assemblies, which mount onto the pilots' night vision goggles, provide flight symbology during night-time operation. Navigation redundancy is accomplished through use of a Kalman filter to blend navigation sensor data into the most accurate solution of aircraft position and motion, even during transient periods or loss of individual navigation sensor inputs.

The ACMS provides comprehensive integration and control of mission management, air vehicle management and pilotage subsystems. Its flexible architecture will accommodate growth well into the next century. Central computers provide control and data handling capability for a wide variety of aircraft systems and equipment. This control and computation ability decreases crew workload and enhances navigation accuracy and mission safety. Overall, the ACMS provides the pilot and co-pilot with added functional capabilities which contribute to a more effective and efficient mission.

Operational status
The RNLAF accepted seven CH-47D aircraft with ACMS in early 1996. Six more ACMS-equipped Chinooks will be delivered to the RNLAF in 1998. Other international customers have expressed interest.

Contractor
Honeywell Inc Defense Avionics Systems.

VERIFIED

Belgian Air Force C-130H avionics upgrade programme

In January 1992, the Belgian Air Force selected Honeywell Defense Avionics Systems and Sabena to design, develop and produce a state-of-the-art cockpit avionics upgrade for the C-130H. The objective for the modernisation programme was to provide additional performance functions and, at the same time, improve the aircraft's availability to carry the Belgian Air Force C-130H fleet well into the next century. In response to these objectives, the Honeywell/Sabena team orchestrated a solution that included: a precision navigation solution; a fully automated digital flight management system to reduce the crew's workload; a digital solution for an integrated autopilot/flight director; electronic flight instruments; a digital, ground map weather radar; and a two-level maintenance support system that draws heavily on the BIT functions resident within the system processor and the individual LRUs.

The system designed for this modernisation programme is referred to as the Integrated Vehicle Mission Management System (IVMMS). A mix of commercial off-the-shelf and military non-development hardware was utilised to keep the IVMMS affordable. Both MIL-STD-1553B and ARINC 429 bussing is used to integrate the various military and commercial hardware. Sufficient I/O for the Flight Management processor has been included to directly tie a myriad of existing COMM/NAV radio sensors and analogue avionics to the system processor. System requirements and qualification test profiles were specifically tailored by the Belgian Air Force to reflect the existing operating conditions experienced by C-130 aircraft accomplishing a worldwide mission scenario. This foresight in tailoring the design requirements to existing conditions rather than standard MIL requirements is responsible for the cost-effective design that directly satisfies the Hercules' mission.

Honeywell is responsible for the avionics design, integrated, test, qualification and production hardware. Sabena is responsible for the installation of the IVMMS hardware and integration of the IVMMS with the other C-130H Hercules systems.

Operational status
The first of 12 Belgian Air Force C-130Hs officially rolled out 31 May 1995. Production deliveries will be complete for the BAF C-130H fleet in 1999.

Contractors
Honeywell Inc Defense Avionics Systems.

VERIFIED

C-130J Communication/Navigation/Identification Management System (CNI-MS)

Honeywell's Communication/Navigation/Identification Management System (CNI-MS) provides C-130J aircrews with the latest technology in an integrated avionics architecture. The CNI-MS architecture includes: dual system processors, solid-state electronics; three mission management units for the crew interface; flat panel, NVIS-compatible displays; and MIL-STD-1553B bussing.

The CNI's operational functions reside in the system processor and include: a fully digital flight management capability; flight planning; database management; lateral and vertical guidance; climb, cruise and descent performance; take-off and landing data; COMM/NAV radio management; and a precision integrated navigation solution that blends GPS/INS, Doppler Velocity System (DVS), and VOR/TACAN nav sensors.

Either CNI processor can manage all of the C-130J's COMM/NAV functions in the case of battle damage or an LRU malfunction. Single CNI operation includes control of all three mission management units, again with individual pages, if desired.

Mission planning data from the customer's ground-based mission planning station is uploaded into the CNI-MS via a 20 Mbyte pocket-size cartridge. Mission planning can also be accomplished in the cockpit by the aircrew with manual entry on any of the three Mission Management Units.

Two complete mission routes can reside in the system processor at any one time, either of which is selectable by the crew. These routes include an origin, destination and alternate, connected by a string of up to 60 waypoints. The routes also include a menu of mission options that can include up to: 10 Computed Altitude Release Points (CARP), four Search And Rescue (SAR) profiles; two rendezvous/intercept points in addition to instrument departures and approach procedures for the origin, destination and alternate airfields.

The aircrew interface to the CNI system is through three individual, mission management units, located in the centre console between the pilot and co-pilot, with easy access to the pilot, co-pilot or an additional crew member occupying the jump seat. The CNI-MUs replace virtually all of the traditional dedicated control heads in the cockpit for the onboard COMM/NAV avionics.

The CNI-MU displays use a flat-panel, thin film electroluminescence technology. Cockpit readability has been judged to be excellent by aircrews. The display is easily readable in direct sunlight.

The displays are also night vision compatible, operating from a level of zero visible light to a brilliance of 100 ft-lamberts.

The CNI-MU units are cross strapped for redundant operation in the event of battle damage or an LRU malfunction.

Particular attention had been directed to the CNI-MU display pages to ensure that the pages are user-friendly for the aircrew. Each major flight management function (examples: take-off and land, COMM/tune, NAV/tune, NAV control, and so on) has a dedicated mode select key for easy access to the most frequently used display pages. Twelve bezel push-buttons are included in this design concept to facilitate data entry from the scratch pad.

Nav radios are automatically tuned by the CNI for the best navigation solution based on present position. COMM radio channels/frequencies for the UHF1,2, VHF1,2 and HF1,2 are preset for the mission plan to include 30 preset channels/frequencies per individual radio.

The CNI-MU contains a full alphanumeric keyboard. Honeywell's design does not require a shift key to convert keystrokes to alphabetic strokes and vice versa.

Crew members can select individual display pages on a given display without affecting what appears on the other cockpit CNI-MU displays.

Six spare card locations are available in each System Processor for customer desired growth fucntions: differential GPS approaches, single card digital map, video processing, voice generation and so on.

Specifications
Dimensions:
(CNI-system processor) 205 × 202 × 317 mm
(CNI-management unit) 146 × 222 × 163 mm
Weight:
(CNI-system processor) 9.1 kg
(CNI-management unit) 3.6 kg
Power:
(CNI-system processor) 60 W, 28 V DC
(CNI-management unit) 30 W, 28 V DC

Operational status
In production.

Contractor
Honeywell Inc Defense Avionics Systems.

VERIFIED

C-130J Digital Autopilot/Flight Director (DA/FD)

The system consists of dual, Automatic Flight Control Processors (AFCP), each with two central processor units for redundant control law computation and mode logic implementation for: autopilot, flight director and autothrottle performance. System inputs to the control surfaces (ailerons, rudder, elevators) are made through three, individually housed, servo drive units containing: servo motor, tachometer, synchro, solenoid engage clutch, drive spline and electrical connector.

Honeywell's digital autopilot/flight director provides a proven solution for the C-130J programme. Early versions of the system flew on C-130E/Hs with the US (RAMTIP and PRAM), Royal Australian and Royal New Zealand Air Forces. An identical hardware version is flying on the Belgian Air Force C-130Hs. Aerodynamic control laws and flight director functions have been thoroughly tested in a series of development programmes for the US Air Force and for foreign governments.

The C-130J version of this autopilot is qualified to both MIL-STD and Federal Aviation Administration requirements in accordance with MIL-STD-810, MIL-STD-454, MIL-STD-461, AC 25.1309-1A, DO-178A and DO-160C. The autopilot interfaces with C-130J sensors and avionics via MIL-STD-1553B and ARINC 429 databusses. The interface to the three servo drive units and the autothrottle is via direct wire, to ensure adequate response time.

Key design features include: digital electronics programmed in Ada; fail-passive operation; dual flight control computers, each with two central processor units for redundant control law and mode logic computation (internal comparison of LRU outputs verify each LRU operation); flight director system internal to the autopilot computer; comprehensive built-in test covering 98 per cent of circuitry (power-up, continuous, maintenance diagnostics); parameter modification permitting selected control law gains to be adjusted under flight test-only conditions for special-purpose C-130 aircraft (air refuelling hardware, external antennas, and so on); autothrottle control for precise speed control; each LRU provides all of the main computing and power amplification functions for the digital autopilot/flight director system (each processor contains dual independent CPUs and ARINC 429/MIL-STD-1553B); comparison of servo outputs for validity and tolerance (disagreement results in system disengagement).

Operational modes for the autopilot include: heading and altitude select; coupled navigation (global positioning system, inertial navigation, Doppler); radio navigation (VOR/LOC, TACAN, GS); altitude hold; turn/pitch knob adjust; coupled flight management system operation; pitch/roll go-around; autothrottle; and speed select.

Specifications

Dimensions: 292 × 203 × 505 mm
Weight: 15.8 kg

Operational status

In production.

Contractor

Honeywell Inc Defense Avionics Systems.

VERIFIED

Digital Automatic Flight Control System (DAFCS) upgrade for the B-52, KC-135 and C-130

The DAFCS comprises three LRUs which replace numerous LRUs of existing systems. These new LRUs are the Digital Amplifier Unit (DAU), the Air Data Sensor Unit (ADSU) and the Status Test Panel (STP).

The DAU is the heart of the autopilot system and is a dual Z-8002 processor design. A separate MIL-STD-1750 processor card can be added later. A 1553B input/output card will be demonstrated as part of the DAU but will not be delivered under this contract.

The ADSU provides the DAU with the necessary sensor information for the airspeed/altitude/Mach hold modes of the autopilot. The STP provides maintenance with an interface during diagnostic testing as well as reading the results of the numerous built-in tests.

Operational status

In service on B-52 and KC-135.

Contractor

Honeywell Inc Defense Avionics Systems.

VERIFIED

EQUIX^{SM} integrated avionics architecture for JSF

Honeywell Defense Avionics Systems will supply avionics for Lockheed Martin's Joint Strike Fighter (JSF) concept demonstration aircraft. Honeywell is allied for the project with Hamilton Standard and Sundstrand Aerospace in an entity called EQUIX^{SM}.

Lockheed Martin's JSF concept demonstration aircraft will be the first application to a military aircraft of the integrated modular avionics architecture Honeywell Air Transport Systems in Phoenix, Arizona, has developed for commercial airliners.

Honeywell will leverage the synergies among its military and commercial avionics divisions to develop components for the vehicle management computer consisting of the processor, high-speed cross-channel datalink, power supply and chassis. Honeywell will also provide an embedded GPS/INS (Global Positioning System/Inertial Navigation System), a radar altimeter, an air data transducer and an engineering test station. Defense Avionics Systems will be supported for this portion of the work by two other divisions of Honeywell's Space and Aviation Control business: Air Transport Systems and Sensor and Guidance products of Minneapolis, Minnesota, and Clearwater, Florida.

Operational status

In development for the Lockheed Martin JSF concept demonstrator aircraft.

Contractor

Honeywell Inc Defense Avionics Systems.

NEW ENTRY

Lasernav special mission management system

The Lasernav special mission management system was designed for aerial photography, surveillance, paradrop, search and rescue, border patrol, maritime patrol and airborne early warning. Added I/O capability provides interfaces with radars, sensors and cameras. The total I/O capability enables the equipment to be fitted in a wide range of aircraft.

The inertial reference system is a derivative of that used in Lasernav. This provides additional analogue outputs, improved performance and the option of an internal GPS receiver. Alignment times of 2½ to 10 minutes give a reasonably quick response. Pure GPS, pure inertial and hybrid outputs are available when the optional GPS receiver and processor module are installed in the inertial reference unit.

Navigation computations, steering and additional I/O are provided by the mission computer which is available in two models. The standard variant provides flight management capability which includes mission planning, flight plan storage, an internal database and fuel management. The advanced version offers camera interfaces, crossfill and automatic fuel flow inputs.

The control and display unit uses a 5 in colour CRT which displays data in a chapter/page format, uses colour to distinguish types of data and can display flight patterns in graphic form. The keyboard utilises full alphanumeric keys, line select keys and function keys. Radio frequency management is also provided via the unit with an optional interface box.

Specifications

Dimensions:
(inertial reference unit) 322 × 324 × 193 mm
(mode select unit) 146 × 38 × 63.5 mm
(standard mission computer) 57 × 197 × 388 mm
(advanced mission computer) 124.5 × 197 × 388 mm
(control and display unit) 146 × 162 × 182 mm
Weight:
(inertial reference unit) 21.3 kg
(mode select unit) 0.45 kg
(standard mission computer) 3.2 kg
(advanced mission computer) 5.0 kg
(control and display unit) 3.2 kg
Power:
(with standard computer) 145 W
(with advanced computer) 160 W

Operational status

In production.

Contractor

Honeywell Inc Defense Avionics Systems.

VERIFIED

Stability Augmentation/Attitude Hold System (SAAHS) for the AV-8B

The SAAHS was the first digital system to be applied to any of the Harrier series of aircraft. It comprises a limited authority stability augmentation system with some conventional autopilot functions to reduce pilot workload. The single-channel system operates on all three axes and is accompanied by a mechanical back-up for reversionary operation. Extensive self-monitoring circuitry switches out the system in the event of a fault, so that it fails passive. Monitoring is accomplished in three ways: equipment, software monitoring of equipment and software monitoring of performance. In addition a comprehensive preflight schedule is provided.

Stability augmentation modes comprise three-axis rate damping in both vertical and cruise flight and in transition between these phases. It also includes rudder/aileron and stabiliser/aileron interconnects to improve turn co-ordination. Autopilot modes include attitude, altitude and heading hold, automatic trim, control column steering and airspeed limit schedules. A rudder pedal shaker alerts the pilot to the onset of potentially dangerous sideslip during the hover.

Principal element of SAAHS is a 6.1 kg digital flight control computer with 2901-bit slice, 16-bit processor working at 470 Kips, with UVEPROM. The flight control programme occupies 20 k words of memory of which 9 k are taken by the control law code and 11 k by the built-in test schedule. The other units comprise a three-axis rate sensor, lateral accelerometer, stick sensor and forward pitch amplifier. In December 1982 the system demonstrated a completely hands-off automatic vertical landing.

Specifications

Weight: 12.3 kg total

Operational status

In service on the US Marine Corps AV-8B.

Contractor

Honeywell Inc Defense Avionics Systems.

VERIFIED

Lasernav special mission management system

Automatic flight control system for the C-5B

Honeywell provided the flight control system for the US Air Force Lockheed C-5A Galaxy. It was claimed to be the first operational flight control system to provide complete Cat. III automatic landing and roll-out guidance with fail-safe performance. Honeywell provides an improved (though still analogue) five-box form, fit and function version of the original flight control system for the C-5B; the computer boxes fit into the same racking as in the C-5A, have the same connectors and are functionally interchangeable with the flight control boxes in the earlier aircraft. Commonality of the two analogue systems, in order to exploit the heavy investment in ground test and avionics maintenance equipment, was a major factor in the decision to go for a minimum change system. The principal purpose of its upgrading is to improve reliability and ease of maintenance. The system incorporates three-axis stability augmentation and autothrottle.

Operational status

In service in the C-5B Galaxy. Depot repair modification for C-5A.

Contractor

Honeywell Inc Sensor and Guidance Products.

UPDATED

IEC 9002 Flight Management Systems (FMS)

The IEC 9002 FMS provides complete LNav and approach capabilities, and accurate navigation in all phases of flight using the 12-channel GPS receiver, which accepts differential GPS corrections.

IEC 9002 features a control display unit with a 4 in diagonal, sunlight-readable, high-resolution, 16-colour LCD and can be enhanced with a kinematic upgrade to provide Cat. III landing accuracy. IEC 9002 provides rapid GPS satellite acquisition (time to first fix 2 minutes). It is equipped with Jeppesen worldwide navigation database including SIDs, STARs, GPS instrument approaches, airports with runways greater than 4,000 ft, high- and low-altitude airways, intersections, VHF navaids and NDBs, and provides for 400 flight plans (company routes or pilot-defined) of 100 waypoints each; 2,000 pilot-defined waypoints.

IEC 9002 is (S)CAT-I DGPS compatible and GLONASS and WAAS upgradable. It is a full-featured GPS flight management system intended for installation in all types of aircraft. The 9002 FMS accepts differential GPS corrections required for (S) CAT-I operations. Accuracies in the 0.5 m range, allowing Cat. III operations, are attained when the IEC 9002 is enhanced with its kinematic upgrade and used with the compatible Model 8000 DGPS ground station.

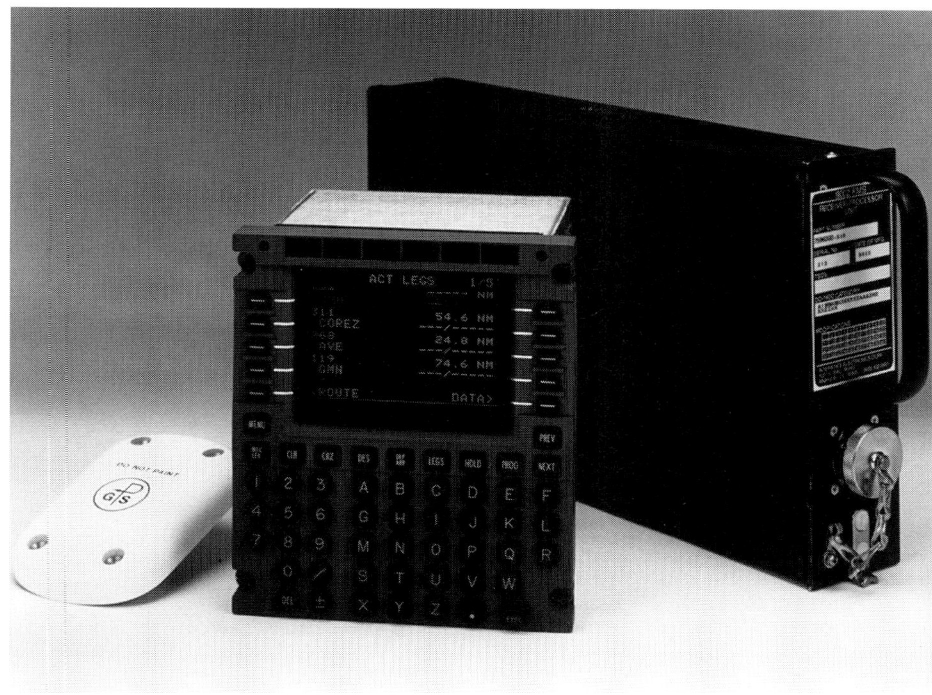

Interstate Electronics Corporation IEC 9002 flight management system ***1998***/0018230

Specifications

Navigational signal: GPS L1 C/A code (SPS)
Receiver type: 12 GPS hardware channels
Time to first fix: 2 min
Dynamics: 900 kt (max) velocity
Accuracy:
100 m 2dRMS (selective availability on)
5 m using differential GPS correction

Operational status

The IEC 9002 has been certified by the FAA to TSO C129 Class A1, and selected by various airlines as an upgrade for Boeing 737-200 aircraft.

Contractor

Interstate Electronics Corporation.

UPDATED

Analogue fully fly-by-wire system for the Pre-Block 40 F-16

The system is entirely analogue in nature, with four channels in each of the three axes providing the necessary redundancy to meet the failure criterion of being able to continue flight after two failures. At the heart of the system is a 20 kg box containing the four flight control computers, each rigorously isolated from electrical contact with any other. Each has its own power supply and individual input signals from the pilot's sidestick controller, sensors and air data system. Each computer processes its signals and provides independent output commands to the servo actuators that move the flying control surfaces.

Fault isolation within the computer box, and subsequent reconfiguration of the computers to compensate for the new situation, are achieved by the use of analogue output voters that interface sensors and computations with movement of the servo-actuators. A 63-function self-test schedule is run automatically as part of preflight checks.

Specifications

Dimensions: 457 × 254 × 203 mm
Weight: 20 kg
Power: 150 W
Digital channels: 4
Independent back-up: 1
Fault tolerance: 2 fail-operate
Analogue inputs: 88
Memory:
(PROM) 16 k
(RAM) 2 k
Language: LSI HOL
Reliability: 2,181 h MTBF

Operational status

In service in F-16A and pre-Block 40 C single-seat and F-16B and D two-seat versions.

Contractor

Lear Astronics Corporation.

VERIFIED

Digital Flight Control System (DFCS) for the F-15E

Whereas the standard F-15 has a Lockheed Martin analogue flight control system, the additional requirements of the F-15E for deep penetration with emphasis on terrain-following and terrain-avoidance, allied with new sensors, called for a digital triplex automatic system of greater performance and reliability. Digital technology permits control laws to be optimised both for air combat and for high-speed low-level flight. Lear Astronics is also responsible for the digital pressure sensors used to provide speed and aerodynamic pressure sensing for the system. The system is based on MIL-STD-1750 instruction set architecture and uses Lockheed Martin Fairchild Systems 9450 microprocessors.

Operational status

In service on F-15E.

Contractor

Lear Astronics Corporation.

VERIFIED

Flight control augmentation system for the KC-135R

Lear Astronics provides the flight control augmentation system for US Air Force Boeing KC-135 tankers retrofitted with General Electric/SNECMA CFM56 turbofan engines. The Lear Astronics flight control augmentation system is essentially a high-authority single-channel yaw damper, based on a rate gyro, that monitors engine rpm differences on the outboard engines and provides a correction signal to the rudder servo if the difference exceeds a given threshold value. The system is nuclear hardened.

Operational status

In production for the Boeing KC-135R.

Contractor

Lear Astronics Corporation.

VERIFIED

Flight control computer for the RAH-66

In 1991, Lear Astronics was selected by Boeing Helicopters to be the flight control computer subsystem contractor for the RAH-66 Comanche helicopter, which is being developed in association with Boeing Defense and Space Group. Lear Astronics has overall responsibility for delivery of an integrated subsystem and for complete systems integration of the triplex computers. Lear Astronics is also developing the controller, flux valve, engineering control panel, flight control panel and back-up computer.

Operational status

In development for the RAH-66 Comanche helicopter.

Contractor

Lear Astronics Corporation.

VERIFIED

F-22A - VMS/IVSC: Vehicle Management System/Integrated Vehicle Subsystem Control

The VMS combines flight and propulsion controls, whilst the IVSC controls the aircraft utilities. The VMS utilises a triplex digital flight control system, with no electrical or mechanical back-up, to provide full carefree handling and enhanced manoeuvrability, with a sidestick fly-by-wire controller.

The IVSC modules control the following services: electric power, hydraulics, fuel systems, integrated warning/caution/advisory functions, diagnostics and health monitoring, auxiliary power, environmental and life support functions.

Equipment modules are included in the avionics common module racks. The Raytheon Systems Company provides common 1750A processor modules for these systems.

Operational status

Installed in F-22A EMD aircraft.

Contractor

Lear Astronics Corporation.

VERIFIED

LTN-400 flight management system

The LTN-400 flight management system consists of the navigation computer, data transfer unit and control/display unit.

The LTN-400 navigation computer incorporates gate arrays, VLSI surface mounting and flash memory devices. The 3.1 Mbyte memory capacity includes provisions for growth up to 5 Mbytes. An advanced microprocessor provides fast and efficient data acquisition. High reliability is achieved through the use of newly developed components which possess greater endurance to stress and shock. Fuel sensor inputs are accommodated in a variety of formats for up to four engines.

The software program provides quick access to a large navigation database, flight planning with SID and STAR procedures and airways, a Kalman-filtered best computed position, auto-scanning DME/DME/DME, en route manoeuvres such as direct-to, heading commands, VNav, holding patterns, three-dimensional approach mode, fuel management and, for certain aircraft types, aircraft-specific performance features, including optional frequency management.

Aircraft parameters are programmed into the module through the control/display unit. Data entered include aircraft identification and the communication formats for all input/output ports, indicating specifications of nav sensors, air data, EFIS displays, flight guidance systems and fuel inputs. Variables can include types, limits and scalings for interface compatibility and safety requirements.

The easily programmed module increases flexibility and simplifies upgrades and sensor complement modifications. The module also provides for quick configuration verification and positive identification of aircraft model for performance data.

The LTN-400 incorporates advanced concepts in

Control/display unit for the LTN-400 flight management system ***1995***

vertical guidance and control. VNav pages provide computed top of descent, target vertical speed and vertical direct-to commands. Vertical waypoints can be defined with altitudes or flight levels and can include at, above and below parameters. Lateral waypoint offsets can be entered as well. The LTN-400 outputs both digital and analogue vertical deviations for flight guidance displays. Digital and analogue pitch commands, analagous to circular arc steering, are output to compatible autopilots for fully coupled descents.

The LTN-400 can utilise a total of 20 different leg types. This allows complex procedures such as heading to altitude, DME arcs, procedural turns and radius-to-fix precision arc legs to be flown.

All pertinent en route data is displayed on the first nav page. This page, coupled with the nav displays on the flight guidance system, provides completely integrated real-time information on flight progress. Line select keys provide access to flight plan manoeuvres such as establishing a parallel course, tracking to/from a pseudo VOR, heading commands, holding patterns, arrivals and approaches. All these manoeuvres are controlled by the LTN-400, while the flight guidance system remains in the flight management system mode.

The control/display unit provides the focal point from which the operator controls functions. The menu operating format provides logical data sequencing and cursor prompting, reducing the most complex flight manoeuvres to a few simple key strokes. The full alphanumeric keyboard and 10 line select keys provide for quick and easy data selection and entry. The use of two character sizes, along with colour displays, facilitates data recognition.

Specifications

Dimensions:
(navigation computer unit) 194 × 56.8 × 388 mm
(control/display unit) 169 × 146 × 200.2 mm
(data transfer unit) 57.2 × 146 × 205 mm
Weight:
(navigation computer unit) 3.45 kg
(control/display unit) 3.54 kg
(data transfer unit) 1.47 kg
Power supply:
27.5 V DC, 60 W (max)
26 V AC, 400 Hz, 1 VA
Temperature range: −55 to +70°C
Altitude: up to 70,000 ft

Contractor

Litton Aero Products.

VERIFIED

AN/ASN-150 Tactical Data Management System (TDMS)

The AN/ASN-150 TDMS controls and displays navigation, communication and armament for a wide variety of fixed-wing aircraft and helicopters. It includes two dual-redundant tactical data processors, communications and armament system controllers, multifunction displays and various display and intercom system control panels. Operator interface with the system is provided by four control/display units, each with an alphanumeric keyboard and LED display.

The TDMS is interfaced with other avionics by a dual-redundant MIL-STD-1553B databus and by discrete interconnections.

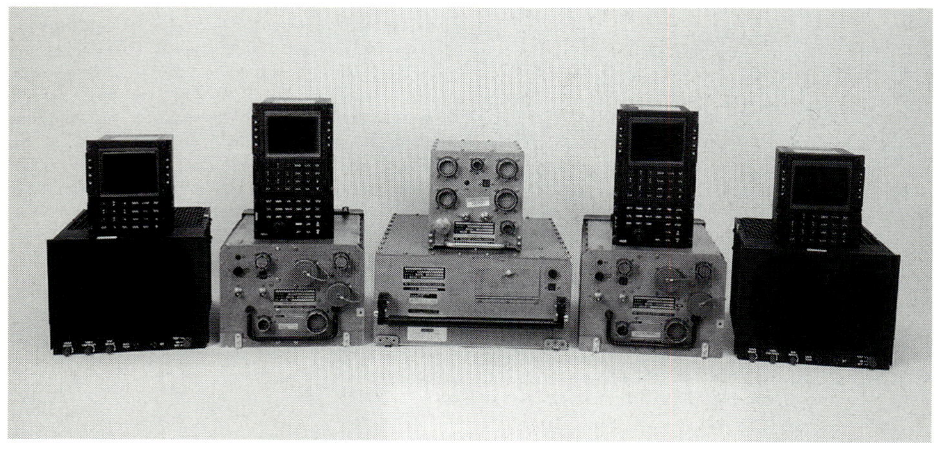

Units comprising the AN/ASN-150 tactical data management system ***1995***

Specifications

Dimensions:
(tactical data processor) 257 × 193 × 387 mm
(communications system controller) 356 × 178 × 443 mm
(armament system controller) 191 × 152 × 262 mm
(multifunction display) 267 × 211 × 381 mm
(control/display unit) 116 × 170 × 178 mm
(intercom system) 127 × 38 × 46 mm
(display and control panel) 86 × 147 × 76 mm
Weight: 123 kg
Reliability: 250 h MTBF

Operational status

In production and in service with US Navy HH-60H and SH-60F, US Naval Reserve SH-2G and US Coast Guard HH-60J. The system is also in use internationally in S-70 variants procured or selected by Greece, Kuwait, China and Thailand. The TDMS has been modified for use in fixed-wing aircraft such as the China S-2T.

Contractor

Litton Guidance & Control Systems.

VERIFIED

AN/ASW-38 automatic flight control set for the F-15

The ASW-38 provides the control functions for the F-15A, B, C and D versions. It provides three-axis command augmentation to improve the fighter's handling qualities over the very wide speed/height envelope. The analogue system incorporates two channels in each of the three axes, also supplied by dual sensors. Three-axis fail-safe trim control and stall inhibit functions are also provided. The set comprises two computers (one for pitch, the other for roll and yaw), a three-axis dual-channel rate sensor box, two-axis dual-channel accelerometer box, dual-channel pressure sensor unit, stick force sensor and pilot's controller.

Specifications

Dimensions:
(computers) 152 × 152 × 381 mm
(rate sensor) 114 × 114 × 178 mm
(acceleration sensor) 102 × 57 × 102 mm
(pressure sensor) 114 × 76 × 127 mm
Weight:
(total set) 17.8 kg
Power: 265 W
Reliability: 1,100 h MTBF over 10,000 flight h

Operational status

In service in the F-15A, B, C and D. No longer in production.

Contractor

Lockheed Martin Control Systems.

VERIFIED

Digital flight control system for the F/A-18

The F/A-18 flight control system is a digital four-channel Fly-By-Wire (FBW) system operating the aileron, stabilator and rudder primary flying controls, leading-edge and trailing-edge flaps and nosewheel steering. The system incorporates 32 servo loops to drive and control these functions. At the same time the conventionally mounted control column, in contrast to the F-16's sidestick controller, has a mechanical link to the tailerons for reversionary pitch and roll control. All FBW computations are accomplished by four digital computers operating in parallel, accepting inputs from linear variable differential transformers sensing control column and rudder pedal movement and analogue motion sensors, and formulating commands to the redundant electrohydraulic servo-actuators driving the control surfaces. The system therefore remains operational after two failures. It has a high degree of integration with other aircraft equipment, communication being accomplished by means of a MIL-STD-1553 databus.

Two special display modes are used in conjunction with the system. The first is a flight control failure matrix, the second provides recovery guidance during spins and both use the same CRT unit. In the first mode, the display is selected to show details of the fault, after the pilot's attention has been drawn to its presence by an indication on the central warning system. The display signals 'X' at the position of the fault on a schematic diagram of the system painted on the screen. In the second mode, the display shows the position of the control column needed to recover from a spin, assumed to exist when yaw rates greater than 15°/s occur simultaneously with speeds of 125 kt or below. This mode is selected automatically, having absolute priority over other displays when this combination of speed and yaw rate occurs. The two modes were originally provided for flight test purposes, but have been retained in production aircraft following recommendations from the Naval Air Test Center and test pilots.

The system comprises two flight control computers, each incorporating two microprogrammable digital processors specially designed for flight control applications, two rate sensor boxes, two acceleration sensor boxes, an air data sensor box, rudder pedal force sensor and pilot's control panel.

Operational status

In service on the F/A-18.

Contractor

Lockheed Martin Control Systems.

VERIFIED

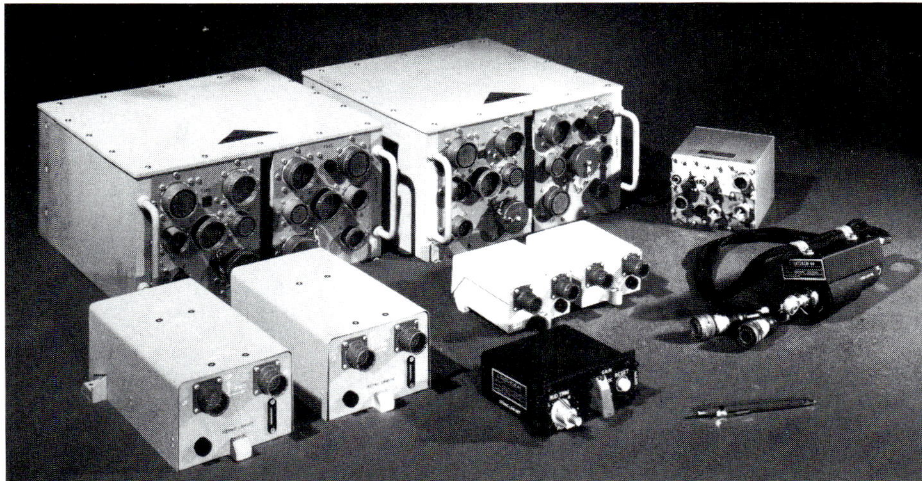

The flight control electronics set for the F/A-18 Hornet

Components for the C-17 electronic flight control system

Electronic Flight Control Set (EFCS) for the B-2

The digital fly-by-wire Electronic Flight Control Set (EFCS) consists of two parts: a quadruple redundant set of Flight Control Computers (FCCs) and quadruple redundant pilot sensor assemblies. Each FCC is a single-channel, real-time digital computer, connected via a MIL-STD-1553 bus controller to the avionics sytems, remote terminals, pilot sensor assemblies, cockpit panels, nosewheel steering and hydraulic system. The computer includes a PACE 1750A processor with a throughput of 2 Mips.

There are three pilot sensor assemblies: yaw pedal, pitch stick and roll stick. The sensor assemblies transform pilot input to an electronic signal which communicates with the FCC.

Specifications

Dimensions:
(FCC) 388.4 × 345.9 × 193.5 mm
(pilot sensor assembly) 215.9 × 66 × 91.4 mm
Weight:
(FCC) 18.6 kg
(pilot sensor assembly) 1.81 kg

Operational status

In service in the Northrop Grumman B-2.

Contractor

Lockheed Martin Control Systems.

VERIFIED

Electronic Flight Control System for the C-17

The C-17 transport is designed to carry men and equipment into austere airfields and drop zones near the front line. Operation of a large aircraft at low speed needs a flight control system that provides stable yet responsive control, augments the basic stability of the aircraft at low speeds and is extremely reliable. These requirements are being met through development of a quadruple redundant digital fly-by-wire control system for both the primary and secondary flight controls of the C-17. The Electronic Flight Control System (EFCS) controls the actuators that drive the movable ailerons, flaps, leading-edge slats, spoilers, horizontal stabilisers, elevators and rudders.

All flight controls are electrically commanded in their normal mode of operation. A mechanical system provides back-up control between pilot controls and the elevator, aileron and lower rudder.

The EFCS is a full-time, full-authority fly-by-wire control system with stability augmentation in all axes. Electronic flight control functions are provided by quadruple redundant sensor, computation and actuation channels. In the fly-by-wire control mode, the sensors relay pilot force signals to the four Flight Control Computers (FCC). The FCCs are programmed, using other sensor information and inputs, to generate flight control signals that control flight path and attitude. In addition, the EFCS provides automatic aileron and stabiliser trim, autopilot, flight director and autothrottle functions. An angle-of-attack limiting system is provided to prevent stall during high-lift configuration.

The EFCS is designed to provide continued operation following both failures and battle damage. This capability is achieved through four-channel architecture which can continue full-function operation on two channels. The channels are separated on the aircraft and the components are located so as to minimise the risk of losing multiple channels through battle damage.

Operational status

In service on the C-17.

Contractor

Lockheed Martin Control Systems.

VERIFIED

Flight Control Electronics Assembly-Upgraded for JAS 39 Gripen

The JAS 39 Flight Control Electronics Assembly-Upgraded (FCEA-U) by Lockheed Martin is a digital, high performance triplex flight control computer which provides full authority, fly-by-wire control of seven primary control surfaces, four secondary control systems, and interfaces with a variety of sensors and I/O devices. The FCEA-U system is provided to Saab Scania AB, Saab Military Aircraft for the Swedish Air Force.

Within each channel of electronics, the FCEA-U contains two high performance 32-bit processors. A Motorola 68040 performs primary flight control law processing and is interfaced via dual port RAM to a Texas Instruments TMS320C30 processor which acts as the I/O processor in addition to computing back-up control laws. A 10 MHz serial cross-channel datalink with built-in error checking is closely coupled with a data acquisition system that can exchange data between channels with less than 150 μs latency.

The electronics are housed in a single, air-cooled, lightweight chassis which is designed with an integrated flexible harness/mother-board and removeable circuit cards for easy maintenance. Reliability of the FCEA-U exceeds 1,500 h.

Specifications

Dimensions: 358 × 454 × 192 mm.

Contractor

Lockheed Martin Control Systems.

VERIFIED

Flight Control System for the V-22 Osprey

The V-22 Osprey Flight Control System (FCS) is a fly-by-wire system composed of triple, dual Primary Flight Control System (PFCS) processors and triple Automatic Flight Control System (AFCS) processors. The Flight Control Computers (FCCs) provide interfaces for the swashplate, elevator, rudder, flap and pylon primary actuators. The FCCs also provide interfaces for the cockpit control force/driver actuator, cockpit control thrust driver actuator and nosewheel steering secondary actuators. The FCC interfaces with the 1553B avionics multiplex bus and also incorporates a dedicated flight control system multiplex bus in each channel. The system is two fail-operational with respect to PFCS functions and one fail-operational with respect to AFCS functions.

Each FCC incorporates dual 1750A processors for the PFCS control functions and one 1750A processor for the AFCS control function. An input/output processor is used for data management and built-in test functions. The PFCS and AFCS processors interface with the I/O bus via a dual-port memory to enhance computerised throughput.

The PFCS electronically connects the pilot's controls to the various control surface actuators for safe operation of the aircraft. The automatic flight control system provides the necessary level of control augmentation required for reliable mission performance.

The PFCS provides basic control of the aircraft by connecting and mixing pilot control inputs during the helicopter, transition and aeroplane modes of operation. All actuators are commanded from the PFCS processors and each dual PFCS processor has failure detection and shutdown logic so that either processor in each channel can independently shutdown the actuator channel controlled by that channel.

The automatic flight control system provides stability and control augmentation and mission related selectable modes of the flight control system. It is a triplex cross-channel monitored system which provides operation after first failure in most triplex sensor paths. On a second failure, the function is shut down in a fail-safe manner.

Automatic flight control system inputs to the primary flight control system are voted at the input interface of the PFCS. The selected inputs are further restricted by rate and authority limiting in the control laws of the primary flight control system.

A non-redundant, dual-computation analogue back-up computer, in conjunction with analogue elements of the PFCS, provides for continued control of the aircraft in the event of a total loss of the digital processing functions. It also directly controls the back-up modes of the pylon actuator and electronic engine controls.

The flight control computer processor monitors the back-up computers continuously in normal operation. Built-in test for the back-up is controlled by the I/O processor during preflight and maintenance modes of operation.

The system consists of:

Flight control panel: this provides pilot/co-pilot control for the automatic flight control system.

Engine control panel: lever type engine condition controls are provided for the V-22's two engines. Illuminated push-button switches are used to select and indicate which engine controller is active. A third lever controls the rotor brakes.

Cockpit interface unit: the Cockpit Interface Unit (CIU) passes cockpit discrete data to the dedicated flight control system multiplexer bus. The CIU also contains a roll rate sensor.

Flight control computers: one flight control computer using 1750A architecture is provided for each channel. Primary and automatic flight control system functions each incorporate separate processing. Within the computer, the systems communicate via a common dual-port memory designed so that the hardware failures on the AFCS bus or memory side do not cause failures of the PFCS functions.

Analogue back-up computer: a non-redundant, dual computation channel analogue back-up computer provides for continued control of the aircraft in the event of a system failure of the digital processing functions.

Conversion actuator electronic interface unit: this provides control of the electric back-up pylon actuator motor.

Flap panel: this allows the pilot/co-pilot to select the operation of the flaps.

Operational status

In production for the V-22 Osprey.

Contractor

Lockheed Martin Control Systems.

VERIFIED

Stability augmentation system for the A-10

Stability augmentation for the US Air Force A-10 in pitch and roll is provided by a two-axis, two-channel system that not only improves handling qualities but makes for more accurate weapon delivery. Each channel in a given axis is independent of the other, except for monitoring cross links, and both are disengaged if significant desynchronisation occurs. Separate engage switches permit single-channel operation, however.

The system also provides turn co-ordination, compensation for the pitching moment caused by operation of the speed brake or recoil from the GAU-8 30 mm cannon. The pitch and yaw monitors can also be self-tested.

The system comprises a single-box computer and pilot's control panel. The computer contains six rate sensors, four of which are solid-state devices manufactured by Lockheed Martin.

Specifications

Dimensions:
(computer) 155 × 196 × 318 mm
(control panel) 146 × 86 × 165 mm
Weight:
(total) 7 kg
Power: 73 W

Operational status

No longer in production. In service in the US Air Force A-10 aircraft.

Contractor

Lockheed Martin Control Systems.

VERIFIED

Standard Automatic Flight Control System (SAFCS) for the EA-6B

The Standard Automatic Flight Control System (SAFCS) for the EA-6B Prowler is a standard computer design suitable for application to a variety of Navy aircraft. This capability is made possible by the inclusion of additional input/output circuitry and multiple memory options. Growth in redundancy up to triplex is possible through the inclusion of a cross-channel databus for exchanging information between computers.

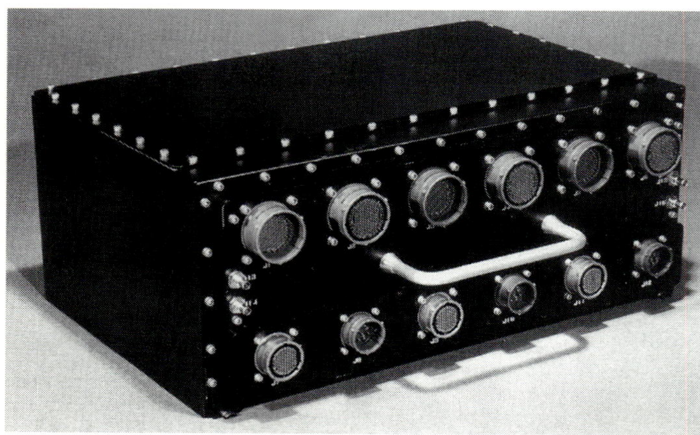

JAS 39 Gripen **1997**/0001442

Flight Control Electronics Assembly-Upgraded (FCEA-U) for JAS 39 Gripen
1997/0001443

The EA-6B SAFCS currently interfaces with analogue sensors and a MIL-STD-1553B avionics databus and drives electrohydraulic and trim actuators. Command augmentation, attitude and directional control servo-loops provide aircraft stability and control.

An extensive built-in test is provided together with non-volatile RAM for in-flight fault data storage. Maintenance information displayed in the front panel simplifies testing and logistic support. It also provides a quick turnaround resulting in higher availability. Use of hybrid circuits and low-power electronics minimises space, power and weight considerations while maximising reliability.

Specifications

Dimensions: 178 × 145 × 267 mm
Weight: 9.1 kg
Power supply: 115 V AC and 28 V DC
Power dissipation: 42 W
Cooling: convection, no forced cooling air
CPU throughput: 429 Kips, DAIS mix
Interfaces: RS-232, 1553B
I/O: 96 analogue inputs, 12 analogue inputs, 128 discrete I/O, 1553 and cross-channel datalink. (Above I/O is not a limitation, but current capability.)
Programs: Jovial and Assembly

Operational status

In service in the EA-6B Prowler. The system has also been adapted for US Air Force A-10 and TR-1 aircraft and US Army AH-64 helicopters.

Contractor

Lockheed Martin Control Systems.

VERIFIED

Tiger IV Avionics

Northrop Grumman has integrated digital weapons systems avionics into an F-5E Tiger IV under a co-operative research and development agreement with the US Air Force. The enhanced F-5 features: improved pilot situational awareness display, advanced multi-mode radar capabilities and a ring laser/GPS navigation unit, integrated via a MIL-STD-1553B databus.

Participating suppliers include: AlliedSignal Equipment and Control Group, standard air data computer; Base10 Systems Inc, stores management system; Honeywell Inc, inertial navigation system; Litton Systems Inc, Guidance and Control Division, airborne video camera, mission computer; Northrop Grumman-ESSD, AN/APG-66 radar; Fairchild Defense OSC, data transfer system; TEAC America Inc, airborne video tape recorders, Dassault Electronique EWS-A RWR.

Operational status

Tiger IV Avionics has completed flight tests evaluation.

Contractor

Northrop Grumman Corporation.

UPDATED

C-141 all-weather flight control system

The C-141 all-weather flight control system upgrade contract was awarded to Chrysler Technologies Airborne Systems (now Raytheon Systems Company) in 1992. The system comprises four subsystems: two digital autopilot computers (replacing: autopilot, autothrottle clutchpack, flight director computers, stall warner); four 6 × 8 in active matrix liquid crystal displays and two display avionics management units (replacing: ADI, HSI, airspeed and altimeters, navigation selector panels); a ground proximity warning subsystem and windshear system (replacing: existing GPWS); a global positioning system that adds GPS and Kalman-Filtered Navigation Solutions (replacing: FSAS).

Operational status

Trial installation completed May 1996, flight testing to December 1996, installations scheduled for August 1997 to February 1999, with 63 aircraft being modified.

Contractor

Raytheon Systems Company.

UPDATED

AMS-850 avionics management system

The AMS-850 avionics management system provides centralised avionics management, including control of EFIS modes, weather radar, TCAS, radio tuning and flight management, in one convenient location. The system offers consistent, intuitive menu-driven operation that facilitates training, minimises entry errors and simplifies avionics management.

An integral part of the Pro Line 4 system, the system takes advantage of Pro Line 4 multifunction displays to provide head-up operation, with navigation maps and tables and other avionics information shown on the MFD. Integration of the AMS-850 as part of the Pro Line 4 system reduces weight and size and simplifies interconnections. Flight management computers are packaged as line-replaceable modules, centrally housed in the Rockwell Collins integrated avionics processing system. Pro Line 4 integration allows the system to perform diagnostic functions, identifying any malfunctioning avionics unit and displaying the information on the MFD. The system also offers complete LRU fault history and detailed analysis of system status.

The AMS-850 avionics management system offers full AFIS capability **1995**

The menu-driven operation of the AMS-850 simplifies flight crew route planning. Flight plans may be entered waypoint by waypoint or selected from up to 99 previously defined and stored routes. En route, terminal area and non-precision approach navigation as well as primary means navigation in oceanic and remote areas are provided by the system database, and new waypoints may be added within a flight plan or appended. When editing a flight plan, pressing the line select key next to the displayed waypoint automatically brings up the edit page on the CDU for consistent operation. Pilot-defined waypoints can be created by plotting a path on the system's present position map using the joystick, and direct-to is also available for straight point to point routeing.

VNav capability is available with the advanced Collins system, fully integrated with the autopilot, navigation and air data systems for smooth operation. The AMS-850 is available with a full alpha keyboard to accommodate a wide range of applications.

The AMS-850 is available with AFIS compatibility, offering flight crews access to flight planning, weather information, SIGMET advisories and messaging capabilities. The AMS-850 also provides straightforward fuel management, computing fuel used, fuel remaining and endurance at the present fuel flow. The system's worldwide database is contained on disks that are updated every 28 days, which contain global VHF navigation and airport information.

Operational status

FAA Supplemental Type Certification on Beechjet 400A awarded in December 1996 and on Starship in October 1997. FAA TSO C129 Class B1 and Technical Order 8110.60 criteria for IFR operation to make non-precision approaches with GPS as primary means of navigation.

Contractor

Rockwell Collins.

UPDATED

AP-105 autopilot

Available for aircraft in the executive jet category, the AP-105 provides three-axis automatic flight control. The system offers a full complement of navigation and vertical mode options, including altitude preselect, and can present all computed steering data on flight director displays for manual control guidance, or to monitor the autopilot performance.

The AP-105 is designed to exceed automatic and manual Cat II approach requirements. Glide slope scheduling using radio altimeter information is incorporated and an all-angle VOR/Loc capture capability is included, plus automatic back course operation. The glide slope function can automatically capture from above or below, either before or after localiser capture. Vertical speed hold, airspeed hold, Mach hold and altitude hold modes can be integrated to fly departure, en route and arrival segments. Automatic mode changing is standard and automatic elevator trim permits ripple free transitions between modes and autopilot disconnects.

The system comprises an autopilot amplifier, controller, mode coupler, mode selector, airspeed sensor flight computer, altitude selector, yaw damper and three servos.

A control wheel synchronisation facility is engaged by depressing a button on the yoke and the autopilot then follows pilot control inputs, taking up the desired attitude when the wheel is released. The rudder channel can be operated in manual flying modes to enhance turn co-ordination, or engaged for all flying modes. In established attitudes the aircraft is gyrostabilised, but displacement of pitch or roll rate manoeuvre controls permits manual override. Attitude rates are used as a basis for all control inputs and automatic lift compensation is provided in the basic control laws.

Specifications

Weight: 23.4 kg total

Operational status

In service, out of production.

Contractor

Rockwell Collins.

UPDATED

APS-65 autopilot

Introduced in 1982, the APS-65 was claimed to be the first digital autopilot offered for turboprop types with dual microprocessor computation. The basic operating modes are: roll hold, pitch hold, heading hold, navigation mode, approach mode, indicated airspeed hold, vertical speed hold, climb and descent, altitude select and go-around. Additionally, when climbing, the autopilot can be programmed to fly the aircraft at optimum efficiency, thus providing fuel savings.

A new control technique for the pilot to use when introducing pitch, altitude, airspeed and vertical speed hold changes has been incorporated. This employs a rocker switch that can be operated to select precise alterations. Altitude can be adjusted in increments of 25 ft, pitch in increments of 1°, airspeed in increments of 1 kt and vertical speed in increments of 200 ft/min. The new technique has been incorporated to reduce pilot workload and permit smoother flying. Operational safety is enhanced through the dual processor configuration, which ensures that no single equipment or software fault can result in exceeding preset limits or a malfunction in more than one axis of control. Built-in monitoring ensures automatic disengagement when a fault is detected and appropriate diagnostics assist in fault detection in any element. The system is lighter and has a lower parts count than comparable analogue systems; higher reliability is attributed to these features.

The autopilot is suitable for the heavier business turboprop types such as the King Air and can be integrated with electromechanical or electronic flight director systems and navigation systems produced by Rockwell Collins.

Specifications

Weight: 10 kg total

Operational status

In production and service. First to be certified was the Beech King Air in April 1983. The APS-65 has now been chosen for 28 types of turboprop business aircraft. The King Air was noteworthy as the first general aviation all-digital turboprop to be certified. Its Rockwell Collins avionics suite included APS-65 autopilot, EFIS-85 electronic flight instrument system, ADS-80/85 air data system and nav/com equipment.

Contractor

Rockwell Collins.

UDPATED

APS-85 digital autopilot

The APS-85 autopilot completes the Pro Line 2 family of digital avionics for general and business aviation. APS-85 is an entirely digital, fail-passive autopilot with dual-redundant flight guidance computers. It has been certified for Cat II operation; Cat IIIA landings will be possible with some growth.

Apart from its digital nature, the APS-85 has a number of features not found on earlier Rockwell Collins general aviation autopilots, including the APS-80. For example it includes climb and descent modes. When selected in the climb mode, the system flies the aircraft according to an airspeed or Mach number schedule appropriate to that type of aircraft and its weight at take-off. In the descent mode the system sets up a rate of descent tailored to the aircraft manual. Three diagnostic modes – report, input and output – are incorporated. In the first, faults in the flight control computers are displayed to the crew. In the second, the system reads out information from outside entering the system, for example from the air data system or control surface position sensors. In the final mode, the system provides particular flight control computer outputs for examination. The diagnostics on the APS-85 are self-contained and do not require additional test equipment.

Specifications

Weight: 14.0 kg total

Operational status

In production and service. It has been certified as the autopilot on the Saab 340, the Raytheon Hawker 800, Falcon 50B, Falcon 20F, Learjet 55, CL-600/601, Astra and DC-8.

Contractor

Rockwell Collins.

UPDATED

Automatic Flight Control and Augmentation System for the Fokker 100

The Rockwell Collins Automatic Flight Control and Augmentation System (AFCAS), Electronic Flight Instrument System (EFIS) and MultiFunction Display System (MFDS) provide the Fokker 100 with advanced operating capabilities in the areas of automatic landing, display integration and engine and aircraft annunciation.

The AFCAS integrates cruise autopilot, yaw stability augmentation, fail-operative automatic landing, fail-operative thrust management and envelope protection functions into a single system. Certified for Category IIIB conditions, the AFCAS will give the Fokker 100 automatic landing capabilities down to a decision height of 50 ft and a runway visual range of 150 m.

The main flight panels of the Raytheon Hawker 800 with the Rockwell Collins APS-85 autopilot mode select panels fitted on the coaming in front of the pilot and co-pilot

Operational status

The Fokker 100 has demonstrated Category IIIB automatic landing capabilities using AFCAS.

Contractor

Rockwell Collins.

UPDATED

CMS-80 cockpit management system

The CMS-80 cockpit management system family consists of hardware and software building blocks. The baseline CMS-80 consists of a full complement of Rockwell Collins avionics plus selected equipments from other manufacturers.

The CMS-80 unclutters the cockpit by removing individual controls which normally crowd the panel. These are replaced with centrally located control and display units which use a standard screen layout and human interface to control all equipment. System interconnection is via dual-redundant MIL-STD-1553B multiplex cables. The CMS-80 can also provide mission computing and navigation integration.

A CMS-80 option is weapons and sight integration and control. This feature has been implemented on the B-406CS Combat Scout demonstrator and the MD-530 NOTAR helicopter. A single keystroke shifts the CDU from avionics to weapons control and all options for guns, rockets and missiles can be selected and managed. Weapons selection and firing can also be accomplished from the handgrips, allowing the pilot to keep his attention focused outside the cockpit.

Various versions of the system have been chosen for the Sikorsky HH-65A and HU-25A helicopters and Lockheed C-130 for the US Coast Guard. It has also been selected for the US Air Force Fairchild A-10A Thunderbolt aircraft.

In December 1981 Delco Electronics ordered the CMS-80 under a $15 million contract to support its commitment to provide 300 sets of Fuel Saving Advisory System (FSAS) equipment to the US Air Force, part of a programme to upgrade the fleet of Boeing KC-135s. Deliveries for this application began in January 1983 and ended in 1986. The FSAS system on the KC-135 is expected to save between 2 and 4 per cent of the fuel used by advising the crew of the most efficient speed, engine pressure ratio, altitude and descent profile.

In 1988, the system provided the baseline for the next-generation FMS-800 flight management system for the German Air Force C-160 Transall autonomous navigation system. In this application the CDU provides crew interface for communications, navigation and IFF, as well as flight management functions.

Operational status

In production and service. The US Army has selected the CMS-80 for Special Operations and Special Electronic Mission Aircraft (SEMA). In addition the CMS-80 has been chosen for the AH-64A Apache and for three US Navy/Marine Corps helicopters including the AH-1W, CH-46 and UH-1N.

Contractor

Rockwell Collins.

UPDATED

FCC-105-1 automatic flight control system

Based on the FCC-105 system, the FCC-105-1 flight control system is applicable to the Sikorsky S-76 and H-76 helicopters and has been in production since 1977. It offers a wide range of features up to full three-axis stability augmentation with turn co-ordination, pitch, roll and yaw attitude and altitude/airspeed hold modes. Each system amplifier has independent pitch, roll and yaw channels contained in individual modules which can be selected separately. This building block principle permits customers to select as many features as necessary to meet their particular requirements.

The system comprises a stability augmentation system amplifier, yaw switch, heading hold amplifier, airspeed switches, control panels, linear electromechanical actuators, indicator panel, rate gyros, cyclic switches and airspeed hold amplifier.

Specifications

Dimensions
(main processor) 190 × 318 × 113 mm
Weight
(main processor) 3.4 kg
Power: 35 W

Operational status

In production for the Sikorsky S-76 and H-76 helicopters.

Contractor

Rockwell Collins.

UPDATED

FCC-110 autostabilisation system

The FCC-110 is a dual-redundant autostabilisation system for the US Army AH-64 Apache helicopter. The system comprises two rate gyros and two analogue computer units. Airspeed sensors and the gyros provide inputs to the computers which calculate the appropriate stabiliser angle for a mission configuration, taking account of longitudinal stability demands. Other features include pitch axis control augmentation and trim features which reduce pilot workload and improve control characteristics during low-altitude high-speed combat operations.

Specifications

Dimensions:
(one unit) 178 × 203 × 279 mm
Weight:
(one unit) 3.75 kg
Power: 40 W per unit

Operational status

In production for the AH-64 helicopter.

Contractor

Rockwell Collins.

UPDATED

FCS-110 flight control system for the Lockheed Martin L-1011

The FCS-110 analogue flight control system, a combined project involving both Rockwell Collins and Lear Astronics, has been installed in the majority of Lockheed Martin L-1011 TriStar airliners. A comprehensive set of autopilot modes, including Cat IIIB automatic landing capability, is incorporated. It is installed in conjunction with the FD-100 flight director system. All work on the autoland system was done by Lear Astronics, which was also responsible for other elements of the system. In conjunction with a Hamilton Standard flight management system, the FCS-110 permits flight path optimisation. Two-dimensional optimisation was certified in 1972 and three-dimensional in 1977.

Operational status

In service in the Lockheed Martin L-1011.

Contractors

Rockwell Collins.
Lear Astronics Corporation.

UPDATED

FCS-240 flight control system for the Lockheed Martin L-1011-500

The FCS-240 advanced digital flight control system was designed to replace the FCS-110 fitted to Lockheed Martin L-1011 aircraft. It is a dual/dual integrated autopilot/flight director system and was initially tailored to the Lockheed Martin TriStar 500 wide-body transport. Compared to its analogue predecessor the number of computing units has diminished from seven to two and benefits are weight savings, improved reliability and lower cost of ownership. Cat IIIB automatic landing capability is included with a comprehensive set of conventional autopilot modes.

Operational status

In production and service in the Lockheed Martin L-1011-500.

Contractor

Rockwell Collins.

UPDATED

FCS-700 flight control system for the 767/757

The fully digital FCS-700 triplex autopilot and flight director system, with fail-operational automatic landing capability, was designed for Boeing 767 and 757 transports.

Development sprung from the FCS-111X experimental flight control system which Rockwell Collins and Boeing evaluated during the late 1970s. The fully digital architecture permits integration of comprehensive self-test and failure protection monitoring.

Autopilot modes include control wheel steering, automatic cruise hold/select modes for heading, altitude, vertical speed and airspeed/Mach number, approach modes with back course capability and automatic approach and landing. The latter facility includes automatic flare control, roll-out guidance and coupled go-around. Computed flight director steering is available in non-coupled flight, including take-off. Turn co-ordination and dutch-roll damping is provided. Dual or triplex control computer configurations can be used.

Steering commands from a navigation computer, using a wide variety of navigation sensors, will permit completely automatic control of preplanned vertical and lateral flight profiles. Large-scale integrated circuit technology is used in ARINC 429 bus interface drives and digital multiplexers and in several areas of each flight control computer, which is based on the Rockwell Collins CAP-6 processor configuration. Computer operating speed is 300 Kops and high-order language programming is used throughout the system.

Maintainability improvements are claimed from the MCDP-701 maintenance central display panel which uses a microprocessor to perform control, display and data management tasks. In-flight system failures are indicated by the maintenance control display panel and fault data is stored for up to 10 flights.

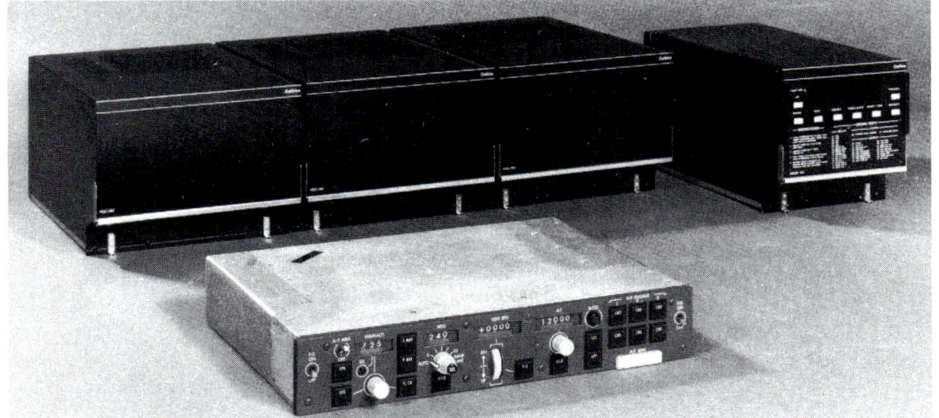

Computers and glareshield controller for the Rockwell Collins FCS-700 automatic flight control system used in the Boeing 757 and 767

Operational status

In production for Boeing 757 and 767 aircraft.

Contractor

Rockwell Collins.

UPDATED

FCS-700A autopilot/flight director system for the 747-400

The FCS-700A is an enhanced derivative of the autopilot on the Boeing 757 and 767 and is designed to provide LRU interchangeability between those two aircraft types and the 747-400. It is a Cat IIIB fail-operative system tailored to the control requirements of the 747-400 and is usable within the complete flight envelope for autopilot and autoland functions including automatic landings with a decision height of zero feet and a runway visual range of 600 m.

The FCS-700A consists of three flight control computers and a mode control panel. Each computer has dual Rockwell Collins Adaptive Processor System (APS) microprocessors for increasing processor capacity to allow for future growth in autopilot capability for MLS.

The system replaces 14 dedicated LRUs on previous 747 aircraft, saving considerable space and weight.

Operational status

In service in Boeing 747-400 aircraft.

Contractor

Rockwell Collins.

UPDATED

FD-108/109 flight director system

Essentially similar systems, the FD-108 uses 4 in (102 mm) attitude director and horizontal situation indicator instruments and the FD-109 uses 5 in (127 mm) instruments. Two categories of each are available, suffixed Y and Z, the latter incorporating air data and additional features. Either flight director system can be integrated with many types of modern autopilots and both versions are used on current production airliners.

The attitude director indicator has a V-bar display and a flat tape background which provides linear pitch information through ±90°. Outputs are provided for annunciations of all submodes and approach progress indications. A conventional compass rose horizontal situation indicator is provided with an electronic digital display of distance to go and a mechanical course readout.

Operational status

In service. No longer in production.

Contractor

Rockwell Collins.

UPDATED

FD-110 flight director system

The FD-110 system uses 5 in (127 mm) attitude director and horizontal situation indicator instruments. Several configurations are available, making them suitable for individual airliner, company and autopilot combinations. Configurations can use cross-pointer or V-bar attitude director indications, various ARINC system interface compatibilities or area navigation and inertial navigation system integration. The equipment is used in many current airliner types.

Operational status
In service.

Contractor
Rockwell Collins.

UPDATED

FDS-84 Pro Line flight director system

Featuring separate 4 in (102 mm) attitude director and horizontal situation indicator displays, the FDS-84 Pro Line system is compatible with Rockwell Collins FCS-80 and FCS-105 autopilot/flight control systems. It is a comprehensive system suitable for high-performance business and executive aircraft and commuter airliners.

The attitude director indicator uses a flat tape attitude display background and has V-bar steering command symbology. A radio altimeter readout is optional in this unit. The horizontal situation indicator uses a compass rose presentation with an electronic readout which can show distance, time to go or groundspeed information. The equipment is fully compatible with ARINC standard VOR, DME, INS and Omega/VLF sensors.

Operational status
In service. No longer in production.

Contractor
Rockwell Collins.

UPDATED

FDS-85 Pro Line flight director system

Based on 5 in (127 mm) ADI-85 attitude director indicator and HSI-85 horizontal situation indicator instruments, and compatible with the Rockwell Collins APS-80 autopilot, the FDS-85 Pro Line system is suitable for high-performance business aircraft and commercial airliners. It is the company's most comprehensive mechanical flight director system. Flat tape attitude indication, V-bar steering commands and electronic readout of distance, time to go, speed or elapsed time are standard. A separate HCP-86 heading/course control panel is used.

Specifications
Dimensions:
(ADI-85) 131 × 131 × 206 mm
(HSI-85) 131 × 112 × 229 mm
(HCP-86) 146 × 38 × 152 mm
Weight:
(ADI-85) 3.2 kg
(HSI-85) 3.3 kg
(HCP-86) 0.28 kg
Power supply:
26 V AC, 400 Hz, 44 VA
28 V DC, 260 mA
5 V AC/DC, 10 W for lighting
Temperature range: −20 to +70°C
Altitude: up to 35,000 ft
Cooling: convection

Operational status
In limited production.

Contractor
Rockwell Collins.

UPDATED

FIS-70 Pro Line flight instrumentation system

Compatible with the Rockwell Collins APS-80 or AP-106A autopilots, the FIS-70 Pro Line comprises the 4 in (102 mm) ADI-70 attitude director indicator and the HSI-70 horizontal situation indicator instruments and is suitable for high-performance turboprop and business jet aircraft. It is essentially a low-cost version of the FDS-84 system and excludes the digital distance/course readouts and radio altimeter display options. All other FDS-84 features are incorporated.

Specifications
Dimensions:
(ADI-70) 106 × 106 × 210 mm
(HSI-70) 106 × 106 × 229 mm
Weight:
(ADI-70) 2.1 kg
(HSI-70) 2.2 kg
Power supply:
26 V AC, 400 Hz, 1.04 A
28 V DC, 0.2 A
28 V DC, 0.34 A for lighting.
Temperature range: −15 to +70°C
Altitude:
(ADI-70) −1,000 to 50,000 ft
(HSI-70) up to 35,000 ft
TSOs:
(ADI-70) C3b, C4c, C52a
(HSI-70) C6c, C52a

Operational status
In service. No longer in production.

Contractor
Rockwell Collins.

UPDATED

FMS-800 flight management system

The FMS-800 flight management system, based on the CMS-80 cockpit management system, integrates the functions of communications, navigation and IFF control, GPS/INS navigation, flight instruments and controls, autopilot, stores and radar for transport, tanker, trainer and utility aircraft. The system automates many of the functions normally carried out by the navigator. It also simplifies the complex tasks of the pilot and co-pilot, permitting them to concentrate on mission planning and execution.

The FMS-800 outputs dynamic data to the flight instruments and automatic flight control system using MIL-STD-1553B, ARINC 429/561 or analogue synchro signals. The system is compatible with existing mechanical flight instruments and analogue autopilots as well as digital systems such as the Rockwell Collins FDS-255 flight display system and APS-85 autopilot. The FMS-800 integrates GPS/INS navigation using a 12 state Kalman filter for airborne alignment and continued high accuracy. GPS and INS stand-alone navigation are also provided.

Workload is reduced and flight accuracy is improved by the automated processing features of the FMS-800. These features include tactical airdrops, intercepts, raster towlines, orbit/rendezvous, search patterns, VNav, FMS non-precision approach, coupled flight director/autopilot guidance and speed commands for precise time of arrival. The FMS-800 also includes embedded dynamic simulation software for ground mission rehearsal and training.

The standard system consists of dual-control display units, dual mission computers which also provide for the interface of non-MIL-STD-1553B avionics, dual-remote readout units and a data loader.

The FMS-800 flight management system is supported by a Windows 95 Mission Planning Station (MPS), which inputs data to the aircraft system via the DR-200 Airborne PC Card Receptacle. The system allows for integration of aircraft mission planning data with data from Jeppesen and Aeronautical Flight Information File (DAFIF) sources.

Operational status
In production. The FMS-800 is being installed in approximately 100 German Air Force C-160 Transall aircraft as part of a US$21 million contract, and is currently flying on the US Air Force/US Army Joint STARS E-8C. The system is currently being integrated on US Air Force B-1, C-5, C-9, CC-130, E-3, E-4, KC-10 and KC-135 aircraft. It has received FAA TSO approvals for installation on the US Army's C-12 aircraft and the US Air Force's KC-10 and C-9 aircraft, meeting military Global Air Traffic Management (GATM) requirements, and permitting aircraft so equipped to operate without civilian airspace restrictions.

Rockwell Collins Flight 2 FMS-800 system has been selected by the Royal Danish Air Force for its C-130 Avionics Upgrade Programme. Other elements to be integrated include: FDS 255 liquid crystal flat panel displays; APS-85 digital autopilot; dual ADS-85 air data systems; FMR-200X colour weather radar.

Contractor
Rockwell Collins.

UPDATED

The FMS-800 flight management system is fitted in the E-8C Joint STARS aircraft **1995**

FMS-4100 Flight Management System

The FMS-4100 Flight Management System (FMS) is designed for the new generation of regional airline aircraft and is available to operators of Canadair and Saab aircraft.

The FMS-4100 is a fully integrated part of the Rockwell Collins Pro Line 4 avionics system standard on the Canadair Regional Jet and Saab 2000. FMS-4100 computers are packaged as LRUs housed in the system's Integrated Avionics Processing System (IAPS). Pro Line integration of the FMS minimises weight and volume requirements and enhances diagnostics capability. The full-sized control/display unit provides a large, seven-colour, sunlight-readable CRT with line select keys and a full alphanumeric keyboard.

The FMS-4100 offers crews standardised cockpit operation, with an ARINC CDU with scratchpad data entry. Preflight and general flight planning are available: a primary flight plan is used for active guidance and a secondary flight plan can be stored as an alternative and activated if required. Up to 1,000 standard routes, each with up to 100 waypoints, can be stored in the system database. This contains navaids, waypoints, NDBs, airports, airport reference points and runway thresholds. The FMS-4100 also includes SIDs, STARs, parallel offsets, high- and low-altitude airways and holding patterns.

Dual FMS-4100 flight management systems in the Canadair Regional Jet aircraft ***1995***

Navigation information is shown on Pro Line 4 MultiFunction Displays (MFDs). Present position and planning maps are displayed on the MFD, in addition to text pages for progress, nav status, position summary, VOR/DME status and Loran status. The FMS-4100 also provides complete radio tuning capability and is both AFIS- and ACARS-compatible.

Operational status

In service.

Contractor

Rockwell Collins.

UPDATED

FMS-4200 Flight Management System

Significant software and hardware upgrades to the Rockwell Collins FMS-4100 Flight Management System have created the new FMS-4200 Flight Management System.

Non-precision GPS approach capability, coupled or advisory VNav guidance, thrust and fuel management and ACARS/AFIS™ compatibility are among the in-flight features of the new system.

The navigational database for the new FMS-4200 has been expanded from 4 Mbytes to 12 Mbytes to facilitate international corporate operations that rely on the memory-intensive demands of the WorldWide Navigational Data Base (WWNDB).

Pro Line 4 multifunction displays are used for the FMS-4100 ***1995***

Through integration with the GPS-4000 Global Positioning System sensor, the new FMS-4200 calculates navigational solutions based on GPS signals from all satellites in view, making GPS-based RNAV non-precision approaches possible worldwide.

New technologies developed for the FMS-4200 include the ability to provide full-profile coupled or advisory vertical navigation guidance for climb, cruise and descent. The advisory VNav system displays situational guidance — vertical speed, required vertical speed and vertical deviation — eyes-up on a multifunction display. Vertical information — top of climb, top of descent, vertical constraints and descent rates — is displayed on the control display unit and the multifunction display. During climb or descent, advisory vertical cues are presented on the primary flight display, including glidescope, vertical speed target and next vertical constraint.

Contractor

Rockwell Collins.

UPDATED

FMS-6000 Flight Management system

The FMS-6000 provides integrated multisensor navigation, flight plan modification and execution, sensor control, multifunction display map support and steering/pitch commands to the flight control system. When integrated with the Rockwell Collins GPS-4000 navigation system, it forms the AVSAT 6000 system, which in turn can be integrated with the Pro Line 4 avionics system to create a complete avionics package.

Operational status

Certified aboard the Canadair Challenger 604.

Contractor

Rockwell Collins.

NEW ENTRY

HFCS-800 helicopter flight control system

The HFCS-800 helicopter flight control system can be configured for single- or dual-pilot operation and offers what the company calls total mission IFR capability. Operating modes, covering operations from start-up to shutdown, have been incorporated, with special emphasis given to autopilot assistance and control at low speeds and in hover.

Two separate subsystems are used, one for Automatic Flight Control Systems (AFCS) and the other for Flight Director Systems (FDS). The AFCS provides automatic stabilisation and control, while the FDS computes automatic path steering for any desired manoeuvre. The subsystems may be purchased jointly or separately and they meet all the requirements of FAA TSO C9A and C52A.

The following list of operating modes provides an indication of the system's performance:

Airspeed hold: maintains indicated airspeed; the pilot may 'beep' airspeed after initial selection. May be used if airspeed is above 35 kt.

Vertical speed hold: holds barometric vertical speed if airspeed is above 60 kt.

Altitude hold: holds barometric altitude if airspeed is above 60 kt.

Heading select: captures selected HSI heading if airspeed is above 35 kt.

Navigation: captures and tracks selected VOR, Loc, Doppler, Omega or RNav course if airspeed is above 35 kt.

Approach: captures and tracks course, providing three-dimensional control from VOR, ILS, Loc (back course), MLS or RNav sensors. May be used above 35 kt airspeed and up to 12° glide slope.

Airspeed/vertical speed: holds both parameters if airspeed is above 35 kt.

Navigation transfer: in two-pilot operation, transfers control to co-pilot's side.

Hover augmentation: holds hover position and radio altitude.

Transition to hover: decelerates aircraft to hover at 50 ft radio altitude.

Go-around: accelerates to climb at 70 kt. May be used after take-off for departure.

Specifications

Weights:

Automatic flight control system

(AFCS computer) 6.43 kg
(AFCS panel) 1.5 kg
(servos) 0.96 kg each (three or four installed)
(feel/trim units) 1.36 kg each (two installed)
(yaw servo) 2.1 kg
(collective servo) 2.1 kg
(vertical gyros) 3.1 kg each (two installed)
(airspeed sensor) 0.5 kg
(yaw rate gyro) 0.9 kg

Flight director system

(flight director computer) 4.9 kg
(flight director panel) 0.6 kg
(attitude control) 1.5.kg
(attitude director indicator) 3.1 kg each (two installed)

Operational status

In production. The system has been selected by the US Coast Guard for installation in the Eurocopter HH-65A (AS 336G1) Dolphin helicopter.

Contractor

Rockwell Collins.

UPDATED

PACER CRAG C/KC-135 avionics upgrade

In October 1995 a Rockwell Collins team was awarded a US$35 million contract to serve as the prime contractor for the C/KC-135 PACER CRAG upgrade. This programme is an avionics-driven upgrade for the aircraft fleet and includes the integration and installation of the FMS-800 flight management system, FDS 255 colour flat panel flight display system, an FMR-200X weather radar and an embedded INS/GPS navigation system to replace the existing compass. The potential value of the contract is in excess of US$250 million, if all options are exercised.

The contract was awarded by the US Air Force through the C/KC-135 Systems Program Office. It was one of the first contracts to be awarded under the US Air Force 'Lightning Bolt' procurement streamlining initiative. This initiative permits the US Air Force to streamline processes and allows the procurement of commercially available products.

The PACER CRAG integration and upgrade is aimed at modernising the C/KC-135 cockpits and, through human factors techniques, reducing crew workload and automating many routine cockpit functions.

US Air Force C/KC-135 PACER CRAG Block 20 cockpit ***1997***/0001434

Rockwell Collins is prime contractor for the programme and has responsibility for design, test and installation of the flight management system, displays, weather radar (FMR-200X) and embedded INS/GPS for compass replacement.

The FMR-200X radar provides weather, windshear and skin paint functions. The skin paint capability allows radar detection, identification and separation maintenance of aircraft during refuelling operations.

Additionally, as part of the PACER CRAG modification, an integrated Traffic alert Collision Avoidance System (TCAS) and Enhanced Ground Proximity Warning System (EGPWS) are being installed.

Operational status

In July 1996, Rockwell Collins delivered the first C/KC-135 PACER CRAG (Compass, Radar And GPS) aircraft with avionics upgrade to the US Air Force, signifying the formal release of the aircraft into the Qualification Test and Evaluation (QT&E) phase of the programme. Ultimately, more than 600 C/KC-135 aircraft will undergo the upgrade.

Contractor

Rockwell Collins.

UPDATED

Performance computer system

Safe Flight's performance computer system is designed to provide increased accuracy in obtaining the best balance between fuel economy and airspeed, while reducing crew workload in the operation of business jets. The system computes and displays the speed appropriate to maximum specific range under the prevailing conditions. Alternatively long-range speed, percentage of maximum specific range being achieved, optimum cruise altitude or several other parameters may be selected as the baseline. Inputs from various aircraft sensors, together with stored performance data, are used in these computations which take into account all external factors such as the effects of fuel burn or change of wind velocity appropriate to the type of aircraft. The computer is based on an Intel 8085 microprocessor and has 28 kbytes of read-only memory and 2.25 kbytes of RAM. The computer can supply data to compatible autothrottles or thrust management systems produced by Safe Flight which are controlled by means of a control/display unit.

Specifications

Dimensions:
(computer) ⅜ ATR short
(control/display unit) 57 × 146 × 121 mm
Weight:
(computer) 3.4 kg
(control/display unit) 0.55 kg
Power supply: 115 V AC, 400 Hz

Operational status

In production and in service.

Contractor

Safe Flight Instrument Corporation.

VERIFIED

AMS-2000 Altitude Management and alert System

The Shadin AMS-2000 Altitude Management and alert System uses information from Mode-C altitude encoders to deliver: time-based altitude alerting, rather than fixed altitude buffers, to notify the pilot 15 seconds before an altitude target or limit is reached – regardless of rate of climb or descent; automatic calculation and display of density altitude; real-time display of instantaneous vertical speed without the inherent lag of the aircraft's static system; calculation and display of aircraft and engine performance percentage.

Specifications

Dimensions: 38 × 80 × 146 mm
Power: 10 to 28 V DC, at 300 ma max
Weight: 0.25 kg

Contractor

Shadin Co Inc.

VERIFIED

AMS-2000 Altitude Management and alert System. ***1997***/0001445

Flight Management Computer System (FMCS)

The Flight Management Computer System (FMCS) is used for commercial air transport applications. Currently it is available in single or dual configuration for the Boeing 737 and in a triple configuration for the Ilyushin Il-96. The system is designed to meet the new airspace requirements specified by ICAO in their instrumentation plans for communication navigation and surveillance for Air Traffic Management. This includes communications with an adaptable datalink to allow each airline to customise its airline communications; navigation with the latest in required navigational performance/actual navigation performance methodology certified to use GPS and surveillance with the building blocks to implement functions such as ATC clearance entry and ADS reporting.

The flight management computer system has a single high-power 32-bit microprocessor and contains options for 4, 8 or 16 Mbytes of on-aircraft loadable EEPROM for storage of the operational flight program and navigation database. Up to two million bytes of memory is dedicated to the operator's navigation database and can be structured according to the customer's specification. A wide range of options is available from a simple listing of navaids and airports to a detailed coverage of the airline's operating routes, SIDs, STARs and gate assignments. Smiths Industries offers airlines a choice of using Jeppesen, Racal or Swissair for the navigation database update service.

The crew interface is via a CRT control and display panel on which 14 by 24 character lines of information can be displayed. Aircraft lateral and vertical profile data is presented and critical information is highlighted by reverse video presentation. The bottom line of the CRT is used as a scratchpad for crew entries.

The system can be coupled to the autopilot and autothrottle for automatic profile tracking and energy management. Fuel savings in the range 4 to 14 per cent are predicted in normal operations.

The system is available to all Boeing 737 operators either in a dual-unit or single-unit installation. The dual unit manages all performance, navigation and approach functions, as well as improving dispatch availability. The dual-unit format can be installed in place of the current single unit. The computer is a passively cooled 4 MCU form factor that is half the weight, uses one fifth the power and is over five times faster than competitive flight management systems. The system has spare memory and processor throughput to accommodate the FMCS functions of the FANS implementation programme.

Specifications

Dimensions:
(display) 267 × 146 × 229 mm
(computer) 320 × 257 × 193 mm (4 MCU)
Weight:
(display) 8.2 kg
(computer) 7.7 kg
Power required:
(total) 75 W
Reliability: 18,000 h MTBF (predicted)

Operational status

In production and in service on Boeing 737-300, 400 and 500 aircraft in single or dual configurations. The triple configuration is standard for the Ilyushin IL-96M. The FMCS is also on the 737-700 and -800 and on the VC-25 US Air Force One and Two aircraft (two Boeing 747-200 aircraft).

Contractor

Smiths Industries Aerospace.

UPDATED

The Smiths Industries flight management computer system is available in single or (as shown here) in dual configurations ***1995***

Fuel Savings Advisory System (FSAS)

The Fuel Savings Advisory System (FSAS) is intended for military transport aircraft. The system includes a Fuel Savings Computer (FSC), Control and Display Unit (CDU) and a Display Interface Control Unit (DICU). The computer has software for both the Lockheed Martin C-141 and C-5 supported in a single computer program. Identical hardware is used in both aircraft.

An aircraft wiring jumper selects aircraft type and version when the computer is installed. A Smiths Industries installation kit is used for rapidity and ease of aircraft modification. Software is written in high-level language with full documentation and support tools. The system has a specified MTBF of 2,000 operating hours.

Operational aspects include climb, cruise, descent and transition coupling to the autothrottle and autopilot. The system also provides advisory information on an alphanumeric CDU that interfaces with the aircraft's INS for enhanced navigation.

The FSAS features flight management, navigation, fuel savings, Tacan and waypoint database, take-off and landing computations, colour graphic tactical aids, moving map displays, enhanced airdrop capability, time of arrival control, altitude alerting, windshear prediction and detection and audio alert tones.

Tactical functions are provided to enhance the mission performance of the aircraft and crew. Performance advisory aids include maximum climb, obstacle clearance, VMO/MMO, buffet margins, rapid descent and optimum approach data, all of which help to improve tactical efficiency and reduce crew workload.

A built-in test system verifies overall system operational status to the flight crew and offers a method of isolating faulty LRUs without support equipment.

Operational status

In service in Lockheed Martin C-141 and C-5 aircraft.

Contractor

Smiths Industries Aerospace.

VERIFIED

LCD altitude selector/alerter

The new S-TEC LCD altitude selector/alerter is a compact, lightweight unit combining the computer and programmer in a single package. It is fully TSOd to C9c standards.

This system allows the pilot to preselect altitudes and rates of climb/descent through use of the autopilot, and also provides an altitude alert mode, decision height mode, altitude readout from the encoder, and barometric calibration in inches of mercury, or in millibars.

Specifications

Power required: 14/28 V DC
Weight: 0.6 kg
Dimensions: 41 × 87 × 172 mm

Operational status

In production.

Contractor

S-TEC Corporation.

UPDATED

LCD altitude selecter/alerter ***1997***/0001449

Single-cue flight director system

S-TEC's single-cue flight director system provides for steering horizon operation both when the autopilot is engaged, and when it is not.

When the autopilot is engaged, the flight director provides a visual display of what the computers are telling the autopilot servos to do. When used during manual flight, it provides the pilot with steering indications that integrate information from several signal sources, dramatically reducing the complexity of the instrument scan. Flight director operation is automatic when the autopilot is engaged.

S-Tec's single-cue flight director is similar in appearance to other systems in order to provide pilots with a measure of conformity, and to reduce error during operation. It contains a fixed delta-shaped symbol, bright orange in colour, which represents the airplane. Pitch and roll attitudes are displayed by a movable attitude field, coloured with a blue sky, and an earthtone ground, separated by a thin white line.

In a half-circle above the attitude field are the bank indexes – the centre or 'level' index, represented by a white inverted triangle, with bank indexes indicated by white vertical lines on either side of the centre index. They represent, 10, 20, 30 and 60° left and right bank angles. Four white lines above, and two below the white horizontal line on the attitude field represent various pitch attitudes: 5° increments in pitch-up to a maximum of 20°; and 10° increments in pitch down attitude, to a maximum of 20°.

The command bars, painted yellow, display computed bank and pitch commands. They rotate around the attitude fields to command climb, descent, left and right bank angles. In manoeuvering, the

Single-cue flight director system ***1997***/0001448

airplane symbol is 'flown into' the command bars, until the two are accurately aligned, which satisfies the command.

When operating in approach mode, operation basically is the same as in other autopilot modes. When the glide slope is captured, the command bars indicate the commands to be satisfied. The airplane symbol is 'flown into' the command bars to satisfy the commands.

Specifications

Power required: 14/28 V DC
Weight: 1.4 kg
Dimensions: 89 × 89 × 176 mm

Operational status

In production.

Contractor

S-TEC Corporation.

UPDATED

Stability and Control Augmentation System (SCAS) for helicopters

The S-TEC SCAS for rotary-wing aircraft provides a new and innovative approach to stabilising the helicopter in flight while enhancing controllability.

The S-TEC SCAS is a full-time series system which augments the pilot's control of the aircraft from hovering flight throughout the normal manoeuvring flight regime all the way to the landing sequence. It improves basic aircraft control harmony and reduces outside disturbances to the desired flight path thus reducing pilot workload and improving the quality of flight. During operation the system is completely transparent to the pilot.

Options to the basic SCAS will include a force trim system which will restrain the cyclic control, allowing the pilot freedom, for short periods, to accomplish other cockpit management duties such as the operation of avionics and other aircraft systems. The force trim system can be converted to parallel trim actuator operation in the upgrade to a complete autopilot system.

This primary system, in addition to being available to improve flight characteristics for the aircraft in which it is installed, becomes the first component of what will be a complete family of modular helicopter flight control systems. As full featured systems are added to the S-TEC rotary-wing product line, the SCAS will become the nucleus of the building block upgradability to a complete flight control system.

Growth to VFR flight control systems will require only the addition of an autopilot circuit card and transducers and the conversion of the force trim system into parallel trim actuators. The upgrade will incorporate full authority autopilot functions of IAS Hold, Altitude Hold, Heading Select/Hold and VOR/LOC/GPS navigation intercept and tracking.

Contractor

S-TEC Corporation.

VERIFIED

ST-180 HSI slaved compass system

The ST-180 system combines a magnetically slaved gyroscopic compass with a VOR/Localiser and glide slope display. The ST-180 was designed specifically for rotary-wing and high performance fixed-wing aircraft. This HSI slaved compass system offers a single convenient display which provides the pilot with all necessary information about the aircraft's position relative to ground-based navigational aids. Simultaneous indications are provided for selected course, course deviation and selected heading. Glide slope deviation is displayed when an active ILS frequency has been selected. The ST-180 HSI Slaved Compass System was developed and approved for helicopter operations.

The ST-180 consists of the Horizontal Situation Indicator (HSI), the remote electric gyro, the magnetic flux sensor and the model slaving panel. The gyro employs electromechanical erection thereby eliminating the performance and reliability degradation frequently experienced with air-erected gyros.

Critical circuits within the HSI are continuously monitored to minimise potential erroneous information. Should the flux sensor fail, the system may be switched to a free-gyro mode. Should a gyro failure be detected, the system may be transitioned into the 'Automatic Emergency Mode' (AEM). In AEM, the compass card is controlled by the Flux Sensor and behaves similarly to a normal wet compass. The pilot can continue to use the HSI.

Specifications

FAA TSO: C34e; C36e; C40c; C6d; C9c; C52a
Input power: 27.5 V DC (supplied by remote gyro)
Compass card accuracy: ±1°
Dimensions: ARINC specification 408 ATI-3
Weight: 1.4 kg

Operational status

In production.

Contractor

S-TEC Corporation.

UPDATED

System 40/50/60/65 autopilots

The basic System 40 rate autopilot consists of a combined programmer/computer/annunciator in a single box, a turn co-ordinator and an electrically operated roll servo. It permits both straight and turning flight, while inputs from a radio beacon provide the means for VOR tracking for en route navigation, localiser tracking for approach and reverse or back course tracking. The system can be upgraded to include a heading mode by adding the optional directional gyro.

The System 50 has all the foregoing functions, but adds an altitude hold mode and elevator trim indicators. The optional directional gyro can be added to provide heading hold. An accelerometer confers pitch stability and automatically disconnects the system in the event of a pitch axis fault. In addition the system incorporates a preflight test feature that tests internal limiter circuitry.

The System 60 is a rate autopilot using rate of change of aircraft motion instead of attitude to generate demands to the flying controls. The System 60 is a single- or two-axis system (60-1 and 60-2) comprising an electric turn co-ordinator, 3 in (76 mm) air-driven directional gyro, mode programmer/annunciator, pitch and/or roll guidance computers and servos, master switch and control wheel disengage switch and altitude transducer.

The System 65 autopilot has the same performance standards as the System 60 but includes automatic elevator pitch trim, liquid crystal display programmer and a remote annunciator.

A liquid crystal display altitude selector/alerter combines the computer and programmer in a single package.

System 60 PSS provides a cost-effective way of adding vertical flight control to virtually any single axis autopilot system. It provides altitude hold, glide slope coupling and vertical speed capabilities. The system does not interface with existing roll axis autopilots, but complements them. Being self-contained the System 60 PSS does not connect to the aircraft pneumatic or vacuum systems.

Operational status

All systems are in production.

Contractor

S-TEC Corporation.

UPDATED

System 55 autopilot

The System 55 is a pure rate-based autopilot. It combines programmer, computer, annunciator and servo amplifier functions into one panel and is the first S-TEC autopilot designed specifically to be integrated into the aircraft radio panel.

The System 55 roll axis has heading select, VOR/Localiser front and back course intercept and tracking. It can be interfaced with RNAV, GPS or Loran systems and all radio couplers are standard.

Operational status

In production.

Contractor

S-TEC Corporation.

VERIFIED

UNS-1B Flight Management System (FMS)

The UNS-1B FMS comprises a Control/Display Unit (CDU), Navigation Computer Unit (NCU) and Data Transfer Unit (DTU) for uploading and downloading navigation database information. The CDU is available with a 127 mm colour display, ten line-select keys, ten function keys and a full alphanumeric keyboard.

The UNS-1B uses position data from long- and short-range navigation sensors to determine the best computed position. Automatic scanning DME/DME/DME positioning with slant range error correction, as well as en route Rho/Theta, is computed. Vertical navigation, three-dimensional approach mode —

including GPS approaches – and holding pattern are also included. All twenty leg types can be flown, including heading to altitude, radial intercept, DME arcs and procedure turns – all manoeuvres required to fly complete SIDs, STARs and holding pattern procedure accurately. The UNS-1B FMS accepts fuel flow data from up to four engines.

Analogue and digital outputs are provided to flight directors, autopilots, EFIS, multifunction displays and radar navigation displays. Digital communications are provided using the ARINC 429/571/561/575 formats.

The UNS-1B has a 3.1 Mbyte database capacity, equivalent to 200,000 waypoints, and can include SIDs, STARs and approaches. It also includes airports, navaids, en route and terminal waypoints and airways in the database. The Jeppesen database is formatted on 3.5 in diskettes and is updated on a 28-day cycle. Pilot-defined data can include 200 routes, with up to 98 waypoints in each for a total of 3,000 waypoints, 100 arrivals/departures, 100 approaches and 100 runways. Built-in batteries prevent memory loss when the unit is removed from the aircraft.

Additional system features include a heading mode for direct control of aircraft heading through the CDU, bank angle commands correlated to altitude and turn anticipation to eliminate overshoots. Options include frequency management for centralised control of aircraft navigation and communication radios, AFIS interface capabilities for airborne ground link information and aircraft specific performance data functions.

The latest variant is UNS-1B plus. It utilises a colour flat-panel control/display. The 2MCU-sized navigation computer unit houses the new ASCB-interface printed circuit board. This board can also be installed in the UNS-1C and UNS-1D models as retrofits for the Challenger 601-3A and Falcon 900 aircraft.

The UNS-1B plus with GPS-1200 sensors is certified for GPS operation in en route, terminal and approach phases of flight (TSO C1156 and C129 Class B1/C1), and meets the requirements for primary means of navigation in oceanic/remote airspace.

UNS-1B with GPS-1000 12-channel receiver, RAIM and Fault Detection and Exclusion (FDE) has received certification by the Civil Aviation Authority of New Zealand for GPS primary means of navigation in remote/oceanic airspace.

Specifications

Dimensions:
(CDU) 169 × 146 × 200.2 mm
(NCU) 194 × 57 × 388 mm
(DTU) 53.8 × 146 × 205 mm
Weight:
(CDU) 3.45 kg
(NCU) 3.45 kg
(DTU) 1.47 kg
Power supply: 27.5 V DC, 60 W (max)
26 V AC, 400 Hz, 1 VA

Operational status

In production.

Contractor

Universal Avionics Systems Corporation.

VERIFIED

UNS-1C Flight Management System (FMS)

The UNS-1C Flight Management System (FMS) is an all-in-one unit that includes an integral 12-channel GPS receiver with real-time and predictive Receiver Autonomous Integrity Monitor (RAIM). It is designed for integration with the advanced all-digital avionics suites of the next-generation aircraft, such as the Lear 45, Falcon 900EX, Falcon 2000, IAI Galaxy and Astra SPX.

The UNS-1C updates the UNS-1B and enhancements include: a 5 Mbyte navigation database, full-colour active matrix liquid crystal flat-panel display and the latest surface mount technology available, providing a more compact design and much higher reliability.

Specifications

Dimensions:
(UNS-1C) 162 × 146 × 228 mm
(data transfer unit) 53.8 × 146 × 205 mm

UNS-1C all-in-one flight management system

1995

Weight:
(UNS-1C) 3.63 kg
(data transfer unit) 1.47 kg
Power supply: 27.5 V DC, 50 W (max)
26 V AC, 400 Hz, 1 A

Operational status

The UNS-1C is available for retrofits offered as a turnkey update for the Dash 8, selected by Horizon Air for its 25 new Dash 8 Series 200 aircraft. It has also been certified on the Bell 430 helicopter.

A more compact 'all-in-one unit' version, designated UNS-ICsp (special package), is available as a special option on the Citation, Ultra and Citation VII.

The UNS-1C is also available in MultiMission Management System (MMMS) versions. The MMMS includes the ability to fly six mission patterns and to accommodate EFIS, Doppler, radar, video and FLIR data.

Contractor

Universal Avionics Systems Corporation.

UPDATED

UNS-1D Flight Management System (FMS)

The UNS-1D is an upgraded UNS-1C with new Control Display Unit (CDU) and new remote Navigation Computer Unit (NCU) with internal 12-channel GPS receiver. The UNS-1D is also available in MultiMission Management System (MMMS) versions. The MMMS includes the ability to fly six patterns and to accommodate EFIS, Doppler, radar, video and FLIR data.

Operational status

Available. Selected for retrofit to US Navy C-9B/DC-9 transport aircraft to enable them to meet CNS/GATM, BRNAV and ACAS requirements.

Contractor

Universal Avionics Systems Corporation.

UPDATED

UNS-1K Flight Management System (FMS)

The UNS-1K is positioned between the higher end UNS-1C and UNS-1D systems and the lower placed UNS-1M navigation management system produced by Universal Avionics Systems Corporation. The UNS-1K utilises the same software, SCN 602, as the UNS-1C and UNS-1D systems and features 10 line select keys and large 4 in colour flat-panel display. The control display unit is housed in a standard 4.5 in tall × 5.75 in wide Dzus-mounted unit 3.25 in deep. The navigation computer unit is 2 MCU in size, and includes an integral 12-channel GPS receiver.

The system will meet emerging Required Navigation Performance/Actual Navigation Performance requirements around the world, including new European B-RNAV requirements.

System features include flight planning with full SID/STAR procedures, airways and approaches. A best computed position is based upon inputs from the integral GPS receiver, auto-scanning DME, and the operator's complement of external navigation sensors. The system will fly all ARINC 424 leg types. En route manoeuvre capabilities include a dedicated direct-to function, FMS heading commands, PVOR tracking, coupled VNAV with computed top of descent and vertical direct-to commands. Procedural holding patterns and approaches along with their transitions and missed approach procedures are contained in the database. The Approach Mode provides IFR-approved, pseudo-localiser, pseudo-glideslope guidance to any airport making all approaches look like an ILS. Fuel Management and optional Frequency Management functions are available and the MultiMission Management System (MMMS) is available on the UNS-1C/-1D models. A standby power-off mode retains flight plan and fuel initialisation data for up to 8 hours. A comprehensive test mode is incorporated to facilitate installation check out and return to service.

The integral 12-channel GPS receiver provides real-time and predictive Receiver Autonomous Integrity Monitoring (RAIM), automatic Fault Detection and Exclusion (FDE), step detection, and manual satellite

UNS-1K flight management system

1998/0018224

deselection capabilities. The UNS-1K is certified for GPS operation en route, terminal and approach phases of flight and meets the GPS navigation operational approvals and Minimum Navigation Performance Specifications (MNPS) for navigation in the North Atlantic Track (NAT) airspace. The system can also be approved under FAA Notice 8110.60 for primary means of navigation in remote/oceanic airspace using GPS alone in conjunction with Universal Avionics' off-line PC-based Flight Planning and RAIM Fault Detection and Exclusion programme. The UNS-1K will also be compatible with Universal Avionics' GLS-1250 GPS landing system (currently in flight test) which will provide future growth by adding GPS precision approach capability.

The UNS-1K will also interface with Universal Avionics' new UniLink air-to-ground datalink. Adding UniLink provides the operator with such capabilities as predeparture clearances, flight plan up/downlinking, oceanic clearances, position reporting, digital ATIS, messages and text weather, through the UNS-1K control display unit. The UniLink UL-600 is housed in a 1 MCU size unit which can provide datalink information through several communications media including VHF, telephony and satcom systems. The UNS-1K with UniLink combine to provide the operator with full capability Communication/Navigation/Surveillance (CNS) avionics suite.

Operational status

Both the UNS-1K and UniLink are available. UNS-1K is certified TSO C129a Class A1/B1/C1 and C115b. Selected for a number of regional aircraft types.

Contractor

Universal Avionics Systems Corporation.

NEW ENTRY

UNS-1M navigation management system

The all-in-one UNS-1M is a full navigation management system, self-contained in a control/display unit, which meets the standards of TSO C-129 A1/B1/C1. It features a flat-panel display, integral 12-channel GPS receiver with Receiver Autonomous Integrity Monitor (RAIM) and full alphanumeric keyboard. Smart auto-scanning DME/DME/DME provides continuous DME updating. Three external long-range sensor inputs accommodate combinations of Omega/VLF, Loran C, inertial and GPS sensors. The UNS-1M provides digital and analogue outputs for autopilot, mechanical HSI and EFIS systems. The unit's three-dimensional approach mode provides guidance on approach and is IFR certified for non-precision GPS, RNav, VOR and VOR/DME approaches. Fuel management utilises two-engine DC analogue sensor inputs to provide real-time fuel data, including gross weight and landing weight; fuel on board, predicted overhead the destination, required to the destination and amount of reserves; usage figures for ground and air n miles/lb, and endurance and range. A worldwide Jeppesen navigation database is stored on, and updated via, a non-volatile flash memory card. The database includes airports, VORs, DMEs, VOR/DMEs, ILS/DMEs, VORTacs, Tacans, NDBs and en route and terminal waypoints.

Specifications

Dimensions: 114 × 146 × 241 mm
Weight: 2.81 kg
Power supply: 19-32 V DC, 35 W (max) at 27.5 V
26 V AC, 400 Hz

Operational status

In production. Selected by Executive Airlines for ATR-42 and ATR-72 aircraft.

Contractor

Universal Avionics Systems Corporation.

VERIFIED

COCKPIT DISPLAYS, INSTRUMENTS AND INDICATORS

Rcgerson Kratos NeoAV Model 550 EFIS Electronic Flight Instrumented System and NeoAv IIDS Integrated Instrument Display System installed in the Bell Model 430 helicopter cockpit **1998**/0018163

BELGIUM

CDU 3.4M flat panel Control Display Unit

The CDU 3.4M series has been developed to meet the difficult specifications found in avionics projects where cost and space are important.

In order to minimise the depth a 4 × 3 in (101.6 × 75.2 mm) electroluminescent panel is used. Appropriate drive electronics ensure good contrast performance under bright sunlight-readable conditions. A large dimming ratio ensures good dimmability for night-time operation.

The unit is NVIS-compatible per MIL-L-85762A.

The CDU 3.4M communicates with onboard systems through a dual redundant MIL-STD-1553 bus interface.

Specifications

Dimensions: 146 × 219 × 178 mm
Weight: 4.5 kg
Power supply: 28 V DC, 45 W
Temperature range: −54 to +55°C
Altitude: up to 50,000 ft
Image dimensions: 101.6 × 75.2 mm
Resolution: 320 × 240 pixels

Operational status

In production, utilised in Belgian Air Force C-130H, Hercules C-130J, P-3C Orion aircraft.

Contractor

BARCO nv/Display Systems.

VERIFIED

MHDD 3.3 Monochrome Head-Down Display

The MHDD 3.3 is a compact high-resolution 2.7 in (68 mm) circular avionics display used in airborne applications. It is a monochrome, sunlight-readable display, with a raster scanning system suitable for fighter aircraft environmental requirements.

The MHDD 3.3 offers a high-brightness display, with very low power consumption, compact, lightweight design and brightness control located on the front of the display. It has full MIL specifications and is convection cooled, requiring no forced air.

Specifications

Dimensions: 82.6 × 82.6 × 176 mm
Weight: 17 kg
Power supply: ± 15 V DC, 20 W (max.)
Temperature range: −40 to +55°C
Altitude: up to 80,000 ft
Image dimensions: 68 mm (2.7 in) circular
Resolution: 384 × 384 pixels (cross)

Operational status

In production, utilised for international airforce programmes such as Belgian Air Force F-16 aircraft on the Carapace passive ECM subsystem. Also selected for several helicopters of the ALAT (France) as the display for the EWR-99/Fruit radar warning receiver display.

MHDD 3.3 monochrome circular head-down display **1998**/0018219

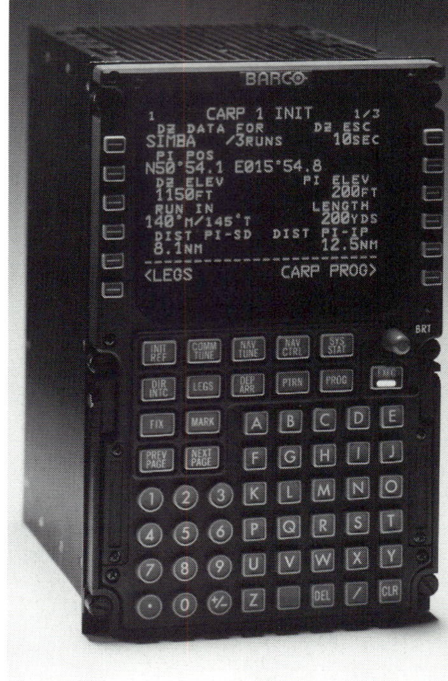

CDU 3.4M flat panel Control Display Unit **1997**/0001347

Contractor

BARCO nv/Display Systems.

UPDATED

MPRD 100 high-resolution flat panel colour displays

The MPRD 100 series of rugged high-resolution flat panel colour displays includes the MPRD 126 display 10.4 in (264 mm) and the MPRD 134 display 13.3 in (338 mm).

The MPRD 100 series is equipped with an LCD panel, incorporating amorphous silicon TFT (Thin Film Transistor).

The MPRD 100 series is light-weight LCD panels, offering superior image quality and excellent off-axis viewability.

It displays video signals as well as signals from graphic boards and scalers.

Due to compact packaging the MPRD 100 Series is suitable for size and weight critical defence applications.

Specifications

MPRD 126
Dimensions: 297 × 241 × 100 mm
Weight: 6.5 kg
Power supply:
28 V DC, 40 W (heater OFF) / 65 W (heater ON)
115 V AC/170 V DC options
Image dimensions: 211 × 158 mm
Resolution: 640 × 480 pixels (VGA)

MPRD 134
Dimensions: 372 × 276 × 125 mm
Weight: < 15 kg
Power supply:
28 V DC, 60 W (heater OFF) / 85 W (heater ON)
115 V AC/170 V DC options
Image dimensions: 270 × 203 mm
Resolution: 1,024 × 768 pixels (XGA)

Operational status

All units are in production.

Contractor

BARCO nv/Display Systems.

VERIFIED

MPRD 200 series flat panel displays

RFD241 16 in flat panel displays

Introduced in 1996, both the FD 241 and RFD 241 use 16 in LCD glass. The units are designed for benign (FD) and rough (RFD) environments. Resolutions of both the FD and RFD 241 displays are 1,280 × 1,024 pixels, with refresh rates of up to 76 Hz.

RFD251 20 in flat panel display

The RFD 251 20 in flat panel display is Barco's largest rugged environment display for use in air-, land- and sea-based platforms. The LCD components and the display electronics have been integrated to improve ruggedness.

The RFD 251 provides the user with a large image surface and a viewing angle of at least 60° in all directions. The high contrast ratio of this display allows it to be used in bright environments. Resolution is 1,280 × 1,024 pixels, and the display is capable of accepting graphic signals from VGA onwards. The RFD 251 display is compatible with standard 19 in racks.

Barco's MPRD 126 fully ruggedised flat panel display **1997**/0001349

Operational status

The RFD 251 is available from 1998.

Contractor

BARCO nv/Display Systems.

UPDATED

MPRD 9000 high-resolution CRT colour displays

The MPRD 9000 series of rugged multiscan, high-resolution CRT colour displays includes the MPRD 9639 display 15 in (14 in V; 356 mm), MPRD 9643 display 17 in (16 in V; 650 mm), MPRD 9651 display 20 in (19 in V; 483 mm), MPRD 9874 display 29 in (27 in V; 686 mm).

The MPRD 9000 Series offers flexibility, expansion capability and ease of integration with low volume and low power. Both display and control functions are available in a compact package. Primary airborne applications include radar, sonar, anti-submarine warfare, maritime patrol, imaging and C³I.

Specifications

MPRD 9639
Dimensions: 387 × 323 × 355 mm
Weight: 24 kg
Power supply:
115/220 V AC, 40-440 Hz, 100 W
28/170/270 V DC options
power factor improvement module (90-264 V AC) option
Image dimensions: 264 × 198 mm
Resolution: 1,280 × 1,024 pixels

MPRD 9643
Dimensions: 435 × 355 × 391 mm
Weight: 27 kg
Power supply:
115/220 V AC, 40-440 Hz, 100 W
28/170/270 V DC options
power factor improvement module (90-264 V AC) option
Image dimensions: 300 × 225 mm
Resolution: 1,280 × 1,024 pixels

MPRD 9651
Dimensions: 482 × 399 × 444.5 mm
Weight: 32 kg
Power supply:
115/220 V AC, 40-440 Hz, 120 W
28/170/270 V DC options
power factor improvement module (90-264 V AC) option
Image dimensions: 300 × 225 mm
Resolution: 1,600 × 1,200 pixels

MPRD 9654
Dimensions: 520 × 419 × 467.9 ± 1.5 mm
Weight: 39.5 kg (max)
Power supply: 115 V or 220 V AC, 40-440 Hz, 28/170/270 V DC options; power factor improvement module (90-264 V AC) option
Image dimensions: 413 × 294 mm
Resolutions: 1,600 × 1,200 pixels

MPRD 9874
Dimensions: 685 × 558 × 583 mm
Weight: 79 kg
Power supply:
power factor improvement module (90-264 V AC)

MPRD 9651 rugged CRT display **1998**/0018218

Barco's RFD251 20 in flat panel display **1997**/0002355

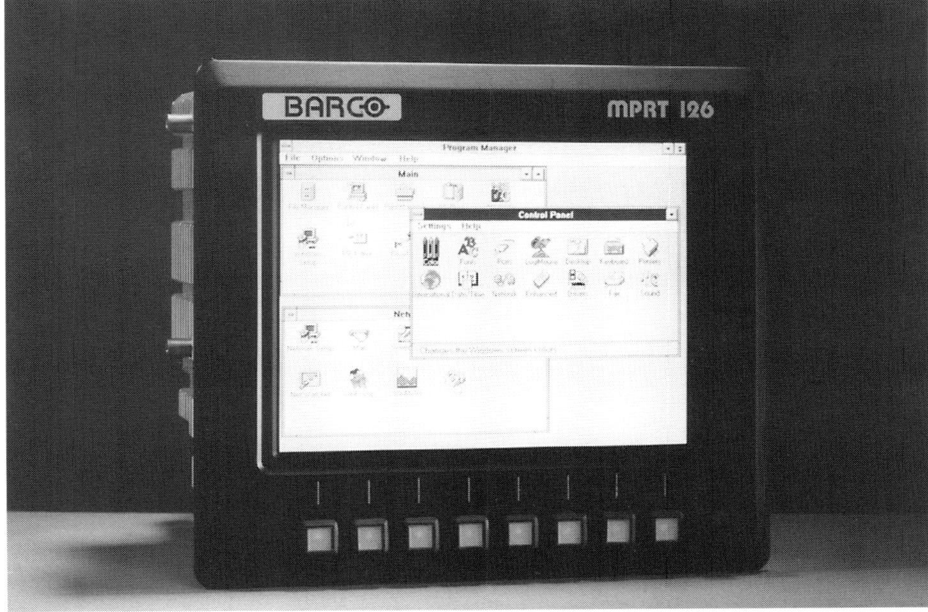

MPRT 126 flat panel colour display (264 mm) with an integrated terminal; offers video window capability inserted in high-resolution graphics **1997**/0001351

Image dimensions: 520 × 390 mm
Resolution: 2,048 × 1,536 pixels

Operational status

All units are in production. The MPRD 9639 15 in MIL-tailored display has been selected for the RAF Nimrod AQS-901 acoustic radar colour update programme. The MPRD 9651 has been selected by Ericsson for the Brazilian SIVAM Erieye Command and Control System.

MPRD 9874 rugged CRT display **1998**/0018217

Contractor

BARCO nv/Display Systems.

UPDATED

MPRT 126 multipurpose rugged AMLCD terminal

The MPRT 126 is a 10.4 in (264 mm) flat panel AMLCD display. It provides a video window capability inserted in high-resolution graphics, together with X-terminal touch-panel capability.

New technologies can be easily qualified and implemented in the MPRT 126 terminal, as they become available. The terminal incorporates either a 486 or pentuim processor.

Specifications

MPRT 126
Dimensions: 286 × 243 × 140 mm
Weight: 7.4 kg
Power supply:
28 V DC, 41 W (heater OFF) / 73 W (heater ON)
115 V AC/170 V DC options
Image dimensions: 211 × 158 mm
Resolution: 640 × 480 pixels (VGA) or 800 × 600

Operational status

Available from October 1997. Selected by Ericsson for integration into the Brazilian SIVAM Erieye Command and Control System, as the touchscreen man/machine interface for the aircraft system programmable workstations.

Contractor

BARCO nv/Display Systems.

UPDATED

RGD 600 ruggedised CRT colour displays

The RGD 600 series of rugged multiscan CRT high-resolution colour displays includes the RGD 639 display 15 in (14 in V; 356 mm), RGD 643 display 17 in (16 in V; 650 mm), RGD 651 display 20 in (19 in V; 483 mm).

The RGD 600 series comprises high-performance, rugged displays. It enables tight budgetary requirements to be met, without sacrificing electro-optical performance.

The RGD 600 series is mechanically compatible with the MPRD 9000 series.

Specifications

RGD 639
Dimensions: 387 × 323 × 335 mm
Weight: 24 kg
Power supply:
115/220 V AC, 40-440 Hz, 90 W
28/170/270 V DC options
power factor improvement module 90-264 V AC option
Image dimensions: 264 × 198 mm
Resolution: 1,280 × 1,024 pixels

RGD 643
Dimensions: 435 × 355 × 391 mm
Weight: 27 kg
Power supply:
115/220 V AC, 40-440 Hz, 100 W
28/170/270 V DC options
power factor improvement module 90-264 V AC option
Image dimensions: 300 × 225 mm
Resolution: 1,280 × 1,024 pixels

RGD 651
Dimensions: 482 × 400 × 445 mm
Weight: 32 kg
Power supply:
115/220 V AC, 40-440 Hz, 120 W
28/170/270 V DC options
power factor improvement module 90-264 V AC option
Image dimensions: 300 × 225 mm
Resolution: 1,600 × 1,200 pixels

Operational status

All units are in production.

Contractor

BARCO nv/Display Systems.

VERIFIED

Barco's flat panel display is available in 16, 13 or 10 in and various levels of ruggedisation; display sources range from VGA to high-resolution graphics and video **1997**/0001350

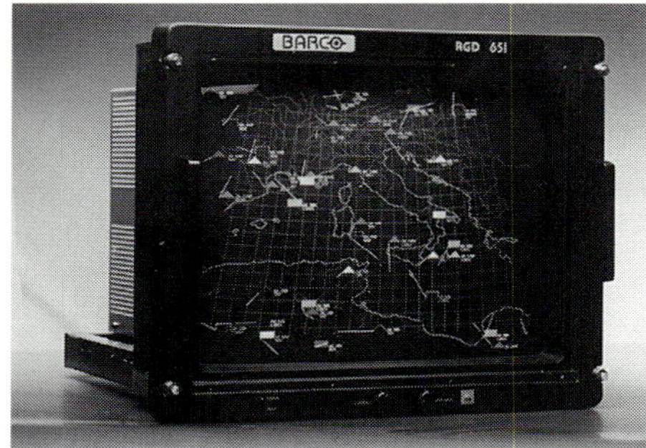

RGD 600 rugged CRT display **1998**/0018216

CANADA

Peripheral vision display

The peripheral vision display is designed to prevent pilot spatial disorientation and improve flight safety.

The system presents the pilot with a line of light projected across the instrument panel. This laser-produced line moves parallel to the earth's horizon as the aircraft changes its attitude. As this line can be seen by the pilot's peripheral vision, he will be aware of changes in aircraft attitude even while his attention is diverted outside or to other cockpit tasks.

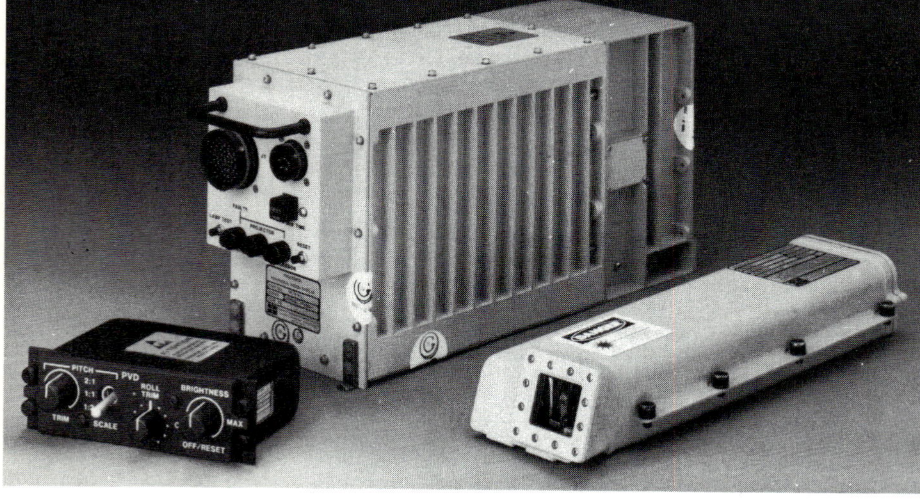

AlliedSignal Aerospace Canada peripheral vision display system

Specifications

Dimensions:
(control panel) 146 × 48 × 123 mm
(processor) 127 × 51 × 318 mm
(laser projector) 102 × 51 × 318 mm
Weight: <9 kg

Operational status

In operational service on a classified USAF programme. Under evaluation at the Defence Research Agency, Farnborough on a Sea King helicopter. Also under evaluation on board the US Air Force F-16, the Canadian Forces CH-113 and the US Navy SH-2F ship motion simulator and vertifuge.

Contractor

AlliedSignal Aerospace Canada.

VERIFIED

CMA-2014 Multipurpose Control and Display Unit (MCDU)

The CMA-2014 MCDU is primarily designed as a cockpit management system. The ARINC 739-compliant MCDU controls all GPS/FMS functions in applications such as the CMA-900 GPS/FMS, can provide control/display functions for other ARINC 739-compatible systems, and is also used for various system maintenance operations. The current system uses a 9-line × 21-character, ARINC 739-compatible display; a 14-line × 24-character colour AMLCD display was due to be available in the third quarter of 1996. Both systems include a full alphanumeric keyboard, line-select keys and nine dedicated annunciators.

Canadian Marconi's CMA-2014 multipurpose CDU **1996**

Specifications

Dimensions: 171 × 146 × 214 mm
Weight: 4 kg
Power supply: 28 V DC, 35 W (max)
5 V AC for integral lighting
Display: bit mapped raster scan CRT, 128 × 96 pixels; 8 lines of 16 characters; 8 colours plus highlighting, image overlays, graphics and multiple display pages
Keyboard: 56 keys including 8 soft keys

Operational status

In production.

Contractor

Canadian Marconi Company.

VERIFIED

CMA-2055 Integrated Instrumentation Display System (IIDS)

The single unit CMA-2055 Integrated Instrumentation Display System (IIDS) reduces the cockpit space required to display a variety of information including engine, transmission, electrical, fuel and hydraulic parameters, caution/warning annunciators and rotor track/balance and vibration monitoring data. With the system, associated parameters are grouped together, thereby increasing the rate of transfer of information to the pilot. The IIDS also stores engine in-flight performance data for subsequent downloading to a ground-based computer for trend monitoring and analysis.

The liquid crystal technology used in the IIDS has excellent resolution, wide viewing angle and high contrast under all lighting conditions. Canadian Marconi's approach in using this technology enables the design of displays that are lightweight, compact and reliable.

Specifications

Dimensions: 305 × 152 × 196.8 mm
Weight: 5.4 kg
Power supply:
40 W nominal
50 W (max)
Display size: 2 screens each 114 × 114 mm
Interfaces: RS-422, RS-232, analogue, discrete
Environmental: DO-160C
Reliability: >10,000 h MTBF

Operational status

In Production. The CMA-2055 is certified and installed on the Boeing MD-900 helicopter.

Contractor

Canadian Marconi Company.

UPDATED

CMA-2056 Multifunction Display System

The CMA-2056 is a high-brightness CRT-based colour/monochrome display system which has been designed for minimum cost, power consumption and weight. All electronics and interfaces are contained in one LRU; no remote symbol generator or electronics unit is required. The CMA-2056 has full graphics capability, is visible in direct sunlight and is ANVIS (Class 1 Type B) compatible.

The CMA-2056 can interface with a wide range of other avionics equipment via the MIL-STD-1553B or ARINC 429 databus. A discrete interface assembly can be added to integrate other avionic systems that are not to MIL-STD-1553B or ARINC 429.

Specifications

Dimensions: 349.7 × 200.7 × 200.7 mm
Weight: 18.88 kg
Power supply: 28 V DC, 100 W (max)
Cooling: convection
Operating modes: colour, with any of 8 colours displayed from a palette of 4,096. Monochrome (green or white) with grey scales
Display size: 152 × 152 mm
Display format: 60 Hz; 525 lines non-interlaced graphics with 512 × 512 × 3 pixels; 525 or 875 lines interlaced
Key switches: integrated front panel keyboard with 18 soft keys and 4 dedicated keys
Interfaces: MIL-STD-1553B RTU, dual RS-343 video inputs, dual RS-232/RS-422 ports, remote keyboard (optional), direct voice input (optional), RGB linear video input (sync on green), RGB linear video output (for remote access), spare capacity for custom interfaces
Environmental: MIL-E-5400 Class 1A
Reliability: 2,000 h MTBF

Operational status

In service. Installed on US Air Force KC-135 Speckled Trout aircraft.

Contractor

Canadian Marconi Company.

UPDATED

Avionic Keyboard Unit (AKU)

The Avionic Keyboard Unit (AKU), originally designed and developed for the AH-64C Apache helicopter, is available as a durable intelligent keyboard for crewstation applications.

The AKU can be incorporated in any digital cockpit or mission architecture and is fully qualified as a MIL-E-5400 Class II device for airborne applications. It is a lightweight, versatile general purpose keyboard, with MIL-STD-1553B serial digital interface or optional ARINC 429 or RS-422 serial interfaces.

Display for the C-130J **1995**

Specifications

Dimensions: 124 × 155 × 63.3 mm
Weight: <1.4 kg
Power supply: 28 V DC
Display: 22 characters in 1 line
Temperature range: −54 to +55°C
Environmental: MIL-STD-810D
Reliability: >7,500 h MTBF

Contractor

Litton Systems Canada Ltd.

VERIFIED

C-130J display suite

Litton is under contract to supply four portrait-oriented smart multifunction displays and up to five 2.3 × 3 in (58.4 × 76.2 mm) graphic monochrome units for each Lockheed Martin C-130J. The colour multifunction display unit is designed for a distributed EDIS architecture and provides for anti-aliased symbology. It operates with a MIL-STD-1553B databus and is NVG-compatible.

Specifications

Dimensions: 198.1 × 274.3 × 200.7 mm
Weight: 8.16 kg
Power supply: 180 W
Display area: 152.4 × 203.2 mm
Resolution: 480 × 640 RGBG quad pixels

Operational status

In development.

Contractor

Litton Systems Canada Ltd.

VERIFIED

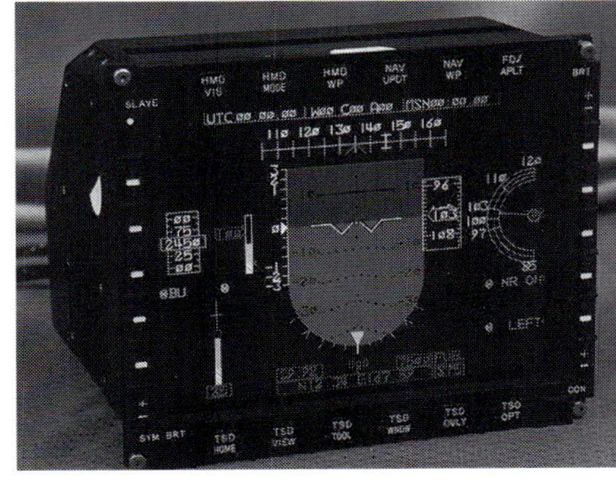

Comanche colour MultiFunction Display **1995**

Comanche colour and mono MultiFunction Displays

The MultiFunction Displays (MFD) for the RAH-66 Comanche comply with MIL-L-85763 for the NVIS Class A mode of operation, using advanced techniques to achieve the Comanche optical specification. There are four MFD per aircraft, with two displays providing high-resolution monochrome for sensor information and two displays that provide colour.

Specifications

Dimensions: 203.2 × 241.3 × 218.4 mm
Weight: 5.72 kg
Power: 130 W excl heaters
Display area: 152.4 × 203.2, 480 × 640 RGBG stripe

Operational status

In development.

Contractor

Litton Systems Canada Ltd.

VERIFIED

Comanche MultiPurpose Display

Litton Canada has designed, developed and produced the monochrome MultiPurpose Display (MPD) to provide graphics for Comanche aircraft and weapon functions. There are two MPDs located in each of the two cockpits, making four displays in all in each Comanche.

Specifications

Dimensions: 129.5 × 200.7 × 149.9 mm
Weight: 2.99 kg
Power: 50 W excl heaters
Display: 91.4 × 170.2 mm, 288 × 544 pixels

Operational status

In development.

Contractor

Litton Systems Canada Ltd.

VERIFIED

Data Entry Display (DED) system

The data entry display set consists of two units: a Data Entry Display (DED) assembly and a power supply. The DED assembly is a sunlight-readable upfront display developed under the F-16 multinational staged improvement programme. The display unit, although graphics capable, usually provides five lines of 24 characters and is used for presenting communications, navigation and IFF data. The DED set has been under high-rate production since 1981.

The purpose of the DED is to take in serial data in a raster scan format and display it on a 76 mm wide by 25 mm high screen. The display screen is composed of three LED display modules, each of which provides an array of 64 × 64 pixels on a 25 × 25 mm viewing surface. The DED is designed to store a serial bit stream of data and display it in raster scan format on a screen 192 pixels wide by 64 pixels high, consisting of three modules. In order to keep the information updating process as independent from internal timing constraints as possible, two identical memories are provided in the DED. When one memory is used to refresh the screen the other memory is available for updating. When the memory which is being updated becomes completely filled, the memory functions are exchanged so that the memory with the latest complete set of data is used to refresh the screen and the other memory is available for updating. Each memory has one bit of storage available for each pixel on the screen.

In addition to the circuitry required to provide the display, the DED also contains independent fault detection and temperature alert discretes and a thermostat to enable in-flight non-interruptive monitoring.

Operational status

In service on the Lockheed Martin F-16C/D.

Contractor

Litton Systems Canada Ltd.

VERIFIED

EH 101 Merlin mission displays

Litton Canada is under contract to supply 625-line CCIR-compatible Active Matrix Liquid Crystal Displays (AMLCDs) to meet the mission video display requirements for the Royal Navy EH 101 Merlin cabin and cockpit. A Merlin set consists of four cabin and two cockpit units.

Specifications

Dimensions:
(cockpit) 254 × 241.3 × 175.3 mm
(cabin) 396.2 × 299.7 × 182.9 mm
Weight:
(cockpit) 7.48 kg
(cabin) 10.43 kg
Power supply:
(cockpit) 130 W
(cabin) 110 W
Display:
(cockpit) 147.3 × 195.6 mm, 100 pixels/in
(cabin) 205.7 × 274.3 mm, 576 × 768 Quad RGBG

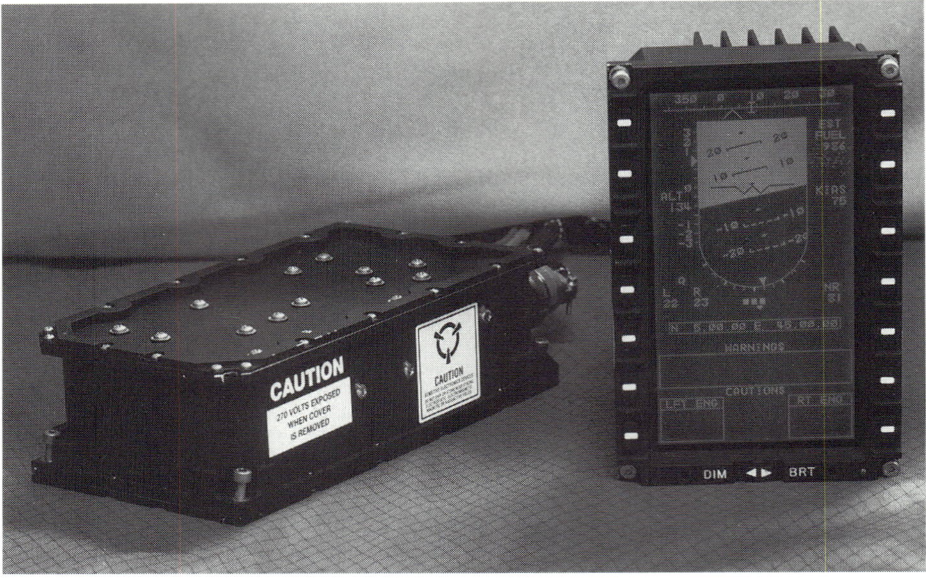

Litton Comanche Multipurpose Display **1995**

Operational status

In production.

Contractor

Litton Systems Canada Ltd.

VERIFIED

General Purpose Control Display Unit (GPCDU)

The General Purpose Control Display Unit (GPCDU) has been developed as a low-cost, rugged avionic user interface for data input and display, control and management of aircraft systems. The GPCDU incorporates a Litton light emitting diode, 10-row by 24-character dot matrix display and a large number of interface options. The GPCDU can also be supplied as a SNU-84-1 compatible keyboard entry and display system or tailored to specific sensor requirements.

Operational status

The GPCDU has been selected for US Air Force RC-135, RF-4C, TR-1A and U-2 aircraft and a number of other airborne platforms, including the AWACS and RF-5E, for other air forces.

Contractor

Litton Systems Canada Ltd.

VERIFIED

Up-Front Display (UFD)

The Up-Front Display (UFD) is a military qualified alphanumeric display which employs Litton's matrix light emitting diode technology. The UFD is designed to operate in the most severe airborne environments. In its standard configuration, the UFD is equipped with dual-redundant RS-422 serial ports, but can be supplied with MIL-STD-1553B or ARINC 429 interfaces.

The UFD is a compact and versatile display device, with built-in scroll and function keys. The flexible UFD can easily be incorporated in an avionic architecture requiring a display repeater and/or data display capability.

Specifications

Dimensions: 83 × 230 × 165 mm
Weight: <2.35 kg
Power supply: 28 V DC, <57 W
Display: 10 lines × 35 characters
Temperature range: −54 to +55°C
Environmental: MIL-STD-810D

Operational status

The UFD has been selected for the AH-64C and AH-64D Apache helicopters.

Contractor

Litton Systems Canada Ltd.

VERIFIED

Warning and caution annunciator display

The warning and caution annunciator display consists of three LED dot matrix annunciation display zones, an LED dot matrix resettable cue display, an edge-lit information panel with key switches and a dual-multiplex databus receiver/transmitter.

The left and right LED modules and the cue display are ANVIS yellow for the display of cautionary messages. The upper zone of the centre LED module is

The Litton up-front display (left) and avionics keyboard (right) have been selected for AH-64C and AH-64D Apache helicopters **1995**

ANVIS green B to display advisory messages and the lower zone is ANVIS red for the presentation of warning messages. The edge-lit information panel has seven push-button control switches for scrolling, display brightness, cue reset and self-test.

Specifications

Dimensions: 115 × 396 × 224 mm
Weight: 6.58 kg
Power supply: 28 V DC, 180 W, 5 V AC

Operational status

In production for US Air Force C-17 aircraft.

Contractor

Litton Systems Canada Ltd.

UPDATED

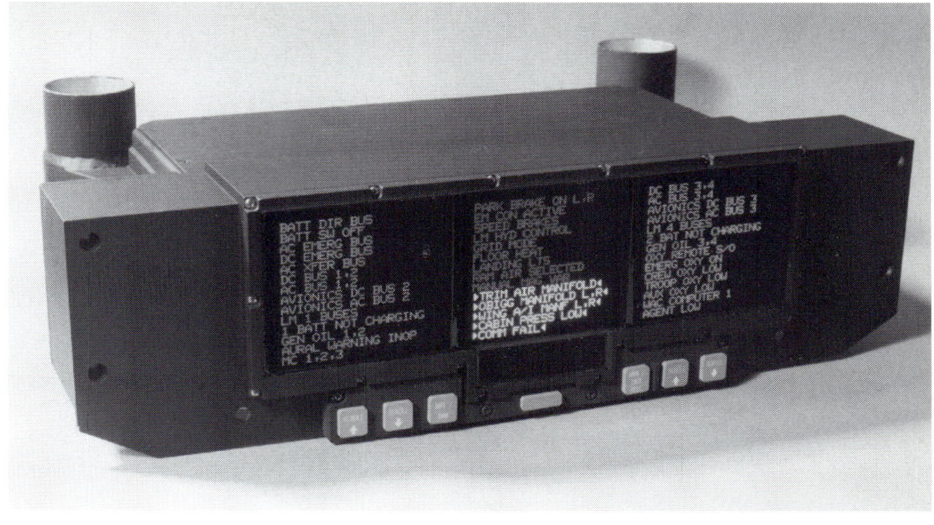

The Litton warning and caution annunciator is in production for the C-17

CHINA, PEOPLE'S REPUBLIC

BGZ-1 combined precision altimeter

The BGZ-1 combined precision altimeter combines the functions of precision altitude measuring, preselected altitude, altitude warning and parallel altitude reporting. It senses the total pressure and static pressure signals independently to implement precision altitude measuring and altitude indicating. It has the functions of preselected altitude and over-altitude warning.

Specifications

Performance data:
Operating range: −1,600 to 89,900 ft
Accuracy: <4%

Preselected altitude:
Operating range: 0-89,900 ft
Resolution: 100 ft
Power supply:
Operating power supply: 27 V DC ±10%, 10 W
Illuminating power supply: 27 V DC, 4 W
Dimensions: 80 × 80 × 220 mm
Weight: 2.3 kg

Contractor

Chengdu Aero-Instrument Corporation.

VERIFIED

BGZ-1 combined precision altimeter
1997/0002356

CZECH REPUBLIC

Airspeed indicators

Mikrotechna Praha a.s manufactures a range of 3 in airspeed indicators, including the models: LUN 1106, LUN 1106.XX-8, LUN 1107.XX-8, LUN 1113.XX-8, LUN 1114; LUN 1115, LUN 1116, LUN 1117, and UL 20 series. All are metal pressure-capsule instruments. All except the UL series utilise pilot-static pressure; the UL series use venturi tubes. In general, options are available within the following specifications:

Specifications

Range: 0-350 kt, or 0-400 mph, or 0-600 km/h
Lighting: none; 5 V DC; 14 V DC; 27 V DC
Case: ARINC 408 3 ATI; round MS-33638 (AS); round MS-33638 (AS) short
Dial layout: single- or dual-scale and range marked to customer requirements
Approvals: TSO C2d

Leading details of particular models:

LUN 1114
LUN 1114 is calibrated 0-300 kt, and indicates maximum allowable airspeed; it is TSO-C46a, and RTCA/DO 160C approved.

UL 20 series
The UL 20 series are designed specifically for use in ultralight aircraft. The UL 10-10 is calibrated −300 to +3,000 m; the UL 20-20 is calibrated −1,000 to +10,000 ft. Both 3 in and 2 in versions are available.

Contractor

Mikrotechna Praha a.s.

UPDATED

Altimeters

Mikrotechna Praha a.s manufactures a range of 3 in altimeters, including the models: LUN 1127, LUN 1128, and UL 10 series. Altimeters can be manually adjusted to variances in barometric pressure; temperature is automatically compensated by a bi-metallic element.

Specifications

Range: −1,000 to +20,000 ft; −1,000 to +35,000 ft (LUN 1127 only)
Lighting: none; 5 V DC, 14 V DC; 27 V DC
Case: ARINC 408; 3 ATI; round MS 33549 (AS)
Dial layout: mbar/in Hg; mbar; in Hg
Approvals: TSO C10b

UL 10 series
The UL 10 series is designed for ultralight aircraft. The UL 10-10 is calibrated −300 to +3,000 m, and the UL 10-20 −1,000 to +10,000 ft; both are also calibrated in mbars. They are 80 mm in diameter, and weigh 0.52 kg. Both 3 in and 2 in versions are available.

LUN 1114 airspeed indicator ***1998***/0018214

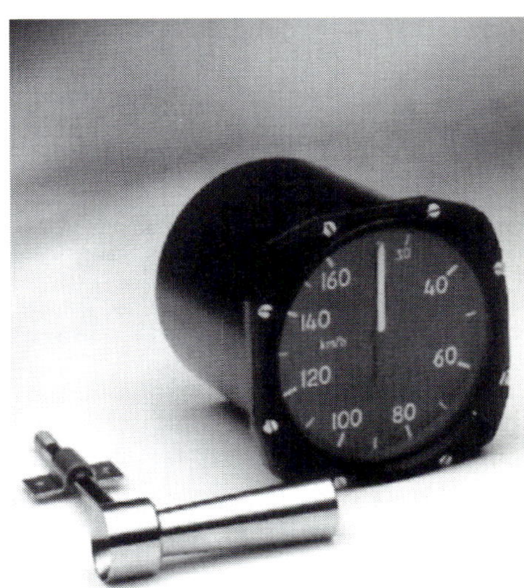

UL 20-10 airspeed indicator with UL 60 venturi tube ***1998***/0018212

UL 10-10 altimeter **1998**/0018213

Contractor
Mikrotechna Praha a.s.

NEW ENTRY

Artificial horizon LUN 1205.XX-8

Mikrotechna Praha a.s. manufactures the 3 in LUN 1205.XX-8 artificial horizon. The instrument provides warning of extreme bank and pitch angles. It also provides failure warning.

Specifications
Bank angle: ±360°
Pitch angle: ±60°
Extreme bank and pitch angles: ±5, ±32°
Power: 115 V AC, 400 Hz gyro; 5 or 27 V DC
Weight: 3.1 kg
Approval: RTCA/DO-160B

Contractor
Mikrotechna Praha a.s.

NEW ENTRY

Artificial horizon LUN 1205.XX-8 **1998**/0018211

Barometric altimeter LUN 1124

The LUN 1124 barometric altimeter indicates the absolute or the relative altitude of the aircraft using barometric pressure. It is equipped with integral lighting.

Specifications
Weight: 1.1 kg
Power supply: 5 V DC
Range: 0-10 km
Scale adjusting range: 790-1,050 mbar

Contractor
Mikrotechna Praha a.s.

UPDATED

Barometric altimeter LUN 1127.XXXX **1997**/0002357

Barometric altimeter LUN 1127.XXXX

The LUN 1127.XXXX barometric altimeter is a three-pointer metal-capsule instrument.

Specifications
Weight: 0.5 kg
Operating range: −1,000 to +35,000 ft, or −1,000 to +20,000 ft
Min increment: 20 ft
Barometric scale: 946-1,050 mbar
Qualifications: TSO-C10b, RTCA/DO-160C

Contractor
Mikrotechna Praha a.s.

UPDATED

Combined airspeed indicator with machmeter LUN 1170.XX-8

The LUN 1170.XX-8 can be used for simultaneous measurement of the indicated and true airspeed, including Mach number.

It can indicate up to three range scales, and can be equipped with white or NVIS integrated lighting.

Specifications
Ranges:
Indicated airspeed: 100-1,200 km/h
True airspeed: 300-1,200 km/h
Mach number: 0.5-0.9
Ceiling: 15 km
Weight: 0.9 kg
Power supply: 5 V or 28 V DC

Combined airspeed indicator with machmeter LUN 1170.XX-8 **1998**/0018210

Contractor
Mikrotechna Praha a.s.

UPDATED

Gyroscopic horizon LUN 1241

The LUN 1241 gyroscopic horizon is a direct-indicating instrument, which indicates pitch and roll angle. The artificial horizon is equipped with an arresting device which can be set by a tie rod within the range +10 to −7°. The main display indicates angles between +20 and −20°. Greater values of pitch, up to ±75° are shown in an indication window. Failure of the gyro supply or the arresting functions is indicated by a warning flag. Internal lighting is provided, including NVIS.

Specifications
Dimensions: ARINC 3 ATI
Weight: 1.6 kg
Power supply: gyro: 27 V DC, or 14 V DC; lighting: 5 V DC/AC, or 28 V DC/ AC
Range: ±70° pitch; 360° roll
Accuracy: ±0.5°, or ±1.0°
Qualifications: TSO C4c and RTCA/DO-160C

Contractor
Mikrotechna Praha a.s.

UPDATED

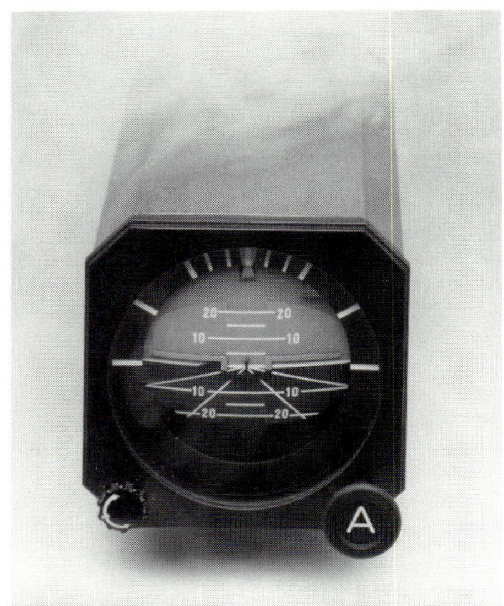

Gyroscopic horizon LUN 1241 **1997**/0001352

Integrated artificial horizon LUN 1208.XX-8

Mikrotechna Praha a.s. manufactures the 5 in LUN 1208.XX-8 integrated artificial horizon. The instrument provides warning of extreme bank and pitch angles, and for indication of the up-down and left and right

Integrated artificial horizon LUN 1208.XX-8 **1998**/0018209

deviations during either en route or the approach phase of flight. It also provides failure warning.

Specifications

Bank angle: ±360°
Pitch angle: ±75°
Extreme bank and pitch angles: ±15, ±32°
Power: 115 V AC, 400 Hz gyro; 5 or 27 V DC
Weight: 3.85 kg
Approval: RTCA/DO-160B

Contractor

Mikrotechna Praha a.s.

NEW ENTRY

Maximum allowable airspeed indicators LUN 1114.XXXX

The LUN 1114.XXXX airspeed indicator is a metal pressure-capsule instrument for indication of actual and maximum allowable airspeed.

Specifications

Range: 0-300 kt
Dimensions: ARINC 3 ATI
Weight: 0.5 kg
Power supply: 5 V or 28 V DC
Qualification: TSO-C46a, RTCA/DO-160C

Contractor

Mikrotechna Praha a.s.

UPDATED

Vertical speed indicators

Mikrotechna Praha a.s. manufactures a number of vertical speed indicators includig the LUN 1144, LUN 1149.24, and UL 30 series. They are metal-pressure capsule instruments.

Specifications

Ranges: 0-2,000; 0-3,000; 0-4,000; 0-6,000 ft/min
Dimensions: ARINC 3 ATI, or MS 33549 (AS) round
Weight: 0.6 kg
Power: 5 or 27 V DC

UL 30 series

The UL 30 series of vertical speed indicators is specifically designed for sport aircraft; both 3 in and 2 in variants are available, with scales calibrated in ft/min or m/s.

Contractor

Mikrotechna Praha a.s.

UPDATED

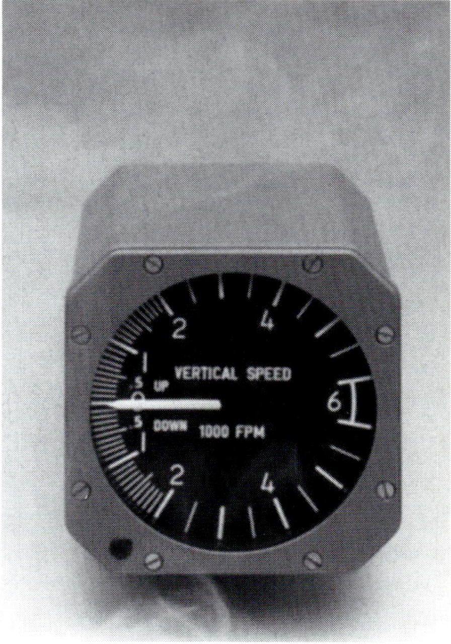

Vertical speed indicator LUN 1149.24
***1998**/0018215*

FRANCE

Attitude director indicators

Sextant Avionique produces a wide range of attitude director indicators for both civil and military use.

The current range provides a wide choice of instruments and features flight director command bars, failure flag indicators and warning lights for decision height alerting. Some instruments include a basic yaw indicator for additional assistance in asymmetric flight.

Operational status

Current military programmes include Transall C-160, Dassault Aviation Atlantique 2, Aerospatiale Nord 262 and Eurocopter SA 341 and SA 342 helicopters. Civil standard attitude director indicators are fitted in Airbus Industrie's A300/A310.

Contractor

Sextant Avionique.

VERIFIED

Digital VOR/DME indicator

The Sextant Avionique VOR/DME instrument takes navigation data derived from suitable ground beacons and processed by a flight management system such as the Collins Pro Line, using an ARINC 429 digital databus. It requires a 28 V DC input and weighs less than 2.1 kg.

Operational status

In production.

Contractor

Sextant Avionique.

VERIFIED

Dracar digital map generator

Designed for modern aircraft, the Dracar digital map generator provides information on terrain for both head-up and head-down displays. In automatic mode, images are computed to take account of aircraft position and show the terrain as it would appear through the front of the aircraft. In manual mode the point of view can be slewed manually to fore and aft, left and right and up and down. Images may be superimposed on the HUD, either as a direct profile comparison with the outside terrain or in conjunction

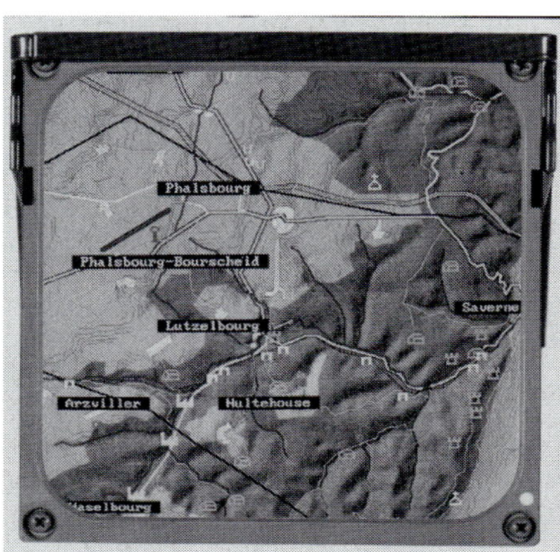

Dracar digital map display

with FLIR imagery. The map generator can also display a plan view of terrain and false colour can be added to provide a ground collision avoidance mode.

The database contains Digital LandMass System (DLMS) data or raster paper charts. The synthesised three-dimensional image can be either standard map imaging with better visual quality or use an expert system for integration of mapping data to optimise the trajectory.

Specifications

Resolution: 512 × 512 pixels
Coverage:
(1:100,000 scale) 500,000 km^2
(1:250,000 scale) 3 million km^2

Operational status

In development.

Contractor

Sextant Avionique.

VERIFIED

Electronic Flight Instrument System for the A310/A300-600 (EFIS)

The electronic flight instrument display designed for the A310 uses six identical 6.25 × 6.25 in (159 × 159 mm) shadow-mask colour CRTs. Each pilot has on the flight panel in front of him a pair of EFIS instruments: a primary flight display CRT which replaces the conventional electromechanical attitude director indicator and a navigation display CRT which supplants the earlier horizontal situation indicator and weather radar display. Each pair of CRTs is driven by a single symbol generator and a third system acts as a hot standby. Also associated with each EFIS pair is a control/display unit. Two more displays on the centre panel represent the Electronic Centralised Aircraft Monitor (ECAM) display, presenting information on aircraft systems in any phase of flight and schematic diagrams of the hydraulic or electrical systems, for example, to supervise their operation. One of them is normally reserved for warnings, the other for systems. The ECAM system operates independently of the pilots' EFIS instruments, being driven by two symbol generators (one operational, the other a hot spare) and a single control unit mounted on the throttle box. All six CRTs are interchangeable, reducing the number of spares required. Frame repetition frequency is 70 Hz.

Sextant Avionique has continued to refine the system. Colour stability and visibility of the displays in high-ambient lighting conditions have been demonstrated, the latter at 100,000 lux instead of the 85,000 lux specified. Nine colours are used on the primary flight display, seven are employed for symbology to avoid confusion and blue and brown represent the sky and earth. A uniformly smooth sky unmarred by the raster scan was achieved by slightly out of focus imaging and has proved particularly acceptable to the eye. The three-dimensional effect familiar to pilots who have used electromechanical primary flight displays has been achieved by masking certain symbols when they would normally be occulted by others mounted further forward in the instruments. Sextant Avionique and Airbus Industrie have devised, therefore, a presentation which is instantly recognisable as that of a typical primary flight display, but one that can also utilise the vast quantity of newly available digital data circulating within an aircraft on the databus. Airbus Industrie and Sextant Avionique have added the following information to the periphery of the instrument: a moving speed scale along the left-hand edge with its standard symbols generated by computer, a selected pressure altitude readout along the right-hand edge, an autopilot and autothrottle mode annunciator along the top edge and radio altimeter heights along the lower edge of the instrument.

For the navigation display, classic compass card symbology has been retained, again in conjunction with the flexibility that digital sensing and processing affords. Its use is innovative in that it creates new symbology to suit each phase of flight, in the form of an electronic map display with five distance scales marking the course to be followed with radio waypoints and their identifying codes. On selection by the pilot, a three-colour weather map can be superimposed on the navigation map, resulting in the saving of panel space that would be required by a separate weather CRT and the more comprehensible integration of weather and navigation information. The warning and system status CRTs on the centre panel show information previously unavailable to flight crews. In addition to special alarms, failure warning is given by presenting the crew with synoptic displays of failed systems, indications of vital actions to be followed to meet the difficulty and the effects on other systems.

The possibility of common mode faults affecting the software in the two sets of flight director equipment is eliminated by using different software for each. The two sets of software were developed by two teams working independently of one another in different localities. Software was implemented in machine code.

Operational status

In production for the Airbus A310 and A300-600.

Contractor

Sextant Avionique.

VERIFIED

Airbus A310 flight deck

Electronic Flight Instrument System for the A320 (EFIS)

In August 1984 Sextant Avionique was selected to develop and build the electronic flight instrument system for the Airbus Industrie A320 airliner. The system for this aircraft is an integration of the EFIS and ECAM CRT suites installed on the earlier A310, although with a number of important differences.

As with the A310 there are six CRT displays, but their disposition on the instrument panels is different. While on the earlier aircraft the flight director and navigation director are stacked vertically with the two ECAM instruments side-by-side on the centre panel, on the A320 the pilots' CRTs are ranged side by side, with the ECAM displays situated one above the other. The CRTs themselves are larger, 7.25 × 7.25 in (184 × 184 mm), compared with 6.25 × 6.25 in (159 × 159 mm), so that more symbology can be displayed without congestion. Each CRT can show at least seven colours. The upper ECAM, for the first time, shows primary engine parameters, replacing the 10 conventional electromechanical dial instruments on the A310. The lower ECAM will normally display systems information, such as electrical or hydraulic circuits, to show the location of faults. Confidence in the electronic display generated by A310 experience has resulted in the decision to delete most of the traditional electromechanical dial and pointer instruments, retaining just a few as a back-up in the most important functions.

As with the A310 system, all six CRTs are identical and interchangeable. However, whereas the A310 employs five symbol generators to drive them, the A320 uses only three and, with more complex functions, they are called display management computers. All three are identical. The CRTs and computers are grouped within very advanced architecture, permitting extensive redundancy. In the event of a CRT, sensor or computer failure the system automatically reconfigures itself, with top priority given to flight commands and engine warning. At the same time there has been a significant weight decrease from 153 to 115 kg. As with the A310 system, elimination of software errors in the pilots' flight director instruments is accomplished by the use of two independent development teams. However, unlike the earlier equipment, software is implemented in high-order language.

Operational status

In production for the Airbus A320.

Contractor

Sextant Avionique.

VERIFIED

Electronic instrument system for the A330/A340

In 1989 Sextant Avionique was selected to develop the electronic instrument system for the Airbus A330 and A340. The system is an integration of the EFIS and ECAM CRT suites. The EFIS system integrates the flight director and the navigation director with a weather radar link. The upper ECAM shows the primary engine parameters while the lower ECAM displays 15 system pages such as electrical, hydraulic, fault location, doors, bleed, brake circuits, auxiliary power unit, secondary engines, and so on.

The system uses six identical 7.25 × 7.25 in (184 × 184 mm) shadow-mask colour CRTs and three display management computers. In the event of a CRT, sensor or computer failure, the system is automatically reconfigured with top priority given to flight commands and engine warnings.

Specifications

Weight: 10 kg
Display dimensions: 184.1 × 184.1 × 355.6 mm

Operational status

In production.

Contractor

Sextant Avionique.

VERIFIED

The electronic instrument system for the A330 and A340 includes six shadow-mask colour CRTs

FCD 34 colour displays

The FCD 34 is a high-resolution shadow-mask full-colour multimode CRT display, using video scanning or stroke and raster images to show information on flight and engine control, navigation and weapon status. With a 3.5 × 4.5 in (89 × 114 mm) screen, the FCD 34 compact units adapt to the instrument panels of combat or training aircraft, either as original equipment or as a retrofit.

Operational status

The FCD 34 has been selected for the French Air Force Mirage 2000D and the Mirage 2000-5 and Mirage 2000 Strike export versions.

Contractor

Sextant Avionique.

VERIFIED

Flat-panel instruments

Designed to replace the previous family of electromechanical instruments, the 3 ATI and 4 ATI size liquid crystal display instruments are capable of

3 ATI AND 4 ATI flat-panel instruments have been selected by over 50 airlines ***1995***

acquiring digital, analogue and pneumatic signals. These indicators are well adapted for use on both helicopters and fixed-wing aircraft flight decks.

The first application for this programme was the TCAS VSI indicator. This indicator combines vertical speed and TCAS information on a 3 ATI instrument. The latest version is used in the navigation instrument display for the French Air Force Tucano.

Operational status

In production and selected by over 50 airlines and manufacturers for the TCAS VSI display.

Contractor

Sextant Avionique.

UPDATED

Gyro horizons

Sextant Avionique gyro horizons have been installed on virtually all commercial transport aircraft. Today, approximately 150 airlines and 30 air forces use Sextant Avionique gyro horizons as stand-by attitude reference instruments.

The Series H3XX gyro horizons are 3 ATI-sized and weigh about 1.5 kg. They make extensive use of modern alloys, have built-in static inverters and gyro speed monitors, and Sextant Avionique patented simplified erection and anti-spin devices. The basic versions include H301 (drum display), H321 (sphere display) and H341 (sphere display with ILS capability). 2 ATI and 4 ATI versions are also in production.

Operational status

In production and in service.

Contractor

Sextant Avionique.

VERIFIED

Icare map display and Mercator Remote Map Reader for the Mirage 2000N (RMR)

Sextant Avionique, under a French Ministry of Defence contract, developed for the Dassault Mirage 2000N aircraft, an integrated electronic map system with two full-colour display units that collects data from the electronic Mercator Remote Map Reader (RMR) radar, and aircraft symbol generation systems.

Operational status

In production. The first Icare systems entered operational use in 1986 with the French Air Force.

Contractor

Sextant Avionique.

UPDATED

Integrated Electronic Standby Instrument (IESI)

The Sextant Avionique solid-state Integrated Electronic Standby Instrument (IESI) is designed to replace conventional pneumatic altimeter, airspeed and gyroscopic horizon indicators with a single LRU

Integrated Electronic Standby Instrument (IESI) ***1997***/0001353

standard 3 ATI box in civil aircraft (commercial transport, commuter and business types).

The IESI incorporates design growth capability to display: SSEC, ILS (localiser and glide path), back course, slip heading, dual baroset scales and Mach number to meet customer requirements.

Two configurations are available: a single LRU configuration to display data from its three internal functions, together with two remote and interchangeable Sextant air data modules; a single LRU configuration dedicated to attitude, which acts as a Standby Horizon Indicator (HSI).

Specifications

Dimensions: 3 ATI case to ARINC 408A standard
Weight: 1.9 kg
Altitude range: up to 51,000 ft
Power supply: 28 V DC (emergency bus)
Certification: TSO-C10c (altitude), -C2d (airspeed), -C4c (attitude), -C113

Operational status

In production from end 1997.

Contractor

Sextant Avionique.

VERIFIED

LCD engine indicator

The LCD engine indicator uses a 4 ATI size liquid crystal display screen with the same technology as Sextant Avionique's 3 ATI VSI/TCAS instrument. The basic engine instrument comprises three screens which display data from the FADEC system, secondary sensors, fuel management computer and APU.

Sextant Avionique flat-panel instrument engine control display

Operational status

In production. Selected by Dassault for the Falcon 2000, and Falcon 50 EX.

Contractor

Sextant Avionique.

UPDATED

Low-airspeed system

Sextant Avionique has acquired the VIMI exclusive licence and has developed the CLASS low-airspeed system as an answer to the problem of low-airspeed measurements in helicopters for flight management, navigation and weapon firing requirements.

Using the helicopter's flight characteristics and, in particular, the measurement of cyclic pitch controls and attitudes, CLASS allows the combination of low airspeed computed data with conventional air data measurements to provide airspeed information. This information is valid throughout the flight range and requires no additional external mobile probes.

Operational status

In production for the Eurocopter Tiger helicopter.

Contractor

Sextant Avionique.

VERIFIED

MFD55/MFD66 liquid crystal multifunction displays

The MFD55/MFD66 multifunction displays are full-colour multimode displays using 5 × 5 in (125 × 125 mm) and 6.13 × 6.13 in (155.7 × 155.7 mm) active matrix liquid crystal displays in the MFD55 and MFD66 respectively to provide good visibility under any ambient light conditions.

The displays are high-resolution units enabling

MFD55/66 liquid crystal multifunction displays ***1995***

display of any kind of video image overlaid with synthetic symbols. The MFD55/MFD66 is fitted with surrounding soft keys.

Specifications

Dimensions:
(MFD55) 160 × 160 × 232 mm
(MFD66) 196 × 192 × 238 mm
Weight: <7 kg
Power supply: 115 V AC, 400 Hz, or 28 V DC, 130 W

Operational status

The MFD55 is in production for Mirage 2000, LCA, and MiG AT. The MFD66 is in production for the Tiger and Rooivalk helicopters and for Nimrod MRA4.

Contractor

Sextant Avionique.

UPDATED

SMD 45 H liquid crystal smart multifunction display

The SMD 45 H was specially designed for helicopter applications and is currently proposed for multiple light helicopter upgrade programmes.

The SMD 45 H is a 'Smart' full-colour multifunction display using a 4 × 5 in active matrix liquid crystal display (AMLCD) providing excellent viewability under any aircraft cockpit ambient light conditions.

The SMD 45 H integrates in one panel-mounted LRU all necessary functions required for stand-alone operation such as systems bus interface, data processor and graphics generator.

The SMD 45 H high-resolution display also provides imagery from various aircraft sensors with synthetic flight symbology overlay.

Specifications

Active screen size: 4 × 5 in
Resolution: 640 × 480 pixels
Brightness and contrast: >150 fL in white
Aircraft interface: ARINC 429 bus (6 in, 4 out) or Mil-STD-1553B Bus (optional)
Power: 28 V DC, 50 W
Overall dimensions:
(front panel) 185 × 134 × 23 mm (without push-buttons)
(case size) 185 × 201 × 166 mm
Weight: 2.8 kg
Video interface: 1 input STANAG 3 350 C
Environmental: DO - 160 - C
NVG-Compatible

Operational status

Fitted on EC 120, EC 135, Dauphin N4 and Ecureuil B3 helicopters.

Contractor

Sextant Avionique.

UPDATED

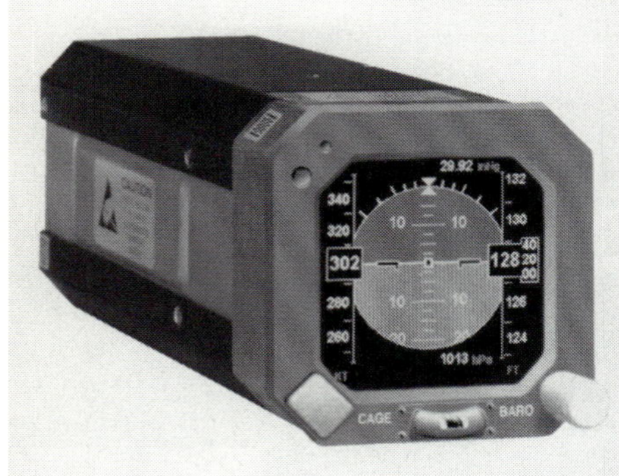

SMD 45 H liquid crystal smart multifunction display
1997/0001354

SMD66 integrated multifunction display

The SMD66 is a multipurpose electronic display developed for use on helicopter and fixed-wing aircraft flight decks. The full-colour high-resolution shadow-mask fully integrated display is capable of providing stroke and raster image for primary flight, navigation and tactical displays and systems, and engine monitoring. The image measures 6 × 6 in (152 × 152 mm). The display features brightness automatic setting up to 8,000 ft candles, to improve legibility, and is compatible with night vision goggles.

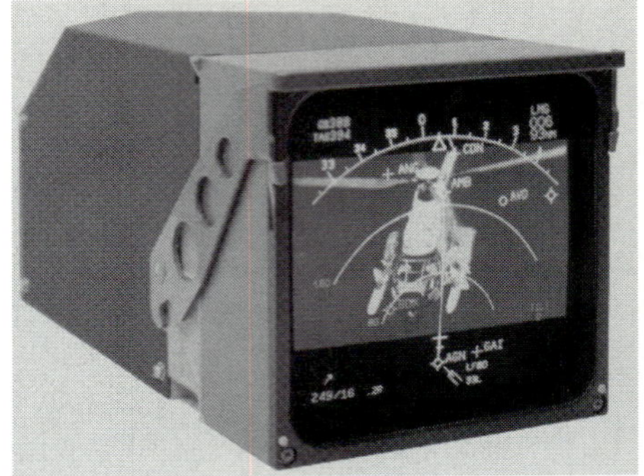

Sextant Avionique SMD66 integrated multifunction display

Operational status

In production for the Super Puma Mk 2. The SMD66 is the last in a family of cockpit displays which equip the Airbus A310, A300-600, A320, A330, A340, Mirage 2000 and Rafale.

Contractor

Sextant Avionique.

VERIFIED

TCAS Resolution Advisory/Traffic Advisory (RA/TA) VSI

Sextant Avionique's Resolution Advisory/Traffic Advisory (RA/TA) VSI combines the vertical speed and Traffic alert Collision Avoidance System (TCAS) information on a 3 ATI instrument using an active matrix type full-colour liquid crystal display. The instrument is capable of acquiring digital, analogue and pneumatic signals, enabling Sextant to propose a unique part number for a given airline. The RA/TA VSI is Sextant Avionique's first application of its flat-panel instrument programme. This flat panel instrument programme aims to replace all the 3 ATI and 4 ATI electromechanical instruments.

Operational status

In production. Over 50 airlines and aircraft manufacturers have chosen the 3 ATI Vertical Speed Indicator/Traffic alert Collision Avoidance System (VSI/TCAS) liquid crystal flat-panel instrument.

Contractor

Sextant Avionique.

UPDATED

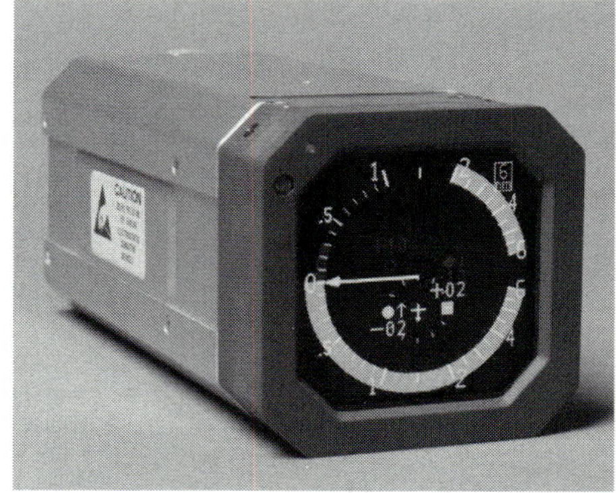

Sextant Avionique VSI/TCAS indicator

ATC-TCAS II control panel

The ATC-TCAS II control panel combines the independent control of two mode 'S' transponders with the LCD display of ATC code, and the selection of the TCAS II modes. It equips the A319/320/321 and A330/340 aircraft families. A version certified on A300-B4/300-600/310 including TCAS test activation is available for general cases of TCAS II retrofit installations.

The Sextant Avionique ATC-TCAS control panel
1997/0001355

Operational status

In production.

Contractor

Sextant Avionique.

VERIFIED

TMV 544 forward view repeater display

Sextant Avionique has developed what it terms a forward view repeater which reproduces the pilot's field of view forward for the benefit of rear-seat occupants. The TMV 544 was developed in response to conclusions that some shortfalls in the instruction of pilots in advanced training aircraft resulted from the instructor's inability to scan the view directly ahead because it was obscured by the pupil's ejection seat headrest. The system is based on a video camera that films the head-up display and outside world from the front compartment and reproduces it on a television monitor in the rear compartment.

Operational status

In production.

Contractor

Sextant Avionique.

VERIFIED

VEMD Vehicle and Engine Management Display

The VEMD is designed for both single- and twin-engined helicopters. It is interfaced to engine and vehicle sensors and linkable to data concentrators as well as FADEC.

It displays the information on two 5 in (127 mm)

VEMD vehicle and engine management display
1997/0001356

diagonal displays. This allows several display modes: normal operation mode (first limitation indication, vehicle information), reversionary mode performance, health monitoring flight report (overlimit detection) and maintenance modes. The dual architecture both in terms of processing and displays, the high level of failure detection and the presence of reversionary modes provide a high level of availability.

The VEMD displays the information on a twin-active matrix colour LCD display unit with a wide viewing angle and provides high readability in any ambient lighting conditions (compatible with filtered NVG). In normal operation the upper matrix displays engine information and the lower one, vehicle information. If one channel fails, the main information is displayed on the remaining display.

The VEMD can also display health monitoring information such as engine cycle count, engine power check, BIT results. This information is stored in a non-volatile memory and is readable on the display in maintenance mode.

Each channel of the dual architecture of the VEMD (display, processing, power supply) performs the acquisition of each parameter and cross-checks its consistency with the other channel. This monitoring added to the self-test of each channel provides a high level of failure detection and low level of erroneous data display.

The VEMD has dual processing and display systems interfaced to engine and vehicle sensors: thermocouples (TOT); frequencies (N1, N2, NR); voltages (torque, fuel quantity, oil pressure, pressure, gearbox oil pressure); resistors (oil temperature, OAT, gearbox oil temperature); voltages (voltmeter, ammeter) pressure sensor as well as numeric interfaces for FADEC (RS-422, ARINC 429, EIA 485). It generates the power supply required for the sensors as well as discrete outputs.

Specifications

Useful screen size: (3 × 4 in) × 2
Lighting in day conditions: 0.4 to 100 Cd/mL
Dimensions: 156 × 212 × 205 mm
Max weight: 3 kg
Power consumption: 44 W
Power supply: 28 V DC
MTBF: 5,000 h
No cooling required

Operational status

Fitted on EC 120, Ecureuil B3, EC 135, Dauphin N4.

Contractor

Sextant Avionique.

VERIFIED

All-attitude indicators

SFIM Industries has developed a range of spherical indicators on which attitude information is displayed to give the pilot heading, roll and pitch information on a single dial without freedom limits around the three axes.

Some versions are fitted with command bars to display signal information from navigation aids or landing systems. A helicopter version is designated 11-2 and features a manual pitch setting mode of ±10°. The various features of the series are:

Specifications

Type	810	811	816
Roll	Yes	Yes	Yes
Pitch	Yes	Yes	Yes
Heading	Yes	Yes	Yes
ILS/VOR/Tacan	Yes	No	Option
To-from	Yes	No	*
Beacon	Yes	No	Option
Sideslip	Yes	Yes	Yes
Failure detection	Yes	Yes	Yes
Warning flag	Yes	Yes	Yes

* separate unit, on option

Type 810 and 811
Dimensions: 97 × 97 × 203 mm
Power supply: 26 or 115 V AC, 400 Hz
28 V DC

Type 816
Dimensions: 81 × 81 × 208 mm
Power supply: 26 V AC, 400 Hz
28 V DC

Operational status

In production. Four types of instruments are available and equip British Aerospace Harrier GR. Mk 7, Dassault Mirage F1 and 2000, Sepecat Jaguar and Dassault Super Etendard, French Navy Westland Lynx helicopters and Soko Galeb, VTI/CIAR Orao and FMA Pucara aircraft.

Contractor

SFIM Industries.

VERIFIED

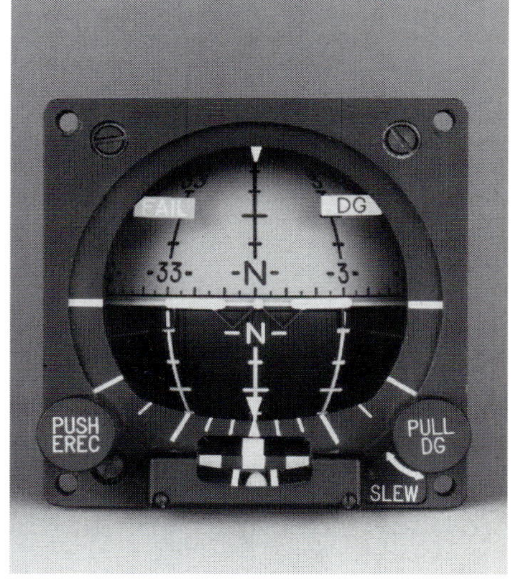

SFIM IS 816-1 attitude indicator

GERMANY

DKG cockpit map displays

The DKG is a cockpit map display for helicopters. The map display is controlled by the navigation system which shifts, zooms and rotates the map according to the flight motion. The equipment software offers full command and control functionality. The map data is stored in semiconductor memory. A sortie-related map can be transferred from the ground-based planning system to the airborne system by means of a solid-state mission data transfer system. The DKG features a 68020 CPU with a 512 × 512 pixels 50 Hz interlaced display, external semiconductor 8 Mbytes exchangeable memory, independent access to two displays and MIL-STD-1553B bus interconnection to the navigation system.

The DKG 3 is a low-cost electronic map display for helicopters and fixed-wing aircraft. It replaces the traditional knee-board and paper maps with a fully electronic multicoloured movable map with additional mission-related information. Position is continuously tracked on the selected map without page breaks by means of a GPS-based navigation component.

Contractor

Daimler-Benz Aerospace AG, Defense and Civil Systems.

UPDATED

EuroGrid European geographic information display system

The digital map generator of EuroGrid comprises the following main functional modules:

(1) The main processor for the control of the EuroGrid system with graphic-overlay management including thematic classes, control and performance of the Main Machine Interface (MMI), and an Internal Fast Serial Link (FSL)
(2) The geographic processor for 2-D map and graphic overlay presentation, output via video link to the displays in the cockpit with the following functions: presentation of basic maps on different scales: 1:100 k, 1:250 k, 1:500 k; presentation of real moving map: translation and rotation in real time; presentation of maps with different zooming factors
(3) A geographic memory for the storage of all necessary map data and the symbol library
(4) The communication processor for the control of two different radio links via software, with the following main tasks: receipt and transmission of tactical messages; management of transmission and reception; administration of data link participants; control of data link channels (VHF/FM and HF); failure management. Communication messages are as follows: tactical, navigation and threat overlays; target hand-over; position requests and answers; text messages
(5) A video memory (video interface) for acquisition and storage of IR and TV images. Presentation on the selected monitors beside the respective map section in which the lines of sight of the image are represented
(6) An elevation processor for the vector map image presentation with the features: shading with or without hypsometric colours; ground avoidance presentation by coloured safe flight areas; terrain profile, presentation of terrain on the helicopter track.

Specifications

Type of system: real time map/graphic elevation processing; 32 bit processing
Architecture: solid-state
Memory technology: raster, vector and charts
Type of map data in main map storage: raster, vector, let down, plates, charts and tactical data
Dimensions: ARINC 600, 6 MCU, 1 cockpit unit
Mass: 10.1 up to 13.1 kg for EuroGrid LRU set (incl MDE/MDT) max
Power: 130 VA for EuroGrid LRU set max
MTBF: 1000 h

Contractor

Daimler-Benz Aerospace AG, Defense and Civil Systems.

NEW ENTRY

Helicopter electronic map display system

Daimler-Benz has received a contract from the German Procurement Agency to develop a digital map display system for the PAH-2 anti-tank helicopter. The digital map display consists of a highly integrated processing unit for image generation and a mass memory unit for map data storage. The maps are displayed with graphical overlays to represent navigation and tactical information on high-resolution colour monitors. The advantage of this system is that all map-based information can be displayed in an endless and continuous form directly tied to the helicopter flight path.

The digital map data is derived from paper maps and aerial or satellite photography; a laser scanner is used for digitisation of the source material. The data sets are subdivided into files for storage in the mass memory and for endless display in the glass cockpit. The displayed information can be laterally moved and rotated. A continuous zoom facility is available for detail enlargement. For display of flight and planning data various graphical symbols representing position, flight path, wind direction and so on, are available. Other information such as groundspeed and altitude can be displayed by means of numeric symbols.

The system can also be employed for graphical communication, relaying flight events directly to the ground operator. Using manual controls the pilot places a symbol on the map which, via datalink, is transferred to the map in the ground station. In the reverse direction the onboard map display system can receive and display information from the ground station.

For future applications the digital map display system will include special obstacle and relief map data to aid helicopter operations at night and under adverse weather conditions.

Operational status

Under development for the Eurocopter PAH-2 Tiger helicopter.

Contractor

Daimler-Benz Aerospace AG, Defense and Civil Systems.

UPDATED

Cockpit warning system

This unit is designed to indicate both warning and caution messages in the aircraft cockpit. Some 16 warning captions provide sunlight readable indications. When illuminated they appear as red letters on a black background and extinguished they remain black. For night operation the brightness can be dimmed by input voltage variation.

When the unit receives a warning signal, the corresponding warning caption is illuminated and also the unit activates the external master warning indication. The external master warning indication turns off when no warning caption is active or can be reset by an external switch.

If required, a timer makes the master warning lamp flash. The master caution inputs can be activated by this unit. They only extinguish when the reset button has been pressed. A lamp test switches on all warning captions in the panel as well as the external master warning lamps.

Every warning caption can be changed individually from the front of the unit without removing the whole system from the instrument panel. Every caption is backlit by two incandescent lamps.

Specifications

Power: 28 V DC nominal, 900 mA
Weight: 0.81 kg

Contractor

Diehl GmbH and Company, Luftfahrt Electronik.

NEW ENTRY

Diehl cockpit warning system ***1998***/0018204

TV tabular display unit

The TV tabular display is capable of displaying various sensors such as a TV camera, FLIR, low-light level TV and reconnaissance cameras and alphanumeric characters. By pressing related buttons on the keyboard the operator can select the preferred sensor out of three which is then internally mixed and displayed with alphanumerics overlaid. The display information on the Display Unit (DU) is also output for recording purposes for future reference or for post-mission analysis.

The DU also generates permanent data output by two 32-bit data words that are transferred to the main computer for selection and control of a computer program. The data words are output in both true and complement form and clocked out by a continuous train of 64 kHz data synch pulses which is also presented in both true and complement form.

The display unit brightness is controlled by an automatic brightness control circuit which guarantees readability of the displayed information during changing light levels, from darkness up to 10^5 lux ambient illumination. In addition to this, the operator can, within certain limits, override the automatic control system by manually setting a brightness and contrast level.

The TV tab DU consists of modules interchangeable between like assemblies which guarantees good serviceability and low maintenance costs.

Specifications

Dimensions: 450 × 210 × 210 mm
Weight: 11.6 kg (max)
Power supply: 200 V AC, 400 Hz, 3 phase, 80 VA
28 V DC

Operational status

In service on the Panavia Tornado.

Contractor

ESW-Extel Systems Wedel.

VERIFIED

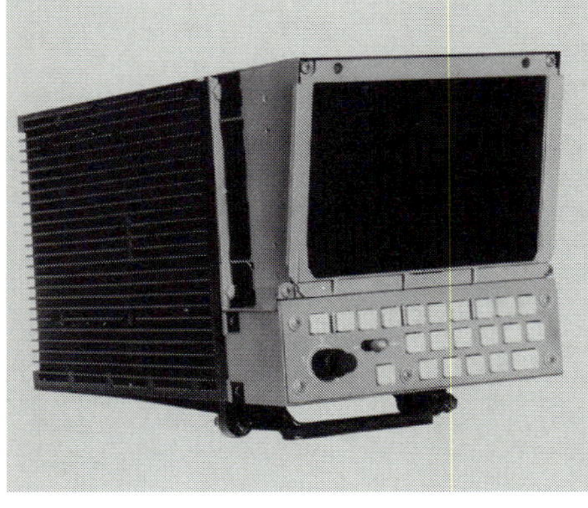

TV tabular display for the Panavia Tornado

GB 609 Control and Display Unit (CDU)

The GB 609 is a Control and Display Unit (CDU) specially devised for avionics systems for entering and displaying different types of data. Its operating concept is determined by software, so that operating strategies can be implemented by the user.

The GB 609 is a smart CDU exhibiting high intelligence and computing capability as a result of its hardware concept, and can manage complex systems in the role of a central computer, for example as a bus control as defined in MIL-STD-1553B. An arithmetic co-processor can be incorporated as an option, accelerating substantially the execution of extensive arithmetic operations.

The panel integrated solid-state electroluminescent display has full graphic capability, call up of prestored masks, 14 lines × 40 characters in alphanumeric mode and three character size standards. It features high contrast and is viewable in direct sunlight. Coloured green, the display is NVG-compatible. The reliability of the ultra-flat and lightweight display, which is insensitive to vibration and shock, is unequalled by any other display technology. Unlike a CRT, it does not have to be changed at regular intervals.

Specifications

Dimensions: 146 × 217 × 185 mm
Weight: 4.5 kg
Power supply: 16-31 V DC, 45 W
Temperature range: −40 to +55°C
Program memory: 0.5 Mbytes EPROM
Data memory: 16 kbytes EEPROM, 64 kbytes RAM

Contractor

Rohde & Schwarz and Co GmbH.

VERIFIED

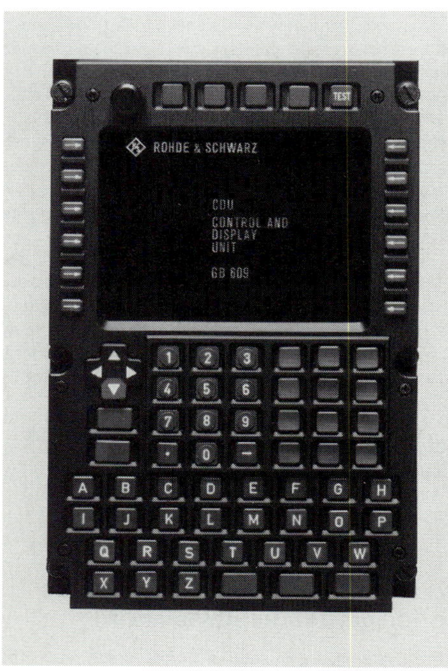

GB 609 control and display unit

INTERNATIONAL

EF 2000 display processor

Alenia Difesa Avionic Systems and Equipment Division is leading a consortium with GEC-Marconi Avionics, Teldix and ENOSA in the full-scale development of a dual-redundant lightweight display processor for the EF 2000. The computer symbology generator contains efficient and very high-performance processing power through distributed architecture, by means of independently functioning intelligent interfaces and concurrent processing operations to drive a head-up display, three multifunction head-down displays and the helmet display.

Design features include all symbology alterable at Operational Flight Program (OFP) level without any hardware change, high-level command interfaces for ease of use and minimum OFP requirements, application of quadruple 68040 parallel processing developed by Teldix giving very high OFP CPU throughput readily upgradable from 3 Mips to over 5 Mips and low-risk dual-redundant STANAG 3910 fibre optic ASIC solution developed from the STANAG 3838 MIL-STD-1553 design. The processor offers maximum use of surface mount components and high-density double-sided circuit modules to minimise space requirements and mass, maximum use of advanced electronic components including ASICs to ensure low operating temperatures and high reliability, modular concept and extensive BIT to minimise test and maintenance support and two spare plug-in module positions connected directly to the internal VME bus to minimise changes during upgrades.

Operational status

In development for the EF 2000.

Contractors

Alenia Difesa, Avionic Systems and Equipment Division, GF Sistemi Avionici.
ENOSA.
GEC-Marconi Avionics, Rochester.
Teldix GmbH.

UPDATED

EF 2000 Multifunction Head-Down Displays (MHDD)

Three Multifunction Head-Down Displays (MHDDs) are installed in the cockpit of the Eurofighter, with six in the two-seat trainer version. The MHDD provides flight, tactical situation and sensor data, as well as vital onboard systems information, on a 6 × 6 in (152 × 152 mm) usable screen area. This combines exceptional brightness and high resolution to give full legibility in full sunlight of raster scan images overlaid with fine stroke graphics. Interface with other aircraft systems is provided by 17 programmable push-button key displays on the unit's bezel. The equipment is a major development of the MPCD fitted in the AV-8B/Harrier GR. Mk 7.

Smiths Industries is the project leader and design authority for the programme.

Operational status

In prototype production for the EF 2000.

Contractors

Smiths Industries Aerospace.
ENOSA.
Alenia Difesa, Avionic Systems and Equipment Division, GF Sistemi Avionici.
VDO Luftfahgerate Werk.

UPDATED

Multifunction Control Unit (MCU)

The Litton Systems Canada/Harris Multifunction Control Unit (MCU) is a full-capability CDU for controlling and managing a diverse range of avionic equipment, including navigation sensors, communication systems and stores. The MCU is available for custom application development with a bundled Ada software toolset.

The MCU includes a host of built-in input/output capabilities, including MIL-STD-1553B, ARINC 429 and RS-422, with other optional digital interfaces available. The MCU incorporates Litton's LED technology, on a 10 line by 24 dot matrix, for readability in high-ambient lighting environments, with optional NVIS-B compliant optics available.

The MCU incorporates a 32-bit, 24 MHz processor/co-processor and a 2 Mbyte memory as standard. Options include a 33 MHz processor and up to 6 Mbytes of flash EEPROM. A full US Air Force approved alphanumeric keyboard is standard, including 22 customisable keys for dedicated functions.

The MCU is supplied with sample application code, providing GPS flight management, planning and control, plus inertial navigation update capability. This application provides for a Trimble GPS as the sole means of navigation and includes a 200 waypoint database and other features.

Specifications

Dimensions: 184.1 × 146 × 165.1 mm
Weight: <4.54 kg
Power supply: 28 V DC to MIL-STD-704A
Reliability: >4,000 h MTBF

Operational status

The MCU has been fitted to the A-10, Tornado and various military helicopter applications.

Contractors

Harris Corporation Electronic Systems Sector.
Litton Systems Canada Ltd.

VERIFIED

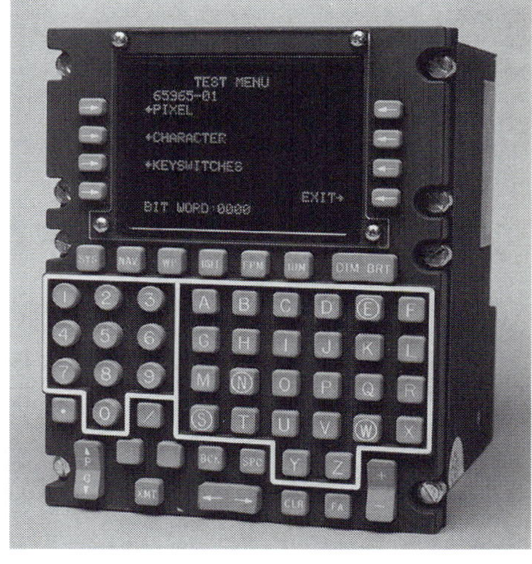

Harris/Litton Canada multifunction control unit
1995

ISRAEL

Cockpit Integration Unit (CIU)

The CIU replaces multiple control panels with a single panel-mounted control display unit. Cockpit management tasks executed via the CIU include radio communication, navigation, stores management and text messages terminal.

Specifications

Keyboard:
Full alphanumeric keyboard with function and soft keys
Sealed for sand/dust environments
Display:
High contrast, sunlight readable
Processor:
32 BIT 386 sx/387 sx instruction set
512 kW memory upgradable to 2 MW
Interfaces:
1553 bus controller or remote terminal
RS-422 half or full duplex communication channels
ARINC 575/429 communication channels
Input/Output discretes
Weight: 5 kg
Dimensions: 181 × 146 × 156 mm
Power: 28 V DC 40 W

Contractor

Elbit Systems Ltd.

UPDATED

Colour Liquid Crystal Display (LCD)

Elbit Systems' colour LCD represents the latest generation of airborne displays. The flat panel display packages high optical performance and reliability into a compact unit with low power consumption. It is suitable for ruggedised airborne and ground applications.

It features a full-colour active matrix LCD; high-resolution graphic display; wide viewing angle; slim design (only 3 in (76 mm) thick); and high contrast in day and night operation.

Cockpit integration unit **1997**/0001357

Specifications
Display size: 97 × 127 mm
Resolution:
640 × 600 pixels (384,000 over display area)
640 × 200 colour triads over display area
Brightness: 50 ft-lamberts (white)
Contrast: 40 (white)
Colours: 8
Viewing angle: ±45° horizontally and vertically for a contrast of 10
Interface: RGB at TTL levels
Power supply: 28 V DC in range of 20-36 V
Power consumption: 20 W
Dimensions: 190 × 190 × 70 mm

Operational status
Operational on MiG-21 and other aircraft.

Contractor
Elbit Systems Ltd.

UPDATED

Engine Performance Indicator (EPI)

Elbit Systems' EPI is an advanced cockpit display that replaces nine electromechanical switches with one electronic display. The display conserves valuable space on the instrument panel for additional displays (MFCDs) required for comprehensive F-5 upgrades. The EPI is based on LCD technology and emphasises human engineering in its use of colour and combination of analogue and digital symbology.

Specifcations
Display size: 110 × 58 mm area
Display technology: direct drive LCD with 4 colours
Power supply: 28 V DC, 40 W
(80 W while heater operates)
Dimensions: 86 × 125 × 273 mm
Weight: 3 kg
Interface: NVG-compatible

Operational status
Operational.

Contractor
Elbit Systems Ltd.

UPDATED

Map Display Generator (MDG)

Designed for two-seat fixed-wing aircraft or helicopters, the Map Display Generator (MDG) has two channels capable of producing a variety of images as selected by the pilot or co-pilot. The system uses two 6 × 6 in (152 × 152 mm) multifunction colour displays. Control of the various map modes is via a keyboard mounted on the display. The MDG is equipped with an optical disk drive and removable optical disk for storing map databases on board the aircraft. Two powerful graphics units simultaneously draw raster and vector maps.

Using the mission planning ground station, the pilot can plan the intended mission in advance, co-ordinating all maps, tactical mission data, flight plans, threats, obstacles and communications. Accumulated data is then downloaded onto the optical disk which the pilot transfers to the MDG disk drive for use during the mission.

The MDG displays paper charts, vectorial maps, digital terrain elevation data and SPOT or Landsat satellite photos. With continuous improvement offline at the ground station and online at the MDG, the image presented on the multifunction colour display is clear, crisp and easily discernible. The system also provides the user with zoom, scale, plan and freeze facilities.

The MDG's computing powers are capable of calculating the line of sight between any two points, point altitude, radar threat area coverage and obstacle clear flight and landing corridors.

Operational status
Selected by the Israeli Air Force for the modernised CH-53 helicopter.

Contractor
Elbit Systems Ltd.

UPDATED

MultiFunction Displays (MFD)

The multifunction display is essentially a monochrome armament control and display panel for combat aircraft with limited cockpit space, showing video information from a variety of sensors on a 5 × 5 in (127 × 127 mm) or 4 × 4 in (102 × 102 mm) CRT. The display CRT is surrounded on the face of the panel by 28 push-buttons for the selection of different functions.

Specifications
Power supply: 28 V DC per MIL-STD-704A at 40 W
Dimensions: 140 × 138 × 344 mm
Weight: 6 kg
Display size: 4 × 4 in (102 × 102 mm) or 5 × 5 in (127 × 127 mm
Phosphor type: P43 (green)
Raster video input: 525/60, 625/50 or 875/60 lines per RS-170, RS-343 or CCIR standards
Brightness: 200 ft-lamberts (685.2 cd/m^2) min
Contrast ratio: 9:1 at 10,000 ft candles sunlight
Video bandwidth: 30 Hz-20 MHz to 3 dB points
Reliability: 1,500 h MTBF per MIL-HDBK-271A

Operational status
In production for Israeli Air Force F-4 Phantoms, IAI Kfir fighters, F-5, T-45 and MiG-21.

Contractor
Elbit Systems Ltd.

UPDATED

JAPAN

Flat-panel display

The flat-panel display uses a high-resolution liquid crystal display now under development for use in advanced aircraft cockpits. It features a full-colour display with adjustable high-brightness back-light, to give exceptional contrast. The system is lighter and thinner, with less power consumption, than conventional CRTs.

Contractor
Japan Aviation Electronics Industry Ltd.

VERIFIED

Central warning display

The central warning display has been developed for commercial aircraft. It is designed to work with two master warning lights which can effectively alert pilots and identify potential problems or hazards.

Koito central warning display for commercial aircraft

The indication panel is a liquid crystal display which can indicate a maximum of 10 warning items in red, amber or green. When more than 10 items occur at the same time, up to 50 additional warnings are listed by a scrolling function. The LCD has a high back-light unit consisting of halogen lamps, allowing the manually or automatically dimmable display to be read under sunlight conditions. The red and amber warning legends can be tailored to customer requirements by a software change. Dual redundancy is incorporated in the power supply, input interface circuit, digital computer and LCD display.

Specifications
Dimensions: 134 × 125 × 250 mm
Weight: 2.5 kg
Display size: 55 × 75 mm
Power supply: 16-30 V DC, 1.3 A at 28 V DC
Temperature range: −40 to +70°C
Altitude: −100 to 25,000 ft
Reliability: >5,000 h MTBF

Contractor
Koito Manufacturing Company Ltd.

VERIFIED

Colour liquid crystal display

The colour liquid crystal display is a bright display which is composed of a high-quality colour filter with excellent colour purity and brilliant back-up illumination with uniform brightness. The display is an eight-colour type using the three primary colour filters of red, green and blue and can easily provide stationary coloured pictures at low cost. In addition, it can increase display information by changing a monochrome liquid crystal display into a multicoloured display.

Contractor
Koito Manufacturing Company Ltd.

VERIFIED

Electroluminescent display

Koito has developed two types of ElectroLuminescent (EL) displays: the AC powder EL display and the thin-film EL display. Both types are thin lightweight surface light emitting devices and are considered ideal for displays due to features such as the uniform brightness on the light emitting surface, high reliability and lack of heat generation.

The AC powder EL display is a flexible type of surface light emitting device which can be bent. It is formed by applying fluorescent and transparent conducting layers on aluminium foil and then laminating a transparent plastic sheet. It is very resistant to excessive vibration and load and is built into passenger aircraft cabin aisles, in aircraft such as the Boeing 767, as floor marker lamps.

In the AC powder EL display the ultra-slim flat light source provides extremely uniform and soft illumination. Composed of organic materials with emphasis on plastic film, it is easily bendable and will provide illumination from a curved surface. Lightweight, tough and durable to vibration and shock, it does not suffer filament breakages as in conventional bulbs. As an EL element, it provides cool light without generating heat.

The thin-film EL display consists of an insulating layer, luminous layer and electrodes applied on the glass substrate by vacuum evaporation. It is far brighter than the AC powder EL display.

It can be built easily into portable apparatus and the high precision of electrode stripes and thin-film combine to provide an easily visible display with high resolution. Since it is an all-solid-state self-luminous element, it has a higher stability and longer life. The response speed of the display is about the same as a CRT display and so it can display letters, numbers, graphs and so on, in real time. The thin-film EL display is available as a dot-matrix or segment display.

Contractor
Koito Manufacturing Company Ltd.

VERIFIED

Radar display

The radar display provides radar information to the crew in the form of symbology and/or video. The radar display consists of a forward indicator unit, aft indicator unit and electronic unit.

Operational status
In production for the F-4EJ Kai aircraft.

Contractor
Shimadzu Corporation.

VERIFIED

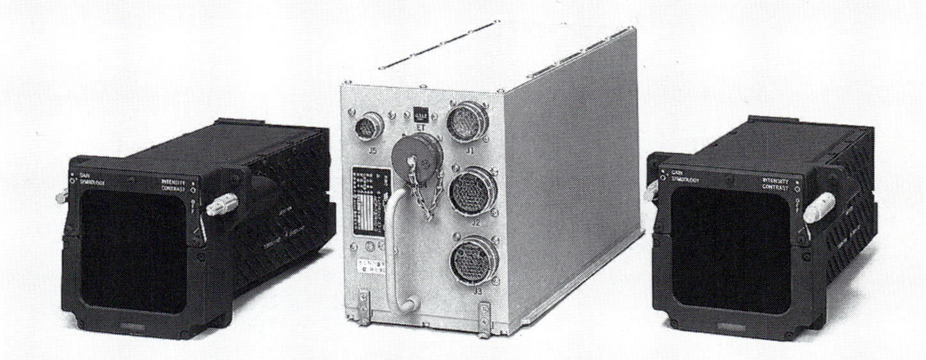

The Shimadzu radar display consists of (left to right) the forward indicator unit, the electronic unit and the aft indicator unit **1995**

3 ATI flat-panel display

The 3 ATI flat-panel display features an active matrix liquid crystal display and an integrated symbol generator. It is an electronic display that provides graphics and/or characters via an ARINC 429 interface. It has highly integrated modules, incorporating customised ASICs for high-speed graphic processing and SMDs for reduced size.

Specifications
Dimensions: 3 ATI
Weight: 1.5 kg
Power supply: 27.5 V DC, 17 W (42 W with heater)
Display size: 57.7 × 57.7 mm, 480 × 480 RGBW Quad dots
Viewing angle: 35° vertical × 90° horizontal
Contrast ratio: >70
Brightness: >100 fL

Operational status
Under development.

Contractor
Tokyo Aircraft Instrument Co Ltd.

VERIFIED

FMPD-10 EFIS package

FMPD-10 is an Electronic Flight Instrument System (EFIS) package both for commercial and military aircraft. The standard package consists of four smart type displays, two display interface processors and two display control panels. The displays show EHSI, EAI, EHSI/EAI composite graphics, WXR graphics, waypoint map, and fuel information on 3.7 × 3.7 in AMLCD glasses, and also provides hover mode data for helicopter application.

Operational status
In production and in service.

Contractor
Tokyo Aircraft Instrument Co Ltd.

VERIFIED

LK-35 series standby altimeter

The standby altimeter is an exceptionally accurate pressure-actuated instrument with a −1,000 to 50,000 ft range and P/N LK-35-1 and P/N LK-35-9 provide two resolver outputs of barometric setting. A three-drum counter supported by ball bearings provides digital indication in tens of thousands, thousands and hundreds of feet, supplemented by a pointer which indicates altitude in hundreds of feet.

A knob on front of the altimeter provides the means for setting barometric pressure which is shown in inches of mercury and MB on two sets of four-digit counters.

Specifications
Altitude range: −1,000 to +50,000 ft
(LK-35-8) −2,000 to +50,000 ft
Barometric range: 22-31.02 inHg
(LK-35-7/LK-35-8) 22-31.99 inHg
(LK-35-9) 22-30.99 inHg
Accuracy certifications: FAA/TSO-10b
Lighting: white (unfiltered)
Electrical input:
(vibrator) 28 V DC
(lighting) 5 V AC or DC
(resolver) 26 V, 400 Hz
Weight:
(LK-35-1, -9) 1.45 kg
(LK-35-2, -4, -5, -7) 1.3 kg
(LK-35-8) 1.5 kg

3 ATI flat-panel display **1995**

FMPD-10 EFIS package **1997**/0001358

LK-35 series standby altimeter ***1996***

Operational status

In production for B737-400, B757, B767, DC-9.

Contractor

Tokyo Aircraft Instrument Co Ltd.

VERIFIED

SFPD-20 Smart Flat-Panel Display (SFPD)

The SFPD-20 Smart Flat-Panel Display (SFPD) is a self-contained 4 × 5 in (102 × 127 mm) display unit that can replace conventional electromechanical instruments or be used in new aircraft with digital databusses like the ARINC 429. The SFPD, having an active matrix liquid crystal display, provides HSI, ADI, engine data and map information.

Specifications

Dimensions: 127 × 101.6 × 165.1 mm
Weight: 2.5 kg
Power supply: 27.5 V DC, 25 W (53 W with heater)
Display area: 92.7 × 74.2 mm
Viewing angle: 50° vertical by 90° horizontal
Contrast ratio: >100
Brightness: >100 ft-lamberts

Operational status

In production for commercial aircraft.

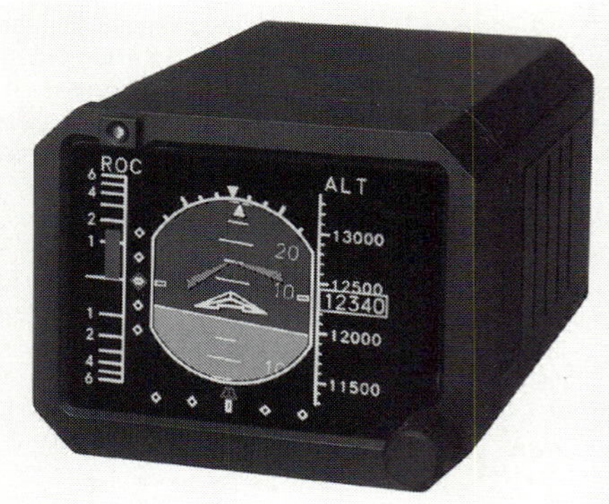

SFPD-20 display ***1995***

Contractor

Tokyo Aircraft Instrument Co Ltd.

VERIFIED

RUSSIAN FEDERATION AND ASSOCIATED STATES (CIS)

VM type altimeters

AeroPribor-Voskhod VM type altimeters are designed on a unified constructional base, and they have a built-in vibrator for reducing friction in the mechanism. Altitude is displayed by a counter calibrated in kilometres. The altimeters are produced in three separate colour options: integral red; red/white; and white dial lighting.

Operational status

Widely used.

Contractor

AeroPribor-Voskhod Joint Stock Company.

AeroPribor-Voskhod Joint Stock Company VM type altimeters ***1998***/0018208

Specifications

	VM-1	VM-1A	VM-2	VM-3	VM-P
Altitude range	−500 to 10,000 m	−500 to 5,000 m	−500 to 15,000 m	0 to 30,000 m	−500 to 30,000 m
Atmospheric pressure range	700-1,080 kPa	700-1,080 kPa	700-1,080 kPa	700-1,080 kPa	700-1,080 kPa
Dimensions	85 × 85 × 190 mm	85 × 85 × 190 mm	85 × 85 × 240 mm	95 × 85 × 190 mm	65 × 65 × 160 mm
Weight	1.2 kg	1.2 kg	1.4 kg	1.6 kg	1.0 kg
Power (all systems):	27 V DC, 1.5 W; 5.5 V AC, 2.2 VA				

NEW ENTRY

AGB-96, AGB-98 and AGR-100 horizon gyros

The AGB-96, AGB-98 and AGR-100 horizon gyros are designed for aircraft pitch/roll indication.

Specifications

Compliance: ENLGS P 8.1.2

	AGB-96	AGB-98	AGR-100
Readiness time	2.0 min	2.0 min	2.0 min
Angular range			
(roll)	±360°	±360°	±360°
(pitch)	±85°	±85°	±85°
Power			
(gyro) all systems:	27 V DC		
(sensors) all systems:	36 V AC, 400 Hz, 0.8 A		
Temperature range all systems:	−20 to +70°C		
Dimensions	105 × 105 × 250 mm	85 × 85 × 250 mm	61 × 61 × 220 mm
Weight	2.2 to 2.5 kg	1.8 to 2.0 kg	1.5 kg

Contractor

AviaPribor

NEW ENTRY

AviaPribor AGR-100, AGB-98, AGB-96 horizon gyros (left to right) ***1998***/0018207

AviaPribor AGR-81 (left) and AGR-29 (right) standby horizon gyros ***1998***/0018206

AGR-29 and AGR-81 standby gyros

The AGR-29 and AGR-81 standby horizon gyros are designed for aircraft pitch/roll indication. The AGR-29M can be used as a remote indicator of pitch and roll from another vertical sensor.

Specifications

Compliance: ENLGS P 8.1.2

	AGR-29	AGR-81
Readiness time	2.0 min	2.0 min
Angular range		
(roll)	±360°	±360°
(pitch)	±360°	±360°
Power		
(gyro)	27 V DC	
(sensors)	36 V AC, 400 Hz, 1.0 A	
(back lighting)	6 V DC	
Temperature range	−30 to +70°	
Dimensions	105 × 105 × 250 mm	85 × 85 × 250 mm
Weight	up to 3.1 kg	2.2 to 2.5 kg

Contractor

AviaPribor

NEW ENTRY

Airspeed and altitude indicators

AviaPribor produces a range of airspeed and altitude indicators that conform with ARINC 429 data interface requirements. The photograph shows some representative examples; the specification below provides representative data.

Specifications

	VMC altimeters	USC airspeed indicators
Measurement range	0-10,000 m	80-800 km/h
Error	±10 to ±30 m	5.5 to 10 km/h
Dimensions	86 × 86 × 250 mm	86 × 86 × 230 mm
Weight	2.0 kg	1.5 kg

Contractor

AviaPribor

NEW ENTRY

IM-3, IM-5, IM-6, IGM multifunction displays

AviaPribor produces a number of multifunction displays to provide: flight/navigation information; radar data; EFIS/EICAS/EIS advisory, caution, warning and status data, and aural warnings.

AviaPribor USC airspeed and VMC altitude indicators ***1998***/0018205

AviaPribor multifunction displays for the Tu-204 and Il-96 aircraft ***1998***/0018203

Specifications

	IM-3	IM-5	IM-6	IGM
Display technology	CRT stroke	CRT raster	CRT raster	AMLCD matrix
Useful screen size	159 × 159 mm	159 × 159 mm	83 × 107 mm	101 × 101 mm
Pixels pitch	0.15/0.3 mm	0.3/0.3 mm	0.2/0.3 mm	0.2 mm
Refresh rate	50/80 Hz	40/80 Hz	40/80 Hz	80 Hz
Colours displayed	8	8 (16)	16	8 (16)
Allowable ambient illumination	70,000 lx	60,000 lx	75,000 lx	75,000 lx
Viewing angle	±53°	±53°	±53°	±40°
Dimensions	203 × 230 × 356 mm	203 × 222 × 356 mm	130 × 145 × 390 mm	130 × 130 × 230 mm
Weight	17.5 kg	12.5 kg	6 kg	4 kg
Power	200 VA	100 VA	60 VA	90 VA

Operational status

The displays illustrated are fitted to the Tu-204 and Il-96 twin and four turbofan airliners. They are also suitable for other aircraft and helicopters.

Contractor

AviaPribor

NEW ENTRY

IRM-1 radio compass indicator

The IRM-1 radio compass indicator is designed to display data derived from the MKS-1 compass and the ADF system.

Specifications

Compliance: ENLGS, p.82.1 g

Interfaces: MKS-1 compass; KR87 digital ADF receiver (AlliedSignal)

IRM-1 radio compass indicator ***1998***/0018202

Heading, selected course and ADF bearings: 360°
Error: ±1.5° heading; ±2.0° ADF bearing
Power: 27 V DC
Temperature range: −20 to +55°C
Dimensions: 85 × 85 × 150 mm
Weight: 1.1 kg

MKS-1 compact compass system

The MKS-1 compact compass system has two operating modes; gyro and manual slaving; automatic magnetic slaving. The system comprises: GK heading gyro: KU-1 slaving unit; ID-6-1 magnetic field sensor.

Specifications

Compliance: AP-23; ENLG-S, p.82.1.d
Warm-up time: 2 min
Gyromagnetic heading measurement error: 2°
Gyro drift: 0.4°/min
Power: 27 V DC
Operating temperature range:
(GK-1) −55 to +55°C
(KU-1) −20 to +55°C
(ID-6-1) −60 to +150°C
Dimensions:
(GK-1) 115 × 200 ×145 mm
(KU-1)55.4 × 55.4 × 94 mm
(ID-6-1) 102 × 60 mm
Weight:
(GK-1) 2.2 kg
(KU-1)0.35 kg
(ID-6-1)0.6 kg

Contractor

AviaPribor

NEW ENTRY

SEI-85 and KISS-1-1M multifunction displays

The ElectroPribor Kazan Plant manufactures more than 700 types of instruments, systems and displays for installation in all types of military and civilian fixed- and rotary-wing aircraft built throughout the CIS. The photograph below shows the flight deck of the Tu-204 aircraft cockpit which incorporates the SEI-85 and KISS-1-1M electronic multifunction displays to present navigation, weather radar and aircraft/engine performance data. Similar displays are used in the Il-96-300 aircraft.

Operational status

In service and in production.

Contractor

ElectroPribor Kazan Plant.

NEW ENTRY

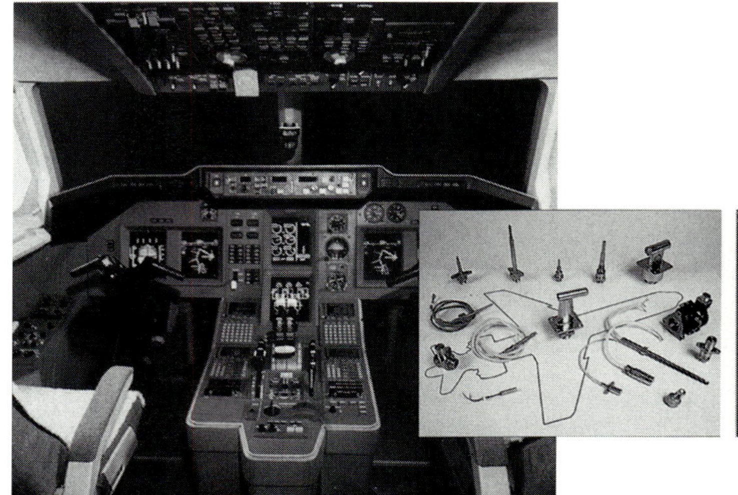

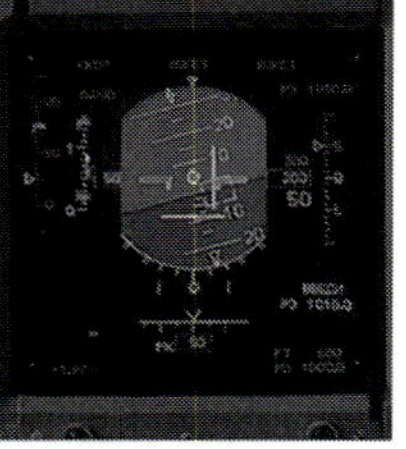

The Tu-204 aircraft cockpit incorporates SEI-85 and KISS-1-1M electronic multifunction display and alerting systems, manufactured by ElectroPribor Kazan Plant ***1998***/0018201

Tachometer Indicators

ElectroPribor Kazan Plant manufactures 14 types of tachometer for measuring engine shaft speed, calibrations can be either in revolutions per minute (rpm) or per cent.

Specifications

ITE-2TB
Operating temperature range: −60 to +60°C
Reading range: 10-110%
Error: +0.5%
Weight: 1.30 kg

ITE-2
Operating temperature range: −60 to +60°C
Reading range: 10-110%
Error: +0.5%
Weight: 0.95 kg

ElectroPribor Kazan Plant tachometers ***1998***/0018200

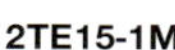

2TE15-1M
Operating temperature range: −60 to +60°C
Reading range: 1,000-15,000 rpm
Error: +75 rpm
Weight: 0.90 kg

Operational status

In service and in production for a wide variety of military and civilian fixed- and rotary-wing aircraft throughout the CIS.

Contractor

ElectroPribor Kazan Plant.

NEW ENTRY

Temperature gauges

ElectroPribor Kazan Plant manufactures 10 types of temperature gauges and thermocouples for measurement of gas flow temperatures in turbojet and turboprop engines, and in the cylinder-head of piston engines.

Specifications

ITG-1
Measuring temperature range: +200 to +1,100°C
Error: +12°C
Weight: 0.76 kg

TBG-1
Measuring temperature range: +300 to +900°C
Error: +7°C
Weight: 0.76 kg

ElectroPribor Kazan Plant temperature gauges ***1998***/0018199

TUT-9
Measuring temperature range: −40 to +300°C
Error: +14°C
Weight: 0.25 kg

Operational status

In service and in production for a wide variety of military and civilian fixed- and rotary-wing aircraft throughout the CIS.

Contractor

ElectroPribor Kazan Plant.

NEW ENTRY

MIKBO series of compact integrated avionic systems

The MIKBO series of compact integrated avionic systems is designed and produced by the Aircraft Instrument-making Establishment MIKBOTRON, which is part of NIIAO, the Institute of Aircraft Equipment.

The MIKBO series is specially designed for use in superlight aircraft. Included in the series are the following systems:

MIKBO-2

MIKBO-2 is in development for use in superlight single-engined aircraft operating in VFR conditions in the daytime, under visual meteorological conditions in the local area and during route and regional transit flights. MIKBO-2 comprises: magnetic compass MKB-90; VHF radio; microcomputer; AMLCD; ASI; engine speed indicator; cylinder head temperature sensor; integrated instrument panel. Total weight <3 kg; dimensions 350 × 150 × 150 mm; power 12 V DC <5 W.

MIKBO-21

MIKBO-21 is designed for superlight single-engined aircraft operating in VFR conditions in daytime under visual meteorological conditions in the local area. MIKBO-21 comprises: ASI YC-150; barometric altimeter CD-10; variometer BP-5; magnetic compass KI-13; cylinder head temperature sensor; and engine shaft speed indicator. Total weight <300 × 180 × 200 mm; power 12 V DC, 1.5 W.

MIKBO-22

MIKBO-22 is in preproduction for use in superlight single-engined aircraft operating in VFR conditions in the daytime, under visual meteorological conditions in the local area and during route and regional transit flights. MIKBO-22 comprises: barometric altimeter BD-10; variometer BP-5; magnetic compass MKB-90; ASI; cylinder head temperature sensor/indicator; engine shaft speed indicator; speed indicator, VHF radio. Weight < 5 kg; dimensions 350 × 230 × 200 mm; 12 or 27 V DC, 13 W.

MIKBO-23

MIKBO-23 is in preproduction for use in superlight twin-engined aircraft operating in VFR conditions in the daytime, under visual meteorological conditions in the local area and during route and regional transit flights. MIKBO-23 comprises: barometric altimeter BD-10; variometer BP-5; magnetic compass MKB-90; ASI; cylinder head temperature sensor/indicator; engine shaft speed indicator; speed indicator, slip indicator; main rotor speed indicator (MIKBO-23AJ only); VHF radio 'ptakha'. Weight 6 kg; 12 or 27 V DC, <13 W.

MIKBO-32

MIKBO-32 is in design for use in superlight and light primary training aircraft certified to AP-23 or FAR-23 operating in VFR conditions from unequipped airfields. MIKBO-32 includes: a multifunction instrument panel with two flat panel displays and databus integrating: VHF radio, ADF, radio altimeter, heading computer, engine and air data sensors, magnetic compass, and CVR/FDR. Weight <30 kg; power <300 W.

MIKBO-43

MIKBO-43 is in design for use in light multipurpose aircraft, including amphibians, operating in IFR conditions from unequipped airfields in all geographical areas, as well as off water in wave heights up to 0.5 m. MIKBO-43 includes the same equipment as MIKBO–32, plus: multifunction weather radar and emergency COSPAS-SARSAT radio beacon. Weight <75 kg; power <700 W.

Contractor

NIIAO Institute of Aircraft Equipment.

NEW ENTRY

MFI multifunction active-matrix liquid-crystal display

The Ramenskoye Design Company MFI AMLCD is designed for presentation of raster enhanced graphic TV and other image data. It forms the display part of the Ramenskoye Design Company's SINUS integrated navigation, flight management and display system.

Specifications

Display size: 130 × 130 mm
Pixels: 864 × 864
Resolution: 0.15 × 0.15 mm
Viewing angles:
(horizontal) ±45°
(vertical) +35 to −10°C
Power: 115 V AC, 400 Hz
Dimensions: 205 × 205 × 205 mm
Weight: <7 kg

Contractor

Ramenskoye Design Company AO RPKB.

NEW ENTRY

MFI active-matrix liquid-crystal display
1998/0018198

PKP-72 and PKP-77 flight directors and PNP-72 compass

The PKP-72 and PKP-77 flight director indicators and the PNP-72 compass are analogue instruments that have found wide application in Russian aircraft.

Specifications

	PKP-72	PKP-77	PNP-72
Roll error	0.7°	0.7°	
Pitch error	0.7°	0.7°	
Yaw error			1.0°
Azimuth			0.5 to 1.5 km

Contractor

Ramenskoye Instrument Engineering Plant.

NEW ENTRY

PKP-72 and PKP-77 flight directors and PNP-72 compass ***1998***/0018197

SOUTH AFRICA

Grintek Avitronics display systems

The Grintek Avitronics' family of displays include MultiFunction Displays (MFD), MultiFunction Keyboards (MFK) and a new Tactical Situation Display (TSD).

The TSD is an intelligent avionics display and control system consisting of two LRUs, the Display Electronic Unit (DEU) and the Display Unit (DU). It is capable of displaying continuously updated map types (digitised paper charts, digitised images, vectorial maps and terrain elevation maps). Each map type can have overlaid information pertaining to the tactical, navigation and weapon status of the aircraft and the operational threat scenario.

The TSD is also capable of displaying external imaging systems (FLIR), video or datalink overlaid with system information. In addition a separate output can be provided for the HUD.

Cockpit-mounted displays are sunlight-readable; provide a wide viewing angle, and are NVG-compatible.

Colour or monochrome displays are available from 3 to 10 in. They feature intelligent or non-intelligent displays, with medium- to very high-performance Graphics Processor Boards (GPB). Future growth will provide three-dimensional graphics capability.

Video modes include: FLIR; VCR video in and out; LLTV; HUD video.

Specifications

Power requirements: 28 V DC (accepts 18-32 V)
Interfaces: Data: 1553B, RS-422, RS-485, ARINC 429
Video: CCIR, STANAG 3350
Environmental:
MIL-STD-810E
MIL-STD-461C
DO 160C

Contractor

Grintek Avitronics, Grintek Electronics Limited.

VERIFIED

Grintek Avitronics display systems
1997/0001359

SWEDEN

EP-12 display system for the JA 37 Viggen

The EP-12 integrated display system for the interceptor version of the Saab JA 37 Viggen collects, processes and displays all flight, navigation, radar and tactical data. It comprises five main units: head-up, radar and tactical displays, waveform generator and power supply. The three displays utilise CRTs with contrast enhancement techniques to provide clear symbology even in bright sunlight. The radar display supports all weather surveillance and interception, while the tactical display shows a track oriented moving map superimposed with navigation and tactical data, updated at between 10 and 50 Hz depending on the amount of data.

Also associated with the EP-12 system is a cassette recorder to sample data at 4 Hz for the purpose of debriefing and training.

Operational status

In service in the Saab JA 37 Viggen. No longer in production.

Contractor

Ericsson Saab Avionics AB.

VERIFIED

EP-17 display system for the JAS 39 Gripen

The cockpit of the Saab JAS 39 Gripen features advanced electronic information presentation on four display units: three head-down and one head-up. Conventional dial and pointer instruments will be retained only as back-ups for some of the more critical parameters. All four displays are computer-controlled, permitting the presentation to be tailored to every type of mission and flight mode. It also provides considerable flexibility and redundancy under emergency conditions by redirecting information between displays.

The three head-down units are the flight data display, the tactical display and the multisensor display. The flight data display provides the pilot with altitude, air data, engine data and so on.

The electronic map on the tactical display shows geographical features and obstacles, such as radio towers and masts, with the tactical situation superimposed. Map scale and display are automatically selected to suit different phases of a mission. The multisensor display presents a computer-processed radar picture or, optionally, infrared or thermal imagery superimposed with target symbology.

The Ericsson Saab Avionics radar display (centre) and tactical display (right) in a Saab JA 37 Viggen

All head-down displays are dimensionally and mechanically identical and measure about 6 × 4.7 in (152 × 120 mm). The large diffractive optics head-up display, presents a large bright picture to the pilot.

The EP-17 has two display processors, each driving two displays and communicating with other aircraft systems by means of MIL-STD-1553B databusses. Each processor contains an Ericsson D-80 computer, the standard computing element in the aircraft. Specially developed high-performance graphics processors provide excellent dynamics in the imagery.

The head-down displays employ raster-generated symbology, while stroke writing is used for the head-up display. Updating is accomplished at 30 Hz.

There are four presentation areas in the EP-17 system:

(1) Head-Up Display (HUD) displaying navigation mode; spherical attitude lines; and landing mode, tactical ILS data; featuring: 20 × 28° FOV; diffraction optics; symbol generation through strokes, with possibility to display raster images;
(2) Flight Data Display (FDD) with flight, system, and weapon data; and HUD back-up data;

The cockpit display system for the JAS 39 Gripen comprises a wide-angle HUD and three head-down displays ***1995***

(3) Horizontal Situation Display (HSD) with tactical overview based on an electronic map displaying air-to-surface mode; air-to-air mode, and emergency checklist data;
(4) MultiSensor Display (MSD) with information from radar and other sensors and air-to-air and air-to-surface mode data.

The three multifunction displays (MFDs) feature three video systems (525/675/875 lines), identical and interchangeable hardware, display area 150 × 120 mm.

Display processing features full software control in computers and symbol generators, anti-aliasing, full colour, five map scales and radar scan conversion.

Recording capabilities included are multiplexed sensor video, MIL-STD-1553B bus data and audio using an HI-8 mm video cassette recorder.

Operational status

In production for the Swedish Air Force Saab JAS 39 Gripen. In May 1996 Ericsson Saab Avionics was contracted to update the PP-1 and PP-2 display processors to PP-12 standard.

Contractor

Ericsson Saab Avionics AB.

UPDATED

SWITZERLAND

Timearc Visualizer multifunction digital display system

The Timearc Visualizer multifunction digital display system is a versatile display system for enhanced situational awareness and mission management.

A large, flat panel, full-colour, high-resolution active matrix (TFT) LCD driven by the Digital Map Display Generator (LRU) is used to combine and use mapping data from multiple raster data sources.

A special Smart Point Track Stick, integrated into the flat panel LCD, allows the user to scroll over the entire activated map and to pinpoint any desired position. A special window shows the digital readout of such a position in latitude/longitude, or any other selected grid system. A pinpointed position an also be used for an instant transfer into the GPS Navigation Management System or for a Direct To Navigation.

The Timearc Visualizer does not include a GPS engine, but most stand-alone GPS navigation systems or GPS sensors can be interfaced with the Visualizer, provided they possess a free serial interface either RS-232/422/485 or ARINC 429.

Special purpose software available includes an integrated Electronic Library System (ELS) and 'Waypoint' or 'Street Find' options. The Visualizer version 3.0 displays moving maps and video in real time. The active matrix LCD is available in two models: the standard LCD, and a high-performance LCD for optimal sunlight readability.

Timearc Visualizer ***1997***/0018196

Specifications

Colour Display (AMLCD)
Dimensions: 280 × 199 × 52 mm
Resolutions: 640 × 480
Colours: 256
Weight: approx 1.9 kg
Electrical:
interfaces: keyboard
Display:
screen size: 214 mm

Data Generator (DMDG)
Dimensions: 96 × 163 × 318 mm
Weight: approx 3.0 kg
Electrical:
input voltage: 10-40 V DC
power consumption: 25 W (max)
interfaces: RS-232/422, ARINC 429 (optional)

Operational status

Widely used by police forces, search-and-rescue organisations, forestry/oil/gas industries in both helicopter and fixed-wing aircraft installations.

Contractor

Flight Components AG.

UPDATED

Altimeters, types 3A and 3H, 2 in

Thommen 2 in altimeters are barometric instruments. The counter-pointer display is actuated by a high-performance mechanism with a high-stability beryllium capsule. A built-in vibrator minimises friction and optimises accuracy in use. Temperature is compensated by a bimetallic element.

All instruments are offered customer-specific (see type designation) and are mounted for easy servicing.

The counter has two moving drums, which show tens of thousands and thousands of feet. Three fixed zeros are printed on the display. The pointer indicates 1,000 feet per revolution on a scale calibrated at intervals of 20 ft. Warning flags are displayed on the counter between 10,000 and 0 ft, respectively below 0 ft.

The barometric counter indicates the pressure as four digits as a single baro in mb or in inches of mercury, or as dual baro in both. Adjustment for QNH/QFE is provided by a knob on the front, built-in stops are included.

An electrical indication of pressure (3H43, 3H42 models only) can be provided by potentiometer geared to the baro-scale mechanism.

The altimeters are installed in lightweight aluminium ARINC, square and round cases.

Altimeters, type 3A and 3H, 2 in ***1998***/0018189

Lighting is offered in different voltages and different colours. All instruments are available in NVG-compatible lighting.

Operational status

The 2 in altimeters are installed as primary or standby instruments in civil and military aircraft.

Contractor

Revue Thommen.

NEW ENTRY

Encoding altimeters, types 3A and 3H, 3 in

Thommen 3 in encoding altimeters are totally self-contained, requiring no electrical power to drive the altitude mechanism. The altimeters incorporate an optical encoder providing height reporting output. The counter-pointer display is actuated by a high-performance mechanism with a high stability beryllium capsule. A built-in vibrator minimises friction and optimises accuracy in use. Temperature is compensated by a bimetallic element.

The counter has three moving drums, which show tens of thousands, thousands and hundreds of feet. The pointer indicates 1,000 ft per revolution on a scale calibrated at intervals of 20 ft. Warning flags are displayed on the counter between 10,000 and 0 ft, respectively below 0 ft.

The barometric-counter indicates the pressure as four digits as single baro in mbar or inHg or as dual baro in mbar/inHg. Adjustment for QNH/QFE is provided by a knob on the front, built-in stops are included.

There is an optical electronic system supplying altitude data in digital form, ICAO coded (Gillham) for SSR pressure altitude transmission in 100 ft increments, to standard accuracy TSO-C88.

Encoding Altimeters are provided with a code ON/OFF flag alarm, which indicates any encoding power malfunction.

An electrical indication of pressure (3H67/3H66 only) can be provided by potentiometer geared to the baro-scale mechanism.

The altimeters are installed in lightweight aluminium ARINC, square and round cases.

Altimeters type 3A and 3H, 3 in ***1998***/0018188

Lighting is offered in different voltages and different colours. All instruments are available in NVG-compatible lighting.

Operational status

The 3 in encoding altimeters are installed as primary or standby instruments in civil and military aircraft.

Contractor

Revue Thommen.

NEW ENTRY

MACH/Airspeed Indicators (MAI), type 5, 3 in

The 3 in MACH airspeed indicators are pneumatically operated instruments deriving indicated airspeed and MACH information from the pilot-static sources. The combined displacements of the airspeed pointer and the MACH-disk determine the MACH number, which is indicated simultaneously with airspeed by the airspeed pointer. The control relays for the outside warning system (Vmo and Mmo) are actuated by optical

MACH/airspeed indicators, type 5, 3 in
1998/0018187

detection devices. All instruments are offered customer-specific.

The MACH airspeed indicators are installed in lightweight aluminium ARINC and square cases.

Lighting is offered in different voltages and different colours. All instruments are available in NVG-compatible lighting.

Operational status

The 3 in MAI are installed as primary or standby instruments in civil and military aircraft.

Contractor

Revue Thommen.

NEW ENTRY

Vertical Speed Indicator (VSI), type 4A16, 3 in

A pointer indicates vertical speed on a fixed dial. Zero adjustment used to compensate internal stresses in the mechanism, is operated by turning the whole mechanism frame in relation to the case/dial assembly.

The Vertical Speed Indicators are installed in lightweight aluminium ARINC, square or round cases.

Lighting is offered in different voltages and different colours. All instruments are available in NVG-compatible lighting.

Operational status

The 3 in Vertical Speed Indicators are installed as primary instruments in civil and military aircraft.

Vertical Speed Indicator (VSI) Type 4A16, 3 in
1998/0018186

Contractor

Revue Thommen AG.

NEW ENTRY

UNITED KINGDOM

CGI-3 compass gyro indicator

The CGI-3 is a panel-mounted instrument containing all the elements, apart from the flux valve and a small annunciator, of a gyromagnetic compass and an RMI. Although it is ideal for installation in helicopters and executive aircraft, the CGI-3 has been fully type tested and cleared for operation in a severe military environment.

The gyro uses a wheel and gimbal assembly and the readout of heading is by means of a rotating dial with 5° graduations which moves in the conventional sense with changes in aircraft heading.

The transistorised slaving amplifier contained within the instrument case provides the necessary energising voltage for a flux valve magnetic detector unit and then amplifies and discriminates the signals received from this unit in order to maintain the indicated heading in accordance with the direction of the earth's magnetic field.

Specifications

Dimensions: 76 × 76 mm
Weight: 2.5 kg

Operational status

In service.

Contractor

British Aerospace Systems & Equipment.

VERIFIED

DG13 - directional gyro indicator

DG13 is a panel-mounted directional gyroscope. Normally used as a standby heading unit, it contains a directional gyro geared to a rotating compass card which provides a heading readout. The compass card is graduated in 5° divisions and is read against a white lubber line at the top of the unit. A yellow 'Set Heading' marker is also provided. Drift rate is a nominal 10°/h, after making allowance for the effect of the Earth's rotation.

Specifications

Dimensions: 83 mm square case, 190 mm length
Weight: 2.5 kg
Power supply: 28 V DC plus 5 V AC/DC for lighting

Operational status

Unit in production. Applications include the Hawk aircraft.

BASE gyro magnetic compass system
1996

Contractor

British Aerospace Systems & Equipment.

VERIFIED

HGU B9, B19 horizon gyro units

The HGUs Type B9 are electrically operated gyroscopic flight instruments which present the pilot with continuous indication of the aircraft attitude in pitch and roll with respect to the natural horizon.

Movement of the aircraft causes the instrument, with its miniature aircraft and roll angle scale, to move in relation to the stabilised horizon bar and roll angle pointer. Thus the horizon gyro unit indicates the degree of roll by means of the pointer and scale, and the attitude of the aircraft with respect to the natural horizon by means of the horizon bar and miniature aircraft.

HGU Type B9 must be used in conjunction with a junction box which provides a junction point for the AC supply to the respective HGU and the DC supply for the potentiometer pick-offs.

HGU Type B19 is an improved form of the electrically operated vertical gyro designed to meet the requirements for attitude information to an autopilot, reduced pendulosity and reduced erection rates.

Specifications

Dimensions:
(mounting face to rear of case) 195 mm
(case diameter) 112 mm
Weight: 2.7 kg
Power supply: 115 V, 400 Hz, 3 phase AC

Operational status

In Service.

Contractor

British Aerospace Systems & Equipment.

VERIFIED

HL8, HL9, HL11, HL12 Horizon Gyro Units (HGU)

The HGU is an electromechanical flight instrument, designed to provide the pilot with a constant visual indication of the pitch and roll attitude of the aircraft relative to the natural horizon.

Pitch attitude is indicated by a gyro-stabilised horizon bar registering against a fixed 'gull wing' datum engraved on the bezel glass; roll attitude is indicated by a pointer fixed on the sky-plate registering against a roll angle scale. To maintain the gyro axis in the vertical, a mechanical erection system is incorporated. The erection system is gravity-controlled and automatically responds to any deviation of the gyro from the vertical axis.

The instrument also incorporates a power failure indicator which carries a warning flag marked off. The flag is visible in the frame located in the upper half of the sky-plate. When the instrument is not operating, the flag indicates off. When the instrument is switched on the flag clears and during normal operation the flag remains in the clear position.

Specifications

Dimensions: 82 × 82 × 216 mm
Weight: 2.15 kg
Power supply: 115 V or 55 V, 400 Hz, single phase.

Operational status

Instrument fitted in Gazelle, Sioux, Whirlwind, Wasp helicopters as well as VC10 and Dove aircraft.

Contractor

British Aerospace Systems & Equipment.

VERIFIED

RAI 4 remote attitude indicator

The RAI 4 remote attitude indicator works on a synchro output from a vertical gyro. The indication is essentially a sphere which moves relative to a fixed aircraft pitch symbol and to a peripheral bank scale for roll. The sphere has 360° of freedom in roll and ±100° in pitch.

Specifications

Dimensions: 4 ATI × 210 mm
Weight: 1.82 kg
Power supply: 115 V AC, 400 Hz, 11 VA
28 V DC plus 5 V AC/DC for lighting

Operational status

In service.

Contractor

British Aerospace Systems & Equipment.

VERIFIED

RL8 - heading repeater

RL8 is a heading repeater series that provides heading information displayed on a servo-driven moving card, with settable markers for desired heading and wind direction. The 3.25 in (83 mm) diameter instrument accepts heading information in standard three-wire synchro form. The rotating card display is calibrated through 360° in 5° increments and the compass cardinal points are designated. Wind direction is indicated by a green cursor, set by pulling out and rotating the knob on the front bezel. Required heading is indicated by a white pointer, set by pushing in and rotating the knob. Once set, both reference indicators rotate with the card display.

Operational status

Unit in production. Applications include Lynx, Sea King and Tornado ADV.

Contractor

British Aerospace Systems & Equipment.

VERIFIED

Airborne radar indicator unit and track-while-scan system

The Caledonian Airborne Systems Ltd airborne radar indicator unit is a direct solid-state replacement for existing CRT-based indicators. It handles a variety of radar, composite and VGA video standards, and provides a high-resolution flicker-free display.

The 10.4 in multifunction display depicted serves as a direct replacement for the AlliedSignal IN-1502A radar indicator, and as the display for the Caledonian Airborne Systems Ltd Track-While-Scan System (TWSS).

The design can, however, be tailored to suit displays from 4 in to 16 in sizes and a number of video standards.

The Track-While-Scan System interfaces with the AlliedSignal RDR-1500B radar and navigation systems to enable on-screen target track and marker placement. TWSS, radar and Scorpio 2000 datalink (see Communications section) system control is by point-and-click using the joystick control. Target, aircraft and marker data are overlaid on the radar indicator unit; messages are presented on the lower screen.

Contractor

Caledonian Airborne Systems Ltd.

NEW ENTRY

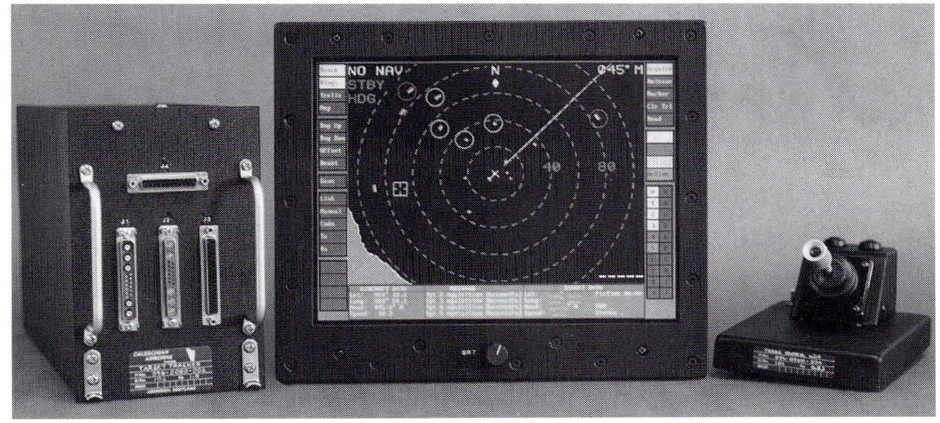

Track-while-scan system processor (left), digital solid-state radar indicator unit (centre) and joystick control unit (right)
1998/0018195

SAHIS Standby Attitude, Heading and rate of turn Indicating System

SAHIS is a Standby Attitude, Heading and rate of turn Indicating System designed primarily for use as an emergency back up in the event of failure of aircraft generated power and/or loss of primary aircraft attitude and heading information.

The system consists of a three axis spherical indicator, mounted in the aircraft's instrument panel, and a gyro unit, mounted remotely. The gyro unit contains a vertical gyro for attitude reference and a directional gyro for heading reference.

The indicator displays pitch, roll and heading information by means of a sphere moving behind a fixed aircraft symbol. There is a roll scale below the sphere and slip and rate of turn indications below the roll scale.

All signals and power inputs to the indicator (except slip) are derived from or via the gyro unit.

The system can operate in either of two modes whereby heading information is derived from the aircraft's inertial navigation system or from the unit's own directional gyro.

Specifications

Freedom:
(heading) unlimited
(roll) unlimited
(pitch) ± 85°
Accuracy:
(heading) ±1°
(roll) ±1°
(pitch) ±1°
Rate of turn: up to 380/min
Dimensions:
(indicator) 81 × 203 × 81 mm
(gyro unit) 110 × 265 × 170 mm
Weight:
(indicator) 1.8 kg
(gyro unit) 5.6 kg

Operational status

SAHIS is currently in production to meet UK RAF requirements on Harrier GR. Mk 7 and Harrier T Mk 10 aircraft.

Contractor

Ferranti Technologies Limited.

NEW ENTRY

Standby Attitude, Heading and rate of turn Indicating System **1998**/0018194

Angle of attack system

The Angle of attack system is designed to measure and display angular airflow direction relative to a fixed datum, such as the horizontal fuselage datum.

The system consists of the FTRP100 Airstream Direction Detector (ADD) and an Angle of Attack Indicator (AAI).

The FTRP100 ADD utilises non-contact inductive resolvers to sense angular position of the detector probe relative to the fixed datum.

The detector probe is equipped with internal heaters for anti-icing and moisture control plus drainage at the base to prevent water ingress. The aerodynamic design and performance creates positive pressure at the input slots, thus preventing sand and dust entering the detector probe tube.

Ferranti Technologies' Angle of attack system
1998/0018223

The rear housing contains either one or two non-contact inductive resolvers which sense the position of the detector probe and can be specified to output information in either digital or analogue format.

The AAI is an easy to read indicator displaying relative angular airflow in increments of one degree from −5 to +25°C. Provision can be made for markers on the display to highlight optimum angle of attack and stall conditions.

Contractor

Ferranti Technologies Limited.

NEW ENTRY

Attitude indicators - FH series

Attitude indicators of the FH series are electrically driven gyroscopic instruments which display aircraft attitude in two axes by a spherical type presentation.

The product range encompasses instruments with 2, 3, and 4 in displays, all with a variety of colour and lighting options and a choice of AC or DC eletrical input. An intergral slip indicator may also be specified.

All instruments are hermetically sealed and feature alternative panel angle options, manual caging and automatic gyro control during accelerated flight.

Operational status

In production for fixed- and rotary-wing aircraft for both military and commercial applications. FH series instruments are specified as primary or standby instrumentation in many aircraft including Jaguar, Tornado, Nimrod, Hawk, Chinook, Super Lynx, Bell 412 and Eurocopter AS 350.

Contractor

Ferranti Technologies Limited.

NEW ENTRY

Ferranti Technologies' attitude indicators FH22, FH30, FH32 and FH40 (left to right) **1998**/0018193

COmbined Map and Electronic Display (COMED)

The maps used with current moving map displays have a major disadvantage in that it is not possible to annotate them with planned route, target information and other tactical data. This has to be marked up on a hand-held map as an adjunct to the moving map. GEC-Marconi Avionics, however, has developed the Combined Map and Electronic Display (COMED) system, in which the map is electronically annotated with intelligence or navigation data appropriate to the particular mission. Using no more space than a conventional moving map, COMED provides a colour topographical map display annotated with dynamic navigation information as required. This can include aircraft track and commanded track, present position, locations of known hostile detection or anti-aircraft devices and tactical information such as the delineation of forward edge of the battle area. In addition, the system can print out alphanumeric information such as time to go to fix point or target.

COMED's CRT and projection facilities permit it to perform other operational tasks. These can include the display of high-resolution high-contrast symbology in raster form from radar, low-light television or FLIR sensors. Electronic countermeasure threats can also be shown. Tabular displays of weapons status, destination co-ordinates, or other tactical information can be shown on tabular 'forms' projected from images stored on the film. The system can be programmed with a library of aircraft and engine checklists. COMED can also act as a back-up primary flight information display, particularly for the horizontal situation and attitude director indicators.

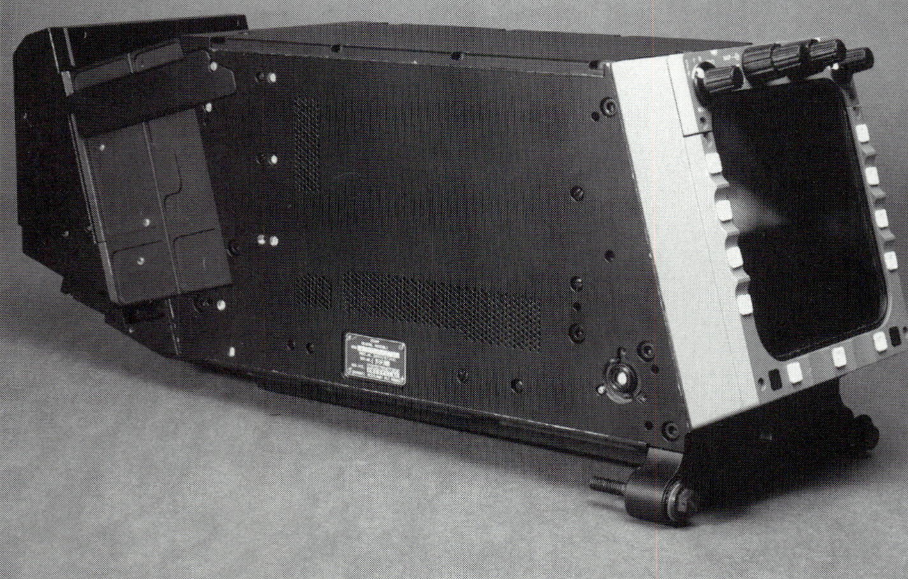

CEDAM, the latest model COMED for the German Air Force ECR Tornado

COMED interfaces with the main aircraft navigation computer via a MIL-STD-1553 serial digital datalink. From a knowledge of film strip layout, aircraft present position and demanded scale, the main computer calculates the appropriate map drive words and transmits them to COMED by the datalink. Within the display the information is converted into a form suitable for driving the map servo resolvers.

COMED uses standard 35 mm colour film up to 57 ft (17 m) long. Typically, this provides coverage of an area of 10,000 sq n miles at a scale of 1:250,000, plus selected target areas at 1:50,000, and sufficient film frames for tabulated displays. Film replacement can be achieved through a side access panel in under a minute.

More than 1,000 systems have been supplied for the F/A-18 Hornets with the US Navy and the armed forces of Australia, Canada and Spain. They are delivered to AlliedSignal Flight Systems Division, the Projected Map Assembly (PMA) forming part of the AlliedSignal horizontal situation indicator, a combined map and electronic display for that aircraft. It projects a coloured topographical moving map image superimposed with data from a CRT image.

Specifications

Dimensions:
(CRT face) 139.7 × 139.7 mm
(unit) 180.3 × 217.9 × 604 mm

Operational status

No longer in production. The system has been installed by the Indian Air Force in its Sepecat Jaguars made under licence by Hindustan Aeronautics. The first production COMED system was delivered during early 1983. In May 1986 CEDAM (the latest variant of COMED) was selected for the Tornado ECR version ordered by the German Air Force.

Contractor
GEC-Marconi Avionics Ltd.

VERIFIED

Digital Map Generator (DMG)

The GEC-Marconi Avionics Digital Map Generator (DMG) has been developed to meet the mission requirements of the next generation of combat aircraft. Housed in a ¾ ATR box, the DMG is suitable not only for these new aircraft but also for a large number of retrofit applications. The experience gained in developing and manufacturing moving map displays for the F/A-18 Hornet, Jaguar, Tornado and Harrier has enabled GEC-Marconi Avionics to optimise the DMG to provide maximum flexibility with minimum pilot workload. The video output of the DMG can be displayed on the normal range of cockpit displays.

The DMG hardware and software can handle both true digital maps and digitised chart information. The digital database can be held in solid-state or in optical disk. Both of these options have been developed. Features on the presentation can be selectively displayed or erased. Hazards such as terrain and obstacles above aircraft altitude can be made to stand out in contrasting colour. Safe areas occasioned by terrain-masking can be depicted and areas may be viewed from different angles and altitudes. Data updating is rapid and simple. The map is displayed on a GEC-Marconi Avionics colour electronic display which incorporates a multifunction keyboard and electronically generated symbology.

The map can be displayed either north up or track up, with the aircraft present position centred or de-centred on the display. Multiple map scales can be accommodated and zoom and declutter facilities are available. Map stabilised overlays for routes and navigational information can be displayed.

Operational status
In production for the Royal Air Force Jaguar, Tornado GR. Mk 4 and Lockheed Martin C-130J.

Contractor
GEC-Marconi Avionics Ltd.

VERIFIED

Display processors/graphics generators

The advent of glass cockpits demands that the display symbol generators have a capability which matches display performance and operational requirements. GEC-Marconi Avionics has adopted a modular approach to graphics generation and can supply boxes which drive either a single display surface in monochrome or a full display suite of multifunction colour displays and a dual-mode head-up display.

A variety of interfaces is offered which can be modified, as necessary, to integrate with existing or new aircraft systems. As well as dedicated display generation, the boxes can have weapon aiming and mission computer functions added easily.

Operational status
In production for the Royal Air Force Tornado GR. Mk 4.

Contractor
GEC-Marconi Avionics Ltd.

VERIFIED

CNI/control and display unit

Developed for Hawk 100/200 aircraft, the CNI/control and display unit provides the common point for the control and data information display for the various radio, navigation and IFF systems. The unit has the primary role of controlling and monitoring the radio systems via a MIL-STD-1553B databus. It has a secondary role, upon request, to perform a reversionary bus controller function without affecting its primary role. The unit can, however, be reconfigured for other applications through software changes via an RS-424 serial datalink.

Operational status
In production.

Contractor
GEC-Marconi Avionics Ltd.

VERIFIED

The CNI/control and display unit has been developed for the Hawk 100/200

Engine monitor panel

The engine monitor panel for the Hawk 100 and 200 displays all necessary fuel and engine information as well as indications of any associated malfunction. The display technology combines the latest LCD technology and LED arrays to provide a flat panel which can be mounted in any cockpit position and still maintain high readability over all ambient light conditions.

The display technology is adaptable to cockpit environments where integration of analogue instruments is required to allow more use of cockpit space.

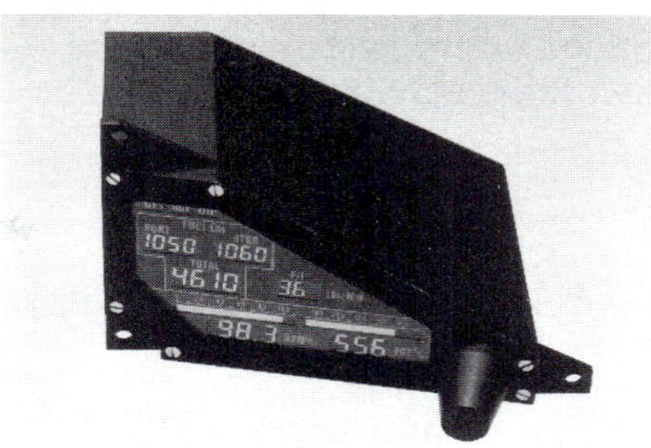

The engine monitor panel for the British Aerospace Hawk 100 and 200

The panel incorporates a fuel 'bingo' facility which allows the pilot to set minimum fuel levels at which an audio warning will be generated. An option to this panel is the incorporation of engine low-cycle fatigue recording for post-flight evaluation.

Operational status
In production.

Contractor
GEC-Marconi Avionics Ltd.

VERIFIED

Head-down displays

With the background of such major projects as Panavia Tornado and British Aerospace Nimrod, GEC-Marconi Avionics has developed a range of multisensor, multifunction television tabulator, E-scope and other displays. The systems comprise a CRT driven by a waveform generator embodying raster scan techniques. The waveform generator produces synthetic symbols, by what the company calls a time-shared digital technique, to give high accuracy and resolution of modulated video signals with a minimum of components; a notable feature is the elimination of the staircase effect in raster graphics. The synthetic video is directly compatible with standard television signals from a variety of sensors, so that video signals can be mixed to produce an overlay and symbology on a pictorial display.

The waveform generator is completely digital in operation and receives data in standard serial digital form. By storing this information it can synthesise continuously video signals to drive one or more displays. The signal range can be modified or extended to meet individual requirements, particularly those suited to raster applications. The display unit provides a high-contrast television picture of sensor and computed data and has a multifunction keyboard permitting the operator to communicate with the aircraft computer. A filter on the face of the CRT improves the contrast under high-ambient illumination and photosensors adjust the brightness of the display in accordance with the ambient lighting.

Operational status
No longer in production. In addition to the Tornado and Nimrod MR 2, other applications include the gunship version of the Lockheed Martin C-130 Hercules.

Contractor
GEC-Marconi Avionics Ltd.

VERIFIED

The GEC-Marconi Avionics displays which equip the Panavia Tornado. Left to right, the electronic head-down display, the TV tab display and the E scope

MED 2060 series monochrome head-down displays

MED 2060 monochrome head-down displays are small high-brightness raster presentation devices designed for use where space is at a premium. Units are currently available with screen diagonals ranging from 105 to 280 mm. Different aspect ratios are selectable and 525- or 625-line variants are available. The displays can be supplied with or without a passive contrast enhancement filter matched to the CRT phosphor. The filter can be either bonded to the CRT or mounted away from its face to provide optimum visibility in specific conditions. All units are compatible with GEC-Marconi Avionics special automatic test set facilities, but can be maintained and serviced using only standard laboratory equipment.

The display can be used as full multifunction displays for FLIR, radar or stores status, or as simple head-up display repeaters in the rear cockpit of tandem-seat trainers.

Additionally, it has facilities to switch automatically between 1:1 and 4:3 aspect ratios, catering for a range of presentations from maps and engine/systems status data to the display of video from the pilot's head-up display and FLIR night vision system. The unit is also fully compatible with night vision goggles.

MED 2060 series monochrome head-down displays

Specifications

Weight: 7.5 kg
Power supply: 115 V AC, 400 Hz or 28 V DC, 50 W

Operational status

In production. The Type MED 2060 series has been selected for the Sea Harrier F/A-2 port and starboard displays, for the BAe Hawk 100/200, a variety of Mirage III and 5 aircraft, Danish SAR S-61 helicopters and for a Far Eastern A-4 Skyhawk retrofit contract.

Contractor

GEC-Marconi Avionics Ltd, Mission Avionics Division, Edinburgh.

UPDATED

Miniature LCD Tacan controller for the Shorts Tucano

The miniature LCD on the Shorts Tucano will be used in conjunction with the lightweight Tacan navigation equipment. Fitted in each cockpit of the Tucano, the LCD will provide normal Tacan control and display two Tacan channels, the one in use and the standby channel. Either unit is able to take control of the Tacan.

Occupying only a 2.5 × 2.25 in (63.5 × 57.1 mm) space, the panel-mounted miniature LCD controller is particularly suited to fixed-wing aircraft and helicopters where panel space is at a premium. The controller features the latest industry standard liquid crystal display technology and is compatible with most modern avionic systems.

Operational status

In production for the Royal Air Force Shorts Tucano.

Contractor

GEC-Marconi Electro Optics Ltd, Sensors Division, Basildon.

VERIFIED

Miniature LCD Tacan controller for the RAF Shorts Tucano

Airborne FM telemetry torquemeter

The airborne FM telemetry torquemeter provides a direct, accurate and reliable measurement of the torque in a helicopter's transmission system. The indicated torque is independent of any correction factors for air temperature, torquemeter shaft temperature, engine and gearbox life and efficiency, fuel characteristics or aircraft operating conditions.

The system consists of four component parts: rotating assembly, fixed antenna and head amplifier, signal processor and indicators. It operates without physical contact between the rotating shaft and the stationary structure and is said to have an overall accuracy of within ±1 per cent of full-scale torque. BITE is provided within the torquemeter equipment to enable system integrity and calibration accuracy to be checked and faults to be located.

Specifications

Power supply: 28 V DC, 1.5 A (max)

Contractor

GKN Westland Aerospace.

VERIFIED

DEWD Dedicated Electronic Warfare Display

Bell Helicopter Textron has selected Meggit Avionics Inc (an operating division of Meggitt Avionics), to develop and produce the Dedicated Electronic Warfare Display (DEWD) for the US Air Force Special Operations CV-22 Osprey tiltrotor aircraft.

The DEWD is a 3 in Air Transport Indicator (ATI) size active matrix liquid crystal display that interfaces with the Suite of Integrated Radio Frequency Countermeasures (SIRFC) system to provide electronic warfare situation awareness to CV-22 operators.

The DEWD is a derivative of the Meggitt Avionics Secondary Flight Display System (SFDS).

Contractor

Meggitt Avionics.

NEW ENTRY

Solid-state air data instruments

Airspeed and Mach/airspeed indicators

With integral pitot and static silicon pressure transducers, microprocessor technology and liquid crystal displays, these units provide display of airspeed or Mach/airspeed complete with all markers and bugs. Repeater variants are also available.

Altimeters

The range of solid-state altimeters is based on silicon pressure transducers and custom liquid crystal displays, with patented features to provide the high standard of human factors required for primary flight

Meggitt Avionics airspeed and Mach/airspeed indicator
1997/0002451

Meggitt Avionics altimeter **1997**/0002450

instruments. Configurations are available for civil and military applications and as primary, primary with reversion to standby, standby or repeater variants.

Contractor

Meggitt Avionics.

VERIFIED

Solid-state attitude indicator

A high-reliability implementation to meet the requirements for a self-contained attitude reference and indicator, the Meggitt Avionics solid-state unit features the latest technology in inertial sensors and active matrix liquid crystal displays. Proprietary software algorithms written in Ada and executed by a high-performance microprocessor proved an accurate attitude reference for either primary or standby applications. Housed in a 3 ATI case, this instrument readily replaces conventional mechanical gyroscope based units with an order improvement in reliability and life cycle costs.

Contractor

Meggitt Avionics.

VERIFIED

Secondary Flight Display System™ (SFDS)

The Meggitt Avionics Secondary Flight Display System™ (SFDS) fulfils the requirement for high-reliability standby flight information — altitude, attitude and airspeed: replacing two or three electromechanical cockpit standby instruments with a single 3 ATI instrument. The system comprises a 3 ATI colour active matrix liquid crystal display that displays flight information in a format compatible with primary electronic flight displays. A second Line-Replaceable Unit, the Air Data Unit, is connected to the appropriate pitot and static ports and includes solid silicon pressure sensors and a microprocessor to provide digital air data to the display.

Operational status

This display has been selected for the US Air Force C-130/C-141B upgrade and South African Air Force C-130 upgrade programmes, and for fitment to the US Air Force C-5 fleet. In January 1998 Boeing selected the Meggitt Secondary Flight Display System™ (SFDS) for all its new-generation 737 aircraft and Meggitt expects it to be fitted throughout the Boeing 7 series aircraft; deliveries start in January 1999.

Contractor

Meggitt Avionics.

UPDATED

Secondary Navigation Display™ (SND) system

The Secondary Navigation Display™ (SND) is a 3 ATI instrument that utilises an active matrix display and interfaces with ADF, VOR and DME radios together with the magnetic flux gate to provide navigation and heading data on a high-quality display. The SND is a complete back-up navigation display.

Contractor

Meggitt Avionics.

VERIFIED

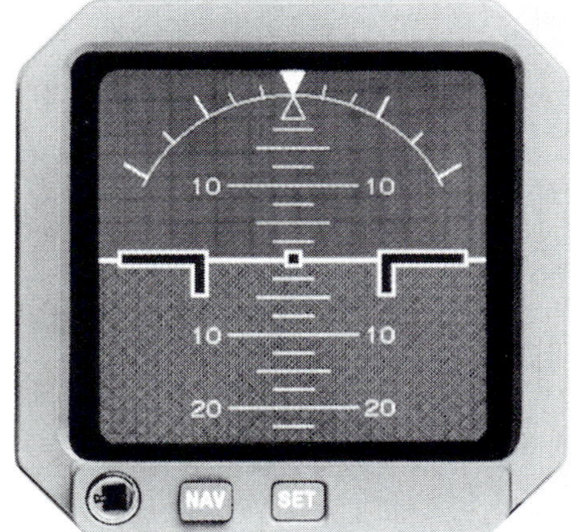

Meggitt Avionics solid-state attitude indicator **1997**/0002452

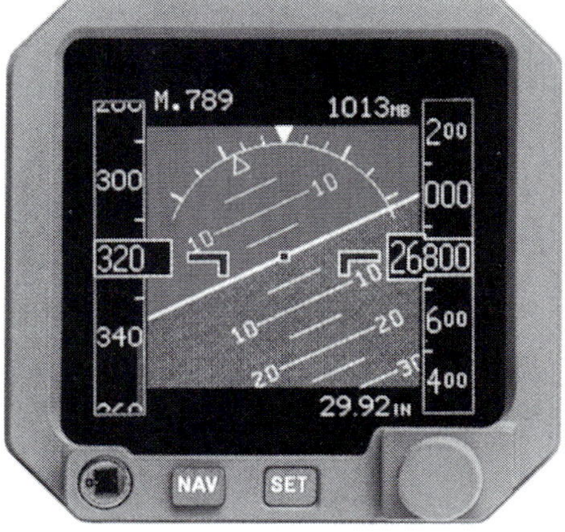

Meggitt Avionics Secondary Flight Display System™ **1997**/0002453

Meggitt Avionics Secondary Navigation Display™ **1997**/0002454

Central warning unit

The central warning unit has been designed and manufactured for use on modern fighter aircraft. The unit houses all the control circuitry and displays to provide 70 illuminated captions, along with audio tone and synthetic voice generation.

Each unit incorporates 28 primary red channels and 42 secondary amber channels. Additional features include day/night mode dimming of caption illumination, test facilities for active and non-active sensor channels, ground activation of certain channels with override facility for test purposes and generation of audio horn tones and synthetic voice.

Various combinations of caption are available in the range, while all variants incorporate a self-illuminating Betalight panel over the switches. Electrical connections are made via a pair of multipin connectors. The unit is capable of total NVG-compatibility.

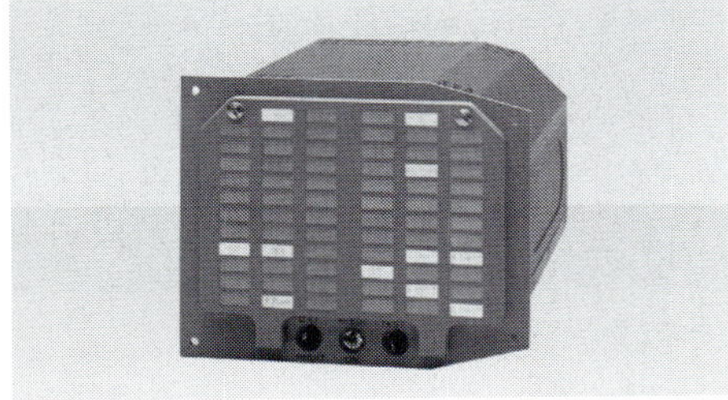

Page Aerospace central warning unit

Specifications

Dimensions: 134 × 165 × 165 mm
Power supply: 22.5-30.5 V DC normal (16-30.5 emergency)
Temperature range: −40 to +70°C
Altitude: up to 40,000 ft

Contractor

Page Aerospace Ltd.

VERIFIED

Standby Master Warning Panel (SMWP)

The Standby Master Warning Panel (SMWP) provides the pilot with essential red primary warning indications in the event of any failure of the standard warning system. It is designed for use in glass cockpits, so the one-piece display matches the appearance of glass instruments.

The display panel has 12 red warning captions, lit by LEDs, which flash until acknowledged. Different captions can be included, to satisfy customer requirements, up to 16 characters in length. Brightness control, to switch between day sunlight-readable and night illumination levels, is achieved by a Dim push-switch mounted directly under the display.

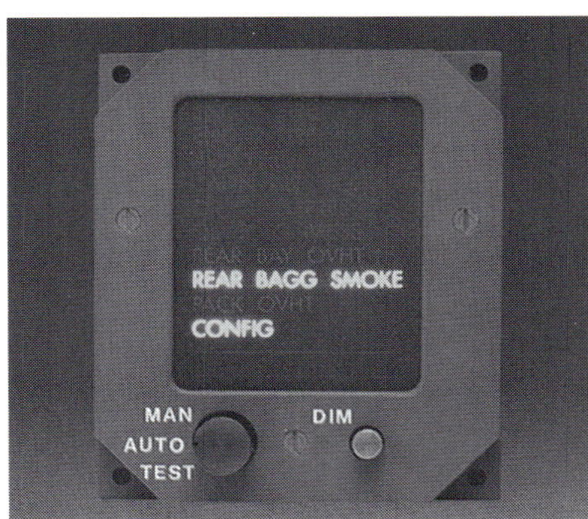

Standby warning panel **1995**

Situated adjacent the Dim switch is a rotary mode switch to enable selection of either automatic or manual operation and test. The test position checks more than 97 per cent of the circuit.

In view of its essential role the SMWP is made fault tolerant, so that no single fault will cause the loss of more than one warning. Even in the event of power supply failure, the unit will still operate in a degraded mode.

Specifications

Dimensions: 87.8 × 100.3 × 152 mm
Weight: 0.73 kg
Power supply: 28 V DC, <22 W (max)
Temperature range: −15 to +55°C
Altitude: up to 45,000 ft
Reliability: >100,000 h MTBF

Contractor

Page Aerospace Ltd.

VERIFIED

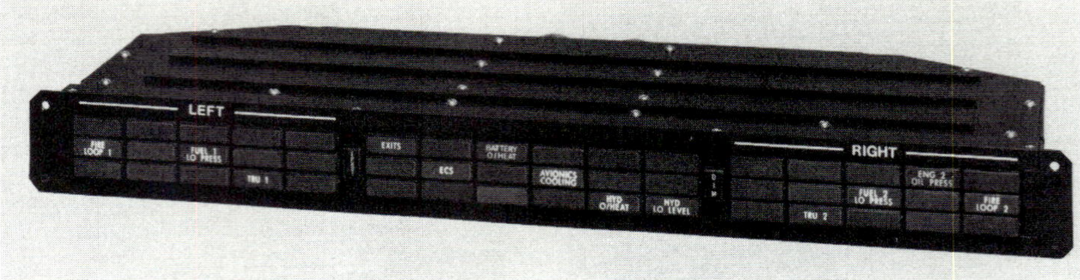

Page central warning panel for the BAe ATP

Two- and three-terminal light modules

Designed for use in single or stacked multichannel configuration, the filament lamp indicator/annunciator modules are a development of a module used extensively in aircraft central warning systems. The modules, available in two- and three-terminal arrangements, provide a low-cost, lightweight and compact alternative to other devices.

Both types of modules incorporate a clip-on caption frame within which a wide range of sunlight-readable blank or engraved caption screens may be contained. The three-terminal module may be used with two screens, one for each lamp, for systems where dual lamping is not essential. In these instances a lamp-holder separator screen may be fitted in a groove provided. Various alternative mounting arrangements facilitate single or multiple panel or stacked installation.

Specifications

Dimensions: 28 × 10.9 × 29 mm
Weight: 0.018 kg
Temperature range: −40 to +55°C
Altitude: up to 60,000 ft

Contractor

Page Aerospace Ltd.

VERIFIED

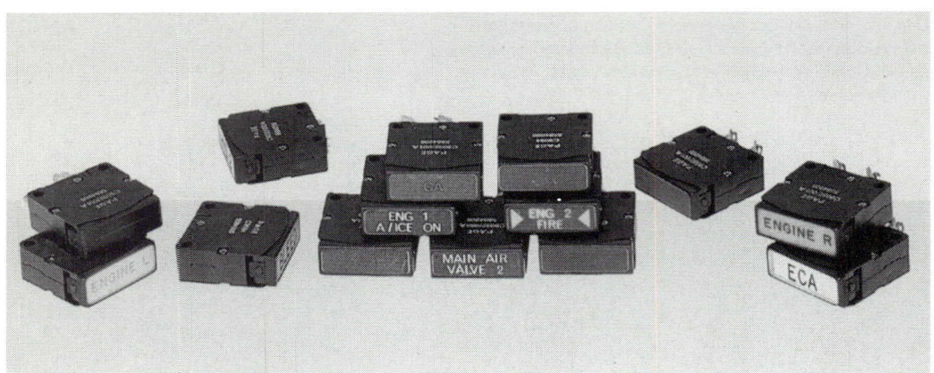

Page Aerospace two- and three-terminal light modules

Voice, tone and display warning systems

Page Aerospace designs and manufactures a wide range of alerting systems for civil and military aircraft. These include centralised and decentralised warning systems, synthesised programmable tone and voice systems, flight mode annunciators, attention alerting devices and indicator modules.

In recent years the company has developed a screen finishing process which results in high levels of sunlight legibility.

For night flying operations the company has developed a night vision goggle-compatible range of screens and bezels for displays and instruments. These screens effectively filter out the infrared content from illuminated cockpit devices and minimise flooding of the goggles.

Other areas of research and development cover high-quality synthesis of audio tones and voice. Units are currently being evaluated by the Defence Evaluation Research Agency, Farnborough and aircraft manufacturers in order to assess the optimum man/machine interface.

Operational status

Systems are now used in the central warning panel installations of the Saab 340, British Aerospace 146 and ATP and Raytheon Hawker 800 aircraft.

Contractor

Page Aerospace Ltd.

UPDATED

EICAS/EIDS Engine Instrument Crew Alerting System/Engine Instrument Display System

The EICAS/EIDS instruments use the latest technology to record engine conditions and provide indications on AMLCDs (Active Matrix Liquid Crystal Displays) that immediately alert flight crews to abnormal values.

Specifications

Dimensions: 134 × 135 × 242 mm
Weight: 3.4 kg
Power: 18-32 V DC, 50 W (nominal), 70 W (maximum)
Display area: 112 × 84 mm
Grey levels: 64

Contractor

Penny & Giles Aerospace Ltd.

VERIFIED

Engine Instrument Crew Alerting System (EICAS)
1997/0002457

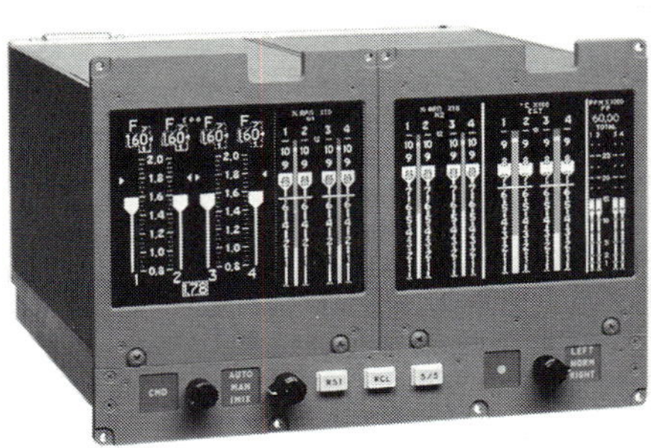
Engine Instrument Display System (EIDS)
1997/0002458

Ice and snow detection system

The ice and snow detection unit monitors ice conditions in low-speed aircraft. By having a fast response time and being totally independent of airspeed, this system is suited to helicopter applications.

The unit uses bleed-air from the compressor to cause air to flow through the detector even at zero airspeed. Information is displayed as liquid water content per cubic metre.

Specifications

Weight:
(detector) 0.9 kg
(meter) 0.34 kg
Airspeed: 0-250 kt
Altitude: sea level to 15,000 ft
Reliability: 10,850 h MTBF

Operational status

A version incorporating snow detection and digital outputs has been selected for the EH 101 helicopter where automatic control of the anti-de-ice systems is provided.

Contractor

Penny & Giles Aerospace Ltd.

VERIFIED

Penny & Giles ice detection system

In-Step digital technology step-motor aircraft engine instruments

B&D Instruments and Avionics Inc, a subsidiary of Penny & Giles Aerospace Ltd, has designed microprocessor-controlled micro-stepping motors to reduce the maintenance cost of aircraft engine instruments.

Both 2 in MS style instruments and 2 in ATI style instruments are available, to display the following engine values: torque, propeller speed (Np), gas generator speed (Ng), propeller oil pressure/temperature, engine oil pressure/temperature and exhaust gas temperature.

Contractor
Penny & Giles Aerospace Ltd.

VERIFIED

In-Step digital technology step-motor aircraft engine instruments **1997**/0002455

In-Step engine gas temperature (EGT) instrument **1997**/0002456

Groundspeed/drift meter Type 9308

The Type 9308 standard aircraft indicator size unit shows groundspeed within 3.5 kt at 100 kt or 5 kt at 300 kt. Drift is registered within 0.5° of true value. If the signal is lost, the last measured groundspeed and drift are displayed. The Type 9308 is compatible with Doppler 71 and 72 sensors.

Operational status
In production.

Contractor
Racal Avionics Ltd.

VERIFIED

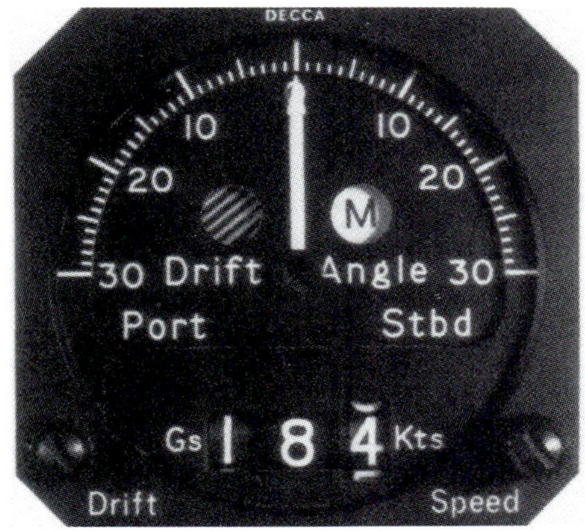

Racal Avionics Type 9308 groundspeed/drift presentation

The position bearing and distance indicator for the Racal Avionics Doppler 80

Hovermeter Type 9306

The Type 9306 standard aircraft indicator size unit displays along heading velocity over the range −10 to 20 kt with an error not greater than 1 kt and vertical velocity range ±500 ft/min with a maximum error of 40 ft/min. The Type 9306 is compatible with Doppler 71 and 72 sensors.

Operational status
In production.

Contractor
Racal Avionics Ltd.

VERIFIED

Hovermeter Type 80564B

The Hovermeter Type 80564B is similar to the Type 9306 in capability and appearance, but is compatible with Doppler 80 and 90 series sensors.

Contractor
Racal Avionics Ltd.

VERIFIED

VU2010 airborne display

The VU2010 display is a colour raster display for airborne applications which meets the exacting requirements of the Royal Air Force and French Air Force E-3 Sentry aircraft. The display is a high-resolution unit with a 20 in diagonal (508 mm) tube and a 0.26 mm pitch shadow-mask. The high performance is obtained by the use of digitally controlled dynamic convergence circuits. Similarly, digitally controlled dynamic focus and active colour purity controls are also incorporated to provide outstanding picture quality. The VU2010 is fitted with a magnetic shield, flashover protection and implosive protection.

Specifications
Dimensions: 464 × 384 × 510 mm
Weight: 53.5 kg
Power supply: 115 V AC, 400 Hz, 3 phase
Reliability: 2,780 h MTBF as MIL-H-217D

Operational status
In production and in service in Royal Air Force, NATO and French Air Force E-3 Sentry AWACS aircraft. Some 56 monitors have been ordered (with an option for a further 19) by the Japanese government under a £2.2 million contract, for installation in Boeing 767-200 AWACS aircraft.

Contractor
Racal Radar Defence Systems Ltd.

VERIFIED

Racal produces the VU2010 colour monitor for the UK, NATO and French E-3 Sentry

Ruggedised and long scale meters

The SIFAM range of ruggedised and long scale meters is available to existing designs or can be modified to user specifications. Movements can also be incorporated into custom enclosures or be supplied for assembly. Meters are available in three sizes: 24, 27, and 37 mm diameter, with fully magnetically shielded movements, sapphire jewels, and tungsten carbide pivots. Accuracy from 1.5 per cent.

Operational status

Selected for Westland EH 101 helicopter.

Contractor

SIFAM Limited.

VERIFIED

SIFAM ruggedised and long scale meters
***1997**/0001361*

SARFIND cockpit display unit

The SARFIND cockpit display unit presents GPS data to the aircrew. It employs an aviation standard sunlight-readable vacuum fluorescent display screen which is able to display up to 40 characters across two lines. The unit is a standard aircraft instrument panel fit. An ARINC version is under development.

The display consists of two screens that scroll at 10 second intervals. Screen 1 gives latitude, longitude, quality of fix and age of the data. Screen 2 gives time of the fix and user identification. The display has a wide viewing angle with filtered display characters. In addition, there is an audio/visual alarm which is activated on receipt of position data.

Contractor

Signature Industries Ltd.

VERIFIED

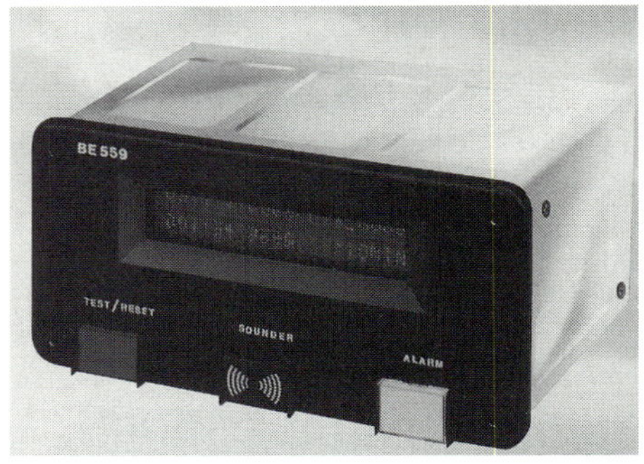

SARFIND cockpit display unit
1995

Colour Skymap and Colour Tracker

Colour Skymap provides a GPS-based colour moving map. Colour Skymap is a 'building block' platform which is capable of coping with forthcoming high-precision and fast update rate requirements for IFR navigation. Colour Tracker is the repeater equivalent of Colour Skymap, and has been specifically engineered to interface with existing GPS and LORAN receivers.

Both products can be installed into a standard 6.25 in radio stack, and will automatically display pilot position within 30 seconds of switch on, relative to geographic features, controlled airspace, airports, danger/restricted areas and other aeronautical data points. The heart of Colour Skymap and Colour Tracker is the high-resolution, ultra-high contrast, active matrix TFT colour LCD, which offers clarity over a wide viewing angle and varying light conditions. This feature, coupled with the ability to rotate the display through 180°, provides complete mounting flexibility within the cockpit, enabling installations to be optimised for left- or right-handed operation.

Colour Skymap and Colour Tracker both come with a worldwide navigational capability; the regions supported are the Americas, Atlantic International and Pacific International. Each front-loading PCMCIA data card contains the regional Jeppesen® data, the Skyforce geographic database and the entire operating software. It is this approach that allows Skyforce to provide periodic operating system enhancements during the normal Jeppesen® updating process.

In addition to their extensive user-defined waypoint storage and flight planning capability, Colour Skymap and Colour Tracker offer many other features, including simplified joystick data entry and aeronautical feature interrogation, user-defined airfields, split screen navigation modes including HSI, an on-screen CDI, auto zoom, extended track, emergency search, emergency airport minimum criteria, map customisation, map configuration (North Up or Track Up), choice of map datum and turning point arrival, airspace and marker annunciation.

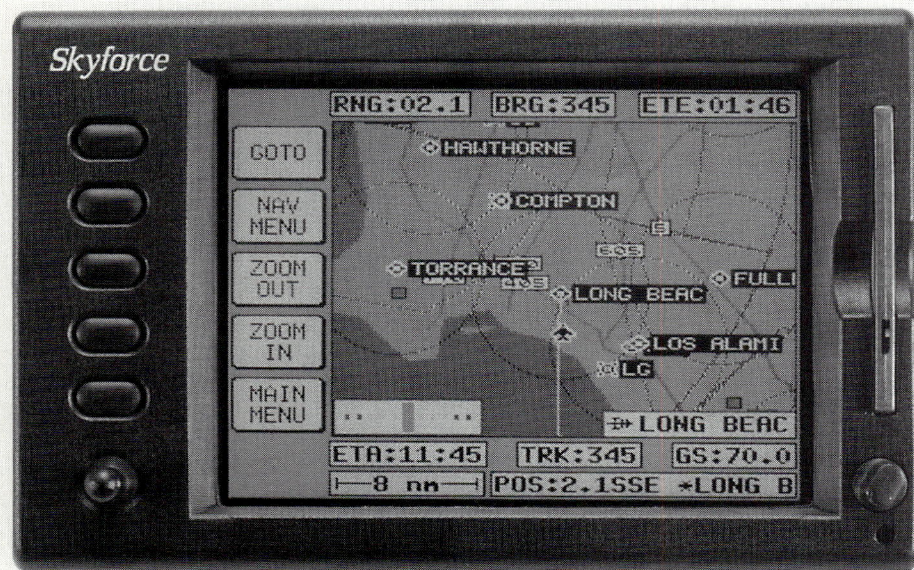

Colour Skymap and Colour Tracker
***1997**/0001362*

Specifications

Screen: 127 mm diagonal active matrix TFT LCD
Dimensions: 158 × 101 × 175 mm
Weight: 1.5 kg

GPS Receiver (Colour Skymap only)
Receiver: 8 channel parallel, simultaneous tracking
Acquisition:
12 s (almanac, position, time and ephemeris known)
43 s (almanac, position and time known)
Reacquire: 1.5 s
Accuracy:
15 m (without S/A)
1-5 m with differential option

Operational status

In service in both the general aviation and military aviation market. IFR approved.

Contractor

Skyforce Avionics.

UPDATED

Observer™ moving map task management system

The Skyforce Observer™ is a fully integrated moving map task management system that has been designed to create a paperless cockpit for an airborne observer in military, paramilitary, emergency or utility services operations. The system is capable of displaying and navigating over a number of highly detailed maps down to street/house level, whether they be in a raster-scanned or digital vector format. The maps are layered, allowing the customer to switch instantly from layer to layer. It is configured for GPS input.

Observer uses a standard 5 in colour display, which can be mounted permanently in the cockpit, or in a semi-permanent mounting, or simply be left free so that it can be passed around the cockpit. It can also be interfaced with other onboard monitors, larger displays, cameras, or datalink systems.

Using a special-purpose user defined facility, customers can introduce databases into the system via a PCMCIA card. These databases can be configured to hold any type of information whether it be text, pictorial or both, thereby allowing the user to store operation orders, daily tasking, or to highlight special waypoints. With a flight logging capability that can be used for post-flight analysis and a search-and-rescue function, Observer can be used for both operational and fleet management tasks.

Contractor

Skyforce Avionics.

NEW ENTRY

Specifications

	Display Head	Control Panel	Processor Unit
Standard configuration	5 in active matrix TFT LCD 320 × 240 pixels with 5 soft keys and joystick	combined PCMCIA card on/off control and display/ keyboard connections	133 MHz Pentium 3.2 Gbytes GPS input video output
Dimensions	160 × 112 × 60 mm	146 × 65.5 × 150 mm	75 × 170 × 230 mm
Weight (approx)	0.65 kg	0.6 kg	1.5 kg
Options	10 in active matrix TFT LCD, 640 × 480 pixels, instead of, or additional to 5 in display	keyboard; split PCMCIA card reader	video input; ARINC 429; MIL-STD-1553B

Skyforce Observer™ moving map task management system ***1998***/0018192

Skymap II™ and Tracker II™

The Skymap II/Tracker II system is a monochrome variant of the Skymap/Tracker product line, which can be knee, yoke, swivel, gimbal, panel or rack mounted. The monochrome reflective supertwist liquid crystal display providing sunlight readability and wide-angle viewing.

Skymap II provides GPS moving map precision navigation worldwide. Tracker II is the repeater equivalent of Skymap II, and has been engineered to interface with existing GPS and Loran receivers.

Skymap II and Tracker II include interchangeable data modules that contain regional Jeppensen data, Skyforce geographic data, and the entire operating software. This architecture permits Skyforce to offer periodic operating system enhancements within the normal Jeppesen updating process. A worldwide navigational capability is available in three cassettes to cover the Americas, Atlantic International and Pacific International. Operator facilities are similar to those listed for Colour Skymap/Tracker.

Specifications

Screen: 127 mm diagonal high contrast reflective supertwist back-lit LCD (128 × 240 pixels)
Dimensions: 158 × 115 × 35 mm
Weight: 0.65 kg

GPS Receiver (Skymap II only)
Receiver: 8 channel parallel, simultaneous tracking
Acquisition: 12 s (almanac, position, time and ephemeris known); 43 s (almanac, position and time known)
Reacquire: 1.5 s
Accuracy: 15 m (without S/A); 1-5 m with differential option

Skymap II™ Version 2.00+
The following additional features were added in summer 1997: Aeronautical data — Jeppesen Plus+; cartographic data version 2.00+; operating system version 2.00 Plus+ to offer a choice of three languages: German, French and English selectable from the set-up menu.

Operational status

Amongst many military and civil installations, Skymap II is fitted to the Hawk aircraft of the Royal Air Force Red Arrows Display Team. At the beginning of 1997, new features added included: HSI, E6B calculator and flight logging capabilities, and map detail was improved. In summer 1997, Skymap II was enhanced to the Skymap II Version 2.00+ standard.

Contractor

Skyforce Avionics.

NEW ENTRY

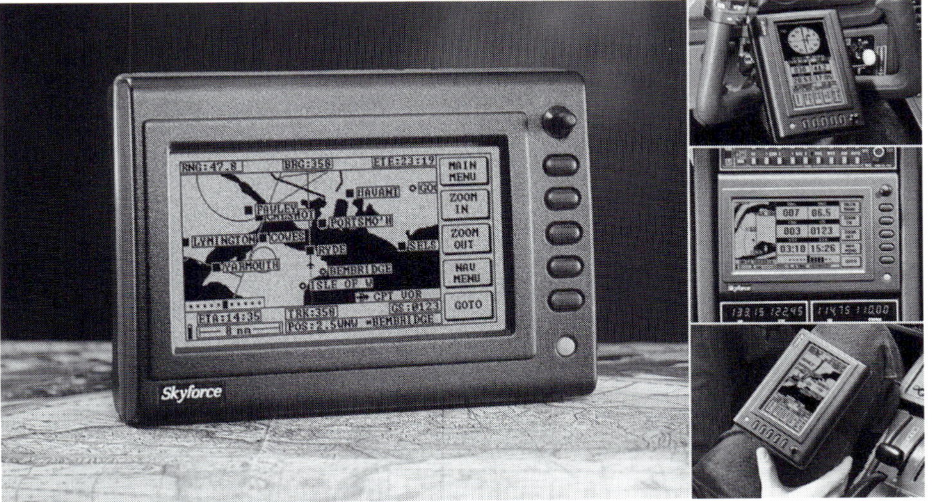

Skymap II and Tracker II display and optional mounting configurations ***1998***/0018190

Ruggedised flat panel LCD aircraft monitors

Skyquest Aviation makes a range of LCD aircraft monitors including the following sizes: 5.0 in, 5.8 in, 10.4 in, 12.1 in, 14.5 in and 17.7 in. Data provided here relates to the 12.1 in product.

The 12.1 in monitor features full PAL or NTSC capability. It is a high-brightness, sunlight-readable display, with switch selectable NVG option. It is suitable for both panel mounting and free standing installation.

Specifications

Viewing area: 12.1 in diagonal
Dimensions: 313 × 243 × 48.4 mm
Contrast ratio: >120:1
Brightness: 870 cd/m² (sunlight-readable)
Resolution: full PAL
Pixels: 800 × 600
TV lines: 730
Grey scales: 64
Colour capability: 262,144 colours
Operating temperature range:
(standard unit) –5 to +60°C

military unit (with heater) −40 to +60°C
military unit (with heater and cooler) −40 to >+60°C
Power: built in 28 V DC

Contractor
Skyquest Aviation.

NEW ENTRY

Some of Skyquest Aviation's LCD monitors.
1998/0018191

300 RNA series Horizontal Situation Indicators (HSI)

The 300 RNA range of Horizontal Situation Indicators (HSIs) is intended for civil and military fixed-wing aircraft and helicopters. Each instrument consists of a mainframe, synchro frame and electronics. Large-scale integrated circuits are used for signal processing and synchros are used to drive the various displays.

The instruments can be used in Nav, Tac, App or ADF modes. The range display is a four-digit electronic module at the top of the instrument face reading up to 999 n miles. Another numeric display is used to indicate the setting of the command track pointer. A knob is provided for selecting the relevant runway QDM on the command track pointer. These instruments provide complete ILS information as well as Tacan and ADF displays.

Operational status
304/305 RNA HSIs are used in the British Aerospace Nimrod and 748 aircraft as integral parts of the SFS6 flight systems. The 309 RNA HSI equips Royal Navy Westland Sea King helicopters and Royal Air Force British Aerospace Hawk trainers. The 306/307 RNA is used in the Sepecat Jaguar and British Aerospace Strikemaster. The 330 series HSI is used in the Royal Air Force's Panavia Tornado aircraft and Chinook helicopters.

Contractor
Smiths Industries Aerospace.

VERIFIED

2100 series MultiPurpose Colour Display (MPCD)

The Type 2100 MultiPurpose Colour Display (MPCD) meets the stringent requirements found in the bubble-canopy cockpits of combat aircraft. It provides a high-resolution full-colour display in all light conditions from low-level night to high-altitude day; special filters ensure full NVG-compatibility. The ruggedised CRT uses a stretched shadow-mask and a periodic focus electron gun to achieve very high screen brightness without loss of picture quality.

The MPCD presents data or imaging in stroke, raster or raster with stroke in frame flyback on a usable screen area of 5 × 5 in (127 × 127 mm). Stroke symbology is displayed as the output of an external graphics generator. Raster modes, in a variety of line standards, come from a range of sources such as electro-optical sensors, radar and video map generators. The hybrid mode enables the precise overlay of stroke symbology on to the raster display.

Operator control is exercised by 20 momentary action keys. Display control is via four rocker switches. Display brightness is determined by the operator and automatically compensates for ambient light measured by sensors on the front bezel.

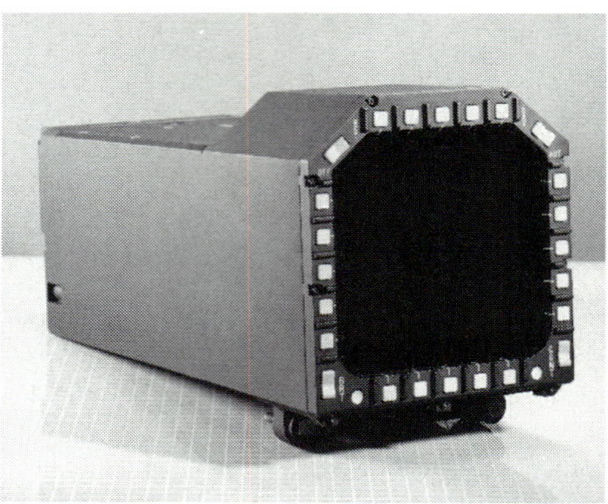

The 2100 series multipurpose display is in production for the AV-8B, F/A-18 and Harrier GR. Mk 7

Specifications
Dimensions: 170 × 180 × 440 mm
Weight: 10.9 kg
Power supply: 200 V AC, 400 Hz, 3 phase, 180 VA
Usable screen area: 5 × 5 in (127 × 127 mm)

Operational status
In production with over 1,500 units delivered. The 2100 Series complies with all relevant UK, NATO and US military standards and specifications. The unit is fitted in the US Marine Corps AV-8B Harrier II, US Navy/US Marine Corps F/A-18 C/D Hornet and Royal Air Force Harrier GR. Mk 7.

A version of the MPCD with a 6 × 6 in (152 × 152 mm) usable screen area is under development for the EF 2000. Other variants are under development or in production for the Tornado GR. Mk 4, RAF Jaguar and Italian/Brazilian AMX.

Contractor
Smiths Industries Aerospace.

VERIFIED

Type 3000 integrated colour display

The Type 3000 is a self-contained colour multipurpose display designed to be integrated into aircraft navigation and attack systems. All primary flight and navigation data is displayed, replacing conventional attitude and horizontal situation indicators, plus a route map representation. In addition, flight plan, system status including fuel information and engine data, radar warning receiver and maintenance information can be displayed under the control of the 17 soft keys around the front panel.

The unit combines a graphics generator, deflection amplifiers, high-voltage power supplies and a full shadow-mask CRT into a single display unit. The display data and control functions are transmitted via the aircraft MIL-STD-1553B databus. The latest large-scale integration and hybrid packaging techniques have resulted in a compact self-contained unit.

Specifications
Dimensions: 162 × 162 × 317 mm
Weight: 11 kg
Power supply: 115 V AC, 400 Hz, 3 phase, 200 VA
Usable screen area: 127 × 127 mm
Reliability: 1,500 h MTBF

Operational status
In production for the British Aerospace Hawk 100 and 200.

Contractor
Smiths Industries Aerospace.

VERIFIED

Active Matrix Liquid Crystal Displays (AMLCD)

Smiths Industries' flat panel Active Matrix Liquid Crystal Displays (AMLCDs) are compact, lightweight, high-resolution colour displays, designed to meet military and commercial requirements, for both new-build and retrofit aircraft and helicopter installations. Smiths Industries has production contracts for AMLCDs in all its display facilities, both in the UK and USA.

3 ATI Flat Panel Instruments
Selected for the JPATS aircraft, 3 ATI flat panel instruments provide a cost-effective, state-of-the-art, means of providing graphics in a high-visibility display. Features include: construction based on five modules (one dispay card and four flexible I/O cards); up to 16 colours at 130 foot Lamberts illumination, reconfiguration by pilot-operated switch, ARINC 429 interface, NVIS option. Applications include air data instruments, engine instruments, system status displays and standby displays.

These displays have been selected by Raytheon for the JPATS aircraft.

3 ATI Self-sensing flat panel instruments
3 ATI self-sensing flat panel instruments are designed for standby instrument applications. A wide range of instruments with integral air data and/or altitude sensors complement glass cockpit systems. Digital input options are available.

5 ATI flat panel instruments
Smiths Industries' 5 ATI design provides a fully self-contained solid-state instrument capable of directly interfacing to either analogue or digital sensors and presenting primary flight and/or engine information using graphics symbology. The 5 ATI flat panel design uses standard mountings, with no additional requirements for remote symbol generators or cooling air – they are, therefore, suited to upgrade and retrofit

***1997**/0001367*

***1997**/0001368*

***1997**/0001369*

Smiths Industries' 3 ATI airspeed indicator display (left), engine instrument display (centre), and standby altimeter and airspeed indicator (right)

EFIS requirements, as well as new-build installations. Features include self-contained interface and graphics processing; analogue and/or digital ARINC 429 or 1553B interfaces; EADI and EHSI formats; ADI/HSI and engine instrument formats; video option (for example HUD), GPS/CNS/ATM compatibility; passive cooling.

Applications include retrofit and OEM installations, standby displays, message displays, video monitor displays.

These displays have been selected for the Royal Australian Air Force Hawk aircraft. Each twin-cockpit aircraft will have up to six identical display units to show flight, navigation, weapon and system symbology, plus digital map and sensor displays.

Multifunction Control Display Unit (MCDU)

Smiths Industries' flat panel MCDU employs a large full-colour AMLCD which complements modern military and civil cockpits. Features include: flexible reconfiguration of front panel keys; passive cooling; alternative ARINC 429, MIL-STD-1553, RS-422 databusses. Applications include: GPS; CNS/ATM; ACARS; weather radar; FMS; Satcom/radio control; video, cockpit and maintenance displays

Selected by the UK MoD for the Nimrod 2000 aircraft upgrade.

5 × 6 in Electronic Display Unit (EDU)

Smiths Industries' 5 × 6 in Electronic Display Unit (EDU) presents engine and utility system parameters on a multicolour, flat panel AMLCD.

MultiFunction Glareshield Display (MFGD)

Smiths Industries' MultiFunction Glareshield Display (MFGD) is designed for glareshield installation to provide the pilot with continuous peripheral awareness weather in head-up or head-down attitude. The unit utilises full-colour AMLCD and is suited to the display of tactical messages in the CNS/ATM environment. Boeing has selected this display as a Para Visual Director (PVD) for the B777.

7 in Cube—Smart Graphic Display Unit (SGDU)

Smiths Industries 7 in Cube — Smart Graphic Display Unit (SGDU) comprises a full-colour AMLCD complete with integral interface and graphics processing. The SGDU offers video capabilities and intelligent processing with anti-aliasing graphic displays of databus information.

Smiths Industries' multifunction control display unit ***1997**/0001364*

Contractor

Smiths Industries Aerospace.

VERIFIED

Advanced displays for Airbus

Over the years 1997-99, Airbus Industrie will certify FANS systems for A319s, A320s, A321s, A330s and A340s incorporating Smiths Industries Datalink Control/Display Units (DCDUs). The DCDU is a development of the company's existing AMLCD technology described previously. The DCDU is a cost-effective way of providing datalink messages to the pilot. There are plans to incorporate this unit into many aircraft and as retrofits to avoid costly modifications to the current display systems.

Contractor

Smiths Industries Aerospace.

VERIFIED

Display Processor/Mission Computer

The wholly modular Display Processor/Mission Computer (DP/MC) can be configured to meet customer requirements. It is based on an extensive library of standard electronic cards, power supplies and ATR short cases. The standardised internal databus allows the unit, and hence the system, to be updated or expanded by replacing or inserting the appropriate cards. This flexibility enables improved components to be incorporated as technology evolves, including faster and more numerous microprocessors and, for example, the adoption of surface-mounted components.

The unit meets military specifications and is suitable for computing and symbol generation applications for head-up displays, weapon aiming, navigation, mission and system control and electronic flight instruments.

Various high-speed 32-bit and 16-bit processor cards are available in the library of modules. Other modules provide global memory, raster and/or stroke display generation for either head-up or head-down displays and extensive analogue, discrete and serial input/outputs. MIL-STD-1553B interfaces are also available as remote terminals and/or bus controllers.

Specifications

Dimensions: ¼ ATR to 1¼ ATR
Example dimensions ¾ ATR short 190 × 193 × 384 mm
Weight: (¾ ATR short) 9 kg

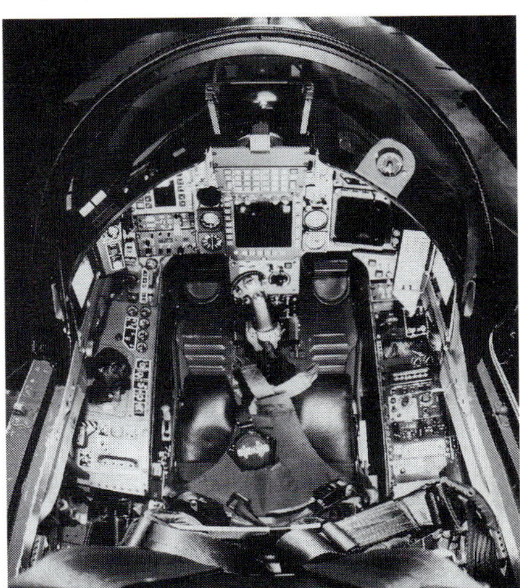

The front cockpit of the Hawk 100/200 with the Smiths Industries HUD with its data entry panel and the colour multipurpose display immediately below ***1995***

***1997**/0001365* ***1997**/0001366*

Smiths Industries' 5 ATI attitude director (left) and map/weather display (right)

Operational status

In production. Latest applications include the display electronic unit for the T-45A Goshawk, bus interface control unit, head-up display and weapon aiming computer and DP/MC for the Hawk 100 and 200, electronic unit for the F-5E avionics update and symbol generator unit for the EH 101 Merlin helicopter.

Contractor

Smiths Industries Aerospace.

VERIFIED

Electronic Instrument System (EIS)

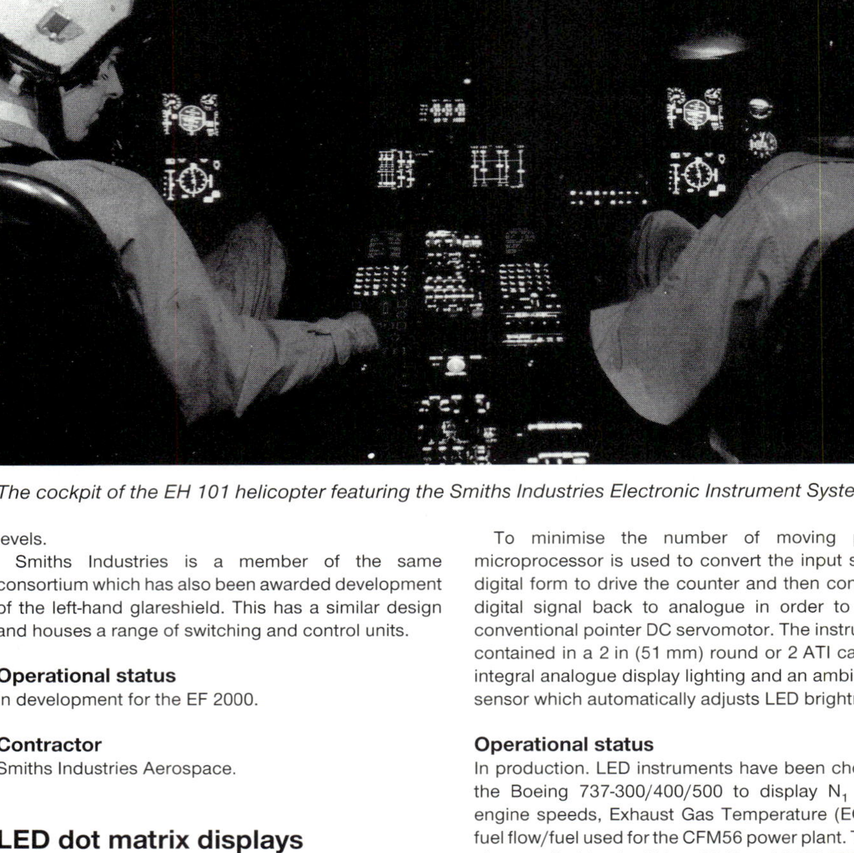

The cockpit of the EH 101 helicopter featuring the Smiths Industries Electronic Instrument System

The Electronic Instrument System (EIS) is a fully integrated cockpit instrument system which displays flight, navigation and aircraft systems data on full-colour shadow-mask CRT displays. The advanced symbology, operational flexibility, enhanced system redundancy and integrity is based on many years of experience in the design, development and production of electronic display systems. Cross monitoring validates both the input and the displayed data. In the event of a failure in any individual unit in the EIS, either pilot can immediately reconfigure the system to display the necessary information at the appropriate crew station.

The EIS for the EH 101 helicopter consists of six display units, three symbol generator units and three flexible display mode selector panels. The software developed by Smiths Industries will produce the different formats for all variants of the aircraft, including composite formats as determined by the reversionary display protocol. The aircraft will be provided with all critical flight, navigation and systems data. Program pins are used to indicate variant and sensor fit.

Specifications

Dimensions:
(display unit) ARINC B
(symbol generator unit) 6 MCU
Weight:
(display unit) 9 kg
(symbol generator unit) 8.4 kg
Power supply: 115 V AC, 400 Hz, single phase

Operational status

In production for the Royal Navy EH 101 Merlin helicopter.

Contractor

Smiths Industries Aerospace.

VERIFIED

EF 2000 glareshield displays

Smiths Industries has the project and technical leadership for development of the EF 2000's right-hand glareshield display. This contains a set of high-intensity LED displays used for standby engine and aircraft attitude information. These displays, which present analogue information generated digitally, have automatic compensation for changing ambient light levels.

Smiths Industries is a member of the same consortium which has also been awarded development of the left-hand glareshield. This has a similar design and houses a range of switching and control units.

Operational status

In development for the EF 2000.

Contractor

Smiths Industries Aerospace.

LED dot matrix displays

In parallel with the development of CRT-based electronic flight instrument system displays, Smiths Industries has produced a family of engine performance indicators using a hybrid DC-servo pointer/LED readout. The servo-driven pointer moves over a conventional circular scale, while an emissive electro-optic drum counter constructed from a dot matrix of high-brightness LED offers a display which is fully compatible with the CRT and light emissive displays now being introduced in civil and military aircraft.

The readout is of the rolling drum type, with values appearing to roll smoothly upwards or downwards. Compared with LED bar instrumentation, this device is much less affected by single LED failures and very accurate interpolation of data is possible due to the apparent motion in the display. A 7 × 5 matrix 4.5 mm high is used, with three blank rows of LEDs between each figure, but a 9 × 5 matrix giving 5.2 mm high characters and with reduced character spacing is an alternative. In place of a mechanical failure flag the LED matrix displays N_1, and N_2 and the word 'Fail' if an indication is outside limits or invalid.

To minimise the number of moving parts a microprocessor is used to convert the input signal to digital form to drive the counter and then convert the digital signal back to analogue in order to drive a conventional pointer DC servomotor. The instrument is contained in a 2 in (51 mm) round or 2 ATI case, with integral analogue display lighting and an ambient light sensor which automatically adjusts LED brightness.

Operational status

In production. LED instruments have been chosen for the Boeing 737-300/400/500 to display N_1 and N_2 engine speeds, Exhaust Gas Temperature (EGT) and fuel flow/fuel used for the CFM56 power plant. They are also supplied for the Airbus A310 and A300-600 aircraft and the Saab JAS 39.

Contractor

Smiths Industries Aerospace.

VERIFIED

LED engine and system displays

Smiths Industries solid-state instrument systems include primary engine displays, aircraft system displays and annunciator panel, and are direct replacements for electromechanical instruments. They are ideal for both new-build and existing aircraft, where they can be retrofitted by easy change of panels and without the need for mechanical or electrical modifications.

Each display system presents information in formats which are familiar to the aircrew who require minimal training in their use. These displays provide significant benefits in terms of weight, power consumption, reliability, ease of maintenance and cost of ownership.

Smiths Industries engine indicators for the display of engine speed (left), exhaust gas temperature (centre) and fuel flow (right)

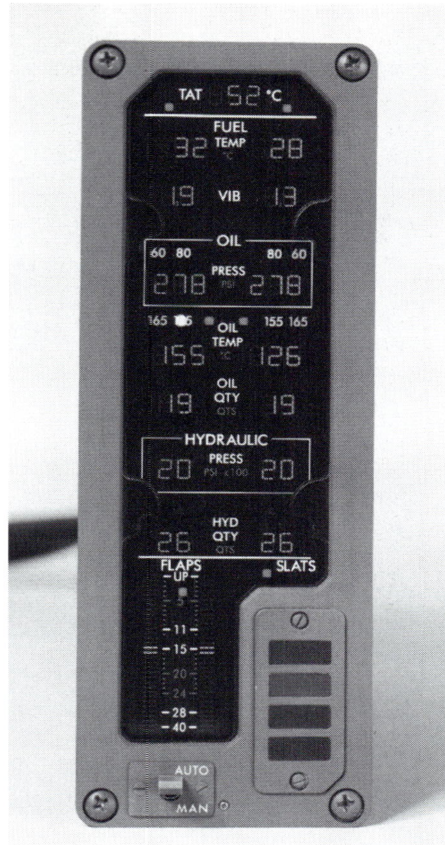

Smiths Industries' LED engine and system display

Primary engine display

The primary engine and aircraft displays are each contained in a single unit which, for ease of maintenance, comprises three modules: a display and associated dimmers, a printed circuit board assembly and a power supply module. Dial markings and legends associated with the displayed parameters are printed on a glass lens which is held in place over the LED displays. Both units are housed in conveniently sealed, lightweight cases with connectors mounted on the rear. Interconnections between the modules and rear connectors are by flexible tape wiring.

For optimum integrity the primary engine display incorporates a separate processor for each engine parameter. Four power supplies are provided to ensure system integrity. The power supplies are configured as two independent pairs so that a failure will result only in partial loss of the parameters of one engine.

Secondary or system display

In the secondary or system display the required reliability is achieved by using multiple processors and power supplies with the parameters to be processed suitably arranged between them.

Brightness is controlled automatically to achieve comfortable viewing over the whole range of ambient lighting conditions on the flight deck, from total darkness to direct sunlight.

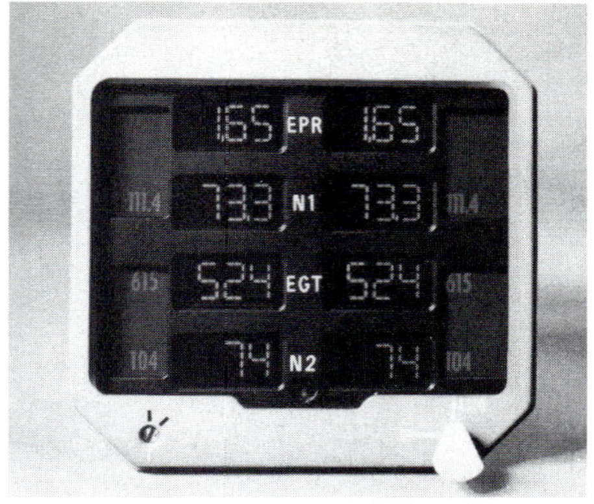

Standby engine display panel for the Boeing 757 and 767

Comprehensive BITE facilities are a feature of all Smiths Industries LED display systems. On power-up, a test sequence is initiated automatically. Test routines can be initiated by either the pilot or the maintenance crew.

Annunciator panel

The overhead annunciator panel is a solid-state replacement for existing discrete annunciator panels. Capable of displaying up to 120 cautionary messages, six warnings and 15 advisory indications, it can be installed in any MD-80 series aircraft without the need for electrical or mechanical modifications. On new-build aircraft in particular it makes possible considerable savings in wiring, connectors and weight. Caution messages are displayed on two dot matrix LED panels. Each panel can display six simultaneous cautions of up to 20 characters in length. Both message displays can be scrolled up or down so that a total of 120 different cautions can be viewed.

Eight push-button switches with integral filament lamps operate in conjunction with the two LED message panels. Each switch is associated with a particular aircraft system. Whenever a caution input is received the corresponding lamp is lit so that a system caution is annunciated even if both LED message panels are full.

Normally, cautions received from all aircraft systems are displayed on the message panels in either chronological or priority order as required. If the pilot prefers, he can use the push-button to select one particular aircraft system so that only cautions associated with that system will be displayed.

The two panels are arranged to display a block of 12 consecutive messages. If a fault occurs on one panel then all messages are automatically made available on the remaining panel while the faulty panel remains blank.

A new development of the annunciator panel is the master warning and caution system selected by for the MD-90. This system, comprising a separate display and control unit, identifies and presents more than 200 messages.

Operational status

Fitted in Boeing 737, MD-80 and MD-90, and British Aerospace 146 and Jetstream 41 aircraft. Also available as a developed system for C-130 aircraft retrofit.

Contractor

Smiths Industries Aerospace.

VERIFIED

LED standby engine display panel

Smiths Industries has been selected to supply a standby engine indicator display panel for the Boeing 757 and 767.

The display of engine parameters is by a light emitting diode in a seven-bar format and incorporates four numeric readouts for each engine — N_1 and N_2, EGT and EPR. Eight engine displays are housed in a 3 ATI case.

Operational status

In production for the Boeing 757 and 767.

Contractor

Smiths Industries Aerospace.

VERIFIED

Para Visual guidance Display (PVD)

The Smiths Industries automatic landing system programme included the development of the Para Visual guidance Display (PVD).

In addition to the Boeing 777, Smiths Industries PVDs have been installed in Hawker Siddley Tridents, Lockheed TriStars, Boeing 767s and 747s and MD-11s. PVD provides the pilot with essential information for ground roll guidance in take-off and landing in low-visibility conditions. The PVD indicator, which is mounted in the cockpit coaming of the 777 in front of each pilot, presents left-right director ('follow me') information by means of a rotating 'barber's pole' in the pilot's peripheral vision when looking forward.

The PVD is certificated as a back-up aid to the 777's automatic landing system and provides the essential visual guidance during take-off should the runway visibility unexpectedly fall below normal operating limits.

The latest implementation of the PVD principle uses a full-colour, Active Matrix Liquid Crystal Display (AMLCD). The use of AMLCD for this application means that in future a range of information may be presented in addition to the basic PVD 'barber pole' pattern. Potential future applications for this type of display include the provision of air traffic control information or aircraft-related system warnings using graphics, text or combined formats.

Operational status

In production for Boeing 777.

Contractor

Smiths Industries Aerospace.

VERIFIED

Flight deck warning system

Ultra has developed a microprocessor-based flight deck warning system which can be adapted for use on a wide range of aircraft types from small business and commuter aircraft to large airliners. The system consists of the central warning electronics, central warning panel and audio warning system.

The main electronics package monitors aircraft systems and sensors to provide the appropriate alerts to the flight crew via the central warning panel. The central warning electronics are made up of a number of rack-mounted printed circuit boards containing microprocessor-control, signal conditioning and input signal circuits. The number of boards and type of packaging can be varied to suit any particular application. Use of a microprocessor results in a system of compact size, low weight and high flexibility. It enables alert signals to be prioritised and reduces spurious alerts through the use of time delays. Changes to the system can be made by simply altering the software. An important feature of the central warning electronics is the self-test facility which can be activated during the preflight phase and is in continual operation during flight. When a failure is detected in the unit, a warning is given to the pilot on the central warning panel and BITE indicators isolate the fault to board level. A failure of the microprocessor does not render the system inoperable as warnings will still be indicated on the flight deck.

The central warning panel is a dedicated display incorporating hidden legend annunciators. It is constructed from identical display modules which can be assembled to suit the application.

Entirely separate from the central warning system, the audible warning system monitors critical aircraft functions and can provide up to 10 warning tones to the flight crew headphones. This high-integrity system has duplicated outputs and dual power supplies to minimise loss of warning under fault conditions.

Operational status

In service on the Avro RJ.

Contractor

Ultra Electronics, Controls Division.

VERIFIED

Aircraft moving coil indicators

Weston Aerospace manufactures a range of moving coil indicators for the aircraft industry, in varying case sizes from 1 in to 2 in, of modular construction, with integral lighting. Indicators are available to display parameters such as: electrical parameters, temperature, pressure, position, torque and speed. Options for indication of 1, 2 or 3 parameters on each instrument are available.

Specifications

Accuracy:
direct reading (mA or V)
±1.5% to ±4% full scale deflection
indirect reading (pressure, temperature) ±4.5% full-scale deflection
Display: black dial/white markings as standard; colour as required

Contractor

Weston Aerospace.

NEW ENTRY

Weston Aerospace aircraft moving coil indicators
1998/0018185

UNITED STATES OF AMERICA

Aerosonic's 2 in and 3 in instruments

Aerosonic produces a range of 2 in (51 mm) and 3 in (76 mm) instruments including: 2 in ASIs, 2 in altimeters, 2 in vertical velocity indicators, 3 in general aviation ASIs, altimeters and vertical speed indicators.

All indicators come in ARINC, military, round or clamp mounting and comply with TSO C-10b and TSO C-88a or military type 32/A requirements. Options include NVIS capability.

Contractor

Aerosonic Corporation.

VERIFIED

3" 3 POINTER

3" COUNTER DRUM

2" 3 POINTER

2" COUNTER DRUM

Aerosonics 2 in and 3 in instruments
1997/0001371

AH-64 caution and warning system

This system provides caution and warning annunciation for the AH-64 Apache helicopter. It consists of a 60 station pilot panel, 24 station co-pilot panel and master caution warning panel. Subsystem inputs are monitored and fault information displayed as applicable on any or all three units. The legend modules are fully sunlight-readable and incorporate front lamp replacement.

Specifications

Dimensions:
(pilot panel) 146 × 206.3 × 177.8 mm
(co-pilot panel) 146 × 95.2 × 177.8 mm
(master caution) 193.5 × 25.4 × 69.8 mm
Weight:
(pilot panel) 4.4 kg
(co-pilot panel) 2.36 kg
(master caution) 0.73 kg

Operational status

In service in the AH-64 Apache helicopter.

Contractor

Aerospace Avionics.

VERIFIED

Caution and warning advisory panel and cockpit lighting controller

The all-solid-state caution and warning advisory panel continuously monitors specific subsystem inputs, displays fault information on an integral 30-channel annunciator panel and drives external warning caution advisory displays. Twenty annunciators are yellow and 10 are red. The unit also provides power to 112 separate external 5.1 V constant current cockpit annunciators. The panel interfaces with an external dimming control to provide discrete dim and a continuously variable lamp power for the annunciators.

The panel is sunlight-readable in 10,000 ft candle

diffuse ambient light and is EMP- and nuclear-hardened.

Specifications

Dimensions: 200 × 72.4 × 279.4 mm
Weight: 3.4 kg
Power supply: 22-29 V DC and 6.5 V DC
Temperature range: −40 to +74°C
Reliability: 15,000 h MTBF

Contractor

Aerospace Avionics.

VERIFIED

Cockpit control panels (C-130J)

These cockpit control panels provide the functional interface between the operator and the applicable subsystems. They provide data, system status, controls and information displays.

Interfaces are available through MIL-STD-1553B databus, ARINC 429 databus or via discrete wire signals. The system is available with Night Vision Imaging Systems (NVIS), non NVIS-compatible lighting, and with or without edge lighting panels.

Examples of current control panels designs include: aerial delivery; air conditioning; automatic flight control system; bleed air; caution and warning; cursor control; fire handles; flight select; fuel management; heading/course select; hoist and winch control; ice protection; landing gear; lighting control; pressurisation; radar control; and reference set/mode select.

Operational status

In production for C-130J.

Contractor

Aerospace Avionics.

UPDATED

Cockpit control panels (C-130J) **1996**

CV-Helo fuel management panel

The microprocessor-based fuel management panel provides fuel management and control during refuelling and fuel transfer operations on board the SH-60F CV-Helo. Fuel flow activity of the auxiliary fuel system is monitored via inputs from system sensors and displayed with indicator lights. The unit automatically controls sequential refuelling and fuel transfer functions.

Specifications

Dimensions: 146 × 120.6 × 165.1 mm
Weight: 2.72 kg
Power supply: 28 V DC
5 V AC for panel edge lighting
Processor: 2 MHz Z80 microprocessor, 1,024 bytes RAM, 4,096 bytes PROM
Reliability: 23,000 h MTBF

Fuel management panel in the US Navy Sikorsky SH-60F helicopter

Operational status

In service in the US Navy Sikorsky SH-60F helicopter.

Contractor

Aerospace Avionics.

VERIFIED

F-18 cockpit lighting controller

The cockpit light controller is a multichannel solid-state device used to control the intensity of incandescent-type lamps used for the instrument lights, console lights, floodlights, master/caution warning lights, and caution/advisory lights in the cockpit of the F/A-18 aircraft.

Specifications

Inputs: 115 V AC, 400 Hz, 1 phase and 28 V DC
Outputs: 5.35 V AC, 7 V AC, 14 V DC

Contractor

Aerospace Avionics.

VERIFIED

Primary lighting control unit for V-22 aircraft

The Primary Lighting Control Unit (PLCU) directs power to the cockpit console lights and to the formation and proprotor tip lights on the V-22 aircraft. The cockpit console is divided into five zones each of which can be controlled independently or as a whole. Each zone contains two subchannels, one for powering electroluminescent panel lights and one for powering incandescent push-button lights. The tip lights and formation lights channels power electroluminescent lights. Dimming commands are received from the Cockpit Management System (CMS) via an ARINC 429 databus. The PLCU also interfaces with the aircraft Secondary Lighting Control Unit (SLCU) and provides status and Built-In-Test (BIT) information for both units to the CMS via the databus.

Specifications

Inputs: 115 V AC, 400 Hz, 1 phase and 28 V DC per MIL-STD-704
Outputs: 0-5 V AC and 0-115 V AC

Contractor

Aerospace Avionics.

VERIFIED

Primary lighting control unit for V-22 **1997**/0001373

ARINC 700 series digital integrated avionics

The ARINC 700 series Communications, Navigation and Identification (CNI) equipment began as part of the US industry-wide momentum in the 1970s to develop new technology systems for the transport aircraft to replace the Boeing 707 and DC-8. AlliedSignal was successful in winning the competition to provide the CNI suite for the Boeing 767 launched by United Airlines in July 1978 and the slightly later Boeing 757. The current ARINC 700 series consists of:

ALA-52A radio altimeter

8 × 10 in AMLCD display unit ***1998***/0018184

DFA-75A automatic direction-finder
RIA-35A ILS receiver
RTA-44A VHF communication transceiver
RVA-36A VHF navigation receiver
SMA-37A distance measuring equipment
TRA-67 Mode S transponder.

Operational status

In production and in service in Boeing 757 and 767 aircraft.

Contractor

AlliedSignal Commercial Avionics Systems.

VERIFIED

8 × 10 in AMLCD display unit

This is currently, AlliedSignal's largest ruggedised military and avionic crew station AMLCD MPD.

Specifications

Grey shades: 64
Resolution: 1,024 × 768 (XGA)
NVG: NVIS Class B (option)
Dimensions: 292 × 241 × 76 mm
Weight: 6.8 kg
Power: 28 V DC, 75 W

Contractor

AlliedSignal Commercial Avionics Systems.

NEW ENTRY

EFS 10 Electronic Flight instrument System

The EFS 10 electronic flight instrument system is designed to interface directly with the Gold Crown III multisensor KNS 600 flight management system and KFC 400 flight control system. It uses a dual-stroke and raster display writing technology to create symbols and backgrounds, allowing each stroke/raster module to concentrate fully on a single display to make it brighter and easier to read.

A typical five-tube EFS 10 installation consists of identical and interchangeable symbol generators for each pilot's EADI and EHSI, plus one for the multifunction display. This provides multiple reversionary options for all displays in the event of a failure. Display units are available in 5 × 6 in (127 × 152.4 mm) horizontal format or 5 × 5 in or 6 × 6 in (127 × 127 or 152.4 × 152.4 mm) square formats.

The EFS 10 includes full-time self-diagnostics and automatic internal testing of key circuits and sensor inputs, which may also be initiated by the pilot. A more comprehensive test routine allows checks down to circuit board level.

Contractor

AlliedSignal Commercial Avionics Systems.

VERIFIED

EFS 40/EFS 50 Electronic Flight instrument Systems

The EFS 40 is a 4 in (101.6 mm) electronic flight instrument system designed for operators of light jet and turboprop aircraft and turbine helicopters. The EFS 50 is a 5 in (127 mm) system. They are modular systems comprising an Electronic Attitude Director Indicator (EADI) and Electronic Horizontal Situation Indicator (EHSI). The package can be installed in two-, three-, four- or five- tube configurations.

The system's basic building block is the EHI. This unit incorporates a navigation mapping capability and modular design with interfaces for analogue and digital inputs. Added functions include a weather map display and joystick controller, plus a wide range of sensor input interfaces. The EHI is available with either a combination control/display unit or display with separate control panel.

The interchangeable ED 461/462 in the EFS 40 measures 106 × 106 × 238.8 mm and weighs 2.275 kg. The ED 551A in the EFS 50 measures 117.3 × 133.4 × 238.8 mm and weighs 3.76 kg. The remote SG 465 EADI/EHSI symbol generator unit fits into a ⅜ ATR short box. Control panels are offered as a built-in mode controller in the ED 461 or as the remote CP 468 mode controller for the ED 462 display unit. Standard interfaces are provided for most of the avionics, flying control and long-range navigation systems in current use in business aviation aircraft.

Contractor

AlliedSignal Commercial Avionics Systems.

VERIFIED

EHI 40 Electronic Horizontal situation Indicator

The EHI 40 is a 4 in (101.6 mm) electronic horizontal situation indicator which has been designed to offer reductions in size, weight and cost. It includes a built-in or separate mode controller, navigation mapping capability and modular design with interfaces for analogue and digital inputs. Added functions include a weather map display and joystick interface, and a wide range of sensor input interfaces. The system consists of a control display unit, symbol generator and associated navigation sensors. The EHI 40 is available with either a combination control display unit or traditional display with separate control panel.

During 1990 AlliedSignal introduced a companion attitude director indicator. The complete 4 in EFIS is designated EFS 40 (see previous item). The system is also available in a 5 in format, designated the EFS 50.

Specifications

Dimensions:
(display unit) 106 × 106 × 238.8 mm
(symbol generator) 193.5 × 57.2 × 320.5 mm
Weight:
(display unit) 2.275 kg
(symbol generator) 3.64 kg
Power supply: 28 V DC

Contractor

AlliedSignal Commercial Avionics Systems.

VERIFIED

Engine performance indicators

AlliedSignal makes a series of engine performance indicators for business, commuter and military aircraft. The indicators provide a continuous analogue display of critical engine variables, such as fan and core rotation speeds, turbine temperature and fuel flow. Configurations with circular or square cross-section are available.

Solid-state circuitry is employed throughout. The indicators are unaffected by the large fluctuations in electrical supply during engine starts.

Specifications

Power supply:
12-34 V DC, 120 mA nominal, 450 mA (max)
5 V DC, 250 mA for lighting
Input signals:
(rotation) monopole pulse
(turbine temperature) chromel-alumel thermocouple per NBS Monograph 125
(fuel flow) second harmonic selsyn mass flow transmitter, 115 V AC 400 Hz reference
Accuracy:
(rpm) (0 to +55° C ambient) ±0.25% at 100% rpm
(−30 to +70° C ambient) ±0.5% at 100% rpm
(temperature) (0 to +55° C ambient) ±5° C at 900° C
(−30 to +70° C ambient) ±10° C at 900° C
(fuel flow) (0 to +55° C ambient) ±1% of full-scale
(−30 to +70° C ambient) ±2% of full-scale
Rpm range: 0-110%
Turbine temperature range: +100 to +1,000° C
Fuel flow range: 0-2,300 lbs/h
Response: 3 s full-scale slew
Temperature range:
(Operating) −30 to +55° C
(short term) −30 to +70° C
(ambient extreme) −65 to +70° C
Shock: 6 *g* for 0.011 s
Vibration: 5-3,000 Hz, 0.02 in amplitude, limited to ±1.5 *g*
Humidity: 95%, +70°C for 6 h, cool to +38°C for 18 h

Operational status

In production. Typical applications are the Raytheon Hawker 125-700, Gates Learjet 35/36 and 55,

The Series 3 and EFS 10 electronic flight instrument systems on the Citation III

Lockheed Martin JetStar 2, Casa CN-235 and Taiwanese XAT-3.

Contractor

AlliedSignal Commercial Avionics Systems.

VERIFIED

FPD 500 Flat Panel Display system

The FPD 500 is an Active Matrix Liquid Crystal Display (AMLCD) that can be used directly to replace the four electromechanical primary flight displays. It has an adaptor designed to readily plug-in to existing electromechanical instrument wiring harnesses. Its form factor matches that of most 5 in instruments.

The FPD 500 can provide the functions for primary and secondary flight displays, navigation, engine displays and other multifunctional use. It can be used for a number of applications including: Attitude Director Indicator (ADI); Horizontal Situation Indicator (HSI); moving-map, radar/windshear display; or TCAS indicator.

Specifications

Dimensions: ARINC 408 5ATI
Weight: 2.61 kg
Power: 28 V DC, or 115 V AC/400 Hz; 42 W typical, 60 W max
Resolution: 192 × 256 DPI (min)
NVG-compatible TSO: DO-160C

Operational status

In production. FAA approved for B 727-200 aircraft.

Contractor

AlliedSignal Commercial Avionics Systems.

VERIFIED

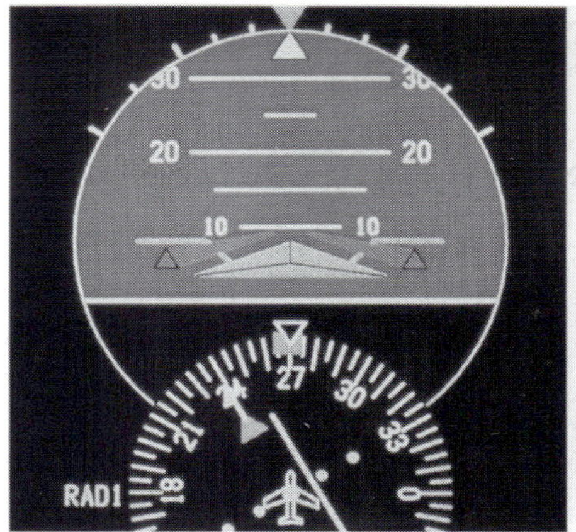

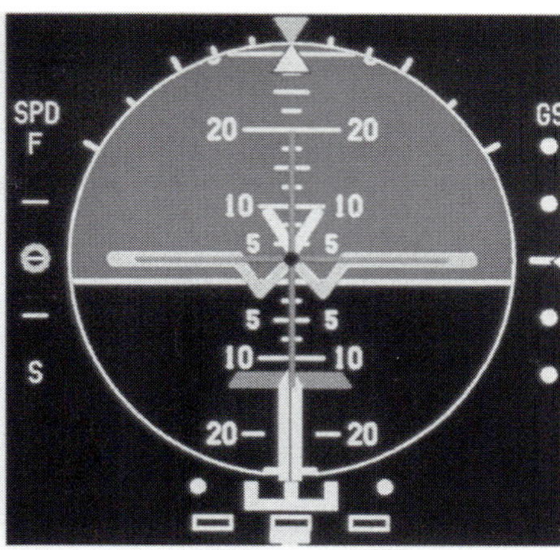

FPD 500 flat panel display system **1997**/0001377

Gold Crown nav/com avionics family

In the late 1960s the Gold Crown range of remote-mounted (that is, panel-mounted flight deck instruments driven by rack-mounted processing amplifiers) equipment was introduced for the larger piston and turbine twins then being developed and to complement the Silver Crown range. In 1972 a more advanced family of digital equipment appeared and began to take a large share of this market. The most recent range is Gold Crown III, unveiled in 1981. Designed for top of the line corporate and commuter twins, turboprops and turbine helicopters, this completely redesigned family incorporates custom LSI and microprocessor technology. The remote-mounted units are some 40 per cent smaller and 30 per cent lighter than their predecessors. Being of solid-state design the systems are substantially more reliable.

AlliedSignal has modified its Gold Crown III avionics to be compatible with night vision goggles. Existing gas discharge displays have been replaced with dichroic liquid displays. This line of equipment is often specified by overseas customers for use in military transport and training aircraft.

The Gold Crown III family comprises the following equipment:

KAA 955 audio amplifier/control
KDF 806 digital automatic direction-finder
KDM 706 distance measuring equipment
KFC 200 flight control system
KFC 250 autopilot/flight director system
KFC 325 autopilot/flight director system
KFC 400 autopilot/flight director system
KHF 950 HF single-sideband transceiver
KHF 990 HF SSB transceiver
KNI 582 indicator
KNR 634 digital navigation receiver
KNS 81 integrated nav/RNav system
KQI 553A pictorial navigation indicator
KTR 908 VHF communications transceiver
KTR 909 UHF communications transceiver
KXP 756 transponder
MST 67A Mode S transponder.

More detailed descriptions of some of these systems can be found under the appropriate section headings elsewhere in Jane's Avionics..

Operational status

In production and service.

Contractor

AlliedSignal Commercial Avionics Systems.

VERIFIED

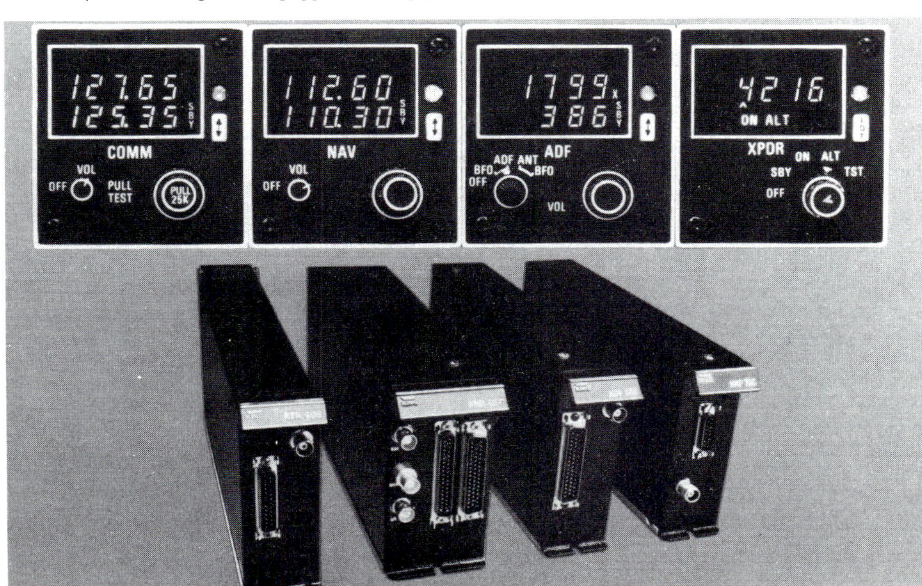

Some of the units in the AlliedSignal Gold Crown III avionics family

Horizontal Situation Indicator (HSI)

Designed for common installation with VSI and mode selector equipment, this HSI is suitable for a wide range of civil/military helicopter or fixed-wing applications. It can display aircraft heading, two simultaneous bearings, distance and course deviation. There are several electrical outputs for use in other aircraft navigation systems.

Specifications

Dimensions: 108 × 127 × 321 mm
Weight: 3.1 kg
Power supply: 115 V AC, 400 Hz, 11.5 VA plus lighting

Operational status

In production.

Contractor

AlliedSignal Commercial Avionics Systems.

VERIFIED

Series III Integrated Digital Avionics System

AlliedSignal announced the Series III integrated digital avionic equipment in 1982. Originally and informally known as the X-line, the system is now designated

Series III, indicating it to be the third range of digital electronic equipment developed by AlliedSignal. Based on ARINC 429 databusses, the system was also designed for compatibility with a new line of EFIS displays being produced by the company.

At the centre of the Series III system is the EFIS display, which is produced in three sizes - 6.25 in square (158 mm), 5 × 6 in (127 × 152 mm) and 4.75 × 5 in (120.7 × 127 mm) - and with what AlliedSignal claims to be a 'unique stroke writing method'. Typical configuration, at least for the higher-performance turbine types, would be two CRTs each for pilot and co-pilot and two shared EICAS displays for systems and navigation.

The current Series III family comprises:

DFS-43 automatic direction-finder
DME-44 distance measuring equipment
TRS-42 transponder.
VCS-40 VHF communication transceiver
VNS-40 VHF navigation receiver

More detailed descriptions of some of these systems can be found under the appropriate section headings elsewhere in Jane's Avionics.

Operational status

In production. An initial installation on board a Cessna Citation III was completed in December 1984, having been certified by the FAA earlier in that year. The five-system suite was chosen by Fokker in May 1985 as the standard communication, navigation and identification package for its Fokker 50 airliner.

Contractor

AlliedSignal Commercial Avionics Systems.

VERIFIED

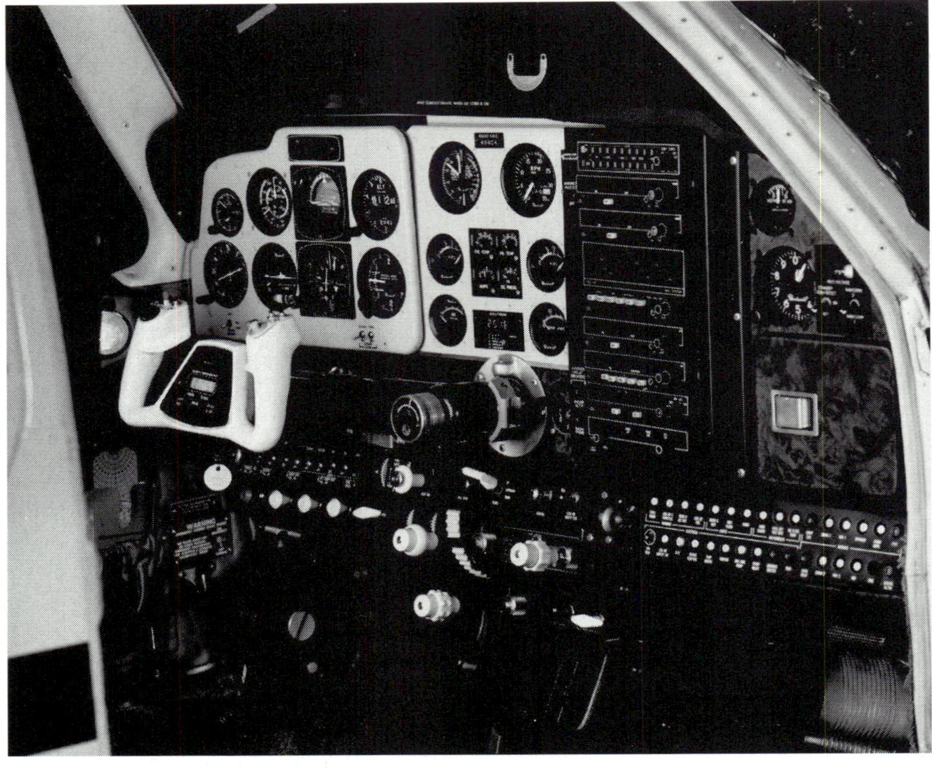

The cockpit of the Beech Bonanza with all AlliedSignal Silver Crown avionics

Silver Crown nav/com avionics family

The Silver Crown designation applies to a family of Communications, Navigation and Identification (CNI) and autopilot equipment for single- and light twin-engine aircraft up to turboprop size and for light helicopters. The various units are self-contained, with processing and presentation or indication mechanisms accommodated together in single panel-mounted boxes without the need for remote units. Initial members of the range appeared in the 1960s, and subsequent additions have made Silver Crown into a very comprehensive suite. In 1980 the series was upgraded by the introduction of some units incorporating digital technology.

The Silver Crown family consists of a large number of alternative units:

KA 134 audio control console/KR 22 marker beacon receiver
KAP 100 single axis autopilot
KAP 150 two-axis autopilot
KAP 200 two-axis autopilot
KCN 90 GPS navigation system
KFC 150 two- or three- axis digital flight control system
KFC 200 two- or three- axis flight control system
KGS 55A compass system
KMA 24/24H audio control system with three light marker beacon receiver or five station intercom
KN 53 navigation receiver
KN 62A digital DME
KN 63 digital DME
KNS 80 integrated navigation system (VOR/Loc receiver, digital DME, digital RNav computer, glide slope receiver)
KNS 81 integrated navigation system (VOR/Loc/glide slope receiver, RNav computer)
KR 86 automatic direction-finder
KR 87 digital automatic direction-finder
KRA 10A radar altimeter
KT 70 Mode S transponder
KT 76A transponder
KT 96 radiotelephone
KX 125 nav/com transceiver
KX 155/165 nav/com transceivers
KY 92 VHF communication transceiver
KY 196/KY 197 communication transceiver.

More detailed descriptions of some of these units can be found under the appropriate section headings elsewhere in Jane's Avionics.

Operational status

In production for and in service with a wide range of light single- and twin-engined aircraft.

Contractor

AlliedSignal Commercial Avionics Systems.

VERIFIED

Vertical Situation Indicator (VSI)

The AlliedSignal Vertical Situation Indicator (VSI) is designed for installation with the company's HSI and a common mode selection panel. It has conventional ADI functions and uses a spheroid pitch/bank background. Roll and pitch trim is provided and indications of glide slope, localiser and navigation warnings are standard. There is also an inclinometer and a rate of turn pointer.

Specifications

Dimensions: 133 × 127 × 314 mm
Weight: 2.7 kg
Power supply: 115 V AC, 400 Hz, 8.5 VA plus lighting

Operational status

In production.

Contractor

AlliedSignal Commercial Avionics Systems.

VERIFIED

Colour AMLCD MultiPurpose Displays (MPDs)

4 × 4 in Colour AMLCD MultiPurpose Display (MPD)

The AlliedSignal 4 × 4 in MPD offers a complete range of formats, and is ideal for new build and retrofit requirements in fighter aircraft. High resolution, low reflectivity and excellent chromaticity is provided. VAPS software is programmable to meet customer requirements, with AMLCD technology. Simultaneous video and graphics capability is offered, with 15 grey shades (option 64), MIL-STD-1553B interfaces, NTSC output, and NVIS Class B compliance.

Specifications

Dimensions: 141 × 141 × 275 mm
Weight: 4.8 kg
Power Supply: 28 V, 95 W

6.25 × 6.25 in Colour AMLCD MultiPurpose Display (MPD)

The AlliedSignal 6.25 × 6.25 AMLCD is in Engineering Manufacturing Development for the US Army Longbow Apache helicopter, where the display system comprises four 6.25 × 6.25 colour AMLCD MPDs and two Colour Display Processors (CDPs). The CDPs can be populated selectively to a maximum of four completely independent channels, using a MIL-STD-1553 bus if desired. Full feature digital map system can be embedded within the CDP. The MPD provides 64 grey shades (growth to 256). Resolution is 512 × 512 colour pixels in Quad RGGB Colour Pixel

Four of AlliedSignal's 6 × 8 in AMLCD display units (two pilot and two co-pilot) in cockpit configuration
1998/0018181

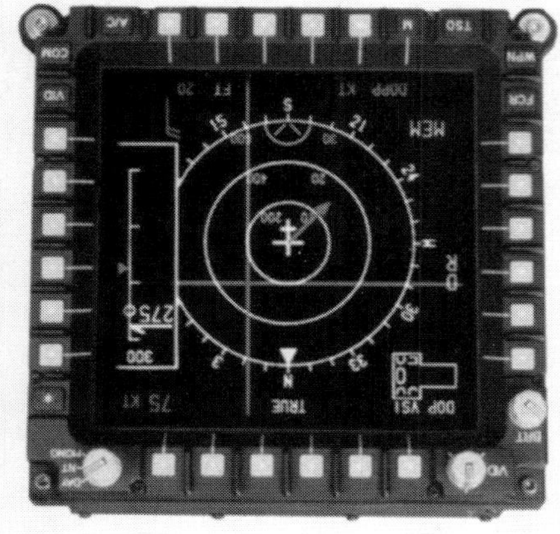

6.25 × 6.25 in colour AMLCD multipurpose display ***1998***/0018180

arrangement. The MPD is NVIS Class B compliant, and tested with Class A goggles in mono green mode.

Specifications

Dimensions: 216 × 216 × 184 mm
Weight: 5.8 kg
Power supply: 115 V 3-phase 400 Hz, 115 W

Operational status

First production displays will be installed on US Army Longbow Apaches in March 1998 and will be standard for all Netherlands and UK Longbow Apaches. A total or more than 4,000 MPDs will be required for the Longbow Apache programme including remanufacture of all US Army AH-64s.

6 × 8 in AMLCD Display Unit

The AlliedSignal 6 × 8 in AMLCD has been designed for use on the C-141 aircraft upgrade, where it acts as either a primary (ADI, HSI, Airspeed, Altitude display) or secondary (Heading, Waypoint, Weather Map) flight display. A high performance graphics processor is used with a 486DX2 general purpose processor; the Virtual Application Prototyping System (VAPS) permits the drawing and simulation of formats in near real time for format changes. Resolution is 480 × 640 colour pixels (RGGB Quad) with 80 colour groups/in. The display is NVIS Class B compliant, with NVIS Class A option. Fitment of 6 × 8 in AMLCDs forms part of the US Air Force C-130 and C-141 integrated avionics upgrade programme.

Specifications

Dimensions: 196 × 246 × 136 mm
Weight: 8.2 kg
Power supply: 28 V, 117 W

Contractor

AlliedSignal Electronic Systems.

UPDATED

Digital map reader

The digital map provides a multitude of navigational perspectives to enhance pilot effectiveness. These perspectives are provided through the use of an AlliedSignal proprietary graphics processor designed to produce three-dimensional images. The Graphics Differential Engine (GDE) makes use of Defense Mapping Agency Digital Terrain Elevation Data (DTED) to produce perspective as well as true three-dimensional views. The digital landmass data is stored in memory in a north up orientation. The GDE does the required algebraic and polynomial manipulations of the data required to create the image electronically. Through data compacting techniques, large areas of coverage are available.

In addition to DTED data, the digital map reader can process and display virtually any type of digitised database. This includes Defense Mapping Agency Digital Feature Analysis Data (DFAD), as well as digitised aeronautical charts, reconnaissance photographs, approach plates and checklists.

GDE was initially implemented in standard CMOS logic, but developments have led to the conversion of the GDE into a VLSI chip.

4 × 4 in colour AMLCD multipurpose display ***1998***/0018182

The system processing is done via a MIL-STD-1750A processor which interfaces with the MIL-STD-1553B multiplexed database. It receives aircraft present position and mode control information and handles all overhead and built-in test functions in the digital map reader. The database is stored on a militarised optical disk.

Along with the two- and three-dimensional perspectives, the digital map reader provides numerous other capabilities. These include look ahead, continuous zoom, freeze frame, sun angle shading, terrain occulting, contour line generation, altitude colour coding, de-clutter and text overlay. In addition, the digital map reader can provide positional correlation algorithms such as SITAN for a complete navigational aid to the flight crew.

The output of the digital map reader is a red/green/blue composite video signal that can provide both variable aspect ratios and variable line rates. Embedded raster text can be added within the digital map reader and/or the composite video output can be fed into a display processor for additional stroke overlays.

Operational status

In development.

Contractor

AlliedSignal Electronic Systems.

VERIFIED

Moving map display for the F-15

The F-15 remote map reader provides a continuous colour or monochromatic high-resolution video signal to a multifunction display. It uses existing 35 mm film strips housed in self-indexing, interchangeable cassettes which automatically engage and align when inserted into the unit. The system can be mounted in the equipment bay, so reducing pressure on cockpit space. The map images appearing on the screen are therefore synthetic, as opposed to the real images projected on to the screen of conventional moving map displays.

The centre of the map image is automatically aligned with the aircraft present position using data from the navigation system. The unit incorporates a built-in test facility.

Specifications

Dimensions: 193 × 191 × 319 mm
Weight: 12.3 kg

Operational status

In service.

Contractor

AlliedSignal Electronic Systems.

VERIFIED

V-22 Osprey advanced colour display system

The system comprises four 6 × 6 in (152 × 152 mm) high-resolution colour displays driven by a pair of three-channel programmable display processors. These use an MIL-STD-1750A computer and feature triple redundant display processing.

Operational status

In development for the V-22 Osprey.

Contractor

AlliedSignal Electronic Systems.

VERIFIED

LM Series compact indicators

The LM Series indicators are a low-cost response to the FAA FAR 121.343 requirements for in-flight recording of engine parameters. Standard LM Series features include back-lighted dial and pointer, 360° dial arc, BIT, smooth pointer movements, modular construction for easy servicing, compact 3.5 in (88.9 mm) long design and 40,000 h MTBF.

The simple mechanical design incorporates only one moving part — the pointer — so there are no brushes or wipers to wear out. Surface-mounted PCB design helps to ensure long, trouble-free service life for the electronic components. The indicators are designed for display of primary and secondary engine parameters in retrofit as well as new production aircraft installations.

Optional features available include flight recorder inputs, high-accuracy redundant digital display, RS-232 interface, output warning signals and exceedance and trend monitoring.

Operational status

LM Series indicators are currently installed in Fokker F27 and Mitsubishi YS-11 aircraft.

Contractor

Ametek Aerospace Products.

VERIFIED

Sentinel instrument system

The Sentinel instrument system is designed to meet the need for fully integrated aircraft systems providing a complete solution from signal interface to cockpit display. The system consists of multiple cockpit display units and a dual-redundant Data Acquisition Unit (DAU). All signals are acquired and transmitted to the displays in digital format by both halves of the DAU which can be mounted in any convenient location in the aircraft.

Typical system applications include primary and secondary engine instruments and caution/advisory panels. Auxiliary functions such as fuel and electrical power management can be incorporated on additional selectable screens or can be displayed on a separate display.

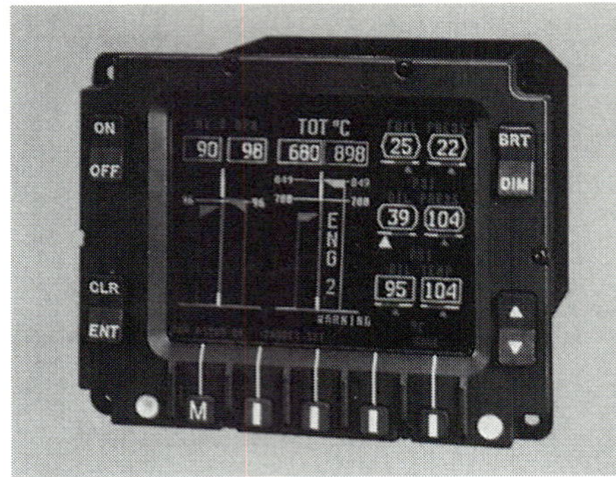

Sentinel instrument system is installed in the Agusta A 109 helicopter ***1995***

The 4.4 × 3.3 in (117.8 × 83.8 mm) active matrix liquid crystal display provides crisp full-colour graphics in a presentation which can be custom-designed for any application. Graphics and fonts can be created and changed without impact on hardware. Individual symbols and fonts can be selected from a large variety of existing styles or can be tailored to meet specific needs.

Bezel-mounted soft keys provide the pilot with access to lower level functions and allow input of information such as barometric pressure or target set points. Menu options direct the pilot to the desired information. Solid-state design provides high reliability to meet demanding requirements. The Sentinel can display up to 256 colours simultaneously, selectable from a palette of 4,096 colours. Colours can be easily selected and matched to a standard chromatic co-ordinate. Easily replaceable fluorescent lamps are used to provide a uniformly lit 125 ft-lambert average white brightness, suitable for viewing in direct sunlight.

Specifications

Weight: 2.49 kg
Power supply: 28 V DC or 115 V AC, 25 W without heater
Temperature range: −40 to +70°C
Display: 111.8 × 83.8 mm, 960 × 234 pixels

Operational status

Currently installed on the Agusta A 109 helicopter.

Contractor

Ametek Aerospace Products.

VERIFIED

Standby engine indicator

The standby indicator provides a display of four engine parameters for each engine. The digital displays are part of a translective LCD, having white characters on a black background. These are easily readable in direct sunlight, as well as at dusk or night, when a separate lighting circuit provides a high-brightness, high-contrast presentation.

A special electronics design eliminates digit toggling during static conditions and provides fast response and sequential counting during dynamic conditions. The display update rate is variable, causing the display to simulate a mechanical counter and giving the flight crew a sense of the rate change of the parameter. When a parameter reaches a programmable limit, the respective display will flash to communicate the warning. If more than one parameter is over limit, they will flash synchronously to avoid confusion.

Loss of signal for any of the inputs is detectable and results in a display of three dashes. Loss of power results in a blank display. BITE permits mechanics to determine if a fault resides outside the indicator. Thus a high mean time between unscheduled removals is achieved.

Specifications

Dimensions: 82.8 × 103.1 × 188.7 mm
Weight: 1.59 kg
Power supply: 10-32 V DC, 2.8 W per channel

Contractor

Ametek Aerospace Products.

VERIFIED

DC-1590 series magnetic Digital Compass

The DC-1590 series magnetic digital compass provides encoded heading data. The display is in red, sunlight-readable, 0.5 in (13 mm) high, three-digit numerals and features leading zero suppression. Brightness is adjustable and, to prevent display jitter, display rate is also fully adjustable. A test switch is provided to test all digit segments. Display readings are in 1° increments. No warm-up time is required.

The compass sensor unit is a high-quality liquid-filled instrument, fully illuminated for night viewing. All-solid-state to ensure reliable performance, it can be installed up to 38 m from the display. A set of East/West offset switches, mounted internally, provides unlimited compensation for magnetic variation. This remains in operation permanently until readjusted, allowing compensated operation to give true heading values. At the same time, a set of East/West offset push-button switches, mounted on the display, allows input compensation for variation changes.

The DC-1590 may be used to drive remote displays, autopilots, plotters and other electronic navigation equipment through a binary encoded decimal output connector.

Specifications

Weight: 4.5-6.8 kg depending on model

Contractor

Arc Industries Inc.

VERIFIED

DC-2200TM series magnetic Digital Compass

The DC-2200TM series magnetic digital compass is a twin display system. One display shows true heading, while the other shows magnetic heading. The operator can at any time add or remove any amount of variation for course correction.

Contractor

Arc Industries Inc.

VERIFIED

MFD 5200 colour AMLCD MultiFunction Display

Designed for the general aviation market, the MFD 5200 colour AMLCD display is a lightweight, low-cost version of those installed in jet transport aircraft. The system consists of a low profile 1.15 kg remote-mounted symbol generator and 1.3 kg colour AMLCD control display unit. The sunlight readable 5 in diagonal display provides situational awareness from onboard systems including navigation, airdata, engine and airframe, as well as environmental elements such as the weather. The basic function of the MFD 5200 is moving map navigation.

Position data input from any GPS or Loran shows as an airplane or helicopter icon over a digitised aviation

MFD 5200 showing moving map and Nexrad weather radar via WxLink
1998/0018179

chart. The flight plan automatically uploads from the Arnav STAR 5000 or FMS 5000 GPS navigators or AlliedSignal KLN 89-89B and KLN 90-90B GPS navigators, reducing pilot workload. It will interface to the next generation BF Goodrich WX-500 Stormscope®, to display thunderstorms, lightening strikes, and building storms in relation to the aircraft position and flight plan. It is FAA approved for depiction of broadcast Nexrad precipitation radar and METAR weather reports. An optional VIPER (Video InPut EncodeR) board converts the MFD 5200 from its digital moving map role, to a composite NTSC host for video or infrared camera display.

Operational status

Certified under a multiple STC, the MFD 5200 is now operating in Sikorsky, Bell, Mooney, Cessna, Piper, and Raytheon aircraft.

Contractor

Arnav Systems Inc.

NEW ENTRY

System 6 avionics system

System 6 is a series of products designed for integration into a complete avionics system. Products include flight management system, multi-engine display, datalink, dispatch terminal, engine monitor and data recorder.

The FMS5000 and FMS7000 flight management systems will complement any avionics suite and operate in conjunction with GPS and Loran C to provide position to within 3 m anywhere in the world. They feature Loran receiver only, GPS receiver only or both Loran and GPS receivers, differential GPS receivers, fuel/air data computer, ARINC 429 interface, single or tandem CDU operations, sunlight readable LED display, anti-reflective optic filter, NVG optic filter, function keys and turn/push knob control, Jeppesen worldwide NavData card and SAR GridNav mission management.

The multifunction display shows position on a moving map and ranges from high-contrast monochrome to full-colour panel mount displays.

The datalink provides the capability to send and receive position reports, encrypted or open messages, text or graphic databases and monitored engine health data. It can track up to 400 ground vehicles and aircraft and integrates VHF, UHF, HF, microwave transmissions and satellite communications. A built-in differential GPS sensor provides high accuracy for position reporting.

The dispatch terminal is a central control facility that can be located on the aircraft or on the ground. Standard IBM PC equipment and System 6 software perform vehicle tracking, data communication, archiving and dissemination of data.

The engine monitor measures up to 70 piston or turbine engine components and environmental conditions on the aircraft. Actual operations and set tolerance exceedances are measured 10 times per second. This information can be shown on the multifunction display, transmitted via the datalink to a dispatch terminal and recorded on the data recorder.

The data recorder is a small, ruggedised aircraft quality computer capable of recording mission plans, position history, engine health and environmental conditions on a PCMCIA storage card for playback on a standard IBM computer.

Units in the System 6 avionics system **1995**

Specifications

Dimensions:
(FMS5000) 51 × 159 × 255 mm
(FMS7000) 57.2 × 139.7 × 127 mm
(MFD 5000 monochrome display) 120.6 × 159 × 226.1 mm
(MFD 5100 colour display) 120.6 × 159 × 226 mm
(MFD 5200 colour display) (see previous entry)
LRU 5010 computer/recorder) 500.8 × 159 × 235 mm
(radio transceiver/tracking unit) 184.2 × 69.9 × 190.5 mm

Operational status

The FMS7000 and datalink have been flight tested in an Erlanger Bell 412 medical transport helicopter.

Contractor

Arnav Systems Inc.

UPDATED

ACA Attitude Director Indicators (ADIs)

ACA 117410 ADI

Operating on the F-5, the 3 in (76 mm), two-axis ACA 117410 ADI provides aircraft attitude information. In the flight director mode it also provides pitch and roll steering information. Glide slope information is presented on a separate pointer. When the flight director is not used, the roll steering pointer displays localiser information. The unit is military qualified and hermetically sealed.

Operational status

In production and in service in the F-5.

ACA 129060 ADI

Presently used on the Black Hawk UH-60A, the 5 in (127 mm) ACA 129060 hermetically sealed military qualified ADI provides command as well as raw data information. The command pitch, roll and collective information is provided by Astronautics 3 cue flight director computer. Eyebrow annunciator lights provide go-around, decision height and marker beacon status. Glide slope, localiser and turn and slip information is also provided.

A similar unit used in the SH-60B Sea Hawk - the ACA 126370 - has demonstrated a MTBF in excess of 5,000 hours.

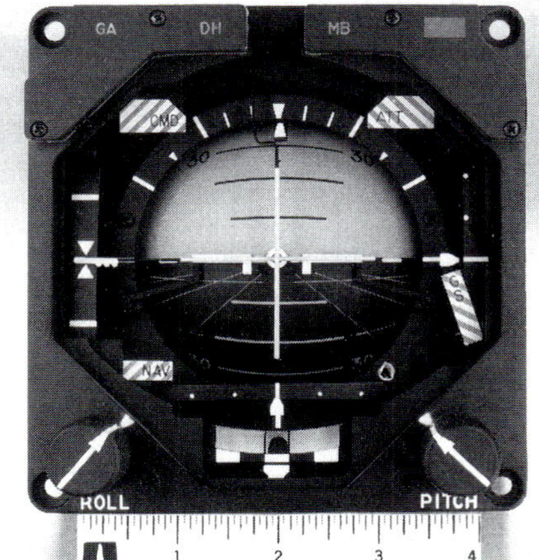

The ACA 129060 ADI as fitted in the Black Hawk UH-60A helicopter

Operational status

In production and in service in the Sikorsky UH-60A Black Hawk.

ACA 131070 ADI

The ACA 131070 4 in (102 mm), two-axis unit is fitted in the Cobra AH-1S. It has cyclic pitch and roll and collective pitch steering pointers. In addition, glide slope, localiser and turn and slip information is presented. The hermetically sealed unit is military qualified.

Operational status

In service in the AH-1S Cobra helicopter.

ACA 135160 ADI

Presently being built for the T-45, the ACA 135160 is a 4 in (102 mm), three-axis ADI. It provides pitch, roll, heading and turn and slip information. Localiser and glide slope information is also provided. The instrument is military qualified.

ACA 137100 ADI

The ACA 137100 3 in (76 mm), three-axis ADI is presently in use on the F-16 and the AMX. It provides pitch, roll and heading as well as turn and slip information. Vertical and horizontal steering pointers provide glide slope and localiser information. The servos are failure monitored and the unit is hermetically sealed and military qualified.

Operational status

In service in the F-16 and Alenia/Embraer AMX.

Contractor

Astronautics Corporation of America.

VERIFIED

ACA Horizontal Situation Indicators (HSIs)

4 in (102 mm) HSI for the Hawk

British Aerospace is the customer for the 4 in (102 mm) HSI which is fitted to export versions of its Hawk light strike/trainer aircraft. Major features include a course bar indicator, to/from indicator, glide slope pointer, two bearing pointers, course set knob, course selection window and digital readout of range which is compatible with ARINC 568 digital input. The instrument meets full MIL-SPEC standards.

Operational status

In production and in service in the British Aerospace Hawk.

ACA 113515 HSI for helicopters

The ACA 113515 HSI has proved to be a popular instrument for helicopters and has course bar indicator, to/from flag, glide slope pointer, two bearing pointers and course set knob with associated course selection window.

Operational status

In production and operational on Bell 212, 214, 412, Sikorsky S-76, S-61, Eurocopter Super Puma and Agusta AB 212, AB 412 helicopters.

ACA 123790 HSI

The ACA 123790 is a 4 in (102 mm) HSI to military specifications fitted in the Bell AH-1S Cobra helicopter. It has similar features to other members of the company's family of instruments, including a range readout compatible with ARINC 582.

The instrument will also accept direct digital input from Doppler navigation.

The Astronautics 4 in (102 mm) horizontal situation indicator for the British Aerospace Hawk

Operational status

In production and in service in the Bell AH-1S helicopter.

ACA 126370 HSI for the SH-60B

The ACA 126370 HSI is a 5 in (127 mm) instrument featuring standard ARINC 407 inputs and outputs with digital interface and extensive built-in test equipment. It was designed for the US Navy Sikorsky Sea Hawk SH-60B LAMPS helicopter programme and is currently in production and operational. An MTBF in excess of 5,000 hours has been demonstrated from operational service.

Operational status

In production and service in the US Navy Sikorsky SH-60B LAMPS helicopter.

ACA 126460 HSI

A 3 in (76 mm) HSI, the ACA 126460 has been supplied for the F-5 and F-16 as well as the B-1B. A slightly different version is used on the AV-8C and a type suitable for use in simulators is also available.

Operational status

In production and operational in the F-5, F-16 and B-1B.

ACA 130500 HSI

Similar in presentation to other company HSIs, the 3 in (76 mm) ACA 130500 is supplied for the US Marine Corps AV-8B. The instrument will accept direct digital input from a Tacan.

Operational status

In production and service in the US Marine Corps AV-8B Harrier.

Contractor

Astronautics Corporation of America.

VERIFIED

Airborne 19 in colour display

Astronautics has designed and produced a large colour display for airborne use. This display is a full militarised high-resolution colour monitor. It has a usable diagonal screen size of 19 in (482.6 mm) and a pitch dot size of 0.31 mm, a feature which contributes to the exceptional resolution. It is equipped with extensive self-test diagnostics for fault isolation and reporting.

The colour display has numerous applications such as for airborne command, warning and control centres, JTIDS operations and control centres and ASW operations. It can easily be adapted to other environments.

Operational status

In production. Qualified for C-130 aircraft and surface ships and submarines.

Contractor

Astronautics Corporation of America.

VERIFIED

Airborne multifunction CRT displays

Astronautics' airborne multifunction (CRT) displays feature:

(1) High-resolution, high-brightness displays viewable in 10,000 ft lambert ambient light
(2) Automatic or manual brightness and contrast control
(3) Raster-scanned stroke and stroke/raster CRT operation
(4) Contrast enhancement filters
(5) Comprehensive – BIT circuitry, and compatibility with Night Vision Goggles (NVG).

Operational status

In service.

Contractor

Astronautics Corporation of America.

NEW ENTRY

Specifications

	P/N 603737	P/N 602700	P/N 133000	P/N 133970	P/N 142000	P/N 133400	P/N 119450
Dimensions	138 × 155 × 475 mm	172 × 147 × 285 mm	168.4 × 200 × 305 mm	176.3 × 190.5 × 336.5 mm	217 × 292.2 × 319.4 mm	251.5 × 228.6 × 330.2 mm	254 × 342.9 × 457.2 mm
Weight	7.0 kg	6.36 kg	6.82 kg	6.82 kg	14.54 kg	18.18 kg	21.82 kg
Power	35 W/28 V DC MIL-STD-704D	30-45 W/28 V DC MIL-STD-704D	30-45 W/28 V DC MIL-STD-704D	30-45 W/28 V DC MIL-STD-704D	175 W/115 V DC 400 Hz	250 W/110 V DC 400 Hz	150 W/115 V DC 400 Hz
Display size	4.0 × 4.0 in	5.0 × 5.0 in	5.0 × 5.0 in	3.75 × 5.00 in	7.0 × 7.0 in	7.0 × 7.0 in	7.0 × 9.0 in
Brightness	200 fl	220 fl	220 fl	220 fl	160 fl	100 fl	120 fl
Scanning line or writing rate	raster 525 or 875 lines	raster 525 or 875 lines	raster 525 or 875 lines	raster 875 lines	stroke 70,000 in/s	stroke/raster 200,000 in/s	stroke 50,000 in/s
MTBF	<3,000 h	2,000 h	3,000 h	3,500 h	2,700 h	1,000 h	853 h

Area navigation control/display for the L-1011 TriStar

This system, installed on forward sections of the L-1011 throttle console so that it can be viewed and operated by both pilots, comprises a centrally mounted electronic automatic chart with identical Control/Display Units (CDUs) situated alongside a control unit and an electronic unit for the chart display. The CDUs present pages of alphanumeric information as required, while the electronic chart shows pictorially the route being flown, with positions of waypoints, destination airfields and navigation beacons marked.

The control unit has a scale selector that can shrink or expand the navigation situation shown on the chart, a slew control that can move the display bodily about the screen or rotate it (the alphanumeric symbols, however, remaining upright), mode selector permitting north up, track up or look ahead display and an RNav switch that can select data from one or other of the navigation computers on the aircraft. The chart can store 1,000 symbols and display 500 of them at any one time, it can also act as a back-up to the CDUs, presenting commanded alphanumerics as required.

The CDUs can display 12 lines of information with 17 characters per line and can access and display data stored in an aircraft computer, for example, route details, flight plans, en route navaid frequencies and terminal area procedures such as standard departures and arrivals.

The system incorporates continuous failure monitoring and self-test and the computing section furnishes a self-test pattern for complete end to end checks.

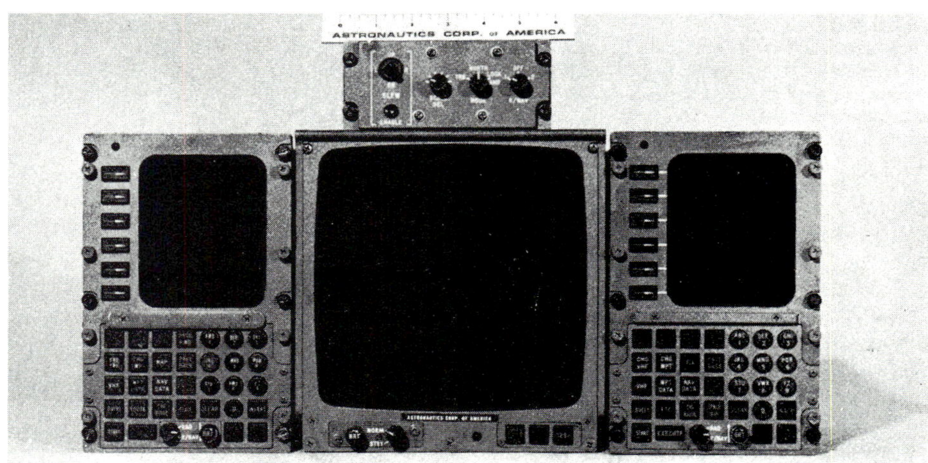

The Astronautics area navigation electronic chart for the L-1011 flanked by two control/display units

Specifications

Dimensions:
(control unit) 147 × 69 × 160 mm
(chart) 216 × 231 × 384 mm
(electronic unit) 124 × 196 × 368 mm
(CDU) 147 × 229 × 305 mm

Weight:
(control unit) 0.68 kg
(chart) 12.25 kg
(electronic unit) 6.35 kg
(CDU) 8.17 kg

Power:
(chart) 200 VA
(electronic unit) 90 VA
(CDU) 90 VA

Operational status

In service in the L-1011 TriStar.

Contractor

Astronautics Corporation of America.

VERIFIED

Colour AMLCD MultiFunction Displays (MFDs)

The Astronautics Corporation of America produces a range of colour AMLCD MFDs for the military and civil

The Astronautics electronic flight instrument in HSI mode **1998**/0018178

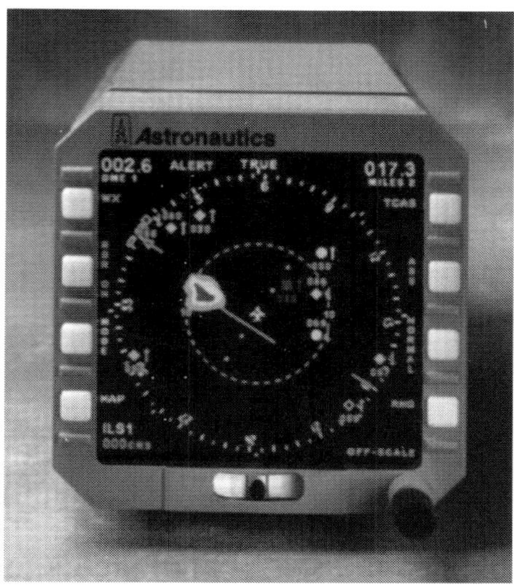

The Astronautics electronic flight instrument in TCAS mode ***1998***/0018177

aircraft markets, including: 3 × 4 in; 3.5 × 4.5 in; 4 × 4 in, 5 × 5 in; 8 × 6 in sizes. All are designed with modular construction, and all modules can be replaced without adjustment. Allmodels are NVG compatible, and provide high-resolution, high-brightness performance.

A specific application is the Astronautics 5 ATI 4 × 4 in Electronic Flight Instrument (EFI). It is pin-programmable as either an ADI or HSI. It also includes software that allows integration of other display formats such as map, TCAS, and colour weather radar. This instrument is TSO'd by the FAA and can be used in military or civil applications.

Operational status

The 5 × 5 in, 4 × 4 in, and 5ATI displays are utilised in A-4, F-5 and OV-10 aircraft; the 8 × 6 in touchscreen display is flying in a DC-10 in-service evaluation programme; and the EFI is flying in the revenue service on a United Airlines DC-10 aircraft.

Contractor

Astronautics Corporation of America.

NEW ENTRY

Display processor for the 530MG helicopter

The display processor designed for the Boeing Helicopters 530MG Defender generates independently programmable symbology for the multifunction displays, control/display unit and the TOW missile display. Heart of the display system is a 16-bit high-speed general purpose processor programmed in high-order language that manages all video, digital and analogue inputs and outputs, performs real-time computations and drives two independent symbol generators each controlling raster and stroke displays.

Specifications

Dimensions: 240 × 145 × 270 mm
Weight: 8 kg
Inputs: (digital) high-speed parallel, (DC analogue) symbol brightness, 525- and/or 625-line video per EIA standard
Outputs: 4 raster video: 1 at 525/875-line, 2 at 525 per RS-330, 1 at 875 per RS-343
Symbology: azimuth and target, direction scales, altitude and rate of climb, pitch, roll, TOW firing symbology, navigation and multifunction display switch annotation.

Operational status

In production and in service in the Boeing 530MG helicopter.

Contractor

Astronautics Corporation of America.

VERIFIED

Engine Performance Indicator (EPI)

The Engine Performance Indicator (EPI) is a dichroic liquid crystal display monitoring and displaying five critical engine parameters. It is designed to achieve full dual redundancy, be of minimum weight, volume and power, easily maintainable and have high reliability. In bright ambient light the EPI is illuminated by a reflective mode. At lower light levels the display is illuminated by a high-intensity source.

The EPI is designed to operate over a temperature range of −55 to +85°C and within an altitude band from sea level up to 70,000 ft. It meets the requirements of MIL-E-5400T and MIL-STD-810C.

Operational status

In production.

Contractor

Astronautics Corporation of America.

VERIFIED

E-Scope radar repeater display for the Tornado IDS

The E-scope radar repeater display portrays video signals from the Raytheon Systems Company terrain-following and attack radar in the Panavia Tornado IDS variant. In the terrain-following mode it shows a hyperbolic shaped symbol representing the ground ahead, above which the aircraft symbol remains if the autopilot is functioning correctly. Topographical information can be shown in the ground-mapping mode. For check purposes, an operator initiated test pattern can also be brought up on the screen.

Operational status

In service in the Panavia Tornado IDS.

Contractor

Astronautics Corporation of America.

VERIFIED

F-16 aft seat HUD monitor

The Astronautics Corporation of America's display unit is a compact high-resolution, high-brightness, lightweight, ruggedised monochromatic raster display presently in use in the rear cockpit of the two-seat F-16B and D aircraft.

The display is configured as a single scan rate unit of 525 lines at 30 Hz frame rate with single aspect ratio of 4:3 on a 4.125 × 5.5 in (105 × 140 mm) active screen size.

The display unit is modular. Each module can be replaced independently without adjustments to the display (including the CRT assembly replacement). Integral BIT continually checks the display circuitry, providing go/no go indications and immediate failure isolation. It also has an internal video test pattern generator.

The display has an illuminated control panel assembly and automatic brightness control circuitry to compensate for cockpit ambient light levels.

Operational status

In production and in service in F-16B and D aircraft.

Contractor

Astronautics Corporation of America.

VERIFIED

Helicopter integrated coloured moving map

The helicopter integrated coloured moving map is a high-performance low-cost system. It comprises the Astronautics modular display processor and multifunction display and provides the pilot with a coloured moving map, night and all-weather capability, nap of the earth capability and workload reduction.

The coloured moving map is stored in the modular display processor and has multiple layers including topography, tactical data, navigation data, flight guidance and obstacles. It has track up and north up modes and a zoom capability from 1:50,000 to 1:1 million.

Options include a helmet-mounted display, data gathering from sensors such as FLIR or CCD and Hands On Collective And Stick (HOCAS).

Contractor

Astronautics Corporation of America.

VERIFIED

Multifunction display

Astronautics' multifunction display is a high-resolution, high-brightness single unit system that accepts composite video data from external sensors and puts it on view on a raster-scanned CRT. Data can be put up as full grey-scale pictures or two-tone green and white alphanumerics. Push-button switches on the front panel select the display required from external sources. Each switch is illuminated for identification in darkness or low-light levels and the display has an automatic brightness control to compensate for changing ambient conditions. There are also manual brightness and contrast controls. Built-in test circuits provide go/no go indications of the unit's health and also other facilities to assist check-out using an external test set. The unit is self-contained, with its own power supplies and cooling provisions.

The system can be used to display television images from the Hughes AGM-65 Maverick air-to-ground missile so that the pilot can guide it to the target, duplicate the Head-Up Display (HUD) scene for a HUD camera and present radar information or data from a FLIR. It is also suitable as a primary ADI or HSI flight instrument.

Specifications

Dimensions: 200 × 168 × 305 mm
Usable screen area: 5 × 5 in (127 × 127 mm)
Weight: 6.59 kg

Operational status

In production. In service in the Agusta A 129 Mangusta and Boeing Helicopters 530MG Defender helicopters.

Contractor

Astronautics Corporation of America.

VERIFIED

Video display for the AH-64A helicopter

The video display for the AH-64A helicopter is a high-resolution, high-brightness CRT for displaying flight, navigation and weapon data. Information is passed to the visual display unit in the form of analogue composite video. Pitch and roll trim controls are provided on the front panel, together with a turn and slip indicator. Built-in test circuitry provides go/no-go status for the VDU and assists in checkout and fault location. The display is in use on the US Army AH-64A Apache helicopter.

Specifications

Dimensions: 186 × 152 × 318 mm
Weight: 6.59 kg

Operational status

In service in the AH-64A Apache helicopter.

Contractor

Astronautics Corporation of America.

VERIFIED

Colour AMLCD multifunction indicators

Avionic Displays Corporation manufactures a range of AMLCD, full colour, NVIS compliant, MultiFunction Display (MFD), military qualified, indicators; leading specifications are outlined below:

Operational status

In production and in service in military aircraft. In February 1998 Avionics Displays Corporation (ADC) and Universal Avionics Systems Corporation (VASC) entered into a business alliance agreement whereby VASC will have exclusive rights to market flat-panel integrated displays manufactured by ADC to corporate and commercial aviation, where they will be integrated with Universal flight management systems, and this new complete line of flat-panel flight displays (4 × 5, 5ATI, 5 × 6 and 8 × 10 in sizes) will provide a new-generation of avionic suites for new and retrofit applications, displaying primary flight and navigation data, multifunction data, and engine data.

Specifications

	Model 104	Model 550	Model 570	Model 640
Dimensions	8.3 × 6.2 in	4.4 × 3.3 in	4.5 × 3.5 in	5.1 × 3.8 in
Design purposes	split screen: HSI, ADI, video engine data warnings	HSI, ADI, video	EFI, HSI, ADI	HSI, ADI, video
Gray scales	64	256		64
Pixels	640 × 480 standard 800 × 600 available 800 × 600 available	640 × 480	640 × 480	640 × 480
Cooling	fan-forced cold wall construction	passive	passive	fan-forced cold wall construction
MTBF	7,500 h	8,200 h	8,000 h	8,200 h

Contractor

Avionic Displays Corporation.

NEW ENTRY

Model 104 AMLCD, MFD, video display unit ***1998***/0018176

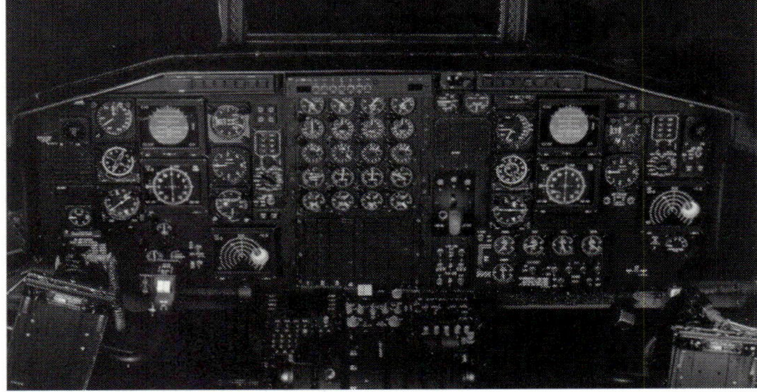

Portuguese Air Force C-130H cockpit upgraded with six Avionic Display Corporation AMLCD indicators ***1998***/0018175

Angle Of Attack (AOA) system

The Angle Of Attack (AOA) system is designed to assist pilots to obtain optimum aircraft performance on descent, approach and landing. It also provides useful data during instrument flying, in-flight turbulence, navigation and low-speed flight. The system consists of the AOA transmitter, AOA control panel indicator and AOA approach indexer.

The AOA transmitter is the foundation of the system. It senses the airflow direction at the side of the aircraft's fuselage. Specially equipped with heaters for de-icing and moisture control, the transmitter has a unique drainage feature to prevent water intake in the air or on the ground.

The AOA indicator is an easy-to-read indicator which displays aircraft lift information on a graded scale from 0 to 1.0. 0 represents zero lift and 1.0 represents 100 per cent lift, or stall. Combined with the system's flap position information, the AOA control panel indicator's display is valid for all flap configurations. The indicator also operates the indexer and the fast-slow pointer on flight director systems.

Mounted on the aircraft's glareshield, the AOA approach indexer is a three-light/three-colour unit that instructs the pilot on the best speed of approach based on the angle of attack.

Operational status

The system is fitted on US Air Force fighter aircraft, US Navy and US Marine Corps carrier-based aircraft and aircraft such as the Astra, Avanti, Beechjet, Citation, Falcon, Gulfstream, Jetstar, Sabreliner, Starship and Westwind.

Contractor

Avionics Specialities.

VERIFIED

TCAS/IVSI collision avoidance display

The collision avoidance/vertical speed indicator (TCAS/IVSI) unit gives collision avoidance messages that are easily understood. The dial of the indicator is fashioned like a traditional IVSI unit for easy recognition. The indicator's pointer and failure flag are prominent on the dial so they can be quickly seen at a glance. Resolution Advisories (RAs) are indicated in red and green segment lamps around the perimeter of the dial. Preventive RAs are indicated in red and corrective RAs in red and green.

Housed in a 3 ATI case, the TCAS/IVSI will replace current vertical speed indicator units. The TCAS/IVSI provides inertial led features plus integral lighting which conforms to most airline lighting requirements. The instrument requires connection to the aircraft's static pressure and electrical systems.

Using the TCAS/IVSI display as part of a collision avoidance system allows combination with equipment from other manufacturers to complete the TCAS II system.

Contractor

Avionics Specialities.

VERIFIED

ADI-330 self-contained Attitude Director Indicator

The ADI-330, with a full electrical erection system, combines the reliability and safety of a case-contained gyro with the convenience and operational features of a remote gyro. Ideally suited for large corporate and transport aircraft, the ADI-330 provides accurate, reliable pitch and roll information under all normal conditions. When used in conjunction with an emergency power supply, it also serves as an efficient, long-running standby attitude reference.

Electrical erection automatically provides 20°/min fast erect during initial power-up, eliminating the need for manual caging or uncaging during preflight checks. The unit also includes a bezel-mounted push-button for fast erect on demand.

The face presentation of the ADI-330 has been designed to be visually integrated into cockpits which use EFIS electronic displays. The style, colours and integral incandescent lighting of the instrument have been chosen to provide an attractive, consistent and familiar attitude information presentation.

Designed for 28 V DC, the unit provides useful attitude information down to 18 V and features power failure monitoring. Estimated MTBF for the ADI-330 is 4,500 operating hours and it is TSO C4c/C4d qualified.

Contractor

BFGoodrich Aerospace Avionics Systems.

VERIFIED

ADI-330/331 self-contained Attitude Director Indicator

The ADI-330/331 is a 3 ATI compact, completely self-contained instrument which features glide slope and localiser cross-pointers, integral inclinometer for slip and skid indication and mechanical erection with manual caging. The ADI-330 accepts analogue glide slope/localiser signals; the ADI-331 accepts ARINC 429 digital glide slope/localiser signals. Direct mechanical linkage eliminates electrical servo response lag.

The ADI-330/331 features 9 minutes of usable attitude information after complete power loss, 18-30 V DC power supply, blue/white internal lighting, Power Off warning and GS/Loc signal validity flags. The design is compatible with EFIS displays and the instrument is qualified to TSO C4c.

Contractor

BFGoodrich Aerospace Avionics Systems.

VERIFIED

ADI-332/333 self-contained Attitude Director Indicator

The ADI-332/333 is a 3 in (76.2 mm) compact, completely self-contained attitude director indicator featuring glide slope and localiser cross-pointers with back course, integral inclinometer for slip and skid indications and mechanical erection with manual caging. The ADI-332 accepts an analogue glide slope/localiser signal; the ADI-333 has an ARINC 429 digital interface. Direct mechanical linkage eliminates electrical servo response lag.

The ADI-332/333 features 9 minutes of usable attitude information after complete power loss, a selectable switch for BC/ILS/OFF modes, 18-30 V DC power supply, blue/white internal lighting, Power Off warning and glide slope/localiser signal validity flags and self-contained lateral and fore and aft acceleration compensation. The design is compatible with EFIS displays and the instrument is qualified to TSO C4c.

Contractor
BFGoodrich Aerospace Avionics Systems.

VERIFIED

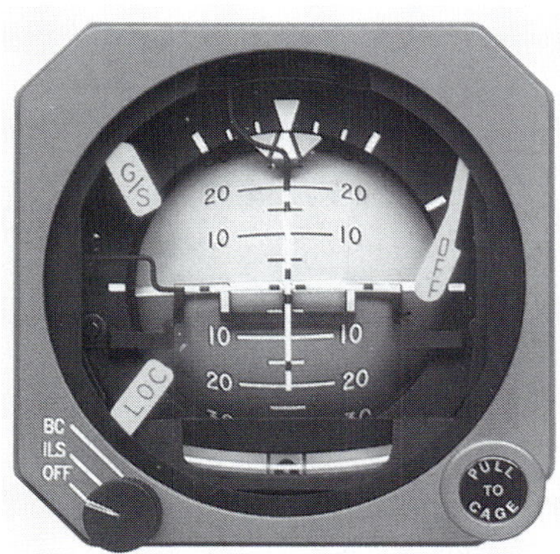

ADI-332/333 attitude director indicator **1995**

ADI-335 standby attitude and navigation indicator

The ADI-335 standby attitude and navigation indicator provides roll and pitch attitude information by electromechanical means. In addition, the instrument contains both en route VOR/DME and ILS navigation displays. Both the navigation modes contain appropriate validity flags and the indicator will interface with an ARINC 429 bus.

The ADI-335 is qualified to TSO C4c and C52a in accordance with DO-160C.

Contractor
BFGoodrich Aerospace Avionics Systems.

VERIFIED

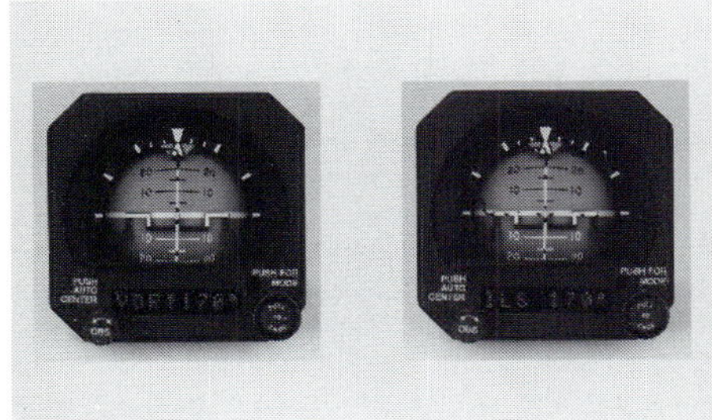

VOR (left) and ILS (right) displays on the ADI-335 standby attitude and navigation indicator

ADI-350 Attitude Director Indicator

The ADI-350 is a self-contained gyro indicator with flight director needles, rate of turn and slip indication. Synchro pick-offs are provided for remote indicators, radar stabilisation and flight control functions. Inputs are needed to operate the flight director needles and the rate of turn indicator.

Specifications
Dimensions: 82.04 × 82.04 × 228.60 mm
Weight: 2.72 kg
Power supply: 115/208 V AC, 400 Hz, 3 phase or 28 V DC

Operational status
In production. Applications include the Northrop Grumman A-6E, KA-6D and EA-6B, Boeing F/A-18, Canadair CP-140, Rockwell B-1B and Bell OH-58D. The ADI-350V has also been selected for use in the upgrade programme for the F-14 Super Tomcat.

Contractor
BFGoodrich Aerospace Avionics Systems.

VERIFIED

AI-803/804 2 in standby gyro horizon

The AI-803/804 is a 2 in standby gyro horizon for high-performance military and general aviation aircraft. The self-contained package eliminates the need for additional electronic components. After complete loss of external power, attitude information remains available for 9 minutes.

Specifications
Dimensions: 2 ATI
Weight: 1.1 kg
Power supply: 115 V AC, 400 Hz or 28 V DC

Operational status
In production and widely deployed in civil and military aircraft.

Contractor
BFGoodrich Aerospace Avionics Systems.

UPDATED

AIM 205 Series 3 in directional gyros

Low cost and light weight characterise the AIM 205 Series of 3 in directional gyros. They are precise flight instruments with a low drift rate, designed to provide the pilot with a constant azimuth reference free from the instability inherent in magnetic compasses. A high-speed electric rotor, mounted in a universal gimbal system, maintains angular momentum to overcome normal bearing friction and establishes a gyroscopically stable datum reference relative to space.

Specifications
Dimensions: 200.7 mm long
Weight:
(205-1) 1.13 kg
(205-2) 1.32 kg
Power supply: 14 V DC, 28 V DC or 115 V AC, 400 Hz, single phase
Temperature range: −30 to +50°C
Environmental: DO-160A

Contractor
BFGoodrich Aerospace Avionics Systems.

VERIFIED

AIM 510 Series 3 in and 4 in attitude gyros

The AIM 510 Series electric 3 and 4 in attitude gyros combine high performance with a ruggedised design, making them suitable for helicopters and high-performance aircraft requiring maximum reliability and long service life. They are available with pick-offs for radar stabilisation or autopilots, with a rotating or fixed dial.

Specifications
Dimensions: 222.25 mm long
Weight: 1.59 kg nominal
Power supply: 14 V DC, 28 V DC or 115 V AC, 400 Hz, single phase
Temperature range: −30 to +50°C
Certification: TSO C4c

Contractor
BFGoodrich Aerospace Avionics Systems.

VERIFIED

AIM 520 Series 2 in attitude gyros

Compact design and light weight characterise the AIM 520 Series electric attitude gyros. The 2 in case makes the series suitable for use as a secondary attitude gyro and for those aircraft where panel space is at a premium.

Specifications
Dimensions: 157.5 mm long
Weight: 0.77 kg nominal
Power supply: 28 V DC
(starting) 0.67 A
(running) 0.45 A
Temperature range: −30 to +50°C
Certification: TSO C4c

Contractor
BFGoodrich Aerospace Avionics Systems.

VERIFIED

AIM 1100 3 in self-contained attitude indicator

The AIM 1100 is designed as a form, fit and function replacement for the AIM Model 305. It is characterised by its lightweight and rugged design. Engineering enhancements include a proprietary bearing design

AIM 1100 3 in self-contained attitude indicator **1998**/0018174

improving rotor life and an improved pointer bar resulting in better performance and reliability under vibration.

It is designed for helicopters and general aviation aircraft operating in high duty cycle environments. It is available in 14 or 28 V DC form with an optional slip/skid indicator and fixed or trimmable pitch airplane symbol.

The AIM 1100 is certified to FAA TSO C4c, RTCA DO-160C, Section 8.0, Vibration Curves S and P.

Contractor

BFGoodrich Aerospace Avionics Systems.

NEW ENTRY

DG-700 Directional Gyro system

The DG-700 provides three operational modes: slaved DG, free DG and compass. The microprocessor-controlled gyro permits such features as extensive built-in test and instantaneous slaving.

Standard features include two isolated synchro outputs and autopilot interlock.

Specifications

Dimensions: 155 × 139 × 167 mm
Weight: 2.7 kg
Power supply: 115 V AC, 400 Hz

Operational status

In production. The DG-700 has been selected for the Royal Air Force Tucano trainer.

Contractor

BFGoodrich Aerospace Avionics Systems.

VERIFIED

GH-3000 electronic standby instrument system

The GH-3000 electronic standby instrument system replaces all three traditional standby instruments with a single, fully digital, flat panel AMLDC display. The 3 ATI size, self-contained inertial measurement cluster eliminates the need for a mechanical gyro. Designed to DO-160C, the GH-3000 requires 28 V DC and weighs 1.59 kg. External systems interfaces provide ILS/VOR/DME/FMS/magnetic heading and a compact remote air data computer at less than 0.62 kg supplies airspeed and altitude.

GH-3000 electronic standby instrument system **1997**/0001379

Specifications

Dimensions: 3 ATI
Weight: 1.59 kg
Power: 28 V DC
Cert: DO-160C

Operational status

TSO certification for conformance to the following standards granted in March 1997: TSO-C2d/-C4c/-C10b/-C34e/-C36e/-C113. Available for production delivery. Certified in Challenger 604, Falcon 50 Gulfstream IV and V, Raytheon Hawker 800XP and Eurocopter AS 365N2.

Contractor

BFGoodrich Aerospace Avionics Systems.

UPDATED

Cockpit 21 for T-45C Goshawk

Cockpit 21 was developed to replace the T-45A's analogue displays with digital displays similar to those found in the US Navy's F/A-18, AV-8B Harrier II and other advanced carrier-based jets. Aircraft equipped with Cockpit 21 are designated T-45C. Since students transitioning to these aircraft from Cockpit 21 will have already mastered cockpit information management skills and situational awareness, they can concentrate on the primary mission of learning how to perform key tactical manoeuvres.

Cockpit 21 in the T-45C Goshawk **1997**/0018164

Smiths Industries Aerospace has performed the full systems integration on Cockpit 21 and supplies the Head-Up Display. Cockpit 21 comprises a Display Processor Unit (PDU), Pilot-Display Unit (PDU), and Data Entry Panel (DEP). The DEP drives five display surfaces including the PDV and four raster Head-Down Diaplays (HDD) (monochrome multifunction displays sourced from Elbit Systems Ltd). The new cockpit provides navigation, weapons delivery, aircraft performance and communication data to both stations in the two-seat cockpit. The cockpit also includes a Global Positioning System/Inertial Navigation Assembly and a multiplex databus that will allow expansion of cockpit capabilities to accommodate changing training requirements.

Operational status

The first Cockpit 21 aircraft was delivered in October 1997 for testing of the production configuration at US Naval Air Station Patxent River. The second T-45C delivered in December 1997 went to US Naval Air Station Meridian for Training Wing US Navy plans call for the existing 72 T-45A Goshawk aircraft with analogue cockpits to be upgraded to the T-45C Cockpit 21 digital configuration. Smiths Industries will provide Boeing with 103 new digital Cockpit 21 assemblies by 2004. Retrofit kits for 84 T-45As are planned.

Contractor

The Boeing Company.

UPDATED

Windshear Alert and Guidance System (WAGS)

WAGS detects changes between air data and inertial data. When this rate of change becomes excessive the pilot will be alerted aurally and visually.

WAGS is active below 1,500 ft AGL. It is a stand-alone box that interfaces with the flight guidance system to provide flight director and thrust setting guidance. The system will couple with the autopilot and autothrottle systems to provide optimal flight path guidance through a hazardous windshear.

Operational status

In service in the MD-88 and MD-11. WAGS can be retrofitted to all MD-80 aircraft.

Contractors

The Boeing Company.
Honeywell Inc Air Transport Systems.

UPDATED

NSD Series Horizontal Situation Indicators (HSI)

Century offers four NSD Series HSI models:

NSD360A slaved HSI has all the features available in an HSI, matched with the simplicity of air gyro operation.

NSD360A slaved HSI with RMI bootstrap provides accurate heading information to a variety of other flight instrument displays.

NSD360A non-slaved HSI provides effective HSI performance.

NSD1000 HSI does not require a remote gyro, but combines the power of a remote gyro with the light weight, reliability and simple installation of a self-contained one-box instrument. Slaving is a standard feature. RMI bootstrap is optional.

The NSD HSIs feature 360° heading presentation, rectilinear course deviation indicator, full-view glide slope indicator, masking glide slope warning flag, 45° tick marks, referencing heading bug, failed gyro warning flag, free gyro mode, gyro caging knob, lost power warning flag, discrete nav warning flag and RNav and Loran compatibility. They also feature autopilot outputs for heading and course, continuously caged heading and course selection knobs, reference aircraft and heading lubber line, diffused incandescent perimeter lighting and course arrow with reciprocal indicators. The slaved models have built-in slaving indicator and automatic magnetic gyro slaving. The NSD1000 includes a built-in electric gyro.

The Century NSD1000 HSI (left) and NSD360A HSI (right) **1995**

Specifications

Dimensions: 85.6 × 85.6 × 220.7 mm
Weight: 2.09 kg
Power supply: 14 or 28 V DC

Contractor

Century Flight Systems Inc.

VERIFIED

Colour control display unit

The Chelton Avionics, Wulfsberg Electronics colour control display unit is designed to operate with the company's GNS-X flight management system. Compared with the control display unit it replaces, it features a larger display area with more lines and characters on a page, pages with titles, page numbers and the number of pages in the section, line select keys (in addition to cursor control), pilot control for parallax adjustment, and large- and medium-size characters. In addition, a dedicated Airborne Flight Information System (AFIS) button is included to speed access to AFIS pages.

Operational status

In production and in service.

Contractor

Chelton Avionics Inc, Wulfsberg Electronics Division.

UPDATED

Chelton Avionics, Wulfsberg Electronics colour control display unit

Onboard Aircraft Server and Information System (OASIS)

The Computing Devices International and Honeywell Onboard Aircraft Server and Information System (OASIS) integrates flight deck, cabin, maintenance and ground-based operations into a seamless communication architecture.

The airborne system is a high performance computing platform and display system capable of hosting multiple software applications. It provides interfaces to flight and cabin crew, and fixed or mobile maintenance terminals, as well as to various aircraft systems, including In-Flight Entertainment (IFE) and cabin management systems. The system concept implements an office LAN environment in each aircraft, with each aircraft existing as a virtual node on an airlines WAN system.

The separate onboard LAN environments are interconnected through an existing communications path (initially the airborne phone system) and establishes new communications paths to the ground such as direct broadcast satellite, wireless LAN system and tethered connection backup.

The airborne system uses COTS hardware and software components and a standard Intel Pentium processor that hosts the Microsoft Windows NT operating system. The airborne system consists of the following components: OASIS server with BITE; OASIS terminal(s); wireless LAN transceiver; wireless LAN antenna; installation kit.

Contractors

Computing Devices International.
Honeywell.

VERIFIED

IP-5110 high-resolution Digital Radar Display

The IP-5110 Digital Radar Display (DRD) converts radar into high-resolution raster scan format. Compatible with a wide variety of radars, the DRD has applications in airborne, shipboard and ground installations. Incorporating two digitisers, the DRD can provide two independent display presentations from the same radar. Graphic displays allow the overlay of ESM and other platform sensor data with radar displays, including ASW and ESM.

Input radar video is digitised at a 20 MHz rate. Overlay graphics are stored in four display planes with 16 levels of intensity. Output video resolution is selectable between 1,024 by 832 lines or 1,024 by 486 lines. Internal video matrix provides switching to video recorders or external monitors.

A trackerball allows the operator to designate radar returns quickly, to determine location in latitude and longitude and to measure range and bearing to or between multiple targets. Target types can be designated with Navy Tactical Data Systems (NTDS) symbols. Up to 64 targets can be designated.

Features include improved radar performance with digital processing; two independent digital scan converters; trackerball control; operation with any radar with pulse rates up to 10 kHz and antenna rotation up to 200 rpm; advanced measurement and display marking up to 64 concurrent symbol markers with independent

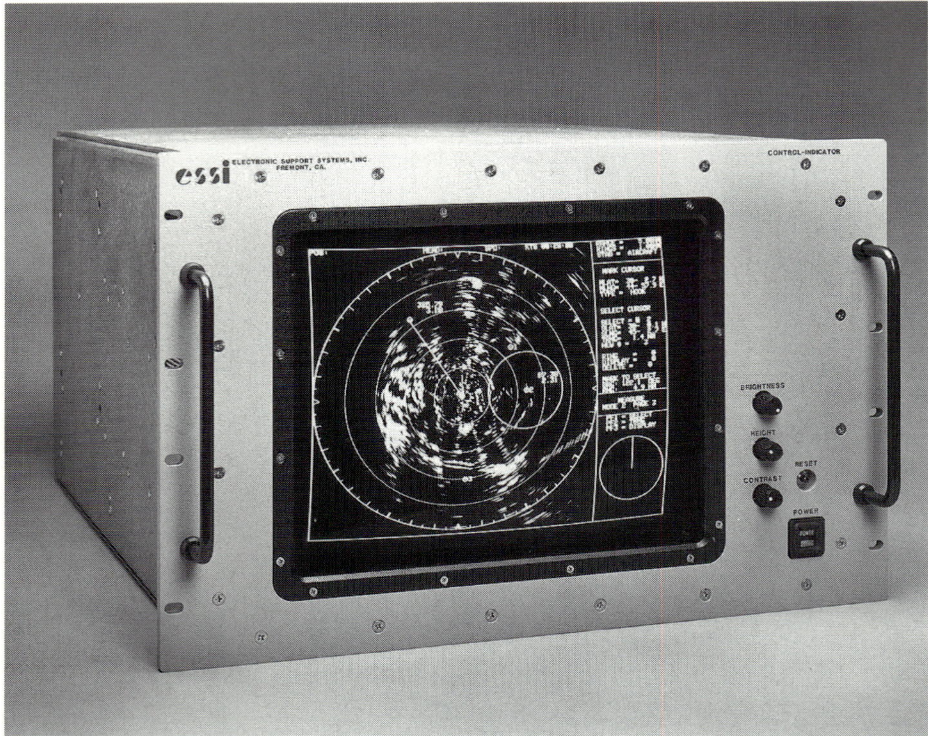

target ring generation; ruggedised and MIL qualified composite video outputs selectable between high and low resolution. The IP-5110 also features programmable overlay graphics; four display planes with independent 16-level intensity control; display ranges from 1.5 n miles to 200 n miles; selectable aircraft, waypoint and ground display stabilisation; freeze and zoom functions; video outputs compatible with VHS video recorders; averaging; peak detecting; weather and fast time constant processing.

Contractor

Condor Systems.

VERIFIED

The IP-5110 high-resolution digital radar display
1995

61000 series temperature exceedance monitoring

The DIAMOND J alumel/chromel 61000 series Thermocouple Engine Instruments with exceedances provide a back-lit analogue pointer with a seven-segment digital display. The seven-segment digital display can be obtained in up to four digits. The signal is processed digitally throughout the instrument to a high-resolution stepper motor. There are 2 in, 2.5 in and 3 in round, or ATI cases available. The pointer imperceptibly steps in increments of 0.45° in a smooth sweeping motion (2 in instrument); 0.225° per step for the 2.5 in and 3 in variants.

The DIAMOND J product line includes instrumentation for both fixed-wing and helicopter applications. The full complement of instruments include single-, dual- and triple-function instruments from one basic design. In many cases, only the dial face and software differentiate one instrument from another.

The DIAMOND J line of microprocessor stepper motor instruments does not require periodic calibration or adjustment during the life of the instrument. The high impedance input of the instrument eliminates the need for in-line balancing resistors. The instrument's only moving parts are contained in the stepper motor/pointer shaft assembly and are calibrated in the electronic circuitry design. The stepper motor has a step increment of 0.45°; this is imperceptible to the human eye at normal distances between the pilot and the instrument panel. Dial expansion is offered to enhance resolution in areas of critical temperature. Accuracy is ±5°C.

Exceedance functions include the following:

(1) Start-up exceedance: the temperature and time over preset parameters are monitored and stored
(2) Flight operations exceedance: engine conditions exceeding preset parameters are monitored and stored
(3) Exceedance display: at any time that an exceedance is detected the display flashes twice per second. The display is normalised when engine conditions are within tolerance, but the exceedance reading can only be eliminated by resetting the instrument out of the instrument panel
(4) Pilot's pre-warning: the display can be used to flash once per second if an exceedance is appoached (above a preset value) to enable the pilot to avoid an exceedance
(5) Data retrieval and reset: on power-up the instrument plays back all data in non-volatile memory; a reset unit must be used to purge data from memory.

Specifications

Power supply: 28 V DC, will operate from 8-36 V DC continuously, as well as accepting 100 V+ spikes
Power consumption: 6 W
Temperature range: −40 to +70°C
Certification: TSO-C43b

An example of DIAMOND J's 61000 series Instruments ***1996***

Contractor

DIAMOND J Inc.

UPDATED

Ice detection system

DNE Technologies has developed an ice detection probe which employs what is claimed to be the most sensitive method of ice accretion detection currently available. The probe operates in a cyclic fashion using the thermal characteristics of the ice which forms on it combined with the heat of fusion effect as the ice is formed. The signal produced may be used to initiate a warning signal or to activate the aircraft's de-icing system automatically. Between detection cycles, the probe is cleared of ice by an integral heater circuit. It is claimed that an ice accretion thickness of 0.12 mm can be detected within 5 seconds. A test facility for airborne confidence checking and for ground servicing purposes is incorporated. The standard ice detector is designed to sense ice formation on fuselage and nacelle air intakes for turbine engines, but other versions are available for carburettor ice detection on piston-engined aircraft and for other ice sensitive areas.

Operational status

DNE Technologies ice detectors are in service on the B-1B, B-2, the F-16C/D and F-117. An ice detector is under development for the F/A-18E/F and a flush-mounted ice detector is also being developed.

Contractor

DNE Technologies Inc.

UPDATED

Eagle-19 colour AMLCD

The Eagle-19 active-matrix liquid crystal display (AMLCD) is designed to replace 17 to 21 in monitors based on CRT technology. It provides flat-panel display solutions suitable for mission-critical crew-station applications in platforms such as AWACS and JSTARS. With full 24-bit colour and a proprietary optical stack that enables wide-angle viewing, the Eagle-19 supports full-motion video without ghosting or smearing and is intended for display of radar, FLIR, ASW applications.

Specifications

Display area: 19.0 in diagonal; 14.8 × 11.9 in
Pixels: 1,280 × 1,024
Colour bit depth: 24 bits per pixel
Contrast ration: >100:1 on axis

Viewing angle: ±55° horizontal; ±45° vertical
Refresh rate: >60 Hz, line sequential
Data input: digital; application-specific analogue interface

Contractor

dpiX, a Xerox New Enterprise Business.

NEW ENTRY

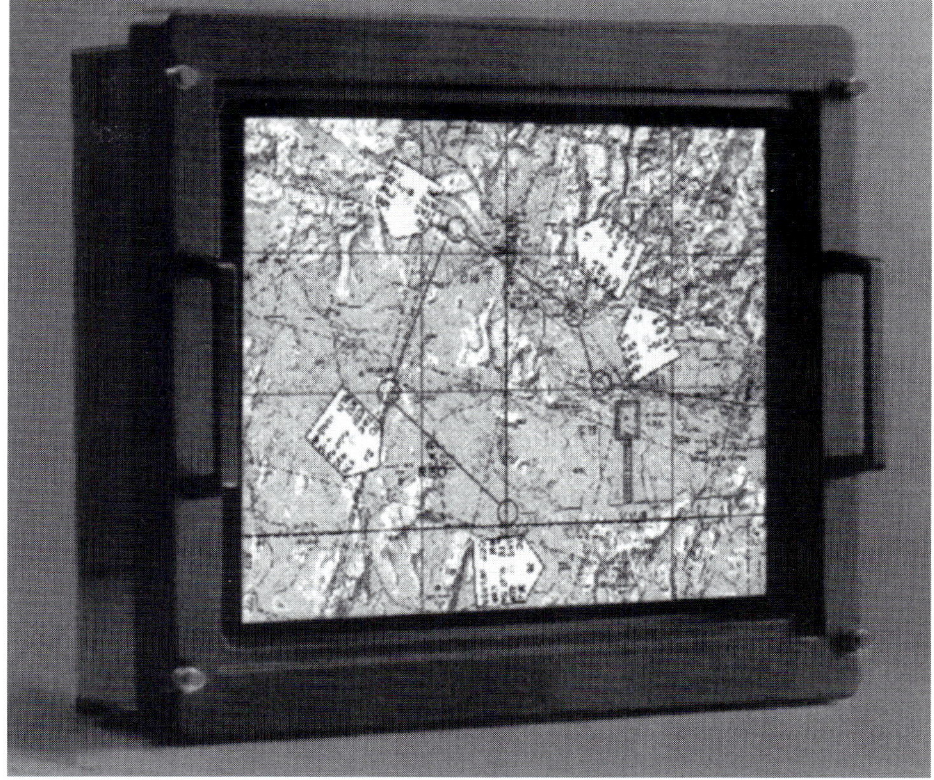

The dpiX Eagle-19 colour AMLCD
1998/0018171

Eventide Airborne Multipurpose Electronic Display (EAMED)

The Eventide Airborne Multipurpose Electronic Display (EAMED) is a powerful compact digital avionics development platform. It is a self-contained 16-bit avionics computer with high memory capacity which includes a CPU board, power supply, video board and CRT and a comprehensive set of input/output peripheral connections, including RS-232C and RS-422. EAMED displays both text and graphics on a high-resolution sunlight-readable screen. The unit fits in a standard panel cutout and hardware meets the DO-160B specification.

EAMED has a wide range of applications, including onboard communications, data acquisition, reconfigurable indicators and interactive situation displays.

Specifications

Dimensions: 76 × 76 × 266.7 mm
Weight: 1.6 kg
Power supply: 11-33 V DC, 15 W

Contractor

Eventide Inc.

VERIFIED

Field Emission Displays (FEDs)

FED Corporation teamed with Honeywell, Harris, and Astronautics Corporation of America for the US Air Force Wright Laboratory research and development contract to develop field emitter displays for military aircraft. The heart of the military display is FED Corporation's recently developed Field Emission Display (FED), essentially a flat CRT in which the three colour electron guns are replaced with an array of emitters each responsible for a single pixel on the screen.

FED technology offers the avionics systems providers significantly easier integration issues compared with AMLCDs. The ability of field emitter displays to operate over wide temperature ranges results in less sensitivity to thermal issues of the emitter display. This translates into lower system design cost and manufacturing cost for integration companies.

Operational status

Development.

Contractor

FED Corporation (consortium leaders).

VERIFIED

Command and control display system

The command and control display system provides the man/machine interface between the mission crew and the sensors and communications systems in the Boeing E-3A AWACS. It maintains the display database, filters and positions the selected data and generates all graphics, alphanumerics and sensor target reports with real-time responses. Situation Display Consoles (SDCs), each comprising a 19 in (482.6 mm) CRT MIL-SPEC colour monitor, data entry, filtering and control panels, trackerball and keyboard with associated processing electronics, provide the mission crew with all display and control features required to carry out surveillance, weapons direction and battle staff functions. Data Display Indicators (DDIs) with monochrome monitors support the communications, maintenance and data processing functions of the mission crew. The E-3A has 14 SDCs and 2 DDIs.

The SDC presents the appropriate colour pictorial representation of the situation required to support the function assigned to the SDC by the operator. Using the high electro-optical qualities of the NDI monitor to achieve high legibility of dense data presentations

The E-3A AWACS command and control display system has 14 SDCs and 2 DDIs ***1995***

under all operating conditions, the pictorial contents range from individual symbols indicating only sensor type and target positions to a combination of symbols and tabular notes that display such information as target type, speed, direction of flight, bearing, mission and altitude. Supporting tabular data, also in colour, is presented in the lower 20 per cent of the display surface. From this data, and from background pictorial information such as maps, landmarks and unsafe areas, the SDC operator can determine the appropriate responses to developing situations. The mission crew can also configure the SDCs in flight to serve as battle staff, surveillance or weapons consoles.

Specifications

Power supply: 115 V AC, 400 Hz, 3 phase
Temperature range: –54 to +55°C
Altitude: up to 40,000 ft
Environmental: MIL-E-5400 Class 1

Operational status

In service in US Air Force Boeing E-3A AWACS and Royal Saudi Air Force aircraft, and through a technology transfer on the NATO E-3 AWACS. Selected for the Japanese Air Self-Defense Force Boeing 767 AWACS fleet.

Contractor

GEC-Marconi Hazeltine Corporation.

VERIFIED

Electronic Flight Instrument System (EFIS) for the MD-80 aircraft

The Electronic Flight Instrument System (EFIS) gives a pictorial presentation of all primary flight instrument data. In addition to displaying conventional ADI and HSI information, these easily installed indicators also display weather information, advisory maps and radio altitude data. Maximum flexibility is achieved with numerous pilot-selectable display formats. The compact, easy-to-read system is programmed with standard commercial display formats and can also be tailored to specific customer requirements.

Specifications
Dimensions: 129 × 154 × 267 mm
Weight: 4.8 kg
Power supply: 115 V AC, 400 Hz, 65 W

Operational status
In production and in service in MD-80 Series aircraft.

Contractor
Honeywell Inc Air Transport Systems.

VERIFIED

The Honeywell MD-80 EFIS

RD-350J Horizontal Situation Indicator (HSI)

The RD-350J Horizontal Situation Indicator (HSI) provides heading, two DME distances, radio navigation information via a displacement bar, indications of selected course and heading and to/from indications for VOR operations. Flags give indication of failures.

A nav mode annunciator takes the form of a rotary display in the centre of the compass card, controlled by VOR/Loc valid, Loc tuned, to/from and back course selected signals. The annunciator displays a symbol to indicate selected modes and can also indicate when radio data is invalid.

Specifications
Dimensions: 127 × 127 × 226 mm
Weight: 4.2 kg
Power supply: 26 V AC, 400 Hz, 6.4 W

Operational status
In production.

Contractor
Honeywell Inc Air Transport Systems.

VERIFIED

RD-700 Series Horizontal Situation Indicators (HSI)

The RD-700 Series has been developed for applications in new aircraft or retrofit installations. High-torque, low-power flag and shutter movements eliminate sticking displays and low-power devices coupled with open card construction result in low heat dissipation and power demands. All indications are conventional in presentation and numerical indicators feature standard veeder readouts. The following list summarises the presentation and displays of each instrument:

RD-700 horizontal situation indicator
Displays all standard navigation radio inputs, compass system and ARINC 561 INS data including digital readout of drift angle and groundspeed.

Specifications
Dimensions: 5 ATI
Weight: 3.9 kg
Power supply: 115 V AC, 400 Hz
or 26 V AC, 400 Hz

RD-700A horizontal situation indicator
All standard navigation radio and compass data are presented together with dual digital DME readout. The RD-700A HSI features new flag, shutter and annunciator mechanisms and improved packaging to enhance reliability. Honeywell claims a 44 per cent reduction in power consumption compared with earlier designs.

Specifications
Dimensions: 5 ATI
Weight: 3.9 kg
Power supply: 115 V AC, 400 Hz
or 26 V AC, 400 Hz

RD-700C horizontal situation indicator
In addition to radio and compass navigation data, this instrument includes ARINC 561 INS data and digital readout of time and distance to waypoint and groundspeed.

Specifications
Dimensions: 5 ATI
Weight: 3.9 kg
Power supply: 115 V AC, 400 Hz
or 26 V AC, 400 Hz

RD-700D horizontal situation indicator
Although very similar to the RD-700C, this presentation does not include the time to waypoint counter.

Specifications
Dimensions: 5 ATI
Weight: 3.9 kg
Power supply: 115 V AC, 400 Hz
or 26 V AC, 400 Hz

RD-700F horizontal situation indicator
Displays all standard data including to/from, drift angle and digital readout of groundspeed and distance to waypoint.

Specifications
Dimensions: 127 × 127 × 216 mm
Weight: 3.9 kg
Power supply: 115 V AC, 400 Hz
or 26 V AC, 400 Hz

RD-700G horizontal situation indicator
In addition to all standard navigation, radio, compass and ARINC 561 INS inputs, this instrument presents digital readout of drift angle, distance to waypoint and groundspeed, although the drift and groundspeed presentation is at the lower area of the indicator rather than the more usual upper region of the instrument.

The Honeywell RD-850 horizontal situation indicator

Specifications
Dimensions: 5 ATI
Weight: 3.9 kg
Power supply: 115 V AC, 400 Hz
or 26 V AC, 400 Hz

RD-700M horizontal situation indicator
In addition to standard information displays, the RD-700M uses 11-position low-power magnetic wheels for dual DME displays. A new thermal design includes a more efficient heat-sink mounting for high-power components and heat sensitive capacitors. Electronic and mechanical sections are segregated for maintenance access, while open-board packaging in the electronic section facilitates troubleshooting.

Specifications
Dimensions: 127 × 127 × 218 mm
Power supply: 115 V AC, 400 Hz
or 26 V AC, 400 Hz

Contractor
Honeywell Inc Air Transport Systems.

VERIFIED

RD-800 Series Horizontal Situation Indicators (HSI)

The RD-800 Horizontal Situation Indicator (HSI) features three digitally driven servoed displays in conjunction with two four-digit gas tube displays showing time and distance to waypoints. Microprocessor control gives improved versatility in navigational data processing.

Specifications
Dimensions: 5 ATI
Weight: 4 kg
Power supply: 115 V AC, 400 Hz
or 26 V AC, 400 Hz

RD-800J horizontal situation indicator
In the RD-800J HSI the readout of true airspeed is provided by conventional counter displays and for ease of interpretation the command bars are colour identified.

Specifications
Dimensions: 5 ATI
Weight: 4 kg
Power supply: 115 V AC, 400 Hz
or 26 V AC,400 Hz

RD-850 Horizontal Situation Indicator
The RD-850 HSI features the most up-to-date applications of instrument technology including microprocessor control. Coloured display elements are included together with distance to go and groundspeed counters. Automatic direction-finder annunciators are fitted in the lower instrument area.

Specifications
Dimensions: 5 ATI
Weight: 4.7 kg
Power supply: 115 V AC, 400 Hz
or 26 V AC, 400 Hz

Contractor
Honeywell Inc Air Transport Systems.

VERIFIED

VIA 2000

In the Versatile Integrated Avionics VIA 2000 concept, Honeywell Air Transport Systems has developed further the Integrated Modular Avionics (IMA) technology developed for the Boeing 777. The VIA 2000 concept consolidates control of multiple avionics functions into cards on a VIA standard chassis, that will fit in existing aircraft.

The VIA 2000 system will be used for display functions on the Boeing 737 and MD-90. For the Boeing MD-95, the VIA 2000 system will be used in a multifunction capacity with flat-panel displays, and with a dual-computer configuration for the flight management system. The VIA 2000 system replaces 24 duplicated captain and first officer instruments on the aircraft instrument panel, as well as nine shared instruments.

The VIA 2000 system also has growth potential for future developments in commercial aviation. It can offer enhanced provisions for using GPS for navigation, and upgrades can be incorporated to bring in the Future Air Navigation System and Aeronautical Telecommunications Network. Provisions can also be included for a fail-active automatic landing option meeting Cat IIIB requirements for approach and landing in low visibility.

Operational status

Selected by Saudia for its MD-90 aircraft, the VIA 2000 system will use six 8 × 8 in flat panel LCDs for the primary flight instruments and to display all systems readouts to the pilots.

A similar configuration of the VIA 2000 system with six 8 × 8 in flat panel LCDs is being introduced as the advanced flight deck for the MD-95 aircraft. Together the six new displays will provide information that required 33 separate instruments in the original MD-95 design.

Contractor

Honeywell Inc Air Transport Systems.

UPDATED

Data Nav V navigation/checklist display system

Honeywell's Data Nav V turns any Honeywell colour weather radar screen into a versatile en route navigation map or aircraft checklist. It allows pilots to select navigation waypoints and instantly create new waypoints by positioning an electronic designator 'bug' anywhere on the radar screen.

The Data Nav systems are designed to add navigation map and/or checklist display capability to all Honeywell colour weather radar indicators. The latest version, Data Nav V, features an ARINC 429 interface with a variety of Flight Management System (FMS) and long-range navigation systems, enhanced on-screen information and simplified pilot operation. Operators can compile their own aircraft-specific checklists on a personal computer and load them quickly and easily into the Data Nav computer, using the new optional Honeywell Standard Checklist software program.

The ARINC 429 interface links the Data Nav V system not only with the Honeywell NZ-2000 FMS system but those of other makes that provide ARINC 429 outputs, as well as with certain other long-range navigation systems. The FMS or long-range navigation system computer will lay out a flight plan, convert it to ARINC 429 digital data format and transmit it to the Data Nav V computer, which generates navigation map, course, heading, lat-lon position and waypoint symbology for display on the Honeywell weather indicator.

In addition, Data Nav V, using FMS or nav system inputs, will display VORTAC locations, track lines, projected course, distance to waypoints and ground speed.

En route, the navigation display creates a moving picture of the aircraft's position as preselected courses are flown. Combined with the radar weather display, it gives a complete visual presentation of selected course and weather conditions ahead.

Data Nav V presents a full menu – up to 200 custom preprogrammed pages of normal and emergency preflight procedures and cockpit checklists, together with performance data, operating notes and other selected data.

The Data Nav V Electronic Checklist simplifies the routine checklist procedure and permits quick retrieval of emergency checklists, reducing the chance of errors and oversight by giving positive indication of a checked procedure. Because the system remembers checklist position, a pilot may instantly switch back and forth between weather/map display and the previously selected checklist.

Data Nav V provides nav map and checklist capabilities for both EFIS and non-EFIS equipped aircraft.

It can be used with a two or four-tube EFIS system, where the Honeywell radar indicator serves as an additional display to enhance overall system capability.

The Honeywell EHSI for corporate aircraft and regional airliners

The Data Nav V system is available in three configurations, with panel-mounted controllers for selecting nav map with weather only, weather and checklists only, or either map/weather or checklist displays. Data Nav V is a direct replacement for the Honeywell Data Nav III system, when using the most common FMS interfaces.

Operational status

Data Nav I, II and III no longer in production. Replaced with Data Nav V, in production.

Contractor

Honeywell Inc Business & Commuter Aviation Systems.

UPDATED

DU-870 CRT Display Unit

Honeywell produces an 8 × 7 in (203 × 177.8 mm) CRT display that provides all the information found on larger displays, yet fits in smaller cockpits. The DU-870 streamlined display is designed for the cockpits of twin-turbine aircraft and is featured in the Primus 1000 and 2000 advanced avionics system.

The DU-870 is the result of aircraft panel size and fit surveys conducted by Honeywell to determine the optimum size, shape and screen area required to meet market needs. The Primus 1000 and 2000 display system architecture provides the ability to view any display format on the DU-870. A micro symbol generator is integrated into the DU-870 to reduce aircraft wiring.

As with the other components of the Primus 1000 and 2000 system, hardware techniques such as surface-mount technology, very large-scale integration and application specific integrated circuits are incorporated to minimise weight and volume. The DU-870 also features self-contained cooling for improved reliability and quiet operation.

Contractor

Honeywell Inc Business & Commuter Aviation Systems.

UPDATED

EDZ-605/805 electronic flight instrument systems

The EDZ-605 and EDZ-805 electronic flight instruments are intended for corporate aircraft and regional airliners. The EDZ-605 is a 5 in (127 mm) system, while the EDZ-805 is a 6 in (152 mm) system. The products are the result of a new symbol generator and software changes. The generator has increased stroke writing and memory capabilities, while software updating has provided a variety of cosmetic improvements to the display.

The Electronic Attitude Director Indicator (EADI) features an enlarged sphere presentation with linear pitch tape, improved single-cue aircraft symbol and stroke filled single-cue command bar. A digital T-bar airspeed and/or angle of attack presentation has also been added. Other improvements include a stroke filled roll pointer, larger roll indices, shorter horizon indices and colour reversal on glide slope, angle of attack, rate of turn and expanded localiser scales. In addition, the 5 in (127 mm) EADI is now available in a truncated sphere presentation.

The Electronic Horizontal Situation Indicator (EHSI) features a stroke filled lubber line, colour reversal on the glide slope scale and weather radar mode annunciation.

The ED-605/ED-805 are compatible with most analogue systems and all digital systems.

Operational status

In production. Chosen for many business aircraft.

Contractor

Honeywell Inc Business & Commuter Aviation Systems.

VERIFIED

EDZ-705 Electronic Flight Instrument System (EFIS)

The EDZ-705 Electronic Flight Instrument System (EFIS) is tailored to meet the needs of specific helicopter applications from executive use to search and rescue missions. The system provides the ability to display standard search patterns as provided by various navigation sources, collective cue symbology which allows the pilot to follow flight director collective cue demands, hover display symbology which allows the pilot to track and maintain target location in relation to the helicopter position in a search and rescue environment and four-axis helicopter flight director mode annunciation.

Since panel space is at a minimum, the EDZ-705 offers 5 × 5 in (127 × 127 mm) displays.

Contractor

Honeywell Inc Business & Commuter Aviation Systems.

VERIFIED

LSZ-860 lightning sensor system

Honeywell's newest lightning sensor, the LSZ-860, is an upgraded version of the LSZ-850. The LSZ-860 senses both visible and high-energy invisible electrostatic and electromagnetic disturbances caused by electrical discharge activity within a 200 n mile radius around the aircraft. When lightning occurs, the system carefully analyses the discharge and creates the proper symbol for display on most Primus colour radar indicators or on most Honeywell EFIS/MFD displays. The system's computer rapidly and accurately determines the rate of vertical lightning in a fixed geographical area and then displays the centre of that area with the lightning rate symbol. Wide bandwidth, extensive signal processing and lightning stroke recognition algorithms ensure a more accurate display. Extraneous signal filtering minimises noise which would otherwise clutter and confuse interpretation.

The LSZ-860 system computes the location and lightning rate for up to 50 thunderstorm areas. The computer tracks the location of each of these areas. To ensure accurate tracking of lightning areas, all displays are both heading and velocity stabilised to keep the symbol over the same ground position regardless of aircraft manoeuvring.

The system gathers information in a full 360° pattern around the aircraft, even in the standby mode. Thus, the lightning sensor system is always ready to present the weather picture in either the full 360° display mode without a radar overlay or in the sector display mode that can include a radar and navigation data overlay.

Three distinct levels of lightning rate are computed for display, each depicted by a unique lightning rate symbol. The symbols represent the vertical lightning rate-of-occurrence. The symbol location is the average position of the lightning that occurred in the previous two minutes, and is displayed inside the radius of the range selected. Each lightning symbol represents the centre of a circular area with a radius dependent on the range selected (8 n mile radius at a range of 25 n mile, 18 n mile radius at a range of 100 n miles and 30 n mile radius at a range of 200 n miles). Whenever any lightning activity is detected at any range, the lightning sensor computer will place a magenta lightning alert symbol at the proper bearing and at the end of the selected range for five seconds.

Specifications

Dimensions: (LP-860) 375 × 194 × 61.5 mm; (AT-850) 294 × 31.8 × 154 mm
Weight: (LP-860) 3.06 kg; (AT-850) 1.14 kg
Power supply: 28 V DC, 28 W

Operational status

LSZ-860 available from January 1998.

Contractor

Honeywell Inc Business & Commuter Aviation Systems.

NEW ENTRY

Primus 1000 integrated avionics system

The Primus 1000 integrated avionics system is designed for mid-sized business jets and regional turboprops. Advanced processing technologies and the integration of key functions result in increased capability, reliability and flexibility while achieving significant reductions in size, weight, power requirements and installation costs.

The Primus 1000 is based on the IC-600 integrated avionics computer which combines a display processor, flight director, Cat II fail-passive autopilot and EICAS processor in a single ½ ATR box. The Primus 1000 utilises an ARINC 429 architecture and is composed of an EFIS electronic display system incorporating from two to five 8 × 7 in (203.2 × 177.8 mm) large format displays, single or dual flight director and an optional single or dual fail-passive autopilot. The other components of the system include AZ-840 micro air data computer, separate vertical and directional gyros or the AH-800 fibre optic AHRS, Primus II radio system and the Primus 650 weather radar system. Standard options are Honeywell's TCAS, MLS, lightning sensor system, Primus 870 turbulence detection weather radar, Primus 700 or Primus 450 weather radar and the Laseref III inertial reference system.

Operational status

The Primus 1000 was selected for the Lear 45 in September 1992 and later for the Citation V Ultra and Embraer EMB-145. A Primus 1000 system selected for the Sino Swearingen SJ30-2 in April 1996 specified dual IC-600 computers, AZ-850 all-digital Micro Air Data Computers (MADC), Primus II digital integrated radios and Primus 650 weather radar.

Contractor

Honeywell Inc Business & Commuter Aviation Systems.

VERIFIED

The Primus 1000 Integrated Avionics System installed in the Embraer-145 aircraft, showing a five display EFIS/Engine Instrument and Crew Advisory System (EICAS) using DU-870 8 × 7 in colour displays

1997/0001387

Primus 2000 advanced avionics system

The Primus 2000 advanced avionics system is designed for twin-turbine aircraft in the business and regional airliner markets. Its small size and weight are the result of using the most advanced technology in components, packaging techniques and systems design. The system incorporates surface-mount technology, very large-scale integrated circuits, application specific integrated circuits and high-density multilayer circuit boards. The Primus 2000 also offers flexibility and maximum growth potential through the use of an advanced architecture built around Honeywell's Avionics Standard Communications Bus (ASCB).

Interconnection between all major systems is accomplished by the ASCB, which provides both the total data handling capacity and requires a lower wire count than the one-way ARINC 429 standard. The bidirectional databus has critical level capability that eliminates the need for numerous dedicated lines between avionics systems. The ASCB architecture enhances system level availability, allowing the ASCB to be the sole means of interconnect for most avionics subsystems.

The Honeywell Primus 2000 integrated avionics system on the Citation III aircraft

The Primus 2000 is composed of an electronic display system incorporating from two to six large screen 8 × 7 in (203 × 177.8 mm) display units and associated control panels, an integrated avionics computer containing electronic display processors, a fault warning computer, a fail-operational/fail-passive automatic flight control system and an optional flight management system, micro air data computers, attitude and heading reference system or inertial reference system, data acquisition units, Primus II radio and Primus 650 weather radar systems and the avionics standard communications bus. Honeywell's flight management system, Traffic alert and Collision Avoidance System (TCAS), Global Positioning System (GPS), Microwave Landing System (MLS), lightning sensor system, Primus 870 turbulence detection radar and laser inertial reference system are offered as standard options.

Operational status

Certified first on the Fairchild Dornier 328. Also now certified on the Cessna Citation X and on the Dassault Falcon 900EX, in a five, 8 × 7 in, full colour EFIS configuration, including two MFDs, two primary flight displays and an EICAS display.

Contractor

Honeywell Inc Business & Commuter Aviation Systems.

VERIFIED

Primus 2000XP integrated avionics cockpit

Honeywell's revolutionary Integrated Avionics Computer IC-800 is standard on the Global Express in a triple configuration. It combines the function of five major subsystems into a compact, lightweight package that is 75 per cent smaller than the combined volume of the units it has replaced.

The Primus® 2000XP integrated cockpit was designed specifically for the Global Express. Its high-accuracy air data system precisely measures speed, altitude, temperature and other critical flight information, while six 8 × 7 in displays keep the pilot fully informed about all aspects of the flight.

Primus 2000XP standard features include:

(1) six-tube Electronic Flight Instrument System (EFIS)/Engine Instrument and Crew Advisory System (EICAS) using DU-870 8 × 7 in colour displays
(2) dual fail-operational autopilot
(3) dual integrated NZ-2000 Flight Management Systems (FMS)
(4) triple long-range Nav source (two FMS and LaserTrak)
(5) Global Navigation Satellite Sensor Unit (GNSSU) Global Positioning System (GPS)
(6) dual full flight regime autothrottle triple Integrated Avionics Computers (IAC)
(7) triple AZ840 Micro Air Data Computers (MADC)
(8) triple LASEREF III Inertial Reference Systems (IRS)
(9) quad Ametek DA-810 Data Acquisition Units (DAU)
(10) Primus 880 colour weather radar
(11) Central Aircraft Information Maintenance System (CAIMS)
(12) dual Primus II Integrated Nav/Com/Ident radios
(13) TCAS II Traffic Alert and Collision Avoidance System.

Optional avionics for the Global Express include: a third integrated FMS; second integrated GPS; third VHF com; lightning sensor system; three or six channel SATCOM, featuring worldwide phone, fax and data satellite communication.

The IC-800 features include: two symbol generators; fault warning computer; tone and voice generator; Radio System Bus (RSB) interface; Avionics Standard Communication Bus (ASCB) controller and interface; flight guidance computer; flight management computer; autothrottle control and power supply.

The exceptional system integration within the IAC is made possible by the lastest technologies, including: Very Large Scale Integration (VLSI); Surface Mount Technology (SMT) and advanced manufacturing techniques to reduce circuit card size. Single cards within two of the IACs contain entire FMS computers. The IACs also house the electronics for automatic flight control, autothrottle and systems monitoring.

Operational status

Pending certification on the Bombardier Global Express in 1998.

Contractor

Honeywell Inc Business & Commuter Aviation Systems.

NEW ENTRY

Primus Epic avionics system

The key to the Primus Epic avionics system is the virtual backplane network developed exclusively for this system. This architectural concept blends the cabinet-based modular capabilities of the Primus 2000 system. The architecture allows a very high degree of system integration and scalability by allowing all data generated by any function to be globally available within the system.

The Primus Epic system hardware is built on the Modular Avionics Unit (MAU) derived from Honeywell technology, developed for the Boeing 777. Computing modules within the MAUs and flat panel displays utilise Honeywell's Digital Engine Operating System (DEOS), which allows the different aircraft and functions to run simultaneously and independently. The full size 8 × 10 in flat panel liquid crystal displays, in a two- to six-display configuration, feature new functionality within a point-and-click Graphical User Interface (GUI) environment, which supports moving maps, ground-based weather, real-time video and electronic pilot manuals.

Pilots may choose traditional interfaces or new cursor control devices, including touchpad, joystick, light pen or tracker ball to interact with on-screen 'soft key' controls. In future, Primus Epic will offer voice command as a control option for some functions.

Operational status

Launch customer will be the Hawker Horizon from Raytheon Aircraft Company. The Horizon is expected to make its first flight in late 1999, with certification in 2001.

Contractor

Honeywell Inc Business & Commuter Aviation Systems.

UPDATED

Colour multifunction display for C-17

Honeywell's 6 × 6 in (152 × 152 mm) colour multifunction display is a high-resolution, shadow-mask CRT display developed for the C-17 military air transport. Honeywell's newest family of full-colour, military-qualified displays includes a number of advanced technologies to provide significant performance advantages. Superior vibration tolerance, sunlight viewability and high-resolution graphics are inherent in the design of the display.

The multifunction display can be used to present primary flight and navigation information, colour weather radar, digital map information, engine instrumentation and tactical display formats. The unit presents stroke, raster or hybrid formats in 16 colours. Raster images are driven from sensor inputs with an RS-170 or RS-343 interface.

Honeywell's multifunction display has been designed as a stand-alone 'smart' unit. The display contains a MIL-STD-1750A processor to convert stored modes into formats using aircraft parameters received over the MIL-STD-1553B bus. An internal vector generator is used to draw the formats on the screen. The unit also includes an expanded built-in test capability and can report failures to the host computer via the 1553B bus.

A separate display processor is not necessary when linking the 'smart' multifunction display directly to the aircraft's mission computer via the 1553B bus. This configuration reduces system weight, reduces potential interface problems and increases system reliability.

Specifications

Dimensions: 203 × 203 × 381 mm
Weight: 16.8 kg

Contractor

Honeywell Inc Defense Avionics Systems.

VERIFIED

Colour and monochrome multifunction display systems for the F-16

Honeywell's 4 × 4 in (102 × 102 mm) multifunction display system for the F-16 multirole fighter provides the pilot with clear, precise displays of navigation, weapons aiming and radar information that is easily viewed in all ambient lighting conditions.

The system consists of a programmable display generator and two or four multifunction display units. The user can choose from monochrome CRT displays or colour AMLCDs.

Honeywell's colour cockpit system greatly increases pilot and co-pilot situational awareness. The high-performance colour displays offer improved recognition of display information, especially in high-density display formats common in modern fighter cockpits.

The multifunction display system, an all-digital system completely compatible with all current F-16 weapons systems avionics, also includes provision for integrating advanced sensors and avionics. The display generator also has outputs available to drive video tape recorders in monochrome or colour.

The display generator uses advanced task-sharing techniques for high-speed and symbol generation. It is a software-based device permitting rapid mode changes and the use of several sets of symbology. Data from the F-16's avionics computer and electro-optical systems are received via a 1553 databus; radar communications are received via a dedicated serial bus. Upgrades to the display generator will include growth to accommodate a real-time moving map.

The multifunction display units utilise advanced techniques to overlay symbology onto background video. Twenty interactive programmable push-button switches located on the bezel provide control of the display formats.

The displays are convection cooled and efficiently packaged, allowing them to be used in a variety of cockpit applications. The 4 × 4 in (102 × 102 mm) multifunction display systems are proven performers, having demonstrated consistent reliabilty and capability in the field.

Operational status

Honeywell's monochrome displays have been in production since 1982. See later entry (Multifunction displays for the F-16C/D) for details of the F-16A/B Mid-Life Update programme displays.

Contractor

Honeywell Inc Defense Avionics Systems.

UPDATED

Colour multipurpose display indicator for F-117A

Honeywell's colour multipurpose display indicator is a new 5 × 5 in high-brightness colour display which uses a taut mask, delta gun CRT.

Presenting stroke, raster or hybrid video displays, the display is normally operated in conjunction with a display processor that provides the required stroke deflections and video signals. The display processor also provides a digital serial interface which can receive display bezel switch information, operational mode and built-in test status. It can operate in a raster-only mode without a display processor, and provides 525- or 875-line rasters in colour or monochrome.

A five-position rotary switch, located on the display bezel, controls the operational mode of the display: off, night, auto, day and equipment override. The day/night/auto mode controls the video gain function of the red-green-blue inputs. In day mode, full contrast and brightness control is provided for use in room ambients to full-sunlight. The auto mode can also be used in daylight operation for automatic adjustment of the contrast and brightness in varying light conditions. The night mode is provided to allow more contrast and brightness resolution in night situations and is operational up to room ambient lighting conditions. The display bezel is designed to accept a detachable filter, such as neutral density, red or green filter, for night operations.

Selection of the equipment override changes the video input from red-green-blue to separate monochrome video input. This feature allows the display to operate in an alternate mode if the display processor channel driving the display should fail.

Operational status

In service. The display is ideal for colour presentation in severe military environments and is used on the US Air Force's F-117A aircraft.

Contractor

Honeywell Inc Defense Avionics Systems.

VERIFIED

Control Display System (CDS) for OH-58D Kiowa Warrior

Honeywell's Control Display System (CDS) provides embedded controls and displays for communication (ETICS), navigation, engine/power-train, Mast-Mounted Sight (MMS) and weapons, as well as Rotorcraft Map System (RMS – digital map) and Video Image cross-Link (VIXL).

The ETICS (Embedded Tactical Information Control System) (see Communications section), is part of the Honeywell CDS, and it embeds integrated Task Force XXI Variable Message Format (VMF) capabilities for command and control, fire support and situational awareness.

The rotorcraft mapping system provides Kiowa Warrior crews with situational awareness by translating data from ETICS and the Honeywell Embedded Global positioning system/Inertial navigation system (EGI) into icons that report the aircraft's position in relation to the terrain and other air and ground vehicles, both friend and foe.

The VIXL transmits near-real-time reconnaissance video images from the Kiowa Warriors mast-mounted sight over standard combat radio links to other Kiowa Warriors or ground command and control centres. In addition, Honeywell is also providing the SINCGARS/SIP (Single-Channel Ground-Air Radio System/System Improvement Program) radio, and EGI system as part of the overall Kiowa Warrior upgrade. The display system itself comprises two monochrome multifunction displays (one for each pilot), and the radio frequency display.

The Kiowa Warrior cockpit, showing the Honeywell Defense Avionics Systems multifunction displays, and the radio frequency display ***1998***/0018170

Contractor
Honeywell Inc Defense Avionics Systems.

NEW ENTRY

Digital map display set for the AV-8B and F/A-18

In July 1985, Honeywell was contracted to design and develop a digital map system for use in the AV-8B Harrier II and F/A-18 night attack programmes. The map display set consists of a digital map computer and a digital memory unit. It provides pilots with a moving map display that includes digitised paper maps and Defense Mapping Agency digital terrain information presented in colour.

Operational status
In service with the AV-8B and F/A-18 aircraft.

Contractor
Honeywell Inc Defense Avionics Systems.

VERIFIED

Monochrome multifunction display for SH-60F helicopter

Honeywell's 6.5 × 8.5 in (165 × 216 mm) monochrome multifunction display, used onboard the US Navy's SH-60F CV-Helo aircraft, presents tactical navigation data, situational plots and sensor information.

The display, which is presently in production, operates in three modes: stroke, raster and hybrid. The raster mode is used to display sensor information such as sonor acoustic plots. Stroke writing may consist of patterns, symbols and alphanumerics. In the hybrid mode, the display presents both raster video and stroke data information. The display operates conventionally in the raster mode, but switches to the stroke mode during the raster's vertical blanking periods.

Display interface is via a 16-bit tactical data processor for stroke operation. For raster performance, the display accepts either RS-170 or RS-343 formats.

Honeywell's mature, field-proven 6½ × 8½ in (165 × 216 mm) multifunction displays are cost-effective solution for airborne requirements designed to meet military specifications, including NVG-compatibility.

Operational status
In production and in service in HH-60, SH-60 and SH-3 helicopters and EA-6B aircraft.

Contractor
Honeywell Inc Defense Avionics Systems.

VERIFIED

Multifunction displays for the F-16C/D

The initial fit multifunction display for the F-16C/D comprises two 4 in (102 mm) square CRT displays with a digital, programmable display generator for each aircraft. The two seat F-16Ds that make up about 15 per cent of the Fighting Falcon fleet have four displays but still feed from the one signal generator. The displays present the pilot with navigation, radar and weapon aiming information. They use raster scan techniques to provide clear and sharp presentations of sensor information, symbol overlays and alphanumerics in all ambient light conditions. The displays are capable of providing both 525-line and 875-line rates. Each display has 24 push-button keys arranged around the bezel for interactive control and data entry.

The programmable display generator uses advanced task sharing techniques for high-speed operation and symbol generation. It is a software-based device with two processors and 64 k words of core memory, permitting rapid changes of mode and the use of several sets of symbology. Data from the F-16's fire control computer is fed to the system via the MIL-STD-1553 digital databus while video information comes from its electro-optical sensors.

The system's digital design permits rapid reprogramming and this flexibility includes the ability to integrate LANTIRN and JTIDS. The display system incorporates comprehensive test and diagnostic routines and there are growth provisions for the generation of colour symbols and for the MIL-STD-1750A instruction set architecture.

Specifications
Dimensions:
(display) 143 × 143 × 308 mm
(display generator) 159 × 194 × 448 mm
Weight:
(display) 5.9 kg
(display generator) 12.7 kg
Display colour: phosphor green P43
Line width: 0.0075 in (0.19 mm)
Usable screen area: 4 × 4 in (102 × 102 mm)
Video inputs: 9 total
Video outputs: 6 total, 2 independent channels, 3 outputs/channel

Operational status
In service on all F-16C/D aircraft. A variation of this display is also being used in the US Navy SH-60F CV-Helo aircraft. An updated colour Active Matrix Liquid Crystal Display (AMLCD) was first produced by Honeywell Defense Aviation in 1996 for the F-16A/B Mid-Life Update for Belgium, Denmark, Norway and Netherlands. The same display has been ordered for the US Air Force's new F-16Cs, and for the F-16Cs being bought by Taiwan.

Contractor
Honeywell Inc Defense Avionics Systems.

VERIFIED

MultiPurpose Colour Display (MPCD) for the F-15C/D MSIP and F-15E

The MultiPurpose Colour Display (MPCD), developed for the F-15 fighter as part of a continuing Multinational Staged Improvement Programme (MSIP) enhancement for the C and D versions, is a very high-resolution 5 × 5 in (127 by 127 mm) shadow-mask system with stroke, raster, or combined stroke/raster writing in any of 16 colours as determined by software in the symbol generator. Raster symbology is a full-colour representation of a colour sensor output. Functions include built-in test for a number of aircraft systems, graphic representation of aircraft stores configuration (replacing the electromechanical armament control panel on earlier F-15s), display of video from electro-optic sensors and weapons and interface for secure tactical information from a JTIDS communications system.

Honeywell has been contracted for a second version of the colour display for the F-15E programme. The F-15E fleet, which incorporates air-to-ground and air-to-air mission capabilities, is equipped with three Honeywell MPCDs on each aircraft. Honeywell is also under contract for the display processor for F-15E aircraft. The equipment not only powers the aircraft's three colour displays, but all the displays on the aircraft.

The display's delta gun shadow-mask tube uses dynamic and static convergence for maximum symbol fidelity. A bandpass filter and 17,000 in/s writing speed facilitates viewing in high ambient light levels, or even direct sunlight. Two CRT versions are available, one having 0.008 in phosphor dot spacing said by Honeywell to be only two-thirds the dot size and spacing of other shadow-mask CRTs designed for airborne applications.

The Honeywell programmable signal data processor interfaces with other equipment and sensors, generating monochrome and colour symbology for other displays. It comprises two elements: a three-channel symbol generator and a general purpose processor. Self-test circuit monitor all critical signals continuously and failures are flagged on a built-in test indicator.

Specifications

Dimensions: 172 × 180 × 387 mm
Weight: 11.8 kg
Screen size: 127 × 127 mm
Linewidth: 0.016 in typical
Video bandwidth: DC 10 MHz
Television scan resolution: 70 line pairs/in

Operational status

In service with F-15C/D MSIP and F-15E fighters. A version of this display is in the F-117A aircraft.

Contractor

Honeywell Inc Defense Avionics Systems.

VERIFIED

Multipurpose stroke display

In the multipurpose stroke display the 6.5 × 8.5 in (165 × 216 mm) CRT is driven by a self-contained display generator, deleting the requirement for peripheral display hardware. The display requires no cooling air. The stroke formatting provides clear and crisp symbology in high ambient light conditions. A variation of this display, characterised by a stroke and raster capability, is now being produced for the US Navy's SH-60F CV-Helo anti-submarine warfare helicopter. The CV-Helo display can simultaneously provide raster formats for TV, FLIR and so on, along with stroke symbology overlay.

Specifications

Dimensions:
(SH-2F) 229 × 279 × 381 mm
Weight:
(SH-2F) <12.7 kg
Power consumption: 70 W nominal
Display colour: phosphor green P43
Usable screen area: 6.5 × 8.5 in (165 × 216 mm)

Operational status

In service in the Sikorsky SH-3, Kaman SH-2F helicopters and EA-6B aircraft.

Contractor

Honeywell Inc Defense Avionics Systems.

VERIFIED

Radar Control Display Unit (RCDU) for UK E-3D

Honeywell Defense Avionics Systems will develop the colour Radar Control Display Unit (RCDU) for the United Kingdom's fleet of seven E-3D Airborne Warning And Control System (AWACS) aircraft under a subcontract to Northrop Grumman Electronic Sensors and Systems Division of Baltimore, Maryland as part of the Boeing Company's United Kingdom AWACS Radar System Improvement Program (RSIP).

The RCDU adapts proven Commercial Off-The-Shelf (COTS) hardware Honeywell Air Transport Systems originally developed and integrated on the Boeing 777, incorporating Honeywell's flat panel, active matrix liquid crystal display technology.

Operational status

In development for the UK E-3D AWACS aircraft.

Contractor

Honeywell Inc Defense Avionics Systems.

NEW ENTRY

Remote frequency display

The operational status of each of the five radios carried aboard OH-58D helicopters is displayed by means of the remote frequency presentation system. It shows, on five lines of liquid crystal readout, the radio number, operator in control, frequency, whether coded or in plain language and two alphanumeric characters for channel identification.

The display receives its information through a serial digital interface with either of the two master controller processor units, the choice of unit being left to the operator. Data for each window remains displayed until updated. If the system does not receive an update within two seconds, the screen goes blank as a warning to the operator. The screen is designed for reflective daylight viewing and is back-lit for night operation. Characters are painted in white on a black background.

Specifications

Dimensions: 146 × 99 × 102 mm
Weight: 10.8 kg
Power:
3 W normal
53 W with heater
Contrast:
20:1 in direct sunlight
16:1 when back-lit
Back-lighting: electroluminescent

Operational status

In production for and in service on the OH-58D.

Contractor

Honeywell Inc Defense Avionics Systems.

VERIFIED

Smart multifunction display

Honeywell has developed a 6 × 6 in (152 × 152 mm) smart multifunction display for use as a primary flight instrument on military aircraft such as the C-17. Functions normally provided by a display controller are incorporated into the display unit, eliminating the need for a peripheral controller. Format generation uses high-level macro commands which may be stored as tables and then compiled along with the associated aircraft dynamic data to generate the actual displays. Multifunction display modes and submodes are completely programmable and are stored in a non-volatile memory within the display. This system allows rapid switching between formats with minimal bus loading because only format designations and dynamic data are sent over the 1553 bus.

The CRT assembly uses a delta gun, taut shadow-mask tube with an aluminised tri-colour phosphor dot triad on a black matrix. Convergence is controlled statically and dynamically over the entire viewing area. The stroke writing rate of this hybrid display is software-controlled with a symbol generation speed of over 35,000 in/s. Synthetically generated raster for background and stroke symbology fill is written at a rate of 180,000 in/s. Display brightness and contrast is automatically controlled by two bezel-mounted light sensors and a remote light sensor.

Specifications

Dimensions: 203 × 203 × 381 mm (max)
Weight: 16.3 kg
Screen size: 152 × 152 mm
Line width: 20 mils (max); 14 mils nominal;
11 mils min
Operating modes: stroke, hybrid (stroke/raster)
Digital interface: MIL-STD-1553B
Raster type: RS-170 (525 lines)

Operational status

In service on the C-17 aircraft.

Contractor

Honeywell Inc Defense Avionics Systems.

VERIFIED

MIAMI ice detection system

The Microwave Ice Accretion Measurement Instrument (MIAMI) permits the accurate measurement of ice accumulating in critical areas of the aircraft.

As an ice warning system, the MIAMI alerts the pilot to the earliest initiation of ice growth. As little as 0.076 mm of ice can be detected, illuminating a warning light on the annunciator panel. Then the preprogrammed microprocessor takes over and both computes icing rate in in/min and indicates in the cockpit digital display the ice thickness.

If required, the system can be used to activate or deactivate the de-icer equipment on aircraft and missile systems and works equally well with all types of de-icing or anti-icing equipment.

The system transducer element consists of a resonant surface waveguide. The resonant frequency of this varies according to the amount of ice accreted and so a relationship between ice thickness and frequency shift can be established. It is said that this type of transducer has the advantage of not requiring an external and frangible probe and the transducer may be profiled to conform to the contour of the mounting surface.

The transducer can be mounted anywhere on the aircraft, including rotors or wings, and the system can be protected against sand or rain erosion. The system is microprocessor-controlled and a single microprocessor unit can control any number of transducers.

Operational status

In production. Installed on the Cessna T303 Crusader.

Contractor

Ideal Research & Development Corporation.

VERIFIED

Avionic display modules

ImageQuest Active Matrix Liquid Crystal Displays (AMLCDs) are designed for avionic applications, and feature amorphous Silicon Thin Film Transistors (TFTs), thermal control of AMLCD and back-light, and VGA format.

Contractor

ImageQuest Technologies Inc.

Specifications

Parameters	4 × 4 in avionic display	6.24 × 8.31 in avionic display	10.4 in avionic touchscreen display
Active display area	101.76 × 101.6 mm	211.2 × 158.4 mm	211.2 × 158.4 mm
Module dimensions	127 × 127 × 76.2 mm	284 × 184.4 × 73.5 mm	300 × 246 × 55 mm
Weight	1.42 kg	3.5 kg	3.41 kg
Pixel configuration	RGB Delta Triad	RGB stripe	RGB stripe
Pixel array	480 × 480	640 × 480	640 × 480
Luminance	200 fL	>200 fL	>50 fL
Grey shades	256	64	16
Power supply	12 to 28 V DC, 29 W	12 to 28 V DC, 29 W	12 V DC, 5 W

VERIFIED

EDM-700 family of Engine Data Management systems

The EDM-700 family is a complete engine data recording system, it records not only Exhaust Gas Temperature (EGT) and Cylinder Head Temperature (CHT) as some systems do, but all 24 engine temperatures and engine pressures, plus RPM and per cent horsepower.

Single- and twin-engined installations are available with Record, 'Snapshot' and Alarm modes of operation.

Contractor
JP Instruments.

NEW ENTRY

EDM-700 Engine Data Management systems
1997/0001441

Display suite for the F-14D update

In 1985 Kaiser was awarded the contract for the entire display suite for the US Navy's F-14D update programme. The multifunction panel display specified by the Navy was Kaiser's Multipurpose Display Repeater Indicator (MDRI) already present in the Navy's inventory for the AV-8B and F-16 aircraft.

In addition to the MDRI, the display suite consists of an advanced Display Processor (DP) and a wide-angle, low-profile HUD. The F-14D complement of equipment includes one HUD, two DPs and three MDRIs.

The F-14D head-up display features an advanced, extremely compact optics design that provides a 30° total field-of-view and dual flat holographic combiners for increased brightness, better see-through vision and raster and cursive capability.

The Display Processor (DP) uses custom gate arrays and a single-card MIL-STD-1750A processor. It can simultaneously drive up to six cockpit displays, including HUDs, multifunction displays, helmet displays and a videotape recorder, all under Mission Computer (MC) control through the primary interface, a dual MIL-STD-1553B bus. The generated functions include alphanumerics, geometric shapes, over 160 prestored formats and new formats definable over the bus. These can be mixed with 525- or 875-line raster video selected from eight possible inputs. In addition, any of the three selected raster formats may be augmented by stroke symbology written in retrace. Stroke only presentations can be in red, yellow and green on display devices utilising the Kaiser Kroma liquid crystal shutter technology.

Four processors handle the input/output and display functions of the DP. Two MIL-STD-1750A CPUs provide system function and display function processing, while a Motorola 658000 microprocessor provides the primary interface with the MC and an Intel 8031 microprocessor provides input/output functions for nine RS-422 serial channels. The software, coded in CMS-2 high-order language, is contained in EEPROMs that are reprogrammable in the aircraft through the IEEE-488 maintenance bus or MIL-STD-1553B bus.

Specifications (DP)
Dimensions: 498.3 × 190.5 × 193.5 mm
Weight: 20.38 kg
Power supply: 115 V AC, 400 Hz, 3 phase, 195 W
Inputs: 8 video, 8 analogue, 52 discrete
Outputs:
(raster) 6 display ports, 1 VTR port, 525/875-line rates, 1:1 or 4:3 aspect ratio
(stroke) 6 independent sets
(hybrid) raster graphics/stroke/video to any display
Interface: dual MIL-STD-1553B
Reliability: 1,700 h MTBF

Operational status
In service in F-14D.

Contractor
Kaiser Electronics.

VERIFIED

Display system for the F-15E

Kaiser is building the head-up/head-down displays for the F-15E aircraft. Each aircraft has two 6 × 6 in (152 × 152 mm) CRT head-down Multifunction Displays (MFDs) in each cockpit and one holographic, wide field of view Head-Up Display (HUD).

By presenting both high-resolution, high-contrast video and fine line, fast writing stroke symbology, the MFD has the capability to fulfil a wide variety of requirements and is compatible with almost any system configuration. High-resolution video from E-O sensors, radar and missiles in either 525- or 875-line rates is fully readable in the high ambient light of the tactical cockpit. Pure stroke modes provide the high information content necessary for such displays as tactical situation displays and JTIDS. The hybrid stroke during retrace mode allows high-resolution stroke symbology to be placed on top of raster displays.

The MFD has nine plug-in subassemblies, all replaceable without harmonisation. It is fully equipped with continuous and initiated BIT. Software cueing on the display surface provides in-flight programmability through bezel-mounted push-buttons.

The HUD employs both raster scan and stroke written symbology to accommodate the FLIR imagery from the Lockheed Martin LANTIRN system and the Raytheon Systems Company AN/APG-70 radar. The Kaiser wide field of view head-up display employs a holographic single combiner glass, permitting a smaller and lighter support structure with less obscuration of forward view.

The holograms for this HUD are made by an associate company, Kaiser Optical Systems Inc.

Specifications
MFD
Dimensions: 345.9 × 190.5 × 196.8 mm
Weight: 10.42 kg
Power supply: 115 V AC, 400 Hz, 3 phase, 180 W
Display: 152 × 152 mm
Resolution: 100 line pairs/in
Reliability: 3,000 h MTBF

Operational status
In service on the F-15E.

Contractor
Kaiser Electronics.

UPDATED

Integrated cockpit display for the F/A-18C/D

The integrated cockpit display in the F/A-18C/D includes the Multifunction Display Indicator (MDI) and Multipurpose Display Repeater Indicator (MDRI).

The F/A-18C/D 5 × 5 in (127 × 127 mm) NVG-compatible colour MDI and symbol generator includes an integral display processor. The processor consists of three Motorola 68020 microprocessors that handle input/output via a MIL-STD-1553 bus and provide display functions to the mission computer. The display functions include a variety of alphanumeric symbols, geometric shapes and large format macros such as a compass rose or pitch ladder. Three fully independent symbol generators are included, allowing the MDI to drive four additional displays with stroke or video symbology, or both.

The MDI suports a variety of formats, including 525-line raster in either 1:1 or 4:3 aspect ratio, 675-line square format, arc scan and wedge for display of high-resolution ground mapping radar and 875- or 1,224-line formats for sensor video. The display generators can handle three independent simultaneous video outputs and can overlay stroke-in-retrace on all of them.

Kroma liquid crystal shutter technology is incorporated into these displays, allowing stroke symbols to be presented in either red, yellow or green. The use of colour improves pilot situation awareness by providing quick differentiation between friend, enemy and unknown in a cluttered tactical situation without sacrificing resolution and brightness. Colour signals are available externally so that each MDI can drive similar remote colour displays, as in the F/A-18 night attack aircraft.

Originally designed for the night attack version of the F/A-18, the 5 × 5 in (127 × 127 mm) MDRI raster/stroke display provides night vision-compatible Kroma colour stroke and monochrome raster displays in a wide variety of formats.

While air-to-air displays are presented in colour stroke, with red, green and yellow symbology available, air-to-ground views are matched to the AN/APG-65 radar by means of special arc-scan, DBS patch and rotating raster forms which preserve in the display the inherently high-resolution available from that radar. Sensor video, such as FLIR, can also be accommodated with normal rasters in 25-, 675- or 875-line format, in either 1:1 or 4:3 aspect ratios. All of these rasters may be augmented by sharp, clear, fine line stroke symbology written in retrace.

Dependent on an external symbol generator for raster sweeps and stroke symbols, the resultant small size and 'cathedral' top allow installation high in the instrument panel either on the left or right sides or, as in the case of the F/A-18, on both sides. Push-buttons around the display surface are software cued to provide in-flight programming.

Specifications

Dimensions:
(MDI) 658.4 × 170.2 × 179.1 mm
(MDRI) 397 × 170.2 × 179.1 mm
Weight:
(MDI) 18 kg
(MDRI) 9.7 kg
Power supply: 115 V AC, 400 Hz, 3 phase
Display: 127 × 127 mm
Resolution: 120 lines/in
Reliability:
(MDI) 1,500 h MTBF
(MDRI) 2,600 h MTBF

Operational status

In service on the US Navy and US Marines F/A-18C/D aircraft and on export F/A-18C/D aircraft.

Contractor

Kaiser Electronics.

VERIFIED

MultiPurpose Colour Display (MPCD) for the F/A-18E/F

The MultiPurpose Colour Display (MPCD) is a smart full colour LCD with a 6.25 × 6.25 in (158.7 × 158.7 mm) quad RGGB colour pixel resolution. Incorporating high brightness, the MPCD provides full-colour map video or high-resolution monochrome sensor video overlaid with full-colour graphics symbology. Colour improves pilot situational awareness by providing improved contrast ratio and improved resolution in full sunlight, enhancing the pilot/aircraft interface and reducing pilot workload.

A variety of video formats, including 525-line mono or GRB rasters in either 1:1 or 4:3 aspect ratios, 675-line square format for display of high-resolution ground mapping radar and 875 or 1,224-line formats for sensor video, is standard.

The MPCD architecture includes two fully independent processors which drive the MPCD as well as the Up-Front Control Display (UFCD), a separate monochrome 3.9 × 4.9 in (99.1 × 124.5 mm) LCD with an infrared touchscreen for pilot interactive control. Each display processor can provide formats using external video, internally generated graphics symbology or both, and comprises an 80960 microprocessor, high-performance graphics generator ASICs and high-speed video digitising and formatting. Primary graphics and control instructions are received via a MIL-STD-1553B bus from the aircraft mission computer. Dual-display generators allow any two independent video inputs, overlaid with internally generated symbology, to be displayed on the MPCD and UFCD.

Display functions include an extensive library of alphanumeric symbols, geometric shapes and large format macros such as a compass rose or pitch ladder. Priority information is colour-coded into symbology, alerts, messages and push-button labels.

A third 80960 processor, embodied in a standard HAC-32 MCM, provides UFCD touchscreen processing and communication, navigation and identification format processing. The HAC-32 processor has significant reserve throughput and memory to accommodate expanded future general purpose processing functions.

Specifications

Dimensions: 196.8 × 196.8 × 424.2 mm
Weight: 11.79 kg
Power supply: 115 V AC, 400 Hz, 3 phase, 246 W
Display: 158.7 × 158.7 mm
Resolution: 82 pixels/in
Viewing angle:
(horizontal) ±20°
(vertical) +35 to −5°
Interface: MIL-STD-1553B, discrete, video out
Reliability: 3,000 h MTBF

Operational status

Selected for the F/A-18E/F aircraft.

Contractor

Kaiser Electronics.

VERIFIED

KN-0001 Attitude Director Indicator (ADI)

The KN-0001 ADI is a self-contained 3 in cockpit display designed for both rotary- and fixed-wing aircraft. The ADI is electronically driven, and displays aircraft pitch and roll attitude through use of an inertial displacement gyro. Attitude information is displayed as an analogue spherical display and is available as synchro outputs. The gyro is driven from an alternating supply. To minimise errors in accelerating flight, erection control of the gyro is automatically stopped when the instrument is subjected to lateral accelerations. The gyro maintains gyro erection through use of a pneumatic erection system, thus eliminating the need for the gyro torquers used in mechanical erection systems, improving reliability and MTBF. Options include red or white lightning and NVG compatibility.

Specifications

Dimensions: 76.2 × 76.2 × 228.6 mm
Weight: 2.16 kg
MTBF: 7,000 h

Contractor

Kearfott Guidance and Navigation Corporation.

NEW ENTRY

KN-0001 Attitude Director Indicator (ADI)
1998/0018169

31400 Pneumatic altimeter

The Kollsman 31400 pneumatic altimeter is a 2 in ATI indicator designed to meet the requirements of TSO C10b. The altimeter is used as a standby indicator for aircraft equipped with EFIS. The indicator measures and displays accurate pressure altitude. A single pointer affixed to a handstaff makes one revolution per 1,000 ft and is read against a dial graduated in 20 ft increments. The handstaff simultaneously drives a counter-drum in synchronism with the pointer.

The instrument provides a digital display with a baro correction range of 28.1 to 31 in mercury or 950 to 1,050 millibars. Operating range of the instrument is −1,000 to +50,000 ft. The indicator receives inputs of static pressure. It is integrally lit with 5 V white or blue white lighting.

Operational status

The indicator is in production and is used in a number of aircraft including the de Havilland Dash 8.

Contractor

Kollsman Inc.

VERIFIED

44929 Digital pressure altimeter

The Kollsman 44929 is a 3 in ATI altimeter meeting the requirements of TSO C10b and TSO C88a. This digital altimeter senses atmospheric pressure changes, displays altitude with high accuracy, and generates an altitude reporting signal. The altimeter is designed specifically for aircraft that do not require static pressure source correction.

The instrument is driven solely by atmospheric pressure acting on dual aneroid diaphragms. No electrical power is required to operate the pneumatic mechanism. The encoding feature of the altimeter is provided by means of an optical encoder which uses light emitting diode light sources and phototransistor detectors. A code disc is driven by the main shaft of the altimeter mechanism and rotates between the light sources and photo detector array to provide the encoded information to the transponder.

Operating range of the instrument is −1,000 to +50,000 ft.

Operational status

In production and used on many types of subsonic aircraft.

Contractor

Kollsman Inc.

VERIFIED

46650 fuel flow/fuel used indicator

The 46650 fuel flow/fuel used indicator family was designed to replace older electromechanical indicators directly. It features a single microcontroller, four printed circuit boards, a DC torque converter which drives the

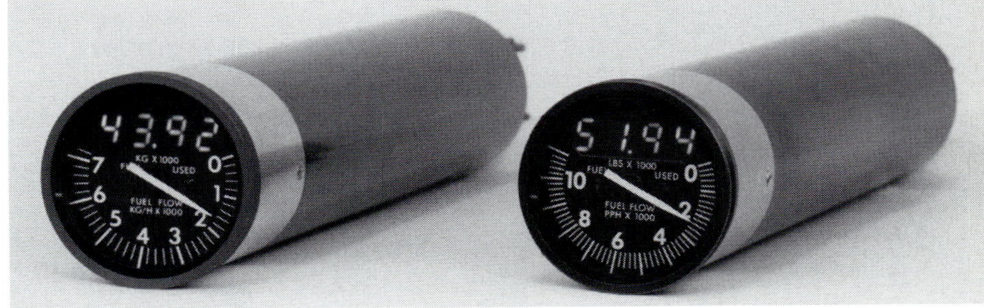

PN 46650 fuel flow/fuel used indicator

pointer and an LED digital display. With a quoted MTBF of over 20,000 hours, the unit is highly reliable. In addition, the circuit boards and major components are easy to assemble, permitting quick repairs.

A pointer, which swings in a 220° arc below the LED display, indicates the total fuel flow. The LED shows the fuel used. The indicator can be used either with old fuel flow transducers or the new motorless transducers.

The 5 V aircraft instrument lighting system controls the bezel lighting. Based on bezel lighting values, the microprocessor controls the intensity of the digital display. Input power is 115 V AC, 400 Hz. In the event of power failure, a non-volatile memory saves the fuel used readings for a minimum of 30 days. A self-test feature verifies indicator integrity.

Operational status
The fuel flow/fuel used indicator is used on Boeing 727, 737, DC-8 and DC-9.

Contractor
Kollsman Inc.

UPDATED

47174-( ) Resolution Advisory/ Vertical Speed Indicator (RA/VSI)

The 47174-( ) Resolution Advisory/Vertical Speed Indicator (RA/VSI) was designed to meet wide applications for Traffic alert and Collision Avoidance Systems (TCAS) installations. It is fully compliant with the TCAS guidelines set out in ARINC 735 and will interface with all TCAS equipment.

A single resolution advisory arc is composed of 52 bi-colour surface-mounted LEDs. These LEDs are individually addressed by an embedded microcontroller and provide a high-resolution and flexible RA display.

The modular design of the instrument satisfies all VSI interfaces. These VSI inputs include an internal pneumatic sensor, ARINC 429, ARINC 565, ARINC 575 and Manchester code. Options are available to accommodate specific VSI interfaces and 28 V DC aircraft power. This flexibility allows an operator to utilise a single indicator that will be universal throughout the whole fleet.

Specifications
Power supply: 115 V AC, 400 Hz or 28 V DC

Operational status
AMR Eagle has selected the Kollsman RA/VSI for its fleet of Saab 340 aircraft.

Contractor
Kollsman Inc.

VERIFIED

48660-( ) EGT indicator

The 48660-( ) solid-state Exhaust Gas Temperature (EGT) indicator has been designed as a direct replacement for older electromechanical indicators. The installation requires no panel or harness modifications. The EGT indicator interfaces with the existing aircraft chromel/alumel Type K thermocouple harness. A single microcontroller processes the thermocouple signal and executes all commands within the indicator. Output for an Airborne Integrated Data System (AIDS) is a standard feature with each indicator.

In the normal operating mode, the EGT indicator displays temperature on both the dial and the three-digit LED display. When sensed temperature drives the pointer into the yellow or caution zone, an amber warning light illuminates to warn of an impending over temperature condition. If the maximum allowable temperature is exceeded, the indicator stores the greatest over temperature value. The blue recall push-button is then illuminated. The duration of the over temperature event is calculated and stored.

The stored information is retrieved by depressing the recall push-button, and the over temperature value and duration are shown on the LED display. When the aircraft over-limit reset switch is activated the blue recall push-button is extinguished. Non-volatile memory, provided through an EEPROM, does not require an internal or external battery when aircraft power is off.

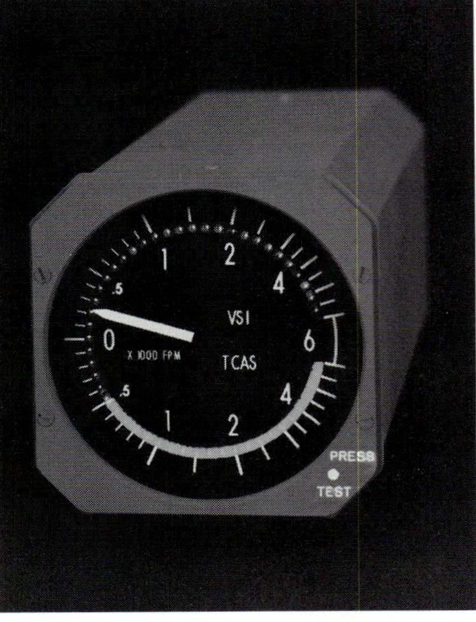

The Kollsman resolution advisory/vertical speed indicator

Operational status
Selected for the Federal Express DC-10-10 and DC-10-30 fleet.

Contractor
Kollsman Inc.

VERIFIED

C-17 Mission computer/display unit

The C-17 computer/display unit, in conjunction with the C-17 multifunction control panel (see next item) and other keyboard units, acts as the single point of control for all primary navigation system modes and sensors to control sensors, display sensor and mission computer data and for input/output control of the mission computer. Display data is available in either 5 × 7 in (127 × 177.8 mm) or 7 × 9 in (177.8 × 228.6 mm) alpha and numeric forms.

Specifications
Dimensions: 122 × 146 × 238 mm
Weight: 4.32 kg
Power supply: 115 V AC, 400 Hz, 84 W
Reliability: >10,000 h MTBF

Operational status
In production for the C-17 aircraft.

The multifunction control panel for the C-17 **1995**

Contractor
Litton Guidance & Control Systems.

VERIFIED

C-17 Multifunction control panel

The C-17 multifunction control panel serves as the control for data displayed on the multifunction displays and the HUDs. It is used to command the operating modes and display formats, as well as for range selection. It also controls and displays the altitude set function.

Specifications
Dimensions: 160 × 146 × 194 mm
Weight: 3.98 kg
Power supply: 115 V AC, 400 Hz, 35.8 W
Reliability: >10,000 h MTBF

Operational status
In production for the C-17 aircraft.

Contractor
Litton Guidance & Control Systems.

VERIFIED

EA-6B Digital display group

The EA-6B digital display group is a derivative of the AN/ASN-123 tactical navigation set developed for ASW. It is employed on the EA-6B aircraft to display electronic warfare situation data and is the primary method of EW system control and operation.

The system comprises one Display Processor (DP), two Digital Display Indicators (DDIs), two Digital Display Indicator Control (DDIC) units and an Interface Computer Unit (ICU). The additional capability to drive these displays and development of a full-colour capability have been completed recently.

The DP interfaces with the aircraft master computer via a MIL-STD-1553 interface. It also interfaces with all the other components of the digital display group and master control panels via discrete lines.

The DDI provides the high-resolution, high-contrast display capability of the EA-6B avionics suite. When used in its three modes of stroke, raster and hybrid, it can display tactical navigation, situation plots and sensor image data. The modes and command data are generated by the DP, while data entry and display control are accomplished through the DDIC.

Operational status
In service on the EA-6B.

Contractor
Litton Guidance & Control Systems.

UPDATED

Full function AMLCD display/ processing systems

Litton Guidance & Control Systems multifunction and full function display systems combine commercial and military components to provide cost effective design. Features include thermal control for cold temperature operation and a backlight of proprietary design that provides very high output.

The full function display system is capable of real-time 2-D and 3-D graphics required for modern digital map system display. Map memory storage can be internal to the full function display system or accessed via the SCSI-2 interface.

The video processor is capable of receiving various sensor inputs (RS-170, RS-343 and PAL) and performing the appropriate resizing to meet the selected resolution. These video inputs can have digital data overlaid, or be viewed in two independent windows. Various chassis sizes are available to incorporate custom input/output channels and to accommodate mechanical constraints for cockpit installation in existing aircraft.

Specifications

Display type: colour active matrix liquid crystal display
Display area: nominal 6 × 8 in; 10 in diagonal
Luminance: 0.05-200 ft L
Colours: 262,144
Pixel resolution: 640 × 480; 600 × 800; 1,024 × 768
NVIS: Class B
MTBF: 3,500 h
Dimensions: 252 × 211 × 152-381 mm (various models)
Weight: <11.4 kg
Power: 28 V DC, <75 W

Contractor

Litton Guidance & Control Systems.

NEW ENTRY

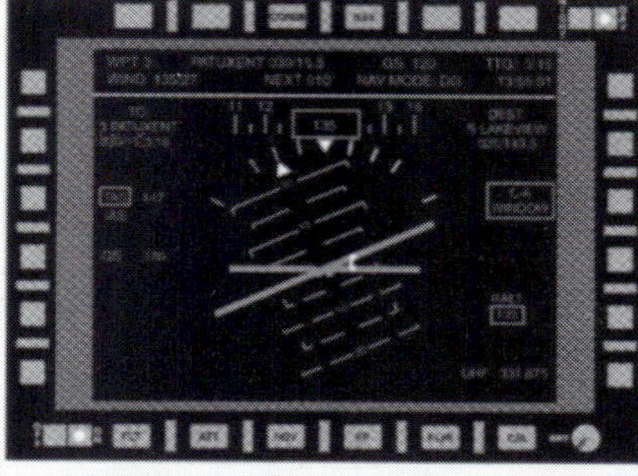

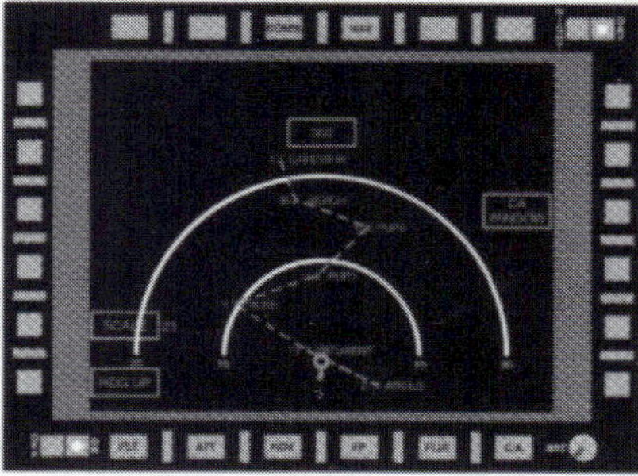

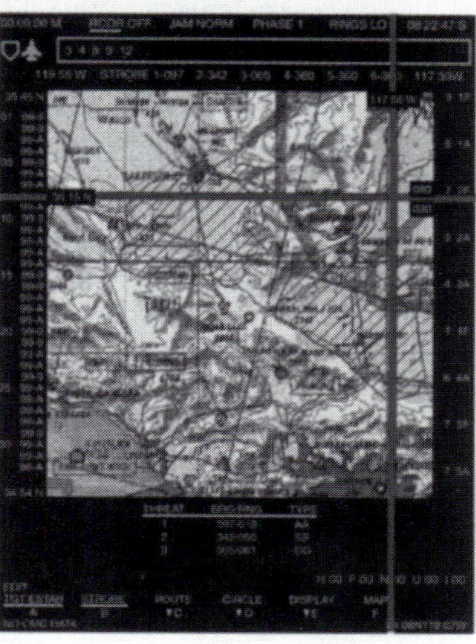

Litton Guidance & Control Systems full function AMLCD display/processing systems **1998**/0018168

Model 1040C Doppler indicator

The Model 1040C Doppler indicator is a pilot-operated control display unit that provides functional control of the Litton AN/APN-217 and digitally displays the Doppler-provided readout of aircraft groundspeed and drift angle.

Major control functions of the system include Doppler On/Off and Land/Sea modes. Test capabilities include the ability to execute and report the result of the Doppler BIT and Doppler self-test velocity. Doppler BIT failures are reported by the malfunction light, while self-test velocity results are indicated on the digital display. Other Doppler indications are memory condition, indicated by a panel light, and invalid data, by dashes on the display. The Model 1040C also performs its own BIT which blanks the display in the event of a failure.

Specifications

Dimensions: 125 × 146 × 75 mm
Weight: 2.2 kg
Power supply: 115 V AC, 400 Hz
28 V DC

Operational status

In production.

Contractor

Litton Guidance & Control Systems.

VERIFIED

The Model 1040C Doppler indicator shows groundspeed (left) and drift angle (right) **1995**

Model 1046 multipurpose indicator

The Model 1046 multipurpose indicator is a solid-state instrument which is compatible with third-generation night vision goggles. It uses six digital displays and two graphic displays to provide the pilot with steering and hover information, making it an ideal instrument for rotary-wing as well as fixed-wing applications.

In the navigation mode the indicator provides a fly-to graphic display as well as command heading and range to checkpoint, groundspeed and estimated time en route. In the hover mode the graphic displays show heading, drift and vertical velocities. Both modes include a system status display.

Specifications

Dimensions: 205 × 83 × 83 mm
Weight: 1 kg
Input signal: ARINC 419 compatible 32-bit data

Operational status

Installed in the US Marine Corps OV-10 and US Navy UH-1N and CH-46 aircraft.

Contractor

Litton Guidance & Control Systems.

VERIFIED

Smart Display Unit (SDU)

The SDU combines the functions of a powerful mission computer, a versatile colour graphics processor and a colour display and control panel in a single lightweight package. The SDU includes a high-resolution 4 × 4 in (101.6 × 101.6 mm) viewable area, active matrix liquid crystal display with integral controller and back-lighting. A 32-bit CPU with 8 Mbytes of memory provides the SDU with the computational power for solving complex tasks in real time.

The SDU is suited for applications such as avionics management, navigation, ECM, tactical datalink, fire control and stores/weapons management, GPS navigation and control, internal digital map and communications control.

The display unit contains a full alphanumeric keyboard, function keys, trackerball and colour active matrix LCD which can be used to superimpose targeting symbology over a FLIR/TV video image or display graphics of system status for the operator. With snapshot of video datalink module installed, the operator can freeze the screen and transfer the whole or a portion of the image to another similarly equipped platform via standard VHF/UHF radios and crypto equipment. Multiple images can be stored for later transmission.

The SDU also provides an independent external display output and inputs for a standard keyboard and trackball which allows it to function as a powerful tactical workstation processor. Full video windowing is provided on the 800 by 600 external display output, which means two video sources can be displayed simultaneously along with a full-colour digital map with tactical symbology.

Specifications

Dimensions: 146 × 171.5 × 180.3 mm
Weight: <3.63 kg
Power supply: 28 V DC, <40 W
Temperature range: −40 to +55°C
Environmental: MIL-E-5400 Class 1A
Display: 101.6 × 101.6 mm, 512 × 512 pixels
Interfaces: MIL-STD-1553B, RS-232, RS-422, RS-170A video, discretes

Operational status

Prototype units have been built and demonstrated.

Contractor

Litton Guidance & Control Systems.

VERIFIED

Bearing/Distance Horizontal Situation Indicators (BDHSI)

The BDHSI-421 was designed to display heading, VOR, ILS, ADF, DME, Tacan or long-range navigation information. Several variations are available that can be configured to the operator's requirements. This indicator is a 4 ATI form factor.

The BDHSI-423 was designed as a successor to the BDHSI-421, displaying heading, VOR, ILS, ADF, DME, Tacan or long-range navigation information, but with the capability of a second bearing pointer. The unit is compatible with any medium- or high-performance fixed-wing aircraft or helicopter.

Operational status

The BDHSI-421 is no longer in production. A variant of this indicator is fitted to UK Royal Air Force Tucano aircraft.

The BDHSI-423 is currently fitted on the Agusta A 109 and Eurocopter BK-117 helicopters.

Contractor

Litton Special Devices.

VERIFIED

Digital Bearing/Distance/heading Indicator

The BDI-300A, a dual-switched RMI with dual-digital DME display is designed to interface directly with

The BDI-302 digital bearing/distance/heading indicator **1995**

new-generation avionics transmitting ARINC 429 data. The indicator is a 3 × 4 in (76.2 × 102.6 mm) ATI form factor.

The BDI-302A, a dual-switched RMI with dual-digital DME display, is designed to interface directly with new-generation avionics, transmitting ARINC 429 or Collins CSDB, RS-422 format. The indicator is a 3 × 4 in (76.2 × 102.6 mm) form factor.

The indicator is compatible with any high-performance fixed-wing aircraft or helicopter.

Specifications

Dimensions:
(BDI-300A) 106.7 × 86.4 × 190.5 mm
(BDI-302A) 106.7 × 86.4 × 177.8 mm
Weight:
(BDI-300A) 0.27 kg
(BDI-302A) 1.4 kg
Temperature range:
(BDI-300A) −15 to +55°C
(BDI-302A) −15 to +70°C
Altitude: up to 15,000 ft
TSO: C-6a, C-66b

Operational status

The BDI-300A is currently fitted to the Gulfstream IV and Canadair Challenger aircraft.

The BDI-302A is fitted on the Dassault Falcon 2000 and the Raytheon Hawker 1000 aircraft.

Contractor

Litton Special Devices.

VERIFIED

Digital radio magnetic indicators

The RMI-303a is a 3 in (76.2 mm) form factor indicator which displays magnetic heading from either of two ARINC 429 sources or from a flux valve. Annunciators are provided to identify the source of the heading information. When in the normal mode, ARINC 429 heading is displayed from the primary heading source. On loss of valid primary heading, the indicator automatically reverts to the alternative ARINC 429 input. If this source also becomes invalid, the indicator automatically reverts to the standby heading mode, receiving heading information from the flux valve. A mode switch is provided to manually select the desired heading source. In addition to the heading display, two bearing pointers are provided. Each pointer is switchable between its respective ARINC 429 ADF or VOR source.

The RMI-321 is fitted in the Falcon 50, Learjet 60 and Yak-142 **1995**

Although designed for the Gulfstream V, the RMI-303A is compatible with any high-performance aircraft.

The RMI-321 is a dual-switched pointer Radio Magnetic Indicator (RMI) in a 3 in (76.2 mm) form factor. The unit is designed to interface with the later generation avionics transmitting signals in ARINC 429, Collins CSDB, sin/cos or ARINC synchro formats for ADF, VOR or heading information.

Specifications

(RMI-321)
Weight: 1.41 kg (max)
Power supply: 27.5 V DC, 0.65 A (max)
Accuracy:
(card) ±1°
(pointer) ±2°
Temperature range: −15 to +70°C
Altitude: −1,000 to 15,000 ft
Environmental: DO-160C

Operational status

Currently fitted in the Dassault Falcon 50, Learjet 60 and Yak-142.

Contractor

Litton Special Devices.

VERIFIED

Horizontal Situation Indicators (HSI)

The HSI-315 is a 3 in (76.2 mm) form factor indicator displaying heading from either ARINC 429 or ARINC synchro input, ADF bearing from a DC sin/cos source and a manually selected course pointer. The unit contains a built-in VOR/Loc converter.

The unit is compatible with any high-performance aircraft.

The HSI-415 is a 3 in (76.2 mm) form factor indicator capable of functioning as an HSI or switchable to perform as an RMI. In the HSI mode, the unit displays aircraft heading, bearing from either ARINC synchro or DC sine/cos source, manually selected course, VOR/Loc deviation, VOR to/from indication and glide slope deviation. In the RMI mode, the course pointer is continuously motorised to a position that will centre the VOR deviation bar and indicate to the station. With the simultaneous display of ADF bearing, the HSI-415 provides all the functions of an RMI.

The unit is compatible with any high-performance fixed-wing aircraft or helicopter.

The HSI-8131 was designed as a 3 ATI form factor instrument compatible with most ARINC analogue directional gyros. It is applicable to all rotary- and fixed-wing aircraft.

Specifications

(HSI-415)
Dimensions: 3 ATI form factor
Weight: 1.59 kg (max)
Power supply: 27.5 V DC, 1.5 A (max)
Temperature range: −30 to +70°C
Altitude: −1,000 to 55,000 ft
Environmental: DO-160B

Operational status

The HSI-315 is currently fitted on the Citation X, Citation V Ultra and Citation II Bravo aircraft.

The HSI-415 is currently fitted on the Learjet 31A, Sikorsky S-76 and Agusta A 109.

The HSI-8131 is no longer in production.

Contractor

Litton Special Devices.

VERIFIED

Liquid Crystal Display indicators

Litton solid-state Liquid Crystal Display (LCD) indicators employ advanced microprocessor circuitry with liquid crystal technology. Interface can be accomplished with analogue and digital signals from inputs of thermocouples, strain gauges, potentiometers and LVDTs. The indicators can be adapted for direct replacement of electromechanical indicators.

The LCDs function over a wide temperature range and work well under harsh environmental conditions. They can be grouped together in one display or installed as separate indicators and can be supplied in round scale formats or vertical bar graphs in various sizes. Digital augmentation is available to provide greater accuracy when required. Displays can be back-lit with electroluminescent, incandescent lighting or LEDs, and can be made NVG-compatible.

Formats available include a compact space-saving design as small as 1.5 in (38 mm), vertical scales with or without digital display for torque, tachometer, pressure or temperature display and large formats available as a single-pointer, dual-pointer or single-bar display augmented with a digital readout.

Operational status

Currently used in military, commercial and business fixed- and rotary-wing aircraft.

Contractor

Litton Special Devices.

VERIFIED

Radio Magnetic Indicators (RMI)

The Model 3100 is a dual-switched Radio Magnetic Indicator (RMI) in a 3 in (76.2 mm) form factor. The VOR pointers will accept AC sine/cos bearing information and the ADF pointer will accept ARINC synchro information. The slaved heading card accepts ARINC synchro data bearing information for selected ADF or VOR stations. The unit is TSO'd and is applicable to any fixed-wing or helicopter application.

The Model 3337 is a dual-switched Radio Magnetic Indicator (RMI) in a 3 in (76.2 mm) form factor. The pointers in VOR position will accept either AC sine/cos or ARINC synchro information. In the ADF position, the pointers will accept either DC sine/cos or ARINC synchro signals. The slaved heading card will accept ARINC synchro signals.

Specifications

Weight: 1.13 kg
Power supply: 28 V DC
(Model 3100) 0.55 A (max)
(Model 3337) 0.75 A (max)
Accuracy:
(card) ±1°
(pointers) ± 2°
Temperature range:
(Model 3100) −30 to +55°C
(Model 3337) −30 to +70°C
Altitude:
(Model 3100) −1,000 to 20,000 ft
(Model 3337) −1,000 to 40,000 ft
Environmental : DO-138

Operational status

The Model 3100 is currently fitted in the Jetstream 61, de Havilland Dash 8 and other aircraft.

The Model 3337 is currently fitted on the ATR-42 and ATR-72, Saab 340, UK Royal Air Force Tucano, Jetstream 41 and corporate and commuter aircraft, as well as helicopters.

Contractor

Litton Special Devices.

VERIFIED

The Model 3100 RMI is fitted in the Jetstream 61 and de Havilland Dash 8 **1995**

Actiview active matrix LCD displays

The Actiview product line provides ruggedised, COTS, flat panel displays to meet a multitude of system architectures and installation requirements. Available with screen sizes ranging from 8.8 to 20.1 in diagonal, each may be acquired as stand-alone or as integrated smart displays.

Features include embedded graphics generation, sensor video processing, multiple line rate video inputs and a full complement of interfaces. Actiview displays offer superior image quality combined with true sunlight readability.

Specifications of Actiview 20

Display area: 399.4 × 319.5 mm
Resolution: 1,280 × 1,024
Luminance: 40 fL
Colours: 262,144
Grey shades: 256 levels
Viewing angle: 120° cone (+60° horizontal and vertical)
Weight: 15.9 kg
Power input: 110 V AC, 60 Hz
Reliability: 10,000 h

Contractor

L.3 Display Systems.

UPDATED

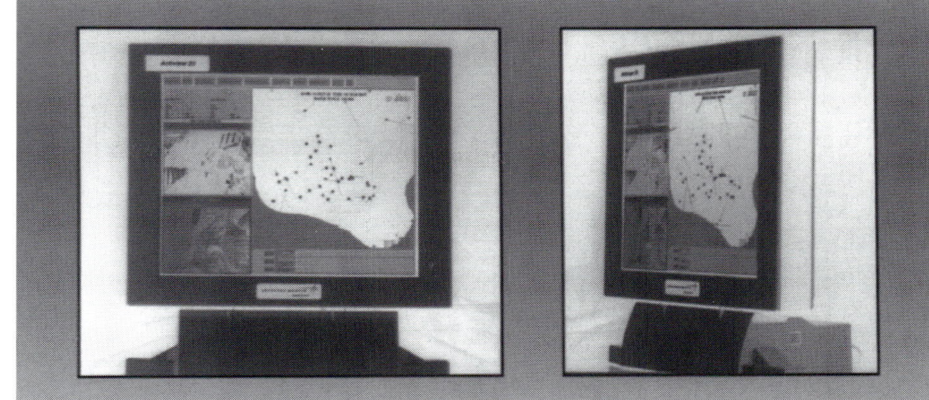

The L.3 Display Systems Actiview AMLCD **1997**/0001389

Cockpit Multifunction Displays and Display Electronics Units (MFD/DEU) for V-22 Osprey aircraft

The MFD/DEUs are ruggedised, full colour beam index CRT cockpit displays and display electronics units for the V-22 Osprey aircraft. The suite comprises four Multi-Function Displays (MFDs) and two Display Electronics Units (DEUs) per aircraft. The displays are NVG-compatible and provide graphics/symbology overlays on digital map, FLIR and radar sensor video inputs. There are two independent channels of high-performance display processing.

Operational status

US Navy/US Marine Corps EMD contract awarded 1994; LRIP contract awarded 1996.

Contractor

L.3 Display Systems.

UPDATED

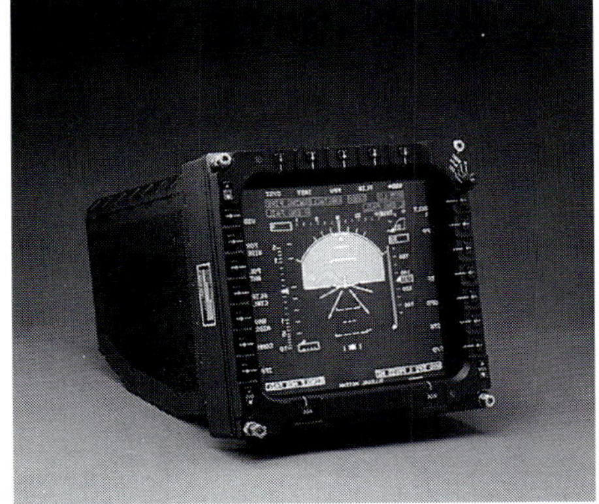

L.3 Display Systems multifunction display for the V-22 Osprey aircraft **1998**/0018167

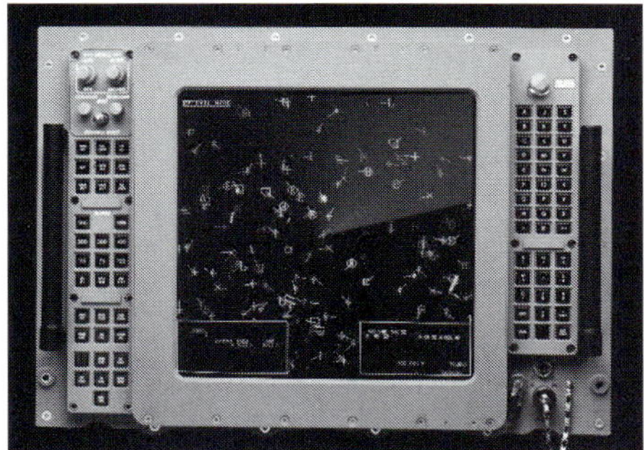

The L.3 Display Systems enhanced main display unit for the E-2C aircraft

Enhanced Main Display Unit for the E-2C

The Enhanced Main Display Unit (EMDU) is the primary information display for the crew of the E-2C Hawkeye AEW aircraft. Each aircraft carries a complement of three EMDUs. The full-colour EMDU displays more than 2,000 radar tracks on an 11 × 11 in (279.4 × 279.4 mm) screen. The EMDU also allows the operators to overlay a map of the search area over their track files and display up to three separate windows containing additional information.

The EMDU utilises beam index CRT technology to provide sharp high-resolution colour images. In addition to the CRT, the EMDU contains processing subsystems that perform graphics generation, radar scan conversion and data input/output functions. The EMDU performs all display oriented processing internally, thus offloading that requirement from the E-2C's L304 central computer.

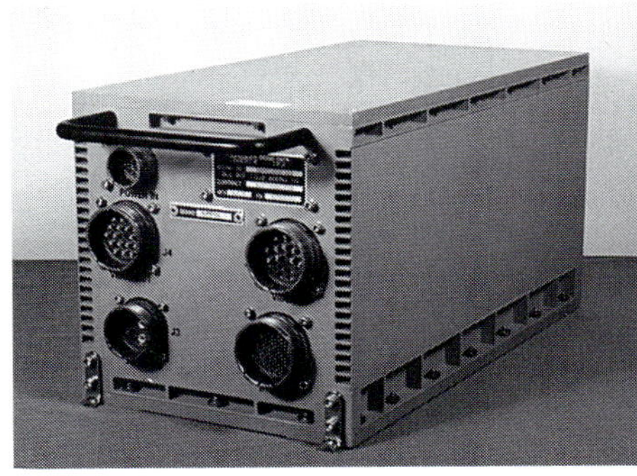

L.3 Display Systems display electronics unit for the V-22 Osprey aircraft **1998**/0018166

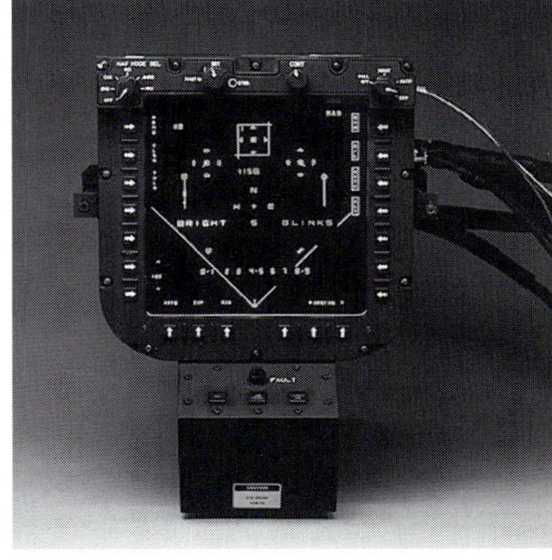

L.3 Display Systems programmable tactical information display for the F-14 aircraft **1998**/0018165

Specifications

Weight: 65.77 kg
Resolution: 900 × 900 pixels
Reliability: 1,100 h MTBF

Operational status

In service in US Navy E-2C aircraft.

Contractor

L.3 Display Systems

UPDATED

Programmable Tactical Information Display (PTID) for F-14 aircraft

The PTID is an 8 × 8 in monochrome CRT display. It features high brightness and contrast; hybrid stroke on raster picture; embedded display processor with dual 1750A CPUs at 30 MHz using ADA software (MIL-STD-2167A); video select mixer and MIL-STD-1553B/F-14-unique I/O. It is now integrated with LANTIRN.

Operational status

Currently in production.

Contractor

L.3 Display Systems.

UPDATED

Integrated tactical display

A multipurpose television and random write alphanumeric, graphic and video display for military applications, this system is essentially a colour version of the AN/ASA-82. It is suitable for stand-alone operation by virtue of a built-in TI-9900 microprocessor, display generator and 8 k 16-bit random access/read-only refresh memory. It presents on a 16 in (406 mm) screen information from electro-optical, radar and acoustic sensors in different formats, each of which can be overlaid with tabular, graphic situation, television raster and raw radar symbology.

Local data storage, an internal refresh function and the ability to program through the microprocessor all help to offload the host computer, increase the quantity of data that can be displayed and allow a more flexible operation of the equipment. The use of colour reduces clutter in high-density displays and increases operator comprehension.

Specifications

Dimensions:
(display) 457 × 368 × 559 mm
Weight:
(display) 33 kg
Display size: 229 × 323 mm
Spot size: 15 mils
Addressability: 1,024 × 1,024
Random write speed: 100,000 in/s
Linearity: 1%

Lockheed Martin integrated tactical display

Operational status

In limited production for some aircraft types with classified electronic warfare functions.

Contractor

Lockheed Martin Fairchild Systems.

VERIFIED

IDME 891 DME and VOR/ILS indicator

The Narco IDME 891 is a remotely channelled DME and VOR/ILS indicator. Designed for use with the Mk 12D or E Nav/Com, Nav 825 navigation receiver, NS 801 RNav and the CP 136M audio panel/marker beacon receiver, the IDME 891 simplifies navigation and requires very little panel space.

The IDME 891 combines VOR/ILS/DME and marker beacon lights all in one package and fits the standard 3 in instrument panel hole. The IDME 891 matches the Narco DME-890 for power, accuracy and lock on capability. The readout is a high-intensity display which indicates in increments of 0.1 n mile in the range mode and 1 kt in the groundspeed mode. Course indication indicator, glide slope, operational flags and marker beacon lights are standard.

Specifications

Weight:
(IDME-891) 1.18 kg
(antenna) 0.21 kg
Power supply: 11-32 V DC
Channels: 220
Display range: 0-160 n miles
Accuracy:
(range) ±0.1 n mile
(groundspeed) ±5 kt or 5% (whichever is greater)

Contractor

Narco Avionics.

VERIFIED

Liquid Crystal Crew Display Units (LCCDU)

The LCCDU provides monaural or binaural capability as required to each position equipped with an LCCDU and provides the microphone interface to the Communications Control Unit (CCU).

Through the full function keyboard, the LCCDU features control of the transmit and receive functions including the control and reassignment of all communication assets, crypto assignments, and Built-In Test (BIT) readouts. Complete status of the communications assets for each operator is available through the eight colour liquid crystal display. Each LCCDU is identical and completely interchangeable with any other position location. The CCU may be programmed in real time if necessary to allow or disallow access of any given communications asset by any given crew position via the LCCDU.

Liquid Crystal Crew Display Unit (LCCDU)
1998/0018162

The integral display intensity control is capable of adjusting the display brightness from off to full intensity without external controls or potentiometers. The LCCDU interfaces with the CCU via a serial RS-422 control data twisted-shielded wire pair.

Options include: tone generation crypto/modem switching; dual crew binaural audio; cross-band or in-band radio relay.

Specifications

Dimensions: 165.1 × 146 × 152.4 mm
Weight: 2.73 kg
Power: 28 V DC, 24 W

Contractor

Palomar Products, Inc.

NEW ENTRY

Model 8000 Multifunction Display (MFD)

The Model 8000 MFD is fully compatible with AN/AAQ-16 night vision systems for either new installation of for retrofit purposes. It is also compatible with Generation II and Generation III night vision goggles.

The Model 8000 is sunlight readable, with a display output greater than 200 fL (up to 400 fL) and a contrast ratio of 6:1 @ 10,000 fc ambient illumination.

Model 8000 Multifunction sunlight readable display
1997/0001393

The front control panel features 14 push-buttons with 14-bit encoded outputs, day/night/off switch, contrast and brightness adjustment, display size, centre and symbol brightness adjustment, Video 1 or 2 select. Front panel functions can be customised for specific requirements.

Specifications

Weight: < 6.4 kg
Largest viewable display area: 122 × 162 mm

Contractor

Palomar Products, Inc.

VERIFIED

T8653 and T8660 taut shadow mask colour CRTs

A patented taut shadow-mask design for this full colour line of CRTs offers very high purity and brightness. Excellent thermal stability means that high beam currents can be achieved without loss of colour purity. This results in a rugged CRT for military aircraft applications that offers a sharp, highly readable display usable in brilliant cockpit sunlight or under demanding night vision conditions.

Specifications

Model:	T8660	T8653
Dimensions:	6″ × 6″	5″ × 5″
maximum weight:	5.0 kg	3.86 kg
maximum length:	372.1 mm	326.4 mm
minimum viewing area:	6.0 × 6.0 in 152.4 × 152.4 mm	5.0 × 5.0 in 127.0 × 127.0 mm
outer width/height:	7.3 × 7.3 in 185.4 × 185.4 mm	6.2 × 6.2 in 157.5 × 157.5 mm

Contractor

Planar Advance Inc.

VERIFIED

AMLCDs

Planar manufactures a range of AMLCD displays for military, industrial and commercial applications. They include ElectroLuminescent (EL) flat panel displays, miniature Active Matrix ElectroLuminescent (AMEL), 1000 lines per in units suitable for Head-Mounted Displays, Active Matrix Liquid Crystal Displays (AMLCD), as well as full colour CAT-based avionic systems.

The Planar 11 × 8 in landscape orientation flat panel display is an amorphous silicon AMLCD full colour display for use in environmentally demanding applications that require maximum readability, rugged construction, and a wide temperature range.

Planar AMLCD **1997**/0001392

B-2 cockpit with Planar displays **1997**/0003322

Specifications

Model: 5.0 × 5.0 in
Active Display Area (mm): 127 × 127
Pixel configuration: RGGB Quad
Display Resolution:
Full Colour: 480 × 480
Monochrome: 960 × 960
Grey levels: 64/256

Viewing angles:
horizontal: >±45°
vertical: >+35°; −15°
NVIS compatibility (optional): NVIS-B MIL-STD-L-85762 Class B

Model: 6.25 × 6.25 in
Active Display Area (mm): 159 × 159
Pixel configuration: RGGB Quad
Display Resolution:
Full Colour: 512 × 512
Monochrome: 1,024 × 1,024
Grey levels: 24/256

Viewing angles:
horizontal: >±45°
vertical: >+35°; −15°
NVIS compatibility (optional): NVIS-B MIL-STD-L-85762 Class B

Model: 6.0 × 8.0-Q/6.0 × 8.0-S/8.0 × 6.0-S in
Active Display Area (mm):
157 × 211/157 × 211/211 × 157
Pixel configuration: RGGB Quad (Q model); RGGB Quad (S model)
Display Resolution:
Full Colour: 600 × 800
Monochrome: 1,200 × 1,600
Grey levels: 64/256

Viewing angles:
horizontal: >±45°
vertical: >+35°; −15°
NVIS compatibility (optional): NVIS-B MIL-STD-L-85762 Class B

Model: 11.0 × 8.0-Q
Active Display Area (mm): 276 × 200
Pixel configuration: RGGB Quad
Display Resolution:
Full Colour: 1,536 × 1,120
Monochrome: 3,072 × 2,240

Contractor

Planar Advance Inc.

UPDATED

ColorGard displays

ColorGard is a family of display products. Based on NuColor technology, the displays offer the combination of a high-resolution and high-brightness monochrome CRT and a NuColor shutter. The shutter is an electrically switchable filter made up of two fast liquid crystal optical switches, plus a combination of colour and neutral polarisers. Colour is produced by sequentially displaying the red, green and blue field information on the monochrome CRT while the shutter is switched to transmit red, green and blue respectively. Alternate fields, viewed through the different coloured filters, create full-colour, very high-resolution images.

The display offers several advantages over conventional shadow-mask displays, including a contrast ratio of 7.4:1 in 10,000 ft-candles diffuse ambient illumination, resolution that depends only on the performance of the fine focus CRT, brilliant colour uniformity created from colour polarisers that utilise light energy from a single high-energy electron beam and a solid mechanical design for full military applications. The display's colour purity is impervious to magnetic field corruption.

ColorGard 19/R display **1995**

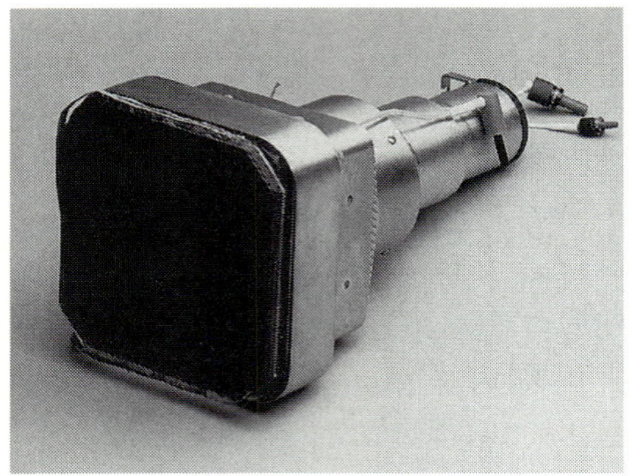

ColorGard 6 display ***1995***

ColorGard 6 offers good colour saturation viewable even in high ambient light. It provides a wide dynamic range of colours and brightness, no shadow-mask restrictions and the high resolution to allow better colour definition and viewability. It is suitable for avionics cockpit displays, terrain-mapping, FLIR imaging and ANVIS and other shipboard and ground-based applications.

ColorGard 19/R meets the needs of users who require a high-quality, full-colur display which has vibrant colours. It is suitable for avionics tracking displays and terrain-mapping and other shipboard and ground-based applications.

Specifications

Dimensions:
(ColorGard 6) 137.2 × 137.2 × 279.4 mm
(ColorGard 19/R cabinet) 558.8 × 431.8 × 584.2 mm
(ColorGard 19/R rack-mount) 482.6 × 444.5 × 538.5 mm
Weight:
(ColorGard 6) 3.17 kg
(ColorGard 19/R cabinet) 38 kg
Power supply:
(ColorGard 19/R) 87-128 V AC, 174-256 V AC, 48-63 Hz, 300 W (max)
Temperature range:
(ColorGard 6) MIL-E-5400 Class 2
(ColorGard 19/R) 0 to +40°C
Altitude:
(ColorGard 6) MIL-E-5400R T Class 2
(ColorGard 19/R) MIL-E-5400R Class 1B
Display:
(ColorGard 6) 114.3 × 114.3 mm
(ColorGard 19/R) 355.6 × 279.4 mm
Resolution:
(ColorGard 6) 960 lines
(ColorGard 19/R) 1,280 × 1,024 pixels

Contractor

Planar Advance Inc.

UPDATED

CDU-900 Control Display Unit

The latest addition to the Rockwell Collins Control Display Unit (CDU) range is the CDU-900. This retains all the features of the CDU-800 series CDU/CDNU, while providing powerful new capabilities. Through utilisation of an Intel 80486 microprocessor and large-scale integration, the CDU-900 provides powerful built-in processing capability and expansion capacity for embedded functions or interfaces to external equipment.

An embedded military P(Y) code GPS receiver/processor is the first expansion module for the CDU-900. The embedded receiver consists of a GPS Embedded Module (GEM) GEM II or GEM III receiver/processor and accompanying adaptors designed to fill two of the three available expansion slots in the CDU-900. Additional expansion modules planned for the CDU-900 include processor and memory modules, various digital, analogue or discrete interface modules for non-MIL-STD-1553B applications and modem/datalink modules. The standard and expansion capabilities of the CDU-900 make it suitable for the integration and control of avionics in both fixed-wing aircraft and helicopters.

Operational status

The CDU-900 with embedded GPS is being integrated with the FMS-800 flight management system for the US Air Force C-5, C-9, E-4B and KC-10. The CDU-900, less the embedded GPS, is being integrated on the US Air Force B-1B and KC-135.

Latest variant of the CDU-900 family is the CDU-900G which provides a complete capability to meet US DoD GPS Integration Guidelines (GIG) for stand-alone GPS navigation, including RNAV and airways.

Contractor

Rockwell Collins.

UPDATED

The Rockwell Collins CDU-900 ***1995***

Control Display Navigation Unit (CDNU)

The Control Display Navigation Unit (CDNU) is a high-power floating point computational engine, keyboard and display system in a single product suitable for a wide variety of integrated flight management and cockpit management applications. The existing Ada software provides a full-feature lateral navigation implementation integrating GPS and air data sensors. Memory and throughput reserves are large, providing significant opportunities for growth. All application software resides in EEPROM, permitting updated software to be installed via the bus interface on the aircraft. There are seven function keys that can be user defined for specific program needs.

Current functions include flight planning with automatic sequencing, guidance computations with associated display driver, intercept of moving waypoints, integrated GPS/air data/heading navigation, internal non-volatile database containing 200 waypoints and access to online identifier database of 20,000 waypoints accessible from a data loader cartridge, integrated control and display of system tests and status, and software updates via the databus with no unit disassembly.

Specifications

Dimensions: 181 × 146 × 165.1 mm
Weight: 4.54 kg (max)
Power supply: 16-32 V DC, 30 W (max)

Operational status

In production for more than 20 US Air Force, Navy, Marine Corps, Army and Coast Guard aircraft types. Deliveries began in 1992.

Contractor

Rockwell Collins.

UPDATED

Digital RMI/DME indicators

Designed to interface with both the current ARINC 700 sensors and the older-generation analogue avionics, this new series of combined Radio Magnetic Indicators and Distance Measuring Equipment (RMI/DME) forms a family of instruments that share many common features including electrical, servo, thermal, packaging and lighting methods. Use of a patent digital encoder/driver module, common to each instrument and each channel within it, permits packaging techniques which greatly improve the effectiveness of the heat transfer arrangements. The encoder/driver module consists of an 11-bit digital encoder, DC motor and associated drive circuits, all under microprocessor control. Each of the six channels is isolated from its neighbour for integrity and each has its own single chip microcomputer. Each channel monitors its own faults and activates its own flag or shutter and output signal.

The members of the digital RMI/DME instrument family comprise:

RMI-733A radio magnetic indicator: a compact lightweight VOR/ADF selectable three-servo instrument. A version designated RMI-733A is a three-servo instrument designed to display information from an ADF receiver.

RDMI-743 radio distance magnetic indicator: this instrument features liquid crystal DME readouts for better reliability and readability. The three-servo unit is selectable to either VOR/ADF or VOR only.

RDMI-743A radio distance magnetic indicator: with a magnetic wheel DME display, the RDMI-743A is a four-servo VOR instrument.

Operational status

All in production.

Contractor

Rockwell Collins.

UPDATED

DLC-800 interactive touchscreen control display

The DLC-800 interactive touchscreen control display is part of the Rockwell Collins Airline Communication And Reporting System (ACARS) product line. It is designed to make data entry for datalink information simpler and error free.

The DLC-800 features a high-resolution flat-panel liquid crystal display with an infrared touch input system and reconfigurable menu formats. The menus can be input from either the DLM-700 or DLM-700B management unit. Other features include highlighted key actions, flashing messages and a wide range of graphic capabilities.

The high-contrast ratio screen is easily read in direct sunlight and at a wide viewing angle for both pilots. Inverse dark video characters on a light screen also enhance readability and message legibility. Larger text sizes can be used to highlight menu titles, while smaller sizes can list detailed information such as flight plans and weather reports.

Aircraft system communication with the DLC-800 is accomplished via ARINC 429. The DLC-800 can be readily tailored to meet individual requirements.

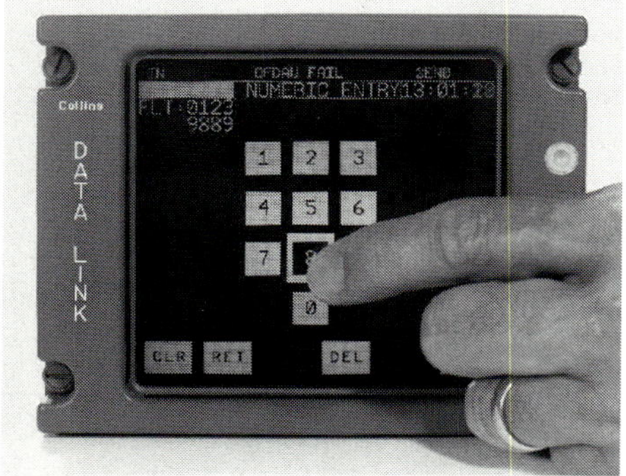

The Rockwell Collins DLC-800 control display shows highlighted keys to confirm correct data entry

Future growth potential includes graphics capability for pictorial and symbol displays such as weather plots. Other growth options include ATC communications such as clearances and route maps and Satcom compatibility.

Specifications

Dimensions: 114 × 146 × 178 mm
Weight: 2.1 kg
Power supply: 115 V AC, 400 Hz, single phase, 25 W

Operational status

Air Canada has ordered the DLC-800 for the Airbus A320.

Contractor

Rockwell Collins.

UPDATED

EFIS-84 Electronic Flight Instrument System

The EFIS-84 is a 4 in electronic flight instrument system offering full capability and high reliability at an affordable cost. The EFIS-84 pairs an EADI and an EHSI to provide the same capabilities as the Collins EFIS-85 and EFIS-86 systems.

The system is designed for applications ranging from helicopters to regional airliners and corporate aircraft. Retrofit of existing 4 in electromechanical systems with the EFIS-84 offers significant operational, reliability and cost of ownership benefits.

The EFIS-84 offers bright clear attitude and navigation information in easy-to-interpret formats. Attitude is displayed in full sky presentation which provides an attitude area more than twice as large as that of electromechanical instruments. A race-track attitude display, with a circular attitude depiction similar to the familiar electromechanical ADI, may also be selected as an installation option. A customer choice of V-bar or cross pointer steering commands is also offered.

A full selection of modes is offered on the EHSI, including the traditional compass rose, arc and map. Significant operational advantages are offered by the extensive map capability, particularly to pilots operating in terminal areas. A key benefit of the EFIS-84 is the ability to display weather radar information integrated with the navigation map. The system is compatible with the Collins WXR-350 or the advanced TWR-850 turbulence weather radar.

System flexibility allows interface with a variety of analogue and digital aircraft sensors, as well as a broad range of long-range navigation options. The optional multifunction display offered as part of the EFIS-84 provides additional system capability, including expanded map display and checklists.

Operational status

In production and in service. Certified on the Beech 1900D regional airline aircraft.

Contractor

Rockwell Collins.

UPDATED

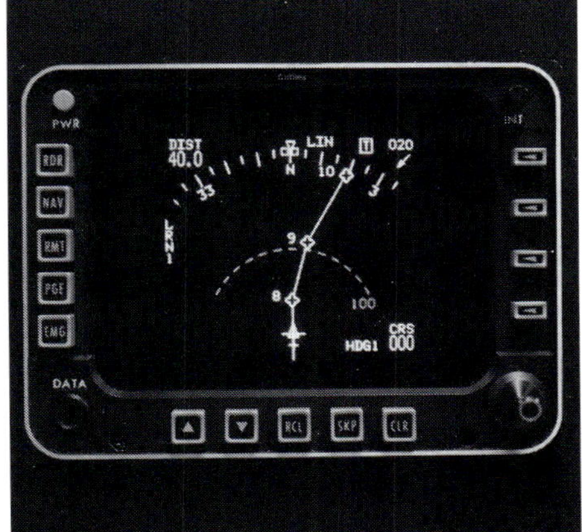

The Rockwell Collins multifunction display offered as an EFIS-84 option

EFIS-85 Electronic Flight Instrument System

A full EFIS-85 system comprises dual 5 × 5 in (127 × 127 mm) EADIs and EHSIs with one set of equipment for each pilot, a shared multifunction display on the centre panel and mode controls. The EADI and EHSI CRTs can display all the information traditionally associated with electromechanical flight director instruments as well as weather radar patterns, navigation maps, performance data and navigation waypoints. Presentation is very flexible. The EHSI-85 can show conventional HSI information on a circular scale or it can expand just the forward sector of the display and show weather maps.

The EFIS-85 CRT incorporates a three-gun assembly, a shadow-mask, a faceplate with phosphor coating and a glass envelope to enclose the elements. The in-line electron gun assembly provides improved convergence and mechanical rigidity and the high-resolution shadow-mask gives four to six times better resolution than that of a domestic television set because the phosphor dots are so much closer together. The displays use both stroke and raster writing. The high-intensity stroke writing of symbols, in conjunction with contrast enhancement filters, enables displays to be read even in full sunlight. Primary colours are red, blue and green with easy synthesis of several derivative colours, including white.

The new advanced map displays are now available with EFIS-85 and -86 electronic flight instrument systems. These include heading up, north up, aircraft centred, north up max view and plan displays. The maps display airports and navaids from navigation systems including the UNS-1A or GNS-1000.

Operational status

In production and in service. The aircraft used for approval trials was a Dassault Falcon 100; certification was announced in December 1982.

Rockwell Collins has certified EFIS-85 on over 50 types of aircraft.

Contractor

Rockwell Collins.

UPDATED

EFIS-86 Electronic Flight Instrument System

In 1983, Rockwell Collins introduced the EFIS-86 Advanced, an electronic flight instrument system with five 6 × 6 in (152 × 152 mm) or 5 × 6 in (127 × 152 mm) CRT displays for top of the range general aviation and commuter aircraft. It comprises dual EADIs, dual EHSIs and a single MFD. The two pilots each have an EADI and an EHSI and share the centrally mounted MFD display.

The CRTs combine raster and stroke writing. The EADI provides conventional ADI information, together with airspeed, airspeed trend and multisource vertical and lateral deviation information. The addition of air data information is facilitated by the increase in the display area of the EFIS-86 which ensures the display remains legible. The large EHSI makes it possible to consolidate all conventional HSI, navigation and weather information directly in front of the pilot, thereby reducing the need to scan other instruments. Three EHSI modes can be selected: full compass rose (as in a conventional HSI), expanded sector display or sector display with weather radar paints.

The system is linked with long-range navigation and flight management systems to show additional information such as an expanded display of HSI, radar and navigation data from the flight instruments either in combination or separately for more detailed examination. The display can store and show preselected lists of waypoints or up to 100 pages of preprogrammed data such as checklists and emergency procedures, written in chapter form for ease of input and retrieval. The display can also act as a standby flight instrument in the event of an EADI or EHSI CRT failure.

An EFIS-86C has been certified on the Canadair Challenger CL-601.

Operational status

In production and in service. EFIS-86 Advanced has been certified in the Dassault Falcon 50 and 200, with airspeed indication on the ADI as the primary speed indication. The aircraft and their EFIS systems have been approved to Cat II operation, the first such approval for an EFIS-equipped business jet. New display symbology has been developed for two Dassault aircraft with speed scale and autopilot mode annunciation. Other systems are flying on the Embraer EMB-120, Gulfstream III, Saab 340 and the Astra Jet.

Contractor

Rockwell Collins.

UPDATED

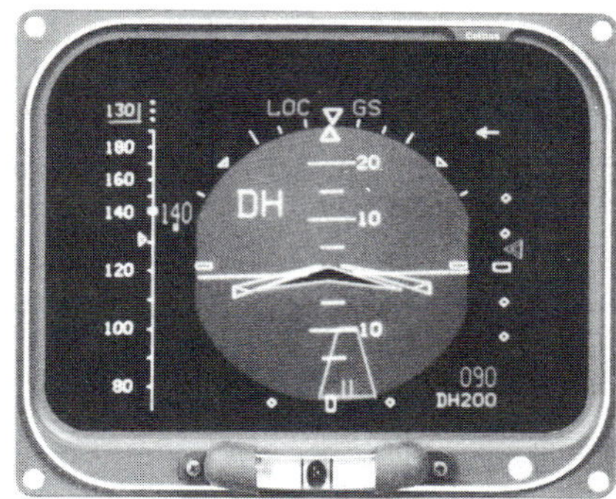

The Rockwell Collins 6 × 5 in EADI with primary airspeed display (above) and weather imagery (below)

EFIS-700 for the Boeing 737, 757 and 767

The EFIS-700 electronic flight instrument system for the Boeing 737, 757 and 767 comprises an EADI and an EHSI for each of the two pilots. Each pair of instruments has an associated mode control panel. The system provides all the functions associated with earlier electromechanical ADI and HSI flight director instruments and, in addition, shows map and flight plan data, weather patterns, radio height, automatic flight control modes, autoland, system status, windshear and flight path information on 7 × 6 in EHSI (178 × 152 mm) and 5 × 6 in EADI (127 × 152 mm) CRTs.

These instruments utilise bright three-gun robust shadow-mask CRT technology permitting no fewer than eight colours. The traditional red, blue and green associated with CRTs is augmented by magenta, yellow, green, cyan and white (not normally used). When used in conjunction with a contrast enhancement filter, the high-resolution CRTs provide bright displays that are readable under all flight deck lighting conditions.

Each of the two pairs of EFIS instruments in an aircraft is driven by its own symbol generator, but a third standby generator is retained as a spare which can be switched in as necessary on failure of a dedicated unit. These symbol generators utilise the Collins RAMP processor, implemented with the latest very large-scale integration and CMOS technology.

The EADI presents primary attitude information, together with pitch and roll steering commands. Secondary data is also shown such as groundspeed, autopilot and autothrottle mode. In order to keep the display uncluttered, information is switched out as soon as it is not needed, for example, instrument landing system and radio height symbols are absent during cruise, appearing only during the final approach.

The EHSI depicts the horizontal position of the aircraft in relation to selected flight data and a map of the navigation features in the vicinity of the aircraft at any given time. Aircraft track, trend vector information and desired flight plan are also displayed. This allows rapid and accurate manual or automatic flight path correction. Other information can be displayed, such as windspeed and direction, vertical deviation from a selected profile and time to the next navigation waypoint. Weather patterns can be superimposed on the navigation picture.

Operational status
Installed in the Boeing 737, 757 and 767.

Contractor
Rockwell Collins.

UPDATED

EFIS-1000 for the Fokker 100

In 1984, Rockwell Collins was chosen to design a third-generation 6 × 7 in (152 × 178 mm) EFIS-1000 system for the Fokker 100. The system utilises six integrated, full-colour CRT displays; each pilot has a Primary Flight Display (PFD) mounted over a Navigation Display (ND) on the instrument panel. These are also two multifunction displays. EFIS control panels mounted on each side of the glareshield allow individual control of CRT brightness, mode of operation and weather radar display. The multifunction display system controller is mounted on the centre pedestal for use by both crew members. Remote light sensors behind the windscreen help to control the automatic brightness function so that the displays are readable under all flight deck lighting conditions.

Operational status
In production and in service in the Fokker 100.

Contractor
Rockwell Collins.

UPDATED

EHSI-74 Electronic Flight Instrument System (EFIS)

EHSI-74 is for the general aviation single- and light twin-turboprop market and comprises a 4 × 4 in (102 × 102 mm) EHSI designed to work in conjunction with the company's ADI-84 attitude director indicator and an information display/radar navigation centre, the Rockwell IND-270 CRT, which is used in conjunction with the company's WXR-270 weather radar. The system is completed by a DCP-270 display control panel.

The IND-270 CRT can present up to 128 pages of easily accessed pilot programmable text and navigation information from a Collins LRN-85 long-range navigation system. The equipment is programmed by a portable data reader that can load the system with performance tables such as cruise/consumption, emergency checklists and other alphanumeric information. Chapter by chapter indexing facilitates input and retrieval. In addition to the information stored within the DCP-270, many pages of data are available from the LRN-85.

The EHSI-74 no longer requires a US STC as do other EFIS in most installations.

Operational status
In production and in service. Certified on the Commander 690, King Air 200, Cessna 441, King Air F90, Bonanza, Beech 100, Mitsubishi MU-2 and dual installation for the Learjet Model 35A. Recently the system was installed in the Chinese Y-7.

Contractor
Rockwell Collins.

UPDATED

Engine Indication and Crew Alerting System for the Boeing 757 and 767 (EICAS)

The Rockwell Collins Engine Indication and Crew Alerting System (EICAS) is standard equipment on the 757 and 767 airliners. The EICAS system comprises two multicolour CRT displays, two computers and a single selector panel. The display unit is identical to the EHSIs on the pilot's display panels, though rotated through 90° in its function as an engine indicating system.

Each aircraft has two EICAS CRTs mounted one above the other on the centre instrument panel where they can be monitored by the two pilots. On the centre panel the top EICAS CRT is programmed to display primary engine information such as engine pressure ratio, fan speed and exhaust gas temperature as electronic symbols representing traditional circular scale and pointer instruments, together with cautionary information, for example, wheel-well overheat or failure of a yaw damper in the flight control system. The lower EICAS display shows lower priority information such as compressor speed, fuel flow and oil temperatures, pressures and tank contents.

In the case of the failure of one EICAS display, priority information automatically switches to the other CRT, and the dual-redundant computer installation permits both CRTs to be driven from one unit.

The multicolour CRT displays in the EICAS configuration measure 7 × 6 in (178 × 152 mm) and are driven by one of the two computers, the other acting as a hot spare.

Layout of the flight panels in the Boeing 757 and 767. Each pilot has a set of two EFIS displays mounted one above the other. The two EICAS displays on the centre panel can be monitored by either pilot

Operational status
In service in Boeing 757 and 767 aircraft.

Contractor
Rockwell Collins.

UPDATED

FDS-255 Flight Display System

The Rockwell Collins Flight Display System (FDS-255) is a 5ATI colour flat-panel display offering an extremely wide viewing angle. The FDS-255 is a key element of the Rockwell Collins Flight2 Systems. It can replace existing instruments, combine functions of existing instruments, or provide the entire display system in a new or retrofit application. The active matrix liquid crystal design and proven Rockwell display circuitry give the user a highly reliable and easily maintainable flight instrument. FDS-255 is programmed in Ada. User specific display formats can be developed and downloaded over the ARINC 429 bus.

Featuring a 'smart head' design, the FDS-255 contains internal graphics generation, analogue and digital input/output and weather radar. In a dual installation, reversionary colour, full-motion video is provided with a video card slot.

The FDS-255 video with 64 grey scales will display FLIR, digital map and TCAS; colour, full-motion video is provided via an optional video card. The wide viewing angle permits easy cross-cockpit viewing.

Because of the highly integrated display design, Rockwell is able to provide 17.7 in^2 of usable viewing area in a 5 ATI format. The FDS-255 has eight standard modes, but can easily be tailored to each customer's specific application.

Flight/navigation operating modes include: ADI, HSI, PFD, ARC, radar-map, windshear, map and hover. TSO versions are also available.

Specifications
Dimensions: 129 × 129 × 222 mm
Weight: 3.8 kg (max)
Power supply: 28 V DC, 50 W typical
(See also Flight2 Systems entry)

Operational status
In production for the US Air Force C/KC-135 Pacer Crag and US Navy P-3. Deliveries began in 1996. Also in production for C-12, C-130, KC-10, MH-60, P-3, VC-10 and various helicopter and commercial aircraft applications.

Contractor
Rockwell Collins.

UPDATED

FDS-255 Flight Display System **1996**

Flight2 Systems for the integrated cockpit

Flight2 Systems are designed to provide enhanced capabilities such as autonomous traffic, weather and terrain awareness; seamless navigation and guidance from take-off to landing, including /G and /E (FMS) approved flight operations; impromptu flight planning; fusion of flight sensor information; and digital automatic communication - all functions associated with and in support of the future Communication, Navigation, Surveillance/Air Traffic Management (CNS/ATM) system.

The Flight2 Systems include flight communication, navigation, positioning, display and flight management and multimode radar systems. The first Flight2 System introduced is the FDS 255 Flight Display System, a high-performance 5 ATI colour flat panel liquid crystal multifunction display system. It consolidates flight information from diverse sources into display presentations appropriate for each phase of flight: integration of Traffic alert Collision Avoidance System (TCAS), radar and navigation information to promote rapid assessment of the flight situation. In addition, Forward Looking InfraRed (FLIR) video and digital map imagery can be displayed without sacrificing any essential flight control or guidance awareness.

The flight display system comprises two LCD multifunction displays, plus a control panel at each crew station. The displays present all flight and navigation data in multiple EADI and EHSI formats, as selected on the control. FDS 255 interfaces include aircraft flight systems, flight management systems, automatic flight control systems, conventional VOR, TACAN, INS, Doppler and GPS navigation systems. In addition, flexibility for free text display will accommodate eyes-up crew messages in the future.

Specifications

Height: 5.08 in
Width: 5.08 in
Depth: 8.75 in
Weight: 3.64 kg
Power: 50 W typical, MIL-STD-704A, 28 V DC
Display type: Colour active matrix 4.22 × 4.22 in, liquid crystal display
Grey shades: 64 video grey shades
Colour capability: 262,000 colours (video); 256 colours (symbols)
Display luminance: 120 fL at full brightness over field of view (200 fL for video)
Display resolution: Over 80 colour groups/inch (vertical); 113 colour groups/inch (horizontal)
NVIS compatibility: Type I, Class A (blue, green) and B (all colours), per MIL-L-85762A
Viewing angle: ±65°H, +40°/–20°V
Reliability: >11,000 hours @ 25°C in an airborne inhabited transport environment per MIL-HDBK-217E, includes backlight
Interfaces: ARINC 429, ARINC 708; analogue, synchro and discrete interfaces
Video input: RS-170 (525 line); or STANAG 3350 (625 line)
RTCA: DO-160C

Contractor

Rockwell Collins.

UPDATED

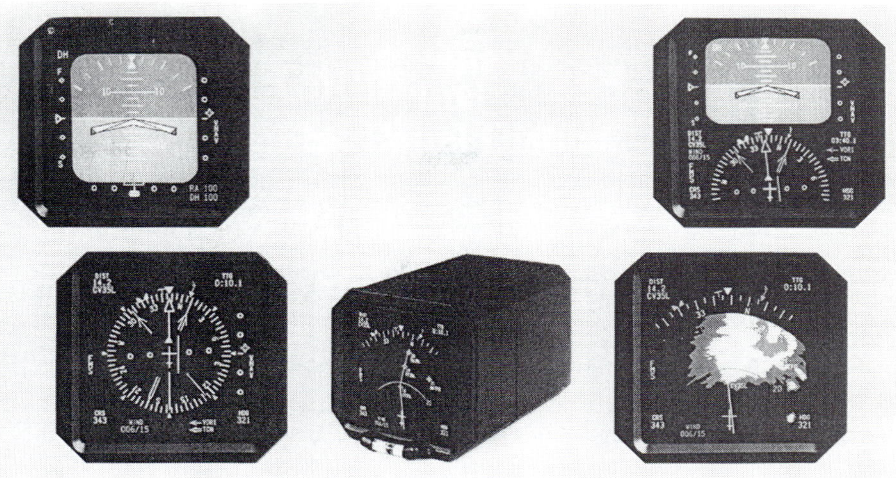

Flight2 systems 5 ATI display formats ***1997***/0001381

The Rockwell Collins displays and flight control system developed for the Boeing 747-400

Integrated display system for the Boeing 747-400

Rockwell Collins has developed an integrated electronic colour display system for the Boeing 747-400. A Rockwell Collins digital flight control and central maintenance computer (CMC) also are standard on the aircraft. The display system features six 8 in (210 mm) square colour CRT displays. Any one of these units can be selected to show either EFIS or EICAS data.

The new technology gives the 747-400 only 38 per cent of the cockpit lights, gauges and switches compared with older 747 aircraft and leads to better aircraft availability and significant reductions in crew workload.

The IDS-7000 integrated display system utilises the third-generation hardware design first certificated in the EFIS-1000 system. Each display unit contains all drive and symbol generation electronics necessary to perform any of the display tasks. Each pilot has a Primary Flight Display (PFD) and Navigation Display (ND) mounted in a side-by-side configuration. All primary air data and heading information is included on the PFD displays. EICAS functions are provided on two displays mounted in an over/under configuration in the centre panel.

The display system also includes three electronics interface units (EIUs) which function as data conversion and collection for EICAS displays, message processing, snapshot information recording, exceedance data recording and CMC-7000 central maintenance computer data collection. The central maintenance computer, in conjunction with the display units, provides ground maintenance crews with maintenance page displays. The display units, EIUs and CMCs are all capable of being programmed on the aircraft without removal of the equipment.

Operational status

In production for the Boeing 747-400.

Contractor

Rockwell Collins.

UPDATED

MFD-68S Multifunction Display

The MFD-68S offers user-defined graphics for primary flight displays, engine instruments, synoptics, text, fuel management, targeting reticles and automated checklists. The 6 × 8 in (152 × 203.2 mm) active matrix liquid crystal display presents video, graphics, video with graphics overlay, split-screen video/graphics and split-screen graphics/graphics.

The MFD-268 series is offered in either a 'smart' or 'video' display to support a wide variety of architectures. The display contains internal graphics generation, analogue and digital input/output interfaces and a MIL-STD-1553B interface. The operational software can be modified over the MIL-STD-1553B bus without removal of the display from the installed position.

The MFD is programmable in the Ada language for local display format generation. A high-level graphics language simplifies tailoring of display formats.

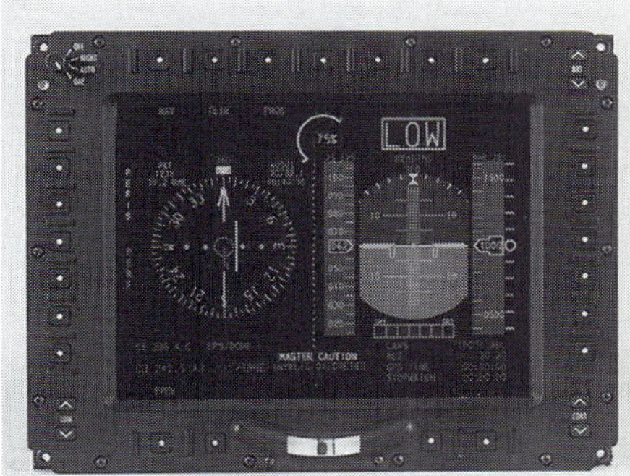

MFD-68S multifunction display ***1995***

Specifications

Dimensions:
254 mm (depth) × 264 mm (width) × 208 mm (length)
Weight: 7.71 kg nominal
Display: 152 × 203.2 mm active area

Contractor

Rockwell Collins.

UPDATED

Micro Line nav/com family

In the early 1970s, Rockwell Collins introduced the first members of a family of panel-mounted navigation and communication units, called Micro Line, for light single- and twin-engined aircraft. The rationale for this step was the lack of any agreed standard or specification for avionics systems appropriate to this type of aircraft, which gave rise to considerable compatibility or interchangeability problems between equipment from different manufacturers. By contrast, the airlines already had the very comprehensive ARINC characteristics to ensure compatibility. Intended for low-budget operation and for aircraft with limited internal space and weight allowance, the various Micro Line units each contained the relevant processing and indication circuits and mechanisms within single boxes, so that they can be mounted as self-contained items on the pilot's panel without the need for any associated remote rack-mounted equipment.

The Micro Line family comprises:
ADF-650A automatic direction-finder
AMR-350 audio control unit/marker beacon receiver
ANS-351 area navigation system
AUD-250 audio control unit
DME-450/451 DME
GLS-350 ILS glide slope receiver
IND-350/351 course deviation indicators
MKR-350 marker beacon receiver
PWC-150 28/14 V power converter
TDR-950 transponder
VHF-251/253 communication transceiver
VIR-351 navigation receiver.

More detailed descriptions of some of these systems can be found under the appropriate section headings elsewhere in *Jane's Avionics.*

Operational status
In service. No longer in production.

Contractor
Rockwell Collins.

UPDATED

Pro Line nav/com family

Introduced in 1970, the Pro Line series avionics was intended for medium and large general aviation piston and turbine-engined twins. The family was originally designated Low-Profile to emphasise the compact size and form factor of individual units but, in 1975, was renamed Pro Line in recognition of its acceptance by professional pilots in regional airlines, and by the defence forces. Unlike the self-contained members of the Micro Line family, Pro Line systems comprise panel-mounted indicators and controls driven by or controlling separate rack-mounted processing and computing boxes. The size and form factor was laid down by Rockwell Collins, there being no industry-wide agreement on packaging for general aviation electronics, in contrast to the highly defined ATR standards governing the characteristics of equipment for airlines. In the military field, Pro Line is found on a wide range of aircraft, not only those equivalent in size and performance to general aviation types, but also on attack and surveillance aircraft such as the A-4, F-5E, C-130 and E-3 AWACS.

The Pro Line family comprises:
346B audio control/isolation and speaker amplifiers
ADF-60 automatic direction-finder
ALT-50/55 radio altimeters
AP-105 autopilot
AP-106A autopilot
APS-80 autopilot
BDI-36 bearing/distance indicator
CTL series control heads
DME-40 distance measuring equipment
FDS-84 flight director system (comprising FD-108 flight director and FIS-70 flight instrument system)
FDS-85 flight director system (comprising FD-109 flight director and associated horizontal situation indicator)
FPA-80 flight profile advisory system
HF-230 HF transceiver
PN-101 pictorial navigation system
TDR-90 transponder
VHF-20A/B communication transceivers
VIR-30A/31A navigation receivers
WXR-220/270/300/350 colour weather radars.

More detailed descriptions of some of these systems can be found under the appropriate section headings elsewhere in *Jane's Avionics.*

Operational status
Production is being phased out in favour of Pro Line II avionics, but equipment remains in wide service. Well over 200,000 Pro Line boxes and controls have been manufactured and sold.

Contractor
Rockwell Collins.

UPDATED

Pro Line II digital nav/com family

Acknowledging the advances in technology, notably in signal processing, since the appearance of Pro Line in 1970, Rockwell Collins decided in the late 1970s to develop a replacement. The result was Pro Line II, the first members of which (VHF communications transceiver, VHF navigation transceiver and DME) appeared in January 1983. Other units were added and there now exists a complete range of Pro Line II equipment.

Pro Line II boxes contain analogue/digital and digital/analogue circuits so that individual units of the earlier family can be exchanged on a one-for-one basis without change to the aircraft wiring or racks. Microprocessors within the new units are programmed to accept either analogue or digital frequency tuning arrangements.

The Pro Line II family consists of:
ADS-82 air data system
AHS-85 attitude/heading reference system
APS-65 autopilot
APS-85/95 autopilot
CTL-22 communication control unit
CTL-32 navigation control unit
CTL-62 automatic direction-finder control unit
CTL-92 transponder control unit
DME-42 DME receiver
EFIS-85/86 electronic flight instrument systems
EHSI-74 electronic flight instrument system
IND-42 DME control unit
MCS-65 compass system
TWR-850 turbulence weather radar system
VHF-21/22 VHF communication transceiver
VIR-32 navigation receiver.

Operational status
In production and service on a wide range of regional types.

Contractor
Rockwell Collins.

UPDATED

Pro Line 4 integrated avionics system

The Pro Line 4 system includes five large side-by-side electronic flight displays that present primary flight, navigation and engine information. Autopilot, flight direction and centralised avionics maintenance functions are provided by the system's integrated avionics processing system. The heart of the Pro Line 4 avionics line is control central, the integrated avionics processing system.

Functions performed by control central include processing of inputs from sensors located throughout the aircraft, and computation of commands and displays for autopilot functions, flight management system and electronic displays. Instead of a number of units performing flight control functions, eight modules in the control central work together to serve as the aircraft central processing system. The small size of the modules allows control central to provide fail/safe integrity, with quadruple redundancy of critical functions. Internal communications between critical functions are protected, eliminating exposed external wires and busses. If one module fails, the system automatically reconfigures to carry on operation without interruption.

Advances in component and manufacturing technology allow circuit boards in control central to perform the functions of up to the eight boards normally used. Surface-mounted devices, automatic insertion of components and manufacture control have allowed a considerable reduction in circuit board size.

As well as control central, Pro Line 4 includes EFIS and EICAS displays, radio tuning units and a TWR-850 Doppler weather radar.

Operational status
In production. Pro Line 4 has been selected for the Beech Starship, Beechjet 400A, Canadair Regional Jet and CL-604, Dassault Falcon 2000 and EX, Embraer CBA-123 and EMB-145 fanjet, IAI Galaxy and Astra SPX, IPTN-250, Learjet 60, and the Saab 2000 regional airliner.

Pro Line 4 is now being retrofitted to early Falcon 50 aircraft to update their avionics capabilities to the standard available on Falcon 2000 and Falcon EX (The

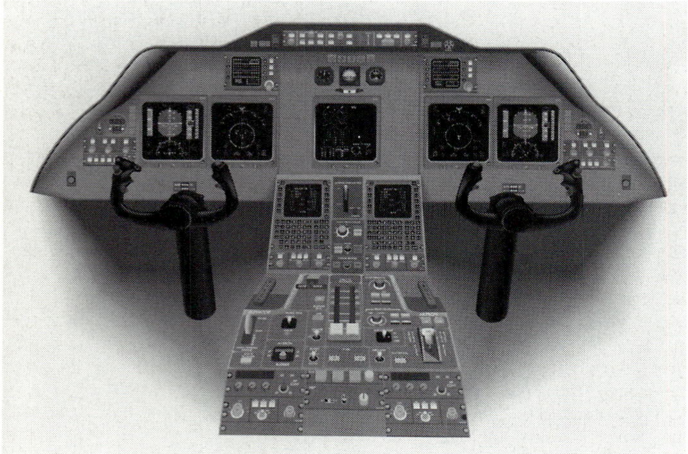

The Rockwell Collins Pro Line 4 has been chosen for the Astra SPX
1997/0001384

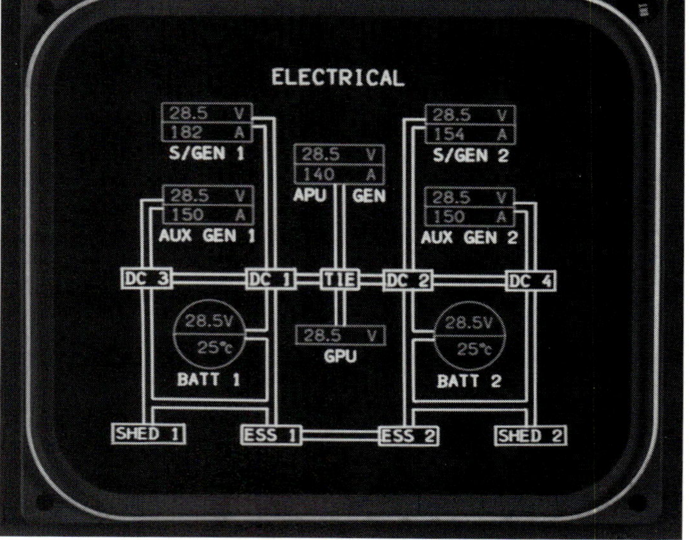

Rockwell Collins Pro Line 4 EICAS

Falcon 50 also includes comprehensive update of communications and navigation facilities and options for TCAS II installation). Recently, a new 7.25 in square active matrix colour LCD has been added to the Pro Line 4 range.

Contractor

Rockwell Collins.

UPDATED

Pro Line 21 integrated avionics system

Advancements in man-machine interface engineering and incorporation of large-format active matrix liquid crystal displays (AMLCDs) are the key elements of the Pro Line 21 avionics system, introduced in 1995 for corporate and regional aircraft.

Pro Line 21 flight decks are custom-configured with two to five adaptive flight displays that utilise a mix of AMLCD formats, including a 10 × 8, 8 × 8, 7.25 × 7.25 and 4 × 5 in. New to corporate jet applications, Pro Line 21 8 × 10 in AFD-3000 adaptive flight displays are nearly 30 per cent larger and approximately two-thirds the weight and depth of standard 8 × 8 in displays on board the Boeing 747-400. In addition to primary flight and navigation information, this LCD technology allows clear presentation of approach plates, terrain maps, real-time video and other highly detailed - and even three-dimensional - graphics that deliver flight operations information to pilots in innovative formats developed and refined by pilots.

Pro Line 21 key subsystems include the Rockwell Collins AVSAT™ satellite-based precision navigation and communication system with GPS-4000 Global Positioning System sensor, an advanced technology AHRS and an advanced flight control system with fail-passive autopilot. Also standard are solid-state weather radar and Pro Line radio sensors, including transponder, TCAS and DME. An optional maintenance system displays current LRU status, fault history and diagnostic data on the multifunction display.

Operational status

Pro Line 21 is now in production for the Raytheon Premier I. It has also been selected for the Bell Boeing 609 civil tilt-rotor, which will feature three 10 × 8 in AMLCDs including two Primary Flight Displays (PFDs) and one MultiFunction Display (MFD).

Contractor

Rockwell Collins.

UPDATED

Series 500 avionics family

Developed for commercial aircraft, the Series 500 avionics family is based on ARINC 500 characteristics and is the final step in the company's range of analogue equipment. Though still analogue in nature, more advanced solid-state circuits together with a small number of components per function (a lower parts count) make for a substantial weight saving and greater reliability over previous equipments. Series 500 boxes are available on a one-for-one replacement basis for earlier systems.

The Series 500 family comprises:

51RV-4 VOR/ILS receiver
51Y-7 (DF-206) automatic direction-finder
54W-1 comparator warning monitor
346D-2/2B passenger address amplifier
490S-1 HF antenna coupler
618M-3 VHF transceiver
621A-6A air traffic control transponder
618T-1/2/3 HF transceiver
860E-4/860E-5 DME systems
860F-4 digital radio altimeter
Datalink system (comprising DLC-700 control unit and 597A-1 management unit)
FD-110 flight director system (comprising FMC-28 flight mode controller, 562A-5F5 flight computer, 329B-8J attitude director indicator, and 331-8K horizontal situation indicator)
ILS-70 instrument landing system receiver.

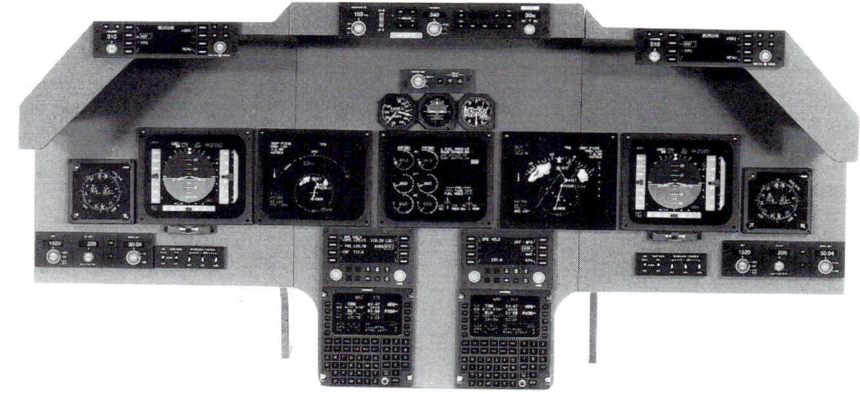

A variety of Pro Line 4 configurations is available to provide flexibility in aircraft applications. Five tubes may be selected to provide primary flight, navigation and engine information to the flight crew

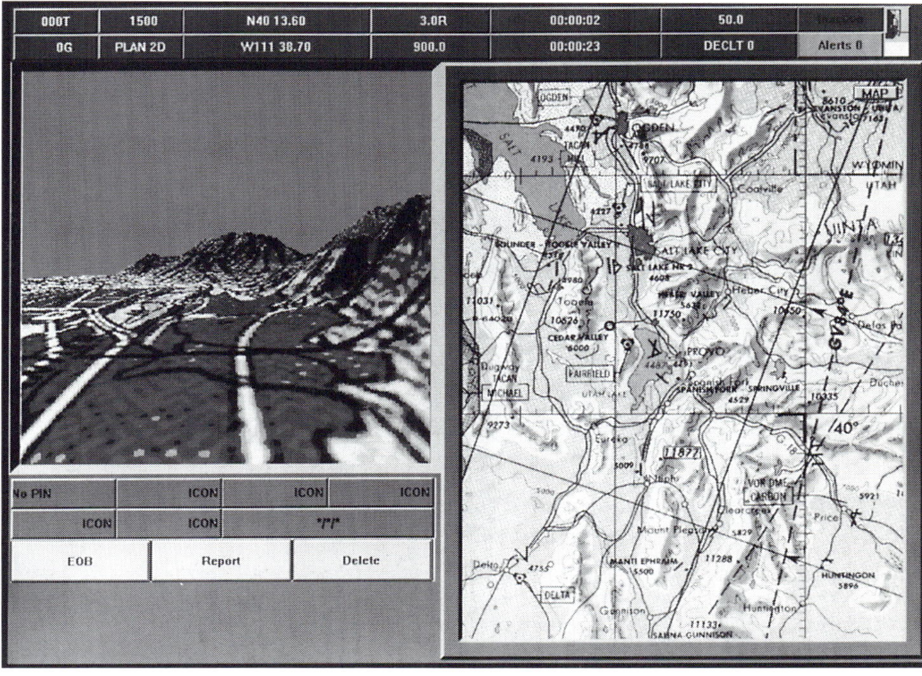

Pro Line 21 real-time situational awareness 3-D information on surrounding airspace and terrain
***1997**/0001386*

Pro Line 21 AMLCDs
***1998**/0018172*

Operational status
In production and in service.

Contractor
Rockwell Collins.

UPDATED

Series 700 digital avionics

Design of the Series 700 digital avionics for commercial aircraft dates back to the early 1970s when US industry in general began planning for the new generation of transports then in prospect. The new aircraft emerged as the Airbus A300 series and the Boeing 767 and 757. The digital systems for these and later projects were defined by the ARINC characteristics of the 700 series. This series was introduced in 1978, launch year of the first of the new transport aircraft, the Boeing 767, and is now offered on the 747-400 and A320. The Series 700 avionics provides centralised fault monitoring capability as specified in ARINC 604.

Together with the usual range of navaids and other electronic devices, the Series 700 introduced an electronic flight instrument system based on television-style CRTs and a similar display suite for engine indication and warning.

The Series 700 family comprises:
ADF-700 automatic direction-finder
DME-700 DME
HF-700 HF radio transceiver
ILS-700 instrument landing receiver
LRA-700 low-range radio altimeter
PAU-700 passenger address amplifier
TPR-720 transponder
VHF-700 VHF radio transceiver
VOR-700 VHF navigation receiver

Operational status
In production and service.

Contractor
Rockwell Collins.

UPDATED

Rockwell Collins series 700 avionics. From left to right the units are the HFS-700, VHF-700, PAU-700, VOR-700, ILS-700, ADF-700, DME-700, LRA-700 and TPR-700

Series 900 avionics system

Series 900 avionics include VHF, ADF, VOR, ILS, DME, altimeter, transponder, weather radar, TCAS and satellite communications systems. They meet industry and FAA requirements for equipment certified to Criticality Level 2. This requires tolerance to aircraft environments including 200 ms extended power interrupts, immunity to high-intensity radiated fields and conformity to the stringent environmental requirements of DO-160C. Conformity to the strict software documentation standards of DO-178C and the higher VHF FM interference levels required by ICAO Annex 10 are also included.

Series 900 is packaged in ARINC standard LRUs. Only the high-power systems require forced-air cooling, due to improvements in heat dissipation techniques. All other systems can be passively cooled.

All software is in Ada. External software loading capability has been enhanced and expanded to allow easy shop or on-aircraft modification.

Series 900 units are interchangeable with Series 700 products. This allows fleet commonality and use of the same top-level test equipment.

Operational status
In production and in service for Airbus A320/330/340 and Boeing 747-400, 757, 767 aircraft.

Contractor
Rockwell Collins.

UPDATED

Standby Instruments for the Boeing 777

Rockwell Collins is supplying flat-panel colour LCD standby indicators for the Boeing 777. These three 3 × 3 in instruments are identical and include the attitude indicator, airspeed indicator and altimeter. They are passively cooled.

Specifications
Dimensions: 76.2 × 76.2 × 215.9 mm
Reliability: >15,000 h MTBF

Operational status
Certified on Boeing 777. In production.

Contractor
Rockwell Collins.

UPDATED

NeoAV Model 500 EFIS Electronic Flight Instrument System

The NeoAV Model 500 is a flat-panel electronic flight instrument system. The equipment consists of a fully self-contained computer-controlled symbol generator using colour active matrix liquid displays. It incorporates a high-speed 32-bit microprocessor-based system with BIT and flight recording capability. Each of the units is interchangeable and networked for maximum redundancy.

Operational status
Certified by the FAA in December 1993 on the Sikorsky S-76C helicopter and selected as standard on this and the S-76B. The NeoAV Model 500 replaces existing HSI/ADI instruments in Federal Express Boeing 727 and DC-10 aircraft. Also selected for the Boeing 707 and 737, Gulfstream GII and Bell Helicopter Model 430 helicopter.

Contractor
Rogerson Kratos, a Rogerson Aircraft Corporation Subsidiary.

NEOAV 500 displays are shown here in a Boeing 727 cockpit **1995**

NeoAV IIDS Integrated Instrument Display Systems

The Rogerson Kratos NeoAV Integrated Instrument Display Systems comprise a family of cockpit displays using advanced Active Matrix Liquid Crystal Display (AMLCD) technology. IIDS can efficiently replace up to 38 conventional instruments, as well as the caution advisory system. Standard features include engine instrument indication, fuel quantity measurement indications, hydraulic systems indications, electrical system indications, caution, warning, and advisory messages, and outside air temperatures. IIDS also has FADEC or EEC compatibility, trend monitoring and weight and balance synoptic indications.

Special helicopter capabilities include mast torque indications and chip detector warning outputs.

Functions are monitored and managed with colour AMLCDs, featuring redundancy and maintenance capabilities. Multiple reversionary display modes operated by bezel-mounted buttons enable the system to be 'fail operational' in case of in-flight malfunctions. Built-In Test Equipment (BITE) is available. Metric or English displays are available, as is built-in non-volatile memory for engine history and maintenance logs. Engine performance instruments are typically presented on one display, while aircraft systems instruments are on a second display. Solid-state components, redundant architecture and derating assure high reliability. IIDS also offers significant reductions in weight, heat and power consumption.

Operational status
IIDS displays have been certified for the Bell 430, Sikorsky S76 and Canadair CL415. Certification is in process for the Casa 212, 235 and 295, Bell 412 and 427, as well as for Zeppelin Airship. Rogerson Kratos IIDS installations will be factory standard on all Bell twin-engined helicopters (Models 212, 412, 427 and 430).

Contractor
Rogerson Kratos, a Rogerson Aircraft Corporation subsidiary.

NEW ENTRY

NeoAV Model 550 EFIS Electronic Flight Instrument System

The NeoAV Model 550 is an Active Matrix Liquid Crystal Display (AMLCD) Electronic Flight Instrument System (EFIS). The system is a microprocessor-based, self-contained unit that includes the symbol generator, Colour Active Matrix Liquid Crystal Display (CAMLCD) and interface hardware/software. The system design also incorporates full-time, Built-In Test (BIT) operated by a high-speed 32-bit microprocessor. The NeoAV 550 EFIS also includes a manual, self-test feature for the pilots and maintenance personnel. Faults are continuously recorded and stored in non-volatile memory for review after the aircraft has landed. Each NeoAV 550 component in the system is interchangeable (single part number) and is networked for maximum redundancy. Both analogue and digital interfaces are available, offering efficient, 'plug in' display interchangeability. Self-contained (bezel-mounted) EFIS control or remote controllers are available as options.

Operational status
NeoAV was certified on the Boeing 727 in 1994. The Bell Helicopter 430 was certified in 1996, follow-on certification will occur on the Bell 412 and Bell 427. Rogerson Kratos EFIS installations will be factory standard on all Bell twin-engined helicopters (Models 212, 412, 427, and 430). Other active certification programme include factory standard installation on the Agusta 109 Power and EFIS upgrades to the US Air Force C18B (B707) and VC-25 (B747-Air Force 1) aircraft.

Contractor
Rogerson Kratos, a Rogerson Aircraft Corporation subsidiary.

NEW ENTRY

The Rogerson Kratos NeoAV IIDS display system was fitted to the Bell 427 helicopter as it entered flight test **1998**/0022211

Bell 430 cockpit showing both the IIDS Integrated Instrument Display System and the Model 550 EFIS Electronic Flight Instrument System **1998**/0018163

Angle of attack computer/ indicator

The angle of attack computer/indicator provides the pilot with a continuous display of aircraft lift information on a decimal scale, with 1.0 representing the stall. The display is valid regardless of bank angle, aircraft weight or wing configurations.

The computer/indicator face is scaled red below 1.1 V, amber from 1.1 V to 1.3 V and black from 1.3 V to maximum speed. It also features a settable bug and index slaved to each other, which can be set for a desired airspeed target between 1.2 V and 1.5 V. Centring and maintaining the angle of attack pointer within the index will result in the selected speed target.

The system computer drives the ADI fast/slow pointer, a 2 in round angle of attack indicator and Safe Flight's speed indexer lights. The panel-mounted angle of attack indicator displays aircraft lift information. An approach reference is provided and the display is valid for all flap positions.

Operational status
The system has been certified on the Raytheon Hawker 800.

Contractor
Safe Flight Instrument Corporation.

UPDATED

Recovery Guidance System

Safe Flight offers recovery guidance, an optional enhancement of the basic windshear warning system. With the combined WindShear Warning/Recovery Guidance System (WSW/RGS), as soon as the warning occurs continuously, computed pitch guidance for recovery is displayed on the flight director command bars. The system also provides pitch guidance for take-off and go-around on the same instrument, to maintain pilot familiarity with use of the system and promote confidence in it. The pilot does not have to change his flight scan or depart from accustomed procedures during the crucial emergency escape manoeuvre. The company maintains that, by following the command bars, the best possible climb profile to maximise the performance capabilities of the aircraft will be achieved.

The system is armed automatically, even if the flight director is turned off, by the windshear warning system's alert output, but only becomes operative, displaying pitch guidance for recovery, when the pilot activates the GA switch. System logic may be programmed to perform these switching and display functions automatically.

Safe Flight's RGS displays pitch attitudes up to, but not in excess of, the stick shaker target. However, the RGS can be programmed to display stick shaker target information on the ADI slow/fast indicator alongside the pitch guidance display on the command bars. With this optional function, when the pilot activates the GA switch for recovery guidance, the slow/fast indicator changes from a speed mode to shaker mode with the slow bar representing shaker target. The pilot then has a continuous visual indication of his margin to stick shaker. Internal system monitoring and a self-test function ensure system reliability.

Operational status
Recovery guidance is presently available in combination with the windshear warning system in a single ¾ ATR box, or where Safe Flight's Speed Command of Attitude and Thrust (SCAT) system is

desired (or already installed), through tie-in of the windshear warning and SCAT system computers.

Contractor

Safe Flight Instrument Corporation.

VERIFIED

Windshear warning system

Safe Flight's airborne windshear warning system provides a voice alert to the crew of high-performance aircraft at the start of an encounter with hazardous low-level windshear. The system is operative during take-off and approach and is a computer-based device which, using conventional sensing elements, resolves the two orthogonal components of a wind gradient with altitude and provides a threshold alert that an aircraft is encountering a potentially hazardous situation. The vectors concerned are horizontal windshear and downdraught drift angle.

Horizontal windshear is derived by subtracting groundspeed acceleration from airspeed rate. The latter term is obtained by passing airspeed analogue data from the airspeed indicator or the air data computer through a high-pass filter. Longitudinal acceleration is sensed by a computer integral accelerometer, the output of which has been summed with a pitch attitude reference gyro to correct for the acceleration component due to pitch. A correction circuit is employed to cancel any errors due to prolonged acceleration. This circuit has a 'dead band', equivalent to 0.2° of pitch, which prevents correction for airspeed rates of less than 0.1 kt/s. Summed acceleration and pitch signals are fed through a low-pass filter, the output from which is summed with the airspeed rate signal to give horizontal windshear.

The vertical computation for downdraught drift angle is developed through the comparison of measured normal acceleration with calculated glide path manoeuvring load. Flight path angle is determined by subtracting the pitch attitude signal from an angle of attack signal sensed by the stall warning flow sensor. This is fed to a high-pass filter and from there to a multiplier to which the airspeed signal has been applied. Thus, the flight path angle rate, corrected for airspeed, provides the computed manoeuvring load term. This is compared in a summing junction with the output of a normal computer integral accelerometer and the failure of the two values to match is the indication of acceleration due to downdraught. The acceleration, when integrated, is the vertical wind velocity and is further divided by the airspeed signal to compute the downdraught angle.

The outputs of both horizontal and vertical channels are determined solely by the atmospheric conditions and ignore manoeuvres that do not increase the total energy of the aircraft. Windshear correctly compensated by increased engine thrust shows no change in airspeed in the presence of an inertial acceleration as thrust is applied. If, however, the shear goes uncorrected, an acceleration or deceleration becomes apparent. Similarly, in the case of the vertical component, a vertical displacement compensated by the crew shows a positive flight path angle rate in a downdraught with less than the computed incremental normal acceleration. Correspondingly, in a downdraught for which the drift angle is allowed to develop, a negative angle rate at a near constant 1 *g* results in the same computation and output. This is important to the crew as it eliminates the possibility that their actions in anticipating or countering windshear might well mask the condition as far as the warning system is concerned.

Both downdraught drift angle and horizontal windshear signals are combined and the resulting output fed through a low-pass filter to the system computer. This provides two output signals: a discrete alert and, through a voice generator, an audio alert. Warning output is set at a threshold of −3 kt/s for horizontal shear and −0.15 rad downdraught drift angle, or for any combination of the two components which, acting together, would provide an equivalent signal level. According to Safe Flight, any wind condition requiring additional thrust equivalent to 0.15 *g* to maintain glide path and airspeed will result in a non-stabilised approach.

A crossover network is employed to sense zero crossovers of the combined windshear warning signal and this is sampled every 25 seconds. If the warning signal does not pass through a band close to zero, the network automatically provides failure indication, alerting the crew to the fact that the unit is inoperative. A self-test function activated by the pilot is also built into the system.

The company has now entered into a licensing agreement with Boeing Commercial Airplanes for the further exploitation of windshear warning technology. This agreement provides for the licensing to Boeing of Safe Flight's existing and pending patents and proprietary data in the areas of windshear detection, alert and escape guidance.

The agreement will facilitate the incorporation of windshear warning capabilities on new models of Boeing commercial aircraft. It is anticipated that the technology will be an added feature on aircraft including the 737-300, 757, 767 and 747-400 models. A version will also be offered for retrofitting on Boeing aircraft currently in service.

Specifications

Dimensions: ¼ ATR
Weight: 2.72 kg

Operational status

In production and service. Certified for Cessna Citation III, Falcon Jet Falcon 50, and Raytheon Hawker 800.

Contractor

Safe Flight Instrument Corporation.

UPDATED

Common Multifunction Display Unit (CMDU)

The Sanders CMDU is designed for flight display systems for military and commercial cockpits. Each CMDU has a 6.2 × 8.3 in (10.4 in diagonal) portrait oriented viewing area and provides for the display of 480 × 640 pixel colour images. Display imagery is locally generated by the Smart Graphics Processor (SGP). The video interface provides access to external digital video signals. The external video can be merged with locally generated graphics allowing for overlays.

The CMDU is: sunlight readable (160 fL green, 200 fL white) and NVG/NVIS compatible with a ±60° viewing angle.

Specifications:

Dimensions: 241 x 191 mm
Weight: 8.3 kg
Power: 180 W
Interface: MIL-STD-1553B
Certifications: DO-160C and DO-178B

Operational status

In service in C-130J aircraft and the Sikorsky S-92 Helibus

Contractor

Sanders, a Lockheed Martin Company.

UPDATED

The F-22 display and control system consists of (left to right) the upfront display, secondary MFD and primary MFD ***1995***

Display and control system for the F-22

Sanders is responsible for the display suite for the F-22. The F-22 display and control system consists of one primary MultiFunction Display (MFD), three secondary MFDs and two upfront displays. Sanders also provides the Graphics Processor Video Interface (GPVI), Airborne Video Tape Recorder (AVTR) and Operational Debrief Station (ODS).

The primary MFD presents a situation display of the air and ground situation.

The secondary MFDs present attack and defensive data, together with stores management information. The upfront display presents CNI data, critical flight data and warnings.

The GPVI resides in the Common Integrated Processor (CIP) and hosts software that generates the tactical situation displays. It both generates display formats and mixes these formats with video from other aircraft sources. The GPVI includes the i960MX processor and two graphics ASICs developed by Sanders. The Graphics Drawing Processor (GDP) ASIC is the high-performance anti-aliasing drawing processor. Using subpixel addressing and advanced anti-aliasing algorithms, the GDP generates complex display formats at better than a 30 Hz rate. The anti-aliasing provides for smooth display dynamics, without artifacts such as stair-stepped lines.

The video merge ASIC controls the merging of external video with the locally generated format and transmits the result over the fibre optic interface to the MFDs. The MFDs also utilise the GDP ASIC to enable local display generation. Thus the MFDs can either display video from the CIP or locally generated formats derived from information received over dual MIL-STD-1553 interfaces.

The GDP is a single 3,000 Mops ASIC which accomplishes on-chip translation, rotation, scaling, anti-aliasing drawing, filling and micro-positioning to better than one-sixteenth of a pixel for smooth instrument movement on an active matrix LCD medium. F-22 systems engineering trade-offs have resulted in design advances such as dual-brightness sensors for automatic compensation for cockpit ambient light level, expanded luminance ranging from 220 ft-lamberts down to 0.1 ft-lamberts for enhanced night operations with minimum canopy reflections, lamps designed to outlast the life of the aircraft and comprehensive BIT for improved safety and maintainability.

Specifications

Dimensions:
(primary MFD) 257.8 × 257.8 × 190.5 mm
(secondary MFD) 213.4 × 213.4 × 190.5 mm
(up front display) 156.2 × 130.8 × 218.4 mm
Weight:
(primary MFD) 7.03 kg
(secondary MFD) 6.12 kg
(up front display) 3.3 kg

Display size:
(primary MFD) 203.2 × 203.2 mm
(secondary MFD) 158.7 × 158.7 mm
(up front display) 76.2 × 101.6 mm

Operational status

In development for the US Air Force F-22 advanced tactical fighter.

Contractor

Sanders, a Lockheed Martin Company.

VERIFIED

Electronic Display Units (EDU) for the B-1B

Sanders has supplied a number of Electronic Display Units (EDUs) for the Rockwell B-1B. The EDU is a random position display capable of presenting graphic information in response to X and Y deflection of signals and an unblank signal Z.

The indicator uses a 12 in (305 mm) diagonal, square CRT with a usable display area of 8 × 8 in (203 × 203 mm). The deflection system is capable of linear response at writing rates of up to 125,000 in/s. Small signal deflection bandwidth is greater than 3 MHz. Positioning moves can be made and settled to within 0.05 per cent of full-screen in less than 35 μs. These displays depict threat and panoramic information which enables the B-1B's defensive systems operator to analyse threat situations quickly and assign countermeasures.

Operational status

No longer in production. In service on the US Air Force Rockwell B-1B.

Contractor

Sanders, a Lockheed Martin Company.

VERIFIED

Miligraphic display system for the P-3 AEW

Sanders has supplied a display system for the Lockheed Martin P-3 AEW. It is a colour graphics terminal with the control electronics and the raster display integrated into a single 19 in (483 mm) rack-mounted unit. The control electronics use dual 68020 microprocessors to generate real-time displays of radar data overlaid on situation maps. The operator keyboard and the display touch control provide interactive operator control of the airborne early warning system. The display uses a high-resolution 19 in (483 mm) shadow-mask colour CRT that has been specially designed to meet the airborne environment and to protect it from varying magnetic fields. The miligraphic display software uses a multitasking executive with multiple windows that is designed to support application programming of real-time display tasks.

The Sanders miligraphic display system for the Lockheed Martin P-3 AEW aircraft

Operational status

Some 300 systems have been sold to the US Customs Service, US Navy and foreign customers.

Contractor

Sanders, a Lockheed Martin Company.

VERIFIED

MultiFunctional Display (MFD)

The MultiFunctional Display (MFD) is the latest addition to the Miligraphic product line. Using the Miligraphic operating system, the 14 in (355.6 mm) flat tension mask CRT display with the 0.28 mm pitch shadow-mask is capable of displaying high-contrast high-visibility red/green/blue colour images. Three multifunctional displays are used on the ES-3A aircraft: one of these is in the cockpit and two are at crew stations. Each MFD is supported by an individual controller housed in a separate common chassis rack. This enables installation of the MFD monitors in limited access areas, with remote installation of the control electronics.

The MFD features contrast of 4:1 with a visor in 8,000 ft-candles ambient light, 91 lines/in resolution and compact construction. The monitor weighs 41.28 kg and the control unit 42.64 kg. The MFD meets MIL-E-5400F, MIL-T-5422F and MIL-STD-461.

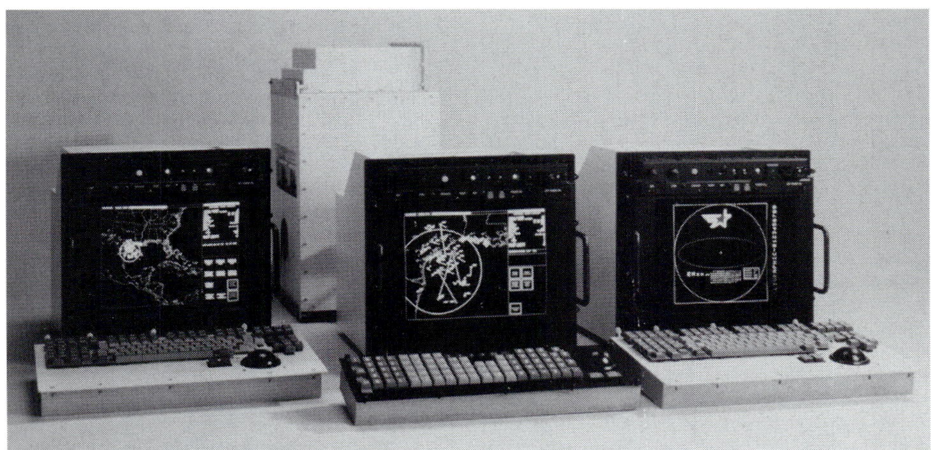

The Sanders control and display system for the Lockheed Martin ES-3A

Operational status

The MFD is used in the Lockheed Martin ES-3A.

Contractor

Sanders, a Lockheed Martin Company.

UPDATED

Primary Flight Display Subsystem (PFDS) for the S-92 Helibus

The Sanders PFDS is designed for flight display systems for military and commercial cockpits.

The S-92 Helibus cockpit can be configured with either four or five Active Matrix Liquid Crystal Displays (AMLCDs) in the PFDS. In the baseline configuration, the left and right displays provide primary flight information, while the two centre displays provide Engine Information and Caution/Advisory Status (EICAS) and navigation information. The optional fifth display provides for sensor information, such as Forward Looking InfraRed (FLIR) or moving map. In either the four or five display suite, the flight crew can designate any display to provide any of the information.

Each CMDU has a 6.2 × 8.3 in (10.4 in diagonal) portrait oriented viewing area and provides for the display of 480 × 640 pixel colour images. Display imagery is locally generated by the Smart Graphics Processor (SGP). The video interface provides access to external digital video signals. The external video can be merged with locally generated graphics allowing for overlays.

An optional Mission Computer (MC) can be integrated with the displays to provide a multitude of additional functionality to the system. The mission computer processor is an R4000 RISC. In addition to the processor, up to eight mission specific VME modules (mission planning, health management and so on.) can be accommodated in the mission computer. Multiple video inputs (analogue or digital) can also be accommodated by the mission computer and converted to digital video compatible with the display.

Specifications

Sunlight readable: 160 fL green, 200 fL white
Viewing angle: ±60°C

NVG/NVIS: compatible
Approvals: DO-160C, DO-178B
Interface: MIL-STD-1553B

Operational status

Designed for the S-92 Helibus. A version of the Mission Computer is in service in the C-130J aircraft.

Contractor

Sanders, a Lockheed Martin Company.

NEW ENTRY

F-15 avionics control panels

SCI designed and manufactured many of the F-15 avionics control panels including the Integrated Communications Control Panel (ICCP), the Take Command Control Panel (TCCP), the Integrated Navigation Aids Control Panel (INACP), the Identification Friend or Foe Control Panel (IFFCP), the Main Communications Control Panel (MCCP) and the Air-to-Air Interrogator Control Panel (AAICP). In addition, SCI has produced a TeleBrief Control Panel (TBCP) for the Israeli F-15. All these avionics control panels were designed into the original F-15 cockpit in 1969 and have been in service with the F-15 for many years.

ICCP

The ICCP is the heart of the system. Using dual-microprocessor control, this unit switches audio paths, controls and displays frequencies for the aircraft radios (including 40 channels of preset non-volatile memory), generates and formats synthesised voice alerting messages, generates audio warning tones and has provision for antenna selection. Built-in test circuitry is claimed to detect 95 per cent of all electronic component failures.

MCCP

A second panel, the MCCP for main communications control, is mounted on a head-up display to provide the pilot with selection of frequency, preset channels and volume control for one of the aircraft radio sets. It also has code selection controls for IFF Mode 3/A operation, a master caution annunciator and two electronic warfare warning displays.

INACP

The ILS segment of the INACP utilises a six wire parallel output to select the ILS frequency. Controls for the ILS operating frequency are adjusted from 108.1 to 111.95 MHz. The Tacan segment utilises a serial digital data train to control the R/T channel and mode of operation. Two rotary knobs control the Tacan channel selection from 0 up to 126. X-Y selection is provided by a separate switch. A volume control is provided to control the identification tone audio. Concentric to the volume switch is a three position mode control selector which selects receive only, receive and transmit or air-to-air modes of operation.

IFFCP

The IFFCP contains three printed circuit card assemblies, controls and indicators necessary to control the functions and modes of the IFF transponder, the IFF reply evaluator and the transponder computer. The master mode control allows selection of standard or low transponder sensitivities or activates all modes in an emergency and controls and displays Mode 1 codes. It has toggle switches for enabling individual IFF modes, a Mode 4A, 4B or out switch, a Mode 4 reply select control which disables the reply signal and routes it through the audio system or display lamp, a lamp indicating a correct reply from an interrogated aircraft and a Mode 4 code hold/zero switch.

AAICP

The AAICP contains the controls necessary to command the functions of the aircraft IFF evaluator and the interrogator computer. The master control switch provides automatic, normal and correct code challenges. The mode switch selects AAI Modes 1, 2, 3, 4A and 4B and an individually controlled four digit code selector.

Specifications

Dimensions:
(AAICP) 47.6 × 146 × 101.6 mm
(ICCP) 146 × 181 × 165.1 mm
(IFFCP) 85.7 × 146 × 101.6 mm
(INACP) 85.7 × 146 × 101.6 mm
Weight:
(AAICP) 0.36 kg
(ICCP) 4.4 kg
(IFFCP) 0.91 kg
(INACP) 1 kg
(MCCP) 0.95 kg

Operational status

In service in the McDonnell Douglas F-15.

Contractor

SCI Systems Inc.

VERIFIED

SCI avionic control panels. Top row (left to right): F-15 ICCP, F/A-18 ICS, F-15 ICCP. Middle row (left to right): F-15 TCCP, F-15 INACP, F-15 IFFCP. Bottom row (left to right): F-15 TBCP, F-15 MCCP, F-15 AAICP

F-16 Voice Message Unit (VMU)

The Voice Message Unit (VMU) is a compact ruggedised voice warning system currently in service as an integral part of the F-16 avionics. The VMU is capable of monitoring twelve 28 V DC discrete and four 5 V DC differential inputs for message activation. The VMU can be arranged to react to aircraft systems sensor activity, delays, priority and a number of occurrences under software control of the VMU microprocessor. The VMU has the capacity for 12 seconds of verbal or other messages which can be broadcast over an intercom or other facilities. As the VMU is microprocessor controlled it can be adapted to monitor and provide voice/tone warnings for most aircraft and vehicle systems.

Specifications

Dimensions: 63.5 × 98.6 × 82.6 mm
Weight: 0.63 kg
Power: (max) 4 W

Operational status

The VMU is currently in service on the F-16 and interfaces with the avionic systems to provide the pilot with an audible indication of aircraft system status which has exceeded preset parameters.

Contractor

SCI Systems Inc.

VERIFIED

V-22 Osprey Control Display Unit (CDU)

The SCI V-22 Control Display Unit (CDU) is a multipurpose multimission-capable CDU that interfaces directly with the V-22 aircraft computer system via a MIL-STD-1553B databus. The CDU provides the pilot with a screen and keyboard input/output menu to control and monitor and display aircraft communications, avionics instrumentation and computer systems. The V-22 CDU has full alphanumeric capabilities with eight programmable CRT bezel switches.

The V-22 CDU is currently designed to provide the pilot with a visual indication of aircraft systems status, using the CDU's menu mode driven software with user-friendly screen displays.

Specifications

Dimensions:
(display unit) 145.8 × 133.3 × 266.7 mm
(keyboard unit) 145.8 × 152.4 × 180.8 mm
Weight:
(display unit) 3.12 kg
(keyboard unit) 2.94 kg
Power supply: 28 V DC, 30 W (max)

Operational status

The SCI CDU has been incorporated into the V-22 cockpit system design and is currently in use on all V-22 Osprey aircraft.

Contractor

SCI Systems Inc.

VERIFIED

The SCI V-22 Osprey control display unit

DigiData fuel/airdata system

DigiData provides fuel management, navigation and Airdata functions including:

E-6B data: Pressure; altitude; density altitude; outside and true air temperature; wind aloft; wind component (speed and direction); indicated and true airspeed; ground speed; Mach; and instantaneous vertical speed.

Fuel flow: Left and right fuel flow; fuel used; fuel remaining; fuel to and reserves at destination; range; cruise efficiency (nautical mileage); and endurance.

Navigation: Heading; ground track; and magnetic variation.

DigiData is connected to aircraft systems to measure Indicated AirSpeed (IAS), pressure altitude, Outside Air Temperature (OAT), heading and fuel flow. From these raw data, TAS and temperature data are calculated by the microprocessor and combined with data from the navigation receiver to calculate automatically wind aloft, fuel needed to destination, specific range and density altitude.

Pressure altitude data is automatically used by most GPS receiver manufacturers to substitute for the fourth satellite range if it is not in view to provide 24-hour, three-dimensional position accuracy with only three satellites, without manual entry of altitude.

The engine fuel flow interface makes the fuel management data totally dynamic, basing it on real-time, winds aloft readings.

With an external Shadin ARINC 429 converter, DigiData can also drive EFIS displays and multisensor navigational management systems.

Specifications

Dimensions: 3.125 (round) × 6½ in (deep) (79 (round) × 165 mm (deep))
Power: 9-35 V DC, 1 W nominal
Weight:
(indicator) 623.7 g
(transducer) 453.6 g
Flow rate: up to 450 gal/h
Max usable fuel: 1,800 gal
Accuracy: ±1%
Operating temperature: −20 to +55°C
Input: pitot pressure, static pressure, outside air temperature, heading synchro, fuel flow
Output: IAS, TAS, ground speed, Mach, P.ALT, D.ALT, OAT, TAT, wind aloft, wind component, fuel flow, fuel used, fuel remaining, cruise efficiency, endurance, left fuel used, right fuel used, IVS, heading, track, magnetic variation
Output format: RS-422/RS-232
ARINC 429 (optional)
Compatible receivers: ARNAV (R-15, R-30, R-40, R-50, STAR 5000, FMS 5000, FMS 7000); Trimble (2000, 2000A, 2100, 3000, 3100); Bendix/King (KLN90, KLN90A, KLN88); II Morrow (604, 612, 614, 618); Northstar; Garmin; Magellen

Contractor

Shadin Co Inc.

VERIFIED

2180 series mini-Control Display Unit (CDU)

The Smiths Industries 2180 series mini-control display unit is a general application CDU which features a seven-colour raster display and has an RS-422 interface. The unit has a full alphanumeric keyboard and a unique colour display which is small in size and offers high resolution, making it ideal for use with navigation and communication systems including GNSS.

Specifications

Dimensions: 95 × 146 × 203 mm
Weight: 2.2 kg

Operational status

In production.

Contractor

Smiths Industries Aerospace.

VERIFIED

The Smiths Industries miniature control and display unit

Specifications

Dimensions: 146 × 181 × 101.6-177.8 mm
Weight: 2.72 kg (approx)
Display: 139.7 mm diagonal
14-lines by 28-characters

Operational status

The 2,000th unit was scheduled for delivery in spring 1998. Current platforms include: A300/310, ATR-42/72, BAe 146/RJ-85, B-727, B-747, C-130, C-141, DC-8, DC-10, King Air, MD-80, Nimrod MRA 4, VC-25 Air Force One & Two.

Contractor

Smiths Industries Aerospace.

UPDATED

2882 Series AMLCD Multifunction Control Display Unit (MCDU)

The 2882 Series flat-panel AMLCD Multifunction Control Display Unit (MCDU) provides a cockpit interface unit with an enhanced display technology, updated ARINC 739 format, advanced electronics and flexible system interface. The unit is designed to facilitate the integration of numerous functions such as GPS, RNav, ACARS, performance advisories and airport maps, and to help conserve space on the flight deck by the elimination of dedicated control heads.

The flat-panel MCDU provides a display of alphanumeric and graphics information, together with a keyboard for crew selection of modes and data entry. A set of dedicated function keys provides the crew with immediate access to particular pages and a set of dedicated alphanumeric keys allows the pilot to enter data. The ARINC 429 digital data enters through the rear connector and is processed by the graphics system processor module where the data is decoded and then displayed on the active matrix LCD. The unit contains standard RS-422 high-speed input and output. An optional MIL-STD-1553 interface unit is available as are NVIS-B and low-temperature operation options.

Smiths Industries 2882 MCDU **1996**

Active Matrix Liquid Crystal Displays (AMLCD)

Smiths Industries has been involved in Active Matrix Liquid Crystal Display (AMLCD) technology since its inception in the aerospace industry. Various display sizes have been integrated into a number of AMLCD products. Due to the multifunction nature of AMLCDs, a variety of applications has been developed. Most notable, digital triple torque, airspeed, altitude, vertical speed, collision avoidance, engine parameters and aerodynamic surface or control position indication can be provided by MIL-STD-1553B or ARINC 429 digital databus input or by integrated interfaces.

Operational status

In production. Smiths Industries Aerospace has been selected by Raytheon Aircraft Company to supply the major cockpit instrument package and other avionics for the US Air Force and Navy Joint Primary Aircraft Training System (JPATS).

The Smiths Industries AMLCD flat panel altimeter

An AMLCD vertical speed indicator

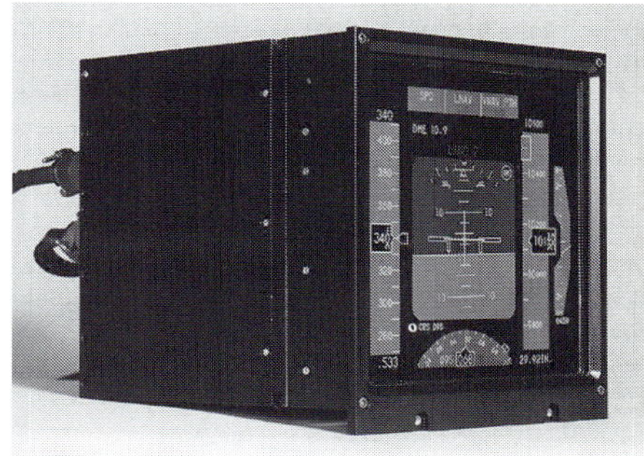

7 in cube AMLCD **1995**

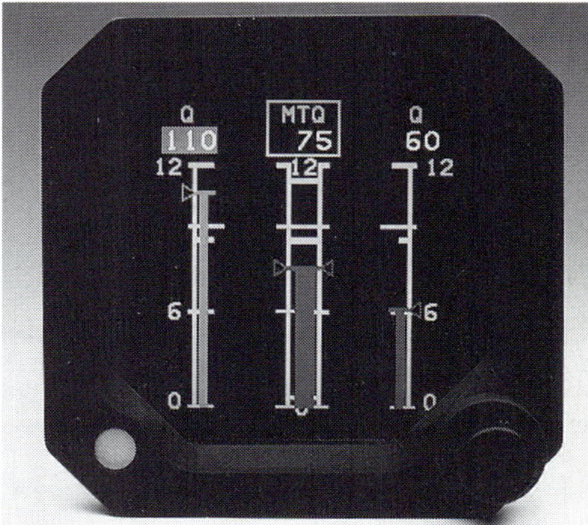

The digital triple torque indicator on the Bell 430 helicopter **1995**

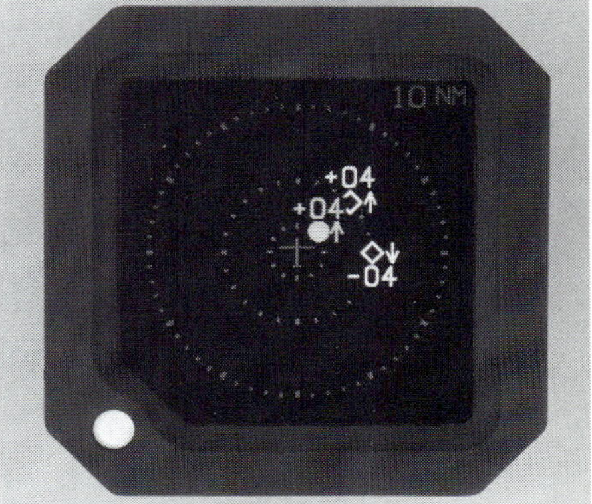

Flat-panel TCAS I display

The twin-cockpit layout includes 12 AMLCDs, in a distributed digital databus (ARINC 429) architecture, to provide the instructor and the student with flight and engine data including altitude, airspeed, vertical speed, fuel quantity, engine parameters, cabin pressure, and collision avoidance information.

Other typical applications of AMLCD include: the triple torque indicator on the Bell 430 helicopter, the para visual display, the engine fuel display for the F/A-18E/F and the V-22 standby flight display.

Smiths Industries AMLCDs have also been selected for the UK MoD Nimrod MRA 4 aircraft upgrade.

In November 1997, Smiths Industries Aerospace shipped its 1,500th AMLCD Multifunction Control Display Unit (MCDU).

Contractor
Smiths Industries Aerospace.

VERIFIED

Smiths Industries control display units

Control Display Units (CDU)

The Smiths Industries control display unit family includes colour and monochrome configurations. CDUs are available in a variety of sizes and interfaces including MIL-STD-1553B. The CDU is the cockpit flight management centre and a CRT is used to display functional menus and the requested data. The control display units are compatible with both ARINC 429 and MIL-STD-1553 databusses. Later models incorporate active matrix colour flat-panel displays.

Produced as keyboard interface units for other avionics systems manufacturers, the Model 2180 Series mini CDU and the Model 2181 Series ARINC CDU are utilised in a variety of navigation systems, such as VLF/Omega and GNSS. The Model 2180 is used primarily by business and rotary aircraft, while the Model 2181 is in service with regional airlines.

The units feature a seven-colour raster display with full alphanumeric keyboards. A total of 128 character fonts or symbols can be displayed on the eight lines, with 14 characters per line. An RS-422 serial port is standard, RS-429 is optional and MIL-STD-1553 bus versions are available.

Operational status
In production. The CDUs serve as remote terminals in a number of aircraft including the F-4, A-6, E-6B, E-3, MH-47, MH-60, A310 and a number of special purpose P-3s.

Contractor
Smiths Industries Aerospace.

VERIFIED

Digital fuel gauging systems

Smiths Industries has had its 2300 Series digital fuel quantity indicators in service for some years on an increasing variety of civil and military aircraft. The system is basic fit on the Boeing 737, 747-200/300 and 777.

Smiths Industries has developed an integrated fuel quantity system in a single LRU. This device performs automatic refuelling, valve control, preselection of fuel quantity and digital data output for EFIS displays.

Operational status
In addition to supplying Boeing, for the 727, 737 and older 747s, Smiths Industries has certified the system for retrofit to a variety of aircraft including Boeing DC-8, DC-9 and DC-10 and the Fokker F28.

Smiths Industries digital fuel gauging repeater and master indicators for the Boeing 747 **1996**

Military retrofits have been completed on the A-4 Skyhawk, C-130 Hercules, C-141 and F-4 Phantom.

Contractor
Smiths Industries Aerospace.

UPDATED

Electronic chronometers

The 2600 Series electronic chronometers utilise two dichroic LCDs to display GMT, chronograph, elapsed time and day and date. Smiths digital clocks utilise a temperature compensated crystal oscillator for extreme accuracy. Trickle current from the aircraft's hot bus maintains functionality with bus power off.

A variant of the 2600 is the 2620 which has been selected for the Boeing 777 and features ARINC 429 input and output, GPS and ASIC technology.

Operational status
In production. Smiths digital clocks are standard fit on the Airbus A320 and A340, BAe Avro 85/100, Boeing 727, 737, 747, 757, 767 and 777 Series, MD-80, MD-11 and MD-90, and Fokker 50/70 and 100, IPTN N-250.

Contractor
Smiths Industries Aerospace.

UPDATED

Graphics generation

Smiths Industries graphics generation consists of independent video, stroke, raster graphics and scan conversions for multiple monochrome and colour displays. The display computers generate calligraphic symbology for the head-up display and a combination of calligraphic hybrid symbology for the head-down multipurpose display. In the event of a mission computer failure, the display computer can assume control of sufficient aircraft avionics to provide all essential information for navigation and aircraft attitude control.

Smiths Industries 2620 digital clock as fitted to Boeing 777 **1996**

The display computer features a MIL-STD-1553B bus controller, advanced BIT and 4,000 hours demonstrated MTBF.

Operational status
In production and in service.

Contractor
Smiths Industries Aerospace.

VERIFIED

TAMMAC Tactical Aircraft Moving MAp Capability

The new US Navy Tactical Aircraft Moving MAp Capability (TAMMAC) is fitted with the Smiths Industries Aerospace Advanced Memory Unit (AMU) which will replace earlier US Navy data storage and mission data loader equipment.

Operational status
The TAMMAC system is scheduled to be fitted initially to five baseline US Navy TAMMAC aircraft: F/A-18, AV-8B, AH-1W, UH-1N, and V-22, with the following aircraft likely to be fitted later: F-14, S-3, CH-53, CH-60, SH-60, and P-3.

Contractor
Smiths Industries Aerospace.

NEW ENTRY

TCAS RA Vertical Speed Indicators

Smiths Industries has developed special vertical speed indicators for use with TCAS I and TCAS II Traffic alert and Collision Avoidance Systems. The units are form, fit and function interchangeable with most existing VSIs. Designed for ARINC 735 compatibility, the units may be used with TCAS II systems now entering service.

The 2074 Series instruments are outwardly similar to the Smiths Industries 2070 Series VSIs already in service with many airlines. The 2074 Series is intended for those applications where traffic advisories will be presented through some other medium such as a

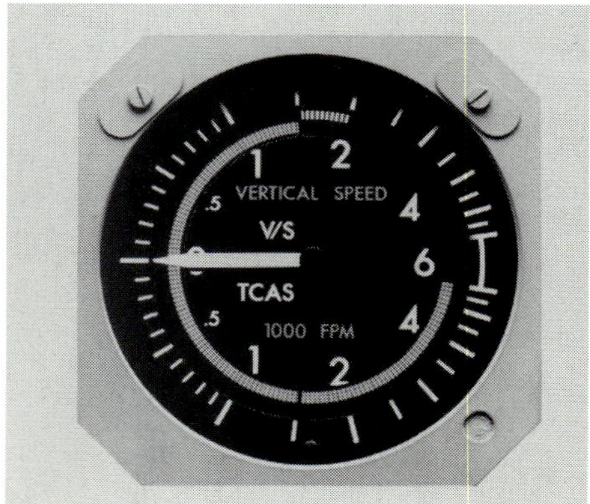

Smiths Industries Model 2074 RA VSI

weather radar. Resolution advisories are presented through red and green LED 'eyebrows'. Green segments indicate those vertical speeds which will maintain safe separation from other aircraft, while red depicts potentially dangerous vertical speeds.

The 2074 flat-panel vertical speed/TCAS I instrument provides TCAS I data in addition to acting as a standard VSI. This instrument has successfully interfaced with both the BFGoodrich and AlliedSignal TCAS I systems. An ARINC 429 channel provides good readability with a full-colour display.

Operational status

The 2074 RA VSI is certified for use with AlliedSignal and Rockwell Collins TCAS. Over 700 units are currently in operation.

Contractor

Smiths Industries Aerospace.

UPDATED

ST-170 Radio Magnetic Indicator (RMI)

The ST-170 radio magnetic indicator was designed for interface with panel-mounted avionics in single-engine, light and medium twin-engine aircraft and helicopters. Magnetic heading information is presented to the pilot on a servo-driven compass azimuth dial against a fixed lubber line. Each of two pointers provides bearing information for a selected ADF or VOR. The system directly interfaces with ADF and navigational receivers without the need for adaptors or external AC inverter. Its light weight and small size permit easy installation almost anywhere in the panel. Rugged all-metal construction renders the ST-170 suitable for all environments including helicopters.

The ST-170 consists of dual-switch RMI display indicator and remote converter electronics unit. A second dual-switch RMI can also be installed and operated from the same converter.

The converter electronics package consists of solid-state plug-in modules. The converter contains two separate and complete converters which will accept standard VOR composite signals. When combined with the magnetic heading information, relative bearing information for the RMI VOR pointer display is produced.

Specifications

Dimensions:
(indicator) 80.6 × 80.6 × 97.3 mm
(converter) 63 × 101.1 × 259.3 mm
Weight:
(indicator) 0.68 kg
(converter) 1.27 kg
Power supply: 28 V DC, 750 mA (max)

Contractor

S-TEC Corporation.

VERIFIED

AN/ASN-165 radar navigation data display set

The AN/ASN-165 is a stand-alone radar indicator system that upgrades the displays on older radars. It replaces the display subsystem on ground mapping, weather avoidance and navigation radars such as the APN-59, APS-133 and APQ-122. The aim is to improve reliability, maintainability and operational performance without replacing the entire radar. The system is designed as a drop-in replacement for the radar displays with little modification to existing structures and cabling in the aircraft. The radar's antenna and receiver/transmitter subsystems remain unchanged.

The AN/ASN-165 consists of a radar data converter, pilot indicator and navigator indicator/control panel. The radar data converter contains two independent digital scan converters that convert the radar data into a high-resolution raster video signal. Navigational data is overlayed on the radar imagery and displayed on the pilot and navigator colour indicators. Aircraft navigational systems are interfaced through a standard ARINC serial bus or dual-redundant MIL-STD-1553B databus.

The AN/ASN-165 radar navigational data display set showing (left to right) the radar data converter, navigator indicator/control panel and pilot indicator

Radar ground map, terrain-avoidance and weather imagery are typical displays. The imagery is displayed with 16 levels of shading using standard RS-170 or RS-343 video format. Multiple colours are used to indicate different weather intensities. The standard video format also permits recording of the radar imagery for mission review and training.

With the navigational interface, aircraft data, such as true heading, groundspeed and track angle error, is displayed on both the navigator's and the pilot's display. The aircraft's present position is also displayed and the data is used to calculate the position of ground targets identified by a movable cursor that can track a fixed point on the ground. The navigational interface also provides the option of stabilising the radar display to true north.

Flight plan waypoints and navigational aids can be displayed, depending on the navigation system used. Additionally, data from EW systems, such as detected threats and threat zones, can be overlayed on the radar image.

Operational status

The AN/ASN-165 is installed on the WC-130, C-130 and US Air National Guard KC-135. Variants of the system are used on the MC-130E, HC-130P/N, C-141B, RC-135 and other Special Operations and reconnaissance aircraft.

Contractor

Systems Research Laboratories.

VERIFIED

AT 3000 Altitude Digitiser

The AT 3000 is an all-solid-state blind encoding altimeter designed to interface with the TRT 250D transponder and most other modern Mode C transponders. The addition of the AT 3000 to the transponder provides an altitude reporting capability conforming to FAR 91-36B. The Federal Aviation Administration has granted a TSO C88 approval for the AT 3000.

Specifications

Dimensions: 42 × 65 × 159 mm
Weight: 0.23 kg
Temperature range: −20 to +55°C
Altitude: −1,000 to 30,000 ft ±50 ft

Operational status

In service.

Contractor

Trimble Navigation Ltd, Avionics Products.

UPDATED

Tri-Nav C VOR indicator with Loran C course deviation

A gas discharge display indicator used in conjunction with most brands of navigation radios with or without glide slope capability, the Tri-Nav C displays one VOR/Loc course or at the flip of a switch the display changes to show deviation from the selected Loran course. The Tri-Nav C presents horizontal deviation information in increments calibrated to 10° full-scale right and left.

Specifications

Dimensions: 82.6 × 82.6 × 114.3 mm
Weight: 0.57 kg

Operational status

In service.

Contractor

Trimble Navigation Ltd, Avionics Products.

VERIFIED

Tri-Nav VOR/Loc indicator

A gas discharge display indicator used in conjunction with most brands of navigation radios, with or without glide slope capability, the Tri-Nav displays two VOR/Loc courses with glide slope deviation.

Specifications

Dimensions: 82.6 × 82.6 × 114.3 mm
Weight: 0.57 kg

Contractor

Trimble Navigation Ltd, Avionics Products.

VERIFIED

HEAD-UP DISPLAYS, HELMET-MOUNTED DISPLAYS AND WEAPON AIMING SIGHTS

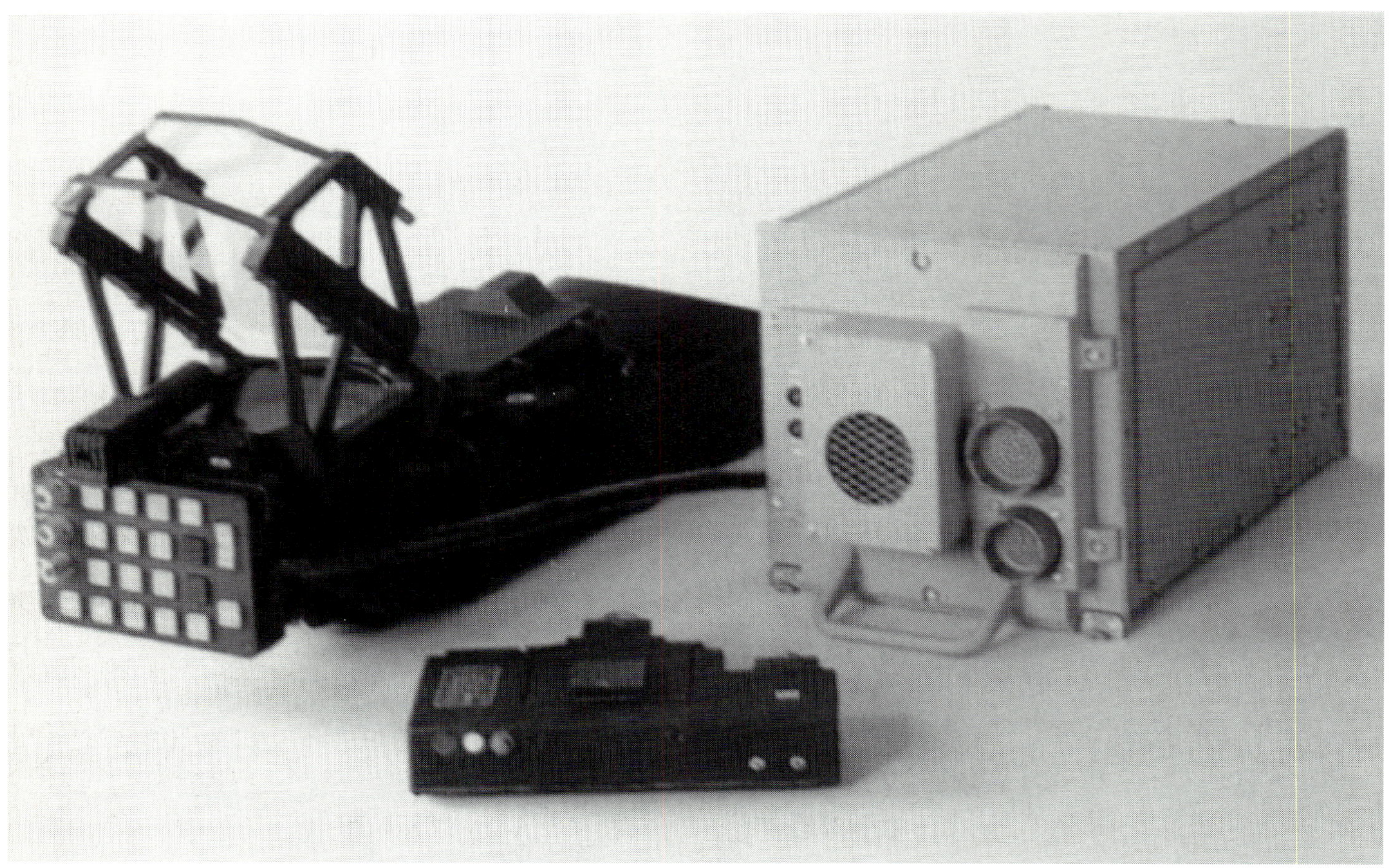

The Smiths Industries' head-up display for the US Navy T-45C Goshawk ***1998***/0018153

FRANCE

Air-to-ground mission helmet-mounted display

In common with Sextant Avionique's other fighter helmet-mounted displays, the air-to-ground mission display features visor projection for pilot comfort and safety, accurate fully qualified and flight test proven head positioning sensing system, integral concept for ejection safety, NBC protection provision, eye protection against laser threats and positive pressure breathing.

The air-to-ground mission display is a binocular system with a 30 × 40° field of view and a 60 mm eye relief. It provides stroke symbology and raster video imagery, and integrated night vision. Head supported weight is 1.8 kg, including the oxygen mask, and it has an optimised centre of gravity.

Contractor

Sextant Avionique.

VERIFIED

CTH 3022 head-up display

The CTH 3022 head-up display for the Rafale is a holographic 30 by 22° field of view HUD that displays computer-generated symbology and FLIR imagery. Information is received from an associated symbol generator unit.

Operational status

Developed for the Rafale aircraft.

Contractor

Sextant Avionique.

UPDATED

Electronic Head-Up Display (HUD) for the A330/A340

The A330/A340 HUD comprises an Optical Display Unit (ODU) and a Head-Up Display Computer (HUDC).

The ODU is installed in the overhead cockpit panel and is stowable behind the pilot's head when not in use. It moves automatically along the mounting tray to the operating position. The combiner glass is collapsible and can move forward in a crash situation. The ODU consists of optical lenses and a combiner which presents collimated symbology superimposed on the outside world, a miniature high-brightness CRT and an automatic brightness control system. The HUD controls are located in the ODU.

The HUDC is located in the electronic bay. It is connected on one side to the display management computer and on the other to the ODU. The HUDC is designed around VLSI circuits. Input parameters received from two distinct channels are monitored so that false information is instantaneously detected. The HUDC includes BITE. The symbol generator is capable of driving two optical display units.

The HUD includes growth potential for enhanced vision systems.

Specifications

Weight:
(ODU) 10 kg
(HUDC) 5 kg
Power supply: 115 V AC, 400 Hz
Field of view: 30 × 24°

Operational status

In production.

Contractor

Sextant Avionique.

VERIFIED

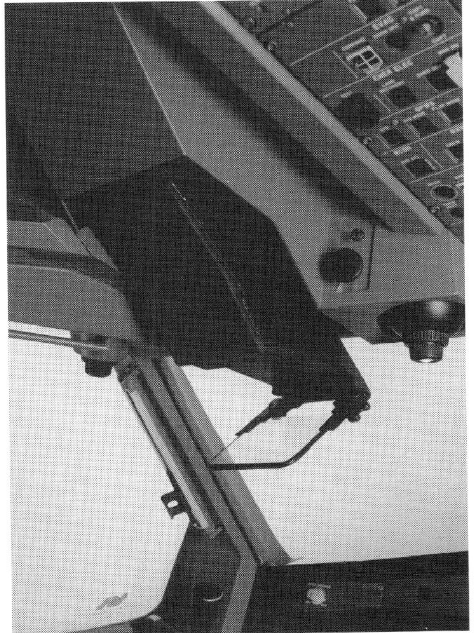

The HUD for the A330/A340

Head-Up Flight Display Systems (HFDS)

Sextant Avionique, previously the avionic division of Thomson-CSF has pioneered Head-Up Displays (HUD) for commercial use since 1975. Its latest development is the Head-Up Flight Display System (HFDS), which is a multipurpose HUD system that can be fitted to any cockpit, whether analogue or digital.

The system aim is to provide assistance to take-off and landing, to improve safety at the minima. The HFDS has been developed as a series:

Series 100, called Integrated HFDS
Series 200, called Manual HFDS
Series 300, called Hybrid HFDS

The Integrated HFDS is intended for aircraft already designed with a 'Fail Op' autopilot.

The Manual HFDS is intended for aircraft with 'fail passive' auto-pilots in order to reach Cat IIIa minima on manual landing with 50 ft decision height. Bombardier and de Havilland offer this system as options on their Global Express and Dash 800-400 aircraft respectively.

The Hybrid HFDS is already in operation on Aerospatiale's Boeing 737-300 and on the MD 82 of Alitalia to enhance the landing capabilities to Cat IIIb

Sextant Avionique's Head-Up Flight Display System for Cat IIIb operation on Aerospatiale's B737-300 test aircraft **1997**/0001450

Sextant Avionique's Head-Up Flight Display System **1997**/0001451

standard with respective decision heights of 35 ft and 20 ft. The system allows a 75 m Runway Visual Range (RVR) for take-off.

Display symbology is matched to the series type and operational configuration.

The HFDS is also able to support Enhanced Vision System (EVS) procedures.

The HFDS comprises 4 LRUs: the projector located in the cockpit above the pilot's head; the holographic combiner with a full 40 x 28° field of view; the control panel; the computer.

Operational status
In production.

Contractor
Sextant Avionique.

VERIFIED

HFDS shipset **1997**/0001452

Helmet-mounted sight for helicopters

The helmet-mounted sight for helicopters features visor projection for pilot comfort and safety, accurate fully qualified and flight test proven head position sensing system, modular concept with a single-size customised helmet, NBC kit compatibility and eye protection against laser threats. It is used in target acquisition and designation and the firing of guns, rockets and air-to-air missiles.

It is a monocular system with a 6° field of view and 600 mm eye relief. The display capability includes a sight reticle and fixed-mode symbology. The head supported weight is 1.5 kg and the helmet has an optimised centre of gravity location. It is NVG-compatible.

Operational status
Deliveries for Eurocopter Tiger helicopter.

Contractor
Sextant Avionique.

VERIFIED

Para-visual display

The para-visual display has been developed as optional equipment for the A320. Mounted on the glareshield, the instrument guides the pilot along the runway centreline during roll-out in poor visibility. Guidance is provided by vertical white and black strips moving to the left or right according to the deviation from runway centreline.

The display uses liquid crystal and is within the pilot's field of view when looking outside. It is connected to the flight guidance system and to the flight warning computer of the aircraft, or equivalent sources. The system comprises the autoland warning and the picture generator.

Specifications
Dimensions: 150 × 155 × 36 mm
Weight: 1 kg
Interface: ARINC 429 HS discrete inputs (autoland warning control)
Angle of view: 60° under day or night conditions

Operational status
Optional equipment for the Airbus A320.

Contractor
Sextant Avionique.

VERIFIED

Smart Head-Up Display (SHUD)

The SHUD is a fully integrated head-up display with a 24° circular field of view. It is able to computerise and draw symbology and to present simultaneously FLIR video and flight information. The SHUD is particularly well adapted to combat aircraft upgrading.

Operational status
Selected by Lockheed Martin for the Argentine Air Force A-4M avionics upgrade programme. Other orders include: the Mikoyan MiG-AT trainers, Spanish Air Force Mirage F1 modernisation, Iryda I-22 trainers of the Polish Air Force. Over 100 SHUDs have been produced.

Sextant Avionique Smart Head-Up Display (SHUD)

Contractor
Sextant Avionique.

UPDATED

T100 and T200 weapons sight for helicopters

The T100 and T200 electromechanical head-up weapons sights show a collimated reticle which can be moved between −10 and +7° in elevation and ±6° in azimuth, angles compatible with the requirements for air-to-air and air-to-ground weapons launch. The device is mounted on the canopy frame of the helicopter, weighs 2 kg and has a field of view of 7.5°.

The sighting system is based on a modular concept. The basic component is the T100 single- or T200 dual-axis monocular sight head which can be used for both day and night weapon firing in conjunction with third-generation microchannel NVG. Sight recording by CCD camera has been validated for both training and operational firing and is available as an option. The physical features of the sight head, such as its low weight, small size, simple and strong high-performance optical system of Angenieux lens, diode array on micro-electronic support, give it the capacity to produce a remarkably high-quality image, both by day and night.

Sextant Avionique T100 helicopter head-up display **1998**/0018160

Sextant Avionique T200 helicopter head-up display **1998**/0018159

The second main component of the sighting system is a control unit combining command and calculation functions. This unit includes the necessary controls for moving the sight head, symbology animation controls, firing tables stored in its internal memory, weapon firing computation system for guns and air-to-air missiles and sighting telescope interface. The links with the missile system and telescope are designed in particular for export requirements and include coupling with Sextant Avionique's Nadir computer for target designation. The system is equipped with a built-in automatic testing facility, with status information being displayed on the sight head.

To meet the weaponry requirements of the French Army Gazelles which are equipped with guns and Mistral missiles, Sextant Avionique proposed a multipurpose sighting and fire-control system comprising the above-mentioned components that have already been qualified for the Gazelle. With this system, the pilot can fire the various weapons, including rockets, without any manipulation other than manual selection of functions on the front panel of the control unit.

Operational status
In production. The system has been installed on a wide variety of helicopters including the Boeing 500/530, Eurocopter BO 105 and BK 117, Sikorsky S-76 and Black Hawk, Bell 206, 406 and 412, Westland WS 70. It has also been selected to equip the Gazelle helicopters of the French Army Light Aviation Corps and the Ecureuil helicopters of the French Air Force, and by Eurocopter for export versions of the Dauphin, Ecureuil, Gazelle and Puma.

Contractor
Sextant Avionique.

VERIFIED

TMV 980A head-up/head-down display

The TMV 980A is an integrated head-up/head-down display system for the Dassault Mirage 2000 multirole fighter. It comprises a digital computer and processor to generate the display symbology and help with flight and weapon aiming computation and three display units: a VE 130 CRT head-up display, a VMC 180 interactive multifunction head-down colour display and a VCM 65 complementary monochrome CRT for electronic support measures information.

The head-up display has a high-resolution, high-brightness CRT with a collimating optical system based on a 130 mm lens providing a wide total field of view. The binocular instantaneous field of view is increased in elevation by the use of a twin-glass combiner which transmits 80 per cent of the light incident upon it. Automatic brightness control, with manual adjustment, permits symbols to be read in an ambient illumination of 100,000 lux. The system provides continuous computation of tracer line in the air-to-air mode and impact and release points in the air-to-ground mode.

The main head-down display presents, in red, green and amber on a 127 by 127 mm CRT radar display, information such as a radar map, synthetic tactical situation and range scales, raster images from television or FLIR sensors and tactical data from the system itself or from an external source.

A helmet-mounted sight may be integrated into the TMV 980A unit to improve target discrimination and off-boresight target designation. Another option is the substitution of the VE 130 head-up display by a VEM 130 system.

Specifications
Weight:
(electronic unit) 9 kg
(head-up display) 13 kg
(VMC 180 head-down display) 14 kg
(VCM 65) 4 kg

Operational status
In service in the Dassault Mirage 2000. No longer in production.

Contractor
Sextant Avionique.

VERIFIED

TMV 1451 electronic head-up display for commercial aircraft

The TMV 1451 head-up display comprises the Optical Head Unit (OHU) and the Head-Up Display Computer (HUDC). The OHU is installed in the glareshield panel either behind the glareshield front panel or in the operational or pull-out position with the combiner appearing in the forward field of vision of the pilot. The HUDC is located in the electronics bay. It is linked on one side to the display management computer and on the other side to the OHU itself. The HUD's control panel is included in the OHU.

The OHU is an electronic head-up display system designed specifically to be installed in the glareshield of commercial aircraft and particularly in the Airbus A320. The head-up display, linked to the existing aircraft systems, can be used for roll-out guidance, visual approach guidance and monitoring of automatic approach and flare. The OHU is linked to the HUDC which generates the symbology to be displayed according to the flight phase.

The OHU consists of optical lenses and combiner which present collimated symbology to the pilot superimposed on the outside world, a miniature high-brightness CRT, an automatic brightness control which adjusts the symbol brightness to the required level and a control panel. Of limited volume, the OHU is designed and installed in such a way that the pilot's lower field of vision is not interrupted.

The head-up display computer is designed around VLSI circuits already in use in the display systems of A310 and A320 aircraft. Input parameters received from two distinct channels are monitored so that false information is instantaneously detected. The HUDC includes BITE. The symbol generator function of the computer is capable of driving two optical head-up displays.

Specifications
Weight:
(OHU) 10.5 kg
(HUDC) 5 kg
Power supply: 115 V AC, 400 Hz
Field of view: 24 × 15°
Reliability:
(OHU) 5,000 h MTBF
(HUDC) 13,000 h MTBF

Operational status
In production.

Contractor
Sextant Avionique.

VERIFIED

Topdeck® avionics suite for military transport retrofits

The Sextant Avionique Topdeck® avionics suite is specially designed for retrofit on surveillance aircraft and military transports.

Topdeck® is built around a dedicated flight management system supporting logistic and tactical missions, an EFIS function presented on four 6 × 8 in liquid crystal displays, two ring laser gyros in an IRS embedding a GPS function, two air data computers and an autopilot/flight director. A Head-Up Flight Display System (HFDS®) is available on option to improve mission capabilities.

Contractor
Sextant Avionique.

VERIFIED

Topowl day/night helmet-mounted sight and display for Tiger helicopter

The main features of Sextant Avionique's Topowl helicopter helmet-mounted sights are: visor projection for pilot safety and comfort; qualified and flight test proven head position system; modular concept with a single-size customised helmet; NBC protection, HID and AEA compatibility and eye protection against laser threats. They are used for piloting, navigation, sensor steering, target acquisition and designation and the firing of guns, rockets and air-to-air missiles, and air-to-ground missiles.

The day/night helmet-mounted sight and display is a binocular system with 100 per cent overlap, a field of view of 40° and eye relief of 60 mm. The integrated display capabilities include stroke symbology and infrared and TV raster video imagery, integrated night vision sensors and integrated cursive symbol generator. The helmet weighs 2.0 kg in the day configuration and 2.2 kg in the night configuration. It has an optimised centre of gravity.

The Topowl day/night helmet-mounted sight and display for the Eurocopter Tiger helicopter
1997/0001453

Operational status
Under development for the Eurocopter Tiger helicopter and NH 90. First units were delivered to Eurocopter in March 1994 for flight test on Tiger PT4. First units for Rooivalk were delivered for test in April 1996.

Contractor
Sextant Avionique.

VERIFIED

Topsight® helmet-mounted display

The Topsight® helmet-mounted display for Rafale is a head-up sight and display fully integrated in a helmet. Symbology is displayed directly on the visor without any optics between the eye and the visor. The helmet is an ergonomically designed integral unit that weighs less than 1.8 kg, including electronics and mask.

Topsight provides both target designation and acquisition. Designation involves locking the weapon system on an identified target. Acquisition allows the pilot, with a simple move of the head, and without having to manoeuvre the aircraft, to bring a target detected by the onboard systems within his field of view, even if the target is outside the field of view of the conventional HUD.

The Topsight® helmet-mounted display for Rafale

Topsight also provides continuous visor display of all parameters needed to carry out the mission, including piloting, navigation, aircraft operation, firing data and system and detection warnings. This means the pilot does not have to look down at the instrument panel during critical phases of the mission.

Operational status

Developed for Rafale. In May/June 1997, Sextant Avionique carried out an operational flight test evaluation of the Topsight helmet-mounted display for the French Air Force and Navy, as part of the overall Rafale development programme. Meanwhile, Topsight is available for modernisation of other fighter aircraft.

Contractor

Sextant Avionique.

UPDATED

VE 110, VE 120 and VE 130 head-up displays for combat aircraft

Sextant Avionique has developed a full family of head-up displays for tactical aircraft of various sizes and sophistication. The displays include advanced software for each phase of a flight, allowing accurate attacks as well as navigation and piloting.

The VE 120, chronologically the first of the family, was configured to the Dassault Super Etendard and Mirage F1.

The VE 110, especially designed for small cockpit tactical or trainer aircraft such as the most recent version of the Dassault Alpha Jet, has the most compact Pilot's Display Unit (PDU).

The VE 130, with a larger PDU, is fitted to versions of the Mirage 2000.

The instantaneous fields of view of the dual-combiner displays are typically about 20° in azimuth and 15 to 18.5° in elevation, depending on aircraft type.

Specifications

VE110
Dimensions:
(PDU) 427 × 123 × 125 mm
(EU) ½ ATR
Weight:
(PDU) 8-10 kg
(EU) 8-9 kg

Operational status

No longer in production.

Contractor

Sextant Avionique.

VERIFIED

VEM 130 and VEH 3020 combined head-up and head-level displays

The VEM 130 is a derivative of the VE 130 which includes an additional capability for raster FLIR image presentation at head level. This display gives a field of view of 14 x 10° and the green phosphor raster display can be set at 525, 625 or 675 lines, 50 or 60 Hz.

The VEH 3020 has been fitted to the French Dassault Rafale demonstrator aircraft. It is a dual head-up and head-level display and also forms a part of the Sextant Avionique integrated multidisplay system designed for the Enhanced Mirage 2000. The VEH 3020 HUD is a dual-raster/cursive type, while the TMM 1410 head-level display is the same as that employed in the VEM 130 system.

Specifications

VEM 130
Dimensions: 645 × 378 × 145 mm (plus combiner glass)
Weight: 23 kg

VEH 3020
Dimensions: 645 × 382 × 158 mm (plus combiner glass)
Weight: 24 kg

Operational status

The VEH 3020 has been selected for the Dassault Mirage 2000-5.

Contractor

Sextant Avionique.

VERIFIED

APX M 334 sights for helicopters

The designation APX M 334 covers a family of roof-mounted sighting systems for helicopters. All are single eyepiece devices and have magnifications of ×3 and ×10 with corresponding 300 and 90 mrad fields of view.

The principal requirement of the M 334 series is to provide an operational system able to detect a target at a range of about 10 km and recognise a tank at 5 km. An optional special TV camera arm allows the recording of observed images by a video recorder.

Variants of the basic system have specific capabilities:

M 334: observation and manually guided missile firing
M 334-04 Athos: observation and firing of manually guided missiles or seeker-head missiles
M 334-25: observation sight with rangefinding and target localisation. When used in conjunction with a navigation system it constitutes the Osloh III system giving target absolute position. When used in connection with an aiming collimator it constitutes the Oshat system for air-to-ground gun and rocket firing
M 397 HOT: observation sight able to fire HOT automatic infrared guided missiles produced by the Franco-German Euromissile partnership.

Operational status

All members of the M 334 family are currently in production and service. A total of 1,500 systems are in operation in 30 countries.

The sight has been fitted to such helicopters as: the Eurocopter Gazelle 341 and 342 and Super Frélon SA 321; Westland Wasp, Lynx and WG 13; Bell 204, 205, 206/OH-58 and 212; Sikorsky SH-3D, Boeing 500 and 530, Agusta A 109, Eurocopter BO 105 and BK 117.

Contractor

SFIM Industries.

VERIFIED

The SFIM Industries APX M 397 sight

Strix and Viviane day/night sights

SFIM Industries has developed a family of day/night nose- and roof-mounted sights for helicopters.

Strix
The Strix sight is fitted with: an IRCCD thermal imager; direct view channel; TV channel; high-rate laser

The SFIM Industries Strix roof-mounted targeting system for the HAP (multirole support) Tiger helicopter
1998/0018158

rangefinder; and micromonitor. The thermal and TV video images and firing symbols are projected onto the sight eyepiece.

The weapon system uses Strix as a data reference system so as to determine its exact line of sight in space. Sighting precision is ensured by automatic target tracking on the TV and thermal channels. The sight is designed to maintain stabilisation and precision of the line of sight throughout the entire operating range of the helicopter and is MIL-STD-1553B compatible to interface with the associated weapon systems. Composite materials are used for the structure so as to obtain the best trade-off between the mass of the integrated sensors and its own mass.

Operational status

Strix has been selected for a roof-mounted installation on the HAP Tiger helicopter, and it is in production for a nose-mounted installation on the Rooivalk helicopter.

Viviane

Viviane is used to provide a day/night capability for Euromissile's HOT missile.

Operational status

In service on the Eurocopter SA 342M Gazelle helicopter of the French Army Light Aviation Corps (ALAT).

Contractor

SFIM Industries.

NEW ENTRY

INTERNATIONAL

AN/AVS-7 ANVIS/HUD system

The AN/AVS-7 ANVIS/HUD system projects flight data into the views of the pilots NVGs. By eliminating the need to shift attention and focus to the cockpit interior, the AN/AVS-7 Aviator's Night Vision System/Head-Up Display (ANVIS/HUD) enhances flight safety and operational effectiveness. Pilots can fly head-up, viewing the situation outside the cockpit, while at the same time receiving all essential flight data. The AN/AVS-7 features a real-time, high resolution, lightweight display unit which is easily mounted on NVGs. It is adaptable to any aircraft platform and has four independent display modes, each with declutter facilities. The pilot and co-pilot can independently select symbols and display modes from controls on the centre console and collectives, using ANVIS compatible control panel and display facilities. BIT provides 95 per cent failure detection and fault isolation.

Day-HUD optics interface directly with the AVS-7 harness to provide day symbology, including expanded symbology for day/night pilot operations, plus maintenance and training.

AN/AVS-7 ANVIS/HUD showing: display unit (top left), signal data converter (top right), converter control (bottom left) and sensors and harness (bottom right) ***1998***/0018157

Specifications

Field of View (FoV): 34°
Display type: stroke (highly stable)
Resolution: 512 × 512 pixels
Interfaces: MIL-STD-1553B, ARINC 429, RS-422, analogue, discrete and synchro
Reliability: >1,000 h
Power: 28 V DC, <5A

Avionics B kit
Dimensions:
(converter control (CC)): 63.5 × 139.7 × 76.2 mm
(display unit (DU)): 88.9 × 38.1 × 39.1 mm
(signal data converter (SDC)): 274.3 × 190.5 × 198.1 mm
Aircraft installation A kit: sensors and harness, as required

Operational status

AN/AVS-7 is built by Tracor Aerospace, Inc in collaboration with Elbit Systems Ltd. AN/AVS-7 is installed on many Israeli Air Force, US Army, US Marine Corps and US Navy rotary- and fixed-wing aircraft. AN/AVS-7 adapts to the following aircraft types: CH-46E, CH-47D, CH-53E, HH-60H, KC-130T, MH-47E, MH-60K, MV-22, UH-1N and UH60A/L.

A derivative system (Gideon) has been ordered by the US Marine Corps and other forces for Bell AH-1W helicopters.

Tracor Aerospace, Inc and Elbit Systems Ltd have together been contracted to upgrade the computer element of 1,100 AN/AVS-7 systems for the US Army, to double the number of inputs and outputs, to transfer data to the Crash Survival Memory Unit (CSMU), and to increase CPU speed, and facilitate programming via a front panel connector.

Contractors

Elbit Sytems Ltd.
Tracor Aerospace, Inc.

NEW ENTRY

Crusader helmet display system

GEC-Marconi Avionics, in association with Gentex and Pilkington Optronics, is developing the Crusader helmet system for both fixed- and rotary-wing aircraft. The team is taking a fully integrated approach and is concentrating on visor projection for fighter/attack applications with full day and night capability. An optical head tracker is used for maximum accuracy and minimum latency. A 40° visor-protected field of view has been demonstrated and is being further developed.

Crusader employs Pilkington Safe-Lite helmet display optics.

Operational status

In development. A derivative of the Crusader helmet display system is in development for the Eurofighter 2000.

Contractor

GEC-Marconi Avionics Ltd, Mission Avionics Division, Edinburgh.
Gentex Corporation.
Pilkington Optronics, Glasgow.

UPDATED

The Crusader helmet display system is under development

Eurofighter helmet display system

GEC-Marconi Avionics and GF Sistemi Avionici have been awarded a £28.5 million contract by British Aerospace Military Aircraft, on behalf of Eurofighter, to develop a fully integrated pilot's helmet display system for the Eurofighter aircraft.

The Eurofighter helmet display system will incorporate latest technological developments in binocular, fast jet, day and night helmets. It will provide to pilots on the visor of the helmet all the necessary flight and target data and provide night vision capability. The helmet will also provide the pilot with the ability to fire missiles at very high off-boresight angles using an optical tracking system.

The helmet will also provide NBC and laser protection, and support direct-voice input communication.

Operational status

Development.

Contractors

Alenia Difesa, Avionic Systems and Equipment Division, GF Sistemi Avionici.

GEC-Marconi Avionics Ltd, Mission Avionics Division, Edinburgh.

UPDATED

VH 100 Head-Up Display

Sextant Avionique and Hamilton Standard have produced a family of head-up displays as a part of advanced weapon control systems in helicopters, in particular for the air-to-ground firing of Stinger or Matra Mistral missiles.

The VH 100 is designed to provide comprehensive navigation and weapon aiming information to the helicopter crew, while being small enough to obscure the crew's vision to the minimum. In weapon aiming it can offer air-to-air rockets and guns and air-to-ground guns and missile symbology and for navigation the display is compatible with NVG. The system comprises a pilot's display unit, an electronics unit and a control panel. The total field of view is 20°.

The system consists of a Pilot Display Unit (PDU), Electronic Unit (EU) and Optional Control Unit (OCU).

The PDU utilises a miniature CRT which can generate symbology visible in a high-brightness environment and optical lenses and combiner to present superimposed symbology over the external scene.

The EU provides power supply and signal processing for the CRT, generation of synthetic symbology, electrical interfacing with weapons system sensors, airframe and fire-control computations and aircraft sensor signal processing.

The CU controls weapons mode selection, symbol luminance adjustment, firing distance selection and weapon elevation offset.

The VH 100 has been flight tested fitted to a Gazelle in France, on OH-58 and Bell 406 helicopters in the USA and BO 105 in Germany. It has been selected for the US Army's OH-58 helicopter Stinger missile sight subsystem programme. Initial flight trials were completed in Autumn 1986.

Specifications

Weight:
(pilot display unit) 2.99 kg
(electronics unit) 1.99 kg
Power supply: 28 V DC, 2.5 A
Field of view: 20°
Reliability: >3,600 h MTBF

Operational status

No longer in production. In service in US Army OH-58 helicopters. Twin VH 100 HUDs have been installed in the C-160 Transall upgrade.

Contractors

Hamilton Standard Division of UTC.

Sextant Avionique.

VERIFIED

Head-Up Display (HUD) for the Eurofighter 2000

GEC-Marconi Avionics, leading a consortium with Teldix, GF Sistemi Avionici and ENOSA, has been awarded thedevelopment contract for the head-up display for the Eurofighter 2000. The HUD forms part of the aircraft displays and controls subsystem and provides the primary display of flight information to the pilot.

The head-up display in development for the Eurofighter 2000

The Pilot's Display Unit (PDU) uses diffractive optics featuring a single-element holographic combiner which consists of two glass elements bonded to produce a flat parallel-sided assembly. An optically powered hologram is recorded on photosensitised gelatine on the spherical interface sandwiched in the assembly and acts as the collimating combiner. The resulting advanced optical system, manufactured using computer-generated holographic techniques, provides new levels of display capability. Additionally, the uncluttered simplicity of the combiner support structure allows virtually a clear out-of-cockpit field of regard. The total field of view is 30° azimuth by 25° elevation and the instantaneous field of view is 30 by 20°. The display is required to operate in three modes; cursive, raster and raster/cursive.

The optical module brightness levels are optimised to operate in a very high ambient light environment, whilst minimising solar reflection and maximising outside world transmission and display uniformity from within the large eye motion box.

Associated with the PDU is a complex HUD control panel, developed by Teldix, which is attached to the aft face of the unit and incorporates LED technology for multifunctional displays.

Operational status

In development.

Contractors

Alenia Difesa, Avionic Systems and Equipment Division, GF Sistemi Avionici.

ENOSA.

GEC-Marconi Avionics Ltd, Mission Avionics Division.

Teldix GmbH.

UPDATED

Head-Up Display (HUD) for the Tornado

Teldix (Germany) is jointly responsible with Smiths Industries (UK) and Ottica Meccanica Italiana (Italy) for the production of the head-up display system for the Panavia Tornado.

The system consists of a Pilot's Display Unit (PDU) and an Electronics Unit (EU). The PDU incorporates a 5 in (127 mm) CRT with deflection amplifiers and an extra high-tension power supply. It also includes the reflector, lens assembly and combining glass. The control panel for the display is on the PDU. The optical system incorporates a standby sight in case of HUD failure and a mounting support for a recording camera.

The EU includes a digital computer for symbol generation, symbol motion and fire-control functions and is programmable to customer requirements. The electronically generated symbols are automatically adjusted by a photoelectric cell to a preset brightness level, to cater for conditions ranging from bright daylight to non-dazzle viewing at night. The BITE circuits allow for the direct detection of a defective assembly.

Operational status

In service.

Contractors

Ottico Meccanica Italiana SpA.

Smiths Industries Aerospace, UK.

Teldix GmbH.

UPDATED

The head-up display for the Panavia Tornado

Heli-TOW multirole scout and weapon sight system

Heli-TOW is a helicopter-mounted day and night anti-armour and scout system. The system consists of a stabilised sight subsystem and a missile launching/guidance subsystem.

For day scout missions the stabilised sight subsystem, also referred to as Helios Observation System (HOS), (see item under Saab Dynamics AB in this section) provides ×3 and ×12 direct view optics to give the gunner or co-pilot a sharp close-up true colour picture to locate and identify the enemy. The stabilised sight allows the gunner to locate and track targets accurately. The FLIR is utilised for night vision and improves target visibility under conditions of battlefield obscurants. In addition to FLIR, sight options include laser rangefinder, laser rangefinder/designator for target designation and gun camera for mission review and reconnaissance. Since the sight can be roof- or nose-mounted, it is compatible with almost all helicopter configurations.

Heli-TOW/Hellfire on a Eurocopter BO 105

The Heli-TOW multirole weapons system can be delivered in various configurations that accommodate TOW or Hellfire missile delivery, rocket or gun interface, or other special applications. The system is a modern, lightweight, low-cost modular system utilising digital electronics to provide precise missile launch and control to target impact. The modular Heli-TOW system was designed to be compatible with almost all helicopter configurations. Helicopter missions can be quickly and easily changed with this multirole system.

With the modular Heli-TOW design, growth potential to upgrade configurations is preplanned. Starting with the day-only HOS, upgrades can easily provide day and night operations with laser rangefinding or laser designation. Basic Heli-TOW can be supplied for day scout and anti-armour TOW operation with limited countermeasures, Heli-TOW 2 day and night capability, or with full countermeasures day and night scout and anti-armour configuration with full TOW 2 thermal tracking capabilities and Heli-TOW/Hellfire for day/night operations with autonomous Hellfire and full TOW 2 thermal tracking capabilities.

The day and night Heli-TOW system can be used with all versions of TOW missiles and laser-guided Hellfire missiles. Autonomous Hellfire guidance capability is provided by a stabilised laser designator, as well as launching missiles in a hand-off mode to co-operative designators.

Operational status

Six countries have ordered and have in operation more than 180 systems for use on five different military helicopter types. These countries are Sweden on the Eurocopter BO 105, Denmark and Singapore on the Eurocopter AS 550 Fennec, Saudi Arabia on the Bell 406 Combat Scout, Italy on the Agusta A 129 Mangusta and Belgium on the Agusta A 109.

Contractors

Saab Dynamics AB.
Systems & Electronics Inc.

VERIFIED

HUD 2020 and HUD 2022 Head-Up Displays

GEC-Marconi Avionics has utilised its C-17 and combat aircraft head-up display technology to develop the HUD 2020 system for corporate jet operators and from it the HUD 2022 for air transport aircraft. In these developments GEC-Marconi Avionics has teamed with Honeywell Business and Commuter Aviation Systems on the HUD 2020, and with Honeywell Air Transport Systems on the HUD 2022.

In the HUD 2020, the Honeywell Business and Commuter Aviation Systems computer produces the HUD symbology and drives it as analogue data, to the GEC-Marconi display optics. Over longer cable lengths required for air transport aircraft, this has proved less satisfactory, and in the HUD 2022, the Honeywell Air Transport Systems computer interfaces directly to the GEC-Marconi unit, which contains both the optics and the symbol generator.

The HUD 2020 system comprises a low-profile overhead-mounted display unit driven by the Honeywell display guidance computer. Cursive only, raster only and combined cursive over raster symbology modes are available.

Multiple processing techniques are incorporated to achieve the system integrity required to enable the system to be used as the primary flight display with less than one critical failure in 10^9 flight hours.

The display guidance computer interfaces with the aircraft avionic systems and generates the symbology displayed on the HUD.

System control is achieved via the existing EFIS control unit resulting in 'seamless' cockpit integration of the system. The display unit comprises an overhead projector unit and a combiner assembly which connect via a one-piece mounting tray which is boresighted to the airframe. This allows the units to be changed without realignment.

The single-piece refractive combiner assembly utilises the latest coating technology which enables a diffractive coating to be fabricated on the external surface of the combiner in a hard material, dramatically reducing the combiner's mass. Stowable when not in use, the combiner will break away under forward pressure for safety. The instantaneous field of view of the system is 30° in azimuth by 25° in elevation.

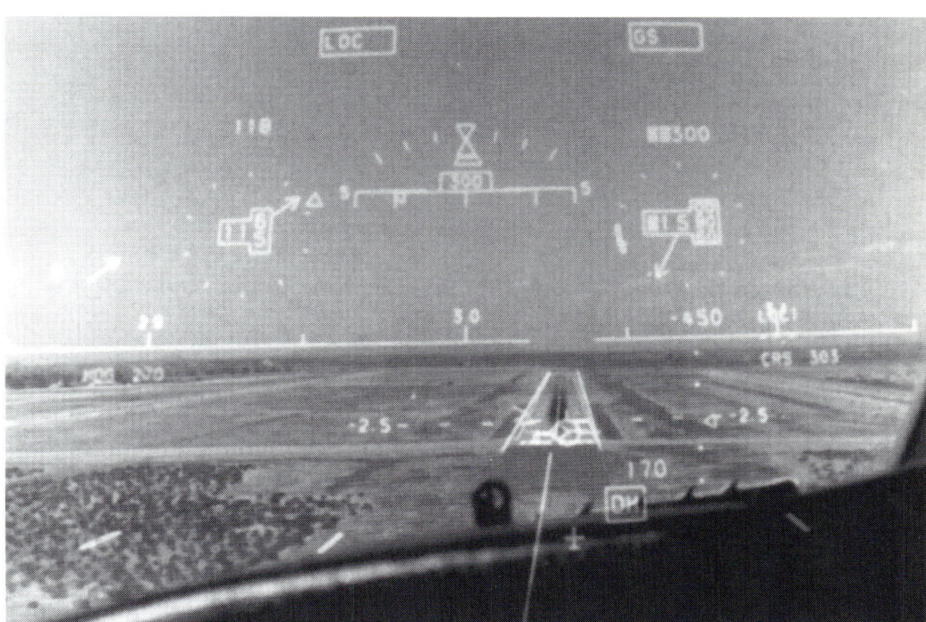

HUD 2020 display for corporate aircraft **1998**/0018156

Operational status

HUD 2020 is in production with 60 on order. The system was first certified to Cat II on a Gulfstream IV aircraft in April 1997, with growth to Cat III and Enhanced Vision Sensor (EVS) approval.

HUD 2022 is still in development, but it is intended to fit the entire Airbus and Boeing air transport range. Selected by American Airlines for its new Boeing 737/800 aircraft. Certification scheduled December 1998.

Contractors

GEC-Marconi Avionics Ltd, Mission Avionics Division.
Honeywell Inc, Air Transport Systems.
Honeywell Inc, Business and Commuter Aviation Systems.

UPDATED

OSIRIS: Tiger anti-tank helicopter mast-mounted sight

The OSIRIS sight is a day and night sight which will detect, lock on and transfer potential target information to the missile seekerhead. OSIRIS is fitted with a high definition TV camera and cockpit display that can be operated in bad atmospheric conditions and throughout the helicopter's operating envelope.

Developed as part of the third-generation European anti-tank missile programme, SFIM Industries is the prime contractor of this stabilised platform sight, being designed for Aerospatiale under a trilateral agreement with ESW Extel Systems Wedel (a subsidiary of Daimler Benz Aerospace Germany) and to GEC-Marconi Radar and Defence Systems.

OSIRIS embodies multispectral optics (bands 0.5-0.7 μm, 0.7-1 μm, and 8-12 μm) and meets stringent stabilisation specifications.

Contractors

SFIM Industries.
ESW — Extel Systems Wedel.
GEC-Marconi Radar & Defence Systems Ltd.

UPDATED

The OSIRIS mast-mounted sight on the rotor head of the Tiger anti-tank helicopter ***1998***/0018351

Virtual Retinal Display (VRDTM) technology in helmet-mounted displays

Microvision Inc, Saab AB (Future Products & Technology Division) and Ericsson Saab Avionics will co-operate to explore the possibilities of advanced visual display systems incorporating Microvision's Virtual Retinal DisplayTM (VRDTM) technology.

Microvision will initially develop two high-resolution helmet-mounted display, technology demonstrators for use in an advanced aircraft simulation system. The demonstrators will be designed to serve as basic multi-use platforms that can be modified or extended to meet the needs of a variety of potential future products.

Ericsson Saab Avionics will contribute its knowledge and expertise in the area of telecommunications, electronic cockpit display systems, and head-mounted displays for military and commercial applications.

Saab and Ericsson Saab Avionics will work with Microvision to investigate commercial development of VRD technology in military aircraft and vehicles, weapons sighting systems and ground troop training.

Operational status

Prototype systems evaluation.

Contractors

Microvision Inc.
Saab AB Future Products & Technology Division.
Ericsson Saab Avionics.

NEW ENTRY

ISRAEL

AN/AVQ-35 Night Vision Goggles/Head-Up Display NVG/HUD system

Elbit Systems' AN/AVQ-35 is an advanced electro-optical system that combines the image seen through the ANVIS with computer-generated graphic and digital symbology. With vital flight information presented on the head-up display, the pilot can fly, and execute the entire mission, head-out of the cockpit.

The AN/AVQ-35 accepts stroke video from the aircraft fixed HUD (in parallel) and projects it through Night Vision Goggles (NVG) using a special combiner. One or two pilots can use the same display through one or two combiners and control panels.

System features allow pilots to fly head-out of the cockpit for entire mission; ease pilot workload and reduce stress by eliminating the need to look inside at the cockpit; increase mission safety; and display sight information for aiming and shooting purposes (if available). The AN/AVQ-35 technology can also be added to second- or third-generation NVGs.

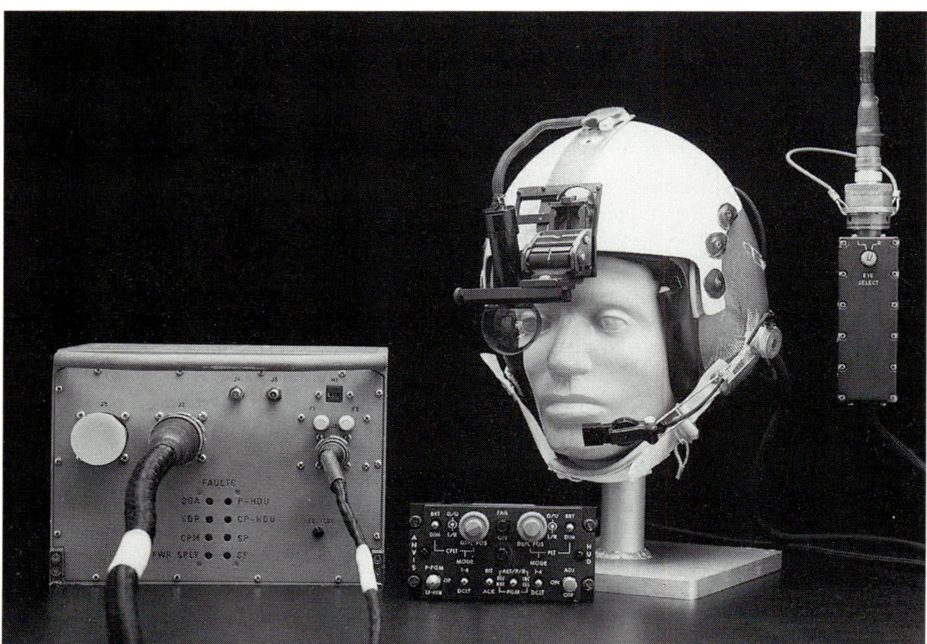

The Elbit Systems/ELOP Day-HUD System ***1996***

Specifications

Weight:
(ECU) 2.25 kg
(CU) 0.36 kg
(HDU) 0.45 kg
Power: 28 V DC, 1 A (MIL-STD-704A)

Operational status

The system is operational in the US Marine Corps AH-1F Cobra helicopter.

Contractor

Elbit Systems Ltd.

UPDATED

Day-HUD

Safe, daytime nap-of-the-earth flight demands that both pilots pay full attention to the external world. The ability to present comprehensive cockpit and navigational information directly in the pilot's field of view dramatically increases flight safety and reduces crew workload. The Day-HUD enables both pilots to fly head-out and receive all vital data, including altitude, height, speed and aircraft condition, flight and platform warnings at eye level. The Day-HUD head assembly adapts to ANVIS-AVS-6 or similar NVG-mountings. It comprises a miniature display source, and an optical combiner which interfaces with the AN/AVS-7 system.

Specifications

Field of view:
(horizontal) 28°
(vertical) 20 °
Exit pupil: 10 ÷ 12 mm

Eye relief: 25 ÷ 40 mm
Weight:
(helmet-mounted element) 0.27 kg
(interface unit) 0.29 kg

Operational status

The US Army is in the process of buying its first lot.

Contractors

Elbit Systems Ltd.
ELOP Electro-Optics Industries Ltd.

UPDATED

Display and Sight Helmet (DASH)

The Elbit Systems' Display and Sight Helmet (DASH) allows the crew of a combat aircraft to direct missiles or sensors on to targets or points of interest by simply looking at that point. Head position, and hence sightline, is computed. DASH slaves all armament systems to the pilot's line of sight. The pilot directs missiles, radar and INS to specific targets by looking at them and receives feedback on his visor on the target that has been acquired. The pilot can also point out a target to a second crew member by looking at it.

DASH presents the head-up display information directly on the pilot's visor so that he is always aware of flight conditions. The position of a target seen by one crew member can be electronically cued in the visor of any other person linked into the system. DASH incorporates the Nightsight integrated avionics system.

Nightsight is an integrated avionics system designed to ensure increased safety and efficiency in tactical night missions. The system's major capabilities and advantages include low-altitude navigation in dusk and darkness, target location, sensor/pilot slaving and cueing target attack, formation flying, flying in adverse weather conditions, map and aerial photo display and full head-out mission capability.

Nightsight's integrated system concept combines the Night Vision Goggles/Head-Up Display (NVG/HUD) and the Line Of Sight (LOS) tracker. Wearing the NVG/HUD, the pilot sees flight symbology collimated with the external view and can fly head-out, realising an added dimension of mission safety and operational capabilities.

Units of Elbit Systems' DASH helmet-mounted sight

The LOS tracker provides active measurement of pilot line of sight and helmet position with high accuracy in all dynamic environments. With pilot LOS slaved to aircraft sensors and vice versa, the result is improved air-to-ground target detection and improved air-to-air interception capability.

Integrating the FLIR pod, NVG/HUD and pilot LOS measurement is a powerful connection, resulting in enhanced overall system performance. The system is designed so that the same hardware drives both day and night displays.

Specifications

Weight: 1.8 kg (total system: 13.1 kg)
Field of view: 22°
Coverage:
(azimuth) ±160°
(elevation) ±70°
(roll) ±60°
Accuracy:
Forward Cone (20°) 6 mrad
Rest of Envelope: 10 mrad
Reliability: 2,000 h MTBF

Operational status

DASH is in service on Israeli Air Force F-4, F-15, F-16 and with foreign customers on F-5, F-16 and MiG-21.

A third-generation helmet display, more compact and lighter than the two previous versions, is undergoing flight tests.

Contractor

Elbit Systems Ltd.

UPDATED

Helmet-Mounted Display Device (HMDD)

The HMDD consists of a head assembly featuring add-on capability to the ANVIS/6 or similar NVG adaptor. It contains a miniature display source, an optical subassembly and a lower electronic adaptor unit which interfaces with the aircraft symbol generator.

The HMDD features a stroke and/or raster display to comply with the existing display source and is adjustable by controls similar to those incorporated in the NVG. It is removable from the helmet, can be folded while on the helmet and is compatible with the aviator's eyeglasses and gas mask. The system is transferable to either eye, to allow use of the dominant eye, and the use of two HMDDs as a binocular display is an option.

The helmet-mounted display device can be used with either eye ***1995***

Specifications

Weight:
(helmet element) 0.11 kg
(interface unit) 0.29 kg
Field of view: (horizontal) 28° × (vertical) 20°

Contractor

ELOP Electro-Optics Industries Ltd.

VERIFIED

Model 849A Head-Up Display (HUD)

The ELOP Model 849A HUD consists of a customer-tailored upfront control panel and an integral high-resolution colour cockpit TV sensor for post-flight debriefing.

The Model 849A is designed for small fixed-wing aircraft and helicopters such as the A-4, L-39, Tucano and Pampa. Ideal for avionics upgrades, it is a compact, low-cost, lightweight, easy to maintain HUD which is capable of displaying cursive symbology.

Specifications

Weight:
(PDU) 4 kg
(PSVS) 7 kg
Power supply: 115 V AC, 400 Hz, 3 phase
28 V DC
Field of view: 20° circular
Environmental: MIL-E-5400T

Operational status

In service in the A-4 Skyhawk, L-39 and IA 63 Pampa.

Contractor

ELOP Electro-Optics Industries Ltd.

VERIFIED

Model 921 Head-Up Display (HUD)

The HUD 921 is capable of displaying high brightness cursive (stroke) symbology with an option of overlaying it on raster display.

Versions of this HUD differ in their combiner-design to accommodate different types of cockpits. Target markets include upgrades for: F-5, IAR 99/109, Kfir, L-39/59, MiG 21/23/27, and Mirage aircraft.

The HUD comprises a customer tailored Up Front Control Panel (UFCP) and can accommodate a high-resolution Colour Cockpit TV Sensor (CCTVS) for post flight debriefing.

Specifications

Modes of operation:
stroke writing
stroke on raster (optional)
Field of view: 24°
Stroke brightness/contrast:
full readability in 10,000 fL ambient light
automatic brightness control
Power supply: 115 V AC, 400 Hz, 3-phase; 28 V DC
Power consumption: 97 W

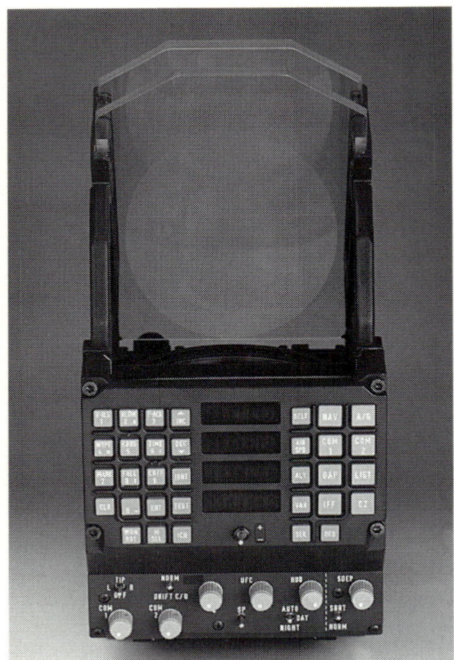

The Model 921 HUD is designed for medium and small cockpits ***1998***/0018155

Weight: 17 kg
Dimensions: 167 × 190 × 560 mm

Contractor
ELOP Electro-Optics Industries Ltd.

UPDATED

Model 959 Head-Up Display (HUD)

The Model 959 HUD comprises a customer-tailored upfront control panel and an integral colour cockpit TV sensor for post-flight debriefing. It is designed for medium and small cockpits in aircraft such as the F-5, F-16, Kfir, MiG-21, MiG-23, Mirage and trainers.

The HUD 959 is capable of displaying high-brightness cursive symbology with an option to overlay it on a raster display. Its shape results from a repackaging of modules from other ELOP HUDs in order to maximise the field of view in a given volume. The Model 959 HUD can function as the mount for a multifunction display or other flight instrument.

Specifications
Weight: 23 kg
Power supply: 115 V AC, 400 Hz, 3 phase
28 V DC
Field of view: 24°
Environmental: MIL-E-5400T

Contractor
ELOP Electro-Optics Industries Ltd.

VERIFIED

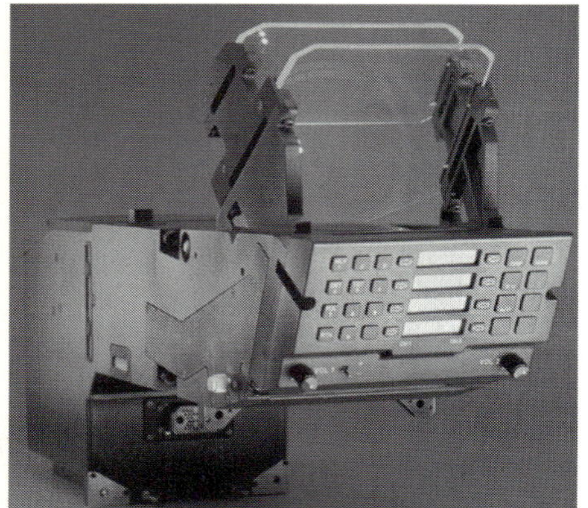

The Model 959 HUD is designed for medium and small cockpits

The Model 967 HUD is third-generation NVG compatible ***1998***/0018154

Model 967 Head-Up Display (HUD)

The Model 967 HUD consists of a customer-tailored upfront control panel and an integral high-resolution colour cockpit TV sensor. It is designed for high-performance aircraft with medium or large cockpits such as the F-4, F-5, F-15, F-16, F/A-18, MiG-29 and Su-27/35.

The Model 967 is capable of displaying high-brightness cursive (stroke) symbology and raster with or without stroke symbology overlaid on it. It is a wide field of view, high-reliability lightweight HUD which is third-generation NVG-compatible.

Specifications
Weight: 21 kg
Power supply: 115 V AC, 400 Hz, 3 phase; 28 V DC; 5 V AC
Field of view: 28°
Stroke brightness/contrast:
2,500 fL automatic brightness control
Environmental: MIL-E-5400T

Operational status
Successfully flown on Israeli Air Force F-16. Selected by Israel Aircraft Industries for its bid for the Turkish Air Force F-4 upgrade programme.

Contractor
ELOP Electro-Optics Industries Ltd.

UPDATED

Model 979 Head-Up Display (HUD)

The Model 979 HUD comprises a customer-tailored upfront control panel and an integral high-resolution colour cockpit TV sensor for post-flight debriefing. It is designed for medium and small cockpits for aircraft such as the F-5, F-15, F-16, IAR 99 and 109, L-39, L-59, Kfir, MiG-21, MiG-23, Mirage, MiG-29, SU-27/35.

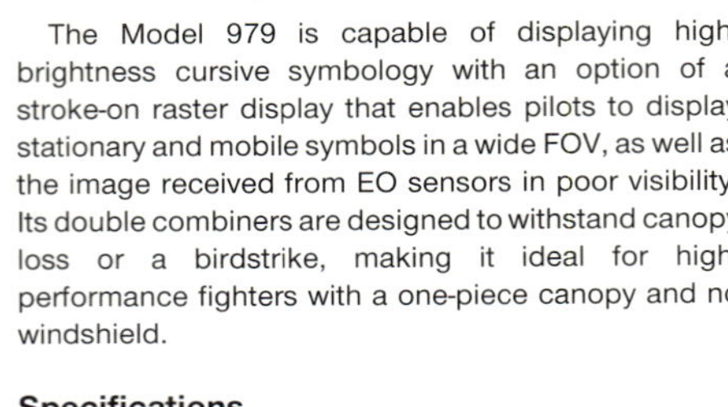

The Model 979 is capable of displaying high-brightness cursive symbology with an option of a stroke-on raster display that enables pilots to display stationary and mobile symbols in a wide FOV, as well as the image received from EO sensors in poor visibility. Its double combiners are designed to withstand canopy loss or a birdstrike, making it ideal for high-performance fighters with a one-piece canopy and no windshield.

Specifications
Weight: 18 kg
Power supply: 115 V AC, 400 Hz, 3 phase
28 V DC
Field of view: 24°
Environmental: MIL-E-5400T

Operational status
In service in the Kfir and F-5. Tested on Israeli Air Force F-16 aircraft.

Contractor
ELOP Electro-Optics Industries Ltd.

VERIFIED

Model 981 Head-Up Display (HUD)

The ELOP Model 981 HUD consists of a Pilot's Display Unit, Display Processor Unit, cockpit TV sensor and power supply unit, all combined into a single package. It features a raster and stroke display and air-to-air and air-to-ground modes. Interface to other equipment is via a MIL-STD-1553B multiplexer bus or analogue, synchro or discrete signals. The integral BIT is automatic or manually initiated.

Specifications
Weight: 43 kg
Power supply: 115 V AC, 400 Hz, 3 phase, 530 W
28 V DC, 30 W
Field of view: (horizontal) 30° × (vertical) 21°

Operational status
In service in the Israeli Air Force Phantom 2000.

Contractor
ELOP Electro-Optics Industries Ltd.

VERIFIED

Model 982/3 Head-Up Displays (HUDs)

The Model 982/3 HUDs comprise a customer-tailored upfront control panel and an integral high-resolution colour cockpit TV sensor for post-flight debriefing. The HUDs are designed for medium and small cockpits in aircraft such as the F-5, IAR 99 and 109, Kfir, L-39, L-59, MiG-21, MiG-23 and Mirage.

The ELOP Model 981 HUD is in service in Israeli Air Force Phantom 2000 aircraft

The Model 982 and 983 HUDs are two models which differ in their combiner design to accommodate different types of cockpits. This results in the reduction of cost and weight, while maintaining high performance. Both HUDs are capable of displaying high-brightness cursive symbology with an option to overlay it on a raster display.

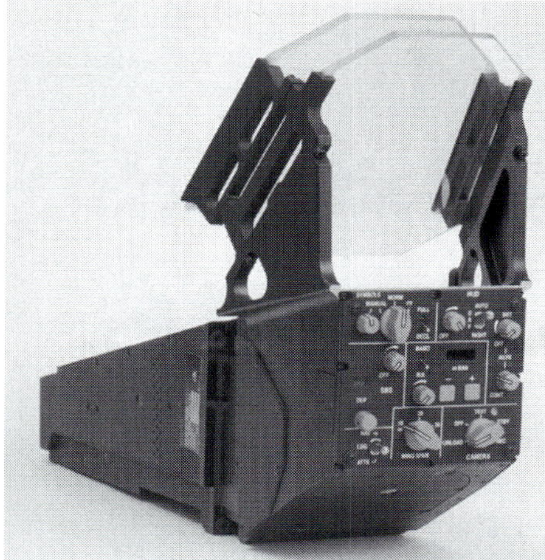

The ELOP Model 989 HUD is installed in the AMX

Specifications

Weight: 19 kg
Power supply: 115 V AC, 400 Hz, 3 phase
28 V DC
Field of view: 24°
Environmental: MIL-E-5400T

Operational status

In service in the Kfir and F-5.

Contractor

ELOP Electro-Optics Industries Ltd.

VERIFIED

Model 989 Head-Up Display (HUD)

The ELOP Model 989 HUD consists of a Pilot's Display Unit (PDU) with integral power supply unit and an Electronics Unit (EU). It is designed for medium-size cockpits in aircraft such as the AMX, MiG-21, MiG-23 and trainers.

The Model 989 features stroke, raster and stroke on raster displays and also operates as a standby sight and has navigation, air-to-air, air-to-ground and test modes. Interface to other equipment is via a MIL-STD-1553B multiplexer bus or through logic discretes. The integral BIT is automatic or manually initiated. Provisions are made for an upfront control panel and cockpit TV sensor.

Specifications

Weight: 18 kg
Power supply: 115 V AC, 400 Hz, 3 phase
28 V DC
Field of view: 24° circular
Environmental: MIL-E-5400T

Operational status

In service in the AMX.

Contractor

ELOP Electro-Optics Industries Ltd.

VERIFIED

ITALY

Head-Up Display (HUD) for the AMX

Alenia has collaborated with OMI to produce the Type 35 HUD for the AMX. OMI designed the Pilot's Display Unit (PDU) and Alenia was responsible for the Symbol Generator Unit (SGU) using its MARA family of data processors. The symbols generated can be changed by software and a display recorder, using either tape or film, can be fitted to the display.

Specifications

Dimensions:
(PDU) 136 × 350 × 650 mm
(SGU) ½ ATR short
Weight:
(PDU) 13.75 kg
(SGU) 8.5 kg
Power supply: 115 V AC, 400 Hz

Operational status

In production for the AMX.

Contractors

Alenia Difesa, Avionic Systems and Equipment Division.
Ottico Meccanica Italiana SpA.

UPDATED

The Alenia/OMI head-up display for the AMX

JAPAN

Head-Up Displays (HUDs)

The Head-Up Display (HUD) provides information such as altitude, airspeed, magnetic heading, attitude, angle of attack and aiming reticle to the pilot during the mission. The information is presented as symbology with a collimated image overlaid in the pilot's view. The HUD consists of the display unit and a symbol generator.

Operational status

In production for Japanese Self Defense Force AH-1S, F-4EJ Kai, F-15J/DJ, T-4 and US-1A aircraft. Also in limited production for the CCV-T2 and C-1 QSOL experimental aircraft.

Contractor

Shimadzu Corporation.

VERIFIED

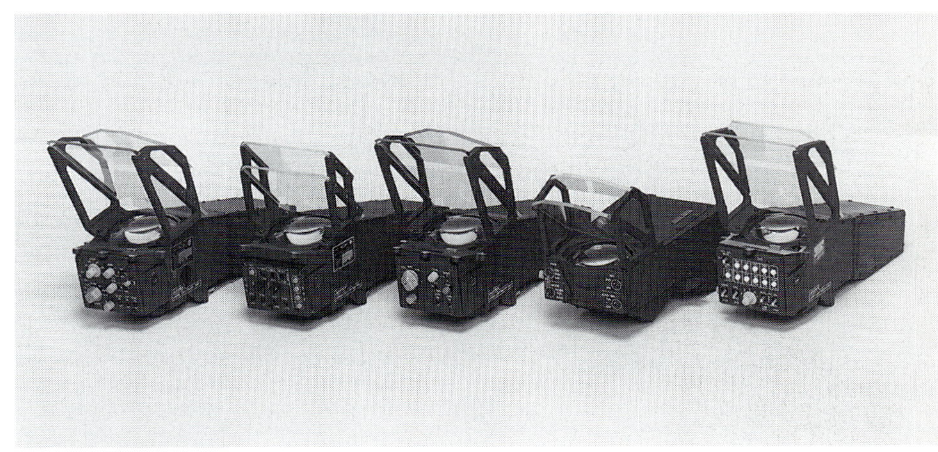

The range of HUDs made by Shimadzu Corporation **1995**

SWEDEN

Helios Helicopter Observation System

The Helios is a roof- or nose-mounted observation system for scout/rescue helicopters. The system has ×3 and ×12 magnification, high-resolution optics (by the UK company Pilkington) and high-stabilisation accuracy. The eyepiece arm protruding into the cockpit in the roof-mounted version, can be stowed close to the roof when not in use and the eyepiece height is adjustable. The system comprises five units: sight head, electronic unit, control panel, control unit and line of sight indicator.

The Helios has provisions for laser rangefinder/designator and thermal imaging system for night vision. It can accept commands from laser warning receiver, helmet sight or avionics systems for accurate target positioning and generate outputs for weapons aiming. A camera port for a film or CCD camera is included to allow recording of the screen.

A full upgrade to a Heli-TOW weapon system capability can be achieved by adding TOW trackers and a Missile Launching and Guidance System (MLGS) to Helios.

Helios can also be converted to a HeliHOT or Heli-Hellfire system, by incorporating goniometers, an MLGS, a Laser Designator and a Hellfire Launching system, depending on which capability is required.

Specification

Dimensions:
(electronic unit) 257 × 194 × 320 mm
(sightline indicator) 60 × 60 × 85 mm
(control panel) 48 × 146 × 70 mm
(control unit) 100 × 130 × 175 mm
Weight:
(observation unit) 46 kg
(electronic unit) 13.3 kg
(sightline indicator) 0.3 kg
(control panel) 0.8 kg
(control unit) 1.1 kg
Power supply: 28 V DC, 290 W
Field of view: 20° (×3 mag), 5° (×12 mag)
Sightline deflection: ±120° (azimuth), ±25° (elevation)
Environmental: MIL-E-5400 T Class 1B

Operational status

In production.

Contractor

Saab Dynamics AB.

VERIFIED

Helios mounted on a Eurocopter AS 355 helicopter ***1995***

SEOS mounted on a Eurocopter BO 105 ***1995***

Stabilised Electro-Optical System (SEOS)

The SEOS is a surveillance and targeting system designed for a variety of operational uses in helicopters and for ships and land vehicles. The system interface architecture allows easy integration with the aircraft management and control system. The design has been optimised for surveillance and target acquisition, as well as for target tracking and weapon aiming.

SEOS comprises a sensor head, control unit, control panel, presentation unit and electronics unit.

The sensor head can be attached to the nose, roof, or rotor-head of a helicopter. Sensors included in the basic version are a FLIR and two TV cameras. Optional sensors include laser rangefinder and laser designator. The picture is electronically transmitted and presented on a TV screen. The combination of sensors, together with advanced image processing, gives the user flexibility in the use of the information.

Operational status

Under development.

Contractor

Saab Dynamics AB.

VERIFIED

UNITED KINGDOM

RGS Series weapon aiming systems

The RGS Series is a comprehensive range of airborne weapon aiming systems, from the simplest of aiming devices to the more sophisticated type of lead computing HUDs. The compact size of the RGS Series allows easy installation in the smallest of cockpits, making it suitable for small turboprop and jet trainers. When cockpit space is particularly restricted, low-profile versions of the sight heads can be provided. All variants of the Series utilise the same sight head/cockpit interface.

Although the primary role of the RGS Series is for use in trainer aircraft, the more sophisticated derivatives can be utilised as alternatives to full HUDs. The Series includes the RGS1 universal manually operated fixed-type sight, RGS2 basic lead computing rate gyro sighting system, RGS2A compact gyro gunsight and weapon delivery computer for enhanced ground attack capability and RGS2R laser-augmented system based on a microprocessor for high-speed low-level attacks.

The RGS1 is a fixed optical sight head which provides a basic weapons delivery system.

Incorporated in the sight head are a combining glass, fixed-plane mirror, movable spherical mirror and injected reticle system. Utilising a 100 mm wide aperture, the RGS1 provides the pilot with binocular viewing, enhancing efficiency even in the most turbulent conditions when otherwise unavoidable head movements would cause loss of the aiming mark. For ease of viewing, the RGS1 displays an aiming mark which is projected at infinity. Compensations for ordnance characteristics, mode of attack and crosswind are made by deflecting the aiming mark in azimuth and elevation. These deflections are made manually using control knobs on either side of the sight head and are displayed, in mils, on indicator scales facing the pilot. A brightness control allows the aiming mark illumination to be varied according to prevailing light conditions.

The RGS2 is a lead computing optical gunsight which comprises an optical sight head, computer and gyro unit and a control unit. The sight head comprises a combining glass, fixed-plane mirror and movable spherical mirror which is suspended in gimbals, controlled by torque motors and displacement transducers. An injected reticle system, with two reticle lamps, is also included. The mirror deflection is very accurately determined by the controlling signals and does not suffer from the temperature dependence and long-term drift encountered in the frequently used open-loop system. The computer and gyro unit houses the system electronic circuitry and a gyro pack. The electronics comprise circuits for weapon aiming calculations and the system power supply. The circuits are built of electronic components assembled on plug-in printed circuit cards. The gyro pack consists of

The twin RGS2 sight head installation in a Saab 105 side-by-side two-seat trainer

three miniature high-precision rate gyros measuring aircraft pitch, yaw and roll rates. Output signals from these gyros control the deflections of the aiming mark in the sight head. The control unit operates the system and incorporates a built-in check function. A facility to set the aiming mark manually is also provided by this unit.

The RGS2A has all the features of the RGS2 but utilises a more powerful weapons aiming computer. This computer provides improved delivery of bombs and rockets by incorporating a CCIP function.

The RGS2R has all the features of the RGS2A but with the integration of a separate laser rangefinder.

The RGS2 Series can be used in a twin-sight head arrangement for training purposes in which the instructor's and trainee's aiming marks are deflected equally and in exact synchronisation. For light attack aircraft, the same principle can be used for slaving a laser rangefinder or an IR missile or radar homing device to the optical sightline of the sight head.

Specifications

Dimensions:
(control unit) 38 × 146 × 84 mm
(computer and gyro unit) 114 × 175 × 275 mm
Weight:
(RGS1 sight head) 2.5 kg
(RGS2 sight head) 3.1 kg
(control unit) 0.8 kg
(computer and gyro unit) 4.5 kg
Power supply:
(RGS1) 28 V DC, 30 W (max)
(RGS2) 115 V AC, 400 Hz, single phase, 70 VA (max)
28 V DC, 60 W (max)

Operational status

The RGS Series has been in operational service since 1982. RGS systems have been fitted to the BAe Hawk, CASA Aviojet C-101, Aermacchi MB-326 and MB-339, Agusta (SIAI-Marchetti) S.211, Northrop Grumman F-5, Fiat G-91R3, Saab 105, MB Vectro 2, Dassault Super Mystère, Pilatus PC-7 and PC-9, Shorts Tucano, Embraer Tucano, IA 63 Pampa, Pucara and A-4 Skyhawk.

Contractor

Avimo Ltd.

VERIFIED

AF500 Series roof observation sights

The AF532 roof-mounted helicopter sight superseded the AF120 sight introduced in 1970. The new sight is half the weight of its predecessor, but confers greatly improved optical performance.

This gyrostabilised, monocular, periscopic telescope is designed for the gunner/observer in reconnaissance helicopters, particularly when scouting targets for anti-tank helicopters.

The device has a built-in interface for a laser designator and rangefinder. It can also be adapted for night vision equipment, helmet sights and weapons, and there are facilities for attaching a television recording camera for training or intelligence gathering. The design of the optical system is such that the varying eye positions in different helicopter installations can be easily accommodated.

The gyrostabilised head protrudes above the roof forward of the rotor mast, while the down-tube and eyepiece extend downwards from the roof so that the eyepiece falls into a comfortable viewing position. The down-tube is adjustable in height and retracts sideways, locking close to the roof when not in use. A control handle is extended by the operator and adjusted in tilt so that it can be used with the right forearm resting on the knee. A horizontal thumbstick is used to steer the sightline and a direction indicator, to show its direction relative to aircraft heading, is mounted on the glareshield in front of the pilot.

The sight provides a stabilised image of the chosen field of ×2.5 magnification for search and ×10 for identification and laser operation. The sightline may be steered through ±30° and ±120° in pitch and yaw planes respectively.

AF580 systems, fitted with GEC-Marconi laser ranger and target designators and integrated with a Collins

The AF532 sight unit installed on the roof of a British Army Westland Gazelle helicopter

Automatic Target Handover System have been evaluated by the US Army in Bell OH-58C Kiowa helicopters. This combination of systems permits the range and bearing of the target to be determined by the observation helicopter and transmitted by datalink to an attack helicopter. In a recent development a thermal image from a separate FLIR has been injected into the sight to permit night observations. The FLIR and the sight are steered by the same controller.

Operational status

In service in British Army Air Corps' Westland Gazelle helicopters. Laser targeting equipment has now been fitted to a number of systems.

Contractor

Ferranti Technologies Ltd.

UPDATED

4500 Series Head-Up Displays (HUDs)

The 4500 Series head-up display system is primarily designed for light attack aircraft and trainers. Designed from the outset to be a dual-mode cursive and raster display, the HUD is modular allowing the mechanical outline to be optimised to fit most aircraft installations.

The display gives a 24° total field of view with a large instantaneous field of view available from the 4.5 in (114 mm) exit lens. P53 phosphor is used on the CRT, giving a very bright display. The upfront control panel gives complete control over the rest of the system and a colour TV camera fitted to record the pilot's view.

The interface unit or Head-up display Electronics Unit (HEU) is a ¾ ATR box containing multiple analogue and discrete synchro interfaces, 1553B R/T or bus control, 68020 or higher-performance processors and symbol generators and graphics processors. Full weapon aiming, mission computations and head-up and head-down display symbol generation are available.

The Type 4510 has both cursive and raster displays, the latter being selected by a switch on the upfront control panel. The 4500 and 4510 are physically identical and were based on the optical and symbol generation technology previously used on the COMED system. The extensive production runs of COMED had removed any design problems, while the weapon aiming, interfacing and air data techniques were derived from a series of complete weapon systems designed and built by GEC-Marconi Avionics.

The 4510 head-up display provides a full suite of navigation symbology with steering and location cues available at all times and generates automatic or selected weapon aiming symbology for all known air-to-air and air-to-ground weapons. Weapon release can be

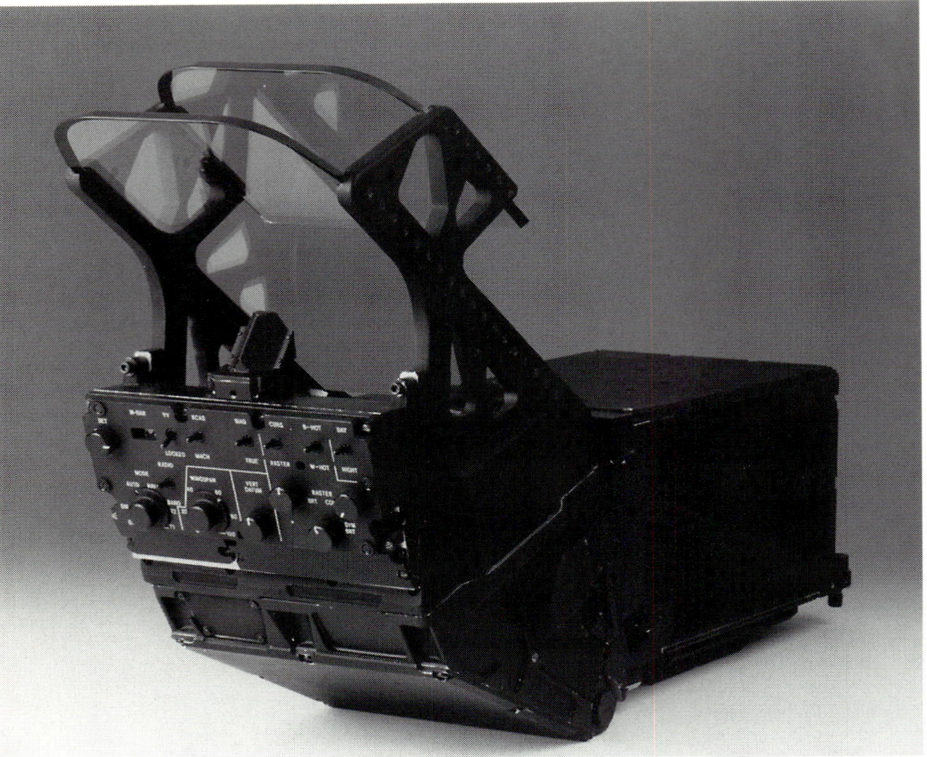

4500 Series HUD ***1997***/0001457

triggered manually or automatically at the pilot's discretion. The control panel allows the pilot to set up the navigation system, select and display modes during flights and then change to a raster display when the mission sensors, such as FLIR, come into play.

Since the introduction of the 4500 Series HUD, systems have undergone rigorous flight trials on the Buccaneer and Harrier Nightbird aircraft at the Defence Evaluation Research Agency, Farnborough. An RAF Jaguar was fitted out with a GEC-Marconi Avionics system and flew early in 1989. The success of these trials was due in part to the flexibility of the highly programmable software in the HEU and the performance of the HUD in both cursive and raster modes.

The HUD is an integral part of various total system options at present being considered in the worldwide retrofit market. The unit can, because of its compact size, meet the installation requirements of many types of attack aircraft.

Specifications

Weight:
(HUD) 12 kg
(HEU) 14 kg
Power supply: 115 V AC, 400 Hz

Operational status

In production and in service for Royal Air Force Jaguar aircraft and for the A-4, C-101, F-5, Mirage III and Mirage 5 aircraft retrofits.

Contractor

GEC-Marconi Avionics Ltd, Mission Avionics Division, Edinburgh.

UPDATED

Type 9000 Head-Up Display (HUD)

To cater for larger attack/fighter aircraft or those aircraft with less installation constraints, GEC-Marconi Avionics has developed the Type 9000 wide-angle conventional HUD with a 6.5 in (165 mm) exit lens. It has self-contained high- and low-voltage power supplies and is capable of cursive and raster display with cursive-in-raster flyback. High-brightness combiners give daylight viewability of the raster FLIR image. A high-resolution colour camera is fitted to record the pilot's view.

The HUD is driven by a Computer Symbol Generator (CSG) which offers full mission computing and faster/cursive symbol generation capability.

The CSG can be fitted with multiple processor cards to ensure all customer systems moding, control and operational requirements can be easily accommodated. The CSG is fully programmable using the ADA software language.

The CSG provides video routeing and mixing of external video with internally generated symbology and can operate as either a bus controller or an R/T on the aircraft 1553 B databus.

The CSG is of modular construction enabling various combination of symbol generation and aircraft interface to be readily configurable. A typical CSG application utilises 12 cards in a 19 slot box providing expansion capability to add extra functions to the unit. Functions which may be added include digital mapping, terrain following and intelligent ground proximity warning.

Operational status

In production for the Royal Air Force Tornado GR. Mk 4.

Contractor

GEC-Marconi Avionics Ltd, Mission Avionics Division, Edinburgh.

VERIFIED

ISIS weapon aiming sights

The Integrated Strike and Interception System (ISIS) has been designed for the aiming of guns, rockets and bombs in the close support role and for guns and missiles in air combat.

The ISIS sight can also use range information from a laser rangefinder and can serve as a low-cost alternative to a CRT head-up display by displaying pointing information from a laser seeker such as the Lockheed Martin Pave Penny or GEC-Marconi Avionics LRMTS. The company claims that adequate accuracy is coupled with low-maintenance cost and high reliability.

Operation is essentially the same for all members of the ISIS family. A two-axis rate gyro generates a lead-angle proportional to the rate of turn and a mirror attached to the rotor reflects the image of an illuminated reticle pattern through a collimating lens onto a combining glass and into the pilot's field of view.

In the ground attack role the reticle is depressed below the armament datum through an angle which compensates for the combined effects of the weapon's ejection velocity, angle of attack of the airframe in the intended release conditions, the gravity drop of the weapon during its flight and the surface wind, if known. These angles may be set up by the pilot to suit the intended attack speed, dive angle and slant range at weapon release, or he may prefer to use a predetermined set of parameters for each type of weapon carried.

In air combat, the unit measures the rate of turn of the sightline and scales it to target range, thus producing a first order lead-angle solution.

To assist steering and enable the pilot to make allowance for sudden manoeuvres by the target, most sights include a second fixed reticle representing the gun axis.

Specifications

ISIS D-195R Mk 3
Dimensions:
(sight head) 243 × 251 × 132 mm
Weight:
(sight head) 4.1 kg
(control unit) 1.36 kg
(gyro interface unit) 1.63 kg
(throttle unit) 0.2 kg

Operational status

In service in the British Aerospace Hawk strike/trainer. The ISIS first entered service in 1968 and current users include the air forces of Argentina, Australia, Austria, Canada, India, Indonesia, Italy, Kenya, Norway, Singapore, UK, Venezuela and the US Marine Corps. ISIS has been fitted to 25 different aircraft types including the Saab 105 (ISIS F-126), Fiat G91Y (ISIS B), Aermacchi MB-326 (ISIS F-126), Boeing A-4S (ISIS D-101), Northrop Grumman CF-5A (ISIS N), Northrop Grumman NF-5A (ISIS F-195R), BAe Hawk (ISIS D-195R), HAL Ajeet (ISIS F-195R), HAL HF24 (ISIS F-124), MiG-21FL, Mostar Galeb and LAS A-4C (ISIS D-126R).

Contractor

GEC-Marconi Avionics Ltd, Mission Avionics Division, Edinburgh.

UPDATED

The D-195R, fitted to the British Aerospace Hawk trainer, is a typical member of the GEC-Marconi Avionics ISIS range of weapon sights

Optical helmet tracking system

GEC-Marconi's optical helmet tracking system is designed to satisfy the requirement for a tracking system to interface with helmet-mounted displays and sights that is fast, accurate and remains uninfluenced by metal structure within the aircraft cockpit.

One or more optical sensors located on the aircraft structure detects the position of several helmet-mounted light emitting diodes (LEDs). A multiplexer energises each LED in turn and the position of the helmet in the cockpit is determined from the sensed relative position of the LED from each sensor. The aircraft-mounted electronics unit can be provided in standalone form or as a card set for integration into an existing unit. The system does not need to be aligned or boresighted after initial installation.

Operational status

Under development.

Contractor

GEC-Marconi Avionics Ltd, Mission Avionics Division, Edinburgh.

UPDATED

AN/AVQ-29 Head-Up Display (HUD) for the A-7D and A-7K

The AN/AVQ-29 HUD was developed from the HUD which equips the F-16C/D in order to give the A-7 a full night attack capability when used in conjunction with an electro-optical sensor.

Operational status

Production complete.

Contractor

GEC-Marconi Avionics Ltd, Mission Avionics Division, Rochester.

VERIFIED

DC electromagnetic helmet tracking system

The GEC-Marconi electromagnetic helmet tracking system has been designed for both fixed- and rotary-wing applications and determines the helmet position and orientation by measuring magnetic fields which it generates in the cockpit. This data can be used to position aircraft sensors and weapon seekers and for referencing Helmet Mounted Display (HMD) Systems.

The system has been developed to overcome many of the disadvantages of competing electromagnetic tracker systems whilst maintaining simplicity of installation. A number of prototype systems have been delivered to both US and European customers by GEC-Marconi Avionics as part of fully integrated HMD systems.

A small transmitter containing three orthogonal coils is rigidly mounted within the cockpit. The coils are pulsed with currents which generate known magnetic fields. A small magnetic sensor, again containing three orthogonal coils, is rigidly mounted to the helmet. These coils sense the direction and magnitude of the generated magnetic fields and, using this data, a processor calculates the helmet position and orientation. The pulsed nature of the transmitted signal leads to this type of system being termed a 'DC Electromagnetic Tracker' to differentiate it from the older 'AC electromagnetic' systems.

The key functional elements are the transmitter, the sensor and an electronics module which is normally installed within the HMD system electronics unit.

This system provides full spherical coverage within the aircraft cockpit offering high accuracy and good dynamic performance. It does not affect and is not affected by other cockpit functions and is significantly less susceptible to cockpit metal than traditional systems, requiring a single generic mapping for all aircraft of a particular type.

Operational status

This system is now in preproduction development for trials with the German armed forces and also for UK technology demonstrator programmes.

Contractor

GEC-Marconi Avionics Ltd, Mission Avionics Division, Rochester.

UPDATED

Head-Up Display and Weapon Aiming Computer (HUDWAC) for the F-5

GEC-Marconi Avionics has developed a HUDWAC for the Northrop Grumman F-5E that, as an avionics upgrade, allows the performance and manoeuvrability of the F-5 to be combined with the advantages of head-up flight.

GEC-Marconi Avionics HUD for the F-5E

The F-5 HUD utilises technology and hardware from other GEC-Marconi Avionics HUD programmes and performs air-to-air and air-to-surface weapon aiming calculations as well as delivering symbology for flight and navigation modes. The HUD comprises a Pilot Display Unit (PDU), an Electronics Unit (EU) and a Weapon Data Input Panel (WDIP).

The PDU fits in an existing mounting tray. The PDU's 25° total and 15.75° vertical by 16.9° azimuth instantaneous fields of view are achieved without impingeing on the F-5's ejection plane. A 16 mm cinematic camera is included in the PDU, but a TV camera is available as an option. The PDU control panel includes switches for controlling symbol brightness, standby sight selection and symbol declutter. Data entry is via a keypad on the same control panel.

The EU will fit in all existing configurations of the F-5E in the same location as the LCOS gyro lead computer which is removed for HUDWAC installation. It contains a power supply, integral cooling fans and the circuit cards for processing, symbol generation and interfacing with the F-5's avionics systems.

The HUDWAC in its baseline configuration interfaces with the F-5's power supply, CADC, AHRS, fire-control system and AN/APQ-153, 157 or 159 radar. The system has flown with the LN-93, with the H423 laser inertial navigation system and also with the LN-39. Optionally, the HUDWAC will interface with the AN/APG-66, 67 or other pulse Doppler radar.

The system software provides weapon aiming calculations for weapons certified for F-5 carriage and is written to display symbology which complies with MIL-STD-1787. The system has the capability to expand and handle the AIM-9P-4 missile line of sight and off-boresight aiming.

The system operates in three modes; navigation, surface attack and air combat.

In navigation and landing submode the display

symbology shows speed, altitude, heading, velocity vector, pitch, bank, Mach number and *g*. Inertial navigation information, when available, is also displayed.

Surface attack modes include Continuously Computed Impact Point (CCIP) and Continuously Computed Release Point (CCRP) displays.

There are three air-to-air modes: missiles; Lead Computing Optical Sight (LCOS) for tracking 20 mm cannon attacks and dogfight, which combines missiles and guns; LCOS and snap-shoot or Continuously Computed Impact Line (CCIL) on the same display.

The system has built-in test facilities to isolate faults to a particular LRU. Predicted MTBF for the HUDWAC is greater than 1,400 hours. The HUDWAC's support equipment includes a semi-automatic system test set which uses a 16-bit microprocessor.

Operational status
In production.

Contractor
GEC-Marconi Avionics Ltd, Mission Avionics Division, Rochester.

UPDATED

Head-Up Display (HUD) for the C-17

The HUD for the C-17 is claimed to be the world's first HUD designed as a critical flight instrument. The unit has a single box, the electronics being built into the optical unit rather than being separate. It has a 30° azimuth by 24° elevation field of view and has twin integral MIL-STD-1750A processors. Two HUDs are fitted to the aircraft, one each for pilot and co-pilot. When not required, the HUD combiner can be folded away below the line of sight.

Specifications
Power supply: 115 V AC, 400 Hz, single phase, 100 W
Reliability: 5,000 h MTBF

Operational status
In production.

Contractor
GEC-Marconi Avionics Ltd, Mission Avionics Division, Rochester.

UPDATED

Head-Up Display (HUD) for the F-16 and A-10

Notwithstanding the capability and growth potential of the standard Lockheed Martin F-16 head-up display, the US Air Force in the mid-1970s issued requirements to industry for a more advanced system. This, in conjunction with improvements in other areas, provided the basis for a substantial upgrading of the F-16s under the designation Multinational Staged Improvement Programme (MSIP). The specific improvement sought was the adoption of the Lockheed Martin LANTIRN to permit the F-16s to operate in all weathers and at night and the provision of a HUD which meets the qualification requirements, including wind blast and birdstrike.

It was soon realised that the field of view needed for this would be far greater than that available with existing head-up display technology. Conventional head-up displays with lateral and vertical fields of about 13.5 and 9° respectively could be expanded to 20 and 15° using standard optics, but this would still be less than the field size the US Air Force had set as its ultimate objective.

GEC-Marconi Avionics started developing a head-up display based on holographic techniques using the principles of diffractive optics and, in July 1980, was awarded a development contract. The system combines the wide-angle display geometry made possible by using hologram technology with a company-developed method of combining raster and cursive symbol writing that greatly reduces the amount of equipment needed for day/night head-up display. The optical train uses three combiner glasses, each of which is a sandwich, the hologram being imprinted on to a gelatin filling. It provides a field of view of 30° laterally and 18° vertically. On the electronics side, it incorporates MIL-STD-1589B high-order language, MIL-STD-1750A airborne instruction set architecture and MIL-STD-1553B digital databus standards.

Installation of the GEC-Marconi Avionics holographic head-up display unit in the F-16 cockpit as part of the LANTIRN system

Operational status
In service in the F-16 with over 900 systems delivered.

Contractor
GEC-Marconi Avionics Ltd, Mission Avionics Division, Rochester.

VERIFIED

Head-Up Display for the F-16C/D

In March 1983, GEC-Marconi Avionics announced a US$50 million order to begin production of a new, wide-angle non-holographic head-up display for the US Air Force F-16C/D fighter programme. This head-up display is based on development work undertaken for the US Air Force's Advanced Fighter Technology Integration (AFTI) programme.

The head-up display provides electronically generated symbols thrown up on a total field of view of 25° which is much wider than that attained with previous head-up displays. The instantaneous field of view (that is, the field seen by the pilot without moving his head) is 21° in azimuth and 13.5° vertical. The system uses the same electronics unit as that developed for the LANTIRN system and the symbology and raster scan pictures are particularly suited to guidance and target acquisition at night or in poor weather. The system is claimed to represent the first applications of MIL-STD-1750A processor architecture, MIL-STD-1553B digital data transmission and Jovial 73 MIL-STD-1589B high-order language.

Specifications
Pilot's display unit
Dimensions: 635 × 163 × 170 mm
Weight: 21.6 kg
Power: 98 W (including 25 W for the standby sight)
Predicted MTBF: >2,000 h

Electronics unit
Dimensions: 337 × 180 × 191 mm
Weight: 14.1 kg
Addressable memory: 64 k words (48 k EPROM, 16 k RAM)
Predicted MTBF: >1,000 h

F-16Cs and Ds have GEC-Marconi Avionics wide-angle head-up displays

Operational status
Still in current production for the US Air Force F-16C/D, with more than 2,200 delivered.

Contractor
GEC-Marconi Avionics Ltd, Mission Avionics Division, Rochester.

VERIFIED

Head-Up Display (HUD) for the F-22

GEC-Marconi Avionics has been selected to develop the HUD for the Lockheed/Boeing F-22 advanced tactical fighter. This will feature a single flat combiner which will ensure maximum clarity while preserving the degree of head freedom to perform operations in a modern fighter aircraft.

The Smart HUD is a critical flight instrument with the display processors and drivers in a single LRU. It uses diffractive optics featuring a single element holographic combiner which consists of two glass elements bonded to produce a flat parallel-sided assembly. An optically powered hologram is recorded on photosensitised gelatine on the spherical interface sandwiched in the assembly and acts as the collimating combiner. The resulting advanced optical system, manufactured using computer-generated holographic techniques, provides new levels of display capability. Additionally, the uncluttered simplicity of the combiner support structure allows virtually a clear out-of-cockpit field of regard. The total field of view is 30° azimuth by 20° elevation and the instantaneous field of view is 24 by 20°.

The optical module brightness levels are optimised to operate in a very high ambient light environment, whilst minimising solar reflection and maximising outside world transmission and display uniformity from within the large eye motion box.

Associated with the PDU is a complex HUD control panel attached to the aft face of the unit which incorporates LED technology and a colour camera system.

Operational status
In development.

Contractor
GEC-Marconi Avionics Ltd, Mission Avionics Division, Rochester.

VERIFIED

Helmet-Mounted Sighting System (HMSS)

The GEC-Marconi Avionics' HMSS is a visor-projected sighting system which comprises an optical subassembly mounted to a virtually unmodified standard aviator's helmet (such as the Helmet-Integrated Systems Limited Mk 10B) with the associated electronics remotely located in the aircraft's avionics bay or similar. Sighting and cueing information is presented to the pilot by means of a high-brightness LED reticle, relayed by a prism and reflected into the pilot's eye by a dichroic patch coating on the inner surface of the clear visor.

The HMSS, originally developed in conjunction with the UK Defence Evaluation Research Agency, has now undergone various stages of upgrade to provide maximum optical performance with minimum obscuration. Used in conjunction with a helmet tracking system, the sight enables the pilot to perform off-boresight missile target acquisition fully exploiting the advantages of modern missile seeker head capabilities. Aircraft sensors can also be pointed using the sight, and with the reticle slaved to the sensor, to cue the pilot.

The GEC-Marconi Avionics helmet-mounted sighting system is under development for agile combat aircraft **1997**/0001458

Specifications
Weight: 0.15 kg
Field of view (circular): 2.0°
Exit pupil: 16 mm at centre field of view
Eye relief: Compatible with prescription spectacles

Operational status
Contract awarded for preproduction units for use on the RAF Jaguar GR1B/T2B as part of the Jaguar '97 upgrade. Further production orders expected.

Contractor
GEC-Marconi Avionics Ltd, Mission Avionics Division, Rochester.

NEW ENTRY

Knighthelm integrated Helmet-Mounted Display (HMD)

Currently under development by GEC-Marconi Avionics is the Knighthelm integrated HMD System. This binocular 40° field of view system incorporates a one-piece display module for both day and night mission requirements. For low-visibility and night

The GEC-Marconi Avionics Knighthelm integrated helmet-mounted display **1997**/0001459

missions, the system displays both intensified images from third-generation image intensifiers and imagery from a Forward-Looking InfraRed (FLIR) sensor. This together with flight and weapon symbology information, is projected onto clear combiners placed in front of each of the pilots' eyes. The combiners may be rapidly flipped up out of the line of sight if required. Pilot comfort and mission capability requirements have been fully integrated and all helmet controls for tactical sensors and weapons may be controlled by Hands on Collective and Stick (HOCAS) or via a control panel. Sensors and weapons are interfaced to the helmet via GEC-Marconi Avionics' advanced helmet tracking system.

Specifications

Weight: 2.2 kg
Field of view: 40°
Eye relief: 30 mm

Operational status

Selected for the development phase of the Integrated Helmet System for the German Tiger helicopter.

Contractor

GEC-Marconi Avionics Ltd, Mission Avionics Division, Rochester.

VERIFIED

Monocular Head-Up Display (MONOHUD) for helicopters

Designed for installation in existing cockpits, the MONOHUD is a fully capable system in miniature. It offers flight and navigation modes to provide the pilot with full flight parameters head-up during low-level flight and landing, as well as computed weapon aiming solutions when used as a sighting system on military aircraft. It has been designed for applications in which size and weight constraints play a major role in the choice of systems. The MONOHUD system consists of a Pilot Display Unit (PDU), an Electronics Unit (EU), a High-Voltage Power Supply Unit (HVPSU) and a control panel.

The PDU can be fitted with minimal modification of the airframe. It allows an unobstructed view of instrumentation and the outside world in both the stowed and operational positions. In the operational position the PDU is situated approximately 8 cm in front of the pilot's eye and thus a 30 × 24° field of view is achieved. The PDU control panel includes manual controls for symbol brilliance and declutter.

The EU is compatible with a comprehensive range of discrete, analogue, digital, synchro, ARINC 429 and MIL-STD-1553B inputs. In addition to the input interface, processor and symbol generator, the ½ ATR electronics unit contains the low-voltage power supply and BIT. Built-in test eliminates the need for external support equipment at organisational and intermediate levels.

GEC-Marconi Avionics Viper I **1997**/0001461

Specifications

Dimensions:
(PDU) 76.2 × 76.2 × 190 mm
(EU) 320 × 124 × 193 mm
(HVPSU) 178 × 153 × 102 mm
(control panel) 57 × 146 × 79 mm
Weight:
(total system) 12.7 kg
Field of view: (vertical) 30° × (azimuth) 24°
Contrast ratio: 1.2:1 against 10,000 ft-lamberts
Environmental: MIL-E-5400T, Class 1B
Reliability: 2,500 h MTBF

Operational status

In production.

Contractor

GEC-Marconi Avionics Ltd, Mission Avionics Division, Rochester.

VERIFIED

Viper Helmet-Mounted Display (HMD) system

The GEC-Marconi Viper family of helmet mounted displays provides a low-cost, lightweight visor-projected display capability by addition of a display module to almost any type of flying helmet. The Viper family comprises two variants; Viper I and Viper II.

Viper I

The Viper I Helmet Mounted Display is a monocular system providing a field of view suitable for off-boresight missile aiming.

The visor reflects dynamic flight and weapon aiming data to the pilot from a high-efficiency miniature CRT display projected via an optical relay assembly. The optical design allows the use of a standard aircrew visor with the addition only of a neutral density reflection coating. This technique enables a high outside world transmission without colouration. The design also ensures that the displayed image is stable and accurate even when the visor is raised.

Although primarily configured for daytime use, the Viper I system is capable of displaying video from a sensor, providing the pilot with an enhanced cueing system after dark or in bad weather.

Specifications

Viper I
Weight: 1.72 kg (on a US standard flying helmet)
Field of view: >20°
Eye relief: >70 mm

Operational status

Prototype flight tested.

Viper II

The Viper II Helmet Mounted Display provides a binocular, fully overlapped, 40° field of view capability. The system is able to receive sensor video (for example, from a Forward Looking InfraRed (FLIR) sensor), and display this together with overlaid flight and weapon aiming symbology.

GEC-Marconi Avionics Viper II **1997**/0001460

The Viper II has a similar optical concept to the Viper I, and by use of a spherical visor, maintains a stable image within the large exit pupil regardless of visor position. Full interpupillary adjustment is also provided.

Specifications

Viper II
Weight: 1.54 kg (on a US standard flying helmet)
Field of view: 40°
Eye relief: 85 mm
Exit pupil: >15 mm

Operational status

Prototype flight tested; preproduction units in development.

Other common Viper I and II features:

a) they fit neatly onto all sizes of standard UK, US Air Force and US Navy fixed-wing aircraft helmets;
b) the HMD incorporates provision for a lightweight video camera head. When the output of this camera is mixed with scan converted helmet mounted diplay symbology, a video output is available for recording purposes and subsequent use on the ground for mission review and training;
c) they provide full space stabilised symbology when coupled to a suitable Head Tracking System (HTS), such as the advanced DC system developed by GEC-Marconi Avionics;
d) the Viper helmet mounted display system includes an Electronics Unit (EU) and a Cockpit Unit (CU). These units can be configured to provide a variety of functions, depending on user requirements and which Viper variant is required. These functions include: display processor, video processing, MIL-STD 1553 bus interface, symbol generation, CRT display drive, high voltage power supply, display control and head tracker interface; further improvements to the Viper family of HMDs are currently in development.

Contractor

GEC-Marconi Avionics Ltd, Mission Avionics Division, Rochester.

VERIFIED

Guardian helmet-mounted display and sight system

Guardian has an advanced helmet-mounted display that is projected onto the user's visor via a miniature cathode-ray tube. The display is easily tailored to suit customer requirements. Depending on the application, flight and/or weapon data can be displayed, using static and moving symbols. Night-Vision Goggles (NVGs) can also be used with the system.

The direction of the user's line-of-sight is determined by clusters of Light-Emitting Diodes (LEDs) positioned on the helmet, their relative positions being continually observed by two miniature cameras mounted on the airframe. The LEDs emit infrared light that is invisible to the user and does not interfere with NVGs. The movement of the user's head is not restricted in any way.

The helmet consists of two parts, namely the inner and the outer. The inner contains the audio and life-support equipment and can be adjusted to any head size. The outer contains the CRT, optical module and electronics. Using the same inner helmet, outer helmets can be configured to cater for differing user

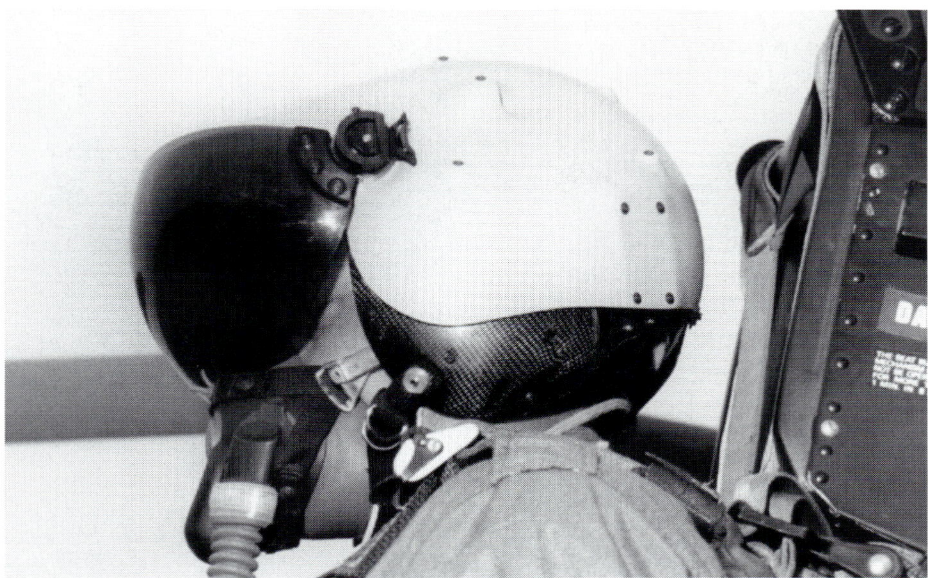

Guardian helmet-mounted display and sight system ***1997**/0001462*

and mission requirements. This simplifies maintenance and logistics, thereby offering definite long-term cost-savings.

Specifications

Accuracy: better than 5 mrad
Measuring envelope
Azimuth: −180° to +180°
Elevation: −90° to +90°
Roll: −90° to +90°
HMD imagery
Angular dimensions: 20° circular (day); 40° circular (night)
Exit pupil: 16 mm
Colour: green
Focused at infinity
Weight
Helmet (without NVGs): <1.6 kg
Total system weight: <15 kg
Power consumption: <120 W

Contractor

Pilkington Optronics, Glasgow.

VERIFIED

Guardian monocular display helmet ***1996***

Safe-Lite display helmet optics

Pilkington Optronics is developing a range of display helmet optical subsystems known as Safe-Lite. The range comprises monocular and binocular designs and includes relay optics, display media, multifunction visors, electro-optical counter-countermeasures and filters. Safe-Lite has applications for both visor-projected and combiner systems.

Operational status

A binocular system was flight trialled in mid-1997.

Contractor

Pilkington Optronics, Glasgow.

UPDATED

LC-40-100-NVG helicopter gunsight

The LC-40-100-NVG is a solid glass, unit power sight for machine guns or cannon firing sideways from a helicopter. The gunner has to aim-off for helicopter speed, range and the other external ballistic conditions. The sight has a special-to-purpose graticule which enables him to do this, so that targets on the ground can be engaged quickly and accurately. The graticule also enables him to adjust fire by burst on target techniques.

The graticule is lit by ambient light in daylight and by an LED array in low light and at night. The brightness of the array can be set by the gunner to suit the conditions and with it he can use night vision goggles.

Optional extras include a laser illuminator, laser rangefinder, tracing camera, dry zeroing device and laser protective filters.

Specifications

Dimensions: 210 × 142 × 114 mm
Weight: 2.5 kg
Power supply: 2 ½ AA or AA lithium thionyl chloride cells
Field of view: 40 × 32 mm
Magnification: ×1

Operational status

In service with Royal Air Force, British Army and other helicopters. The sight has been adopted as standard equipment by Fabrique Nationale, Giat and other weapon manufacturers.

Contractor

Ring Sights Defence Ltd.

VERIFIED

LC-40-100-NVG helicopter gunsight, shown on a door-mounted M134 Minigun ***1998**/0018161*

Type 1502 Head-Up Display (HUD)

The Type 1502 HUD is of modular construction. The unit is highly reliable and is easy to maintain; cost, volume and weight have been reduced without any compromise over optical accuracy and capability. The Type 1502 is suitable for installation in many new build attack aircraft or for retrofit, particularly where space or volume is limited.

The total field of view is 25°, achieved by using a 140 mm diameter exit lens, truncated fore and aft, and dual combiner glasses. Stroke symbology, raster video imagery or hybrid formats can be displayed. A video camera and an electronic variable standby sight can be incorporated as options. A customised data entry panel is an integral part of the HUD and enables the pilot to control the aircraft's nav/attack system and HUD mode. The Type 1502 is precision hard-mounted on the aircraft and requires no on-aircraft harmonisation.

Operational status

In service/production for the F-5E/F and Hawk 100 and 200.

Contractor

Smiths Industries Aerospace.

VERIFIED

The Smiths Industries' Type 1502 head-up display

Head-Up Display (HUD) for the AV-8B

The HUD which equips the US Marine Corps AV-8B, features a 4.5 in diameter (113 mm) exit lens which produces a vertical field of view of 22°. The unit also has a precision dual-combining glass, an electronically depressible standby sight and a built-in test system. The unit weight, 13.6 kg, is some 20 per cent less than traditional designs, without loss of mechanical strength, due to modern design and manufacturing techniques.

Compensation for windscreen distortion has been applied electronically and optically. The CRT is protected from damage from sunlight by infrared and ultraviolet filters and the brightness, display accuracy and deflection amplifier performance are all monitored by the built-in test system.

The standby sight is a precision LED matrix on a ceramic substrate, with variable brightness. HUD symbology and the outside world view through the HUD are recorded by a video camera which views through a periscope arrangement. A MTBF of 1,700 hours is being achieved in service.

In 1985, Smiths was awarded a contract by McDonnell Douglas to modify the existing HUD on the AV-8B to provide a complete night attack capability.

Operational status
No longer in production. In service in the US Marine Corps AV-8B Harrier.

Contractor
Smiths Industries Aerospace.

VERIFIED

Head-Up Display (HUD) for the Jaguar

From inception the Jaguar system has featured a Smiths Industries' electronic HUD. The system, which comprises a pilot's display unit, waveform generator, pilot's control panel and extra high-tension unit, is designed to provide the pilot with accurate analogue and alphanumeric symbol displays of primary flight data and navigation and weapon aiming information.

The pilot's display unit consists of the optical system and the CRT assembly. The optical system consists of a collimating lens assembly and a combining glass. The 100 mm f0.97 lens assembly provides a 25° total field of view. The installation provides an instantaneous field of view in the region of 18° in azimuth and 16° in elevation. All glass surfaces are treated with an anti-reflective coating to reduce spurious reflections.

Operational status
No longer in production. In service on Jaguar.

Contractor
Smiths Industries Aerospace.

VERIFIED

Head-Up Display (HUD) for the night attack AV-8B Harrier

Refractive optics are used on the head-up display for the AV-8B night attack HUD to give an unusually wide instantaneous field of view of 20° horizontally and 16° vertically. The large diameter collimating lens has been truncated on the fore and aft edges to save weight and to place the lens nearer the pilot's eyes in order to achieve this performance.

The Smiths HUD for the Harrier GR. Mk 7 and night attack variant of the AV-8B

Conventional stroke symbology can be overlaid during the raster flyback period on a raster picture derived from electro-optic sensors such as FLIR. Brightness of the two displays can be controlled independently.

Operational status
In service in the Royal Air Force Harrier GR. Mk 7 and night attack variants of the AV-8B.

Contractor
Smiths Industries Aerospace.

VERIFIED

Head-Up Display (HUD) for the Saab JA 37 Viggen

The Smiths Industries' pilot's display unit for the Saab JA 37 Viggen is a fully line-replaceable unit consisting of an electronic or front module, optical module and control panel.

The optical module consists of a combining glass assembly and a high-accuracy 100 mm lens system, providing a total field of view of 28°. A standby sight permits reversionary weapon aiming in the event of a failure.

The electronic module forms the main structure of the pilot's display unit and houses the deflection amplifiers, power supplies, BITE, CRT assembly, brightness control, combiner servo-amplifier and all the peripheral electronics circuitry.

Operational status
No longer in production. In service on the Swedish Air Force Saab JA 37 Viggen.

Contractor
Smiths Industries Aerospace.

VERIFIED

Low-Profile Head-Up Display (LPHUD)

The LPHUD was designed to meet the needs of modern cockpit design and the high-performance man/machine interface. It matches the capabilities of new sensors, avionics and displays.

The optical design of the lenses has reduced the instrument panel height for the LPHUD by 40 per cent. Valuable panel space is therefore freed to allow active matrix liquid crystal displays to be sited in their optimum position. A video camera and electronic standby sight can be incorporated.

The low-profile HUD frees valuable panel space **1995**

Operational status
Under development for new aircraft or retrofit applications in high-performance military aircraft.

Contractor
Smiths Industries Aerospace.

VERIFIED

Training Head-Up Display (HUD) system

The training head-up display is a compact low-weight system suitable for a wide range of light attack and training aircraft. Proven off-the-shelf equipment has been combined with significant growth provision for additional computing, interfacing and graphics generation.

The HUD has a total field of view of 25°. A display electronics unit responds to analogue, synchro, discrete and databus interfaces to generate HUD symbology for navigation, flight and weapon aiming data. In addition, its expansion capability has been exploited in order to provide the raster graphics for four head-down displays. A data entry panel is integral with the HUD for system control and input. A video recording system records information for post-flight debriefing. The baseline configuration features F/A-18 style symbology.

Operational status
In production for the Cockpit 21 installation of the US Navy T-45C Goshawk.

Contractor
Smiths Industries Aerospace.

UPDATED

The Smiths Industries' head-up display for the US Navy T-45C Goshawk **1998**/0018153

UNITED STATES OF AMERICA

Model 131A Head-Up Display (HUD)

The Model 131A HUD was produced for the US Army Bell AH-1G light attack helicopter.

Specifications

Exit aperture: 3 in (76 mm)
Field of view: 20°
Instantaneous field of view: 8.5° monocular (15.5° binocular) at 18.33 in from combiner glass
Standby reticle: fixed, red colour
Computer: microprocessor, 500 ns/instruction full arithmetic capability
Memory: 4 k 16-bit words ROM, 256 k 12-bit words RAM

Operational status

In service in the US Army Bell AH-1G helicopter.

Contractor

Astronautics Corporation of America.

VERIFIED

Head-Up Display (HUD) for the F-15A/B/C/D

The F-15A/B/C/D HUD consists of a display unit and signal data processor. The HUD accepts inputs from various sensors and processes and formats the display.

The HUD subsystem is being upgraded as part of the F-15 cockpit reliability and supportability enhancement programme. The display will have increased field of view, better combiner optics, higher writing speed and enhanced BIT. The signal data processor will have increased symbology capability, a helmet-mounted tracker option and enhanced BIT.

The F-15 improved HUD subsystem provides the pilot with navigation, attack and targeting and primary flight information necessary for effective flight control and weapons management. When complete, the upgraded system will be offered as an option for aircraft modification and upgrades for the F-15A/B/C/D.

Operational status

Over 1,300 HUDs have been delivered for F-15A, B, C and D aircraft over a period of 17 years. The Boeing Company has been awarded a $3 million contract for the full-scale design and development of the improved HUD subsystem.

Contractor

The Boeing Company.

VERIFIED

The HUD subsystem in the F-15A/B/C/D is being upgraded

Mast-Mounted Sight (MMS) for the AHIP helicopter

In October 1984, the US Army ordered the MMS for its Army Helicopter Improvement Programme (AHIP) Bell OH-58D helicopters. Mounted on the rotor mast above the rotor itself, the sight unit acts like the periscope of a submarine, permitting the aircraft to conceal itself by using natural cover but allowing the crew to peer over the cover to observe enemy dispositions and forces.

The MMS contractor, integrates stabilisation platform and electrical systems with a sensor suite purchased from Northrop Grumman. The sensors include a low-light television, a thermal imaging sensor and laser rangefinder/designator for Copperhead shells and Hellfire anti-tank missiles. A feature of AHIP is the Automatic Target Hand over System by which targeting information from one aircraft can be transmitted to another or to ground-based weapons. AHIP itself is the precursor to the much more ambitious RAH-66 light battlefield helicopter system.

Mounted over the main rotor drive shaft, the performance of such a system would be severely degraded by the high vibration levels in this area were it not for the soft mount devised by The Boeing Company which provides a high degree of isolation. Performance of the anti-vibration system is such that target bearing can be measured to within 20 mrad. The overriding need to keep down weight above the rotor has produced a sensor package weighing only 73 kg. Equipment bay systems add another 41 kg.

The television camera has a silicon-vidicon dawn to dusk capability with an 8° field of view for target acquisition and a 2° field for recognition. The FLIR sensor has a 120 element common module detector array and two fields of view of 10 and 3°. Television and laser systems share a common optical path to minimise the number of components. A video tracker and digital scan converter in the data processor together permit

The mast-mounted sight on an OH-58D helicopter

the incorporation of other features such as autotrack, frame-freeze and point-track.

In the cockpit are two multifunction displays for the presentation of video from the MMS and flight guidance and communication and navigation information. The sensors themselves are contained within a 650 mm carbon-epoxy sphere with a sightline 810 mm above the plane of rotor rotation. Data is channelled into the cockpit area via cables through a 23 mm tube running inside the drive shaft.

The MMS is applicable to a variety of helicopter platforms, including the Boeing 500 Series, the Sikorsky H-76, the Bell 406, the Agusta A 129 and the Eurocopter BK 117. The MMS can be used to provide these platforms with both AHIP scout features and light attack capability, using Hellfire and/or TOW missiles. The Apache helicopter which does not have a mast-mounted system, the Lockheed Martin TADS/PNVS system being nose-mounted, is regarded as a potential future application.

Operational status

In service in OH-58D helicopters.

Contractor

The Boeing Company.

UPDATED

The mast-mounted sight display is on the left of this Bell OH-58D helicopter cockpit. The view is adjusted by controls on the stick on the pilot's right-hand side

Spasyn visual target acquisition system

The Spasyn (space synchro) visual target acquisition system is an advanced ultra-light helmet-mounted sighting system. Spasyn measures the location and orientation of an object in space by means of radiating directional magnetic fields in three orthogonal co-ordinates. Applied to a helmet-mounted sight, this closed-loop transducing system provides a full six degrees of freedom in measuring the pilot's line of sight to an accuracy of better than 0.5° within a head motion envelope of 0.14 m^3. The system enables the pilot to achieve visual acquisition of a given target and lock on automatic tracking systems while remaining in the head-up position and free to perform evasive manoeuvres or other in-flight tasks.

A small transmitter mounted on the cockpit frame emits magnetic field vectors which induce voltage signals in a sensor mounted in the helmet. These signals are processed by a computer in the system's electronics unit to define the sensor's translation and orientation which correspond to the operator's line of sight angles. The sensor mounted on the helmet weighs approximately 14 g and does not encumber the wearer.

The Spasyn system can be used in both fixed-wing aircraft and helicopters to command weapons and sensors or for land and sea platforms.

Specifications

Dimensions:
(sensor) 12.6 mm cube
(radiator) 25.4 mm cube
(electronics unit) 333 × 193 × 122 mm
Weight:
(sensor) 14 g
(radiator) 28 g
(electronics unit) 10 kg
Accuracy:
(nominal head position) 0.1°
(max error condition) 0.5°

Contractor

The Boeing Company.

UPDATED

HGS-2000 Head-Up Guidance System

Flight Dynamics has pioneered the application of the head-up display for commercial, military and corporate transport aircraft with the development of a Cat IIIa HGS. The system comprises five main components: the overhead unit integrating a CRT and lens assembly; a combiner containing the holographic element; a high-integrity computer; a drive electronics unit; a pilot's control panel.

The HGS is aimed at improving safety and performance by offering a system capable of operating to lower weather minima and as an economic alternative to Cat III automatic landing systems. In July 1984, Flight Dynamics received FAA approval for its HGS for manually flown Cat IIIa approaches down to a runway minimum range of 700 ft and decision height of 50 ft in the Boeing 727. This was the first manual system certified by the FAA for the demanding low-visibility environment. In August 1987, the HGS was also certified for windshear detection and recovery guidance and in 1990 was approved by the FAA for low-visibility take-offs down to 300 ft RVR.

The system projects flight guidance symbology, focused at infinity, on the holographic combiner. Along with basic flight information such as airspeed, altitude, course and heading, the HGS displays inertial flight path and acceleration, providing the sensitivity and accuracy required for Cat III operations. Safety is also improved, as industry studies have demonstrated that projected flight path and precise energy management greatly improves the pilot's situational awareness and aircraft control, particularly in difficult or unexpected conditions.

The combiner provides the pilot with a full 30 × 24° field of view, a feature especially useful in high cross-wind conditions. By comparison, military fighter HUDs typically have 20 × 15° fields of view. The holographic technology improves both the reflectivity of the projected symbology and the transmissivity of real-world details as seen by the pilot.

Flight Dynamics' holographic head-up guidance system

Flight Dynamics is currently evaluating the use of the HGS to obtain even lower take-off and landing minima. The commercial transport Industry has also expressed interest in combining a fail-passive autoland system with the fail passive HGS to achieve Cat IIIb capability. This hybrid landing system should be certifiable to 300 ft runway visual range and would combine the benefits of an automatic landing system with the projected flight path and head-up advantages of the HGS.

Alaska Airlines, the first carrier to equip its 23 aircraft Boeing 727 fleet with the HGS, received FAA operational approval to operate in revenue service down to 50 ft decision height in late 1988 and the system has now been cleared for 300 ft RVR take-offs.

In October 1989, Alaska Airlines conducted the world's first manually flown Cat IIIa landings with passengers on board.

Most recently, Flight Dynamics is proposing integration of infrared sensor imagery to improve pilot situational awareness during climb-out and descent phases of flight.

Operational status

The Flight Dynamics line of Head-Up Guidance Systems has logged more than 1 million flight hours and is currently operated worldwide on eight aircraft types for business and scheduled airline operators, providing low-visibility take-off and landing as well as flight regime guidance. The HGS displays flight information and guidance on a holographic combiner in the pilot's normal head-up field of view. Aircraft equipped with the Flight Dynamics HGS include the Falcon 2000, Saab 2000, Boeing 727 and 737, de Havilland Dash 8, Canadair Regional Jet, Dornier 328 and the Lockheed Martin C-130J.

Latest in the series of Flight Dynamics head-up guidance systems in the HGS-4000, for which FAA supplemental type certification is expected late in 1998, before delivery to Delta Airlines as launch customers for its new generation Boeing 737-600/700/800 aircraft.

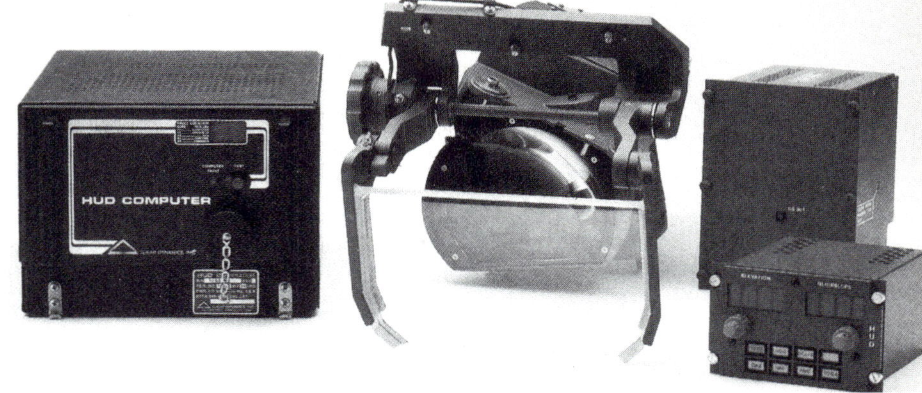

Components of the holographic guidance system

Contractor

Flight Dynamics.

UPDATED

FV-2000 Head-Up Display (HUD)

The FV-2000 HUD incorporates both expanded (conformal) and compressed (non-conformal) pitch scales and has the ability to display a velocity vector or flight path marker. The velocity vector feature shows the aircraft's actual ground track and projects the touchdown point. It also shows the immediate effects of windshear. The accelerate/speed cue and all of the navigational and flight information is also depicted on an easy-to-read display. The system has been certified as a primary flight display and can be retrofitted in aircraft with either electromechanical instruments or EFIS, creating fleet commonality for flight departments.

The system comprises the optical unit, combiner glass, control panel and HUD computer and weighs 9.98 to 14.25 kg. Each Flight Visions FV-2000 HUD is specifically designed for seamless integration into each aircraft type for which it is certified. The control panel is mounted in the cockpit pedestal or panel. Other features include TCAS, GPWS, runway overlay, raster capability and an RS-170 video port. The processor is certified for use outside the pressure vessel. The FV-2000 contains self-diagnostics through the control panel and maintenance pages are accessible through an RS-232 port.

Flight Visions is developing an enhancement of the FV-2000, designated FV-2000E, together with an Enhanced Vision Sensor (EVS) to give pilots improved capability to operate in poor conditions.

The current FV-2000 system already incorporates design elements to accept enhanced vision technologies including FLIR and Millimetre Wave Radar (MWR). After new EVS sensors become available, existing systems can be updated by replacing the FV-2000 remote-mounted computer with an FV-2000E computer. The FV-2000's overhead optical unit already has raster capability to display all EVS available.

The FV-2000E head-up display symbol generator is a new product that combines enhanced display generation modules from the FV-3000 with the less expensive computing engine of the FV-2000. It was designed for users that do not need the processing power and flexibility of the FV-3000 but still need a HUD symbol generator. Flight Vision also offers the FV-2000E with the civil head-up display product line.

Flight Visions claim that the FV-2000E will give operators access to Cat I airports in Cat II and Cat III conditions.

Operational status

FAA certified: Falcon 50, Gulfstream III and IV, Learjet 55, Citation 550, all Beech King Air models and the Bell 230 helicopter. Certifications for about 20 more aircraft types are expected. The FV-2000 will be available for most turbojet-powered business aircraft.

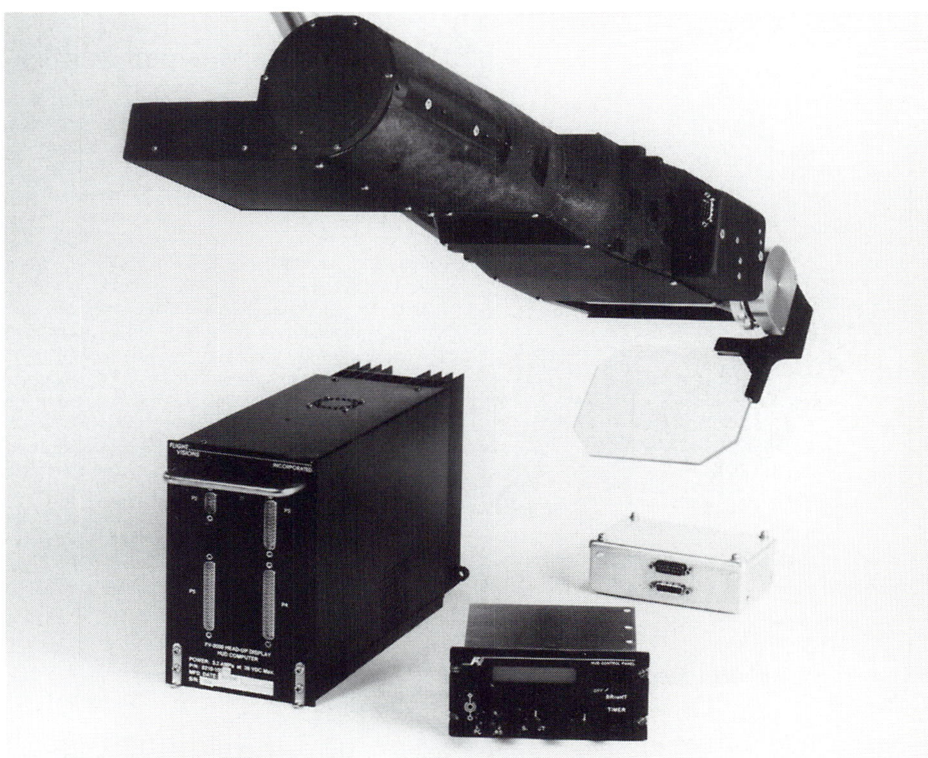

Flight Visions HUD components (top to bottom, clockwise order): straight mounted overhead optical unit, grid amp, control panel, HUD computer ***1997***/0001464

The FV-2000 HUD installed in a Falcon 50

The FV-2000E, display processor together with the Sparrow Hawk™ HUD, has been selected by israel Aircraft Industries for upgrade of seven Kfir aircraft, with options for a further 12 systems.

Contractor

Flight Visions Inc.

VERIFIED

FV-3000 Modular Mission Display Processor (MMDP)

The FV-3000 MMDP forms the core of modern retrofit or new aircraft avionics packages. It can drive both Head-Up Displays (HUD) and Multi-Function Displays (MFD) while providing spare power for mission processing.

Containing two 50 MHz 68040 microprocessors and a dual-redundant MIL-STD-1553B databus, the FV-3000 MMDP generates both flight and military displays. It has the capability to input RS-170 video, synchronise it with stroke-generated graphics and output the combined image for display on a HUD. The FV-3000 MMDP also offers: air-to-ground weapon delivery processing, air-to-air gunnery and missile processing, primary navigation processing, primary MIL-STD-1553B databus, primary system control, upfront control panel switch processing, HOTAS switch processing.

The FV-3000 MMDP meets both military and civilian specifications while designed to meet the high demands of fighter aircraft. It has the capability to detect and isolate faults through a combination of operational and intermediate-level testing. In accordance with MIL-STD-810E, the FV-3000 can be fitted in jet fighter aircraft.

FV-3000 Modular Mission Display Processor (MMDP) **1997**/0001463

Specifications

Dimensions: 152.4 × 228.6 × 406.4 mm
Weight: 7.3 kg
Power supply: +28 V DC MIL-STD-704D
200 watts

Operational status

Czech L-159.

Contractor

Flight Visions Inc.

VERIFIED

Sparrow Hawk™ weapon delivery system

Sparrow Hawk is a complete weapon delivery system providing a powerful and cost-effective solution for aircraft that perform a broad range of missions. Sparrow Hawk is designed for new fighter aircraft and for upgrading existing aircraft. The Sparrow Hawk is designed to be customised in order to offer fleet commonality to ease pilot training and/or transition across aircraft types.

Air-to-air functions include LCOS, hotshoot and AIM-9; air-to-ground functions include gun and rocket dive attack, dive and level bombing with both instantaneous and delayed release, combination rocket or gun and bomb attacks in the dive toss mode and toss bombing. A 25° wide field of view Head-Up Display (HUD) provides attack flexibility on a sharp and crisp stroke on raster head-up display with electronic boresight.

The Sparrow Hawk HUD has been designed to be fully customised in instrument interface, symbology and upfront controls. At a total system weight of 7.8 kg, it is one of the lightest weapon delivery systems available. It can provide weapon delivery with a wide range of sensors, from basic attitude heading reference systems and no radar to full inertial reference and radar. A wide range of interfaces are available, from analogue and synchro to ARNIC 429 to MIL-STD-1553B. Sparrow Hawk features extensive BIT for easy maintainability and can be almost completely diagnosed in the aircraft. Internal data logging and continuous in-flight testing keeps track of system health. Sparrow Hawk is Gen. II and Gen. III NVG-compatible and features raster display capability, integrated colour HUD camera and built-in standard training features.

Flight Visions' FV-2000/Sparrow Hawk Primary Flight Display HUD is mounted at the top of the instrument panel in the Pilatus PC-7 Mk II turboprop military trainer **1996**

Operational status

Sparrow Hawk is in service on the Czech Republic Air Force's L-159 aircraft and the Ecuadorian Navy's Bell-412 helicopters; it has been selected by Israel Aircraft Industries for the upgrading of Kfir aircraft.

Contractor

Flight Visions Inc.

VERIFIED

Helmet-Mounted Cueing System (HMCS)

Honeywell's HMCS design aim is to provide the pilot with the 'first-shot' advantage. The HMCS features: integrated electronics including sight, display and symbol generation, and auto-brightness sensor; also integrated is a colour camera and Honeywell's advanced magnetic head-tracker system. The symbol-set is fully programmable.

Honeywell's advanced visor display helmet-mounted cueing system **1998**/0018152

The HMCS is suitable for single or dual cockpit use, with HGU-86, HGU-55 and HGU-53 helmet shell options.

Specifications

Symbols: stroke/raster or mixed operation
Field of view: 20°
Exit pupil: >25 mm
Eye relief: 69 mm
Accuracy: <2 mrad RMS accuracy
Weight: 1.27 kg

Contractor

Honeywell Inc, Sensor and Guidance Products.

NEW ENTRY

Integrated Helmet and Display Sighting System (IHADSS)

The latest version of the Honeywell electro-optical system is the IHADSS which is in production for the AH-64A helicopter. Both the pilot and co-pilot/gunner are provided with helmet units and controls to allow independent and co-operative use of the system. The IHADSS sight component provides off-boresight line of sight information to the fire-control computer for slaving weapon and sensor to the pilot's head movements. Real-world sized video imagery from the slaved and gimballed infrared sensor is overlaid with targeting as well as flight information symbology and projected on a combiner glass immediately in front of the pilot's eye.

Honeywell IHADSS helmet sight and display

The IHADSS allows night nap-of-the-earth flight at below treetop altitudes, without reference to cockpit instruments, and rapid target engagement.

Specifications

Dimensions:
(sensor surveying unit × 4) 111.76 × 129.54 × 78.74 mm
(sight electronics unit) 177.8 × 177.8 × 289.56 mm
(display electronics unit) 139.7 × 177.8 × 289.56 mm
(display adjust panel × 2) 152.4 × 182.88 × 76.2 mm
Weight:
(helmet × 2) 1.4 kg each

(sensor surveying unit × 4) 0.57 kg each
(sight electronics unit) 7.26 kg
(boresight reticle unit × 2) 0.23 kg each
(helmet display unit × 2) 0.57 kg each
(display electronics unit) 6.35 kg
(display adjust panel × 2) 1.58 kg each
(total weight) 23.45 kg
Power supply: 115 V AC, 400 Hz, 3 phase, 460 W
Field of view: (horizontal) 40° × (vertical) 30°
Coverage:
(azimuth) ±120°
(elevation) ±70°
Accuracy: 5-10 mrad RMS
Slew rate: 120°/s

Operational status

Currently in service on the Boeing AH-64A Apache and Agusta A 129 helicopters.

Contractor

Honeywell Inc, Sensor and Guidance Products.

UPDATED

Agile Eye Plus helmet

Kaiser Electronics has developed Agile Eye Plus, a second-generation helmet-mounted display that provides significantly greater capability than the original Agile Eye. The field of view of Agile Eye Plus has been increased from 12 to 18° to improve situational awareness, and a raster display capability has been added to the stroke display mode so that information from TV cameras and other sources such as FLIR can be projected. A retractable visor replaces the clip-in visor of Agile Eye.

Agile Eye Plus projects head-up display information on to the visor of the pilot's helmet superimposing, in a monocular presentation, information over the field of view without obscuring vision. The system includes two declutter modes so that the pilot can include only those items of information he needs to see. A magnetic head tracker mounted inside the canopy tracks the pilot's head movements and continuously updates display data to correspond to the direction of his head. Mounted in the helmet are a 0.5 in diameter Hughes CRT; a small high-voltage power supply; a system of mirrors for projecting the image on to the visor. The Agile Eye Plus helmet weighs just over 1 kg. The system also includes a display driver unit in the cockpit and a display processor/tracker unit in the avionics bay.

The third generation of Agile Eye, Agile Eye Mk III, has now been released. This features a tracking sensor, detachable display unit, interchangeable day/night visor, visor/mask seal adjustment, high-voltage quick disconnect and 600 kt windblast protection.

Specifications

Agile Eye Mk III
Weight:
(helmet-mounted display) 1.45 kg
(cockpit control panel) 2 kg
Power supply: 115 V AC, 400 Hz
5 V AC, 400 Hz
Field of view: 20° monocular, right eye
Field of regard: unlimited
Reliability: >2,000 h MTBF

Contractor

Kaiser Electronics.

VERIFIED

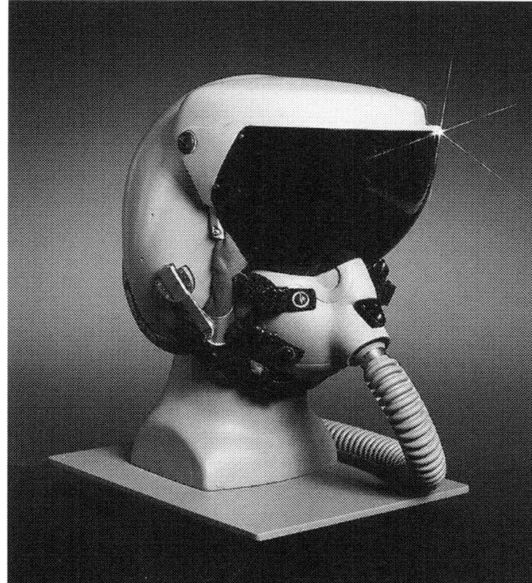

Kaiser Agile Eye Plus helmet ***1995***

Specifications

Weight:
(control panel) 0.45 kg
(display electronics unit) 17.8 kg
(helmet-mounted display) 2.04 kg
Power supply: 115 V AC
Field of view: 40° (vertical) by 60° (horizontal)
Colours: red, green and blue
Modes of operation: day, night, boresight

Contractor

Kaiser Electronics.

VERIFIED

Colour helmet-mounted display

The colour helmet-mounted display provides a high-quality, high-brightness, wide field of view, full-colour helmet display in a system configuration which is easy to use and compatible with aircraft and simulator installations. The system is a suitable flight research tool for experimentation in guidance, control and display flight research.

Head-Up Display (HUD) and weapon aiming system for the MB-339C

Kaiser supplies the head-up display and weapon aiming system for the MB-339C. The three-unit system comprises a computer/symbol generator, sweep driver unit and pilot's display unit. All or part of the same system is used in the MT-4, F-4EJ, C-1 and CCV programmes in Japan and the Taiwanese AT-3. The Kaiser HUD fitted to the Italian MB-339C, known in this application as Sabre, has dual-flat holographic combiners and a flexible upfront control panel.

Operational status

In service in the MB-339C.

Contractor

Kaiser Electronics.

VERIFIED

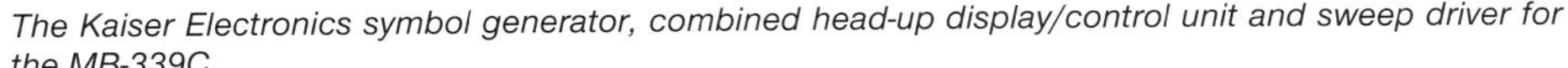

The Kaiser Electronics symbol generator, combined head-up display/control unit and sweep driver for the MB-339C

Head-up Display (HUD) for the A-10

The updated version of the Kaiser HUD for the US Air Force A-10 incorporates inertial navigation data from the A-10's inertial navigation system. The HUD contains a MIL-STD-1553 multiplex databus which can compute total velocity vectors and other algorithms.

Operational status

In service in the US Air Force A-10.

Contractor

Kaiser Electronics.

VERIFIED

Head-Up Display (HUD) for the AH-1S

The HUD for the US Army Bell Helicopter AH-1S Cobra helicopter is a lightweight low-profile conventional CRT that superimposes aiming information for the multibarrelled gun and TOW anti-tank missile on to the pilot's forward field of view. The programming flexibility and spare capacity of the microprocessor-controlled symbol generator permits other functions to be incorporated, such as the derivation of flight commands for nap-of-the-earth flying and laser tracking and pointing information.

Operational status

In service in the US Army Bell AH-1S helicopter. Over 1,000 HUDs have been delivered.

Contractor

Kaiser Electronics.

UPDATED

Head-Up Display (HUD) for the AH-1W

Both the control/display subsystem and Full-Function Signal Processor (FFSP) are produced by Kaiser. The HUD is identical to that of the US Army's Bell AH-1S, which Kaiser also supplies. In contrast, the FFSP is an entirely new design, which features dual 68000 processors and software designed to DoD-STD-1679A requirements. The processor is programmable and includes the capabilities for vectors, circles, arcs and rotation.

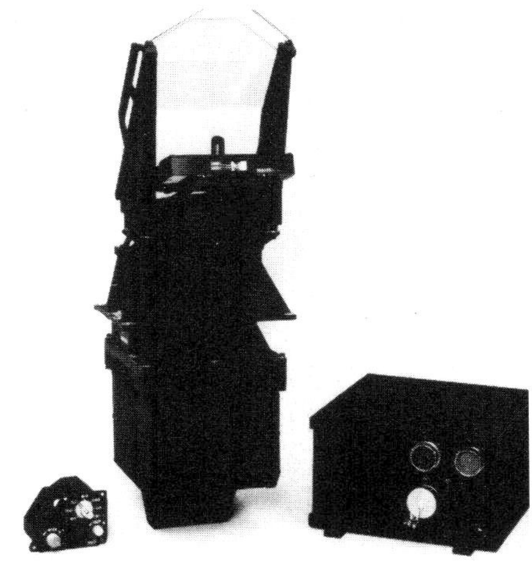

The pilot's display unit for the Kaiser Electronics head-up display in the AH-1S Cobra helicopter

Operational status

The US Marine Corps has ordered 44 Bell AH-1Ts and is expected to retrofit its fleet to the AH-1W configuration.

Contractor

Kaiser Electronics.

VERIFIED

Head-Up Display (HUD) for the JAS 39

Sweden awarded a production contract for a HUD incorporating diffraction-optics technology, because the advantages of this technology are claimed to be a key factor in providing the JAS 39 with its capability in the air-to-air, air-to-ground and reconnaissance roles.

Diffractive-optics HUDs have two principal advantages. First, by comparison with conventional systems, they have a much wider field of view, typically 30 × 20° compared with 20 × 15°, and are thus more suited to the new generation of combat aircraft in which weapon aiming symbology can make large angles with the flight vector. The wide field will also be useful in night operations to display data from electro-optical sensors such as FLIR.

Secondly, the combiner glass on which the symbology is superimposed on the outside world acts as a mirror reflecting only a narrow band of light. The transmission index is about 85 per cent compared with 50 to 70 per cent for refractive HUDs. The symbology is also bright enough to stand out in direct sunlight without having to operate the CRT at such high-power levels that its life is shortened.

The system also provides resistance to glare, reflections and spurious sun images. The latter is particularly important since bright sunlight can create hot spots on the display that prevents the pilot from seeing the symbology. The design, based on proprietary technology using holography and lasers, employs a single-combiner glass, eliminating the bulky support structure necessary in HUDs that support two or more.

Operational status

In production and in service in the JAS 39 Gripen.

Contractor

Kaiser Electronics.

UPDATED

Helmet Integrated Display Sight System (HIDSS)

The HIDSS is a second-generation binocular, wide field of view helmet-mounted display system designed for the RAH-66 Comanche helicopter. Critical for night pilotage, accurate delivery of weapons and improved situational awareness, the lightweight, high-resolution HIDSS utilises a two-piece modular helmet design, advanced optics and precision magnetic tracking to provide head-up, eyes-out operation. Driven by the high-performance SEM-E expanded display unit, HIDSS combines Gen.II FLIR video, raster graphics imagery and growth to stroke symbology for a day and night, all-weather helmet-integrated display system that is adaptable to a variety of helicopter platforms and missions.

Specifications

Weight:
(control panel) 0.5 kg
(helmet-mounted display) 2 kg
(expanded display electronics unit) 14.97 kg
Power supply: 270 V DC, 335 W
Field of view: 35 × 22° (18° overlap)
Field of regard:
(azimuth) ±180°
(elevation) ±90°
(roll) ±180°
Reliability: 1,000 h MTBF

Operational status

Under development for the US Army RAH-66 Comanche helicopter.

Contractor

Kaiser Electronics.

VERIFIED

Holographic Head-Up Display (HHUD) for the F-4E

A Wide Field Of View (WFOV) HHUD is now in production for the F-4E aircraft under a contract from an international customer. The F-4E HHUD features a single curved holographic combiner which offers a 20 × 30° total and instantaneous field of view. The HHUD operates in raster and/or cursive modes and the display processor, which utilises a MIL-STD-1750A processor, is packaged with the HUD.

Operational status

Some 100 systems have been produced for the F-4E.

Contractor

Kaiser Electronics.

UPDATED

Low-profile Head-Up Display (HUD)

Designed to deliver a 23.5 × 30° wide picture in the demanding tactical cockpit environment, the low-profile HUD for the F-14D uses modern holographic combiners to provide the highest see-through transmission and contrast. The high contrast is made even more valuable through the inclusion of a circularly polarised filter which suppresses the reflections so common in many HUDs. Used in the F-14D, the low-profile HUD provides crisp stroke symbology to display navigation, steering, flight situation and attack cues in the brightest daylight or, at the flick of the day/night switch, at subdued brightness to allow smooth adjustment of the critical night-time brightness levels.

This performance is packed into a unit of cross-section 102 × 152 mm, enabling cockpit designers to preserve forward visibility in instrument panel designs with large displays underneath and alongside the HUD. A TV recording camera is chin-mounted to view the same scene as the pilot, while eliminating the need for scan conversion.

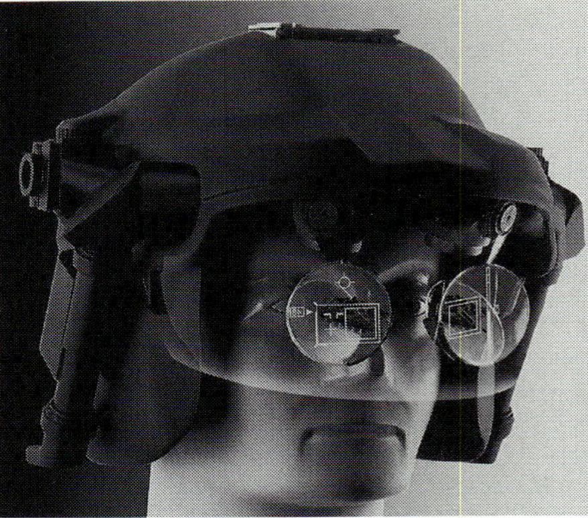

HIDSS is being developed for the RAH-66 Comanche helicopter **1995**

The unit is completely self-contained. All nine shop-replaceable assemblies, including the combiner, are replaceable without harmonisation. BIT isolates 90 per cent of possible failures to these subassemblies, including the low- and high-power supplies.

Specifications

Dimensions: 674.9 × 165.1 × 274.8 mm
Weight: 17.5 kg
Power supply: 115 V AC, 400 Hz, 3 phase, 81 W
Field of view: 23.5° (elevation) × 30° (azimuth)
Reliability: 3,300 h MTBF

Operational status

In service with the Northrop Grumman F-14D.

Contractor

Kaiser Electronics.

VERIFIED

Wide Eye helmet integrated display

Wide Eye is a fully integrated binocular helmet display system with retractable combiners for day and night use by helicopters at low level in all weather. It incorporates dual one inch CRTs with stroke/raster and hybrid capabilities and image intensifiers. The optical subsystem is detachable and remains with the aircraft. The system consists of the headgear, display electronics unit, tracker and boresight reticle control unit.

Operational status

Weight:
(helmet) 1.8 kg
(system) 10.25 kg
Power supply: 115 V AC, 200 W
28 V DC, 20 W
Alignment time: 5 s
Field of view: 52 × 35°
Accuracy: <4.9 mrad

Contractor

Kaiser Electronics.

VERIFIED

AN/ASG-29 Lead-computing optical sight

The AN/ASG-29 lead-computing optical sight has been specially produced for the US Air Force F-5E. The AN/ASG-29 has a family relationship with the AN/ASG-26A but comprises only two units: a pilot's display and lead computer. The system provides guidance for air-to-air and air-to-ground weapons delivery.

Specifications

Weight:
(sight head) 6.9 kg
(lead computer) 7.9 kg
(mounting base) 0.8 kg
Field of view:
(azimuth) 12°
(elevation) 14°
Instantaneous field of view: 7.5°
Collimating lens aperture: 4 in (102 mm)
Reliability: over 300 h MTBF claimed

Operational status

In service in the F-5E, but no longer in production.

Contractor

Lockheed Martin Ocean Radar and Sensor Systems.

VERIFIED

Advanced helmet-mounted sight

Polhemus designs, develops and manufactures helmet-mounted sights specifically for tactical military aircraft, based on a company-patented method of measuring sight angles by the use of magnetic sensors. A fourth-generation system, marketed under the name Magnetrak, is now in production (see next entry).

A source mounted in the cockpit generates a magnetic field some distance around it, which is sensed by a sensor installed under the pilot's helmet visor. Signals are processed by a computer in an associated electronics unit and converted into angles representing the direction of the pilot's line of sight in relation to a particular datum, usually the aircraft's reference frame.

Coincidence between the helmet aiming axis and the pilot's line of sight is attained by a small helmet-mounted optical generator which projects a virtual image located at infinity so that the pilot does not have to refocus his eyes on to the helmet visor. The pilot then aligns the aiming reticle with the target. The reticle can also be used to provide cueing information which permits the pilot to be signalled or directed from an external source of information, for example, from a radar, FLIR or radar warning device or by a weapons systems operator. The same system, in conjunction with a more elaborate display, can provide the pilot with flight director symbology or imagery and information from other aircraft sensors. The complete system, irrespective of complexity, comprises a magnetic field source, helmet-mounted sensor, memory unit, electronics unit and helmet visor display unit.

Contractor
Polhemus Inc.

VERIFIED

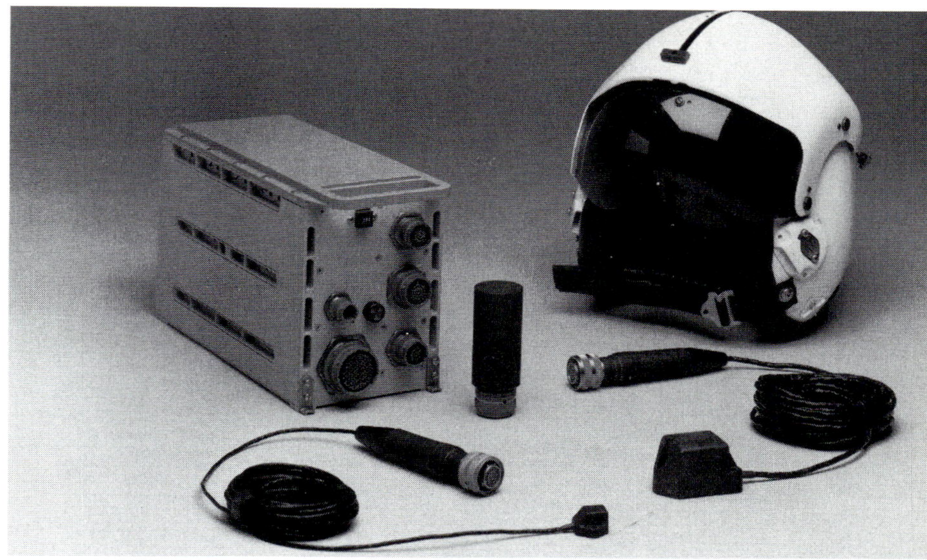

The Polhemus Magnetrak helmet sight. Left to right: system electronics unit and connecting cables, aircraft memory unit, magnetic field sensor, helmet with visor display and source unit

Magnetrak helmet-mounted sight

The line of sight data acquired by the sensor/source arrangement is passed into the aircraft weapon aiming system using a MIL-STD-1553B databus link. Conversely, target data detected by one of the aircraft's sensors such as the radar or thermal imager, can be relayed to the Magnetrak system, to give target cueing information on the pilot's visor. Thus the pilot's own line of sight can be directed to a possible target for visual identification before attack.

The components of the Magnetrak system include a small three-axis sensor and source, system electronics unit and memory unit, the latter supplying mapping data to the former. There is also a visor display consisting of a parabolic visor, (and cover) and an integral LED reticle generator which projects a collimated cross-hair image on to the visor, allowing the pilot's eye to focus on target and reticle simultaneously. Discrete dots around the cross-hair give a cueing facility. The visor fits any standard military helmet without modification to the shell. The electronics unit is ½ ATR in format.

The Magnetrak offers a resolution of 0.1° and covers 360° movement in azimuth and roll axes and ±90° in elevation. The motion box, inside which the head must remain for accurate system performance, is 410 × 254 × 150 mm either side of a central point.

Specifications
Dimensions:
(source) 61 × 35 × 35 mm
(sensor) 28 × 23 × 18 mm
(electronics unit) 124 × 174 × 283 mm
(memory unit) 109 × 33 mm
Weight:
(source) 0.16 kg
(sensor) 19 g
(electronics unit) 5.13 kg
(memory unit) 0.17 kg
Power supply: 115 V AC, 400 Hz, single phase, 0.7 A

Operational status
In production.

Contractor
Polhemus Inc.

VERIFIED

Tiger-Paws weapon system upgrade

Sierra Technologies has developed the Tiger-Paws upgrade for the F-5A/B/E/F avionics refurbishment. This integrates major avionics enhancements by replacing outdated analogue electronics with digital equipment already in use in many US and European military aircraft. Key items in the digital core include the MIL-STD-1553B databus, a smart Head-Up Display/ Weapon Aiming Computer (HUDWAC), ring laser gyro inertial navigation system and the standard central air data computer. For the radar-equipped F-5E/F, the digital core can be augmented by replacing the analogue pulse radar with a pulse Doppler coherent radar modified for compatibility with the aircraft.

Growth potential afforded by the digital system facilitates integration of a wide variety of options such as video, data transfer, stores management, radar warning receiver, multifunction displays, laser ranger and designator, FLIR, ECM, HOTAS, chaff and flares, Sidewinder control and incorporation of existing systems into a reliable modern digital system enhanced by very high MTBF.

Cockpit design can emulate the F-16 or other specified aircraft for logistics, training and operational compatibility. Installation of the Tiger-Paws digital system can be accomplished locally with kits, if required.

Similar digital enhancements are available for the T-38 Talon supersonic trainer.

Operational status
In 1991 Sierra received a contract, valued at $20.7 million, to modify 15 Royal Norwegian Air Force F-5A/B aircraft with the Tiger-Paws digital core and other enhancements. This contract was completed in August 1994 and the aircraft are currently in service as F-16 lead-in trainers.

Contractor
Sierra Technologies Inc.

VERIFIED

Joint Helmet-Mounted Cueing System (JHMCS)

Vision Systems International LLC (VSI), a joint venture of Elbit Systems, EFW and Kaiser Electronics has been selected by The Boeing Company in conjunction with Lockheed Martin Tactical Aircraft Systems to equip various US fighter aircraft that the two companies produce with the JHMCS.

The JHMCS is designed for compatibility with US fighters and supports high-off boresight weapons engagement under high-G forces.

Operational status
Development, based on Elbits' DASH Helmet-Mounted Display (HMD) and Kaisers' Agile-Eye, will take several years.

Contractor
Vision Systems International.

UPDATED

STORES MANAGEMENT

A two seat Super Hornet, designated F/A-18F2, launching AMRAAM ***1998***/0018150

BELGIUM

BT14 Series weapons panels

The BT14 Series is a command system for controlling weapons on fixed-wing aircraft and helicopters. The Series includes the BT14-2 for aircraft with two stations and the BT14-6 for aircraft with four to six stations. The BT14-6 RE is a repeater for the BT14-6 and is dedicated to the back seat of a fixed-wing aircraft.

The BT14-2 allows weapons to be realeased in case of emergency by the use of a jettison switch. A weapon selector is provided for the use of mixed weapon loads. The command system includes a rocket mode selector for single-shot or ripple firing.

The BT14-6 allows weapons to be released singly, in ripple or sequentially from each side of the aircraft to reduce trim disturbances to a minimum. A weapon selector is provided for the use of mixed weapon loads. The system includes a bomb fuzing selector to allow activation of nose or tail fuze on impact, delayed activation of the tail fuze or for weapons to be dropped inert with no fuze activation. Six push-buttons in the BT14-6 are used to select the weapon station. A test button allows a check on the correct functioning of all lights in the panel.

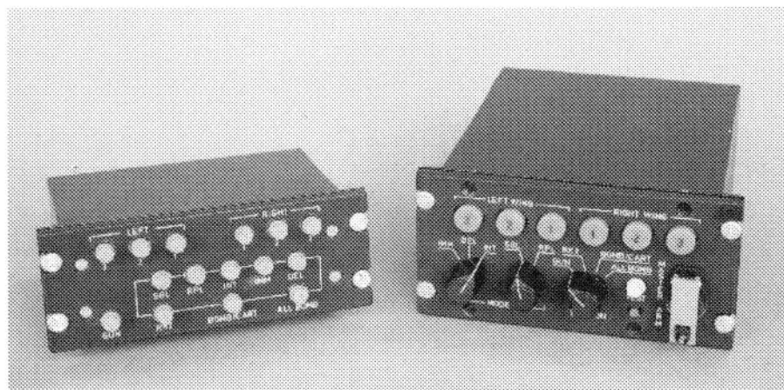

The FN HERSTAL BT14-6 RE repeater (left) and BT14-6 firing control panel (right)

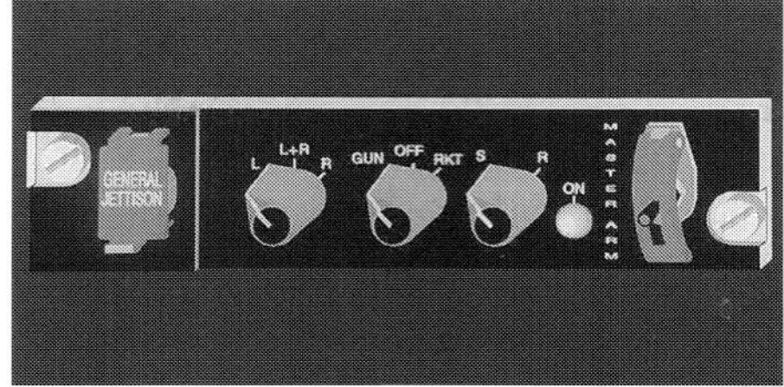

The BT14-2 firing control panel **1995**

Specifications

Dimensions:
(BT14-2) 145 × 37 × 100 mm
(BT14-6) 145 × 75 × 160 mm
(BT14-6 RE) 145 × 57 × 100 mm

Weight:
(BT14-2) 0.5 kg
(BT14-6) 1.2 kg
(BT14-6 RE) 0.4 kg

Contractor

FN HERSTAL.

VERIFIED

PC16 pod control unit

The PC16 pod control unit monitors and controls aircraft gunpod functions. The PC16 features a burst controller which limits the number of rounds fired in a burst. Once the preset figure is reached, firing is automatically interrupted until the trigger is released. It includes a bar graph indicator which shows, on two 10 LED linear indicators, the total ammunition content of each pod, with each LED representing 10 per cent of the full load. When ammunition reaches less than 10 per cent of the full load, amber spare ammunition indicator lamps illuminate.

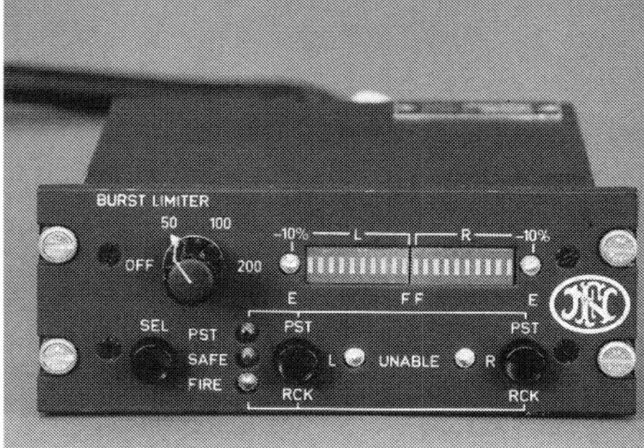

The FN HERSTAL PC16 pod control unit

The PC16 incorporates a number of safety features. These include a non-reversible sequencing selector to prevent wrong or dangerous use of the system, a positive locking selector to prevent the gun being fired during the landing sequence, a control test which tests the system during the pre-take-off checks, a firing auto control to allow the gun to be fired only when certain conditions are met and auto-recocking to bring the gun back into operational condition after a misfire.

The system has a number of dimmable status lights. Information from these can, as an option, be fed into the aircraft intercom system to provide audio signals.

Contractor

FN HERSTAL.

VERIFIED

GERMANY

Gun control unit for the Tornado

Base Ten builds the electronic control units for the Mauser 27 mm cannon in the Tornado aircraft of the British, German, Italian and Saudi Arabian air forces, governing rate of fire, auto-recocking and purging. It provides a rounds remaining indication to the pilot and has comprehensive built-in test equipment.

Operational status

In service in Panavia Tornado aircraft.

Contractor

Base Ten Systems Electronics GmbH.

VERIFIED

Sidewinder control unit for the Tornado

Base Ten has further developed the standard US system used to control the Sidewinder air-to-air missile, to meet UK requirements for Royal Air Force Tornados GR. Mk 1s. The microprocessor-based missile unit interfaces with the stores management system to provide semi-automatic control of the missile. Features include both normal and off-boresight search and lock, simultaneous missile running, rapid acquisition, automatic selection of optimum missile in the presence of obscuration by the airframe and extensive built-in test equipment.

The current UK missile unit has been developed to enable it to control four missiles. Base Ten's parent company in the USA is providing a new control unit for the operation of four Sidewinders on US Air Force A-10s.

Operational status

The four-missile variant has been supplied to the Royal Air Force, the German Air Force and Royal Saudi Air Force for the Tornado IDS.

Contractor

Base Ten Systems Electronics GmbH.

UPDATED

SMS-50 Stores Management System

The SMS-50 is designed to offer low-cost control of a wide range of weapon types in a single unit for helicopter and fixed-wing activities. Weapon types include gun-pods, rocket systems, air-to-air missiles, bomb release mechanisms and practice bomb dispensers. Meeting a requirement for a stand-alone stores management system, the SMS-50 embodies many features common to other Base Ten systems.

At the core of the SMS-50 is a microprocessor-based Stores Control Unit (SCU) which oversees all control functions in the system. Most operations are carried out under software control, but emergency jettison uses a hard-wired sequencer.

Pilot interaction is completed through the Cockpit Control Unit (CCU) and the SMS-50 can support two CCUs in a variety of configurations. Store and pylon interfaces are wired directly from the SCU.

All three units in the system feature dual-channel architecture to maintain full performance characteristics in the event of any single failure. They maintain immunity against unintended release and prevention of intended release.

CCU switches and indicators are arranged into three main functional groups: store and pylon selectors and indicators, weapon mode selectors and indicators, and system function selectors and indicators.

The SCU processes all pilot selections from the CCU and inputs from the aircraft safety critical switches. All pilot selections are verified before appropriate CCU indications are made. The SCU is responsible for sending arming and single or multiple release signals to pylons. Store status is indicated at all times on the CCU.

A number of optional features is available. Inventory loading and validation is accomplished via the CCU and checked against actual stores present. Stores mode selection allows single, multiple, burst and multiple firing options to be put into use. Stores selection enables different weapons packages to be activated. Stores fuzing helps weapon fuzing to take place in different modes appropriate to the store type. Selective jettison enables stores on one or more pylons to be selected and jettisoned in sequence. Emergency jettison enables all aircraft stores to be safely jettisoned in a predetermined sequence. Safe mode enables the pilot to reset the SMS and all associated stores to a safe state. Comprehensive BIT operates during flight to check functions essential to mission success and safety. Interruptive BIT operates on the ground in an interactive mode to detect all safety critical failures.

Specifications

Dimensions:
(CCU) 76 × 146 × 163 mm
(SCU) 105 × 193 × 321 mm

Weight:
(CCU) 1.5 kg
(SCU) 6 kg

Operational status

Ordered by two overseas customers for Tucano aircraft.

Contractor

Base Ten Systems Electronics GmbH.

VERIFIED

Stores Management System (SMS) for the Tornado

Base Ten builds the Stores Management System (SMS) for the IDS interdictor/strike versions of the Panavia Tornado operated by the German and Italian air forces and the German Navy.

The Base Ten stores management system for the Tornado. Left to right: pylon decoder units, pilot's control unit, weapon processor, navigator's control unit and pylon decoder units. Foreground: pylon decoder unit

The Tornado SMS consists of a weapon processor, pilot's control unit, navigator's control unit and a number of pylon control units. Recently, the system has been updated to provide a MIL-STD-1553B databus capability.

Operational status

In service in German Air Force and Navy and Italian Air Force Panavia IDS Tornado aircraft.

Contractor

Base Ten Systems Electronics GmbH.

VERIFIED

INTERNATIONAL

Weapons interface unit for the Tornado

Teldix, leading an international consortium with GEC-Marconi Avionics and Alenia Difesa, has been awarded a series production contract for the weapons interface unit for the German Avionic Structure Programme as well as for the Royal Air Force Tornado mid-life update. Two weapons interface units will be installed in each aircraft and will provide the interface for transferring high-frequency signals between the aircraft subsystems and the weapon stations. The weapons interface unit, together with the upgraded stores management system which is being delivered by GEC-Marconi Avionics, will give the aircraft MIL-STD-1760 weapon carriage capability.

Operational status

In production.

Contractors

GEC-Marconi Avionics Ltd.

Alenia Difesa, Avionic Systems and Equipment Division, GF Sistemi Avionici.

Teldix GmbH.

UPDATED

ISRAEL

Stores management system

Elbit Systems' stores management system is computerised, flexible, easy to manage and has dual-mode operation for maximum integrity against the accidental release of stores. Computed information is shown on an armament control and display panel which is essentially a multifunction CRT unit. The pilot prepares several weapon release schedules according to his preflight briefing so that, during the attack phase, he merely has to select the appropriate one for the task. The system comprises four units: an armament control and display panel, stores management computer, video electronic unit and an optional video recorder.

The stores management system replaces the switches and selectors which would normally control guns, weapons and fuzes, weapon data, stores release programming, weapon data insertion panel and stores emergency jettison. Preflight selections comprise: master mode for the aircraft's weapon delivery and navigation system, which takes in air-to-air combat, air-to-ground missiles, gun firing, selective jettison and navigation; CCIP, dive/toss, toss and direct weapon delivery mode; type of weapon to be used; single weapon, pair or salvo release mode; release interval and nose, tail or nose and tail fuze mode.

The CRT continuously displays the number, location and type of stores being carried on the mission and is updated as stores are released or jettisoned. It also

Elbit Systems' stores management system display showing the INV (inventory) mode, with store types and positions on the aircraft. The pattern indicates that four AIM-9 Sidewinder missiles and four Mk 82 and SN 82 bombs are carried on wing pylons and an ECM pod on the fuselage centreline station

shows the information appropriate to the particular release programme and warns of incorrect programming, equipment failure or beyond limits conditions. The system computer can also control a videotape recorder for post-mission analysis or debriefing after a training flight.

Specifications

Computer

Dimensions: 321 × 194 × 172 mm
Weight: 13.5 kg
Redundancy: two independent computing systems, dual power supply
Memory: 58 k EPROM, 3 k RAM, 1 k non-volatile RAM
Input/output: 160 discrete inputs, 120 discrete outputs

Armament control and display panel

Dimensions: 172 × 147 × 267 mm
Weight: 7 kg
Scanning: 525 or 875 lines
Bandwidth: 20 MHz
Contrast ratio: 7:1

Operational status

In production and in service in the Kfir, MiG-21, A-4 Skyhawk, F-4, F-5, and in other aircraft exported from Israel.

Contractor

Elbit Systems Ltd.

UPDATED

ITALY

Stores Management System (SMS) for the AMX

Alenia has developed the stores management system for the AMX aircraft which performs the automatic control of selection, arming, release and jettison of the aircraft's external stores.

The system consists of seven LRUs and includes a weapon processor unit containing an advanced computer which manages all the logic operations and controls the other units. There are also five station units which control guided and unguided weapons and a weapons control panel on which the crew selects the various functions and which displays the stores status. The system can be easily configured for different weapons.

From this project, Alenia is working on a more advanced stores management system which uses the cockpit multifunction display to show and control the various functions and which has a MIL-STD-1760A interface.

Specifications

Dimensions:
(weapon processor unit) ¾ ATR short
(weapons control panel) 190 × 90 × 180 mm

Weight:
(weapon processor unit) 15 kg
(weapons control panel) 5 kg
(Station unit × 5) 3 kg
Power supply: 28 V DC

Operational status
In production for the AMX.

Contractor
Alenia Difesa, Avionic Systems and Equipment Division.

UPDATED

The Alenia stores management system fitted to the AMX showing the weapons control panel (left), the weapon processor unit (centre) and three station units (right)

SOUTH AFRICA

Stores Management System (SMS)

The SMS consists of a Weapons Computer (WEAC) and a Stores Power Unit (SPU). The WEAC is a remote terminal to the mission computer and a bus controller to the weapons MIL-STD-1553B bus. Full video and power and data switching are performed by the SMS and interfaced to the rest of the aircraft systems. MIL-STD-1760A is fully implemented, permitting the coupling of any compatible weapon. The SMS is typically implemented as a dual WEAC/dual SPU system, with complete intercommunication permitting full dual redundancy.

Application software is executed on a 32-bit RISC processor. All input/output is managed by the processor. A PC-based applications software development environment is fully implemented and validated. Extensive BIT ensures failure detection to SRU level of better than 95 per cent.

Specifications
Dimensions:
(WEAC) 330 × 129 × 197 mm
(SPU) 330 × 129 × 197 mm
Weight:
(WEAC) 6 kg
(SPU) 8 kg
Power:
(WEAC) 50 W
(SPU) 70 W
Reliability: 1,000 h MTBF

Operational status
Flight tested.

Contractor
Advanced Technologies & Engineering Co (ATE).

VERIFIED

SPAIN

SCAR armament control system

The SCAR (Sistema de Control de Armamento) armament control system provides full weapon selection and control facilities for up to six underwing carriage stations and a centrally located gunpod. The system's functions include selection of weapon station for use in a mission, selection of weapon type fitted at a station, mode selection for bomb arming, time interval selection for bomb release in ripple mode, air-to-air and air-to-ground gunnery selection, emergency mode selective jettison, weapon availability and status, stores selected and stores remaining indication and status and gun ammunition counter indication.

Designed principally for the CASA C-101, the SCAR system provides weapon selection and control facilities for bombs, rockets, guns and combined stores dispenser/launcher units and also supplies signals for control of weapon sighting and gun camera systems.

The equipment comprises a central processor, usually located in the aircraft's nose bay, a main unit, monitoring unit and emergency unit installed in the cockpit. In the case of trainer aircraft, additional monitoring and emergency units, identical to those in the attack aircraft, are mounted in the student's cockpit.

The main unit is used for all control and selection functions, with the exception of emergency jettison for which the emergency unit is used. Monitoring units indicate the status and type of stores available and the quantities of stores or ammunition remaining or expended. In the training role, the functions of the student's portion of the system can be completely inhibited by the instructor pressing a button. An error light indicates a system malfunction or a mis-selection of a store or weapon. In the case of mis-selection, the system, with the exception of emergency release, is automatically rendered inoperative. The emergency unit is hard-wired to ensure complete reliability and contains a test-lamp push-button for status checking of all indicator lamps in the system.

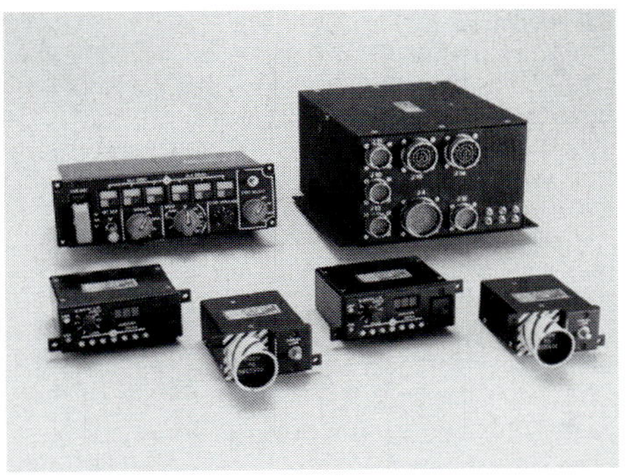

The SCAR armament control system with (rear left) the main unit, (rear right) the central processor and (front) two monitor units and two emergency units

SCAR is a microprocessor-controlled system employing digital techniques and TTL integrated circuitry. Operational parameters can be modified via software changes and an EAROM memory ensures that stored data is retained even when the system is unpowered. The system is of modular design and conforms to the recommendations of MIL-STD-1472 and MIL-HDBK-244. It can be readily adapted to different types of aircraft by changing module input/output interfaces. An interface to a centralised WAC, MFD, HUD control/display system has been incorporated via discrete disturbance-free output lines. Compatibility with previous self-supporting systems is maximised as the new interface is provided through the central processor connector previously dedicated to the front cockpit monitoring unit and redundant in the new configuration.

The new interface lines provide not only all status information included in the substituted monitoring unit, including gun rounds, but also ripple interval status and certain control capabilities for remote gunnery selection (HOTAS interface). Power requirements are 2.5 A maximum for an illuminated panel system or 1.6 A maximum for a non-illuminated system, both at 28 V DC.

Specifications
Dimensions:
(central processor) 180 × 210 × 120 mm
(main unit) 200 × 70 × 101 mm
(monitor unit) 100 × 42 × 75 mm
(emergency unit) 70 × 40 × 81 mm
Weight:
(central processor) 3.3 kg
(main unit) 1.62 kg
(monitor unit) 0.46 kg
(emergency unit) 0.28 kg

Operational status
In service.

Contractor
Tecnobit SA.

VERIFIED

SWEDEN

JAS 39 Gripen stores management unit

The stores management unit distributes and interfaces signals between the various weapons and tactical systems in the aircraft. It is based on a 486 computer module system. The unit is partly operated from the main mode selector, status panel and target acquisition panel. The unit also handles the preparation of stores, functions for firing and releasing stores in co-operation with the system computer, stores monitoring, control of the gun and control of countermeasures.

The stores management unit features modular design, LRU concept to facilitate maintenance, chassis integrated power supply for optimal heat dissipation, SMT for increased packing density and built-in redundancy.

Operational status

In production for the JAS 39 Gripen.

Contractor

Ericsson Saab Avionics AB.

UPDATED

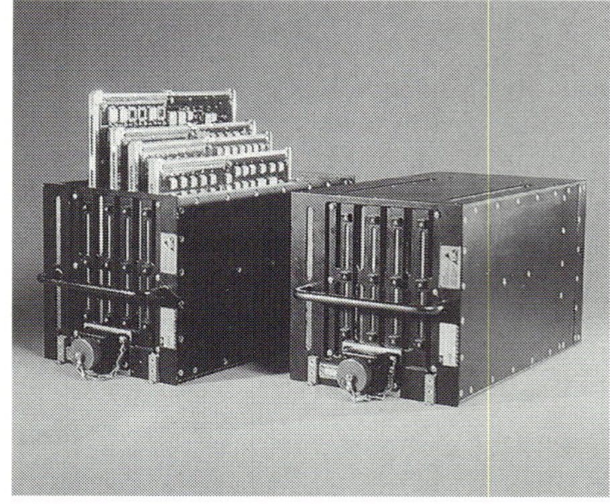

The stores management unit (left) and aircraft interface unit (right) for the JAS 39 Gripen
1995

UNITED KINGDOM

SMS 2000 series stores management systems

Computing Devices offers several variants of the SMS 2000 family of airborne stores management systems derived from the SMS 2100 system fitted to early export marks of the BAe Hawk aircraft. The SMS 2100 is particularly suitable for fixed-wing aircraft with up to nine weapon stations, plus internal cannon. The SMS 2112 and 2120 are currently standard equipment on Mks 60, 100 and 200 export Hawk aircraft. The SMS 2200 is a dual-channel system for close integration with the mission avionics and is optimised for naval ASW aircraft and helicopters; a variant is standard equipment on the EH 101 helicopter for the Royal Navy and Italian MMI. The SMS 2600 is a simple system for small aircraft with, typically, four weapon stations using modules from the SMS 2120. Its capabilities can be expanded to match the SMS 2120.

The SMS 2000 family controls the complete range of stores carried by helicopters and light- and medium-sized aircraft, including torpedoes, mines, buoys, rockets, air-to-air and air-to-ground guided missiles, weapon and equipment pods and drop tanks. A system typically comprises a digital stores management computer, remote switching units mounted in the pylons or sponsons and cockpit control panels for those aircraft requiring dedicated crew interfacing.

The system permits in-flight preselection of a weapon delivery package to suit particular targets or conditions. Additional facilities are provided to control the release modes and, where necessary, ground spacing optimisation in relation to aircraft speed or fuzing optimisation in relation to altitude. Weapon failures for whatever reason are indicated on the SMS panel(s) and the presentations are optimised for ease of assimilation by the aircrew. Each system consists of a Weapon Processing Unit (WPU), Weapon Control Panel (WCP) and Stores Interface Units (SIUs) or pylon interface units.

The SMS computer maintains the centre of gravity and/or lateral balance within the aircraft's handling constraints. The system contains, as a minimum, duplex emergency jettison control, so that no single defect can prevent safe jettison when it is demanded. Multiple interlocks ensure the highest levels of safety; the SMS 2200 has two channels in a hot standby configuration further to ensure availability of all functions. Avionics databus interfaces to MIL-STD-1553B are fitted to some configurations, allowing an enhanced interface with a nav/attack computer.

Modular construction allows the SMS 2000 to be upgraded in service; current models include a high degree of BIT. Fault location is further enhanced by the use of Computing Devices automatic armament test equipment.

Specifications

Dimensions:
(SMS 2100 WPU) 220 × 130 × 300 mm
(SMS 2200 WPU) 190 × 197 × 356 mm
(SMS 2600 WPU) 20 × 130 × 300 mm
(SMS 2100/2200/2600 WCP) 145 × 165 ×165 mm
(SMS 2200 SIU) 170 × 57 × 235 mm
(SMS 2100/2600 pylon interface unit) 170 × 57 × 235 mm
Weight:
(SMS 2100 WPU) 9 kg
(SMS 2200 WPU) 13 kg, (SMS 2600 WPU) 6 kg
(SMS 2100/2200/2600 WCP) 2 kg
(SMS 2200 SIU) 2.2 kg
(SMS 2100/2600 pylon interface unit) 1.5 kg
Power supply: 28 V DC
(SMS 2100) 35 W
(SMS 2200) 59 W
(SMS 2600) 29 W

Operational status

The SMS 2112/2120 are in service and in production. SMS 2100, SMS 2112 and SMS 2120 are standard equipment on BAe Hawk Mks 50, 60, 100 and 200 aircraft.

The SMS 2200 is developed and entering production. A variant has been selected for the EH 101 helicopter for the Royal Navy and the Italian MMI.

The SMS 2600 is in development.

Contractor

Computing Devices Company Ltd.

VERIFIED

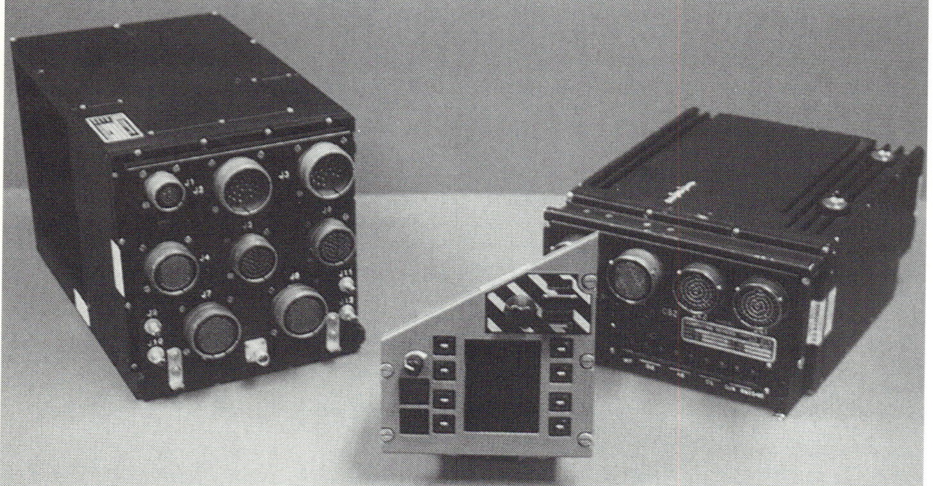

The weapon processing units and weapon control panel of the SMS 2200 and SMS 2100 stores management systems
1995

SMS 3000 series stores management systems

SMS 3000 stores management systems are based largely on the experience gained with the company's SMS 2000 Series, but offer greater flexibility in heavier aircraft.

The SMS 3000 Series has a dual mode of operation which has a two-fold intent: maximising safety by reducing the risk of accidental release, and improving mission reliability. This involves the employment of a dual-redundant channel system which operates in serial form to release stores in the safety mode. When in combat, one of two parallel systems can be used to effect release. There is provision for expansion by means of a range of modular stores interface units to match the increased processor capacity.

Operational status

In service in all Panavia Tornado ADV aircraft.

Contractor

Computing Devices Company Ltd.

VERIFIED

Panavia Tornado ADV aircraft have the Computing Devices SMS 3000 stores management system

SMS 4000 series stores management systems

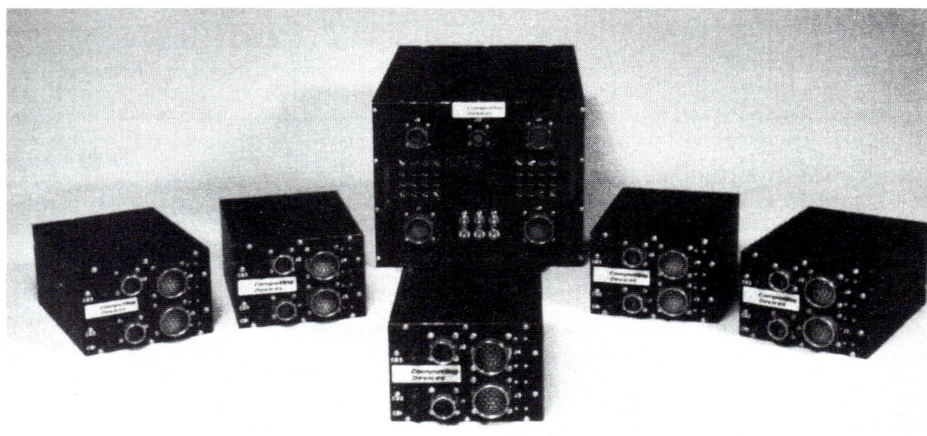

Units of the Computing Devices SMS 4000 stores management system

The SMS 4000 Series stores management systems cover the full range of stores applicable to heavier aircraft, especially to those based around the MIL-STD-1553 multiplex serial databus and MIL-STD-1760 interface standard.

The SMS 4000 hardware and software architecture has been optimised to meet the five key needs for modern weapon management. These are full implementation of MIL-STD-1760/STANAG 3837AA for the standard interface of stores, compatibility with existing non-standard stores such as Sidewinder, integration with MIL-STD-1553 and high-speed databus avionics where no dedicated SMS control panels are implemented, use of high-level languages such as Ada and use of standard databusses for communication between separate SMS equipments.

The SMS 4000's four main functional elements are safety critical processing, non-safety critical processing, critical signal switching and high/low-bandwidth signal networking. It is expected that at least two of these functions would be incorporated into any SMS 4000 configuration. Processing equipment is designed for installation in conditioned bays; critical switching and networking equipments are designed for harsher environments, including installation in removable launchers, dispensers and pylons.

Separation of safety critical and non-safety critical processing is a key feature of the SMS 4000 architecture, enabling achievement of the high data integrity required by MIL-STD-1760. Physical separation of the two types of processing - critical decisions from targeting data processing - is sometimes required. The optimum implementation also depends on any avionics software support environment required by the user. The SMS processor (s) manages all weapon aspects up to and including release/launch/firing, maintaining high-integrity levels at all phases.

Specifications

Dimensions:
(distribution unit) 290 × 210 × 130 mm
(armament control panel) 515 × 190 × 216 mm
(store/station equipment) 300 × 100 × 100 mm
Weight:
(distribution unit) 5 kg
(armament control panel) 10.5 kg
(store/station equipment) 4 kg
Power:
(distribution unit) 30 W
(armament control unit) 115 W
(stores/station equipment) 30 W

Operational status

In development. Advanced systems have been delivered for government development of MIL-STD-1760. A variant of the SMS 4000 has been selected for the EF 2000 armament system processors and signal distribution unit, and is nearing development completion. Safety critical databus-controlled power switching units of the SMS 4000 range have been selected for fitting to the EF 2000 decoy dispensers.

Contractor

Computing Devices Company Ltd.

VERIFIED

Stores Management System (SMS) for the Tornado GR. Mk 1

The SMS developed by GEC-Marconi Avionics for the Royal Air Force Panavia Tornado GR. Mk 1 aircraft can handle 26 types of weapon, controlling arming, fuzing, firing or release functions from its 35 outputs. The system also initiates the timed and sequenced emergency and selective jettison of all weapons and other stores such as fuel and equipment pods. This is a controlled output of 29 pulses over 2.2 seconds. A comprehensive bogus weapon inventory is also included for training purposes.

Among the principal design aims of the Tornado stores management system are: meeting the specified rates of safety, jettison availability and mission success, an unprecedented degree of electromagnetic compatibility hardness and a significant reduction in crew workload. The system is engineered to provide a minimum of interconnect wiring between the 12 units.

To achieve these targets, the company developed a dual-channel digital architecture system employing mini-computers and a digital data transmission system to provide the required integrity. Wherever possible, CMOS circuitry is used on account of its low power consumption, to eliminate the need for cooling and to meet the electromagnetic compatibility specification. Both channels are fully synchronised and perform in a consolidated mode until channel unserviceability is detected, when authority is given to the serviceable channel to proceed independently.

The level of redundancy ensures that the channels only close down for safety critical failures and that the emergency jettison system is not only separated from processor control but made duplex to avoid any consolidation requirements.

Weapon control is split into defensive and offensive weapons. The pilot's control panel takes the defensive weapons, and the navigator's control panel has the others. Crew workload is minimised by eliminating from the weapon package selection process both safety critical and non-applicable options.

Operational status

In service on Royal Air Force Panavia Tornado GR. Mk 1 aircraft.

Contractor

GEC-Marconi Avionics Ltd.

UPDATED

Stores management system for the Tornado GR. Mk 4

GEC-Marconi Avionics is developing the stores management system for the Tornado GR. Mk 4 as an upgrade to that which is currently used on the Royal Air Force Tornado GR. Mk 1. It involves two upgraded LRUs within the system and further enhances the features of the current Tornado stores management system. Major enhancements are the use of Ada high-order language in conjunction with MC68020 processing capability. MIL-STD-1553B interfaces have been introduced to cater for the new generation of weapons now becoming available. The stores management system now possesses a back-up bus control expansion capability.

The navigator's control panel is being upgraded with a custom switch and display interface offering a multifunction soft key capability with further reduced crew workload features and enhanced night vision capability.

Operational status

In production for the Royal Air Force Panavia Tornado GR. Mk 4.

Contractor

GEC-Marconi Avionics Ltd.

UPDATED

Weapon control system for the Harrier

The Harrier weapon control system is digitally based and makes extensive use of integrated circuitry. Logic circuits are of diode-transistor type and the number of relays has been reduced to the minimum required to handle the heavy switching currents which occur at the time of weapon release.

When stores are loaded on to the aircraft, thumbwheel switches are set on the control panel to provide the pilot with an indication of the external stores status; the role selection switches are also set for the type of stores carried. Two modes of operation, manual or automatic, are possible. In the manual mode, release intervals of 20, 40, 80, 160 and 320 milliseconds are selectable from a rotary switch on the weapon control system panel.

For automatic release, the auto/manual switch interfaces the weapon control system with the aircraft's nav/attack computer providing it with data on stores status, types selected and spacing required. Release interval then comes under computer control. In either mode, the pilot may use individual or 'clear aircraft' overriding jettison control switches to jettison stores separately or as a single salvo in an emergency.

During rocket or guided weapon release or gun firing, the weapon control system ensures that power is supplied to the engine igniter circuits to safeguard against flame-out.

In two-seat Harriers a monitor unit in the rear cockpit duplicates the front cockpit jettison facilities and enables the control panel to be monitored during operation.

The system's capacity permits mixed loads to be carried and controlled on each pylon. The range of stores includes single or twin bombs, flares, one or two rocket launchers, two gun pods, external fuel tanks or guided weapons. Bomb fuzing signals can be transmitted to the five pylons.

Specifications

Dimensions:
(monitor) 114 × 210 × 160 mm
(weapon control system unit) 114 × 210 × 321 mm
Weights:
(monitor) 2.38 kg
(weapon control system unit) 4.65 kg

Operational status

In service on the Harrier. More than 200 units have been procured.

Contractor

GEC-Marconi Electro Optics Limited, Airadio Division, Portsmouth.

VERIFIED

Weapon control system for the Jaguar

The GEC-Marconi weapon control system for the Sepecat Jaguar is a development of that provided for the Harrier (see previous entry) and employs the same digital solid-state electronics technology. In the single-seat Jaguar the system is divided into two major elements: the main unit housing the control logic and the cockpit-mounted control unit which contains the indicator lamps, the pilot-operated armament switches and the individual jettison switches. In the two-seat variant a monitor unit is fitted in the rear cockpit for training purposes.

As well as providing all the functions of the Harrier system, the Jaguar unit has greater reliability as the result of wider use of solid-state logic. It provides fully duplicated release and jettison of individual stores, dispenses with autoselector switches on tandem store carriers and provides a more advanced method of rocket firing control. It may be operated as an autonomous system or used in conjunction with the Jaguar nav/attack computer.

Specifications

Dimensions:
(control unit) 152 × 146 × 157 mm
(logic unit) 155 × 349 × 425 mm
(monitor) 85 × 146 × 130 mm
Weights:
(control unit) 1.76 kg
(logic unit) 14.54 kg
(monitor) 0.56 kg

Operational status

In service.

Contractor

GEC-Marconi Electro Optics Limited, Airadio Division, Portsmouth

VERIFIED

SGS-10 general purpose stores management system

SGS-10 is a low-cost and flexible stores management system which meets the requirements of the latest generation of light combat aircraft armed with modern weapons. It can drive up to nine weapon stations and can be configured to suit a variety of stores.

The modular concept enables a wide range of options and variants to be offered so that systems can be selected to meet the needs of retrofit programmes and the requirements of new build aircraft. All variants offer the highest standards of safe operation while innovative designs provide cost-effective system solutions.

The baseline system configuration achieves a high degree of safety without the added cost and complexity of full duplication. It is controlled by a single microprocessor featuring duplex safety critical interfaces and an innovative software technique. A back-up capability, independent of the primary electronics and power supplies, is provided to handle release of certain self-defence weapons in a reversionary mode. Independent emergency jettison circuits are included. An integral part of the design is a control and display panel which features software-configured LED displays. This SGS-10 baseline system configuration has been selected for the Sea Harrier F/A-2 programme.

Operational status

In service on Sea Harrier F/A-2.

Contractor

Smiths Industries Aerospace.

VERIFIED

UNITED STATES OF AMERICA

ARmament Control And Delivery System (ARCADS)

The AlliedSignal ARCADS has been developed to provide symmetrical release external stores management for the US Army Bell AH-1J light attack helicopter. It permits selective programming and controlled release of stores from between one and four external stations and can be indexed and programmed for up to 12 separate stores. Control functions comprise stores selection, quantity for release, rate of release, single, pair or all mode of release and emergency jettison. The system displays to the flight crew the type and quantity of stores that have been selected, the desired rate of release (slow or fast), the selected mode of release (single, pair or all), stores low quantity, stores empty and co-pilot override.

A complete ARCADS comprises seven main units: cockpit stores control panel, emergency jettison control unit, emergency jettison switch, two wing rocket delivery units and two wing adaptor cable assemblies. The units are of modular construction and modules are readily replaceable to allow for changes of weapon complement. The system contains safety interlocks and built-in safeguards to ensure symmetrical release of stores.

Specifications

Dimensions:
(stores control panel) 146 × 114 × 165 mm
(emergency jettison select) 38 × 146 × 129 mm
(wing rocket delivery) 137 × 193 × 51 mm
Weight:
(stores control panel) 2.4 kg
(emergency jettison select) 0.54 kg
(wing rocket delivery) 2.36 kg

Operational status

In production and in service in the US Army Bell AH-1J helicopter.

Contractor

AlliedSignal Flight Systems.

VERIFIED

Integrated Conventional Stores Management System/Global Positioning System (ICSM/GPS) for the B-52H

In early 1993, The Boeing Company, Product Support Division in Wichita was awarded a $25.1 million contract by the US Air Force Aeronautical Systems Division for work on the B-52H bomber ICSM/GPS. The Conventional Enhancement Modification (CEM) programme provides standard electrical and software interfaces for all future weapons specified for the B-52H, using the MIL-STD-1760 protocol. The system permits the crew to load smart missiles with details of route and target, and also allows the weapon load of a B-52 to be changed from one flight to the next with only a software change loaded into the offensive avionics suite using a tape cassette. Integration of GPS provides precision navigational capability which greatly improves weapon delivery accuracy.

Operational status

Initial operational capability for the B-52H was achieved in December 1994.

Plans to add the ICSM/GPS capability for additional B-52Hs are in hand.

Contractor

Boeing, Product Support Division.

UPDATED

AN/ASQ-165 Armament Control Indicator Set (ACIS)

The AN/ASQ-165 ACIS is an airborne ordnance stores management system developed for the US Navy Sikorsky SH-60B LAMPS III helicopter. The system controls one or two Mk 46 Mod 0/1/2 torpedoes and up to 25 air-launched sonobuoys. Interfaces for control functions and for the ACIS microprocessor-control system are provided for two BRU-14A bomb racks, the sonobuoy launcher and a MIL-STD-1553 databus which interfaces in turn with an AN/AYK-14 airborne computer.

The system consists of a C-10488/ASQ-165 Armament Control Indicator (ACI) for the airborne tactical operator or the helicopter pilot and a CV-3531/ASQ-165 Armament Signal Data Converter (ASDC) which contains logic, interlock, driver and relay circuits. The microprocessor controls all tasks such as inventory usage, functional status and built-in self-test, with the exception of the jettison function which is hard-wired. Jettison functions are redundant and the circuitry ensures that no single failure will cause inadvertent release or inability to jettison. Communication with all parts of the system apart from the jettison function is provided by the databus.

The ACI control panel functions include: master arm voltage switching for positive armament safety; torpedo selection, arming and launch; torpedo search depth, mode, ceiling and course programming; three torpedo programme status indicators; sonobuoy selection, arming and launch; and jettison left store, right store, all sonobuoys or all ordnance.

The AN/ASQ-165 precludes unintentional launching from single point failures. The system features low-power Schottky TTL 54-LS series logic and tri-state, sunlight-readable switches and indicators. With the exception of some launch and jettison circuits, the self-test facility is quoted as being able to detect more than 98 per cent of faults. The system is compatible with automatic test equipment at both unit and module level. Failure rates for the ACI and the ASDC are stated as better than one in 25,700 hours and 3,850 hours respectively.

Specifications

Dimensions:
(ACI) 219 × 146 × 170 mm
(ASDC) 193 × 149 × 457 mm
Weight:
(ACI) 2.04 kg
(ASDC) 11.05 kg

Operational status

No longer in production. In service in US Navy Sikorsky SH-60B LAMPS III helicopters.

Contractor

Fairchild Defense OSC.

VERIFIED

AN/AWG-15 Armament Control System (ACS) for the F-14A

The AN/AWG-15 ACS for the US Navy F-14A/A+ was designed and developed for weapon stores control to be implemented from a control indicator by an aircrew member or in conjunction with an Airborne Missile Control System (AMCS) computer. The system comprises a Control Indicator (CI), a Power Switching Unit (PSU) and a set of Signal Command Decoders (SCDs), one of which is located at each weapon station.

The ACS provides the preparation and control functions before releasing Phoenix, Sparrow and Sidewinder missiles or bombs, rockets and flares. It also controls and fires the M61-A1 gun.

For air-to-air functions, the system provides missile options, control/speed gate selection, inventory/ready control, weapon/station control and release by the pilot or navigator. For air-to-ground functions, using bombs, rockets or flares, it provides weapon and station selection, inventory/ready control, attack, delivery, fuzing and weapon mix mode selection and release by the pilot. It also provides jettison command functions. Protective logic circuitry is incorporated to prevent inadvertent weapon firing or release.

The system has a high level of inbuilt redundancy with BIT for 90 per cent fault detection and automatic isolation of faulty circuits, sub-units or components.

Specifications

Dimensions:
(CI) 318 × 292 × 254 mm
(PSU) 203 × 178 × 419 mm
(SCDA) 64 × 102 × 203 mm
(SCDB) 51 × 102 × 152 mm
Weight:
(CI) 6.13 kg
(PSU) 9.09 kg
(SCDA) 0.85 kg
(SCDB) 0.5 kg

Operational status

In September 1986, Fairchild Defense was awarded US$23.1 million by the US Naval Air Systems Command to modify the Navy's AN/AWG-15E weapon fire-control system in its F-14 aircraft. The upgraded system is designated AN/AWG-15F and over 470 aircraft were modified; electronics in the control indicator and part of the power switching units were replaced by items of more modern technology.

Earlier, in May 1985, Fairchild Defense was awarded a full-scale development contract by Northrop Grumman for the AN/AWG-15 Technology Improvement Programme (TIP) and a prototype of the new system was flight-tested in August 1986.

The first upgraded units were delivered to the navy at the end of 1987, and deliveries completed in 1989.

No longer in production. In service in US Navy F-14A and F-14A+ Tomcats.

Contractor

Fairchild Defense OSC.

UPDATED

Armament Control System (ACS) for the A-10A

The A-10A ACS provides the pilot with stores status information and programmed control over the operation and release modes of all of them. The ACS comprises three separate types of LRU. The Armament Control Panel (ACP) permits the pilot to communicate with the system, the Interstation Control Unit (ICU) carries out signal processing and the Station Control Units (SCUs) act as the interfaces between the system and the weapon stations. A complete system consists of one control panel, one interstation and 11 station control units.

The ACS allows the pilot to select the weapon stores at any of the 11 weapon stations at any time. The GAU-8A gun and camera may be selected for operation, together with several types of weapon fuzing option, by means of a switch. Single pairs, ripple-single and ripple-pairs are the possible weapon-release mode choices. Selective rack and missile jettison are the choices available with the emergency jettison control. A built-in timing and sequencing system generates the appropriate time intervals between releases, ensuring safe release intervals for weapons and the appropriate weapon station priorities. Video signal channelling assists in the release of electro-optical weapon stores. A front panel displays the current status of each of the 11 stations and the quantity of ammunition remaining in the GAU-8A gun. The system can operate over a temperature range from −55 to +71°C. Microprocessor-controlled test equipment is used for flight line and intermediate level support.

Specifications

Dimensions:
(ICU) 241 × 388 × 216 mm
(ACP) 165 × 235 × 192 mm
(SCU types A and B) 95 × 142 × 71 mm
Weight:
(ICU) 14.7 kg
(ACP) 4.88 kg
(SCU type A) 1 kg
(SCU type B) 0.63 kg

Operational status

In service on the US Air Force A-10. No longer in production.

Contractor

Fairchild Defense OSC.

VERIFIED

The US Navy F-14 has the AN/AWG-15 armament control system

Tactical Weapons Management System (TWMS)

The Rockwell Collins Tactical Weapons Management System (TWMS) was originally designed in 1985 for the US Army JOH-58C scout helicopter programme. Since then, the TWMS, also known as Quick Draw, has been expanded and enhanced to address a wide range of weapons for attack and scout/attack helicopters. This system not only provides the crew with control and display capability for communication and navigation management but also acts as the control point for the weapon systems. It is capable of managing multiple rocket, gun and missile types, including air-to-air Stinger, TOW and Hellfire.

Head-up management of weapons and communications and navigation avionics is provided by the Hands On Throttle And Stick (HOTAS) feature. Such fingertip control improves weapons delivery and mission safety and effectiveness. Weapons stores management, as well as weapons selection and sequencing, is also provided.

There is no need to enter weapon type information manually into the system because it is automatically fed the data as weapons are attached to the stores stations.

Units of the Tactical Weapons Management System: from left to right, the weapons relay unit, weapons management unit, master arm/emergency panel and control display unit

TWMS can be installed as a stand-alone system or integrated with the CMS-80 cockpit management system.

Operational status

Originally designed for the JOH-58C helicopter, the TWMS is now being developed for the Bell 406CS helicopter. A Bell helicopter 406 Combat Scout equipped with the TWMS has successfully completed live firing tests with guns and rockets against ground targets and the system is also flying on board the MG 530 NOTAR.

Contractor

Rockwell Collins.

UPDATED

AN/AYQ-9(V) weapons control and management system for F/A-18

The AYQ-9(V) weapons control and management system provides fully computerised control of stores release to optimise accuracy and minimise crew workload. Although available for a variety of attack aircraft, it is standard equipment on the US Navy F/A-18 aircraft, and is claimed to be the most advanced system of its kind. Functions in the F/A-18 include maintaining an inventory of stores types, locations, quantity, status and special conditions such as a hung store or locked pylon for correct sequencing and display on the cockpit CRT; preparation and activation of store and store release mechanism; deactivation of stores after completion of attacks and self-testing to diagnose failures in the system.

Fifty types of weapons on the F/A-18 can be controlled by the AYQ-9(V) including AMRAAM, Sidewinder, Sparrow, Maverick, Harpoon, SLAM and HARM missiles, several types of nuclear, free-fall, retarded and smart bombs, FLIR cameras and the aircraft guns. Eleven types of mines can also be carried.

The AN/AYQ-9(V) system consists of the processor and control unit, together with decoders at each of eight weapons stations. It interfaces with the AN/AYK-14 mission computer via a MIL-STD-1553B digital databus and a dedicated -1553B armament control bus connects the processor with the various decoders.

The AN/AYQ-9(V) provides for the operational readiness assessment of each store and weapons station, prepares the suspension and release equipment (including power-up, alignment and fuzing) and activates the stores sequencing, arming and release, including the off-boresight guidance of Sparrow and Sidewinder missiles.

The system also provides a stores inventory facility by type, number, location and so on, to permit correct release and also for the display of requisite data to the pilot. There is an extensive built-in test capability.

Stores Management Upgrade F/A-18E/F

The latest aircraft version, the F/A-18E/F, made its first flight on 29 November 1995 with Smiths Industries' Stores Management Upgrade System aboard. The weapons system in the F/A-18 interfaces with the aircraft's mission computers via a dual MIL-STD-1553 digital databus and controls 88 different weapons and stores. The system for the new F/A-18E/F is lighter, more reliable and more easily maintainable than the earlier system. Each aircraft will have a stores management processor unit and nine control units.

Operational status

In test on the F/A-18E/F Super Hornet at US Navy Air Station Patuxent River. So far the F/A-18E/F has fired: AIM-7 Sidewinder; AIM-120 AMRAAM; ALE-47 flares; ALE-50 towed decoy; it has also launched: SLAM missile; Harpoon missile; a ripple of 10 Mk 82 bombs; Mk 83 bombs; 480 gallon tanks, from both the aircraft's centreline position and wing stations; and dual loads of CBU-100 (Rockeyes).

Contractor

Smiths Industries Aerospace.

UPDATED

A two seat Super Hornet, designated F/A-18F2, launching AMRAAM ***1998***/0018150

F/A-18 E/F stores management system ***1998***/0018151

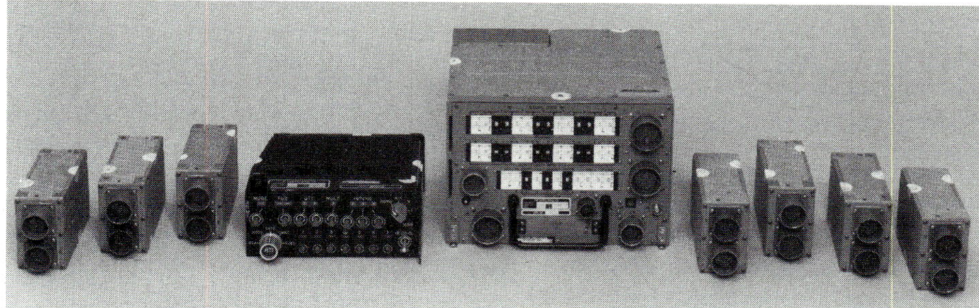

The Smiths Industries' AN/AYQ-13 weapons control and management system for the AV-8B

AN/AYQ-13 weapons control and management system for the AV-8B

The AN/AYQ-13 weapons control and management system is a development of the AYQ-9 system optimised for the US Marine Corps AV-8B and the Royal Air Force Harrier GR. Mk 7 aircraft.

As well as the primary functions for the AYQ-9 system, the AYQ-13 interfaces with and controls a 25 mm gun and provides for the control of stores at seven stations. Additionally, the system allows for the emergency jettison of all or a selection of the stores being carried.

For the AV-8B, the AYQ-13 is compatible with 45 types of stores including Maverick and Sidewinder missiles, laser-guided, cluster, practice, anti-tank, fire and conventional bombs, rocket launchers, decoy dispensers, the aircraft gun and external fuel tanks.

Additionally the AYQ-13 maintains an inventory of stores carried by type, location and quantity: it also takes account of special conditions such as minimum release interval, hung store, weapon gone and rack locked. This is to provide correct release sequencing and the required data for the cockpit display. The system controls the preparation of the suspension and release equipment, power-up, alignment and fuzing and the dynamic power management of the SMS: it also provides for instant deactivation of stores on landing or when a change of circumstances in flight demands. The AN/AYQ-13 has a comprehensive built-in test facility and a pilot training mode for simulation of weapons release.

The system comprises nine weapon-replaceable assemblies, seven station control units, an armament control panel and a stores management processor.

Operational status

In service in US Marine Corps AV-8B aircraft. Smiths Industries Aerospace is to provide an update for the AN/AYQ-13 system under programme Open Systems Core Avionics Requirements (OSCAR), which will include COTS hardware, together with software updates. The update is in development.

Contractor

Smiths Industries Aerospace.

VERIFIED

AN/AYQ-15 Weapons Control and Management System (WCMS) for the F-14D

The AN/AYQ-15 WCMS has a high degree of commonality with the WCMS on board the F/A-18. The stores management processor in the F-14 is similar to that in the F/A-18 and differs only in the removal of one card and the addition of two others to make up the F/A-18 processor. The AN/AYQ-15 controls AIM-54 Phoenix, AIM-7 Sparrow, AIM-9 Sidewinder and AIM-120 AMRAAM missiles, fuel tanks, bombs and the M61-A1 gun.

The F-14D WCMS consists of 15 Weapon Replaceable Assemblies (WRA): the microprocessor-controlled Stores Management Processor (SMP), 10 station decoders/encoders, a gun decoder, a missile power relay unit and two fuel tank jettison units.

The SMP interfaces through the MIL-STD-1553B bus with the aircraft's AN/AYK-14(V) mission computer and the cockpit displays. The various WRAs communicate via a 1553B armament control bus controlled by the SMP. The software within the SMP permits the easy introduction of new stores and weapons in the future.

Operational status

In production and in service in the US Navy's F-14D aircraft.

Contractor

Smiths Industries Aerospace.

VERIFIED

The Smiths Industries' IAMS mounted above the TOW missile sight on the MD 530 helicopter instrument panel

Integrated Armament Management System (IAMS)

The Smiths Industries' IAMS was designed to move helicopter weapons management into the digital computer age.

IAMS provides significant advances in aircraft survivability and ease of operation and maintenance. Utilising a single-chip microprocessor with solid-state memory for system control, the system permits the pilot to carry out any one of four preprogrammed attacks for each of the weapon types on board. This can be done without having to remove hands from the flight controls.

IAMS is composed of two major subassemblies: the armament control panel, containing controls and displays along with the microprocessor, and the stations interface unit containing the high-current switching and weapons interface.

The system is applicable to all types of new light attack fixed-wing aircraft and helicopters, or for modifications or retrofits, and is suited for single or multicrew cockpits.

Operational status

IAMS is currently installed in the Sikorsky H-76 Eagle and UH-60A Black Hawk and on The Boeing Company 500 Defender helicopter.

Contractor

Smiths Industries Aerospace.

VERIFIED

Pylon Interface Unit (PIU)

The PIU supplies the weapons control and management functions for stores recognition, target acquisition and tracking, weapons selection, aiming, fuzing, and release for the Longbow Apache. The Load Maintenance Panel provides the ground maintenance crew with the capability to position the pylons for loading, simulation of airborne conditions for aircraft ground testing, and entry of rocket type loaded within each zone of the rocket pod.

The PIU is an intelligent MIL-STD-1553B remote terminal with a dedicated MIL-STD-1750 processor programmed with Ada software. It consists of six electronic subassemblies designed to reduce supportability costs.

Operational status

Smiths Industries Aerospace has been awarded two contracts by Boeing Helicopter Systems, Mesa, Arizona as part of the Apache modernisation programme. The contracts include supplying four Pylon Interface Units and one Load Maintenance Panel for each US Army Apache to the AH-64D configuration.

Contractor

Smiths Industries Aerospace.

VERIFIED

Weapons control and management system for Nimrod MRA.4

The Nimrod MRA.4 weapons control and management system will provide integrated control of air-to-surface weapons and include provisions for future weapons systems.

Smiths Industries Aerospace is also providing navigation and flight management systems, and components of the utilities system, to which the weapons control and management system will interface.

Operational status

In development. Deliveries are scheduled to begin in 1999.

Contractor

Smiths Industries Aerospace.

NEW ENTRY

Smiths Industries Aerospace upgrades the US Army Apache to the AH-64D configuration providing the weapons control and management system function
1996

ADDENDA

COMMUNICATIONS

DENMARK

TT-3024A Inmarsat-C aeronautical capsat

The compact TT-3024A aeronautical Inmarsat-C/GPS transceiver is designed for automatic data reporting and message transfer of position reports, performance data and operational mesages on a global basis, from sea level to 55,000 ft and from 70° north to 70° south.

The TT-3024A operates through the established network of Inmarsat and GPS satellites with interconnection to the international telex, fax and packet switched data networks, offering fast and reliable transfer of information, 24 hours per day.

The integrated GPS receiver calculates position, altitude, speed and heading every second, used for automatic Doppler compensation and transfer of position status reports to any predefined air traffic control or other receiver at specified intervals.

Data collecting equipment may be connected to the TT-3024A via an RS-422/423 port, enabling all selected data to be transferred as data reporting packages every 2 minutes or whenever a special event occurs.

Operational messages, weather and flight plan information as well as passenger messages may be transferred to/from any telex or data subscriber via the international networks.

Specifications

General specifications: the TT-3024A receiver meets or exceeds all Inmarsat specifications for the Inmarsat-C aero system and all relevant GPS specifications
Antenna: integrated Inmarsat-C/GPS omnidirectional antenna, RHC polarised
Figure-of-merit (G/T): −23 dB/K at 5° elevation
EIRP: 12 dBW min at 5° elevation
Transmit frequency: 1,626.5-1,646.5 MHz
Receive frequency: Inm-C 1,530.0-1,545.0 MHz, GPS 1,575.42 MHz
Channel spacing: 5 kHz
Modulation: 1,200 symbols/s BPSK
Data rate: 600 bits/s
TX message channel: TDMA and FDMA, interleaved code symbol
Position reporting: built-in 5-channel GPS receiver with Lat/Long/alt/speed/track calculation, update rate 1 s
Position accuracy: C/A code with 96 m spherical error probability
Initial stabilisation: 15 min at max Doppler shift correction and GPS almanac update
Solid-state storage: 256 kbyte RAM memory
Onboard message/data interface: RS-422/423, 110-9,600 bps and Centronics parallel
Navigational interface: RS-422/423 V.10 Special for interface to onboard navigational systems
Roll and pitch: min ±25° from level flight
Altitude: MSL-55,000 ft
Airspeed: full Doppler compensation to min 620 kt
Power: floating 10.5-32 V DC, 9.5 W Rx, 80 W Tx
Dimensions:
(antenna) 114 × 274 × 98 mm
(LNA/HPA) 161 × 213.9 × 49.5 mm
(electronics assembly) ¼ ATR short
Weights:
(antenna) 0.75 kg
(LNA/HPA) 2.3 kg
(electronics assembly) 2.5 kg

Operational status

Full US FAA and Inmarsat approval for aeronautical use.

Contractor

Thrane & Thrane A/S.

NEW ENTRY

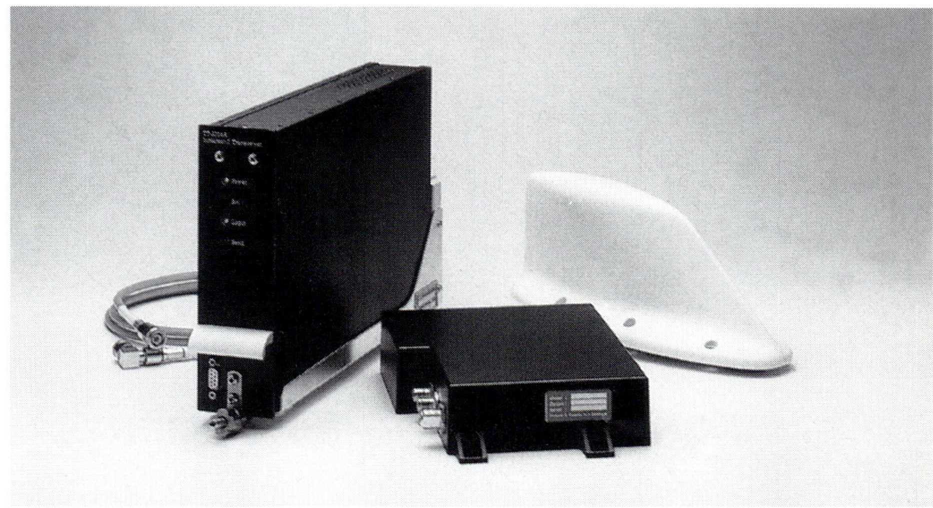

TT-3024A Inmarsat-C aeronautical capsat **1998**/0022245

TT-3608F Aero-C capsat printer unit

The TT-3608F printer unit for the Aero-C capsat transceiver TT-3024A, is designed as a compact and ruggedised printer for use in aeronautical installations.

The TT-3608F is connected to the capsat transceiver through a single two-wire RS-485 cable which may be extended up to 300 m. The TT-3608F may be used both as the general message printer for the capsat system as well as a local printer for the terminal/PC connected to the transceiver.

The TT-3608F includes a receive mail indication and LOGIN status, enabling 'non-visible' installation of the capsat transceiver electronics.

An RS-232/422 front panel connector allows direct interface to any PC compatible or dedicated terminal for message preparation and status/system set-up.

The TT-3608F may be located at the rear of the centre console in the aircraft cockpit, enabling easy access and in-flight message reception.

Up to eight TT-3608F printers may be connected to the Aero-C transceiver for distributed printing.

Specifications

Printer: 112 mm 9 × 7 dot thermal printer with 2 motors, 40/80 char per line, 1 line/s, selectable printing orientation
Indication: LED indicators for receive mail, system LOGIN and low paper
Capsat interface: two-wire RS-485 interface
Terminal/PC interface: RS-232/RS-422 serial interface from front panel connector
Power: 10.5-32 V floating DC, 2.5 W standby, 7 W printing
Altitude: MSL-55,000 ft
Dimensions: 117.5 × 213.9 × 57.5 mm
Weight: 1.3 kg

TT-3608F Aero-C capsat printer unit **1998**/0022246

Contractor

Thrane & Thrane A/S.

NEW ENTRY

TT-5000 series Inmarsat Aero-I system

The Thrane & Thrane compact, lightweight, low-power TT-5000 series Aero-I system functions as a high-quality multiple channel communication centre, for telephone calls, fax prints, data transfers or email messages to any destination.

The Aero-I system is designed to be an integrated part of the CNS/ATM system, offering three simultaneous channels (two voice/fax or data channels and one packet data channel). The built-in Cabin Telephone Unit (CTU) provides up to four handsets and two fax/phone/modem ports.

Integration of an optional Navigational Reference System (NRS) utilising 12-channel GPS data and 3-D flux gate technology, makes the TT-5000 series totally independent of aircraft systems. However, aircraft navigational systems can also be connected to create an integrated aircraft system.

Specifications

General specifications: the TT-5000 system meets or exceeds current and proposed Inmarsat specifications for the Inmarsat Aero-I system
Figure-of-merit (G/T): −19 dB/K min
EIRP: >16.5 dBW total
Coverage volume: >85%
Operating frequencies:
(receive) 1,530.0-1,559.0 MHz
(transmit) 1,626.5-1,660.5 MHz
(GPS) 1,575.42 MHz
Channel spacing: 2.5 kHz
Antenna system: Jet Sat-97 mechanically steered antenna, or CAL ANT-30 electronically steered antenna
Navigational interfaces: stand-alone with NRS (option) or ARINC 429 IRS bus
Altitude: MSL to 55,000 ft
Airspeed: full antenna tracking and Doppler compensation to min Mach 1.0

TT-5033A SDU
Channels: 3 (2 voice/fax/modem data, 1 packet data). Voice 4.8 kbytes/s, fax and modem data 2.4 kbytes/s, packet data 0.6/1.2 kbytes/s
Dimensions: 2 MCU ARINC 600
Weight: 3.6 kg
Power: +28 V DC, 20 W

TT-5010A HPA
Dimensions: 260 × 243 × 74 mm
Weight: 4 kg
Power: +28 V DC, 25-100 W
Output power: 18 W linear

TT-5012A DLNA
Dimensions: 254 × 193 × 50 mm
Weight: 2.2 kg

TT-5620A handset
Display: 2 × 12 character LCD
Key pad: 21 keys
Interface: 4 wire and RS-485
Power: +12 V DC, 0.2 A from SDU

TT-5622A handset
WH-10 AlliedSignal

TT-5008A NRS antenna
Dimensions: 38 × 243 × 129 mm
Weight: 0.68 kg
Power: 12 V DC, <2.5 W

Contractor
Thrane & Thrane A/S.

NEW ENTRY *TT-5000 series Inmarsat Aero-I system* ***1998**/0018940*

COCKPIT DISPLAYS, INSTRUMENTS AND INDICATORS

UNITED STATES OF AMERICA

Cockpit control panels

G6990-51 TCAS panels
Gables TCAS panels are available for all Boeing commercial aircraft types in both keyboard and rotary layouts. The latest variant to be certified is the G6990-51 for Boeing New Generation aircraft.

G7130-03 ATC/TCAS control panel
The G7130-03 control panel has been certified for the Falcon-900 aircraft, compatible with the Rockwell Collins 822-0078-005 radio.

G7400 series VHF control panels
Continuing VHF control panel programmes include six variants of the G7400 series for Boeing 737, 757 and 767 aircraft. Included in this series is the G7406, which is capable of tuning both ARINC 566A and 716/750 series radios. The G7404 radio tuning panel is now available in three variants for use on Boeing 747-400 aircraft. Gable's radio tuning panels have also been certified on Boeing 737 New Generation aircraft. The G7409 control radio panel has been selected for the Boeing MD-10 aircraft.

G7400 series control panels are also certified on the Russian Aviation Register.

G7500-01 VHF navigation control panel
The G7500-01 panel is standard on Boeing 737 New Generation aircraft.

Contractor
Gables Engineering Inc.

NEW ENTRY

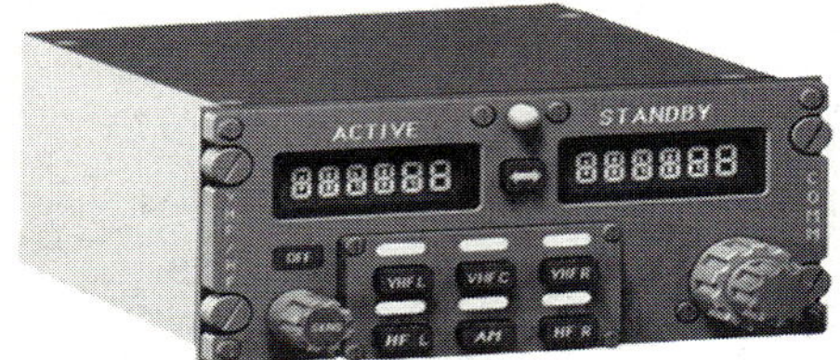

Gables' G7404 radio tuning panel ***1998**/0022250*

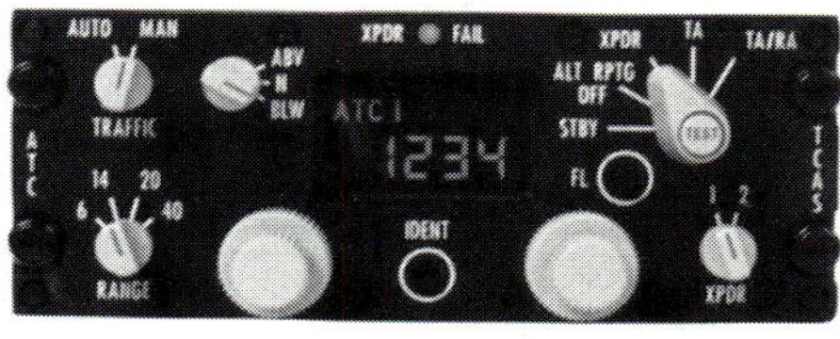

Gables' G7130-03 ATC/TCAS general aviation control panel ***1998**/0022248*

Gables' G7409 radio control panel ***1998**/0022249*

HEAD-UP DISPLAYS, HELMET-MOUNTED DISPLAYS AND WEAPON AIMING SIGHTS

UKRAINE

SURA Helmet-Mounted Target Designation System (HMTDS)

The SURA Helmet-Mounted Target Designation System (HMTDS) is designed to be used either independently, or in combination with other aircraft systems, concerned with the operation and aiming of guided weapon and gun systems.

The HMTDS generates target designation signals for weapons in proportion to the angles of turn of an operator's (pilot's) head, as well as collimating the image of an aiming mark and initiating one-time commands to his field of view.

The HMTDS helmet position sighting system is implemented using small IR emitting diodes mounted on the pilot's helmet, together with two scanning units (designated scanning unit A and scanning unit B) mounted one each side of the head-up display (HUD). This concept facilitates use of the system in a variety of aircraft with minimum requirement for aircraft modification. It also meets safety requirements for pilot ejection and emergency evacuation, and makes it possible to install the system on three standard sizes of pilot's helmet. Units of the HMTDS comprise:

(1) the helmet-mounted sighting device, attached to the pilot's helmet, using a special bracket
(2) scanning units A and B, mounted each side of the HUD
(3) an electronics unit that processes sensor information, and interfaces it to other aircraft weapon and communications systems.

Specifications
Target designation angles:
(horizontal) +60 to −60°
(vertical) +45 to −30°
Target designation accuracy: <3 mrad
Sensor data format: asynchronous binary numerical code

Power: 115 V AC, 400 Hz, <150 VA; +27 V DC
Weights:
(helmet-mounted sighting device) 0.36 kg
(scanning units) 0.8 kg
(electronics unit) 6.0 kg
Temperature limits: −40 to +60°C
Warm up time of HMTDS: <1 min
Continuous operation: limited to 3 h, followed by 25 min cycle time

Operational status
SURA HMTDS has been developed from the SHCH-3UM series of helmet-mounted target designation systems used on the MiG-29 and Su-27 aircraft.

Contractor
Arsenal Central Design Office.

NEW ENTRY

SURA electronic unit and scanning units
1998/0022251

SURA scanning unit installation, either side of the HUD
1998/0022252

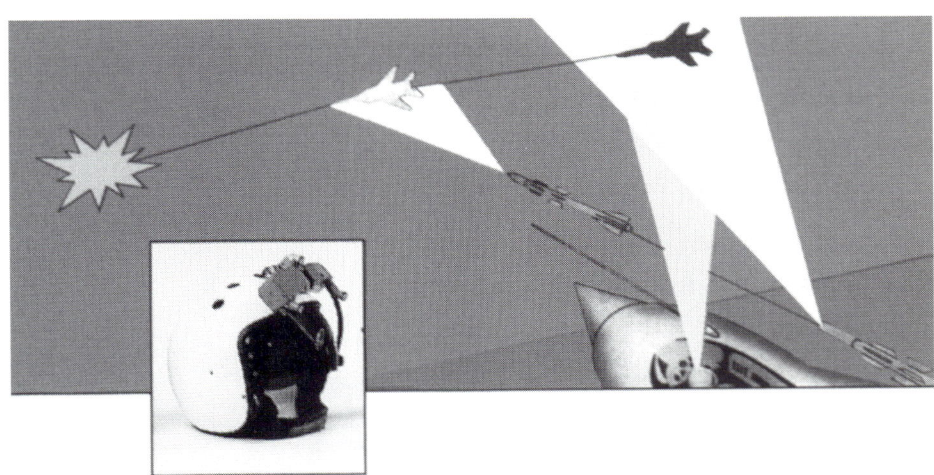

SURA principle of operation, and helmet-mounted unit
1998/0022253

Contractors

Australia

Aerospace Technologies of Australia Ltd
226 Lorimer Street
Port Melbourne
Victoria 3207
Australia
Tel: (+61 3) 647 31 11
Fax: (+61 3) 646 22 53

British Aerospace Australia Ltd
14 Park Way
Technology Park
The Levels
South Australia 5095
Australia
Tel: (+61 8) 82 90 88 88
Fax: (+61 8) 82 90 88 00

Belgium

Barco NV
Theodor Sevenslaan 106
B-8500 Kortrijk
Belgium
Tel: (+32 56) 23 32 11
Fax: (+32 56) 23 32 79
Website: http://www.barco.com

Delft Sensor Systems OIP NV
Westerring 21
B-9700 Oudenaarde
Belgium
Tel: (+32 55) 33 38 11
Fax: (+32 55) 31 68 95
Website: http://www.oip.be/oip

FN Herstal S.A
33 Voie de Liège
B-4040 Herstal
Belgium
Tel: (+32 41) 40 81 11
Fax: (+32 41) 40 88 99

Canada

Allied Signal Aerospace Canada
240 Attwell Drive
Etobicoke
Rexdale
Ontario M9W 6L7
Canada
Tel: (+1 416) 675 14 11
Fax: (+1 416) 675 40 21

Array Systems Computing Inc
1120 Finch Avenue West
8th Floor
North York
Ontario M3J 3H7
Canada
Tel: (+1 416) 736 09 00
Fax: (+1 416) 736 47 15
Website: http://www.array.ca

CAE Electronics Ltd
CP 1800
Saint Laurent
Quebec H4L 4X4
Canada
Tel: (+1 514) 341 67 80
Fax: (+1 514) 341 76 99

CAL Corporation
1050 Morrison Drive
Ottawa
Ontario K2H 8K7
Canada
Tel: (+1 613) 820 82 80
Fax: (+1 613) 820 83 14

Canadian Marconi Company
600 Dr Frederik Philips Boulevard
Ville Saint-Laurent
Quebec H4M 2S9
Canada
Tel: (+1 514) 748 30 46
Fax: (+1 514) 748 31 84

Computing Devices Canada Ltd
PO Box 8508
Ottawa
Ontario K1G 3M9
Canada
Tel: (+1 613) 596 70 00
Fax: (+1 613) 820 50 81

DRS Flight Safety and Communications
A DRS Technologies Canada Company
Kanata
Ontario K2K 2C9
Canada
Tel: (+1 613) 592 34 30
Fax: (+1 613) 592 44 86

Hughes Elcan Optical Technologies Ltd
Leitz Road
Midland
Ontario L4R 5B8
Canada
Tel: (+1 705) 526 54 01
Fax: (+1 705) 526 58 31

Litton Systems Canada Ltd
25 City View Drive
Etobicoke
Ontario M9W 5A7
Canada
Tel: (+1 416) 249 12 31
Fax: (+1 416) 245 03 24

Lockheed Martin Canada
6111 Royalmount Avenue
Montreal
Quebec H4P 1K6
Canada
Tel: (+1 514) 340 83 10
Fax: (+1 514) 340 84 48

MacDonald Dettweiler & Associates Ltd
Airborne Radar Division
13800 Commerce Parkway
Richmond
British Columbia V6V 2J3
Canada
Tel: (+1 604) 278 34 11
Fax: (+1 604) 278 12 85

Optech Systems Inc
100 Wildcat Road
North York
Toronto
Ontario M3J 2Z9
Canada
Tel: (+1 416) 661 59 04
Fax: (+1 416) 661 41 68

Pelorus Navigation Systems Inc
5418 - 11 Street NE
Calgary
Alberta T2E 7E9
Canada
Tel: (+1 403) 730 55 55
Fax: (+1 403) 730 55 11
Website: http://www.pelorus.ca

Wescam Inc
45 Innovation Drive
Flamborough
Ontario
Canada L9H 7L8
Tel: (+1 905) 689 22 31
Fax: (+1 905) 689 66 27

Chile

DTS Ltda
Rosas 1444
Santiago
Chile
Tel: (+56 2) 697 09 91
Fax: (+56 2) 699 33 16

China, Peoples Republic

Chengdu Aero-Instrument Corporation
PO Box 229
Cheng Du
Sichuan 610091
China
Tel: (+86 28) 740 90 18
Fax: (+86 28) 776 94 04
Website: http://www.caic-china.com

China Leihua Electronic Technology Research Institute
PO Box 3
Neijang
Sichuan 641003
China
Tel: (+86 0832) 202 38 21
Fax: (+86 0832) 202 48 22

China National Aero-Technology Import & Export
5 Liangguaching Road
PO Box 647
East City District
Beijing
China
Tel: (+86 1) 401 77 22
Fax: (+86 1) 401 53 81

China National Electronics Import & Export Corporation
A23 Fuxing Road
Beijing
China
Tel: (+86 1) 821 95 50
Fax: (+86 1) 821 23 52

China Precision Machinery Import & Export Corporation
17 Wen Chang Hutong
PO Box 845
Beijing
China
Tel: (+86 1) 65 18 96
Fax: (+86 1) 831 16 57

Shanghai Avionics Corporation
432 Gui Ping Lu
Shanghai 200233
China
Tel: (+86-21) 436 77 07
Fax: (+86-21) 470 01 50

Southwest China Research Institute of Electronic Equipment
PO Box 429
Chadianzi Western Suburb
Chengdu 610036
Sichuan
China
Tel: (+86 28) 751 42 43
Fax: (+86 28) 751 42 43

Czech Republic

Elektrotechnika-Tesla Kolin, a.s.
Havlickova 260
CZ-280 00 Kolin IV
Czech Republic
Tel: (+42 321) 72 46 13
Fax: (+42 321) 72 46 05

Mikrotechna Praha a.s.
Barrandova 409
CZ-143 00 Praha 4 -Modrany
Czech Republic
Tel: (+42 2) 61 31 32 10
Fax: (+42 2) 402 56 35

VZLU-SPEEL Ltd
Automatic Research and Test Institute - Special Electronics
Beranovych 130
CZ-199 05 Praha 9
Czech Republic
Tel: (+42 2) 82 34 76
Fax: (+42 2) 858 45 83

Denmark

Jorgen Andersen Ingeniorfirma A/S (JAI)
Productionswej 1
DK-2600 Glostrup
Copenhagen
Denmark
Tel: (+45 44) 91 88 88
Fax: (+45 44) 91 32 52
Website: http://www.jai.dk

Per Udsen Company Aircraft Industry A/S
Fabrikvej 1
DK-8500 Grenaa
Denmark
Tel: (+45 86) 32 19 88
Fax: (+45 86) 32 14 48

TERMA Elektronik A/S
Hovmarken 4
DK-8520 Lystrup
Denmark
Tel: (+45 86) 22 20 00
Fax: (+45 86) 22 27 99

Thrane & Thrane A/S
Tobaksverjen 23
DK-2860 Soborg
Denmark
Tel: (+45 39) 55 88 00
Fax: (+45 39) 55 88 88

France

AATON
2, rue de la Paix
BP3002
F-38001 Grenoble Cedex
France
Tel: (+33 4) 76 42 95 50
Fax: (+33 4) 76 51 34 91

Alkan
rue du 8 Mai 1945
BP 23
F-94460 Valenton
France
Tel: (+33 1) 45 10 86 00
Fax: (+33 1) 43 89 10 61

CEIS TM - LCD Division
4 avenue D Daurat
Centreda II 2e étage
BP 48
F-31702 Blagnac Cedex
France
Tel: (+33 5) 61 16 32 30
Fax: (+33 5) 61 16 32 31
Website: http://ourworld.compuserve.com/homepages/ceistm.lcd

Compagnie Industrielle des Lasers (CILAS)
route de Nozay
BP 27
F-91460 Marcoussis
France
Tel: (+33 1) 64 54 48 06
Fax: (+33 1) 64 54 48 19

Dassault Electronique SA
55 quai Marcel Dassault
BP 301
F-92214 Saint-Cloud Cedex
France
Tel: (+33 1) 49 11 80 00
Fax: (+33 1) 46 02 57 58

ELECMA Division Electronique de la SNECMA
Site de Villaroche
BP 42
F-77552 Moissy Cramayel Cedex
France
Tel: (+33 1) 60 59 71 23
Fax: (+33 1) 60 59 84 44

Enertec SA
Data Acquisition and Recording Division of Schlumberger Industries
1 rue Nieuport
BP 54
F-78141 Vélizy-Villacoublay Cedex
France
Tel: (+33 1) 30 70 30 70
Fax: (+33 1) 30 70 30 80

GIE Radar ACT-ACM Rafale
La Clef de Saint Pierre
1 boulevard Jean Moulin
F-78852 Elancourt Cedex
France
Tel: (+33 1) 34 59 60 00
Fax: (+33 1) 34 59 62 36

Matra BAe Dynamics
20-22 rue Grange Dame Rose
BP 150
F-78141 Vélizy Cedex
France
Tel: (+33 1) 34 88 30 00
Fax: (+33 1) 34 88 22 88

Monit'air
ZAC des Pres Rouz
81, rue Alain-Fournier
F-38920 Crolles
France
Tel: (+33 4) 76 08 14 39
Fax: (+33 4) 76 08 89 04

Mors SA
Centre d'Affairs Paris Nord
Tour Continental
F-93153 Le Blanc-Mesnil
France
Tel: (+33 1) 49 39 32 32
Fax: (+33 1) 49 39 01 05

Rockwell-Collins France SA
6 avenue Didier Daurat
BP 8
F-31701 Blagnac Cedex
France
Tel: (+33 1) 61 71 77 00
Fax: (+33 1) 61 71 51 69

SAGEM SA, (Société d'Applications Générales d'Electricité et de Mécanique)
Defence & Security Division
61 rue Salvador Allende
F-92751 Nanterre Cedex
France
Tel: (+33 1) 40 70 67 33
Fax: (+33 1) 40 70 64 54
Website: http://www.sagem.com

Satori
Aéroport du Bourget
BP 151 - Batiment 66
F-93352 Le Bourget Cedex
France
Tel: (+33 1) 48 62 73 00
Fax: (+33 1) 48 64 98 56

Sextant Avionique
Aerodrome de Villacoublay
ZAL Bréguet
BP 200
F-78141 Vélizy-Villacoublay Cedex
France
Tel: (+33 1) 46 29 70 00
Fax: (+33 1) 40 94 02 51
Website: http://www.sextant-avionique.com

SFIM Industries
13 avenue Marcel Ramolfo-Garnier
F-91344 Massy Cedex
France
Tel: (+33 1) 69 19 66 00
Fax: (+33 1) 69 19 69 19

TEAM Télécommunications, Electronique Aéronautique et Maritime
10, place Vauban
Silic 127
F-94523 Rungis Cedex
France
Tel: (+33 1) 49 78 66 00
Fax: (+33 1) 49 78 66 99

Thomson-CSF Communications
Parc d'Activités Kléber
160 boulevard de Valmey
BP 82
F-92704 Colombes Cedex
France
Tel: (+33 1) 41 30 30 00
Fax: (+33 1) 41 30 33 57
Website: http://www.tcc.thomson-csf.com

Thomson-CSF Applications Radar
6 rue Nieuport
BP 86
F-78143 Vélizy-Villacoublay Cedex
France
Tel: (+33 1) 30 67 88 00
Fax: (+33 1) 39 46 34 11

Thomson-CSF Optronique
rue Guynemer
BP 55
F-78283 Guyancourt Cedex
France
Tel: (+33 1) 30 96 70 00
Fax: (+33 1) 30 96 75 50

Thomson -CSF Radar Contre-Mesures
Le Clef de Saint Pierre
1 boulevard Jean Moulin
F-78852 Elancourt Cedex
France
Tel: (+33 1) 34 59 60 00
Fax: (+33 1) 34 59 62 36

Thomson Marconi Sonar SAS
BP 157
F-06903 Sophia Antipolis Cedex
France
Tel: (+33 ?) 92 96 30 00
Fax: (+33 ?) 92 96 46 30

Germany

Aerodata Flugmesstechnik GmbH
Hermann-Blenk-Strasse 36
D-38108 Braunschweig
Germany
Tel: (+49 531) 235 90
Fax: (+49 531) 235 91 58
Website: http://www.aerodata.de

Alcatel
Air Navigation Systems GmbH
Lorenzenstrasse 10
PO Box 40 07 49
D-70407 Stuttgart
Germany
Tel: (+49 711) 82 13 46 61
Fax: (+49 711) 82 13 44 05

Base Ten Systems Electronics GmbH
Am Söldnermoos 10
D-85399 Hallbergmoos
Germany
Tel: (+49 811) 55 98 0
Fax: (+49 811) 55 98 258

Bavaria Keytronic Technologie GmbH
PO Box 500272
Boschstrasse 23
D-22761 Hamburg
Germany
Tel: (+49 40) 89 69 90
Fax: (+49 40) 890 30 14

Becker Avionic Systems
Baden Air Park
PO Box 34
D-76549 Hugelsheim
Germany
Tel: (+49 7229) 30 52 28
Fax: (+49 7229) 30 52 17

Bodenseewerk Gerätetechnik GmbH
Control and Navigation Division
PO Box 101155
D-88641 Überlingen
Germany
Tel: (+49 7551) 89 0
Fax: (+49 7551) 89 28 22

Daimler-Benz Aerospace AG
Defense and Civil Systems, LFK-
Lenkflugkorpersysteme GmbH
Postfach 801149
D-81663 Munich
Germany
Tel: (+49 89 607) 292 33
Fax: (+49 89 607) 255 15

Daimler-Benz Aerospace AG
Defense and Civil Systems, Sensor Systems, Airborne Systems
Worthstrasse 85
D-89077 Ulm
Germany
Tel: (+49 731) 392 37 74
Fax: (+49 731) 392 41 08

Daimler-Benz Aerospace AG
Defense and Civil Systems
D-88039 Friedrichshafen
Germany
Tel: (+49 7545) 801
Fax: (+49 7545) 844 11

DLE Diehl GmbH & Co. Luftfahrt Electronik
Heinrich-Diehl-Strasse 2
D-90552 Röthenbach/Pegnitz
Germany
Tel: (+49 911) 957 25 60
Fax: (+49 911) 957 24 44

Walter Dittel GmbH
Erpftingerstrasse 36
PO Box 1261
D-86899 Landsberg 1
Germany
Tel: (+49 8191) 335 10
Fax: (+49 8191) 33 51 49

DLR Deutsches Zentrum für Luft- und Raumfahrt e.V.
D-51170 Cologne
Germany
Tel: (+49 22 036) 01 0
Fax: (+49 22 036) 73 10

ESG Elektroniksystem- und Logistik-GmbH
Einsteinstrasse 174
D-81675 Munich
Germany
Tel: (+49 89) 921 60
Fax: (+49 89) 92 16 26 31

ESW-Extel Systems Wedel
Gesellschaft für Ausrüstung GmbH & Co. KG
Industriestrasse 23-33
D-22876 Wedel
Germany
Tel: (+49 4103) 60 55 84
Fax: (+49 4103) 60 45 03

EuroAvionics Navigationssysteme GmbH & Co
Am Lindenberg 8
D-71263 Weil der Stadt - Hausen
Germany
Tel: (+ 49 7033) 53 82 0
Fax: (+ 49 70 33) 53 82 82

Honeywell Regelsysteme GmbH
PO Box 2010
D-63475 Maintal
Germany
Tel: (+49 6181) 40 15 99
Fax: (+49 6181) 40 14 67

LITEF GmbH
Lörracher Strasse 18
PO Box 774
D-79115 Freiberg
Germany
Tel: (+49 761) 490 10
Fax: (+49 761) 490 14 80

Nord-Micro Electronik Feinmechanik AG
Victor Slotosch Strasse 20
D-60388 Frankfurt/Main
Germany
Tel: (+49 6109) 30 30
Fax: (+49 6109) 366 98

Racal-Heim GmbH
Huttenstrasse 38
D-51469 Bergisch Gladbach
Germany
Tel: (+49 2202) 93 68 30
Fax: (+49 2202) 93 68 35

Rohde & Schwarz GmbH & Co KG
PO Box 801469
D-816714 Munich
Germany
Tel: (+49 89) 41 29 22 32
Fax: (+49 89) 41 29 32 08

Teldix GmbH
PO Box 10 56 08
D-69046 Heidelberg
Germany
Tel: (+49 62 21) 51 20
Fax: (+49 62 21) 51 23 05

VDO Luftrtgeräte Werk GmbH
An der Sandelmuhle 13
D-60439 Frankfurt/Main
Germany
Tel: (+49 69) 580 50
Fax: (+49 69) 580 53 99

Zeiss-Eltro Optronic GmbH
Carl Zeiss Strasse 22
D-73446 Oberkochen
Germany
Tel: (+49 7364) 20 27 94
Fax: (+49 7364) 20 46 01

India

Hindustan Aeronautics Ltd
Corporate Office
PO Box 5150
15/1, Cubbon Road
Bangalore-560 001
India
Tel: (+91 80) 286 67 01
Fax: (+91 80) 286 75 33

Israel

Controp Precision Technologies Ltd
PO Box 611
Hod Hasharon
IL-45105 Israel
Tel: (+972 9) 740 70 62
Fax: (+972 9) 740 53 76

ECI Telesystems Ltd
42 Hamagsimin Street
IL-49130 Petach Tikva
Israel
Tel: (+972 3) 926 69 66
Fax: (+972 3) 926 65 0

Elbit Systems Ltd
Advanced Technology Center
PO Box 539
IL-31503 Haifa
Israel
Tel: (+972 7) 831 53 15
Fax: (+972 7) 855 00 02
Website: http://www.elbit.co.il

Elisra Electronic Systems Ltd
48 Mivtza Kadesh Street
IL-51203 Bene Beraq
Israel
Tel: (+972 3) 617 51 11
Fax: (+972 3) 617 58 50
Website: http://www.elisra.com

Elop Electro Optics Industries Ltd
PO Box 1165
IL-76110 Rehovot
Israel
Tel: (+972 8) 938 64 33
Fax: (+972 8) 938 62 37

IAI ELTA Electronics Industries Ltd
100, Yitzhak Hanassi Blvd
PO Box 330
IL-77102 Ashdod
Israel
Tel: (+972 8) 857 23 33
Fax: (+972 8) 856 39 30

Israel Aircraft Industries (IAI) Ltd
Ben-Gurion International Airport
IL-70100 Israel
Tel: (+972 3) 935 85 14
Fax: (+972 3) 935 85 12

Israel Aircraft Industries (IAI) Ltd
Lahav Division
Ben Gurion Airport
Il-70100 Tel Aviv
Israel
Tel: (+972 3) 935 31 63
Fax: (+972 3) 935 36 87

Israel Aircraft Industries (IAI) Ltd
Tamam Division, Electronics Group
PO Box 75
Industrial Zone
IL-56100 Yahud
Israel
Tel: (+972 3) 531 50 03
Fax: (+972 3) 531 51 40

Opgal Optronic Industries Ltd
PO Box 462
Industrial Zone
IL-20101 Karmiel
Israel
Tel: (+972 4) 995 39 31
Fax: (+972 4) 995 39 46

Rada Electronic Industries Ltd
12 Medinat Hayedhudim Street
PO Box 2059
Herzlia
IL-46120
Israel
Tel: (+972 9) 954 21 82
Fax: (+972 9) 955 51 76

Rafael Electronic Systems Division
PO Box 2250 (80)
IL-31021 Haifa
Israel
Tel: (+972 4) 879 52 32
Fax: (+972 4) 879 40 93
Website: http://www.rafael.co.il

Rafael Missile Division
PO Box 2250
IL-31021 Haifa
Israel
Tel: (+972 4) 990 85 03
Fax: (+972 4) 990 62 57

Rokar International (A Tracor Company)
Science Based Industry Campus
Mount Hotzvim
PO Box 45049
IL-91450 Jerusalem
Israel
Tel: (+972 2) 532 98 88
Fax: (+972 2) 582 25 22

RSL Electronics Ltd
Ramat Gabriel Industrial Zone
PO Box 21
Migdal Ha'emek
IL-10550
Israel
Tel: (+972 6) 54 75 10
Fax: (+972 6) 54 75 20

Tadiran Com Ltd
Marketing Department
26 Hashoftim Street
PO Box 267
IL-58102 Holon
Israel
Tel: (+972 3) 557 42 41
Fax: (+972 3) 557 45 89

Tadiran Electronic Systems Ltd
29 Hamerkava Street
PO Box 150
IL-58101 Holon
Israel
Tel: (+972 3) 557 72 40
Fax: (+972 3) 557 72 70

Tadiran Spectralink Ltd
29 Hamerkava Street
IL-58101 Holon
Israel
Tel: (+972 3) 557 31 14
Fax: (+972 3) 557 73 84

Italy

Agusta Sistemi SpA
Via Caldera 21
I-20153 Milan
Italy
Tel: (+39 2) 45 27 51
Fax: (+39 2) 48 20 47 51

Alenia Difesa
Avionic Systems & Equipment Division
Viale Maresciallo Pilsudski, 92
I-00197 Rome
Italy
Tel: (+39 6) 80 77 81
Fax: (+39 6) 808 82 80

Alenia Difesa, Avionic Systems and Equipment Division, FIAR (Fabbrica Italiana Apparecchiature Radioelettriche)
Via GB Grassi, 93
I-93-20157 Milan
Italy
Tel: (+39 2) 35 79 05 03
Fax: (+39 2) 35 79 00 74

Alenia Difesa, Avionic Systems and Equipment Division, Officine Galileo SpA
Via Albert Einstein 35
Campi Bisenzio
I-50013 Florence
Italy
Tel: (+39 55) 895 03 21
Fax: (+39 55) 895 06 19

Elettronica SpA
Via Tiburtina Km 13.700
I-00131 Rome
Italy
Tel: (+39 6) 415 41
Fax: (+39 6) 419 28 69

Elmer SpA
Vialle dell'Industria 4
I-00040 Pomezia
Rome
Italy
Tel: (+39 6) 91 09 11
Fax: (+39 6) 91 09 12 03

Litton Italia SpA
Via Pontina Km 27.800
I-00040 Pomezia
Rome
Italy
Tel: (+39 6) 91 19 21
Fax: (+39 6) 912 25 17

Marconi SpA
via Campo nell'Elba, 3
I-00138 Rome
Italy
Tel: (+39 6) 88 69 31
Fax: (+39 6) 88 63 968

Ottico Meccanica Italiana SpA (OMI)
Via della Vasca Navale 79
I-00146 Rome
Italy
Tel: (+39 6) 54 78 81

Japan

Japan Aviation Electronics Industry Ltd
21-2 Dgenzaka 1-chome
Shibuya-Ku
Tokyo 150
Japan
Tel: (+81 3) 37 80 27 11
Fax: (+81 3) 37 80 27 33

Japan Defense Agency
Boei Cho
7-45 Akasaka 9-chome
Minato-ku
Tokyo 107
Japan
Tel: (+81 3) 34 08 52 11
Fax: (+81 3) 34 08 64 80

Mitsubishi Electronics Corporation
Corporate Communications Dept
6-3, Marunouchi 2-chome
Chiyoda-ku
Tokyo 100-86
Japan
Tel: (+81 3) 32 10 21 21
Fax: (+81 3) 32 10 80 51

NEC Corporation
7-1 Shiba 5-chome
Minato-ku
Tokyo 108-01
Japan
Tel: (+81 3) 34 54 11 11
Fax: (+81 3) 37 98 15 10

Shimadzu Corporation
1 Nishinokyo-Kuwabaracho
Nakagyo-ku
Koyoto 604
Japan
Tel: (+81 3) 32 19 58 44
Fax: (+81 3) 32 19 55 10
Website: http://www.shimadzu.co.jp/

Tokimec Inc
Electronics Systems Division
16-46 Minami-Kamata 2-chome
Ohta-ku
Tokyo 144
Japan
Tel: (+81 3) 37 37 86 41
Fax: (+81 3) 37 37 86 68

Tokyo Aircraft Instrument Co. Ltd
35-1 Izumi-Honcho 1-chome
Komae-shi
Tokyo 201
Japan
Tel: (+81 3) 34 89 11 25
Fax: (+81 3) 34 89 11 82

Toshiba Corporation
1-1 Shibaura 1-chome
Minato-ku
Tokyo 105-01
Japan
Tel: (+81 3) 34 57 31 19
Fax: (+81 3) 34 57 05 45

Netherlands

Delft Sensor Systems
DIEO BV
Rontgenweg 1 - PO Box 5083
NL-2600 GB Delft
Netherlands
Tel: (+31 15) 269 80 50
Fax: (+31 15) 269 80 98

Hollandse Signaalapparaten BV
Signaal Special Products
PO Box 241
NL-2700 AE Zoetermeer
Netherlands
Tel: (+31 79) 344 59 99
Fax: (+31 79) 344 59 38

Signaal USFA
Meerenakkerweg 1
NL-5652 AR Eindhoven
Netherlands
Tel: (+31 40) 250 36 03
Fax: (+31 40) 250 37 77
Website: http://www.usfa.nl

Norway

Eidsvoll Electronics AS
Gruemyra
N-2080 Eidsvoll
Norway
Tel: (+47 63) 96 42 30
Fax: (+47 63) 96 20 48

Kongsberg Aerospace
PO Box 1003
N-3601 Kongsberg
Norway
Tel: (+47 32) 73 82 00
Fax: (+47 32) 73 85 86

Navia Aviation AS
PO Box 50
Manglerud
N-0612 Oslo
Norway
Tel: (+47 23) 18 02 26
Fax: (+47 23) 18 02 12
Website: http://www.naviaav.com/

Poland

ATM Inc
Grochowska 21a
PL-041-86 Warsaw
Poland
Tel: (+48 22) 610 60 73
Fax: (+48 22) 610 41 44

Russian Federation and Associated States (CIS)

AeroPribor-Voskhod Joint Stock Company
19 Tkatskaya Street
105318 Moscow
Russian Federation
Tel: (+7 369) 10 81
Fax: (+7 369) 76 56

All-Russia Research Institute of Radio Equipment
Shkiperski protok 19
199106 St Petersburg
Russian Federation

All-Russia JSC Nizhegorodskaya Yarmarka
PO Box 648
420032 Kazan
Republic of Tartarstan
Tel: (+8432) 55 71 63
Fax: (+8432) 55 34 85

AviaAvtomatika, Design Bureau of the Pribor Joint Stock Company
47 Zapolnaya Street
305000 Kursk
Russian Federation
Tel: (+7 122) 264 68
Fax: (+7 122) 244 15

AviaPribor
5, Aviatzionny Per
125319 Moscow
Russian Federation
Tel: (+7 095) 152 48 74
Fax: (+7 095) 152 26 31

Elara Cheboksar Instrument-Making Plant, Joint Stock Company
Moskovsky pr 40
428034 Cheboksary
Russian Federation

ElectroPribor Kazan Plant
20 N. Ershov Street
Kazan
420045 Tartarstan
Russian Federation
Tel: (+7 8432) 76 40 01
Fax: (+7 8432) 38 89 83

ElectroPribor Voronezh Plant Joint Stock Company
20-letiya Oktyabrya 59
394006 Voronezh
Russian Federation
Tel: (+7 0732) 36 58 36
Fax: (+7 0732) 77 85 25

Geophizika-NV
Matrosskaia Tishina Street
House 23, Building 2
107016 Moscow
Russian Federation
Tel: (+7 095) 269 27 42
Fax: (+7 095) 268 01 42

GosNIIAS (State Research Institute for Aviation Systems)
7 Victorenko Street
125319 Moscow
Russian Federation
Tel: (+7 095) 157 70 47
Fax: (+7 095) 157 31 27

Joint Stock Company Pribor
47 Zapolnaya Street
305040 Kursk
Russian Federation
Tel: (+7 122) 255 72
Fax: (+7 122) 229 12

Joint Stock Company Tambovsky Zavod Electropribor
Morshanskoye shosse 36
392000 Tambov
Russian Federation
Tel: (+7 0752) 37 73 03

Manufacturing Stock Company AVECS
17 I-ya Ul
Yamskova Polya
Moscow
Russian Federation
Tel: (+7 812) 257 08 46
Fax: (+7 812) 257 77 32

Moscow Institute of Electromechanics and Automatics
125167 Moscow
Russian Federation

Moscow Scientific Research Institute of Instrument Engineering MNIIP
34 Kutuzov Avenue
121170 Moscow
Russian Federation
Tel: (+7 095) 249 07 04
Fax: (+7 095) 148 79 96

NAVIS
12, Tufeleva Roscha Street
109280 Moscow
Russian Federation
Tel: (+7 095) 274 63 04
Fax: (+7 095) 274 00 77

NIIAO Institute of Aircraft Equipment
Tupoleva 18
Zhukovsky-2
140160 Moscow
Russia Federation
Tel: (+7 095) 556 58 44
Fax: (+7 095) 556 23 28

Phazotron Scientific & Production Company
1 Electrichesky Pereulok
123557 Moscow
Russian Federation
Tel: (+7 095) 253 56 13
Fax: (+7 095) 253 04 95

Pirometr St Petersburg Joint Stock Company
Bolshaya Monetnaya 16
197101 St Petersburg
Russian Federation
Tel: (+7 812) 233 74 59
Fax: (+7 812) 233 83 06

Production Association Urals Optical & Mechanical Plant (PA UOMZ)
33-B Vostochnaya Street
620100 Yekaterinburg
Russian Federation
Tel: (+7 3432) 24 18 63
Fax: (+7 3432) 24 16 80

Production Association RadioPribor
2 Fatkullin Street
Kazan
420022 Tartarstan
Russian Federation
Tel: (+7 8432) 37 09 01
Fax: (+7 8432) 37 40 93

Ramenskoye Design Company AO RPKB
Gurjev Street 2
Ramenskoye
140103 Moscow
Russian Federation
Tel: (+7 095) 556 23 93 (Moscow)
Tel: (+7 096) 24 63 39 32 (Ramenskoye)
Fax: (+7 096) 463 19 72

Ramensky Instrument Engineering Plant
39 Mikhalevich Street
Ramenskoye
140100 Moscow
Russian Federation
Tel: (+7 095) 501 41 11
Fax: (+7 095) 556 43 28

Scientific-Technical Centre Osnova, Leninetz Holding Company
212 Moskovskiy Prospect
St Petersburg
Russian Federation
Tel: (+7 812) 293 35 62
Fax: (+7 812) 293 12 60

TekhPribor State Enterprise
1a Korpusnoi Proezd
196084 St Petersburg
Russian Federation
Tel: (+7 296) 97 38
Fax: (+7 812) 95 72

Ulyanovsk Instrument Design Office
10a Krylova Street
432001 Ulyanovsk
Russian Federation

Vympel State Machine Building Design Bureau
90 Volokolamskoje sh.
123424 Moscow
Russian Federation
Tel: (+7 095) 491 02 39
Fax: (+7 095) 490 22 22

Zenit Foreign Trade Firm, State Enterprise P/C S.A Zverev Krasnogorsky Zavod
8 Rechnaya Street
Krasnogorsk
143400 Moscow
Russian Federation
Tel: (+7 095) 561 33 77
Fax: (+7 095) 562 82 75

South Africa

ADS (Altech Defence Systems)
PO Box 432
Mount Edgecombe
4300 KwaZulu-Natal
Republic of South Africa
Tel: (+27 31) 508 11 11
Fax: (+27 31) 59 53 60

Advanced Technologies & Engineering Co (IATE)
226A Old Pretoria Road
Halfway House 1685
Republic of South Africa
Tel: (+27 11) 314 21 70
Fax: (+27 11) 314 21 51

Analysis Management & Systems (Pty) Ltd
PO Box 1980
Halfway House 1685
Republic of South Africa
Tel: (+27 11) 315 10 02
Fax: (+27 11) 315 16 45
Website: http://www.am.co.za

Grintek Avitronics, Grintek Electronics Limited
PO Box 8492
Hennopsmeer 0046
Republic of South Africa
Tel: (+27 12) 672 60 00
Fax: (+27 12) 672 62 22

Grintek Comms, Grintek Electronics Limited
PO Box 1463
Pretoria 0001
Republic of South Africa
Tel: (+27 12) 810 10 00
Fax: (+27 12) 803 60 39

Grintek System Technologies GST, Grintek Electronics Limited
PO Box 912-561
Silverton 0127
Republic of South Africa
Tel: (+27 12) 421 62 14
Fax: (+27 12) 421 62 16
Website: http://www.grinaker.co.za

Irene Commercial Enterprise (Pty) Ltd
PO Box 520
Irene 1675
Republic of South Africa
Tel: (+27 12) 671 16 05

Plessey South Africa Limited, Plessey Defence Systems, Avionics Division
64/74 White Road, Retreat 7945
PO Box 30451
Tokai 7966
Republic of South Africa
Tel: (+27 21) 710 29 11
Fax: (+27 21) 710 23 50

RDI (Pty) Ltd
PO Box 118
New Germany
Kwazulu-Natal 3620
Republic of South Africa
Tel: (+27 31) 719 57 11
Fax: (+27 31) 723 57 07

Reutech Systems (Pty) Ltd
Midrand
Republic of South Africa
Tel: (+27 11) 652 55 55

Spain

ELT SA
Polig Ind La Mina
Parcela 11
E-28770 Colmenar Viejo
Madrid
Spain
Tel: (+34 1) 846 03 01
Fax: (+34 1) 846 03 02

ENOSA
C/Joaquin Rodrigo 11
E-28300 Aranjuez
Madrid
Spain
Tel: (+34 1) 894 88 00
Fax: (+34 1) 891 80 56

INDRA
Ctra De Loeches 9
Torrejon de Ardoz
E-28850 Madrid
Spain
Tel: (+34 1) 396 81 97
Fax: (+34 1) 396 80 02

Technobit SA
Polignano Industrial
E-13300 Valdepenas
Ciudad Real
Spain

Sweden

CelsiusTech Electronics AB
SE-17588 Jarfälla
Sweden
Tel: (+46 8) 58 08 40 00
Fax: (+46 8) 58 03 22 44
Website: http://www.celsiustech.se

Ericsson Microwave Systems AB
Defence and Space Systems
Bergfotsgatan 2
SE-431 84 Mölndal
Sweden
Tel: (+46 31) 747 00 00
Fax: (+46 31) 27 78 91
Website: http://www.emw.ericsson.se

Ericsson Saab Avionics AB
Airborne Equipment Division
Box 1017
SE-551 11 Jönköping
Sweden
Tel: (+46 36) 19 46 00
Fax: (+46 36) 19 45 00

Ericsson Saab Avionics AB
EW and IFF Systems Division
SE-164 84 Stockholm
Sweden
Tel: (+46 8) 757 30 00
Fax: (+46 8) 752 81 72

Forsvarets Forskningsanstalt
Ergopingsvagan 126
SE-172 90 Stockholm
Sweden
Tel: (+46 8) 706 30 00
Fax: (+46 8) 706 30 75

GP&C Sweden AB
PO Box 4207
SE-171 04 Solna
Sweden
Tel: (+46 8) 627 64 34
Fax: (+46 8) 627 64 49
Website: http://www.ssc.se/

Polytech AB
Box 20
SE-640 32 Malmköping
Sweden
Tel: (+46 157) 217 42
Fax: (+46 157) 213 48

Saab Dynamics AB
SE-581 88 Linköping
Sweden
Tel: (+46 13) 18 60 00

Switzerland

Flight Components AG
Bitzibergstrasse 5
PO Box 57
CH-8184 Bachenbulach
Switzerland
Tel: (+41 1) 861 12 00
Fax: (+41 1) 861 17 15

LH Systems GmbH
Heinrich-Wild-Strasse
CH-9435 Heerbrugg
Switzerland
Tel: (+41 71) 727 31 77
Fax: (+41 71) 727 46 91
Website: http://www.leica.com

Revue Thommen
Postcheck 40-321-1
CH-4437 Waldenburg
Switzerland
Tel: (+41 61) 965 22 22
Fax: (+41 61) 961 81 71

Turkey

ASELSAN Inc, Microelectronics, Guidance and Electro-Optics Division
PK 30 Etlik
TR-06011 Ankara
Turkey
Tel: (+90 312) 847 53 00
Fax: (+90 312) 847 53 20
Website: hhtp://www.aselsan.com.tr

MIKES Microwave Electronic Systems Incorporated
Cankin yolu 5, km
Akyurt
TR-06750 Ankara
Turkey
Tel: (+90 312) 847 51 00
Fax: (+90 312) 847 51 14

Ukraine

Ukraine Research Institute of Radio Equipment
Kiev
252072 Ukraine

Arsenal Central Design Office
8 Moskovskaya Str
Kiev-10 'GSP'
252601 Ukraine
Tel: (380 44) 291 49 78
Fax: (380 44) 293 15 09

United Kingdom

Ampex Great Britain Limited
Beechwood
Chineham Business Park
Basingstoke
Hants RG24 8WA
UK
Tel: (+44 1256) 81 44 10
Fax: (+44 1256) 81 44 74
Website: http://www.ampex.com

Avalon Electronics Ltd
Langhorne Park House
High Street
Shepton Mallet
Somerset BA4 5AQ
UK
Tel: (+44 1749) 34 52 66
Fax: (+44 1749) 34 52 67

Avimo Ltd
Lisieux Way
Taunton
Somerset TA1 2JZ
UK
Tel: (+44 1823) 33 10 71
Fax: (+44 1823) 27 44 13

British Aerospace Systems & Equipment
Clittaford Road
Southway
Plymouth
Devon PL6 6DE
UK
Tel: (+44 1752) 72 20 59
Fax: (+44 1752) 72 31 24
Website: http://www.bae.co.uk

Caledonian Airborne Systems
Ninian Road
Aberdeen Airport
Aberdeen AB21 0PD
UK
Tel: (+44 1224) 72 22 74
Fax: (+44 1224) 72 28 96

Cargo Aids Ltd
Brett Drive
De La Warr Road
Bexhill-on-Sea
East Sussex TN40 2JP
UK
Tel: (+44 1424) 21 66 11
Fax: (+44 1424) 21 66 36

Chelton (Electrostatics) Ltd
Fieldhouse Lane
Marlow
Buckinghamshire SL7 1LR
UK
Tel: (+44 1628) 47 20 72
Fax: (+44 1628) 48 22 55

Computing Devices Company Ltd
Castleham Road
St Leonards-on-Sea
East Sussex TN38 9NJ
UK
Tel: (+44 1424) 85 34 81
Fax: (+44 1424) 85 15 20

Defence Evaluation Research Agency
Farnborough
Hampshire GU14 6TD
UK
Tel: (+44 1252) 39 55 21
Fax: (+44 1252) 39 51 20

Enterprise Control Systems Ltd
31 The High Street
Wappenham
Northants NN12 8SN
UK
Tel: (+44 1327) 86 00 50
Fax: (+44 1327) 86 00 58

Ferranti Technologies Limited
Cairo Mill
Waterhead
Oldham
Lancashire OL4 3JA
UK
Tel: (+44 161) 624 02 81
Fax: (+44 161) 624 52 44

The Flight Data Company Ltd
The Lodge
Harmondsworth Lane
West Drayton
Middx UB7 0LQ
UK
Tel: (+44 181) 759 34 55
Fax: (+44 181) 564 90 64
Website: http://www.fdata.demon.co.uk

GEC-Marconi Limited
The Grove
Warren Lane
Stanmore
Middlesex HA7 4LY
UK
Tel: (+44 181) 954 23 11
Fax: (+44 181) 954 78 08
Website: http://www.gec.com

GEC-Marconi Avionics Ltd, Displays Division; and Flight Systems Division; and Mission Avionics Division
Airport Works
Rochester
Kent ME1 2XX
UK
Tel: (+44 1634) 84 44 00
Fax: (+44 1634) 82 73 32

GEC-Marconi Avionics Ltd, Displays Division
1 South Gyle Crescent
Edinburgh EH12 9HQ
Tel: (+44 131) 332 24 11
Fax: (+44 131) 314 82 37

GEC-Marconi Avionics Ltd, Radar Systems Division
Crewe Toll
Ferry Road
Edinburgh EH5 2XS
Tel: (+44 131) 332 24 11
Fax: (+44 131) 343 06 90

GEC-Marconi Avionics Ltd, Radar Systems Division
Foxhunter Drive
Linford Wood
Milton Keynes MK14 6LA
UK
Tel: (+44 1908) 22 00 44
Fax: (+44 1908) 31 71 37

GEC-Marconi Radar and Defence Systems Ltd, Defence Systems Division
The Grove
Warren Lane
Stanmore
Middlesex HA7 4LY
UK
Tel: (+44 181) 954 23 11
Fax: (+44 181) 954 78 08

GEC-Marconi Radar and Defence Systems Ltd, Electronic Systems Division
Browns Lane
The Airport
Portsmouth
Hampshire PO3 5PH
Tel: (+44 1705) 22 60 00
Fax: (+44 1705) 22 60 01

GEC-Marconi Radar and Defence Systems Ltd, Navigation and Electro-Optic Systems Division
Silverknowes
Edinburgh EH4 4AD
Tel: (+44 131) 332 24 11
Fax: (+44 131) 343 50 50

GEC-Marconi Radar and Defence Systems Ltd, GEC-Marconi S3I Ltd
Command and Information Systems (CIS) Division
PO Box 133
Chobham Road
Frimley
Camberley
Surrey GU16 5PE
Tel: (+44 1276) 633 11
Fax: (+44 1276) 69 54 98

GEC-Marconi Electro Optics Ltd, Sensors Division
Christopher Martin Road
Basildon
Essex SS14 3EL
UK
Tel: (+44 1268) 52 28 22
Fax: (+44 1268) 88 31 40

GEC-Marconi Electro Optics Ltd, Sensors Division, Airadio Division
Browns Lane
The Airport
Portsmouth
Hampshire PO3 5PH
UK
Tel: (+44 1705) 22 60 00
Fax: (+44 1705) 22 71 33

GEC-Marconi Electro Optics Ltd, GEC-Marconi Infrared Limited
PO Box 217
Millbrook Industrial Estate
Southampton SO15 0EG
UK
Tel: (+44 1703) 70 23 00
Fax: (+44 1703) 31 67 77

GKN Westland Aerospace Ltd
Kings Building
Castle Street
East Cowes
Isle of Wight PO32 6RH
UK
Tel: (+44 1983) 29 41 01
Fax: (+44 1983) 29 10 06

Graseby Dynamics Ltd
459 Park Avenue
Bushey
Watford
Hertfordshire WD2 2BW
UK
Tel: (+44 1923) 22 85 66
Fax: (+44 1923) 24 02 85

Helmet Integrated Systems Ltd
Moat Factory
Wheathampstead
St Albans
Hertfordshire AL4 8QT
UK
Tel: (+44 1582) 83 42 11
Fax: (+44 1582) 83 42 10

ITT Defence Ltd
Jays Close
Viables Estate
Basingstoke
Hampshire RG22 4BW
UK
Tel: (+44 1256) 31 16 00
Fax: (+44 1256) 84 05 56

Lockheed Martin Solartron Systems
Hale Road
Hertford
Hertfordshire SG13 9DX
UK
Tel: (+44 1992) 53 41 11
Fax: (+44 1992) 53 41 47

Lockheed Martin UK Government Systems
PO Box 41
North Harbour
Portsmouth
Hampshire PO6 3AU
UK
Tel: (+44 1705) 56 30 74
Fax: (+44 1705) 56 30 05

LucasVarity
Lucas Aerospace
Brueton House
New Road
Solihull B91 3TX
UK
Tel: (+44 121) 627 57 24
Fax: (+44 121) 500 64 05

M/A COM Ltd
Humphrys Road
Woodside Estate
Dunstable
Bedfordshire LU5 4SX
UK
Tel: (+44 1582) 47 12 00
Fax: (+44 1582) 47 22 77

Meggitt Avionics
6 Manor Court
Barnes Wallis Road
Fareham
Hampshire PO15 5TH
UK
Tel: (+44 1489) 57 90 90
Fax: (+44 1489) 57 90 13

MS Instruments plc
Electron House
Farwig Lane
Bromley
Kent BR1 3RE
UK
Tel: (+44 181) 290 02 00
Fax: (+44 181) 464 65 96

Navstar Systems Ltd
Mansard Close
Westgate
Northampton NN5 5DL
UK
Tel: (+44 1604) 58 55 88
Fax: (+44 1604) 58 55 99
Website: http://www.telecom.com/navstar

Normalair Garrett Ltd
Electronics Division
PO Box 27
Yeovil
Somerset BA20 2YD
UK
Tel: (+44 1935) 751 81
Fax: (+44 1935) 276 00
Website: http://www.x-cd.com/normalair-garrett/

Octec Ltd
The Western Centre
Western Road
Bracknell
Berkshire RG12 1RW
UK
Tel: (+44 1344) 86 10 51
Fax: (+44 1344) 86 09 83
Website: http://www.octec.co.uk

Page Aerospace Ltd
Forge Lane
Sunbury-on-Thames
Middlesex TW16 6EQ
UK
Tel: (+44 1932) 78 76 61
Fax: (+44 1932) 78 03 49

Penny & Giles Aerospace Ltd
6 Airfield Way
Christchurch
Dorset BH23 3TT
UK
Tel: (+44 1202) 48 17 71
Fax: (+44 1202) 48 48 46

Penny & Giles Data Systems
The Mill
Wookey Hole
Wells
Somerset BA5 1BB
UK
Tel: (+44 1749) 67 54 54
Fax: (+44 1749) 67 10 92

Photo-Sonics International Ltd
5, Thame Park Business Centre
Wenman Road
Thame
Oxford OX9 3FR
UK
Tel: (+44 1844) 26 06 00
Fax: (+44 1844) 26 01 26

Pilkington Optronics Ltd
1 Linthouse Road
Govan
Glasgow G51 4BZ
UK
Tel: (+44 141) 440 43 93
Fax: (+44 141) 440 40 53

Pilkington Optronics Ltd
Pilkington PE Ltd
Glascoed Road
St Asaph
Clwyd LL17 0LL
UK
Tel: (+44 1745) 58 80 00
Fax: (+44 1745) 58 42 58

Pilkington Optronics Ltd
Thorpe Road
Staines
Middlesex TW18 3HD
UK
Tel: (+44 1784) 41 22 00
Fax: (+44 1784) 41 22 06

Racal Electronics plc
Racal Corporate Communications Centre
Western Road
Bracknell
Berkshire RG12 1WL
UK
Tel: (+44 1344) 38 80 65
Fax: (+44 1344) 38 80 62
Website: http://www.racal.com

Racal Acoustics Ltd
Hailsham Drive
Waverley Industrial Park
Harrow
Middlesex HA1 4TR
UK
Tel: (+44 181) 427 77 27
Fax: (+44 181) 427 03 50

Racal Avionics Ltd
614/615 Spur Road
North Feltham Trading Estate
Middlesex TW14 0SU
UK
Tel: (+44 181) 890 48 98
Fax: (+44 181) 890 44 03

Racal Radar Defence Systems Ltd
Manor Royal
Crawley
West Sussex RH10 2PZ
UK
Tel: (+44 1293) 52 87 87
Fax: (+44 1293) 54 28 18

Racal Radio Ltd
PO Box 3621
Bracknell
Berkshire RG12 1WJ
UK
Tel: (+44 1334) 38 70 00
Fax: (+44 1334) 38 74 03

Racal Recorders Ltd
Hardley Industrial Estate
Hythe
Southampton
Hampshire SO45 3ZH
UK
Tel: (+44 1703) 84 32 65
Fax: (+44 1703) 84 82 19

Racal Survey Ltd
Burlington House
118 Burlington Road
New Malden
Surrey KT3 4NR
UK

Raytheon Systems Limited, Electronic Systems Division
The Pinnacles
Elizabeth Way
Harlow
Essex CM19 5BB
UK
Tel: (+44 1279) 42 68 62
Fax: (+44 1279) 41 04 13

Raytheon Systems Limited, Systems Integration Division
1 Jenner Road
Manor Royal Industrial Estate
Crawley
West Sussex RH10 2GA
UK
Tel: (+44 1293) 40 19 00
Fax: (+44 1293) 40 19 01

Ring Sights Defence Ltd
PO Box 22
Borden
Hampshire GU35 9PD
UK
Tel: (+44 1420) 47 22 60
Fax: (+44 1420) 47 83 59

Rolls Smiths Engine Controls Ltd (RoSEC)
PO Box 31
Derby
Derbyshire DE24 8BJ
UK
Tel: (+44 1332) 24 76 66
Fax: (+44 1332) 24 75 00

SIFAM Ltd
SIFAM Meters Division
Woodland Road
Torquay
Devon TQ2 7AY
UK
Tel: (+44 1803) 61 38 22
Fax: (+44 1803) 61 39 26
Website: http://www.sifam.com

Signature Industries Ltd
Tom Cribb Road
Thamesmead
London SE28 0BH
UK
Tel: (+44 181) 316 44 77
Fax: (+44 181) 854 51 49

Simrad Optronics Limited
3 Meadowbrook Industrial Estate
Maxwell Way
Crawley
West Sussex RH10 2SA
UK
Tel: (+44 1293) 56 04 13
Fax: (+44 1293) 56 04 18

Skyforce Avionics
5 The Old Granary
Boxgrove
Chichester
West Sussex PO18 0ES
UK
Tel: (+44 1243) 78 37 63
Fax: (+44 1243) 78 39 92

SkyQuest Aviation
Unit A11
Fairoaks Airport
Chobham
Surrey GU24 8HX
UK
Tel: (+44 1276) 85 53 34
Fax: (+44 1276) 85 57 25

Smiths Industries Aerospace
765 Finchley Road
Childs Hill
London NW11 8DS
UK
Tel: (+44 181) 458 32 32
Fax: (+44 181) 458 43 80

Smiths Industries-Newmark
48 Vulcan Way
New Addington
Croydon
Surrey CR9 0BD
UK
Tel: (+44 1689) 83 44 41
Fax: (+44 1689) 84 59 70

Stewart Hughes Ltd
School Lane
Chandlers Ford
Eastleigh
Hampshire SO5 3YG
UK
Tel: (+44 1703) 24 20 00
Fax: (+44 1703) 24 20 01
Website: http://www.shl.co.uk

Techtest Limited
Street Court
Kingsland
Leominster
Herefordshire H26 9QA
UK
Tel: (+44 1568) 70 87 44
Fax: (+44 1568) 70 87 13

Thomson Marconi Sonar Limited
Wilkinthroop House
Templecombe
Somerset BA8 0DH
UK
Tel: (+44 1963) 37 05 51
Fax: (+44 1935) 44 22 00

Ultra Electronics Controls Division
417 Bridport Road
Greenford
Uxbridge
Middlesex UB6 8UA
UK
Tel: (+44 181) 813 44 44
Fax: (+44 181) 813 43 51

Ultra Electronics Noise & Vibration Systems
1 Cambridge Business Park
Cowley Road
Cambridge CB4 4WZ
UK
Tel: (+44 1223) 42 66 99
Fax: (+44 1223) 42 66 96

Ultra Electronics Sonar and Communication Systems
419 Bridport Road
Greenford
Middlesex UB6 8UA
UK
Tel: (+44 181) 813 45 67
Fax: (+44 181) 813 45 68

Vosper Thornycroft Controls Ltd
HSDE Division
Tewin Road
Welwyn Garden City
Hertfordshire AL7 1LR
UK
Tel: (+44 1707) 36 80 00
Fax: (+44 1707) 32 75 63

W Vinten Ltd
Vicon House
Western Way
Bury St Edmunds
Suffolk IP33 3SP
UK
Tel: (+44 1284) 75 05 99
Fax: (+44 1284) 75 05 98

Weston Aerospace Limited
124 Victoria Road
Farnborough
Hampshire GU14 7PW
UK
Tel: (+44 1252) 54 44 33
Fax: (+44 1252) 37 12 16
Website: http://www.solartron.com

United States of America

II Morrow Inc
2345 Turner Road SE
Salem
Oregon 97302
USA
Tel: (+1 503) 391 34 11
Fax: (+1 503) 364 21 38

Accurate Automation Corporation
7001 Shallowford Road
Chattanooga
Tennessee 37421
USA
Tel: (+1 423) 894 46 46
Fax: (+1 423) 894 46 45
Website: http://www.accurate-automation.com

ADC Avionic Displays Corporation
6090-A Northbelt Parkway
Norcross
Georgia 30071
USA
Tel: (+1 770) 242 74 66
Fax: (+1 770) 242 75 33

Aerosonic Corporation
1212 North Hercules Avenue
Clearwater
Florida 34625
USA
Tel: (+1 813) 461 30 00
Fax: (+1 813) 447 59 26

Aerospace Avionics Inc
1000 MacArthur Memorial Highway
PO Box 1000
Bohemia
New York 11716
USA
Tel: (+1 516) 467 55 00
Fax: (+1 516) 467 59 39

AIL Systems Inc
415 Commack Road
Deer Park
New York 11729
USA
Tel: (+1 516) 595 50 00
Fax: (+1 516) 595 59 45

Alliant Defense Electronics Systems, Inc
a wholly owned subsidiary of Alliant Techsystems, Inc
PO Box 4648
Clearwater
Florida 34618
USA
Tel: (+1 813) 572 31 84
Fax: (+1 813) 572 24 33

Alliant Defense Electronics Systems, Inc.
Lundy Product Group
13133 34th Street, North
Clearwater
Florida 34618
USA
Tel: (+1 813) 572 19 00

AlliedSignal Aerospace
2525 West 190th Street
Torrance
California 90504
Tel: (+1 310) 512 10 16
Fax: (+1 310) 512 32 51

AlliedSignal Inc, Commercial Avionics Systems
400 N Rogers Road
Olathe
Kansas 66062
USA
Tel: (+1 913) 768 30 00
Fax: (+1 913) 768 29 40

AlliedSignal Inc, Electronic Systems, Communications Systems
1300 East Joppa Road
Baltimore
Maryland 21286
USA
Tel: (+1 410) 583 43 53
Fax: (+1 401) 583 44 99

AlliedSignal Inc, Electronic Systems, Inertial & Sensor Products
699 Route 46 East
Teterboro
New Jersey 07608
Tel: (+1 201) 393 41 32
Fax: (+1 201) 288 82 70

AlliedSignal Engine Systems & Accessories
717N Bendix Drive
South Bend
Indiana 46620
USA
Tel: (+1 219) 231 37 17
Fax: (+1 219) 231 33 35

AlliedSignal Ocean Systems
15825 Roxford Street
Sylmar
California 91342
USA
Tel: (+1 818) 367 01 11
Fax: (+1 818) 367 04 03

Ametek Aerospace
50 Fordham Road
Wilmington
Maryland 01887
USA
Tel: (+1 978) 988 41 01
Fax: (+1 978) 988 49 44
Website: www.AMETEK.com

Amherst Systems Inc
30 Wilson Road
Buffalo
New York 14221
USA
Tel: (+1 716) 631 06 10
Fax: (+1 716) 631 06 29

Ampex Data Systems Corporation
500 Broadway
Redwood City
California 94063
USA
Tel: (+1 415) 367 41 05
Fax: (+1 415) 367 31 06

Anaren Microwave Inc
6635 Kirkville Road
E. Syracuse
New York 13057
USA
Tel: (+1 315) 432 89 09

Andrew SciComm
2908 National Drive
Garland
Texas 75041
USA
Tel: (+1 972) 840 49 00
Fax: (+1 972) 271 93 79
Website: http://www.andrew.com

Arc Industries Inc
PO Box 033498
Indialantic
Florida 32903
USA
Tel: (+1 407) 254 19 17
Fax: (+1 407) 254 85 44

ARGO Systems Inc
A subsidiary of The Boeing Company
324 N. Mary Avenue / PO Box 3452
Sunnyvale
California 94088
USA
Tel: (+1 408) 524 17 78
Fax: (+1 408) 524 20 23

ARINC Incorporated
2551 Riva Road
Annapolis
Maryland 21401
USA
Tel: (+1 410) 266 46 51
Fax: (+1 410) 266 23 29

Arnav Systems Inc
PO Box 73730
Puyallup
Washington 98373
USA
Tel: (+1 206) 848 60 60
Fax: (+1 206) 848 35 55

Artex Aircraft Inc
PO Box 1270
10714 S. Township Road
Canby
Oregon 97013
USA
Tel: (+1 503) 266 39 59
Fax: (+1 503) 266 33 62
Website: http://www.earthworld.com/artex

Astronautics Corporation of America
PO Box 523
4115 N Teutonia Avenue
Milwaukee
Wisconsin 53201
USA
Tel: (+1 414) 447 82 00
Fax: (+1 414) 447 82 31
Website: http://www.astronautics.com

Avionics Specialities Inc
Charlottesville-Albermarle Airport
PO Box 6400
Charlottesville
Virginia 22906
USA
Tel: (+1 804) 973 33 11
Fax: (+1 804) 973 29 76

AVTECH Corporation
3400 Wallingford Avenue North
Seattle
Washington 98103
USA
Tel: (+1 206) 634 25 40
Fax: (+1 206) 634 30 11

AYDIN Corporation
Radar & EW Division
32 Great Oaks Boulevard
PO Box 1371
San Jose
California 95119
USA
Tel: (+1 408) 629 11 00
Fax: (+1 408) 224 55 92

AYDIN TELEMETRY
47 Friends Lane
PO Box 328
Newtown
Pennsylvania 18940
USA
Tel: (+1 215) 968 42 71
Fax: (+1 215) 968 32 14
Website: http://www.aydin.com

Ball Aerospace & Technologies Corporation (BATC)
PO Box 1235
10 Long Peaks Drive
Broomfield
Colorado 80038
USA
Tel: (+1 303) 939 40 00
Fax: (+1 303) 460 23 15
Website: http://www.ball.com/aerospace/batchp.htm

BFGoodrich Aerospace Aircraft Integrated Systems
100 Panton Road
Vergennes
Vermont 05491
USA
Tel: (+1 802) 877 29 11
Fax: (+1 802) 877 41 13

BFGoodrich Aerospace Avionic Systems, Inc.
5353 52nd Street SE
Grand Rapids
Michigan 49588
USA
Tel: (+1 616) 949 66 00
Fax: (+1 616) 285 42 24

The Boeing Company, McDonnell Aircraft and Missile Systems
PO Box 516
St Louis
Missouri 63166
USA
Tel: +1 (314) 234 41 87
Fax: (+1 314) 233 64 55

The Boeing Company, Communications & Information Management Systems
3370 Miraloma Avenue
PO Box 4921
Anaheim
California 92803
USA
Tel: (+1 714) 762 81 11

The Boeing Company, Information and Communication Systems
20403-68th Avenue S.
Kent
Washington 98032
USA
Tel: (+1 253) 773 28 16
Fax: (+1 253) 773 39 00

The Boeing Company, Product Support Division
PO Box 7730
Wichita
Kansas 67277
USA
Tel: (+1 316) 526 39 02
Fax: (+1 316) 526 76 01
Website: http://www.boeing.com

Century Flight Systems Inc
Municipal Airport
PO Box 610
Mineral Wells
Texas 76068
USA
Tel: (+1 940) 325 25 17
Fax: (+1 940) 325 25 46
Website: http://www.centuryFlight.com

Chelton Avionics Inc, Wulfsberg Electronics Division
6400 Wilkinson Drive
Prescott
Arizona 86301
USA
Tel: (+1 520) 708 15 00
Fax: (+1 520) 445 55 81

Cincinnati Electronics Corporation
7500 Innovation Way
Mason
Ohio 45040
USA
Tel: +1 513 573 6100
Fax: +1 513 733 6741

Computing Devices International
8800 Queen Avenue South
Bloomington
Minnesota 55431
USA
Tel: (+1 612) 921 60 80
Fax: (+1 612) 921 69 66

Concurrent Computer Corporation
2 Crescent Place
Oceanport
New Jersey 07757
USA
Tel: (+1 908) 870 45 00
Fax: (+1 908) 870 59 67

Condor Systems
2133 Samaritan Drive
San Jose
California 95124
USA
Tel: (+1 408) 371 95 80
Fax: (+1 408) 371 95 89

Cubic Communications Inc
9535 Waples Street
San Diego
California 92121
USA
Tel: (+1 619) 643 58 10
Fax: (+1 619) 643 58 03
Website: http://www.cubiccom.com

Cubic Defense Systems Inc
9333 Balboa Ave
PO Box 85587
San Diego 92186
USA
Tel: (+1 619) 277 67 80
Fax: (+1 619) 505 15 39

Daedalus Enterprises Inc
PO Box 1869
Ann Arbor
Michigan 48106
USA
Tel: (+1 313) 769 56 49

Delco Electronics Corporation
Delco Systems Operations
6767 Hollister Avenue
Goletta
California 93117
USA
Tel: (+1 805) 961 59 03
Fax: (+1 805) 961 54 16

Diamond J, Inc
PO Box 9526
Wichita
Kansas 67277
USA
Tel: (+1 316) 945 10 10
Fax: (+1 316) 945 22 44

DNE Technologies Inc
50 Barnes Park North
PO Box 30
Wallingford
Connecticut 06492
USA
Tel: (+1 203) 265 71 51
Fax: (+1 203) 265 91 01

dpix, A Xerox New Enterprise Business
3406 Hillview Avenue
Palo Alto
California 94304
USA
Tel: (+1 650) 842 96 00
Fax: (+1 650) 842 98 08
Website: http://www.dpix.com

DRS Precision Echo Inc
3105 Patrick Henry Drive
Santa Clara
California 95054
USA
Tel: (+1 408) 988 05 16
Fax: (+1 408) 727 74 91
Website: http://www.precisionecho.com

DRS Technologies, Inc.
5 Sylvan Way
Parsippany NJ 07054
USA
Tel: (+1 973) 898 15 00
Fax: (+1 973) 898 47 30
Website: http://www.drs.com

EDO Corporation
Marine and Aircraft Systems
14/04 111th Street
College Point
New York 11356
USA
Tel: (+1 718) 321 40 00
Fax: (+1 718) 321 01 19

Electronics & Space Corporation
8100W Florissant Avenue
St Louis
Missouri 63136
USA
Tel: (+1 314) 553 45 29
Fax: (+1 314) 553 45 55

Endevco
30700 Rancho Viejo Road
San Juan Capistrano
California 92675
USA
Tel: (+1 714) 493 81 81
Fax: (+1 714) 661 72 31

Eventide Inc
One Alsan Way
Little Ferry
New Jersey 07643
USA
Tel: (+1 201) 641 12 00
Fax: (+1 201) 641 16 40

Fairchild Defense OSC
20301 Century Boulevard
Germantown
Maryland 20874
USA
Tel: (+1 301) 428 60 00
Fax: (+1 301) 428 69 75

Flight Dynamics
16600 SW 72nd Avenue
Portland
Oregon 97224
USA
Tel: (+1 503) 684 53 84
Fax: (+1 503) 684 01 69

Flightline Electronics Inc
7525 County Road 42
PO Box 750
Fishers
New York 144553
USA
Tel: (+1 716) 924 40 00
Fax: (+1 716) 924 57 32

Flight Visions Inc
Aurora Municipal Airport
43 W752 Route 30
PO Box 250
Sugar Grove
Illinois 60554
USA
Tel: (+1 708) 466 43 43
Fax: (+1 708) 466 43 58

FLIR Systems Inc
16505 SW 72nd Avenue
Portland
Oregon 97224
USA
Tel: (+1 503) 684 37 31
Fax: (+1 503) 684 32 07
Website: http://www.flir.com

Gables Engineering, Inc.
247 Greco Avenue
Coral Gables
Florida 33146
USA
Tel: (+1 305) 774 42 80

Garmin International
1200 East 151st Street
Olathe
Kansas 66062
USA
Tel: (+1 913) 397 82 00
Fax: (+1 913) 397 82 82

GEC-Marconi Hazeltine Corporation
164 Totowa Road
PO Box 975
Wayne
New Jersey 07474
USA
Tel: (+1 201) 633 60 00
Fax: (+1 201) 633 64 31

General Atronics Corporation
1200 East Mermaid Lane
Wyndmoor
PA 19038
USA
Tel: (+1 215) 233 41 00
Fax: (+1 215) 233 99 47
Website: http://www.generalatronics.com

Gentex Corporation
PO Box 315
Carbondale
Pennsylvania 18407
USA
Tel: (+1 717) 282 35 50
Fax: (+1 717) 282 85 55

GTE Government Systems Corporation
77 A Street
Needham Heights
Maryland 02194
USA
Tel: (+1 617) 449 20 00
Fax: (+1 617) 455 30 30

Hamilton Standard Division of UTC
One Hamilton Road
Windsor Locks
Connecticut 06096
USA
Tel: (+1 203) 654 60 00
Fax: (+1 203) 654 26 20

Harris Corporation
Government Aerospace Systems Division
PO Box 94000
Melbourne
Florida 32902
USA
Tel: (+1 407) 737 40 00
Fax: (+1 407) 729 76 75

Honeywell Inc, Commercial Aviation, Air Transport Systems
21111 N 19th Avenue
Phoenix
Arizona 85027
USA
Tel: (+1 602) 436 22 03
Fax: (+1 602) 436 53 00

Honeywell Inc, Commercial Aviation, Business & Commuter Aviation Systems
21111 N 19th Avenue
Phoenix
Arizona 85036
USA
Tel: (+1 602) 436 23 11
Fax: (+1 602) 436 22 52

Honeywell Inc, Defense Avionics Systems
9201 San Mateo Boulevard NE
PO Box 9200
Albuquerque
New Mexico 87113
USA
Tel: (+1 505) 828 50 00
Fax: (+1 505) 828 55 00

Honeywell Inc, Sensor and Guidance Products
2600 Ridgway Parkway NE
PO Box 312
Minneapolis
Minnesota 55440
USA
Tel: (+1 612) 951 53 77

Icom Inc
2380 116th Avenue NE
Bellevue
Washington 98004
USA
Tel: (+1 425) 454 81 55
Fax: (+1 425) 454 15 09

Ideal Research & Development Corporation
1810 Parklawn Drive
Rockville
Maryland 20852
USA
Tel: (+1 301) 468 20 50
Fax: (+1 301) 230 08 13

ImageQuest Technologies Inc
48611 Warm Springs Blvd
Fremont
California 94539
USA
Tel: (+1 510) 249 05 00
Fax: (+1 510) 249 05 50

Interstate Electronics Corporation
1001 East Ball Road
PO Box 3117
Anaheim
California 92803
USA
Tel: (+1 714) 758 05 00
Fax: (+1 714) 758 41 48
Website: http://www.iechome.com

ITT Aerospace
Communications Division
Fort Wayne
Indiana
USA
Tel: (+1 219) 487 63 84

ITT Defense & Electronics, Avionics Division
100 Kingsland Road
Clifton
New Jersey 07014
USA
Tel: (+1 201) 284 01 23
Fax: (+1 201) 284 33 34

ITT Defense & Electronics, Electro-Optical Products Division
7635 Plantation Road NW
PO Box 7065
Roanoke
Virginia 24019
USA
Tel: (+1 703) 563 03 71
Fax: (+1 703) 366 90 15

JP Instruments
Box 7033
Huntington Beach
California 92615
USA
Tel: (+1 714) 557 54 34
Fax: (+1 714) 557 98 40

Kaiser Electronics
2701 Orchard Parkway
San Jose
California 95134
USA
Tel: (+1 408) 432 30 00
Fax: (+1 408) 432 84 40

Kaman Aerospace International
Bloomfield
Connecticut 06002
USA
Tel: (+1 860) 243 73 19
Fax: (+1 203) 243 63 67

Kearfott Guidance & Navigation Corporation
150 Totowa Road
Wayne
New Jersey 07474
USA
Tel: (+1 973) 785 60 75
Fax: (+1 973) 785 59 05

Kollsman Inc
220 Daniel Webster Highway
Merrimack
New Hampshire 03054
USA
Tel: (+1 603) 889 25 00
Fax: (+1 603) 889 7966

L.3 Communication Aviation Recorders
PO Box 3041
Sarasota
Florida 34230
USA
Tel: (+1 941) 377 08 11
Fax: (+1 941) 377 55 91
Website: http://www.L-3ar.com

L.3 Communications Display Systems
1355 Bluegrass Lakes Parkway
Alpharetta
Georgia 30004
USA
Tel: (+1 770) 752 54 86
Fax: (+1 770) 752 55 25

Lear Astronics Corporation
3400 Airport Avenue
PO Box 442
Santa Monica
California 90406
USA
Tel: (+1 310) 915 60 00
Fax: (+1 310) 915 83 84

Litton Aero Products
21050 Burbank Boulevard
Woodland Hills
California 91367
USA
Tel: (+1 818) 226 20 00
Fax: (+1 818) 226 21 99

Litton Amecon
5115 Calvert Road
College Parkway
Maryland 20740
USA
Tel: (+1 301) 864 56 00
Fax: (+1 301) 864 59 55

Litton Applied Technology
4747 Hellyer Avenue
PO Box 7012
San Jose
California 95150
USA
Tel: (+1 408) 365 40 30
Fax: (+1 408) 365 40 40

Litton Electron Devices
1215 S 52nd Street
Temple
Arizona 85281
USA
Tel: (+1 602) 968 44 71
Fax: (+1 602) 966 90 55

Litton Guidance & Control Systems
19601 Nordhoff Street
Northridge
California 91324
USA
Tel: (+1 818) 717 69 64
Fax: (+1 818) 349 68 17

Litton Marine Systems (Sperry Marine)
1070 Seminole Trail
Charlottesville
Virginia 22901
USA
Tel: (+1 804) 974 20 00
Fax: (+1 804) 974 22 59

Litton Laser Systems Division
2787 South Orange Blossom Trail
Noopka
Florida 32703
USA
Tel: (+1 407) 295 40 10
Fax: (+1 407) 297 48 48

Litton Special Devices
750 West Sproul Road
Springfield
Pennsylvania 19064
USA
Tel: (+1 215) 328 40 00
Fax: (+1 215) 328 40 16

Lockheed Martin Aircraft & Logistic Centers
107 Frederick Street
Greenville
South Carolina 29607
USA
Tel: (+1 864) 422 62 41
Fax: (+1 864) 422 62 76

Lockheed Martin Control Systems
600 Main Street
Johnson City
New York 13790
USA
Tel: (+1 607) 770 26 37
Fax: (+1 607) 770 25 67
Website: http://www.lmco.com/controlsystems/

Lockheed Martin Electronics & Missiles
5600 Sand Lake Road
Orlando
Florida 32819
USA
Tel: (+1 407) 356 53 51
Fax: (+1 407) 356 20 80

Lockheed Martin Fairchild Systems
Ridge Hill
Yonkers
New York 10710
USA
Tel: (+1 914) 964 28 28

Lockheed Martin Fairchild Systems
300 Robbins Lane
Syosset
New York 11791
USA
Tel: (+1 516) 349 23 28
Fax: (+1 516) 349 25 89

Lockheed Martin Federal Systems
9500 Godwin Drive
Manassas
Virginia 22110
USA
Tel: (+1 703) 367 44 40
Fax: (+1 703) 367 32 36

Lockheed Martin Federal Systems
1801 State Route 17c
Owego
New York 13827
USA
Tel: (+1 607) 751 45 24
Fax: (+1 607) 751 32 59

Lockheed Martin Government Electronic Systems
199 Borton Landing Road
Moorestown
New Jersey 08057
USA
Tel: (+1 609) 722 27 08
Fax: (+1 609) 722 39 47

Lockheed Martin IR Imaging Systems
2 Forbes Road
Lexington
Massachusetts 02173
USA
Tel: (+1 617) 862 62 22
Fax: (+1 817) 863 34 96
Website: http://www.lmco.com

Lockheed Martin Ocean, Radar & Sensor Systems
Electronics Park
Syracuse
New York 13221
USA
Tel: (+1 315) 456 38 28
Fax: (+1 315) 456 38 81
Website: http://www.lmco.com/orss

Lockheed Martin Tactical Aircraft Systems
Lockheed Boulevard
Fort Worth
Texas 76101
USA
Tel: (+1 817) 763 40 86
Fax: (+1 817) 763 47 97

Lockheed Martin Tactical Defense Systems
1210 Masillon Road
Akron
Ohio 44315
USA
Tel: (+1 330) 796 84 58
Fax: (+1 330) 796 32 74

Lockheed Martin Tactical Defense Systems
PO Box 64525
St Paul
Minnesota 55164
USA
Tel: (+1 612) 456 22 10
Fax: (+1 612) 456 27 36

Lockheed Martin Tactical Defense Systems
PO Box 85
Litchfield Park
Arizona 85340
USA
Tel: (+1 602) 925 70 00
Fax: (+1 602) 925 70 80

Lockheed Martin Western Development Labs
3200 Zanker Road
San Jose
California 95134
USA
Tel: (+1 408) 473 40 36
Fax: (+1 408) 473 40 97

Magellan Systems Corporation
960 Overland Court
San Dimas
California 91773
USA
Tel: (+1 909) 394 50 00
Fax: (+1 909) 394 70 50
Website: http://www.magellanangps.com

Metrum -Datatape Inc
4800 East Dry Creek Road
Littleton
Colorado 80122
USA
Tel: (+1 303) 773 47 00
Fax: (+1 303) 773 47 62
Website: http://www.metrum-datatape.com

Microvision Inc
Seattle
Washington
USA
Tel: (+1 206) 623 70 55

Motorola Inc
Government & Systems Technology Group
PO Box 9040
8220 E Roosevelt Road
Scottsdale
Arizona 85252
USA
Tel: (+1 602) 441 40 79
Fax: (+1 602) 441 23 68

Narco Avionics Inc
270 Commerce Drive
Fort Washington
Pennsylvania 19034
USA
Tel: (+1 215) 643 29 05
Fax: (+1 215) 643 01 97

NavCom Defense Electronics Inc
4323 Arden Drive
El Monte
California 91731
USA
Tel: (+1 626) 579 86 91
Fax: (+1 626) 444 76 19

NavSymm Positioning Systems
85 West Tasman Drive
San Jose
California 95134
USA
Tel: (+1 800) 486 63 38
Fax: (+1 408) 428 79 72

Northrop Grumman Corporation
1840 Century Park East
Los Angeles
California 90067
USA
Tel: (+1 310) 553 62 62
Fax: (+1 310) 556 45 61

Northrop Grumman Corporation, Electronic Sensors & Systems Division (ESSD)
PO Box 451, A275
Baltimore
Maryland 21203
USA
Tel: (+1 410) 765 26 44
Fax: (+1 410) 765 45 97

Northrop Grumman Corporation, Electronic Sensors & Systems Division (ESSD), Norden Systems Inc
Norden Place
PO Box 5300
Norwalk
Connecticut 06856
USA
Tel: (+1 203) 852 50 00
Fax: (+1 203) 852 76 98

Northrop Grumman Corporation, Electronics & Systems Integration Division
800 Hicks Road
Rolling Meadows
Illinois 60008
USA
Tel: (+1 847) 259 96 00
Fax: (+1 847) 870 57 13

Northrop Grumman Corporation, Electronics & Systems Integration Division
1111 Stewart Avenue
Bethpage
New York 11714
USA
Tel: (+1 516) 575 51 19
Fax: (+1 516) 575 36 91

Northstar Technologies
30 Sudbury Road
Acton
Maryland 01720
USA
Tel: (+1 978) 897 66 00
Fax: (+1 978) 897 82 64
Website: http://www.northstarcmc.com

Palomar Products Inc
2051 Palomar Airport Road
Carlsbad
California 92009
USA
Tel: (+1 760) 931 33 00
Fax: (+1 760) 931 51 98

Photo-Sonics Inc
820 S Mariposa Street
Burbank
California 91506
USA
Tel: (+1 818) 842 21 41
Fax: (+1 818) 842 26 10

Planar Advance Inc
13950 SW Karl Braun Drive
PO Box 4001
Beaverton
Oregon 97075
USA
Tel: (+1 503) 614 41 11
Fax: (+1 503) 614 41 01

Polhemus Inc
1 Hercules Drive
PO Box 560
Colchester
Vermont 05446
USA
Tel: (+1 802) 655 31 59
Fax: (+1 802) 655 14 39

Racal Communications Inc
5 Research Place
Rockville
Maryland 20850
USA
Tel: (+1 301) 948 44 20
Fax: (+1 301) 948 60 15

Raytheon Systems Company
2000 E. Imperial Highway
PO Box 92426
Los Angeles
California 90009
USA
Tel: (+1 310) 334 71 07
Fax: (+1 310) 334 03 85

Raytheon Systems Company
2000 E. El Segundo Boulevard
PO Box 902
El Segundo
California 90245
USA
Tel: (+1 310) 616 10 22
Fax: (+1 310) 616 63 30

Raytheon Systems Company
One South Los Carneros
Goleta
California 93117
USA
Tel: (+1 805) 967 55 11
Fax: (+1 805) 964 04 70

Raytheon Systems Company
Greenville Division
Majors Field
PO Box 6056
Greenville
Texas 75403
USA
Tel: (+1 903) 457 93 34
Fax: (+1 903) 457 44 13

Raytheon Systems Company
PO Box 405
2501 South Highway 121
Lewisville
Texas 75067
USA
Tel: (+1 972) 462 65 00
Fax: (+1 214) 462 65 08
Website: http://www.raytheon.com/rtis

Raytheon Systems Company
Intelligence, Information and Aircraft Integration Systems
1301 E. Collins Boulevard
Richardson
Texas 75081
USA
Tel: (+1 972) 470 20 00
Fax: (+1 972) 301 59 90

Recon/Optical Inc
550 West Northwest Highway
Barrington
Illinois 60010
USA
Tel: (+1 847) 381 24 00
Fax: (+1 847) 381 24 25

Rockwell Collins
Collins Commercial Avionics
400 Collins Road NE
Cedar Rapids
Iowa 52498
USA
Tel: (+1 319) 395 10 00
Fax: (+1 319) 395 97 46

Rockwell International Corporation
Communications Systems Division
3200 East Renner Road
PO Box 833807
Richardson
Texas 75082
USA
Tel: (+1 214) 705 39 50
Fax: (+1 241) 705 33 98

Rockwell International Corporation
Defense Systems
2201 Seal Beach Boulevard
PO Box 4250
Seal Beach
California 90740
USA
Tel: (+1 310) 797 33 11
Fax: (+1 310) 797 58 28

Rodale Electronics Inc
603 Chestnut Street
Garden City
New York 11530
USA
Tel: (+1 516) 222 00 50
Fax: (+1 516) 222 02 23

Rogerson Kratos
a Rogerson Aircraft Corporation Subsidiary
403 S. Raymond Avenue
Pasadena
California 91109
USA
Tel: (+1 714) 660 06 66
Fax: (+1 714) 660 79 55
Website: http://www.rogerson.com

Safe Flight Instrument Corporation
20 New King Street
White Plains
New York 10604
USA
Tel: (+1 914) 946 95 00
Fax: (+1 914) 946 78 82

Sanders
A Lockheed Martin Company
65 Spit Brook Road
Nashua
New Hampshire 03060
USA
Tel: (+1 603) 885 43 21
Fax: (+1 603) 885 36 55
Website: http://www.sanders.com

Sargent Fletcher Inc
9400 E. Flair Drive
El Monte
California 91731
USA
Tel: (+1 818) 402 29 15
Fax: (+1 818) 579 91 83

SCI Systems Inc
8600 South Memorial Parkway
PO Box 4000
Huntsville
Alabama 35802
USA
Tel: (+1 205) 882 48 00
Fax: (+1 205) 882 46 52

Shadin Corporation Inc
14280 North 23rd Avenue
Minneapolis
Minnesota 55447
USA
Tel: (+1 612) 557 65 00
Fax: (+1 612) 557 91 96

Sierra Technologies Inc
Sierra Research Division
485 Cayuga Road
PO Box 222
Buffalo
New York 14225-0222
USA
Tel: (+1 716) 631 65 25
Fax: (+1 716) 631 61 16

Sigtronics Corporation
949 N. Cataract Avenue
San Dimas
California 91773
USA
Tel: (+1 909) 305 93 99

Smiths Industries Aerospace & Defense Systems Ltd
4141 Eastern Avenue SE
Grand Rapids
Michigan 49518
USA
Tel: (+1 616) 241 86 43
Fax: (+1 616) 241 73 18

Smiths Industries Aerospace
7-11 Vreeland Road
Florham Park
New Jersey 07932
USA
Tel: (+1 201) 514 40 17
Fax: (+1 201) 822 08 74

SPS Signal Processing Systems
(A Division of Global Associates, Ltd)
13112 Evening Creek Drive South
San Diego
California 92128
USA
Tel: (+1 619) 679 62 28
Fax: (+1 619) 679 64 00
Website: http://www.sps-globalus.com

S-TEC Corporation
One S-TEC Way
Municipal Airport
Mineral Wells
Texas 76067
USA
Tel: (+1 817) 325 94 06
Fax: (+1 817) 325 39 04
Website: http://www.s-tec.com

Sunair Electronics Inc
3101 SW 3rd Avenue
Fort Lauderdale
Florida 33315
USA
Tel: (+1 954) 525 15 05
Fax: (+1 954) 765 13 22
Website: http://www.sunairhf.com

Systems & Electronics Inc
201 Evans Lane
St Louis
Missouri 63121
USA
Tel: (+1 314) 553 46 78
Fax: (+1 314) 553 49 50
Website: http://www.seistl.com

Systems Research Laboratories (SRL)
2800 Indian Ripple Road
Dayton
Ohio 45440
USA
Tel: (+1 513) 426 60 00
Fax: (+1 513) 426 19 84

TEAC Airborne Video Products
TEAC America Inc
7733 Telegraph Road
Montebello
California 90640
USA
Tel: (+1 213) 726 03 03
Fax: (+1 213) 727 76 21

Teledyne Controls
8640 154th Avenue NE
Redmond
WA 98052
USA
Tel: (+1 425) 861 69 06
Fax: (+1 425) 885 15 43

Teledyne Controls
Helicopter Systems
12333 W. Olympic Boulevard
Los Angeles
California 90064
USA
Tel: (+1 310) 442 42 38
Fax: (+1 310) 442 43 24

Teledyne Electronic Tech
Microwave Production
1274 Terra Bella Avenue
Mountain View
California 94043
USA
Tel: (+1 415) 962 69 44
Fax: (+1 415) 962 68 45

Telephonics Corporation
815 Broad Hollow Road
Farmingdale
New York 11735
USA
Tel: (+1 516) 755 70 00
Fax: (+1 516) 755 70 46
Website: http://www.telephonics.com

Telex Communications Inc
9600 Aldrich Avenue South
Minneapolis
Minnesota 55420
USA
Tel: (+1 612) 884 40 51
Fax: (+1 612) 884 00 43
Website: http://www.telex.com

Titan Linkabit
3033 Science Park Road
San Diego
California 92121
USA
Tel: (+1 619) 552 95 00
Fax: (+1 619) 552 96 45

Tracor Aerospace Inc
6500 Tracor Lane
Austin
Texas 78725
USA
Tel: (+1 512) 929 28 81
Fax: (+1 512) 929 23 80

Tracor Aerospace Electronic Systems Inc
305 Richardson Road
Lansdale
Pennsylvania 19446
USA
Tel: (+1 215) 822 29 29
Fax: (+1 215) 822 91 65

Trimble Navigation Ltd
Avionics Products
2105 Donley Drive
Austin
Texas 78758
USA
Tel: (+1 512) 432 04 00
Fax: (+1 512) 836 94 13
Website: http://www.trimble.comavionics

Trimble Navigation Ltd
Military Systems
645 North Mary Avenue
Sunnyvale
California 94086
USA
Tel: (+1 408) 481 80 00
Fax: (+1 408) 481 20 82

TRW Electronics Systems & Technology Division
Space & Electronics Group
One Space Park
Redondo Beach
California 90278
USA
Tel: (+1 310) 812 41 05

TRW Electromagnetic Systems Division
Systems Integration Group
495 Java Drive
PO Box 3510
Sunnyvale
California 94088
USA
Tel: (+1 408) 752 23 34
Fax: (+1 408) 752 26 40

Universal Avionics Systems Corporation
3260 East Lerdo Road
Tucson
Arizona 85706
USA
Tel: (+1 520) 295 23 00
Fax: (+1 520) 295 23 90

US Air Force, Materiel Command, Wright Laboratory
Wright-Patterson Air Force Base
Ohio 45433
USA
Tel: (+1 513) 255 27 25
Fax: (+1 513) 476 40 22

Versatron Corporation (a division of Wescam Inc)
103 West North Street
Healdsburg
California 95448
USA
Tel: (+1 707) 433 30 00
Fax: (+1 707) 433 71 10

Vickers Inc
Systems Monitoring Division
24 E. Glenolden Avenue
Glenolden
Pennsylvania 19036
USA
Tel: (+1 610) 583 94 00
Fax: (+1 610) 583 39 85

Vision Systems International
2711 Orchard Parkway
San Jose
California
USA
Tel: (+1 408) 433 97 20
Fax: (+1 408) 432 84 49

Watkins Johnson Company, Telecommunications Group
700 Quince Orchard Road
Gaithersburg,
Maryland 20878
USA
Tel: (+1 301) 948 75 50
Fax: (+1 301) 921 94 79
Website: http://www.wj.com

Zeta
2811 Orchard Parkway
San Jose
California 95134
USA
Tel: (+1 408) 434 36 00
Fax: (+1 408) 433 02 05

Alphabetical index

The prefixes 'Mark', 'Model', 'Series', 'System' and 'Type' have been ignored in the placing of items in this index.

A

B

C

F

G

H

I

J

K

L

M

N

O

T

U

V

W

X

Y

Z

Manufacturers index

The prefixes 'Mark', 'Model', 'Series', 'System' and 'Type' have been ignored in placing items in this index.

Alphabetical list of advertisers